ANALYTE REFERENCE CHART

Analyte*	Expected adult reference or therapeutic range: Conventional units		Expected adult reference or therapeutic range: SI units		Method of analysis (page)	Clinical correlation (page)
Creatinine:					1247	403, 509
Serum	6.4-10.4 mg/L	males	57-92 μmol/L	males		
	5.7-9.2 mg/L	females	50-81 μmol/L	females		
Urine	1-2 g/day	males	8.8-17.7 mmol/day	males		
	0.8-1.8 g/day	females	7.1-16 mmol/day	females		
Digoxin (S)	0.5-2 ng/mL		0.6-2.4 nmol/L		1352	488
Drug screen (U)	Negative		—		1355	897
Estriol (serum and urine)	Varies with gestational age				1140	686, 789
Ferritin (S)	20-200 μg/L		45-450 pmol/L		1279	633
Folic acid (folate assay) (S)	1.9-14 ng/mL		4.3-31.7 nmol/L		1402	656
Gamma-glutamyl transferase (S)	8-37 U/L	males	$11.7\text{-}67 \times 10^{-8}$ katal/L		1120	420, 594
	5-24 U/L	females	$6.7\text{-}42 \times 10^{-8}$ katal/L			
Gentamicin (S)	2-9 μg/mL		4.3-19.4 μmol/L		1365	962
Glucose (S)						
Hexokinase	700-1100 mg/L		3.9-6.1 mmol/L		1032	520
Oxygen electrode	650-1100 mg/L		3.6-6.1 mmol/L			
Glycohemoglobin (B)	4.5%-8.5%		—		1281	520
Haptoglobin (S)	600-2700 mg/L		7.1-31.8 μmol/L		1284	611
Hemoglobin (B)						
Hb A_2	1.5%-4%		—		1286	611, 832
Hb F	<0.9%		—		1294	
Hb S	0		—		1286	
Hemoglobin, electrophoresis (B)	—		—		1286	611
High-density lipoprotein (HDL) cholesterol (S)	Age and sex dependent, p. 579				1207	550
Human chorionic gonadotropin (HCG, hCG)	Varies with gestational age				1147	686
Nonpregnant serum			<10 mU/mL			
Nonpregnant urine			<50 mU/mL			
5-Hydroxyindoleacetic acid (5-HIAA) (U)	1.8-6.0 mg/24 hr		9.4-31.4 μmol/24 hr		1153	468, 864
Immunoelectrophoresis (S)	—		—		1297	718
Immunoglobulin (Ig) (S) quantitation	Age dependent				1304	718
IgA	1.4-3.5 mg/mL		8.7-21.9 μmol/L			
IgD	0-0.14 mg/mL		0-0.76 μmol/L			
IgE	<300 ng/mL		<1.5 nmol/L			
IgG	8.0-16.0 mg/mL		53-106 μmol/L			
IgM	0.5-2.0 mg/mL		0.56-2.2 μmol/L			
Insulin (S)	<1042 pg/mL (<25 μU/mL)		<181 nmol/L		1157	460, 520
Iron (S)	400-1600 μg/L		7.16-28.6 μmol/L		1063	633
Ketones (acetoacetate) (S)	5-30 mg/L		49-294 μmol/L		1037	520
Lactate dehydrogenase (LD) (S)	63-155 (L-P)U/L	male	$1.1\text{-}2.6 \times 10^{-6}$ katal/L		1124	488
	90-320 (P-L)U/L	male	$1.5\text{-}5.3 \times 10^{-6}$ katal/L			
	62-131 (L-P)U/L	female	$1.0\text{-}2.2 \times 10^{-6}$ katal/L			
	90-320 (P-L)U/L	female	$1.5\text{-}5.3 \times 10^{-6}$ katal/L			
Lactate dehydrogenase isoenzymes (S)	LD_1 18%-33% LD_2 28%-40% LD_3 18%-30% LD_4 6%-16% LD_5 2%-13%		—		1127	488, 953
Lactic acid (S)	<180 mg/L		<2 mmol/L		1040	387, 520
Lead (B)	100-200 μg/L		483-965 nmol/L		1369	639
Lecithin-sphingomyelin (LS) ratio (amniotic fluid)	>2 indicates fetal maturity		—		1182	686
Lipase (S)	2-7.5 lipase units/mL (66-248 U/L)		$1.1\text{-}4.1 \times 10^{-6}$ katal/L		1130	460, 550
Lipoprotein electrophoresis (S or P)	—		—		1213	550
Lithium (S)	—		0.4-1.0 mmol/L		1377	864
Low-density lipoprotein (LDL) cholesterol (S)	Very age and sex dependent, p. 576				—	550
Magnesium						
Serum	15.8-25.5 mg/L		0.65-1.05 mmol/L		1065	439
Urine	24-255 mg/24 hr		(1-10.5 mmol/24 hr)			
Melanins (U)	Not detectable		—		1391	996

*B, Whole blood; P, plasma; S, serum; U, urine.

(Continued on back endsheet)

CLINICAL CHEMISTRY
theory, analysis, and correlation

CLINICAL CHEMISTRY

theory, analysis, and correlation

LAWRENCE A. KAPLAN, Ph.D.
Associate Professor,
Pathology and Laboratory Medicine,
University of Cincinnati,
Cincinnati, Ohio

AMADEO J. PESCE, Ph.D.
Professor,
Experimental Medicine, Pathology,
and Laboratory Medicine,
University of Cincinnati Medical Center,
Cincinnati, Ohio

With 84 contributors

with 920 illustrations

The C. V. Mosby Company

ST. LOUIS • TORONTO • PRINCETON
1984

A TRADITION OF PUBLISHING EXCELLENCE

Editors: Don Ladig/Rosa Kasper
Manuscript editor: Carl Masthay
Book design: Staff
Cover design: Nancy Steinmeyer
Production: Linda Stalnaker

Printed in the United States of America

The C.V. Mosby Company
11830 Westline Industrial Drive, St. Louis, Missouri 63146

Library of Congress Cataloging in Publication Data

Main entry under title:
Clinical chemistry.

Bibliography: p.
Includes index.
1. Chemistry, Clinical. I. Kaplan, Lawrence A., 1944- . II. Pesce, Amadeo J. [DNLM: 1. Chemistry, Clinical. QY 90 C6415]
RB40.C58 1984 616.07′56 83-21983
ISBN 0-8016-2705-2

C/VH/VH 9 8 7 6 5 4 3 2 1 03/D/309

Contributors

LENOX B. ABBOTT, Ph.D.
Postdoctoral Fellow,
The Cleveland Clinic Foundation,
Cleveland, Ohio

ROY L. ALEXANDER, Jr., Ph.D.
Assistant Professor, Department of Clinical Pathology,
St. Louis University Hospital,
St. Louis, Missouri

SUSAN BASSION, Ph.D.
Section Head, Immunology,
Scott and White Clinic,
Temple, Texas;
Assistant Professor,
Departments of Pathology and Microbiology,
Texas A & M Medical School,
College Station, Texas

JOHN D. BAUER, M.D.
Hematology Laboratory,
St. Peter's Hospital,
St. Peters, Missouri

HELEN K. BERRY, M.A.
Professor of Pediatric Research,
University of Cincinnati College of Medicine;
Director, Inborn Errors of Metabolism,
Department of Pediatrics,
Children's Hospital Medical Center,
Cincinnati, Ohio

GORDON L. BILLS, M.D.
Chief Resident,
Department of Pathology and Laboratory Medicine,
University of Cincinnati Medical Center,
Cincinnati, Ohio

GAYLE BIRKBECK, M.S.
Director, Hematology/Immunology,
Department of Laboratory Medicine,
The Christ Hospital,
Cincinnati, Ohio

LARRY D. BOWERS, Ph.D.
Associate Professor,
Division of Clinical Laboratories,
Department of Laboratory Medicine and Pathology,
University of Minnesota,
Minneapolis, Minnesota

W. FRASER BREMNER, M.D., Ph.D.
Associate Professor of Medicine,
Department of Medicine,
University of Cincinnati Medical Center,
Cincinnati, Ohio

JOHN M. BREWER, Ph.D.
Professor, Department of Biochemistry,
University of Georgia,
Athens, Georgia

MARGE A. BREWSTER, Ph.D.
Associate Professor,
Departments of Pathology and Biochemistry,
University of Arkansas for Medical Sciences,
Little Rock;
Clinical Biochemist,
Department of Pediatric Pathology,
Arkansas Children's Hospital,
Little Rock, Arkansas

BERNDT B. BRUEGGER, Ph.D.
Supervisor, Chemistry Laboratory,
Valley Children's Hospital,
Fresno, California

CHARLES RALPH BUNCHER, Sc.D.
Professor, Division of Biostatistics and Epidemiology,
University of Cincinnati Medical Center,
Cincinnati, Ohio

Sr. ELIZABETH ANN BYRNE, M.S., C.L.S.
Director, M.L.T. Program,
Assistant Professor, Biology Department,
College of Mount St. Joseph,
Mount St. Joseph, Ohio

R. NEILL CAREY, Ph.D.
Clinical Chemist, Department of Pathology,
Peninsula General Hospital Medical Center,
Salisbury;
Associate Clinical Professor,
Department of Medical Technology,
Salisbury State College,
Salisbury, Maryland

JOHN F. CHAPMAN, Dr.P.H.
Assistant Professor, Department of Pathology,
School of Medicine,
University of North Carolina at Chapel Hill;
Associate Director, Clinical Chemistry Laboratories,
Department of Hospital Laboratories,
The North Carolina Memorial Hospital,
Chapel Hill, North Carolina

I-WEN CHEN, Ph.D.
Professor of Radiology,
Eugene L. Saenger Radioisotope Laboratory,
Department of Radiology,
University of Cincinnati College of Medicine,
Cincinnati, Ohio

BRADLEY E. COPELAND, M.D., A.B.
Professor of Clinical Pathology,
Department of Pathology and Laboratory Medicine,
University of Cincinnati School of Medicine;
Chief, Laboratory Service,
U.S. Veterans Administration Medical Center,
Cincinnati, Ohio

VICKI N. DAASCH, Ph.D.
Postdoctoral Fellow in Clinical Immunology,
Department of Laboratory Medicine,
University of North Carolina at Chapel Hill;
Department of Hospital Laboratories,
North Carolina Memorial Hospital,
Chapel Hill, North Carolina

JAMES E. DAVIS, Ph.D.
Quillen Building
Concord Place,
E.I. du Pont de Nemours & Co., Inc.,
Wilmington, Delaware

JOSEPH R. DiPERSIO, Ph.D., Diplomate of ABMM
Director, Microbiology,
Department of Laboratory Medicine,
The Christ Hospital,
Cincinnati, Ohio

E. CHRISTIS FARRELL, Jr., Ph.D.
Director, Clinical Chemistry Laboratory,
Ohio Valley Hospital,
Steubenville, Ohio

MARIANO FERNANDEZ-ULLOA, M.D.
Associate Professor of Radiology (Nuclear Medicine),
Eugene L. Saenger Radioisotope Laboratory,
University of Cincinnati Medical Center,
Cincinnati, Ohio

M. ROY FIRST, M.D.
Associate Professor,
Division of Nephrology,
Department of Internal Medicine,
University of Cincinnati Medical Center,
Cincinnati, Ohio

ROBERT S. FRANCO, Ph.D.
Assistant Professor, Experimental Medicine;
Director, Diagnostic Hematology Laboratory,
Division of Hematology/Oncology,
Department of Internal Medicine,
University of Cincinnati Medical Center
Cincinnati, Ohio

CHRISTOPHER S. FRINGS, Ph.D.
Director, Clinical Chemistry and Toxicology,
Medical Laboratory Associates; Clinical Professor,
Pathology and School of Allied Health Sciences,
University of Alabama in Birmingham,
Birmingham, Alabama

CARL C. GARBER, Ph.D.
Associate Professor,
Department of Pathology and Laboratory Medicine,
University of Wisconsin Hospital,
Madison, Wisconsin

DAVID L. GARVER, M.D.
Professor of Psychiatry, Pharmacology and Cell Biophysics,
Department of Psychiatry,
University of Cincinnati,
Cincinnati, Ohio

NANCY GAU, M.T. (A.S.C.P.)
Supervisor, Clinical Chemistry Laboratory,
Department of Pathology and Laboratory Medicine,
University of Cincinnati Medical Center
Cincinnati, Ohio

JACK GAULDIE, Ph.D.
Associate Professor,
Department of Pathology;
Director, Regional Clinical Immunology Laboratory,
McMaster University Medical Center,
Hamilton, Ontario,
Canada

STEPHEN M. GENDLER, Ph.D.
Division of Biochemistry,
Two Blum Pavilion,
Michael Reese Hospital,
Chicago, Illinois

LEWIS GLASSER, M.D.
Chief, Hematopathology Laboratories,
Professor, Department of Pathology;
Arizona Health Sciences Center,
Tucson, Arizona

F. MICHAEL HASSAN, B.S., C.(A.S.P.)
Assistant Director, Toxicology Laboratory,
Department of Pathology and Laboratory Medicine,
University of Cincinnati Medical Center,
Cincinnati, Ohio

LINDA A. HEMINGER, B.S. (C.N.M.T.)
Eugene L. Saenger Radioisotope Laboratory,
Department of Radiology,
University of Cincinnati Medical Center,
Cincinnati, Ohio

ROBERT HOFF, M.D.
Chief Resident,
Department of Pathology and Laboratory Medicine,
University of Cincinnati Medical Center,
Cincinnati, Ohio

DAVID C. HOHNADEL, Ph.D., F.A.C.B.
Director, Clinical Chemistry Laboratory,
The Christ Hospital,
Cincinnati, Ohio

PETER HORSEWOOD, Ph.D.
Research Associate, Department of Pathology,
McMaster University Medical Center,
Hamilton, Ontario,
Canada

PAUL E. HURTUBISE, Ph.D.
Associate Professor,
Department of Pathology and Laboratory Medicine,
University of Cincinnati Medical Center,
Cincinnati;
Director, Diagnostic Immunology Laboratory,
Division of Laboratory Medicine,
University of Cincinnati Hospital,
Cincinnati, Ohio

STEPHEN N. JOFFE, M.D.
Professor of Surgery,
University of Cincinnati Medical Center,
Cincinnati, Ohio

THADDEUS E. KELLY, M.D., Ph.D.
Professor of Pediatrics;
Chief, Medical Genetics,
University of Virginia Medical Center,
Charlottesville, Virginia

ELIZABETH J. KICKLIGHTER, M.D.
Resident, Department of Pathology and Laboratory Medicine,
University of Cincinnati Medical Center,
Cincinnati, Ohio

MARY ELLEN KING, Ph.D.
Assistant Professor, Department of Pathology,
Medical College of Virginia,
Richmond, Virginia

LEONARD I. KLEINMAN, M.D.
Professor and Director of Newborn Services,
Department of Pediatrics,
State University of New York at Stony Brook,
Stony Brook, New York

ANTHONY KOLLER, Ph.D.
Division of Biochemistry,
Two Blum Pavilion,
Michael Reese Hospital,
Chicago, Illinois

RICHARD J. KOZERA, M.D.
Professor of Clinical Internal Medicine,
University of Cincinnati Medical Center,
Cincinnati, Ohio

BIDY KULKARNI, Ph.D.
Director, Reproductive Endocrinology Laboratory,
Department of Obstetrics and Gynecology,
Cook County Hospital, Chicago;
Associate Professor of Obstetrics and Gynecology,
Chicago Medical School,
North Chicago, Illinois

BOLESLAW H. LIWNICZ, M.D., Ph.D.
Associate Professor of Pathology and Laboratory Medicine,
University of Cincinnati Medical Center,
Cincinnati, Ohio

REGINA G. LIWNICZ
McMicken College of Arts and Sciences,
University of Cincinnati,
Cincinnati, Ohio

JOHN M. LORENZ, M.D.
Assistant Professor,
Department of Pediatrics,
University of Cincinnati Medical Center,
Cincinnati, Ohio

JOHN A. LOTT, Ph.D.
Professor, Department of Pathology;
Director, Outpatient Clinical Laboratory,
The Ohio State University,
Columbus, Ohio

HAROLD MARDER, M.D.
Philadelphia Children's Hospital and Medical Center,
Philadelphia, Pennsylvania

HARRY R. MAXON, M.D.
Professor of Radiology,
Eugene L. Saenger Radioisotope Laboratory,
University of Cincinnati Medical Center,
Cincinnati, Ohio

MICHAEL D.D. McNEELY, M.D.
Island Medical Laboratories,
Victoria, British Columbia,
Canada

CHARLES L. MENDENHALL, M.D., Ph.D.
Professor, Department of Internal Medicine,
University of Cincinnati Medical Center;
Chief, Digestive Disease Section,
Veterans Administration Medical Center,
Cincinnati, Ohio

W. GREGORY MILLER, Ph.D.
Associate Professor, Department of Pathology,
Medical College of Virginia,
Richmond, Virginia

GERALD MORIARTY, M.D.
Assistant Professor, Department of Neurology,
University of Cincinnati Medical Center,
Cincinnati, Ohio

ROBERT L. MURRAY, Ph.D.
Department of Biochemistry,
Lutheran General Hospital,
Park Ridge, Illinois

BÉLA NAGY, Ph.D.
Associate Professor, Department of Neurology and Department of Pharmacology and Biophysics,
University of Cincinnati Medical Center,
Cincinnati, Ohio

HERBERT K. NAITO, Ph.D.
Head, Section of Lipids, Proteins, and Metabolic Diseases,
Division of Laboratory Medicine and Division of Research,
The Cleveland Clinic Foundation,
Cleveland, Ohio

RONALD OVERNOLTE, Ph.D.
Product Manager,
Hamilton Company
Reno, Nevada

SUMAN PATEL, Ph.D.
Postdoctoral Fellow, Clinical Chemistry,
Department of Pathology,
The Ohio State University,
Columbus, Ohio

GERARDO PERROTTA, M.T. (A.S.C.P.)S.H.
Assistant Director, Hematology Laboratory,
Department of Pathology and Laboratory Medicine,
University of Cincinnati Medical Center,
Cincinnati, Ohio

NATHAN A. PICKARD, Ph.D.
Assistant Director of Clinical Chemistry,
Department of Pathology,
St. Vincent's Hospital and Medical Center,
New York, New York

ALPHONSE POKLIS, Ph.D.
Director, Forensic and Environmental Laboratory;
Associate Professor, Department of Pathology;
Assistant Professor, Department of Pharmacology,
St. Louis University School of Medicine,
St. Louis, Missouri

DAVID G. RHOADS, Ph.D.
Consultant, David G. Rhoads Associates, Inc.
Wilmington, Delaware

WOLFGANG RITSCHEL, Ph.D.
Professor of Pharmacokinetics,
College of Pharmacy,
University of Cincinnati Medical Center,
Cincinnati, Ohio

PAUL T. RUSSELL, Ph.D.
Associate Professor,
Department of Obstetrics and Gynecology,
University of Cincinnati Medical Center,
Cincinnati, Ohio

TIMOTHY J. SCHROEDER, M.S.
Staff Technologist,
Department of Pathology and Laboratory Medicine,
University of Cincinnati Medical Center,
Cincinnati, Ohio

ARNOLD L. SCHULTZ, Ph.D.
Laboratory Service,
Veterans Administration Medical Center,
Denver, Colorado

G. BERRY SCHUMANN, M.D.
Associate Professor of Pathology,
Director, Cytology Division,
University of Utah College of Medicine,
Salt Lake City, Utah

JOHN E. SHERWIN, Ph.D.
Director, Clinical Chemistry,
Valley Children's Hospital,
Fresno, California

LAWRENCE M. SILVERMAN, Ph.D.
Assistant Professor of Pathology,
Clinical Chemistry Laboratory,
University of North Carolina at Chapel Hill,
Chapel Hill, North Carolina

THOMAS SKILLMAN, M.D.
Professor, Department of Internal Medicine,
The Ohio State University,
Columbus, Ohio

STEVEN J. SOLDIN, Ph.D., F.A.C.B.
Associate Biochemist and Director,
Therapeutic Drug Monitoring,
Department of Biochemistry,
Hospital for Sick Children,
Toronto, Ontario,
Canada

MATTHEW I. SPERLING, B.S.
Eugene L. Saenger Radioisotope Laboratory,
Department of Radiology,
University of Cincinnati Medical Center,
Cincinnati, Ohio

BERNARD E. STATLAND, M.D., Ph.D.
Director of Clinical Chemistry;
Director, Department of Laboratory Medicine,
University Hospital,
Boston, Massachusetts

JOSEPH SVIRBELY, M.S.
Staff Technologist,
Department of Pathology and Laboratory Medicine,
University of Cincinnati Medical Center,
Cincinnati, Ohio

M. WILSON TABOR, Ph.D.
Assistant Professor,
Department of Environmental Health,
University of Cincinnati Medical Center,
Cincinnati, Ohio

STEPHEN G. THOMPSON, Ph.D.
Senior Research Scientist,
Ames Division, Miles Laboratories, Inc.,
Elkhart, Indiana

REGINALD TSANG, M.D.
Professor of Pediatrics, Obstetrics and Gynecology;
Director, Division of Neonatology,
Department of Pediatrics,
University of Cincinnati Medical Center,
Cincinnati, Ohio

STEPHEN G. WEBER, Ph.D.
Assistant Professor,
Department of Chemistry,
University of Pittsburgh,
Pittsburgh, Pennsylvania

ROBERT E. WEESNER, M.D.
Assistant Professor, Department of Internal Medicine,
University of Cincinnati Medical Center;
Director, GI Endoscopy Services,
Veterans Administration Medical Center,
Cincinnati, Ohio

DAN WEINER, Ph.D.
Adjunct Associate Professor,
Division of Biostatistics and Epidemiology,
University of Cincinnati Medical Center,
Cincinnati, Ohio

WILLIAM C. WENGER
Predoctoral Fellow, Department of Pathology,
The Ohio State University,
Columbus, Ohio

PER WINKEL, M.D., Doc.Med.Sci.
Visiting Professor, Department of Laboratory Medicine,
University Hospital,
Boston, Massachusetts

LINDA L. WOODARD, M.T. (A.S.C.P.)
Clinical Chemistry Laboratories,
North Carolina Memorial Hospital,
Chapel Hill, North Carolina

To our wives,
Judith Kaplan and *Anna Pesce,*
from whom we took two years of their lives,

and our mentors,
Samuel Natelson and *George F. Grannis*

Foreword

Clinical chemistry is the application of the science of chemistry to the understanding of the human in health and disease. Much of the progress in basic chemistry in the past has been motivated largely by an intense interest in the nature of disease in the human. For example, the alchemist's search for the panacea, to cure all human ills, resulted in the identification of many of the common elements and basic principles of chemistry. However, the term "clinical chemistry," to designate this specific discipline, did not come into general use until 1949, with the organization of the American Association of Clinical Chemists (now called "American Association for Clinical Chemistry") and the subsequent publication of clinical chemistry journals in the United States, Scandinavia, England, Canada, Germany, Italy, and other countries.

Since the early 1950s, substantial progress has taken place in clinical chemistry, not only in highly sophisticated instrumentation, but also in the scope of the science. Instruments not generally available 30 years ago are in common use today in the clinical chemistry laboratory. Examples include automated instrumentation for assaying more than 20 blood components on over 100 small blood specimens per hour, highly sophisticated computers for controlling these devices and analyzing the data, gas chromatography/mass spectrometry assemblies, high-performance liquid chromatography equipment, and atomic absorption and infrared spectrometers. Even the analytical gravimetric balance has been replaced largely by the rapidly weighing electronic balances with digital readout. pH, Pco_2 and Po_2 measurements, which were difficult chores in the fifties, are now carried out routinely with relatively inexpensive instruments, with a degree of accuracy and precision not considered possible 30 years ago. Many components of biological materials that were considered outside the range of capability of the clinical chemistry laboratory are now assayed daily. For example, immunoassay by radiometric, fluorimetric, and enzymatic techniques are performed for the biologically active polypeptides, tumor markers, and other components normally present in minute quantities in the serum such as steroids and iodothyronines. The blood of patients receiving therapeutic drugs is analyzed regularly, even when concentrations are very low. Toxicological sections of the clinical chemistry laboratory have developed extensively, using the modern techniques just mentioned.

With the expansion of the role of the laboratory of clinical chemistry in public health care has come a corresponding increase in their economic importance. Clinical chemistry is now a multibillion-dollar industry. Most major industrial chemical and pharmaceutical complexes are now directly involved in providing instrumentation and supplying reagents for use in the clinical laboratory. Several of these companies operate service laboratories for private physicians and provide reference laboratories or, by contract, operate hospital laboratories. The private-service laboratories have moved from being one- or two-person establishments to national multimillion-dollar enterprises.

The growth of clinical chemistry in the past 20 years has added a major industry to the United States and the world. This growth has been so rapid that adequate time to develop books on the subject, which present all the newer ramifications and could serve as a sophisticated basis for training personnel, has not been available. The text presented by Kaplan and Pesce is designed to fill this void. They accomplished it by assembling the contributions of 84 experts in the field who contributed their talents to this enterprise. In this manner the subject could be addressed in depth and its subject matter kept current.

This book presents the basic analytical principles and integrates them with the newer instrumentation and techniques. The physiology and anatomy of the various organs of the human are then presented, with emphasis on the use of the findings of the clinical chemistry laboratory in the evaluation of their function in disease. Finally, it presents detailed procedures for carrying out the numerous diagnostic tests. In this regard it serves as a detailed laboratory manual.

This text is an excellent source book for those engaged in the practice of clinical chemistry. It is also an excellent manual for the training of medical technologists, clinical chemists, and clinical pathologists. It can serve well the practicing physician who wishes to use the clinical chemistry laboratory's findings in the diagnosis and management of the patient.

Samuel Natelson, Ph.D.
Department of Environmental Medicine,
College of Veterinary Medicine,
University of Tennessee,
Knoxville, Tennessee

Preface

Clinical Chemistry: Theory, Analysis, and Correlation is designed to serve primarily as a teaching text for medical technologists and medical laboratory technicians. However, this book also contains the depth of information required in a bench reference for clinical chemists and laboratory supervisory personnel.

Clinical chemistry is a multidisciplinary field that draws upon diverse disciplines including pharmacology, toxicology, physiology, immunology, and hematology. The teaching and application of information from these areas to analytical biochemistry is complicated by the volume of knowledge required. We have provided a synthesis of the most relevant and current information from the associated disciplines and thus a complete survey of the clinical chemistry field.

The wide range of information covered in *Clinical Chemistry* includes the use of many terms unfamiliar to students. For example, the term "analyte," used to describe any substance that can be measured, may not be found in current dictionaries. Therefore definitions of such terms have been included in glossaries in the beginning of each chapter.

The format of *Clinical Chemistry* is designed to separate the large body of facts into three categories:

1. *Laboratory techniques:* the principles of analytical techniques.
2. *Pathophysiology:* The ways in which the laboratory data generated are related to disease or organ dysfunction.
3. *Methods of analysis:* An in-depth survey of the measurement of commonly analyzed biochemical substances.

The chapters within each of the three major sections are organized according to a consistent pattern. Our intent is to facilitate both the presentation of the information by instructors and the retrieval of information by students and clinicians. In addition, to facilitate teaching and learning we have grouped together chapters that explore common themes, such as endocrinology, chromatography, or data processing.

The chapters in Section One, Laboratory Techniques, are directed primarily to baccalaureate-level medical technology students. However, the depth of information provided is intended to be sufficient to meet the requirements of advanced and graduate clinical chemistry students, as well as laboratory supervisors and clinical chemists.

Each of the general information chapters in Laboratory Techniques is organized according to a logical progression of ideas.

1. The scientific principles upon which a technique is based.
2. The application of a technique to a specific class of analytes or process.
3. The limitations, such as sensitivity, specificity, and accuracy of each technique—that is, the factors that govern when a certain technique should or should not be used.

Section Two, Pathophysiology, is composed of chapters covering the interpretation of laboratory data obtained from the methods of analysis described in Section Three. To recognize erroneous analytical data, the clinical chemist must have an understanding of pathophysiological mechanisms. It has been our experience that superior medical technologists or technicians are those who have a thorough understanding of pathophysiology and can therefore understand the reason for ordering certain tests and can determine whether the tests ordered are appropriate for the analyte in question. Accordingly, the analytes are discussed in their *usual clinical contexts* by disease or organ dysfunction. For example, rather than providing a chapter on carbohydrates, we wrote the two chapters on diabetes and genetic disease to include discussions on abnormal levels of carbohydrates in the clinical context in which they are most likely to be encountered.

Within each chapter of Pathophysiology, the following organizational plan has been used, where appropriate:

1. A brief description of the anatomy of the organ.
2. A discussion of the normal biochemistry and physiology of the organ or system.

3. A discussion of pathological conditions, their clinical symptoms, and their effects on function and biochemistry.
4. Function and challenge tests that require laboratory analysis.
5. A change of analyte with disease section in which the analytes are correlated with various disease states.

We have organized the Pathophysiology chapters in this manner to guide the student to a clear comprehension of the biochemical nature of the disease state as compared to the normal, nondiseased state. Thus changes in analyte concentration can be more readily understood. The pathophysiology section therefore highlights the biochemical changes associated with each disease state.

Section Three, Methods of Analysis, is introduced by several chapters that present the following:

1. Classifications of biochemicals by type and a brief description of how biochemical class can affect the laboratory approach of measurements.
2. Enzyme analyses, including the variables to be controlled.
3. A description of isoenzymes and how they are analyzed.
4. A discussion of pharmacokinetic interpretation of data.
5. Interferences in spectrophotometric analysis, composed of a discussion of the nature of those interferences and how they can be reduced.
6. Additional notes.

After these introductory chapters are detailed methods of analysis grouped together as much as possible by common chemical class. The methods cover a wide range of analysis, including toxicology, endocrinology, and the most recent advances in analytical technique. The methods for the analytes included in Section Three are not intended to provide access to all possible methods that may be used in the clinical chemistry laboratory. We have chosen to include the methods that demonstrate most effectively the principles and techniques of analytical chemistry discussed in Section One and that compose those analyses most commonly performed in the clinical chemistry laboratory.

The format used for each analyte is as follows:

1. Appropriate cross-referencing to chapters in Section Two.
2. A concise chemical description and classification of the compound. Since *The Merck Index,* ninth edition, can rapidly provide chemical information on a wide variety of biochemicals, whenever possible we have listed *The Merck Index* number for each analyte.
3. A description of the methods, including principles of analyses and the current usage of available procedures.
4. A discussion of why one or more procedures is used as a reference method or why it was chosen as the *preferred* method. By comparing the benefits and disadvantages of each method, such as accuracy and precision, the student and clinical chemist can understand more fully the usefulness of each assay.
5. Specimen information, including which type of body fluid is appropriate, specimen stability and preservation, and important interferences.
6. A presentation of methods of analysis for the analyte. Since the mid-1970s many analytes have been most commonly measured by automated procedures. In addition, these automated procedures invariably utilize prepackaged reagents available from instrument manufacturers or commercial vendors. Therefore, rather than list a *manual* method for such analytes, we have chosen to *critique* those methods currently available. A comparison of variations of the preferred method to one another, as well as to a reference method (if available), is given with respect to precision, final reagent composition and concentration, and sample fraction (sample volume divided by total volume). This information allows the comparison of current or proposed procedures to the reference or preferred method.

For analytes that are commonly measured by nonautomated, manual procedures, a *complete* description of a working assay has been included. The preferred manual method includes the following:

1. The principle of the reaction or analysis.
2. Reagent preparation, stability, and storage.
3. The procedure in sufficient detail for ready adaptation by most laboratories.
4. Sample calculation as needed.
5. Reference ranges using this and other commonly used methods.

A concerted effort is being made in clinical chemistry toward the use of SI units (Système International d'Unités) for the reporting of results of laboratory analysis. The SI units are in either moles per unit volume (such as liter) or mass per unit volume (mass per liter) rather than the historical form of mass percent (mass per deciliter). In an attempt to begin this difficult but necessary task, all concentration units are reported as mass per liter or milliliter as well as moles per liter or, for enzymes, in accepted SI units. Because the use of deciliters is still in widespread use, it is our hope that by our expressing mass concentrations per liter readers will be easily able to convert concentrations to mass per deciliter and vice versa. Since the current literature uses SI units, the text will allow students to familiarize themselves with these units and enable them to more readily use current journals in their fields.

The amount of material available for these chapters is enormous. To present the necessary information concisely, we have used many figures and summary tables. These figures and tables demonstrate certain patterns of information, allowing greater ease in understanding. For those who desire to have a more profound understanding, numbered references or a general bibliography at the end of each chapter lists appropriate reading material.

The enterprise that produced this text was not undertaken by a single or even several persons. Indeed, the editors of the text have enlisted the aid of scores of gifted clinicians, laboratorians, and educators, and it is to these people that we owe the success of this book.

We would like to give a special note of appreciation to Drs. David C. Hohnadel, Carl C. Garber, and R. Neill Carey, and to Joseph Svirbely for their help in reviewing and revising several chapters. We would also extend our thanks to the Document Processing Area of the Department of Pathology and Laboratory Medicine, Mary Grannen and Liz Wendelmoot, who typed large portions of the manuscript. Our special thanks also go to Julie Hodde who directed the artists drawing many illustrations and to Beverly Etter, Manager of Medical Illustrations of the Biomedical Communications Department, who provided us with superb artistic support.

We also wish to thank our reviewers: Bethany Wise, M.S., M.T. (A.S.C.P.), School of Allied Medical Professions, Ohio State University; Edna Mains, B.S., M.T. (A.S.C.P.), Medical Technology Program, University of Colorado; Joseph Svirbely, M.S., Department of Environmental Health, University of Cincinnati; Anne Sullivan, M.S., M.T. (A.S.C.P.), Department of Medical Technology, University of Vermont; Sharon A. Jackson, M.Ed., M.T. (A.S.C.P.), Department of Medical Technology, University of Florida; Janet von Laufen and Barbara Smith Michael, Program in Medical Technology, The University of Texas Health Science Center at Houston; and Lester Hardegree, Jr., M.Ed., M.T. (A.S.C.P.), Armstrong State College, Savannah, Georgia. Their suggestions have been invaluable during the development of this book.

We are indebted to Don Ladig of The C.V. Mosby Company, who believed in our ability to deliver this text, and to his associates Rosa Kasper and Cindy Bendet. We would like to thank the Book Editing Department, especially Carl Masthay and Peggy Fagen. Last, we extend our thanks to Dr. Roger D. Smith, Director of the Department of Pathology and Laboratory Medicine, and Dr. Colin Macpherson, who produced the academic environment in which this text could be undertaken.

Lawrence A. Kaplan
Amadeo J. Pesce

Contents

APPENDIXES

Section one

Laboratory techniques

Chapter 1 Basic laboratory principles and calculations

balances Mechanical or electronic instruments used to measure mass accurately.
beakers Laboratory utensils used to contain liquids or solids.
buret (burette) Laboratory utensil used to deliver a wide range of volumes accurately.
centrifuge Instrument used to separate materials from solution by application of increased gravitational force to rotate or spin samples rapidly.
chemical purity Degree of purity or homogeneity as designated by various scientific agencies, such as the American Chemical Society and National Bureau of Standards.
desiccant Material used in a desiccator to absorb water from the air.
desiccator Large container used to store material in a water-free environment.
dilution Process of preparing less concentrated solutions from a solution of greater concentration.
Erlenmeyer flask Laboratory utensil used to contain liquids.
funnel Laboratory utensil used to transfer liquids or solids into a container; also used for extraction of liquids.
graduated cylinder Laboratory utensil used to measure a volume of liquid.
metric system A system of measurement of weights, distances, and volumes.
pipet (pipette) Laboratory utensil used to transfer a specific or varying volume of liquid.
Système International d'Unités (SI) An internationally accepted system of measurements.
thermometer A device, physical, electronic, or optical, that is used to measure temperature.
volumetric flask Laboratory utensil used to contain a specific volume of liquid.
water purity Three levels of purity, based upon the amount of biological and dissolved organic and inorganic material present in the water.

Table 1-1. Basic quantities and units of the SI (Système International d'Unités)

Quantity	Basic unit	Symbol
length	meter	m
mass	kilogram	kg
time	second	s
electric current	ampere	A
temperature	kelvin	K
luminous intensity	candela	cd
amount of substance	mole	mol

Part I: Basic laboratory principles

ROY L. ALEXANDER*
JOHN D. BAUER

UNITS OF MEASUREMENT

Metric system

The metric system was introduced in France at the end of the eighteenth century and legally adopted by that country in 1795. The basic units are the meter, defined as one 10-millionth of the earth's quadrant, and the kilogram, defined as the weight of one cubic decimeter of water. Other systems of weights and measures used at that time were based on human anatomical measurements and agricultural units.[1-3]

The first attempt at an international standardization of weights and measures occurred in France in 1875 at the Meter Convention, where the first General Conference of Weights and Measures (Conférence Générale des Poids et Mesures, CGPM) was held. The International Bureau of Weights and Measures was established. Seventeen nations including the United States and Great Britain took part in this conference. The centimeter, gram, second (CGS) system was adopted by the CGPM in 1881 and changed to the meter, kilogram, second (MKS) system in 1900. Later, the ampere, candela, kelvin, and mole were added. It is interesting that the MKS system was not universally accepted by all the countries that helped to establish it.

International system of units

In 1954, the CGPM adopted a more coherent† system that defined six basic units from which all other units could be derived. This system was called the International Sys-

*Discussions of units of measurements, measurement of mass, thermometry, centrifuges and heating baths, and laboratory water.

†Coherent implies that derived SI units are the product or quotient of basic units; that is, no factor is required to convert basic units to derived units. For example, concentration can be expressed as kilogram per cubic meter or mole per cubic meter.

tem of Units (Système International d'Unités) and given the official abbreviation *SI*. The CGPM introduced in 1971 a seventh basic unit, the mole, defined as the unit for *amount of substance* (Table 1-1). The United States Congress passed the Metric Conversion Act in 1975, establishing the Metric Board to plan and coordinate the adoption of the metric system in this country. The Metric Conversion Act recognized the SI but made its adoption voluntary. The SI has been introduced in 21 countries and legalized in England, France, Germany, Finland, Norway, and Sweden. In 1977, the World Health Organization (WHO) endorsed the SI and recommended its adoption throughout the world. The recommendations of the WHO and the application of SI units in medicine have been reviewed.[4,5]

SI-derived units

Two or more basic units may be combined by multiplication or division to form SI-derived units (Table 1-2). Basic and derived units may be too small or too large for convenient use, and prefixes that form decimal multiples or submultiples of the units are permitted (Table 1-3). A few non-SI units have been retained because of difficulties encountered in converting them to SI units and their widespread usage. Non-SI units relevant to clinical chemistry and their symbols are time, expressed in minutes (min), hours (h), or days (d), and volume, expressed as liter (L). The CGPM has approved *l*, *ℓ*, or *L*, however, *L* is the official abbreviation in the United States.

Table 1-2. SI-derived units used in medicine

Derived quantity	Derived unit	Symbol
area	square meter	m^2
volume	cubic meter	m^3
speed	meter per second	m/s or $m \cdot s^{-1}$
substance concentration	mole per cubic meter	mol/m^3 or $mol \cdot m^{-3}$
pressure	pascal	Pa
work energy or quantity of heat	joule	J
Celsius temperature	Celsius degree	°C
activity (radionuclide)	becquerel	Bq
power	watt	W
electric charge or quantity	coulomb	C
electric potential	volt	V
resistance	ohm	Ω
conductance	siemens	S

Table 1-3. SI prefixes

Prefix*	Factor	Symbol	Prefix*	Factor	Symbol
atto	10^{-18}	a	deka	10^1	da
femto	10^{-15}	f	hecto	10^2	h
pico	10^{-12}	p	kilo	10^3	k
nano	10^{-9}	n	mega	10^6	M
micro	10^{-6}	μ	giga	10^9	G
milli	10^{-3}	m	tera	10^{12}	T
centi	10^{-2}	c	peta	10^{15}	P
deci	10^{-1}	d	exa	10^{18}	E

*It is recommended that only one prefix be used.

SI units in the clinical laboratory

The adoption of SI units in the clinical laboratory will change the laboratory report form. Whenever the molecular weight of the analyte is known, its concentration is to be expressed in mol/L or submultiples rather than mass/L. When the molecular weight is not known, as of specific proteins or their mixtures, concentration should be reported as mass/L. The CGPM does not recommend the use of pH, and hydrogen-ion concentration should be reported as nmol/L. The elimination of the pH scale has not been accepted by most scientific disciplines, and this proposal is still undergoing study.

Enzymes

The SI unit of enzyme activity is defined as the amount of enzyme that will catalyze the transformation of 1 mole of substrate per second in an assay system. The Joint Commission on Biochemical Nomenclature of the International Union of Biochemistry (IUB) and the International Union of Pure and Applied Chemistry (IUPAC) have recommended that this unit be called the *katal*.* The CGPM has not approved this nomenclature. Whether or not this approval is forthcoming, it seems very likely that the international unit (IU)† will not be replaced by the katal in the near future. Although in many laboratories the IU has replaced traditional units of activity, one IU of activity may differ from the next depending on how the standard conditions of the assay are defined (pH, temperature, type of buffer and its concentration, substrate concentration, activator, and coenzyme concentrations, if required).

The adoption of SI units in the clinical laboratory will result in some inconvenience to the laboratory and medical staff and initially some expense to the laboratory. Physicians and laboratory personnel will have to learn new reference values. Errors might occur during this transition and they could endanger the health of the patient, however, this has not been reported to have occurred in Canada, England, or Europe where SI units have been adopted. To some extent, SI units have been used in the laboratory for many years. For example, equations for cal-

*The katal (kat) is defined as the amount of enzyme that produces a reaction rate of 1 mole per second in a method of assay.

$$10^{-9}\ kat/L\ (1\ nanokat/L) = 0.06\ IU/L$$

†In 1964 the IUB defined the international unit as the amount of enzyme that will catalyze the transformation of 1 micromole of substrate per minute under standard conditions.

culating the anion gap are based on mole concentrations of serum cations and anions.[6] Likewise, one equation for calculating the serum osmolarity makes use of mole concentrations of sodium, glucose, and urea nitrogen.[7] It is possible that the replacement of mass concentration with mole concentration will reveal other important interrelationships involving analytes in body fluids.

It does not appear that the replacement of the international unit by the katal will offer any advantages in the measurement of enzyme activity. Also, there is much opposition toward adoption of the pascal unit for measuring blood pressure or partial gas pressures in blood.[8] Conversion to SI units would involve considerable expense in the restandardization of many automated instruments, the revision of report forms, and the reprograming of laboratory computers and other data-handling equipment.

Despite these objections, there are some signs that the metric system and the SI are gaining ground in the United States. A number of scientific journals in this country require that manuscripts report data in SI units or follow traditional units with the appropriate SI unit.

A number of organizations in this country have recommended the use of SI units in reporting laboratory data and published suggestions and guidelines for implementing SI units in the clinical laboratory.[9-11]

MEASUREMENT OF MASS

Laboratory balances are essential to the operation of the clinical chemistry laboratory. An analytical balance is needed for the preparation of primary standards from the National Bureau of Standards (NBS) standard reference materials and for the gravimetric assay of fecal lipids. The gravimetric calibration of semiautomatic and automatic pipets affords one possible method for checking their precision and accuracy. In the toxicology laboratory, the analytical balance is needed to prepare primary drug standards and for weighing quantities of drugs used in recovery studies. Laboratory balances are mechanical or electronic in design. Both are available in a number of models with different capacities, readabilities, and precision (Table 1-4).

Mechanical balances

Trip balance. The trip balance consists of two pans of equal mass suspended from the ends of a beam that is supported at its center of gravity by a knife-edge fulcrum (Fig. 1-1, *A*). The material being weighed is placed in the left pan. The final weight is obtained by adjustment of the position of a rider or small weight on an extension of the beam called the *balance arm bridge*. The balance arm bridge is calibrated in 0.1 g increments up to a maximum of 200 g.

Single-pan balance. The single-pan balance is a modified trip balance with arms of unequal length. The fulcrum is located close to the weighing pan (Fig. 1-1, *B*). The right arm of the beam consists of two or three balance arm bridges that support counterbalancing weights. Trip balances are useful for weighing masses quickly when a weight to the nearest 0.1 g is satisfactory, as in the preparation of strong bases or buffers.

Table 1-4. Characterization of types of balance in relationship to their operating ranges

Type of balance	Weighing range, g	Scale-reading limit, g	Reproducibility, g
Precision balances			
Electronic	30 000	1	±0.5
	15 000	0.1	±0.05
	3 200	0.1	±0.05
	1 200	0.01	±0.005
	320	0.001	±0.0005
Mechanical	20 000	2	±1
	10 000	1	±0.5
	5 000	0.1	±0.05
	2 200	0.01	±0.01
	160	0.001	±0.001
Analytical balances			
Electronic	160	1 mg	±1 mg
	160	0.1 mg	±0.1 mg
	160	0.01 mg	±0.02 mg
Mechanical	160	0.1 mg	±0.05 mg
	160	0.01 mg	±0.01 mg
Microbalances			
Electronic	3.3	10 μg	±10 μg
	3.12	1 μg	±1 μg
	3.10	0.1 μg	±0.3 μg
Mechanical	20	0.001 mg	±0.001 mg

From Richterich, R., and Colombo, J.P.: Clinical chemistry, New York, 1981, John Wiley & Sons, Inc.

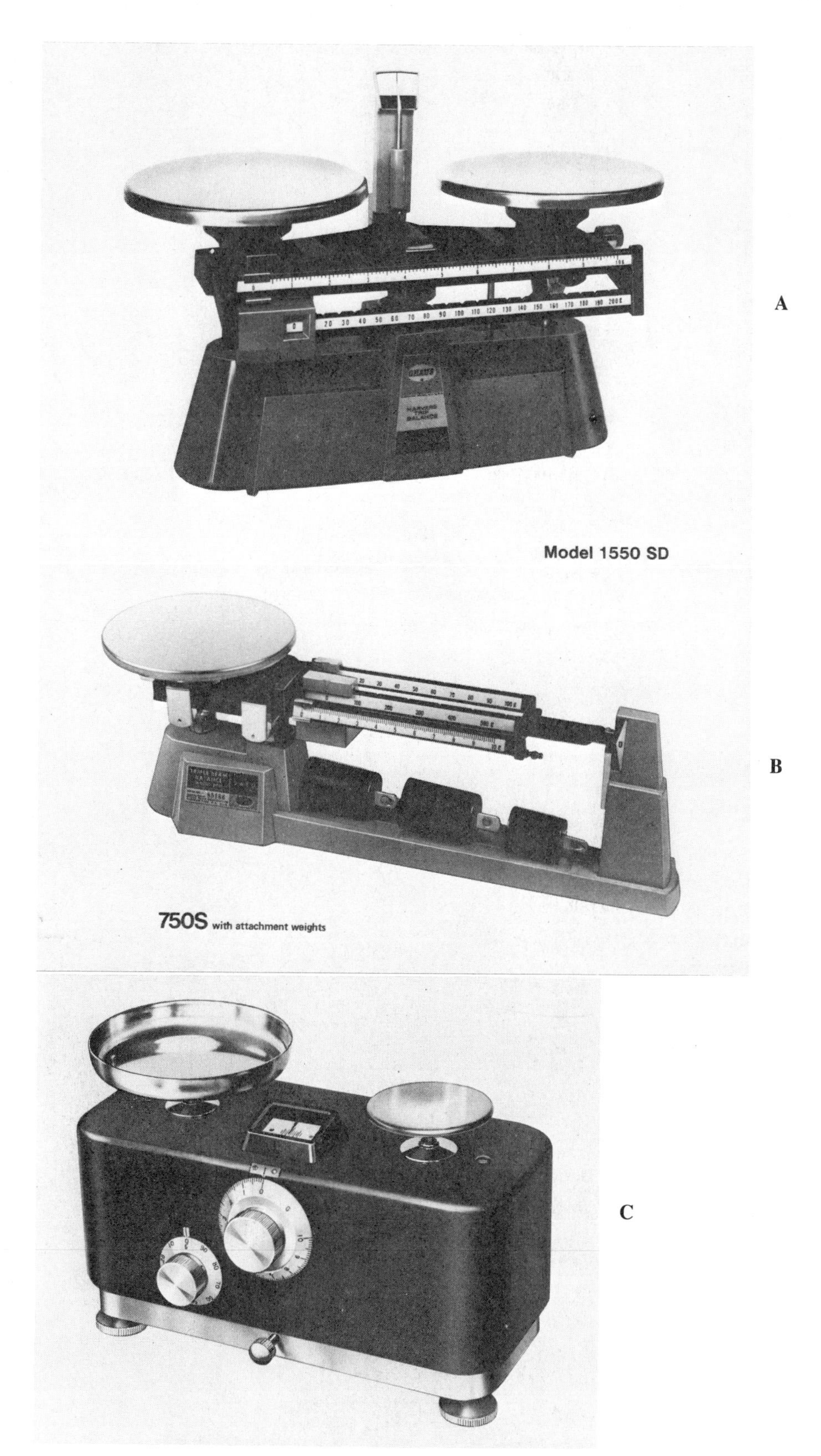

Fig. 1-1. A, Double beam trip balance. **B,** Single pan trip balance. **C,** Torsion balance. (**A** and **B,** Courtesy Ohaus Scale Corp., Florham Park, N.J.; **C** and **D,** courtesy The Torsion Balance Co., Clifton, N.J.)

Continued.

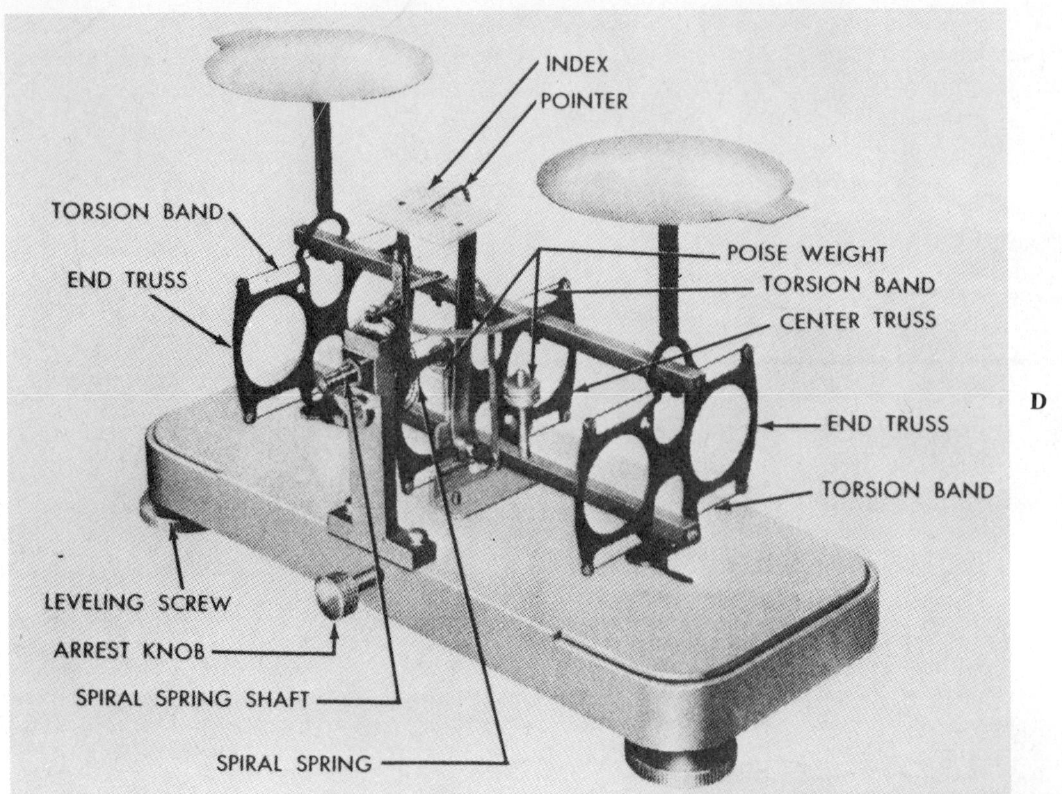

Fig. 1-1, cont'd. D, Torsion balance showing internal design.

Torsion balance. A torsion balance can usually be found in the pharmacy and has application in the chemistry laboratory when a readability is desired intermediate to that of the trip and analytical balances. Torsion balances do not have a knife-edge fulcrum. Metal bands serve to support the weight of the beam. As the balance oscillates, the movement of the pans is restricted by the torque on these bands (Fig. 1-1, *C*). Torsion balances cannot support as much weight as the knife-edge balances but can have a readability of 2 mg and provide a more accurate, rapid means of weighing small samples of material. By substitution of helical quartz fibers for metal bands, ultramicrotorsion balances can be constructed that will have great accuracy with readabilities of 0.1 to 0.001 microgram. These balances are more likely to be found in a research environment.

Analytical balance. Analytical balances generally have much greater readability, precision, and accuracy than trip or macrotorsion balances. Before 1960, most analytical balances were a two-pan design with a knife-edge fulcrum. The sensitivity of these balances varies inversely with the load. In 1946, Erhard Mettler built an improved analytical balance in Switzerland that was the forerunner of the modern substitution balance. The substitution balance is a single-pan balance with unequal arms. Suspended above the pan are a series of weights that are counterbalanced by a

Fig. 1-2. Analytical balances. **A,** Single-pan mechanical balance.

single weight located at the opposite end of the beam (Fig. 1-2, *A* and *B*). Because the load on either side of the knife-edge is always constant, the sensitivity of a substitution balance does not vary. The material to be weighed is placed inside a tared container or weighing paper on the pan and weights to the nearest 0.1 g are removed from the beam by a dial control lever. Weights less than 100 mg are read from an optical scale attached to the end of the beam opposite the pan. A light source coupled with appropriate lenses and mirrors projects the optical scale (0 to 100 mg) on a screen located in the front of the balance. Weights to the nearest 0.1 or 0.01 mg are read with a vernier. The total weight to the nearest 0.1 g is indicated by a digital register located near the screen. The substitution balance has a number of advantages over the two-pan analytical balance other than constant sensitivity. Errors attributable to unequal arm length are eliminated because both sample and weights are compared on the same arm. The operator does not handle the weights, and the time required for a weighing is reduced to about 30 seconds. Substitution balances are either air or liquid damped.

Top-loading balances. Top-loading balances are modified torsion or substitution balances (Fig. 1-2, *C* and *D*) and are much faster and easier to use than other balances. Weighings can be completed in a few seconds; however, precision is not so good as that obtained with analytical balances and ranges from ±20 to ±1 mg, depending on the design.

Electronic balances

The electronic balance was introduced in the late 1960s. There are available electronic balances that equal the precision and accuracy of all types of mechanical balances and are beginning to replace the latter in research and clin-

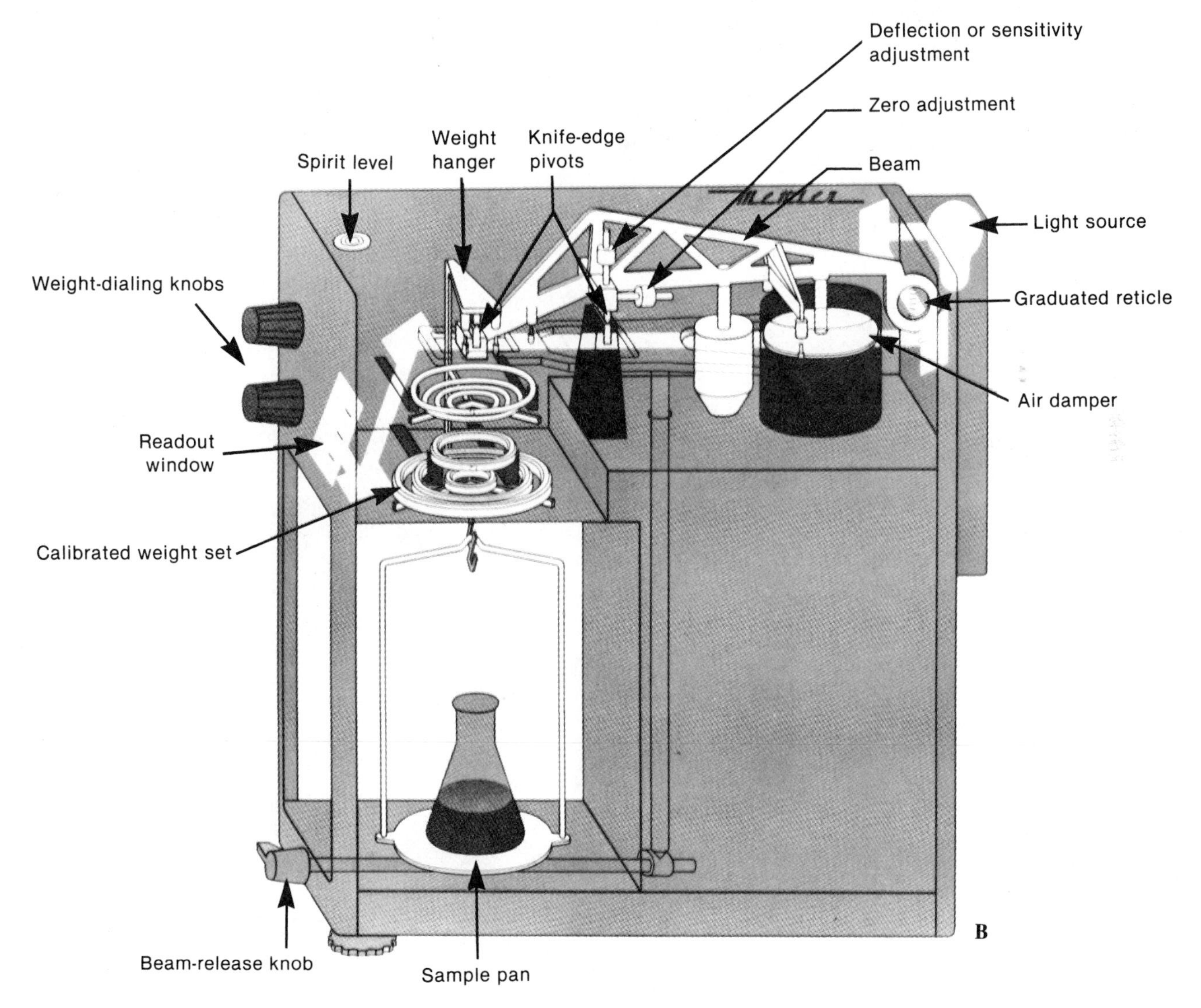

Fig. 1-2, cont'd. B, Single-pan balance showing internal design. (**A** to **E,** Courtesy The Mettler Instrument Corp., Hightstown, N.J.)

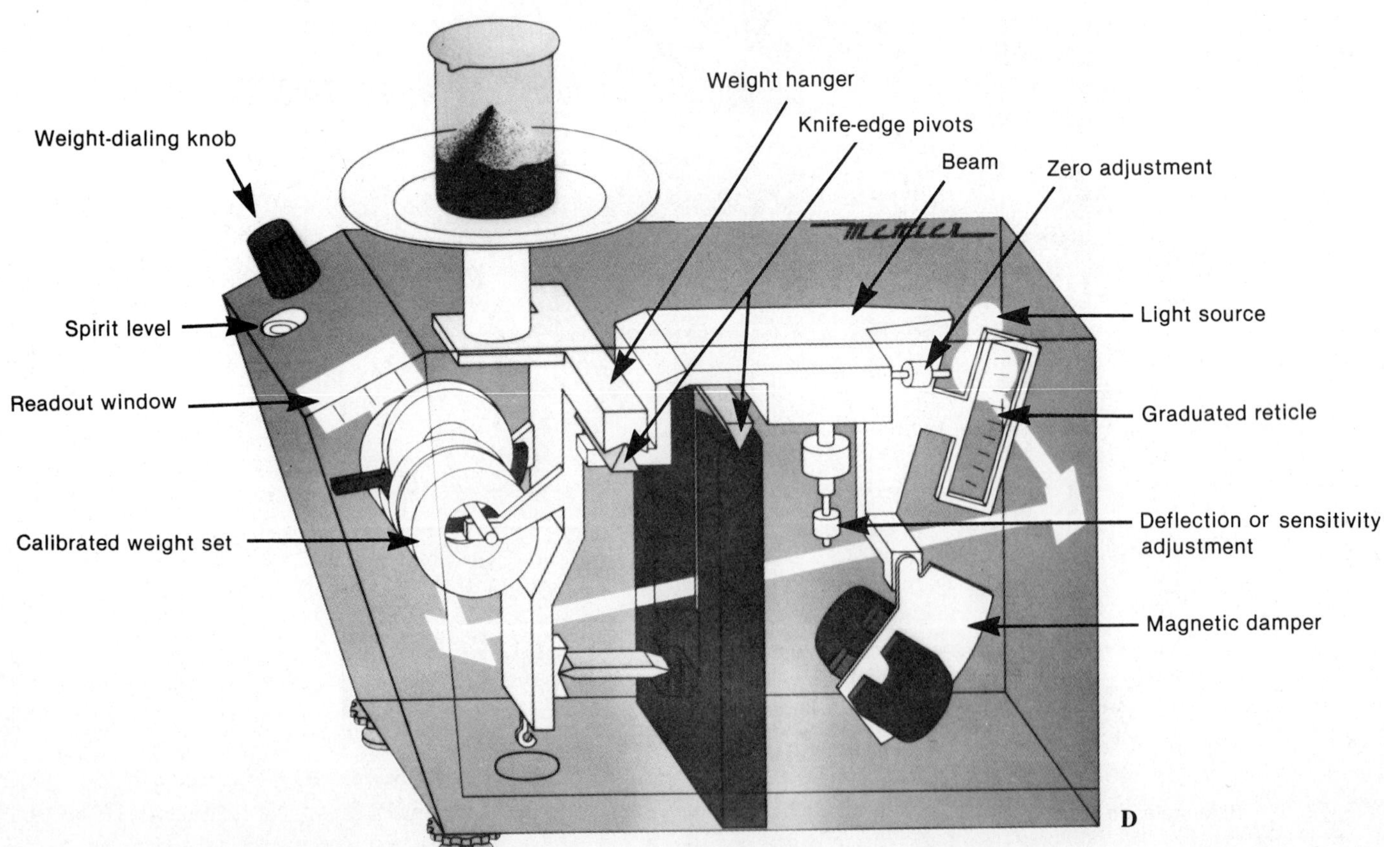

Fig. 1-2, cont'd. C, Top-loading mechanical balance. **D,** Top-loading balance showing internal design.

E

Fig. 1-2, cont'd. E, Electronic top-loading balance.

ical laboratories. The electronic balance is a single-pan balance that employs an electromagnetic force to counterbalance the load placed on the pan (Fig. 1-2, *E*). The pan is attached directly to a coil suspended in the field of a permanent magnet. A current is passed through the coil producing an electromagnetic force that keeps the pan in a constant position. When a load is placed on the pan, a photoelectric cell scanning device attached to the lever arm changes position and transmits a current to an amplifier that increases the current flow through the coil and restores the pan to its original position. This current is proportional to the weight of the load on the pan and produces a measurable voltage that is converted by a microprocessor to a numeric display or data output. These balances can be interfaced with data-processing equipment to provide calculations such as weight averaging and statistical analysis of multiple weighings. Electronic balances are either top loading or analytical in design and permit weighings to be made in 5 seconds or less.

Operation and maintenance

Balances should be located in an area of the laboratory that is free of drafts and vibration and has a relatively constant temperature. Some electronic balances have a built-in electronic vibration damper. Excessive vibration can be detected when the variation of the pointer or oscillation of numbers in the last decimal place of the digital display is observed. The analytical balance should be located in a relatively isolated part of the chemistry laboratory away from areas where corrosive chemicals are used or stored. The balance should be placed on a heavy, dedicated weighing table that can be purchased commercially. The balance pan and surrounding area should be kept clean. Chemicals should be weighed on weighing paper, in plastic boats, or in weighing bottles and never placed directly on the pan. Before the balance is used, its level should be checked by adjustment of the foot screws and centering of the bubble in the spirit level. The optical zero should be checked before each weighing.

Performance

Mechanical and electronic balances that are maintained and used according to the manufacturer's instruction manual may not require service more than once a year, depending on how frequently the balance is used. Most mechanical and electronic analytical balances have internal weights that meet the tolerances for class S weights established by the National Bureau of Standards.[12] The performance and reliability of an analytical balance (often erroneously called "calibration") can be checked if a set of class S weights is available. The College of American Pathologists requires that approved laboratories have access to such weights. There are two adjustments that can be made inside the balance, one to correct the zero point and the other to correct optical scale deflection. If the balance cannot be zeroed with the external zero control knob, the horizontal zero counterweight located near the main knife-edge (Fig. 1-2) can be rotated to bring the zero into an adjustable range. The full deflection of the optical scale can be checked by placement of a weight on the pan

Table 1-5. Individual National Bureau of Standards tolerances for class S weights

Nominal mass	Individual tolerance (mg)	Maintenance tolerance (mg)
1, 2, 3, 5, 10, 20, 30, 50 mg	±0.014	±0.014
100, 200, 300, 500 mg	±0.025	±0.05
1, 2, 3, 5 g	±0.054	±0.11
10, 20, 30 g	±0.074	±0.148
50 g	±0.12	±0.22
100 g	±0.25	±0.5

equivalent to a full-scale reading, usually 100 mg or 1000 mg. If the full-scale reading is more or less than the nominal value of the weight, the vertical-sensitivity counterweight located near the main knife-edge (Fig. 1-2) can be rotated to correct the reading. It may be necessary to adjust the optical zero between adjustments of the full-scale reading. The pan should be arrested whenever these counterweights are rotated. Once the optical scale range is adjusted, the internal balance weights can be checked when one weighs the individual class S weights. A 100 g weight should weigh 100 g ± 0.5 mg. One should check other class S weights to determine whether their apparent weight lies within the NBS maintenance tolerance limits (Table 1-5). If the class S weights weigh greater or less than the maintenance tolerance values, the balance requires service.

Some electronic top-loading and analytical balances have an internal weight built into the balance for checking performance. Information is available concerning the maintenance and performance of balances.[13,14]

LABORATORY SUPPLIES

Laboratory ware or apparatuses are used for three primary functions; storage, measurement, and confinement of reactions. The most widely used laboratory utensils are made of glass, though many of these articles are available in a variety of plastics. Generally, most commonly used glassware can be replaced by the cheaper, more durable plastic products. Glassware is most frequently chosen, however, because of its chemical stability and clarity. For certain specific applications, such as gas and high-performance liquid chromatography, glass is the preferred material because of its chemical stability. On the other hand, alkaline reagents should not be stored in glass products, since these reagents can dissolve many types of glass.

Types of glasses

There are many grades of glassware commercially available that vary in their tensile strength and heat or light resistance. Since soft glass is generally not recommended for general laboratory work, most glass utensils in the chemistry laboratory are constructed from highly resistant borosilicate glass. These are available under the brand names of Kimax (Kimble Glass Company, Vineland, N.J.) and Pyrex (Corning Glass Works, Corning, N.Y.).

Borosilicate glass has a low alkaline-earth content and is free of many contaminants such as heavy metals. This type of glass can be heated to approximately 600° C and will not soften until approximately 820° C. Table 1-6 list additional types of commonly used glass.

Types of plastics

Plastics used for laboratory utensils are constructed from polymerized organic monomers. The properties and the types of plastics available depend on the nature of the monomer and the final polymer forms used to prepare the plastic materials. The most commonly used plastics include the polyolefins (polyethylene, polypropylene), polytetrafluoroethylene (Teflon, a fluorinated hydrocarbon produced by du Pont), polystyrene, polycarbonate, and polyvinylchloride.

The polyolefins are noted for their strength and resistance to elevated temperatures. Teflon is almost totally chemically inert and is resistant to a wide range of temperatures. Polycarbonate glassware is very clear and is ideal for graduated cylinders. Teflon, polycarbonate, and some of the polyolefin plastics are autoclavable. Polyvinylchloride plastics are soft and flexible materials used frequently to construct tubing. Table 1-7 reviews some physical properties of commonly encountered plastics and their resistance to a number of chemicals.

Table 1-6. Types of commonly used glass and their properties

Glass	Properties	Purpose
Kimax or Pyrex	Relatively inert borosilicate glass	All-purpose
Vycor	96% silicate; high resistance to heat and cold shock	For use with temperatures up to 900° C; also has good optical properties
Corning Boron Free Glass	Soft, low-boron glass (<0.2%) highly resistant to alkali; poor heat resistance	For use with highly alkaline solutions
Low-actinic	Glass contains substances giving it an amber or red tint to reduce quantity of light passing through	For use with light-sensitive reagents (such as carotene, bilirubin, or vitamin A)
Corex	Aluminum-silicate glass, six- to tenfold stronger than conventional borosilicate glass; highly resistant to scratching, alkaline etching, and so on	For glassware to be used in stressful conditions

Laboratory utensils

Flasks. Some types of flasks used in the laboratory are shown in Fig. 1-3. The most commonly used flask in the clinical chemistry laboratory is the Erlenmeyer flask, which is available in sizes ranging from 25 mL to several liters. Small round-bottom flasks are often used to evaporate samples to dryness.

Beakers. Beakers are wide, straight-sided cylindrical vessels available in a wide range of volumes from 5 mL to several liters.

Table 1-7. Chemical resistance and physical properties summary of various plastics*

	Types of resins†									
Classes of substances (20° C)	LDPE	HDPE	PP, PA	PMP	FEP, TFE, ETFE	PC	PSF	PVC bottles‡	PS	Nylon
Chemical resistance										
Acids, dilute or weak	E	E	E	E	E	E	E	E	E	F
Acids,§ strong and concentrated	E	E	E	E	E	N	G	E	F	N
Alcohols, aliphatic	E	E	E	E	E	G	G	E	E	G
Aldehydes	G	G	G	G	E	F	F	N	N	F
Bases	E	E	E	E	E	N	E	E	E	F
Esters	G	G	G	G	E	N	N	N	N	E
Hydrocarbons, aliphatic	F	G	G	F	E	F	G	E	N	E
Hydrocarbons, aromatic	F	G	F	F	E	N	N	N	N	E
Hydrocarbons, halogenated	N	F	F	N	E	N	N	N	N	G
Ketones	G	G	G	F	E	N	N	N	N	E
Oxidizing agents, strong	F	F	F	F	E	N	G	G	N	N
Physical properties										
Maximum-use temperature (°C)	80	120	135 (PP) 130 (PA)	175	205 (FEP) 150 (ETFE)	135	165	70#	—	—
Brittleness temperature (°C)	−100	−100	0 (PP) −40 (PA)	20	−270 (FEP) −100 (ETFE)	−135	−100	−30	—	—
Sterilization‖										
Autoclaving	No	No	Yes	Yes	Yes	Yes¶	Yes	No#	—	—
Gas	Yes	Yes	Yes	Yes	Yes	Yes	Yes	Yes	—	—
Dry heat	No	No	No	Yes¶	Yes	No	Yes¶	No	—	—
Chemical	Yes	Yes	Yes	Yes	Yes	Yes	Yes	Yes	—	—

*Modified from 1983-1984 Nalgene Labware Catalog, Nalge Co., Division of Sybron Corp., Rochester, N.Y.

†**Resin codes:** *ETFE,* Tefzel ETFE (ethylene tetrafluoroethylene); *FEP,* Teflon FEP (fluorinated ethylene propylene); *HDPE,* high-density polyethylene; *LDPE,* low-density polyethylene; *PA,* polyallomer; *PC,* polycarbonate; *PMP,* polymethylpentene (''TPX''); *PP,* polypropylene; *PS,* polystyrene; *PSF,* polysulfone; *PVC,* polyvinyl chloride; *TFE,* Teflon TFE (tetrafluoroethylene).

Chemical resistance classification: *E,* 30 days of constant exposure cause no damage. Plastic may even tolerate it for years. *G,* Little or no damage after 30 days of constant exposure to the reagent. *F,* Some effect after 7 days of constant exposure to the reagent. Depending on the plastic, the effect may be crazing, cracking, loss of strength, or discoloration. Solvents may cause softening, swelling, and permeation losses with LDPE, HDPE, PP, PA, and PMP. The solvent effects on these five resins are normally reversible; the part will usually return to its normal condition after evaporation. *N,* Not recommended for continuous use. Immediate damage may occur. Depending on the plastic, the effect will be a more severe crazing, cracking, loss of strength, discoloration, deformation, dissolution, or permeation loss.

‡For polyvinyl chloride tubing, see the current Nalgene Labware Catalog.

§Except for oxidizing acids. For oxidizing acids, see ''Oxidizing agents, strong.''

‖Sterilization—*Autoclaving:* Clean and rinse item with distilled water before autoclaving. Certain chemicals that have no appreciable effect on resin at room temperature may cause deterioration at autoclaving temperatures unless removed with distilled water beforehand. *Gas:* Ethylene oxide. *Dry heat:* At 160° C. *Chemical:* Benzalkonium chloride, formalin, ethanol, and so on.

¶Sterilizing reduces mechanical strength. Do not use polycarbonate vessels for vacuum applications if they have been autoclaved.

#Except for the polyvinyl chloride in tubing, which will withstand temperatures to 121° C and can be autoclaved. Refer to ''The Use and Care of Plastic Labware'' in the current Nalgene Labware Catalog for detailed information on sterilization.

Interpretation of chemical resistance. This summary is a general guide only. Because so many factors can affect the chemical resistance of a given product, you should test under your own conditions. If any doubt exists about specific applications of Nalgene products, please contact Technical Service, Nalgene Labware Department, Nalge Company, Box 365, Rochester, NY 14602 or call (716)586-8800. Telex 97-8242.

Effects of Chemicals on plastics. Chemicals can affect the strength, flexibility, surface appearance, color, dimensions, or weight of plastics. The basic modes of interaction that cause these changes are (1) chemical attack on the polymer chain, resulting in reduction in physical properties, including oxidation; reaction of functional groups in or on the chain; and depolymerization; (2) physical change, including absorption of solvents, resulting in softening and swelling of the plastic; permeation of solvent through the plastic; dissolution in a solvent; and (3) stress cracking from the interaction of a ''stress-cracking agent'' with molded-in or external stresses. The reactive combination of compounds of two or more classes may cause a synergistic or undesirable chemical effect. Other factors affecting chemical resistance include temperature, pressure, and internal or external stresses (such as centrifugation), length of exposure, and concentration of the chemical. As temperature increases, resistance to attack decreases.

CAUTION. Do not store strong oxidizing agents in plastic labware except that made of Teflon FEP. Prolonged exposure causes embrittlement and failure. Although prolonged storage may not be intended at the time of filling, a forgotten container will fail in time and result in leakage of contents. Do not place plastic labware in a flame or on a hot plate.

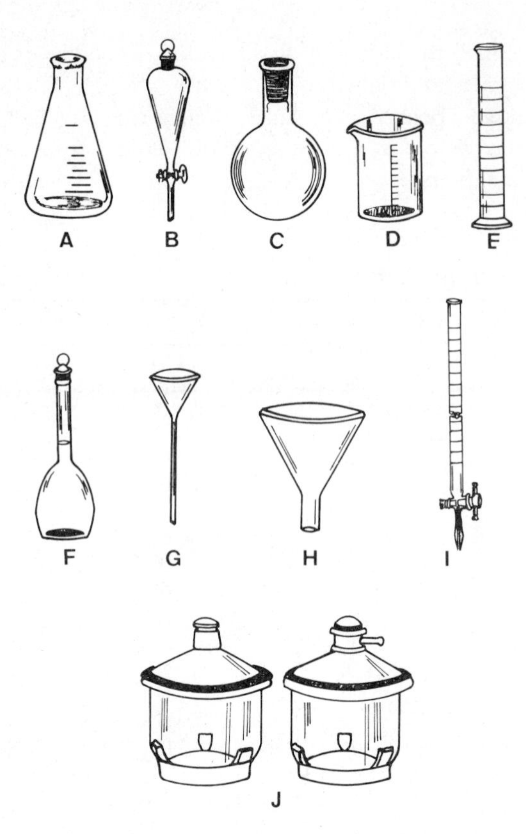

Fig. 1-3. Examples of commonly used laboratory utensils. **A,** Erlenmeyer flask. **B,** Separatory funnel. **C,** Round bottom flask. **D,** Beaker. **E,** Graduate cylinder. **F,** Volumetric flask. **G,** Long-stem funnel (filtering). **H,** Powder funnel. **I,** Buret. **J,** Desiccators.

Flasks and beakers are used for general mixing and preparation of liquid reagents. Because of their narrower mouths, Erlenmeyer flasks contain liquids better than beakers do and thus are more suited for the storage of liquids.

Graduated cylinders. Graduated cylinders are narrow, straight-sided vessels that are used to measure specific volumes (Fig. 1-3 and see below). Graduated cylinders are available in sizes ranging from 10 mL to several liters and are used to deliver (TD) the volume indicated at specific temperatures. They are graduated into subdivisions of approximately 100 portions of the total volume of the flask.

Volumetric flask (Fig. 1-3). Volumetric flasks are used to contain (TC) an exact volume when the flask is properly filled to the indicator line at a specified temperature. Volumetric flasks cannot be used to deliver this volume however.

Burets. Burets are long, graduated tubes with a stopcock at one end (Fig. 1-3). These devices are used to accurately deliver known volumes of liquid into a container. By measuring from graduated line to graduated line, one can deliver fractional volumes (that is, less than 1 mL) of liquid with a high degree of accuracy.

Funnels. Most commonly used funnels are employed to transfer liquids or solids into containers. Filtering funnels (Fig. 1-3) are usually 58 or 60 degree–angle funnels with either short or long, thin stems. These funnels are used with filter paper to remove particles from solution. Many funnels have ridges to increase the surface area available for filtering purposes.

Powder funnels (Fig. 1-3) have wide-mouthed stems that allow solids to easily pass through, and thus these funnels are used for transferring solids into a container or flask.

Separatory funnels are constructed with a ground-glass stoppered opening at one end and a stopcock opening at the other end. These devices are used for manual liquid-liquid extractions of relatively large volumes of samples. The lower phase is separated from the upper phase through the stopcock.

Desiccators. Desiccators are used to dry, or keep dry, solid or liquid materials. Desiccators usually have an area at the bottom where a desiccant, or water absorbing material, is placed (Fig. 1-3). A shelf is placed on top of the desiccant upon which the material to be stored can be set. The top of the desiccator has a wide, flat, ground glass lip that fits snugly against an opposing lip of the bottom part of the desiccator. Stopcock grease is usually placed on the surface of the lips to provide an airtight seal. Many desiccators also have a stopcock outlet on the upper portion to allow the desiccator to be evacuated. A laboratory will often have at least two, sometimes three, desiccators so that one may store material dry at ambient, 4° and −20° C temperatures.

Some materials used as desiccants are listed in Table 1-8.

Pipets. Most pipets are constructed from glass though disposable, plastic, serological pipets are available. Pipets are discussed in more detail below.

Glassware washing

Clean glassware is essential for any analytical procedure. Described below is an example of a type of washing procedure for various laboratory glassware and quality control check for this important function.

1. Pipets
   a. Pipets are soaked first in a 10% sodium hypochlorite (Clorox or Dazzle) solution. Immediately after use, pipets are placed tip down in this solution in a plastic jar (place cotton on bottom to prevent tips from breaking).
   b. Pipets are then placed into a stainless steel or plastic basket and washed with hot water and detergent (Alcotab) in the pipet washer. After all the detergent has been rinsed away, remove pipet basket and allow to drain 15 minutes.
   c. Place pipets in acid solution (No-Chromix, available from Fisher Scientific, in H_2SO_4) for 30

Table 1-8. Some common drying agents (desiccants)

Desiccant	Properties	Uses
Anhydrous $CaCl_2$	High capacity, slow acting, alkaline in nature, works well below 30° C	Most conditions, very inexpensive
Anhydrous $MgSO_4$	Neutral, rapid action	Most conditions, inexpensive
Anhydrous Na_2SO_4	Neutral, high capacity, works only below 32° C, slow action	Can remove large volumes of water
Anhydrous $CaSO_4$	Extremely rapid in action, chemically inert, limited capacity to absorb water (6% to 10% weight in water)	More expensive than $MgSO_4$ and Na_2SO_4; sold commercially as Drierite; can be easily regenerated by heating at 230° to 240° C for 3 hours
Al_2O_3 (activated alumina)	Can absorb 15% to 20% of its weight in water	Can be repeatedly reactivated by heating at 175° C for 7 hours

minutes. Hang pipet basket on hook for 15 minutes to drain the acid.

d. Rinse pipets thoroughly with tap water in the pipet rinser for 30 minutes and with three changes of deionized (class I or II) water.
e. Pipets are dried in the oven at 110° C for 30 minutes.

2. Other glassware
   a. Glassware soaking. After use, glassware is rinsed by the technologist before being placed in the soaking basin of tap water to which 8 ounces (approximately 240 mL) of Clorox has been added.
   b. Automatic glassware washer. As much as possible, miscellaneous glassware is washed in an automatic glassware washer (such as the Heinicke washer, Heinicke Instruments Co., Hollywood, Fla.). Glassware is loaded with beakers, flasks, tubes, and so on, with their openings down on the appropriate racks or in baskets. For small, light items, a cover basket is placed on top. For the first load of the day, the washer is filled and 3 tablespoons of detergent is added. Wash and rinse times are as follows:

Wash	4 minutes
Rinse	3 minutes
Deionized (class I or II) rinse	3 minutes

   After removal, glassware is dried in the oven 30 minutes. Plastic items are never placed in the oven. At the end of the day, the washer is drained. Once a week, the washer is run through a cycle with Heinicke mineral solvents to prevent lime deposits from forming.
   c. Manual glassware washing. Oddly shaped, very large or small, or plastic items may require handwashing. A solution of 2 tablespoons of detergent per basin of tap water is used with a brush. Glassware is then rinsed with tap water thoroughly, followed by five rinses with deionized water, and may then be dried in the oven for 30 minutes. Plastic items are never placed in the oven.
3. Glassware check. Daily, as items are removed from the oven:
   a. Select at random a dried flask, beaker, and so on before it is put in the glass cabinet. Examine item for water spots.
   b. If water spots are present, *all* glassware washed that day must be checked and rewashed, rinsed, and dried.
   c. Check off on quality control sheets as OK, or S (for spotted).
4. Glassware in the laboratory is checked weekly for:
   a. Appearance of water spots, which indicate inadequate rinsing.
   b. Draining of deionized water from glassware as a continuous film (clean) rather than in little drops (not clean).
   c. Detergent removal with a dilute (20 mg/L) aqueous solution of sodium sulfobromophthalein (Bromsulphalein, BSP) dye. The solution is used to rinse a beaker, and the formation of any pink color indicates residual detergent. If pink color appears, rewash last load *and* contact supervisor.
   d. Glassware that is damaged or chipped or has unreadable markings, which is then discarded.

Chemicals

Chemicals are obtainable in many degrees of purity. In addition, many are analyzed so that the type or amount of impurities are known. The highest grade or most pure chemicals are obtainable from the National Bureau of Standards. However, only very few such compounds are available for the clinical chemistry laboratory. These are termed "standard, clinical type."

Additional standards of purity for certain chemicals have been specified by the International Union for Pure and Applied Chemistry (IUPAC). These include atomic-weight standards (grade A); ultimate standards (grade B); primary standards, which are commercially available and have less than 0.002% impurities (grade C); working standards, which are commercially available and have less than 0.05% impurities (grade D); and secondary substance (grade E), which are defined or standardized using a primary standard (grade C) as the reference material.

For ordinary analytical work, chemicals obtainable from various vendors come in many grades. The usual material suitable for analytical work is termed "reagent grade." These meet specifications defined by the American Chemical Society (ACS). These specifications establish the maximum amounts of various types of impurities that can be present in each chemical. Some manufacturers will sell a certified or very pure material when specifications have not been set by the ACS. Usually these chemicals have the impurities listed and are analyzed by lot; that is, each batch of material has been checked for the most interfering types of impurities. For some chemicals, particularly pharmaceuticals, the specifications of purity are set in *The National Formulary, The United States Pharmacopeia,* and *The Food Chemical Codex*.

Less pure chemicals and reagents are available and are purchased because of their lower cost. Grading of these types of chemicals include terms such as "practical grade" and that of the lowest quality "technical grade." These should not be used in analytical work.

For many analyses, particularly those involving spectroscopy and chromatography, it is necessary to purchase chemicals that are more pure than those certified as reagent grade. Such chemicals will specify the maximum amount of impurity or interference. For example, a spectral grade of *n*-butanol will have an absorbance guaranteed to be less than 0.05 at 260 nm when purchased. Compounds with desired purity for ultraviolet and infrared spectroscopy work must be purchased with these specifications. Also available are chromatographic solvents that exceed reagent-grade requirements. Two types of purity analyses are done: one to ensure minimum spectral or detector interference and the other to ensure minimal residual contamination after extraction and evaporation of the solvents in the procedure. For example, pesticide-grade solvents for use in the gas chromatographic analysis of chlorinated organic compounds (including tranquilizers) must have less than 1 part in 100 billion of chlorinated organic impurities. Similar types of high-purity chemicals exist for high-performance liquid chromatography. Most often these chemicals are solvents purified by distillation in glass apparatuses.

Gases, particularly those used for gas chromatography and atomic absorption analysis, must also be of the highest purity. Helium purity must be 99.9999% for use with the gas chromatograph mass spectrometer. Thus careful attention must be paid to the specifications of gases.

There are a number of chemicals that are teratogenic and carcinogenic agents. These are specified to have these properties in the most modern catalogs. They should also be listed in the laboratory safety manual.

VOLUMETRIC MEASUREMENTS

Accurate volumetric measurements and transfers are central to the quantitative analysis of the clinical chemistry laboratory. Such measurements are made by manual pipets including graduated and volumetric ones, micropipets, dispensing devices, automated pipets, syringes, burets, volumetric flasks, and graduated cylinders. The reliability of these devices can vary considerably, and not all of these devices are suitable for the most accurate work. The maximum allowable error or tolerance limits for these devices are set by the National Bureau of Standards. Class A devices are more accurate, whereas class B ones are less accurate, though they may be as precise. Originally, the class A glassware was defined as NBS certified. Currently, volumetric glassware that meets federal specifications or equivalent is termed "class A." The College of American Pathologists (CAP) inspection requirements specify that volumetric pipets be of certified accuracy, that is, class A. If not, they must be checked by a gravimetric, colorimetric, or some other verification procedure. In addition, automatic pipets and diluting devices must be checked for accuracy and reproducibility. Volumetric glassware used to make solutions must be of the class A category. Thus for most laboratories the preparation of solutions proceeds with the use of class A glassware. One should remember that accurate delivery of volumes can be accomplished *only* with clean glassware. If liquid does not drain cleanly, that is, if beads of liquid cling to the sides of a pipet, volumetric flask, or buret, one must assume an inaccurate volume. The glassware should be set aside and replaced by a clean one.

Manual pipets

There are three types of manual pipets: to contain (TC), to deliver/blow out (TD), and to deliver (TD). *To-contain,* or *rinse-out, pipets* must be refilled or rinsed out with the appropriate solvent after the initial liquid has been drained from the pipet. These pipets contain or hold an exact amount of liquid, which must be completely transferred for accurate measurement. The more common names given to these pipets include the Micro-Folin, dual purpose, Sahli hemoglobin, Kirk Micro, White-Black Lambda, transfer micro, measuring micro, and Lang-Levy. None of these to-contain pipets meets class A specifications.

To-deliver/blow-out pipets are filled and allowed to drain, and the fluid remaining in the tip is blown out. These pipets thus transfer or deliver an exact amount of liquid and are not rinsed out. The common names given to these pipets are Ostwald-Folin, serological, serological long tip, and serological large-tip opening. They can be readily identified by the two frosted bands near the mouthpiece of the pipet (Fig. 1-5). Serological pipets are long glass tubes of uniform diameter. They have volume graduations that extend to the delivery tip of the pipet. Thus the last blown-out drop of liquid is also included in the delivery volume. These pipets come in designs with large-tip openings for delivery of viscous fluids and long tapered tips.

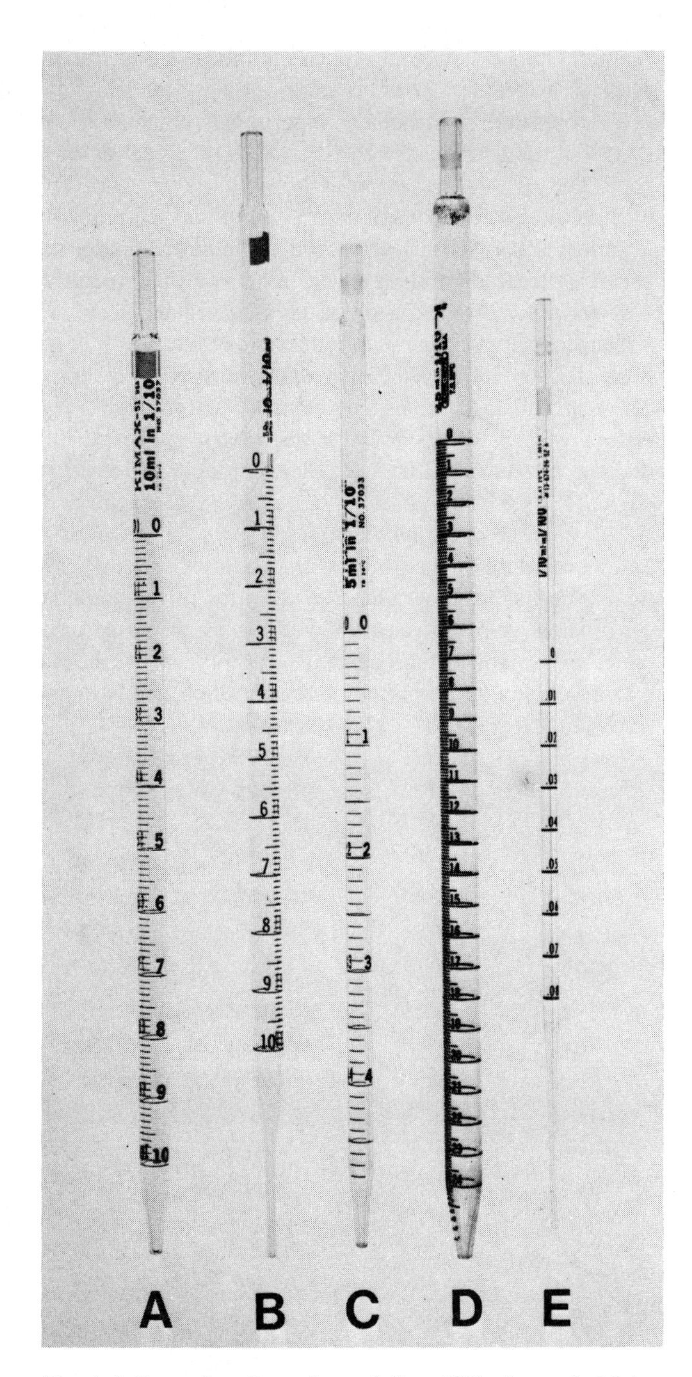

Fig. 1-4. Examples of transfer to-deliver (TD) pipets. **A,** Mohr. **B,** Mohr long tip. **C,** Serological. **D,** Serological large opening. **E,** Serological long tip.

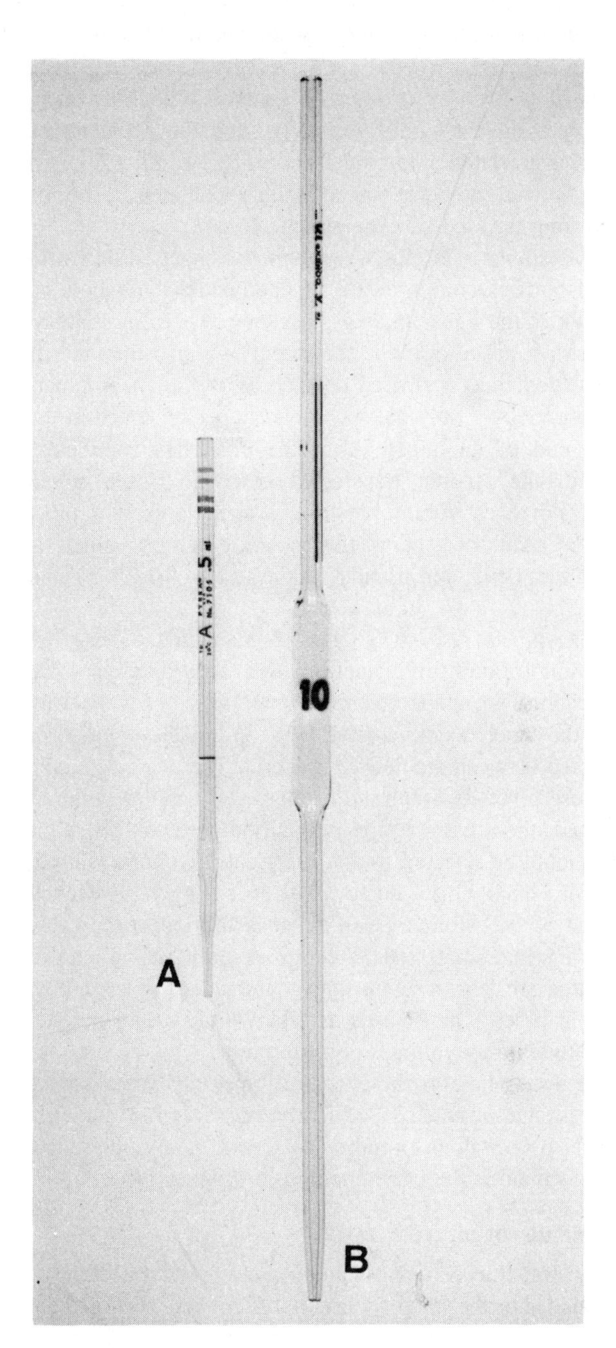

Fig. 1-5. Examples of to-deliver (TD) pipets. **A,** Ostwald-Folin. **B,** Class A volumetric.

Table 1-9. Accuracies of manual pipets

Type of pipet	1.0 mL	5.0 mL	10.0 mL	25.0 mL
NBS standard	—	0.01	0.02	0.025
Class A volumetric	0.006	0.01	0.02	0.03
Mohr	0.01	0.02	0.03	0.10
Mohr long tip	0.02	0.04	0.06	—
Serological	0.01	0.02	0.03	0.10
Serological large opening	0.05	0.10	0.10	0.20
Serological long tip	0.02	0.04	0.06	—

To-deliver pipets are filled and allowed to drain by gravity. To ensure complete draining of the liquid, one must set the flow rates according to maximum NBS flow rates. The pipet must be held vertically, and the tip must be placed against the side of the accepting vessel. Common names given to these pipets include volumetric, Mohr, Mohr long tip, and bacteriological.

An example of a class A volumetric pipet is shown in Fig. 1-5. It is composed of an open-ended bulb holding the bulk of the liquid, a long glass tube at one end that has the mark (line) to describe the extent to which the pipet is to be filled, and a tapered delivery portion. These pipets hold and deliver only the specific volumes indicated at the upper end of the pipet. After the pipet has completely drained, the amount transferred is equal to the stated value. For each volume desired, a specific size of pipet must be used. In general, the accuracy of these pipets is better than 0.6% and usually on the order of 0.2% (Table 1-9).

Mohr pipets also meet class A standards. These are glass tubes of uniform diameter with a tapered delivery tip. Graduations are made at uniform intervals but well away from the tapered delivery tip (Fig. 1-4). The solution is delivered between the desired markers. The listed accuracy for these pipets is for the *full* volume. If smaller volumes are used, the accuracy proportionally decreases. These pipets should be selected so that the greatest volume is used; therefore maximum accuracy will be achieved. A second version of the Mohr type of pipet is the long tip, which easily fits into small vials. According to the manufacturer, long tips are less accurate than the standard tapered tips. A Mohr pipet is never used as a blow-out type of pipette, but delivers only volumes point to point.

The accuracies for these several types of manual pipets are presented in Table 1-9. Clearly, the class A volumetric transfer pipette is much more accurate than any other. The least accurate is the large-opening serological pipet.

Techniques of pipetting

The first rule of using manual pipets is that nothing is aspirated into the pipet by mouth. A rubber bulb or other device is used to fill the pipet above the mark. The pipet is grasped by the thumb and middle finger. The index finger, placed over the upper opening, is used to control the flow of liquid (Fig. 1-6). With the pipet filled above the mark, the tip of the pipet is wiped free of adhering fluid with a tissue, such as Kimwipe. The liquid is allowed to drain to the mark, and the pipet is transferred to the receiving vessel where the liquid is allowed to drain with the pipet held in a vertical position, tip against the side of the vessel.

Micropipets

Micropipets contain or deliver volumes ranging from 1 to 500 microliters (μL). Another term often used is "lambda pipet" since the term "lambda" was usually given to a volume of 1 microliter.

The original, most popular type of micropipet was the Lang-Levy pipet depicted in Fig. 1-7. The pipet has a tapered delivery tip and a constriction at the other end. The pipet is filled to the constriction. These pipets are probably the most accurate ever developed for microvolumes, but because of the difficulties in cleaning and their fragility, they are not commonly used in the clinical laboratory.

An inexpensive, disposable micropipet consists of capillary tubing with a line demarking a specific volume. These are filled to the line by capillary action. The liquid is delivered by positive pressure (blown out) by a medicine dropper or equivalent device. These are not very accurate devices.

The most common type of micropipet used in the laboratories was introduced by the Eppendorf Company, and this name has become almost generic for a large number of to-deliver types of pipettes working by the same principle. These are piston-operated devices. Disposable and exchangeable tips are placed on the barrel of the pipet, and

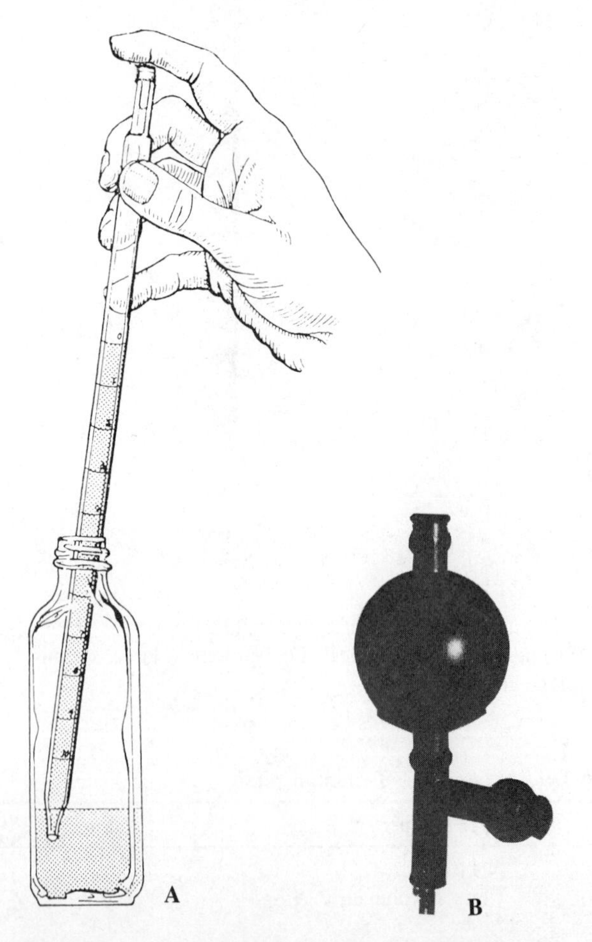

Fig. 1-6. A, Proper pipetting technique as described in text. **B,** Example of rubber pipetting bulb used to aspirate sample into pipet.

these receive and dispense liquid (Fig. 1-8, *A*). The piston is depressed to a stop position on the device, the tip is placed in the liquid, and then slowly the piston is allowed to rise back to the original position (Fig. 1-8, *C*). This fills the tip with the desired amount of liquid. The tips are not usually wiped because the plastic surface is considered nonwettable, but are drawn along the wall of the vessel from which the fluid is to be taken so that any adhering liquid is removed. The tip is then placed on the wall of the receiving vessel and the piston depressed, first to the stop position, allowing the liquid to drain, and then to a second stop position beyond the first one. This ensures the full dispensing of the fluid similar to that of the blow-out type of pipet. The manufacturer claims to have 99% sample recovery with this device, a standard deviation (for volumes of 10 to 500 μL) of less than 1%, and a reproducibility for multiple pipetting of 0.6% to 0.3%. For volumes under 10 μL the errors become significantly greater.

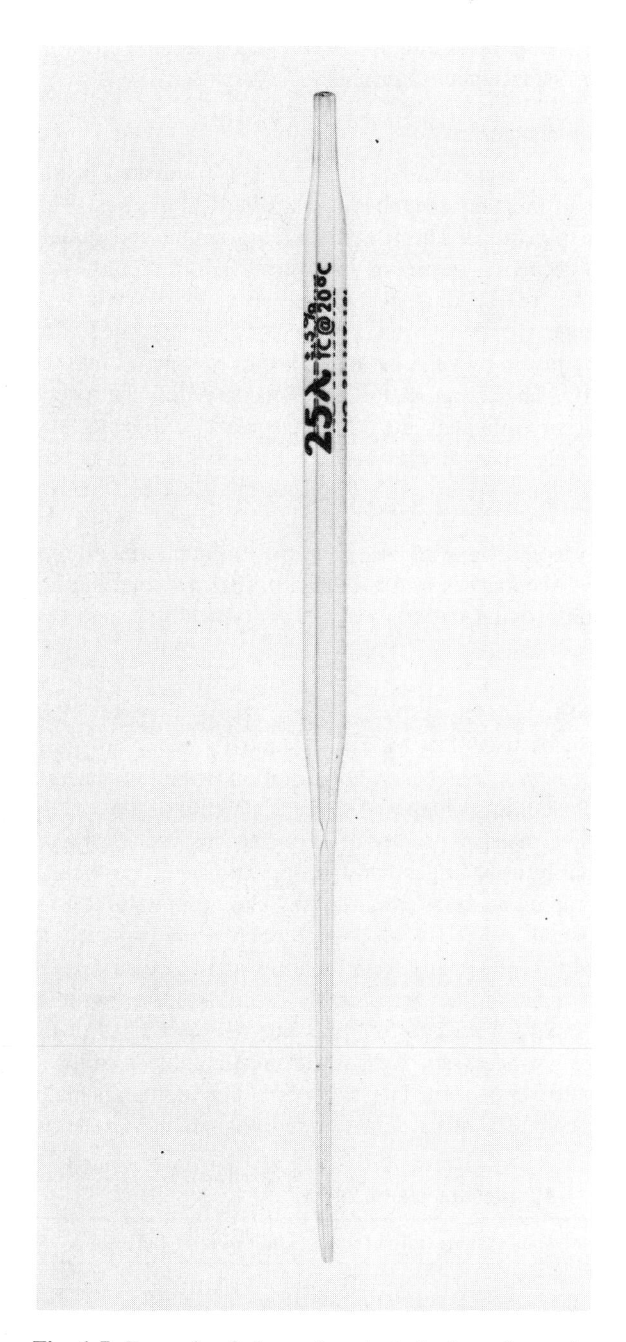

Fig. 1-7. Example of glass micropipet, the Lang-Levy pipet. This pipet is most often used as a to-contain (TC) pipet.

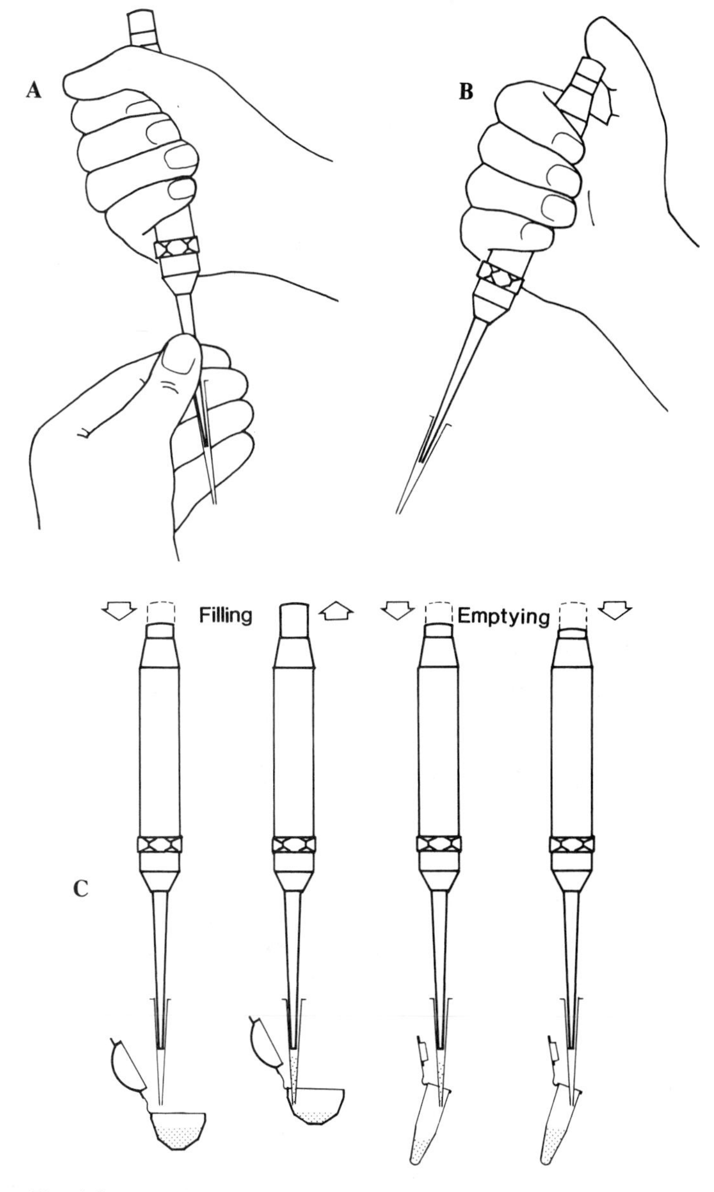

Fig. 1-8. Steps in using Eppindorf type of micropipet. **A,** Attaching proper tip size for range of pipet volume and twisting tip as it is pushed onto pipet to give an airtight, continuous seal. **B,** Holding pipet before use. **C,** Detailed instructions for filling and emptying pipet tip. Follow manufacturer's complete instructions for care and use of micropipets.

For these devices the pipet tips are disposable so that cleaning is not necessary. Various types of such tips are available; most are made of polypropylene, since this surface is considered nonwettable. For reagents that will react with plastics, micropipetters such as the SMI micropipetter system are available with glass capillaries.

Devices with adjustable volumes, working by the same principles as the Eppendorf style of pipet, are also available. These claim to deliver set volumes to 1%; however, in our opinion one should calibrate these devices at several volumes before using them.

Automated dispensers

Automated dispensers are frequently used by the laboratory to add repeatedly a specific volume of reagent or diluent to a solution. Many types of dispensers are available. The Oxford Repipette is a typical example (Fig. 1-9). A long tube leading from the dispenser is placed in the reagent bottle. The dispenser is composed of a plunger, valve system, and dispensing tip. Typically, once the dispensing device is primed with liquid, pressing on the plunger dispenses a selected amount of liquid. When the plunger is returned to the original position, usually by spring action, the dispenser chamber is refilled. The manufacturers claim accuracy of 1% and a reproducibility of 0.1% for these devices at full deflection of the plunger.

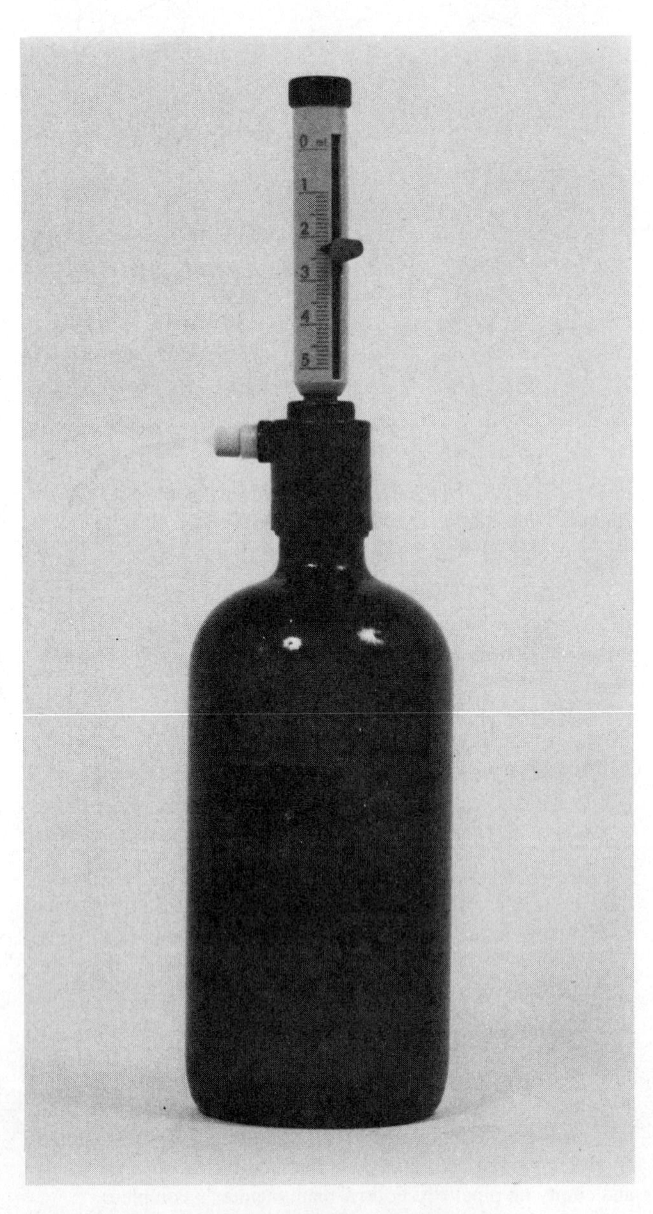

Fig. 1-9. Example of manual dispenser or repipettor.

Such devices often require constant cleaning to prevent material from disturbing the piston action and not allowing proper displacement to occur.

Diluter-dispensers

The diluter-dispenser is often used in automated instruments to prepare a number of samples for analysis. This device pipettes a selected aliquot of sample and diluent into a receiving vessel or instrument. Most of these devices are of the dual piston type (Fig. 1-10). When an electronic signal is given, two pistons are activated. The smaller piston by valvular action aspirates samples into the pipet tip. The second piston simultaneously fills the second syringe or equivalent device with solvent or diluent. At a second signal the valves are changed, and diluent is sent through the pipet tip. This displaces the sample and rinses the pipet tip. The automatic diluter is now prepared for the next sample. Many ratios of sample to diluent may be selected. Accuracy is to 1% of full syringe volume, and repeatability is on the order of 0.05% of a full-syringe volume.

Syringes

Syringes used for accurate volumetric work are glass cylinders with a precision-bored hole in which is placed a very close fitting plunger. The dispensing tip of the syringe is a metal needle, usually of very fine diameter. Syringes are mainly used in gas and high-performance liquid chromatographic analyses because they can accurately transfer very small volumes of solvent. Such solvents, such as methanol, are usually very volatile and are usually not readily quantitatively transferable by glass pipets. Syringes range in size from 1 to 500 μL. Needles on the end of the syringe are necessary for gas chromatography in order to pierce the septa of the injection ports. Manufacturers claim accuracy of 1% of the total syringe volume and a repeata-

Table 1-10. Accuracies of burets

Total volume (mL)	Limit of error (mL)
5	0.01
10	0.02
50	0.05
100	0.10

bility of 1% of the dispensed volume for those of sizes greater than 5 μL. For those less than 5 μL in size, 2% accuracy is the best that is achievable. Many syringes are not calibrated because internal standards are used in gas and high-performance liquid chromatographic analyses to quantify the samples and correct for transfer errors.

Burets

A buret is one of the most accurate devices for the dispensing of volume. It consists of a long glass cylinder with graduations corresponding to accurate volumes, a stopcock or other device to stop the flow of liquid, and a tapered dispensing tip. Burets meet the class A requirements for volumetric accuracy. A buret is filled with liquid, and an amount of fluid is allowed to flow into a receiving device for waste material. The tip is wiped, and the liquid is allowed to flow into the waste-receiving device with the buret tip on the side of the device until the meniscus reaches a desired graduation. The reading on the buret is recorded. By opening the stopcock, one can allow a specific amount of liquid to flow from the buret into the titrating or receiving vessel. The accuracy requirements of burets to meet NBS standards are presented in Table 1-10.

It is clear from this table that the error of this device is on the order of 0.2% to 0.1%.

Volumetric flasks

Volumetric flasks are essential for the accurate preparation of solutions. Class A specifications are required for such usage. The typical volumetric flask consists of a large bulbous lower portion with a flat bottom and a long slender neck. A line or an equivalent mark on the neck describes the position at which the meniscus must be located to achieve the stated volume. The flask is calibrated to contain, *not* deliver, a given volume. Thus they cannot be used as transfer devices. The top of the volumetric flask is capped by a tight-fitting, ground-glass stopper. This allows the flask to be inverted without loss of liquid. Under no circumstances should the flask be heated because this may distort its shape and volume.

To dissolve a solute in a given amount of solvent, several procedures are appropriate. In any event, the solute must be transferred into the volumetric flask and dissolved in a small volume of solvent in the flask. The solute must be completely dissolved before one brings it to final volume. The reason is that many solutes have appreciable volume changes when placed in solution. Once dissolved, small portions of solvent are added with swirling until the meniscus is reached. The volumetric flask is then stoppered and inverted with swirling motions 15 times for 4 to 5 minutes. This should result in uniform mixing.

Volumetric flasks are the most accurate of all devices described. Table 1-11 describes the accuracy of class A volumetric flasks.

One should keep in mind that these figures are accurate only at the temperature specified on the flask. This temperature and other NBS specifications are usually imprinted directly on the glassware (Fig. 1-11).

Graduated cylinders

As their name implies graduated cylinders have marks that allow for filling to specified volumes. The cylinders

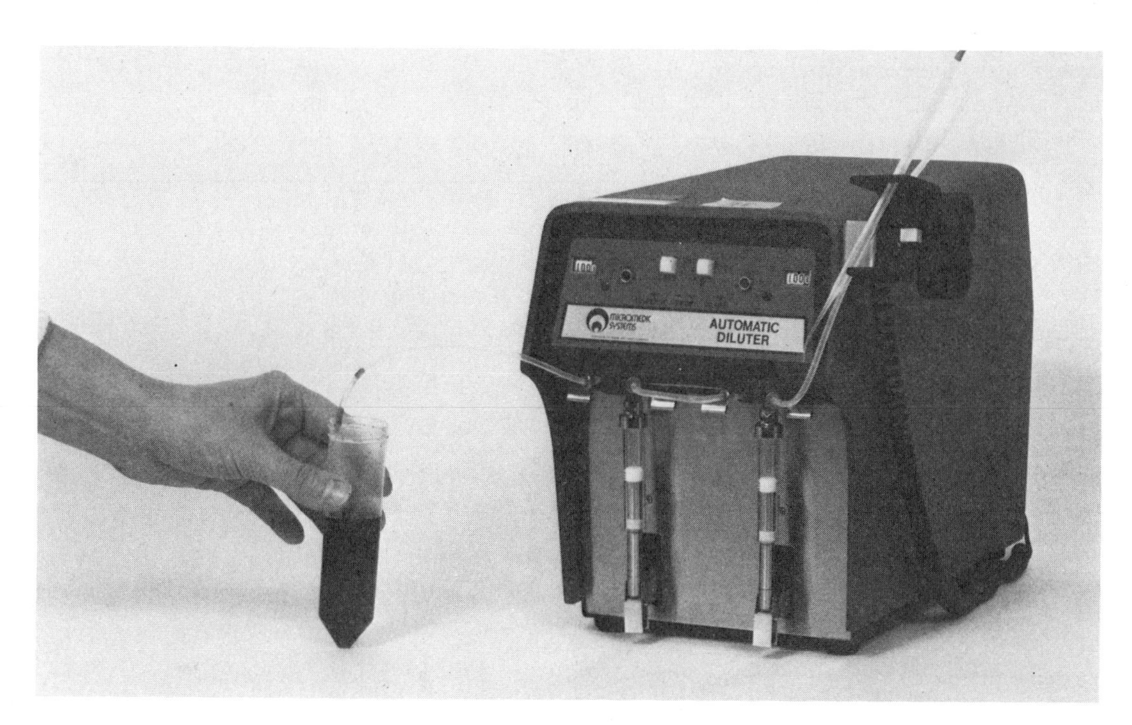

Fig. 1-10. Example of dual-syringe type of diluter-dispenser.

are calibrated to deliver the amount of fluid described at the marking. They are most often used to deliver fractional volumes of liquid. One can purchase cylinders that meet class A specifications. These specifications and some of the properties are described in Table 1-12.

Quality control procedure for micropipets, dispensers, and diluters

Micropipets. The following protocol establishes a spectrophotometric method for the assessment of accuracy and precision of manual micropipets used in the clinical chemistry laboratory. The method employs absorbance measurements of diluted, stock green food coloring solutions by a spectrophotometer and the mathematical calculation of delivered volumes for each pipet.

Method. Selected manual micropipets with disposable tips are used to deliver measured aliquots of concentrated green dye to class A volumetric flasks. Aliquots are diluted to the mark with doubly distilled water. Most dilutions will employ 10 mL volumetric flasks (refer to Table 1-13 for the appropriate flask size to be used with each pipet). Whenever possible, the same disposable tip should be used with each pipet.

The spectrophotometer is set to wavelength 620 nm at 25° C and zeroed with use of doubly distilled water. Absorbances are measured for each pipet dilution according to standard operating procedure. This yields the value $A_{dilution}$.

Calculations. The delivered volume for each micropipet is calculated from the measured absorbance of each dilution using a conversion factor (CF) based upon the mean absorbance value ($A_{1:50\ standard}$) of a manual 1/50 dilution of the stock green food coloring (pipet 1 mL of green dye into a 50 mL volumetric flask; dilute with water up to mark). The conversion factor and pipet volume are calculated according to the following relations:

$$V_{pipet} = A_{dilution} \times CF \qquad \textit{Eq. 1-1}$$

Table 1-11. Accuracies of volumetric flasks

Capacity (mL)	Limit of error (mL)	Percent error
25	0.03	0.1
50	0.05	0.1
100	0.08	0.08
250	0.11	0.04
500	0.15	0.03
1000	0.30	0.03

Table 1-12. Specifications and accuracies of graduated cylinders

Volume (mL)	Subdivision (mL)	Tolerance (mL)	Percent error
10	0.1	0.1	1
25	0.2	0.18	0.9
50	1	0.26	0.5
100	1	0.40	0.4
250	2	0.8	0.4
500	5	1.3	0.26

Table 1-13. Volumetric dilutions for check of micropipet accuracy

Eppendorf pipet volume (μL)	Volumetric flask (mL)	Dilution
4	10	2500
5	10	2000
10	10	1000
20	10	500
25	10	400
40	10	250
50	10	200
75	10	133.3
100	100	1000
250	100	400
500	100	200

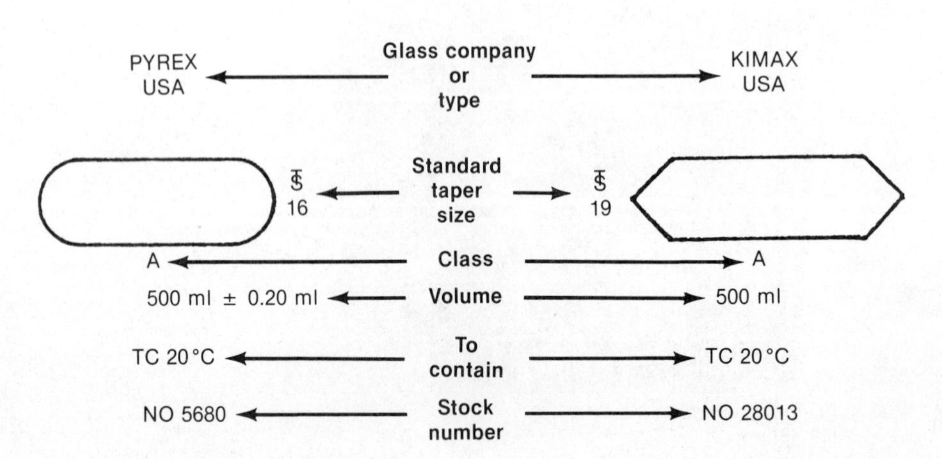

Fig. 1-11. Example of NBS specifications found imprinted on class A volumetric flasks.

where

$$CF = \frac{V_{flask}\ (mL) \times 20}{A_{1:50\ standard}} \quad Eq.\ 1\text{-}2$$

V_{flask} is the volume of the receiving flask, and 20 is a volume factor. Equation 1-1 allows the calculation of the actual delivered volume of the micropipet after one measures the absorbance of the diluted solution and multiplies the reading by the concentration factor, CF.

The concentration factor in equation 1-2 depends only on the volume of the flask (in milliliters) in which the diluted solution is prepared, and on the absorbance of the standard, manually diluted 1:50 solution.

Quality control schedule. Accuracy determinations should be made three to four times a year for each manual micropipet employed in the laboratory. These values should be plotted on a Levy-Jennings plot to determine drifts or sudden shifts of the performance of any volumetric pipet in use. Mean volumes and standard deviations may be established when one follows the accumulation of values over 3 to 4 weeks. During the accumulation of base-line data, a greater than 3% coefficient of variation from the labeled volume for each pipet should be used to identify pipet malfunction.

Routine maintenance. The Eppendorf or other piston type of micropipets have a fixed stroke length, which must be maintained. In addition, there are seals to prevent air from leaking into the pipet during its motion. These must be greased to maintain proper operation. Refer to the manufacturer's guidelines for this type of maintenance.

Accuracy and precision check for automatic diluters. Weekly, take a stock dye solution of known optical density, prepare a manual dilution similar to the one made by the diluter, and read at a specific wavelength. For example, if the automatic diluter effects a 1:50 dilution, use a 1 mL volumetric pipet to transfer an aliquot of the stock dye solution into a 50 mL volumetric flask. Carefully dilute to 50 mL with water and mix. Using the stock solution, run 20 consecutive dilutions on the automatic diluter. Read all dilutions in a spectrophotometer. Compute $\bar{x}$ (mean), SD (standard deviation), and CV (coefficient of variation) of the 20 dilutions. The CV should be equal to or less than 1%. The mean ($\bar{x}$) should be within limits established by the laboratory, as usually within 2% of the normal dilution. If either the deviation from the mean or the CV are not within limits, check settings and repeat the procedure. Green food dye can be used as stock solution, taking absorbance readings at 620 nm on the spectrophotometer.

Accuracy check for dispensers. Daily, take a clean, unscratched graduated cylinder whose volume is as small as possible and can contain several dispensings of the dispenser. NBS graduated cylinders are available for this purpose. Adjust setting and prime dispenser and ensure that no air bubbles are present. Carefully dispense an aliquot down the side of the cylinder and allow the liquid to drain. Read the volume contained. If the correct volume was delivered, check the maintenance sheet. If it was not, readjust the dispenser settings to correct for the error. Repeat dispensing, calculating the new volume delivered by subtraction. Repeat until correct volume (within an established error, such as 2% to 4%) is delivered.

THERMOMETRY

Types of thermometers

Water baths and metal heating devices must be maintained at relatively constant temperature when kinetic or other temperature-sensitive assays are performed. Liquid-in-glass, thermistors, and electronic digital thermometers are used to monitor the temperature of heating baths.

The temperature of heating baths should be checked every day and recorded for maintenance and quality control purposes. Thermometers that are used to monitor these heating baths should be calibrated at regular intervals, usually 6 months to 1 year. It has been recommended[15] that the thermometer have an accuracy range of one half that of the desired range for the bath. For instance, if the desired accuracy for the heating bath is $\pm 0.1°$ C, the thermometer should have a maximum uncertainty of $\pm 0.05°$ C.

Calibration of liquid-in-glass thermometers

For calibration standards, one should use thermometers that have been certified by the NBS or calibrated against an NBS-certified thermometer. As part of the NBS Standard Reference Material (SRM) program, certified thermometers are available that can be used to calibrate thermometers in the range of 24° to 38° C.[16] High-quality liquid-in-glass thermometers are available that have ranges greater than the SRM thermometers and have been calibrated against NBS thermometers by the manufacturer. Liquid-in-glass thermometers are available for partial or total immersion. Partial-immersion thermometers are used to measure the temperature of water baths, heating blocks, and ovens. The immersion depth is engraved on the stem and is usually located about 76 mm from the bulb. Total-immersion thermometers are generally used to check refrigerator and freezer temperatures but can be substituted for partial-immersion thermometers if they are calibrated at the same immersion depth that they will be used in the laboratory.

Calibration procedure: SRM or certified thermometers

1. The mercury column should be checked to make sure that it is not separated and that bubbles are not present. If either exists, the bulb should be cooled in a mixture of sodium chloride and ice until the mercury retreats into the bulb. After cooling, it may be necessary to swing the thermometer in a circular arc in order to force all the mercury in the column into the

bulb. The bulb should then be allowed to warm to room temperature in an upright position.

2. The ice point is a fixed reference point and should be checked and recorded each time a calibration is made. Clear, shaved ice or ice made from distilled water should be used. The ice is placed in a Dewar flask with enough distilled water to form a slush with no floating ice. The bulb should be rinsed with distilled water before being placed in the ice bath.
3. If the ice point is higher or lower than the SRM or certified value, other corrections should be adjusted accordingly, that is, higher or lower by the same amount. A small 10-power telescope will permit precise readings.

Calibration procedure: noncertified thermometers

1. The mercury column should be checked for a separation or bubbles, as indicated above. If the column of mercury is continuous, the ice point should be determined.
2. The heating bath should be adjusted to the temperature required for the test application.
3. The correction should be determined by comparison of the reference with the noncertified thermometer. The thermometers should be gently tapped before readings are taken so that the mercury column is prevented from sticking.
4. If a total-immersion thermometer is being calibrated for use as a partial-immersion one, it should be immersed in the heating bath to the same depth used for test applications.
5. Electronic thermometers that employ a thermistor probe may be calibrated by use of SRM thermometers or thermometers calibrated against an NBS-certified thermometer.

In the calibration of all types of thermometers, it is important that a liquid bath be used that is maintained at a constant level. It is recommended that the volume of the bath fluid be at least 100 times greater than the volume of the thermometers being calibrated so that a uniform temperature can be maintained throughout the bath.[17] The reference thermometer and thermometer being calibrated should be far enough apart so that the flow of liquid around them is not disturbed. Liquid-in-glass thermometers have a temperature response time of several minutes, whereas thermistors respond in a few milliseconds. This response time must be taken into consideration when calibrations are carried out.

Available from the NBS is a gallium melting point cell that provides a fixed point (melting point of gallium, 29.772° C) for calibrating electronic thermometers that have a thermistor probe. An electronic thermometer calibrated at 29.772° C can be used to standardize liquid-in-glass thermometers in the 20° to 40° C range. The application of the gallium melting point standard for thermometry has been reviewed.[18]

Manual and automated chemistry analyzers are available that use concealed thermistor probes for monitoring the heating-bath temperature. Some of these instruments have flow-through or other cuvette designs that make the measurement of the temperature in the cuvette very difficult using thermistor probes. A technique has been devised for monitoring cuvette temperatures by measurement of the change in absorbance of a temperature-sensitive solution.[19] The temperature of the cuvette is obtained when this absorbance is compared with an absorbance-temperature calibration curve. This approach to thermometry is claimed to be capable of detecting changes of 0.1° C.

CENTRIFUGES AND HEATING BATHS

Types of centrifuges and their uses

Laboratory centrifuges are available in floor or table models with a number of different designs. Centrifuges used most frequently in chemistry have either horizontal or fixed-angle (45 to 52 degree) rotor heads. These centrifuges operate at speeds up to 6000 revolutions per minute (rpm), generating relative centrifugal force (RCF) up to 7300 G. Microhematocrit centrifuges for determining red blood cell volumes operate at 11,000 to 15,000 rpm with an RCF up to 14,000 G. Ultracentrifuges are generally used for research; however, a small air-driven ultracentrifuge (Beckman Instruments, Spinco Division, Palo Alto, CA 94304) is available and operates at 90,000 to 100,000 rpm and generates a maximum RCF of 178,000 G. This type of centrifuge has been used to separate chylomicrons from serum, fractionate lipoproteins, perform drug-binding assays, and prepare tissue for steroid hormone receptor assays.

Centrifuges are used in the chemistry laboratory primarily to separate clotted blood or cells from serum or plasma and to clarify body fluids. Although the relative centrifugal force necessary to carry out these separations is not critical, a force of at least 1000 G for 10 minutes will give a good separation. Serum separation tubes are available that contain a silicone gel with a specific gravity intermediate to that of serum and blood coagulum. When these tubes are centrifuged, the gel is displaced up the side of the tube forming a stable, inert barrier between the serum and the clot. The RCF needed to displace the gel is 1000 to 1300 G for 10 minutes. RCF less than 1000 G may result in an incomplete displacement of the gel.

Refrigerated centrifuges have internal refrigeration units that maintain temperatures ranging from −15° to 25° C during centrifugation. This permits centrifugation at higher speeds and protects the specimens from the heat generated by the rotor. The temperature of these centrifuges should be checked daily, and the thermometer checked annually for accuracy.

Maintenance of centrifuges

The speed of a centrifuge is expressed in revolutions per minute (rpm), and the centrifugal force generated is ex-

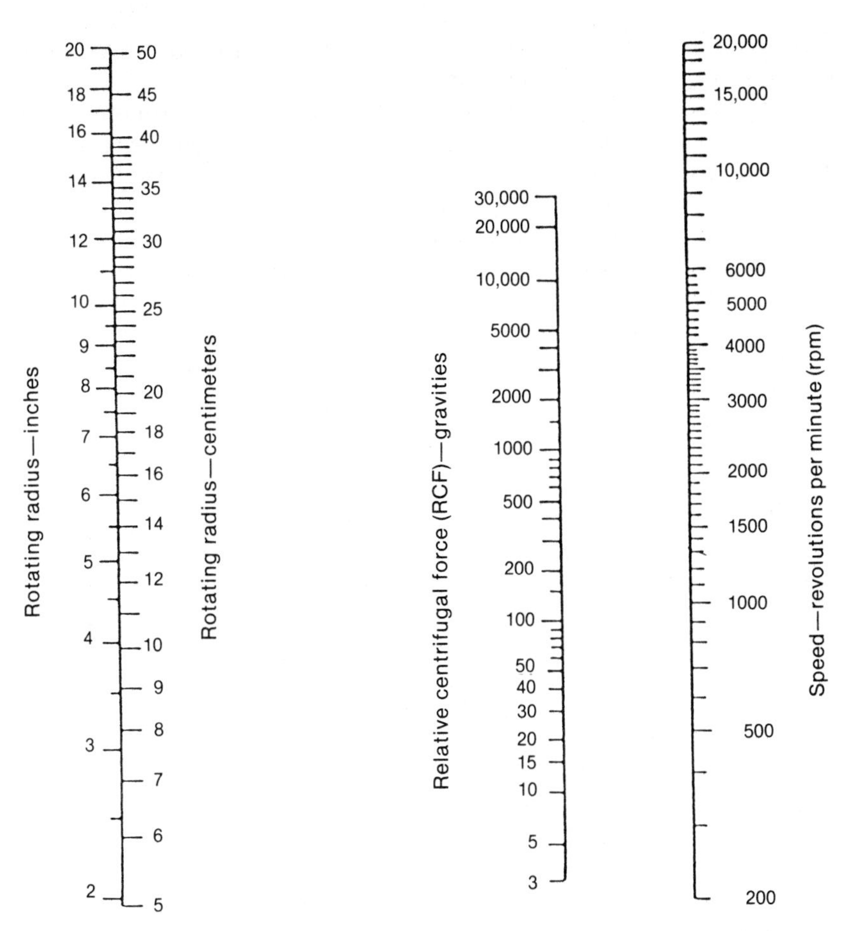

Fig. 1-12. Nomograph for relating relative centrifugal force, RCF, to revolutions per minute, rpm.

pressed in gravities as relative centrifugal force (RCF). The number of revolutions per minute is related to the relative centrifugal force by the following formula:

$$RCF = 1.12 \times 10^{-5} \times r \times (\text{rpm})^2$$

r is the radius of the centrifuge expressed in centimeters and is equal to the distance from the center of the centrifuge head to the bottom of the tube holder in the centrifuge bucket. Fig. 1-12 shows a monograph for determining the relative centrifugal force from the number of revolutions per minutes and the radius. The College of American Pathologists (CAP) recommends that the number of revolutions per minute of a centrifuge used in a chemistry laboratory be checked every 3 months. This is most easily accomplished with a photoelectric tachometer or strobe tachometer. Most centrifuges have a hole or plastic window located in the centrifuge lid that permits a stroboscopic check of rpm. Some floor model centrifuges do not have this window, but do have a digital tachometer that displays the rpm while the centrifuge is running. The rpm of these centrifuges cannot be checked without taking the lid off the centrifuge or the cowling from around the motor. This creates a hazardous situation; consequently, it is advisable to have factory-trained personnel check these centrifuges.

The CAP also recommends that the timer and speed control be checked on a quarterly basis with any corrections posted near the controls.

Instructions for lubrication, maintenance, and replacement of brushes are given in the manufacturer's service manual. It is important that the centrifuge cups in the rotor head be balanced before centrifugation. This will permit maximum relative centrifugal force and minimize breakage of tubes and wear on the motor and bearings. Tubes should be capped during centrifugation to prevent release of infectious material inside the centrifuge by aerosol formation. If breakage occurs with release of potentially infectious material, the centrifuge bowl should be cleaned with a germicidal disinfectant, and the rotor head and buckets should be autoclaved. Further information on the maintenance and care of centrifuges is available in the manufacturer's and other manuals.[13,20]

Heating baths and ovens

Laboratory ovens are used to dry chemicals, extracts, media used in electrophoresis, and glassware. For most purposes, temperature control at ±1° C is satisfactory. Thermometers that monitor oven temperature should be checked for accuracy annually. A daily record of the temperature will indicate malfunction of heating elements or thermistor controls and maintain quality control of operation.

Water baths are circulating or noncirculating in design. Noncirculating water baths generally maintain temperature within a range of ±1° C or better. For certain applications such as enzyme, blood gas, pH, or electrophoretic assays, a narrower range is required (see Chapter 49), and these baths have an external or internal circulating pump to maintain thermal equilibrium in the bath. Some of these pumps are coupled with a refrigeration unit to permit temperature control below room temperature. The bath liquid should be type II or type III water. The addition of a bactericidal agent such as a solution of 1:1000 thimerosal (Merthiolate) will reduce the frequency with which the bath and its components must be cleaned.

Metal heating blocks are somewhat less efficient in maintaining a relatively constant temperature and usually operate within ±0.5° C. Blocks that are incorporated into the cuvette compartment of a spectrophotometer will operate with a greater accuracy, usually ±0.2° C or better. The temperature of water or metal heating baths should be recorded daily with a thermometer calibrated against an NBS-certified thermometer.

LABORATORY WATER

Significant error can be introduced into clinical laboratory assays by the presence of inorganic or organic impurities in laboratory water. Interference by these impurities may be relatively easy to detect in assays for iron, lead, and trace metals by analysis of a blank, but it is more difficult to recognize where enzymatic, fluorometric, or chromatographic procedures are concerned. The requirement for certification of clinical laboratories by state and federal agencies has resulted in the establishment of well-defined specifications for water purity (Table 1-14).

Three levels of water purity have been recommended by the National Committee for Clinical Laboratory Standards and the College of American Pathologists.[21,22] Type I water should be used in all the quantitative chemistry procedures, in the preparation of buffers, controls, and standards, and in electrophoresis, toxicology screening tests, and high performance liquid chromatography. Type II water is suitable for qualitative chemistry methods and can be used in most procedures carried out in hematology, immunology, and microbiology. Type III water can be used as a water source for producing type II and type I water and for washing and rinsing glassware. The final glassware rinse should be made with either type II or type I water, depending on the utilization of the glassware.

Storage of water

Type I water will absorb carbon dioxide rapidly and must be used immediately after purification. Type II and type III water can be stored in glass or polyethylene bottles but should be used as soon as possible because of possible contamination with airborne microorganisms. Sterile bottled water is usually equal to or better than type II water and has a shelf life of about 3 years if unopened.

Methods of purifying water

Water originating from rivers, lakes, springs, or wells contains a variety of inorganic, organic, and microbiological contaminants. The contaminants cannot be completely removed by any single purification system, but a combination of several methods can be used to produce water of type I purity or better (Fig. 1-13).

Distillation. Distillation of water in glass will remove nonvolatile organics, inorganic impurities, and microbiological organisms. Volatile impurities such as ammonia, carbon dioxide, chlorine, and low-boiling organic compounds will be present in the distillate. Distilled water meets the specifications of type II or type III water.

Deionization. Deionization is accomplished when water is passed through basic and acidic ion-exchange beds or a mixture of the two. This treatment results in the exchange of hydroxyl and hydrogen ions located on the surface of

Table 1-14. Recommendations for reagent water used in the clinical laboratory (Commission on Inspection and Accreditation for Water Quality Control, College of American Pathologists)

Specifications	Type I	Type II	Type III
Resistivity (megohm/centimeter at 25° C)	10 (in-line)	2.0	0.1
Silicate (mg/L, as SiO_2)	0.05	0.10	1.00
pH	NA*	NA	5.0-8.0
Microbiological content (colony-forming units/mL, maximum)	10	10^3	NA
Particulate matter (>0.2 μm)	<500/L	NA	NA
Organic compounds	Activated carbon treatment	NA	NA

*NA, Not applicable.

the resin for cation and anion impurities. The resins eventually lose their exchange capacity but can be regenerated by treatment with acid and alkali. Water containing less than 1 million colony-forming units per liter can be deionized to produce type II water. Further treatment with activated charcoal and membrane filtration to remove organic impurities, particulate matter, and microorganisms is necessary to produce type I water.

Reverse osmosis. Reverse osmosis is the passage of water under pressure through a semipermeable membrane made of cellulose acetate, aromatic polyamides, cellulose acetobutyrates, or other materials. This treatment removes approximately 90% of dissolved solids and 98% of organic impurities, insoluble matter, and microbiological organisms. Reverse osmosis removes about 10% of ionic impurities but does not remove dissolved gases. This method of purification will produce type III water and is frequently used to treat water before its passage through ion-exchange resins.

Filtration. Filtration of water through semipermeable membranes with pore sizes of about 0.2 μm will remove insoluble matter, emulsified solids, pyrogens, and microorganisms. Sterile water can be produced by filtration through membranes with a pore size of 0.3 μm (300 nm) or less.

Adsorption. Organic impurities can be removed from water by adsorption on activated charcoal, clays, silicates, or metal oxides. A combination of deionization, adsorption, and filtration will produce type I water.

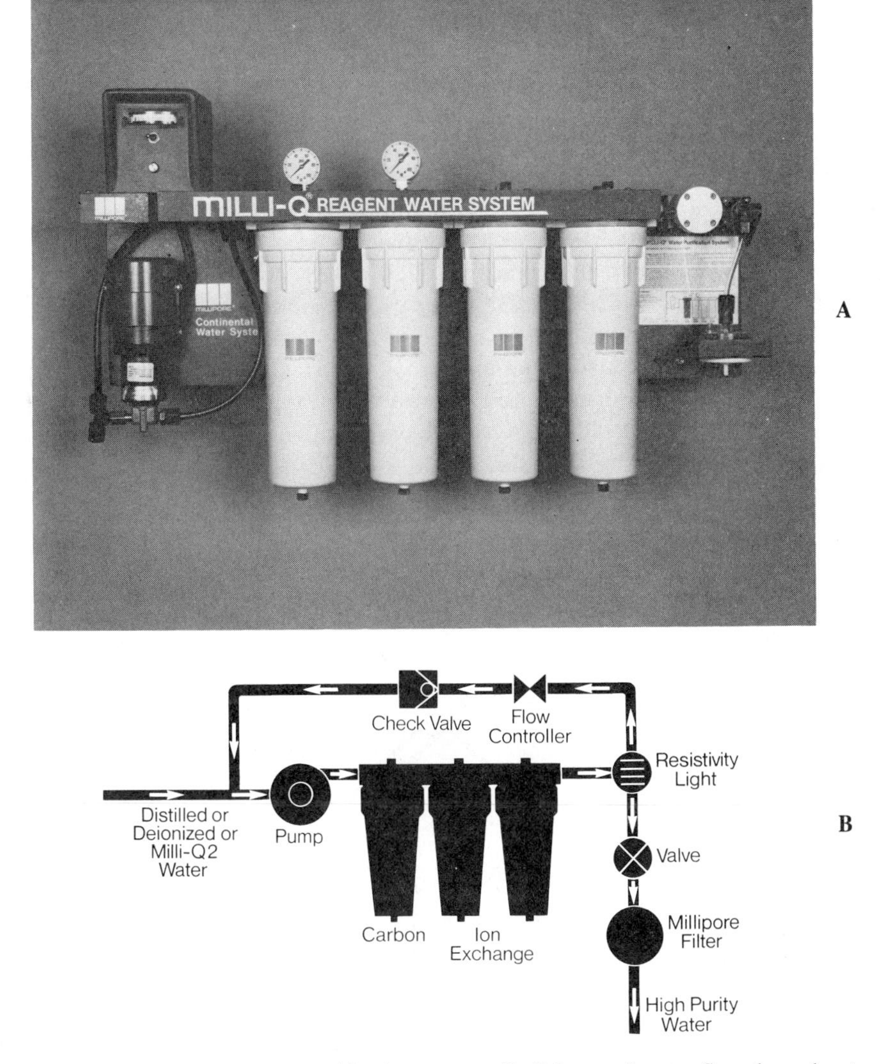

Fig. 1-13. A, Type I water purification system. **B,** Scheme of water flow through a type I purification system. (Courtesy Millipore/Continental Water System, Bedford, Maine.)

Tests for purity of water

Measurement of the conductivity or specific resistance (resistivity) of water is used to indicate the purity of water with regard to inorganic ionized substances. Ion-exchange tanks usually have a conductivity light located in line at the outlet of the tanks that is adjusted to go out when the resistance of the water goes below 200,000 ohms/cm indicating that the capacity of the tanks has been exceeded. Systems that supply type I water usually have an ohmmeter located in the line outlet that has a scale graduated in the range of 0.5 to 18 megohm/cm. Type I water should have a resistivity of at least 10 megohm/cm, equivalent to a total dissolved solid concentration of less than one part per million. Procedures for measuring the resistivity, silicate, and microbiological content of water are available.[21,22]

Water-purity checks

Water-resistance test. This test is a measure of dissolved ionized substances in the water. As the purity of the water increases and the amount of dissolved ionized materials decreases, the ability of the water to conduct an electrical current decreases and the resistance increases.

$$\frac{\text{Purity} \uparrow}{\text{Dissolved ionized material} \downarrow} = \frac{\text{Resistance} \uparrow}{\text{Conductivity} \downarrow}$$

The Beckman Solu-bridge (Beckman Instruments, Inc., Fullerton, California) is an example of a portable flowmeter. Only a flow-through conductivity meter can be used to measure the purity of type I water. When the Beckman flowmeter is used according to the manufacturer's instructions, type I water should have a resistance between 15 and 18 megohms/cm. This check is to be performed daily to weekly.

Weekly organic material check

1. Rinse a 1000 mL volumetric flask with distilled, deionized water three times. Collect about 500 mL of this water in the flask.
2. Add 1 mL of concentrated sulfuric acid.
3. Add 0.4 mL of 0.01 N potassium permanganate (see below).
4. Stopper and mix well by swirling.
5. Let stand for 1 hour at room temperature.
6. If the purple permanganate color does not disappear completely after 1 hour, there is no organic contamination.
7. If the color disappears, this is an indication of the presence of organic contaminants. Notify the supervisor.

Reagents

1. Potassium permanganate, 0.01 N. Dissolve 0.316 g of potassium permanganate in 1 liter of boiling distilled water in a 2-liter Erlenmeyer flask. Cover with a beaker and simmer for one-half hour. Cool to room temperature. Store in a dark-colored stoppered bottle. Date bottle. Prepare fresh every 6 months.
2. Sulfuric acid, concentrated.

Monthly bacteriological contamination test

1. Obtain a 10 mL sterile culture tube from the bacteriology laboratory.
2. Let approximately 30 to 50 mL of water flow from the deionizer to waste. Quickly open sterile tube and fill halfway with water from deionizer. Close tube and give to bacteriology laboratory for organism check.
3. Record all readings and results on monthly water quality control chart and list any actions taken on water that fails any one of these checks.

SAFETY IN THE CLINICAL LABORATORY*

General considerations

Safety in the clinical laboratory has received increased emphasis in recent years, partly as a result of the manifold regulations of the Occupational Safety and Health Administration (OSHA). Regulations concerning the physical layout of the laboratory and emergency exits will not be discussed here except to note that all personnel should know the exit routes and should be familiar with the locations of the various safety devices—fire extinguishers, emergency showers, eye washers, fire blankets, and other equipment such as respirators and goggles—and with their use and operation. This information should be carefully explained to new employees and to all employees after the acquisition of new emergency equipment. (If any information is not understood, laboratory personnel should not hesitate to ask the supervisor for clarification.) Personnel should also recognize the various auditory or visual signals

*From Bauer, J.D.: Clinical laboratory methods, ed. 9, St. Louis, 1982, The C.V. Mosby Co.

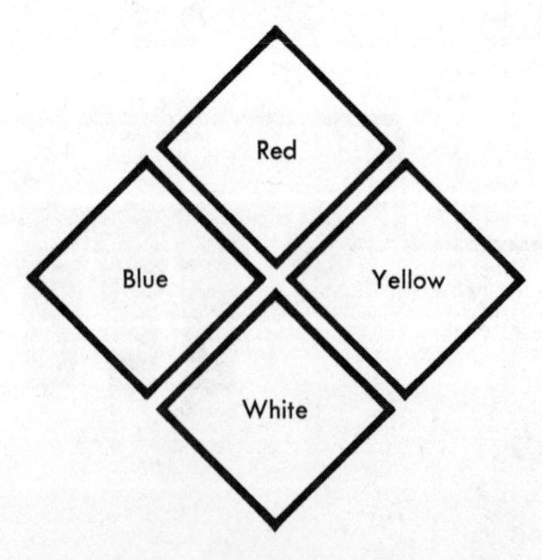

Fig. 1-14. Identification system of the National Fire Protection Association. (From Bauer, J.: Clinical laboratory methods, ed. 9, St. Louis, 1982, The C.V. Mosby Co.)

for fire or other emergencies and should fully cooperate in all emergency drills.

Safety manual. Every laboratory should have a safety manual covering all safety practices and precautions, including regulations concerning the proper use of equipment and the handling of toxic or infectious materials. The material here deals in a general way with factors most common to clinical laboratories.

Warning signs. Consistent warning signs should be used to indicate the presence of hazards. The Hazards Identification System developed by the National Fire Protection Association has been adopted by some laboratories. This system consists of four small, diamond-shaped symbols grouped into a larger diamond shape (Fig. 1-14). The left diamond is blue and is used to identify health hazards. The top diamond is red and indicates a flammability hazard. The diamond on the right is yellow and warns of a reactivity-stability hazard (meaning that materials capable of explosion or violent chemical change are present). The bottom diamond is white and is used to provide special hazard information. It may indicate the presence of radioactivity (Fig. 1-15, *A*), special biohazard (Fig. 1-15, *B*), corrosives, or other dangerous elements (Fig. 1-15, *C*). The severity of each hazard included in the symbol is suggested by numbers from 0 to 4, the greatest degree being 4. This simple system can provide safety information for all personnel and aid in any fire fighting as well.

Personal behavior. Smoking, eating, and drinking should be prohibited in all work areas. These activities should be carried out only in designated rest areas completely separated physically from work areas. Foot-operated drinking fountains may be allowed just outside the work areas. Although the greatest danger from eating or smoking in the laboratory is the possibility of infection from laboratory specimens, there is also the possibility that food or tobacco could contaminate test materials. For the same reasons application of cosmetics should not be done in work areas. No matter what type of work is done in the laboratory, the hands should always be washed before eating, smoking, or leaving the laboratory.

Long hair must be secured so that it will not come in contact with specimens, work areas, reagents, or media or become entangled in laboratory equipment. In the microbiology laboratory it may be advisable to use the same precautions that food service workers use in regard to hair containment, since contamination may occur from shedding of long hair and beards.

Volatile flammables

Storage. All refrigerators must be clearly marked on the outside as to the type of material to be placed in them, indicating whether they are explosion proof. (Under no circumstances should food or drinks be placed, even temporarily, in a refrigerator in the work area; these should be left only in designated refrigerators in the rest area.) Volatile solvents should be labeled as to their hazardous qualities and should be kept in explosion-proof refrigerators. These solvents should be kept in these refrigerators only in as small a quantity as is consistent with daily use, preferably not more than 1 week's supply. (There is some question as to the safety of storing flammable solvents such as ethyl ether under refrigeration instead of on an open shelf in a well-ventilated area. Opening a refrigerator door can release evaporation vapors, resulting in ignition and explosion.) Large amounts of such solvents must be kept in a fireproof, specially ventilated area with explosion-proof switches and lighting and a 4 in. sill or ramp at the bottom of the door to prevent the escape of heavier-than-air vapors. Flammable liquids in quantities of more than 2 L should be stored only in clearly labeled, approved red safety storage containers with pressure-relieving, spring-closing covers and flame-arrester screens. Maintaining only the minimal amount of flammables necessary is a good practice.

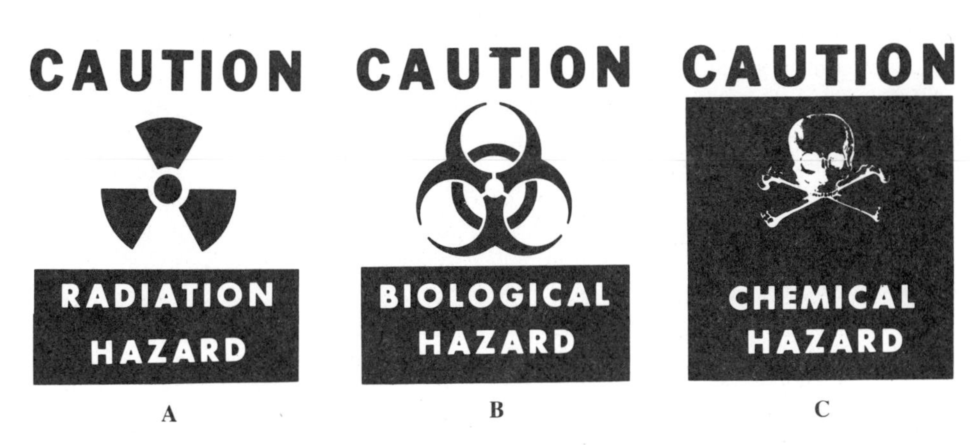

Fig. 1-15. Special hazard warnings. (From Bauer, J.: Clinical laboratory methods, ed. 9, St. Louis, 1982, The C.V. Mosby Co.)

Handling. All flammable and toxic liquids should be handled in a space with good ventilation, preferably in a fume hood. Evaporation of such liquids must be carried out in a fume hood using a water bath. Extreme caution should be taken with very volatile flammable liquids such as ethyl ether, since the vapors of these liquids can be ignited by any ignition source, even a hot plate or oven. In the evaporation of larger quantities of volatile liquids, glass beads or other antibump granules should be added to prevent bumping and the sudden release of large amounts of vapors. The fume hood function may be checked easily by noting the disposal of smoke or fumes produced after cautiously bringing close together two applicator sticks whose cotton tips have been dipped in strong ammonia and hydrochloric acid, respectively (rinse the tips in running water before discarding).

Compressed gas

Handling and storage. Large cylinders (200 ft^3) of compressed gases often are used in the laboratory. The most commonly used gases are oxygen with acetylene or propane (for atomic absorption or flame spectroscopy), hydrogen or helium (for gas-liquid chromatography), nitrogen, and carbon dioxide. The cylinders are color coded, but the contents must be clearly marked on the cylinder. (Otherwise it should be returned unused to the supplier marked "Contents Unknown,") All cylinders must be securely fastened to the wall or bench or placed in floor holders so that they cannot overturn. A fall can rupture the outlet valve, and the cylinder may then act like a torpedo and inflict serious injury. When transported, the cylinders must always be secured on a dolly or hand truck, and the protective caps must be on when the cylinders are not connected to equipment.

It is advisable to place a large tag on the tank indicating the date that the cylinder was put into use. Empty cylinders must be marked as such. (Note, however, that a small amount of gas must be left in the cylinder and valves closed to prevent negative pressure and drawing in of foreign contaminating materials.) Extra cylinders should be stored away from the laboratory in a secure, upright position, preferably in a locked, ventilated, fire-resistant space with the empty cylinders well separated from the full ones.

Precautions. The reduction valves for the different types of gases are not interchangeable, and no attempt should be made to interchange them. Laboratory personnel should not attempt to force or free stuck or frozen cylinder valves. When in use, all connections to the cylinders should be tested for leaks with soapy water. Very small leaks of oxygen or nitrogen are of little consequence (except for the waste of gas), but leaks of hydrogen, acetylene, or other flammable gases are not to be tolerated. When one is shutting down flammable gases, the gas is first turned off at the main take-in valve and the gas allowed to burn out; then the reduction valves are closed. It should be remembered that propane is heavier than air and that a little leaking gas can flow along the top of the bench to be ignited by a flame elsewhere. Many flame photometers use small cylinders of propane. These are very convenient and usually cause little difficulty. Because of potentially hazardous characteristics, only properly trained persons should handle compressed gases.

Corrosives

Handling. Caution must be observed when using toxic or corrosive solutions such as those of mercury salts or caustic acids or alkalis. When handling these chemicals, laboratory personnel should wear goggles or a face mask. Strong stirring should be avoided or done in a hood. If considerable stirring is needed, a magnetic stirrer can be used to avoid splashing. In preparing reagents, caution should always be taken to *slowly* add acid to water. If a small amount of any reagent gets into an eye, it should immediately be rinsed well (for at least 15 min) in the eye washer while one holds the eyelids apart with the fingers and rotates the eyeball. Contact lenses should not be worn in the laboratory. They prevent proper washing of the eyes in the event a harmful solution does reach them, which could result in serious, permanent damage. Plastic lenses may be damaged by organic vapors.

Spills. All spills should be cleaned up at once. If the material is toxic or corrosive, rubber gloves and a plastic apron should be donned before cleaning up. All potentially infectious materials such as serum or other body fluids should be flooded with a 5% solution of sodium hypochlorite or other disinfectant before being wiped up (wearing gloves); the waste should be disposed of with other infectious materials, not with ordinary wastes. Dilute solutions of acids or bases may be neutralized with solid sodium bicarbonate. For stronger acids, after dilution with water, solid sodium carbonate may be used with caution. For strongly basic solutions, a 1:5 dilution of acetic acid can be used. The "Space Clean-Up Kit" (Mallinckrodt Chemical Co., St. Louis) is satisfactory for laboratory use. It contains rubber gloves, towels, a scoop, and various chemicals for neutralizing and absorbing acid or alkaline spills, as well as material for absorbing organic solvents. Similar kits can be made with gloves, towel and other tools, sodium bicarbonate, an organic acid such as citric acid, and a quantity of absorbing material such as celite or diatomaceous earth. For such kits only the cheapest grade of technical chemicals need be used. These kits should be kept in strategic places in the laboratory, conspicuously labeled to indicate their use.

Mercury

A particularly troublesome substance to recover is metallic mercury. Although it is not used so much as formerly (with Van Slyke or Natelson apparatus), it may still be used occasionally. Spilled mercury tends to break up into very fine droplets, which are difficult to pick up. Even after collection disposal is a problem, since metallic mercury should not be incinerated or buried. If mercury is used in the laboratory, the "Mercury Absorption/Disposal Kit" (Aldrich Chemical Co., Milwaukee) may be helpful. It contains material that will readily absorb mercury droplets, producing a less toxic substance that may be disposed of by burial. Older benches or floors may contain fine cracks in which minute mercury globules may be lodged. These are very difficult to remove, but the cracks should be cleaned if possible, since mercury is somewhat volatile at room temperature. Rubbing powdered sulfur or sodium polysulfide into the cracks may help to change the mercury to the less volatile sulfide.

Some colorimetric methods, particularly for chloride, use mercury salts in a reagent. The disposal of large amounts of spent reagent in automated methods may be a problem, since the solution should not be disposed of in the sewer. The waste material may be collected in large plastic containers and made slightly acid with acetic acid, if necessary, and with thioacetamide (about 10 g/L) added. This is stored in a ventilated area (small amounts of hydrogen sulfide may be liberated), and over a period of time the mercury will be precipitated as mercuric sulfide. The supernatant may then be decanted and disposed of in the sewer. The mercuric sulfide may be disposed of by burial.

Azides

Another troublesome chemical, sodium azide, was used as a preservative in the past, though its use has largely been discontinued. Azides form explosive salts with a number of metals such as copper and iron. These salts are readily detonated by mechanical shock. Although the amount used as a preservative is relatively small, continued use with disposal through the sewer can result in a buildup of the metallic salts in the sewer pipes. These salts are extremely explosive; even use of a wrench on such a drain line may result in a violent explosion. It is difficult to remove the azides from the pipes. One suggestion is to close the lower end of a section of pipe and allow a 10% solution of sodium hydroxide to stand at least 16 hr in the pipe; then rinse copiously with water (a minimum of 15 min). It is best to avoid the use of solutions containing azides, particularly since they are also said to be carcinogenic.

Carcinogenic hazards

The use of chemicals that are possibly carcinogenic is another hazard in the laboratory. Many of these are aromatic amines. One, benzidine, has been used frequently in the laboratory for testing hemoglobin (plasma hemoglobin and occult blood). Generally this can be replaced by the compound 3,3′,5,5′-tetramethylbenzidine dihydrochloride, which is much safer. With proper precautions some potentially carcinogenic compounds can be used occasionally, if necessary. Precautions should include performing the procedure in an isolated area or in a good fume hood, wearing rubber gloves and (if dusty material is handled) a respirator, careful cleanup of the work area, rinsing the glassware with a strong acid or an organic solvent before placing it in the regular washing cycle, and using disposable equipment as much as possible. No pipetting should be performed by mouth.

Procedures and equipment

Whenever a potentially toxic or dangerous chemical is used in a procedure, the directions in the procedure manual should include warnings of the noxious properties of that chemical. For example, it may seem superfluous to add "Caution: Corrosive" whenever the use of concentrated sulfuric acid is mentioned, or "Caution: Highly Flammable" whenever the use of diethyl ether is mentioned, but these warnings may be helpful to someone who is not familiar with the procedure or to the worker who may become careless through the routine use of a dangerous chemical.

Cracked or chipped glassware should not be used (being more likely to break during use), but should be discarded. Flasks or beakers used with corrosive or toxic substances should be rinsed well with water or alcohol (depending on the solubility) before being placed in the collection container for soiled glassware. No pipetting by mouth should be allowed. Although it may seem safe to orally pipet distilled water or saline, it is best to use a pipetting device at all times; thus an attempt to pipet a dangerous substance by mouth is less likely. For simple dilutions with water or saline, the use of a good quality buret with a Teflon stopcock (or automatic dispensing bottle) is often convenient.

Hepatitis

The presence of hepatitis viruses in blood, tissue, urine, and feces from infected individuals constitutes a hazard to laboratory personnel. Samples from known (or suspected) hepatitis patients should be noticeably marked as such. Food, beverage, and smoking restrictions, good personal hygiene, use of disposable gloves and disinfectants, and prevention of aerosols must be stringently applied. Employee education in regard to this problem is important.

Safeguards

Disposable, plastic, Pasteur type of pipets are useful for transferring serum or other samples from the collection tubes to the test tube or sample containers of many types of automated apparatus. Samples should not be poured, since this could contaminate the fingers and the outside of the tubes. The empty sample cups may first be placed in the rack, the serum samples then added to the cups with the pipets, and the rack placed in the analyzer. After analysis the cups can be dropped directly from the rack into the waste container.

Centrifuges should not be operated without the covers being completely closed. (Many of the newer models cannot be operated unless this is done.) Potentially infectious, volatile, or toxic substances should not be centrifuged in open tubes, as centrifugation may result in the formation of infectious aerosols or the volatilization of the liquid. In using older centrifuges, do not attempt to open the cover until the rotor has completely stopped. In those with hand-operated brakes, do not brake sharply. Be certain that the tubes are properly balanced and that the cushions are at the bottoms of the holders. Always use tubes of the correct size and strength for the desired application. If a tube breaks in the centrifuge, immediately turn the centrifuge off, and after donning rubber gloves and other protective clothing (if necessary), clean up. Allow any droplets in the chamber to settle for about 15 min; then wash it with an appropriate solution (dilute acid for strongly alkaline solutions, sodium bicarbonate solutions for acid solutions, and 5% sodium hypochlorite for potentially infectious material).

Radioactivity precautions

All precautions against eating, drinking, or smoking are particularly applicable to the radioisotope laboratory. Although the amount of radioactivity associated with the kits for radioimmunoassays is small, it can present some hazards. Radioisotopes often used for labeling are ^{125}I and ^{131}I. If either is ingested, the radioactive compound may be broken down in the body and radioactive iodine absorbed and concentrated in the thyroid gland. Thus the thyroid can receive a much larger dose of radiation than would be expected from a random distribution in the body. All personnel performing work with radioactive material should wear film badges at all times when in the isotope laboratory.

The radioactive material used in radioimmunoassays is generally so small that it can be harmlessly flushed down the sewer with copious amounts of water. Plastic beads or other resins used to absorb some of the radioactivity should not be incinerated. Burial is satisfactory since the radioactivity gradually decays. All radioactive material should be stored in special refrigerators or cabinets conspicuously labeled with appropriate signs indicating the presence of radioactive elements. Generally the level of radioactivity used in a radioimmunoassay laboratory that purchases the material in the form of kits and does not prepare its own will not be sufficiently high to require lead shielding for storage.

The procedures using radioactive materials should be carried out in a separate room with a good ventilating system. The door into the room should be conspicuously marked "Radioactive Material—Authorized Personnel Only." All waste containers should be clearly marked as containing radioactive material so that the waste will not be incinerated.

Microbiological hazards

Precautions and decontamination. All work with potentially highly infectious specimens should be done in clean air hoods. If contaminated material is accidently dropped, pour disinfectant over the contaminated area, cover with paper towels, let stand about 30 min, and then clean up. Use rubber gloves for the cleaning operations. All discarded material should be covered with disinfectant-soaked paper towels after it is placed in the disposal container. The container should later be autoclaved.

Precautions for handling material (blood, feces, or any body fluid or tissue) likely to contain highly infectious agents (viruses, bacteria, fungi, or parasites) should include covering all cuts or skin breaks with tape, wearing protective clothing and rubber gloves, and working under a safety hood if possible. Aerosol formation must be prevented. Screw-capped centrifuge tubes in an enclosed centrifuge should be used. Leaky or spilled specimens require thorough disinfecting of outer surfaces. Clearing, cleaning, and disinfecting work surfaces, as well as thorough hand washing, reduce chances of infection. Overfilled stool specimen containers should be handled with gloved hands only, and soiled requisitions must be autoclaved, discarded, and replaced.

Electrical equipment

Safeguards. All electrical apparatus should have three-pronged grounding plugs, and sufficient grounded outlets should be available. The only exceptions may be small items such as clocks, which are totally enclosed in plastic, and microscope lamps, which are similarly enclosed. No attempt should be made to adjust any piece of electrical apparatus while it is plugged in, unless it is absolutely necessary (e.g., adjusting the lamp position in a spectrophotometer; this should be done with insulated tools). Under no circumstances should hands be placed inside an electrical instrument with the current on. Rings or jewelry along the forearms should not be worn. Preferably, rubber gloves should be worn for delicate adjustments. All electrical cords should be as short as possible and kept out of any areas where water or other solutions might contact them. Personnel should know the locations of the fuse boxes or circuit breakers for the different outlets. If electrical equipment fails to work properly (and particularly if smoke appears or sparks are seen), the apparatus should be disconnected and the maintenance department notified.

Quality control

Safety precautions, like all other aspects of laboratory function, must undergo a sort of "quality control." Inspection of safety equipment operation and availability must be regular. It must be determined whether warning signs are displayed where needed and whether decontamination and disposal practices are adequate, in addition to whether adherence to other safety procedures and rules is being practiced. Maintaining safe working standards may require extra effort, but risk minimization is "part of the job."

Part II: Calculations in clinical chemistry

ELIZABETH ANN BYRNE

DILUTION

In several areas of the medical laboratory, one must dilute blood or body fluids to prepare a measurable concentration. Accurate preparation of these dilutions is mandatory for reporting the actual concentrations of body-fluid constituents. Diagnosis, prognosis, and therapy depend on these test results.

Dilution can be defined as expressions of concentrations. Dilutions express the amount, either volume or weight, of a substance in a specified total final volume. A 1:5 dilution contains 1 volume (weight) in a *total* of 5 volumes (weights), that is, 1 volume and 4 volumes.

Expression of a 1:5 dilution can be stated as the common fraction ⅕. This fraction enables one to calculate the actual concentration of a diluted solution.

Example. A 100 mg/mL nitrogen standard is diluted 1:10. The concentration of the resulting solution is 100 × 1/10 = 10 mg/mL.

The most commonly used equation for preparing dilutions is

$$V_1 \times C_1 = V_2 \times C_2 \qquad \textit{Eq. 1-3}$$

V_1 is the volume, C_1 is the concentration of solution 1, and V_2 and C_2 are for the resulting dilution. These may be expressed as % (weight/volume, w/v), or molarity, or normality concentration. Similarly V_2 and C_2 are related. This basic equation can also be expressed as

$$\frac{V_1}{V_2} = \frac{C_1}{C_2} \qquad \textit{Eq. 1-4}$$

The most common error in setting up any equation of this type is *not placing* the related volumes or concentrations in the proper place and having the units cancel out, leaving the final, uncanceled units.

One helpful practice for successfully solving laboratory calculations is to label all numbers in an equation with their respective units of measurement. It may take an extra minute but will save many minutes of reviewing calculations when the final result appears illogical or incorrect. A problem that does *not* properly cancel out units cannot be successfully solved. A second helpful practice is to reduce fractions to their least common denominators before one calculates the results.

Example. Prepare 500 mL of 0.5 M NaCl (molecular weight of NaCl = 58.5 g/L).

$$500\text{ mL} \times \frac{\text{Liter}}{1000\text{ mL}} \times \frac{0.5\text{ mol}}{\text{Liter}} \times \frac{58.5\text{ g}}{\text{mol}} =$$

$$\frac{0.5 \times 58.5\text{ g}}{2} = \frac{29.25\text{ g}}{2} = 14.6\text{ grams}$$

14.6 g of NaCl diluted to 500 mL = 0.5 M NaCl

Example. Preparation of 250 mL of 0.1 M HCl from stock 1 M HCl.

Using $C_1 \times V_1 = C_2 \times V_2$

Where V_1 is the unknown $\quad V_2 = 250$ mL

$C_1 = 1.0$ mol/L $\quad C_2 = 0.1$ mol/L

$1.0\text{ mol/L} \times V_1 = 250\text{ mL} \times 0.1\text{ mol/L}$

$V_1 = 25$ mL

Measure 25 mL of 1 M HCl; dilute to 250 mL with distilled water. This diluted solution has a 0.1 M HCl concentration. (Mathematical reasoning indicates that a 1:10 dilution of stock 1 M HCl results in a 0.1 M concentration, and 25 mL diluted to 250 ml equals a 1:10 dilution.).

Another application of dilutions

So that a 24-hour urine creatinine concentration could be assayed, the specimen had to be diluted 1:5 before measurement. Calculate the 24-hour excretion if the 24-hour urine volume is 1800 mL and the measured creatinine concentration is 260 mg/L:

Total excretion = Total urine volume × Concentration × Dilution

$$\text{Total excretion} = 1800\,\frac{\text{mL}}{24\text{ hrs}} \times 260\,\frac{\text{mg}}{\text{L}} \times 5 \times \frac{1\text{ L}}{1000\text{ mL}}$$

Total excretion = 2340 mg/24 hr or 2.34 g/24 hr

Exercises

Calculate the concentrations (answers are in the appendix at the end of this chapter).

1. 10 M NaOH, which is diluted 1:20 = ____ M?
2. 2 M HCl, which is diluted 1:5 = ____ M?
3. 1000 mg/L glucose, diluted 1:10 and then 1:2 = ____ mg/L

Serial dilutions are those in which all the dilutions after the first one are the same. Exceptions to this general description of preparation of serial dilutions are included with certain techniques in serology, such as the antistreptolysin titer.

Serial dilution example. To determine the anti–RH_o (D) titer, serum is diluted 1:5 by addition of 0.2 mL of serum to 0.8 mL of saline solution, in tube 1. Tubes 2 through 8 contain 0.5 mL of saline as diluent. Dilution is performed by transferal of 0.5 mL of tube 1 to tube 2, mixing, and then transferring 0.5 mL through the tubes to tube 8. The concentration of serum in these tubes decreases by a factor of 2 with each dilution: 1:5, 1:10, 1:20, 1:40, 1:80, 1:160, 1:320, and 1:640.

4. For an ABO titer, tube 1 contains 0.9 mL of diluent, tubes 2 to 8 contain 0.5 mL of diluent, 0.1 mL of serum is added to tube 1, and serial dilutions using 0.5 mL are carried out in the remaining tubes. If the last tube showing agglutination with A cells is tube 6, what is the anti-A titer of the serum?

5. All tubes for serial dilution contain 0.5 mL of saline solution, 0.5 mL of serum is added to tube 1, and 0.5 of mL is transferred through the row of tubes. Sheep cells are added to the tubes, and agglutination is demonstrated through tube 7. What is the titer of sheep cell agglutinins?

WEIGHTS AND CONCENTRATIONS

Definitions and examples

Percent concentrations. Percent concentrations are generally expressed as parts of solute per 100 parts of total solution; hence the expression *percent,* or *per one hundred.* The use of percent concentration is derived historically from the early pharmaceutical chemists. Although these terms are still commonly used in the United States, major organizations (AACC, CAP) are attempting to use unified SI units. Concentrations in SI units are described in moles per liter when the molecular weight of the substance is known, or as weight (mass) per milliliter, or weight per liter, when the molecular weight is not known. Throughout the text SI units are used where possible.

The three basic forms of concentration are as follows:

Weight per unit weight (w/w). Both solute and solvent are weighed, the total equaling 100 grams.

Example. 5% w/w of NaCl contains 50 g of NaCl + 950 g of diluent.

Volume per unit volume (v/v). The volume of liquid solute per the total volume of the solute and solvent is expressed.

Example. 1% of HCl (v/v) contains 1 mL of HCl per 100 mL (or 1 deciliter) of solution.

Weight per unit volume (w/v). The most frequently used expression, concentrations of w/v are reported as grams percent (g%) or grams/dL, as well as mg/dL and μg/dL. The use of weight percent to describe concentration is being discouraged by professional organizations and, with few exceptions, is not used in this book. With SI units this would be in terms of weight per microliters (μL), milliliters (mL), or liters (L).

Example. To prepare 100 mL of 100 g/L of NaCl, weigh 10 g of NaCl and dilute to volume in a 100 mL volumetric flask.

Molarity. Molarity expresses concentration as the number of moles per liter of solution. One mole is the molecular weight of the substance in grams. A millimole is 1/1000 of a mole. A molar solution contains one gram-molecular weight of a substance per liter.

$$1 \text{ mole (mol)} = 1000 \text{ mmoles}$$
$$1 \text{ mmole} = 1000 \text{ μmoles}$$
$$1 \text{ μmole} = 1000 \text{ nmoles}$$

Examples. 1 M NaOH (molecular weight, or MW = 40.0 g/mole) contains one gram-equivalent molecular weight per liter, or 40 g diluted to 1000 mL with distilled water. A millimolar (1 mM) solution, or 0.001 molar (0.001 M), contains 1 mmole per liter. 1 millimole of NaOH is 1/1000 of 40 g, that is, 0.040 g (or 40 mg). When diluted to 1000 mL, the concentration of the solution will be 0.001 M.

Normality. Normality expresses concentration in terms of equivalent weights of substances. Equivalent weights are determined by the valence, which reflects the number of combining or replaceable units. A 1 normal (1 N) solution contains 1 equivalent weight per liter. The equivalent weight of an element or compound is equal to the molecular weight divided by the valence.

Normality and *molarity* relationships can be readily calculated if their definitions are understood.

Examples

1 M HCl = 1 N HCl, since 1 mole of H^+ or Cl^- react for every mole of HCl.

1 M H_2SO_4 = 2 N H_2SO_4, since 2 moles of H^+ (that is, equivalents) react for every mole of H_2SO_4.

1 M H_3PO_4 = 3 N H_3PO_4, since 3 moles of H^+ react for every mole of H_3PO_4.

1 M $CaCl_2$ = 2 N $CaCl_2$, since 2 Cl^- can react for every mole of $CaCl_2$.

1 M $CaSO_4$ = 2 N $CaSO_4$, since 2 mole volume electrons are available for reaction with either Ca^{++} or $SO_4^=$

Equivalent weights are known as the number of grams of an element (or compound) that will react with another element (or compound). This so-called law of combining weights is operable for all chemical compounds.

To simplify chemistry procedures and reports, factors can be used to express a quantity of one compound as an equivalent quantity of another compound. This process can be termed *equivalency*.

Example. Calculate the amount of urea if a patient's urea nitrogen level is 800 mg/L. The formula for urea is $NH_2{-}CO{-}NH_2$, and its molecular weight is 60 g/mol. The molecular equivalent weight for nitrogen in the mole is 14 g/mol × 2 molecules = 28. The *urea/nitrogen factor* is determined by the following equation:

$$\frac{\text{MW of urea (60)}}{\text{MW of nitrogen (28)}} = \frac{x \text{ g of urea}}{1 \text{ g of urea nitrogen}}$$
$$28\,x = 60$$
$$x = 2.14 \text{ (factor)}$$

This factor states that 2.14 g of urea would equivalently represent 1 g of urea nitrogen. So 800 mg/L urea nitrogen × 2.14 equals 1712 mg/L of urea. Laboratory results today are reported as urea nitrogen, since historical methods for this particular test are based on measurement of the urea nitrogen.

Competent laboratory personnel should be able to convert mg/dL to mEq/L. Electrolyte equivalents can be calculated from the equation:

$$\text{mg/dL} \times 10 = \text{mg/L}$$

Since mg/mEq weight is the millimolar weight in milligrams divided by the valence

$$\frac{\text{mg/L}}{\text{mg/mEq}} = \text{mEq/L}$$

or

$$\frac{\text{mg/dL}}{\text{mg/mEq}} \times 10\,\frac{\text{dL}}{\text{L}} = \text{mEq/L}$$

Example. What is the mEq/L concentration of serum chloride reported as 250 mg/dL? Since the millimolecular weight of chloride is 35.5 (that is, 1 mmol = 35.5 mg), the milliequivalent weight of chloride is

$$\frac{MW}{Valence} = \frac{35.5 \text{ g/mole}}{1}$$

$$\frac{250 \text{ mg/dL}}{35.5 \text{ mg/mEq}} \times 10 \frac{dL}{L} = 70 \text{ mEq/L}$$

Specific gravity. Specific gravity can be used to determine the mass (weight) of solutions. It relates the weight of 1 mL of the solution and the weight of an equal volume of pure water at 4° C (1 g). One practical use of specific gravity is in preparation of dilutions from concentrated commercial acids, the equation being:

Specific gravity × Percent assay = Grams of compound per milliliter

Example. Concentrated HCl has a specific gravity of 1.25 g/mL and is assayed as being 38% HCl. What is the amount of HCl per milliliter?

$$1.25 \text{ g/mL} \times 0.38 = 0.475 \text{ g of HCl per mL}$$

One common error is neglecting to change the percent assay to its proper decimal; in the above example, 38% = 0.38!

Exercise

6. Using the above example, how many milliliters are needed to prepare 1 liter of a 0.1 N HCl solution if the molecular weight of HCl = 36.5?

Water of hydration. Some salts are available in forms both anhydrous (no water) and hydrated (with water molecules). The form of the available salt, including the water of hydration, is listed on the manufacturer's label. To prepare accurate weight concentrations of these salts, calculations must include the molecules of water present in the compound. This is most easily done by calculation of the percentage of the compound that is in the anhydrous form. With this percentage, the weight of the hydrated form can be corrected to that of the anhydrous form.

The advantage of using molar concentrations is that the water of hydration does *not* have to be accounted for in the calculations. For example, 1 mol of $CuSO_4$ = 160 g, and 1 mol of $CuSO_4 \cdot 5\ H_2O$ = 250 g. One gram-equivalent molecular weight of each compound will contain 1 mol of $CuSO_4$; that is:

$$250 \text{ g of } CuSO_4 \cdot 5\ H_2O = 1 \text{ mole of } CuSO_4 = 160 \text{ g of } CuSO_4$$

Example. How many grams of $MgCl_2$ are there in 1 g of $MgCl_2 \cdot 3\ H_2O$?

Mg	24	Mg	24
Cl_2	71	Cl_2	71
	95 MW	$3\ H_2O$	54
			149 MW

$$\frac{95}{149} = \frac{x}{1}$$

$$149\,x = 95$$

$$x = 63.7\%$$

One gram of $MgCl_2 \cdot 3\ H_2O$ contains 0.637 g of $MgCl_2$.

Mole fraction. Mole fraction refers to the ratio of the amount of a component to the total mixture of components. Mole fraction is a derived unit that is expressed as either a percent or as a decimal.

Example. What percent Mg is contained in $MgCl_2 \cdot 3\ H_2O$?

$$Mg = 24 \quad Cl = 35.5 \quad MgCl_2 \cdot 3\ H_2O = 149 \qquad \frac{24}{149} = 16.1\%$$

Mg is 16.1% of the molecule $MgCl_2 \cdot 3\ H_2O$.

Example. To determine mole percent calcium in calcium carbonate (MW = 100), 1 mole of $CaCO_3$ contains 100 g, comprising 40 (Ca) + 12 (C) + 48 (3 × O), of which 40 is calcium. The mole fraction of calcium in 1 liter of 1 mole of calcium carbonate equals 40%.

Example. How much $CuSO_4 \cdot 5\ H_2O$ must be weighed to prepare 1 liter of a solution containing 80 mg of $CuSO_4$?

Total MW of $CuSO_4 \cdot 5\ H_2O$ = 250
MW of $CuSO_4$ = 160

The proportion of $CuSO_4 \cdot 5\ H_2O$ that is $CuSO_4$ is 160/250 = 0.64. Thus, 1 gram of $CuSO_4 \cdot 5\ H_2O$ contains 1 g × 0.64 = 0.64 g of $CuSO_4$. The rest, 0.36 g, is water.

Therefore

$$\frac{80}{0.64} = 125 \text{ mg of } CuSO_4 \cdot 5\ H_2O$$

Examples of calculations

a. What is the normality of concentrated HCl that has a specific gravity of 1.19 g/mL and a 38% assay?

Specific gravity × Percent = Grams/milliliter

$$1.19 \times 0.38 = 0.452 \text{ g/mL} = 452 \text{ g/L}$$

36.5 g of HCl per liter = 1 normal

$$\frac{452 \text{ g/L}}{36.5 \text{ g/Eq}} = 12.4 \text{ Eq/L}$$

b. If 24.5 g of H_2SO_4 (MW = 98 g/mol) are dissolved in a 1 liter solution
 (1) what is its molarity?
 (2) what is its normality?
 (answer 1) 1 mol of H_2SO_4 = 98 g; therefore 24.5 g equals

$$\frac{1 \text{ mol}}{98 \text{ g}} = \frac{x}{24.5 \text{ g}}$$

$$x = 0.25 \text{ mol}$$

$$0.25 \text{ mol in 1 liter} = 0.25 \text{ mol/L} = 0.25 \text{ M}$$

(answer 2) The valence of H_2SO_4 equals 2; therefore the equivalent weight of H_2SO_4 is expressed as follows:

$$\text{Equivalent weight} = \frac{\text{Molecular weight}}{\text{Valence}} = \frac{98 \text{ g}}{2}$$

Equivalent weight = 49 g

To solve for the number of equivalents in 24.5 g

$$\frac{1 \text{ equivalent}}{49 \text{ g}} = \frac{x}{24.5 \text{ g}}$$

$$x = 0.5 \text{ equivalents}$$

0.5 equivalents in 1 liter = 0.5 Eq/L = 0.5 N

c. If the molecular weight of $CaCO_3$ is 100 g/mol and the atomic weight of calcium is 40 g/mol, how many grams are needed to prepare:
(1) 1 liter of 0.1 M $CaCO_3$?
(2) 100 mL of 10 mg/dL of Ca^{++} using $CaCO_3$?
(3) 50 mg/L of $CaCO_3$?
(4) 0.2 mEq/L of Ca^{++} using $CaCO_3$?
(Answer 1) 1 mol of $CaCO_3$ = 100 g; therefore 0.1 mol = 10 g, since 0.1 molar = 0.1 mol/L = 10 g/L
(Answer 2) The percentage weight of Ca^{++} in $CaCO_3$ is

$$\frac{\text{Atomic weight } Ca^{++}}{\text{Molecular weight } CaCO_3} = \frac{40}{100} = 40\%$$

In 100 ml, 10 mg of Ca^{++} is needed or

$$\frac{10 \text{ mg}}{\% \text{ Ca in } CaCO_3} = \frac{10 \text{ mg}}{40\%} = \frac{10 \text{ mg}}{0.4} = 25 \text{ mg of } CaCO_3$$

Therefore 25 mg of $CaCO_3$ = 10 mg Ca^{++}
(Answer 3) For 1 liter, 50 mg of $CaCO_3$ is needed.
(Answer 4) 1 equivalent weight of $CaCO_3$ equals

$$\frac{\text{Molecular weight of } CaCO_3}{\text{Valence}} = \frac{100 \text{ g/mol}}{2 \text{ equivalents/mol}} = 50 \text{ g/equivalent}$$

To convert to milliequivalents

$$1 \text{ mEq} = \frac{1 \text{ Eq}}{1000}$$

$$\therefore \quad \frac{50 \text{ g}}{1 \text{ Eq}} \times \frac{1 \text{ Eq}}{1000 \text{ mEq}} = 50 \text{ mg/mEq}$$

To calculate amount to prepare 1 liter of 0.2 mEq

$$\frac{1 \text{ mEq wt}}{50 \text{ mg}} = \frac{0.2 \text{ mEq wt}}{x}$$

$$x = 10 \text{ mg of } CaCO_3$$

Exercises

7. 3 M $CaCl_2$ (MW 111.1) = ____ N $CaCl_2$
8. 2 N H_3PO_4 (MW 98) = ____ M H_3PO_4
9. 2 M H_2SO_4 (MW 98) = ____ N H_2SO_4
10. 250 mL of 5% NaCl contains ____ g of NaCl
11. How much $CuSO_4 \cdot 5\ H_2O$ must be weighed to prepare 100 mL of 5% $CuSO_4$? (MW $CuSO_4$ = 159.61; MW H_2O = 18)
12. What percent of $CuSO_4 \cdot 5\ H_2O$ is water? ____%

CALCULATIONS BASED ON PHOTOMETRIC MEASUREMENTS (BEER'S LAW)

Refer to Chapter 3 for a description of the relationship between percent transmittance, absorbance, and concentration.

Colorimetry

Colorimetry is the measurement of the kind and amount of light absorbed or transmitted by a solution. These measurements of absorbance or transmittance are logarithmically related. Beer's law (see below) reflects the relationships between the absorbance and concentration of a known standard solution with that of solutions with unknown concentrations, the patients' samples. Beer's law states that the absorbance of a solution is directly related to its concentration. If Beer's law is true, then:

$$C_u = \frac{A_u}{A_s} \times C_s \qquad \textit{Eq. 1-5}$$

C_u and A_u represent concentration and absorbance of the unknown samples, whereas C_s and A_s reflect that of the standard solution. When preparing a colorimetric method for clinical chemistry analysis, one must be sure that Beer's law is followed or this formula cannot be used. In other words, this formula can only be used if the absorbance and concentration are directly related, that is, if the absorbance doubles with a doubling of concentration. A *standard curve* can be employed to determine graphically concentration values. Standard-curve preparations are described on p. 39.

Absorbance and transmittance

Absorbance measures the amount of light that is blocked, or absorbed by a solution. Absorbance is also termed "optical density" (OD), a term found in the older literature and not in common use today.

Transmittance measures the amount of light that passes through a solution. The transmittance is usually expressed as a percentage, or $\%T$. The $\%T$ scale is linear, as noted on a colorimeter readout scale.

As discussed in Chapter 3, the absorbance and percent transmittance are logarithmically related, since absorbance is a logarithmic function. Interconversion of absorbance and percent transmittance is commonly expressed by the following formulas:

$$A = -\log \frac{\%T}{100} \qquad \textit{Eq. 1-6}$$

This equation can be algebraically converted to the following form:

$$A = -(\log \%T - \log 100)$$
$$A = -(\log \%T - 2)$$
$$A = -\log \%T + 2$$
$$A = 2 - \log \%T \qquad \textit{Eq. 1-7}$$

One can obtain absorbance from a hand calculator using this formula by punching in the numbers for $\%T$, converting to the log form, placing a minus sign, and adding 2.

Examples. Determining concentrations using absorbance (A, or OD) readings.

a. If absorbance of an unknown is 0.25 and the concentration of a standard is 4 mg/L with an absorbancy of 0.40, the concentration of the unknown can be calculated using:

$$C_u = \frac{A_u}{A_s} \times C_s$$
$$C_u = \frac{0.25}{0.40} \times 4 \text{ mg/L}$$
$$C_u = 2.5 \text{ mg/L}$$

b. To calculate the concentration of glucose if the following information is known:

$$C_s = 2000 \text{ mg/L}, A_s = 0.40, A_u = 0.25$$

Using the same formula above

$$C_u = \frac{0.25}{0.40} \times 2000 \text{ mg/L}$$
$$C_u = 1250 \text{ mg/L}$$

c. To calculate glucose concentration of unknown (C_u) if the $\%T$ is given.
If the 1000 mg/L glucose standard (C_s) reads 49% T, and the unknown reads 55% T, the $\%T$ must be converted to absorbancy, since absorbancy is linearly proportional to concentration:

$$A_s = 2 - \log \%T \qquad A_u = 2 - \log \%T$$
$$A_s = 2 - 1.690 \qquad A_u = 2 - 1.740$$
$$A_s = 0.31 \qquad A_u = 0.26$$
$$49\%\ T = 0.31\ A = A_s$$
$$55\%\ T = 0.26\ A = A_u$$
$$C_s = 1000 \text{ mg/L}, A_u = 0.26, A_s = 0.31$$

Using the formula as in (a) and (b) above:

$$C_u = \frac{0.26}{0.31} \times 1000 \text{ mg/L} = 839 \text{ mg/L}$$

Molar extinction coefficient

Molar extinction coefficients are used in the clinical laboratory to calculate concentrations and activities of enzymes in international units (U) and to determine the purity of dissolved substances. Specific applications are checking standard solutions, such as hemoglobin or bilirubin. The molar extinction coefficient, or molar absorbance coefficient, or ϵ, is defined as the absorbance at a given wavelength of a 1 M solution of the substance in a 1 cm cuvette at 25° C. It is related to absorbance by the formula

$$A = \epsilon cl \qquad \textit{Eq. 1-8}$$

where A = absorbance at a specified wavelength, c = concentration of substance being measured, in moles/L, and l = the path length in cm.

A suitable bilirubin standard, as a 1 M solution in chloroform, would have a theoretical absorbance of 60,700 (mean) ± 800 liters · moles^{-1} · cm^{-1} at 453 nm, when measured in a 1 cm cuvette at 25° C. Logical reasoning suggests that if this standard were diluted to 1:60,700 the absorbance would read 1.

Example. 1 M bilirubin standard is diluted to 1:60,700 and then 1:2, with the final dilution being 1:121,400. The absorbance of this dilution reads 0.495 nm in a 1 cm cuvette. What is the extinction coefficient of this bilirubin standard?

$$\epsilon = \frac{0.495}{1 \text{ mol/L}/121{,}400(1 \text{ cm})}$$
$$\epsilon = 60{,}093 \text{ liters} \cdot \text{mol}^{-1} \cdot \text{cm}^{-1}$$

A major application of ϵ is the measurement of concentrations of substances. If the ϵ of a substance is known and a 1 cm cuvette is used, Beer's law formula is simplified to the following:

$$c = \frac{A}{\epsilon}$$

Example. The ϵ of NADH at 340 nm is 6.22×10^3 liters · mol^{-1} · cm^{-1}. If the absorbance of NADH at 340 nm reads 0.350, what is the concentration?

$$c = \frac{0.350}{6.22 \times 10^3 \text{ L} \cdot \text{mol}^{-1}}$$
$$c = 5.6 \times 10^{-5} \text{ mol/L}$$

Exercises

13. NADH has a molar absorptivity of 3.3×10^3 at a wavelength of 366 nm. Calculate the concentration of a solution that has an A (absorbance) of 0.175 at 366 nm.

14. A chemistry technologist is checking a bilirubin standard. What would be the molar absorptivity of a 1 M solution diluted to 1:60,700 and reading 0.70 in a 7 mm cuvette?

15. The chemistry technologist has a solution of NADH with a concentration of 0.05×10^{-3} mol/L. Calculate the molar absorptivity if it measures 0.300 at 334 nm.

BUFFERS

Buffers resist changes in acidity by forming a weakly ionized acid or base with the added H^+ or OH^- ions. For example, when HCl is added to a solution of Na^+Ac^- (sodium acetate) plus H^+Ac^- (acetic acid), the H^+ of HCl will react with the Ac^- forming more HAc, which is only slightly ionized. The acetate–acetic acid effectively buffers by removing H^+ from the solution.

The Henderson-Hasselbalch equation is utilized to express acid-base relationships. There are several forms of this equation that will not be delineated at this time but can be used for calculating acid-base problems. The simplest equation is

$$pH = pK + \log \frac{\text{Concentration of conjugate base}}{\text{Concentration of weak acid}} \quad \textit{Eq. 1-9}$$

The pK value is dependent on a specific set of conditions; these are degree of dissociation, temperature, and pH. The pK for the bicarbonate buffer system in serum or plasma is 6.10 at 37° C. Chemical reference books, such as the *Handbook of Chemistry and Physics,** contains pK values. As capable medical technologists we should grasp the basic calculations of the Henderson-Hasselbalch equations, even though laboratory instruments provide direct "readout" values on patients' acid-base tests.

Examples

a. Calculate the pH of an acetate buffer composed of 0.20 M sodium and 0.05 M acetic acid. (The pK for acetic acid is 4.76.)

$$\begin{aligned} pH &= pK + \log \frac{[\text{Salt}]}{[\text{Acid}]} \\ &= 4.76 + \log \frac{0.20}{0.05} \\ &= 4.76 + \log 4 \\ &= 5.3621 \\ pH &= 5.36 \end{aligned}$$

b. Now for a complicated example. Prepare an acetate buffer whose concentration is 0.2 M and has a pH of 5.0. (The pK of acetic is 4.76; the molecular weight of acetic acid is 60 and is 82 for sodium acetate.)

$$\begin{aligned} pH &= 4.76 + \log \frac{[\text{Salt}]}{[\text{Acid}]} \\ \log \frac{[\text{Salt}]}{[\text{Acid}]} &= 5.0 - 4.76 \\ [\text{Salt}]/[\text{Acid}] &= \text{antilog } 0.24 \\ [\text{Salt}]/[\text{Acid}] &= 1.7 \end{aligned}$$

The number 1.7 is the ratio of the moles per liter of salt to the moles per liter of acid. Any molar concentrations of salt to acid yielding a ratio of 1.7 will result in a 5.0 pH acetate buffer.

Note. The problem specifies a concentration of 0.2 M solution, or HAc + Ac = 0.2 M.

If

$$\text{Ac/HAc} = 1.7$$

then

$$\text{Ac} = 1.7 \text{ HAc}$$

or

$$\text{HAc} + 1.7 \text{ HAc} = 0.2 \text{ M}$$

and

$$\begin{aligned} 2.7 \text{ HAc} &= 0.2 \\ \text{HAc} &= 0.074 \text{ mol/L of acid} \\ \text{MW} \times \text{M} &= \text{g/L} \\ 60 \times 0.074 &= 4.44 \text{ g/L of acid needed} \end{aligned}$$

To calculate the weight of salt needed, use:

$$\begin{aligned} \text{Moles/liter of salt} &= \text{Total moles} - \text{Moles/liter of acid} \\ &= 0.2 - 0.074 \\ &= 0.126 \text{ moles of salt per liter} \end{aligned}$$

As done for the acid:

$$\begin{aligned} \text{MW of salt} \times \text{M} &= \text{Grams of salt per liter} \\ 82 \times 0.126 &= 10.33 \text{ g of salt per liter} \end{aligned}$$

When 4.44 g of acid and 10.33 of salt are dissolved in a total volume of 1 liter, the resulting buffer concentration is 0.2 M at a pH of 5.0.

ENZYME CALCULATIONS

Expressing enzyme activity in international units has been generally accepted since its recommendation by the International Union of Biochemistry in the early 1960s. One international unit, U, of an enzyme is defined as the amount that will catalyze the transformation of 1 μmole of the substrate per minute under standard conditions. Activity is expressed in terms of enzyme units per liter of serum, or milliunits per milliliter, in the following relationship:

$$1 \text{ U/L} = \mu\text{moles/minute/liter of serum}$$

Explanation of the basic equation for the conversion of absorbance data to international units will not be attempted in this portion of laboratory calculation. Suffice it to state that any change in factors such as temperature or volume must be accounted for in the following basic equation (see Chapter 49):

$$\text{U/L} = \frac{\Delta A/\text{min} \times V_t \times 10^6 \ (\mu\text{mol/mol}) \times l}{\epsilon \times V_s} \quad \textit{Eq. 1-10}$$

$\Delta A/\text{min}$ = absorbance change per minute
V_t = Total reaction volume including sample, reagent, and diluent
V_s = Serum volume
l = Cuvette path length
ϵ = Extinction coefficient

The factor of 10^6 μmol/mol is added to convert the answer to μmol/min/L (U/L).

Example. What is the lactate dehydrogenase (LD) activity of 0.1 ml of serum + 3 mL of substrate if the NADH being formed showed a 0.002 ΔA per minute at 340 nm? ϵ for NADH = 6.22×10^3 L·mol^{-1}·cm^{-1}.

*Weast, R.C., editor: Handbook of chemistry and physics, Cleveland, Ohio, CRC press, yearly updated editions.

Using the above formula:

$$U/L = \frac{0.002\ (10^6\ \mu mol/mol)\ (3.1\ mL)\ 1\ cm}{1\ min\left(6.22 \times 10^3 \frac{L}{mol \cdot cm}\right)(0.1\ mL)}$$

$$U/L = 9.9$$

Example. Calculation of international units per liter of alkaline phosphatase activity using *p*-nitrophenol standard requires attention to all factors of the formula:

$$U/L = \frac{\Delta A/min \times V_t \times 10^6\ (\mu mol/mol) \times l}{\epsilon \times V_s}$$

If ΔA of sample = 0.070, the ϵ for *p*-nitrophenol is 50,000 liters/mol·cm; timing = 15 minutes; V_t = 5.5 mL; V_s = 0.5 mL

$$U/L = \frac{0.070\left(10^6 \frac{\mu mol}{mol}\right)(5.5\ mL)\ (1\ cm)}{15\ minutes\left(50{,}000 \frac{L}{mol \cdot cm}\right)(0.5\ mL)}$$

$$U/L = \frac{0.070 \times 5.5 \times 1000}{50 \times 0.4}$$

$$220\ U/L = \text{Alkaline phosphatase activity}$$

STANDARD CURVES

Preparation of standard curves on graph paper is an essential way of examining data for validity. Often calculations or computers do not reveal abnormalities of the system, but calculate averages of results. Therefore graphing of data is a very important way to validate assays.

Previously in this chapter, Beer's law was defined as the direct relationship of the absorbance and concentration of a solution. This means that if a 2% solution reads 0.1 *A*, then a 4% concentration will read 0.2 *A*, and an 8% solution will read 0.4 *A*. To repeat, most solutions obey Beer's law; that is, concentration and absorbance are directly proportional only over specified ranges of concentrations.

Graphs

Fig. 1-16. Absorbances of glucose standard concentrations plotted on linear paper results in a straight line, confirming that Beer's law is followed for the concentrations up to 3000 mg/L.

Fig. 1-17. Plotting the %*T* values of the same glucose concentrations used in Fig. 1-18 on linear paper produces a semicurved line; %*T* values are *not* linear absorbance versus concentrations. (Recall the logarithmic relationship of absorbance and %*T*.)

Fig. 1-18. Plotting %*T* values on semilog paper results in a straight line, which can be used to interpolate the concentrations of glucose from %*T* values.

Exercises (Figs. 1-16 and 1-18)

Find glucose concentrations for the following readings using Figs. 1-16 and 1-18:

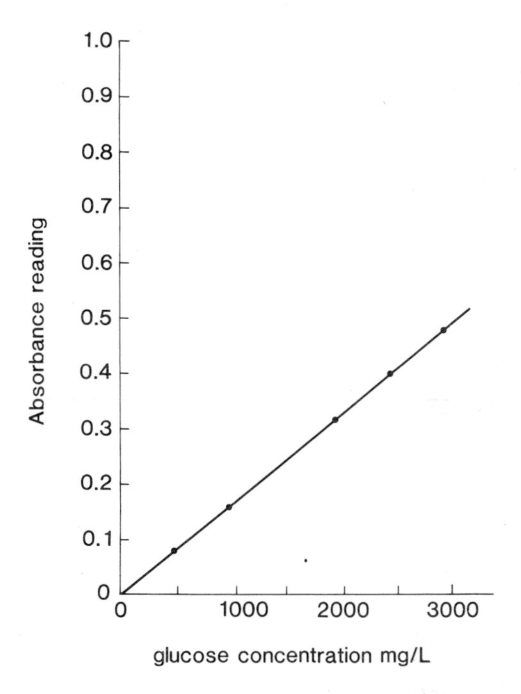

Fig. 1-16. Standard curve for glucose analysis: absorbance versus concentration on linear-linear graph paper.

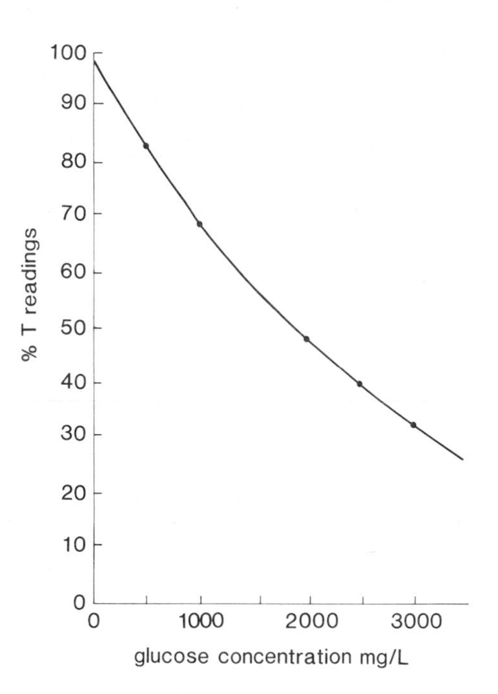

Fig. 1-17. Standard curve for glucose analysis: percent transmittance versus concentration on linear-linear graph paper.

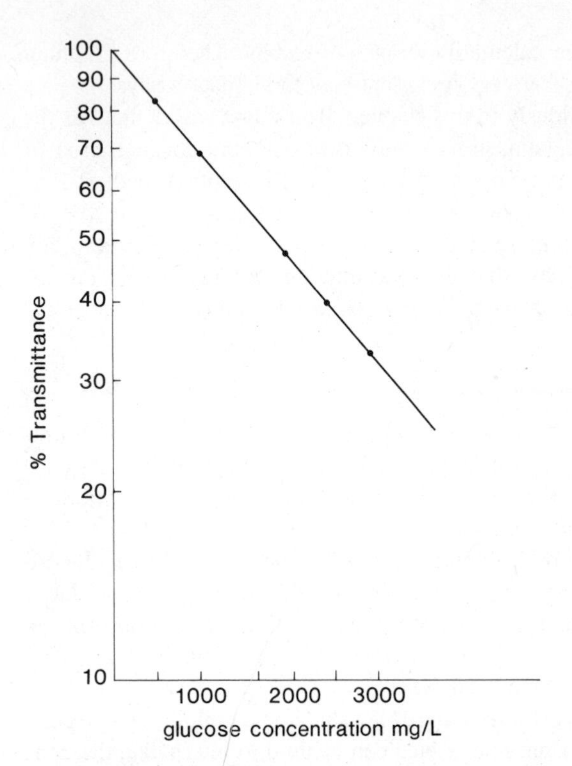

Fig. 1-18. Standard curve for glucose analysis: percent transmittance versus concentration on log-linear graph paper.

16. 0.3 *A* = ___ mg/L
17. 0.39 *A* = ___ mg/L
18. 49% *T* = ___ mg/L
19. 52% *T* = ___ mg/L

Exercises using one known standard value to determine concentrations of unknowns

What are the glucose concentrations of the following patients' samples if the 2000 mg/L standard reads 0.32 *A*?

We can employ either:

$$C_u = \frac{A_u}{A_s} \times C_s$$

or

$$\frac{C_u}{C_s} = \frac{A_u}{A_s}$$

20. 0.22 *A* = ___ mg/L
21. 0.14 *A* = ___ mg/L
22. 0.46 *A* = ___ mg/L

ADDITIONAL EXERCISES

23. The hemoglobin standard solution contains 200 g/L. What amounts would be used to prepare 6 mL of the following concentrations?

200 g/L? 150 g/L? 100 g/L? 50 g/L?

24. a. What fraction of urea is nitrogen? Urea is $CO(NH_2)_2$.

Atomic weights:
C = 12
O = 16
N = 14
H = 1

b. What percent of urea is nitrogen?

25. There is available concentrated HCl having a 38% assay and a specific gravity of 1.170.
a. What is the weight of HCl present in 1 mL?
b. For the preparation of 100 mL of 10% wt/vol HCl, ___ mL of HCl would be diluted to total volume of ___ mL.

26. Normal saline solution is 0.85% concentration. What is its molarity? (NaCI molecular weight is 58.5)

27. If a protein standard reads 0.48 *A*, and a patient's sample reads 0.36 A, what is the patient's protein concentration? Select one of the following answers:
a. Twice the standard concentration
b. Equal to the standard concentration
c. Three fourths of the standard concentration
d. Not enough data for calculation

28. A patient in diabetic coma has high blood glucose levels; so serum from this patient is diluted 1:2 and again 1:2 before it is readable from the glucose chart as 1900 mg/L. What is the actual concentration in (a) mg/L; (b) mg/dL, and (c) g/L?

29. How many mEq/L are there in a solution containing 27.7 mg/dL of potassium? (Atomic weight of K = 39.)

30. How many milliliters of 0.4 N NaOH can be made from 20 mL of 2 N solution?

31. What is the normality of a solution containing 40 mEq per 50 mL?

32. What is the dilution of serum in a tube containing 200 μL of serum, 500 μL of saline, and 300 μL of reagent?

33. Calculate the alkaline phosphatase activity in U/L for the following:

ϵ of standard (*p*-nitrophenol) = 5×10^4 $L \cdot mol^{-1} \cdot cm^{-1}$
ΔA_{sample} = 0.150
V_{sample} = 0.2 mL
V_{total} = 2.2 mL
Timing = 15 minutes

34. What is the extinction coefficient for the following:

Solution concentration = 1.2 molar
Dilution of solution = 1/121,400
A reading in a 1 cm cuvette = 0.6

35. A 0.01 M Na_2HPO_4 solution needs to be prepared (MW of Na_2HPO_4 = 141.98). Only the hydrated salt, $Na_2HPO_4 \cdot 7\ H_2O$ (MW = 267.98) is available. How many grams will be needed to prepare 250 mL?

36. If a 50 mg/mL solution of Na_2HPO_4 needs to be prepared, how many grams of the hydrated salt need be weighed to make a 1 liter solution?

37. A medical technologist desires to prepare a 50 mL of a 10 mg/mL solution of NADH (MW = 663.44). To do this accurately, the technologist will first prepare a stock solution containing approximately 50 mg/mL. The absorbance of a 1:1000 dilution of the stock solution will be measured at 340 nm and from the known molar absorbance of NADH at this wavelength (6.22×10^3), the actual concentration will be calculated. A suitable dilution will then be made to prepare the 50 mL of the desired 10 mg/mL solution. Presume that the technologist, following these directions, has prepared a dilution of the stock solution with an absorbance of 0.075. Calculate the concentration of NADH in this stock solution as mmol/L and mg/L. What dilution should be made to prepare the 10 mg/mL of NADH solution?

38. A patient's serum calcium level is 3.5 mEq/L. The expected normal range is 90 to 110 mg/L. Is the patient's calcium level lower than, within, or higher than the expected normal range?

39. A sodium concentration is reported as 3500 mg/L. What is the concentration in mEq/L? (atomic wt. of sodium = eq. wt. = 23 g/mol or 23 mg/mmol)

40. If the cyanmethemoglobin standard, with a concentration of 200 g/L, reads 0.426 *A*, and a patient's blood sample reads 0.297 *A*, what is the concentration of hemoglobin in the sample?

41. A 200 mg/L urea nitrogen standard reads 0.30 *A* and a patient's sample reads 0.40 *A*. The concentration of the standard compared to the patient's level is:

a. higher.
b. twice as much.
c. 3/4 as much.
d. 4/3 as much.

42. A glucose standard of 2000 mg/L reads 0.4 *A* and a patient's sample reads 1.0 *A*. The technologist should:

a. report result as 500 mg/dL.
b. repeat test before reporting.
c. repeat test on diluted sample.
d. prepare fresh glucose standard.

APPENDIX: ANSWERS TO PROBLEMS

1. 0.5 M
2. 0.4 N
3. 50 mg/L
4. 1:320
5. 1:128
6. 7.68 mL
7. 6
8. 0.667
9. 4
10. 12.5
11. 7.82
12. 36.05
13. $c = 53 \times 10^{-6}$ mol/L
14. $\epsilon = 60{,}700\ L \cdot mol^{-1} \cdot cm^{-1}$
15. $6.0 \times 10^3\ L \cdot mol^{-1} \cdot cm^{-1}$
16. 1900 mg/L
17. 2400 mg/L
18. 1920 mg/L
19. 1740 mg/L
20. 1330 mg/L
21. 870 mg/L
22. 2870 mg/L
23. 200 g/L = 6 mL + 0 mL of diluent
150 g/L = 4.5 mL + 1.5 mL of diluent
100 g/L = 3.0 mL + 3 mL of diluent
50 g/L = 1.5 mL + 4.5 mL of diluent
24. **a.** 28/60 = 7/15
b. 46.6%
25. **a.** 0.445 g
b. 22.5 mL will be diluted to 100 mL
26. 0.145 mol/L
27. c
28. **a.** 7600 mg/L
b. 760 mg/dL
c. 7.6 g/L
29. 7.1 mEq/L
30. 100 mL
31. 0.8 N
32. 1:5
33. 2.2 U/L
34. 6.07×10^4
35. 0.67 g
36. 94.4 g
37. Concentration of NADH in stock solution is 90.4 mmol/L or 60 mg/mL. Take 8.3 mL of the stock and dilute to 50 mL to prepare the 10 mg/mL of solution.
38. 70 mg/L: lower than the expected range
39. 152 mEq/L
40. 139 g/L
41. 3/4 as much
42. c

REFERENCES

1. Encyclopedia Britannica, ed. 15, 1974, section on weights and measures.
2. Langevin, L.: The introduction of the metric system, UNESCO publication: Impact of Science on Society **XI:**77, 1961.
3. Lines, J.G.: S.I. units: another view, Clin. Chem. **25:**1331-1333, 1979.
4. The SI for the health professions, Geneva, 1977, World Health Organization.
5. Lippert, H., and Lehmann, H.P.: SI units in medicine, Baltimore and Munich, 1978, Urban & Schwarzenberg.
6. Gabow, P.A., and Kaehny, W.D.: The anion gap: its meaning and clinical utility, Kidney **12:**5-8, March 1979.
7. Krupp, M.A.: Fluid and electrolyte disorders in current medical diagnosis and treatment. In Krupp, M.A., and Chatton, M.J., editors: Physician's handbook, Los Altos, Calif., 1982, Lange Medical Publications, p. 21.
8. Rose, J.C.: Pressures on the millimeter of mercury, N. Engl. J. Med. **298:**1361-1364, 1978.
9. Statement from Quadrennial International Symposium on Measurable Properties (Quantities) and Units in Clinical Pathology and Clinical Chemistry, Gaithersburg, Md., August 5 and 6, 1976, Am. J. Clin. Pathol. **71:**465-468, 1979.
10. The National Committee for Clinical Laboratory Standards Position Paper (PPC-11): Quantities and units (SI), Clin. Chem. **25:**657-658, 1979.
11. Metrication and SI units, Committee on Hospital Care, American Academy of Pediatrics, Pediatrics **65:**659-664, 1980.
12. Lashor, T.W., and Macurdy, L.B.: Precision laboratory standards of mass and laboratory weights, National Bureau of Standards Circular 547, Washington, D.C., 1954, United States Department of Commerce.
13. Laboratory Instrument Maintenance Manual, ed. 3, Skokie, Ill., 1982, College of American Pathologists.
14. Preventive Maintenance for Precision Balances, Engineering Service, Veterans Administration, Dept. of Medicine and Surgery, G-29, Part 56, May 1974, DM & S Suppl. MP-3.
15. Guide to the selection of accuracy classes of thermistor thermometers and the verification of their accuracy, NCCLS Proposed Standard, PS1-4, Villanova, Pa., 1979, National Committee for Clinical Laboratory Standards.
16. Mangum, B.W.: Standard reference materials 933 and 934: The National Bureau of Standards' Precision thermometers for the clinical laboratory, Clin. Chem. **20:**670-672, 1974.
17. Standard for temperature calibration of water baths, instruments and temperature sensors. NCCLS Approved Standard, AS1-2, Villanova, Pa., 1977, National Committee for Clinical Laboratory Standards.
18. Bowers, G.N., Jr., and Inman, S.R.: The gallium melting-point standard: its application and evaluation for temperature measurement in the clinical laboratory, Clin. Chem. **23:**733-737, 1977.
19. Bowie, L., Esters, F., Bolin, J., and Gochman, N.: Development of an aqueous temperature-indicating technique and its application to clinical laboratory instrumentation, Clin. Chem. **22:**449-455, 1976.
20. Preventive Maintenance for Centrifuges, Engineering Service, Veterans Administration, Dept. of Medicine and Surgery, G-29, Part 16, Dec. 1973, DM & S Suppl. MP-3.
21. Specifications for reagent water used in the clinical laboratory, NCCLS Approved Standard, ASC-3, Villanova, Pa., 1980, National Committee for Clinical Laboratory Standards.
22. Reagent Water Specifications, Commission on Laboratory Inspection and Accreditation, Skokie, Ill., 1978, College of American Pathologists.

BIBLIOGRAPHY

Alexander, J.J., and Steffel, M.J.: Chemistry in the laboratory, New York, 1976, Harcourt Brace Jovanovich, Inc.

Brewer, J.M., Pesce, A.J., and Ashworth, R.B.: Experimental techniques in biochemistry, Englewood Cliffs, N.J., 1974, Prentice-Hall, Inc.

Campbell, J.B., and Campbell J.B.: Laboratory mathematics: medical and biological applications, ed. 3, St. Louis, 1984, The C.V. Mosby Co.

Christiansen, H.E., and Lugibyhl, T.T., editors: Suspected carcinogens, Dept. of Health, Education, and Welfare, Washington, D.C., (NIOSH) 75-188, 1975, U.S. Government Printing Office.

Diehl, H.: Quantitative analysis: elementary principles and practice, Ames, Iowa, 1974, Oakland Street Sciences Press.

Duckworth, J.K., and Von Boechman, J.: In Commission on Inspection and Accreditation: Clinical Laboratory Improvement Seminar, Skokie, Ill., 1980, College of American Pathologists.

Flury, P.A.: Environmental health and safety in the hospital laboratory, Springfield, Ill., 1978, Charles C Thomas, Publisher.

Halper, H.R., and Foster, H.S.: Laboratory regulation manual, vol. 3, Germantown, Md., 1970, Aspen Systems Corp.

Handbook for analytical quality control in water and wastewater laboratories, United States Environmental Protection Agency, EPA-600/4-79-019, March 1979, Chapter 4.

Ibbot, F.A.: In Henry, R.J., Cannon, D.C., and Winkleman, J.W., editors: Clinical chemistry principles and techniques, ed. 2, Hagerstown, Md., 1974, Harper & Row, Publishers.

Microbiological methods for monitoring the environment, United States Environmental Protection Agency, EPA-600/8-78-017, Dec. 1978, pp 32-37.

Richterich, R., and Colombo, J.P.: Clinical chemistry, New York, 1981, John Wiley & Sons, Inc.

Chapter 2 Collection and handling of patient specimens

Nathan A. Pickard

accuracy The extent to which the measurement is close to the true value.

adsorption The attachment of one substance to the surface of another.

aerosol A mist produced by the atomization of a liquid.

anaerobic Occurring only in the absence of molecular oxygen.

analyte Any substance that is measured; usually applied to a component of a biological section.

anticoagulant A substance that can suppress, delay, or prevent coagulation of blood.

calculus An abnormal inorganic mass occurring within the body, usually composed of mineral salts.

chromogen A substance that absorbs light, producing color.

clot Semisolid mass of coagulated blood cells and proteins.

coagulation Process by which blood cells are trapped within a net of precipitated blood proteins to form a clot.

diurnal Occurring during the day.

in vitro Within a glass; observable in a test tube.

in vivo Within the living body.

menses The monthly flow of blood from the genital tract of women.

phlebotomy The opening of a vein for the collection of blood.

plasma The noncellular portion of blood.

relative centrifugal force (RCF) A method of comparing the forces generated by various centrifuges taking into consideration the speed of rotation and radius from the center of rotation.

serum The noncellular portion of blood from which fibrinogen and other clotting proteins have been removed.

stasis Stoppage of the flow of blood.

Quality control in the clinical laboratory begins before the sample is collected from the patient. Accuracy arises from ensuring that the appropriate specimen is collected, the correct collection vessel is used, and the pertinent collection variables are considered. Once the sample is collected and delivered to the laboratory, errors may arise during the period before the analysis of the sample. This chapter reviews these considerations, highlighting those areas where the most common or significant errors may arise.

The problem of sample collection and processing is a complex one because there is no simple system that will meet the requirements for all analytes that the clinical chemistry laboratory measures. Thus one method of collection and storage for a particular analyte may be invalid for another, and so, inappropriately collected specimens are a common problem. This occurs because of a lack of awareness of the availability of many methods of collecting samples. In addition, it is important to determine the *type* of specimen required for a particular analyte before one begins to obtain a specimen. For example, the collection of urine for a toxic-drug screen is an appropriate specimen whereas a serum sample is not. It is a proper function of the laboratory to provide this information to both physicians and nurses.

The laboratory must remind hospital staff and patients as to the proper method of collecting and storing urine and stool samples for clinical analysis. Instructions should be provided in writing and attached to all specimen-collection containers. In addition, an instruction or warnings concerning the presence of preservatives, dietary restrictions, and other precautions should be clearly labeled.

TYPES OF SPECIMENS

Blood

Blood is, of course, a suspension of cells in a protein-salt matrix. The noncellular portion of blood contains a series of proteins, some of which are involved in the coagulation process. This fluid is called *plasma*. When the coagulation process is allowed to proceed to completion, the noncellular fluid, which can be separated from the clotted material, is termed *serum*.

Blood used for biochemical analysis is collected either from the veins, the arteries, or the capillaries. For most testing, the site of phlebotomy has no analytical or physiological significance, so that venous blood is utilized because of the ease of collection. For a limited number of analytes such as blood gases and lactic acid, significant differences arise between arterial and venous samples.[1,2]

Most testing is performed on the liquid or serum fraction of the blood that has been allowed to clot. The assumption is made that the distribution of constituents between the cellular and extracellular compartments of blood is roughly equal. Therefore one can extrapolate the concentration of an analyte in blood from that measured in serum. This assumption is usually valid. For exceptions, see the effects of hemolysis on concentration described below.

For some analytes it is necessary to inhibit the blood-clotting process using an anticoagulant. Other analytes require the addition of a preservative for accurate results. A later section will describe the various anticoagulants and preservatives available and their mechanisms of action.

Urine

Urine is commonly analyzed by the clinical chemistry laboratory. Both timed (such as 24-hour) collections, where quantitative analyses over a period of time are desired, and random or spot collections, where the concentration of the analyte in the random sample is desired, are used. Just as in the case of blood samples, there are certain constituents in urine that require stabilization or preservation. This can be accomplished by procedures such as refrigeration or the addition of acid or base.

Other body fluids

The laboratory is also called upon to perform a variety of testing on cerebrospinal fluid, amniotic fluid, and other body fluids and drainages. In all instances it is essential to ensure that the proper specimen is collected and contamination is avoided. Simple observation of a specimen, such as that of a bloody spinal tap, can be used to rule out use of the specimen. In addition, biochemical analyses may be performed on stool samples, gastric aspirates, and renal or biliary calculi. Stool samples are usually collected over 48- to 72-hour periods. Renal calculi collection may be a random event depending on when the stone is passed.

COLLECTION VESSELS

Syringes

Although syringes are not used as much as they were in the past, they are still required to collect samples for blood gas, spinal fluid, and amniotic fluid analysis. For blood-gas and pH analysis there are some who believe that only glass syringes and not plastic syringes should be used, though there is less than universal agreement.[3] The laboratory must determine whether there is a significant risk of contamination, collection problems such as excessive syringe friction, or other factors, such as adsorption, that would preclude the use of either plastic or glass syringes for a particular test.

Vacutainers

Syringe collection has been largely replaced by the use of evacuated blood collection tubes, as exemplified by the Vacutainer system (Becton Dickinson & Co., Paramus, NJ 07652). The tubes may be siliconized to minimize the risk of hemolysis and to prevent the blood from adhering to the wall of the tube (Fig. 2-1). The Vacutainers now come presterilized by irradiation and are available in a variety of sizes from 2 to 30 mL, with the 10 mL tube being the most commonly used tube. Table 2-1 lists the different types of phlebotomy tubes available according to the anticoagulants and preservatives present. Also indicated are their mechanism of action, stopper color, and specific ap-

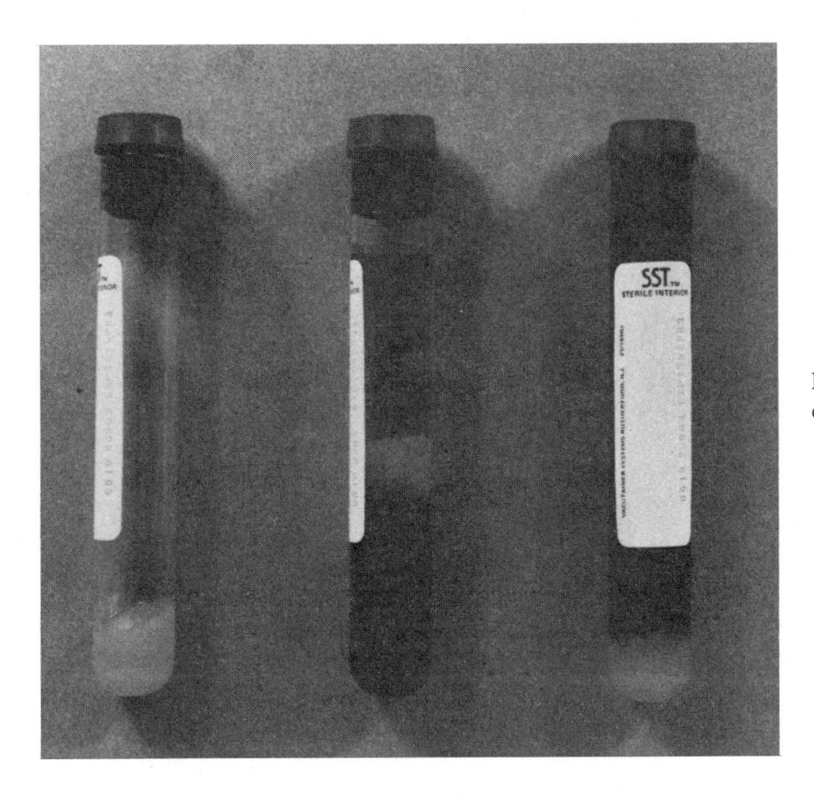

Fig. 2-1. Examples of evacuated phlebotomy tubes containing silicon barrier material.

Table 2-1. Phlebotomy tubes and agents

Samples	Type of anticoagulant	Stopper color	Chemical basis	Application
Phlebotomy Vacutainer tubes				
Plasma	Citrate	Blue	Binds calcium	Coagulation
Plasma	EDTA	Lavender	Binds calcium	Hematology
Plasma	Heparin	Green	Inhibits thrombin	Chemistry
Plasma	Oxalates	Black	Binds calcium	Coagulation
Antiglycolytic agents				
Serum	Iodoacetate	Gray	Inhibits glyceraldehyde-3-phosphate dehydrogenase	Glucose, lactic acid
Partial plasma	Fluoride	Gray	Inhibits enolase	Glucose
Special tubes				
Serum	—	Royal blue	Contaminant free	Trace elements Heavy metals
Serum	—	Brown	Lead free	Lead
Serum	Serum separator	Gray/red	Gel barrier	Chemistry
Serum	Thrombin	Orange	Increased rate of clotting	Stat. chemistries

plication. Specially made needles are used with the Vacutainer systems, with a 20-gauge needle being recommended for most patients and the 21- or 22-gauge needles recommended for pediatric patients and those with poor or traumatized veins. The use of needles outside this range increases the likelihood of hemolysis.[4]

The stoppers currently available from manufacturers have, in many cases, been reformulated to exclude TBEP, a common constituent of rubber shown to interfere with therapeutic drug-monitoring results. They are now recommended for most methodologies.[5]

Serum separators

Older types of Vacutainers required very careful separation of plasma or serum from the cellular components of blood. Often, in order to completely remove contaminating cells, several centrifugation steps were necessary. An advancement over this procedure has been the introduction of serum-separator materials. These are described later in the section on the serum-separation process.

It is the laboratory's responsibility to determine or verify the manufacturer's claim that a specific tube is suitable for a given analysis and that neither the tube itself nor the stopper has an effect on the analyte of interest either by containing a contaminant, absorbing the substance of interest, altering protein binding, or inhibiting an enzyme reaction. Any effect could depend on the specific procedure that is being used by the laboratory. This is of concern not only for routine biochemical determinations but also for tests for metals, drugs, and hormones.[6,7] There can be no single comprehensive reference to these problems, and only a review of the literature and experimental study for the analyte of interest by a specific method will minimize the chance of introducing an analytical error or artifact.

Other collection containers

Special collection containers are also available to facilitate the collection of urine and stool samples, with consideration being taken for the need to add a preservative, when indicated, to stabilize a constituent in the sample. It is also helpful to weigh the container before collection so that minimal manipulation will be required upon receipt of the sample in the laboratory. For the determination of bilirubin one may need to provide a light-shielding container, such as dark glass or aluminum foil wrapping, because of bilirubin's photosensitivity. Patients should also receive clear instructions as to the proper method for collecting the sample.

In addition, contamination of the collection vessel can be a problem. Testing for analytes such as heavy metals requires the use of an acid-washed container to avoid contamination of the sample.

COLLECTION VARIABLES

Diurnal variation

The first item that must be considered in evaluating the clinical significance of results is the time of day the sample was taken. Some analytes demonstrate significant diurnal variation. Some common examples are listed.

Cortisol	Corticosteroids
Iron	Glucose
Estriol	Triglycerides
Catecholamines	

Within-day variation for these substances may be as much as 30% to 50%. Unless there is a specific contraindication, it is best to collect samples as soon as the patient wakens.

Posture

The posture of the patient at the time of collection can have a significant effect upon protein and protein-bound substances in the serum.[8] Some serum components known to demonstrate postural variation are the following:

Total protein	Iron
Albumin	Calcium
Lipids	Enzymes

When patients go from the supine to the standing position, these serum constituents increase their concentration 5% to 15%. This effect is probably attributable to the movement of water out of the intravascular compartment upon standing. Again, the only ideal ways to minimize this effect are (1) drawing all blood samples from supine patients and (2) establishing reference ranges using blood drawn from supine, healthy volunteers.

Stasis

Prolonged use of a tourniquet also may elevate a number of laboratory results. The application of a tourniquet for an extended period results in a stasis or pooling of blood above the constriction. Tourniquets should always be avoided during the collection of samples for blood-gas analysis or for lactic acid determinations.[4] The resultant slowing of blood flow increases blood pH, decreases Po_2, increases Pco_2, and increases the lactate concentration because of its production by anaerobic metabolism. In addition, prolonged use of a tourniquet will increase the serum concentration of protein and protein-bound substances such as those already described. Significant elevations may be seen with as short as a 3-minute application of a tourniquet.[8]

Hemolysis

During sample collection and until the serum is isolated from the red blood cells, care must be taken to minimize the opportunities for hemolysis. Hemolysis may arise because of the use of too large or too small a needle, moisture in a syringe, vigorous mixing of the blood, rapid expansion of the blood into the tube, or the separation process. Whatever the cause, the net effect is to increase falsely the serum concentration of analytes present in high concentration within the red blood cells.[4] On the other hand, for those substances that exist at lower concentrations in the red cells than outside, hemolysis will result in a dilution effect on the serum constituents. Common analytes whose concentrations are significantly affected by hemolysis are as follows:

Total protein	Triglyceride
Albumin	Norepinephrine
Lipids	Renin
Iron	Aldosterone
Calcium	Potassium
Enzymes	Magnesium
Bilirubin	Inorganic phosphorus
Cholesterol	

This phenomenon does not require visible hemolysis and may result simply from prolonged cell-serum contact before separation. Leaking of potassium and enzymes may occur without the visible leakage of hemoglobin. In addition to the alterations in concentration of certain substances because of hemolysis, the hemoglobin itself may cause a methodological interference. This has been documented by certain procedures for the measurement of albumin, angiotensin II, calcium, carotene, cortisol, iron, insulin, and lipase.[9,10] The effect of chromagens such as hemoglobin in the serum may be lessened in spectrophotometric measurements by the utilization of serum blanking or bichromatic analysis.[11]

Preservation

Once the blood sample is collected, if requested for certain analytes such as blood-gas analysis or lactate, which would be affected by the red blood cell glycolytic activity, it should be placed immediately on ice. Erroneous results will occur within minutes if the sample is not placed on ice. Blood-gas samples once placed on ice will be stable for approximately 1 hour.[12] Samples for plasma catecholamines, ammonia, and acid phosphatase may also require stabilization by placement in an ice bath.

The stress of phlebotomy may also affect laboratory results. Anxiety of the patient may result in changes in catecholamine levels and blood-gas results through direct hormonal effects and hyperventilation. Every effort should be made to calm the patient before collecting the sample. In addition, during the phlebotomy, errors may arise from improper swabbing of the venipuncture site, traumatizing the arm or drawing too close to an intravenous site.[13]

TRANSPORTATION

Consideration must be taken of the use of an automated delivery system, pneumatic or other, as to its effect on the validity of laboratory results. Of greatest concern is the degree to which the samples are subjected to vibrations that would affect the integrity of the red blood cell membranes so as to cause the leakage of potassium, lactate dehydrogenase, acid phosphatase, aspartate aminotransferase, and so on.[14] The laboratory personnel should check the system's literature and obtain references and written commitments from the manufacturer of the system being considered before contracting for such a system. The chance of damage to the red blood cells is lessened by systems that employ carriers that travel at slower speeds and use only totally filled tubes.[15]

LABORATORY CRITERIA FOR UNACCEPTABLE SAMPLES

For the reasons described above, some samples cannot be used for the proper analysis of certain analytes. The clinical laboratory is responsible for establishing criteria for specimen rejection. Documentation of rejected samples should be kept and monitored to establish any patterns that could be corrected. Following are conditions that cause blood specimens to be rejected.

Inadequate sample identification. Each hospital must determine the minimum amount of patient information to be included on each laboratory slip and specimen. This usually includes the patient's name, address, room, identification number, age, and sex. Phlebotomists must visually and verbally verify the identity of the patient, comparing the name on the wristband, test requisition, and labels. The tube and laboratory slip should be rechecked for verification of its identity upon receipt in the lab. Differences between the name on the laboratory slip and the name on the sample container are grounds for sample rejection.

Inadequate volume of blood collected into an additive tube or syringe. The amount of additive added to a Vacutainer tube presumes the tube will be completely filled with blood. For example, preheparinizing a syringe and then collecting an insufficient volume of blood may result in erroneous results.

Use of an improper collection tube. In general, serum is the preferred sample for most biochemical analyses. Sodium fluoride tubes designed for glucose samples are unsuitable for most other procedures. Chelating agents are unacceptable for many enzymatic methods. Heparin is the anticoagulant least likely to affect clinical chemistry procedures, though this is largely method dependent.

Hemolysis. Visible hemolysis (greater than 200 mg/L of hemoglobin) is unacceptable when one is testing for these analytes if certain methods are used. The degree of interference is dependent on the extent of hemolysis, analyte concentration, and methodology. The following analytes can be significantly affected:

Potassium (+)*
Lactate dehydrogenase (+)
Aspartate aminotransferase (+)

*"+" denotes a positive interference.

Acid phosphatase (+)
Creatine phosphokinase (+)
Iron (+)
Magnesium (+)
Inorganic phosphorus (+)
Haptoglobin (−)*
Bilirubin (−)

Improper transportation. Samples for blood gases, lactic acid, ammonia, and other procedures where there is a significant sample lability should not be analyzed if not transported to the laboratory on ice and delivered within a specified time.

Interferents. The presence of potential analytical interferents such as icteric serum, lipemia, turbidity, drugs, diet, or isotope exposure, which may interfere with a specific analytical procedure, is a basis for rejecting a sample.

SAMPLE PROCESSING

Clotting

Blood will normally clot within 20 to 60 minutes, faster if the blood collection tube contains a clotting activator, slower if the sample is on ice. Clot formation should be completed before centrifugation, or fibrin formation will cause clotting problems within the laboratory instrumentation and hamper pipetting. The time between sample collection and receipt by the laboratory should not exceed 45 minutes. Tubes of blood during transport to the laboratory should be kept in the stopper-up position to promote clot formation and reduce agitation of the sample. In addition, this position minimizes the possibility of hemolyzing the sample and contaminating the sample because of substances released from the stoppers.[16]

Centrifugation

To separate the liquid portion of the blood from the cellular components, centrifugal force generated by a centrifuge is used. The first step is to close all tubes of blood. This may be either the original rubber stopper of the Vacutainer or some other suitable tube closure. Open tubes of blood should never be centrifuged because an aerosol, produced from the heat and vibration, is generated by the centrifuge. Aerosols will increase the risk of infection to the laboratory staff. Centrifugation of open tubes of blood also may result in evaporation of the sample.

The most common type of centrifuge used in the clinical laboratory is the horizontal head. Other types of centrifuges include the angle head, ultracentrifuge, and special-purpose centrifuge.

In the horizontal-head centrifuge the samples are placed in holders that are in a vertical position when the rotor is stationary. As the rotor spins, the holders swing to a horizontal position with respect to the axis of rotation. The angle-head centrifuge has holders in a fixed position, usually at an angle of 52 degrees to the shaft around which it spins. Because of the nature of their construction, horizontal-head centrifuges can attain speeds up to roughly 3000 revolutions per minute, about 1700 *g*, without producing excessive heat caused by friction between the head and air. On the other hand, angle-head centrifuges, because of their smoother design, produce less heat and may attain speeds in the range of 7000 rpm (about 9000 *g*), though this is not required for routine clinical use. Samples spun in a horizontal-head centrifuge result in the cells traveling the entire length of the sample to reach the bottom. The cells may remix with the serum (plasma) as the tubes drop from the horizontal to vertical position when the centrifuge slows to a stop. This is more of a problem when the centrifuge brake is used rather than letting the head gradually coast to a stop. Remixing does not occur when an angle-head rotor is used where the cells have less of a distance to travel—the efficiency of separation is greater and the sample cups do not change position on deceleration.[17] With the use of separator or barrier tubes, horizontal centrifugation may give a more uniform barrier between the fluid and cell phases and thus a more discrete separation.

In general, complete separation of the serum from cells is achieved with centrifugation at 1000 to 1200 *g* for 10 (± 5) minutes.[18] Manufacturer's directions should be followed when utilizing special collection tubes or serum separator devices that may require different conditions.

Most centrifuges display the approximate speed of the centrifuge in rpm (revolutions per minute). A more meaningful expression is the RCF, or relative centrifugal force, which can be understood more clearly and takes into account the radius of the centrifugal head and thus the force on the tubes. There are nomograms available that graph the RCF, rpm, and radius in centimeters to allow conversion of the terms (Fig. 1-12, p. 23). Alternatively the RCF may be evaluated by the following expression:

$$\text{RCF} = 1.118 \times 10^{-5} (R)r^2$$
$$R = \text{Rotating radius (cm)}$$
$$r = \text{Speed of rotation (rpm)}$$

Example. If the rotating radius is 12 cm and the speed of the centrifuge is 2000 rpm, the resultant RCF would be 1.118×10^{-5} (12) 2000^2, or 537 *g*.

In those cases where a sample is being tested for an unstable analyte (such as lactate), the heat normally generated by the centrifuge would cause sample deterioration. To prevent this from happening, one must use a refrigerated centrifuge to maintain sample integrity. The samples that are being analyzed for substances unstable at cold temperatures (such as lactate dehydrogenase isoenzymes) should never be spun in a refrigerated centrifuge.

Serum-separator devices

There are many types of serum-separator devices commercially available that are easy and safe to use if the laboratory follows the manufacturer's specifications and recognizes any limitations stated in the directions or

*"−" denotes a negative interference.

documented in the scientific literature. Otherwise, it is the laboratory's obligation to verify the use of these devices, determining whether they have any effect on laboratory tests as done by their specific procedures. Serum-separator devices may be broadly separated into two major types, those used during centrifugation and those used after centrifugation.

Devices used during centrifugation may be either integrated gel-tube systems or devices inserted into the collection tube just before centrifugation. Integrated gel-tube systems contain a gel that starts at the bottom of the tube. During centrifugation, because of its viscosity and density, the gel floats to a position above the cells and below the serum. These tubes have the advantage of avoiding the need to remove the stopper before centrifugation and so saving time and ensuring that aerosol production and evaporation do not occur. Depending upon the type of barrier system used, it is generally safe to store serum on the barrier for up to 48 hours, if not longer, though this should be verified by the laboratory.[19]

Other separation systems

There are also devices that can be added to the blood-collection tubes before centrifugation. One such device is a container that is placed on top of the tube, generally also serving as a tube closure. Other separation aids include a variety of nongel devices, which may be beads, crystals, disks, filters, or fiber plugs made of glass, plastic, fiber, or felt. In all cases the centrifugal force (at least 500 *g*) causes the materials, which are of an intermediate density, to form a barrier between the serum and the cells. Precautions should be taken to avoid hemolysis, evaporation, or aerosols. These devices also function, as the gel devices, on the basis of their density to form barriers between the cells and serum. They are more permeable than the gel devices so that the serum should be removed from the tube within 1 hour of centrifugation.[20]

There are devices that may be added after centrifugation to separate the clot from serum. Generally, they are plunger type of filters, having a plastic tube with a filter tip at the end in a plastic or rubber base. After centrifugation of the samples these devices are inserted with the filters passing through the serum, stopping just above the surface of the cells, avoiding contact with the cells, which can cause hemolysis. The device should then be withdrawn slightly to produce a small air gap below the filter, separating the filter from the cells. This minimizes the potential for leakage of constituents from the cells, through the filter into the serum.

SPECIMEN STABILITY AND STORAGE

Removal of cellular components

After the serum is isolated from the blood cells, care must be taken to maintain the integrity of the analyte in the sample. Consideration should be taken to ensure that the serum is essentially free of red blood cells in particular if there is an analysis of potassium, LDH, AST, or lactate. If it is necessary, the sample should be protected from light.

Stabilization of sample

Certain samples will require stabilization if the analysis is not performed immediately. Depending on the analyte it may need refrigeration, freezing, or deep freezing. Other analytes such as lactate dehydrogenase are more stable at room temperature and deteriorate on cooling. Some circumstances, as when processing samples for cardiac enzymes, may require saving several aliquots of serum on one patient if several tests are ordered with differing storage requirements. Therefore one might want to refrigerate an aliquot for creatine phosphokinase and save an aliquot at room temperature for lactate dehydrogenase. In addition, some blood and urine tests need acid or alkaline stabilization, which may make that aliquot unsuitable for other testing. Urines for vanillylmandelic acid and catecholamine analysis need acid stabilization, whereas porphyrins are more stable in alkaline urine.[21]

Storage

If sample analysis is not going to be completed the same day as it was collected, the sample should be covered and refrigerated or frozen as necessary until it can be analyzed. If it is necessary, the serum should be removed from contact with any serum-separator device or barrier and stabilized. Excessive exposure of the serum in the aliquot tube or in open sample cups to room air should be minimized to avoid evaporation with a resulting concentration of analytes in the serum. In addition, there is a significant loss of carbon dioxide dissolved in the sample to room air, falsely lowering the measured carbon dioxide result.[22]

The speed with which a serum evaporates is directly proportional to the room temperature, air flow in the laboratory, the duration of exposure, and the surface area of the serum exposed to room air. Greater errors arise by the use of microsample cups, which hold less than 500 µL of serum and have a large surface area.[23] Samples in microcups should probably be analyzed within 10 to 15 minutes after transferal of serum to the cup and should be kept covered as much as possible. Some investigators have recommended the layering of an organic solvent over the serum to minimize sample evaporation and the loss of carbon dioxide to room air.[24]

Other storage problems

Many other potential problems may occur after separation of the serum or urine, especially if the sample is saved for an extended period of time. Medical technologists should be aware of the possibility of bacterial or fungal growth. Errors may arise from recentrifugation of the sample because of hemolysis and as well the rimming of clots, though such rimming occurs less frequently now with newer siliconized blood collection tubes.[4] There has also been a recent report that placing of wood applicator sticks in the serum for extended periods of time may produce analytical artifacts. Potassium, calcium, and glucose concentrations increase after only 3 minutes of contact with wood applicator sticks.[25]

It would have been impossible to attempt to list every potential source of error in the collection and handling of patient specimens; they are endless and constantly expanding. Only continuing education, reading of the scientific literature, and critical analysis of manufacturer's claims to product applications will minimize the risk of introducing errors into a patient sample before the sample is analyzed. Analytical errors to be discussed in a later chapter may be significant enough without adding the avoidable problems discussed in this chapter.

REFERENCES

1. Cohen, J.J., and Kassirer, J.P.: Acid/base, Boston, 1982, Little, Brown & Co.
2. Klimt, C.R., Prout, T.E., and Bradley, R.F.: Standardization of the oral glucose tolerance test: report of the Committee on Statistics of the American Diabetes Association, June 14, 1968, Diabetes **18:**299-310, 1969.
3. Shapiro, B.A., Harrison, R.A., and Walton, J.R.: Clinical application of blood gases, ed. 2, Chicago, 1977, Year Book Medical Publishers, Inc.
4. Calam, R.R.: Reviewing the importance of specimen collection. J. Am. Med. Technol. **39:**297-298, 1977.
5. Kessler, K.M., Kewal, J., and Narayanan, S.: Effect of blood collection system in percent of free quinidine in serum, Clin. Chem. **26:**1004, 1980.
6. Anand, V.D., White, J.M., and Nino, H.V.: Some aspects of specimen collection and stability in trace element analysis of body fluids, Clin. Chem. **21:**595-602, 1975.
7. Missen, A.W., and Gwyn, S.A.: Another source of contamination from sample containers, Clin. Chem. **24:**2063, 1978.
8. Statland, B.E., Winkle, P., and Bokelund, H.: Factors contributing to intra-individual variation of serum constituents. 4. Effects of posture and tourniquet application on variation of serum constituents in healthy subjects, Clin. Chem. **20:**1513-1519, 1974.
9. Frank, J.J., Bermes, E.W., Bickel, M.J., and Watkins, B.F.: Effect of in vitro hemolysis on chemical values for serum, Clin. Chem. **24:**1966-1970, 1978.
10. Young, D.S., Pestaner, L.C., and Gibberman, V.: Effects of drugs on laboratory tests, Clin. Chem. **21:**133D-1334D, 1975.
11. Teitz, N.: Fundamentals of clinical chemistry, Philadelphia, 1976, W.B. Saunders Co.
12. Kelman, G.R., and Nunn, J.F.: Nomograms for correction of blood pO_2, pCO_2 and pH and base excess for time and temperature, J. Appl. Physiol. **21:**1484-1490, 1966.
13. Bodansky, O.: Diagnostic biochemistry and clinical medicine: facts and fallacies, Clin. Chem. **9:**1-18, 1963.
14. Stige, M., and Jones, J.D.: Evaluation of pneumatic tube system for delivery of blood specimens, Clin. Chem. **17:**1160-1164, 1971.
15. Pragay, D.A., and Edwards, L.: Evaluation of an improved pneumatic tube system for transportation of blood specimens, Clin. Chem. **20:**57-60, 1974.
16. Pragay, D.A., Brinkley, S., Rejent, T., and Gotthelf, J.: Vacutainer contamination revisited, Clin. Chem. **25:**2058, 1979.
17. Hicks, R., Schenken, J.R., and Steinrauf, M.A.: Laboratory instrumentation, Hagerstown, Md., 1974, Harper & Row, Publishers.
18. Calam, R.R., Benoit, S., Du Bois, J.A.: Proposed standards for the handling and processing of blood specimens: National Committee for Clinical Laboratory Standards, vol. 1, no. 16, pp. 504-505, 1981.
19. Narayanan, S., et al.: Control of blood collection and processing variables to obtain extended (greater than 5 day) stability in serum constituents, Clin. Chem. **25:**1086, 1979.
20. Seckinger, D.L., Antonio Vasquez, D., Rosenthal, P.K., and Heller, Z.H.: Evaluation of a new serum separator, Clin. Chem. *28:*157-159, 1982.
21. Pisano, J.J., Crout, R.J., and Abraham, D.: Determination of 3-methoxy-4-hydroxymandelic acid, Clin. Chim. Acta **7:**285-291, 1962.
22. Gambino, S.R., and Schreiber, H.: The measurement of carbon dioxide concentration with the autoanalyzer: a comparison with three standard methods and a description of a new method for preventing loss of carbon dioxide from open cups, Am. J. Clin. Pathol. **45:**406-411, 1966.
23. Burtis, C.A., Begovich, J.M., and Watson, J.S.: Factors influencing evaporation from serum cups, and assessment of their effect on analytical error, Clin. Chem. **21:**1907-1917, 1975.
24. Bandi, Z.L.: Estimation, prevention and quality control of carbon dioxide loss during aerobic sample processing, Clin. Chem. **27:**1676-1681, 1981.
25. Joseph, T.P.: Interferences from wood applicator sticks used in serum, Clin. Chem. **28:**544, 1982.

Chapter 3 Spectral techniques

Christopher S. Frings
Jack Gauldie

absorbance Defined as 2 − log %T, it is directly proportional to concentration of absorbing species if Beer's law is followed.

absorption spectrum The range of electromagnetic energy that is used for spectroanalysis, including both visible light and ultraviolet radiation; also graph of spectrum for a specific compound.

absorptivity Absorbance divided by the product of the concentration of a substance and the sample path length.

angle of detection The angle at which scattered light is measured in nephelometry.

band pass The range of wavelengths that reaches the exit slit of a monochromator; usually referred to as the range of wavelengths transmitted at a point equal to half the peak intensity transmitted.

Beer's law The concentration of a substance is directly proportional to the amount of radiant energy absorbed.

blank A solution comprised of all the components including solvents and solutes except the compound to be measured. This solution is used to set I_o, the original intensity.

cuvette The receptacle in a photometer in which the sample is placed.

electronic transition The change in the orbital position of an electron of an atom or molecule caused by the absorption of a photon of light. The electron usually goes from the ground or the lowest energy level to some higher one with a consequent higher energy state (increased energy) of the molecule. Basis of fluorescence phenomena.

emission wavelength The wavelength of light (λ_{em}) that is used to monitor decay of excited molecules into fluorescence; usually refers to the wavelength of output photons by a fluorometer.

excitation wavelength The wavelength of radiant energy (λ_{ex}) that is absorbed by a molecule and causes it to be raised to a higher energy state; usually refers to the wavelength of incident energy of a fluorometer.

filter An optical device (usually glass) that allows only a portion of polychromatic, incident light to pass through. The amount of transmitted light is related to the band pass of the filter.

fluorescence The light emitted by an atom or molecule after absorption of a photon. This light is at longer wavelengths (less energy) than the absorbed light.

grating An optical device consisting of a reflecting, ruled surface that disperses polychromatic light into a uniform, continuous spectrum. Dispersion of light is attributable to interference phenomena at the ruled surface.

hollow cathode lamp A lamp consisting of a metal cathode and an inert gas. When an electric current is passed through the cathode, the metal is sputtered free and, after colliding with the gas in the lamp, emits a line spectrum of specific wavelengths related to the metal of the cathode.

infrared radiation The region of the electromagnetic spectrum extending from about 780 to 300,000 nm.

internal standard An element or compound added in a known amount to yield a signal against which an instrument or an analyte to be measured can be calibrated.

light scattering The interaction of light with particles that cause the light to be bent away from its original path (cause of turbidity).

line spectrum Discontinuous emission spectrum of elements in which the emitted light bands cover a very narrow (0.1 nm) range of energies.

molar absorptivity **(ϵ)** The absorbance of light, at a specific wavelength, divided by the product of concentration in moles per liter and the sample path length in centimeters. Molar absorptivity is expressed as L/mol·cm.

monochromatic Light of one color (wavelength). In practice this refers to radiant energy composed of a very narrow range of wavelengths.

monochromator Device used to isolate a certain wavelength or range of wavelengths. Usually refers to prisms or gratings.

nephelometry A technique that measures the amount of light scattered by particles suspended in a solution.

photodetector A device that responds to light (photons) usually in a manner proportional to the number of photons striking its light-sensitive surface. Commonly a current that is proportional to the incident light intensity is generated.

photometer An instrument that measures light intensity; composed of a source of radiant energy, filter for wavelength selection, cuvette holder, detector, and a readout device.

photon A particle consisting of a discrete packet of radiant energy.

polychromatic Light of many colors (wavelengths), usually referring to white light, or that encompassing a defined portion of the spectrum.

Rayleigh scatter The reflection of light at different angles by particles suspended in a solution. This scattering occurs when the wavelength of light is greater than the size of the particles.

refraction A process by which the path of incident light is bent after passing obliquely from one medium to another of different density.

refractive index The ratio of the speed of light in two different mediums; usually the reference medium is air.

refractometer An instrument for measuring the refractive index (refractivity) of various substances, especially of solutions.

spectrophotometer An instrument that measures light intensity, composed of a source of radiant energy, an entrance slit, a monochromator, exit slit, cuvette holder, detector, and readout device. Measurements in these instruments can be made over a continuous range of available spectrum.

stray light Radiant energy reaching the detector and consisting of wavelengths other than those defined by the filter or monochromator.

ultraviolet radiation The region of the electromagnetic spectrum from about 180 to 390 nm.

visible light The radiant energy in the electromagnetic spectrum visible to the human eye (approximately 390 to 780 nm).

wavelength The linear distance traversed by one complete wave cycle of electromagnetic energy.

Table 3-1. Electromagnetic spectrum

	Gamma rays	X rays	Ultraviolet (UV)	Visible	Infrared (IR)	Microwaves
Wavelength (nm)*	0.1	1	180	390	780	400×10^3

*This is the wavelength interval where the lowest type of respective radiant energy occurs.

LIGHT AND MATTER[1-3]

Properties of light and radiant energy

Electromagnetic radiant energy is a form of energy that can be described in terms of its wavelike properties. Electromagnetic waves travel at high velocities and do not require the existence of a supporting medium for propagation.

The wavelength, λ, of a beam of electromagnetic radiant energy is the linear distance traversed by one complete wave cycle and is usually given in nanometers (nm, 10^{-9} meters). The frequency, ν, is the number of cycles occurring per second and is obtained by the relationship

$$\nu = \frac{c}{\lambda}$$

The velocity, c, varies with the medium through which the radiant energy is passing ($c = 3 \times 10^{10}$ cm/sec when measured in a vacuum).

Radiant energy can be shown to behave as if it were composed of discrete packets of energy called "photons." The energy of a photon is variable and depends on the frequency or wavelength of the radiant energy. The relationship between the energy, E, of a photon and frequency is given by the formula

$$E = h\nu$$

h is Planck's constant and has a numerical value of 6.62×10^{-27} erg·sec. The equivalent expression involving wavelength is

$$E = \frac{hc}{\lambda}$$

This equation shows that shorter wavelengths have a higher energy than longer wavelengths.

The electromagnetic spectrum covers a very large range of wavelengths, as shown in Table 3-1. The areas of the electromagnetic spectrum that are commonly used in the clinical laboratory are the ultraviolet (UV) and visible regions. The visible region is generally specified as being between 390 and 780 nm, whereas the ultraviolet spectrum usually referred to in the clinical chemistry laboratory falls between 180 and 390 nm. Sunlight or light emitted from a tungsten filament is a mixture of radiant energy of different wavelengths that the eye recognizes as "white." The breakdown of the visible region into color absorbed and color reflected is shown in Table 3-2. If a solution absorbs radiant energy (light) between 400 and 480 nm (blue), it will *transmit* all other colors and appear yellow to the eye. Therefore yellow is the complementary color of blue. If white light is focused on a solution that absorbs energy between 505 and 555 nm (green), the transmitted light and thus the solution will appear purple (blue + red). If a red light is shined on a red solution, red light will be transmitted because this solution cannot absorb red light. On the other hand, if green light is focused on the red solution, no light is transmitted, since the solution absorbs all light but red. The human eye responds to radiant energy between 390 and 700 nm, but laboratory instrumentation permits measurements at both shorter wavelengths—ultraviolet (UV)—and longer wavelengths—infrared (IR)—of the spectrum.

Interactions with light and matter

Absorption process. When an atom, ion, or molecule absorbs a photon, the added energy results in an alteration of state, and the species is said to be excited. Excitation may involve any of the following processes:

1. Transition of an electron to a higher energy level
2. A change in the mode of vibration of the molecule's covalent bonds
3. Alteration of its mode of rotation about the covalent bonds

Table 3-2. Colors and complementary colors of visible spectrum*

Wavelength† (nm)	Color absorbed†	Complementary or solution color transmitted
350-430	violet	yellow blue
430-475	blue	yellow
475-495	green blue	orange
495-505	blue green	red
505-555	green	purple
555-575	yellow green	violet
575-600	yellow	blue
600-650	orange	green blue
650-700	red	blue green

*If a solution absorbs light of the color listed in the second column, the observed color of the solution, that is, the transmitted complementary light, is given in the third column.
†Because of the subjective nature of color, the wavelength ranges are only approximations.

Each of these transitions requires a definite quantity of energy; the probability of occurrence for a particular transition is greatest when the photon absorbed supplies this exact quantity of energy.

The energy requirements for these transitions vary widely. Usually elevation of electrons to higher energy levels requires greater energy absorption than those needed to cause vibrational changes. Rotational alterations usually have the lowest energy requirements. Therefore absorption of energy in the microwave and far infrared regions results in shifts in the rotational energy levels, since the energy of the radiant energy is insufficient to cause other types of transitions. Changes in vibrational levels are caused by absorption in the near infrared and visible regions. Promotion of an electron to a higher energy level occurs after energy absorption in the visible, ultraviolet, and x-ray regions of the spectrum. The energy content of the electrons of covalent bonds varies with the nature of the bonds. The energy of a photon of light needed to excite an electron will therefore vary with the bond, and each type of bond will have its own characteristic pattern of optimal wavelength of light that can be absorbed by that bond. Table 3-3 gives the electronic absorption bands for a number of organic groups.

The absorption pattern of a complex organic molecule containing tens or thousands of bonds must therefore de-

Table 3-3. Electron absorption bands for representative chromophores

Chromophore	System	λ_{max}	ϵ_{max}	λ_{max}	ϵ_{max}	λ_{max}	ϵ_{max}
Ether	$-O-$	185	1000				
Thioether	$-S-$	194	4600	215	1600		
Amine	$-NH_2$	195	2800				
Thiol	$-SH$	195	1400				
Disulfide	$-S-S-$	194	5500	255	400		
Bromide	$-Br$	208	300				
Iodide	$-I$	260	400				
Nitrile	$-C\equiv N$	160	—				
Acetylide	$-C\equiv C-$	175-180	6000				
Sulfone	$-SO_2-$	180	—				
Oxime	$-NOH$	190	5000				
Azido	$>C=N-$	190	5000				
Ethylene	$-C=C-$	190	8000				
Ketone	$>C=O$	195	1000	270-285	18-30		
Thioketone	$>C=S$	205	strong				
Esters	$-COOR$	205	50				
Aldehyde	$-CHO$	210	strong	280-300	11-18		
Carboxyl	$-COOH$	200-210	50-70				
Sulfoxide	$>S-O$	210	1500				
Nitro	$-NO_2$	210	strong				
Nitrite	$-ONO$	220-230	1000-2000	300-400	10		
Azo	$-N=N-$	285-400	3-25				
Nitroso	$-N=O$	302	100				
Nitrate	$-ONO_2$	270 (shoulder)	12				
	$-(C=C)_2-$ (acyclic)	210-230	21,000				
	$-(C=C)_3-$	260	35,000				
	$-(C=C)_4-$	300	52,000				
	$-(C=C)_5-$	330	118,000				
	$-(C=C)_2-$ (alicyclic)	230-260	3000-8000				
	$C=C-C\equiv C$	219	6,500				
	$C=C-C=N$	220	23,000				
	$C=C-C=O$	210-250	10,000-20,000			300-350	weak
	$C=C-NO_2$	229	9,500				
Benzene		184	46,700	202	6,900	255	170
Diphenyl				246	20,000		
Naphthalene		220	112,000	275	5,600	312	175
Anthracene		252	199,000	375	7,900		
Pyridine		174	80,000	195	6,000	251	1,700
Quinoline		227	37,000	270	3,600	314	2,750
Isoquinoline		218	80,000	266	4,000	317	3,500

From Willard, H.H., Merritt, L.L., and Dean, J.A.: Instrumental methods of analysis, ed. 4, Princeton, N.J., 1965, D. Van Nostrand Co., Inc.

scribe the cumulative sum of the absorption of *all* the individual covalent bonds.

The absorption of radiant energy by a solution can be described by means of a plot of the absorbance as a function of wavelength. This graph is called an *absorption spectrum* (Fig. 3-1). The absorption spectrum reflects the sum of the energy transitions particular for a molecule at each wavelength of light. The interaction of the type described by Fig. 3-1 is attributable to the photon energy causing an electron in the molecule to jump into a higher orbital. When this happens, the molecule has gained energy. These absorption spectra are often helpful for qualitative identification purposes. This is particularly true for low-energy absorptions such as those found in the infrared region.

Irrespective of the amount of energy absorbed, an excited species tends to return spontaneously to its unexcited, or ground, state. To accomplish this, the excited electron must give up its excess energy, and this is usually dissipated as heat, as shown as follows:

$$A^{\circ} + h\nu \rightarrow A^{*} \rightarrow A^{\circ} + \text{Heat}$$

A° is the absorber in an unexcited energy state, A^{*} is the absorber in an excited state, and $h\nu$ (h is Planck's constant; ν is the frequency) is the energy of the incident photon having a wavelength $\lambda = \frac{c}{\nu}$ (where c is the velocity of light).

Emission process. Some elements and compounds can be excited in such a fashion that when the electrons return from the excited state to the ground state the energy is dissipated as radiant energy. The radiant energy may consist of one or more than one energy level and therefore may consist of different wavelengths. This principle is used in flame photometry and fluorometric methods and will be further discussed under these topics.

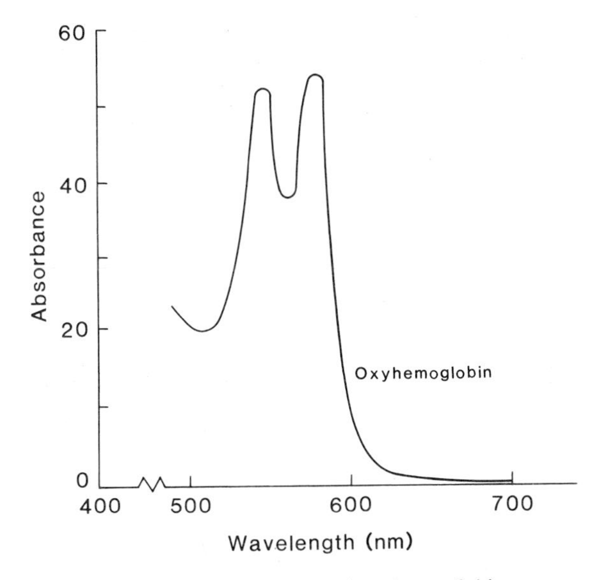

Fig. 3-1. Absorption spectrum of oxyhemoglobin.

ABSORPTION SPECTROSCOPY

Radiant-energy absorption

Consider a beam of radiant energy with an original intensity I'_o impinging on and passing through a square cell (whose sides are perpendicular to the beam) containing a solution of a compound that absorbs radiant energy of a certain wavelength (Fig. 3-2). The intensity of the transmitted radiant energy, I_s, will be less than I'_o. Some of the incident radiant energy may be reflected by the surface of the cell or absorbed by the cell wall or the solvent. Therefore these factors must be eliminated if one is to consider *only* the absorption of the compound of interest. This is done by use of a blank or reference solution containing everything but the compound to be measured. The amount of light passing through the blank solution is set as the new I_o (relative to the reference cell and solution). The transmittance for the compound in solution is defined as the proportion of the incident light that is transmitted:

$$\text{Transmittance} = T = I_s/I_o$$

Usually this ratio is described as a percentage.

$$\text{Percent } T = \%T = I_s/I_o \times 100\%$$

The concept of transmittance is important because it is only transmitted light that can be measured.

As the concentration of the compound in solution increases, more light is absorbed by the solution and less light is transmitted. The decrease in percent transmittance varies inversely and logarithmically with concentration. However, it is more convenient to use absorbance, A, which is directly proportional to concentration. Therefore:

$$A = -\log I_s/I_o = -\log T = \log \frac{1}{T}$$

To convert T into $\%T$, the denominator and numerator are multiplied by 100:

$$A = \log \frac{1}{T} \times \frac{100}{100} = \log \frac{100}{\%T}$$

This can be rearranged to:

$$A = \log 100 - \log \%T$$

or

$$A = 2 - \log \%T$$

It is important to remember that absorbance is *not* a directly measurable quantity but can only be obtained by mathematical calculation from transmittance data.

The relationship between absorbance and percent transmittance is shown in Fig. 3-3 in which the linear $\%T$ scale

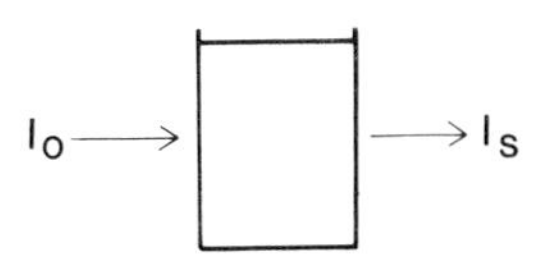

Fig. 3-2. Transmittance of radiant energy through a cuvette. I_o is the incident radiation; I_s is the transmitted radiation.

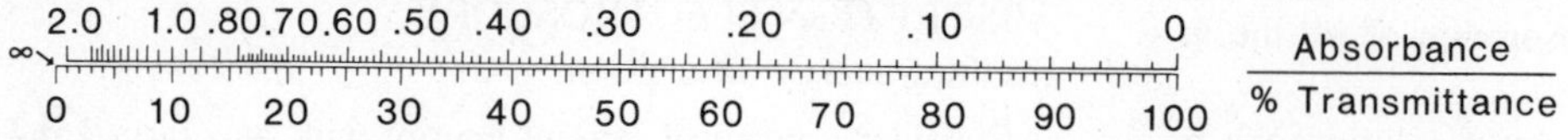

∞ 2.0 1.0 .80 .70 .60 .50 .40 .30 .20 .10 0 Absorbance

0 10 20 30 40 50 60 70 80 90 100 % Transmittance

Fig. 3-3. Scale showing relationship between absorbance and percent transmittance.

runs from 0 to 100%, whereas the logarithmic absorbance scale runs from infinity to 0.

Beer-Lambert law

The Beer-Lambert law (most commonly referred to simply as "Beer's law") states that the concentration of a substance is directly proportional to the amount of radiant energy absorbed or inversely proportional to the logarithm of the transmitted radiant energy. If the concentration of a solution is constant, and the path length through the solution that the light must traverse is doubled, the effect on the absorbance is the same as doubling the concentration, since twice as many absorbing molecules are now present in the radiant energy path. Thus the absorbance is also directly proportional to the path length of the radiant energy through the cell.

The mathematical relationship between absorbance of radiant energy, concentration of a solution, and path length is shown by Beer's law:

$$A = abc$$

A is absorbance; a, absorptivity; b, light path of the solution in centimeters; and c, concentration of the substance of interest.

This equation forms the basis of quantitative analysis by absorption photometry or absorption spectroscopy. Absorbance values have no units. The absorptivity is a proportionality constant related to the chemical nature of the solute and has units that are reciprocal of those for b and c.

When c is expressed in moles per liter and b is expressed in centimeters, the symbol ϵ, called the *molar absorptivity*, is used in place of a and is a constant for a given compound at a given wavelength under specified conditions of solvent, pH, temperature, and so on. It has units of L/mole·cm. The higher the molar absorptivity, the higher the absorbance for the same mass concentration of two compounds. Therefore, in selecting a chromogen for spectrophotometric methods, the chromogen with a higher molar absorptivity will impart a greater sensitivity to the measurement.

Once a chromogen is proved to follow Beer's law at a specific wavelength, that is, a linear plot of A versus c with a zero intercept (Fig. 3-4, *A*), the concentration of an unknown solution can be determined by measurement of its absorbance and interpolation of its concentration from the graph of the standards. In contrast, when $\%T$ is plotted versus concentration (on linear graph paper), a curvilinear relationship is obtained (Fig. 3-4, *B*). Because of the linear relationship between absorbance and concentration, it is possible to relate unknown concentrations to a single standard by a simple proportional equation. Therefore:

$$\frac{A_s}{A_u} = \frac{C_s}{C_u}$$

and

$$C_u = \frac{A_u}{A_s} \times C_s$$

where C_u and C_s are concentration of the unknown and standard, respectively; A_u and A_s are the absorbance of the unknown and standard.

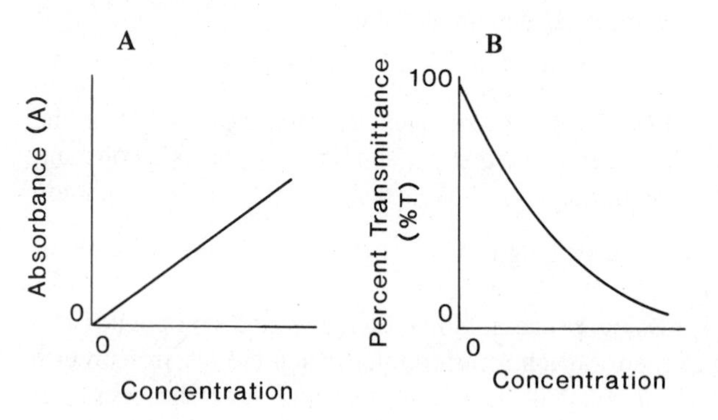

Fig. 3-4. Relationships of absorbance, **A**, and percent transmittance, **B**, to concentration.

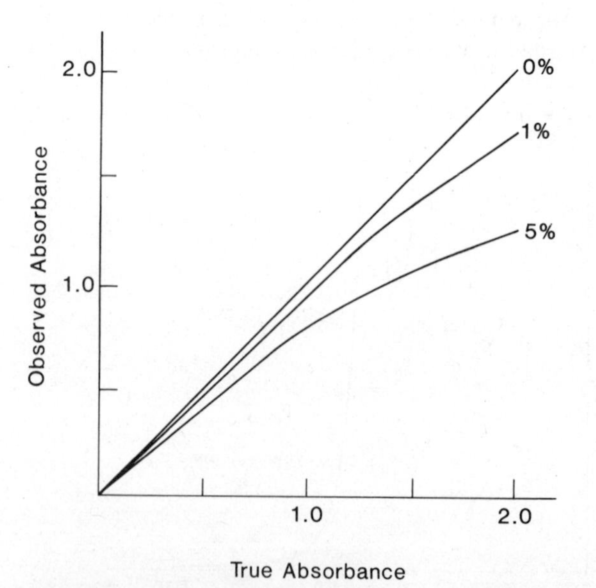

Fig. 3-5. Effect of stray radiation on true absorbance. (From Frings, C.S., and Broussard, L.A.: Clin. Chem. **25:**1013-1017, 1979.

The above equation is *only* valid if the chromogen obeys Beer's law and both standard and unknown are measured in the same cell. The concentration range over which a chromogen obeys Beer's law must be determined for each set of analytical conditions.

Beer's law is an ideal mathematical relationship that contains several limitations. Deviations from Beer's law, that is, variation from linearity of the absorbance versus concentration curve (Fig. 3-5), occur when (1) very elevated concentrations are measured, (2) incident radiant energy is not monochromatic, (3) the solvent absorption is significant compared to the solute absorbance, (4) radiant energy is transmitted by other mechanisms (stray light), and (5) the sides of the cell are not parallel. If two or more chemical species are absorbing the wavelength of incident radiant energy, each with a different absorptivity, Beer's law will not be followed. If the absorbance of a fluorescent solution is being measured, Beer's law may not be followed.

''Stray radiation'' (''stray light'') is radiant energy that reaches the detector at wavelengths other than those indicated by the monochromator setting. All radiant energy that reaches the detector with or without having passed through the sample will be recorded. Fig. 3-5 shows the effects of ''stray light'' on Beer's law. As the amount of ''stray light'' increases (or monochromicity decreases), deviation from Beer's law also increases (that is, linearity decreases).

Instrumentation

Single-beam spectrophotometer. The major components of a single-beam spectrophotometer are shown in Fig. 3-6. The apparatus needed can be divided into five basic components: (1) a stable source of radiant energy, (2) a wavelength selector, (3) a device to hold the transparent container (cuvette), which contains the solution to be measured, (4) a radiant-energy detector, and (5) a device to read out the electrical signal generated by the detector. If a filter is used as the wavelength selector, monochromatic light at only discrete wavelengths is available, and the instrument is called a ''photometer.'' If a monochromator is used (that is, a prism or grating, see below) as the wavelength selector, the instrument can provide monochromatic light over a continuous range of wavelengths and is called a ''spectrometer'' or ''spectrophotometer.'' Spectrophotometers are often double-beam instruments having two cuvette holders, one for the sample and the other for the blank.

Sources of radiant energy. A tungsten-filament lamp is useful as the source of a continuous spectrum of radiant energy from 360 to 950 nm (Fig. 3-7). Tungsten iodide lamps are often used as sources of visible and near-ultraviolet radiant energy. The tungsten halide filaments are longer lasting, produce more light at shorter wavelengths, and emit higher intensity radiant energy than tungsten filaments do.

Fig. 3-6. Components of a spectrophotometer. *A*, Source of radiant energy; *B*, entrance slit; *C*, wavelength selector; *D*, exit slit; *E*, cuvette and cuvette holder; *F*, detector; *G*, read-out device.

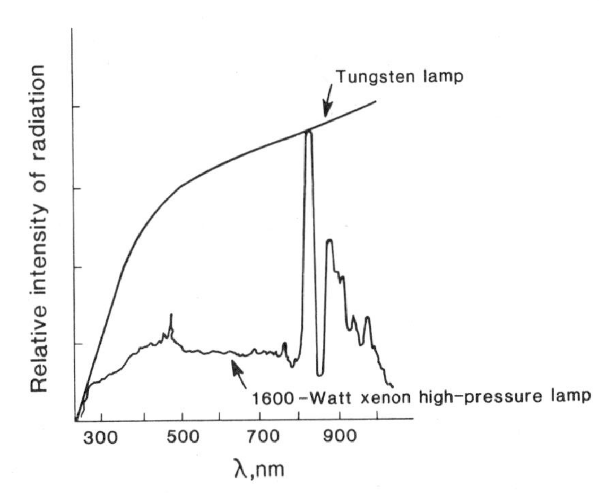

Fig. 3-7. Intensity of radiant energy versus wavelength for a tungsten filament and a 1600-watt xenon light source. Tungsten lamp intensity has been magnified approximately a hundredfold to place it on the same scale as the xenon lamp. (Modified from Brewer, J.M., et al., editors: Experimental techniques in biochemistry, Englewood Cliffs, N.J., 1974, Prentice-Hall, Inc.)

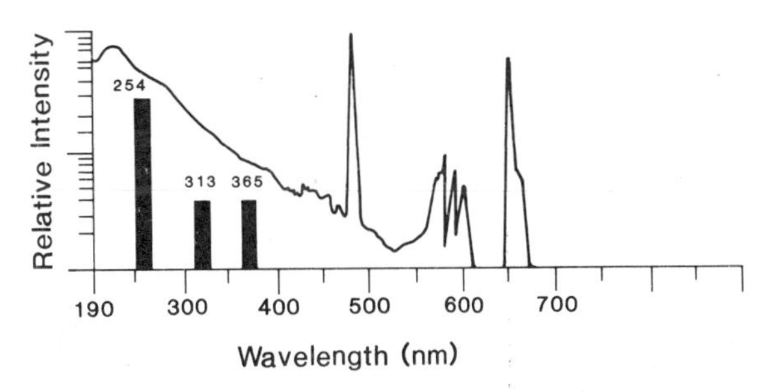

Fig. 3-8. Intensity of radiant energy versus wavelength for a mercury lamp, *solid bars,* and a deuterium lamp, *continuous line*. For illustrative purposes, the intensity of the mercury emission lines has been reduced several hundredfold, and only those lines (wavelegths are numbers above bars) in the ultraviolet region of the spectra have been depicted.

Hydrogen and deuterium discharge lamps emit a continuous spectrum and are used for the ultraviolet region of the spectrum (220 to 360 nm) (Fig. 3-8). The deuterium lamp has more intensity than the hydrogen lamp. Mercury-vapor lamps emit a discontinuous or line spectrum (313, 365, 405, 436, and 546 nm) (Fig. 3-8). This is useful for wavelength calibration purposes but is not used in many spectrophotometers. The mercury lamp is used in photometers or spectrophotometers employed for high-performance liquid chromatography.

It is important to understand that the amount of light emitted from a light source is not constant over a continuous range of wavelengths. Thus a typical lamp has a complex transmittance spectrum with maxima and minima (Figs. 3-7 and 3-8). Lamps of different types, and even different manufacturers, can vary. Therefore care must be taken in choosing a lamp for a particular analysis, since the amount of light emitted at the desired wavelength may be too little or too much. For example, hydrogen or deuterium lamps, used for ultraviolet analysis, have a maximum output of ultraviolet light in the 250 to 300 nm range. The output of radiant energy at longer wavelengths, that is, greater than 340 nm, is considerably less and can be too weak for many analyses.

Wavelength selectors. Isolation of the required wavelength or range of wavelengths can be accomplished by a filter or monochromator. Filters are the simplest devices, consisting of only a material that selectively transmits the desired wavelengths and absorbs all other wavelengths. In a monochromator, radiant energy from the source lamp is dispersed by a *grating* or *prism* into a spectrum from which the desired wavelength is isolated by mechanical slits.

Filters. There are two types of filters: (1) those with selective transmission characteristics, including glass and Wratten filters, and (2) those based on the principle of interference (interference filters). The Wratten filter consists of colored gelatin between clear glass plates; glass filters are composed of one or more layers of colored glass. Both of these types of filters transmit more radiant energy in some parts of the spectrum than in others. When glass filters are available, they are preferred because they are more durable and less susceptible to damage by heat and sunlight.

Interference filters work on a different principle. The principle is the same as that underlying the play of colors from a soap film, namely, interference. When radiant energy strikes the thin film, some is reflected from the front surface while some of the radiant energy that penetrates the film is reflected by the surface on the other side. The latter rays of radiant energy have now traveled farther than the first by a distance two times the film thickness. If the two reflected rays are in phase, their resultant intensity is doubled; whereas, if they are out of phase, they destroy each other. Therefore, when white light strikes the film, some reflected wavelengths will be augmented and some destroyed, resulting in colors. Interference filters can be purchased commercially and can usually isolate a more narrow range of wavelengths than glass filters can.

Monochromators. Monochromators can give a much narrower range of wavelength than filters and are easily adjustable over a wide spectral range. The dispersing element may be a prism or a grating.

Dispersion by a prism is nonlinear, becoming less linear at longer wavelengths (over 550 nm). Therefore, to certify wavelength calibration, one must check three different wavelengths. Prisms give only one order of emerging spectrum and thus provide higher optical efficiency, since the entire incident energy is distributed over the single emerging spectrum.

A grating consists of a large number of parallel, equally spaced lines ruled upon a surface. The original etched grating can be quite expensive. However, copies can be made that are usually less expensive than well-made prisms. Dispersion by a grating is linear; therefore, only two different wavelengths must be checked to certify the wavelength accuracy.

Band pass. Except for laser optical devices, the light obtained by a wavelength selector is not truly monochromatic, that is, of a single wavelength, but is comprised of a range of wavelengths. The degree of monochromicity is defined by the following terms. *Band pass* is that range of wavelengths that pass through the exit slit of the wavelength-selecting device. The *nominal wavelength* of this light beam is the wavelength at which the peak intensity of light occurs. For a wavelength selector such as a filter or a monochromator whose entrance and exit slits are of equal width, the nominal wavelength is the middle wavelength of the emerging spectrum. For filters, the nominal wavelength is usually the number etched on the filter, whereas for monochromators, the nominal wavelength is that number corresponding to the instrument setting.

The range of wavelengths obtained by a filter producing a symmetric spectrum is usually noted by its *half-band width* (or *half band pass*). This describes the wavelengths obtained between the two sides of the transmittance spectrum at a transmittance equal to one half the peak transmittance (Fig. 3-9). For monochromators, the degree of monochromicity is described by the *nominal band width,* which corresponds to those wavelengths centered about the peak wavelengths and transmitting 75% of the total radiant energy present in the emerging beam of light. For monochromators with variable exit slits, the band pass will also vary.

Slits. There are two types of slits present in monochromators. The first, at the entrance, focuses the light on the grating or prism where it can be dispersed with a minimum of stray light. The second, or exit, slit determines the band width of light that will be selected from the dispersed spectrum. By increasing the width of the exit slit, the band

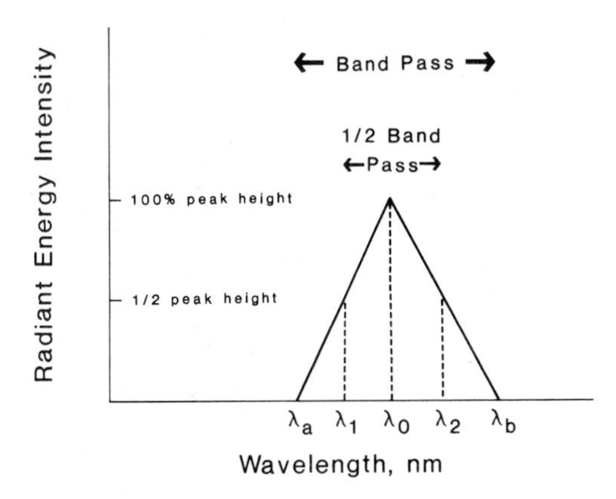

Fig. 3-9. Schematic depicting idealized energy distribution of radiant energy emerging from exit slit of wavelength selector. For a filter, or a monochromator with entrance and exit slits of equal width, a symmetric distribution of transmitted energy occurs, as shown.

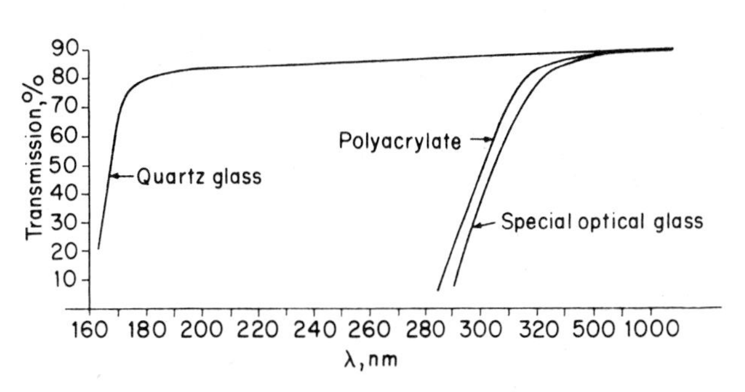

Fig. 3-10. Transmission characteristics of several types of optical materials used for cuvettes. (From Keller, H.: In Richterich, R., and Colombo, J.P., editors: Clinical chemistry, New York, 1981, John Wiley & Sons.)

width of the emerging light is broadened, with a resultant increase in energy intensity but a decrease in spectral purity. In diffraction-grating monochromators, the exit slit may be of fixed width resulting in a constant band pass. In contrast, prism monochromators have variable exit slits.

The purpose of both slits in filter photometers is only to make the light parallel and reduce stray radiation.

Cuvettes. The receptacle in which a sample is placed for spectrophotometric or photometric measurement is called a "cuvette" or "cell." Glass cuvettes are satisfactory for use in the range of 320 to 950 nm. For measurement below 320 nm it is necessary to use quartz (silica) cells. Such cells can be used at higher wavelengths also. Fig. 3-10 shows the transmission pattern of several types of cuvettes. Cuvettes having a square cross section and those having a circular cross section (that is, test tubes) are available. Greater accuracy is achieved by square cuvettes with parallel sides made of "optical glass." Although cuvettes usually have internal dimensions (that is, path lengths) of 1 cm, cuvettes with other dimensions are available.

Detectors

Barrier layer (photovoltaic) cells. These detectors consist of a plate of copper or iron upon which a semiconducting layer of cuprous oxide or selenium is placed. This layer is covered by a light-transmitting layer of metal that serves as a collector electrode. Upon illumination through the transparent electrode, an electron flow in the semiconducting layer is induced, and this flow can be sensed by an ampmeter. These detectors are rugged, relatively inexpensive, and sensitive from the ultraviolet region up to about 1000 nm. No external power is required, and the photocurrent produced is essentially directly proportional to the radiant energy intensity.

Barrier layer cells exhibit the fatigue effect, which means that upon illumination, the current rises above the apparent equilibrium value and then gradually decreases. Therefore it is best to wait 30 seconds between readings.

Photomultiplier tubes. A photomultiplier tube is an electron tube that is capable of significantly amplifying a current. The cathode is made of a light-sensitive metal that is capable of absorbing radiant energy and emitting electrons in proportion to the radiant energy that strikes the surface of the light-sensitive metal. These surfaces vary in their response to light of different energies (wavelengths) and so also the sensitivity of the photomultiplier tube (Fig. 3-11). The electrons produced by the first stage go to a secondary surface where each electron produces between four and six additional electrons. Each of these electrons from the second stage goes on to another stage, again producing four to six electrons. As many as 15 stages (or dynodes) are present in today's photomultiplier tubes (Fig. 3-12). Photomultiplier tubes have rapid response times, do not show as much fatigue as other detectors, and are very sensitive.

Instrument performance

The sensitivity of response of a spectrophotometer is a combination of lamp output, L, efficiency of the filter or monochromator to transmit light, M, and response of the photomultiplier, R. This may be expressed as follows:

$$\text{Sensitivity} \sim L \times M \times R$$

Since these factors are all functions of wavelength, it is clear that the instrument must be reset when one changes wavelengths. This resetting most often takes the form of adjusting the blank solution to read 100% T (0 absorbance) by changing the photomultiplier gain.

A series of recommendations on instrument specifications has been proposed which covers many aspects of in-

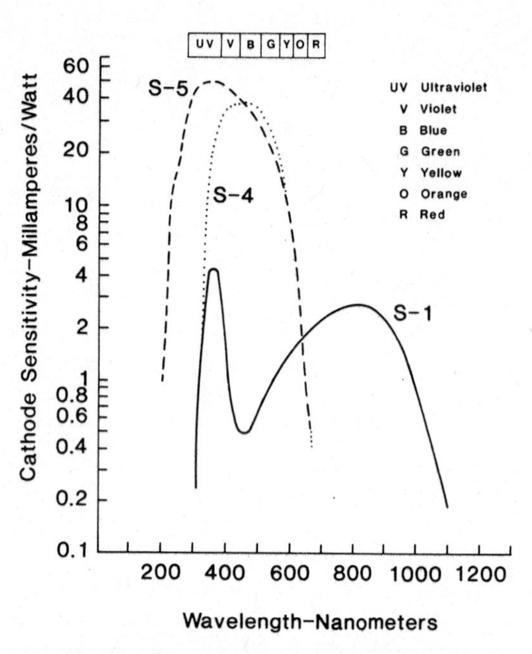

Fig. 3-11. Response of cathode of several photomultiplier tubes to energy of different wavelengths. Sensitivity is expressed as milliamperes of current generated per watt of incident radiation.

Table 3-4. Guidelines for photometric enzyme instruments

Parameter	Error or range (95% confidence, ±2 SD)
Carry-over	
Sample to sample	<0.3%
Temperature accuracy	±0.1° C
Equilibration time	20 sec
Sample handling	
Accuracy	1%
Precision	0.5%
Size	50 μL or less
Reagent handling	
Mixing time	≤10 sec
Photometric performance (at a rate of 0.01 *A*/min)	
Initial absorbance 0-1 *A*	<3%
Initial absorbance 1-2 *A*	<5%
Wavelength accuracy	±2 nm
Bandwidth	<8 nm
Wavelength range	Variable
Absorbance range	0-2.0 *A*
Linearity	<2%
Cell path/placement	<0.6%
Absorbance drift (10 to 60 min)	<2%
Absorbance accuracy	<2%
Absorbance reproducibility	
Low 0-1 *A*	±2%
High 1-2 *A*	±4%

From Instrumentation Guidelines Study Group, Subcommittee on Enzymes, Clin. Chem. **23:**2160-2162, 1977.

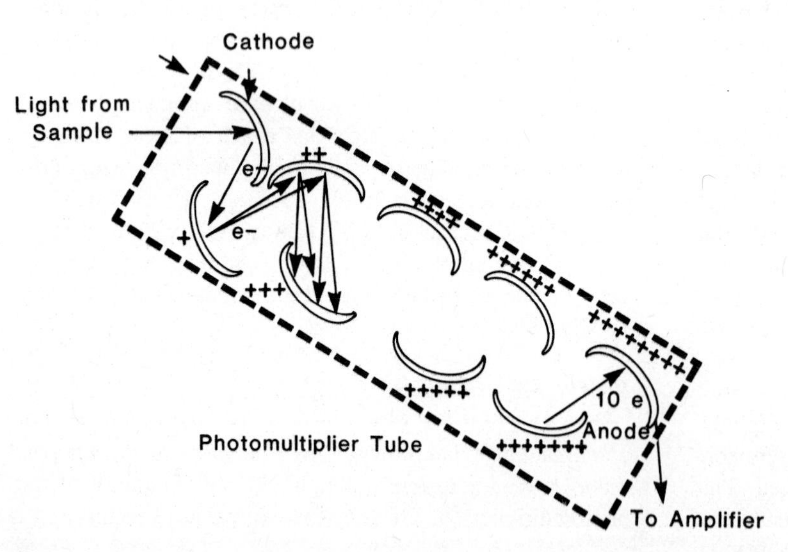

Fig. 3-12. Schematic of photomultiplier tube. Each dynode (electrode used to generate secondary emissions of electrons) is represented by a crescent. Light impinges upon cathode and frees an electron. Electron is drawn towards first dynode (stage) by applied voltage. Secondary electrons are released and pass on to successive dynodes, which are at increasingly higher voltages, as depicted by the + symbols. Increasing numbers of secondary electrons are generated at each stage. In this diagram a tenfold amplification of the initial signal is produced at the anode. Many photomultiplier tubes increase signal several thousandfold. (From Simonson, M.G.: In Kaplan, L.A., and Pesce, A.J., editors: Nonisotopic alternatives to radioimmunoassay, New York, 1981, Marcel Dekker, Inc.)

strumentation used for photometric analysis.[9] These specifications are listed in Table 3-4.

Selection of optimum conditions and limitations

When establishing a new spectrophotometric procedure, it is important to record the absorption spectrum of the material being measured. This absorption spectrum should be recorded in relation to either water or a reagent blank, depending on the actual method of analysis chosen. Examples of such spectra are presented in the Methods Section of this text. This spectrum will help in the determination of the best wavelength for the spectrophotometric analysis. The optimal wavelength for a specific analysis will depend on several factors, including the absorption maxima of the chromogen, the slope of the absorption peak, and the absorption spectra of possible interfering chromogens.

An example of an absorption spectrum is seen in Fig. 3-13. According to Beer's law, the higher the molar absorptivity, the greater the absorption at a given concentration and wavelength and the higher the sensitivity of the analysis will be. In this spectrum there are three peaks of absorption (highest absorption coefficient): λ_1, λ_2, and λ_3 nm. The absorptivity at λ_2 is too low and can be ruled out immediately as being of any analytical use.

If an absorption peak is too *narrow,* as at λ_1, any small error in the setting of the spectrophotometer at this wavelength will result in a large change in absorbance. With spectrophotometers using manually set wavelengths, this can cause large run-to-run imprecision and analytical error. With filter photometers one would require a high-quality, accurate filter to ensure accuracy when monitoring at a narrow absorption peak.

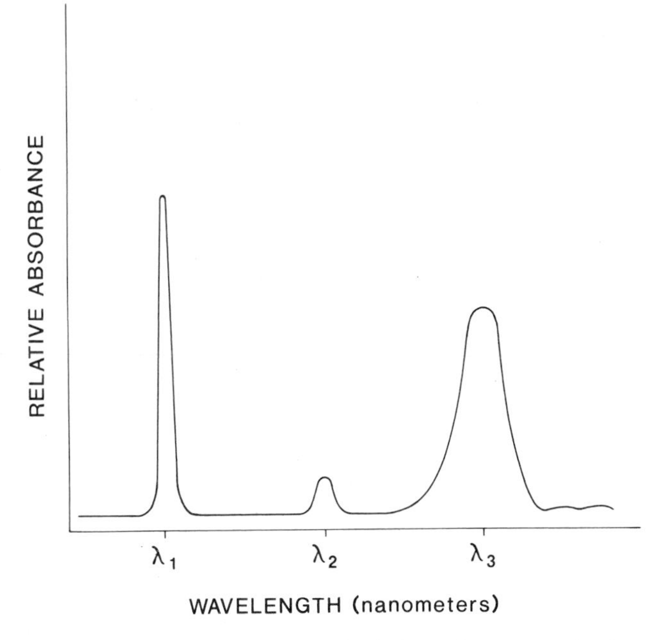

Fig. 3-13. Schematic of an idealized absorption spectra. λ_1, λ_2, and λ_3 represent the absorption bands of a chromophore.

These problems can be avoided by use of a wider absorption peak (λ_3). Therefore small changes in wavelength adjustment will result in small changes in absorptivity with resulting high precision and accuracy.

The sensitivity of many methods may be improved by use of absorption bands at shorter wavelengths (such as the ultraviolet), since very often these are more intense. However, often there is additional nonspecific absorption from buffers or other chemical moieties in the solution at shorter wavelengths. Therefore appropriate blanks must be used to obtain accurate measurements. In the case of some techniques where the analyte is purified before analysis, detection at short wavelengths (ultraviolet) is feasible and provides optimal sensitivity.

Knowledge of where the commonly interfering chromogens absorb light will also help determine the wavelength of choice.

A general rule for selecting the optimal wavelength at which to monitor a spectrophotometric reaction is to choose an absorption peak of the greatest possible molar absorptivity with a relatively broad peak. In addition, this peak should be as far as possible from absorption peaks of commonly interfering chromogens.

Quality control checks of spectrophotometers

There are several quality control checks that should be performed to certify that spectrophotometers are functioning within specifications. These checks are wavelength accuracy, linearity of detector response, stray radiation (stray light), and photometric accuracy. Details of the spectrophotometer performance checks can be found in references 5 and 10.

Wavelength accuracy. If the wavelength calibration of an instrument changes, the measured absorbance will change. The magnitude of the absorbance error attributable to inaccurate wavelength calibration is dependent on the relative location of the point on the absorption spectrum of the chromophore to be measured. That is, the absorbance error relative to the wavelength error is greater when the absorbance measurement is on the slope of the absorbance band than when the absorbance measurement is on or near the peak of the absorbance band. Maintenance of wavelength calibration is especially important for analyses such as spectrophotometric enzyme assays.

The most accurate method of checking the wavelength accuracy involves the replacement of the source lamp with a radiant energy source that has strong emission lines at well-defined wavelengths. Useful radiant energy sources are (1) the mercury vapor lamp, which has strong emission lines at 313, 365, 405, 436, and 546 nm, and (2) the deuterium or hydrogen lamp, which has useful emission lines at 486 and 656 nm (Fig. 3-8). Spectrophotometers equipped with a hydrogen or deuterium radiant energy lamp have built-in sources for checking wavelength accuracy.

A second method for checking wavelength calibration involves the use of rare-earth glass filters such as holmium oxide and didymium. Holmium oxide has strong absorption lines at approximately 241, 279, 287, 333, 361, 418, 453, 536, and 636 nm. Didymium has much broader absorption bands at approximately 573, 586, 685, 741, and 803 nm. Because of the possibility of filter deterioration, this wavelength accuracy should be periodically checked.

A third method for checking wavelength calibration involves the use of solutions. A solution of a stable chromogen can be used as a secondary wavelength calibration standard to determine whether the wavelength accuracy of an instrument has changed *after* the wavelength accuracy has been certified by a primary wavelength calibration standard such as a mercury or deuterium lamp. Disadvantages of chemical solutions are that the absorption peaks are generally broad and spectral shifts may result from contamination, aging, or preparation errors.

Irrespective of the method, calibration at two wavelengths is necessary for grating instruments, and calibration at three wavelengths is necessary for prism instruments.

Linearity of detector response. A properly functioning spectrophotometer must exhibit a linear relationship between the radiant energy absorbed and the instrument readout. Instrumental linearity is a prerequisite for spectrophotometric accuracy and for analytical accuracy as well. Solid glass filters may be used to check instrumental linearity. The most common method for certifying linearity of detector response is through the use of solutions of varying concentrations of a compound known to follow Beer's law. Some compounds used for this purpose are oxyhemoglobin at 415 nm, *p*-nitrophenol at 405 nm, cobalt ammonium sulfate at 512 nm, copper sulfate at 650 nm, and green food coloring at 257, 410, and 630 nm.

The absorbances of solutions containing increasing concentrations of one such compound are plotted against the known concentration. A nonlinear plot of absorbance versus concentration indicates either an error in dilution or an instrumental problem. Besides a faulty detector, "stray radiation," or too wide a slit width may cause a nonlinear response.

Stray radiation. An increase in "stray radiation" is often observed at the extreme ends of the spectral range where detector response or source energy is at its lowest. Stray radiation usually causes a negative deviation from Beer's law. Methods to detect stray radiation use filters or solutions that are highly transmitting over a portion of the spectrum but are essentially opaque below an abrupt "cutoff" wavelength. Several solutions have been used to check for stray radiation including Li_2CO_3 below 250 nm, NaBr (0.1 mol/L) below 240 nm, and acetone below 320 nm. The exact wavelength at which the cutoff occurs is a function of concentration, cell path length, and temperature; so the wavelengths reported may vary somewhat. Many filters can detect stray radiation. If solutions or filters that transmit no radiant energy at the measurement wavelength are used, the measured transmittance would be the amount of stray radiation present. Multiplication of this transmittance by 100 would give the percentage of stray radiation. An instrument malfunction is indicated whenever the amount of stray radiation exceeds 1%.

Action taken to eliminate stray radiation includes changing the light source, verifying wavelength calibration, sealing light leaks, realigning of instrument components, and cleaning optical surfaces.

Photometric accuracy. When one performs analyses that do not use chemical standards, absorbance accuracy is essential. An absorbance standard should have a constant, stable absorbance at a suitable wavelength that is insensitive to the spectral band width of the instrument and to variations in the geometry of the light beam. Such standards should be easy to use and are readily available. The National Bureau of Standards (NBS) has a set of three neutral-density glass filters (SBM 930) that have known absorbances at four wavelengths for each filter. These filters are not completely stable and must be recalibrated by the NBS periodically.

Potassium dichromate solution, cobalt ammonium sulfate solution, and potassium nitrate solution have been used as standards for checking photometric accuracy. Standard solutions are subject to absorbance changes with time, temperature, and pH, which make them unsuitable as long-term calibration standards for photometric accuracy.

ATOMIC ABSORPTION[11,12]

Atomic absorption spectrophotometry is widely used in the clinical laboratory for the determination of calcium, magnesium, lithium, copper, zinc, and other metals.

Principle

Vaporized atoms in the ground state absorb light at very narrowly defined wavelengths. These absorption bands are on the order of 0.001 to 0.01 nm in width and thus the entire absorbtion spectrum of atoms is called a "line spectrum." If these atoms in the vapor state are excited, they can return to the ground state by emitting light of the same discrete wavelengths as the line spectrum. In atomic absorption spectrophotometry the ionic form of the element is not excited in the flame but is dissociated from its chemical bonds and, by attracting free electrons produced by the combustion process, is placed in the atomic ground state. In this form it is capable of absorbing light at the specific wavelengths of its line spectrum. The series of reactions may be written as follows:

$$A^{++} + 2e^{-} \rightarrow A^{o}$$

Ionized metal plus electrons produces ground-state metal atom.

$$A^{o} + h\nu \rightarrow A^{*}$$

Ground-state metal and light produces excited-state atom.

$$A^{*} \rightarrow A^{o} + h\nu$$

Excited-state atom releases photon and returns to ground state.

A^{++} is a metal ion in solution, A^{o} is an atom having ground-state electronic energy, A^{*} is an atom in the excited state, and $h\nu$ is a photon.

In atomic absorption, a beam of radiant energy containing the line spectrum of the element to be measured is passed through a flame containing the vaporized metal to be determined. The source emitting such radiant energy is called a *hollow-cathode lamp*. The wavelength of the absorbed radiant energy is the same as that which would be emitted if the element were excited. With the aid of a monochromator, a measurement is made of the attenuation of one of the wavelengths of the incident light. This attenuation is caused by the photons interacting with ground-state atoms in the flame. The Beer-Lambert law is valid for relating the concentration of atoms in the flame and transmission or absorption of light. Only a small percentage of atoms in the flame are excited, and most atoms are in a form capable of absorbing radiant energy emitted by the hollow-cathode lamp.

Instrumentation

Fig. 3-14 shows the salient components of an atomic absorption spectrophotometer.

Hollow-cathode lamp. The hollow-cathode lamp is the most practical means for generating the "line spectrum" of the required spectral purity. The lamps have a hollow or cuplike cathode that is lined with the pure metal of the element to be determined or with an appropriate alloy. A separate lamp is used for each element except for a few instances where the cathode can be constructed in such a manner that a single lamp serves for two or three elements (such as calcium and magnesium). The lamp is filled with an inert monatomic gas, usually argon or neon, at low pressure. The inert gas selected for the lamp can vary with the analyte to be measured. For example, lead and iron can be better analyzed with neon-filled lamps, whereas lithium analysis is better performed with argon-filled lamps. For other measured elements, the choice of inert gas is not critical. Quartz or a special glass that allows transmission of the proper wavelength is used as the window. A current is supplied to the cathode and metal atoms are continually released (sputtered) from the inner surface of the cathode, filling the lamp with an atomic vapor. Atoms in this vapor undergo electronic excitation by collisions with the inert gas, and the resulting excited atoms emit their characteristic radiant energy when returning to the ground-state electron level. This results in radiant energy with the correct wavelength for absorption by ground-state atoms in the flame.

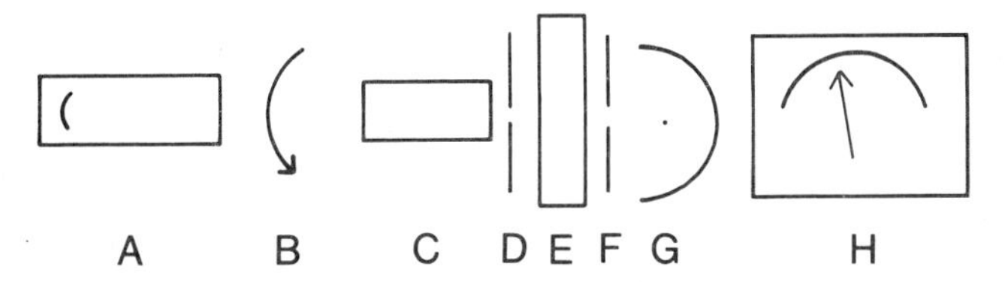

Fig. 3-14. Essential components of atomic absorption spectrophotometer. *A*, Hollow cathode lamp; *B*, chopper; *C*, flame and burner assembly; *D*, entrance slit; *E*, wavelength selector; *F*, exit slit; *G*, detector; *H*, read-out device.

Burner. In atomic absorption spectrophotometry, the sample solution must be converted into a fine spray or aerosol while being introduced into the flame. This process is called "nebulization." The nebulizer is usually considered part of the burner. Within the flame, solvent evaporates from the aerosol leaving microscopic particles that disintegrate under the influence of heat to yield atoms. This phenomenon is termed "atomization." Acetylene is the commonly used fuel in the burner. Temperatures of 2300° C are usually achieved in flame atomic absorption.

Two kinds of burners have been used in most clinical applications. One is the total consumption burner. With this burner, the gases and the sample mix within the flame. Relatively large droplets are produced in the flame and can cause signal noise. The flame in this type of burner can be made hotter, causing molecular dissociations that may be desirable for some chemical systems. The second type of burner is the premix burner (laminar-flow burner). In this burner system, larger droplets from the atomization go to waste and not into the flame, and so there is less signal noise. The path length through the premix burner is longer than that of the total consumption burner, and so there is greater sensitivity. The flame temperature is not as hot as that of the total consumption burner.

Monochromator and detector. Monochromators (grating or prisms) and photomultiplier tubes can isolate a pure radiant energy signal and measure the intensity of that signal, respectively. Extraneous radiant energy, both from other wavelengths of the line spectrum and from light generated by the flame, is kept from reaching the photomultiplier tube by the monochromator. The photomultiplier tube converts the radiant energy from the hollow cathode, which was *not* absorbed in the flame, into a signal and amplifies this signal to drive a recorder or meter.

Flameless atomic absorption. The purpose of the flame is to convert the sample into an atomic vapor. The flame can be replaced by other atomization processes. One process applicable to mercury analysis uses chemical reactions to convert mercury into an atomic vapor. The sample is decomposed by digestion with acids, then a reducing agent is added to convert mercury to the elemental state, and finally a stream of gas is bubbled through the apparatus pushing mercury vapor into a sealed cell with quartz windows in the optical beam. Absorbance measurements are made at 253.7 nm.

In another atomization technique, the sample is dried on a tantalum ribbon, small boat, or some other support. The sample is vaporized in an inert atmosphere when an electric current is passed through the support to create an instantaneous temperature of approximately 2100° C. The other technique for vaporization of metals is the use of a graphite furnace. These atomizers are in the space normally occupied by the flame. The higher temperatures achieved by flameless atomic absorption are necessary to vaporize heavier metals. They also have a greater sensitivity than do flame atomic absorption instruments.

Sources of error

Chemical, ionization, matrix, and burner interferences can occur in atomic absorption measurements. Additional factors that may cause variable behavior from sample to sample or between unknowns and standards include temperature, solvent composition, salt content, viscosity, and surface tension.

Chemical interference. With some elements the presence of certain anions in the sample results in the formation of compounds that are not completely dissociated in the flame. The result is a decrease in the number of ground-state atoms present in the flame. The most common example of chemical interference in atomic absorbance is the formation of a tight complex of calcium with anions, especially phosphate ions. The effect of tightly complexing anions can be minimized or eliminated when lanthanum is added to the sample to displace calcium from the complex and form a more stable complex with phosphate than calcium does.

Ionization interference. When atoms in the flame become ionized (A^+) instead of remaining in the ground state (A^o), they will not absorb the incident light. This is termed "ionization interference," and this effect will result in an apparent decrease in analyte concentration. Ionization interference can be corrected when one adds an excess of a substance that is more easily ionized, thus providing free electrons. The excess free electrons thus shifts the reaction

$$A^+ + e^- \rightarrow A^o$$

to the formation of ground-state atoms. Ionization interference is minimized by operation of the flame at the lower temperatures of acetylene-air combustion.

Matrix interference. Differences in the matrix between the sample and the standard can result in errors. Protein is sometimes included in the standards when the serum dilution factor is small. The matrix effects are minimized as compositional differences between the standard and the sample become negligible. Calcium standards must contain physiological concentrations of sodium, since sodium will cause a negative interference.

Burner problems. The most critical component in the atomic absorption spectrophotometer is the flame and its associated nebulizer. A steady flame is essential, and controlled gas flows are required for both oxidant and fuel. A clean burner head is essential for precise and accurate analysis.

Emission interference. A number of the analyte atoms introduced into the flame will become excited and emit a photon to return to the ground state. The light emitted is, of course, at the same wavelengths as the incident light being measured. The emitted light enhances the signal being received by the photodetector. The increased signal is translated as a decreased absorption and therefore a falsely lower concentration of analyte. This interference can be eliminated by use of a "chopper" (Fig. 3-14) to create a pulsed beam of incident light from the hollow-cathode lamp. The light caused by emission interference, however, is a constantly produced beam. This steady emission can be electronically differentiated from the pulsed beam of transmitted light and thus eliminated as a source of interference.

FLAME PHOTOMETRY[13,14]

Flame photometry is widely used in the clinical laboratory to determine sodium, potassium, and lithium concentrations in biological fluids.

Principle

Atoms of some metals, when given sufficient heat energy as supplied by a hot flame, will become excited and reemit this energy at wavelengths characteristic for the element as described above. The reactions undergone by ions in the flame are as follows:

$$A^+ + e^- \rightarrow A^o$$
$$A^o + \text{Heat} \rightarrow A^*$$
$$A^* \rightarrow A^o + h\nu$$

A^* represents the excited atom in the flame, A^o an atom having ground-state electron energy, and $h\nu$ a photon. Alkali metals are relatively easy to excite in a flame. Lithium produces a red emission; sodium, a yellow emission; and potassium, a red-violet color in a flame. These colors are characteristic of the metal atoms that are present as cations in solution.

The intensity, I, of the characteristic wavelength of radiant energy produced by the atoms in the flame is directly proportional to the number of atoms excited in the flame, which is directly proportional to the concentration $[A^+]$ of the substance of interest in the sample.

$$I = k[A^+]$$

The actual number of atoms present in the excited state is a small fraction of the total number of atoms present in the flame.

Instrumentation

Fig. 3-15 shows the salient components of a flame photometer. Natural gas, propane, and air are used to produce the flame in most flame photometers encountered in the clinical laboratory. Since it is essential that the flame temperature be held constant, regulators are required to maintain a constant flow of gas. An atomizer is needed to convert the sample into fine droplets before introduction of the sample into the flame. The filter, entrance and exit slits, and detectors have a function similar to those discussed previously for spectrophotometers.

Direct and internal standard instruments

The two types of flame photometers are (1) the direct type and (2) the internal standard type. In the direct type, the intensity of the characteristic spectral emission is used as a measure of concentration. In the internal standard type, the intensity of the spectral emission of the element to be determined is balanced against that of an added element that acts as an internal standard. All flame photometers used in the clinical laboratory are of the internal standard type, and the internal standard is lithium or cesium. The internal standard is normally absent from biological fluids and emits light at a wavelength sufficiently removed from the elements to be measured (such as sodium and potassium) so that it does not interfere with the measurement of these elements. The sample and standards are diluted with a constant concentration of internal standard. The light emission of the internal standard provides a reference signal against which the emission of the other elements can be measured. The flame photometer makes a comparison of the emission of the desired element with the emission of the reference element. Small variations in sample-aspiration rate, atomization rates, solution viscosity, flame temperature, and flame stability will affect the amount of excited atoms present in the flame. With direct standardization these variables will affect the intensity of the emission and thus the calculated result. With internal standards, fluctuations of these variables will affect both the internal standard and the element to be measured *simultaneously* and to the same degree. Although the intensity of emitted light for both the internal standard and the analyte will vary, the *ratio* of the two will not. Thus the use of internal standardization allows more precise and accurate determinations.

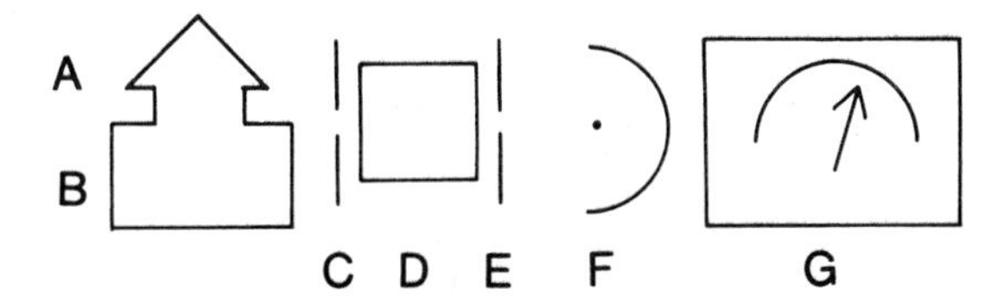

Fig. 3-15. Essential components of a flame photometer. *A*, flame; *B*, burner head and atomizer; *C*, entrance slit; *D*, wavelength selector (filter); *E*, exit slit; *F*, detector; *G*, read-out device.

An internal standard can also be used to minimize the effect of *mutual excitation*. This phenomenon is the result of photons emitted by excited sodium atoms, in turn exciting potassium atoms. This can result in falsely elevated potassium values unless concentrations of sodium and potassium in the standards closely approximate the concentrations of these analytes in unknowns—a difficult achievement for urine specimens. The internal standard (lithium or cesium) acts to absorb the radiation and buffer the potassium atoms from the effect of mutual excitation.

Limitations

Mechanisms that may be responsible for interferences in flame photometry include the following:

1. *Alteration of light emission because of altered flame temperature*. The higher the flame temperature, the less the error that is produced by this phenomenon.
2. *Inadequate selectivity of wavelength*. The magnitude of the background error caused by photoemission by extraneous elements is a function of the filter or monochromator. Accurate selection of the emission lines of sodium or potassium can be achieved with a monochromator; it may or may not be achieved with a filter.
3. *Differences in the viscosities of the standards and samples*. Differences in viscosities can result in differences in aspiration rates into the flame and cause interference in measurements. Inclusion of wetting agents in the internal standard used to dilute standards and samples helps to minimize this source of error.
4. *Molecular emission bands*. Certain radicals, such as cyanide, emit spectral bands that overlap spectral lines of metallic ions and therefore produce erroneous enhancement. It is difficult to evaluate the magnitude of errors because of this factor.

FLUOROMETRY[15,16]

Principle

Fluorescence may be considered as one of the results of the interactions of light with matter. When light impinges upon matter, it can simply pass through, as in a transparent solution, be scattered by the interaction, or absorbed. When light is transmitted there is no loss of energy. When light is scattered, there is no change of energy; the light is the same wavelength before and after it interacts with matter. But when light is absorbed, there is conversion of the light energy into any one of a number of forms, including radiationless transitions (converting the energy into heat) and others such as fluorescence and phosphorescence

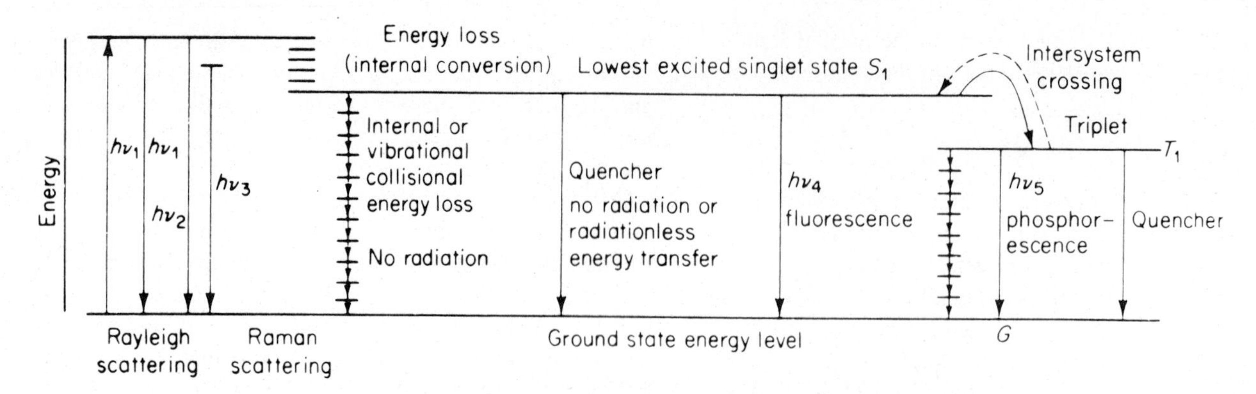

Fig. 3-16. Schematic showing conversion of light energy into different forms of molecular and radiant energy. (From Pesce, A.J., et al., editors: Fluorescence spectroscopy, New York, 1971, Marcel Dekker, Inc.)

where photons are emitted (Fig. 3-16). Absorption of the light can be used, of course, to determine the concentration of compounds as is done in absorption spectroscopy. If the absorbed light is reemitted, the emitted photons can be used to quantitate the amount of the light-emitting compound (fluor). Quantitation is also possible with use of scattered light, since the amount of scattered light is related to the number and size of particles in solution. Methods that use light scattering are termed "nephelometry" and "turbidity."

Fluorescent light is the result of a photon of radiant energy being absorbed by a molecule. Once the molecule absorbs a photon, it has an increased energy level, and because the molecular energy is greater than that of its environment, it will seek to eject the excess energy. When the energy is lost as an ejected photon, the result is fluorescence or phosphorescence emission.

For fluorescence to occur, there has to be a good probability that the energy of the excited state can be converted to the ground state by the ejection of a photon. In general, not all compounds fluoresce; indeed, only a very few fluoresce. Of those that do fluoresce, not every single photon absorbed will be converted to fluorescent light. Some excited compounds will lose energy by radiationless transitions, that is, by transfer of the energy to the solvent. For the same amount of light absorbed, the molecules with higher fluorescence efficiency will have brighter or more intense fluorescence. In solutions, when a molecule returns to the ground state by emitting a photon, there is usually considerably less energy in the emitted photon compared to the one initially absorbed. That is, the emitted fluorescent light is at a longer wavelength than the exciting or absorbed radiation (Fig. 3-17).

Instrumentation

The basic components of a fluorometer are similar to an absorption spectrophotometer. The major difference is the introduction of a set of filters or a monochromator before and after the cell to isolate the emitted light. A diagram of a fluorometer is presented in Fig. 3-18.

The principal components are the exciting light, filters or monochromators to separate the exciting light from the emitted light, and a sensitive detector. Most often, the measurement of fluorescent light is made at 90 degrees to the exciting light. This is done to maximize the sensitivity of the instrument by minimizing the amount of excitation light that can reach the photodetector. The detector is a photomultiplier or similar device that can quantitate the very small fluorescent light signal and thus achieve the desired level of sensitivity. Because the spectrum of absorption and emission varies from one compound to another, the instrument must be optimized for every analyte measured. This is done by adjustment of the exciting wavelength to achieve the maximum absorption of photons, which usually means setting the instrument to the absorption maxima of the compound. By the same token, the wavelength of maximum emission of the fluorescent photons must also be ascertained, and this is the wavelength at which the fluorescence signal is most often recorded.

Limitations

The fluorescence signal of a compound is subject to many variables including (1) solvent, (2) pH, (3) temperature, (4) absorbance of the solution, and (5) presence of interfering or specifically quenching compounds. Standardization is not usually done by an absolute procedure as in absorption spectroscopy. The reason is that the fluorescence varies depending on (1) the intensity of the incident light upon the sample, (2) the amount of light intercepted by the detector as controlled by the slits, (3) the band width of light analyzed, and (4) the efficiency of the detector. The quantum yield or efficiency of light emission of a photon is constant if solvent, pH, temperature, and so on, are kept constant. But in general the instrument will not be constant on a daily basis. Therefore relative fluorescence yield is used for most measurement. For a reagent

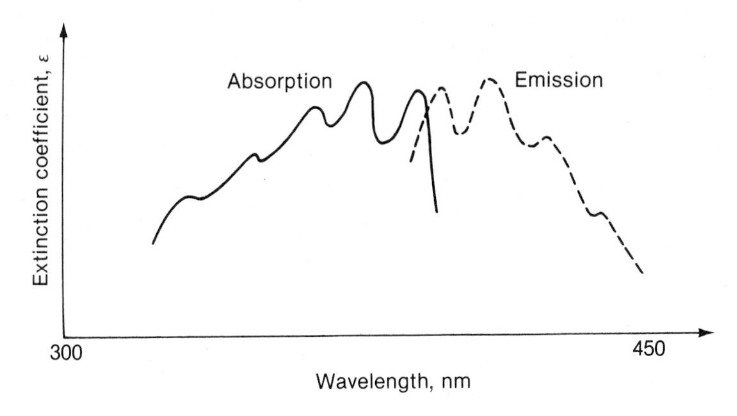

Fig. 3-17. Absorption (excitation) and emission (fluorescence) spectra of a fluorescent compound. (From Pesce, A.J., et al., editors: Fluorescence spectroscopy, New York, 1971, Marcel Dekker, Inc.)

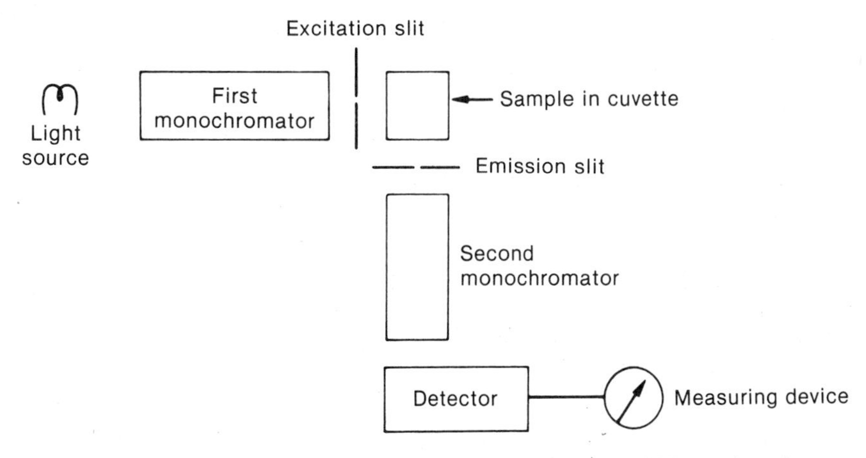

Fig. 3-18. Essential components of fluorometer. (From Brewer, J.M., et al., editors: Experimental techniques in biochemistry, Englewood Cliffs, N.J., 1974, Prentice-Hall, Inc.)

blank in a fluorometric assay, only the 0 or null fluorescence can be set. There is no equivalent to the 100% scale of absorption. Therefore the electronic signal will vary from instrument to instrument for the same concentration of analyte. For a series of fluoresent standards to form a curve that is linear with concentration, the absorbance of the solutions should not exceed 0.1. Above this absorbance, all portions of the solution are not uniformly illuminated; that is, the initially illuminated layer of the solution will absorb more light than the final layer and thus the initial layer will fluoresce more than the final layer. With dilute solutions (such as an absorbance less than 0.1) this does not occur.

NEPHELOMETRY AND TURBIDITY[17-19]

Principle

Interaction of light with particles. To understand the principle of the various assays, we must first examine the concept of light scattering. When a collimated (that is, parallel, nondivergent) beam of light strikes a particle in suspension, some light is reflected, some scattered, some absorbed, and some transmitted. Nephelometry is the measurement of the light scattered by a particulate solution. Turbidity measures light scattering as a decrease in the light transmitted through the solution.

In considering nephelometry, the question of how light is scattered by a *homogeneous* particle suspension must be examined. Three types of scatter can occur. If the wavelength, λ, of light is much larger than the size of the particle ($d < 0.1\lambda$), the light is symmetrically scattered around the particle, with a minimum in the intensity of the scatter occurring at 90 degrees to the incident beam, as described by Rayleigh (Fig. 3-19, *A*).

If the wavelength of the incident light is much smaller than the size of the particle ($d > 10\lambda$), most of the light appears to be scattered forward because of destructive out-of-phase backscatter, as described by the Mie theory (Fig. 3-19, *B*).

If, however, the wavelength of light is approximately equal to the size of the particles, more light appears scattered in a forward direction than in a backward direction (Fig. 3-19, *C*), as described by Rayleigh-Debye scatter.[17]

One of the most common uses of light-scattering analyses is the measurement of antigen-antibody reactions.

A SMALL PARTICLES
-LIGHT SCATTERED SYMMETRICALLY
BUT MINIMALLY AT 90° (RAYLEIGH)

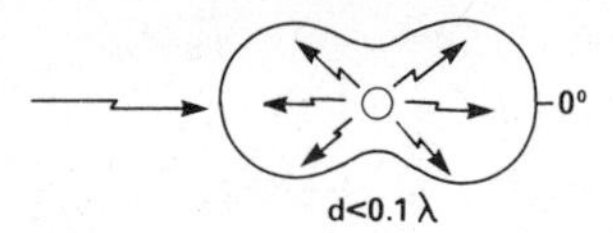

B VERY LARGE PARTICLES
-LIGHT MOSTLY SCATTERED FORWARD
(MIE)

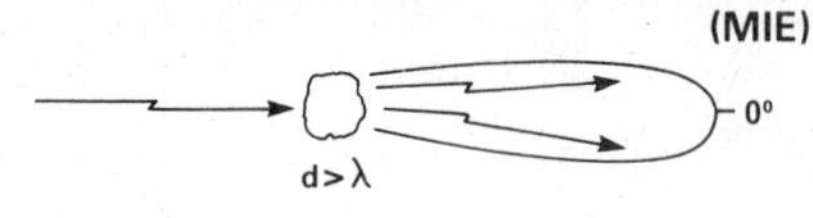

C LARGE PARTICLES
-LIGHT SCATTERED PREFERENTIALLY
FORWARD (RAYLEIGH-DEBYE)

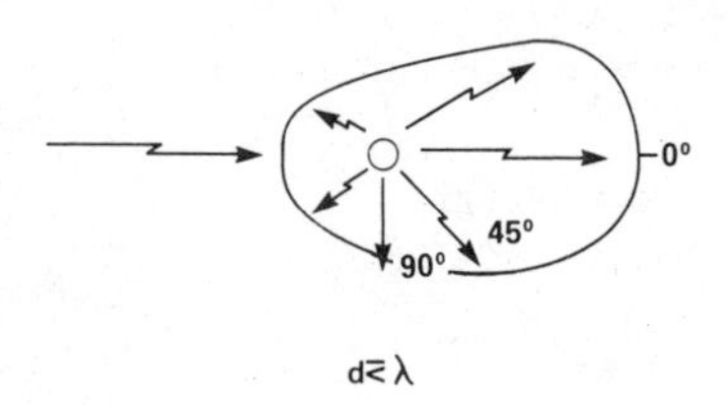

d≥λ

Fig. 3-19. Effect of particle size on scattering of incident light in a homogeneous solution. (From Gauldie, J.: In Kaplan, L.A., and Pesce, A.J., editors: Nonisotopic alternatives to radioimmunoassay, New York, 1981, Marcel Dekker, Inc.)

TURBIDIMETRY

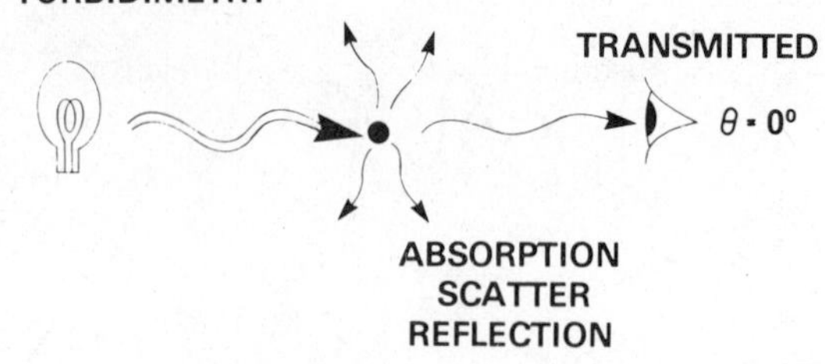

Fig. 3-20. Schematic diagram of turbidity measurements. *θ*, Angle of detection. (From Gauldie, J.: In Kaplan, L.A., and Pesce, A.J., editors: Nonisotopic alternatives to radioimmunoassay, New York, 1981, Marcel Dekker, Inc.)

NEPHELOMETRY

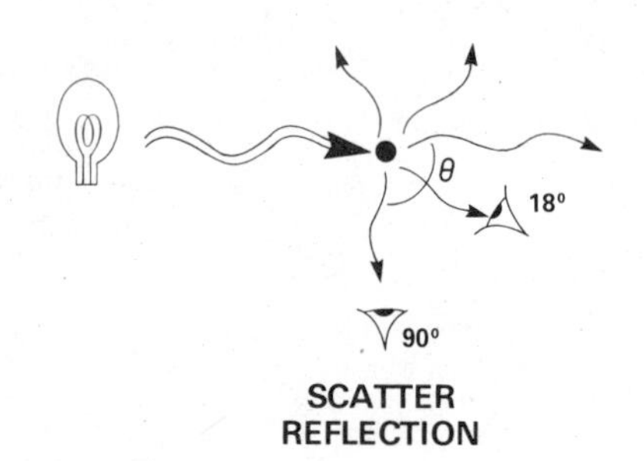

Fig. 3-21. Schematic diagram of nephelometric measurement. *θ*, Angle of detection. (From Gauldie, J.: In Kaplan, L.A., and Pesce, A.J., editors: Nonisotopic alternatives to radioimmunoassay, New York, 1981, Marcel Dekker, Inc.)

Since most antigen-antibody complex systems are heterogeneous with particle diameters of 250 to 1500 nm and the wavelengths used in most light-scattering analyzers are 320 to 650 nm, the scatter seen is essentially Rayleigh-Debye, with the blank scatter being primarily described by Rayleigh scatter. Thus the ability to detect light scatter in a forward direction (θ = 15 to 90 degrees) would lead to greater sensitivity for nephelometric determinations. Such is the case in the newer rate and laser nephelometers.

Detection of scattered light

Turbidity. Turbidimetry is a measure of the reduction in the light transmission caused by particle formation, and it quantifies the residual light transmitted (Fig. 3-20). The instrumentation required for turbidity measurements ranges from a simple manual spectrophotometer available in most laboratories to a sophisticated discrete fast analyzer. Because this technique measures a decrease in a large signal of transmitted light, the sensitivity of turbidimetry is limited primarily by the photometric accuracy and sensitivity of the instrument. These instruments can be used for many other assays such as enzymes or those based on color development.

Nephelometry. Nephelometry, on the other hand, detects a portion of the light that is scattered at a variety of angles (Fig. 3-21). The sensitivity of this method is primarily dependent on the blank or background scatter of the samples because the instruments are detecting a small increment of signal at a scatter angle, θ, on a supposedly black or null background. Ideally, no light is detected in the absence of a scattering species, and so subsequent scatter is measured against this black background. The signal is magnified by the use of a photomultiplier, so that the detection range is increased. However, such measurements require the committed use of a nephelometer, which has limited use in other assays.

Instrumentation

Schematic layout of instruments. A schematic layout of the basic components of a nephelometer is shown in Fig. 3-22. Typical systems consist of a light source, a collimating system, a wavelength selector such as a filter (the last two items are unnecessary with laser light sources), a sample cuvette, a stray light trap, and a photodetector.

Light source. Fluoronephelometers such as the Technicon instrument use a medium-pressure mercury-arc lamp as a light source, which serves both for nephelometry and fluorometry. The relatively high intensity light and short-wavelength emission bands make this a good source. Other light sources range from simple low-voltage tungsten-filament lamps to sophisticated low-power lasers. Lasers pro-

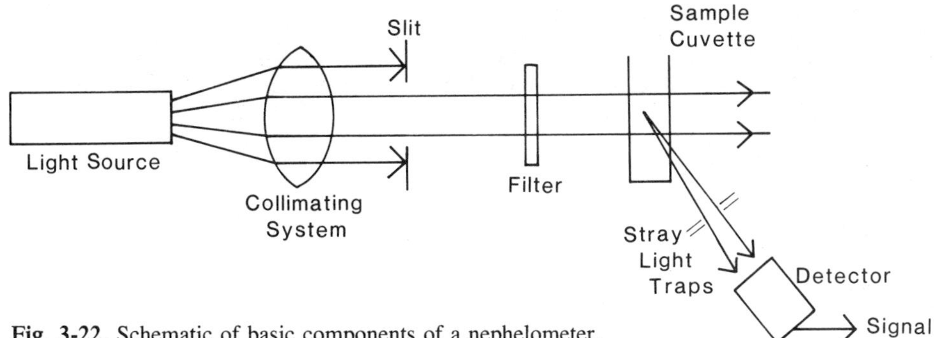

Fig. 3-22. Schematic of basic components of a nephelometer.

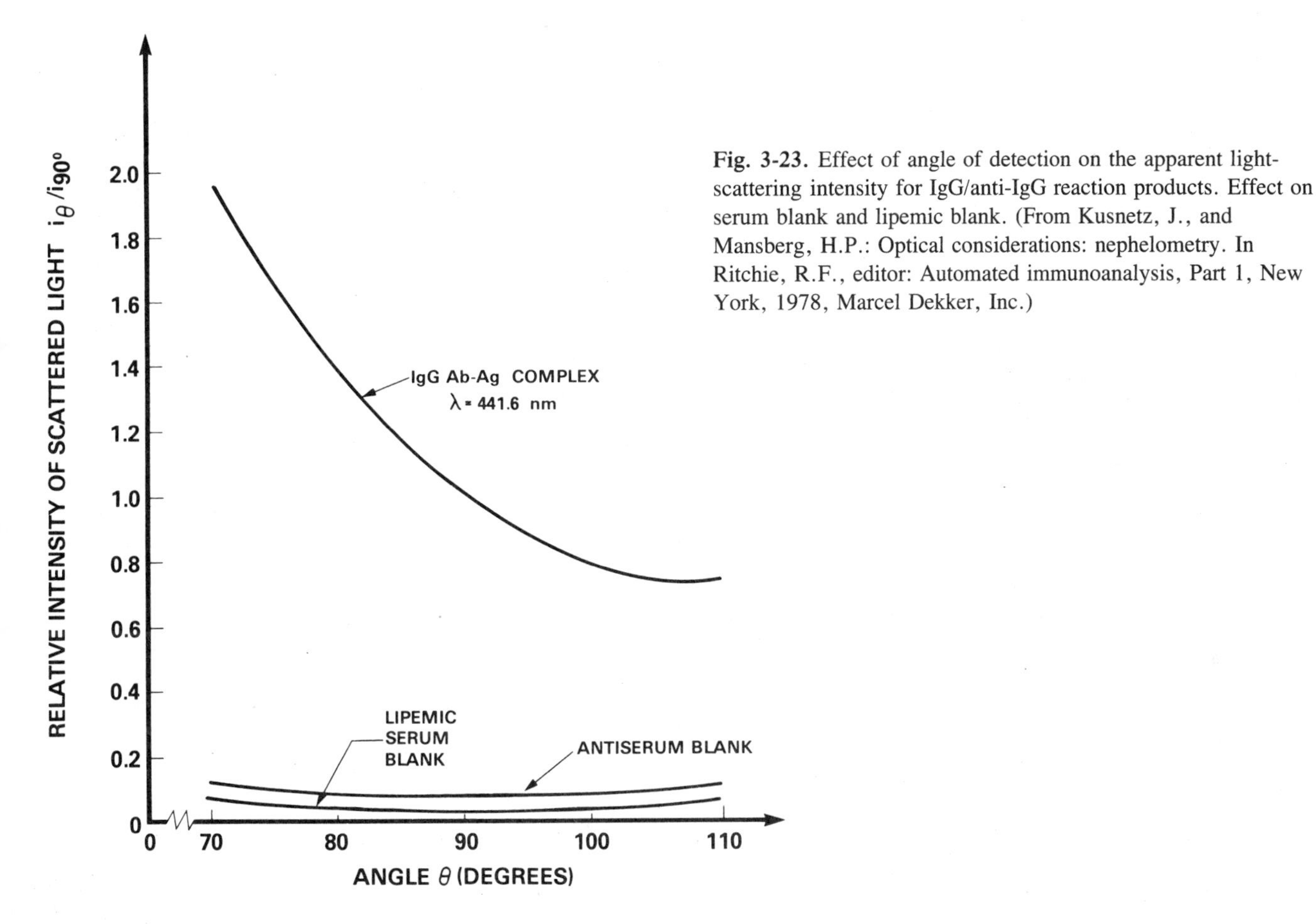

Fig. 3-23. Effect of angle of detection on the apparent light-scattering intensity for IgG/anti-IgG reaction products. Effect on serum blank and lipemic blank. (From Kusnetz, J., and Mansberg, H.P.: Optical considerations: nephelometry. In Ritchie, R.F., editor: Automated immunoanalysis, Part 1, New York, 1978, Marcel Dekker, Inc.)

duce stable, highly collimated, and intense beams of light (typically 1 milliradian divergence) that require no additional optical collimators as do other light sources. In optical systems using laser light, it is easier to reduce stray light, which contributes to background scatter, and to mask the transmitted beam thus allowing measurement of forward scatter. The increase in light intensity achievable with lasers also results in an improvement of signal-to-noise ratio, but this is limited somewhat by detector saturation. A disadvantage of laser sources is their cost and safety and restricted availability of limited fixed wavelengths. Since particle size may continually change during the course of reaction analysis, as during immune precipitate formation, light scatter at a single wavelength may change while the average light scatter over a number of wavelengths remains relatively constant. The Beckman Auto-ICS System employs a broad-band filter for selection of a wavelength region from a normal tungsten lamp source to overcome this problem, which is obviously more acute in rate methods when the size of the particle is changing most rapidly.

In all cases, the photodetector system must be matched to the wavelength or wavelengths of the scattered light which, for nephelometry and turbidimetry, corresponds to the incident light wavelength or wavelengths.

Angle of detection. Since particles the size of antigen-antibody complexes appear to scatter light more in the forward direction, there is an increased signal-to-noise ratio

Table 3-5. Automated clinical instruments available for light-scattering analysis*

	Technicon AIP	Hyland PDQ (OISC 120)	Behring Auto-laser LN	Beckman Automated ICS	Reaction rate analyzer†	Centrifugal fast analyzer‡
Nephelometer (N) (angle) or turbidometer (T)	N (90°)	N (31°)	N (5°-12°)	N (70°)	T	T
End-point (E) or kinetic (K) analysis of reaction	E + blank	E + blank	E + blank	K	K or E	K or E
Light source (wavelength, nm)	Mercury arc (357 nm)	Laser (632.8 nm)	Laser (632.8 nm)	Tungsten lamp (400-550 nm)	Variable: ultraviolet to 700 nm	Variable: ultraviolet to 700 nm
Calibration	Standard curve	Standard curve	Standard curve	Single point, reagent lot	Standard curve, reagent lot	Standard curve, reagent lot
Samples/hour	120	120	120-240	30-50	20-40	300
Reaction time	10 min	1 hour	1 hour (no PEG)	30 sec	1-2 min (kinetic)	1-2 min (kinetic)
Sensitivity	5 mg/L	1 mg/L	—	1 mg/L	—	—
Precision (CV§)						
Within-run	2%-5%	2%-5%	—	2%-5%	—	—
Between-run	5%-10%	5%-10%	—	5%-10%	—	—
Capital outlay	+ +	+ + +	+ + +	+ + +	+ + +	+ + + +
Original monitoring capability‖	0	+ +	±¶	+ + +	±‖	±‖
Other type of analysis	+ +	+	+	+	+ + +	+ + + +

*Adapted from Deverill, I., and Reeves, W.G.: J. Immunol. Methods **38:**191-204, 1980.
†For example, those by Abbott (Irving, Tex.); du Pont de Nemours (Wilmington, Del.); Gilford (Oberlin, Oh.); LKB/Clinicon (Rockville, Md.); Pye-Unicam (Cambridge, Eng.); and Vitatron (Lexington, Mass.).
‡For example, those by Aminco (Rotachem) (Silver Springs, Md.); ENI (GemENI) (Fairfield, N.J.); Instrumentation Laboratory (Multistat III) (Lexington, Mass.); Roche (COBAS Bio) (Nutley, N.J.); Baker (CentrifiChem) (Pleasantville, N.Y.).
§*CV*, Coefficient of variation; *PEG*, polyethylene glycol.
‖Drug monitoring by hapten inhibition of light scatter.
¶Method allows drug monitoring but no information available as to method and so on.

as the detector is placed nearer the transmitted path (0 degrees). Fig. 3-23 shows the magnitude of the difference expected in an IgG–anti-IgG system.[20]

The blank signal, described best by Rayleigh scatter (Fig. 3-19, *A*), is not as affected by an altered angle of detection. Thus, although most early nephelometers detect light scattered at 90 degrees, for reasons of manufacturing ease and limited low-angle measurement capability, the detection of forward light scatter should provide theoretically greater sensitivity. The newer instruments tend to operate with lower detection angles, optimized in many cases to give the highest signal-to-noise ratio for the particular instrument's optics. Several of the systems seen in Table 3-5 detect forward light scatter and possess the increased sensitivity expected. Obviously, detection at 0 degrees is not possible because of the high intensity of the transmitted beam, but some laser-equipped fast analyzers using a mask to block the transmitted beam are able to operate at very low angles. Instruments employing low-angle detectors tend to have increased sensitivity over the 90-degree type of instruments.

Limitations: turbidimetry versus nephelometry

Although the principle of nephelometry, detection of a small signal (amplifiable) on a black background, should lend this method high sensitivity, the sophistication and specifications of the instruments available do not achieve this promise. Turbidimetry, detection of a small decrease in a large signal, should be limited in sensitivity; however, current instruments have excellent discrimination and can quantify small changes in signal, thereby allowing turbidimetric measurements to achieve high sensitivity.

Turbidimetry and nephelometry have similarities to absorption spectrophotometry, and many sources of interference and errors are common to all these systems. Many techniques, discussed in Chapter 52, to minimize absorption interferences are also applicable to turbidimetry and nephelometry. Because of the uniqueness of nephelometric measurements, especially in the case of antibody-antigen reactions, some specific applications are discussed below.

Endogenous color and choice of wavelength. Basic light-scattering theory predicts that the intensity of scattered light increases as shorter wavelengths of incident light are used. Most immunological assay reactions employ serum protein reactions requiring the choice of a wavelength at which neither the proteins nor colored serum components absorb appreciably. Since proteins absorb strongly below 300 nm and serum has an absorption peak at 400 to 425 nm because of porphyrins, instruments tend to operate in the 320 to 380 or 500 to 650 nm ranges.

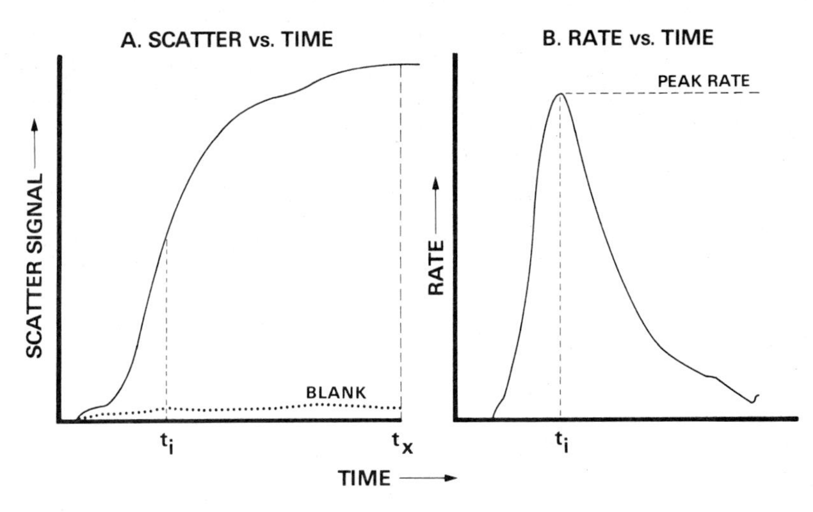

Fig. 3-24. Kinetic analysis of light scattering. *Curve A,* Intensity of scattered light signal versus time; *curve B,* rate of change of scattered light signal versus time.

Reduction of the protein concentration by dilution will decrease background absorption. Most immunochemical reactions measured by nephelometry use high-affinity antibodies that allow for large dilutions of protein and consequent improvement of sensitivity.

Comparison of sensitivity. Sensitivity in nephelometers is largely controlled by the amount of background scatter from sample and reagents. Since background scatter can be high relative to specific scatter, instruments do not reach their full potential of sensitivity. This limitation, coupled with the higher wavelengths generated in laser instruments, accounts for the fact that laser instruments show no great increase in sensitivity over conventional nephelometers.

Sensitivity in turbidimetric measurements is dependent on the ability of the detector to resolve small changes in light intensity. Using low wavelengths and high-quality spectrophotometers with their high-precision detection systems, sensitivity in turbidimetry is usually adequate for many measurements and in many cases compares well to nephelometry. Theoretically, with additional refinements, nephelometry ultimately should provide higher sensitivity than turbidimetry.

End-point versus kinetic analysis. Examination of light scattered as a function of time, after there is mixture of an antibody and antigen, shows that after an initial delay there is an almost linear increase in scatter followed by a slower attainment of plateau scatter. The secondary reaction occurs much slower than the first because larger particles form and begin to flocculate and they distort the scatter intensity seen at forward angles. Both turbidity and nephelometry measurements behave in this manner.

There are two basic ways of measuring light scatter caused by this reaction—end-point analysis and rate analysis. End-point analysis requires blank (reagent) determinations and a reasonable time must elapse for final measurement. Fig. 3-24 shows the forward scatter developed at 70 degrees of a rate nephelometric analyzer.[19] Comparing the two graphs, one can see the differences between an end-point analysis (blank value versus reading at $t = x$) and a rate or kinetic analysis (increase in scattered intensity over a set time interval). The kinetic approach, which electronically subtracts any blank signal, thus does not require a separate reagent blank to be run. Both kinetic and end-point analysis can be applied equally to turbidimetry and nephelometry.

Table comparing several instruments. Both nephelometry and turbidimetry can be automated, and both techniques can be analyzed by either rate or end-point methodology. The choice of method is therefore made by extraneous considerations such as availability, cost, and versatility of instrumentation.

Table 3-5, adapted from the review by Deverill and Reeves,[18] shows the major features of the automated instruments available. The Beckman Automated ICS system, in my opinion (J.G.), represents the best of the nephelometers. The choice of the fast analyzers is usually determined more by their versatility than by turbidimetric considerations.

Each of the instruments has advantages and drawbacks, and the selection depends on the service provided or specific needs of the consumer. The Beckman, Hyland, and Behring nephelometers are also available as less expensive manual instruments for laboratories with low throughput needs.

The nephelometric and turbidimetric assays have demonstrated excellent correlation with standard radial immunodiffusion or Mancini assays. Correlation coefficients for immunoglobulins range from 0.89 to 0.99, with good scatter around the regression line. Automation of analysis results in interassay coefficients of variation of 2% to 8%. In the ranges of analyte normally encountered, the sensitivity of the instruments is generally in the milligram-per-liter range.

REFRACTIVITY

Principle

When a beam of light impinges upon a boundary surface, it can be reflected, absorbed, or, if the material is transparent, pass into the boundry and emerge on the other side. When light passes from one medium into another, the path of the light beam will change direction at the boundary surface if its speed in the second medium is different from that in the first (Fig. 3-25). This bending of light is called "refraction."[21]

Since the degree of refraction of a light beam is dependent on the difference in the speed of light between two different mediums, the *ratio* of the two speeds has been expressed as the *index of refraction,* or *refractive index.* The relative ability of a substance to bend light is called "refractivity." The expression of a refractive index, *n,* is always relative to air with the convention that n of air $=$ 1. The measurement of the refractive index is the measurement of angles, since the light is bent at an angle proportional to the relationship of n in the medium through which the light is passing.

$$\frac{n}{n_1} = \frac{\sin \theta}{\sin \theta_1}$$

The refractivity of a liquid is dependent on (1) the wavelength of the incident light, (2) the temperature, (3) the nature of the liquid, and (4) the total mass of solid dissolved in the liquid. If the first three factors are held constant, the refractive index of a solution is a direct measure of the total mass of dissolved solids.

Applications

Refractometry has been applied to the measurement of total serum protein concentration.[22] The assumption of this analysis is that the serum matrix (that is, the concentration of electrolytes and small organic molecules) remains essentially the same from patient to patient. Since the mass of protein is normally so much greater than for other serum constituents, small variations of these other substances have no significant effect upon the refractive index of serum. Refractometers are calibrated against "normal" serum, and total protein concentrations are read directly from a scale.

Refractometry is also used to estimate the specific gravity of urine samples. The refractive index is linearly related to the total mass of dissolved solids and thus to specific gravity. This remains valid over most of the range normally encountered for urine (that is, up to 1.035 g/mL).

Interference

When the concentration of small molecular weight compounds or particulate matter greatly increases, positive interference results. This interference occurs in the presence of hyperglycemia, hyperbilirubinemia, azotemia (increased serum urea), lyophilized samples, and hyperlipidemia.

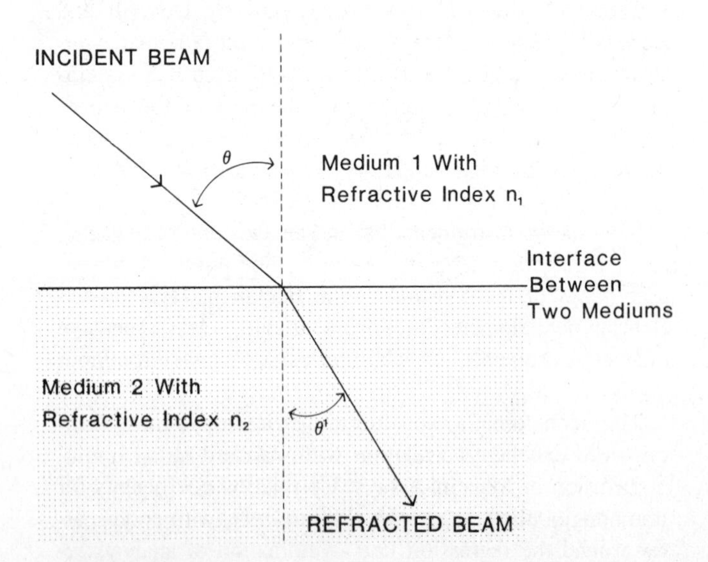

Fig. 3-25. Schematic illustrating bending of light when it passes from a medium of one density into a medium of a different density, with an angle of deflection, θ^1.

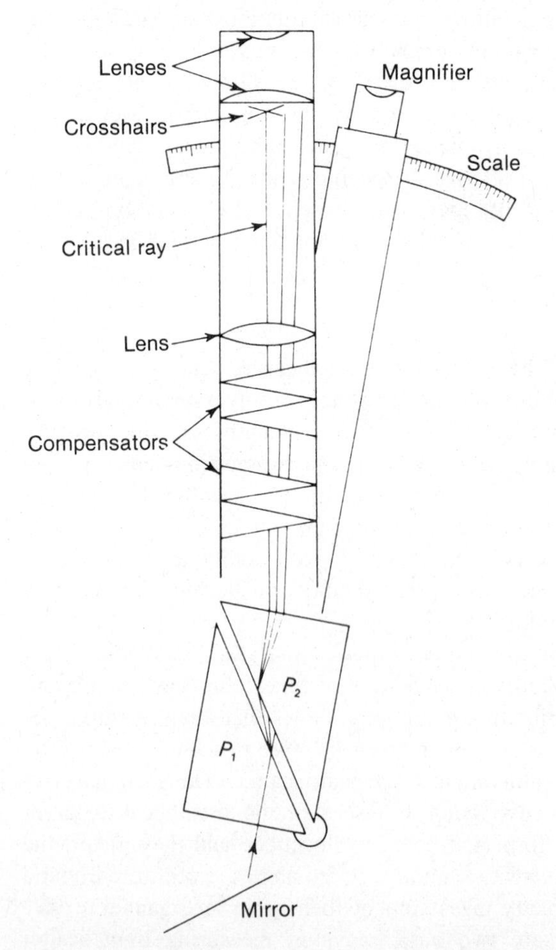

Fig. 3-26. Schematic of an Abbé refractometer. (From Shugar, G.J., Shugar, R.A., and Bauman, L.: Chemical technicians' ready reference book, New York, 1973, McGraw-Hill Book Co.)

Hemolysis will also result in false-positive values for total serum protein.

Instrumentation

Most clinical refractometers are based on the Abbé refractometer (Fig. 3-26), marketed by American Optical Corporation. This refractometer consists of two prisms and a series of lenses. Light passes through the first prism where the light beam is dispersed. The dispersed light passes into and through the thin layer of the liquid sample where it is refracted. The light beam passes through a second prism where the light is again dispersed and upon leaving is again refracted. The boundary at the edge of the refracted light beam is aligned perpendicularly to the scale for reading of serum protein concentrations or specific gravity. The scale for reading serum protein (g/dL or g/L) is established by calibration of the instrument against a "normal" serum solution.

This type of refractometer is extraordinarily simple, having no moving or electrical parts. Thus it is easily reproducible, measuring protein with a precision of ±1% and an accuracy of ±1 g/L. The sample size is on the order of 50 μL.

More complex refractometers are used to monitor column effluents for high-performance liquid chromatography (HPLC) analysis. See Chapter 5.

REFERENCES

1. Ditchburn, R.W.: Light, New York, 1962, John Wiley & Sons, Inc.
2. Clayton, R.K.: Light and living matter: the physical part, New York, 1970, McGraw-Hill Book Co.
3. Jaffe, H.H., and Orchin, M.: Theory and applications of ultraviolet spectroscopy, New York, 1966, John Wiley & Sons, Inc.
4. Willard, H.H., Merritt, L.L., and Dean, J.A.: Instrumental methods of analysis, ed. 4, Princeton, N.J., 1965, D. Van Nostrand Co.
5. Frings, C.S., and Broussard, L.A.: Calibration and monitoring of spectrometers and spectrophotometers, Clin. Chem. **25**:1013-1017, 1979.
6. Brewer, J.M., Pesce, A.J., and Ashworth, R.B., editors: Experimental techniques in biochemistry, Edgewood Cliffs, N.J., 1974, Prentice-Hall, Inc.
7. Keller, H.: Optical methods of measurement. In Richterich, R., and Colombo, J.P., editors: Clinical chemistry, New York, 1981, John Wiley & Sons, Inc.
8. Simonson, M.G.: The application of a photon-counting fluorometer for the immunofluorescent measurement of therapeutic drugs. In Kaplan, L.A., and Pesce, A.J., editors: Nonisotopic alternatives to radioimmunoassay, New York, 1981, Marcel Dekker, Inc.
9. Instrumentation Guidelines Study Group, Subcommittee on Enzymes, Clin. Chem. **23**:2160-2162, 1977.
10. Alexander, L.R., and Barnhart, E.R.: Photometric quality assurance instrument check procedures, Atlanta, Ga., 1980, U.S. Department of Health and Human Services, Centers for Disease Control, Bureau of Laboratories.
11. Rubeska, I.: Atomic absorption spectroscopy, Cleveland, 1969, Chemical Rubber Company Press.
12. Robinson, J.W.: Atomic absorption spectroscopy, ed. 2, New York, 1973, Marcel Dekker, Inc.
13. Winefordner, J.D., editor: Spectrochemical methods of analysis, New York, 1977, John Wiley & Son, Inc., Wiley Interscience.
14. Dvorak, J., Rubeska, I., and Rezak, Z.: Flame photometry: laboratory practice, Cleveland, 1971, Chemical Rubber Company Press.
15. Pesce, A.J., Rosen, C.G., and Pasby, T.L., editors: Fluorescence spectroscopy, New York, 1971, Marcel Dekker, Inc.
16. Wehry, F.L., editor: Modern fluorescence spectroscopy, New York, 1976, Plenum Press.
17. Ritchie, R.F., editor: Automated immunoanalysis, Parts 1 and 2, New York, 1978, Marcel Dekker, Inc.
18. Deverill, I., and Reeves, W.G.: Light scattering and absorption developments in immunology, J. Immunol. Methods **38**:191-204, 1980.
19. Gauldie, J.: Principles and clinical applications of nephelometry. In Kaplan, L.A., and Pesce, A.J., editors: Nonisotopic alternatives to radioimmunoassay, New York, 1981, Marcel Dekker, Inc.
20. Kusnetz, J., and Mansberg, H.P.: In Ritchie, R.F., Automated immunoanalysis, Part 1, New York, 1978, Marcel Dekker, Inc.
21. Glover, F.A., and Gaulden, J.D.S.: Relationship between refractive index and concentration of solutions, Nature **200**:1165-1166, 1963.
22. Rubini, M.D., and Wolf, A.V.: Refractometric determination of total solids and water of serum and urine, J. Biol. Chem. **225**:869-876, 1957.
23. Shugar, G.J., Shugar, R.A., and Bauman, L.: Chemical technicians' ready reference handbook, New York, 1973, McGraw-Hill Book Co.

Chapter 4 Chromatography: theory and practice

M. Wilson Tabor

adsorption Process whereby one substance adheres to another because of attractive forces between surface atoms of the two substances. (See Fig. 4-13.)

analyte The substance or component in a sample for which the analysis is being conducted or that is being measured.

band A chromatographic zone, that is, a region where the separated substance is concentrated.

capacity factor The ratio of the elution volume of a substance to the void volume in the column. (See Equation 4-8 and Fig. 4-9, *A*.)

chromatogram A series of separated bands or zones detected either visually, as in some paper chromatographic or thin-layer chromatographic separations, or indirectly by a detection system. In the latter case, the detection system usually outputs an electrical signal, which is graphically plotted through time, to display the series of separated bands or zones.

chromatography A method of analysis in which the flow of a mobile phase (gas or liquid) containing the sample promotes the separation of sample components by a differential distribution between this phase and a stationary phase. The stationary phase may be a solid, or a liquid coated or bonded on a solid.

dipole The attractive force of compounds having centers of both positive and negative charges that is the result of an unequal sharing of bonding electrons between two elements of the compound with large differences in electronegativity. (See Fig. 4-11, *B*.)

dispersive The attractive force, sometimes termed "van der Waals' forces," of compounds that results from the induction of a temporary dipole within an individual compound. (See Fig. 4-11, *A*.)

distribution isotherm A graphic representation of the interaction of a solute in solution (gas or liquid mobile phase) and an adsorbent (stationary phase) at a given temperature. (See Fig. 4-4.)

eddy diffusion (A) Diffusion of solute molecules during the chromatographic process attributable to an uneven rate of movement of the solute molecules through the column because of differences in the path lengths individual solute molecules have to travel. (See Equations 4-6 and 4-7.)

efficiency A measure of chromatographic performance usually related to the sharpness of the peaks in the chromatogram and quantitated by the number, N, of theoretical plates of a column. (See Equation 4-3, Figs. 4-6 and 4-7, and Table 4-1.)

electrostatic attraction The attractive force of compounds having formal positive or negative charges. (See Fig. 4-11, *D*.)

eluotropic series Series of solvents or solvent mixtures arranged in the order of their ability to elute a solute from an adsorbent.

elution Removal of a solute from a stationary phase by passage of a suitable mobile phase.

elution volume (V_e) The volume of mobile phase required to elute a solute from a chromatographic column. (See Equation 4-8 and Fig. 4-9, *A*.)

emulsion For a liquid-liquid extraction, a mixture formed when one of the immiscible liquid phases becomes dispersed as fine droplets in the other immiscible liquid phases.

equilibrium concentration distribution coefficient (K_D) The ratio of the concentration of a sample component in one phase to its concentration in a second phase at equilibrium. The two phases may be two immiscible liquids or the mobile phase and the stationary phase. (See Equation 4-1 and Fig. 4-3.)

height equivalent to a theoretical plate (HETP) The number obtained by dividing the column length by the theoretical plate number. (See Equation 4-4 and Fig. 4-7.)

hydrogen bonding The attractive force of compounds where a hydrogen atom covalently linked to an electronegative element, like oxygen, nitrogen, or sulfur, has a large degree of positive character relative to the electronegative atom thereby causing the compound to possess a large dipole. (See Fig. 4-11, C.)

longitudinal diffusion (B) Diffusion of solute molecules in the direction of flow of the mobile phase. (See Equations 4-6 and 4-7.)

mass transfer (C, C_m, C_s) The rate at which solute molecules transfer into and out of the stationary (C_s) phase from the mobile (C_m) phase. (See Equations 4-6 and 4-7.)

pK_a The pK of an acid is the pH at which it is half dissociated.

partition Process by which a solute is distributed between two immiscible phases. (See Fig. 4-15.)

peak A term applied to a chromatogram showing a maximum of concentration between two minima, such as a band or zone

polarity (P') The attractive forces encompassing the total interaction of solvent molecules with sample molecules and of solvent or sample molecules with the stationary phase.

R_f A ratio used in paper chromatography and thin-layer chromatography that is the distance from the origin to the center of the separated zone divided by the distance from the origin to the solvent front. (See Equation 4-11.)

resolution (R or R_s) The degree of separation between two components by chromatography. (See Equations 4-2, 4-10, and 4-12 and Fig. 4-5.)

retention time (t_R) The time that has elapsed from the injection of the sample into the chromatographic system to the recording of the peak (band) maximum of the component in the chromatogram. (See Fig. 4-7.)

selectivity (α) The ratios of the capacity factors for two substances measured under identical chromatographic conditions; sometimes termed "separation factor," or "chromatographic selectivity." (See Equation 4-9 and Fig. 4-9, *B*.)

theoretical plate number (N) A number defining the efficiency of the chromatographic column. (See Equation 4-3 and Fig. 4-7.)

void volume (V_o) The interstitial volume of the chromatographic column, that is, the volume of mobile phase imbibed in the pores and around the stationary phase in a column. (See Equation 4-8 and Fig. 4-9, *A*.)

The need for fast, reproducible, and accurate analyses for many classes of analytes present in small amounts in the clinical laboratory is being met today largely as a result of the developments in chromatography during the past two decades. Chromatography is a collective term referring to a group of separation processes whereby a mixture of solutes, dissolved in a common solvent, are separated one from the other by a differential distribution of the solutes between two phases. One phase, the solvent, is mobile and carries the mixture of solutes through the other phase, the fixed or stationary phase. Chromatographic methods encompass a number of variations in technique in which the mobile phase ranges from liquids to gases, and the stationary phase ranges from sheets of cellulose paper to capillary glass tubes as fine as a human hair that are internally coated with a covalently bonded complex or complex organic polymers. A cursory examination of the scientific literature shows that both the numbers of chromatographic methods published and their applications have been growing exponentially. For example, see the *Analytical Chemistry* compendium reviews of Fundamentals published by the American Chemical Society in April of every year. Recent reviews have included clinical chemistry,[1] gas chromatography,[2] liquid chromatography,[3] and thin-layer chromatography.[4]

Modern chromatography began in 1906 when Michael S. Tswett detailed his separation of chlorophylls using a column of calcium carbonate (chalk).[5] Although it was not Tswett's first publication on separation methods, this one contained a discussion of the physicochemical basis of chromatography—the mechanism of adsorption—and the introduction of a system of nomenclature that is now universally applied to chromatography (''color writing'', from the Greek words *khrōma, khrōmato-*, 'color,' and *graphē*, 'writing').

Additional details of the history of chromatography are in chapters by Heftman,[6] and Strain and Svec.[7] Histories of thin-layer and gas chromatography have been reviewed more extensively by Pelick et al.[8] and Ettre,[9] respectively. The Analytical Chemistry Divison of the American Chemical Society has published a book on the history of analytical chemistry, containing several chapters related to chromatography.[10] These are but a few of the many published accounts of the colorful history of chromatography.

BRANCHES OF CHROMATOGRAPHY

Chromatography is the separation of components in a mixture by their distribution between a fixed and a moving phase, that is, the stationary phase and the mobile phase. These components are made to undergo a differential migration by a combination of the flow of the mobile phase, a liquid, or a gas, and their interactions with the stationary phase. The fixed-phase interactions include adsorption, partition, size exclusion, and electrostatic attraction.

Chromatographic methods are generally classified according to the physical state of the solute carrier phase, that is, the mobile phase. These branches are represented in Figs. 4-1 and 4-2 as solution and gas chromatography, referring to the respective liquid and gaseous states of the mobile phase. In Fig. 4-1 these branches are further classified according to how the stationary phase matrix is contained for a particular chromatographic method. For example, solution chromatography is divided into flat and column methods, depending on whether the stationary phase is mechanically supported as a thin-layer sheet or is packed into a column. The flat method of support may involve use of a sheet of paper, such as cellulose, or a thin layer on a mechanical backing, such as glass or plastic.

Column methods are classically referred to as ''liquid chromatography.'' Furthermore, ''column methods'' is a phrase generally used to subdivide solution chromatography wherein the stationary phase is packed into a glass or metal tube. However, it is noted that gas chromatography is strictly a column method for containment of the stationary phase. For gas chromatography one contains the stationary phase in the column in one of two ways: by coating the stationary phase onto particles, that is, a packed column, or by covalently bonding the internal walls of the column with the stationary phase. This latter method is found in capillary columns.

The main divisions of chromatography, based on mobile phase, may also be subdivided according to mechanism of solute interaction with the stationary phase (Fig. 4-2). Two mechanisms, adsorption and partition, are the most commonly encountered for both solution and gas mobile-phase separations. Adsorption chromatography (liquid-solid, L/S, or gas-solid, G/S) is a process whereby solutes of a sample are separated by their differences in attraction to the stationary versus the mobile phase. Partition chromatography (liquid-liquid, L/L, or gas-liquid, G/L) is a process whereby the solutes of a sample are separated by differences in their distribution between two liquid phases (L/L) or between a gas and a liquid phase (G/L). In both

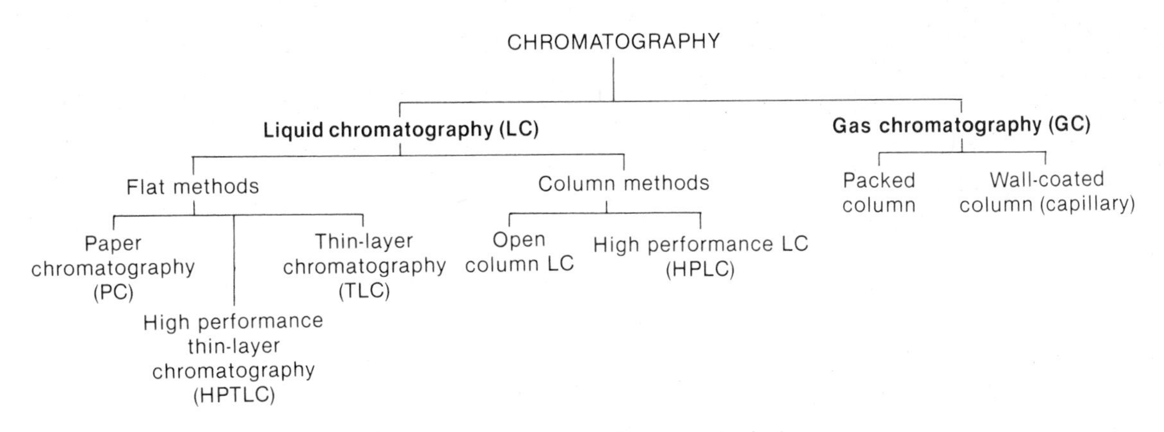

Fig. 4-1. Branches of chromatography according to mobile phase and physical apparatus.

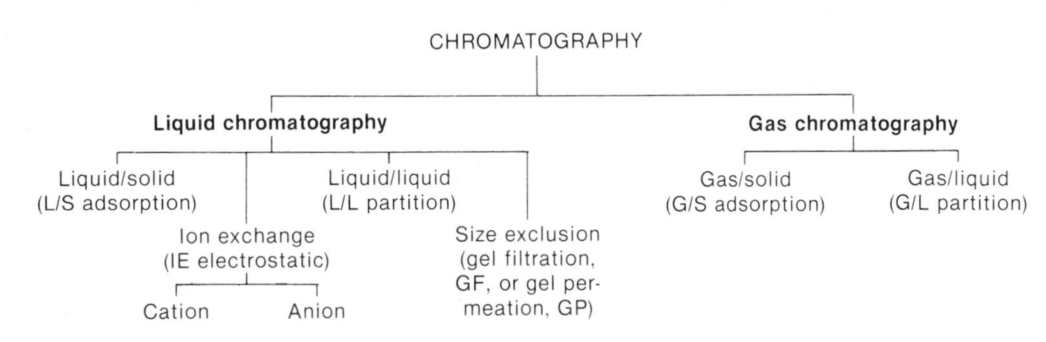

Fig. 4-2. Branches of chromatography according to mechanism of separation on stationary phase.

cases, the stationary phase is liquid and the mobile phase is a liquid or a gas.

Other mechanistic divisions of solution chromatography are ion exchange and gel permeation. Ion-exchange chromatography (IE) uses an insoluble matrix containing covalently linked ionic groups for the stationary phase, which can reversibly exchange either cations or anions with the mobile phase. Gel filtration (GF) chromatography refers to a stationary phase of solvent-swollen hydrophilic gel in the form of porous beads that is used with an aqueous-solvent mobile phase. Gel permeation (GP) chromatography refers to a stationary phase of solvent-swollen hydrophobic gel in the form of porous beads that is used with an organic-solvent mobile phase. Both GF and GP chromatography are sometimes referred to as "molecular exclusion (or inclusion) chromatography." Solutes in a sample are separated based upon their size relative to the size of the pores in the stationary phase. Small solute molecules penetrate the pores of the gel particles, that is, inclusion occurs, and are retarded relative to the large solute molecules, which do not penetrate the pores of the gel particles, that is, exclusion occurs.

The boundaries between these mechanistically different types of chromatography are not finite, since for some chromatographic separations more than one mechanism may be operating. For example, in gel filtration chromatography, adsorptive interactions between the solute molecules and the stationary phase are common, in addition to the prevailing size-exclusion mechanism. More discussion of these mechanisms is presented below, after a brief discussion of chromatographic theory and principles.

GENERAL PRINCIPLES

The theoretical basis of chromatography is well developed, with both solution and gas-phase methods sharing the same foundation.[11] Only a few general concepts of this theory are discussed in this section. For more extensive discussions, refer to representative books[12-20] and reviews,[21-24] and the *Analytical Chemistry* compendium reviews[1-4] on specialized chromatographic techniques for additional reference leads.

The separation of a mixture containing two or more components is an operation with the goal of producing fractions, each of which has an increased concentration of one component relative to the other components contained in the original mixture. The physicochemical basis of the separation process can be described in two distinct equilibriums[13]: phase equilibriums and distribution equilibriums. Phase equilibrium refers to the physical state equilibriums existing between the gas, liquid, and solid states. Separation techniques such as sublimation and the various types of distillation are based on phase equilibriums. Distribution equilibrium refers to the differences in solubility and adsorption of a component between two immiscible phases. Chromatographic separation techniques are principally based on distribution equilibriums.[13,15] The theoretical basis and predictive capabilities of separation equilibriums are grounded in the physicochemical principles of thermodynamics and intermolecular interactions.[13]

Separation equilibrium begins with a consideration of the distribution of a solute *S* between two immiscible phases, upper phase *(u)* and lower phase *(l)*, at constant temperature and pressure. The ratio of the solute concentrations in the two phases determines the separation, which can be defined by an equilibrium concentration distribution coefficient, K_D, for the molar concentration, C_u and C_l, of solute in the upper and lower phases, respectively.

$$K_D = \frac{C_u}{C_l} \qquad \textit{Eq. 4-1}$$

Molar concentration is defined in the classical sense as the number of moles of solute per unit volume of solvent. The distribution coefficient is sometimes referred to as a partition ratio. In practical terms, a K_D of 1.0 means that 50% of the solute is distributed in the upper phase and 50% is distributed in the lower phase (Fig. 4-3, *A*). Likewise a K_D of 9.0 means that 90% of the solute is distributed in the upper phase and 10% is in the lower phase (Fig. 4-3, *B*). The principle of thermodynamic distribution coefficients will be further considered in the discussion of partition chromatography.

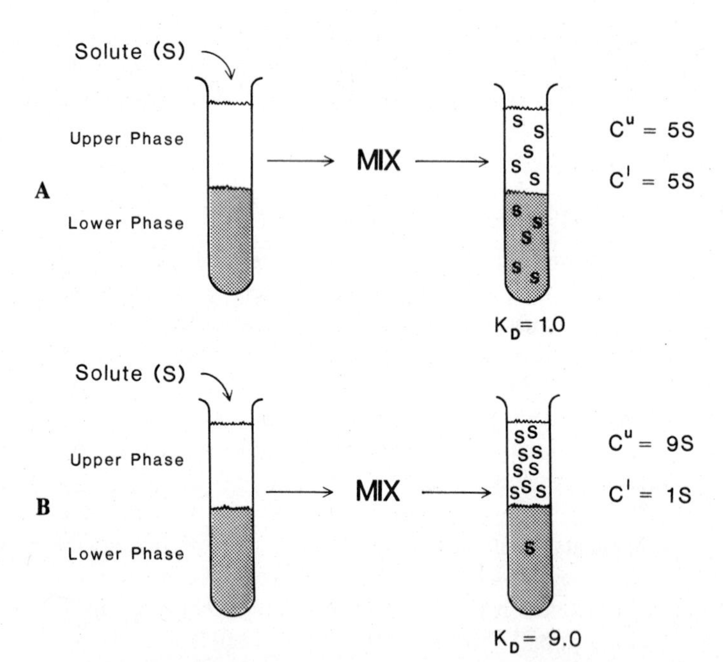

Fig. 4-3. Separation of a solute, S, by partition into two different solvent systems. In the first system, **A**, solute has a distribution coefficient, K_D, of 1.0, indicating an equal partitioning between upper and lower phases after mixing. In second system, **B**, solute has a K_D of 9.0, indicating a partitioning of nine parts of the solute in the upper phase and one part of the solute in the lower phase after mixing. C_u, Upper-phase concentration; C_l, lower phase concentration.

For a more generalized application of distribution coefficients to chromatography, let $C_l = C_m$, where C_m refers to the amount of the solute distributed into a unit amount of mobile phase, and let $C_u = C_s$, where C_s refers to the amount of the solute distributed into a unit amount of stationary phase. In this more generalized case, the distribution equilibriums for a solute being separated in a given chromatographic system at constant temperature, that is, isothermal, can be graphically illustrated by a plot of solute concentration in the mobile phase, C_m, versus the concentration in the stationary phase, C_s. This is expressed as an adsorption-distribution isotherm (Fig. 4-4, *A*). The slope of the isotherm is equal to the distribution equilibrium coefficient, K_D.

The resulting shape of a distribution isotherm plot (Fig. 4-4, *A*), is dependent on several factors. Since solute movement between the mobile phase and the stationary phase is a thermodynamic equilibrium process, the distribution equilibrium coefficient, K_D, is both temperature and pressure dependent. These three variables are related by the standard entropy (S_o, that is, a measure of the amount of disorder within a system) and enthalpy (ΔH_o, that is, heat of the adsorption-desorption process) of the overall solute–stationary phase interaction. However, temperature and pressure are not contributing factors in the situation under discussion, since one earlier assumption was that these variables were constant for the distribution process. This assumption is usually made for most routine chromatographic procedures. Also, it should be noted that this assumption and the overall thermodynamic basis of solute-solvent interaction theory[12-14,17] hold only for dilute solutions of the solute in the mobile phase. This is the type of situation existing for several chromatographic techniques, most notably in conventional gas chromatography and in other types of chromatography, such as HPLC (high-performance liquid chromatography), when dealing with trace amounts of solute components in the sample. For these separations, the distribution isotherms approach linearity (Fig. 4-4, *A1*), and the solute-concentration profiles resemble a discrete circular spot (Fig. 4-4, *B1*), or a symmetric bell-shaped (that is, gaussian) peak (Fig. 4-4, *C1*). But it must be noted that one cannot predict behavior of any given sample in any given chromatographic system.

Linearity of distribution isotherms, as in Fig. 4-4, *A1*, is the exception rather than the norm. In most cases, concave or convex distribution isotherms are observed for solute–stationary phase interactions (Fig. 4-4, *A2* and *A3*, respectively). Several factors influence the degree to which nonlinearity is observed. At higher concentrations of solute in the mobile phase, nonlinearity results, since the thermodynamic basis of solute-solvent interactions only holds for dilute solutions. Another factor is the complexity of the sample, that is, the presence of multiple solutes. Interaction of these components with each other will cause a deviation from linearity.

If dilute solutions are used or there are no solute-solute interactions, a Langmuir or convex isotherm (Fig. 4-4, *A2*) may be observed. This isotherm shows that the limited number of adsorptive sites on the stationary phase become occupied with solute with increasing solute concentration, thereby losing their capacity to adsorb in proportion to the overall increase in solute concentration. Therefore the resulting solute-elution profile (Fig. 4-4, *B2* and *C2*) is characterized by a sharp front or leading edge that is indicative of a high solute concentration. The rear boundary of the solute elution pattern decreases asymmetrically from the peak.

For situations where the solute is poorly adsorbed to the stationary phase, preferring the mobile phase, the anti-Langmuir or concave isotherm (Fig. 4-4, *A3*) is observed. In this situation, the resulting solute-elution profile (Fig. 4-4, *B3* and *C3*) is characterized by sloping, that is, low-solute concentration, front boundaries and sharp, that is, high-solute concentration, rear boundaries. Also the elution volume for the solute is a function of sample size.

RESOLUTION

The ultimate goal of any given chromatographic technique is to separate the components of a given sample within a reasonable period of time. The purpose of such a separation is to detect or quantitate a particular component or group of components of interest in pure form. The ability to resolve the components one from the other and the degree to which this resolution is accomplished are measures of the adequacy of the chromatographic separation.

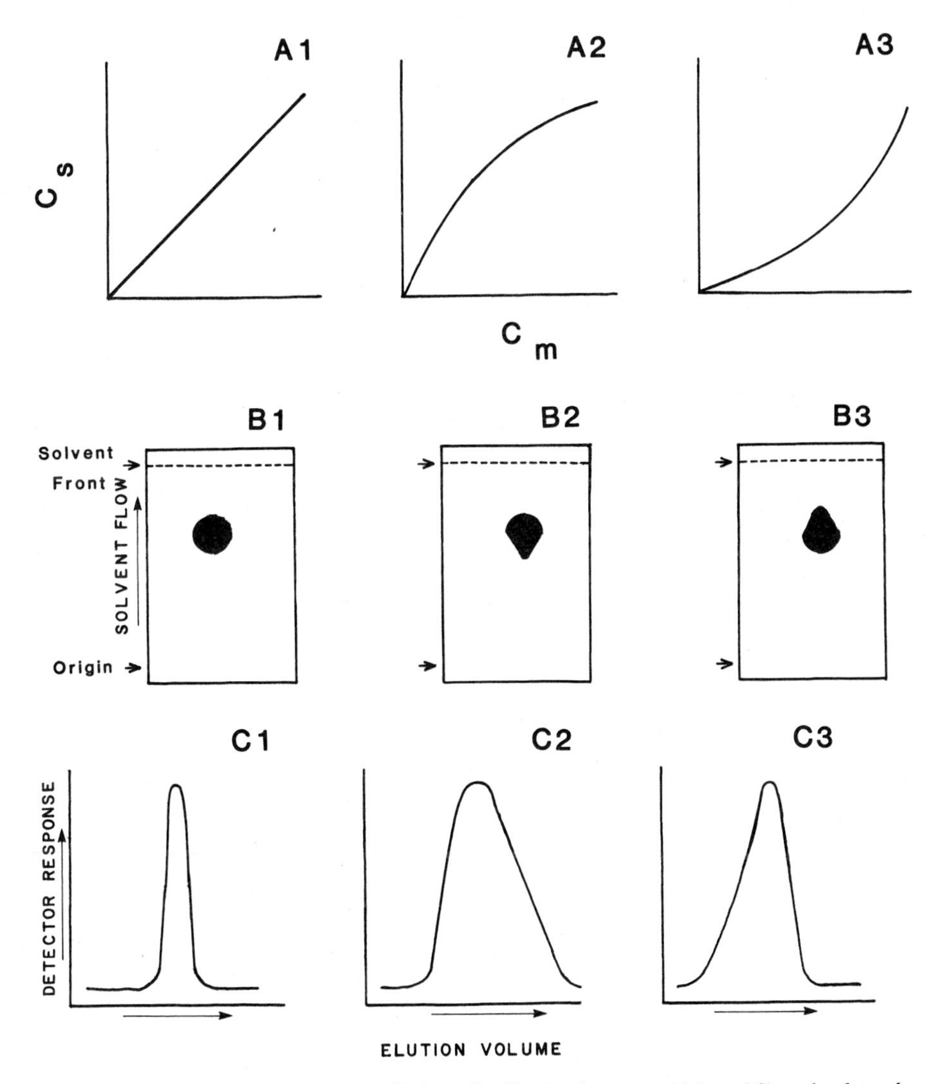

Fig. 4-4. Relationship between equilibrium distribution isotherm (A1 to A3) and solute shape, position, and elution profile in a PC/TLC chromatogram (B1 to B3) or a GC/HPLC chromatogram (C1 to C3). C_s, Concentration of solute in stationary phase; C_m, concentration of solute in mobile phase.

The question of what is adequate resolution for a given sample has been detailed by Snyder.[25] For the purpose of the present discussion, this question can be answered only if the analyst defines the objectives of the chromatographic separation. Generally, objectives for the analyst in a chemical laboratory depend on the following questions: (1) Is a particular substance present in a sample; that is, should a qualitative analysis be followed? or (2) how much of a particular substance is present in a sample; that is, should a quantitative analysis be followed? If unexplained components are detected on separation of a sample, the objective of the analyst may be to purify, by chromatography, sufficient quantities of the unknown for structural elucidation by other techniques such as mass spectrometry. In the following discussion of the theory of resolution, the principal emphasis and corresponding illustrations will be with column, gas, or liquid techniques rather than flat methods.

Note that conventionally the concentrations of solutes in a chromatographic system are plotted out versus time or distance. The bands or zones separated are usually termed a "peak."

An actual chromatographic separation of a three-component mixture by high-performance liquid chromatography is shown in Fig. 4-5, indicating important parameters for assessment of resolution. The quantity, *R*, for any two components is defined as the distance, *d*, between the peak centers of two peaks divided by the average base width, *W*, of the peaks, that is:

$$R = \frac{d_2 - d_1}{{}^1\!/_2(W_1 + W_2)} \qquad \text{Eq. 4-2}$$

For this calculation, both the distance, *d*, and the peak width, *W*, are measured in the same units. This method of calculating resolution is based upon the assumption that the distribution isotherms, for the components being separated, approach linearity (Fig. 4-4, *A1*) under specified conditions. The set of specified conditions are mobile phase, stationary phase, and solute-concentration range.

A resolution value of 1.25 or greater is required for good quantitative or qualitative chromatographic analyses. If the resolution is 0.4 or less, the peak shape does not clearly show the presence of two or more components. The actual value of the resolution depends on two factors: width of the peak and the distance between the peak maxima for column separations; diameter of the circular spots and the distance between these spots for flat-method separations.

These determinant factors of resolution also are indicative of the efficiency of the chromatographic process. Efficiency is decreased by the broadening of a solute band as it migrates through the stationary phase. If broadening occurs to any significant extent during the chromatographic process, the resulting peaks will be wide or the resulting spots will be diffuse. The separation of components is then poor, and the sensitivity with which they can be detected is reduced. An example of high versus low efficiency of separation is illustrated in Fig. 4-6 for both column and flat-method separations.

Solute-band broadening occurs during the actual chromatographic process and may be described as follows for a column separation. The sample, in a small volume of solvent, is introduced into the mobile phase at a point near the inlet end of the column. Once entering the column, the sample begins to disperse by thermal diffusion processes,

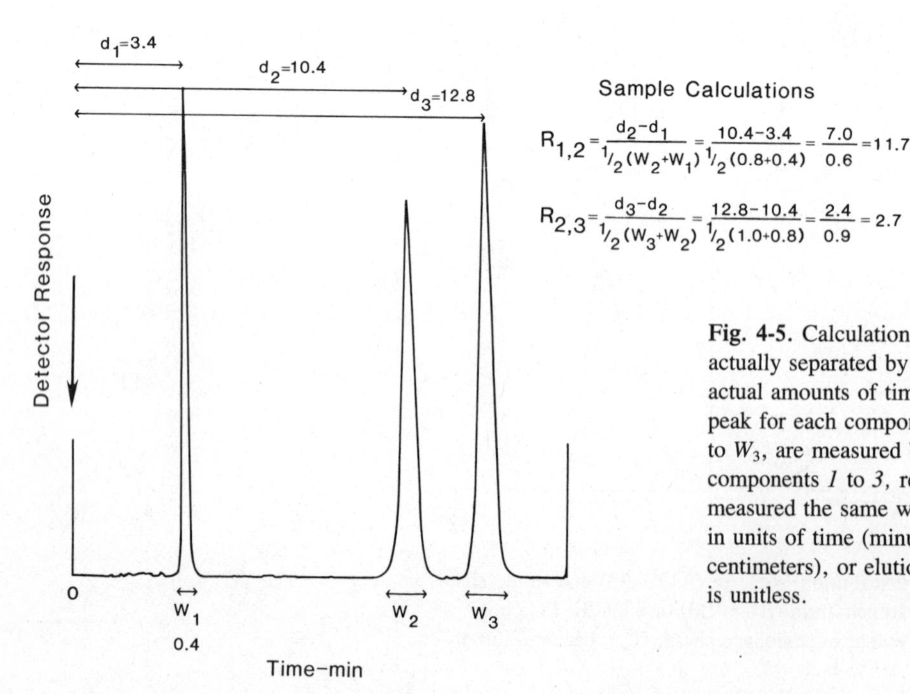

Sample Calculations

$$R_{1,2} = \frac{d_2 - d_1}{{}^1\!/_2(W_2 + W_1)} = \frac{10.4 - 3.4}{{}^1\!/_2(0.8 + 0.4)} = \frac{7.0}{0.6} = 11.7$$

$$R_{2,3} = \frac{d_3 - d_2}{{}^1\!/_2(W_3 + W_2)} = \frac{12.8 - 10.4}{{}^1\!/_2(1.0 + 0.8)} = \frac{2.4}{0.9} = 2.7$$

Fig. 4-5. Calculation of resolution of sample components actually separated by HPLC. The distances d_1 to d_3 are the actual amounts of time from injection (↓) to apex of eluting peak for each component, *1* to *3*, respectively. Peak widths, W_1 to W_3, are measured by triangulation at base of each peak for components *1* to *3*, respectively. Both *d* and *W* must be measured the same way from the time of injection, that is, in units of time (minutes or seconds), length (inches or centimeters), or elution volume (milliliters). Resolution, *R*, is unitless.

which continue as it passes through the column. The longer the time a solute band spends in a column, that is, the longer the retention time, the greater the opportunity for thermal diffusion. An increased retention time will cause more band dispersion. The result is broader but gaussian (that is, symmetric as in Fig. 4-4, *C1*) elution peaks for the more retained solutes. This process is similar in flat methods of chromatography resulting in larger diameter but symmetric spots, as in Fig. 4-4, *B1*, for the more retained solutes. Several additional factors contribute to solute-band broadening.[12,14,26-28] These include nonuniform regions of stationary-phase, particle-sized distribution and nonuniformity of column packing, which may result in nonuniform passage of solute molecules. In the latter case, some molecules spend more time in the separation system than others do; that is, they have a longer path. In addition, localized nonequilibriums during mass transfer of solutes between mobile and stationary phases can also occur. Both of these processes result in broader, nongaussian dispersion of solute bands, such as asymmetric eluting peaks or trailing spots as in Fig. 4-4, *B2* and *C2*, respectively.

Therefore one way to maximize column efficiency, that is, decrease the extent of band broadening, is to use a well-packed chromatography column that contains a stationary-phase packing that is not only small but also uniform with regard to size distribution of particles. Columns meeting these criteria are readily obtainable today from numerous commercial sources of chromatography materials.

Theoretical plates

A numerical measurement of column efficiency can be obtained by calculation of the number of theoretical plates, N, for a given column. A theoretical plate is a microscopic segment of a column where a perfect equilibrium is assumed to exist between the solute in the mobile and stationary phases. The theory originated with Martin and Strain[29] in their mathematical treatment of the chromatographic process. The number of theoretical plates can be calculated for a column directly from the resulting chromatogram (Fig. 4-7) by the following:

$$N = 16\left(\frac{t_R}{W}\right)^2 \qquad \textit{Eq. 4-3}$$

In this expression, W, is the base width of the chromatographic peak and t_R is the retention time of the solute, that is, the time from introduction of the sample onto the column to the apex of the eluting solute peak. Both t_R and W are measured in the same units—time or distance. (Note that N is dimensionless.) A large number of theoretical plates indicates relatively narrow peaks, that is, an efficient column.

Related to the number of theoretical plates is the column

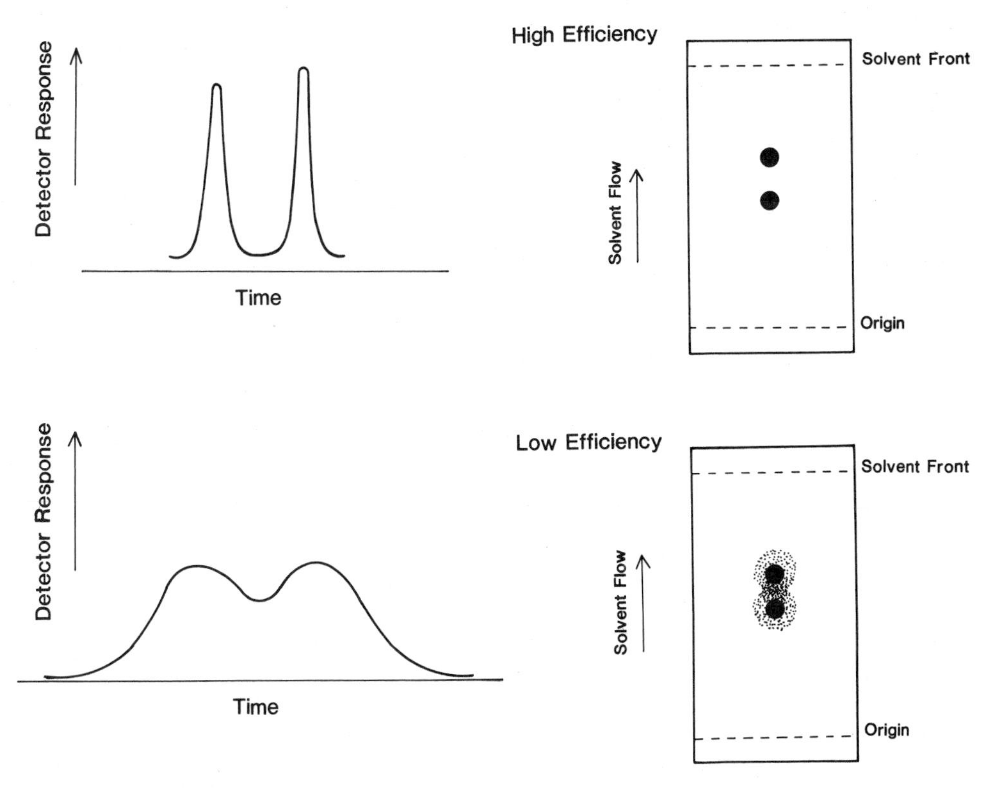

Fig. 4-6. Model chromatograms exemplifying high-efficiency separations and low-efficiency separations.

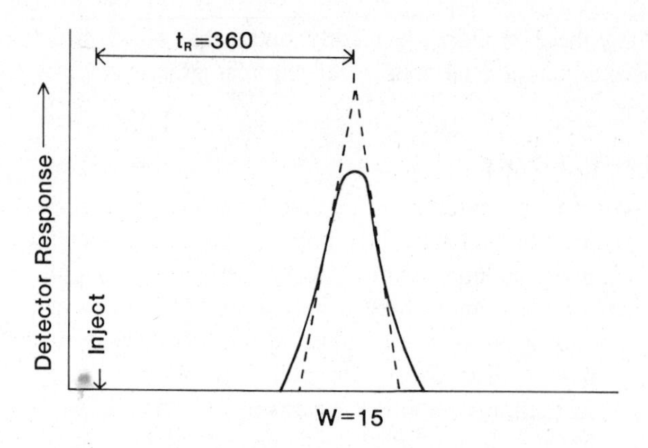

Fig. 4-7. Calculation of number of theoretical plates, N, from an HPLC or gas chromatogram. Elution or retention time, t_R, is measured from time of injection to apex of eluting peak for component. Peak width, W, is measured at the base of the peak by triangulation, as shown. Both are measured in same units. Column length is calculated here as

$$N = 16\left(\frac{t_R}{W}\right)^2 = 16\left(\frac{360}{15}\right)^2 = 9216$$

height equivalent to a theoretical plate[29] (HETP), which can be calculated by the following:

$$\text{HETP} = \frac{L}{N} \qquad \textit{Eq. 4-4}$$

In this equation, L equals the column length, usually in millimeters. Maximum column efficiency is obtained when HETP is as small as possible.

In addition to the factors affecting column efficiency, N is affected by the flow rate of the mobile phase. The mobile-phase linear velocity, μ, can be calculated by:

$$\mu = \frac{L}{t_o} \qquad \textit{Eq. 4-5}$$

In this equation, L is column length and t_o is the time it takes a discrete portion of the solvent to flow through the column. A plot of HETP versus μ shows the experimental relationship of these two variables in gas chromatography (GC) (Fig. 4-8, *A*) and in liquid chromatography (LC) (Fig. 4-8, *B*). At the minimum HETP, the optimum flow velocity, μ_{opt}, is obtained. For gas chromatography (Fig. 4-8, *A*) this plot is the well-known van Deemter plot,[30] and these two parameters are related by the equation:

$$\text{HETP} = A + \frac{B}{\mu} + C\mu \qquad \textit{Eq. 4-6}$$

In this expression, the van Deemter equation, the terms A, B, and C are constants for a given system and refer to the contributions of eddy diffusion, longitudinal diffusion, and the sum of stationary- and mobile-phase mass transfer to HETP. For liquid chromatography Fig. 4-8, *B*, a minimum in HETP is seldom observed, but is related by the equation:

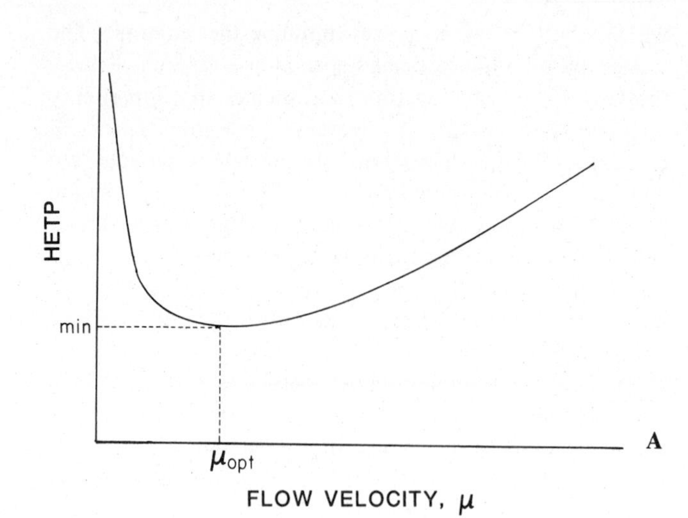

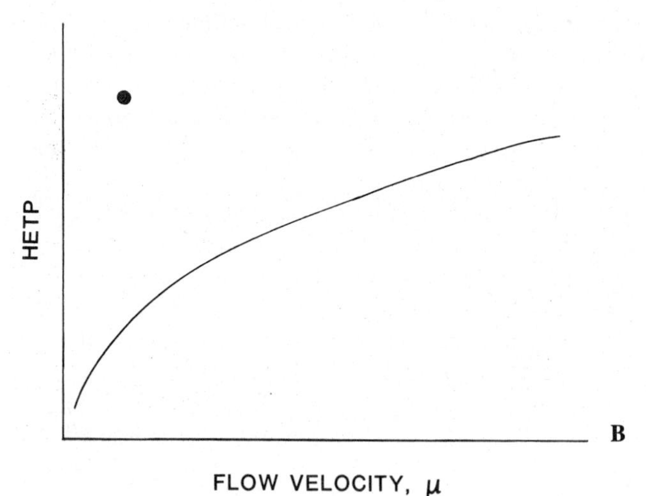

Fig. 4-8. Dependence of theoretical plate height, HETP, on mobile-phase velocity, μ, in gas chromatography, **A**, and in liquid chromatography, **B**.

$$\text{HETP} = \frac{B}{\mu} + \left(\frac{1}{A} + \frac{1}{C_m\mu}\right)^{-1} + C_s\mu \qquad \textit{Eq. 4-7}$$

In this equation, B, A, C_m, and C_s are constants. The terms in the equation containing these constants refer to the longitudinal diffusion, eddy diffusion, mobile-phase mass transfer, and stationary-phase mass transfer, respectively.

These relationships of HETP and μ point out the fundamental difference between gas chromatography and liquid chromatography with regard to solute-band spreading. This arises from the 10^4 to 10^5 greater solute diffusivities in gas versus liquid mobile phases.[27]

At high mobile-phase velocities, HETP varies linearly with μ for gas chromatography (Fig. 4-8, *A*), but in liquid chromatography HETP tends to level off at high values for μ. As can be seen, at comparable velocities the efficiency, or HETP, of the LC system is much less than that observed in GC. In practice GC mobile-phase velocities are usually between one and two orders of magnitude greater than LC. For LC the lower velocities are necessary because of the active role the mobile phase plays in the chromatographic separation process. Therefore, to improve separation efficiency in LC, one uses smaller stationary-phase particles, usually one tenth to one hundredth the size of those used in GC.

Most LC stationary phases are in a particle-size range of 3 to 10 μm. To achieve a reasonable flow rate of liquid mobile phase through a column containing such particles, pressures to 2500 psi or greater are required. These pressures are easily achieved by the pumping systems used in the current high-performance liquid chromatography techniques.[16,28,31]

Retention

Resolution is dependent on factors in addition to theoretical plates. One of these is the ratio of the volumes of mobile and stationary phases in the column, that is, the capacity factor, k', which can be calculated from the chromatogram (Fig. 4-9, *A*) by the equation:

$$k' = \frac{V_e - V_o}{V_o} = \frac{t_R - t_o}{t_o} \qquad \textit{Eq. 4-8}$$

The volumes in this equation are the void volume, V_o, of the column, that is, the volume of the mobile phase in the column, and the elution volume, V_e, of a solute retained by the stationary phase and undergoing chromatography. The HPLC chromatogram for Fig. 4-9, *A*, was obtained by injection of a sample containing two solutes, the first of which was not retained by the stationary phase and the second of which was retained, undergoing chromatography. The volumes, V_o and V_e, were then measured from the injection point to the apex of the peak of each component. As indicated above, the capacity factor can be calculated also by the measurement of the times, t_o and t_R, from sample injection to the apex of the peak of the component not retained and of the peak of the component retained, respectively.

Small values of k' indicate that the sample components are little retained by the stationary phase and elute close to the unretained peak. Large values of k' indicate not only that the sample components are well retained by the stationary phase, but also that long analysis times are required. For this latter situation, one must remember that solute-band broadening increases with residence time on the column because of an increased diffusion of the solute. Therefore the resulting peaks on elution will be wide and diffuse, decreasing sensitivity and making detection difficult.

The k' value for a particular solute is constant for any given chromatography system at constant mobile-phase compositions and stationary-phase size and composition. Within these limits, the capacity factor varies neither with

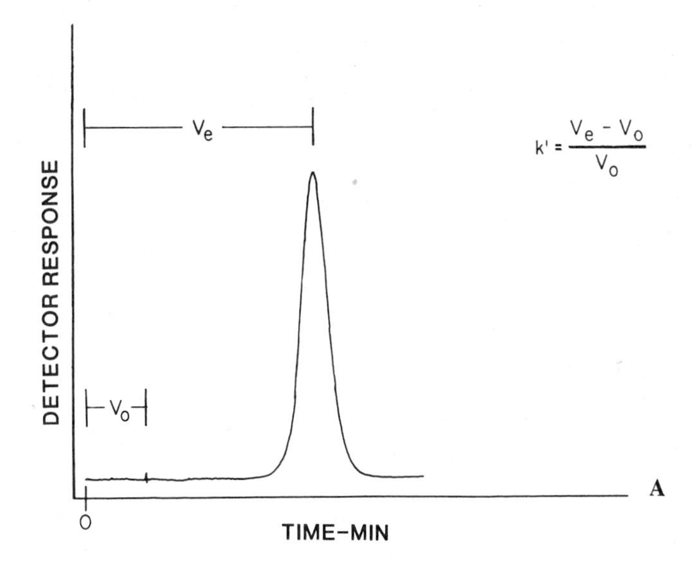

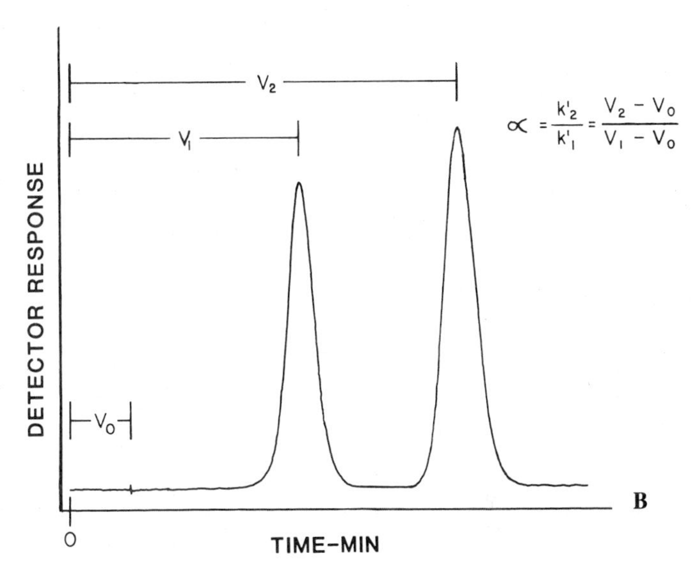

Fig. 4-9. A, Calculation of capacity factor, k', from an HPLC chromatogram. Sample was retained by stationary phase and underwent chromatography. **B,** Calculation of selectivity factor, α, from HPLC chromatogram. Solutes A and B were retained by stationary phase and underwent a chromatographic separation as indicated.

flow rate of the mobile phase nor with column dimensions, that is, length and diameter.

Selectivity

Another parameter on which resolution is dependent is the selectivity factor, α (alpha), a term that describes the ability to separate two solutes. It is the ratio of the capacity factors for the two solutes being examined for the degree of their resolution achieved on separation by a chromatographic method.

$$\alpha = \frac{k'_2}{k'_1} \qquad \textit{Eq. 4-9}$$

The capacity factors for two solutes are determined as shown in Fig. 4-9, *B*, from which the resultant selectivity of the system can be calculated. The selectivity of any given column for the sample is a function of the thermodynamics of the solute-exchange process between the mobile phase and the stationary phase.[13] Therefore, to affect selectivity, one can change the chemical composition of either the mobile phase or the stationary phase to increase the preference of the solute for one phase or the other.

In liquid chromatography, changes in mobile-phase chemistry, that is, increases or decreases in polarity, are usually made to improve selectivity. However, in gas chromatography, changes in stationary-phase chemistry, that is, with a column packed with a more or less polar stationary phase, are used for selectivity improvement.

Improving peak resolution

General equation for resolution. With the definition of the factors affecting resolution, a more fundamental equation can now be written:

$$R = \left(\frac{N^{0.5}}{4}\right)\left(\frac{\alpha - 1}{\alpha}\right)\left(\frac{k_2'}{1 + k_1'}\right) \qquad \textit{Eq. 4-10}$$

Therefore the resolution of any two solutes in a given chromatographic system is a function of three factors: (1) theoretical plates *(N)*, a column-efficiency factor; (2) a selectivity factor that varies with α; and (3) a capacity factor that varies with k'. This relationship is under the assumption that a given pair of solute bands passes completely through the column of stationary phase and is eluted from the column.

For flat methods of chromatography, such as thin-layer chromatography, a similar relationship can be derived.[13] However, a major difference exists between flat and column methods of chromatography. In a column method of chromatography, the solutes in a given sample pass completely through the bed of the stationary phase. But in a flat method of chromatography, the separation process is stopped when the mobile phase has reached the end of the bed of the stationary phase, thereby resulting in the solute bands having migrated through only a portion of the bed (Fig. 4-10). For this type of separation, solute retention is measured in terms of the R_f value, which is the distance, d_2, migrated by the solute divided by the distance, d_1, migrated by the mobile phase, that is, solvent front. This relationship can be expressed as:

$$R_f = \frac{d_2}{d_1} \qquad \textit{Eq. 4-11}$$

In consideration of this major feature of flat methods of chromatography, the resolution equation for the separation of two solutes becomes as follows:

$$R = \left(\frac{N^{0.5}}{4}\right)\left(\frac{\alpha - 1}{\alpha}\right)\left(\frac{k_2'}{(1 + k_2')^{2/3}}\right) \qquad \textit{Eq. 4-12}$$

The terms of this equation are as previously defined.

The question of the practical significance of these two resolution equations can now be addressed. Specifically, how do these equations relate resolution to the actual experimental conditions of the chromatographic separation and the physical design of a particular chromatographic device? The three fundamental parameters N, α, and k' of resolution can be adjusted more or less independently of each other. The dynamics or rates of the various physical processes that occur during the separation determine N. Both α and k' are thermodynamic or equilibrium properties of the chromatographic system.

Variation of *N* to optimize resolution. Experimentally one can change N by adjusting or varying a variety of parameters or conditions. Remember that a doubling of the value for N will increase the resolution by a factor of 1.4. (Note that in the resolution equation R is proportional to the square root of N). Experimental parameters that can be changed to optimize N are summarized in Table 4-1.

Variation of capacity factor to optimize resolution. Another way to increase resolution is to vary the capacity factor, k', or the separation factor, α, or both. One can change these factors experimentally by altering the physical chemistry of the mobile phase, the stationary phase or the sample. Temperature is one parameter that is varied to affect a change in the capacity factor. Since the solute–stationary phase interactions are temperature dependent, a change in this parameter will affect solute retention. This is especially important in gas chromatography, but to a much lesser extent for other chromatographic techniques.

Consider a series of gas chromatographic optimization experiments on the same sample at differing isothermal, that is, constant temperature, conditions. By lowering the temperature, one usually improves the resolution when compared to the first experiment. Whereas in an experiment where the temperature is increased, analysis time will

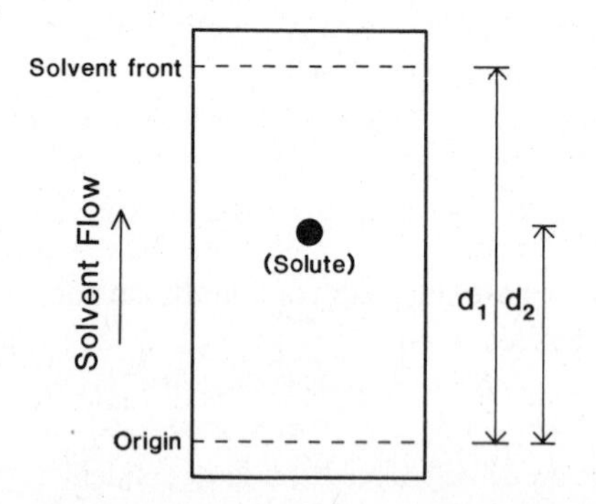

Fig. 4-10. Calculation of R_f (solvent-front ratio) value for a sample component from a paper or thin-layer chromatogram. The distance d_1 from origin to solvent front and distance d_2 from origin to center of spot of separated component are measured, as shown, in terms of same units, that is, length (centimeters or inches). Resulting R_f value, d_2 divided by d_1, is unitless.

be decreased and peak shape will be improved. However for this case, resolution will be reduced. Therefore the final isothermal conditions for a GC analysis must be a compromise.

A second way to vary the capacity factor is by affecting changes in the strength of the stationary phase. For partition chromatography, increasing the percentage of liquid-phase coating on the matrix support will effect an increase in k'. In gas-liquid chromatography (GLC), efficiency decreases sharply when the liquid-phase coating exceeds 30% of the matrix support.[27,32] Most GLC analyses are done on columns containing 2% to 10% ratios of liquid to stationary phase. Since retention time is proportional to the amount of liquid phase present, lower percentage ratios mean lower analysis times. However, higher percentages of stationary phase for partition chromatography mean higher resolution. Therefore the final selection of the percentage of stationary phase must be a compromise between analysis time and resolution.

Additionally, one can vary the capacity factor, k', to improve resolution by changing the solvent strength of the mobile phase, but this is applicable only to liquid chromatographic methods. In gas chromatography the mobile phase is an inert gas like helium or nitrogen, but in liquid chromatography the mobile phase is an active component of the chromatographic procedure, ranging from very nonpolar, such as hexane, to polar, such as water, in solvent strength.

For the adsorption and partition modes of liquid chromatography, scheme I, the chromatographic process involves a continual distribution of the solute molecules between the mobile and the stationary phases. Any shift in the equilibrium concentration distribution coefficient (K_D) affects the capacity factor, that is, the elution volume per time of a particular solute in the given sample. One way to shift this coefficient is to increase or decrease the polarity of the mobile phase relative to the stationary phase.

An example is an HPLC separation of a sample using a silica gel stationary phase. If one does not know the proper mobile phase to be used, a good solvent for an initial separation would be a 50/50 (vol/vol) mixture of hexane and methylene chloride containing 0.1% isopropanol, to maintain consistency in the stationary-phase adsorbent activity. If the components of the sample elute with relatively large k' values, the mobile phase can be made more polar by increasing the percentage of methylene chloride relative to hexane. This change will effect a decrease in the k' of the components by increasing the solubility of the solutes in the mobile phase relative to the stationary phase. However, if the components elute with relatively small k' values, the polarity of the original mobile phase can be decreased by an increase in the percentage of hexane relative to methylene chloride. The net effect of this change is to shift the distribution coefficient to a larger value indicating a more favorable interaction of the sample components with the stationary phase. The net result is an increase in the k' values of the solutes.

Likewise, similar solvent-strength changes can be made for other forms of adsorption-partition liquid chromatography. For example, in ion-exchange chromatography the ionic strength of the mobile phase may be varied. In partition chromatography using a hydrocarbon-bonded stationary phase, such as octadecylsilyl high-performance liquid chromatography, the aqueous mobile phase may be decreased in polarity by the addition of methanol. Numerous other examples of mobile-phase alterations could be given. However, suffice to state that this is a powerful technique for varying the k' of solutes being separated.

Variation of selectivity to optimize resolution. After optimization of the k' values in liquid chromatography, the selectivity, or separation factor, α, then can be adjusted to maximize the resolution. In general, the separation factor is varied by changes in the chemistries of the mobile phase, the stationary phase, or the sample. These changes may be accomplished as follows.

Derivatization. The chemistry of the sample can be altered in a variety of ways to affect selectivity. The most common method is derivatization. Knapp[33] has summarized an extensive variety of derivatization reactions for samples to be separated by gas chromatography. Not only is derivatization used in gas chromatogaphy to improve sample volatility or solute detection limits, but also it is used to improve selectivity. This technique accomplishes an improvement in peak shape to a more gaussian nature for individual solutes by negating charge effects and making solute molecules more uniform in their adsorption and partition characteristics. In liquid chromatography, derivatization techniques are generally used to improve detection,[28,30] with the most extensive collection of methods presented in a book by Frei and Lawrence.[34]

Ion-suppression reagents. A more common method of change in sample chemistry to improve selectivity in liquid chromatography is the use of ion-suppression reagents, that is, ion-pair chromatography.[28,31,35,36] For this partic-

Table 4-1. Experimental parameters affecting N

Parameters	Direction of change necessary to increase N
Mobile-phase flow rate	Decrease
Column length	Increase
Average particle size of stationary phase	Decrease
Particle-size distribution	Decrease
Volume of sample introduced	Decrease
Viscosity of sample introduced	Increase
Viscosity of mobile phase	Increase
Temperature (as it controls mobile-phase viscosity)	Decrease
Extracolumnar effects (that is, dead volumes, excessive connective tubing)	Decrease

ular method, ionic components are made neutral by the addition of an ion-suppression reagent to the mobile phase. The process is described by the following:

$$\text{Sample}^{\pm} + \text{Reagent}^{\mp} \rightleftharpoons \underset{\text{Ion pair}}{R^{\mp} S^{\pm}} \qquad \textit{Eq. 4-13}$$

If the sample is acidic, such as carboxylic acids or sulfonic acids, a quaternary ammonium compound, such as tertiary butyl ammonium hydroxide, is commonly used as the ion pair reagent. If the sample is basic, like many drugs, an organic sulfonic acid, such as pentane sulfonate, is used. The resulting neutral ion-pair species of sample-reagent molecules are then separated by liquid chromatography, usually a partition method. However, controversy does exist as to the exact chromatographic mechanism of separation,[28,31,35-39] but in many cases the use of ion-pairing reagents does improve the resolution of ionic species.

Alterations in mobile-phase chemistry. Another method for improving the separation factor in liquid chromatography is by alterations in the mobile-phase chemistry. Previously, solvent polarity changes were used to improve the capacity factor, k'. For improving the separation factor, solvents of similar polarity are interchanged without affecting the overall polarity of the mobile phase but resulting in changes in the solvent selectivity. For example, the resolution of an HPLC separation initially developed with a mobile phase of methylene chloride/hexane, 50/50 by volume, could be improved by substitution of tetrahydrofuran for methylene chloride without changing the solvent strength.[17,40-44] The reason for improvement in resolution would be that tetrahydrofuran yields quite different selectivities toward certain classes of compounds, such as lipids, than methylene chloride does. In this case of solvent substitution of an ether for an organic halide, the functional groups of the mobile phase have been altered. Changes in solvent functionality without changing solvent strength is a powerful technique for improving selectivity, thereby resolution, in liquid chromatography. Depending on the sample and chromatographic and detector requirements, the choice of solvent or solvents available to effect changes in selectivity is extensive, as detailed by Schneider[45] in his compendium of properties of 911 solvents. A further discussion of the properties of solvents is presented in a later section.

Alterations in stationary-phase chemistry. As adjustments in the functionality of the mobile phase are used to improve the separation factor in liquid chromatography, similar changes in the functionality of the stationary phase are commonly used to improve the separation factor in gas chromatography. For example, many polyester stationary phases, such as ethylene glycol succinate, can be replaced by the cyanopropyl silicone stationary phases, such as Silar 10C, allowing for a change in functionality with little or no change in polarity. For relatively nonpolar stationary phases, the hydrocarbon phases, such as Apiezon N, can be readily replaced by the alkyl silicone phases, such as OV-1, for a functionality change. These changes in stationary phase afford an improvement in selectivity, thereby leading to an optimization in resolution. Details as to the initial choice of phase and their chromatographic mechanisms are discussed in the following sections.

POLARITY

In previous sections, the importance of solvent strength and stationary-phase strength were discussed in relationship to their effect on resolution. Both forces affect chromatographic resolution by the capacity factor and the separation factor as described earlier in equations 4-8 and 4-9. These parameters are constituted by the principle of "polarity."

The role of polarity is central to the interaction of molecules in the liquid or gaseous state and is a major determinant property in the overall chromatographic process. The concept of polarity can be interpreted several different ways[28,40-44] but generally is considered to encompass the total interaction of solvent molecules with sample molecules, and of solvent or sample molecules with the stationary phase. The physicochemical basis for polarity is the interaction of attractive forces that exist between molecules. These four attractive forces are more specifically referred to as dispersive, dipolar, hydrogen-bonding, and dielectric interactions. As illustrated in Fig. 4-11 and discussed below, all four of these interactions involve the attraction of induced, partial, or formal positive and negative charges.

Dispersion interactions of molecules, sometimes termed "van der Waals' forces," refer to the induced attraction between two molecules. The basis for this interaction is that the electrons, normally in random motion on a given atom in a molecule, may assume an asymmetric configuration at any given instant thereby causing at that instant a temporary dipole in the molecule (Fig. 4-11, *A*). This temporary separation of opposite charges in one molecule induces the polarization of electrons in an adjacent molecule, thereby causing the two molecules to be attracted to each other by electrostatic interactions. The formation of temporary dipoles in molecules is the physical basis for existence in the liquid state of many compounds composed of elements with small differences in electronegativity. For example, the elements carbon and hydrogen exhibit a relatively small difference in electronegativity, and many compounds composed of these elements, such as hydrocarbons like pentane, exist as liquids because of these interactions. Generally, dispersive interactions are an important determinant in polarity only when other forces are lacking or when some elemental constituents of molecules are electron-rich species (such as halogens in halohydrocarbons).

Some molecules possess permanent rather than temporary *dipoles* (Fig. 4-11, *B*). These compounds have centers

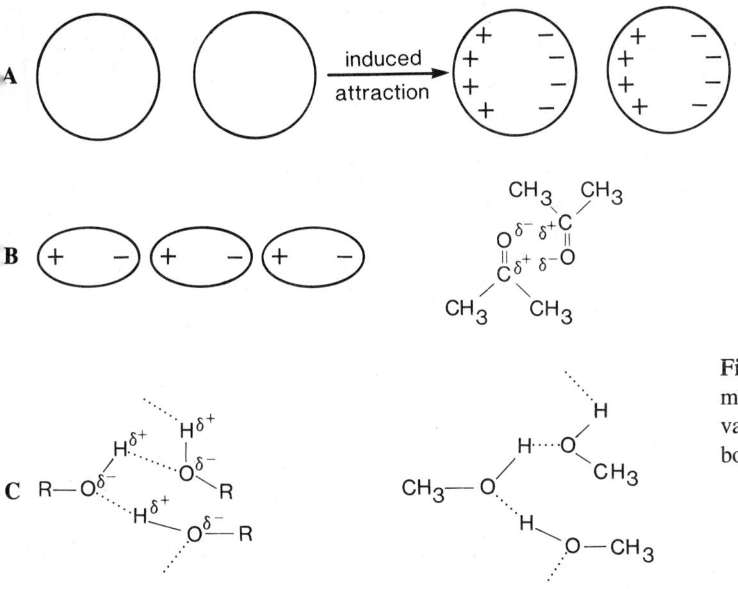

Fig. 4-11. Illustration of physicochemical interactions between molecules that constitute concept of polarity: **A,** dispersive or van der Waals' interactions; **B,** dipole interactions; **C,** hydrogen bonding; **D,** electrostatic interactions.

of positive and negative charge that are the result of an unequal sharing of bonding electrons between two elements with large differences in electronegativity within the same molecule. This overall molecular dipole is enhanced by the presence of elemental nonbonding electron pairs within a compound. Elements like oxygen, sulfur, halogens, and nitrogen possess nonbonding electrons when covalently linked to other atoms in compounds. The resulting permanent dipole is directional, with one end of the molecule being partially positive while the other is partially negative.

One special category of dipolar molecules is composed of those whose hydrogen is covalently linked to an electronegative element like oxygen, nitrogen, or sulfur. In these molecules the hydrogen has a large degree of positive character relative to the electronegative atom to which it is bonded. Because of the small size of the hydrogen atom relative to other atoms, this positive end of the dipole can approach close to the negative end of a neighboring dipole. The force of attraction between the two is quite large, about 10 times that of normal dipolar interactions. This special case of dipole-dipole interactions is termed *"hydrogen bonding"* and is one of the most important types of weak attractive forces. This bonding is illustrated in Fig. 4-11, *C,* for the alcohol methanol.

The fourth type of attractive force important to the overall concept of polarity is the *electrostatic* or *dielectric interaction*. In this case, the solute molecule or the stationary phase is a charged ionic species having either a formal positive or a formal negative charge. A small counterion, such as H^+ or Cl^-, is present but is generally separated from the charged solute or stationary phase because of solvation by the mobile phase (Fig. 4-11, *D*). These ionic species increase the dipolar character of the solvent by an enhancement of polarization. Dielectric interactions are quite strong and favor the dissolution of ionic or ionizable sample molecules in strongly dipolar solvents like water or methanol.

The polarity of a solute molecule is the result of the four attractive forces described above. The polarity of a molecule will affect its interactions with the mobile and stationary phases. The more polar molecules have this property primarily because of strong dipoles, an ionic character, an ability to form strong hydrogen bonds, or a combination of the three forces. The less polar, that is, nonpolar, molecules have dispersive forces as a primary basis of interaction with a very weak ability to interact through dipolar, hydrogen bonding, or dielectric forces. The practical aspects of these interactive forces and the degree of polarity or nonpolarity form the basis for the mechanisms of chromatography. The role of polarity in these mechanisms is discussed below for each of the three chromatographic constituents—the solute, the solvent or mobile phase, and the stationary phase.

Solvent polarity and solvent strength

Solvent strength in liquid chromatography is a measure of the ability of the mobile phase to compete with the solute molecules for active sites, that is, interaction or attraction sites, on the stationary phase. When the stationary phase is silica gel, the active sites are the highly polar hydroxyl groups (-Si-OH). Therefore, in this case, solvent strength increases with solvent or mobile phase polarity. However, when the stationary phase is nonpolar, such as a polydivinylbenzene like XAD-2, the solvent strength decreases with increases in solvent or mobile phase polarity. The strength of a solvent is directly related to its polarity.

Solvent polarity has been described and quantitated in various ways.[40-44,46] The four most common solvent polarity classification schemes are (1) the *Hildebrand solubility parameter,* which is based on thermodynamic properties of the compound in question[46]; (2) the *Rohrschneider polarity scale* of P' values, which is based on measurement of solvent properties through the use of model solutes[41]; (3) the *eluotropic series,* which ranks solvents in order of their eluting power for removing solutes from a polar stationary phase like alumina.[47] This series was further defined to state that eluting power of solvents was proportional to their dielectric constant[48]; and (4) the *solvent selectivity grouping* of Snyder and Karger, which is a blend of the three previous approaches and incorporates specific solubility parameters for dispersion, dipole, and hydrogen-bonding interactions. These forces are derived from measurements of physical and thermodynamic properties of the solvents.[40,43,44] A listing of the solvent selectivity groups is given in Table 4-2 along with representative classes of compounds and specific examples for each group. Solvent selectivity groups were originally designated by roman numerals[40,43,44] but now are designated by arabic numerals.

Solvents within any one group exhibit the same types of attractive forces. For example, the compounds of group 1 (I) are all strong hydrogen-bonding acceptors and weak hydrogen-bonding donors and have intermediate dipole moments. The solvents in group 5 (V) are compounds whose dipole interactions predominate over any hydrogen-bonding interactions. However, the solvents in group 0 are compounds in which only dispersion interactions, that is, temporary dipoles, are the predominate force. These solvents do not have permanent dipoles, nor do they interact through hydrogen bonding.

The overall degree of the interactive forces of the solvent is quantitated in the polarity index, P.[40,43,44] A list of solvents commonly used in chromatography is presented in Table 4-3 with values for their polarity index and their solvent selectivity group classification. It is noted that solvents within any one group may vary widely in their overall degree of polarity. For example, the solvents listed for group 6 (VIA) vary more than three polarity units. In

Table 4-2. Solvent classification by selectivity groups

Solvent group	Representative classes or examples of compound
0	*n*-Alkanes, cyclohexane, saturated fluorochlorohydrocarbons
1 (I)	Aliphatic ethers, trialkylamines
2 (II)	Aliphatic alcohols
3 (III)	Pyridine derivatives, tetrahydrofuran, amides
4 (IV)	Glycols, benzyl alcohol, acetic acid
5 (V)	Methylene chloride, ethylene chloride, bis-2-ethoxyethyl ether
6 (VIA)	Aliphatic ketones and esters, nitriles, dioxane, sulfoxides
7 (VIB)	Nitrocompounds, phenyl alkyl ethers, aromatic hydrocarbons, carbon tetrachloride
8 (VII)	Halobenzenes, diphenyl ether
9 (VIII)	Fluoroalkanols, chloroform, water

Table 4-3. Representative solvent polarity indices

Solvent	Polarity index	Solvent group
Isooctane	−0.4	0
Hexane	0.0	0
Carbon tetrachloride	1.7	7
Dibutyl ether	1.7	1
Triethylamine	1.8	1
Toluene	2.3	7
Chlorobenzene	2.7	8
Diphenyl ether	2.8	8
Diethyl ether	2.9	1
Benzene	3.0	7
Ethyl bromide	3.1	6
Methylene chloride	3.4	5
1,2-Dichloroethane	3.7	5
Bis-2-ethoxyethyl ether	3.9	5
n-Propanol	4.1	2
Tetrahydrofuran	4.2	3
Chloroform	4.3	9
Ethyl acetate	4.3	6
Isopropanol	4.3	2
2-Butanone	4.5	6
Dioxane	4.8	6
Ethanol	5.2	2
Nitroethane	5.3	7
Pyridine	5.3	3
Acetone	5.4	6
Benzyl alcohol	5.5	4
Methoxyethanol	5.7	4
Acetic acid	6.2	4
Acetonitrile	6.2	6
Dimethylformamide	6.4	3
Dimethyl sulfoxide	6.5	6
Methanol	6.6	2
Nitromethane	6.8	7
Water	9.0	9

group 2 (II), the polarity index varies more than two units. Even though these variations are relatively large, one must remember that solvents within the same group exhibit the same kinds of interactions.

This information can be used as a basis for solvent selections and can be used to vary the mobile-phase selectivity, α, in order to improve resolution. The first important step in solvent selection is to maximize solute solubility in a given solvent. For a solvent to be an effective mobile phase in chromatography, it should dissolve the sample over the range of expected solute concentrations.

The most important factor affecting solute solubility is solvent polarity. One can vary this systematically over a wide range of P' values by using a combination of two solvents, since solvent polarities are additive according to:

$$P'_{\text{(of combined solvents A + B)}} = \phi_A P'_A + \phi_B P'_B \qquad \textit{Eq. 4-14}$$

In this expression, ϕ_A and ϕ_B are the volume fractions of solvents A and B, which have solvent polarities of P'_A and P'_B, respectively. The two solvents should differ in P' values so that maximization of sample solubility can be achieved. It is important that the two solvents are miscible in all proportions. Usually a solvent that is too weak (A) to dissolve the sample is mixed with a solvent that is too strong (B). The two solvents are now chosen from the solvent groups (Table 4-3) according to the compatibility of interactive forces with sample solutes and stationary-phase functionalities. An intermediate mixture, such as 50%:50% by volume, will usually provide maximum solubility for the sample. Adjustments in polarity are made accordingly.

For example, to calculate the P' for solvent mixtures of isooctane (solvent A) and chloroform (solvent B), their individual polarity values are obtained from Table 4-3. Any mixtures of these two solvents would range in polarity from -0.4 to 4.3, that is, weakest (100% isooctane) to strongest (100% chloroform). A 50%:50% mixture of these two solvents would have a P' of $([0.5 \times -0.4] + [0.5 \times 4.3])$, or 1.95. Also, for each 10% change in solvent composition, the polarity of the mixture would change by 0.47 units; that is, $0.1(P'_B - P'_A)$ or $0.1(4.3 - [-0.4]) = 0.47$.

Once a suitable solvent mixture has been found to dissolve the sample, it is then applied to a chromatographic system like a silica HPLC column. As previously discussed, the retention index, k', for the solutes in the sample is maximized to values between 2 and 8. This is accomplished by adjustment of the solvent polarity by small changes in the relative proportions of the two solvents.

Resolution is then maximized for the solute separation by adjustments in chromatographic selectivity, α. This is accomplished by exchange of one of the mobile-phase solvents for a solvent in another solvent selectivity group (Table 4-3), but with the same polarity of the overall mixture being kept. For example, chloroform, $P' = 4.3$, could be exchanged with isopropanol, $P' = 4.3$, in the previously described isooctane-chloroform mixture. The net effect of this exchange would be in going from a solvent that is a strong hydrogen-bonding donor, chloroform, to a solvent that is both a strong hydrogen-bonding donor and hydrogen-bonding acceptor. Other examples of such changes were given earlier in the discussion on selectivity. Further discussion of solvent selection and optimization are given by Snyder[28,43,44] and in Chapter 5 on high-performance liquid chromatography.

In principle, this approach is a general guide to solvent choice both for dissolving the sample and for use as a mobile phase in liquid chromatography (HPLC, TLC, and so on). It does, however, require knowledge of the forces of interaction exhibited by the sample and those forces involved in the chromatographic separation process. The latter is discussed below in the sections on the role of stationary-phase polarity and the mechanisms of chromatography.

Stationary-phase polarity and selectivity

The stationary phase is the fundamental component of the chromatographic system to effect the separation of solute molecules in a given sample. This role of the stationary phase is dependent on its selectivity, which, in turn, is determined by the polarity of the phase. The forces constituting stationary-phase polarity are the same interactions responsible for mobile-phase (solvent) polarity: dispersion, dipole, hydrogen bonding, and dielectric.

The relative strength of stationary phases in terms of polarity is more difficult to ascertain than the relative strength of the polarity for liquid mobile phases, as previously described. Most of the attention to this problem has been to the stationary phases used in gas chromatography wherein the two major variables determining the separation are the stationary phase itself and the temperature of the column. In this mode of chromatography, studies of the affinity of solute molecules of varying polarities has led to a classification system for stationary phase according to polarity.[49]

The system called "retention index classification" was originally developed by Rohrschneider[49] and subsequently modified by McReynolds[50] to rank stationary phases according to their relative polarities. Details on the Rohrschneider and McReynolds classifications can be found in Chapter 6.

Using the McReynolds system for retention indices, several hundred stationary phases have been classified for gas chromatography. The choice of a suitable stationary phase could be overwhelming for the analyst.

Most gas chromatographic separations can be accomplished in the clinical laboratory by use of one of the "preferred" stationary-phase types, as recommended by a committee.[51] The recommended phase types and examples of each are listed in Table 4-4. Although these six primary

Table 4-4. Preferred stationary phases

Phase type	Examples
Dimethylsilicone	OV-1, OV-101, SE-30, SP-2100
50% Phenylmethylsilicone	OV-17, SP-2750
Trifluoropropylmethylsilicone	OV-210, SP-2401
Polyethylene glycol	Carbowaxes
Polyesters	DEGS, EGSS-X, EGA
3-Cyanopropylsilicone	Silar 10C, SP-2340, Apolar 10C

From Hawkes, S., Grossman, D., Hartkopf, A., et al.: J. Chromatogr. Sci. **13:**115-117, 1975.

classes of stationary phases are preferred for most gas chromatographic separations, a secondary list of 24 additional phases was recommended[51] to provide a broader choice for specific polarity and temperature requirements not met by the initial list. Further discussions of stationary-phase selection for gas chromatography are presented in Chapter 6.

MECHANISMS OF CHROMATOGRAPHY

The mechanisms by which a chromatographic method can separate sample components are generally based on the interactions that constitute the concept of polarity, with the addition of physical interaction because of the size and shape of the solute molecules. As discussed below, this latter interaction is principally the mechanism for gel permeation chromatography though molecular size plays a minor role in other modes of chromatography. The other broad mechanistic classes of chromatography are adsorption, partition, and ion exchange. Each is briefly discussed in the following section.

Adsorption

Adsorption chromatography is the oldest mechanism encountered in the history of modern chromatography. Tswett correctly described the mechanism of adsorption as the physicochemical basis of chromatography in his 1906 publication.[5] However, this was only a beginning, for since this publication many books, reviews, and research papers have been published to further detail the theoretical basis of the adsorption mechanism. For example, reviews in references 6 and 21 to 24 may be consulted for leading references to this literature, whereas references 20, 27, and 28 are examples of books that treat the mechanism of adsorption in chromatographic separations for TLC, GC, and LC, respectively.

The interactions of solute or mobile phase molecules at the surface of a solid particle form the basis of the adsorption mechanism. There are fundamentally two types of adsorbents, nonpolar and polar. The latter category includes those that are acidic, that is, having electron-accepting surfaces, and those that are basic, that is, having electron-donating surfaces.

The nonpolar adsorbents have limited application in gas-solid chromatography[27] and are characterized by stationary phases,[52] such as graphitized thermal carbon black, and polymers, such as polyethylene and Teflon. The mechanism for adsorption of the solute to this nonpolar stationary phase is principally by dispersive interactions. Retention is determined by the adsorption energy and the surface volume of the stationary phase. The adsorption energy for GSC is determined by the temperature and the pressure of the separation process. A decrease in temperature or an increase in pressure increases adsorption. The converse is also true; that is, an increase in desorption can be accomplished by an increase in temperature or a decrease in pressure. A more extensive treatment of the thermodynamics of this mechanism is given by Karger et al.[13] The most important applications of nonpolar adsorption phase gas-solid chromatography (GSC) are in the separation of permanent gases, such as the sulfur gases[53] and the low molecular weight hydrocarbons.[54]

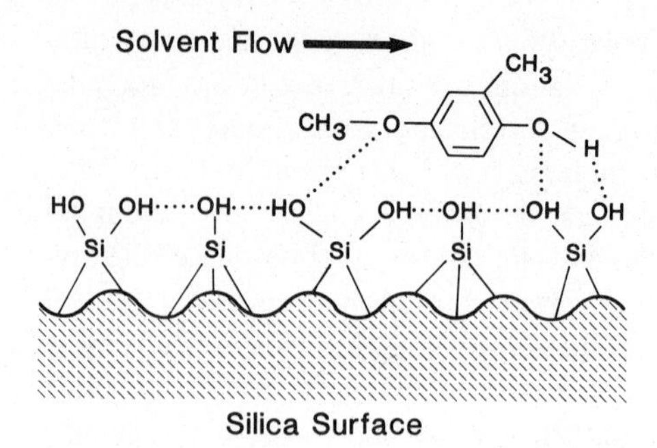

Fig. 4-12. Mechanism of separation of a metabolite of methylanisole by silica gel chromatography. Hydrogen bonds, ...; covalent bonds, ——.

Polar adsorbents are most widely used in liquid-solid chromatography (LSC), with applications in both flat and column methods. Limited applications are found in GSC. The most common stationary phases are silica and alumina for LSC, and silica or porous glass and aluminosilicates, that is, zeolites or molecular sieves, for GSC. This latter group also has application in LC for gel permeation chromatography where the separation is principally through the mechanism of molecular exclusion. This is discussed later in the section on the mechanism of gel permeation.

The principal polar adsorbents used in liquid chromatography are silica and alumina, accounting for more than 95% of the applications in HPLC[28,31] and TLC.[20,55,56] Both hydrogen bonding and dipole interactions between the solute and the stationary phase, by the surface hydroxyl (silica and acid-washed alumina) or the oxygen anionic (base-washed alumina) groups, constitute the mechanism of separation by this method (Fig. 4-12). The number and topographic arrangement of these groups, along with the total surface area, determine the activity and strength of the adsorption. Retention of solutes on

Table 4-5. Selected groups of solutes in order of increasing retention in normal-phase and reversed-phase chromatography

Reversed phase	Solute type	Normal phase
Most retained ↓	Fluorocarbons	Least retained ↑
	Saturated hydrocarbons	
	Unsaturated hydrocarbons	
	Halides and esters	
	Aldehydes and ketones	
	Alcohol and thiols	
Least retained	Acids and bases	Most retained

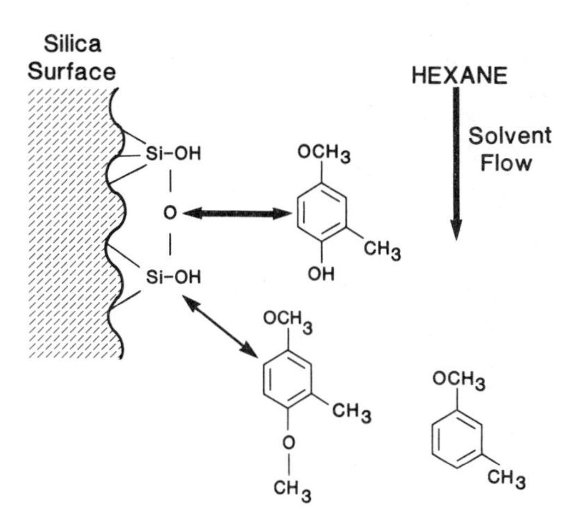

Fig. 4-13. Mechanism of adsorption chromatography by separation of 3-methylanisole and two of its biochemical metabolites. The most polar sample components, such as 3-methyl-4-hydroxyanisole, are retained the most by polar silica gel stationary phase, *heavy arrow*. Sample components of intermediate polarity, such as 2,5-dimethoxytoluene, are retained to a much lesser degree, *light arrow*, whereas relatively nonpolar components, such as 3-methylanisole, are not retained and prefer the nonpolar mobile phase, hexane.

these phases increases with increasing polarity of the compound class (Table 4-5). Retention of a solute molecule requires displacement of adsorbed solvent molecules (Fig. 4-13). Adjustments in solvent polarity, as previously described, ultimately determine the strength of adsorption of the solute to the stationary phase and the retention characteristics of the system.

Adsorption chromatography offers many advantages for use in liquid chromatographic separations. These advantages are as follows: (1) An extensive literature is available to the investigator for the separation of many types and classes of compounds by thin-layer chromatography (TLC). For example, see the *Journal of Chromatography* reviews on TLC.[57] These methods are readily transferable to adsorption HPLC. (2) The flexibility, speed, and low cost of TLC allow its use in experimental development, particularily for selection of mobile phases. Once the optimum separation has been achieved, the transfer of the method to the HPLC is straightforward. (3) TLC has a great value for use in preliminary investigation of samples of unknown constituents, particularly when one considers the advantages noted above. (4) Adsorption chromatography, particularly with silica gel, has been widely used for the separation of drugs in both the HPLC and TLC modes of chromatography. The silica gel support is superior to other LC stationary phases for the separation of structural isomers, though other phases are superior to silica gel for the separation of a homologous series of compounds with similar polarity.[28]

Partition

Partition chromatography is based on the separation of solutes by differences in their distribution between two immiscible phases. In liquid-liquid chromatography, the phase support is usually coated with a polar substance (normal phase), with separations being accomplished by use of an immiscible mobile phase. A common normal-phase partition system would be silica coated with a monolayer of water and the use of a relatively nonpolar solvent system. Separations in this system are based upon solute polarity, with the least polar compounds eluting first and the most polar substances being retained the longest (Fig. 4-14). A similar separation system operates in paper chromatography where the cellulose is coated with an aqueous monolayer and immiscible solvents are used as the mobile phase.

Other types of polar stationary-phase coatings are utilized in liquid-liquid chromatography (LLC). For example, coatings of β,β'-oxydipropionitrile and the polyethylene glycols on silica, diatomaceous earth, or micropheres have had wide application in the separation of organic solutes.[3] In these cases, relatively nonpolar mobile phases like hexane or isooctane modified with 10% to 20% of a more polar solvent like methylene chloride or tetrahydrofuran are used for affecting the desired separation. For the normal-phase separation in LLC, the mechanism of solute retention is generally considered to be a partitioning of the solute molecules between the liquid mobile phase and the liquid-coated stationary phase. The principle interacting forces between solutes and these phases are dispersion, dipole and hydrogen bonding, depending on the chemical nature of each of the constituents of the chromatographic system. A more extensive discussion of polar-phase liquid/liquid chromatography, with examples, is given in Chapter 5.

In 1969, Halasz and Sebestian[58] introduced a variation in the stationary phase for LLC where the silica support was chemically modified to produce a monolayer of a nonpolar organic substituent (Fig. 4-15). These chemically

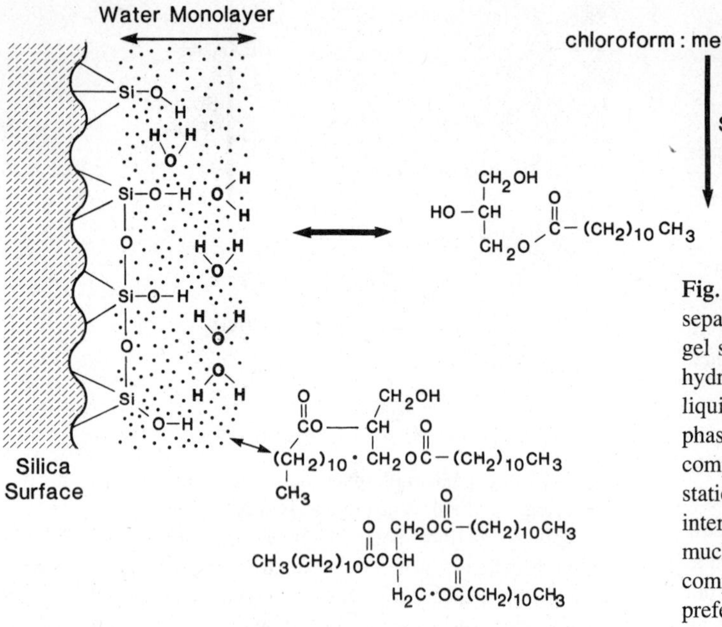

Water Monolayer

chloroform : methanol (5:15)

Solvent Flow

Silica Surface

Fig. 4-14. Mechanism of liquid/liquid chromatography by separation of mono-, di-, and triglycerides of lauric acid. Silica gel stationary phase has monolayer of water strongly held by hydrogen bonding. Solute molecules are partitioned between liquid mobile phase (chloroform:methanol) and liquid stationary phase, that is, water monolayer. Most polar sample components, the monoglycerides, are retained most by polar stationary phase, *heavy arrow*. Sample components of intermediate polarity, such as the diglycerides, are retained to a much lesser degree, *light arrow*, whereas relatively nonpolar components, such as the triglycerides, are not retained and prefer relatively nonpolar stationary phase.

Si — OH

+ Cl Si–$(CH_2)_7$ CH_3 →

Si-O-Si-$(CH_2)_7$ CH_3

Silica Surface

Octanylchlorosilane → Octanylsilane

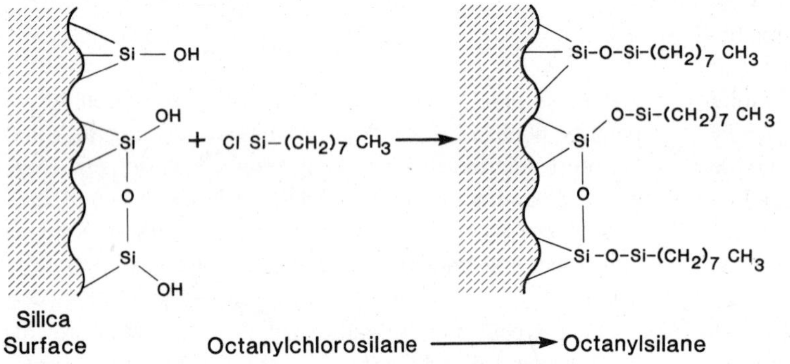

Fig. 4-15. Chemical preparation of bonded, stationary phase (reversed phase). Organochlorosilane reacts with nucleophilic hydroxyl (OH) groups of silica gel, forming siloxane covalent bond (Si-O-Si).

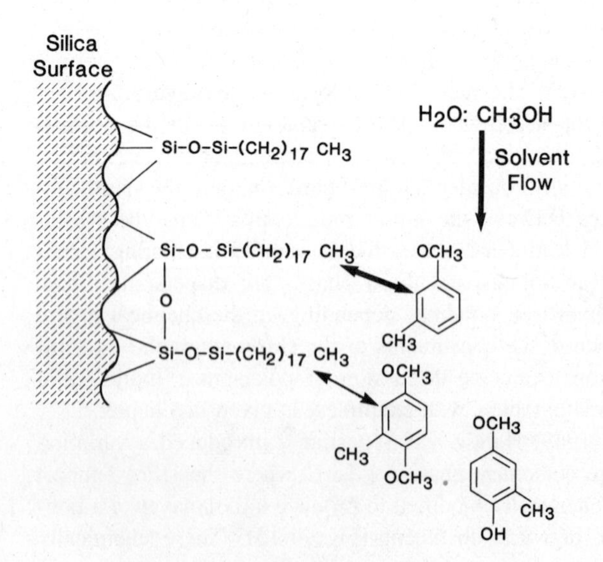

Fig. 4-16. Mechanism of bonded-phase (reversed phase) chromatography by separation of 3-methylanisole and two of its biochemical metabolites. Least polar (most nonpolar) sample components, such as 3-methylanisole, are most retained by nonpolar octadecylsilane stationary phase, *heavy arrow*. Sample components of intermediate polarity, such as 2,5-dimethoxytoluene, are retained to a much lesser degree, *light arrow*, whereas relatively polar components, such as 3-methyl-4-hydroxyanisole are not retained and prefer the polar mobile phase, water:methanol.

bonded stationary-phase supports are available with a variety of functional groups.[3,28,31] The most commonly used bonded phases are hydrocarbon phases such as octadecyl or octyl groups bonded to silica. The recent literature (summarized in reference 3) suggests that 70% or more of the HPLC separations are accomplished through these stationary phases.

The organic nature of the bonded phases impart a nonpolar character to the stationary phase. Therefore the mobile phases commonly used are highly polar, like water, methanol, or acetonitrile (Table 4-3). Solutes are separated by their relatively nonpolar character, that is, the most polar eluting first, whereas the nonpolar solutes are retained longer (Fig. 4-16). From this type of separation characteristic, the use of bonded phases in LLC is termed "reversed-phase chromatography."

The mechanism of solute retention for reversed-phase LC has not yet been clearly delineated.[28] The mechanism proposed by Horvath et al.[59,60] is an adsorption mechanism whereby the solute and mobile-phase molecules compete for sites on the organic surface of the bonded phase. In this case, the forces of interaction would be similar to those postulated for adsorption chromatography, as described in the previous section. However, one alternative retention mechanism for reversed-phase chromatography is that the sample molecules partition from the mobile phase into a "liquid" phase, defined by the organic coating bonded to the silica plus associated molecules of the mobile phase.[61] Again, in this case, the forces of interaction between sample constituents and the two phases would be dispersion, dipole, and hydrogen bonding. A further discussion, with examples, of reversed-phase LC is given in Chapter 5.

Another example of chromatography where a partition mechanism operates is in the case of gas-liquid chromatography. The forces of interaction between solute molecules and the liquid-coated stationary phase are as previously discussed for LLC. However, for GC, the mobile phase serves as an inert carrier for the sample constituents, whereas in LLC the mobile phase is an active, interacting component in the partition mechanism.

Ion exchange

This mode of liquid chromatography was originally introduced for the separation of amino acids and other ionic species in physiological fluids. Substantial progress in biochemistry has been made during the past 25 years because of the application of ion-exchange chromatography to the isolation and purification of proteins for their characterization. Numerous examples of the use of ion-exchange chromatography can be found in continuing series, such as *Methods in Enzymology*[62] and *The Enzymes*,[63] in reviews,[64] and in books.[65]

This mode of chromatography utilizes stationary phases that possess formal positive or negative charges. The most common retention mechanism is the exchange of sample ions, *A*, and mobile-phase ions, *B*, with the charged groups, *R*, of the stationary phase:

$$A^- + R^+B^- \rightarrow B^- + R^+A^- \quad \textbf{Anion exchange} \qquad \textit{Eq. 4-15, A}$$
$$A^+ + R^-B^+ \rightarrow B^+ + R^-A^+ \quad \textbf{Cation exchange} \qquad \textit{Eq. 4-15, B}$$

In the first case, anion exchange is occurring, whereas cation exchange is shown for the second. For ion-exchange chromatography, sample ions compete with mobile-phase ions for ionic sites on the stationary phase. The sample ions that interact weakly with the stationary phase will be retained least, whereas those that interact strongly will be retained the most and will elute later. The principle force of these interactions is electrostatic, that is, the attraction of opposite charges.

The cation-exchange stationary phases most commonly used are those containing a sulfonic acid ($-SO_3^-H^+$) group where the H^+ counterion can be exchanged with positively charged ions of the solute molecules or mobile phase. A weaker cation-exchange stationary phase is the carboxylic acid group ($-COO^-H^+$), which is used most commonly in open-column LC for the purification of proteins. For the anion-exchange stationary phases, the quaternary ammonium group ($-CH_2-N^+-[CH_3]_3Cl^-$) is most commonly used. In this case the chloride anion can be exchanged with anionic solutes or anions in the mobile phase. A weaker anion-exchange stationary phase is the protonated ammonium group ($-N^+H[R_2]Cl^-$).

To effect a separation of sample constituents, the extent of ionization of sample molecules is controlled by variations in pH of the mobile phase. Since the solutes are predominately weak acids, HA, or weak bases, B, a change in pH will shift the following ionization equilibriums either to the right or to the left:

$$\textbf{pH}\downarrow \qquad \textbf{pH}\uparrow$$
$$HA \rightleftharpoons H^+ + A^- \qquad \textit{Eq. 4-16, A}$$
$$BH^+ \rightleftharpoons H^+ + B \qquad \textit{Eq. 4-16, B}$$

An increase in ionization leads to an increased retention of the sample.

Factors, other than pH, controlling solute retention in ion-exchange chromatography are (1) charge strength of the solute ion, (2) ionic strength of the mobile phase, and (3) charge strength of the counterion on the stationary phase. One can decrease the retardation of solutes by increasing the ionic strength of the mobile phase, decreasing the strength of the counterion, such as use of Na^+ instead of H^+ for cation-exchange phases, or by adjusting the pH of the mobile phase in a manner to decrease dissociation of either the solute, the counterion on the packing, or both.

Gel permeation

In contrast to the previous mechanisms and modes of chromatography, gel permeation chromatography (GPC) separation is strictly based on molecular size. The station-

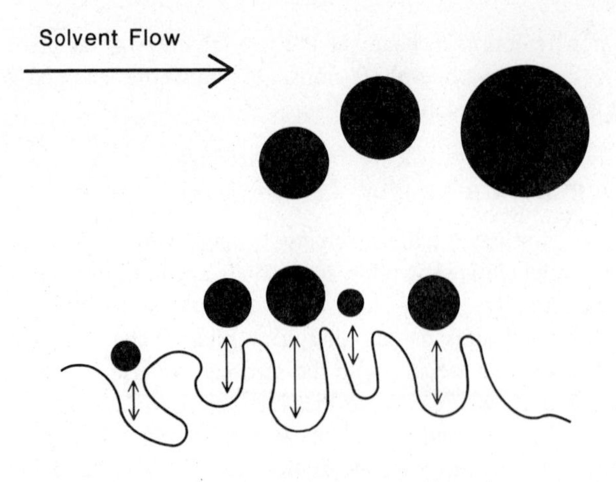

Fig. 4-17. Mechanism of size-exclusion chromatography. Stationary phase, in form of porous beads, contains pores of varying diameter. Mobile phase outside and inside pores is same, except that liquid inside is immobilized. When a sample containing solutes varying from small to large molecules elutes through column, small molecules penetrate all pores and are retained, thus being eluted later than large molecules, which move only in mobile phase. Molecules of intermediate size penetrate only some pores, thereby being retained to a lesser degree than small ones are.

ary phase for GPC contains pores of a particular average size and if the sample molecules are too large to enter the pores, they are not retained by the stationary phase. Small sample molecules permeate deeply into the pores and are retained. They ultimately diffuse from the pores and are swept away by the flow of the mobile phase. Intermediate-sized sample molecules enter the pores to some extent but are not retained as easily as the small sample molecules because of a lack of deep penetration into the pores. This mechanism is illustrated in Fig. 4-17.

The major advantage of this mode of chromatography is that the liquid chromatographic method can be used to separate virtually any sample, as long as it is soluble in a mobile phase. Additionally, it is applicable to soluble species having an average molecular weight of 50 to greater than 10 million. Since molecular size is the property of interest, representative calibration curves should be obtained by use of calibration standards of known molecular weight. Likewise, stationary-phase choice is based upon expected molecular weight range of the solute molecules in the sample and a compatibility with the mobile phase. A mobile phase for GPC should be chosen first on the basis of sample solubility, and then a compatible stationary phase is selected. Most stationary phases are compatible with aqueous or proton-donating (such as methanol) solvents. However, stationary phases are available that are compatible only with organic solvents. Lists of the available phases for GPC are tabulated in the manufacturers' literature,[66,67] in reviews,[64,68] or in books.[28,31,69]

SAMPLE PREPARATION FOR CHROMATOGRAPHY

Nature of problem

Few chromatographic analyses are conducted on the sample as submitted to the clinical laboratory. For any given sample, the goal of the chromatographic analysis is either a qualitative or a quantitative determination of its component or components. To achieve this objective, one should separate the component or components of interest as a discrete zone or zones with the same peak or spot distribution and k' or R_f value as the standards under identical chromatographic conditions. However, the complexity of a biological-sample matrix usually renders the chromatographic separation ineffectual by (1) interaction of sample impurities with the stationary phase, causing a reduction in the resolving power of the system; (2) saturation of most chromatographic detector systems, tending to raise the noise level and thereby decreasing sensitivity; and (3) interaction of the component of interest with other matrix components, leading to irreproducibility of the separation from sample to sample. To minimize these sample matrix effects in the chromatographic separation, a strategy for separation of the analyte from interfering components (sample clean-up) is required.

Any separation method employed in the laboratory must meet the criteria of yield, separation, capacity, and cost effectiveness. The advantages of having high yield in any sample manipulation step are obvious, but if recovery is quantitative with little separation, the method is unsatisfactory. The corollary is also true; that is, if the separation is excellent but there is a low yield, the method is of little value. Many separation methods are readily applied on a large scale where large amounts of sample are available, but others are only applicable to small-scale separations. The criterion of cost effectiveness includes time, equipment, reagents, and labor, which may render a separation method impracticable.

The strategy of preparation for chromatographic analysis should include consideration of whether the objective of the analysis is to qualitatively detect or to quantitate the substance under investigation.

Mechanical methods for initial isolation of analyte

The type of sample matrix received by the analyst in a clinical chemistry laboratory varies from a simple homogeneous-appearing liquid like perspiration to a complex heterogeneous solid like feces. However, the most commonly received sample matrices are urine and blood (or plasma). The initial step in analyte preparation for chromatography will vary according to matrix.

Solid samples, such as tissues or feces, are first disrupted or treated for preparation of a homogeneous solution or suspension from which the analyte can be isolated. Homogenization of tissues in a blender such as a Polytron (Brinkmann Instruments, Inc., Westbury, N.Y.), with an appropriate solvent may solubilize the desired analyte. The use of a Potter-Elvehjem tissue grinder is also effective. Tissue can also be extracted in a mortar with a pestle and a small amount of solvent. In addition to these grinding or shearing techniques, solid samples can be disrupted by sonication in solvent, or hydrolyzed by acid, base, or enzymes.[16,62,63]

A common procedure applied to solid or liquid samples is lyophilization, that is, freeze-drying, or evaporation under reduced pressure. This technique is generally applied to samples containing heat-labile analytes, or in situations where a large volume of sample, such as a 24-hour urine, is to be analyzed.

A relatively new procedure for preparing homogeneous powders of feces for subsequent extraction has been developed for the investigation of drug metabolism. It involves grinding the sample with a stainless steel ball mill in the presence of anhydrous sodium sulfate.[70,71]

Liquid samples may also require an initial treatment for removal of analytes sequestered by matrix components. Mild-base hydrolysis has been used to release sequestered polychlorinated biphenyls from blood lipid components.[72] Whole blood can be diluted with sterile water to disrupt blood cells osmotically before analyte isolation. Another initial treatment applied to blood or urine samples is to remove proteins and other macromolecules through precipitation. Some of the more commonly used protein-precipitating agents are trichloroacetic acid or barium sulfate.

In many other mechanical methods of matrix disruption, such as homogenization in a solvent or buffer, centrifugation is commonly employed to remove cell debris, particulate matter, or other large contaminants. An alternative method for the removal of insolubles is filtration, either through an inert material, such as glass wool, or through a membrane, such as a Millipore.

Chromatographic methods for initial isolation of analyte

An increasingly common method for the initial isolation of components of interest from aqueous solutions, like urine or blood, is the use of XAD-2 resin chromatography.[7,74] This stationary phase of polydivinylbenzene has a large surface area and is of a nonionic character, making it capable of adsorbing many classes of organic compounds from aqueous solution, principally by dispersive and dipole interactions. The adsorbed organics are eluted from the XAD-2 by organic solvents like methanol, acetone, diethyl ether, hexane, methylene chloride, or combinations of these solvents.[73] The XAD-2 method has been applied most often to urine- or blood-screening methods for drugs of abuse and their metabolites, but they can also be applied to isolate trace amounts of compounds.[75]

Another recently introduced chromatographic technique for initial analyte isolation is the use of small columns of octadecylsilyl-bonded phase.[76,77] The analytes from a relative large volume of sample are adsorbed from aqueous solution by forces similar to those operating in the XAD-2 procedure. Desorption is accomplished when a small volume of methanol is passed through the reversed-phase cartridge; the sample may then be processed for any mode of chromatography. Many additional resins, of the type just described, are currently available for the isolation of compounds of interest to the clinical chemist.[78]

Other types of chromatographic methods have been used in the preparation of samples for analysis. Ion-exchange chromatography has been widely used to isolate charged analytes. Anion exchange, first suggested by Horning and Horning,[79] has widely been used for the isolation of acidic constituents from biological fluids.[80] Although DEAE-Sephadex is the most widely used ion-exchange stationary phase for sample cleanup,[80] other anion exchangers such as AG1X[81,82] and Dowex 3[83] have been used.

Extraction methods for analyte isolation

Liquid-liquid and liquid-solid partition methods have been widely used for both primary and secondary extraction steps in a wide variety of clinical chemistry analyses before the chromatographic quantitation step. The reasons for the use of extraction procedures are numerous, including the isolation of the analyte from large quantities of contaminating materials and its concentration into a small volume of solvent, making detection easier. Liquid-liquid extraction procedures are easily accomplished, usually permitting the workup of multiple samples simultaneously.

The success of an extraction step depends on knowledge of the polarity of the analyte. This information is used to select an extracting solvent that will effectively remove the analyte from the sample. A general rule of solvent selection is that compounds tend to favor solvents having the same polarity interaction forces. It is important that the chosen solvent be immiscible with the sample matrix.

Other points to consider in solvent selection include the following. The solvent must be chemically compatible with the analyte; that is, no chemical reaction should be possible between the two. The solvent must be compatible with all subsequent operations after the extraction. For example, a high boiling solvent would be difficult to remove, and so the analyte solution would be difficult to concentrate. The solvent should not introduce any contaminants that would make the analysis difficult. Many laboratory supply companies offer common solvents of high-purity grades such as the three following types: (1) *HPLC-grade solvents,* which are compatible with most detector systems and do not contain particulate matter, which would foul the HPLC equipment; (2) *pesticide-grade solvents,* which are compatible with electron-capture GC detectors in not introducing any contaminating substances; and (3) *lipo-grade-grade solvents,* which do not contain any greases or other substances that would interfere with the analysis of lipids. These are but a few of the types of quality solvents available. If the solvent is not available in the required purity, a purification of the solvent must be done before use in any sample cleanup procedure. Most commonly, a distillation of the solvent will suffice, but sometimes more extensive purification measures are required. Methods of more rigorous purification procedures for most solvents are described in Weissberger.[84] Even with the use of the highest quality solvents commercially available or prior purification of solvents, impurities may still be a problem. The most common contaminant is plasticizers, usually coming from cap liners and other plastic materials.[85] These contaminants, various alkyl phthalates, can interfere with some analyses, particularly when electron-capture gas chromatography is used.

Once a decision concerning the solvents for extraction has been made, the actual operations in extraction must be considered. In general, a repeated series of extractions with smaller volumes of solvent will be more efficient than a single extraction with a large volume. For solid samples, the solvent may be introduced during the mechanical disruption step as previously mentioned. The cycle of grinding, sonication, and so on, is repeated several times with several volumes of solvent. However, doing so sometimes does not effectively extract the desired analyte. In this case, the pulverized solid sample may have to be extracted with a Soxhlet extractor or a continuous infusion extractor. Both of these methods are more efficient than manual operations for extracting substances from a solid matrix. However, the requirements for these methods include a reasonably volatile extracting solvent and stability of the analyte at the boiling point of the solvent.

To extract an analyte having ionizable groups, it is best to first solubilize the solid sample in an aqueous solution. The pH of the solution is then adjusted below the pK_a of acidic components or above the pK_a of basic components with the addition of acid or base, respectively, to convert the analyte, 95% or greater, into its extractable (nonionized) form. A nomogram relating pK_a values of acids to percent ionization at various pH values has been published by Hopgood.[86] If the pK_a of the analyte is not known, a lowering of the pH of the aqueous solution to a pH of 2.0 by the addition of acid is usually sufficiently low to permit the extraction of most acidic analytes. Likewise, raising the pH to 12 usually is sufficiently high to permit the extraction of most basic analytes of unknown pK_a.

For liquid-liquid extraction, an increase in the ionic strength of the aqueous layer will enhance the ease of extraction of the analyte causing it to favor the extracting solvent. An ionic neutral salt, such as sodium chloride or potassium bromide, is commonly used for this purpose.

One of the problems commonly encountered in liquid-liquid extractions is the formation of emulsions, that is, one of the immiscible phases becoming dispersed as fine droplets in the other. To avoid emulsion formation, several precautions can be taken during the actual extraction process: (1) If the two liquid layers have a large contact surface, avoid vigorous mixing of the phases. The use of gentle agitation will accomplish the extraction. (2) Filter all finely divided particulate matter before extraction. (3) Use solvent pairs with large differences in density.

If an emulsion does form, one may try several steps that will possibly break it. (1) Try to get the dispersed droplets to achieve coalescence by mechanically disrupting their surfaces. Stirring with a glass rod or filtration through a loose bed of glass wool sometimes will break it. (2) If the densities of the two solvents are sufficiently different, centrifugation will sometimes effect separation. (3) Cooling or freezing the mixture sometimes causes a coalescence of droplets. (4) An increase in ionic strength, by the addition of salt or a small amount of an alcohol, such as ethanol or 2-ethylhexanol, may cause a decrease in the forces stabilizing the emulsion. (5) A change in the ratio of the two solvents by addition of more extraction solvent or a partial evaporation of solvent may break the emulsion. (6) Filtration through phase-separation filter paper will break many emulsions commonly encountered. In the majority of the cases, one of these procedures will be successful in breaking the emulsion.

Examples of solvent extraction procedures include the following. Folch et al.[87] developed a procedure, widely used today, for the extraction of lipids from biological tissues. In this procedure, the sample is homogenized with a 2:1 solvent mixture of chloroform and methanol. The homogenate is filtered, and the extract is washed with water. The chloroform layer is removed and is prepared for analysis. Sunshine et al.[88] have described a procedure whereby basic organic drugs and their metabolites are extracted from potassium carbonate–saturated urine with ether. The extract is adjusted to pH 8.5 and the drugs are extracted into ether. The ether extract is then analyzed by thin-layer chromatography. Finkle et al.[89] have described a comprehensive extraction procedure for isolating basic, acidic, and neutral compounds from blood and urine for GC analysis. The procedure has been applied to forensic toxicology samples for the detection of poisons, drugs, and their metabolites. Many other examples of extraction procedures can be found in almost every issue of the following representative journals: *Clinical Chemistry, Journal of Chromatography,* and *Clinica Chimica Acta.*

Processing of sample extracts

Many analyte extracts are too dilute for direct chromatographic analysis or for derivatization reactions before chromatography and are usually concentrated by evaporation of the solvent.

Any solvent-evaporation procedure must be conducted with care to avoid loss of the analyte. Such a loss of analyte can occur if traces of water are present in the extract. These can be removed by use of an anhydrous salt such as sodium carbonate or sodium sulfate. Alternative desiccating salts such as calcium oxide or magnesium sulfate can also be used. Other purposes for drying an extract may be to conduct subsequently a derivatization procedure, such as acetylation or silylization, or to remove traces of water, which may interfere with the chromatography step.

To concentrate an extract, one must take care to avoid losses of the analyte. The analyte may be lost to the concentration vessel by irreversibly binding to the walls during concentration. This can be avoided by prior silylization of the glassware. Some substances are sufficiently volatile to form azeotrope mixtures with the solvent during evaporation. To avoid this, many gentle concentration methods or apparatuses are available. Micro-Synder or Kurderna-Danish concentrators evaporate solvent under mild conditions. If the analyte is heat sensitive, evaporation of the solvent under a stream of purified inert gas, such as nitrogen or argon, can be employed. In this case, one can warm the vessel to a range of 35 to 50 degrees C to expedite the evaporation process. The use of a rotary evaporator under reduced pressure is also a gentle method for solvent evaporation.

Another method for concentrating the analyte is the back-extraction of the compound of interest from the solvent. For example, MacGee has published a variety of methods[90] whereby the analyte, in the original extracting solvent, is back-extracted into a small volume of analyte-derivatizing solvent before gas chromatography. Methods of this type expedite the analysis, since solvent-evaporation steps are not required. Other examples of analyte cleanup procedures for preparing samples for chromatography are detailed in reviews by Ko and Petzold,[91] and Coutts and Jones,[92] and in books by Clarke,[93] Schirmer,[94] and Sunshine.[95] Additional examples are given in Chapters 5 and 6 on HPLC and GC.

REFERENCES

1. Evenson, M.A., and Carmack, G.D.: Clinical chemistry, Anal. Chem. **51**:35R-79R, 1979; and Evenson, M.A.: Clinical chemistry, Anal. Chem. **53**:214R-233R, 1981.
2. Cram, S.P., Risby, T.H., Field, L.R., and Yu, W.-L.: Gas chromatography, Anal. Chem. **52**:324R-360R, 1980; and Risby, T.H., Field, L.R., Yancy, F.J., and Cram, S.P.: Gas chromatography, Anal. Chem. **54**:410R-428R, 1982.
3. Walton, H.F.: Ion exchange and liquid column chromatography, Anal. Chem. **52**:15R-27R, 1980; and Majors, R.E., Barth, H.G., and Lochmuller, C.H.: Column liquid chromatography, Anal. Chem. **54**:323R-363R, 1982.
4. Zweig, G., and Sherma, J.: Paper and thin-layer chromatography, Anal. Chem. **52**:276R-289R, 1980; and Sherma, J., and Fried, B.: Thin-layer and paper chromatography, Anal. Chem. **54**:45R-57R, 1982.
5. Tswett, M.: Absorption analysis and the chromatographic method: application in the chemistry of chlorophyll, Berichte der deutschen botanischen Gesellschaft **24**:384-393, 1906. Translated by Strain, H.H., and Sherman, J., in Tswett, M.: Absorption analysis and chromatographic methods, J. Chem. Ed. **44**:238-242, 1967.
6. Heftman, E.: History of chromatography. In Heftman, E., editor: Chromatography: a laboratory handbook of chromatography methods, ed. 3, New York, 1975, Van Nostrand Reinhold Co.
7. Strain, H.H., and Svec, W.A.: Differential methods of analysis. In Heftman, E., editor: Chromatography: a laboratory handbook of chromatography methods, ed. 3, New York, 1975, Van Nostrand Reinhold Co.
8. Pelick, N., Bolleger, H.R., and Mangold, H.: In Giddings, J.C., and Keller, R.A., editors: Advances in chromatography, III, New York, 1966, Marcel Dekker, Inc.
9. Ettre, L.S.: The development of chromatography, Anal. Chem. **43**:20A-21A, 25A, 27A-31A, 1971.
10. Latiinen, H.A., and Ewing, G.W., editors: A history of analytical chemistry, Washington, D.C., 1977, Analytical Chemistry Division of American Chemical Society.
11. Giddings, J.C.: Reduced plate height equation: a common link between chromatographic methods, J. Chromatogr. **13**:301-304, 1964.
12. Giddings, J.C.: Dynamics of chromatography, I, Part I, Principles on theory, New York, 1965, Marcel Dekker, Inc.
13. Karger, B.L., Snyder, L.R., and Horvath, C.: An introduction to separation science, New York, 1973, John Wiley & Sons, Inc.
14. Dean, J.A.: Chemical separations methods, New York, 1969, Van Nostrand Reinhold Co.
15. Blackburn, T.R.: Equilibrium, a chemistry of solutions, New York, 1969, Holt, Rinehart & Winston, Inc.
16. Tsuji, K., and Morozowich, W.: GLC and HPLC determination of therapeutic agents, Part I, New York, 1978, Marcel Dekker, Inc.
17. Snyder, L.R.: Principles of absorption chromatography, New York, 1968, Marcel Dekker, Inc.
18. Zweig, G., and Sherman, J.: Handbook of chromatography, I and II, Boca Raton, Fla., 1972, CRC Press, Inc.
19. Heftman, E.: Chromatography: a laboratory handbook of chromatographic and electrophoretic methods, ed. 3, New York, 1975, Van Nostrand Reinhold Co.
20. Stahl, E.: Thin-layer chromatography: a laboratory handbook, New York, 1969, Springer-Verlag.
21. Giddings, J.C., and Keller, R.A., editors: Advances in chromatography, New York, Marcel Dekker, Inc., a continuing series of reviews and techniques.
22. Lederer, M., editor: Chromatographic reviews, Journal of Chromatography, Elsevier Scientific Publishing Co., Amsterdam; reviews appear on a regular basis as a part of this journal publication.
23. Library series, Journal of Chromatography, Elsevier Scientific Publishing Co., Amsterdam.
24. Keller, R.A., editor: Journal of Chromatographic Science, Evanston, Ill., Preston Technical Abstract Co.
25. Snyder, L.R.: A rapid approach to selecting the best experimental conditions for high speed liquid column chromatography. Part I. Estimating initial sample resolution and the final resolution required by a given problem, J. Chromatogr. Sci. **10**:200-212, 1972.
26. Giddings, J.C.: Non-equilibrium and diffusion: a common basis for theories of chromatography, J. Chromatogr. **2**:44-52, 1959.
27. Grob, R.L., editor: Modern practice of gas chromatography, New York, 1977, John Wiley & Sons, Inc.
28. Snyder, L.R., and Kirkland, J.J.: Introduction of modern liquid chromatography, ed. 2, New York, 1979, John Wiley & Sons, Inc.
29. Martin, A.J.P., and Synge, R.L.M.: A new form of chromatogram employing two liquid phases. 1. A theory of chromatography; 2. Application to the micro-determinations of the higher monoamino-acids in proteins, Biochem. J. **35**:1358-1368, 1941.
30. Van Deemter, J.J., Zuiderweg, F.J., and Klinkenberg, A.: Longitudinal diffusion and resistance to mass transfer as causes of nonideality in chromatography, Chem. Eng. Sci. **5**:271-289, 1956.
31. Johnson, E.L., and Stevenson, R.: Basic liquid chromatography, Palo Alto, Calif., 1978, Varian Associates.
32. McNair, H.M., and Bonelli, E.J.: Basic gas chromatography, ed. 5, Palo Alto, Calif., 1969, Varian Associates.
33. Knapp, D.R.: Handbook of analytical derivatization reactions, New York, 1979, John Wiley & Sons, Inc.
34. Frei, R.W., and Laurence, J.F.: Chemical derivatization in liquid chromatography, New York, 1977, Elsevier Scientific Publishing Co.
35. Persson, B.A., and Karger, B.L.: High performance ion pair partition chromatography: the separation of biogenic amines and their metabolites, J. Chromatogr. Sci. **12**:521-528, 1974.

36. Gloor, R., and Johnson, E.L.: Practical aspects of reverse-phase ion-pair chromatography, J. Chomatogr. Sci. **15**:413-423, 1977.
37. Kissinger, P.T.: Comments on reverse-phase ion-pair partition chromatography, Anal. Chem. **49**:883, 1977.
38. Scott, R.P.W., and Kucera, P.: Some aspects of ion-exchange chromatography employing adsorbed ion exchangers on reversed-phase columns, J. Chromatogr. **175**:51-63, 1979.
39. Wahlund, K.G., and Beijersten, I.: Stationary phase effects in reversed-phase liquid chromatography of acids and ion pairs, J. Chromatogr. **149**:313-329, 1978.
40. Keller, R.A., and Snyder, L.R.: Relation between the solubility parameter and the liquid-solid solvent strength parameter, J. Chromatogr. Sci. **9**:345-459, 1971.
41. Rohrschneider, L.: Solvent selection in absorption liquid chromatography, Anal. Chem. **46**:470-473, 1974.
42. Saunders, D.L.: Solvent selection in absorption liquid chromatography, Anal. Chem. **46**:470-473, 1974.
43. Snyder, L.R.: Classification of the solvent properties of common liquids, J. Chromatogr. **92**:223-230, 1974.
44. Karger, B.L., Snyder, L.R., and Eon, C.: An expanded solubility parameter treatment for classification and use of chromatographic solvents and absorbents: parameters for dispersion, dipole and hydrogen bonding interactions, J. Chromatogr. **125**:71-88, 1976.
45. Snyder, L.R.: Solvent selection for separation processes. In Perry, E.S., and Weissberger, A., editors: Techniques of chemistry: separation and purification, ed. 3, vol. 12, New York, 1978, John Wiley & Sons, Inc.
46. Hildebrand, J.H., and Scott, R.I.: The solubility of non-electrolytes, ed. 3, New York, 1964, Dover Publications, Inc.; and Hildebrand, J.H., and Scott, R.I.: Regular solutions, Englewood Cliffs, N.J., 1962, Prentice-Hall, Inc.
47. Trappe, W.: Die Trennung von biologischen Fettstoffen aus ihren natürlichen Gemischen durch Anwendung von Adsorptionssäulen, I. Mitteilung: Die eluotrope Reihe der Lösungsmittel, Biochem. Z. **305**:150-161, 1940.
48. Barton, A.F.M.: Solubility parameters, Chem. Rev. **75**:731-753, 1975.
49. Rohrschneider, L.: Eine Methode zur Charakterisierung von gaschromatographischen Trennflüssigkeiten, J. Chromatogr. **22**:6-22, 1966.
50. McReynolds, W.O.: Characterization of some liquid phases, J. Chromatogr. Sci. **8**:685-691, 1970.
51. Hawkes, S., Grossman, D., Hartkopf, A., et al.: Preferred stationary liquids for gas chromatography, J. Chromatogr. Sci. **13**:115-117, 1975.
52. Kiseler, A.V., and Yashin, Y.I.: Gas adsorption chromatography, New York, 1969, Plenum Press.
53. Thornsberry, W.L.: Isothermal gas chromatographic separation of carbon dioxide, carbon oxysulfide, hydrogen sulfide, carbon disulfide, and sulfur dioxide, Anal. Chem. **43**:452-453, 1971.
54. Halasz, I., and Heine, E.: Packed capillary columns in gas chromatography, Anal. Chem. **37**:495-498, 1965.
55. Zlatkin, A., and Kaiser, R.E.: HPTLC: high performance thin-layer chromatography, Journal of Chromatography Library Series, 9, New York, 1977, Elsevier Scientific Publishing Co.
56. Touchstone, J.C., and Dobbins, M.F.: Practice of thin layer chromatography, New York, 1978, John Wiley & Sons, Inc.
57. Macek, K., Hais, I.M., Kopecky J., et al.: Bibliography of paper and thin-layer chromatography and survey of applications, J. Chromatogr., Suppl. vol. 5, 1976.
58. Halasz, I., and Sebestian, I.: New stationary phase for chromatography, Angew. Chem., Int. Ed. **8**:453-454, 1969.
59. Horvath, C., Melander, W., and Molnar, I.: Solvophobic interactions in liquid chromatography with nonpolar stationary phases, J. Chromatogr. **125**:129-156, 1976.
60. Horvath, C., and Melander, W.: Liquid chromatography with hydrocarbonaceous bonded phases: theory and practice of reverse phase chromatography, J. Chromatogr. Sci. **15**:393-404, 1977.
61. Kirkland, J.J.: High speed liquid partition chromatography with chemically bonded stationary phases, J. Chromatogr. Sci. **9**:206-214, 1971.
62. Colowick, S.P., and Kaplan, N.O., editors: Methods in enzymology, New York, Academic Press, Inc.; a continuing series on biochemical techniques.
63. Boyer, P.D., editor: The enzymes, ed. 3, New York, Academic Press, Inc.; a continuing series on biochemical techniques.
64. Majors, R.E.: Recent advances in high performance chromatography packings and columns, J. Chromatogr. Sci. **15**:334-351, 1977; and Brown, P.R., and Krstulovic, A.M.: Ion-exchange chromatography. In Perry, E.S., and Weissberger, A., editors: Techniques of chemistry: separation and purification, vol. 12, New York, 1978, John Wiley & Sons, Inc.
65. Helfferich, F.: Ion exchange, New York, 1972, McGraw Hill Book Co.
66. Sourcebook for successful HPLC, Milford, Mass., 1982, Waters Associates.
67. Chromatography, electrophoresis, immunochemistry and HPLC, Richmond, Calif., 1982, Bio-Rad Laboratories.
68. Anderson, D.M.W.: Gel permeation chromatography. In Simpson, C.F., editor: Practical high performance liquid chromatography, Philadelphia, 1978, Heyden & Sons Inc.
69. Determann, H.: Gel chromatography, ed. 2, New York, Berlin, 1969, Springer-Verlag.
70. Smith, C.C., Tabor, M.W., and Wolfe, G.J.: Metabolism studies on WR-158,122, an antimalarial drug, in bile duct–cannulated rats, Toxicologist **2**:30, 1982.
71. Smith, C.C., Khalil, A., and Tabor, M.W.: Fractionation of urinary and fecal metabolites of the antimalarial drug WR-158,122 following oral doses in rats and rhesus monkey, Toxicologist **3**:52, 1983.
72. Que Hee, S.S., Ward, J.A., Tabor, M.W., and Suskind, R.R.: Screening method for Aroclor 1254 in whole blood, Anal. Chem. **55**:157-160, 1983.
73. Weissman, N., Lowe, M.L., Beattie, J.M., and Demetriou, J.A.: Screening method for detection of drugs of abuse in human urine, Clin. Chem. **17**:875-881, 1971.
74. Stolman, A., and Pranitis, P.A.: XAD-2 resin drug extraction methods for biologic samples, Clin. Toxicol. **10**:49-60, 1977.
75. Yanasaki, E., and Ames, B.N.: Concentration of mutagens from urine by adsorption with the non-polar resin XAD-2: cigarette smokers have mutagenic urine, Proc. Natl. Acad. Sci. USA **74**:3555-3559, 1977.
76. Shackleton, C.H.L., and Whitney, J.D.: Use of Sep-Pak® Cartridges for urinary steroid extraction: evaluation of the method for use prior to gas chromatographic analysis, Clin. Chim. Acta **107**:231-243, 1980.
77. Heikkinen, R., Fotsis, T., and Adlercreutz, H.: Reversed-phase C_{18} cartridge for extraction of estrogens from urine and plasma, Clin. Chem. **27**:1186-1189, 1981.
78. Dressler, M.: Extraction of trace amounts of organic compounds from water with porous organic polymers, J. Chromatogr. **165**:167-206, 1979.
79. Horning, E.C., and Horning, M.G.: Metabolic profiles: gas-phase methods for analysis of metabolites, Clin. Chem. **17**:802-809, 1971.
80. Jellum, E.: Profiling of human body fluids in healthy and diseased states using gas chromatography and mass spectrometry, with special reference to organic acids, J. Chromatogr. **143**:427-462, 1977.
81. MacGee, J.M., Roda, S.M.B., Elias, S.V., et al.: Determination of delta-aminolevulinic acid in blood plasma and urine by gas-liquid chromatography, Biochem. Med. **17**:31-44, 1977.
82. Mrochek, J.E., Butts, W.C., Rainey, W.T., and Burtis, C.A.: Separation and identification of urinary constituents by use of multiple analytical techniques, Clin. Chem. **17**:72-77, 1971.
83. Nakamura, E., Rosenberg, L.E., and Tanaku, K.: Microdetermination of methylmalonic acid and other short chain dicarboxylic acids by gas chromatography: use in prenatal diagnosis of methylmalonic acidemia and in studies of isovaleric acidemia, Clin. Chim. Acta **68**:127-140, 1976.
84. Riddick, J.A., and Bunger, W.B.: Organic solvents. In Weissberger, A., editor: Techniques of chemistry, vol. 2, ed. 3, New York, 1978, John Wiley & Sons, Inc.
85. DeZeeuw, R.A., Jonkman, J.H.G., and van Mansvelt, F.J.W.: Plasticizers as contaminants in high purity solvents: a potential source of interference in biological analysis, Anal. Biochem. **67**:339-341, 1975.
86. Hopgood, M.F.: Nomogram for calculating percentage ionization of acids and bases, J. Chromatogr. **47**:45-50, 1970.

87. Folch, J., Lees, M., and Sloan-Stanley, G.M.: A simple method for the isolation and purification of total lipids from animal tissues, J. Biol. Chem. **226:**497-509, 1957.
88. Bastos, M.L., Kananen, G.E., Young, R.M., Monforte, J.R., and Sunshine, I.: Detection of basic organic drugs and their metabolites in urine, Clin. Chem. **16:**931-940, 1970.
89. Finkle, B.A., Cherry, E.J., and Taylor, D.M.: AGLC based system for the detection of poisons, drugs, and human metabolites encountered in forensic toxicology, J. Chromatogr. Sci. **9:**393-419, 1971.
90. Kossa, W.C., MacGee, J., Ramachandran, S., and Webber, A.J.: Pyrolytic methylation/gas chromatography: a short review, J. Chromatogr. Sci. **17:**177-187, 1979.
91. Ko, H., and Petzold, E.N.: Isolation of samples prior to chromatography. In Tsuji, K., and Morozowich, W., editors: GLC and HPLC determination of therapeutic agents. Part I, New York, 1978, Marcel Dekker, Inc.
92. Coutts, R.T., and Jones, G.R.: Significance of analytical techniques in drug metabolism studies. In Jenner, P., and Testa, B., editors: Concepts in drug metabolism, vol. 10A, New York, 1979, Marcel Dekker, Inc.
93. Clarke, E.G.C.: Isolation and identification of drugs in pharmaceuticals, body fluids and post-mortem material, London, 1969, The Pharmaceutical Press.
94. Schirmer, R.E.: Modern methods of pharmaceutical analysis, vol. I, Boca Raton, Fla., 1982, CRC Press, Inc.
95. Sunshine, I.: Manual of analytical toxicology, Boca Raton, Fla., 1971, CRC Press, Inc.

Chapter 5 Liquid chromatography

Larry D. Bowers

α Chromatographic selectivity.

bonded phase A chromatographic support in which the stationary phase is covalently bound to the surface of the support.

capacity factor The retention of a compound normalized for the size of the column (see equation 5-6).

effluent Mobile phase that has exited from the column.

eluate A compound or mixture that has been separated in and exited from the column.

eluent Mobile phase.

gradient elution An elution system where the solvent composition is varied during the run.

H Height equivalent to one theoretical plate (HETP).

isocratic elution Elution with a solvent mixture of constant composition.

k' See capacity factor.

L Length of the column.

mobile phase The mixture of solvents that is percolated through the column.

μ Solvent velocity in the column ($= L/t_o$).

N Number of theoretical plates.

normal phase A chromatographic mode in which the mobile phase is less polar than the stationary phase.

permeability A measure of the ease with which the mobile phase can be forced through the column.

R_s Resolution.

reversed phase A chromatographic mode in which the mobile phase is more polar than the stationary phase.

selectivity The relative retention of two compounds in a chromatographic system (see equation 5-7.).

stationary phase The portion of the separation system that is immobilized in the column.

support The particles on which the stationary phase is held.

t_o Time required to elute an unretained substance.

t_R Retention time.

V_O Volume of solvent required to elute an unretained compound; also called "void volume."

V_R Retention volume.

Liquid chromatography is a form of separation science in which a liquid mobile phase is percolated through a column or thin layer of particles. Fig. 5-1 shows a schematic diagram of a column chromatograph used in liquid chromatography. The liquid mobile phase is taken from the reservoir and moved through the column by some type of driving force. Initially the force was gravity, but a pump has become a necessary component to drive the liquid through the column. A method of introducing the sample into the chromatographic system is also required. The most important constituent of a chromatographic instrument is the column. The column is packed with small particles on which specific sites or a layer of solvent (called the "stationary phase") is held. The differential equilibration of the analyte or analytes between the mobile and stationary phases results in their separation. All chromatographic modes (see Chapter 4) can be used in liquid chromatography. Finally, we can either collect the column effluent for further analysis or analyze the effluent with an on-line detector such as a photometer. Aliquots of the liquid phase are easy to collect if subsequent analysis is desired. The recording of any parameter that allows the analyte or analytes to be monitored as a function of elution volume or time is called a "chromatogram" (see Fig. 5-2).

Liquid chromatography is well suited for use in the clinical laboratory. Since the retention of a compound is determined by equilibriums, the position of the peak in the chromatogram (that is, the retention volume) can be helpful in identification if it coelutes with a known compound. In addition, one can obtain quantitative information by measuring the height or area of the peak. The separation of the components of the mixture allows quantitation of several compounds in a single analysis, which is useful in the analysis of drugs or intermediates in a metabolic pathway (such as porphyrins). Liquid chromatography has the additional advantage that the relatively polar compounds present in body fluids readily dissolve in commonly used mobile-phase solvents. This is in contrast to gas chromatography, which requires volatile analytes. Proteins and peptides are readily separated by liquid chromatography.

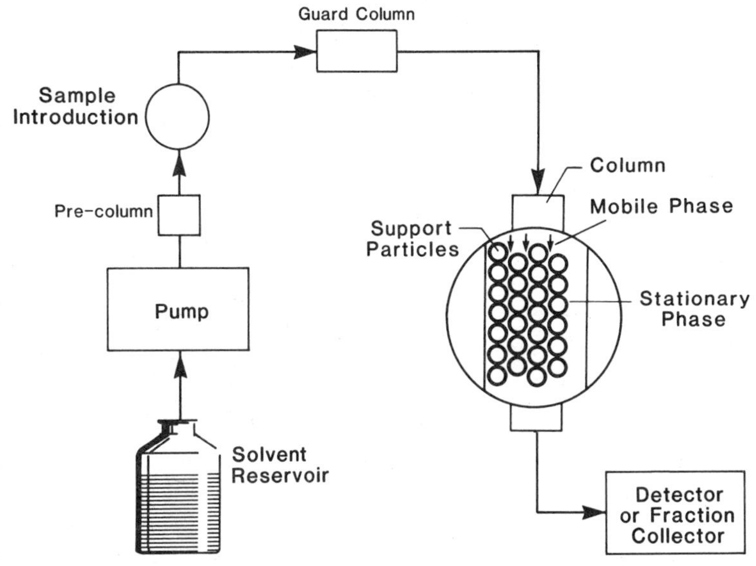

Fig. 5-1. Schematic diagram of a column liquid chromatographic system. (From Bowers, L.D., and Carr, P.W.: Quantitative Aspects of HPLC Workshop, Minneapolis, Minn., 1983.)

Despite the fact that liquid chromatography was discovered before gas chromatography, it was used in only a very small number of analytical applications in the clinical laboratory. The reason was that classical liquid chromatography typically required hours or days to complete a separation whereas gas chromatography required only minutes. Liquid chromatography was used preparatively somewhat more frequently but still lacked popularity because of its sensitivity to solvent and sample impurities and its labor intensiveness. With the advent of high-performance liquid chromatography (HPLC) in the early 1970s, liquid chromatography was able to achieve speed and resolution comparable to gas chromatography (GC). The introduction of covalently bonded stationary phases resulted in the further popularization of HPLC. Instrumental developments, such as microprocessor automation, have made both analytical and preparative chromatography easier to perform and more appealing. At present, HPLC is recognized as a true complement to GC in chromatographic analysis.

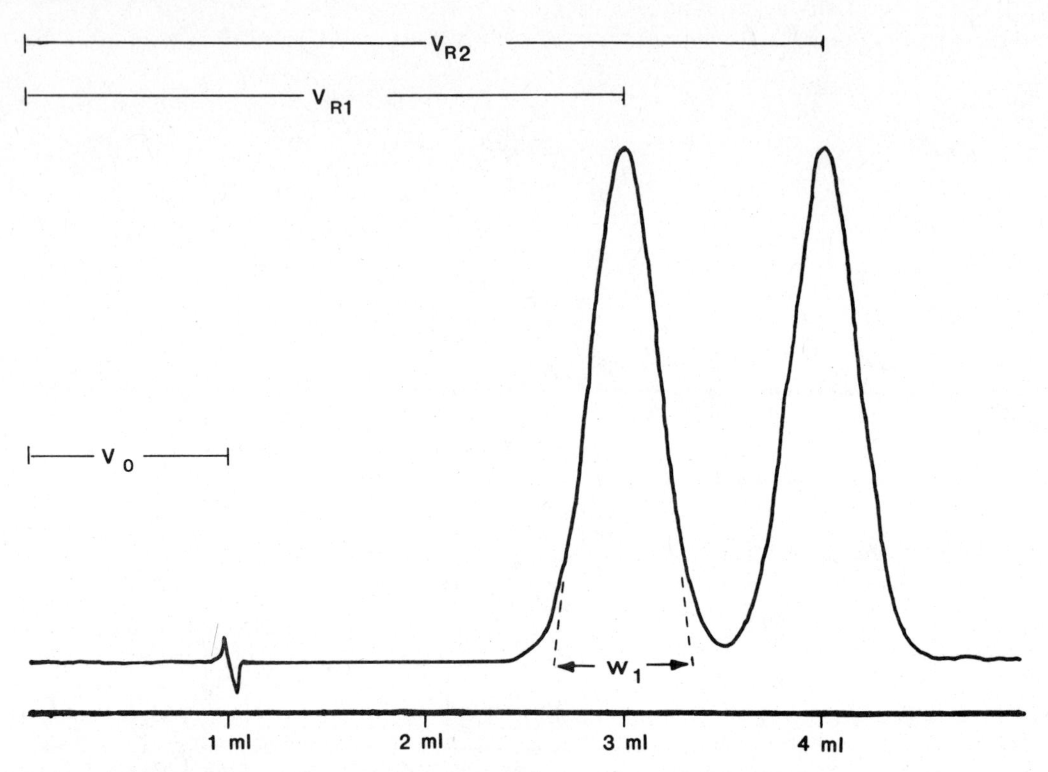

Fig. 5-2. Typical chromatogram obtained with a column liquid chromatographic system. V_o, Void volume; V_{R1}, peak 1 retention volume; V_{R2}, peak 2 retention volume.

RESOLUTION, EFFICIENCY, AND SPEED OF ANALYSIS

The object of any chromatographic technique is to separate, or resolve, the species of interest from other compounds of interest or from interferences in the sample matrix. As analysts, we must also be concerned about the speed of the analytical scheme, including any sample preparation steps, the separation itself, and calculation and reporting of results. Not surprisingly, speed of analysis and resolving power are related. The evolution of HPLC was based on an understanding of the factors that affect resolution and the optimization of those factors.

Resolution

The relative separation of two chromatographic peaks is measured by a parameter known as *resolution*. It is dependent on the positions of the centers of the peaks that correspond to each compound and on the width of the peak between the points where it is indistinguishable from the background signal. If, as shown in Fig. 5-2, the peak tracing reaches the background level before rising for the second peak, base-line resolution has been achieved. We calculate a numerical value for resolution (R_s) by dividing the difference in retention volumes (V_R), or times, by the average width, W, of the peaks.

$$R_s = \frac{V_{R2} - V_{R1}}{\frac{(W_1 + W_2)}{2}} \qquad \textit{Eq. 5-1}$$

Note that the value for resolution has no units. In Fig. 5-2, the resolution is 1.5 (see also box below).

From a separation standpoint, the important aspect of this resolution is that if the first peak were collected up to the minimum in the valley between the two peaks the compound in that peak would be 100% free of the compound making up the second peak. Table 5-1 shows the relative impurity in the first peak when both peaks are the same size for various resolution values. Snyder and Kirkland have covered this topic more completely.[1] Resolution of peaks will also affect quantitation as is discussed later.

Example. In Fig. 5-2, the retention volume of peaks 1 and 2 are 3 and 4 mL respectively. The width of the peaks at the base line is 0.6 and 0.7 mL respectively.

$$R_s = \frac{V_{R2} - V_{R1}}{\frac{(W_1 + W_2)}{2}}$$

$$R_s = \frac{4 - 3}{\frac{(0.6 + 0.7)}{2}}$$

$$R_s = \frac{1}{0.65} = 1.5$$

Similar results would be obtained if retention times and peak widths in time units were used.

Table 5-1. Effect of resolution on various peak parameters*

Resolution	Purity†	% error in area‡	% error in peak height‡
0.6	90	−>25	~15
0.8	95	−10	1
1.0	98	−3	<1
1.25	99.5	−<1	0
1.5	100	0	0

*Assuming a peak-height ratio of 1 to 1 for the two components.
†Purity of major peak, assuming collection is stopped at the lowest point of the valley between the peaks.
‡Error for the smaller peak.

Resolution must therefore be controlled by the factors that govern the peak width, namely, efficiency and relative peak retention. Peak retention is determined by the capacity factor and selectivity, as discussed in Chapter 4.

$$R_s = \frac{\sqrt{N}}{4}(\alpha - 1)\left(\frac{k'}{1 + k'}\right) \qquad \textit{Eq. 5-2}$$

R_s is the value of the resolution, N is a measure of efficiency called the *number of theoretical plates,* α is the relative retention of the two compounds called the *selectivity,* and k' is the *capacity factor* of the first peak. These terms have been defined in the previous chapter; so this discussion will be limited to their adjustment in liquid chromatography. The effect of changes in each of these parameters on resolution is illustrated in Fig. 5-3.

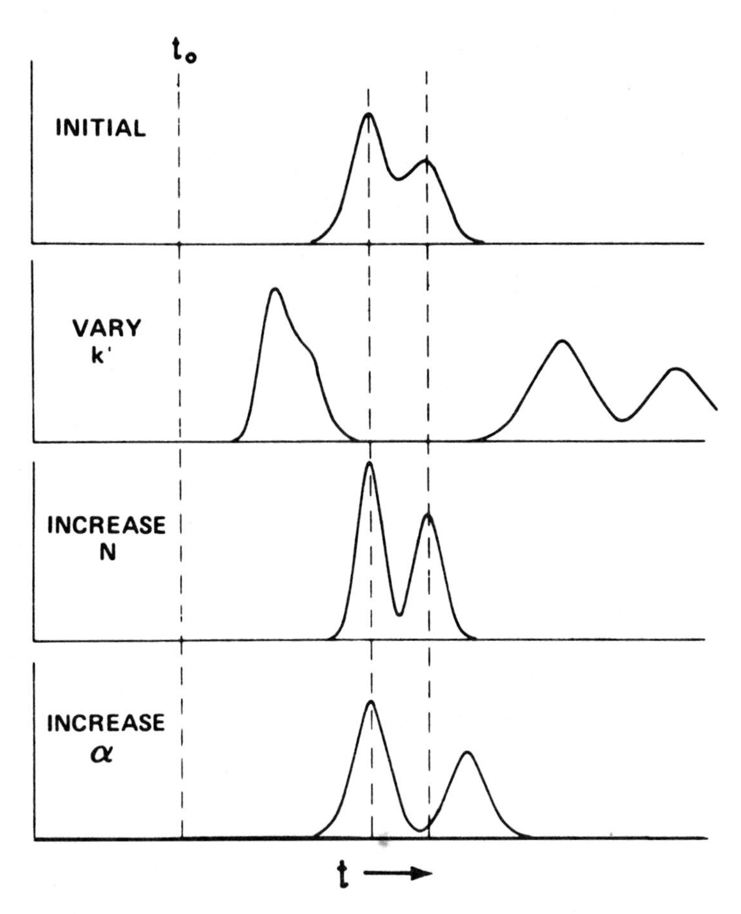

Fig. 5-3. Effect of varying k', N, and α on resolution. t, Time. (From Snyder, L.R., and Kirkland, J.J.: Introduction to modern liquid chromatography, ed. 2, New York, 1979, John Wiley & Sons, Inc.)

Chromatographic efficiency

The number of theoretical plates is a measure of peak width determined from the ratio of the retention volume and the standard deviation (σ) of the "gaussian" chromatographic peak squared:

$$N = \left(\frac{V_R}{\sigma}\right)^2 \qquad \textit{Eq. 5-3}$$

Since the width of a gaussian curve at half the peak maximum is 2.354σ, the peak width at half-peak height ($W_{1/2}$) is commonly measured and substituted in the equation:

$$N = 5.54\left(\frac{V_R}{W_{1/2}}\right)^2 \qquad \textit{Eq. 5-4}$$

to calculate the number of theoretical plates available in the chromatographic system. (See example at right.)

Originally this term was associated with phenomena occurring in the chromatographic column. Thus one could optimize the efficiency of the column by adjusting the flow rate, increasing the temperature, and so on. The Knox equation:

$$H = A_{\mu}^{1/3} + \frac{B}{\mu} + C\mu \qquad \textit{Eq. 5-5}$$

relates H, the height equivalent to a theoretical plate, to the flow velocity, μ. H is defined as the length of the column divided by N. The constants A, B, and C are numbers calculated from formulas derived from the mechanisms associated with the movement of a compound through the column. One outcome of this theory is that as

Example. From Fig. 5-2, the width at half-height for peak 2 is 0.4 mL.

$$N = 5.54\left(\frac{4\text{ mL}}{0.4\text{ mL}}\right)^2$$

$$N = 554\text{ plates}$$

This is not a very efficient column. Modern HPLC columns have about 10,000 plates. What would the width of peak 2 be on such an HPLC column? (Answer: 0.094 mL) the diameter of the column packing material decreases, H decreases. By decreasing H, N can be increased in the same length of column. Hence, 10, 5, and even 3 μm totally porous particles have been used in HPLC to maximize efficiency. It is doubtful that particles smaller than 3 μm will result in increased efficiency.

For the practicing chromatographer, four final points about efficiency are worthy of mention. First, a new column should always be tested upon receipt to be sure that reasonable efficiency is observed on that column. This also serves as a bench mark for measuring the decline in column performance as it is used. Second, the column should be tested under the same flow conditions the manufacturer uses. The buyer should be aware that a column tested at 0.1 mL/min to obtain a high number of theoretical plates may not be the best column to use at reasonable flow rates. Third, N is in reality a measure of *system* efficiency. Thus a poorly designed detector can make the best column look bad. To achieve the high efficiencies reported by some column manufacturers, the entire system must be optimized. Finally, note from equation 5-2 that to enhance resolution twofold, one must increase N fourfold. Thus adjustment of N is normally used to fine tune a relatively good separation. References 2 to 5 contain a more detailed treatment of system efficiency.

The retention of a compound is often measured by its *capacity factor, k′*. The capacity factor is a normalizing factor that allows retention on different-sized columns, or columns operated at different flow rates, to be compared because the k', by definition, is related to the equilibrium of the analyte between the mobile and stationary phases. For liquid chromatography:

$$k' = \frac{V_R - V_o}{V_o} \qquad \textit{Eq. 5-6}$$

where V_R is the retention volume and V_o is the void volume of the column. In Fig. 5-2, the capacity factor for the first peak is 2. The latter corresponds to the retention volume of a compound that does not interact with the column and therefore is not retained. In terms of resolution, k' values in excess of five do not increase resolution much and can, in fact, slow analysis and deteriorate the limit of detection. In liquid chromatography, k' is adjusted mainly by changes in mobile-phase composition though it is also inversely related to the temperature.[2,6]

The difference in retention of two compounds as measured by the ratio of their capacity factors is called the *selectivity,* or α:

$$\alpha = \frac{k'_2}{k'_1} = \frac{V_{R2} - V_o}{V_{R1} - V_o} \qquad \textit{Eq. 5-7}$$

The selectivity of the chromatographic system depicted in Fig. 5-2 for compounds 1 and 2 is 1.5. Note that if α is large there will be a large difference in the retention volumes of the two compounds and we can perform the separation with few plates and little retention. Selectivity is frequently adjusted by changes in solvent composition. One of the major advantages of liquid chromatography is the wide range of selectivity achievable by varying the composition of the mobile phase. For example, an ion-exchange separation depends on the number of charges on the analytes. We can change the selectivity by varying the mobile phase pH, ionic strength, or salt used (NaCl versus LiCl). Prior selection of a mobile-phase system requires an understanding of the separation mechanisms and a great deal of experience. A change in the stationary phase can also be used to adjust selectivity, since the equilibrium achieved between the stationary and mobile phases is the basis of any chromatographic separation. If the retention of the analytes is adequate ($1.5 \leq k' \leq 6$), one can improve the resolution most readily by changing the selectivity.

General elution problem

The chromatographer is sometimes required to separate compounds that, though structurally related, behave quite differently in the separation system chosen. In Fig. 5-4, a separation of the bile acid conjugates demonstrates the problem. When the mobile-phase composition is adjusted to achieve resolution of peaks 2 and 3 (taurochenodeoxycholate and taurodeoxycholate), peak 8 (glycolithocholate) is retained for over 30 minutes. Since there are no other peaks near peak 8, the excessive base line present between peaks 7 and 8 is a waste of valuable analysis time. The alternative to this is to change the mobile-phase composition *during* the chromatographic run. This can be done as a single change from one mobile phase to another *(step gradient)* or as a continuous change in any of a variety of shapes (such as *linear, segmented linear,* or *exponential gradient*). A complete treatment of gradient elution can be found in reference 7. A mobile phase with a constant composition is referred to as *isocratic*. In HPLC, the ability to vary mobile-phase composition can be purchased as a part of the solvent-delivery system. It generally increases the cost of the system significantly.

Speed of analysis

To this point, we have only discussed retention in terms of mobile phase volume, V_R. The reasons for this are that (1) the volume of mobile phase is a direct reflection of cost and (2) changes in the flow rate do not affect the V_R. In contrast, the retention time, t_R, is a function of flow rate and retention volume, that is:

$$t_R = \frac{V_R}{F} \qquad \textit{Eq. 5-8}$$

where F is the flow rate in milliliters per minute. For example, if the diameter of the column were doubled, its volume, and thus the retention volume, would increase by a factor of 4. We could keep t_R constant by increasing the

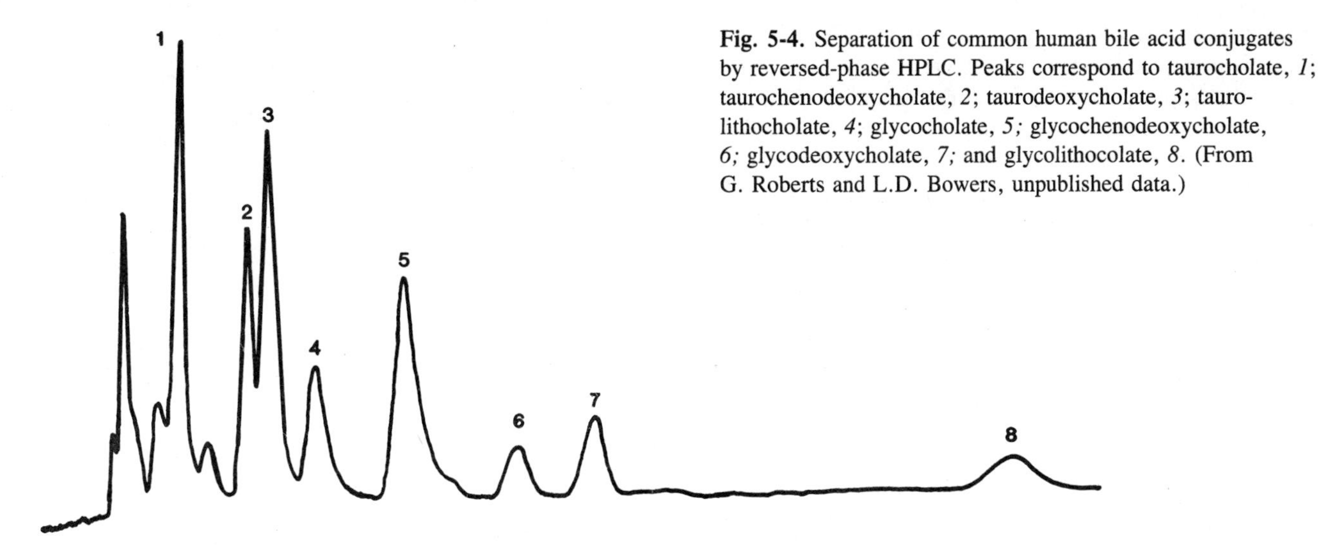

Fig. 5-4. Separation of common human bile acid conjugates by reversed-phase HPLC. Peaks correspond to taurocholate, *1*; taurochenodeoxycholate, *2*; taurodeoxycholate, *3*; taurolithocholate, *4*; glycocholate, *5*; glycochenodeoxycholate, *6*; glycodeoxycholate, *7*; and glycolithocolate, *8*. (From G. Roberts and L.D. Bowers, unpublished data.)

flow rate by a factor of 4, but solvent consumption would increase accordingly.

The optimization of analysis speed depends on a number of factors. The most important relationships in achieving a rapid separation are given in the box below. Guichon has discussed these factors in great detail.[8] The minimum time possible for a separation is the product of the time required for an analyte to pass one plate and the number of plates required for adequate resolution. The smaller the height of a theoretical plate, H, the faster the separation. Retention time is also increased by the need for more resolution, by small selectivities, α, and by large capacity factors, k'. Again, to obtain the fastest separations, a mobile phase must be selected to maximize α at relatively small values of k'. Under optimal conditions for a separation requiring 3000 plates, an analysis time of 100 seconds is feasible with HPLC. One of the problems in translating theory to practice in the clinical laboratory is that the analyte may have to be separated from an unknown metabolite or interferent, and hence α is not known during the development of the separation. In this case, a slower separation time is acceptable and the selectivity of the detector must be relied upon to indicate a potential problem.

Practical considerations in speed of analysis

$$t_R = N \cdot \frac{H}{\mu} \cdot (1 + k')$$

$$L = NH$$

$$\Delta P = NH\mu \cdot \frac{\eta}{K_o}$$

μ, Solvent velocity in the column ($= L/t_o$); η, solvent viscosity; K_o, column permeability. (See reference 8 for more detail.)

QUANTITATION

Approaches

Quantitation in liquid chromatography is achieved when one relates either the peak height or the peak area to the concentration of analyte in the sample. The chromatographic trace is a recording of the concentration of the analyte or analytes as sensed by the detector. At the peak maximum:

$$C_{max} = \frac{C_s V_s}{V_R} \sqrt{\frac{N}{2\pi}} \qquad \textit{Eq. 5-9}$$

C_s and V_s are the concentration and volume of sample injected, V_R is the retention volume, and N is the number of theoretical plates. The greater the peak height for any given concentration, the more sensitive the method will be. Thus minimizing the retention volume and maximizing the number of theoretical plates will result in the most sensitive assay. The factors that decrease the retention volume are a small column void volume and a small capacity factor. Large plate counts are favored by low flow rates and small support-particle diameters. Also note that unlike that of gas chromatography, liquid chromatograhic sensitivity is improved by injection of larger sample volumes. The peak height can be related to the concentration in the sample by a sensitivity factor, S. The sensitivity factor will change if the retention volume changes, making day-to-day operation without standardization difficult but not impossible.

The second approach to quantitation of an analyte is to measure the peak area. Although other methods of integration are available and are described in references 1 to 3, microprocessor-based integrators have become so inexpensive that they are the most reasonable means of measuring peak areas. Most liquid chromatography detectors are concentration dependent; that is, they measure a concentration in the flow cell. (This is in contrast to most GC detectors, which are mass-flow dependent.) The peak area obtained from a concentration-dependent detector is inversely proportional to flow rate. Therefore there can be significant variation in peak area if the flow rate changes during a chromatographic run. In addition, less resolution is required for an equal degree of accuracy when peak heights are used rather than peak areas. The reason is that the overlap of peaks, even with a resolution of 1.0, affects peak area but does not affect the peak maximum.[1] A rough rule of thumb is to use peak *heights* when there are interfering peaks or when maximum accuracy is required, but to use peak *area* when precision is the main requirement. For peaks that are only barely above the base-line noise, peak heights should always be used.

Standardization

Standardization in liquid chromatography can be done in any of three ways: external standardization, internal standardization, and standard addition. For external standardization, a calibration curve is constructed for the peak height (or peak area) values obtained with known concentrations of analyte and a constant injection volume. The slope of the curve is the sensitivity factor, *S*, in peak-height units per concentration unit (such as mm/mM). The concentration of the unknown is then simply its peak height divided by the sensitivity factor. Notice that if the calibration curve is linear, the sensitivity factor can be obtained from a single standard. It would be important in this case to check several control specimens to verify the validity of the sensitivity factor. The principal sources of error in external calibration are variable losses in the preparative steps before LC analysis and sample injection variability. It is thus important to treat the standards and samples in the same way. It should be possible to achieve 1% precision with external calibration, but up to 5% is commonly observed.

Internal standardization utilizes a compound that is usually structurally similar to the analyte to correct for losses and injection imprecision. The same amount of internal standard is added to each sample before sample pretreatment and chromatography. The calibration curve is then constructed for the ratio of the peak heights (or areas) of the standard and the internal standard at various standard concentrations (Fig. 5-5). Again, unknown concentrations can be measured by use of the sensitivity factor or interpolation of the value from the calibration curve. It is hoped that any losses or variations that occur affect the analyte and internal standard equivalently. In practice, this is difficult to achieve. In addition to desired improvements in accuracy and precision, an internal standard can be used as a quality control check, since its peak height should be the same in all chromatograms. There are a number of requirements that must be met in the selection of an internal standard, and these are summarized in the list at right. It should be appreciated that in some cases internal standards do not improve precision and accuracy and may deteriorate precision because of the imprecision involved in measuring two peaks. The precision that can be attained with HPLC has been studied and sources of error have been analyzed.[9]

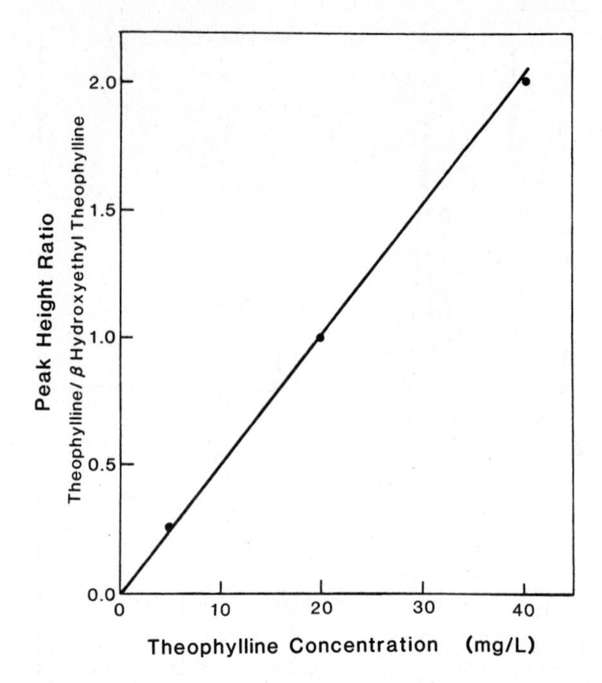

Fig. 5-5. Calibration curve for theophylline using internal standard technique.

The final calibration method, standard addition, requires two analyses to be performed on each sample and thus is not so popular as the other methods are. After a sample is analyzed once, a known amount of the compound of interest is added in a very small amount of liquid so that no dilution occurs, and the sample is reanalyzed. To correct for extraction variability, the addition should be made to the biological fluid and the entire analytical scheme re-

Requirements for internal standard (IS) selection and use

1. The internal standard must be completely resolved from all peaks in the sample.
2. It must be eluted near the analyte, with $k' \pm 30\%$ being preferable.
3. It must behave similarly to the analyte in pretreatment if losses are to be corrected. This may require more than one internal standard.
4. It must have a peak height or area approximately equal to a standard in the concentration range desired.
5. It must not normally be present in the sample.
6. It should be commercially available in pure form.
7. It should be added as a liquid.

peated. Quantitative data can be obtained from the ratio of the peak height (or area) before and after addition of the standard. Addition of the standard can also help to verify the identity of the peak, since the standard must coelute for the unknown to be the same compound. The converse is not true, however, since more than one compound may elute at the same retention volume.

Many additional problems are associated with quantitation when gradient elution is used. Their discussion is beyond the scope of this material. Very careful evaluation of a quantitative gradient elution method is strongly recommended.

SELECTION OF A CHROMATOGRAPHIC MODE

In Chapter 4 the various mechanisms of chromatography were discussed including ion exchange, steric exclusion (gel permeation), adsorption, and partition. As mentioned previously, all these modes are available in liquid chromatography. The selection of the "best" mode is a problem of significant magnitude for the chromatographer. The choice is made based on an understanding of the mechanism of each mode and its strengths and weaknesses, and frequently it is accomplished by experience. Fig. 5-6 illustrates one method of selecting a chromatographic mode using mechanistic considerations as illustrated below.

One of the first considerations of the chromatographer is the size of the molecules to be separated. If the molecules are relatively large (> 100,000 MW), steric exclusion (SEC) is a logical first choice. If, on the other hand, the molecular weights of the compounds of interest are less than 1000, steric exclusion is probably not the mode of choice. Column packings for steric exclusion have pores with a narrow range of diameters. Molecules that fit into the pores have a larger volume (pores and space between support particles) in which to distribute than a larger molecule, which cannot enter the pores. Therefore molecules that are larger than the pores, and thus sterically excluded, are eluted first, whereas those that penetrate all of the pore volume are eluted last. Molecules of intermediate size are eluted between the two extremes according to the fraction of pore volume to which they have access. It is assumed in steric exclusion that there are no interactions between the solute molecules and the support. For this reason, hydrophobic polymers are separated on polar steric exclusion packing materials, whereas hydrophilic macromolecules such as proteins must be separated on a packing with almost no charged groups present. Molecular weight and quantitative information can be obtained with steric exclusion, though the most frequent use is probably protein purification. Another example is the removal of interfering

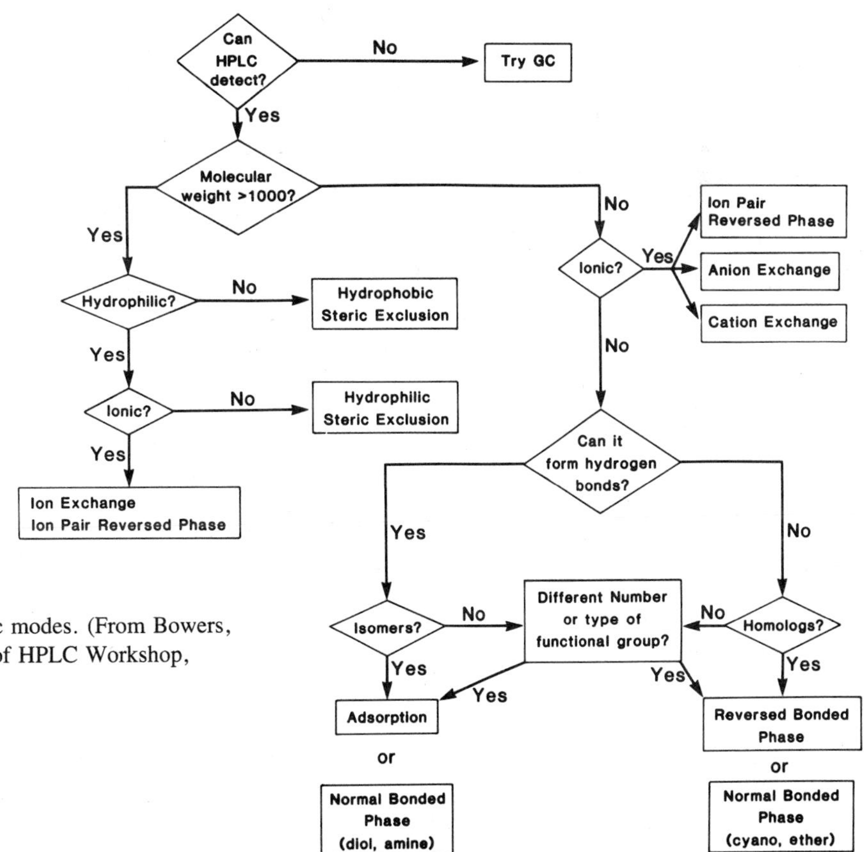

Fig. 5-6. Selection guide for chromatographic modes. (From Bowers, L.D., and Carr, P.W.: Quantitative Aspects of HPLC Workshop, Minneapolis, Minn., 1983.)

endogenous glucose from the enzyme amylase before the colorimetric measurement of amylase activity on the du Pont ACA. Glucose, a small molecule, is trapped in the steric exclusion column while the amylase is eluted into the reaction pack and quantified.

Another key consideration in choosing a separation mode is the existence of an ion or the presence of ionizable groups on the molecule. Ion exchange would be a logical choice for molecules that can be charged regardless of molecular size. Ion exchange is based on the interaction of analyte charges with an oppositely charged group bound to the chromatographic support. One can vary retention by varying pH and ionic strength, or even adding organic solvents such as methanol and ethanol. The greater the number of charges on the analyte, the greater the retention. Increasing ionic strength, and with it the number of charged groups (such as Na^+) competing with the analyte for the exchange sites on the support, reduces retention. The limitation of ion-exchange chromatography is that it does not exhibit high theoretical plate counts, and thus separations are relatively slow. Mobile-phase manipulations in bonded-phase partition chromatography called "ion pairing" (see p. 110) can in many cases accomplish the same separation more efficiently.

If the molecule has a molecular weight less than 1000 and is not ionizable, another characteristic of the compound must be used to achieve the separation. One such characteristic is the ability of the compound to form hydrogen bonds. Adsorption chromatography is based on the interaction of the analyte with a three-dimensional binding site on the support matrix, which may involve hydrogen bonding. Thus the structure and polarity of the solute are important in determining retention. For example, it would be relatively easy to separate *p*-dinitrobenzene from *o*-dinitrobenzene using adsorption chromatography because of the structural differences. On the other hand, separating caproic (C-6) acid from caprylic (C-8) acid would be very difficult because the part of the molecule that interacts with the stationary phase is identical. Compounds are eluted from the column because of competition between the analyte and the solvent for the binding site. A solvent that elutes compounds more rapidly and therefore competes better for the chromatographic sites is called a "strong solvent." Water would be a strong solvent for silica adsorbents because it interacts strongly with the silanol (SiOH) groups responsible for the adsorption mechanism. A solvent that does not compete well for sites is called a "weak solvent." Hexane would be a weak solvent for silica adsorbents. One can adjust solvent strength by using mixtures of solvents. Snyder[10] has developed a scale of solvent strength called the *eluotropic series* and has extended the solvent strength theory to binary and ternary[11] mixtures. Interestingly, solvent mixtures of the same strength can show differences in selectivity because Snyder's theory considers only the adsorption process whereas in reality the solute can also interact with the solvent. One final consideration in adsorption chromatography is the control of the "activity" of the adsorbent. Silica contains several types of silanol groups, which interact differently with various types of compounds. A small amount of water, methanol, acetonitrile, isopropanol, or other strong solvent is used to block the strongest sites and give a more reproducible adsorption surface. This problem has caused some chromatographers to avoid adsorption systems, probably unnecessarily. A separation that involves nonpolar to intermediate polarity compounds that can form hydrogen bonds or have isomeric components will probably be easily achieved with adsorption chromatography.

The interactions between solute and stationary phase in partition chromatography are not nearly so well defined as those for adsorption systems. Any chemical forces that exist between molecules can be used in partition-based separations, including hydrogen bonding, van der Waals' forces, ion-ion interactions, and so on (see Chapter 4). The basis of partition systems is the distribution of a solute between two liquid solvent layers, one stationary and the other mobile. It is essentially analogous to thousands of extractions taking place in a column. In classical partition chromatography, a polar liquid such as β,β′-oxydipropionitrile (β,β′-ODPN) was coated onto the support particles and an immiscible solvent such as hexane was used as the mobile phase. Since solvent-solute interactions involve the entire molecule, a typical partition system as given above would separate a homologous series (C-6, C-7, and C-8 carboxylic acids) and some positional isomers. A partition system with a polar stationary phase and a nonpolar mobile phase is called *normal phase* because it was the first type of system developed. Later, a system with a nonpolar stationary phase, such as squalane, and a polar mobile phase, such as a water-acetonitrile mixture, was developed and called *reversed phase*. Both types of liquid-liquid partition chromatography had many problems related to the finite solubility of all solvents in each other. For example, the β,β′-ODPN would slowly dissolve in the hexane and slowly change the amount of stationary phase, which in turn would change the retention volumes of the analytes. The temperature had to be very closely controlled, and gradient elution was impossible with partition chromatography. These problems made partition chromatography difficult to use in the clinical laboratory.

All of this changed with the development of chemically bonded stationary phases. These materials are prepared by covalent bonding of an organic moiety onto the surface of the support particle, usually silica. The organic groups include nonpolar functions such as octadecylisilane (ODS or C-18) or octylsilane (C-8) and polar groups such as cyanopropyl (CN), aminopropyl (NH_2), or glycidoxypropyl (diol) silanes. The advantages of bonded phases are that (1) polar and ionic compounds are readily separated, (2) the stationary phase does not strip off, (3) gradient elution can be used, and (4) the columns are easy to use and take care of. The main disadvantage of bonded phases is that at pH's below 2 the bonded group is cleaved from the support and at pH's above 8 the silica support particles dissolve. Since most separations for clinical laboratory applications can be performed within the workable pH window, bonded phases are very popular. It has been estimated that 90% of all separations are now performed on reversed-phase (such as C-18) chemically bonded columns. A separate section has been included to discuss the variety of separations feasible with reversed-phase systems.

As stated on p. 107, selection of a chromatographic system depends on an understanding of the strengths and weaknesses of the various separation modes and some experience. It is possible that all separations of interest in a laboratory could be performed on one type of column (such as reversed phase). This, however, would require considerable mastery of mobile phase modification and might result in a separation inferior to that obtainable with relative ease on another type of column. In summary, there is no "best" way to solve a chromatographic problem.

BONDED REVERSED-PHASE CHROMATOGRAPHY

The most popular type of bonded-phase support is the octadecylsilane (ODS or C-18) reversed-phase (RP) column packing. The reason for this popularity is the ease with which polar biological molecules are separated. The mobile phase in reversed-phase chromatography is a polar solvent such as water, and therefore the polar or ionic molecules are readily soluble in the mobile phase. The stationary phase is a nonpolar octadecyl-bonded phase. If a molecule has a nonpolar part, as shown for tyrosine in Fig. 5-7, it can interact with the nonpolar stationary phase while the polar part interacts with the polar mobile phase. Molecules with a small nonpolar area do not interact very

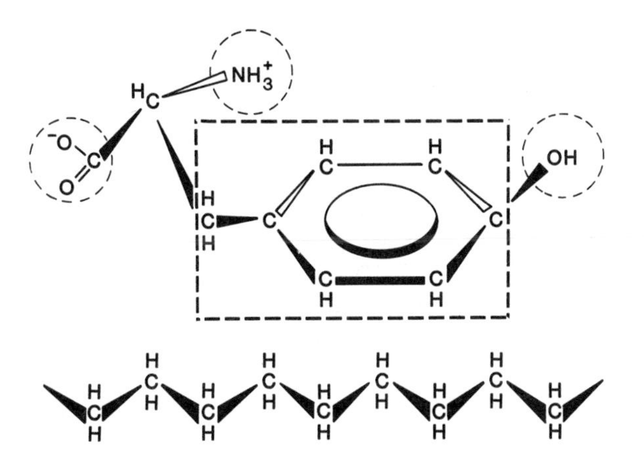

Fig. 5-7. Retention in reversed-phase chromatography is result of interaction of nonpolar portion of compound such as tyrosine *(enclosed in box)* with nonpolar stationary phase. Hydrophilic groups *(circled)* tend to decrease retention.

strongly with the stationary phase, but they spend most of their time in the mobile phase and thus are rapidly eluted from the column. Molecules with a large nonpolar area interact strongly with the stationary phase and may not be eluted with water. In this case, a less polar solvent such as methanol, acetonitrile, or tetrahydrofuran would be mixed with the water so that the mobile phase can compete with the stationary phase for the nonpolar part of the molecule. The interactions involved are not competition for a specific site as in adsorption chromatography, but rather competition in terms of solubility. Thus nonpolar molecules can be eluted from the column if a sufficient amount of organic solvent is added to the eluent. A detailed theory of reversed-phase chromatography can be found in references 12 to 14. In summary, polar molecules are eluted first and nonpolar molecules are eluted last. The selectivity and elution volume are determined by the mobile-phase conditions including pH, ionic strength, and percentage of organic solvent.

Stationary-phase considerations

As noted above, the separation obtained for any group of analytes is a function of both the stationary phase and the mobile phase. The analyst has the ability to vary mobile-phase conditions but is dependent on manufacturers for the stationary phase, particularly for bonded-phase packings. If one is to do reproducible chromatography, the behavior of the stationary phase toward nonpolar, polar, and ionic compounds must be the same from column to column and lot to lot. In addition, the durability of the column is important, particularly since columns cost $250 to $350 each. Significant improvements in these features have occurred in recent years, but there are some limitations in producing a column with absolutely reproducible reversed-phase column packings.

The preparation of an octadecylsilane (ODS) stationary phase is usually accomplished by reacting the silanol (SiOH) groups on the silica gel with an octadecylsilane such as octadecyldimethylchlorosilane. The resulting surface contains octadecyl groups bound to the surface by siloxane (Si-O-Si) bonds as shown in Fig. 5-8. Most manufacturers of columns use silanes with only one chloride group and the resulting stationary phase is called "monomeric." Because of the stereochemistry of the silica gel surface, only about one third of the silanol groups can react with the ODS groups. The remaining silanol groups are polar and can interact with polar analytes, changing the selectivity of the stationary phase. Trimethylchlorosilane can react with about an additional 20% of the surface silanol groups with a resultant increase in nonpolar character of the support (Fig. 5-8). This process is called "end capping" and is used in many commercially available packing materials. As might be expected, differences in the surface morphology of the silica gel, reaction conditions in the ODS-bonding step, and the presence or absence of end

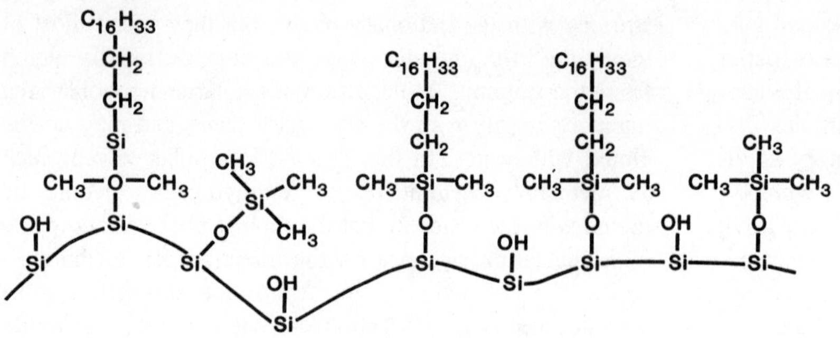

Fig. 5-8. Schematic representation of a silica-based octadecyl reversed-phase support that has been end capped. Note presence of residual silanol groups on surface.

capping make columns purchased from different manufacturers perform differently. In fact, variations from lot to lot of packing material may be quite noticeable in the separations obtained for certain analyses. Thus it is not surprising that in adapting a method to a laboratory one may require significant changes in the mobile phase if a C-18 column from a manufacturer other than that in the original report is used.[15] Choice of a reversed-phase column still requires trial and error. Several good reviews of columns in biomedical applications have appeared.[16,17]

Mobile-phase considerations

As mentioned above, the real power in liquid chromatography arises from the fact that changes in mobile-phase composition can have major affects on selectivity and thus on resolution. In reversed-phase systems, there are two types of changes that can be made: (1) type of organic solvent, (2) additions to the mobile phase that affect its pH, ionic strength, or complexing ability. It has been recognized for some time that the type of organic solvent used has an effect on retention and selectivity.[14] In some cases, a change from methanol to acetonitrile actually alters the elution order of the compounds. Unfortunately at this time the use of solvents to vary selectivity is largely empirical[18] and their exact role is poorly defined. Their relative strength is well defined with solvent strength and therefore the ability to elute solutes, increasing in the following order: methanol < DMSO < ethanol ≤ acetonitrile < tetrahydrofuran < dioxane < isopropanol. As a rule of thumb, when one is adjusting the retention of a solute, a 10% increase in the fraction of organic solvent in water causes a two- or threefold decrease in k' value.

Ion suppression. Mobile-phase modifications that change retention by introducing a second chemical equilibrium process in the mobile phase have been used since the inception of reversed-phase chromatography.[19] The first approach was control of pH to affect retention. If, for example, ascorbic acid was to be separated by HPLC, there would be a strong influence of pH on retention. At pH's above the pK_a of ascorbic acid, the acid would be deprotonated and charged and therefore would not partition itself strongly into the nonpolar stationary phase. Retention would be relatively low. On the other hand, at pH's below the pK_a the acid would be protonated and uncharged and so would be much more strongly retained. In the area about one pH unit on either side of the pK_a, the retention changes rapidly as a function of pH, as shown in Fig. 5-9. The use of pH control to increase retention for acids has been termed "ion suppression." It is a very useful method of adjusting retention behavior. Buffers are normally used to control the pH. It is important to remember that an acid-base pair is only a good buffer within one pH unit of the pK_a. Table 5-2 lists some useful buffers for reversed-phase HPLC.

Paired-ion chromatography. Since silica-based reversed-phase materials can only operate in the pH range of 2 to 8, many basic compounds and strong bases cannot be handled by ion suppression. In these cases, the charge on the analyte can be neutralized by addition to the mobile phase of an oppositely charged ion that will form an ion pair with the analyte. This neutral pair will be more strongly retained than the analyte alone. The more hydrophobic the counterion, the greater the increase in retention. For example, the catecholamines, which are weak bases, have their retention increased by using ClO_4^-, pentane sulfonate, and octane sulfonate. If equimolar amounts of each were tested, octane sulfonate–modified mobile phases

Table 5-2. Useful buffers for reversed-phase high-performance liquid chromatography

Buffer pair	pK_a
Phosphoric acid/dihydrogen phosphate	2.12
Chloroacetic acid/chloroacetate	2.87
Succinic acid/monohydrogen succinate	4.23
Acetic acid/acetate	4.77
Piperazine phosphate	5.33
Monohydrogen succinate/succinate	5.65
Tetramethylethylenediamine phosphate	6.13
Dihydrogen phosphate/monohydrogen phosphate	7.20
Tris(hydroxymethyl)-aminomethane	8.19

would have the greatest retention as a result of the nonpolar octyl group. Perchlorate ion–containing mobile phases would have the least increase in retention. In addition to the hydrophobicity of the ion-pairing agent, its concentration in the mobile phase also is important. As the concentration is increased from zero, there is a linear relationship between increased retention and concentration as shown in Fig. 5-10. At higher concentrations of counterion, the retention remains constant despite increased concentrations of pairing ion. Thus selection of the type and concentration of pairing ion, as well as the pH and concentration and type of organic modifier or modifiers, allows rather precise adjustment of the retention of compounds of interest. Strong acids can have their retention modified by addition of tetraalkylammonium salts such as tetraethylammonium perchlorate or amine buffers. The theory associated with ion-pair chromatography (IPC), though still controversial, is discussed in references 19 to 21. The advantages of ion-pair chromatography are threefold: (1) both ionic and polar compounds can be separated in one chromatographic system, (2) it is more efficient (that is, N is greater) than ion-exchange separations, and (3) the separation can be controlled by mobile-phase modifications. The major disadvantage of IPC is that in some cases the ion-pairing reagents permanently modify the column.

Other mobile-phase modifiers. It should be apparent that anything that is added to the mobile phase that makes the analyte less polar will increase its retention on a reversed-phase column. An additive that makes the analyte more polar reduces retention. The addition of silver ions to the mobile phase results in the selective complexation of compounds containing double bonds, making them more hydrophilic. Thus selectivity between compounds differing only in the number of double bonds (such as prednisolone and cortisol) can be achieved by use of argentation chromatography (that is, Ag^+). The use of cupric, zinc, nickel, and cadmium ions, often in the form of a hydrophobic chelate, has also been reported with great success. For example, the selectivity, α, for the *cis*- and *trans*-dicarboxylic acids fumeric acid and maleic acid, respectively, increased from 0.9 to 2.5 upon addition of a Zn(II)-4-dodecyldiethylenetriamine chelate to the mobile

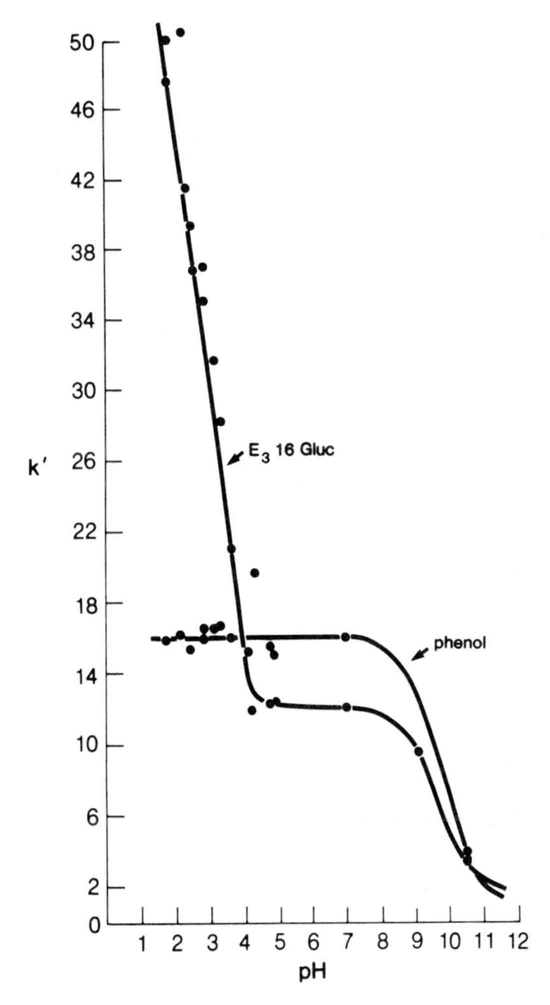

Fig. 5-9. Change in k' as a function of pH for estriol-16α-glucuronide and phenol. Decrease in k' at pH 2 is attributable to ionization of glucuronic acid; decrease at pH 10 is attributable to ionization of phenolic group. (From C. Oliphant and L.D. Bowers, unpublished data.)

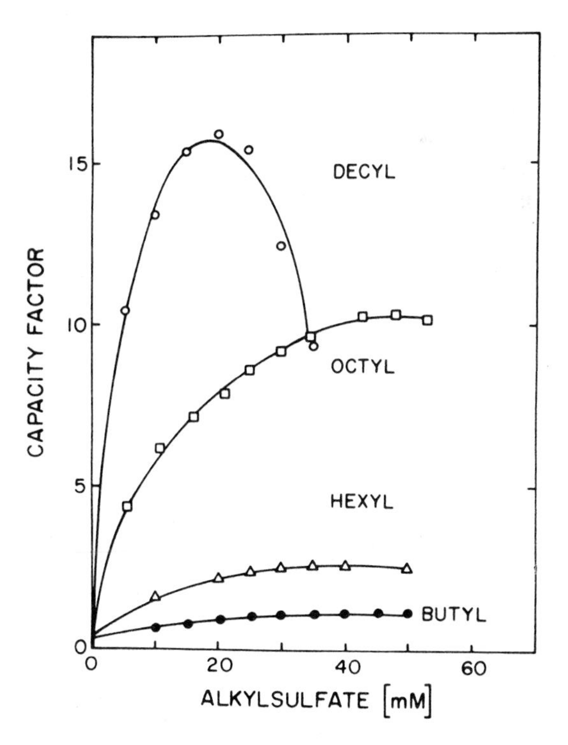

Fig. 5-10. Variation of capacity factor, k', of epinephrine as a function of ion-pairing reagent concentration for several *n*-alkylsulfonates. (From Horvath, C., et al.: Anal. Chem. **49**:2295, 1977.)

phase. This was the result of the formation of a complex between the dicarboxylic acid and the metal chelate. In summary, anything the chromatographer can do to change the retention of one compound relative to another improves selectivity. The advantage of liquid chromatography is that mobile-phase modifications are relatively easy to make and are limited in selectivity only by the chemical insights of the chromatographer.

INSTRUMENTATION

As mentioned previously, modern HPLC requires relatively sophisticated instrumentation to achieve difficult separations in less than 10 minutes. There are four basic components in the separation system itself: a solvent-delivery system to provide the driving force for the mobile phase, a sample introduction system, the column, and the detector (refer to Fig. 5-1). In addition, a recorder or integrator must be used to display or calculate the results, as discussed earlier.

Solvent-delivery systems

The most common delivery system is based on the reciprocating piston pump. Other types including pneumatic amplification, syringe, and diaphragm pumps are mainly of historical interest and are discussed in references 1 to 3.

In the first pumps used in HPLC, solvent delivery occurred during less than half the cycle time. This meant that flow through the column was erratic. The stoppage of flow and the compression of the solvent that occurred when the pump head refilled also resulted in a signal at the detector, which limited the detection of analyte. In the jargon of chromatography, the stoppage is known as a "pulse." Subsequently, manufacturers have used a variety of designs to minimize pulsation. These range from a very rapid refill stroke relative to the delivery stroke, to two or more pump heads that are out of phase such that one head is always delivering solvent. There are also mechanical devices known as "pulse dampeners" that work quite well under isocratic conditions with all but the most crude pumps. It must be emphasized, however, that pulses can be minimized in a reciprocating piston pump; they can never be eliminated. An excellent monograph on the care and maintenance of HPLC equipment has been published.[22]

Gradient elution can be attained by mixing of the solvents after they have passed through the pumps or by mixing of them before they enter the pump. A well-maintained system of either type can function well, and selection of one or the other is based on gradient reproducibility, gradient shapes available, cost, and operator preference.

Sample-introduction systems

The most widely used method of introducing a sample into the chromatographic system is the fixed-loop injection valve. A sample aliquot is loaded into an external loop of stainless steel tubing. The valve is then rotated so that the sample loop is flushed onto the column by the mobile phase from the pump. Returning the valve to the original position allows loading of the next sample. Fixed-loop valves can be used in two ways: partial-fill method and full-loop method. In the latter the entire loop (such as 20 μL) is filled with sample and injected. It is the most precise method. One should recognize, however, that accurate results require flushing of the loop with five to ten loop volumes before loading. In the partial-fill mode, the sample loop is not filled with sample. In this case, the precision of the injection volume is determined by the loading syringe.

Columns and connectors

The column is of course the most important part of the separation system. The packing material for columns has been discussed at length. Packing a high-efficiency column is still an art. The column tube itself is usually of no. 316 stainless steel with an inside diameter of between 2.1 and 4.6 mm and a length of 50 to 1500 mm. Column end fittings are required at both ends to connect the column to the other system components and to hold the packing material inside. The frits used should have pores less than one fourth the average diameter of the packing material. In addition to the analytical (or preparative) column, which actually performs the separation, two types of protector columns might be used. A *guard column* (Fig. 5-1) is located between the injector and the analytical column and is 1/15 to 1/25 the volume of the latter. It is packed with a material similar to that of the main column. Its function is to collect any particulate matter or any strongly retained components of the sample and therefore protect the expensive analytical column. A *precolumn* is positioned between the pump and the injection valve. It is always packed with silica, the purpose of which is to saturate the mobile phase with silicate and thus prevent dissolution of the packing material in the analytical column. This has reportedly allowed operation of silica columns at pH's of 10, far beyond the normal pH for dissolution of silica. Rabel[23] has discussed the use of protector columns at length.

The analytical column has been described previously. The high efficiencies obtained with current HPLC columns result from the use of small (3, 5, or 10 μm), totally porous particles. To achieve efficient columns and reasonable operating pressures, the range between the largest and smallest particle must be as small as possible. Both spherical and irregularly shaped particles are available. Columns packed with spherical particles seem to give lower operating pressures and are more durable.

In addition to the columns, the connections made between system components are critical because the fittings should introduce as little peak spreading as possible. They should be of the zero dead volume (ZDV) or at least low dead volume (LDV) type.

Table 5-3. Performance characteristics of commonly used detectors

Ideal detector characteristic	Fixed wavelength	Variable wavelength	Fluorescence	Electrochemical	
				Oxidation	Reduction
Selective?	No	No	Yes	Yes	Yes
Sensitivity to flow-rate changes?	Possibly	Possibly	No	Yes	Yes
Limit of detection?	10^{-10} g/L	10^{-9} g/L	10^{-11} g/L	10^{-12} g/L	10^{-9} g/L
Cell volume?	8-10 μL	8-10 μL	20 μL	≤ 1 μL	
Compatible with gradient?	Yes	Yes	Yes	?	

Detection systems

The final component in the chromatographic system is responsible for detecting the compounds as they elute from the column. Ideally it would respond to any compound, would detect picograms or less of the analytes, would be immune to any solvent-related phenomena, and would respond linearly to a wide range of concentrations. Unfortunately, liquid chromatography does not have such a detector. In the clinical laboratory, three types of detectors have been used—photometers, fluorometers, and electrochemical devices. The selection of the appropriate detector will depend on the required selectivity and sensitivity. The performance characteristics of the common LC detectors are summarized in Table 5-3.

Absorbance detection. Ultraviolet and visible photometers are used in the same manner as other photometers, with quantitation based on Beer's law. The main difference is that because of the small size of the flow-through cell (5 to 10 μL) intense light sources are required to get a reasonable amount of light through the cell. The most common source is the mercury-arc lamp, which has an intense emission at 254 nm. A relatively inexpensive dual-beam, fixed-wavelength filter photometer can thus be constructed and is simple, easy to use, and relatively rugged. Absorbance changes as small as 10^{-5} absorbance units can be detected. Fortunately, most compounds that contain an aromatic ring absorb at 254 nm, but the detector is far from universally sensitive. For compounds that do not absorb at 254 nm, there are two alternatives. Most manufacturers make available other lamps or lamp/phosphor systems that have selected emissions in the wavelength range from 280 nm to 660 nm (such as 280, 313, 365, 405, 436, 546, 578, 660 nm). More recently, zinc and cadmium hollow-cathode lamps, which have emissions at 214 and 229 nm respectively, have been used. The use of a lower wavelength allows more compounds to be detected.

The second alternative is a variable-wavelength photometer where the wavelength is chosen by a monochromator from the spectra emitted by a deuterium or tungsten-halide lamp in the ultraviolet or visible region respectively. The advantage here is that the wavelength of maximum absorption can be selected. On the other hand, the diminished intensity of these light sources generally makes the limiting base-line noise a factor of five to ten worse than a fixed-wavelength detector. A second advantage is that these detectors can be operated at 190 nm where even sugar moieties such as glucose absorb light. Unfortunately, since just about everything absorbs light below about 205 nm, it is very difficult from a practical viewpoint to use this region of the spectrum. Even the widely used reversed-phase solvents, acetonitrile and methanol, have an absorbance of 1.0 at 190 and 205 nm respectively. Of course, as they are diluted, the absorbance decreases, but they can absorb a great deal of the available light in the far ultraviolet and thus cause detection-limit problems. One final characteristic of most photometers is that they respond to refractive-index changes that occur because of pressure changes (pulses), temperature, or solvent changes (such as a gradient). Well-designed photometers and flow cells, to a large extent, eliminate this problem (which can affect the detection limit of the detector and its utility in gradient work) and are probably worth the greater capital investment.

Fluorescence detection. The technique of fluorescence is based on the ability of a molecule to emit light after it has been excited by light radiation. For a description of fluorescence spectroscopy, refer to Chapter 3. The main differences in fluorometer performance between manufacturers arise from differences in the lamp intensity and the detection efficiency. Commercially available instruments use emission from either deuterium or xenon arc lamps for exciting light. Since the fluorescent intensity is directly proportional to the excitation light intensity, there has been a great deal of interest recently in laser sources. Fluorescence is a highly sensitive detection method for those compounds that fluoresce. Since many molecules do not fluoresce, numerous methods for derivatizing compounds have been developed. Epinephrine and other amine compounds are reacted with dimethylaminonaphthalenesulfonyl (dansyl) chloride or *o*-phthalaldehyde to produce highly fluorescent compounds.

Electrochemical detection. Although a variety of electrochemical detector types have been reported, amperometric detectors are the most widely used. In an amperometric cell, the column effluent flows past an electrode to which a voltage has been applied. If the voltage is sufficiently large, the analyte molecules at the interface between the electrode and the solution can either accept electrons and be reduced or give up electrons and be oxidized. The net movement of electrons causes a current to flow, and the current is proportional to the concentration of analyte at the interface, as in the following:

$$i = hnFA(C - C_o) \qquad \text{Eq. 5-11}$$

where i is the current, h is a measure of the ability of the analyte to reach the electrode surface, n is the number of electrons involved in the oxidation or reduction, F is Faraday's constant, A is the electrode area, and C and C_o are the analyte concentrations in solution and at the electrode surface, respectively. The reason that several detector designs have been developed is in an effort to have h, and therefore the current, as large as possible. The two most popular designs, the thin layer or channel cell and the wall jet cell, are shown in Fig. 5-11. Although there are theoretical advantages and disadvantages, these two types of cells are comparable in performance. In addition to the variety of cell designs, many electrode materials have been used, including carbon paste, glassy carbon, gold, and mercury. The reasons for the number of electrode materials are their compatibility with HPLC solvents and the voltages that can be applied to them. Carbon paste and glassy carbon have a potential range useful for oxidation of compounds such as the catecholamines. Mercury has a range most useful for reductions.

As discussed in Chapter 9, the oxidation or reduction of a compound occurs in a distinct potential range. Thus electrochemical detection can be quite selective. For oxidative (or anodic) processes, amperometric detection is very sensitive. As little as 10 pg of epinephrine can be detected. For cathodic reactions, a number of problems associated with oxygen reduction restrict the limit of detection to the nanogram range. To make the reductive mode more sensitive, all oxygen would have to be removed from the mobile phase. Another advantage of amperometric detectors is the small cell volume, which can be smaller than 1 μL.

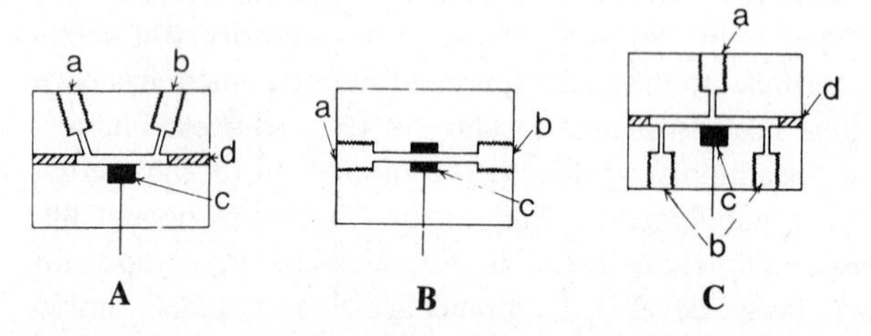

Fig. 5-11. Schematic diagram of thin layer or channel, **A**, tubular, **B**, and wall-jet, **C**, electrochemical flow cell designs. (From Weber, S.G.: I & EC Product Research and Development **20**:593, 1981.)

SUMMARY

Liquid chromatography has become an integral part of the analytical capability of the clinical laboratory. The evolution of high-performance liquid chromatography was motivated, in part, by the interest expressed by clinical chemists in separating and measuring the concentrations of drugs and their metabolites in body fluids. These early applications took advantage of the speed of HPLC for compounds that were easily measured by the standard detectors (254 nm). It has become apparent, subsequently, that HPLC has several limitations for clinical application: (1) a significant amount of sample clean-up may be required; (2) there is no sensitive detector that responds to all compounds; (3) there are limited applications of selective detectors that are highly sensitive. The sample preparation problem is being addressed in a number of ways, including off-line preparation with solid-phase extractants such as Bond-Elut, and Sep-Paks, or automated systems such as the du Pont Prep I, and on-line extractions such as the Technicon Fast-LC system. Although there is nothing at present to fill the universally sensitive detector void, on-line derivatization systems to enhance detection are being investigated. Recently two commercial derivatization systems have been introduced, the Kratos Post Column Reaction System and the Varian System I-III post-column reactor. Refer to Lawrence and Frei's monograph[23] and to Knapp's compilation of methods[24] to appreciate the advantages and disadvantages of derivatization. A great deal of research being done will have an impact on the limits of detection in HPLC, and before long it should be possible to reach the picogram range for a number of compounds of clinical interest. HPLC has come a long way from being a tool limited to the drug analysis laboratory. As summarized in Table 5-4, liquid chromatography has permeated every area of the laboratory for preparative, qualitative identification, and quantitative separations. The table is not meant to be comprehensive, but a representative sampling in terms of equipment requirements and areas of application.

Users of high-performance liquid chromatography have enjoyed improved column reliability and efficiency to the point that commercially available columns are as good as "research-grade" columns. The user must be aware of the inevitable differences between different types, and even different lots, of packing materials, but even this situation has improved. Solvent-delivery systems that are both sophisticated and reliable are available for routine use. Presently available detection systems are also reliable and relatively easy to maintain. Improvements in the performance of instrumentation, particularly detectors, will be required before any further enhancement of speed is obtained. Nevertheless, HPLC is at present a valuable asset to the clinical laboratory.

Table 5-4. Sample of HPLC usage in the clinical laboratory

Analyte	Sample matrix	Chromatographic mode	Mobile phase*	Detection system†
Antiarrhythmic drugs (procainamide, lidocaine, quinidine, disopyramide, propranolol)	Serum	Reversed phase	I	UV (254), fluorescence
Anticonvulsant drugs (phenobarbital, phenytoin, ethosuximide, primidone, carbamazepine)	Serum	Reversed phase	I	UV (200)
Amino acids	Urine	Reversed phase	G	Fluorescence, dansyl derivatives
Bile acids	Serum	Reversed phase	G	Fluorescence, enzyme reduction (NADH)
Bilirubin	Serum	Reversed phase	I	Vis (403)
Carbohydrates (from fungal infections)	Serum	Normal bonded phase (NH_2)	I	UV (200)
Chloramphenicol	Serum	Reversed phase	I	UV (278)
Cholesterol	Serum	Reversed phase	I	UV (205)
Cortisol	Serum	Reversed phase	I	Fluorescence
Creatinine	Serum	Reversed phase	I	UV (238)
Creatinine kinase isozymes	Serum	Ion exchange	G	Chemiluminescence
1,25-Dihydroxyvitamin D	Plasma	Normal bonded phase (NO_2)	G	CPB
Epinephrine	Plasma, urine	Reversed phase	I	Electrochemical
Estriol	Urine	Reversed phase	I	Fluorescence
Gentamicin	Serum	Reversed phase	I	Fluorescence, *o*-phthalaldehyde derivative
Hemoglobin A_{1c}	Red blood cell hemolysate	Ion exchange	G	Vis (403)
Homovanillic acid	Urine	Reversed phase	I	Fluorescence, *o*-phthalaldehyde derivative
5-Hydroxyindole acetic acid	Urine	Reversed phase	I	Electrochemical
17α-Hydroxyprogesterone	Serum	Reversed phase	I	UV (254)
Metanephrines	Urine	Reversed phase	I	Fluorescence, *o*-phthalaldehyde derivative
Norepinephrine	Plasma, urine	Reversed phase	I	Electrochemical
Porphyrins	Urine	Reversed phase	G	Fluorescence
Prednisone, prednisolone	Serum	Reversed phase	I	UV (254)
Proteins	Serum	Ion exchange	I	UV (254), *p*-bromophenacyl derivative
Serotonin	Serum	Reversed phase	I	Electrochemical
Tricyclic antidepressants	Serum	Adsorption	I	UV (254)
Vitamin A and E	Serum	Reversed phase	I	UV (290)

*Type of elution system used: *I*, isocratic; *G*, gradient.
†*CPB*, Competitive protein binding; *UV*, Ultraviolet photometry; *Vis*, visible light; numbers in parentheses indicate wavelength.

REFERENCES

1. Snyder, L.R., and Kirkland, J.J.: Introduction to modern liquid chromatography, ed. 2, New York, 1979, John Wiley & Sons, Inc., pp. 16-82.
2. Willard, H.H., Merritt, L.L., Dean, J.A., and Settle, F.A.: Instrumental methods of analysis, ed. 6, New York, 1981, Van Nostrand Reinhold Co., pp. 430-453; 495-564.
3. Johnson, E.L., and Stevenson, R.: Basic liquid chromatography, Palo Alto, Calif. 1978, Varian Associates.
4. Karger, B.L., Snyder, L.R., and Horvath, C.: An introduction to separation science, New York, 1973, John Wiley & Sons, Inc.
5. Giddings, J.C.: Dynamics of chromatography, New York, 1965, Marcel Dekker, Inc.
6. Majors, R.E.: In Grushka, E., editor: Bonded stationary phases in chromatography, Ann Arbor, Mich., 1974, Ann Arbor Science Publishers.
7. Snyder, L.R.: Gradient elution. In Horvath, C., editor: High performance liquid chromatography: advances and perspectives, vol. 1, New York, 1980, Academic Press, Inc., pp. 207-316.
8. Guichon, G.G.: In Horvath, C., editor: High performance liquid chromatography: advances and perspectives, vol. 2, New York, 1981, Academic Press, Inc., pp. 1-56.
9. van der Wal, S.J., and Snyder, L.R.: Precision of high-performance liquid chromatographic assays with sample pretreatment: error analysis for the Technicon Fast-LC system, Clin. Chem. **27:**1233-1240, 1981.
10. Snyder, L.R.: Principles of adsorption chromatography, New York, 1963, Marcel Dekker, Inc.
11. Snyder, L.R., Glajch, J.L., and Kirkland, J.J.: Theoretical basis for a systematic optimization of mobile phase selectivity in liquid-solid chromatography, J. Chromatogr. **218:**299-326, 1981.
12. Horvath, C., Melander, W., and Molnar, I.: Solvophobic interactions in liquid chromatography with nonpolar stationary phases, J. Chromatogr. **125:**129-158, 1976.
13. Horvath, C., and Melander, W.: Liquid chromatography with hydrocarbonaceous bonded phases: theory and practice of reversed phase chromatography, J. Chromatogr. Sci. **15:**393-404, 1977.
14. Atwood, J.G., and Goldstein, J.: Testing and quality control of a reversed phase column packing, J. Chromatogr. Sci. **18:**650-654, 1980.
15. Karger, B.L., and Giese, R.W.: Reversed phase liquid chromatography and its application to biochemistry, Anal. Chem. **50:**1048A-1073A, 1978.
16. Majors, R.E.: Practical operation of bonded-phase columns in high-performance liquid chromatography. In Horvath, C., editor: High performance liquid chromatography: advances and perspectives, vol. 1, New York, 1980, Academic Press, Inc., pp. 75-111.
17. Glajch, J.L., Kirkland, J.J., Squire, K.M., and Minor, J.M.: Optimization of solvent strength and selectivity for reversed-phase chromatography using an interactive mixture-design statistical technique, J. Chromatogr. **199:**57-79, 1980.
18. Karger, B.L., LePage, J.N., and Tanaka, N.: Secondary chemical equilibria in high-performance liquid chromatography. In Horvath, C., editor: High performance liquid chromatography: advances and perspectives, vol. 1, New York, 1980, Academic Press, Inc., pp. 113-206.
19. Horvath, C., Melander, W.R., Molnar, I., and Molnar, P.: Enhancement of retention by ion-pair formation in liquid chromatography with nonpolar stationary phases, Anal. Chem. **49:**2295-2305, 1977.
20. Melin, A.T., Askemark, Y., Wahlund, K.G., and Schill, G.: Retention behavior of carboxylic acids and their quaternary ammonium ion pairs in reversed phase chromatography with acetonitrile as organic modifier in the mobile phase, Anal. Chem. **51:**976-983, 1979.
21. Walker, J.Q., Jackson, M.T., Jr., and Maynard, J.B.: Chromatographic systems: maintenance and troubleshooting, New York, 1957, Academic Press, Inc.
22. Rabel, F.M.: Use and maintenance of microparticulate high performance liquid chromatography columns, J. Chromatogr. Sci. **18:**394-408, 1980.
23. Lawrence, J.F., and Frei, R.W.: Chemical derivatization in liquid chromatography, Amsterdam, New York, 1976, Elsevier/North-Holland, Inc.
24. Knapp, D.R.: Handbook of analytical derivatization reactions, New York, 1979, John Wiley & Sons, Inc.—Interscience.

Chapter 6 Gas chromatography

Alphonse Poklis

active sites Places, usually on the stationary phase, that reversibly bind the compound to be separated.

capacity ratio The ratio of sample weight in the stationary phase to that in the mobile phase.

chemical ionization The component molecule to be analyzed is mixed with an ionized gas, such as methane or isobutane. A positive charge is transferred to the molecule, the M + 1 charged molecule and its fragments are separated by the mass spectrometer, and their size and relative abundance are measured. (''M'' is 'mass.')

chromatogram A plot of the response of the detector versus time. This reflects the separation of the components of a solution and their relative amounts.

corrected retention time The amount of time a compound is retained on a column minus the gas holdup time.

derivative A molecule chemically altered from the original one. Usually in gas chromatography it refers to chemical groups added to increase the volatility of the initial compound.

detector A device that responds to the presence of a compound, usually in a proportional or linear manner.

diffusivity The ability of a molecule to diffuse or spread because of the thermal energy inherent in the molecule.

effective plate (number of effective plates) The number of partitions that are practically available on a column.

efficiency The ability of a column or chromatographic system to produce narrow elution peaks.

electron-capture detector A device that releases beta particles into the carrier gas stream, producing low-energy electrons that are collected on electrodes and measured. Halogen atoms collect the low-energy electrons, decreasing the current in a manner proportional to their concentration. This type of detector is very sensitive and specific for compounds with chemical groups of high electronegativity, such as halogens.

electron impact Fragmentation of molecules into specific charged fragments by collision with high-energy electrons.

flame ionization detector A device in which eluted components are mixed with hydrogen and burned in the air to produce a flame, which ionizes these components. A pair of electrodes measures the number of ions.

flow regulator A system of valves that is set to yield a desired gas pressure and thus control the rate of gas movement (flow) through a gas chromatographic system.

free acid The nonsalt form of an acid, such as, for carboxylic acids, the form R-COOH.

free base The nonsalt form of a base, such as, for amines, the form R-NH_2.

gas chromatography A physical technique that separates components based on their distribution between a gas and a stationary phase.

gas holdup time The amount of time it takes for the carrier gas to move from the injection port to the detector, analogous to the void volume of liquid chromatography.

gas-liquid chromatography A separation technique in which the stationary phase is a liquid.

gas-solid chromatography A separation technique in which the stationary phase is a solid.

HEEP (height equivalent to effective plate, h) A term to establish column efficiency. Separation is usually done on a column, which can be of variable length. This term defines the effective number, N, of plates per unit of column length, L. Equation: $h = L/N$.

HETP (height equivalent to theoretical plate, H) A term to establish column efficiency. Although similar to HEEP, this term uses the uncorrected retention time for the calculation.

injection port A device usually having a septum and a heating block to volatilize the compounds to be separated. This is placed before the column.

Kovats' index This index relates the logarithm of the retention time of a compound, regardless of its chemical nature, to those of the *n*-paraffins.

liquid phase The nonvolatile fluid that coats the immobile support medium. These fluids have the property of acting as solvents for the compounds to be separated.

mass fragment A degraded portion of a molecule containing one or more charges.

mass spectrometer A device that may be considered a gas chromatographic detector. Compounds are fragmented into specific groups of charged molecules, which are separated into their mass and charge components, and their relative abundance is measured.

mass transfer The movement of mass from one phase to another.

McReynolds' constant A constant describing a system that classifies the stationary phase in terms of its ability to separate various compounds.

mobile phase The moving component, such as gas, of a chromatographic system. The compound to be separated equilibrates between this material and the stationary phase.

negative chemical ionization Similar to chemical ionization except that the gas is oxygen or hydrogen, which produces primarily negative ions of the form M − 1. (''M'' is 'mass.')

nitrogen-phosphorus detector A device similar to a flame ionization detector but into which alkaline metals are introduced. When nitrogen- or phosphorus-containing compounds are burned, the rate of release of alkaline metal vapor and thus of current flow is increased. It has high specificity for nitrogen or phosphorus.

nonvolatile liquid A fluid that does not vaporize or have a form that is readily gaseous in nature.

nonpolar Usually applied to molecules that have a hydrophobic affinity, that is, ''water hating.'' Nonpolar substances tend to dissolve in nonpolar solvents.

overloading When too much of a compound is presented for adsorption by the stationary phase, nonequilibrium between the two phases occurs.

partition coefficient The ratio of the concentration of a compound in the stationary phase to that in the mobile phase.

phase ratio The ratio of mobile-phase (gas) volume to stationary-phase (column) volume. Usually indicated by the term ''β.''

plate A chromatographic term that refers to a single partitioning unit of the chromatographic system.

polar Usually applied to molecules that have a hydrophilic affinity, that is, ''water loving.'' Polar substances tend to dissolve in polar solvents.

relative peak sharpness The ratio of the corrected retention time of a compound to the peak width at its base.

relative retention time The ratio of the corrected retention times of the reference compound to the sample compound.

resolution The ability of a chromatographic system to separate two adjacent peaks.

retention index A system relating the retention time to a standard.

retention time The amount of time a compound is retained on a column.

Rohrschneider constant Similar to the McReynolds constant.

selected-ion chromatogram A technique in which only mass fragments of a preselected size are recorded and quantified by the mass spectrometer.

separator A device that removes large portions of the carrier gas and concentrates the solutes before entrance to the mass spectrometer.

septum A device that separates the chromatographic column from the laboratory environment. Usually it is a small disk of silicone rubber through which the solution to be separated is injected into the column.

silanization The chemical process of converting the SiOH moieties of a stationary phase to the ester form.

sorbent A material that has the property of interacting with the compound of interest, usually to make it bind.

stationary phase The nonmoving component in a chromatographic system. The compound to be separated equilibrates between this material and the mobile phase.

thermal compartment The temperature-regulated oven in which the chromatographic column is placed.

thermal-conductivity detector A device that measures the difference between the heat conductivities of the carrier gas and that of the sample-gas effluents. A sample carried in gas increases the heat conductivity.

Van Deemter's equation Relates the HETP (height equivalent to the theoretical plate) to the linear velocity of the carrier gas.

Gas chromatography (GC) is a physical technique that separates two or more compounds based on their distribution between two phases, a stationary and a mobile one. The stationary phase may be a liquid or a solid, and the mobile phase is a gas that percolates over the stationary phase. When separation of sample components is accomplished by use of a mobile gas phase and a stationary phase consisting of a thin layer of nonvolatile liquid held on a solid support, the technique is called gas-liquid chromatography (GLC). Gas-solid chromatography (GSC) employs a solid sorbent as the stationary phase. Any compound that can be volatilized or converted to a volatile derivative can theoretically be analyzed by GC. Independent of the type of mobile or stationary phase, separation is achieved by the difference in partitioning of the various molecules of the sample between the two phases.

Gas chromatographic separation is illustrated by the following example: a sample containing the components to be separated is injected into a heated block in which they are immediately vaporized and swept by a stream of carrier gas through a column of stationary phase. The components are adsorbed onto the stationary phase at the head of the column and then are gradually desorbed by fresh carrier gas. The partitioning between the two phases occurs repeatedly as carrier gas sweeps the components toward the column outlet.

As the components are eluted, they enter a detector where their presence is converted to an electric signal, which is then measured, usually by a strip chart recorder, which produces a series of peaks charted versus time (Fig. 6-1). The appearance time, height, width, and area of these chromatogram peaks may be measured to yield valuable qualitative and quantitative data.

MOLECULES THAT CAN BE SEPARATED BY GAS CHROMATOGRAPHY

Theoretically any compound that can be vaporized or converted to a volatile derivative may be analyzed by gas chromatography. Compounds as small as carbon monoxide and methane and as large as 800 daltons have been successfully analyzed. Compounds larger than 800 daltons lack sufficient volatility. Generally a compound must be stable as a vapor so as to produce a single identifiable chromatographic peak. Unstable compounds may be converted to stable, volatile derivatives. However, if a compound degrades to known products or a consistent number of products, the resultant pattern of multiple compounds may be used as a means of tentative identification.

Inorganic compounds, or the inorganic salts of organic acids and bases, lack sufficient volatility for gas chromatographic analysis. Thus the technique is generally applied to analysis of organic molecules in their neutral or free-base or free-acid forms. Before chromatographic analysis, compounds are generally isolated and concentrated by means of solvent extraction and evaporation to dryness. The residues containing the analytes are dissolved in small amounts of volatile organic solvents. The solvent-analyte solution is then chromatographed. The analyte vapor should not interact with the solvent. The solvent should have greater volatility and much less affinity for the stationary phase than the analyte compounds do, thereby eluting far ahead of the analyte and not interferring with the chromatogram.

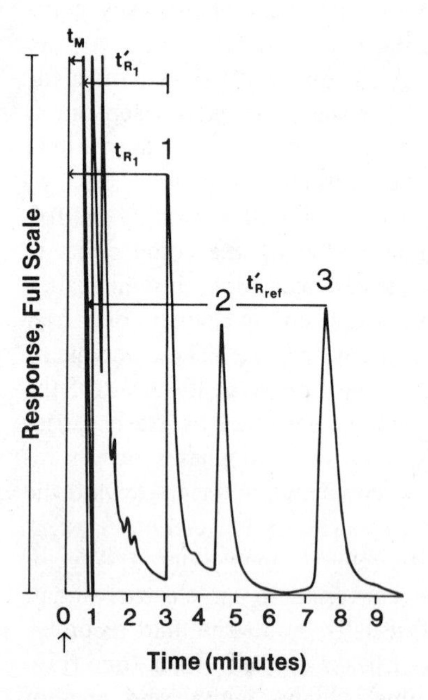

Fig. 6-1. Example of gas chromatogram showing detector response versus time. Vertical arrow (time = 0) indicates time of injection, and initial peak is "dead time," t_M. t_R designates uncorrected retention time, whereas t'_R is corrected retention time. $t'_{R_{ref}}$ is corrected reference time for internal standard or reference compound. Relative retention time for peak 1: $r = t'_{R1}/t'_{R_{ref}}$. (Modified from Mackell, M.A., and Poklis, A.: J. Chromatogr. **235**(2):445, 1982.)

THEORY OF GAS CHROMATOGRAPHIC SEPARATION

The following is a brief discussion of the basic theory necessary to understand and apply gas-liquid chromatography. For a more complete treatment of the nomenclature of chromatography[1] and the many complex variables that influence gas chromatographic separations, one should read references 1 to 4.

Partition coefficient

Each component travels through the column at a rate determined by its partition coefficient, K_D, as defined in equation 6-1.

$$K_D = C_s/C_m \qquad \textit{Eq. 6-1, A}$$

C_s is the concentration of component in the stationary phase, and C_m is the concentration of component in the mobile phase.

This partition coefficient can also be examined as a time function rather than a concentration function.

$$K_D = \frac{\text{Time sample is in stationary phase}}{\text{Time sample is in mobile phase}} \qquad \textit{Eq. 6-1, B}$$

Thus time and concentration in each phase are related to each other. If a compound is not adsorbed at all by the stationary phase, $K_D = 0$. The more time spent adsorbed to the stationary phase, the higher the K_D. The components then elute from the column in order of their partition coefficients. Those with the smallest K_D elute first; those with the largest, last.

Retention time

The length of time a compound spends on a column is called the *retention time*. For a given operating condition, such as column, temperature, carrier gas, and flow rate, retention time is constant whether a compound is pure or in a mixture. The time between sample injection to the peak of the eluted compound is called the "uncorrected retention time," t_R (Fig. 6-1). The transit time required for the carrier gas to move from the point of injection to the end of the column is called the "gas holdup time," or "dead time," t_M. It arises from the internal volume of the injector column and detector and is equivalent to the void volume in HPLC. A compound that does not partition into the stationary phase ($K_D = 0$) will be eluted from the column at t_M.

Because the "dead time" is a characteristic of both the particular gas chromatograph used for analysis (injector and detector volumes) and the column volume, it is the same for all sample components and is of no significance in identification. The difference between the uncorrected retention time and the dead time is called the "corrected retention time," t'_R.

$$t'_R = t_R - t_m \qquad \textit{Eq. 6-2}$$

The corrected retention time is a characteristic of a compound analyzed at specific conditions of column size, stationary phase, carrier-gas flow rate, and other chromatographic factors. It is theoretically independent of the constructional design of the gas chromatograph. The ratio of the corrected retention time to the dead time is called the "capacity ratio," k.

$$k = t'_R/t_M \qquad Eq. 6\text{-}3$$

Although in theory t'_R is a constant for a given set of chromatographic conditions, in practice it is difficult to reproduce from day to day and from laboratory to laboratory. To minimize the influence of operational variables, retention behavior is often expressed by relative retention, r. A reference substance, which may be a component of the sample or externally added, is chromatographed with the sample. The relative retention is the ratio of the corrected retention times of the reference to sample compound.

$$r = (t'_R)_{ref}/(t'_R)_{sample} \qquad Eq. 6\text{-}4$$

Although the relative retention is independent of column length and carrier gas flow, it is not independent of temperature. To assure the reproducibility of an analysis between different instruments or laboratories, one should calculate relative retention using only corrected retention times, and when reporting chromatographic data, one should cite information on the sample t'_R, liquid phase, and temperature.

Partitioning in GC systems

The chromatographic behavior of a solute's molecules is not specific with respect to their molecular skeleton or to their particular functional group. Therefore, except for members of a homologous series, such as compounds with the same basic structure differing only by the number of methylene groups, the K_D or corrected retention time, t'_R, of a solute in a specific system must be experimentally determined; it cannot be easily predicted. Both the partition coefficient and retention time of a solute are directly related and depend on the affinity of solute for the stationary phase. The old rule "like dissolves like" offers a simple guide to solute affinity. Polar solutes will have greater partition coefficients, K_D, hence longer retention times on hydrophilic (polar) phases than hydrophobic (nonpolar) stationary phases. Likewise, hydrophobic solutes exhibit greater K_D's and longer retention times on the nonpolar rather than the polar stationary phase. If both a hydrophobic and a polar solute are chromatographed together on a polar stationary phase, the hydrophobic solute will have less affinity for the phase than the polar solute ($K_{D\ hydrophobic} < K_{D\ polar}$) and will be eluted before the polar solute ($t'_{R\ hydrophobic} < t'_{R\ polar}$).

The K_D of solute is expressed as a concentration (amount/volume) ratio and therefore may be written as:

$$K_D = [W_s/V_s]/[W_m/V_m] = \frac{V_m}{V_s} \cdot \frac{W_s}{W_m}$$

W_s = Weight of the sample in the stationary phase
W_m = Weight of the sample in the mobile phase
V_s = Volume of the stationary phase
V_m = Volume of the mobile phase

The ratio of the mobile-phase (gas) and stationary-phase (column) volumes (V_m/V_s) is called the phase ratio (β). The ratio of sample weights (W_s/W_m) in each phase is equal to the capacity ratio, k. Therefore the partition coefficient can be expressed as:

$$K_D = \beta k$$

The phase ratio and capacity ratio are characteristics for a particular column. However, their product, K_D, is independent of the particular column. Therefore the higher the phase ratio (smaller the volume of stationary phase), the smaller the capacity ratio (less time for elution, t'_R). In general, the smaller the capacity ratio (shorter t'_R), the more difficult it is to achieve a particular separation. The amount of solute analyzed on column without overloading is dependent on the amount of stationary phase that influences β. The greater the amount of stationary phase, the smaller the β and the larger the k (greater t'_R).

Temperature dependence

Temperature is the most important single parameter in a gas chromatograph separation. This is attributable to the great dependence of the partition coefficient, K_D, on temperature. The basic relationship between K_D and temperature is given by the equation:

$$\log K_D = \frac{\Delta H}{2.3R \cdot T_c} + \text{Constant} \qquad Eq. 6\text{-}7$$

ΔH is the partial molar heat of solution of the solute in the liquid state, R is the gas constant, and T_c is the column temperature.

This equation demonstrates that in a given system, the higher the column temperature, the lower the K_D, which means a lower capacity ratio, k (shorter retention time). Therefore the retention times of solute molecules may be readily altered when one changes the column temperatures. Roughly, a 30° C decrease in column temperature will approximately double the retention time. Conversely, a 30° C increase in column temperature will approximately halve the retention time. The influence of temperature on separation is discussed later.

Column performance

The ability of a column to produce optimum separations is measured by two quantities: "efficiency," which is the ability to produce narrow peaks, and "resolution," which is the ability to separate two adjacent peaks.

The components of a mixture elute as narrow peaks from an efficient column. The narrower the peaks, the better the chance for resolution of the mixture into its individual components. The peak width is a result of many variables; simply stated, it is related to the affinity of a compound for the stationary phase. The greater the affinity, the greater the retention time and the wider the peak. The relative peak sharpness, Q, may be defined as the ratio of the corrected retention time to the peak width at the base, W_b.

$$Q = t'_R/W_b \quad \textit{Eq. 6-8}$$

In distillation technology, efficiency is measured in separation plates per foot. Borrowing this terminology, efficiency in gas chromatography is expressed as the "number of effective plates," N. It may be related to the relative peak sharpness by the equation:

$$N = 16Q^2 = 16\,(t'_R/W_b)^2 \quad \textit{Eq. 6-9}$$

The number of effective plates depends on the column length, L; therefore, to permit comparisons of column efficiencies independent of their length, one expresses N as a function of length. The "height equivalent to one effective plate" (HEEP, h) is defined as:

$$h = L/N$$

If the uncorrected retention time, t_R, is used in place of t'_R, efficiency is expressed as the "height equivalent to one theoretical plate" (HETP, H). This term defines not only the efficiency of the column, but also the overall efficiency of the system, because of the gas holdup time of the injector, column, and detector. Since the gas holdup time varies in different instruments, the column efficiency is best expressed as above, by the number of effective plates.

The relationship of the HETP to the linear gas velocity, μ, is complex. The factors effecting gas flow are expressed by Van Deemter's equation[2]:

$$H = A + B/\mu + C\mu \quad \textit{Eq. 6-10}$$

A is the eddy diffusion term, which relates the effect of support-particle diameter and the column-packing procedure to the distance that a streamline of carrier gas persists before its velocity is drastically changed by the support. B is the longitudinal gas diffusion term, which relates peak broadening to the effect of diffusion in the flowing gas along the direction of flow. C is the resistance-to-mass-transfer term, which relates the diffusion processes in the gas and liquid phases.

The relationship of plate height to linear velocity, when graphed, corresponds to a hyperbola (Fig. 6-2). The A term is the intercept of the slope. At low velocities, the plate height is controlled by the longitudinal diffusion, B, of the sample in the carrier gas. Eventually the curve reaches a minimum, which is the optimum velocity (μ_{opt}) for the smallest plate height. The optimum velocity is ideal for only one compound; however, similar compounds have closely related optimum velocities, and a single flow rate is suitable for their separation. The resistance to mass transfer, C, is the limiting slope at high velocity. The component molecules need time to diffuse from the liquid phase into the gas phase. At higher flow rates the gas sweeps the diffusing molecules from the liquid before all have emerged; thus the peak is broadened.

Resolution is expressed when the distance between two peak maxima is related to the average width of the peaks.

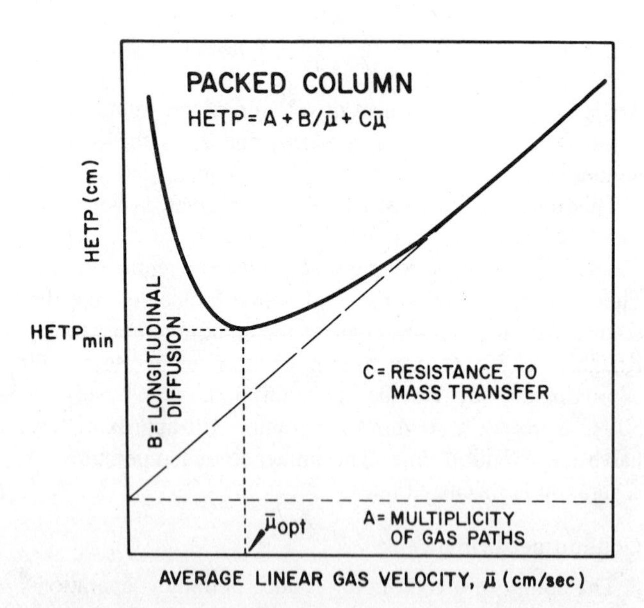

Fig. 6-2. Relationship between height equivalent to a theoretical plate (HETP) and average linear gas velocity ($\bar{\mu}$). $HEPT_{min}$ is smallest theoretical plate height obtained at optimal gas velocity. (From Ettre, L.S.: Basic relationships of gas chromatography, Norwalk, Conn., 1977, Perkin-Elmer Corp.)

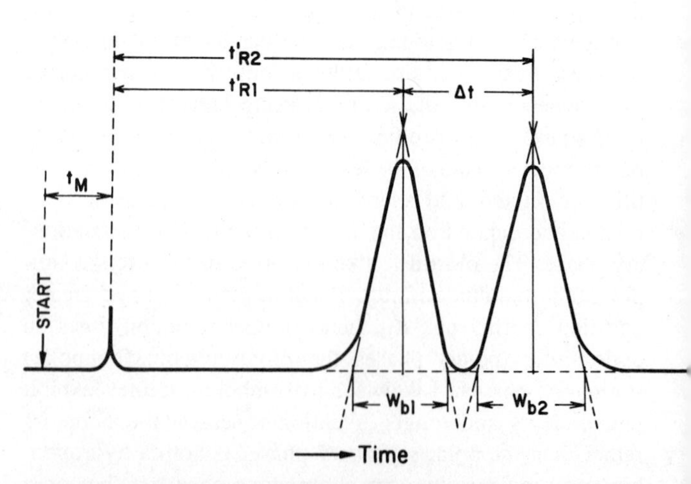

Fig. 6-3. Schematic showing resolution of two closely eluting compounds from gas chromatogram. t'_{R1} and t'_{R2} are corrected retention times for compounds 1 and 2, respectively. W_{B1} and W_{B2} are width at base of each peak for compounds 1 and 2. Resolution of these two compounds may be calculated with equation 6-11 in the text. (From Ettre, L.S.: Basic relationships of gas chromatography, Norwalk, Conn., 1977, Perkin-Elmer Corp.)

Peak resolution, R, is defined as follows:

$$R = \frac{t'_{R2} - t'_{R1}}{{}^1/_2(W_{b1} + W_{b2})} \qquad \textit{Eq. 6-11}$$

W_b is defined in equation 6-8 and shown in Fig. 6-3.

If $R = 1$, there is 94% separation of the two peaks. If $R = 1.5$, the separation is practically complete. Resolution changes with the square root of column length; therefore attempting to increase resolution by use of a longer column is not a practical approach. Changing the stationary phase or analytical conditions is more efficient.

MOBILE-PHASE CONSIDERATIONS

The most commonly used carrier gases are presented in Table 6-1. The carrier gas must be inert so as not to react with the sample components. Large quantities of relatively pure gas must be commercially available because appreciable amounts of carrier gas are utilized for analysis. Common impurities in carrier gases are moisture, oxygen, and hydrocarbons. Each of these contaminants may adversely affect various detectors, producing unstable recorder baseline or extraneous peaks. In certain situations, carrier-gas impurities may interact with sample components and prevent their analysis. For example, prepurified-grade nitrogen contains up to 200 ppm of oxygen. If high column temperatures are necessary to separate compounds that are readily oxidized, the oxygen impurity in nitrogen carrier gas may degrade the compounds on the column and prevent their detection or may produce multiple extraneous peaks of the degradation products. In such a situation, helium, which contains less oxygen contamination, should replace nitrogen as the carrier gas.

The choice of carrier gas does influence column performance (efficiency and resolution) and time required for analysis (retention time). As presented by the Van Deemter equation on p. 122, the "height equivalent of a theoretical plate" (HETP) is related to the linear gas velocity (μ) of the carrier gas. This interaction is highly complex, but the following brief discussion presents the basic effects of carrier gas upon separation.

In a flowing system (column) where the gas is compressible, the density, pressure, and velocity of the gas are different at each point in the column. The carrier gas is compressible, and the value of the carrier-gas velocity must be corrected to average conditions. The average linear gas velocity ($\bar{\mu}$) is determined by two factors: (1) the time necessary for an unretained solute to pass through the column (dead time, t_M) (p. 120) and (2) the length of the column, L. These factors determine $\bar{\mu}$ by the following equation:

$$\bar{\mu} = L/t_M \qquad \textit{Eq. 6-12}$$

In a given column, the optimum linear gas velocity ($\bar{\mu}_{opt}$) is proportional to the diffusivity of the vapors of the substances being chromatographed. However, the kinetic theory of gases states that the diffusivity in gas is inversely proportional to the square root of the molecular weight or density of the gas (see Table 6-1). Therefore the $\bar{\mu}_{opt}$ will be lower for high-density gases (low solute diffusivity) such as nitrogen and argon, and higher for low-density gases (high solute diffusivity) such as helium and hydrogen. Therefore in the same column different values of $\bar{\mu}_{opt}$ will be obtained for different gases. Fig. 6-4 shows the relationship between the $\bar{\mu}$ of the two different carrier gases and the resultant HETP (see Van Deemter's equation 6-10 and Fig. 6-2). Nitrogen has a minimum HETP of

Table 6-1. Common carrier gases

Gas	Molecular weight	Density (g/L)	Impurities (ppm)
Argon	39.944	1.784	—
Helium	4.007	0.177	Hydrocarbons (1 to 100)
Hydrogen	2.018	0.089	—
Nitrogen	28.014	1.251	Oxygen (20)

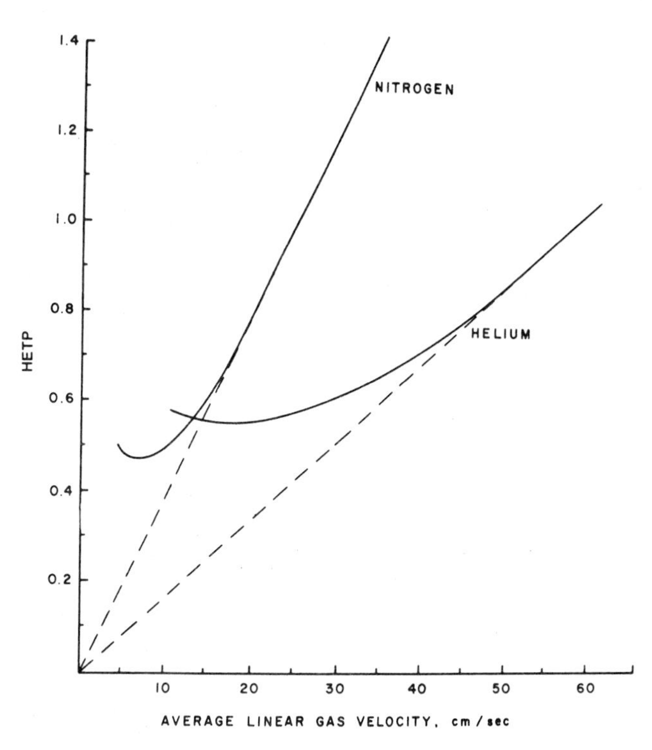

Fig. 6-4. Relationship between HETP (heat equivalent to a theoretical plate) and average linear gas velocity for two different carrier gases: nitrogen and helium. (From Ettre, L.S.: Practical gas chromatography, Norwalk, Conn., 1973, Perkin-Elmer Corp.)

0.465 at a $\bar{\mu}_{opt}$ of 7 cm/sec, whereas helium has a minimum HETP of 0.55 at a $\bar{\mu}_{opt}$ of 17.5 cm/sec. This demonstrates that in the same length of column, a high-density gas (nitrogen or argon) will produce better efficiency (more theoretical plates per unit length, lower HETP) than a low-density gas (helium or hydrogen).

The only parameter of resolution, R, that is dependent upon $\bar{\mu}$ is h, the height equivalent to one effective plate. The relationship of R to h is given by the following equation.

$$R = \frac{1}{4} \cdot \frac{L}{h_2} \cdot \frac{k_2}{1 + k_2} \cdot (r - 1/r) \qquad \textit{Eq. 6-13}$$

L is the length of the column, h_2 is the height equivalent to one effective plate for the second compound, k_2 is the capacity ratio ($t'_R W_m/t_M$) for the second compound, and r is the relative retention for the second compound.

Hence, for a given system, the maximum resolution will be obtained at minimum h, which is determined at the $\bar{\mu}_{opt}$ of the carrier gas. Also, for a given length of column, better resolution (greater R) will be obtained by the carrier gas producing the lowest h value. Also notice that resolution is a function of the capacity ratio, k, which is related to the interaction of the solute (affinity) with the stationary phase. The capacity ratio is dependent on column temperature (see p. 121). Therefore temperature will influence resolution. The column temperature is usually a compromise between better resolution (when lowered) and higher speed of analysis (when raised).

For a low density gas, the diffusivity of the solute and the $\bar{\mu}_{opt}$ in a given column are greater than those of a high-density gas. Therefore, shorter times of analyses (smaller t_R) are obtained with helium or hydrogen than with nitrogen or argon. The relationship between time of analysis and resolution is given by the equation:

$$t_R = 16R^2\,(r^2/r - 1) \cdot [(k_2 + 1)^3/k_2^2] \cdot h/\bar{\mu} \qquad \textit{Eq. 6-14}$$

At a given degree of resolution for two compounds, their retention times will depend on the ratio of h to $\bar{\mu}$. The differences between observed h values for high- and low-density gases in a given column are generally small; however the $\bar{\mu}_{opt}$ values are appreciable. Therefore the higher $\bar{\mu}_{opt}$ values obtained for low-density gases will give rise to shorter retention times.

STATIONARY-PHASE CONSIDERATIONS

Gas-solid stationary phases

In gas-solid chromatography (GSC) the column is packed with an adsorptive solid material on which the sample components are partitioned by adsorption on the surface of the solid. This material should possess a large surface area per unit volume to ensure rapid equilibrium between the stationary and gas phases. It should possess uniform particle size and pore structure and be strong enough to resist breakdown during handling and column packing. Theoretically the smaller the particle size of support, the greater the efficiency of the column. However, the smaller the particles, the greater the resistance to flow and the greater the necessary carrier-gas pressure.

The most common chromatographic solids for adsorption phases are made from diatomaceous earth (kieselguhr). The processed white kieselguhr is sold under many trade names: Chromosorb W, Celite, Gas Chrom, and Anakron. The diatomite may also be crushed, blended, pressed into brick, and processed such that mineral impurities form oxides and silicates, which give the material a pink color. It is marketed as crushed firebrick, or Chromosorb P. This material has greater density and is less fragile than the white material. The pore size of the pink material is only 2 μm compared to 9 μm for the white. Therefore greater efficiency is obtained with the pink material.

Each support possesses individual properties that may enhance or hinder its use for a particular application. The white material is slightly alkaline and will interact with acidic compounds. Its surface, however, is nonadsorptive, a property that favors its application to analysis of polar compounds. The pink material adsorbs polar compounds; thus it is best suited for the separation of nonpolar molecules like hydrocarbons.

An additional type of solid stationary phase consists of porous polymer beads, which allow the analyte molecules to partition directly from the gas phase into the amorphous polymer. Porapak, a polymer of ethylvinylbenzene cross-linked with vinylbenzene, is the most popular polymer phase. The material may be modified by copolymerization with various polar monomers to produce beads of varying polarity. Porapak columns are thermally stable up to 250° C. At temperatures above 250° C the column material will be degraded and be eluted, a phenomenon called "column bleed." These degradation products can be observed by the detector. Water and highly polar molecules are rapidly eluted from the polymer. Porapak is especially useful for base-line separation of aqueous samples containing low molecular weight alcohols, esters, halogens, hydrocarbons, ketones, and mercaptans (Table 6-2).

Gas-liquid-solid supports for GLC

The stationary phase in gas-liquid chromatography (GLC) is a thin film of liquid held on an inert platform or support. In capillary chromatography, the liquid is coated on the walls of the tubing. In packed columns, the liquid is held in a thin-layer film across the surface of an inert support (Fig. 6-5). Many materials that act as stationary phases for GSC are also supports for the liquid phase in GLC. Both the pink and white solid phases described above are popular liquid supports. Although the support should be inert and not influence separation, both pink and white materials have "active sites" because of metallic impurities and silanol (-SiOH) and siloxane (SiOSi-) groups, which form hydrogen bonds with polar com-

Table 6-2. Examples of commonly used stationary phases and their applications

Stationary phase	Structures	Activity	Temperature (°C min/max)	Application	Specific compounds
Silicone OV-1 (100% methyl)	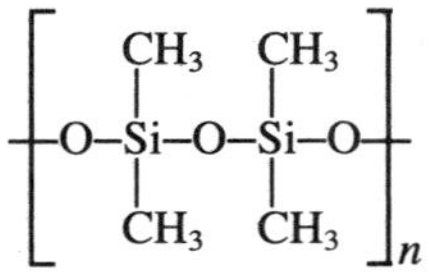 $[-O-Si(CH_3)_2-O-Si(CH_3)_2-O-]_n$	Nonpolar	100/350	Bacteria, drugs	Fatty acid methyl esters, benzodiazepines
Silicone OV-17 (50% phenyl)	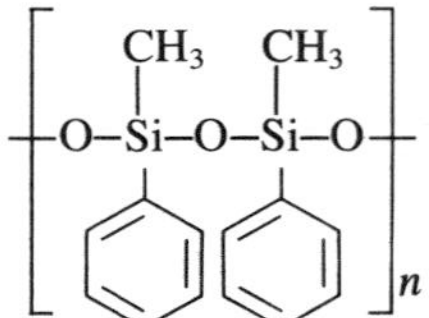 $[-O-Si(CH_3)(C_6H_5)-O-Si(CH_3)(C_6H_5)-O-]_n$	Intermediate polarity	20/350	Drugs, steroids	Tricyclic antidepressants, barbiturates, cholesterol
Silicone OV-210 (50%, 3,3,3-trifluoropropyl)	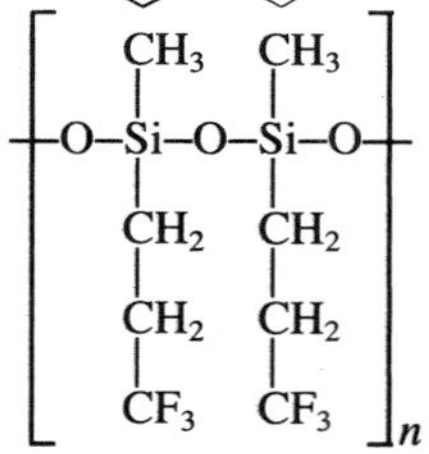 $[-O-Si(CH_3)(CH_2CH_2CF_3)-O-Si(CH_3)(CH_2CH_2CF_3)-O-]_n$	Polar	20/300	Drugs, pesticides	Basic drugs, lindane, aldrin, DDT
Silicone OV-225 (25% cyanopropyl, 25% phenyl)	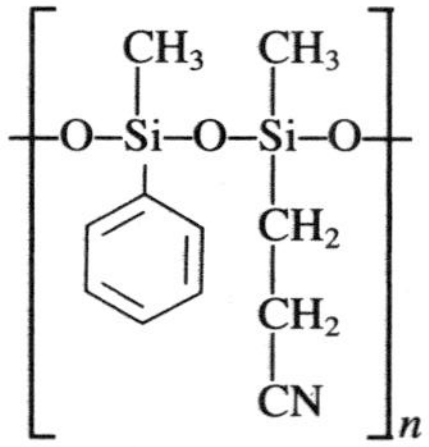 $[-O-Si(CH_3)(C_6H_5)-O-Si(CH_3)(CH_2CH_2CN)-O-]_n$	Polar	20/275	Steroids	TMS derivatives of 17-ketosteroids
10% Apiezon L/ 2% KOH	Undefined mixture of high-boiling hydrocarbons	Nonpolar	50/225	Amines	Amphetamine
NPGS (neopentyl glycol succinate)	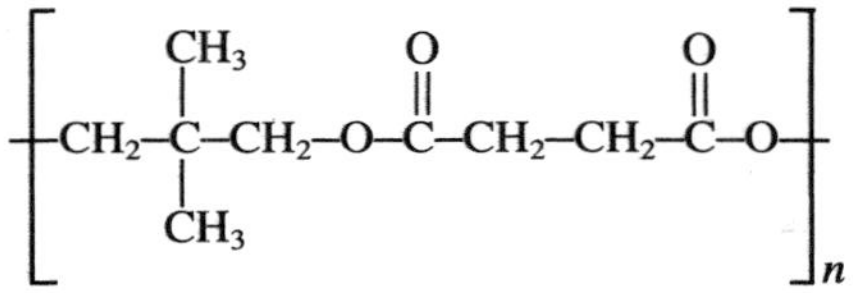 $[-CH_2-C(CH_3)_2-CH_2-O-C(=O)-CH_2-CH_2-C(=O)-O-]_n$		50/240	Volatile fatty acids	Acetic through caproic acids
Carbopack B/5%	$(-CH_2-CH_2-O-)_n$	Polar		Alcohols, aldehydes, ketones	Methanol, ethanol, acetaldehyde acetone

Continued.

Table 6-2. Examples of commonly used stationary phases and their applications—cont'd

Stationary phase	Structures	Activity	Temperature (°C min/max)	Application	Specific compounds
DEGS (diethylene glycol succinate)	$\left(-CH_2-CH_2-O-CH_2-CH_2-O-\overset{O}{\overset{\Vert}{C}}-CH_2-CH_2-\overset{O}{\overset{\Vert}{C}}-O-\right)_n$	Polar	20/200	Bacteria	Fatty acid methylesters
EGA (ethylene glycol adipate)	$\left(-CH_2-CH_2-O-\overset{O}{\overset{\Vert}{C}}-CH_2-CH_2-CH_2-CH_2-\overset{O}{\overset{\Vert}{C}}-O-\right)_n$		100/210	Amino acids	NBTFA* derivatives of amino acids
Chromosorb 102 (styrene divinyl benzene polymer)	$\left(-CH=C-CH=CH-C_6H_4-CH=CH-\right)_n$		<250° C	Alcohols, aldehydes	Methanol, ethanol, acetaldehyde
Porapak Q (ethylvinyl benzene + divinyl benzene polymer mixture)	$\left(-CH_2=CH_2-C_6H_4-CH=CH-\right)_n$		<250° C	Low molecular weight	Chlorinated hydrocarbons

**NBTFA*, Nitroblue tetrazolium fatty acid.

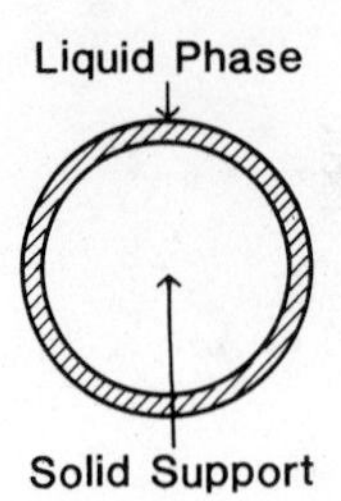

Fig. 6-5. Schematic of solid support particle for gas chromatography with liquid stationary-phase coating.

pounds. This interaction gives rise to distorted peaks (nongaussian) in the resultant GLC chromatogram. These "active sites" may be removed by acid washing of the mineral impurities from the support and by conversion of the silanol groups to silylesters (silanization) of dimethyldichlorosilane or hexamethyldisilazone. Silanization reduces surface activity but also reduces the surface area of the support so that no more than 10% (v/w) of the liquid stationary phase to total column weight may be applied. In certain instances, special additives are mixed with the liquid phase to block the "active sites" of untreated support material. Two such examples are the incorporation of stearic acid in silicone oil used in separation of fatty acids and the addition of potassium hydroxide to polar liquid phases used to separate amines. Informative data describing preparation, applications, and limitations of support materials are readily available from commercial manufacturers and suppliers.

Liquid phases

The universal popularity of GLC as a separation method is attributable to the large variety of liquid phases with differing solution properties and therefore different affinities for various classes of analytes. The range of liquids used as stationary phases is limited only by their volatility, thermal stability, and ability to wet the support. No single stationary phase will achieve all desired separations. Commercial suppliers typically offer 100 to 200 liquid phases; however, many of these phases are duplicates sold under various trade names or are so similar in character as to have little difference in their separation abilities. In fact, few laboratories require the use of more than a half dozen different liquid phases. Eighty percent of a wide range of organic compounds may be successfully separated using only four to seven phases: OV-101, OV-17, Carbowax 20M, OV-225, DEGS, OV-275, and OV-210.[5,6] Examples of liquid phases, characteristics, and applications are presented in Table 6-2.

Liquid phases may be generally classified into five categories: (1) Nonpolar ones that are hydrocarbon liquids such as squalane, silicone greases, Apiezon L, and silicone gum rubber. Generally, compounds are eluted from these phases in order of increasing boiling point. (2) Intermediate polarity phases that include polar or polarizable groups attached to a long nonpolar skeleton such as esters of high molecular weight or alcohols such as diisodecylphthalate. Both polar and nonpolar compounds are separated by these phases, with the more polar ones being eluted first. (3) Polar phases, which contain a high concentration of polar groups such as carbowaxes. These phases differentiate between polar and nonpolar compounds by interacting strongly only with polar compounds, separating these from the earlier eluting, less polar compounds. (4) Hydrogen-bonding phases, which contain many hydrogen atoms readily available for hydrogen bonding, such as glycol phases. Polar compounds have greater affinity for the stationary phase and are eluted more slowly. (5) Special-purpose phases that can be prepared to utilize a specific chemical interaction between the sample and the stationary phase. These phases are usually applied to a specific application. An example of a special-purpose phase is silver nitrate dissolved in glycol to enhance separation of unsaturated hydrocarbons by charge-transfer interactions.

Each liquid phase has a specific temperature range for efficient use (Table 6-2). The maximum temperature at which a phase may be used is determined by its volatility. Beyond this temperature the phase is lost because of decomposition or volatilization and is carried into the detector producing extensive background noise (column bleed). A column may be heated above the maximum temperature for brief periods of time as in temperature programming, but the maximum temperature must never be exceeded for isothermal (constant-temperature) analysis. However, below the minimum temperature the increased viscosity or solidification of the liquid renders analysis irreproducible.

The amount of stationary phase in the column is expressed in percent by weight of the liquid phase on the support. In general, analytical columns contain 3% to 10% liquid phase. Deviations from these values may occur in specific applications: very low liquid loads for high molecular weight compounds and high loads for small, highly volatile compounds such as hydrocarbons containing 1 to 4 carbon atoms. The amount of stationary phase directly affects the sample capacity and efficiency of the column. The greater the amount of liquid phase, the larger the amount of sample that may be chromatographed. The influence of the thickness of liquid phase on efficiency is discussed under column efficiency.

DERIVATIZATION

Often it is desirable to modify chemically a molecule so that newly formed products have properties that are preferable to their precursors. One may need to derivatize a compound to make it volatile and stable as a gas and thus be analyzable by GC. Derivatives are also prepared to achieve increased sensitivity, selectivity, or specificity for a given separation. Derivatives may be eluted from the column sooner, have less tailing, produce sharper peaks, provide stability to thermally labile compounds, and increase resolution. Derivatization involves a chemical reaction between some functional group on the sample molecule (usually a polar group, which reduces volatility or interacts with the stationary phase to increase retention time) and a smaller molecule (derivatizing agent), which forms a new product of increased volatility with a smaller partition coefficient (K_D). The derivatization may be carried out before sample injection or may occur in the injection port of the chromatograph (''on column'' or ''flash derivatization''). A few derivatization techniques are briefly presented, but for a more complete discussion consult the literature.[7,8]

A popular GC derivatization technique is the replacement of an active hydrogen by a trimethylsilyl (TMS) group. The resultant ''silyl'' derivatives are usually less polar and more volatile and display greater thermal stability than their parent compounds. Silylizing reagents react vigorously with water or alcohol-containing solvents; therefore the conversion reactions are carried out in anhydrous solvents such as acetonitrile or tetrahydrofuran. TMS reagents are flammable, and some are highly corrosive. They should be handled with care.

$$\underset{\textbf{Alcohol}}{ROH} + \underset{\textbf{Trimethylchlorosilane}}{(CH_3)_3SiCl} \longrightarrow ROSi(CH_3)_3 + HCl$$

Esterification is often used for GC analysis of compounds containing a carboxylic acid group. Methyl esters possess the greatest volatility and hence are most popular. Alkylation reactions with quaternary alkylammonium hydroxides or dimethylformamide–dialkyl acetals have become popular as ''flash-derivatizing'' reagents. Fig. 6-6 presents the derivatization reaction of tetramethylammonium hydroxide and barbiturate drugs. To increase sensitiv-

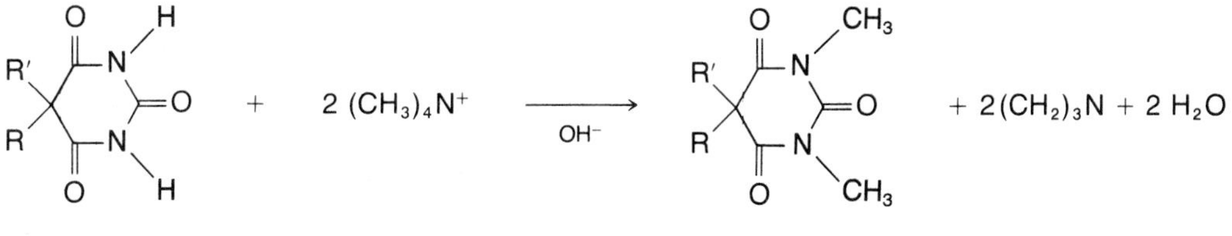

Fig. 6-6. Tetramethyl derivatization of barbiturate drugs.

ity, derivatizing reagents that produce halogen- or nitrogen-containing compounds are used with electron-capture detectors.

SELECTION OF A SEPARATION SYSTEM

Choosing the mobile phase

The mobile phase or carrier gas has one major function in gas chromatography: to carry the vaporized sample through the column and into the detector. As described above, selection of the proper carrier gas is predicated upon three considerations: (1) the operating principles of the detector through which the gas will be continuously flowing, (2) the presence of impurities in the carrier gas, and (3) the desired speed of analysis and performance of the column. Compounds that are negligibly partitioned into a stationary phase cannot be separated from each other. Similarly, compounds with too great an affinity for the stationary phase will have unacceptably long retention times or may be irreversibly retarded.

Stationary-phase selection

Liquid phases with the same physical properties as the sample will retain the sample and generally effect a separation. Simply stated, "like substances retain like substances." However, this general rule does not aid in determining which specific stationary phase is potentially the best for a particular separation. Several approaches to the choice of liquid-phase selection for a desired separation are briefly presented.

There are many sources of irreproducibility that can affect GC analysis. These include variations in assay conditions and variations in stationary-phase packaging (lot to lot, company to company, and so on). For reproducible identification of peaks of interest regardless of exact assay conditions, approaches have been developed to standardize the reporting of relative retention times by their conversion to indices or constants. These values can then be used to compare data between analyses, within a laboratory or between laboratories.

Kovats index. If the components of the sample are known, the most likely stationary phase that will effect a separation may be selected by use of the Kovats retention index.[9-11] Retention indices relate the retention time of a compound, regardless of its chemical nature, to those of the *n*-paraffins (straight-chain hydrocarbons) being eluted directly before and after it. Chromatographing the *n*-paraffins on a given column under set conditions yields a nearly linear relationship between the log of their retention time (t_R) and a number of carbon atoms in each paraffin, as shown in Fig. 6-7. Each *n*-paraffin is given a retention index, I, that equals 100 times the number of carbon atoms. The retention index of all other compounds is calculated from the following relationship (equation 6-15):

$$I = 100z + 100\,[(\log t_{Rx} - \log t_{Rz})/(\log t_{R(z+1)} - \log t_{Rz})] \quad \textit{Eq. 6-15}$$

z equals the number of carbon atoms in the unknown compound; t_{Rx} is the retention time of the unknown substance x; t_{Rz} is the retention time of the *n*-paraffin being eluted immediately before x; and $t_{R(z+1)}$ is the retention time of the *n*-paraffin being eluted immediately after x. For example, assume the data presented in Table 6-3 and graphically represented in Fig. 6-7 were obtained by chromatography from a series of *n*-paraffins and benzene on a given liquid phase. The retention index, I, for benzene ($z = 6$) on that liquid phase is calculated to be 650.

$$I = 100(6) + 100(0.052/0.104) = 600 + 50 = 650$$

Table 6-3. Sample calculation of a retention index

Compound	Carbon atoms	Symbol	t_R (min)	Log t_R (min)	Retention index (I)
Hexane	6	z	14.96	1.175	600 (by definition)
Benzene	6	x	16.86	1.227	650 (by experiment)
Heptane	7	$z + 1$	19.01	1.279	700 (by definition)
Octane	8	—	24.15	—	800 (by definition)
Nonane	9	—	30.76	—	900 (by definition)

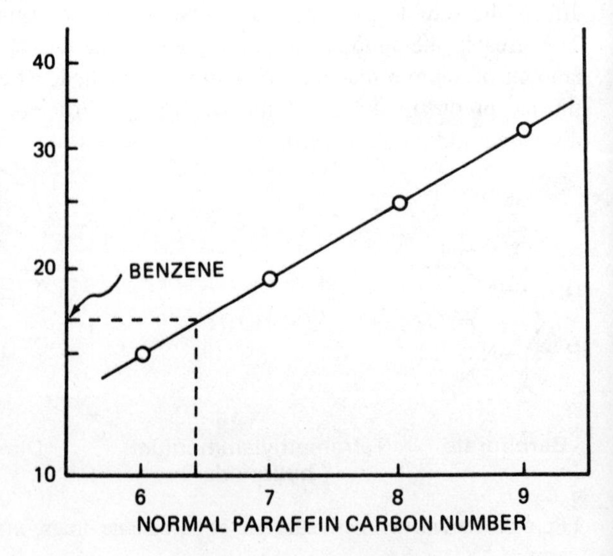

Fig. 6-7. Linear relationship between log of retention time and number of carbon atoms in a series of paraffin hydrocarbons. (From Rowland, F.W.: The practice of gas chromatography, Palo Alto, Calif., 1974, Hewlett-Packard Co.)

Table 6-4. Reference compounds used to determine McReynolds constants

Reference compound	Abbreviation	I*	Organic compound expected to display similar behavior on liquid phase
Benzene	x'	650	Aromatics, olefins
Butanol	y'	590	Alcohols, phenols, weak acids
2-Pentanone	z'	627	Aldehydes, esters, ketones
Nitropropane	u'	652	Nitrogenous and nitrile compounds
Pyridine	s'	699	Nitrogenous aromatic heterocyclics, bases

*Absolute value of retention indices observed on squalane.

The calculated I applies only to that particular stationary phase and temperature conditions. However, the effects of flow rate of the mobile phase and the percent loading (quantity of liquid phase) will change the retention time proportionally for all n-paraffins and thus the calculated I values are unchanged. Extensive tabulation of retention indices of numerous compounds on liquid phases are available in the *ASTM Gas Chromatography Data Compilation Catalog AMD 25A*[12] and a supplement catalog *AMD 25A S-1*.[13] If several compounds of a different chemical nature are to be separated simultaneously, the compounds of interest are located in the table and their respective I values for various stationary phases noted. A difference of at least 30 I units between the compounds will indicate that a particular phase will efficiently separate them. The I values are based on peak apex only and give no indication of peak width. Therefore I values do not indicate the resolution of the compounds. However, since the retention times increase with the retention index, I, the retention indices indicate the order at which compounds will be eluted from the column. A recent clinical application of I values has been the qualitative identification of drugs on standard liquid phases under both isothermal and temperature-programmed conditions.[14,15]

Rohrschneider and McReynolds constants. Rohrschneider and McReynolds constants are related systems that classify stationary phases in terms of their separating power.[16,17] The I values of a set of reference compounds of varying polarity are determined on the liquid phase being tested and compared against the I values for the same compounds obtained on a reference liquid phase. Squalane, a nonpolar liquid, is used as the reference phase. The constants are then calculated as indicated in the equations below.

$$\text{Rohrschneider constant (X)} = 1/100\,(I_{\text{test phase}} - I_{\text{squalane}}) \qquad \textit{Eq. 6-16}$$

$$\text{McReynolds constant (X')} = I_{\text{test phase}} - I_{\text{squalane}} \qquad \textit{Eq. 6-17}$$

Table 6-4 presents data related to five reference compounds used to determine McReynolds constants. Both Rohrschneider and McReynolds constants may be used for two purposes: to select a liquid phase for a particular application and to classify liquid phases as to how similar or different they are in the ability to perform chromatographic separations.[18] An example of the use of McReynolds constants for selection of a liquid phase for a given application would be the separation of saturated and unsaturated fatty acid methyl esters. If one has available liquid phases 1(DEGS) and 2(Carbowax 20M) presented in Table 6-5, an examination of their McReynolds constants permits one to choose the best phase for the separation. For the ability to separate saturated and unsaturated fatty acid esters, one considers the constants x' (olefinic compounds, unsaturated esters) and z' (esters) (Table 6-4). The McReynolds constants for x' and z' (in bold face in Table 6-5) are much higher for DEGS than for Carbowax 20M; therefore DEGS is better suited for the separation of the fatty acid esters.

The characterization of liquid phases as to their ability to perform separations by use of McReynolds constants is demonstrated by examination of phases 3(Emulphor ON-870), 4(Triton X-100), and 5(XE-60) in Table 6-5. All liquid phases are commercially available, and one can determine differences or similarities in separating power as illustrated in Table 6-5. Liquid phases 3 and 4 are almost identical in their ability to separate the reference compounds. Therefore, if used as a liquid phase, they would yield very similar chromatograms. Liquid phase 5 is similar to phases 3 and 4 for constants x' and y'. Therefore separation of compounds characterized by these constants (Table 6-5) on phase 5(XE-60) would be essentially the same on phases 3(Emulphor EN-870) and 4(Triton X-100). The constants z' and u' are higher for phase 5, and the separation of the compounds listed in Table 6-4 for these constants would be better on XE-60 than on the other two phases. However, although keto compounds (constant z')

Table 6-5. McReynolds constant of various liquid phases

Liquid phase	McReynolds constant*				
	x′	y′	z′	u′	s′
1. DEGS	**496**	746	**590**	837	835
2. Carbowax 20M	**322**	536	**368**	572	510
3. Emulphor ON-870	202	395	251	395	344
4. Triton X-100	203	399	268	402	362
5. XE-60	204	381	340	493	367

*x', Benzene; y', butanol; z', 2-pentanone; u', nitropropane; s', pyridine. For bold-faced type, see text above.

are better separated on phase 5, the separation of alcohols (constant y') would be practically identical on phases 3, 4, and 5. When using packed columns, one finds that differences in McReynolds constants of 20 or less are insignificant. Differences of 100 McReynolds units indicate significantly better separating ability of one phase compared to the other.

COMPONENTS OF GAS CHROMATOGRAPH

Basically a gas chromatograph consists of six components (Fig. 6-8): (1) a pressurized carrier gas with ancillary pressure and flow regulators, (2) a sample injection port, (3) a column, (4) a detector, (5) an electrometer and signal recorder, and (6) thermostated compartments encasing the column, detector, and injection port.

Carrier gas

The efficiency of a gas chromatograph depends on a constant flow of carrier gas. The carrier gas from a pressurized tank flows through a toggle valve, a flowmeter (range 1 to 1000 liters/min), metal restrictors, and a pressure gauge (1 to 4 atmospheres). The flow is adjusted by a needle valve mounted at the base of the flowmeter. The gas moves more slowly at the head of the column than at the outlet because of a pressure drop in the column. Thus the flow rates are measured as the gas exits from the column. This is done with a soap-film flowmeter. A simple sidearm buret with a rubber bulb filled with soap solution is connected to the detector outlet. One determines the flow rate by noting the time required for a film (bubble) to pass between two calibrated volume marks on the buret.

Carrier gas should be inert, dry, and pure. The most common carrier gases are inert, but they may contain contaminants that affect column performance and the response of ionization detectors. Hydrocarbon gases and water are removed from the carrier gas by a molecular sieve trap between the gas cylinder and the chromatograph.

Sample-injection port

Most GC analyses are performed on nonaqueous, liquid samples that are injected by a glass microsyringe by insertion of a needle through a septum into a heated block where the sample is vaporized and swept by carrier gas into the column. The pressure inside the injection port is usually well above atmospheric pressure, and the stream of carrier gas sweeps away the sample and aids in vaporization. Thus a sample may be vaporized at temperatures below its atmospheric boiling point. However, the injection port temperature is usually set at 25° to 50° C higher than the boiling point of the highest boiling components in the sample. This assures that immediate vaporization will occur and the components will not be diluted by carrier gas and will enter the head of the column as a single band. The time required for vaporization is dependent on the amount and volatility of the sample. Dilute samples vaporize faster than concentrated samples. High-boiling or temperature-sensitive compounds may be diluted with volatile solvents, which lower injection temperatures significantly.

Since heated metal may catalyze the degradation of many biological compounds, many injection ports are equipped with a glass liner or a glass column that extends through the injectors flush to the septum. The latter approach is called "on-column injection." For maximum efficiency it is imperative that the sample be the smallest possible volume (0.5 to 10 μL) consistent with detector sensitivity and be injected as a single, uniform band ("slug injection"). Insertion, injection, and withdrawal of the needle should be performed quickly and smoothly. Gaseous samples are injected by a gas-tight syringe or a calibrated bypass loop. The loop consists of a glass system of three stopcocks, between two of which a standard volume of gas is trapped and introduced into the carrier gas stream when the stopcocks are switched.

A septum separates the chromatographic column from the laboratory environment. Septums are small disks of sil-

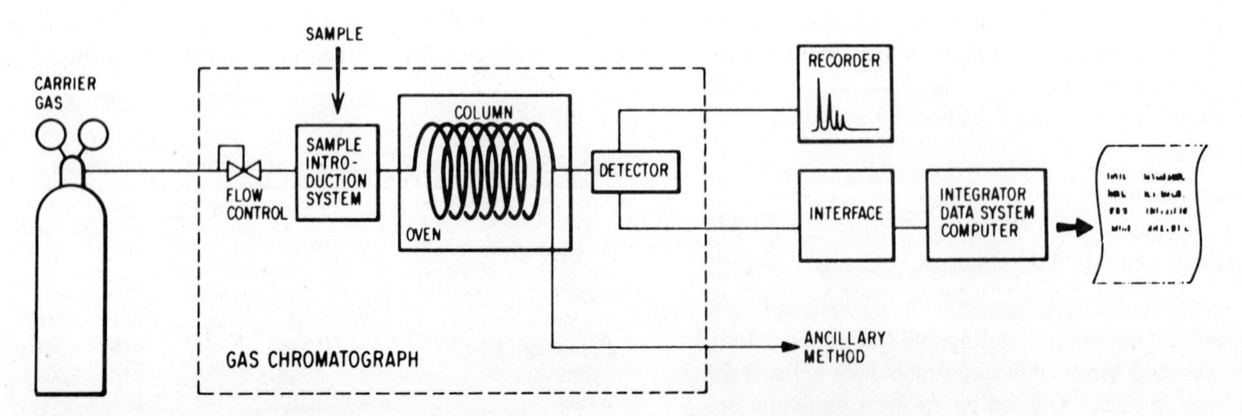

Fig. 6-8. Basic components of a gas chromatographic system. (From Ettre, L.S.: Practical gas chromatography, Norwalk, Conn., 1973, Perkin-Elmer Corp.)

icone rubber, and numerous types are available, depending on analytic requirements. Silicone rubber septums may absorb certain types of samples. In such instances, special septums (such as Teflon R coated) will alleviate this problem. Low molecular weight solvents used in the manufacture of septums may be released as the injection port is heated. This "bleed" of solvent may produce unwarranted peaks (ghost peaks) in chromatograms, and it increases the background level of the detector. Low-bleed septums from which the solvents have been extracted are available. Repeated injections through the septum will gradually destroy its mechanical strength, causing leakage. As a result, the retention time and sensitivity decrease as the carrier gas and part of the sample are released back through the septum into the atmosphere. This problem is easily avoided by regular insertion of new septums.

Various specialized injection systems are commercially available. If large numbers of similar analyses are to be performed, automatic sampling units are commonly used.

Column tubing

The column tubing is a container for the stationary phase (packing material) and directs the carrier gas flow. It should be inert and not affect the separation by reaction with the stationary phase or the sample. Depending on the gas chromatograph employed, the columns may be shaped as a U-tube or coiled in an open spiral or flat pancake shape.

Stainless steel and copper columns are often used for analyses requiring temperatures greater than 250° C. However, for the analysis of drugs, steroids, or other biological compounds, metal columns may absorb these analytes or catalyze their degradation. Therefore glass is the tubing of choice for the majority of clinical analyses. However, glass is fragile and inflexible, and if not properly handled, the columns are easily broken during transport or installation. Recently nickel has been recommended as a substitute for glass. Nickel tubing has been effectively used in the analysis of specific drugs, pesticides, and cholesterol, which previously required glass tubing.[19] However, the application of nickel tubing to the broad range of biological compounds has not yet been established. Until such time, glass tubing should remain the primary support when one is performing a clinical analysis.

Inside column diameters vary from capillary to larger dimensions. Most clinical analyses are performed on columns of 2 or 4 mm inner diameter (ID). Columns of 4 mm ID contain four times the stationary phase as 2 mm ID columns of the same length and therefore possess a greater sample capacity. However, the same separation will require higher temperatures and a longer time of analysis on the wider column. In addition, columns should be only as long as necessary to affect the desired separation. A short column provides a short analysis time, low temperatures, long column life, and less background in the detector. Columns of 0.7 to 2 m (2 to 6 feet) are sufficient for most chemical separations.

Thermal compartment

Precise control of column temperature is imperative in gas chromatography. The column oven is controlled by a system that is sensitive to changes of 0.01° C and maintains the column temperature to ± 0.1° C of the desired temperature. The column oven, injection block, and detectors should have separate heaters and controls. Analysis may be performed at a constant oven temperature (isothermal) or the temperature may be varied during the analysis (temperature programming). The temperature change during analysis can be programmed to vary with time according to predetermined, reproducible patterns, giving linear, convex, or concave curves when column temperature is plotted against time (Fig. 6-9). Temperature programming is often used in separating a complex mixture, the components of which have widely varying affinity for the stationary phase. Initially the column temperature is set low to permit separation and elution of the compounds with little affinity for the stationary phase. The temperature is then raised to elute compounds of higher stationary-phase affinity. Many chromatographs are equipped with specialized oven controls that uniformly raise the column temperature after each sample injection.

Detectors

As the carrier gas exits from the column, a detector senses the separated components of the sample and pro-

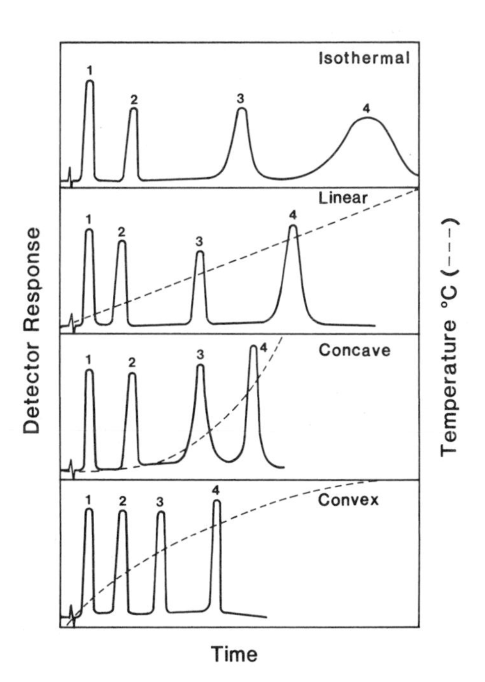

Fig. 6-9. Schematic diagram of theoretical separation of four compounds showing varying elution patterns with different temperature programming.

Table 6-6. Detectors and appropriate gases

Detector	Carrier gas	Detector gas
Thermal conductivity (TCD)	Helium, hydrogen	—
Flame ionization (FID)	Helium, nitrogen	Air and hydrogen
Nitrogen-phosphorus (NPD)	Helium, nitrogen	1. Air and hydrogen 2. Air and 8% hydrogen in helium
Electron capture	Nitrogen	5% methane in argon
	5% methane in argon	—

vides a corresponding electrical signal. Any physical device that accomplishes this may be used as a detector; however, only a few have obtained widespread application. For proper operation or optimum response, each type of detector requires a specific carrier gas (Table 6-6). The most widely used detectors are discussed below.[20,21]

Thermal-conductivity detector (TCD). A thermal-conductivity detector measures the difference in ability to conduct heat (thermal conductivity) between pure carrier gas and the carrier with sample mixture. A sample carried in the gas increases the thermal conductivity. Usually four heat-sensing elements, thermistors or wires, are mounted in a brass or stainless steel heat sink and connected to form the arms of a Wheatstone bridge (Fig. 6-10). An electric current is passed through the wires composing the bridge. Two filaments in opposite arms of the bridge are cooled by carrier gas (reference) and the other two by the column effluent (sample) (Fig. 6-9). The heat lost over both sets of wires is balanced by adjustment of the flow rate of the pure carrier gas. Emerging components from the column increase the rate of cooling of the sample wires because of the increased thermal conductivity of the gas mixture. This changes the electrical resistance of the sample wire pattern, making the Wheatstone bridge out of balance. This imbalance causes a response on the recorder. Important variables in optimum TCD response are carrier gas, flow rate, filament current, and detector temperature. TCD's lack selectivity, since any compound cooling the wire will cause a response. They are not as sensitive as other detectors, with minimum detection ranging from 0.1 to 0.5 μg of analyte per microliter.

Flame-ionization detector (FID). In a flame-ionization detector, eluted components in the carrier gas are mixed with hydrogen and burned in air to produce a very hot flame to ionize organic compounds. A pair of electrodes, charged by a polarizing voltage, collects the ions and generates a current proportional to the number of ions collected. The resultant current is amplified by an electrometer, producing a response on the recorder. The response of an FID is directly proportional to the number of carbons in a molecule bound to hydrogen or other carbon atoms. It is insensitive to water, carbon monoxide, carbon dioxide, and most inorganic compounds. The FID is the most popular detector for the determination of organic compounds. Sensitivity depends on chemical structure; therefore the detector response must be determined for each compound analyzed. At optimal conditions the minimal detectable quantity of organic compound is 1 ng. A cross section of an FID detector is represented in Fig. 6-11.

Nitrogen-phosphorus detector (NPD). A nitrogen-phosphorus detector is similar to an FID except that ions of an alkali metal (rubidium) are introduced into the hydrogen flame. When a compound containing nitrogen or phosphorus is burned in the flame, the rate of release of alkali metal vapor is increased. The alkali metal vapor readily ionizes in the flame and increases the current flow, which results in enhanced sensitivity for nitrogen and phosphorus. The optimum response is greatly dependent

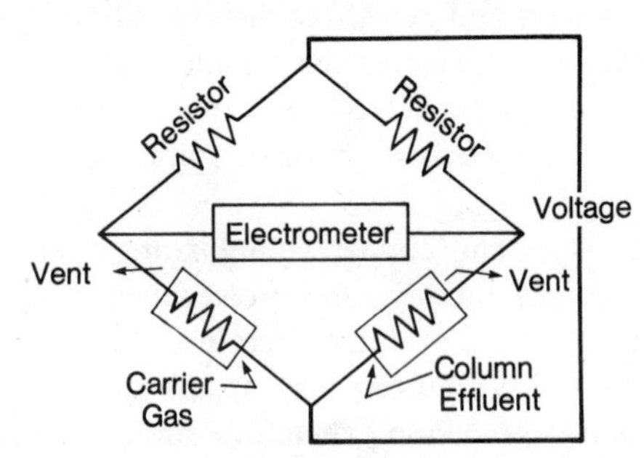

Fig. 6-10. Schematic diagram of thermal conductivity detector. (From Werner, M., Mohrbacher, R.J., and Riendeau, C.J.: In Baer, D.M., and Dito, W.R.: Interpretation of therapeutic drug levels, Chicago, 1981, American Society of Clinical Pathologists.)

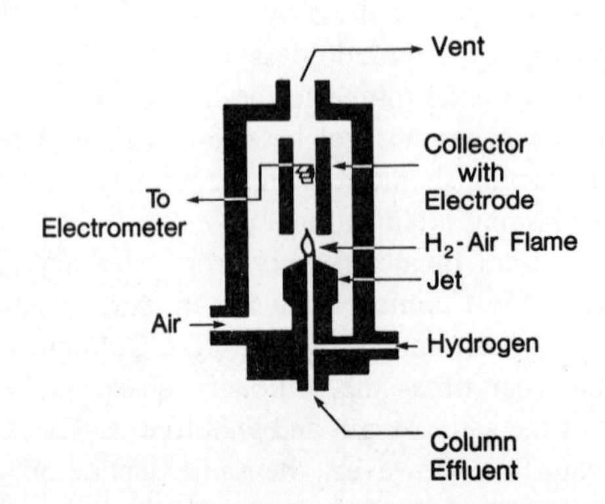

Fig. 6-11. Schematic diagram of flame ionization detector. (From Werner, M., Mohrbacher, R.J., and Riendeau, C.J.: In Baer, D.M., and Dito, W.R.: Interpretation of therapeutic drug levels, Chicago, 1981, American Society of Clinical Pathologists.)

on the flow of hydrogen. The selective interaction of alkali metal ions with these compounds is complex and poorly understood. However, the sensitivity to organonitrogen compounds and lack of response to other organics make the NPD highly advantageous for the analysis of biological samples. At optimal conditions, the minimal detectable quantity of nitrogenous organic compounds is less than 1 ng. A cross section of an NPD detector is presented in Fig. 6-12.

Electron-capture detector (ECD). In an electron-capture detector, a radioactive isotope releases beta particles that collide with the carrier gas molecules, producing many low-energy electrons. The electrons are collected on electrodes and produce a small, measurable, ''standing current.'' As sample components that contain chemical groups with high electron affinity (electrophilic species), particularly halogen atoms, are eluted from column, they capture the low-energy electrons generated by the isotope to form negatively charged ions. The detector measures the loss of cell current because of the recombination of the electrons. Three techniques are used for the collection of the electrons: (1) direct current (DC), (2) pulsed method, and (3) linear method. The DC method is the application of a constant voltage to the cell electrodes, and the electrons are collected continuously to produce a steady current. The sensitivity of the method is less than other ECD methods, since both negative ions and free electrons are collected by the electrodes. The reduction in current is less than if only free electrons were collected. The pulsed method is the application of a voltage in continuous pulses of short duration; therefore the heavy negative ions do not have time to respond and only free electrons are captured. Between pulses, the electron concentration in the detector builds up to levels exceeding those of the DC method. Thus the pulse method has greater sensitivity. The DC and pulsed methods inherently produce a nonlinear response over a wide range of sample concentrations. Such a response is attributable to the finite amount of beta radiation

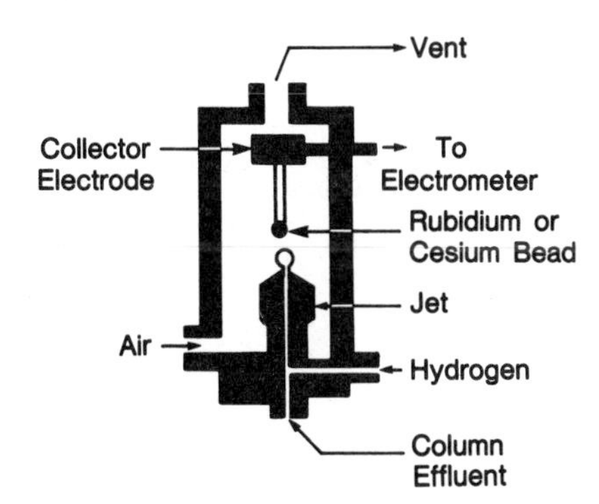

Fig. 6-12. Schematic diagram of alkali-metal flame detector. (From Werner, M., Mohrbacher, R.J., and Riendeau, C.J.: In Baer, D.M., and Dito, W.R.: Interpretation of therapeutic drug levels, Chicago, 1981, American Society of Clinical Pathologists.)

emitted by the detector source per unit time. Since a decrease in current is measured, once a concentration of eluting solute captures a majority of available low-energy electrons, only small changes in current (detector response) will be observed with increasing concentrations of solute. The linear range is usually 400 to 500 times the detection limit of a solute for a tritium source and 100 times for a nickel-63 source. However, the linearized method uses electronic modifications that operate the detector in a pulsed mode such that constant cell current is produced. The linear range is thus expanded to ranges of 10,000:1 for a nickel-63 source. The sources of beta particles in an ECD are usually tritium or nickel-63. The ECD is the most sensitive detector available, since as little as 1 picogram of halogen-containing compound may be measured. Laboratories using electron-capture detectors must be licensed by the Nuclear Regulatory Commission and are subject to all regulations concerning employee safety and possible environmental contamination as set forth by the commission.

Mass spectrometer (MS) as a detector. The mass spectrometer is a specialized gas chromatographic detector that provides extremely sensitive detection (picogram quantities) and is the ultimate specific identification technique currently available to the analyst. The mass spectrometer is often considered a separate instrument rather than a component of a gas chromatograph because samples may be directly inserted into the instrument (direct-probe technique) or it may be coupled with other chromatographic instruments such as a high-performance liquid chromatograph. The primary difficulty in combining a gas chromatograph with a mass spectrometer is that the GC operates at a positive pressure whereas the MS operates under a high vacuum. This limitation is overcome by use of large vacuum pumps (700 L/sec) that maintain a vacuum despite the large flow of gas from the GC. Additionally, a ''separator'' inserted between the effluent of the GC column and the ionization chamber of the MS removes large portions of the carrier gas, concentrating the analyte and reducing the total volume of gas entering the MS. Two types of separators are available, the ''membrane'' and the ''jet.'' The membrane separator, made of silicone rubber, is mounted between the effluents of the GC and the vacuum of the MS. The organic solutes are more soluble in the membrane than the carrier gas and after being dissolved in the membrane are pumped away by the vacuum system and transferred into the ion source. The jet separator operates on the principle of differential mobility of sample molecules and carrier gas. The effluent from the GC enters a vacuum through a narrow jet. The core of the molecular beam contains a higher concentration of sample molecules than the perimeter does. A pump removes the lighter gas molecules, whereas the heavier sample molecules in the core are collected by a small orifice and transferred into the MS.

The generation of a spectrum by the mass spectrometer involves three steps: ionization, mass filtration, and detection. At present there are three modes of ionization used in the analysis of biological specimens: (1) electron impact (EI), (2) chemical ionization (CI), and (3) negative chemical ionization (NCI). In the electron-impact mode the sample molecules in the gas phase are bombarded with high-energy electrons (70 eV) so that they produce a charged molecule or shatter the molecule into ionic fragments. The charged fragments are separated and detected according to their atomic masses. The mass spectrum is a display of the different masses of the charged fragments and their relative abundance. The mass spectrum generated by electron impact is characteristic of the molecule analyzed and is often used to establish an unequivocal identification of an unknown drug. The electron-impact spectrum of 3,4-methylenedioxyamphetamine (MDA), a hallucinogenic drug of abuse, isolated from the urine of a fatal overdose case[22] is presented in Fig. 6-13. The identification of MDA is based upon the characteristic mass to charge (*m/e*) fragment pattern and relative ion abundance (height) of these fragments compared to the major ion fragment at ion mass 44 (Table 6-7).

In the chemical ionization technique, the sample molecules are mixed with an ionized gas such as methane or isobutane. A positive charge (proton) is transferred to the sample molecule as a result of the ion-molecule collisions. In the electron beam, methane CH_4 becomes transformed to CH_5^+:

$$CH_4 + \text{Electron beam} \rightarrow CH_4^{+}\bullet$$
$$CH_4 + CH_4^{+}\bullet \rightarrow CH_5^+ + \text{Other products}$$

The CH_5^+ can then interact with the analyte of interest to form a positive M + 1 ion.

$$R + CH_5^+ \rightarrow RH^+ + CH_4$$

This can then break down to form fragments:

$$RH^+ \rightarrow \text{Fragments}$$

By controlling experimental conditions, one can minimize fragmentation of the sample molecule, and the dominant ion produced equals the molecular weight (M) of sample plus a proton (M + 1, Fig. 6-14). The electron-impact and chemical-ionization mass spectra of cocaine are compared in Fig. 6-14. Unlike the fragmented EI spectrum, the CI spectrum of cocaine obtained using methane and ammonia as reagent gases shows only a single abundant ion corresponding to the protonated molecular ion (*m/z* 304). The CI mass spectrum using methane displays a base peak (fragment of highest abundance) at *m/z* 182, which results from the loss of benzoic acid from the protonated cocaine molecule. In negative chemical ionization, oxygen or hydrogen are mixed with the sample molecules to produce a dominant negative ion of M − 1. The detection of compounds containing halogen (fluorine, chlorine) atoms may

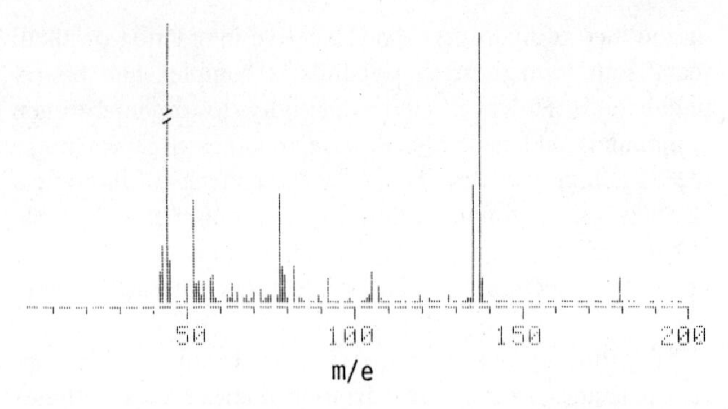

Fig. 6-13. 3,4-Methylenedioxyamphetamine (MDA) electron impact spectrum, *m/e*. Ion mass 44 base peak and an ion mass peak at 179. (From Poklis, A., Mackell, M.A., and Drake, W.K.: J. Forensic Sci. **24**(1):70, 1979.)

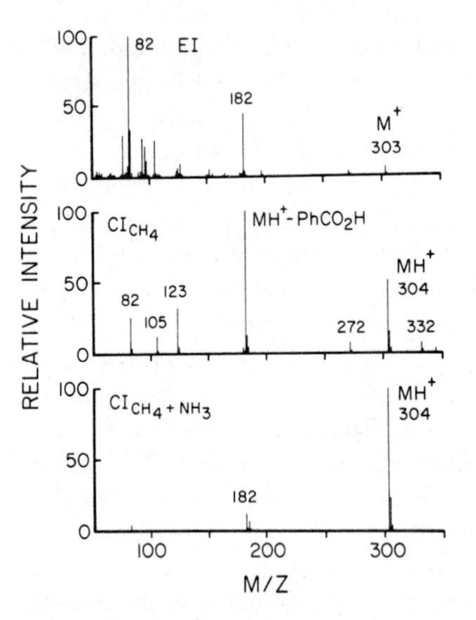

Fig. 6-14. Mass spectra, M/Z, of cocaine. (From Foltz, R.L., Fentiman, A.F., and Foltz, R.B.: GC/MS assays for abused drugs in body fluids, Washington, D.C., 1980, U.S. Dept. of Health and Human Services.)

Table 6-7. Ion mass and mass abundance of MDA* standard and MDA extracted from decedent's urine†

Ion mass, *m/e*	Ion abundance of MDA standard	Ion abundance of urine compound‡
44	100.0	100.0
51	12.7	12.1
77	16.4	12.5
78	6.4	4.5
79	6.4	4.5
105	4.3	3.6
135	13.1	13.6
136	35.6	31.0
179	5.2	3.1

*3,4-Methylenedioxyamphetamine (MDA).
†From Poklis, A., Mackell, M.A., and Drake, W.K.: J. Forensic Sci. **24**:70-75, 1979.
‡Similarity index, 0.9957; molecular weight, 179.2.

be greatly enhanced by negative chemical ionization. The chemical ionization technique determines the molecular weight of the sample and, because of the paucity of the fragment ions, is ideally suited for high-sensitivity, quantitative analysis. Once produced, the ions are directed and separated by mass by a magnetic field.

Two types of chromatograms may be produced when the MS is used as a GC detector: the "total chromatogram" and the "selective ion chromatogram." In the total chromatogram, the total ion current of the various compounds in the GC effluent is displayed. The chromatogram has the appearance of any other separation of sample components as produced by the conventional GC detectors, such as the FID or NPD. The mass spectrum of each peak is collected and stored by a computerized data system. The spectrum of each separated component may then be produced on a printout of the data system. All ionized compounds being eluted from the GC are detected. The selective ion chromatogram is produced when a specific mass fragment or ion is monitored. Depending on the capabilities of the instrument, several specific ions may be monitored. Only ionized or fragmented compounds produce the specific charged mass unit and are detected and displayed in the chromatogram. Therefore in GC separation of a complex mixture the total chromatogram will display all components present, whereas selective ion monitoring (SIM) will display only those components that produce a specific chosen ion. The m/z 182 ion in the electron-impact spectrum in Fig. 6-14 could be used for selective ion monitoring.

Because of the complexity of operation and the resultant spectral lines, mass spectrometers are connected to computerized systems for operation and data collection. For qualitative analysis, the computer automatically stores spectra of sample compounds and searches a stored spectral library of known compounds for comparison (Table 6-7).

Readout. Strip chart recorders are the most common readout devices in gas chromatography. Recorder sensitivity is usually 1 to 10 mV, with a full scale response of 1 second or less. Quantitative determinations of separate compounds are performed in two ways: peak-height or peak-area measurements. Both the peak height and peak area of the detector response to the effluent sample are proportional to its concentration. Peak-height measurements are useful in repetitive analysis when performed by the same operator in a fixed system, when extensive calibration is necessary, or when only partial resolution of compounds makes peak-area determinations difficult. In general, peak-area measurements are more precise. Peak area may be determined by several manual or instrumental means. Electronic integration of the peak area produces both the most precise and the most accurate measurements. Today, detectors may be connected to microprocessor units or a computerized data system that automatically records the response, identifies the sample components, integrates the signals, performs calculations, stores all data, and prints out the analytical results in final form.

REFERENCES

1. Ettre, L.S.: The nomenclature of chromatography. I. Gas chromatography, J. Chromatogr. **165**:235-256, 1979.
2. Ettre, L.S.: Practical gas chromatography, Norwalk, Conn., 1973, Perkin Elmer Corp.
3. Rowland, F.W.: The practice of gas chromatography, ed. 2, Avondale, Pa., 1974, Hewlett-Packard Co.
4. Grob, R.L.: Modern practice of gas chromatography, New York, 1977, John Wiley & Sons, Inc.–Interscience.
5. Delley, R., and Friedrich, K.: System CG72 von bevorzugten Trennflüssigkeiten für die Gas-chromatographie, Chromatographia **10**:593-598, 1971.
6. Hawkes, S., Grossman, D., Hartkopf, A., et al.: Preferred stationary liquids for gas chromatography, J. Chromatogr. Sci. **13**:115-117, 1975.
7. Siggia, S.: Instrumental methods of organic functional group analysis, New York, 1972, John Wiley & Sons, Inc.–Interscience.
8. Ahuja, S.: Derivatization in gas chromatography, J. Pharm. Sci. **65**:163-181, 1976.
9. Kovats, E.: Gas-chromatographische Charakterisierung organischer Verbindungen. Teil 1: Retentionsindices aliphatischer Halogenide, Alkohol, Aldehyde und Ketone, Helv. Chim. Acta **41**:1915-1932, 1958.
10. Kovats, E.: The Kovats retention index system, Anal. Chem. **36**:31A-41A, 1964.
11. Lorenz, L.J., and Roger, L.B.: Specification of gas chromatographic behavior using Kovats indices and Rohrschneider constants, Anal. Chem. **43**:1593-1599, 1971.
12. American Society for Testing and Materials: Gas chromatographic data compilation catalog AMD 25A, Philadelphia, Pa., 1967.
13. American Society for Testing and Materials: Gas chromatographic data compilation catalog, suppl. 25A S-1, Philadelphia, Pa., 1971.
14. Perrigo, B.J., and Peel, H.W.: The use of retention indices and temperature-programmed gas chromatography in analytical toxicology, J. Chromatogr. Sci. **19**:219-226, 1981.
15. Moffat, A.C.: Use of SE-30 as a stationary phase for the gas-liquid chromatography of drugs, J. Chromatogr. **113**:69-95, 1975.
16. Supina, W.R., and Rose, L.P.: The use of Rohrschneider constants for classification of GLC columns, J. Chromatogr. Sci. **8**:214-217, 1970.
17. McReynolds, W.O.: Characterization of some liquid phases, J. Chromatogr. Sci. **8**:685-691, 1970.
18. Ettre, L.S.: Basic relationships of gas chromatography, ed. 2, Norwalk, Conn. 1979, Perkin-Elmer Corp.
19. Fenimore, D.C., Whitford, J.J., Davis, C.M., and Zlatkis, A.: Nickel gas chromatographic columns: an alternative to glass for biological samples, J. Chromatogr. **140**:9-16, 1977.
20. David, D.J.: Gas chromatographic detectors, New York, 1974, John Wiley & Sons, Inc.
21. Sevcik, J.: Detectors in gas chromatography, New York, 1975, Elsevier/North-Holland, Inc.
22. Poklis, A., Mackell, M.A., and Drake, W.K.: Fatal intoxication from 3,4-methylenedioxyamphetamine, J. Forensic Sci. **24**:70-75, 1979.
23. Foltz, R.L., Fentiman, Jr., A.F., and Foltz, R.B., editors: GC/MS asays for abused drugs in body fluids, Rockville, Md., 1980, National Institute of Drug Abuse.

Chapter 7 Isotopes in clinical chemistry

I-Wen Chen

becquerel (Bq) Système International d'Unités (SI) unit of radioactivity corresponding to a decay rate of one per second (1 Bq = 1 sec^{-1} = 2.7 × 10^{-9} Ci) (see ***curie***).

chemiluminescence Production of light photons by an interaction of the sample material with the solute or solubilizer added to the scintillation solution in liquid scintillation counting.

circuit, anticoincident A circuit used in the pulse-height analyzer of a radioactive-particle counter for setting window width (see ***window***). It transmits a pulse arriving at its input from the lower discriminator only if there is no pulse arriving from the upper discriminator at the same time (see ***discriminator***).

circuit, coincident A circuit used in a liquid scintillation counter to eliminate the electronic noise. It determines if a pulse from one photomultiplier tube is accompanied by a corresponding pulse from the other within the allowed time interval (see ***coincidence resolving time***).

circuit, summation A circuit used in a liquid scintillation counter to sum all coincident pulses to improve counting efficiency for low-energy beta-particle emitters.

coincidence resolving time A time interval within which the output pulses from each photomultiplier tube of a liquid scintillation counter have to arrive at the coincident circuit in order to be counted.

curie (Ci) Unit of radioactivity. One curie is defined as an activity of a sample decaying at a rate of 3.7 × 10^{10} disintegrations per second (dps).

decay constant A constant unique to each radioactive nuclide (see ***nuclide***), representing the proportion of the atoms in a sample of that radionuclide undergoing decay in unit time.

decay factor The fraction of radionuclides remaining after a time, t.

discriminator(s) Device(s) used in the pulse-height analyzer of a radioactive-particle counter for setting upper (upper-level discriminator) and lower (lower-level discriminator) voltage limits for counting.

electron capture One mode of radioactive decay in which the neutron-poor nuclides decay by capturing electrons from orbits closest to the nucleus to transform a proton to a neutron. The orbital vacancy created by the electron capture is filled by the electron from a higher orbit, resulting in emission of characteristic x rays.

electron volt (eV) Basic unit of energy commonly used in radiation, defined as the amount of energy acquired by an electron when it is accelerated through an electrical potential of 1 volt.

half-life ($t_{1/2}$) Time required for a given number of radionuclides in the sample to decrease to one half its original value.

isobar Nuclides (see ***nuclide***) with the same atomic mass number but different atomic number.

isotope Nuclides (see ***nuclide***) with the same atomic number but different atomic mass number.

isotopic abundance Amounts of isotopes present in an element.

nucleon A collective term for protons and neutrons within the nucleus.

nuclide A nucleus with a particular atomic number and atomic mass number.

rad (radiation absorbed dose) A measure of local energy deposition per unit mass of material irradiated. One rad is equal to 100 ergs of absorbed energy per gram of absorber.

rem (roentgen equivalent, man) A unit of biological dose as a result of exposure to ionizing radiation. In the case of x, gamma, or beta radiation, rems are equal to the absorbed dose in rads; in the case of alpha radiation, however, the dose in rems equals the dose in rads multiplied by 20 because only 0.05 rad of alpha radiation is needed to produce the same biological effect as 1 rad of x, gamma, or beta radiation.

roentgen (R) A unit of x rays or gamma rays representing the quantity of ionization produced by photon radiation in a given sample of air. One R equals that quantity of photon radiation capable of producing 1 electrostatic unit of ions of either sign in 0.001293 g of air.

specific activity Activity of the radionuclide per unit mass of the radioactive sample, expressed as Ci per μg, μCi per μmole, and so on.

specific ionization Number of ion pairs produced per unit path length of ionizing radiation.

transmutation A radioactive decay process that results in a change in nuclear constitution, such as electron capture decay (see ***electron capture***).

wavelength shifter The secondary scintillator added to scintillation liquid for shifting the wavelength of light emitted by the primary scintillator for more efficient detection by the photocathodes of photomultiplier tubes in liquid scintillation counting.

window The voltage limit set by the upper-level and lower-level discriminators (see ***discriminator(s)***) of the pulse-height analyzer of a radioactive-particle counter for differential counting.

The use of isotopes, both stable and radioactive, has provided a great store of information in the field of medical sciences, much of which could not have been obtained in any other way. The usefulness of isotopes depends on the fact that isotopes of an element have identical chemical properties but different isotopic properties, such as radioactivity and increased mass, and on the fact that the isotopic properties and chemical properties of an element are independent of each other. Therefore substitution of an atom in the molecule of a substance by other isotopes will not chemically alter that substance, and the isotopic properties of the isotopes incorporated into that substance will remain unchanged. The isotopic properties will make that substance more easily identifiable. For example, thyroxine is a thyroid hormone containing four atoms of iodine. One or all of these iodine atoms may be replaced by radioactive iodine without appreciable alteration of its chemical properties, and the radioiodine atoms incorporated into the thyroxine molecules will maintain their characteristic radioactivity. The radioiodine-labeled thyroxine molecules can be identified and quantified easily by virtue of their radioactivity. Radioiodine-labeled thyroxines are used in various thyroid-function tests.

The application of radioisotopically labeled compounds in clinical chemistry has greatly expanded since the advent of radioimmunoassays, in which the quantity of antigen bound to antibody is determined by measurement of radioactivity. Because of its inherent sensitivity, specificity, and wide applicability, radioimmunoassay has been playing an important role in the diagnosis of various disorders since its discovery in 1956 by Berson and Yalow.

Radioactive elements are in many respects more useful tracers than stable isotopes are, because the analytical methods for their measurement are exceedingly sensitive. However, recent advances in mass spectrometry have greatly improved the sensitivity, specificity, and quantitative accuracy of stable isotope measurements, and the application of stable isotopes in clinical chemistry is expected to increase in the future. Isotope-dilution mass spectrometric methods have been used for determinations of many analytes, including serum glucose and cholesterol. In this chapter, some basic principles involved in the measurements of isotopes are discussed.

BASIC STRUCTURE OF ATOM

Fundamental particles of atom

An atom is the smallest unit of matter that still exhibits the chemical properties of an element. The primary building blocks of atoms are the electron, the proton, and the neutron and are termed *elementary particles*. According to the planetary model of the atom developed by Rutherford in 1911, the atom consists of a central, small, positively charged body (the nucleus composed of protons and neutron) around which the negatively charged electrons move in defined orbits. Although the Rutherford model is oversimplified, one can use it to explain many atomic phenomena satisfactorily. The planetary model of an atom of carbon is illustrated in Fig. 7-1.

The nucleus of carbon contains six protons and six neutrons. Since complete atoms are electrically neutral, six orbiting electrons are present in the carbon atom to match the six protons in the nucleus. They move around the nucleus in a series of orbits or shells, at varying distances from the nucleus, much as the planets of the solar system travel in different orbits at varying distances from the sun. The orbits, or shells, are called "K," "L," "M," and so on, starting from the inner one. Only two electrons can be accommodated in the K shell; the L shell of the carbon atom contains the remaining four.

The differences between the atoms of different elements depend on the number of protons and neutrons contained in an atomic nucleus, which determines both the mass and charge of the nucleus, and the number and arrangements of the orbital electrons, which determine the chemical properties of elements.

Atomic nomenclature

There are several important terms that are helpful in understanding atomic structure:

nucleon A collective term for protons and neutrons within the nucleus.
atomic number, Z The number of protons in the nucleus.
atomic mass number, A The total number of nucleons within the nucleus.
neutron number, N The number of neutrons within the nucleus.
nuclide A nucleus with particular Z and A numbers.
element, E A nucleus with a given Z number.
isotope Nuclides with the same Z but different A numbers (various nuclear species of the same element).
isobar Nuclides with the same A but different Z numbers (different elements with the same atomic mass).

The atomic mass number is represented as a left superscript and the atomic number as a left subscript to the

Fig. 7-1. Planetary model of carbon atom.

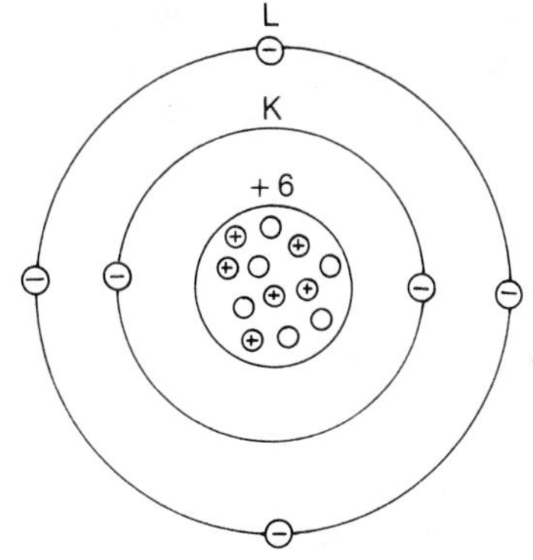

chemical symbol. Thus an element, E, is written as $^{A}_{Z}E$. The most abundant, naturally occurring, stable isotope of carbon has six protons and six neutrons in the nucleus, as shown in Fig. 7-1. The atomic number, Z, is therefore 6; the atomic mass number ($A = Z + N$) is 12; and the whole atom may be described as $^{12}_{6}C$. The other naturally occurring but less abundant isotope of carbon is $^{13}_{6}C$, which contains seven neutrons in the nucleus. $^{12}_{6}C$ and $^{13}_{6}C$ are both stable isotopes of carbon, neither of which is radioactive. The best-known radioactive isotope of carbon is $^{14}_{6}C$, which contains six protons and eight neutrons. Examples of other groups of isotopes of an element commonly used in clinical chemistry are $^{1}_{1}H$, $^{2}_{1}H$, and $^{3}_{1}H$ and $^{125}_{53}I$, $^{127}_{53}I$, and $^{131}_{53}I$. $^{1}_{1}H$ is the most abundant, naturally occurring isotope of hydrogen and has one proton but no neutron in the nucleus. $^{2}_{1}H$ is a stable isotope of hydrogen and is known as *deuterium* because its nucleus contains two nuclear particles, one proton and one neutron. $^{3}_{1}H$, called *tritium,* is a radioactive isotope of hydrogen, the nucleus of which is formed by a combination of a proton and two neutrons. All isotopes of hydrogen have a single circling electron and therefore have identical chemical properties; however, their physical properties are different. For example, they will have different boiling and freezing points. In the case of tritium, its nucleus is unstable and, as is discussed later, will undergo radioactive transitions to become a different and stable nucleus—the nucleus of helium. The naturally occurring stable isotope of iodine is $^{127}_{53}I$. The other two isotopes of iodine mentioned here are radioactive isotopes with different numbers of neutrons in their nucleus, as indicated by their atomic mass numbers. In many cases the atomic-number subscript is redundant because the atomic number and the chemical symbol both identify the chemical species. Therefore, except in some equations describing nuclear reactions, the subscript is normally omitted (such as ^{14}C, ^{3}H, ^{125}I).

PRINCIPLES OF RADIATION AND RADIOACTIVITY

Nuclear radiation

The release of energy or matter during the transformation of an unstable atom to a more stable atom is termed *nuclear radiation.* The numbers and arrangement of protons and neutrons in the nucleus of an atom determine whether the nucleus is stable or unstable.

Nuclear stability. There are favored neutron-to-proton ratios among stable nuclides. The ratio is equal to or close to unity for the light nuclides. When the atomic mass number exceeds 40, no stable nuclides exist with equal numbers of neutrons and protons because, as the number of protons increases, the repulsive coulombic forces between the protons increase at a greater rate than the attractive nuclear force does. Therefore the addition of extra neutrons is necessary to increase the average distance between protons within the nucleus to reduce the coulombic force. For heavy nuclei the neutron/proton ratio is 1.5 or greater. For example, the heaviest stable isotope of lead, ^{208}Pb, has a neutron/proton ratio of 1.53.

Fig. 7-2 illustrates the relationship between the neutron and proton numbers of the stable nuclides. An imaginary line, called the *line of stability* and represented by a dashed line in the graph, can be obtained from the neutron/proton plot, and the stable nuclides are clustered around this line. Nuclides deficient in protons lie below the line of stability and are unstable. Nuclides deficient in neutrons lie above the line and are also unstable. The graph also illustrates the fact that as nuclides become heavier, more neutrons are required to maintain stability.

In addition to the favored neutron/proton ratio, the stable nuclides tend to favor even numbers. For example, 168 out of approximately 280 known, stable nuclides have even numbers of both protons and neutrons, reflecting the tendency of nuclides to achieve stable arrangements by pairing up nucleons in the nucleus.

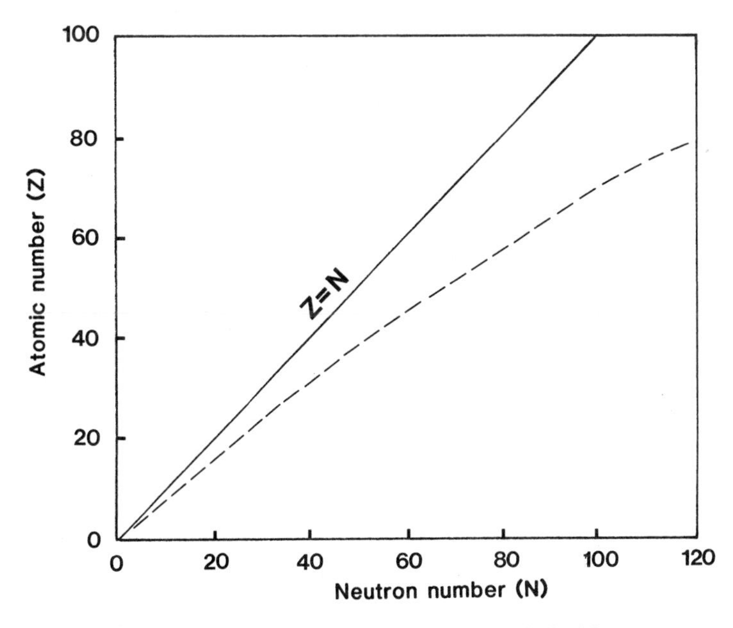

Fig. 7-2. Neutron-proton ratios for stable isotopes, *dashed line.*

Modes of radioactive decay. Unstable nuclides are generally transformed into stable nuclides by one of the radioactive-decay processes described below.

Decay by alpha-particle emission. An alpha (α) particle consists of two neutrons and two protons and is essentially a helium nucleus. Heavy nuclides that must lose mass to achieve nuclear stability frequently decay by alpha-particle emission because alpha-particle emission is an effective way to reduce the mass number. The emission of one alpha particle removes two neutrons and two protons from the nucleus, resulting in the reduction of an atomic number by 2 and a mass number by 4. Very heavy radioactive nuclides that decay with alpha-particle emission are of little interest in clinical chemistry. An example of alpha-particle decay is given below:

$$^{226}_{88}Ra \rightarrow ^{222}_{86}Rn + ^{4}_{2}He \text{ (alpha particle)}$$

Decay by beta-particle emission. Beta (β) particles are either negatively charged electrons (negatrons, β^-), or positively charged electrons (positrons, β^+). Proton-deficient nuclides lying below the line of stability, shown in Fig. 7-2, usually decay by negatron emission because this mode of decay transforms a neutron into a proton, moving the nucleus closer to the line of stability. Neutron-deficient nuclides lying above the line of stability usually decay by positron emission since this mode transforms a proton into a neutron. Schematically these modes of radioactive decay can be represented by the following equations:

$$n \rightarrow + \beta^- + \nu + \text{Energy} \quad \text{(Negatron emission)}$$
$$p^+ \rightarrow + \beta^+ + \nu + \text{Energy} \quad \text{(Positron emission)}$$

The beta particles and neutrino (ν) ejected from the nucleus carry away the energy released in the decay process as kinetic energy. The neutrino is a particle with no mass or electrical charge and virtually does not interact with matter. The only practical consequence of its emission from the nucleus is that it carries away some energy released in the decay process.

In beta-particle decay processes the mass number does not change because the total number of nucleons in the nucleus remains the same. Such decay processes are known as "isobaric transitions." However, the atomic number increases by 1 in the negatron emission and decreases by 1 in the positron emission, resulting in a transmutation of elements (conversion of one element to another). Examples of decay by beta emission are as follows:

$$^{3}_{1}H \rightarrow ^{3}_{2}He + \beta^- + \nu \quad \text{(Negatron emission)}$$
$$^{11}_{6}C \rightarrow ^{11}_{5}B + \beta^+ + \nu \quad \text{(Positron emission)}$$

Decay by electron capture. In addition to the decay by positron emission, the neutron-deficient nuclides may decay by electron capture to transform a proton to a neutron. Thus the electron capture is sometimes called *inverse negatron decay*. It is also an isobaric transition leading to a transmutation of elements. In the electron-capture process the electron is captured from orbits closest to the nucleus, that is, the K and L shells (K and L capture; see Fig. 7-1). The orbital vacancy created by the electron capture is quickly filled by the electron from a higher orbit, resulting in emission of a characteristic x ray:

$$p^+ + e^- \rightarrow n + \nu + \text{X ray} + \text{Energy}$$

The daughter nucleus formed by this mode of decay is frequently in an excited or metastable state and may further undergo decay by gamma (γ)-ray emission, as described below.

Decay by gamma-ray emission. In some cases, the isobaric transitions previously mentioned (negatron emission, positron emission, electron capture) result in a daughter nucleus in an excited or metastable state, which means that it possesses excess energy above its minimum possible ground-state energy. Such an excited or metastable nuclide decays promptly to a more stable nuclear arrangement by the emission of gamma rays, electromagnetic radiation of very short wavelength. Examples of such modes of decay are as follows:

$$^{131}_{53}I \xrightarrow{\beta^-} ^{131}_{54}Xe^* \xrightarrow{\gamma} ^{131}_{54}Xe$$
$$^{68}_{31}Ga \xrightarrow[\text{electron capture}]{\beta+} ^{68}_{30}Zn^* \xrightarrow{\gamma} ^{68}_{30}Zn$$
$$^{125}_{53}I \xrightarrow[\text{capture}]{\text{electron}} ^{125}_{52}Te^* \xrightarrow{\gamma} ^{125}_{52}Te$$

Note that gamma emission is not accompanied by any change in mass number, proton number, or neutron number. This is called an *isomeric transition*.

Extranuclear radiation

In a series of events occurring in some electron-capture decay processes discussed above, characteristic x rays, as well as gamma rays, are emitted. X rays and gamma rays are both electromagnetic waves and are physically indistinguishable. However, they are distinguished by their origin of emission: x rays are extranuclear and gamma rays are nuclear in origin. Other extranuclear radiation is emitted in the form of Auger (pronounced o-zháy) electrons. A photon of the characteristic x rays discussed above may collide with one of the other orbital electrons. When the photon's energy is greater than the electron's binding energy, the electron is ejected. The ejected electron is termed an *Auger electron*. This process reduces energy levels by emitting matter, as compared to the emission of a photon in x-ray production.

Radiation energy

In any of the radioactive decay processes mentioned previously, a fixed amount of energy is released with each disintegration. Most or all of the released energy will appear as the kinetic energy of the emitted particles or photons. The basic unit of energy commonly used in radiation

*The nuclide is in an excited or metastable state.

is the electron volt (eV). One electron volt is defined as the amount of energy acquired by an electron when it is accelerated through an electrical potential of 1 volt. Basic multiples are the kiloelectron volt (keV; 1 keV = 1000 eV) and the megaelectron volt (MeV; 1 MeV = 1000 keV = 1,000,000 eV). In general, the energy of beta particles emitted from radionuclides in clinical use ranges from 18 keV to 3.6 MeV, and that of gamma rays from 27 keV to 2.8 MeV.

Decay of radioactivity

Decay constant, decay factor, and half-life. Radioactive decay is a spontaneous process; that is, it is not possible to predict when a given radioactive atom will decay, and the probability of decay can be given only on a statistical basis. For a sample containing N radioactive nuclei, the number of nuclei decaying at any given moment (dN/dt) can be given by the following:

$$\frac{dN}{dt} = -\lambda N \qquad \textit{Eq. 7-1}$$

In this equation, λ is the decay constant of the radioactive nuclide, and the minus sign, $-$, indicates that the number of radioactive nuclides is decreasing with time. Each radionuclide has a characteristic decay constant that represents the proportion of the atoms in a sample of that radionuclide undergoing decay in unit time. The decay constant, λ, is measured in units of $(\text{time})^{-1}$. Therefore the equivalence $\lambda = 0.05\ \text{sec}^{-1}$ means that, on the average, 5% of the radionuclides are disintegrating per second. On integration of equation 7-1, we obtain

$$N = N_0 e^{-\lambda t} \qquad \textit{Eq. 7-2}$$

where N_0 is the number of radionuclides present at time $t = 0$, and e is the base of the natural logarithm. Therefore the number of radionuclides remaining after a time, $t(N)$, is equal to the number of radionuclides at a time, $t = 0$ (N_0), multiplied by the factor ($e^{-\lambda t}$). This factor is the fraction of radionuclides remaining after a time, t, and is termed the *decay factor*. The decay factor, $e^{-\lambda t}$, is an exponential function of time, t; that is, a constant fraction of the number of radionuclides present in the sample disappears during a given time interval. A given time interval is customarily expressed as the time required for a given number of radionuclides in the sample to decrease one half its original value. This time interval is termed the *half-life* ($t_{1/2}$). The half-life of a radionuclide is related to its decay constant as follows:

$$t_{1/2} = 0.693/\lambda \qquad \textit{Eq. 7-3}$$

This equation is derived as follows:

In Eq. 7-2, $N = N_0/2$ after time $t_{1/2}$.

$\therefore N_0/2 = N_0 e^{-\lambda t_{1/2}}$ $\qquad \therefore \ln 2 = \lambda t_{1/2}$

$\therefore 2 = e^{\lambda t_{1/2}}$ $\qquad \therefore 0.693 = \lambda t_{1/2}$, or $t_{1/2} = 0.693/\lambda$

A plot of the decay factor versus the number of half-lives elapsed, t, on a semilogarithmic graph paper gives a straight line, as shown in Fig. 7-3. One can obtain this straight line by connecting two points on the curve, that is, at $t = 0$, decay factor $= 1.0$; at $t = 2t_{1/2}$, decay factor $= 0.25$. This graph can be used to determine the decay factor of any radionuclide at any given time, provided that the elapsed time is expressed in terms of number of radionuclide half-lives elapsed. For example, the half-life of ^{125}I is 60 days. If it is desired to calculate the residual radioactivity at 360 days, the number of half-lives must first be calculated:

$$\frac{360 \text{ days}}{60 \text{ days/half-life}} = 6 \text{ half-lives}$$

From Figure 7-3 one can see that the decay factor at 6 half-lives is approximately 0.016. When this value is multiplied by the initial amount of radioactivity, it gives the residual activity at 360 days.

One can also determine the decay factor from tables of such factors, which are available from most radiopharmaceutical companies or instrument manufacturers. An example of such a table for ^{125}I is shown in Table 7-1. For example, the decay factor for ^{125}I 28 days after the manufacturer's calibration date is 0.724. This number, multiplied by the initial amount of radioactivity, gives the level of radioactivity left at 28 days.

If the desired time, t, does not appear in the table, one can express it as a sum of times, $t = t_1 + t_2 + \ldots$, that do appear in the table, and the desired decay factor, DF, can be calculated according to principles based on the properties of exponential functions, as follows:

$$DF(t_1 + t_2 + \ldots) = DF(t_1) \times DF(t_2) \times \ldots \qquad \textit{Eq. 7-4}$$

For example, the decay factor for ^{125}I after 88 days can be calculated as follows:

$$DF(88 \text{ days}) = DF(60 \text{ days}) \times DF(20 \text{ days}) \times DF(8 \text{ days}) = 0.500 \times 0.794 \times 0.912 = 0.362$$

Occasionally, one receives radionuclides in precalibrated quantities. For example, a shipment of $Na^{125}I$ may be received 8 days before the calibration date. To determine its present radioactivity, it is therefore necessary to calculate the decay factor for 8 days before the calibration date, that is, DF(-8 days). As will be discussed in the

Table 7-1. Decay factors for ^{125}I

	Days				
Days	**0**	**4**	**8**	**12**	**16**
0	—	0.955	0.912	0.871	0.831
20	0.794	0.758	0.724	0.691	0.660
40	0.630	0.602	0.574	0.548	0.524
60	0.500	0.477	0.456	0.435	0.416

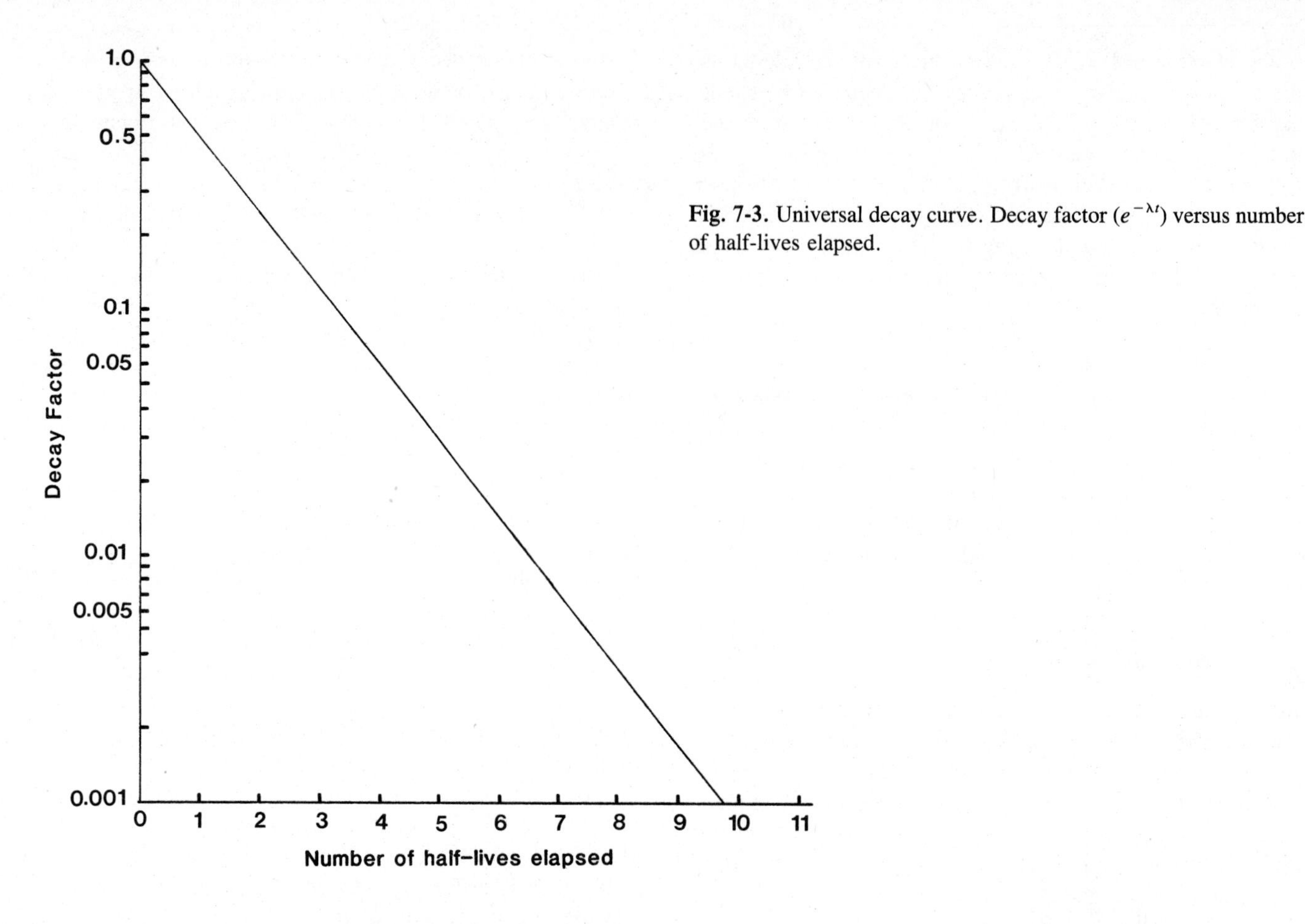

Fig. 7-3. Universal decay curve. Decay factor ($e^{-\lambda t}$) versus number of half-lives elapsed.

next section, the decay factor also applies to radioactivity versus time. According to another of the properties of exponential functions:

$$DF(-t) = 1/DF(t) \qquad \textit{Eq. 7-5}$$

Therefore

$$DF(-8 \text{ days}) = \frac{1}{0.912} = 1.10$$

Units of radioactivity. The average rate of decay of a sample (equation 7-1), namely, the average number of nuclides disintegrating per second (dps) or per minute (dpm), is the activity of the sample and is used to determine the amount of radioactivity present in the sample.

Radioactivity is measured in curie (Ci) units. One curie is defined as an activity of a sample decaying at a rate of 3.7×10^{10} dps (2.22×10^{12} dpm), which is very close to the activity of 1 g of ^{226}Ra (3.656×10^{10} dps/g). In fact, the curie was originally defined as the activity of 1 g of ^{226}Ra. The basic multiples of the curie are as follows:

1 curie (Ci) = 10^3 millicurie (mCi) = 10^6 microcurie (μCi) = 10^9 nanocurie (nCi) = 10^{12} picocurie (pCi)

In clinical chemistry the amounts of radioactivity used are usually in the range of nCi to μCi; occasionally, pCi quantities are measured. The use of SI units in radioactivity measurements have been introduced. The basic unit of this system is the becquerel (Bq), in which 1 Bq is one disintegration per second. Thus

$$1\ \mu Ci = 3.7 \times 10^4\ Bq$$

This system has not gained widespread acceptance in the United States.

In equation 7-2, N_0 and N are the numbers of radionuclides present at times 0 and t, respectively. These quantities are extremely difficult to measure. However, the effects of the nuclear disintegrations can be measured more easily with use of one of the radioactive detectors described on pp. 145 to 147. In this way, the total number of disintegrations per second occurring within the radioactive sample, or the radioactivity, at any given time can be estimated. Since the radioactivity A is proportional to the number of atoms N, equation 7-2 can be written as

$$A = A_0 e^{-\lambda t} \qquad \textit{Eq. 7-6}$$

Therefore the decay constant, the decay factor, and the half-life are also applicable to activity versus time. In equation 7-4 the residual activity of 5 mCi of ^{125}I after 88 days is reduced to 5 mCi × 0.362 = 1.81 mCi, whereas in equation 7-5 a shipment of sodium iodide (^{125}I) labeled "5 mCi at 1200 hr GMT on July 28, 1982," has an activity of 5 mCi × 1.10, or 5.5 mCi at 12 hr GMT on July 20, the date the reagent is to be used.

Table 7-2. Radiation characteristics of radionuclides commonly used in clinical chemistry

Nuclide	Half-life	Main radiation		Specific activity*	
		Type	Energy (keV)	mCi/μg	mCi/μmole
^{3}H	12.3 years	β^-	18	9.7	29
^{14}C	5760 years	β^-	158	0.0044	0.062
^{32}P	14.3 days	β^-	1700	285	9120
^{35}S	87.1 days	β^-	167	42.8	1500
^{51}Cr	27.8 days	EC†	γ320	92	4690
^{59}Fe	45 days	β^-/γ	460/1099	49.1	2900
^{57}Co	270 days	EC†	γ122	8.5	480
^{125}I	60 days	EC†	γ35	17.3	2200
^{131}I	8.1 days	β^-/γ	807/364	123	16,100

*Carrier free.
†Electron capture.

Table 7-3. Basic properties of radiation

Radiation	Charge	Energy range	Approximate range of travel in		Relative specific ionization*
			Air	Water	
Particles					
α	+2	3-9 MeV	2-8 cm	20-40 μm	2500
β^-	−1	0-3 MeV	0-10 m	0-1 mm	100
β^+	+1	0-3 MeV	0-10 m	0-1 mm	100
Electromagnetic					
X rays	None	1 eV − 100 keV	1 mm − 10 m	1 μm − 1 cm	10
Gamma rays	None	10 keV − 10 MeV	1 cm − 100 m	1 mm − 10 cm	1

*The number of ion pairs produced per unit path length relative to that of gamma rays.

It is often necessary to know the specific activity of a radioactive sample. This is the activity of the radionuclide per unit mass of the radioactive sample and thus is measured in μCi per μg, μCi per μmole, and so on, or submultiples thereof. The specific activity of the nuclide is inversely proportional to the half-life of the radionuclide and is an important factor in determining the sensitivity of radioimmunoassay. A list of nuclides commonly used in clinical chemistry and some of their radiation characteristics are presented in Table 7-2. The low decay rate of ^{3}H and ^{14}C with long half-lives gives them low specific activity and makes them less suitable for radioimmunoassay work. The specific activities of nuclides listed in Table 7-2 are calculated under the assumption that all nuclides present are radioactive (carrier free). This assumption does not always hold true. For example, the isotopic abundance of available ^{131}I preparations seldom exceeds 20%; that is, only about 20% of iodine atoms present in the ^{131}I preparation are ^{131}I; the rest are ^{127}I (stable iodine). Therefore the actual specific activity of ^{131}I preparations is only about one fourth of the theoretical specific activity shown in Table 7-2. Similarly, the specific activity of molecules depends on the isotopic abundance of the radionuclides present in the molecules. A thyroxine molecule contains four iodine atoms, and thus the specific radioactivity of radioiodine-labeled thyroxine preparations depends not only on the kind of radioactive iodines used for labeling but also on how many of the stable iodines are replaced by the radioactive iodine.

Properties of radiation and interaction with matter

Understanding of the properties of radiation and the mechanism of the energy loss of radiation as it passes through matter is important because the operation of every detecting device for any type of radiation depends on one or more of the particular properties of the radiation being measured and the interactions of radiation with matter. Furthermore, the safe manipulation of radioactive substances requires a knowledge of the nature of radiation and its ability to penetrate matter. The harmful effects of radiation on tissues are highly dependent on the ability of the radiation to ionize matter and on the energy of the incident radiation.

Radioactive decay produces a stream of alpha or beta particles or electromagnetic radiation. Particles may be negatively charged (β^-) or positively charged (α and β^+). Electromagnetic radiation may possess a relatively small quantity of energy (x rays) or a large amount of energy (gamma rays). Such differences in the nature of radiation create important differences in its modes of interaction with matter. Various properties of radiation are presented in Table 7-3.

The interactions of radiation with matter result in the transfer of energy from a radioactive nucleus to the surrounding material. This transfer is accomplished through processes of excitation and ionization; therefore radiation emitted from radionuclides is frequently termed *ionizing radiation*.

Excitation occurs when orbital electrons are perturbed from their normal arrangement by absorbing energy from the incident radiation. Ionization occurs when the energy absorbed is sufficient to cause an orbital electron to be ejected from its orbit, creating an ion pair (a free electron and a positively charged atom or molecule). This ionizing ability of radiation is best expressed by the number of ion pairs produced per unit path length, that is, the specific ionization. The relative specific ionizations of various forms of radiation are presented in Table 7-3.

Alpha particles. Alpha particles are helium nuclei and thus have two protons and two neutrons and a charge of +2. The relatively large mass causes the particles to travel slowly in essentially a straight line during the ionization process. In addition, because of their double charge, they produce a great deal of ionization that results in loss of their energy quickly in a short distance. Therefore alpha particles are weakly penetrating and can be stopped completely by very thin layers of solid materials. For this reason, they are less hazardous externally. However, if they get into the body, they will irradiate the tissues around them intensely, causing a serious health hazard. Alpha-emitting nuclides are seldom used in medicine.

Beta particles. There is no known difference between the negatron particle and the electron except for their origin. In general, pure beta particle–emitting nuclides have a continuous spectrum of energy ranging from zero to a maximum of E_{max}. E_{max} is equivalent to the total energy available from the nuclear decay and is characteristic of each radionuclide. Fig. 7-4 shows the beta-particle energy spectrum of ^{14}C. When gamma emission is also present, sharp peaks superimposed on the continuous spectrum can be observed. As in the case of alpha particles, negatron particles also lose their energy by causing excitation and ionization.

Positrons are antiparticles of electrons. As in the case of negatron particles, the energetic positron particles can lose their energy by excitation and ionization. When the positron particle has lost most of its kinetic energy and comes into contact with a free electron in the matter, they completely annihilate each other and two 0.55 MeV gamma photons are formed in opposite directions. In this way the total mass of both particles is completely converted into energy. The energy equivalent of one electron or positron mass is 0.51 MeV.

Electromagnetic radiation. Electromagnetic radiations encountered in the field of medicine include gamma ray, x ray, and the annihilation radiation mentioned previously. Except for possible differences in energy, these photons are indistinguishable and engage in the same type of interactions with matter. Because photons have no mass, are uncharged, and travel with the velocity of light, they might travel through matter for a considerable distance without any interaction and then lose all or most of their energy in a single interaction. Photons can interact with matter in several different ways, depending on their energies and

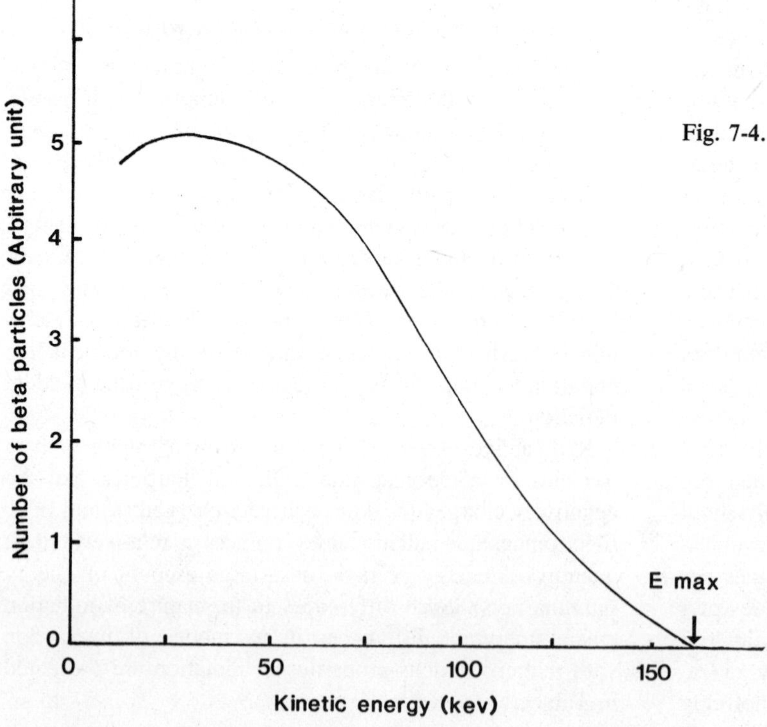

Fig. 7-4. Beta-particle spectrum for carbon 14.

the properties of the material with which they interact.

Photoelectric effect is especially important for photons with low energy (below 0.5 MeV). The photon interacts directly with one of the orbital electrons in the matter (a photon-electron interaction), and the entire photon energy is transferred to the electron. Some transferred energy is used to overcome the binding energy of the electron, and the remaining energy is carried by the electron as kinetic energy. The ejected electron (photoelectron) in turn transfers its kinetic energy to many other electrons in its path. The photoelectric effect is especially pronounced if the atomic number of the absorbing material is high.

Compton effect, or *Compton scattering,* occurs primarily with photons of medium energy (0.5 to 1 MeV). In this process a collision between a photon and an electron results in the transfer of only a portion of the photon energy to the electron. The scattered photon with reduced energy emerges from the site of interaction in a new direction. The ejected electron (Compton electron) and the scattered photon lose more energy by subsequent interactions.

In *pair production* the entire energy of the photon is consumed in the production of a positron and an electron in the region of a strong electromagnetic field, such as that surrounding the nucleus. Therefore pair production is involved only with photons having energies greater than 1.02 MeV. Photon energy in excess of 1.02 MeV is carried by the positron and electron as kinetic energy. The positron and electron formed can produce secondary ionization. When the positron loses most of its kinetic energy, annihilation radiation will result, as described in the interaction of positrons with matter.

ISOTOPE DETECTION

Radioactive isotopes

Radioactivity measurements depend on the ability of radionuclides to produce ionized or excited atoms within the sensitive volume of the detector. Two basic types of radiation detectors are in common use: gas ionization and scintillation. The latter is capable of detecting both negatron and gamma radiation and providing information regarding the type and energy of the radiation and hence is currently the most commonly used detector in the field of medicine. Therefore the following discussion is devoted primarily to scintillation detectors; other detectors are described only briefly.

Ionization of gases. The Geiger-Müller counter and ionization chambers are examples of gas ionization detectors. In the Geiger-Müller counter the radiation is detected through ionization produced within a suitable gas. Because the ions produced are accelerated by the relatively high voltage applied between the electrodes of the detector, considerable secondary ionization occurs, leading to a large output pulse (electron multiplication). The major advantage of this type of detector, as compared to ionization chambers, is its ability to detect low levels of radiation. The ionization chamber functions on a similar principle. However, because a lower voltage is used, electron multiplication does not occur in the ionization chamber, and the output signal is relatively small. Both detectors are widely used in survey meters for measuring exposure of personnel and locating a spilled radionuclide.

Scintillation detectors. Scintillation counting is based on the principle that a charged particle (alpha or beta) entering the detector, or an electron excited in the detector after an interaction with an incoming photon (gamma ray), will dissipate its energy within the scintillator contained in the detector by various processes of interaction mentioned previously. A portion of the energy absorbed by the scintillator is emitted as photons in the visible or near-ultraviolet region of the electromagnetic spectrum. Scintillators or fluors are substances capable of converting the kinetic energy of an incoming charged particle or photon into flashes of light (scintillation). The light emitted by the scintillator is converted into photoelectrons at the photocathode of a phototube. The multiplication (amplification) of the initial number of photoelectrons by a photomultiplier tube and amplifiers results in measurable electrical current pulses, the heights of which are related to the energy of the radiation, and their number to the activity of the radionuclide.

Crystal scintillation detectors. The most commonly used fluor for detecting gamma radiation by scintillation is a single crystal of sodium iodide containing small amounts of thallium (about 1%) as the activator. Fig. 7-5 is a block diagram of the common types of thallium-activated sodium iodide crystal scintillation detectors. The crystal is usually in the shape of a well, and the sample to be counted is allowed to sit within the well. The sodium iodide crystal is very hygroscopic. It is encapsulated in a metal can (such as aluminum) to prevent it from absorbing atmospheric moisture, except for one face (usually the bottom face) of the crystal well, which is covered by a transparent material such as Lucite and is optically coupled to the transparent face of a photomultiplier tube.

A gamma ray emitted from the sample placed in the crystal well is highly penetrating and therefore can pass through the glass or plastic wall of the test tube containing the radioactive sample and enter into the crystal. As the gamma ray passes into the crystal, it produces excitation or ionization, and light photons are emitted. About 20 to 30 light photons are produced for each electron volt of energy absorbed. The photons pass through the transparent crystal and strike the photocathode of the photomultiplier tube to cause a release of electrons from the cathode. The energy required to release one photoelectron from the photocathode is about 300 to 2000 eV.

In addition to the conversion of the light photons emitted by the fluor into a pulse of detectable electrons, the photomultiplier also amplifies the minute amount of current produced from the photocathode to a level that can be

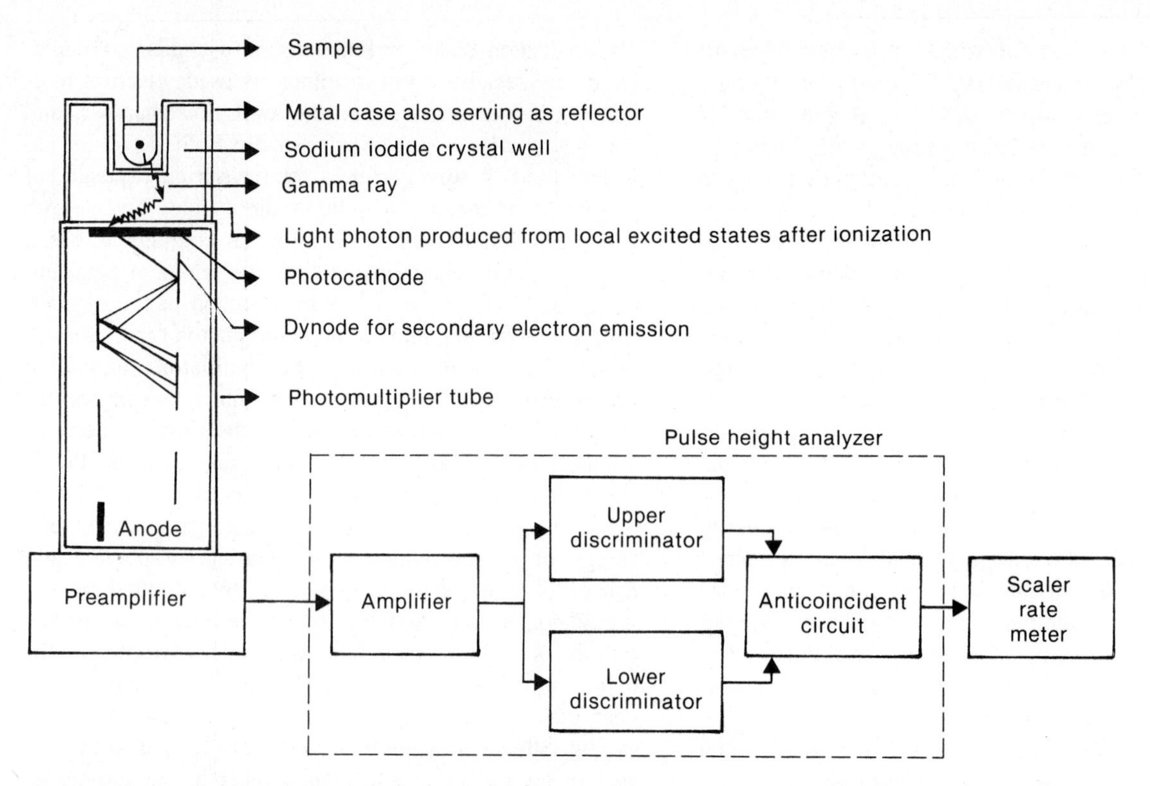

Fig. 7-5. Block diagram showing principle components of typical crystal scintillation counter.

effectively handled in conventional electronic amplifier circuits. This is achieved by a process of electron multiplication. As illustrated in Fig. 7-5, a series of metal plates, termed *dynodes,* are spaced along the length of the photomultiplier tube. The dynode surface is coated with a material capable of emitting secondary electrons when struck by an accelerated electron. Each dynode is maintained at a potential higher than the preceding one. The initial photoelectrons are accelerated toward the first dynode and strike it to produce secondary electrons, which are then accelerated toward a second dynode. About three or four electrons are released from the dynode for each striking electron. This process is repeated until an amplification of about 10^8 is achieved.

The current output of the photomultiplier tube is amplified, and the resulting voltage pulse is shaped for optimal counting by conventional electronic circuitry such as that shown in Fig. 7-5. The preamplifier reduces the distortion of the electrical signal produced by the photomultiplier tube. The preamplifier output is further amplified by the amplifier to give a voltage of up to 10 V.

The height of the voltage pulse produced by the amplifier is proportional to beta- or gamma-ray energy deposited within the detector. Each radionuclide has a characteristic spectrum of energies (pulse height) such as that shown for ^{14}C in Fig. 7-4. The function of the pulse-height analyzer is to sort out the pulses according to their pulse height and to allow those pulses that lie within a restricted range (the photopeak) to reach the rate meter for counting. This is accomplished by means of *discriminators.* A lower discriminator sets the lower limit and an upper discriminator sets the upper limit of the energy range to be counted. The lower discriminator excludes all voltage pulses below the lower limit; the upper discriminator excludes voltage pulses above the upper limit. The energy interval represented by the difference between the two discrimination levels is called the *window width.* Only the pulses with energy within the preset discriminator window pass through the anticoincident circuit and are counted because the anticoincident circuit will transmit a pulse arriving at its input from the lower discriminator only if there is no pulse arriving from the upper discriminator at the same time. This mode of operation is termed *differential counting.* The discriminators can also be set to count every pulse that exceeds the setting of the base control. This mode of operation is termed *integral counting.* In some counters the discriminator controls consist of a lower level discriminator, termed the *base control,* and window-width control. The value of the base control and the window width gives the upper level of the energy window selected.

A scintillation detector equipped with two or more pulse-height analyzers (multichannel analyzers) can be used to simultaneously count two or more radionuclides, either in the same sample or in different samples, provided that there is sufficient energy difference between them so that a certain portion of the energy of one radionuclide can be detected free from the second radionuclide. For example, the major photopeak of ^{125}I occurs at 27 keV and that of ^{131}I at 364 keV. In addition to the 364 keV photopeak, a minor ^{131}I photopeak occurs at 32 keV. For counting a mixture of these two isotopes, one analyzer channel (A) is centered at 27 keV and the other channel (B) at 364 keV, with the window width of about 20 to 40 keV. Channel B gives the true count for ^{131}I because ^{125}I does not contribute counts to channel B. Counts from channel A, however, represent the sum of the true counts for ^{125}I and the ^{131}I spillover. One can estimate the extent of the ^{131}I spillover by counting the pure ^{131}I standard in both channels.

In order to save counting time, crystal scintillation counters equipped with up to 20 detectors have been developed in recent years. With such a counter, the counting time can be reduced to as much as one twentieth of that required by a single detector instrument. However, it is absolutely necessary to make sure that all detectors in a multidetector instrument perform in an equivalent fashion.

Liquid scintillation detectors. Liquid scintillation detectors are primarily used for counting beta particle–emitting radionuclides such as ^{3}H, ^{14}C, and ^{32}P. Unlike that of gamma photons, the penetration of negatron particles is so short that they cannot penetrate the wall of the sample container for interaction with crystal scintillators. In liquid scintillation counting, the sample is dissolved or suspended in a solution or ''cocktail'' consisting primarily of a solvent such as toluene, a primary scintillator such as 2,5-diphenyloxazole (PPO), and a secondary scintillator such as 1,4-bis-2(5-phenyloxazolyl)-benzene (POPOP). The beta particles from the radioactive sample dissolved in the scintillation cocktail ionize and excite the molecules of the solvent. The excitation energy is transferred to the primary scintillator, which emits light photons when the excited electrons return to the ground energy level. The wavelength of light emitted by the primary scintillator is frequently too short (about 350 to 400 nm) for efficient detection by the photocathodes of photomultiplier tubes. The secondary scintillator absorbs the photons emitted by the primary scintillator and reemits them at a longer wavelength (about 430 nm). Thus the secondary scintillator is also termed a *wavelength shifter*.

A typical arrangement of the principle components of a liquid scintillation counter is shown in Fig. 7-6. The light photons produced in the sample vial are detected and amplified by the photomultiplier tubes in the same manner as for the crystal scintillation counter. In the liquid scintillation detector, however, a second photomultiplier tube, a coincidence circuit, and a summation circuit are incorporated to eliminate the electronic noise associated with the photomultiplier tube and to improve counting efficiency for low-energy beta-particle emitters.

Noise pulses are random events, and the probability of two photomultiplier tubes producing noise pulses simultaneously is relatively small. In contrast, the beta particle produces a burst of photons, and two photomultiplier tubes will receive photons almost simultaneously. The output pulses from each photomultiplier tube are fed into a coincidence circuit to check if a pulse from one photomultiplier tube is accompanied by a corresponding pulse from the other within the allowed time interval (termed the *coincidence resolving time,* usually about 20×10^{-8} sec). Pulses within the resolving time produce a coincident signal that is electrically sent to the coincidence gate. Most noise pulses do not meet the coincident resolving-time requirement and are excluded. The summation circuit is incorporated to sum all coincident pulses to obtain the true pulse height. The summed coincident pulses are amplified, sorted, and counted in a manner similar to that for the crystal scintillation counter.

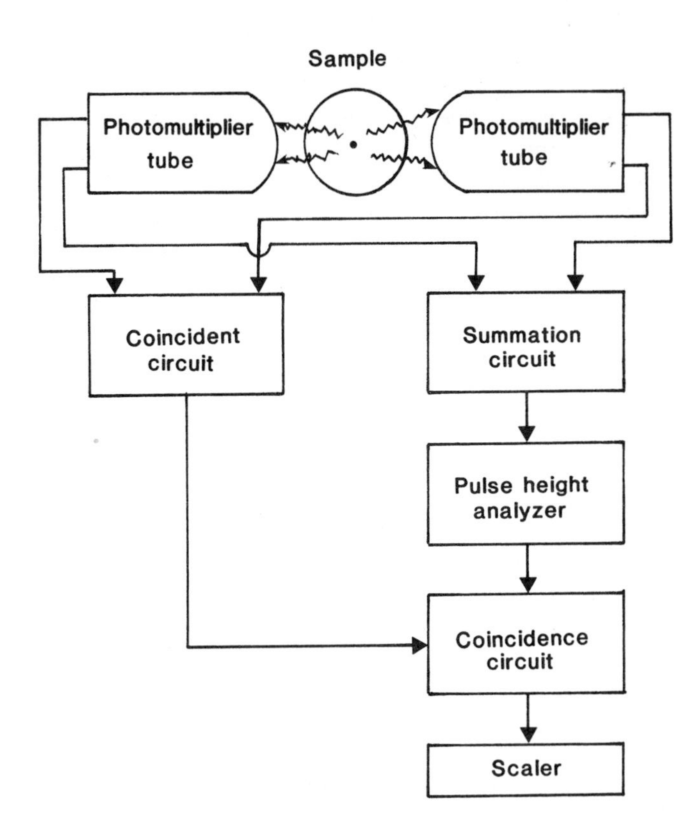

Fig. 7-6. Schematic diagram of liquid scintillation counter.

In liquid scintillation counting, proper energy transfer can not occur unless the sample is in contact with the scintillation solution to give a colorless, transparent, homogeneous solution. Some radioactive samples are not soluble in the scintillation solution, and so it may be necessary to add one or more substances to obtain a homogeneous scintillation mixture. Solubilizers such as methylbenzethonium chloride (Hyamine 10 ×) are used to facilitate dissolution of the sample in the scintillation solution or jelling agents such as aluminum stearate to enhance the counting efficiency by stabilizing the sample suspension in liquid scintillators. Many commercial liquid scintillation cocktails of nonpolar media (toluene or xylene) contain some type of surfactant such as the Tritons (polyoxyethylene ethers and other surface-active compounds) to maintain aqueous samples in colloidal suspensions so that the aqueous samples can be counted at high efficiency. Nonvolatile, radioactive materials are also counted on solid supports such as filter paper disks or glass fibers immersed in a scintillation solution. The disadvantage of this counting method is the relatively low counting efficiency because of self-absorption (impurity quenching).

Quenching is basically a process that results in the reduction of the overall photon output of the sample. Impurities present in the radioactive sample may compete with the scintillators for energy transfer, that is, the energy is lost to a non–light producing process. This phenomenon is termed *impurity quenching*. Water in aqueous samples or a support medium such as a filter disk may cause impurity quenching. Colored substances such as hemoglobin may absorb the light photons produced by the scintillation process before they can be detected by the photomultiplier tubes or may change the wavelength of the light photons to a value not suitable for efficient detection by the photocathodes of photomultiplier tubes. Quenching in liquid scintillation counting is detected and corrected by efficiency determination. The efficiency of the measurement is defined as the ratio of the observed counts per minute, cpm, to the absolute units of disintegrations per minute, dpm; thus

$$\text{Efficiency} = \frac{\text{cpm}}{\text{dpm}}$$

Since the quenching characteristics of each sample are different, the efficiency must be determined for each sample. By knowing the counting rate (cpm) and the counting efficiency of a sample, one can calculate the absolute radioactivity (dpm) of the sample. Several methods for efficiency determination have been developed, but only those most frequently used are discussed.

Internal standardization. Internal standardization is one of the oldest methods and is the most accurate method for efficiency determination when properly carried out. In this method the sample is counted before and after the introduction of a calibrated standard of the measured radionuclide. The difference between the count rates before and after the spike, divided by the calibrated activity of the spike in disintegrations per unit time, is termed the *counting efficiency*. The disadvantages of this method are the time-consuming manipulation of the sample and the loss of sample for recount after the introduction of the spike.

Sample-channels ratio. The sample-channels ratio method is based on a downward shift of the pulse-height spectrum of the photon as a result of the quenching-induced decrease in the pulse height of many energetic decays (Fig. 7-7). The degree of the shift is related to the extent of quenching or counting efficiency and is expressed by the change in the ratio of the sample counts obtained from two different discriminator window settings (channels). As shown in Fig. 7-7, one channel is usually set to measure the entire isotope spectrum (L_1 to L_3), and the

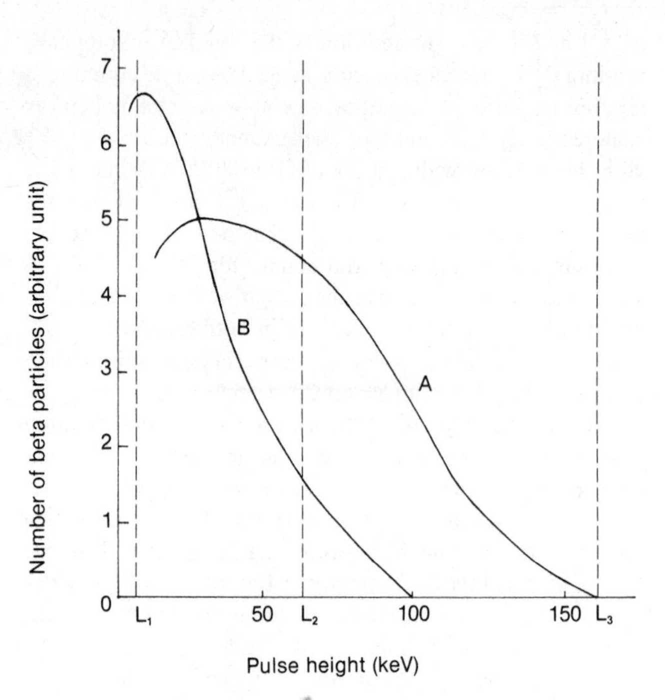

Fig. 7-7. Unquenched, *A*, and quenched, *B*, pulse-height spectra of carbon-14. L_1 to L_3 denote discrimination levels.

second channel is restricted to only a portion of the spectrum (L_1 to L_2). In this method, channel ratios (L_1 to L_2)/(L_1 to L_3) of a set of artificially quenched standards of known efficiencies are determined and plotted against counting efficiency to obtain a quench curve. The efficiency of any unknown sample can be determined from its channel ratio and the quench curve. This method requires no additional sample manipulation, unlike internal standardization, and is suitable for handling a large number of samples through automation. This method, however, may give large errors in highly quenched samples or samples with low count rates.

External standards. Unlike the internal standard method, the known activity in the external standard method is provided by an external source of gamma radiation, such as ^{226}Ra, placed at a fixed position adjacent to the sample vial. The external gamma-ray source generates electrons through the Compton collision process in the scintillation solution. The Compton electrons transfer energy to the solution and cause scintillation in the same way as beta particles do in the scintillation medium. The energy spectrum produced by Compton electrons is also affected by the presence of quenching materials, as in a typical beta-particle spectrum. The sample is counted twice, once in the absence of and once in the presence of an external standard. As in the sample-channels ratio method, a set of quenched standards of known efficiencies is used to obtain a correlation curve between the sample-counting efficiency and the count rate of the external standard. The counting efficiency of a sample can be determined from the count

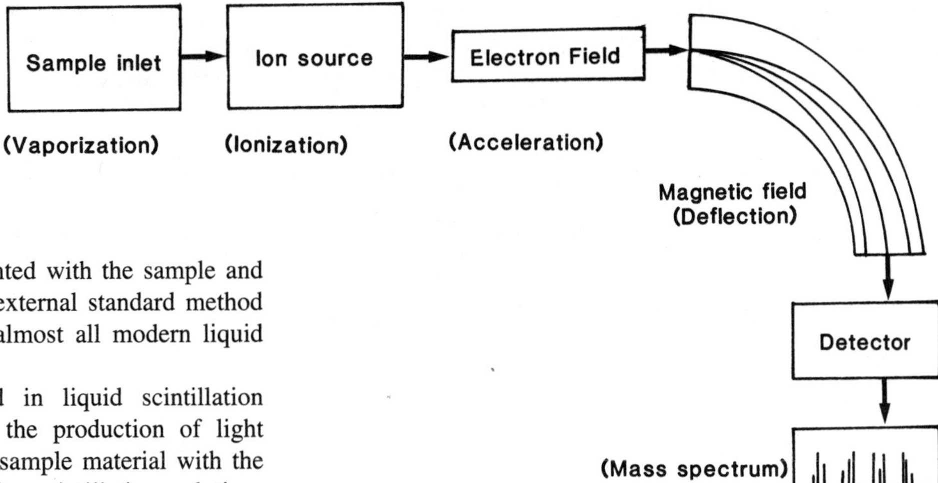

Fig. 7-8. Schematic of a mass spectrometer.

rate of the external standard counted with the sample and from the correlation curve. The external standard method has become an integral part of almost all modern liquid scintillation detectors.

Another problem encountered in liquid scintillation counting is *chemiluminescence,* the production of light photons by an interaction of the sample material with the solute or solubilizer added to the scintillation solution. Chemiluminescence gives rise to single photons and can be excluded by the coincidence circuit of the liquid scintillation counter. However, when chemiluminescence reactions are of sufficient intensity, non–beta particle coincidence pulses may be generated to interfere with the beta-particle scintillation counting. The chemiluminescent effect will eventually disappear but may take several hours or longer, especially at low temperatures. The presence of chemiluminescence can be monitored by repeated counting of the sample. Some modern instruments are capable of automatically monitoring and correcting the chemiluminescent effects.

As with crystal scintillation counting, mixed-isotope counting is also possible with a liquid scintillation counter with multichannel analyzers. A common example of dual-label counting involves a mixture of tritium, with a maximum beta-particle decay of 18.6 keV, and ^{14}C, with a maximum beta-particle decay of 156 keV.

Counting statistics. Since radioactive decay is essentially a random process, it is unlikely that successive measurements on a given sample will result in the same number of counts. However, radioactive decay obeys a *Poisson distribution*. The Poisson distribution density formula can be applied in the calculation of the precision of measurement at a given count rate. If a single measurement of total counts, N, is made, then precision of this measurement in terms of the coefficient of variation, CV, can be estimated to be as follows:

$$\text{CV} = \frac{\sqrt{N}}{N} \times 100$$

For example, at a total count of 100, CV = 10%; at 1000, CV = 3.2%; at 10,000, CV = 1%. This approximation is applicable only when the background count is negligible compared with the sample count. When significant background counts are present, the formula for the standard deviation of a difference must be used:

$$\text{CV} = \frac{\sqrt{N + B}}{S} \times 100$$

where B is the background count and S is the sample count $(N - B)$. Thus a total count of 100 in the presence of a background count of 10 gives the following:

$$\text{CV} = \frac{\sqrt{100 + 10}}{100 - 10} \times 100 = 11.7\%$$

Precision can be increased by prolongation of the counting time, but a 1% CV (that is, 10,000 counts) is satisfactory in most applications.

Stable isotopes

The amounts of stable isotopes present in an element *(isotope abundance)* are determined primarily by use of a mass spectrometer. The principle of a mass spectrometer is illustrated in Fig. 7-8. A mass spectrometer consists of four principle components: a sample inlet for vaporization of the sample; an ion source for ionization of the vaporized sample by, for example, electron bombardment; a mass separation unit for separation of resulting positively charged ions by means of electrical and magnetic field according to mass; and a detector-recorder.

In general, radioactive isotopes rather than stable ones are used whenever there is a choice between the two because the radioactivity measurement is usually easier, less expensive, and more sensitive than the measurement of stable isotopes. However, the latest commercially available isotope-abundance mass spectrometers, especially when coupled with the gas chromatograph (gas chromatography–mass spectrometry, or GC-MS) or with another mass spectrometer (tandem mass spectrometry, or MS-MS), are accurate, sensitive, specific, and relatively easy to operate and are expected to become as acceptable as scintillation counters in a routine clinical laboratory.

Stable isotopes are nonradioactive and suitable for use as tracers in humans, especially in infants, children, and pregnant women. In the case of nitrogen and oxygen, no suitable radioactive isotopes are available, and the stable isotopes ^{15}N and ^{18}O are used exclusively. In addition to being useful tracers, stable isotopes have also been used in the quantitative analysis of various substances in recent years. For example, isotope-dilution mass spectrometry has been utilized in definitive methods for calcium, lithium, potassium, chloride, cholesterol, certain steroid hormones, glucose, and other clinically important compounds.

RADIATION HEALTH SAFETY

Although the quantity of radioactivity handled in the clinical chemistry laboratory is usually very small, a basic knowledge of radiation safety is vital to every laboratory worker who has frequent contact with radioactive substances because the biological effects of long-term exposure to very low doses of ionizing radiation are still largely unknown and may prove to be hazardous to health.

Radionuclides commonly used in the clinical laboratory are either beta-particle emitters, such as ^{14}C and ^{3}H, or gamma-ray emitters, such as ^{125}I and ^{57}Co. Both forms of radiation produce their biological effects by producing ionization and excitation along their paths in the tissue. However, beta radiation is less penetrating than gamma radiation; thus beta-particle emitters are considered to be more hazardous in terms of internal radiation and less hazardous in terms of external radiation than gamma-ray emitters. Therefore the primary concern with beta-particle emitters is to prevent the entry of radioactive materials into the body through inhalation, ingestion, or absorption by the skin; however, with gamma-ray emitters other factors such as shielding, exposure time, and exposure distance are also important in the consideration of radiation safety.

Monitoring

Regular monitoring of both personnel and work areas is an important radiation safety procedure. It is necessary to measure periodically the radiation exposure doses of personnel to ensure that radiation doses received are below the recommended limits. The following three basic units are used to measure radiation exposure and dose.

The roentgen, *R*, is a unit of x rays or gamma rays and measures the quantity of ionization produced by photon radiation in a given sample of air. One R equals that quantity of photon radiation capable of producing 1 electrostatic unit of either sign in 0.001293 g of air.

The rad (radiation absorbed dose) is a measure of local energy deposition per unit mass of material irradiated by any ionizing radiation. One rad is equal to 100 ergs of absorbed energy per gram of absorber.

The rem (roentgen equivalent, man) is a unit of biological dose as a result of exposure to ionizing radiation. It is equal to the absorbed dose in rads when the radiation exposure involves x, gamma, or beta radiation. In the case of alpha radiation, the dose in rems equals the dose in rads multiplied by 20 because only 0.05 rad of alpha radiation is needed to produce the same biological effect as 1 rad of x, gamma, or beta radiation. The recommended maximum permissible dose to the whole body is 0.5 rem per year for the general public and 5 rem per year for radiation workers. Film badges are probably the most commonly used and cost-effective way of monitoring personnel. The photographic film becomes progressively optically dense when exposed to ionizing radiation and thus may be used to monitor the radiation dose received by the wearer. Since most clinical laboratory personnel involved in radioimmunoassays routinely handle ^{125}I-labeled compounds, it is advisable to monitor possible accumulation of radioactive iodine in the thyroid glands. Arrangements should be made to have the radioactive content of each worker's thyroid taken at least twice a year or after each radioiodination experiment. It is necessary to keep all records of radiation exposure of all workers handling radioactive materials for at least 5 years. Each laboratory should have a portable radiation detector, such as a portable Geiger-Müller survey meter, to monitor radioactivity in an area in which radioactive materials are routinely handled. Monitoring of beta radiation usually requires taking samples of the work area with swabs and using a liquid scintillation counter to determine the presence of radioactivity.

Contamination control

Internal radiation exposure is controlled only by the prevention of the entry of radioactive materials into the body. This requires strict adherence to the general rules for radiation safety. No smoking, eating, drinking, application of cosmetics, or storing of food is allowed in work areas. Mouth pipetting of radioactive materials should never be done. All persons working in radioactive areas must wear the designated protective clothing (a standard laboratory coat is satisfactory in a clinical laboratory involved in radioimmunoassays) and disposable gloves. Radioactive materials must be properly labeled, stored, and used only at specially designated areas. Work involving the possible generation of volatile, radioactive substances, such as radioiodination, should be performed in an exhaust hood. The working surface should be covered by a layer of disposable absorbent material. In addition to the proper operating technique, cleanliness and good housekeeping are essential to prevent and minimize the spread and buildup of contamination.

Persons contaminated by radioactive materials should be quickly decontaminated to prevent the possible transfer of radioactivity to internal organs by absorption through the skin. Facilities for decontamination, such as a shower and an eyewash station, should be available in each laboratory. Absorbent materials should be used to remove spilled radioactive material. The contaminated area should then be scrubbed with soap and water. It is a good practice to cover the contaminated area immediately with a piece of paper to prevent spreading of the radioactivity to other parts of the laboratory.

Waste disposal

The level of radioactivity of the radioactive waste material generated in clinical chemistry laboratories involved in radioimmunoassays is usually very low. The radioactive waste materials in most cases are of such low specific activity or of such a low concentration that they may be disposed of directly by release to air, water, or ground. Radioactive waste disposal is regulated by the Nuclear Regulatory Commission (NRC) of the United States. Recent changes in the disposal regulation adopted by the NRC allow users of most in vitro radioassay kits to dispose of the waste in the same manner as for nonradioactive trash. All radiation labels should be removed and destroyed before the disposal. The NRC proposes to allow a total of 7 Ci of water-soluble radioactivity to be disposed of each year through sewer systems in the following groups:

1 Ci—undefined waste
1 Ci—carbon-14
5 Ci—tritium

Liquid scintillation media containing less than 0.5 μCi per gram of material may be incinerated. Some states (NRC agreement states) are approved by the NRC to regulate the use, safety, and disposal of radioactive material within the state provided that the regulations are more restrictive than the NRC regulations. It is important therefore to be familiar with the state regulations regarding radioactive materials if your laboratory is located in an NRC agreement state.

BIBLIOGRAPHY

Sorenson, J.A., and Phelps, M.E.: Physics in nuclear medicine, New York, 1980, Grune & Stratton, Inc.

Peng, C.T., Horrocks, D.L., and Alpen E.L.: Liquid scintillation counting: recent applications and development, vol. 1, Physical aspects; vol. 2, Sample preparation and applications, New York, 1980, Academic Press, Inc.

Rinehart, K.L., Jr.: Fast atom bombardment mass spectrometry, Science **218:**254, 1982.

Heal, A.V.: Safety and disposal changes that affect regulations in radioassay labs, Lab World, pp. 50-53, Dec. 1981.

Chapter 8 Electrophoresis

John M. Brewer

Ampholyte A trade name for a mixture of substances with a range of isoelectric points that have high buffering capacities at their isoelectric points.

amphoteric A substance that can have a positive, zero, or negative charge, depending on conditions.

anion Negatively charged particle or ion.

boundary Edge of a zone, as of a macromolecule solution next to the solvent.

buffer A mixture of proton-donating and proton-accepting substances whose function is to keep the proton concentration (the pH) constant or nearly so. An example is a mixture of acetic acid and sodium acetate.

cation Positively charged particle or ion.

co-ion An ion of the same charge as the one under consideration. Generally a much smaller ion.

conductivity The readiness of a conductor to carry a current. In an ionic solution, the sum of the product of the charge concentrations and charge mobilities.

convection Mass or bulk movement of one part of a solution relative to the rest, usually because of density differences.

counterion An ion of opposite charge to the one under consideration. Generally a smaller ion.

disk electrophoresis A stacking or isotachophoretic step followed by zone electrophoresis, usually on a polyacrylamide gel.

discontinuous solvent A solution consisting of at least two separate regions that have different ions in them.

effective mobility The actual mobility of a substance under certain conditions. Generally less than the "mobility" because of a lower charge or resistance by a supporting medium.

electric field An influence measured in volts (or volts per centimeter) that is manifested by the behavior of a charged particle in it.

electrical neutrality A condition in which total positive charges equal total negative charges.

electrodes Substances in contact with a conductor. The substances are connected to a source of an electric field.

electrolyte Ionic substances, usually of low molecular weight, added to provide as constant and uniform an ionic environment for electrophoresis as possible.

electro-osmosis Tendency of a solution to move relative to an adjacent stationary substance when an electric field is applied.

electrophoresis Movement of charged particles because of an external electric field.

frictional coefficient A measure of the resistance a particle offers to movement through a solvent.

gel A network of interacting fibers, or a polymer that is solid but traps large amounts of solvent in pores or channels inside.

ionic strength The sum of the concentrations of all ions in a solution, weighted by the squares of their charges.

isoelectric Condition of zero net charge on an amphoteric substance.

isoelectric focusing The ordering and concentration of substances according to their isoelectric points.

isoelectric point The pH at which a substance has a zero net charge.

isotachophoresis The ordering and concentration of substances of intermediate effective mobilities between an ion of high effective mobility and one of much lower effective mobility, followed by their migration at a uniform velocity.

joule heating Heating of a conductor by the passage of an electric current.

mobility The velocity a particle or ion attains for a given applied voltage. A relative measure of how quickly an ion moves in an electric field.

molecular sieving Separation of molecules on the basis of their effective sizes.

polyacrylamide Polymer of acrylamide and usually some cross-linking derivative.

polyelectrolyte Substance with many charged or potentially charged groups.

resolving power Ability to separate closely migrating substances.

SDS Sodium dodecyl sulfate. A detergent and an especially effective protein denaturant.

stacking Ordering or arranging and concentrating macromolecules according to their effective mobilities.

zeta potential The potential produced by the effective charge of a macromolecule, usually taken at the boundary between what is moving with the macromolecule and the rest of the solution.

zone A particular region or space within a larger one, generally distinguished by some property, such as its occupancy by a protein.

DEFINITIONS

A particle with one or more charges on it can be shown to have an electrical field associated with it.[1] This is seen when another charged particle, called a "test particle," is brought near: a strong force is exerted on the test particle. This fact is important for two reasons.

Since it has this strong force, a charged particle tends to polarize nearby atoms or molecules, slightly altering the normal placement of electrons and protons. A consequence of this polarization is that charged particles in water solution (ions) interact with water molecules to become hydrated (Fig. 8-1). Every ion has a cluster of water mole-

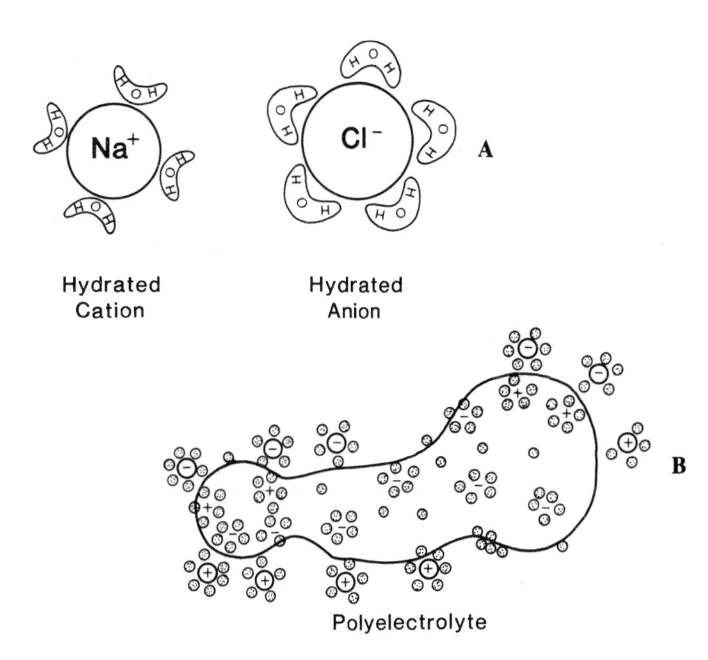

Fig. 8-1. State of charged particles in water solution. **A,** Small ions (Na^+ and Cl^-) with associated water molecules. **B,** Macromolecule with water molecules *(stippled smaller circles)* associated with charged and polar groups. Hydrated co-ions and counterions are also shown as larger circles around plus or minus signs.

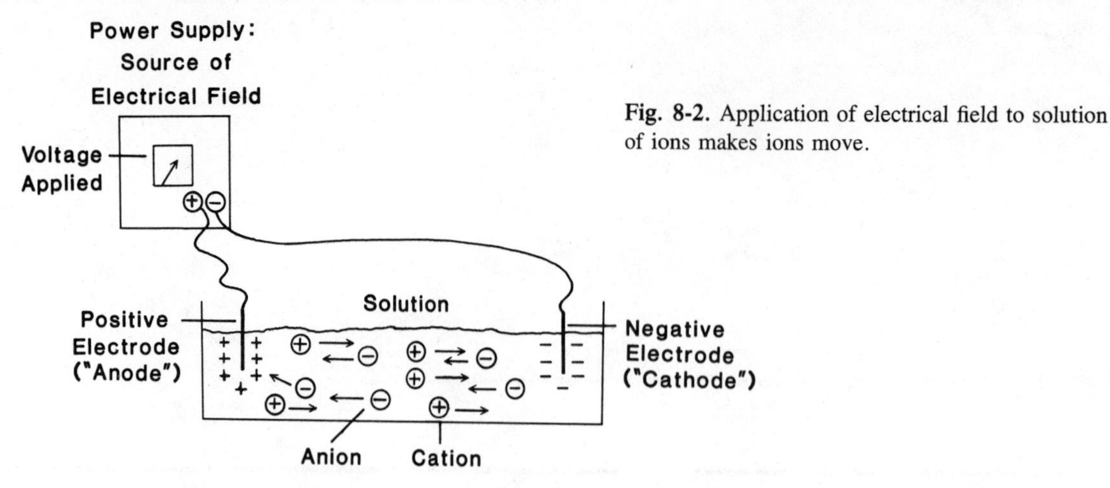

Fig. 8-2. Application of electrical field to solution of ions makes ions move.

cules about it, even though water molecules themselves have no charge.[2] If other charged particles are about, these will also interact, attracting or repelling each other, depending on the signs of their charges.

The strength of the force is important for another reason. It can cause charged particles, even very large ones, to move unless they are attached to something much larger.

Movement of charged particles because of an *external* electrical field is called "electrophoresis."[3] Since charged molecules can be made to move, different molecules can be separated if they have different velocities in an electrical field. Therefore electrophoresis is a separation technique, just as chromatography and ultracentrifugation are, and there are similarities between these techniques.

The electric field is applied to a solution through oppositely charged "electrodes" placed in the solution (Fig. 8-2). A particular ion then travels through the solution toward the electrode of opposite charge. Thus positively charged particles (cations) move to the negatively charged electrode (cathode) while negatively charged particles (anions) migrate to the positively charged electrode (anode).

APPLICATION OF ELECTRICAL FIELD TO SOLUTION CONTAINING A CHARGED PARTICLE

Forces on a particle

The basis on which a separation can be made is the interaction between an applied external electric field, V (in volts or volts per centimeter), and the charge on a particle, Q. The force on the charged particle is the product:

$$F_{elec} = QV \qquad \textit{Eq. 8-1}$$

This electrical force, F_{elec}, when exerted on the particle, will cause it to move. However, a particle moving in a solvent with a certain velocity will experience resistance because of the viscosity of the solvent.[3,4] For low velocities (v), the resistance, which is itself a force, is proportional to the velocity:

$$F_{resistance} = fv \qquad \textit{Eq. 8-2}$$

The proportionality constant, f, is called the "frictional coefficient."

The frictional coefficient depends on the viscosity of the solvent and the size and shape of the particle. The greater the viscosity, the slower the movement. The bigger or more asymmetric the particle, the slower its movement through the solvent. The frictional coefficient of a large particle such as a protein is a characteristic property of the particle.

Mobility of a particle

When an electric field is applied to a charged particle, it will begin to migrate. Its migration will be opposed by the frictional force (F_{resist}). The electrophoretic and frictional forces oppose each other, and the particle's velocity increases until the forces are equal. At this point one can say that:

$$F_{resist} = F_{elec};\ \text{or}\ fv = QV;\ \text{or}\ v = \frac{QV}{f};\ \text{or}\ \frac{v}{V} = \frac{Q}{f} \qquad \textit{Eq. 8-3, A}$$

The velocity, v, a particle attains for a given electrical field, V, is determined by two properties of the particle—its charge and its frictional coefficient. Consequently, the value of v/V is also a characteristic property of the particle and is important enough to be given its own name. It is called the "mobility" of the particle.

$$\text{Mobility of particle} = U = \frac{v}{V} = \frac{Q}{f} \qquad \textit{Eq. 8-3, B}$$

Effect of pH on mobility

Each ion has a particular charge and mobility. However, a *solution* of a substance whose pK is near the pH of the solution contains a *population* of species of the substance, some with its particular charge and some without. The fraction of species with the charge will depend on the pK of the substance and the pH of the solution. When the pH is equal to the pK_a of a weak acid, only 50% of the particles will be charged. At 1 pH unit below the pK_a, 90% will be uncharged. At 1 pH unit above the pK_a, 90% will be in the charged state. One can also say that the effective (average) charge of a substance varies with the pH. Hence its "effective mobility" varies with the pH.

This is particularly true for substances such as proteins. Proteins are clearly "amphoteric" substances; that is, they contain acidic and basic groups. Their overall (net) charge will be highly positive at low pH's, zero (isoelectric) at a particular higher pH, and negative at still more alkaline pH's. Since mobility is directly proportional to the magnitude of the charge, Q (see equation 8-3), the effective mobility of a protein is very much a function of the pH.

The most important practical consequence of this is that electrophoresis solutions must be buffered to maintain a constant pH. The buffer pH is chosen to give an optimal net charge for maximal separation. For proteins, pH's in the 7 to 9 range are generally used. The buffer is used to maintain this pH and thus the net protein charge throughout the electrophoretic process. Buffers are ionic substances themselves and so take part in any electrophoretic process, a fact that must also be considered.

Electrolytes

Interactions between charged groups naturally include those between charged groups on a single molecule. A single molecule with many such groups is called a "polyelectrolyte." These interactions also include those between different "polyelectrolytes" and those between polyelectrolytes and the generally smaller "counterions" (those of opposite charge) and "co-ions" (same charge) present in solution[5] (see Fig. 8-1).

Normally one determines velocities by measuring how far a boundary of a substance moves in a given time.[6] This is a discontinuous system wherein a substance, such as a protein, is put in one limited region or "zone" of the system and made to move into another region or zone. Therefore much of the solution will not have protein in it at any particular time.

If the protein and its associated counterions are not present to carry the current in a particular region, other ions must be present to carry the current. For this reason it is a common practice to add an excess, usually about 0.1 M, of low molecular weight buffer to the solution through which the protein must travel. The buffer and salt ions (electrolytes) provide a constant electrical environment so that the overall movement of the protein will be as constant as possible and be minimally influenced by other protein molecules.

Ion movement and conductivity

In any electrical system the current is proportional to the voltage:

$$V = R \cdot i \qquad \textit{Eq. 8-4}$$

R is the resistance and i is the current.

The proportionality constant can be the resistance, R, or the conductivity, c:

$$V = \frac{i}{c} \qquad \textit{Eq. 8-5, A}$$

That is:

$$R = \frac{1}{c} \qquad \textit{Eq. 8-5, B}$$

Since the system described above consists of ion flows, the more affirmative quantity, conductivity, is used. Each separate ion carries a specific fraction of the current[2]; that is, each ion contributes to the overall conductivity. This contribution is determined by the concentration and charge of the particular ion and by its mobility:

$$c = CUQ \qquad \textit{Eq. 8-6}$$

U, the mobility of a particular ion, is defined for an ionic species with a whole charge. It does not change, but Q, the effective charge on the ion, can change with pH. The product UQ is the effective mobility. The concentration of the substance, C, is usually in moles per liter.

However, the total conductivity of a solution is the sum of the contributions of each dissolved ionic species. This is described by equation 8-7.

$$c_{\text{total}} = C_1U_1Q_1 + C_2U_2Q_2 + \ . \ . \ . = \sum_i C_iU_iQ_i \qquad \textit{Eq. 8-7}$$

It should be clear that an ion with a higher effective mobility will carry a greater fraction of the current.

It follows from equation 8-5 that increasing the conductivity at a fixed voltage increases the current. This increases the electrical heat (joules) generated in the system, since heating is proportional to the square of the current. Excessive heating produces convective disturbances in the solutions, which distort the electrophoretic patterns and may also denature macromolecules.

The voltage, conductivity, and current are related (equation 8-5). If the conductivity is increased by an increase in the salt concentration while the current is kept constant, the voltage must decrease. Such a decrease in voltage reduces the electrical force, F_{elec}, on charged particles, slowing the movement of the macromolecules. This increases the time needed for a given separation, and the resolution decreases because of increased diffusion. Since increasing the conductivity at either a fixed voltage or current has deleterious effects, optimal results are achieved when one keeps the concentration of ions and therefore the conductivity at moderate values.

FACTORS AFFECTING MOBILITIES OF MACROMOLECULES

Charge and conformation

The clinical laboratory usually deals with polyelectrolytes, substances with many charged or potentially charged groups on them. Proteins, for example, have aspartyl, lysyl, and other side chains that can carry charges.

The strength of the electrical fields produced by each charged group makes interactions between groups on the same molecule possible. Attraction of unlike-charged groups (salt effect), repulsion of like-charged groups, and hydrogen-bonding effects can occur. Such interactions can change the pK_a's of one or both interacting groups. For this to happen, the groups must be close, either in the linear sequence or because of the conformation of the macromolecule. And just as the conformational folding of the macromolecule can change the pK's of groups brought together, the charges on these groups can affect the conformation of the macromolecule.

The net charge of a polyelectrolyte is determined by the total number of charged groups within the polyelectrolyte and its conformation. A solution of polyelectrolytes consists of a population of species of different charge distributions and different conformations. These species are all in equilibrium with each other. If the rates of interconversion between species are all fast relative to the average electrophoretic velocity of the polyelectrolyte, the observed mobility will be an average one, sometimes called a "constituent" mobility.[7] If the rates of interconversion are relatively slow, a chemically homogeneous macromolecule would appear to be a heterogeneous mixture of macromolecules with different mobilities.

In some cases the electrolyte ions bind strongly and specifically to the macromolecule. For example, bovine serum albumin can bind several chloride ions. This changes the net charge of the macromolecule and therefore its mobility.

Ionic "atmosphere" and zeta potential

Counterions, ions of opposite charge, naturally tend to hover in the vicinity of the charged groups of macromolecules. However, they do not actually neutralize the charges on the macromolecule but are instead located at a spectrum of distances from the charged groups of the macromolecule, forming a "double layer" of charge about the macromolecule, called an "ionic atmosphere."

The macromolecule moves with its entourage of hydration and hydrated counterions (Fig. 8-3). These reduce the effective charge of the macromolecule to a level given by the "zeta potential." The zeta potential is the potential (voltage) produced by the effective charge of the macromolecule at the "surface of shear." The surface of shear is the boundary between the entire macromolecule complex in solution (hydration layer and embedded counterions) and the material that is staying behind (the solvent).

The effectiveness of interaction of small ions with macromolecules is proportional to the square of the charge on the small ions.[4] Electrophoretic solutions are often described by their "ionic strengths," in which case the concentration of each ion, C_i, is multiplied by the square of its effective charge, Q_i. For a solution, the ionic strength, I, is written:

$$I = {}^1\!/_2 C_1 Q_1^2 + {}^1\!/_2 C_2 Q_2^2 + \ldots = {}^1\!/_2 \sum C_i Q_i^2 \qquad \textit{Eq. 8-8}$$

For univalent electrolyte solutions, such as NaCl, the ionic strengths and concentrations are the same.

Thermal energy: relaxation effect

Macromolecules do not move in a continuous fashion or in a straight line. The reason is that random thermal motion of the macromolecules is superimposed on the motion produced by the electric field. Thermal energy will cause a macromolecule to move in "jumps," irrespective of any electric field. At each jump, the counterion atmosphere is left somewhat behind. The counterions (or their replace-

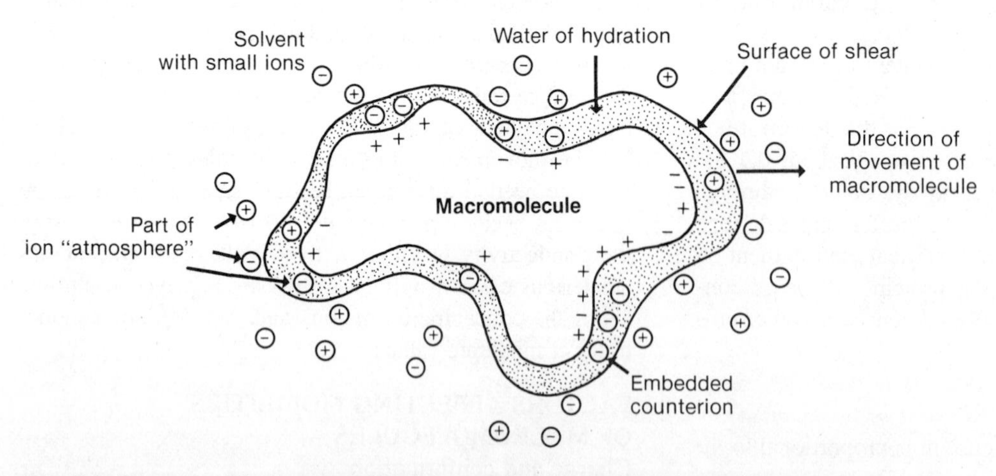

Fig. 8-3. Zeta potential of macromolecule is average effective electrical field strength ("potential") produced by charges on macromolecule and any charged particles embedded in solvent carried along ("water of hydration") with macromolecule. *Stippled area,* Water of hydration. Zeta potential is measured at "surface of shear": the boundary between water of hydration and rest of solvent.

ments from other parts of the solution) then move to catch up or reposition themselves, but this takes a little time. This is called a "relaxation effect." It also tends to lower the mobility of the macromolecule, since the retarded or misplaced counterions momentarily produce a field in the opposite direction from the applied field.

Electrophoretic effect

Since ions in water solution are hydrated, the counterions of the electrolyte moving in the opposite direction carry a lot of solvent along with them. The macromolecule is thus moving against a *flow* of solvent, and so its mobility is further reduced. This "electrophoretic effect" increases the frictional drag on the protein because its speed relative to the solvent, part of which is moving in the opposite direction, is greater. Since equal charges are moving in opposite directions, this effect produces little *net* flow of solvent in either direction.

SUPPORT MEDIA

It often is desirable to separate a mixture of proteins into completely separate zones. The narrower (thinner) the original zone of a mixture of macromolecules, the less migration distance necessary to achieve separation. Use of narrower zones means that complete separation can be effected in less time. This provides an additional benefit; blurring or remixing of the separated zones as a result of diffusion is also reduced.

The major technical difficulty in using narrow (thin) zones of relatively concentrated macromolecules is a mechanical one. Such zones will be considerably more dense than the solvent; thus the zones would "fall" through the solvent faster than the macromolecules would electrophorese. This is called "bulk flow" or "convection." This problem is especially severe when one is trying to electrophorese large particles such as cells. The conventional answer to this problem is to use a supporting medium.

Functional basis

The supporting medium must allow as free a penetration of the material to be separated as possible and yet cut off bulk flow (convection). Most media do this by offering a restricted pore size for electrophoretic movement of the macromolecules. A capillary tube has the same effect.

Electro-osmosis

The supporting medium should not adsorb the molecules, since this will inhibit or stop migration. The usual interaction problem encountered is not from actual adsorption of the material. Electrostatic interactions, from charged groups attached to the medium, are more common.

Agar, which is often used as a supporting medium in electrophoresis, is a mixture of agarose and agaropectin. The agaropectin has a relatively large number of carboxyl groups in it, which at neutral pH have counterions. If a voltage is applied, the counterions will move but the carboxyl groups attached to the polysaccharide matrix will not. The counterions carry enough solvent with them to produce a *net* flow of solvent in *one* direction. This is called "electro-osmosis," or sometimes "endosmosis."

Electro-osmosis is a very general effect.[2] It is more pronounced when charged groups are present in the supporting medium, but it always occurs to some extent. If any two different substances (phases) are brought into contact, an electrical potential (voltage) develops because the "chemical potentials" of the two phases will usually be different. (There is no experimental difference between a chemical and an electrical potential.) If an external voltage is applied, there will always be a tendency for one phase to move relative to the other.

Types of supporting media

The supporting medium can be a solution such as a sucrose-density gradient, but, in general, insoluble materials are used. Some are self-supporting, whereas others are mechanically supported by the apparatus. Paper or sheets of plastic such as cellulose acetate have enough mechanical strength to allow electrophoresis on sheets hung or stretched over rods.

Support media can also be classified as particulate or continuous. Particulate support media include glass beads, Sephadex, or cellulose fibers. Continuous supports include polyacrylamide, starch, and agarose gels.

Gels are jellylike solids in which all the solvent is included. For example, starch suspensions are heated and cooled. The starch fibers interact, tangling with each other but trapping the solvent so that large gaps or pores exist between the fibers. These gaps or pores are available for macromolecular movement. Similar gels can be made from

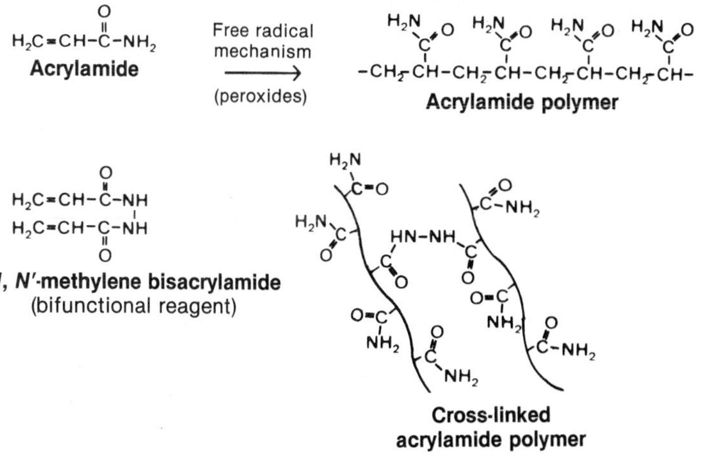

Acrylamide

Free radical mechanism (peroxides)

Acrylamide polymer

N, N'-methylene bisacrylamide (bifunctional reagent)

Cross-linked acrylamide polymer

Fig. 8-4. Polyacrylamide gels are produced by polymerizing a mixture of acrylamide and a bifunctional (cross-linking) acrylamide derivative. Derivative shown is that in common use.

agar, agarose, or some chemical polymers. Gels can also be made by polymerization of acrylamide with a small percentage of a bifunctional acrylamide derivative that cross-links the acrylamide polymers (Fig. 8-4).

Molecular sieving

The porosity, or average pore sizes, of some media are fixed, but the porosities of other media can be controlled. For example, by changing the gel concentrations of starch or agar, one can vary the pore size. If the average pore size is near the average diameter of the macromolecules being electrophoresed, the supporting medium will produce "molecular sieving" effects.

The average pore size of polyacrylamide gels cast at 5% to 10% concentrations is comparable to the effective diameters of many "globular" (relatively compact) proteins of 15,000 to 250,000 molecular weights. These gels will filter such solutions, adding the possibility of separation of macromolecules on the basis of size and mobility.* The molecular sieving effects can produce enhanced resolution, that is, narrower zones of macromolecules.

ENHANCED-RESOLUTION TECHNIQUES

Discontinuous solvents and voltage gradients

The value of electrophoretic techniques depends on their resolving power, that is, their ability to separate different molecules from each other. One series of techniques that increases the resolving power is based on the observation that mobilities are different for each ion; therefore Na^+ ions in solution migrate differently from Ca^{++} ions.

One can layer solutions containing ions of different mobilities on top of each other so that they are in series. When a solution of ions of low effective mobility and one of higher effective mobility are placed in an electric field so that the slower ions follow the faster, the layered ions move at the same velocity. The reason is that if the faster ions moved away from the slower ions, a gap between the solutions would occur that would contain no ions; thus no current would be conducted.

The current must be the same in both solutions. To keep the slower ions moving at the same pace as the faster ions, the voltage driving the slower ions must be greater (Fig. 8-5). The difference in voltage prevents diffusion between the two solutions.[11]

If a third ion solution of intermediate effective mobility is placed between the two layered ion solutions described above, this ion will remain in place between the other two ions. If the intermediate ion has a low charge relative to a high molecular weight, as many proteins do, the zone the intermediate ion occupies will shrink (Fig. 8-6) so that it carries the same current as the other two solutions do.[11] Thus this molecule is "stacked" between the other two. One can expand the technique to separate a mixture of

*Chapter 15, reference 8; Chapter 4, reference 9.

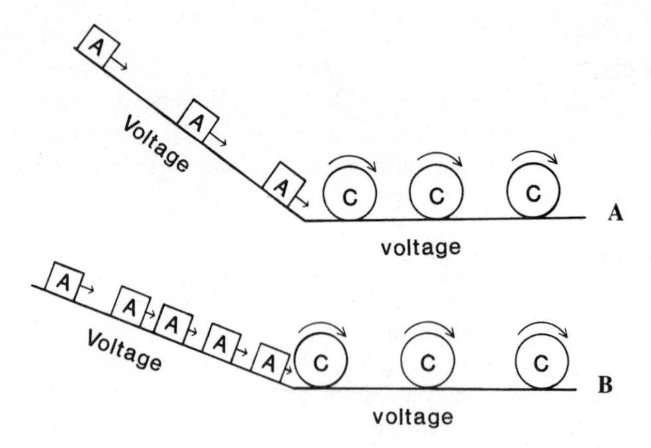

Fig. 8-5. **A,** A much greater driving force (voltage) is required to make ions of low effective mobility (*A* ions) keep up with ions of high effective mobility (*C* ions). In **B,** concentration of *A* ions is greater and so conductivity of *A* solution is greater than that in **A.** Voltage keeping *A* ions up to *C* ions is less. This is represented by slope under *A* ions being lower than that in **A.**

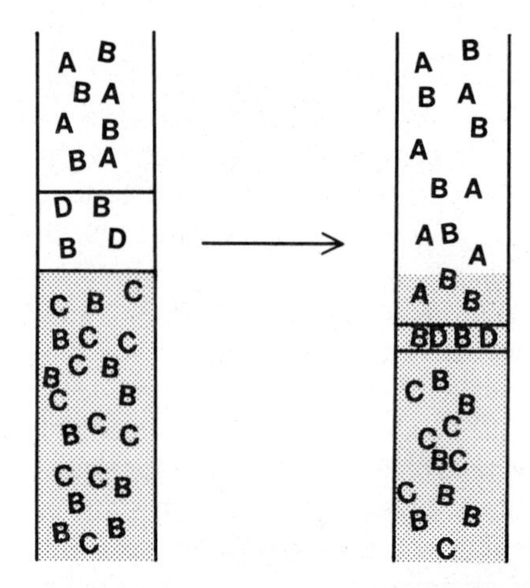

Fig. 8-6. Movement of *A* and *D* ions into region vacated by *C* ion *(stippled region)* is accompanied by changes in concentration and pH. The latter is not shown. *D* ions are assumed to be much larger, with a relatively low charge.

molecules. If this intermediate solution contains many different ions of intermediate effective mobilities, such as a series of proteins, they will arrange themselves in order of their effective mobilities and concentrate themselves according to their charges when the electric field is applied. The restricted diffusion and concentration produce enhanced resolution.

There are three major applications of the use of discontinuous systems to produce enhanced resolution: isotachophoresis, isoelectric focusing, and disk electrophoresis.

Isotachophoresis

Isotachophoresis involves ranking macromolecules in order of decreasing effective mobilities.

A "leading ion," corresponding to a chloride ion in the standard disk electrophoretic system, and a "terminating ion" (such as glycinate) are required. A counterion for buffering the system is also necessary.

The profile of distribution of materials in an isotachophoretic separation is a series of abrupt boundaries, or "steps," that can be detected from their temperatures, conductivities, or refractive index gradients. The lengths of the zones are proportional to the concentration of the substances separated.

The method works well for low molecular weight materials such as simple ions, amino acids, and nucleotides. This could have clinical application in situations in which chromatographic procedures are not available.

The major problem with the technique is the fact that the different zones are adjacent. Overall, the separation effectiveness of the method lies between those of ordinary zone electrophoresis in continuous buffers and disk electrophoresis. Other problems with the technique include electro-osmosis, temperature gradients, convection, and limitations on the range of usable pH's. Currently there are no widespread clinical applications of this technique.

Isoelectric focusing

In principle, isoelectric focusing is believed to be related to isotachophoresis. However, in isoelectric focusing the leading and trailing ions are an acid and base. At the interface, there will be a pH gradient between these two solutions. If a third solution that comprises a special series of molecules (ampholytes) is layered between the acid and the base, a pH gradient can be made throughout this third or intermediate solution. Ampholytes are extremely heterogeneous mixtures of alphatic polyamino-polycarboxylic acids. When placed in an electric field between acid and base solutions, ampholytes migrate according to their effective mobilities. However, there is a pH gradient between the two electrodes. The ampholytes will migrate until they reach a pH at which each type will be electrically neutral; that is, the positive and negative charges on the molecules will be equal. The importance of this migration is that other ions such as proteins will also migrate in the pH gradient until they reach a point at which they are electrically neutral, or isoelectric.

The main advantage of isoelectric focusing is its resolving power, that is, its ability to separate molecules with very close isoelectric points. This is enhanced by use of a high field strength (voltage) and a shallow pH gradient. Resolution to within 0.02 pH units is possible. Focusing in analytical gels can be done in 1 to 2 hours. However, it is important to remember that given enough time, the pH gradient breaks down, and so there is a limit to the time a sample can be kept focused. More complete descriptions of this technique are given by Righetti* and Catsimpoolas.* An application of the technique, an analysis of Pi-system variants, is described by Fagerhol and Laurell.[13]

*Chapter 19 in reference 8.

Disk electrophoresis

Disk electrophoresis essentially involves adding ordinary zone electrophoresis to isotachophoresis. A supporting medium, usually a polyacrylamide gel, is almost always used. Its great usefulness is derived from its sensitivity or resolving power. This resolving power comes partly from the initial isotachophoresis step, partly from separations based on molecular sieving, and partly from the restriction of diffusion of macromolecules. The resolution is also a function of the sensitivity of the method used for detection of a protein zone.

The major limitation of the method is that the ion systems and usable pH's are limited. The method is technically exacting, but for high-resolution analysis it is still one of the best techniques available.[8] It is often used for analysis of serum proteins.

METHODS INVOLVING SEPARATIONS BASED ON MOLECULAR SIZE

Size measurements

Gradient gels. Determination of macromolecular size can be achieved by use of gradient gels. This method of measuring molecular size is based on the relationship between the effective size of a macromolecule (Stokes radius) and its ability to penetrate a gel. A gradient of decreasing pore size in a gel can be formed by successive addition of acrylamide solutions of decreasing concentration. At pore sizes smaller (high concentrations of polymer) than its Stokes radius, a macromolecule cannot penetrate the gel and is effectively excluded. A polyelectrolyte placed in an electric field in such a gel will migrate to its exclusion limit. One can separate a series of polyelectrolytes of varying molecular size by using this technique. If appropriate standards are used, the molecular weight or Stokes radius can be calculated.

SDS gel electrophoresis. Sodium dodecyl sulfate (SDS) is a detergent with a long-chain aliphatic hydrocarbon terminating in a sulfate group. It denatures most proteins by binding in large amounts. For most proteins the amount bound is nearly constant, about 1.4 g of SDS per gram of protein. This is on the condition that all disulfide (−SS−) bonds and any other cross-links between polypeptides are broken. If there are no cross-links, the SDS-protein complex assumes a rodlike shape in solution. The uniform shape and the high negative charge from the sulfate of SDS means that the complexes migrate proportionally to the logarithm of the molecular weight of the polypeptide in molecular sieving media. Since protein polymers are broken up, the molecular weights obtained are subunit molecular weights.[14,15]

*Chapter 9 in reference 9.

Methods involving separations based on size and charge

Two-dimensional electrophoresis. The resolving power of electrophoresis is theoretically very high. In practice, the enormous variation in concentration of different macromolecules in cells and other biological samples makes it hard to detect the less abundant species. For example, the albumin in serum forms a zone so thick that many serum proteins cannot be detected at all.[16] The use of two-dimensional techniques, in which molecules are spread over the surface of a plate rather than along a column, helps overcome this problem. The best resolution is achieved when molecules are separated in one direction according to a different molecular property than in the other direction. In the case of electrophoresis, one can achieve separation according to charge in one dimension and according to molecular size in the second dimension. The resolving power is extraordinary—1100 different proteins from an *Escherichia coli* lysate were detected.[17] Recently, an entire issue of the journal *Clinical Chemistry* was devoted to two-dimensional electrophoresis.[18] Its potential applications in clinical work are only starting to appear. The information available in the gels is so great that computerized scanning and mapping techniques are necessary.[18,19]

It is convenient to make the first-dimension separation by charge by employing isoelectric focusing. Usually the focusing step is carried out on a gel column, after which the column is mechanically attached or embedded in a slab for the second-dimension separation. Separation in the second dimension by use of the slab can be done by isotachophoresis, ordinary gel electrophoresis, gel-gradient electrophoresis, or SDS-gel electrophoresis. Since SDS-gel electrophoresis separates on the basis of subunit molecular weights, it is the most popular. This is the "O'Farrell technique."[17]

Table 8-1. Common effects of electrophoretic parameters on separation

Parameter	Effect on electrophoresis
pH	Changes charge of analyte and hence effective mobility. Can affect structure of analyte, such as denaturing or dissociating a protein.
Ionic strength	Changes voltage or current; so increased ionic strength usually reduces migration velocity and increases heating.
Ions present	Can change migration velocity if interaction is strong. Can cause tailing of bands.
Current	Too high a current causes overheating.
Voltage	Migration velocity proportional to voltage.
Temperature	Temperature gradients in support media cause bowed bands. Overheating can denature (precipitate) proteins (see "Current"). Lower temperatures reduce diffusion but also reduce migration velocity; no effect on resolution.
Time	Separation of bands (resolution) increases linearly with time, but dilution of bands (diffusion) increases with the square root of time.
Medium	Major factors are endosmosis and pore-size effects, which affect migration velocities.

SELECTION OF METHODS AND CONDITIONS

The method chosen will depend partly on the information available and that which is desired. Knowledge of the isoeletric point and the molecular weight of the compound to be examined can help determine the optimal conditions. A pH is desired that will provide the maximum separation without destroying the properties of the sample. Very acidic or basic conditions pose problems for any system, since an increasing fraction of the current is carried by protons or hydroxyl ions and will result in poorer separation. A summary of the effects of the various parameters on electrophoresis is provided in Table 8-1.

Support media

The choice of a supporting medium is based on many considerations. Slabs or flat-surface media allow greater amounts of material to be electrophoresed. These media are also useful when one is comparing different samples. Use of gel cylinders can provide great sensitivity, whereas thin-layer methods provide even greater sensitivity.

Paper and cellulose acetate. Paper is especially favored for separation of low molecular weight substances in specialized biochemical laboratories. The main advantages of these materials are their thinness and mechanical strength. A thinner support means greater sensitivity because less material will produce a detectable spot or zone.

Cellulose acetate is somewhat stronger than paper and a good deal more chemically uniform. Hence it is used more often in clinical work. Adsorption of material to paper or to any electrophoretic support medium leads to losses of material and to "tailing" of zones. Cellulose acetate is more inert in this respect.

Gels. Gels can be cast with varying thicknesses to increase or decrease capacity. They also offer the possibility of molecular sieving effects because their porosity can be controlled when their composition is changed. On the other hand, their mechanical strength tends to be lower.

Starch and agar. Starch gels are not as extensively used as agarose, acrylamide, or even Sephadex. The starch solution must be heated to 100° C and then degassed, a process that is awkward. The starch gels tend to provide greater resolution than agar or agarose does. However, the inconvenience of preparing starch gels has limited their use.

Agar and agarose (agar without the agaropectin) are easier to handle. Because agarose has lower electro-osmosis and exhibits fewer problems with adsorption, it is preferred over agar. The pore size of agarose is much greater than that of polyacrylamide. This is the reason why agarose is used in most immunoelectrophoretic techniques, since antigen and antibody must be able to diffuse freely through the gel. Another advantage of agarose is that it may be poured after reheating to only about 50° C; so some proteins, such as antibodies, can be mixed in without denaturing. Precast agarose gels for a variety of separations are available commercially. These gels are used for separation of isoenzymes, hemoglobins, glycoproteins, and so on.

Polyacrylamide. Polyacrylamide gels are rarely used in clinical laboratories but are a common research tool. They are clear and easy to prepare and exhibit reasonable mechanical strength over the acrylamide concentration range normally employed for proteins. In addition, they have a low endosmosis and a pore size well suited for the separation of the "average" proteins, RNA molecules, and smaller restriction fragments of DNA. A major use of polyacrylamide gels is for the separation of alkaline phosphatase isoenzymes.

Conditions

Horizontal versus vertical position. Electrophoresis is done horizontally or vertically; there is no theoretical reason for preferring one or the other procedure. Horizontal electrophoresis places less mechanical stress on the support, whereas vertical electrophoresis supporting media are often supported between glass plates.

If horizontal electrophoresis is used and the surface of the medium is open to air, evaporation of the solvent can cause problems. As a result of evaporation, salt concentrations rise, usually unevenly along the support, leading to nonuniform current flow and heating. This can lead to problems of uneven migration, especially at the sides of a horizontal, flat electrophoresis bed. If one buffer reservoir is higher than the other, convective flow of the buffer through the supporting medium can occur. Therefore the electrophoresis apparatus should always be level.

Sample application. Sometimes samples are simply applied to the surface of the supporting medium and allowed to soak in. Sometimes special slots or holes are cast or cut in a gel. Occasionally the sample may be polymerized into the gel or cast with the gel. Injection of the sample is rarely used, except with isotachophoresis. If electrophoresis takes place in stages, the gel containing part or all of the sample may be cut out and reattached, sometimes with an agarose "glue," to another gel for the next stage in the separation. Often the sample is layered onto the surface of a gel. The sample is then usually made more dense than the solvent with sucrose or glycerol.

Current and voltage considerations. Electrophoresis can be carried out at constant voltage, constant current, or constant power. Selection of any of these modes often depends on the type of power supply available. Since diffusion increases with the square root of time, it is best to complete the electrophoresis as quickly as possible. This requires use of the maximum voltage. However, the voltage is always limited by the efficiency of cooling of the apparatus. Some workers claim that temperature gradients of more than 0.1° C across a gel or other support lead to noticeable distortions of macromolecule zones. For many clinical applications, temperature control does not appear to be necessary and separations are carried out at ambient temperatures.

The conductivity of any electrophoresis system will change with time because the ionic composition will change as a result of movement (electrophoresis) of the sample along the system. Such changes are minimal in continuous systems, such as high-voltage paper electrophoresis, and application of constant voltage is satisfactory for these systems. For isotachophoresis, a constant velocity of zone migration is desired, and constant current is used. For other electrophoretic systems, heating is usually the limiting factor, and so constant power (wattage) should be used. Disk electrophoresis and isoelectric focusing fall into this category. "Pulsed" power supplies provide no advantage.[20]

Separation time. In the case of isotachophoresis, the electrophoresis is stopped when the trailing ion emerges. Isoelectric focusing is complete after the current has dropped to a stable value, since the gradient is at equilibrium. The time to stop a disk or ordinary zone electrophoresis separation is usually indicated by the position of the "tracking dye." Dyes, such as bromphenol blue, that have high mobilities are employed. They are usually added with the sample. Since some proteins bind such dyes, their apparent mobilities may be changed.

LOCATING THE ANALYTE

Analysis involves determining where the substances to be identified or quantified are on the support. This can be accomplished by measurement of a physical property of a molecule, such as light absorption or refractive index, or by a chemical reaction such as staining.

Generally, measurements of physical properties lack specificity, sensitivity, or resolution. For example, not all proteins absorb strongly at 280 nm.

Staining

Staining often achieves the desired goals of resolution, sensitivity, and specificity (Table 8-2). Since the zones of material broaden by diffusion after electrophoresis is stopped, the first step in the analytical procedure is to eliminate diffusion. One can do this in paper electrophoresis by drying the paper or in the case of autoradiography by freezing the gels.

Proteins

In the case of proteins in gels, the protein often is denatured, that is, precipitated in the gel matrix. One often

Table 8-2. Commonly-used stains for various substances

Substance	Stain	Comments
Proteins	Ponceau S*	Less sensitive than amido black, but more specific for proteins Most widely used stain
	Bromophenol blue† Light green SF†	Low sensitivity, but can be used with ampholyte gels
	Coomassie brilliant blue R250*	Can detect less than 0.2 μg of protein; can be used with ampholyte gels
	Silver stain (silver reduced onto oxidized macromolecules)‡	At least 10 to 50 times more sensitive than others; different proteins give different colors, for unknown reasons.
	Stains-All (a cationic carbocyanine dye)§	General sensitivity, including phosphoproteins
	Amido black 10B ("buffalo black")‖	Very sensitive stain
Lipoproteins	Sudan black B¶	
	Oil red O#	
	Coomassie brilliant blue R250#	Used with SDS gels
Glycoproteins	PAS (periodic acid–Schiff)**	Best for neutral glycoproteins; 2 to 3 μg of carbohydrate detectable
	Stains-All§	Best for sialic acid–rich glycoproteins
Nucleic acids	Stains-All§	Best for RNA, DNA, and mucopolysaccharides
	Silver stain§	
	Ethidium bromide††	Fluorescent bands with DNA; less than 10 ng detectable
Enzymes		
Dehydrogenases	NADH (fluorescence)‡‡	
	Nitro blue tetrazolium chloride‡‡	
Esterases Cholinesterases	Beta-naphthyl esters and tetrazotized O-dianisidine‡‡	
Phosphatases	1-Naphthylphosphate and fast blue B‡‡	

*Righetti, P.G., and Drysdale, J.W.: Isoelectric focusing, New York, 1976, Academic Press, Inc.
†Chapter 19, reference 8.
‡Merril, C.R., Goldman, D., Sedman, S.A., and Ebert, M.H.: Science **211**:1437, 1981.
§Green, M.R., Pastewka, J.V., and Peacock, A.C.: Anal. Biochem. **56**:43, 1973.
‖Wilson, C.M.: Anal. Biochem. **96**:263, 1979.
¶Swahn, B.: Scand. J. Clin. Lab. Invest. **4**:98, 1952.
#Weller, H.: Klin. Wochenschr. **36**:563, 1958.
Matthieu, J.M., and Quarles, R.H.: Anal. Biochem. **55:313, 1973.
††Brunk, C.F., and Simpson, L.: Anal. Biochem. **82**:455, 1977.
‡‡Gabriel, O.: Methods in enzymology, vol. 22, New York, 1971, Academic Press, Inc., p. 578.

does this by soaking the gels in dilute acetic acid. Trichloroacetic acid is more effective. Addition of sulfosalicylic acid further improves the denaturating ability of the staining solution.

Sometimes heat must also be applied to make the proteins insoluble. However, some resist denaturation by all these conventional procedures and remain soluble in the stain. If detergents such as sodium dodecyl sulfate (SDS) or other solubilizing agents are present, they will interfere with precipitation. Inclusion of methanol in the acid solutions helps remove such substances before staining.

Choice of stain

Many types of stains are employed, depending on the need. Sometimes it is desirable to stain "everything," such as all proteins. A dye called "Stains-All" is suitable.[21] A "silver stain," which is reactive to both proteins and nucleic acids, may be an alternative choice.[22] Proteins, after electrophoresis on cellulose acetate, are most often stained with Ponceau S.[23]

Stains and staining procedures are often specific for one chemical group. The ninhydrin stain for amino groups, often used after paper electrophoresis of peptides, is an example of this. Glycoproteins are detectable by treatment with periodic acid (for oxidation), and color is developed with a dye (fuchsin) in the presence of a reducing agent (sulfite). This periodic acid–Schiff stain (PAS) treatment oxidizes carbohydrate groups to aldehydes, which react with the dye to form a Schiff base. The sulfite reduces the Schiff base, making the stain permanent. There is also a specific stain for phosphoproteins. A fairly complete list of stains is given by Righetti and Drysdale.[23] (See Table 8-2 for a list of commonly used stains.)

Once the stain has been introduced, usually by soaking of the support in the stain solution, excess stain must be removed. This can be done electrophoretically or, most

Table 8-3. Commonly encountered problems in electrophoresis

Problem	Likely cause	Corrective action
No migration	Instrument not connected	Check electrical circuits.
	Wrong pH; electrodes connected backwards	Check isoelectric point of protein and pH of buffer; check electrode polarity.
Bowed electrophoretic pattern on edges of support	Overheating or drying out of support	Humidify chamber; check buffer ionic strength.
Tailing of bands	Chemical reaction: subunit dissociation or adsorption to support	Use different support; try different pH.
	Salt in sample	Check sample for salt; dialyze sample against electrophoresis buffer.
	Buffer co-ion effect	Use different buffer co-ion.
Holes in staining pattern	Analyte present in too high a concentration	Apply less concentrated sample.
Very thin, sharp bands	Molecular weight of sample very high for support pore size	Use support with larger pore size.
	Sulfhydryl oxidation and aggregation	Run sample with sulfhydryl reducing compound or at lower pH.
Very slow migration	High molecular weight	Use support with larger pore size.
	Low charge	Change pH so that charge increases.
	Ionic strength too high	Check conductivity; dilute buffer.
	Voltage too low	Increase voltage.
Sample precipitates in support	pH too high or low	Run at different pH.
	Too much heating	Use lower wattage or external cooling.

Table 8-4. Table of compounds commmonly separated by electrophoresis

Class of compound	Stain	Support medium
Amino acid	Ninhydrin	Paper, cellulose acetate
Serum protein	Ponceau S	Cellulose acetate
	Coomassie blue 250	Polyacrylamide (with or without isoelectric focusing)
Lipoproteins	Oil red O	Agarose
Glycoproteins	PAS (periodic acid–Schiff)	Agarose
Nucleic acids	Ethidium bromide (fluorescent)	Agarose
Hemoglobins	Silver stain	
	o-Dianisidine	Cellulose acetate agar
	Ferricyanide	
	Peroxide	
Isoenzymes		
Lactate dehydrogenase	Fluorescent NADH or tetrazolium	Agarose
Creatine kinase	Fluorescent NADH or tetrazolium	Agarose
Alkaline phosphatase	1-Naphthylphosphate + fast blue B or 5-bromo-4-chloroindolyl phosphate	Polyacrylamide, cellulose acetate
Immunoglobulins	Coomassie blue 250	Agarose
Specific antigens by immunological electrophoretic techniques (such as Laurell rocket)	Amido black 10B	Agarose

commonly, by diffusion. Electrophoretic removal is fast but can result in distortion of the stained zones. Diffusion involves changing the solvent or use of a destainer, which removes free stain.

Many enzymes are identified with the use of colored or fluorescent substrates or products (zymograms), even in gels or other support media. For example, alkaline phosphatase hydrolyzes *para*-nitrophenylphosphate to *para*-nitrophenol, which has a yellow color at pH 8. Soaking a gel in such assay solutions produces colored bands where the enzymes are. Prolonged soaking in such solutions leads to zone broadening from diffusion of the colored substrate or product. A list of such zymograms is given in Righetti and Drysdale.[23]

Some commonly encountered problems in electrophoresis and their most likely causes are listed in Table 8-3, as well as suggested corrective action.

CLINICAL APPLICATIONS

The most common uses of electrophoretic techniques in the laboratory today are listed below:

1. Specific protein electrophoresis (see Chapters 10, 50, 55, and 59)
   a. Quantitative analysis of specific serum protein classes such as gamma globulins and albumins
   b. Identification and quantitation of hemoglobin and its subclasses
   c. Identification of monoclonal proteins, such as Bence Jones gamma globulins in either serum or urine
2. Isoenzyme analysis. Separation and quantitation (see Chapter 50) of enzymes such as creatine kinase, lactate dehydrogenase, and alkaline phosphatase into their respective molecular subtypes
3. Immunoelectrophoresis (see Chapter 10). Most often used to determine qualitatively the elevation or deficiency of specific classes of immunoglobulins. In addition, the technique can be used to semiquantitate serum proteins such as transferrin and complement component C3

The two-dimensional procedures will undoubtedly replace some or all of the preceding procedures, but at this time mostly exploratory work is being done. All the procedures involve measurement of alterations in an electrophoretic pattern relative to a "normal" control. One can often use these to diagnose specific diseases[16,24] (Fig. 8-7).

Electrophoresis is sometimes used in assays of genetic defects. The two-dimensional techniques now being developed have tremendous potential in that area. Table 8-4 lists compounds normally separated by electrophoresis.

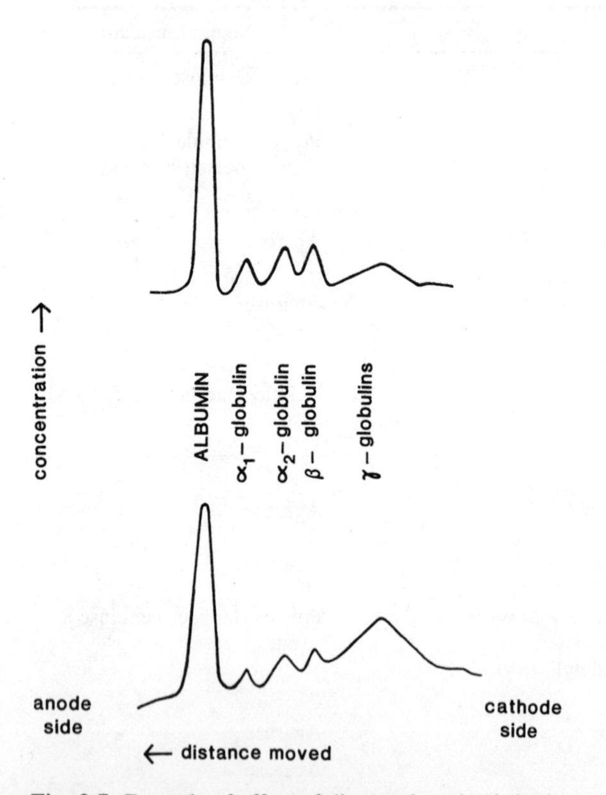

Fig. 8-7. Example of effect of disease, hepatic cirrhosis, on blood serum protein electrophoretic pattern. *Upper profile,* Distribution characteristic of healthy people.

ACKNOWLEDGMENT

Supported by a grant from the National Institutes of Health, GM 27742.

REFERENCES

1. Richards, E.G.: An introduction to physical properties of large molecules in solution, Cambridge, 1980, Cambridge University Press.
2. Moore, W.J.: Physical chemistry, ed. 4, Englewood Cliffs, N.J., 1970, Prentice-Hall, Inc.
3. Van Holde, K.E.: Physical biochemistry, Englewood Cliffs, N.J., 1971, Prentice-Hall, Inc.
4. Tanford, C.: Physical chemistry of macromolecules, New York, 1961, Academic Press, Inc.
5. Bier, M.: Electrophoresis, New York, 1959, Academic Press, Inc.
6. Tiselius, A.: Electrophoresis, methods in enzymology IV, New York, 1957, Academic Press, Inc., p. 3.
7. Cann, J.R.: Interacting macromolecules, New York, 1970, Academic Press, Inc.
8. Righetti, P.G., Van Oss, C.J., and Vanderhoff, J.W.: Electrokinetic separation methods, New York, 1979, Elsevier/North Holland, Inc.
9. Deyl, Z., Everaerts, F.M., Prusik, Z., and Svendsen, P.J.: Electrophoresis—a survey of techniques and applications, J. Chromatogr. **18** (series), New York, 1979, Elsevier Scientific Publishing Co., Inc.
10. Ornstein, L.: Disc electrophoresis, Ann. N.Y. Acad. Sci. **121**:321, 1964.
11. Brewer, J.M., and Ashworth, R.B.: Disc electrophoresis, J. Chem. Educ. **46**:41, 1969.
12. Jovin, T.M., Dante, M.L., and Chrambach, A.: Multiphasic buffer systems output PB 196085 to 196092 and 203016, Springfield, Va., 1970 National Technical Information Service.
13. Fagerhol, M.K., and Laurell, C.B.: The Pi system-inherited variants of serum alpha 1-antitrypsin, Prog. Med. Genet. **7**:96, 1970.
14. Laemmli, U.K.: Cleavage of structural proteins during the assembly of the head of bacteriophage T4, Nature **227**:680, 1970.
15. Wyckoff, M., Rodbard, D., and Chrambach, A.: Polyacrylamide gel electrophoresis in sodium dodecyl sulfate–containing buffers using multiphasic buffer systems, Anal. Biochem. **78**:459, 1977.

16. Henry, J.B.: Clinical diagnosis and management by laboratory methods, ed. 16, Philadelphia, 1979, W.B. Saunders Co.
17. O'Farrell, P.H.: High-resolution two-dimensional electrophoresis of proteins, J. Biol. Chem. **250**:4007, 1975.
18. Clinical Chemistry **28,** 1982 (entire issue).
19. Lemkin, P.F., and Lipkin, L.E.: GELLAB: a computer system for 2D gel electrophoresis analysis. II. Pairing spots, Comput. Biomed. Res. **14**:355, 1981.
20. Allington, R.W., Nelson, J.W., and Aron, G.G.: ISCO Applications Research Bulletin, no. 18, Lincoln, Neb., 1975, Specialities Co.
21. Green, M.R., and Pastewka, J.V.: Identification of sialic acid–rich glycoproteins on polyacrylamide gels, Anal. Biochem. **65**:66, 1975.
22. Merril, C.R., Goldman, D., Sedman, S.A., and Ebert, M.H.: Ultrasensitive stain for proteins in polyacrylamide gels shows regional variation in cerebrospinal fluid proteins, Science **211**:1437, 1981.
23. Righetti, P.G., and Drysdale, J.W.: Isoelectric focusing, New York, 1976, Academic Press, Inc.
24. Annino, J.S., and Giese, R.W.: Clinical chemistry, principles and procedures, ed 4., Boston, 1976, Little, Brown & Co.

Chapter 9 Immunological reactions

Susan Bassion

affinity Measure of the binding strength of the antibody-antigen reaction.

antibodies Proteins that combine specifically with antigens.

antigenic determinant That portion of an antigen involved in a reaction with an antibody.

antigens Substances that induce an immune response.

avidity Measure of the binding strength of antibodies to multiple antigenic determinants on natural antigens.

B cells B lymphocytes, which transform to plasma cells and produce antibodies.

constant region Noncombining region of immunoglobulin polypeptide chains.

cross-reactivity Binding of an antibody to an antigen other than the one initiating the immune response.

Fab fragment Portion of immunoglobulin molecule made by papain degradation and containing light chain and portion of heavy chain.

Fc fragment Portion of immunoglobulin molecule containing most of heavy chain after papain degradation.

flocculation Precipitation reaction producing large, loosely bound precipitate.

haptens Low molecular weight substances that can induce an immune response only when coupled to high molecular weight immunogenic molecules.

heavy chain Portion of immunoglobulin molecule comprised of a polypeptide chain of about 50,000 daltons.

hypervariable regions Amino acid sequences in the variable region that have an increased likelihood of variation.

idiotype Portion of immunoglobulin molecule conferring unique character; most often including its binding site.

immunoglobulins (Ig) Proteins with antibody activity.

immunopotency The ability of an antigen to elicit an immune response.

J chain Portion of IgM molecule possibly holding structure together, thus "joining chain."

lattice formation The cross-linked, three-dimensional structure formed by the reaction of multivalent antigens with antibody.

light chain Portion of an immunoglobulin molecule composed of a polypeptide chain of about 22,000 daltons.

nonself Refers to molecules and organs that are not part of an organism.

plasma cells Immunoglobulin-producing cells.

precipitation A reaction in which antigen and antibody are in proper ratios so that a large lattice or matrix is formed by the reaction. This matrix becomes insoluble and precipitates, forming a precipitin line.

secretory piece Polypeptide chain attached to IgA and necessary for secretion into mucosal spaces.

self Refers to all the macromolecules and structures of an organism.

valency The effective number of antigenic determinants on an antigen molecule.

variable region N-terminal portion of immunoglobulin polypeptide chain whose amino acid sequence can change; this region includes the antigen-combining site.

zone of equivalence Region of antibody-antigen reaction in which concentrations of both reactants are equal.

Immunological reactions can occur between two types of substances, *antigens* and *antibodies*. This chapter examines these substances and the interactions between them.

ANTIGENS

Antigens, or *immunogens,* are defined as substances that induce an immune response. The immune response produced may be an antibody (humoral) response or production of sensitized cells (cellular response). Usually both humoral and cellular responses are stimulated.

Factors affecting antigenicity

Many factors determine the antigenicity of a molecule. The nature and dosage of an antigen, the route of administration, the organism immunized, and the sensitivity of the detection method are important factors in the evaluation of antigenicity. Many other conditions must be satisfied in order for a molecule to be immunogenic. These conditions are discussed below.

Chemical nature. The first antigens investigated were bacteria and red blood cells. Studies showed that such complex macromolecular structures were composed of many different proteins, carbohydrates, and lipids. Subsequent investigations have proved that immunogens are found in several chemical classes. Proteins, polysaccharides, glycolipids, nucleic acids, and polynucleotides are capable of inducing an immune response.

Size. There is no absolute size requirement, but size is of considerable importance in determining the antigenicity of a molecule. The most potent immunogens are macromolecules with molecular weights greater than 100,000 daltons. The A and B polypeptide chains of insulin (2500 daltons) and of glucagon (3600 daltons) are immunogenic in guinea pigs. Nevertheless, most molecules with molecular weights less than 10,000 daltons are weakly immunogenic, if at all.

Haptens. Substances with low molecular weights can induce an immune response when coupled to higher molecular weight immunogenic *carrier* molecules. Such incomplete antigens, or *haptens,* do not elicit an immune response by themselves but do react with antibody. Many low molecular weight compounds have been shown to act as haptens, including monosaccharides, lipids, peptides, hormones such as ACTH and prostaglandins, toxins such as arsphenamide, and drugs such as barbiturates and sulfonamides.

Complexity. A molecule must exhibit a certain degree of chemical complexity in order to be antigenic. Synthetic amino acid homopolymers, composed of repeating units of a single amino acid, have been shown to be poor immunogens; copolymers of two or three amino acids are much better immunogens. Increasing immunogenicity follows increasing complexity. For example, the addition of aromatic amino acid residues such as tyrosine to synthetic amino acid copolymers increases their immunogenicity.

Antigenic determinants. The portion of an antigen involved in the reaction with an antibody is called an *antigenic determinant.* An antigen may contain more than one type and number of antigenic determinants; the number of antigenic determinants per antigen varies with the size and complexity of the molecule (Fig. 9-1). The effective number of antigenic determinants on an antigen is its *valency.* This is the number of antibody molecules that can be bound to an antigen at the same time (Fig. 9-1). Antibodies recognize the three-dimensional shape of an antigenic determinant (conformational antigenic determinant) as well as the basic amino acid structure (sequential antigenic determinant). An antigenic determinant sometimes comprises as few as four amino acid residues. The combining site of an antibody molecule reacts with an antigenic determinant in the complementary "lock-and-key" manner of protein-enzyme interactions. The affinity of the binding of antibody to antigenic determinant is directly proportional to the closeness of fit. The capacity of a portion of an antigen to serve as an antigenic determinant is termed its *immunopotency*.

Conformation and accessibility. The tertiary structure or spatial folding of molecules is a significant factor in their immunogenicity. Antibodies to native proteins do not react with denatured molecules. Antibodies to native proteins are directed primarily to conformational rather than sequential antigenic determinants.

In addition, accessibility or exposure to the environment is an important factor in determining immunogenicity. The terminal side chains of polysaccharides, the portions of a polysaccharide molecule that stick out from the main part of the molecule, are the most immunopotent regions of polysaccharide antigens. Accessibility of an antigen to the environment also depends on the solubility of an antigen in aqueous medium. The more soluble an antigen is, the greater the probability of interactions with it. The influence of charge on immunogenicity may be a manifestation of accessibility. Charged or hydrophilic residues are more in contact with the environment than hydrophobic residues, which tend to be sequestered in the interior of molecules.

Foreignness. The immune system is capable of distinguishing *self* from *nonself* in such a way that, under normal circumstances, a vigorous immune response is produced only to substances recognized as foreign. The more distant the evolutionary relationship between antigen and host, the more immunogenic the molecule. Thus guinea pig albumin will not evoke an immune response when in-

Fig. 9-1. Antigen X contains many different antigenic determinants numbered *A* to *Q* in this schematic representation. Antibody molecules when combined with antigen X bind to different sites. Maximum number of molecules of antibodies bound in this figure is 3; therefore valence is 3.

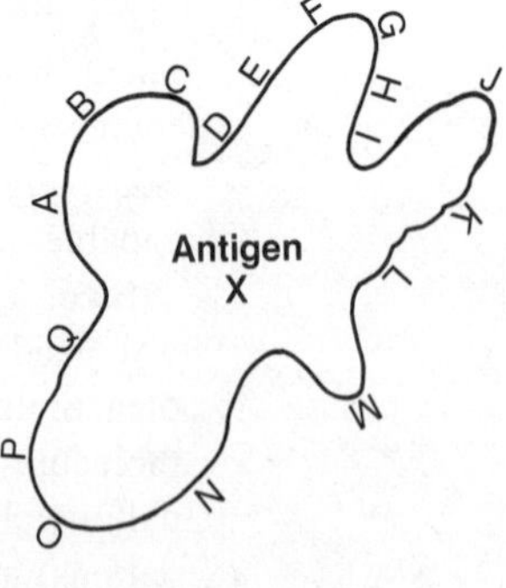

Antibody

Antibody

Antigen
X

Antibody

Antigenic determinants A to Q

Valence = 3

jected into another guinea pig. The same guinea pig albumin will evoke a strong immune response, however, when injected into a different or more complex (higher) vertebrate such as a rabbit or a monkey. Similar proteins produced by animals that are closely related in evolutionary terms have remarkably similar amino acid sequences. As animals diverge from each other in evolutionary development, the differences in the amino acid sequence of analagous proteins increase.

Genetics. It has recently been shown that the ability to recognize an antigen and the strength of the immune response produced may be under strict genetic control. Some strains of mice injected with synthetic polypeptides are capable of producing a vigorous immune response. Other mice, with closely related, but nonidentical, genetic backgrounds, may be poor responders or nonresponders.

In summary, many factors influence the immunogenicity of an antigen: chemical nature, size, molecular complexity, conformation, accessibility, foreignness, and genetics.

ANTIBODIES

The proteins that combine specifically with antigens are termed *antibodies*. Antibodies are produced by a subset of lymphocytes called *B lymphocytes* and by their progeny, *plasma cells*. B lymphocytes, through their production of antibodies, are responsible for the phenomenon of humoral immunity. Proteins with antibody activity are also called *immunoglobulins*. Immunoglobulins are an extremely heterogeneous group of molecules that constitute approximately 20% of the plasma proteins. Immunoglobulins are heterogeneous with respect to their antigen specificity, amino acid sequence, migration within an electrical field, and functions. There may be as many as 10,000 different molecules circulating in a person that can be classified as immunoglobulins.

Structure

The discovery that electrophoretically homogeneous proteins found in the serum of patients with multiple myeloma were structurally homogeneous and were very closely related to normal immunoglobulin was an important advance in the study of immunoglobulin structure. Such myeloma proteins could be isolated in large quantities and chemically characterized. These studies produced an understanding of the precise structure of the antibody molecule.

H and L chains. Antibodies are glycoproteins composed of 82% to 96% polypeptide and 4% to 18% carbohydrate. All immunoglobulin molecules have a common structure of four polypeptide chains. Two identical large or *heavy chains* (H chains) and two identical small or *light chains* (L chains) are held together by noncovalent forces and by covalent interchain disulfide bonds (Fig. 9-2).

The carbohydrate portion of the immunoglobulin molecule is covalently bonded to amino acids in the polypeptide chains. The carbohydrates are usually found bound to the C-terminal half (Fc) of the molecule. Their function is poorly understood. They may be involved in transporting the molecule or protecting it from metabolic degradation.

Fab and Fc fragments. Enzymatic digestion of immunoglobulin molecules has provided further evidence of their structure (Fig. 9-2). Digestion with papain splits the molecule on the N-terminal side of the disulfide bond, yielding three fragments of approximately equal size. Two of these fragments are identical and retain the antigen-binding capacity associated with an intact immunoglobulin

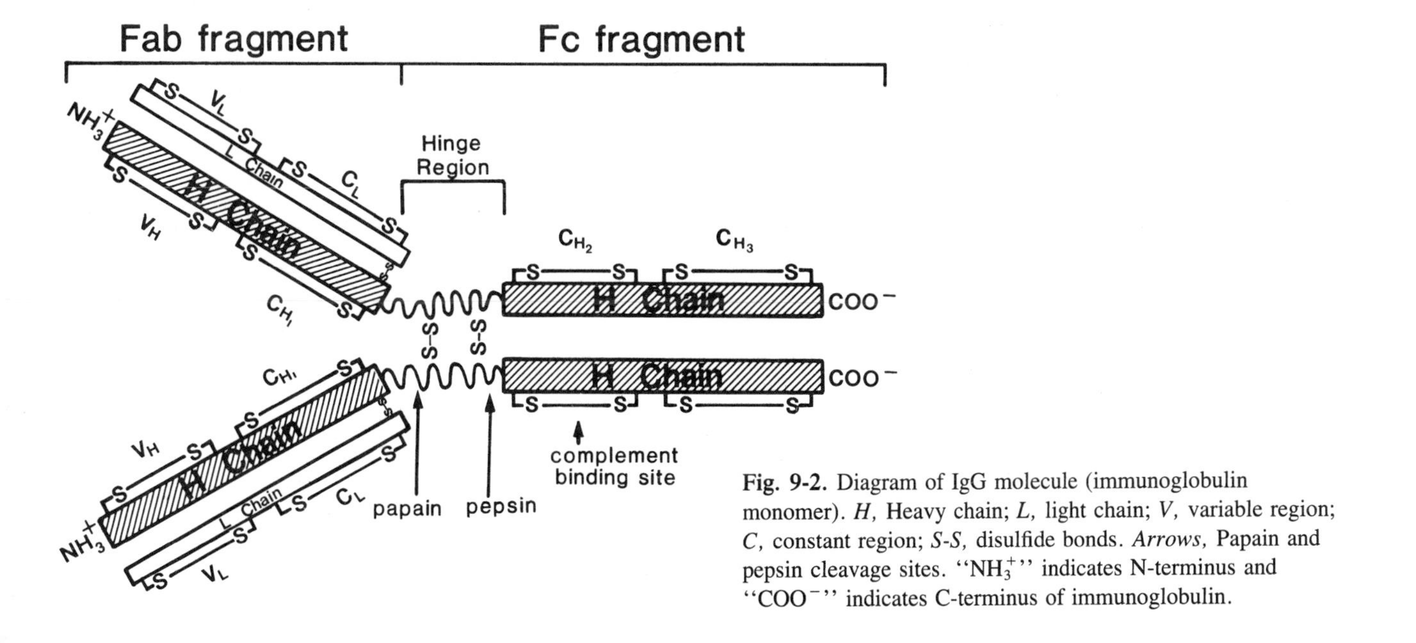

Fig. 9-2. Diagram of IgG molecule (immunoglobulin monomer). *H,* Heavy chain; *L,* light chain; *V,* variable region; *C,* constant region; *S-S,* disulfide bonds. *Arrows,* Papain and pepsin cleavage sites. "NH_3^+" indicates N-terminus and "COO^-" indicates C-terminus of immunoglobulin.

molecule. The fragments with the antibody-combining site *(Fab fragments)* are composed of the entire light chain and a portion of the heavy chain. The third fragment has no antigen-binding activity and is crystallizable *(Fc fragment)*. The Fc fragment retains the other biological activities associated with immunoglobulin molecules: interaction with the complement system and binding to tissue. The Fc fragment is composed of the C-terminal half of the heavy chain.

Digestion with pepsin cleaves the antibody molecule on the C-terminal side of the disulfide bond. This digestion results in the Fab_2' fragment composed of the two Fab fragments linked by the disulfide bond. The remainder of the molecule undergoes extensive degradation.

V and C regions. Each polypeptide chain is composed of *domains,* or peptide sequences of uniform size (100 to 110 amino acid residues) that contain intrachain disulfide bonds. The domain of the N-terminal or antibody-combining site is more variable in its amino acid sequence than the rest of the polypeptide chain and is called the *variable region (V region)*. The remainder of the polypeptide chain is composed of domains that are similar in immunoglobulin molecules of the same and other species. These domains are termed *constant regions (C regions)*. Light chains are composed of one variable and one constant region (V_L and C_L). Heavy chains are composed of one variable and three or four constant regions (V_H and $C_{H1\text{-}H4}$) (Fig. 9-2).

The specific amino acid sequences of the variable regions of the light and heavy chains of an antibody molecule confer a unique character on the antibody and are termed the *idiotype* of the molecule. The idiotype is determined by the antigenic determinant to which the antibody is directed. The character of the idiotype permits the complementary fit of the antigenic determinant to the antibody-combining site.

Light-chain types. There are two types of light chains found in immunoglobulin molecules, kappa and lambda. Kappa and lambda light chains differ structurally in the amino acid sequence of their constant regions. A given antibody molecule always has two identical kappa light chains or two identical lambda light chains. An antibody molecule can never have both a kappa and a lambda light chain together. More immunoglobulin molecules have kappa than lambda light chains. In human serum the ratio of kappa to lambda antibody molecules is approximately 2:1.

Heavy-chain types. There are five types of heavy chains in humans, based on structural differences in their constant regions. These structural differences permit functional differences. The heavy-chain types are designed gamma (γ), alpha (α), mu (μ), delta (δ), and epsilon (ϵ). The heavy-chain types vary in molecular weight. The gamma, alpha, and delta heavy chains are composed of three constant regions. The mu and epsilon heavy chains have four constant regions. The heavy-chain type determines the class of immunoglobulin. In humans there are five immunoglobulin classes, corresponding to the five heavy-chain types: immunoglobulin G (IgG), immunoglobulin A (IgA), immunoglobulin M (IgM), immunoglobulin D (IgD), and immunoglobulin E (IgE) (Table 9-1).

In addition, some immunoglobulin classes have subclasses based on additional amino acid differences in their constant regions. IgG has four subclasses, and IgA and IgM have two subclasses each. The biological properties and concentrations of the subclasses may differ.

IgG

IgG molecules are monomers of the basic immunoglobulin subunit. They are composed of two kappa or two lambda light chains and two gamma heavy chains. IgG molecules may therefore be represented as $\gamma_2\lambda_2$ or $\gamma_2\kappa_2$. Approximately 75% of serum immunoglobulin is IgG. The frequency of IgG subclasses varies as follows: IgG_1, 60% to 70%; IgG_2, 14% to 20%; IgG_3, 4% to 8%; and IgG_4, 2% to 6%. There is some evidence that antibodies produced to certain antigens may be restricted in their sub-

Table 9-1. Properties of human immunoglobulin classes

Properties	IgG	IgA	IgM	IgD	IgE
H-chain	γ	α	μ	δ	ϵ
Subclasses	1-4	1 and 2	1 and 2	None	None
L-chain	κ and λ	κ and λ	κ and λ	κ and λ	κ and λ
Form	Monomer	Monomer and dimer	Pentamer (some monomer may circulate)	Monomer	Monomer
Formula	$\gamma_2\kappa_2$ or $\gamma_2\lambda_2$	$\alpha_2\kappa_2$ or $\alpha_2\lambda_2$	$\mu_{10}\kappa_{10}$ or $\mu_{10}\lambda_{10}$	$\delta_2\kappa_2$ or $\delta_2\lambda_2$	$\epsilon_2\kappa_2$ or $\epsilon_2\lambda_2$
J-chain	No	On dimer	On pentamer	No	No
Molecular weight (approximate)	150,000	Monomer 160,000 Dimer 400,000	900,000	180,000	190,000
Complement fixation (classical pathway)	$G_3>G_1>G_2$	No	M_1 and M_2	No	No
Crosses placenta	Yes	No	No	No	No
Concentration in serum	8-16 mg/mL	1.4-3.5 mg/mL	0.5-2 mg/mL	0-0.14 mg/mL	≤300 ng/mL

classes. Polysaccharide antigens tend to produce IgG_4 antibodies. Antinucleoprotein antibodies are found primarily in the IgG_1 and IgG_3 subclasses.

IgG molecules cross the placenta and are responsible for protection of the newborn. IgG molecules also fix to the surface of effector cells, which are then capable of antibody-mediated cytotoxic reactions important in protecting the host. IgG molecules bind or "fix" complement, a complex of serum proteins that assists in the lysis or elimination of foreign particles. Complement proteins are bound to the IgG molecule in the midpoint of the heavy chain, near the disulfide bond in the CH_2 domain. This area is called the "hinge region" because the two Fab fragments of the molecule open to expose the complement-binding site (see Fig. 9-2). Molecules in the IgG subclasses fix complement proteins with unequal ability. IgG_3 is most active, followed by IgG_1, IgG_2, and IgG_4.

IgM

IgM constitutes approximately 10% of serum immunoglobulins and exists primarily as a pentamer of the basic immunoglobulin structure (Fig. 9-3). The five immunoglobulin monomers are held in a circle by disulfide bonds between H chains of the subunits. In addition, the IgM molecule contains a polypeptide chain called the *joining, or J, chain* that may help in maintaining its structure. The J chain is a small glycoprotein (molecular weight 15,000 daltons) that is covalently bonded to the H chains of the molecule.

IgM is the predominant immunoglobulin in the initial immune response to an antigen. It is the most efficient immunoglobulin in fixing complement. This efficiency is a result of its pentameric structure. The polymeric structure of IgM would also be expected to increase its valency. The presence of 10 Fab units conveys on the IgM molecule a theoretical valency of 10. This means that an IgM molecule should be able to bind 10 antigen molecules simultaneously. Although this value has been computed in some experimental systems, it is not normally observed. Steric hindrance may be responsible for this disparity.

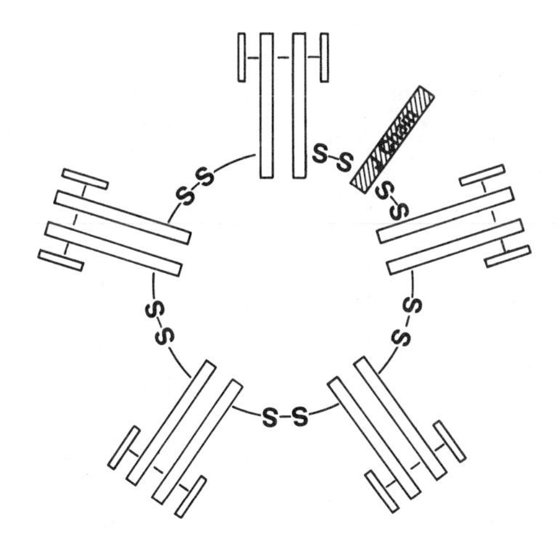

Fig. 9-3. Diagram of pentameric IgM molecule showing intermonomeric disulfide bridges (S-S) and location of J chain.

IgA

IgA constitutes approximately 15% of the serum immunoglobulin, but it is the predominant immunoglobulin in body secretions such as saliva, tears, sweat, human milk, and colostrum. In serum, IgA exists in both monomeric and polymeric forms. Polymeric serum IgA possesses the J chain. Secretory IgA exists as a dimer of the basic immunoglobulin unit combined with a J chain and an additional polypeptide chain called the *secretory piece*. The secretory piece is bound to dimeric secretory IgA by strong noncovalent linkages. It is important in secretory transport of the molecule and in its protection from proteolytic digestion in the gut. Secretory IgA provides the first line of defense against local infections and is important in the processing of food antigens in the gut.

IgD

IgD is a monomer of the basic immunoglobulin unit and is present in human serum in trace amounts. In addition, it is expressed on lymphocyte cell surface membranes. The main function of IgD is unknown. It may be involved in lymphocyte differentiation.

IgE

IgE is also a monomeric immunoglobulin. It is present in human serum in very low concentrations. IgE, which binds to cells by means of its F_c portion, is responsible for the physiologic manifestations of allergy.

ANTIGEN-ANTIBODY REACTIONS

Antigen-antibody reactions were first recognized by bacteriologists who also surmised that such reactions exhibited specificity. Bacteriologists noticed that the serum of patients recovering from infectious diseases could agglutinate the organism responsible for their disease but not unrelated organisms. Serum from persons not exposed to the disease or serum from patients before they contracted the disease could not agglutinate the same organisms. From such evidence, scientists proposed the existence of antibody molecules, the specificity of their interactions with antigens, and the importance of such interactions in host defense.

The following sections consider the forces involved in antigen-antibody binding, the specificity of the reaction, and the mechanism of the reaction.

Binding forces

The strength of the binding of an antigen to an antibody depends on the complementarity of fit of the antigenic determinant to the idiotype of the antibody. It also depends on the sum of weak, noncovalent, intermolecular forces. Such weak, short-range forces can operate between antigen and antibody if their closeness of fit brings them into proximity with one another. The attraction between antigen and antibody molecules involves electrostatic attraction, hydrogen bonding, van der Waals forces, and hydrophobic interactions.

In solution at physiological pH, charged polar groups on the amino acid residues of proteins can be strongly attracted to one another. These electrostatic forces are the strongest and most important contributors to noncovalent attraction between antigen and antibody.

Hydrogen bonding between the amino and carboxy groups of peptide bonds also contributes to the attractive forces. Hydrogen bonds are weaker than electrostatic forces, but their numbers make them an important factor.

Van der Waals forces are the weakest forces involved. They can function only within a very small radius because of their low power. The close approximation of antigenic determinant to idiotype induces charge fluctuations within the atoms of the molecules. At very close distances the nucleus of one atom can be attracted to the external orbit electrons of a second atom. These van der Waals forces contribute to binding strength.

The final component of the attractive forces involves hydrophobic bonding between apolar groups in solution. Hydrophobic bonding functions by the exclusion of polar water molecules to bring hydrophobic molecules together. Such interactions also serve to attract polar water molecules to hydrophilic amino acid residues on protein molecules. Antibody molecules have increased numbers of hydrophobic amino acid residues such as alanine, leucine, tyrosine, tryptophan, and methionine in their antibody-combining sites, where they enhance bonding to hydrophobic residues in antigenic determinants.

Antibody affinity

The strength of the binding of a single antigenic determinant to an antibody is a function of the closeness of fit and is called *antibody affinity*. Antibody affinity is an expression of the attraction between molecules of antibody and antigen. It is a function of the sum of the short range, noncovalent, intermolecular forces.

Binding of antigen to antibody is a reversible reaction. The equilibrium of the reaction favors antigen-antibody association if the "fit" between molecules is good and the forces binding the molecules together are relatively strong and stable. The strength of the association between antigen and antibody is represented by the association constant, which may be derived as follows:

$$Ag + Ab \underset{k_2}{\overset{k_1}{\rightleftharpoons}} Ag \cdot Ab \qquad \textit{Eq. 9-1}$$

$$\frac{[Ag \cdot Ab]}{[Ag][Ab]} = \frac{k_1}{k_2} = K_a \text{ (Association constant)} \qquad \textit{Eq. 9-2}$$

To study these reactions, one places solutions of a small antigen, or hapten, on either side of a semipermeable membrane. As the hapten diffuses across the membrane, the reaction proceeds to the right of equation 9-1; that is, hapten and antibody associate to form complexes. Eventually, equilibrium is reached. At equilibrium, a constant number of antibody molecules react with hapten to form complexes, and a constant number of complexes dissociate to form free hapten and free antibody. One rate constant, k_1, expresses the tendency of the reaction to move toward the right, or the tendency for association. The other rate constant, k_2, expresses the tendency of the reaction to move to the left, or the tendency for dissociation. These reaction-rate constants differ for each antigen and antibody pair. The concentrations of antigen, antibody, and complex at equilibrium is described by equation 9-2. The equilibrium constant, K_a, expresses the tendency of the reaction to favor association between antigen and antibody.

Heterogeneity of immune response

Analysis of the binding of a simple hapten containing a single antigenic determinant shows variation in binding strength of antibody molecules. Immunization with a single antigenic determinant produces a variety of antibodies with different antibody-combining sites and with a range of antibody affinities. This is termed the *heterogeneity* of the immune response. The immune system responds to three-dimensional antigenic determinants that present more than one area of contact (Fig. 9-4). Antibodies can bind to different configurations of a single antigenic determinant. Thus different antibody molecules can be produced against the varied configurations of a single determinant. Such different configurations present differing distributions of charged and hydrophobic residues, resulting in varying closeness of fit with an antibody.

Antibody avidity

In natural situations in vivo a variety of antibody molecules are generated in response to a large number of multivalent antigenic stimuli. Thus there are two areas of complexity: (1) multiple antibodies generated to different conformations on a single antigenic determinant and (2) multiple antigenic determinants on a single natural antigen. Both of these generate the *diversity* of the immune response. The measure of binding strength of antibodies to multiple antigenic determinants on natural antigens is termed *avidity*. Avidity is a measure of the stability of the antigen-antibody complex. It is partially dependent on the affinity of each antibody for its complementary antigenic determinant. There is an enhanced effect, however, with multivalent antigens. The sum of the binding is greater than its individual parts. This is caused by the reversible nature of antigen-antibody bonds. It is also caused by the divalent nature of IgG molecules or the multivalent nature of IgM molecules. If the single bond between antigen and antibody dissociates, the antigen escapes. If an antigen has two antigenic determinants, each of which is bound by an-

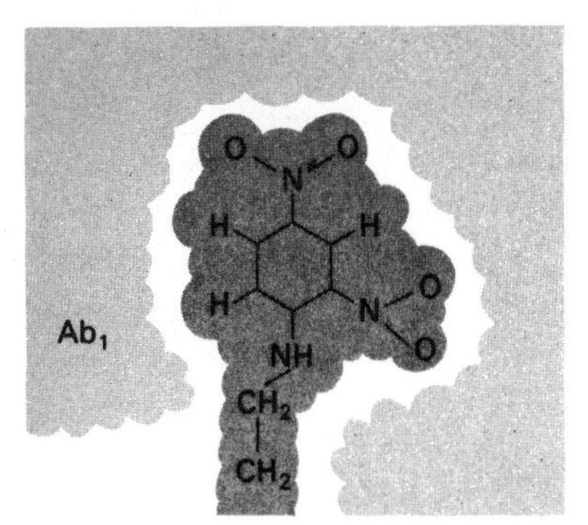

(a) High affinity

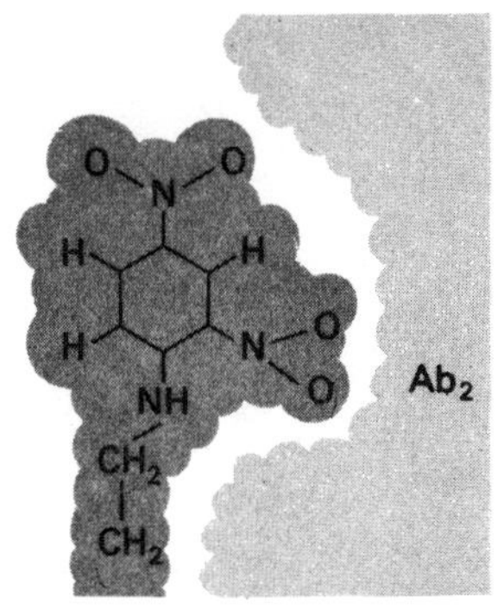

(b) Moderate affinity

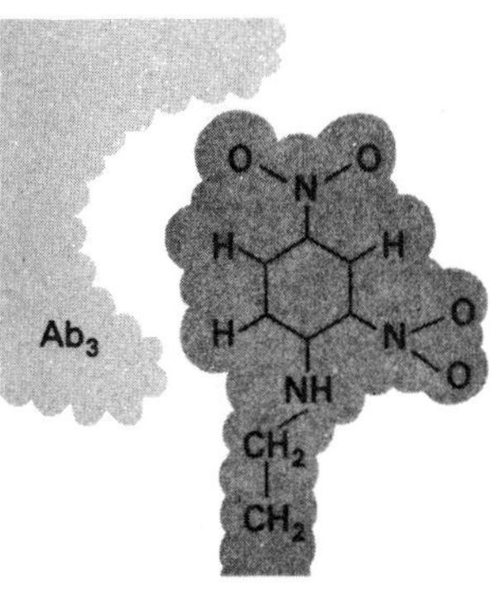

(c) Low affinity

Fig. 9-4. Binding of antibodies present in same antiserum with different affinities to same hapten (dinitrobenzene linked to amino group of lysine). **(a)** Antibody$_1$ fits with nearly whole hapten and is thus of high affinity. **(b)** Antibody$_2$ fits with less of molecule and not so closely and has a moderate binding affinity, whereas **(c)** low affinity antibody$_3$ is complementary in shape to so little of hapten surface that its binding energy is very little above that occurring between completely unrelated proteins. Only a portion of antibody combining site is shown. (From Roitt, I: Essential immunology, ed. 4, Oxford, England, 1980, Blackwell Scientific Publications.)

tibody, the antigen is kept in place until the broken bond reforms (Fig. 9-5). Thus avidity is a measure of the stability of the multivalent antigen–multivalent antibody complex.

Cross-reactivity

Cross-reactivity of antigen and antibody is a by-product of the heterogeneity of the immune response. As stated previously, immunization with a simple hapten produces a variety of antibodies of differing antibody affinities. Some of these antibodies will combine with chemically related and structurally similar haptens. Cross-reactivity usually involves low-affinity antibodies that exhibit a less precise fit to antigens. The reactivity of an antibody to a different antigen may also indicate that the two antigens in question share a previously unknown but common antigenic determinant. Thus cross-reactivity may result from similar or identical antigenic determinants in different antigens. Such cross-reactivity is often observed with antibodies produced to such drugs as penicillin. The reactivity of an antibody with penicillin may be very high, but metabolic derivatives containing the basic drug structure may also react with an antibody produced to the complete drug.

In cases of prolonged antigenic challenge, such as that occurring in natural infection, animals exhibit a natural selection for high-affinity antibodies. As the heterogeneity of the antibody response narrows, the specificity of antigen-antibody reactions increases. This adaptation of the immune response promotes effective protection of the host against infection.

Genetic basis of antibody diversity

The heterogeneity of the immune response or the variety of antibodies produced to a single antigen is known to be genetically determined. The varible (V) regions of light and heavy chains of the antibody molecule encode for antibody specificity. The amino acid sequence of the V region is genetically controlled. Recent evidence suggests that there are positions in the amino acid sequence of the

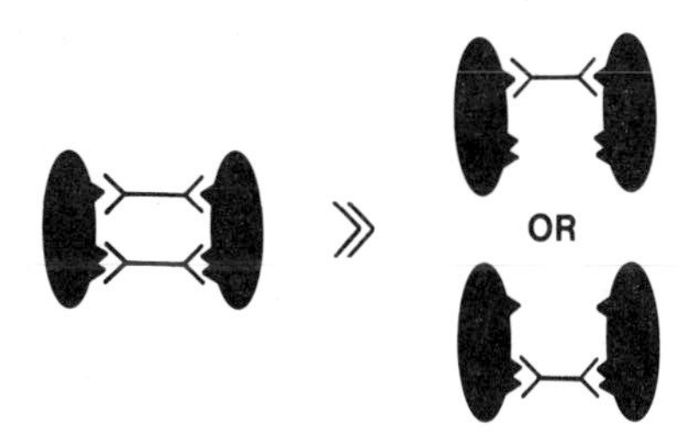

Fig. 9-5. Multivalent bonding of antigen-antibody increases bonding strength. A single bond created by divalent antibody molecules between a single antigenic determinant on two adjacent antigens is much weaker than binding created by two divalent antibodies bound simultaneously to two unique antigenic determinants on two adjacent antigens. Strength and complexity of this multivalent bonding is described by the term "affinity."

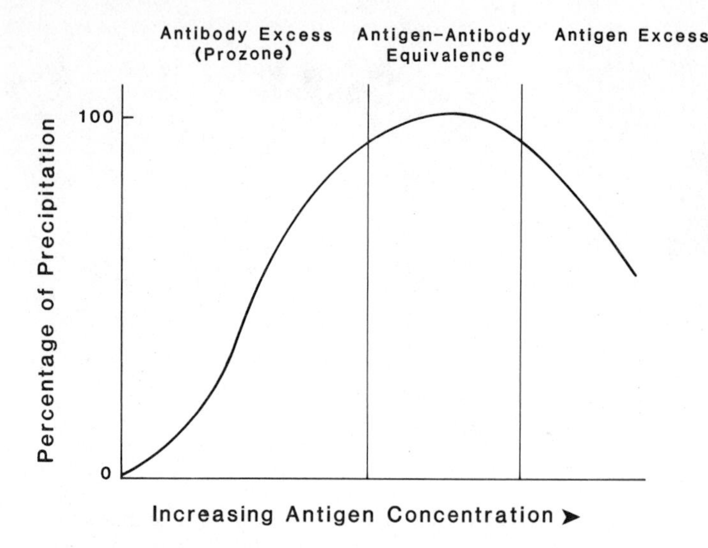

Fig. 9-6. Quantitative precipitin curve in which amount of antibody antigen complex that precipitates is plotted as function of antigen concentration.

V region that have an even more increased likelihood of amino acid variation. These *hypervariable regions,* scattered throughout the amino acid sequence of the variable region, are brought into proximity to each other by the natural folding of the antibody molecule. The natural folding, or tertiary structure, of the antibody molecule is also genetically determined and is also dependent on the amino acid sequence. The sequence dictates possible attraction between polar amino acid residues as well as the possibility of intrachain disulfide bonds. The approximation of hypervariable regions by the folding of the antibody molecule results in the formation of the antibody-combining site.

ANTIGEN-ANTIBODY PRECIPITATION REACTIONS

The primary reaction of antigen with antibody is usually detected by secondary manifestations of the reaction. The nature of the secondary manifestations is dependent on experimental conditions, the class of antibody involved in the reaction, the number of antigenic deteminants on the antigen, and the size and solubility of the antigen. The reaction of antibody with soluble molecules possessing multiple antigenic determinants that permit cross-linking are detected by *precipitation* of the complex out of solution. The term *flocculation* may be used to describe a precipitation reaction that produces a large, loosely-bound precipitate. The reaction of antibodies with large, particulate, multivalent antigens is detected by *agglutination* of the antigen. These reactions are considered separately.

Precipitation curve

When a known quantity of antibody is present in solution in a series of tubes to which increasing amounts of antigen are added and allowed to react, precipitation occurs in some of the test tubes. When the amount of precipitate is measured and correlated with the amount of antigen present, one obtains a curve similar to that in Fig. 9-6.

No precipitate is formed in the first area of the curve, and examination of the fluid phase at this time shows no detectable antigen, without attached antibody (free antigen). Unattached (free) antibody can be detected, however; this phase of the reaction is called the *antibody-excess phase*. As increasing amounts of antigen are added, the amount of precipitate detectable at the bottom of the test tubes increases until a point of maximum precipitation is reached. At this point, no free antigen or free antibody can be detected in the fluid. This is called the *zone of equivalence*. As the amount of antigen added continues to increase, the amount of precipitate detected diminishes. Examination of the fluid phase of the reaction at this time shows no free antibody, but increasing amounts of free antigen. This area of the curve is called the *antigen-excess* phase.

Lattice theory

Antigen-antibody complexes precipitate out of solution because of the multivalent nature of both molecules. The reaction of antigens possessing multiple antigenic determinants and antibodies with two (IgG) or more (IgM) antibody-combining sites produces a lattice of interlocking molecules. Antibody molecules can cross-link antigenic sites on the same or different molecules of antigen. As the size and complexity of the lattice increases, the lattice becomes insoluble and precipitates out of solution (Fig. 9-7).

In the antibody-excess zone, a single molecule of antigen binds to each antibody molecule. The excess of antibody ensures that each molecule of antigen can encounter a free antibody molecule. The absence of cross-linking produces small soluble complexes.

As the antigen concentration increases and the zone of equivalence is entered, complexes of increasing size with increasing levels of cross-linking are formed. Such large, complex lattices precipitate out of solution.

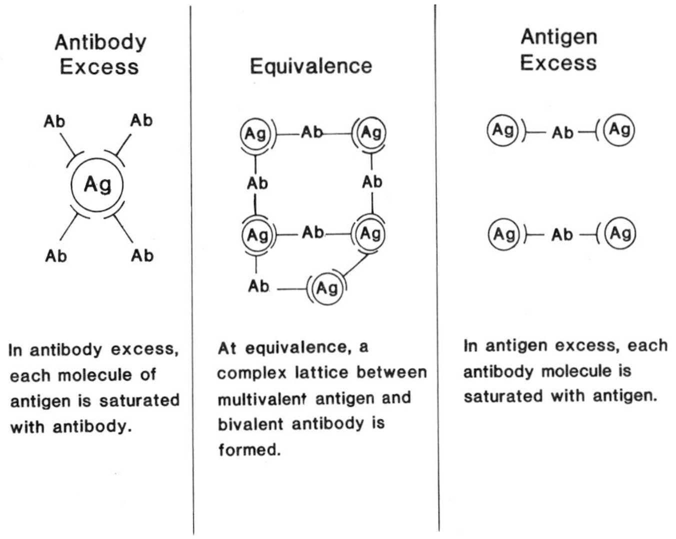

Fig. 9-7. Representation of sizes of molecular complexes formed at varying ratios of antigen and antibody.

Antibody Excess
Equivalence
Antigen Excess
In antibody excess, each molecule of antigen is saturated with antibody.
At equivalence, a complex lattice between multivalent antigen and bivalent antibody is formed.
In antigen excess, each antibody molecule is saturated with antigen.

As the antigen concentration continues to increase, the zone of antigen excess is reached. In this area, smaller complexes are again seen. The size of the lattice decreases because there is sufficient antigen to permit binding of a free antigen molecule to each antibody-combining site. This is called the *prozone* area. At high concentration of antigen, false-negative precipitation reactions can occur as a result of antigen excess. Obviously, detection of antigen by antibody precipitation requires optimal concentration of both reactants. Formation of lattices best suited for precipitation occurs at equal equivalent concentration of antigen and antibody or at slight antigen excess.

Other factors affecting precipitation

The precipitation of antigen-antibody complexes out of solution can be affected by factors other than the ratio of antigen concentration to antibody concentration. Different antibody molecules can precipitate the same antigen to varying degrees. The efficiency of the antibody depends on its affinity and specificity. The charge and shape of the antigen-antibody complex is also important. Excessively charged complexes are difficult to precipitate. More energy is required to displace polar water molecules from the complexes and to bring the charged molecules into proximity. The best precipitates involve protein antigens with molecular weights ranging from 40,000 to 160,000 daltons. Proteins in this size range are easily cross-linked by multivalent antibody molecules. Polysaccharide antigens, denatured proteins, and viruses produce broader precipitation curves. Their size produces steric hindrances to optimal precipitation. Precipitation can also be affected by temperature, pH, and ionic concentration. Such factors influence antigen-antibody interactions on a molecular level.

Precipitation reactions in gel

Precipitation reactions are frequently carried out in a gel-support matrix composed of agar or the more purified polysaccharide, agarose. The agar prevents convective mixing of antigen and antibody and thereby ensures establishment of concentration gradients of the two reactants. Precipitation in agar is only a moderately sensitive technique when compared to newer advances such as radioimmunoassay, but it is widely employed because of its ease and versatility. In addition, some modification of precipitation reactions in gel permit the study of antigenic relationships among different compounds. The following section is a discussion of two gel-precipitation reactions, double immunodiffusion and radial immunodiffusion.

Double immunodiffusion. In double-immunodiffusion reactions, or *Ouchterlony tests,* agar or agarose is poured onto a solid support such as a glass slide or petri dish. Wells are then cut into the agar. Antigen and antibody solutions are placed into separate wells. The solutions then diffuse toward one another in the gel in a radial fashion during room-temperature incubation (Fig. 9-8). With diffusion into the agar, the solutions establish concentration gradients that diminish with distance from the well. At the point of antigen-antibody equivalence at the interface of the diffusing fronts, a precipitation line is formed (Fig. 9-8). The positioning and the shape of the line are dictated by the concentration of the reactants and by their size. The line will be closer to the well with the reactant of lower concentration because the distance travelled is directly proportional to concentration. The rate of diffusion is also inversly proportional to molecular size. High molecular weight compounds such as IgM diffuse more slowly than lower molecular weight substances such as IgG. The precipitation line that is formed at the interface of the two concentration gradients will be concave to the higher molecular weight compound whose diffusion rate is slower. Because precipitation occurs at antigen-antibody equivalence to slight antigen excess, an inappropriate ratio of antigen to antibody will result in failure to form a precipitate.

The presence of different antigenic determinants on the

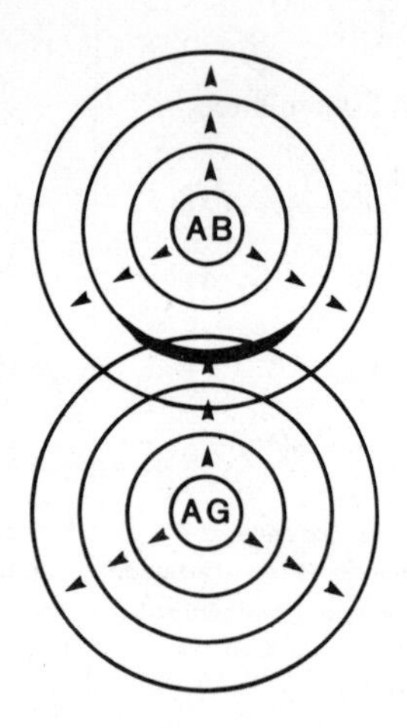

Fig. 9-8. Depiction of protein gradients in radial immunodiffusion. Concentric circles represent decreasing protein concentrations. Both antigen (AG) and antibody (AB) diffuse radially from application wells. Precipitation, *heavy black arc,* occurs at point of antigen-antibody equivalence. Precipitin line is closer to well of lower concentration and concave toward reagent of higher molecular weight.

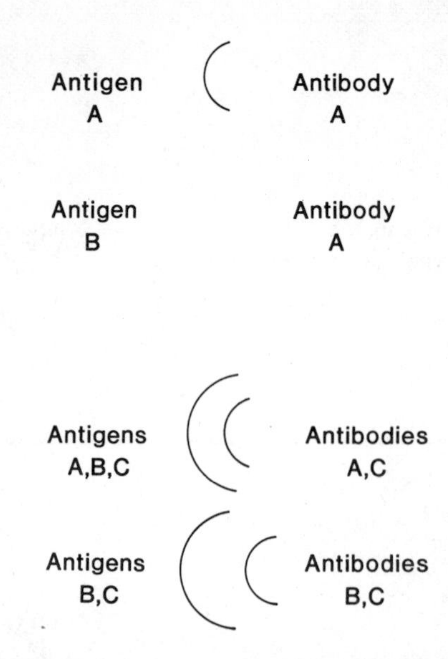

Fig. 9-9. Precipitation occurs at equivalence point of antigen with corresponding antibody. Multiple precipitation lines are seen with multiple antigens and corresponding multiple antibodies.

same or on different molecules can be detected by the production of more than one precipitation line if the antiserum used contains antibodies against the multiple antigens (Fig. 9-9). Two antigens in the antigen solution will create two independent concentration gradients as they diffuse into the agar. Precipitation will occur at the point of equivalence of each antigen with its corresponding antibody. In this way the components of an antigen mixture can be studied.

Ouchterlony testing also permits analysis of the relationship between two antigenic mixtures. The antibody solution is placed in a center well, which is surrounded by several wells into which antigen solutions are placed. Placing two antigen solutions in adjacent wells permits conclusions about their antigenic relationship. This is done by the study of precipitation lines obtained by reactions between their two concentration gradients.

Unrelated antigens that have corresponding antibodies in the antibody solution will form separate precipitation lines corresponding to their distinct components. Radial diffusion from the adjacent wells causes superimposition of the two gradients, but because their concentration gradients are independent, two separate and distinct lines are formed (Fig. 9-10). The double spurs formed at the two antigen interfaces are the hallmark of the *reaction of nonidentity.* The two antigens or antigen mixtures have nothing in common in relationship to the antibody solution used.

Two identical antigen solutions that are tested with their

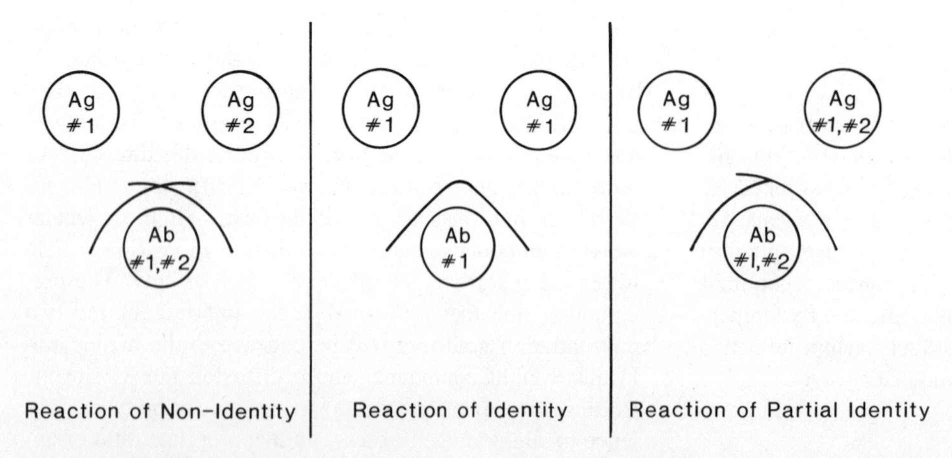

Fig. 9-10. Double immunodiffusion patterns. *Ag,* Antigen; Ab, antibody; *heavy line,* precipitin line; *circles,* application wells.

common antibody result in a *reaction of identity* (Fig. 9-10). Precipitation lines form at their separate but adjacent points of equivalence against antibody. Because diffusion through the gel is radial, an area of shared and common antigen concentration between the two adjacent wells is formed. For this reason, there is fusion of the two separate precipitation lines. This fusion or reaction of identity indicates that the antigen solutions are identical with respect to the antibody employed or that the antigen solutions have one antigenic determinant in common. It does not imply molecular identity. If antigen concentrations in the adjacent wells differ, a common fused precipitation line is still formed at the point of average concentration of the two antigen fronts.

A *reaction of partial identity* is formed when adjacent wells share some but not all the antigens detected by the antibody solution. A reaction of partial identity is a reaction of nonidentity superimposed on a reaction of identity. (Fig. 9-10). A common fused line is formed by reaction of the shared antigen. A second identical equivalence point between the novel antibody creates a superimposed precipitation line that extends into the common area between the adjacent wells. The second antigen creates a concentration gradient that is not contributed to by the adjacent well. The spur indicates that the second antigen solution lacks an antigenic determinant present in the first antigen solution that is recognized by the antibody solution. The spur always points to the well of the antigen that is monospecific with respect to the antibody.

A complex relationship of antigen and antibody solutions is presented and discussed in Fig. 9-11.

Radial immunodiffusion. Radial immunodiffusion, the Mancini technique, is a precipitation reaction carried out by application of antigen solution to a gel into which a monospecific antibody solution has been impregnated. It is an adaptation of gel-precipitation reactions that permits quantitation of antigen. The antibody-gel solution is applied to a solid support. Wells are then cut into the agar and dilutions of antigen are placed in the wells. The antigen diffuses out radially into the agar. This diffusion produces a concentration gradient that is inversely proportional to the distance from the well. Antibody concentration within the gel is constant. At the point where antigen and antibody concentrations are equivalent, precipitation occurs. Because diffusion from the well is radial, the precipitation appears as a ring around the well. The square of the diameter of the ring (mm^2) is directly proportional to antigen concentration. The precipitation reaction is not a static but a dynamic one. The precipitation ring first forms close to the well at the initial point of antigen-antibody equivalence. As antigen continues to diffuse from the well, antigen excess causes conversion of the precipitate to soluble complexes, which resolubilize and continue to diffuse outward. A new ring is formed at

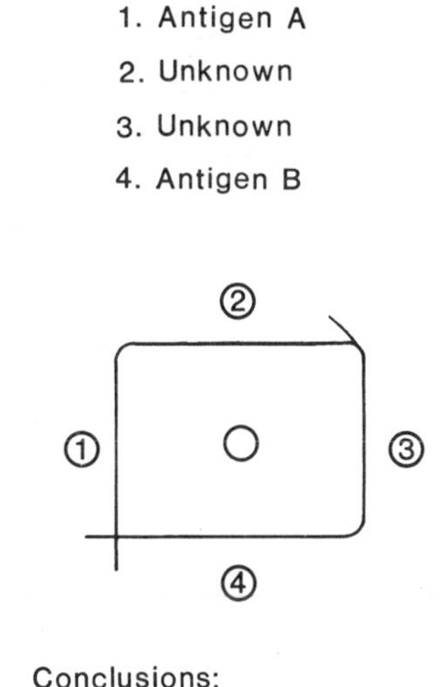

Conclusions:
2 = antigen A
3 = antigen A and antigen B

Well Content
1. Antigen A
2. Unknown
3. Antigen B+unknown
4. Unknown

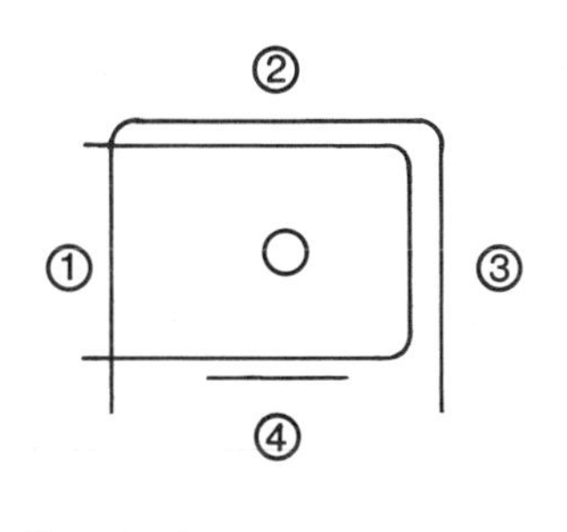

Conclusions:
#2 = antigen A+antigen B
#3 = antigen A+antigen B
#4 = antigen B+antigen C

Well Content
1. Unknown
2. Unknown
3. Antigen A+unknown
4. Antigen B

Conclusions:
1 = contains no antigens recognized by antisera or reagent concentrations are not correct
Antigen A and B are the same
#2 = contains antigen A plus antigens C and D
#3 = contains A and E

Fig. 9-11. Interpretations of double immunodiffusion patterns. Center well contains an antiserum to several possible antigens. Surrounding wells contain test antigens. *Heavy lines,* Precipitin reaction.

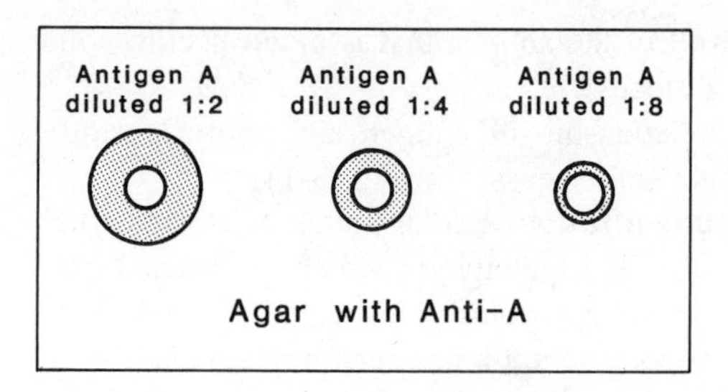

Fig. 9-12. Radial immunodiffusion patterns. Band of precipitation, *stippled area,* extends as a disk from center of each circular well. Area of precipitation is proportional to concentration.

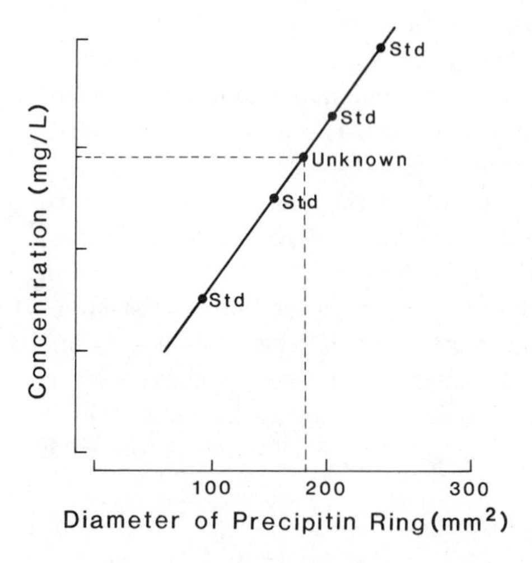

Fig. 9-13. Graph of concentration of antigen expressed as mg/L versus square of diameter of precipitin ring. *Std,* Standard.

a new point of antigen-antibody equivalence. The final distance traveled at end-point equilibrium is a function of antigen concentration. The thickness of the ring is a function of the final concentration of antigen-antibody complexes at the equivalence point (Fig. 9-12).

With constant sample, volume, temperature, pH, incubation time, and antibody concentration, an unknown antigen concentration can be determined. This is accomplished when one compares the square of the diameter of the precipitation ring of the unknown with rings obtained by several dilutions of a standard antigen solution. When concentration of the standard is plotted against ring area, the concentration of the unknown can be easily determined (Fig. 9-13).

BIBLIOGRAPHY

Antigens

Goodman, J.W.: Antigenic determinants and antibody combining sites. In Sela, M., editor: The antigens, vol. 8, New York, 1975, Academic Press, Inc., pp. 127-187.

Williams, C.A., and Chase, M.W.: Methods in immunology and immunochemistry, vol. 6, Preparation of antigens and antibodies, New York, 1967, Academic Press, Inc.

Antibodies

Natvig, J.B., and Kunkel, H.G.: Human immunoglobulins: classes, subclasses, genetic variants, and idiotypes, Adv. Immunol. **16:**1-59, 1973.

Nisonoff, A., Hooper, J.E., and Spring, S.: The antibody molecule, New York, 1975, Academic Press, Inc.

Antigen-antibody interactions

Gill, T.J.: Methods for detecting antibody, Immunochemistry **7:**887-1000, 1970.

Rose, N.R., and Friedman, H., editors: Manual for clinical immunology, ed. 2, Washington, D.C., 1980, American Society for Microbiology.

Weir, D.M., editor: Handbook of experimental immunology, ed. 3, Oxford, 1978, Blackwell Scientific Publications, Ltd.

Williams, C.A., and Chase, M.W.: Methods in immunology and immunochemistry, vol. 3, Reactions of antibodies with soluble antigens; vol. 4, Agglutination, complement, neutralization and inhibition, New York, 1971, 1977, Academic Press, Inc.

Chapter 10 Immunochemical techniques

Paul E. Hurtubise
Susan Bassion
Jack Gauldie
Peter Horsewood

Enzyme immunoassay
- Principles of enzyme immunoassays
- Sample requirements and preparation
- Reagents
- Instrumentation
- Interfering substances
- Common pitfalls
- Test sensitivity and coefficient of variation

agglutination Clumping or aggregating together by specific antibody of a particle, such as a red blood cell or latex bead to which the specific antigenic determinant is attached.

agglutinin Specific antibody that causes agglutination.

anti-antibody An antibody with specificity for immunoglobulins reacting in test assay systems.

antibody absorption The process of removing or tying up undesired antibodies in an antiserum reagent by allowing it to react with undesired antigens.

antibody reagent A high titer, high-affinity IgG class of antibody prepared in animals.

antigen reagent A stabilized solution containing a known amount of an antigen; to be used as a standard.

antigenic determinant Portion of an antigen involved in the reaction with a single antibody.

cold agglutinin An agglutinin that reacts better at temperatures less than body temperature; best reaction is usually at 4° C.

complement A group of serum proteins activated as a result of an antibody-antigen reaction. When the reaction is on the surface of a red blood cell, the activated complement can lyse the cell.

complement fixation A term applied to a set of assays in which complement is activated or "fixed" by a test reaction system.

Coombs' test A type of agglutination reaction. A direct Coombs' test measures the presence of antibody on cells; an indirect test measures its presence in serum.

counterimmunoelectrophoresis An assay in which antigen and antibody migrate toward each other under the influence of an electric field. The presence of antigen is observed by the formation of a precipitin line.

cryoglobulin Protein that precipitates at temperatures less than body temperature; precipitates maximally at 4° C.

fluoroimmunoassay Any immunoprocedure that uses a fluorescent molecule as the indicator label.

hemolysin Anti–sheep red blood cell antibody.

heterogeneous enzyme immunoassay Any technique that uses two phases, usually liquid and solid, to separate reacted from unreacted components.

immobilization The fixation of antigen or antibody onto a solid support such as a plastic tube or microliter plate.

immunodiffusion Random, spreading movement of antibody or antigen, or both, in a support medium.

immunoelectrophoresis An immunoprecipitation technique in which antigens are separated from each other by migration in an electric field, followed by reaction with antibody by immunodiffusion.

indicator phase The portion of an immunochemical reaction that can be measured.

inhibition assay A term for those types of immunoassays in which an excess of antigens prevents or inhibits the completion of either the initial or indicator phase of the reaction.

lattice formation The cross-linked three-dimensional structure formed by the reaction of multivalent antigens with antibody.

monoclonal antibody A monospecific antibody that is produced by a single plasma cell or a single clone of plasma cells of a lymphocyte myeloma hybrid.

monospecific An antibody that will react with only one type of antigen molecule.

nephelometric assay Measurement of antigen or antibody by determination of the amount or rate of formation of antibody-antigen aggregate by the amount of light scattered.

nephelometric inhibition assay (NINIA) Measurement of haptens by inhibition of formation of antibody-antigen lattice.

nephelometry Measurement of light-scattering properties of large particles (such as antigen-antibody complexes) in solution.

Ouchterlony double diffusion A version of the original gel diffusion technique invented by Oudin in which antigen and antibody in separate wells are allowed to spread (diffuse) toward each other.

polyclonal antibody Heterogeneous antibodies with diverse affinities produced by a large number of plasma cells.

prozone phenomenon Apparently lower reactivity or nonreactivity of antibody solution at low dilutions.

radial immunodiffusion (Mancini technique) Measurement of antigen concentration by allowing antigen to spread (by diffusion) into agarose containing the desired monospecific antibody. The area of the immunoprecipitin ring is proportional to antigen concentration.

"rocket" (Laurell) immunoelectrophoresis Assay system in which the antigen, under the influence of an electric field, migrates into agarose-containing antibody, with a resultant immunoprecipitation reaction. The precipitin lines appear rocket shaped.

"sandwich" assay A term applied to a solid-phase immunoassay in which the first layer is immobilized antibody, the second is antigen, and the third is labeled antibody.

specificity Property of an antibody molecule that restricts its reactivity to a defined molecule or group of molecules.

titer Maximum dilution of a specific antibody that gives a measurable reaction with a specific antigen; usually expressed as the reciprocal of that dilution.

In the previous chapter, Chapter 9, the molecular nature of antigens and antibodies is described, as well as the general characteristics of the antigen-antibody reaction. This chapter deals with many techniques that use the antigen-antibody reaction as the basis to detect, characterize, or quantitate constituents in blood and other body fluids submitted to the laboratory for analysis. These constituents can range from small molecular weight drugs and their metabolites to large molecular weight proteins, such as IgM and alpha$_2$-macroglobulin. Most frequently, the patient's sample contains the antigen (analyte), and antibody is added as the reagent to detect or measure the antigen. In contrast, in cases of infectious disease, serological determinations, and autoimmune antibody testing, the patient's sample is the source of antibody, and its measurement is of clinical importance. For these determinations, antigen of known composition is used. These may be soluble or tissue based. However, this latter form of testing is usually performed in the immunology section of the laboratory. Since this chapter is directed primarily to those techniques used in the clinical chemistry laboratory, it will concentrate on the procedures that detect antigen in the patient's sample.

REAGENTS

Antibody as reagent

Reagent antibodies are usually prepared in animals, such as rabbits or goats, by repeated exposures to substances foreign to the animal. The type and magnitude of the antibodies produced are the result of highly complex processes and are controlled by numerous factors, such as the genetic makeup of the responding animal, the various interactions of the animal's immune cells, the complexity of the antigen, and the foreignness of the immunizing material to the animal. When all these factors are present in an appropriate manner, the animal's response to a specific antigen usually produces antibody protein of sufficient magnitude that it can be detected easily in the serum of the immunized animal. This serum contains a large number of antibodies that will specifically interact with surface structures on the antigen. Some of these surface structures are the major determinants of the antigen molecule and cause the production of the largest amount of antibodies, whereas some of these features are minor and cause a lesser amount of antibodies to be produced. Since many different antibodies are present and they are attributable to expansion of several classes of antibody-producing cells, the antiserum thus produced is a *polyclonal* reagent antiserum. For example, an antibody against the protein human serum albumin (abbreviated anti-HSA) is a reagent that has multiple antibodies to antigenic determinants or specific molecular configurations that are characteristic and specific for the surface of HSA.

It is important to demonstrate that this anti-HSA will not react with other serum proteins, such as IgG and transferrin. If this anti-HSA is to be used as a reagent in the clinical laboratory, its *specificity* (that is, its reactivity with only HSA) must be verified, particularly in the same immunological test system used to generate patient results. For every reagent antibody, the specificity of its immunochemical reactivity is the single most important factor in the success or failure for any immunologic technique used in the clinical laboratory.

More recently, a technology has been developed utilizing hybridization of a single antibody-producing cell (plasma cell) with tumor cells. These hybrid cells can be isolated and produce antibody directed to a single antigenic determinant. The antibodies have a single homogeneous primary structure. The products of the new isolation technique are called *monoclonal antibodies*. Although the use of monoclonal antibodies is still relatively limited, they are expected to play a larger role in the future. An example of the use of a monoclonal antibody in a competitive binding assay is presented in Chapter 11. Therefore, in this chapter, antibodies as reagents will be understood to be polyclonal in nature.

Selection of antibody as a reagent in an immunological procedure requires information about its characteristics such as its strength (titer), affinity, and specificity. The amount of antibody that is available for reactivity in a specific immunologic method is termed the "titer of the antibody." The titer is the reciprocal of the maximum dilution of the antibody that gives a detectable reaction for a specific method. The titer of the reagent antibody is often different for each kind of immunological procedure. For example, anti-HSA may react in an immunoprecipitation technique at a maximal dilution of 1:32, but the same antiserum may react at a maximal dilution of 1:6400 in a radioimmunoassay procedure. Occasionally the amount of reagent antibody present may be expressed in weight, that is, milligrams per milliliter. This expression of antibody amount is determined by precipitation techniques and is often helpful in determining the amount of reagent needed.

Affinity refers to the strength of binding of an antibody with an antigen in an antibody-antigen reaction. Reagent antibodies generally fall into two categories, those of high affinity and those of low affinity. The antibodies in the reagent may be a mixture of both, but one should use reagents where high-affinity antibodies are predominant. The high affinity indicates that there will be a strong union with the antigen that is not readily reversible and that this union will not be influenced greatly by alteration of the conditions of the antibody-antigen reaction. Low-affinity antibodies do not bind well with the antigen and can be influenced by temperature, pH, and ionic strength with consequent changes of the reaction resulting in dissociation of the antibody-antigen reaction. Most commercial-reagent antibodies are of the high-affinity type. However, if one is preparing reagents, they should be tested to be certain they are of the appropriate, preferably high, affinity.

Specificity refers to the ability of the antibody to restrict its reaction to a specifically defined group of molecules. Since these reagents are really a collection of antibodies, they are directed to multiple antigenic determinants on a single antigen and thus could have multiple reactivities. Some antigenic determinants may be shared on different molecules. For example, *human* serum albumin has many determinants that can also be found on *bovine* serum albumin and on other animal albumins as well. Therefore human anti-HSA might interact with other animal sera if they are part of the reagents of the reaction. This interaction with other species of albumins could significantly interfere with the results. Other reagent antibodies may react with antigenic determinants that may be common to several molecular forms of plasma proteins. For example, a reagent antibody directed to the IgG molecule should only recognize the IgG molecule, but there may be antibody in the reagent that would also react with light chains of that IgG molecule. Since light chains are common to all the immunoglobulin classes, that is, IgA, IgM, and IgD, the reagent antibody would then react with all immunoglobulin molecules. It would not be appropriate to say that it was recognizing only the IgG molecule. The problem of cross-reactivity with other serum proteins can usually be controlled by a technique termed *antibody absorption,* which binds or removes from the reagent that population of antibody reacting inappropriately with other molecules of the test solution. Absorption is necessary for virtually all antibody reagents of the polyclonal type. It is often accomplished by addition of the undesired reacting antigen or by preparation of pure antibody by affinity columns.

Specificity of the antibody is especially important when one is dealing with reagent antibodies that are directed to drugs or metabolites of these drugs. For example, antibody directed to a small molecule, such as gentamicin, is usually produced by chemical binding of many of these molecules to a protein or polypeptide backbone, such as poly-L-lysine. This material is subsequently antigenic in the animal, and a good reacting antibody against the small molecule can be produced. Once the reagent antibody is produced, however, problems related to specificity frequently arise. Gentamicin by itself would be considered an antigenic determinant, but there may be close analogs of other molecular configurations that could interact with the antibody to gentamicin, such as tobramycin and kanamycin. This potential problem is uncovered by determination of the interaction of the antibody with as many closely related analogs as possible. Specificity of the reagent antibody is extremely important in enzyme immunoassays and radioimmunoassays that are most frequently used to measure the presence of small molecules such as drugs and hormones. However, often there is a residual reaction between the reagent antibody and a closely related compound. This reaction between antibody and the undesired antigen is termed *cross-reactivity*. There are times, however, when the cross-reactivity with very similar antigenic determinants cannot be avoided. For example, antibody directed to the small molecule trinitrophenol will cross-react with dinitrophenol. To the antibody, both of these small molecular weight entities look very similar. The only way to establish the specificity of the antibody is to determine the relative affinity of the reagent antibody to presumptive cross-reacting molecules at concentrations likely to occur in patients. Particularly in the case of antibody reagents used in therapeutic drug monitoring, the cross-reactivity with the metabolites of the drug and other drugs is given by the manufacturer.

Since reagent antibody is protein, all precautions to preserve this molecule should be taken. The reagent should be kept free of bacterial contamination and should be stored in the refrigerator (4° C) if to be used within several days. Long-term storage usually is adequate at −20° C. The reagent antibody is very sensitive to heat denaturation. Quality controlling the reagent antibody is one of the most important steps needed to ensure the success of immunological procedures in the clinical laboratory.

Antigen as analyte

Numerous naturally occurring molecules or antigens that are protein, glycoprotein, or lipoprotein in nature can be detected and measured easily in biological fluids, if specific reagent antibodies are available. In addition, many small molecules, such as drugs and hormones, can be measured. The following box lists examples of these large and small molecules that are frequently measured by immunological techniques. To ensure accurate detection and precise measurement of these molecules using immunological techniques, close attention by the technologist to the proper handling and storage of the biological fluid containing these antigens is necessary.

The most common biological fluids available to the laboratory for analysis are serum, urine, and cerebrospinal fluid. Antigens present in each of these different kinds of fluids are subject to degradation depending on (1) the nature of the antigen, (2) its concentration, (3) its susceptibility to various enzymes in the fluid, and (4) its relative stability at various storage temperatures (such as room temperature, 4° C, −20° C, and −70° C). Knowledge of the proper conditions to preserve the antigen before measurement is important for valid results. Preservation or handling conditions must be such that the antigen molecule is unaltered so that the reagent antibody can react with the appropriate antigenic determinants on the molecule. An example of varying degradation rates of an antigen in different kinds of biological fluids is characterized by attempts to measure the fourth component (C4) of the complement system. The C4 of serum is stable and can be measured accurately up to a week after receipt of the serum if the specimen is stored at 4° C before analysis. However, the C4 of cerebrospinal fluid is very labile and is usually present at very low concentrations. If this kind of sample is stored more than 8 hours at 4° C before analysis, the C4 will have been degraded rapidly and be unmeasurable. Thus spinal fluid must be frozen and stored at −70° C if the C4 is not to be degraded before measurement. Another example is that of antigen denaturation in urine specimens. Because most urine specimens are acidic, immunological measurement of various proteins is often suspect. Proteins are degraded in an acid pH, and many antigenic determinants on these proteins are lost. Beta$_2$-microglobulin, a protein found in urine and used to estimate renal tubular dysfunction, is rapidly destroyed if the pH of urine is less than 6.0. Quantitation of specific proteins in urine samples requires immediate neutralization of the acid pH at the time of collection. The problems associated with specific protein measurement and antigen degradation are not so acute when small molecules are measured, but it is always good laboratory practice to store biological fluids at 4° C if the analysis is to be done on the same day and in a frozen state if the analysis is to be done much later.

Examples of molecules in biological fluids frequently measured by immunological techniques

Large molecules	*Small molecules*
Immunoglobulins (IgG, IgA, IgM, IgD, IgE)	Digoxin and digitonin
Complement components (C3, C4, factor B)	Antibiotics
Coagulation factors (factor VIII, fibrinogen)	Cytotoxic drugs
Lipoproteins	Prostaglandins
Acute-phase proteins (alpha-1-antitrypsin, C-reactive protein)	Hormones
Albumin	Theophylline
Selected urine and cerebrospinal-fluid proteins	Anticonvulsant drugs
Viral antigens	Antiarrhythmic drugs

It should be emphasized that the immunological reactivity of a molecule may not be related to its biological activity. The importance of this distinction is illustrated by the immunological measurement of alpha$_1$-antitrypsin and C3, the third component of complement. Alpha$_1$-antitrypsin is a potent inhibitor of the proteolytic enzyme trypsin, and its production is under genetic control. Certain genetic variances occur between individuals in which the amount of the molecule is estimated to be at normal levels by immunochemical techniques, but its enzyme-inhibiting capability is greatly impaired. Immunochemically the genetic variants react as well as the normally functioning molecule does; however, there is a great biological difference. Another example is that of C3, which is a reasonably stable molecule in serum, but in other body fluids, such as joint fluid and cerebrospinal fluid, the C3 molecule may not be present as an intact molecule. When levels of C3 are obtained by immunological methods, they can appear to be normal or increased when, in fact, there is a low level of the complete molecule. This discrepancy is attributable to the reaction of antibody with the breakdown products of C3, which retain the appropriate antigenic determinants with which the anti-C3 can react. Examples of immunological reactivity without biological activity occur frequently and demonstrate that normal levels of molecules assayed by immunological methods do not necessarily describe their functional activity.

Qualitative and quantitative measurements of antigen in biological fluids require a highly specific reagent antibody and a known reference standard of antigen with which to compare the reactivity of this antibody with the antigen in the patient's biological fluid. For the most part, standards are supplied in immunological test kits. If these test kits are approved by the Food and Drug Administration (FDA), the technologist is reasonably assured that the reagent antibody is detecting the antigen, as stated by the manufacturer. However, it is good practice when using immunological methods to evaluate the test system with reference antigen obtained from other sources. The World Health Organization (WHO) supplies reference antigen for many of the serum proteins as primary standards. Secondary standards have been developed by the College of American Pathology (CAP) in collaboration with the Centers for Disease Control (CDC) and are easily available. These reference materials, when used in the immunological methods, provide the technologist with a level of confidence that they are providing valid results on measurement of antigen in biological fluids.

ANTIBODY-ANTIGEN REACTION

In Chapter 9, the basic mechanism of the antibody-antigen reaction is described in detail and should be reviewed before the descriptions of the various immunological techniques presented below are read. However, it is appropriate to reiterate several key points regarding the antibody-antigen reaction. The antibody-antigen reaction can be considered a two-phase reaction. The first phase involves the immediate union of the antibody with various antigenic determinants on the antigen. If the conditions of pH, ionic strength, temperature, and mixture of antibody-antigen are optimal, this phase of the reaction occurs within seconds to minutes after mixture of the reagent antibody with the antigen in the biological fluids. This first phase of the reaction, however, is not always detectable and requires the development of the second or *indicator phase* of the antibody reaction to proceed. The indicator phase can take multiple forms in the variety of immunological techniques used in the laboratory. In immunoprecipitation techniques, additional time is required for large aggregates of antibody-antigen complexes to form. When the aggregates are no longer soluble, they precipitate. The indicator phase can require from hours to days to occur. In agglutination assays, the indicator phase of the antibody-antigen reaction is usually visualization of the clumping of the particle on which the antigenic determinant resides. This phase may occur within minutes to hours after the initial phase of the antibody-antigen reaction. In some immunological techniques, the indicator phase of the antibody-antigen reaction may require utilization of other reagents, such as complement, radioisotope, an enzyme label, or a fluorescent molecule. In each immunological technique described below, the indicator phase, or second step of the antibody-antigen reaction, is the distinguishing characteristic that separates one technique from another. The relative sensitivity of each of these assays reflects the relative sensitivity of each of the indicators used to demonstrate that an antibody-antigen reaction has occurred. The immunoprecipitation technique is generally considered to be the least sensitive, whereas the enzyme immunoassays and radioimmunoassays are considered to be the most sensitive techniques for detection of antigen in biological fluids (refer to Table 10-1 outlining each immunological method).

IMMUNODIFFUSION (OUCHTERLONY)

This technique is commonly used to determine if antibody or antigen is present in a test solution. It can also be used to establish if there are changes in antigenic structure or as a rough criterion of purity of either antigen or antibody.

Principles

The principles of this technique are discussed in Chapter 9.

Sample requirements and preparation

The samples to be tested should have the test antigen or antibody in a concentration suitable to observe the reaction. Usually, several dilutions of antibody and antigen must be tested to observe the desired result. Serum or plasma is often suitable with appropriate dilution. Urine and cerebrospinal fluid often must be concentrated. If sam-

Table 10-1. Summary of immunological techniques*

Technique	Assay end point	Equipment needed	Assay sensitivity	Time needed for assay results	Common analytes	Comments
Immunodiffusion (Ouchterlony)	Precipitation (qualitative)	Agar template	45 μg/mL	8-72 hours	Bacterial, viral, or fungal antigens	Most frequently used to screen for presence of antigen
Immunoelectrophoresis	Precipitation (qualitative)	Agar template Buffer tanks Electrodes Power-supply unit	500 μg/mL	12-24 hours	Serum, urine, and cerebrospinal-fluid protein	Used to assay complex mixture of analytes in biological fluids
Counterimmunoelectrophoresis	Precipitation (qualitative)	Agar template Buffer tanks Electrodes Power-supply unit	3 μg/mL	2-3 hours	Bacterial, viral, or fungal antigens	Commonly used to screen for antigens associated with infectious agents; more rapid than immunodiffusion
Two-dimensional immunoelectrophoresis	Precipitation (qualitative)	Buffer tanks Cooling blocks Electrodes Power-supply unit	500 μg/mL	8-10 hours	Serum proteins	Used to examine subtle differences in proteins (such as C3, factor VIII, and alpha-1-antitrypsin)
Radial immunodiffusion (Mancini)	Precipitation (quantitative) CV, 10%-15%	Calibrated device to measure precipitation diameter Graph paper	50 μg/mL	12-24 hours	Serum and CSF proteins	Most commonly used immunological technique to measure serum and CSF proteins.
Laurell rocket immunoelectrophoresis	Precipitation (quantitative) CV, 8%-12%	Buffer tanks Electrodes Power-supply unit Calibrated device to measure precipitated "rocket" height Semilog graph paper	50 μg/mL	4-8 hours	Serum and CSF proteins	More rapid than radial immunodiffusion
Immunonephelometry	Light-scattering aggregates of antigen-antibody complexes (quantitative) CV, 3%-8%	Nephelometer must be equipped with microcomputer if rate of formation of light-scattering aggregates is measured.	1 μg/mL	½ to 1 hour	*Direct mode:* Serum and CSF proteins *Inhibition mode:* Phenobarbital, theophylline and phenytoin	Popularly accepted technique to quantitate protein in direct mode; in many labs, this technique has replaced radial immunodiffusion
Direct and indirect agglutination	Agglutination of bacteria or red blood cell–containing antigen (semiquantitative)	None	15 μg/mL	1-5 minutes	Antibodies to bacterial antigens (e.g. febrile agglutinins) and red blood cell antigens	Techniques commonly used by serology laboratory and blood bank; not often used in chemistry lab
Agglutination inhibition	Inhibition of agglutination (semiquantitative)	None	15 μg/mL	2-5 minutes	Detect antigens such as pregnancy hormones (HCG)	Rapid test procedure often used to screen urine of pregnant women for HCG

*Abbreviations: *CSF*, cerebrospinal fluid; *CV*, coefficient of variation; *HCG*, human choriogonadotropic hormone.

Continued.

Table 10-1. Summary of immunological techniques—cont'd

Technique	Assay end point	Equipment needed	Assay sensitivity	Time needed for assay results	Common analytes	Comments
Immunodiffusion (Ouchterlony)	Precipitation (qualitative)	Agar template	45 µg/mL	2-4 hours	Detect complement-fixing antibodies to bacterial, viral, and fungal antigens	Worldwide, most commonly used serological procedure; sensitivity of assay approaches radioimmunoassay; assay difficult to perform
Complement fixation	Lysis of red blood cells or inhibition of red blood cell lysis (semiquantitative)	Spectrophotometer Graph paper	10 µg/mL			
Fluoroimmunoassay (direct or indirect)	Fluorescent intensity of labeled reagent antibody (quantitative) CV, 5%-10%	Fluorometer Graph paper	100 ng/mL	1-2 hours	Measure proteins in serum and CSF	Not widely accepted as a method to measure proteins
Enzyme immunoassay (ELISA, sandwich)	Color reaction between enzyme and substrate (quantitative) CV, 8%-15%	Spectrophotometer or microtiter-plate ELISA reader Graph paper Microtiter plates	1 ng/mL	1-24 hours	Serum proteins (such as IgE) Bacterial, viral, and fungal antigens Antibodies to infectious agents	Excellent assay for screening for presence of small amounts of antigen or antibody
Enzyme immunoassay (competitive binding)	Color reaction between enzyme and substrate (quantitative) CV, 8%-15%	Microtiter plates Spectrophotometer or microtiter-plate ELISA reader Graph paper or microcomputer	1 ng/mL	2-4 hours	Small amounts of antigen (such as hormones, drugs, viral antigens)	

ples are concentrated, they must not contain too high a salt concentration or improper pH, which would prevent the reaction from occurring.

Reagents

If an antibody is to be tested for purity, it is desirable to have the pure antigen and a series of solutions containing antigens capable of reacting with possible contaminating antibodies. In contrast, if antigen is to be tested for purity, one should use antibodies that can react with possible antigen contaminants.

Instrumentation

Essentially there is no instrumentation, though some laboratories prefer an indirect light apparatus.

Common pitfall

Improper dilution of antigen or antibody may occur.

Interpretation of results

See Chapter 9.

Limitations

The Ouchterlony technique requires a concentration of antigen greater than 45 μg/mL and thus is not as sensitive as other techniques. It is not as discriminatory as immunoelectrophoresis and resolves only a few antigens compared to other techniques. Large molecules do not diffuse readily into the gel, and the technique resolves these poorly.

IMMUNOELECTROPHORESIS

The immunological technique immunoelectrophoresis is used primarily as a qualitative procedure to evaluate the electrophoretic and immunological characteristics of proteins and glycoproteins in serum, urine, or cerebrospinal fluid. Immunoelectrophoresis (IEP) is a more sophisticated technique than serum protein electrophoresis and is used routinely to characterize qualitative abnormalities of specific proteins or to analyze the composition of a complex mixture of proteins in biological fluids. Immunoelectrophoresis is not considered a screening method.

Principles

Immunoelectrophoresis is performed as a two-stage procedure. The first stage involves the separation of antigenic material in biological fluids by their differential migration in an electrical field. Generally, serum proteins can be separated into five discrete regions (albumin, α_1, α_2, β, and γ), this stage of the procedure is termed "zone electrophoresis." The second stage of this technique is the immunological characterization of each of the separated proteins by immunodiffusion procedures. In this process, antibody diffuses into the gel and the antigen-antibody reaction is visualized by precipitation. The immunological characterization of each of these proteins is dependent on the reagent antisera used to immunoprecipitate the proteins. Only if the antibody to a specific protein is present will there be precipitation. Thus, selection of which protein is to be observed is dictated by the antibody.

In this technique, a solid surface, such as a glass slide, is covered with agarose in a buffered solution (pH 8.2), and the antigen or sample well is cut into the center of this slide. A long, parallel cut in the agarose is made so that antibody in the second stage can diffuse perpendicularly to the separated proteins. Fig. 10-1 illustrates the technique of immunoelectrophoresis. The biological fluid to be analyzed is placed at the application point (antigen well). Because of the alkaline pH of the agarose buffer, most biological molecules assume a net negative charge and in an electrical field will migrate toward the positive pole. If serum samples are used, proteins are separated into five distinct regions when a potential difference at 3.3 V/cm is established across the agarose plate for 30 to 60 minutes. Albumin travels the farthest toward the positive pole. After the albumin are the α_1, α_2, β, and γ regions, respectively. After the electrophoresis, antisera are then placed immediately in the parallel antibody trough and allowed to diffuse toward the electrophoretically separated proteins. The density, position, and shape of the resultant immunoprecipitin bands are then interpreted as a means of describing each of the precipitated proteins.

Permanent records of the immunoelectrophoretic patterns can be obtained by staining of the precipitin bands with protein stains, such as amido black or Coomassie blue. Some laboratories obtain permanent records by photographing the immunoprecipitin bands in indirect lighting in lieu of staining.

Sample requirements and preparation

Serum. Most immunoelectrophoretic procedures are established to characterize various serum proteins that are greater than 500 μg/mL. The system can be used to detect albumin, the immunoglobulins, and about 30 other serum proteins that are present in at least these concentrations. No further preparation of the serum specimen is required. The sample should be stored in a frozen state to preserve antigens if the procedure is to be delayed more than a day after receipt of the specimen. Plasma is considered an inappropriate sample for immunoelectrophoresis because of the high concentrations of fibrinogen in an unclotted sample.

Urine. As mentioned above, immunoelectrophoretic techniques are used to characterize proteins in concentrations greater than 500 μg/mL. Urine usually contains dilute concentrations of protein and therefore must be concentrated to bring them into the detectable range for the immunoelectrophoretic procedure. Concentration procedures usually involve semipermeable membranes of sufficient porosity to allow water and salts to pass but not proteins. Methods that concentrate the urinary salts, such as lyoph-

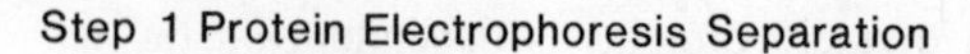

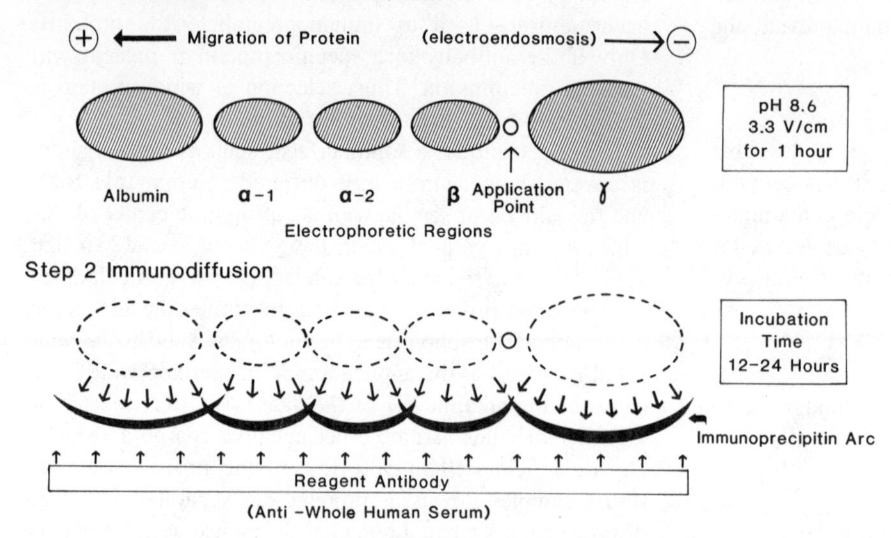

Fig. 10-1. Immunoelectrophoresis technique using normal human serum and anti–whole human serum. Usually many more than five immunoprecipitin arcs are seen.

ilization, are not acceptable. Many of these proteins can be detected by immunoelectrophoretic procedures if the urine is concentrated 50 to 100 times. It is important that urine specimens be properly stored in a frozen state if any delay in immunoelectrophoretic analysis is anticipated.

Cerebrospinal fluid. Cerebrospinal fluid is also a dilute protein solution that requires concentration 50 to 100 times its original volume to allow the proteins present to be analyzed by this procedure. Since the major proteins of clinical significance in a cerebrospinal fluid observed by this procedure are albumin, transferrin, and immunoglobulins, it is important that they be concentrated to detectable ranges for immunoelectrophoretic analysis. Again, storage of the sample in a frozen state is important if the analysis is to be delayed more than 1 day.

Reagents

In immunoelectrophoresis, three quality reagents are needed: agarose, buffered solution, and specific reagent antibody. A high-quality agarose as the inert support medium through which proteins will migrate during the electrophoretic stage of this procedure is important. Agarose, when compared to agar, does not develop surface charges, which can interfere with the migration of proteins during the electrophoresis. The phenomenon of electroendosmosis is not so noticeable when agarose is used.

Immunoelectrophoresis is usually carried out in a buffered medium (pH 8.6). Maintaining the alkaline pH of this buffer is important in the performance of a good immunoelectrophoretic analysis. Barbital or barbituric acid is frequently added to the buffering solution, not only because it has good buffering properties, but also it is an excellent bacteriostatic reagent. Close attention to the freshness of a buffer solution each time immunoelectrophoresis is performed is necessary.

The final reagent that requires close attention is the reagent antibody used to define the various migrated proteins. These antibodies should perform well in an immunoprecipitation reaction and should be titered to sufficient dilution to obtain the maximum precipitation band for those proteins in lowest concentration that can be detected by immunoelectrophoretic analysis.

Instrumentation

Four specific pieces of equipment are needed for the performance of immunoelectrophoresis: power supply, buffer tanks, proper wicking material, and agar-cutting template. The power supply used in the electrophoretic stage of this procedure should be able to supply a constant voltage over a several-hour period at a rate of 3 to 5 V/cm (Fig. 10-2). Buffer tanks should be of sufficient size to contain an adequate amount of buffer to prevent changes in pH during the electrophoresis stage of this technique. Electrodes should be made of platinum because of its high conductance and inertness to chemical degradation. The wicking material should be porous and inert and should allow free passage of buffer from the tank to the agarose-containing plate. This wicking material also should be of low electrical resistance. Material that has a high electrical resistance can cause overheating of the system and produce

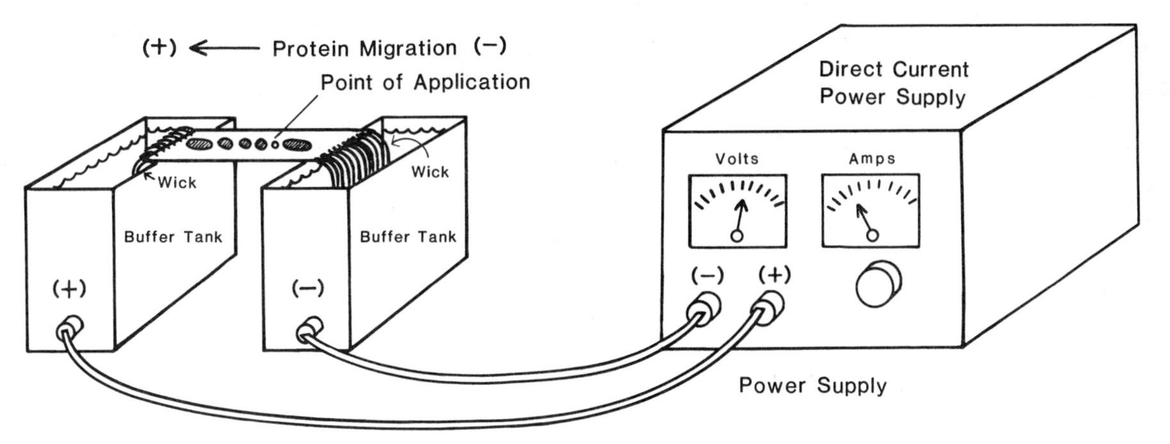

Fig. 10-2. Major equipment components needed to perform immunoelectrophoresis techniques.

inaccurate results. The agar-cutting template should be built with razor-sharp cutting blades set in a rigid form such that uniform wells and troughs can be cut into the agarose. Close attention should be paid to maintaining sharp cutting edges in this template.

Common pitfalls

Immunoelectrophoresis is a multicomponent system that requires close attention to each of the major components for successful application of this technique. Common pitfalls are listed below.

1. Spent buffer solution, causing improper migration of the proteins in electrophoretic separation. It can cause overheating in the agarose, which may result in some denaturation of the protein and destruction of antigenic determinants. Spent buffer occurs when the buffer solution is used for too many electrophoretic runs.
2. Improper concentration of specimens, resulting in no precipitation, insoluble aggregated proteins, or too high a concentration to observe.
3. Inappropriate attachment of the wick to the agarose plate and buffer tanks, resulting in intermittent or nonuniform flow of electricity across the agarose plate during the electrophoretic procedure. Improper attachment of the wick to the agarose plate is a frequent source of errors in the technique.
4. Dull cutting blades in the template, causing ragged cutting of the agarose plate with consequent artifactual influence on the diffusion properties of the reagent antibody.
5. Improper dilution of the antisera, usually with loss of precipitation reaction.
6. Improper ionic strength of the buffer. The ionic strength of the buffer solution used should be sufficiently low so that when the migration of the antigenic material is induced the electrical charge is conferred mostly to those antigens rather than to the buffer solution.
7. Overheating, usually caused by improper wicking, spent buffer solutions, and too long electrophoretic migration, with a resulting denaturation of the proteins.
8. Immunodiffusion for too long a time, resulting in immunoprecipitation lines in inappropriate positions, making interpretation questionable. Antisera and antigen concentrations should be titrated to a point where the maximum precipitation line can be observed between 14 and 24 hours.
9. Improper staining and photographic procedures.

Interpretation of results

Immunoelectrophoretic analysis of biological fluids requires a firm understanding of electrophoretic separation of proteins and immunodiffusion analysis. Fig. 10-3 illustrates the use of this technique in the interpretation of normal and abnormal serum. Analysis of these immunoprecipitin bands can provide much useful clinical information. The presence of monoclonal or oligoclonal immunoglobulins, the presence or absence of various proteins, alpha$_1$-antitrypsin, and so on, can be used for the diagnosis of disease processes.

COUNTERIMMUNOELECTROPHORESIS

Counterimmunoelectrophoresis (CIE) is also called "double electroimmunodiffusion" (double EID). This technique is frequently used for the detection of single antigens present in a patient sample. It has the advantage of being much more rapid than immunodiffusion techniques.

Principles

In immunodiffusion techniques described earlier, antigen and antibody are allowed to come into contact and precipitate purely by passive diffusion processes. This diffusion process of antigen coming together with antibody can be accelerated if one allows the antigen and the antibody to migrate more quickly toward each other in an electrical field. Fig. 10-4 illustrates this technique. The an-

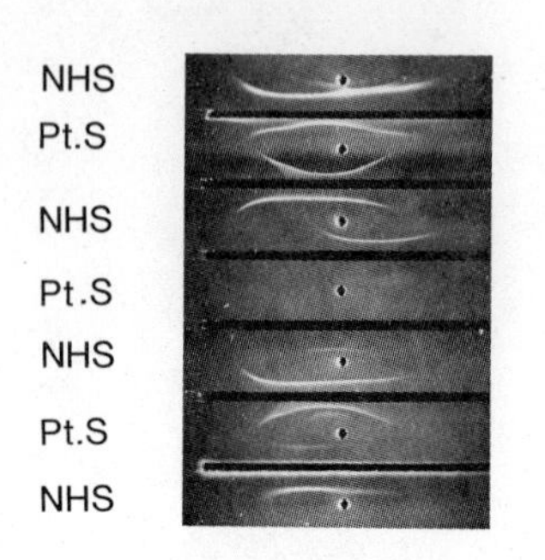

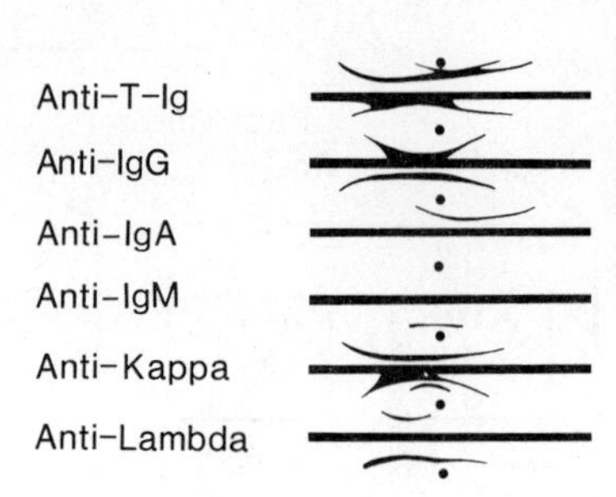

Fig. 10-3. Examples of immunoelectrophoresis patterns. *On left,* Actual gel. *On right,* scheme drawn from photograph. *NHS,* Normal human serum; *Pt.S,* serum from a patient with IgG(K) myeloma; *Anti–T-Ig,* antisera to all (total) human immunoglobulins; *Anti-IgG,* antisera to human immunoglobulins.

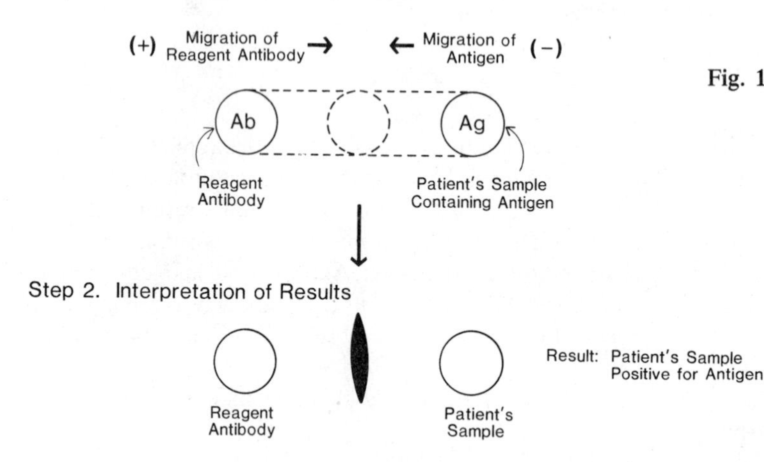

Fig. 10-4. Counterimmunoelectrophoresis.

tibody-containing well is positioned closest to the positive pole, whereas the antigen-containing well is positioned toward the negative pole. If the buffer is selected appropriately, the antibody will migrate toward the negative pole. The antigen must carry a net negative charge so that it can travel toward the positive pole. If the electrical current and antigen and antibody concentrations are all in proper proportion, a precipitation line will form halfway between the two wells. Counterimmunoelectrophoresis is a screening procedure. It is used primarily for the detection of bacterial and viral antigens present in biological fluids.

Sample requirements and preparation

A biological fluid most frequently tested by this technique is serum, though urine and cerebrospinal fluid may also be used. Proper storage of the fluids to preserve the antigen is necessary if the analysis is to be delayed. Occasionally, the antigen may be of low concentration and may require that the biological fluid be concentrated so that the antigen can be detected. This is particularly true of urine and cerebrospinal fluid specimens. It should be noted that since this technique is used primarily for bacterial or viral antigen detection, precaution should be taken by the technologist in dealing with these specimens, and they should be treated as potentially infectious materials.

Reagents

The *reagent antibody* used in counterimmunoelectrophoresis should migrate in an electrical field toward the negative pole. In general, these antibodies are of the IgG class, since IgM migrates primarily toward the positive pole. The antibody should also be a good precipitating antibody and have singular specificity for the antigen to be detected.

Instrumentation

All the equipment listed in the immunoelectrophoretic section is required, except that the agar-cutting template configuration is different. This template should be able to cut antigen and antibody wells that are properly spaced in the agarose plate such that an appropriate antibody-antigen reaction can occur.

Common pitfalls

1. Improper handling and storage of specimens.
2. Antigen to be detected does not migrate toward the positive pole.
3. Improper distance between the antigen- and antibody-containing wells. If the distance is too great, the antibody-antigen reaction will not occur in sufficient intensity for an immunoprecipitation line to occur. If the wells are too

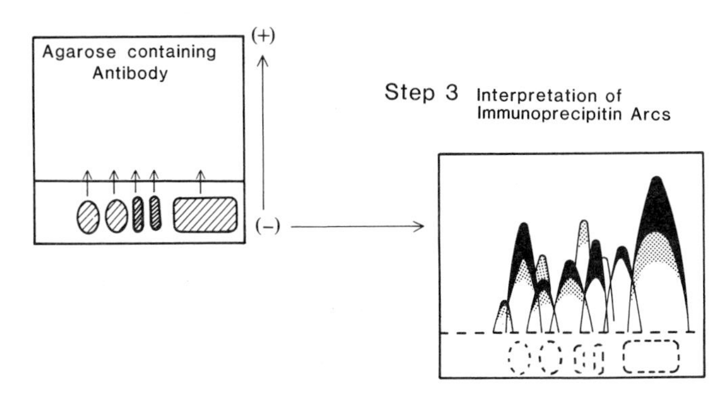

Fig. 10-5. Two-dimensional immunoelectrophoresis. Pattern partially representative of normal human serum versus anti–whole human serum.

close, the antibody-antigen reaction may occur within one or the other wells and therefore not be visible in the agar between the wells.

4. Improper concentration of specimen.
5. Multiple specificities of the reagent antibody used.

Interpretation of results

A positive test result is considered when a clear immunoprecipitin band is found between the antibody- and antigen-containing wells. Serial dilution of the test specimen can be used to determine the titration of the antigen present in the test specimen. Counterimmunoelectrophoresis is a screening procedure, and therefore confirmation of positive test results might be considered appropriate.

TWO-DIMENSIONAL IMMUNOELECTROPHORESIS

This method is used as a research technique for analytical separation of closely related antigens with similarly related electrophoretic properties. It is used in special situations to identify the heterogeneity of certain proteins such as factor VIII (hemophiliac factor) or distinguish genetic variants of some proteins, such as alpha$_1$-antitrypsin. In both of these situations, analysis of the fine details of these proteins may be important in some clinical situations. Even though this procedure is an excellent tool to evaluate subtle differences of biological material, it is a difficult procedure that is expensive and time consuming. It is not likely to be used in the routine laboratory.

Principles

This procedure is a two-stage technique that uses electrophoresis in both stages. The antigen mixture, usually serum, urine, or cerebrospinal fluid, is first electrophoresed in a neutral agarose to separate components by net charge. After this first stage, the agarose strip containing the separated protein is transferred to a larger support plate. Liquid agarose (pH 8.6) containing antibody against the protein to be analyzed is poured on the plate, making contact at one end with the neutral agarose containing the antigens. Fig. 10-5 illustrates this procedure. After the antibody-containing gel has hardened, wick material is placed at each end of the plate, and electric current is applied in a direction perpendicular to the first electrophoretic separation so that the antigens migrate into the antibody-containing agarose. This second stage of electrophoresis may take as long as 20 hours in order for a good interaction with the antibody and antigen to be obtained so that definable immunoprecipitin arcs are formed. After the current on the second electrophoresis is stopped, the plate is washed in saline solution to remove excess protein not associated with the visible precipitation. Excess fluid is pressed out with filter paper, and then the plate is fixed and stained. The plate can be retained as a permanent record.

Sample requirements and preparation

The type of sample required and the sample processing are similar to those described in the immunoelectrophoresis section. Special attention should be paid to those samples in which the protein constituents may be labile. Analysis of factor VIII requires plasma be used rather than serum because this protein is easily consumed by clotting.

Concentration of the analyzed antigen is important in this assay. Since the antigen is electrophoresed in two dimensions, a significant amount of antigen dilution occurs. Samples often must be concentrated before analysis.

Reagents

High-quality agarose. The agarose should be essentially neutral at a pH 8.6 (see p. 188). The agarose should have a low melting point so that it can be poured at a temperature that will not denature the reagent antibody (less than 56° C).

Reagent antibody. The reagent antibody should be a good precipitating antibody and should not migrate appreciably in the agarose at a pH 8.6.

Instrumentation

Along with the routine electrophoresis apparatus, special cooling accessories may be appropriate. Since electrophoresing takes a long time or is performed at high voltages, heat generation is an important factor that should be controlled by frequent replacement with refrigerated buffer, or use of ice packs under and on top of the chamber.

Common pitfalls

1. Test sample is not adequately concentrated.
2. Inadequate cooling during electrophoresis such that agarose may be overheated and cause denaturation of sample.
3. Improper dilution of reagent antibody.
4. When pouring the agarose containing antibody before the second stage of electrophoresis, extreme care should be taken to ensure a good contact of gel with the gel from the first stage to further ensure that all protein will migrate into the antibody-containing gel.
5. Inadequate electrical current supplied in both stages of the electrophoresis.
6. Improper saline washing, fixation, or staining of the completed product.

Test sensitivity

This procedure is slightly more sensitive than immunoelectrophoresis and can detect protein as low as 500 μg/mL. The advantage of this technique lies in its ability to separate protein antigens with subtle differences. It is a qualitative procedure but can be used in a semiquantitative manner if appropriate standards are analyzed under the same conditions as the test sample.

Interpretation of results

The immunoelectrophoresis pattern resembles a series of immunoprecipitation arcs, or ''peaks,'' that stretch the entire length of the antibody-containing agarose. Identification of each of these peaks in the test sample requires establishment of the pattern of each protein's precipitation characteristics. This can be obtained by prior analysis of the isolated protein or by use of a monospecific reagent antibody for each of the proteins. Interpretation of the patterns requires a high degree of skill.

RADIAL IMMUNODIFFUSION

This technique is one of the most commonly used to quantify antigens because of its simplicity and accuracy.

Principles

The principles are discussed in Chapter 9.

Sample requirements and preparation

The assay is only valid over a rather narrow range of antigen concentrations. In part, this is dictated by the antibody concentration in the gel and by the ease of diffusion of the antigen into the gel. For some antigens, several dilutions are necessary to achieve the optimal range. Urine and cerebrospinal fluid generally must be concentrated before they can be quantified by this technique. Excess salt must be removed from these specimens if they have been concentrated by lyophilization.

Reagents

Monospecific antibody to the desired antigen is the only crucial reagent.

Instrumentation

A ruler or similar device is needed.

Common pitfalls

If the antibody measures more than one component, a double ring may be seen. Temperature should be kept constant. If the antigen is too large or aggregated, the resulting diffusion pattern will not be quantitative.

Test sensitivity

Test sensitivity is dependent on antigen size, since the greater the size, the poorer the diffusion. For most serum proteins, the assay is sensitive to 50 μg/mL.

LAURELL "ROCKET" IMMUNOELECTROPHORESIS

This technique is a quantitative method that is used to measure antigens in biological fluids and provides data for individual antigens similar to that obtained by radial immunodiffusion. The advantage of this technique over radial immunodiffusion is that the time needed to produce results is 4 to 6 hours rather than 18 to 72 hours. This shortened time is attributable to the forced migration of antigen into the agarose containing antibody rather than passive diffusion of the antigen, as in radial immunodiffusion. The ''rocket'' technique is usually more expensive to perform than radial immunodiffusion because of the need for electrophoretic equipment and the increased quantity of reagent antibody and agarose used. This technique is best used when the antigen to be measured is in concentrations of 50 to 20,000 μg/mL. Antigens most frequently measured are the serum proteins.

Principles

In this technique, the antigen combines with the antibody during electrophoresis of the antigen into an antibody-containing medium such as agarose. The pH of the agarose-antibody medium is usually 8.6, so that the antibody has little or no net negative charge and will not migrate when current is passed through the gel. The antigen to be measured must have a net negative charge so that it will migrate into the gel when the electrophoresis is performed. The agarose-antibody medium is usually prepared with circular wells cut at one end of the gel and filled with measured amounts of antigen (standards) and the test specimen. Electrical current is applied to the gel plate in such a way as to cause migration of the antigen to the center of the plate. As the antigen migrates and combines with the antibody in the gel, cones or rocket-shaped bands of precipitate are formed. Under the influence of the electrical current, the unbound antigen within the rocket-shaped band of precipitate migrates into the precipitate and causes the precipitate to dissolve in antigen excess. The leading edge of the rocket reforms ahead of the antigen in the direction the antigen is migrating. The amount of antigen within the leading edge is successively diminished until equivalence of the antigen-antibody complex is reached. At this point, a stable precipitate will form and will no longer migrate. One can determine the concentration of antigen in the test sample by taking the length of the rocket from the edge of the circular well to the edge of the precipitin band and comparing it with the lengths of rockets obtained from reference standards run at the same time in the same gel (Fig. 10-6). The length of the rocket in the test sample is proportional to the log concentration of the antigen in the reference standard.

The time of the electrophoresis is important in this technique if accurate measurements are to be obtained. Electrophoresis must continue until the edge of the rocket reaches equilibrium, otherwise a rounded or blunted apex is observed, which indicates equilibrium has not been reached. The usual electrophoresis time is 4 to 6 hours, and if the proper-shaped rocket fails to form, the test sample should be diluted and run again.

Sample requirements and preparation

As in the other immunological techniques, care must be taken to ensure that the antigen to be measured is not degraded or denatured. These precautions are especially true when one is using the "rocket" technique to measure complement components such as C3 and C4 and the immunoglobulin IgM. Breakdown of the protein molecules can cause several peaks within the same rocket to form, and then the difficulty facing the technologist is to decipher which peak is the one to measure for accurate quantitation. In some disease states, breakdown of the protein C3 and C4 occurs in vivo, and knowing how to interpret these several peaks may be of clinical significance, but if the test sample has not been properly handled and stored before testing, the observation of several peaks would be difficult to interpret.

Reagents

Agarose. See p. 188.

Reagent antibody. The reagent antibody should have the same properties as those used in the two-dimensional immunoelectrophoresis (p. 192).

Instrumentation

This technique requires an electrophoresis chamber and an excellent electrical power source. Because prolonged electrophoresis cooling of the electrophoresis chamber is necessary, an appropriate cooling apparatus is needed.

Common pitfalls

1. Improper antigen migration. For an accurate measurement of antigen concentration with this technique, the

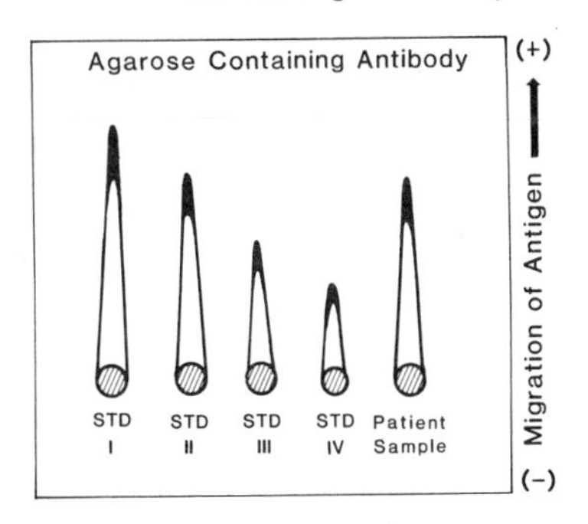

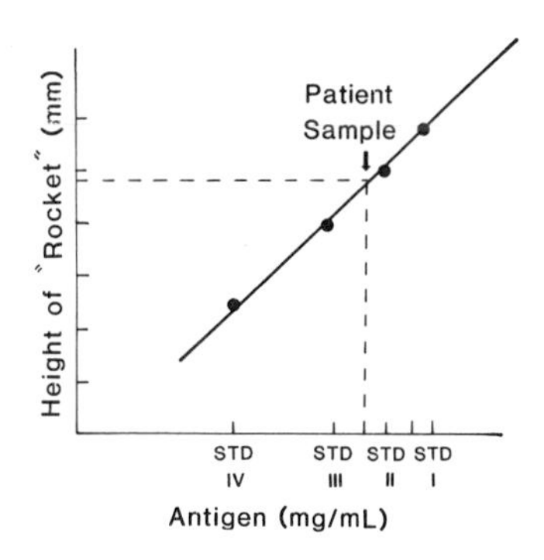

Fig. 10-6. "Rocket" immunoelectrophoresis. *STD*, Standard.

antigen must be able to assume a net negative charge at a pH of 8.6. If the antigen's isoelectric point is greater than 7.5, the antigen will not migrate adequately into the agarose-antibody medium to obtain peaks of high resolution.

2. Improper wicking. As in all electrophoretic procedures, proper placement of the wicks at the edge of the gel plate such that all of the gel has a constant electrical current is necessary. If the contact point of the wick with gel is improper, the antigen will not migrate in a consistent manner throughout the plate, giving false peaks.

3. Overheating during electrophoresis. Since the electrophoresis time is prolonged in this technique, close attention to cooling of the gel plate during electrophoresis is essential. Overheating can cause the gel to dry, the antigen to denature, and the antigen to migrate improperly.

4. Improper measurement of peak height.

Test sensitivity and coefficient of variation

The Laurell rocket immunoelectrophoresis technique has a sensitivity of approximately 50 μg/mL for serum proteins found in biological fluids. The upper limits of detection are usually around 20,000 μg/mL for these antigens. Within-run variations are between 5% and 10%, whereas between-run variations, when the same antigen is measured, are 10% to 15%.

IMMUNONEPHELOMETRY

Nephelometry

Principles. When an antibody and an antigen combine in solution, small aggregates that can scatter light quickly form, causing a turbid appearance to the solution. These aggregates then slowly associate to form a larger matrix, which eventually gives rise to the precipitate formed, as seen in immunoprecipitation assays such as double diffusion (Ouchterlony) or radial immunodiffusion (Mancini).

The light-scattering properties of the aggregates were demonstrated initially by Pope and Healey in 1938 and later confirmed by Gitlin and Edelhoch. These early indications of a useful assay procedure measuring the turbidity of a solution proved to be of clinical usefulness with the work of Schultze and Schwick, who, in 1959, reported that for some plasma proteins the intensity of scattered light was a measure of the amount of precipitate formed, as long as the reaction was carried out in antibody excess.

These advances were not actively pursued until the early 1970s, when light-scattering techniques, such as nephelometry, became useful tools in the clinical laboratory. Workers, such as Ritchie et al. and Killingsworth and Savory, using the newly developed Technicon Automated Immunoprecipitation System (AIP), reported on a rapid, high-throughput, automated system for the quantitation of specific proteins in biological fluids that was comparable to the frequently used radial immunodiffusion assay in sensitivity and specificity.

Although the method suffered some initial instrument-instability difficulties, the latter half of the 1970s saw the emergence of more stable and reliable instrumentation. This included laser nephelometers, rate or kinetic nephelometers, and light-scattering assays carried out on a number of discrete fast analyzers.

There are now available fast (seconds to hours) and precise methods for the specific measurement of plasma proteins, such as albumin, immunoglobulins, complement components, and acute-phase reactants, as well as hormones and therapeutic drugs.

In Chapter 9, the structure of the aggregates created by early lattice formation of immune complexes is shown. These aggregates from primary reactions occur within seconds to minutes, whereas the secondary interactions leading to precipitation occur over a period of hours.

Light-scattering assays measure this early second-order reaction, presumably between antigen and high-affinity antibody, in which there is formed a micelle of protein large enough to scatter light but not large enough to precipitate.

Although the initial reaction of antigen with antibody is fast, the build-up of small light-scattering complexes takes time. This reaction, aggregate formation, can be greatly enhanced with the addition of the water-soluble polymer polyethylene glycol (PEG) (MW, 6000) at concentrations of 2% to 4%. The polymer causes a severalfold increase in light scatter while decreasing the reaction time tenfold.

Sample requirements and preparation. The most frequently measured biological fluids are serum and cerebrospinal fluids, though urine may also be measured.

Reagents. Reagent antibody is of very high titer and affinity and specially clarified by microfiltration to minimize scattering.

Instrumentation. Refer to Chapter 3 for a complete description.

Common pitfalls. One common pitfall occurs when the reaction is not in antibody excess. In these circumstances the amount of precipitate formation will be recorded as a falsely low value. The rate-nephelometric determinations are preferred to end-point determinations because they minimize the background contributions. For example, it is possible to detect a 5% increase in scattering over background using a rate measurement. In contrast, such a difference would not be measureable as an end-point change. In addition, end-point methods are influenced by colored solutions. These absorb the scattered light, tending to yield lower values. However, even kinetic measurements are lower in highly colored solutions.

Because the rate method requires constant agitation to make uniform particles, the mixing efficacy must be constantly monitored.

Limitations. The difficulty of determining whether a given amount of precipitate (light scatter) is occurring in antigen or antibody excess is one that makes all the methods suspect in measuring the concentration of molecules, such as immunoglobulins, where one expects five- to tenfold variations to occur in pathological conditions. Although each instrument manufacturer has devised ways of recognizing antigen excess, the foolproof method is to carry out the measurements at two different dilutions, or have available a serum protein electrophoresis of the sample for comparison.

It may be possible, using kinetic assays, to differentiate the early reaction rate seen in antibody excess from that seen in antigen excess, but the microprocessor software needed has not yet been made available.

Light-scattering inhibition immunoassay.

Principles. The antibody population that reacts first with antigen represents only a small portion of the total antibody content of a polyclonal antiserum. The concentration of this high-affinity antibody dictates the rapidity and intensity with which the system approaches the optimum scatter for a particular antigen concentration. Addition of an unknown but small amount of antigen to the antibody followed by back-titration of residual antibody by a standard antigen can be used to quantitate very small amounts of antigen and, as has been described for IgE, allows detection down to 100 ng/mL. This level of protein determination begins to overlap with radioimmunoassay, though it does not reach the limits of detection of RIA.

Nephelometric-inhibition immunoassay (NINIA) of haptens. If a hormone or drug, such as digoxin, which by itself is nonantigenic, is covalently linked to a large carrier protein, such as keyhole limpet hemocyanin (KLH), the resulting conjugate acts as an antigen and will cause the formation of antibodies that will recognize both KLH and digoxin (hapten) (Fig. 10-7).

If the digoxin is coupled to a different molecule, such as bovine serum albumin (BSA), the conjugate will react with the anti-digoxin-KLH antiserum. Precipitation with the BSA-digoxin (developer) antigen can be attributable only to antihapten (digoxin) specificities. Pauling et al. in 1944 had demonstrated that precipitation from antihapten activity could be inhibited by addition of the free hapten, and it is on this principle that the nephelometric-inhibition immunoassay is based. The technique was first described for progesterone by Cambiaso et al. (1974), and they referred to the term NINIA as an acronym for the approach. With appropriate manipulation of the parameters, one can adjust the number of haptenic groups for adequate precipitation while allowing maximum sensitivity for inhibition by free hapten.

Addition of sample containing free hapten to the reaction mixture neutralizes available antibody and causes a decrease in the signal generated by a standard hapten-conjugate preparation (Fig. 10-8).

Methods have been developed for rapid analysis of drugs occurring in milligram-per-liter amounts, such as phenytoin, phenobarbital, and theophylline.

Sample requirements and preparation. The most frequently measured biological fluid is serum. The most measured compounds are drugs such as theophylline and phenytoin.

Reagents. The requirements for reagent antibody are similar to those for any nephelometric technique. Reagent antigen is composed of the drug to be monitored covalently bound at several sites to a protein, usually albumin.

Instrumentation. Same as used for nephelometry.

Common pitfalls. Similar to those described for nephelometry, except that the reaction is titered to be always in antigen excess.

New developments

Particle-enhanced light scattering. Since the amount of light scatter is dependent on the size and amount as well as the refractive index of the scattering species, an increase in any of these parameters should result in greater sensitivity. This potential has been realized when either antigens or antibodies are coupled to various inert carrier particles, but, because of availability and better control of coupling conditions, polystyrene latex beads have become the particles of choice. This type of assay is essentially a type of agglutinating procedure (see next column). The method also offers faster rates of signal generation and economy of reagents. The latex-fixation test for detection of rheumatoid factor is the classical example of this type of assay, though it depends on visual observation of agglutination and is thus only semiquantitative. More recently, several particle-enhanced methods have been developed for both direct and inhibition assays, and a wide variety of light-scatter detection techniques have been employed. Repre-

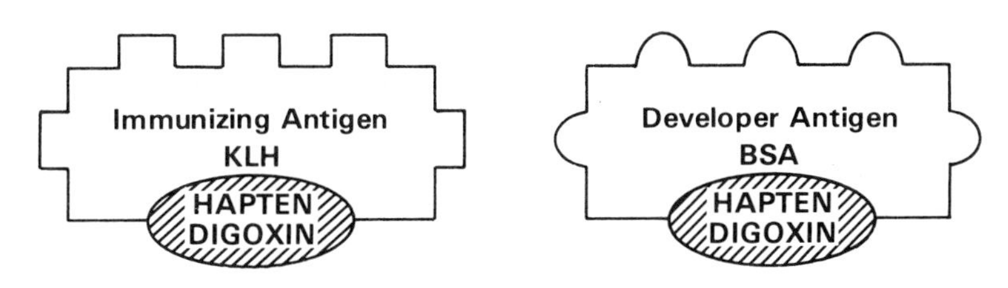

Fig. 10-7. Formation of hapten (digoxin) complexes for use as immunizing antigen (keyhole limpet hemocyanin–digoxin) and developer antigen (bovine serum albumin–digoxin) for nephelometric-inhibition immunoassay (NINIA) approach to measurement of small molecules.

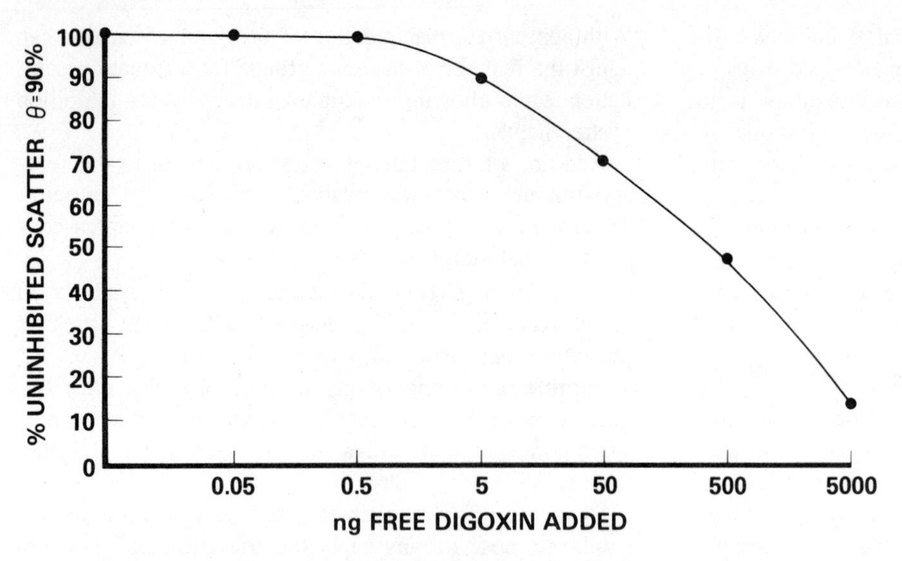

Fig. 10-8. Digoxin standard curve using nephelometric-inhibition immunoassay.

sentative of such assays is the particle-counting immunoassay (PACIA) of Cambiaso et al. (1977), whereby the number of nonagglutinated particles remaining are counted using a light-scatter instrument (Technicon Autocounter). Using inhibition of agglutination of antigen-coated beads by antibody-coated beads, these workers have produced an assay for digoxin in the low microgram-per-liter range.

Once the many controlling parameters, such as particle size, type, wavelength, and detection method, have been resolved more fully, it is to be expected that this technique, now in its infancy, will mature to become a major tool in immunoassay.

Monoclonal antibody reagents. The performance of light-scattering assays is greatly dependent on the quality of the antiserum used. With conventional polyclonal reagents there is a continual need to monitor and adjust antiserum titer, specificity, and affinity.

Such variability is overcome with the use of newly developed monoclonal antibody preparations from "immortalized" hybrid plasmacytoma cell lines. However, unless the antigen has a number of identical antigenic sites, monoclonal antibody cannot cause matrix formation and will cause little or no light scatter. If an appropriate mixture of monoclonal antibodies can be made, complexes will form, causing measurable light scatter. This blending of monoclonal antibodies or the use of monoclonals in particle-enhanced light-scatter assays will ensure constancy of reagent production and give these assays further stability and specificity.

AGGLUTINATION ASSAYS

Agglutination is the clumping and sedimentation of particulate antigen after primary reaction with antibody. It was first noted by the reaction of bacteria after incubation with infected patient serum. Agglutination of red blood cells after incubation with serum led to the discovery of ABO blood groups. Agglutination has been extensively employed as a laboratory test because of its ease and versatility. It is, however, only a semiquantitative procedure. Repeat testing shows reproducibility of results only within fourfold dilutions. Agglutinating antibodies *(agglutinins)* may be directed against naturally occurring antigens on the surface of cells *(active* or *direct agglutination)* or against substances that have been applied to the surface of cells or inert particles *(passive* or *indirect agglutination).*

Principles

Agglutination reactions are dependent on the formation of antibody bridges by bivalent (IgG) or multivalent (IgM) antibody between antigen particles with multiple antigenic determinants. Large particles, such as red blood cells or bacteria, contain many different antigens, as well as antigens that appear hundreds of times on the cell or particle surface. Thus it is possible for antibody molecules to bind to more than one site on a single particle or to bind to equivalent sites on different particles. Such binding is called "cross-linking." Antigens with a single antigenic determinant would not permit cross-linking and would therefore not agglutinate. Antibody can bind to antigenic determinants on the same or different particles, creating a high molecular weight lattice that clumps together and falls out of suspension. Because of its size and multivalency, IgM is said to be 750 times more efficient at agglutination than IgG. The effectiveness of IgM in agglutination is the reason that IgM molecules are called *complete antibodies.* IgG molecules, because they sometimes cannot be effective at cross-linking, are called *incomplete antibodies.* Agglutination reactions are generally used to detect antibody in serial dilutions of serum containing antibody directed to particulate antigens. *Reverse agglutination* can be used to detect soluble antigen by adsorption of antibody onto cell or particle surfaces. Agglutination reactions are read by the naked eye or with the aid of magnification. Reactions are scored 1+ to 4+ by extent of agglutination. The titer of the serum is the reciprocal of the highest dilution giving visible (1+) agglutination.

Factors influencing agglutination reaction

Many factors influence agglutination. These include particle charge, antibody type, electrolyte concentration, viscosity of the medium, reactant concentrations, location and concentration of antigenic determinants, and time and temperature of incubation. These factors are considered separately.

Particle charge. Red blood cells, bacteria, and inert particles such as latex have a net negative surface charge, which is called the *zeta potential*. The surface charge of red blood cells is caused by sialic acid residues on the cell membrane. These charges must be overcome to permit the cross-linking that will result in agglutination. Enzyme treatment of red blood cells with papain or ficin makes red cells more easily agglutinable. Such treatment may function by clearing the negative residues from the cell surface to permit more contact between them.

Antibody type. IgM antibodies are more efficient at agglutination because their size permits more effective bridging of the gap between cells caused by charge repulsion.

Electrolyte concentration and viscosity. The ionic strength of the medium used for the agglutination reaction can assist in reducing the negative surface charge of particles. This can be accomplished by addition of charged low molecular weight molecules, such as albumin, to the medium. The pH of the medium should be near physiological conditions. At neutral pH, high electrolyte concentrations act to neutralize the net negative charge of particles. Increasing the viscosity of the medium with polymerized molecules, such as dextran, also assists in bringing the charged particles together.

Antigenic determinants. As stated earlier, antigens with multiple antigenic determinants are necessary for agglutination. A monovalent antigen would not permit cross-linking. The placement of the antigenic determinants on the particle can also affect agglutinability. Antigenic determinants that are sparsely distributed will not be so easily cross-linked as antigenic determinants that are densely distributed. Antigenic determinants can also be inaccessible to antibody binding because they are buried within cell membranes. IgG antibodies are frequently unable to react with hidden antigenic determinants because such antigens increase the distance that antibody molecules have to bridge. Enzyme treatment of red blood cells, in addition to decreasing negative surface charge, may also increase agglutinability of red blood cells by exposing hidden antigenic determinants.

Temperature and time of incubation. At increased antigen concentration, results in the presence of antibody will be rapid. Agitation of the antigen suspension with antibody solution will increase reaction rate by increasing the surface area exposed to antibody. At lower antigen concentrations, reaction time can be shortened by centrifugation, which increases contact between the antigenic particles and the antibody. Concentration of reactants can therefore influence reaction time. Temperature of incubation is also an important variable. Some antigens are best bound by antibody at 37° C. These antigens include microbes. Some antigens react optimally with antibody at 4° C. These *cold agglutinins* include antibody to the i antigen of red blood cells. Optimal temperature for the agglutination reaction will vary with different antigen-antibody systems. Optimal binding temperature affects the behavior of these antibodies in vitro.

Direct agglutination

Direct agglutination involves detection of antibody to intrinsic antigens on the surface of red blood cells, microbes, or other particulate antigens. The titer of the serum reflects the concentration of the predominant antibody in the serum. Infection by bacteria that contain many different types of antigenic determinants produces antibody against a variety of these different antigens. The titer of the serum will reflect the level of the antibody in the greatest concentration.

Direct agglutination tests are frequently used in the immunologic diagnosis of microbial infections. Early detection of a high titer or documentation of a significant rise in titer are important tools in diagnosis. Antibodies to *Brucella* (brucellosis), *Salmonella* (typhoid fever), and *Proteus* (Rocky Mountain spotted fever) are detected in this way. Direct agglutination tests are also used in the typing of human red blood cells in the blood bank.

Different bacterial antigens may give different patterns of agglutination. Antibodies to bacterial flagella cause cross-linking of the flagella themselves. These antibodies cause formation of a loose, rapidly formed agglutinate. Antibodies to antigens in the body of the bacterium cause cross-linking of the organisms themselves. This results in a granular compact precipitate that develops more slowly.

Indirect agglutination

Indirect agglutination involves reaction of antibody with antigens that have been passively transferred onto the surface of particles (Fig. 10-9). Red blood cells, especially those from human, sheep, and turkey, are employed. Inert particles such as latex (0.81 μm in diameter) and bentonite (clay) are also used. Polysaccharide and some protein antigens, such as albumin and purified protein derivative

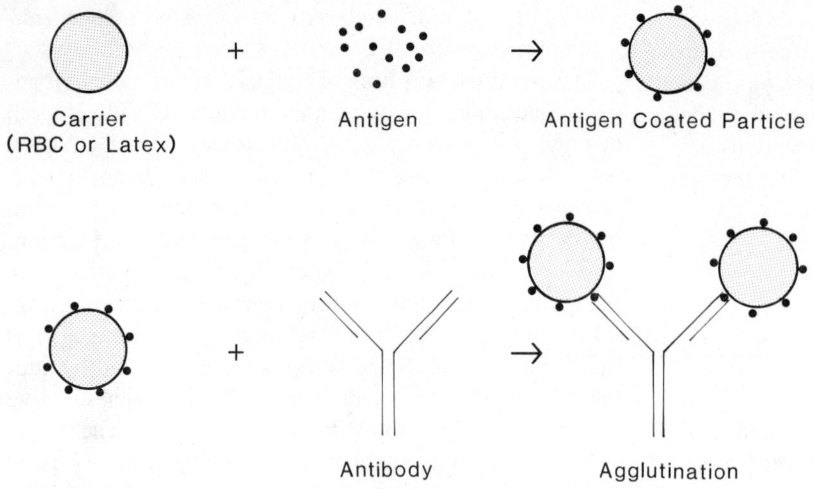

Carrier
(RBC or Latex)
Antigen
Antigen Coated Particle
Antibody
Agglutination

Fig. 10-9. Passive (indirect) agglutination reaction. Antigen is adsorbed onto surface of carrier particle, which is then agglutinated by antigen-specific antibody.

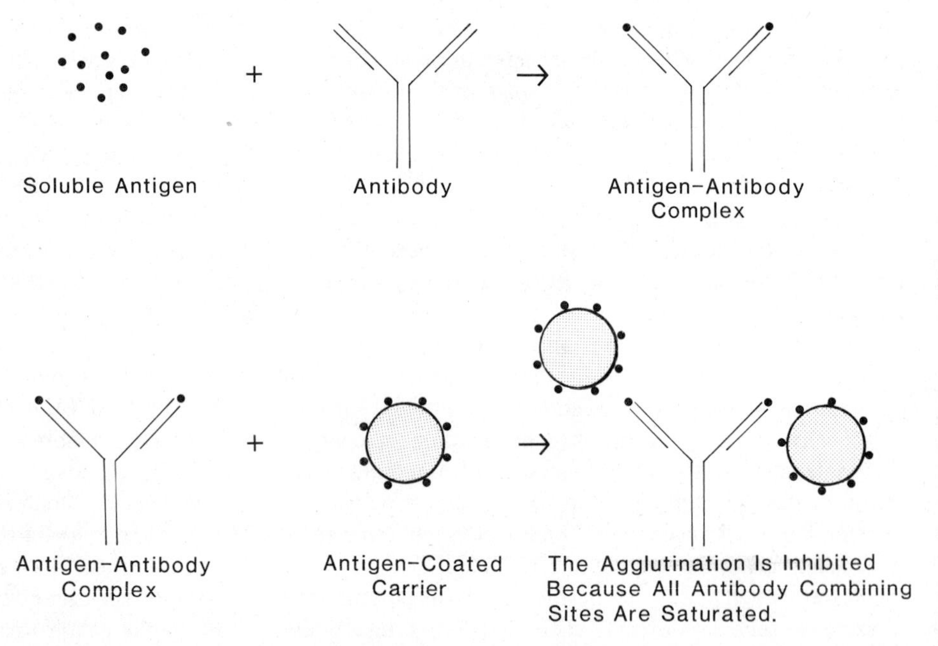

Fig. 10-10. Agglutination-inhibition reaction. Same reaction as that shown in Fig. 10-9 but inhibited by soluble antigen.

(PPD), are easily adsorbed onto the particle surface. Other antigens require pretreatment of particles for adsorption. Treatment of cells is done by incubation with tannic acid or chromium chloride, which modifies the surface of the cells. Treatment probably affects cell surface charge. Covalent bonding of antigen to cell surface by bifunctional molecules such as bisdiazobenzidine (BDB) or glutaraldehyde is also performed. Such molecules bind to the cell surface and then bind the antigen. Antigens are often applied to the surface of the cells at the highest concentration that does not produce nonspecific agglutination. This concentration is determined in independent experiments performed before antibody testing is done. Because the concentration of antigen can be artificially controlled, the sensitivity of indirect agglutination tests is somewhat greater than that of direct agglutination tests. Indirect agglutination procedures are used in the diagnosis of syphilis. The VDRL test (Venereal Disease Research Laboratory) employs cholesterol crystals coated with cardiolipin antigen. Detection of rheumatoid factor, useful in the diagnosis of rheumatoid arthritis, utilizes agglutination of latex particles coated with human IgG.

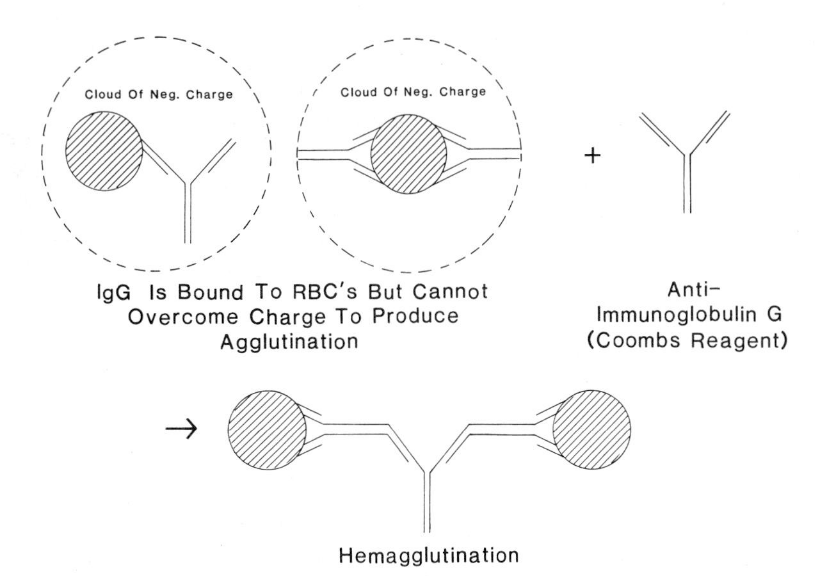

Fig. 10-11. Direct Coombs' test for antibody to red blood cells (RBC).

Agglutination inhibition

Agglutination inhibition is an adaptation of agglutination reactions that permit detection and quantitation of soluble antigen. The test concentrations of antibody and particles are carefully controlled to prevent antibody excess. Results are assessed by the competitive inhibition of agglutination by prior incubation of antibody with unknown antigen solution (Fig. 10-10). When antibody is incubated with the test antigen solution, antigen is bound to available antibody-combining sites. This antibody is then added to particulate antigen suspensions. The failure to agglutinate indicates that enough antibody-combining sites have been saturated with a soluble form of the same antigen so that insufficient antibody-binding sites are available for binding and cross-linking of the particulate antigen. Quantitation of the soluble antigen can be done by assessment of the degree of inhibition in serial dilutions of the antigen solution.

Confirmation of pregnancy is performed by an agglutination-inhibition assay. Human chorionic gonadatropin (HCG) is present in the urine of pregnant women. In the pregnancy test, patient urine is incubated with antibody to HCG. Latex particles coated with HCG are then added. If agglutination is inhibited, HCG is present in the urine and the patient from whom the specimen was obtained is pregnant. If no HCG is present in the urine, agglutination will occur and the patient is not pregnant.

Antiglobulin testing

Antiglobulin testing is a modification of agglutination reactions to permit detection of an incomplete (IgG) antibody, which may not produce agglutination even after binding to the particle surface. IgG is less effective at agglutination than IgM because its smaller size is less effective at bridging antigen particles. Addition of unagglutinated IgG-coated particles to an anti-immunoglobulin, or *Coombs' reagent,* bridges the gap between particles by bivalent binding (Fig. 10-11). This enables cross-linking to achieve agglutination. The *direct Coombs' test* is used in blood banks to determine the presence of IgG antibodies to red blood cells.

The *indirect Coombs' test* is a variation of the antiglobulin test and is used to detect free antibody to red blood cells in patient serum (Fig.10-12). Serum is screened against a panel of red blood cells of known and varied antigenicity. Agglutination of cells to which patient serum and antiglobulin reagent have been added indicates the presence of antibody in patient serum to an antigen present on the agglutinated cells.

Sample requirements

The reaction will measure components of plasma, serum, or cerebrospinal fluid. Urine must be buffered because of its usual acidity. Care must be taken to consider the possible role of complement of the plasma or serum samples to be tested. Specific directions regarding the need for inactivation or dilution to minimize the effect of this blood component should be followed.

Reagents

As cited above, particularly for red blood cell agglutination techniques, the factors influencing the reactions are particle charge, type of antibody, electrolyte concentration, and viscosity. These requirements will vary according to the type of assay. Red blood cells, when used fresh, have a shelf life of about 2 weeks. Therefore many manufacturers have developed fixed red blood cells (usually stabilized by tannic acid or glutaraldehyde) or latex beads to overcome the need to prepare the reagents every few weeks or to extend the reagent shelf life.

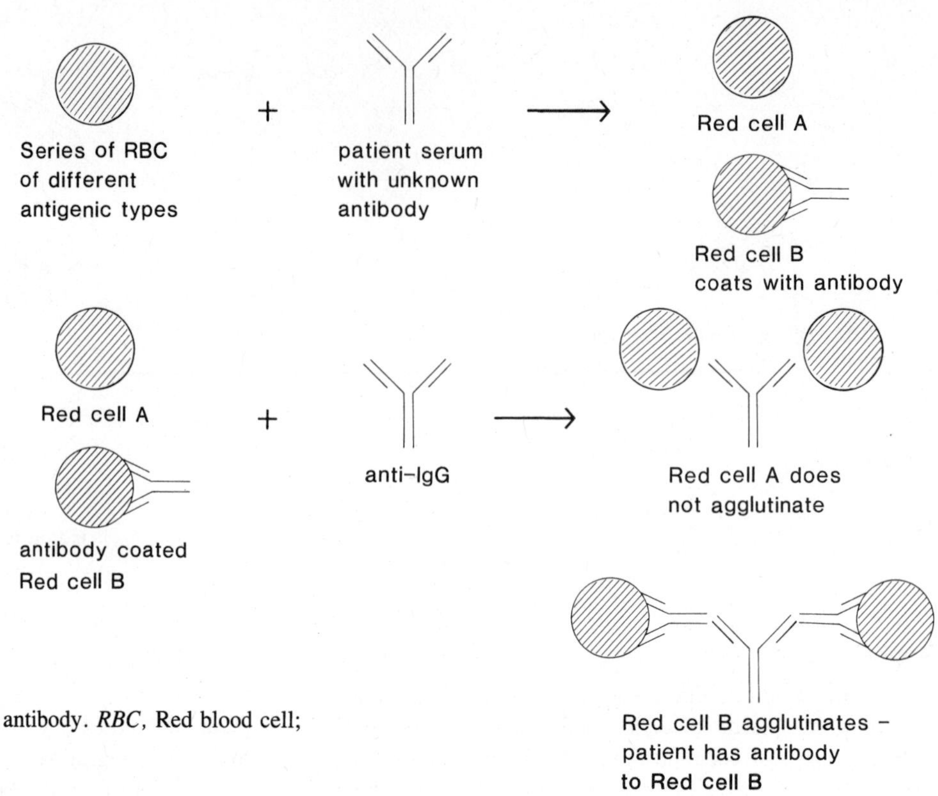

Series of RBC of different antigenic types + patient serum with unknown antibody →

Red cell A

Red cell B coats with antibody

Red cell A + anti-IgG →

antibody coated Red cell B

Red cell A does not agglutinate

Red cell B agglutinates – patient has antibody to Red cell B

Fig. 10-12. Indirect Coombs' test for antibody. *RBC,* Red blood cell; *anti-IgG,* anti–human Ig antibody.

Instrumentation

The great advantage of this technique is the simplicity of instrumentation. Results can be recorded by eye or with the aid of mirrors or magnifying glasses.

Common pitfalls

Antigen excess will often result in a prozone phenomenon with consequent false-negative results. Expired red blood cells or other reagents can result in lack of agglutination.

Limitations

The technique is only semiquantitative and thus allows estimates of the true value within a factor of two. For screening assays, such as pregnancy testing, the rapid assays may not have adequate sensitivity, and a slower, more sensitive reaction may be required to achieve the necessary sensitivity.

COMPLEMENT-FIXATION ASSAYS

The complement-fixation tests are probably the most sensitive of the immunologic procedures that were developed early in the history of immunology. Tests more sensitive than complement fixation, such as radioimmunoassay and enzyme immunoassay, have now been developed, but complement fixation tests are still important, especially in the diagnosis of fungal, viral, and parasitic infections and in the quantitation of functional complement levels (total hemolytic complement) and complement components.

Complement proteins

The term *complement* is used to denote a series of plasma proteins that are activated in sequence after antigen-antibody reactions. Not all classes and subclasses of immunoglobulin are capable of activating complement or can activate it to the same degree (see Table 9-2). The antigen-antibody complex used in complement-fixation tests must therefore involve an antibody that is capable of activating, or *fixing,* complement. As with agglutination reactions, IgM is more efficient in fixing complement than IgG. Because of its pentameric structure, IgM molecules are believed to be a thousandfold more efficient in fixing complement than monomeric IgG molecules when the two immunoglobulins are compared on a molar basis.

The first protein in the complement sequence is bound to a site in the interior of the immunoglobulin molecule in the second constant domain of the heavy chain (C_{H2}). This area is inaccessible on an unreacted or unbound antibody. After interaction and binding with antigen, conformational changes in the immunoglobulin molecule cause the hinge region to open and the complement-binding site to be exposed. The first complement protein binds to this region. Subsequent complement proteins are then activated and

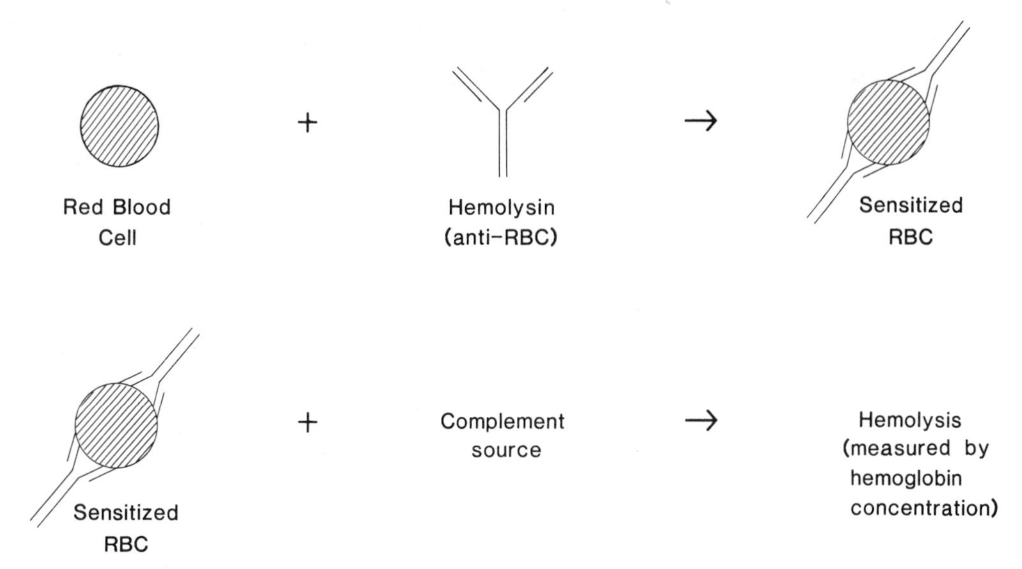

Fig. 10-13. Complement fixation—one-stage testing. For measurement of total hemolytic complement, test serum is added as source of complement. For measurement of complement components as complement source, one uses test serums, to which are added purified complement components; for example, to measure complement component 3, one adds all components, except C3, in excess to test serum. Therefore reaction is limited only by concentration of C3 in test serum.

can bind to the membrane of the cell to which the antigen-antibody complex is bound. Binding of the complete sequence of nine complement proteins results in small defects in the membrane of the cells. Cytoplasmic cell contents are lost through these holes, and extracellular fluid is admitted. This results in hypotonic swelling of the cell and ends in cell lysis. If the cells used in the test are red blood cells, cell lysis results in release of hemoglobin into the medium. The amount of hemoglobin, which can be quantitated spectrophotometrically, is directly proportional to the amount of complement fixed to the surface of the cells.

One-stage testing

When used to assess total complement levels (the level of complement activity in the serum that reflects the quantities of all nine major complement proteins) or to assess separately the levels of the nine individual major complement proteins, a one-stage test system is used. A constant volume of red blood cells, usually derived from sheep, is added to a constant amount of anti–sheep red blood cell antibody, or *hemolysin*. A source of complement is added. For total complement measurements, dilutions of the unknown serum are used. For measurement of complement components, serum with an added excess of all other complement components but the component to be measured is added. The degree of hemolysis is measured in a spectrophotometer and is directly proportional to the amount of complement available for fixation in the test serum (Fig. 10-13).

Some immunologically mediated diseases cause a decrease in total serum complement levels or in the levels of individual complement components. In addition, hereditary deficiencies of certain components of the complement system have been described.

Two-stage testing

Two-stage complement-fixation testing is used to measure antigen or antibody. In the first reaction, antigen and antibody are incubated with a known amount of complement. In the second stage, the residual complement activity in this solution is determined by an indicator system. The indicator is a suspension of sheep red blood cells that are coated with hemolysin. The degree of hemolysis of the indicator cells is inversely proportional to the amount of complement fixed in the first reaction (Fig. 10-14).

The first reaction can be adapted so that the complement-fixation reaction measures antigen or antibody. To determine antigen levels, one uses a constant volume of antibody. To determine antibody levels, a constant amount of antigen is used.

Complement-fixation tests to determine the presence and titer of an antibody in an unknown serum frequently employ antigens that have been fixed to the surface of red blood cells. In this test, a sequence of two complement-mediated hemolysis tests are performed and measured (Fig.10-15). Antigen bound on red blood cells is incubated with test serum. A known amount of complement is then added, and the degree of hemolysis is measured. After the

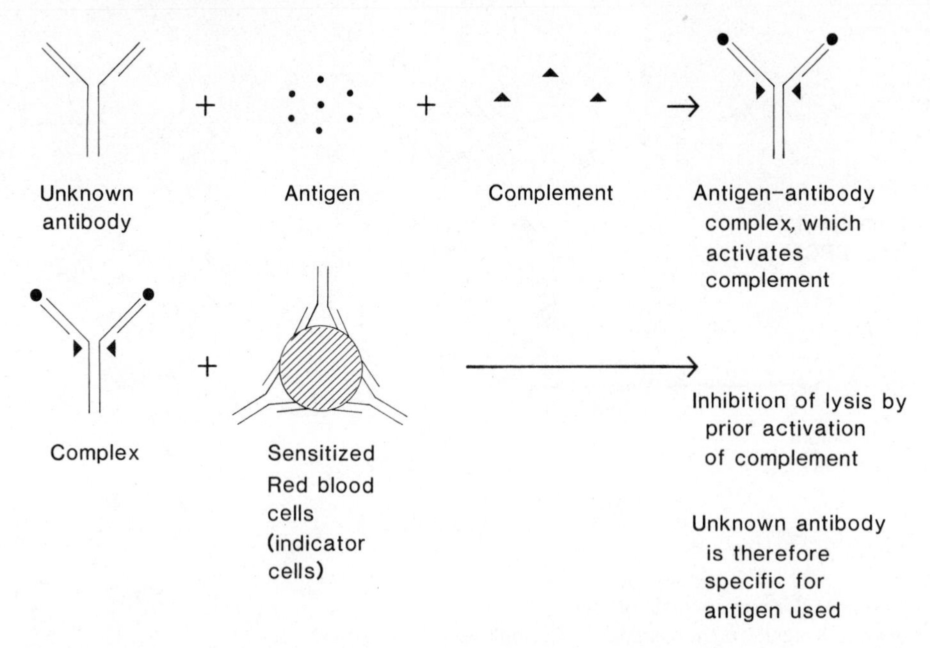

Fig. 10-14. Complement fixation—two-stage testing. Antigen or antibody can be measured by holding constant all but the variable to be tested, in this case unknown antibody.

measurement, hemolysin-treated red blood cells are added. The residual complement in the test mixture will lyse the hemolysin-treated indicator cells. The difference in complement activity between the two reactions is inversely proportional to the antibody level present in the original reaction.

Sample requirements and preparations

Plasma is not an appropriate sample for testing because the clotting proteins or the anticoagulants can bind complement. Urine is not an appropriate sample unless the analyte is extracted or the specimen is highly diluted.

Serum and cerebrospinal fluid may be analyzed, but certain precautions are necessary. The samples cannot be hemolyzed. For detection of antibody, the endogenous complement of the sample must be inactivated. Usually this is accomplished by heating of the specimen for 15 to 30 minutes at 56° C. Antigens, if obtained from serum or other fluids, must also be free of endogenous complement. For unknown reasons some antigens react with complement without the need for specific antibody and thus cannot be measured.

Reagents

Exogenous complement must be prepared daily and cannot be stored. The complement activity varies significantly from batch to batch and must be standardized daily.

Instrumentation

Only a spectrophotometer is necessary.

Limitations

Complement-fixation tests are sensitive to many variables. They are inhibited by anticomplementary activity (factors that inactivate or interfere with any of the complement proteins) in serum, including factors such as circulating immune (antigen-antibody) complexes, lipemic sera, aggregated immunoglobulins, and heparin.

It is critical to keep all components in the test constant except the one that is to be measured. Red blood cell number, concentration of complement, and concentration of antigen should be rigorously defined for antibody determinations.

Tests are influenced by the instability of some complement proteins, variability in red blood cells, variation between lots of hemolysin, and the narrow range of optimal reactivity for many reagents. The need for fresh reagents and the great variability make this assay one of the most difficult to be performed by a laboratory.

The Centers for Disease Control (CDC) evaluates complement-fixation reagents and has developed standardized procedures for complement-fixation tests.

INDICATOR-LABELED IMMUNOASSAYS

This section describes several immunoassays that use indicator molecules attached to either the reagent antibody or antigen, to demonstrate that an antibody-antigen reaction has taken place. These indicator molecules can be enzymes, fluorescing molecules, or radioactive compounds. The two immunoassays described below use enzymes or fluorescein as the indicator. Assay formats that use radioactive compounds are described in the competitive-binding

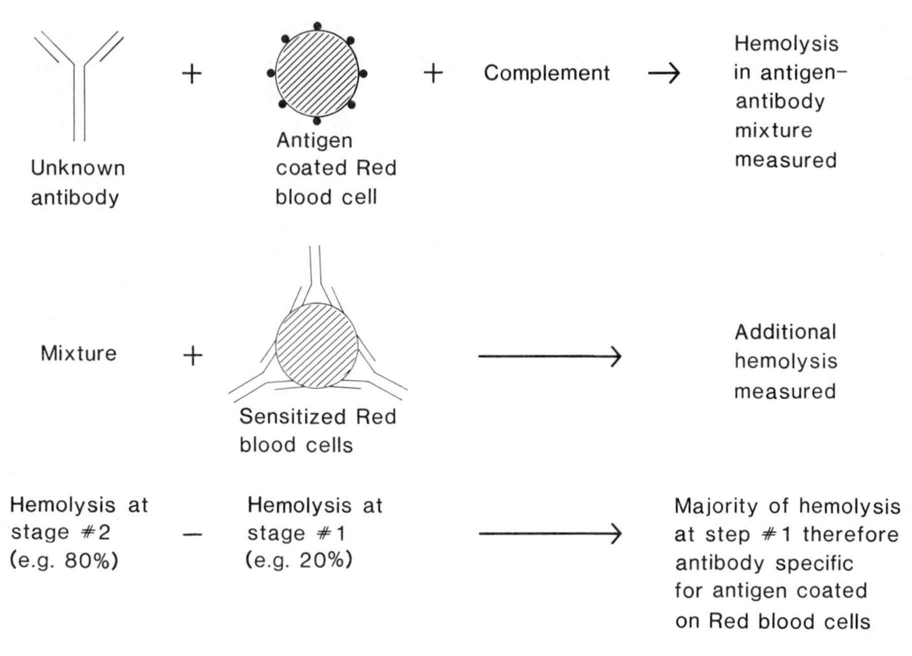

Fig. 10-15. Complement fixation using antigen adsorbed onto red blood cells. Testing for specificity of unknown antibody.

assay section of this book. Indicator-labeled immunoassays are very popular in the laboratory today for many reasons. These assays use minimal amounts of reagents, are rapid in producing test results, have maximum sensitivity compared to other immunoassays, and are suited to doing large numbers of tests at a single time. Enzyme and fluorescent immunoassays have an additional advantage over radioimmunoassays in that they do not require all the special precautions necessary with the handling of radioisotopes.

Indicator-labeled immunoassays are usually designed to detect the first stage of an antibody-antigen reaction rather than the second stage as described in the immunoprecipitation and agglutination immunoassays. First-stage detection has several advantages in that the rate of reaction is faster, the obligatory requirement for electrolytes to support the detection stage is not necessary, and reagent antibody can be monovalent rather than polyvalent. These types of immunoassays are usually quantitative procedures in the clinical chemistry laboratory and are sensitive to the microgram and nanogram range. Qualitative indicator-labeled immunoassays are very popular in the serology laboratory for the detection of antibodies to infectious organisms and for the characterization of autoimmune antibodies such as antinuclear antibody (ANA) and antithyroid antibodies (ATA).

FLUOROIMMUNOASSAY

Two types of quantitative fluorescein-labeled immunoassays are used routinely in the clinical laboratory. These assays are used primarily for the measurement of serum proteins in biological fluids, but are sensitive enough to be adapted for the measurement of hormones and drugs. One type of immunoassay system measures the direct uptake of labeled antibody by the antigen after the antigen is immobilized on a solid phase. The amount of labeled-antibody uptake is directly proportional to the amount of antigen present. This format is termed *direct fluoroimmunoassay*. The other type of immunoassay measures the residual amount of fluorescent-labeled antibody that can be reacted with a known amount of antigen fixed to a solid phase after first reacting with the antigen in the test specimen. This format is called *indirect fluoroimmunoassay*. The indirect format is often more sensitive than the direct method for the detection of small amounts of antigens. Common to both the indirect and direct format is the use of a fluoresceinated antibody.

In both assays, the indicator-reagent antibody is usually conjugated or covalently linked with a fluorochrome molecule. The fluorochrome molecule is a chemical that can absorb electromagnetic energy of short-wavelength light (200 to 400 nm) and then instantaneously emit light at a longer wavelength of the visible spectrum (400 to 700 nm). Intensity of the emitted visible light is the measurable indicator in this assay. The most popular fluorochrome is fluorescein-isothiocyanate, often abbreviated FITC, which can be conjugated easily to free amine groups on the reagent antibody.

The indirect assay system is available commercially in several kit forms that can be used to quantitate many serum proteins in biological fluids. It has not had the popular success of nephelometric immunoassays but is equivalent in sensitivity and precision.

Principle of direct assay

Several formats are available for the direct fluoroimmunoassay, but the procedure is generally a two-step assay. The first step is the removal of the antigen from the biological fluid by immobilization onto a solid surface. This can be done either by incubation of the biological fluid that contains the antigen with excess specific antibody that is attached to inert polyacrylamide beads or by adsorbance of the antigen onto a specially prepared adsorbent material. The solid phase is then washed extensively to remove all unattached material. The second step is to incubate the bound antigen with FITC-labeled reagent antibody. The antibody forms a complex with the antigen present on the solid surface. (If antibody is used in the first stage to immobilize the antigen, the complex that is formed is antibody-antigen-FITC-labeled antibody.) The solid-phase complex is then washed extensively to remove unreacted labeled antibody, and the amount of FITC present is measured in a fluorometer or spectrophotofluorometer. Standard concentrations of antigen are treated in the same manner and compared with the amount of fluorescence in the patient's sample. The relative amount of fluorescence is directly proportional to the concentration of antigen bound to the solid phase (Fig. 10-16).

Principle of indirect assay

The initial step of this assay is to react excess FITC-labeled specific antibody with the antigen in test solution. After incubation, the solution contains some unreacted antibody and antibody-antigen complexes. The solution is then incubated with a special immunoadsorbent that will bind the unreacted antibody in the solution. The immunoadsorbent contains a known amount of antigen that will react with the excess FITC-labeled antibody that has not reacted with the antigen in solution. The fluorescence of the bound labeled antibody is measured in a fluorometer. The intensity of the fluorescence is inversely related to the amount of antigen present in the test solution (Fig. 10-17).

Sample requirements and preparation

As previously described, all precautions should be taken to preserve the antigens by proper collection and storage before measurement. Serum and cerebrospinal fluid are the usual samples that are collected.

Reagents

Antibody. The most important reagent in this type of immunoassay is the reagent antibody. Several points are important concerning this key reagent. First, the specificity of the reagent antibody is crucial. The antibody should be monospecific and not contain any other antibody specificity that might react with other antigens in the test mixture. These contaminating antibodies can raise the background of fluorescence to a level that will dramatically reduce the sensitivity of the assay system. Second, the reagent should be properly labeled with the fluorescein molecule. The antibody should be purified to a great extent to ensure the fluorescein molecule will label only the antibody molecule and not other proteins that are present in the unfractionated

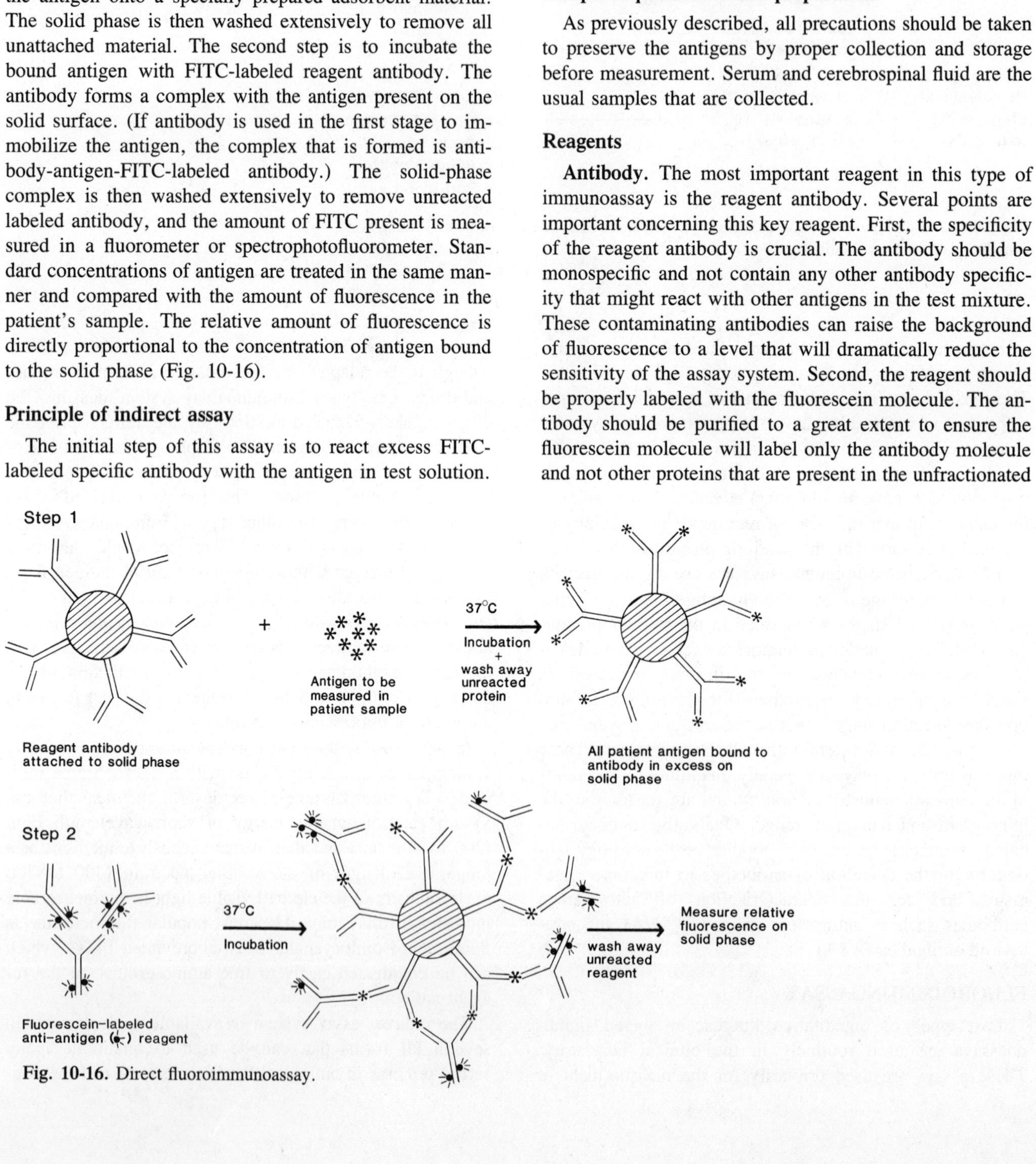

Step 1
Reagent antibody attached to solid phase
+
Antigen to be measured in patient sample
37°C
Incubation + wash away unreacted protein
All patient antigen bound to antibody in excess on solid phase

Step 2
Fluorescein-labeled anti-antigen (✱) reagent
37°C
Incubation
wash away unreacted reagent
Measure relative fluorescence on solid phase

Fig. 10-16. Direct fluoroimmunoassay.

antiserum. Fluorescein labeling of other protein in the mixture along with the reagent antibody can significantly raise the background of fluorescence and cause decreased assay sensitivity. Monospecificity and proper labeling of the reagent antibody are keys to a successful immunofluorescent assay. Third, the antibody should be conjugated with a sufficient amount of fluorochrome to maximize the sensitivity of this reagent. Care should be taken not to overconjugate the antibody with fluorochrome. This will cause the labeled antibody to aggregate freely into large complexes that can nonspecifically adsorb to the solid-phase elements of this assay. This nonspecific adsorption also greatly reduces the sensitivity of the assay by causing increased fluorescence, which is not attributable to specific antibody-antigen interaction.

Buffer solutions. The buffers used in this assay must not only be able to facilitate the antibody-antigen reaction but also potentiate the fluorescing characteristic of the fluorochrome. FITC absorb and emit light best in an alkaline solution at a pH greater than 7.6. The buffer solution used to support the antibody reaction is usually at a pH of 7.2, whereas the buffer solution used in the measurement of the fluorochrome is at a pH of around 8.0.

Instrumentation

Both the indirect and direct fluoroimmunoassays require a fluorometer or spectrophotofluorometer to obtain accurate reading of the fluorochrome-labeled antibody. The more sensitive the required measurement, the more sophisticated the instrument needed to detect and record the fluorometric signal accurately. Thus for very sensitive measurements devices such as platen counters are required.

Common pitfalls

1. Improper labeling of antibody with fluorochrome. Too much conjugation or too little conjugation of the reagent with the fluorochrome can cause the assays to function improperly.
2. Buffers. Often, buffers are improperly prepared or become contaminated with bacteria and quickly become unable to buffer in the appropriate pH range.
3. Improper storage of reagent. The fluorochrome-labeled antibody should not be repeatedly frozen and thawed. This manipulation causes the reagent to become denatured and aggregated. Even prolonged storage under appropriate conditions may cause aggregates to form. These can be removed by periodic filtering of the reagent to be used in the assay.
4. Instrument malfunction.
5. Improper storage of test specimen.

Test sensitivity and coefficient of variation

Fluoroimmunoassays have a sensitivity for measurement to a lower limit of approximately 100 ng/mL and an upper limit of 200 mg/mL. The between-run variation for the direct assay is between 5% to 10%, whereas the indirect assay is usually between 4% to 8%. The variation observed within the same run is reported to be 3% to 6%.

ENZYME IMMUNOASSAY

Enzymes are biological catalysts that convert a substrate into some product. In participating in this reaction, the enzyme is not consumed. A single enzyme can convert millions of substrate molecules into product molecules. This enormous amplification means a very small quantity of en-

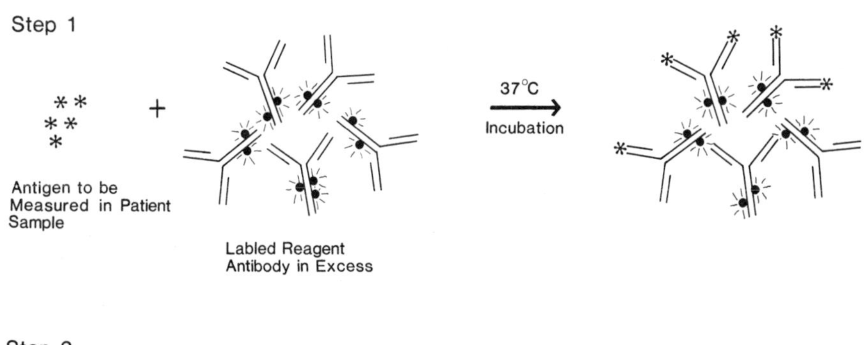

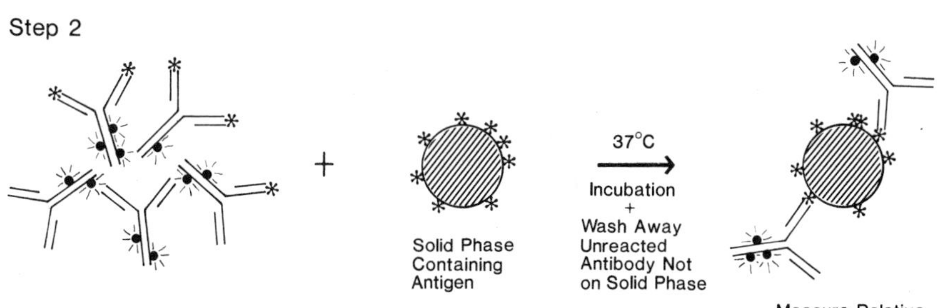

Fig. 10-17. Indirect fluoroimmunoassay.

zyme can be detected. When an enzyme is used to label an antibody or an antigen, very minute quantities of the antibody-antigen complex can be detected. Numerous applications of enzyme immunoassays have been developed in the past decade such that these assays rival radioimmunoassay as the preferred method for detection of small quantities of antigens in biological fluids. Enzyme immunoassays are not restricted by the stringent safeguards and requirements that are associated with the handling of radioisotopes of the radioimmunoassay systems. Other advantages of enzyme immunoassays over radioimmunoassays include longer shelf life of the reagents, and the elimination of the radiation-counting instruments.

Numerous configurations for enzyme immunoassays have been described in the literature and are available in commercial kits. This section will describe the assays that are considered under the category of heterogeneous immunoassays.

Homogeneous immunoassays are described in the sections on competitive-binding assays. The heterogeneous enzyme immunoassays are those procedures that require the physical separation of the antibody-antigen complex from the unbound constituents so that measurement of enzyme activity associated with either the complex or unbound constituents can be made. Homogeneous enzyme immunoassays do not require this physical separation for final measurement because either the enzyme or substrate is part of the reaction. Heterogeneous assays can be subdivided into those assays in which the antibody is labeled with enzyme and those in which the antigen is labeled. One type of assay for antigen involves coating of a surface with antibody, followed by reaction with the test antigen and developing the extent of the reaction with enzyme-labeled antibody. This is frequently referred to as "sandwich" type of methodology. A second configuration frequently used is a competitive type in which the antigen is labeled with enzyme. The labeled antigen competes with unlabeled antigen in the test sample for sites on an antibody that is immobilized on a solid surface, such as polystyrene, latex, or iron.

The appropriate enzyme selected to label an immunochemical reagent must have certain qualities. The enzyme should be inexpensive to obtain and be purified to a high degree. It should also have a specific activity associated with it (see Chapter 49). The enzyme should not be present in biological fluids in concentrations sufficient to interfere with the detection of the antigen. Contaminating enzyme in the biological fluid would produce reasonably high backgrounds of substrate conversion such that the sensitivity of the assay would be greatly decreased. This last condition is not an essential for heterogeneous assays but is crucial for homogeneous assays. Several enzymes fulfill most of the above requirements and have been successfully used in enzyme immunoassays. These are horseradish peroxidase, alkaline phosphatase, and glucose oxidase. Selection of the appropriate enzyme to label the immunochemical reagents is often empirical, and each has a distinct advantage or disadvantage associated with it.

The enzyme is conjugated (coupled) to the antibody or the antigen by use of bifunctional reagents, that is, those that have two reactive sites. The most frequently used reagent to combine the enzyme to the appropriate antibody or antigen molecule is glutaraldehyde. Care must be taken to preserve the enzyme activity so that the assay retains its sensitivity for low concentrations of the test sample. Loss of activity results in loss of sensitivity.

Principles of enzyme immunoassays

Sandwich technique (antibody labeled). This technique can be used to measure either antigen or antibody. Usually there are three layers. For antigen measurement, these are antibody in the first layer, antigen (test sample) second layer, and antibody labeled with enzyme third layer. For antibody measurements, the three layers are antigen first layer, antibody (test sample) second layer, and second antibody labeled with enzyme third layer. Each of these assay systems are discussed in detail.

For the antigen-measuring system, two different molecules of antibody must bind to the antigen. Thus only large antigens, such as proteins, can be measured by this system. Fig. 10-18 is a schematic of this assay. In the first step, antibody of the desired specificity is immobilized to a solid surface, which may be the wells in a microtiter plate or a plastic test tube. The microtiter well or test tube is washed to remove all unreacted materials and may then be coated with other material (proteins) to minimize nonspecific reactions with subsequent possible false-positive results. In the second step, the fluid containing the antigen is reacted with the immobilized antibody. All nonreacting material is washed away. One should note that, for quantitative results, the amount of antigen added in the test sample should not exceed the antibody-binding capacity. The third step is to react the enzyme-labeled antibody with the antigen that has now been immobilized by the antibody on the solid phase. All unreacted enzyme-labeled antibody is then washed away and substrate with appropriate cofactors is added so that the enzyme on the antibody can then convert the substrate to the product. The amount of product is then measured by a color change or color reaction. The intensity of the color is directly proportional to the amount of antigen that has been immobilized on the solid surface by reaction with the antibody. This format is used most often to detect polyvalent antigens such as various serum proteins found in biological fluids. These polyvalent antigens contain multiple antigenic determinants that can react with the antibody on the solid phase and the enzyme-labeled antibody as well.

For an antibody-measuring system, the antigen must be first immobilized on an insoluble matrix, such as a plastic surface or bead. Most often microtiter plates are used. Fig. 10-19 is a scheme of this assay. The appropriate immobi-

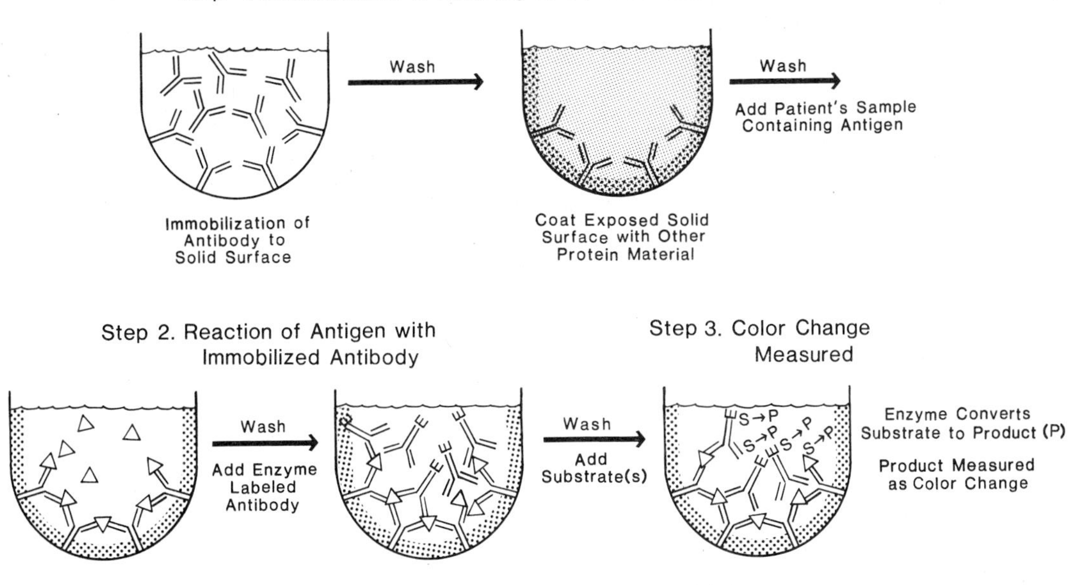

Fig. 10-18. Enzyme immunoassay. Sandwich technique with antibody label.

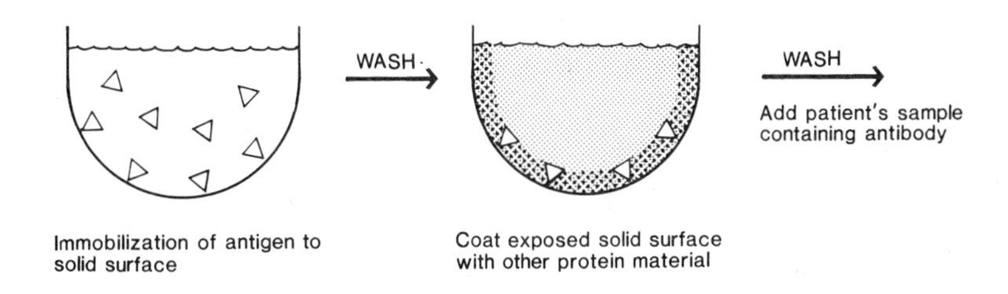

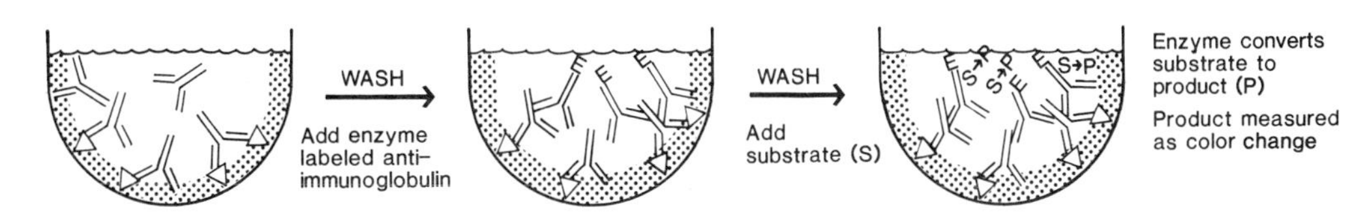

Fig. 10-19. Enzyme immunoassay. Detection of IgE specific for an allergen.

lization with retention of antigenic reactivity is the first step of this procedure. In the second step, the biological fluid containing presumptive antibody toward the immobilized antigen is allowed to react. Any antibody present will be bound to the antigen immobilized on the solid phase. After separation of the unreacted components by washing of the support surface, the presence of antibody is detected and quantitated by addition of enzyme-labeled anti-antibody that is directed to the class specificity of the antibody under consideration. Finally, the enzyme reaction is quantitated. This format is used to measure IgE in serum specific for an allergen (the so-called RAST test). It is important to note that if the antibody concentration exceeds the number of antigenic sites on the surface, the reaction will not be quantitative. Also the enzyme-labeled anti-antibody must be present in excess over the antibody.

Competitive-binding assays. The simplest competitive-binding assay uses labeled antigen (Fig. 10-20). The enzyme-labeled antigen is mixed with the test solution containing an unknown amount of the antigen. The solution containing the labeled and unlabeled antigen is allowed to react with a limited amount of antibody bound to a solid

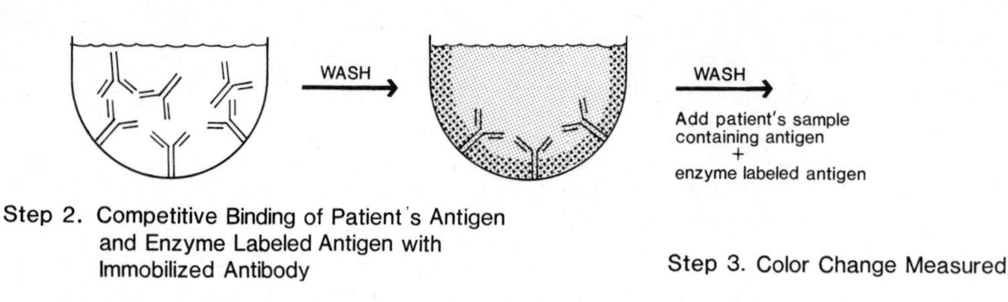
Step 1. Immobilization of Antibody to Solid Surface
WASH
WASH
Add patient's sample containing antigen
+
enzyme labeled antigen
Step 2. Competitive Binding of Patient's Antigen and Enzyme Labeled Antigen with Immobilized Antibody
WASH
Add substrate (S)
Step 3. Color Change Measured
Enzyme converts substrate to product (P)
Product measured as color change

Fig. 10-20. Enzyme immunoassay. Competitive binding.

matrix. One removes unbound antigen (both labeled and unlabeled) by washing and measures the amount of labeled antigen by determining the amount of enzyme bound to the solid surface. This assay is always performed in antigen excess. The concentration of the antigen present in the test sample is inversely proportional to the intensity of the color measured. This format of enzyme immunoassay can be used to detect hapten groups or small molecular antigens found in biological fluids such as drugs and hormones.

Sample requirements and preparation

All the precautions required to preserve antibodies and antigens should be taken when one is using enzyme immunoassays. It is particularly important to establish that the antigen or antibody does not bind nonspecifically to the surface. In some systems where antibody is immobilized or the serum is mixed with coupled antibody, rheumatoid factor may react, yielding false-high values.

Reagents

Enzyme-labeled reagents. Enzyme-labeled reagents must be selected so that the coupled enzyme product will have a high specific activity.

Microtiter plates and test tubes. Since the assays described above are performed primarily with the use of a solid phase (an immobilization of the analyte and reagents into a solid matrix), care should be taken to use the proper type of plastic that will hydrophobically attach these reagents so that, once bound, they will not be removed by wash solutions. The quality of plastic varies from lot to lot and from manufacturer to manufacturer. Recently, several manufacturers have developed plastic microtiter plates and test tubes that are used specifically for enzyme immunoassays.

Substrate. Whichever enzyme is used in these assays, the appropriate substrate specific for the enzyme should be selected to maximize the catalytic activity of the enzyme. Care should be taken to add sufficient amount of substrate so that if large quantities of enzyme are bound to the solid matrix, full color development can occur and substrate is not depleted.

Buffers. Close attention should be paid to the ionic strengths, pH, and composition of the buffer used to promote the enzyme's activity. Frequently, specific cations are required for appropriate enzyme activity.

Instrumentation

Measurement of color changes that are a result of enzyme conversion of substrate into product is usually accomplished by use of a spectrophotometer. More recently, the use of microtiter plates, where the volume of fluid in which the color change has occurred is very small (such as 100 to 200 μL), has made it necessary to use special spectrophotometers. These are referred to as "microtiter plate readers." The drawback to the microtiter plate readers, however, is that they are frequently not so sensitive as a standard spectrophotometer is.

Interfering substances

As indicated above, care should be taken to eliminate substances that would interfere with the enzyme's activity, such as enzyme inhibitors, antibodies, or complement which might react binding an artificial matrix.

Common pitfalls

1. Factors that must be considered in successful measurement of the enzyme activity in these assays are tight control of temperature, pH, ionic strengths of the buffer, and various cofactor concentrations necessary for the en-

zyme to convert substrate into product. Finally, since enzymes are protein and can be subject to rapid denaturation under improper incubation conditions, close attention must be paid to preserve the enzyme activity.

2. Improper storage of immunochemical reagents. Care should be taken to preserve the enzyme activity of the labeled reagents.

3. Inappropriate plastic used to manufacture the microtiter plates or test tubes.

4. Substrate depletion. Often, if a high quantity of enzyme-labeled reagents are present on the solid matrix, substrate can be depleted very rapidly. One must take care to have sufficient substrate so that full color development can occur. The upper limits of linearity should be carefully defined.

5. Improper pH and ionic strengths of buffer.

6. Nonspecific adsorption of reactants to plastic surface. This nonspecific adsorption can be minimized by use of proteinaceous material, such as gelatin or bovine serum albumin, after initial adsorption of the reagents to the solid phase.

Test sensitivity and coefficient of variation

Numerous formats are described in the use of the enzyme immunoassays. The sensitivity and coefficient of variation of each of these immunoassays is dependent on the format selected and the instrumentation used to measure the color change. Enzyme immunoassays are generally considered to be sensitive to the nanogram level of detection. An acceptable enzyme immunoassay should have a coefficient of variation of less than 10% in the nanogram range.

SUMMARY

In this chapter, numerous immunological techniques are described and are summarized partially in Table 10-1. Each of these procedures was developed to meet a specific need to identify or quantitate an antigen present in a patient's sample. The immunoprecipitation techniques are the least sensitive but have the highest degree of specificity in determining the presence of an antigen. Immunonephelometry is the most sensitive of the quantitative assays using precipitation as an end point and is fast growing to be the most popular form of assay to quantitate serum and cerebrospinal fluid protein. Recent advances in instrumentation have made this technique the most precise of the immunological techniques for quantitation of antigen and is the method of choice over the radial immunodiffusion and "rocket" immunoelectrophoresis techniques. Immunoelectrophoresis and two-dimensional immunoelectrophoresis are strictly qualitative tools. The former is very commonly employed to assess the quality of an antigen present in a patient's sample (such as polyclonal IgG versus monoclonal IgG, or C3 breakdown versus C3 present in native form).

The agglutination and complement-fixation techniques are procedures used mostly in the serology and blood bank laboratories, but the agglutination-inhibition technique is used widely in the chemistry laboratory to screen for the presence of human chorionic gonadotropin (HCG) in urine of women suspected of being pregnant. This assay format is now available to the general public in the home pregnancy test kits available at the local drug store.

Indicator-labeled immunoassays are becoming more popular in the clinical laboratory as quantitative tools because they extend the sensitivity of antigen detection into the range of nanogram quantities. Enzyme immunoassays are the most popular for reasons previously cited. The assay formats are many, and numerous kits are commercially available to measure minute quantities of serum proteins, hormones, and drugs.

Selection of the appropriate immunological technique to be used for detection of an antigen in the laboratory is dependent on many variables: technologist's skill, instrumentation, volume of samples to be tested, availability of test format in kit form, quality control reference samples of antigen, ease of the technique, and cost to perform the assay. Whatever assay format is selected, the crucial factors that must always be considered are the specificity of the reagent antibody, understanding of the antigen's structure, and sample preservation.

BIBLIOGRAPHY

General

Carpenter, P.L.: Immunology and serology, ed. 3, Philadelphia, Pa., 1975, W.B. Saunders Co.

Hudson, L., and Hay, F.C.: Practical immunology, ed. 2, Oxford, England, 1980, Blackwell Scientific Publications.

Nakamura, R.M.: Immunopathology: clinical laboratory concepts and methods, Boston, Mass., 1974, Little, Brown & Co.

Natelson, S., Pesce, A.J., and Dietz, A.A., editors: Clinical immunochemistry: chemical and cellular basis and application in disease, Washington, D.C., 1978, The American Association for Clinical Chemistry.

Rose, N.R., and Friedman, H., editors: Manual of clinical immunology, ed. 2, New York, 1980, American Society of Microbiology.

Weir, D.M., editor: Handbook of experimental immunology, ed. 3, Oxford, England, 1978, Blackwell Scientific Publications.

Immunodiffusion

Crowle, A.J.: Immunodiffusion, ed. 2, New York, 1973, Academic Press, Inc.

Ouchterlony, O.: Handbook of immunodiffusion and immunoelectrophoresis, Ann Arbor, Mich., 1968, Ann Arbor Science Publishers, Inc.

Rose, N.R., and Bigazzi, P.E., editors: Methods in immunodiagnosis, New York, 1973, John Wiley & Sons, Inc.

Immunoelectrophoresis

Grabar, P., and Brutin, P.: Immunoelectrophoresis analysis, New York, 1964, American Elsevier Publishing Co., Inc. (English translation).

Penn, G.M., and Batya, J.: Interpretation of Immunoelectrophoretic patterns, Chicago, Ill., 1978, American Society of Clinical Pathologists.

Ritzmann, S.E., and Daniels, J.C., editors: Serum protein abnormalities: diagnostic and clinical aspects, New York, 1975, Little, Brown & Co.

Counterimmunoelectrophoresis

Blumberg, B.S., Sutnick, A.I., and London, W.T.: Australia antigen as a hepatitis virus, Am. J. Med. **48:**1, 1970.

Palmer, D.F., Kaufman, L., Kaplan, W., and Cavallaro, J.J.: Serodiagnosis of mycotic diseases, Springfield, Ill., 1977, Charles C Thomas, Publisher.

Prince, A.M., and Burke, K.: Serum hepatitis antigen (SH): rapid selection by high voltage immunoelectrophoresis, Science **169:**593-595, 1970.

Two-dimensional immunoelectrophoresis

Arvan, D.A., and Shaw, L.M.: Two-dimensional immunoelectrophoresis technique and applications in the clinical laboratory, Separation Science **8:**123, 1973.

Ressler, N.: Two-dimensional electrophoresis of protein antigens with an antibody-containing buffer, Clin. Chim. Acta **5:**795, 1960.

Weeke, B.: Crossed immunoelectrophoresis. In Axelsen, N.H., Kroll, J., and Weeke, B., editor: A manual of quantitative immunoelectrophoresis, Scand. J. Immunol. **2**(Suppl. 1):47-56, 1973.

Radial immunodiffusion

Fahey, J.L., and McKelney, E.: Quantitative determination of serum immunoglobulins in antibody-agar plates, J. Immunol. **94:**84, 1965.

Mancini, G., Vaerman, J.P., Carbonara, A.D., and Heremans, J.F.: A single radial immunodiffusion method for the immunological quantitation of proteins. In Peeters, H., editor: Protides of the biological fluid, 11th colloquium, Amsterdam, 1964, Elsevier Publishing Co.

Ritzman, S.E., Fisher, C.L., and Nakamura, R.M.: Quantitative immunochemical procedures. In Ritzman, S.E., and Daniels, J.C., editors: Serum protein abnormalities: diagnostic and clinical aspects, Boston, Mass., 1975, Little, Brown & Co.

Laurell rocket immunoelectrophoresis

Laurell, C.B.: Quantitative estimation of proteins by electrophoresis in agarose gel containing antibodies, Annal. Biochem. **15:**45, 1966.

Weeke, B.: Rocket immunoelectrophoresis. In Axelsen, N.H., Kroll, J., and Weeke, B., editors: A manual of quantitative immunoelectrophoresis, Scand. J. Immunol. **2**(Suppl. 1):37-46, 1973.

Immunonephelometry

Cambasio, C.L., Leek, A.E., de Steenwinkel, F., Billon, J., and Masson, P.L.: Particle counting immunoassay (PACIA). I. A general method for the determination of antibodies, antigens and haptens, J. Immunol. Methods **18:**33, 1977.

Cambiaso, C.L., Riccomi, H., Masson, P.L., and Heremans, J.F.: Automated nephelometric immunoassay. II. Its application to the determination of hapten, J. Immunol. Methods **5:**293-302, 1974.

Finley, P.R., Dye, J.A., Williams, J., and Lichti, D.A.: Rate-nephelometric inhibition immunoassay of phenytoin and phenobarbital, Clin. Chem. News **27:**405-409, 1981.

Galvin, G.P., Looney, C.E., Leflar, C.C., et al.: Particle enhanced photometric immunoassay systems, Proc. New Directions in Clin. Lab. Assays, La Jolla, Calif., 1982.

Gitlin, D., and Edelhoch, H.: A study of the reaction between human serum albumin and its homologous equine antibody through the medium of light scattering, J. Immunol. **66:**67-77, 1951.

Killingsworth, L.M., Savory, J., and Teague, P.O.: Automated immunoprecipitin technique for analysis of the third component of complement (C′3) in human serum, Clin. Chem. **17:**374-377, 1971.

Nishikawa, T., Kubo, H., and Saito, M.: Competitive nephelometric immunoassay methods for antiepileptic drugs in patient blood, J. Immunol. Methods **29:**85-89, 1979.

Pauling, L., Pressman, D., and Grossberg, A.: The serological properties of simple substances. VII. A quantitative theory of the inhibition by haptens of the precipitation of heterogeneous antisera with antigens, and a comparison with experimental results for polyhaptenic simple substances and for axoproteins, J. Am. Chem. Soc. **66:**784-792, 1944.

Pope, C.G., and Healey, M.: A photoelectric study of reactions between diphtheria toxin and antitoxin, Br. J. Exp. Pathol. **19:**397-410, 1938.

Ritchie, R.F., Alper, C.A., Graves, J., Pearson, N., and Larson, C.: Automated quantitation of proteins in serum and other biological fluids, Am. J. Clin. Pathol. **69:**151-159, 1973.

Schultze, H.E., and Schwick, G.: Quantitative immunologische Bestimmung von Plasma Proteines, Clin. Chim. Acta **4:**15-25, 1959.

Sternberg, J.C.: A rate nephelometer for measuring specific proteins by immunoprecipitin reactions, Clin. Chem. **23:**1456-1464, 1977.

Agglutination

Bell, C.A., editor: A seminar on antigen-antibody reaction revisited, Washington, D.C., 1982, American Association of Blood Banks.

Lennette, E.H., Valows, A., Hausler, W.J., Jr., and Truant, J.P., editors: Manual of clinical microbiology, ed. 3, New York, 1980, American Society for Microbiology.

Williams, C.A., and Chase, M.W.: Methods in immunology and immunochemistry. vol. 3, Reactions of antibodies with soluble antigens; vol. 4, Agglutination, complement, neutralization and inhibition, New York, 1977, Academic Press, Inc.

Complement fixation

Kabat, E.A., and Mayer, M.M.: Experimental immunochemistry. Springfield, Ill., 1961, Charles C Thomas, Publisher.

Stansfield, W.D.: Serology and immunology, New York, 1981, McMillan Publishing Co., Inc.

Fluoroimmunoassays

Blanchard, G.C., and Gardner, R.: Two immunofluorescent methods compared with a radial immunodiffusion method for measurement of serum immunoglobins, Clin. Chem. **24:**808, 1978.

Nakamura, R.M., and Dito, W.R.: Nonradioisotope immunoassays for therapeutic drug monitoring, Lab. Med, **11:**807-817, 1980.

Soini, E., and Hemmila, I.: Fluoroimmunoassay: present status and key problems, Clin. Chem. **25:**353, 1979.

Enzyme immunoassays

Engvall, E., and Perlmann, P.: Enzyme-linked immunosorbent assays (ELISA): quantitative assay of immunoglobin G, Immunochemistry **8:**871, 1971.

Greenwood, H.M., and Schneider, R.S.: Current development of enzyme immunoassay. In Natelson, S., Pesce, A.J., and Dietz, A.A., editors: Clinical immunochemistry, Washington, D.C., 1978, The American Association for Clinical Chemistry.

Maggio, E.T., editor: Enzyme immunoassay, Cleveland, Ohio (Boca Raton, Fla.), 1980, CRC Press, Inc.

Chapter 11 Competitive-binding assays

Stephan G. Thompson

antibody An immunoglobulin that is employed as a binding protein in a competitive protein-binding assay (immunoassay) usually of the IgG class, with the IgG antibody having two specific binding sites.

antigen A high molecular weight substance that elicits an antibody response and is reversibly bound by its specific antibody. A macromolecular ligand in this chapter is considered as an antigen. Antigens have specific sites or determinants on their surface to which antibodies bind.

antiserum The serum of an animal that contains antibodies to an antigen or immunogen. Since most antigens or immunogens have multiple antigenic sites, the antiserum usually contains a mixture of different populations of antibodies to the same antigen or immunogen.

affinity The attraction that a binding protein and its ligand have for one another. Affinity can be expressed by a constant, K_a, referred to as the association constant, in liters per mole with

$$K_a = \frac{[Ab{:}L]}{[Ab][L]}$$

for a binding reaction in which an antibody is the binding protein.

competitive-binding assay An analytical procedure based on the reversible binding of a ligand to a binding protein. In proportion to its concentration, the ligand competes with a labeled derivative for binding to the limited number of available binding sites.

cross-reactivity The ability of substances similar to the ligand to bind to the binding protein. If such molecules are present in the reaction mixture, they may compete for binding with the ligand and cause an overestimation of its concentration.

detection limit The smallest concentration of a ligand that can be statistically distinguished from a zero level in an assay. The detection limit is also referred to as the sensitivity of an assay.

enzyme-linked immunosorbent assay (ELISA) A heterogeneous protein-binding assay that can employ an enzyme-labeled ligand that competes with the ligand in the sample for antibody immobilized to a solid phase.

enzyme-multiplied immunoassay technique (EMIT) A homogeneous enzyme immunoassay in which a low molecular weight ligand is labeled with an enzyme whose activity is inhibited when the ligand is bound by a specific antibody. Competitive binding of unlabeled ligand to the antibody relieves the inhibition and generates a dose response in proportion to the ligand concentration. This assay format does not require a step to separate the antibody-bound label from the free label.

fluorescence immunoassay (FIA) Competitive-binding assays that employ a fluorophore or fluorogen as the label.

fluorogen A nonfluorescent molecule that becomes fluorescent when modified by a chemical or enzymatic process.

heterogeneous assay A competitive-binding assay in which it is necessary to separate physically the protein-bound, labeled ligand from the unbound ligand, before measurement of the signal generated by the label.

homogeneous assay A competitive-binding assay in which it is not necessary to separate the protein-bound and free ligand fractions since the signal of the label is modulated by protein binding.

immunoassay A competitive-binding assay in which the binding protein is an antibody.

immunogen A substance that stimulates an antibody response when administered to an appropriate animal. Immunogens include macromolecular antigens and the otherwise nonantigenic haptens coupled to a macromolecular carrier.

immunometric assay Competitive and noncompetitive protein-binding assays in which the antibody rather than the ligand is labeled with a radioisotope or other suitable label.

label An atom or molecule attached to either the ligand or binding protein and capable of generating a signal for monitoring the binding reaction.

ligand A molecule or part of a molecule that is reversibly bound by the binding protein in a competitive-binding assay. It usually is the analyte but can also be a cross-reactant.

nonisotopic binding assays Assays with labels other than a radioactive or stable isotope.

prosthetic group Nonprotein components of enzymes and other functional proteins that are required for functional activity of the protein.

sensitivity The degree of response to a change in the ligand concentration in an assay. Also refers to and is synonymous with the detection limit.

specificity The degree to which a binding protein binds its particular ligand and does not bind structurally similar compounds.

DEFINITION OF COMPETITIVE-BINDING ASSAYS

Competitive-binding assays are a group of in vitro analytical methods based on noncovalent, reversible binding of a small molecule or ligand to a specific binding protein. The binding assay for small molecules is most often described by the reaction:

$$\text{Binding protein} + \text{Ligand} \rightleftharpoons \text{Binding protein:Ligand} \qquad \textit{Eq. 11-1}$$

where the binding protein has an affinity for the small molecule (ligand) that interacts with it. In general, only one binding protein can attach to the small molecule. This reaction can be considered simply as one molecule of ligand reacting with one protein-binding site. The important molecular feature of binding proteins that enables them to be used in quantitative assays is their ability to bind compounds with a high specificity and affinity. Some proteins with this ability are specific binding proteins found in serum in addition to antibodies. Examples of specific binding proteins are listed in Table 11-1.

Antibodies that bind small molecules are usually immunoglobulin G (IgG) produced in animals sensitized to the specific ligand. The methods of raising such antibodies are discussed in Chapters 9 and 10 and in more detail in other texts listed in the bibliography at the end of this chapter.

Small molecules that by themselves cannot provoke an immune response but can elicit antibodies when coupled with larger molecules are termed *haptens*. Because such small molecules are involved in the competitive-binding assays where antibody is used, the ligand is often referred to as a hapten.

The ligand in equation 11-1 is the analyte to be quantified. Often these are drugs (digoxin, theophylline), hormones (cortisol, T_4), vitamins (B_{12}), and so on. For most competitive-binding reactions, the ligand includes both the analyte and a labeled derivative of the analyte. Both must bind to the specific binding protein for the analyte to be measured. The final complex of binding protein and ligand is usually stable and, in favorable circumstances, dissociates slowly.

Table 11-1. Specific binding proteins present in serum or tissues

Binding protein	Ligand
Antibodies	Varied antigens
Corticosteroid binding globulin (CBG)	Cortisol, corticosterone
Estrogen receptor	Estrogen
Intrinsic factor	Vitamin B_{12}
Thyroglobulin (TBG)	Thyroxine (T_4) Triiodothyronine (T_3)
Vitamin D receptor	1-α,25-Dihydroxyvitamin D_3

The reaction described above is more complex for large molecules, such as proteins, because these usually have many different possible binding sites (antigenic determinents). Thus the molecule can have, one, two, and so on, binding proteins attached at the same time. In addition, such sites are usually quite different from each other, so that the population of antibodies generated in response to large molecules is more heterogeneous. For large molecules, assays other than those of the competitive binding type may be used for quantitation. These are discussed in the previous chapter.

QUANTITATION OF ANTIGEN-ANTIBODY REACTIONS: LAW OF MASS ACTION

The antigenic determinants of high molecular weight antigens and low molecular weight haptens can be considered similar when one is discussing the behavior of antibody-binding reactions. Those reactions involving haptens are, however, a simplification of the complexes that can be formed through the interactions of antibodies with high molecular weight molecules.

The law of mass action describes some aspects of the phenomenon that occurs when molecules bind to one another. This is best illustrated when one examines the concentration of an antibody and its ligand or specific binding partner under equilibrium conditions. The bimolecular reaction shown below:

$$Ab + L \underset{k_{-1}}{\overset{k_1}{\rightleftharpoons}} Ab{:}L \qquad \textit{Eq. 11-2}$$

can be rearranged for calculation of the equilibrium association constant

$$K_a = \frac{k_1}{k_{-1}} = \frac{[Ab{:}L]}{[Ab][L]} \qquad \textit{Eq. 11-3}$$

where k_1 and k_{-1} are the respective rate constants for association and dissociation of the bound complex; $[Ab]$ is the concentration of the unbound or free antibody at equilibrium; $[L]$ is the equilibrium concentration of unbound ligand (a term denoting the antigen, hapten, or other substance); and $[Ab{:}L]$ is the equilibrium concentration of the ligand-antibody complex. The K_a, also referred to as the affinity constant, is defined in reciprocal molar concentrations, M^{-1} or liters per mole (L/mol). This is the volume into which a mole of the binding protein can be diluted to yield 50% binding of the ligand. The larger the K_a, the greater the affinity of the antibody for the ligand. It follows that for a constant amount of antibody in the reaction, less ligand is required for a high-affinity antibody to bind 50% of the ligand than is required for 50% binding by a low-affinity antibody.

If Ab_T is the total concentration of antibody in the system where the concentration of unbound antibody, Ab, is

$$Ab = Ab_T - (Ab{:}L) \qquad \textit{Eq. 11-4}$$

and if the terms B and F describe the bound and free ligand, respectively, equation 11-3 can be rewritten as:

$$K_a = \frac{B}{(Ab_T - B)F} \qquad \textit{Eq. 11-5}$$

If any two of the three unknowns in equation 11-5: K_a, B, and F, are known, the third one can be calculated. This equation can be rearranged so that

$$\frac{B}{F} = K_a\,(Ab_T - B)$$

or *Eq. 11-6*

$$\frac{B}{F} = -K_a\,(B) + K_a \cdot Ab_T$$

This is equivalent to the equation describing a straight line

$$y = mx + b$$

where B is on the abscissa (x), B/F is on the ordinate (y), and $-K_a$ is the slope (m). The y intercept (b) is $K_a \cdot Ab_T$, and the x intercept is equivalent to the concentration of antibody in the assay. The graphic description of this equation is called a Scatchard plot, an example of which is shown in Fig. 11-1.

Antiserum produced by an animal that has been immunized with a particular antigen usually contains a mixture of different populations of antibody to that antigen. The slope of the Scatchard plot in this case is curved, reflecting the relative concentrations of different populations of an-

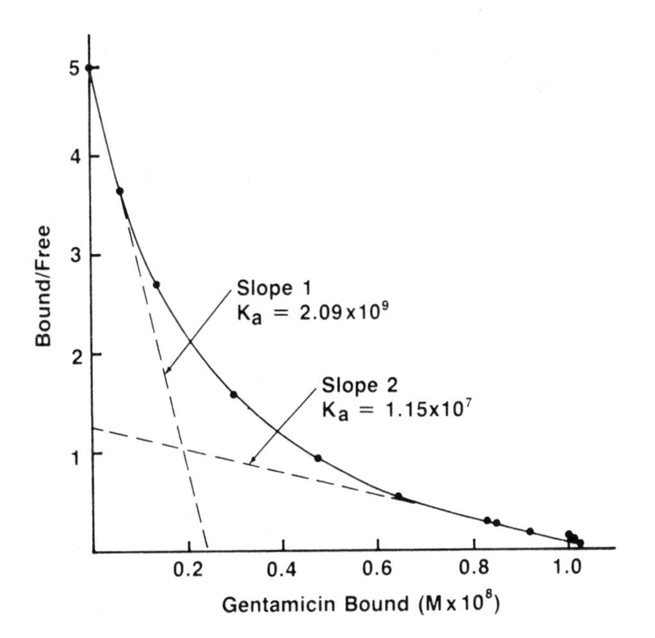

Fig. 11-1. Scatchard plot of gentamicin binding to gentamicin rabbit antiserum. Curvature of plot illustrates presence of mixture of various populations of antibodies with different affinities for gentamicin. Slope 1 and slope 2 indicate association constants (K_a) for high and low affinity binding sites, respectively.

tibodies, with different K_a's, in the heterogeneous mix of an antiserum. Fig. 11-1 shows that two straight lines can be extrapolated from the curve. The point of intersection with the x axis approximates the amount of each concentration of antibody-binding sites. The high-affinity population of antibodies to gentamicin in a gentamicin antiserum has a K_a of 2.09×10^8 L/mol with 2.4×10^{-9} M binding sites. The low-affinity population has a K_a of 1.15×10^7 L/mol with 1.1×10^{-8} M binding sites. These different populations of antibodies vary in their ability, that is, affinity, to bind the ligand and even how they might bind it, that is, recognizing different sites on its surface. The degree of heterogeneity (or homogeneity) of an antiserum for a ligand can be quantitated with the use of the Sips equation:

$$\frac{B}{Ab_T} = \frac{(K_oE)^a}{1 + (K_oE)^a} \qquad \textit{Eq. 11-7}$$

and

$$\log [B/(Ab - B)] = a \log E + a \log K_o \qquad \textit{Eq. 11-8}$$

where Ab_T is the concentration of total antibody and K_o is the average affinity constant, and a is equal to the heterogeneity index.

When the binding protein or antibody is homogeneous, meaning that there is only one population of uniform binding sites for a ligand, the shape of the plot is linear. This usually occurs when the analysis of a binding reaction is performed with monoclonal antibodies (see Chapter 10). These antibodies are homogeneous in that they reflect uniform affinity and specificity for the antigen.

Many experimental techniques have been developed to measure antigen-antibody and hapten-antibody interactions. They are dependent on the ability to detect the signal generated by the labeled antigen and also discriminate between the bound and free ligand. Examples include equilibrium dialysis, fluorescence quenching, fluorescence enhancement, fluorescence polarization, radioimmunoassays, and modified Farr assays. These methods are all applicable to the low molecular weight ligands. Equilibrium dialysis and fluorescence polarization are not suitable for measurement of the interaction between a high molecular weight ligand and its antibody because of the large size of the ligand.

Low-affinity binding proteins and antibodies typically have association constants in the order of 10^5 to 10^7 L/mol, whereas binders suitable for immunoassays and other competitive protein-binding assays must have association constants between 10^8 and 10^{11} L/mol. A higher association constant enables one to design assays with sensitivity down to 10^{-9} or to 10^{-12} M, provided that the label itself is detectable at such low concentrations.

COMPETITIVE BINDING

The competitive-binding assay can be imagined as the addition of increasing amounts of unlabeled ligand to re-

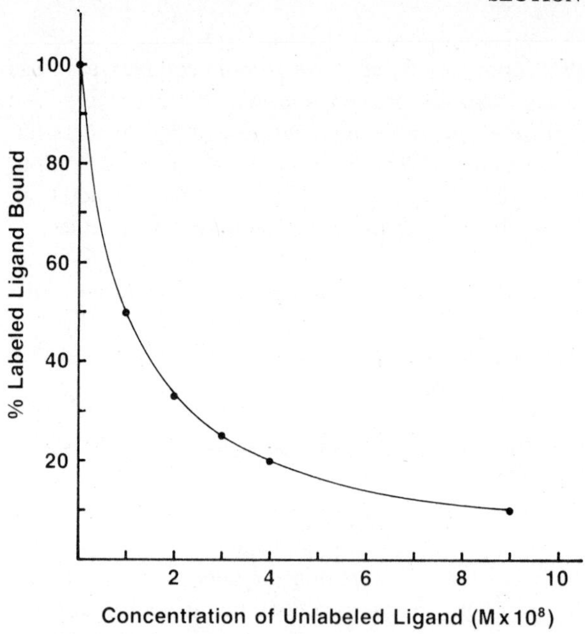

Fig. 11-2. Linear dose-response curve for data shown in Table 11-2.

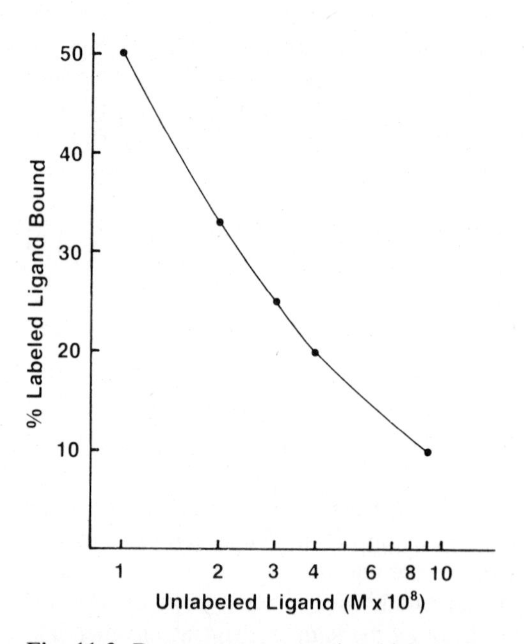

Fig. 11-3. Dose-response curve in which response is plotted as function of log of ligand concentration.

Table 11-2. Calculation of displacement reaction

Moles per liter of reactant $\times 10^8$			
Antibody	Labeled ligand	Unlabeled ligand	% labeled ligand bound
1	1	0	100
1	1	1	50
1	1	2	33
1	1	3	25
1	1	4	20
1	1	9	10

action mixtures containing known, constant amounts of labeled ligand and specific binding protein. In this case, labeled ligand, L^*, and antibody, Ab, are added together in equimolar amounts. Presuming that all the L^* is bound, the reaction becomes:

$$L^* + Ab \rightleftharpoons Ab{:}L^*$$

Two things happen with the addition of increasing amounts of unlabeled ligand: (1) unlabeled ligand competes with the labeled ligand for antibody-binding sites, and (2) there is an excess of the total ligand (L and L^*) in solution. The concentration of antibody-binding sites is therefore limiting with respect to total ligand, thus modifying the above reaction as follows:

$$L + AbL^* \rightleftharpoons AbL + AbL^* + L^* + L$$

Less L^* is antibody bound (as $Ab{:}L^*$), and more L^* is free as the amount of L increases. The amount or percentage of L^* in the bound form can be calculated from the amount of L and L^* present. Examples of such calculations are shown in Table 11-2.

When the percentage of labeled ligand bound (Table 11-2) is plotted as a function of the concentration of the unlabeled ligand, it yields a dose-response curve shown in Fig. 11-2. The term "dose-response curve" applies to a plot of binding versus increasing amounts of ligand. The curvature of the dose response in Fig. 11-2 is attributable to the logarithmic increase in the percentage of L that is bound when the concentration of L (dose) in the assay increases arithmetically. Thus the decrease in bound L^* is also logarithmic. Conversion of the concentration of L to a log value as shown in Fig. 11-3 makes the relationship become more linear.

Table 11-3. Techniques to separate bound from free labeled ligand

Technique	Principle
Absorbents	
Nonspecific	Low molecular weight ligands such as drugs and steroids are absorbed by particles such as charcoal.
Specific	Antibodies to the ligand or to the ligand-binding antibody are immobilized to a solid matrix such as glass or plastic. The immobilized antibody-ligand complex is separated from the unbound ligand by decantation or washing.
Electrophoresis	The protein-bound ligand moves at a lower rate through the electrophoretic medium than does the free ligand.
Gel filtration (size exclusion)	The protein-bound ligand moves at a faster rate through a column than the unbound ligand, since it is excluded from the porous gel.
Precipitation	
Ammonium sulfate	The antibody-bound ligand is precipitated by ammonium sulfate (Farr technique).
Double antibody	The antibody-bound ligand is precipitated by the addition of a second antibody specific for the antibody in the antibody-ligand complex.

SEPARATION TECHNIQUES

Figs. 11-2 and 11-3 show that to derive a dose-response or standard curve one must know the amount of labeled ligand that is antibody bound as a function of the amount of unlabeled ligand added. A variety of techniques have been developed to measure either the bound or free forms of the labeled ligand.

Some of these techniques require that the labeled ligand bound by antibody be physically separated from the free labeled ligand. These assays are called "heterogeneous assays." Table 11-3 lists some methods used to separate the bound and free labeled ligand. Other immunoassay approaches that do not require physical separation of bound and free labeled ligand are called "homogeneous assays." The signal of the labels in the homogeneous assays is altered upon binding to the specific binding protein; thus the bound and free labeled ligands can be directly distinguished from one another.

LABELED LIGAND

Types of labels

There are three common types of markers used to label ligands. These are radioisotopes, enzymes, and fluorophores. They can be used in both homogeneous and heterogeneous competitive-binding assays. The type of label, assay to which it is suited, and detection system are presented in Table 11-4.

Factors determining choice of label

Radioisotopes. Radioisotopic labels can be used only with heterogeneous immunoassay systems, since binding by antibody does not change the radioactive emission. In general, the desired sensitivity of the assay limits the type of radioactive labels to certain specific isotopes that have high specific activity, high energy output, and manageable half-lives and are easily obtained. The radioisotope must be readily incorporated or coupled to the ligand molecule, and its emission must be easily detected. Isotopes that meet these requirements are listed in Table 11-5. Consideration of all the factors, especially the need for high specific activity and ease of incorporation, has made ^{125}I the label of choice for most radioassays.

Enzymes. Enzymes can be used as labels but differ from radioisotopes in that the binding reaction can modify their activity. Again the enzymes must have high specific activity (that is, conversion of many moles of substrate to product per minute per mole of enzyme) and also be easily coupled to the ligand without losing significant activity.

Table 11-4. Some labels for competitive-binding assays and their means of detection

Label	Assay: Homogeneous	Assay: Heterogeneous	Detector
Enzymes			
Chromogenic substrates	Yes	Yes	Spectrophotometer
Fluorogenic substrates	Yes	Yes	Fluorometer
Enzyme substrates	Yes	Yes	Spectrophotometer, fluorometer
Enzyme prosthetic groups	Yes	Yes	Spectrophotometer, fluorometer
Fluorophores	Yes	Yes	Fluorometer
Metals	No	Yes	Atomic absorption spectrophotometer
Radioactive	No	Yes	Radioactivity counter

Table 11-5. Radioisotopes used in competitive protein-binding assays

Isotope	Emission	Maximum specific activity* (Ci/g)	Half-life	Counter†
^{3}H	Beta	9.6×10^{3}	12.3 years	LS
^{14}C	Beta	4.5	5730 years	LS
^{32}P	Beta	2.85×10^{5}	14.2 days	LS
^{125}I	Gamma	1.74×10^{4}	60.0 days	Crystal
^{57}Co	Gamma	8.48×10^{3}	270 days	Crystal

*The curie (Ci) is a unit of radioactivity equal to 3.7×10^{10} disintegrations per second.
†Beta-particle emitters are counted in liquid scintillation (LS) counters by the release of photons from organic phosphors in solution. Gamma-ray emitters are counted in detectors with a sodium iodide crystal that contains fluor from which photons are released.

Table 11-6. Fluorophores used as labels in competitive-binding assays

Fluorophore	Excitation wavelength (nm)	Emission wavelength (nm)	ϵ*	Fluorescence quantum yield†
ANS‡	385	471	12,300	0.80
DNS§	340	480-520	3,400	0.30
Fluorescein	490	520	72,000	0.85
Rhodamine	550	585	50,000	0.70
Umbelliferone	380	450	20,000	0.50

*ϵ is the absorbance of a 1 molar solution through a 1 cm light path.
†The fluorescence quantum yield is relative to the quantum yield of acridine, which is 1.0.
‡ANS, 1-Anilino-8-naphthalene sulfonic acid.
§DNS, Dimethylaminonaphthalene sulfonic acid.

Enzymes that have been used successfully include peroxidase, alkaline phosphatase, glucose-6-phosphate dehydrogenase, and glucose oxidase.

Fluorophores. Fluorophores as labels have requirements similar to enzyme labels in that they can be attached to the ligand and still can fluoresce with a high degree of efficiency. It is also helpful if the absorption (excitation) and emission wavelengths are well separated so that light scatter does not contribute to the fluorescence seen at the emission wavelength. Examples of fluorophores used as labels for competitive-binding assays and their properties are shown in Table 11-6. Of these, fluorescein and umbelliferone are the most commonly used. Fluorescein has a high extinction coefficient and quantum yield, whereas the largest difference between the excitation and emission wavelengths is 70 nm, as seen with umbelliferone.

AFFINITY OF PROTEIN-LIGAND INTERACTIONS

The binding proteins in competitive-binding assays, particularly antibodies, are usually characterized and described with respect to their concentration (titer) and affinity for the ligand. The affinity of an antibody is the force with which the antibody binds the ligand and is expressed as the K_a in liters per mole as was described on pp. 211 and 213. The K_a is the reciprocal of the dissociation constant, K_d, which can be derived from equation 11-2:

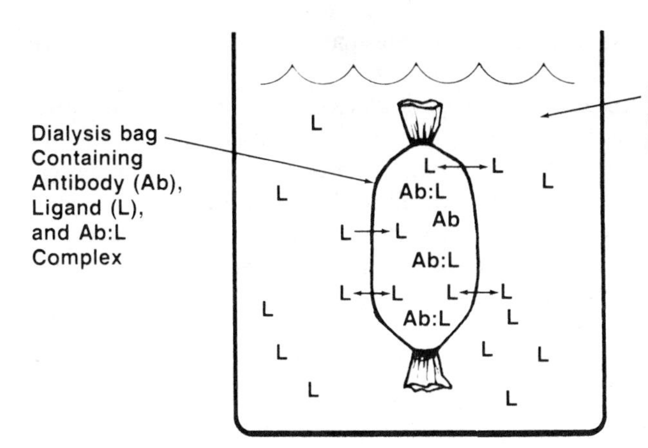

Fig. 11-4. Diagram of equilibrium dialysis experiment. Semipermeable dialysis bag containing antibody, *Ab,* or other binding protein is placed in a solution that contained one of various concentrations of ligand, *L*. Ligand crosses membrane barrier and binds to antibody. Unless ligand itself can be monitored at low concentrations by fluorescence or other means, it will be labeled with an isotope or other detectable molecule.

$$K_d = \frac{k_{-1}}{k_1} = \frac{[Ab][L]}{[Ab:L]} \qquad \textit{Eq. 11-9}$$

and is expressed in molarity (moles per liter). The K_d of an antibody with a K_a of $5 \times 10^9\ M^{-1}$ is $1/K_a$, or 2×10^{-10} M. This is the concentration of ligand and antibody required to obtain 50% binding of the ligand. At 50% binding B/F equals 1; therefore K_d equals $[Ab]$.

The determination of antibody affinity requires that the equilibrium concentrations of antibody-bound or antibody-free ligand be measured. When the ligand is small and dialyzable, one of the most frequently used methods for determining antibody affinity is equilibrium dialysis. This procedure is diagrammatically shown in Fig. 11-4. The binding protein of interest, such as an antibody, is placed in a dialysis bag that is in turn placed in a solution containing the corresponding ligand. Assuming that the ligand can easily pass through the semipermeable membrane of the dialysis bag, the equilibrium concentration of *unbound* ligand inside and outside the bag is the same. Since the antibody within the bag is also binding the ligand, the concentration of ligand within the bag includes both bound and free ligand. Subtraction of the concentration of free ligand found outside the dialysis bag, $[L]$, from the concentration inside the bag, $[Ab:L] + [L]$, yields the concentration of bound ligand $[Ab:L]$. $[Ab:L]$ is also equal to the concentration of occupied combining sites. The concentration of unbound-antibody combining sites, $[Ab]$, is equal to the total antibody concentration minus the bound-antibody concentration. The concentration of antibody is not always known, and so the K_a is calculated for $[Ab:L]$ and $[L]$ at different concentrations of L.

A result of an analysis to determine the K_a for monoclonal antibody to the antibiotic gentamicin is provided in Fig. 11-5. A constant amount of ^{125}I-gentamicin was used to monitor the binding of unlabeled gentamicin to the antibody within the dialysis bag. Contrast this with the plot shown in Fig. 11-1, which illustrates two populations of antibodies with respective high and low affinities for gentamicin. In either case the K_a is given by the negative slope. The concentration of antibody-binding sites is estimated from the intercept on the x axis.

The sensitivity of a binding-reacting assay is a function of the affinity of the binding protein for its ligand. Consequently, for 50% binding to occur, a ligand present at a concentration of 1×10^{-7} M would require an antibody that had a K_a of 10^7 L/mol, whereas a ligand present at a

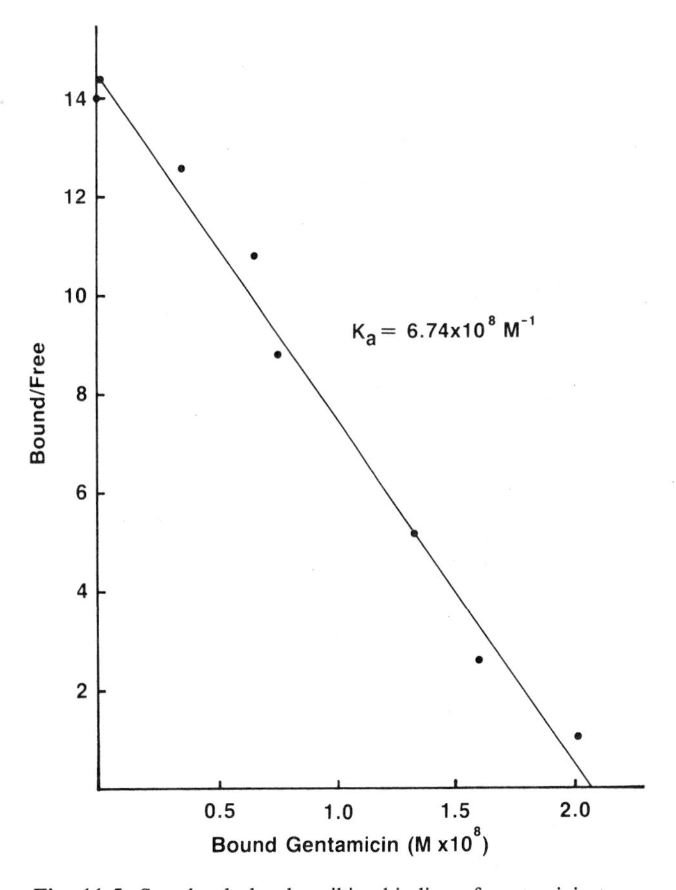

Fig. 11-5. Scatchard plot describing binding of gentamicin to gentamicin monoclonal antibody. Linearity of plot indicates a homogeneous grouping of antibody-binding sites.

concentration of 1×10^{-10} M would need an antibody that had a K_a of 1×10^{10} L/mol.

Ideally the affinity of the binding protein for the labeled and unlabeled ligands would be the same in a competitive-binding assay. Usually this is not the case. In some instances the label or the labeling procedure will alter the immunochemical properties of the ligand to the extent that antibody does not bind it as well as it does the unlabeled compound. The converse is also true: antibodies are often made against haptens in which the chemical bridge used to couple the ligand to the protein carrier is the same bridge used for coupling the ligand to the label. In these cases the antibodies may have higher affinity for the labeled ligand with the chemical bridge than it does for the unmodified ligand. One who designs a competitive-binding assay must consider differences in relative affinity of the antibody for the labeled and unlabeled ligand.

CROSS-REACTIVITY

The specificity of a binding protein for its ligand is measured by its ability to bind only the ligand and not other substances. Cross-reacting molecules are those that are so similar to the ligand in structure that they are capable of being bound by the antibody. The greater the chemical difference between the ligand and a potential cross-reactant, the less likely the cross-reactant will be bound by antibody to the ligand. Differences in antibody binding of ligand and cross-reacting substances are therefore a function of differences in affinity. These differences are reflected by responses to cross-reactants in competitive-binding assays. Table 11-7 presents the relationship between the K_a of the antibody for its ligand and two cross-reactants and the concentration of each that is required in the assay to deliver the same binding response. The concentrations of cross-reactant_a required for 50% binding to antibody_1 is 10 times the concentration necessary to bind 50% of ligand_1. Similarly, 10,000-fold less ligand_2 is required to achieve 50% binding to antibody_2 than is necessary to bind 50% of cross-reactant_d. Table 11-7 shows that the lower the K_a, the more antibody is required to bind 50% of the ligand or cross-reactant, further illustrating the relationship between sensitivity and K_a.

Table 11-7. Cross-reactant binding as a function of antibody affinity

Antibody	Bound species	K_a	Concentration (M) required for 50% binding*
1	Ligand_1	1×10^8	2×10^{-8}
	Cross-reactant_a	1×10^7	2×10^{-7}
	Cross-reactant_b	5×10^7	4×10^{-8}
2	Ligand_2	1×10^{10}	2×10^{-10}
	Cross-reactant_c	2×10^8	1×10^{-8}
	Cross-reactant_d	1×10^6	2×10^{-6}

*When 50% of the ligand or cross-reactant is bound, $B/F = 1$. Since $K_a = \frac{B}{F[Ab]}$, when $B/F = 1$, the $K_a = \frac{1}{[Ab]} = \frac{1}{K_D}$.

Ideally, antibodies or other binding proteins that participate in competitive-binding reactions are very specific for the ligand with essentially no cross-reactivity with closely related molecules. Realistically this is rare, since the antibodies present in a heterogeneous antiserum bind the ligand with different affinities and orientations and are therefore also likely to bind structurally similar molecules. One of the advantages of the use of monoclonal antibodies is the potential for selecting very specific antibodies that have essentially no cross-reactivity with other compounds. Examples of the cross-reactivity antiserums and a non-cross-reactive antibody are shown in Figs. 11-6 and 11-7. The dose-response curves seen in Fig. 11-6 show that caffeine, which is structurally similar to the antiasthmatic drug theophylline, displaces the bound label from antiserum to theophylline only at much higher concentrations. The degree of caffeine cross-reactivity with theophylline antibodies, determined by the "classical" approach to cross-reactivity, is calculated when one divides the concentration of ligand (in this case theophylline), which displaces 50% of the label (indicated in Fig. 11-6) by the concentration of cross-reactant (caffeine), which also causes a 50% displacement of bound label according to the following equation:

$$\frac{\text{[Ligand] at 50\% displacement}}{\text{[Cross-reactant] at 50\% displacement}}(100) = \text{\% cross-reactivity} \qquad \textit{Eq. 11-10}$$

There is 12.3% caffeine cross-reactivity for the antiserum shown in Fig. 11-6, but only 1.2% cross-reactivity shown with the theophylline monoclonal antibody in Fig. 11-7. Table 11-8 summarizes these results. A competitive-binding assay that uses the monoclonal antibody to theophylline is much more specific than that based on an antiserum; consequently the former is less prone to caffeine interference.

DATA REDUCTION

Earlier in this chapter the displacement reaction was discussed and also the general principle of competitive-binding assays, which is described by the reaction:

$$L^* + Ab + L \rightleftharpoons (Ab{:}L^*) + (Ab{:}L)$$

Table 11-8. Caffeine cross-reactivity with theophylline antiserum or monoclonal antibodies

	Concentration (*M*) when 50% of label is bound		% cross-reactivity
	Theophylline	Caffeine	
Antiserum	1.29×10^{-7}	1.05×10^{-6}	12.3
Monoclonal antibody	6.15×10^{-8}	5.09×10^{-6}	1.2

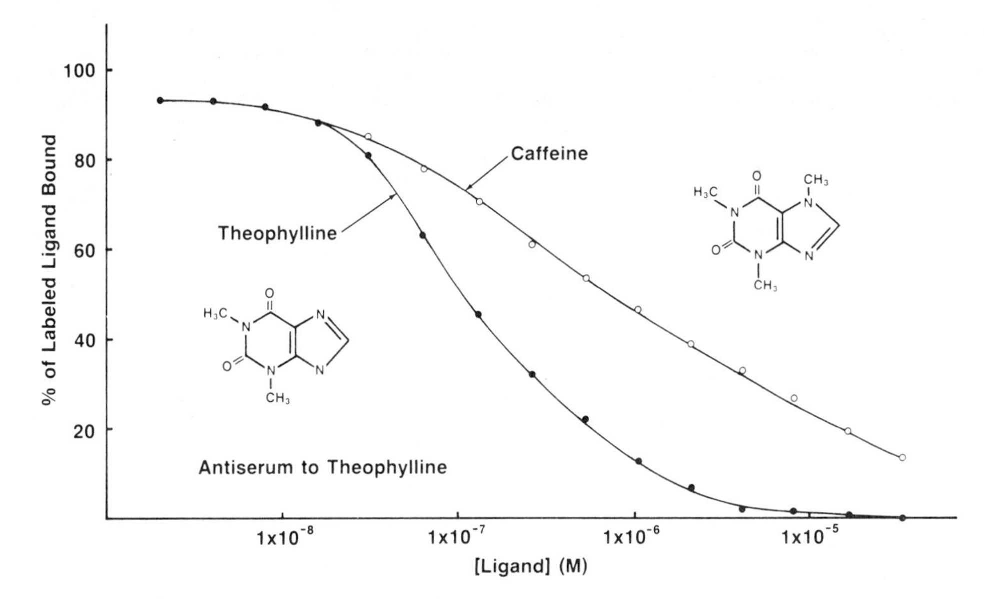

Fig. 11-6. Cross-reactivity of caffeine with an antiserum to antiasthmatic drug theophylline in a substrate-labeled fluorescent immunoassay. Degree of cross-reaction is determined at concentrations of theophylline and caffeine required for 50% of dose response. This is equivalent to 46.5% of bound label. Refer to Table 11-8 for cross-reactivity data.

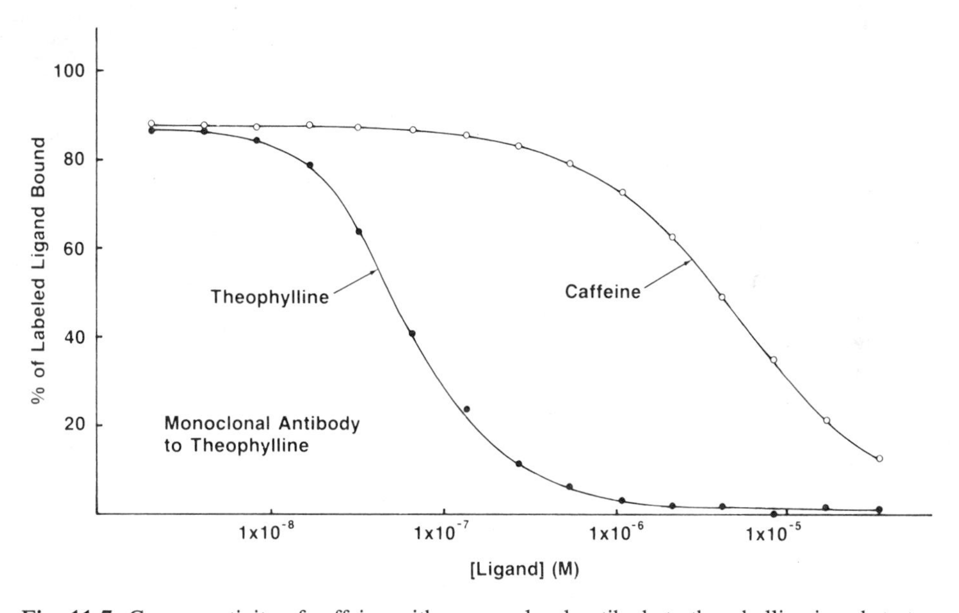

Fig. 11-7. Cross-reactivity of caffeine with a monoclonal antibody to theophylline in substrate-labeled fluorescent immunoassay. Cross-reactivity is determined at 43.2% of bound label.

where *Ab* is the antibody or another binding protein providing a limiting number of ligand-binding sites. The concentration of the labeled ligand, L^*, is constant and in excess of the binding-site concentration. The ligand, L, will compete with L^* for the available binding sites. As was shown in Table 11-2, the amount of L^* bound by the antibody is inversely proportional to the concentration of L in the assay. The dose response is quantitated when one measures either the bound L^* or the free L^* that has been displaced by L. Radioimmunoassay (RIA) is a heterogeneous assay that will serve as an example to illustrate the most common methods for generating dose-response curves and reducing data to forms usable for quantifying concentrations of analytes.

The data shown in Table 11-9 was generated by use of a double antibody RIA for the aminoglycoside antibiotic amikacin, where ^{125}I-amikacin was the labeled ligand. The primary antibody was rabbit antiserum to amikacin. Goat antiserum to rabbit antibodies is included in the assay to precipitate the antibody-bound labeled ligand (^{125}I-amika-

Table 11-9. Data for amikacin radioactivity*

Dose of Amikacin ($M \times 10^{-8}$)	Total bound cpm *(TB)*	Specifically bound cpm *(B)*†	%*B*‡	*B/F*§	B/B_0‖	Logit B/B_0
0	14019	13588 (B_0)	64.0	1.78	1.00	—
1.07	10694	10264	48.4	0.94	0.76	1.13
2.14	9235	8805	41.5	0.71	0.65	0.61
4.28	7184	6754	31.8	0.47	0.50	−0.01
8.56	5360	4930	23.2	0.30	0.36	−0.56
17.12	3925	3495	16.5	0.20	0.26	−1.06
Unknown 1	8912	8482	40.0	0.67	0.62	0.51
Unknown 2	6910	6480	30.5	0.44	0.48	−0.09
Unknown 3	4340	3910	18.4	0.23	0.29	−0.91

*Total counts per minute (cpm), *T,* of ^{125}I-amikacin in each reaction are 21,225; nonspecifically bound (NSB) counts per minute are 430.
†Specifically bound cpm = *B* = Total bound cpm − NSB.
‡% bound = $(B/T)100$.
§$B/F = \dfrac{B}{TB - B}$
‖B_0 = *B* at zero dose of drug.

Table 11-10. Concentrations of unknown sample as determined with different data-reduction techniques

Unknown sample	Amikacin ($M \times 10^{-8}$) %*B* versus log dose	*B/F* versus dose	Logit B/B_0 versus log dose
1	2.38	2.40	2.33
2	4.75	4.60	4.93
3	14.06	14.40	13.62

Fig. 11-8. Three dose-response curves drawn from amikacin radioimmunoassay data presented in Table 11-9.

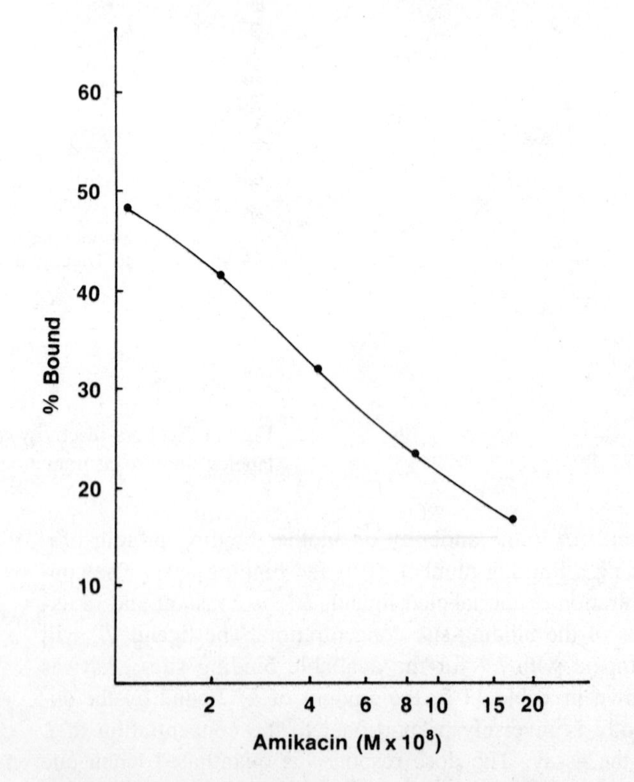

Fig. 11-9. Amikacin radioimmunoassay dose-response curve with percentage of bound amikacin plotted as function of log of amikacin concentration (see Table 11-9).

cin). The precipitate is collected by centrifugation, and the pellet in the bottom of the tube is counted for radioactivity after the supernatant has been removed. The quantity "nonspecifically bound (NSB) counts per minute" refers to the radioactivity nonspecificity bound to the bottom of the tube in the absence of antibody to amikacin.

Table 11-9 shows how the amikacin RIA data are reduced to generate the four most common dose-response curves. In addition to the bound counts per minute *(TB),* the *y* axis can be drawn as *%B*, B/F, B/B_0, or logit B/B_0. The logit transformation is the following:

$$\text{logit } (B/B_0) = \ln \frac{B/B_0}{1 - B/B_0} \qquad \textit{Eq. 11-11}$$

The bound counts per minute, *%B*, and B/F can be plotted as a function of the arithmetic dose as is shown in Fig. 11-8. Changing the arithmetic dose to a log dose and plotting the *%B* as a function of the log of the amikacin concentration yield the slightly sigmoid dose-response curve seen in Fig. 11-9. All these amikacin dose-response curves are inversely proportional to the concentration of unlabeled ligand in the reaction. The log dose versus the logit B/B_0 like that shown in Fig. 11-10 has been the most accepted empirical method for linearizing competitive protein-binding dose-response curves. Automatic data reduction and processing of the log-logit transformation is easily done with computerized systems, whereas the plots shown in Figs. 11-8 and 11-9 can be easily done either manually or with a calculator. All the standard curves demonstrated in

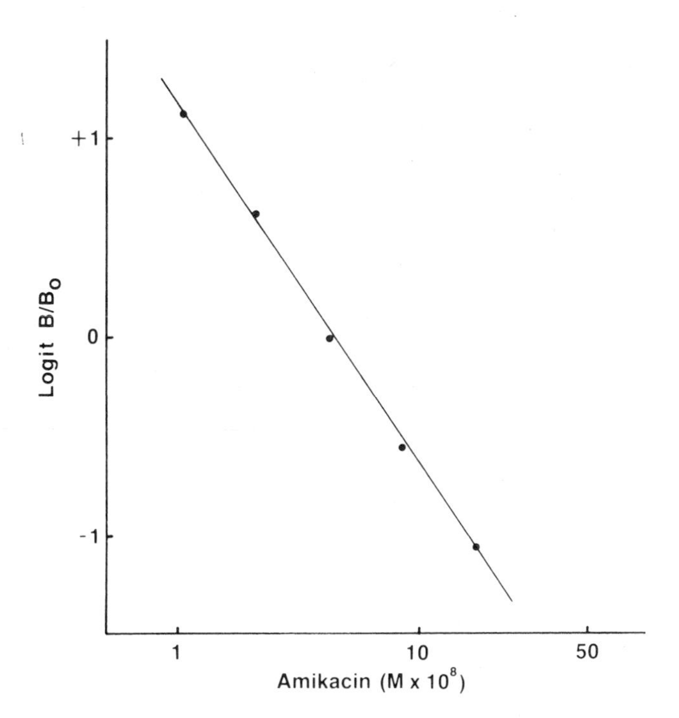

Fig. 11-10. Logit (B/B_0) versus log of amikacin concentration for radioimmunoassay data presented in Table 11-9.

Figs. 11-8 to 11-10 can be used for estimation of the concentration of the unknowns listed in Table 11-9. These results are shown in Table 11-10. One can see from Table 11-10 that the interpolated concentrations of the unknown samples are close and relatively independent of the plotting method. Of the three methods, the best linear fitted line over the range of the standard curve is the logit B/B_0 versus log dose, whereas the shallow response seen at higher concentrations of amikacin in the B/F versus dose plotting routine is more prone to errors in precision. The reason is that slight changes in the B/F response will cause greater changes in the interpolated dose when the standard curve is shallow whereas the same changes in the region of a steep slope (such as seen at lower concentrations) will cause less variation. Since the logit B/B_0 versus log dose plotting method can be routinely linearized for rapid data reduction and processing of the dose responses of unknown concentrations of analyte, this is the most popular method for generating standard curves.

EXAMPLES OF SPECIFIC COMPETITIVE-BINDING ASSAYS

Radioimmunoassay

Although radioimmunoassay (RIA) was used to describe the various dose-response curves (Table 11-9; Figs. 11-8 to 11-10) that can be generated with competitive-binding reactions, a brief description of the principles of RIA will be useful to introduce the basic principles of other competitive-binding assays. In RIA, the ligand and a constant amount of radioactively labeled ligand compete for a limited number of antibody-binding sites. The concentration of antibody is usually sufficient to bind between 30% and 80% of the labeled material. For example, the antibody to amikacin used in the earlier example bound 64% of the ^{125}I-amikacin in the absence of unlabeled ligand. Addition of unlabeled ligand to the assay results in a net increase in the total ligand (labeled plus unlabeled) but, because of competition for antibody-binding sites, a decrease in the proportion of labeled ligand that will be bound by the antibody. In RIA, if one counts the radioactivity bound to the antibody after the separation step, the dose-response curve will have a negative slope like that shown in Fig. 11-11, *A*. As the concentration of unlabeled ligand increases and the antibody-binding sites approach saturation, the slope levels off. When the unbound radioactively labeled ligand is monitored, the dose-response curve has a positive slope (Fig. 11-11, *B*) with the same shape as that in Fig. 11-11, *A*.

Radioimmunoassay is applicable to the measurement of both low molecular weight and high molecular weight ligands, provided that the labeling procedure or the labeled ligand conjugate itself does not adversely affect the immunoreactivity of the ligand. Some ligands (including proteins) lack the tyrosine residues required for radioactive labeling with the isotopes ^{125}I or ^{131}I. In these cases the ligand must be derivatized with tyrosine before labeling. Derivatization of the ligand with tyrosine, or even nonisotopic labels, could yield a labeled ligand that has lost some or all of its immunoreactivity. Another consequence of conjugation is that some ligands may be unstable when labeled. These problems are overcome sometimes by labeling of the antibody rather than the ligand since the structural differences between IgG molecules are substantially less than the differences between ligands. Losses in immunoreactivity are less likely to occur when the antibody is labeled. IgG has many tyrosines potentially available for labeling with radioactive iodide. The labels are often conjugated to sites on the antibody distal to the binding regions, with such conjugation allowing the antibody to bind the ligand without interference by the label.

Assays based on the use of a labeled antibody are called "immunometric" assays. One that uses a radiolabeled antibody is an immunoradiometric assay (IRMA). With this assay format the ligand in the sample competes with the ligand attached to a solid surface for the binding sites of the labeled antibody (Fig. 11-12). The amount of labeled antibody bound to the solid surface is determined after excess label and sample have been removed. A representative dose-response curve is also shown in Fig. 11-12. The immunometric format is considered again below in the discussion of ELISA techniques.

Table 11-11. Enzyme labels for immunoassays

Enzyme	Source	Molecular weight	Specific activity (units/mg)*
Alkaline phosphatase	Calf intestine	100,000	400
β-Galactosidase	*Escherichia coli*	540,000	400
Glucose oxidase	*Aspergillus niger*	160,000	200
Glucose-6-phosphate dehydrogenase	*Leuconostoc mesenteroides*	130,000	250
Peroxidase	Horseradish	40,000	900

*A unit of enzyme activity represents the conversion of 1 μmol of enzyme substrate to product per minute.

Enzyme-linked immunoabsorbent assay

The enzyme-linked immunoabsorbent assays (ELISA) are heterogeneous nonisotopic assays that usually have an antibody immobilized onto a solid support (Table 11-3) and the ligand is labeled with the enzyme. Table 11-11 shows some enzymes used for ELISA (or other enzyme immunoassays) and some of their properties. These enzymes are useful for labels because they satisfy the following criteria: (1) The enzymes have a high specific activity. The amplification seen with an enzyme is related to the amount of substrate converted to product during the time of incubation. Enzymes with the highest specific activities are potentially capable of giving the greatest amplification. Assays using such enzymes may have greater sensitivity and may be able to detect lower concentrations of ligand.

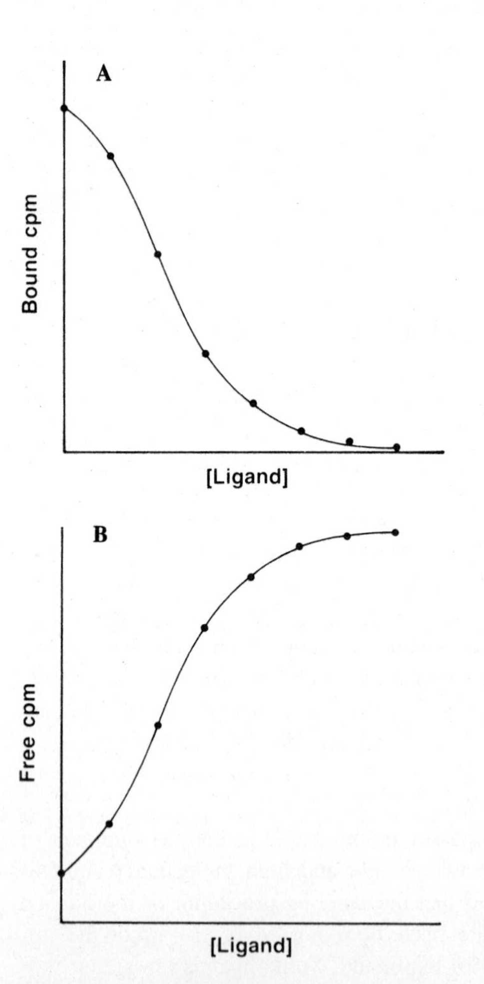

Fig. 11-11. Radioimmunoassay dose-response curves illustrating slopes for antibody-bound, **A**, and free, **B**, radioactivity in counts per minute (cpm) versus dose.

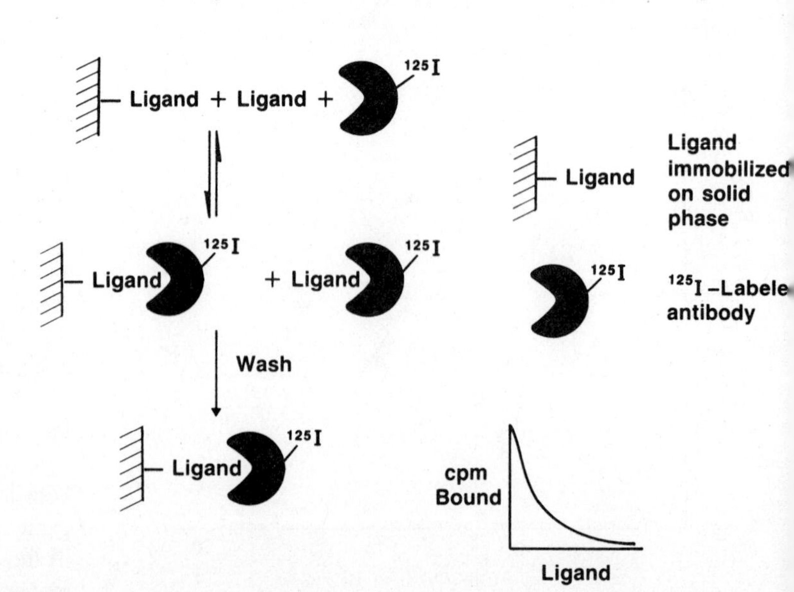

Fig. 11-12. Principle of competitive immunoradiometric assay (IRMA) and typical dose-response curve.

(2) The labels are stable during the assay and under refrigerated storage conditions. (3) The enzymes are not present in the biological fluid or tissue sample to be analyzed. (4) The enzymes still retain most of their activity when attached to the ligand (or antibody).

Alkaline phosphatase and horseradish peroxidase are both inexpensively available in a highly purified form. For this and the reasons listed above, these two enzymes are most often used as labels for ELISA. The enzymes listed in Table 11-11 can use either chromogens or fluorogens as substrates, with the respective products yielding an absorbance or fluorescence signal. The advantage of a chromogenic substrate is that its products can be detected visually. Fluorescent products can be detected at 100 to 1000 times lower concentrations than that of chromophores. The incubation time can be greatly shortened when one is using fluorogenic substrates and the sensitivity for the measured ligand is potentially greater.

Many configurations for ELISA have been devised. Some are based on competitive reactions, whereas others are direct immunometric "sandwich" assays. The "sandwich" ELISA is discussed in Chapter 10. The two basic formats for the competitive assays and the shape of the respective typical dose-response curves describing the signal remaining on the solid phase are shown in Figs. 11-13 and 11-14.

Of the two, the configuration in which antibody has been immobilized onto the solid surface (Fig. 11-13) has been the most frequently described. This is analogous to the configuration in an RIA because the ligand in the sample competes with the enzyme-labeled ligand for the limiting amount of antibody-binding sites fixed to the solid phase. After the binding reaction has taken place, the solid phase is washed with buffer to remove the unbound labeled ligand so that it does not contribute to the signal. The amount of enzyme bound to the solid phase is proportional to the absorbance (or fluorescence or other signal) of the product formed after the addition of substrate, which is inversely proportional to the concentration of unlabeled ligand. This method is applicable to both low and high molecular weight analytes.

If, instead of the antibody, the ligand is attached to the solid surface as shown in Fig. 11-14. Only those binding sites not occupied by the ligand in solution will bind the immobilized ligand. The solid phase–labeled antibody complex is washed before addition of the substrates. Immunometric ELISA's with immunoradiometric assays can have the same basic advantages when compared to ELISA's in which the ligand is labeled.

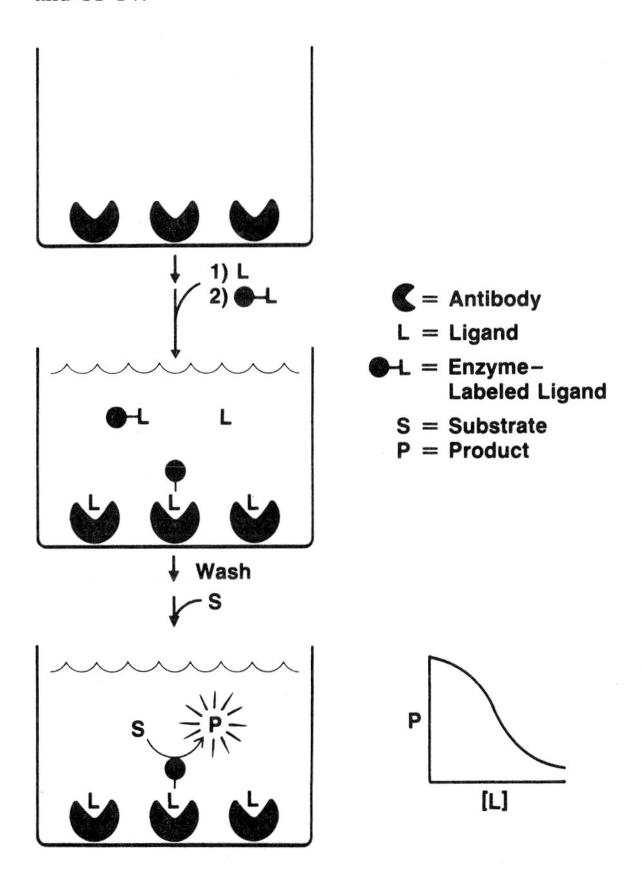

Fig. 11-13. Principles of competitive enzyme–linked immunosorbent assay (ELISA) with ligand labeled with an enzyme and typical dose-response curve.

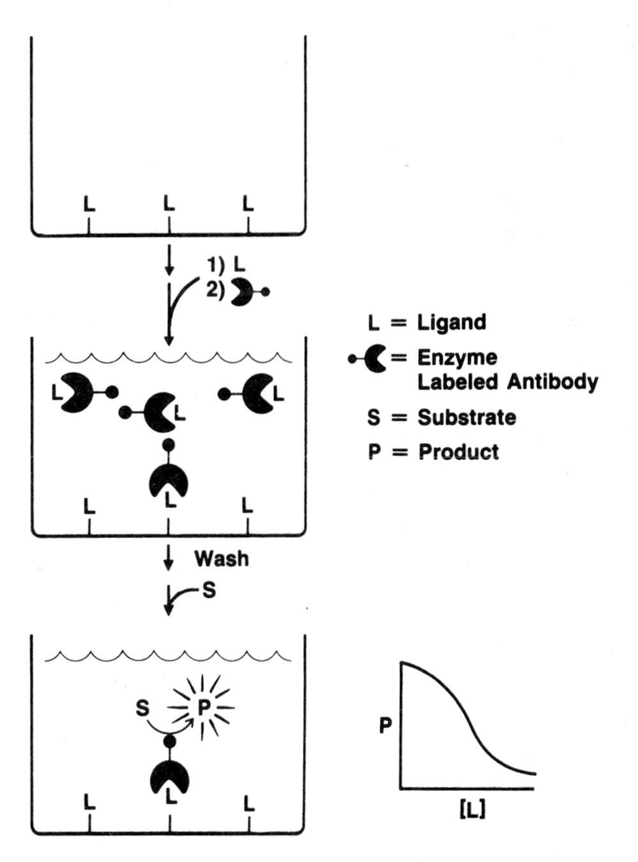

Fig. 11-14. Principles of competitive ELISA where antibody is labeled with enzyme. This is an immunometric assay.

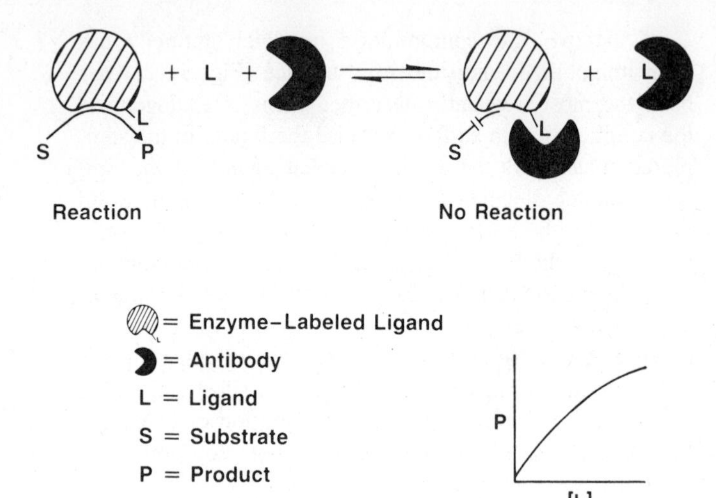

Fig. 11-15. Principle of homogeneous enzyme immunoassay (enzyme-multiplied immmunoassay technique, EMIT) and typical dose-response curve.

Homogeneous-enzyme immunoassay

The first of the homogeneous immunoassays to be described here is an enzyme immunoassay for low molecular weight ligands. This assay format is referred to as the enzyme-multiplied immunoassay technique (EMIT). Binding of antibody to the enzyme-labeled ligand changes the enzymatic activity of the label, so that the antibody-bound label can be distinguished from the unbound labeled ligand.

In most instances, binding of antibody to the labeled ligand sterically inhibits the enzyme by preventing the enzyme substrate access to the catalytic site (Fig. 11-15). A typical enzyme-inhibition profile is shown in Fig. 11-16. Increasing amounts of gentamicin monoclonal antibody inhibits the activity of the enzyme label, glucose-6-phosphate dehydrogenase (Fig. 11-16). The dose-response curve for the homogeneous gentamicin enzyme immunoassay is shown in Fig. 11-17. The unlabeled ligand competes with enzyme-labeled gentamicin for the limiting number of antibody-binding sites. Less labeled gentamicin is therefore bound by the antibody resulting in an increase in catalytic activity. Usually this response is reflected by a change in the rate of enzyme activity with two readings of each reaction being taken 30 seconds apart. The change in absorbance (ΔA) is used to calculate the dose response. The ΔA at zero dose (ΔA_0) is routinely subtracted from all standards, controls, and specimens before one plots the standard curve and calculates the unknown concentrations.

The homogeneous enzyme immunoassay for thyroxine (T_4) is based on the antibody-mediated reversal of the inhibition of the enzyme malate dehydrogenase by its conjugated ligand, thyroxine. Free thyroxine in the reaction mixture prevents reversal of the inhibition by antibody; consequently malate dehydrogenase activity is reduced as the concentration of thyroxine increases.

Substrate-labeled fluorescent immunoassay

Another homogeneous competitive protein-binding assay is the substrate-labeled fluorescent immunoassay (SLFIA). The assay is based on a label that is a fluorogenic enzyme substrate. When the label is hydrolyzed by a specific enzyme, it yields a fluorescent product. Binding of the labeled ligand by specific antibody prevents the enzyme from hydrolyzing the substrate label. Since fluorescence will not be produced by antibody-bound label, the bound label can be distinguished from unbound.

β-Galactosidase and its substrate, β-galactosyl–umbelliferone, are the enzyme-and-label combination shown in Fig. 11-18. This combination has been used to develop a variety of SLFIA for therapeutic drugs and proteins, demonstrating its applicability to ligands with low and high molecular weights, respectively. The substrate, β-galactosyl–umbelliferone, when covalently attached to the ligand, is nonfluorescent under the conditions of the assay until it is hydrolyzed by β-galactosidase to yield the fluorescent product, umbelliferone ligand.

Inhibition of the enzyme-catalyzed fluorescence production from the fluorogenic substrate label conjugated to theophylline in the presence or absence of unlabeled theophylline is shown in Fig. 11-19. The fluorescence resulting from hydrolysis of the label decreases as the concentration of antibody in the reaction increases. The difference between the two inhibition curves reflects the magnitude of the dose response at any one antibody level.

In a competitive-binding reaction assay, the substrate-labeled ligand and sample ligand compete for the limited number of antibody-binding sites as illustrated in Fig. 11-20. As the concentration of ligand in the sample increases, the fraction of β-galactosyl–umbelliferone–ligand not bound by the antibody also increases. It is hydrolyzed by the enzyme, resulting in fluorescence production proportional to the concentration of ligand in the sample. A typical SLFIA dose-response curve is shown in Fig. 11-21.

Similar dose-response curves have also been demonstrated for high molecular weight ligands such as IgA, IgG, and IgM.

Fluorescence polarization immunoassay (FPIA)

Fluorescence polarization is a technique that has often been used to study protein-ligand binding interactions but has only recently been routinely applied to clinical immunoassay systems. These assays are based on the amount of polarized fluorescent light detected when the fluorophore label is excited with plane-polarized light. The degree of fluorescence polarization is dependent on the rate of rotation of the fluorophore-ligand complex in solution. Small molecules rotate freely, and consequently fluorescent light emitted by the molecule is relatively depolarized, whereas large molecules like proteins rotate more slowly, resulting in a greater degree of polarization of emitted fluorescent light. When an antibody binds a low molecular weight ligand labeled with a fluorophore, as shown in Fig. 11-22,

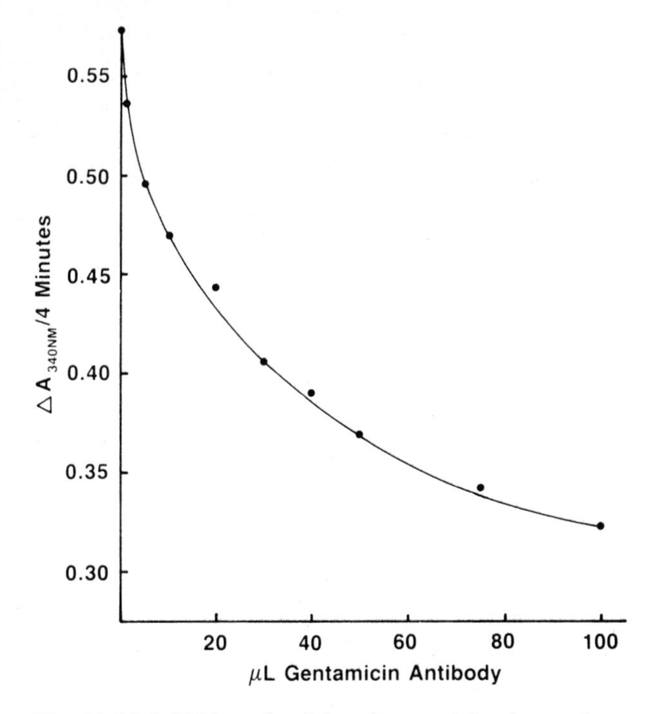

Fig. 11-16. Inhibition of activity of gentamicin–glucose-6-phosphate dehydrogenase conjugate by monoclonal antibody to gentamicin.

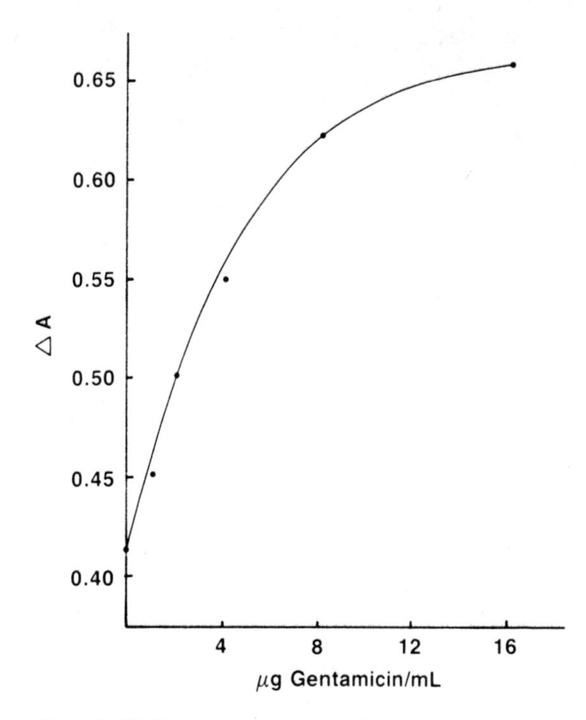

Fig. 11-17. Dose-response curve for homogeneous gentamicin enzyme immunoassay.

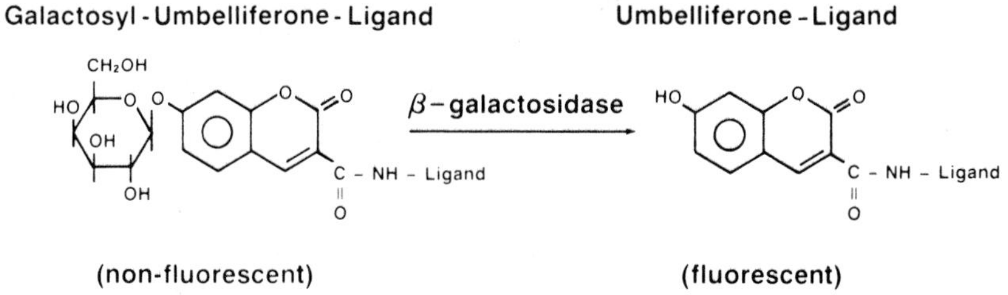

Galactosyl - Umbelliferone - Ligand

Umbelliferone - Ligand

β-galactosidase

(non-fluorescent)

(fluorescent)

Fig. 11-18. Fluorogenic substrate label, β-galactosyl–umbelliferone, before and after hydrolysis by β-galactosidase.

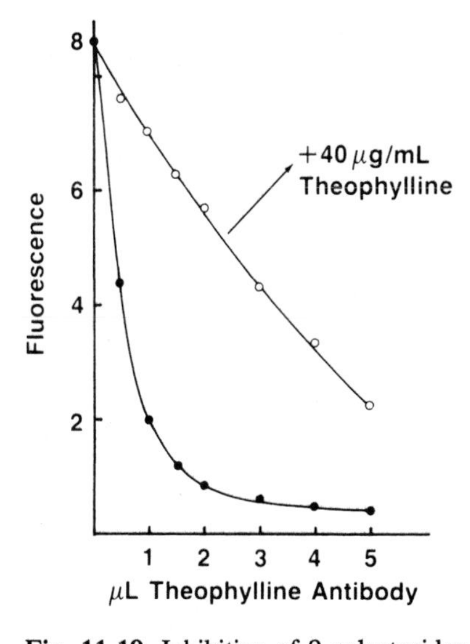

Fig. 11-19. Inhibition of β-galactosidase-mediated hydrolysis of β-galactosyl–umbelliferone–theophylline by monoclonal antibody to theophylline in absence (●) or presence (○) of a diluted 40 μg/mL theophylline standard.

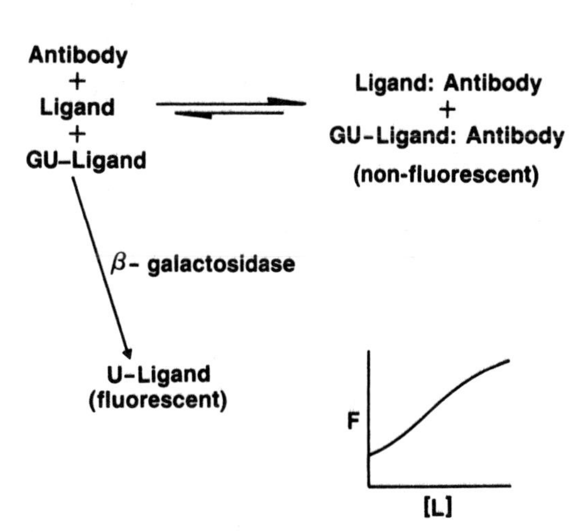

Fig. 11-20. Principle of substrate-labeled fluorescent immunoassay (SLFIA). *GU-Ligand,* β-Galactosyl–umbelliferone–ligand.

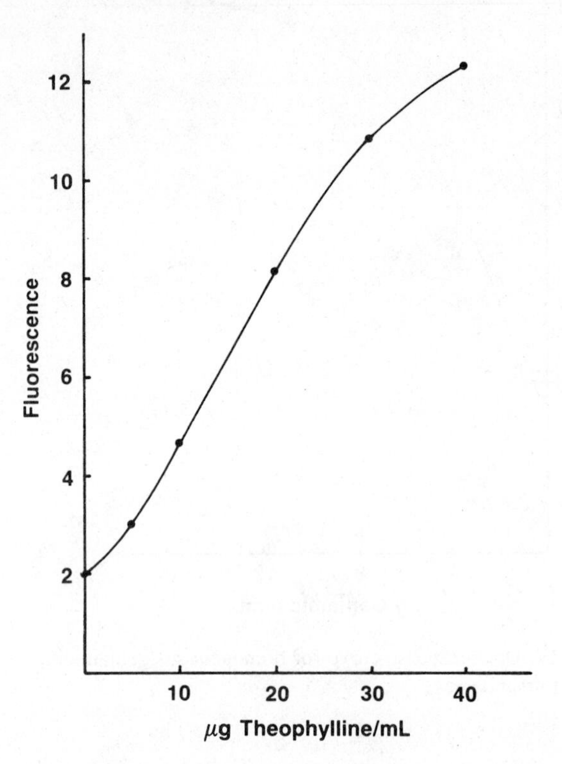

Fig. 11-21. Dose-response curve for theophylline SLFIA. This assay contains monoclonal antibody to theophylline.

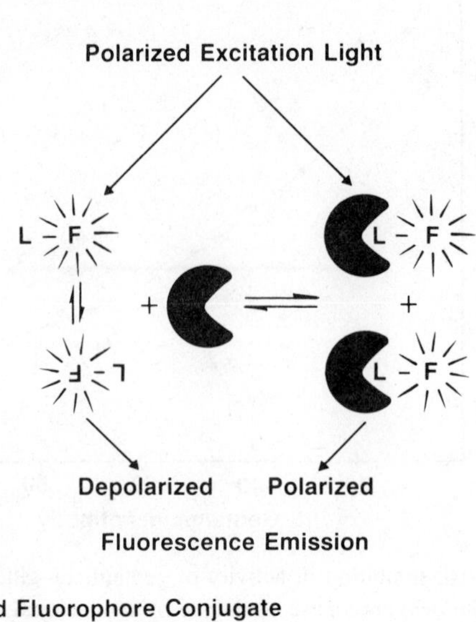

Fig. 11-22. Fluorescence polarization of a fluorophore-labeled ligand free in solution and when bound by an antibody. Unbound fluorescent conjugate rotates in solution so that its fluorescence emission is depolarized, whereas bound label has less rotation, resulting in a polarized fluorescence emission.

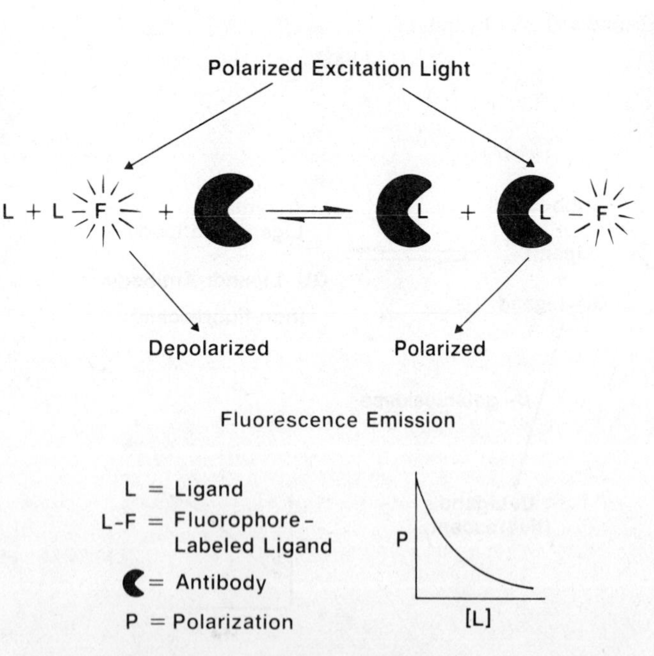

Fig. 11-23. Principle of fluorescence polarization immunoassay (FPIA) and typical dose-response curve.

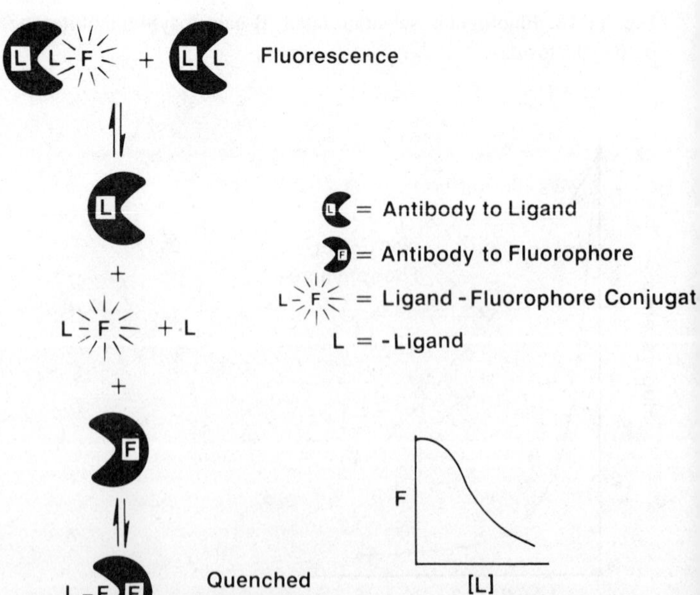

Fig. 11-24. Principle of fluorescence protection immunoassay and typical dose response. Antibodies to ligand and fluorophore are designated by *L* and *F*, respectively.

the fluorescence polarization of the labeled ligand is increased, since rotation of the labeled ligand-antibody complex is much slower than that of labeled ligands alone. Unlabeled ligand will compete with the labeled ligand for antibody-binding sites so that the amount of polarized fluorescent light resulting from a competitive-binding reaction is inversely proportional to the concentration of unlabeled ligand in the reaction. The principle of the fluorescence polarization immunoassay and a typical dose response are shown in Fig. 11-23.

Fluorescence polarization immunoassays have been limited to the measurement of low molecular weight ligands such as drugs and some hormones. The reason is that the fluorescence polarization of a labeled high molecular weight ligand is already large, and no measurable change in polarization is seen when the macromolecular ligand is bound by antibody.

Fluorescence protection immunoassay

Another homogeneous fluorescent immunoassay, the fluorescence protection immunoassay, is dependent on the fluorescence quenching of a fluorophore by antibody to the fluorophore. Thus, when a macromolecular ligand such as IgG is labeled with fluorescein, binding of this conjugate to an antifluorescein antibody almost completely quenches its fluorescence. Such interaction is prevented, however, when antibody to the ligand is allowed to bind the ligand before the addition of excess fluorescein antibody to the reaction. The principles of the fluorescence protection assay and a typical dose-response curve are shown in Fig. 11-24. In the absence of unlabeled macromolecular ligand, antibody to IgG binds the fluorescein-IgG conjugate and sterically hinders the fluorescein antibody from binding the labeled ligand, with a maximum fluorescent signal being produced. As the concentration of the unlabeled ligand, IgG, increases, a competitive-binding reaction results, with less antibody to IgG binding the labeled conjugate. This allows the antibody to the label to bind the conjugate and quench, that is, diminish, its fluorescence. Quenching is inversely proportional to the concentration of ligand in the reaction. The fluorescence protection immunoassay has been successfully demonstrated for the measurement of levels of macromolecules such as IgG. Variations have also been applied to measurement of endogenous antibody to cardiolipin, as well as for T_3 uptake. Since the competitive approach requires that a ligand-antibody complex be formed for optimum inhibition of antifluorophore binding, the assay is readily applied to the measurement of macromolecules that have more than one antigenic determinant and to the monovalent, low molecular weight haptens.

Fluorescence energy-transfer immunoassay

When some fluorophores are brought into proximity to other absorptive molecules where the fluorescence spectrum of the former (donor) overlaps the absorption spectrum of the latter (acceptor), there can be energy transfer between the two molecules. This phenomenon results in quenching of the donor fluorescence by the acceptor. Fluorescein and rhodamine are a suitable energy-transfer pair if they are brought within 6 nm of one another. Under these conditions, the acceptor rhodamine, which absorbs light maximally at 520 nm, will quench the fluorescence of the donor fluorescein. This principle has been applied to competitive immunoassays as is illustrated in Fig. 11-25. If the ligand is labeled with fluorescein and its binding partner, the specific antibody, is labeled with the acceptor rhodamine, binding of the fluorescein-ligand conjugate by rhodamine-labeled antibody will quench the fluorescence of the fluorescein label. Competition for the antibody-binding sites by unlabeled ligand prevents the interaction, thus resulting in fluorescence increases in proportion to the concentration of unlabeled ligand, as also shown in Fig. 11-25.

This homogeneous immunoassay technique can be applied to the measurement of both low and high molecular weight analytes. It has already been used to determine the concentrations of thyroxine, morphine, thyroxine-binding globulin, albumin, and IgG.

Prosthetic group–labeled immunoassay (PGLIA)

Another homogeneous immunoassay that uses a reactant label is the prosthetic group–labeled immunoassay. Prosthetic groups are nonprotein components of some enzymes (or other proteins) that are required for biological activity of the enzyme (or protein). These groups are very tightly bound by the protein component, apoprotein or apoenzyme. With some enzymes, the prosthetic group and apoenzyme can be disassociated to render the enzyme inactive. Combining the apoenzyme and prosthetic group

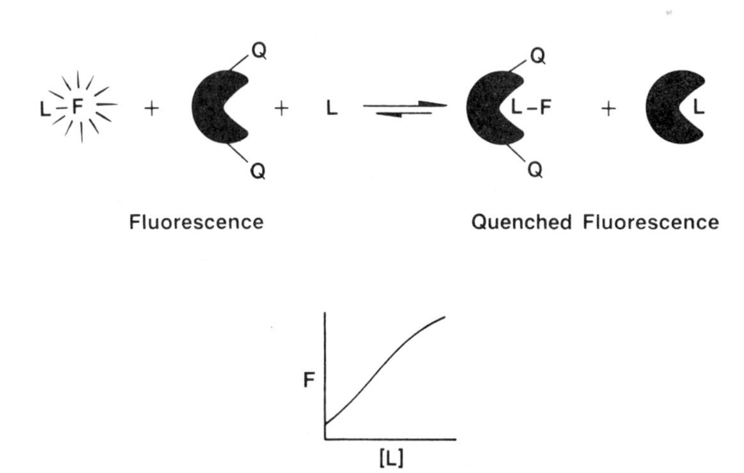

Fig. 11-25. Principle of fluorescence energy transfer immunoassay and typical dose-response curve. Acceptor molecules, *Q*, attached to antibody quench the fluorescence of antibody-bound fluorophore–labeled ligand, *L-F*.

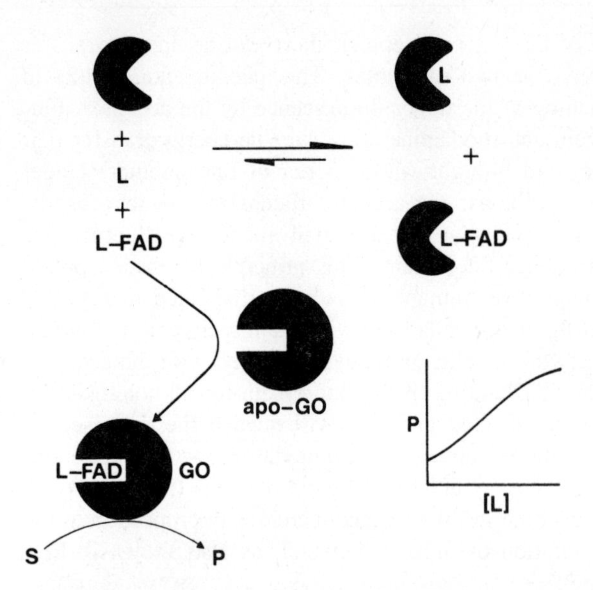

Fig. 11-26. Principle of prosthetic group–labeled immunoassay (PGLIA) and typical dose-response curve. Ligand, *L,* competes with flavin adenine dinucleotide–labeled ligand *L-FAD,* for antibody-binding sites. Unbound L-FAD is free to react with apo–glucose oxidase (apo-GO) to generate an active enzyme.

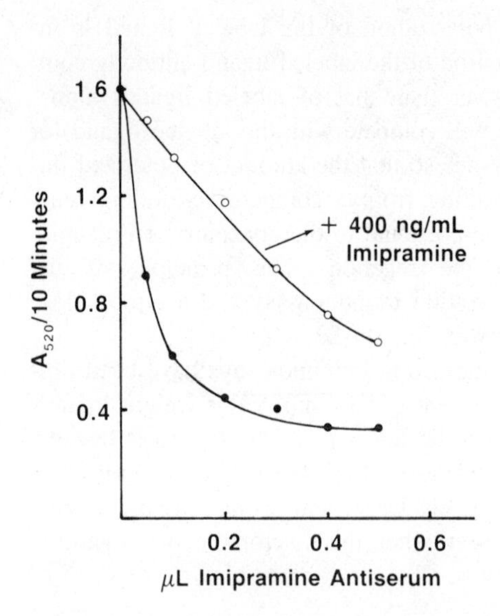

Fig. 11-27. Inhibition of formation of active glucose oxidase by rabbit antiserum to tricyclic drug imipramine in absence and presence of diluted 400 ng of imipramine per milliliter standard.

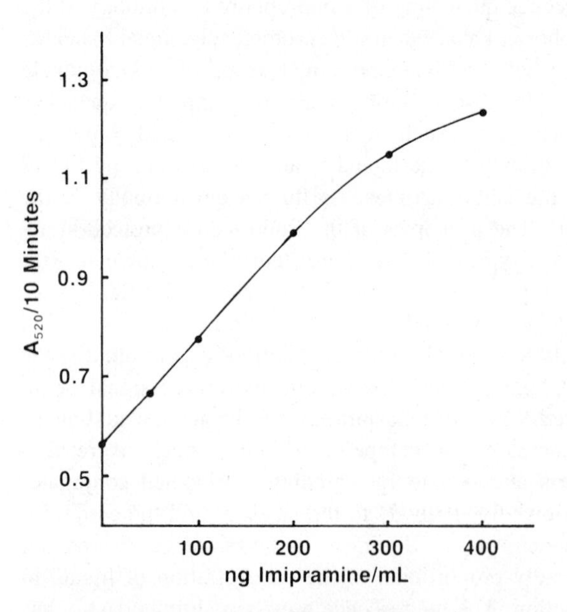

Fig. 11-28. Dose-response curve for imipramine prosthetic group–labeled immunoassay (PGLIA).

can regenerate the enzyme to yield an active protein. The prosthetic group–labeled immunoassay is dependent on the ability of a ligand labeled with the prosthetic group flavin adenine dinucleotide (FAD) to combine with apoglucose oxidase and regenerate glucose oxidase activity. Fig. 11-26 illustrates the principle of this assay. As shown in Fig. 11-27, binding of a constant amount of imipramine-FAD by antiserum to the antidepressant drug imipramine prevents the FAD-labeled conjugate from combining with excess apoglucose oxidase in the reaction mixture, thereby reducing the amount of active enzyme regenerated (as monitored by the color produced in the detection system). Inhibition of reactivation by ligand-specific antibody is reversed in a competitive protein-binding assay by the addition of unlabeled ligand. The dose response of a prosthetic group–labeled immunoassay for imipramine is shown in Fig. 11-28. The absorbance generated in this assay is proportional to the concentration of imipramine in the sample. The prosthetic group–labeled immunoassay has also been used to measure the concentration of macromolecular ligands such as IgG in biological fluids.

ADVANTAGES AND DISADVANTAGES OF DIFFERENT APPROACHES TO COMPETITIVE PROTEIN-BINDING ASSAYS

Before I describe the advantages and limitations of the competitive-binding assays that have been described above, it is useful to discuss some factors to be considered when one is designing or choosing a particular assay format. These include both convenience and performance factors.

Factors that should be considered by the user for convenience purposes are as follows:

1. Number of pipetting steps
2. Incubation time
3. Rate or end-point assay
4. Need for additional equipment such as incubators, centrifuges, or specialized detectors
5. Automation possible
6. Sample volume requirement
7. Sample pretreatment (extraction, and so on)
8. Temperature-control requirement

Table 11-12. Some characteristics of various competitive-binding assays

Immunoassay	Homogeneous or heterogeneous	Detection limit (M)	Amplification	Low/high molecular weight ligands
RIA (radioimmunoassay)	Heterogeneous	1×10^{-12} to 1×10^{-14}	No	Both
ELISA (enzyme-linked immunosorbent assay)	Heterogeneous	1×10^{-11}	Yes	Both
EMIT (enzyme-multiplied immunoassay technique)	Homogeneous	5×10^{-10}	Yes	Low
SLFIA (substrate-labeled fluorescent)	Homogeneous	1×10^{-9}	No	Both
FPIA (fluorescence polarization)	Homogeneous	5×10^{-9}	No	Low
Fluorescence protection	Homogeneous	3×10^{-9}	No	Both
Fluorescence energy transfer	Homogeneous	1×10^{-10}	No	Both
PGLIA (prosthetic group labeled)	Homogeneous	5×10^{-10}	Yes	Both

9. Cost
10. Operator time
11. Radioactive-waste disposal
12. Applicable to a variety of analytes
13. Stability of reagents

Obviously, the ideal assay would require very little sample with no pretreatment, a short incubation time at room temperature, few or no pipetting steps (if automated), and no additional equipment other than the detector or an automated instrument. It would be able to be done quickly (stat.), cheaply, with very little operator hands-on time, and have a rapid data reduction capability. The reagents would be stable for at least a year.

Some of the most important performance factors are as follows:

1. Good assay response over the range of the standard curve.

2. A low background signal—that signal caused by either nonspecific interactions of the reagents in the assay or contributions by other substances in the standards or samples—must be kept to a minimum to maximize the slope and range of the standard curves.

3. High antibody affinity increases the slope of the dose-response curve. It determines the sensitivity or detection limit of an assay (the lowest concentration in the assay that is significantly different from zero). Sensitivity is a function of the slope of the dose-response curve and the experimental error (accuracy and precision of the assay), in addition to the activity and detectability of the label.

4. Good precision and accuracy are also important requirements for competitive-binding assays. Accuracy is simply the ability of the assay to measure the true concentration of the ligand in a sample. Accurate measurements are dependent on the accuracy of the claimed concentrations of standards and the ability for correct interpolation of the true shape of the dose-response curve between the standards. Obviously, a linear dose response, or a linear transformation of a nonlinear one, enables one to make a reasonably accurate interpolation between the standards. Precision is the measure of variation of multiple determinations of the same sample in the same assay run (intra-assay precision) and in different assay runs (interassay).

5. Specificity in competitive-binding assays is determined by the ability of the antibody to discriminate between the ligand and similarly structured substances. In most cases, the screening and selection process for suitable antisera or monoclonal antibodies plays a very important part in determining antibody specificity. This is true for both native ligands (such as proteins) and the synthetic immunogens prepared to produce antibodies to low molecular weight substances. Since specificity is a function of antibody affinity for the ligand, the various assays that have been described will have similar cross-reactivity for related compounds. For example, if caffeine cross-reacts 10% in a theophylline radioimmunoassay, it will also cross-react in an enzyme-linked immunoabsorbent assay.

Table 11-12 compares some characteristics of the previously described competitive-binding assays. The heterogeneous assays RIA and ELISA have the most potential for sensitivity, with detection limits of a picomolar or less in RIA. In addition to the high and specific activities of the radiolabels or enzyme labels, potential interferences that can be caused either by substances in the sample or by impurities in the reagents are removed by the separation or wash steps. These procedures reduce the background signal so that a greater volume of sample can participate in the assay, thus increasing the sensitivity.

The short half-life of the radioactive labels used in the clinical laboratory (such as ^{125}I) limits the use of RIA reagents to a range of 6 to 12 weeks. This is short in comparison to the non–isotope labeled assays, which often are stable for a year or longer. In addition, radioisotopes may be hazardous and are subject to governmental regulation in their use and disposal.

ELISA and the other nonisotopic immunoassays are neither regulated nor potentially hazardous. Since enzymes are biochemical amplifiers, the systems that employ enzyme labels are capable of producing a greatly amplified signal, depending on the specific activity of the enzyme and the incubation time for conversion of substrate to product. The use of fluorogenic substrates instead of chromogens enables these assays to be 100 to 1000 times more sensitive. Their use in the case of ELISA, where the washing steps have removed potential background fluorescence and other interferences present in the sample or reagents, can be particularly advantageous.

The washing steps employed when one is performing an ELISA make this format less convenient than the homogeneous assays. With ELISA techniques there is also the chance for steric hindrance in the ligand-antibody binding reaction because of the introduction of the enzyme label. Coupling of the label to the ligand, or binding of the ligand-enzyme conjugate to the antibody, could lower the activity of the label, thus reducing the sensitivity of an ELISA. Reproducible application of antibody to the solid phase reduces the error often seen in competitive ELISA's and other solid-phase heterogeneous competitive assays.

The homogeneous enzyme immunoassay (enzyme-multiplied immunoassay technique, EMIT), though limited by the absorbance of the product, is still sensitive to subnanomolar levels of ligand. Like ELISA, the sensitivity may be increased by use of a fluorogenic substrate, enabling the amount of sample to be increased in addition to lowering the sensitivity limit inherent in fluorescence measurements. The EMIT is subject to possible interferences by hemolyzed, lipemic, and icteric samples. The EMIT is performed by use of a rate measurement, making it difficult to do manually, but it is quickly and easily done with automated equipment. It normally has a very short incubation time. The assay is not suitable for the higher molecular weight ligands, since binding by antibody usually will not cause sufficient enzyme inhibition for unlabeled ligand to provide a dose response.

The fluorescence assays are also sensitive at subnanomolar concentrations of ligand. The described assays are all homogeneous assays; therefore they avoid errors that can be introduced by the separation steps of heterogeneous systems. These assays are also prone to interferences by hemolyzed, lipemic, and icteric samples. In addition, other possible sample interferences in fluorescence immunoassays include light scattering from lipids and particulates, fluorescence quenching, and background fluorescence from the presence of endogenous fluorophores. The fluorescence polarization immunoassay is not applicable to determination of the concentration of high molecular weight ligands, is sensitive to the depolarized scattered light of particulates in the assay, and requires sophisticated instrumentation. Some labeled ligands may be nonspecifically bound by endogenous proteins, resulting in an increased polarization background. Each fluorescent assay can be adapted to a heterogeneous assay format.

The prosthetic group–labeled immunoassay has the same potential for amplification of the signal because an active enzyme is generated in the assay. Unlike other homogeneous enzyme immunoassays, antibody binds a ligand-label conjugate necessary to generate an active enzyme, rather than binding a ligand coupled to an enzyme that is already active. Therefore the prosthetic group–labeled assay should have lower background activity, resulting in a more sensitive assay. The sensitivity of the prosthetic group–labeled immunoassay can also be improved by the use of a fluorescent readout.

BIBLIOGRAPHY

Ligand binding and competitive protein-binding assays

Haurowitz, F.: Immunochemistry and the biosynthesis of antibodies, New York, 1968, John Wiley & Sons, Inc.

Odell, W.D., and Doughaday, W.H., editors: Principles of competitive protein binding assays, Philadelphia, Pa., 1971, J.B. Lippincott Co.

Pressman, D., and Grossberg, A.L.: The structural basis of antibody specificity, Reading, Mass., 1973, W.A. Benjamin, Inc.

Travis, J.C.: Fundamentals of RIA and other ligand assays, Anaheim, Calif., 1977, Scientific Newsletter, Inc.

Immunoassays (under a variety of formats)

Hakamura, R.M., Dito, W.R., and Tuckler, E.S., III: Immunoassays: clinical laboratory techniques for the 1980's (laboratory and research methods in biology and medicine, vol. 4), New York, 1980, Alan R. Liss, Inc.

Kaplan, L.A., and Pesce, A.J., editors: Nonisotopic alternatives to radioimmunoassay: principles and applications, New York, 1981, Marcel Dekker, Inc.

Radioimmunoassay

Parker, C.W.: Radioimmunoassay of biologically active compounds, Englewood Cliffs, N.J., 1976, Prentice-Hall, Inc.

Walker, W.H.C.: An approach to immunoassay, Clin. Chem. **23:**384-402, 1977.

Enzyme immunoassay

Maggio, E.T., editor: Enzyme-immunoassay, Boca Raton, Fla., 1980, CRC Press, Inc.

Wisdom, G.B.: Enzyme-immunoassay, Clin. Chem. **22:**1243-1255, 1976.

Substrate-labeled fluorescent immunoassay

Burd, J.F.: The homogeneous substrate labeled fluorescent immunoassay. In Langone, J.J., and Van Vunaskis, H., editors: Methods in enzymology, vol. 74, Immunochemical techniques, Part C, New York, 1981, Academic Press, Inc.

Burd, J.F., Carrico, R.J., Fetter, M.C., et al.: Specific protein binding reactions monitored by enzymatic hydrolysis of ligand fluorescent dye conjugates, Anal. Biochem. **77:**56-67, 1977.

Burd, J.F., Wong, R.C., Feeney, J.E., Carrico, R.J., and Boguslaski, R.C.: Homogeneous reactant–labeled fluorescent immunoassay for therapeutic drugs exemplified by gentamicin determination in human serum, Clin. Chem. **23:**1402-1408, 1977.

Fluorescence polarization immunoassay

Dandliker, W.B., Kelly, R.J., and Dandliker, J., Farquhar, J., and Levin, J.: Fluorescence polarization immunoassay: theory and experimental method, Immunochemistry **10:**215-227, 1973.

Jolley, M.E., Sharpe, S.D., Wang, C.-H.J., et al.: Fluorescence polarization immunoassay: monitoring aminoglycoside antibiotics in serum and plasma, Clin. Chem. **27:**1190-1197, 1981.

Fluorescence protection immunoassay

Zuk, R.F., Rowley, G.L., and Ullman, E.F.: Fluorescence protection immunoassay: a new homogeneous assay technique, Clin. Chem. **25:**1554-1560, 1979.

Fluorescent energy-transfer immunoassay

Ullman, E.F., Schwarz, Berg, M., and Rubenstein, K.E.: Fluorescent excitation transfer immunoassay, J. Biol. Chem. **251**:4172-4178, 1976.

Van Der Werf, P., and Chang, C.-H.: Determination of thyroxine binding globulin (TBG) in human serum by fluorescence excitation transfer immunoassay, J. Immunol. Methods **36**:339-347, 1980.

Fluorescence immunoassay

Visor, G.C., and Shulman, S.G.: Fluorescence immunoassay, J. Pharm. Sci. **70**:469-475, 1981.

Prosthetic group–labeled immunoassay

Morris, D.L., Ellis, P.B., Carrico, R.J., et al.: Flavin adenine dinucleotide as a label in homogeneous colorimetric immunoassays, Anal. Chem. **53**:658-665, 1981.

Chapter 12 Measurement of colligative properties

James E. Davis

"acetone" bodies The acetone and other ketones that are present in the serum of patients with diabetic ketoacidosis.

activity The effective concentration of the molecules of a solution.

boiling-point elevation A phenomenon in which addition of solute molecules raises the temperature at which the solution will boil (in equilibrium between liquid form and gas form). For water, this is degrees per mole of solute per kilogram.

colligative property A characteristic to which all the molecules of a solution contribute, regardless of their individual composition or nature.

colloid A large molecule, usually in aqueous solution. Normally the term is applied to protein solutions.

colloid osmotic pressure (COP) The osmotic pressure generated by that portion of a solution having high molecular weight (greater than 30,000 daltons).

crystalloids The uncharged solute molecules of a solution.

dew point The temperature at which condensation of water from the vapor phase occurs.

diffusion Mixing or movement of molecules as a result of their random motion.

Donnan effect The distribution of ions caused by having a high molecular weight ion on one side of a semipermeable membrane.

freezing-point depression A phenomenon in which the addition of solute molecules to a solution lowers the temperature at which the solution will freeze (that is, the equilibrium between solid and liquid form).

molality The number of moles of solute per kilogram of water or solvent.

molarity The number of moles of solute per liter of water or solvent.

oncotic pressure Another term for colloid osmotic pressure.

osmolal gap The difference in serum between the observed and calculated osmolalities. The calculated osmolar values include sodium concentration multiplied by 2, plus glucose and blood urea nitrogen.

osmolality The measurement of the number of moles of particles per kilogram of water.

osmometry The measurement of a colligative property of a solution in which the number of moles of a substrate per unit volume (concentration) are determined.

osmosis Water flow across a semipermeable membrane.

osmotic pressure The hydrostatic pressure required to prevent a change in volume when two solutions of different concentrations are placed on opposite sides of a semipermeable membrane.

plasma expander Usually a high molecular weight dextran that is administered intravenously to increase the oncotic pressure of a patient.

Seebeck effect The phenomenon of a voltage difference when two ends of a specially made wire are at two different temperatures.

semipermeable membrane A barrier that allows one type of molecule, such as water, to pass but does not allow another type of molecule, such as protein, to pass.

thermistor A device in which the conduction of electrons is temperature dependent. It is derived from the words *"thermal resistor."*

thermocouple A device that generates a voltage (Seebeck effect) when the two ends of a wire are at different temperatures.

ultrafiltrate The solution remaining after passage through a semipermeable membrane. Usually it contains only low molecular weight solutes.

vapor pressure depression The phenomenon in which the addition of a solute molecule to a solvent will decrease the amount of solvent in equilibrium between the vapor phase and the liquid phase.

COLLIGATIVE PROPERTIES

Osmosis

Osmosis is neither simply a mixing of two fluids nor simply a diffusion. Diffusion is the mixing of molecules as a result of random motion caused by thermal kinetic energy (brownian motion). For example, if an albumin solution were carefully overlayed with water, the albumin molecules would randomly move back and forth across the original interface boundary. Because there are more albumin molecules in the albumin solution, the odds are great that an albumin molecule will cross into the water side. Thus albumin will diffuse into the water layer until the solutions become homogeneous, that is, when the odds are equal that an albumin molecule will diffuse one way or the other across the original boundary because the concentration is equal on both sides. The term *osmosis* specifically applies to water flow across a semipermeable membrane such as a cell wall. It can occur with any fluid, but water is the most important to this discussion. A semipermeable membrane allows some particles (molecules, ions, or aggregates of molecules) to pass through it, whereas it inhibits others; hence it is *semi*permeable (permeable to some particles). The simplest example of a semipermeable membrane is a dialysis membrane, which is usually made of cellophane. It has very small pores through which water and some small molecules and ions pass. However, large molecules, such as proteins, cannot pass. To demonstrate, place an albumin solution in a section of dialysis tubing and tie the ends of the tubing. If the tubing is placed in a beaker of water, the albumin molecules cannot move out of the membrane, but water molecules will move in and affect dilution of the albumin. As a result, the tubing will swell as water flows into the albumin solution inside the tubing. The pressure inside will increase because more water is added to a fixed volume. If the tubing does not burst from this pressure, an equilibrium will be maintained between the water flowing in and the water being forced out by the internal pressure. The buildup and maintenance of hydrostatic pressure by this process is called *osmotic pressure*.

Perhaps the most graphic example of osmosis is lysis of red cells by water. So much water flows into the more concentrated intracellular fluid that the cell swells and bursts. Cells can also shrink if exposed to a fluid of high osmolality. In this case the water in the cell flows out of the cell to the concentrated solution outside. A laboratory example of the sensitivity of this process is the measurement of mean corpuscular volume (MCV) by a cell counter. If the diluent is not isotonic, that is, of equal osmotic pressure, the cells will swell or shrink, giving an erroneous MCV and hematocrit value because the latter is calculated from the mean corpuscular volume.

Osmolality

Usually the term *molarity* is used to characterize concentration, that is, the number of moles of solute per liter of water. *Molality* is the number of moles of solute per kilogram of water. Because a liter of water has a mass of 1 kg, the difference between these two expressions of concentration is usually small. Only for concentrated solutions is the difference appreciable. In practice it is the difference between adding material to a liter of water (molality) and adding water to material to make a liter of solution (molarity).

Molality is the term best suited to osmometry because it gives a simpler theoretical formula for osmotic pressure than molarity does. The term *osmolality* is used to identify the number of moles of particles per kilogram of water.

Because the osmolality of a solution does not depend on the kind of particles but only on the number of particles, it is called a *colligative* property. That means that isolated particles act together without regard to their type to produce the osmotic effect. For example, a solution that is 1 millimolal in sodium chloride is 2 milli*osmolal*. The reason is that sodium chloride separates into sodium and chloride ions. Each kind of ion represents a particle that contributes to the osmolality. Furthermore, a 1 millimolal calcium chloride solution is 3 milliosmolal because each molecule ionizes to give one calcium ion and two chloride ions. It is important to note that the charge on an ion has only a secondary effect on the osmolality, whereas the charge has a primary effect on the ionic strength. To be technically correct, presume that the osmolality is equal to the molal "activity." The activity is the effective concentration and depends on the ionic strength. Corrections for activity are between 5% and 10% for physiological solutions and are necessary for accurate calibration of osmometers. Usually it is not necessary to be concerned with such calculations because the calibration standards have already taken this into account. Changes in the formulation of calibration standards that are used to make theoretical corrections for activity can lead to changes in the clinical reference interval (normal range).

Osmometry

Osmometry is a measure of concentration, not of a particular molecule, but of molecules and ions in general. Physiologically it relates to the balance between the intracellular and extracellular water. In this chapter the clinical importance and use of osmometry is discussed. The techniques for measuring the osmolality are reviewed, with examples of instrumentation used.

Osmolal gap

There are just a few substances in plasma that contribute significantly to the osmolality, and they are mostly small molecules and ions. For example, plasma usually contains 40 g of albumin per liter, but the number of moles of albumin is very small (only about 50 μmol). In contrast, plasma contains about 150 mmol of sodium ion and 100 mmol of a corresponding anion such as chloride. This is only 5.8 g of sodium chloride per liter. Thus sodium chloride contributes 3000 times more to osmolality than a similar mass of albumin does.

Many formulas have been used to calculate the approximate osmolality of serum or plasma. Most such formulas are an attempt for accuracy with simplicity in calculation. A formula that requires measurement of many substances is not very useful clinically. The calculated osmolality can be used for comparison with the measured osmolality; the difference is called the *osmolal gap*. An abnormal osmolal gap is an important clue to abnormal concentrations of unmeasured substances in the blood. Because the formula predicts the plasma osmolality so well, there is little new information to be gained from *routine* measurements of the osmolality. However, in those special situations described in the next section, the measurement is informative and worthwhile.

The formulas below for the calculation of the osmolality are "zero-order approximations" because they consist of only the most important contributors:

Historical units

$$\text{Calculated osmolality (mOsm/kg)} = 2 \cdot \text{Na (mEq/L)} + \frac{\text{Glucose (mg/dL)}}{18} + \frac{\text{BUN (mg/dL)}}{2.8} \quad \textit{Eq. 12-1}$$

S.I. units*

$$\text{Calculated osmolality (mOsm/kg)} = 2 \cdot \text{Na (mmol/L)} + \text{Glucose (mmol/L)} + \text{BUN (mmol/L)} \quad \textit{Eq. 12-2}$$

The S.I. units formula is very straightforward. The factor 2 in both equations counts the cation (sodium) once and the corresponding anion once. Glucose and BUN are undissociated molecules and are counted once each. All other components are ignored. In the historical-units formula the dividing factors represent the respective molecular weights and conversion from deciliters to liters.

Notice that these formulas use molarity rather than molality. This approximation fortuitously compensates for

*Système International d'Unités, or simply "metric" units.

some of the other serum components and theoretical corrections that were ignored.

The osmolal gap is defined as:

$$\text{Osm/kg} = \text{Measured Osm/kg} - \text{Calculated Osm/kg} \qquad \textit{Eq. 12-3}$$

The average osmolal gap is near zero. However, the calculated osmolality has not been corrected for the actual water content of plasma (lipids and protein take up some of the volume). Such a correction would be nearly constant and therefore would not improve the quality of information from the formula, though it would change the reference interval (normal range). Because the work of making such corrections does not improve the clinical utility of the osmolal gap or of osmolality in general, such corrections are not used.

CLINICAL USE OF OSMOMETRY

There are two major uses of osmometry: (1) detection of unmeasured substances in the plasma and (2) assessment of renal concentrating ability.

Plasma osmolality

Only a few substances can be ingested in amounts sufficient to affect the plasma osmolality. Table 12-1 lists the substances and concentrations necessary to clearly affect the osmolal gap, that is, to increase the gap to 10 or more milliosmoles per kilogram. The most common substances are alcohols. If ethanol is measured specifically and does not agree with the osmolal gap within 10 mOsm/kg, with the molar conversion to molal being ignored, then there is reason to suspect that another of the substances listed in Table 12-1 is also present. Table 12-1 shows that trichloroethane can be at near-lethal levels in the blood without being readily detected by osmometry.

Although osmometry has long been recommended as a means to detect alcohol, it should be noted that vapor-pressure osmometers are not useful for the detection of alcohol. This insensitivity occurs because alcohol forms part of the solvent and thus its vapor pressure.

An increase in the osmolal gap will also reflect an increase in the anion gap in patients with metabolic imbalance. This change can occur because of an increase in the presence of "ketone bodies." The term *acetone* is often used to refer to those substances associated with ketoacidosis caused by diabetes. In fact, many substances present in ketoacidosis are ketones or ketone metabolites and not specifically acetone. The term "acetone" is used because these compounds react with sodium nitroprusside, a common test for the presence of acetone and other ketones as well.

Urine osmolality

Renal concentrating ability is a sensitive measure of kidney function. In the glomerulus, an ultrafiltrate (an aqueous solution containing no large molecules) is extracted from the bloodstream and passed to the renal tubules. The osmolality of the ultrafiltrate is nearly equal to that of the plasma. As it passes through the nephron, most of the ultrafiltrate, including water, is resorbed into the bloodstream. The urine that is delivered to the bladder is usually more concentrated than the plasma, typically one to three times more concentrated.

A random urine specimen is sufficient to demonstrate the kidney's ability to concentrate urine, that is, if the osmolality of the random urine specimen is greater than 600 mOsm/kg. However, if the random urine specimen is dilute, no conclusion about concentrating ability can be made. A definitive follow-up test involves overnight water restriction. After the morning void, at least one urine specimen should exceed 850 mOsm/kg. Patients who are compulsive water drinkers may need continuous observation to assure that no water has been ingested.

One should note that the specific gravity as estimated by the refractive index can be used to measure urine concentration. However, osmometry is less affected by the presence of protein or radiocontrast dyes.

Serum or plasma osmolality

Sample-collection technique is important to obtain a valid specimen for measurement of osmolality. For example, stasis during phlebotomy should be avoided. In addition, the sensitivity of osmometry is sufficient to measure physiologic water shifts between the supine and upright

Table 12-1. Substances affecting plasma osmolality

Substances	Toxic or lethal concentration		Corresponding increase in osmolality (mOsm/kg)
	Historical units (mg/dL)	S.I. units (mmol/L)	
Ethanol	350	80	80
Isopropanol	340	60	60
Methanol	80	24	24
Ethyl ether	180	24	24
Trichloroethane	100	9	9
Acetone*	55	10	10

*Includes other ketones or ketone metabolites.

Table 12-2. Estimated effect of anticoagulants on osmolality (compared to serum)

Anticoagulant	Full tube (mOsm/kg)	Half-full tube (mOsm/kg)
Heparin	+0	+0
EDTA (disodium salt)	+15	+30
Fluoride-oxalate (sodium fluoride–potassium oxalate)*	+150	+300
Iodoacetic acid (lithium salt)	+5	+10

*This hyperosmolal state accounts for the hemolysis usually observed in the plasma of these samples.

positions. Thus a sample from a fasting, hospitalized patient will give the most uniform results, not because lipemia or other effects are avoided but because it is more likely that the patient was still in bed at the time of the morning phlebotomy rounds.

Serum and heparinized plasma have similar osmolality values. The contribution to the osmolality by fibrinogen in plasma is small, and it is important only in the measurement of colloid osmotic pressure. Freezing-point depression techniques can use whole blood and are not affected by lipemia or hemolysis. Anticoagulants other than heparin increase the measured osmolality. Table 12-2 shows the estimated effect caused by various anticoagulants. On occasion, the kind of anticoagulant used can be verified by measurement of the osmolality of the plasma.

PRINCIPLES OF MEASUREMENT

Osmolality is a colligative property; thus any of four measurements that depend on colligative properties may be used. The measurements are (1) osmotic pressure, (2) boiling-point elevation, (3) freezing-point depression, and (4) vapor-pressure depression. Osmotic-pressure measurement has been used only in a special form of osmometry called ''colloid osmotic pressure'' and is discussed separately in a later section of this chapter. Boiling-point elevation is not useful for clinical samples because proteins will coagulate, causing gross changes to the sample composition. Of the remaining two, freezing-point depression is the more common technique. Vapor-pressure depression as measured by the dew point is a relatively new technique and is more commonly used in pediatric labs.

Freezing-point depression

The use of salt to melt ice and snow is a well-known practice. This is an example of freezing-point depression; that is, the salt increases the osmolality, thereby lowering the freezing point of the solution compared to that of the pure solvent (ice or snow). ''Salting'' is useful over a limited temperature range, and colder temperatures require more salt to reduce the freezing point adequately. When the temperature is colder than the depressed freezing point, the ice and snow will not melt. The temperature at which ice and the water solution are in equilibrium is a function of the salt concentration. More precisely, the temperature at equilibrium is a function of the number of particles in solution. The temperature is depressed 1.86° C for each mole of particles dissolved per kilogram of water. Because the osmolality of blood is about 0.3 Osm/kg (300 mOsm/kg), the freezing point is −0.62° C. Precise measurement of this temperature requires a sensitive thermometer. A thermistor (thermal resistor) is made from a mixture of transition metal oxides such as manganese, cobalt, and nickel. These materials are semiconductors, and the number of electrons in the conduction band (valence electrons of metal lattice capable of conducting a current) depends on the temperature. They become better conductors as the temperature rises. The conductance or resistance of the metals can be related to the temperature and hence to the osmolality.

The sequence of steps involved in the measurement of the freezing point depression is as follows:

1. The sample is cooled by a bath containing an antifreeze solution that is maintained at about −5° C by a conventional refrigerator or a thermoelectric cooler.
2. The sample is supercooled; that is, its temperature falls below the equilibrium freezing point. This occurs because pure ice crystals are slow to form.
3. The crystallization process is induced by vigorous stirring. Once ice crystals begin to form, additional water molecules are rapidly added to the ice crystals. However, heat is released in the freezing process just as it is absorbed in the melting process. The heat that is released from the formation of the ice crystals raises the temperature of the sample until the rapid freezing stops and an equilibrium temperature is established.
4. The temperature is measured at the plateau, that is, at the temperature at which the heat removed by the cooling bath is matched by the heat released from the freezing process. The temperature at this equilibrium is the freezing point of the solution and, of course, is inversely related to osmolality. The plateau's temperature is detected electronically by the thermistor, and the temperature reading is converted to milliosmoles per kilogram and displayed. At this time, before complete freezing, the thermistor is removed from the sample because the bath is still removing heat from the sample. The whole sample will eventually freeze and, in the process, could cause mechanical damage to the thermistor probe.

The nature of a colligative property is such that all solute molecules have the same effect on osmolality. In this case, the number of osmoles of solute is directly related to

Table 12-3. Characteristics of clinical osmometers*

Manufacturer	Model	Technique†	Sample size (μL)	Precision‡	Measurement time (sec)
Advanced Instrument, Inc. (Needham Heights, Mass.)	3D11	FP	200	1.5%	50-70
Fiske Associates, Inc. (Needham Heights, Mass.)	OS	FP	250	1.5%	90
	OR	FP	50		30
Precision Systems, Inc.	5002	FP	200	1.5%	180
(Sudbury, Mass.)	μOsmette	FP	50		
Wescor, Inc. (Logan, Utah)	5100C	VP	8	2.8%	90
Instrumentation Lab, Inc. (Lexington, Mass.)	186	COP	300	—	60
Wescor, Inc. (Logan, Utah)	4100	COP	300	—	

*All models are manually loaded with sample and have automated measurement and reporting. Variations in sample size, automated sampling, and printing are available.
†*FP*, Freezing-point depression; *VP*, vapor pressure; *COP*, colloid osmotic pressure.
‡From College of American Pathologists survey.

the extent of the freezing-point depression. One osmole of solute lowers the freezing point by 1.86° C. Therefore osmolality can be calculated directly by the formula:

$$\text{Freezing-point depression} = 1.86° \text{ C per mole of solute per kilogram (Osmolality)} \qquad \textit{Eq. 12-4}$$

However, it is more practical to calibrate the osmometer using saline solutions. Calibration also corrects for systematic or procedural effects, such as the increase in concentration of the sample because of the removal of pure water (as ice) before measurement of the temperature.

A number of factors must be considered to achieve high precision in freezing-point depression osmometry. These include the bath temperature, fluid composition, and amount of fluid. The fluid composition and volume change as moisture condenses from the room air. The thickness of the sample container and the amount of sample must be standardized. Of course the probe needs to be wiped to minimize carryover from one sample to the next. This is especially important between samples of widely differing osmolality, such as standards and urine samples. Samples can be remeasured, but great care must be taken to warm the sample (for example, by holding the sample cup in one's hand) until *all* of the ice is melted, otherwise the sample will freeze prematurely. Any sample droplets on the side of the cup should be joined with the main body by tipping of the cup to coalesce the droplets.

The most common freezing-point osmometers are listed in Table 12-3, along with several key characteristics.

Vapor-pressure depression

The temperature at which the atmosphere is saturated with solvent is measured by a thermocouple. A thermocouple generates a voltage (Seebeck effect) in the following way. When the temperature of one end of a wire is increased, the number of electrons in the conduction band increases. The excess of electrons created at the hot end will flow toward the cold end, giving it a negative charge, which will inhibit further electron flow. The voltage difference between the ends depends on the temperature of the ends. However, to complete the electrical circuit, another conductor needs to be attached to the hot and cold ends of the original wire. If the metal of this wire is the same as that of the original wire, the voltage between the hot and cold ends will be the same as the original; hence the voltage difference will be zero. Thus to get a useful voltage difference, one must have two wires of different materials so that the degree to which electrons are elevated to the conduction band in the second wire is different from that in the original wire.

Thermocouples also exhibit the Peltier effect, which is the opposite of the Seebeck effect. An electronic current through the thermocouple transfers heat from one junction to the other. One junction cools, whereas the other heats. The vapor-pressure osmometer passes an electric current through the thermocouple in the measurement chamber, causing it to cool. When its temperature falls low enough, water (solvent) begins to condense on it. The electronic current is discontinued, and the thermocouple comes to an equilibrium temperature at which the water condensing on it is matched by the water evaporating from it. This equilibrium temperature is measured by the Seebeck voltage, which is linearly related to the osmolality.

The sequence of steps involved in measurement of the vapor-pressure depression is as follows:

1. The sample is sealed in a chamber. An equilibrium is allowed to occur in which the air quickly changes humidity until its humidity is in equilibrium with the sample.
2. The thermocouple cools until its temperature is below the dew point. The electric current is turned off,

and the junction temperature rises as vapor condenses on it.

3. The plateau temperature (that is, the temperature at which an equilibrium exists between condensation and evaporation and their respective heating and cooling effects balance each other) is measured.

The vapor pressure of the sample is directly proportional to the thermocouple voltage. Again, it is more practical to calibrate this type of osmometer than to apply theoretical factors. Systematic or procedural effects that must be controlled include the sample volume, the size and composition of the sample absorbent disk, the time delay between sample application and sealing of the chamber, cleanliness of the chamber, and changes in room temperature.

Table 12-3 lists the single maker of a clinical vapor pressure osmometer, together with the comparison characteristics.

COLLOID OSMOTIC PRESSURE

Definitions

Osmotic pressure is a colligative property and hence reflects osmolality. This is strictly true for a semipermeable membrane that is permeable to water only. The difficulty in finding such a membrane has kept the measurement of osmotic pressure from being used as a technique in the assessment of osmolality in clinical samples. However, measurement of the osmolal contribution of only a portion of a group of molecules is practical and useful. This group is responsible for the colloid osmotic pressure (COP). This property is measured by use of membranes that are permeable to small molecules. Small molecules, less than 30,000 daltons of molecular weight, are called *crystalloids* if they are uncharged and *ions* if they are charged. Large molecules are called *colloids*. Hence colloid osmotic pressure measures only the contribution made to osmolality by large molecules. An alternate term is the *oncotic* pressure.

Clinical use of colloid osmotic pressure

The major use for the measurement of the colloid osmotic pressure is detection of conditions leading to pulmonary edema.

In this condition, there is an accumulation of water in the lungs, which interferes with oxygen and carbon dioxide exchange. The actual diagnosis can be obtained from x-ray measurements. Two measurements are needed to predict pulmonary edema: left ventricular pressure and colloid osmotic pressure. As long as the COP is greater than the pulmonary blood pressure (as measured by the "pulmonary artery wedge pressure"), pulmonary edema is unlikely. If heart failure is not present, that is, if the pulmonary blood pressure is normal, then COP measurements *alone* will predict the probability of pulmonary edema.

Knowledge of the albumin or total protein content of the plasma permits the calculation of the COP. However, the formula is inaccurate when used for acutely ill patients. The calculation is complicated when the patient has received dextrans, or "plasma expanders." Measurement of COP is useful for acutely ill patients with heart failure and for those having dextran plasma expanders because their COP cannot accurately be calculated. There is little reason to recommend COP measurements for screening.

Measurement of colloid osmotic pressure

The colloid osmotic pressure is measured with a microporous filter or membrane that contains pores or channels whose diameter is carefully controlled. Physiological saline is placed in a sealed chamber on one side of the membrane and the sample on the other. Saline flows into the sample until the back pressure stops further flow. This back pressure, or negative pressure, is sensed by a pressure gauge. In addition to the osmotic pressure, an additional pressure is created by the Donnan effect. This effect arises because the large molecules are proteins that are electrically charged. At the physiological pH, most proteins are negatively charged. Because the sample is electrically neutral, there are positive charges equal in number to the negative charges on the proteins. These positive charges are mostly in the form of sodium ions. Charged sodium ions diffuse through the membrane, whereas the corresponding negatively charged proteins do not. This leads to a separation of electrical charges. Because of the charge separation, additional small, negatively charged molecules will be attracted across the membrane. As a result the number of particles that diffuse will be larger than that resulting from simple osmosis, and the pressure across the membrane will be larger. Because the net charge on proteins changes with pH, the measured COP will also change with pH.

Customarily, COP is reported in millimeters of mercury (mm Hg). In practice a maximum pressure occurs 30 to 90 seconds after the sample is placed into the instrument. This value is chosen because the pressure decays with time as a result of imperfections in the membrane that slowly allow large molecules to diffuse to the saline side, thus reducing the true pressure.

To achieve high precision and good reliability, care must be taken with the delivery of the sample into the instrument. A syringe is normally used to put the sample into the measurement chamber. Forcing the sample into the chamber can damage or change the calibration of the pressure sensor. Enough sample must flow through the measurement chamber to wash out the previous sample. When the measurement is complete, the sample should be flushed out with saline solution. Saline should always be kept in the measurement chamber, otherwise damage to the membrane may result or the base-line reading (zero) may shift. Special microsample techniques are used to minimize the amount of sample used to wash out the measurement chamber. Two makers of colloid osmometers are listed in Table 12-3.

NONCLINICAL USE OF OSMOMETRY

Because osmometry does not depend on the particular molecules in solution, it is exceptionally useful for quality control and troubleshooting of reagents. For example, if an inappropriate amount of water were used in the reconstitution of reagents, the osmolality would show that an error existed. In addition, if the wrong reagent were used, the osmolality would show a change. The pharmaceutical industry has long used osmolality as a part of its quality control. The clinical laboratories seem to use this fine resource less frequently than they should.

BIBLIOGRAPHY

Barlow, W.K.: Volatiles and osmometry, Clin. Chem. **22:**1231, 1976.

Dorwart, W.V., and Chalmers, L.: Comparison of methods for calculating serum osmolality from chemical concentrations and the prognostic value of such calculations, Clin. Chem. **21:**190-194, 1975.

Eklund, J., Granberg, P.O., and Halberg, D.: Clinical aspects of body fluid osmolality, Nutr. Metab. **14**(suppl.):74, 1972.

Hendry, E.B.: The osmotic pressure and chemical composition of human body fluid, Clin. Chem. **8:**246, 1962.

Johnston, R.B., Jr.: Osmolality in serum and urine. In Meites, S., editor: Standard methods of clinical chemistry, vol. 5, New York, 1965, Academic Press, Inc., p. 159.

Warhol, R.M., Eichenholz, A., and Mulhausen, R.O.: Osmolality, Arch. Intern. Med. **116**:743, 1965.

Wolf, P.L., Williams, D., Tsudaka, T., and Acosta, L.: Methods and techniques in clinical chemistry, New York, 1972, John Wiley & Sons, Inc. –Interscience.

Chapter 13 Electrochemistry

Stephen G. Weber

activity The property of a species that determines its effect on chemical equilibriums.

amperometry The measurement of current at a single applied potential.

anode The electrode at which oxidation occurs.

auxiliary electrode The electrode through which current flows, completing the circuit with the working electrode.

cathode The electrode at which reduction occurs.

charge A property of matter that if possessed by a species, causes that species to ''feel'' a potential.

charge transfer The event in which charge moves from one side of an interface to the other.

conductance See conductivity.

conductivity The property of a material that dictates the amount of current that will pass through the material because of a specified potential gradient.

coulometry The measurement of charge.

current The flow of charge. A current of one ampere is the flow of one coulomb per second.

electrode potential The electrostatic potential that defines the energy of electrons entering or leaving an electrode.

Faraday's constant The number of coulombs in a mole of electrons, designated F, which is 96,485 coulombs per mole.

free energy The theoretical maximum amount of energy available from a system out of equilibrium as the system goes to equilibrium.

half-cell reaction An electrically and chemically balanced reaction that explicitly contains electrons as a reagent. A half-cell reaction cannot proceed without a source or sink of electrons, such as another half-cell reaction.

input impedance A frequency-dependent measure of the resistance to the flow of current attributable to an applied voltage. A high-input impedance is required for electrochemical potential measurements.

ion-selective electrodes Potentiometric electrodes that develop a potential in the presence of one ion (or class of ions) but not in the presence of a similar concentration of other ions.

normal hydrogen electrode (NHE) A reference electrode that is assigned a value of 0 volts. The potential of the reaction

$$2H^+ + 2e^- \rightleftharpoons H_2$$

is measured at a platinum black electrode. The H_2 pressure is 1 atmosphere and the pH = 0.00.

oxidation The process in which a chemical species loses electrons.

polarography Voltammetry at a dropping mercury electrode.

potential The energy of a charged species has as one contribution the product of its charge times the potential. The gradient (slope) of a potential causes charge to move. The movement of one coulomb of charge from a potential of V to a potential of $V - 1$ volts requires one joule of energy.

potentiometer A voltage-measuring device.

reduction The process in which a chemical species gains electrons.

reference electrode An electrode that has a well-established potential and is used as a reference against which other potentials may be measured.

resistance The frictional force opposing charge flow.

salt bridge A device that allows charge to flow as ions between two separate containers of electrolyte that are also connected by metallic conductors.

saturated calomel electrode (SCE) A reference electrode commonly used in polarography. The potential of this electrode is based on the reaction

$$Hg_2Cl_2 + 2e^- \rightleftharpoons Hg^0 + 2\ Cl^-$$

in the presence of saturated KCl.

transference number For an ion in solution, the transference number of the ion is the fraction of the total current passing through the solution that is being carried by that ion.

valence orbitals The region around a molecule or ion that contains or may contain the electrons involved in chemical bonding and intermolecular forces.

volt The unit of electrostatic potential: 1 volt = 1 joule per coulomb.

voltammetry The measurement of current as a function of potential.

voltmeter A voltage-measuring device.

working electrode The electrode at which the process of interest occurs.

The field of electrochemistry deals with the interactions between chemical species and electric fields or electrons. The most dramatic events that one might see involving electrochemistry include changes in chemical properties. Consider the difference in mechanical strength between iron and rust, the difference in taste or smell between chlorine and sodium chloride, or more subtle differences such as the difference in oxygen-binding capabilities of hemoglobin (Fe^{2+}) and methemoglobin (Fe^{3+}), the difference in fluorescence between NAD^+ and NADH. All these differences are caused by a difference in the number of electrons in the *valence orbitals* of the ion, atom, or molecule under consideration. Recalling that the electrons and vacancies in valence orbitals are responsible for chemical bonding, one is not surprised that changing the number of valence electrons alters chemical properties.

In making chemical measurements one may take advantage of the fact that electrons are required to cause some chemical change. If the electrons come from an electrical circuit, the flow of electrons, as they cause their chemical change, may easily be measured, since flow of charge is *current*. Furthermore, in certain cases, the energy of the electrons required to cause a particular chemical change may be measured as an *electrode potential*. Finally, the fact that charged species create potentials and move in electric fields leads to other possibilities for measurement. Several electrochemical techniques and their properties are shown in Tables 13-1 and 13-2.

UNITS, CONVENTIONS, AND DEFINITIONS

At least some of the difficulty in understanding electrochemical phenomena is certainly attributable to misunderstandings of various terms. We will first discuss electrical potential, charge and current, conductivity, and finally activity.

Potential

Potential is an expression of energy. An important energy is the *free energy*. The free energy of a species is a

Table 13-1. Some electrochemical techniques in which electrodes are an integral part

Technique	Brief description	Typical application
Potentiometry	Measurement of potential of one electrode with respect to another. No permanent change of solution occurs.	Measurement of H^+, Na^+, K^+, and Ca^{2+} in serum.
Voltammetry	The measurement of current flowing into or out of an electrode in a solution. Measurements are made as a function of electrode potential. Extremely minute changes in the solution occur (<1% over several hours of measurement).	"Scouting" solutes for the ability to oxidize or reduce them and thereby determine them electrochemically.
Polarographic techniques	Voltammetry at a special electrode, the dropping mercury electrode. More sensitive polarographic techniques apply the potential to the electrode as a series of pulses. Extremely minute changes in the solution occur.	Quantitative determination of inorganic and organic species, such as Cu^{2+}, Pb^{2+}, Cd^{2+}, Zn^{2+}, tetracycline; 10^{-8} M detection limit is typical.
Stripping	A two-step technique. First, reduce metal ions to metal, which will dissolve in the mercury and can be concentrated there. Then, using voltammetry, observe the current, which is caused by oxidation of the metal amalgam back to the ions. Alterations in metal ion concentration may occur if solution is very small (<1 mL).	Trace-metal analysis, such as Pb^{++}, Cu^{++}, and As^{++}
Amperometry	The measurement of current at an electrode that is kept at a single constant potential. Small changes in solution may occur: in end-point detection, these are extremely small; in liquid chromatography, several percent.	End-point detection in titration, such as chloride. Detection in liquid chromatography at 10^{-8} to 10^{-9} M (in chromatographic effluent) is typical detection limit.
Coulometry	The measurement of charge (the integral of current over time) with an electrode at a constant potential. The attempt is made to alter completely the redox form of a solute; for example, all Cl^- in a sample goes to AgCl.	Titration, such as chloride. Detection in liquid chromatography at 10^{-8} M (in chromatographic effluent) is typical detection limit.

Table 13-2. Some electrochemically related techniques in which electrodes serve only to create a potential gradient

Technique	Brief description	Typical application
Conductance	The measurement of the ability of a solution to conduct electricity.	Measurement of presence of ions, as for water purity, detection in liquid chromatography.
Electrophoresis	The separation of charged species by their motion in an electric field. The field is applied across a buffered medium.	Separation of proteins.
Isotachophoresis	The separation of charged species. The sample is applied between two buffers. The components of the sample separate into contiguous zones based on mobility.	Separation of ions.

measure of its ability to enter into chemical reactions. Species move toward a lower free energy, and at equilibrium the free energy of a species is the same for all its forms. Part of the free energy of a charged species is equal to the charge on that species, z, multiplied by the potential at the point in space where the charge exists, ϕ, and a constant called *Faraday's constant, F. F* has a value of 96,485 coulombs per mole of electrons. Thus this portion of the free energy, G_ϕ, of a species is represented as $zF\phi$.

$$G = zF\phi \qquad \textit{Eq. 13-1}$$

The potential, ϕ, has units of energy per charge; one *volt* is one *joule per coulomb*. If one had one coulomb of charge (1.036×10^{-5} moles of electrons), it would require one joule of energy to move that coulomb from a place where the potential was zero volts to one where the potential was -1 volt. It would require one joule of energy to move that coulomb from a place where the potential was 10 volts to one where the potential was 9 volts. A characteristic of electrical potentials is that the important parameter is the difference between two potentials, not the absolute value of the potentials. Thus all batteries have two poles, and a certain potential exists between the two poles. In a 12-volt car battery the positive pole has a potential that is 12 volts higher than the negative pole. In fact, it is quite meaningless to speak of a single potential; when one hears of single potentials, there will always be some implied reference point. The following example shows a common reference point. The standard electrical outlet in this country is said to provide 120 volts. The

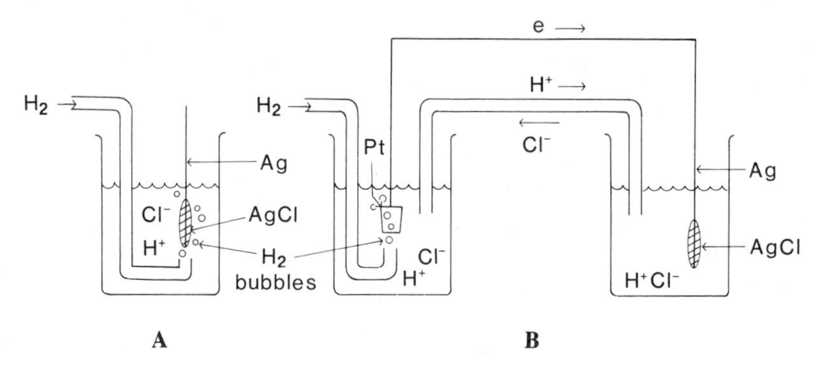

Fig. 13-1. A, Experimental apparatus in which hydrogen could reduce silver chloride (AgCl). **B,** Similar experimental apparatus in which hydrogen reduces AgCl, but here electrons from H_2 pass through external circuit to reduce AgCl. Charge is balanced by ionic-charge transport through salt bridge.

reference point is earth, or ground. If you were to drive a metal post into the ground, attach a wire to it, and then (carefully!) measure the potential difference between it and one of the slots in a home electrical outlet, a value of about 120 volts (the root-mean-square of the sine wave of potential) would appear. The other slot, when measured, would be near zero volts.

In electrochemical measurements one needs a device that can maintain a constant potential under a variety of conditions. Such a device is used as a reference against which potential measurements are made and is called a *reference electrode*. Thus any "potential measurement" is really a measurement of the potential difference between two electrodes. In analytical and physical investigations one is almost always concerned with changes at one of the electrodes, not at both. Thus, to measure reproducibly a change in potential of a single electrode, one must measure the change in potential difference between two electrodes, one of which is at a constant potential. This is the function of the reference electrode.

There are several reference electrodes in common use. One of the earliest, and the one against which oxidation-reduction potentials are commonly measured, is the *normal hydrogen electrode* (NHE). The platinum electrode used in this reference electrode is in equilibrium with electrons that are moving back and forth between H^+ (aqueous solution, pH = 0.00) and H_2 (gas at 1 atmosphere of pressure). A common reference in *polarography* and in other electroanalytical techniques is the *saturated calomel electrode* (SCE). Here a pool of mercury metal is in equilibrium with calomel (Hg_2Cl_2) in the reaction

$$Hg_2Cl_2 + 2e^- \rightarrow 2Hg^0 + 2Cl^- \qquad \textit{Eq. 13-2}$$

The aqueous solution is saturated with Hg_2Cl_2 and KCl. The silver–silver chloride reference electrode is also commonly used. A silver wire covered with a layer of AgCl is in contact with a solution of sodium or potassium chloride. Thus the reaction that controls the energy of the electrons in the silver wire is

$$AgCl + e^- \rightleftharpoons Ag^0 + Cl^- \qquad \textit{Eq. 13-3}$$

The reactions shown above for the reference electrodes are typical *half-cell reactions*. As the term implies, a half-cell reaction is incomplete; one needs two half-cell reactions to form a complete cell. Notice that the electron is explicitly included in the reaction scheme. Electrons are not kept on the shelf like other reagents, they must come from other half-cell reactions. One half-cell produces electrons while the other half-cell consumes them. For example, one might wish to reduce AgCl by using the hydrogen half-cell as the electron source, as follows:

$$H_2 \rightarrow 2H^+ + 2e^- \qquad \textit{Eq. 13-4, A}$$

$$2e^- + 2AgCl \rightarrow 2Ag^0 + 2Cl^- \qquad \textit{Eq. 13-4, B}$$

$$H_2 + 2AgCl \rightarrow 2Ag^0 + 2H^+ + 2Cl^- \qquad \textit{Eq. 13-4, C}$$

Experimentally one can do this in two ways: the electrons may go directly from H_2 to the silver ion, as by bubbling H_2 gas across AgCl on a silver wire, or one may require the electrons to travel through a wire to achieve the same result; for example, in one beaker place the reagents for a hydrogen half-cell with its platinum electrode attached to a platinum wire, and in another beaker place a silver half-cell with its silver wire; then connect the two wires and connect the two solutions with a *salt bridge*. The salt bridge conducts electricity through the solution by the motion of ions in the solution. Notice that the salt bridge completes the electrical circuit. These two systems are compared in Fig. 13-1.

Notice how charge neutrality is maintained in both systems. In Fig. 13-1, *A*, the neutral AgCl is reduced so that the silver ion becomes an uncharged silver atom and the chloride ion is set free to roam about in the solution (equation 13-3). On the other hand the hydrogen has gone from a neutral to a positive ion (equation 13-4, *A*); thus the charges from H^+ and Cl^- cancel. In Fig. 13-1, *B*, one can see that the salt bridge is necessary for the purpose of charge neutralization. In the silver half-cell, negative chlorides are produced, and in the hydrogen half-cell, positive protons are produced. The ions move through the salt bridge to keep charge neutrality in each half-cell. Thus the salt bridge completes the electrical circuit. In both *A* and *B* in Fig. 13-1, we are allowing chemical change to occur and electrical current to flow. Rather than allowing the hydrogen to actually reduce the silver ion, one could measure the potential of the hydrogen to reduce the silver ion. An analogy will help make the distinction clear. Consider that in place of the hydrogen-bound electrons there is a closed, pressurized container of fluid, such as a tank of compressed air. The silver ion, or electron-receiving system will be replaced by a container at room pressure. Now there are two ways to treat the pressurized container. One can ask it to do work, such as blowing up a balloon, in analogy to Fig. 13-1, *B,* in which the hydrogen half-cell is reducing the silver ion. On the other hand, one may ask what is the potential of the air tank to do the work of blowing up balloons. One can determine this by measuring the pressure of the cylinder. Analogously, one can measure the potential difference between the two metal electrodes, in this way measuring the potential of the system to do work.

The potential difference between the two half-cells would be measured as in Fig. 13-2. The measuring device, a *voltmeter,* or *potentiometer,* allows only the merest trickle of current to flow (about 1 picoampere, pA, equivalent to about 10^{-17} moles of electrons per second). This allows one to make the potential measurement without changing the relevant concentrations to any degree (to observe a 0.1%-concentration change in a 100 mL solution of 10^{-3} molar Ag^+ would take about 300 years!). The description of the potential difference may take many forms and has historically been the subject of much confusion. Today, however, the convention used is very rational, thus easy to

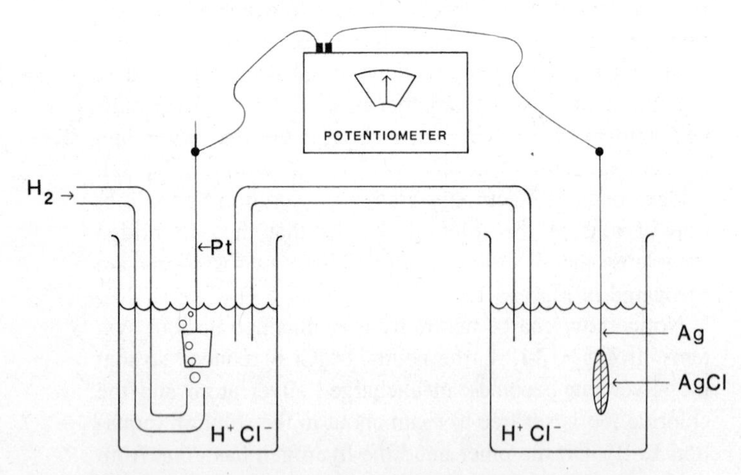

Fig. 13-2. As in Fig. 13-1, apparatus involves reduction of AgCl by H_2. In this apparatus, however, a voltmeter, through which very little current flows, is used to measure potential of cell.

remember. Briefly stated, *the half-cell with the more positive potential is a stronger chemical oxidizer than the other half-cell.* Any oxidation-reduction (redox) reaction involves two pairs of species, or two couples. One is an oxidizer, or oxidizing agent, and one is a reducer, or reducing agent. In a redox reaction the oxidizing agent is reduced while the reducing agent is oxidized. *Oxidation* implies the removing of electrons from a species, whereas *reduction* is the opposite. One may use the mnemonic "LEO (the lion, says) GER; Lose Electrons Oxidized, Gain Electrons Reduced." A species that is reduced has its formal charge reduced, since it has taken on electrons that are negative ($2H^+$ reduced to H_2; Cl_2 reduced to $2Cl^-$; Fe^{3+} reduced to Fe^{2+}). Conversely, a chemical species that is oxidized has its formal charge increased (H_2 oxidized to $2H^+$, $2Cl^-$ to Cl_2).

Returning the focus to the convention itself, let us look at a specific example. Consider the potential-difference measurement between a solution of unit activity of Ag^+ (such as silver nitrate) with a silver wire electrode. ("Activity" is discussed on p. 246; for now, consider the concentration to be near 1 molar.) The half-cell reaction is written as a reduction, by convention:

$$Ag^+ + e^- \rightarrow Ag^0$$

This is connected by means of a salt bridge to a solution of H^+ at unit activity (pH 0.00) with an electrode of platinum and with H_2 gas at 1 atmosphere of pressure. The half-cell reaction is

$$2H^+ + 2e^- \rightarrow H_2$$

The measured potential difference between the silver electrode and the platinum electrode at 25° C will be + 0.799 volts (with silver being positive with respect to platinum). Knowing this, consider the following question: Which of the following pairs of species, all at unit activity, will spontaneously react?

1. H_2 and Ag^0
2. H^+ and Ag^+
3. H^+ and Ag^0
4. H_2 and Ag^+

Oxidation-reduction reactions are under scrutiny, and one must remember that electrons are transferred from one species to another; thus while one species is reduced, the other is oxidized. Therefore (1) and (2) may be rejected as possibilities, H_2 and Ag^0 are both the reduced species, whereas H^+ and Ag^+ are both the oxidized species. Now the convention is that the half-cell with the more positive potential is the better oxidant. The silver half-cell has the more positive potential; it is the better oxidizer; it will therefore take electrons from the hydrogen half-cell. Of the silver half-cell, Ag^+ can take electrons; of the hydrogen half-cell, H_2 can give electrons; thus only reaction 4 will occur spontaneously.

Let us try another example. If the standard potential for the Zn/Zn^{2+} couple is -0.763 V and the standard potential for the Cu/Cu^{2+} couple is +0.340 V, which of the following pairs, all at unit activity, will spontaneously react?

1. Cu^{2+}, Zn^0
2. Cu^{2+}, Zn^{2+}
3. Cu^0, Zn^0
4. Cu^0, Zn^{2+}

Again, since an exchange of electrons must occur, (2) and (3) can be ruled out. The Zn^0/Zn^{2+} couple has the more negative potential, and therefore it is the better reducing agent; thus (1) occurs because Zn^0 is the species that has the capability to perform reductions.

Many compendiums have been published with lists of reduction potentials. The potentials are usually reported as if the potential difference between a particular half-cell has been measured against a reference hydrogen half-cell; thus the hydrogen half-cell used above is *arbitrarily* assigned a potential of 0.000 V. The other common reference electrodes have the following potentials versus the hydrogen half-cell: SCE (standard calomel electrode), +0.241 volt; Ag/AgCl (KCl saturated), +0.197 volt. To switch from one reference electrode scale to another, one may use a diagram such as that in Fig. 13-3.

It is important to realize that the relative values of half-cell potentials are independent of the reference scale. If one wants to use the SCE as the reference potential, it must be assigned a value of 0.000 V. One can do this by subtracting 0.241 V from its potential against the NHE. To keep the relative values of the half-cell potentials on the SCE scale the same, one must therefore subtract 0.241 V from *all* potentials against the NHE to arrive at a set of potentials against the SCE. Fig 13-3 does this by showing parallel scales shifted by 0.241 V for the two reference electrodes.

In Fig. 13-3, what is the standard potential for the reduction of NAD^+ (nicotinamide adenine dinucleotide) to $NADH + H^+$ using an SCE as a reference electrode if the potential for the couple is 0.060 versus the hydrogen electrode? The potential on the SCE scale just above 0.060 on the NHE scale is −0.181 V (0.060 V − 0.241 V). Thus this is the E^0 for the NAD couple with the SCE reference.

Current and charge

Current is the flow of charge. This charge is measured in coulombs; one mole of electrons has a charge of 96,485 coulombs. This number of coulombs is called a *faraday*. Current is measured in *amperes;* one ampere is the flow of one coulomb per second. Charge is caused to flow, creating a current when a potential difference exists between two points in space. *Ohm's law* is a simple expression of this fact:

$$E = IR \qquad \text{Eq. 13-5}$$

E is the potential difference in volts, I is the current that flows, in amperes, and R is called the resistance, measured in ohms. Current is defined as the quantity or amount of electrical charge passing a fixed point per unit of time. The *resistance* is a measure of the frictional force opposing charge flow.

In electrochemistry there are two distinctly different types of charge flow. One is conduction of electrons through metal electrodes and wires, whereas the other is the conduction of charged species, ions, in solution. In an electrochemical cell, ions flow through a salt bridge from one half-cell to the other. Unlike the resistance to electron conduction through conducting materials, the resistance to ionic flux in ionic conduction cannot be ignored. Very often in electroanalytical chemistry, matters are simplified if the species of interest do not themselves carry the current in the cell. This is accomplished if a large excess of other ions is present in the cell. Thus in the determination of ascorbic acid (vitamin C), by measuring the current from its oxidation at a carbon electrode, one would add to the sample an excess of an indifferent or supporting electrolyte, say potassium chloride or an acetate or phosphate buffer. Virtually any salt, acid, or base may be used as long as it does not interfere with the analysis.

If ions carry current in solution and electrons carry current in wires and electrodes, then at the interface between the two there must be a process in which an electron forms or changes the charge of an ion. This event, called *charge transfer,* is at the heart of much of electrochemistry. Recall that all the half-cell reactions, which were introduced in the discussion of potential, involved a change in charge on the chemical species as they gained or lost electrons. At that time we were concerned about the energetics of the process. If one were making a measurement of current in an electrochemical cell, because of the known stoichiometry of the half-cell reaction of interest, one would also be measuring the flow of the analyte to the electrode. This flow is concentration dependent, leading to the possibility of quantitative analysis.

In any electrochemical cell, oxidation must occur at one electrode and reduction at the other. If the electrons in an electrode are reducing a species at that electrode, the electrode is called a *cathode* and the current is called *cathodic*. The oxidizing current of the other electrode, an *anode,* is called *anodic*.

Although most electroanalytical procedures involve potential or current measurements, which depend on the specifics of charge transfer, *conductance* is one that does not.

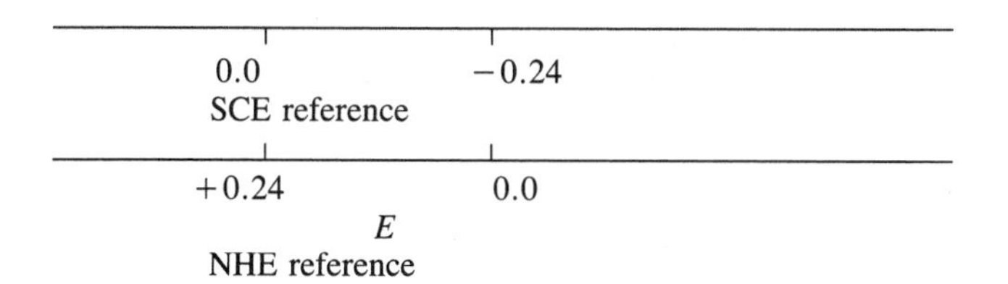

Fig. 13-3. SCE (saturated calomel electrode) reference scale versus NHE (normal hydrogen electrode) reference scale. *E*, potential difference (of electrode) in volts.

The objective of a conductance measurement is to measure the ability of a solution to conduct electricity. This ability of a solution to conduct a current depends on the concentration and mobility of ions in that solution. Ions conduct current through a solution by moving in an electric field. The viscous forces of the fluid surrounding the ion limit the rate at which an ion can move in a given field. The larger the ion and the more viscous the fluid, the lower the mobility of the ion. Increasing the charge on an ion increases its mobility.

Since any ionic solution is made up of at least two ions (a cation and an anion), one might ask how much of the current is carried by each ion. The fraction of the total current carried by an ion is called the *transference number* for that ion. For a simple two-ion solution, such as Na^+ and Cl^-, the transference number is determined by the relative mobilities of the two ions; the ion with the higher mobility will carry more of the current. The mobility of ions is most often determined by the conductance technique itself or by electrophoresis. Analytically speaking, conductance measurements are exceedingly nonselective, since all ions carry charge and one cannot discriminate among various ions using conductance. Conductance methods are most often used where clean, nonionic solutions exist as a matrix, or where changes in ionic content are important. Detection of ions after they are separated by ion-exchange chromatography and testing distilled water for ionic contaminants are examples of conductance measurements giving information in clean systems, whereas measurement of rates of enzyme reactions or bacterial growth are examples of determinations that are based on measuring changes in conductance with time.

Activity

Activity is a very important concept in chemistry and biochemistry, yet it is one for which it is difficult to give an intuitive feel. Activity and concentration are related by an activity coefficient

$$a = \gamma c$$

where a is activity, γ is the activity coefficient, and c is the concentration. The activity of a species represents its ability to participate in or influence chemical equilibriums. The activity coefficient may be profoundly altered by changes in the surroundings of a chemical species. Broadly speaking, things that tend to stabilize, or "tie up," a species will lower its activity coefficient and thus its ability to participate in reactions. For example, increasing the albumin concentration in serum reduces the activity of bilirubin, since the albumin binds bilirubin. It is a characteristic of potentiometric measurements that activity, not concentration, is the parameter measured. To distinguish between the two expressions, the convention used in this book will be that brackets "[ ]" represent concentration and parentheses "( )" activity.

POTENTIOMETRY

Energetics

In the measurement of the potential difference between two half-cells, there are several places that may contribute to the overall measured potential difference. In Fig. 13-2 one can see that the following potential differences may occur:

1. Between the Pt electrode and the HCl solution
2. From one end of the salt bridge to the other
3. Between the HCl solution and the AgCl, Ag electrode

Here we have neglected possible potential differences through the solutions themselves, since most analytical procedures are carried out under conditions where these potentials are small. To measure one potential difference only, one must keep the other two constant. We have already seen above the case in which one electrode-and-solution (half-cell) potential difference, and the salt-bridge (junction) potential difference are held constant. This results in a reference electrode. The remaining electrode-and-solution potential difference can then be measured. Such a measurement is called "potentiometry." It is not often used in this way in clinical chemistry because of its lack of selectivity. However, the simple relationship that such systems obey, the Nernst equation, is of fundamental importance in chemistry. The Nernst equation gives the electrode potential for a half-cell as a function of the standard potential for the redox couple present in solution and the concentrations of involved species.

For the reaction

$$\alpha A + \beta B + \ldots + ne^- \rightarrow \pi P + \sigma S + \ldots$$

the Nernst equation is

$$E_{\text{half-cell}} = E^0 + \frac{RT}{nF} \ln \frac{(A)^\alpha (B)^\beta \ldots}{(P)^\pi (S)^\sigma \ldots} \qquad \textit{Eq. 13-6}$$

In this equation R is the gas-law constant (8.316 joule/mol · K), T is the absolute temperature (0° C ≈ 273 K), F is the faraday ($F \approx$ 96,500 coulombs/mol). The value of RT/F at room temperature is 0.0257 volts. The number of electrons transferred in the reaction is n. (A) indicates the activity of chemical species A, α moles of which enter into the reaction with β moles of B to produce π moles of P and σ moles of S. E^0 is the standard potential for the half-cell, the potential that would be measured in a half-cell in which all the involved chemical species had unit activity. These potentials are compiled in many handbooks.

Example A: A platinum electrode in equilibrium with a solution of Fe^{2+} and Fe^{3+}, in 1 M HCl, will have a potential for the reaction

$$Fe^{+2} \rightleftharpoons Fe^{+3} + 1e^-$$

In this reaction, *one* iron(II) reacts to give *one* iron(III); thus the stoichiometric coefficient for each species is 1.

The number of electrons transferred is 1; thus $n = 1$. The standard potential that one would look up is that for the reduction of Fe^{3+} to Fe^{2+} (in 1 M HCl).

$$E = E^0_{Fe^{3+},Fe^{2+}} + \frac{RT}{nF} \ln \frac{(Fe^{3+})^1}{(Fe^{2+})^1}$$

At room temperature ($T = 293$ K), one then has

$$E = 0.770 + 0.0257 \ln (Fe^{3+})/(Fe^{2+})$$

It is often convenient to use base-10 logarithms. Since $2.303 \log x = \ln x$, one has

$$E = 0.770 + 0.059 \log (Fe^{3+})/(Fe^{2+})$$

Since the potential, E, depends on the ratio of activities of Fe^{3+} and Fe^{2+}, the measurement of potential in this way is best suited to measurements where both the activities are important, as in a titration.

Example B: A platinum electrode in equilibrium with a solution of H^+ and with H_2 gas will have a potential for the reaction

$$2H^+ + 2e^- \rightleftharpoons H_2$$

In this reaction the stoichiometric coefficient for H^+ is 2, for H_2 is 1, and n is 2. Therefore the proton activity is squared, and the H_2 activity is to the first power:

$$E = E^0_{H^+,H_2} + 2.303 \frac{RT}{nF} \log \frac{(H^+)^2}{(H_2)^1}$$

Since

$$E^0_{H^+,H_2} = 0$$

one has

$$E = 0 + \frac{0.059}{2} \log \frac{(H^+)^2}{(H_2)}$$

If the pressure of H_2 is constant and equal to 1 atmosphere, it has (by definition) an activity of 1, therefore:

$$E = 0 + \frac{0.059}{2} \log (H^+)^2$$

Recalling that

$$\log x^m = m \log x$$

and that

$$\text{pH} = -\log (H^+)$$

one can see that

$$E = -0.059 \text{ pH}$$

Thus the hydrogen half-cell could be used as a pH-sensitive electrode. It is impractical, however, and more practical alternatives exist.

Example C: As a final example, let us consider the silver–silver chloride half-cell in which the potential is a function of a species that is not itself oxidized or reduced. The half-cell reaction is

$$AgCl + e^- \rightleftharpoons Ag^0 + Cl^-$$

It is most conveniently thought of as two reactions that are in equilibrium with one another:

$$AgCl \rightleftharpoons Ag^+ + Cl^-$$

$$\frac{Ag^+ + e^- \rightleftharpoons Ag^0}{AgCl + e^- \rightleftharpoons Ag^0 + Cl^-}$$

The second reaction is the redox reaction, and one can write

$$E = E^0_{Ag^+,Ag} + 0.059 \log \frac{(Ag^+)}{(Ag)}$$

The activity of a solid is 1, therefore

$$E = E^0_{Ag^+,Ag} + 0.059 \log (Ag^+)$$

Because of the first reaction, the activity of Ag^+ is governed by the activity of Cl^-. The solubility product is the relevant equilibrium constant:

$$K_{sp} = (Ag^+)(Cl^-)$$

from which one has

$$(Ag^+) = K_{sp}/(Cl^-)$$

thus

$$E = E^0_{Ag^+,Ag} + 0.059 \log K_{sp}/(Cl^-)$$
$$E = \text{Constant} - 0.059 \log (Cl^-)$$

Thus the electrode potential depends on the activity of Cl^-.

Now let us return to a consideration of Fig. 13-2. A more useful configuration than the one just discussed is to hold the two electrode-solution interfaces at constant potential differences (use two reference electrodes) and allow the salt-bridge potential difference to be altered. In this case the salt bridge does not simply connect two half-cells, it is a separate phase, a membrane, crystal, glass, or other material, with very distinctive properties. Such electrode

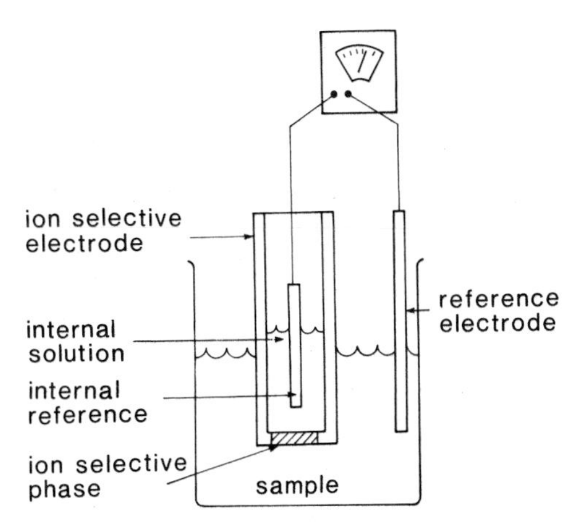

Fig. 13-4. Diagram of an ion-selective electrode. Measurement is of potential difference created across ion-selective phase.

systems are known as *ion-selective electrodes*. The potential that develops across a membrane (or other active phase) can be made to respond to the activity or concentration of a particular species. The properties of a membrane that make it selectively develop a potential gradient because of a difference in concentration of a certain ion are its ability to "dissolve" the ion and its ability to allow the ion to move relatively freely through the membrane. An oversimplified but useful way to visualize the ion-selective electrode is to consider the concentration-cell analogy. The ion-selective electrode consists of a reference electrode, a solution, and an ion-selective phase that closes the system (Fig. 13-4). This is dipped into a sample solution containing a reference electrode. The potential difference between the two reference electrodes is measured. If one makes two half-cells that respond to some ion, the difference in half-cell potentials will be governed by the ratio of the activities of the ions in the two half-cells. For example, with two hydrogen half-cells exposed to 1 atmosphere of H_2:

$$E_{cell} = E_1 - E_2$$
$$E_2 = 0.059 \log (H^+)_2$$
$$E_1 = 0.059 \log (H^+)_1$$
$$E_1 - E_2 = 0.059 \log \frac{(H^+)_1}{(H^+)_2}$$

In practice it is almost impossible to make use of such a phenomenon because there are many species present in real samples that can cause cell-potential change. If, however, the salt bridge allows only one type of ion to "pass," only concentration cells for that ion can be made. Differences in concentrations of other ions cannot be communicated through the system. This is the reason that in an ion-selective electrode one separates a pair of reference electrodes by a *selectively permeable* membrane or crystal.

Ion-selective electrodes

There are several different media that can be used to make an ion-selective electrode. Solid materials include single crystals, composites of crystals and inert substances, glasses, ceramics, and plastics, whereas the liquids used are water-insoluble liquids in which are dissolved species that can bind certain ions. Materials that are used to give ion-selective electrodes their particular selectivity are shown in Table 13-3.

Table 13-3. Composition of the ion-selective electrode for electrodes of clinical interest

Analyte	Composition
H^+	Glass
Na^+	Glass
K^+	Polyvinylchloride/valinomycin, potassium tetraphenylborate
NH_4^+	Glass
Ca^{2+}	Polyvinylchloride/*o*-nitrophenyl octylether, sodium tetraphenylborate
Cl^-	Silver sulfide–silver chloride crystals
F^-	Lanthanum fluoride crystal

Crystalline lanthanum fluoride (LaF_3) and silver sulfide (Ag_2S) are widely used to determine F^- and Ag^+ respectively. The latter may also be used to determine sulfide and other anions that precipitate silver ion (such as Cl^-). Glasses and ceramics, since they consist of various oxygenated species (such as borate, alumina, and silica) that require counterions have application in the selective detection of certain cations, most importantly H^+, Na^+, and K^+. The liquid-membrane electrodes contain a complexing species that can transport ions through the hydrophobic membrane phase.

Gas-sensing electrodes

Except for the oxygen electrode, which is amperometric, the other gas-sensing electrodes used in clinical chemistry are potentiometric sensors separated from the fluid being analyzed by a thin, gas-permeable membrane. Both the CO_2 and the NH_3 electrodes measure changes in pH, which occur because of the following reactions:

$$(CO_2)_{sample} \rightleftharpoons (CO_2)_{membrane} \rightleftharpoons (CO_2)_{electrode} \overset{H_2O}{\rightleftharpoons} H_2CO_3 \rightleftharpoons H^+ + HCO_3^-$$

$$(NH_3)_{sample} \rightleftharpoons (NH_3)_{membrane} \rightleftharpoons (NH_3)_{electrode} \overset{H_2O}{\rightleftharpoons} NH_4^+ + OH^-$$

The actual measurement is a pH measurement. The applications of these electrodes are covered in Chapter 14.

Instrumentation

Since potentiometric measurements are, in principle, carried out with no current being drawn, the voltmeter or potentiometer used to measure a potential difference between two half-cells must have a high internal resistance. By Ohm's law, the current that flows through the measuring device will equal the potential difference between its two inputs (that is, the potential difference being measured) divided by the electrical resistance between those two poles:

$$i = E_{meas}/R_{internal}$$

Thus, the higher the internal resistance of the potentiometer, the lower the current and the better the measurement. In a measuring instrument, this internal resistance is called the "input resistance." The most commonly used instruments, such as pH meters, employ high input resistance operational amplifiers as the input stage in the measuring device. The phrase *high input impedance* is used more generally to indicate a high internal resistance for both AC and DC signals.

Most high input impedance potentiometers in use may be used to measure potential difference directly, or the internal log-conversion circuit may be used to convert the measured potential to an activity. Since the potential-activity conversion is temperature dependent, the instrument's sensitivity changes with temperature, and there is generally a temperature-correction feature on such instruments. Of course, calibration of the instrument is required. The most accurate measurements are made with instruments for which a two-point calibration is required (''slope'' and ''intercept''). For certain applications (not generally in the clinical lab) a single-point calibration is sufficient. For pH measurements, buffers with accurately known pH values, which can be used as standards, are commercially available. Since activity is a function of a solution's ionic strength, for the highest accuracy the standards and the sample should have the same ionic strength.

AMPEROMETRY AND COULOMETRY

Amperometry is the measurement of current, whereas coulometry is the measurement of charge. The charge measured is the integral of the current over time. Since both methods require current to flow, the chemical basis for their use is similar.

Thermodynamics and kinetics

We have seen above that if there is a potential difference between two electrodes, the possibility exists for the occurrence of current flow because of interfacial charge transfer. One can reduce this current flow by interposing a high-impedance resistor in between the two electrodes and measuring the potential. Alternatively, one can allow the current to flow and measure it. Analytical measurements based on measuring current are called *amperometry*. Electroanalytical amperometric techniques are carried out so that the current is proportional to the concentration of some species present in solution. *Coulometric* techniques oxidize or reduce *all* of a particular species. These techniques measure the total quantity (not concentration) of that species present in the solution.

To perform amperometry, one must control the oxidizing or reducing power of the electrode, called the *working electrode*, which is oxidizing or reducing the species of interest in the solution. A potentiostat is used for this. Fig. 13-5 schematically demonstrates the potentiostat. There are three electrodes in the measuring circuit. Two of them act as a simple potential measuring circuit; the working electrode potential, *W*, is measured with respect to the reference electrode potential, *R*. One sets the potentiostat so that it maintains some potential, *E*, in the working electrode. This user-set potential is constantly compared to the measured potential. When current is drawn, the working-electrode potential tends to change. The *auxiliary* electrode (Fig. 13-5) is responsible for completing the current-carrying circuit with the working electrode. When the measured potential differs from the desired potential, the auxiliary electrode pushes (or pulls) current through the working electrode. The current is measured after it passes through the working electrode. Thus one has control over the oxidizing and reducing power of the working electrode. One can visualize the operation of the potentiostat with an example. Consider a set of working, reference, and auxiliary electrodes in a flowing stream of electrolyte. This flowing stream may be emanating from a liquid chromatograph. The working-electrode (mercury) potential has been set at -1.1 V versus the silver–silver chloride reference electrode. The flowing stream contains acetic acid as a supporting electrolyte. The potential on the electrode is stable, and there is no current passing between the auxiliary and working electrodes. Now a reducible substance, such as an *N,N*-diethyl-*N*-nitrosamine, at a low concentration, passes through the cell. Since this substance is reducible at -1.1 V (versus silver–silver chloride), some electrons leave the mercury electrode to reduce the nitrosamine. The electrode's potential becomes more positive, a result that causes a difference between the set and measured potentials. Current flows in the electrical circuit to replace the lost electrons. The electrons come from the auxiliary electrode's oxidizing something in solution. It is not selective. Generally solvent or supporting electrolyte is oxidized, as in the following:

$$2H_2O \rightarrow 4e^- + 4H^+ + 2O_2$$

Having regained its electrons, the working electrode has its potential return to the previous value. All this occurs many times quite rapidly and invisibly. The current that has been measured is the amperometric response of the electrode to the nitrosamine.

The initial estimate of the potential required to carry out

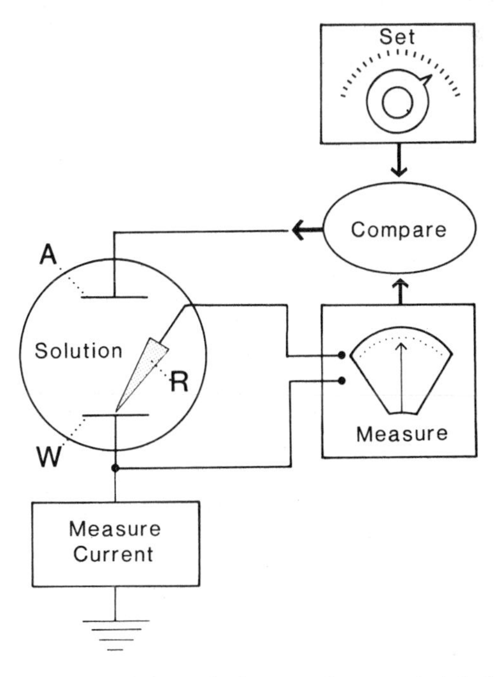

Fig. 13-5. Schematic diagram of a potentiostat. *A*, Auxiliary electrode; *R*, reference electrode; *W*, working electrode.

a particular oxidation or reduction is made with the aid of standard potentials. As was seen earlier, the half-cell with the more positive potential is the oxidizing half-cell. Analogously, to oxidize Fe^{2+} to Fe^{3+}, one must apply to the working electrode a potential more positive than the standard potential of the Fe^{+3}/Fe^{+2} half-cell. For half-cells that involve complexing or precipitating species, or are acids or bases, one must use the Nernst equation to determine the appropriate potential, since the potential will depend on the activity of complexing agent, precipitating agent, or H^+, respectively. This is shown in the reduction of oxygen to hydrogen peroxide at pH 7:

$$E = E^0_{O_2,H_2O_2} + \frac{0.059}{2} \log \frac{(O_2)(H^+)^2}{(H_2O_2)}$$

For this calculation, let (O_2) and (H_2O_2) be equal, remembering that log (H^+) is pH; thus

$$E = E^0_{O_2,H_2O_2} - 0.059 \text{ pH (volts)}$$
$$E_{pH\ 7} = E^0_{O_2,H_2O} - 0.413 \text{ (volts)}$$
$$E_{pH\ 7} = 0.682 - 0.413 = 0.169 \text{ V}$$

Thus, at pH 7, to reduce O_2 to H_2O_2, one must apply a potential somewhat more negative than 0.169 V versus the NHE, whereas to oxidize H_2O_2 to O_2, one requires a potential that is more positive than 0.169 V versus the NHE.

One can use the Nernst equation to indicate how much different from 0.169 V one wants to be. In potentiometry the species' activities determine the electrode potential according to the Nernst equation. In amperometry and coulometry one uses the Nernst equation to predict a species' activity at the electrode surface at a given potential. From the previous example one can see that at pH 7 and when $(H_2O_2) = (O_2)$ the working electrode would be in equilibrium with this species if it were set to 0.169 V versus the NHE. To drive the (O_2) to a low activity, that is, to reduce O_2 efficiently, one must apply a potential more negative than 0.169 V versus the NHE. Let us arbitrarily say that a (O_2) that is 0.1% of the (H_2O_2) is sufficiently low. Then $(O_2)/(H_2O_2)$ in the Nernst equation is 0.001, and one has

$$E_{pH\ 7,\ (O_2)} = 0.1\% \text{ of } (H_2O_2)$$
$$= E^0_{O_2,H_2O_2} - 0.59(7) + \frac{0.059}{2} \log 0.001$$
$$= 0.169 + 0.0295 \cdot (-3)$$
$$= 0.080 \text{ V versus the NHE.}$$

As a rule of thumb, 0.059 V per decade of concentration ratio is the potential change required for a one-electron-per-species process, 0.029 V per decade for a two-electron-per-species process, and so on.

A key feature of amperometry and coulometry is the requirement for mass transport. These techniques allow current to flow. Since the current comes from the transformation of one species to another and the transformation occurs *at the surface* of the working electrodes, there is a requirement that the species being destroyed be transported to the surface of the electrode. This transport can be brought about by convection (such as stirring) or diffusion. In amperometry, the current depends not only on the concentration of species being oxidized or reduced, but on the rate of mass transport. Thus, for precise analytical results, the mass transport must be carefully controlled.

Let us consider an example, the detection of Ag^+ by amperometry. This is important in the determination of chloride in body fluids. Since Cl^- is precipitated by Ag^+, in the presence of Cl^- the concentration of Ag^+ in solution will be low (governed by K_{sp} for AgCl). If Ag^+ is continuously added to a sample containing Cl^- in the manner of a titration, the point will be reached where the (Cl^-) is sufficiently low so that a measurable concentration of Ag^+ will exist in solution. Ag^+ may be reduced to Ag^0 ($E^0 = +0.799$). Thus an electrode at a potential somewhat more negative than this will reduce Ag^+ to Ag^0. This electrode can be used to detect the Ag^+ by the measurement of current: amperometry. The detection of a specified current attributable to the Ag^+ represents the end point of the titration. For this end point to be reproducible, one must control the stirring, or mass transport rate, carefully. The number of moles of Cl^- originally present in the solution is equal to the number of moles of Ag^+ used to titrate it. If the silver ion is consumed at a constant rate, the time taken for the titration will be related to chloride concentration.

The magnitude of the current in amperometry is related to the mass-transport rate. A highly sensitive amperometric device is the electrochemical detector used in liquid chromatography. This device is capable of quantitative detection of very small quantities (nanograms to picograms) of oxidizable and reducible substances. Many important endogenous biochemicals (catacholamines, some indoles, thyroid hormones, some amino acids, NADH, and others) and exogenous compounds (drugs and toxins) can be determined amperometrically after separation by liquid chromatography.

Bulk measurements

Whereas amperometry measures current caused by a species, coulometry measures the total charge consumed in the exhaustive oxidation or reduction of some species. The amount of species reacted is governed by Faraday's law:

$$Q = nFm \qquad \textit{Eq. 13-7}$$

where Q is the charge in coulombs, n is the number of electrons in the redox event, F is the faraday, and m is the quantity, in moles, of species reacted. The most common use of coulometry is in coulometric titrations. Here, as in volumetric titrations, one seeks to convert completely one form of a species (such as pyruvate, $CH_3\overset{\overset{O}{\|}}{C}COO^-$) to another form (lactate, $CH_3{-}\underset{\underset{H}{|}}{\overset{\overset{OH}{|}}{C}}COO^-$). The conversion is carried out by the electrochemical generation of reagents. Thus one may employ fairly unstable reagents, since they are only generated in the presence of analyte where they quickly react. An excellent example of this is the coulometric titration of ammonia, NH_3. Ammonia may be oxidized by hypobromite ion, BrO^-. This ion is a powerful oxidant and quite unstable. However, a suitably oxidizing, working electrode may be used to oxidize Br^- (very stable) to BrO^-. The BrO^- then reacts with NH_3 to form N_2:

$$3H_2O + 3Br^- \xrightarrow{-6e^-} 3BrO^- + 6H^+$$
$$3BrO^- + 2NH_3 \rightarrow N_2 + 3Br^- + 3H_2O$$

The net reaction is

$$2NH_3 - 6e^- \rightarrow N_2 + 6H^+$$

For every mole of ammonia reacting, 3 moles of electrons are consumed. The total charge change for m moles of NH_3 is $Q = 3F \cdot m$ coulombs. The number of moles converted can be calculated from the total number of moles of electrons. Experimentally the number of coulombs passed is measured, and the number of moles of ammonia can be calculated from equation 13-7.

Another example of a coulometric titration is the titration of Cl^- with Ag^+. We discussed above the detection of the excess Ag^+ at the end point. The Ag^+ in practice is generated from a silver anode at which the reaction

$$Ag^0 - e^- \rightarrow Ag^+$$

occurs. This yields the Ag^+ required for precipitation of the Cl^- ion in the sample.

Coulometry is also employed successfully in liquid chromatograhic detection and gas chromatograhic detection. The liquid chromatographic detector uses an electrode with a large surface area so that as solution passes through the detector, all the species capable of being oxidized or reduced have a chance to contact the electrode and react. The gas chromatographic detector is more complex. Solute molecules in the gaseous effluent are combusted in a flame. The resulting fragmented molecules are dissolved in a suitable electrolyte and coulometrically determined. Only certain fragmented molecules can be determined in this way; thus the detectors are *selective* for sulfur-containing, nitrogen-containing, and halogen-containing compounds.

A word about the applicability of amperometry and coulometry is in order. These electrochemical techniques have a degree of selectivity, though the selectivity is not so great as one might like. The selectivity resides in the ability to choose the oxidizing or reducing power of the electrode. Thus, by careful choice of the optimal working electrode potential, only a few species will react, whereas others will be unreactive at the same potential. Thus, given a solution of ascorbic acid (vitamin C) and a supporting electrolyte, one can select an electrode potential at which the ascorbic acid will be oxidized. At more negative potentials the ascorbic acid will not be oxidized. However, if a serum sample is allowed to react at the same potential as used above to oxidize the ascorbic acid, one would obtain current from the oxidation of many other substances (in particular uric acid), leading to falsely elevated results. If one can create a system in which there is more selectivity, these measurements are much more successful. Thus the precipitation of Cl^- with Ag^+ is specific, and the measurement of Ag^+ after titration of the Cl^- by amperometry is indirectly specific. In general, the separation of one species from another (and from undesired components of a sample) by chromatography, followed by electrochemical detection, is an extremely powerful analytical technique.

The apparatus used for amperometry consists of three components: the potentiostat, the current-to-voltage convertor, and the cell. The potentiostat has been discussed. The current-to-voltage convertor is a device that changes the current through the working electrode to a voltage. Voltages are much more conveniently measured, for example, by recorders and computers. The cell may take many forms. Details may be found in many texts on analytical chemistry (see the bibliography below).

Coulometric instruments exist in two forms. When coulometric detection after chromatography is used, the potentiostatic apparatus is used, and an integrator after the current-to-voltage convertor determines the number of coulombs produced. For titration a constant current source is used so that the generation of titrant (such as BrO^-) is linear with time.

One important aspect of cells is the material used to make the working electrode. For oxidation, various types of carbon (pyrolytic graphite, glassy carbon, carbon paste, wax-impregnated graphite, and compositions of plastics and graphite powder, particularly Kel-F) are used, as well as metals, particularly platinum and gold. For reductions the most common electrode is mercury, and this also occurs in many forms: bulk mercury, mercury films on graphite, and mercury films on gold. It is very important that a rational choice of an electrode in any investigation be made. Proper use of that electrode can be accomplished only if one knows what sort of electrode is being used in a particular application and why.

BIBLIOGRAPHY

Skoog, D.A., and West, D.M.: Principles of instrumental analysis, ed. 2, Philadelphia, 1980, W.B. Saunders, Co., pp. 500-652.

Bauer, H.H., Christian, G.D., and O'Reilly, J.E.: Instrumental analysis, Boston, 1978, Allyn & Bacon, Inc., pp. 1-138.

Willard, H.H., Merritt L.L., Jr., Dean, J.A., and Settle, F.A.: Instrumental methods of analysis, ed. 6, New York, 1974, D. Van Nostrand Co., pp. 628-800.

The following companies publish useful information on the use of electrodes in analysis:

Beckman Instruments, Inc., 2500 Harbor Blvd., Fullerton, CA 92634.

Bioanalytical Systems, Purdue Research Park, West Lafayette, IN 47906.

Orion Research, Inc., 840 Memorial Dr., Cambridge, MA 02139.

Chapter 14 Electrode-based measurements

Stephen G. Weber

activity coefficient A proportionality constant, γ, relating activity, a, to concentration, $a = \gamma c$.
detection limit The concentration at which the coefficient of variation for the measurement is 50%.
ionic strength $\frac{1}{2} \sum z_i^2 c_i$; one half the sum, over all ions in a solution, of the ionic charge squared times the ion concentration.
pH $-\log (H^+)$; minus the base 10 logarithm of the hydrogen ion activity of a solution.
selectivity coefficient The fractional degree to which an ISE (ion-selective electrode) responds to ion j, where the response to ion i is 1.0.
solutes In a solution, the molecules that are dissolved in the solvent; the minor constituent in a solution.
solvent The major constituent of a solution.
standardization The process by which the measured quantity produced by an instrument (such as voltage) is related to a chemically meaningful property, such as concentration.
standard state A real or hypothetical state of matter in which molecules are assigned an activity coefficient of 1.

One reason that there is a large number of analytical techniques is that each technique has certain advantages. Among the advantages possessed by potentiometric sensors are their ability to make measurements in situ (because they are "probes"), their specificity, the fact that any lack of specificity can be understood and quantitated, and their ability to make activity measurements rather than concentration measurements. Their major disadvantage is the necessity of putting the measurement system into the sample (as opposed to just observing the sample as in photometry), an action that causes interference problems. Most of these attributes are shared by amperometric sensors. With certain potentiometric and amperometric sensors, one can greatly improve performance by preceding the chemical measurement with a separation process. The separation gives the overall analytical system more selectivity, that is, more freedom from interferences. The specific systems in which this is used are discussed from p. 254 on. Before specific discussions of various electrode systems are discussed, it is important that the distinction between activity and concentration be clear.

ACTIVITY

The concept of activity was introduced by G. N. Lewis to describe nonideal behavior of *solutes* in solution. It is simplest to consider the case of a nonionic substance, and then the extension to ionic substances can be made. A measure of the degree to which some solute is soluble in various *solvents* is related to the partial pressure of the solute in equilibrium with the various solutions. The more soluble the solute in a particular solvent, the lower will be its vapor pressure above the solution. It is also true that a decrease in the concentration of the solute in the solution will decrease the partial pressure of the solute above the solution. Thus the partial pressure of a solute above a solution depends on its concentration and the degree to which it is solvated (that is, made soluble because of being surrounded by molecules of solvent). The activity, a, of the solute is defined as the ratio of the partial pressure of the solute in a particular solvent at a particular concentration to the partial pressure of the solute above a solution in a so-called *standard state*. It is customary to define the standard state for a thermodynamic frame of reference. One common standard state is that of pure solute. Another common standard state is that in which one has a fictitious solution of concentration 1 M in which the solute in the solution is surrounded only by solvent molecules. In the latter case the standard state changes with the solvent, whereas with the former it does not. In each case, when the solute is in the standard state, the concentration equals the activity, which equals 1 (the concentration is in a "mole fraction" for the former state and "molar" for the latter state). Thus these 4 situations are deemed "standard." A proportionality constant, γ, the *activity coefficient,* relates concentration and activity:

$$a = \gamma c \qquad \textit{Eq. 14-1}$$

In the standard state, $\gamma = 1$. Note that now the two properties that control the solute's vapor pressure or activity are separated, the concentration information is in c (an increase in c increases a), whereas the solvation information is in γ (because of a decrease in solubility, an increase in γ increases a).

Consider a pair of solutions that have identical concentrations but have different activities of a solute. Why might their activities differ? Clearly the activity coefficients must differ, thus the solvating ability of the two solvents must differ. The one with the lower activity coefficient, solution 1, has the lower activity, and solute is more soluble in solvent 1 than in solvent 2. The following relationships show how this affects the activity of solute in the solutions:

$$\gamma_1 < \gamma_2$$
$$a_1 = \gamma_1 c_1, \; a_2 = \gamma_2 c_2$$
$$c_1 = c_2$$
$$\frac{a_1}{a_2} = \frac{\gamma_1}{\gamma_2}$$

Solution 1 also has a lower vapor pressure of solute above it than solution 2. The vapor pressure is one measure of the solute's escaping tendency (how much the solute wants to get out of the solution to be in the gas phase). The ability of a solute to enter into other chemical reactions is regulated in the very same way. Thus, if there were a chemical reaction (say a binding equilibrium) in which the solute in both solutions could participate and the activity of each of the participants (except for the solute) in the reaction did not differ between solutions, the reaction would proceed to a larger degree in solution 2, where the activity of the solute is larger. Remember that the concentrations of solute in the two solutions were the same, but the reaction proceeds to a different degree in each solution. The activity describes the tendency of solutes to interact in chemical equilibriums with other species. This indicates the importance of activity measurements.

For ionic solutes, a major determining factor in the solute's activity coefficient is the solution's *ionic strength.* The ionic strength increases as the total (solute plus all other) ionic concentration increases. A measure of Ca^{2+} ion *concentration* tells how much Ca^{2+} is in the sample. A measure of Ca^{2+} *activity* tells one about the chemical "pressure" (that is availability) that the Ca^{2+} exerts in its various equilibriums. It is often the latter that is physiologically important.

POTENTIOMETRIC SENSORS

Ion-selective electrodes

Ion-selective electrodes (ISE's) are used in the clinical lab for the measurement of H^+, Na^+, K^+, Ca^{2+}, F^-, and Cl^-. Each electrode has a unique ion-selective phase that is intended to make the electrode respond to only one ion. Of course in practice this is not absolutely the case. Interfering ions do exist for all ion-selective electrodes.

An important property of these electrodes is that they are not absolutely specific. Fortunately the interference problem is well understood so that experiments can easily be done to determine the extent of the interference of a particular ion. The extent of the interference is given by a *selectivity coefficient,* K_{ij}. For an electrode that is meant to be specific to ion i (charge z) in the presence of both ions i and j (also of charge z) a potential given by equation 14-2 will develop:

$$E = \alpha + \beta \log (a_i + K_{ij}a_j) \qquad \textit{Eq. 14-2}$$

Here α is a constant that depends on the specific electrode, and β is a constant that is usually close to the theoretical value of $\beta = RT/zF$, a_i is the activity of ion i, and a_j is the activity of ion j. Note that a selectivity coeffiecient of 1 would mean that the electrode would respond equally well to i and j. For the operation with the least interference, a low selectivity factor is required.

For more than one interferent, where the interfering ion is not necessarily of the same charge as i, one must use the more general expression:

$$E = \alpha + \beta \log (a_i + \sum_j K_{ij}a_j^{z_i/z_j}) \qquad \textit{Eq. 14-3}$$

H^+. Perhaps the most often performed measurement in chemical labs is that of pH. The usual pH ion-selective electrode consists of a glass membrane containing a solution of constant pH and an internal reference electrode. The glass consists of mostly Na_2O and SiO_2 with some CaO. The outer surfaces of the glass membrane become hydrated, and the negatively charged oxides at the surface reach an equilibrium with the cations in the solution. For a pH electrode the equilibrium is such that H^+ is the preferred cation. When the H^+ activities inside and outside the electrode are not the same, the number of charges on the oxide of the inner and outer surface will not be the same, and a potential will develop across the membrane. This will be measurable as the potential difference between the internal and external reference electrodes (Fig. 13-4). If the only potential difference measured between the two reference electrodes were the potential difference across the membrane, a measurement of that potential difference would suffice to determine the pH of the test solution. Unfortunately this is not the only potential difference measured. There is an asymmetry potential, there are usually junction potentials at each electrode, and there are diffusion potentials. The asymmetry potential is reflected in the following thought experiment. Using a standard pH electrode, which consists of a reference electrode in a solution that is isolated from the outside by a pH-sensitive glass membrane, measure the difference in potential between a reference electrode and the pH electrode in a buffered solution. Now remove the pH electrode from the solution and turn the membrane around so that its inner surface is now the outer surface and vice versa. Measure the potential again. There will be a difference in the potential-difference measurement; that difference is the asymmetry potential. Junction potentials arise at the interface between a reference electrode and the analytical solution. Many reference electrodes consist of an electrode material in contact with a solution of constant composition. This solution must not be contaminated, and it is best if it very slowly leaks out into the analytical solution. The slow leak is allowed by use of porous materials or perhaps by allowing fluid to leak between a cylinder and a sleeve. These slow leaks are called "liquid junctions." For a variety of reasons, such as differences in ion mobility (called a "diffusion potential") or adsorption, there is generally a liquid-junction potential or junction potential. The liquid-junction potential can depend on the analytical solution. Variations in the diffusion potentials in the ion-selective membrane can occur. This is caused by variations in the ions carrying current through the membrane.

Since all these potentials other than the analytically derived potential exist and it becomes impossible to determine pH from a potentiometric measurement directly, *standardization* is required. The relevant criterion for suitable performance then becomes the fact that all the potentials except the analytically interesting one should be the same in the standard and in the sample. It is for this reason that potentiometric standards should reflect as accurately as possible the composition of the sample.

Interferences. Nonspecific interferences can occur in pH measurements on whole blood because of the alteration of the reference electrode's junction potential. One can avoid this by creating a large bath containing the reference electrode, placing the whole-blood sample into a container that interfaces with the pH-electrode surface, and allowing the whole-blood reference electrode to interface at the surface at the end of a long capillary tube. The blood does not come into contact with the liquid junction because the bath separates the two.

Specific interferences do not occur in the physiological

region. In strong base (pH ≥ 10, depending on glass composition), cations such as Na^+ begin to interfere. In very strong acid there are also difficulties in making accurate pH measurements.

Care and use. The pH electrode's glass surface must be in equilibrium with its surroundings in order to yield reproducible results. Thus it is best to keep the electrode stored in an ionic, aqueous solution. The electrode, since it is a probe, can carry contamination from one solution to another; thus it should be cleaned between measurements. A rinse in distilled water followed by patting (not rubbing, which establishes electrostatic charges on the electrode surface and causes interference) of the electrode surface to remove excess water is usually sufficient to prevent gross contamination of the sample.

Standardizaton of a pH electrode is best done with standards that mimic the sample. One especially wishes to avoid large shifts in the junction potential. A two-point standardization is appropriate for most clinical measurements. pH standards are commerically available. The manufacturer's instructions should be followed when using these standards. Be aware that the E-pH relationship is temperature dependent; thus not only should the temperature during the measurement be controlled, but also the pH of the standard at the operating temperature should be known. The two points of pH at which standardization is to occur should bracket the expected pH of the sample. It is important to use two standards. This is the minimum number that can be used to establish the slope and the intercept of the line that relates measured potential, E, to pH. One should be patient when calibrating and using the pH electrode; the measured potentials must be stable in order to obtain the most accurate results.

Na^+. The sodium electrode is currently used to measure both the activity and the concentration of Na^+ in blood and urine. The activity measurements can be made on whole blood or plasma directly. Standards that match the blood's ionic strength are used, so that this component of the activity is controlled. However, there are two effects that may still cause an ion-selective electrode measurement to differ from a flame-emission photometric measurement. The Na^+ is predominantly dissolved in the water of the blood. The blood plasma is normally 93% water; the remainder is predominantly protein. Lipemic plasma can also cause the total water content in a given sample to be decreased. Variations in the water content of the serum or plasma will cause the total quantity of Na^+ to vary in a given volume of serum or plasma. A flame-emission photometric measurement is based on the total quantity of Na^+ per volume of fluid. The ion-selective electrode responds only to the Na^+ activity, and neither γ nor c in the aqueous portion of the sample is altered significantly by the addition of small amounts of lipid to the sample. At equilibrium the activity of Na^+ will be the same in all phases (as required by thermodynamics). Thus an activity measurement is insensitive to small changes in lipid content of undiluted serum. The case for changes in protein content is not so clear cut; there are conflicting reports. It is best at this stage to consider that it is possible that proteins "bind" Na^+ (and K^+) and the activity coefficient of Na^+ may change as the concentration of various fractions of protein change.

A concentration measurement can be made by ion-selective electrodes that will then yield data consistent with flame-emission photometry. This is done by dilution of the sample severalfold in a buffer. In this way small changes in water content of the serum are no longer significant. For example, if the water contents of two otherwise identical sera were made to be 93% and 90% and the Na^+ activities were the same in the two cases, then the ISE results made on whole blood, serum, or plasma would lie in agreement with each other, but flame-emission photometric results would be 3.3% lower for the second serum. If each of these sera were diluted 1:20 before the determinations, the water content in the more concentrated diluted serum will be 99.65% and in the other, 99.5%. The difference is now negligible, and the difference in Na^+ content is also negligible. Furthermore, the concentration of protein is reduced twentyfold, with a reduction of its influence on the activity coefficient of Na^+. By dilution, then, agreement between flame-emission photometry and ISE results may be obtained. However, in many instances the knowledge of Na^+ activity and the rapidity and ease with which whole blood measurements can be made argue strongly for the use of the nondilutional ISE techniques.

Electrode glass. The Na^+ electrode operates on a principle that is the same as that of the H^+ electrode. The difference is in the contents of the glass membrane itself.

Interferences. The measurement of Na^+ by ISE suffers from no interferences of any significance in normal blood-derived samples. There is a potential for H^+ interference in highly acidic samples that may be derived from urine. The user should consult the manufacturer's specifications carefully to be aware of potential interference problems.

Care and use. The care and standardization procedures are similar to those for the pH electrode.

K^+. The measurement of K^+ is, like that of Na^+, done profitably by ISE. Differences similar to Na^+ measurements between flame-emission photometry and ISE exist for K^+ measurements. There may be a negative bias for urine $[K^+]$ >50 mmol/L in ISE results as compared to flame-emission photometry. This problem can be avoided by dilution of the urine samples that have such high values.

Electrode. Although there are K^+-sensitive glasses, none are specific enough for clinical use. The K^+ electrode employed routinely in clinical chemistry is based on the cyclic molecule valinomycin. This molecule specifically binds K^+. An electrode membrane's composition is usually a polyvinylchloride matrix in which there is valinomycin, K^+, and tetraphenylborate as counterion.

Interferences. Like the Na^+ electrode the K^+ electrode is almost interference free in reasonably normal samples, though the information supplied with the electrode should be consulted. There is an effect of ammonia on the K^+ electrode's response. Depending on operating conditions ammonia may increase or decrease the apparent K^+ level in blood or urine. The effect, even under conditions of very high ammonia levels in both blood and urine, is at most a few percent.

Care and use. See the section on the pH electrode and the manufacturer's literature.

Ca^{2+}. Calcium electrodes may be based on phosphate type of ion exchangers or neutral carriers. Measurements of Ca^{2+} in serum yield results (on a concentration basis) that are much lower than the results one obtains using flame methods. In a healthy human a typical total Ca^{2+} concentration as measured by atomic absorption is 2.5 mEq/L. A measurement by an ion-selective electrode yields about 1.1 mEq/L. The measurement by ISE is sensitive to Ca^{2+} activity, and the activity is lower than expected on purely ionic considerations because of the association of Ca^{2+} with proteins and other smaller anions. Thus the ISE measures the ionized, or unbound, uncomplexed calcium fraction. These associations are pH and temperature dependent, and so these parameters must be well controlled in measurements of Ca^{2+} activity.

Cl^-. The active membrane in a Cl^- ISE is usually a composite of Ag_2S and AgCl. The electrode responds to other anions (such as Br^-, F^-, CN^-, OH^-, and S^{2-}) as well as Ag^+, but these others are rarely present in interfering quantities in physiological fluids.

F^-. The fluoride electrode is based on the conducting LaF_3 crystal. The only interfering anion is OH^-, and the selectivity coefficient is such that the interference is not significant at any but the most alkaline pH's, that is, above 10.

Gas-sensing potentiometric electrodes

The CO_2 and NH_3 electrodes are ion-selective electrodes that are placed behind a gas-permeable membrane.

CO_2. A typical CO_2 electrode such as that introduced by Severinghouse consists of a pH electrode that has a flat surface, a thin layer of a weak bicarbonate buffer, and a silicone membrane covering the glass-electrode surface. The silicone is a hydrophobic polymer; thus it does not absorb water, and polar constituents of serum cannot go through the membrane. Nonpolar constituents can go through the membrane.

The total CO_2 in a serum or whole blood sample is in an equilibrium in which several species take part:

$$CO_2 + H_2O \rightleftharpoons H_2CO_3 \rightleftharpoons H^+ + HCO_3^- \rightleftharpoons H^+ + CO_3^=$$

At equilibrium, the partial pressure of CO_2 is regulated by the total quantity of CO_2 present in the sample and by pH as long as the reactions shown above are the only ones that occur. The CO_2 in a sample can permeate the silicone membrane, and it can change the pH of the weak buffer inside the membrane (Fig. 14-1).

The major advantage of such electrodes is that the measurement system is isolated from the sample, and even the reference-electrode junction is inside the silicone membrane. There are very few interferences in such a system, but nonpolar weak acids can interfere, since they can dissolve in and therefore permeate the silicone membrane. Because the equilibrium between CO_2 and H_2CO_3 is not rapid and because of diffusion of CO_2 through the membrane and the internal filling solution, the electrode responds slightly more slowly than a non–gas sensing electrode. Electrodes that respond to carbonate or bicarbonate have been made using quaternary amines as an ion-exchange species incorporated into a membrane. These respond to other small anions as well, but if one places the ion-exchange ISE behind a silicone-rubber membrane, the anionic interferences are obviated. Furthermore, since this type of ISE does not respond to nonpolar weak acids, this interference is also eliminated.

Ammonia. The ammonia sensor works on similar principles as the CO_2 electrode and suffers from analogous interferences; thus nonpolar, volatile amines can interfere. The ammonium ion can also be measured with an ion-selective electrode based on nonactin. This responds to K^+ as well, and thus one must still employ a membrane (such as microporous polytetrafluoroethylene) to separate the sample and the electrode.

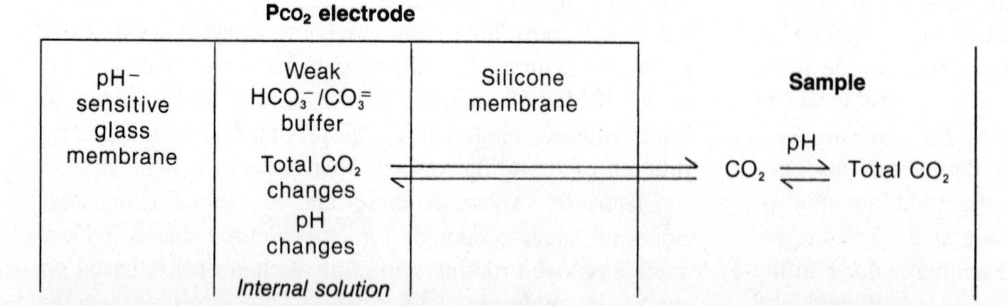

Fig. 14-1. Ion-selective electrode for measurement of carbon dioxide. Carbon dioxide in sample, whose partial pressure (activity) is regulated by total CO_2 and pH, goes through membrane to alter total CO_2 on inside of membrane. This changes pH of internal solution.

Care. Gas-sensing electrodes require a minimum of attention. The most severe test for such an electrode is the determination of the time taken for the electrode to respond to a significant diminution of analyte concentration. If the electrode responds slowly, the membrane may need to be changed. A rapidly and reproducibly responding electrode is desired.

Use of potentiometric electrodes. A high-input impedance voltmeter is used to measure the potential difference between a sensor and a reference electrode. Potentiometric electrodes yield a voltage output that is a logarithmic function of analyte activity. The slope of the relationship between the voltage and the logarithm of the activity can often be predicted from equilibrium considerations and is related to the charge on the *analyte* (and to the stoichiometry of the species that is formed as a result of the chemical interaction of the analyte with the electrode). Thus the lanthanum fluoride–based ISE responds directly to fluoride, whose charge, z, is -1:

$$(0.059/z_F)\ \log\ (F^-)_{sample} \propto \Delta E_{measured}$$

becomes

$$-0.059\ \log\ (F^-)_{sample} \propto \Delta E_{measured}$$

Thus a tenfold increase in F^- activity in the sample leads to a 0.059 V decrease (at room temperature) in measured potential. Standardization of ISE's should be done with standards that mimic the sample to a high degree. The use of total ionic strength activity buffers (TISAB) is a useful way to achieve this. These buffers are used in standard preparations to cause the activity coefficients of the various standards to be the same and as close as possible to the sample. For gas-sensing electrodes, gas standards can be used for standardization.

Calibration. The specific calibration procedures used vary with the type of electrode. A general idea of how to proceed is given here. An electrode should be tested for its linearity of log ion activity versus potential. The negative of the log of the ion activity is termed *pIon,* such as pNa or pCl, by analogy with pH. A series of standards should be prepared in an ionic strength buffer, and the concentrations should vary over a wide range. Measurements of the potential developed by the ISE after the potential reading has stabilized should be made for each standard. A plot of E versus pIon should be linear over some region of pIon, typically for concentrations in the 10^{-2} to 10^{-5} M region. The slope of the linear portion should be noted. Well-behaved electrodes obey an equilibrium-based equation as shown above.

Once it has been established that the electrode shows a linear relationship between E and pIon, one may rely upon a pair of standards that bracket the expected sample activities for routine standardization.

Most ISE's employed in the clinical lab are currently incorporated into automated or semiautomated equipment. It is important to realize that this packaging in no way decreases the attention that should be paid to the integrity of standards, controls, and samples. Electrode-based systems operate practically flawlessly when they work well. Periodic attention to junctions, membranes, electrical contacts, and solid-state electrode surfaces is required for best performance.

Enzyme-based electrodes

Although not yet widely used in clinical labs, enzyme-based electrodes are the subject of much research. Comprehensive coverage of this subject is beyond the scope of this chapter, but by seeing one illustrative example, appreciation for the possibilities for these electrodes can be had. The basic idea is to use an enzyme as a reagent in such a way that an analyte is converted into a species that is detected by an ISE. The classical example of this is the urea sensor based on the enzymatically catalyzed hydrolysis of urea by urease:

$$H_2NC(=O)NH_2 \xrightarrow[H_2O]{Urease} 2NH_3 + CO_2$$

$$\downarrow$$

$$2NH_4^+ + H_2CO_3 \rightleftharpoons H^+ + HCO_3^-$$

By suitable means, urease can be held near the active surface of a cation or ammonium-ion ISE, an ammonia-sensing gas electrode, or a CO_2-sensing gas electrode. The amount of product (NH_3 or other) produced is related to the amount of urea present in the sample.

CURRENT-MEASURING SENSORS

Amperometric sensors measure the current produced by an electrolytic process that occurs because of the presence of an analyte species. For systems of analytical interest the current is directly proportional to the concentration of the species in the solution.

Oxygen electrode

The oxygen electrode, invented by Clark in 1956, consists of a platinum (Pt) cathode, an anode or reference electrode, typically silver coated with silver chloride, a buffered solution between them, and a membrane to separate the system from the sample. O_2 diffuses through the membrane into the buffer. The Pt cathode is held at a potential more negative than the Ag/AgCl electrode by several hundred millivolts. O_2 can then be reduced at the Pt electrode, and as long as the internal-solution pH remains constant, the current generated by the O_2 reaction will be a linear function of O_2 partial pressure. As with other membrane-based electrodes, attention must be given to the membrane if response times are slow. After long periods of use, silver metal may be plated onto the Pt causing poor electrode behavior. One may fix this by polishing the Pt with a fine (0.1 to 0.05 μm) polishing compound and a polishing cloth.

The oxygen electrode responds linearly to O_2 dissolved

in the blood. Standardization is accomplished by use of gaseous standards that are saturated with water vapor.

Coulometric titration of Cl^- with amperometric detection

The Büchler-Cotlove chloride titrators give rapid, accurate determinations of Cl^- in the sample by the coulometric generation of Ag^+. The sample is diluted in a solvent that contains nitric acid, acetic acid, and gelatin. This solution is put into contact with a set of two pairs of electrodes: a generator pair and a detector pair. A constant current is passed between the generator pair, the anode of which is silver. The oxidation of silver yields Ag^+, which precipitates as AgCl. The concentration of Ag^+ in solution is kept very low (and can be calculated from $[Cl^-]$ and $K_{sp,AgCl}$), since, as it is produced by the anode, it precipitates. The detector pair is acting as a $[Ag^+]$-measuring device; when Ag^+ is present in solution, a current will flow between them, and this current is proportional to the $[Ag^+]$ in solution. While the Cl^- is being precipitated, the detector pair sees very little Ag^+, but when the Cl^- has all been precipitated, the end point has been reached, and the Ag^+ generated by the generators remains in solution where it is detected. The presence of a significant current passing between the detector pair is used to turn off the generator pair and stop a clock that has been running since the experiment began. Since one Ag^+ precipitates with one Cl^-, the number of moles of Cl^- in the sample is equal to the number of moles of Ag^+ generated. This is given by Faraday's law:

$$Q = \int_0^T i(t)dt = nF \text{ (moles)}$$

Q is the number of coulombs passed, n is the number of electrons transferred per redox event (for Ag/Ag^+, $n = 1$), F is Faraday's constant, and *(moles)* is the number of moles of material oxidized or reduced in the passage of Q coulombs of charge. If the generator current is constant, then

$$\int_0^T i(t)dt = iT$$

where i is the constant current and T is the time to the end point. Note that the titration time is proportional to the number of moles of Cl^- precipitated:

$$T = \frac{nF \text{ (moles)}}{i}$$

Thus the number of seconds recorded on the clock at the end point is proportional to the concentration of Cl^- in the sample for a given sample volume taken.

Samples may be analyzed serially by this technique. In fact, the technique works best if a titration has been performed just before the titration of interest; thus the use of the technique is well suited to many samples. The major errors are attributable to poor pipetting technique, since the coulometric analysis is inherently precise. Interferences as by CN^-, Br^-, I^-, and NH_3 exist, but the concentrations are low enough to be neglected. It is possible that control sera may indicate improper performance. Simple cleaning of the reaction vessel, polishing, or perhaps replacing dirty or corroded silver electrodes will usually return the instrument to its original functioning status.

Electrochemical detectors for liquid chromatography

Amperometry does not possess the inherent selectivity of the ISE's, and consequently the usefulness of an amperometric detector is strongly coupled to the selectivity of ancillary analytical methods. Liquid chromatography is a powerful separation technique that can be used to separate oxidizable or reducible species, and electrochemical detectors are sensitive and selective tools for the detection of the separated species. The electrochemical detectors also yield low (that is, good) detection limits.

There are different requirements for an electrochemical detector, depending on whether an oxidation or reduction is required. For oxidations the working electrode is generally a form of carbon or platinum, whereas for reductions mercury or mercury-coated metal electrodes, or gold electrodes, are more useful. For reductions one must remove oxygen from the liquid chromatographic mobile phase so that the reduction of O_2 does not obscure the analytical signal.

The electrochemical detection is capable of detection limits in the 10 pg range for oxidations and the 100 pg

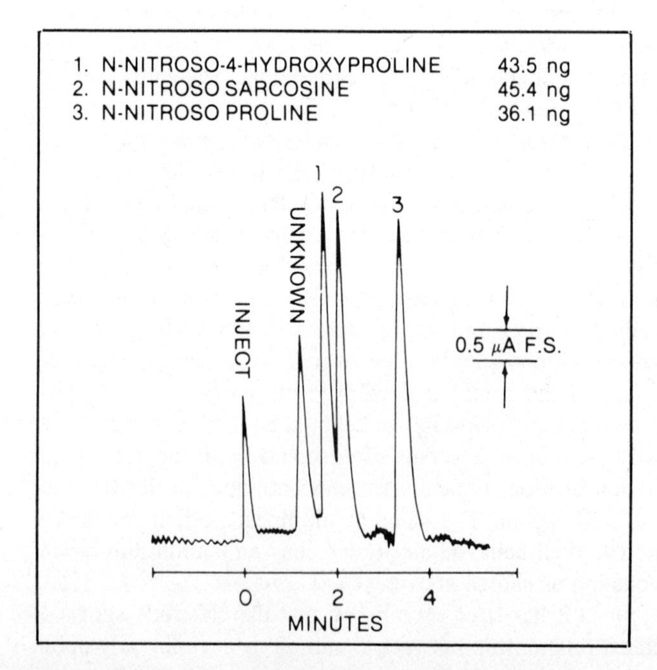

Fig. 14-2. Amperometric detection of *N*-nitrosamines at a mercury electrode after separation, on a reversed-phase column. (Courtesy EG & G Princeton Applied Research, Princeton, N.J.)

range for reductions. Of course, these are only intended as a guide because more accurate limits can be defined for particular compounds.

The electrochemical detectors respond to a variety of organic functional groups, notably phenols (including catechols), quinones, aromatic amines, thiols, alkyl sulfides, sulfones, phenothiazines, hydrazines, hydrazones, and nitrogen-containing compounds. It also responds to many inorganic species, particularly complexes of transition metals.

One can determine *N*-nitrosamines by liquid chromatography with electrochemical detection (LCEC) using a mercury electrode held near −0.8 volt versus SSCE (NaCl-saturated calomel electrode). The separation of several *N*-nitrosamines is shown in Fig. 14-2. The chromatogram is actually a plot of detector current versus time. The detector reduced the *N*-nitrosamines in a four-electron reduction to the corresponding hydrazine. Only when the compound is in the solution near the electrode can it be reduced. As solution is continually being swept through the detector, the current increases, reaches a peak, and decreases as the concentration of the compound does the same in the detector cell. At constant working-electrode potential and solution-flow rate, the peak current and area are proportional to analyte concentration and amount, respectively.

The use of multiple working electrodes is proving to be a useful way to increase detector selectivity. This is exemplified in Fig. 14-3. Here a mixture separated by liquid chromatography passes through porous working electrodes. The cell is designed so that all the analyte that can be reduced is reduced (or oxidized) in the cell. Detectors that have this property are called "coulometric." In this example, three compounds in the mixture are oxidized at the first electrode. A second electrode downstream from the first is set at a potential more negative than that of the first electrode. If the oxidation of the analytes is chemically reversible, it is possible to detect oxidized analyte (by reducing it) at the second electrode. Fig. 14-3 shows that, in this case, only acetaminophen satisfies this criterion and is detected at the second electrode.

The use of chemical derivatives that are electroactive extends the scope of electrochemical detectors. The *ortho*-phthalaldehyde derivatives of primary amines are electroactive, as well as being fluorophores. Fig. 14-4 shows the separation and amperometric detection of amino acids in 10 minutes. The *o*-phthalaldehyde derivatives were prepared (by a simple addition of reagent to the sample) before injection onto the column.

The use of LCEC detection, especially for oxidations (with no oxygen interference), is a simple procedure with some simple requirements. The detector (electrode) participates in the detection, and the electrode surface can become fouled. This is not a major problem in most applications, but should this difficulty occur, a physical (as by polishing) or electrochemical treatment can recover the electrode's performance. Solvents for LCEC detection must contain ions (to conduct electricity). This has meant that most applications are in a water-based, reversed-phase chromatography.

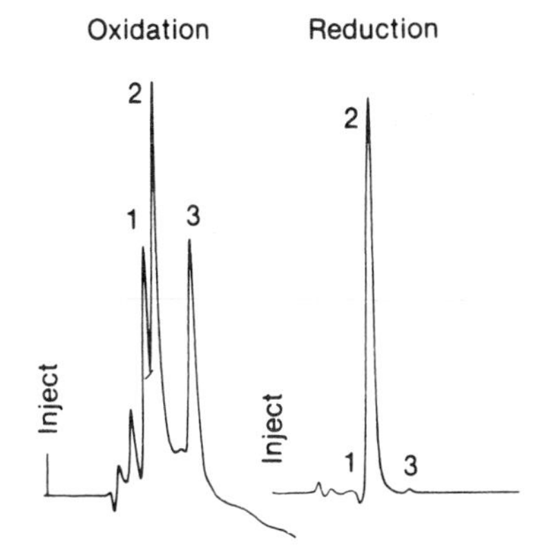

Fig. 14-3. Detection of compounds after chromatographic separation. Acetaminophen is oxidized at first electrode, *left chromatogram,* and resulting molecule is reduced at second electrode, *right chromatogram.* (Courtesy Environmental Sciences Association, Inc., Bedford, Mass.)

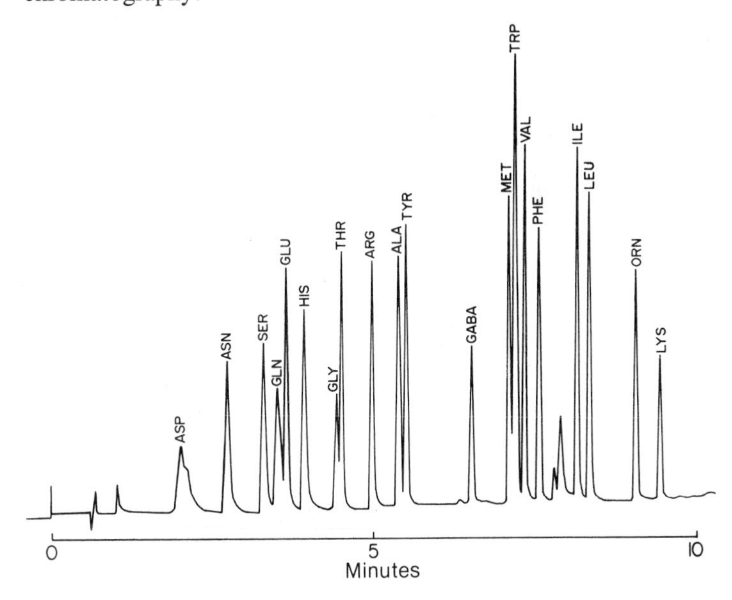

Fig. 14-4. Gradient separation of *ortho*-phthalaldehyde-mercaptoethanol derivatives of 20 amino acids using a 3 μm reversed-phase column. A thin-layer amperometric detector was used, with glassy-carbon working electrode at a potential of +0.700 V against Ag/AgCl reference. Electrochemically detectable functional group of each *ortho*-phthalaldehyde-mercaptoethanol derivative consists of a substituted isoindole, which is readily oxidizable. (Courtesy Bioanalytical Systems, West Lafayette, Ind.)

BIBLIOGRAPHY

Apple, F.S., Koch, D.D., Graves, S., and Ladenson, J.H.: Relationship between direct-potentiometric and flame photometric measurement of sodium in blood, Clin. Chem. **28:**1931, 1982.

Cammann, K.: Working with ion-selective electrodes, Berlin, 1979, Springer-Verlag.

Covington, A.K.: Ion-selective electrode methodology, Boca Raton, Fla., 1979, CRC Press, Inc.

Czaban, J.D., Cormier, A.D., and Legg, K.D.: Establishing the direct-potentiometric 'normal' range for Na/K: residual liquid junction potential and activity coefficient effects, Clin. Chem. **28:**1936, 1982.

Durst, R.A.: Ion-selective electrodes in science, medicine, and technology, Am. Sci. **59:**353, 1971.

Fisher electrode handbook, Pittsburgh, 1981, Fisher Scientific, Co.

Freiser, H.: Ion-selective electrodes in analytical chemistry, New York, 1978, Publishing Corp.

Guide to ion analysis, Cambridge, Mass., 1983, Orion Research, Inc.

Ion-selective electrodes, Russell Technicial Data Bull. RU-20, Fife, Scotland, Russell pH Limited.

Kobos, R.K., Parks, S.J., and Meyerhoff, M.E.: Selectivity characteristics of potentiometric carbon dioxide sensors with various gas membrane materials, Anal. Chem. **54:**1976, 1982.

Electrochemical detectors

Allison, L.A., and Shoup, R.E.: Anal. Chem. **55:**8, 1983.

Kissinger, P.T., Felice, L.J., Miner, D.J., Preddy, C.R., and Shoup, R.E.: Detectors for trace organic analysis by liquid chromatography principles and applications. In Hercules, D.H., Hieftje, G.M., Snyder, L.R., and Evenson. M.A., editors: Contemporary topics in analytical and clinical chemistry, vol. 2, New York, 1978, Plenum Publishing Corp.

Shoup, R.E., editor: Recent reports on liquid chromatography/electrochemistry, West Lafayette, Ind., 1982, Bioanalytical Systems, Inc.

Weber, S.G., and Purdy, W.C.: Electrochemical detectors in liquid chromatography: a short review of detector design, I & EC Product Research & Development **20:**593, 1981.

Chapter 15 Automation

James E. Davis

analog A measurement derived directly from an instrument's continuous signal, that is, voltage, and usually presented in graphic form.

automation Use of a machine designed to follow repeatedly and automatically a predetermined sequence of individual operations.

bar coding A computer-driven sample-recognition system that allows both the specimen and the ordered analyses to be identified and relayed to the automated analyzer.

bulk reagents Those that must be measured before being added to a reaction mixture to attain the desired proportion. Usually a reservoir contains the reagents for more than one analysis.

carry-over Contamination of a specimen by the previous one.

centrifugal analyzer Uses centrifugal force to mix the sample aliquot with reagent, and a spinning rotor to pass the reaction mixture through a detector.

computation Calculation of a desired result from the signal or readout of an instrument, and it can be electronically automated by use of either digital or analog conversion.

continuous flow Instruments constantly pumping reagent and sample through tubing and coils, forming a continuous stream.

dead volume The volume in a sampling container that must be present for proper sample aliquoting but is not consumed.

digital Relating to data available in the form of discrete units or the calculations using such data.

discrete Term applied to instruments that compartmentalize each sample reaction.

dwell time The minimum time required for an instrument to obtain a result as calculated from the initial sampling of the specimen.

flow injection Placement of a sample into the stream of a continuous-flow analyzer.

incubation The time allowed for a chemical reaction or process to proceed.

mixing Process by which individual components of a chemical assay are formed into a homogeneous solution.

proportioning Addition of individual components of a chemical assay in proper ratios or amounts.

readout Visualization of the result of an instrumental analysis.

selective instruments An instrument capable of performing multiple tests on a sample, but performing only those programmed.

sensing A system or device that monitors changes in the reaction mixture that are related to analyte concentration.

simultaneous analyzers Automated analyzers capable of more than one analysis per sample at the same time.

test repertoire The number of different tests available either at one time or with changing of reagents or instrument components.

throughput The maximum number of individual samples or test analyses that can be practically performed per hour by an assay system, with the required dwell time being taken into account.

unit reagents Premeasured reaction chemicals packaged so that only one package (unit) is used per sample test.

DEFINITIONS

In this chapter the reasons for automation and the ways to achieve it are considered. Examples from the major instrument categories are examined.

In this section the concept of automation is examined in a historical perspective. It is a fact that more tests are being performed per capita every year. This has come about for several reasons: the repertoire of available tests has increased, more patients survive acute illnesses and trauma, thus continuing to need additional tests, and results are generated more rapidly and reported in time to affect medical decisions.

Until recently, the test work load has increased at a 15% annual rate, doubling every 5 years. This kind of growth has been fueled and sustained by improvements in productivity, part of which derives from automation. A major result of automation has been the almost unlimited access to laboratory analyses by physicians. The salutary effect of this phenomenon on patient care and the cost to the health system is still to be determined.

Automation, as applied to clinical chemistry, can be defined as the self-moving or mechanical transfer of a specimen within a complex, industrial assembly to a succession of self-acting machines each of which completes a specified stage in the total analytical process from crude sample to analytical result. A second definition is the application of fully automatic procedures in the efficient performance and control of operations involving a sequence of complex standardized or repetitive processes on a large scale. Automation in the laboratory may be considered in the broader context of the need to obtain information about the state of a patient in a hospital or outpatient setting. One way to visualize how this information is obtained is shown in the block diagram of Fig. 15-1. The patient is examined by a physician, and there is a decision to obtain specific laboratory information. A specimen is collected, transferred to the laboratory, received, processed by the laboratory, and analyzed to obtain the desired results, which are then transferred back to the requesting physician. In

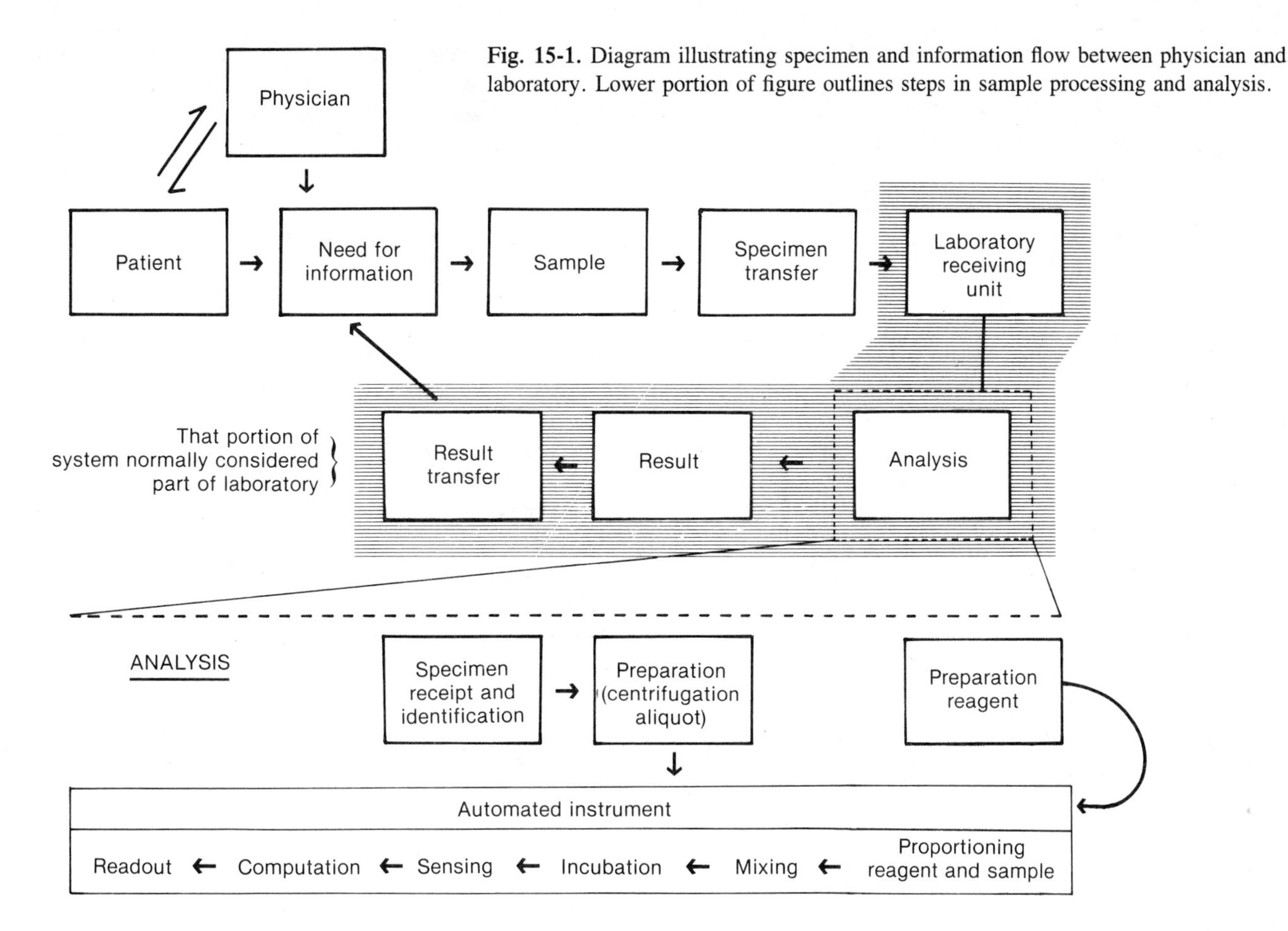

Fig. 15-1. Diagram illustrating specimen and information flow between physician and laboratory. Lower portion of figure outlines steps in sample processing and analysis.

general, laboratory automation has focused on those steps from the reception of the sample in the laboratory to the transfer of information back to the requesting physician. Most laboratories have concentrated on purchasing instruments that will automate the steps of analysis. The steps of analysis are outlined in the lower part of Fig. 15-1 and discussed in detail below. Although the term "automation" implies the processing of large numbers of samples, the principles inherent in automation can be applied to the analysis of single samples as well. For example, a blood-gas analyzer is a single-sample analyzer but commonly is highly automated.

The goal of automation is often so desirable that the equipment, and its mode of performing certain types of actions, often restrict the types of analyses that can be performed. For example, the reference method for glucose analysis requires a protein precipitation followed by the hexokinase reaction. The inability of current technology to automate easily the precipitation step has eliminated this method of assay as a procedure available with automated glucose analyzers.

AUTOMATION OF CHEMICAL ANALYSES

To understand how patient samples are processed by automated procedures, it is necessary to divide the process of analysis, as might be performed in a manual assay, into a series of stages or steps. Commonly the following series of steps is performed during the course of an analysis: (1) obtaining a patient sample in the proper form, (2) mixing an aliquot of the sample with a series of reagents in an ordered sequence with defined amounts, (3) monitoring or sensing the result of the reaction, (4) quantitating the extent of the reaction, and (5) providing an appropriate readout or permanent record. Automation may be applied to any or all of these steps necessary to perform a specific analytical procedure. The automation of each one of these steps is now discussed in some detail.

Sample preparation and identification

Remarkably little automation has been applied to this step. By and large, the specimens are manually labeled, centrifuged, and divided into aliquots if tests for more than one work station have been requested. The reason for continued manual processing is the need to determine if the sample meets acceptable criteria for analysis (such as proper phlebotomy tubes or absence of debris).

The manual performance of these procedures can often lead to errors in the generation of accurate patient results. Mislabeling of patient samples leads to laboratory information being erroneously transferred. Similarly, the necessity of dividing the sample into aliquots for separate work stations also can lead to improper patient identification.

Various approaches are being tried to avoid or minimize these manual processing steps. For example, use of computer-based *bar coding* of patient samples allows electronic identification of the sample and the tests requested. The use of single instruments capable of measuring at one work station a large number of analytes from an individual specimen has been advocated to delete the need for sample aliquoting. By using whole blood rather than serum as the sample, one can also avoid the centrifugation steps. The type of sample capable of being analyzed by an instrument is usually determined by the chemistry of the analysis rather than the mechanics of automation. Thus most analyzers are capable of processing most biological fluids, such as serum, plasma, urine, and cerebrospinal fluid, but not whole blood. A few instruments, notably those employing ion-selective electrodes, are designed for use with whole blood.

In the future, instruments combining several of these system approaches for processing samples will find their place in the laboratory.

Reagent preparation

Bulk reagents can be manually prepared, though an increasing number are used as concentrates or lyophilates. Thus simple dilution or reconstitution of the reagent with water is all that is required. However, this simple step can also lead to analytical errors because of improper dilution or processing. These errors can be avoided by purchasing reagent in a form suitable for instrument use without *any* processing. Unit test-reagent preparation (where sufficient reagent is present for the performance of a single test) has been automated in two ways. The first is the dry-film or impregnated-paper technique. The dry-chemical techniques use either paper or series of thin films impregnated with the desired reagent. The analytical reactions take place when the sample is placed on the dry reagent (see Eastman Kodak Ektachem, Table 15-1 and p. 269). In this type of reagent, preparation consists in wetting this reagent with water, buffer, or sample. The second kind of unit test reagent is a container or test tube containing premeasured liquids or powders to which water, buffer, or sample is added. Unit test reagents tend to be more expensive than bulk reagents, a consideration that can be important for many laboratories.

Proportioning of samples and reagents

Most chemical reactions require the combining of reagent and sample in exact amounts to yield specific, final concentrations of analyte and reagents. Since the reagents, as just described, are prepared in predetermined amounts, the ratio, or proportion, of reagent to sample must be kept constant to achieve reproducible and accurate final reagent concentrations. Thus the addition of sample to reagent is termed *proportioning*. The case of unit test reagents is considered first. In these systems, the reagents are already proportioned in the required amounts; therefore only the sample must be proportioned. The dry-film reagent may have the sample added volumetrically by syringe, pipet, or peristaltic pump, or by saturation addition. The latter technique requires some explanation. The film is exposed to an excess of the sample, and the pores of the film will allow only a fixed amount of the sample to be absorbed. This fixed amount of sample required to wet the film represents the proportioning mechanism. In some cases of saturation addition, the rate of the diffusion of the sample into the film may also affect the proportioning step. In the case of bulk reagents, proportioning is always by volumetric addition. There are three automated dispensing methods in common use. Syringes or volumetric overflow devices are used in discrete test analyzers where sample and reagents are volumetrically added to a test tube or container. The second mechanism is the continuous-flow technique used where sample and reagents are proportioned by their relative flow rates. Typically, peristaltic pumps are used to move the reagents through tubing, and the flow rate is controlled by the cross-sectional area (diameter) of the pump tubing. Usually the sample and reagent streams are allowed to flow continuously through the tubing where mixing and incubation are also accomplished. Additional discussion on this technique will be found under the Technicon autoanalyzers. The third type uses electric valves to control the time the reagents can flow. The flow rate is controlled by the air pressure applied to the reagent container and the flow resistance in the tubing to the reaction vessel.

In almost all systems the sample is introduced into the analyzer with a thin, stainless steel probe. This probe is passed into a sample, aspirates a defined quantity of sample, and moves from the sample to dispense the aliquot into an appropriate vessel. Since the same probe is used repeatedly for sequential samples, there is the potential for contamination of a specimen by a preceding one. This is called *sample carry-over*. Various techniques have been used to minimize the interaction between samples. These include (1) aspiration of a wash liquid (such as saline solution or water) between sample aspirations and (2) a backflush of the probe. The latter technique has the wash liquid flow through the probe in the direction opposite to the aspiration and then into a waste container.

Measurement of the degree of sample carry-over can be achieved by the following procedure. Samples containing low and high levels of analyte are analyzed in the following order: low (L_1); high (H); and a repeat aliquot of the low (L_2). The degree (%) of sample carry-over or sample interaction between the high and low concentration samples can be described by the following equation:

$$\text{Percent carry-over} = 100\,(L_2 - L_1)/H$$

One should determine the percentage of carry-over at several levels of analyte to establish those levels of analyte where significant contamination of subsequent samples will occur.

Sample carry-over, seen most commonly with continuous-flow analyzers, occurs because the stream exhibits laminar flow. The fluid at the walls of the tubing is stationary while the fluid in the center flows at the highest velocity. Thus an abrupt change between samples will become a gradual change because some of the previous sample is left behind at the walls of the tubing. This material eventually diffuses back into the center of the tube where it is carried onward. Carry-over affects the test results by contaminating the current sample with a proportional part of the previous sample. The amount of contamination or carry-over that is permitted determines the throughput. If less carry-over is permitted, a longer time must be allowed for the previous sample to be flushed out. A longer time spent on each sample reduces the number of samples that are processed per hour.

Chromatographic theory has been applied to reduce the amount of carry-over. This is usually accomplished by the use of narrow-bore tubing so that the distance for the sample to diffuse from the wall to the center of the tubing is small and the time is short. This reduction in the tubing size results in a considerable increase in flow resistance; that is, the pressure increases. The pressure could be lowered by reduction of the flow rate; however, this makes carry-over from dead volumes more severe. A dead volume is a nook or cranny that is not readily flushed out by the flowing stream. Recently, instruments have been designed to reduce the carry-over sufficiently so that adequate throughput can be achieved. This form of continuous-flow analysis is usually called *flow-injection analysis*.

Almost all continuous-flow analyzers have streams that are *air segmented* to reduce the carry-over. Air bubbles are injected into the flowing stream so that the fluid on the walls is forced to move along with the fluid in the center of the tubing. The family of instruments from the Technicon Instruments Corporation (Tarrytown, N.Y.) is closely identified with this technique.

Mixing

Those instruments using the dry-film technique mix sample and reagents by the diffusion of sample into the reagents. Most dry-film reagents are premixed during manufacturing, though some are mixed by diffusion, which becomes possible only when the film is wet. The discrete test analyzers can mix the reagent and sample by (1) motion of a test tube or container, (2) stirring by paddle or stick, (3) agitation by air bubbles or ultrasonic waves, or (4) convection resulting from forceful addition of sample into the container. Mixing in continuous-flow analyzers is accomplished by convection in a mixing coil. When the tubing is wrapped in a tight coil, the distance that the fluid flows is different between the inside and outside of the flow path within the coil. This difference causes the fluid to tumble between the inside and outside of the coil, affecting the mixing process.

Incubation

Automated incubation is merely a delay station where the test mixture is allowed to react. This is performed, in most cases, under conditions of a specified, constant temperature, which is most frequently achieved by heating blocks and air or water baths. These constant temperature devices are monitored electronically by thermocouples. In a continuous-flow system, incubation is accomplished when the length of the tubing is increased. In this case, the delay is the volume of tubing divided by the flow rate. For example, a 4.2 mL incubator with a 1.2 mL/min flow rate would give a 3.5 min incubation time. Discrete analyzers accomplish incubation by allowing the reaction mixture to dwell in a chamber (test tube or cuvette) for a specified time. A similar approach is used for the dry-film analyzers.

It should be noted that many test methods require an addition of a second reagent, possibly followed by additional incubation. The automated means for doing this are not different from those just discussed above.

Sensing

The techniques of automation are not dependent on the forms of sensing, whether optical, thermal, or electrical. There are two major approaches to automated sensing: in situ and external. The term "in situ" refers to measurement in the vessel where the reaction has taken place. The term "external" is applied to systems of measurement where the sample is transferred from its original position in the reaction vessel to the sensing device. The dry-film tests are measured in situ by reflectance photometry or by means of integral electrodes (electrometric). Discrete test instruments use both in situ and external sensing mechanisms. The proportional flow instruments use external sensors; this is by definition, since it is difficult to achieve what is meant by "in its original position" for a flowing stream. Currently there is no preference between in situ and external sensing. External sensing generally exposes the sensing chamber to many samples so that care must be taken to eliminate carry-over from one sample to the next. Optical or electrode surfaces may also be contaminated by components from the samples. On the other hand, in situ sensing makes special demands on the test chamber. If the test container is disposable, it is impractical to calibrate it for optical or electrical characteristics. Thus such containers must be manufactured with very good reproducibility. However, disposable containers are meant to eliminate the mechanical complexity required to wash and recertify the sensing chamber. It is most likely that in the future sensing will be done in situ because this approach decreases the mechanical complexity of the instrument.

Chemical reactions can be monitored either at one time point or at many. Commonly, single-point monitoring is used for end-point analysis in which the reaction has gone to completion. Multiple-point monitoring is used for kinetic analysis. Discrete analyzers are most easily adapted for multiple time-point monitoring, whereas continuous-flow systems are not.

Computation

Automated computation has taken two forms: *analog* and *digital*. Analog computations use the electrical signal such as a voltage or current from a sensor (such as a phototube) and quantify the signal by comparing it to a reference signal. For example, a "blank" reaction mixture will give a 100% T (blank transmittance) resulting in a certain electronic signal. A test standard will give a lower percentage of T and thus a decreased electronic signal. The analog computer compares the two signals and takes the logarithm of the result. The final result is related to the quantitation of the reaction.

Some reaction signals, by their very nature, are in a form of discrete numbers. Two examples are individual photon-counting events and counting of radioactive decay. This signal, consisting of a number of individual events, can be monitored by a digital computer that can process this signal. Digital processing is restricted to certain arithmetic functions (such as subtraction or addition) unless the computer is programmable.

For a digital computer to process signals from many types of sensing devices in automated instruments (such as the spectrophotometer and ion-selective electrodes), an analog to digital converter is necessary. This converts the voltage or current signal into a digital form, which can be processed by the digital computer.

There are no straightforward rules as to which is the best form of computation. The decision is usually based on economics. However, if any part of the signal processing is done digitally, virtually all the processing is done digitally. Perhaps the major exception is the analog conversion of transmittance to absorbance, which is done for reason of analytical performance.

Readouts

The simplest visualization form of an instrument readout is the use of light-emitting diodes (LED) or a television monitor (cathode-ray tube, CRT) to express the data in numbers. This readout is usually converted to a hard-copy such as a paper-tape printout. This data must be manually transferred to laboratory slips or other permanent records. This manual step, of course, is subject to transcription errors.

More sophisticated, automated systems collate the test results for each patient. Usually the results are printed directly on the report slip. The hallmark of the Technicon SMA was the bar chart, which was an analog representation of the analytical results.

When the results are put into a laboratory computer, the instrument readout can be directly interfaced or connected to the computer. Although analog connections are possible, the majority are digital.

CONCEPTS OF AUTOMATION: DEFINITIONS

Test repertoire

Economics suggest that instruments perform more than one kind of test; that is, the investment in automation and the labor in sample loading is not duplicated. Following this logic to an extreme would require that an instrument be capable of performing every conceivable kind of test. This has not been possible. However, the characteristics of the analyses requested are such that six chemistry tests account for 50% of the work load and the next 14 tests account for another 40% of the total work load. Automated instruments have been designed to perform the most frequently ordered tests, those ordered less frequently, and the infrequently ordered specialty tests. Thus the most highly automated instruments are usually reserved for the most frequently performed analyses, whereas less automated analyzers are used for less frequently ordered tests. Automation is usually not essential for rarely ordered tests.

The automation of tests may also be done on the basis of *type* of analysis rather than test volume (number of samples); that is, an automated radioimmunoassay instrument will process RIA's for many different analytes regardless of numbers of specimens per analysis.

The *immediate test repertoire* can therefore be defined as the number of tests that can be performed at any one time, whereas the *total test repertoire* includes the total number of different tests that can possibly be performed on the instrument, that is, by changing reagents and a few components.

Selective

Instruments that are capable of performing multiple tests are selective if the particular tests to be performed on an individual sample can be specified and if no sample and no reagent are consumed by tests that were not requested. For example, the Technicon SMAC is a nonselective instrument, since all tests in the immediate repertoire are performed on each sample regardless of the exact tests requested. A discrete analyzer, such as the ACA, performs only the test requested. Selective analyzers have also been termed "random-access analyzers" for their ability to process different test combinations for each individual specimen.

Discrete

Instruments that compartmentalize each sample reaction are discrete analyzers. Typically the sample aliquot and reagent are contained in a single cuvette that is physically separated from all other cuvettes. This contrasts with the continuous-flow instrumentation in which all samples physically flow through the same sample path.

Continuous flow

Instruments that continuously pump reagent through tubing and coils to form a flowing stream and continuously pump sample into that stream are called "continuous-flow analyzers." The proportioning of sample and reagent is accomplished by control of the respective volumetric flow rates. For example, the sample might be pumped at 0.1 mL/min and the reagent at 0.9 mL/min, giving a 1 to 10 dilution of sample in the final reaction mixture.

Batch analyzer

Instruments that perform the same test simultaneously on all samples presented to it are termed "batch analyzers." The type of test can vary widely, but usually only a limited number of samples are processed per analysis. Examples of a batch analyzer are the centrifugal analyzers such as the Roche COBAS and CentrifiChem. All batch analyzers are selective analyzers.

Centrifugal analyzer

A centrifugal analyzer is an instrument that uses centrifugal force to mix the sample aliquot with reagents. The sample and reagent are loaded into discrete compartments of a centrifuge rotor. When the rotor begins to spin, the sample and reagent are thrown against the periphery of the rotor. Radial dividers keep the samples separate. A light beam shines through each sample as the sample rotates over a light source and its absorbance (or fluorescence) is measured. The resultant data are generated in rapid sequence and are processed by a digital computer.

Dwell time

The dwell time is the minimum time required to obtain a result after the initial sampling of the specimen. Some instruments can give results in as little as 15 seconds for single tests such as glucose. Commonly, instruments that perform multiple tests on a single sample have longer dwell times, ranging from 60 seconds to 10 minutes. Certain test procedures, such as kinetic analysis for enzyme activity or radioimmunoassays, that require long incubations will, of course, have longer dwell times. Dwell time is extremely important when significant or life-threatening physiological changes can take place rapidly. Thus blood-gas determinations (pH, P_{CO_2}, and P_{O_2}) need instruments with dwell time on the order of seconds. On the other hand, a "dwell time" of a week for a vitamin assay would be clinically acceptable.

Throughput

The throughput is the maximum number of samples or tests that can be processed in an hour. For simultaneous analyzers, one can calculate the total test throughput by multiplying the number of samples processed per hour by the number of tests performed on each specimen. For discrete analyzers, the sample throughput will obviously depend on the number of different tests requested on each sample. In addition, the time required per test can vary widely (that is, from less than 30 seconds to approximately 10 minutes). In general, the more tests ordered per sample, the slower the sample throughput on a discrete analyzer. Thus it is more difficult to give a simple, accurate value for the sample throughput for a discrete analyzer. The calculation for throughput does take into account the dwell time; that is, the fact that no results are produced until the dwell time has elapsed. The desired throughput of an instrument is usually matched to the number of samples that need to be processed in a given time period. Generally the throughput of an instrument is chosen so that a maximum number of samples can be processed in a timely fashion. For example, a higher throughput (and more costly) instrument may be required to process samples from a clinic so that results can be made available before the patients return home. In general, an automated analyzer is chosen on the basis of its ability to process the bulk of the routine work load in time for routine clinical decision making.

Stat testing

The word "stat." is an abbreviation of the latin word *statim,* meaning 'immediately.' There are many valid clinical reasons why a test result is needed immediately. Some instruments are not well adapted to processing stat tests. For example, instruments (batch instruments) that require reagent change to perform another kind of test may not have the reagents on line to perform a stat test. The interruption in the processing of the current samples and time required to change to the proper reagents may not be acceptable or practical.

Instruments that are to be used for stat testing need not necessarily have a high throughput but should have a short dwell time. Most stat instruments are not multiple-test, simultaneous analyzers, but dedicated instruments that analyze no more than a half-a-dozen high-frequency tests simultaneously. The Beckman ASTRA 8 is an example of such an instrument.

Cost

The resources consumed in producing a patient's test result represents the cost. It consists of labor, which is the monetary representation of the time spent processing the sample, preparation and maintenance of the instrument, reagents (which are calculated from the cost of the chemicals used for the test and a proportionate part of that for start-up), calibration and quality control, consumables (comprising the cost of sample containers, paper, and so on), and capital (which is a proportionate amount of the life of the instrument consumed and hospital overhead including the cost of items such as lab slips and maintenance).

COMMON AUTOMATED INSTRUMENTS

This section describes a number of instruments that are common in the hospital laboratory or illustrate a category of instrument type. A comparison of operational parame-

Table 15-1. Comparison of operational features for several automated analyzers

	SMA-12	SMAC	ASTRA-8	CentrifiChem	ACA-II	Ektachem	Abbott ABA-100	Hitachi 705	Micromedic Concept 4	Automated HPLC
Type[a]	C,S	C,S	D,S,SL	B,D,SL	D,SL	D,SL	B,D	D,SL	B,D	B,D
Sample volume (μL)[b]	1600-3300	400	120	5-50	20-800	10	5-25	5-20	2-200	2-200
Minimum volume (μL)[c]	1800-3500	450	175	20	30	25	50	55-75	Prediluted	2
Sample identification[d]	None	Keyboard	Keyboard	None	Photo image	Keyboard	None	None	None	Keyboard
Reagent[e]	Bulk, A	Bulk, A	Bulk, A	Bulk, A	Unit, M	Unit, M	Bulk, A	Bulk, A	Bulk, M	Bulk, A
Proportioning[f]	Flow ratio	Flow ratio	Vol Add	Vol Add	Prepackaged, Vol Add	Prepackaged, saturation	Vol Add	Vol Add	Vol Add	Vol Add
Minimization of matrix interferences[g]	Dialysis	Dialysis	Dilution	Dilution	Dilution	Filtration	Dilution	Dilution	NA	Extraction
Mixing	Coils	Coils	Flow turbulence	Centrifugal force	Vibrational (mechanical)	Diffusion	Flow turbulence	Flow turbulence	Flow turbulence	None
Sensing[h]	Spec, flame	Spec, electrodes	Spec, electrodes	Spec	Spec, electrodes available	Reflectance spec, electrodes	Spec	Spec, electrodes available	Gamma-ray counter	Spec, fluorometric, electrochemical
Test repertoire available[i]	12	20	9	1	50	16	1	16	1	1
Total repertoire[j]	23	23	20	>50	60	16+	>50	>50	10-15	>50
Stat capability[k]	NR	NR	Yes	NR	Yes	Yes	Yes	NR	NR	Yes
Dwell time[l]	10 min	12 min	1¼ min	5-30 min	8 min	5-6½ min	1-10 min	15 min	35-100	10-100 min
Throughput, samples per hour[m]	60	150	72	100+	97	270	115	180	50-100	1-8
Throughput, tests per hour[n]	720	3000	648	100+	97	270	115	180	50-100	1-8
Readout	Graphic (analog)	Digital	Digital	Digital	Digital	Digital	Digital	Digital	Digital	Graphic

[a] *B*, Batch analyzer; *C*, continuous flow; *D*, discrete analyzer; *S*, simultaneous analyzer; *SL*, selective.
[b] Sample volume needed to perform a test (discrete analyzer) or simultaneous profile.
[c] Sample volume plus dead space in sample cup.
[d] Means of identifying sample for final printout.
[e] *A*, Available from alternative sources; *M*, only available from manufacturer.
[f] Manner of adding reagent, diluent, and sample; *Vol Add*, volumetric addition.
[g] Means of minimizing protein interference; *NA*, not applicable.
[h] *Spec*, Photometric; *flame*, flame emission or absorption spectroscopy; *electrodes*, ion-selective electrodes.
[i] The number of tests available at one time, without a change of reagents or instrument module.
[j] Total number of analytes for which reagents are commercially available (as of 1984).
[k] *NR*, Not recommended.
[l] Approximate time between sampling and availability of result.
[m] Number of samples capable of being processed per hour. Data listed are for one simultaneous profile or a *single* test per sample (approximate reaction time for variable analyses = 1 minute) for discrete analyzers.
[n] Calculated by multiplication of maximum number of tests per sample available times number of samples capable of being processed per hour.

ters for some automated instruments is shown in Table 15-1. Obviously this table is not meant to be all inclusive but to demonstrate the features available for common types of instruments.

Technicon AA and SMA

The Technicon Auto-Analyzers (AA) and Technicon Sequential Multiple Analyzers (SMA and SMAC, that is, SMA with computer) are a family of instruments based on a continuous-flow principle invented by Leonard Skeggs in the late 1950s. His idea was to continuously mix reagent and sample flow streams. The volume ratio of sample and reagents is controlled by the relative flow rate, which is controlled by use of larger or smaller tubing in a peristaltic pump. His key contribution was the injection of air bubbles into the flowing streams. These bubbles effectively scrubbed the walls of the tubing so that carry-over of sample was minimized.

The first AA performed two tests: glucose and blood urea nitrogen. It processed 20 samples per hour, and the reagent was consumed at 5 mL/min. The SMA's reduced the sample and reagent consumption while simultaneously increasing the throughput. The SMAC can process 150 samples per hour and uses reagent flow rates on the order of 0.75 mL/min.

A wide range of test methods are available or have been reported. However, the most common use of the SMA and SMAC has been for profiles in the medium and large laboratories. As much as any instrument, the SMA's have been a leader in the automation of the clinical chemistry laboratory. Their dominance has only recently been challenged by the computer-controlled, selective analyzers. Because of the considerable start-up time, the SMA's are not well adapted to off-hour or stat testing. While they are running, the dwell time is under 10 minutes for 6 to 20 test results. The SMA's use bulk reagent that can be purchased from Technicon or from alternative commercial sources. Consisting primarily of water, the reagent for Technicon equipment has a relatively low cost. The reagent is usually stable at room temperature.

Eastman Kodak Ektachem

The Eastman Kodak Ektachem is a discrete, selective analyzer that uses dry-film technology. The addition of 10 μL of serum sample provides the solvent (water) necessary to rehydrate the dry reagents. The film is composed of multiple layers, some of which serve to ultrafilter the sample (remove protein) whereas others serve to provide reactive reagents. The colored reaction products are measured by reflectance on the side opposite the sample addition. The reflected light is converted into concentration units (of the reaction products) by the Williams-Clapper formula (analogous to Beer's law for converting light transmission into absorbance units).

The samples are automatically processed by dispensing of a drop of serum (10 μL) from the sample cup onto the slide. The slides are automatically dispensed from cartridges, each containing a different method. The slides are incubated while the color forms. Finally the reflectance measurements are performed. The instrument is capable of rate ("kinetic") and enzyme measurement. The system also permits electrometric measurements for the determination of sodium and potassium. The Ektachem uses, by definition, unit reagent. The reagent slides, which must be stored at refrigerated temperature, are available only through the manufacturer. The reagent cost per test is relatively high.

Centrifugal analyzers

There are several manufacturers of these discrete batch analyzers. The instrument represents an example of technological transfer from the Oak Ridge National Laboratory to the private industrial sector. The key element is a centrifuge whose rotor contains 16 to 32 cuvettes. The sample and reagent are mixed together by the action of centrifugal force when the rotor is first accelerated. The cuvettes rotate past a photometer that permits absorbance readings at a preselected wavelength. These analyzers have been especially useful for rate ("kinetic") and enzyme measurements and for end-point analysis as well.

They are batch instruments usually performing only one kind of analysis at a time. Their reagent consumption is small, a feature that makes them economical to operate. Although the theoretical test throughput can be high, in practice it is considerably lower because of the need for subsequent runs for reanalysis of samples that give unsatisfactory results such as excess enzymatic activity.

Many of these analyzers have automatic sampling devices that take aliquots of the samples and reagents and place them in separate compartments of the rotor. Most sampling devices are separate from the analyzer, though the Roche COBAS has a sampler integrated into the analyzer. Some instruments use disposable rotor-cuvettes, whereas others wash the rotor for reuse. Although all the centrifugal analyzers can perform spectrophotometric and turbidimetric analysis, newer instruments (such as COBAS and Instrumentation Laboratories' Multistat) have nephelometric and fluorometric capabilities.

The centrifugal analyzers use bulk reagents available from a wide variety of commercial suppliers. The flexibility of these instruments allows for easy adaptation of "in-house" reagents, or commercially supplied reagent, for routine analysis.

Du Pont ACA Discrete Clinical Analyzer

The du Pont ACA Discrete Clinical Analyzer is a discrete, selective analyzer that serves as a general chemistry analyzer in small-sized hospitals and a specialty analyzer in the medium- and large-sized hospitals. Some 50 test methods are available, ranging from routine and stat.

chemistries to immunochemistries to therapeutic drug monitoring to coagulation. It processes 97 tests or samples per hour with a dwell time of 8 minutes. Any combination of tests can be run on any sample at any time. Its relatively large sample-volume requirements make it impractical for pediatric situations. The reagents are prepackaged in plastic packs, which also serve as the reaction cuvettes.

Sample processing begins by automatic sample aspiration. The sample and buffer is then automatically dispensed into the pack. The reagents, which were isolated in separate compartments of the pack, are released and mixed with the sample and buffer. There are stages where two separate reagent additions can take place. The contents of the pack are then measured photometrically when the plastic pack is formed into a 1 cm path-length cuvette. The end-point methods are measured bichromatically, except for several methods that use two packs to permit blank absorbance measurement at the desired wavelength. The rate ("kinetic") and enzyme methods are measured at a single wavelength over a 17-second interval.

The reagent packs are available only from du Pont de Nemours (Wilmington, Del.) at this time and must be stored at refrigerator temperatures. The cost per test is relatively high.

Beckman ASTRA

The ASTRA-4 and ASTRA-8 are discrete, selective analyzers that process the high-volume tests, glucose, creatinine, BUN, Na, K, Cl, and CO_2. The ASTRA-8 will process up to 72 samples per hour and up to 648 tests per hour. The test repertoire is defined by eight analytical modules, which can be selected from a larger set including calcium and a number of enzymes. The modules make innovative use of photometric and nonphotometric methods of analysis that are monitored in a kinetic mode.

The ASTRA-8 has been a popular replacement for the SMA 6/60 because of its low cost, simplicity of operation, and reduced maintenance requirements.

The ASTRA's use bulk reagent that is available from alternative reagent suppliers and from the manufacturers as well.

Abbott ABA

The ABA-100 is a discrete batch analyzer that can process up to 120 tests per hour. The newer version, the VP, is more automated and provides greater throughput than its predecessors. This family of instruments is noted for using bichromatic-photometric measurements. The absorbance is measured at two wavelengths and the absorbance difference is used to calculate the concentration. The advantage is removal of the effects of perturbations in the optical path that act similarly at both wavelengths. Thus effects of cuvette imperfections and turbidity are reduced. The bichromatic principles also can be applied to rate measurements where the rate of change of the bichromatic measurement is related to the analyte concentration.

The ABA uses bulk reagents obtainable from a wide number of commercial suppliers and from the manufacturer as well.

Hitachi 705

The Hitachi 705 is a discrete, selective analyzer having a throughput of 180 tests per hour. It can have a 16-test repertoire available at any one time and, in addition, offers an ion-selective electrode accessory. It is representative of a group of new random-access instruments, which include Technicon's RA 1000 (240 tests per hour) and Coulter's DACOS (450 tests per hour). These analyzers have been designed for microprocessor (microcomputer) control, and as a result the test methods can be readily changed. Perhaps the key feature of these analyzers is the use of a single photometer that measures multiple cuvettes at several wavelengths and multiple times. Previous analyzers used multiple photometers to process multiple tests or processed one sample at a time. This technique differs from that of centrifugal analyzers, which process only one test type at a time.

All these random-access analyzers employ bulk reagents available from a wide number of suppliers and from the manufacturers as well. It is relatively simple to adapt new procedures for use on most random-access analyzers.

Micromedic Concept 4

The Micromedic Concept 4 is a discrete analyzer for RIA (radioimmunoassay). It is based on the use of special test tubes whose inner surfaces have been coated with antibody. The sample and reagent are added to each tube, which can be incubated at room temperature or at 45° C. After an incubation of up to 17 hours, the free ligand (sample analyte and radiolabeled analyte) is aspirated, and the bound ligand attached to the antibody on the test tube is measured in a gamma-ray counter. Because the precision of the measurement depends on the number of gamma rays counted, among other things, it has been necessary to measure the counts over a minute or more. To increase the throughput, the newer instruments employ multiple counting chambers. The system is designed to use racks of 10 tubes each and is capable of processing up to 200 antibody-coated tubes per assay. The precoated tubes are obtainable only from the manufacturer.

Squibb Gammaflo

The Squibb Gammaflo is an air-segmented, continuous-flow analyzer for radioimmunoassay. The photometer of the conventional chemistry analyzer has been replaced by two stages. The first stage separates fluid from the air and passes the liquid to a resin-charcoal column for separation of the free and bound ligand. The second stage is a stop-flow gamma-ray counter. A microcomputer is used for data reduction and system control.

Becton Dickinson ARIA II

The Becton Dickinson ARIA II is a continuous-flow analyzer for radioimmunoassay that sequentially pumps assay mixtures through a reusable chamber containing antibody covalently bound to solid support media. In this system the bound ligand is chemically stripped from the antibody and passed to a flow-through gamma-ray counter. The free ligand can also be counted.

TRENDS IN AUTOMATION

It is of course risky to predict the future, but there are some trends that will affect the kind of automation employed in the clinical laboratory over the next 5 years. Increasing economic pressure to reduce operating costs will make instruments that improve productivity attractive. These will tend to be selective analyzers to conserve sample and costly reagents and be capable of performing a large number of tests. The areas needing improvement are sample preparation and identification and reduction of manual transcription of results. To this end, laboratory computers reading bar codes and transcribing data will become common, and instruments will be relieved of tasks better handled on a lab-wide, computer basis, yet interaction between the two will become more sophisticated. The reaction volumes will continue to decrease so that the reagent cost will be small. Spectrophotometric instruments will continue to dominate. Electrometric methods will be used for fewer than a dozen test types.

The new instruments will be hygienic and mechanically simple and require minimal daily maintenance. They will have a relatively large repertoire of tests so that the number of work stations in the laboratory will be reduced. The capital cost will increase because of features that reduce the labor and operative costs. The communication between the operator and the instrument will become much more convenient.

ECONOMIC JUSTIFICATION

When the laboratory wishes to acquire a new piece of automated equipment, it is usually necessary to present to the hospital administration an economic justification that is based on a return-on-investment (ROI) concept. Even if the hospital is a nonprofit organization, the return on investment is a useful concept for the allocation of funds. This is true because the decision that provides the best return on investment will permit the largest number of additional services. In other words, a poor investment will squander limited resources.

In order to understand the concept of return on investment, it is important to calculate the "payback." The cost of the instrument is divided by the annual revenues in excess of expenses to give the number of years necessary to repay the investment. A rule of thumb is a 2- to 3-year payback for an investment to be worthy of consideration. If the payback is longer than 2 to 3 years, there are presumably other investment opportunities that are more desirable. Thus in the laboratory it may be more economical to send certain tests to outside laboratories even though the laboratory could "do it cheaper."

The payback concept is simple but can be misleading when considered by itself. Thus the payback concept is, at best, a screening device to identify investments for further study. Suppose a certain instrument provides a 2-year payback, but a higher-capacity instrument provides a 3-year payback. A more rigorous analysis relating both time and money is required to decide which is actually the best investment. Nonetheless, an instrument with a 10-year payback is exceedingly unlikely to be a good investment. Notice that an *essential* piece of equipment *must* be purchased even if the payback shows it to be a poor investment. However, even in that event a good economic analysis can show which of several instruments will provide the essential service at the smallest economic cost.

Although it is possible to calculate revenues associated with a test, because of the mechanisms of reimbursement to the hospital by third-party payers, the exercise may be meaningless. Thus it is more reasonable to look at an investment based on cost differences between possible instruments. For example, if two instruments consume $10,000 per year in deionized water, there is no relative advantage of one instrument over the other. So the analysis need not be burdened by including the cost for deionized water.

Suppose the laboratory owns a batch analyzer that performs a profile or panel of six tests on each sample. Further suppose that the instrument is old and takes too much time to start up each day. A new instrument is being considered that is *selective* in the tests that it performs. However, the reagent cost is 2¢ per test more than the current instrument (Table 15-2).

This table shows increased costs for the new instrument. The purchase of the new instrument would never pay back the investment relative to continued use of the existing instrument. Of course, one assumes that the current instru-

Table 15-2. Cost comparison of two instruments

Item	Instrument	
	Current	New
Start-up	Slow	Fast
1.5 hr/day × $10/hr × 300 days/yr	$4,500/yr	—
Reagent-cost differential		
250 samples/day × 6 tests/sample × $0.02/test × 300 days/yr	—	$9,000/yr
	$4,500/yr	$9,000/yr
Net savings with new instrument		(−$4,500/yr)

Table 15-3. Cost comparison of two instruments with new reimbursement conditions

Item	Instrument	
	Current	**New**
Start-up	Slow	Fast
1.5 hr/day × $10 hr × 300 days/yr	$4,500/yr	—
Reagent-cost differential		
250 samples/day × 6 tests/sample × $0.13/test × 300 days/yr	$58,500/yr	
660 tests × $0.15/test × 300 days/yr		$19,700/yr
	$63,000/yr	$29,700/yr
Net savings with new instrument		$33,300/yr

ment with its excessive start-up time (and maintenance) does provide adequate service, that is, accurate and timely results.

Now suppose that the third-party payers refused to reimburse for "profiles" and that the physicians begin to order discrete tests. It is the laboratory's prerogative to run the test on the profiling instrument but report only the requested test or tests. Now consider the relative costs, assuming, for example, that the 250 profiles of the previous example would be equivalent to 150 glucose, 120 K, 120 Na, 105 urea, 90 creatinine, 50 CO_2, or 660 tests (Table 15-3).

Suppose the new instrument costs $100,000 and the proceeds from selling the old instrument (after moving expenses and so on) are $10,000. The investment is $90,000, and it pays back in 2.7 years ($90,000 ÷ $33,300/yr).

Notice that there is no need to consider the revenues. Presumably, in time, the reduced billings, because of discrete tests being ordered rather than profiles, will result in an increased reimbursement rate to cover the laboratory costs. If the laboratory can reduce its costs attributable to the change in pattern of tests ordered, that is all to the good. Furthermore, it has been of no consequence whether the current instrument has been fully amortized. We mean by this term, the remaining value of the instrument for accounting purposes. For example, a $21,000 instrument might have an expected life of 7 years. For accounting purposes, the utility of the instrument is consumed at $3,000 per year. At the end of 7 years there is no utility or value left in the instrument; that is, its "book value" is zero. Through good fortune or careful maintenance the instrument may be very useful after 7 years, but its purchase cost has been paid by charges allocated to each result produced by the instrument.

BIBLIOGRAPHY

Coakley, W.A.: Handbook of automated analysis: continuous flow technique, New York, 1981, Marcel Dekker, Inc.

Ferris, C.D.: Guide to medical laboratory instruments, Boston, 1980, Little, Brown & Co.

Kinney, T.D., and Melville, R.S., editors: Evaluation of the uses of automation in the clinical laboratory, D.H.E.W. Publication 79-501, vol. 147, Washington, D.C., 1979.

Price, C.P., and Spencer, K., editors: Centrifugal analyzers in clinical chemistry, New York, 1980, Praeger Scientific.

Ritchie, R.F., editor: Automated immunoanalysis, New York, 1978, Marcel Dekker, Inc.

Tompkins, W., and Webster, J., editors: Design of microcomputer-based medical instrumentation, Englewood Cliffs, N.J., 1981, Prentice-Hall, Inc.

Chapter 16 Laboratory computers

David G. Rhoads

accumulator The "scratch pad" of the computer's central processing unit (CPU). Data written there can be accessed and processed very rapidly.

address The physical location where specific pieces of information are stored.

algorithm The series of operations required to do a simple task. For example, an algorithm might be set up to calculate the mean from a list of numbers.

analog computers Devices that use a continuum of voltage or some other continuously variable parameter as the fundamental principles of their operation.

ASCII Acronym for American Standard Code for Information Interchange. It is the code for the numeric values assigned to each character.

batch A job submitted to a batch-oriented computer. The machine puts the job in a queue with many other jobs. When it is ready, it performs the job and prints out the results.

baud Usually treated as equivalent to the unit of bits per second. Approximately 10 bits of serially transmitted data are required to form one ACSII character.

bit The fundamental unit of information for the digital computer. It has two possible states, on and off. These are translated by the computer to be a 1 or a 0, respectively.

bootstrapping The process of starting the computer after it has been turned off. It is derived from the phrase "pulling oneself up by one's bootstraps." A series of programs (usually two or three) of increasing complexity are used to start the computer. Initially, a small program reads in the next one, which reads in the third one, and so on.

bug A mistake or problem in the code of a program. In some systems, DDT (for dynamic debugging technique) is used to eliminate bugs.

byte A group of eight bits. A byte in most computer systems will hold one ASCII character.

card reader An input-output (I/O) device that reads computer cards. The computer cards in common use either have punched holes or are marked sense cards.

code The set of instructions written to perform a task in the computer. A code may be written in any language compatible with the computer and its other software.

compilation The process of converting program code into instructions that the computer uses to perform the task at hand. It is performed in a separate operation before one runs the program.

CPU (central processing unit) The physical component that supervises operation of the computer. It directs the operations of the computer on the basis of the instructions from the program.

data base The collection of data used in the computer. In a clinical laboratory application, this could correspond to all the data on each test, the names of physicians, the results of the tests, and so on. In a business application it might correspond to the information on each customer or all the data on payroll.

data reduction The processing of raw data into a meaningful result. An example of data reduction is the calculation of a mean and standard deviation from a month's worth of quality control data.

data types Different types of data that a computer may use, set up for purposes of efficiency. The simplest is *logical* or *boolean,* in which a variable has only the value of "true" or "false." Other data types include *integers* (5, 2486, and so on), *real* (5.5, -0.82, $9892.2 = 9.8922 \times 10^3 = 9.8922 \times 10^{**} = 9.8922$ EØ3), and *alphanumeric* or string data (R2D2; Good morning, Mrs. Robinson; words).

digital computers Devices that use "on/off" or binary logic as the fundamental principle of operation.

directory The index of file names and locations in a data-storage device such as a disk; it corresponds to the index of a book. In a laboratory computer, directory entries would include names of files and computer programs stored on the disk.

file A region in which data is stored on a device that has a directory or index of file names; it corresponds to a file folder or a chapter in a book. A file is composed of one or more records. In a laboratory computer there might be a file of all the tests offered by the laboratory along with their names, costs, and reference ranges.

floppy disk An input/output device similar in concept to the ***hard disk.*** However, its speed is slower and its storage capacity much less. It has the advantages of being relatively inexpensive and portable.

hard copy Paper containing printout from a computer.

hard disk (or Winchester drive) A high-speed, high-capacity data storage device. It is capable of random access of the data. It has a magnetic surface on which data are stored and resembles a heavy-duty phonograph record.

hardware The generic name for the physical equipment used in computer systems. Hardware includes the CPU, memory, components for the interfaces, and all the input/output devices.

high-level language A computer language constructed along more or less conventional thought patterns. Frequently one instruction in a high-level language will require a dozen or more instructions in a machine-level language. High-level languages are usually much slower in operation but are much easier to program than machine-level languages.

instructions Commands to the computer that specify what it is to do.

instruction set The allowed set of instructions in whatever language is being used.

interfaces The connection between different elements of a computer system or between different computers.

interpretation Similar to compilation but performed as an integral part of the process of running the program. The code must be interpreted each time the computer uses a section of it. Programs usually take much longer to run in an interpretive environment than in a compiled environment.

I/O Input and output.

I/O devices (also known as peripheral devices) Pieces of equipment used for input and output of information.

machine-level language A language in which the code is equivalent to the instruction set of the CPU. This code is very fast, usually less than 1 μsec per instruction. The permissible operations are limited, and the code is machine specific; that is, it cannot easily be transferred to another model of computer.

magnetic tape drives I/O devices that read and write on tape. The tape used closely resembles that used to record music.

mainframes Large, fast, sophisticated computers. They are the largest computers normally used in the medical environment and typically would be used for hospital administrative functions.

memory Stores data and programs for use by the computer.

microcomputer A computer consisting of a CPU, a substantial amount of memory, and a fair number of I/O ports. Prices range from $100 to $10,000.

microprocessor A central processing unit in a single electronic chip. They are used to control the operation of equipment.

minicomputers Moderately large computers ranging in price from $10,000 to $1,000,000. Many laboratory information systems are minicomputers.

modem A device that allows connection of a computer to a telephone line. Common rates of data transfer are 300 and 1200 baud.

network A system of several computers joined together to communicate and facilitate sharing of data among them.

nibble Half a byte, or 4 bits.

OCR (optical character reader) An I/O device capable of reading characters from paper. Usually there are fairly severe restrictions on the form and location of the characters in the text.

operating system A program that continually monitors the operation of the computer while it is running and performs the following tasks: controls access to I/O devices, controls which job will run at any given time, times the computer operation, and controls the interrupts (requests by various devices to the CPU for time). It may also be called a "monitor" or an "executive."

parallel interface The type of interface in which the bits that form each character are transmitted simultaneously along many wires lying side by side in a ribbon cable.

peripheral device See ***I/O devices.***

printer An I/O device that generates printed copy.

program A series of instructions designed to perform tasks on a computer.

RAM (random-access memory) Memory to which one has random access and in which one can both read and write frequently and easily.

random access The ability to obtain any element of an array of information without having to look at all the elements between it and the beginning. An example of random access is the ability to play a particular selection on a phonograph record without having to play all the record up to that point.

real time To perform the necessary operations in a period closely associated with the event. This is in contrast to collecting the data and then performing the operations at some later time. An example of real-time operations is the Technicon SMAC, in which the computer continually collects and analyzes the absorbance data coming from the photometers.

record A unit of storage on a storage device, frequently ranging from 128 to 1024 or more bytes in length; it corresponds to a piece of paper in a file folder or to a page in a book. In the example used for *file,* a record might contain the data on a single test.

ROM (read-only memory) Memory to which one has random access. However, it is designed to hold a given set of information permanently but only to be read. Special equipment is required to write or erase this type of memory.

RS-232C The IEEE (Institute of Electrical and Electronics Engineers) standard for a serial interface. It specifies only the electronic nature of data transmission and says nothing about the data that is transmitted.

sequential access The ability to obtain access to any element of an array. However, one has to pass over all the other elements between it and the beginning. A cassette tape is an example of a device with sequential access.

serial interface The type of interface in which the bits forming each ASCII character are transmitted sequentially along a single wire.

software The generic name for data and programs for computers.

terminal An I/O device that has two-way communication capability with the computer. It has a keyboard and either a screen or a facility to print text. Other names for certain types of terminals are CRT (cathode ray terminal) or VDT (video display terminal).

time sharing To have more than one person using the computer during any given time. A requirement for effective time sharing is that most users at any time are running jobs that require only limited amounts of computer resources. If anyone were doing a lot of calculations or disk I/O, the rest of the jobs would be slowed down significantly.

Winchester drive See ***hard disk drive.***

word A group of bits on which the computer operates. The size of a word ranges from 8 to 80 bits and is defined by the hardware and software used in any particular system.

BACKGROUND

Computers are capable of prodigious feats. They can store vast amounts of information and have it available for virtually instant access. Computers are being used to control the operation of equipment easily and accurately. It is clear that computers have changed the way society functions because of the revolution in the approach to the collection, distribution, and manipulation of information.

Numerous farsighted institutions have installed computers in the hope of improving their flow of information. The initial cost of the largest of these systems has been a million dollars or more. After installation, the performance of these computers ranged from terrific to terrible. In the worst cases, not even substantial additional expense was sufficient to guarantee success.

Computers come in many sizes and are used in many ways other than the handling of massive amounts of data. A major clinical laboratory application since 1970 has been instrument control. The ACA, introduced in 1971, was one of the first of this type of computer. In the past few years, most new instruments that have been introduced contain a microprocessor if not a full-fledged computer. In general, these new instruments have been successful because they manage many functions, such as instrument operation, data collection, and processing, all at the same time.

In contrast, laboratory information systems (LIS), the other major laboratory computer application, have been installed (as of 1982) in less than 10% of the clinical chemistry laboratories in the United States. This lack of widespread acceptance is an indicator of the anticipated risks and expense.

OBJECTIVES

The objectives of this chapter are the following:

1. To develop a general understanding of how a computer works. The concepts to be developed include hardware, software, interfacing, computer speed, and the different types of computer compatibility.
2. To describe some applications of computers in instruments used in clinical chemistry laboratories.
3. To define some factors in the various stages of the plan to purchase a laboratory information system. These include justification, selection and purchase, and implementation.

HOW COMPUTERS FUNCTION

Hardware

General concepts. There are two main types of computers, analog and digital. In practice the digital computer dominates the computer market because it is much more powerful than the analog computer. The digital computer is the subject of the remainder of this chapter.

All digital computers use the same general principle in their design, whether they are a $100 microcomputer or a $10 million-plus supercomputer.[1,2] Details, of course, vary tremendously. Generally, as the size of the machine increases, the word size increases, along with speed, memory capacity, complexity of job performed, and the number of input/output devices attached. In order to understand these systems it is essential that one learn what these various elements are and how they interact.

A computer system has two major conceptual components, the physical machine itself (hardware) and the instructions and data used in the operation of the machine (software). The whole system is a logical device capable of performing sequences of operations tirelessly for extended periods of time. Normally it will follow all its instructions exactly, even to the slightest mistake.

Central processing unit. Computer hardware consists of three major elements: the central processing unit (CPU), data storage capability (memory), and input/output (I/O) devices. The CPU is the component that controls the operation of the machine. Physically, the CPU is a unit of electronic circuitry. In recent models it is a single chip etched with thousands of carefully organized circuits. A CPU on a single chip is known as a microprocessor[3] (Fig. 16-1). Logically, it supervises what happens within the computer; the other chips are in fact "slaves" to the CPU. The computer program is fed to the CPU one instruction at a time. The CPU then has the machine perform the electronic function specified by the instruction. If an arithmetical or logical operation is called for, the CPU usually does it itself. If storage of data in memory is required, it sets up and supervises the storage of that data in memory. If printout of data is specified, the CPU activates the printer and directs that the data be passed to it. Whatever functions are going on within the computer are under the control of the CPU as directed by the program. Interestingly enough, the CPU is usually the cheapest major component of the computer. A CPU board for a microcomputer can be purchased for less than $300. Memory and I/O devices are usually much more expensive.

Memory. Computer memory is that part of the machine

Fig. 16-1. This is a 16-bit CPU (central processing unit) microcomputer board. The 8086 CPU chip and 8089 I/O (input/output) processor chips are large chips in upper left center of board. Two chips in upper right with white tape are ROM (read-only memory) chips, which contain bootstrap instructions. Most of remaining chips on board are used to facilitate access to memory and I/O devices.

in which data is stored.[4] Data include both programs and results. Physically the memory is a large array of electronic circuits of which individual elements are capable of maintaining either an "on" or an "off" condition for an extended period of time. The "on" state corresponds to a 1, the "off" state to a 0. These abilities to "remember" data in the form of an on/off condition and to manipulate them are the fundamental concepts of the digital computer.

The simplest memory element, the on/off condition, is termed a "bit." Groups of 8 bits are called "bytes" (rhymes with "bites"). Words are variable in length and depend on the electronic structure of the machine and the software. They range from 8 to 80 bits in length and include sizes such as 8, 12, 16, 32, 36, 64, and 80 bits. The on/off property of computers makes them binary machines. As a result, many attributes are in powers of 2. One kilobyte equals 1024 bytes, slightly more than the 1000 bytes one would expect. Similarly, 64 kilobytes, a common amount of memory, equals 65,536 bytes. A 10-megabyte disk drive would hold 10,485,760 bytes.

A single character, such as the *A* at the beginning of this sentence, takes up 1 byte of memory in many common implementations. A full page of text, 80 characters per line for 50 lines, requires 4000 bytes. A 5-megabyte hard disk, the smallest hard disk commonly available, would store 1250 such pages. It is clear that the amount of data stored by such a machine can be vast indeed.

There are two major types of computer memory in common use in the early 1980s. Somewhat confusingly, they are termed RAM (random-access memory) and ROM (read-only memory). In fact, both are capable of random access, that is, the capability of accessing (finding) any element of memory without having to traverse (examine and reject) many other elements. The difference between RAM and ROM is that with the former one may easily both store and retrieve data. With ROM the data are stored more or less permanently, and normally the users can only read the data. The trade-off between them is that the contents of RAM are lost when the power is turned off, whereas the contents of ROM memory are stable indefinitely, even without power. (Video-arcade games all store their games in ROM.) The technology of memory is changing rapidly. New types are being developed to have the advantage of both RAM and ROM; that is, they maintain their integrity when the power is off and are easily written into and read by the average user.

Logically, computer memory has two major properties: the content of memory, and the *addresses* at which the data are stored. The content is obvious. The addresses are not because they are not "visible" property. Without the addresses the computer has no way of knowing where its data are located. Fortunately the various computer languages make it easy to keep track of the addresses. This is done symbolically. These symbolic names implicitly specify memory locations of data and programs. The connection between the symbolic name and the memory location is made by the programmer's code, which is converted by a compiler or interpreter program into terms the machine can understand. Usually the programmer will neither know nor care about the exact location for any given variable. All the programmer cares about is that the datum is stored where it can be accessed as necessary.

Input/output devices. I/O devices provide for transfer of information between the user and the CPU and memory. The two major classes of I/O devices are on-line data storage devices (disk drives and magnetic tape drives) and everything else (terminals, printers, interfaces, card readers, and so on).

On-line storage devices are technically I/O devices. However, access can be so fast (50 msec access time for a hard disk) that their role is becoming that of memory, a way in which the apparent amount of data stored in memory may be vastly increased. The speed of access to the data stored on some devices is fast enough that the user may not notice the delay. On slower devices, however, the delay is extensive. For a review of disk storage technology, see White.[5] The major forms of on-line storage devices are the following.

Hard-disk drives. Hard-disk drives, as a form of on-line storage, are the fastest and potentially the largest. Sizes of a single hard-disk drive can range from 5 to 300 megabytes. They are random access devices. Typical time required to access data is about 50 msec. However, these drives are not cheap. In 1983, 5 and 10 megabyte drives for microcomputers cost about $2000 to $4000. The larger drives are proportionately more expensive. Some large computer installations have banks of many hard-disk drives on-line simultaneously, with a total hard-disk storage capacity measured in the billions of bytes.

Floppy (flexible) disk drives. An inexpensive way to store data is by use of floppy disks, which resemble 45 rpm records in square jackets. Their sizes are 3½, 5¼, and 8 inches in diameter. An 8-inch, double-density floppy disk can store about 600 kilobytes per side. Smaller disks will store less data. Access time is about 10 times longer than for the hard disk. However, their advantages are that they are much less expensive and are very portable. A cheap, easy way to transfer data between computer locations is to mail a floppy disk. The cost of a floppy disk that stores 1.2 megabytes ranges from $5 to $10.

Magnetic tape. This form of storage is characterized by vast storage capacity, especially in the larger tapes. Tape sizes range from cassettes to reels 30 cm in diameter. It is another inexpensive form of data storage. The major problem with the use of magnetic tape is that it can take relatively long periods of time to retrieve data. Unless the data is well-organized, a full reel can wind for minutes to retrieve a specific item. Tapes are used primarily because they store so much data very cheaply.

Other types of I/O devices. Terminals, which provide for input into the computer from a keyboard. Messages from the computer are either printed on paper or displayed on a monitor (TV screen). Other names for terminals with a video display include CRT (cathode ray terminal) or VDT (video display terminal).

Printers, which print data on paper at speeds ranging from 15 to 1200 lines per minute. Some devices now available are even faster; they function using ink-drop or laser technology. Inexpensive printers (less than $3000) come in two general categories—dot matrix printers and letter-quality printers. Dot matrix printers have a row of small wires in the print head. The machine sweeps them across the paper, telling each wire when to strike the paper to form the various patterns for each character. Often the quality of the result leaves much to be desired; however, some newer models are approaching letter quality. With a letter quality printer, a typewriter-like character strikes the paper in a single blow and forms a much cleaner image. The trade-off is that the letter-quality printer is more expensive and also substantially slower. Good dot-matrix printers (120 characters per second, 7 to 9 dots per character) cost less than $1000 in 1983. A good letter-quality printer (40 to 60 characters per second) cost less than $2500.

Modems, which provide for connection to a telephone line. Speeds for data transmission are 300 to 1200 bits per second (approximately 30 to 120 characters per second, respectively) but can be made faster with special equipment. A 300-bit-per-second modem in 1983 cost about $300; faster ones cost more. See the section on interfacing for more details.

Card reader, which reads punched or marked sense cards.

Optical character reader (OCR), which reads symbols (usually letters and numbers written on paper).

Interfacing

Protocol for transmission. Interfacing is the process of connecting various devices one to another so that they may communicate. Two computers, or a computer and a peripheral device such as a line printer, may be interfaced. The two elements of the interface are (1) the electronic connection between the devices, which includes not only the connecting cables but also the protocol to translate the electronic signals between the internal machine form and the form transmitted over the cable; and (2) the structure of the data that are transmitted. A scheme for the interfaces between various components of a microcomputer is shown in Fig. 16-2.

The two types of interface protocols are serial and parallel transmission of data. In the serial type the signals for each character are transmitted in sequence along a single line.

In a parallel interface all the signals representing a sin-

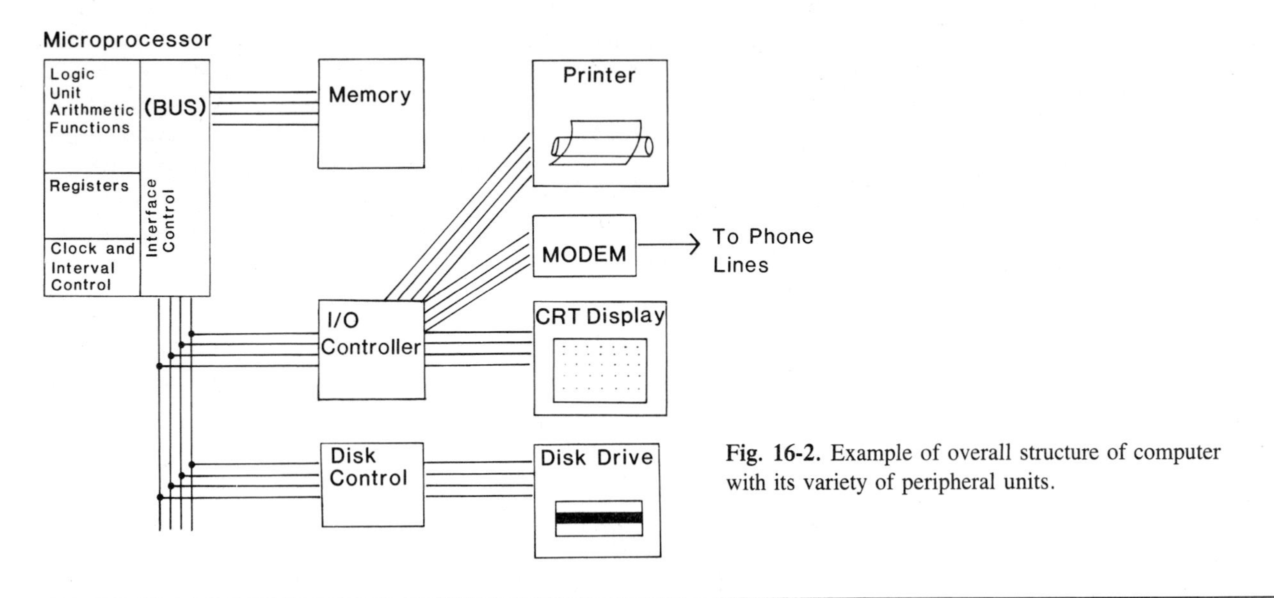

Fig. 16-2. Example of overall structure of computer with its variety of peripheral units.

SAMPLE OF A DATA STRING

```
Position in string  1   5   10   15   20   25   30   35   40   45   50
                    ....+....+....+....+....+....+....+....+....+....+
       Data string  James M. Infarction 123-45-6789 Dr. Goodhealth
```

Fig. 16-3. In this arbitrary example of structure of data in a string of characters, it is apparent that patient's name is from characters 1 to 20, identification number from characters 22 to 33, and name of attending physician from characters 34 to 50. The only way computer can identify which fields go with which data is by prior definition in program.

gle character are transmitted simultaneously along the many lines of a ribbon cable. The parallel transmission rate is much faster than the serial rate. One frequent restriction for the parallel interface is the need to have the communicating devices located within a few meters of one another.

One may interface two devices using phone lines. The usual transmission rates are 300 to 1200 bits per second, though faster rates are possible with special equipment or dedicated phone lines. The use of the phone lines requires a modem to translate the phone line signal to the internal form of the computer.

Data structure. The structure of the transmitted data depends solely on the application and the programs. In the simplest case, text such as that you are reading now is transmitted. In situations with high-security requirements, complicated encoders are applied to the data to scramble it so that unauthorized persons are prevented from gaining access to it. In the case of transmission of laboratory data, the programmers alone determine order and format. An example of this is shown in Fig. 16-3. Characters 1 to 20 specify the patient's name, characters 22 to 33 the ID number, and so on. The only way the computer knows that these datum elements correspond to patient name and ID number is that the program specifies the sequence of expected data. Remember that the transmitter and the receiver must "agree" on both the protocol and the content in order for the communication to be successful.

Software

General concepts. Computer software is defined as the set of instructions and data used by the machine to perform various tasks. The instructions are followed in sequential order, except when the machine is told to branch to some other address in the list. They instruct the machine to perform various operations on the data, such as numerical calculations, logical decisions, input and output from various devices, and transfer of data between various locations within the memory. The upper box on p. 280 gives examples of how a simple addition is expressed in several languages. The data may be laboratory results, specimen requirements and normal ranges for various tests, or the parameters required to run the printer or operate the disk drive, to mention a few of the myriad possibilities.

Computer languages. The set of instructions used to perform a given job is called a program. The simplest form of program from the operational standpoint is written in machine-level language. In this case the instructions are equivalent to the logical and mathematical operations of the CPU. Individual assembly-language instructions reference individual instructions programmed into the CPU.

Different expressons of C = A + B

Interpretation: Add contents of memory location A to the contents of memory location B. Store the result in memory location C. The "=" expression is an operational term and does not imply algebraic equivalence.

FORTRAN	C = A + B
BASIC	Let C = A + B
COBOL	Add A to B giving C
PASCAL	C := A + B;

A machine-level language

LDA A	Load contents of memory location A into accumulator
ADD B	Add contents of memory location B to the contents of the accumulator
STR C	Store contents of accumulator in memory location C

BASIC program to calculate statistics

```
 10 REM-First initialize the variables.
 20 Let S = 0
 30 Let S2 = 0
 40 Let S3 = 0
 50 REM
 60 REM-Now to have a loop
 70 REM to read 10 items.
 72 REM Calculate their sum,
 74 REM their sum of squares,
 76 REM and to count them.
 80 For I = 1 to 10
 90 Print "Enter next result"
100 Input R
110 Let S = S + R
120 Let S2 = S2 + R * R
130 Let N = N + 1
140 Next I
150 REM-Statement 140 was the last in the loop.
160 REM-Now to calculate the mean,
162 REM standard deviation,
164 REM and coefficient of variation.
170 Let M = S / N
180 Let S3 = SQRT ((S2 -(S*S/N))/(N-1))
190 Let C = S3*100/M
200 REM
210 REM-Now to print out the results.
220 Print "Mean = ",M
230 Print "Standard deviation = ",S3
240 Print "Coefficient of variation = ",C
250 Print "Number of items = ",N
260 End
Note: The statements beginning with "REM" are
comment statements that are intended to docu-
ment the program internally. Operations in
the program are not caused by their presence.
```

Machine and assembly language programs can run very fast and efficiently; however, they are relatively difficult to program.

High-level languages are much easier to program and are the ones usually encountered in the laboratory. BASIC, FORTRAN, PASCAL, and MUMPS are most frequently used in the clinical laboratory. These languages are expressed in terms relatively easy to understand (see the lower box in the left column for a sample program written in BASIC). However, a fairly large program (called a compiler or an interpreter) is required to translate the instructions into a form that can be understood by the computer.

If the same job is programmed in both assembly language and in a high-level language, the assembly-language program will usually run five to 10 times faster than that of the high-level language. However, the program written in the high-level language will take about one fifth the time to write in comparison to the one in assembly language. The choice of time is therefore a trade-off between computer time and the programmer's time. If a certain portion of code is used intensively, that portion may be written in assembly language and then combined later with the high-level program to give a more efficient program.

The reason that operations in the high-level language take longer is that small subprograms are needed to perform the tasks specified by the main program, which frequently are more complex than those in the instruction set for the CPU. If the routine were written in assembly language, little shortcuts could be taken. However, the high-level languages usually employ routines that have been generalized and hence take longer. An example of this is the FORTRAN addition on the almost obsolete PDP-8A (Digital Equipment, Maynard, Mass.). The machine-level instruction that adds two 12-bit words takes about 2 μsec. The FORTRAN instruction that adds two 36-bit words, each with mantissa and exponent, takes about 2 msec, thus 1000 times slower.

Operating systems. It is essential to realize that software not only is used in the applications of computers (to be discussed below) but also to control the operations of the computer. The program that does this is called the "operating system." Its role is to perform the generalized functions of the system. These include establishing access to the I/O devices, transferring data among the devices, and controlling the timing and access of the devices to the CPU. Another important function occurs in multiple-user or time-sharing systems, that is, the control and management of access by each user to the system.

Software development. Software is developed as persons write computer programs. Programming has three major phases: (1) the planning stage (systems design or analysis), (2) the execution stage (writing the code), and (3) fixing the problems (debugging).

One aspect of programming that facilitates acceptance of the system is its "friendliness." One expects a friendly system to accept data easily, to do what is expected of it, and to be tolerant of the user's errors.

Computer speed

Much ado has been made about computer speed. The issue is far from simple. It is essential to realize that the system, functioning as a whole, is the determinant of computer speed. The major contributing factors to the problem are (1) CPU speed, (2) CPU instruction set, (3) operating system efficiency, (4) speed of the I/O devices, and (5) applications program. In many applications the speed of most of these various components makes little difference. If one is largely doing word-processing applications on a microcomputer, there is little point in getting a fast, powerful machine. The rate-limiting steps are typing the text in and printing it out, hardly a process pushing a system capable of 100,000+ instructions per second. On the other hand, if one is doing lots of calculations (''number-crunching''), a fast, powerful system becomes more important. In this situation one should pay more attention to the software because of the different speeds of the different systems.

Compatibility between microcomputers

One of the difficult problems with microcomputers is the lack of compatibility between apparently similar equipment. The reason for the incompatibility may be that vendors want to improve their product as the field advances, or perhaps they believe that full implementation of a standard is really not necessary because of expense. In any case, differences exist between systems. These differences can be considered to include (1) differences in instruction sets of the CPU's, (2) differences in computer languages, (3) physical and electronic differences in transferable media (floppy disks and magnetic tapes), and (4) differences in the size or complexity of an application exceeding the capacity of a smaller system.

There are three general levels of compatibility. The simplest is the direct use of hardware or software from one system in another without modification. This situation occurs in two general cases: In the first, a manufacturer produces a product so popular that in effect it establishes a standard that others emulate. For example, several manufacturers have attempted to ''clone'' the IBM-PC (IBM, Armonk, N.Y.). The second of these cases occurs when various interested parties get together to establish a standard. The RS 232C serial interface is such an example.

The second level of compatibility occurs when a program or a piece of equipment can be shifted to another system with only minor changes. An example of this might be a program written in PASCAL on the IBM-PC. To run the same program on an Apple II (Apple Computer, Cupertino, Calif.), one would have to transmit the program to the Apple and then recompile it. To change from one printer to another on an otherwise identical system, one may have to reconfigure one's software according to the ''menus'' built into the programs.

The third level of compatibility occurs when extensive changes are required to shift between systems. For example, much work is required to translate a program written in FORTRAN to PASCAL. Another example occurred when the popular word processing program, Wordstar (MicroPro, San Raphael, Calif.), was translated from one assembly language to another.

One of the major difficulties in transferring programs from one system to another is the incompatibility of the medium (disk or tape) on which it is stored. Clearly, it would be most desirable to place a floppy disk into the drive and read it immediately. However, there are innumerable formats in which data may be recorded on the disk, not to mention the different disk sizes. Special programs must be used for reading disks written in different formats. Fortunately, standards for floppy-disk protocols are being developed.

APPLICATIONS

The best applications of computers are those involving manipulations of complex or extensive data. Those tasks that are only to be done a few times usually do not warrant the effort and expense required to set them up. The major uses of computer include the following:

1. To perform repetitive tasks. Classic examples of this include billing for patient tests and logging specimens into a lab.
2. To perform complex calculations rapidly. Examples of this are linear-regression analyses associated with enzyme analyses on some instruments, RIA data reduction, and some newer quality control schemes.
3. To collect, organize, store, and distribute large amounts of data. Examples are the collection of patient test results, preparation of a cumulative report, and collection and analysis of quality control results.
4. To operate machines in a highly reproducible manner. Many scientific and clinical instruments are now run by computers.

This section is a discussion first of instrument operation, then general clinical laboratory applications that involve data collection from various instruments and their output as patient reports, and finally some specialized applications, such as diagnostic programs.

Applications for laboratory instrumentation

Analytical instruments. A substantial number of instruments introduced into the clinical laboratory of the 1980s are computer or microprocessor controlled. ''Intelligent'' instruments range from spectrophotometers and chromatographic systems to discrete analyzers and large multichannel analyzers. In many instruments the computer controls each step, including aspiration of sample and reagents, collection of data, calculation, and printout of results. The paragraphs that follow provide examples of how computers are used in several instruments.

Examples of simple computer-controlled instruments are the newer spectrophotometers with microprocessors. The operations that are controlled in several instruments include control of the wavelength scan, the intervals at which observations are made, collection of data, data manipulations such as subtraction of a previous blank or comparison with a previous scan, and output of the data, either by transmission to an interfaced computer or printout of results, either by listing or in a graph.

More complex computer-controlled instruments are the gas and liquid chromatographic systems. In the simplest cases the computer merely operates the instrument by controlling the temperature or the solvent gradients and cooling or purging the instrument in preparation for the next run. More sophisticated instruments control sample injection, program gradient changes, collect the data, and calculate the results.

The most sophisticated chromatographs are the gas chromatographs coupled to mass spectrometers. In these instruments, not only are all the physical operations controlled by the instrument, but the machine also collects vast amounts of data from the mass spectrometer. At the end of the run, it can compare its scans with previous scans in a large "data base" in order to identify the unknown compounds. In a slightly different application, it can determine the concentration of various analytes, even in the presence of other materials that co-chromatograph with the desired analyte.

Discrete analyzers. The du Pont ACA-3 (E.I du Pont de Nemours, Inc., Wilmington, Delaware) is an excellent example of how a computer has been incorporated into a discrete analyzer. (See Fig. 16-4 for the computer architecture of the ACA-3.) Virtually all the manual operations previously performed for the ACA-1 and ACA-2 have been automated. These include filter balance, absorbance check, pack calibration, and determination of concentrations for immunochemical procedures. The only remaining manual operations are preparation and loading of specimens and test packs, plus preventive maintenance and trouble-shooting.

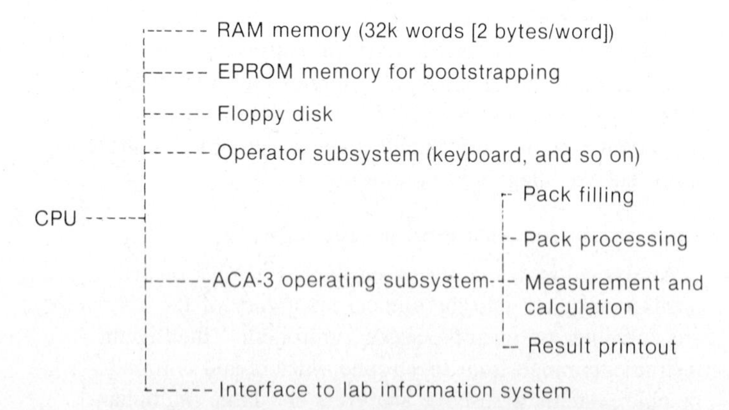

Fig. 16-4. Architecture of ACA-3. EPROM memory for bootstrapping is a type of read-only memory that, using special equipment, can be erased and reprogrammed (EP).

Error checks included in the software are (1) check for absorbance being either too high (greater than 2) or too low (less than 0) (A flag); (2) check for out-of-range temperature (T flag); (3) check for gross filling errors (P flag); and (4) check for decode-head errors (D flag).

The software for the ACA-3 contains data-reduction schemes for various immunochemical assays, such as EMIT procedures for several therapeutic drugs, and nephelometric procedures for IgG and IgM. These calculations, which involve fitting a four-parameter model to the results of up to six standards, are quite involved.

Quality control applications

One of the major roles of laboratory computers is to collect and monitor quality control (QC) data. It is a natural application because of the ease with which large amounts of data are collected and manipulated in the various calculations.

In some cases the computer is interfaced to the laboratory instrument so that the QC data and patient results may be easily collected. The computer then stores the QC data for later use. In the simplest programs the computer simply compares the current value with the acceptable range supplied by the user. If the result is outside this range, it is starred so that the user may choose to reject the run.

More complex QC schemes have been proposed.[6] Some of these proposed schemes are only practicable, with the aid of a computer. Algorithms have been developed to indicate, based on the pattern of QC results, the type of error being encountered and whether it is an accuracy or precision problem. Systems such as these are commercially available for microcomputers.

Another procedure used to check results is the delta-check method.[7,8] The delta check involves comparing the current result for a given patient with the previous result. If the comparison exceeds specified limits, the result is flagged for checking. In one study,[9] the delta-check method determined an error-incidence rate of 1.2% to 4% of the specimens assayed. Up to 20% of the specimens were flagged by this check. Most delta-check flags occurred because of real changes in the patient and were caused by administration of an analyte, hemodialysis, or major changes in the patient's condition. Note that delta-check methods will not detect errors in which the comparison result does not exceed the specified threshold.

Applications in laboratory information systems

General concepts. Laboratory information systems (LIS) are sophisticated computer systems designed to han-

Laboratory Information System Features. Listed are some of the features available on laboratory information systems (LIS). The features listed are included in at least one system. This is not meant to be a complete listing of all available features and does not imply the merit of any given feature. That is left to the judgment of the prospective user.

I. Sources of Data
  A. Patient demographic information (Name, ID number, location, tests ordered, physician's name, and so on)
    1. CRT's in laboratory specimen reception area
    2. CRT's at nursing stations (when installed)
    3. Hospital computer (when interfaced to LIS)
  B. Specimen test results
    1. CRT's in laboratory
      a. Manual tests—results typed in
      b. Hematology cell count—CRT used as counter
      c. Radioimmunoassay—raw data input from CRT or counter; LIS calculates results
      d. Adjacent to automated instruments—operator verifies data from instrument before release
    2. Interfaces to automated instruments
      a. Chemistry instruments (many automated discrete and continuous-flow analyzers)
      b. Hematology instruments (multichannel blood analyzers, cell counters)
      c. Coagulation analyzers
      d. RIA gamma counters
    3. Card readers for marked sense cards for microbiology or urinalysis sections (optional form of input)
  C. Lab director information—mostly input during installation of LIS from CRT or by vendor
    1. Descriptive data for each test
      a. Test name
      b. Specimen requirements
      c. Reference ranges
      d. Panic values (if any)
      e. Price of test (if billing option selected)
      f. Where test is done
      g. QC specimens used for control of test
      h. QC limits for control specimens
    2. Persons using LIS
      a. Role
        (1) Lab computer manager
        (2) Lab supervisors
        (3) Lab technicians
        (4) Clerks
        (5) Attending physicians
        (6) Nursing staff
      b. Name
      c. Access code (used to limit access to various elements of system)

II. Output of data
  A. Patient reports
    1. Output device and location
      a. Bulk printed on fast line printer
      b. Telephone reports available from clerk at CRT in lab office
      c. Stat reports sent directly to slow printer in critical care areas (ICU, ER, and so on)
      d. CRTs at nursing stations (when installed).
    2. Content
      a. Patient demographic data
      b. Test results
        (1) Single set of tests
        (2) Cumulative report
      c. Graphic display of numerical result
      d. Starred if result abnormal or in critical range.
    3. Report format is variable with customer deciding which format to use
  B. QC reports
    1. Results for automated instruments output to CRT adjacent to instrument—allows operator to verify results
    2. Summary report output on line printer
      a. Summary of data
      b. Calculated values from data (mean, standard deviation, and so on)
    3. Graphic display of results
    4. Delta and discrepancy tests—this may be optional
  C. Laboratory technologist data
    1. Worklists
    2. Specimen labels
    3. Phlebotomy lists and labels
  D. Lab supervisor data
    1. Listing of uncompleted tests
      a. By test
      b. By section
    2. Listing of status of tests on a given patient
    3. CAP workload report

III. Storage of data
  A. Storage of patient reports
    1. On-line storage—done on hard-disk drives
      a. Limited by amount of storage space on disks
      b. Limited by number of patient names—this may only be a factor on smaller systems
    2. Off-line storage
      a. Storage on magnetic tape
      b. Storage on microfiche—institution may have an outsider prepare microfiche from tape
      c. Amount of storage in these forms is limited only by the volume of storage space available
  B. QC results
    1. Raw data stored for periods of 30 to 365 days

Continued.

IV. Sections of hospital served by the computer
   A. Within laboratory
      1. Chemistry
      2. Hematology
      3. Urinalysis
      4. Serology
      5. Coagulation
      6. Microbiology (optional if available)
      7. Blood bank (when available, this package allows a wide range of computerized functions, from storage of typing and cross-matching data to a relatively complete record-keeping facility)
   B. Outside laboratory
      1. Admission/transfer/discharge—this optional feature is used to keep track of the locations of patients during their hospital stay
      2. Pharmacy—this optional feature has substantial requirements for additional hardware and software

V. Other features
   A. Hardware service
      1. One-year warranty
      2. Service contract in later years
      3. Some vendors offer 24-hour service
   B. Software service
      1. Twenty-four-hour service by means of modem to headquarters
   C. Source code is available to user at a cost only after signing a nondisclosure agreement (optional)
   D. Word processing features (this feature may be used for preparation of anatomical pathology reports, ward manuals, and procedure manuals if wanted, though it is not universally available)

dle the problems of collecting, organizing, and reporting data from the clinical laboratory. An account of how one LIS was used is given by Nisbet and others.[10]

Justification of purchase. The reason to purchase an LIS is to solve problems that a given laboratory may have. Early in the decision-making process, it is essential to define the problems that point toward purchase of the computer and then determine if there is some way they can be solved without the use of a computer. If there is, it is advisable to do it that way, for that solution will be much cheaper than purchasing a computer.

The LIS will pay for itself in two ways. First, it will generate more income because it is more efficient than a manual method for billing tests. The other identifiable way net income will be increased is that the LIS will perform many clerical tasks of the laboratory, such as filing, and may replace several clerks. Other savings are more difficult to quantify. Among them are many efficiencies that occur because the LIS makes the data much more available than before. One laboratory did a follow-up study after purchase of the LIS and found that the payback period was about 4 years. The calculation was based on the initial cost of the system and the increased revenue from a more complete billing for tests.[11]

Selection consideration. It is essential that one carefully check the proposed instrument system before purchase to ensure that it will meet one's needs. This involves two major tasks: (1) determination of one's needs, and (2) determination of the capabilities of the proposed instrument system. (Normally these would be done whenever one buys any instrument.) As one investigates the various possible systems, one should look at them as a system as a whole and not as an IBM or DEC (Digital Equipment Corp., Maynard, Mass.) computer plus someone else's software. The design of the whole system is what will make it good, regardless of the manufacturer. If one buys the whole system from one vendor, that vendor can be held responsible for its performance. For an account of the process one laboratory went through to select an LIS, see Marshall.[12]

Several capabilities that should be considered are the following:

Does the system have enough capacity to handle not only the current workload but also the expected growth in workload over its expected lifetime (5 to 7 years)? Capacity is perceived by the user as response times to simple commands. Response times of 1 second or less under peak-load conditions indicate sufficient capacity. Times of 5 seconds or more indicate a severely overloaded machine.

Increased workload for the computer will come from two major sources: (1) an increase in the work load of the laboratory and (2) an increase in the applications for the computer. The work load increase can be projected from its recent history. The increase in applications depends on the desire or ability to write new programs or to purchase new software. (It may be necessary for the management of the LIS to restrict severely additional applications in order that the efficiency of the system in its primary task is not compromised.)

Does the system have the features needed of it? The system should support the appropriate parts of the laboratory (chemistry, hematology, and so on), record the necessary demographic data (name, ID number, address, and so on) in the form used by the institution, monitor quality control, and perform other features listed on pp. 283 and 284. If any necessary items are not included, one must buy additional software by hiring either a programmer or an outside contractor. Programming can be expensive. It is worth determining in great detail whether the proposed system meets all of one's requirements. One should determine in advance how the system will allow one to make these changes. Some systems are much more flexible than others.

Does one have the capability to change the system as one's requirements change? The 5- to 7-year lifetime of the computer is sufficient time for numerous changes to occur in the clinical laboratory. Consider the major changes in instrumentation and in tests offered that have occurred during the past 5 years. This dynamic situation makes it mandatory that one's LIS be sufficiently flexible to accommodate the inevitable changes.

Changes are to be expected in at least two major forms: (1) modification of file contents and sizes to accommodate changes in test offerings, and (2) new or modified programs to allow one to perform new or updated applications that may not be anticipated at the time of purchase. If the user takes responsibility for programming the changes, the user must have access to at least the file structure of the system and preferably to the program code. Without these, it is essentially impossible to write any programs that will interact successfully with the file structure of the LIS.

These and other items are considered with great thoroughness in the excellent ASTM manual *Computer Automation in the Clinical Laboratory*.[13]

Costs. Perhaps the most apparent cost of a computer system is the cost to purchase the computer and the software. The cost of microcomputer-based laboratory systems for both hardware and software will range from $10,000 to $50,000. Such a system will retrieve data from several instruments, usually only in chemistry and hematology. It will print out patients' reports and perform some QC functions. Larger systems provide an extensive service for all areas of the hospital laboratory, including chemistry, hematology, coagulation, microbiology, blood bank, and anatomical pathology. These machines, which are interfaced to 10 or more instruments and have 10 to 80 terminals, will cost $150,000 and up. The approximate cost is $1000 per hospital bed.

Implementation

Personnel. People, in the form of users and operators, are major elements in any computer system. If the user is able to work smoothly and easily with the system, the likelihood of success is much greater. The keys to success lie in several areas: training, involvement in planning and implementation, and removal of threats to various persons.

The importance of training goes virtually without saying. One item that should be considered for the training program is that persons be given a fair amount of background information in addition to what they need to know to do their routine tasks. The reasons for this are that (1) when things go wrong they will know better how to describe and cope with the problems, and (2) through better understanding of what is happening, they will be better able to utilize the systems.

Because there are so many ways a computer system can go wrong, it is essential that all the groups that will be involved with the system are thoroughly consulted throughout the selection and implementation process. This not only develops a sense of involvement and excitement but also helps identify potential problems at an early stage. It also helps to develop the cooperative attitude that helps to ensure the success of the system.

Anytime substantial changes are made in any environment, some people are going to feel threatened. In the case of computer purchases it may occur because they are afraid of losing their jobs or because they are afraid of machines. Major efforts should be made to minimize these threats.

Security. The security of an LIS is important. There are two aspects of security that require special attention: (1) physical protection of the machine and its programs, and (2) release of the data to authorized persons only.

The first problem is a matter of restricting access to the computer room and the program code. It is essential that unauthorized persons not be allowed near the machine because of the incalculable mischief they could do to the machine or to the data stored nearby.

The second problem is much tougher. In essence, it is to protect patient privacy and confidential laboratory data. A properly designed system will have access only for those people who really need it.

Machine failure. Failure of the computer (a "crash") is always possible. This may occur because some component fails or the user has discovered some new way (usually much to his or her chagrin) to abuse the machine so that it fails. For a system that has been installed in many institutions, the chances of the latter is remote; the former is much more likely. The most likely components to fail are those that involve mechanical systems, such as disk drives and line printers. When an electronic component does fail, the usual way to approach repair is to replace all the computer boards that might be involved. Some computer installations cope with the problem of machine failure by duplicating the most vulnerable components. When the original fails, the duplicate is brought on line to replace it.

In most failures (crashes), at least some data are lost. For most centers, anywhere from an hour's to a day's worth of data will be lost if the entire system fails. If a disk system fails, it is possible that a great deal of data will be lost. As a result, back-up disks should be prepared at frequent intervals to protect against such a circumstance.

For an LIS, back-up disks or tapes are either maintained on-line or copied afresh every night. In other instrument systems, it is worthwhile to make or keep extra copies of disks or tapes to protect against the inevitable failure.

In addition to back-up computer systems, a back-up

noncomputer system should always be available in case of an extended computer failure. All the old manual systems (log sheets, work sheets, laboratory slips, charting, and so on) should be available for use if needed.

Computer environment. Computer systems are very sensitive to variations of temperature and humidity. Like any electrical device, they put out a fair amount of heat. Certain components are more sensitive to temperature than others; disk drives are most sensitive. After that come the various computer chips. Some intolerant systems require temperatures within a few degrees of 24° C. Others may allow up to 10° or more variation. Humidity must be kept low enough so that it is noncondensing but high enough so that static electricity is not a problem. These environmental problems will frequently appear in the form of a memory error or a disk-read error.

ACKNOWLEDGMENTS

I am indebted to David Chou, M.D., for numerous helpful conversations and to Peter F. Schatzki, M.D., of the McGuire Veterans Administration Medical Center, Richmond, Virginia, for his support of this project. I would also like to express my gratitude to Nancy Ninneman and Gerry Maass for the time they spent with me.

REFERENCES

1. Levine, R.D.: Supercomputers, Sci. Am. **246**(1):118-135, 1982.
2. Toong, H.D., and Gupta, A.: Personal computers, Sci. Am. **247**(6):87-107, 1982.
3. Toong, H.D.: Microprocessors, Sci. Am. **237**(3):146-161, 1977.
4. Hodges, D.A.: Microelectronic memories, Sci. Am. **237**(3):130-145, 1977.
5. White, R.M.: Disk storage technology, Sci. Am. **243**(2):138-148, 1980.
6. Westgard, J.O., Barry, P.L., and Hunt, M.R.: A multi-rule Shewhart chart for quality control in clinical chemistry, Clin. Chem. **27**:493-501, 1981.
7. Ladenson, J.H.: Patients as their own controls: use of the computer to identify "laboratory errors," Clin. Chem. **21**:1648-1653, 1975.
8. Whitehurst, P., DiSilvio, T.V., and Boyadjian, G.: Evaluation of discrepancies in patient results: an aspect of computer-assisted quality control, Clin. Chem. **21**:87-92, 1975.
9. Wheeler, L.A., and Sheiner, L.B.: A clinical evaluation of various delta check methods, Clin. Chem. **27**:5-9, 1981.
10. Nisbet, J.A., Owen, D.J.M., Owen, J.A., and Walduck, A.G.: A comprehensive laboratory computer system working round the clock, Clin. Chim. Acta **91**:89-101, 1979.
11. Hospital recoups $128,000 in lost lab charges, Mod. Health Care, **11**(12):46, 1981.
12. Marshall, C.: Campaign to computerize a clinical laboratory, Lab World, pp. 27-31, Aug. 1981.
13. Standard Guide for Computer Automation in the Clinical Laboratory, ASTM Manual E792-81, Philadelphia, 1981, American Society for Testing and Materials.

Chapter 17 Laboratory statistics

Carl C. Garber
R. Neill Carey

accuracy Estimate of nonrandom, systematic bias between populations of data or between a population of data and the true value.

central tendency The value about which a population is centered. The mean, the median, and the mode are all used to describe the central tendency of a population.

coefficient of variation (CV) A relative standard deviation calculated by multiplying by 100%, the standard deviation divided by the mean.

confidence interval A range around an experimentally determined statistic that has a known probability of including the true parameter.

correlation coefficient A statistic that measures the distribution of data about the estimated linear-regression line.

degrees of freedom The number of independent observations in a data set. It is the number of observations minus the number of restrictions for a set of data.

F test A statistical test used to determine whether there are differences between two variances.

histogram A graphic display of data in which the frequency of a certain value (or range of values) is plotted against a scale of all values.

mean Arithmetic average of a set of data.

median A value or interval of a population occurring in the middle of a population, half of which falls above and half below the median.

mode The value or interval of a population occurring with the greatest frequency.

nonparametric statistics Statistics that do not assume a normal or symmetric (that is, gaussian) distribution.

null hypothesis The working hypothesis of a statistical test stating that there is no difference between the statistics of two different populations.

parametric statistics Statistics that assume that a population has a symmetric distribution (such as gaussian or log-normal).

precision Estimate of amount of random variation in a population of data.

random error Error that affects reproducibility of a method (precision).

range The difference between the highest and lowest values in a population.

standard deviation Square root of a variance.

statistic A number that describes a property of a set of data or other numbers.

statistics The plural of statistic; also the science that deals with the use and classification of numbers or data.

systematic error Nonrandom error that affects the mean of a population of data and defines the bias between the means of two populations.

t test A statistical test used to determine whether there are differences between two means or between a target value and a calculated mean.

variance A statistic used to describe the distribution or spread of data in a population.

Daniel[1] has defined statistics as ". . . a field of study concerned with 1) the organization and summarization of data, and 2) the drawing of inferences about a body of data when only a part of the data is observed."

A statistic is a number that describes a property of a set of other numbers. A familiar example is the average, or arithmetic mean, of a set of data. Often, only a few simple statistics are required to describe an entire population of data; thus statistics are the essence that is left after one "boils down" the experimental data. (Notice that *statistic* is singular, *statistics* as a science is singular, and a *group of statistics* is plural.) Statistical descriptions of data sets are useful for several purposes:

1. To assess random variation in a population of data

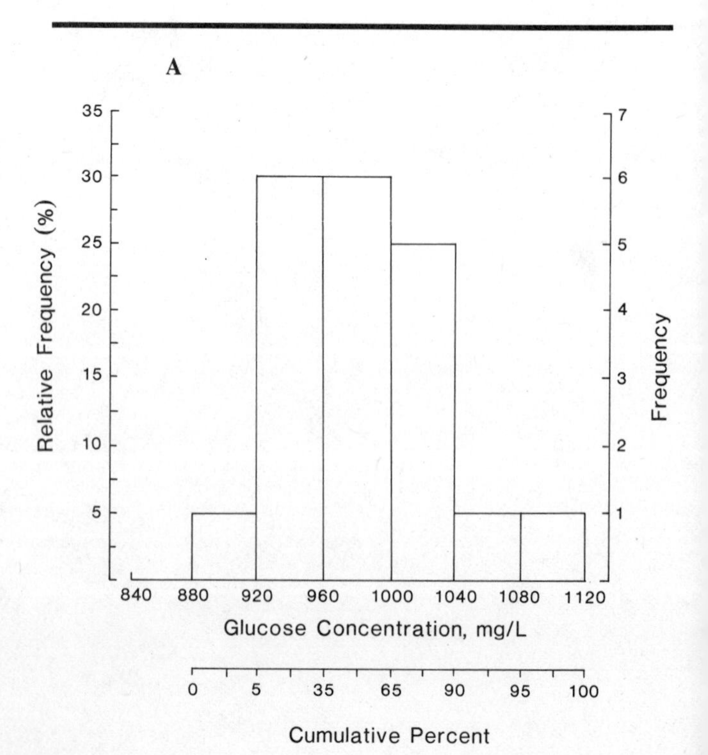

2. To compare the amounts of random variation among populations of data
3. To test for systematic differences between populations of data
4. To assess the degree of correlation between populations of data

Each of these uses of statistics is briefly described in this chapter.

The importance of statistical methods in clinical chemistry cannot be overemphasized. The reader is encouraged to study statistics textbooks, such as references 1 to 6, to obtain a basic understanding for intelligent use of statistics. The reader is also cautioned about the necessity for sound experimental design to yield data amenable to unambiguous statistical treatment and interpretation.

STATISTICAL DESCRIPTIONS OF A POPULATION

No chemical measurement is exact because random variation (random error) is inherent in all laboratory measurements. For example, if the glucose concentration of a single blood specimen is repeatedly measured using the same analyst, method, reagents, and instrument, a population of results (observations) will be obtained. This variation has many causes that one can minimize but not eliminate. Some of these are heterogeneity of the specimen itself with time, variation in the technique of the analyst, heterogeneity of reagents with time, and instrument variation.

Population distributions

Conceptually, perhaps the simplest way to describe a population of data is to construct a histogram (also called a "frequency diagram" or "frequency distribution"). It shows the frequency with which a particular value, or range of values, is obtained versus the scale of all values. Fig. 17-1, *A*, is a histogram of the glucose results obtained in the repeated measurement of the same specimen mentioned above. The horizontal axis is glucose concentration, divided into small convenient ranges or "bins." The vertical axis is the frequency (or relative frequency) with which results from each bin are obtained. When relative frequencies are used, each bin frequency is presented as a percentage of the total number of samples. The histogram's horizontal axis can also represent cumulative percentiles (cumulative percentage of the population up to and including each bin) as well as concentration units. When the number of observations, *N*, is small and the bins are relatively wide, the histogram has a choppy appearance. As *N* increases and the bins are made narrower, the shape

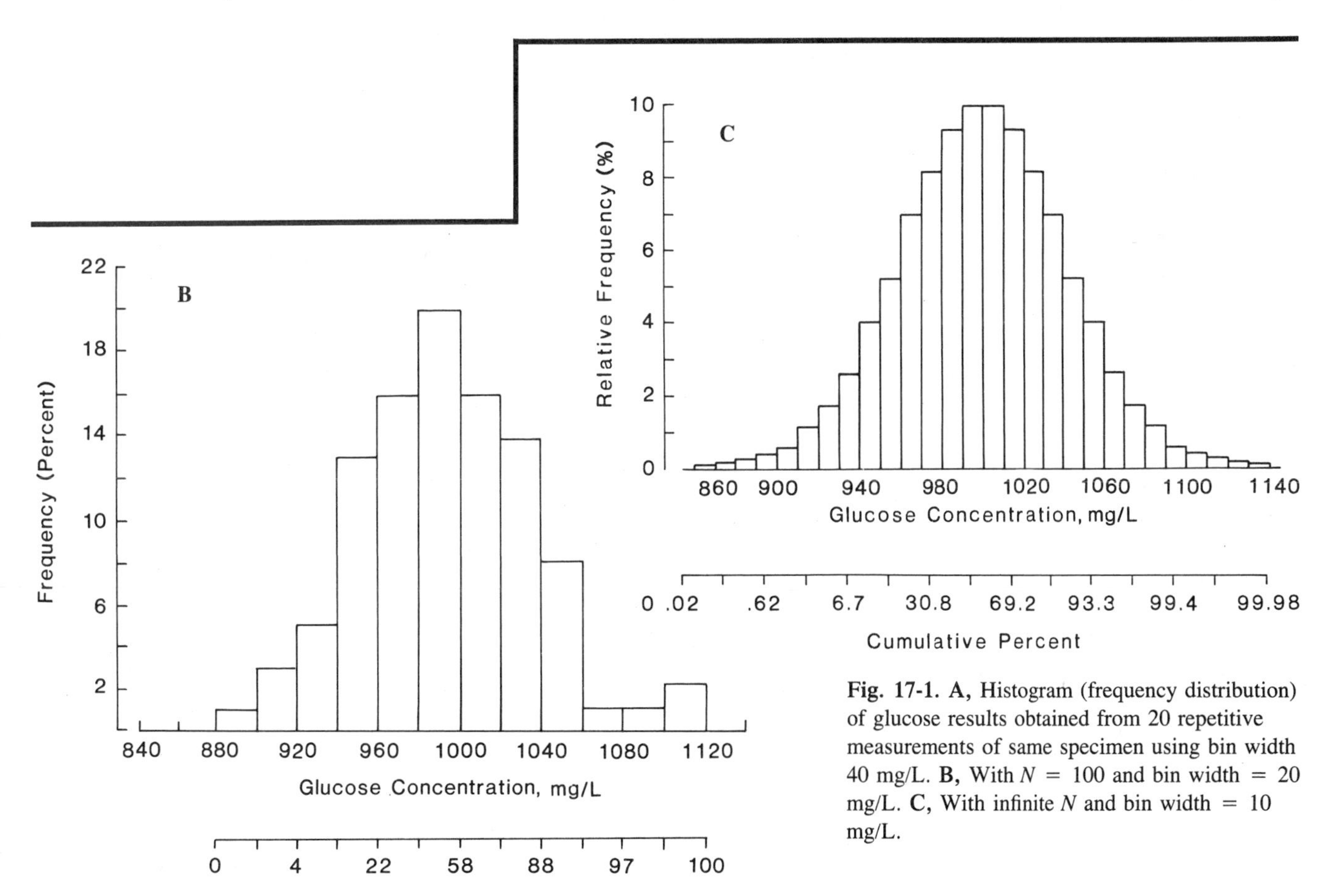

Fig. 17-1. A, Histogram (frequency distribution) of glucose results obtained from 20 repetitive measurements of same specimen using bin width 40 mg/L. **B,** With $N = 100$ and bin width = 20 mg/L. **C,** With infinite N and bin width = 10 mg/L.

of the histogram becomes smoother and the histogram becomes more truly representative of the population. As N increases further, the histogram takes on the appearance of a continuous function. In the histogram of hypothetical glucose data in Fig. 17-1, *B*, one can see that the population is centered around 1000 mg/L, few of the observations are less than 920 mg/L, and few are greater than 1080 mg/L. One can also assess the general spread of the data.

If enough data are represented in the histogram and the data are truly random (that is, each result was affected by random processes alone), one can use the histogram to predict the probability of obtaining future results above or below a given value. Fig. 17-1, *C*, shows that there is about a 2.3% chance that a future glucose result from this population will be less than 920 mg/L, and a 2.3% chance that a future glucose result will be greater than 1080 mg/L.

Central tendencies

There are several terms that define the values about which a population is centered. It is customary in everyday life to talk about the "average." The simplest average is the arithmetic mean, which one obtains by adding up all the observations and dividing by the number of observations, N. The mean, $\bar{x}$, is given by equation 17-1, where x_i is an individual observation.

$$\bar{x} = \frac{\sum x_i}{N} \qquad \textit{Eq. 17-1}$$

The arithmetic mean is representative of the central tendency of the population only if the histogram is symmetric about the mean, as in Fig. 17-2. Fig. 17-2 also represents the "normal" or gaussian, distribution, which is obtained when repeated measurements are made on the same specimen with only the small inherent random errors of the method present as in Fig. 17-1. Many statistical tests assume that data are "normally" distributed. If the entire population of data is used to calculate the mean, this calculated mean is the true mean of the population, indicated by μ. If a subset of the population of data is used, the symbol, $\bar{x}$, denotes that the mean is an estimate of the true mean based on the subset of data (or limited data).

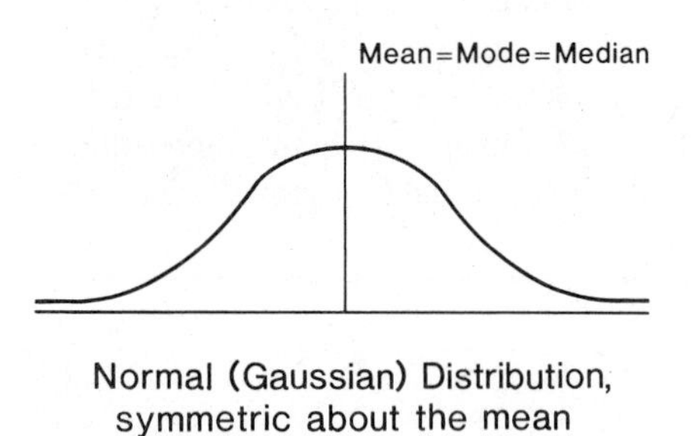

Fig. 17-2. Normal (gaussian) distribution, symmetric about mean.

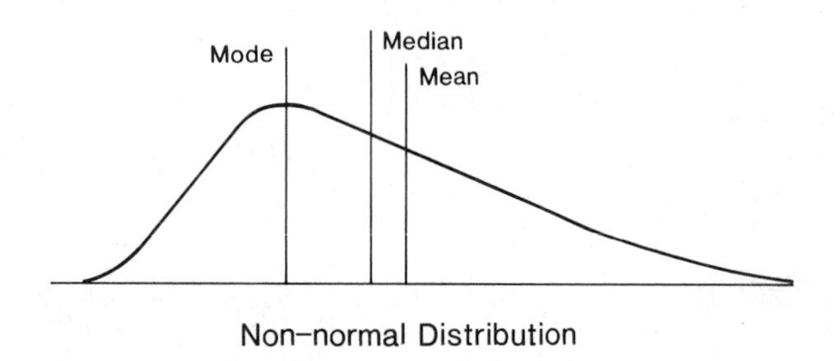

Fig. 17-3. Nonnormal distribution.

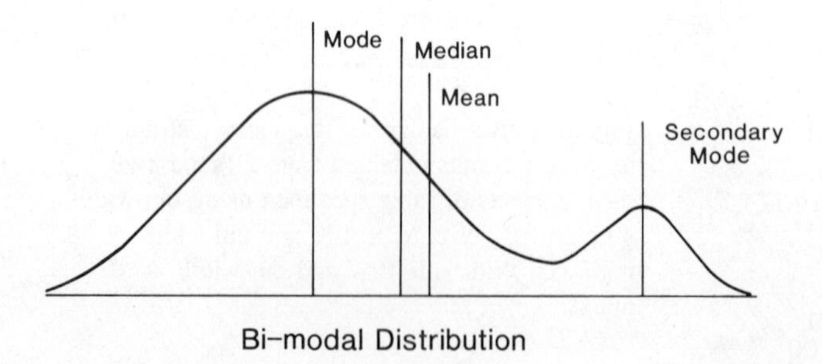

Fig. 17-4. Bimodal distribution.

When a nonsymmetric distribution is obtained, the arithmetic mean does not describe the center of the distribution satisfactorily. One alternative is to find the most frequently obtained value, called the "mode." A distribution whose mode differs from the arithmetic mean is shown in Fig. 17-3. Another term describing the center of the distribution is the "median." The median is the value in the middle of the distribution. Half the observations are greater than the median and half are less; the median is the observation at the 50th percentile. Mode and median are useful terms to describe the location of distributions that are skewed, for example, with a long tail of higher values, as in Fig. 17-3. An extreme example of the nonsymmetric distribution, the bimodal distribution, is demonstrated in Fig. 17-4. This distribution would not be expected for repeated testing of the same specimen. Differences between the arithmetic mean, median, and mode alert the analyst to the presence of nonsymmetric distributions (see p. 296).

Measures of variation

Indicators of the variation of the measurements or spread of the distribution are given by the range, the variance, and the standard deviation. The range is the difference between the lowest and highest values in the data set. The range is especially useful for indicating the spread of the data when N is small. It makes no assumptions about the shape of the distribution.

One calculates the variance by adding the squares of the differences between the individual results and the mean, and dividing by $N - 1$, as in equation 17-2. The standard deviation, s, is the square root of the variance.

$$\text{Variance} = s^2 = \frac{\sum (x_i - \bar{x})^2}{N - 1} \qquad \textit{Eq. 17-2}$$

Example 1. Calculation of common statistics from a sample population of repeated glucose measurements.

Construct a frequency histogram of the population below using bins of 20 mg/L. Calculate the arithmetic mean, mode, median, variance, standard deviation, and coefficient of variation (CV).

Population: 1090, 1020, 1040, 970, 1050, 980, 1000, 960, 1030, 970, 960, 970, 1040, 990, 1050, 940, 950, 970, 1000.

Answer:

x_i (mg/L)	$x_i - \bar{x}$	$(x_i - \bar{x})^2$
1090	92.5	8556.25
1020	22.5	506.25
1040	42.5	1806.25
970	−27.5	756.25
1050	52.5	2756.25
980	−17.5	306.25
1000	2.5	6.25
960	−37.5	1406.25
1030	32.5	1056.25
970	−27.5	756.25
970	−27.5	756.25
960	−37.5	1406.25
970	−27.5	756.25
1040	42.5	1806.25
990	− 7.5	56.25
1050	52.5	2756.25
940	−57.5	3306.25
950	−47.5	2256.25
970	−27.5	756.25
1000	2.5	6.25

$\Sigma x_i = 19950$ $\Sigma(x_i - \bar{x})^2 = 31775$

$$\text{Mean} = \bar{x} = \frac{\Sigma x_i}{N} = \frac{19950}{20} = 997.5 \text{ mg/L}$$

$$s^2 = \frac{\Sigma(x_i - \bar{x})^2}{N - 1} = \frac{31775}{19} = 1672$$

Median = 980 mg/L

$s = 40.9$ mg/L

Mode = 970 mg/L

$$\%\text{CV} = \frac{s}{\bar{x}} \cdot 100 = \frac{40.9 \text{ mg/L}}{997.5 \text{ mg/L}} \cdot 100 = 4.1\%$$

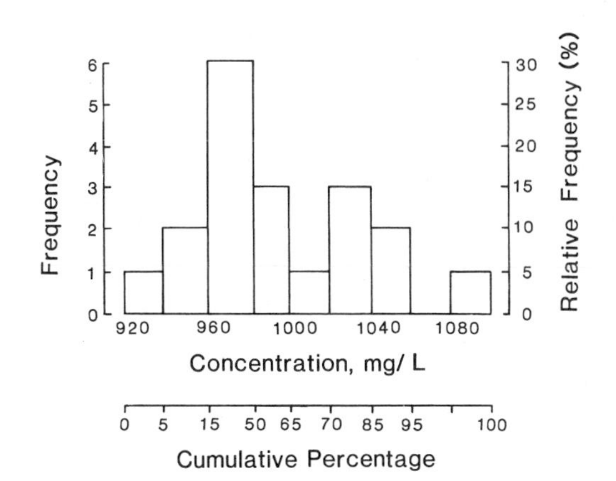

The denominator of equation 17-2 is $N - 1$ rather than N because there are only $N - 1$ degrees of freedom for the variance, since $\bar{x}$ has been used to calculate the variance. Calculation of mean and standard deviation is demonstrated in Example 1 (p. 291). If the true standard deviation of the population has been calculated from all the data in the population, the Greek sigma, σ, is the symbol for standard deviation. The letter s (or sometimes *SD*) is used when a subset of the population has been used to calculate the standard deviation.

The coefficient of variation, % CV, is simply 100 times the quotient of the standard deviation divided by the mean. It is given in units of percentage.

$$\%CV = 100 \frac{s}{\bar{x}} \qquad \textit{Eq. 17-3}$$

CV's are often used to express random variation of analytical methods in units independent of analytical methodology. It is often assumed that the coefficient of variation of an analytical method is independent of concentration; unfortunately this is not the case.

Confidence intervals

If a population is normally distributed, it is completely described by two statistics, the arithmetic mean and the standard deviation. Confidence in the validity of the description increases as N increases. Statistics that assume a gaussian distribution of a population are often called "parametric statistics." Fig. 17-5 shows the probability distribution for a population described by the gaussian curve. One can see that 68.2% of the observations are within $\pm$ 1.0 s of the mean, 95.5% are within $\pm$ 2.0 s, and 99.7% are within $\pm$ 3.0 s. These intervals containing a stated portion of the data are called "confidence intervals." They are the basis of statistical quality control "rules" for run acceptance and rejection decisions. Suppose the same specimen from Example 1 is analyzed for glucose with every batch of patient specimens. Its mean and standard deviation for glucose are well known. Nearly all the data, 99.7%, are within $\pm$ 3 s of the mean (that is, between 875 and 1120 mg/L); so there is only a 0.3% probability of obtaining a result more than $\pm$ 3 s from the mean because of random chance alone. Any result obtained this far from the mean is extremely suspect. Such a result is not likely to be from the same population as the previous data and indicates that the method's performance has changed.

If the measurement of the glucose concentration in Example 1 is inexact, so is the mean of a group of measurements on the specimen, though the mean is more precise than a single measurement. If several means are obtained from different groups of measurements of the same specimen, the individual means are distributed about the grand mean. The random variation in the population of means is described by the standard error of the mean, $s_{\bar{x}}$, given by equation 17-4.

$$s_{\bar{x}} = \frac{s}{\sqrt{N}} \qquad \textit{Eq. 17-4}$$

Confidence limits for means of groups of measurements may be set up about the mean using $s_{\bar{x}}$ in a manner analogous to those above. In the glucose example, if three measurements were made each time, then:

$$s_{\bar{x}} = 40.9/1.732 = 23.6 \text{ mg/L}$$

Thus 99.7% of the means of triplicate determinations of glucose in the specimen will be within 3 × 23.6 mg/L of the grand mean, or between 926.7 mg/L and 1068.3 mg/L.

The true mean glucose concentration of the example patient specimen cannot be known exactly unless an infinite number of measurements are made. However, one can use the standard error of the experimentally determined mean to develop a confidence interval that has a known probability of including the true mean. The interval is described by equation 17-5, where μ is the true mean.

$$\mu = \bar{x} \pm ts_{\bar{x}} \qquad \textit{Eq. 17-5}$$

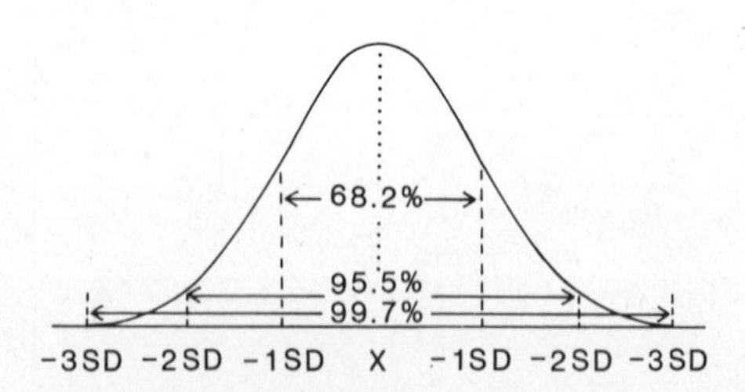

Fig. 17-5. Gaussian curve.

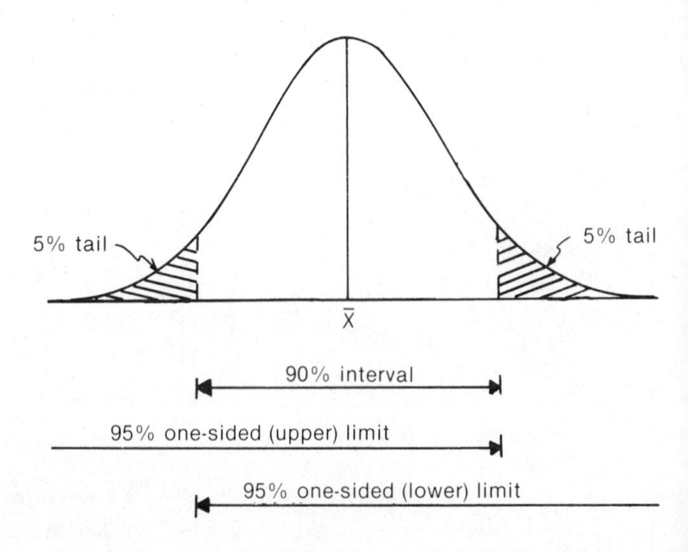

Fig. 17-6. One-sided versus two-sided *t* values.

The t value is taken from a t table and depends on the number of degrees of freedom (one less than the sample number, or $N - 1$) and the desired probability, p, that the true mean is outside the confidence interval because of chance alone. Thus $p = 0.05$ implies a 95% confidence ($100 [1 - p]\%$) that the interval includes the true mean. The t values describe the same probability distribution as in Fig. 17-5. The difference is that in Fig. 17-5 one assumes that the true population mean and standard deviation are known. The t table makes allowances for the decreasing confidence in the estimated values of these parameters as N decreases. Because a confidence interval is being constructed to include the true mean, the 0.05 probability that the true mean is beyond the calculated limits must be spread over both ends or tails of the distribution (Fig. 17-6). Thus a two-sided, $p = 0.05$, t value is used. If one is only interested in stating that the true mean is greater than some limit, a one-sided t value would be used. The t value for a two-sided interval at $p = 0.05$ is the same as the t value for a one-sided limit at $p = 0.025$.

Sources of variation

In reality, random variation has many components that can be related to the actual laboratory setting. Potential sources of variance can be tested by performance of an analysis of variance (ANOVA) experiment. A general discussion of ANOVA is beyond the scope of this chapter; however, a pertinent example is described. A procedure for determining a clinical chemistry method's within-run, run-to-run, day-to-day, and total variance has been proposed by the National Committee for Clinical Laboratory Standards (NCCLS).[6] Total variance is given by equation 17-6, in which σ_t is total standard deviation, σ_{wr} is within-run standard deviation, σ_{br} is between-run, within-day standard deviation, and σ_{bd} is day-to-day standard deviation.

$$\sigma_t^2 = \sigma_{wr}^2 + \sigma_{br}^2 + \sigma_{bd}^2 \qquad \textit{Eq. 17-6}$$

(Notice in equation 17-6 that standard deviations do not add up directly; they add up as variances). In the experiment, a suitably stable specimen is analyzed in duplicate in each of two runs per day for 20 days. The resulting data are reduced mathematically to estimates of each of the quantities in equation 17-6.[6]

STATISTICAL COMPARISONS OF POPULATIONS

One can use the information in the previous section to define certain analytical parameters used in clinical chemistry. These are the terms "accuracy" and "precision." Accuracy is used to describe the ability of an analytical method, after replicate analyses, to obtain the "true" or correct result. The closer the mean of N replicate analyses comes to the "true" result, the more accurate a method is. Precision defines the reproducibility of a method. After replicate analyses, the narrower the distribution of results, the more precise a method will be.

The definitions of accuracy and precision are demonstrated in Figs. 17-7 and 17-8. Fig. 17-7, *A*, shows the results of replicate anaylses on an identical sample by three different methods. All three methods have the same mean value and thus the same relative degree of accuracy. However, the distributions of the results by each method differ from each other, with method A having the narrowest distribution (standard deviation) and thus the best precision, whereas method C has the widest distribution of results and the poorest precision. Fig. 17-8 shows two different methods whose distribution of replicate analyses, that is, precisions, are equal but whose means (that is, relative accuracy) are *not* equal. The different means indicate a nonrandom bias between the methods.

Thus the concepts of accuracy (or bias) and precision are independent of each other. This can be conceptualized

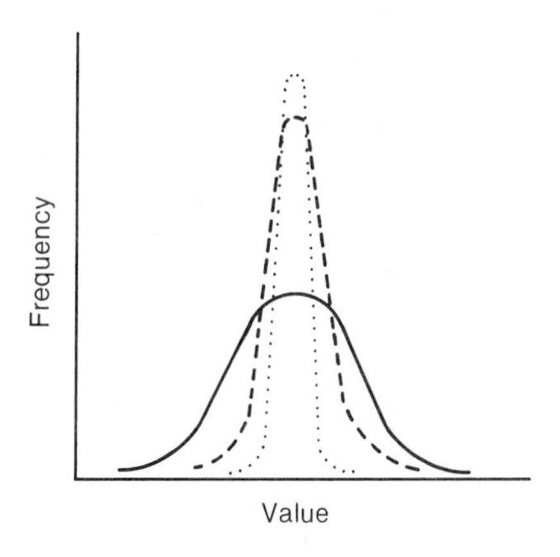

Fig. 17-7. Frequency distributions for 3 methods having same means, but different distributions, σ. *A* (.....) has narrowest distribution, whereas distribution of *C* (——) is wider than that of *B* (– – –).

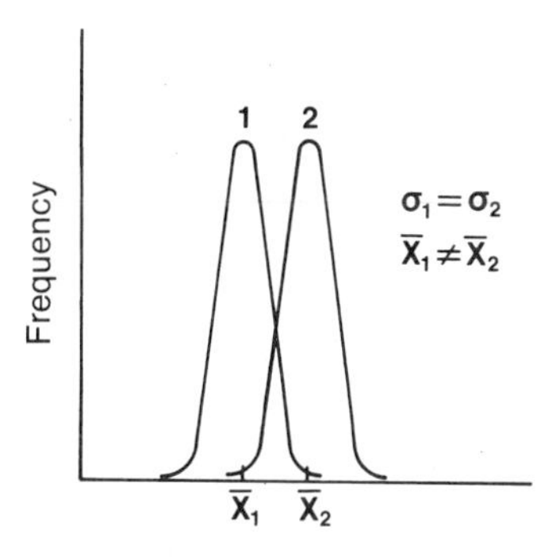

Fig. 17-8. Frequency distributions for replicate analysis by two different methods, *1* and *2*. Both methods are equally precise ($\sigma_1 = \sigma_2$) but are biased from each other ($\bar{X}_1 \neq \bar{X}_2$).

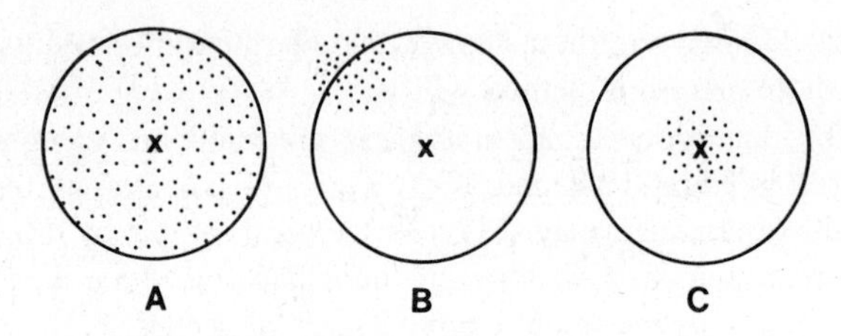

Fig. 17-9. X's of these targets denote true value for a sample. **A** to **C** denote three replicate analyses by three different methods: **A,** imprecise but accurate; **B,** precise but inaccurate; **C,** accurate and precise.

two-dimensionally by the targets in Fig. 17-9. Repeated attempts to hit the middle of the target (''true'' or accurate value) are indicated by dots. A method can be accurate but not very precise, as in Fig. 17-9, *A*, in which the average value falls on the mean of the target. Or a method can be precise but not very accurate, as shown in Fig. 17-9, *B*, in which the values fall close together but are grouped far from the middle. It is the goal of clinical chemists to design methods that are both accurate and precise (Fig. 17-9, *C*).

When the same specimen is analyzed repeatedly by two different analytical methods, the means and standard deviations calculated for the two methods are nearly always different. Unless the specimen is analyzed an infinite number of times by both methods, it is impossible to be absolutely certain that the means or standard deviations really are different. However, one can perform statistical tests on the estimated parameters to determine how likely it is that there is a significant difference between the standard deviations or the means.

Comparisons of random variation (precision)

The *F* test is used to determine whether an observed difference in variances is ''statistically significant.'' One performs it by dividing the larger variance (s_1^2) by the smaller variance (s_s^2), as shown in equation 17-7.

$$F = \frac{s_1^2}{s_s^2} \qquad \textit{Eq. 17-7}$$

The calculated *F* value is compared with a critical *F* value (Table 17-1) to test the null hypothesis, which states that there is no difference in the true variances of the two methods. If the calculated *F* value exceeds the critical *F* value for a given probability (such as $p = 0.05$), there is only the given probability, that is, less than 5%, that the calculated *F* value occurred because of chance alone. In this case, one can assume with 95% confidence that there is a statistically significant difference in the variances. Thus one can reject the null hypothesis with 95% confidence and state that the method with the larger estimated variance really is less precise than the method with the smaller estimated variance. If the calculated *F* value does not exceed the critical *F* value, the data do not demonstrate that a real difference exists at the given probability level; that is, there is not a statistically significant difference.

For example, method A has a standard deviation of 30 mg/L ($N = 21$), and method B has a standard deviation of 40 mg/L ($N = 31$). Is method A really more precise?

$$F = \frac{s_1^2}{s_s^2} = \frac{s_B^2}{s_A^2} = \frac{40^2}{30^2} = 1.78$$

From Table 17-1, with 30 degrees of freedom for the numerator and 20 degrees of freedom for the denominator, the critical *F* value ($p = 0.05$) is 2.04. The data do not

Table 17-1. Critical values of *F* for $p = 0.05$ and selected degrees of freedom *(df)*

df for denominator	Degrees of freedom *(df)* for numerator						
	5	10	15	20	30	60	∞
1	230.00	242.00	246.00	248.00	250.00	252.00	254.00
2	19.30	19.40	19.40	19.40	19.50	19.50	19.50
3	9.01	8.79	8.70	8.66	8.62	8.57	8.53
4	6.26	5.96	5.86	5.80	5.75	5.69	5.63
5	5.05	4.74	4.62	4.56	4.50	4.43	4.36
6	4.39	4.06	3.94	3.87	3.81	3.74	3.67
7	3.97	3.64	3.51	3.44	3.38	3.30	3.23
8	3.69	3.35	3.22	3.15	3.08	3.01	2.93
9	3.48	3.14	3.01	2.94	2.86	2.79	2.71
10	3.33	2.98	2.85	2.77	2.70	2.62	2.54
15	2.90	2.54	2.40	2.33	2.25	2.16	2.07
20	2.71	2.35	2.20	2.12	2.04	1.95	1.84
30	2.53	2.16	2.01	1.93	1.84	1.74	1.62
60	2.37	1.99	1.84	1.75	1.65	1.53	1.39
∞	2.21	1.83	1.67	1.57	1.46	1.32	1.00

Modified from Barnett, R.N.: Clinical laboratory statistics, ed. 2, Boston, 1979, Little, Brown & Co.

show that the difference in standard deviations is statistically significant.

Comparisons of means (accuracy or bias)

The t distribution was used previously to describe intervals that have a stated confidence of containing the true mean. The t test is also used to test for statistically significant differences between two experimental means or between an experimental mean and a stated value.

Testing the statistical significance between two means actually involves testing the degree of overlap of their respective probability distributions. If there is little or no overlap, the populations are demonstrated to be different. If there is significant overlap, one cannot be sure that there is any difference. (For further discussion of hypothesis testing, see references 1 to 5). A t value is calculated from equation 17-8, in which subscripts 1 and 2 denote the respective populations.

$$t = \frac{\bar{x}_1 - \bar{x}_2}{\sqrt{\frac{s_1^2}{N_1} + \frac{s_2^2}{N_2}}} \qquad \text{Eq. 17-8}$$

This t value is compared to the two-sided critical t value for the desired probability level and $(N_1 + N_2 - 2)$ degrees of freedom (if $s_1 \cong s_2$).[1] If the absolute value of the calculated t exceeds the critical t value, the difference between the two means is statistically significant.

Table 17-2. Critical values of t for selected probabilities (p) and degrees of freedom (df)

	Two-sided intervals or tests		
	$p = 0.10$	$p = 0.05$	$p = 0.01$
	One-sided limits or tests		
df	$p = 0.05$	$p = 0.025$	$p = 0.005$
1	6.31	12.70	63.70
2	2.92	4.30	9.92
3	2.35	3.18	5.84
4	2.13	2.78	4.60
5	2.01	2.57	4.03
6	1.94	2.45	3.71
7	1.89	2.36	3.50
8	1.86	2.31	3.36
9	1.83	2.26	3.25
10	1.81	2.23	3.17
12	1.78	2.18	3.05
14	1.76	2.14	2.98
16	1.75	2.12	2.92
18	1.73	2.10	2.88
20	1.72	2.09	2.85
30	1.70	2.04	2.75
40	1.68	2.02	2.70
60	1.67	2.00	2.66
120	1.66	1.98	2.62
∞	1.64	1.96	2.58

Condensed from Davies, O.L., and Goldsmith, P.L.: Statistical methods in research and production, ed. 4, New York, 1972, Hafner Press.

For example, assume that a stable control material is included with the patient specimens in each daily batch of glucose measurements. In April the mean was 1110 mg/L, with a standard deviation of 25 mg/L, with $N = 30$. In May the mean was 1090 mg/L, with a standard deviation of 20 mg/L, with $N = 31$. Are the means really different (from two different populations) or can the difference be explained by random variation within a single population?

$$t = \frac{1110 - 1090}{\sqrt{\frac{25^2}{30} + \frac{20^2}{31}}} = 3.44$$

$$df = 30 + 31 - 2 = 59$$

The critical t value for $p = 0.05$ and 59 df is 2.00 (Table 17-2, two-sided interval). Since the calculated t value exceeds the critical t value, one is more than 95% certain of a real difference between the populations.

A special case of comparison of means is the paired-sample t test. It is used to minimize the sources of variation that can cause ambiguity in testing for differences. For example, if glucose method X were being compared to glucose method Y and one compared the means using different random patient specimen populations for each method, the extraneous variation of the populations could mask true differences. To eliminate this variance, the same specimens are analyzed by both methods. This procedure is described in Chapter 20.

Testing the difference between an experimental mean and a stated (given, assigned, known, and so on) value involves testing to see if the stated value is included in the confidence interval around the experimental mean. If it is not, the null hypothesis is rejected, and there appears to be a difference between the stated value and the experimental mean value. Equation 17-9 is used to calculate the t value, in which μ_0 is the stated value.

$$t = \frac{\bar{x} - \mu_0}{s/\sqrt{N}} \qquad \text{Eq. 17-9}$$

There are $N - 1$ df.

Using the above example with quality control data for April and May, assume that the stated glucose concentration in the quality control specimen by the National Bureau of Standards is known to be 1120 mg/L. Is the mean for April (1110 mg/L) significantly different from 1120 mg/L?

$$t = \frac{1110 - 1120}{25/\sqrt{30}} = -2.19$$

The critical t value for $p = 0.05$ and 29 df is 2.04. Thus April's mean of 1110 mg/L is statistically significantly different from the assigned glucose concentration of 1120 mg/L.

NONPARAMETRIC STATISTICS

The statistical tests described to this point are termed "parametric statistics" because they assume a gaussian or other symmetric distribution of the data. Many sampling populations do not meet this criterion, and the analyst needs techniques for describing these populations statistically. This is especially true for distributions of concentrations of some analytes in healthy subjects. The seriousness of this problem depends on the use to be made of the statistics derived from these populations. Statistical procedures involving the dispersion of data in an unknown or nonnormal distribution may require the use of data transformations to yield a normally distributed population (see Chapter 18) or the use of nonparametric statistics.

Nonparametric or distribution-free statistics require no assumption of the distribution. In this sense, nonparametric statistics may be considered more general. Advantages of nonparametric statistics are that (1) they may be used when the population distribution is unknown, (2) computational procedures are simple and can be performed quickly, and (3) they may be applied to ranked or semiquantitative data and to quantitative data as well.

However, disadvantages in the use of nonparametric statistics are that (1) many data are wasted and so if the data can be summarized by parametric statistics, this is preferred; (2) some applications may be laborious and tedious for large data sets; and (3) they are less powerful than the corresponding parametric tests. Nonparametric statistical procedures are not commonly used in clinical chemistry. They are described here as an introduction to the concepts rather than to make you a practitioner.

The simplest nonparametric procedure is to rank the data in order from lowest (value = 1) to the highest (value = N). The range of the data set is the difference between the lowest and highest value. The middle value (or average of the two central values if N is even) is the median and indicates the central tendency of the data set. A confidence interval of probability, P, may be established for the ranked data by exclusion of the lower [100(1 − P)/2%] of the data and similarly the upper [100(1 − P)/2%] of the data.

A nonparametric test analogous to the t test is the *sign test*. It essentially tests the median rather than the mean of a data set. All the data in a single data set can be compared to some stated (critical) value. Data points higher than the stated value are assigned a plus value (+), lower points are assigned a minus value (−), and zeros are assigned to those values equal to the critical value. The sign test can also be used to compare the results of two methods (A to B). If the B value is higher than A for a given sample, that sample is assigned a plus value. If the B value is less than A, it is assigned a minus value, and zeros are assigned to those samples in which A = B.

If the null hypothesis is true and the median equals the stated value (or method A and method B are equivalent), the probability for plus values is equal to the probability for minus values.

$$P(+) = P(-) = 0.5 \qquad \textit{Eq. 17-10}$$

To test whether the null hypothesis holds, one totals the sum of the occurrences of the less frequent sign. This total is compared to a critical total for N and the desired p.[4] If the total is less than the critical total number of occurrences for the less frequent sign, the null hypothesis is rejected. Alternatively, one can compute the actual probability of obtaining the observed frequency of plus (+) and minus (−) signs, assuming that the null hypothesis is true. This is really the probability that the observed frequency is part of the population of frequencies that would be obtained in a random sampling if the probability for plus signs is equal to the probability for minus signs. One can compare the calculated probability to the chosen level of significance to determine whether the null hypothesis holds.[1]

The actual probability, P, of observing m or less occurrences of the less frequent sign can be calculated by use of equation 17-11, in which n is the number of signs remaining after eliminating zeros, k is the number of the less frequently observed sign, p is the assumed probability, and q is $(1 - p)$.

$$P = \sum_{k=0}^{m} \frac{n!\, p^k q^{n-k}}{k!\,(n-k)!} \qquad \textit{Eq. 17-11}$$

For example, in a group of 10 samples, if only one minus sign is observed with nine plus signs, the frequency for only one or less minus sign to occur is the sum for the frequency for no minus-sign values plus that for only one minus-sign value:

$$P = \frac{10!\,(0.5)^0(0.5)^{10}}{0!\,10!} + \frac{10!\,(0.5)^1(0.5)^9}{1!\,9!}$$

$$P = (0.5)^{10} + 10(0.5)^{10} = 0.011$$

Thus the null hypothesis may be rejected because there is only a 1.1% probability of observing one or less minus sign if there are equal chances for plus and minus signs.

An extension of the sign test is the Wilcoxon Signed Rank test. This test and the Wilcoxon Rank Sum Test are analogous to the parametric paired t test. They are sensitive not only to the sign of the differences but also to the magnitudes of the differences.[5]

LINEAR REGRESSION AND CORRELATION

The degree of correlation or association between two variables can be tested by statistical methods. For example, one can determine the degree of correlation between creatinine and pH in urine specimens, or the degree of correlation between measurements of an analyte made by two different methods. Fig. 17-10 and Table 17-3 are examples using estriol concentrations obtained from plasma samples by an HPLC method and an RIA (radioimmu-

Table 17-3. Estriol concentrations of 30 specimens measured by HPLC and RIA methods

Sample no.	RIA (X)	HPLC (Y)	X*Y	X*X	Y*Y
1	5.1	8.5	43.35	26.01	72.25
2	5.4	2.8	15.12	29.16	7.84
3	3.6	3.2	11.52	12.96	10.24
4	6.8	3.6	24.48	46.24	12.96
5	6.6	7.4	48.84	43.56	54.76
6	6.5	7.1	46.15	42.25	50.41
7	11.4	10.0	114.00	129.96	100.00
8	5.3	6.2	32.86	28.09	38.44
9	4.9	3.3	16.17	24.01	10.89
10	9.8	8.4	82.32	96.04	70.56
11	5.5	4.1	22.55	30.25	16.81
12	6.3	4.6	28.98	39.69	21.16
13	4.9	2.9	14.21	24.01	8.41
14	3.5	2.5	8.75	12.25	6.25
15	6.5	4.4	28.60	42.25	19.36
16	11.1	7.1	78.81	123.21	50.41
17	7.3	5.2	37.96	53.29	27.04
18	6.2	6.9	42.78	38.44	47.61
19	14.6	17.8	259.88	213.16	316.84
20	11.9	15.2	180.88	141.61	231.04
21	8.8	11.2	98.56	77.44	125.44
22	16.0	19.9	318.40	256.00	396.01
23	21.0	20.7	434.70	441.00	428.49
24	3.4	2.4	8.16	11.56	5.76
25	8.0	6.2	49.60	64.00	38.44
26	8.2	6.5	53.30	67.24	42.25
27	17.1	12.4	212.04	292.41	153.76
28	11.4	7.4	84.36	129.96	54.76
29	11.0	11.5	126.50	121.00	132.25
30	12.7	13.8	175.26	161.29	190.44
SUM	260.8	243.2	2699.09	2818.34	2740.88

*$\overline{X}$ = 8.69; $\overline{Y}$ = 8.11.

noassay) method. The RIA data are presented as the independent variable, X (that is, the method being used as the accepted or standard method), and the HPLC data are presented as the dependent variable, Y.

Fig. 17-10 appears to disclose a linear relationship between the results obtained by the HPLC and RIA methods. It is possible to determine the equation of the line that best fits the data. The general form of this equation, the regression line, is given in equation 17-12:

$$Y = \alpha + \beta X + \epsilon \qquad \textit{Eq. 17-12}$$

where α denotes the true value of the intercept of the line and β, the true value of the slope. The variable X is used to make a prediction of Y and is presumed to be a preset or nonrandom variable. In practice α and β are unknown and must be estimated as a and b, respectively.

It is unrealistic to assume that every data point will fall exactly on the regression line. A term, ϵ, is included in the above model to represent the amount any individual, Y, falls off the regression line $\alpha + \beta X$. In practice this term represents all the variability in each measurement caused by unclean glassware, temperature variation, electrical surges, biological variability, nonlinearity, chemical interferences, and so on. To put it another way, one of the assumptions underlying a regression analysis is that for any fixed, accurately known values of X, there is a corresponding normal distribution of Y values. This assumption is shown graphically in Fig. 17-11. Notice that the regression line goes through the means of the distributions.

The regression line estimated from limited data is given in equation 17-13, where $\hat{Y}$ ("y hat") denotes the predicted mean value of Y for a given value of X.

$$\hat{Y} = a + bX \qquad \textit{Eq. 17-13}$$

One may determine visually the values for a and b from a scatter plot of the data, aligning a ruler as well as possible along the data. The most rigorous method to determine the regression coefficients is the method of least squares. By this approach, the sum of the squares of the difference $(Y_1 - \hat{Y}_1)$ for each X_1 is minimized mathematically, hence the term "least squares." Formulas for the estimated regression coefficients, a and b, under these conditions are as follows:

$$b = S_{xy}/S_{xx} \qquad \textit{Eq. 17-14}$$

where

$$S_{xy} = \sum xy - N\overline{x}\overline{y} \qquad \textit{Eq. 17-15}$$

and

$$S_{xx} = \sum x^2 - N\overline{x}^2 \qquad \textit{Eq. 17-16}$$

and N is the number of samples. A mathematically equivalent equation for b is

$$b = \frac{N \sum x_i y_i - \sum x_i \sum y_i}{N \sum x_i^2 - (\sum x_i)^2} \qquad \textit{Eq. 17-17}$$

The equation for a is then:

$$a = \overline{Y} - b\overline{X} \qquad \textit{Eq. 17-18}$$

Calculations of the estimated regression coefficients for the estriol data presented in Table 17-3 are as follows:

$$b = \frac{30(2699.09) - (260.8)(243.2)}{30(2818.34) - (260.8)^2} = 1.06$$

and

$$a = 8.11 - (1.06)8.69 = -1.10\ \mu g/mL$$

Thus the estimated regression line can be written as follows:

$$\hat{Y} = -1.10\ \mu g/mL + 1.06X$$

The regression line is graphed in Fig. 17-10. As one can see, very few of the data points fall exactly on the line. But as stated before, the sum of the squares of these deviations about the regression line is minimized to yield the line of best fit through the data. A measure of the variability of the data points about the line is the estimated

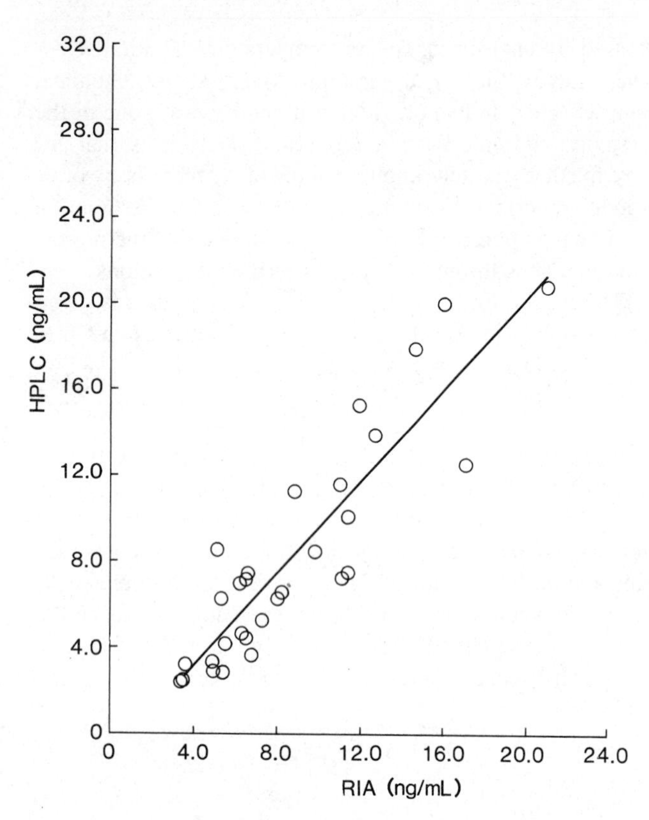

Fig. 17-10. Graph of estriol method comparison data. High-performance liquid chromatography versus radioimmunoassay.

standard deviation about the regression line of the differences between the observed and predicted Y values, $s_{y/x}$, otherwise known as the standard error of the estimate.

$$s_{y/x} = \sqrt{\frac{\sum(y_i - \hat{Y}_i)^2}{N - 2}} \qquad \textit{Eq. 17-19}$$

The $(N - 2)$ degrees of freedom in the denominator come from the fact that two regression coefficients, a and b, had to be calculated from the data in order to compute Y; thus there are two restrictions placed on the N observations. For computational purposes, the following formula is usually used:

$$s_{y/x} = \sqrt{\frac{S_{yy} - bS_{xy}}{N - 2}} \qquad \textit{Eq. 17-20}$$

For the estriol example,

$$s_{y/x} = \sqrt{\frac{[2740.88 - 30(8.11)^2] - 1.06[(2699.09) - 30(8.69)(8.11)]}{28}}$$

$$s_{y/x} = 2.3\ \mu\text{g/mL}$$

Inference on estimated slope, *b*

A measure of the variability of the estimated slope, b, can be obtained by computation of the standard deviation of b, s_b.

$$s_b = s_{y/x}/\sqrt{S_{xx}} \qquad \textit{Eq. 17-21}$$

It can be shown that a $100(1-p)\%$ confidence interval for β is as follows:

$$\beta = b \pm ts_b \qquad \textit{Eq. 17-22}$$

where t is obtained from a two-sided t table for $N - 2$ degrees of freedom and the desired level of significance. In the estriol example, $s_{y/x} = 2.3\ \mu$g/mL and $S_{xx} = 547.6$. Thus a 95% confidence interval for the true slope, β, is given by

$$\beta = 1.06 \pm 2.05\ (2.3)/\sqrt{547.6}$$

$$\beta = 1.06 \pm 0.20$$

to give a lower limit of 0.86 and an upper limit for the slope of 1.26. The fact that the interval brackets the ideal value for the slope of 1.0 implies that the slope of the line is not significantly different from 1.0 at the $p = 0.05$ level of significance.

Inference on estimated intercept, *a*

A confidence interval of a can be constructed in a fashion similar to that for b. The estimated standard deviation of a, s_a, *is computed as follows:*

$$s_a = s_{y/x}\,(\textstyle\sum x^2/NS_{xx})^{1/2} \qquad \textit{Eq. 17-23}$$

Thus a $100(1 - p)\%$ confidence interval for α is given by:

$$\alpha = a \pm ts_a \qquad \textit{Eq. 17-24}$$

Substitution of the appropriate values for the estriol example yields an interval from -3.07 to $0.83\ \mu$g/mL. Since the ideal value for the intercept, zero, is contained within this interval, the true value of the intercept is not significantly different from the ideal value.

Correlation and regression

The estimated Pearson product-moment correlation coefficient, r, provides a measure of how closely the data points lie to the estimated regression line. The correlation coefficient can take on values from -1 to $+1$ and will be equal to $+1$ or -1 only if every datum point lies exactly on the regression line. Thus a value of r close to $+1$ or -1 is indicative of a strong (positive or negative) linear relationship between X and Y. The sign of r is the same as the sign of b. The formula for r is given by equation 17-25.

$$r = S_{xy}/\sqrt{S_{xx}S_{yy}} \qquad \textit{Eq. 17-25}$$

Care should be exercised in the interpretation of r. Consider the four hypothetical data sets graphed in Fig. 17-12. Although a high value of r might be obtained for each of the four data sets, only the first one is properly seen as a truly linear relationship between x and y. It is thus recommended that X and Y data always be graphed. In addition, it has been shown that the value of r is sensitive to the scatter of the data and the range of data (see Chapter 20). Although the scatter of the data is inherent in the precision of the method being studied, it is possible to in-

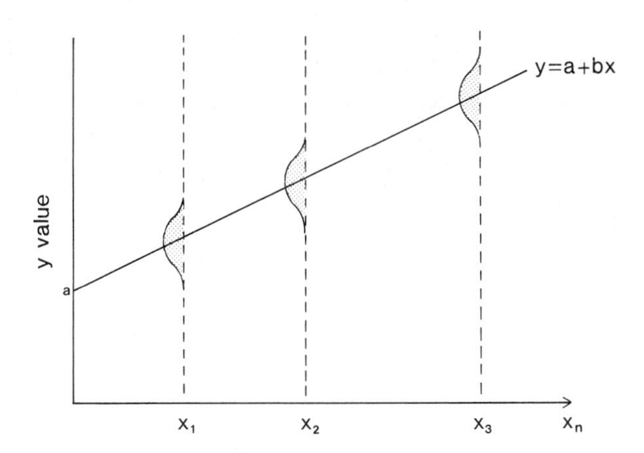

Fig. 17-11. Normal distribution of y values corresponding to a known x value.

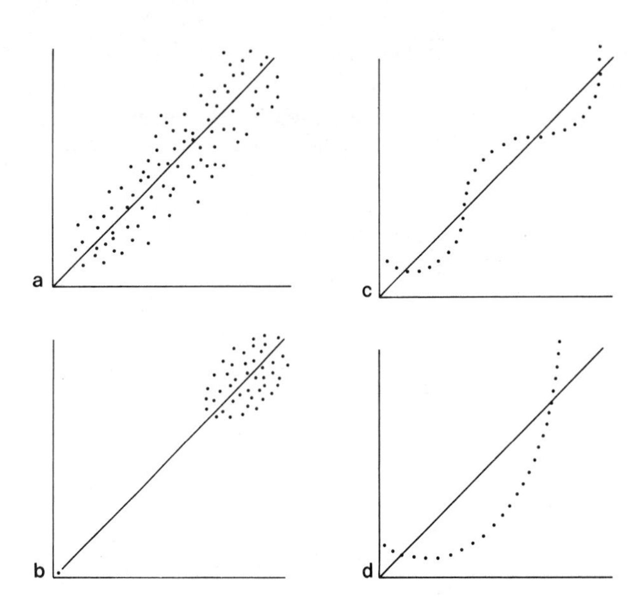

Fig. 17-12. Hypothetical data giving high values for r (estimated Pearson product-moment correlation coefficient).

crease the value of r just by extension of the range of data. This is seen in equation 17-26:

$$r = \frac{1}{s_x}\sqrt{s_x^2 + s_{y/x}^2} \qquad \textit{Eq. 17-26}$$

where s_x is the standard deviation of the X population, indicating the spread in the X data. As s_x becomes very large, $s_{y/x}$ becomes negligible and r approaches 1.0.

There are statistical approaches available[1] to determine the significance of the value of r, but the practical meaning of r has in general been somewhat elusive. It appears to be more useful in population studies rather than in comparison of method studies. In any case, it is essential to keep in mind that the demonstration of a ''relationship'' between data sets proves neither cause and effect nor clinically acceptable agreement.

The nonparametric form of the correlation coefficient is the Spearman rank correlation coefficient. It tests for the ranking of the X and Y data together. It essentially tests whether the X and Y for each datum point are at the same relative positions in their respective distributions. It ignores their absolute differences from their respective means.[1]

In summary, one should be aware of the assumptions made on the properties of data being analyzed by linear regression to avoid inappropriate use and interpretation of this statistical method.

To use linear regression between two data sets, X and Y, it is assumed that:

1. The observed Y's are normally distributed at any given X (see Fig. 17-11).
2. X is known without error.
3. The variance of observed Y's for a given X is constant for all X's.
4. The mean Y at a given X is a linear function of X.

In general it is not recommended that least-squares analysis be used in situations in which the X value is not known without error. The X values should, ideally, be predefined or preselected. The application of linear regression analysis to nonpreselcted data results in an estimated slope that is biased low (nonaccurate) and an intercept that is biased high. The regression line pivots around the $\overline{X},\overline{Y}$ point. It is important to avoid inappropriate application of linear regression to avoid biased results. If the range of data is wide enough, the problem introduced by using imprecise, nonpreselected X values may be minimized. In the estriol example, since $r = 0.90$ (which is less than the recommended cut-off value of 0.99 for accurate linear regressions), one might suspect the regression coefficients to be biased in relation to the true values of the parameters α and β. In such cases, other statistical approaches should be used (see Chapter 20).

Prediction of Y for a value of X

It is possible to compute a $100(1 - p)\%$ confidence interval for the mean value of Y for a given X. For example, for samples having an estriol concentration of 12 µg/mL by the RIA method, an approximate confidence interval, CI, for the mean of the corresponding HPLC assay is given in equation 17-27:

$$\text{CI} = Y(\text{at } X = 12) \pm ts_{y/x}\left(1/N + \frac{(12 - \bar{x})^2}{S_{xx}}\right)^{1/2} \qquad \textit{Eq. 17-27}$$

$$\text{CI} = [-1.10 + 1.06(12)] \pm 2.05(2.3)\sqrt{\frac{1}{30} + \frac{(12 - 8.69)^2}{547.6}}$$

This gives limits extending from 10.53 to 12.71 µg/mL.

A confidence interval for a single assay value (as opposed to the mean of all values) can be computed in a similar fashion by replacement of $1/N$ under the radical with $(1 + 1/N)$. The variability for a single measurement (patient sample) is greater than that for the average. A word of caution: If r is less than 0.99 in a method comparison study, the slope will be biased low in relation to the true slope, and the confidence interval as calculated above for Y may not correct for this bias.

Inverse prediction: calibration problem

Consider a situation in which one has fitted a regression line $Y = a + bX$ and is interested in estimating a value, X_0, corresponding to a fixed value, Y_0. This situation arises when known corresponding peak heights or areas (Y) are measured from a chromatogram. Later, a peak height (Y_0) is measured for a sample containing an unknown concentration of the analyte. Based on the fitted regression line, it is then possible to obtain an estimate of the corresponding unknown concentration (X_0). One obtains the estimate by solving the equation $Y_0 = a + bX_0$ for X_0. Thus equation 17-12 becomes as follows:

$$\hat{X}_0 = (Y_0 - a)/b \qquad \textit{Eq. 17-28}$$

An approximate 95% confidence interval (or more correctly the fiducial interval, see reference 3) for X_0 can be found in equation 17-29:

$$X_0 = \hat{X}_0 + \frac{(\hat{X}_0 - \overline{X})}{1 - g} \pm \frac{\frac{t \cdot s_{y/x}}{b}\{(\hat{X}_0 - \overline{X})^2/S_{xx} + (1 - g)/N\}^{1/2}}{1 - g} \qquad \textit{Eq. 17-29}$$

where

$$g = \frac{t^2(s_{y/x})^2}{b^2 S_{xx}} \qquad \textit{Eq. 17-30}$$

The width of the confidence interval for X_0 is greatly affected by how far X_0 is from X, thus illustrating the importance of a good statistical design. How "significant" or well determined b is also has a pronounced effect on the width of the confidence interval. To see this, one can rewrite the equation for g as follows:

$$g = \left(\frac{t}{b/s_b}\right)^2 \qquad \textit{Eq. 17-31}$$

If b is large in relation to its standard deviation, the denominator of g will be large and consequently g will be small. The smaller the value of g, the narrower the limits for X. Values of g less than approximately 0.20 are indicative of a slope (b) that is highly statistically significant, that is, well determined.

For example, consider the estriol data presented in Table 17-2 and suppose that one wishes to estimate the concentration of estriol by the RIA method $(\hat{X}_0)$ that corresponds to a concentration of 8 μg/mL by the HPLC method:

$$\hat{X}_0 = \frac{8 - (-1.10)}{1.06} \text{ or } 8.58$$

with corresponding 95% limits of 7.76 to 9.42 μg/mL.

REFERENCES

1. Daniel, W.W.: Biostatistics: a foundation for analyses in the health sciences, ed. 2, New York, 1978, John Wiley & Sons, Inc.
2. Box, G.E.P., Hunter, W.G., and Hunter, J.J.: Statistics for experimenters, New York, 1978, John Wiley & Sons, Inc.
3. Draper, N., and Smith, H.: Applied regression analysis ed. 2, New York, 1981, John Wiley & Sons, Inc.
4. Davies, O.L., and Goldsmith, P.L.: Statistical methods in research and production, ed. 4, New York, 1972, Hafner Press.
5. Massart, D.L., Dijkstra, A., and Kaufman, L.: Evaluation and optimization of laboratory methods and analytical procedures, New York, 1978, Elsevier/North Holland, Inc.
6. National Committee for Clinical Laboratory Standards: NCCLS tentative standard EP5-T, tentative guidelines for user evaluation of precision performance of clinical chemistry devices, Subcommittee for User Evaluation of Precision of the Evaluation Protocols Area Committee, Villanova, Pa., 1983.
7. Barnett, R.N.: Clinical laboratory statistics, ed. 2, Boston, 1979, Little, Brown & Co.

Chapter 18 Reference values

Charles Ralph Buncher
Dan Weiner

gaussian A particular symmetric statistical distribution; also called the "normal distribution."

log-normal A symmetric gaussian population distribution obtained by a plot of the log of the data or interval.

log-normal distribution A sample of values with a long tail to the right can often be made to act like a gaussian distribution when one uses the logarithms of values.

negative predictive value The probability that a laboratory result falling within the normal range reflects the true absence of disease; defined as true negatives divided by the addition of true negatives and false negatives.

normal A term with many meanings including those persons in the nondiseased population and a gaussian distribution (see this chapter for a discussion).

positive predictive value The probability that a laboratory result exceeding the upper limit of normal actually reflects the presence of disease; defined as true positives divided by the addition of true positives and false positives.

predictive value Probability that a laboratory result accurately reflects the true presence or absence of disease. It is dependent on the actual prevalence of the disease.

prevalence The number of persons who have a disease in a given population at any one point in time, or more commonly the rate of such disease, which is also called the "disease frequency."

reference range The limits of laboratory values of a population without disease; should be further defined by other variables such as gender, age, heritage, and other epidemiological factors.

sensitivity A term used to describe the probability that a laboratory test is positive (that is, greater than the upper limit of normal) in the presence of disease; defined as true positives divided by the addition of true positives and false negatives.

specificity Used to describe the probability that a laboratory test will be negative (that is, within the normal range) in the absence of disease; defined as true negatives divided by the addition of true negatives and false positives.

standard deviation A measure of variability; in the gaussian distribution, two standard deviations above and below the mean encompass the central 95.5% of the population data, and one standard deviation above and below encompasses 68.3% of the data.

DEFINITION OF NORMAL

This chapter is a discussion of what are considered "normal" values in laboratory medicine and the statistical methods that are used to work with these values. For example, let us assume that some laboratory value, which shall be called the "serum analyte value," was measured. After one has measured the serum analyte value, the next step is to inquire whether this is a "normal" or "abnormal" value. Does the value suggest a healthy person, a condition that can be called normal, or does it indicate that the person's condition is abnormal, or in other words, diseased? If life were simple, the distribution of such values would be as in Fig. 18-1. In this situation each value would be considered to be normal if in the left-hand curve (*A* to *B*) and abnormal if in the right-hand curve (*C* to *D*). If the value were found to be outside the range of normal values, this would be a sign that this person's condition was abnormal. There are almost no laboratory tests that fit this ideal model.

A more common situation is that shown in Fig. 18-2, in which there is some overlap of normal and abnormal values. In this situation there are three choices for each laboratory value: (1) the value may be strictly in the reference range of normal (*E* to *F*), (2) the value may be strictly in the reference range of abnormal (*G* to *H*), or (3) the value may be in the intermediate range (*F* to *G*), in which case a probability statement can be made allocating that determination as more likely to be normal or more likely to be abnormal. Many situations appear as in Fig. 18-3. In this case the area of overlap of normal and abnormal is the most common outcome.

At this point, one needs to discuss what is really meant by the word "normal." Few disease conditions are such that a person's condition can be considered normal at one moment and abnormal at the next. This is especially true of chronic diseases in which persons move slowly from the range of normal into the range of abnormal, often over an interval of many years. Moreover, clinical studies show that the progression is not one of a uniform change from normal to abnormal. Monitoring these changes can be more important than measuring their current level, but that issue is not discussed in this chapter.

A single laboratory parameter is rarely sufficient to decide whether a person's condition is normal or not. Usually one laboratory test contributes to the diagnosis rather than determines the clinical situation. Moreover, for simplicity in this discussion an elevation of serum analyte will show disease, though a lack of a substance or lower concentration might also indicate disease.

In this text the term "reference range" applies to the values of laboratory tests on analytes, determined on healthy (nondiseased) populations, and is equivalent to the less desirable term "normal range."

Another complication is that the term "normal" has many different meanings in the English language. In one meaning, normal distinguishes a healthy person from the abnormal or unhealthy. In another meaning, "normal" means 'usual' as in a usual value for a serum cholesterol. Many would say that Americans in general have high levels of serum cholesterol, and even though a value may be considered usual, it is not necessarily the level associated with good health.

A third definition of the word "normal" is a set of laboratory results for a test that are distributed in the same manner as a particular bell-shaped statistical distribution, often called the "gaussian distribution." Although this symmetric bell-shaped distribution is of great importance in statistics, there is no reason to believe that a particular laboratory analyte has this distribution. In fact, it is clear that most do not. This is another reason for not applying the term "normal" to a set of laboratory values. There are mathematical ways of making the laboratory measurements conform more closely to this distribution, such as use of the logarithm of the concentration rather than the actual measurement. Yet many statistical determinations depend on this distribution, and despite its name, it is a part of statistical science and should not be confused with the term "healthy." In fact, many persons consider any symmetric curve that has a central peak to be a normal distribution. Statisticians reserve this terminology for a particular set of such curves with specific mathematical properties.

ARE DATA NORMALLY DISTRIBUTED?

Probability plot

An important statistical question that must be faced in a clinical laboratory is whether a set of data can be considered to be normally (gaussian) distributed. There are a number of statistical techniques for testing this assumption. One effective and simple alternative is to graph the data on normal probability paper.

Consider 60 values from parathyroid hormone determinations. It is useful to know whether these data or their logarithms can be considered to be normally distributed. The first thing to do is to arrange the data in increasing

Table 18-1. Parathyroid hormone data

x (pg/mL)	Frequency	Cumulative frequency	F(x)
≤160	6	6	0.100
180	1	7	0.117
190	1	8	0.133
200	1	9	0.150
230	4	13	0.217
240	1	14	0.233
290	1	15	0.250
320	1	16	0.267
330	3	19	0.317
350	1	20	0.333
360	2	22	0.367
380	5	27	0.450
400	2	29	0.483
410	2	31	0.517
430	1	32	0.533
450	1	33	0.550
470	4	37	0.617
500	1	38	0.633
520	1	39	0.650
530	2	41	0.683
560	1	42	0.700
570	2	44	0.733
580	2	46	0.767
600	1	47	0.783
650	2	49	0.817
670	1	50	0.833
730	3	53	0.883
800	1	54	0.900
920	1	55	0.916
1150	1	56	0.933
1280	1	57	0.950
1370	2	59	0.983
1520	1	60	1.000

numeric order. Then one should calculate the percentage of the data that are at a given value or less, which is called the "cumulative proportion of data" and denoted with $F(x)$. The data are shown in Table 18-1.

Next the data are plotted on a piece of normal probability paper. This paper is designed and drawn so that data that conform to the normal distribution will be graphed as a straight line. The horizontal axis is scaled to be a normal (gaussian) distribution, whereas the vertical axis is an ordinary arithmetic scale. We now turn to the data and, for each value of parathyroid hormone, find the point equal to that value (level) on the vertical scale and equal to the cumulative proportion of the whole distribution, $F(x)$, on the horizontal scale (plotted as a percentage). These points have been graphed in Fig. 18-4.

A word of caution is necessary at this time. One must be aware that there is neither a 0% nor a 100% on the normal probability paper. Although 99.99% appears to be close to 100%, in actuality, on this type of paper it is not at all close. Therefore one can never plot either a 0% or a 100% point on this type of paper.

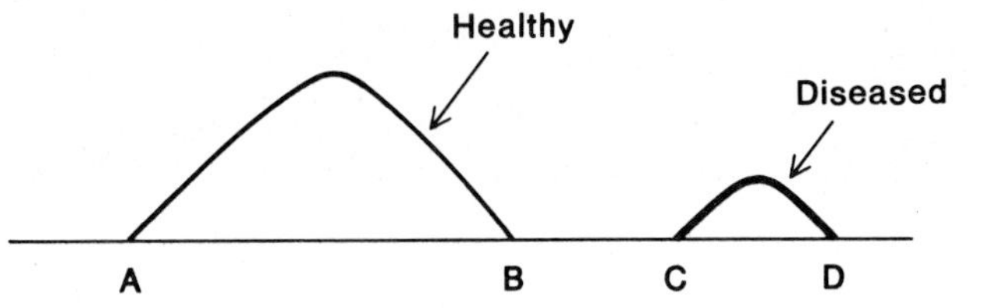

Fig. 18-1. Perfectly separated distributions of healthy and diseased populations. This clear separation does not occur in reality.

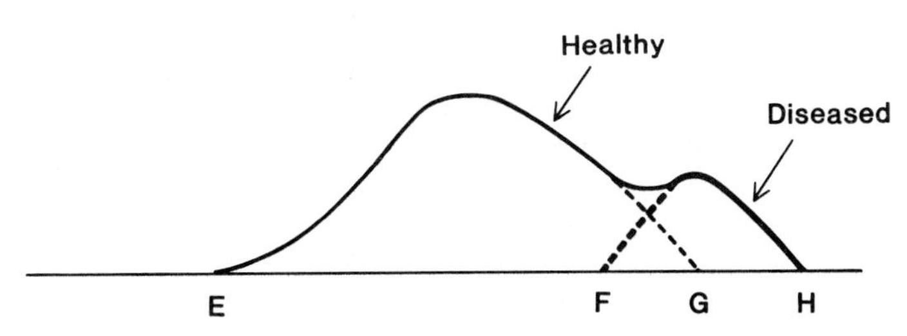

Fig. 18-2. Usual distributions of healthy and diseased populations in which an overlap between the two occur.

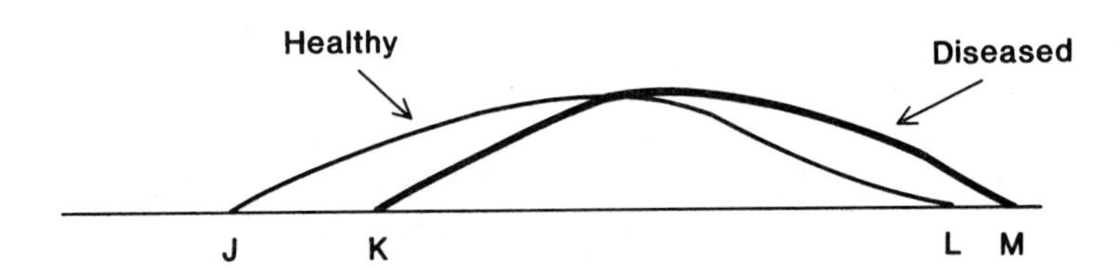

Fig. 18-3. Situation in which degree of overlap between healthy and diseased populations is most common outcome.

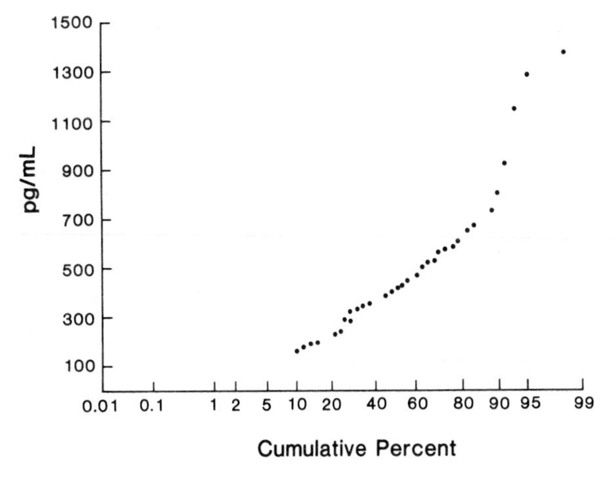

Fig. 18-4. Data for parathyroid hormone (PTH) levels plotted on normal probability paper: concentration of PTH is on y axis, and cumulative proportion for each value is on x axis.

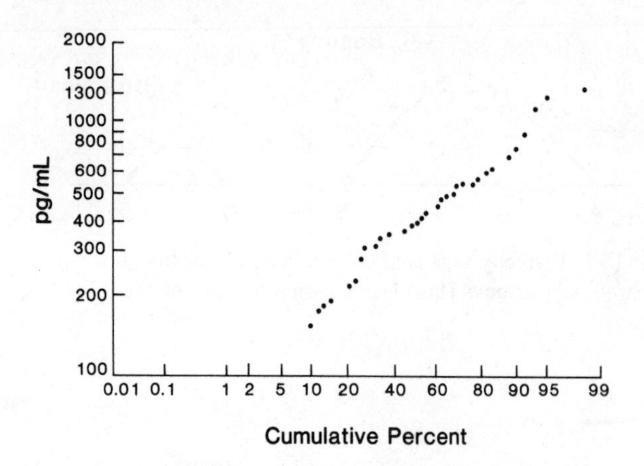

Fig. 18-5. Data for parathyroid hormone (PTH) levels plotted on log-normal probability paper: log of PTH concentration is on *y* axis, and cumulative proportion for each value is on *x* axis.

One can now look at Fig. 18-4 and decide whether the data appear to be in a straight line. In this instance the data do not lie on a straight line because the points at the higher values bend upward. An alternative would be to take the same data set and graph it on log-normal probability paper. On this graph paper the horizontal axis is the same as before, but the vertical axis is drawn on a logarithmic scale. Thus a straight line will indicate a log-normal distribution of the data. This is shown in Fig. 18-5. This simple test has shown that these parathyroid hormone data can be easily described by a log-normal statistical distribution. One can then use the mean value and geometric standard deviation to describe the distribution mathematically and to obtain precise percentile values, such as the 2.5 and 97.5 percentiles. These points could be read directly from the graph.

Percentiles

The mean value of a normal distribution is also the 50th percentile, and so the mean value can be read directly from this graph. In the same manner, it is known that from the 16th percentile (approximately) to the 50th percentile is 1 standard deviation, and likewise from the 50th percentile to the 84th percentile is 1 standard deviation. Thus on arithmetic graph paper one can subtract the value corresponding to the 16th percentile from the 84th percentile and divide by 2 to obtain the standard deviation. The same procedure can be done with log-normal probability paper, but the standard deviation must be interpreted as a geometric standard deviation. Alternatively one could have subtracted the value corresponding to the 2.5 percentile from the 97.5 percentile and divided this difference by 4 to obtain the standard deviation. If one understands this calculation, it becomes apparent that the scale of the horizontal axis is really in standard deviations.

One last point concerns the occasional observation of a graph that appears to consist of two lines joined together, one line in the lower values and a different line in the upper values. This type of result is obtained when a bimodal distribution is graphed and should prompt the researcher to find a way to separate the data into two different normal distributions based on two subpopulations.

Statistical tests for normality

There are other numerical tests that one can use to test whether the results from a laboratory determination can be considered to have a gaussian distribution. These include a chi-squared goodness-of-fit test and a Kolmogorov-Smirnov (K-S) nonparametric test of the cumulative distribution.[1] Sometimes, as in our example, a transformation of the data gives a useful fit though the original scale did not. Common transformations include the logarithmic, the square root, and other power functions. A biostatistician can be of help in finding an acceptable transformation.

The chi-squared (χ^2) goodness-of-fit test is based on the following equation:

$$\chi^2 = \sum_{i=1}^{M} \frac{(O_i - E_i)^2}{E_i}$$

where O_i refers to the observed frequency for the group (bin, cell), E_i is the expected frequency in group *i*, and *M* is the total number of groups. If the observed frequencies are in exact agreement with the theoretical distribution, χ^2 will be zero. If the observed frequencies are not in satisfactory agreement with the theoretically expected frequencies, χ^2 will be large. Thus, if the calculated χ^2 is greater than the critical value of χ^2, the null hypothesis (stating that there is no difference between the observed and the theoretical distributions) is rejected. The critical value is a function of the degrees of freedom and the level of significance selected.

The chi-squared statistic may be used to test whether a sample distribution fits any type of theoretical distribution, discrete or continuous, including the gaussian distribution.

The Kolmogorov-Smirnov (K-S) test can also be used to determine whether the observed distribution of data is consistent with some particular type of distribution, such as the normal curve. To do this test, the difference, *D*, between the observed cumulative frequency and the theoretical cumulative frequency is calculated for each succeeding value in the data set. The maximum difference, D_{max}, is compared to a critical value found in a table for the K-S test. If the distribution of the observed data is in agreement with the theoretical curve, D_{max} will be small. If there are large differences between the cumulative frequencies, D_{max} will be large. The null hypothesis, stating that there is no difference between the two distributions, will be rejected for D_{max} greater than a critical value, which is a function of the number of observations and the level of significance. Table 18-7 illustrates the procedure for using this test.

It is important to keep in mind the two assumptions for the K-S test: first, the requirement for randomized samples, and, second, that the theoretical distribution be continuous. Because most laboratory data have been rounded off and hence have been made more discrete, the attempt to relate discrete noncontinuous data to a continuous theoretical model may result in normal data failing the K-S test because of the fact that the numbers have been rounded off; such data will then be considered nonnormal, or nongaussian. One should avoid the indiscriminant use of the K-S test, and we recommend that the simpler graphic approaches be used to determine the nature of a data distribution.

NUMBER OF SAMPLES NEEDED FOR REFERENCE RANGE

A particularly troubling problem to many people working in clinical laboratories is the question of how many persons must be sampled before a statistically valid reference range is obtained. If this question as stated is brought directly to a statistician, it sounds similar to the question "How high is up?" The training of the statistician requires the laboratory worker to specify (1) the size of error that a laboratory is willing to accept, (2) the frequency of various other errors, (3) a difference that is clinically important, and (4) the variability in the measurements. With these pieces of information the statistician can calculate an answer to the question of sample size. Although these types of statistical calculations give mathematically definitive answers, it is possible to provide a simpler response. The classic statistical answer is that the more observations there are, the better.

PROPER SAMPLING OF POPULATION FOR REFERENCE RANGE

When laboratories attempt to establish a reference range, there is a tendency to obtain specimens from persons who are easily available, such as laboratory personnel, nurses, and medical students. These groups are useful in the early stages of testing only. One problem is that they differ from diseased persons in many ways other than that they are healthy. For example, they tend to be younger, more likely to be more health conscious, and less likely to have other diseases.

A series of recommendations is being developed by the International Federation of Clinical Chemistry (IFCC) on the "theory of reference values." The first recommendation[2] provides a series of definitions and lists factors to be controlled when one is planning a study to estimate reference values for a procedure. The reference population should be defined in terms of age, gender, race, and possibly genetic background. Other physiological factors that must be defined within the chosen population during the study are phlebotomy specifications (posture, time of day, state of fasting), obesity, smoking, medications, alcohol, and so on.

The clinical situations in which a reference range may be used can affect the actual cut-off point (upper limit of healthy) of a reference range. The first use of a range is to differentiate healthy from nonhealthy persons and determine the specific cause of the illness. The second application of a reference range is the monitoring of patients for progression of disease or for drug therapy after an illness has been identified, in which case the comparison of a person's laboratory data against previous results is the concern. In the case of monitoring a drug level, the renal and hepatic status of the patient can affect the acceptable range of values.

The last purpose of laboratory testing may be to determine prognostic or risk factors for a person. This prognostic indicator (such as estrogen receptors for breast cancer) can determine the type of treatment chosen for a patient. In this instance, the laboratory data for the person will be compared to a reference range obtained from the population at large, rather than to some specifically defined population.

For most purposes, when the literature suggests that the data or the logarithms have a gaussian distribution, about 40 representative observations will provide sufficient data from which to get a reasonable level for normal versus abnormal, even if this does not yield the most precise values for a reference range. If no prior information exists, a larger sample size is essential.[3-5] Clearly, a large sample size specifies more precisely the boundary between normal and abnormal. A sample size of 40 is made for those who want a local reference range and have a limited number of determinations available. A standard reference range brought from outside may not be applicable to local conditions. The reference range may be based on a gaussian distribution by use of the mean plus or minus 2 standard deviations as estimates of the 2.5 and 97.5 percentiles. These percentiles cover the central 95% of the observations and are regarded as the normal range. Alternatively, one could obtain nonparametric estimates of these limits. One ranks the observations and then determines the 2.5 and 97.5 percentiles from these ranked data. Remember that the implication of either type of reference range is that some "normal" but unusual persons will be called "unhealthy" because their values are outside the reference range.

One way to test the recommendation that 40 samples can give reasonable results is to return to the example of parathyroid hormone data and randomly select 40 values from the 60 values that are presented. One can do this by tossing a fair die (one of two dice) once for each of the 60 values and using only those results for which the toss results in a 1 or 2 or 3 or 4. This will provide a random sample of about 40 values from the distribution. Then one should reconstruct the two graphs using only the 40 obser-

Table 18-2. Random samples from a normal (gaussian) distribution with mean of 100 and standard deviation of 20

First sample	Second sample	Cumulative mean	Cumulative standard deviation
111	94	102.50	12.021
101	92	99.50	8.583
144	141	113.83	23.198
76	70	103.62	27.281
109	100	103.80	24.156
85	103	102.17	22.510
87	105	101.29	21.124
71	101	99.38	21.071
66	121	98.72	22.008
93	103	98.65	20.881
98	67	97.18	20.975
109	97	97.67	20.188
108	80	97.38	19.789
98	103	97.61	19.072
61	96	96.33	19.577
114	86	96.56	19.287
118	103	97.38	19.077
91	97	97.19	18.555
95	106	97.37	18.107
98	110	97.70	17.749

vations. Finally, one can read the mean value and 2.5 and 97.5 percentiles to see whether they are very different from the results obtained using all 60 observations. Alternatively, one can use a table of random numbers to generalize the sample.

Another test of a reference range based on small numbers of results is provided in Table 18-2. In this table, 40 observations are obtained from a known statistical distribution that will have a mean value of 100 and a standard deviation of 20 if the sample size becomes very large. Observations are obtained two at a time, and then the mean value and standard deviation are calculated after each new pair of observations. By the design, we know that the mean value will be 100 and the standard deviation will be 20, if a very large sample is created. The coefficient of variation will be 20% (= 20/100) and the 95% confidence limits will be 60.8 and 139.2 when the data are numerous. This example shows how rapidly a well-behaved set of data will conform to theory and also that there is still some deviation between the theory and the results actually obtained. In this typical situation, use of a sample size of 40 was satisfactory but resulted in some error compared to the larger population. The larger the number of observations, the smaller the deviation from the underlying distribution.

In a real laboratory situation, in addition to the consistent and fixed standard deviation that can be considered, there are many other sources of variation caused by undefined factors. There are also variations caused by temperature, technician, age of reagents, and characteristics of patients that will change the standard deviation or the mean value in this calculation. A small coefficient of variation indicates that the determinations are well-controlled, whereas a larger coefficient of variation suggests the need for a larger sample size.

If the laboratory has a lot more than 40 points available from which to calculate a reference range, the next suggestion is to try to find out whether there are subgroups in the data. For example, do the men differ from the women in their average values? Do the old differ from the young? Do hypertensive persons differ from normotensive persons? When there are only a small number of observations, it is difficult to discriminate between two subgroups. On the other hand, as the data become more numerous, one can distinguish different subgroups of patients. Similarly, one can compare the determinations carried out by technician A with those done by technician B to see whether there are many differences. One can compare daytime determinations with evening determinations, and so forth.

At this time we should point out that a large number of determinations, such as 200 or more, is very likely to result in some statistically significant differences between subgroups. At this point one must evaluate the practical importance of the differences observed. For example, if men are consistently 10 points higher than women when the average is 100, it appears that separate reference ranges must be created for men and women. On the other hand, if the difference is only 1%, what does one do? If the sample size is sufficiently large, this 1% difference will eventually be found to be statistically significant. It is a real difference not caused by chance variation, even though very small. Individual variation will be much larger than this difference between the groups. Therefore this difference is not what we would call ''clinically significant,'' which is to say that there is so much variability anyway that the 1% difference between men and women, even though found consistently, is not of sufficient importance that one should create two separate ranges. Rather, in this instance one should use the more simple and efficient procedure of having a single range for all persons. Each laboratory should consult with an appropriate laboratory user to determine what difference can be considered clinically significant.

SENSITIVITY AND SPECIFICITY[6]

Suppose a laboratory has developed a new test for a serum analyte and you believe that the biochemistry is such that diabetic persons will be positive on this test and nondiabetic persons will be negative. Suppose the test is a color reaction, so that only qualitative results are available, either a positive color reaction or a negative reaction. A natural first step is to take a series of diabetic persons and test each with the serum analyte (SA).

In usual terminology, the proportion of diabetic (diseased) persons who are positive on this test is called the *sensitivity* of the laboratory test. Using the notation in Table 18-3, this would be given by the fraction $a/(a + c)$. The other group, labeled c, consists of persons who are called negative when they should in fact be positive on the test. This group consists of false negatives, and the proportion $c/(a + c)$ is the false-negative rate. Obviously one would like the sensitivity to be as close to 100% as possible.

For comparison one would take a group of healthy persons who are known not to have diabetes and test them on the SA test. The proportion of the nondiabetic persons who are found to be negative on the test, or $d/(b + d)$, is the *specificity* of the test. Then there are b false-positive reactions to the test, and $b/(b + d)$ is the false-positive rate. One would like specificity to be as close to 100% as possible.

A typical problem with laboratory tests is the sensitivity of the test at various stages in the disease. In general one wishes to have a test that is positive at the earliest possible time in a disease for which one can make a confirmed diagnosis and then begin useful treatment. Thus a test that was specific for Hodgkin's disease and was diagnostic several months before current tests would be of great value. On the other hand, a perfect blood test that would be diagnostic of all epithelial cancers 3 years before symptoms would produce nothing but frustration because one would still have to wait years to make the specific cancer diagnosis before treatment could be started.

Let us reexamine Table 18-3 to determine how the clinicians will react to the results coming out of the SA test. These results say that $a + b$ persons are positive for the disease, whereas $c + d$ persons are negative. Obviously the clinicians must do additional testing and a clinical work-up for the purpose of confirming or denying a diagnosis on the $a + b$ persons who came up as positive in the SA test. An important statistic is the percent *positive accuracy,* or positive predictive value, of the SA tests, which is $a/(a + b)$. If a is much larger than b, the percent positive prediction will be close to 100% and everyone will be happy; on the other hand, if b is larger than a, the majority of persons found to be positive on the test will in fact, after an extensive diagnostic work-up, be found to be false positives.

Suppose the SA test is perfected and is tried out on a series of 100 diabetic persons. The clinicians find that 90 of these persons react positively, though in another series of 100 other hospital patients confirmed not to be diabetic, 95 of them are negative on the SA test. Then it can be proudly proclaimed in a medical journal that the sensitivity of the test is 90% and the specificity is 95% and Table 18-4 can be published. However, someone might calculate that the positive accuracy is 90 (true positives) divided by 95 (all those positive on the test), or 95%, which would imply that the test is an excellent one for these characteristics. On the other hand, there is a basic error in this thinking. In our example, half of those persons tested were diseased, whereas in ordinary clinical practice perhaps 10% of the persons undergoing the SA test will be diseased. Table 18-5 shows more reasonable data for this type of test, though maintaining the same sensitivity and specificity, and now it is observed that the positive predictive accuracy is two thirds. One can recalculate that if the disease frequency (the prevalence of disease) is only 5%, a majority of persons who are found to be positive on the test will be false positives so that the positive predictive accuracy is less than 50% (Table 18-6).

An even more extreme situation exists if one wishes to screen unselected, asymptomatic persons with the SA test rather than selected persons seeking a diagnosis in a hospital. Suppose that one suggests that every person coming

Table 18-3. Sensitivity and specificity*

	Diseased	Not diseased	Total
Test positive	a	b	$a + b$
Test negative	c	d	$c + d$
Total	$a + c$	$b + d$	N

*Sensitivity = $a/(a + c)$; specificity = $d/(b + d)$.

Table 18-4. Sensitivity and specificity: disease frequency, 50%

	Diabetic persons	Other persons	Total
Test positive	90	5	95
Test negative	10	95	105
Total	100	100	200

Sensitivity = 90/100 = 90%; specificity = 95/100 = 95%; positive accuracy = 90/95 = 95%.

Table 18-5. Sensitivity and specificity: disease frequency, 10%

	Diabetic persons	Other persons	Total
Test positive	90	45	135
Test negative	10	855	865
Total	100	900	1000

Sensitivity = 90/100 = 90%; specificity = 855/900 = 95%; positive accuracy = 90/135 = 67%

Table 18-6. Sensitivity and specificity: disease frequency, 5%

	Diabetic persons	Other persons	Total
Test positive	90	95	185
Test negative	10	1805	1815
Total	100	1900	2000

Sensitivity = 90%; specificity = 95%; positive accuracy = 90/185 = 49%.

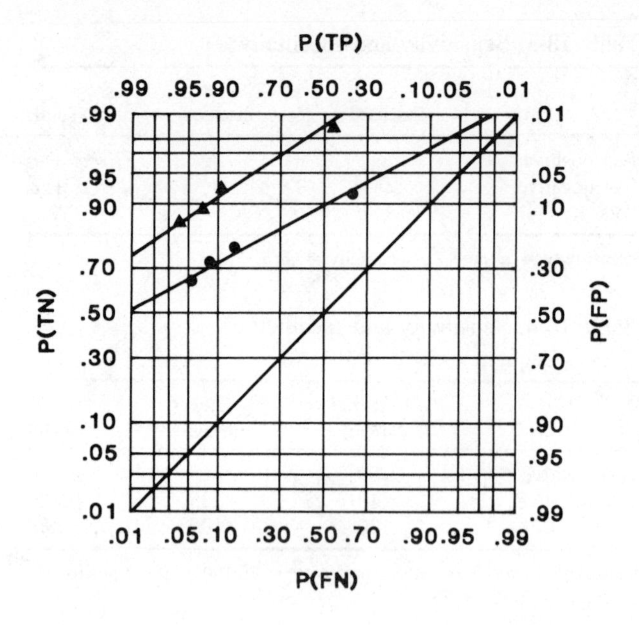

Fig. 18-6. Example of a receiver-operator curve (ROC) to determine optimal cut-off point for a test. Horizontal and vertical axes are normal probability scales. Goal is to find a decision point as close to upper-left corner as possible. ▲, ROC for more accurate test; ●, ROC for less accurate test; *45° diagonal line*, results obtained by coin tossing alone (probability of health, 50%).

into a hospital, or even worse, every person at a local supermarket, be screened with the SA test to see whether they have the disease (diabetes). In this situation let us say that only 1% of the persons tested have the disease. The positive predictive accuracy is then 15%; fully 85% of the people who turn up positive on the SA test will be found after a diagnostic work-up to be free from the disease! This example serves to illustrate that the positive predictive accuracy of a test depends on the disease prevalence of a population being tested. If the rate of false positives is unacceptable, the specificity of the test must be made greater, such as more than 98%. Alternatively the test should be applied only to populations with a high prevalence of the disease.

So far, we have been discussing only qualitative tests rather than those tests that have a numeric measurement, such as quantitative tests. With quantitative tests one has the ability to choose different decision points along the quantitative scale. Therefore one can adjust the sensitivity and specificity of a given test. The same is true to a lesser extent with ordinal tests (+ vs + + vs + + +; normal-slight-moderate-severe, and so on), in which there will be an ordered listing of positivity with three or more levels. Thus, by raising the numeric or ordinal level at which one says a person is diseased, one automatically becomes more restrictive. The result is that the test will have greater positive predictive accuracy. On the other hand, the price of greater accuracy in the subgroup of those found positive is poorer accuracy in those found negative because the marginally ill persons in the sample, who have laboratory results in that interval, will now be declared not diseased.

There is no easy choice concerning the optimal combination of sensitivity and specificity other than the general goal to have them both as near to 100% as possible. The choice to a great extent depends on the nature of the disease and the nature of the confirmatory diagnostic procedure. When the confirmatory procedures are easy to do, there can be more false-positive results. If the confirmatory procedures are more difficult to do or if they are not conclusive and will have to be repeated or if they carry an important element of danger, one would like to keep false-positive results to a minimum. Usually one chooses different levels of a quantitative test and works out the sensitivity and specificity at each of these levels.

One good way of seeing the results of various cut-off levels is to create a receiver-operator-characteristic (ROC) curve. An example of such a curve is given in Fig. 18-6. In this type of curve, one plots the proportion of false positives (or specificity in reverse) against the proportion of true positives (sensitivity). The left vertical axis uses as a scale the probability of obtaining a true-negative (TN) result. Since the probability of obtaining a false-positive (FP) result is the complement of the probability of TN, one can also show $P(\text{FP})$ at the same time on the right vertical axis. They are related by the equation

$$1 - P(\text{TN}) = P(\text{FP})$$

The lower horizontal axis is the probability of a false negative (FN), and the upper horizontal axis is the probability of a true positive (TP). They are related by the equation

$$1 - P(\text{FN}) = P(\text{TP})$$

The points (triangles and dots in the figure) for a test are obtained by examination of various levels of the cut-off value that is used to define healthy and diseased persons. Increasing the sensitivity of a test will increase the probability of finding a true positive but will also increase the probability of finding a false positive. The goal is to get as close to the upper left corner as possible. The diagonal shows the results obtained when one tosses a fair coin. Additional information can be found in Swets and Pickett.[7]

EXAMPLE CALCULATION OF KOLMOGOROV-SMIRNOV TEST

See Table 18-7.

REFERENCES

1. Daniel, W.W.: Biostatistics: a foundation for analysis in the health sciences, ed. 2, New York, 1978, John Wiley & Sons, Inc.
2. Gräsbeck, R., Siest, G., Wilding P., et al.: Provisional recommendation on the theory of reference values (1978). I. The concept of reference values, Clin. Chem. **25**:1506-1508, 1979.
3. Bezemer, P.D., Netelenbos, J.C., Mulder, C., et al.: Determining reference ('normal') limits in medicine: an application, Statistics in Medicine **2**:191-198, 1983.
4. Dybkaer, R.: The theory of reference values, VI. Presentation of observed values related to reference values, Clin. Chim. Acta **127**:441F-448F, 1983.
5. Reed, A.H., Henry, R.J., and Mason, W.B.: Influence of statistical method used on the resulting estimate of normal range, Clin. Chem. **17**:275-284, 1971.
6. Galen, R.S., and Gambino, S.R.: Beyond normality: the predictive value and efficacy of medical diagnoses, New York, 1975, John Wiley & Sons, Inc.
7. Swets, J.A., and Pickett, R.M.: Evaluation of diagnostic systems—methods from signal detection theory, New York, 1982, Academic Press, Inc.

Table 18-7. Illustration of Kolmogorov-Smirnov test of a gaussian distribution for serum sodium data

	1	2	3	4	5	6	7
Na (X_i)	**Observed frequency (O_i)**	**Cumulative total**	**Observed cumulative frequency (OCF_i)**	Z_i	**Expected cumulative frequency (ECF_i)**	D	D'
134	1	1	.0043	−2.72	.0033	.0010	−.0033
135	4	5	.0213	−2.30	.0107	.0106	−.0064
136	7	12	.0511	−1.87	.0307	.0204	−.0094
137	16	28	.1191	−1.45	.0735	.0456	−.0224
138	24	52	.2213	−1.02	.1539	.0674	−.0348
139	30	82	.3489	−0.60	.2743	.0746	−.0530
140	39	121	.5149	−0.17	.4325	.0824	−.0836
141	37	158	.6723	0.26	.6026	.0697	−.0877
142	31	189	.8043	0.68	.7517	.0526	−.0794
143	26	215	.9149	1.11	.8665	.0484	−.0622
144	12	227	.9660	1.53	.9370	.0290	−.0221
145	6	233	.9915	1.96	.9750	.0165	−.0090
146	2	235	1.000	2.38	.9913	.0087	+.0002
147	0			2.81	.9975		
	Total 235						

$\bar{X} = 140.4$ $D_{max} = .0877$ $D_{crit(.99)} = .1063$
$s = 2.352$ $D_{crit(.95)} = .0887$
$D_{crit(.90)} = .0796$

Conclusion: $D_{max} < D_{crit(.95)}$. Therefore there is no reason to reject the null hypothesis that the sodium data are normally distributed.

Kolmogorov-Smirnov (K-S) test procedure:

Step 1. Record the observed frequency for each group or cell. Use a theoretical mean and standard deviation or those calculated from the data.
Step 2. Calculate the cumulative total of samples beginning at the lowest values.
Step 3. For each cell calculate the observed cumulative frequency (OCF).
Step 4. Calculate the standard deviation units (observed value minus the overall mean divided by the standard deviation) for each cell (Z_i).
Step 5. Using a table of areas* from the gaussian distribution (normal curve), find the area for each cell corresponding to Z_i. These areas are equivalent to the cumulative frequencies for the gaussian distribution that will be used as the expected cumulative frequency (ECF).
Step 6. Calculate the difference *(D)* between the observed cumulative frequency (OCF_i) and the expected cumulative frequency (ECF_i) for each of the cells. In addition find the difference *(D')* between the *i*-1st observation (OCF_{i-1}) and the *i*-th expectation (ECF_i).
Step 7. The maximum (in absolute value) of the differences (D_{max}) calculated in step 6 is compared to a table of critical values (D_{crit}) for this K-S test based on *N* observations and the desired level of statistical significance.*

*See Daniel, W.W.: Biostatistics: a foundation for analysis in the health sciences, ed. 2, New York, 1978, John Wiley & Sons, Inc.

Chapter 19 Quality control

Bradley E. Copeland

average difference test A statistical procedure that indicates whether two averages are really different or whether the observed difference is a matter of chance or inherent variability. Also called the *t* test. According to Youden (personal communication), *t* was selected as a nonspecific algebraic term with no acronymic meaning.

control limits Numerical limits within which a control sample must fall to be part of the normal distribution of values, usually target average ± 2 USD (usual standard deviation).

external quality control A program in which an external agency provides unknown samples for analysis. The results are submitted for an independent evaluation and are returned to the participant with an evaluation of "acceptable" or "not acceptable" performance.

inherent variability The characteristic of repeated measurements on the same material to vary by chance around an average.

internal quality control A program that (1) verifies the validity of laboratory observations and (2) is planned and carried out as part of the daily regular routine within the laboratory.

monthly average The average value of the daily quality control values for a 1-month period.

monthly standard deviation The standard deviation calculated with the daily quality control values for a single month.

out of control The condition in which a quality control sample is outside the numerical control limits. An out-of-control quality control sample must be retested and the validity of the analysis verified.

quality control pool serum A quantity of liquid serum that is preserved in small portions, such as 5 to 10 mL, by freezing or lyophilization and whose constituents remain stable for at least 1 year.

regional quality control program A group of 10 to 500 laboratories that jointly purchase a large amount of serum control material so that comparative results can be established.

shift A change in an analytical system that happens abruptly and continues at the new level.

significant difference Statistically, a difference that is shown to be outside the expected variability limits; medically, a difference that is large enough to influence a medical decision; operationally, a statistically significant difference that is not medically significant but one that analysts and supervisors believe to be large enough to require investigation.

standard deviation (SD) An important descriptor of a varying population of measurement data. The average is the central point. Around the average; within ±1 SD 67% of the total population fall; within ±2 SD 95% of the total population fall; and within ±3 SD 99.7% (for all practical purposes the total population). This single concept is the cornerstone of statistics.

usual standard deviation (USD) The average of 6-monthly standard deviation values. This represents the usual precision capability of an instrument or method.

survey specimen A preserved sample similar to a human specimen that is prepared by an independent agency and submitted as an unknown to a group of participating laboratories.

systematic bias *Constant*—a constant difference between the true value and the observed value, regardless of the concentration level. *Proportional*—a difference between the true value and the observed value, which changes as the concentration level changes; that is, if the concentration level doubles, the proportional bias doubles. For example, bias at 50 mg = 2 mg; at 100 mg = 4 mg; and at 200 mg = 8 mg.

target average *Temporary*—the average of the 40 initial values collected on a new serum pool. Duplicate values from two vials per day for 10 days. *Final*—the average of the temporary target average, the first month's average, and the second month's average.

trend A gradual change in one direction of the results of repeated analyses of a sample.

PURPOSE

The first comprehensive daily quality control program was developed by two medical technologists, Freier and Rausch[3] in 1958 in the Clinical Pathology Laboratory of the University of Minnesota, where they were colleagues of Dr. Gerald Evans and Dr. Ellis Benson, two of the foremost pioneers in the field of clinical chemistry.

The purpose of quality control of analytical testing is to ensure the reliability of each measurement performed on a patient sample.

There are two requirements for all quality control programs: (1) the programs should lead to decisions regarding the reliability of the analytical data, and (2) the quality control decisions should be related to the medical purposes for which the analyses are being done.[1, 2]

The final judgment regarding the effectiveness of a quality control system is made on the basis of the question, Have consistently precise and accurate biochemical data been provided for both short- and long-term medical decision making?

Quality control actions should end with decisions regarding not only the analytical significance but also the medical significance of the quality control data. Decisions should be recorded, dated, and signed. The practical decision-making system presented in this chapter has been developed with ease of operation in mind. Additional quality control tools are also presented. Because control samples and patient samples are analyzed by the same system, data from both reflect changes that occur in the system itself. Thus with two sources of data—control samples and patient samples—and several ways of analyzing each set of data, it is possible for the analyst to select an operationally effective combination of quality control tools. There is no one quality control system that is best. Each laboratory should determine and use the quality control tools that best meet its needs.

QUALITY CONTROL POOL

Types of material available

The type of quality control material used in the laboratory is based on the laboratory's needs. The majority of decision making, that is, the daily bench-level work of medical technologists, involves the question of whether a particular set of patient analyses is valid. Because the quality control material is analyzed along with patient specimens, large amounts (liters) of control material are needed each year. There are currently several ways in which a laboratory can obtain sufficient quantities of quality control material. These are (1) frozen, pooled, patient specimens (serum and urine), (2) commercial lyophilized pool material (serum or urine), and (3) commercial stabilized low-temperature liquid serum pools.

Most often serum or urine pools are used for quality control material, but one can use plasma for specialized purposes. Serum is more frequently used than plasma because it is more readily available and is less likely to have precipitated material.

The qualities of cost, clarity, stability, validation, and lyophilization error are compared in Table 19-1.

The following statements relate to all quality control serum pools. First, all pooled serum material should be free of hepatitis B. The Food and Drug Administration (FDA) requires commercial products to be free of hepatitis B. Noncommercial frozen pools should not be used if there is HBsAg present. Second, all control material requires refrigerator or freezer space for storage of a 1- to 2-year supply. The annual or biennial purchase requires careful planning for storage area. Often commercial distributors will supply quantities on a quarterly basis of a purchased 1- or 2-year supply. One would prefer to change control lots only once a year, which is the present practice in most laboratories.

The approach used for urine quality control pools are similar to those for serum. That is, either pooled, frozen, patient specimens or commercially available lyophilized material is most commonly used. For convenience, most laboratories use the commercial lyophilized pools.

The first regional quality control program was developed by Preston in 1970 for all the hospital laboratories of Colorado. After that breakthrough, more than 25 professional groups and manufacturers now offer participation in large regional quality control pool programs in which 100 to 500 laboratories use the same batch of pooled serum. There is both a cost advantage and a scientific advantage when one compares results as a participant in one of these programs. Names of the professional regional quality control pool programs can be obtained from the College of American Pathologists (CAP), North Skokie, Illinois. Manufacturers' representatives should be contacted for information about commercial quality control regional programs. The regional programs are open to all and cover the entire United States.

Table 19-1. Comparison of quality control materials

Criteria	Frozen	Lyophilized	Low-temperature liquid
Cost	Low, if not manipulated* Medium, if manipulated	High	High
Clarity	Clear, if carefully collected	Turbid	Clear
Stability	12 months	18 to 24 months	18 to 24 months
Validation	Compare with accurately measured materials (NBS and CAP†)	Regional and manufacturer's peer group analysis available, or by NBS and CAP	Regional and manufacturer's peer group analysis available
Lyophilization error	Absent	Present	Absent

*That is, if additional analyte is added.
†*NBS,* National Bureau of Standards; *CAP,* College of American Pathologists.

Target average value of quality control pools (TA)

The "target values" of the quality control pool are the estimated concentrations of each analyte within the pool. Each laboratory must establish a target value for each analyte by the regular procedures performed by that laboratory.

When establishing target values for a new lot of quality control material, it is assumed that the analytical systems of the laboratory are performing optimally during the data collection. This is normally assured by analysis of the new lot of quality control material in parallel with the current quality control program in operation. If the analytical data from the current quality control material indicates satisfactory control performance of the methods, one assumes that the data for the new lot is valid. When setting up a quality control system for the first time, one accepts the current methodology as valid. Future methodology decisions will be based on experience with medical usefulness, significant change limits, external quality control comparisons, and accuracy materials.

A simple outline for establishing target values for quality control pools is listed below.

1. Procure a 1-year supply of quality control test material.
2. It is preferable to plan a 6-week lead time to allow for (a) analyses to be run (2 weeks); (b) data to be calculated and evaluated (1 week); (c) methodology decisions to be made (1 week); (d) and 2 weeks for safety because not all planning is perfect. It is also advisable to retain 20 or 30 vials of each expiring pool for use during the subsequent year as reference material for problem solving.
3. Always reconstitute the lyophilized material carefully. There are idiosyncrasies of solubilization involved here. Follow the label directions. Mixing too quickly or vigorously may interfere with the solubilization of the lyophilized material. This is especially true for enzymes and other proteins that can be denatured and inactivated by such procedures. Date and initial each sample vial on reconstitution. If a frozen liquid pool is used after thawing, mix six times by inversion because the protein and other compounds become concentrated at the bottom of the tube during freezing.
4. For each constituent, analyze duplicate samples from each of two separate vials for 10 days (20 vials, 40 measurements). An alternative procedure is to reconstitute one vial and analyze in duplicate on each of 20 consecutive days.
5. Calculate average and standard deviation of the 40 analytic values $n = 40$ (Table 19-2).
6. Use the average $\pm$ 2 usual standard deviations as the acceptable control limit for the new lot of quality control material.
7. When a new quality control pool is introduced, the best estimate of the average value (target average) is the preliminary set of 40 values from 20 vials, which are made as described above. This initial average is the temporary target average (TTA). A final target average can be established at the end of the second month. At that time, there are three average values (that is, a temporary target value and two monthly averages) that one can average to calculate the final target average.

Usual standard deviation (USD)

Every method has a characteristic inherent variability. The *usual standard deviation* (USD) is calculated when one averages a series of 6-monthly standard deviations. This is a valid estimate of the usual day-to-day variability of individual measurements. The usual standard deviation is used to establish the daily control limits around the *target average*. The usual standard deviation is an important performance characteristic of every method.

The usual standard deviation has other uses: (1) to establish the statistical significance of the difference between the target average and the monthly average and (2) to establish the statistical significance of the difference between the monthly standard deviation and the usual standard deviation (see p. 322).

The decisions using the usual standard deviation as a starting point are based on the usual procedures of parametric statistics. To facilitate prompt decision making,

Table 19-2. Example of calculation of temporary target average (TTA), target average (TA), and usual standard deviation (USD) from observed data*

Temporary target average (TTA)

Duplicate samples, two vials per day, 10 days; n = 40 measurements of 20 vials per 10 days. The average includes the variability effect of sampling, different vials, and different days. For potassium, mEq/L.

Day	Vial 1		Vial 2	
	Sample A	Sample B	Sample A	Sample B
1	6.1	6.1	6.2	5.9
2	6.2	6.2	6.0	6.0
3	5.7	5.8	6.0	6.0
4	5.9	5.8	5.9	5.8
5	6.0	6.0	6.0	6.0
6	5.9	6.0	6.0	6.0
7	5.9	6.0	6.0	6.0
8	5.9	5.8	6.0	5.9
9	6.0	6.1	6.1	6.2
10	6.0	6.1	6.1	6.1

Total, all observations = 245.7 mEq/L
Number of observations = 40.0

$$\text{Temporary target average} = \frac{\text{Total}}{n} = \frac{245.7}{40} = 6.14 \text{ mEq/L}$$

Final target average

Temporary target average, 6.14 mEq/L
First month, 6.07 mEq/L
Second month, 6.12 mEq/L
Total of averages, 18.33 mEq/L
Number of averages, 3

$$\text{Average of averages} = \frac{\text{Total of avg.}}{\text{Number of avg.}} = \frac{18.33}{3} = 6.11 \text{ mEq/L}$$

This is the *target average,* which replaces the temporary target average in the decision-making process.

Usual standard deviation (USD)

Month	Monthly standard deviation (mEq/L)
April	0.13
May	0.11
June	0.13
July	0.10
August	0.15
September	0.11

Total of monthly SD's, 0.73 mEq/L
Number of monthly SD's, 6

$$\text{Average of monthly SD's,} \frac{\text{Total}}{\text{Number}} = \frac{0.73}{6} = 0.12 \text{ mEq/L}$$

This is the usual standard deviation (USD).

*Laboratory Service, Chemistry Section, Veterans Administration Medical Center, Cincinnati, Ohio.

one rounds off the numerical data. The usual framework is the 95% confidence limit with a 2 SD distribution. See Chapter 17 for more details.

Control limits for each level of control pool

The target average plus and minus two times the usual standard deviation is the control limit for each pool sample.

Introduction of quality control sample into analyte-measuring system

Daily preparation and analysis of quality control samples is a regular responsibility of the analyst. The quality control pools are thus analyzed as "known" controls by the medical technologists during analysis of patient samples. The frequency of analysis of the quality control material is established by each laboratory for each method. For example, for some continuous-flow multichannel instruments, controls usually are run every 20 samples.

Most laboratories use two different pools, one "normal" and one "abnormal." A "normal" pool contains constituents at concentrations within the nondiseased reference range, whereas an "abnormal" pool contains the analytes at concentrations outside the reference range. For some tests, laboratories may employ three pools, low, normal, and high, when medically significant decisions are made at each level. The following discussion assumes that only two levels are used.

The medical technologist must use the data from each quality control analysis to make a decision as to the validity of the patient analysis data. This decision should be recorded permanently with date and name of analyst, for example "in control" or "out of control," plus the initials or signature of the analyst.

Calibration and calibration materials

Controls may not be used as calibrators. Controls and calibrators must be different for each has a separate and important function.

Each method must have a calibration system that is independent of the control system. The calibrator has an assigned value that is established by the manufacturer or the user by a reference method. The calibrator is used to standardize the method or instrument. Calibrator materials come in a variety of forms such as aqueous and serum matrices. Once a material is used to calibrate, it may not be used as a control, and vice versa.

The differences between an aqueous and serum matrix can include the surface tension, which can affect sample pipetting, the interactions between analytes and proteins, and the effect of the volume fraction occupied by protein on the actual concentration of certain analytes (such as sodium). When creating spiked pools for accuracy determination, care should be taken to obtain pooled patient serum that contains no known possible interferent, such as lipemia, bilirubinemia, or hemolysis. Furthermore, it is best to spike with only one analyte at a time to minimize possible interference by other analytes.

Calibrators are usually purchased in lots large enough to last 12 to 18 months. It is recommended that a new lot of calibrator material be tested 6 weeks before use to detect systematic bias between "present calibrator" and "new calibrator." Although the FDA requires manufacturers to use reference methods to assign calibrator values, there are frequently significant differences between calibrator lots.

A practical system for new calibrator verification is as follows:

1. Ten-day verification period.
2. Each day, insert 2 aliquots from one vial of new calibrator as an unknown in the regular daily run (n = 10 vials; 20 values). Calculate the average for each analyte. Compare each average with the value assigned by the manufacturer. This will predict the average change in the quality control pool average value anticipated when the new calibrator is introduced. A change in the quality control pool average greater than 1.0 usual standard deviation is statistically significant, and a decision must be made as to which calibrator is truly accurate.
3. Each day during this 10-day period, reset the instruments using manufacturer's assigned values for new calibrator.
4. Run 2 aliquots of each quality control pool and four patient samples (patients n = 40).
5. Calculate the average and standard deviations for each quality control pool level measured with the new calibrator (n = 20 values).
6. Compare these averages with current target average for the pool. This step will predict the new average values for the quality control pool.
7. Apply the average difference test (t) to the patient sample data using the parallel value from the current calibrator and the values from the new calibrator to calculate a series of 40 differences and the average difference (n = 40).
8. Calculate the average of differences between values from new calibrator and corresponding values from calibrator in use. Calculate the standard deviation of the differences and then the standard deviation of the average difference ($SD_{avg\ diff}$ = SD_{diff} divided by the square root of n).
9. If the average difference is greater than two times the standard deviation of the average difference, there is a statistically significant difference present between the patient values measured in relation to the two calibrators.
10. Some significant differences may be large enough to be of medical importance, and some may be of operational significance. There are many statistically significant differences that are neither medically nor operationally important (that is, no action is required).
11. Before institution of a new calibrator, a decision should be made as to whether the manufacturer's value for any analyte should be changed. Changes and data supporting these changes should be recorded, dated, and signed by supervisors.

Keep in mind that the overall goal is to maintain a year-in, year-out consistent level of analytical accuracy. The above verification procedure will identify possible systematic bias caused by analytical errors in the manufacturer's assay procedure.

DAILY DECISIONS

The medical technologist generates a set of quality control data along with each set of patient analyses.

Daily control problems are particularly likely to occur when new batch lots of solution are introduced and immediately after regular maintenance has been done.

It is preferable to check out new batch lots of commercial reagents as soon as they are received rather than just before they are used. It is also good practice to schedule maintenance so that one can establish a test set of controls and run a few patient samples from a previous batch before the next regular daily run is processed. Often maintenance leads to an out-of-control situation. A record of all solution changes, all instrument repairs, and all maintenance procedures must be kept to help in subsequent troubleshooting or in supervisory planning of maintenance. This documentation is required by the Joint Committee on Accreditation of Hospitals (JCAH) and the College of American Pathologists (CAP) accreditation programs.

It should be emphasized that the frequency of analysis of calibrators and quality control samples is highly dependent on the specific instrument being used. Newer instruments claim calibration stabilities of weeks, whereas others have considerably less stability. Thus the frequency of control testing must be tailored to an individual instrument as well as to the experience and need of an individual laboratory.

Most quality control decisions are made on a daily basis. Many instrument runs are made in which the controls are tested first. Therefore there is an immediate identification of problems, and no patient samples are quantified until the instrument or system is running in control. For other procedures that are run in a batch process, the controls and unknown sample values are only available at the end of the analytical run. Daily bench-level quality control testing can be used to detect only systematic errors and a decrease in precision and not random errors, which occur unpredictably. These are observed when significantly abnormal results are repeated.

Batch analysis decisions using two controls

1:2 SD rule

1. When both controls are within ±2 SD. Technologist decision: approve batch and release analyte results.
2. When both controls are outside ±2 SD. Technologist decision: quarantine all results and report to supervisor.

Technologist check of systems: (1) check for calculation error, (2) repeat standards and controls, (3) prepare and analyze a new dilution of control material, (4) check reagent solutions, (5) check standard solutions, (6) check instrument. When trouble is identified, repeat all samples.

3. When one control is outside ±2 SD and the second control is within ±2 SD. Technologist decision: quarantine all results.

If no obvious source of error is identified by the above-mentioned check of systems, release values near the in-control value.

The acceptable control range is based on 95% confidence limits (±2 SD). Thus five out of every 100 times a pool is analyzed, results exceeding the 95% limits will be obtained *even though the instrument is functioning optimally*. To ensure that one out-of-control quality control value is not the result of chance, repeat the analysis of both pools. This approach is most feasible for automated or semiautomated analysis. If the repeated controls are within acceptable limits, continue the analysis by releasing all patient results.

If five consecutive quality control values are outside the 1 SD limit and inside the 2 SD limit on either side of the plus or minus side of the average, study the system for a correctable bias. Results can continue to be reported, but prompt review within 2 or 3 days is recommended.

Batch analysis decisions using multirule Shewhart plan[4]

The multirule Shewhart procedure recommended by Westgard[5,6] is as follows: *Decision:* An analytic run is out of control when (1) one control observation exceeds control limits set at ±3 SD from the average (a 1:3 SD rule; that is, one control value outside the 3 SD limit establishes the out-of-control situation); (2) two consecutive control observations exceed control limits set at ±2 SD from the average (a 2:2 SD rule); (3) one control observation exceeds the +2 SD limit and a second control observation exceeds the −2 SD limits (called the range:4 SD rule); (4) four consecutive control observations exceed the mean +1 SD or the mean −1 SD (4:1 SD).

Comparison of the 1:2 SD rule and the multirule Shewhart plan. These rules are similar logical extensions of the 1:2 SD rule, which is recommended here.

Rule 1 (Shewhart). Accept a control value outside 2 SD and inside 3 SD; 1:2 SD rule does not accept values outside 2 SD.

Rule 2 (Shewhart). Accept one control value outside 2 SD, but stop the acceptance when there is a consecutive value outside 2 SD. 1:2 SD stops the acceptance with one value outside 2 SD and inside 3 SD.

Rule 3. The Shewhart range: 4 SD would be handled identically by the 1:2 SD rule because both values are outside 2 SD. This rule emphasizes that wide shifts are occurring and that the procedure should be shut down for revalidation.

Rule 4. The 4:1 SD rule detects shifts appropriately and would be part of a daily or weekly 1:2 SD review, depending on how many batches were processed in a day. This Shewhart rule is a big improvement over the "four consecutive values above or below the average rule" that has been suggested by some as an operational out-of-control guideline. Although statistically valid, this suggested rule leads to an excessive number of "no action" decisions.

In summary, the four Shewhart rules recommended by Westgard et al.[5,6] and the one sample:2 SD rule recommended here as a basic control decision plan illustrate different but comparable decision pathways. The work of Westgard et al.[5-7] using simulated data to predict (1) the probability of rejecting a valid control value and (2) the probability of detecting a statistically significant change in the average represents an interesting exploration of the advantages and disadvantages of multirule control systems.

Each individual laboratory is responsible for defining its system for quality control decisions and is also responsible for making this system a written part of its laboratory manual.

Continuous analysis decisions

A control specimen may be introduced at intervals according to the tendency of the continuous-analysis instrument to drift or according to other factors deemed important. A set of two controls for every 10 samples is rigorous. A set of two controls for every 20 samples is generally acceptable. A calibrating material is inserted after each tenth (rigorous) or at least each twentieth sample. This allows prompt correction of specific operational drift occurrences.

All samples analyzed between two sets of "in control" quality control samples, often termed a "run," may be reported to the patient's physician.

Samples that have been analyzed between two sets (normal and abnormal equal one set) of controls, one control value of which is outside 2 SD and within 3 SD, may be reported to the patient's physician. If one control value is outside 3 SD, the samples should be repeated after one follows the system check noted above.

Stat analysis

For certain procedures, each emergency test performed must have a complete set of controls analyzed. For discontinuous stat testing, the frequency of control analysis is not dependent on the number of samples analyzed, but the time between analytic runs. Depending on the stability of the instrument and the nature of the test, it is a reasonable decision to analyze a control sample every 20 to 60 minutes. Other instruments, such as the recently developed blood-gas instruments, may be automatically recalibrated every hour. Controls must be measured at least at the beginning of every 8-hour shift.

Response to out-of-control decisions

Once a decision is made to reject a set of analyses, an effort must be made to determine the cause of the improper analysis. Most laboratories have alternative or back-up methods of analysis. Therefore a decision must be made as to how much time should pass in the troubleshooting mode before one uses the alternative systems. This decision will vary with the type of analyte, the work flow, and the medical situation requiring the result (stat versus routine). The extreme examples would be a stat potassium analysis versus a 72-hour fecal fat determination. In the case of potassium, a delay of more than 30 minutes in providing a back-up stat analysis may affect medical decision making, whereas several days' delay in the fecal fat analysis would not be crucial to patient care. These decisions should be made in consultation with the supervisory staff.

Thus there is an immediate decision pattern for daily controls that is followed by the individual analyst. When controls are outside the quality control limits, indicating unacceptable performance, one must activate a second level of decision making, which often includes a laboratory supervisor. Again, whenever possible, a structured pattern of immediate response should be planned ahead of the actual problem situation.

Every quality control decision should be recorded with the data and the signature of the person who made the decision. This is a good practice for several reasons: (1) it gives credit for an important function, namely, decision making; (2) it provides easy follow-up for supervisors; and (3) it meets CAP, JCAH, and Medicare requirements for documented control records. The record should include acceptable quality control limits and a place in which responses to out-of-range quality control values are noted. These response-to-out-of-control notes should include date, analyte, response to correct the problem, and initials of medical technologist[8] (Fig. 19-1). It may be convenient to prepare check-off graphs for each of the multiple constituent levels in an instrument system that provides multiple control values during a single day. This record is useful in predicting the need for

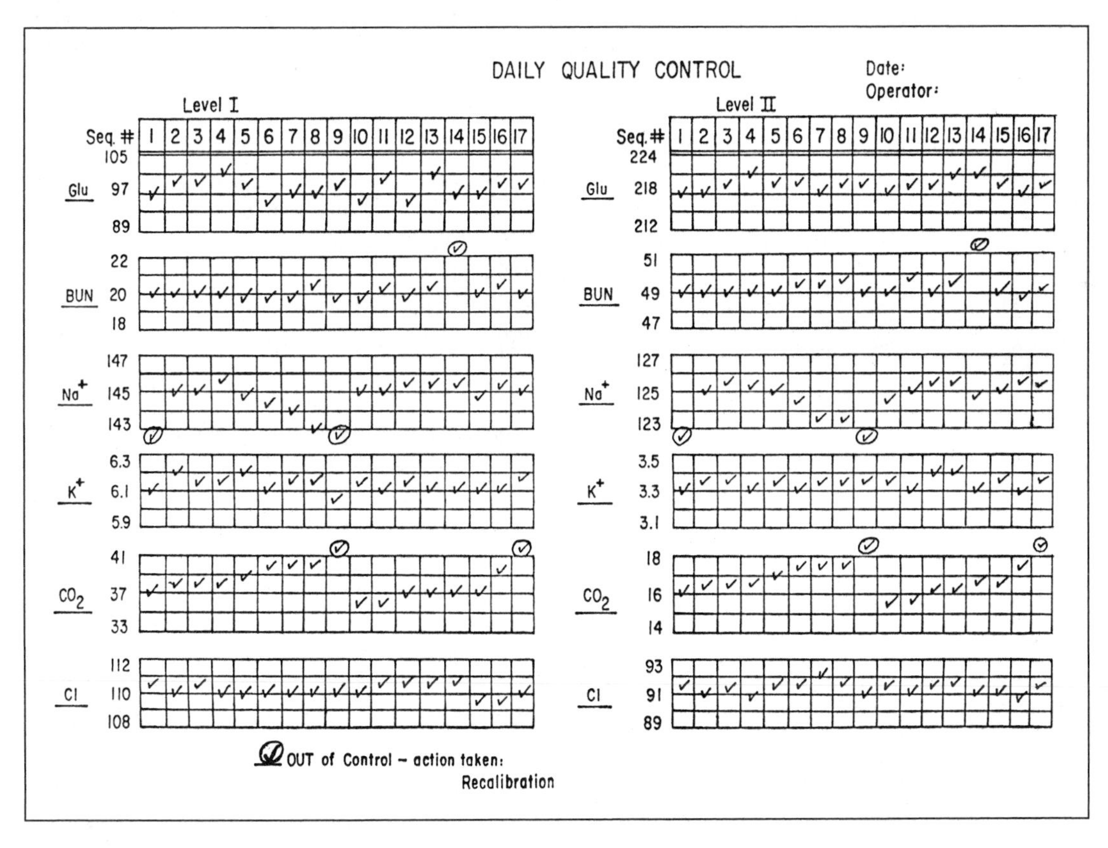

Fig. 19-1. Multiple analyte daily quality control check-off record. As each control is reported, it is quickly logged in on data sheet. Notes of out-of-control values and action taken are included. A daily value for quality control calculation is selected by use of a random number table basis. (Form developed by Rosvoll, R.V.: In Copeland, B.E., Rosvol, R.V., and Casella, J.M.: Quality control workshop manual, Chicago, 1978, American Society of Clinical Pathologists Commission on Continuing Education.)

maintenance, repair of worn parts, or replacement of deteriorating solutions.

Use of patient data in daily decision making

The results of most patient sample analyses usually fall within the reference ("normal") ranges established for each compound. Thus when one is analyzing a set of patient samples either by continuous or batch operation, the results fall into a familiar pattern, that is, most within the nondiseased-patient reference range and a few elevated and a few low values. Deviations from this pattern of results should alert the medical technologist that a shift in the system may be occurring and that the patient analysis may be invalid. It is unlikely to have a consecutive series of two or three patient values that are greatly abnormal. For example, three consecutive patients with potassium values above 6 mEq/L should trigger a response to recheck these analyses, even though the controls are within the limits.

An example of this situation is shown in Table 19-3. The data in patient set A show a typical set of results along with acceptable quality control data. In patient set B, however, there is an increased number of abnormal patient results from specimen 4 on, though the quality control data are within acceptable limits. This pattern suggests that the patient samples 4 through 8 are not valid. If sequence B listed in the table was from a continuous operation, one can assume a shift in the calibration at approximately patient sample 4. If the sequence was from a batch operation, one might assume the introduction of some systematic error into the analytical system, such as a pipettor malfunction. In either situation, the instrument should be examined by the medical technologist and recalibrated, followed by reanalysis of the patient samples, until the results are consistently repeatable.

Table 19-3. Use of patient data in daily quality control

Sample number	Patient set A	Patient set B
Control I	4.4—in control	4.4—in control
Control II	6.9—in control	6.8—in control
1	3.8	4.1
2	4.6	3.9
3	5.0	5.7
4	4.3	6.1
5	4.2	6.5
6	3.6	5.8
7	4.7	6.4
8	4.0	6.2
9	4.6	5.1
10	3.9	4.7
Control I	4.3—in control	4.6—in control (at 2 SD)
Control II	6.8—in control	7.0—in control (at 2 SD)

Would you wait until the quality control samples after the tenth sample to make a judgment about the system? No. After about the third or fourth patient sample with an extremely high or low value, quality control could be moved ahead and trouble detection should begin. Keep in mind that occasionally by chance a series of specimens from very ill patients may fall in consecutive order. Repeated testing will usually solve this problem.

Daily communication with physicians

When a physician doubts the accuracy of a particular test, this is an emergency situation for the laboratory, the physician, and the patient.

When a physician states that he does not believe a particular analytical value, he is exercising an important aspect of his responsibility to the patient, namely, the correlation of all data input into a diagnosis of the patient's condition and plan of treatment. The physician's decision to question the validity of a given measurement is based on his knowledge of the patient's past history, present condition, current therapy, and expected progress. Each laboratory must be ready to confirm any measurement or observation that it has made during the previous 1 to 2 weeks. Each physician must be responsible for immediately alerting the laboratory that a particular value does not fit the clinical picture in his or her opinion. At this point, neither the physician nor the laboratory knows whether the values in question are correct or incorrect. A quick confirmation response promotes good patient care and builds confidence in the laboratory.

In the laboratory, the most common sources of such problems are (1) a mix-up in data recording; (2) a confusion in sample numbering; (3) an error in sample labeling; or, worst of all, (4) the collection of a sample from the wrong patient. On the other hand, it is my experience that half the questioned reports will be confirmed as valid and will represent specific but unexpected changes in a patient's condition.

WEEKLY REVIEW OF DAY-TO-DAY VALUES

This portion of the quality control program assumes that the methods of analyses are basically accurate and precise and that each day's quality control data fell within the criteria described previously. However, *short*-term changes in a method can occur, affecting both accuracy and precision. Such variables as new lots of reagent or standards, poor or improper preventive maintenance, deterioration of instrument, reagents, or standards, and so on must be considered as sources of problems. The review of past daily quality control data, either on a weekly or monthly basis, is designed to monitor short-term changes in the analytical systems.

Selection of daily quality control value for statistical analysis

How does one select the daily control value for inclusion in long-range data analysis? First, if a single batch is run, there will be only a single value for evaluation. Second, when multiple batches are run or a continuous system is used, there will be multiple quality control analyses. Which value should be selected for the day-to-day analysis? There are two acceptable procedures. A value should be chosen at random from the daily series. One can prepare a random-number series for each month using a random-number table. The standard deviation of these daily values will approximate the inherent day-to-day variability of a single patient analysis. Another acceptable practice is to choose the first quality control analysis of each daily run.

Why select one value from each day? If all the daily results for each day are averaged and the daily average is used to calculate a monthly average and standard deviation, this standard deviation will be the inherent variability of the daily average rather than the inherent variability of a single quality control analysis, which is also the variability of a single patient analysis. Because each patient sample is analyzed only once, the variability of the average of n measurements on the same day is not appropriate as the descriptor of the inherent variability of the single-patient measurement. Hence the preference for the use of a single analysis from each day. This is the estimate of the variability of single measurements on a day-to-day basis. The daily decisions based on analysis of single quality control samples and the daily decisions that the physician makes on the single analysis of the patient sample are related to the variability of the individual measurement on a day-to-day frame of reference.

Alternatively, many laboratories average the n values obtained for each quality control pool for each analyte for each day and plot that average on a quality control chart (such as a Levy-Jennings graph). The standard deviation of the analysis obtained in this manner will be less than the standard deviation when one uses only a single, random value. Using this method, narrower control limits are obtained and more statistically significant deviations from the average are observed; however, frequently these are not operationally or medically significant.

Levy-Jennings plots[9]

The data obtained from daily analysis of quality control pools can be plotted to give a visual presentation of the data. The most common visual analysis is the Levy-Jennings plot. The established target average ±2 USD is drawn on the y axis, and the days of the month are indicated on the x axis (Fig. 19-2). By using large pieces of paper one can obtain extended graphs covering 6 months. Thus cumulative information showing patterns of quality control results can be obtained.

Trends or shifts from the average target value are known as "biases." Biases can be either positive or negative. Levy-Jennings plots should be observed on a routine basis by supervisory personnel looking for trends or shifts in the data. Normally one notices a trend or a shift within 6 to 7 days after it begins.

An example of a Levy-Jennings plot is shown in Fig. 19-2. If the data slowly deviate up or down from the target value, this is called a *trend* (days 6 to 10, Fig. 19-2). If there is a sudden jump of data points from one average to a new average, this is called a *shift* (days 13 to 20, Fig. 19-2). One can also note changes in the precision of the procedure by observing the scatter or distribution of data points. Thus days 23 to 31 show an obvious decrease in precision, that is, greater dispersion of data points, than was obtained in days 1 to 6. Changes in accuracy and precision can be more formally demonstrated using the procedures described below.

When a systematic bias (that is, accuracy change) or a change of precision is noticed, the supervisory personnel should decide what appropriate action should be taken. This usually includes continuing the analysis while checking reagent, standard, and instrumentation. If the bias becomes severe, so that data points repeatedly exceed the 2 SD limit, a decision is required to discontinue performing the analysis while seeking the source of the error. In multichannel instruments this may mean shutting down the defective channel and running analyses by a backup method.

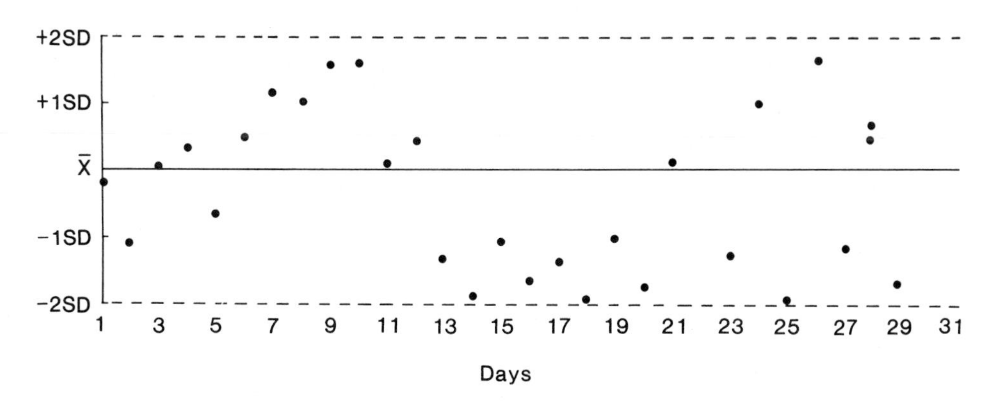

Fig. 19-2. Levy-Jennings plot for recording daily quality control values. Established target average, $\overline{X}$, is given as solid horizontal line. Value of +2 SD and −2 SD are also represented as dashed horizontal lines. Values for each day are plotted sequentially along x axis.

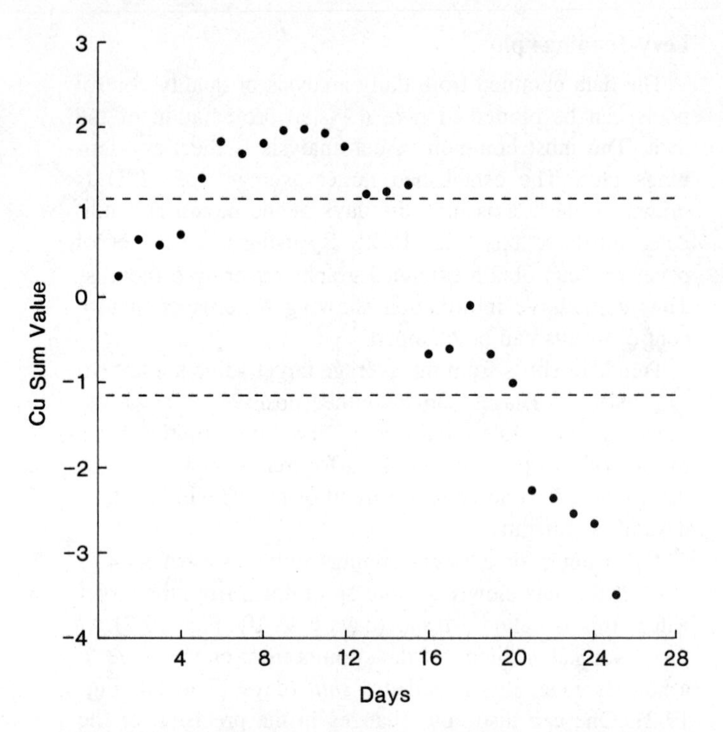

Fig. 19-3. A plot of cumulative sum (Cu Sum) of differences between observed and target value (*y* axis) on successive days (*x* axis). In this circumstance, a difference of 1.1 concentration units was 3 SD. Trends and shifts are readily observed.

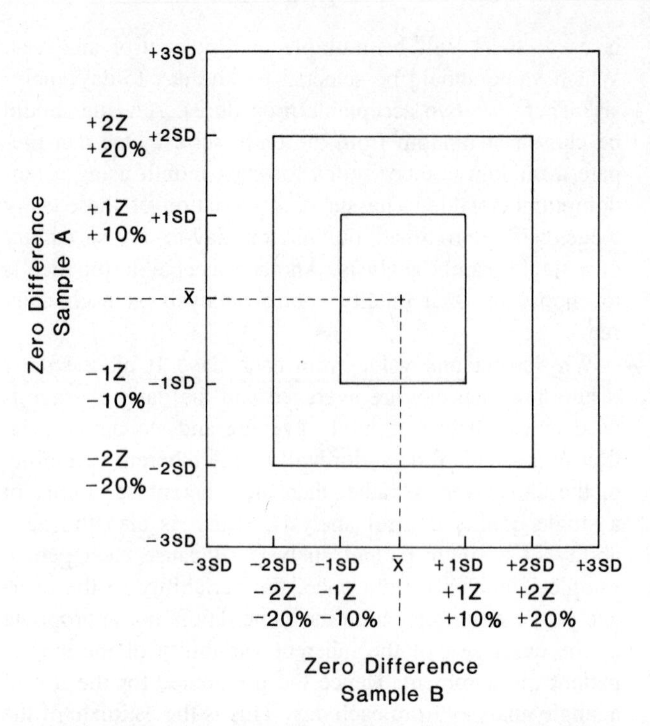

Fig. 19-4. Youden plot of external quality control for long-term record of performance. Plan of action:
1. Each constituent will have a pair of values (samples *A* and *B*) from each survey; absolute values will vary, but point of reference is true value established by external evaluator.
2. Convert each of the values of the set to one of following: descriptions of value's deviation from target value: percentage, standard deviation interval (SDI), or Z score, according to your preference.
3. Plot successive control coincidence points. If any points are outside ±2 SDI or ±2 Z or ±20%, immediately look for sources of systematic constant bias.
4. After four points are established, decision-making starts.
5. If the points cluster within ±2 SDI or ±2 Z or ±20% range, there is no need to study for sources of systematic bias. Points outside ±2 SDI, ±2 Z or ±20% should be investigated.
6. Usually with a stable internal quality control system, external quality control results plotted on Youden plot as described (a) will be within 2 SD, 2 Z, or 20% square and (b) will gather usually in one area.
7. A group of four results in same vicinity within ±2 SD square is good performance. Avoid fallacy of trying to achieve central point. Consistency is preferable to moving one's average values if they are within 2 SD. Any effort to move within 1 SD box is futile and should be discouraged, since this depends upon chance.
Other methods for expressing this data are (1) a tabular chart, a graph that plots values above and below a central zero line, or (2) a 45-degree identity graph with 45-degree lines as zero difference flanked by ±2 SD, 2 Z, or ±20% lines in Fig. 19-12.

Cu Sum plots

Cu Sum analysis operates on the basis that there should be a random scatter of quality control points around the average value; the cumulative sum of positive and negative differences of individual control values from the mean value will approach zero when the system is in control. A Cu Sum or cumulative sum of differences from the mean in excess of 3 USD indicates an out-of-control situation that requires attention. It can be seen that a shift or a trend will quickly produce an out-of-control Cu Sum (Fig. 19-3). The Cu Sum also relates to certain rules regarding the number of consecutive quality control values that are on the same side of the average. Five or six consecutive values all above or all below the average are statistically an out-of-control situation. Even if the values are within the 1 SD range, this finding indicates a statistically significant change, though this situation is usually not medically or operationally significant. However, five or six values between the 1 SD and the 2 SD levels should be investigated because this usually indicates an operationally significant systematic-bias situation.[5,6]

*Negative and positive features of Cu Sum compared to the Levy-Jennings.*The Cu Sum may be better at detecting small bias changes in a system than the visual inspection of a Levy-Jennings plot. On the other hand, detection of small changes that are neither medically nor operationally significant is not effective if it occupies a supervisor's time

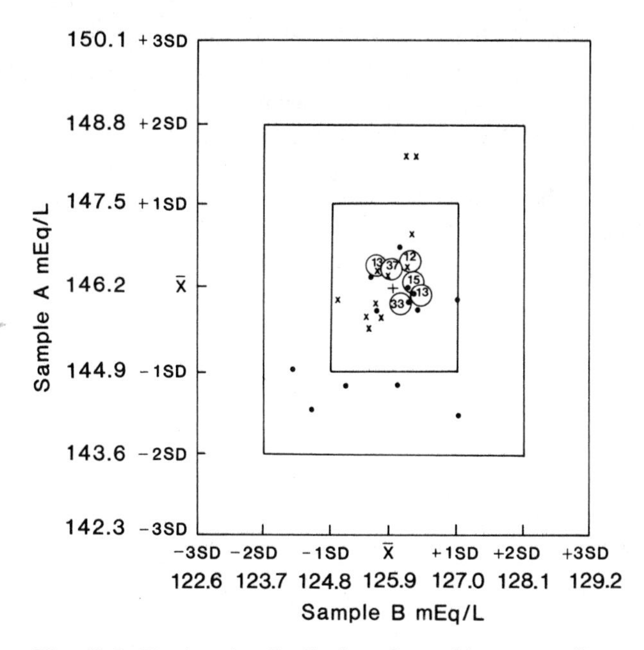

Fig. 19-5. Youden plot distribution of monthly mean sodium values by instrument and method from 6-month experience of regional pool having 226 members in Great Lakes and southeastern USA areas. •, Flame emission; *X*, Ion-specific electrode. Note systematic bias between two types of measurements. Circled numbers indicate number of responders with that value.

or interrupts regular test reporting until a ''not operationally significant'' decision is made by the supervisor.

In my opinion, Levy-Jennings and Cu Sum are both valid techniques for quality control evaluation and decision making. The same data base is being used in two different ways. To establish a conclusive preference, one would have to operate for at least a year using both systems simultaneously in order to tabulate the number of (1) statistically significant changes detected, (2) medically important changes detected, and (3) operationally important changes detected. It would then be possible to formulate a decision as to which system was most effective in a given situation.

This is the type of study that would be required to answer the question, Would using both Levy-Jennings and Cu Sum together be more effective than using either system alone? Up to now, no long-term (1-year) study has been done. Recommendations here and elsewhere represent my practice and opinion that the program of two control-sample levels at 2 SD and 95% limits is reliable and efficient.

Two-way convergence plot (Youden)

When two control-pool samples are used, a useful technique for quality control evaluation is the two-way convergence plot[10] (Fig. 19-4). In essence the diagram shows the convergence points of two quality control samples. One value is plotted on the horizontal axis and one on the vertical axis. The average values for the two quality control samples intersect at the center. The ±1 SD and ±2 SD lines intersect to make two corresponding squares representing an inner 1 SD and an outer 2 SD zone. The simultaneous plot point of the two daily quality control values will indicate an in- or out-of-control situation. Trends and shifts will also show up clearly in this graphic approach.

Another application for the convergence plot of data handling is the representation of a comparison of the results of different laboratories analyzing the same two quality control pool samples. This is characteristic of the regional quality control pool programs and gives each laboratory a clear idea of the relationship of its results to the other laboratories using the same two quality control pools. In this case, a laboratory whose convergence point is outside the 2 SD square should consider immediate investigation to determine the source of its systematic bias with respect to other laboratories. Systematic bias of this sort may be a reflection of instrument or method bias. Nevertheless, it is important to study such observed deviations and to record the results of the study. Fig. 19-5 illustrates the use of a Youden plot. The sodium values from the Great Lakes Regional Quality Control Program are graphed by method. The systematic bias between the two types of measurements becomes obvious.

MONTHLY REVIEW OF DAY-TO-DAY VALUES

Before the tenth of the month, the average and standard deviations of the past month's daily quality control analyses should be evaluated for change in average from the target average and change in the standard deviation from the usual standard deviation.

Monthly average compared to target average

The monthly average is compared to the target average (TA), which is the best available estimate of the true average value of the quality control sample under the actual conditions of analysis in your laboratory.

The first analysis is to determine if the average for the month changed from the target average. One can do this in the following manner, using Fig. 19-6 as a cumulative record.[10]

Step 1. Record the monthly average in column 1.

Step 2. Subtract the target average from the monthly average. Record the difference in the adjacent column.

Step 3. Make a statistical decision based on the following statistical approximation: *if the difference between the target average and the monthly average is greater than the usual standard deviation, the difference is a statistically significant one at approximately the 5% level of probability.* This means there is a one-twentieth chance of being incorrect in this decision.[11] This

QUALITY CONTROL

METHOD	GLUCOSE									
POOL	CAP 11									
	$\overline{TA}$ 105[a] mg/dl			4.0[b] USD			$\overline{TA}$			USD
MONTH	$\bar{X}$	QCD	$\bar{X}-\overline{TA}$	SD	QCD	SD-USD	$\bar{X}$	QCD	$\bar{X}-\overline{TA}$	SD	QCD	SD-USD
INITIAL	105	✓		5.0	✓	+1.0						
JAN	106	✓	+1.0	4.5	✓	+0.5						
FEB	104	✓	-1.0	3.8	✓	-0.2						
MAR	99	*	-6.0[c]	6.3	*	+2.5[d]						

Fig. 19-6. Software form for cumulative monthly records, including data and decisions. *a,* Target average (TA) calculated from (1) initial, (2) January, and (3) February. *b,* Usual standard deviation (USD) calculated from previous 6-month experience. *c,* Difference of monthly $\overline{X}$ from TA of greater than 1 USD is statistically significant; it is not medically significant (NMS); it is operationally significant (OS). *d,* Difference greater than 0.5 USD is statistically significant; it is not medically significant (NMS); it is operationally significant (OS). *Decision:* Glucose method needs investigation during April.

is a commonly accepted probability. The monthly average difference decision is based on the following assumptions: The monthly average is $\bar{x}$, the target average is $\bar{y}$, and it is assumed that (1) n is about 20 or greater and (2) the SD of $\bar{x}$ and $\bar{y}$ are equal to USD/$\sqrt{n}$. The sum of the variances of $\bar{x}$ and $\bar{y}$ equals the variance of the average difference $\bar{x} - \bar{y}$.

The USD (usual standard deviation) is equal to variance by the relationship

$$(\text{USD}/\sqrt{n})^2 = \text{Variance} = (\text{SD}_{\bar{x}})^2 = (\text{SD}_{\bar{y}})^2$$

To test whether the $\bar{x}$ is different from $\bar{y}$, the standard deviation of the average is used:

$$\text{SD}_{\text{average difference}} = \sqrt{n}\text{USD}/\sqrt{n}$$

Then the quantity of the ratio of the measured difference in averages $\bar{x} - \bar{y}$ and $\text{SD}_{\text{average difference}}$ is calculated as follows:

$$\bar{x} - \bar{y} \,/\, (\sqrt{2} \cdot \text{SD}_{\text{average difference}}) = t$$

For these observations, $t > 2.0$ indicates a significant difference. Necessarily, the decision rule is an approximation.

Decision A. If the difference is not statistically significant, no action is taken. Place a check mark in column 2, labeled QCD (quality control decision).

Decision B. If the difference is statistically significant, place an asterisk in column 2. In this format, it is easy to pick up the significant differences that must be studied. The next action is to make a decision regarding the medical significance of the observed change in the average. In other words, will this change in the observed average cause physicians to misinterpret the tests performed on samples from their patients?

If the answer is yes, there is an immediate emergency situation that must be corrected before further analytical values are sent to physicians.

If the answer is no, place the letters NMS (not medically significant) in the difference box and refer to Decision C.

Decision C. If the difference is statistically significant but is not medically significant, is the difference operationally significant? The persons actually using an instrument always have a perception as to when the instrument, system, or method is operating properly. At the monthly conference, one can arrive at a decision using information from the above-mentioned persons. If the decision is made that the difference is "operationally significant," record OS (operationally significant) in the difference box. A plan should then be drawn up and carried out to investigate the cause of the observed difference. Since this is not an emergency matter, one can schedule the investigation to be carried out before the next monthly review.

A useful action at this point is to review the daily values of the Levy-Jennings plot for the month under study to see if there is a detectable daily point at which a shift occurred. Such a point is usually detectable. It can serve as a focal point for investigating concurrent events such as instrument-maintenance procedures, solution changes, introduction of new standards, and other possible variables. If a break point is not detectable in the values from the month under study, one may find that consulting the daily values for the previous month may reveal a shift that took place at the end of the previous month. A shift occurring late in the previous month may not be large enough to influence the average value, but when the shift or trend continues, the mean of the following month will be found to be significantly changed.

If the review of the system or instrument shows no obvious reason for the shift, the laboratory may make the decision to "watch" the procedure for another month to see whether the change is resolved.

Comparison of monthly standard deviation with usual standard deviation

1. Subtract the current monthly standard deviation from the usual standard deviation. Record in column 5 (SD-USD) of Fig. 19-6.

Table 19-4. Guidelines for usual standard deviation values*

Analyte	Method	Average SD† (normal level)	Average SD (abnormal level)	Significant change limit
Glucose	All	31 (500-1000 mg/L)	54 (1100-3300 mg/L)	90 mg/L
Urea N	All	6 (100-300 mg/L)	15 (350-860 mg/L)	20 mg/L
Creatinine	12/60 Technicon	0.7 (9-18 mg/L)	1.0 (19-73 mg/L)	2.1 mg/L
Cholesterol	12/60 Technicon	53 (750-2600 mg/L)		150 mg/L
Potassium	12/60 Technicon	0.07 (3.0-4.9 mEq/L)	0.1 (5.0-8.0 mEq/L)	0.2 mEq/L
Sodium	6/60 Technicon	1.3 (133-160 mEq/L)	1.3 (121-132 mEq/L)	3.9 mEq/L
Calcium	12/60 Technicon	1.6 (70-109 mg/L)	2.0 (110-140 mg/L)	4.8 mg/L

*From Kurtz, S.R., Copeland, B.E., and Straumfjord, J.V.: Am. J. Clin. Pathol. **68:**463-473, 1977.
†In this table, average standard deviation (SD) is equivalent to the usual standard deviation (USD). Normal and abnormal levels indicate quality control pools.

2. *If the difference is greater than one-half the usual standard deviation, the monthly standard deviation is statistically different from the usual standard deviation.* The monthly standard deviation decision is based on the following assumptions: $n = 16$ or greater, and the F-ratio test is at a 95% level of certainty.
3. If the difference is not statistically different, place a checkmark in the QCD column; if it is statistically different, place an asterisk in the QCD column.
4. At this point, the decision can be made as to whether the monthly standard deviation has increased to the point at which it is a medically significant change.

One can use the *significant change limit* to evaluate the monthly standard deviation. Table 19-4 lists suggested significant change limits for several analytes. Ask the question, Is the significant change limit using the monthly standard deviation useful for medical purposes? For example, a usual standard deviation for potassium of 0.07 mEq/L will give a significant change limit of 0.2 mEq/L, which is satisfactory for the required medical decisions.

On the other hand, a monthly standard deviation of 0.2 mEq/L would result in a significant change limit of 0.6 mEq/L. This would require the patient's potassium level to change by more than 0.6 mEq/L before the change in the patient's value would be significant. This would not be satisfactory for medical decision making. Conservatively a doubling of the precision value may be expected to interfere with medical decision making.

Steps to correct a "medically significant" change in the monthly standard deviation should be taken on an emergency, "right-now" basis.

If the decision is that the statistically different monthly standard deviation was not medically significant, an "operational decision" should be made.

An *operationally significant change* requires action, but one can plan it and it need not be taken immediately. One can usually take corrective steps within 10 to 14 days.

Medical importance of a stable usual standard deviation and use of significant change limit

The day-to-day standard deviation of individual measurements is the characteristic frame of reference for medical decisions. The physician assumes that the day-to-day precision is maintained at the same level from month to month and year to year. The physician makes many clinical decisions on the basis of day-to-day differences. The precision of the measurement procedure directly influences the medical interpretation of a day-to-day change. The physician needs to know which changes can be regarded as changes in the patient as opposed to changes that represent the inherent day-to-day variation in the measurement system.

The *significant change limit* is a decision-making tool that can be given to physicians to help them distinguish day-to-day changes that are caused by the inherent variability of the instruments and methodology from changes that are caused by changes in the patient. As an approximation, *the significant change limit is three times the usual standard deviation.* The significant change limit is based on the assumption that the usual standard deviation represents day-to-day variability for individual measurements. The calculation is dependent on the quantity SD_{diff} (p. 315), which is related to the usual standard deviation (USD) by the following formula:

$$SD_{diff} = \sqrt{2(USD)^2} = 1.4 \text{ USD}$$

The 95% confidence limits of true difference are as follows:

$$2\ SD_{diff} = 2.8 \text{ USD}$$

For practical use, significant change is 3 USD.

Changes in patient analyte values smaller than the significant change limit are in fact changes that cannot be differentiated from inherent method variability. Changes greater than the significant change limit represent a real change in the patient. For example, if the usual standard deviation for cholesterol is 50 mg/L, the significant change limit is 150 mg/L. A change from 2000 to 2200 mg/L would exceed the significant change limit and would represent a real change in the patient. A change from 2000 to 1900 mg/L would not exceed the significant change limit and would most likely represent the inherent variability of the method. To facilitate consistent decisions by the attending physician, it is important to maintain a consistent level of precision from month to month and year to year.

Software form for recording monthly quality control data and monthly quality control decisions

Fig. 19-6 is a convenient cumulative form (software) that was developed over 20 years of quality control experience at the New England Baptist Hospital (Boston) and the New England Deaconess Hospital (Boston).[10] Its purpose is to permit easy decision making at the monthly conference and to provide a convenient format for a scanning review of previous months. Two symbols are used in the narrow quality control decision (QCD) columns. A check mark means no statistical difference; an asterisk means there is a statistically significant difference.

When the check mark appears, no added comments are needed. When an asterisk appears, there must be further decisions. These are recorded as follows in the space above the appropriate numbers: *MS*, medically significant; *NMS*, not medically significant; *OS*, operationally significant; *NOS*, not operationally significant; *W*, watch. These decisions are also recorded in a report of the monthly quality control review. In addition, hypotheses as to causes of changes and plans to test the hypotheses are recorded. For example, glucose-low bias corresponds to opening a new bottle of calibrator on February 15, 1983. Action: compare new bottle with a new lot number of standard.

An important item to add to the monthly summary is the date when a new lot of calibrator material is introduced.

Use of patient population average in monthly evaluation

By comparing the results obtained from a large number of patient analyses, one can also evaluate the constancy of a laboratory's analytic values. If the analytical system is constant, the premise is that the distribution of results for a large patient population (most of whom are nondiseased) will be constant. Therefore a statistically significant change in the daily or monthly mean of such a population reflects the introduction of a bias in the method when consecutive comparisons of the average population values are made. This analysis, of course, is invalid if there is a change in the patient population. When this type of data review involves large numbers of data, it usually is performed by computer analysis.

Procedure for quality control using average of patient values[11]

1. Calculate the mean ± 2 SD for your patient population.
2. Eliminate values outside the average ± 3 SD, and recalculate the average value for groups of at least 16 to 25 patient values. (There may be several groups per day, or it may require several days to accumulate a group.)
3. The average of each group of values should fall within the following limits: mean of ± 2 SD of the average. The standard deviation of the average of patient's values, that is, the inherent variability of the average of these values, is established by the following formula: the standard-deviation average equals standard deviation of individual patients divided by the square root of the number, n, in the group (16 or 25). For example, the standard deviation of persons in a patient population equals 20 mg/L at $n = 16$. The standard-deviation average of values equals 20 mg/L divided by the square root of 16, or 20 divided by 4, which equals 5.0 mg/L. One deliberately chooses 16 or 25 to simplify the square-root calculation.
4. For the example system above, the control limits for patient population samples would be patient average ± 2 SD. Assume the patient average equals 160 mg/L. For example, control limits $160 \pm 2(5.0)$, or 160 ± 10 mg/L. Barnett states that a shift in the average of 1.0 SD will be detected 99% of the time using this system.[11]

What is acceptable precision?

Often the question is asked, How precise should the measurements be? There are three points to consider: (1) What is the medical use for which these measurements are being made? (2) Will the significant change limit allow the physician to make useful medical decisions? and (3) Is the instrument or method being operated with the optimum expertise and the optimum carefulness?

If the answers to questions 2 and 3 are yes, the usual standard deviation obtained in that framework is valid and acceptable performance. These two questions emphasize two specific points: first, medical measurements should be medically useful, and, second, each method and instrument should be operated with optimum care.

A summary selection of the usual standard deviations of frequently performed tests obtained in the university hospitals of the United States is presented in Table 19-4.[12]

As a rule, if the method's day-to-day standard deviation is less than 1.5 times the usual standard deviation, this is good performance. If the standard deviation is greater than two times the average, the method needs attention. The last column in Table 19-4 indicates when immediate action is necessary to reduce the standard deviation. It is specifically recommended that the usual standard deviation for sodium, chloride, and carbon dioxide not exceed 2 mEq/L and that for potassium it not exceed 0.2 mEq/L.

It should be understood that in many instances, as the

number of persons doing a procedure increases, the standard deviation may also increase. Therefore small laboratories may have an advantage over large laboratories for those tests that are done in a daily batch, since fewer persons contribute to the day-to-day inherent variability. When comparing the standard deviations from different laboratories, never compare the standard deviation of individual values day to day with the standard deviation of a series of daily average values. The standard deviation of a series of averages from multiple daily values will have much less variability than the standard deviation of individual measurements on a day-to-day basis. Compare like with like or an incorrect decision will be made about the quality of work.

LONG-TERM DECISIONS ON METHODOLOGY RELATED TO ACCURACY AND PRECISION

Consistency goals and methodology evolution

Daily, weekly, monthly, and yearly quality control programs verify that the method in use maintains (1) consistency, (2) precision, and (3) accuracy. Quality control programs are established to eliminate aberrations of accuracy or precision year in and year out. The month-by-month and year-by-year quality control review guarantees that the method retains the same degree of accuracy and precision over long periods. A patient returns to the physician's office over a period of years. Consistency is important in order to maintain comparability of measurements so that observed differences represent patient changes and not changes in laboratory systematic bias.

The purpose of long-range quality control evaluation is to verify that the current method retains desired qualities of accuracy and precision.

The technology used in clinical chemistry is not static. New methods of analysis and instrumentation are constantly appearing. After some years of use, a new method can be perceived as generally being more accurate, more precise, or more cost effective than older methods and may become accepted as the new state of the art. For example, according to the 1978 CAP quality control summary, 58% of participating laboratories used the diacetylmonoxime method for measurement of urea. In 1983, the percentage of laboratories using that method had decreased to 23%; by then urea was largely being quantitated by enzymatic methods. There have been similar shifts toward different methods for the measurement of glucose, sodium, and potassium. Similarly, the precision and sensitivity of many analyte measurements have improved in the past 5 years. Long-range quality control allows a laboratory to compare itself to others and determine if it is using the most accurate and precise methods available.

Although every new method should undergo vigorous validation before introduction, only continual evaluation can determine the long-range validity of a procedure under working conditions. When changing to a new instrument or method, it is important to avoid positive or negative systematic bias changes that may change the reference value ranges.

Another important use of long-term external quality control is to establish the interchangeability of the results from different laboratories if a patient is transferred from one hospital to another. The goal of having a continual and accurate and interchangeable biochemical record of a person can be achieved by laboratories comparing themselves to each other using regional pools and external quality control programs as reference points and by correlation with reference methods and reference laboratories. This goal can be advanced by the increased acceptance of standard procedures as proposed by the NBS, FDA, and NCCLS (National Committee for Clinical Laboratory Standards).

Internal quality control programs for precision and consistency control

The day-to-day internal quality control program is the most useful tool to maintain long-term consistency control and long-term precision control. The number of quality control analyses in each monthly average is 20 to 25. This number makes it possible to form statistical decisions that detect small changes before they become operationally significant or medically significant. All regional quality control programs and most commercially available quality control pools offer a computerized reporting system. The monthly reports incorporate the values obtained by all users of the pools. These reports usually list the individual laboratory's monthly mean, standard deviation, and coefficient of variation. Some reports also provide the cumulative mean, standard deviation, and coefficient of variation.

The quality control report is designed so that each laboratory can compare its data to its own peer group (that is, same instrument and method), as well as to other instruments or methods. By comparison of each laboratory's data with peer and other methods, one can detect possible laboratory bias in a method as well as problems in precision.

Medically or operationally significant deviations from the peer mean or standard deviation should be used as an indication of either an improper target value or a true bias in the method. The alerted supervisory staff should look for additional evidence, either continued bias in succeeding months or bias in other external quality control data (see p. 326). A laboratory's standard deviation that is 1.5 times greater than the average standard deviation of a group of regional pool members indicates that the laboratory's procedure has a statistically different precision value.

Another useful way of presenting the internal quality control monthly means in order to detect long-term drifts in the method is to use a modified Youden plot. The Youden plot was described previously as a technique to compare visually the laboratory's mean value for an analyte

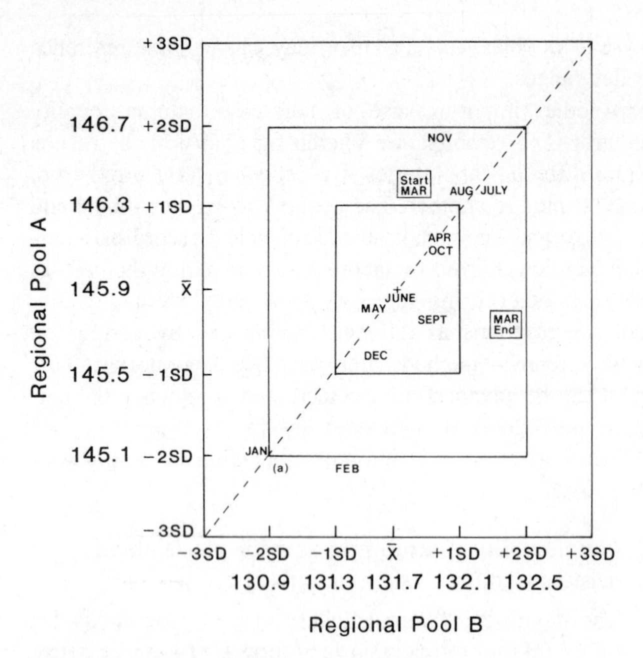

Fig. 19-7. Regional pool-participant monthly-average values for sodium (mEq/L) plotted with use of Youden two-level concept. Pool was started in March. Notice that January shows a value out of control for both levels *(a)*. February shows a value out of control for level *A*. Systematic bias was corrected in March (end). Coordinates are established using temporary target average (TTA) as mean *($\bar{X}$)* and standard deviation of the average. One obtains this by dividing the usual standard deviation (USD) by the square root of *n,* where *n* equals the usual number of days on which these analyses are run in a month's time. *Example:* Test run for 25 days in month of June. SD for plotting equals USD divided by square root of number of days 5.

with the values obtained by other laboratories using similar or different methods. For long-term evaluations, the laboratory's assigned target values and standard deviation values for the two levels of pool are placed on the *x* and *y* axes. One then plots each month's calculated coincident mean on this graph by placing the abbreviation of the month on the proper coordinates (Fig. 19-7). The cumulative monthly means should randomly fall within the +2 standard-deviation range. Movement of the coincident average point along the 45-degree operational line *(dashed)* indicates systematic bias; that is, both averages are affected in the same way. One can detect significant long-term drifts from the target value and, along with the monthly evaluation, use them to make decisions as to the importance of small successive changes in the monthly mean.

External quality control programs for accuracy control

To provide independent validation of the internal quality control programs, external surveys have been developed to provide unknown samples for analysis by participating laboratories. Because the analyst has no idea of the target value, operator bias is eliminated. Thus these programs, if properly used, can give a valid estimation of the inherent accuracy of a system. These programs involve the testing of a wide range of analytes, including hormones, drugs, and enzymes, in addition to the usual biochemical analytes in both serum and urine. Some programs that appear to provide laboratories successfully with information on accuracy and precision are listed in Table 19-5.

Table 19-5. List of external quality control programs

Name	Type of survey	Address
College of American Pathologists	Physician's office General chemistry Enzyme Blood gas Therapeutic drugs Toxicology Instrumentation	7400 North Skokie Boulevard, Skokie, IL 60077
Centers for Disease Control	Chemistry Therapeutic drug monitoring Toxicology	Centers for Disease Control Atlanta, GA 30333
American Association of Clinical Chemists	Therapeutic drug monitoring	1725 K Street NW Washington, DC 20006
American Association of Bioanalysts	Chemistry	818 Olive, Suite 918, St. Louis, MO 63101
Burroughs-Wellcome	Chemistry	Wellcome Diagnostics, 3030 Cornwallis Road, Research Triangle Park, NC 27709

College of American Pathologists. The CAP Comprehensive Chemistry Survey involves approximately 7000 participating laboratories. This survey covers most common analytes (such as electrolytes, glucose, and urea) and several hormones as well. The results of a typical survey for magnesium are shown in Fig. 19-8. The report includes the following information: (1) constituent and the reporting laboratory's method and units of measure; (2) comparative method, if available; (3) results submitted by the laboratory ("your result") (these results should have been obtained in a manner identical to that used for patient specimens; that is, the quality control sample should not be analyzed more than once or treated in a more careful fashion); (4) the mean and standard deviation of all participants using a method similar to yours; and (5) your standard-deviation index, which is calculated when you subtract your result from the group mean and divide it by the group standard deviation. The standard-deviation index (SDI) is thus an indication of distance of the participant value from the group mean value. The standard-deviation index is also

CONSTITUENT UNIT OF MEASURE YOUR LABORATORY'S METHOD/SYSTEM COMPARATIVE METHOD	EVALUATION STATISTICS COMPARATIVE STATISTICS SPECIMEN NUMBER	CODE	YOUR RESULT	MEAN	S.D.	NO. OF LABS.	YOUR SDI
MAGNESIUM	C-07	01	2.9	2.91	.19	147	-.1
MG/DL	C-08	01	2.8	2.78	.19	147	+.1
ATOMIC ABSORPTION	C-09	01	2.6	2.66	.18	147	-.3
PERKIN ELMER A A SPECT							
ATOMIC ABSORPTION	C-07	07		2.89	.21	229	.0
ALL A.A. SPECTROPHOTOM	C-08	07		2.75	.21	227	+.2
	C-09	07		2.63	.20	228	-.2

Fig. 19-8. Survey results for magnesium from the College of American Physiologists (CAP) comprehensive chemistry survey. Survey lists analyte (magnesium) units of measurement, laboratory's method, and reference method (all atomic absorption spectrophotometer). Results of laboratory's analysis of three samples sent by CAP for that survey (C-07, C-08, and C-09) are compared to group mean and to reference group.

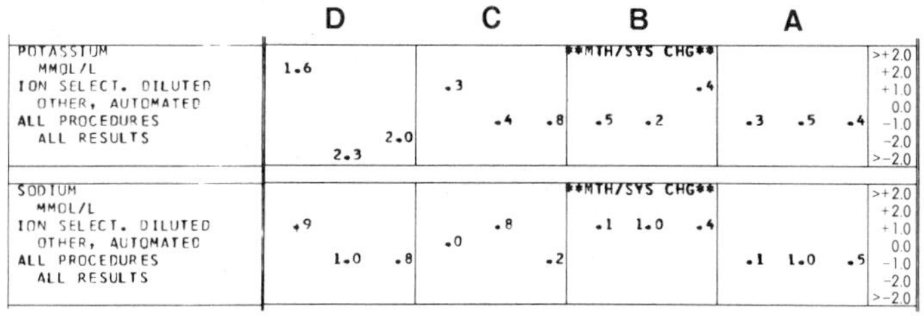

	D	C	B	A	
POTASSIUM			**MTH/SYS CHG**		>+2.0
MMOL/L	1.6				+2.0
ION SELECT. DILUTED		.3	.4		+1.0
OTHER, AUTOMATED					0.0
ALL PROCEDURES		.4 .8	.5 .2	.3 .5 .4	-1.0
ALL RESULTS	2.0				-2.0
	2.3				>-2.0
SODIUM			**MTH/SYS CHG**		>+2.0
MMOL/L					+2.0
ION SELECT. DILUTED	.9	.8	.1 1.0 .4		+1.0
OTHER, AUTOMATED		.0			0.0
ALL PROCEDURES	1.0 .8	.2		.1 1.0 .5	-1.0
ALL RESULTS					-2.0
					>-2.0

Fig. 19-9. Graph of CAP survey results of 1 year for sodium and potassium. Results are plotted as standard deviation intervals from target or average value of other participants using the reporting laboratory's method. *A* to *D*, Different surveys.

given in a visual presentation to show the pattern of deviation from the group mean (Fig. 19-9). The previous three survey standard deviation indices are also presented to provide a visual concept of systematic bias development. Information is also provided comparing the submitting laboratory's results to the designated reference method for that analyte.

Fig. 19-9 shows the CAP survey results for 1 year from a single laboratory for sodium and potassium, which are analyzed simultaneously. The survey results for both analytes are excellent except for the last set of results for potassium. The entire year's results suggest that the out-of-range potassium results in survey D most likely reflect a random error and not a systemic, chronic bias.

The mean, standard deviation, and coefficient of variation for each method or system are also presented. The summary report for glucose for the 1982 set C-8 is shown in Table 19-6. It is important to note the difference in mean and standard deviation values between methods and as well between the numerous instruments within a method system. Each laboratory should compare its results with its closest peer group (same method, same instrument). The difference between the individual laboratory result and average value of the peer or "reference" group provides an indication of a possible bias. When one is choosing a new method or instrument, this list, showing the standard deviations and average values, may be very helpful in selecting a system that is precise and accurate.

The difference between the mean values of the various groups reflect true methodological biases and not random laboratory error. If a laboratory's result is biased statistically from its peer group, that is, outside ±2 SD, this will be pronounced "not acceptable" performance. When this occurs, there should be a review of the quality control for that day and repeat testing of some of the remaining sample or of the follow-up survey validated sample. The study, conclusions drawn, and decisions made should be recorded on the survey report itself. If repeated survey results for the same constituent show a bias in the nonacceptable range, the method or instrument should be changed. If a peer group is consistently biased from other groups, the question must be raised as to which group is correct. A search of the literature for method comparisons and the use of reference material can often help answer this important question. The CAP survey has been instrumental in clearly showing statistically significant method and instrument biases. Many of these have been corrected by the FDA, CDC, instrument makers, and reagent suppliers. Proved biases should be corrected by modification of existing procedures or replacement by newer ones. It is important to consider the biases that will affect medical decision making.

There is a difference of opinion among authorities as to how frequently external quality control testing should be done. Quarterly (3-month) intervals are considered satisfactory by the Secretary of Health and Human Services

Table 19-6. Example of an all-participant summary for glucose from 1982 College of American Pathologists Survey

Constituent method and system	Specimen C-7			
	Number of laboratories	Mean	Standard deviation (SD)	Coefficient of variation (CV)
Glucose (mg/dL)				
Glucose oxidase (colorimetric)				
All results	1285	47.5	4.8	10.1
Bio-Dynamics/bmc	153	45.3	6.2	13.7
Sclavo	57	48.4	2.9	6.0
Other, manual	39	50.0	6.7	13.4
All manual procedures	303	47.1	6.2	13.1
American Monitor KDA	262	46.4	2.8	6.1
Centrifugal analyzers	77	44.3	4.7	10.6
Beckman ASTRA 8	29	46.4	2.4	5.1
Gilford Impact 400, and so on	24	51.4	4.4	8.5
I.L. 919	65	43.9	2.4	5.5
Technicon AAI or AAII	29	50.0	3.1	6.3
Autoanalyzer I and II	29	50.0	3.1	6.3
Technicon SMA 6/60	45	52.2	4.8	9.1
Technicon SMA II	28	49.9	3.6	7.2
Technicon SMA 12/60	127	50.9	3.7	7.2
All Technicon SMA's	202	51.0	3.9	7.6
Technicon SMAC	69	48.1	3.6	7.5
CentrifiChem Baker	45	42.5	3.7	8.7
Other, automated	87	45.9	4.9	10.7
All multiconstituent analyzers	968	47.7	4.4	9.1
Orthotoluidine				
All results	984	50.6	4.0	7.9
Manual, in-house reagent	78	49.8	4.0	8.0
Data Medical Associates	77	50.2	4.4	8.7
Dow	207	49.6	3.4	6.9
Harleco	70	50.2	3.8	7.6
Hycel	45	52.4	4.3	8.2
Pfizer Diagnostics	25	48.3	4.7	9.6
Pierce	43	49.8	3.4	6.9
Sigma	27	51.2	3.5	6.9
Stanbio	24	51.5	4.7	9.0
Other, manual	41	48.7	4.3	8.8
All manual procedures	734	50.1	4.1	8.2
Coulter 22	61	51.9	1.7	3.3
Harleco/IL Clinicard	72	49.3	2.1	4.3
Hycel Super 17 or Mark 17	80	54.8	2.5	4.6
All multiconstituent analyzers	247	51.8	3.3	6.4
Glucose oxidase (O_2 electrode)				
All results	1271	46.9	2.9	6.1
Beckman ASTRA 8	688	46.4	2.5	5.3
Beckman Specific Analyzer	416	47.7	3.0	6.2
YSI Glucose Analyzer	30	48.9	2.9	5.9
Other, automated	30	46.8	2.5	5.4
All multiconstituent analyzers	1183	46.9	2.8	5.9
Beckman ASTRA 4	79	46.1	2.8	6.0
All electrolyte analyzers	80	46.1	2.8	6.0
Hexokinase (ultraviolet monitoring)				
All results	2332	51.4	4.9	9.6
Worthington	63	56.0	4.8	8.6
Other, manual	40	55.6	4.3	7.8
All manual	147	55.7	4.6	8.3
Abbott ABA 50	54	49.0	2.2	4.5
Abbott ABA 100	86	49.3	2.9	6.0
Abbott ABA 200	23	50.1	7.5	15.0
Abbott VP	158	47.2	2.5	5.3
Aminco RotoChem	25	51.9	5.7	10.9

Table 19-6. Example of an all-participant summary for glucose from 1982 College of American Pathologists Survey—cont'd

Constituent method and system	Specimen C-7			
	Number of laboratories	Mean	Standard deviation (SD)	Coefficient of variation (CV)
Centrifugal analyzers	437	47.4	3.0	6.3
Chemetrics	61	57.0	4.6	8.1
Chemetrics II	134	53.0	4.3	8.0
Du Pont ACA	367	48.8	2.3	4.8
ElectroNucleonics GemENI	168	47.8	2.5	5.2
Gilford Impact 400, and so on	141	57.6	2.5	4.4
Gilford 102, and so on	82	57.0	3.4	6.0
Gilford 103, System 5, 202	89	56.4	3.5	6.2
Hycel HMA 1600	27	58.9	6.6	11.2
I.L. multistat III	100	46.1	2.2	4.8
Technicon SMA II	115	51.8	3.2	6.2
Technicon SMA 12/60	64	55.0	4.6	8.3
All Technicon SMA's	198	53.2	4.1	7.7
Technicon SMAC	163	51.6	2.4	4.7
CentrifiChem, Baker	120	47.2	2.1	4.5
Other, automated	111	53.5	4.8	8.9
All multiconstituent analyzers	2187	51.1	4.9	9.5
Hexokinase (colorimetric)				
All results	710	50.6	4.8	9.6
All manuals	53	55.6	5.8	10.5
Centrifugal analyzers	146	47.0	2.6	5.6
Chemetrics II	29	51.8	5.8	11.2
Du Pont ACA	207	48.6	2.4	5.0
ElectroNucleonics GemENI	77	47.5	2.5	5.2
Gilford Impact 400, and so on	39	57.4	3.2	5.6
Gilford 103, System 5, 202	26	57.3	3.1	5.3
I.L. Multistat III	31	45.6	3.2	7.1
Technicon SMA II	51	51.4	3.3	6.3
Technicon SMA 12/60	22	54.3	3.3	6.1
All Technicon SMA's	81	52.6	3.7	7.0
CentrifiChem Baker	31	47.3	2.0	4.2
Other, automated	25	52.1	4.2	8.0
All multiconstituent analyzers	655	50.2	4.6	9.1
All procedures and all results	6908	49.5	4.8	9.6

(HHS), JCAH, CAP, and the American Association of Bioanalysts (AAB).

The quarterly programs conducted by the CAP and the CDC assume that daily, weekly, and monthly internal quality control programs are in regular operation in order to detect both spurious daily values and shifts and trends of the average. It is also assumed that the external quality control unknown specimens are analyzed when the analytical system is "in control" and therefore is representative of the system.

On the other hand, the Burroughs-Wellcome survey (every 2 weeks) and the American Association for Clinical Chemistry (AACC) TDM Program (every month) recommend and provide unknown samples more frequently. These programs also assume that daily, weekly, and monthly internal quality control systems are being used.

In the assessment of accuracy, the regional pool systems conducted by commercial companies and by user groups provide more powerful estimates of interlaboratory accuracy bias (since these comparisons are based on averages of 20 to 25 measurements) than the external monthly or quarterly sample programs, which supply one to three samples and request only one or two measurements per sample. Comparison data with peer groups are published on a monthly or quarterly basis.

In addition, the CAP program provides for purchase of additional lyophilized *survey validated samples,* which are delivered after the survey results have been reported to the individual laboratory. Therefore one can study all methods or instruments that show a systematic bias to identify and correct the source of the bias. This survey-validated serum pool is a very valuable tool for accuracy control.

American Association for Clinical Chemistry. The therapeutic drug monitoring (TDM) program sponsored by the AACC is designed specifically to monitor analysis of therapeutic drugs. The TDM program sends out three ly-

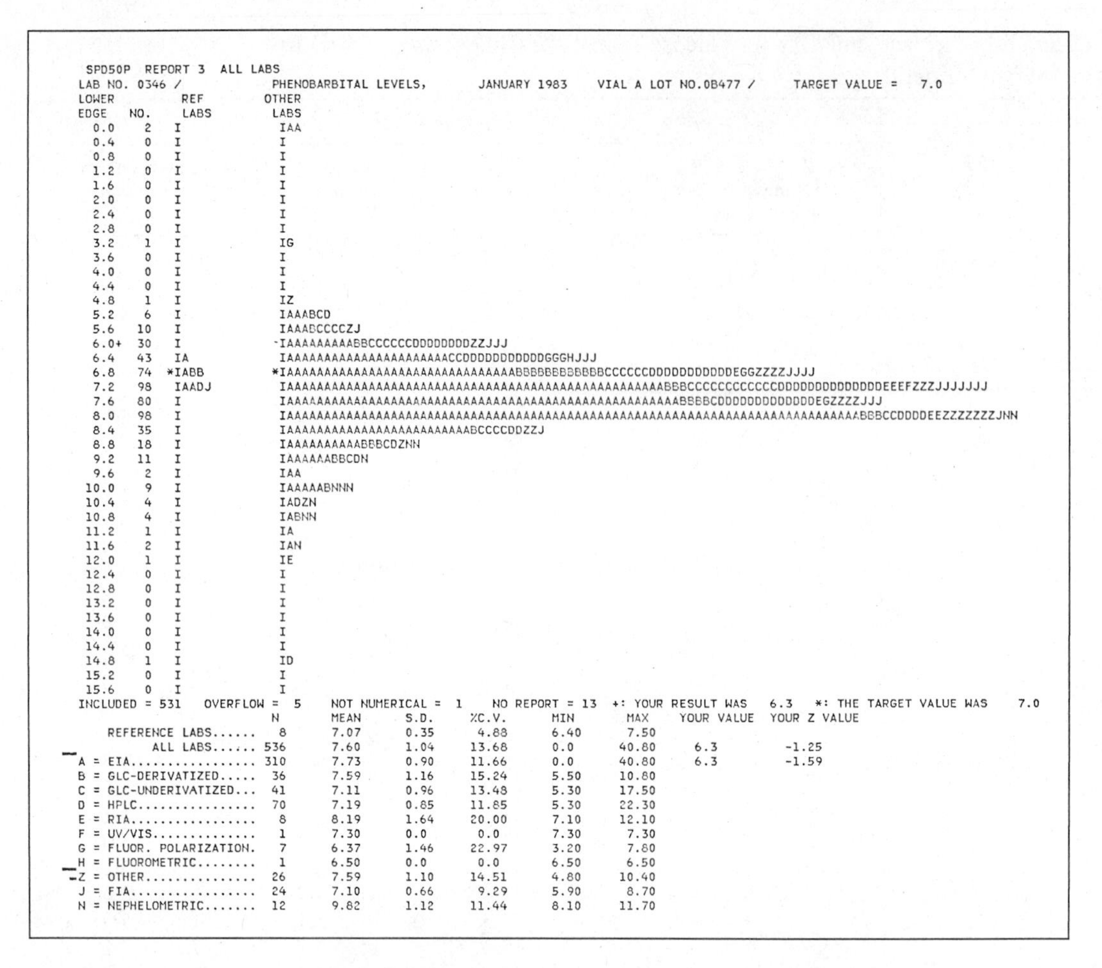

```
SPD50P  REPORT 3  ALL LABS
LAB NO. 0346 /            PHENOBARBITAL LEVELS,        JANUARY 1983    VIAL A LOT NO.0B477 /      TARGET VALUE =   7.0
LOWER          REF        OTHER
EDGE   NO.     LABS        LABS
  0.0    2   I              IAA
  0.4    0   I              I
  0.8    0   I              I
  1.2    0   I              I
  1.6    0   I              I
  2.0    0   I              I
  2.4    0   I              I
  2.8    0   I              I
  3.2    1   I              IG
  3.6    0   I              I
  4.0    0   I              I
  4.4    0   I              I
  4.8    1   I              IZ
  5.2    6   I              IAAABCD
  5.6   10   I              IAAABCCCCZJ
  6.0+  30   I             -IAAAAAAAAABBCCCCCCDDDDDDDDZZJJJ
  6.4   43   IA             IAAAAAAAAAAAAAAAAAAAAAACCDDDDDDDDDDDGGGHJJJ
  6.8   74  *IABB          *IAAAAAAAAAAAAAAAAAAAAAAAAAAAAAAAAABBBBBBBBBBBCCCCCDDDDDDDDDDDEGGZZZZJJJJ
  7.2   98   IAADJ          IAAAAAAAAAAAAAAAAAAAAAAAAAAAAAAAAAAAAAAAAAAAAAAAAAAAAABBBCCCCCCCCCCCDDDDDDDDDDDDDEEEFZZZJJJJJJ
  7.6   80   I              IAAAAAAAAAAAAAAAAAAAAAAAAAAAAAAAAAAAAAAAAAAAAAAAAAAAABBBBCDDDDDDDDDDDDDDEGZZZZJJJ
  8.0   98   I              IAAAAAAAAAAAAAAAAAAAAAAAAAAAAAAAAAAAAAAAAAAAAAAAAAAAAAAAAAAAAAAAAAAAAAAAAAAAAAABBBCCDDDDEEZZZZZZJNN
  8.4   35   I              IAAAAAAAAAAAAAAAAAAAAAAAAABCCCCDDZZJ
  8.8   18   I              IAAAAAAAAAABBBCDZNN
  9.2   11   I              IAAAAAABBCDN
  9.6    2   I              IAA
 10.0    9   I              IAAAAABNNN
 10.4    4   I              IADZN
 10.8    4   I              IABNN
 11.2    1   I              IA
 11.6    2   I              IAN
 12.0    1   I              IE
 12.4    0   I              I
 12.8    0   I              I
 13.2    0   I              I
 13.6    0   I              I
 14.0    0   I              I
 14.4    0   I              I
 14.8    1   I              ID
 15.2    0   I              I
 15.6    0   I              I
INCLUDED = 531   OVERFLOW =  5    NOT NUMERICAL =  1    NO REPORT = 13   +: YOUR RESULT WAS   6.3   *: THE TARGET VALUE WAS    7.0
```

	N	MEAN	S.D.	%C.V.	MIN	MAX	YOUR VALUE	YOUR Z VALUE
REFERENCE LABS......	8	7.07	0.35	4.88	6.40	7.50		
ALL LABS......	536	7.60	1.04	13.68	0.0	40.80	6.3	-1.25
A = EIA.................	310	7.73	0.90	11.66	0.0	40.80	6.3	-1.59
B = GLC-DERIVATIZED.....	36	7.59	1.16	15.24	5.50	10.80		
C = GLC-UNDERIVATIZED...	41	7.11	0.96	13.48	5.30	17.50		
D = HPLC................	70	7.19	0.85	11.85	5.30	22.30		
E = RIA.................	8	8.19	1.64	20.00	7.10	12.10		
F = UV/VIS..............	1	7.30	0.0	0.0	7.30	7.30		
G = FLUOR. POLARIZATION.	7	6.37	1.46	22.97	3.20	7.80		
H = FLUOROMETRIC........	1	6.50	0.0	0.0	6.50	6.50		
Z = OTHER...............	26	7.59	1.10	14.51	4.80	10.40		
J = FIA.................	24	7.10	0.66	9.29	5.90	8.70		
N = NEPHELOMETRIC.......	12	9.82	1.12	11.44	8.10	11.70		

Fig. 19-10. Histogram of results of phenobarbital levels from therapeutic drug monitoring (TDM) survey sponsored by American Association for Clinical Chemistry. In this particular survey, target value of 7.0 was established when a known amount of drug was added. Values from all reporting laboratories (536) are divided into increments of 0.4 units (μg/mL), number of laboratories reporting each result, and method used by letter are presented as a frequency histogram. In this particular circumstance, although target value was 7.0, mean value obtained by all laboratories was 7.6. This laboratory reported a result of 6.3. Since standard deviation was 1.04 for all laboratories, Z value was

$$\frac{6.3 - 7.6}{1.04} = -1.25$$

or 1.25 times usual standard deviation. By this laboratory's method (A = EIA), however, SD was 0.90; therefore Z value was

$$\frac{6.3 - 7.73}{0.90} = -1.59$$

which is acceptable, on a comparative basis.

ophilized serum specimens 12 times a year. The TDM report includes the following information (Fig. 19-10): (1) presentation of all results for each analyte in a histogram form, which also gives the distribution of data for all instrument and technique reporting; (2), your value and the target value, which is the calculated concentration of the weighed-in drug; (3) the mean, standard deviation, and coefficient of variation of each group, including the minimum and maximum values of each laboratory; and (4) the participating laboratory's value as compared to an all-laboratories value and to its peer group. The extent of participating laboratory bias from these groups is designated by the *z* value, which is similar to the standard-deviation index in the CAP survey. In addition, the mean standard deviation and coefficient of variation of reference laboratories and all participants are also presented.

The TDM program is useful because of its large data base (more than 500 participating laboratories) and the fre-

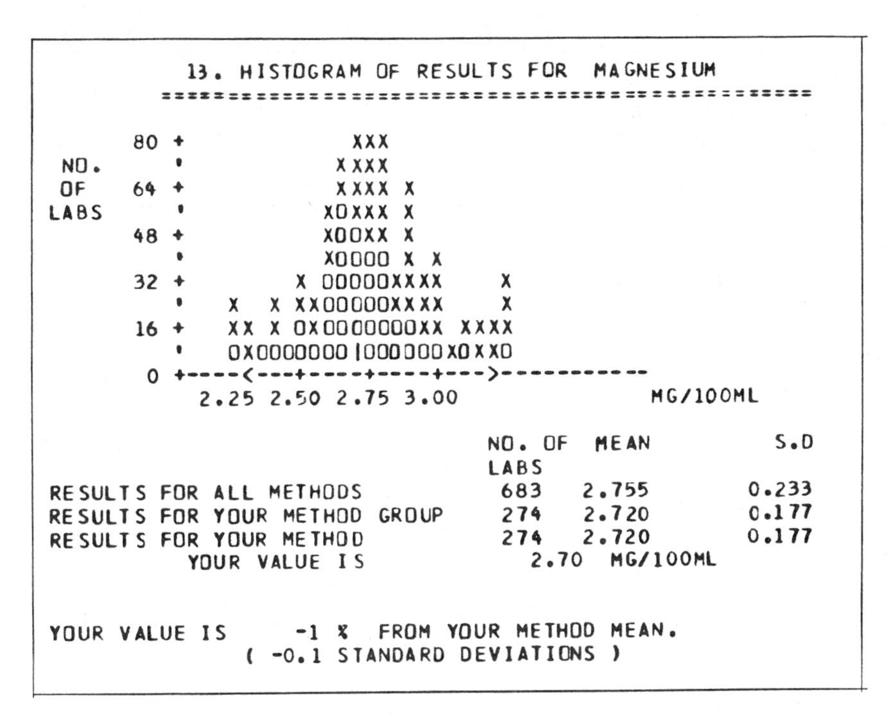

Fig. 19-11. Histogram for magnesium results from a Burroughs-Wellcome Survey. All method results, *X*, results of peer method group, *O*, and individual reporting laboratory's result, your value, *vertical line* |, are presented. An evaluation in percentage and standard deviations are given.

quency of sampling data. This allows rapid accumulation of information covering 12 different time points, which increases the validity of decisions based on an apparent bias. Again the significance of a bias between a participating laboratory and other groups must be determined. Because the TDM program lists an accurate "weight" target value, such decisions can be reached rigorously. The AACC quality control program was extended in 1982 to include hormone analysis.

Burroughs-Wellcome Survey. A third type of program available is one provided by the Burroughs-Wellcome Company. This program surveys many common serum analytes, including some enzymes, hormones, and drugs. As in the other programs, this program lists all the results obtained for a particular analyte and the results of the peer group closest to each participating laboratory as well. As in the TDM program, the results of all data, produced by methods similar to the participating laboratory and its peer group, are presented in a histogram form (Fig. 19-11). The mean and standard deviation of each group are also listed. The data of your laboratory are compared to your peer group mean as the percentage of deviation from the mean and as the number of multiple standard deviations from that mean.

An interesting feature of the Burroughs-Wellcome program is the use of repeat analysis of samples assayed previously. One can use the results of the duplicate analysis to estimate the precision of the laboratory for each analyte throughout a sampling period (6 months). At the end of each sampling period, each analyte is given a ranking compared to the results of all other laboratories based on the cumulative bias and precision. Each laboratory is then given an overall rank in the entire program. A presentation of these data to a laboratory staff can serve as a simple but effective overview of the laboratory's performance. It might be added, however, that the ranking by number of a series of laboratories, all of whom lie within ±2 SD around a mean value, gives a false sense of satisfaction to some and a false sense of discouragement to many.

A shortcoming of the Burroughs-Wellcome program, like many others, is the difficulty of properly choosing the best peer group for comparison of data. Nevertheless, the sufficiently large data base and frequent sampling intervals (biweekly) make this group a useful approach for long-term evaluation of methods.

CAP Enzyme Chemistry Survey. Because of the absence of pure preparations of human enzymes, it is difficult to assign absolute values for enzyme concentrations. Measurement of enzymes by their catalytic properties has led to a wide number of assay procedures. This has made it difficult to compare laboratories in terms of accuracy and precision.

A unique approach to this problem has been the CAP Enzyme Chemistry Survey. The participating laboratory receives, on a quarterly basis, a set of 3 lyophilized samples. This set consists of linearly related pools of serum with increasing amounts of enzymes. Thus laboratories might differ in the absolute values for enzymatic activity, but the relative recovery of activity and the linearity of the recovery should be similar for all laboratories using similar methods and temperatures. This program also gives each participating laboratory an estimate of their short- and long-term precision. This program offers the best means of comparing a laboratory's accuracy and precision for enzyme analysis.

Multiple external-control sample-recording procedure for long-term decision making with respect to accuracy and precision

To make objective decisions as to the validity of an ongoing procedure or instrument, it is important to accumulate sufficient data over a long period. If a specific pattern of bias or imprecision persists over an extended time, the conclusion that the performance is unacceptable may be reached. The degree of bias or imprecision and its medical significance must be considered before such a decision is reached. Small biases or imprecision, though real, may not warrant immediate action but could be used as justificaton for eventual change.

A visual way of comparing the results of cumulative, long-range external quality control data is shown in Fig. 19-12. On the *x* axis is plotted the target values for the laboratory's peer group. The *y* axis records the results obtained by the laboratory. In this graph, *all* external quality control data for an analyte are thus plotted. An equal distribution of data points falling randomly above and below the line of identity are evidence that no bias exists for a method. One can also arbitrarily delineate 1 and 2 SD limits at various levels by using the usual standard deviation data for each analyte and connecting the points, as shown in Fig. 19-12. If the laboratory method for an analyte has been accurate over a long time, the values will fall close along the 45-degree line of identity, the line representing equality of laboratory target values. In nearly all the points fall above or below the identity line; this finding possibly indicates a long-term bias. If the majority (approximately 67%) of points falls within ±1 SD of the line of identity, with most of the remaining points falling between ±1 to 2 SD, this possibly indicates an acceptably accurate and precise procedure.

Responsibility for systematic bias

The burden for elimination of systematic bias problems should not rest solely on the shoulders of the user but should also be shared by the manufacturer and distributors of equipment and reagents. Careful perusal of the national surveys in clinical chemistry will quickly reveal instrument systems and reagent systems that show the presence of significant systematic bias. This information is useful when choosing new equipment. The FDA has developed an excellent reporting system for immediate communication (by a toll-free telephone line) of instrument or reagent defects to a central clearinghouse operated by the U.S. Pharmacopeia. Reports bring rapid response from manufacturers.

ACCURACY CONTROL WITH DEFINITIVE METHODS, REFERENCE METHODS, PRIMARY STANDARDS, AND REFERENCE MATERIALS

The control of accuracy is an important ongoing activity as a part of each laboratory's program to maintain a con-

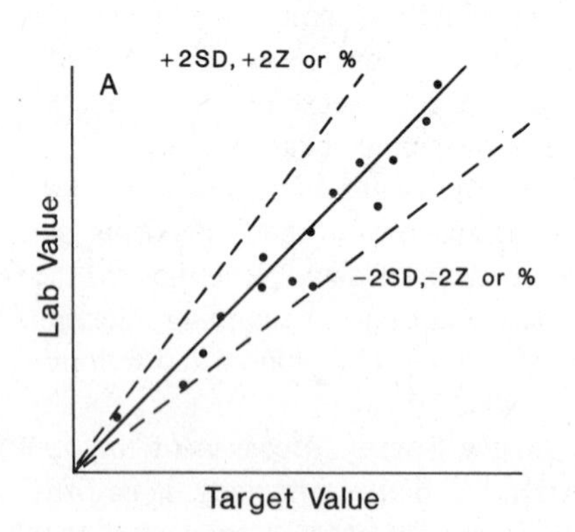

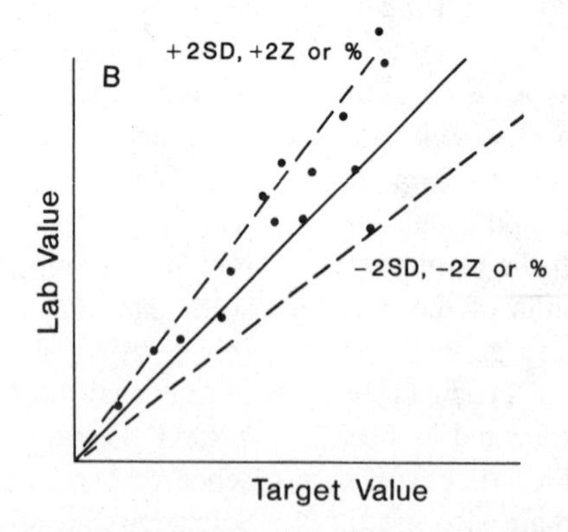

Fig. 19-12. Line of identity plot for external quality control long-term record of performance. External quality control data from CAP and Burroughs-Wellcome quality control programs plotted as lab value versus target value. *A,* Method that compares well with peer groups; *B,* method that shows positive proportional bias compared to peer groups. Line of identity is at 45 degrees from horizontal. One expects variation above and below line of identity because of inherent variability of all methods. One can use line of identity as a zero-difference line as in Youden plot, and divergent lines above and below may be labeled ±2 SDI, ±2 Z or ±% (which may be 10% or 20% or other %). In addition, axes may be labeled with absolute concentration values and divergent lines labeled percent above or below.

sistent base line of analytical measurements. Precision control comes first, since accuracy control studies require a knowledge of inherent variability of the method and a capability of maintaining a consistent average from month to month.

Practical plan for accuracy control

There are several circumstances in which it is necessary to investigate the accuracy of a method. These are as follows: (1) when introducing a new method into the laboratory, (2) when the method in use is questioned because of accumulated external quality control data (see previous section), and (3) when clinicians voice concern as to the clinical validity of the results.

There are two approaches to an investigation of the accuracy of a method. The first approach is to analyze patient samples in parallel by two methods: (1) the method under investigation and (2) by a reference or definitive method. The reference or definitive method analyses may be performed by another laboratory. Twenty comparisons are sufficient for the initial study. By use of appropriate statistical tests, usually the average difference test (t test), one can establish the presence or absence of a statistically significant bias between the methods.

The second approach is to analyze a reference material with a known concentration level by the method under investigation. The reference material provides a specific target value for the method it is to be compared with. Repeated analysis of this material will indicate a bias if one exists. Duplicate values on 2 days with the system in control would be sufficient for calculation of an average value that might indicate a difference (reference value minus the average of the four analyses) greater than 1 USD for the method, suggestive of a true bias.

Definitive and reference methods (Table 19-6)

A *definitive* method[14] is the scientifically most accurate way to measure a particular chemical substance. This involves instrumentation of the most sophisticated type and separation procedures that usually isolate the analyte in high purity before its concentration is measured. These methods are available in institutions such as the U.S. National Bureau of Standards. Because of the complex nature of proteins, peptides, hormones, and enzymes, definitive methods are not available for these analytes at the present time.

A *reference* method is a method that has been compared to the definitive method and found to give analytical concentration values within $\pm 1\%$ or 2% of the true definite value in the hands of a series of expert laboratories. Thus the reference method has a demonstrated record of transferability and accuracy. The equipment and methodology are such that these methods are usually available in a university hospital–level laboratory. If a definitive method is not available for comparison, the reference method is established by consensus among authorities in the field.

A standard-setting body such as NCCLS or the standards committee of a professional society may make this selection or recommendation, or it may be made by a group of knowledgeable individuals specializing in a given field who form a consensus recommending a reference method.

A *derived* method is any method that has been compared to a reference method and has been shown to give comparable results within a range acceptable to the FDA, which licenses these products, or within a range acceptable to the user. Information on these method comparisons and evaluations are available in the medical literature.

Definitive methods and reference methods are established by groups of interested persons who represent different disciplines and different organizations that include the measurement experts from the U.S. National Bureau of Standards, members of methodology and standardization committees of professional organizations, experts from industry, and representative experts from governmental agencies with a legal responsibility, such as the CDC and the FDA. In the United States, the NCCLS has been a final meeting place for these expert groups. Publications of the NCCLS represent the best state-of-the-art consensus agreements in the United States. Other similar international consensus organizations are now being formed on the basis of the NCCLS model.

Primary crystalline standards are always required by definitive and reference methods. The U.S. National Bureau of Standards provides 14 primary crystalline standards that may be used to prepare primary liquid standards. The analytes are glucose, cholesterol, urea, creatinine, uric acid, sodium, potassium, chloride, phosphorus, calcium, magnesium, lithium, albumin, and hydrogen ion. Aqueous sealed vials prepared from the above NBS primary standards are also available from the CAP. When NBS primary standards are used, a method's accuracy may be said to be ''traceable to'' the NBS primary standard.

Prominent sources for information on reference methods are the U.S. National Bureau of Standards, Analytical Division, Methodology Section, Gaithersburg, Md.; Clinical Chemistry Section, Communicable Disease Center, Atlanta, Ga.; National Committee for Clinical Laboratory Standards, Villanova, Pa.: Standards Committee of the American Association Clinical Chemists; and the Standards Committee of the College of American Pathologists.

Reference materials

There are now available four protein-based, reliable accuracy control materials. Each of these reference materials is useful for investigating the accuracy of a method. Target concentrations assigned by definitive methods are the most accurate values obtainable by state-of-the-art technology and thus are preferable when they are available.

1. National Bureau of Standards Human Serum SRM #909 (Immunology Section, Gaithersberg, Md.). The

NBS provides a human serum pool assayed by definitive methods for eight constituents: glucose, cholesterol, uric acid, calcium, chloride, potassium, lithium, and magnesium.

2. Seronorm. A commercially available reference pool is sold under the name of "Seronorm" (Accurate Chemical Company, Bethpage, Long Island, NY.). This lyophilized serum pool has six constituents assayed by definitive methods and 18 constituents assayed by reference methods. Other analytes are analyzed by derivative methodology. Analyzed values from four highly sophisticated laboratories are presented, two in Scandinavia and two in the United States.

3. College of American Pathologists; national survey-validated pools. These pools are from regular CAP Comprehensive Chemistry Surveys and thus have well-documented target values.[33] Since these target values are the mean reported values of all methods, they are not viewed as definitive values but are rather good approximations of the actual concentration. Studies conducted on six vials for seven analytes by the NBS indicate that the definitive method value and the average value of all methods were the same (that is, all within −2% and a majority within −1%). In addition to the all-methods value, these materials also have estimates for particular methodology, instrument, and reagent combinations.[10] The CAP also has a special collaborative project to provide specific values for serum protein constituents, which has resulted in the CAP-CDC Reference Preparation of Serum Proteins (RPSP). The reference preparation contains albumin, alpha-l-glycoprotein, alpha-1-antitrypsin, alpha-2-macroglobulin, ceruloplasmin, haptoglobin, transferrin, C3 complement, C4 complement, IgG, IgA, and IgM. The assigned values are based on analyses by acknowledged experts in each protein measurement.

4. Some regional pools submit their material for definitive method analysis to the National Bureau of Standards and thus may be used as an accuracy control.

Selection of a laboratory for accuracy control assistance

When engaging in an investigation of the accuracy of a method using another laboratory as a reference point, it is important to be completely confident of the quality of its analytical work. One should always obtain information as to the accuracy and precision of the analytical data. It is useful to ask if a definitive or reference method is being used.

How to decide on an accuracy standard

Choose one or more of the following reference averages.

1. A national external survey: (a) average of a particular instrument or reagent system, (b) all-methods average, or (c) average of a method selected for reference.

2. A regional pool program classified by instrument: (a) average of a particular instrument or reagent system, (b) all-methods average, or (c) average of a method selected as a reference point.

3. Definitive-method value: (a) NBS serum standard reference material, (b) Seronorm, or (c) special laboratory performing a definitive method.

4. Reference-method value: (a) Seronorm or (b) special laboratory performing a reference method.

5. Local-leader reference value parallel testing with (a) university hospital, (b) major clinic, or (c) clinical investigator with local importance.

The choice of an accuracy reference point requires an independent decision for each laboratory and for each method. As mentioned previously, within the laboratory consistency from day-to-day and year-to-year is the first and most important consideration.

The decision on an accuracy reference point depends on local needs and local opinions. For example, the wishes of the attending physicians should be considered. They may wish to have their values approximate those at a local university hospital or to have results approximate those of a local clinical investigator whose advice is influential in patient care. There are usually one or two clinicians in each medical community whose opinions are highly respected regarding the clinical decision making related to laboratory measurements. The opinion of these persons should be included in the discussion of accuracy goals.

Theoretically, everyone should attempt to approximate the definitive method value because, with a single reference point, all results would be interchangeable between institutions and between time periods. This may not be practical in some cases, and it is impossible for analytes that do not as yet have definitive methods, such as enzymes, hormones, and drugs. Each laboratory should make the selection and should record this decision in its procedure manual.

YEARLY HEALTHY HUMAN VALUE STUDY

An essential part of a quality control program is the yearly healthy human value study. The average value for each analyte for each year should be compared with the preceding year. If the difference between the average for these 2 years exceeds the usual standard deviation of the procedure, an inaccuracy caused by systematic bias may be found. A practical yearly program is to analyze serum from 12 males and 12 females within the 20 to 30 year age group.

Eliminating bias in healthy human values

Each average value and standard deviation includes the inherent variability of the method as expressed by the usual standard deviation of the method day to day and also contains any systematic bias present in the system. Therefore it is recommended that the healthy human values study be done each year after about 6 months of experience has been obtained with the quality control pool. In addition, collection of this data should not be done immediately after a new reference calibrator has been introduced, since this is also a time when systematic bias may be introduced. It usually requires 2 or 3 months to be sure that a new calibrator or a new pool is bias free.

Acquisition of a bias-free set of age-related healthy humans

By selecting a series of persons within the same decade (10-year period), one can compare the average values from year to year to detect systemic bias. To provide an understanding of age-related changes, first find an age-related article in the literature and then compare the decade that the laboratory has studied with the corresponding decade reported in the literature. The difference between the obtained analyte average value of the selected decade and the average from the literature for the same decade can be considered to be the systematic bias between the two measurement systems. One may apply this bias to the other decades in the article to obtain the predicted value for different age groups.

Thus by studying healthy persons in a single decade (I have found the age 20 to 30 decade most convenient), it is possible to make the published literature relevant to the laboratory reference values. Male-female differences can be evaluated by selection of 11 or 12 persons of each sex for the total of 22 or 24 nondiseased persons.

RECORD KEEPING AND SOFTWARE FORM DEVELOPMENT

The best reason for developing convenient and durable forms is to permit a convenient ongoing monthly evaluation of the measurement system. One can study problems and remedy out-of-control situations in a timely fashion.

Cincinnati General Hospital | Clinical Chemistry Laboratory

Maintenance log – IL 513 blood gas Year ____

First shift

Required maintenance	**Jan**	**Feb**	**Mar**	**Apr**	**May**	**June**	**July**	**Aug**	**Sept**	**Oct**	**Nov**	**Dec**			
Monthly:															
1. Clean PO_2 cathode and check															
electrolyte solution															
2. Change PO_2 membrane															
3. Clean sampler															
4. Clean waste sensor															
5. Change CO_2 absorber															
6. Change pump windings															
Initials of technologist															
Six months:															
1. Change pH tip seal															
2. Change belly band															
3. Change KCl tube															
4. Change bath water															
5. Change sampler needle seal															
6. Clean pH junction															
Initials of technologist															

Fig. 19-13. Example of preventive maintenance record for a specific instrument.

A second reason for good records is found in the requirements of the several national bodies that accredit medical laboratories such as the Health Care Financing Administration (HCFA) for Medicare, CAP, and the Joint Committee for Accreditation of Hospitals (JCAH).

These agencies require that daily quality control be run and that records be kept that demonstrate that quality control decisions were made. This requires documentation of quality control decisions on a daily, monthly, and year-end basis. The daily decision is recorded as "in" or "out" of control with signature or initials. With the monthly decisions, a record should be kept and signed by those attending the monthly review conference. The yearly evaluation requires similar documentation.

PREVENTIVE MAINTENANCE

Instrument maintenance principles and documentation

All instrument maintenance should be done as recommended by the manufacturer's instrument manual as a minimum, with additional maintenance added by the user as necessary. Any errors in the instrument manual should be brought to the manufacturer's attention and changed with the manufacturer's authorization.

Need to date and sign maintenance records (Fig. 19-13)

The maintenance records should be audited once a month for completeness and for a record of instrument down time and actions taken. The audit should be recorded and signed by those who perform it.

Program of quality control for materials and reagents

All prepared reagents should be dated when received and when put into use. All reagents prepared within the laboratory should have on the label a storage temperature and an outdate period. When using prepared reagents, one should record the lot numbers on the daily work sheet or in a log book so that the problems involving reagents can be identified by reagent lot numbers. Reagent lot variation is a frequent source of out-of-control values.

REFERENCES

1. Jones, R.J., and Palulonis, R.M.: Laboratory tests in medical practice, Chicago, 1980, American Medical Association.
2. Elion-Gerritzen, W.E.: Medical significance of laboratory results in relation to analytic performance, doctoral thesis, Rotterdam, 1978, Erasmus University.
3. Frier, E.F., and Rausch, V.L.: Quality control in clinical chemistry, Am. J. Med. Technol. **24:**195-207, 1958.
4. Shewhart, W.A.: Economic control of quality of manufactured products, New York, 1931, D. Van Nostrand Co., Inc.
5. Westgard, J.O.: Better quality control through microcomputers, Diagnostic Medicine **61:**741, 1982.
6. Westgard, J.O., Barry, P.L., Hunt, M.R., et al.: Performance characteristics of rules for internal quality control: probabilities of false rejection and error detection, Clin. Chem. **27:**493-501, 1981.
7. Westgard, J.O., Groth, T., Aronsson, T., et al.: Performance characteristics of rules for internal quality control: probabilities for false rejection and error detection, Clin. Chem. **23:**1857-1867, 1977.
8. Copeland, B.E., Rosvol, R.V., and Casella, J.M.: Quality control workshop manual, Chicago, 1978, American Society of Clinical Pathology Commission on Continuing Education.
9. Levy, S., and Jennings, E.R.: The use of control charts in the clinical laboratory, Am. J. Clin. Pathol. **20:**1059, 1950.
10. Youden, W.J.: The sample, the procedure, and the laboratory, Anal. Chem. **32:**23A-37A, 1960.
11. Barnett, R.N.: Clinical Laboratory Statistics, ed. 2, Boston, 1979, Little, Brown & Co.
12. Kurtz, S.R., Copeland, B.E., and Straumfjord, J.V.: Guidelines for clinical chemistry quality control based on the long-term experience of 61 university and tertiary care referral hospitals, Am. J. Clin. Pathol. **68:**463-473, 1977.
13. Velapoldi, R.A., Paule, R.C., Schaffer, R., et al.: A reference method for the determination of sodium in serum, National Measurement Laboratory, National Bureau of Standards, sp. pub. no. 260-60, Washington, D.C., 1978, U.S. Government Printing Office.
14. Velapoldi, R.A., Paule, R.C., Schaffer, R., et al.: A reference method of the determination of potassium in serum, National Measurement Laboratory, National Bureau of Standards, sp. pub. no. 260-63, Washington, D.C., 1979, U.S. Government Printing Office.
15. Velapoldi, R.A., Paule, R.C., Schaffer, R., et al.: A reference method for the determination of chloride in serum, National Measurement Laboratory, National Bureau of Standards, sp. pub. no. 260-67, Washington, D.C., 1978, U.S. Government Printing Office.
16. Schaffer, R., White, E., and Welch, M.J.: Accurate determination of serum glucose by isotope dilution mass spectrometry—two methods, Biomed. Mass Spectrom. **9:**395-402, 1982.
17. U.S. Food and Drug Administration: Proposed performance standard for in vitro diagnostic devices used in quantitative measurement of glucose in serum or plasma, an FDA Medical Devices Standards publication technical report, U.S. Department of Health, Education and Welfare, Washington, D.C., 1980, U.S. Government Printing Office.
18. Velapoldi, R.A., Paule, R.C., Schaffer, R., et al.: A reference method for the determination of lithium in serum, National Measurement Laboratory, National Bureau of Standards, sp. pub. no. 260-69, Washington, D.C., 1978, U.S. Government Printing Office.
19. Schaffer, R., Sniegoski, L.T., Welch, M.J., et al.: Comparison of two isotope dilution/mass spectrometric methods for determination of total serum cholesterol, Clin. Chem. **28:**5-8, 1982.
20. Abell, L.L., Levy, B.B., Brodie, B.B., and Kendall, F.E.: Extraction method for measuring total cholesterol in serum, J. Biol. Chem. **195:**357, 1952.
21. Björkhem, I., Bergman, A., Falk, O., et al.: Accuracy of some routine methods used in clinical chemistry as judged by isotope dilution–mass spectrometry, Clin. Chem. **27:**733-735, 1981.
22. Moore, L.I., and Machlan, L.A.: Isotope dilution mass spectrograph method for measurement of calcium, Anal. Chem. **44:**2291, 1972.
23. Cali, J.P., Mandel, J., Moore, L., and Young, D.S.: A referee method for determination of calcium in serum, National Measurement Laboratory, National Bureau of Standards, sp. pub. no. 260-636, Washington, D.C., 1978, U.S. Government Printing Office.

24. Duncan, P.H., Gochman, N., Cooper, T., et al.: Development and evaluation of a candidate reference method for uric acid in serum, Atlanta, Ga., 1980, Centers for Disease Control, U.S. DEW—PHS Clinical Chemistry Division.
25. Durst, R.A.: Standard reference materials: standardization of pH measurements, National Bureau of Standards spec. pub. 260-53, Washington, D.C., 1975, U.S. Government Printing Office.
26. Itano, H., Fogarty, W.M., and Alford, W.C.: Molar absorptivity constant for cyanmethemoglobin, Am. J. Clin. Pathol. **55**:135-140, 1971.
27. Cannan, R.K.: National Research Council, Cyanmethemoglobin standard, Am. J. Clin. Pathol. **25**:376-380, 1955.
28. Eilers, R., et al.: Experience with certification of the NBS cyanmethemoglobin standard.
29. Cohen, A., Hertz, H.S., Mandel, J., et al.: Total serum cholesterol by isotope dilution/mass spectrometry: a candidate definitive method, Clin. Chem. **26**:854-860, 1980.
30. Duncan, I.W., Mather, A., and Cooper, G.R.: The procedure for the proposed cholesterol reference method, Center for Environmental Health, Atlanta, Ga., 1982, U.S. Centers for Disease Control.
31. Velapoldi, R.A., Paule, R.C., Schaffer, R., et al.: A reference method for the determination of sodium in serum, National Bureau of Standards spec. pub. 260-60, Washington, D.C., Aug. 1978, U.S. Government Publishing Office.
32. Doumas, B.T., Bayse, D.D., Carter, R.J., et al.: A candidate reference method for determination of total protein in serum. I. Development and validation, Clin. Chem. **27**:1642-1650, 1981.
33. Doumas, B.T., Bayse, D.D., Borner, K., et al.: A candidate reference method for determination of total protein in serum. II. Test for transferability, Clin. Chem. **27**:1651-1654, 1981.
34. Gilbert, R.: Definitive method comparison with all methods averages from CAP chemistry survey, Am. J. Clin. Pathol. **70**:450-470, 1978.

Chapter 20 Evaluation of methods

R. Neill Carey
Carl C. Garber

accuracy The agreement between the mean estimate of a quantity and its true value.[21]

allowable error (E_A) The amount of error that can be tolerated without invalidating the medical usefulness of the analytical result. Allowable error is defined as having a 95% limit of analytical error; only one sample in 20 can have an error greater than this limit.[2]

assigned value The value assigned either arbitrarily (as by convention) or from preliminary evidence (as in the absence of a recognized reference method).[21]

bias A systematic component of analytical error, estimated from a comparison-of-methods experiment.[4] Also known as the difference between two quantities. A measure of inaccuracy.

comparative method The analytical method to which the test method is compared in the comparison-of-methods experiment. This term makes no inference about the quality of the comparative method.[2]

comparison-of-methods experiment An evaluation experiment in which a series of patient samples are analyzed by both the test method and comparative method. The results are assessed to determine whether differences exist between the two methods.[2]

confidence interval The numerical interval that contains the population parameter with a specified probability.[4]

constant systematic error (CE) An error that is always in the same direction and magnitude, even as the concentration of analyte changes.[3]

error The difference between a single estimate of a quantity and its true value. If a good estimate of the true value is not available, the difference may have to be expressed as the deviation from an assigned value. Note that errors as defined cannot be classified as ''random'' or ''systematic'' without sets of measurements being considered.[21]

ideal value The value of a parameter under conditions of zero error.

imprecision The standard deviation or coeffficient of variation of the results in a set of replicate measurements. The mean value and number of replicates must be stated, and the design used must be described in such a way that other workers can repeat it. This is particularly important whenever a specific term is used to denote a particular type of imprecision, such as between-laboratory, within-day, or between-day.[21]

inaccuracy The systematic error, estimated from the mean of a set of data relative to the true value or estimated from other approaches, that indicates the difference between the observed value and true or assigned value.

interference The effect of a component on the accuracy of measurement of the desired analyte.[21]

interference experiment An evaluation experiment that estimates the systematic error in a method resulting from interference or lack of specificity.[2]

linear regression An approach to choose a single line through a data set that "best" describes the relation between two subsets or two methods. This mathematical technique minimizes the sum of the squares of the differences between the observed y values and the y values predicted by the regression line for a given x value. One uses this approach assuming that there are no errors in the data by the x method.

medical decision level (X_C) A concentration of analyte at which some medical action is indicated for proper patient care. There may be several medical decision levels for a given analyte.

parameter A number that describes a feature of a population. This is in contrast to a statistic, which is an estimate of a parameter derived from a sample of the population.

precision The agreement between replicate measurements.[21]

proportional systematic error (PE) An error that is always in one direction and whose magnitude is a percentage of the concentration of analyte being measured.[2]

random analytical error (RE) An error, either positive or negative, whose direction and exact magnitude cannot be predicted; imprecision.[2]

recovery experiment An evaluation experiment that estimates proportional systematic error.[2] The amount of analyte recovered is divided by the amount of analyte added to a sample and the ratio is expressed as the percentage of recovery. The deviation of the percentage of recovery from 100% is the proportional error.

replication experiment An evaluation experiment that estimates random analytical error.[2] Measurements are made on aliquots of a stable sample over specified periods of time, as within a run, within a day, or over a period of days.

sample The appropriately representative part of a specimen used in the analysis. This sample should be called a "test sample" when it is necessary to avoid confusion with the statistical term "random sample from a population."[21]

standard error of the estimate The standard deviation of the differences, $s_{y/x}$ between the observed y values and the y values predicted by the regression line for a given x. This statistic measures the dispersion or spread of the data around the regression line.

systematic analytical error (SE) An error that is always in one direction; inaccuracy.[2]

test method The method that is chosen for experimental testing or study by means of method evaluation.[2]

true value A term considered to have self-evident meaning requiring no definition. In practice the true value is closely approximated by the definitive (method) value and somewhat less closely by the reference (method) value.[21]

total error (TE) A combination of the random and systematic analytical errors that estimates the magnitude of error that might occur in a single measurement.

variance The square of the standard deviation.

Over the last decade, the quantitative analytical methods used in clinical laboratories have become more reliable and more standardized. A growing proportion of analytical procedures are supplied by commercial manufacturers. The emphasis of the hospital clinical chemist has shifted away from methods development and toward selection of the method that suits the laboratory situation best among the commercially available methods and the subsequent evaluation of the selected method.

The process of method evaluation has also been evolving.[1] It has been recognized that a method's performance can be objectively judged as acceptable only if its errors are small enough to be acceptable for medical use. The common theme in the protocols developed by Westgard et al.,[2] the Food and Drug Administration (FDA),[3] and the National Committee for Clinical Laboratory Standards (NCCLS)[4,5] is to measure the errors in terms of analyte concentration units and to compare them to medically allowable error.

PURPOSE OF METHOD EVALUATION

Laboratory requirements

New analytical methods are usually developed for improvement of accuracy and precision over existing methods, for automation, for reduction of reagent or labor cost, or for measurement of a new analyte. In any case, the method's analytical performance in a clinical laboratory setting must be verified experimentally, even if the new method is believed to be an improvement over all previous methods. The extent of the experiments and the interpretation of the data vary, depending on the purpose of the evaluation and who performs the evaluation, but the basic approach and basic experimental design are similar for all evaluations.

The process of evaluating a method is different from the process of routine quality control of a method after it has been introduced into daily use. Routine (daily) quality control (see Chapter 19) is the process by which increases in the size and frequency of the analytical errors of a method in routine use are detected so that the method can be "repaired." Routine quality control detects errors only when they exceed the error that was present in the method when the control ranges were established. Routine quality control is not able to determine the magnitude of the inherent errors of the method or to decide whether they are acceptable. Method-evaluation experiments are required to assess the inherent analytical errors of the method and relate them to medical requirements. Routine quality control can detect only that the method's performance has deteriorated.

Manufacturer requirements

When a manufacturer develops a new method and prepares to market it, the manufacturer is required by the FDA to make claims about the analytical performance of the method, specifically about precision and accuracy.[6] These claims must be supported by experimental method evaluation data. It is essential that these claims be realistic and conservative. The level of performance of the method in most users' laboratories must be at least as good as that claimed by the manufacturer. However, potential customers often compare claims made by different manufacturers as they select methods, and so the claims must be competitive. Extensive experimental data will be required for the manufacturer to develop defensible claims. Protocols for manufacturers to follow for experimental testing of methods and statistical treatment of data to produce defensible performance claims have been developed by NCCLS.[4]

Method-evaluation studies are also performed by some other organizations to verify manufacturer's claims of analytical performance for a method. The FDA has indicated its intention to test whether the performance of any commercial method ("in vitro diagnostic device") meets the performance claimed by the manufacturer and has published its testing protocol.[7] NCCLS is also developing testing protocols for claims verification.[5] It is not cost effective to repeat all of the experiments performed by the manufacturer to develop the claims. Instead, fewer tests are done, and conservative statistical tests are used to determine whether the actual performance is worse than the claimed performance. The claim will be rejected only if there is a very high probability (such as 95%) that the method's actual performance is worse than its claimed performance.

Most method evaluations are performed by "users" (laboratory personnel) in hospitals and commercial laboratories. These evaluations are performed to determine whether the performance of a method meets the requirements for the medical applications intended by the user. The method may be commercial, newly "home grown," or a method the user has seen in the literature and is setting up in one's own laboratory. The user needs to perform the evaluation as efficiently as possible and, if performance is unacceptable, to detect it with a minimum of experimental

Table 20-1. Recommended medically allowable errors, E_A

Chemistry analyte	Decision level, X_C	Barnett	Tonks	Gilbert	Cotlove	1976 Aspen Conference		Elion-Gerritzen	Ross
						Group	Individual		
Albumin	35 g/L	5			3	1	1		
Bicarbonate	20 mmol/L	20			16				
	30 mmol/L	20			16				
Bilirubin	10 mg/L	4	2						
	200 mg/L	30	60						
Calcium	85 mg/L	6	4			15	15		
	110 mg/L	5	5.5	5.5	3.2	2	2	5	
Chloride	90 mmol/L	4	24	3.6	1.8	2.7	2		
	110 mmol/L	4	26	4.4	2.2	3.3	2.4	4.8	
Cholesterol	2500 mg/L	400	250	400	350	500	100	350	
Creatinine	20 mg/L	3	2	4					
Cortisol	50 μg/L					30	13		
	300 μg/L					180	78		
Globulin	35 g/L	5	50						
Glucose	500 mg/L	100	120		90	31	22		
	1200 mg/L	100	200	130		74	53		
	2000 mg/L							310	
Iron	1000 μg/L			150		360	260		
Magnesium	20 mg/L			3	1.4				
Osmolality	270 mOsm/kg		15						
pH	7.35	0.008							
	7.45	0.012							
P_{CO_2}	35 mm Hg	6							
	50 mm Hg	6							
P_{O_2}	30 mm Hg								12
	80 mm Hg	10							8
	195 mm Hg								31
Phosphorus	45 mg/L	4	4.5	5	4.5	3.1	2.6	7.7 (at 78 mg/L)	
Potassium	3.0 mmol/L	0.5	0.26	0.24	0.28	0.17	0.13	0.47	
	6.0 mmol/L	0.5	0.59	0.5	0.28	0.34	0.26	—	
Protein, total	70 g/L	6	4.9	7	4.4	2.8	2.1	—	
Sodium	130 mmol/L	4.0	2.3	3.8	1.0	1.0	1.0	0.5	
	150 mmol/L	4.0	2.7	4.4	1.0	1.2	1.2	—	
Thyroxine	30 μg/L					4	2	—	
	130 μg/L					18	10	—	
Triglycerides	1600 mg/L	300				1180	420	—	
Urea nitrogen	270 mg/L	40	27	50	30	54	33	50	
Uric acid	60 mg/L	10	6	9	11	6	4		
Enzymes (E_A given as percentage of measured value)									
Acid phosphatase (ACP)			20%						
Alkaline phosphatase (ALKP)			20%			13%	1.8%		
Amylase (AML)			20%						
Aspartate amino transferase (AST, GOT)			10%						
Creatine kinase (CK)						26%	13%		
Lactate dehydrogenase			20%					22.2%	

work. Thus the user can reject the method without performing all the time-consuming studies that would be required for acceptance.

Medical requirements

The decision to accept or reject a candidate laboratory method should be based on the ability of the method to meet the requirements of the final user, the physician who is interpreting the results of a laboratory test on a patient. The error of the test result is excessive if it causes a misdiagnosis. The greatest chance for misdiagnosis caused by analytical error of a test result occurs at the concentration at which a medical diagnosis is made; this concentration is termed the "medical decision-level concentration." For example, a fasting glucose concentration below 500 mg/L may be diagnostic of hypoglycemia.[8] For each decision-level concentration, a performance standard may be formulated, consisting of the decision-level concentration, X_C, and the allowable error, E_A. Allowable error is stated in concentration units, so that errors of the test method may be judged by comparison with clinically allowable error. One interprets the method evaluation data by using the data to estimate the error of the method at the medical decision-level concentration and then comparing this estimate with the allowable error. If the method's error exceeds allowable error, performance is not acceptable. If it is less than allowable error, performance is acceptable.

The amount of error present in the single measurement of an analyte is different each time the analyte is measured because a portion of the error is purely random. Thus the magnitude of error for a measurement on a given patient specimen cannot be known exactly, and one cannot predict the absolute maximum error a method could ever make on the analysis of a single patient specimen. However, one can calculate an estimate of the upper limit of the error such that there is only a 5% chance (or a lesser amount if desired) that the actual error would exceed the upper limit. Similarly, one can define allowable error as a 95% upper limit of error. There will be only a 5% chance that the error might be larger than this defined amount and hence could cause a misdiagnosis.

Exact performance standards (for allowable error) do not exist for most analytes. Performance standards have been proposed for the analytes measured most often, but mostly one must use one's own professional judgment and input from clinicians to establish the performance standard for a particular analyte. There are some guidelines for general performance standards. Several of these have been combined in Table 20-1. The medical decision-level concentrations (X_C) for Table 20-1 are those suggested by Barnett.[9,10] Tonks proposed that allowable error be either one fourth of the normal range, or 10%, whichever is less.[11] Allowable standard deviations and coefficients of variation from all references have been converted into 95% limits of allowable error when multiplied by 1.96. (A multiplier of 2 may be used for convenience; see below.) For enzymes, the limit is expanded to 20%.[12,13]

The Aspen Conference[14] sponsored by CAP also recommended the use of intraindividual and interindividual biological variations for determining the goals for the precision of a method used for group testing. The analytical coefficient of variation is denoted as CV_A, where

$$CV_A = \frac{1}{2}\sqrt{(CV_{intra})^2 + (CV_{inter})^2}$$

and CV_{intra} is the biological variation observed within an individual and CV_{inter} is the biological variation observed between individuals.[15] To enable the physician to monitor intraindividual changes, the method must be even more precise:

$$CV_A = \frac{1}{2}\, CV_{intra}$$

Solid guidelines for performance exist only for the most commonly measured analytes. It is often observed that the least quantitative aspect of clinical laboratory testing is the medical interpretation of laboratory data. One survey of clinicians' interpretation of therapeutic drug-monitoring data revealed a range of an order of magnitude in their limits for allowable errors.[16] Laboratory workers need advice about medically allowable error from the clinician users of the test in their own hospitals. A continuing dialogue with clinicians helps make their interpretations and expectations more realistic and informs the chemist of the analytic performance needed medically.

Other factors, such as turnaround time, affect the medically allowable error. Clinicians can sometimes accept increased error if turnaround time is fast. For example, the relatively insensitive, rapidly performed "slide test" for urinary human chorionic gonadotropin (HCG) is used for emergency pregnancy testing instead of the radioimmunoassay for HCG in blood, which requires several hours.

SELECTION OF METHODS

Evaluation of need

The quality of service achievable by a laboratory is set by selections of personnel, equipment, and analytical methods. The many considerations involved in the process

Steps in selection process
Determine need
Define requirements
Application
Methodological
Performance
Review literature
Select candidate method

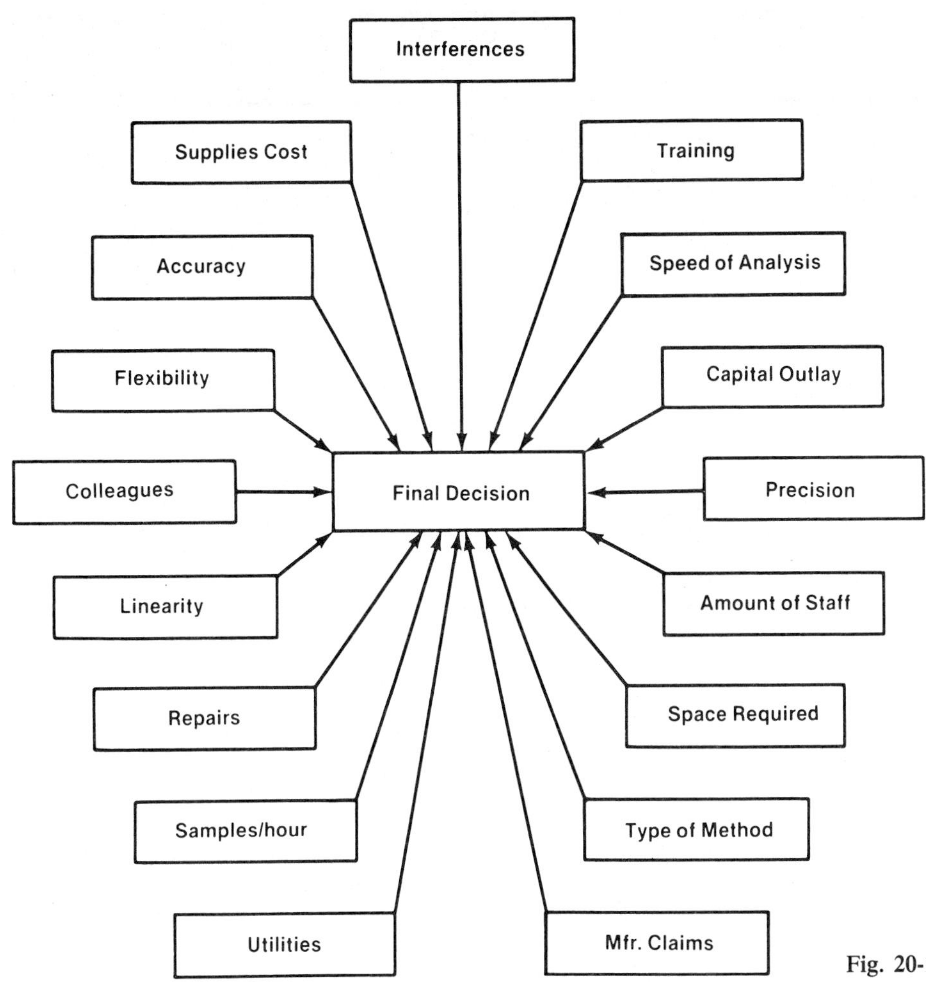

Fig. 20-1. Factors in method selection.

of method selection are shown in Fig. 20-1. Unless this process is well organized, method selection can be a traumatic and costly experience. The box on p. 342 provides a logical sequence to follow in method selection.

Often the decision to set up a new method or instrument is based on a medical requirement that a new test must be provided on site by the laboratory. A change in the methodology of a presently offered test may also be dictated by advances in laboratory practice. For example, during the 1970s most laboratories converted their neocuproine or alkaline ferricyanide glucose methods to the highly specific enzymic glucose methods, which are based on glucose oxidase or on the coupled hexokinase/glucose-6-phosphate dehydrogenase procedure. The need for a new method or device may also be dictated by the age and lack of operational reliability of the present analyzer.

Application characteristics

After the need for a new method (or analyzer) has been determined, all the practical features required of the method are defined. These are termed "application characteristics." They include requirements of sample size, turnaround (dwell) time, sample throughput rate, specimen type, automated calibration, on-line quality control review, self-diagnostics, laboratory space required, reagent storage facilities required, availability and skill of laboratory staff, time available for training, cost per test, and safety and environmental hazards. Emphasis will be placed on sample size for pediatric applications, on turnaround time and interrupt features for stat applications, and on the sample throughput rate for high-volume screening applications. It is essential that a candidate method meet these fundamental requirements before one considers it any further.

Cost per test has been defined as an application characteristic. This item could be considered separately in light of present emphasis on rising medical costs. The factors affecting the direct costs should be considered for comparison purposes between the candidate methods. These include the depreciated capital cost, reagent and supplies cost, service and repair cost, and labor cost. Much of this information is available from the manufacturer in terms of initial equipment cost, estimated reagent and supplies consumption and cost, estimated productivity, and service costs (by contract or per visit). Other information, such as

Estimation of direct costs

Laboratory information
- 8000 patient tests per year
- 50% estimated test yield (samples, quality control, repeats, dilutions, troubleshooting)
- 16,000 total number of assays
- 4 CAP units per sample = 64,000 total units
- Five-year depreciation of capital
- $750 set-up costs (change laboratory bench)

Manufacturer information
- $11,000 Cost of equipment
- $1200 Cost of reagents and supplies (for 16,000 assays)
- $1100 Service contract (two visits)
- $500 Replacement parts

Calculate equipment costs

Capital and setup: $11,750 ÷ 5 yr = $2350/yr ÷ 8000	= $0.294/sample
Service and repair: $1100 + $500 = $1600/yr ÷ 8000	= $0.20/sample
Subtotal	= $0.494/sample

Calculate labor costs (assume $10/hour including benefits)

64,000 CAP units ÷ 60 units/hour × $10/hour ÷ 8000 samples	= $1.333/sample
Total direct costs	= $1.827/sample

the expected work load and anticipated modifications in the productivity based on the internal quality control procedures, are available from within the laboratory. One can use CAP work-load recording units to make an initial estimate of the direct labor costs. Users must adjust this information to their own laboratory situation. The example in the box above illustrates how the above information may be combined to arrive at a total direct cost per test.

Method characteristics

The next step in the selection process is the definition of ideal methodological characteristics that will enable the selected method to have a good chance for success in the user's laboratory. These characteristics include preferred methodology, which will potentially have the necessary chemical specificity (freedom from interferences) and chemical sensitivity (ability to detect small quantities or small changes in the analyte's concentration). The ability to use primary aqueous standards for calibration (freedom from matrix effects) is also important. The choice of reagents, temperature, reaction time, measurement time, and measurement approach (such as one-point, two-point, or multipoint kinetic methods) are all characteristics of a method and should be defined. A source of recommended principles for clinical chemistry methods has been developed by NCCLS.[18]

Analytical performance characteristics

The method should also be defined in terms of its analytical performance capabilities. Overall goals for analytical performance have been discussed in terms of allowable error based on the medical application of the test. Other aspects of performance that must be defined are working range of the method (linearity), stability of the reagents, ability of the analyzer to detect reagent depletion in the case of enzyme substrates, expected reference range, amount of error caused by interfering substances, precision (within-run, between-run, between-day, and total), and accuracy of the method (determined by comparison of results to those obtained by a reference or standard method). The manufacturer is now required to provide information about precision and accuracy. However, it is essential that the new method be tested to see whether, in fact, it does perform to manufacturer's specifications. More importantly, the selected method must be evaluated experimentally to determine if the method's actual performance in the user's laboratory is good enough to meet the needs for the medical application in the user's institution. The manufacturer's claimed performance should be considered only as a starting point for determining the actual performance in the user's laboratory setting.

Next, one should review the technical and professional literature to determine what methods are available and to obtain some information about their application, methodological, and performance characteristics. It is also useful to confer with colleagues as to their experience and recommendations.

The final step in the selection process involves putting all the above-garnered information together to arrive at a final choice. The use of a rating scheme enables one to obtain a more objective overall rating of the candidate methods.[19,20] The rating scheme can be customized by use of appropriate weighting factors for the characteristics that are more important.

LABORATORY EVALUATION OF A METHOD

The goal of a method evaluation study is not to test all methods to determine the method with the smallest error but to determine whether the selected method has acceptably small analytical errors. The process of method evaluation involves estimation of the magnitude of analytical error for a single patient specimen. The laboratory experiments performed to obtain data for estimating the errors are chosen because they give quantitative estimates of random and systematic errors with a minimum of experimental work. The error estimates obtained may be invalid, however, if certain underlying assumptions are not true. These assumptions include operator familiarity with the method's procedure, stability of calibrators, controls, reagents, and linearity of response throughout the working range.

Familiarization

It is essential that the operators of the method become thoroughly familiar with the details of the method and instrument operation before the collection of any data that will be used to characterize the method's performance. This familiarization period has been addressed by NCCLS[4,5] and may include training by the manufacturer. It should be of sufficient duration so that one can perform all aspects of the method or instrument operation comfortably at its completion. Obviously the length of time for device familiarization varies with the complexity of the method or analyzer.

Stability

Verification of the stability of reagents, calibrators, and control materials, especially those prepared in house, can be a lengthy procedure. The matter is simplified considerably for commercially prepared materials. One can use the expiration dating of the manufacturer during the method evaluation because serious stability problems will be detectable through unacceptable analytical performance of the method. For in-house preparations, it is necessary to document these characteristics. One should perform preliminary studies with crossover analyses comparing the results for fresh calibrators and ''old'' calibrators using fresh reagents to test the stability of calibrators. This should be done several times, and the differences for each specific age of calibrator should be averaged to reduce the effects of different preparations. Similarly, one can test the stability of reagents by periodically (daily, weekly, or monthly, depending on the anticipated decay rate) preparing new reagents and testing them against the older reagents. The older reagents should be stored under specified conditions for the subsequent measurements. One can test the observed differences by use of a t test (see Chapter 17).

Linearity

The International Federation of Clinical Chemistry (IFCC) has defined the analytical range in a qualitative sense, stating that it is ''the range of concentration or other quantity in the specimen over which the method is applicable without modification.''[21] When the limits of linearity are studied experimentally, the range of concentrations included should at least encompass the limits claimed by the manufacturer. The absolute minimum number of different concentrations that must be measured is three, in a mathematical sense, but at least five different concentrations are recommended, evenly spaced throughout the range of interest. Duplicate measurements should be made on each concentration sample. Ideally, an initial linearity study should be conducted by use of aqueous standards throughout the range to identify the capabilities of the method in an ideal specimen matrix. This should be followed by the analysis of an analyte dilution series of samples containing the biological matrix (such as serum or urine). The relative performances between aqueous and matrix samples will provide important information about the impact of the biological matrix on the method. It may be difficult to prepare specimens in a biological matrix with a range of analyte concentrations from zero to the limit of linearity. For analytes not normally present in the matrix, such as drugs, the analyte is simply added to an analyte-free specimen to the desired maximum concentration, and a dilution series is prepared by use of analyte-free serum or urine. One can also approximate serum matrices by diluting stock aqueous pools of analyte with human serum albumin, enzyme inactivated serum, or Plasmonate (Cutter Laboratories, Inc., Berkeley, Calif.). One can also use a patient specimen containing the analyte at a concentration known to exceed the linearity of the method and then construct a dilution series using the analyte-free materials. The accuracy of the volumetric dilutions is very important, and serial dilutions are not recommended because of the propagation of errors through the subsequent samples. Rather, each sample should be prepared by direct dilution from the original high sample or pool. Finally, all the data points should be plotted for visual inspection of linear performance. The actual result of the analysis of each dilution is plotted against the percentage of high pool present in each dilution. The straight portion of the resulting curve represents the linear portion of the assay. In the case of methods with curvilinear response (such as radioimmunoassay procedures), the results obtained from the recommended curve-straightening algorithms should be plotted to show linearity of results.

Random and systematic error

In general, errors that affect the performance of analytical procedures are classified as either random or systematic. Factors contributing to random error are those that affect the reproducibility of the measurement. These include instability of the instrument, variations in the temperature, variations in the reagents and calibrators, variabilities in handling techniques such as pipetting, mixing, and timing, and variabilities in operators. These factors superimpose their effects on each other at different times. Some cause rapid fluctuations, and others occur over a longer time. Thus random error has different components of variation that are related to the actual laboratory setting. The within-run component of variation (σ_{wr}) is caused by specific steps in the procedure, such as pipetting precision, and short-term variations in the temperature and stability of the instrument. Within-day, between-run variation (σ_{br}) is caused by differences in recalibration that occur throughout the day, longer-term variations in the instrument, small changes in the condition of the calibrator and reagents, changes in the condition of the laboratory during the day, and fatigue of the laboratory staff. The between-day component of variation (σ_{bd}) is caused by variations in the instrument that occur over periods of days, changes in calibrators and reagents (especially if new vials are opened each day), and changes in staff from day to day. One can combine these components in such a way as to produce an estimate of the total variance of a method (σ_t).

$$\sigma_t^2 = \sigma_{wr}^2 + \sigma_{br}^2 + \sigma_{bd}^2$$

Terms used to indicate random error include *precision, imprecision, reproducibility,* and *repeatability*. In each case, they refer to the random dispersion of results or measurements around some point of central tendency.

Systematic error describes the error that is consistently low or high. If the error is consistently low or high by the same amount, regardless of the concentration, it is called constant systematic error (Fig. 20-2). If the error is consistently low or high by an amount proportional to the concentration of the analyte, it is called "proportional systematic error."

Factors that contribute to constant systematic error are independent of the analyte concentration because this type of error is of a constant magnitude throughout the concentration range of the analyte. This type of error is caused by an interfering substance that gives rise to a false signal. The false signal could be high or low; that is, the error could be positive or negative. A reaction between the interfering substance and the reagents caused by a lack of specificity is an example of a constant systematic error. Another cause of systematic error is an interfering substance that affects the reaction between the analyte and the reagents. This type of error is seen in enzymatic methods using oxidase-peroxidase–coupled reactions in which the hydrogen peroxide intermediate is destroyed by endogenous reducing agents such as ascorbic acid. An interfering substance may also inhibit or destroy the reagent so that it remains in suboptimal amounts for the reaction with the analyte. A nonchemical source of constant systematic error may be caused by improper blanking of the sample or the reagents.

Proportional error may be caused by incorrect assignment of the amount of substance in the calibrator. If the calibrator has more analyte than is labeled, all the unknown determinations will be low in proportion to the amount of error in the calibration (and vice versa). Erroneous calibration is the most frequent cause of proportional error. Proportional error may also be caused by a side reaction for the analyte. The percentage of analyte that undergoes a side reaction will be the percentage of error in the method.

EXPERIMENTS TO ESTIMATE MAGNITUDE OF SPECIFIC ERRORS

In designing experiments to be performed to determine the analytical errors of a method, it is imperative that they be carefully conceived to avoid ambiguous conclusions. The aim of this section is to describe specific experiments

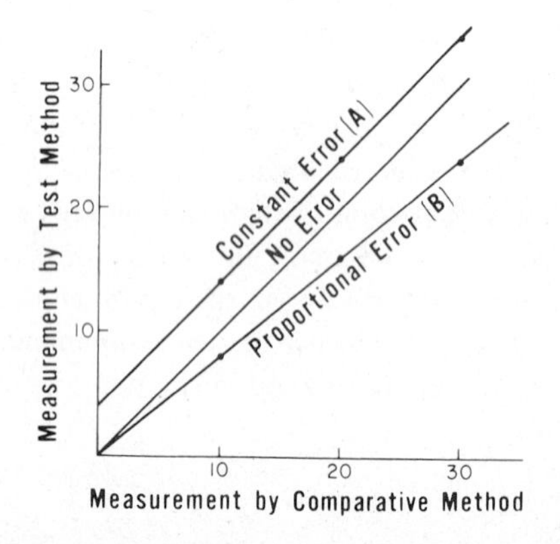

Fig. 20-2. Constant and proportional errors. (From Westgard, J.O., de Vos, D.J., Hunt, M.R., et al.: Am. J. Med. Technol. **44**:290, 1978.

TYPE OF ANALYTIC ERROR	EVALUATION EXPERIMENTS	
	PRELIMINARY	FINAL
RANDOM ERROR	REPLICATION WITHIN RUN PURE MATERIALS REAL SAMPLES	REPLICATION RUN TO RUN REAL SAMPLES
CONSTANT ERROR	INTERFERENCE	COMPARISON WITH COMPARATIVE METHOD
PROPORTIONAL ERROR	RECOVERY	

Fig. 20-3. Specific evaluation experiments for estimating specific types of analytical error. (From Westgard, J.O., de Vos, D.J., Hunt, M.R., et al.: Am. J. Med. Technol. **44**:290, 1978.)

that will enable one to estimate the magnitude of a specific error. One can then compare the size of the error to the allowable error to determine the acceptability of the method. This approach is used for all the types of errors described previously. Each type of error is considered individually before combinations of errors are considered. Fig. 20-3 presents an organization of experiments to be performed for specific error determinations arranged in such a way that the easy experiments can be done first. The more extensive (and expensive) final studies are performed only if the errors estimated by these preliminary experiments are acceptable.

Random error estimated from replication studies

The within-run replication experiment is the simplest type of study and should be one of the first performed to assess the performance of a new method. Because it allows assessment of precision over a very short time, the results cannot be extrapolated to indicate long-term performance. The short-term performance must be judged acceptable before one goes to the trouble to study the long-term performance of the method.

The replication study should first be performed with an aqueous solution of calibrator or standard and then repeated with samples whose matrix is as similar as possible to that of the intended patient samples. The concentrations to be studied should be at or near the medical-decision concentrations for the analyte. This is where the laboratory data will be interpreted most critically; thus one must be certain of the method's performance at those concentrations.

An estimate of random error is developed by consideration of repeated analyses of the same specimen. Sixty-eight percent of the results are within ± 1.0 standard deviation of the test mean (SD, or s_{TM} in this chapter), and 95% of the results are within 1.96 SD of the mean. The 95% limit for random error is 1.96 times the standard deviation. (1.96 SD is taken as the 95% limit of random error throughout this chapter. You may wish to round to 2 SD for simplicity, with an accompanying 2% difference from the rigorous estimate of random error. The standard deviation is seldom known accurately to within 2%; thus this rounding would probably not degrade the quality of the estimate of error significantly.) If the estimate of random error is less than allowable error, random error is acceptable. Example calculations are shown on p. 355 and in Chapter 17.

Constant error estimated from interference studies

The interference study measures the "constant" error caused by the presence of a substance suspected of interfering with the test method. A sample is spiked with a suspected interfering substance. The volume of this addition should be small, less than 10% of the sample volume, so that the disruption of the matrix is minimal. To compensate for the dilution of the spiked sample, a base-line sample should be prepared by addition of the solvent, used for the interferent, to another aliquot of the sample. The two samples should then be measured, at least in duplicate. The difference between the results in the two samples is attributable to an interference caused by the added substance.

An alternate scheme for studying the effects of hemolysis involves taking two blood samples. One is centrifuged and analyzed directly (base-line sample), and the red blood cells in the other blood tube are physically traumatized to rupture the cell membranes to yield an elevated amount of serum hemoglobin. After centrifugation, this hemolyzed sample is analyzed. The difference between the two samples is attributable to the effects of hemolysis. Mild, moderate, or severe hemolysis may be simulated, depending on the volume of red cells traumatized. This approach is more consistent with the actual problems encountered in the laboratory than the approach in which pure hemoglobin is added to a sample. It is not valid if red blood cells contain the analyte.

One may study the effects of lipemia by dividing a lipemic sample into two portions and analyzing one directly but centrifuging the other with an ultrahigh speed centrifuge to remove the lipoproteins before analysis. The difference in results is attributable to the effects of lipemia. Alternatively, one may also prepare turbid specimens for each decision-level concentration by adding small amounts of Intralipid (Cutter Laboratories, Inc., Berkeley, Calif.) to nonlipemic specimens of appropriate analyte concentrations to obtain slightly, moderately, and grossly appearing lipemia. Base-line concentrations are prepared by addition of equal volumes of water to the original specimens.

Pools with increased amounts of unconjugated bilirubin are produced from a stock solution of bilirubin prepared by the dissolving of pure bilirubin in *N,N*-dimethylsulfoxide to 2500 mg/L. Clear, nonicteric patient sera are spiked to the desired bilirubin concentration. Base-line specimens are prepared as above.

The choice of substances to be tested is almost infinite. For all absorbance-measuring methods, the effects of hemolysis, icterus, and lipemia should be determined. Other substances that have been reported to affect similar methods should be tested. The accuracy of the pipetting is important in terms of knowing how much interfering substance has been added. More importantly, the pipetting should be precise so that the base-line and spiked samples reflect the same extent of dilution. Again, it is important that the concentration of the analyte in the sample be near the medical-decision levels. The concentration of the suspected interfering substance should be added in such an amount that the final concentration is at the maximum physiologically expected amount. If no errors are caused at this high concentration, one can assume that lower concentrations will not adversely affect the performance of the method. If an error is too large at the maximum concentration of interfering substance, it may be appropriate to test the interference at lower concentrations. A slightly icteric sample may be acceptable, but a grossly icteric one may not. It is recommended that these interference studies be conducted on the comparative method (see next column) at the same time as a check on the experimental technique.

Calculation of the constant error from the interference experiment data is shown on p. 356. The overall average difference (bias) is called a "constant error" (CE) because it is independent of the analyte concentration. This constant error is compared directly to the allowable error for the appropriate decision level. If the constant error is less than allowable error, the constant error caused by the interferent is judged acceptable. This decision is based on clinical limits instead of a statistical test of significance. The standard deviation of the interference values is a measure of the uncertainty of the estimated constant error.

Proportional error estimated from a recovery experiment

Another preliminary study is the recovery experiment. This procedure involves the addition of a known amount of analyte to an aliquot of sample. As in the interference experiment, the sample is divided into two aliquots. One aliquot is spiked with analyte. An equivalent amount of diluent is added to the second; it is termed the *base-line sample*. The two samples are then analyzed. The base-line sample provides the original amount of analyte. The difference between the spiked sample and the base-line sample indicates the amount "recovered." The amount "added" is calculated from the volumetric details of the addition step and the concentration of the "stock" solution of the analyte. The volume of analyte added to the sample should be less than 10% to avoid major disruption of the sample matrix. Pipetting accuracy is critical because the amount of added analyte is calculated from the volumetric dispensings. The concentration of the sample and the amount added should be such that they test the performance of the method near the medical-decision levels of the analyte. In some instances a very small amount of analyte is added to the sample, and the amount recovered is lost in the randomness of the method. Thus it is advisable to make two to four measurements on each sample to reduce the effects of the imprecision of the method. Analysis of these samples with the comparison method is recommended as a check on the experimental technique. The calculation of recovery is illustrated with an example on p. 356. Recovery is defined as the ratio of the amount recovered to the amount added and is given as a percentage. The difference between the calculated percentage of recovery and its ideal value (100%) is the percent of proportional error. The standard deviation of the percentage of recovery is a measure of the uncertainty of percentage of proportional error. One cannot compare percent of proportional error directly to allowable error to decide acceptability because it is not in concentration units. One can convert proportional error to concentration units at the medical-decision level, as shown on p. 357. If the proportional error is less than the allowable error, the proportional error is acceptable. Again, the decision is based on medical requirements rather than statistical tests of significance.

If within-run random error, constant error, and proportional error are acceptable, the day-to-day replication and comparison-of-methods experiments are performed.

FINAL-EVALUATION EXPERIMENTS

The final-evaluation experiments take the most time to perform and potentially yield the most definitive information about the test method's day-to-day performance on real patient specimens.

Between-day replication experiment

The between-day replication experiment is an expansion of the within-run experiment over a period of many days, usually 20. This period must be long enough to allow the random effects occurring over several days to influence the long-term estimate of random error. This experiment and the comparison-of-methods experiment described below are usually combined for better efficiency in the study.

A material known to be stable for the time of the experiment is used. Usually a frozen serum or plasma pool or a lyophilized control product is used. Aliquot-to-aliquot variation of the material must be minimal because it will appear to be day-to-day variance of the test method.

Random error is estimated as 1.96 times the total standard deviation and compared to allowable error as described previously for the within-run study.

Comparison-of-methods experiment

The comparison-of-methods experiment determines the systematic error of the test method, using real patient specimens. A group of patient specimens are analyzed by both the test method and a comparative method, a method known to be accurate and precise. Systematic differences between the two methods are interpreted as errors of the test method if the results of the comparison method are known to have little or no error (negligible random and systematic errors). Thus the comparative method should be of the highest quality possible so that errors will not be erroneously assigned to the test method.

Methods may be classified as three types in terms of the quality of their performance. These are definitive, reference, and routine methods. See Chapter 19 for a description of definitive and reference methods.

In practice, the comparative method is often the method in routine use in the laboratory but is not of reference quality. It is useful to see how results from the test method compare with those of the routine method, but differences between the two methods should be interpreted cautiously unless the quality of the comparative method is known to be high.

At least 40 and preferably 100 or more patient specimens should be analyzed. They should include the variety of disease states that will be encountered by the test in routine use. Analyte concentrations of the specimens should be evenly distributed throughout the analytical range; otherwise regression analysis of the comparison data will be inaccurate. Hemolyzed, lipemic, and icteric specimens should be included, if not proscribed by the manufacturer of the test method, and only if they do not cause errors by the comparative method. If included in the study, they should be identified. Specimens must be carefully selected from the routine work load to be an efficient representation of the patient mix; preanalysis by the routine method is usually necessary.

Specimens are analyzed in duplicate by each method. Results should be examined carefully and plotted daily. Any specimen whose duplicates do not agree closely by either method should be reanalyzed in duplicate by both methods in the next run. Any specimen with large differences between results by two methods should be similarly reanalyzed. If the large difference is confirmed, one should investigate the patient for disease or diseases present and the specimen for other analytes (possible interferents) to determine the cause of the large difference. Immediate follow-up is essential to avoid unanswerable questions about outliers later.

The test and comparative methods should be run at the same time, or as closely as possible to each other. If this is not possible, specimens must be stored in a manner that guarantees analyte stability.

The comparison-of-methods experiment is usually combined with the between-day replication experiment. Patient specimens should be evenly spread over at least five runs, and preferably all 20 runs, to ensure that day-to-day effects have a chance to influence the data (and to ensure that day-to-day effects are "fully confounded" in statistical parlance). Both methods must be maintained in quality control during the period.

***t*-test statistics: bias, s_d.** The systematic differences between the test and comparative methods are most easily estimated from the comparison-of-methods data by the bias. The bias is the difference between the average result by the test method and the average result by the comparative method. Bias can indicate the magnitude of the systematic error between the two methods. (Each patient specimen must be analyzed by both methods for bias to be valid.) Bias is given by equation 20-1, in which y_i and x_i are the analyte concentrations of the individual specimens by the test method and comparative method, respectively, and N is the number of paired results compared.

$$\text{Bias} = \frac{\sum(y_i - x_i)}{N} \qquad \textit{Eq. 20-1}$$

The standard deviation about the bias, called the "standard deviation of the differences," s_d, is calculated in a manner analogous to the standard deviation from the replication experiment. One may view it as an indicator of the random error between the two methods.

$$s_d = \sqrt{\frac{\sum(y_i - x_i - \text{Bias})^2}{N - 1}}$$

The statistical significance of the bias, that is, whether it really differs from zero, or no bias, is determined by use of the t test. A t value is calculated according to the formula

$$t = \frac{\text{Bias}\sqrt{N}}{s_d}$$

The t value is the ratio of a systematic-error term (bias) to a random-error term (s_d). If the bias increases relative to the standard deviation of differences, there is less of a probability that the observed bias is caused by random variations and more of a probability that there really is a systematic difference between the test and comparative-method mean values. For example, in a comparison of glucose methods there were 101 specimens, the bias was 30 mg/L, and the t value was 2.11. The critical t value for $p = 0.05$ and for 100 degrees of freedom (obtainable from a statistics textbook) is 1.99. (The two-sided critical t value is used because the bias could be either positive or negative.) The calculated t value exceeds the critical t value; therefore a statistically real bias exists between the two methods. (For a more thorough discussion of the t test, see Chapter 17.)

The acceptability of the systematic error, as estimated by the bias, is judged by comparison with allowable error. If bias is less than allowable error, the systematic error is acceptable. If bias exceeds allowable error, the systematic error is not acceptable. Decisions about acceptability should never be based on the t value alone. A large bias and large s_d may combine to give an insignificant t value, even though the bias is unacceptably large.

Westgard and Hunt[22] have shown that bias can give inaccurate estimates of systematic error if proportional error is present. Both proportional and constant errors are combined in the bias. Proportional error also increases s_d. Bias should not be used as an estimator of systematic error unless proportional error is absent, or the mean analyte concentration as measured by the comparative method is very near the decision-level concentration (X_C) and the data are well distributed around X_C. Otherwise the bias will be weighted toward the side of X_C that has the most samples with large individual biases.

Correlation coefficient. The statistic most frequently cited in reports of comparison-of-methods experiments is the correlation coefficient. It is calculated according to equation 20-2, in which r is the correlation coefficient.

Statistic	Range of Concentrations Studied 0-1.5 mg/dl	0-2.5 mg/dl	0-4.5 mg/dl
r	0.773	0.878	0.950
bias	0.17	0.17	0.17
s_d	0.30	0.29	0.31
$s_{y/x}$	0.29	0.29	0.31
a	0.17	0.17	0.20
b	1.025	1.007	0.966

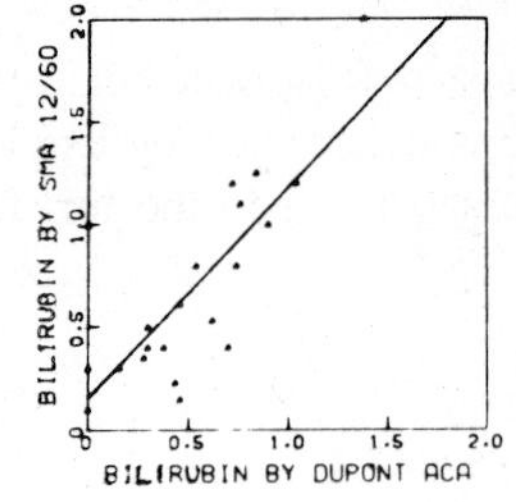

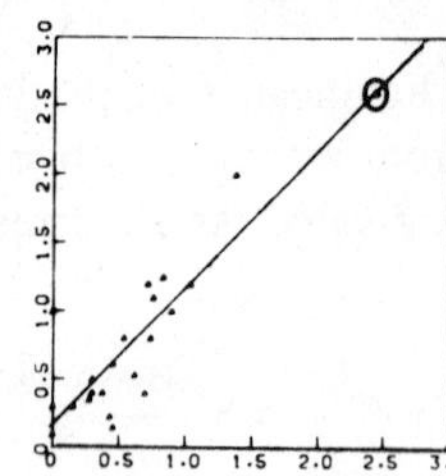

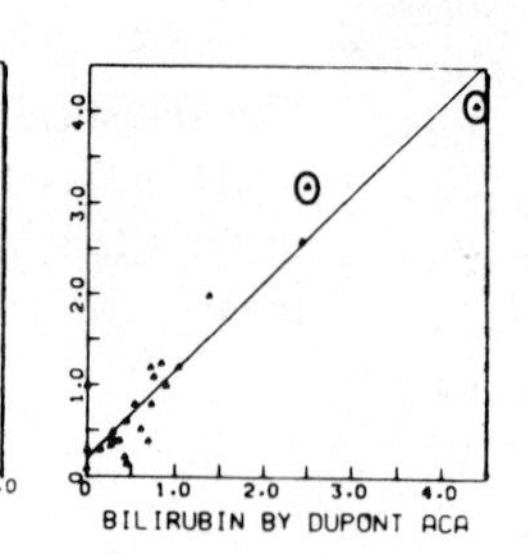

Fig. 20-4. Effect of range of data on correlation coefficient, *r*. (From Westgard, J.O., de Vos, D.J., Hunt, M.R., et al.: Am. J. Med. Technol. **44**:552, 1978.)

$$r = \sqrt{\frac{N\sum x_iy_i - \sum x_i\sum y_i}{[N\sum x_i^2 - (\sum x_i)^2][N\sum y_i^2 - (\sum y_i)^2]}} \quad \text{Eq. 20-2}$$

It is the ratio of the covariance (variation in which x and y move proportionally together away from their means) to the total variance of x and y. An r value of zero indicates that there is no correlation between the methods. A value of $+1$ indicates perfect positive correlation.

The correlation coefficient is often considered with the linear regression statistics. It is treated separately here because of its frequent abuse in method evaluation reports. Westgard and Hunt[22] demonstrated that the correlation coefficient is extremely sensitive to the range of analyte concentrations of the patient specimens in the comparison-of-methods experiment. In a comparison of bilirubin methods over a range of 0 to 45 mg/L, a correlation coefficient of 0.950 was obtained. When data pairs with bilirubin concentrations above 15 mg/L were eliminated, the correlation coefficient dropped to 0.773. This is shown in Fig. 20-4.

Decisions about the acceptability of the analytical performance of a method should never be based on the value of the correlation coefficient alone.

Linear regression statistics. If the test method and comparative method do correlate with each other, an $X:Y$ plot of results resembles a straight line, which can be described by the linear regression expression:

$$Y_i = a + bx_i \quad \text{Eq. 20-3}$$

where Y_i is the calculated value on the straight line corresponding to the actual comparative method result, x_i. The proportionality between the methods is given by the slope, b, whose ideal value (no errors) is 1. Constant error is indicated by the y intercept, a. Random error between the methods is indicated by the standard error of the regression, $s_{y/x}$ (also called the standard error of estimate and the standard deviation of the residuals). Linear regression statistics are calculated by equations 20-4 to 20-6.

$$b = \frac{N\sum x_iy_i - \sum x_i\sum y_i}{N\sum x_i^2 - (\sum x_i)^2} \quad \text{Eq. 20-4}$$

$$a = \bar{y} - b\bar{X} \quad \text{Eq. 20-5}$$

$$s_{y/x} = \sqrt{\frac{\sum(y_i - Y_i)^2}{N - 2}} \quad \text{Eq. 20-6}$$

An estimate of systematic error at X_C, the decision-level concentration, may be obtained from the linear regression statistics by substitution of X_C for x_i in equation 20-3, to calculate Y_C, the concentration the test method would measure for a specimen whose true analyte concentration is X_C. The systematic error, SE, is calculated by subtraction of X_C from this Y_C:

$$SE = |Y_C - X_C| = |a + bX_C - X_C|$$

This estimate of error will be valid only if the following limitations of linear regression are observed.

The data must be carefully examined for nonlinearity, and the range of data must be limited to the linear range. Nonlinearity at higher concentrations (falling off at high concentrations) will lower the slope, increase the y intercept, and increase $s_{y/x}$.

The data must be carefully examined for outliers. The importance of daily examination and plotting of comparison-of-methods data cannot be overemphasized. Outlier specimens (for which $y_i - Y_i > 3.5\ s_{y/x}$) must be detected immediately and reanalyzed by both methods so that the data can correct or confirm the outlier. The linear regression line is "pulled" toward the outlier, with the greatest effects caused by outliers at the extremes of the data. Confirmed outliers should be investigated for their causes. A confirmed outlier really is representative of the true analytical performance of the method. The systematic error of

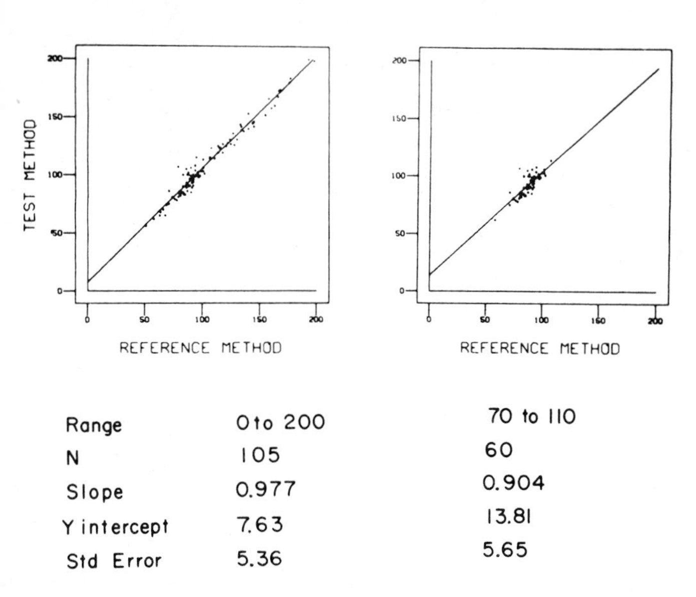

Range	0 to 200	70 to 110
N	105	60
Slope	0.977	0.904
Y intercept	7.63	13.81
Std Error	5.36	5.65

Fig. 20-5. Effect of range of data on linear regression statistics. (From Westgard, J.O., and Hunt, M.R.: Clin. Chem. **19**:49, 1973.)

TRUE VALUE
OBSERVED VALUES
PRECISION — RANDOM ANALYTIC ERROR
ACCURACY (1) — SYSTEMATIC ANALYTIC ERROR
(2) — TOTAL ANALYTIC ERROR

Fig. 20-6. Total analytical error. (From Westgard, J.O., Carey, R.N., and Wold, S.: Clin. Chem. **20**:825, 1974.)

the test method should be calculated both with the outlier included in the data set and with it excluded. If errors are acceptable with the outlier excluded and excessive with it included, extreme caution should be exercised. There are statistical tests for removal of outliers,[23] but no more than one outlier should be excluded in a set of 40 patient comparisons. If more than one outlier is present per 40 patient comparison samples, the test method should be rejected until a cause for the outliers can be found and corrected.

The range of analytic concentrations must be wide. The effects of a narrow range of data on the least-squares statistics are seen in Fig. 20-5. Methods-comparison data often fail to meet one additional assumption of linear regression calculations. This assumption that is violated requires that the x data (comparison) be known without error. Actually, in a methods-comparison experiment, random errors do affect the results of the comparison method. When the range of the data is sufficiently large, the effects of the failure to know the x values without error becomes negligible.[24,25]

Waakers[24] has suggested that the correlation coefficient be used to decide whether the range of data is sufficient for using the traditional least-squares calculation. If the correlation coefficient is greater than 0.99, the traditional least-squares approach will calculate a slope whose mathematical error will be less than 1%. If the correlation coefficient is less than 0.99, the slope will be falsely low and the y intercept will be too high. Cornbleet and Gochman[25] have suggested another decision limit. If the ratio of the analytical standard deviation of the comparative method to the standard deviation of the comparison-of-methods experiment specimen population by the comparative method is less than 0.2, the least-squares calculation will be appropriate. If these tests on the data fail, one should use another regression approach, such as those discussed by Cornbleet and Gochman.

Calculation of systematic error by use of linear regression statistics is demonstrated on p. 358.

ESTIMATION OF TOTAL ERROR

Estimates of random error and systematic error are combined to estimate the total error of the test method. This is the most severe criterion for the test method to meet. The rationale for the total error concept is shown in Fig. 20-6. The horizontal line is error in concentration units, and the vertical line is located at the "true-value" concentration, the medical decision-level concentration, or zero error. The vertical distance from the horizontal line represents the probability of obtaining a test method result at any given amount of error (difference from X_C). The bell-shaped curve shows the distribution of test-method data obtained from repeated analyses of a patient specimen whose true analyte concentration is X_C. The distance from the mean of that curve to the true value is the systematic error. The dispersion around the mean of the data is the random error, which has a 95% limit of 1.96 times the standard deviation. There will be instances in which the combined error will be exactly equal to the systematic error, other times when the combined error for a given result will be less than the average systematic error by some amount because of the random error of the method, and other times when the combined error will be greater than the systematic error, again by some amount caused by the random error of the method. The physician has no way of

Table 20-2. Point-estimate criteria for acceptable performance

Type of error	Criteria
Random (RE)	$1.96 \cdot s_{TM} < E_A$
Constant (CE)	$\text{Bias} < E_A$
Proportional (PE)	$\frac{\lvert \bar{R} - 100 \rvert}{100} \cdot X_C < E_A$
Systematic (SE)	If $\bar{X} = X_C$, $\lvert \bar{Y} - \bar{X} \rvert < E_A$ or $\lvert Y_C - X_C \rvert = \lvert a + bX_C - X_C \rvert < E_A$
Total (TE)	$RE + SE = 1.96\ s_{TM} + \lvert a + bX_C - X_C \rvert < E_A$

knowing what the various components of error are or when they will cause a larger error. Therefore it is essential to consider the worst-case combination, and define this as total error (TE):

$$TE = RE + SE$$

If total error is less than allowable error, the method's overall performance is acceptable. Calculation of total error is demonstrated on p. 358.

Equations for estimating the magnitudes of the various errors and the criteria for judging their acceptability are summarized in Table 20-2.

CONFIDENCE-INTERVAL CRITERIA FOR JUDGING ANALYTICAL PERFORMANCE

To this point it has been assumed that the error estimated by each of the previous equations is absolutely accurate. However, if the same experiment were to be repeated in as identical a manner as possible, a slightly different estimate of error would probably be obtained. Exact measurements of random and systematic errors cannot be obtained from the limited numbers of specimens analyzed in the procedures recommended above.

In the approach developed by Westgard, Carey, and Wold,[26] 95% upper and lower limits of error are calculated. If the 95% upper limit of an error is smaller than the allowable error, one can be at least 95% sure that the method's performance is acceptable. If the 95% lower limit is greater than the allowable error, one can be at least 95% sure the method's performance is not acceptable, and no further testing is indicated. The method should be rejected or modified to improve its analytical performance. When the lower 95% limit is less than allowable error and the 95% upper limit of error exceeds allowable error, one cannot decide whether the method is unacceptable or acceptable, and more data are required to make a definitive decision.

Calculations of confidence-interval estimates of each type of error are demonstrated on pp. 355 to 358.

Confidence-interval criterion for random error

In the calculation of random error, the true value of the standard deviation is not known. One can estimate upper and lower confidence limits of the standard deviation by multiplying the observed standard deviation by the appropriate one-sided 95% factors. These factors (see Table 20-4, p. 355) are referenced to $N - 1$ degrees of freedom.

$$s_{TM_u} = s_{TM} \times A_u$$
$$s_{TM_l} = s_{TM} \times A_l$$

where A_u and A_l are the factors for computing the upper and lower one-sided limits of the standard deviation.

The upper confidence limit of random error is 1.96 times the upper confidence limit of the standard deviation, and the lower confidence limit of random error is 1.96 times the lower confidence limit of the standard deviation.

$$RE_u = 1.96 \times s_{TM_u}$$
$$RE_l = 1.96 \times s_{TM_l}$$

Confidence-interval criterion for constant error

Upper and lower limits for constant error are estimated from the interference study data. Again, an upper limit is derived such that there is a 95% probability that the true error is less than that limit.

$$CE_u = CE + \frac{t \cdot s_d}{\sqrt{N}}$$

and

$$CE_l = CE - \frac{t \cdot s_d}{\sqrt{N}}$$

The t value is a one-sided, 95% t. Fig. 20-7, *A*, shows the upper 95% limit of constant error, leaving only a 5% chance that the error exceeds this upper limit, CE_u. Similarly, Fig. 20-7, *B*, shows the lower 95% limit of constant error, CE_l. A one-sided t is used only when one is concerned with an upper limit on the constant error without regard for how small the constant error is (and vice versa for a lower limit). (A two-sided t is used to answer the question, "Is there a difference?" without regard to whether the difference is positive or negative, as in the t test used in the interpretation of the comparison-of-methods experiment.)

Confidence-interval criterion for proportional error

A similar approach is used to estimate the upper and lower limits for recovery. A one sided t value is used in the equation below for $N - 1$ degrees of freedom and a 95% limit.

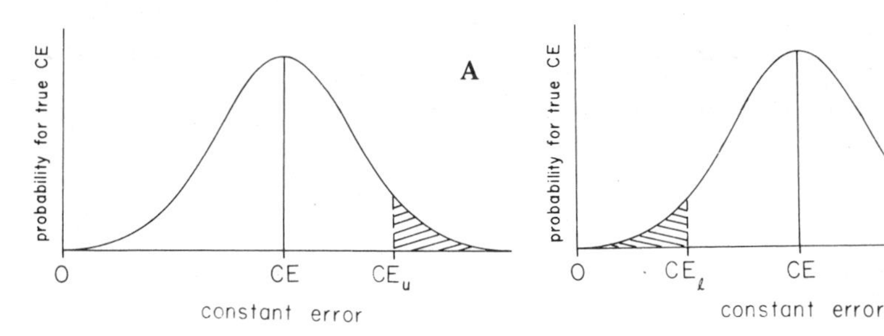

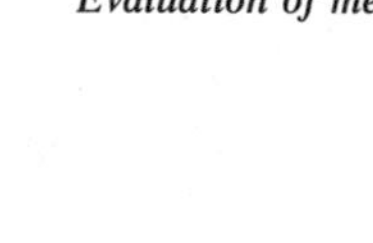

Fig. 20-7. A, One-sided 95% upper limit of constant error, CE_u, **B,** One-sided 95% lower limit of constant error, CE_l.

$$\overline{R}_u = \overline{R} + \frac{t \cdot s_R}{\sqrt{N}}$$

and

$$\overline{R}_l = \overline{R} - \frac{t \cdot s_R}{\sqrt{N}}$$

where $\overline{R}$ is the mean recovery and s_R is the standard deviation of the recovery (see p. 357). The upper confidence limit of the percent of proportional error is the confidence limit that is the greater distance from the ideal value of 100%, and the lower confidence limit of the percent of proportional error is the confidence limit that is closer to 100%. These percentages are converted to concentration units at the critical concentration or concentrations to make the comparison with allowable error.

$$PE_u = X_C \left| \frac{\overline{R}_u \text{ or } \overline{R}_l - 100}{100} \right|_u$$

and

$$PE_l = X_C \left| \frac{\overline{R}_l \text{ or } \overline{R}_u - 100}{100} \right|_l$$

Confidence-interval criterion for systematic error

Fig. 20-8 shows the profile of a confidence interval around a least-square regression line. The expression for the limits, w, of this interval is given as

$$w = ts_{y/x}\left[\frac{1}{N} + \frac{X_C - \overline{X}}{\sum(X_i - \overline{X})^2}\right]^{1/2} \qquad \textit{Eq. 20-7}$$

This equation is similar to those used to calculate the limits for constant and proportional error in terms of the component $ts_{y/x}$. The component under the square-root sign becomes zero if X_C equals the mean of the patient data by the comparative method. As one progresses from the mean, this term begins to contribute to the widening of the limits. One can calculate the denominator of this second term from the standard deviation of the patient population by the comparative method (s_x) as follows:

$$\sum(x_i - \overline{x})^2 = s_x^2(N - 1)$$

In this situation, one cannot know exactly what the regression line is, and for a given X_C, the corresponding Y_C could be as large as $(Y_C + w)$, or as small as $(Y_C - w)$. The limit that is farther from the ideal value is used to estimate the upper limit of systematic error, and the limit closer to the ideal value is used to estimate the lower limit of systematic error:

$$SE_u = [(Y_C \pm w) - X_C]_u \qquad \textit{Eq. 20-8}$$

and

$$SE_l = [(Y_C \pm w) - X_C]_l \qquad \textit{Eq. 20-9}$$

Equations 20-7 to 20-9 may not provide a valid estimate of the confidence limits of the linear regression line if the precision of the test method is not reasonably constant throughout the concentration range of the patient specimens included in the comparison-of-methods experiment.

Confidence-interval criterion for total error

As described before, total error is the worst-case combination of the random and systematic error. Since both the random and systematic errors have variances included in the equations used to calculate them, they must be combined vectorially. Their variances are combined as shown below:

$$TE_u = \sqrt{RE_u^2 + w^2} + SE_u$$

and

$$TE_l = \sqrt{RE_l^2 + w^2} + SE_l$$

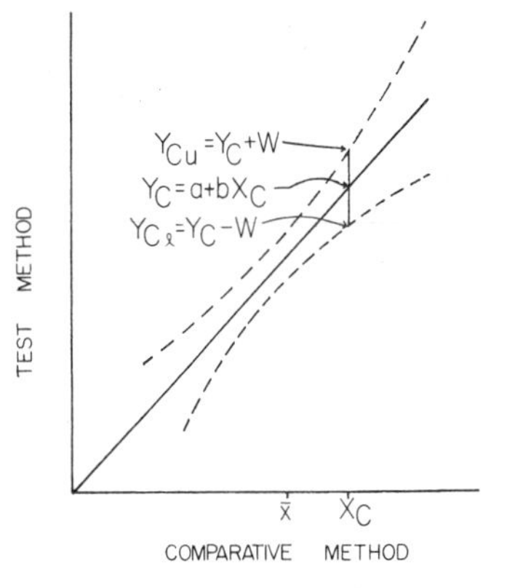

Fig. 20-8. Confidence interval around regression line. (From Westgard, J.O., de Vos, D.J., Hunt, M.R., et al.: Am. J. Med. Technol. **44:**727, 1978.)

Table 20-3. Confidence-interval criteria

Error	Acceptable performance	Unacceptable performance
Random (RE)	$1.96\ s_{TM_u} < E_A$	$1.96\ s_{TM_l} > E_A$
Constant (CE)	$\lvert Bias \rvert + \frac{ts}{\sqrt{N}} < E_A$	$\lvert Bias \rvert - \frac{ts}{\sqrt{N}} > E_A$*
Proportional (PE)	$\frac{\lvert \bar{R}_u \text{ or } \bar{R}_l - 100 \rvert_u}{100} \cdot X_C < E_A$	$\frac{\lvert \bar{R}_l \text{ or } \bar{R}_u - 100 \rvert_l}{100} \cdot X_C > E_A$*
Systematic (SE)	$\lvert (a + bX_C \pm w) - X_C \rvert_u < E_A$	$\lvert (a + bX_C \pm w) - X_C \rvert_l > E_A$*
Total error (TE)	$\sqrt{RE_u^2 + w^2} + SE < E_A$	$\sqrt{RE_l^2 + w^2} + SE > E_A$*

*If the ideal value is between the two limits, it is possible that there is no error and hence the lower limit of error might be zero. Thus in these cases, CE_l is redefined as zero, or PE_l is redefined as zero, or SE_l is redefined as zero, whatever the case might be. If $SE_l = 0$, TE_l becomes RE_l.

If the upper 95% limit of the total error is less than the allowable error, one can be 95% sure that the method performs acceptably. If the lower 95% limit of the total error exceeds allowable error, one can be 95% sure the method does not perform acceptably and it should be modified or rejected.

One should note that whenever the ideal value (zero-error condition) is between the upper and lower limits of the estimated error, there is a chance that the true error might be zero. Thus in these situations the lower limit of error is simply zero. The upper limit of error remains as calculated above. This situation can arise for the constant, proportional, or systematic error estimates, but not of course for random error. If the lower limit of systematic error is zero, the lower limit of the estimate of total error is equal to the lower limit of the estimate of random error because there may not be a systematic error present.

Equations for calculating confidence intervals of the various errors and criteria for judging their acceptabilities are summarized in Table 20-3.

OTHER EVALUATION PROTOCOLS

Protocols for manufacturers to follow to produce defensible performance claims have been developed by NCCLS.[4] These protocols involve more laboratory experimental work than described previously. Claims are stated by statistical tolerance limits:

> For example, one could be 99% confident that 95% of patient samples will agree within the limit of 8 mg/dL of the values determined by a selected analytical method. This would be a tolerance limit for total error. A similar tolerance limit statement can be made for imprecision.[4]

Use of a tolerance limit for the upper limit of an error estimate increases the probability that a user who repeats the manufacturer's experiments will obtain an error estimate less than the claimed error.

In the total-error protocol, a tolerance limit is developed for the maximum difference between the test and comparative method for 95% of patient specimens analyzed by both methods. This is a larger estimate of total error than the one presented earlier in this chapter because it includes the random error of the comparative method and gives more weight to differences caused by interferents in the patient specimens. The random-error portion of the total-error tolerance limit is derived from $s_{y/x}$ instead of s_{TM}.

The NCCLS has proposed a protocol for user evaluation of precision and precision-claims verification.[5] It requires duplicate measurements on at least two different materials in a run, two runs per day, for a period of 20 days. An analysis of variance calculation is used to determine within-run, within-day, and day-to-day components of variance. These are combined to estimate the total standard deviation.

The FDA has developed a protocol for use to verify conformance of in vitro diagnostic devices' performance to that claimed by the manufacturers. Performance is tested separately at 10%, 50%, and 90% of a method's claimed linear range.[7] A chi-square test is used to compare the total standard deviation observed in an analysis of variance experiment to the manufacturer's claimed standard deviation. In the comparison-of-methods experiment, 20 patient specimens with analyte concentrations within 10% of each of the above concentrations (60 patient specimens total) are selected and run in duplicate by the test and comparative methods. The bias is calculated for the 20 specimens at each level, and it is compared to the bias claim by use of a one-sided *t* test at each level. Since data used to make a decision at each concentration are within 10% of the decision-level concentration, the problems of the bias and *t* test described earlier in this chapter are eliminated. It is assumed that if bias is within a manufacturer's claim at these three levels, it is also within claimed limits at all intermediate levels. This method does not use data as efficiently as one that uses a linear regression approach, but it avoids the problems caused by violating the underlying assumptions of linear regression.

DISCUSSION

There are situations, as in the study of different enzyme methods, in which suitably close agreement is not expected or possible because of different reaction conditions or different definitions of enzyme units. In these cases, rather than conclude that the method is unacceptable, a new clinical base line of information is necessary, and a new reference range is needed. Specific disease-related data should be obtained to provide new clinical information for interpretation of test method results.

Evaluation of a method for a "new" analyte previously not measured in the user's laboratory is an analogous situation. Since there is no comparative method on site, accurate estimates of systematic error are harder to obtain. Reliance on published evaluation reports increases. The conclusions of these reports must be reviewed cautiously after the analysis of one's own experimental data is completed. If an analyte is not commonly measured, emphasis shifts to experiments to estimate specific errors. Accurate recovery studies are essential. The interference studies are expanded to include a broader range of chemicals that could interfere with the measurement reactions. Patient specimens that have been analyzed in another laboratory may be analyzed for comparison purposes, but the reliability of the systematic-error estimate may be decreased by specimen instability and lack of user control of the other laboratory's procedure. However, if the other laboratory is the reference laboratory to which the user has previously referred specimens for measurement of this analyte, the comparison really is being made to present practice.

Smaller laboratories often do not have the resources for exhaustive method-evaluation studies but fortunately are usually not among the first to evaluate a new method. Usually some evaluation reports have been published. Even when a method's performance is well documented by published evaluation studies, the user should still evaluate random error and perform the comparison-of-methods experiment to verify acceptable performance in his own laboratory. Again a reference range study should be performed.

EXAMPLE CALCULATIONS AND DECISION MAKING FOR A PERFORMANCE EVALUATION STUDY FOR GLUCOSE

I. ESTIMATION OF RANDOM ERROR FROM REPLICATION DATA

A. Statistics calculations

(y_i = results from the method being tested)

1. Mean:

$$\bar{Y} = \frac{\sum y_i}{N}$$

2. Standard deviation:

$$s_{TM} = \sqrt{\frac{\sum (y_i - \bar{Y})^2}{N-1}}$$

or

$$s_{TM} = \sqrt{\frac{\sum y_i^2 - N\bar{Y}^2}{N-1}}$$

3. Coefficient of variation:

$$CV = \frac{s_{TM}}{\bar{Y}} \cdot 100\%$$

4. Example: glucose

$$N = 21,\ \sum y_i = 24{,}990,\ \text{and}\ \sum y_i^2 = 29{,}765{,}480$$
$$\bar{Y} = 1190\ \text{mg/L}$$
$$s_{TM} = 37\ \text{mg/L}$$
$$CV = 3.1\%$$

B. Point estimate of random error (RE)

$$RE = 1.96 \cdot s_{TM}$$

If RE < E_A, performance is acceptable.

Example: glucose, E_A = 100 mg/L at X_C = 1200 mg/L

$$\bar{X} = 1190\ \text{mg/L}$$
$$s_{TM} = 37\ \text{mg/L}$$

RE = 1.96 × 37 = 73 mg/L; RE is acceptable.

C. Confidence-interval estimate of random error (RE_u, RE_l)

$$s_{TM_u} = s_{TM} \times (A_{.95}) \ldots \text{(see Table 20-4)}$$
$$= 37 \times 1.358 = 50.2\ \text{mg/L}$$
$$s_{TM_l} = s_{TM} \times (A_{.05})$$
$$= 37 \times 0.7979 = 29.5\ \text{mg/L}$$

$$RE_u = 1.96 \times s_{TM_u} = 98\ \text{mg/L}$$
$$RE_l = 1.96 \times s_{TM_l} = 58\ \text{mg/L}$$

Since $RE_u < E_A$, we are 95% certain that random error is acceptable.

Table 20-4. Factors for computing one-sided confidence limits for standard deviation

Degrees of freedom ($N-1$)	$A_{.05}$	$A_{.95}$
1	.5103	15.947
5	.6721	2.089
10	.7391	1.593
15	.7747	1.437
20	.7979	1.358
25	.8149	1.308
30	.8279	1.274
40	.8470	1.228
50	.8606	1.199
60	.8710	1.179
70	.8793	1.163
80	.8861	1.151
90	.8919	1.141
100	.8968	1.133

From Natrella, M.G.: Experimental statistics, National Bureau of Standards Handbook 91, Washington, D.C., 1963, U.S. Government Printing Office.

II. ESTIMATION OF CONSTANT ERROR FROM AN INTERFERENCE STUDY FOR A GLUCOSE METHOD

A. Sample preparation

1. 1.00 mL of serum A + 0.10 mL of water
2. 1.00 mL of serum A + 0.10 mL of 1000 mg/L of creatinine standard
3. 1.00 mL of serum A + 0.10 mL of 3000 mg/L of creatinine standard

B. Results

	Creatinine added (mg/L)	Glucose measured (mg/L)	Interference (mg/L)	Average interference (CE) (mg/L)	s
1.	—	1200, 1220, 1190	—	—	s
2.	91	1240, 1240, 1230	+40, +20, +40	+33	11.5
3.	273	1310, 1340, 1290	+110, +120, +100	+110	10

C. Formulas for calculations

$$\text{Concentration added} = \text{Concentration of standard} \times \frac{\text{Volume standard}}{\text{Total volume}}$$

$$\text{Interference} = \text{Concentration (test)} - \text{Concentration (base line)}$$

D. Point estimate of constant error (CE)

$$\text{CE} = \text{Interference}$$

If CE $< E_A$ performance is acceptable.

Example: For glucose, E_A = 100 mg/L at X_C = 1200 mg/L

In the presence of 91 mg/L of creatinine, CE = 33 mg/L; CE is acceptable.

In the presence of 273 mg/L of creatinine, CE = 110 mg/L; CE is not acceptable.

E. Confidence-interval estimate of constant error (CE_u, CE_l)

1. In the presence of 91 mg/L of creatinine, CE = 33 mg/L, s = 11.5, N = 3, and t for ($N-1$), or 2 degrees of freedom, 95% one-sided limit is 2.92.

$$CE_u = CE + t\frac{s}{\sqrt{N}}$$

$$= 33 + \frac{2.92 \times 11.5}{\sqrt{3}} = 52 \text{ mg/L}$$

$$CE_l = CE - t\frac{s}{\sqrt{N}}$$

$$= 33 - \frac{2.92 \times 11.5}{\sqrt{3}} = 14 \text{ mg/L}$$

Since $CE_u < E_A$ (52 < 100 mg/L), the constant error caused by creatinine of 91 mg/L is acceptably small.

2. In the presence of 273 mg/L of creatinine, CE = 110, s = 10, N = 3

$$CE_u = 110 + \frac{2.92 \times 10}{\sqrt{3}} = 127 \text{ mg/L}$$

$$CE_l = 110 - \frac{2.92 \times 10}{\sqrt{3}} = 93 \text{ mg/L}$$

Since $CE_l < E_A < CE_u$, we cannot be 95% sure the method is acceptable or 95% sure the method is not acceptable for glucose analysis in the presence of 273 mg/L of creatinine. More data should be obtained to narrow the confidence limits to one side of E_A or the other.

III. ESTIMATION OF PROPORTIONAL ERROR FROM A RECOVERY STUDY FOR A GLUCOSE METHOD

A. Sample preparation

1. 2.0 mL of serum A + 0.1 mL of water
2. 2.0 mL of serum A + 0.1 mL of 10,000 mg/L of glucose standard
3. 2.0 mL of serum B + 0.1 mL of water
4. 2.0 mL of serum B + 0.1 mL of 10,000 mg/L of glucose standard

B. Results (mg/L)

	Glucose added	Glucose measured	Glucose recovered	Percent of recovery*
1.	—	510, 530, 540	—	—
2.	476	970, 1000, 980	460, 470, 440	96.6%, 98.7%, 92.4%
3.	—	1240, 1200, 1210	—	—
4.	476	1690, 1660, 1640	450, 460, 430	94.5%, 96.6%, 90.3%

*Average recovery ($\bar{R}$) = 94.8%; SD of Rec. (s_R) = 3.09%; SD of Avg. Rec. = 1.26%.

C. Formulas for calculations

$$\text{Concentration added} = \text{Concentration of standard} \times \frac{\text{Volume of standard}}{\text{Total volume}}$$

$$\text{Concentration recovered} = \text{Concentration (test)} - \text{concentration (base line)}$$

$$\%\ \text{Recovery} = \frac{\text{Concentration recovered}}{\text{Concentration added}} \times 100$$

D. Point estimate of proportional error (PE)

$$\text{PE}(\%) = \bar{R} - 100$$

$$\text{PE (concentration units)} = \left|\frac{\bar{R} - 100}{100}\right| \cdot X_C$$

If PE < E_A, performance is acceptable.
Example: For glucose, E_A = 100 mg/L at X_C = 1200 mg/L

$$\bar{R} = 94.8\%$$

$$\text{PE}(\%) = 94.8 - 100 = 5.2\%$$

$$\text{or PE} = \left|\frac{94.8 - 100}{100}\right| \cdot 1200 \text{ mg/L} = 62 \text{ mg/L}$$

PE is acceptable.

E. Confidence interval estimate of proportional error (PE_u, PE_l)

$\bar{R}$ = 94.8%, s_R = 3.09%, N = 6, and t for $(N-1)$, or 5 degrees of freedom, and 95% one-sided limit is 2.02.

$$\bar{R}_u = \bar{R} + \frac{ts_R}{\sqrt{N}}$$

$$= 94.8 + \frac{2.02 \times 3.09}{\sqrt{6}} = 97.3\%$$

$$\bar{R}_l = \bar{R} - \frac{ts_R}{\sqrt{N}}$$

$$= 94.8 - \frac{2.02 \times 3.09}{\sqrt{6}} = 92.3\%$$

The limit that deviates more from the ideal recovery of 100% is used to estimate the upper limit of proportional error, $\text{PE}_u\%$.

$$\text{PE}_u\% = |92.3 - 100| = 7.7\%$$

and for the lower limit

$$\text{PE}_l\% = |97.3 - 100| = 2.7\%$$

To relate $\text{PE}_u\%$ or $\text{PE}_l\%$ to E_A, convert them to concentration units at X_C.

$$\text{PE}_u = \frac{\text{PE}_u\%}{100} \cdot X_C$$

$$= \frac{7.7\%}{100} \times 1200 = 92 \text{ mg/L}$$

$$\text{PE}_l = \frac{\text{PE}_l\%}{100} \cdot X_C$$

$$= \frac{2.7\%}{100} \times 1200 = 32 \text{ mg/L}$$

Since $\text{PE}_u < E_A$, one can be 95% sure proportional error is acceptably small for this glucose method.

IV. ESTIMATION OF SYSTEMATIC ERROR FROM A COMPARISON-OF-METHODS STUDY FOR GLUCOSE

A. In the comparison of an automated glucose oxidase method *(y)* versus the manual glucose national reference method *(x)*, the following statistics were obtained:

$Y = 0.973x - 57$ mg/L, $s_{y/x} = 37$ mg/L, $N = 82$, $\bar{x} = 1723$, $\bar{y} = 1619$, $s_x = 571$ mg/L, where s_x is the standard deviation of the x values for the 82 samples, and $r = 0.9941$

B. Point estimate of systematic error (SE)

1. Consider bias:

$$\text{Bias} = |\bar{y} - \bar{x}| = 104 \text{ mg/L}$$

The bias provides an estimate of systematic error at the mean of the data. However, if $\bar{x}$ is not equal to the X_C of interest, there must be no proportional error between methods for the bias to provide an accurate estimate of systematic error at these other concentrations. If proportional error is present, use linear regression statistics or calculate the bias for those samples close to X_C.

2. Consider linear regression:

$$\text{SE} = |Y_C - X_C|$$

where $Y_C = a + bX_C$

For $X_C = 1200$ mg/L of glucose,

$$\begin{aligned} Y_C &= 0.973 \times 1200 - 57 \text{ mg/L} \\ &= 1111 \text{ mg/L} \\ \text{SE} &= 1111 - 1200 = 89 \text{ mg/L} \end{aligned}$$

Since SE $< E_A$, systematic error is acceptable.

C. Confidence-interval estimate of systematic error (SE_u, SE_l)

$$\begin{aligned} Y_{C_u} &= Y_C + w \\ Y_{C_l} &= Y_C - w \end{aligned}$$

where

$$w = ts_{y/x}\sqrt{\frac{1}{N} + \frac{(X_C - \bar{x})^2}{\sum(x_i - \bar{x})^2}}$$

where w is the width of the confidence interval around the regression line (Fig. 20-8). The value for t, obtained from a 95% one-sided t table and $N-2$ degrees of freedom, has the value of 1.66.

$$w = 1.66 \times 37\sqrt{\frac{1}{82} + \frac{(1200 - 1723)^2}{571^2 \times 81}}$$

where

$$\begin{aligned} \sum(x_i - \bar{x})^2 &= s_x^2(N - 1) \\ w &= 9 \text{ mg/L} \end{aligned}$$

thus

$$\begin{aligned} Y_{C_u} &= 1111 + 9 = 1120 \text{ mg/L} \\ Y_{C_l} &= 1111 - 9 = 1102 \text{ mg/L} \end{aligned}$$

The limit that deviates more from the ideal value for Y_C (ideally, $Y_C = X_C$) will be used to estimate the upper limit of systematic error.

$$\begin{aligned} \text{SE}_u &= |(Y_C \pm w) - X_C|_u \\ &= |1102 - 1200| = 98 \text{ mg/L} \end{aligned}$$

and

$$\begin{aligned} \text{SE}_l &= |(Y_C \mp w) - X_C|_l \\ &= 1120 - 1200 = 80 \text{ mg/L} \end{aligned}$$

Since $\text{SE}_u < E_A$, we can be 95% sure that the systematic error is acceptable between these methods.

V. ESTIMATION OF TOTAL ERROR

A. Point estimate of total error (TE)

$$\begin{aligned} \text{TE} &= \text{RE} + \text{SE (see Fig. 20-6)} \\ &= 1.96\, s_{TM} + |Y_C - X_C| \\ \text{TE} &= 7.3 + 89 \text{ mg/L} = 162 \text{ mg/L} \end{aligned}$$

Since TE $> E_A$, the total error of the new glucose method is not acceptable.

B. Confidence-interval estimate of total error (TE_u, TE_l)

$$\text{TE}_u = \sqrt{\text{RE}_u^2 + w^2} + \text{SE}$$

and

$$\text{TE}_l = \sqrt{\text{RE}_l^2 + w^2} + \text{SE}$$

Notice that the variances of the uncertainty in repetitive measurements and the uncertainty in the regression line are added, and the square root of the sum is taken to estimate the overall uncertainty, which is then combined with the point estimate of systematic error to yield the appropriate limit for total error. Thus

$$\text{TE}_u = \sqrt{98^2 + 9^2} + 89 = 187 \text{ mg/L}$$

and

$$\text{TE}_l = \sqrt{58^2 + 9^2} + 89 = 148 \text{ mg/L}$$

Because $\text{TE}_l > E_A$, we are at least 95% sure that the total error is not acceptable. Although both components of error—random and systematic—are fairly large, the errors that are easier to reduce or eliminate are systematic errors. However, in this example, both RE_u and SE_u are nearly 100 mg/L. Steps should be taken to reduce both types of error before the automated method is acceptable.

REFERENCES

1. Westgard, J.O.: Precision and accuracy: concepts and assessment by method evaluation testing, CRC Crit. Rev. Clin. Lab. Sci., pp. 283-330, 1981, Boca Raton, Fla., CRC Press, Inc.
2. Westgard, J.O., de Vos, D.J., Hunt, M.R., et al.: Concepts and practices in the selection and evaluation of methods, Am. J. Med.; Technol.; Part I. Background and approach, **44:**290-300, 1978; Part II. Experimental procedures, **44:**420-430, 1978; Part III. Statistics, **44:**552-571, 1978; Part IV. Decision on acceptability, **44:**727-742, 1978; Part V. Applications, **44:**803-813, 1978.
3. Proposed establishment of product class standard for detection or measurement of glucose, Federal Register **39:**126-24136, 1974.
4. National Committee for Clinical Laboratory Standards: NCCLS tentative standards EP2-T to EP4-T, Protocol for establishing performance claims for clinical chemical methods, Instrumentation Evaluation Subcommittee of the Evaluation Protocols Area Committee, Villanova, Pa., 1979.
5. National Committee for Clinical Laboratory Standards: NCCLS proposed standard EP5-P, Proposed guidelines for user evaluation of precision performance of clinical chemistry devices, Subcommittee for User Evaluation of Precision of the Evaluation Protocols Area Committee, Villanova, Pa., 1982.
6. Labeling requirements and standards development for in-vitro diagnostic products, Federal Register **21** CFR:809-10, 1974.
7. Lee, H.T., Daniel, A., and Walker, C.D.: Conformance test procedures for verifying labeling claims for precision, bias, and interferences in in-vitro diagnostic devices used for the quantitative measurement of analytes in body fluids, Bureau of Medical Devices Biometrics Report, U.S. Food and Drug Administration pub. no. 8202, Silver Spring, Md., 1982, U.S. Government Printing Office.
8. Berkow, R., editor: The Merck manual of diagnosis and therapy, ed. 13, Rahway, N.J., 1977, Merck Sharp & Dohme Research Laboratories, p. 1302.
9. Barnett, R.N.: Medical significance of laboratory results, Am. J. Clin. Pathol. **50:**671-676, 1968.
10. Barnett, R.N.: Analytic goals in clinical chemistry: the pathologist's viewpoint. In Elevitch, F.R., editor: Proceedings of the 1976 Aspen Conference on Analytic Goals in Clinical Chemistry, Skokie, Ill., 1977, College of American Pathologists, pp. 20-24.
11. Tonks, D.: A study of the accuracy and precision of clinical chemistry determinations in 170 Canadian laboratories, Clin. Chem. **9:**217-233, 1963.
12. Tonks, D.: A quality control program for quantitative clinical chemistry estimations, Can. J. Med. Technol. **30:**38-54, 1968.
13. Cotlove E., Harris, E., and Williams, G.: Biological and analytic components of variation in long-term studies of serum constituents in normal subjects. III. Physiological and medical implications, Clin. Chem. **16:**1028-1032, 1970.
14. Elevitch, F.R., editor: CAP Aspen Conference 1976: analytical goals in clinical chemistry, Skokie, Ill. 1977, College of American Pathologists.
15. Elion-Gerritzen, W.E.: Analytic precision in clinical chemistry and medical decisions, Am. J. Clin. Pathol. **73:**183-195, 1980.
16. Ross, J.W.: Blood gas internal quality control, Pathologist **34:**377-379, 1980.
17. Gerson, B.: Quality control in therapeutic drug monitoring: intralaboratory precision and medical requirements, Ther. Drug Monit. **2:**225-232, 1980.
18. National Committee for Clinical Laboratory Standards: NCCLS Proposed standard PSC-5, Methodological principles for establishing principal assigned values to calibrators, Villanova, Pa., 1977.
19. Tremblay, M.M.: Evaluation of instruments in biochemistry, Can. J. Med. Technol. **41:**65-76, 1979.
20. Shaikh, A.H.: A systematic procedure for selection of automated instruments in the clinical laboratory, Am. J. Med. Technol. **45:**710-714, 1979.
21. Buttner, R., Borth, R., Boutwell, J.H., et al.: Approved recommendation (1978) on quality control in clinical chemistry. Part I. General principles and terminology, Clin. Chim. Acta **98:**129F-143F, 1979.
22. Westgard, J.O., and Hunt, M.R.: Use and interpretation of common statistical tests in method-comparison studies, Clin. Chem. **19:**49-57, 1973.
23. American Society for Testing and Materials: ASTM Standard E178-68: Standard recommended practice for dealing with outlying observations, Philadelphia, Pa, 1968.
24. Waakers, P.J.M., Hellendoorn, H.B.A., Op De Weegh, G.J., and Heerspink, W.: Applications of statistics in clinical chemistry: a critical evaluation of regression lines, Clin. Chim. Acta **64:**173-184, 1975.
25. Cornbleet, P.J., and Gochman, N.: Incorrect least-squares regression coerffiicients in method-comparison analysis, Clin. Chem. **25:**432-438, 1979.
26. Westgard, J.O., Carey, R.N., and Wold, S.: Criteria for judging precision and accuracy in method development and evaluation, Clin. Chem. **20:**825-833, 1974.
27. Natrella, M.G.: Experimental statistics, National Bureau of Standards Handbook 91, Washington, D.C., 1963, U.S. Government Printing Office.

Section two

Pathophysiology

Chapter 21 Physiology and pathophysiology of body water and electrolytes

Leonard I. Kleinman
John M. Lorenz

acidosis Abnormally low body fluid pH. *Respiratory*—caused by an abnormally high P_{CO_2}; *metabolic*—caused by an abnormally low bicarbonate concentration.

active transport The passage of ions or molecules across a cell membrane by an energy-consuming process. This energy is generated by cellular metabolism.

adipsia Absence of thirst.

aldosterone A mineralocorticoid hormone secreted by the adrenal cortex that influences sodium and potassium metabolism.

alkalosis Abnormally high body fluid pH. *Respiratory*—caused by an abnormally low P_{CO_2}; *metabolic*—caused by an abnormally high bicarbonate concentration.

anabolism The process of assimilation of nutritive matter and its conversion into living substance.

angiotensin A vasoconstrictive polypeptide produced by the enzymatic action of renin on angiotensinogen. A converting enzyme in the lung removes two C-terminal amino acids from the inactive decapeptide, angiotensin I, to form the biologically active octapeptide, angiotensin II.

angiotensinogen A serum globulin produced in the liver that is the precursor of angiotensin.

anion An ion that carries a negative charge.

anorexia Diminished appetite for food.

antidiuretic hormone A nonapeptide hormone of the neurohypophysis that acts on the collecting tubule of the kidneys to allow increased water reabsorption and therefore decreased free water excretion by the kidney. Also known as *vasopressin.*

arrhythmia Irregularity of the heartbeat.

ascites The accumulation of fluid in the peritoneal cavity.

asphyxia Interference with oxygen delivery to and carbon dioxide removal from the tissue.

baroreceptor A nerve ending that responds to change in pressure.

Bartter's syndrome Primary juxtaglomerular cell hyperplasia with secondary hyperaldosteronism.

catabolism The breaking down of complex chemical compounds in the body into simpler ones, often with the production of energy.

cation An ion that carries a positive charge.

cirrhosis Progressive disease of the liver characterized by damage to hepatic parenchymal cells, with nodular regeneration, fibrosis, and disturbance of normal architecture.

colloid As used in this chapter, this term applies to the large molecules in the body to which the capillary endothelium and cell membrane are impermeable.

colloid osmotic pressure The effective osmotic pressure of plasma and interstitial fluid across the capillary endothelium, largely the result of the presence of protein.

dehydration Abnormal decrease in total body water (see Table 21-4). *Hypernatremic*—net loss of sodium and water from the body, with net water loss exceeding net sodium loss; *hyponatremic*—net loss of sodium and water from the body, with net sodium loss exceeding net water loss; *normonatremic*—net loss of sodium and water from the body in equal extracellular proportions; *simple*—net loss of body water alone with no net sodium loss.

diabetes insipidus The chronic excretion of very large amounts of hypoosmotic urine caused by inability to concentrate urine because of the lack of antidiuretic hormone production, secretion, or effect. *Pituitary*—caused by inadequate antidiuretic hormone (ADH) synthesis or secretion; *nephrogenic*—caused by unresponsiveness of the renal tubules to ADH.

distension receptor A nerve ending that responds to stretch.

edema An increase in the interstitial fluid volume.

extracellular water (ECW) Water external to cell membranes. *Anatomical*—all body water external to cell membranes; *physiological*—plasma plus body water into which small solutes can freely diffuse; excludes the transcellular portion of the anatomical extracellular water; includes the plasma and interstitial fluid (see Fig. 21-1).

flux The rate of transfer of fluid or solute across an interface.

free water Water containing no solute.

Gibbs-Donnan equilibrium The steady-state distribution of permeable ions and transmembrane potential that results across a semipermeable membrane when an impermeant ion exists in unequal amounts on either side of the membrane and solvent movement across the semipermeable membrane is exactly opposed (see Fig. 21-11).

hydrophilicity The property of attracting or associating with water molecules.

hyperaldosteronism A disorder caused by excessive secretion of aldosterone, and characterized by hypokalemic alkalosis, muscular weakness, hypertension, polyuria, polydipsia, and normal or elevated plasma sodium concentration.

hyperchloremia An abnormally high plasma chloride concentration.

hyperkalemia An abnormally high plasma potassium concentration.

hypernatremia An abnormally high plasma sodium concentration.

hyperosmotic Denoting an effective osmotic pressure higher than that of plasma.

hypertonic Denoting a theoretical osmotic pressure higher than that of plasma.

hypochloremia An abnormally low plasma chloride concentration.

hypokalemia An abnormally low plasma potassium concentration.

hyponatremia An abnormally low plasma sodium concentration. *Dilutional*—hyponatremia caused by an excess of water (relative to sodium) in the extracellular compartment.

hypoosmotic Denoting an effective osmotic pressure lower than that of plasma.

hypothalamus A group of nuclei at the base of the brain in approximation to the floor and walls of the third ventricle.

hypotonic Denoting a theoretic osmotic pressure lower than that of plasma.

hypovolemia An abnormally low circulating blood volume.

indicator As used in this chapter, a substance that can be used to determine the volume of a body water compartment.

indicator space The body water compartments into which the indicator in question diffuses.

insensible water loss Evaporation of water through the skin or from the respiratory tract.

interstitial fluid (ISF) Extravascular, extracellular water (see Fig. 21-2).

intracellular water (ICW) Water inside the cells of the body; water within cell membranes.

ischemia A decrease in arterial blood flow below the minimum level necessary to meet the metabolic demands of the tissue or organ in question.

juxtaglomerular cells Smooth muscle cells lining the glomerular end of the afferent arterioles in the kidney that are in opposition to the macula densa region of the early distal tubule. These cells synthesize and store renin and release it in response to decreased renal perfusion pressure, increased sympathetic nerve stimulation of the kidneys, or decreased sodium concentration in fluid in the distal tubule.

macromolecule A molecule of colloidal size, notably proteins, nucleic acids, and polysaccharides.

nephrotoxin A substance that causes dysfunction or death of renal parenchymal cells with some degree of specificity.

obligatory water requirements The minimum volume of water necessary to replace insensible water loss and to excrete the existing renal solute load when the urine is maximally concentrated.

oliguria Abnormally low urine output, that is, less than 400 mL/day in an adult.

osmolality Osmotic concentration defined as osmoles of solute per kilogram of solvent.

osmolarity Osmotic concentration defined as osmoles of solute per liter of solution.

osmole The total number of moles of a solute in solution after dissociation. For example, 1 mole of NaCl in solution dissociates into 1 mole of Na^{+} and 1 mole of Cl^{-}. Therefore there are 2 osmoles for every mole of NaCl.

osmosis The movement of water across a semipermeable membrane from a solution with low solute particle concentration to a solution with high solute particle concentration (see Fig. 21-5).

osmotic pressure The force necessary to exactly oppose osmosis into a solution across a semipermeable membrane.

paresthesia An abnormal spontaneous sensation, such as burning, pricking, numbness, and so on.

passive diffusion Movement of ions or solute across a membrane and down a chemical or concentration gradient without energy consumption or a carrier process.

plasma The extracellular, intravascular fluid of the body (see Fig. 21-2).

polyanionic Possessing multiple negative charges.

polydipsia Excessive fluid intake secondary to extreme thirst. *Psychogenic*—polydipsia secondary to a psychiatric disorder, without a demonstrable organic lesion.

polyuria Excessive urine output, that is, more than 1 to 2 liters per day in the adult.

pseudohyperkalemia Abnormally high plasma potassium concentration in a sample obtained from a patient in the absence of true elevation of plasma potassium concentration in that patient.

renin An enzyme produced, stored, and secreted by the juxtaglomerular cells of the kidney, which acts on circulating angiotensinogen to form angiotensin I.

semipermeable Permeable to certain molecules but not to others; usually permeable to water.

syndrome of inappropriate antidiuretic hormone secretion (SIADH) A grouping of findings, including hypotonicity of the plasma, hyponatremia, and hypertonicity of the urine with continued sodium excretion, that is produced by excessive ADH secretion and improves with water restriction.

total body water (TBW) All water within the body, both inside and outside the cells, including that present in the gastrointestinal and genitourinary tracts.

transcellular water That portion of extracellular water that is enclosed by an epithelial membrane and whose volume and composition is determined by the cellular activity of that membrane.

transmembrane potential The magnitude of an electrical energy gradient across the wall of a cell.

transudation The passage of a fluid through a membrane with nearly all the solutes the fluid contains remaining in solution or suspension.

volume of distribution The size of the compartment throughout which an indicator would have to be evenly distributed in order for a known amount of that indicator to produce a given concentration of indicator in that compartment.

water intoxication An increase in free water in the body; results in dilutional hyponatremia.

Water is the most abundant constituent of the human body, accounting for approximately 60% of the body mass in a normal adult. Water is important not only because of its abundance but also because it is the medium in which body solutes, both organic and inorganic, are dissolved and metabolic reactions take place. This chapter is devoted to a discussion of body water and its inorganic solutes, the stable *milieu intérieur* (interior environment) of Claude Bernard. This milieu intérieur is maintained in a dynamic steady state by the constant expenditure of energy derived from cellular metabolism. The discussion in this chapter focuses on (1) the description of the dynamic steady-state compartmentalization of body fluid and its inorganic solutes, (2) the physiological mechanisms involved in the maintenance of this compartmentalization, and (3) the pathophysiological events that occur during certain clinical states that alter the milieu intérieur.

BODY WATER COMPARTMENTS

Definitions (Figs. 21-1 and 21-2)

Total body water (TBW) includes water both inside and outside of cells and water normally present in the gastrointestinal and urinary tracts. Total body water can be theoretically divided into two main compartments. The *anatomical extracellular water* (ECW) includes all water external to cell membranes and constitutes the medium through which all metabolic exchange occurs. *Intracellular water* (ICW) includes all water within cell membranes and constitutes the medium in which chemical reactions of cell metabolism occur. This compartment is heterogeneous and discontinuous—the interior of each cell is separated from the extracellular water and from the interior of every other cell by the semipermeable cell membrane.

The anatomical ECW is functionally subdivided into *physiological extracellular water* and *transcellular water*. The physiological ECW is the portion of the anatomical ECW whose volume is accessible to direct measurement and includes *plasma* (intravascular water) and *interstitial fluid* (ISF). The ISF includes extravascular, extracellular water into which ions and small molecules diffuse freely from plasma and is the fluid that directly bathes the cells of the body. In addition, there are potential spaces in the body (pericardial, pleural, peritoneal, and synovial) that are normally empty except for a few milliliters of viscous lubricating fluid and are considered to be part of the ISF compartment. Transcellular water includes water in extracellular compartments enclosed by an epithelial membrane, whose volume and composition is determined by the cellular activity of that membrane. These heterogeneous compartments include the aqueous humor in the eye, the cerebrospinal fluid, and water within the gastrointestinal, genitourinary, and nasorespiratory tracts. The volume of the transcellular water portion of the anatomical ECW is not included in conventional measurements of "extracellular water."

Determination of body water compartment volumes

The most widely applicable in vivo method of determining compartment volume is based on the *indicator dilution* principle: the volume of a compartment is determined from the concentration that results when a known quantity of some suitable substance (or indicator) is dissolved in the fluid. Mathematically this is expressed as:

$$\text{Compartment volume} = \frac{\text{Quantity of indicator in compartment}}{\text{Concentration of indicator in compartment}}$$

Isotopes of water, antipyrine, and *N*-acetylaminopyrine are all reliable indicators for determining TBW volume. In contrast, no ideal indicator exists for the ECW compartment. There is no known substance whose volume of distribution includes only and all the ECW compartment. Useful ECW indicators include saccharides (inulin, raffinose, and mannitol) and ions (radiosulfate, thiosulfate, radiochloride, bromide, and thiocyanate). The relative magnitudes of their volumes of distribution in the adult are as follows: inulin $<$ raffinose $<$ thiosulfate, radiosulfate, mannitol $<$ corrected bromide, thiocyanate, radiochloride. The relatively wide range of volumes of distributions of these ECW indicators is in large part caused by differences in their abilities to diffuse into the ISF of dense connective tissue, cartilage, and bone and into transcellular water. Since the "true" volume of the ECW compartment cannot be measured, results obtained with ECW indicators are referred to as *indicator spaces* rather than ECW space.

Plasma volume is accurately measured using Evan's blue dye (which binds to albumin) or radioiodinated serum albumin. The interstitial fluid volume must be calculated as the difference between the volume of distribution of one of the ECW indicators and the plasma volume. The indicator dilution technique is not applicable to the ICW compartment. ICW volume is calculated as the difference between TBW volume and the volume of distribution of one of the ECW indicators.

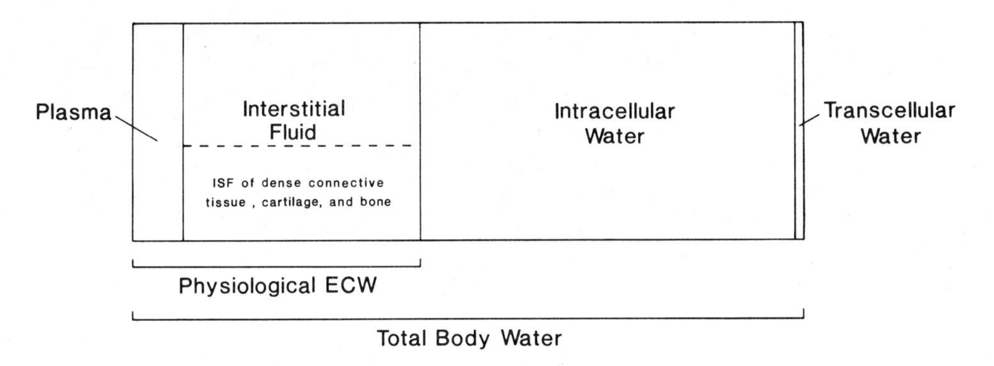

Fig. 21-1. Body water compartments. Note that anatomical extracellular water *(ECW)* includes physiological extracellular water and transcellular water. *ISF,* Interstitial fluid.

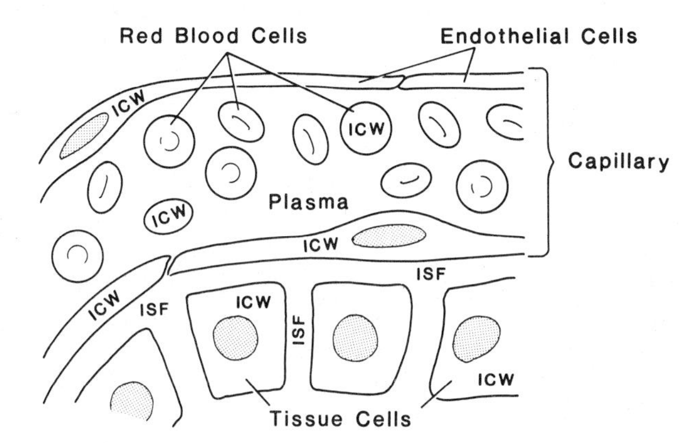

Fig. 21-2. Diagram of plasma, interstitial fluid, *ISF,* and intracellular water, *ICW,* in tissue at microscopic level.

Table 21-1. Compartment volumes

	Percentage of body weight	Percentage of total body water	Volume in 70 kg man
Total body water	60%		42 L
Extracellular water*	20%	33%	14 L
Plasma	5%	8%	3.5 L
Interstitial fluid†	15%	25%	10.5 L
Intracellular water†	40%	67%	28 L

*Volume of distribution of thiosulfate.
†Calculated using the volume of distribution of thiosulfate as the extracellular water volume.

Volume of body water compartments (Table 21-1)

TBW is 65% of body weight in average adult men and 55% of body weight in women. This difference between men and women is largely the result of differences in body fat. As a percent of total body weight, TBW varies inversely with body fat content—from approximately 70% in very thin persons to 50% in very obese persons.

Volumes of distribution of ECW indicators range from 16% for inulin to 26% for radiochloride corrected for intracellular distribution. The thiosulfate space has been considered by some to be the best measure of the physiological ECW volume. It is 20% of body weight and one third of TBW in the average adult.

Intracellular water calculated using the volume of distribution of thiosulfate is equal to 40% of body weight and two thirds of TBW in the average adult. ICW calculated in this manner includes transcellular water, which has been estimated to be 1% to 3% of body weight.

Maturational changes in body water compartment volumes (Fig. 21-3)

The fraction of body weight that is water and the proportion of total body water that is ECW and ICW do not remain constant during growth. When expressed as a percentage of body weight, TBW gradually decreases during intrauterine gestation and early childhood, reaching a value approximating that in the adult by about 3 years of age. During this time, ECW (expressed as a percentage of body weight) decreases and ICW (expressed as a percentage of body weight) increases. Thus ECW becomes a lesser and ICW a greater proportion of TBW. Plasma volume remains constant at 4% to 5% of body weight throughout life. Of course, the absolute volumes of TBW, ECW, ICW, and plasma all increase with growth.

Composition of body water compartments (Table 21-2)

Plasma compartment. The plasma compartment is the only compartment whose composition is directly measurable. Notice that the concentration of ions in the plasma is lower than that in *plasma water*. The reason is that plasma is composed of water and macromolecules. Ions are present only in the water phase. The term "plasma water" is used to indicate this aqueous fraction in distinction to the remainder, which is composed of protein, lipid, and other macromolecules. The concentration of ions in plasma is lower than that in plasma water because plasma contains

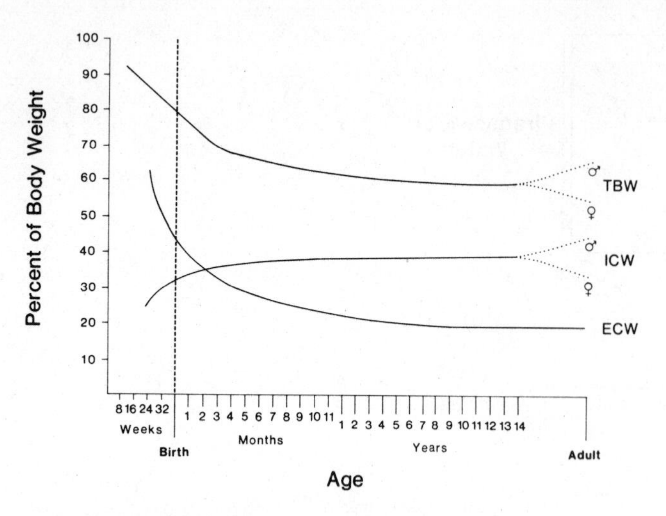

Fig. 21-3. Changes in body water compartments (expressed as a percentage of body weight) with age. *TBW,* Total body water; *ICW,* intracellular water; *ECW,* extracellular water. (Adapted from data of Friis-Hansen, B.: Acta Paediatr. Scand. **46** (suppl. 110):1, 1957.)

both the plasma water (in which plasma ions are dissolved) and the macromolecule fraction (in which no ions are dissolved). Plasma water represents only 93% of total plasma volume. Consequently, the concentration of ions in plasma is 93% of that in plasma water. It should be emphasized that, although the concentration of ions in plasma is that which is conventionally measured and reported, the chemical activities of ions and small molecules are a function of their concentrations in plasma water. It is the chemical activities of ions that determine their distribution between various compartments. If there were an abnormally increased amount of macromolecules in the plasma (such as lipids), the measured plasma concentration of ions would be low even though the concentration of ions in plasma water, and the resultant chemical activities of the ions, might be normal. In addition to protein, plasma contains high concentrations of sodium and chloride, moderate concentrations of bicarbonate, and low concentrations of calcium, magnesium, phosphate, sulfate, and organic acids.

The sum of all the charges of positively charged ions (cations) must be equal to the sum of all the charges of negatively charged ions (anions) to maintain electrical neutrality in the plasma. Most commonly in clinical medicine, however, the plasma concentrations only of sodium, potassium, chloride, and bicarbonate are measured. The sum of these *measured* cations exceeds that of the *measured* anions. Therefore the sum of unmeasured plasma anions must be greater than that of the unmeasured cations. The difference between the sum of measured cations and the sum of measured anions is known as the *anion gap* and is calculated either as $[Na]+[K]-[Cl]-[HCO_3]$ or as $[Na]-[Cl]-[HCO_3]$. The latter is frequently used because the plasma potassium concentration is relatively constant and often subject to measurement error with hemolysis. The anion gap is normally 15 if potassium is included in the calculation and 12 if it is not. Because total plasma cation concentration must equal total plasma anion concentration and decreases in unmeasured cations have little effect in the calculation, an increased anion gap is usually indicative of an increase in the concentration of one or more of the unmeasured anions (Fig. 21-4). A decrease in the anion gap suggests the opposite possibility. The most frequent use of the anion gap clinically is in the differential diagnosis of metabolic acidosis (see Chapter 22).

Interstitial fluid compartment. The interstitial fluid cannot normally be sampled in sufficient amounts for chemical analysis. The major difference between the ISF and plasma is the presence of protein in the plasma and its relative absence from the ISF. Although the concentrations of freely diffusible solute in ISF might be expected to be equal to those in plasma water, this is true only for uncharged solutes. The presence of impermeant polyanionic protein molecules in plasma leads to the Gibbs-Donnan equilibrium (see p. 374), which results in plasma water cation concentrations slightly greater than those in ISF and

Table 21-2. Composition of body compartments

	Plasma (mEq/L)	Plasma water (mEq/L)	Interstitial fluid (mEq/L H_2O)	Intracellular water (mEq/L H_2O)
Cations	153	164.6	153	195
Na^+	142	152.7	145	10
K^+	4	4.3	4	156
Ca^{++}	5	5.4	(2-3)	3.2
Mg^{++}	2	2.2	(1-2)	26
Anions	153	164.6	153	195
Cl^-	103	110.8	116	2
HCO_3^-	28	30.1	31	8
Protein	17	18.3	—	55
Others	5	5.4	(6)	130
Osmolarity (mOsm/L)		296	294.6	294.6
Theoretical osmotic pressure (mm Hg)		5712.8	5685.8	5685.8

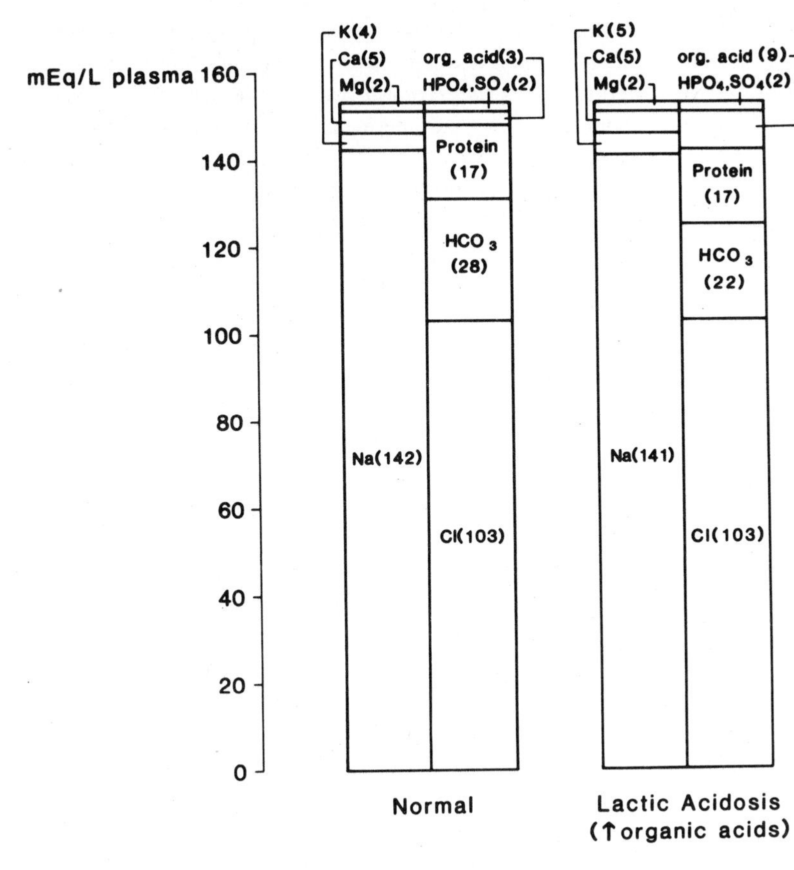

Fig. 21-4. Increased anion gap because of increase in unmeasured anion. *Numbers in parentheses,* Concentration of ions in units of mEq/L plasma. Notice that sum of cations (left-hand side of each bar graph) is always equal to sum of anions (right-hand side of each bar graph), both under normal conditions and in presence of lactic acidosis. Sum of concentrations of unmeasured anions (organic acids, HPO_4, SO_4, and proteins) is larger than sum of concentrations of unmeasured cations (Ca and Mg). During lactic acidosis, difference between unmeasured anions and cations becomes greater because production of lactic acid increases concentration of organic acids.

plasma water anion concentrations slightly less than those in ISF. Values for ISF ion concentrations given in Table 21-2 are theoretical approximations based on Gibbs-Donnan equilibrium calculations.

Intracellular water compartment. Solute concentrations in cell water cannot be directly determined. The ICW compartment is nonhomogeneous; important differences exist in intracellular solute concentrations between different cell types. However, certain features of all cell fluids are quantitatively similar and distinguish ICW from ECW. The major cations of ICW are potassium and magnesium, and the concentration of sodium is always low; the major anions of cell fluids are protein, organic phosphates, and sulfates, whereas chloride and bicarbonate concentrations are low. The profile presented in Table 21-2 is for muscle cells.

Osmotic pressure and osmolarity of body fluids

Osmotic pressure is an important factor determining the distribution of water among the body water compartments. When a membrane permeable to water but impermeable to solute is placed between two fluid compartments of unequal solute concentrations, water moves through the membrane from the solution with higher water (that is, lower solute) concentration to the solution with lower water (that is, higher solute) concentration (Fig. 21-5). This movement of water is called *osmosis*. Osmosis can be opposed by application of a pressure across the semipermeable membrane in the opposite direction. The amount of pressure required to exactly oppose osmosis into any solution across a semipermeable membrane separating it from pure (without solute) water is the *osmotic pressure* of the solution. The osmotic pressure of a solution can then be thought of as its water attractability. The *theoretical* osmotic pressure of a solution is proportional to the number of solute particles per unit volume of solution. The *osmole* (abbreviated *osmol* or *Osm*) is a measure of the total number of solute particles per unit volume of solution. One mole of a nonionic solute (such as urea) equals 1 osmole. However, when a solute dissociates into two or more ions in solution, 1 mole of the ionic solute will equal 2 or more osmoles (depending on the number of ions into which the solute molecule dissociates). For example, 1 mole of NaCl equals 2 osmoles because the molecule dissociates into sodium and chloride ions in solution. Osmolarity is a measure of the total number of solute particles per unit volume of solution and, in biological systems, is usually expressed in units of milliosmoles (1/1000 of an osmole, or mOsm) per liter of water. (*Osmolality* is the total number of solute particles per unit *weight* of solution, or

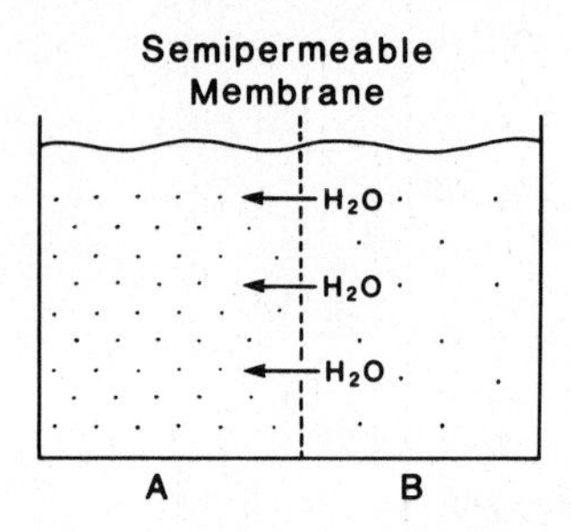

Fig. 21-5. Osmosis. Phenomenon of net water movement across a semipermeable membrane, separating two solutions of unequal solute concentration, from solution of lower solute concentration, *B*, to solution of higher solute concentration, *A*.

milliosmoles per kilogram of water. In the dilute solutions of the normal human body, osmolarity and osmolality are essentially equal.) Therefore the *theoretical* osmotic pressure (and water attractability) of a solution is proportional to its osmolarity. The theoretical osmotic pressure of a solution at body temperature is calculated as follows:

$$\text{Theoretical osmotic pressure (mm Hg)} = 19.3\ (\text{mm Hg/mOsm/L}) \times \text{Osmolarity (mOsm/L)}$$

This calculation assumes none of the solute particles in solution are able to cross the membrane. The osmolarity and theoretical osmotic pressure of each of the body water compartments are listed in Table 21-2.

When a membrane is permeable to a solute, the solute exerts no osmotic pressure across the membrane—it does not contribute to the *effective* osmotic pressure of the solution. The *effective* osmotic pressure of a solution then is dependent on the total number of solute particles in solution *and* the permeability characteristics of the particular membrane in question. The higher the permeability of a membrane to a solute, the lower the effective osmotic pressure of a solution of that solute at any given osmolarity. For example, cell membranes are much more permeable to urea than to sodium and chloride. Therefore the effective osmotic pressure of a solution of urea across the cell membrane would be much less than a solution of NaCl of the same osmolarity.

Osmolarity is usually measured by solution freezing point or vapor pressure determinations (see Chapter 12). This technique of course does not distinguish between solutes that can or cannot cross cell membranes. It is clear that the osmolarity of body compartment water is a measure only of its *theoretical,* not effective, osmotic pressure.

A solution with an *effective* osmotic pressure greater than that of plasma is said to be *hyperosmotic* with respect to plasma. A solution with a *theoretical* osmotic pressure greater than plasma is said to be *hypertonic*. Hypoosmotic and hypotonic solutions are respectively those with effective and theoretical osmotic pressures less than those of plasma.

The effective osmotic pressure of plasma and ISF across the capillary endothelium by which they are separated is referred to as their *colloid* osmotic pressure. The capillary endothelium is freely permeable to most solutes in plasma and ISF, and these therefore contribute to theoretical, but not effective, osmotic pressure. The capillary endothelium is impermeable to large protein molecules *(colloids)* under usual circumstances. It is these colloids that are responsible for the effective osmotic pressure of plasma and ISF.

REGULATION OF BODY FLUID COMPARTMENT OSMOLARITY AND VOLUME

Extracellular compartment

Regulation of the extracellular water osmolarity and volume depends on the independent control of each of these variables by the hypothalamus, the renin-angiotensin-aldosterone system, and the kidney.

Water metabolism and hypothalamus (Fig. 21-6). The regulatory centers for water intake and water output are located in separate areas of the hypothalamus in the brain. Neurons in each of these areas respond to increases in ECW osmolarity, to decreases in intravascular volume, and to angiotensin II. Increased ECW osmolarity stimulates these neurons directly by causing them to shrink (increased osmolarity of ISF bathing any cell will cause water to move out of the cell into the ISF—see p. 374). A decrease in intravascular volume causes a reduction in activity of distension receptors located in the atria of the heart, the inferior vena cava, and the pulmonary veins and a reduction in activity of baroreceptors in the aorta and the carotid arteries. Relay of this information to the central nervous system stimulates neurons in the water-intake and water-output areas of the hypothalamus. Circulating angiotensin II seems to act directly to stimulate neurons located in these water-control areas of the hypothalamus. Stimulation of neurons located in the water-intake area produces the conscious sensation of thirst and thereby stimulates water intake. Stimulation of neurons located in the water-output area results in the release of antidiuretic hormone (ADH) from the axons (which are located in the posterior pituitary gland) of the same neurons. ADH stimulates water reabsorption in the collecting ducts of the kidney, which results in the formation of hypertonic urine and decreased output of free water (water without solute). This elegant integration of the control mechanisms governing water intake and output assures maintenance of appropriate water balance.

Water and sodium metabolism and renin-angiotensin-aldosterone system (Fig. 21-7). The renin-angiotensin-aldosterone system functions as a neurohormonal regulating mechanism for body sodium and water content, arterial blood pressure, and potassium balance. Renin is a proteolytic enzyme synthesized, stored, and secreted by cells in the juxtaglomerular bodies of the kidney. Renin secretion is increased by decreased renal perfusion pres-

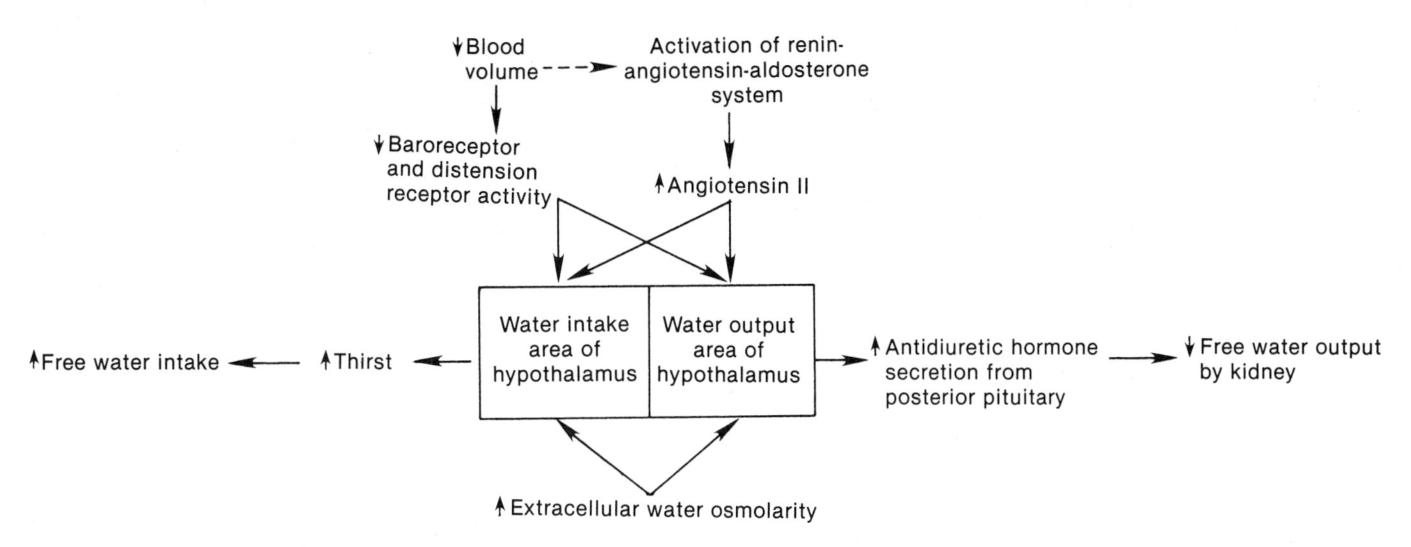

Fig. 21-6. Hypothalamic regulation of water balance.

sure, stimulation of sympathetic nerves to the kidneys, and decreased sodium concentration in the fluid of the distal tubule. Renin activates angiotensinogen (a polypeptide synthesized in the liver) to angiotensin I. Angiotensin I is converted to angiotensin II in the lung and kidney. Angiotensin II is a potent vasoconstrictor. In addition, angiotensin II stimulates aldosterone secretion by the adrenal cortex, as well as thirsting behavior and ADH secretion, as discussed previously. Aldosterone stimulates sodium reabsorption in the distal tubule and collecting tubule of the kidney. As a consequence of this sodium reabsorption, water is retained by the body. Aldosterone also decreases saliva sodium concentration, which (at least in animals) stimulates sodium appetite.

Factors and mechanisms involved in the control of water and sodium excretion by the kidney are discussed in Chapter 23.

Control of extracellular water osmolarity. ECW *osmolarity* is regulated by the hypothalamic control of water intake (regulatory thirst) and renal excretion of free water (Fig. 21-8). Increased ECW osmolarity produces thirsting behavior and increased secretion of ADH by the mechanism discussed previously. Increased water intake and decreased renal free water excretion results in a positive wa-

Fig. 21-7. Renin-angiotensin-aldosterone system.

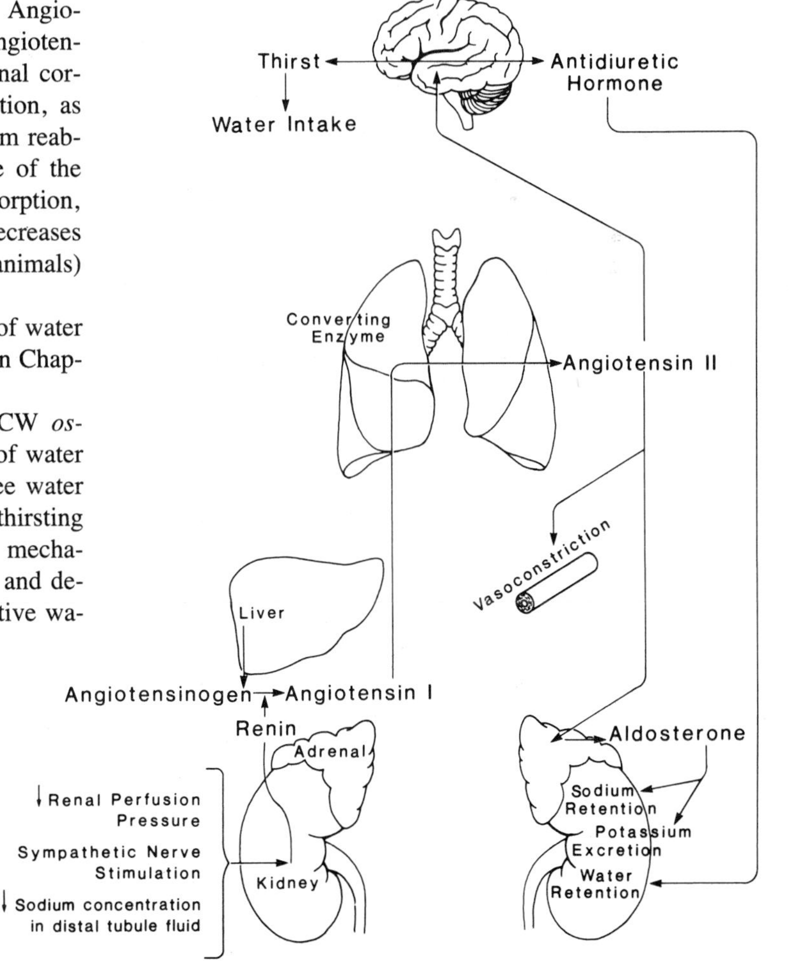

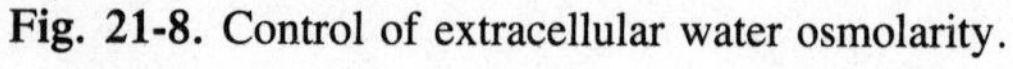

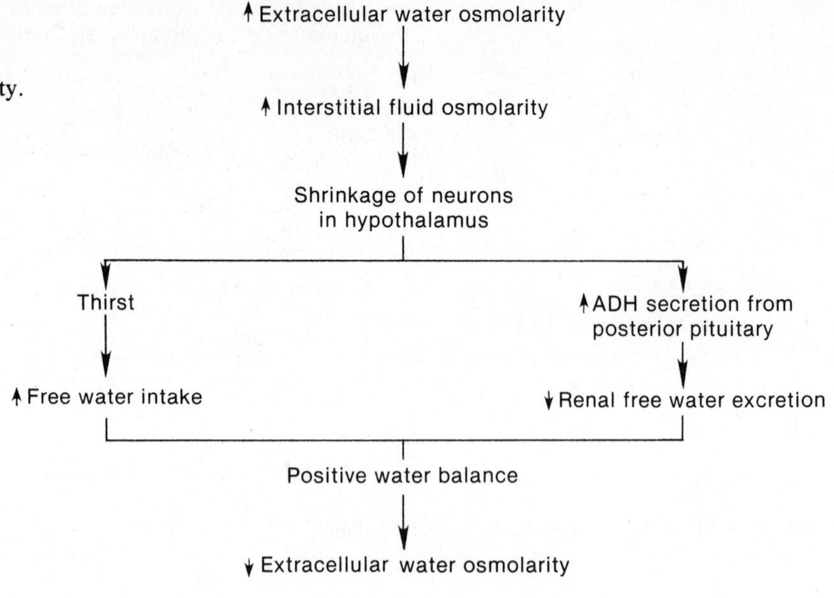

Fig. 21-8. Control of extracellular water osmolarity.

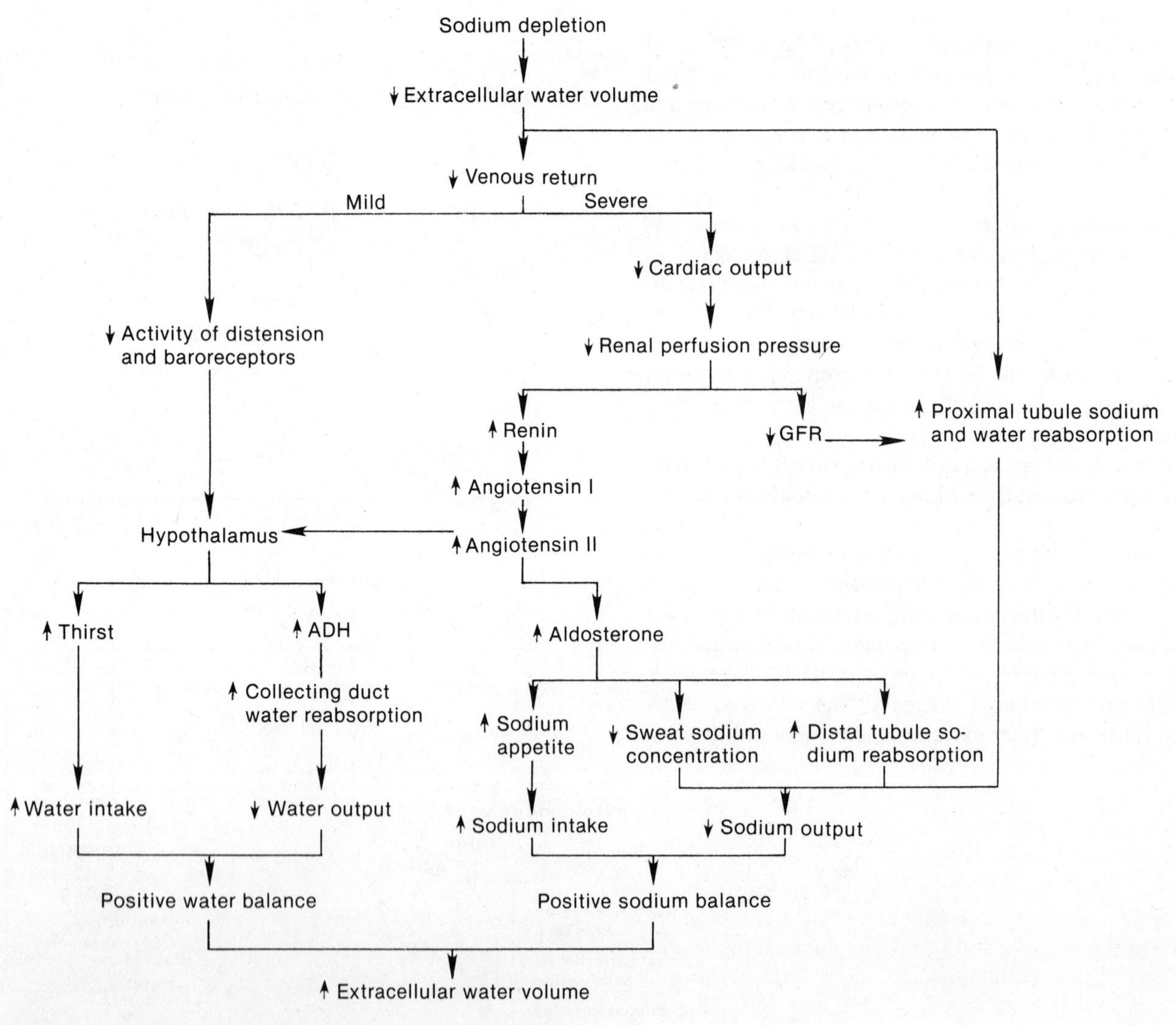

Fig. 21-9. Control of extracellular sodium content and volume.

ter balance, which restores ECW osmolarity to normal. The opposite sequence of events with decreased ECW osmolarity results in the absence of regulatory thirst and inhibition of ADH secretion. In the absence of ADH, hypotonic urine is excreted by the kidney, and free water excretion is increased. In the absence of water intake, increased free water excretion results in negative water balance, and ECW osmolarity is restored to normal.

Control of extracellular water volume. Control of ECW *volume* depends on control of both total ECW sodium content and water balance (Fig. 21-9). A moderate reduction in intravascular volume of 5% to 10% results in decreased return of blood to the heart, decreased activity of distension receptors and baroreceptors in the heart and great vessels, and reflex stimulation of thirst and ADH secretion. A positive water balance (that is, a net increase in body water) results. If there is greater blood volume depletion and further reduction of venous return and cardiac output, renal perfusion pressure decreases and sympathetic nerve stimulation of the juxtaglomerular cells increases, and the renin-angiotensin-aldosterone system is activated. Angiotensin II stimulates regulatory thirst and ADH secretion (independently of distension and baroreceptor cells), and aldosterone stimulates renal sodium reabsortion and sodium appetite. In concert, these processes result in net sodium and water retention and thus restoration of ECW volume.

Expansion of the ECW volume results in the opposite sequence of events with net loss of water and sodium and restoration of ECW balance to normal.

Plasma and interstitial fluid compartments

Water and solute distribution between the plasma and ISF compartments depends on an intact capillary endothelial surface and is controlled passively by the interaction of fluid, osmotic, and electrochemical forces. The capillary endothelium functions as a continuous endothelial tube, with numerous 4 to 5 nm diameter intercellular channels. It is freely permeable to water and small solutes and relatively impermeable to protein.

Water distribution. Water distribution across the capillary endothelial surface is controlled by the balance of forces that tend to move water from the plasma to the ISF (filtration forces) and forces that tend to move water from the ISF into the plasma (reabsorption forces).

Filtration forces. The major filtration force is plasma hydrostatic pressure in the capillary (P_{pl}), which is on the order of 25 mm Hg. A much weaker filtration force is the ISF colloid osmotic pressure (Π_{ISF}). Since the protein concentration in ISF is negligible, Π_{ISF} is low—on the order of 5 mm Hg. Another weak filtration force is a small *negative* ISF hydrostatic pressure, P_{ISF}, which is on the order of −6 mm Hg.

Reabsorption forces. The major reabsorption force is the colloid osmotic pressure exerted across the capillary endothelium by plasma proteins (Π_{pl}). This force is approximately 28 mm Hg.

The net result of these four forces can be described by Starling's equation:

$$\text{Filtration pressure} = (P_{pl} + \Pi_{ISF}) - (\Pi_{pl} + P_{ISF})$$

or

$$\text{Filtration pressure} = (P_{pl} - P_{ISF}) - (\Pi_{pl} - \Pi_{ISF})$$

When filtration pressure is positive, *filtration* of water from the plasma to the ISF occurs. When filtration pressure is negative, *reabsorption* of water from the ISF to the plasma occurs. As mentioned before, ISF hydrostatic pressure is negligible, as is ISF colloid osmotic pressure (because of the relative absence of protein from ISF). As a broad generalization, plasma hydrostatic pressure (which tends to drive water out of the capillary) exceeds plasma colloid osmotic pressure (which tends to draw water into the capillary) at the arteriolar end of the capillary so that filtration occurs. As plasma moves along the capillary and filtration occurs, plasma hydrostatic pressure decreases and plasma protein concentration (and therefore plasma colloid osmotic pressure) increases along the course of the capillary so that reabsorption of water occurs toward the venous end of the capillary. This is depicted schematically in Fig. 21-10. Overall, filtration exceeds reabsorption; therefore water must be returned to the plasma from the ISF compartment by way of the lymphatic system to prevent edema (defined as an abnormal increase in ISF volume).

Solute distribution. The small differences in the concentrations of the various extracellular solutes across the capillary endothelium are caused by the presence of protein molecules in the plasma that are polyanionic (that is, have multiple negative charges) at body pH and to which the capillary endothelium is relatively impermeable. This results in the Gibbs-Donnan equilibrium (Fig. 21-11): the presence of impermeant polyanionic macromolecules restricted to one side of a membrane permeable to solvent

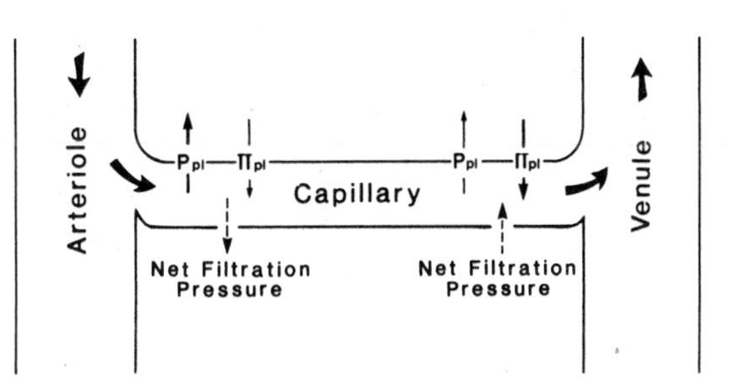

Fig. 21-10. Starling's hypothesis of water distribution between plasma and interstitial fluid compartments. Thickness of arrows representing plasma hydrostatic pressure, P_{pl}, and plasma oncotic pressure, Π_{pl}, indicate their relative magnitudes. *Dashed arrows,* Direction of net filtration pressure.

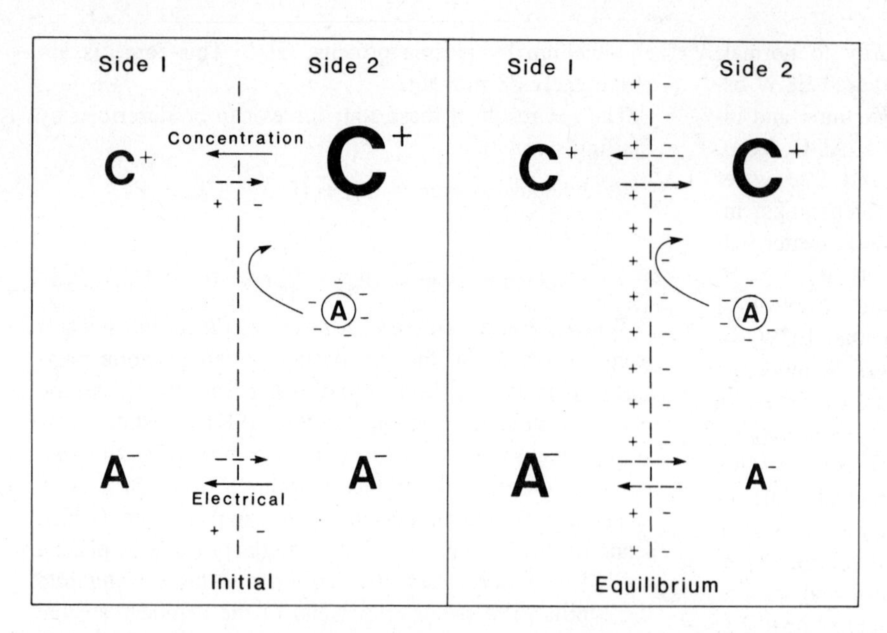

Fig. 21-11. Gibbs-Donnan equilibrium. Distribution of diffusible and nondiffusible ions and development of an electrical potential gradient across a membrane when nondiffusible, polyvalent anion (Ⓐ) with diffusible cation (C^+) is added to one side of membrane in solution of diffusible cation (C^+) and anion (A^-). Initially, diffusible cation moves down its concentration gradient from side 2 to side 1. This movement generates an electrical potential gradient across membrane (side 2 negative with respect to side 1). Diffusible anion moves down this electrical potential gradient from side 2 to side 1. At equilibrium, concentration of diffusible cation will be greater on side 2 than side 1 (as indicated by size of symbols), while concentration of diffusible anions will be greater on side 1 than side 2. No *net* movement of diffusible ions occurs across membrane because no net electrochemical gradients exist. Concentration gradient for each ion is balanced by an equal but oppositely directed electrical gradient.

and ions will establish a characteristic distribution of the permeable ions, with the following results:

1. The concentrations of cations are greater in the compartment containing the macromolecule.
2. The concentrations of anions are lower in the compartment containing the macromolecule.
3. The sum of the concentrations of the diffusible ions is greater in the compartment containing the macromolecule.
4. Osmotic pressure is higher in the compartment containing the macromolecule (because of the macromolecule itself as well as its associated counterions).

In the case of calcium and magnesium, an additional factor is that in the plasma approximately 45% of calcium and 25% of magnesium is bound to protein and is therefore nondiffusible.

Intracellular compartment

Water and solute distribution across the cell membrane between the ISF and ICW depends on the integrity of the cell membrane and on osmotic and electrochemical forces; all of these factors are sustained by cell metabolism. The cell membrane behaves as though it were an oil film with numerous 0.7 nm diameter pores. It is highly permeable to water but differentially permeable to solute. The permeability of the cell membrane to a solute is directly related to the lipid solubility of the solute and inversely related to its hydrophilicity (water attractability) and molecular size. Other factors being constant, membrane permeability is greater to anions than to cations.

Cell volume. Cell volume is controlled by ISF osmolarity. Osmolarity inside the cell must equal osmolarity outside the cell because the cell membrane is highly permeable to water and no fluid pressure gradient can be maintained across animal cell membranes. The osmotic content of the intracellular compartment is maintained constantly by cell metabolism. Therefore, osmotic equilibrium can be maintained in the face of a change in ISF osmolarity only by the movement of water between the intracellular compartment and interstitial space. A decrease in ISF osmolarity must cause movement of water into cells and an increase in intracellular volume. Conversely, an increase in ISF osmolarity must cause movement of water out of cells and a decrease in intracellular volume.

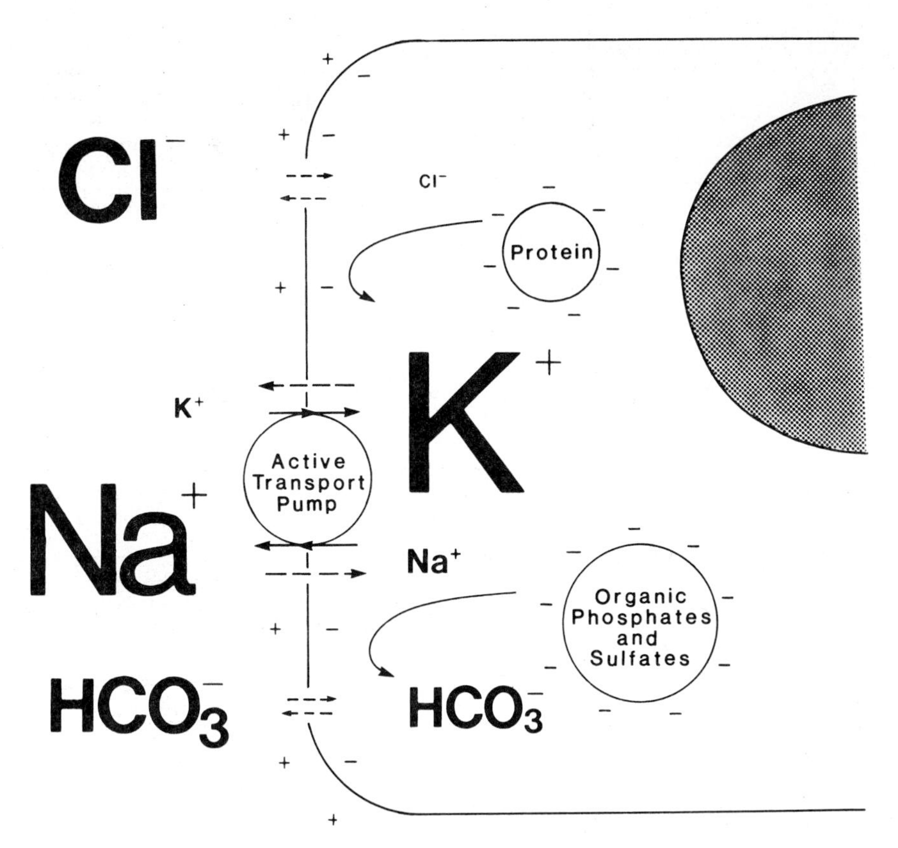

Fig. 21-12. Na^+-K^+-ATPase pump. Concentration of sodium intracellularly is less than in interstitial fluid because of active transport of sodium *out* of cell. On the other hand, potassium concentration is less in interstitial fluid than it is intracellularly because of active transport *into* cell. Chloride concentration is less intracellularly than in interstitial fluid because of electrical potential difference (inside negative) across cell membrane. *Dashed arrows,* Passive diffusion down an electrochemical gradient; *curved lines,* cell membrane impermeable to protein and organic phosphates and sulfates.

Cell solute content. The solute content of the ICW depends on cellular metabolism. Cell metabolism maintains the content of intracellular impermeant macromolecules at a constant level; catabolism is balanced by anabolism. The presence of impermeant polyanionic molecules (protein and organic phosphates and sulfates) in ICW results in the Gibbs-Donnan effect (see p. 374). This results in an osmotic gradient from ISF to ICW that must be opposed. Some other solute must be maintained extracellularly (against the Gibbs-Donnan effect) in order to balance osmotically the intracellular colloid osmotic pressure. This is accomplished by an energy-dependent ionic transport system in the cell-membrane (Fig. 21-12). The active transport of sodium out of cells in exchange for potassium by this Na-K-ATPase pump in concert with the differential permeability of the cell membrane to sodium and potassium (greater to potassium than to sodium) generates electrochemical gradients that result in a relative excess of ions in ISF to which the cell membrane is relatively permeable (which is the opposite of what would be expected with the Gibbs-Donnan effect alone). This increase in total diffusible ions in ISF compared to ICW establishes the osmotic pressure in the ISF, which balances the colloid osmotic pressure produced intracellularly by macromolecules. Thus maintenance of cellular solute content depends not only on the constant cellular content of impermeant macromolecules and rates of active ion transport by the cell but also on differential membrane permeability to ions. All these factors are dependent on normal cellular metabolism. With disruption of cellular metabolism (as during asphyxia), the Gibbs-Donnan effect is unopposed by the Na-K pump, and solute and water enter the cell.

WATER METABOLISM

Water balance

Extracellular water osmolarity is maintained constant at 285 to 298 mOsm/L as a consequence of the dynamic balance between water intake and water excretion, which is controlled by the mechanisms discussed previously. Average daily water turnover in the adult is approximately 2500

Table 21-3. Water balance in average adult under various conditions

Intake (mL/day)				Output (mL/day)			
	Normal	Hot environment	Strenuous work		Normal	Hot environment	Strenuous work
Drinking water	1200	2200	3400	Urine	1400	1200	500
Water from food	1000	1000	1150	Insensible water			
Water of oxidation	300	300	450	Skin	400	400	400
				Lungs	400	300	600
				Sweat	100	1400	3300
				Stool	200	200	200
Total	2500	3500	5000		2500	3500	5000

Table 21-4. Changes in total body water volume and distribution, total body sodium content, and plasma sodium concentration with dehydration and overhydration

	Total body water	Extracellular water	Intracellular water	Total body sodium	Plasma sodium concentration
Dehydration					
Hypernatremic	↓	sl ↓	↓	nl or sl ↓	↑
Normonatremic	↓	↓	nl	↓	nl
Hyponatremic	↓	↓ ↓	↑	↓ ↓	↓
Overhydration					
Water intoxication	↑	↑	↑	nl	↓
Extracellular water volume expansion					
Normonatremic	↑	↑	nl	↑	nl
Hyponatremic	↑	↑	↑	sl ↑	↓

nl, Normal; *sl*, slightly.

mL; however, the range of water turnover possible is great and depends on intake, environment, and activity (Table 21-3).

Under normal conditions, approximately one half to two thirds of water intake is in the form of oral fluid intake, and approximately one third to one half is in the form of oral intake of water in food. In addition, a small amount of water (150 to 350 mL/day) is produced by oxidative metabolism. Oral intake of fluid is the only source of water that is regulated in response to changes in ECW volume and osmolality.

Routes of water excretion include urinary water loss, insensible water loss, sensible perspiration (sweating), and gastrointestinal water loss. The kidney is the principal organ regulating the volume and composition of the body fluids. Urine volume varies over a wide range in response to changes in ECW volume and osmolality. Solute excretion is regulated independently.

Loss of water by diffusion through the skin and through the respiratory tract is known as *insensible water loss* because it is not apparent. It is the only route by which water is lost without solute. Normally, half of insensible water loss occurs through the skin and half through the respiratory tract. The magnitude of cutaneous insensible water loss is a function of body surface area; therefore it is disproportionately greater in infants and children for weight. Insensible water loss varies directly with ambient temperature, body temperature, and activity and inversely with ambient humidity.

Sensible perspiration is negligible in a cool environment but may be quite large with increases in ambient temperature, body temperature, or exercise. Sodium and chloride are the major ionic components of sweat, but sweat is almost invariably hypotonic to plasma. An increase in ECW osmolality causes a decrease in the rate of sensible perspiration.

Net water loss from the gastrointestinal tract is normally small (approximately 150 mL/day). However, the *flux* of water and electrolytes between the gastrointestinal tract and ECW compartment is quite large. Therefore if reabsorption is impaired, water and electrolyte losses from the gastrointestinal tract can be great, as in cases of diarrhea. With the exception of saliva (which is hypotonic), the total solute concentration of most gastrointestinal secretions is similar to ISF.

Water imbalance

Disorders of water balance (dehydration and overhydration) result from an imbalance of water intake and output or sodium intake and output (Table 21-4).

Dehydration

Deficit of water. *Simple dehydration* (defined as a decrease in total body water with relatively normal total body sodium) may result from failure to replace obligatory water losses or failure of the regulatory or effector mechanisms that promote conservation of free water by the kidney (see the box below). Failure of water intake to meet obligatory water losses may be caused by lack of access to water free of solute, inadequate intravenous fluid administration (caused by failure to appreciate the magnitude of insensible water loss), or rarely *adipsia* (absence of thirst) secondary to a neurological deficit in the hypothalamus. Failure to conserve free water may be caused by pituitary diabetes insipidus or nephrogenic diabetes insipidus. *Pituitary diabetes insipidus* is a syndrome that results from failure of the neurohypophysial system to produce or release an adequate quantity of antidiuretic hormone to achieve homeostatic renal conservation of free water. *Nephrogenic diabetes insipidus* is a syndrome in which the renal tubules are totally unresponsive to antidiuretic hormone. Simple dehydration is by definition associated with hypernatremia and hyperosmolarity because water balance is negative and sodium balance is normal. The increase in ECW osmolarity as water is lost from the body results in movement of water out of the ICW compartment. Therefore simple dehydration results in contraction of both the ECW and ICW compartments.

Deficit of water and sodium. More commonly, *dehydration* results from a net negative balance of both water and sodium. In this case, water balance may be more negative than, equal to, or less negative than sodium balance (Table 21-4). If water balance is more negative than sodium balance, the result is *hypernatremic* or *hyperosmolar* dehydration; if it is equally negative, *normonatremic* or *isosmolar* dehydration results; and if it is less negative, *hyponatremic* or *hyposmolar* dehydration results. Hypernatremic dehydration is most common. It can result from food and water deprivation. With usual salt and water intake, it can result from excessive sweating, osmotic diuresis (with glucosuria or mannitol administration), tubular diuretic therapy, vomiting, or diarrhea. Normonatremic dehydration results if the water loss in these conditions is replaced by low-sodium liquids. Hyponatremic dehydration occurs in salt-wasting renal disease, nonoliguric chronic renal failure with salt-wasting, and adrenocortical insufficiency.

The degree of extracellular volume contraction for a given sodium deficit and the associated change in intracellular volume is different for each of these types of dehydration (Table 21-4). The degree of extracellular volume contraction is least with hypernatremic dehydration because the increase in ECW osmolarity causes water to move out of the cell; contraction of ICW volume occurs. Thus the total body water deficit is "shared" by the extracellular and intracellular compartments. The degree of extracellular volume contraction is intermediate with normonatremic dehydration, since no water moves out of or into cells because there is no change in ECW osmolarity. There is no change in ICW volume. The degree of ECW volume depletion is largest with hyponatremic dehydration because the decrease in ECW osmolarity causes water to move into cells. ICW volume is actually increased.

Symptoms of dehydration. The signs and symptoms of dehydration include thirst, dry mucous membranes, decreased skin turgor, decreased urine output and increased urine osmolarity (except when caused by failure of the kidney to conserve free water), increased blood urea nitrogen, and increased hematocrit. With increasing severity, weakness, lethargy, hypotension, and shock may occur.

Overhydration

Excessive water. *Water intoxication* (defined as an increase in total body water with normal total body sodium) results from the intake of water beyond the ability of the kidney to excrete it or impairment of the ability of the kidney to excrete free water. Excessive water consumption is rare but does occur with *psychogenic polydipsia* (excessive water consumption resulting from a psychiatric disturbance) and sometimes with organic lesions in the anterior hypothalamus. Impairment of the kidney to excrete free water results from antidiuretic hormone secretion in excess of that required to maintain normal ECW osmolarity. This condition is known as the syndrome of inappropriate antidiuretic hormone secretion (SIADH). Causes of SIADH are listed on p. 378. It may be caused by the perception of low blood volume by atrial volume receptors in the absence of a true decrease in blood volume. Decreased return of blood to the heart and redistribution of blood within the vascular compartment (without overall decrease in blood volume) may occur with asthma, pneumothorax, pneumonia, positive-pressure ventilation, chronic obstructive pulmonary disease, pneumonitis, right-sided heart failure, and diseases of the spinal cord or peripheral nerves. This "regional hypovolemia" is sensed by atrial volume receptors and results in antidiuretic hormone excretion that is excessive for ECW osmolality. Antidiuretic hormone excess may result from secretion of ADH from the posterior pi-

Causes of pure water (simple) dehydration

- Decreased water intake
  - Lack of access
  - Adipsia—absence of thirst
  - Iatrogenic—insufficient intravenous fluid administration
- Increased free water output
  - Pituitary diabetes insipidus—decreased output of ADH by the hypothalamus
  - Nephrogenic diabetes insipidus—inability of the kidney to respond to ADH

Causes of syndrome of inappropriate antidiuretic hormone secretion (SIADH)

Increased secretion of ADH by hypothalamus secondary to "regional hypovolemia"
- Asthma
- Pneumothorax
- Bacterial or viral pneumonia
- Positive pressure ventilation
- Chronic obstructive pulmonary disease
- Right-sided heart failure
- Disease of spinal cord or peripheral nerves (Guillain-Barré syndrome, poliomyelitis)

Increased secretion of ADH by hypothalamus in absence of appropriate osmolar or volume stimuli
- Central nervous system disorders (intracranial hemorrhage, hydrocephalus, skull fracture, severe asphyxia, brain tumors, cerebrovascular thrombosis, meningitis, encephalitis, seizures, acute psychoses, and cerebral atrophy)
- Hypothyroidism
- Pain, fear
- Anesthesia or surgical stress
- Drugs such as morphine, barbiturates, cyclophosphamide, vincristine, and carbamazepine

Ectopic, autonomous secretion of ADH
- Bronchogenic carcinoma
- Adenosarcoma of pancreas
- Lymphosarcoma
- Duodenal adenocarcinoma
- Pulmonary tuberculosis
- Pulmonary abscess

tuitary in the absence of either appropriate osmolar or volume stimuli with many central nervous system disorders (such as increased intracranial pressure or infection), hypothyroidism, pain, fear, or certain drugs. It can also occur as the result of the autonomous secretion of antidiuretic hormone by tumor cells and even by nonmalignant lung tissue.

With water intoxication, the *dilutional* hyponatremia and hyposmolarity of the ECW results in water movement into the cells. Therefore water intoxication produces expansion of the ECW and ICW compartments (Table 21-4).

The symptoms of water intoxication are related to the degree and rate of fall in sodium. With an acute fall in serum sodium to 120 to 125 mEq/L, nausea, vomiting, seizures, and coma can occur.

Excessive water and sodium. *Expansion of the extracellular compartment* usually results from retention of sodium *and* water. This occurs with oliguric renal failure, nephrotic syndrome, congestive heart failure, cirrhosis, and primary hyperaldosteronism. In these conditions, water excess is associated with normal or low serum sodium and osmolality (Table 21-4). Hypernatremia is uncommon with water excess. If the serum sodium is normal, the increase in total body water will be limited to the ECW. With hyponatremia, the increase of TBW will be shared by the ECW and ICW compartments.

SODIUM METABOLISM

Sodium balance

In a normal adult, the total body sodium is about 55 mEq/kg of body weight, but about 30% is tightly bound in the crystal structure of bone and thus nonexchangeable. Thus only 40 mEq/kg is exchangeable among the various compartments and accessible to measurement. The exchangeable sodium is distributed primarily but not solely in the extracellular space (Fig. 21-13). About 97% to 98% of the exchangeable sodium is found in the ECW space and only 2% to 3% in the ICW space. Approximately 16% of exchangeable sodium is in plasma, 41% is in ISF that is readily accessible to the plasma compartment, 17% is in ISF of dense connective tissue and cartilage, 20% is in ISF of bone, and 3% to 4% in the transcellular water compartment. Total bone sodium (exchangeable plus nonexchangeable) accounts for 40% to 45% of the total body sodium. The concentration of sodium in the various fluid compartments is depicted in Table 21-2. As discussed previously, the difference in sodium concentration between plasma and ISF is caused by the Gibbs-Donnan equilibrium. The difference in sodium concentration between ISF and ICW is caused by the active transport of sodium out of the cell (Fig. 21-12).

The amount of sodium in the body is a reflection of the balance between sodium intake and output. Sodium intake of course depends on the quantity and type of food intake of the person. Under normal conditions, the average adult takes in about 50 to 200 mEq of sodium per day. Sodium output occurs through three primary routes, the gastrointestinal tract, the skin, and the urine.

Under normal circumstances, loss of sodium through the gastrointestinal tract is very small. Fecal water excretion is only 100 to 200 mL/day for a normal adult, and fecal sodium excretion only 1 to 2 mEq/day. However, one should bear in mind that although fecal losses of water and electrolytes are normally small, the total volume of gastrointestinal fluid secreted is large, averaging about 8 L/day. Almost all of this volume is normally reabsorbed. However, with impaired gastrointestinal reabsorption, losses of water and electrolytes are large. The volume and electrolyte content of various gastrointestinal secretions are shown in Table 21-5. Notice that most of the secretions have sodium contents much higher than that of the feces. Thus with severe diarrhea or with gastric or intestinal drainage tubes, sodium losses through the gastrointestinal tract may exceed 100 mEq/day.

The sodium content of sweat averages about 50 mEq/L but is somewhat variable. The sweat sodium concentration is decreased by aldosterone and increased in cystic fibro-

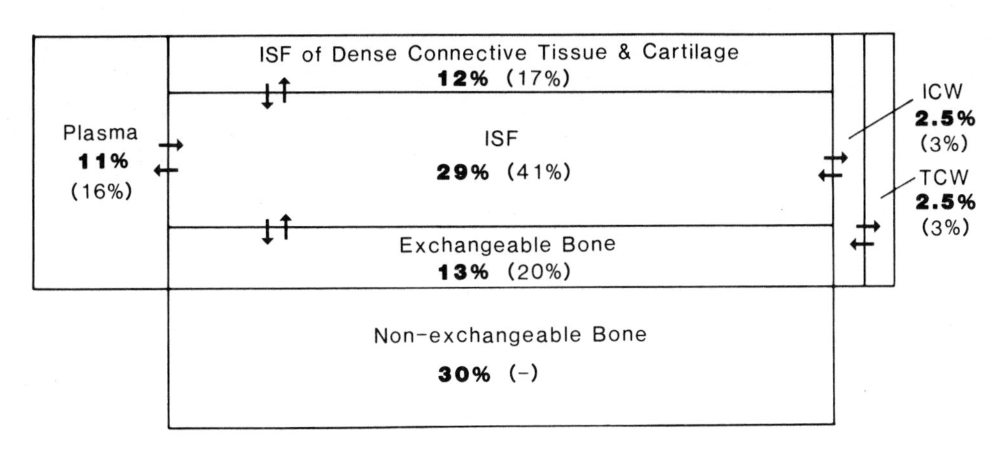

Fig. 21-13. Distribution of sodium among body compartments. *Bold numbers,* Percentages of *total body* sodium in various compartments. *Numbers in parentheses,* Percentages of *exchangeable* sodium in various compartments. *ISF,* Interstitial fluid; *ICW,* intracellular water; *TCW,* transcellular water.

Table 21-5. Electrolyte composition and volume of various gastrointestinal secretions in a normal adult

Fluid	Volume secreted (mL/day)	Electrolyte concentration (mEq/L)			
		Na^+	K^+	Cl^-	HCO_3^-
Gastric juice*	2500	8-120	1-30	8-100	0-20
Bile	700-1000	134-156	4-6	83-110	38
Pancreatic juice	>1000	113-153	2-7	54-95	110
Small bowel	3000	72-120	3.5-7	69-127	30
Ileostomy	100-4000	112-142	4.5-14	43-122	30
Cecostomy	100-300	48-116	11-28	35-70	15
Feces	100	<10	<10	<15	<15

*Electrolyte composition of gastric juice varies, depending on acidity. The higher the acidity, the lower the sodium concentration, the higher the chloride concentration, and the lower the bicarbonate concentration. The average sodium concentration is approximately 100 mEq/L. (From Lockwood, J.S., and Randall, H.T.: Bull. N.Y. Acad. Med. **25**:228-243, April 1949.)

sis. The amount of sweat is of course highly variable, increasing in hot environments, during exercise, and with fever. Under certain circumstances, sweat production can exceed 5 L/day, accounting for a loss of more than 250 mEq of sodium. Under normal conditions, in a cool environment, sodium losses through the skin are small. With extensive burns or exudative skin lesions there is great loss of sodium and water.

The major route of sodium excretion is through the kidney. Furthermore, the urinary excretion of sodium is carefully regulated to maintain body sodium homeostasis. That is, as intake of sodium increases or decreases, renal sodium excretion increases or decreases. The details of the mechanisms of renal sodium excretion are discussed in Chapter 23; only a brief description is given in this chapter.

Sodium is freely filtered by the glomerulus. About 70% of the filtered sodium is reabsorbed by the proximal tubule, about 15% by the loop of Henle, about 5% by the distal convoluted tubule, 5% by the cortical collecting tubule, and another 5% by the medullary collecting duct so that normally less than 1% of the filtered sodium is excreted. When there is an increase in the sodium intake, there is a temporary expansion of the extracellular volume. This results in an increase in the glomerular filtration rate and the amount of sodium filtered and a decrease in the amount of sodium reabsorbed by the proximal tubule and the collecting duct. In addition, there is a decrease in the circulating levels of aldosterone, resulting in a decrease in the reabsorption of sodium in the distal convoluted tubule and cortical collecting duct. All these factors increase the excretion of sodium and return the extracellular volume to normal. The opposite occurs when there is a decrease in the sodium intake.

Overall regulation of body sodium, and therefore extracellular volume, is illustrated for the case of sodium depletion in Fig. 21-9. The opposite effect occurs with sodium excess.

Disorders of sodium balance

Sodium excess. Sodium accumulates in the body when the intake of sodium exceeds output because of some ab-

Clinical conditions resulting in excess body sodium
Cardiac failure Liver disease Renal disease—nephrotic syndrome Hyperaldosteronism Pregnancy

normality of the sodium homeostatic mechanism. Some major clinical causes of sodium retention are presented in the box above.

Since sodium is distributed in the extracellular space, an increase in total body sodium is usually accompanied by an increase in ECW volume. An abnormal increase in ECW volume, particularly an increase in the interstitial space, produces tissue swelling known as edema. Thus those clinical conditions associated with sodium retention are frequently characterized by the presence of edema. Clinically, edema is characterized by swelling and puffiness of the body.

Congestive heart failure. When the heart begins to fail as a pump, a series of pathophysiological mechanisms are called into play (Fig. 21-14). The failing heart does not pump as much blood to the kidney, resulting in less sodium filtration, greater reabsorption, and consequently, less excretion. The greater venous back pressure generated from the failing heart causes fluid to move from the vascular space to the interstitial space, decreasing the effective blood volume, particularly on the arterial side. The low blood volume (hypovolemia) stimulates the secretion of aldosterone and ADH, which further enhance salt and water retention.

Liver disease. In various diseases of the liver there is venous obstruction, which results in production of excessive amounts of hepatic lymph and increased portal venous and sinusoidal pressure. These in turn lead to leakage of fluid out of the hepatic vascular space into the peritoneal space (ascites), which lowers the effective blood volume. The lowered arterial blood volume leads to salt and water retention by mechanisms similar to those for heart failure.

Renal disease. If the kidneys are damaged to such a degree that the glomerular filtration rate is greatly reduced and sodium excretion is thereby compromised, sodium retention will occur. Renal damage of this degree may occur as the result of renal ischemia, nephrotoxins, chronic renal infection, glomerulonephritis, or obstruction of the renal outflow tract. In addition, one kidney disorder, the nephrotic syndrome, is characterized by decreased serum al-

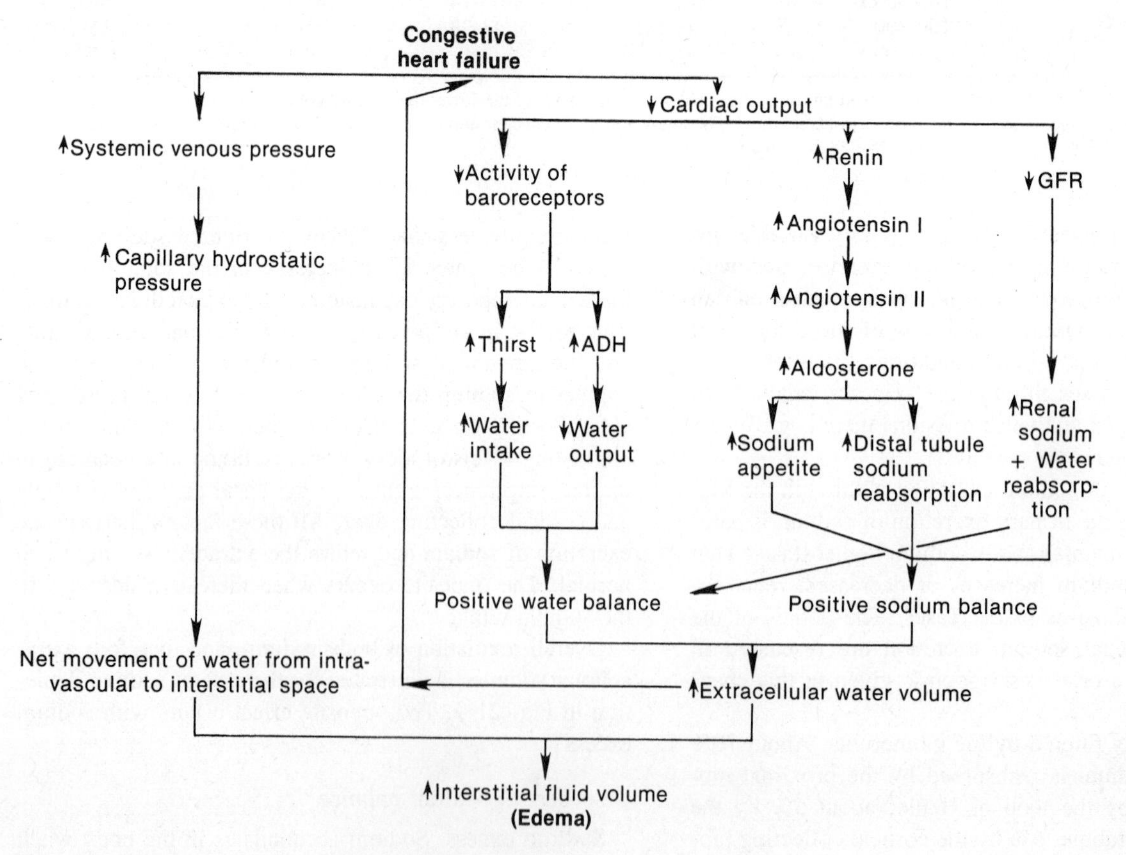

Fig. 21-14. Pathophysiology of congestive heart failure.

bumin, resulting in low plasma colloid osmotic pressure and therefore movement of fluid from the vascular space to the ISF space. This in turn results in effective vascular hypovolemia with consequent salt and water retention, as previously discussed.

Pregnancy. The reasons for sodium accumulation during pregnancy are still unclear, but there is no question that most women accumulate between 500 and 800 mEq of sodium during a normal pregnancy. There is some suggestion that the sodium accumulation may be a resetting of the normal homeostatic mechanism regulating body sodium and water.

Sodium depletion. Sodium depletion occurs when the output of sodium exceeds the intake (see the box below). As discussed previously, only small amounts of sodium are lost in the feces under normal conditions. However, under conditions of severe diarrhea, or drainage of gastrointestinal secretions, gastrointestinal sodium excretion can be quite large, and if this is not replaced by increased input, sodium depletion will result. Moreover, since the gastrointestinal route may not be available for input, the intravenous replacement of water and electrolytes may be necessary.

Similarly, losses of sodium through the skin are normally relatively small, but when the volume of sweat becomes large (as occurs with exercise, fever, or a hot environment), when the concentration of sodium in sweat is abnormally high (as with cystic fibrosis), or when there is abnormal transudation of fluid and electrolytes through the skin (as occurs with extensive burns), the amount of sodium lost through the skin may be substantial and sodium depletion might occur.

When the tubules of the kidney are unable to reabsorb sodium because of disease or hormonal abnormalities, sodium loss can be excessive. For example, aldosterone deficiency, caused by disease of the adrenal gland or abnormalities in the aldosterone-regulating system, would lead to decreased reabsorption of sodium in the distal convoluted tubule and total body sodium depletion. Inhibition of tubular sodium reabsorption by a diuretic would also lead to body sodium depletion.

In SIADH (see p. 377) there is water retention and hypotonic expansion of the ECW and ICW spaces. This in turn inhibits sodium reabsorption in the proximal nephron (and also perhaps the distal nephron), leading to body salt depletion.

Clinical conditions that can result in deficits of body sodium

Gastrointestinal losses—vomiting, diarrhea, fistulas, drainage tubes
Excessive sweating—exercise, fever, hot environment
Renal disease
Adrenal insufficiency—hypoaldosteronism
Diuretic therapy
Osmotic diuresis—diabetes mellitus
Burns
SIADH

Abnormalities of plasma sodium concentration

Changes in total body sodium are not necessarily associated with similar changes in plasma sodium concentration. That is, with salt retention, plasma sodium concentration is not necessarily increased. In fact, plasma sodium is frequently decreased in sodium-retentive states. Similarly, salt depletion is not necessarily associated with decreased plasma sodium concentrations. Sodium concentration reflects the relative balance of extracellular sodium and water.

Hyponatremia (low serum sodium) occurs when there is a greater excess of extracellular water than of sodium or a greater deficit of sodium than of water. Some causes of hyponatremia are listed in the box below. Notice that in many cases there is an excess of total body sodium.

The symptoms of hyponatremia depend on the cause, magnitude, and rate of fall in serum sodium. With acute, pronounced hyponatremia caused by water intoxication, nausea, vomiting, seizures, and coma occur. Symptoms are less fulminant with chronic hyponatremia caused by salt depletion in excess of water depletion. With progres-

Clinical conditions associated with hyponatremia

Water excess greater than sodium excess
- Heart failure*
- Liver disease*
- Nephrotic syndrome*
- Renal failure*
- Inapproriate ADH secretion
- Psychogenic polydipsia (excessive fluid intake)
- Essential hyponatremia (reset ''osmostat'')

Sodium deficit greater than water deficit
- Certain gastrointestinal abnormalities—vomiting, diarrhea, fistulas, and intestinal obstruction
- Burns
- Diuretic therapy
- Adrenal insufficiency—hypoaldosteronism

Movement of sodium from extracellular to intracellular water space
- Adrenal insufficiency—hypoaldosteronism
- Sick cell syndrome—shock

Pseudohyponatremia—hyperglycemia, hyperlipidemia, hyperglobulinemia

*Hyponatremia is dilutional, that is, secondary to excessive water retention. Total body sodium may even increase.

Clinical conditions associated with hypernatremia

Sodium excess greater than water excess
 Ingestions of large amounts of sodium
 Administration of hypertonic NaCl or $NaHCO_3$
 Primary hyperaldosteronism
Water deficiency greater than sodium deficiency
 Excessive sweating*—exercise, fever, hot environment
 Burns*
 Hyperventilation
 Diabetes insipidus
  ADH deficiency
  Nephrogenic—kidney unresponsive to ADH
 Osmotic diuresis*—diabetes, mannitol infusion
 Diminished fluid input—diminished thirst
 Essential hypernatremia—reset ''osmostat''
 Certain diarrheal states and vomiting*

*Total body sodium is decreased. Serum sodium concentration is increased because the magnitude of water loss exceeds the magnitude of sodium loss.

sively severe degrees of chronic hyponatremia, constant thirst, muscle cramps, nausea, vomiting, abdominal cramps, weakness, lethargy, and finally delirium and impaired consciousness occur.

Hypernatremia (high plasma sodium) occurs when there is greater deficit of extracellular water than of sodium or a greater excess of sodium than of water. Some causes of hypernatremia are listed in the box above. Notice that in many cases there is actually a deficit of total body sodium.

Hypernatremia usually occurs as a chronic process secondary to loss of water in excess of sodium. Symptoms are therefore those of dehydration (see p. 377).

POTASSIUM METABOLISM

Potassium balance

About 98% of the total body potassium is found in the intracellular water space, reaching a concentration there of about 150 to 160 mEq/L. In the extracellular water space, the concentration of potassium is only 3.5 to 5 mEq/L. Total body potassium in an adult male is about 50 mEq/kg of body weight and is influenced by age, sex, and, very importantly, muscle mass, since most of the body potassium store is contained in muscle.

The concentration of potassium in the various fluid compartments is listed in Table 21-2. The difference in potassium concentration between plasma and interstitial fluid is attributable to the Gibbs-Donnan equilibrium. The difference in potassium concentration in interstitial fluid and intracellular fluid is caused by the active transport of potassium into the cell (Fig. 21-12). Factors that enhance potassium transport into the cell and thereby increase the ratio of intracellular to extracellular potassium are insulin, aldosterone, alkalosis, and beta-adrenergic stimulation. Factors that decrease potassium transport into the cell or enhance leakage out of the cell include acidosis, alpha-adrenergic stimulation, and tissue hypoxia.

The amount of potassium in the body is a reflection of the balance between potassium intake and output. Potassium intake depends on the quantity and type of food intake of the person. Under normal conditions, the average adult takes in about 50 to 100 mEq of potassium per day (about the same amount as sodium). Potassium output occurs through three primary routes, the gastrointestinal tract, the skin, and the urine.

Under normal conditions, loss of potassium through the gastrointestinal tract is very small, amounting to less than 5 mEq per day for an adult. The concentration of potassium in the sweat is less than that of sodium, and so potassium losses through the skin are usually quite small.

The major means of potassium excretion is urination. Furthermore, the kidney is capable of regulating the excretion of potassium to maintain body potassium homeostasis. The details of the mechanisms of renal potassium excretion are discussed in Chapter 23, and only a brief description is given here.

Potassium is freely filtered by the glomerulus. Almost all the filtered potassium is reabsorbed in the proximal tubule and ascending limb of Henle's loop, so that only about 10% of the filtered potassium reaches the distal convoluted tubule. However, in this region of the nephron, potassium is secreted into the tubular fluid. The amount of potassium appearing in the final urine therefore is largely determined by the amount of potassium secreted by the distal tubule. Factors that enhance distal tubular potassium secretion include aldosterone, increased sodium and water delivery to the distal tubule, high-potassium diet, and alkalosis.

Disorders of potassium balance

Potassium excess. Potassium accumulates in the body when the intake of potassium exceeds output because of some abnormality of the potassium homeostatic mechanisms. Some major clinical conditions causing potassium retention are presented in the next box on p. 383. It should be noted that under most conditions the normal kidney is capable of excreting a great deal of potassium, and a high potassium intake leads to potassium retention only when kidney function is compromised.

Potassium depletion. Potassium depletion occurs when potassium output exceeds intake. As discussed previously, only small amounts of potassium are lost in the feces under normal conditions. As is the case for water and sodium, however, gastrointestinal potassium excretion during diarrhea or drainage of gastrointestinal secretions can be quite large (see Table 21-5). Some major clinical conditions causing potassium depletion are presented in the second box on p. 383.

Causes of potassium retention

Increased potassium intake
- High-potassium diet
- Oral potassium supplementation
- Intravenous potassium administration
- Potassium penicillin in high doses
- Transfusion of aged blood

Decreased potassium excretion
- Renal failure
- Hypoaldosteronism—adrenal failure
- Diuretics that block distal tubular potassium secretion; triamterene, amiloride, spironolactone
- Primary defects in renal tubular potassium secretion

Note that alkalosis results in total body potassium depletion. With alkalosis, potassium moves from the extracellular to the intracellular space. In the cells of the late distal tubule of the kidney, this increase in intracellular potassium stimulates potassium secretion and therefore increases renal excretion of potassium.

Abnormalities of plasma potassium concentration

Abnormalities in plasma potassium concentration can occur, not only because of abnormalities in total body potassium but also because of shifts of potassium between the extracellular and intracellular compartments. Although similar shifts may occur with sodium, the effect of intracellular to extracellular shifting on plasma concentration is

Causes of potassium depletion

Decreased potassium intake
- Low-potassium diet
- Alcoholism
- Anorexia nervosa

Increased gastrointestinal losses
- Vomiting
- Diarrhea
- Fistulas
- Gastrointestinal drainage tube
- Malabsorption
- Laxative or enema abuse

Increased urinary losses
- Increased aldosterone
  - Primary aldosteronism
  - Adrenal hyperplasia
  - Bartter's syndrome
  - Birth control pills
- Adrenogenital syndrome
- Renal disease
  - Renal tubular acidosis
  - Fanconi syndrome
- Diuretics
  - Thiazides
  - Loop diuretics—ethacrynic acid, furosemide
  - Carbonic anhydrase inhibitors—acetazolamide
- Alkalosis

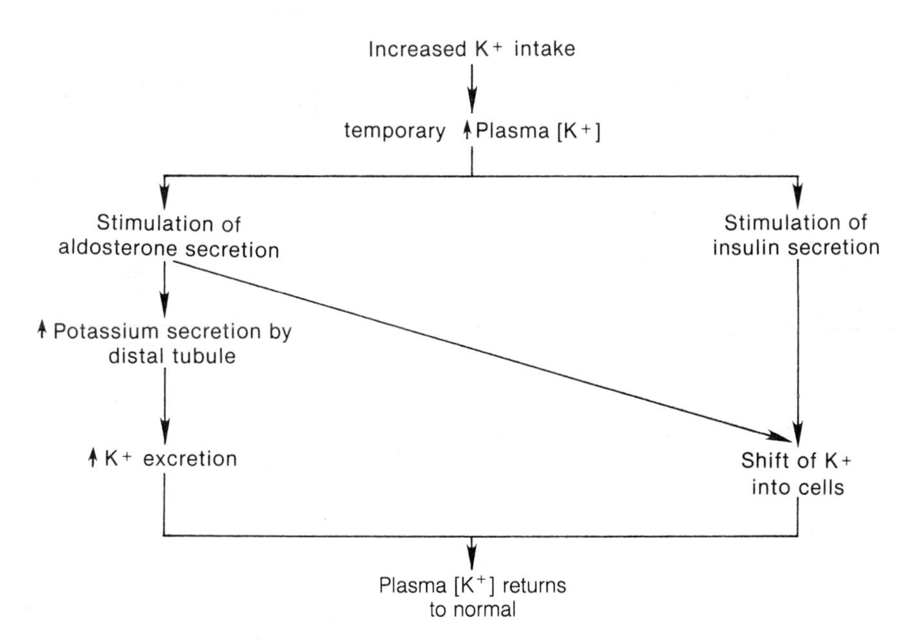

Fig. 21-15. Control of plasma potassium concentration.

Causes of hyperkalemia

Pseudohyperkalemia
  Hemolysis
  Leukocytosis
Intracellular to extracellular shift
  Acidosis*
  Crush injuries
  Tissue hypoxia*
  Insulin deficiency*
  Digitalis overdose*
High potassium intake (see box, p. 383, left)
Decreased potassium excretion (see box, p. 383, left)

*May be associated with total body potassium depletion.

more pronounced for potassium since 98% of the total potassium is intracellular. For example, if only 2% of the intracellular potassium were to shift to the extracellular space, the plasma potassium concentration would increase 100%. Fortunately the plasma potassium concentration is held fairly constant despite large fluctuations in potassium intake. This homeostatic mechanism is depicted in Fig. 21-15 for increased potassium intake. The opposite effect occurs with decreased potassium intake. Clinical conditions associated with elevated plasma potassium (hyperkalemia) are listed in the box above. Actual plasma potassium may be normal, but measured plasma potassium may be elevated (pseudohyperkalemia) if the blood sample is hemolyzed or if there is leakage of potassium from white blood cells under conditions of leukocytosis (elevated white blood cell number). In addition, vigorous arm exercise, tight application of the tourniquet, or squeezing of the area around the venipuncture site may result in cellular potassium release and spurious elevation of plasma potassium concentration.

True hyperkalemia can result from movement of potassium out of the cell into the extracellular water space, increased intake, or decreased output. Hyperkalemia caused by potassium shifts may in fact be associated with total body potassium depletion. This is the case in diabetic ketoacidosis. Hyperkalemia caused by increased intake or decreased output is usually associated with total body potassium excess.

The clinical signs and symptoms of hyperkalemia include changes in the electrocardiogram, cardiac arrythmia, muscular weakness, and paresthesias. The greatest danger of hyperkalemia is that patients may die of cardiac arrythmia or standstill.

Low plasma potassium concentration (hypokalemia) can be caused by movement of potassium into the cell from the extracellular water space, increased output, or decreased intake (see box in right column). Hypokalemia

Causes of hypokalemia

Extracellular to intracellular potassium shift
  Alkalosis
  Increased plasma insulin*
  Diuretic administration
Decreased potassium intake
Increased gastrointestinal losses
Increased urinary losses
} See box, p. 383, right

*May be associated with total body potassium excess.

caused by potassium shifts may in fact be associated with increased total body potassium. Hypokalemia caused by increased excretion or decreased intake is usually associated with total body potassium depletion.

Signs and symptoms of hypokalemia are numerous and include anorexia, nausea, vomiting, abdominal distension, muscle cramps or tenderness, paresthesias, electrocardiographic changes, arrythmias, inability to concentrate the urine with resultant polyuria and polydysia, lethargy, and confusion.

CHLORIDE METABOLISM

Chloride balance

Chloride is the major anion in the extracellular water space. In a normal adult, the total body chloride is about 30 mEq/Kg of body weight. Approximately 88% of chloride is found in the ECW space and 12% in the ICW space. Approximately 14% of total body chloride is in the plasma, 27% in interstitial fluid that is readily accessible to plasma, 17% in ISF of dense connective tissue and cartilage, 15% in ISF of bone, and 5% in the transcellular space. The concentration of chloride in the various fluid compartments is depicted in Table 21-2. Note that the concentration of chloride in the interstitial fluid is greater than that in plasma water, whereas the concentrations of sodium and potassium in interstitial fluid is less than that in plasma water. These differences between plasma and interstitial fluid are caused by the Gibbs-Donnan equilibrium. Chloride is passively distributed across the cell membrane. The difference in chloride concentration between interstitial fluid and intracellular water is caused by the electrical potential difference across the cell membrane (Fig. 21-12). Since the inside of the cell is negative compared to the outside, the concentration of chloride outside the cell will be higher than that inside.

The amount of chloride in the body is a reflection of the balance between chloride intake and output. Chloride intake is dependent on the quantity and type of food intake of the person. The chloride content of most foods parallels that of sodium. Under normal conditions the average adult

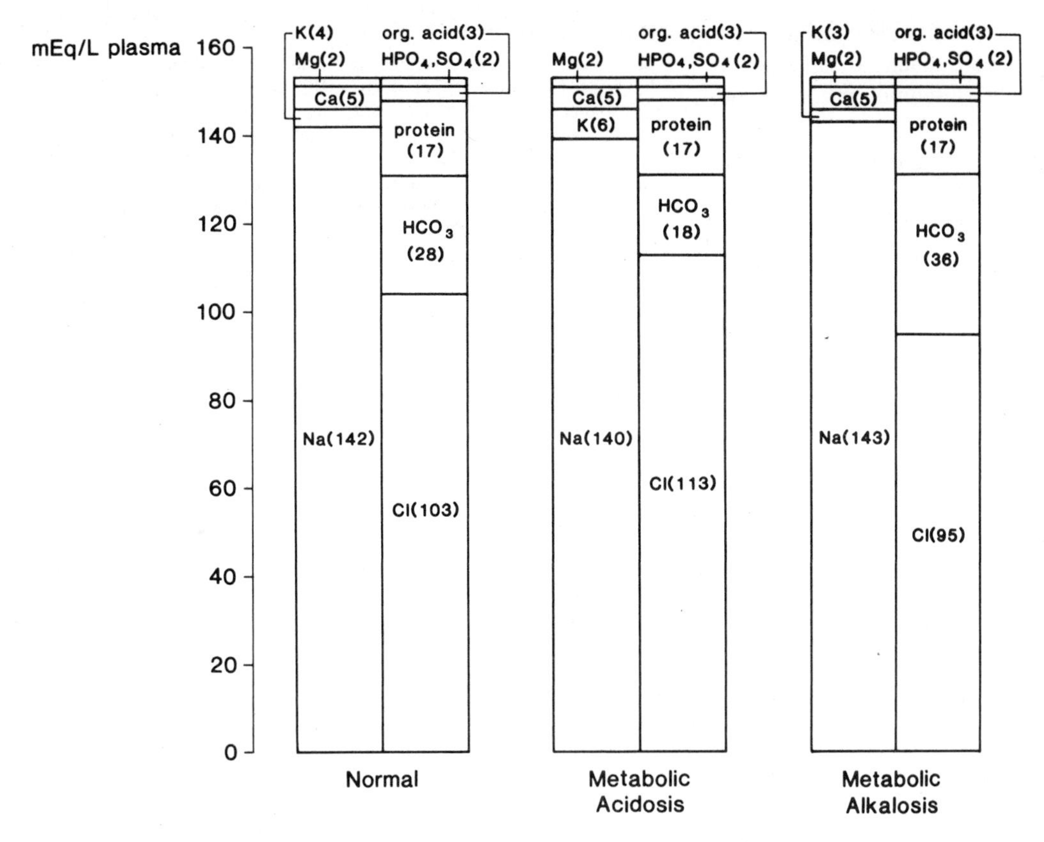

Fig. 21-16. Concentration of electrolytes in plasma (mEq/L) with metabolic acidosis and metabolic alkalosis compared to normal. In example of metabolic acidosis shown, there is no increase in organic acids, only loss of bicarbonate. Metabolic acidosis may be attributable to an increase in organic acids (see Fig. 21-4). In these cases chloride may not be increased. Note that extracellular potassium concentration is elevated in metabolic acidosis and lower in metabolic alkalosis. Under all conditions concentration of anions equals concentration of cations.

takes in about 50 to 200 mEq of chloride per day. Chloride output occurs through three primary routes: the gastrointestinal tract, the skin, and the urinary tract.

Under normal circumstances, loss of chloride through the gastrointestinal tract is very small. Fecal chloride excretion for a normal adult is only 1 to 2 mEq/day. The concentrations of chloride in gastrointestinal secretions are shown in Table 21-5. With severe diarrhea or with gastric or intestinal drainage tubes, chloride losses through the gastrointestinal tract may exceed 100 mEq/day.

The chloride composition of sweat averages about 40 mEq/L but is somewhat variable. As in the case of sodium, the concentration of chloride in sweat is decreased by aldosterone and increased in cystic fibrosis. Under conditions of excessive sweating, chloride losses through the skin can exceed 200 mEq/day. However, under normal conditions chloride losses through the skin are small.

The major route of chloride excretion is through the kidney. Details of the mechanisms of renal chloride excretion are discussed in Chapter 23, and only a brief description is given in this chapter.

Chloride is freely filtered by the glomerulus. About 60% to 65% of the filtered chloride is reabsorbed by the proximal tubule, about 20% to 25% by the loop of Henle, and about 10% to 15% by portions distal to Henle's loop, so that normally only about 1% of the filtered chloride is excreted.

Disorders of chloride balance

Chloride excess. Chloride accumulates in the body when the intake of chloride exceeds output because of some abnormality of the chloride homeostasis mechanism. For the most part, the causes of chloride retention are the same as those of sodium retention. Therefore the pathophysiology of chloride excess is in most cases similar to that of sodium (see box on p. 380). However, there is one clinical condition in which chloride excess may not be associated with sodium excess: metabolic acidosis. The two major extracellular anions are chloride and bicarbonate. In metabolic acidosis, extracellular bicarbonate is depleted; if electrical neutrality is to be maintained, extracellular chloride concentration (and therefore chloride content) must increase (Fig. 21-16). The increase in chloride concentration is caused by the reabsorption by the tubules of the kidney of a relatively greater proportion of sodium with chloride than with bicarbonate.

Chloride depletion. Chloride depletion occurs when the output of chloride exceeds intake. For the most part, the causes of chloride depletion are the same as those of sodium depletion (see box on p. 381, left). However, in one clinical condition, metabolic alkalosis, there may be chloride depletion without sodium depletion. Metabolic alkalosis is associated with bicarbonate excess, which in the presence of a normal sodium content requires the loss of chloride to maintain electrical neutrality (Fig. 21-16).

Abnormalities of plasma chloride concentration

As for sodium, changes in total body chloride are not necessarily associated with similar changes in plasma chloride concentration. That is, with body chloride retention the plasma chloride concentration will remain normal if there is a proportional increase in extracellular water and will decrease if there is a relatively greater increase in extracellular water. Similarly, plasma chloride concentration may remain normal or even increase with chloride depletion, depending on the concomitant change in extracellular water.

In most cases, the causes of hypochloremia and hyperchloremia are the same as those of hyponatremia and hypernatremia (see pp. 381 and 382). The major clinical exceptions to the usual parallel changes in plasma sodium and chloride concentrations occur during chronic metabolic acidosis and alkalosis. With metabolic acidosis, hyperchloremia may be unassociated with hypernatremia; with metabolic alkalosis, hypochloremia may not be accompanied by hyponatremia. The reasons for this are those previously discussed for chloride excess and depletion.

Symptoms are not directly attributable to hypochloremia or hyperchloremia. Rather, symptoms that occur in patients with an abnormal serum chloride concentration are caused by the associated abnormality in serum sodium or pH.

BIBLIOGRAPHY

1. Albert, S.N.: Blood volume. In Blahd, W.H., editor: Nuclear medicine, New York, 1971, McGraw-Hill Book Co., pp. 529-545.
2. Andreoli, T.E., Grantham, J.J., and Rector, F.C., editors: Disturbances in body fluid osmolality, Bethesda, Md., 1977, American Physiological Society.
3. Aukland, K., and Nicolaysen, G.: Interstitial fluid volume: local regulatory mechanisms, Physiol. Rev. **61:**556, 1981.
4. Bauer, F.K.: Radioisotope dilution methods: measurement of body composition. In Blahd, W.H., editor: Nuclear medicine, New York, 1971, McGraw-Hill Book Co., pp. 513-527.
5. Brenner, B.M., and Stein, J.H., editors: Acid-base and potassium homeostasis (Contemporary Issues in Nephrology, vol. 2), New York, 1978, Churchill Livingstone.
6. Brenner, B.M., and Stein, J.H., editors: Sodium and water homeostasis (Contemporary Issues in Nephrology; vol. 1), New York, 1978, Churchill Livingstone.
7. Brozek, J., and Henschel, A., editors: Techniques for measuring body composition, Washington, D.C., 1961, National Academy of Sciences—National Research Council.
8. Cohen, J.J.: Disorders of potassium balance, Hosp. Practice **14**(1):119, 1979.
9. Collins, R.D.: Illustrated manual of fluid and electrolyte disorders, Philadelphia, 1983, J.B. Lippincott Co.
10. Donnan, F.G.: The theory of membrane equilibria, Chem. Rev. **1:**73, 1924.
11. Edelman, I.S., and Leibman, J.: Anatomy of body water and electrolytes, Am. J. Med. **27:**256, 1959.
12. Edelman, I.S., Olney, J.M., James, A.H., Brooks, L., and Moore, F.D.: Body composition: studies in the human being by the dilution principle, Science **115:**447, 1952.
13. Fitzsimons, J.T.: Thirst, Physiol. Rev. **52:**468, 1972.
14. Friis-Hansen, B.: Changes in body water compartment during growth, Acta Pediatr. **46**(supp. 110):1, 1957.
15. Gamble, J.L.: Chemical anatomy, physiology and pathology of extracellular fluid: a lecture syllabus, Cambridge, Mass., 1954, Harvard University Press.
16. Goldberger, E.: A primer of water, electrolyte and acid-base syndromes, Philadelphia, 1980, Lea & Febiger.
17. Guyton, A.C., Granger, H.J., and Taylor, A.E.: Interstitial fluid pressure, Physiol. Rev. **51:**527, 1971.
18. Hays, R.M.: Antidiuretic hormone, N. Engl. J. Med. **295:**659, 1976.
19. Humes, D.H., Narins, R.G., and Brenner, B.M.: Disorders of water balance, Hosp. Practice **14**(3):113, 1979.
20. Levy, M.: The pathophysiology of sodium balance, Hosp. Practice **13**(11):95, 1978.
21. MacKnight, A.D., and Leaf, A.: Regulation of cell volume, Physiol. Rev. **57:**510, 1977.
22. Maxwell, M.H., and Kleeman, C.R., editors: Clinical disorders of fluid and electrolyte metabolism, New York, 1980, McGraw-Hill Book Co.
23. Oh, M.S., and Carroll, H.J.: The anion gap, N. Engl. J. Med. **297:**814, 1977.
24. Scribner, B.H., and Burnell, J.M.: Interpretation of the serum potassium concentration, Metabolism **5:**468, 1956.
25. Schrier, R.W., editor: Symposium on water metabolism, Kidney Int. **10:**1-132, 1976.
26. Starling, E.H.: On the absorption of fluids from the connective tissue spaces, J. Physiol. (London) **19:**312, 1896.

Chapter 22 Acid-base control and acid-base disorders

John E. Sherwin
Berndt B. Bruegger

acidemia A condition of decreased pH of the blood.

acidosis A pathological condition resulting from accumulation of acid in the blood or loss of base from the blood.

alkalemia A condition of increased pH of the blood.

alkalosis A pathological condition resulting from accumulation of base or loss of acid from the body.

alveoli Small outpouchings of walls of alveolar space through which gas exchange takes place between alveolar air and pulmonary capillary blood.

anion gap The concentration of undetermined anions, calculated as the difference between the measured cations and the measured anions.

apnea Cessation of breathing.

base excess or deficit The difference between the titratable acids and bases of the blood at a pH of 7.4, a P_{CO_2} of 40 mm Hg, and a temperature of 37° C.

Boyle's law At fixed mass and temperature, a change in volume, V, is inversely proportional to the change in pressure, P; or $P \times V$ = Constant.

bradycardia Slowing of the heartbeat to less than 60 beats per minute.

carbamino reaction A covalent chemical reaction between CO_2 and a primary amino group ($-NH_2$) of proteins to form a stable, bound form of CO_2.

carbonic anhydrase An enzyme that catalyzes the reaction between CO_2 and water to form carbonic acid (H_2CO_3).

chloride shift Exchange of Cl^- for HCO_3^- in red blood cells in peripheral tissues in response to P_{CO_2} of blood. The shift reverses in the lungs.

conjugate acid Protonated cationic form of a corresponding weak acid.

conjugate base The deprotonated anion form of a corresponding weak acid.

dead spaces Spaces in the respiratory tract not directly involved in exchange of gas.

Gay-Lussac's law At constant pressure, a change in the temperature of the gas will cause a proportional change in volume. V/T = Constant. Also known as Charles' law.

Henderson-Hasselbalch equation Describes the relationship between pH, the pK_a of a buffer system, and the ratio of the conjugate base and a weak acid.

hypercapnia A condition of excess carbon dioxide in the blood.

hypochloremic alkalosis A metabolic alkalosis resulting from increased blood bicarbonate secondary to loss of chloride from the body.

hypoxia A condition of low oxygen content in the blood.

ideal gas law Relates pressure, *P*, volume, *V*, temperature, *T*, the number of moles of gas, *n*, and temperature and a gas constant, *R*, so that $PV = nRT$.

isohydric shift The series of reactions in red blood cells in which CO_2 is taken up and oxygen is released without the production of excess hydrogen ions.

metabolic acidosis Pathological loss of base in the body.

metabolic alkalosis Pathological accumulation of base in the body.

metabolic component The bicarbonate concentration of plasma.

oxygen saturation A term that defines the fraction of total hemoglobin (Hb) in the form of HbO_2 at a defined Po_2. Percentage of saturation = $100(HbO_2)/(HbO_2 + Hb)$.

P_{50} The partial pressure of oxygen at which hemoglobin is half-saturated with bound oxygen.

partial pressure The pressure exerted by a gas, whether it is alone or mixed with other gases. The partial pressure of a gas is denoted by the letter *P* preceding the symbol for that gas; for example, the partial pressure of CO_2 is Pco_2.

relative base deficit A term that describes the lowered HCO_3^-/H_2CO_3 ratio caused by an increase in Pco_2. The HCO_3^- (base) is low relative to the Pco_2.

relative base excess Name for the elevated HCO_3^-/H_2CO_3 ratio caused by a decrease in Pco_2. The HCO_3^- (base) is elevated relative to the Pco_2.

respiratory acidosis Pathological retention of CO_2 in the body caused by respiratory change.

respiratory alkalosis Pathological decrease in carbonic acid caused by respiratory change.

respiratory component "αPco_2," or the acid component, which is immediately modified by respiratory status.

surfactant An agent that decreases surface tension. Applies to agents that coat pulmonary alveolar surfaces.

ventilation The exchange of air between the lungs and ambient air.

ACID-BASE CONTROL

Acids and bases

Definitions. The simplest definition for an acid is a substance that releases protons, or hydrogen ions (H^+), whereas a base is simply defined as a substance that accepts protons, or H^+. Both acids and bases are further defined by their degree of affinity for H^+. A strong acid has little affinity for H^+ and so readily dissociates an H^+, whereas a weak acid has some affinity for H^+ and thus less readily dissociates an H^+. A strong base has a high affinity for H^+; a weak base has a low affinity for H^+. If one molecule differs from another by only a proton, they are called a conjugate acid-base pair. Physiological examples of a weak acid and its conjugate base are carbonic acid (H_2CO_3) and bicarbonate (HCO_3^-). The equilibrium reaction is as follows in equation 22-1:

$$H_2CO_3 \rightleftharpoons H^+ + HCO_3^- \qquad \textit{Eq. 22-1}$$

This equilibrium lies to the right at physiological pH, since bicarbonate, as the conjugate base, has a weak affinity for the hydrogen ion.

Dietary and metabolic sources of acids and bases. Two primary types of acids are dealt with in physiological states: fixed acids and volatile acids. Fixed acids are nongaseous acids such as phosphate ($HPO_4^=$) and sulfate (HSO_4^-) or organic acids such as lactic acid, acetoacetic acid, and beta-hydroxybutyric acid. The physiologically important volatile acid is carbonic acid (H_2CO_3). The volatility of carbonic acid arises from its ability to dissociate into water and carbon dioxide (CO_2), which can be released as a gas. The complete reaction scheme for carbonic acid is as follows:

Eq. 22-2

$$CO_2\ (\text{gas}) \rightleftharpoons CO_2\ (\text{dissolved}) \underset{-H_2O}{\overset{+H_2O}{\rightleftharpoons}} H_2CO_3 \rightleftharpoons H^+ + HCO_3^-$$

At one end of the equilibrium is carbon dioxide, which can be considered the anhydrous form of H_2CO_3, and at the other end is HCO_3^-, the conjugate base of H_2CO_3. Although the reaction of CO_2 and water to form H_2CO_3 will occur spontaneously, the enzyme carbonic anhydrase facilitates this reaction in vivo.

Carbohydrates, lipids, and proteins are metabolized by oxidation reactions that generate acids, which must be neutralized. In anaerobic conditions such as respiratory distress or strenuous exercise, carbohydrates are metabolized to lactic and pyruvic acids, which accumulate until normal oxygenation is achieved. These acids can be further metabolized to the ultimate oxidation product, carbon dioxide, when aerobic metabolism is resumed. Triglycerides are metabolized to fatty acids, which can be further metabolized to ketone bodies, acetoacetic acid, and beta-hydroxybutyric acid under anaerobic conditions. Ultimately these lipid metabolites are further oxidized to carbon dioxide. Proteins are hydrolyzed to amino acids, which are then converted to carbon dioxide. Those proteins composed of sulfur-containing amino acids are catabolized in part to the salt of sulfuric acid. Nucleic acids and some lipids contain phosphorus and are metabolized to salts of phosphoric acid.

pH and hydrogen ion. Water weakly dissociates into H^+ and OH^-. The equilibrium constant for the dissociation is 10^{-14} M. Very small values are used when one discusses the concentration of H^+ in water; for example, in a neutral solution there is an equal concentration of H^+ and OH^-, or 10^{-7} M of each. Rather than describing the concentration of H^+ in moles per liter, an easier convention to use is pH, which is defined as the negative logarithm of the concentration of H^+. Therefore a neutral solution has a pH of 7 $[-\log(10^{-7})]$. An acid solution has an H^+ concentration greater than 10^{-7} M and therefore has a pH of less than 7, whereas a basic (or alkaline) solution has an H^+ concentration less than 10^{-7} M and thus a pH greater than 7. Normal human whole blood maintains a slightly alkaline pH in a range of 7.35 to 7.45, which corresponds to an H^+ concentration of 4.5×10^{-8} M to 3.5×10^{-8} M. Remember, as the concentration of acid (H^+) increases, the pH decreases.

Buffers

Equilibrium. Before studying the mechanisms by which the body maintains a constant pH and neutralizes the acids produced during metabolism, one should review the concept of buffers. A buffer is a chemical solution designed to resist changes in pH after the addition of strong acids or bases. A buffer commonly consists of a weak acid (HA) and its conjugate base (A^-). Equation 22-3 shows the equilibrium reaction of such a hypothetical buffer solution:

$$HA \rightleftharpoons H^+ + A^- \qquad \textit{Eq. 22-3}$$

with an equilibrium equation as follows:

$$K_a = \frac{[H^+][A^-]}{[HA]} \qquad \textit{Eq. 22-4}$$

where K_a is the dissociation constant for the weak acid of the buffer solution. One can rearrange the equation as follows:

$$[H^+] = K_a \frac{[HA]}{[A^-]} \qquad \textit{Eq. 22-5}$$

and then taking the negative logarithm of both sides of the equation

$$-\log[H^+] = -\log K_a - \log \frac{[HA]}{[A^-]} \qquad \textit{Eq. 22-6}$$

and, finally, rearranging it to the Henderson-Hasselbalch equation of

$$pH = pK_a + \log \frac{[A^-]}{[HA]} \qquad \textit{Eq. 22-7}$$

When the concentration of A^- equals the concentration of HA, the pH is equal to the pK_a:

$$pH = pK + \log \frac{[A^-]}{[HA]} = pK_a + \log 1 = pK + 0 \qquad \textit{Eq. 22-8}$$

At this point, the solution is at its maximum buffering capacity. Adding strong acid or strong base to the solution at maximum buffer capacity will change the ratio of $[A^-]/[HA]$, but since that ratio is a log function, there will be little change in the pH until large enough quantities of the strong acid and base are added to substantially change the ratio of $[A^-]/[HA]$. A buffer is considered most effective within ± 2 pH units of its pK_a.

Physiological buffers. Buffering capacity is dependent on the concentration of the buffer and the relationship between the pK_a of the buffer and the desired pH. For maximal blood buffering, the pK_a of the buffers should therefore be near physiological pH, that is, pH 7.4. The physiologically important buffers that maintain the narrow pH range observed in the body are hemoglobin, bicarbonate, phosphate, and proteins. Table 22-1 lists the pK_a and concentrations of these buffer systems and their relative buffering capacities.

Bicarbonate buffer system. The reactive equation for the bicarbonate–carbonic acid buffer system (equation 22-1) is as follows:

$$pH = pK_a + \log \frac{[HCO_3^-]}{[H_2CO_3]} \qquad \textit{Eq. 22-9}$$

and an *apparent* pK_a of 6.33 at 37° C. Instead of being at the maximum buffer capacity with a 1:1 ratio of HCO_3^- to H_2CO_3, the bicarbonate–carbonic acid buffer system in the blood is at a ratio of 20:1 to produce a pH to 7.4:

$$pH = pK_a + \log \frac{[HCO_3^-]}{[H_2CO_3]} = 6.33 + \log 20 = 7.4 \qquad \textit{Eq. 22-10}$$

This 20:1 ratio is primarily maintained by the lungs, which expel CO_2 derived from carbonic acid ($H_2CO_3 \rightleftharpoons H_2O + CO_2$) produced during the metabolism of nutrients to CO_2.

In equation 22-9 there are three unknowns: pH, $[HCO_3^-]$, and $[H_2CO_3]$. Although pH is measurable, there is no direct measure of $[HCO_3^-]$ or $[H_2CO_3]$. In order to use this equation, the term "H_2CO_3" is replaced by an

Table 22-1. Physiologically important buffers and their concentrations, pK_a, and buffering capacity

Buffer	pK_a	Concentration (mmol/L)	Relative buffering capacity (mEq/L)
Bicarbonate	6.1	25	1
Hemoglobin	7.2	53	40
Phosphate	6.8	1.2	0.3
Protein	—	—	8

analyte that is measurable. The concentration of H_2CO_3 is proportional to the amount of dissolved CO_2 (equation 22-2). Thus one can replace $[H_2CO_3]$ with the term "αpCO_2," where α (the Bunson coefficient) is the solubility coefficient of CO_2. The pK_a term in the Henderson-Hasselbalch equation is modified to reflect the equilibrium between CO_2 (dissolved) and HCO_3^-. The modified Henderson-Hasselbalch equation describing this equlibrium is as follows:

$$pH = pK'_a + \log \frac{[HCO_3^-]}{[\alpha pCO_2]} \quad \textit{Eq. 22-11}$$

The apparent pK'_a in human plasma is 6.1 M at 37° C. The solubility coefficient for CO_2 in plasma at 37° C is 0.031 mmol × L^{-1} × mm Hg^{-1}. The preponderance of base to acid in this buffer system reflects the demands put on the body by its metabolism, which primarily produces acids. This buffer system is designed to eliminate the primary metabolic waste product, CO_2. The CO_2 component of the buffer system, that is, H_2CO_3, or $CO_2 + H_2O$, is eliminated by the lungs.

The total CO_2 content (Tco_2) of plasma is described as:

$$Tco_2 = CO_2\text{ (dissolved)} + [HCO_3^-] + [H_2CO_3]$$

One can disregard the $[H_2CO_3]$ term, since it is so small (one twentieth the $[HCO_3^-]$) and replace CO_2 (dissolved) with αCO_2. Thus the equation can be reduced to:

$$Tco_2 = \alpha Pco_2 + [HCO_3^-] \quad \textit{Eq. 22-12}$$

Two of the three unknowns are readily determined, allowing calculation of the third (see p. 396).

Hemoglobin. The major buffer of blood is hemoglobin, which is localized in the red blood cells. Hemoglobin (Hb) takes up free H^+ so that the following reaction proceeds to the right:

$$CO_2 + H_2O \rightleftharpoons H_2CO_3 \rightleftharpoons HCO_3^- + H^+ \quad \textit{Eq. 22-13}$$
$$KHb \rightarrow HHb + K^+$$

Hemoglobin and serum proteins have high concentrations of histidine residues. The imidiazole group of histidine (Fig. 22-1) has a pK_a of approximately 7.3. It is this combination of high concentration and the appropriate pK_a that makes hemoglobin the dominant buffering agent at physiological pH. The bulk of the CO_2 formed in peripheral tissues is transported in the plasma portion of blood as HCO_3^- with the H^+ bound to hemoglobin within the erythrocyte (Fig. 22-2). In tissue during exercise, during which both acidity and the Pco_2 are increased, carbonic CO_2 accounts for about 2 mmol/L of CO_2, as compared to only about 1 mmol/L in arterial blood.

Additionally, significant amounts of CO_2 are transported as a protein-bound moiety. CO_2 reacts nonenzymatically with the accessible amino groups of proteins to form a *carbamino* group

$$O{=}C{=}O + H_2N{-}\text{Protein} \rightarrow {}^-O{-}\overset{O}{\overset{\|}{C}}{-}\overset{H}{\overset{|}{N}}{-}\text{Protein} \quad \textit{Eq. 22-14}$$

Carbamino group

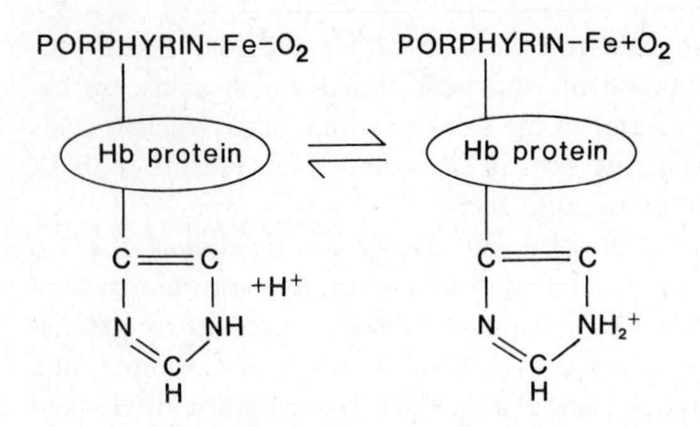

Fig. 22-1. Effect of hemoglobin oxygenation on buffering action of imidazole group of histidine. Binding affects pK_a of imidazole ring making the ring more acidic with release of an H^+.

Approximately 0.5 mmol/L of CO_2 is transported in this fashion, and there is only a very small difference between arterial and venous concentrations. The carbamino reaction with hemoglobin in the red blood cells tends to slightly reduce the buffering capacity of Hb, since a proton is released during the reaction.

The observed arteriovenous difference of total CO_2 content is almost entirely the result of formation of bicarbonate in the red blood cell. In the lungs, where deoxygenated hemoglobin becomes oxygenated and the CO_2 is expelled, the H^+ is released from the deoxygenated hemoglobin (HbH), since oxygenated hemoglobin (O_2Hb) is a stronger acid than the deoxyhemoglobin. In the lungs then, the reaction in equation 21-13 proceeds to the left with the release of H^+ and its reaction with the transported HCO_3^- to form CO_2 that the lungs can now release (Fig. 22-2). The overall equation linking the oxygenation process to buffering is:

$$H^+ + O_2Hb \rightleftharpoons HbH^+ + O_2 \quad \textit{Eq. 22-15}$$

The forward reaction occurs in tisues in which there is a relatively high H^+ and a relatively low O_2 concentration, whereas the reverse reaction occurs in the lungs, in which the H^+ concentration is relatively higher.

Phosphate and proteins. Phosphate is a minor buffering component of the blood, with the following equilibrium reaction occurring:

$$H_2PO_4^- \rightleftharpoons H^+ + HPO_4^{-2} \quad \textit{Eq. 22-16}$$

with a pK of 6.8. Phosphate buffer is an important buffer in urine that has relatively little protein, hemoglobin, or bicarbonate. The phosphate buffers in the blood are inor-

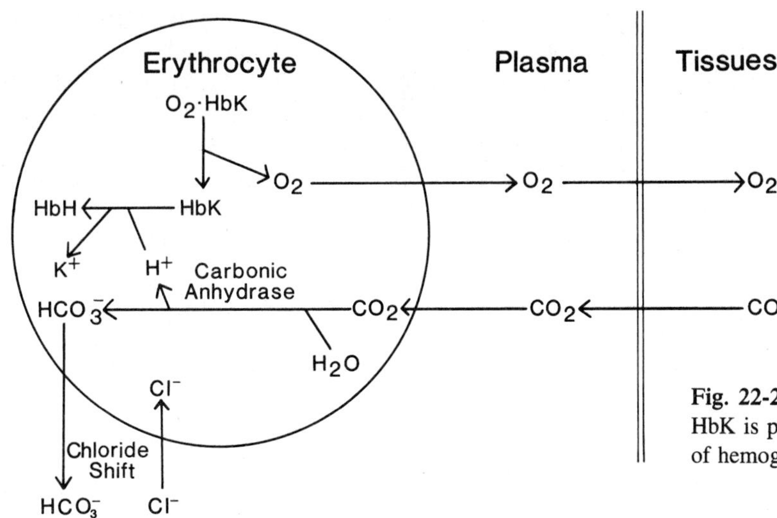

Fig. 22-2. Hemoglobin buffering action in peripheral tissues. HbK is potassium salt of hemoglobin. HbH is protonated form of hemoglobin.

ganic phosphates, in contrast to the intracellular state in which both inorganic and organic phosphates exist as buffers.

In addition, plasma proteins also act as buffers in the blood. The proteins have prosthetic groups that bind or release H^+ and also amino acids such as lysine, arginine, histidine, glutamate, and asparate. This buffering effect is minor compared to that of the bicarbonate system or hemoglobin system (Table 22-1).

Oxygen and carbon dioxide homeostasis

Partial pressure. In a mixture of gases, the total pressure is equal to the sum of the partial pressures of each gas. The composition of air is expressed in units of partial pressure for each gas making up the mixture rather than the percentage of each gas. The two gases of concern in blood gas analysis, oxygen and carbon dioxide, are expressed in terms of partial pressure of oxygen (Po_2) and carbon dioxide (Pco_2), respectively. Partial pressure is a measure of quantitation in a gas phase, but one can extend it to the quantitation of O_2 and CO_2 in a liquid such as blood. The partial pressures of O_2 and CO_2 in a theoretical gas phase that is in equilibrium with the dissolved O_2 and CO_2 in the blood defines the Po_2 and Pco_2 of blood; therefore Po_2 and Pco_2 are used to refer to the gas composition of blood as well as to the air. Historically, the units of Po_2 and Pco_2 have been torr, or millimeters of mercury (mm Hg), which are still used by a majority of laboratories in the United States. The international unit for partial pressure is the "pascal," or Pa (1 mm Hg = 133.3224 Pa).

Table 22-2 shows the composition of atmospheric air, alveolar air (air inside the lung), and expired air. Humidity makes a substantial contribution to the composition of air in the lungs, which in turn alters the partial pressures of the other gases. The correction for gas-volume caused by water vapor is an essential consideration because, at the alveolar surface, the inspired air is saturated with water at 37° C, which amounts to 47 mm Hg.

Gas-law relationships. In understanding the relationship of gases to their occupied volume, a number of basic gas laws need to be reviewed. Boyle's law is based on the observation that, for a fixed mass and temperature, a change in volume, V, of a gas will result in an inverse and proportional change in the pressure, P, exerted by the gas. Boyle's law for an ideal gas can be expressed by the following relationship:

$$P \times V = \text{Constant} \qquad \textit{Eq. 22-17}$$

This equation is applicable to the gases of air at room temperature and atmospheric pressure.

When pressure rather than temperature, T, is held constant, a change in the volume of the gas will produce a proportional change in the temperature of the gas. This is known as Gay-Lussac's law and is expressed by the following relationship:

$$\frac{V}{T} = \text{Constant} \qquad \textit{Eq. 22-18}$$

The laws of Boyle and Gay-Lussac combine to form the *ideal gas law:*

$$PV = nRT \qquad \textit{Eq. 22-19}$$

Table 22-2. Composition of air (partial pressure expressed in units of mm Hg)

Air	N_2	O_2	CO_2	H_2O	Total pressure
Atmospheric air	598.0	158.0	0.3	3.7	760
Alveolar air	573.0	100.0	40.0	47.0	760
Expired air	566.0	115.0	32.0	47.0	760

Thus at 37° C and normally encountered atmospheric pressures, the total pressure, *P*, is equal to the sum of the partial pressures (P_{O_2}, P_{CO_2}, P_{N_2}, and P_{H_2O}). Table 22-2 shows a total pressure of 760 mm Hg; however, extreme examples of lower air pressure (hypobarism) and higher air pressure (hyperbarism) do exist, as in space travel and deep-sea diving. Correction of blood gas data for changes in the total atmospheric pressure is important in these instances, and correction of calibration gases for variation in total pressure is necessary.

Two other concepts used in discussing the oxygen content of blood are *oxygen saturation* and P_{50}. Oxygen saturation refers to the degree to which the hemoglobin molecules are bound with oxygen. Oxygen saturation is the percentage of the total hemoglobin present as oxygenated hemoglobin versus the total hemoglobin content. The term P_{50} denotes the partial pressure of oxygen at which the hemoglobin is 50% saturated with oxygen.

Classification of ventilation. For an understanding of the exchange of gases by the body and the corresponding regulation of the bicarbonate–carbonic acid buffer system, an introduction into the concept of ventilation is necessary. First, ventilation should be differentiated from respiration, in that *ventilation* is the mechanical process of moving air in and out of the lungs; *respiration* is the exchange of gases between the atmosphere and the body. One concept that should be understood before going into the categories of ventilation is that of *dead space*. Not all areas of the respiratory tract contain capillaries of the pulmonary circulation that are properly situated to exchange gases between the blood in the capillaries and the air in the lung space. The areas that are not directly involved in the exchange of gas are known as "dead spaces." The portion of the respiratory tract that takes air into the small, thin-walled air spaces, called *alveoli,* is part of the dead space. Even alveoli, in which the exchange of gases between air and the capillaries of the pulmonary circulation occur, are considered dead space in those instances in which perfusion by the capillaries of the pulmonary circulation is lacking.

Ventilation is classified as total ventilation, dead space ventilation, and alveolar ventilation. Total ventilation consists in the mechanical process of bringing air through the entire respiratory tract. Dead-space ventilation is restricted to air movement in the dead spaces of the respiratory tract. Alveolar ventilation is restricted to air exchange at the surface of those alveoli that are actively perfused by the capillaries of the pulmonary system. In this classification, total ventilation is the sum of the dead-space ventilation and the alveolar ventilation.

Physiology of ventilation. The actual process of inspiring air into the lungs involves two major sets of muscles, the diaphragm and the intercostal muscles, which are located between the ribs. During inspiration of air, the diaphragm contracts, increasing the longitudinal dimension of the thoracic cavity and thereby increasing its volume. An additional smaller volume increase arises from the contraction of the external intercostal muscles, pulling the bottom portion of the ribs upward, raising the sternum, and increasing the anteroposterior dimension of the thoracic cavity. Expiration of air is accomplished either by the relaxation of the muscles involved in inspiration, or, as occurs during exercise, the internal intercostal muscles may pull the bottom part of the ribs and the sternum down while the anterior abdominal wall muscles induce the abdominal viscera to force up the diaphragm. The normal respiration rate is 13 to 16 times per minute.

The walls of the lung contain elastic connective tissues that would collapse the lung were it not for the surface tension between the surface of the lung and the wall of the thoracic cavity. The surface tension of the inner walls of the alveoli, on the other hand, has a tendency to collapse the alveoli after expiration, when the alveoli are deflated. This surface tension is reduced by the presence of a phospholipid *surfactant* that lines the alveolar walls in a thin film and allows the alveolar walls to be easily reinflated. Premature babies without sufficient surfactant lining the alveolar walls can have respiratory difficulties because of the tendency of alveoli to collapse. It is for this reason that the lecithin/sphingomyelin ratio in amniotic fluid is performed to assess fetal lung development (See Chapter 35).

In the newborn the lungs are about the same size as the thoracic cavity and therefore need not stretch to fill the cavity. The thoracic cavity grows faster than the lungs as the child develops, and the surface tension between the lungs and the inner wall of the thoracic cavity prevents the lungs from separating from the thoracic cavity and collapsing. With advanced age, the lungs become less elastic and the chest wall becomes more rigid, reducing the capacity for ventilation.

Gas exchange. The transfer of gas in the alveoli is a concentration-dependent phenomenon. Inspired (room) air has a relatively high P_{O_2} (158 mm Hg) and a low P_{CO_2} (0.3 mm Hg). Pressures of oxygen and carbon dioxide in capillary blood in the lungs are 50 and 40 mm Hg, respectively. Since the P_{O_2} in blood is lower than that of inspired air and the P_{CO_2} of blood is higher than the P_{CO_2} of room air, the gases diffuse from higher to lower concentration

Table 22-3. Reference values for adult blood gas parameters in arterial and venous blood

Parameters	Arterial	Venous
pH	7.35-7.45	7.33-7.43
P_{CO_2}	35-45 mm Hg	38-50 mm Hg
P_{O_2}	80-100 mm Hg	30-50 mm Hg
HCO_3^-	22-26 mmol/L	23-27 mmol/L
Total CO_2	23-27 mmol/L	24-28 mmol/L
O_2 saturation	94%-100%	60%-85%
Venous anion gap	5 to 14 mmol/L	
Base excess	−2 to +2 mEq/L	

areas. That is, CO_2 gas moves in the direction from the capillaries to the alveolar air space, whereas O_2 moves from the alveoli to the capillaries. Reference values for adult blood gas parameters in arterial and venous blood are summarized in Table 22-3. It is of interest to note that because of its great water solubility CO_2 exchanges more rapidly and more efficiently than does O_2. In respiratory acidosis, this phenomenon of differential gas diffusibility can result in low blood Po_2 but relatively normal CO_2.

Control of ventilation. Ventilatory control regulates the carbonate-bicarbonate buffer system but in turn is controlled by the resulting pH of cerebrospinal fluid and plasma. Control of ventilation is localized in a respiratory center of the brain, and its chemoreceptors are influenced by the pH of the cerebrospinal fluid. One set of chemoreceptors is located on the anterior surface of the medulla oblongata. The other chemoreceptors influenced by the changes in pH of arterial blood are located in the carotid and aortic vessel. A rise in Pco_2 will result in a fall in pH, which in turn stimulates the chemoreceptors and thereby initiates a rise in the respiration rate that will result in the release of more CO_2 from the blood into the lungs.

Acid-base balance

The maintenance of a constant pH by the body is important because changes in pH will alter the functioning of enzymes, the cellular uptake and regulation of metabolites and minerals, the conformation of biological structural components, and the uptake and release of oxygen.

In the body, physiological buffers can be thought to act in the following manner. Fixed acids enter into the blood and are immediately neutralized by the bicarbonate buffering system.

$$\begin{array}{l} H^+ A^- \text{ [fixed acid] } + \\ \qquad HCO_3^- \rightleftharpoons H_2CO_3 + A^- \text{ [unmeasured anion]} \\ \qquad\qquad\qquad \downarrow\uparrow \\ \qquad\qquad\quad H_2O + CO_2 \end{array} \quad \textit{Eq. 22-20}$$

However, the volatile acid, CO_2, is neutralized by the hemoglobin buffering system, since all the buffering systems are at equilibrium with each other. It is this overall equilibrium that gives the blood the relative buffering capacities described in Table 22-1.

Thus one of the important buffer systems required to maintain the pH of the blood is the carbonic acid–bicarbonate buffer system. Although this system has relatively low buffering capacity (see Table 22-1), it plays a large role in maintaining blood pH because it acts as the immediate buffer when fixed acids enter the blood. The lungs, by controlling the Pco_2 and thus the ratio of bicarbonate to carbonic acid, are the primary means by which this buffer system is rapidly adjusted to maintain that constant pH.

Changes in respiration rate will alter the bicarbonate–carbonic acid ratio and the pH. To understand this process, one must reconsider equations 22-2 and 22-11. A drop in gas exchange will cause a decrease in the release of CO_2 from the blood in the lungs. The increased blood CO_2 results in the formation of more bicarbonate (and then shifts equation 22-2 to the right), though the increase in bicarbonate is less than the increase in Pco_2. Thus there is a decrease in the bicarbonate-carbonate ratio and a decrease in the pH (equation 22-11). If the ventilation rate were to stay constant and the metabolic release of acid were to increase, the same effect would be observed. In this case H^+ reacts with HCO_3^- to form CO_2, which is released in the lungs. There is an immediate decrease in the concentration of bicarbonate with essentially no change in Pco_2, resulting in a decreased bicarbonate/αPco_2 ratio and a decreased pH. If the ventilation rate increases, more CO_2 is released from the blood at the lungs and the bicarbonate–carbonic acid ratio and pH increase. The ventilation rate can range from zero to 15 times normal, which allows a significant degree of regulation of the bicarbonate–carbonic acid ratio. Thus when the rate of ventilation is increased, excess acid in the form of CO_2 is quickly removed. Similarly, when the rate of ventilation is decreased, acid (CO_2) is added to neutralize excess alkaline (HCO_3^-).

The other important blood buffer, hemoglobin, is vital in regulating the blood pH by buffering and transporting the CO_2 from the tissues to the lungs. The major function of hemoglobin is transporting oxygen through the blood to the cells of the body. There is a complex interaction between the degree of oxygenation of hemoglobin and the pH, Pco_2, and total CO_2 (Tco_2) of blood. Oxygenated hemoglobin is a stronger acid than deoxygenated hemoglobin, and so in the lungs hemoglobin will release H^+ as it becomes oxygenated (equation 22-13). This shifts the dissociation reaction of carbonic acid to the left (equations 22-13 and 22-21), decreasing the bicarbonate level, increasing the level of carbonic acid and also its anhydrous form CO_2, and so increasing the Pco_2 of the blood. The rate at which this reaction proceeds is enormously increased by the presence in the red blood cells of the enzyme *carbonic anhydrase*. It is the action of this enzyme that allows the rapid transfer of CO_2 into and out of red blood cells, with consequent buffering by hemoglobin. This process is summarized by the following series of reactions and Fig. 22-3.

$$\text{Gas exchange in the lungs } \underset{\text{(gas)}}{CO_2} \rightleftharpoons \underset{\text{(dissolved)}}{dCO_2} \xrightarrow[\text{}]{\text{Carbonic anhydrase}} H_2CO_3 \rightleftharpoons H^+ + HCO_3^- \quad \textit{Eq. 22-21}$$

$$Tco_2 = dCO_2 + HCO_3^- \text{ or } Tco_2 = \alpha Pco_2 + HCO_3^-$$

In the lungs, ventilation will eliminate this increased Pco_2 by releasing the CO_2 from the blood and thereby returning the ratio of bicarbonate to carbonic acid to 20.

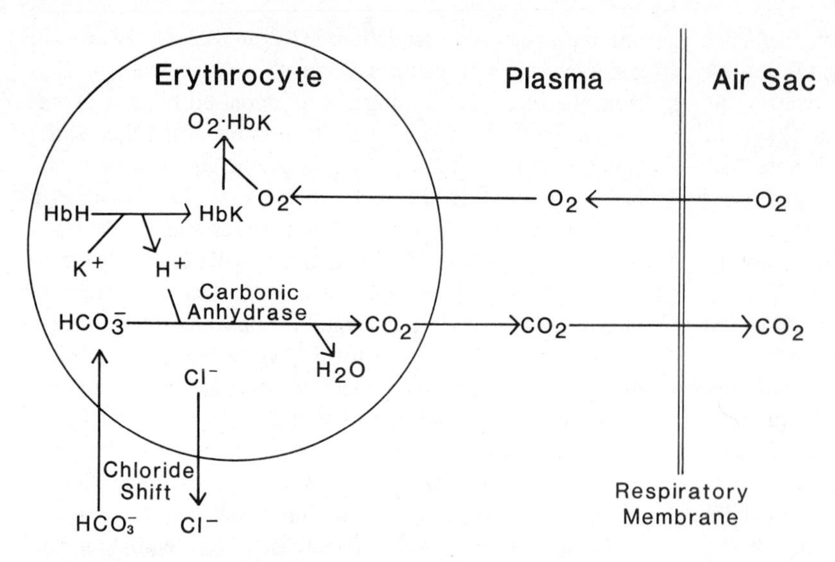

Fig. 22-3. Transfer of CO_2 in lungs from erythrocytes to air sacs. *HbK,* Potassium salt of hemoglobin; *HbH,* protonated form of hemoglobin.

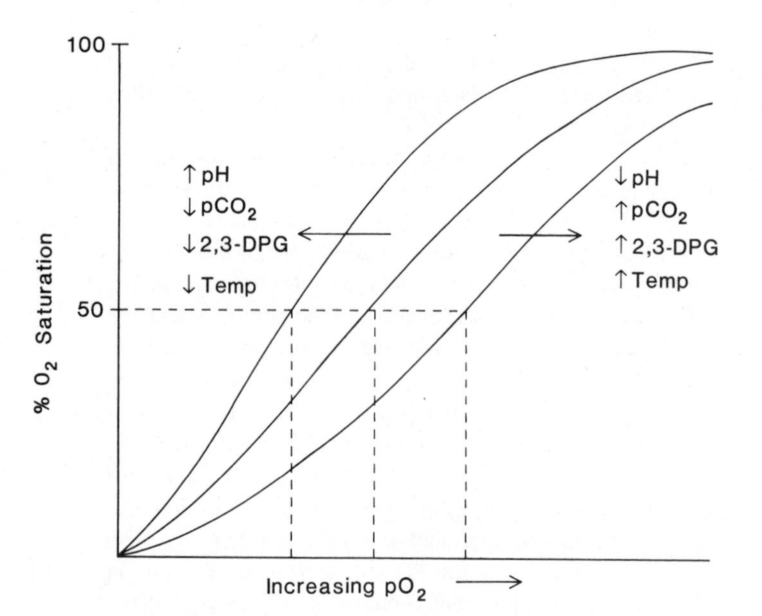

Fig. 22-4. Hemoglobin-oxygen dissociation curves and factors that shift the curve right and left. A shift of curve right or left changes level of Po_2 at which hemoglobin is 50% saturated (P_{50}).

Oxygenated hemoglobin is transported in the blood to cellular sites that have relatively low Po_2 tension and are releasing metabolic products, such as CO_2 and organic acids, into the blood, thus raising the Pco_2 and Tco_2 and lowering the pH. The relatively low Po_2 of peripheral tissues causes the dissociation of O_2 from HbO_2 and consequent delivery of O_2 to the cells. The oxygen release produces the weak acid deoxygenated hemoglobin, which counteracts, in part, the increasing Pco_2 and the decreasing pH. The high CO_2 pressure in the cells drives the CO_2 along a concentration gradient into the red blood cells. Carbonic anhydrase rapidly converts the CO_2 into H^+ and HCO_3^- (Fig. 22-2). Deoxygenated hemoglobin is a weaker acid than oxygenated hemoglobin. It neutralizes the H^+ to raise the pH and causes the dissociation reaction of carbonic acid to proceed to the right to increase the level of bicarbonate and decrease Pco_2.

The dissociation of oxygen from hemoglobin as a function of the Pco_2 is shown in Fig. 22-4 as a graph of the percentage of O_2 saturation of hemoglobin versus Po_2. The sigmoid shape of the curve indicates that at critical levels of Po_2 near the P_{50} there is a strong increase or decrease in the percentage of O_2 saturation, with a minimal shift in Po_2. In an area of the body in which there is a drop in Po_2 below that P_{50} on the sigmoid curve, the hemoglobin will release a larger portion of O_2 than at a Po_2 level above the P_{50}. Similarly, in areas of high O_2, such as the lungs, the hemoglobin will be essentially saturated with O_2.

One factor that has an effect on the position of the oxygen dissociation curve is 2,3-diphosphoglycerate (2,3-DPG), an intermediate in glycolysis. By interacting with the N-terminal amino groups of the hemoglobin molecule itself, 2,3-DPG induces the release of oxygen from hemoglobin. This is reflected in the shift to the right of the O_2 dissociation curve (Fig. 22-4). With a shift to the right, the critical Po_2 level that causes a 50% saturation of hemoglobin by oxygen is increased so that areas of active metabolism, which contain increased 2,3-DPG levels, do not require Po_2 levels as low as areas without increased 2,3-DPG for significant O_2 release from hemoglobin. As is seen in Fig. 22-4, the hemoglobin-oxygen dissociation curve that is shifted to the right requires a higher Po_2 to

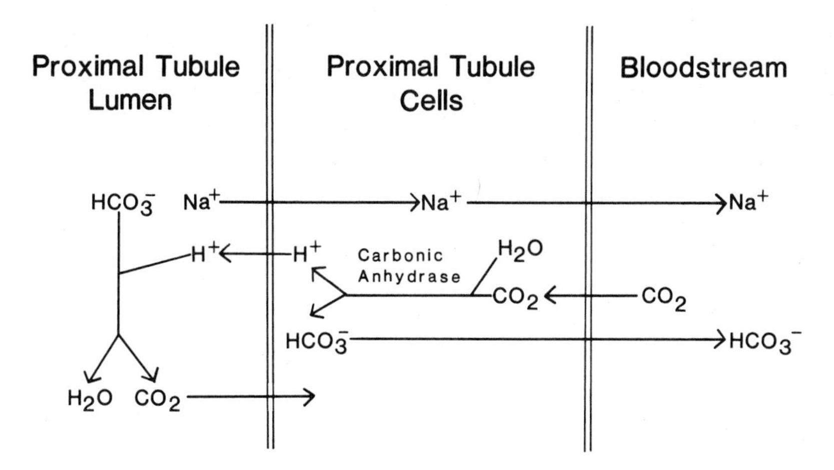

Fig. 22-5. Kidney reabsorption of bicarbonate with excretion of H^+.

be 50% saturated than the curve that is not shifted. Other metabolites that react with the N-terminal group of the β chain and the ϵ-amino groups of hemoglobin, causing a shift in the O_2 saturation curve, include glucose (glycohemoglobin) and CO_2.

This type of shift is observed in patients with increased Pco_2 and decreased pH, since both will shift the oxygen dissociation curve to the right. This effect of pH and Pco_2 is known as the "Bohr effect." Even under normal conditions, tissues with active metabolism are increasing the levels of Pco_2 and H^+, which cause the oxygen dissociation curve to shift to the right, thereby lowering the affinity of hemoglobin for oxygen. The net result is an enhanced release of O_2 to the tissues. During exercise, concentrations of H^+, Pco_2, and 2,3-DPG all increase in the areas of the body under stress, thereby shifting the oxygen dissociation curve to the right and enhancing oxygen release to those areas requiring oxygen. (See Chapter 33 for additional detail on 2,3-DPG and the Bohr effect.)

The decrease in affinity of hemoglobin for oxygen by increasing levels of Pco_2 and H^+ also results in more deoxyhemoglobin, which acts as a buffer to neutralize the Pco_2 and H^+, as mentioned earlier. Two phenomena occur in the erythrocyte during this buffering action of hemoglobin. One is the *isohydric shift* and the other is the *chloride shift*. Increased Pco_2 leads to the formation of bicarbonate (HCO_3^-) and H^+. Most of the H^+ ions are bound by the deoxygenated hemoglobin, and the rest are buffered by the proteins and phosphate buffer in the plasma. Since all the H^+ formed is buffered, there is essentially no change in the pH. This buffering phenomenon is referred to as the "isohydric shift." The HCO_3^- formed in the red blood cells as a result of uptake of H^+ by the hemoglobin diffuses out of the cells into the plasma. To preserve the electrical neutrality of the red cell, as HCO_3^- diffuses out of the cell, Cl^- diffuses into the erythrocytes from the plasma. This increase in the erythrocyte Cl^- is termed the *chloride shift*. Thus the plasma chloride concentration in venous blood (where HCO_3^- is formed in red blood cells) is *lower* than in arterial blood, in which, as CO_2 is expelled from the lungs, Cl^- again shifts out of red blood cells into plasma (see Figs. 22-2 and 22-3). The substitution of the polyvalent hemoglobin anions by the diffusing monovalent chloride anions increases the osmolality of the erythrocyte, leading to diffusion of water into the erythrocyte and slightly increasing the mean cell volume of venous red blood cells over the mean cell volume in arterial blood.

Although the intact respiratory system acts as an immediate regulator of the $HCO_3^-/\alpha Pco_2$ system, long-term control is exerted by renal mechanisms (see Chapter 23 for details). The kidneys excrete nonvolatile acids such as sulfuric, hydrochloric, phosphoric, and some of the organic acids into the urine. Hydrogen ions are excreted by the kidneys into the urine by being buffered against HPO_4^{-2} and ammonia, which is derived from deamination of the amino acid glutamine. Sodium is the cation exchanged for excreted hydrogen ions by the kidney. The kidney also affects the bicarbonate–carbonic acid buffer system by regulating the excretion of bicarbonate (Fig. 22-5). The kidney reabsorbs almost all filtered bicarbonate at plasma bicarbonate concentrations below 25 mEq/L. Only when bicarbonate levels become elevated above 25 mEq/L will bicarbonate be excreted into the urine. The reabsorbed bicarbonate is neutralized by the reabsorbed sodium ions, which have been exchanged for the hydrogen ions excreted in the urine (Fig. 22-5).

ACID-BASE DISORDERS

Definitions

Acid-base disorders resulting in a blood pH of less than 7.4 are termed *acidoses*, and those resulting in a blood pH greater than 7.4 are termed *alkaloses*. Acid-base disorders, whether acidosis or alkalosis, are most readily classified in terms of their immediate cause. Thus acidosis and alkalosis are described as either of respiratory or metabolic origin.

These classifications should always be considered in terms of the modified Henderson-Hasselbalch equation:

$$pH = pK_a + \log \frac{[HCO_3^-]}{[\alpha Pco_2]} \qquad \textit{Eq. 22-22}$$

The term "αPco_2" represents the acid component that is directly and immediately modified by respiratory status. Thus the term αPco_2 is called the *respiratory component*. The concentration of bicarbonate is most immediately affected by changes in the hydrogen-ion concentration caused by production of metabolic acids other than CO_2 (that is, fixed acids) and by physiological processes that directly change the concentration of serum bicarbonate. Thus the bicarbonate concentration of plasma is called the *metabolic component* of acid-base status. Acid-base homeostasis is accomplished when one controls one or both of these components.

It should be noted again that the pH of plasma depends on the *ratio* of the concentration of bicarbonate to αPco_2 rather than the absolute concentration of these components. If the concentration of bicarbonate doubles, and so does the αPco_2, the ratio remains constant and no change in pH occurs. Thus it is possible to classify metabolic and respiratory acidoses or alkaloses as base (HCO_3^-)-deficit or base-excess disorders.

Base-deficient disorders

Metabolic acidosis. In base-deficient disorders the pH is below normal. Such disorders occur if metabolic processes result in the accumulation of abnormal amounts of organic acids other than carbonic acid. Examples of acids that accumulate are lactic acid, beta-hydroxybutyrate, and acetoacetic acid. Metabolic organic acids entering plasma react with plasma bicarbonate to form H_2CO_3, which immediately is converted to CO_2 gas, which in turn is rapidly eliminated from the body by the lungs. Since the Pco_2 is in equilibrium with air, the net result is an immediate decrease in bicarbonate concentration with essentially no loss of Pco_2. This leads to a lowered bicarbonate/αPco_2 ratio and a lowered pH, or metabolic acidosis.

In contrast to this accumulation of acids is the pathological loss of base from the body. In severe diarrhea, bicarbonate ion is lost as part of the watery stool, resulting in a base deficit ($\downarrow [HCO_3^-]$, $\downarrow$ bicarbonate/αPco_2 ratio). These types of disorders are termed *metabolic acidoses*.

Respiratory acidosis. A relative base-deficient disorder can result from a decrease in the bicarbonate/carbonic acid ratio as a result of an increase in carbonic acid. This occurs if the lungs are not able to expel the metabolically produced CO_2 from the blood. This disorder is termed *respiratory acidosis*. The increase in Pco_2 results in an increase in the concentration of bicarbonate as the CO_2 is buffered by hemoglobin. However, the rise in bicarbonate concentration is less than the increase in Pco_2, resulting in a relative base deficit and a decrease in the bicarbonate/αPco_2 ratio, which results in a below-normal serum pH. Any condition associated with a lower-than-normal blood pH is referred to as *acidemia*.

Base-excess disorders

Metabolic alkalosis. In base-excess disorders, the pH is above normal. If such a disorder is caused by an increase in bicarbonate, with little or no change in carbonic acid, the disorder is termed *metabolic alkalosis*. Such a disorder occurs when excess amounts of bicarbonate of soda are ingested or administered or when there is an increased renal absorption of bicarbonate as in hypochloremic alkalosis.

Respiratory alkalosis. If the disorder is caused by a decrease in carbonic acid, as when respiration is overly stimulated, the disorder is termed *respiratory alkalosis*. In this condition, rapid ventilation greatly decreases the Pco_2 of blood, with minimal changes in bicarbonate concentrations. This results in a relative excess of bicarbonate so that the bicarbonate/αPco_2 ratio increases. This increased ratio yields a higher plasma pH. Any condition associated with a blood pH above normal is an *alkalemia*.

Laboratory aids in diagnosis of acid-base disorders

Measured parameters. The laboratory provides analytic results for a number of chemical constituents of the blood when requested to assess the patient's acid-base status. The pH is measured with an electrochemical cell that is in contact with the blood sample through an H^+ selective glass membrane (see Chapter 13). To measure Pco_2, one uses a modified pH electrode that has a membrane permeable to CO_2 gas but not liquid, through which CO_2 diffuses and alters the pH of a weak bicarbonate buffer solution. The change in pH caused by the diffusion of CO_2 is detected by a modified pH electrode and correlated to the change in the level of CO_2. The oxygen tension of the blood is measured by an electrode that also has a gas-permeable membrane that covers an electrochemical cell. The electrochemical cell has an adjusted cathode potential that is depolarized by the diffusing O_2, causing an increased current to flow that is proportional to the Po_2 level of the blood (see Chapter 14).

Calculated parameters. The remainder of the blood acid-base parameters are not measurable but are calculated from equations 22-11 and 22-12. One of the calculated parameters the laboratory provides is bicarbonate, which uses the measured parameters pH and Pco_2 in the Henderson-Hasselbalch equation:

$$HCO_3^- = (\alpha Pco_2) \text{ antilog } (pH - pK_a) \qquad \textit{Eq. 22-23}$$

An example of this calculation is a normal blood gas analysis with a pH of 7.4 and a Pco_2 of 40 mm Hg:

$$HCO_3^- = (.03)(40) \text{ antilog } (7.4 - 6.1)$$
$$HCO_3^- = 23.9 \text{ mmol/L}$$

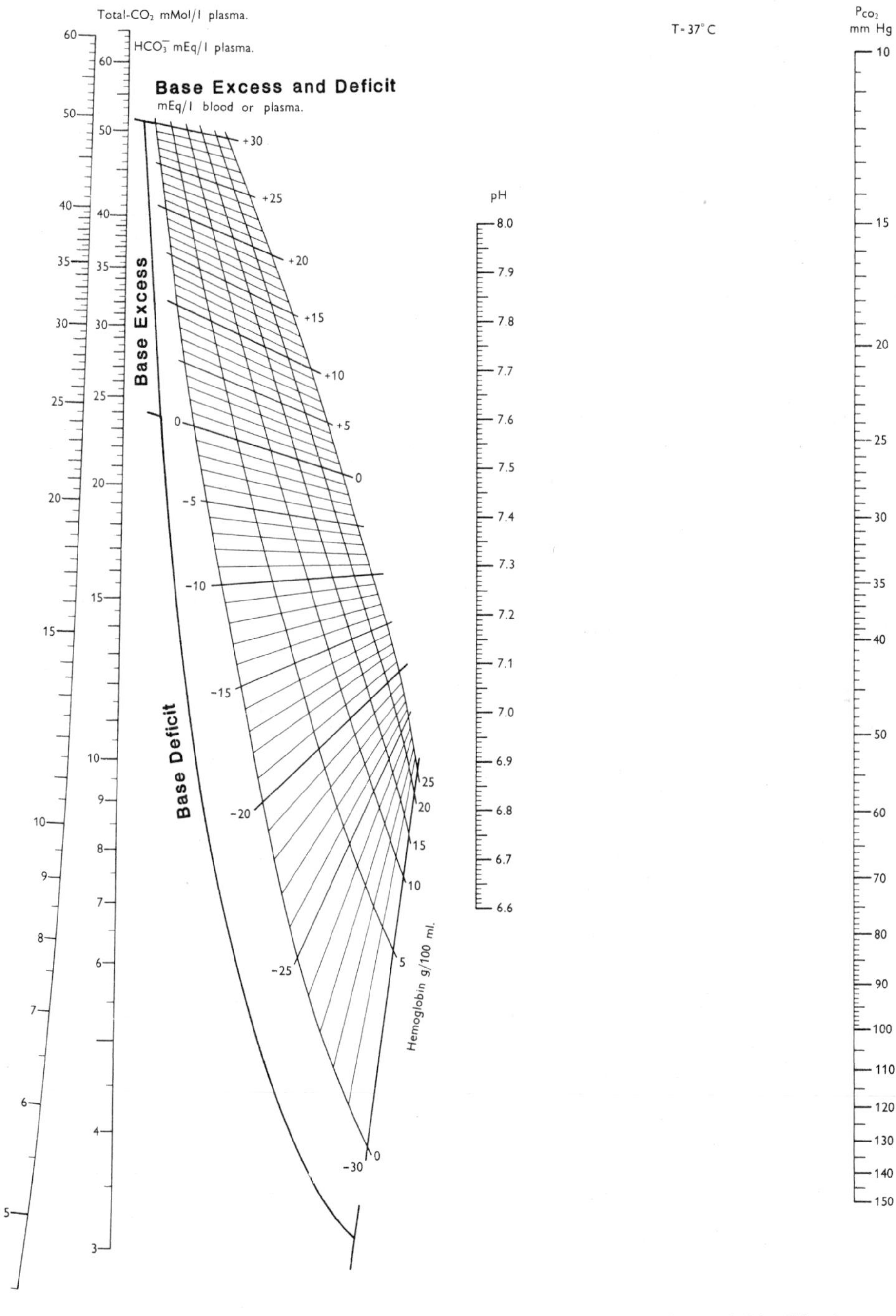

Fig. 22-6. Nomogram of relationship between P_{CO_2}, pH, base excess, hemoglobin, bicarbonate, and total CO_2. A straight line through a value of pH and of P_{CO_2} will connect with a calculated value of HCO_3^- and total CO_2. Base excess or deficit can be derived from that straight line if hemoglobin level is known. (Modified from Radiometer Corp., Copenhagen, Denmark.)

In this example, $pK_a = 6.1$, the apparent dissociation constant of carbonic acid, and α is the solubility coefficient of CO_2 in plasma (.0301). Nomograms have also been developed to derive bicarbonate levels from pH and Pco_2 (Fig. 22-6). Bicarbonate is useful in the assessment of the degree to which metabolic and renal control are involved in the acid-base status of the patient.

Total CO_2 is measured when a sample of plasma or serum is added to a reaction vessel with sufficient acid to shift the equilibrium of the reaction

$$(CO_2)\text{ gas} \rightleftharpoons (CO_2)\text{ dissolved} \rightleftharpoons H_2CO_3 \rightleftharpoons H + HCO_3^-$$

far to the left, thereby converting all forms of carbon dioxide into gaseous carbon dioxide. An electrode is then employed to measure the Pco_2 of the acidified plasma in the reaction chamber. Alternatively, the total CO_2 is measured colorimetrically or gasometrically with the Natelson microgasometer. Total CO_2 is also calculated from the measured parameters of pH and Pco_2 by the following equation:

$$Tco_2 = Pco_2 + HCO_3^- + H_2CO_3 \qquad \textit{Eq. 22-24}$$

The concentration of H_2CO_3 is small enough to be ignored. The bicarbonate concentration is calculated from the measured Pco_2 and pH as described above in equation 22-23. An example of this equation is illustrated below for a pH of 7.4 and a Pco_2 of 40 mm Hg. The bicarbonate is taken from the previous example.

$$Tco_2 = (.03)(40) + 23.9 = 25.1 \text{ mmol/L} \qquad \textit{Eq. 22-25}$$

Nomograms have been created to easily derive Tco_2 from pH and Pco_2 (Fig. 22-6).

Base excess is another calculated parameter that is used to assess the metabolic component of the patient's acid-base disturbance. The term "base excess" is used to describe clinical situations in which there is an excess of bicarbonate (positive base excess) or a deficit of bicarbonate (negative base excess). We use the term "*base deficit*" for a negative base excess as being a more accurate description of the physiological condition. Base excess in the blood at a pH of 7.40, Pco_2 of 40 mm Hg, a normal hemoglobin of 150 g/L, and a temperature of 37° C is zero. The hemoglobin concentration is important because the blood-buffering capacity is greatly dependent on this quantity. The addition of a base, such as bicarbonate, raises the buffer content of the blood and results in a positive base excess. The loss of base, as occurs in diarrhea or the addition of acids, lowers the blood buffer content and results in a base deficit. To calculate base excess, the measured parameters, pH and Pco_2, and either an assumed or measured hemoglobin level are used in a relationship developed by Siggaard-Andersen (Fig. 22-6 and equation 22-26). The calculation of the base excess or deficit is useful in the management of patients with acid-base disturbances because it permits estimation of the number of milliequivalents of sodium bicarbonate or ammonium chloride to be administered to correct the patient's pH to normal. In practice the base excess is only a crude estimate, since as the patient's condition improves, changes in respiration and metabolism will invalidate the original base excess. It is for this reason that blood gas status is closely monitored by collection of sequential blood specimens for analysis.

$$\text{Base excess} = (1.0 - 0.0143 \text{ Hgb}) \cdot (HCO_3^-) - (9.5 + 1.63 \text{ Hgb})(7.4 \text{ pH}) - 24 \qquad \textit{Eq. 22-26}$$

where Hgb is the hemoglobin concentration in g/dL.

Oxygen saturation indicates the amount of oxygen bound to hemoglobin and is used to determine the effectiveness of respiration or oxygen therapy. Oxygen saturation is calculated by use of the measured parameters of pH and Po_2 and the equation for a nomal oxygen dissociation curve. A nomogram to derive O_2 saturation from pH and Po_2 values is presented in Fig. 22-7. Oxygen saturation is also measured directly by use of the difference in the wavelengths of maximum absorbance for oxyhemoglobin and deoxyhemoglobin. This measurement is accomplished by a co-oximeter. Table 22-3 contains reference values for the calculated blood gas parameters.

An additional calculated parameter that also is indicative of the effect of metabolism on acid-base status is the *anion gap*. If the total measured cations are subtracted from the total measured anions as shown below, the difference is the anion gap, or the amount of unmeasured anions present. Or more simply:

$$\text{Anion gap} = (Na^+) - (Cl^-) - (HCO_3^-) \qquad \textit{Eq. 22-27}$$

Usually the only electrolytes measured are sodium, potassium, chloride, and bicarbonate (as total CO_2), though other anions exist in blood, such as phosphates, ketones, lactic acid, proteins, and sulfates, as counterions to the cations. Since these other anions are not measured, there is an apparent excess, or gap, of measured cations over measured anions. Increases in the amounts of these unmeasured anions plus the accompanying Na^+ ion will increase the apparent gap. Normally the anion gap averges 17 mEq/L. The anion gap increases with production of organic acids in lactic acidosis, ketoacidosis, dehydration, and after ingestion of acids. Diabetic ketoacidosis is the most common cause of an elevated anion gap. If diabetes is ruled out, other causes of the acidosis must be sought, such as lactic acidosis, renal tubular acidosis, sepsis, and toxic acidosis.

Acidosis

Metabolic acidosis

Etiology. Increased organic acid production in the blood results in metabolic acidosis, which develops from a number of causes. Uncontrolled diabetes results in accumulation of acetoacetic acid and hydroxybutyric acid, which are derived from the excessive oxidation of fatty acids (see

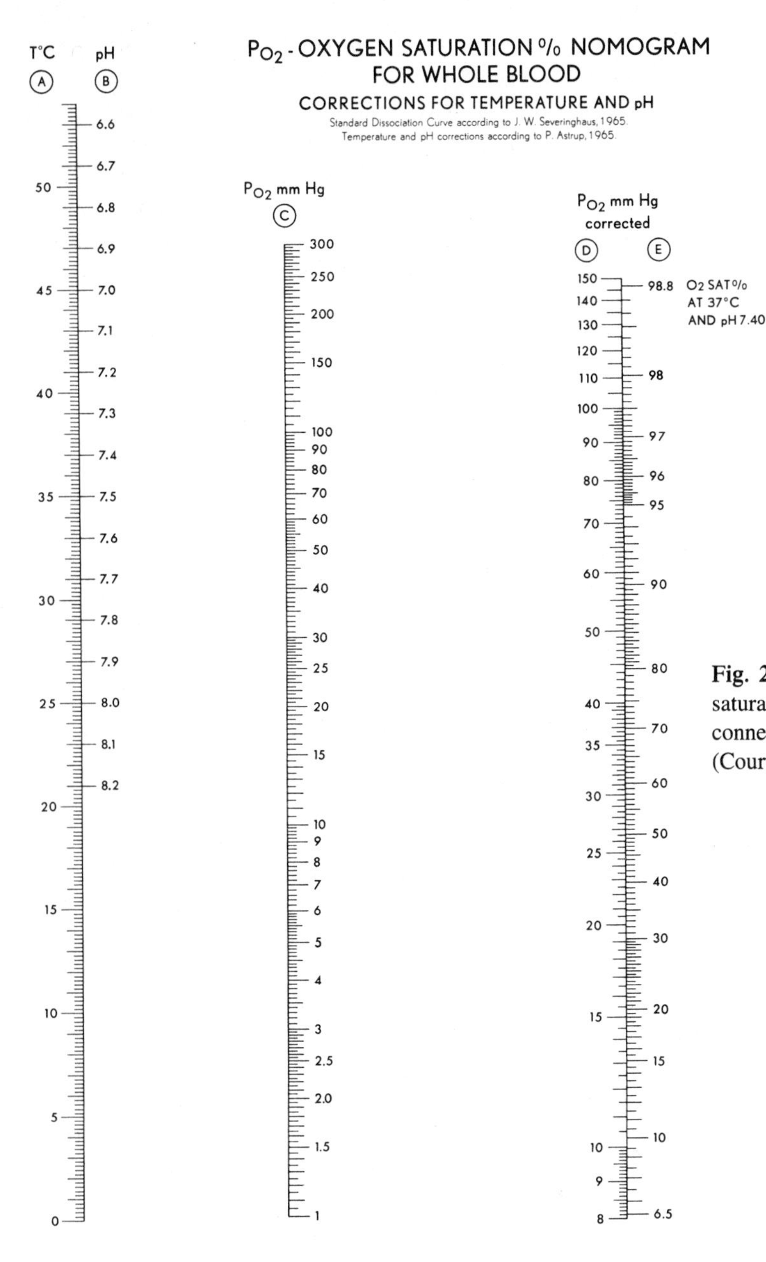

Fig. 22-7. Nomogram of relationship between pH, Po_2, and O_2 saturation. A straight line through a value of pH and of Po_2 will connect with a calculated value of O_2 saturation at 37° C. (Courtesy Radiometer Corp, Copenhagen, Denmark.)

Chapter 30). Fasting or fad diets also lead to increased levels of these acids.

Lactic acid increases as a result of increased anaerobic metabolism caused by strenuous muscular exercise or systemic infections. Lactic acidosis also results from local tissue hypoxia (low tissue Po_2), which is caused by dehydration, poor perfusion as a result of circulatory collapse, or cardiac failure.

Renal tubular acidosis results from a failure of the kidney to acidify the urine by exchanging H^+ for Na^+. This renal insufficiency is acquired as a result of infection or is congenital, as in cases of severe de Toni-Fanconi syndrome. Liver disease that impairs the formation of urea and ammonia will also result in a metabolic acidosis because of retention of H^+.

Salicylate intoxication initially induces respiratory alkalosis because of hyperventilation, which is a result of a stimulatory effect of the drug on the respiratory center. The ingested drug is converted to an acid before excretion, and these large quantities of acid ultimately result in metabolic acidosis. Other poisons that are ingested as acids or as compounds that will lead to acid metabolites are methyl alcohol (converted to formic acid), ethylene glycol (converted to oxalic acid), paraldehyde, and ammonium chloride. These compounds initially cause respiratory alkalosis, followed by metabolic acidosis. Infusion of large quantities of isotonic sodium chloride results in a metabolic acidosis because the high sodium load competes with hydrogen ions for renal excretion. Metabolic acidosis is also caused by the ingestion of carbonic anhydrase inhibitors, such as acetazolamide or sulfonamides, which interfere with the rapid formation of bicarbonate from CO_2 and water that is used by the erythrocyte in the hemoglobin buffer system as well as renal tubule cells (Figs. 22-2, 22-3, and 22-5).

A metabolic acidosis may also be caused by a decreased bicarbonate concentration. Diarrhea and colitis lead to losses of intestinal fluids, which contain high concentrations of bicarbonate. The resultant reduction of the HCO_3^-/H_2CO_3 ratio causes acidemia.

Physiological response. The acidoses described above result from the presentation to the body of an acid load that is compensated at least in part by retention of bicarbonate. When acidemia occurs as a result of acute metabolic acidosis, the body attempts to immediately correct this acidemia by hyperventilation. This hyperventilation lowers the P_{CO_2}, and to a smaller extent the HCO_3^-, and at least partially increases the ratio of HCO_3^-/H_2CO_3, thereby returning the pH toward normal. This mechanism of correcting the pH during acidosis is known as "compensatory respiratory alkalosis." The result is a lowering of the P_{CO_2} and the HCO_3^- concentration and a more normal pH.

In metabolic acidosis that does not involve renal dysfunction, the kidney will excrete the organic acids and exchange the H^+ for Na^+ in the distal tubule, resulting in a more acid urine. This renal compensatory mechanism becomes effective over longer periods of time and will eventually normalize both the blood pH and the bicarbonate. This correction occurs only when the underlying cause of the acidosis has been eliminated. Part of the renal response to chronic acidosis is the excretion of ammonia by the renal tubular cells. This excretion of ammonia into the presumptive urine allows additional H^+ to be excreted and thus reduces the H^+ load in blood.

Laboratory findings. The findings seen in metabolic acidosis, summarized in Table 22-4, include a decrease in both pH and HCO_3^-. Initially the P_{CO_2} may be normal, but it will decrease as a result of the respiratory response to the acidemia. A base deficit (negative base excess) will also be present.

Table 22-4. Classes of acid-base disorders, with corresponding effects on selected blood-gas parameters

Disorder	pH	P_{CO_2}	HCO_3^-	Base excess
Metabolic acidosis	↓	N	↓	↓
Respiratory acidosis	↓	↑	N	N
Metabolic alkalosis	↑	N	↑	↑
Respiratory alkalosis	↑	↓	N	N

N, Initially normal.

Lactic acidosis occurs during increased anaerobic metabolism caused by tissue hypoxia, which may result from strenuous exercise, systemic infections, or circulatory collapse. The laboratory findings are those of a metabolic acidosis with a decreased pH, an initially normal P_{CO_2} and P_{O_2}, a decreased O_2 saturation (the hemoglobin saturation curve is shifted to the right), a decreased bicarbonate and total CO_2, an increased anion gap, a negative base excess, and increased potassium and lactic acid concentrations. As the body attempts to correct the metabolic acidosis, the P_{CO_2} decreases.

In cases of toxic drug ingestion, such as methanol, ethylene glycol, or paraldehyde poisoning, the patient develops a metabolic acidosis. Laboratory findings include a decreased pH, an initially normal P_{CO_2} and P_{O_2}, a decreased O_2 saturation, bicarbonate, and total CO_2, an increased anion gap (caused by the ingested poisons or their metabolites), and a negative base excess. As the body attempts to compensate for the acidosis, the P_{CO_2} decreases initially and the bicarbonate slowly increases toward normal as the kidney reabsorbs increasing amounts of bicarbonate.

Treatment. Initial medical treatment is to correct the cause of the acidemia if possible, such as insulin treatment of diabetes. If the pH falls below 7.2, there is sometimes a deleterious effect on the cardiovascular system, and the base deficit may have to be corrected immediately. This is frequently accomplished by administration of bicarbonate, which corrects the base deficit by raising the HCO_3^-/H_2CO_3 ratio. In all cases the cause of the metabolic acidosis must ultimately be corrected.

Respiratory acidosis

Etiology. Respiratory acidoses are caused by disorders that interfere with the normal ability of the lungs to expel CO_2. These disorders include pulmonary edema, bronchoconstriction, pneumonia, emphysema, apnea, and bradycardia. Morphine ingestion and barbiturate poisoning cause an immediate respiratory despression, resulting in respiratory acidosis.

Respiratory distress syndrome (RDS), which is common in premature infants, results in a respiratory acidosis because these infants lack the surfactant in their lungs that allows the alveoli to exchange gases normally. Respiratory distress is also seen in some adults who experience systemic shock or oxygen toxicity. Initially, the observed blood gas parameters are decreased pH, increased P_{CO_2} and bicarbonate, decreased P_{O_2}, and decreased oxygen saturation. Base excess and anion gap are initially within normal limits.

Physiological response. The physiological response to respiratory acidosis is primarily one of increased renal excretion of acids, the retention of sodium and bicarbonate, and if possible, hyperventilation. If a response compensates for the respiratory acidosis and results in its correction, the acidosis is referred to as "compensated respiratory acidosis." This may be viewed as the development of a metabolic alkalosis that compensates for the respiratory acidosis. In chronic respiratory acidosis, the pH becomes more normal but a base excess remains. When the respiratory disorder is corrected, the normal respiratory response to the acidosis removes the excess CO_2, and a transient metabolic alkalosis may result. Generally this alkalosis does not require treatment.

Laboratory findings. Respiratory disorders frequently lead to an increase in plasma CO_2 concentration with a smaller increase in HCO_3^- and a concurrent decrease in the ratio of HCO_3^-/H_2CO_3, which results in respiratory acidosis. However, as a result of the low oxygen levels in the tissue, coexisting metabolic lactic acidosis can develop. This metabolic acidosis results in an increased anion gap, and the decreasing bicarbonate results in a negative base excess. Table 22-4 reviews these findings.

As a result of the compensatory response, often very elevated concentrations of HCO_3^- with almost normal pH are seen. In chronic respiratory disease the concentration of HCO_3^- and pH are near normal, though the Po_2 may be rather depressed.

Medical treatment. Medical treatment is primarily aimed at correction of the respiratory disorder and ventilation of the patient with gases containing higher Po_2 and lower Pco_2 by use of mechanical respirators. However, initial correction of the acidemia may be achieved by injection of sodium bicarbonate.

Alkalosis

Metabolic alkalosis

Etiology. Occasionally, excessive, chronic ingestion of bicarbonate of soda for gastrointestinal distress results in an increased concentration of blood bicarbonate and a resultant metabolic alkalosis. Similarly, treatment of peptic ulcers with ingestion of large quantities of alkali antacids will also produce metabolic alkalosis. More commonly, metabolic alkalosis arises from the loss of acid. Prolonged vomiting leads to loss of gastric hydrochloric acid. This in turn raises the pH of the blood, since the loss of the chloride anion during the vomiting results in increased renal retention of bicarbonate to neutralize the sodium reabsorbed by the proximal tubule. This condition is known as *hypochloremic alkalosis*. Diseases such as hyperaldosteronism, Cushing's syndrome, and corticosteroid administration, which affect the ability of the kidney to regulate electrolyte balance, will also raise the blood pH. In the distal tubule, Na^+ is retained at the expense of K^+ and H^+. In these diseases, the resultant hypokalemia causes a release of K^+ by cells into the blood and a concurrent balanced movement of H^+ from the blood into the cells, thereby leading to a rise in the pH of the blood.

Physiologic response. To compensate for the increase in the HCO_3^-/H_2CO_3 ratio during metabolic alkalosis, the respiratory system slows in order to raise the Pco_2 and the carbonic acid concentration in the blood. The Pco_2 rises more rapidly than the HCO_3^-, thereby decreasing the pH. This mechanism of readjusting the pH during metabolic alkalosis is termed *compensatory respiratory acidosis*. The result is a more normal pH in the presence of an elevated concentration of HCO_3^-.

If the alkalosis persists, the body will attempt to correct the alkalosis by increasing the renal excretion of the excess bicarbonate, except in those instances in which the proximal tubule of the kidney actually increases the reabsorption bicarbonate, as in hypokalemia, dehydration, or hypochloremia.

Laboratory findings. During metabolic alkalosis, the ratio of HCO_3^-/H_2CO_3 increases as a result of a rise in the concentration of blood bicarbonate. Because of the physiological response to the alkalemia, additional laboratory findings are an increased Pco_2 and an alkaline urine containing titratable bicarbonate (Table 22-4).

Treatment. Treatment of metabolic alkalosis involves administration of NaCl or KCl, depending on the degree of hypokalemia, and perhaps also administration of NH_4Cl if the alkalosis is severe and persistent. The Cl anion of NH_4Cl compensates for the chloride deficit, which may have led to excessive retention of bicarbonate initially. This permits the kidney to begin to excrete the excess bicarbonate and correct the alkalosis.

Respiratory alkalosis

Etiology. Hyperventilation is the chief cause of respiratory alkalosis. The conditions resulting in hyperventilation include hysteria, excessive crying, pregnancy, salicylate intoxication, impairment of the central nervous system's control of the respiratory system, asthma, fever, a pulmonary embolism, and excessive use of a mechanical respirator.

Physiological response. The kidneys respond to the alkalosis by excreting increased amounts of bicarbonate under conditions of lower Pco_2 that occur during respiratory alkalosis. In response to the alkalosis, the proximal tubules of the kidney decrease the reabsorption of bicarbonate. This renal response to respiratory alkalosis is termed *compensatory metabolic acidosis*.

Laboratory findings. Hyperventilation leads to increased loss of CO_2 from the blood at the alveolar surface, which causes the HCO_3^-/H_2CO_3 ratio to increase as carbonic acid is lost. Because of the physiological response to the alkalemia, additional laboratory findings include decreased Pco_2 and an alkaline urine containing titratable bicarbonate (Table 22-4).

Treatment. One corrects respiratory alkalosis by lowering the respiration rate with drugs, such as sedatives, or by having the patient breathe air with a higher CO_2 content. One easily accomplishes this by having the patient breathe in a restricted environment, as into a paper bag, which raises the Pco_2 of the air and the blood. The increased Pco_2 returns the HCO_3^-/H_2CO_3 ratio to normal and corrects the respiratory alkalosis.

Table 22-5. Common disorders of acid-base balance and effects on selected blood-gas parameters

Disorder	pH	P_{CO_2}	P_{O_2}	HCO_3^-	Base excess	Anion gap	O_2 saturation
Respiratory distress syndrome	↓	↑	↓	N	N	N	↓
Lactic acidosis	↓	N*	N	↓	↓	↑	↓
Diabetic ketosis	↓	N*	N	↓	↓	↑	↓
Emphysema	↓	↑	↓	N	N	N	↓
Methanol poisoning	↓	N*	N	↓	↓	↑	↓
Renal failure	↓	N*	N	↓	↓	↑	↓

**N, Initially* normal.

CHANGE OF ANALYTE IN DISEASE
(Table 22-5)

An analysis of the laboratory data observed in some specific common disorders will be useful to consolidate the subjects covered in this chapter.

Diabetic ketoacidosis in patients with uncontrolled diabetes is an example of metabolic acidosis. Laboratory findings include a decreased pH, acidemia, an initially normal P_{CO_2}, a normal P_{O_2}, a decreased O_2 saturation, a decreased bicarbonate and total CO_2 (T_{CO_2}), an increased anion gap, a negative base excess, and increased potassium, ketones, and lactic acid (caused by the disturbed carbohydrate and fat metabolism). As the body compensates for the acidosis, the P_{CO_2} and the bicarbonate concentration decrease.

Emphysema is a disease of impaired respiration that frequently results in respiratory acidosis. Laboratory findings include a decreased pH and P_{O_2}, an increased P_{CO_2} and potassium, a decreased oxygen saturation, and initially a normal anion gap, base excess, bicarbonate, and T_{CO_2}. As the body compensates for the acidosis, the bicarbonate and T_{CO_2} rise. As in the respiratory distress syndrome, the low P_{O_2} may result in a metabolic acidosis caused by a rise in blood lactate because of increased anaerobic metabolism.

Hemoglobinpathies, such as sickle cell anemia, can lead to unusual oxygen-saturation kinetics caused by the abnormal hemoglobin molecule. Some laboratory findings associated with hemoglobinopathies are decreased oxygen saturation and P_{O_2} levels, which result in increased anaerobic metabolism and thereby metabolic acidosis. Persistence of the hypoxemia results in decreased bicarbonate and total CO_2, an increased anion gap and blood lactate, and a negative base excess. The respiratory response to this acidosis is hyperventilation, which decreases the P_{CO_2}. The renal response to this acidosis is to increase the reabsorption of bicarbonate, which tends to return the HCO_3^-/H_2CO_3 ratio to normal.

Renal failure leads to metabolic acidosis with the associated laboratory findings of a decreased pH, initially normal P_{CO_2} and P_{O_2}, decreased oxygen saturation, increased potassium level, and decreased bicarbonate level and T_{CO_2}. As the anion gap increases because of organic acid production and retention, the base excess becomes negative. The respiratory compensation for the metabolic acidosis will lead eventually to a decrease in the P_{CO_2}.

BIBLIOGRAPHY

Beeler, M.F.: Interpretations in clinical chemistry: self-instructional units, Chicago, Ill., 1978, American Society of Clinical Pathologists.

Davenport, H.W.: The ABC of acid-base chemistry, rev. ed. 6, Chicago, Ill., 1974, University of Chicago Press.

Harper, W.A., Rodwell, V.W., and Mayes, P.A.: Review of physiological chemistry, ed. 17, Los Altos, Calif., 1979, Lange Medical Publications.

Natelson, S., and Natelson, E.A.: Principles of applied clinical chemistry, vol. 1, New York, 1975, Plenum Press.

Soloway, H.B.: How the body maintains acid-base balance, Diagn. Medicine, pp. 32-41, Feb. 1979.

Spearman, C.B.: Egan's fundamentals of respiratory therapy, St. Louis, Mo., 1982. The C.V. Mosby Co.

Spence, A.P., and Mason, E.B.: Human anatomy and physiology, Menlo Park, Calif., 1979, The Benjamin/Cummings Publishing Co.

Thimmig, R.: A practical course in the techniques, applications and maintenance of the radiometer acid-base laboratory, Cleveland, 1977, The London Co.

Young, D.S., Pestaner, L.C., and Gibberman, V.: Effects of disease on clinical laboratory tests, Clin. Chem. **26**(4):supplementary issue, 1980.

Chapter 23 Renal function

M. Roy First

absorption (active and passive) Process of uptake of substance into tissues or cells. Active absorption requires the expenditure of energy to move substances against a concentration gradient, whereas, in the case of passive absorption, substances move from higher to lower concentrations.

aldosterone A steroid hormone produced in the adrenal cortex that acts on the distal tubules to stimulate sodium reabsorption and potassium and hydrogen excretion.

angiotensin I A polypeptide of 10 amino acids that is an intermediate in the formation of angiotensin II from angiotensinogen.

angiotensin II An eight–amino acid polypeptide that is the most potent vasoconstrictor known.

antidiuretic hormone (ADH) Also called vasopressin; a pituitary hormone that acts at the collecting duct to increase absorption of water, resulting in the formation of a more concentrated urine.

anuria Lack of formation of urine; no urine.

ascending limb Straight portion of loop of Henle in which the presumptive urine flows up toward the convoluted distal tubule. Fluid decreases in osmolality because of loss of chloride (plus Na^+).

Bence Jones protein Light-chain portion of a monoclonal immunoglobulin produced in excess and excreted into urine in patients with multiple myeloma.

bladder A sac used to collect formed urine before voiding.

Bowman's capsule Structure consisting of glomeruli and extended opening of the proximal tubule.

carbonic anhydrase The enzyme at the brush border of the proximal tubule that catalyzes the reaction: $H_2O + CO_2 \rightarrow H_2CO_3$.

cast Protein aggregates outlined in the shape of renal tubules secreted into the urine.

clearance A theoretical concept expressing that volume of plasma filtered at the glomeruli per unit of time from which an analyte would be completely removed and placed in final urine. It is usually expressed as milliliters of plasma per minute.

collecting tubule The last portion of the nephron, connecting the distal convoluted tubule and the larger collecting ducts, which, in turn, empty into the ureter. The final concentrating processes under the influence of the hormone ADH occur here.

countercurrent mechanism The process by which two streams, flowing in opposite direction, exchange material. In the kidney, urine and blood form opposing flows, and this mechanism allows reabsorption of substances.

creatinine clearance An estimate of the glomerular filtration rate (GFR) obtained by measurement of the amount of creatinine in plasma and its rate of excretion into urine.

descending limb Straight portion of the loop of Henle in which forming urine flows down from the convoluted proximal tubule. This portion of the loop of Henle is freely permeable to water, which leaves the presumptive urine.

distal tubule Convoluted tubule connecting the ascending loop of Henle with the collecting tubule. It has secretory and reabsorptive functions as part of the final urine formation and acidification process.

diuretic A drug whose action promotes the increased excretion of salt and water, thus increasing the flow of urine.

filtered load The amount of a substance presented to the tubules for reabsorption.

glomerular filtration rate (GFR) The rate in milliliters per minute at which plasma substances are filtered through the glomeruli into the proximal tubule.

glomeruli Cluster of small blood vessels in the kidney projecting into the expanded end (capsule) of the proximal tubule. Functions as a filtering mechanism of the nephron.

hematuria Presence of blood or red blood cells in the urine.

loop of Henle A U-shaped tubule connecting the proximal and distal convoluted tubules. It functions in reducing the volume of tubule fluid.

micturition Urination.

nephron The functional unit of the kidney containing Bowman's capsule, the proximal and convoluted tubules, ascending and descending limits of the loop of Henle, and the collecting tubules.

oliguria Formation of small amounts of urine.

proteinuria Presence of protein in urine.

proximal tubule Convoluted tubule beginning at the glomeruli and connecting with the descending loop of Henle. Has secretory and reabsorptive functions as part of the mechanism for urine formation.

pyuria The presence of pus (an inflammation fluid with leukocytes and dead cells) in the urine.

renal cortex Refers to outer part of the kidney that contains mostly glomeruli and convoluted tubules.

renal medulla Refers to the inner part of the kidney that contains mostly collecting ducts and the loops of Henle.

renal threshold The plasma concentration of a substance above which it will be present in urine.

renin An enzyme formed by the juxtaglomerular apparatus in the kidney; it converts plasma angiotensinogen to angiotensin I.

specific gravity The ratio of the weight in grams per milliliter of a body fluid compared to water.

titratable acid Combination of hydrogen ion with phosphate present in final urine.

urethra A membranous tube through which urine passes from the bladder to the exterior of the body.

urine The aqueous liquid and dissolved substances excreted by the kidney.

The clinical laboratory plays a major role in the diagnosis and assessment of diseases of the kidney. Abnormalities of renal function may be discovered by clinical laboratory tests long before symptoms of the disease have developed, thereby triggering a search for the cause of these abnormalities. The main role of kidney function tests in clinical medicine is as a means for assessing the severity of the renal disease and following its progression.

ANATOMY OF KIDNEY

Gross anatomy

The kidneys are paired organs located in the posterior part of the abdomen on either side of the vertebral column. Each kidney is 12 to 13 cm in length, 5 to 7 cm wide, and 2.5 cm thick. In the adult each kidney weighs about 150 g. The kidneys are bean shaped and are enclosed in a capsule of fibrous tissue. Underneath the capsule lies the cortex, which has a granular appearance and contains all the glomeruli. The inner portion of the kidney is the medulla. The collecting ducts form a large portion of the medulla, giving it the striated appearance. A vertical section through the kidney is shown on the left-hand side of Fig. 23-1. The medulla is arranged in a series of pyramids, the apex of which is called the "papilla." The papillae empty into the minor calyces, which in turn unite to form two or three larger tubes, the major calyces, which then unite to form a funnel-shaped sack, the renal pelvis. The pelvis occupies a medial indentation of the kidney known as the hilum. In the hilum the major blood vessels join the kidney; the renal vein and its branches pass anterior to the pelvis to empty into the inferior vena cava, whereas the renal artery arises from the aorta and lies behind the pelvis.

The urinary system is illustrated on the right-hand side of Fig. 23-1. The renal pelvis receives newly formed urine from the ends of the collecting ducts at the tips of the papillae. The renal pelvis rapidly diminishes in caliber and merges into the ureter. Each ureter descends in the abdomen alongside the vertebral column to join the bladder. The bladder provides temporary storage for urine, which is eventually voided through the urethra to the exterior.

Microscopic anatomy

Each kidney is made up of approximately one million functional units or *nephrons*. The component parts of the nephron are illustrated in Fig. 23-2. The nephron begins with the glomerulus, which is a tuft of capillaries that is formed from the afferent (incoming) arteriole and is drained by a smaller efferent (outgoing) arteriole. The glomerulus is surrounded by Bowman's capsule, which is formed by the blind, dilated end of the renal tubule. The proximal convoluted tubule runs a tortuous course through the cortex, entering the medulla and forming the descending limb of the loop of Henle and the ascending limb of the loop of Henle. The thick ascending limb of the loop of Henle reenters the cortex, where it again becomes tortuous, forming the distal convoluted tubule. The end of the distal convulted tubule marks the end of the nephron. The merging of two or more distal tubules marks the beginning of a collecting duct. As the collecting duct descends through the cortex and medulla, it receives the effluent from a dozen or more distal tubules. The collecting ducts join and increase in size as they pass down the medulla. The ducts of each pyramid coalesce to form a central duct, which empties through the papilla into a minor calyx.

RENAL PHYSIOLOGY

The primary function of the kidney is to regulate the body's internal environment, maintaining it in a constant state. The kidney is the chief regulator of all the substances of body fluids and is primarily responsible for maintaining homeostasis, or an equilibrium of fluid and electrolytes in the body. The kidney has five main functions:

1. Urine formation
2. Regulation of fluid and electrolyte balance
3. Regulation of acid-base balance

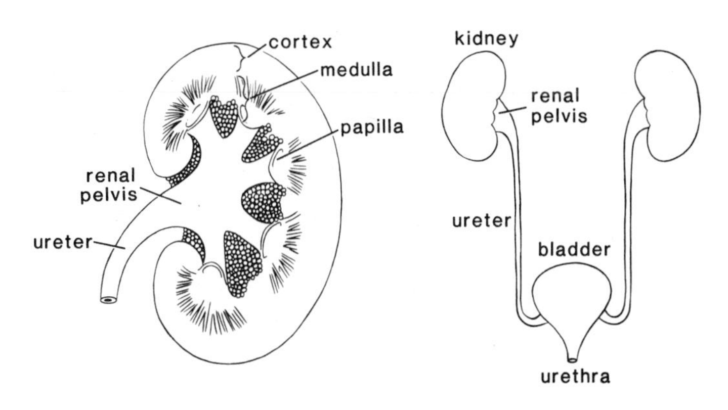

Fig. 23-1. Gross anatomy of kidney and urinary system.

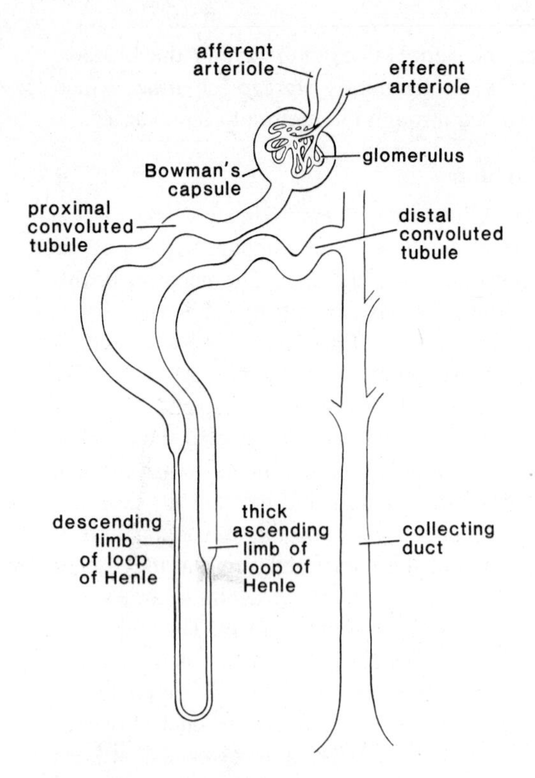

Fig. 23-2. Components of nephron.

4. Excretion of waste products of protein metabolism
5. Hormonal function

The kidney is able to carry out these complex functions by virtue of the fact that approximately 25% of the volume of blood pumped by the heart into the systemic circulation is circulated through the kidneys; thus the kidneys, which constitute about 0.5% of total body weight, receive one fourth of the cardiac output.

Urine formation

The removal of potentially toxic waste products is a major function of the kidneys and is accomplished through the formation of urine. The basic processes involved in the formation of urine are filtration (which occurs in the glomeruli), reabsorption, and secretion (which takes place in the renal tubules). The kidneys filter large volumes of plasma, reabsorb most of what is filtered, and leave behind for elimination from the body a concentrated solution of metabolic wastes that we call ''urine.'' It is through these processes that the kidneys regulate the amounts of various substances in the body and are able to maintain the internal environment for proper functioning of cells throughout the body. In healthy individuals, the kidneys are highly sensitive to fluctuations in diet and in fluid and electrolyte intake and compensate by varying the volume and consistency of the urine.

Glomerular filtration. As just mentioned, approximately 25% of the volume of blood pumped into the systemic circulation passes through the kidneys. The normal cardiac output is 4 to 6 L/min. Thus 25%, or 1000 to 1500 mL, of blood passes through the kidneys each minute. The glomerulus has a semipermeable basement membrane; that is, it allows free passage of water and electrolytes but is relatively impermeable to larger molecules. The glomerular capillaries differ from other capillaries in the body in that the hydrostatic pressure within them is approximately three times greater than the pressure in other capillaries. As a result of this high pressure, substances are filtered through the semipermeable membrane into Bowman's capsule at a rate of approximately 130 mL/min; this is known as the ''glomerular filtration rate'' (GFR). Cells and the large molecular size plasma proteins are unable to pass through the semipermeable membrane. Therefore the glomerular filtrate is essentially plasma without the proteins. The GFR is an extremely important parameter both in the study of kidney physiology and in the clinical assessment of renal function. In normal persons about 130 mL/min of filtrate is formed. Therefore, over 187,000 mL of filtrate is formed per day. Normal urine output is around 1500 mL per day, which is only about 1% of the amount of filtrate formed; therefore the other 99% must be reabsorbed.

Proximal tubule. The proximal tubular cells perform a variety of physiological tasks. Approximately 80% of salt and water are reabsorbed from the glomerular filtrate in the proximal tubule. All the filtered glucose and most of the filtered amino acids are normally reabsorbed here. Low molecular weight proteins, urea, uric acid, bicarbonate, phosphate, chloride, potassium, magnesium, and calcium are reabsorbed to varying extents. A variety of organic acids and bases, as well as hydrogen ions and ammonia, are secreted into the tubular fluid by these cells. In the case of glucose, phosphate, and bicarbonate, a renal plasma threshold and tubular maximal reabsorptive (T_m) mechanism exists. This is best understood by referring to the reabsorptive mechanism for glucose. Under normal conditions, no glucose is excreted in the urine; all that is filtered is reabsorbed. As the plasma concentration of glucose is increased above some critical level, termed the ''renal plasma threshold,'' the Tm for glucose is exceeded, and glucose appears in the urine. The higher the plasma concentration of glucose, the greater is the quantity excreted in the urine.

Most of the metabolic energy consumed by the kidney is used to promote active reabsorption of these substances. Active reabsorption can produce net movement of a substance against a concentration or electrical gradient and therefore requires energy expenditure by transporting cells. Active reabsorption of glucose, amino acids, small molecular weight proteins, uric acid, sodium, potassium, magnesium, calcium, chloride, and bicarbonate are regulated by the kidney according to the levels in the blood and the body's needs. Passive reabsorption occurs when a substance moves downhill by simple diffusion as the result of

Table 23-1. Filtration, reabsorption, and excretion by kidney

Component	Amount filtered per day	Amount excreted per day	Percentage reabsorbed
Water	180 L	1.5 L	99.2
Sodium	24,000 mEq	100 mEq	99.6
Chloride	20,000 mEq	100 mEq	99.5
Bicarbonate	5,000 mEq	2 mEq	99.9
Potassium	700 mEq	50 mEq	92.9
Glucose	180 g	0	100
Albumin	3600 mg	180 mg	95.0

an electrical or chemical concentration gradient, and no cellular energy is involved in the process. Water, urea, and chloride are reabsorbed in this way. Urea is an example of a waste product whose molecules do not diffuse easily through the pores of the tubular membrane. As water is reabsorbed, the concentration of urea in the tubules rises and forms a concentration gradient, which is totally responsible for the reabsorption of urea.

Other poorly absorbed substances that are concentrated in the urine are phosphates, sulfates, nitrates, and phenols. These are all waste products that would be harmful to the body in high concentration. Substances such as creatinine, inulin, mannitol, and sucrose are not reabsorbed at all by the tubules. An idea of the magnitude and importance of these reabsorptive mechanisms can be gained from Table 23-1.

Tubular secretion, which transports substances into the tubular lumen, that is, in the direction opposite to tubular reabsorption, may also be an active or passive process. Substances that are transported from the blood to the tubules and excreted in the urine include potassium, hydrogen ions, ammonia, uric acid, and certain drugs such as penicillin.

Loop of Henle. The descending limb of the loop of Henle is highly permeable to water, which moves passively across the tubular wall when an osmotic gradient exists between the tubular fluid and the interstitial fluid. In the medulla, the loop of Henle descends into an environment that is increasingly hypertonic as the papilla is approached. The reabsorption of water in response to this gradient raises the osmotic concentration of the tubular fluid as it flows deeper into the medulla. No known active-transport process exists in the descending limb. The ascending limb is relatively impermeable to the passage of water, but actively reabsorbs sodium and chloride. This segment of the nephron is often called the "diluting segment" because the removal of salt with little water from the tubular contents dilutes the salt and osmotic concentration of the tubular fluid. The addition of the reabsorbed sodium and chloride with little water to the medullary interstitial fluid compartment raises the osmotic and salt concentration of that compartment. This phenomenon is known as the countercurrent mechanism. The ascending thick limb of the loop of Henle transfers sodium chloride actively from its lumen into the interstitial fluid. The tubular fluid within its lumen becomes hypotonic, and the interstitial fluid becomes hypertonic. A series of successive steps results in sodium chloride being trapped in the interstitial fluid of the medulla. As the isotonic fluid in the descending limb reaches the area into which the ascending limb is pumping out sodium, it becomes slightly hypertonic because of the movement of water into the hypertonic interstitium. The first step repeats itself, and again, as more sodium and chloride are added to the interstitium by the ascending limb, more water is drawn out of the descending limb.

Distal tubule. A small fraction of the filtered amount of sodium, chloride, and water is reabsorbed in the distal tubule. The distal tubule responds to antidiuretic hormone (ADH) so that its water permeability is high in the presence of the hormone and low in its absence. Potassium can be reabsorbed and secreted here. Aldosterone stimulates both sodium reabsorption and potassium secretion in the distal tubule. Hydrogen, ammonia, and uric acid secretion, and bicarbonate reabsorption occur, but there is little transport of organic substances. This segment of the nephron also has a low permeability to urea.

Collecting duct. ADH controls the water permeability of the collecting tubules throughout its length. In the presence of the hormone, the hypotonic tubular fluid entering the duct loses water. In the medulla, the high osmotic concentration of the interstitium provides a force for continued water reabsorption, and the osmotic concentration of tubular fluid continues to increase along the length of the tubule. Sodium and chloride are reabsorbed by the collecting tubule, with the transport of sodium being stimulated by aldosterone. Potassium, hydrogen, and ammonia are also transported by the collecting duct. When ADH is present, the rate of water reabsorption exceeds the rate of solute reabsorption and the concentration of sodium and chloride rises. The collecting duct is relatively impermeable to urea.

Regulation of fluid and electrolyte balance

The kidney is primarily responsible for maintenance of the fluid and electrolyte composition of the body. The kidney maintains the organism in a delicate balance for a wide variety of substances, with the amount of each substance excreted being exactly equal to the amount ingested. Disruption of this balance will result in a wide array of biochemical disorders.

Water. Water is the most abundant component of the human body, constituting approximately 60% of body weight. Body water, and therefore body weight, remains fairly constant from day to day in normal persons despite wide fluctuations in fluid intake. The concentrating and diluting mechanisms of the kidney contribute largely to the homeostasis of total body water and the rate of urine flow can normally be varied within a very wide range.[1] One of the most remarkable properties of the human kidney is its ability to elaborate urine that is either more concentrated or more dilute than the plasma from which it is derived. When the human body needs to conserve water, the concentrating mechanism operates maximally, and urine osmolality increases to about 1200 mOsm/kg. Conversely, when there is excess water in the body, urine flow increases, and the diluting mechanism reduces urine osmolality to as low as 50 mOsm/kg. It is this capacity of the kidney to form urine of greatly varying osmolality that enables the kidney to regulate the solute concentration and hence the osmolality of body fluids within narrow physiological limits despite wide fluctuations of intake of salt and water.[1, 2] Water balance is primarily controlled through voluntary intake, which is regulated through the thirst center in the hypothalamus, and urinary loss. The control of urinary water loss is the major automatic mechanism by which body water is regulated. Factors affecting the urinary water loss include ADH, aldosterone, arterial pressure receptors, and nervous reflexes.

In dehydrated states the urine is concentrated by reabsorbing water without solute. Conversely, urine is diluted by reabsorbing solute without water. The countercurrent mechanism of the kidney is responsible for its ability to concentrate or dilute urine. This takes place in the loop of Henle. When the extracellular fluid volume is reduced, ADH is secreted by the hypothalamus, and this promotes water reabsorption in the distal tubule and collecting duct.

Sodium. Sodium is the main cation found in extracellular fluid. The normal serum sodium level is 136 to 145 mEq/L. Sodium is freely filtered through the glomerulus and actively reabsorbed by the tubules. Sodium reabsorption is very important because it affects the regulation of several other electrolytes. Active reabsorption of the sodium ion in the proximal tubule results in passive transport of chloride and bicarbonate, as well as passive reabsorption of water. In normal persons, daily urinary sodium excretion fluctuates widely according to the dietary intake, thereby keeping the body sodium content remarkably constant. In the normal person, over 99% of the filtered load of sodium is reabsorbed by the kidneys. The sodium reabsorption by the nephron is controlled by the renin-angiotensin-aldosterone system. With sodium loss and resultant extracellular fluid volume contraction, there is a reduced renal plasma flow and glomerular filtration rate. Very slight reductions in glomerular perfusion and filtration lead to the release of the enzyme renin from the juxtaglomerular cells. Once in the circulation, this enzyme cleaves renin substrate (an alpha$_2$-globulin), angiotensinogen, splitting off a decapeptide, angiotensin I. Another enzyme normally present in the plasma converts this compound to the biologically active octapeptide, angiotensin II.[3,4] Angiotensin II preserves body fluid volume by the following mechanisms: (1) it is a powerful vasoconstrictor and further reduces the glomerular filtration rate and thereby the filtered load of sodium; (2) it stimulates the adrenal cortex to produce aldosterone, which in turn stimulates renal tubular reabsorption of filtered sodium; and (3) it causes thirst.

Chloride. The concentration of chloride in the extracellular fluid parallels that of sodium and is influenced by the same factors. However, chloride reabsorption is passive in the proximal tubule and probably active in the distal nephron.

Potassium. Potassium is the chief cation of the intracellular fluid. Maintenance of a normal potassium level is essential to the life of the cells. The serum potassium level is measured as a reflection of the total body potassium; the normal value is 3.5 to 5 mEq/L. The normal person maintains potassium balance by excreting daily an amount of potassium equal to the amount ingested minus the small amount eliminated in the feces and sweat. Once again, renal function is the major mechanism by which the body potassium is regulated. Potassium is freely filtered at the glomerulus and active tubular reabsorption occurs throughout the nephron except for the descending loop of Henle. Only about 10% of the filtered potassium enters the distal tubule. The distal tubule and collecting ducts are able to both secrete and reabsorb potassium, and it is by changes in potassium handling in the distal tubule that changes in potassium excretion are achieved.[5] Two mechanisms that control the net potassium excretion are operative in the distal tubule. The first of these is dependent on the cellular potassium content. When a high-potassium diet is ingested, the potassium concentration in the body cells, including the renal tubular cells, increases. This high concentration enhances the gradient for potassium secretion into the lumen and increases potassium excretion. The second important factor linking potassium secretion to potassium balance is the hormone aldosterone, which, besides stimulating tubular sodium reabsorption, simultaneously enhances potassium secretion in the distal tubule.[6] Aldosterone production by the renal cortex is directly stimulated by an increase in the extracellular potassium concentration.

Calcium. Calcium is reabsorbed in the proximal tubule under the hormonal influence of parathyroid hormone (PTH).[7,8] The maintenance of calcium homeostasis is dependent on the balance between calcium intake and calcium loss. The body loses calcium in the urine, through the gastrointestinal tract, and in the sweat. Calcium balance is achieved largely by the control of calcium absorption rather than by the regulation of calcium excretion. The percentage of ingested calcium absorbed decreases as the dietary calcium content increases, and so the amount absorbed can remain relatively constant. The slight increase in absorption that occurs on a high-calcium diet is reflected in an increased renal excretion.[9]

Phosphorus. Most of the serum phosphorus is present in the freely diffusible ionized form. Unlike calcium, the absorption of phosphorus from the gastrointestinal tract is not controlled according to bodily needs. Over a wide range of dietary intakes, roughly two thirds of ingested phosphorus is absorbed into the bloodstream. The maintenance of the phosphorus balance is achieved largely through renal excretion.[9] Proximal tubular reabsorption of inorganic phosphate is normally about 90% of the filtered load. Parathyroid hormone depresses the renal tubular reabsorption of inorganic phosphate. In progressive chronic renal failure there is a progressive increase in the serum phosphorus level.[10]

Magnesium. The major portion of the body magnesium is in bone and in soft tissues. The filtration of magnesium at the glomerulus and its reabsorption from the proximal tubules parallels that of calcium and is also under the influence of parathyroid hormone. A moderate elevation of the plasma magnesium concentration occurs in patients with advanced chronic renal failure.[11]

Acid-base balance (see Chapter 22)

Each day acid waste products are produced in the body and, if they were not disposed of efficiently, would accumulate and cause cellular damage. Body pH is controlled by three systems: acid-base buffers, the lungs, and the kidneys.

Excretion of hydrogen ions. In subjects on a normal diet, about 50 to 100 mEq of hydrogen ions are generated each day.[12] To prevent a progressive metabolic acidosis, these hydrogen ions are excreted in the urine. The renal excretion of hydrogen is achieved by tubular secretion from the cell into the lumen.[13] Secreted hydrogen ions are generated in the cells of the proximal and distal tubule and the collecting duct. Carbonic anhydrase (CA) is required for the hydration of carbon dioxide. This enzyme is essential for normal urinary acidification because the uncatalyzed hydration reaction is too slow to provide the requisite number of hydrogen ions:

$$H_2O + CO_2 \xrightarrow{(CA)} H_2CO_3 \rightarrow H^+ + HCO_3^-$$

The kidneys' role in the maintenance of the acid-base balance revolves around the regeneration of bicarbonate. This is accomplished by the renal tubular secretion of hydrogen ions into the urine. The passage of bicarbonate ions from the tubular cells into the blood occurs at the same rate as bicarbonate is consumed by the metabolic processes.[14] To maintain electrochemical neutrality, the movement of a hydrogen ion into the lumen can be counterbalanced by the movement of a sodium ion from the lumen into the tubular cell (and hence into the blood) or the movement of an anion (such as phosphate or an organic acid) from the tubular cell into the lumen. The following four mechanisms exist to handle the hydrogen ions that have been secreted into the tubular fluid.

Reaction with filtered bicarbonate ions. Bicarbonate is completely filterable at the glomerulus. In the tubular lumen, the excreted hydrogen ion combines with the filtered bicarbonate to form carbonic acid, which decomposes to water and carbon dioxide, the latter then diffusing into the cell where it is used to generate another hydrogen ion. In this way, one bicarbonate ion is regenerated for every hydrogen ion that is secreted into the tubular lumen. The bicarbonate ion is reabsorbed into the blood as sodium bicarbonate. The reabsorption of sodium bicarbonate by this mechanism conserves most of the filtered bicarbonate. The renal threshold for bicarbonate is 28 mM/L; at a plasma level below this, all filtered bicarbonate is reabsorbed.

Reaction with filtered buffers to form titratable acids. Besides being able to conserve all the filtered bicarbonates, the kidneys can also contribute new bicarbonate to the plasma. The mechanism by which new bicarbonate is added to the blood is fundamentally the same as that for bicarbonate reabsorption, namely, tubular acid secretion. Inorganic monohydrogen phosphate is present in the tubular lumen as the disodium salt. The secreted hydrogen ions react with the filtered phosphate, the released sodium combines with the bicarbonate and is reabsorbed as sodium bicarbonate, and dihydrogen phosphate is excreted:

$$Na_2HPO_4 + H^+ \rightarrow NaH_2PO_4 + Na^+$$
$$Na^+ + HCO_3^- \rightarrow NaHCO_3 \text{ (reabsorbed)}$$

The combination of hydrogen ions with phosphate is called "titratable acidity." The rate of excretion of titratable acid is limited by the filtered buffer load and cannot increase greatly in acidosis. The limiting gradient against which hydrogen ions can be pumped in forming urine of maximum acidity is about 1000:1, a value that corresponds to a blood pH of 7.4 and a urine pH of 4.4.

Reaction with secreted ammonia to form ammonium ion. The glomerular filtrate does not contain ammonia. This compound is synthesized in renal tubular cells by deamination of glutamine in the presence of glutaminase.[15] The ammonia diffuses into the tubular fluid where it reacts with a secreted hydrogen ion to form an ammonium ion. Once again this results in the addition of new bicarbonate to the blood. The most important renal adaptation to acidosis is the increased excretion of ammonium ions.[15]

$$\text{Glutamine-NH}_3 \xrightarrow{\text{Glutamase}} \text{Glutamic acid} + NH_3$$
$$NH_3 + H^+ + NaCl \longrightarrow NH_4Cl + Na^+$$

Excretion as free hydrogen ions. Only negligible quantities of hydrogen ions are handled in this way by the kidneys.

Nitrogenous waste excretion

One of the major functions of the kidney is the elimination of nitrogenous products of protein catabolism. In the presence of impaired renal function, restoration of water and electrolyte balance can be accomplished relatively easily, but the impaired capacity to excrete nitrogenous waste products becomes dominant in causing symptoms. The enormous reserves of the kidney for excretion of the products of protein catabolism are indicated by the fact that symptoms of renal failure do not occur until renal function is reduced to less than one third of normal.[16]

Urea. As amino acids are deaminated, ammonia is produced. The development of toxic levels of ammonia in the blood is prevented by the conversion of ammonia into urea. This takes place in the liver. The concentration of urea in the blood is measured as the blood urea nitrogen (BUN). Urea production and the BUN are increased when more amino acids are metabolized in the liver. This may occur with a high protein diet,[17] tissue breakdown, or decreased protein synthesis. On the other hand, urea production and the BUN are reduced in the presence of a low protein intake and severe liver disease. Urea production exceeds renal urea excretion in normal persons. The difference is chiefly attributable to the degradation of urea by intestinal bacteria.[18] Urea is a small molecular weight compound that is readily filterable. However, urea excretion is not determined solely by glomerular filtration. Approximately 40% to 50% of the filtered urea is normally reabsorbed by the tubules. The reabsorption of urea tends to follow passively that of sodium. Thus, in states of volume depletion in which sodium reabsorption is increased, urea absorption is also enhanced.

As we have described above, a number of factors may influence the BUN level while the glomerular filtration rate remains constant. For this reason the BUN is a poor indicator of renal function and should not be relied upon for that purpose. Its extensive use follows its easy chemical determination and has become hallowed by tradition. It can be used to follow the course of an individual patient provided that the protein intake during the period under observation is reasonably constant.[16]

Creatinine. Serum creatinine levels and urinary creatinine excretion are a function of lean body mass in normal persons and show little or no response to dietary changes or alteration in electrolyte balance.[19] Creatinine is derived from the nonenzymatic interconversion of creatine in skeletal muscle (see reaction in next column) and is released into the plasma at a relatively constant rate. The amount of creatinine per unit of muscle mass is constant, and thus the spontaneous breakdown is constant. As a result, the plasma creatinine concentration is very stable, varying less than 10% per day in serial observations in normal subjects.[20] Since the serum creatinine is a direct reflection of muscle mass, the level is higher in males than in females. Creatinine is freely filtered at the glomerulus and is not reabsorbed by the tubules. A small amount of the creatinine in the final urine is derived from tubular secretion. Because of these properties of creatinine, the creatinine clearance can be used to estimate the glomerulus filtration rate. The normal serum creatinine level will range from 7 to 14 mg/L, depending on age, sex, and muscle mass.

```
CH3–NCH2–COOH        CH3–N–CH2
     |                   |   \
    C=NH      ——→        |    C=O  + H2O
     |                   |   /
    NH2                HN=C–NH
  Creatine             Creatinine
```

Uric acid. Uric acid is derived from the oxidation of purine bases. Plasma levels are quite variable, being higher in males than in females. Plasma urates are completely filterable, and both proximal tubular resorption and distal tubular secretion occur. With advanced chronic renal failure there is a progressive increase in the plasma uric acid level.

Hormonal function

The kidneys have important metabolic and endocrine functions. The kidney serves as a target organ for a number of hormones; it influences the metabolism of various other hormones; and finally it acts directly as an endocrine organ, secreting regulatory hormones.[21] In this section we shall deal with the kidney as an endocrine organ.

Vitamin D metabolism. The role of the kidney in vitamin D metabolism is the production of the major biologically active hormone 1,25-dihydroxycholecalciferol.[22-24] This hormone is formed by hydroxylation of 25-hydroxycalciferol produced in the liver. The enzyme responsible for the final conversion to 1,25-dihydroxycholecalciferol is present only in the mitochrondria of the renal cortex.[21]

Renin. The role of renin in fluid and electrolyte balance through the renin-angiotensin-aldosterone axis has been discussed previously. Mechanisms that have been postulated to stimulate renin release include a fall in pressure in the afferent arterioles, reduced sodium chloride load, renal prostaglandins, and sympathetic nerve activity at the juxtaglomerular apparatus.[21]

Erythropoietin. The kidneys play a major role in the production and release of erythropoietin, a glycoprotein hormone that stimulates red blood cell production. It appears that the kidneys do not produce the actual hormone, for none can be extracted from renal tissues. It is believed that the kidney synthesizes an enzyme, erythrogenin, that acts on a plasma protein substrate produced by the liver, much as renin catalyzes the formation of angiotensin in the plasma. The central role of the kidneys in erythropoietin production explains the anemia of chronic renal failure.[25]

Prostaglandins. Prostaglandins are synthesized in the renal medulla. Although they have important primary and secondary effects on the kidney, their exact role remains unknown. Infusion of exogenous prostaglandins causes an increase in the renal blood flow and in salt and water excretion. They have also been shown to stimulate renin release.[26] Since prostaglandins are vasodilators, it is possible that they play a role in blood pressure regulation. The importance of these physiologic actions of the prostaglandins remains to be determined.

Protein conservation

Under normal physiological conditions, the kidney maintains the homeostasis of the body proteins. In man, 180 liters of plasma, each containing 70 g of protein, are filtered each day by the glomerulus.[27] Without an efficient conservation mechanism, body protein stores would be depleted very rapidly. Yet normal urine contains less than 200 mg of protein per day, only a minute percentage of the 12,600 g passing through the glomerulus daily.[27] The importance of the kidney in conserving protein has been appreciated since the time of Richard Bright, 140 years ago.[28] Since then, proteinuria has been recognized as an important characteristic of renal disease. The normal person excretes between 150 and 200 mg protein per day.[27] Most plasma proteins, except those of very high molecular weight, have been found in the urine. Albumin is less than 20 mg/day.[27] Many proteins of nonserum origin are also found in the urine. One of these, uromucoid or Tamm-Horsfall mucoprotein, is the predominant protein in normal urine, with about 40 mg excreted daily.[29] This high molecular weight mucoprotein is excreted by the cells of the distal tubule and collecting ducts and forms the matrix of all casts seen in the urine.

In the past, it had been taught that normal urine contains no protein. This appears to have been a consequence of the fact that the tests used had been relatively insensitive to very small concentrations of protein found in normal urine.[27] Today, commercially available dipsticks (Albustix) are in widespread use. One often hears that these dipsticks are "too sensitive" because the urine of healthy subjects may contain protein when tested by this method. They give trace reactions with 100 mg of protein per liter and 1+ reactions with 300 mg/L. Thus, a healthy subject excreting 150 mg of protein per 1500 mL of urine will have a trace reaction. Thus the dipsticks are not "too sensitive"; they merely reflect that healthy subjects do excrete very small amounts of protein in the urine. Dipsticks have been shown to be a very accurate estimation of protein excretion. Highly buffered alkaline urine is the only situation under which they give a false-positive reaction; Bence Jones protein, which occurs in multiple myeloma, may give a false-negative reaction.

As mentioned above, the kidney acts as an important conservation mechanism for body protein stores. Only a minute portion of the protein passing through the glomerulus appears in the final urine. The glomerular capillary wall acts as a molecular sieve and retains most of the plasma proteins passing through it. Under normal circumstances, protein molecules of relatively small size, such as lysozyme and insulin, pass the glomerular filter in relatively large amounts; protein molecules of intermediate size, such as albumin, transferrin, and IgG, pass through in relatively small amounts; and high molecular weight proteins, such as IgM, do not pass the normal glomerular basement membrane except perhaps in trace quantities.[27] Only a very small percentage of the filtered protein appears in the final urine. Filtered proteins are absorbed by the tubular cells and are returned to the circulation.

PATHOLOGICAL CONDITIONS OF KIDNEY

There are a number of syndromes that singly or in combination call attention to the possibility of renal disease. These have been elegantly described by Coe.[30]

Acute glomerulonephritis (AGN)

Acute glomerulonephritis is an acute inflammation of the glomeruli. This results in oliguria (a reduction in urine volume), hematuria, increased BUN and serum creatinine levels, decreased glomerular filtration rate, edema formation, and hypertension. The presence of red blood cells in the urine (hematuria) alone is insufficient evidence of acute glomerulonephritis, for blood might arise from elsewhere in the kidney or from the urinary tract. When red blood cell casts are present in the urine, this in an indication of glomerular inflammation and is a finding of great importance. Other abnormalities present in acute nephritis include proteinuria and anemia.

Nephrotic syndrome (NS)

The nephrotic syndrome has been classically defined as a clinical entity characterized by massive proteinuria, edema, hypoalbuminemia, hyperlipidemia, and lipiduria.[31] This syndrome is a clinical entity having multiple causes and is characterized by increased glomerular membrane permeability that is manifested by massive proteinuria and excretion of fat bodies. Protein excretion rates are usually in excess of a range of 2 to 3 g/day in the absence of a depressed glomerular filtration rate. Hematuria and oliguria may be present. The causes of the nephrotic syndrome are illustrated in the box on p. 412. As a result of the massive loss of serum proteins, primarily albumin, into urine, the plasma protein concentration is decreased, with a concomitant reduction in plasma oncotic pressure. This results in fluid movement from the vascular to interstitial space with consequent edema formation.

Causes of nephrotic syndrome
Associated with various forms of glomerulonephritis
Associated with generalized disease processes
Amyloidosis
Carcinoma
Systemic lupus erythematosus
Diabetic glomerulosclerosis
Polyarteritis nodosa
Associated with mechanical or circulating disorders
Renal vein thrombosis
Constrictive pericarditis
Associated with infection
Syphilis
Malaria
Subacute bacterial endocarditis
Associated with toxins and allergens
Penicillamine
Gold salts
Bee sting
Serum sickness
Miscellaneous
Severe preeclampsia
Transplant rejection

Tubular disease (TD)

Selective disorders of renal tubular function may occur so that such function is depressed disproportionately to the reduction in the glomerular filtration rate. Defects of tubular function may result in depressed secretion or reabsorption, or impairment of urine concentration and dilution.

Renal tubular acidosis (RTA) is the most important clinical disorder of tubular function. There are two main types of RTA: (1) proximal RTA, which occurs as a result of reduced proximal tubular bicarbonate reabsorption and causes hyperchloremic acidosis and (2) distal RTA, in which there is an inability of the tubular cells to create and maintain the usual pH difference between tubular fluid and blood.[32] Failure of either the proximal or distal secretory mechanisms occurs in several clinical states. Failure of the proximal mechanism causes acidosis because more bicarbonate is passed on to the low-capacity distal mechanism than it can reabsorb. The loss of alkali in the urine causes the blood (and other body fluids) to become acidotic. In these patients the urine pH can still be reduced to less than 5.0 by reduction of the filtered load of bicarbonate to the point that the unaffected distal mechanism can reabsorb the excess bicarbonate leaving the proximal tubule. A defect in the distal hydrogen-ion secretory mechanism is recognized by the inability of the kidney to reduce the urine pH <5.0, even when the filtered load of bicarbonate is reduced.

Defects in potassium and uric acid secretion may result in elevated serum potassium and uric acid levels out of proportion to the reduction in the glomerular filtration rate.

Reabsorptive disorders of the proximal tubules may result in hypouricemia, hypophosphatemia, aminoaciduria, and renal glucosuria. The Fanconi syndrome is a grouping of renal defects most commonly including glucosuria, aminoaciduria, hypophosphatemia, and renal tubular acidosis. Tubular proteinuria may occur as a result of a defect in tubular handling of proteins. In tubular proteinuria, less than 2 g/day are excreted. This condition is discussed later in this chapter.

Disorders of urine concentration and dilution occur in all renal disease as the glomerular filtration rate falls appreciably, but occasionally they may become extreme and dominate the clinical presentation.[30]

Urinary tract infection (UTI)

Infection of the urinary tract may occur in the bladder (cystitis) or involve the kidneys (pyelonephritis). Diagnosis is made by the presence of a urine bacterial concentration of more than 100,000 colonies/mL. In urinary tract infection there is an increased number of white blood cells in the urine. The presence of white blood cell casts is indicative of pyelonephritis. An increased number of red blood cells may also be present in the urine.

Vascular diseases

Hypertension (HT). Long-standing and severe hypertension may result in progressive renal damage and chronic renal insufficiency (hypertensive nephrosclerosis). On the other hand, hypertension might be a manifestation of the sodium and water retention that occurs in chronic renal failure, acute glomerulonephritis, and the nephrotic syndrome (volume-dependent hypertension), or it may occur as a result of increased renin release from chronically damaged kidneys (renin-dependent hypertension).

Arteriolar disease. Disease of the small arteries of the kidneys (arteritis) may occur in association with generalized disease processes affecting the kidney, such as systemic lupus erythematosus, polyarteritis nodosa, and progressive systemic sclerosis (scleroderma). These diseases may result in the clinical and biochemical abnormalities seen in acute glomerulonephritis, the nephrotic syndrome, or chronic renal insufficiency.

Renal vein thrombosis (RVT). Thrombosis of the renal veins results in massive proteinuria and the nephrotic syndrome. Hypertension, edema, hematuria, and impaired renal function might accompany the proteinuria.

Diabetes mellitus (DM)

Diabetes mellitus results in a wide variety of abnormalities in kidney function according to the stage of the disease. The early phases of the disease are manifested by the presence of pronounced glucosuria, and polyuria and nocturia as a result of the osmotic diuresis caused by the glucose load. In the juvenile diabetic (onset of insulin-dependent diabetes before 20-years of age), disease of the small arteries and involvement of the kidneys is the leading cause of death. The natural history of diabetic nephropathy in the juvenile diabetic results in the development of proteinuria approximately 17 years after the diagnosis has been made, followed by the development of hypertension 1 to 2 years later and chronic renal insufficiency after a further year.[33]

Urinary tract obstruction (UTO)

Obstruction in the urinary tract may occur in the upper or lower portions. Lower urinary tract obstruction is characterized by residual urine in the bladder after micturition (urination), or urinary retention, whereas the presence of upper tract obstruction is established by the demonstration of a dilated collecting system above a constricting lesion.[29] Lower urinary tract obstruction is characterized by a slow urinary stream, difficulty in emptying the bladder, hesitancy in initiating micturition, and dribbling. Chronic renal damage may result from obstruction and incomplete bladder emptying, and symptoms of chronic renal insufficiency may develop. With complete obstruction, oliguria or anuria will occur. Symptoms of urinary tract infection may be superimposed on the obstructed urinary tract. Urinary tract obstruction may occur as a result of congenital disorders of the lower urinary tract, neoplastic lesions (benign prostatic hypertrophy, carcinoma of the prostate or bladder, or lymph nodes compressing the ureters), or from acquired disorders such as retroperitoneal fibrosis, renal calculi, or urethral strictures.

Renal calculi (RC)

Renal calculi, or stones, are seen in combination with renal colic, hematuria, and symptoms of urinary tract infection or obstruction. Kidney stones may occur as a result of recurrent urinary tract infections by urease-producing organisms, or when the urine is supersaturated by large quantities of calcium, uric acid, cystine, or xanthine.

Causes of acute renal failure

Prerenal
- Hypovolemia
- Cardiovascular failure

Renal
- Acute tubular necrosis
- Glomerulonephritis
- Vasculitis
- Malignant nephrosclerosis
- Vascular obstruction
  - Arterial
  - Venous

Postrenal
- Obstruction of lower urinary tract
- Rupture of bladder

Acute renal failure (ARF)

In acute renal failure there is an abrupt deterioration in renal function. Acute renal failure can be classified as follows:

1. Prerenal (occurring before blood reaches the kidney) because of hypovolemia or poor perfusion as a result of cardiovascular failure.
2. Renal (occurring within the kidney) because of acute tubular necrosis (ATN), which is the most frequently observed cause of acute renal failure, or because of other renal diseases causing rapid deterioration in renal function including arterial or venous obstruction.
3. Postrenal (occurring after urine leaves kidney) because of obstruction.

The causes of acute renal failure are listed in the box at left. Acute renal failure is usually accompanied by oliguria (low urine output) or anuria (no urine output), though nonoliguric acute tubular necrosis might occur. Acute renal failure is associated with varying degrees of proteinuria, hematuria, and the presence of red blood cell casts and other casts in the urine. Serum urea nitrogen and creatinine levels increase rapidly, and metabolic acidosis becomes evident. Depending on the cause, acute renal failure might progress to chronic renal failure or lead to recovery of renal function. Most cases of acute tubular necrosis recover once the offending cause has been treated or removed.

Chronic renal failure (CRF)

Chronic renal failure is a clinical syndrome resulting from progressive loss of renal function. The symptomatic manifestations of chronic renal failure result not only from simple excretory failure, but also from the onset of regulatory failure of certain substances, such as sodium or water; from biosynthetic failure, such as inadequate production of erythropoietin resulting in anemia; and from the excessive production of certain normal substances in response to the chemical derangements that occur in chronic renal failure, such as excessive production of parathyroid hormone.[34] In chronic, progressive renal disease there are four stages. In the first stage there is no more than diminution in renal reserve. At least 50% of normal function must be lost before the concentration of urea nitrogen or creatinine in the plasma rises above the normal range. The second stage of progressive renal disease is that of mild renal insufficiency. The third stage is the development of frank renal failure with advancing anemia, acidosis, and other clinical and biochemical manifestations. The fourth and final stage is that of uremia when all the conse-

Classification of causes of chronic renal failure

- Primary glomerular diseases
  - Chronic glomerulonephritis of various types
  - Systemic lupus erythematosus
  - Polyarteritis nodosa
- Renal vascular disease
  - Malignant hypertension
  - Renal vein thrombosis
- Inflammatory disease
  - Chronic pyelonephritis
  - Tuberculosis
- Metabolic disease with renal involvement
  - Diabetes mellitus
  - Gout
  - Amyloidosis
- Nephrotoxins
  - Analgesic nephropathy
  - Chronic heavy metal poisoning
- Obstructive uropathy
  - Calculi
  - Prostatic hypertrophy
  - Congenital anomalies of lower urinary tract
- Congenital anomalies of kidneys
  - Hypoplastic kidneys
  - Polycystic kidney disease
- Miscellaneous
  - Chronic radiation nephritis
  - Balkan nephropathy

quences of renal failure become overt.[34] A classification of the causes of chronic renal failure is shown in the box above.

RENAL FUNCTION TESTS

The kidney performs many physiologic and excretory functions. By performing a relatively small number of tests, one can deduce accurately the functional state of the kidney.[35] In evaluating renal function, the clinician is interested in two different things: first, whether any significant impairment of renal function is present, and, second, the quantitative assessment of some particular renal function in order to make a specific disease diagnosis.[35] In this section, renal function tests are considered in terms of evaluation of glomerular function, evaluation of tubular function, and urinalysis.

Tests of glomerular function

Glomerular function is most conveniently measured by the creatinine clearance test. Clearance is a theoretical concept defined as that volume of plasma from which a measured amount of substance could be completely eliminated into the urine per unit of time. This depends on the plasma concentration and excretory rate, which, in turn involves the glomerular filtration rate (GFR) and renal plasma flow (RPF). The creatinine clearance is a renal function test based on the rate of excretion by the kidneys of metabolically produced creatinine. The amount of creatinine produced by endogenous protein metabolism is relatively constant, being directly proportional to the body surface area. The amount present in the urine is dependent on renal excretion. Creatinine is freely filtered at the glomerulus and is not reabsorbed by the tubules. A small amount of the creatinine in the final urine is derived by tubular secretion. Because of these properties, the creatinine clearance can be used to estimate the GFR. Generally, a 24-hour urine collection is performed. However, shorter collection periods are acceptable. Precise timing is critical to this test. The bladder is emptied at the beginning of the test period, whereafter all urine passed during the timed collection is kept in a single container. A sample of blood is drawn during the urine collection period. The creatinine clearance is calculated from the following formula:

$$\text{Creatinine clearance (mL/min)} = UV/P$$

where U is urinary creatinine (mg/L), V is volume of urine (mL/min), and P is plasma creatinine (mg/L).

The normal range for creatinine clearance corrected to a surface area of 1.73 m^2 is 90 to 120 mL/min. The creatinine clearance usually parallels true GFR. However, at low filtration rates, creatinine clearance becomes increasingly inaccurate because of the relatively high proportion of the secreted creatinine fraction.[36] The creatinine clearance is lower in women, the elderly, and smaller persons. A number of simple formulas have been derived for prediction of the creatinine clearance from the serum creatinine; these formulas take into account the age, sex, and body weight of the person.[37]

Inulin clearance is the method of choice when precise determination of the glomerular filtration rate is required.[36] The glomerular capillary wall is freely permeable to inulin, and it is not reabsorbed, secreted, or metabolically altered by the renal tubule. The clearance of endogenous creatinine may exceed that of inulin by up to 30% in normal man.[36] The main disadvantages in the measurement of inulin clearance are that it has to be injected intravenously and there is the technical difficulty of laboratory analysis when compared with creatinine measurement.

Urea clearance may also be employed as a measure of the glomerular filtration rate. It is freely filtered at the glomerulus, and approximately 40% to 50% is reabsorbed in the tubules. Thus, under usual conditions, urea clearance values parallel the true glomerular filtration rate at about 60% of it. However, several factors may adversely influence the measurement of urea clearance. As mentioned earlier, the blood urea level varies considerably according to the diet and other factors, and the measurement of urea clearance is dependent on a high rate of urine flow.

From a practical point of view, creatinine clearance is

used in clinical medicine as an assessment of the glomerular filtration rate. It is important to understand that there is a large margin of reserve in renal function; over two thirds of the GFR may be lost in the course of chronic renal disease with few clinical symptoms and biochemical abnormalities.[35] For a person whose normal serum creatinine is 7 mg/L, an increase to 14 mg/L, which is still defined as within the normal range for serum creatinine, is indicative of a fall in the GFR to 50% of normal.

Tests of tubular function

As the tubular urine progresses from Bowman's space in the glomerulus to the tip of the collecting duct, many chemical transformations occur. Although information regarding the functional state of the kidneys may be obtained by studying any of the several urinary functions, in practice relatively few are used for clinical assessment of tubular function.[35] These tests include concentration-dilution studies and acid-base balance studies.

Concentration-dilution studies. The countercurrent mechanism in the loop of Henle is responsible for generation of a hypertonic urine. The mechanism also allows for generation of a dilute urine. The determination of reduction in concentrating and diluting ability of the kidney can provide the most sensitive means of detecting early impairment in renal function, since the full range of ability to concentrate urine and conserve water requires an adequate glomerular filtration rate, proper distribution and amount of renal plasma flow, adequate tubular mass, and healthy tubular cells that are able to pump salt against a sizable electrochemical gradient.[35] Release of ADH from the hypothalamus controls water reabsorption in the distal tubule and collecting ducts.

The urinary *specific gravity* and *osmolality* are used as measures of the concentrating and diluting ability of the tubules. There is a general proportionality between specific gravity and osmolality, and as long as the urine does not contain appreciable amounts of protein, sugar, or exogenous material such as contrast dye, a specific gravity of 1.032 will correspond to an osmolality of 1200 mOsm/kg.[35]

Impairment of renal concentrating ability is a relatively early manifestation of chronic renal disease and becomes evident before changes in other function tests appear. Nonetheless, it remains a nonspecific test for reduced renal function, and any disease resulting in chronic renal failure, as well as diuretics, diabetes insipidus, and so forth, may impair renal concentrating ability. The test is performed after 15 hours of fluid deprivation, and urine is then collected on the hour for 3 hours. Dehydration maximally stimulates endogenous ADH secretion. Under these conditions the urine osmolality should be at least three times that of plasma (286 mOsm/kg). A specific gravity of 1.025 or more, or an osmolality of 850 mOsm/kg or above, in one of the specimens is accepted as evidence of normal concentrating ability. Maximum concentrating ability can also be tested by the injection of ADH. A patient with completely normal concentrating ability is unlikely to have serious kidney malfunction of any type.[35] As chronic renal disease progresses, tubular ability to concentrate urine slowly decreases until the urine has the same specific gravity as the plasma ultrafiltrate—1.010. Clinically the fixation of specific gravity is manifested by nocturia and polyuria.

To test the urinary diluting capacity, one uses the following procedure. The patient empties the bladder and is then given 1000 to 1200 mL of water. Urine specimens are then collected every hour for the next 4 hours. Under these circumstances, the urinary specific gravity should fall to 1.005 or less, or an osmolality of less than 100 mOsm/kg. In the patient with chronic renal disease who is unable to dilute the urine, there is a danger of fluid overload with this test.

In diabetes insipidus, which might arise from inadequate ADH production or from insensitivity of the renal tubules to ADH, the distal tubular walls are impervious to water. As sodium is reabsorbed, the fluid left behind may be very dilute. In this disease the base-line urine might have a specific gravity of less than 1.005 and an osmolality of 50 mOsm/kg.

Acid-base balance. As has been discussed earlier, the kidney plays a major role in acid-base balance. Bicarbonate is reclaimed, and hydrogen ions are secreted. The urine pH can be reduced to about 4.5. In chronic renal failure, whole nephrons are lost as the disease advances, while the remaining nephrons remain intact and are able to reduce urinary pH maximally. Evaluation of urinary acidification after the administration of ammonium chloride can be used as an estimate of the remaining tubular mass.[35] This test should not be performed in a patient who is already acidotic. In renal tubular acidosis there is an inability to produce a maximally acid urine. After the administration of ammonium chloride, the minimum pH that can be achieved is 5.5.

Urinalysis. Urinalysis is an indispensable part of the clinical pathology of renal disease. It may reveal disease anywhere in the urinary tract. The standard urinalysis includes the following observations: appearance of the specimen, pH, specific gravity, protein semiquantitation, presence or absence of glucose and ketones, and a microscopic examination of the centrifuged urinary specimen. The importance of the urinalysis is indicated in Table 23-2. The importance of proteinuria, glucosuria, hematuria, pyuria, red blood cell casts, and white blood cell casts has been dealt with previously. Microscopic examination of the centrifuged urinary sediment should be done on a freshly voided specimen. The specimen must be mixed well before one puts it into the centrifuge tube. A centrifuged specimen of urine from a healthy person contains fewer than 3 red blood cells per high-power field, fewer than 5

Table 23-2. Association of pathological conditions affecting the kidney, and clinical and biochemical abnormalities

	AGN	NS	TD	UTI	HT	RVT	DM	UTO	RC	ARF	CRF
Hypertension	+ +	+	0	0	+ +	±	±	0	0	+	+
Edema	+	+ +	0	0	0	+	+	0	0	+	+
Oliguria/anuria	+	±	0	0	0	±	0	+	0	+	+
Polyuria	0	0	+	0	0	0	+	0	0	0	0
Nocturia	0	±	+	±	0	0	+	±	0	0	+
Frequency	0	0	0	+	0	0	0	±	±	0	0
Loin pain	0	0	0	+	0	+	0	+	+	0	0
Anemia	+	0	0	0	0	0	0	0	0	0	+ +
Blood urea nitrogen	+	0	0	0	±	±	±	±	0	+	+
serum creatinine	+	–	–	–	±	±	±	±	0	+	+
GFR	+	0	0	0	±	±	±	±	0	+	+
Serum potassium	±	±	0	0	0	0	0	0	+	+	+
Serum phosphorus	±	0	0	0	0	0	0	0	0	+	+
Serum calcium	0	+	0	0	0	0	0	0	0	+	+
Serum uric acid	0	0	+	0	±	0	±	0	±	+	+
Acidosis	0	0	+	0	0	0	0	0	0	+	+
Proteinuria	+	+ + + +	+	±	±	+ +	+	0	0	±	±
Hematuria	+ +	+	±	+	0	+	0	0	+ +	+	±
RBC casts	+	0	0	0	0	0	0	0	0	±	0
Pyuria	±	0	0	+ +	0	0	0	±	±	0	0
WBC casts	0	0	0	+	0	0	0	+	0	0	0
Glucosuria	0	0	+	0	0	0	+ +	0	0	0	0

AGN, Acute glomerulonephritis; *ARF,* acute renal failure; *CRF,* chronic renal failure; *DM,* diabetes mellitus; *HT,* hypertension; *NS,* nephrotic syndrome; *RC,* renal calculi; *RVT,* renal vein thrombosis; *TD,* tubular disease; *UTI,* urinary tract infection; *UTO,* urinary tract obstruction.
O = absent; ± = variable; + = present.
GFR, Glomerular filtration rate; *RBC,* red blood cell; *WBC,* white blood cell.

Table 23-3. Characteristic urine microscopic findings in renal disease

Condition	Protein	Red blood cells (per high-power field)	White blood cells (per high-power field)	Bacteria	Casts (per low-power field)
Normal	0-Trace	0-3	0-5	0	Hyaline, occasionally
Glomerulonephritis	1-2$^+$	>20	0-10	0	Granular red blood cells
Nephrotic syndrome	4$^+$	0-10	0-5	0	Oval fat bodies; hyaline
Pyelonephritis	0-1$^+$	0-10	>30	+ +	Granular white blood cells

white blood cells per high-power field, and only an occasional hyaline cast.

Casts are protein conglomerates outlining the shape of the renal tubules in which they were formed. Hyaline casts are composed almost exclusively of protein. Cellular elements may be trapped within hyaline casts, resulting in the formation of granular casts. When there is heavy proteinuria, accumulation of protein within tubular cells leads to a fatty degeneration of the cells and desquamation into the urine; these appear in the urine as oval fat bodies. In acute pyelonephritis, white blood cells may aggregate in the tubules to form pus casts. Red blood cell casts are important markers of glomerular inflammation and should be diligently searched for when any form of glomerular nephritis is suspected.

Microscopic examination of the urinary sediment is completed by a search for bacteria in the urine and for crystals. The presence of crystals in the urine may be a clue to the diagnosis of a specific type of renal calculus. The characteristic urine microscopic findings in renal disease are indicated in Table 23-3.

CHANGE OF ANALYTE IN DISEASE

The changes that occur in anlytes have been discussed in the section on pathological conditions of the kidney and summarized in Table 23-2. In this section the following question is examined from a different perspective: What does the finding of a biochemical abnormality, or group of abnormalities, mean in terms of aiding in the diagnosis of the pathological condition in the kidney?

Serum electrolytes

Sodium. Sodium is the major cation in the extracellular fluid and normally has a serum concentration of 136 to 145 mEq/L. Sodium and its attendant anions are the major contributors to serum osmolality.[38] Its measurement is applicable to the management of patients with renal disease in many situations.

Hyponatremia, one of the most common electrolyte problems encountered clinically, occurs either with or without concomitant hypoosmolality. Hyponatremia with hypoosmolality occurs in psychogenic polydipsia, in which the person ingests an enormous quantity of water during a short period of time and overwhelms the kidney's ability to excrete water. Hyponatremia with an increased extracellular fluid volume occurs in patients with chronic renal insufficiency because of the kidney's inability to excrete water without solute (''free water'') and is seen in patients with cirrhosis, nephrotic syndrome, and congestive heart failure. Hyponatremia with a decreased extracellular fluid volume occurs with excessive fluid loss through sweating or through the gastrointestinal tract after replacement with hypotonic fluid, or by potent diuretic agents, or by the administration of electrolyte-free solutions intravenously. When neither overt volume depletion nor volume excess is present, and renal function is normal, hyponatremia can occur with adrenal insufficiency and the syndrome of inappropriate ADH secretion.

Hypernatremia, by definition a relative water deficit, occurs in patients with hypotonic fluid loss through the skin, gastrointestinal tract, or urinary tract. It can also result from pure water deprivation secondary to severely restricted water intake, as occurs in the unconscious or the elderly, or after the infusion of hypertonic sodium solution. Hypernatremia also occurs in diabetes insipidus whenever the oral fluid intake cannot keep pace with the urinary losses.

Chloride. The concentration of chloride in extracellular fluid parallels that of sodium and is influenced by the same factors. Chloride imbalances occur concurrently with sodium imbalances. Hyperchloremia occurs in association with renal tubular acidosis.

Potassium. Potassium is the major cation of the intracellular fluid, with only 2% being in the extracellular fluid.

Hypokalemia is usually associated with overt potassium depletion as a result of excessive losses of potassium-rich fluids. Potassium loss may be renal or extrarenal. Increased renal excretion of potassium occurs with diuretic agents, prolonged use of corticosteroids, primary or secondary aldosteronism, and Cushing's syndrome. Hypokalemia from extrarenal potassium losses usually occurs in the gastrointestinal tract and is seen with prolonged vomiting, diarrhea, fistulas of the intestinal tract, and villous adenomas of the colon.

Hyperkalemia, an acute medical emergency, is usually caused either by increased cellular breakdown exceeding the normal renal excretory capacity, or by impaired renal excretion alone.[38] Hyperkalemia may result (1) from increased intake of potassium, as occurs with dietary excess or intravenous potassium administration in the patient with compromised renal function, or from cellular breakdown as occurs with extensive burns or rhabdomyolysis (acute muscle necrosis); (2) with decreased potassium excretion, as occurs in acute or chronic renal failure, secondary to potassium-sparing diuretics, adrenal insufficiency, and hypoaldosteronism; (3) with transcellular redistribution of potassium, as occurs with acute acidosis, diabetic ketoacidosis, familial hyperkalemic periodic paralysis, and certain drugs.

Urinary electrolytes

Sodium. Urinary sodium determinations are diagnostically useful in three clinical settings: (1) In volume depletion, the measurement of urinary sodium excretion is helpful in determining the route of sodium loss. A low urinary sodium concentration (less than 10 mEq/L) indicates an extrarenal sodium loss, whereas the presence of a high concentration of sodium in the urine is indicative of renal salt wasting, or adrenal insufficiency. (2) In the differential diagnosis of acute renal failure, the urinary sodium excretion will be less than 10 mEq/L in patients with volume depletion who have no intrinsic renal disease, and usually more than 30 mEq/L in patients with acute tubular necrosis.[39] Additional diagnostic indices may aid in this differentiation. In volume depletion a urine to plasma osmolality of more than 1.1 and a urine to plasma urea ratio of more than 10 is observed, compared with values of less than 1.05 and less than 10 respectively in acute tubular necrosis.[39] (3) In hyponatremia, a low urinary sodium concentration (less than 10 mEq/L) indicates avid renal sodium retention, which may be attributable to either severe volume depletion, or the sodium-retaining state seen in cirrhosis, the nephrotic syndrome, and congestive heart failure. When hyponatremia is associated with the urinary sodium excretion that equals or exceeds the dietary sodium intake—and there is no evidence of volume depletion, expansion, or recent diuretic administration—it is likely that the syndrome of inappropriate ADH secretion is present.[38] In the three situations mentioned above, a random urinary sodium concentration can rapidly supply valuable diagnostic information.

Chloride. The measurement of urinary chloride is of clinical value only in patients with persistent metabolic alkalosis who are not receiving diuretics.[38] In patients with chloride-responsive alkalosis caused by gastric losses and diuretics, the urinary chloride concentration is less than 10 mEq/L. In patients with chloride-resistant alkalosis caused by hyperadrenocorticism with stimulation of renal bicarbonate reabsorption, the urinary chloride level approximates the dietary intake.

Potassium. Urinary potassium levels are helpful in the evaluation of patients with unexplained hypokalemia.[38] The finding of a urinary potassium concentration more than 10 mEq/L indicates that the kidney is responsible for the potassium loss, whereas a urinary potassium concentration of less than 10 mEq/L in the presence of hypokalemia strongly suggests that the gastrointestinal tract is the route of potassium loss.

Anion gap

In the normal serum there is a balance between the cations (sodium, potassium, calcium, magnesium) and anions (chloride, bicarbonate, proteins, organic acids, sulfate, phosphate). Because all these substances are not measured routinely in the laboratory, from a practical point of view the anion gap is referred to as the difference between the sodium concentration on the one hand and the chloride with bicarbonate on the other. The normal anion gap is about 12 mEq/L, ranging from 8 to 16 mEq/L. An increased anion gap occurs in renal failure because of the retention of sulfate, phosphate, and organic acid anions.

Urea, creatinine, and uric acid

With progressive renal insufficiency there is retention in the blood of urea, creatinine, and uric acid. Normally the ratio of serum urea nitrogen to serum creatinine is between 10:1 and 20:1. The usual case of renal failure has a similar ratio. Ratios higher than 20:1 occur in disease states of extrarenal origin, namely, prerenal azotemia, gastrointestinal bleeding, or excessive protein intake with marginally adequate renal function. On the other hand, urea production and the BUN are reduced in the presence of a low protein intake and in severe liver disease. Uric acid concentration in the blood rises in advanced chronic renal failure, but this rarely results in the production of classical gout. The serum urate level increases only modestly until the glomerular filtration rate falls below 20 mL/min, a result of the progressive, adaptive increase in uric acid excretion by residual nephrons, and some degree of intestinal uricolysis.[40] In a patient with a long history of gouty arthritis and the presence of tophi (urate deposits in the skin) around the elbows and ears, the elevated uric acid level may be indicative of chronic gouty nephropathy being the cause of chronic renal insufficiency, rather than the elevated uric acid simply being a manifestation of the renal impairment.

Calcium and phosphorus

In chronic renal failure there is impaired excretion of phosphate, and progressive hyperphosphatemia occurs. This results in a fall in the plasma calcium concentration (hypocalcemia), giving rise to secondary hyperparathyroidism. The elevated parathyroid hormone level causes calcium resorption from bone, and normocalcemia or hypercalcemia may result. However, hypocalcemia is more prevalent in uremia, both as a result of the reciprocal fall in the plasma calcium concentration as the plasma phosphate level rises and because of reduced calcium absorption in the gut as a result of impaired 1,25-dihydroxycholecalciferol production.[21-24] Hypocalcemia is also present in the nephrotic syndrome, as a result of the hypoalbuminemia. However, the ionized serum calcium level remains normal in this condition.

Proteinuria

Proteinuria may be of two types: (1) *Glomerular proteinuria,* in which large quantities of high molecular weight proteins enter the glomerular filtrate and ultimately appear in the urine. Heavy proteinuria (more than 2 g/day) results from increased glomerular permeability, and the protein loss may be so great as to result in the nephrotic syndrome.[30] (2) *Tubular proteinuria,* in which the amount of protein filtered by the glomeruli is not increased, and the low molecular weight proteins, which are normally filtered, appear in larger quantities in the final urine because tubular reabsorption is incomplete. Impaired tubular reabsorption of filtered proteins results in modest increases (1 to 3 g/day) in the urinary excretion of low molecular weight proteins and albumin.[27] Physiological increases in protein excretion occur during the maintenance of an upright posture, after strenuous exercise, and in normal pregnancy.[27] Differentiation of the type of proteinuria can be made in the laboratory by the measurement of ''tracer'' proteins, with albumin being a measure of glomerular proteinuria and $beta_2$-microglobulin being a measure of tubular damage.[41]

Enzymes in urine

Enzymes may appear in urine because of filtration, secretion, or tissue damage.[41] Lysozyme is ubiquitous to all cells of the body. This protein has a molecular weight of 14,000 and is polycationic. It is freely filtered at the glomerulus and largely reabsorbed by the tubules, with only small amounts appearing in the final urine. Increased amounts of lysozyme occur in the urine after tubular damage and strenuous exercise.[32] Amylase is an enzyme with a molecular weight of 50,000. It comes from the salivary glands and the pancreas. The clearance of amylase is about 3% that of creatinine.[32] Increased amylase clearance occurs in acute pancreatitis.

Hemoglobin and hematocrit

Anemia is a common feature of chronic renal failure, and its severity reflects the extent of renal impairment.[42] Progressive anemia usually occurs when the glomerular filtration rate falls below 25 mL/min. The anemia of chronic renal failure is attributable to the following factors: (1) reduced erythropoietin production as renal mass decreases, (2) inhibitors of erythropoiesis present in the serum of the uremic patient, (3) reduced red blood cell survival in advanced renal failure, and (4) iron deficiency secondary to blood loss as a result of the hemostatic defect characteristic of renal failure.[43]

REFERENCES

1. Kokko, J.P.: Renal concentrating and diluting mechanism, Hosp. Pract. **14:**110-116, 1979.
2. Rose, B.D.: Clinical physiology of acid-base and electrolyte disorders, New York, 1977, McGraw-Hill Book Co., Inc., pp. 336-340.
3. Peart, W.S.: Renin-angiotensin system, N. Engl. J. Med. **292:**302-306, 1975.
4. Rose, B.D.: Pathophysiology of renal disease, New York, 1981, McGraw-Hill Book Co., Inc., pp. 579-585.
5. Suki, W.N.: Disposition and regulation of body potassium: an overview, Am. J. Med. Sci. **272:**31-41, 1976.
6. Gonick, H.C., Kleeman, C.R., Rubini, M.E., and Maxwell, M.D.: Functional impairment in chronic renal disease: studies of potassium excretion, Am. J. Med. Sci. **261:**281-290, 1971.
7. Sherwood, L.M., Mayer, G.P., Ramberg, C.F., et al.: Regulation of parathyroid hormone secretion: proportional control by calcium, lack of effect of phosphate, Endocrinology **83:**1043-1051, 1968.
8. Agus, Z.S., Gardner, L.B., Beck, L.H., and Goldberg, M.: Effects of parathyroid hormone on renal tubular reabsorption of calcium, sodium, and phosphate, Am. J. Physiol. **224:**1143-1148, 1973.
9. Massry, S.G., Friedler, R.M., and Coburn, J.W.: Excretion of phosphate and calcium, Arch. Intern. Med. **131:**828-859, 1973.
10. Slatopolsky, E., Robson, A.M., Elkan, I., and Bricker, N.S.: Control of phosphate excretion in uremic man, J. Clin. Invest. **47:**1865-1874, 1968.
11. Cantiguila, S.R., Alfrey, A.C., Miller, N., and Butkus, D.: Total body magnesium excess in chronic renal failure, Lancet **1:**1300-1302, 1972.
12. Lennon, E.J., Lemann, J., and Litzow, J.R.: The effect of diet and stool composition on the net external acid balance of normal subjects, J. Clin. Invest. **45:**1601-1607, 1966.
13. Rose, B.D.: Clinical physiology of acid-base and electrolyte disorders, New York, 1977, McGraw-Hill Book Co., Inc., pp. 327-328.
14. Malnic, G., and Giebisch, G.: Mechanism of renal hydrogen ion secretion, Kidney Int. **1:**280-296, 1972.
15. Pitts, R.S.: Control of renal production of ammonia, Kidney Int. **1:**297-305, 1972.
16. First, M.R., Chronic renal failure, Garden City, N.Y., 1982, Medical Examining Publishing Co., pp. 15-16.
17. Addis, T., Barrett, E., Poo, L.J., and Yuen, D.W.: The relation between serum urea concentration and the protein consumption of normal individuals, J. Clin. Invest. **26:**869-874, 1947.
18. Walser, M.: Urea metabolism in chronic renal failure, J. Clin. Invest. **53:**1385-1392, 1974.
19. Bleiler, R.E., and Schedl, H.P.: Creatinine excretion: variability and relationship to diet and body size, J. Lab. Clin. Med. **59:**945-955, 1962.
20. Barrett, E., and Addis, T.: The serum creatinine concentration of normal individuals, J. Clin. Invest. **26:**875-878, 1947.
21. Stein, J.H.: Hormones and the kidney, Hosp. Pract. **14:**91-105, 1979.
22. DeLuca, H.S.: The kidney as an endocrine organ involved in calcium homeostasis, Kidney Int. **4:**80-88, 1973.
23. Haussler, M.R., and McCain, T.A.: Basic and clinical concepts related to vitamin D metabolism and action, N. Engl. J. Med. **297:**974-983, 1041-1050, 1977.
24. Lumb, G.A., Mawer, E.B., and Stanbury, S.W.: The apparent vitamin D resistance of chronic renal failure: a study of the physiology of vitamin D in man, Am. J. Med. **50:**421-441, 1971.
25. Eschbach, J.W., Adamson, J.W., and Cook, J.B.: Disorders of red blood cell production in uremia, Arch. Intern. Med. **126:**812-815, 1970.
26. Dunn, M.J., and Hood, B.L.: Prostaglandins and the kidney, Am. J. Physiol. **233:**169-184, 1977.
27. Pesce, A.J., and First, M.R.: Proteinuria: an integrated review, New York, 1979, Marcel Dekker, Inc., pp. 80-99.
28. Bright, R.: Cases and observations illustrative of renal disease accompanied with secretion of albuminous urine. In Barlow, G.H., and Babington, J.P., editors: Guy's hospital report, vol. I, London, Highley, 1836, pp. 338-400.
29. Perlmann, G.E., Tamm, I., and Horsfall, F.L.: An electrophoretic examination of a urinary mucoprotein which reacts with various viruses, J. Exp. Med. **95:**99-104, 1952.
30. Coe, F.L.: Clinical and laboratory assessment of the patient with renal disease. In Brenner, B.M., and Rector, F.C., editors: The kidney, vol. I, Philadelphia, 1981, W.B. Saunders Co., pp. 1135-1180.
31. Pesce, A.J., and First, M.R.: Proteinuria: an integrated review, New York, 1979, Marcel Dekker, Inc., pp. 100-143.
32. Morris, R.C.: Renal tubular acidosis, N. Engl. J. Med. **281:**1405, 1969.
33. First, M.R., and Pollak, V.E.: Renal insufficiency in the diabetic patient with heart disease. In Scott, R.C., editor: Clinical cardiology and diabetes, vol. 3, Part 2, Mount Kisco, N.Y., 1981, Futura Publishing Co., Inc., pp. 63-92.
34. First, M.R.: Chronic renal failure, Garden City, N.Y., 1982, Medical Examination Publishing Co., Inc., pp. 1-5.
35. Ware, F.: Renal function tests: a guide to interpretation, Hosp. Med. **9:**77-92, 1981.
36. Carrie, B.J., Golbetz, H.V., Michaels, A.S., and Myers, B.B.: Creatinine: an inadequate filtration marker in glomerular diseases, Am. J. Med. **69:**177-182, 1980.
37. Cockcroft, D.W., and Gault, M.H.: Prediction of creatinine clearance from serum creatinine, Nephron **16:**31-41, 1976.
38. Harrington, J.T.: Evaluation of serum and urinary electrolytes, Hosp. Pract. **17:**28-39, 1982.
39. Oken, D.E.: On the differential diagnosis of acute renal failure, Am. J. Med. **71:**916-920, 1981.
40. Steele, T.H., and Rieselbach, R.E.: The contribution of residual nephrons within the chronically diseased kidney to urate homeostasis in man, Am. J. Med. **43:**876-886, 1967.
41. Pesce, A.J., and First, M.R.: Proteinuria: an integrated review, New York, 1979, Marcel Dekker, Inc., pp. 54-79.
42. Kasanen, A., and Kalliomaki, J.L.: Correlation of some kidney function tests with hemoglobin in chronic nephropathies, Acta Med. Scand. **158:**213-219, 1957.
43. First, M.R.: Chronic renal failure, Garden City, N.Y., 1982, Medical Examination Publishing Co., Inc., pp. 17-18.

Chapter 24 Liver function

John E. Sherwin

ACTH Adrenocorticotropic hormone

ATP Adenosine triphosphate

biliary canaliculi Fine channels running between liver cells.

cirrhosis Liver disorder characterized by loss of normal microscopic architecture with fibrosis. Cirrhosis has a variety of causes, such as obstructive biliary cirrhosis caused by obstruction of major intra- or extrahepatic bile ducts.

CoA Coenzyme A

Crigler-Najjar syndrome A familial form of nonhemolytic jaundice caused by the absence of glucuronide transferase activity from the liver. Associated with increased serum unconjugated bilirubin, kernicterus, and nervous system disorders.

cytochrome P-450 A series of proteins within cells that are one-electron carriers and whose active centers are heme groups. These are involved in hydroxylation reactions of drugs and other xenobiotics. The 450 refers to the position of the "Soret" absorption band.

detoxification The process of changing the chemical structure of a foreign substance or poison to make it less poisonous or more readily eliminated.

Dubin-Johnson syndrome A familial form of chronic, nonhemolytic jaundice caused by a defect in the hepatic excretion of conjugated bilirubin.

focal necrosis Death of cells in a small area of tissue.

Gilbert's disease A benign, hereditary form of hyperbilirubinemia and jaundice caused by a defect in the hepatic uptake of unconjugated bilirubin from serum.

gluconeogenesis The formation of glucose from lactate or amino acids by means of the Cori cycle.

glycogenesis The biochemical formation of glycogen from glucose.

glycogenolysis The biochemical degradation of glycogen to form glucose.

hepatitis Inflammation of liver produced by a variety of infections, toxins, and other causes, such as obstructive hepatitis caused by obstruction of biliary tract.

hepatobiliary Relating to liver and biliary ducts.

hepatocellular disease Diseases in which the liver cells are destroyed.

hepatocyte A parenchymal liver cell that performs all the functions ascribed to the liver.

jaundice A syndrome characterized by hyperbilirubinemia and deposition of bilirubin pigment in skin and mucosal membranes, giving a yellow appearance to the patient skin (also called "icterus").

kernicterus Literally "nuclear jaundice," resulting from deposition of unconjugated bilirubin in nuclei of brain and nerve cells, resulting in cell destruction and encephalopathy.

ketone bodies Compounds having carbonyl groups, usually referring to acetoacetic acid, acetone, and beta-hydroxybutyric acid (though the latter compound is not a ketone).

LCAT Lecithin-cholesterol acyltransferase

mRNA Messenger ribonucleic acid

MSH Melanocyte-stimulating hormone

NADH Nicotine adenine dinucleotide, reduced

NADPH Nicotine adenine diphosphonucleotide, reduced

neonatal jaundice (physiological jaundice) A disorder of newborns characterized by increased serum levels of unconjugated bilirubin and caused by transient immaturity of liver.

oncofetal proteins Any protein produced by embryological tumors, such as alpha-fetoprotein.

P_i Inorganic phosphate

parenchymal cells A general term indicating the functional elements of an organ (see *hepatocyte*).

periportal fibrosis The deposition of fibers or fibrous material in the cells lining the portal blood vessels of the liver.

porphyrias A group of disorders caused by disturbances of porphyrin metabolism characterized by increased formation and excretion of porphyrins and their precursors.

Reye's syndrome Acute, often fatal encephalopathy and fatty degeneration of the liver, seen primarily in children.

TSH Thyroid-stimulating hormone

Wilson's disease Hepatocellular degeneration, also associated with a change in the iris and lens of the eye and caused by a defect in copper metabolism.

xenobiotics Any organic compound that is foreign to the body, such as drugs and organic poisons.

ANATOMY OF LIVER

The liver is the largest organ of the body and is responsible for the production of the majority of the endogenous energy sources used by the body. The liver is divided into two primary lobes and is located in the abdominal cavity just below the diaphragm. The right lobe is the largest, being approximately five to six times larger than the left lobe. The wedge-shaped liver normally weighs approximately 1400 g in adults. The liver is soft and friable, with a specific gravity of 1.05. It is covered by a fibrous coat and a serous coat, which are derived from the peritoneum. During embryological development the liver is formed from the ventral surface of the foregut. Morphological development of the liver is nearly completed by the third month of gestation. The liver almost fills the abdominal cavity at this point, and it remains relatively large, even after birth. Although the morphological development is completed relatively early, functional development is not completed until much later.[1] One of the best examples of this delay in functional development is the lack of glucuronyl transferase in infants born prematurely, which results in neonatal jaundice. This functional development explains the need for age-related reference values for such analytes as bilirubin, cholesterol, lactate dehydrogenase, alkaline phosphatase, and alpha-fetoprotein (Table 24-1).

The functional units of the liver are the lobules, which are about 1 mm in diameter. Each lobule has a blood supply from the hepatic artery and portal vein. The parenchymal cells of the lobule are arranged about the hepatic vein in a plate that is generally perpendicular to the portal vessel tracts, which contain the hepatic artery and the portal vein. The lobule is arranged in a fashion such that the cells are surrounded by sinusoid capillaries, blood vessels, and ducts (Fig. 24-1). The sinusoids are lined with intensely phagocytic cells, the Kupffer cells, which engulf and lyse damaged blood cells. As nutrients are supplied to the parenchymal cells, they are metabolized, and the metabolites and wastes are secreted into the central vein or into the

Table 24-1. Size and function of the liver with age

Age	Mean weight (g)		Distance below costal margin (cm)	Bromosulfophthalein clearance (5 mg/kg/45 min) % of sample remaining	Asparate aminotransferase (AST) (U/L)	Alanine aminotransferase (ALT) (U/L)	Total cholesterol (mg/L)	Total bilirubin (mg/L)	Direct bilirubin (mg/L)
	Male	Female							
Neonate									
birth	124	125					500-1200	120	0-10
1-5 days			1.8	10-20	16-74	1-25			
Infant							700-1900		
1-2 months	300	240	1.5	4	9-67	16-36		10	0-10
Preschool									
2 years	400	390	1.1				1350-2500		
3 years	460	450	1.0		6-30				
4 years	510	500	0.6	5		7-23		10	0-10
5 years	555	550	1.0		5-25				
School									
8 years	665	685	0.9		26-28		1350-2500		
10 years	770	810	1.0	5	24-27	2-15		10	0-10
Adolescence									
12 years	950	960	1.2		22-23		1350-2500		
14 years	1150	1180	0.9	5	20-22	2-15		10	0-10
Adult	1630	1415		5	4-40	4-35	1300-2700	10	0-10

From Johnson, T.R.: Johnson, T.R., Moore, W.M., and Jeffries, S.E., editors: Children are different: development physiology, ed. 2, Columbus, Ohio, 1978, Ross Laboratories, pp. 155-160.

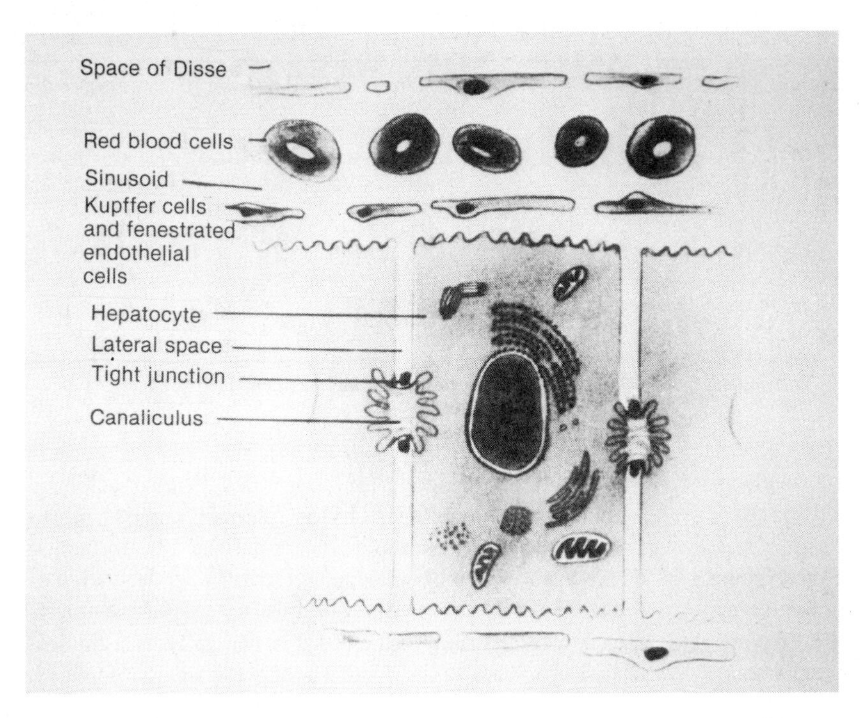

Fig. 24-1. Anatomic relationships of blood, lymphatic, hepatocellular, and biliary systems in liver. (From Soloway, R.D.: In Dermers, L.M., and Shaw, L.M., editors: Evaluation of liver function, Baltimore, Md., 1978, Urban & Schwarzenberg, Inc.)

biliary canaliculi for collection in the hepatic duct and storage in the gallbladder or into the lymphatic vessels, which carry the interstitial fluid of the liver into the general lymphatic circulation.

The subcellular structure of the parenchymal cells, the *hepatocytes,* is that of a relatively undifferentiated cell. These polygonal cells are bounded by a cell membrane that surrounds cytoplasm with a nucleus, mitochondria, lysosomes, Golgi apparatus, endoplasmic reticulum, and vacuoles. Therefore these cells are capable of protein synthesis, energy metabolism, transport and secretion, storage, and other general catabolic reactions, such as conjugation and hydroxylation. Thus this cell type is responsible for the bulk of liver metabolic function.

NORMAL FUNCTION OF LIVER

The liver is the principal organ of the body for the metabolism of carbohydrates, proteins, lipids, porphyrins, and bile acids. The liver is capable of synthesizing most body proteins, except for the gamma globulins and hemoglobin. Remember, however, that the liver of neonates does have the ability to synthesize hemoglobin. It is also the major site for storage of iron, glycogen, lipids, and vitamins. Glycogen represents 5% to 8% of the liver mass and only 1% of the muscle mass, though the total amount of body glycogen stores is greater in muscle because of the greater total muscle mass. Additionally, the liver plays an important role in the detoxification of xenobiotics (molecules foreign to the body) and excretion of metabolic end products such as bilirubin, ammonia, and urea.

Carbohydrate metabolism and liver function

Polysaccharides are a form of energy storage. The liver is capable of producing glycogen, the principle storage polysaccharide, by glycogenesis and degrading glycogen by glycogenolysis. The predominance of the glycogen synthetic reactions or the glycogen degradation reactions depends on the metabolic status of the person. The metabolic pathway for the conversion of glucose to glycogen is reviewed in Chapter 30. Monosaccharides such as fructose and galactose are converted to glucose, phosphorylated, and used for glycogen synthesis. Liver glycogen in the normal adult is not a static storage pool because it functions as the source of glucose for the rest of the body, except for muscle. Since the rate of energy consumption in most tissues is relatively constant, the adenosine triphosphate (ATP) needed for the maintenance of cellular integrity and function is normally supplied by oxidative metabolism of mitochondrial nicotine adenine dinucleotide NADH). The NADH is the result of cellular oxidation of glucose or fatty acids. Under conditions of stress, increased body energy requirements must be met by increased glucose use, that is, glycogenolysis and glycolysis. Additional glucose required for the body in times of stress is derived from increased secretion of glucose by the liver. This is accomplished by the liver increasing the rate of

Table 24-2. Concentration of plasma proteins separated by electrophoresis as a function of age

Age	Total protein (g/L)	Albumin (g/L)	α_1-globulin (g/L)	α_2-globulin (g/L)	β-globulin (g/L)	γ-globulin (g/L)
At birth	46-70	32-48	1-3	2-6	3-6	6-12
Newborn, 1 week	44-76	29-55	0.9-25	3-4.6	1.6-6	3.5-13
3 months	45-65	32-48	1-3	3-7	3-7	2-7
3 to 4 months	42-74	28-50	0.7-3.9	3.1-8.3	3.1-8.3	1.1-7.5
1 year	56-72	39-51	1-3.4	2.8-8.0	3.8-8.6	3.5-7.5
	54-75	37-57	1-3	5-11	4-10	2-9
2 years to adult	53-80	33-58	1-3	4-10	3-12	4-14
4 years	59-80	38-54	1-3.4	3.5-8.0	5-10	3.5-13

From Meites, S., editor: Pediatric clinical chemistry, ed. 2, Washington, D.C., 1981, American Association of Clinical Chemists.

glycogen degradation and by increasing the rate of gluconeogenesis. Gluconeogenesis is an important source of blood glucose when liver glycogen has been depleted. Liver glycogen depletion will occur within several hours in a fasting person. Gluconeogenesis is not simply a reversal of glycolysis, since several of the glycolytic enzymes such as pyruvate kinase, phosphofructokinase, and hexokinase are not reversible. Lactate and amino acids serve as the precursors for the gluconeogenic pathway. Liver use of blood lactic acid (of muscle origin) is an important factor in clearance of this analyte from serum.

The net result of these metabolic pathways is to provide a constant supply of glucose to the blood for export to peripheral tissues to meet their energy requirements. Blood glucose is homeostatically controlled by a variety of hormones, such as insulin, glucagon, glucocorticoids, thyroxine, ACTH, and growth hormone as well as by substrate concentrations of lactate, amino acids, glycogen, and oxygen.

Protein metabolism in liver

The majority of serum proteins are synthesized in the liver, with two exceptions in the adult: gamma globulin and hemoglobin. In the infant the liver retains the ability to synthesize hemoglobin. As in carbohydrate synthesis, extensive impairment of liver function is required before decreased protein synthesis inhibition is unequivocally demonstrated. The normal concentrations of the major electrophoretic subgroups of proteins are quite variable and depend at least in part on the person's age (Table 24-2).

The direct synthesis of small peptides occurs enzymatically without an RNA template. For example, the tripeptide glutathione is synthesized by two enzymes:

$$\text{1. Glutamate + Cysteine + ATP} \xrightarrow[\text{Gamma-glutamyl cysteine synthetase}]{Mg^{++}\ K^{+}} \text{Gamma-glutamyl cysteine + ADP + } P_i$$

$$\text{2. Gamma-glutamyl cysteine + Glycine + ATP} \xrightarrow{\text{Glutathione synthetase}} \text{Glutathione + ADP + } P_i$$

The synthesis of longer oligopeptides is also achieved without an RNA template by an alternative series of enzymatic reactions. Amino acids are activated by formation of the adenosine monophosphate (AMP) acyl derivative of the amino acid. Intermediates in this oligopeptide synthesis are covalently bound to the enzyme generally by a thiolester linkage. Thus in the synthesis of these smaller peptides, the enzyme acts as a template during synthesis.

High molecular weight polypeptides are synthesized in the rough endoplasmic reticulum present in the cytoplasm. This synthesis is directed by the genome through messenger RNA synthesized in the nucleus and transported to the endoplasmic reticulum. A brief summary of the events in protein synthesis follows.

Messenger RNA (mRNA) is synthesized in the nucleus by transcription of DNA and exported to the endoplasmic reticulum, where it serves as the template for a specific protein sequence. Ribosomal RNA subunits are synthesized in the nucleolus and also exported to the endoplasmic reticulum. The 30S ribosomal ribonucleoprotein subunit attaches to the mRNA, and individual amino acids are activated by covalent linkage to transfer RNA (tRNA). These activated amino acids then interact with the 30S ribosomal subunit complexed with mRNA. The tRNA's align themselves (and thus the amino acids) in a specific order dictated by the mRNA template. Elongation of the peptide occurs by association of the 50S ribosomal subunit, which binds additional activated amino acids and places them in proximity with the growing peptide chain so that the enzymatic formation of the peptide bond is energetically favored.

Many liver enzymes exhibit half-lives of several weeks, though structural proteins are stable nearly indefinitely. Plasma proteins synthesized by the liver exhibit quite varied rates of synthesis and degradation (that is, turnover rates). Under normal conditions, the rate of synthesis of each protein equals its rate of degradation, since its concentration in plasma remains constant. Many proteins synthesized by the liver are excreted into extravascular fluid to carry out specific functions. These include: nutrition, blood pressure control (oncotic), transport, and other func-

Table 24-3. Representative plasma proteins and their properties

Electrophoretic fraction	Protein	Approximate concentration (g/L)	Principal function
α_1	Prealbumin	1.0	
	Albumin	40-50	Binds Ca^{++}, T_4, bilirubin
	Antitrypsin	2-4	Inhibits a number of proteolytic enzymes
	High-density lipoproteins (HDL)	3-8	Transport of cholesterol from peripheral tissue to liver
	Retinol binding	3-6	Transport of retinol
	Thyroxine binding	1-2	Transport of thyroxine
α_2	Haptoglobins (3 types)	400-800	Transport of free hemoglobin from destroyed red blood cells
	Lipoprotein (VLDL)	200	Transport of cholesterol and triglycerides
	Ceruloplasmin	20-60	Transport of copper; increases use of iron as ferroxidase
	Prothrombin (bovine)	10	Proenzyme of thrombin
	Angiotensinogen	—	Precursor of angiotensin
	Erythropoietin	<5	Erythropoietic hormone
β_1	Lipoprotein (LDL)	400-1000	Transport of triglycerides and other lipids
	Plasminogen	300	Profibrinolysin
	Fibrinogen	300	Coagulation factor I
$\beta_1\beta_2$	Complement transferrin	1-700	Lysis of foreign cells Transport of iron
β_2	Glycoproteins	30	Unknown
γ	Blood group globulins and immunoglobulins	700-1500	Contains various antibodies, blood globulins, complement C1 and C2, and so on, synthesized by B cells.

Modified from Orten, J.M., and Neuhaus, O.W., Human biochemistry, ed. 10, St. Louis, 1982, The C.V. Mosby Co.

tions. Table 24-3 lists representative liver proteins of each type found in plasma, along with some of their properties.

One of the most important serum proteins produced in the liver is albumin. Present in concentrations of 40 to 50 g/L, albumin represents 50% to 60% by weight of all plasma protein. This molecule has an extraordinarily wide range of functions, including nutrition, maintenance of oncotic pressure and primary serum transport of Ca^{++}, unconjugated bilirubin, free fatty acids, drugs, and steroids. Its multifactorial role in human physiology makes it an important analyte in the monitoring of liver disease.

Metabolic pools of amino acids are present in the liver, and it is from these pools that the amino acids are drawn for the synthesis of proteins. On degradation of the protein, the bulk of the constituent amino acids are returned to these intracellular pools. The released amino acids are also used in gluconeogenesis, transamination, or deamination reactions or reincorporated into new proteins. Important transamination reactions are catalyzed by the enzymes alanine aminotransferase (ALT or SGPT) and aspartate aminotransferase (AST or SGOT). In the normal person who is in nitrogen equilibrium or negative nitrogen balance, the amino groups of excess amino acids in serum are converted to ammonia or urea for excretion (Fig. 24-2).[2] There is also excretion of some amino acids unchanged in the urine. Positive nitrogen balance leads to a diminution of amino acids pools and thus decreased urea excretion. The carbon skeletons of the amino acids are used for the metabolic reactions listed above.

Urea, creatinine, ammonia, and uric acid represent 70% to 75% of the serum nonprotein nitrogen, of which urea accounts for 60% of the total. These compounds represent, in addition to amino acids, the principal sources of excreted nitrogenous compounds. The majority of the metabolism of nonprotein nitrogen occurs in the liver. Urea is produced in the liver because arginase, the enzyme that converts arginine to urea and ornithine, is restricted to the liver (Fig. 24-2). Although blood ammonia concentration is normally quite low, 500 μg/L, it is an important intermediate in amino acid synthesis. Sources of ammonia include hepatic oxidation of glutamate to oxoglutarate, the transamination and oxidative deamination of amino acids and catecholamines, and bacterial breakdown of urea in the gut. The fixation of ammonia results from glutamate and glutamine synthesis and from carbamyl phosphate synthesis. This carbamyl phosphate may be used to synthesize orotic acid and ultimately pyrimidines for nucleic acids or to synthesize urea, which is the principal pathway for the excretion of excess nitrogen.[2]

Lipid biosynthesis and transport in liver function

Lipids, for the sake of this discussion, include only free fatty acids, triglycerides, glycerophosphatides, sphingolipids, cholesterol, and cholesterol esters. General chemical structures for these lipids are shown in Chapter 48. Lipids are not only synthesized by the liver but are also absorbed from the lumen of the small intestine by the intestinal mucosa. These absorbed lipids are bound to apo-

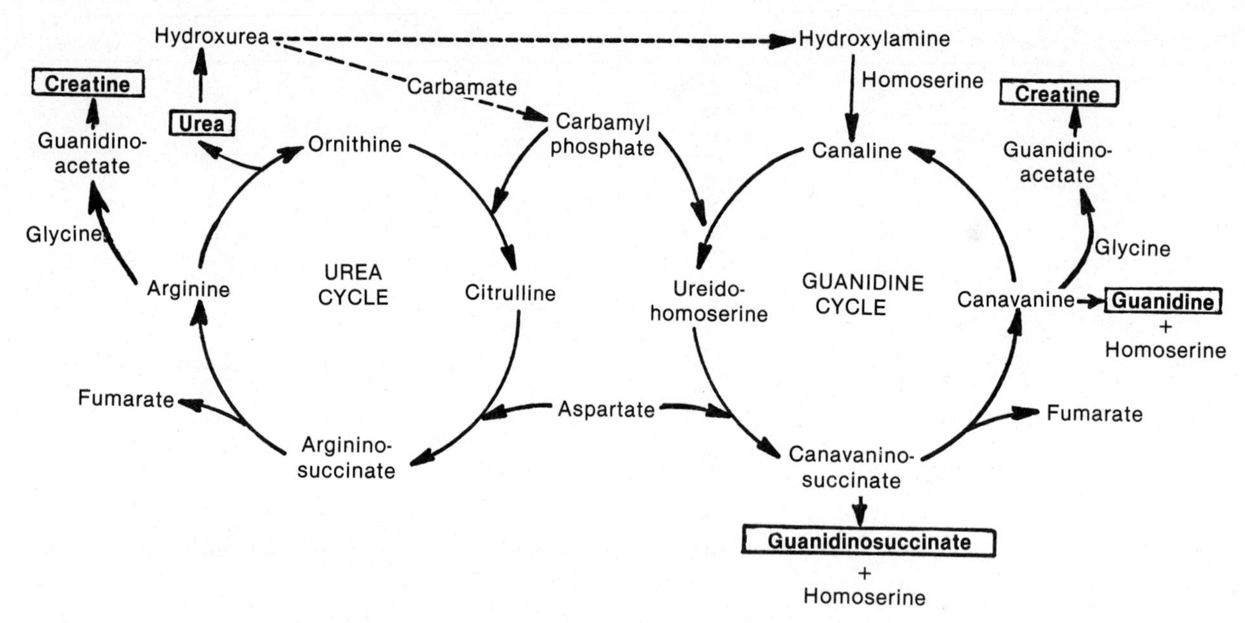

Fig. 24-2. Reutilization of urea nitrogen. Urea is oxidized to hydroxyurea. Hydrolysis yields carbamate and hydroxylamine. Carbamate reacts with ATP to form carbamyl phosphate. This enters both cycles to form citrulline and ureidohomoserine, respectively. Condensation with aspartate forms argininosuccinate and canavaninosuccinate. Argininosuccinate is acted upon by a lyase to form fumarate and arginine. Canavaninosuccinate can be reduced to form guanidinosuccinate and homoserine, a reaction that is not reversible, or it can continue in cycle to form fumarate and canavanine by action of a lyase. Canavanine can be reduced to form homoserine and guanidine. This reaction is irreversible. Canavanine can react with glycine to form guanidinoacetate and then creatinine by methylation. In this reaction, canaline is formed and completes cycle. With arginine, transamidination to glycine forms ornithine, which completes cycle. (From Natelson, S., and Sherwin, J.E.: Clin. Chem. **25**:1343, 1979.)

proteins synthesized in the liver and are transported in the blood as lipoproteins. The chief source of fatty acids is the diet, but a significant portion of the fatty acids normally required are also synthesized by the liver.

Lipids are synthesized in the liver in response to excess carbohydrate intake. The excess carbohydrate is converted to acetyl coenzyme A (acetyl CoA) and a cytoplasmic enzyme system converts it to the fatty acid palmitate, using nicotine adenine diphosphonucleotide (NADPH) and ATP, which are also produced from glucose metabolism. Additionally, a system of enzymes localized in the mitochondria catalyses fatty-acid chain elongation, using acetyl CoA. Triglycerides are formed in the liver, using the fatty acids and glycerol derived from the excess carbohydrate.

The liver also serves as a major site of breakdown of free fatty acids present in serum. The fatty acids are broken down to acetyl CoA, which can then be oxidized to CO_2 by the citric acid cycle. However, a small portion of acetyl CoA is converted to "ketone" bodies, such as acetoacetate, beta-hydroxybutyrate, and acetone. In normal persons, these products are present in blood to the extent of only about 30 mg/L. In the presence of excess mobilization of fatty acids, as in diabetic ketoacidosis or in alcohol intoxication, limiting amounts of NAD/NADP result in an increased hepatic synthesis of the ketone bodies.

Cholesterol is also synthesized in the microsomes of liver from acetyl CoA. Approximately 70% of the total cholesterol in plasma is esterified with fatty acids. These cholesterol esters are formed in plasma by enzymatic transfer of the fatty acid from the 2 position of phosphatidyl choline to cholesterol. This reaction is catalyzed by lecithin cholesterol acyltransferase (LCAT), also produced by the liver.

Bile acids are produced from cholesterol by the cells lining the biliary canaliculi and the ductules of the liver.[3] The bile acids are the final excretory metabolite of cholesterol. They also serve as aids in the digestion of dietary lipids (see Chapter 31). Approximately 80% of the available cholesterol is converted into the four major bile acids (Table 24-4). Cholic acid and chenodeoxycholate are the primary bile acids and are present in a five- to tenfold excess over the secondary bile acids, deoxycholic acid and lithocholic acid, which are produced by metabolism of the primary hepatic bile acids by intestinal bacteria.

Bile acids collect in the biliary canaliculi and ductules. They then flow through the bile ducts to the gallbladder.

Table 24-4. Names and structures of bile acids, their relative contribution to the total bile acid pool, and normal serum concentrations*

Name	Structure	Relative content (%) in bile	Normal serum concentration (mmol/L)
Cholylglycine	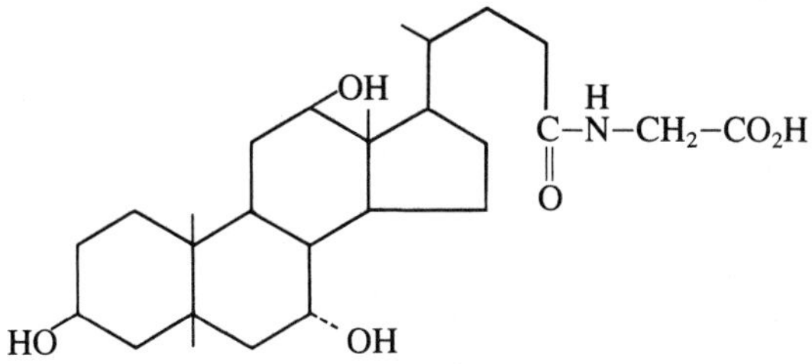 OH; H; $C-N-CH_2-CO_2H$; O; HO; OH	38	0.2–0.9
Deoxycholylglycine	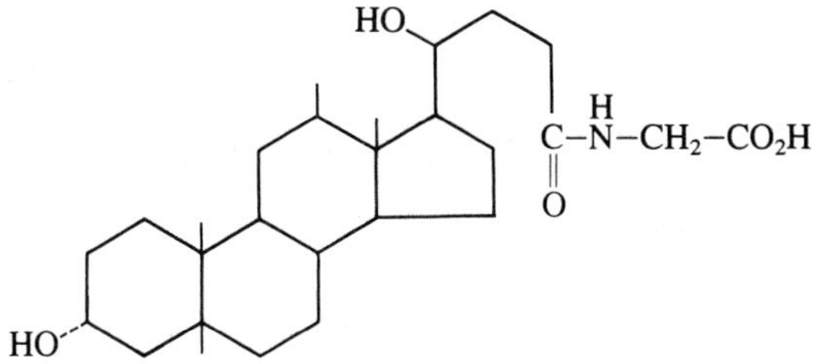 HO; H; $C-N-CH_2-CO_2H$; O; HO	20	0.08–0.7
Lithocholylglycine	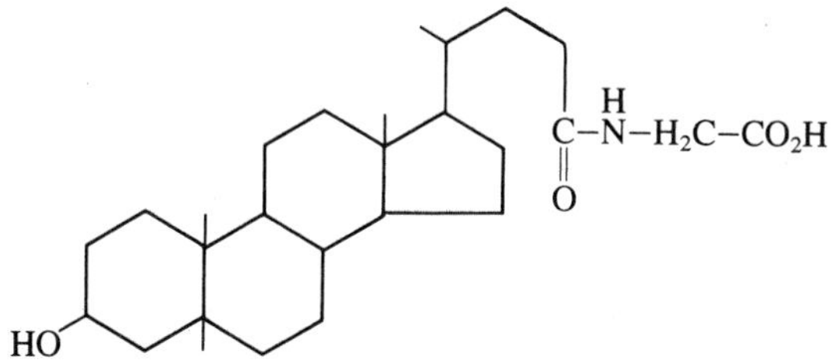 H; $C-N-H_2C-CO_2H$; O; HO	4	0.07–0.3
Chenodeoxycholylglycine	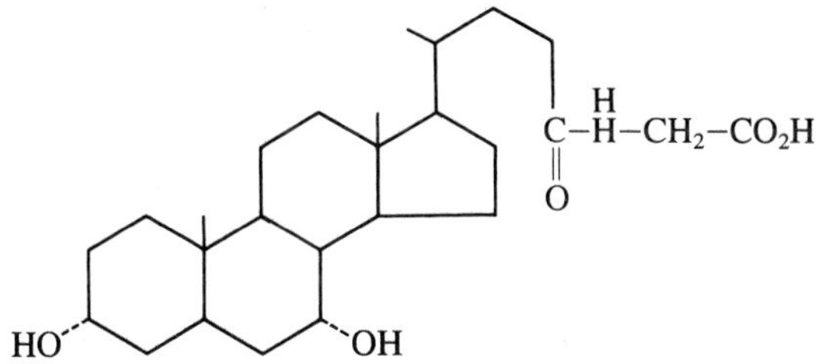 H; $C-H-CH_2-CO_2H$; O; HO; OH	38	0.05–0.2

*From Shaw, L.M.: Lab. Management **20:**56-63, 1982.

The bile acids are transported in the bile to the intestinal lumen, where they emulsify ingested lipids so that they are absorbable. This emulsification of the lipids then permits the intestinal mucosa to digest the lipids and absorb the liberated triglycerides and cholesterol. Absorbed and synthesized triglycerides and cholesterol are then condensed with hepatic apoproteins to form lipoproteins, which are transported to the liver for degradation or to adipose tissue for storage. Free fatty acids are transported to the liver from the intestine or from adipose tissue bound to albumin. More than 90% of the secreted bile acids are reabsorbed and returned to the liver through the portal circulation.[3] The sequence of events in the degradation of lipoproteins is considered in Chapter 31.

Liver function as a storage depot

The liver is an important site for the storage of several metabolites, including iron, glycogen, amino acids, and some lipids and vitamins. Approximately 1 to 1.5 g of iron are stored in the liver, spleen, and bone marrow. The remaining 3 to 4 g of body iron are physiologically active and are present in hemoglobin, myoglobin, the cytochromes or other enzymes. The daily iron needs of the body are small because the body conserves iron very efficiently. In the newborn, the iron requirement is very small because of the low turnover of fetal red blood cells. The iron requirement increases during adolescence to a range of 10 to 40 mg/day. The normal adult requires only 5 to 20 mg/day.

Nutritional iron is absorbed primarily in the intestine. To be absorbed, ferric iron (Fe^{+3}) must be converted to ferrous iron (Fe^{+2}). Immediately after absorption, the Fe^{+2} is reconverted to Fe^{+3} and temporarily stored in the intestinal mucosa as the ferritin complex. The ferritin apoprotein is synthesized in the liver. The sequence of events that result in subsequent reduction of ferritin to apoferritin and Fe^{+2} are unclear. Release of Fe^{+2} into the plasma is followed rapidly by oxidation to Fe^{+3} and complexation with transferrin, a hepatic synthesized alpha$_1$-globulin. In the normal adult, transferrin is 25% to 30% saturated with Fe^{+3}. In the liver, transferrin releases iron, and a new ferritin-Fe^{+3} complex in liver cells is formed. Ferritin is the primary storage form of iron, in which the apoferritin binds iron as a colloidal hydrous ferric oxide. However, a small amount of iron is stored as hemosiderin, which is an insoluble cellular inclusion of Fe^{+3} complexed with ferritin. Hemosiderin granules serve as a storage form for iron when there are insufficient levels of apoferritin. The ratio of iron to protein is much greater in hemosiderin than in ferritin. The adult liver contains about 700 mg of iron. Additional details of iron metabolism are found in Chapter 33.

Glycogen is also stored in the liver. Only liver glycogen is available for maintenance of a constant blood glucose, since only the liver (and kidneys) contain the enzyme glucose-6-phosphatase, which converts glucose-6-phosphate to glucose. Glycogen is stored in the hepatocytes as a semicrystalline inclusion body. These storage granules are associated with the hydrolytic enzymes, which release glucose from the terminuses of glycogen. As a result of the highly branched structure of glycogen, approximately 10% of the glucose of glycogen is available for immediate enzymic release.

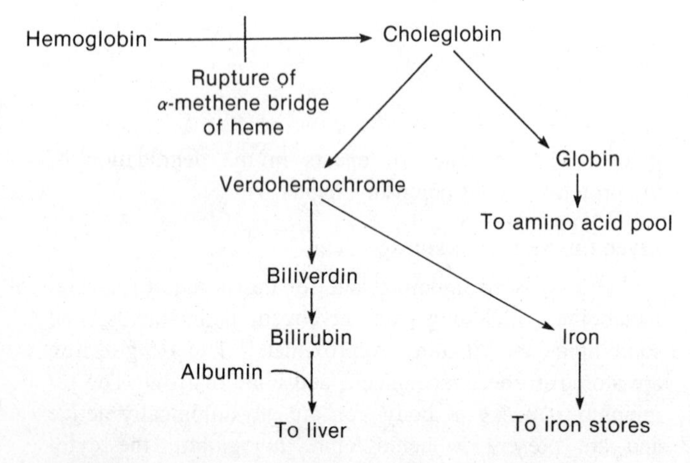

Fig. 24-3. Formation of bilirubin from hemoglobin occurs primarily in reticuloendothelial tissues.

Lipid storage, primarily as triglyceride, occurs preferentially in adipose tissue located subcutaneously. Lipid is also stored in the liver, where it functions as an energy reservoir. Under normal circumstances the liver functions as a temporary storage site for lipids as they are synthesized in the liver or absorbed from the intestine after a meal.

Bile pigment formation

About 126 days after emergence from the reticuloendothelial tissue, the senescent erythrocytes are phagocytized and the hemoglobin is released. The heme portion of hemoglobin is converted to bilirubin, with the release of iron and the globin proteins. The liberated iron is bound by transferrin and returned to the iron stores of the liver or bone marrow, and the globin is degraded to its constituent amino acids. The conversion of heme to bilirubin requires 2 to 3 hours (Fig. 24-3).

Bilirubin, bound to albumin, is transported from the reticuloendothelial cells to the hepatocytes. Albumin has two types of binding sites for bilirubin. One has a high affinity constant of about 1×10^8, and the other has a weaker binding constant of 1×10^6. Bilirubin at the lower affinity binding site is displaced by free fatty acids, barbiturates, and salicylates. Bilirubin bound to albumin is apparently not cytotoxic and remains in solution as it is transported to the liver. A small amount of bilirubin (less than 20 μg/L) is free in the plasma. In the liver, bilirubin is transported through the cellular microvilli of the sinusoids to the hepatocytes. The bilirubin is dissociated from albumin and taken up into the hepatocytes by specific proteins. Within the hepatocyte, bilirubin glucuronide is formed by reaction with uridine diphosphoglucuronate (UDP) in the presence of UDP-glucuronyl transferase. Conventional methodology indicates that 8 to 10 mg/L of unconjugated bilirubin is present in normal adult serum and that normal adult serum contains 0 to 2 mg/L of conjugated bilirubin. Controversy exists as to the relative contribution of the monoglucuronide and diglucuronide to fresh bile pigment. It seems likely that the formation of diglucuronide bilirubin represents the normal conjugation reaction but that in disease states the accumulation of monoglucuronides of bilirubin occurs as a result of accumulation of unconjugated bilirubin in the face of a limited supply of glucuronate.

After formation of the bilirubin-diglucuronide, it is excreted by the hepatocyte into the biliary canaliculi. Any disease resulting in a decreased secretion of the conjugated bilirubin will result in an increased serum concentration of this analyte. As part of the bile, these water-soluble conjugates are secreted into the lumen of the small intestine. The bilirubin conjugates are hydrolyzed by a beta-glucuronidase, and the regenerated bilirubin is converted to

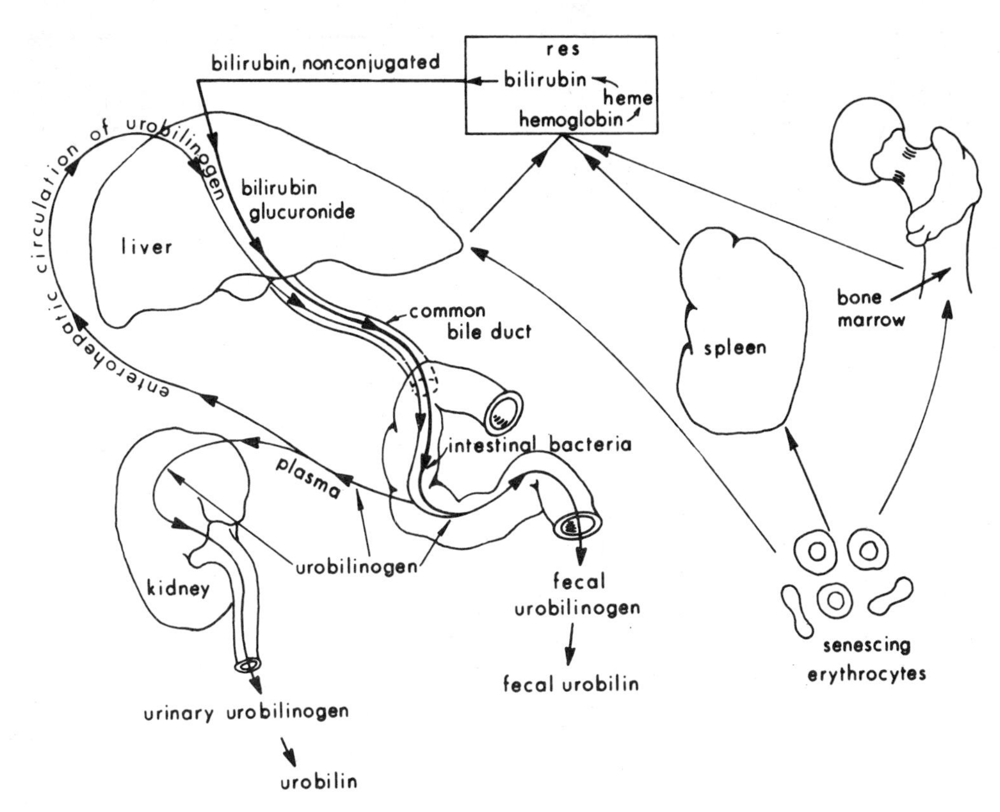

Fig. 24-4. Bilirubin metabolism. (From Bauer, J.D., editor: Clinical laboratory methods, ed. 9, St. Louis, 1982, The C.V. Mosby Co.)

d-urobilinogen and further reduced to *l*-urobilinogen and *l*-sterocobilinogen by anaerobic bacteria of the intestinal lumen. The urobilinogens are reabsorbed from the intestine and recirculated in the extrahepatic circulation, where they are ultimately excreted in the urine. Stercobilinogen is not reabsorbed from the intestine but is a normal constituent of the feces. These bilinogens oxidize spontaneously in air to the corresponding bilins and thereby contribute to the color of both urine and feces (Fig. 24-4). See Chapter 33 for additional details on bilirubin metabolism.

Metabolic end-product excretion and detoxification

The human has two mechanisms for detoxification of foreign (xenobiotic) materials, such as drugs and poisons, and toxic metabolic products, such as ammonia and bilirubin. The first is to reversibly bind the compound to a protein so that the material is inactivated. This is true of bilirubin when it is bound to albumin and of lead when it is bound to hemoglobin. The second mechanism is to modify chemically the compound so that it is readily excreted, such as conversion of ammonia to urea or bilirubin to bilirubin-glucuronide.

The detoxification of exogenous compounds is usually accomplished by hydroxylation or conjugation of the parent compound. This is mediated by one of the cytochrome P-450 enzymes or by conjugation with sulfate or carbohydrate. These reactions are localized in the microsomes of the liver. The microsomal hydroxylation system consists of NADPH, a cytochrome reductase, and a hemoprotein complex, cytochrome P-450. The cytochrome P-450 complex is not a single enzyme but represents a class of enzymes that is induced by exposure to drugs such as phenobarbital, benzopyrenes, and aromatic amines. The hydroxylation of drugs generally results in loss of physiological activity and an increase in the solubility of the drug as well. The increased solubility results in rapid elimination of the hydroxylated drug in the urine.

The esterification of phenolic drugs or steroid hormones with sulfate or carbohydrate presents a significant mechanism for the detoxification and elimination of these compounds. The sulfated compounds, being more water soluble than the parent compounds, are excreted directly in the urine. Carbohydrate conjugates are commonly excreted into the intestinal lumen as part of the bile. Hydrolysis of the conjugates occurs at this point, and some reabsorption of the drug may also occur.

LIVER-FUNCTION ALTERATIONS DURING DISEASE

Morphological changes in response to disease are common and generally are associated with biochemical alterations as well.[4] In neonatal hepatic biliary atresia, portal

Possible causes of acute hepatic failure in children

Ischemia
- Acute circulatory failure
- After cardiac surgery
- Ligation of hepatic artery

Infections
- Hepatitis A
- Hepatitis B
- Herpes simplex
- Cytomegalovirus
- Coxsackievirus
- Rubella
- Adenovirus
- Leptospirosis
- Yellow fever

Poisons
- Arsenic
- Beryllium
- Iron
- Yellow phosphorus
- Hydrocarbons
- *Amanita phalloides*
- Pyrrolizidine alkaloids (venoocclusive disease)
- Aflatoxin
- Hypoglycin A (Jamaican vomiting sickness)

Drugs
- Acetaminophen
- Alcohol
- Halothane
- Tetracycline
- Erythromycin estolate
- Isoniazid
- Methotrexate
- Imipramine compounds
- Corticosteroids
- Azothiaprine
- 6-Mercaptopurine
- Others

Metabolic
- Reye's syndrome
- Tyrosinosis
- Wilson's disease
- Isovaleric acidemia
- Ornithine transarbamylase deficiency
- Others

Tumors
- Histiocytoses
- Lymphosarcoma
- Hodgkin's disease

From Lloyd-Still, J.: In Demers, L.M., and Shaw, L.M., editors: evaluation of liver function, Baltimore, 1978, Urban & Schwarzenburg, Inc.

ductal proliferation in all areas is usually present. Periportal fibrosis is also common. Acute hepatic failure in the pediatric patient may be caused by any of the conditions listed in the box above. Alterations of liver morphology are usually limited to focal necrosis and periportal fibrosis. Chemotherapy or parenteral nutrition is associated with periportal inflammation and reversible portal fibrosis.[5,6] In the adult patient, much the same alterations are present in acute disease states. Neoplastic disease of the liver may result in the production of oncofetal proteins, such as alpha-fetoprotein, as a result of the proliferation of relatively undifferentiated cells.[7] Because the tumor occupies space, some necrosis and fibrosis is generally present and may also result in biochemical changes. After partial hepatectomy, liver regeneration occurs. During this period of regrowth the regenerating tissue also produces fetal proteins. Care must be taken not to misinterpret the production of these fetal proteins as evidence of tumor growth.

Jaundice

Jaundice is a general condition that results from abnormal metabolism or retention of bilirubin, which causes a yellow discoloration of the skin, mucous membranes, and sclera. An organ-system approach to the classification of jaundice leads to identification of three principal types of jaundice: prehepatic, hepatic, and posthepatic. Prehepatic jaundice is the result of acute or chronic hemolytic anemias.

Hepatic jaundice includes disorders of bilirubin metabolism and transport defects such as Crigler-Najjar disease, Dubin-Johnson syndrome, and Gilbert's disease, as well as neonatal physiological jaundice and diseases resulting in hepatocellular injury or destruction.

The specific diseases of bilirubin metabolism each represent a defect in one of the steps in the hepatic processing of serum bilirubin (Fig. 24-5). Thus *Gilbert's disease* is caused by a defect in the transport of bilirubin from plasma albumin into the hepatocyte. Although levels of unconjugated bilirubin are elevated in this familial disorder, conjugated bilirubin is not. Impairment in the conjugation step by UDP-glucuronide caused by a deficiency in the enzyme UDP-glucuronyl transferase will also lead to a large increase in unconjugated bilirubin. When this enzyme deficiency is congenital, it is known as *Crigler-Najjar disease*. However, this impairment is most frequently encountered as *neonatal or physiological jaundice*.[8] This enzyme activity is one of the last liver functions to be activated in prenatal life, since unconjugated bilirubin formed in the fetus is cleared by the placenta. However, in premature births, infants are sometimes born without the enzyme activity

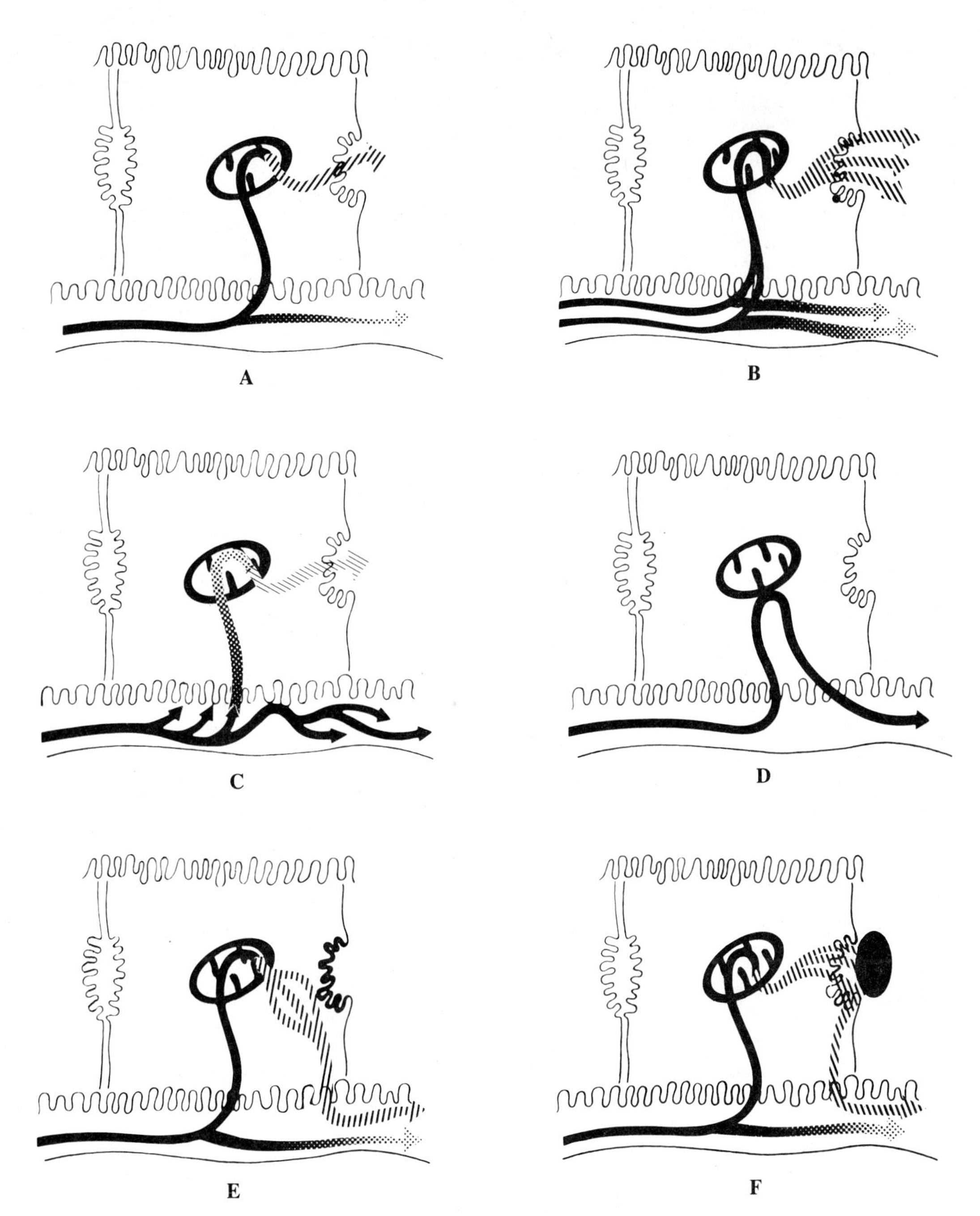

Fig. 24-5. Mechanisms of hyperbilirubinemia. (a) Normal bilirubin metabolism with hepatocyte uptake of unconjugated bilirubin, *dark arrow,* and microsomal conjugation and excretion of conjugated bilirubin, *striped arrow*. (b) Hemolytic jaundice in which increased bilirubin production results in increased excretion of conjugated bilirubin and a rise in excess (exceeding liver capacity) unconjugated bilirubin in blood. (c) Gilbert's disease in which decreased hepatic uptake results in large increase in blood levels of unconjugated bilirubin. (d) Physiological jaundice in which microsomal conjugating system is not functional, resulting in a large increase in unconjugated bilirubin. Congenital deficiency is the Crigler-Najjar syndrome. (e) Dubin-Johnson syndrome in which there is a biochemical defect preventing secretion of conjugated bilirubin, resulting in a backflow into blood. (f) Intrahepatic or extrahepatic obstruction in which a physical block prevents secretion of conjugated bilirubin. Hepatocellular disease results in a pattern similar to a combination of (c) and (d). (From Leevy, C.M., editor: Evaluation of liver function, ed. 2, Indianapolis, Ind., 1974, Lilly Research Laboratories.)

present. This leads to a rapid buildup of unconjugated bilirubin, which can be life threatening. The unconjugated bilirubin, being much more lipid than water soluble, readily passes into the brain and nerve cells and is deposited in the nuclei of these cells resulting in ''kernicterus'' (literally, 'nuclear jaundice'). Kernicterus often results in cell damage and death. Neonatal jaundice is a temporary condition until the enzyme glucuronyl transferase is produced by the newborn liver.[8] Treatment of this condition is to monitor the newborn's blood frequently for dangerously high levels of unconjugated bilirubin (about 200 mg/L). At this point the infant is treated with ultraviolet light to destroy the bilirubin as it passes thrugh the capillaries of the skin, or by exchange transfusion.[8]

The last step in hepatic processing of bilirubin is the postconjugation step of excretion of the bilirubin-glucuronide from the hepatic microsomes into the canaliculi. Impairment of this process, called the *Dubin-Johnson syndrome,* results in large increases in the conjugated bilirubin fraction of serum and a urine positive for bilirubin.

Hepatic jaundice also encompasses the disorders characterized by hepatocellular damage or necrosis, such as hepatitis and cirrhosis. Posthepatic jaundice is generally the result of biliary obstructive disease resulting from spasms or strictures of the biliary tract, ductal occlusion by stones, or compression by neoplastic disease. Since the hepatic functions of transport and conjugation of bilirubin are normal in these diseases, the major increase in serum bilirubin involves the conjugated fraction. Unable to be properly excreted by the liver in these disorders, the conjugated bilirubin fraction increases in serum and also results in the appearance of bilirubin in the urine. The laboratory findings for bilirubin and its metabolites in these diseases are summarized in Table 24-5.

Hepatitis

Hepatitis is a general term meaning 'inflammation of the liver' and is used to describe diseases resulting in hepatocellular damage. Hepatitis is usually caused by infections or toxic agents. Viral hepatitis is the most common cause of acute hepatocellular disease. Three types of hepatitis virus have been recognized: type A, type B, and non-A, non-B hepatitis. The clinical symptoms of these three types are similar, and the majority of hepatitis cases are anicteric because, despite the hepatic necrosis, the liver retains sufficient residual functional capacity to handle the hemoglobin load from normal hemoglobin turnover. Other viruses known to cause hepatitis include cytomegalovirus (CMV), coxsackievirus B, and herpes simplex.

Serum alanine aminotransferase (ALT) and aspartate aminotransferase (AST) rise rapidly during the early course of hepatitis because of hepatic necrosis. In patients who develop jaundice, the rise in the transaminases precedes the increase in bilirubin, which persists for 2 to 8 weeks. AST is usually more elevated than ALT. Serum alkaline phosphatase and gamma-glutamyl transferase (GGT) are elevated during the early cholestatic portion of the disease and remain elevated until the disease has resolved. Diagnosis of the type of viral hepatitis is accomplished by measurement of the specific hepatitis antigen during the prodromal phase of the illness. Specific antibodies are detectable for several weeks after the antigen is no longer detectable. Chronic hepatitis is generally the result of a persistence of hepatitis B infection and is associated with elevation of serum bilirubin and with minimally but persistently elevated serum transaminases (with ALT now greater than AST) caused by hepatic necrosis, and occasionally elevated alkaline phosphatase.

Drug-induced hepatic damage

The most common cause of drug-induced hepatic damage is chronic excessive ingestion of alcohol. Liver damage is expressed as fatty infiltration of the liver, alcoholic hepatitis, and cirrhosis. Laboratory findings associated with fatty infiltration are elevation of gamma-glutamyl transferase, a mild elevation of the transminases, an increase in the globulin and a decrease in the albumin fractions of the serum proteins, and an increase in bromosul-

Table 24-5. Concentrations and changes in concentration of bilirubin and its metabolites in normal persons and those with jaundice

Condition	Serum		Urine		Feces pigment
	Total bilirubin	Conjugated bilirubin	Bilirubin	Urobilinogen	
Normal	2-10 mg/L	0-2 mg/L	Negative	0.5-3.4 mg/day	Brown
Prehepatic jaundice	Increased	Normal	Negative	Increased	Normal
Hepatic jaundice					
Hepatocellular disease	Increased	Increased	Positive	Decreased (normal)	Light brown
Gilbert's disease	Increased	Normal	Negative	Decreased (normal)	Normal
Crigler-Najjar syndrome	Increased	Decreased	Negative	Decreased	Light brown
Dubin-Johnson syndrome	Increased	Increased	Positive	Decreased (normal)	Light brown
Posthepatic obstructive jaundice	Increased	Increased	Positive	Decreased	Light brown

fophthalein (BSP) or indocyanine green retention. The only difference between acute hepatic damage caused by alcohol ingestion and alcoholic hepatitis in terms of laboratory findings is that bilirubin elevations are more common and the transaminases are more elevated, and AST is generally higher than ALT because of mitochrondrial damage in alcoholic hepatitis. Alcoholic liver cirrhosis is generally not differentiated from the preceding types of damage by laboratory test results but requires liver biopsy. See Chapter 32 for a detailed review of alcoholic disease.

Other types of drugs that induce hepatic damage include such drugs as barbiturates, tricyclic antidepressants, and antiepileptics. These drugs are typified by an increase in serum GGT. Drug withdrawl permits liver regeneration and reversal of this damage. Chemotherapeutic drugs such as vincristine, vinblastine, actinomycin D, and 5-fluorouracil will typically result in an elevation of the serum transaminases and lactate dehydrogenase (LD) because of hepatic tissue damage and enzyme release.

Reye's syndrome

Reye's syndrome typically occurs in children between 2 and 13 years of age. The liver has fatty infiltration, and encephalopathy occurs because of accumulation of ammonia. Laboratory findings include an elevated blood ammonia, elevated serum transaminases, and prolonged prothrombin time because of hepatic necrosis and cholestasis.

Congenital deficiency syndromes with altered liver function

Porphyrias. The porphyrias include five types that represent congenital enzyme deficiencies in the synthesis of the heme moiety of hemoglobin and other heme proteins, such as myoglobin and the cytochromes. This leads to increased excretion of specific porphyrin metabolites, which vary with the enzyme defect. Chapter 33 summarizes porphyrin metabolism and discusses the enzyme defects. The five types of porphyria are acute intermittent porphyria, acquired congenital hepatic porphyria, erythropoietic porphyria, and erythropoietic protoporphyria. Diagnosis of the specific porphyria type depends on the demonstration of the accumulation of porphyrins in the serum, urine, or feces. Acute intermittent porphyria is characterized by accumulation of urinary porphobilinogen. Congenital cutaneous hepatic porphyria is characterized by chronic fecal excretion of uroporphyrins and coproporphyrins. During acute episodes, urinary excretion of delta-amino levulinic acid, porphobilinogen, uroporphyrin, and coproporphyrin is increased. Acquired cutaneous hepatic porphyria is differentiated from the congenital variety because only increased urinary excretion of uroporphyrin and coproporphyrin is demonstrated. The erythropoietic porphyrias are differentiated by the demonstration of urinary coproporphyrin I and uroporphyrin I in patients with congenital erythropoietic porphyria. Congenital protoporphyria is diagnosed by demonstration of increased fecal excretion of coproporphyrins and protoporphyrins with normal urinary content of porphyrins.

Wilson's disease. Wilson's disease is an autosomal recessive disorder of copper metabolism that results in jaundice followed by liver cirrhosis as a result of the accumulation of copper in the liver. Wilson's disease is characterized by a low serum concentration of ceruloplasmin, glycosuria, phosphaturia, aminoaciduria, and elevated urinary copper concentrations with excretion greater than 50 μg/day.

Hemochromatosis. Hemochromatosis is another genetic disorder of metal metabolism. In this disease, iron accumulates in the liver, and the resulting cirrhosis is similar to alcoholic cirrhosis. The laboratory tests of value are serum iron (elevated), iron-binding capacity, which is low, and ferritin, which is elevated. These are diagnostically useful if the patient has normal dietary iron intake.

Liver tumors and other hepatic disorders

Congenital hepatic fibrosis, hepatic cysts, and liver abscesses are generally best diagnosed by liver biopsy. However, nonspecific changes in liver enzymes and bromosulfophthalein (BSP) retention can accompany these disorders. Liver tumors frequently cause altered liver function as a result of tissue compression during tumor growth and infiltration. This results in an increase in serum alkaline phosphatase, 5′-nucleotidase, and especially GGT. BSP and indocyanine green retentions are frequently prolonged. The demonstration of an elevation of serum alphafetoprotein is diagnostic of hepatic tumor in the presence of an abnormal liver scan.

Alpha$_1$-antitrypsin deficiency

Alpha$_1$-antitypsin deficiency is an inborn error of protein metabolism that results in emphysema and liver cirrhosis. Numerous alpha$_1$-antitrypsin phenotypes have been identified carrying varying risks of disease.[9] The tissue damage may well be caused by hydrolytic damage to structural protein by trypsinlike enzymes that cannot be neutralized with alpha$_1$-antitrypsin. Since alpha$_1$-antitrypsin represents approximately 80% of the alpha$_1$ fraction of serum proteins on electrophoresis, severe deficiency is often diagnosed by the absence of this fraction in the electrophoretogram. Phenotyping of the various alpha$_1$-antitrypsin deficiencies is best done by isoelectric focusing.

Other genetic disorders that are associated with liver disease include the lipid-storage diseases and the glycogen-storage diseases. These are discussed in detail elsewhere (Chapter 44).

LIVER-FUNCTION TESTS

Liver function tests are generally used to identify liver disease in the absence of jaundice or when the jaundice is the result of hemolytic disease and there is a question of

Table 24-6. Change of serum analyte with disease

	Alkaline phosphatase	GGT	5′-Nucleotidase	AST	ALT	Bile acids	Albumin	NH_3
Acute hepatitis (viral, and so on)	↑	↑	↑	↑ ↑ ↑	↑ ↑	↑ ↑	N	N, ↑
Alcoholic (drug) hepatis	N, ↑	↑ ↑ ↑	↑	↑	↑	↑	N	N, ↑
Chronic hepatocellular	N, ↑	N, ↑	N, ↑	↑	↑	↑	↓	N, ↑
Cirrhosis	N, ↑	N, ↑	N, ↑	N, ↑	N, ↑	↑	↓	N, ↑
Reye's syndrome	N			↑	↑			↑ ↑
Hepatomas	↑ ↑	↑ ↑ ↑ ↑	↑ ↑	↑	↑		N, ↓	N
Choleostatic disease	↑	↑ ↑	↑ ↑ ↑	↑	↑	↑	N	N

GGT, Gamma-glutamyl transferase; *AST,* aspartate aminotransferase; *ALT,* alanine aminotransferase; NH_3, ammonia.
N, Normal; ↑, elevated; ↓, lowered.

the existence of complicating liver disease. Liver function is tested by injection of a dye intravascularly and observation of the retention of the dye in the serum. It is essential that a dye be chosen that is excreted into the bile rather than filtered by the kidney. Thus dyes such as bromosulfophthalein (BSP) and indocyanine green (IG) are acceptable, whereas phenolsulfonphthalein (PSP), which is excreted preferentially in the urine, is not. The use of BSP or IG retention in liver function tests depends on hepatic blood flow, biliary duct function, and liver cell function. Therefore these tests cannot distinguish between hepatocellular disease and obstructive liver disease. In the normal person the percentage of retention is less than 5% at 45 minutes. An increase in the retention is consistent with liver dysfunction resulting from hepatocellular disease, biliary obstructive disease, and space-filling lesions such as tumors in the liver.

CHANGE OF ANALYTE IN DISEASE

(Tables 24-5 and 24-6)

Enzymes

In this section, seven serum enzymes are described and their value in the differential diagnosis of the liver disease examined. The enzymes discussed are alkaline phosphatase, gamma-glutamyl transferase, aspartate aminotransferase, alanine aminotransferase, 5′-nucleotidase, leucine aminopeptidase, and lactate dehydrogenase. Numerous other enzymes have been identified that are useful in the evaluation of liver function. However, the seven enzymes listed are those generally used. Examples of the temporal

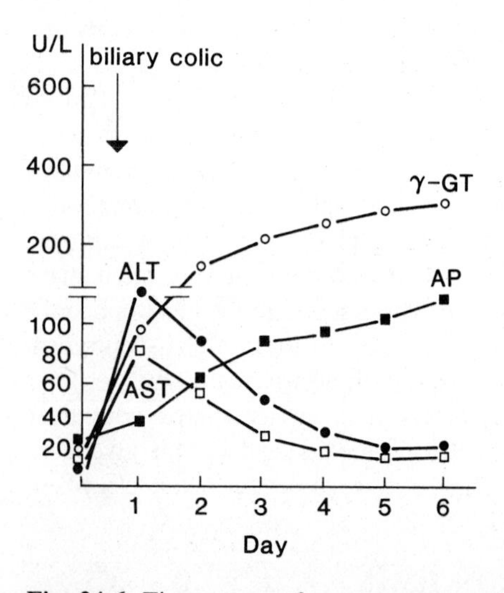

Fig. 24-6. Time course of serum enzyme activities in obstructive jaundice. (From Schmidt, E., and Schmidt, F.W.: Brief guide to practical enzyme diagnosis, 1977, Boehringer-Mannheim Diagnostics, Houston, Texas.)

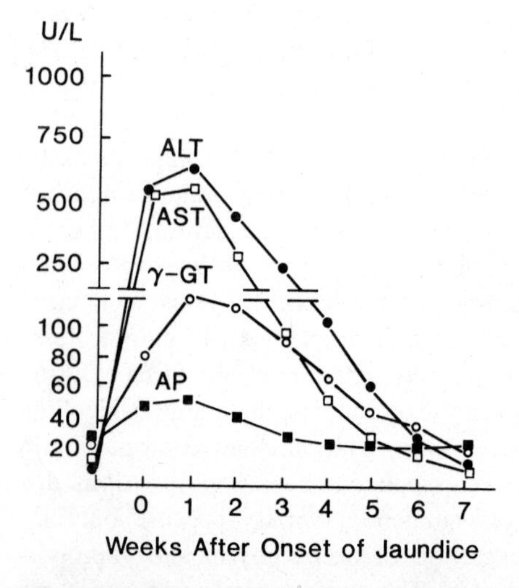

Fig. 24-7. Time course of serum enzyme activities in acute viral hepatitis. (From Schmidt, E., and Schmidt, F.W.: Brief guide to practical enzyme diagnosis, 1977, Boehringer-Mannheim Diagnostics, Houston, Texas.)

changes in activities for several of these enzymes in hepatic disease are shown in Figs. 24-6 to 24-8.

Alkaline phosphatase (EC 3.1.3.1) Alkaline phosphatase is actually a group of enzymes that hydrolyze monophosphate esters at an alkaline pH. The general reaction is as follows:

$$R{-}O{-}\overset{\overset{\textstyle O}{\|}}{\underset{\underset{\textstyle OH}{|}}{P}}{-}OH + HOH \rightarrow ROH + HO{-}\overset{\overset{\textstyle O}{\|}}{\underset{\underset{\textstyle OH}{|}}{P}}{-}OH$$

Their pH optima are generally about 10. The natural substrates for alkaline phosphatase are not known. The enzyme has been identified in most body tissues and is generally localized in the membranes of cells. Alkaline phosphatase activity is highest in the liver, bone, intestine, kidney, and placenta, and as many as eight different isozymes of alkaline phosphatase have been identified. Since alkaline phosphatase normally contains significant amounts of sialic acid, it is possible that some of these multiple enzyme forms are the result of different degrees of sialation. Nonetheless, the enzyme produced by the placenta is known to have a protein composition different from the other enzyme forms.

Measurement of serum alkaline phosphatase is useful in differentiating hepatobiliary disease from osteogenic bone disease, since the enzyme forms present in serum originate primarily from these tissues. Alkaline phosphatase activity increases greatly (10 times) as a result of membrane-localized enzyme synthesis after extrahepatobiliary obstruction such as cholelithiasis or gallstones. Intrahepatic biliary obstruction is also accompanied by an increased serum activity, but the degree of increase is smaller (two to three times). Liver disease resulting in parenchymal cell necrosis does not elevate serum alkaline phosphatase unless associated with damage to the canaliculi or biliary stasis.

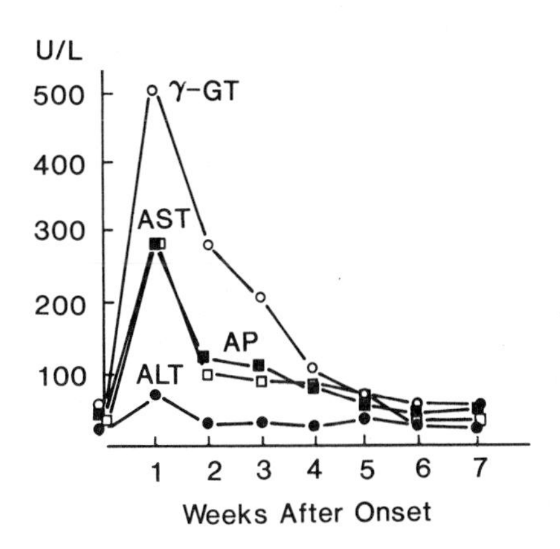

Fig. 24-8. Time course of serum enzyme activities in acute alcoholic hepatitis. (From Schmidt, E., and Schmidt, F.W.: Brief guide to practical enzyme diagnosis, 1977, Boehringer-Mannheim Diagnostics, Houston, Texas.)

Interpretation of serum alkaline phosphatase measurements is complicated by the fact that enzyme activity from sources other than liver is sometimes released in the absence of liver disease. The most common of these disorders causing elevation of alkaline phosphatase are bone diseases, such as Paget's disease, rickets, and osteomalacia. Serum alkaline phosphatase activity also rises during puberty because of rapid bone growth. This rise occurs in the absence of any organic disease process. Serum alkaline phosphatase is also elevated during the third trimester of pregnancy because alkaline phosphatase is released from the placenta. Differentiation of these sources of alkaline phosphatase is best done by isoenzyme analysis combining electrophoresis and heat-stability analysis. The isoenzyme from the placenta is stable at 65° C for up to 30 minutes.

Gamma-glutamyl transferase (EC 2.3.2.2). Gamma-glutamyl transferase (GGT) is a membrane-localized enzyme that plays a major role in glutathione metabolism and in resorption of amino acids from the glomerular filtrate. It is also important in the absorption of amino acids from the intestinal lumen. Glutathione (gamma-glutamyl cysteinylglycine) in the presence of GGT and an amino acid or peptide transfers glutamate to the amino acid by means of a peptide bond on the gamma-carboxylic acid, thereby forming cysteinylglycine and the corresponding gamma-glutamyl peptide:

$$\begin{array}{c} \text{Glutathione} \\ + \\ \text{L-amino acid} \end{array} \xrightarrow{\text{GGT}} \begin{array}{c} \text{Gamma-glutamyl peptide} \\ + \\ \text{Cysteinylglycine} \end{array}$$

Although the GGT activity is highest in renal tissue, serum GGT is generally elevated as a result of liver disease. Serum GGT is elevated earlier than other liver enzymes in diseases such as acute cholecystitis, acute pancreatitis, acute and subacute liver necrosis, and neoplasms of multiple sites at which liver metastases are present. GGT has been advocated as a screening test for alcohol abuse.[10] Since GGT is a hepatic microsomal enzyme, chronic ingestion of alcohol or drugs such as barbiturates, tricyclic antidepressants, or anticonvulsants induces microsomal enzyme production. These drug-induced elevations precede any change in other liver enzymes, and if drug ingestion is stopped at this point, the liver changes are generally reversible. One advantage of serum GGT compared to alkaline phosphatase is that serum GGT values fall in the less diseased reference range in Paget's disease, rickets, and osteomalacia and in children and pregnant women without liver disease. Thus GGT permits differentiation of liver disease in conditions in which serum alkaline phosphatase is elevated. Since the prostate contains significant GGT activity, serum activity is higher in healthy men than in women. Serum GGT is most useful in the diagnosis of cholestasis caused by chronic alcohol or drug ingestion, mechanical or viral cholestasis, liver metastases; in bone disorders in which alkaline phosphatase is elevated but

GGT is normal; and in skeletal muscle disorders in which the transaminase AST is elevated but GGT is normal.

5′-Nucleotidase (EC 3.1.3.5). 5′-Nucleotidase (NTP) is a microsomal and cell membrane–localized enzyme that catalyses the hydrolysis of nucleoside-5′-phosphate esters. The serum enzyme has an apparent pH optimum of 7.5:

$$\text{Nucleoside-5}'\text{-monophosphate} \xrightarrow{\text{NTP}} \text{Nucleoside} + P_i$$

NTP is similar to GGT in its diagnostic values. Serum NTP is increased in hepatobiliary diseases such as gallstone obstruction of the bile duct, cholestasis, biliary cirrhosis, or obstructive disease caused by neoplastic growth. Serum NTP is not generally elevated in drug-induced liver damage. Therefore it is useful in conjunction with GGT in following the course of chemotherapy for liver neoplasms.[7] Since NTP is not elevated in bone disease, it, like GGT, is useful in differentiating hepatic causes of alkaline phosphatase increase from other causes such as bone disease, pregnancy, and normal childhood growth.

Leucine aminopeptidase (EC 3.4.11.1). Leucine aminopeptidase (LAP) is another liver enzyme that has relatively broad specificity. It was initially considered that serum LAP was an effective enzyme for the differentiation of hepatic cell damage, in which LAP is elevated, from biliary obstructive disease, in which LAP activity is normal. However, this has not been substantiated by further studies, and for the most part serum LAP values do not provide information that is not available when other liver enzyme analyses are used.[11] For example, serum LAP activity is increased in infants with jaundice caused by hepatocellular damage but is normal in patients with hemolytic or physiological jaundice. LAP activity is also elevated in viral hepatitis, cirrhosis, hepatic neoplasms, acute pancreatitis, liver abscess, and extrahepatic biliary obstruction. Since LAP is found in most tissues, it is not surprising that slight elevations of serum LAP activity are seen in nephrotic syndrome, pregnancy, osteitis deformans, and congestive heart failure.

Lactate dehydrogenase (EC 1.1.1.27). Lactate dehydrogenase (LD) occurs in many tissues. LD catalyzes the interconversion of pyruvate and lactate:

$$\text{NAD} + \text{Lactate} \xrightarrow{\text{LD}} \text{Pyruvate} + \text{NADH}^+$$

The activity is highest in the kidney and heart and lowest in the lung and serum. LD is localized in the cytoplasm of cells and thus is extruded into the serum when cells are damaged or necrotic.

When only a specific organ, such as the liver, is known to be involved, the measurement of total LD can be useful. Total LD is increased in viral or toxic hepatitis, extrahepatic biliary obstruction, acute necrosis of the liver, and cirrhosis of the liver. However, in conditions in which multiple organs are involved, the measurement of total LD is less useful than the measurement of LD isoenzymes. LD_5 and LD_4 account for the primary activity of liver LD, whereas LD_1 and LD_2 cause the predominant activities of heart and kidney. Since red blood cells also contain significant LD_1, analysis of hemolyzed serum specimens should be avoided. In the conditions listed above, LD electrophoresis indicates that the increased total LD is caused by the release of LD_4 and LD_5 into the serum.

Aspartate aminotransferase (EC 2.6.1.1) and alanine aminotransferase (EC 2.6.1.2). The transaminases aspartate aminotransferase (AST) and alanine aminotransferase (ALT) catalyse the conversion of aspartate and alanine to oxaloacetate and pyruvate respectively:

$$\text{L-Asparate} + \alpha\text{-Ketoacid} \xrightarrow{\text{AST}} \text{Oxaloacetate} + \text{L-Amino acid}$$

$$\text{L-Alanine} + \alpha\text{-Ketoacid} \xrightarrow{\text{ALT}} \text{Pyruvate} + \text{L-Amino acid}$$

ALT is most prevalent in liver, whereas AST is present in heart, skeletal muscle, and liver to nearly the same extent. Serum activity of AST and ALT increases rapidly during onset of viral jaundice and remains elevated for 1 to 2 weeks. In toxic hepatitis, ALT and AST are also elevated, but LD is elevated to an even greater extent as a result of hepatic cell necrosis. Patients with chronic active hepatitis also exhibit increased ASL and ALT.

Acute liver necrosis is accompanied by significant increases in the activity of both ALT and AST. The increase in ALT activity is usually greater than the increase in AST activity. In cirrhotic liver disease, serum transaminase activities are generally not elevated above 300 units/L regardless of the cause of the cirrhotic disease. The elevations of serum ALT and AST seen in Reye's syndrome are directly attributable to hepatic damage, and the increase in ALT is generally greater than the increase in AST. Neoplastic disease also elevates serum transaminase activity.

Measurement of ALT and AST is valuable in the diagnosis of liver disease. However, these laboratory tests are best used with other enzyme assays such as LD and creatinine kinase, as well as other measures of liver and kidney function, such as blood urea, creatinine, ammonia, and bilirubin.[12] This is important in establishing a diagnosis because ALT and AST are present in tissues other than liver, and serum activity of these enzymes can reflect organic disease in tissues other than liver. ALT and AST serum activities are elevated in myocardial infarction, renal infarction, progressive muscular dystrophy, and numerous diseases that only secondarily affect the liver, such as Gaucher's disease, Niemann-Pick disease, infectious mononucleosis, myelocytic leukemia, diabetic ketoacidosis, and hyperthyroidism.

Other hepatic analytes

Bilirubin. Serum bilirubin analysis is helpful in the differentiation of the cause of jaundice. *Prehepatic jaundice* results in a large increase in unconjugated bilirubin because of the increased release and metabolism of hemoglobin after hemolysis. No increase or only a slight increase in conjugated bilirubin is observed because the transport of bilirubin into the liver and the formation of the glucuronide conjugate becomes rate limiting. Additionally, because of the increased levels of conjugated bilirubin excreted by the liver, urinary urobilinogen and fecal urobilin concentrations are elevated, but urinary bilirubin (which is only the freely soluble, conjugated form) is absent. In contrast, *posthepatic obstructive jaundice* is characterized by large increases in serum-conjugated bilirubin. The accumulation of bilirubin is the result of decreased biliary excretion after the conjugation of bilirubin in the liver rather than a result of an increased bilirubin load caused by hemolysis. Excretion of bilirubin metabolites is low, and urinary bilirubin can usually be demonstrated. *Hepatic jaundice* presents an intermediate pattern wherein both conjugated and unconjugated serum bilirubin are increased and conjugated bilirubin is present in the urine. However, the fecal concentration of urobilin is generally decreased.

Cholesterol. Serum cholesterol comprises two forms, free cholesterol and esterified cholesterol. About 70% of the total cholesterol is esterified with fatty acids. Since this esterification takes place in the liver, intrahepatic disease or biliary obstruction is characterized by an increase in the free cholesterol and occasionally a shift in the serum free fatty acid profile, though the total cholesterol usually remains unchanged. In chronic disease associated with cell destruction, the total cholesterol may fall below the reference range.

Bile acids. Bile acid secretion and production is altered in liver disease.[3,12] In the healthy adult, serum contains 1 to 2 μg/mL of bile acids. In hepatobiliary disease, serum bile acid concentrations may rise as much as a thousandfold. Diseases that also can cause a significant rise include hepatitis, cirrhosis, drug-induced liver disease, and hepatoma. Serum bile acid concentrations are normal in Gilbert's disease, hemochromatosis, and polycystic liver disease. Measurement of serum bile acids is useful in the diagnosis of minimal liver dysfunction when other biochemical parameters are still unchanged.

Triglycerides. Serum triglycerides should be measured in a fasting sample, and increases are relatively nonspecific.[12] For example, liver dysfunction resulting from hepatitis, extrahepatic biliary obstruction, or cirrhosis is associated with an increase, but so are such disorders as acute pancreatitis, myocardial infarction, renal failure, gout, pernicious anemia, and diabetes mellitus. Free fatty acids exhibit a similar nonspecificity. They are decreased in chronic hepatitis and as well in chronic renal failure and cystic fibrosis. Serum free fatty acid concentrations are elevated in Reye's syndrome, hepatic encephalopathy, and chronic active hepatitis but also in myocardial infarction, acute renal failure, hyperthyroidism, and pheochromocytoma.

Serum proteins in evaluation of liver function. A healthy functioning liver is required for the synthesis of the serum proteins, except for the gamma globulins. The liver has the ability to increase protein output approximately twofold during diseases associated with protein loss. Therefore it is not surprising that total protein measurements are not altered until an extensive impairment of liver function has occurred.

Albumin is decreased in chronic liver disease and is generally accompanied by an increase in the beta and gamma globulins as a result of production of IgG and IgM in chronic active hepatitis and of IgM and IgA in biliary or alcholic cirrhosis, respectively. Immunoelectrophoresis may facilitate the identification of these subclasses of gamma globulin. However, a decrease in serum albumin is not specific for liver disease, since decreases are also seen in malabsorption, malnutrition, renal disease, alcoholism, and malignant diseases.

The $alpha_1$ fraction of the serum protein globulin is decreased in chronic liver disease, and when this fraction is absent or nearly so, it suggests that $alpha_1$-antitrypsin deficiency may be the cause of the liver disease. Serum $alpha_2$ globulin and beta globulin are increased in obstructive jaundice. Lipoprotein metabolism is altered in liver disease. The increase in $alpha_2$ globulin and beta globulin in obstructive jaundice is largely associated with interferences with normal lipoprotein metabolism. Thus it is important that phenotyping of a lipid disorder not be done in the presence of liver disease. As a result of the increased serum cholesterol associated with liver disease, the use of high density lipoprotein cholesterol to assess risk of coronary heart disease is obviated in patients with alcoholic liver disease, biliary obstruction, and acute liver necrosis.

Coagulation factors are produced by the liver and can decrease significantly in the presence of liver disease. Plasma fibrinogen is normally present in a concentration of 2 to 4 g/L, whereas in serum the concentration is negligible as a result of the clotting process. In liver disease, a decrease in plasma fibrinogen is usually an indication of severe liver disease and is associated with decreased concentrations of other clotting factors, most notably prothrombin. Since prothrombin synthesis occurs in the liver and requires the fat-soluble vitamin K, prothrombin time may be increased in biliary obstructive disease, liver cirrhosis or necrosis, hepatic failure, Reye's syndrome, liver abscess, vitamin K deficiency, and hepatitis. The response of the prothrombin time to exogenous vitamin K is of use in the differentiation of intrahepatic disease from extrahepatic obstructive disease with decreased absorption of vitamin K.

Urea and ammonia in evaluation of liver function. Blood ammonia is increased in infants relative to adult concentrations because of the continuing development of hepatic circulation after birth. Hyperammonemia results infrequently from congenital defects of the urea cycle. The most common of these inborn errors of metabolism is ornithine transcarbamylase deficiency. A much more frequent cause of hyperammonemia in infants is hyperalimentation.[4] Reye's syndrome is frequently diagnosed by an elevated blood ammonia in the absence of any other demonstrable cause.

Adult patients exhibit elevated blood ammonia in the terminal stages of liver cirrhosis, hepatic failure, and acute and subacute liver necrosis. Urinary ammonia excretion is increased in acidosis and decreased in alkalosis, since ammonia salt formation is a significant mechanism for excretion of excess protons. Damage to the renal distal tubules, as occurs in renal failure, glomerulonephritis, hypercorticoidism, and Addison's disease, results in decreased ammonia excretion.

Since urea is synthesized in the liver, liver disease without renal impairment results in a low serum urea nitrogen, though the urea/creatinine ratio may remain normal.[13] An elevated serum urea nitrogen does not necessarily imply renal damage, since dehydration may result in a urea nitrogen as high as 600 mg/L, and infants receiving high-protein formula may exhibit a urea nitrogen level of 250 to 300 mg/L. Naturally, renal disease such as acute glomerulonephritis, chronic nephritis, polycystic kidney, and renal necrosis result in an elevated urea nitrogen.

REFERENCES

1. Johnson, T.R.: Development of the liver. In Johnson, T.R., Moore, W.M., and Jefferies, S.E., editors: Children are different: developmental physiology, ed. 2, Columbus, Ohio, 1978, Ross Laboratories, pp. 155-160.
2. Natelson, S., and Sherwin, J.E.: Proposed mechanism for urea nitrogen reutilization: relationship between urea and proposed guanidine cycles, Clin. Chem. **25:**1343-1344, 1979.
3. Demers, L.M.: Serum bile acids in health and in hepatobiliary disease. In Demers, L.M., and Shaw, L.M., editors: Evaluation of liver function, Baltimore, 1978, Urban & Schwarzenberg, Inc., pp. 33-50.
4. Lloyd-Still, J.D.: Disorders of the liver in childhood. In Demers, L.M., and Shaw, L.M., editors: Evaluation of liver function, Baltimore, 1978, Urban & Schwarzenberg, Inc., pp. 171-200.
5. Dahms, B.B., and Halpin, T.C., Jr.: Serial liver biopsies in parenteral nutrition-associated cholestasis of early infancy, Gastroenterology. **81:**136-144, 1981.
6. Benjamin, D.R.: Hepatobiliary dysfunction in infants and children associated with long-term total parenteral nutrition: a clinico-pathologic study, Am. J. Clin. Pathol. **76:**276-283, 1971.
7. Deeble, T.J., and Goldberg, D.M.: Assessment of biochemical tests for bone and liver involvement in malignant lymphoma patients, Cancer **45:**1451-1457, 1980.
8. Stoner, J.W.: Neonatal jaundice, Am. Family Physician **24:**226-232, 1981.
9. Jeppson, J.O., Franzen, B., Cox, D.W., et al.: Typing of genetic variants of alpha-1-antitrypsin by electrofocusing, Clin. Chem. **28:**219-225, 1982.
10. Shaw, L.M.: The GGT assay in chronic alcohol consumption, Lab. Management **20:**56-63, 1982.
11. Ellis, G., Goldberg, D.M., Spooner, R.J., and Ward, A.M.: Serum enzyme tests in diseases of the liver and biliary tree, Am. J. Clin. Pathol. **70:**248-258, 1978.
12. Friedman, R.B., Anderson, R.E., Entine, S.M., and Hirschberg, S.B.: Effects of diseases on clinical laboratory tests, Clin. Chem. **26:**1D-476D, 1980.
13. Nanji, A.A., and Blank, D.: The serum urea nitrogen/creatinine ratio and liver disease, Clin. Chem. **28:**1398-1399, 1982.

BIBLIOGRAPHY

Halsted, J.A.: The laboratory in clinical chemistry: interpretation and application, Philadelphia, 1976, W.B. Saunders Co.

Meites, S., editor: Pediatric clinical chemistry: a survey of reference (normal) values, methods, and instrumentation, with commentary. ed. 2, Washington, D.C., 1981, American Association of Clinical Chemists.

Natelson, S.: Principles of applied clinical chemistry, vol. 1, New York, 1979, Plenum Press.

White, A., Handler, P., and Smith, E.: Principles of biochemistry, ed. 5, New York, 1973, McGraw-Hill Book Co.

Chapter 25 Bone disease

Reginald Tsang
Harold Marder

cholecalciferol The parent vitamin D compound.
calcidiol Cholecalciferol hydroxylated at the carbon-25 position in the liver.
calcitriol The active vitamin D metabolite; cholecalciferol hydroxylated at both the carbon-1 and carbon-25 positions.
cortical bone Dense compact bone that provides structural support.
diaphysis Shaft of a long bone.
epiphysis End of a long bone.
metaphysis Region in which diaphysis and epiphysis converge.
osteoblasts Cells that synthesize bone matrix.
osteoclasts Cells that resorb bone.
osteocytes Mature bone cells that have limited function and are encased in bone matrix, the composition of which they help to maintain.
osteoid Bone matrix.
osteomalacia Bone containing normal amounts of osteoid but deficient amounts of mineral.
osteopenia The roentgenographic appearance of subnormally mineralized bone.
osteoporosis A generalized reduction in bone mass involving both mineral and osteoid.
rickets Osteomalacia in childhood.
trabecular bone Interlacing delicate spicules of bone that are predominantly involved in mineral homeostasis

BONE STRUCTURE AND FUNCTION

The complexity of the skeletal system has been only recently appreciated. The association of bone with fossils and grave sites contributed to an old hypothesis that bone merely provided static structural support for the body's dynamic organ systems. This theory was suported by the histological appearance of bone. Over 90% of the cells in bone are encased in calcified tissue and separated by great distance from a vascular supply. This gives the cells the appearance of inactivity. However, it is now clear that the skeletal system is a dynamic organ. Albright introduced the concept of constant bone remodeling[1]. He proposed that there were cells specifically committed to bone resorption (osteoclasts) and bone formation (osteoblasts), but the majority of bone cells (osteocytes) were believed to be dormant. Major revisions in Albright's original hypotheses have produced the modern theory of skeletal system function—that bone has two interdependent roles: provision of support and maintenance of mineral homeostasis. Both functions are successfully achieved by continuous bone remodeling. Disturbances in the balance and nature of bone formation and resorption produce the common bone diseases.

This chapter is a brief discussion of the structure and function of bone and an examination in some detail of the hormonal balance that controls bone metabolism. In addition, some common consequences of alterations in this balance are considered.

Bone structure

Macroscopically, the major bones are classified as long bones or flat bones[2]. Long bones are confined to the limbs and consist of a shaft (*diaphysis*), two ends (*epiphyses*), and a region in which the two converge (*metaphysis*) (Fig. 25-1, *A*). Seen in cross section, the diaphysis is lined by dense compact (*cortical*) bone, whereas the metaphysis contains interlacing bony spicules that resemble the structure of a sponge (*trabecular* or *cancellous bone*) Fig. 25-

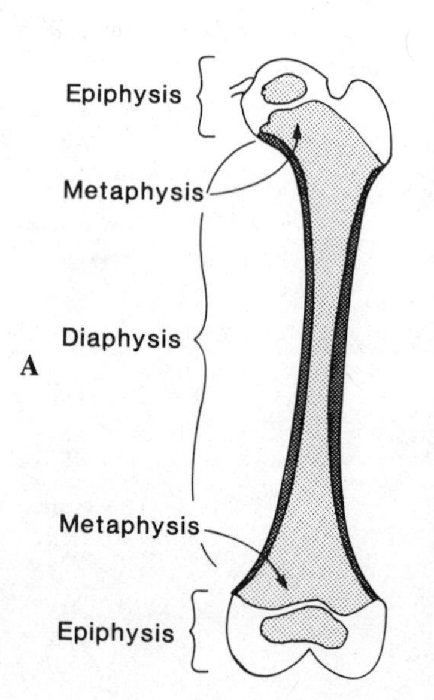

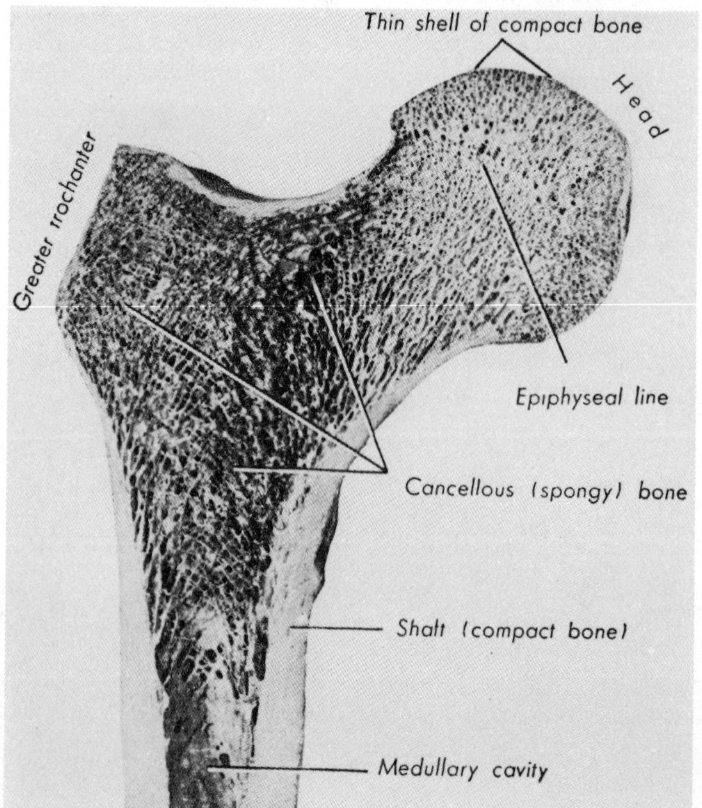

Fig. 25-1. A, Parts of a long bone. **B,** Long bone in cross section; note predominance of trabecular, cancellous bone in diaphysis. (From Copenhaver, W.M., Kelly, D.E., and Wood, R.L., editors: Bailey's textbook of histology, ed. 17, Baltimore, 1978, The Williams & Wilkins Co.)

1, *B*). Flat bones, typified by the bones of the skull, consist of two thin layers of cortical bone that enclose a layer of trabecular bone.

Trabecular and cortical bone differ in microscopic and macroscopic appearance (Fig. 25-2). Their structures offer a clue to their specialized functions. The spicules of trabecular bone provide a large surface area for bone synthesis and resorption and provide a reservoir of minerals for the maintenance of mineral homeostasis. As bone matures, individual trabeculas grow circumferentially, and the relative proportion of bone matrix to bone surface is increased, resulting in the formation of dense cortical bone, which provides the strength needed for structural support (Fig. 25-3).

Bone contains three major types of mature cells.[3] *Osteoblasts,* found along surfaces of both coritcal and trabecular bone, synthesize bone matrix. As osteoblasts become imbedded in bone matrix, they differentiate into mature *osteocytes*. Osteocytes synthesize small amounts of matrix continuously to maintain bone integrity, and they are able to resorb bone (osteocytic osteolysis) in exceptional circumstances when normal mineral homeostasis is altered. *Osteoclasts* contain enzymes that demineralize and digest bone matrix. The moth-eaten appearance of bone resorption is evident in histological sections of areas in which osteoclasts are numerous.

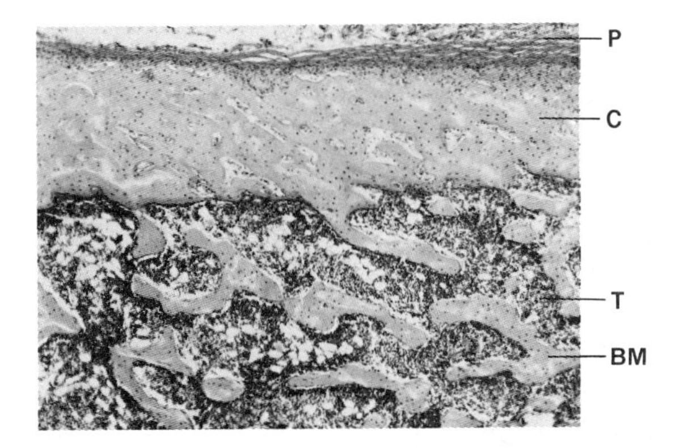

Fig. 25-2. Histological preparation of a rat tibia illustrating microscopic appearance of cortical, *C,* and trabecular, *T,* bone. Bone marrow cells, *BM,* surround spicules of trabecular bone. *P,* Periosteum. (40 ×; cyanuric chloride stain; courtesy Larry Flora, Procter & Gamble Co., Cincinnati, Ohio.)

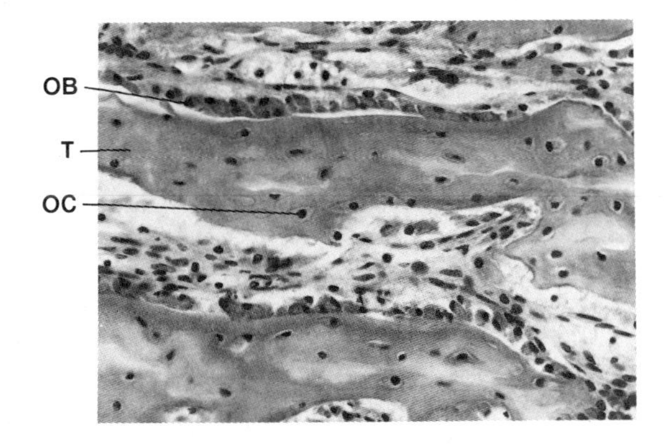

Fig. 25-3. Trabecular bone, *T,* is converted to cortical bone by circumferential formation of new bone. *OB,* Osteoblast; *OC,* osteoclast. (From Ham, A.W., and Cormack, D.H.: Histology, ed. 8, New York, 1979, J.B. Lippincott Co.)

Only a small portion of bone is cellular; calcified matrix predominates. This matrix is primarily composed of collagen fibers. Approximately one fourth of the amino acids present in collagen are either proline or hydroxyproline, neither of which is present to any great extent in other tissues. When collagen is metabolized, hydroxyproline-containing oligopeptides are excreted in the urine, and the amount present correlates with the amount of bone turnover. The mineral elements of bone consist mostly of crystals of calcium and phosphate arranged either amorphously or as hydroxyapatite, $Ca_{10}(PO_4)_6(OH)_2$. A wide range of other elements may be present, including sodium, magnesium, copper, zinc, lead, and fluoride.

Bone function

Bone contains 99% of the body's calcium, 85% of the phosphorus, and 66% of the magnesium.[4] Although only a small percentage of these ions may be rapidly mobilized, they provide the initial mineral source when serum concentrations fall. The large surface area and excellent blood supply of trabecular bone permit a quick response to perturbations in plasma mineral concentrations. In contrast, the abundant calcified matrix of cortical bone provides the strength to support body weight. Despite this segregation of structure and function, disturbances in both often coexist. Examples include vitamin D deficiency, which causes both hypocalcemia and easily fractured bone, and immobilization, which causes bone resorption, osteoporosis, and hypercalcemia. Under normal circumstances, such disturbances do not occur because, in bone remodeling that occurs throughout the body, bone formation and bone resorption are ''coupled'' resulting in equal amounts of bone formation and resorption.[5] Most bone diseases result from alterations in coupling, either anew or secondary to hormonal imbalance, which produces either excessive bone formation or excessive resorption.

Bone remodeling results from both nonhormonal and hormonal stimuli. Bone possesses piezoelectric (pressure-electricity) properties, which permit remodeling in response to mechanical stress. This produces extra strength where structural demand is great.[6] When subjected to mechanical pressure, bone produces a change in electrical potential, which causes current flow. This flow creates ion fluxes, which, in turn, increase either osteoblastic or osteoclastic activity. Interestingly, parathyroid hormone also produces depolarization of osteoclast cell membranes. From these findings, it has been suggested that the common pathway for mechanical and hormonal control of bone remodeling is membrane depolarization.[7]

BIOCHEMISTRY AND PHYSIOLOGY

Vitamin D

Attention has been focused on vitamin D since its discovery, in 1925, led to the elimination of the widespread problem of nutritional rickets. This substance was considered a vitamin because rachitic patients were cured with oral supplementation of vitamin D. Major advances in the understanding of the mechanism of action of vitamin D have led to the current concept that it is a hormone.[8] The major source of vitamin D is not the diet, but its production in skin after exposure to sunlight.[9] It is then transported in the bloodstream to the liver and kidneys for activation. It subsequently localizes at sites of activity in intestine and bone because of the presence of specific cellular receptors in these organs. Finally, as in other hormone systems, the plasma level of activated vitamin D is rigidly controlled by feedback regulation. Although many questions remain unanswered, vitamin D can now be regarded as one of the three major hormones controlling calcium and phosphorus homeostasis and bone mineralization. The references at the end of this chapter list some recent reviews of this subject.[8,10,11]

Biochemistry and metabolism. Vitamin D_3, subsequently referred to as cholecalciferol, is produced in skin through the conversion of 7-dehydrocholesterol by ultraviolet light[8,9] (Fig. 25-4). Cholecalciferol is not the active hormone. When administered intravenously to rats, there is a 12-hour lag phase before intestinal calcium transport increases.[12] This observation led to the search for active metabolites of cholecalciferol (Fig. 25-5). In 1968, calcidiol (25-hydroxycholecalciferol) was purified.[13] When administered to rats, the lag phase was reduced to 4 hours. Subsequently, calcitriol (1,25-dihydroxycholecalciferol) was identified and was shown to act more rapidly and effectively than calcidiol in promoting both intestinal cal-

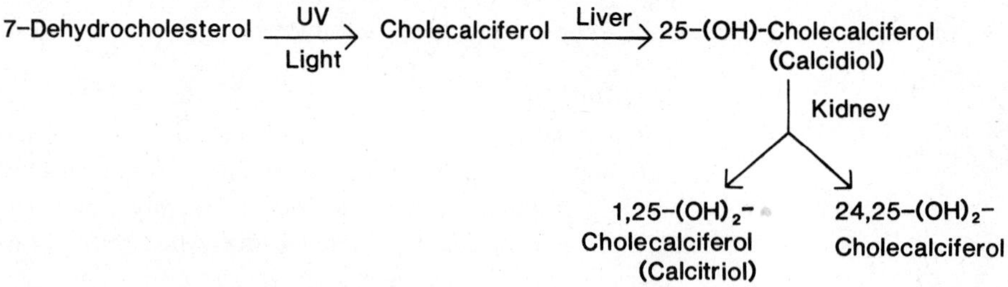

7-Dehydrocholesterol —UV Light→ Cholecalciferol —Liver→ 25-(OH)-Cholecalciferol (Calcidiol)

Kidney

1,25-$(OH)_2$-Cholecalciferol (Calcitriol)

24,25-$(OH)_2$-Cholecalciferol

Fig. 25-4. Conversion of 7-dehydrocholesterol to activated vitamin D by ultraviolet (UV) light and by liver and kidney.

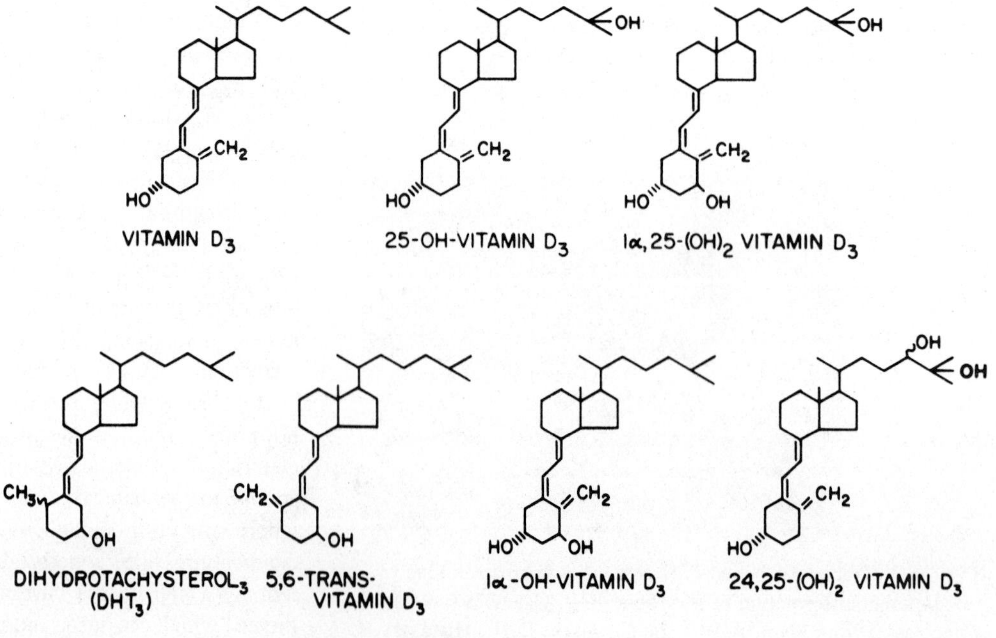

VITAMIN D_3 — 25-OH-VITAMIN D_3 — 1α,25-$(OH)_2$ VITAMIN D_3

DIHYDROTACHYSTEROL$_3$ (DHT_3) — 5,6-TRANS-VITAMIN D_3 — 1α-OH-VITAMIN D_3 — 24,25-$(OH)_2$ VITAMIN D_3

Fig. 25-5. Some common metabolites of cholecalciferol.

cium transport and bone mineral release.[14,15] This is the active metabolite of cholecalciferol.

After cholecalciferol is produced in the skin, it is transported to the liver, in which hydroxylation occurs at the C-25 position (Fig. 25-4) to produce calcidiol. Calcidiol is the most abundant vitamin D compound in plasma, with normal concentrations of approximately 30 mg/mL. Production is not under strict metabolic control, as the plasma concentration varies with dietary intake [16] and sunlight exposure. [17] Calcidiol is then transported to the kidney, in which hydroxylation occurs at the 1α position to produce calcitriol (Fig. 25-4). Plasma calcitriol concentrations are relatively low (approximately 30 pg/mL). Although normal plasma concentrations vary with age,[18] they are probably under strict feedback control.[19] The details of this control are discussed below. If plasma calcitriol concentrations are sufficient, calcidiol is hydroxylated in the kidney at the C-24 position to yield 24,25-dihydroxycholecalciferol (Fig. 25-4). Attempts to discover a role for this 24-hydroxylated metabolite are in progress. It is currently regarded by most investigators as a waste product of vitamin D metabolism. The kidney is generally held to be the only site of calcitriol production. From recent information, however, it has been suggested that extrarenal synthetic sites exist. Elevated plasma calcitriol concentrations have been found in an anephric patient with sarcoidosis.[20] Furthermore, osteogenic sarcoma cells, which metabolically resemble osteoblasts, can synthesize calcitriol in in vitro culture systems.[21] Finally, the placenta in the pregnant woman appears to be a source of calcitriol production.[22]

Other compounds of interest (Fig. 25-5) include ergosterol (vitamin D_2), a plant sterol that is the substance commonly used as a food additive and metabolically resembles cholecalciferol, and dihydrotachysterol and 1α-hydroxyvitamin D_3, synthetic substances that are activated by 25-hydroxylation alone.

Cholecalciferol and its metabolites pass through the circulation attached to vitamin D–binding protein.[24] Certain tissues contain an intracellular protein that serves as a specific receptor for calcitriol. This protein, located in the cytosol of the kidney, intestine, bone, and selected other tissues, functions like other established cellular steroid hormone receptors. It facilitates entry of calcitriol into cells and transports the calcitriol to the cell nucleus, where the hormone, in theory, directs protein synthesis to achieve its desired effect. For example, intestinal cells produce a calcium-binding protein in response to calcitriol stimulation, which is believed to be involved in intestinal calcium absorption.

Mechanisms of action. The function of calcitriol is discussed in many review articles[8,10-12,23,24] (see following box). The three major target organs of calcitriol are intestine, bone, and kidney. Calcitriol facilitates both calcium and phosphate absorption in the intestine. Phosphate transport accompanies calcium transport, but it is also increased by an unknown, calcium-independent mechanism. Calcitriol works cooperatively with parathyroid hormone to increase bone resorption by increasing osteoclast activity. This may be considered paradoxical, since vitamin D is believed to enhance bone mineralization. However, the net effect of the action of calcitriol at bone and intestine is to increase available blood calcium and phosphorus concentrations, which subsequently facilitate mineralization of newly formed bone matrix. Calcitriol also stimulates alkaline phosphatase activity in osteogenic sarcoma cells in vitro.[25] As suggested in a previous section of this chapter, alkaline phosphatase may promote bone mineralization directly. Calcitriol increases the renal reabsorption of both calcium and phosphorus, but since 99% of filtered calcium is normally reabsorbed, the overall effect of alterations in plasma calcitriol concentrations on renal calcium reabsorption is small.

The target organs for calcitriol

Intestine—increased calcium and phosphorus absorption
Bone—enhanced parathyroid hormone–induced bone resorption
Kidney—increased reabsorption of calcium and phosphorus

Calcitriol is probably not the only active vitamin D metabolite. The inability to demonstrate specifically that calcitriol increases bone mineralization has led to an investigation of the actions of other metabolites on bone. The basic disturbance in osteomalacia, an illness produced by vitamin D deficiency in adulthood, is undermineralized bones. From evidence in patients with both nutritional and renal osteomalacia it has been suggested that calcidiol may have a more pronounced effect on bone mineralization than calcitriol[26-28] and may thus have a broader role than that of a precursor hormone.

Regulation of vitamin D metabolism. The regulation of vitamin D metabolism is easily understood once the function of calcitriol is known (see box on p. 444). Although plasma calcidiol levels are poorly controlled,[18] there appears to be relatively strict control of plasma calcitriol concentrations.[21] Initially, decreased calcitriol production was observed in animals fed a calcium-enriched diet. This decrease occurred only in animals with intact parathyroid glands, a finding that supports the thesis that parathyroid hormone (PTH) was the major stimulus for calcitriol formation. This thesis has been subsequently proved directly by intravenous infusion of PTH. PTH administration is now used as a clinical tool to assess the ability of the kidney to produce cal-

Primary stimuli for calcitriol synthesis
Decreased serum calcium concentration Increased parathyroid hormone secretion Decreased intracellular phosphorus concentration

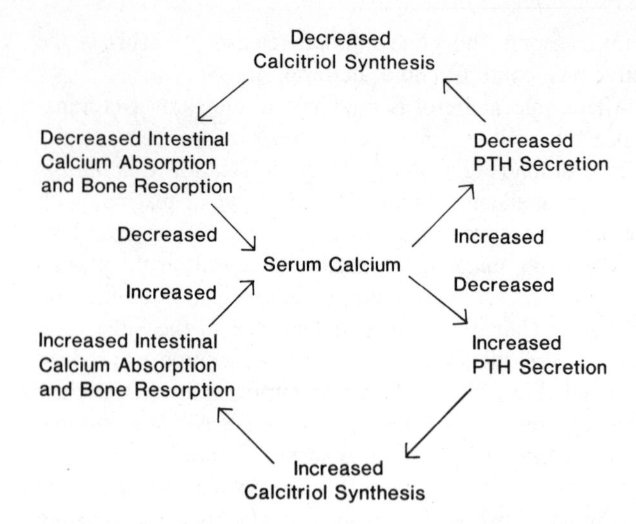

Fig. 25-6. Interrelationships of serum calcium concentrations and parathyroid hormone, *PTH,* and calcitriol.

citriol.[29] PTH may stimulate renal 1-hydroxylase, the enzyme that hydroxylates calcidiol at the C-1 position, indirectly by lowering intracellular phosphorus concentrations. Since phosphorus depletion increases calcitriol synthesis in normal or parathyroidectomized animals,[10] decreased intracellular phosphorus levels may be the ultimate common stimulus for calcitriol synthesis.

Understanding the metabolic control of vitamin D allows one to comprehend control of serum calcium and phosphorus concentrations (Fig. 25-6). When the serum calcium concentration falls, PTH is secreted and acutely restores normal serum calcium concentrations by stimulating osteoclasts to resorb bone. Within hours, calcitriol production is increased, which causes enhanced intestinal calcium absorption, which subsequently restores the serum calcium concentration and indirectly the PTH concentration to normal. Elevations in serum calcium concentration produce the opposite effect, that is, a reduction in both serum PTH concentrations and calcitriol synthesis.

Plasma calcitriol concentrations are altered by aging and pregnancy; they are elevated during adolescence and decline in old age.[20,30] Pregnancy and subsequent lactation are associated with elevated serum estrogen or prolactin concentrations; both hormones increase calcitriol synthesis.[31,32] It should be noted that the adolescent growth spurt, pregnancy, and lactation all increase requirements for calcium. Thus the elevated serum calcitriol concentrations represent an appropriate response to a physiological need.

Parathyroid hormone

In 1908, MacCallum and Voegtlin showed that removal of the parathyroid glands resulted in lowered serum calcium concentrations.[33] It was not until 1925, however, that Collip purified extracts of the parathyroid glands and demonstrated that tetany resulting from parathyroidectomy could be prevented by these extracts.[34] One major advance in understanding parathyroid physiology occurred when measurements of the hormone could be made in blood. These measurements, first by bioassay and later by radioimmunoassay techniques,[35] allowed an appreciation of the important relationships of parathyroid hormone and calcium homeostasis in normal and disturbed physiological processes.

Biochemistry and metabolism. Parathyroid hormone is synthesized in the parathyroid glands. The precursor hormone for parathyroid hormone is ''preproparathyroid hormone.'' This precursor is sequentially converted in the gland, first to proparathyroid and then to parathyroid hormone, which is released into the circulation. Normally, both parathyroid hormone and its fragments circulate in blood; however, in disease states, precursor forms of parathyroid hormone also may be present. The parathyroid hormone molecule has two terminals, known as the amino-terminal (N terminal) and the carboxy terminal (C terminal). The N-terminal section of the hormone is the biologically active section, whereas the C-terminal section is inactive. Thus fragments of parathyroid hormone bearing the N terminal generally are active, whereas those bearing the C terminal are inactive. Finally, catabolism of parathyroid hormone predominantly occurs in the liver and the kidney.[36]

Mechanisms of action. Parathyroid hormone acts on two major target organs, bone and kidney, to produce three major effects: increase in serum calcium concentrations, decrease in serum phosphorus concentrations, and increase in the active hormonal form of vitamin D (cholecalciferol). In bone, parathyroid hormone predominantly mobilizes calcium and phosphorus to the extracellular fluid, thus raising serum calcium and phosphorus concentrations. At the other target organ, the kidney, parathyroid hormone causes increased calcium retention, increased phosphorus excretion, and increased conversion of 25-hydroxycholecalciferol (calcidiol) to 1,25-dihydroxycholecalciferol (calcitriol). Calcitriol, in turn, as described earlier, predominantly causes increased intestinal calcium and phosphorus absorption. The effects of parathyroid hormone on the kidney are mediated through the formation of cyclic adenosine monophosphate (cAMP), and urinary levels of this substance rise when parathyroid hormone production is increased. The resultant effect of parathyroid

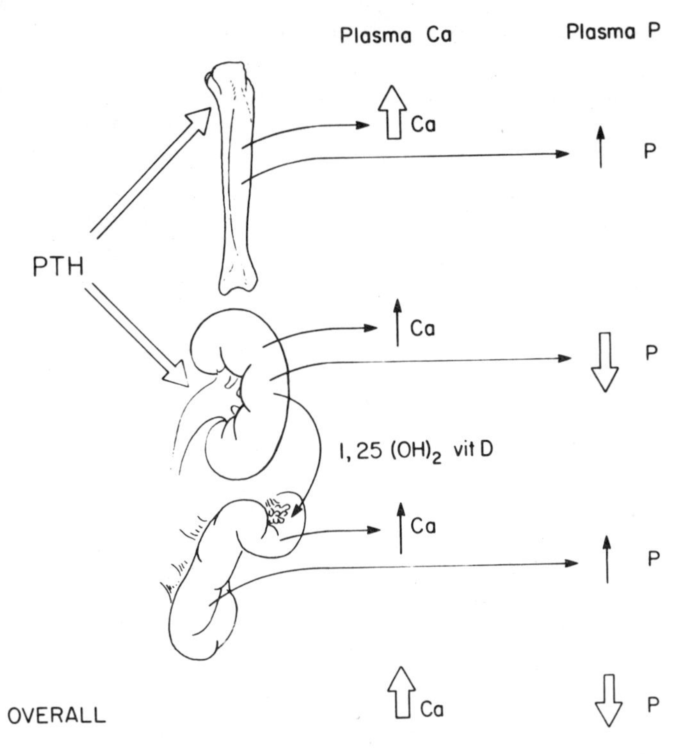

Fig. 25-7. Normal parathyroid hormone physiology. Parathyroid hormone action increases serum calcium concentrations predominantly through its bone and kidney effects but reduces plasma phosphorus concentrations, *P*, through increasing renal phosphorus excretion. (From Tsang, R.C., Noguchi, A., and Steichen, J.J.: Pediatr. Clin. North Am. **26**:223, 1979.)

hormone on the bone, kidney, and indirectly the intestine is to increase calcium concentrations in the blood. Although phosphorus concentrations may be elevated through parathyroid actions on bone and indirectly the intestine, the effect on increased renal phosphorus excretion overwhelms the other effects and, overall, results in decreased serum phosphorus concentrations (Fig. 25-7).

Regulation of parathyroid hormone production. Parathyroid hormone production is predominantly regulated by the concentration of calcium in the blood bathing the parathyroids. Increased serum calcium concentrations decrease parathyroid hormone production, whereas decreased serum calcium concentrations increase its production. Since parathyroid hormone in turn affects calcium metabolism directly, this inverse relationship of calcium concentrations to parathyroid hormone production is one of the major mechanisms by which calcium is finely regulated in the blood.[37]

Calcitonin

Calcitonin, discovered in 1962, is generally regarded as one of three hormones (along with vitamin D and parathyroid hormone) responsible for the control of calcium and phosphorus homeostasis. This function was initially suggested when Copp discovered that the serum calcium concentrations of dogs decreased after their thyroid and parathyroid glands were perfused with hypercalcemic blood.[38] Subsequent work confirmed the thyroidal origin of calcitonin both in animals[39] and humans.[40] Despite great research efforts, a definitive role for calcitonin in calcium homeostasis has not yet been clarified. Interpretation of experiments to define the functions of calcitonin has been difficult because pharmacological rather than physiological doses of the hormone have been used. Furthermore, neither calcitonin deficiency nor excess is clearly associated with bone disease or alteration of serum calcium homeostasis. The reader is referred to a discussion of this subject for further information.[41]

Localization, biochemistry, and metabolism. Immunofluorescent and immunochemical studies have confirmed that calcitonin is produced by the parafollicular C cells of the thyroid gland.[40] The pituitary gland, gastrointestinal tract, and liver may also produce the hormone.[41] Calcitonin is a polypeptide secreted in a precursor form with a molecular weight of 15,000 daltons that is cleaved into the active calcitonin molecule, which has a molecular weight of 3500 daltons.[41] Calcitonin molecules produced by different animal species contain 32 amino acids, but there are large differences in specific amino acid composition. Rat calcitonin most closely resembles human calcitonin, with a difference of only two amino acids. Salmon calcitonin, the only preparation licensed for human use in America, differs from human calcitonin in 16 amino acids. This has potential clinical significance because 40% to 70% of the patients who chronically receive calcitonin develop anti–salmon calcitonin antibodies.[42]

Normal serum calcitonin concentrations are less than 100 pg/ml when assayed by most radioimmunoassay techniques. Exact normal values must be established in individual laboratories because anticalcitonin antibodies differ in their binding specificity. Calcitonin alters cell function by increasing intracellular cyclic AMP production.[43] Newer assays employ bone or kidney tissue, which contains calcitonin receptors, and either displacement of calcitonin tracer or production of cyclic AMP is measured.[44,45] These assays currently are not sensitive enough for commercial application. Calcitonin is rapidly excreted, with a half-life of 10 minutes after intravenous administration.[46] Excretion is predominantly by the kidney, and serum calcitonin concentrations are increased in patients with renal failure.[47]

Biological effects. The biological effects of calcitonin may be divided into those related to calcium and phosphorus homeostasis and those related to gastrointestinal function. Intravenous calcitonin administration causes a prompt decline in the serum calcium concentration. This occurs because of effects of calcitonin on both bone and kidney. Pharmacological doses of calcitonin inhibit osteoclastic bone resorption,[48] an effect demonstrable within 15 minutes after calcitonin administration. Urinary hydroxyproline excretion declines in parallel with the inhibition of bone resorption. Calcitonin also decreases the renal reabsorption of calcium, phosphorus, sodium, potassium, and magnesium.[47] Although renal intracellular cyclic AMP levels are increased by calcitonin, there is no evident change in urinary cyclic AMP excretion.[49] It should be stressed that these described effects on both bone and kidney have been produced with pharmacological calcitonin concentrations. Data on the effects of physiological concentrations are currently unavailable.

Calcitonin also affects gastrointestinal function, resulting in decreased gastrin and gastric acid secretion and increased small intestine secretion of sodium, potassium, chloride, and water.[41] Again, these effects are observed with pharmacological doses of the hormone. Since calcitonin has inhibitory effects on bone, kidney, and the gastrointestinal tract, it has been hypothesized that calcitonin may be a substance that specifically inhibits cell function.[50]

Regulation of calcitonin secretion. Calcitonin secretion is influenced by serum calcium concentrations, gastrointestinal hormones, and sex steroids. Calcitonin release is stimulated by hypercalcemia and inhibited by hypocalcemia.[51] Large amounts of gastrin, glucagon, and cholecystokinin also increase calcitonin secretion, but physiological levels of these hormones produce no such changes.[51] Estrogen administration also raises serum calcitonin concentrations. Serum calcitonin concentrations may be higher in pregnant and lactating women than in controls, and men have higher circulating calcitonin concentrations than women.[51]

Magnesium

Magnesium is predominantly an intracellular cation. Sixty percent of the body's magnesium is in bone, and the remainder is equally divided between muscle and other soft tissue. Only 1% of the body's magnesium is in blood. Magnesium is essential for activation of the molecule adenosine triphosphate, which provides the basic energy for cell function. Magnesium also is intimately related to the intracellular synthesis of RNA and DNA.[52]

Metabolism. Magnesium is absorbed through the intestinal tract, with absorption rates ranging from 44% on an ordinary diet to 76% on a low-magnesium diet. There is an efficient conservation of magnesium in the kidney so that in magnesium deficiency extremely low magnesium-excretion rates occur. Parathyroid hormone (PTH) appears to cause increases in serum magnesium concentrations, possibly by its effects on mobilizing magnesium from bone. Acutely, increase in the serum magnesium concentration results in suppression of the parathyroids, thus theoretically preventing further parathyroid hormone increase and completing a "feedback" loop for magnesium-parathyroid interrelationships.[52,53] This feedback mechanism is thus similar to that for calcium-parathyroid interrelationships (see p. 444).

Although acute lowering of serum magnesium appears to increase serum parathyroid hormone concentrations, chronic magnesium deficiency results in *decreased* release of PTH. In addition to this impairment in parathyroid function, magnesium deficiency can decrease the response of target organs to PTH. Thus magnesium deficiency would lead to hypoparathyroidism and, secondarily, hypocalcemia. Hypocalcemia can also accompany magnesium deficiency because calcium release from bone is decreased. Under normal circumstances, magnesium and calcium undergo an exchange in bone related to their release into the circulation. Lowered magnesium content in bone would result in lowered interchange with calcium and lowered release of calcium from bone.[53,54]

BONE DISORDERS

Disorders of calcium, phosphorus, vitamin D, or parathyroid hormone homeostasis frequently produce *osteopenia,* a general term for the roentgenographic appearance of a subnormal amount of mineralized bone. Many illnesses are associated with osteopenia[55,56] (see accompanying box). The bone histopathological condition of osteopenia can reveal decreased osteoid (bone matrix) formation, decreased osteoid mineralization, or increased

A partial differential diagnosis of osteopenia

- Osteoporosis
  - Aging (senile)
  - Postmenopausal
  - Juvenile
  - Immobilization
  - Cushing's syndrome
  - Multiple myeloma
  - Leukemia
  - Turner's syndrome
  - Alcoholism
  - Chronic liver disease
- Osteomalacia
  - Vitamin D deficiency
  - Chronic gastrointestinal disease
  - Anticonvulsant induced
  - Vitamin D dependency
  - Vitamin D resistance (hypophosphatemic)
  - Chronic acidosis
  - Fanconi's syndrome
  - Chronic renal failure
  - Phosphorus and calcium deficiency
- Osteitis fibrosa
  - Primary hyperparathyroidism
  - Chronic renal failure
- Paget's disease

Table 25-1. Common serum abnormalities associated with metabolic bone disease

Disease	Ca	P	PTH	Alk PO_4	Calcidiol	Calcitriol
Osteoporosis	NL	NL	NL	NL	NL	LO
Osteomalacia	LO	LO	HI	HI	NL or LO	NL or LO
Osteitis fibrosa	HI or NL	LO or HI	HI	HI	NL	HI or LO

NL, Normal; *LO,* decreased; *HI,* increased; *Alk PO_4,* alkaline phosphatase.

bone resorption.[57] These histological categories correlate with the clinical diagnoses of *osteoporosis, osteomalacia,* or *osteitis fibrosa,* respectively. Clinically, osteopenia results in the crush-fracture syndrome[55] in adults and either fractures or growth failure in children. Trabecular bone is more frequently affected than cortical bone, and so fractures most often occur in vertebrae, the femoral neck, and the distal ends of the long bones, where trabecular bone is abundant. Whereas specific diagnoses of osteopenic bone are best made histologically, occasionally, characteristic laboratory abnormalities will permit differentiation between osteoporosis, osteomalacia, and osteitis fibrosa (Table 25-1).

Osteoporosis

Osteoporosis is defined histologically as a generalized reduction in bone mass, including both osteoid and mineral. This may be produced by either suppressed bone formation or increased bone resorption. In either case, the "coupling" of bone formation and bone resorption is altered.[58] Over 6 million Americans currently suffer from acute problems secondary to weakened vertebrae, and approximately 10% of females over 50 years of age have bone loss severe enough to result in fractures.[59] The estimated annual cost of caring for this illness exceeds 1 billion dollars. Osteoporosis is frequently diagnosed in older men and women (senile osteoporosis) and postmenopausal women. It occurs only rarely in childhood as a idiopathic illness and can also accompany certain systemic diseases, including Cushing's syndrome, hematological malignancies, Turner's syndrome, alcoholism, and chronic liver disease (see box on p. 446).

Senile osteoporosis. Progressive bone loss normally occurs during aging. This process begins at 50 years of age in women and 65 to 70 years of age in men and results in a loss of 0.5% of total bone mass per year and approximately 20% in a lifetime.[60] Patients with senile osteoporosis experience accelerated losses of 1% to 2% per year, with symptoms of osteoporosis beginning when 30% of bone mass is lost. It has been suggested that osteoporosis is a natural part of the aging process manifested earlier in those persons who have accrued less skeletal mass during early adult life. The causes of senile osteoporosis are largely unknown. One hypothesis states that chronic inadequate dietary calcium intake results in diminished skeletal mass and eventual osteoporosis.[61] Hormonal alterations that occur during senescence undoubtedly potentiate bone loss (see box below). Decreased serum calcitriol concentrations found in the elderly[62] probably result from a blunted synthetic response to parathyroid hormone.[63] In addition, serum parathyroid hormone concentrations increase[64] and serum calcitonin concentrations decrease[65] with aging. The net effect of these hormonal alterations is diminished intestinal calcium absorption and increased bone resorption.

Postmenopausal osteoporosis. Postmenopausal osteoporosis, which occurs in females at a younger age than senile osteoporosis does, is probably caused by estrogen deficiency. Affected women have diminshed intestinal calcium absorption and lower serum calcitriol concentrations than their normal age-matched peers.[62] Although serum PTH concentrations are normal when compared to controls with normal serum calcitriol concentrations, they are low when viewed in the context of calcitriol deficiency.[62] Estrogen supplementation increases intestinal calcium absorption and serum calcitriol and PTH concentrations.[66] These data have been interpreted to indicate that estrogen deficiency produces postmenopausal osteoporosis by causing bone resorption, which releases calcium into the extracellular space and which, in turn, suppresses PTH secretion, calcitriol synthesis, and intestinal absorption of calcium (Fig. 25-8).

Osteomalacia

Osteomalacia is diagnosed when bone contains normal quantities of osteoid that fail to mineralize. When seen in the growing child, osteomalacia is termed *rickets.* The terms *rickets* and *osteomalacia* are used interchangeably in this chapter.

Calcium-regulating hormone abnormalities associated with aging

- Decreased serum calcitriol concentration and calcitriol secretory reserve
- Increased serum parathyroid hormone concentration
- Decreased serum calcitonin concentration

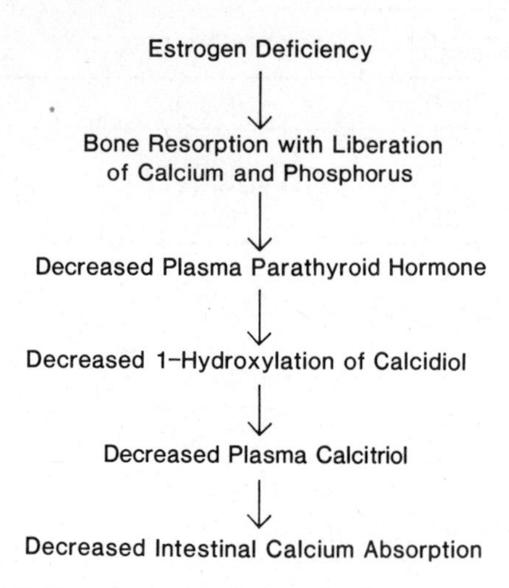

Fig. 25-8. Hypothesized pathogenesis of postmenopausal osteoporosis.

Vitamin D–deficient osteomalacia. Osteomalacia is caused by disturbances in either vitamin D metabolism or the bioavailability of phosphorus. Alterations of vitamin D metabolism range from conditions of insufficient intake or production of cholecalciferol to disturbances in its activation by the liver and kidneys. Generally one can predict the biochemical response to a deficiency of calcitriol (Fig 25-4 and box on p. 443). Intestinal calcium absorption will decrease and produce hypocalcemia. This will stimulate PTH release (secondary hyperparathyroidism), which will mobilize calcium from bone and increase phosphorus excretion by the kidney. Initially, serum calcium concentrations will be maintained at the expense of bone resorption, but as minerals are depleted, hypocalcemia occurs. Hypophosphatemia occurs because of increased urinary phosphorus losses. Thus the characteristic serum abnormalities associated with calcitriol deficiency are hypocalcemia, hypophosphatemia, and hyperparathyroidism (Table 25-2). Osteomalacia also results from phosphorus deficiency. In this situation, low intracellular phosphorus concentrations should stimulate calcitriol synthesis, which will increase both intestinal and renal phosphate absorption. Serum calcium and PTH concentrations should be unaffected (Table 25-2).

The major causes of osteomalacia are listed in the box on p. 446 and their associated biochemical abnormalities are summarized in Table 25-2. Historically the most common cause of osteomalacia was *vitamin D deficiency* caused by a combination of insufficient sunlight exposure and inadequate dietary intake of vitamin D–containing foods.[67] Serum calcidiol concentrations, which reflect the adequacy of vitamin D in the body, are low in osteomalacia. Supplementation of foods with vitamin D has virtually eliminated the problem in industrialized countries, but it may still be seen in underdeveloped nations, particularly among dark-skinned people because skin pigment decreases the production of cholecalciferol, which normally occurs after ultraviolet light exposure.

Osteomalacia secondary to gastrointestinal disorders. Patients with gastrointestinal disease, particularly those with hepatobiliary disease, often develop osteomalacia. Vitamin D is fat soluble and requires bile acids for absorption.[68] Patients with hepatobiliary disease have low serum calcidiol levels that appear to be caused in part by defective intestinal cholecalciferol or ergosterol absorption,[68] impaired calcidiol production by the liver,[69] and enhanced calcitriol metabolism.[70] Osteopenia also may be seen after gastric surgery, though the pathogenesis is not understood.[71]

Osteomalacia secondary to anticonvulsant medication. Rickets may occur in children receiving anticonvulsant medications, such as phenytoin (Dilantin) and phenobarbitol, that induce the hepatic microsomal mixed-oxidase enzyme system.[72,73] This enzyme system, when stimulated, converts calcidiol to polar inactive metabolites, which results in calcidiol deficiency.

Vitamin D–dependent osteomalacia (types I and II). After foods were fortified with vitamin D, it became apparent that normal antirachitic doses of ergosterol failed to heal the rickets of a small subpopulation of rachitic patients. One group of such patients had the classical signs and symptoms of vitamin D deficiency, including early infantile hypocalcemia, hypophosphatemia, and tetany, but required up to 100 times the normal intake of vitamin D

Table 25-2. Biochemical abnormalities associated with rickets

	Serum calcium	Serum phosphorus	Parathyroid homone	Calcidiol	Calcitriol
Vitamin D deficiency	LO	LO	HI	LO	LO or NL
Vitamin D dependency					
I	LO	LO	HI	HI	LO
II	LO	LO	HI	HI	HI
Vitamin D resistance	NL	LO	NL	NL	NL or LO
Dietary phosphorus deficiency	NL	LO	NL	LO	HI

NL, Normal; LO, decreased; *HI*, increased.

to heal their rickets. This group of patients with *vitamin D–dependent rickets* has recently been subclassified into two groups with distinct pathophysiological bases. Patients with vitamin D–dependent rickets type I have the classical biochemical abnormalities of vitamin D–deficient rickets but their serum calcidiol concentrations are normal and they lack circulating calcitriol.[74] The presumed defect is an abnormal or absent renal 1α-hydroxylase enzyme. The osteomalacia of these patients heals when physiological doses of calcitriol are administered.[75] Patients with vitamin D–dependent rickets type II have normal calcidiol and calcitriol levels and are resistant to physiological doses of calcitriol, which suggests an end-organ resistance to calcitriol.[76] Recently, fibroblasts from these patients, examined in vitro, failed to translocate calcitriol from the cytosol to the nucleus, an essential step in the intracellular action of calcitriol.[77] Both types of vitamin D–dependent rickets are inherited in an autosomal recessive pattern.

Vitamin D–resistant osteomalacia. Patients with *vitamin D–resistant rickets* lack most of the usual biochemical markers associated with rachitic patients. Serum calcium and PTH concentrations are normal, but hypophosphatemia is severe. The major defect in these patients is subnormal reabsorption of phosphorus by the renal tubules.[78] Because low intracellular phosphorus concentrations are a major stimulus for calcitriol synthesis, serum calcitriol concentrations should be elevated in this disorder; however, when measured, serum calcitriol concentrations have been found to be low or low-normal.[74,79,80] This finding suggests a potential second defect in this condition, that is, dysfunction of the renal lα-hydroxylase enzyme. Vitamin D–resistant rickets may be inherited in a sex-linked recessive or an autosomal dominant pattern. Traditionally, patients with vitamin D–resistant rickets have been treated with cholecalciferol and phosphate supplements.[81] There appears to be a better response with combined calcitriol and phosphate therapy.[80,82]

Other causes of osteomalacia. Chronic acidosis causes osteomalacia, hypercalciuria, and hyperphosphaturia because of neutralization of acids by bone with subsequent release of bone mineral. Patients with renal Fanconi's syndrome have diminished proximal tubule reabsorption of bicarbonate (resulting in chronic acidosis), phosphorus, glucose, and amino acids. Osteomalacia may be severe because of the chronic acidosis and severe hypophosphatemia. In addition, as part of the proximal tubulopathy, there may be subnormal activity of the renal 1α-hydroxylase enzyme.[83]

Sufficient substrate must be supplied in the diet for proper bone mineralization. Delayed bone mineralization commonly occurs in very low birth weight premature infants who are fed normal infant formulas or breast milk,[84,85] both of which contain insufficient quantities of calcium and phosphorus for the rapid bone mineralization of premature infants.

Histopathology of renal osteodystrophy

- Predominant osteitis fibrosa
  - Normal serum calcium level—calcitriol responsive
  - Pretreatment hypercalcemia—exacerbated by calcitriol
- Predominant osteomalacia
  - Small amount of fibrosis present—calcitriol responsive
  - Pure osteomalacia—hypercalcemia with calcitriol treatment
- Mixed osteitis fibrosa and osteomalacia—calcitriol responsive
- Mild—calcitriol responsive

Osteitis fibrosa

Osteitis fibrosa is the histopathological bone lesion produced by excessive parathyroid hormone secretion. It is primarily seen in two conditions, primary hyperparathyroidism and chronic renal failure. Bone disease is of lesser significance in primary hyperparathyroidism, since surgical removal of the involved parathyroid glands cures the disease. The pathophysiological condition of the secondary hyperparathyroidism associated with chronic renal failure is more complex and less amenable to treatment. Thus uremic patients frequently suffer from severe bone disease.

The complex bone abnormality associated with chronic renal failure is termed *renal osteodystrophy*. There are two distinct histopathological forms of renal osteodystrophy, osteomalacia and osteitis fibrosa (see box above), which frequently coexist in the same patient. Osteomalacia is probably caused by decreased synthesis of calcitriol secondary to renal parenchymal disease. Serum concentrations of calcitriol and 24,25-dihydroxyvitamin D are decreased in both children and adults with chronic renal failure, and calcidiol concentrations are normal.[86-89] The pathogenesis of hyperparathyroidism in patients with renal disease is less clearly understood and may be caused by phosphate rentention, decreased calcitriol synthesis, or a combination of the two[89,90] (Figs. 25-9 and 25-10).

Paget's disease

Paget's disease is a disorder of bone metabolism characterized by increased osteoclastic bone resorption followed by disordered bone formation.[91,92] The incidence of this disease is difficult to determine, since the majority of affected patients are asymptomatic. In autopsy series of persons over 40 years of age, 3% of this group is affected. The cause of the disease is unknown.

The histological pattern of patients with Paget's disease proceeds through three stages. In the early phase of the illness, resorption predominates and the bone marrow is

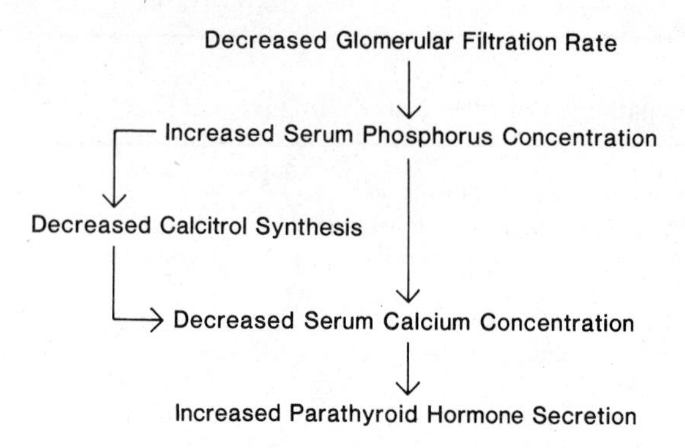

Fig. 25-9. Pathogenetic mechanism of secondary hyperparathyroidism in renal failure according to "phosphate theory."

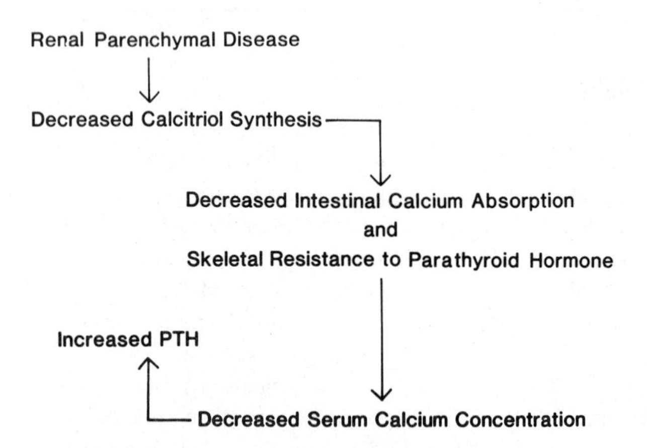

Fig. 25-10. Pathogenetic mechanism of secondary hyperparathyroidism in renal failure according to "vitamin D theory."

replaced with a highly vascular fibrous connective tissue. In the second phase of the disease, bone formation predominates. The pagetic bone is coarse-fibered, dense trabecular bone. In the final phase the rate of bone resorption declines and continued bone formation produces hard, dense bone. The large amount of bone resorption that initially occurs produces greatly elevated urinary hydroxyproline concentrations, and the subsequent rapid rate of bone formation results in dramatically elevated serum alkaline phosphatase concentrations. Serum calcium and phosphorus concentrations are normal. However, pathological fractures occur and are treated by immobilization of the patient. Hypercalcemia frequently occurs, since immobilization increases the rate of bone resorption. Paget's disease is treated successfully with calcitonin, which inhibits osteoclastic bone resorption.[93]

CHANGE OF ANALYTE IN DISEASE

Vitamin D. Serum concentrations of vitamin D metabolites may be altered in a variety of disease states (see box above). Decreased concentrations result from deficient intake, defective metabolic regulation, or increased excretion. Serum calcidiol concentrations are low in patients who have both an insufficient exposure to sunlight and a low intake of foods that contain vitamin D. Patients who receive anticonvulsant drugs convert calcidiol into biologically inactive polar metabolites.[94] Production of calcidiol is impaired in patients with liver disease.[95] Inactive or absent renal 1α-hydroxylase activity and secondary low serum calcitriol concentrations are associated with vitamin D–dependent rickets type I[96] postmenopausal and senile osteoporosis,[97,98] hypoparathyroidism,[99] pseudohypoparathyroidism,[100] vitamin D–resistant rickets,[101] and chronic renal failure.[102] Patients with nephrotic syndrome have low serum concentrations of both calcidiol and calcitriol because of urinary losses of both metabolites and as well the serum protein (vitamin D–binding protein) to which they are attached.[103,104]

Diseases and conditions associated with changes in serum concentrations of vitamin D metabolites

- Calcidiol (25-hydroxycholecalciferol) deficiency
  - Nutritional osteomalacia
  - Anticonvulsant-induced osteomalacia
  - Liver disease
  - Nephrotic syndrome
- Calcitriol (1,25-dihydroxycholecalciferol) deficiency
  - Vitamin D–dependent rickets type I
  - Postmenopausal and senile osteoporosis
  - Hypoparathyroidism
  - Pseudohypoparathyroidism
  - Vitamin D–resistent rickets
  - Nephrotic syndrome
- Caldidiol (25-hydroxycholecalciferol) excess
  - Sunlight exposure
  - Vitamin D intoxication
- Calcitriol (1,25-dihydroxycholecalciferol) excess
  - Childhood
  - Pregnancy and lactation
  - Sarcoidosis
  - Hyperparathyroidism

High serum calcidiol concentrations result from either increased exogenous intake or increased endogenous production secondary to an unusually large sunlight exposure.[105] High serum calcitriol concentrations occur in physiological states of increased calcium requirements such as growth[106] and pregnancy and lactation.[107] High serum calcitriol concentrations are also seen in sarcoidosis, in which an extrarenal source of calcitriol production has been implicated,[108] and in hyperparathyroidism, in which the serum concentrations of parathyroid hormone, a major stimulus for calcitriol production, are elevated.

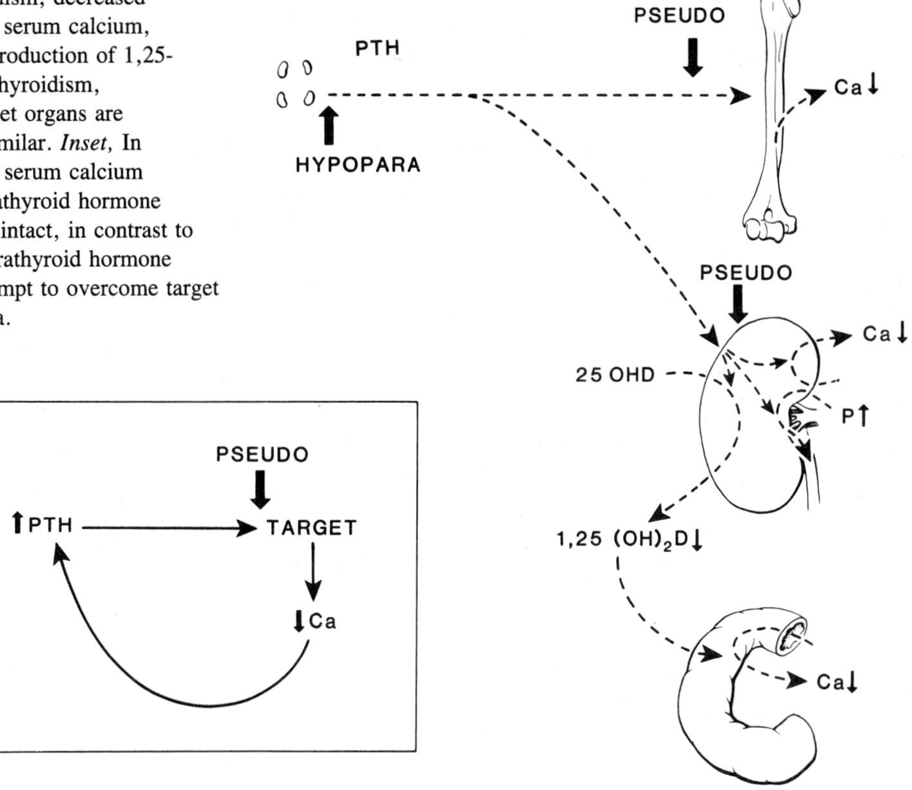

Fig. 25-11. In idiopathic hypoparathyroidism, decreased parathyroid hormone results in decreased serum calcium, increased serum phosphorus, decreased production of 1,25-dihydroxyvitamin D. In pseudohypoparathyroidism, although there is sufficient hormone, target organs are unresponsive and biochemical result is similar. *Inset,* In pseudohypoparathyroidism, resultant low serum calcium concentrations serve as a stimulus to parathyroid hormone production. Since parathyroid glands are intact, in contrast to idiopathic hypoparathyroidism, serum parathyroid hormone concentrations will be *elevated* in an attempt to overcome target organ resistance and rectify hypocalcemia.

Parathyroid hormone

Primary hypoparathyroidism. "Idiopathic hypoparathyroidism" describes the condition of decreased production of parathyroid hormone. In "pseudohypoparathyroidism," production of parthyroid hormone is intact, but there is target organ resistance to parathyroid hormone; in other words, parathyroid hormone, though present, does not exert its physiological actions because the target organs are not responsive. In current terminology, there may be a "receptor defect" for parathyroid hormone. Another way of describing idiopathic hypoparathyroidism would be "hormone-deficient hypoparathyroidism" and pseudohypoparathyroidism would be described as "hormone-sufficient, receptor-deficient hypoparathyroidism"[109] (Fig. 25-11).

In view of the major physiological actions of parathyroid hormone, hypoparathyroidism would be expected to present classically with hypocalcemia and hyperphosphatemia usually in childhood. However, not all the clinical signs and symptoms are directly related to these biochemical changes, and it is possible that other unexplored functions of parathyroid hormone and its interactions with target organs may be disordered or other genetic or immunological problems may coexist.[110]

Secondary hypoparathyroidism. Hypoparathyroidism may result from other disorders. Inadvertent surgical removal of the parathyroids may occur during thyroidectomy, since the parathyroids are closely associated with the thyroid gland and may be difficult to distinguish from thyroid tissue. In this circumstance, postoperative tetany occurs and signals the deficiency of parathyroid hormone. Since magnesium is important for parathyroid hormone secretion, magnesium deficiency may result in hypoparathyroidism. An interesting "physiological" hypoparathyridism occurs in infants. In utero, calcium is transferred actively across the placenta, and serum calcium concentrations in the fetus are extremely high. These high serum calcium concentrations appear to inhibit fetal parathyroid function. Inhibited parathyroid function persists for a short interval after birth and appears to be the major cause of hypocalcemia in the first 3 days of life, especially in the premature infant.[100]

The diagnosis of hypoparathyroidism is made from the clinical presentation of lowered serum calcium and elevated serum phosphorus concentrations. Parathyroid hormone concentrations will be low in hypoparathyroidism but elevated in pseudohypoparathyroidism. In the latter condition, since there is a target organ resistance to parathyroid hormone and subsequent hypocalcemia, parathyroid hormone production is continuously stimulated in a generally futile attempt to overcome the target resistance (Fig. 25-11). To further distinguish idiopathic hypoparathyroidism from pseudohypoparathyroidism, parathyroid hormone infusion is administered. After the infusion, serum calcium and urinary phosphorus and cyclic adenosine monophosphate concentrations are measured. Patients with pseudohypoparathyroidism may have varying degrees of "block" in response to parathyroid hormone, at the bone site or at various "levels" in the kidney.[111] Patients with hypoparathyroidism are treated with supplements of calcium salts and calcitriol.

Primary hyperparathyroidism. Hyperparathyroidism is often described as being related to hyperplasia or adenoma of the parathyroids. However, it is not clear whether these entities are totally separate, because histological features of both may be found within one person. In contrast to hypoparathyroidism, which usually begins in childhood, hyperparathyroidism is usually discovered in adulthood.[112] As expected from the physiological action of parathyroid hormone, excess concentrations of the hormone result in increased serum calcium concentrations and decreased serum phosphorus concentrations. Demineralization occurs as a consequence of the bone lytic action of parathyroid hormone and is associated with areas of extensive resorption (osteitis fibrosa, see p. 449). Many clinical problems are associated with the high serum calcium concentrations. The major organ systems adversely affected by hypercalcemia are the nervous system and the kidney.[108]

Secondary hyperparathyroidism. Conditions that are associated with chronic hypocalcemia will result in chronic stimulation of the parathyroids and secondary hyperparathyroidism. The two major factors resulting in chronic hypocalcemia of nonparathyroid cause are vitamin D–metabolite deficiencies and high phosphorus loads. Any deficiency of vitamin D or its major metabolites will result in decreased intestinal absorption of calcium and hypocalcemia. The initial response to this hypocalcemia will be secondary hyperparathyroidism, which helps maintain serum calcium concentrations in the normal range. High phosphorus loads occur with infusion of phosphorus-containing fluids, ingestion of high phosphorus-containing milk (such as cow's milk given to new born infants), or retention of phosphorus by failing kidneys.[113] High serum phosphorus concentrations result in a secondary decrease in serum calcium concentrations. With decreased serum calcium concentrations there is compensatory secondary hyperparathyroidism.[114]

The diagnosis of primary hyperparathyroidism is based on the findings of high serum calcium, low serum phosphorus, and high serum parthyroid hormone concentrations. However, not all hyperparathyroid patients will have increased serum calcium concentrations or increased serum parathyroid hormone concentrations. Ionized calcium measurements in blood may provide additional diagnostic help, since the ionized calcium fraction is the physiologically active calcium. C terminal–recognizing assays for parathyroid hormone appear to give greater distinction between normal and hyperparathyroid patients. In secondary hyperparathyroidism, serum calcium concentrations are low or normal, since the parthyroid overactivity results from an initial decline in serum calcium. Serum phosphorus concentrations would be low except in situations of phosphorus overload, when it would be high. Serum parathyroid hormone concentrations should be elevated in secondary hyperparathyroidism.[110]

Calcium

Hypercalcemia. Hypercalcemia from hyperparathyroidism has been described previously. Since the parathyroid glands are not inherently abnormal, the consequence of hypercalcemia would be suppression of the parathyroids and low serum parathyroid hormone concentrations[110,115] (see box below).

Endocrine and tumor-related hypercalcemia. Hypercalcemia may occur with disorders of endocrine organs other than the parathyroids. Overproduction of thyroid hormone (thyrotoxicosis) and underproduction of cordicosteroids (Addison's disease or abrupt withdrawal of steroid hormones) are associated with hypercalcemia. A wide variety of tumors appear to produce parathyroid hormone–like substances with osteoclast-stimulatory activity, which results in hypercalcemia. Such tumors include malignancies with bone metastases, lung cancers, hepatomas, renal cancers, adrenal pheochromocytomas, fibrosarcomas, and leukemias.[110]

Vitamin A– and vitamin D–related disorders. Excessive intake of vitamin A or vitamin D may result in hypercalcemia. In sarcoidosis, elevated calcitriol concentrations appear to be the cause of the hypercalcemia. Vitamin A in high doses appears to have a direct effect on bone resorption. Excess intake of these vitamins is usually a result of food fads or accidental ingestion. Therapy for vitamin D–resistant rickets or hypoparathyroidism with high doses of vitamin D is a common cause of hypercalcemia. Idiopathic hypercalcemia of infants is believed to be related to dis-

Causes of hypercalcemia

Hyperparathyroidism
Thyrotoxicosis
Addison's disease
Withdrawal of steroids
Tumors
Vitamin D and vitamin A intoxication
Sarcoidosis
Idiopathic hypercalcemia of infancy
Immobilization
Subcutaneous fat necrosis in infants
Thiazide diuretics
Milk-alkali syndrome
Benign familial hypercalcemia

ordered vitamin D metabolism, possibly increased sensitivity to vitamin D.[116] The condition may be associated with supravalvular aortic stenosis and mental retardation.

Iatral causes. Immobilization of patients, especially male adolescents, results in rapid mobilization of calcium from bone and resultant hypercalcemia. In infants born after traumatic deliveries, a curious condition of hypercalcemia can occur in association with extensive trauma related to subcutaneous fat necrosis. It is believed that mobilization of calcium from the traumatized areas during healing results in increased serum calcium concentrations. Use of thiazide diuretics is classically associated with hypercalcemia, presumably because of their augmentation of the calcium-mobilizing effect of parathyroid hormone on bone. Excessive ingestion of milk and alkali in the treatment of peptic ulcer (the ''milk-alkali syndrome'') also results in hypercalcemia.[110]

Familial form. There is a benign form of familial hypercalemia that is inherited as a dominant trait. Mild hypercalcemia (less than 130 mg/L) occurs and apparently is without adverse effects.[110]

Hypocalcemia. The causes of hypocalcemia are currently classified in relation to the major hormone or biochemical involved: vitamin D, parathyroid hormone, calcitonin, calcium, magnesium, and phosphate (see box below).

Causes of hypocalcemia

Vitamin D
- Decreased solar exposure and endogenous synthesis
- Decreased intestinal intake (malabsorption, dietary deficiency)
- Altered hepatic metabolism of vitamin D (hepatic disease, anticonvulsants)
- Decreased renal synthesis of calcitriol (vitamin D dependency, renal failure)

Parathyroid
- Hypoparathyroidism (primary and secondary)
- Pseudohypoparathyroidism

Calcitonin
- Calcitonin or mithramycin infusion

Calcium
- Intestinal malabsorption
- Acute pancreatitis
- Infusion of agents complexing calcium
- Alkalosis decreasing ionized calcium

Magnesium
- Magnesium deficiency (see box, p. 455, top right)

Phosphorus
- Renal failure
- Phosphate infusion
- Cow's milk formulas

Vitamin D deficiency, which has been covered earlier, occurs as a result of reduced synthesis or intake of the parent vitamin D, altered hepatic metabolism of vitamin D, and decreased renal synthesis of calcitriol, the final active metabolite of vitamin D.

Hypoparathyroidism (primary and secondary) and pseudohypoparathyroidism have been discussed previously. Calcitonin or mithramycin infusions will decrease calcium transport from bone to extracellular space and result in hypocalcemia. Intestinal malabsorption of calcium may lead to hypocalcemia. Acute pancreatitis is associated with fatty acid–calcium complex precipitates in the pancreas and hypocalcemia. Decreased blood ionized calcium occurs with infusion of agents complexing calcium (citrate and acid-citrated blood for transfusion, or EDTA), or alkalosis, which shifts the fraction of calcium that is ionized to that which is protein bound. Hypomagnesemia results in hypocalcemia mostly related to the adverse effect of hypomagnesemia on parathyroid function. Conditions whereby phosphorus concentrations in blood are elevated, as in renal failure (see earlier), phosphate infusion, or infants, receiving cow's milk formulas with their high phosphate content will result in decreased serum calcium concentrations because calcium is shifted from the extracellular space into bone and soft tissues and probably because there is a blunted bone response to the effects of parathyroid hormone.[117]

Changes in parathyroid hormone: vitamin D–axis analytes in hypercalcemia and hypocalcemia. Changes in the serum measurements for phosphorus, the parathyroid hormone–vitamin D axis, and vitamin D status help in evaluation of the causes of hypercalcemia and hypocalcemia. If the parathyroid–vitamin D axis is intact, two effects are seen: (1) cyclic AMP production by the kidney is active; renal cyclic AMP is best determined as ''nephrogenous'' cyclic AMP, which takes into account the cyclic AMP not produced in the kidney,[118] and (2) 1,25-dihydroxycholecalciferol (calcitriol) production is also active (Tables 25-3 and 25-4).

In the *hypercalcemic* disorders (Table 25-3), serum phosphorus concentrations are decreased in hyperparathyroidism because of the phosphaturic effects of parathyroid hormone; in the remaining causes of hypercalcemia, little effect on serum phosphorus is evident. In hyperparathyroidism, serum parathyroid hormone concentrations and the parathyroid hormone–vitamin D axis analytes are increased. In hypercalcemia for other causes, serum parathyroid hormone is suppressed. In turn the parathyroid hormone–vitamin D axis may be suppressed, except for sarcoidosis, in which elevation of serum calcitriol concentrations appears to be a primary problem.

Vitamin D status is best assessed through measurement of serum 25-hydroxycholecalciferol (calcidiol) concentrations. Thus serum 25-hydroxycholecalciferol concentrations are elevated in vitamin D intoxication, with or with-

Table 25-3. Parathyroid hormone–vitamin D axis analytes in hypercalcemia

Disorder	Serum Phosphorus	Parathyroid hormone (PTH)	Parathyroid hormone–vitamin D axis: Nephrogenous cyclic AMP	Parathyroid hormone–vitamin D axis: Calcitriol (1,25-dihydroxy-cholecalciferol)	Vitamin D status: Calcidiol (25-hydroxy-cholecalciferol)
Hyperparathyroidism	LO	HI	HI	HI	NL
Vitamin D disorders					
Vitamin D intoxication	NL	LO	LO	HI	HI
High calcitriol in sarcoidosis	NL	LO	LO	NL	NL
Sensitivity to vitamin D: Idiopathic hypercalcemia of infancy	NL	LO	LO	NL	NL
Non–parathyroid hormone, non–vitamin D disorders					
Malignancy	NL	LO	HI/LO	LO	NL
Immobilization	NL	LO	LO	LO	NL
Thyrotoxicosis	NL	LO	NL	LO	NL

NL, Normal; *LO,* decreased; *HI,* increased.

Table 25-4. Parathyroid hormone–vitamin D axis analytes in hypocalcemia

Disorder	Serum phosphorus	Parathyroid hormone (PTH)	Parathyroid hormone–vitamin D axis: Nephrogenous cyclic AMP	Parathyroid hormone–vitamin D axis: Calcitriol (1,25-dihydroxycholecalciferol)	Vitamin D status: Calcidiol (25-hydroxycholecaliferol)
Parathyroid disorders					
Hypoparathyroidism	HI	LO	LO	LO	NL
Pseudohypoparathyroidism	HI	HI	LO	LO	NL
Vitamin D disorders					
Vitamin D deficiency	LO	HI	HI	NL or LO	LO
Hepatic disease alpha-anticonvulsants	LO	HI	HI	NL or LO	LO
Renal					
Vitamin D–dependent rickets	LO	HI	HI	LO	NL
Osteodystrophy	HI	HI	—	LO	NL
Resistance to 1,25-dihydroxycholecalciferol	LO	HI	HI	HI	NL
Mineral disorders					
Calcium malabsorption	NL	NL or HI	—	—	–
Hypomagnesemia	NL	NL or HI	—	—	NL
High phosphate load	HI	NL or HI	—	—	NL

NL, Normal; *LO,* decreased; *HI,* increased.

out elevations in the serum 1,25-dihydroxycholecalciferol (calcitriol) concentrations.

In *hypocalcemia* (Table 25-4) related to parathyroid disorders, hyperparathyroidism, or pseudohyperparathyroidism, serum phosphorus is elevated because of decreased urinary phosphorus excretion. The parathyroid hormone–vitamin D axis is generally hypofunctioning, except for pseudohyperparathyroidism, in which serum parathyroid hormone concentrations will be elevated because of target organ resistance to the hormone.

In the vitamin D disorders, serum phosphorus concentrations are generally low because one of the major actions of vitamin D is to raise serum phosphorus concentrations. However, in renal osteodystrophy (renal failure) serum phosphorus concentrations are elevated because of decreased renal phosphorus excretion. The parathyroid hormone–cyclic AMP axis in this circumstance may be increased because of hyperparathyroidism secondary to hypocalcemia; however, serum 1,25-dihydroxycholecalciferol concentrations will remain decreased because of deficiency of vitamin D or blocks in vitamin D metabolism. In the condition of increased resistance to 1,25-dihydroxycholecalciferol, high serum concentrations of the metabolite are found, analogous to elevated parathyroid hormone concentrations in pseudohyperparathyroidism. Serum 25-hydroxycholecalciferol measurements will be low in vita-

Causes of hypermagnesemia

Magnesium sulfate therapy
Magnesium-containing antacids and purgatives
Renal failure

min D deficiency or when there is a block in 25-hydroxylation of the vitamin D, but normal in vitamin D disorder caused by metabolic blocks beyond the liver step of hydroxylation.

In the mineral disorders causing hypocalcemia, little effect on the parathyroid hormone–vitamin D axis has been reported. Secondary hyperparathyroidism can be a consequence of hypocalcemia. In hypomagnesemia, however, hypoparathyroidism can occur secondary to magnesium deficiency.

Magnesium

Hypermagnesemia. Excess of magnesium is usually a consequence of increased medicinal intake of magnesium. Magnesium salts ($MgSO_4$) are used in the treatment of hypertension induced by pregnancy (''preeclampsia''). The mother will become hypermagnesemic, as will her infant. Reduced magnesium excretion in severe renal failure may occur, and the use of medicines that contain magnesium (antacids, purgatives) in this situation may result in hypermagnesemia[119,120] (see box above).

Hypermagnesemia and magnesium deficiency. Severe magnesium deficiency in humans is uncommon, possibly because of the highly developed capabilities of conservation of magnesium in the body. Decreased intake of magnesium caused by gastrointestinal disorders (steatorrhea, malabsorption syndromes, gut resections) can cause magnesium deficiency. Specific intestinal malabsorption of magnesium also occurs and can cause hypomagnesemia in infancy. Protein-calorie malnutrition is often associated with magnesium depletion. Increased urinary magnesium losses may result from generalized renal disease or a specific renal defect in reabsorption of magnesium. Dialysis of patients may result in magnesium depletion if a low magnesium content dialysate is used. High rates of production of aldosterone (hyperaldosteronism), hyperparathyroidism, and diabetes mellitus cause increased urinary magnesium losses. Alcoholism, intensive diuretic therapy, and treatment with the antibiotic gentamycin result also in increased urinary magnesium losses[121,122] (see box in top of next column).

Magnesium deficiency is often associated with hypocalcemia, and the signs and symptoms of magnesium deficiency normally are the signs of hypocalcemia. Although serum magnesium concentrations can be low, since magnesium is predominantly an intracellular mineral, serum measurements may not reflect intracellular concentrations. Red cell magnesium concentrations have been advocated as a measure of intracellular magnesium status. Alternatively, magnesium ''retention'' studies have been suggested to assess the degree of magnesium ''depletion.'' A magnesium-''sufficient'' person would excrete large amounts of a magnesium load, whereas a ''deficient'' person would retain magnesium.[119,121,122]

Causes of hypomagnesemia

Decreased intake of magnesium
- Steatorrhea
- Malabsorption syndromes
- Gut resections
- Specific intestinal malabsorption of magnesium
- Protein-calorie malnutrition

Increased loss of magnesium
- Renal tubular loss
- Dialysis with low magnesium dialysate
- Hyperaldosteronism
- Hyperparathyroidism
- Diabetes mellitus
- Alcoholism
- Diuretic therapy
- Aminoglycoside therapy

Calcitonin

Abnormal serum calcitonin concentrations are only rarely encountered (see following box). Serum measurements are most useful in patients suspected of having medullary thyroid carcinoma, a malignancy of the thyroid C cells. This cancer is frequently seen in different members within families and is often associated with a tendency for

Diseases associated with abnormal serum calcitonin concentrations

Excess
- Medullary thyroid carcinoma
- Bronchogenic carcinoma
- Zollinger-Ellison syndrome
- Renal failure

Deficiency
- Thyroid agenesis
- Thyroidectomy
- Osteoporosis

other malignancies (termed *multiple endocrine neoplasia syndrome type II*).[123] Serum calcitonin measurements are useful both in the screening of family members who are potentially at risk of developing the disease and in the follow-up examination of previously treated patients suspected of recurrent metastatic disease. Serum calcitonin elevations are produced by a wide variety of other neoplasias, the most frequent one being bronchogenic carcinoma.[124] Since gastrin is a potent stimulus for calcitonin secretion, serum calcitonin concentrations are elevated in Zollinger-Ellison syndrome, a pancreatic tumor of gastrin-secreting cells. Finally, calcitonin excretion is decreased in patients with renal failure, and that decrease results in secondary elevation of serum concentrations of calcitonin in these patients.

Since the thyroid gland is usually the sole source of calcitonin production, athyroid patients lack circulating calcitonin.[125] Calcitonin levels are also decreased in some patients with osteoporosis. This may be caused by altered regulation of calcitonin synthesis or release.[126,127]

Sources of alkaline phosphatase

Osteoblasts
Bile canalicular cells
Placenta
Leukocytes
Proximal renal tubule cells
Active mammary gland

Bone diseases associated with abnormal serum alkaline phosphatase concentrations

Deficiency
- Hypophosphatasia
- Achondroplasia
- Severe malnutrition
- Scurvy

Excess
- Ostoblastic sarcoma
- Ostemomalacia
- Paget's disease
- Acromegaly
- Hyperparathyroidism
- Growing children

Alkaline phosphatase

In clinical practice, alkaline phosphatase determinations measure a group of enzymes that catalyze the hydrolysis of phosphate esters in alkaline medium.[128,129] Although alkaline phosphatase is produced by many tissues (see the first box on this page), only the portion produced by bone and liver is detected in serum from nondiseased persons. The causes of abnormal serum "bone" alkaline phosphatase concentrations are listed in the second box. Alkaline phosphatase is produced by osteoblasts and, as previously discussed, lowers bone pyrophosphate levels, which probably facilitates mineralization. Alkaline phosphatase synthesis is deficient in hypophosphatasia, a rare hereditary illness associated with undermineralized bones and pathological fractures, and achondroplasia, an inherited disorder of endochondral bone growth. Production is also decreased with generalized malnutrition or scurvy. Far more common than decreased concentrations are diseases associated with elevated serum alkaline phosphatase concentrations. Such elevations signify increased osteoblastic activity, as seen in osteoblastic sarcoma, rickets, Paget's disease, and acromegaly. The elevated levels associated with hyperparathyroidism result from secondary bone mineralization rather than parathyroid hormone–induced osteoclastic activity. One should exercise caution when considering the pathological significance of alkaline phosphatase increases in childhood, since growth is an important physiological cause of such elevations.[130] "Liver" alkaline phosphatase elevations reflect biliary obstruction and do not occur to any great extent with pure hepatocellular disease.

Hydroxyproline

Collagen, which is present predominantly in bone and skin, is the sole source of the amino acid hydroxyproline, which together with proline comprises approximately one third the total amino acid content of collagen. Collagen digestion, associated with either bone or skin breakdown, results in elevated urinary hydroxyproline concentrations (see box below).[131]

Conditions associated with elevated urinary hydroxyproline concentrations

Paget's disease	Acromegaly
Osteomalacia	Rheumatoid arthritis
Neoplastic bone disease	Osteoporosis
Hyperthyroidism	Aseptic bone necrosis
Osteomyelitis	Chronic renal failure
Burns	

REFERENCES

Bone structure and function

1. Albright, F., and Reifenstein, C.C., Jr.: The parathyroid glands and metabolic bone disease, Baltimore, 1948, Williams & Wilkins, p. 5.
2. Grant, J.C.B., and Basmajian, J.V.: Grant's method of anatomy, ed. 7, Baltimore, 1965, The Williams & Wilkins Co, pp. 5-8.
3. Ham, A.W., and Cormack, D.H.: Histology, ed. 8, Philadelphia, 1979, J.B. Lippincott Co., pp. 377-463.
4. Potts, J.T., and Deftos, L.J.: Parathyroid hormone, calcitonin, vitamin D, bone and bone mineral metabolism. In Bondy, P.K., and Rosenberg, L.E., editors: Duncan's diseases of metabolism, Philadelphia, 1974, W.B. Saunders Co., pp. 1275-1430.
5. Baylink, D.J., and Lin, C.C.: The regulation of endosteal bone volume, J. Periodontol. **50:**43-49, 1979.
6. Bassett, C.A.L.: Electromechanical factors regulating bone architecture, In Fleisch, H., Blackwood, H.J.J., and Owen, M., editors: Calcified tissues, Proceedings of Third European Symposium, New York, 1966, Springer-Verlag New York, pp. 79-89.
7. Mears, D.C.: Effects of parathyroid hormone and thyrocalcitonin on the membrane potential of osteoclasts, Endocrinology **88:**1021-1028, 1971.

Biochemistry and physiology

8. DeLuca, H.F.: The kidney as an endocrine organ for production of 1,25-dihydroxyvitamin D_3, a calcium-mobilizing hormone, N. Engl. J. Med. **289:**359-365, 1973.
9. Hollick, M.F., Frommer, J.E., McNeill, S.C., et al.: Photometabolism of 7-dehydrocholesterol to previtamin D_3 in skin, Biochem. Biophys. Res. Commun. **76:**107-114, 1977.
10. Haussler, M.R., and McCain, T.A.: Basic and clinical concepts related to vitamin D metabolism and action, N. Engl. J. Med. **297:** 974-983, 1041-1050, 1977.
11. Avioli, L.V.: Hormonal aspects of vitamin D metabolism and its clinical implications, Clin. Endocrinol. Metab. **8:**547-577, 1979.
12. DeLuca, H.F.: Mechanism of action and metabolic fate of vitamin D, Vitam. Horm. **25:**315-367, 1967.
13. Blunt, J.W., DeLuca, H.F., and Schnoes, H.K.: 25-hydroxycholecalciferol: a biologically active metabolite of vitamin D_3, Biochemistry **7:**3317-3322, 1968.
14. Tanaka, Y., and DeLuca, H.F.: Bone mineral mobilization activity of 1,25-dihydroxycholecalciferol, a metabolite of vitamin D, Arch. Biochem. Biophys. **146:**574-578, 1971.
15. Myrtle, J.F., and Norman, A.W.: Vitamin D: a cholecalciferol metabolite highly active in promoting intestinal calcium transport, Science **171:**79-82, 1971.
16. Haddad, J.G., and Stamp, T.C.B.: Circulating 25-hydroxyvitamin D in man, Am. J. Med. **57:**57-62, 1974.
17. Avioli, L.V., and Haddad, J.G.: Vitamin D: current concepts, Metabolism **22:**507-531, 1973.
18. Chesney, R.W., Rosen, J.F., Hamstra, A.J., and DeLuca, H.F.: Serum 1,25-dihydroxyvitamin D levels in normal children and in vitamin D disorders, Am. J. Dis. Child. **134:**135-139, 1980.
19. Chesney, R.W., Rosen, J.F., Hamstra, A.J., et al.: Absence of seasonal variation in serum concentrations of 1,25-dihydroxyvitamin D despite a rise in 25-hydroxyvitamin D in summer, J. Clin. Endocrinol. Metab. **53:**139-142, 1981.
20. Barbour, G.L., Coburn, J.W., Slatopolsky, E., Norman, A.W., and Horst, R.L.: Hypercalcemia in an anephric patient with sarcoidosis: evidence for extrarenal generation of 1,25-dihydroxyvitamin D, N. Engl. J. Med. **305:**440-443, 1981.
21. Howard, G.A., Turner, R.T., Sherrard, D.J., and Baylink, D.J.: Human bone cells in culture metabolize 25-hydroxyvitamin D_3 to 1,25-dihydroxyvitamin D_3 and 24,25-dihydroxyvitamin D_3, J. Biol. Chem. **256:**7738-7740, 1981.
22. Franceschi, R.T., Simpson, R.J., and DeLuca, H.F.: Binding proteins for vitamin D metabolites: serum carriers and intracellular receptors, Arch. Biochem. Biophys. **210:**1-13, 1981.
23. DeLuca, H.F., Vitamin D metabolism and function, Arch. Intern. Med. **138:**836-847, 1978.
24. DeLuca, H.F.: The vitamin D system in the regulation of calcium and phosphorus metabolism, Nutr. Rev. **37:**161-193, 1979.
25. Manolagas, S.C., Burton, D.W., and Deftos, L.J.: 1,25-dihydroxyvitamin D_3 stimulates alkaline phosphatase of osteoblast-like cells, J. Biol. Chem. **256:**7115-7117, 1981.
26. Bordier, P., Rasmussen, H., Marie, P., et al.: Vitamin D metabolites and bone mineralization in man, J. Clin. Endocrinol. Metab. **46:**284-294, 1978.
27. Fouriner, A., Bordier, P., Gueria, J., et al.: Comparison of 1α-hydroxycholecalciferol and 25-hydroxycholecalciferol in the treatment of renal osteodystrophy: Greater effect of 25-hydroxycholecalciferol on bone mineralization, Kidney Int. **15:**196-204, 1979.
28. Mason, R.S., Lissner, D., Wilkinson, M., and Rosen, S.,: Vitamin D metabolites and their relationship to azotaemic osteodystrophy, Clin. Endocrinol. **13:**375-385, 1980.
29. Eisman, J.A., Pounce, R.L., Ward, J.D., and Moseby, J.M.: Modulation of plasma 1,25-dihydroxyvitamin D in man by stimulation and suppression tests, Lancet **2:**931-933, 1979.
30. Gallagher, J.C., Riggs, L.B., Eisman, J., et al.: Intestinal calcium absorption and serum vitamin D metabolites in normal subjects and osteoporotic patients: effect of age and dietary calcium, J. Clin. Invest. **64:**729-736, 1979.
31. Kumar, R., Cohen, W.R., Silva, P., and Epstein, F.H.: Elevated 1,25-dihydroxyvitamin D levels in normal human pregnancy and lactation, J. Clin. Invest. **63:**342-344, 1979.
32. Whitsett, J.A., Ho, M., Tsang, R.C., et al.: Synthesis of 1,25-dihydroxyvitamin D_3 by human placenta in vitro, J. Clin. Endocrinol. Metab. **53:**484-488, 1981.
33. MacCallum, W.B., and Voegtlin, C.: On the relation of tetany to the parathyroid glands and to calcium metabolism, J. Exp. Med. **2:**118-151, 1908.
34. Collip, J.B, The extraction of a parathyroid hormone which will prevent or control parathyroid tetany and which regulates the level of blood calcium, J. Biol. Chem. **63:**395-438, 1925.
35. Berson, S.A., Yallow, R.S., Aurbach, G.D., and Potts, J.T.: Immunoassay of bovine and human parathyroid hormone, Proc. Nat. Acad. Sci. (Washington) **49:**613-617, 1963.
36. Tsang, R.C., Noguchi, A., and Steichen, J.J.: Pediatric parathyroid physiology, Pediatr. Clin. North Am. **26:**223-249, 1979.
37. Munson, P.L.: Studies on the role of the parathyroids in calcium and phosphate metabolism, Ann. N.Y. Acad. Sci. **60:**776, 1955.
38. Copp, D.H., Cameron, E.C., Cheney, B.A., et al.: Evidence for calcitonin—a new hormone from the parathyroid that lowers blood calcium, Endocrinology **70:**638-649, 1962.
39. Foster, G.V., Baghdzintz, H., Kumav, M.A., et al.: Thyroid origin of calcitonin, Nature **202:**1303-1305, 1964.
40. McMillan, P.J., Hooker, W.M., and Deftos, L.J.: Distribution of calcitonin-containing cells in the human thyroid, Am. J. Anat. **140:**73-79, 1974.
41. Austin, L.A., and Heath, H., III: Calcitonin physiology and pathophysiology, N. Engl. J. Med. **304:**269-278, 1981.
42. Hosking, D.J., Dinton, L.B., Cadge, B., and Martin, T.J.: Functional significance of antibody formation after long-term salmon CT therapy, Clin. Endocrinol. **10:**243-252, 1979.
43. Heersche, J.N.M., Marcus, R., and Aurbach, G.D.: Calcitonin and the formation of 3′,5′-AMP in bone and kidney, Endocrinology **94:**241-247, 1974.
44. Ham, J., Williams, J.C., and Ellison, M.L.: Radioreceptor assay for the detection of biologically active forms of calcitonin, J. Endocrinol. **81:**152-153, 1979.
45. Goltzman, D., and Tischler, A.S.: Characterization of the immunochemical forms of calcitonin released by a medullary thyroid carcinoma in tissue culture, J. Clin. Invest. **61:**449-458, 1978.
46. Huwler, R., Born, W., Ohnhaus, E.E., and Fischer, J.A.: Plasma kinetics and urinary excretion of exogenous human and salmon calcitonin in man, Am. J. Physiol. **236:**15-19, 1979.
47. Ardaillou, R.: Kidney and calcitonin, Nephron **15:**250-260, 1975.
48. Deftos, L.J.: Calcitonin. In Gray, C.H., and James, V.H.T., editors: Hormones and Blood, New York, 1979, Academic Press, Inc.
49. Bijvoet, O.L.M., van der Sluys Veer, J., deVries, H.R., and van Koppen, T.J.: Natriuretic effect of calcitonin in man, N. Engl. J. Med. **284:**681-688, 1971.
50. Deftos, L.J., and First, B.P.: Calcitonin as a drug, Ann. Intern. Med. **95:**192-197, 1981.

51. Cooper, C.W.: Recent advances with thyrocalcitonin, Ann. Clin. Lab. Sci. **6:**119-129, 1976.
52. Alkawa, J.K.: Magnesium: its biologic significance, CRC series on cations of biological significance, Boca Raton, Fla., 1981, CRC Press, Inc., pp. 1-129.
53. Wacker, W.E.C., and Parisi, A.F.: Magnesium metabolism, New Engl. J. Med. **278:**658-662, 712-717, 772-776, 1968.
54. Tsang, R.C.: Neonatal magnesium disturbances, Am. J. Dis. Child. **124:**282, 1972.

Bone disorders

55. Mundy, G.R.: Differential diagnosis of osteopenia, Hosp. Pract. **13:**65-72, 1978.
56. Chase, L.: Osteopenia, Am. J. Med. **69:**915-922, 1980.
57. Parfitt, A.M., Oliver, I., and Villanueva, A.R.: Bone histology in metabolic bone disease: the diagnostic value of bone biopsy, Orthop. Clin. North Am. **10:**329-345, 1979.
58. Ivey, J.L., and Baylink, D.J.: Postmenopausal osteoporosis: proposed roles of defective coupling and estrogen deficiency, Metab. Bone Dis. Rel. Res. **3:**3-7, 1981.
59. Avioli, L.V., Postmenopausal osteoporosis: prevention versus cure, Fed. Proc. **40:**2418-2422, 1981.
60. Wallach, S.: Management of osteoporosis, Hosp. Pract. **13:**91-98, 1978.
61. Draper, H.H., and Scythes, C.A.: Calcium, phosphorus, and osteoporosis, Fed. Proc. **40:**2434-2438, 1981.
62. Gallagher, J.C., Rigg, B.L., Eisman, J., et al.: Intestinal calcium absorption and serum vitamin D metabolites in normal subjects and osteoporotic patients: effect of age and dietary calcium, J. Clin. Invest. **64:**729-736, 1979.
63. Slovik, D.M., Adams, J.S., Neer, R.M., et al.: Deficient production of 1,25-dihydroxyvitamin D in elderly osteoporotic subjects, New Engl. J. Med. **305:**372-374, 1981.
64. Gallagher, J.C., Riggs, B.L., Jerpbak, C.M., and Arnaud, C.D.: Effect of age on serum immunoreactive parathyroid hormone in normal and osteoporotic women, J. Lab. Clin. Med. 95:373-385, 1980.
65. Shamonki, I.M., Fumar, A.M., Tataryn, I.V., et al.: Age-related changes of calcitonin secretion in females, J. Clin. Endocrinol. Metab. **50:**437-439, 1980.
66. Gallaher, J.C., Riggs, B.L., and DeLuca, H.F.: Effect of estrogen on calcium absorption and serum vitamin D metabolites in postmenopausal osteoporosis, J. Clin. Endocrinol. Metab. **51:**1359-1364, 1980.
67. Avioli, L.V.: Hormonal aspects of vitamin D metabolism and its clinical implications, Clin. Endocrinol. Metab. **8:**547-577, 1979.
68. Kooh, S.W., Jones, G., Reilly, B.J., and Fraser, D.: Pathogenesis of rickets in chronic hepatobiliary disease in children, J. Pediatr. **94:**870-874, 1979.
69. Daum, F., Rosen, J.F., Roginsky, M., et al.: 25-Hydroxycholecalciferol in the management of rickets associated with extrahepatic biliary atresia, J. Pediatr. **88:**1041-1043, 1976.
70. Heubi, J.E., Tsang, R.C., Steichen, J.J., et al.: 1,25-Dihydroxyvitamin D_3 in childhood hepatic osteodystrophy, J. Pediatr. **94:**977-982, 1979.
71. Eddy, R.L.: Metabolic bone disease after gastrectomy, Am. J. Med. **50:**442-445, 1971.
72. Hahn, T.J., Hendin, B.A., Scharp, C.R., et al.: Serum 25-hydroxycalciferol levels and bone mass in children on anticonvulsant therapy, N. Engl. J. Med. **292:**550-554, 1975.
73. Winnacker, J.L., Yeager, H., Saunders, J.A., et al.: Rickets in children receiving anticonvulsant drugs, Am. J. Dis. Child. **131:**286-290, 1977.
74. Scriver, C.R., Reade, T.M., DeLuca, H.F., and Hamstra, A.J.: Serum 1,25-dihydroxyvitamin D levels in normal subjects and in patients with hereditary rickets and bone disease, N. Engl. J. Med. **299:**976-979, 1978.
75. Fraser, D., Kooh, S.W., Kind, H.P., et al.: Pathogenesis of hereditary vitamin D–dependent rickets: an inborn error of vitamin D metabolism involving defective conversion of 25-hydroxyvitamin D to 1-α,25-dihydroxyvitamin D, New Engl. J. Med. **289:**817-822, 1973.
76. Brooks, M.H., Bell, N.H., Love, L., et al. Vitamin D–dependent rickets type II: resistance of target organs to 1,25-dihydroxyvitamin D, N. Engl. J. Med. **298:**996-999, 1978.
77. Eil, C., Liberman, U.A., Rosen, J.F., and Marx, S.J.: A cellular defect in hereditary vitamin D–dependent rickets type II: defective nuclear uptake of 1,25-dihydroxyvitamin D in cultured skin fibroblasts, N. Engl. J. Med. **304:**1588-1591, 1981.
78. Scriver, C.: Rickets and the pathogenesis of impaired tubular transport of phosphate and other solutes, Am. J. Med. **57:**43-49, 1974.
79. Chesney, R.W., Mazess, R.B., Rose, P., et al.: Supranormal 25-hydroxyvitamin D and subnormal 1,25-dihydroxyvitamin D: their role in X-linked hypophosphatemic rickets, Am. J. Dis. Child. **134:**140-143, 1980.
80. Drezner, M.K., Lyles, K.W., Haussler, M.R., and Harrelson, J.M.: Evaluation of a role for 1,25-dihydroxyvitamin D_3 in the pathogenesis and treatment of X-linked hypophosphatemic rickets and osteomalacia, J. Clin. Invest. **66:**1020-1032, 1980.
81. West, C.D., Blanton, J.C., Silverman, F.N., and Holland, N.H.: Use of phosphate salts as an adjunct to vitamin D in the treatment of hypophosphatemic vitamin D refractory rickets, J. Pediatr. **64:**469-477, 1964.
82. Glorieux, F.H., Marie, P.J., Pettifor, J.M., and Delvin, E.E.: Bone response to phosphate salts, ergocalciferol, and calcitriol in hypophosphatemic vitamin D–resistant rickets, N. Engl. J. Med. **303:**1023-1031, 1980.
83. Chesney, R.W., Rosen, J.F., Hamstra, A.J., and DeLuca, H.F.: Serum 1,25-dihydroxyvitamin D levels in normal children and in vitamin D disorders, Am. J. Dis. Child. **134:**135-139, 1980.
84. Steichen, J.J., Tsang, R.C., Greer, F.R., et al.: Elevated serum 1,25-dihydroxyvitamin D concentrations in rickets of very low-birth-weight infants, J. Pediatr. **99:**293-298, 1981.
85. Rowe, J.C., Wood, D.H., Rowe, D.W., and Raisz, L.G.: Nutritional hypophosphatemic rickets in a premature infant fed breast milk, N. Engl. J. Med. **300:**293-296, 1979.
86. Chesney, R.W., Hamstra, A.J., Mazess, R.B., et al.: Circulating vitamin D metabolite concentrations in childhood renal diseases, Kidney Int. **21:**65-69, 1982.
87. Mason, R.S., Lissner, D., Wilkenson, M., and Posen, S.: Vitamin D metabolites and their relationship to azotemic osteodystrophy, Clin. Endocrinol. **13:**375-385, 1980.
88. Juttmann, J.R., Buurman, C.J., De Kam, E., et al.: Serum concentrations of vitamin D in patients with chronic renal failure: consequences for the treatment with 1α-hydroxy-derivatives, Clin. Endocrinol. **14:**225-236, 1981.
89. Slatopolsky, E., Rutherford, W.E., Hruska, K., et al. How important is phosphate in the pathogenesis of renal osteodystrophy? Arch. Intern. Med. **138:**848-852, 1978.
90. Massry, S.G., and Ritz, E.: The pathogensis of secondary hyperparathyroidism of renal failure: is there a controversy? Arch. Intern. Med. **138:**853-856, 1978.
91. Singer, F.R., Schiller, A.L., Pyle, E.B., and Krane, S.M.: Paget's disease of bone. In Avioli, L.V., and Krane, S.M., editors: Metabolic bone disease, vol. 2, New York, 1978, Academic Press, Inc., pp. 490-575.
92. S. Wallach, editor: Symposium on Paget's disease, Clin. Orthop. **127:**2-110, 1977.
93. Singer, F.R.: Human calcitonin treatment of Paget's disease of bone, Clin. Orthop. **127:**86-93, 1977.

Analyte changes in disease

94. Hahn, T.J., Hendin, B.A., Scharp, C.R., et al.: Serum 25-hydroxycholecalciferol levels and bone mass in children on anticonvulsant therapy, N. Engl. J. Med. **292:**550-554, 1975.
95. Kooh, S.W., Jones, G., Reilly, B.J., and Fraser, D.: Pathogenesis of rickets in chronic hepatobiliary disease in children, J. Pediatr. **94:**870-874, 1979.
96. Fraser, D., Kooh, S.W., Kind, H.P., et al.: Pathogenesis of hereditary vitamin D–dependent rickets: an inborn error of vitamin D metabolism involving defective conversion of 25-hydroxyvitamin D to 1-alpha,25-dihydroxyvitamin D, N. Engl. J. Med. **289:**817-822, 1973.

97. Gallagher, J.C., Riggs, B.L., and DeLuca, H.F.: Effect of estrogen on calcium absorption and serum vitamin D metabolites in postmenopausal osteoporosis, J. Clin. Endocrinol. Metab. **51**:1359-1364, 1980.
98. Slovik, D.M., Adams, J.S., Neer, R.M., et al.: Deficient production of 1,25-dihydroxyvitamin D in elderly osteoporotic subjects, N. Engl. J. Med. **305**:372-374, 1981.
99. Drezner, M.K., Neelon, F.A., and Jowsey, J.: Hypoparathyroidism: a possible cause of osteomalacia, J. Clin. Endocrinol. Metab. **45**:114, 1977.
100. Drezner, M.K., Neelon, F.A., and Haussler, M.: 1,25-Dihydroxycholecalciferol deficiency: the probable cause of hypocalcemia and metabolic bone disease in pseudohypoparathyroidism, J. Clin. Endocrinol. Metab. **42**:621, 1976.
101. Lyles, K.W., and Drezner, M.K.: Parathyroid hormone effects on serum 1,25-dihydroxyvitamin D levels in patients with X-linked hypophosphatemic rickets: evidence for abnormal 25-hydroxyvitamin D-1-hydroxylase activity, J. Clin. Endocrinol. Metab. **54**:638-644, 1982.
102. Juttmann, J.R., Buurman, C.J., De Kam, E., et al.: Serum concentrations of vitamin D in patients with chronic renal failure: consequences for the treatment with 1-α-hydroxy derivatives, Clin. Endocrinol. **14**:225-236, 1981.
103. Malluche, H.H., Goldstein, D.A., and Massry, S.G.: Osteomalacia and hyperparathyroid bone disease in patients with nephrotic syndrome, J. Clin. Invest. **63**:494-500, 1979.
104. Goldstein, D.A., Haldimann, B., Sherman, D., et al.: Vitamin D metabolites and calcium metabolism in patients with nephrotic syndrome and normal renal function, J. Clin. Endocrinol. Metab. **52**:116-121, 1981.
105. Avioli, L.V., and Haddad, J.G.: Vitamin D.: current concepts, Metabolism, **22**:507, 1973.
106. Chesney, R.W., Rosen, J.F., Hamstra, A.J., and DeLuca, H.F.: Serum 1,25-dihydroxyvitamin D levels in normal children and in vitamin D disorders, Am. J. Dis. Child. **34**:135-139, 1980.
107. Kumar, R., Cohen, W.R., Silva, P., and Epstein, F.H.: Elevated 1,25-dihydroxyvitamin D levels in normal human pregnancy and lactation, J. Clin. Invest. **63**:342-344, 1979.
108. Barbour, G.L., Coburn, J.W., Slatopolsky, E., et al.: Hypercalcemia in an anephric patient with sarcoidosis: evidence for extrarenal generation of 1,25-dihydroxyvitamin D, N. Engl. J. Med. **205**:440-443, 1981.
109. Parfitt, A.M.: The spectrum of hypoparathyroidism, J. Clin. Endocrinol. **34**:152, 1972.
110. Tsang, R.C., Noguchi, A., and Steichen, J.J.: Pediatric parathyroid physiology, Pediatr. Clin. North Am. **26**:223-249, 1979.
111. Tsang, R.C., and Brown, D.R.: The parathyroids. In Kelley, V., editor: Practice of pediatrics, vol. 1, New York, 1979, Harper & Row, Publishers, Inc., pp. 1-16.
112. Paloyan, E., Paloyan, D., and Pickleman, J.R.: Hyperparathyroidism today, Surg. Clin. North Am. **53**:211, 1973.
113. Bricker, N.S.: On the pathogenesis of the uremic state: an exposition of the ''trade-off'' hypothesis, New Engl. J. Med. **286**:1093, 1972.
114. Tsang, R.C., and Venkataraman, P.: Pediatric parathyroid and vitamin D–related disorders. In Kaplan, S.A., editor: Clinical pediatric and adolescent endocrinology, Philadelphia, 1982, W.B. Saunders Co., pp. 346-365.
115. David, N.J., Verner, J.V., and Engel, F.L.: The diagnostic spectrum of hypercalcemia, Am. J. Med. **33**:88, 1962.
116. Taylor, A.B., Stern, P.H., and Bell, N.H.: Abnormal regulation of circulating 25-hydroxyvitamin D in the Williams syndrome, N. Engl. J. Med. **306**:972-975, 1982.
117. Juan, D.: Hypocalcemia differential diagnosis and mechanisms, Arch. Intern. Med. **139**:1166-1171, 1979.
118. Broadus, A.E.: Nephrogenous cyclic AMP, Recent Prog. Horm. Res. **37**:667, 1981.
119. Tsang, R.C.: Neonatal magnesium disturbances, Am. J. Dis. Child. **124**:282, 1972.
120. Randall, R.D., Cohen, M.D., Spray, C.C., et al.: Hypermagnesemia in renal failure, Ann. Intern. Med. **61**:73-88, 1964.
121. Alkawa, J.K.: Magnesium: its biologic significance, CRC series on cations of biological significance, Boca Raton, Fla., 1981, CRC Press, Inc., pp. 1-129.
122. Randall, R.E., Rossmeisl, E.C., and Bleifer, K.H.: Magnesium deficiency in man, Ann. Intern. Med. **50**:257-272, 1950.
123. Grace, K., Spiler, I.J., and Tashjian, A.H., Jr.: Natural history of familial medullary thyroid carcinoma: effect of a program for early diagnosis, N. Engl. J. Med. **299**:980-985, 1978.
124. Silva, O.L., Brode, L.E., and Doppman, J.L.: Calcitonin as a marker for bronchogenic cancer, Cancer **44**:680-684, 1979.
125. Carey, D.E., Jones K.L., Parthemore, J.G., and Deftos, L.J.: Calcitonin secretion in congenital nongoitrous creatinism, J. Clin. Invest. **65**:892-895, 1980.
126. Milhaud, G., Benezech-Lefevre, M., and Moukhtar, M.S.: Deficiency of calcitonin in age-related osteoporosis, Biomedicine **29**:272-276, 1978.
127. McTaggert, H., Ivey, J.L., Sisom, K., et al.: Deficient calcitonin response to calcium stimulation in post-menopausal osteoporosis, Lancet **1**:475-477, 1982.
128. Rosen, S.: Alkaline phosphatase, Ann. Intern. Med. **67**:183-203, 1967.
129. Kaplan, M.: Alkaline phosphatase, N. Engl. J. Med. **286**:200-201, 1972.
130. Root, A.W., and Harrison, H.E.: Recent advances in calcium metabolism, J. Pediatr. **88**:1-18, 1976.
131. Niejadlik, D.C.: Hydroxyproline, Postgrad. Med. **51**:214-216, 1972.

Chapter 26 Pancreatic function

Michael D.D. McNeely

acinar From the latin word *acinus,* which means a 'berry' or 'grape.' In anatomy, the term refers to a small saclike dilatation.

ampulla of Vater A muscular sphincter in the duodenum through which the pancreatic and biliary ducts enter the gastrointestinal lumen.

cholangiopancreatography After careful manipulation of a special tube through the duodenum, x-ray contrast material is injected into the bile and pancreatic ducts so that an x-ray visualization can be obtained.

cholecystography A radiological technique in which contrast material that collects in the gallbladder and bile duct is administered to the patient. An x-ray examination is then used to observe the anatomy of the biliary tract.

dextran A linear glucose polymer of variable molecular weight in which carbon 1 of each unit is linked to carbon 6 of the next unit.

glucogenesis The biochemical process of glucose synthesis.

glycogenolysis The biochemical process of glycogen breakdown into glucose.

hypoglycemia Low blood glucose concentration (often considered less than 500 mg/L).

islets of Langerhans Islets are clusters of cells, and the islets of Langerhans are clusters of cells within the pancreas that provide its endocrine function.

laparotomy Surgically opening the abdominal cavity.

oncofetal antigen A protein found in fetal and neoplastic tissue.

pancreatic duct A conduit that traverses the pancreas to convey exocrine secretions to the duodenum.

peritonitis Inflammation (usually caused by bacterial infection) of the lining of the abdominal cavity.

proteolytic Having the ability (usually of an enzyme) to break protein molecules into peptides.

sialitis Inflammation of salivary glands or ducts.

vagus nerve The tenth cranial nerve, which carries motor, sensory, and autonomic nerve fibers to the neck, thorax, and abdomen.

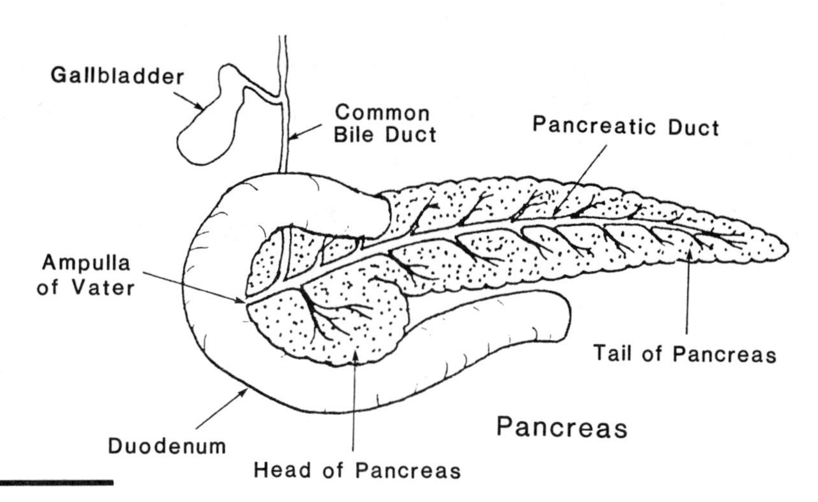

Fig. 26-1. Diagram of pancreas beside duodenum. Pancreatic duct extends throughout organ to convey exocrine enzymes into duodenum. Common bile duct enters duodenum beside or close to pancreatic duct. Endocrine islets are scattered throughout entire organ.

ANATOMY

The pancreas is a soft, variegated, linear strip of tissue approximately 15 cm long. It lies across the posterior wall of the abdomen and is surrounded on three sides by the duodenum. Its head lies to the right, in association with the midportion of the duodenum and its body and tail run toward the left to lie under the spleen. (Fig. 26-1). It receives a vigorous vascular supply from the aorta. The pancreatic duct runs throughout its length and extends branches throughout the organ. The duct ultimately empties into the duodenum through a muscular sphincter called the ''ampulla of Vater.'' In some persons this sphincter is shared with the bile duct. In other persons this sphincter is separate but in close association with that of the bile duct.

The pancreas serves two different functions. It is an *endocrine* gland that synthesizes the hormones glucagon, insulin, and gastrin. At the same time it serves as an *exocrine* gland by providing digestive enzymes in a bicarbonate fluid to facilitate the duodenal digestion of food.

The endocrine functions are invested in a collection of cells grouped in a characteristic configuration known as the ''islets of Langerhans'' (Fig. 26-2). These islets contain beta cells, which synthesize insulin, alpha cells, which synthesize glucagon, and delta cells, which produce gastrin.

The exocrine substances are produced by cells in acinar groupings. The secretory regions of these cells are arranged so that the digestive chemicals are emptied into the pancreatic ductal system (Fig. 26-3). The duct epithelium itself contributes to the fluid by secreting both water and electrolytes.

ENDOCRINE FUNCTIONS

As mentioned above, the endocrine functions of the pancreas are invested in the islets of Langerhans, which are composed of alpha cells, beta cells, and delta cells.[1] The pancreatic endocrine hormones are listed in Table 26-1.

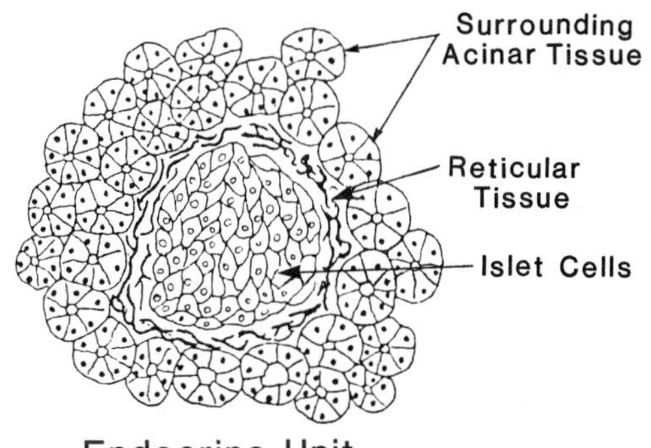

Fig. 26-2. Islet cells are collected together in a cluster separated from acinar tissue by a thin layer of reticular tissue. On routine stains, islet cells appear to be similar but can be distinguished by special staining techniques. Islet cells release their hormones directly into circulation.

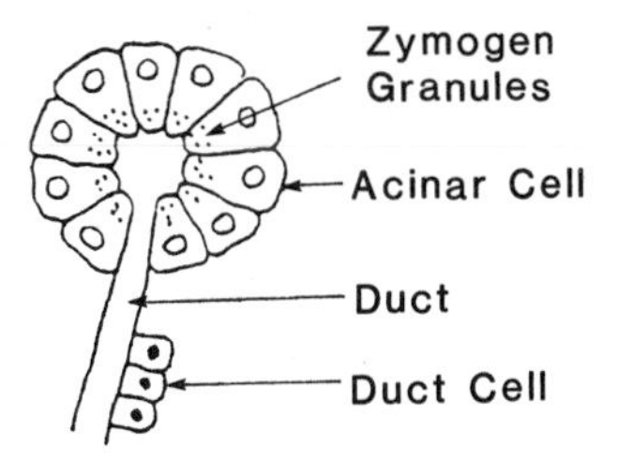

Fig. 26-3. Exocrine acinus terminates with a collection of acinar cells. They contain zymogen granules, which are loaded with proteolytic enzymes. Along wall of duct are located special cells that contribute fluid and bicarbonate.

Table 26-1. Pancreatic endocrine hormones

Cell of origin	Hormone	Molecular weight (daltons)	Factors that initiate hormone release	Factors that inhibit hormone release	Hormone action
Alpha cell	Glucagon	3485	Hypoglycemia, starvation, exercise, stress, pancreozymin, gastroinhibitory polypeptide	Most factors that stimulate insulin release	Increases plasma glucose by stimulating hepatic glycogenolysis and gluconeogenesis and adipose tissue lipolysis
Beta cell	Proinsulin	11,500	Glucose, leucine, arginine, histidine, phenylalanine, sulfonylureas, adrenocorticotropic hormone, glucagon, growth hormone, α-adrenergic blockage, β-adrenergic stimulation, vagal stimulation, secretin, cholecystokinin	α-Adrenergic stimulation, β-adrenergic blockade, thiazide diuretics, phenytoin, diazoxide	Insulin precursor
	Insulin	5734			Stimulates membrane transport, alters membrane-bound enzymes, stimulates protein synthesis, inhibits protein degradation, stimulates messenger RNA synthesis, stimulates DNA synthesis
Delta cell	Gastrin	2098	See Chapters 27 and 40		Role in pancreas not understood
	Somatostatin	1640			Not fully understood
F cell	Pancreatic polypeptide	4226	See Chapters 27 and 42		Not fully understood

Alpha cells

Alpha cells make up 20% to 30% of all the cells in the islet. Similar cells have been identified in the duodenal mucosa.

The alpha cells are known to produce the hormone glucagon. Glucagon is made up of 29 amino acids in a straight chain with a molecular weight of 3485 daltons. A glucagon of gastrointestinal tract mucosal origin is also known to exist. This is immunoreactive material that has been called "gut glucagon," "enteroglucagon," or "glucagon-like immunoreactive material."

Insulin-induced hypoglycemia is followed by increases in the concentration of glucagon in the pancreas and peripheral blood. Similarly, starvation, the ingestion of protein, the administration of certain amino acids (arginine or alanine), exercise, and stress also stimulate the secretion of pancreatic glucagon. Pancreozymin and gastroinhibitory polypeptide will increase glucagon. Hyperglycemia will cause suppression of glucagon secretion.

Glucagon stimulates hepatic glycogenolysis, which converts glycogen into glucose. Glucagon also stimulates glucogenesis by promoting the hepatic uptake of amino acids. These mechanisms increase the plasma glucose concentration. Glucagon is also capable of stimulating the release of insulin from the pancreatic beta cells independently of the increases in blood glucose. It is presumed that the physiological role of glucagon is to smooth the metabolic effects of glucose absorption and insulin release.

Beta cells

The beta cells make up 60% to 70% of the pancreatic islet cells.

The beta cells are responsible for synthesizing insulin. They do this by producing a large, preproinsulin molecule that is stored in beta granules. This large molecule is enzymatically broken down to liberate active insulin, C peptide, and a variety of intermediate products.

The function of insulin is covered in greater detail in Chapter 30.

Delta cells

The delta cells are minor components (2% to 8%) of the islets. They are also present in the stomach and scattered within the pancreatic ductal epithelium. Delta cells are capable of producing gastrin.

EXOCRINE FUNCTIONS

The exocrine function of the pancreas is to synthesize and convey potent digestive enzymes into the duodenum.[2] The action of these enzymes is discussed in detail in Chapter 27.

The hormone cholecystokinin-pancreozymin is the main stimulus for the release of digestive enzymes from the pancreatic acinar cells.

A number of proteolytic enzymes are formed (Fig. 26-4). These are synthesized as zymogen granules in the acinar cells. When cholecystokinin-pancreozymin or vagal stimulation is received, the granule is released into the lumen of the pancreatic duct. The proteolytic enzyme precursors trypsinogen, chymotrypsinogen, proelastase, and procarboxypeptidase are released in this way into the ductal lumen, from which they are then conveyed into the

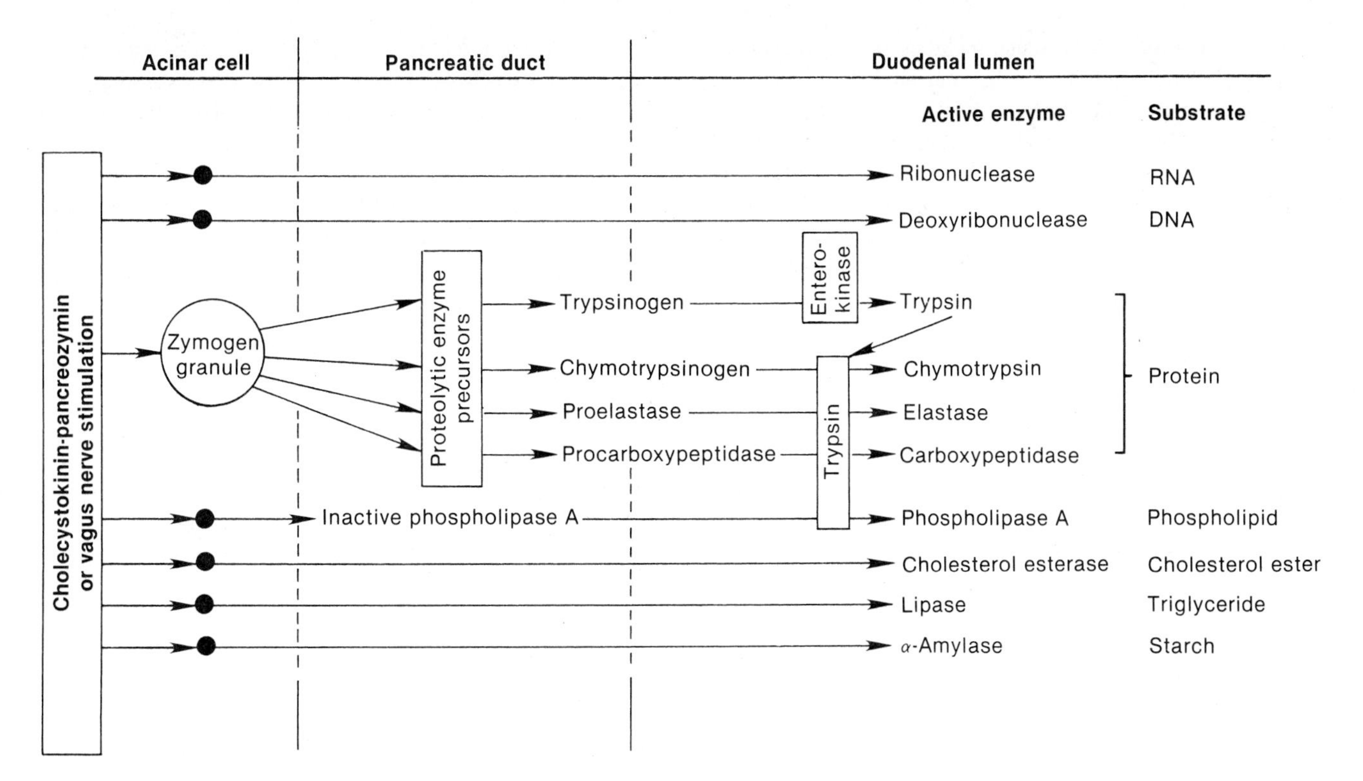

Fig. 26-4. Pancreatic exocrine enzymes.

duodenum. Here an enzyme known as "enterokinase" is released by the duodenal mucosal cell after stimulation by bile salts and proteases. Enterokinase splits the lysine-isoleucine bond in the trypsinogen molecule. The resulting amino acid sequence refolds itself to form an active site. This newly created molecule is known as "trypsin." Trypsin has the ability to cause catalysis of the trypsinogen. This reaction is very slow and is probably of little physiological consequence. Trypsin does act on the other zymogens to create the active proteolytic enzymes chymotrypsin, elastase, and carboxypeptidase A and carboxypeptidase B. Each of the proteolytic enzymes has a different function in protein breakdown.

Also secreted into the pancreatic duct are ribonuclease and deoxyribonuclease. These are secreted in the active form.

Alpha amylase is also synthesized and released in a similar fashion. It will split starch into limit dextrans and maltose.

Lipids are acted on by the enzymes lipase, phospholipase A, and cholesterol esterase. Phospholipase A requires trypsin activation.

The hormone secretin is elaborated by the S cells of the duodenum and jejunum. It stimulates the pancreatic duct to produce water and bicarbonate. These aid in washing the enzymes into the duodenum and provide the appropriate pH environment for enzyme activity.

PATHOLOGICAL CONDITIONS

Pancreatitis[3-5]

Pancreatitis (inflammation of the pancreas) occurs in two clinical forms: acute and chronic. Two types of acute pancreatitis are recognized.

The acute edematous variety occurs in 80% of cases. The remaining 20% are hemorrhagic and are much more serious, being characterized by parenchymal necrosis and hemorrhage. Hemorrhagic pancreatitis has a high mortality. It is usually impossible to distinguish between the two forms when they first occur with severe abdominal pain.

Acute hemorrhagic pancreatitis can cause cardiovascular shock and hemorrhage into the gland. Up to 10 L of fluid may be lost into extracellular spaces in 24 hours. Pancreatic enzymes become activated and cause local damage and microcirculatory effects. Serum calcium may fall precipitously.

Pancreatitis is caused by infections, obstruction to the duct, vascular disorders, toxins, trauma, and metabolic derangements. The common factor is possibly an increase in membrane permeability. Autodigestion of pancreatic cells by proteolytic enzymes is the key event, and it is believed that trypsin activation takes place first. This causes activation of the kinin system, the coagulation cascade, and platelet aggregation. Widespread effects on blood vessels (including those of the lung) contribute to the serious physiological disruption that occurs during pancreatitis.

Current medical management includes the replacement of the fluid deficit, relief of pain, and an attempt to suppress pancreatic stimulation by use of nasogastric decompression. Pharmacological intervention is being investigated, but its efficacy is not yet proved. Attention is given to the respiratory system since respiratory failure is a common problem. It is very important to confirm the diagnosis of acute pancreatitis so that treatment is started early and inappropriate treatment is avoided.

A chronic form of pancreatitis also occurs. In this disorder, inflammation of the pancreatic tissue ebbs and wanes, causing intermittent discomfort and chronic pancreatic insufficiency. Usually the clinical picture is not dramatic enough to make diagnosis easy. Diagnosis is ultimately confirmed by a variety of visualization techniques, including the barium meal, endoscopic retrograde cholangiopancreatography, 2-scale ultrasonography, and angiography.

Cancer of pancreas

Adenocarcinoma.[6,7] The incidence of adenocarcinoma of the pancreas has increased during the past 20 years and now accounts for 5% of all cancer deaths in the United States. The overall 5-year survival rate is less than 1%. The principal reason for this dismal prognosis is the inability to diagnose the disease at an early stage. Approximately 90% of all pancreatic cancers have undergone extension or have metastasized by the time of diagnosis. Most pancreatic cancers are localized in the head of the gland. When enlarged in this site, the cancer can obstruct the bile duct and cause obstructive jaundice. Cancers in the body and tail of the pancreas are extremely difficult to diagnose. The most common symptoms are pain, jaundice, and weight loss, which develop insidiously and progressively. The results of laboratory tests are often normal and are never specific. The serum amylase may be elevated, or the patient may be mildly anemic. Pancreatic function tests may reveal abnormalities but are of no diagnostic value. A variety of oncofetal (tumor) antigens have been examined with the hope that they may shed some light on this disease, but they have not yet been proved to do so. Currently the diagnosis is made using direct visualization techniques, radiological studies, and laparotomy.

Endocrine tumors.[8] A group of unusual hormone-synthesizing tumors of the pancreas are recognized. They are derived from cells with the common ability of amine-precursor uptake and decarboxylation (APUD). Various tumors are recognized.

Insulinomas are the most common of these tumors. They produce a characteristic fasting hypoglycemic syndrome. One can identify them by subjecting the patient to a 72-hour fast and discovering hypoglycemia (glucose less than 400 mg/L) and inappropriate plasma insulin concentrations (insulin in microunits per milliliter divided by glucose in milligrams per liter, greater than 0.03).

The glucagonoma is a glucagon-secreting tumor that, when it metastasizes, produces a syndrome of diabetes, characteristic dermatitis, and anemia. The diagnosis is made when elevated plasma glucagon and plasma insulin are found.

Somatostatinomas do not produce a characteristic syndrome.

Pancreatic polypeptide islet cell tumors have been found in patients with MEA I syndrome.

Carcinoid tumors have been described in the pancreas.

Gastrinomas produce the Zollinger-Ellison syndrome, described in Chapter 27.

Vipomas (Verner-Morrison syndrome) cause a voluminous (5 L/day), explosive, watery diarrhea associated with hypokalemia and hypochlorhydria. The syndrome is caused by excess vasoactive intestinal polypeptide (VIP).

Pancreatic tumors have been known to produce ectopic ACTH and parathyroid hormone.

Pancreatic insufficiency[9]

Pancreatic insufficiency can be caused by repeated attacks of acute pancreatitis or long-standing chronic pancreatitis. In such cases the pancreatic parenchyma is gradually destroyed. When 75% of the gland has been disabled, detectable changes will occur in the digestive process. Because of this tremendous reserve, pancreatic insufficiency becomes apparent only very late in the disease. Diabetes mellitus can also occur, since the islets become damaged. Evaluation of pancreatic function is made by stimulation of the pancreas and measurement of the maximum output of pancreatic enzymes. Therapy is directed toward correcting nutritional deficiencies, avoiding alcohol, correcting the diabetes mellitus, and correcting the enzyme defects with oral pancreatic enzymes. The results of pancreatic insufficiency are discussed in Chapter 27.

Cystic fibrosis[10]

Cystic fibrosis is an inherited defect in exocrine gland function in which secretions from the liver, gallbladder, duodenum, pancreas, lungs, salivary glands, and genital tract are abnormal. It is characterized clinically by chronic lung disease and general malnutrition, secondary to malabsorption. It may present in the neonatal period as intestinal obstruction with undigested meconium. With modern treatment, the life-span of persons with this disease has increased far beyond the 60% mortality before 10 years of age, which was usual a decade ago.

A diagnosis of cystic fibrosis requires that at least two of the following four criteria are met: (1) characteristic lung disease, (2) pancreatic insufficiency, (3) increased concentration of sweat chloride, and (4) a positive family history.

The demonstration of an increased sweat chloride is the most characteristic diagnostic test. Sweat-chloride values must exceed 60 mmol/L. Normal values vary with age, and abnormally high values may occasionally be found in other conditions, including glucose-6-phosphatase deficiency, glycogen-storage disease, hypothyroidism, renal insufficiency, and malnutrition. The only acceptable test for measuring sweat chloride is pilocarpine iontophoresis sweat stimulation followed by gravimetric quantitation of the amount of sweat produced and coulometric measurement of the chloride.

Diabetes

Diabetes is a generalized disorder of glucose metabolism that is generally attributed to an insufficiency of insulin. This is covered in greater detail in Chapter 30.

Hypoglycemia.

Pancreatic tumors that produce insulin are one cause of hypoglycemia. This diagnosis of is often very difficult and is covered in greater detail in Chapter 30.

PANCREATIC FUNCTION TESTS[11]

General

Pancreatic assessment has been greatly enhanced by the development of intubation techniques that allow a variety of tubes to be introduced into very specific regions of the gastrointestinal tract.

Secretin-stimulation test[12-13]

In this test a double-lumen tube is positioned under fluoroscopic guidance so that one tip is situated in the duodenum and the other is left in the stomach to remove gastric juice. Two basal collections of endogenous duodenal contents are obtained 10 minutes apart. Secretin is then administered intravenously as 1 unit per kilogram of body weight. Secretin administered in this way will stimulate the pancreas to produce water and bicarbonate. For assessment of this response, three 10-minute samples are removed from the duodenum, labeled appropriately, and put on ice.

The volume of all samples should be measured. A normal postsecretin response will be a significant increase in volume (1.5 mL/kg/30 min). Bicarbonate is also measured on all samples; a peak bicarbonate concentration will be at least 90 mmol/L. Low fluid volumes are characteristic of pancreatic duct obstruction. Low bicarbonate concentration suggests chronic pancreatitis. Tumors of the pancreas may also cause decreased flow. Normally the amylase activity will increase, but enzyme behavior is not diagnostically useful.

Pancreozymin-stimulation test[12-13]

Pancreozymin is administered intravenously immediately after the collection of the samples for the secretin-stimulation test. This material has not yet been fully approved and may be difficult to obtain. In addition, the normal response is not well characterized. The enzymatic content of the stimulated pancreatic juice is used to assess pancreatic function.

Cholecystokinin-stimulation test

The use of cholecystokinin is restricted to experimental work and is not used routinely in gastrointestinal function testing at this time.

Lundh test[14]

In the Lundh test, trypsin determinations are carried out on duodenal aspirations after a standard meal.

The patient fasts for 12 hours before the test, and medications that affect gastrointestinal motility are withheld. A tube is inserted under fluoroscopic guidance into the third portion of the duodenum. A test meal consisting of 8 g of corn oil, 15 g of casein, and 40 g of glucose in warm water with flavoring is given. The total volume is about 300 mL. The timed meal is taken through a straw over 15 minutes. The patient is then placed on his right side and the tube is allowed to drain into an iced container. The drainage continues for 2 hours. The volume of material collected is measured, and an aliquot is titrated to determine the hydrogen-ion content. The secretion of hydrogen ion by normal persons is 12 to 20 mmol/min/mL. Those with pancreatic insufficiency secrete less than 10 mmol/min/mL.

An intact duodenal mucosa must be present to mediate the hormone. A vagotomy or duodenal bypass surgery will also diminish the response. The test will be abnormal in 90% of patients with chronic pancreatitis and 79% of those with carcinoma of the pancreas.

CHANGE OF ANALYTE IN DISEASE (Table 26-2)

Amylase[15-16]

Amylase is the enzyme secreted by the exocrine pancreas for the purpose of breaking down starch into limit dextrins and maltose. Eighty percent of all patients with acute pancreatitis experience an elevation of serum amylase within 24 hours. Such elevations are between two and 20 times the upper limit of normal. If the attack is self-limiting, the value will return to base line within 2 days to 1 week. As a result, serum amylase determinations provide a convenient approach to the diagnosis of acute pancreatitis. Hypoglycemia can depress amylase values, and therefore the blood must be drawn before intravenous fluid therapy is initiated. Rarely, acute hemorrhagic pancreatitis will be so severe than all pancreatic tissue is destroyed and the serum amylase level is normal. Seventy percent of all persons receiving morphine will display a raised serum amylase up to three times normal. The reason is that opiates are known to cause constriction of the sphincter of Oddi and back pressure into the pancreatic duct. Also to be avoided are choline, methocholine, chlorothiazide, pancreozymin, and secretin. Patients with chronic pancreatitis

Table 26-2. Test changes in disease

	Serum amylase	Urine amylase	Serum lipase
Acute pancreatitis	↑ to ↑ ↑ ↑ in 80%	↑ ↑ ↑	↑ to ↑ ↑ in 90%
Opiate administration	↑ in 75%	↑	↑
Cholinergic drugs	↑ many	↑	↑
Chronic pancreatitis	N or ↑	N or ↑	N or ↑
Carcinoma of pancreas	↑ rare	↑ rare	↑ 50%
Pseudocyst	Prolonged ↑	↑	↑
Macroamylasemia	Prolonged ↑	N	N
Salivary gland disease	↑	↑	N
Tubo-ovarian abscess	↑ or N	↑ or N	N
Sympathetic pleural effusion	↑ or N	↑ or N	N
Intestinal obstruction	↑ or N	↑ or N	↑ or N
Abdominal trauma	↑ or N	↑ or N	↑ or N
Mesenteric thrombosis	↑ or N	↑ or N	↑ or N
Peritonitis	↑ or N	↑ or N	↑ or N
Cholecystography	↑ or N	↑ or N	↑ or N
Common duct obstruction	↑ or N	↑ or N	↑ or N
Perforated ulcer	↑ or N	↑ or N	↑ or N
Dissecting aortic aneurysm	↑ or N	↑ or N	↑ or N
Bronchogenic carcinoma	↑ or N	↑ or N	↑ or N
Splenic hemorrhage	↑ or N	↑ or N	↑ or N
Uremia	↑ usually	↓ or N	N

N, Normal; ↑, elevated; ↓, lowered.

have normal levels of amylase between attacks, and even during attacks amylase elevations may not occur. Carcinoma of the pancreas is not generally associated with a raised amylase unless the disease has caused some form of destructive lesion. Pancreatic pseudocysts occasionally cause prolonged elevations of serum amylase.

A number of other conditions also cause an increased serum amylase, including mumps, parotitis, sialitis, tubo-ovarian abscess, sympathetic pleural effusion, intestinal obstruction, abdominal trauma, intra-abdominal surgery, mesenteric thrombosis, peritonitis, cholecystography, common duct obstruction, perforated duodenal or gastric ulcer, dissecting aortic aneurysm, bronchogenic carcinoma, cerebral trauma (rare), splenic hemorrhage, and uremia.

Amylase is a relatively small protein (molecular weight 45,000 daltons) and is therefore filtered readily into the urine. The enzyme may be found in increased concentrations in the urine for longer periods of time than in the serum. A 2-hour urine collection for amylase content is an excellent test for detecting pancreatitis. The test will be increased by the same conditions that alter the serum amylase level. In addition, urine amylase has been artificially increased by saliva surreptitiously added to the collection vessel.[17]

An interesting condition that has recently been recognized is macroamylasemia.[18] In this condition, very high levels of serum amylase are found but pancreatitis is not present. The elevated values persist indefinitely. The amount of amylase found in the urine is low or normal. Macroamylase is a giant molecule that results from the complexing of normal amylase to high molecular weight proteins (usually immunoglobulins).

Some recent enthusiasm for the application of the amylase-to-creatinine clearance ratio has been generated on the basis that this procedure gives a more specific indication of pancreatitis than serum or urinary measurements alone. The use of this ratio has not been widely accepted because experience has not proved it to be of any substantial benefit except in highlighting macroamylasemia.[19]

Recently, techniques have become available for separating amylase isoenzymes.[20] These methods demonstrate P type (from pancreas), S type (from salivary glands), and other unusual forms.

Lipase[21]

Serum lipase rises as fast as amylase and remains elevated for a much longer period of time (7 to 10 days). It is not elevated by as many conditions as serum amylase and therefore should be considered much more specific. Unfortunately, analytical difficulties have prevented the widespread application of serum lipase in the routine diagnosis of pancreatitis. The test is elevated in 50% of patients with carcinoma of the pancreas and is occasionally found to be elevated in people with cirrhosis of the liver without pancreatitis. It rises with opiates and cholinergics because they affect the sphincter of Oddi. Serum lipase can be detected in 90% of patients with acute pancreatitis using appropriate analytical techniques. Urinary lipase cannot be detected.

Methemalbumin[22]

Detection of methemalbumin in the serum and ascitic fluid of patients with suspected acute pancreatitis is strong evidence for the hemorrhagic variety of this condition. Methemalbumin is formed by the action of pancreatic digestive enzymes on hemoglobin. Other disorders that cause blood loss into the abdominal cavity also result in methemalbumin.

Nonpancreatic enzymes in pancreatic disease

A variety of "nonpancreatic" serum enzyme changes occur in pancreatic disease. None is specific.

If a pancreatic neoplasm causes obstruction to bilary outflow, typical hepatic enzyme changes will be noted. Early enthusiasm for the use of the serum leucine aminopeptidase determination as a clue to pancreatic tumor has waned because this method has not stood the test of time.

Lactate dehydrogenase increases caused by LD_2 and LD_3 are sometimes caused by malignancy of the pancreas.

In pancreatitis, 5′-nucleotidase, alanine aminotransferase, aspartate aminotransferase, and gamma glutamyl transferase are frequently elevated.

Lactoferrin

It has been reported that the measurement of lactoferrin and trypsin in pancreatic juice can differentiate between patients with pancreatic cancer and chronic pancreatitis.[23] The intubation must be made directly from the pancreatic duct.

Insulin

The primary reason for measuring insulin in serum is to detect the presence of an insulinoma. This tumor will be characterized by an inappropriately increased serum insulin in association with fasting hypoglycemia. "Inappropriate" means that the insulin (microunits per milliliter) divided by the plasma glucose (milligrams per liter) is greater than 0.03.

Serum insulin also increases in generalized acute illnesses, hyperthyroidism, some diabetic states, acromegaly, adrenal hyperfunction, type IV hyperlipoproteinemia, cirrhosis of the liver, and renal failure.

Low levels are found in diabetes mellitus, pheochromocytoma, malnutrition, and cystic fibrosis.

A common dilemma is whether hyperinsulinemia is caused by a tumor or is the result of self-administration. Resolution is achieved by measurement of the C-peptide portion of the proinsulin molecule, which is present only with endogenous insulin.

Insulin antibodies

Insulin antibodies are present in most persons who have received pharmaceutical insulin. Some patients have exceedingly high values, which may greatly alter their insulin requirements. It is important to measure both total and unbound antibodies.

REFERENCES

1. Unger, R.H.: Alpha- and beta-cell interrelationship in health and disease, Metabolism **23:**581-593, 1974.
2. Hadorn, B.: The exocrine pancreas. In Anderson, C.M., and Burke, V., editors: Pediatric gastroenterology, Oxford, 1975, Blackwell Scientific Publications, Ltd., pp. 289-327.
3. Keith, R.G., Poncelet, P., Mullens, J.E., et al.: Symposium on pancreatitis, Can. J. Surg. **21:**56-74, 1978.
4. Webster, P.D., and Spainhour, J.B.: Pathophysiology and management of acute pancreatitis, Hosp. Pract. **9**(12):59-66, 1974.
5. Geokas, M.C., Van Lancker, J.L., Kadell, B.M., and Machleder, H.I.: Acute pancreatitis, Ann. Intern. Med. **76:**105-117, 1972.
6. Hermann, R.E., and Cooperman, A.M.: Cancer of the pancreas, N. Engl. J. Med. **301:**482-485, 1979.
7. Dimagno, E.P.: Pancreatic cancer: a continuing diagnostic dilemma, Ann. Intern. Med. **90:**847-848, 1979.
8. Friesen, S.R.: Tumors of the endocrine pancreas, N. Engl. J. Med. **306:**580-590, 1982.
9. Thomson, A.B.R., Dwyer, J., Gee, M., and Holt, T.: Management of pancreatic insufficiency in the adult, Mod. Med. Canada **34:**692-695, 1979.
10. Di Sant'Agnese, P.A., and Farrell, P.M.: Neonatal and general aspects of cystic fibrosis. In Young, D.S., and Hicks, J. M., editors: The neonate, New York, 1976, John Wiley & Sons, Inc., pp. 199-217.
11. Brooks, F.P.: Testing pancreatic function, N. Engl. J. Med. **286:**300-303, 1972.
12. Rick, W.: Der Secretin-pankreozymin-test in der Diagnostik der Pankreasinsuffizienz, Internist **11:**110-117, 1970.
13. Wormsley, K.G.: Further studies of the response to secretin and pancreozymin in man, Scand. J. Gastroenterol. **6:**343-350, 1971.
14. Lundh, G.: Pancreatic exocrine function in neoplastic and inflammatory disease: a test of pancreatic function, Gastroenterology **42:**275-280, 1962.
15. Webster, P.D., and Zieve, L.: Alterations in serum content of pancreatic enzymes, N. Engl. J. Med. **267:**604-607, 654-658, 1962.
16. Salt, W.B., II, and Schenker, S.: Amylase—its clinical significance: a review of the literature, Medicine (Baltimore) **55:**269-289, 1976.
17. Robison, J.C., Gitlin, N., Morrelli, H.F., and Mann, L.J.: P Factitious hyperamylasuria: a trap in the diagnosis of pancreatitis, N. Eng. J. Med. **306:**1211-1212, 1982.
18. Berk, J.E., and Fridhandler, L.: Advances in the interpretation of hyperamylasemia. In Glass, G.B.J., editor: Progress in gastroenterology, New York, 1977, Grune & Stratton, Inc., p.p. 873-894.
19. Warshaw, A.L. and Fuller, A.F.: Specificity of increased renal clearance of amylase in diagnosis of acute pancreatitis, N. Engl. J. Med. **292:**325-328, 1975.
20. Berk, J.E., Amylase in diagnosis of pancreatic disease, Ann. Intern. Med. **88:**838-839, 1978.
21. Ticktin, H.E., Trujillo, N.P., and Evans, P.E.: Diagnostic value of a new serum lipase method, Gastroenterology **48:**12-17, 1965.
22. Geokas, M.C., Rinderknecht, H., Walberg, C.B., and Weissman, R.: Methemalbumin in the diagnosis of acute hemorrhagic pancreatitis, Ann. Intern. Med. **81:**483-486, 1974.
23. Fedail, S.S., Harvey, R.F., and Salmon, P.R.: Trypsin and lactoferrin levels in pure pancreatic juice in patients with pancreatic disease, Gut **20:**983-986, 1979.

Chapter 27 Gastrointestinal digestive disease

Michael D.D. McNeely

achlorhydria Literally, ''without hydrochloric acid.'' Refers to lack of acid production by the stomach.

anticholinergic A drug that opposes the action of the cholinergic nervous system.

chyme The semisolid end product of gastric action on food. Chyme consists of mucus, gastric secretions, and broken-down food.

diarrhea Increased frequency of bowel movements with increased water content.

dipeptide A molecule composed of two amino acids.

disaccharide A molecule composed of two monosaccharides, with the general formula $C_n(H_2O)_{n-1}$, such as $C_{12}H_{22}O_{11}$.

esterification The combination of a carboxylic acid group with an alcohol.

gastrointestinal hormones Substances that are produced by gastrointestinal cells and then travel through the bloodstream to act on a separate site. These hormones include cholecystokinin, secretin, glucagon, gastric inhibitory polypeptide, vasoactive intestinal polypeptide, bombesin, somatostatin, motilin, chymodenin, bulbogastrone, entero-oxyntin, and pancreatic polypeptide.

gluten A protein found in wheat and wheat products.

hydrolysis The splitting of a molecule into fragments by the addition of water.

hydrophobic Literally ''water hating.'' Refers to substances that do not dissolve in or are not miscible with water.

intubation The procedure of introducing a tube-shaped instrument into the body, usually through an anatomical opening, such as the mouth.

long-chain triglyceride Three fatty acids of at least 18 carbon atoms linked to a glycerol molecule.

monosaccharide Carbohydrates consisting of a single basic unit with the general formula $C_n(H_2O)_n$.

pancreatic exocrine enzymes Enzymes required for digestion. Often released in a precursor form. These enzymes include trypsinogen, chymotrypsinogen, proelastase, procarboxypeptidase, ribonuclease, deoxyribonuclease, amylase, lipase, phospholipase A, and cholesterol esterase.

pancreatic hormones Endocrine hormones mainly concerned with carbohydrate intermediary metabolism and including glucagon, insulin, and gastrin.

polypeptide A molecule composed of more than three amino acids.

polyunsaturate A fatty acid containing more than one double bond between its carbon atoms.

pyrexia Fever. A body temperature above 37.5° C.

The gastrointestinal tract is a muscular tube lined with epithelial cells and extending 10 meters from the mouth to the anus. Along its course its structure is modified to suit particular requirements for the digestion and absorption of food.

The old view of the gastrointestinal tract describes it as an inert conduit, across which digested food molecules are allowed to pass into the bloodstream. Today physicians realize that the absorptive surface of the intestine is an extremely elaborate organ covered with minute microvilli that are invested with complex enzyme systems. This intricate microstructure creates an extremely efficient and highly selective absorptive mechanism.[1] In addition, the gastrointestinal tract is controlled by an elaborate hormonal and neural regulatory network.[2]

To complement this enhanced physiological knowledge, new diagnostic techniques, including imaging procedures, fiberoptic intubation, and chemical analyses, have introduced a new era of gastrointestinal diagnoses.

ANATOMY AND GENERAL FUNCTIONS

The gastrointestinal tract has five distinct regions: mouth, stomach, duodenum, jejunum-ileum, and large bowel (Fig. 27-1).

The mouth contains teeth, tongue, salivary glands, and an elaborate swallowing mechanism. It is responsible for tasting, grinding, and lubricating food. The swallowing mechanism propels the food down the esophagus, through the thoracic cavity, and into the stomach.

The stomach is a rough-surfaced, muscular bag coated with a protective mucus layer. The vigorous churning action of the stomach is responsible for mixing and breakdown of food. Hydrochloric acid and the enzyme pepsin are also secreted and continue the digestive breakdown. These actions convert food into chyme.

The chyme enters the duodenum, into which bile and pancreatic enzymes are secreted. Enzymatic degradation of the basic food materials is initiated in the duodenum and continues as the food material enters the small intestine.

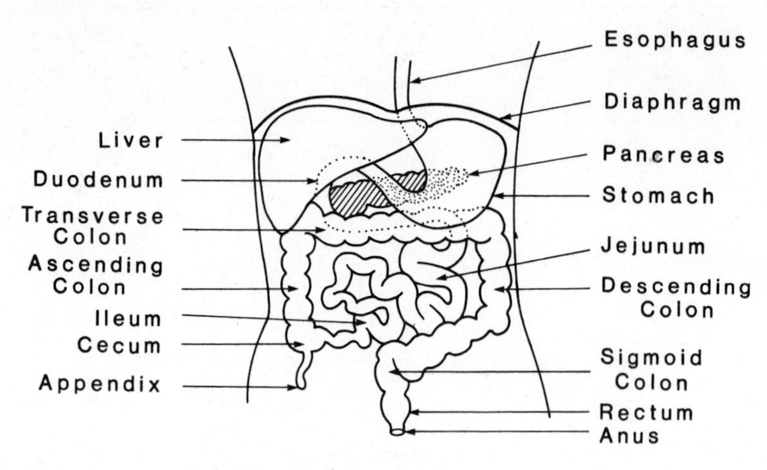

Fig. 27-1. Diagram of gastrointestinal tract.

The small intestine is 4 meters in length. Its absorptive capacity is enhanced by its microvillous substructure. The small intestine consists of two parts—the jejunum proximally and the ileum distally. Here the fragmented food materials are finally broken down and absorbed into the bloodstream.

When the nutrients have been absorbed, the residual matter enters the large intestine, wherein a process of selective water and electrolyte balance occurs. The digestive process terminates with the formation of feces.

The entire absorptive surface of the gastrointestinal tract is drained by the portal venous blood vessels. These convey the newly absorbed materials directly to the liver so that they may be immediately converted into usable forms.

DIGESTION

Digestion is the chemical process of rendering food into a form that is absorbable (assimilated) by the body. The digestive process begins in the mouth and is generally completed in the proximal portion of the small intestine. The digestive process for various foods is summarized in Table 27-1.

Digestive action of mouth

Food is tasted in the mouth by the combined action of the taste receptors on the dorsal surface of the tongue and on the palate, pharynx, and tonsils. Four fundamental tastes are recognized: sweet, salt, sour, and bitter. The taste of food also depends on its smell. The integration of these various neural inputs produces the sensation known as "taste." The taste of a food substance is important to protect us from disagreeable foods, to encourage eating, and to initiate a complex psychoneurogenic reflex that acts on the rest of the gastrointestinal tract.

Part of this reflex stimulates the production of saliva from the three pairs of salivary glands: parotid, mandibular, and sublingual. These glands produce viscid, water-based, mucin-containing secretions that act as a lubricant. They also release salivary amylase to initiate the digestion of starch.

Food is masticated by the complex interaction of the teeth, tongue, and mouth. The resulting bolus of food is then propelled to the stomach.

Digestive action of stomach

The stomach is a thin-walled, muscular sac that one can roughly divide into three zones (Fig. 27-2). The very top part of the stomach is known as the "fundus." The main portion of the stomach is known as the "body." The outlet of the stomach is known as the "antrum" and is segregated from the duodenum by the pyloric region, which contains a strong, muscular sphincter.

The gastric mucosa is covered with numerous coarse folds known as "rugae." The rugae assist in mixing food substances during the churning action of the stomach.

The gastric mucosa contains four types of cell. Mucous cells are found throughout the entire stomach and secrete mucus to protect the surface from attack by acid and enzymes. Also found in all parts of the stomach are the surface epithelial cells, which are also capable of secreting mucus but have been designed to proliferate rapidly and

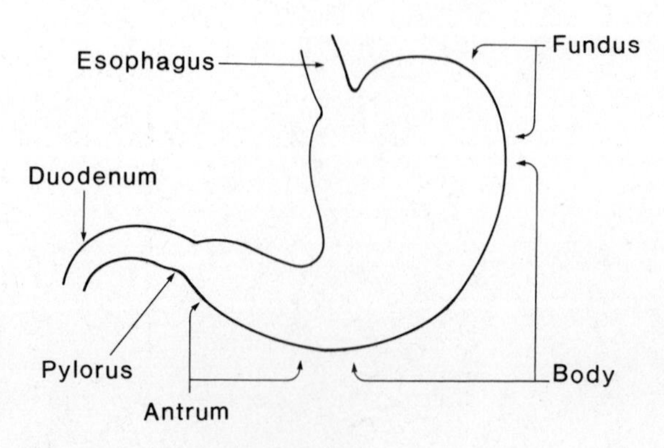

Fig. 27-2. Diagram of stomach.

Table 27-1. Digestion

Food material	Digestive action	End product
Starch	Pancreatic amylase	Disaccharides (mainly maltose)
Disaccharides	Mucosal disaccharidases	Monoglycerides
Monoglycerides	None	
Protein	1. Gastric hydrochloric acid and pepsin	Partial degradation into large polypeptides
	2. Pancreatic trypsin, chymotrypsin, and carboxypeptidase	Polypeptides, dipeptides, and aminoacids
Long-chain triglycerides	Emulsification with bile, hydrolysis by lipase	Fatty acids and glycerol

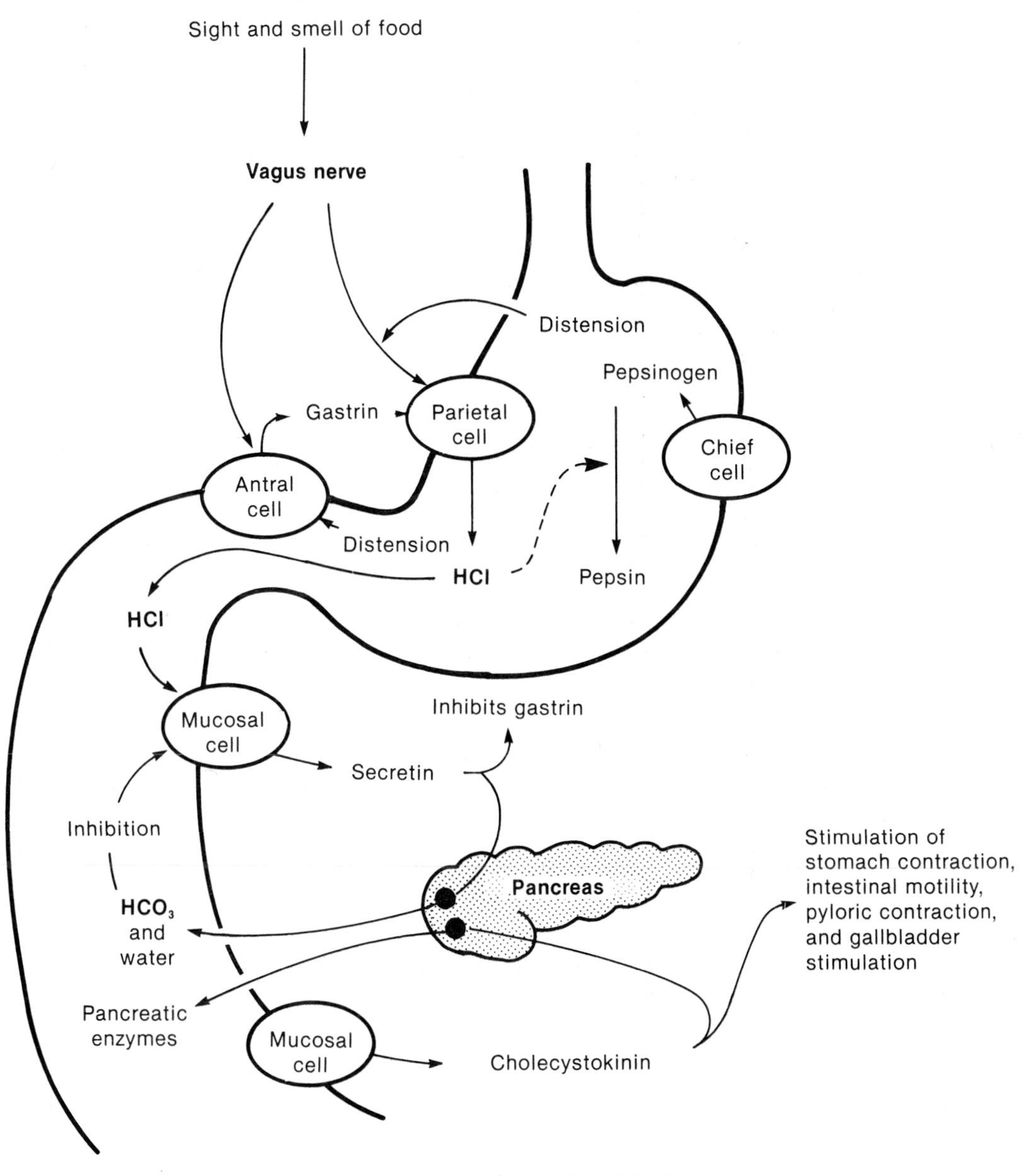

Fig. 27-3. Scheme demonstrating various stimuli of stomach and duodenum.

be shed, thereby allowing a continually viable surface for the stomach. Parietal cells produce hydrochloric acid and intrinsic factor. Chief cells produce the enzyme pepsinogen. These last two cell types are found throughout the body of the stomach.

The antral cells secrete mainly mucus but also some pepsinogen. The hormone gastrin is synthesized and stored in the G cells of the antrum.

There are three phases of gastric activity. The first of these is the cephalic phase, which is initiated by the sight and smell of food. This sensation triggers a direct vagus nerve message from the brain to the stomach to initiate the digestive process.

Next is the gastric phase of digestion, in which a variety of mechanisms interplay to create the digestive milieu (Fig. 27-3).

1. The vagus nerve directly stimulates the parietal cells to release hydrochloric acid.
2. The antral cells are also stimulated by the vagus nerve to secrete gastrin. Gastrin in turn stimulates the parietal cells to produce more hydrochloric acid.
3. Local distension of the gastric antrum also stimulates the production of gastrin and thus the secretion of hydrochloric acid.
4. Cholinergic (vagus) reflexes are further enhanced by distension of the fundus of the stomach.
5. The chief cells contain receptors that respond to the acid environment by initiating self-stimulation of the enzyme precursor pepsinogen. Pepsinogen is rapidly converted into its active form (pepsin) at pH 3. Lipase and other enzymes are also liberated, but these enzymes are of little consequence in the digestive process.

The combined action of antral contractions, pyloric sphincter activity, and chemical secretions render the food into a much degraded, mucus-containing solution known as "chyme."

Stomach activity subsides with time, and the fragmented material is then permitted to pass into the duodenum.

Digestive action of duodenum

The next step in digestion occurs in the duodenum. As chyme enters this portion of the intestine, several gastrointestinal hormones are released by both neural and local stimulation (Table 27-2). These hormones enter the portal blood system and act primarily on various regions of the gastrointestinal tract. It is appropriate to review the gastrointestinal hormones.[3]

Cholecystokinin. In response to the presence of fat or protein in the duodenum, *cholecystokinin* is released. Cholecystokinin consists of 33 amino acid residues, with the five C-terminal residues being identical to gastrin. There are two molecular forms of the hormone. The entire physiological action is invested in 10 amino acids of the C-terminal part of the polypeptide chain. It is produced by the mucosa of the upper small intestine.

Cholecystokinin stimulates enzyme secretion by the pancreas and is a strong stimulant of gallbladder contraction. It also causes release of bicarbonate and insulin from the pancreas; stimulates intestinal motility and contraction of the stomach and pyloric sphincter; stimulates the flow

Table 27-2. Intestinal hormones

Hormone	Number of amino acids	Source	Stimulating factor	Function
Cholecystokinin	33	Mucosa of upper small intestine	Amino acids, fatty acids, hydrochloric acid, and food in duodenum	Stimulates Pancreatic enzyme secretion Gallbladder contraction Contraction of stomach and pylorus Intestinal motility
Secretin	27	Throughout gut mucosa but concentrated in duodenum	Acid in duodenum	Stimulates pancreatic secretion of water and bicarbonate. Stimulates gastric pepsin secretion Stimulates the pyloric sphincter Inhibits gastrin secretion and stomach motility
Glucagon	29	Pancreatic and intestinal mucosa	Arginine, alanine, stress	Stimulates gluconeogenesis, raises blood glucose
Gastric inhibitory polypeptide (GIP)	43	Duodenal mucosa	Glucose and fat	Cholecystokinin-like activity
Vasoactive intestinal polypeptide (VIP)	28	Wide distribution throughout guy		Vasodilatation and hypotensive effects Inhibits histamine, pentagastrin acid release, and pepsin secretion Stimulates electrolyte and water secretion from pancreas Stimulates bile flow
Bombesin	14			Stimulates pancreatic secretion and gastrin release

of bile; stimulates blood flow in the superior mesenteric artery; opens the sphincter of Oddi; and encourages the absorption of sodium, potassium, and chloride from the jejunum and ileum.

Cholecystokinin release is stimulated by amino acids, fatty acids, hydrochloric acid, and food in the duodenum. Different amino acids have a variable ability to release cholecystokinin. Hydrogen ion is a weak releaser.

Secretin.[3] The presence of acid in the duodenum causes its mucosal cells to release *secretin* into the circulation.

Secretin is composed of 27 amino acid residues, 14 of which are identical to a sequence found in glucagon. Secretin-releasing cells are found throughout the gut mucosa but are concentrated in the duodenum.

Secretin is released most vigorously by hydrogen ion and thus the intraluminal pH is of vital importance. It is released when the pH falls below 4.5. The hormone is not released by vagal stimulation; thus an endocrine loop is established in which duodenal pH becomes acidic, releasing secretin, which then stimulates the pancreas to produce HCO_3. This bicarbonate then neutralizes the acid and shuts off the release of secretin.

Secretin acts to stimulate pancreatic secretion of water and bicarbonate. It acts on the stomach to stimulate pepsin secretion, stimulates the pyloric sphincter, and inhibits gastrin secretion and stomach motility. It also augments the action of cholecystokinin on gallbladder contraction.

Glucagon.[3] Three hormones with *glucagon* activity have been reported. The first is classic pancreatic glucagon, and the other two have been found in gastrointestinal mucosa. The hormone is composed of 29 amino acids. The primary structure of all glucagons is identical, and 14 of the amino acids occupy the same position in glucagon and secretin. Macroglucagons have also been reported.

There are many mechanisms for the release of pancreatic glucagon (see Chapter 40).

Gastric inhibitory polypeptide (GIP).[3] Gastric inhibitory polypeptide arises in the duodenal mucosa and has cholecystokinin-like activity, though it can inhibit gastric acid secretion. The molecule contains 43 amino acids. Fifteen are in the same position as glucagon and nine in the same position as secretin. Most of the GIP arises in the duodenum, with lesser amounts in the jejunum.

Gastric inhibitory polypeptide shows a biphasic response to food with an initial pH stimulated by glucose with a later plateau in response to fat.

Vasoactive intestinal polypeptide (VIP).[3] Vasoactive intestinal polypeptide hormone contains 28 amino acid residues and is chemically related to secretin, glucagon, and GIP. It is widely distributed throughout the entire length of the gut, with the largest amount being found in the small intestine.

The hormone has potent vasodilator and hypotensive effects. It dilates pulmonary vessels and reduces the effects of smooth muscle constrictor agents. In the gastrointestinal tract, VIP inhibits histamine and pentagastrin-stimulated acid secretion, inhibits pepsin secretion, and relaxes gastric muscle. It stimulates electrolyte and water secretion by the pancreas and increases bile flow. In the small intestine, it stimulates secretion and elicits muscle contraction. The hormone stimulates lipolysis and glycogenolysis and increases insulin release. It inhibits food-stimulated gastrin release and gastric acid secretion. The physiological role of the hormone and its mechanisms of release are unknown.

Bombesin.[3] Bombesin is a 14–amino acid peptide that has been noted to stimulate pancreatic secretion and the release of gastrin. There is some evidence that cholecystokinin may be released by bombesin. It is possible that bombesin released from the duodenum will in turn release antral gastrin.

Somatostatin.[3] Somatostatin, or growth hormone releasing–inhibiting hormone, has been isolated from the hypothalamus. It suppresses the secretion of growth hormone and impairs stimulation of thyrotropin secretion. It also inhibits the secretion of insulin and glucagon by direct action on the pancreatic islet cells. The hormone has also been demonstrated to inhibit both fasting and meal-stimulated gastrin release. Somatostatin-positive cells are present in high density in the mucosa of the stomach and upper part of the gut. This suggests that somatostatin may be involved in the regulation of secretory activity of endocrine and exocrine glands of the gastrointestinal system.

Motilin.[3] Motilin is a polypeptide containing 22 amino acids. It probably arises in the small intestine. It is released in response to alkalinization of the duodenojejunal region and causes an increase in the motor activity of the stomach.

Chymodenin.[3] Chymodenin has not been completely characterized. It causes a dramatic output of pancreatic juice rich in chymotrypsin, with little effect on other pancreatic enzymes.

Bulbogastrone.[3] Bulbogastrone is a chemically unidentified substance with the ability to decrease gastric acid secretion in response to acidification of the duodenal bulb.

Entero-oxyntin.[3] The chemical nature and site of origin of entero-oxyntin is unknown, but it is postulated that it stimulates gastric acid release when protein is introduced to the small intestine.

Pancreatic polypeptide.[3] A polypeptide has been separated from pancreatic extracts. It appears to have a mediating effect on a number of gastrointestinal functions, including acid secretion, secretin-induced pancreatic secretion, and relaxation of part of the gastrointestinal tract.

Hormone summary. The result of this intricate hormonal feedback activity is the secretion into the duodenum of bile salts, bicarbonate, and the enzymes amylase, lipase, and a variety of protein-degrading enzymes. The action of these agents on the primary food substances is now considered. (Additional discussions are in Chapter 40.)

Carbohydrate digestion[4]

Carbohydrates are present in the diet as monosaccharides, disaccharides, or complex polysaccharides. Only the polysaccharides require extensive digestive action in the duodenum.

Starch is the most common complex polysaccharide. It has a branching structure based on 1,4-carbohydrate or 1,6-carbohydrate linkages. Amylase is capable of hydrolyzing starch into oligosaccharides and ultimately into disaccharides. The dominant disaccharide produced from starch is maltose. Thus as the food leaves the duodenum, monosaccharides and disaccharides from the diet and disaccharides resulting from the action of amylase are passed to the jejunum and ileum, wherein absorption takes place.

Protein digestion[5]

Dietary protein is partially degraded in the stomach by hydrochloric acid and pepsin. In the duodenum, trypsin, chymotrypsin, and carboxypeptidase secreted by the pancreas act on the partially degraded protein to yield polypeptides, dipeptides, and amino acids. These tiny molecules then pass into the ileum and jejunum for assimilation.

Fat digestion[6]

Fat digestion is more complex than digestion of other basic food substances. Most dietary fats are long-chain triglycerides (palmitic, stearic, oleic, and linoleic acids). The stomach decreases the particle size of the fatty substances by its churning action. In the duodenum, fats are emulsified by the detergent action of bile. Emulsification allows the pancreatic enzyme, lipase, to attack the otherwise water-insoluble lipids. Lipase causes stepwise hydrolysis, which first forms a diglyceride, then a monoglyceride, and finally a fatty acid and glycerol (see Chapter 31).

The bile salts, which are so important for fat digestion, are synthesized in the liver from cholesterol and are conjugated to taurine or glycine. The main bile salts are conjugates of cholic and chenodeoxycholic acid.

ABSORPTION

Absorption is the process whereby digested food substances enter the body.[7] Having traversed the duodenum, these digested food substances enter the jejunum and ileum, wherein the final absorptive process takes place.

The intestinal mucosa is thrown into many folds, which assist in propelling its contents distally. The surface of each fold is drawn up into fingerlike projections known as villi (Fig. 27-4). Each villus increases the absorptive surface many times. Electron microscopic studies have shown that each villus is covered by hairlike projections known as microvilli. There are 200 million microvilli per centimeter of epithelium. Thus the intestine is given a massive absorptive surface area measuring 500 m^2.

It was once believed that enzymes were secreted by the small intestine to produce a digestive juice known as "succus entericus." It is now realized that the main enzymatic action occurs in intimate association with the epithelial surface. Rather than being a purely passive sieve through which food substances are permitted to pass, the intestinal mucosa contains a highly selective mechanism for the absorption of each nutrient.

Although there are regional differences in the ability of

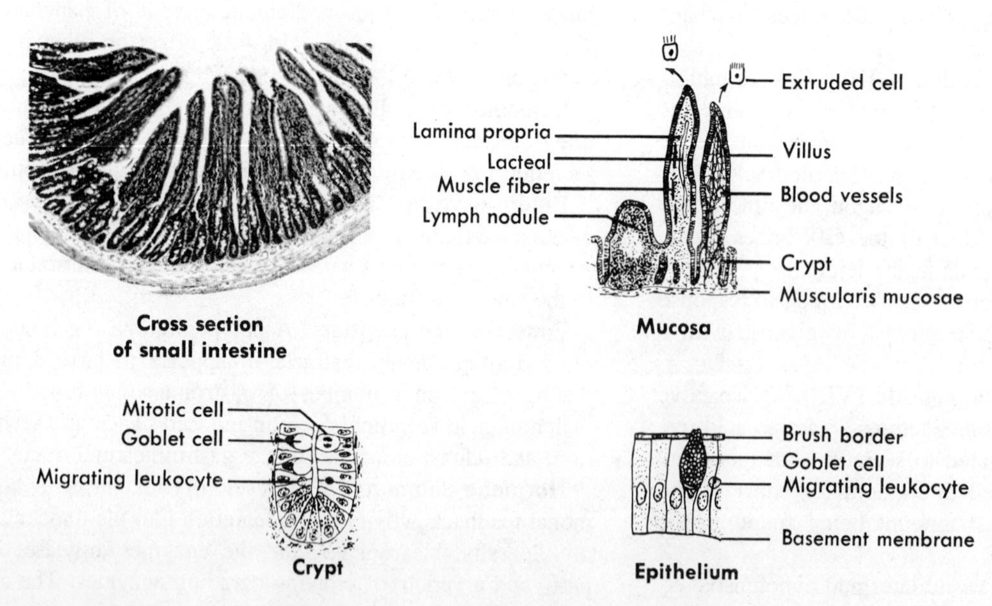

Fig. 27-4. Structures of functional components of small intestine. (From Arey, L.B.: Human histology: a textbook in outline form, ed. 4, Philadelphia, 1974, W.B. Saunders Co.)

the intestine to absorb different food substances, these details are not considered here.

Carbohydrate absorption

The digestive process degrades carbohydrate into monosaccharides and disaccharides. The monosaccharides—glucose, galactose and fructose—are absorbed by specific active transport mechanisms. That is, they are conveyed from the intestinal lumen into the bloodstream against a concentration gradient. Energy is required for this to occur.

The disaccharides are split into monosaccharides by the enzymatic activity of disaccharide enzymes located on the microvilli. For example, the milk sugar, lactose, is split by the enzyme lactase into its component sugars, glucose and galactose. These are then actively absorbed. The disaccharide sucrose is split by the enzyme sucrase into glucose and fructose. Maltose, which is the common product of starch hydrolysis, is split by a surface maltase into two molecules of glucose.

Protein absorption

The digested products of protein are small polypeptides, dipeptides, and amino acids. Dipeptides are absorbed more rapidly than amino acids because of special transport mechanisms. Proteins are not absorbed directly. A very large number of specific absorptive mechanisms designed for various types of amino acid are located in the mucosal surface.

Fat absorption (see also Chapter 31)

The successfully digested fat enters the intestine as a micelle. The fatty acids and monoglycerides enter the intestinal epithelial cells by diffusion, wherein they then interact with a binding protein. Long-chain fatty acids (LCFA) of 16 to 18 carbons are reesterified to triglycerides and then bound to apolipoproteins to form chylomicrons. These tiny lipid droplets are released into the lymphatic system and transported to the thoracic duct before entry into the bloodstream. Medium-chain fatty acids (8 to 10 carbons) are not reesterified and rapidly enter the portal bloodstream bound to albumin.

Vitamins D, E, A, and K[8,9]

Vitamins D, E, A, and K are not water soluble and must therefore be absorbed with lipids. Thus they depend on normal lipid absorption. Vitamin D absorption is modified by calcium intake and metabolism.

Water and sodium absorption

The control over water absorption is not fully understood, but it is believed that bulk flow after sodium absorption is the mode for water transport in the intestine. Sodium is absorbed by an active-transport mechanism that is linked to the absorption of amino acids, bicarbonate, and glucose.

Calcium

Calcium transport is under the influence of Vitamin D and parathyroid hormone and is regulated by a calcium-binding protein in the mucosal cells.[10]

Iron absorption

Iron absorption occurs by a most interesting process.[11] Gastric acid is required to convert the iron to the absorbable ferrous form. Iron then enters the mucosal cells and is transported across these cells before being picked up by carrier proteins in the circulation. The epithelial cells transport iron at a limiting rate. If the absorption of iron into the cells exceeds this limiting rate, the cells are generally shed before the iron is absorbed. In normal circumstances, the epithelium-shedding process prevents overabsorption of iron. This mechanism can be overcome by vastly excessive iron ingestion.

Formation of stool

Having passed through the ileum and jejunum, the intestinal contents enter the large bowel. Very little absorption of nutrients occurs in this region. It is here that water is actively absorbed and returned to the circulation. In addition, the balance of electrolytes is regulated. Progressive dehydration of undigested food substances along with the action of the bacteria (which normally inhabit the colon) causes the formation of feces.

PATHOLOGICAL CONDITIONS

Malnutrition

Malnutrition is caused by an abnormal food intake. On a worldwide basis, malnutrition is one of the leading causes of death. In North America, overnutrition is the most common malnutrient state. Obesity contributes to the mortality of all diseases but is closely related to diabetes, hypertension, cardiovascular disease, and emotional disorders. Undernutrition may be caused by the lack of food, bizarre diets, malabsorption, and hypermetabolic states. There is increasing recognition of the malnourished condition of our elderly population. Alcoholics, drug addicts, or mentally impaired persons may suffer from various forms of undernutrition. Chronically ill patients may suffer from anorexia, the loss of appetite. Occasionally, persons will adopt bizarre diets that lack the basic nutritional requirements. Disease of the gastrointestinal tract may prevent nutrients from being absorbed. Malignancy, pyrexia, and endocrine abnormalities may consume food energy at a faster rate than assimilation can occur.[12]

Stomach pathological conditions

Ulcers.[13] The cause of ulcer disease is multifactorial and relates to genetic and psychological makeup. For example, there is increased incidence of ulcer disease in persons with the blood group O. Some workers have claimed that an increased serum pepsinogin occurs in association with a high duodenal ulcer risk.[14] There has been a decrease in the incidence of ulcer disease in the last 20 years, but the age of the afflicted is now older. Duodenal and gastric ulcers are different disorders.

The pathogenesis of duodenal ulcer has been studied extensively.[15] Patients with duodenal ulcers, as a group, have increased capacity to secrete acid and pepsin. They have an increased responsiveness to gastrin and increased gastric acid and therefore deliver an increased gastric load to the duodenum.

The pathophysiology of gastric ulceration has four major hypotheses. What seems to be clear is that gastric ulcers are not associated with increased acidity. The first theory is that gastritis evolves into gastric ulcers, since diffuse gastritis is commonly associated. It is not known, however, whether this condition is primary or secondary. Antral stasis has been promoted as another cause. In this theory, pyloric stenosis (narrowing) or antral stasis causes a delay in gastric emptying. This in turn causes antral distension, which in turn stimulates gastrin secretion. The gastrin increases stomach acid, which leads to ulcers. As logical as this sequence of events may be, workers have not observed either a delayed gastric emptying or an increased acid secretion in this disorder. Another theory claims that the gastric mucosal barrier becomes deranged. In this theory, chemical agents such as salicylates, bile salts, and alcohol disrupt the mucosal barrier and allow back diffusion of hydrochloric acid, which results in tissue damage.

The final theory is the bile reflux theory. A disturbance in antral duodenal motility is postulated to be the primary event. This leads to the reflux of duodenal contents into the stomach. The duodenal contents contain bile acids and lysolecithin, which cause the breakdown of the stomach mucosal barrier. This permits gastritis to occur, and the mucosa is then susceptible to ulcers.

An ulcer diagnosis is generally made on clinical grounds, with roentgenographic and endoscopic examinations being of prime importance. The response to therapy is also very helpful. Treatment is generally with antacids or anticholinergic agents such as cimetidine. Surgery, which usually includes a vagotomy and antral drainage, is generally considered a second line of therapy and is used for repeated ulceration.

After gastric surgery, several physiologically based complications can occur shortly after eating. Some persons with gastric surgery will absorb glucose at an abnormally fast rate. This triggers an extremely rapid release of insulin, which may ''overshoot'' and cause hypoglycemia.

Another problem is caused by the rapid entry of osmotically active food particles into the intestine. This causes a fluid and electrolyte shift into the intestine. Hypovolemia and transient hypokalemia cause a generalized nausea and dizziness. The activation of the kallikrein-kinin system has an important but not fully understood role in this ''dumping'' syndrome.

Pyloric obstruction. In pyloric obstruction, the outlet of the stomach is constricted by the contraction of an ulcer, malignancy, or congenital abnormality. Obstruction is characterized by vomiting (often projectile), abdominal distension, and loss of hydrochloric acid, leading to severe hypochloremic metabolic alkalosis.

Cancer.[16] The incidence of stomach cancer is declining in America, but it remains high in Russia and in Japan (54% of all cancers). It appears most often in the seventh and eighth decades of life, and the 5-year survival remains at 15%. Over half of all gastric cancers are found in the pylorus or antrum. Surgery in combination with radiotherapy or chemotherapy is used to treat the lesion.

Zollinger-Ellison syndrome. The Zollinger-Ellison syndrome is an extreme form of peptic ulcer disease caused by a gastrin-secreting tumor of the pancreas (gastrinoma).[17] The unrelenting gastrin release stimulates hypersecretion of hydrochloric acid by the stomach.[18] The typical clinical presentation (not seen in all patients) is recurrent peptic ulceration. Seventy-five percent of patients with this syndrome have ulcers in the duodenal bulb or immediate postbulbar area. The tumors are often very small and can be difficult to identify. Sixty percent of tumors metastasize, and multiple tumors are common. Some tumors (10%) arise in the duodenal wall. The excess secretion of hydrochloric acid accounts for most of the clinical manifestations of the syndrome. The large amount of gastric acid entering the duodenum interferes with the digestion of fat and leads to steatorrhea. The prolonged secretion of gastrin causes hypertrophy of the stomach, with parietal cell hyperplasia. The secretion of pancreatic bicarbonate is increased in compensation for the acid load delivered to the duodenum. Often the intestine becomes ulcerated. The proximal intestinal lining often displays abnormal villi, submucosal edema, and hemorrhage. Since gastrin also inhibits salt and water absorption by the intestine, diarrhea will occur in 50% of patients. The diarrhea is enhanced by the fact that very large volumes of gastric contents are presented to the intestine. The Zollinger-Ellison syndrome is associated with hyperparathyroidism in 20% of patients. Other endocrine abnormalities that appear less commonly include pituitary, adrenal, ovarian, and thyroid tumors. This cluster of endocrine adenomas and carcinomas is known as the ''multiple endocrine neoplasia syndrome (MEN) I.'' It may occur with autosomal dominant inheritance as described originally by Werner or may occur sporadically.[19]

A fasting serum gastrin concentration four times the upper limit of normal in the absence of achlorhydria or renal failure strongly suggests the Zollinger-Ellison syndrome. This criterion is not met in 40% of cases.

Provocative testing has been used. The serum gastrin is measured after administration of (1) intravenous secretin 1 to 2 units/kg, (2) intravenous calcium gluconate, or (3) a standard meal. The secretin test, with a postinjection increase of gastrin of 110 pg/mL, is the most reliable.[20,21] A negative secretin response occurs in 5% of patients. Thus the sensitivity of the test is 95% and specificity virtually 100%.

Thus an overall approach to this diagnosis should be to screen appropriate patients with fasting serum gastrin measurements. Those with values less than 100 pg/mL can be discounted as having Zollinger-Ellison syndrome. Those with values over 100 pg/mL should receive the secretin test.

Pernicious anemia. Pernicious anemia is a disease that consists of gastric achlorhydria, gastric atrophy, and failure to secrete intrinsic factor. The intrinsic-factor deficiency prohibits absorption of vitamin B_{12}. This leads to the sequelae of mucosal epithelial insufficiency, degeneration of the posterior columns of the spinal cord, and macrocytic anemia. It is covered in greater detail in Chapter 33.

Malabsorption syndromes

In malabsorption syndromes the gastrointestinal tract is impaired so that it cannot absorb a variety of nutrient materials. This is generally the result of a disorder that causes damage to the mucosal lining. Patients suspected of having a generalized malabsorption syndrome should be evaluated with serum iron, vitamin B_{12}, albumin, and calcium determinations. In addition, immunoglobulin determinations can be useful to rule out IgA deficiency, a condition that permits parasitic infestations to occur. One can use the D-xylose absorption test as a screening procedure for generalized malabsorption.

The other category of malabsorption syndrome should in fact be called "maldigestion." In this group of disorders one of the important factors for the digestive process is in some way impaired. This is most commonly caused by some form of pancreatic insufficiency (see Chapter 26) or surgical procedure. The most common maldigestion syndrome leads to fat malabsorption and steatorrhea.

Steatorrhea. Steatorrhea is a clinical syndrome caused by the malabsorption of dietary fat. The indigested fat travels into the large bowel, and the stools contain an excess amount of lipid and are characteristically pale, bulky, and greasy with a repugnant odor. It is important to distinguish clinically the stools produced in steatorrhea from those produced in diarrhea. When steatorrhea is suspected, testing should be undertaken to estimate the actual amount of fat in the stool. When excess fat has been identified, the specific cause is sought. A deficiency of any factor important for lipid digestion and absorption can cause steatorrhea and include the Zollinger-Ellison syndrome, increased duodenal acid (postgastrectomy syndromes), abnormal bile output, pancreatic insufficiency, intestinal mucosal impairment, and disease of the large bowel that has caused an interruption of bile-salt enterohepatic circulation.

Celiac disease. Celiac disease is an extremely important cause of malabsorption. In this condition persons appear to have an abnormal immunological response to the presence of gluten in the diet. Up to 90% of celiac patients have circulating antibodies to gluten. The response to gluten is shedding of the microvillous mucosal surface of the intestine. This drastically reduces the absorptive surface area and causes malabsorption. Celiac disease may present in very subtle ways and may be definitively diagnosed only by the response to a gluten-free diet.[22]

Lactose intolerance and other carbohydrate malabsorption disorders. The most common carbohydrate malabsorption disorder is lactose intolerance.[23] All infants have the intestinal enzyme mechanism necessary to break the milk sugar disaccharide lactose into its components glucose and galactose, thereby allowing their absorption to occur. In those population groups who characteristically feed on animal milk throughout their lives, these enzyme mechanisms persist into adulthood. However, in those groups who are historically not milk drinkers, the enzyme system regresses (African blacks and Orientals). If persons lacking the lactase enzyme ingest milk or milk products, they will fail to split the lactose in the proper fashion. This unabsorbed sugar will create an osmotic force that pulls fluid into the intestinal lumen. This causes cramping, bloating sensations, and diarrhea. Moreover, large-bowel bacteria can metabolize the sugar to produce gas. Although most people with lactose intolerance are aware of their problem and avoid milk products, there are persons with milder forms who experience discomfort in much more subtle ways. The diagnosis of lactose malabsorption is made by use of the lactose tolerance procedures discussed later in this chapter. Malabsorption syndromes of other disaccharides have been reported but are extremely rare. The malabsorption of monosaccharides is seen only in extreme impairment of the mucosal surface.

Carcinoid syndrome[24]

A syndrome consisting of vascular "flushing," diarrhea, a carcinoid tumor of the bowel, occasional tricuspid insufficiency, and rarely, pellagra, is called the "carcinoid syndrome." Carcinoid tumors are the most common of small bowel tumors and are located predominantly in the distal ileum. The remainder of the extra-appendicular gastrointestinal carcinoid tumors are found in the rectum and stomach. These tumors metastasize most commonly to the regional lymph nodes, liver, and skeleton. Primary carcinoid tumors of the appendix are common but rarely metastasize, whereas those that arise from other parts of the gastrointestinal tract do metastasize. The tumors produce serotonin and kinins in vast excess. It is these hormonal substances that are responsible for the characteristic clinical syndrome. One can detect the presence of the disorder by measuring serotonin or its metabolites.

Large intestine disease

Diarrhea.[25] Diarrhea is defined as the excessive production of feces usually by the overabundance of water in the stool. The causes of diarrhea are many. Diarrhea should be clinically distinguished from steatorrhea.

Severe diarrhea causes sodium and water depletion. Potassium is also lost. The acid-base disturbances caused by diarrhea are variable. However, the most common disorder is acidosis, which results from the increased fecal loss of bicarbonate. In chronic, mild diarrhea, hypokalemic alkalosis may be found.

There are three main reasons for diarrhea: solute malabsorption, secretion of fluid into the intestine, and motility disturbance.

Solute malabsorption is caused by the ingestion of poorly absorbed substances, "dumping," or intestinal malabsorption, such as lactose intolerance.

The secretion of fluid occurs in a number of conditions. Passive secretion will occur if the epithelial permeability is increased by obstruction or inflammation. The secretion of anions will occur through the activity of 3′,5′-cyclic adenosine monophosphate as stimulated by cholera toxin, endotoxin, prostaglandins, bile acids, and certain tumor products (such as vasoactive intestinal polypeptide). Another secretory mechanism is the replacement of absorptive epithelium by crypt epithelium (as occurs with viral gastroenteritis).

Motility disturbances are caused by cathartics and nervous tension. These will increase the motility and decrease the transit time and therefore the absorptive efficiency.

Cancer.[25] Malignancy of the colon and rectum account for over half the malignancies of the entire gastrointestinal system. The 5-year cure rate for these lesions runs between 25% and 50%. The cure rate is directly proportional to how early the lesion is detected and is therefore correlated with its proximity to the anus. Thus early detection and screening is the most desirable approach to curing this often fatal disorder. Digital and sigmoidoscopic examination of the rectum is supplemented by screening for the presence of occult blood. Roentgenological studies are only useful in screening patients with high risk of cancer. The recent development of colonoscopy will become the preferred method of examining high-risk patients.

Blind-loop syndromes.[26] A variety of inflammatory and anatomical disorders of the gastrointestinal tract may cause regional outpouchings to occur in the large bowel. These pockets can trap intestinal material and allow bacterial overgrowth. If this happens, the overabundant bacteria can cause excessive breakdown of bile conjugates. When these materials have been deconjugated, they cannot be reabsorbed by the body and are lost in the feces. This may be the cause of diarrhea. The bile acid breath test sheds light on the condition.

Common bowel disorders. Rarely a chemical diagnostic problem, the most common disorders of the bowel are associated with abnormal motility. Such symptoms as bloating, cramps, and excess flatus production are common clinical problems.

Hyperalimentation[27]

In recent years, techniques for providing nutrition to persons who are otherwise unable to eat have been highly refined. Such techniques employ tube feeding or intravenous alimentation. In the case of those who are undergoing tube feeding, it is important that the material not cause an osmotic load on the gastrointestinal tract and hence produce diarrhea.

The assessment of persons receiving such artificial nutrition is not highly refined. In general, the tests must first allow assessment of whether complications have occurred or undernutrition of one or more substance is present. Patients should be monitored carefully by clinial assessment. Biochemical monitoring of the urine is useful and includes measurements of the urine osmolality to evaluate hydration, sodium, and potassium measurements to indicate electrolyte load, a urea concentration as a rough guide to overall nitrogen balance, and ketones and glucose to indicate poor carbohydrate control and caloric loss. Blood analyses are also useful but should be interpreted with caution, since the intravenous solution administered at the time of the blood collection can drastically influence the results by lipemia or hyperosmotic forces.

GASTROINTESTINAL FUNCTION TESTS

There are a number of gastrointestinal function tests that have been designed for critical evaluation of various gastrointestinal physiological functions.

Tests of gastric acidity

Tests of gastric acidity are required to screen for the ability of the parietal cells to produce hydrochloric acid. The discovery of achlorhydria (anacidity) is strong evidence for the presence of pernicious anemia and will rule out peptic ulcer disease, casting suspicion on gastric ulcer or gastric carcinoma. The presence of acid in the stomach is very strong evidence against pernicious anemia. Acid detection must be carried out to determine the significance of raised serum gastrin determinations.

The only suitable test for the presence of gastric acid is intubation and withdrawal of stomach juice. A pH measurement may then be made directly and should be less than 3. Anacidity in only confirmed by a pH over 6.

A resin-dye exchange method known as the Diagnex Blue test (E.R. Squibb Sons, New York) has been popular but is currently not recommended. In this method the patient ingests a carbacryl-cation exchange resin in which some of the hydrogen ions have been replaced by the dye azure A. Stomach acid, if present, will displace the dye. The dye is then absorbed and excreted into the urine. Thus the presence of stomach acid is signified by colored urine.

There are too many false positives and false negatives to allow this test to be recommended.

Gastric stimulation tests

Tests that evaluate the capacity of the stomach to produce acid have been popular in the past but are now considered to have limited diagnostic utility.

Gastric analysis. The gastric analysis test involves draining stomach secretions for a base-line period to determine basal or unstimulated acid production. Next, a parietal cell stimulant is administered and gastric juice collected to evaluate maximal secretory ability. Several different stimuli have been used. A test meal was popular at one time but is not reproducible and interferes with the desired measurements. Histamine has been used.[28] An antihistamine (diphenhydramine, or Benadryl) must be given in advance to ablate the systemic effects of the histamine. Until recently, betazole hydrochloride (Histalog) has been widely used.[28] No antihistamine is needed when this preparation is given. Currently, pentagastrin, the active 5-amino acid portion of gastrin, is the recommended stimulant.[29,30]

Thus the protocol is (1) collect residual gastric fluid from a fasting patient by intermittent suction on a nasogastric tube positioned within the stomach by fluoroscopy, (2) collect basal secretions for four 15-minute periods, (3) administer pentagastrin intramuscularly in a dose of 5 μg/kg of body weight, and (4) collect further stomach secretions for six consecutive 15-minute time periods. All collections are then evaluated for appearance, blood, bile, pH, volume, millimoles of H^+ per liter, millimoles of H^+ per volume, and millimoles of H^+ per hour for each collection.

The pH measurement is useful. Basal pH over 6 is almost certainly caused by anacidity, less than 3 indicates normal or excessive parietal cell function; intermediate values (3 to 6 mmol/hour) are not diagnostic and merely indicate a balance between hydrochloric acid and buffer. After stimulation, pH values should fall to less than 2. Failure to do so indicates inadequate parietal cell function, which may be found in pernicious anemia, gastric carcinoma, hypochromic anemia, rheumatoid arthritis, and myxedema.

Next, the basal acid output (BAO) should be computed by averaging of the millimole-per-hour output for the closest three basal collections.

The maximal acid output (MAO) is also calculated as the mean of the two highest, poststimulation values in millimoles per hour.

Normal adult men have a BAO of 2.2 to 2.7 mmol/hour, with 5 mmol/hour being the absolute upper limit. The MAO for men under 30 years of age is 14 to 42 mmol/hour, and 3 to 33 mmol/hour for men over 30 years of age. The values for women are approximately 50% of those for men. Detailed tables of normal values have been published[31] and are reviewed in Table 27-3.

Table 27-3. Pentagastrin stimulation test

	Basal acid output (BAO) (mmol/hour)	Maximal acid output (MAO) (mmol/hour)
Normal adult men		
Under 30	2.2-2.7	14-42
Over 30	2.2-2.7	3-33
Normal adult women	1-1.5	7-20
Zollinger-Ellison syndrome	10-100 (or more)	40%-60% above BAO
Ulcer predisposition		
Likely		> 35
Highly likely		> 45
Low risk		< 11

The Zollinger-Ellison syndrome is characterized by high BAO (Table 27-3). The MAO is generally only 40% to 60% higher than BAO, since the stomach is close to maximal stimulation. Indeed, a BAO/MAO greater than 60% is virtually pathognomonic for the Zollinger-Ellison syndrome.

If the MAO is over 35 mmol/hour, an ulcer predisposition is likely. In persons over 45 years of age it is highly likely. An MAO less than 11 mmol/hour in a man under 30 years of age suggests a very low peptic ulcer risk.[31]

Gastric analysis is used by some surgeons to indicate the nature of surgery to be used for ulcer treatment.

Hollander insulin test.[32] The Hollander insulin test is used to assess whether a surgical vagotomy has successfully denervated the stomach. In this procedure the patient is given regular insulin (0.15 units/kg of body weight intravenously) to render him hypoglycemic (plasma glucose less than 300 mg/L). Vagus stimulation is a normal response to hypoglycemia. Those with an intact gastric vagus will release acid in response to hypoglycemia. A successful denervation will cause an MAO less than 0.05 mmol/hour, and the pH will remain over 3.5. The test is not often relied on, since clinical evaluation is generally sufficient.

Gastrin stimulation test. The principal of the gastrin-stimulation test was described earlier.[20,21]

Blood for serum gastrin is collected from a fasting patient. Secretin (2 units/kg as an intravenous bolus) is then administered and serum gastrin collected at 2, 5, 10, 15, 30, and 60 minutes.

An absolute increase of gastrin greater than 110 pg/mL is positive for the Zollinger-Ellison syndrome.

Schilling test

The Schilling test of vitamin B_{12} absorption is an elegant evaluator of both gastric and intestinal function.[33] Radioactively tagged vitamin B_{12} is taken orally by the pa-

Table 27-4. Schilling test

Group	Absorption of intrinsic factor–bound vitamin B_{12}	Absorption of unbound vitamin B_{12}	Ratio of bound to unbound
Normal	>15%	>15%	0.5-1.5
Gastric lesion group	>10%	<5%	>2
Pernicious anemia			
Gastrectomy			
Congenital absence of functional intrinsic factor			
Ileal disease group	<15%	<15%	0.5-1.5
Tropical sprue			
Nontropical sprue			
Regional enteritis			
Intestinal resection			
Neoplasm or granulomatous reaction (rare)			
Selective vitamin B_{12} malabsorption (Imerslund syndrome)			
Fish tapeworm *(Diphyllobothrium latum)*			
Bacterial overgrowth (blind-loop syndrome)			
Chronic vitamin B_{12} deficiency			

tient. If absorption is successful, the radioactive material will be excreted in the urine. The vitamin B_{12} is administered with and without intrinsic factor to determine whether that substance is absent. More specifically, the test is carried out in the following way.

A fasting patient is given an oral preparation containing radioactively labeled vitamin B_{12}. One hour later, an intramuscular injection of unlabeled vitamin B_{12} is given. This "cold" B_{12} saturates the body's binding sites for B_{12}, preventing the "labeled" oral dose from becoming stored. Urine is collected for 24 hours. The amount of radioactivity in the urine is determined as a percentage of the original dose.

After several days, the test may be repeated. This time, the radioactively labeled vitamin B_{12} is given with intrinsic factor. Again, a 24-hour urine collection is carried out and the amount of vitamin B_{12} excreted is determined. There are three possible results (Table 27-4).

If both stages of the test show greater than 15% excretion, vitamin B_{12} absorption is normal and any deficiency of serum vitamin B_{12} must be the result of a dietary problem.

If the absorption of vitamin B_{12} given alone is low but normal when administered in association with intrinsic factor, one can presume a deficiency of intrinsic factor. Such a situation is found in classic pernicious anemia, gastrectomy, and the rare occurrence of nonfunctional intrinsic factor.

If vitamin B_{12} excretion in the urine is low after administration of both preparations, an absorptive defect of the terminal ileum is presumed to be present. This may be found in tropical sprue, nontropical sprue, regional enteritis, intestinal resection, neoplasm or granuloma, selective ileal vitamin B_{12} malabsorption (Imerslund syndrome) (rare), fish tapeworm *(Diphyllobothrium latum)*, bacterial overgrowth, and chronic vitamin B_{12} deficiency.

A two-capsule test has been devised (Dicopac, Amersham Corporation, Arlington Hts., Ill.). In this preparation, two forms of vitamin B_{12} are radioactively labeled using different isotopes of cobalt (cobalt 57, cobalt 58). The advantage of the procedure is that one can give both forms of vitamin B_{12} at essentially the same time. Absorption may then be determined by use of a single 24-hour urine collection. Special counting techniques to subtract the influence of one isotope on the other are required.

Pancreatic challenge tests

The elucidation of malabsorptive disorders depends on the evaluation of pancreatic secretory ability. This is best performed using the pancreatic stimulation tests that are described in Chapter 26.

Fat absorption tests

The definitive test of fat absorption is the quantitative measurement of fat in timed collections of feces obtained while the patient is maintained on a diet containing a known amount of fat. Because the collection is extremely difficult for the patient, a variety of alternative approaches have been promoted. Unfortunately, none of these entirely replaces the diagnostic ability of the quantitative fecal fat measurement.

Fat screening.[34] Fat screening is carried out first by evaluation of the weight and appearance of the stool. A pale, frothy appearance is virtually diagnostic of excessive fat. More reliable than this is the application of a small amount of the fecal material onto a standard microscopic slide, followed by staining with a fat-specific stain.

Trained observers are able to identify excessive fat in 80% of persons with fat malabsorption.

Quantitative fecal fat estimation.[35] This procedure is performed by collecting feces for 3 consecutive days. In the 2 days preceding the collection and during the period of collection the person must take a diet containing approximately 100 g of medium-chain triglycerides.

The easiest way to ensure this diet is to ask the patient to take his normal food but to supplement it with four glasses of whole milk each day and two tablespoons of corn oil with each meal.

It has been found that patients most easily accept the collection of feces into plastic bags. The plastic bags may then be closed with a tin tie and held in a preweighed, 5-gallon paint can. On arrival in the laboratory the can and contents are weighed and thus the weight of the collection is determined by the difference. The chemical analysis is then carried out on a thoroughly mixed aliquot of this 3-day collection.[36]

Serum fat estimations. Another rough screening test for fat absorption is the fat tolerance test.[37] A fasting blood sample is collected. The patient is then given a high-fat meal consisting of 6 ounces of corn oil. Serum is then collected at 3 and 6 hours. Estimations of total fat in these samples are then made. The post-fat specimens should exhibit a minimum of 50% increase in fat content over the fasting value. This test has poor diagnostic ability.

Isotope tests. Radioactively labeled, medium-chain triglyceride is administered to patients in whom fat malabsorption is suspected.[37] After a suitable time interval blood is collected and its radioactivity determined. It is assumed that the radioactivity that finds its way into the bloodstream is a result of the successful digestion and absorption of the radioactive fat. Unfortunately, radioactive iodine tags are not suitable for this procedure because they must be linked to unsaturated fats and because the size of the iodine tag gravely distorts the triglyceride molecule, making it susceptible to incidental breakdown. Thus the absorption of the radioactive iodine from these materials might not indicate successful fat malabsorption. A ^{14}C-labeled triglyceride has been tested and is measurable in the bloodstream or in the feces.[38] This is difficult because of the low activity of this isotope and some fear of its long half-life. A breath test based on the administration of ^{14}C-labeled triolein has been promoted and may eventually be the best test available for fat absorption, though it has not yet found widespread acceptability.[39]

D-Xylose absorption test[40]

D-Xylose is an aldopentose that is absorbed by the small intestine. It is postulated that D-xylose is absorbed in a passive fashion; its successful absorption is therefore a reflection of the integrity of the surface area of the small intestine. Once absorbed into the bloodstream there is a small amount of metabolic alteration, but at least 50% is excreted in the urine within the next 24 hours. It has been shown that the amount of D-xylose excreted into the urine over a 5-hour period is closely correlated with the amount of D-xylose absorbed in the gastrointestinal tract.

A variety of options is available for the performance of this test. The patient is instructed to fast overnight but is encouraged to drink an ample amount of water during this time.

Two doses have been advocated. Most authors suggest that 25 g of D-xylose be dissolved in approximately 300 to 500 mL of water as a suitable dose. Smaller subjects are given 1 g/kg of body weight to a total of 25 g. We have found that a 5 g dose is adequate.[41] Such a dose is less likely to cause abdominal cramps and is probably more sensitive, since some persons with absorptive defects are capable of absorbing sufficient amounts of the larger dose.

After administration of the sugar, urine is collected for a 5-hour period. A quantitative assay for D-xylose is carried out on this sample. Normally at least 25% of the administered dose will appear in the urine over a 5-hour period if renal function is normal.

For children who cannot be relied on to collect a urinary sample, blood collections at 1 and 2 hours may be substituted. Most persons demonstrate plasma levels greater than 300 mg/L in one of the samples. In children, values above 100 mg/L should be considered within the normal limits.

Low levels of urine or plasma xylose suggest an absorptive defect in the jejunum. Low levels are also seen in ascites, vomiting, delayed gastric emptying, improper urine collection, and high-dose aspirin therapy and with neomycin, colchicine, indomethacin, atropine, and impaired renal function. Normal levels are seen in persons who have absorptive defects occurring in a skip pattern. Such a disease distribution allows a sufficient amount of normal mucosa to remain and absorb an apparently normal amount of D-xylose.

Lactose tolerance

As mentioned earlier in the chapter, some persons do not have a fully developed mucosal lactase enzyme system. This deficiency prevents the normal absorption of lactose, resulting in gastrointestinal bloating, cramps, and diarrhea. The lactose tolerance test is useful for the quick identification of such persons.

In this test, 50 g of lactose dissolved in water is administered orally to the patient, who is observed carefully for the onset of symptoms.

The standard protocol is to collect a base-line specimen and 5, 10, 30, 60, 90 and 120-minute specimens of blood. The blood sample is analyzed for the presence of glucose. Glucose will be present in the bloodstream if lactose has been successfully cleaved and absorbed. The galactose moiety of the lactose is converted quickly into glucose by the liver. Normal persons will demonstrate a glucose rise of greater than 200 mg/L over the base-line sample. Those with lactase deficiency will exhibit notable abdominal discomfort and will have less than a 100 mg/L increase in the serum glucose concentration.

Recently it has been shown that the most reliable method of determining lactose absorption is to measure the amount of hydrogen appearing in exhaled breath after the oral administration of lactose.[42] Lactase-deficient persons will not absorb lactose, and it will find its way into the large bowel, where bacteria will metabolize it. Hydrogen is one of the by-products of this bacterial action. Hydrogen passes quickly into the bloodstream and is removed in the exhaled breath. Special-purpose gas chromatographs can detect the presence of postlactose hydrogen. A normal person allows no lactose to enter the colon and therefore has less than 10 parts per million of hydrogen in the exhaled breath. Persons with lactase deficiency demonstrate at least 50 ppm of hydrogen. Intermediate amounts of hydrogen in the breath can be caused by large doses of lactose and are of questionable significance.

The definitive diagnosis is made by tissue enzyme assays carried out on biopsy samples of the intestinal mucosa.

CHANGE OF ANALYTE IN DISEASE (Table 27-5)

Gastrin[21]

The normal fasting serum gastrin concentration is between 30 and 100 pg/mL. Increased serum gastrin may be caused primarily by hyperplasia of the G cells of the antrum, abnormal acid production, such as achlorhydria with compensatory hypergastrinemia or isolated retained antrum after gastrectomy, gastrinomas, and renal disease (see box on the next page).

Low gastrin concentrations are observed in normal persons, in persons with hypothyroidism, and after the administration of oral acid, streptozotocin, and phenformin.

Minimal elevations up to 500 pg are not diagnostic of any specific disorder and are occasionally found in normal persons. Such values are sometimes produced by the ingestion (or even by thought or smell) of food, insulin administration, malignant carcinoma of the stomach, pheochromocytoma, hyperthyroidism, hyperparathyroidism, peptic ulceration, gastritis, cirrhosis of the liver, renal failure, and rheumatoid arthritis. Some cases of Zollinger-Ellison syndrome and pernicious anemia may be found with serum gastrin values in this range, but this is not usual. Values between 500 and 1000 pg/mL are significant. Such results are associated with food ingestion, insulin administration, pheochromocytoma, hyperparathyroidism, renal failure, pernicious anemia, and Zollinger-Ellison syndrome.

Results over 1000 pg/mL are probably caused by either the Zollinger-Ellison syndrome, pernicious anemia, or rarely, parietal cell antibody–positive chronic atrophic gastritis. Most cases of the Zollinger-Ellison syndrome have values greater than 2000 pg/mL.

If the Zollinger-Ellison syndrome is considered, a fast-

Table 27-5. Change of analyte and function tests in disease

Disease	Fecal fat	Lactose tolerance	S-carotene S-vitamin A	S-vitamin B_{12} S-folate	Shilling test	D-Xylose absorption	Stool occult blood	Carcinoembryonic antigen	5-Hydroxyindoleacetic acid (5-HIAA)	Pancreatic enzyme testing	Stool examination
Steatorrhea	↑↑	N	↓	N, ↓	AB	AB	Neg	N	N	AB	Foul smelling, greasy
Celiac disease	N, ↑	N	N, ↓	N, ↓	N,AB	N,AB	Neg	N	N	N	Variable
Lactose intolerance	N	AB	N	N	N	N	Neg	N	N	N	Loose in association with abdominal cramps
Carcinoid syndrome	N	N	N	N, ↓	N,AB	N	Neg, pos	N, ↑	↑↑	N	Loose in association with cutaneous flushing
Functional diarrhea	N	N	N	N	N	N	Neg	N	N	N	Loose
Bowel carcinoma	N	N	N	N, ↓	N	N	Neg or pos	N, ↑	N	N	Change in bowel habits
Inflammatory bowel	N	N	N	N, ↓	N	N	Pos	N, ↑	N	N	Loose, bloody

N, Normal; ↑, elevated; ↓, lowered; *AB,* abnormal; *S,* serum.

Serum gastrin concentrations in disease
Normal (30 to 100 pg/mL)
Low (<30 pg/mL)
Normal persons, hypothyroidism, administration of oral acid, streptozotocin, phenformin
Minimal elevations (100 to 500 pg/mL)
Normal persons (occasionally), ingestion of food, insulin administration, malignant carcinoma of the stomach, pheochromocytoma, hyperthyroidism, hyperparathyroidism, peptic ulceration, gastritis, cirrhosis of the liver, renal failure, rheumatoid arthritis, Zollinger-Ellison syndrome (unusual), pernicious anemia (unusual)
Significant increase (500 to 1000 pg/mL)
Food ingestion, insulin administration, pheochromocytoma, hyperparathyroidism, renal failure, pernicious anemia, Zollinger-Ellison syndrome
Dramatic increase (>1000 pg/mL)
Zollinger-Ellison syndrome, pernicious anemia, parietal cell antibody–positive chronic atrophic gastritis

ing serum gastrin should be obtained. Results under 100 pg/mL essentially rule out the Zollinger-Ellison syndrome. Results greater than this in persons with normal renal function warrant stimulation testing as described on p. 476.

Malabsorption testing

Screening approach. Screening for persons with malabsorption syndromes in North American society is best done clinically. The population at greatest risk to occult malabsorption is the elderly. Laboratory screening for malabsorption is not very sensitive; however, measurement of serum albumin, calcium, vitamin B_{12}, and a peripheral smear looking for evidence of macrocytosis and iron-deficiency anemia constitute a reasonable general laboratory screen for malabsorption. If necessary, more specific tests for iron deficiency can be carried out. Measurements of the B vitamins are not generally available but would be of particular use in the elderly and in the indigent alcoholic population.

Persons who are considered to be suffering from specific malabsorption syndromes should be tested accordingly. Those with steatorrhea and suspected fat malabsorption should first have their feces examined visually. Next, one should carry out a rapid slide evaluation of a stool sample looking for meat fibers and excess fat. A carotene determination is easily performed and reflects gross abnormalities of fat absorption. A D-xylose absorption test will indicate whether significant generalized absorptive problems are present. Protein malabsorption is difficult to assess biochemically, and only when there is serious amino acid malabsorption will the serum albumin be depressed.

Carotene.[43] Beta-carotene, a naturally-occurring pigment, is a common constituent of vegetables and fruit. Collectively, these pigments are called "carotenoids." The carotenoid chemical structure is two 6-member carbon rings joined by a polyunsaturated 18-member carbon chain. It is hydrophobic and insoluble in water. Consequently, it must be absorbed into the body in association with fats. Therefore in clinical conditions in which fat malabsorption is impaired, the absorption of carotenoid pigments will be reduced and the serum carotene will be lower than normal. The normal range for serum carotene is 500 to 2500 mg/L.

Serum carotene is decreased in low carotene diets (low vegetable diets), abetalipoproteinemia, Tangier lipoprotein abnormality, liver failure, and 86% of patients suffering from clinically significant fat malabsorption.

A normal serum carotene concentration is found in 14% of patients with fat malabsorption. The concentration of carotene depends on a balance between the degree of malabsorption and the oral carotene intake.

Serum carotene is increased in hypothyroidism, diabetes mellitus, types I, II, and III hyperlipoproteinemias, and increased dietary carotene.

It should be pointed out that carotene is found in most vegetables. Therefore even those persons who are ingesting large amounts of lettuce but no carrots may develop carotenemia. Indeed, the most common form of observed carotenemia is caused by extreme weight-loss diets in which the person restricts calories and increases green vegetable ingestion. Starvation itself will add to the clinical phenomenon of hypercarotenemia.

Hypercarotenemia is a bizarre syndrome in which the skin may acquire a carrotlike hue. It is generally seen in food faddists.

A carotene-absorption test, which involves 3 days of carotene administration in hope of seeing a significant rise in the serum carotene level, has been promoted as a test of fat absorption. It has not proved to be diagnostic and is not recommended.

Vitamin A.[44] Vitamin A is an alcohol derived from beta-carotene by hydrolytic cleavage at the midpoint of the 18-carbon-polyene chain. Vitamin A is found only in animal tissue and animal products such as milk and egg yolk. Because it is chemically similar to carotene, it is also hydrophobic and must be absorbed into the body along with fat. Thus its presence in serum is also a reasonable estimate of the ability of the body to absorb fat. Serum vitamin A concentrations are less dependent on diet than are serum carotene concentrations. Reduced serum vitamin A concentrations are seen in association with fat malabsorption and liver disease.

Because the serum vitamin A determination is significantly more difficult to perform than the serum carotene and because its diagnostic ability is not significantly greater, it has never achieved popular acceptance as a screening test for fat malabsorption.

Trypsin. The measurement of trypsin in stool has been advocated as a screening test of pancreatic insufficiency.[45] In infants and young children the determination of trypsin is much more reliable than in adults. The reason is that there is less colonic degradation of pancreatic enzymes. The simplest test of tryptic activity is to apply a smear of fecal material to a thin film of gelatin. If trypsin is present, an enzymatic breakdown of the film will be seen. The test will be abnormal in all patients with significant pancreatic insufficiency, except in unusual cases of isolated defects of amylase and lipase. Some normal infants, however, will fail to produce sufficient trypsin to degrade the gelatin layer. Although such tests have some validity in screening, a strong clinical suggestion of pancreatic insufficiency warrants specific testing, as outlined in Chapter 26.

Fecal fat determinations.[35] Persons taking a 100 g fat diet will excrete no more than 5 g of fecal fat per day. More than 10 g per day is certain evidence of fat malabsorption.

Failure to adhere to the diet may invalidate the results. Low fat intake will mask minimal fat malabsorption. Grossly excessive fat intake will raise the fecal content above 5 g.

Tests related to specific disorders

Occult blood in stool. Because the survival of persons with cancer of the bowel depends on early diagnosis, reliable methods for detecting carcinoma of the bowel are extremely important.[46] A number of color reagents that react to trace amounts of hemoglobin in feces are available. Most rely on the ability of hemoglobin and its derivatives to act as peroxidases and catalyze the reaction between hydrogen peroxide and a chromogenic, organic compound.

Benzidine has been used but is carcinogenic and therefore not currently recommended as a laboratory-prepared reagent.

Various commercial systems for stool occult blood measurement are available.[47] These have been evaluated by Morris,[48] who recommends the Hemoccult test (Smith-Kline & French Laboratories, Philadelphia) as the most suitable. The fundamental problem in any test of fecal blood loss is to find a reagent that is suitably sensitive so that it will detect the presence of all gastrointestinal malignancies while being specific enough so that investigators are not burdened with large numbers of false-positive reactions.

The Hemoccult test is based on the guaiac-peroxide reaction and is not made positive by insignificant blood loss or meat ingestion. Moreover, it is suitable for patient use because it provides a convenient sample-handling envelope.

Those laboratories wishing to compose their own reagent system may employ the method of Woodman.[49]

Whatever method is employed, the evaluation of occult blood in three separate stool aliquots is recommended for optimum diagnostic detection.

Carcinoembryonic antigen. Carcinoembryonic antigen (CEA) is a glycoprotein that is abundant in entodermally derived tissues (gastrointestinal mucosa, pancreas, lung).[50] In the gastrointestinal tract it is located in the glycocalyx of the cell. It is found in fetal tissues at 12 weeks of gestation, and peak levels are observed at 22 weeks. The value decreases near the end of the third trimester. It is not known how carcinoembryonic antigen enters the circulation. After removal of a CEA-producing tumor, the material disappears within 2 to 4 weeks, most likely because of hepatic breakdown. The molecular weight is 160,000 to 300,000 daltons. It is a twisted rod that is approximately 50% carbohydrate. Its amino acid sequence is known.

Numerous glycoproteins have been found that have immunological cross reactivity with CEA. They include fetal sufoglycoprotein, normal colonic antigen (NCA), and normal glycoprotein (NGP). This latter antigen is also known as colonic antigen Be, colonic antigen X, colonic CEA II, and colonic cancer antigen III.

There are three possible clinical uses of CEA: for screening, for diagnosis, and for management of colonic carcinomas.[51]

The CEA has been advocated as a method of screening for the presence of occult carcinomas of the gastrointestinal tract. Unfortunately, this is a very difficult proposition because CEA is increased in a variety of colonic disorders, some of which predispose to colonic carcinoma. For example, CEA is increased in 20% of colorectal polyps. Thus it is not specific enough to be recommended as a screening procedure.

A variety of studies have been carried out to determine whether the CEA determination is useful to confirm the diagnosis of colonic cancer in individual patients. Unfortunately, the test is not sufficiently reliable to assist in the diagnosis of this potentially fatal disorder. For example, if a person has a 50% chance of colonic cancer, a positive CEA will increase the chance to 80%. Thus the CEA is not recommended as a preoperative indication of whether a malignancy is present. Indeed, the consensus statement of the U.S. Cancer Institute is as follows:

> We cannot recommend, based upon the available data, that CEA be used independently to establish a diagnosis of cancer. However, in a patient with symptoms, a value greater than five to ten times the upper limit of referenced normal should be strongly suggestive for the presence of cancer . . . . Further diagnostic efforts are indicated.[54]

The application of the CEA test in monitoring the follow-up of persons who have been treated for colonic cancer is a well-established role for the test. For this purpose the CEA should be measured at the time of surgery to establish a base-line level. Six weeks must then be allowed to pass before the next measurement is taken. If the value is elevated, a rising titer over the next few weeks must be noted before residual cancer is considered. If the value is low at the 6-week measurement, one can presume that the original lesion has been removed in its entirety and further measurements at regular intervals should be carried out. Some authors believe that a rising CEA titer requires a second-look operation. However, although some patients will have a rising titer caused by surgically manageable local spread, others will exhibit inoperable metastases.

The CEA concentration approximates the tumor burden and correlates approximately with the clinical course of metastatic disease. Unfortunately, despite many excellent and supportive reports for the use of this test, it has not proved consistently reliable in individual cases.

The normal range for CEA is 0 to 5 ng/mL. However, rising titers significant of disease may be seen within this range. A normal CEA does not exclude an underlying neoplasm.

Mild elevations of 5 to 10 ng/mL may be caused by malignancy or by heavy smoking, chronic liver disease, chronic chest disease, inflammatory bowel disease, and chronic renal failure.

When the value reaches 10 to 15 ng/mL, malignancy is quite likely, since less than 5% of the CEA-inducing, nonmalignant conditions produce results in this range. Values over 15 ng/mL strongly suggest malignancy.

5-Hydroxyindoleacetic acid. The presence of cutaneous flushing and diarrhea is a sufficiently common syndrome to warrant the frequent request for tests that will identify the possibility of the carcinoid syndrome. The most commonly performed biochemical test for this purpose is the 5-hydroxyindoleacetic acid determination.[53] This test is performed because the most common biochemical product of this tumor is serotonin, which is formed by the conversion of tryptophan to 5-hydroxytryptamine (serotonin), which is ultimately converted to 5-hydroxyindoleacetic acid.

The amount of 5-hydroxyindoleacetic acid (5-HIAA) found in the urine is highly method dependent. There are many screening procedures for this substance that are very nonspecific and therefore not recommended as stand-alone tests. An appropriate approach is to perform a screening procedure for all requests for 5-HIAA; those that exhibit an elevated value should be subjected to a somewhat more specific test, such as that of Goldberg.[53]

Normally, up to 15 mg of 5-HIAA is excreted per 24 hours. In the carcinoid syndrome more than 25 mg is usually measured. False-normal results are produced by a number of drugs, including *p*-chlorophenylalanine, corticotropin, ethanol, imipramine, isoniazid, monoamine oxidase (MAO) inhibitors, methenamine, methyldopa, and phenothiazines. Reduction of an elevated value is also seen in renal disease and in phenylketonuria.

In the carcinoid syndrome, results are usually between 25 and 1000 mg/day. False-positive results have been reported in nontropical sprue, intestinal obstruction, pregnancy, sleep deprivation, and oat cell carcinoma of the lung, and in ingestion of avocados, bananas, eggplants, pineapples, plums, and walnuts. Drugs that are known to cause an increase in 5-HIAA value are acetanilid, ephedrine, mephenesin, nicotine, phenacetin, phenobarbital, phentolamine, rauwolfia, reserpine, methocarbamol, and glycerol guaiacolate cough medicines.

Carbon-14 bile acid breath test. The use of breath analysis in gastroenterology has been extensively reviewed by Newman.[54] The most common radioactive breath-testing procedure in current use is the ^{14}C bile acid breath test.[55] In this procedure, glycine-1-^{14}C-cholate is administered orally. If the patient has an intestinal blind loop or other source of bacterial overgrowth, deconjugation of the tracer will occur, allowing ^{14}C to enter the bloodstream. Here it is metabolized to $^{14}CO_2$. Breath CO_2 is trapped in an alkaline solution, and the subsequent detection of radioactivity is a sensitive indication of the presence of bacterial overgrowth.

Tests of celiac disease. The detection of circulating gluten antibodies has been suggested as a means of identifying persons with celiac disease.[56] The test is not sufficiently sensitive or specific to be recommended for routine testing.

IgA. A very good test to perform in persons with chronic undiagnosed diarrhea is the serum concentration of immunoglobulin A (IgA). Approximately one in 800 persons suffers from idiopathic deficiency of IgA. Such persons do not have a normal immunodefense of their mucosal membrane system. They are therefore susceptible to infestations of parasites.[57] If a person is discovered to be suffering from a deficiency of IgA, an intestinal intubation with vigorous mucosal biopsy is warranted in search of such parasites as *Giardia lamblia*.

Urine oxalate. The metabolism of oxalate is very complex. A major component of its biochemistry is the enterohepatic circulation of glyoxalate materials secreted in the bile and subsequently reabsorbed by the bowel. In persons who have had gastrointestinal surgery, suffer from chronic inflammatory disease of the bowel, or have chronic abnormalities of the gastrointestinal tract, it is possible to have excessive reabsorption of oxalate material, which has been split from the excretory products. Such persons secrete large amounts of oxalate in the urine and are prone to the formation of oxalate renal calculi. Thus the evaluation of urinary oxalate is recommended in anyone who has a chronic anatomical or physiological intestinal abnormality.[58]

Indican. Some intestinal bacteria, including (*Escherichia coli* and the *Bacteroides* species) contain the enzyme tryptophanase, which can metabolize dietary tryptophan to indole. This product is then absorbed, converted to indican in the liver, and excreted in the urine.[59] Thus the estimation of indican in a 24-hour specimen of urine has been suggested as a means of diagnosing small bowel bacterial contamination. An increased urinary indican is not always found, however, since the predominant bacteria may not contain the appropriate enzyme. Furthermore, other small bowel diseases causing malabsorption may encourage indican excretion as a result of absorbed dietary tryptophan. Therefore estimations of urinary indican correlate only roughly with abnormalities in the gastrointestinal tract, and the test is not recommended for regular use in the diagnosis of gastrointestinal pathological conditions. The test is commonly requested by naturopathic physicians who ascribe their own particular interpretation to the results. It is of interest that this was the test that identified the first case of Hartnup's disease.

Serum folate.[60] Folate determinations are widely available and are often used in the assessment of malabsorption. The folate value obtained in red cells reflects a chronic situation, whereas the serum level reflects the day-to-day nutritional impact of folate. Increased serum folate is seen occasionally in pernicious anemia. Decreased serum folate is seen in a series of disorders in which the absorptive capacity of the small intestine is impaired, including Whipple's disease, myelofibrosis, hypothyroidism, celiac sprue, abetalipoproteinemia, vitamin B_6 deficiency, regional enteritis, ileitis, ulcerative colitis, acute and subacute necrosis of the liver, malignancy, and Down's syndrome.

Vitamin B_{12}. A recent article[61] has called to question the validity of serum vitamin B_{12} determinations. This article reveals the fact that many currently available tests of vitamin B_{12} detect the presence of nonphysiologically active analogs of vitamin B_{12}. Thus they may give false normal vitamin B_{12} results in B_{12}-deficient persons. Most current methods respect this potential problem.

Low vitamin B_{12} concentrations (less than 150 pg/mL) are found in pernicious anemia, gastric disease, thyroid disease, ileal disease, malabsorption, and vegetarian diets.

Increased vitamin B_{12} is seen with vitamin B_{12} administration, liver tissue damage, leukemia, and lymphoma.

Tests in Crohn's disease. Serum tests for inflammatory disease of the intestine would be very useful but have not yet been developed for routine use.

Some workers believe that serum lysozyme determinations are useful for distinguishing Crohn's disease from ulcerative colitis.[62] There is no general agreement on this fact. Others[63] have used the serum $beta_2$ microglobulin as a marker that correlates with the activity of Crohn's disease.

REFERENCES

1. Moog, F.: The lining of the small intestine, Sci. Am. **245:**154-176, 1981.
2. Track, N.S.: The gastrointestinal endocrine system, Can. Med. Assoc. J. **122:**287-291, 1980.
3. Rayford, P.L., Miller, T.A., and Thompson, J.C.: Secretin, cholecystokinin and newer gastrointestinal hormones, N. Engl. J. Med. **294:**1093-1101, 1157-1164, 1976.
4. Dawson, A.M.: The absorption of disaccharides. In W.I., Card, and B. Creamer, editors: Modern trends in gastroenterology, vol. 4, London, 1970, Butterworth & Co. (Pubs), Ltd., pp. 105-124.
5. McColl, I., and Sladen, G.E.C., editors: Intestinal absorption in man, New York, 1975, Academic Press, Inc.
6. Wilson, F.A., and Dietschy, J.M.: Differential diagnostic approach to clinical problems of malabsorption, Gastroenterology **61:**911-921, 1971.
7. Ingelfinger, F.J.: Gastrointestinal absorption, Nutrition Today, pp. 2-10, March 1967.
8. Wiss, O., and Gloor, U.: Absorption, distribution, storage and metabolites of vitamin K and related quinones, Vitam. Horm. **24:**575-586, 1966.
9. DeLuca, H.F., and Suttie, J.W., editors: The fat soluble vitamins, Madison, Wisc., 1970, University of Wisconsin Press.
10. Holdsworth, C.D.: Calcium absorption in man. In McColl, I., and Sladen, G.E.G., editors: New York, 1975, Academic Press, Inc., pp. 223-262.
11. Martiner-Torres, C., and Layrisse, M.: Nutritional factors in iron-deficiency: food iron absorption, Clin. Hematol. **2:**339-352, 1973.
12. Pearson, W.N.: Assessment of nutritional status: biochemical methods. In Beaton, G.H., and McHenry, E.W., editors: New York, 1966, Academic Press, Inc., pp. 265-315.
13. Mirsky, I.A.: Physiologic, psychologic, and social determinants in the etiology of duodenal ulcer, Am. J. Dig. Dis. **3:**285-314, 1958.
14. Grossman, M.I.: Elevated serum pepsinogen I: a genetic marker for duodenal ulcer disease, N. Engl. J. Med. **300:**89, 1979.
15. Grossman, M.I., Guth, P.H., Isenberg, J.I., et al.: A new look at peptic ulcer, Ann. Intern. Med. **84:**57-67, 1976.
16. Shahon, B.B., and Horowitz, S.: Cancer of the stomach: analysis of 1152 cases, Surgery **39:**204-221, 1956.
17. Deveney, C.W., Deveney, K.S., and Way, L.W.: The Zollinger-Ellison syndrome—23 years later, Ann. Surg. **188:**384-393, 1978.
18. Walsh, J.H., and Grossman, M.T.: Gastrin, N. Engl. J. Med. **292:**1324-1334, 1377-1384, 1975.
19. Johnson, G.J., Somerskill, W.H.K., and Anderson, V.E.: Clinical and genetic investigation of a large kindred with multiple endocrine adenomatosis, N. Engl. J. Med. **277:**1379-1386, 1967.
20. Ippoliti, A.F.: Zollinger-Ellison syndrome: provocative diagnostic tests, Ann. Intern. Med. **87:**787-788, 1977.
21. Deveney, C.W., Deveney, K.S., Jaffe, B.M., Jones, R.S., and Way, L.W.: Use of calcium and secretin in the diagnosis of gastrinoma, Ann. Intern. Med. **87:**680-686, 1977.
22. Strober, W.: The pathogenesis of gluten-sensitive enteropathy, Ann. Intern. Med. **83:**242-256, 1975.
23. Gray, G.M.: Congenital and adult intestinal lactose deficiency, N. Engl. J. Med. **294:**1057-1058, 1976.
24. Kuehn, P.G., Coley, G.M., and Christine, B.: Carcinoid syndrome: a study of 16 cases, Hartford Hosp. Bull. **28:**305-307, 1973.
25. Jeejeebhoy, K.N.: Symposium on diarrhea 1. Definition and mechanisms of diarrhea, Can. Med. Assoc. J. **116:**737-738, 1977.
26. Gilbertson, V.A.: The earlier diagnosis of adenocarcinoma of the large intestine, Cancer **27:**143-149, 1971.
27. Walravens, P.A.: Keeping TPN on course with lab monitoring, Diagn. Med. **4**(3):38-43, 1981.
28. Ward, S., Gillespie, I.E., Passaro, E.R., et al.: Comparison of Histalog and histamine as stimulants for maximal gastric secretions in human subjects and in dogs, Gastroenterology **44:**620-626, 1963.
29. Abernethy, R.J., Gillespie, I.E., Lowrie, J.H., et al.: Pentagastrin as a stimulant of maximal gastric acid response in man: a multicentre pilot study, Lancet **1:**291-295, 1967.
30. Kirsner, J.B., and Ford, H.: The gastric secretory response to Histalog: one-hour basal and Histalog secretion in normal persons and in patients with duodenal ulcer and gastric ulcer, J. Lab. Clin. Med. **46:**307-311, 1955.

31. Blackman, A.H., Lambert, D.L., Thayer, W.R., et al.: Computed normal values for peak acid output based on age, sex and body weight, Am. J. Dig. Dis. **15:**783-789, 1970.
32. McNeely, M.D.D.: Gastrointestinal function. In Sonnenwirth, A.C., and Jarett, L., editors: Gradwohl's clinical laboratory methods and diagnosis, St. Louis, 1980, The C.V. Mosby Co., p. 520.
33. Herbert, V.: Detection of malabsorption of vitamin B_{12} due to gastric or intestinal dysfunction, Semin. Nucl. Med. **2:**220-234, 1972.
34. Drummey, G.D., Benson, J.A., and Jones, C.M.: Microscopical examination of the stool for steatorrhea, N. Engl. J. Med. **264:**85-87, 1961.
35. Massion, C.G., and McNeely, M.D.: Accurate micromethod for estimation of both medium- and long-chain fatty acids and triglycerides in fecal fat, Clin. Chem. **19:**499-505, 1973.
36. Schwartz, L., Woldow, A., and Dunsmore, R.: Determination of fat tolerance in patients with myocardial infarction: method utilizing serum turbidity changes following a fat meal, J.A.M.A. **149:**364-366, 1952.
37. Silver, S.: Radioactive isotopes in medicine and biology, Philadelphia, 1962, Lea & Febiger.
38. Kaihara, S., and Wagner, H.N., Jr. :Measurement of intestinal fat absorption with carbon-14 labeled tracers, J.. Lab. Clin. Med. **71:**400-411, 1968.
39. Meeker, H.E., Chen, I.W.: Connell, A.M., and Saenger, E.L.: Clincal experiences in ^{14}C-tripalmitin breath test for fat malabsorption, Am. J. Gastroenterol. **73:**227-231, 1980.
40. Benson, J.A., Culver, P.J., Ragland, S., et al.: The D-xylose absorption test in malabsorption syndromes, N. Engl. J. Med. **256:**335-339, 1957.
41. Santini, R., Sheehy, T.W., and Martínez-de-Jesús, J.: The xylose tolerance test with a five-gram dose, Gastroenterology **40:**772-774, 1961.
42. Newcomer, A.D., McGill, D.B., Thomas, P.J., and Hofmann, A.F.: Prospective comparison of indirect methods for detecting lactase deficiency, N. Engl. J. Med. **293:**1232-1235, 1975.
43. Onstad, G.R., and Zieve, L.: Carotene absorption: a screening test for steatorrhea, J.A.M.A. **221:**677-679, 1972.
44. Parkinson, C.E., and Gal, I.: Factors affecting the lab management of human serum and liver vitamin A analysis, Clin. Chim. Acta. **40:**83-90, 1972.
45. Erlanger, B.F., Kokowsky, N., and Cohen, W.: The preparation and properties of two new chromogenic substrates of trypsin, Arch. Biochem. Biophys. **95:**271-278, 1961.
46. Winawer, S.J., Sherlock, P., Schottenfeld, D., and Miller, D.G.: Screening for colon cancer, Gastroenterology **70:**783-789, 1976.
47. Christensen, F., Anker, N., and Mondrop, M.: Blood in feces: a comparison of the sensitivity and reproducibility of five chemical methods, Clin. Chim. Acta **57:**23-27, 1974.
48. Morris, D.W., Lee, C.S., and Hansell, J.R.: Presentation at the annual meeting of the American Gastroenterological Society, San Francisco, 1974.
49. MaKarem, A.: Hemoglobins, myoglobins and haptoglobins. In Henry, R.J., Cannon, D.C., and Winkelman, J., editors: Clinical chemistry principles and technics, New York, 1974, Harper & Row, Publishers.
50. Gold, P., and Freedman, S.O.: Specific carcinoembryonic antigens of the human digestive system, J. Exp. Med. **122:**467-481, 1965.
51. Meeker, W.R.: The use and abuse of the CEA test in clinical practice, Cancer **41:**854-862, 1978.
52. National Institutes of Health Consensus Conference: Carcinoembryonic antigen: its role as a marker in the management of cancer, Br. Med. J. **282:**373-375, 1981.
53. Goldenberg, H.: Specific photometric determination of 5-hydroxyindoleacetic acid in urine, Clin. Chem. **19:**38-44, 1973.
54. Newman, A.: Breath-analysis tests in gastroenterology, Gut **15:**1-15, 1974.
55. Hofman, A.F., and Thomas, P.J.: Bile acid breath test: extremely simple, moderately useful, Ann. Intern. Med. **79:**743-744, 1973.
56. Baker, P.G., Barry, R.E., and Read, A.E.: Detection of continuing gluten ingestion in treated coeliac patients, Br. Med. J. **1:**486-488, 1975.
57. Jones, E.A.: Immunoglobulins and the gut, Gut **13:**825-835, 1972.
58. Stauffer, J.Q., Humphreys, M.H., and Weir, G.J.: Acquired hyperoxaluria with regional enteritis after ileal resection, Ann. Intern. Med. **79:**383-391, 1973.
59. Holdsworth, C.D.: Intestinal and pancreatic function. In Brown, S.S., Mitchell, F.L., and Young, D.S., editors: Chemical diagnosis of disease, New York, 1980, Elsevier/North Holland, Inc., p. 721.
60. Rosenberg, I.H., and Godwin, H.A.: The digestion and absorption of dietary folate, Gastroenterology **60:**445-463, 1970.
61. Kolhouse, J.F., Kondo, H., Allen, N.C., Podell, E., and Allen, R.H.: Cobalamin analogues are present in human plasma and can mask cobalamin deficiency because current radioisotope dilution assays are not specific for true cobalamin, N. Engl. J. Med. **299:**785-792, 1978.
62. Falchuk, K.R., Perrotto, J.L., and Isselbacher, K.J.: Serum lysozyme in Crohn's disease, Gastroenterology **69:**893-896, 1975.
63. Gahrton, G., Zech, L., Robert, K.H., and Bird, A.G.: Serum levels of β2-microglobulin: a new marker of activity in Crohn's disease, N. Engl. J. Med. **301:**440-441, 1979.

Chapter 28 Cardiac disease and hypertension

W. Fraser Bremner

actin One of the proteins involved in myocardial and arterial smooth muscle contraction.

angina Pain of cardiac origin, most commonly caused by coronary atherosclerosis. Angina can occur with exertion (angina of effort) or at rest (angina decubitus) or as a result of arterial spasm (vasospastic angina).

atherosclerosis Hardening of the arteries, a process that begins early in life and results in gradual deposition of lipid, fibrin, and calcium in the wall of arteries with, if the affected person lives long enough, eventual occlusion of the vessel. The most common cause of death in Western countries.

atrioventricular node A specialized part of the conduction system that transmits electrical impulse from the atria to the ventricles. The node delays the passage of the impulse, thus permitting mechanical contraction of the atria to occur before the ventricles contract. This node is a common site of heart block if damaged, in which case the electrical impulse from the sinus node fails to reach the ventricles, which may beat independently.

atrium The chamber of the heart that collects blood from the veins and contracts to expel the blood into the ventricles, thereby filling the ventricles. There are two atria; the right collects blood from the systemic veins and fills the right ventricle; the left collects blood from the pulmonary veins and fills the left ventricle.

captopril A drug used to treat hypertension. It acts by inhibiting angiotensin-converting enzyme, thereby reducing the concentration of circulating angiotensin.

cardiac failure Failure of the heart to maintain the circulation, with resultant accumulation of salt and water and inadequate perfusion of essential organs.

cardiomyopathy A heterogeneous group of disorders that have in common a toxic or genetic insult that affects contracting myocardial cells directly.

diastole The period of the cardiac cycle during which the ventricles are not contracting. (Pronounced dy-ás-to-lee.)

diastolic pressure The blood pressure measured at the brachial artery during diastole.

disopyramide An antiarrhythmic drug that is given by the intravenous, intramuscular, or oral route.

Embden-Meyerhof pathway The pathway of anaerobic glycolysis that will convert glucose or glycogen to lactate.

embolus A portion of a blood clot that breaks off from the main clot and travels through the circulation to settle in the lungs (pulmonary embolus) or in one of the systemic arteries (systemic embolus), depending on where the embolus originates.

fibrillation An irregular contraction of the cardiac tissue. In atrial fibrillation the atria are primarily affected but the heart is still able to maintain an output, even though the ventricles beat irregularly. In ventricular fibrillation, ventricular contraction is totally irregular. This condition is incompatible with life.

glycoside A generic term for a group of drugs originally from the foxglove and including digoxin. These drugs improve the contractility of the failing heart and slow the rate of ventricular contraction in atrial fibrillation.

His bundle Part of the cardiac conduction system that relays the electrical impulse from the atrioventricular node to the Purkinje system.

infarction A dead piece of tissue caused by a loss of blood supply. Thus myocardial infarction or heart attack occurs when one of the coronary arteries is occluded and the tissue distal to the occlusion dies.

ischemia Reduction of blood supply to tissue to the point at which it cannot function normally.

Krebs citric acid cycle The pathway of intermediary metabolism that will accept broken-down products from the Embden-Meyerhof pathway and oxidize, decarboxylate, and reduce them, with the production of a relatively large amount of adenosine triphosphate.

myosin One of the contractile proteins involved in cardiac and arterial smooth muscle contraction.

procainamide An antiarrhythmic drug.

propranolol A beta-blocker drug, that is, a drug that will competitively antagonize catecholamine uptake at beta-receptor sites.

Purkinje system Part of the cardiac conduction system that ramifies throughout the ventricles and distributes electrical impulses to the ventricular muscle mass.

quinidine An antiarrhythmic drug.

renin-angiotensin system An important system involved in maintaining the blood pressure. The renin-angiotensin system is of importance in many young hypertensive patients or patients with hypertension caused by renal damage, in which it is often the major system supporting the blood pressure.

saralasin The competitive antagonist of angiotensin. It is used intravenously to assess the role of the renin-angiotensin system in maintenance of blood pressure.

sinus node The center that generates the cardiac electrical impulse and is in turn controlled by the autonomic nervous system.

systole The period of the cardiac cycle during which the ventricles contract. (Pronounced sís-to-lee.)

systolic blood pressure Blood pressure measured at the brachial artery during ventricular contraction.

technetium-99m (***^{99m}Tc***) An isotope injected intravenously to measure left ventricular output or to detect an acute myocardial infarction. (*m* means 'metastable'.)

thallium-201 An isotope injected intravenously to delineate part of the ventricular muscle mass that is ischemic or a scar caused by an old infarction.

troponin-tropomyosin A regulatory protein complex that prevents cardiac contraction from occurring. Calcium annuls this repression.

ventricles The main pumping chambers of the heart that receive blood from the atria and expel the blood into the pulmonary artery and aorta. The left ventricle is more massive and powerful than the right ventricle because the pressure in the systemic circulation is higher than that in the pulmonary circulation; thus a stronger pump is required for the systemic side of the circulation.

verapamil A drug used to treat certain arrhythmias and angina. The drug acts by competitive antagonism of calcium uptake through the slow calcium channel in cardiac conduction tissue and arterial smooth muscle tissue.

Part I: Cardiac disease

ANATOMY OF HEART[1]

The pumping action of the heart is the prime factor in the maintenance of the body's circulation (Fig. 28-1). The heart is a muscular organ composed of four chambers; the two atria act primarily as collecting chambers and the two ventricles act as the major pumps in the system. The right atrium collects blood from the systemic circulation and pumps it into the right ventricle, which then contracts and expels the blood into the lungs. In the lungs, deoxygenated blood is reoxygenated and carbon dioxide, the main metabolite from the body cells, is exhaled. Oxygenated blood drains from the pulmonary vascular tree into four pulmonary veins, which empty into the left atrium. The left atrium pumps this blood into the left ventricle, which then contracts and propels the blood into the systemic circulation. The blood then enters the aorta, the largest artery in the body. The first aortic branches are the coronary arteries, which feed the blood to the heart to maintain its essential supply of oxygen and nutrients. Further branches from the aorta supply blood to the brain, liver, gut, kidneys, and so on.

The cardiac cycle consists of two phases, *diastole* and *systole*. In diastole the valves between the atria and ventricles (the tricuspid valve on the right side and the mitral valve on the left side) are open and blood flows into the ventricles. In systole the ventricles contract and the valves are closed; the outlet valves (the pulmonary valve on the right side and the aortic valve on the left side) open and permit the ventricles to eject the blood into the pulmonary artery and the aorta, respectively. In the normal cardiac cycle both atria contract, followed by simultaneous contraction of both ventricles. Two coronary arteries feed blood to heart muscle.

The left ventricle is the major pumping chamber of the heart, but the integrity of the entire system is vital to the maintenance of the systemic circulation. A sensitive and

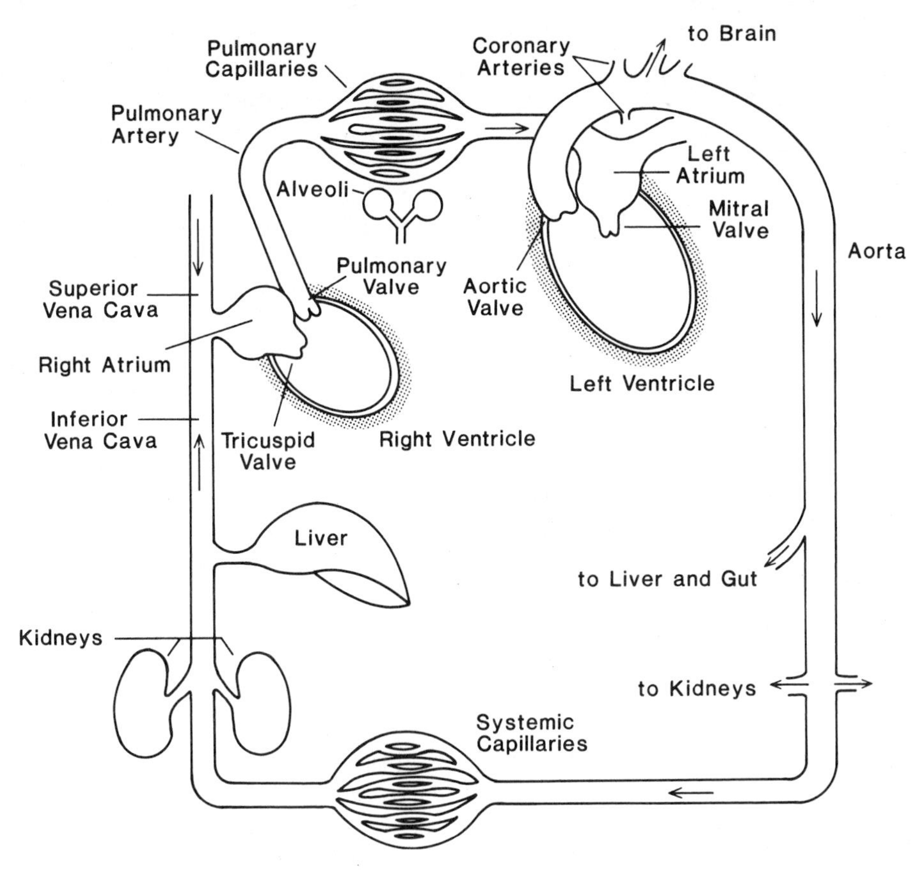

Fig. 28-1. Diagram of heart and circulation. Atria and ventricles are shown in exploded form for greater clarification. In reality they share a common wall and are contiguous. Atria collect blood returning from heart, either from systemic circulation through superior and inferior venae cavae or from lungs through pulmonary veins. They transfer this blood to ventricles, which contract and expel blood into arteries, the vessels that take blood from heart. Pulmonary artery delivers blood to lungs for reoxygenation and removal of carbon dioxide, and freshly oxygenated blood returns to heart to go to essential organs of body, particularly brain, heart, kidney, and gut.

efficient control system is obviously required to regulate the pumping sequence of these four chambers and to adjust their outputs in response to the body's needs. Cardiac output is determined primarily by the heart rate, by the systemic blood pressure, and by the contractile force developed in the wall of the left ventricle.

The control center that normally regulates contraction is the sinus node. This node is composed of a group of cells in the right atrium that possess the property of inherent automaticity; that is, they will discharge spontaneously and regularly and send impulses through the atria to a subsidiary pacemaker site, the atrioventricular node, and then into ventricular muscle. The Purkinje fibers in the left ventricle compose the conduction system for this tissue. The conduction system of the heart has certain inbuilt mechanisms. If the sinus node fails, subsidiary centers further down the conduction system take over and maintain electrical impulse to cardiac muscle. These subsidiary centers generally have slower rates than the sinus node does and are normally suppressed by the dominant sinus node.

CARDIAC FUNCTION[1]

The primary function of the myocardial cell is contraction. The basis of contraction involves the interaction between a fibrous protein, actin, and a globular protein, myosin, whereby the myosin protein moves along the actin molecule, thereby contracting the cell (Fig. 28-2). In the quiescent situation, diastole, the contractile response is inhibited by the presence of a protein complex composed of troponin and tropomyosin. Contractility is triggered by the arrival of the cardiac electrical impulse, resulting in an influx of calcium that allows contraction when it is bound to troponin. In effect, the inhibitory protein complex is pulled out of the way, thereby permitting myosin to interact with actin in systole. The arrival of calcium at the contraction site is regulated at the cell membrane by electrical nerve impulses.

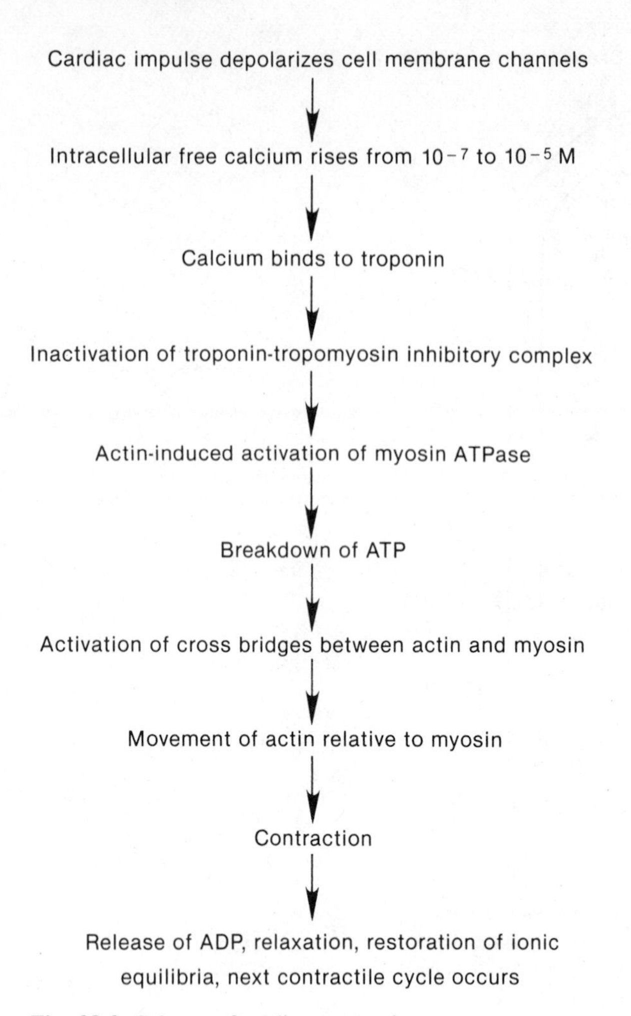
Cardiac impulse depolarizes cell membrane channels
Intracellular free calcium rises from 10^{-7} to 10^{-5} M
Calcium binds to troponin
Inactivation of troponin-tropomyosin inhibitory complex
Actin-induced activation of myosin ATPase
Breakdown of ATP
Activation of cross bridges between actin and myosin
Movement of actin relative to myosin
Contraction
Release of ADP, relaxation, restoration of ionic equilibria, next contractile cycle occurs

Fig. 28-2. Scheme of cardiac contraction.

Once contraction is completed, the breakdown products of adenosine triphosphate (ATP) are released from the myosin, another ATP molecule is bound, and the myosin detaches from the actin. The molecules are then ready for another contraction cycle to occur. The actual amount of calcium that enters the cell during each contraction is small and is used to activate a larger concentration of calcium in the sarcoplasmic reticulum, the complex network of interconnecting (anastomosing) intracellular channels that surround and interweave with the myofibrils of the cell.

Cardiac muscle is highly aerobic because it is extremely active. Delivery of the large quantity of oxygen needed to fuel the heart requires a rich capillary bed. In fact, each myocardial cell is in close apposition to two to four capillaries. Normally, some 75% of the oxygen passing through the heart is extracted from the bloodstream, and changes in metabolic demand, particularly changes in heart rate, are met by alterations in coronary blood flow. The blood flowing through the coronary artery beds depends on regulation of arterial tone, which is mediated by a complex system of multiple feedback processes involving both neural and humoral inputs. The major site of regulation is at the level of the arteriole. Thus, as the heart speeds up in response to an increase in muscle activity, which will occur with exercise, there is reflex neural stimulus to produce arteriolar relaxation. There is preferential shunting of blood to myocardial tissue, whereas less relevant organs, such as the gut, receive less blood. There is also evidence of local release of vasodilator components from cardiac tissue, which presumably play a significant role in the perfusion of myocardial tissue. One should appreciate the incredible efficiency of the heart as a pump. The prime function of the heart is to pump deoxygenated blood to the lungs and oxygenated blood to the systemic tissues, and it does so with an exquisite capability for balancing supply and demand for the lifetime of the person.

MYOCARDIAL METABOLISM

Energy metabolism

Energy is liberated from various substrates that the heart uses by several pathways. These include the Embden-Meyerhof glycolytic pathway, the pentose phosphate shunt, fatty acid oxidation, and the Krebs citric acid cycle (see Chapter 30). The energy produced in the breakdown of substrates is then transported through the electron-transport system of the mitochondria to produce adenosine triphosphate (ATP). This is a chemical form of stored energy that is used by myocardial tissue to perform work. Since performance has to be matched to work demand with speed and efficiency, the heart requires an effective storage method to maintain the energy of ATP. This is achieved through storage of the high-energy phosphate bond of ATP as creatine phosphate, which acts as a backup store for rapid regeneration of ATP when levels fall as a result of increased demand (Fig. 28-3).

Normally the heart is strictly an oxygen-dependent organ. Since it possesses the enzymes of the Embden-Meyerhof pathway and Krebs cycle, in theory 1 mole of glucose metabolized down this pathway will produce 38 moles of ATP (see Chapter 30).

The heart uses many substrates normally present in plasma, and its uptake of most of these substances is proportional to the arterial concentration once the extraction threshold is exceeded. In general terms, the heart uses free fatty acids as its predominant fuel. It also picks up significant quantities of glucose and lactate, as well as lesser amounts of pyruvate and ketone bodies, and small amounts of amino acids. Most of the energy for cardiac function is obtained from the breakdown of metabolites through the citric acid cycle and oxidative phosphorylation. These enzyme pathways are found principally in the mitochondria, which make up some 35% of the total volume of cardiac muscle.

Metabolic processes of myocardial tissue have been classified into three phases, those of energy liberation, conservation, and utilization. In the energy liberation stage, the major cardiac substrates are broken down to

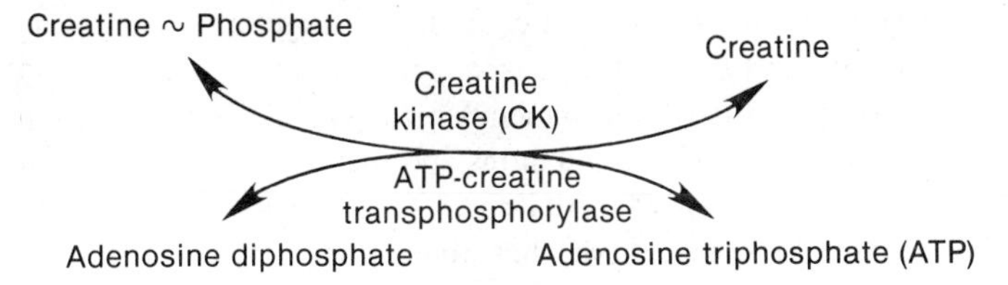

Fig. 28-3. Transfer of high-energy phosphate from creatine phosphate to adenosine diphosphate is accomplished by enzyme creatine kinase (CK). This is a reversible reaction, regulated by concentration of adenosine triphosphate (ATP). Phosphate can be transferred from ATP to creatine to form creatine phosphate by the enzyme ATP creatine transphosphorylase. The symbol ~ denotes a high-energy bond, that is, a bond that when it undergoes hydrolysis produces energy that can be converted to work. There are two such high-energy bonds in ATP.

CO_2 and H_2O, with the release of high-energy electrons by the biochemical pathways. The energy produced is conserved chemically as adenosine triphosphate. Adenosine triphosphate is the energy source for myocardial contractility, for the maintenance and restoration of ionic gradients across the cell membrane, and the source of energy for a host of intermediary cell metabolic processes. During energy utilization, adenosine triphosphate is hydrolyzed to adenosine diphosphate and inorganic phosphate. The adenosine diphosphate is then reconverted to adenosine triphosphate by picking up a high-energy phosphate group from the electron-transport chain or from the reservoir of creatine phosphate. Creatine phosphate is present in relatively small amounts in cardiac muscle, but it provides a small reserve of immediately accessible high-energy phosphate bonds according to the reaction shown in Fig. 28-3.

Enzymes in myocardial cells[2]

Several cardiac enzymes are of clinical importance because detection of their release into the bloodstream has come to provide a sensitive and specific indicator of myocardial cell damage and death. The enzymes, if they are to be of clinical relevance, must be soluble and persist in the bloodstream long enough to be detected. Enzymes measured routinely in the clinical laboratory in assessment of myocardial disease are discussed here in respect to their function in normal myocardial metabolism.

Creatine kinase (CK).[2] Creatine phosphate concentrations drop when energy utilization acutely increases with increased cardiac work and can also fall in hypoxia, hypertrophy, or heart failure. The enzyme responsible for production of creatine phosphate is creatine kinase (CK). It is also termed creatine phosphokinase, has a molecular weight of 85,000 daltons, and exists in several isoenzymatic forms. Mitochondrial and myofibrillar isoenzymes have been described, but they have as yet no relevance to pathological conditions. The three isoenzymes of importance are those found in the cytosol. These isoenzymes are given the abbreviations MM, MB, and BB. In fetal tissue the predominant isoenzyme is the BB isoenzyme, but with growth this is replaced by MB and MM until eventually MM is the significant isoenzyme. Creatine kinase is a dimer, composed of the two subunits M (muscle) and B (brain). They are easy to distinguish electrophoretically, since at pH 8.6 the MM isoenzyme stays at the origin whereas the BB form moves toward the anode with a mobility comparable to that of albumin. The MB isoenzyme falls in between these two. If one looks at the comparative activity of the enzyme in various tissues, skeletal muscle has by far the highest activity, possessing some 50,000 times the concentration of serum CK. The predominant isoenzyme is the MM isoenzyme, with only traces of MB and BB isoenzymes. The MB component, however, is increased in certain muscle disorders, particularly the Duchenne type of dystrophy and in polymyositis. The activity measured in international units is around 2000 U/g, which compares with the cardiac muscle activity of 500 U/g. Some 70% of the cytoplasmic isoenzyme is the MM variety, and 30% is the MB variant. Brain cytoplasm, on the other hand, contains some 200 U/g, and 100% of the isoenzyme is the BB variety. In normal serum, at least 95% is of the MM type and is probably largely the result of leakage from skeletal muscle, particularly during physical activity. Serum CK activity in healthy persons shows an asymmetric distribution skewed toward higher values. In addition, values are lower in women than in men, are lower in morning than in evening, and tend to be lower in hospitalized patients, possibly because bed rest reduces the amount of enzymes released from muscle.

Lactate dehydrogenase (LD). Lactate dehydrogenase[2] is a ubiquitous tissue enzyme that catalyzes the reduction of pyruvate to lactate using nicotinamide adenosine diphosphate (NAD). LD has a molecular weight of about 140,000 daltons and is a tetramer composed of four subunits of molecular weight 35,000 daltons each. The subunits are of two forms, H (heart) and M (muscle), which are polymerized to form the five isoenzymes of LD. The H subunit contains relatively more aspartate and glutamate than the M subunit, which is richer in arginine and methionine so that the polymer HHHH is most electronegative and the polymer MMMM is most electropositive. Electrophoresis readily separates the five isoenzymes. The principal isoenzyme in heart (HHHH) is maximally active in the presence of low concentrations of pyruvate but is inhibited by excess pyruvate. In contrast, the major muscle isoenzyme (MMMM) is maximally active with a higher pyruvate concentration and is less inhibited by excess pyruvate. This parallels the metabolic demands of the two tissues. The heart metabolizes fatty acids and carbohydrates at a fairly constant rate with complete oxidation of pyruvate through the Krebs citric acid cycle. The heart thus has a low tissue concentration of pyruvate and lactate. In contrast, muscle, with sudden demands for increased energy during exercise, has to deal with sudden increases in tissue pyruvate and lactate caused by anaerobic metabolism. This is true of muscle involved in dynamic exercise, whereas the postural muscles, which function at a relatively constant level of activity similar to that of the heart, tend to have LD isoenzymes identical to that of cardiac tissue. LD is found in the cytosol of all human cells and thus would have little diagnostic specificity were it not for the presence of isoenzymes that show differences from tissue to tissue. The half-life of lactate dehydrogenase depends on the particular isoenzyme; isoenzyme 1 (HHHH) has a half-life of about 100 hours but isoenzyme 5 (MMMM) has a half-life of only 10 hours. The significance of this is evident in the discussion of the release of enzymes during myocardial infarction.

All five LD isoenzymes occur in all tissues, but the relative proportions for each tissue are significantly different. Heart and red cells contain mostly LD_1 and LD_2, whereas skeletal muscle and liver contain LD_5 and to a lesser degree LD_4. Normal serum contains mostly LD_2, with lesser amounts of LD_1 and the other enzymes. If the enzymes are released from cardiac tissue into serum, one frequently sees a change in the ratio of LD_1 to LD_2.

Aspartate aminotransferase (AST, SGOT).[2] Aspartate aminotransferase (AST) catalyzes transfer of an amino group between aspartic acid and pyruvate to form oxaloacetate (alpha-ketoglutarate) and alanine (Fig. 28-4). The enzyme permits transfer of amino moieties in intermediary metabolism and permits amino acids, aspartic acid, and glutamic acid to be broken down in the Krebs cycle, since the reaction is reversible. The enzyme exists in two structurally different forms, one being found principally in the cytoplasm and the other in the mitochondria. It is the cytosol form that is found in serum, whereas the mitochondrial isoenzyme, though present in higher concentrations inside the cell, does not normally circulate in serum. One would therefore expect that AST mitochondrial isoenzyme would be particularly valuable in the diagnosis of myocardial infarction, but the isoenzyme assay has not found widespread use. The molecular weight of the cytoplasmic isoenzyme is 120,000 daltons and that of the mitochondrial equivalent, 100,000 daltons. The half-life is probably about 20 hours. Liver has the highest enzyme activity, with some 85 U/g of tissue, whereas heart muscle possesses 75 U/g and skeletal muscle about 50 U/g.

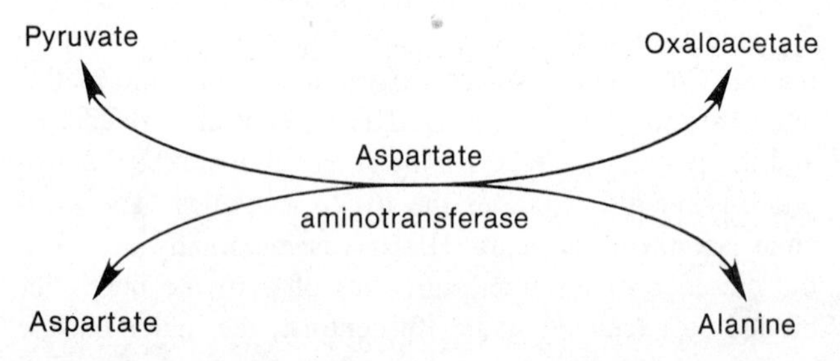

Fig. 28-4. Aspartate amino transferase catalyzes transfer of amino group between aspartate acid and pyruvate with production of oxaloacetate and alanine. This reaction permits several amino acids to participate in intermediary metabolism, such that their carbon chain can be introduced into Krebs citric acid cycle.

There are many enzymes in cardiac tissue that will appear in the serum after death of myocardium. However, they lack tissue specificity and so have little or no value in the confirmation or assessment of a myocardial infarction. However, these enzymes can be useful in assessment of secondary involvement and damage to other organs, such as the liver (see Chapter 24).

Other important myocardial proteins[2]

Myosin and actin. Myosin and actin proteins have already been mentioned in regard to cardiac function on pp. 491 and 492. They form the major part of the contractile apparatus of the myocardial cell. Myosin molecules compose the thick fibrous elements of the contractile unit. They have a molecular weight of 500,000 daltons and are large, about 1.6 μm in length. Myosin is stimulated by both magnesium and calcium and has the ability to split ATP. Actin is a thinner protein and is normally present as a double alpha helix. It has a molecular weight of 47,000 daltons. In addition, there are two regulatory proteins, troponin and tropomyosin. These normally act as inhibitors of contraction until calcium comes along in sufficient concentration to annul this inhibition.

Myoglobin. The heart is critically dependent on oxygen, consuming between 6.5 and 10 ml/100 g of tissue per minute at rest. There is a dramatic increase in oxygen requirements with exercise. Normally, some 70% of the oxygen reaching the heart is extracted. The myocardium contains 1.4 mg of myoglobin per gram of tissue. Myoglobin's oxygen dissociation curve is different from that of hemoglobin and thus facilitates entry of oxygen into the cell and is essential for maintaining cardiac function. If one uses, for example, the average situation in which the oxygen saturation in blood leaving the heart is 30%, the Po_2 is around 20 mm Hg in the coronary sinus. If the same Po_2 exists in the myocardium, this will be sufficient to maintain myoglobin oxygen saturation at 85% by virtue of the hyperbolic dissociation curve of myoglobin. Myoglobin has the capability to function as an intracellular oxygen reservoir, but the actual amount of storable oxygen is only enough for a few beats (less than 10). Myoglobin has a molecular weight of 17,000 daltons and is small enough to be excreted in the urine.

PATHOLOGICAL CONDITIONS

Pathological processes affecting the heart are generally classified as follows:

1. Ischemic heart disease, which is one of the major health problems in developed countries.
2. Diseases of the heart muscle. This is a heterogeneous group of conditions described under the generic term "cardiomyopathy." The term implies a direct toxic effect on or intrinsic abnormality of myocardial cell tissue.
3. Damage to the heart valves, causing them to become constricted or incompetent (leaky). This is a common complication of rheumatic heart disease.
4. Abnormalities of the cardiac conduction system or arrhythmias.
5. Congenital abnormalities (about 0.8% of full-term births).

Ischemic heart disease

Ischemia[1] is defined as the situation in which an organ has an inadequate blood supply to maintain its essential functions. In the case of myocardial tissue, function is strictly dependent on blood supply because, being a highly aerobic organ, the heart has little reserve if blood supply becomes deficient.

Causes. There are many causes of myocardial ischemia, but the most common cause by far is coronary atherosclerosis. This is a lifelong process in which the arteries supplying blood to the heart gradually narrow as a result of deposition of lipid and associated material in the arterial wall, with encroachment on the lumen until the point is reached at which blood supply is inadequate for the needs of myocardial function. Usually cardiac insufficiency is noted only on exertion, but in more severe cases it occurs at rest. There is good evidence that coronary vasospasm can also produce ischemia. Coronary vasospasm is a pathological condition wherein the coronary artery wall is abnormal in that it responds in a supersensitive fashion to normal vasoconstrictor inputs, be they emotional, neural, or hormonal. The effect is that the wall constricts in an abnormal and prolonged fashion so that the tissue becomes ischemic. If the spasm relaxes, there may be no symptoms and no damage, but if it is prolonged, damage may result in myocardial infarction or death. Other less common causes of myocardial ischemia are inflammation of the coronary arteries and thromboses (blood clots), severe anemia, and severe hypotension.

Effects of occlusion on myocardium

Biochemical consequences. It should be noted that cessation of blood flow produces a complex series of consequences. Not only is there serious hypoxia because tissue oxygen concentration drops drastically, but also there is the problem of removal of toxic cellular metabolites from ischemic tissue. Sudden occlusion of the coronary artery leaves myocardial cells with only a few seconds of aerobic metabolism as they rapidly use what oxygen supplies remain in the microvasculature. Once the oxygen supply has been exhausted, oxidative phosphorylation is unable to continue because there is no receptor for electrons to be transferred to. Myocardial metabolism then switches to use of glycogen or glucose in the anaerobic Embden-Meyerhof pathway with ATP production, instead of the aerobic Krebs cycle. Since the Krebs cycle is unable to function, the end product of anaerobic glucose metabolism, pyruvate, is reduced to lactate, and lactate accumulation is one of the earliest and most dramatic signs of myocardial ischemia. Experimental data indicate that a complete occlusion of a coronary artery will double lactate production in the artery distal to the block by 30 seconds and double it again by 60 seconds. The reduction of pyruvate to lactate recycles $NADH^+$ and NAD^+, permitting the Embden-Meyerhof pathway to continue functioning.

As a compensatory mechanism to maintain the level of ATP production, the myocardial uptake of glucose increases, and glycogenolysis is activated. This in turn increases the intracellular accumulation of pyruvate, which is converted to lactic acid. Intracellular accumulation of lactic acid results in intracellular acidosis, and this affects the enzyme phosphofructokinase, which is a rate-limiting step in glycolysis. Thus even glycolysis can be inefficient in ischemic conditions. Free fatty acids are another source of substrate, but under ischemic conditions they are not oxidized and accumulate inside myocardial cells.

As ischemia progresses, creatine phosphate reserves are used up, adenosine triphosphate levels fall, and cardiac tissue becomes more acidic as lactate and other acidic intermediates of glycolysis accumulate. Up to 15 to 20 minutes after an ischemic incident, the tissue will recover if it is reperfused, and in the light of the lack of any evidence of structural damage, the implication is that the cells are able to maintain their structural integrity up to this point. However, by some 20 minutes of occlusion, over 60% of the adenosine triphosphate has been used up and the amount of lactate in wet myocardial tissue is 12 times its normal aerobic level. In addition, all cellular glycogen has been used.

Structural changes. Once all the glycogen has been used, dramatic ultrastructural changes develop. Mitochondria become severely swollen, the myofibrils become excessively stretched, and the sarcolemma develops areas of separation. Cell membrane damage is also evident. If, at this point, the obstruction is relieved and the myocardial tissue is reperfused with blood, cell lysis and contracture develop. The integrity of the myocardial cell is rapidly disrupted, and it is unable to tolerate the arrival of fresh blood. Experimental studies indicate that a major cause of the cellular damage that ensues is excessive calcium influx into the cells and deposition of the calcium, particularly in the mitochondria.

If there is significant reduction in high-energy compounds necessary to maintain the membrane ionic equilibriums, then cells will accumulate sodium and lose potassium. In particular, there is excessive influx of chloride ions and the cells swell.

Biochemical changes as a consequence of cell death: release of cellular constituents. The distinguishing point between reversible and irreversible ischemic injury is considered to be the inability of the cell to maintain membrane integrity. The cell membrane is a vital component in the preservation of myocardial cell function and structure, and it is damage to this structure that results in the release of intracellular contents. Of these contents, it is release of enzymes that is particularly significant in the evaluation and confirmation of irreversible ischemic injury. There is much clinical and experimental evidence to indicate that if these enzymes are not released into the circulation then the cells have survived an ischemic insult, which may then be classified as reversible. Enzyme release indicates irreversible damage to myocytes and places the patient in the context of having a myocardial infarction rather than simply an anginal or ischemic episode. The enzymes that are released after ischemic irreversible injury are soluble enzymes such as creatine kinase (CK), aspartate aminotransferase (AST), lactate dehydrogenase (LD), aldolase, myokinase, and alanine aminotransferase (ALT). Myoglobin is an attractive protein, in some respects, for assessment of cardiac cell death because, as a result of its small molecular weight, it is rapidly released.

Most enzymes released early in the infarction are cytoplasmic. Mitochondrial enzymes, if soluble, are released also, but there is usually some delay before they appear in plasma. Once the damage to the membrane has occurred, the rate of appearance of an enzyme in the circulation appears to be flow dependent. Thus an area of damaged myocardium with poor blood perfusion will release enzymes to the general circulation much more slowly than enzymes released from cells that die in areas where recirculation flow is better. Another factor affecting enzyme release is the size of the enzyme molecule. CK, with a molecular weight of 80,000 daltons, is released before AST, with a molecular weight of 120,000 daltons, which in turn is released before LD, with a molecular weight of 140,000 daltons. Enzyme release also depends on the pattern of the circulatory system within the myocardium. The subendocardial zone of the left ventricle has the poorest blood supply, and so cells will die most rapidly in this area in a myocardial infarction. However, the low blood flow of collateral (parallel) arteries will delay enzyme release into the general circulation. Better-perfused areas of the heart release cellular enzymes into the general circulation more rapidly. These differences in collateral flow probably help to reduce the correlation between quantity of myocardial tissue lost and the loss as estimated from enzyme release.

Clinical presentation. Patients who have an inadequate supply of blood to the heart often complain of a constricting central chest pressure or pain (angina), which comes on with exertion and is relieved by rest. In angina of effort, the pain usually lasts a few minutes and then clears. If the narrowing of the artery is sufficiently severe or there is superimposed spasm, pain may occur at rest, waking up the patient or interfering with even minimal activity. Eventually the occlusive process may be sufficiently severe to reduce the blood supply to the cardiac tissue to such a point that the tissue actually dies. If frank cell necrosis occurs, the patient has sustained a myocardial infarction or heart attack. In a typical acute myocardial infarction, the onset of symptoms, usually a cardiac type of pain, is taken as the start of the infarct. In the typical case of infarction, the pain often radiates down the left arm or to the other parts of the body. Shortness of breath, sweating, and nausea are also typical accompanying signs. Occasionally the ischemia of the cardiac tissue is sufficiently severe to produce a change in the rhythm of the heart (arrhythmias, see p. 497) that is incompatible with survival. Many patients with coronary atherosclerosis suffer sudden death, the majority of which events are considered to be arrhythmic in nature and caused by myocardial ischemia. Generally, after coronary occlusion, the majority of cells that are dead or are going to die have reached the point of no return by 6 hours. Most damage occurs within the first hour.

Cardiac failure

The term "cardiac failure" means that the heart is unable to maintain an adequate circulation to the essential organs of the body, particularly the kidneys, the liver, and the brain. When the heart is damaged as a pump, the extremely complex neural and hormonal integration of cardiac function and blood flow to the various parts of the body goes badly awry. There is frequently an excessive outpouring of catecholamines with an abnormal amount of peripheral vasoconstriction, together with excessive pooling of fluid in the venous side of the circulation and in the interstitial tissue of the body.

Causes. The major causes of cardiac failure are ischemic heart disease and hypertensive disease. Additional causes include viral infection of the myocardium, deposition of iron in myocardial cells (hemochromatosis), and cardiac amyloid deposition. There is a large group of patients who experience heart failure with no obvious cause. These patients are said to have idiopathic cardiac failure.

Clinical presentation. There is significant systemic derangement in the regulation of the circulation. Frequently, blood flow to the kidneys is inadequate, and this initiates renal compensatory mechanisms, creating excess renin production from the kidney and overactivity of the renin-angiotensin system, which in turn produces abnormal vasoconstriction. Abnormal renal function is manifest as abnormal retention of fluid with deposition of fluid in the peripheral tissues, producing edema. If there is an abrupt decrease of left ventricular function, as can occur in a myocardial infarction, there is an acute rise in the left atrial pressure. This rise in pressure is transmitted back into the lungs and can overwhelm the normal balance between the hydrostatic and the oncotic pressures, which prevents transfer of fluid into the pulmonary tissues. Deposition of fluid in lung tissue can produce acute pulmonary edema, in which case the subject becomes extremely breathless. Rise in pressure can be transmitted to the venous side of the systemic circulation with resulting hepatic congestion and deranged liver function. Long-standing cardiac failure can produce hepatic cirrhosis if the hepatic congestion is sustained long enough. Thus the liver may be unable to metabolize adequately antidiuretic hormone and aldosterone, resulting in further derangements of the circulation. Typically, there is a sharp rise in blood urea nitrogen levels because the earliest organ to suffer usually is the kidney. Inadequate blood supply to the brain can produce disorientation, cerebral hypoxia, and in severe cases structural damage.

Cardiomyopathy

Cardiomyopathy is a disorder of contracting myocardial cells caused by direct damage to myocardial cells. Although there are some specific forms of cardiomyopathy, most clinical cases of cardiomyopathy are idiopathic; that is, the cause is unclear and the presentation is of a heart with dilitation of all four chambers and cardiac failure. Some cases of cardiomyopathy are sequelae to severe viral infection.

Many cases of cardiomyopathy present with frank cardiac failure, and at this stage there are often no specific findings to indicate the cause. Such is not the case, of course, in lead cardiomyopathy or hemochromatosis, in which excess deposition of these metals is demonstrated by special staining techniques. The biochemical findings in most cases of cardiomyopathies are nonspecific and reflect the major clinical presentation, that is, cardiac failure.

Arrhythmias

Although the structure of the heart as a pump is relatively simple, the regulatory system that balances one side of the heart against the other and regulates cardiac function in relation to the needs of the organs of the body is necessarily much more complex and therefore, more at risk of damage. The structure of the cardiac conduction system is discussed briefly on p. 491.

The anatomical changes that result in cardiac arrhythmias are relatively nonspecific, frequently being more related to the disease process affecting the heart muscle. If the heart is damaged, this damage frequently distorts the transmission of the cardiac impulse. If a chamber of the heart dilates, the conduction and muscle tissues forming that chamber can develop the capability of producing abnormal and self-sustaining irregular activity. Stretching of cardiac tissue is likely to produce arrhythmias. Myocardial infarction is particularly prone to be complicated by serious rhythm abnormalities. As an area of cardiac tissue becomes ischemic and dies, the area that is not quite dead develops the potential of either producing noncontrolled activity itself, or of conducting the electrical impulses in an abnormal fashion. In general terms, rhythm abnormalities result from either abnormal automaticity, that is, from groups of cells that develop the capability of spontaneously producing electrical activity when it is not desired, or from the development of abnormal conduction pathways. The anatomical changes that are found are usually caused by an underlying disease process, with arrhythmias being a secondary result of that process. There are, however, some conditions that affect the conduction system directly and produce significant derangement with the development of arrhythmias. In elderly subjects it is not uncommon to find diffuse fibrosis of the conduction system, which renders its function abnormal. Thus the sinus node, instead of firing in a regular fashion and altering its firing rate in response to neural input, can develop significant delays in firing or produce abnormally fast rates of firing. Similar damage is seen in the atrioventricular node and in the ventricular conduction system.

In functional terms the rhythm abnormalities resulting from the changes described above are classified as bradycardias (resulting in heartbeat rates less than 60 per minute), or as tachycardias (producing heartbeat rates faster than 100 per minute). Atrial fibrillation is a fairly common rhythm abnormality in which the atria beat in an irregular, chaotic and rapid fashion. The atrial rate may be around

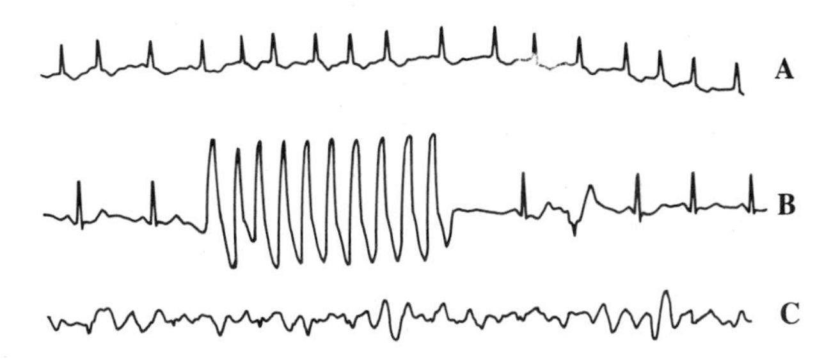

Fig. 28-5. Representative cardiac arrhythmias. These are three rhythm strips. **A,** Atrial fibrillation. In this situation, atria are beating so fast and so irregularly that they appear only as a base-line undulation. Because the atrioventricular node transmits only a portion of these electrical impulses, ventricles beat in a rapid and irregular fashion, since transmission mimics irregular activity in atria. Thus QRS complexes, which represent ventricular contraction, show no obvious relation to one another, and contraction is said to be irregularly irregular. **B,** Ventricular tachycardia. In this situation, ventricles are beating independently of atria. QRS complexes are fast and often bizarre because ectopic focus in ventricle frequently uses an abnormal conduction pathway for transmission of electrical impulse. **C,** Ventricular fibrillation. Here, ventricles have multiple sites of electrical activity, with multiple pathways of transmission through ventricles, so that net result is chaotic electrical activity and no organized sequential contraction of ventricles. In this condition, there is virtually no cardiac output, and it is not compatible with survival.

300 beats per minute, but only a fraction of these impulses are transmitted to the ventricles because the atrioventricular node is unable to conduct much more than 180 beats per minute. This rhythm abnormality is relatively common in the early stages of myocardial infarction and is frequently a complicating factor in rheumatic heart disease, particularly in mitral stenosis, in which the narrowed mitral valve causes a back-pressure effect in the left atrium, which stretches and develops this particular rhythm abnormality. In ventricular tachycardia an automatic focus develops in one of the ventricles that triggers off a very rapid beating rate, which results in a fall in cardiac output because the rate is so fast that the ventricles cannot fill properly. This is a serious rhythm abnormality and requires rapid intervention. In ventricular fibrillation, multiple foci in the ventricles develop the capability of automaticity; the net effect is to render the ventricle unable to pump blood effectively. This rhythm abnormality is incompatible with survival and is only stopped by electric shock treatment (Fig. 28-5). Chronic arrhythmias are controlled by drugs, whose serum levels should be routinely monitored.

Congenital and valvular heart disease

The multitude of congenital abnormalities that have been described will not be discussed. In general terms, all components of the heart are affected by maldevelopment. Most causes are unknown, with one important exception being infection of the mother during the first trimester of pregnancy by rubella.

Of the acquired valvular diseases of the heart, the one large group is that caused by rheumatic carditis. In susceptible subjects affected by the hemolytic streptococcus, the body develops a reaction against myocardial tissue, particularly the valves, which become damaged and undergo deformation. The most common valves affected are the mitral valve and the aortic valve. The valves may become stenosed or incompetent or show a mixture of these lesions.

FUNCTION TESTS

Electrocardiogram

The electrocardiogram[3] is a noninvasive recording of the transmission of the electrical impulse through the heart. Fig. 28-6 is an illustration of the electrical impulse as it appears on the surface electrocardiogram. The components of importance are lettered from *P* through *U*. The electrocardiogram is a most effective means of assessing cardiac rhythm abnormalities at the bedside. However, it has some deficiencies.

Generally, the changes seen in a patient occur in a sequence. Thus in early stages of a myocardial infarction there is severe ischemic damage and ST elevation is seen. The next change that generally occurs is that repolarization becomes abnormal and the T wave becomes inverted instead of being upright. The third and only diagnostic change for myocardial infarction is the development of Q waves. Q waves that are wide and deep are described as pathological Q waves to distinguish them from the small normal Q waves. Pathological Q waves represent tissue death, and the appearance of new Q waves is considered diagnostic for myocardial infarction.

Unfortunately, the electrocardiogram has some limitations.[4, 5] Some conduction abnormalities alter the electrical tracing so that myocardial infarction is obscured by the conduction abnormality. Also, myocardial infarcts can occur in areas of the ventricle that are not easily recorded by the surface electrocardiogram. Myocardial infarctions in the high lateral region of the left ventricle are frequently missed by standard electrode placements, and unfortunately many myocardial infarctions are too small in terms

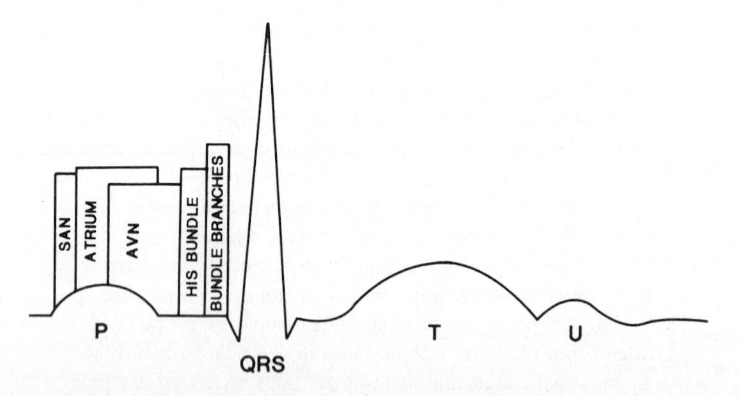

Fig. 28-6. Diagram of surface electrocardiogram. P wave includes activation of sinoatrial node *(SAN)*, atria, and arrival of impulse at atrioventricular node, *AVN*. Electrical impulse then passes through His bundle and bundle branches into ventricular muscle mass proper. QRS complex represents depolarization of ventricular muscle mass starting with depolarization of ventricular septum, illustrated as a small Q wave. T wave represents ventricular repolarization, and U wave is repolarization of cardiac conduction system.

of amount of tissue damage to show up on a surface electrocardiogram with frank changes in Q wave. In addition, there are problems in interpreting the ST- and T-wave changes because they are often altered in myocardial ischemia as distinct from myocardial infarction. For this reason the electrocardiogram, if it is equivocal or not diagnostic, is not sufficiently reliable as a means of ruling out a myocardial infarction. As a means of diagnosing myocardial ischemia, the electrocardiogram is even less reliable. Epidemiological surveys have indicated that some 5% of the adult population have electrocardiographic appearances that are indicative of myocardial ischemia.

Myocardial imaging techniques[1]

The use of radionuclide techniques to assess myocardial function is now common. They permit assessment of cardiac output and definition of wall-motion abnormalities that may occur in ischemia. In general, they are of two forms. One type involves injection of a bolus of technetium-99m pyrophosphate into a vein and observing its passage through the heart. Thallium-201 is a potassium analog that is taken up by myocardial tissue. If an area of the left ventricle is ischemic or scar tissue is there, the thallium is not taken up in that area, which appears as a blank spot.

DRUG THERAPY[6,7]

Two groups of drugs having a direct effect on cardiac tissue are of particular relevance to the laboratory. These are the cardiac glycosides and the antiarrhythmic drugs. There are, in addition, drugs that will affect aspects of myocardial performance, such as oxygen consumption. These drugs include vasodilator drugs, which will reduce cardiac work by reducing systemic blood pressure (such as nitroprusside and hydralazine) or by reducing filling pressure by venous dilatation (such as isosorbide dinitrate and nitroglycerin). In general terms, however, these effects on the heart are not measured and are of secondary importance to their more direct effects on the vascular system.

Glycosides

The cardiac glycosides, of which digoxin is the most commonly used in the United States, hold a unique place in the treatment of cardiac disease. These compounds will increase contractility in the failing heart and are particularly useful in treating heart failure. They also slow conduction of the electrical impulse from the filling chambers of the atria to the pumping chambers of ventricles at the atrioventricular node. Atrial fibrillation is a rhythm abnormality in which the filling chambers beat very rapidly and erratically; this irregular beat is transmitted to the ventricles, which do not fill adequately, so that cardiac output falls (Fig. 28-5). The glycosides slow the ventricular rate in this rhythm abnormality, resulting in improvement in cardiac output. The problem with glycosides is that the therapeutic to toxic ratio is very low. They are particularly prone to cause upsets in the electrical activity of the heart and can induce fatal arrhythmias. Since they have long half-lives in the body, they tend to have cumulative effects, necessitating careful monitoring of blood levels to reduce the possibility of toxic side effects. To this end, much effort has been devoted to defining the range of serum levels in the steady state that are relatively safe and efficacious. The aim of therapy in many patients is to maintain a level regularly within this range so that the maximum benefit from the drug is obtained with minimal toxicity. Assays for various glycosides are available, but they are glycoside-specific; the one most commonly used in this country is for digoxin. We must emphasize that these assays are often misused, since the range that is regarded as safe refers to the steady-state situation. Consequently, use of the assay when patients are being loaded with initial doses of digoxin can be misleading. The range depends not only on the actual assay technique used but also on the experience of the particular institution in which it is in use. Most assay procedures indicate that an upper level of 2.2 to 2.5 ng/mL is useful as a cutoff level. Above this range, toxicity becomes significantly more common. Because there is significant individual variation in the absorption and elimination of digoxin, the assay has come to be widely used as a means of monitoring safety of therapy. Elimination of digoxin is largely by the renal route. The drug is strongly protein bound, and elimination is accordingly slow (biological half-life, 32 hours). Elimination is also altered by cardiac failure, and it is of course reduced in the elderly. The digoxin assay has some other uses. For example, it is used to monitor compliance if a patient is suspected of not taking medication. The assay has become more important recently because of demonstrated interactions between digoxin and other cardiac drugs, such as quinidine and verapamil. These last two drugs, if administered concomitantly with digoxin, can significantly increase digoxin serum levels and induce toxic side effects. Toxicity of digoxin is significantly potentiated by the presence of hypokalemia. Since diuretic therapy for heart failure can induce hypokalemia and digoxin is frequently given to improve cardiac function in this condition, there is a high risk of potential problems.

Lidocaine

Lidocaine is an anesthetic compound that is given intravenously as a bolus or infusion to suppress ventricular irregularities and to prevent the induction of life-threatening arrhythmias, such as ventricular tachycardia. One can also administer the drug intramuscularly, but absorption is somewhat variable. Lidocaine is metabolized by the liver and the half-life in normal persons is approximately 25 minutes. This is prolonged in patients with heart failure or hepatic congestion. Lidocaine is a particularly attractive drug because its side effects are mainly cerebral and there is little or no effect on myocardial function and cardiac conduction. Thus it is one of the safest drugs to use in the context of acute myocardial infarction when cardiac function is changing as the infarct evolves. For this reason, it is most favored in coronary care units as the prophylactic drug of choice in this situation. Although there may be premonitory indications that cerebral toxicity is developing, such as the production of slurred speech, numbness around the mouth, or flickering of the eyelids in many patients, toxicity may be manifested by major convulsive seizures with no prior warning. Accordingly, it becomes important to try to prevent toxicity from lidocaine by use of therapeutic monitoring. Toxicity is particularly common in older subjects. However, since it is being used for life-threatening rhythm abnormalities, the excitement of the moment often prevails even though there is a significant risk of toxicity.

Quinidine and other drugs

Quinidine is another drug used to suppress arrhythmias. In most patients it is used orally, though there are intravenous and intramuscular preparations available. However, in the early days of its use, intravenous and intramuscular routes were associated with a high degree of toxicity and cardiovascular collapse and death, therefore quinidine is much more commonly used as an oral prophylactic medication. Some side effects of quinidine use, such as diarrhea, are unpredictable, but of more importance is its capacity to induce severe arrhythmia in its own right. Indeed, many antiarrhythmic drugs have this potential, and for this reason, estimation of blood levels is often done so that one may use them as safely as possible. Besides quinidine, there are other drugs, such as *procainamide* and *disopyramide,* which are used both intravenously and orally and for which there are assay procedures available to help when one is choosing for a patient a dose regimen that is both effective and as safe as possible. The side effects to be guarded against are primarily electrophysiological, such as induction of arrhythmia, heart block, or significant myocardial depression. There is increasing evidence of significant drug interactions in the cardiovascular field, particularly with quinidine and verapamil, since they have recently been shown to increase digoxin levels significantly. Thus, if these drugs are being used together, serum assays are mandatory to ensure that no toxicity is developing. Many antiarrhythmic drugs have the capability of inducing a particular potentially lethal arrhythmia, which is a variant of ventricular tachycardia. Anxiety about this potential complication has led many clinicians to use blood levels to guide usage.

CHANGE OF ANALYTE IN DISEASE

The only analytes that are used as diagnostic indicators of cardiac disease are myoglobin and the enzymes creatine kinase (CK), lactate dehydrogenase (LD), and aspartate aminotransferase (AST). The use of the major structural proteins, actin and myosin, may become more important in the future. These serum markers are used *only* for the diagnosis of acute myocardial infarction (AMI). Currently there are no biochemical markers unique for the other cardiac diseases. In this section the clinical value and limitation of these analytes in the assessment of acute myocardial infarction is discussed.

Myoglobin

The value of myoglobin assays in acute myocardial infarction is limited by its early and rather brief appearance in serum after a myocardial infarction. The presence of large amounts of myoglobin in skeletal muscle reduces the specificity of this analyte in cases involving possible muscle trauma.

Generally, the biological half-life of myoglobin is around 10 hours, but in some patients there is a delay in peak myoglobin levels. Myoglobin offers the promise of early detection of myocardial infarction, which might be useful if certain interventions were planned but definitive diagnosis had to be made first. Unfortunately, myoglobin is not as specific as CK-MB for myocardial infarction, since serum levels are elevated in patients with myopathies or who are traumatized or undergo surgery. The protein is rapidly excreted in urine, and as may be expected, reduced renal clearance results in elevations of serum myoglobin. Practical experience with myoglobin has not led to an increased ability to detect acute myocardial infarction. Thus this analyte is infrequently used to confirm or rule out acute myocardial infarction.

Enzymes and isoenzymes in acute myocardial infarction

After the onset of symptoms in acute myocardial infarction there is in most patients a so-called time window, during which the enzymes released from the damaged myocardial tissue are found to be elevated in serum. This temporal relationship is unique for each analyte and varies somewhat between persons, though a typical pattern has been determined (Fig. 28-7). Usually, 4 to 6 hours are required after the onset of chest pain before CK-MB becomes elevated in the serum of patients with AMI.[8] Detection times as short as 2 hours or as long as 15 hours have been reported.[9] This activity peaks at 12 to 24 hours and usually returns to base-line levels within 24 to 48 hours. A typical time course for CK and lactate dehydrogenase isoenzymes is for CK-MB to peak first, with LD_1 exceeding LD_2 5 to 20 hours later. The average peak times for enzyme release after infarction in a study of 47 patients were 18 hours for CK-MB and total CK, 24 hours for aspartate aminotransferase (AST), and 48 hours for LD.[10] This time course represents the classic temporal sequence of enzyme changes and is often helpful in distinguishing uncomplicated AMI from extension or reinfarction.

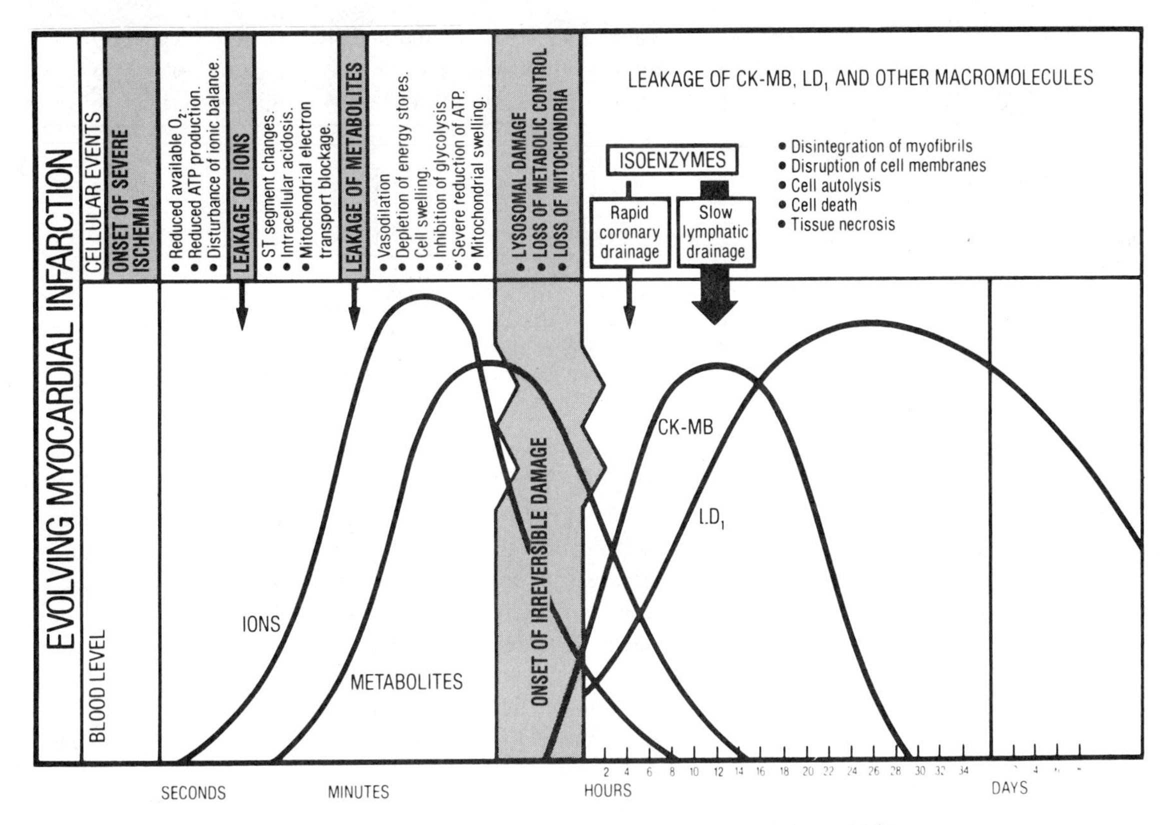

Fig. 28-7. Time course of biochemical events in a typical acute myocardial infarct (AMI). (Adapted from Usategui-Gomez, M.: Lab. World **32:**49-52, 1981.)

Whereas this complete pattern is highly sensitive and specific for infarction, slightly different temporal patterns have been reported in up to 25% of patients with AMI.[11] These workers also report that CK-MB levels are higher and persist longer in patients with cardiogenic shock after AMI.

It follows from the above discussion that proper interpretation of enzyme values in the diagnosis of AMI requires that samples be collected at appropriate time intervals. Since the temporal sequence of events is critical in assessment of the biochemical changes after AMI, the laboratory should normally not analyze a single isolated specimen. Most sources recommend that samples be drawn on admission and at 6 hours, 12 hours, and 24 hours when AMI is suspected.[8,9,12] The absolute minimum number of samples recommended for interpretation is two, obtained 12 and 24 hours *after the onset of symptoms*. Negative results for CK-MB in samples obtained before 12 hours or after 24 hours should not be used to exclude the diagnosis of AMI. A sample collected in this fashion may reflect the period before or after CK-MB elevation occurs in some patients.[13] In the latter case, LD isoenzymes may be helpful, since an increase in LD_1 with concomitant reversal of the normal LD_1/LD_2 ratio, termed the "LD flip," generally occurs at 12 to 24 hours after infarction and may persist for several days. Elevated AST values can also persist for about 96 hours after infarction, and although less specific than LD_1 elevations, can also provide helpful information when the time of onset is unknown or hospital admission is delayed.[11] Note that normal values for total CK should not be used to rule out AMI or to exclude a request for CK-MB analysis. Numerous cases have been documented in which elevated CK-MB levels are accompanied by normal total CK levels during the initial course of an AMI.[13] In most of these cases, however, there is usually a distinct rise and fall of the total creatine kinase values, though none ever exceed the upper limit of normal. However, since approximately five times more CK-MM than CK-MB is released from myocardial tissue, and as a result of the longer half-life of the former, these cases are rather rare. When the time of clinical onset is known, normal results from CK-MB, total CK, and LD in properly collected samples are highly reliable for ruling out AMI.[9]

Isoenzymes in conditions other than myocardial infarction (see following two boxes)

Frequently the kinetics of enzyme clearance in conditions other than myocardial infarction are different from that seen with myocardial infarction and are often chronically elevated. These "false-positive" situations emphasize the importance of testing multiple samples obtained at appropriate intervals. Current research indicates that CK

Elevated serum creatine kinase-MB in conditions other than AMI

Acute myocardial infarction
Severe angina
Chronic atrial fibrillation
Coronary insufficiency
Crush syndrome
Pericarditis
Defibrillation
Insertion of pacemaker
Coronary angiography
Open heart surgery
External cardiac massage or cardiopulmonary resuscitation
Carbon monoxide poisoning
Malignant hyperthermia
Muscular dystrophy, such as Duchenne's syndrome
Poliomyositis
Prostatic surgery or infarction
Dermatomyositis
Reye's syndrome
Malignancy

Causes of increased LD_1/LD_2 ratio ("flip")

Acute myocardial infarction
Acute renal infarction
Hemolysis caused by
  Prosthetic heart valves
  Hemolytic anemias
  Megaloblastic anemias
  Sample during processing
Malignancy

isoenzymes undergo developmental expression that one can duplicate in certain pathological states. This alteration is most commonly observed in skeletal muscle, in which the adult form of creatine kinase is predominantly the MM form. However, fetal muscle is predominantly BB until the sixteenth week of gestation,[14] at which time the expression of the gene coding for the M subunit is significantly increased. Diseases of skeletal muscle characterized by blood-perfusion regeneration are often associated with increased levels of the fetal isomeric forms. Thus Duchenne's muscular dystrophy, polymyositis, and some forms of rhabdomyolysis are examples of skeletal muscle diseases in which increased levels of CK-MB are frequently found in serum in proportion to the degree of fiber regeneration.[14] Again, the time course of CK-MB or BB clearance does not resemble the pattern observed in myocardial infarction. However, the levels of CK-MB are frequently in the range associated with myocardial infarction. Since the absolute amount of creatine kinase in skeletal muscle is about 5 to 10 times that observed in cardiac tissue, the actual elevations of total creatine kinase observed in serum in skeletal muscle abnormalities are frequently dramatically higher than those observed in myocardial infarction. Also, the LD isoenzyme pattern does not reflect myocardial origin.

Table 28-1. Positive and negative predictive values (PV) for serum CK-MB and LD ratio change ("flip") with varying prevalence of acute myocardial infarction

	Test					
	CK-MB*		LD "flip"†		Both‡	
Prevalence	PV+	PV−	PV+	PV−	PV+	PV−
50%	96	100	100	91	100	96
5%	51	100	100	99.5	100	99.7

*Sensitivity = 100%; specificity = 95%.
†Sensitivity = 90%; specificity = 100%.
‡Sensitivity = 95%; specificity = 100%.

In addition to the pathological conditions discussed above, elevations of CK isoenzymes, LD isoenzymes, and AST are obviously present in any situation that increases leakage from myocardial cells. Examples include trauma such as surgery or flail chest and diseases involving the myocardium, such as myocarditis, congestive heart failure, and arrhythmias. Recently, elevations of serum CK-MB have been reported in endurance-trained athletes after competitive events. Muscle biopsy specimens have revealed the source of these serum CK-MB elevations to be increased intracellular content of CK-MB. Cardiac function as judged by scintigraphy does not seem to be impaired in these persons.

Although CK-MB is of great value in a coronary care unit setting, our experience and that of others suggests that it is of lesser use after cardiac surgery. For procedures involving coronary bypass grafting, valve replacement, or repair of congenital defects, CK-MB is usually elevated postoperatively. The release of CK-MB occurs intraoperatively, and by 18 hours after operation the CK-MB has been reported back to normal levels in 74% of cases. Neither the level of CK-MB activity nor the total CK activity reliably indicates the occurrence of AMI in the postoperative period for cardiac surgery. Similar types of conclusions have been reported previously for LD, AST, and total CK, though a recent study found a good correlation of LD isoenzymes with postsurgery morbidity despite the fact that hemolysis may produce the same LD isoenzyme profile in the absence of AMI. The highest levels of CK-MB tend to occur with aortic valve replacement. At present one can only speculate on the contributing effects of hypothermia, fibrillation, defibrillation, and post-pump sequelae after cardiac operations. Establishing the diagnosis of an AMI after a cardiac operation involves documenting appropriate electrocardiographic changes, but the appearance of Q waves postoperatively may merely reflect an old myocardial infarction and not the occurrence of acute myocardial infarction. Myocardial scans have been reported to correlate with the occurrence of AMI, but again both false-positive and false-negative results are known to occur.[8]

Sensitivity and specificity

Clearly, the interpretation of serum enzymes in the diagnosis of AMI becomes obscured by the causes of false-positive and false-negative results. Galen and Gambino[15] have used the predictive value model to express mathematically the clinical use of laboratory tests in various patient populations. Table 28-1 lists both positive and negative predictive values of CK-MB and LD ratio change ("flip") for high- and low-prevalence populations. The use of CK-MB in a coronary care population, with a 50% prevalence of disease assumed, clearly increases the predictive values of elevated enzyme levels. In comparison, in a low-prevalence population (5%) in which many conditions outlined in the top box on p. 502 may be seen, the usefulness of this same enzyme test to assess AMI is greatly diminished.

It is the responsibility of the laboratory not only to ensure accurate and precise enzyme determinations but also to inform the physician of the various clinical and physiological situations that must be considered in effecting appropriate interpretation.

Part II: Hypertension[16,17]

Hypertension is defined as a persistent elevation of blood pressure that continues in the resting, basal state.

In earlier days, treatment of hypertension was limited to severe hypertension, partly because the treatment modalities available were limited and had severe side effects. It is now known that benefit accrues in treating mildly hypertensive subjects in terms of prolongation of life and prevention of hypertensive complications such as stroke.[18-20] With the advent of a large number of different therapies, the treatment of hypertension has been revolutionized over the last 20 years and is now available for consideration in a significant proportion, perhaps 10% to 20%, of the American adult population. Although the clinical manifestations of hypertension are those of end-organ damage, more recently an understanding of the role of arterial function in hypertension has been developed.

ARTERIAL ANATOMY AND FUNCTION[21]

The arterial wall has become an important topic for research and clinical study because of its involvement in hypertension and atherosclerosis. Although other disease entities, such as vasculitis, affect the arterial wall, these

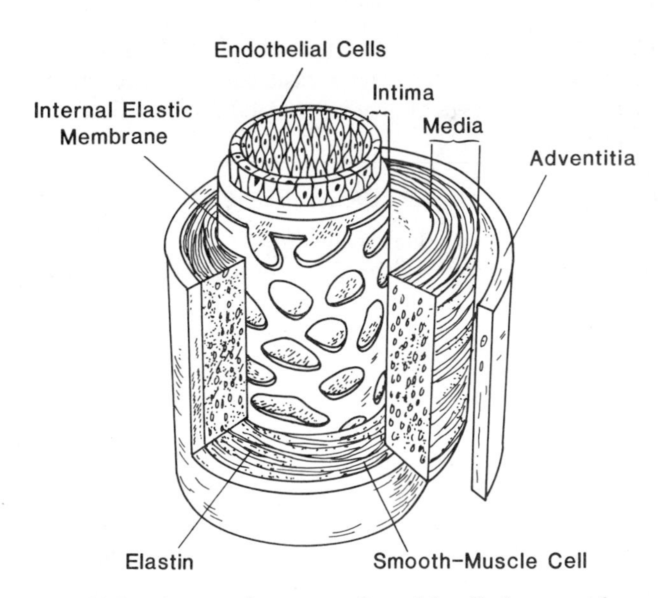

Fig. 28-8. Diagram of structure of arterial wall. Innermost layer is composed of a column of thin endothelial cells. Next comes some connective fibrous tissue and an elastic layer, the internal elastic membrane. Together, these structures form tunica intima. Tunica media is composed of a mixture of elastin and smooth muscle cells, with relative concentrations of each depending on function of artery. Outermost layer is adventitial layer, which is composed largely of collagen.

particular conditions are the major disease processes that affect the arterial wall in Western societies. Significant manifestations of hypertension are those of end-organ damage, particularly involving the kidneys, the heart, and the brain.

The arterial wall is divided into three layers (tunicae), the *intima, media,* and *adventitia* (Fig. 28-8). The part of the intima that faces the bloodstream is a sheet of thin endothelial cells that are important in resisting the shearing force of the bloodstream. The endothelial cells also appear to regulate influx and efflux of materials to the inner muscle layer, and damage to the endothelial layer is considered to be particularly significant in initiating atherosclerotic changes. Next to the endothelial cell lining is fibrous and connective tissue and an internal sheath, composed of elastin, which is the internal elastic membrane. The media is composed of elastin and smooth muscle cells and gives the artery its strength and elasticity. The relative amounts of one component to the other varies, depending on the size of the artery and its particular function. The arteries nearest the heart are thick-walled vessels capable of resisting the pulsatile force of cardiac contraction. These large arteries then branch and subbranch, and as the subbranches approach the tissues they become largely composed of muscle cells because it is at this level that blood flow to the tissues is regulated. Muscle cells are exquisitely capable of contracting and dilating in response to neural and hormonal stimuli that are fed into the arterioles from the central nervous system and from the organs they supply.

There is, however, also some regulatory effect on the arteries closer to the heart, and occasionally, if the artery is abnormal, this aspect of control of blood flow can become significant.

REGULATION OF ARTERIAL FUNCTION[21]

The vascular tone (that is, the degree of vascular contraction) of the arteries depends on two important regulatory mechanisms. The first is sympathetic nervous system input, which acts at alpha and beta receptors. Alpha receptors, if activated, produce vasoconstriction, and beta receptors induce vasodilatation. Vasoconstriction is mediated by an increased passage of calcium into the cell, whereas vasodilatation is modulated through beta-site stimulation (Fig. 28-9).

Of the many humoral effectors interacting with specific receptors on the arterial smooth wall cell, the most important is angiotensin II, which is the most powerful vasopressor agent known to man. Production of angiotensin II is outlined in Fig. 28-10. Briefly, a renal hormone *renin* is produced from a specific group of cells, termed the *juxtaglomerular apparatus,* and this hormone interacts with the glycoprotein *angiotensinogen,* produced by the liver, to produce angiotensin I, which in turn is converted by a ubiquitous enzyme *angiotensin-converting enzyme* (ACE) to angiotensin II. This system is particularly important because many hypertensive persons, particularly young ones and those who have a renal cause of their hypertension, have elevated levels of activity of this system. This has both diagnostic and therapeutic possibilities. One can determine if renal disease is the cause of hypertension in some patients by measuring the components of the renin-angiotensin system. Renin released from the juxtaglomerular cells is regulated by five major factors. These include the pressure within the renal afferent arteriole, which is the most impotant regulator of renin secretion. Plasma sodium concentration within the macula densa (see Chapter 23), input from renal sympathetic nerves, circulating angiotensin II concentration, and plasma potassium concentration also exert significant effects. The enzyme that converts the inactive angiotensin I to the active vasopressor agent angiotensin II by chopping off two terminal amino acids is widespread throughout the body. However, the highest levels occur in the lung; some 33% of angiotensin I is converted in a single passage through the lung vascular bed. Angiotensin II is inactivated by a group of circulating and tissue-bound enzymes, termed *angiotensinases*. The metabolite angiotensin III appears to have a significant effect on the adrenal cortex, stimulating the production of aldosterone. More recently, there have become available

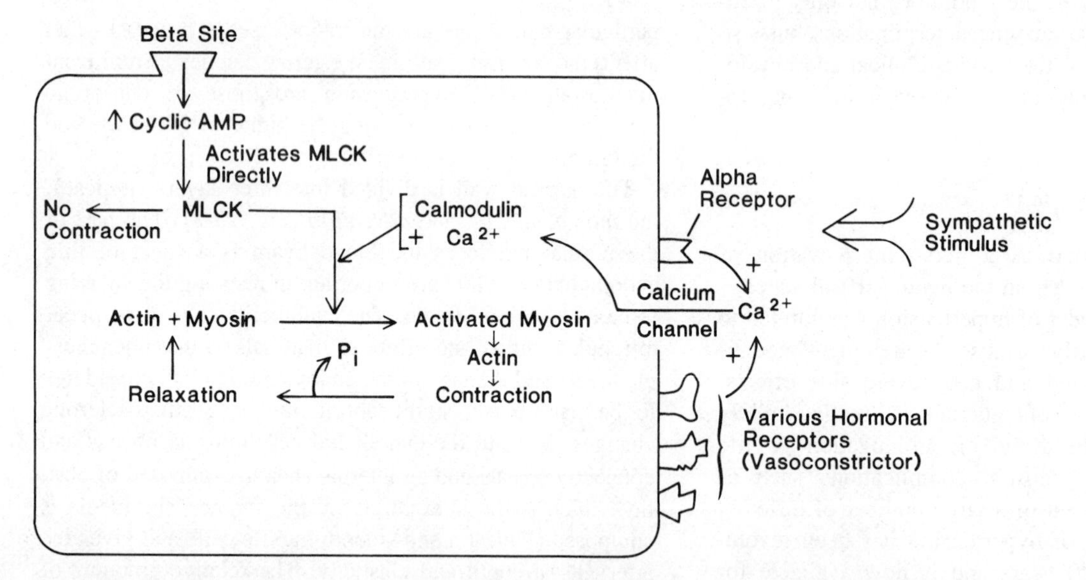

Fig. 28-9. Diagram of contraction in arterial tissue. Arterial cell is critically dependent on inflow of calcium to induce and modulate contraction. Many stimuli, including neural and various hormonal stimuli, regulate inflow of calcium. Calcium combines with an intracellular protein, calmodulin, and this complex, in turn, activates myosin light-chain kinase, an enzyme that phosphorylates myosin and so activates contraction. Contraction is broken by a phosphatase enzyme, and cycle is then ready to be repeated. If beta site on cell is stimulated, levels of cyclic adenosine monophosphate are increased, and this activates myosin light-chain kinase directly. However, without prior combination of calmodulin with calcium, this activation pathway does not activate myosin, and so there is, in effect, no contraction. Thus arterial cell can be seen as a contractile cell where there is a balance between positive and negative stimuli. *MLCK,* Myosin light-chain kinase. P_i, Inorganic phosphate.

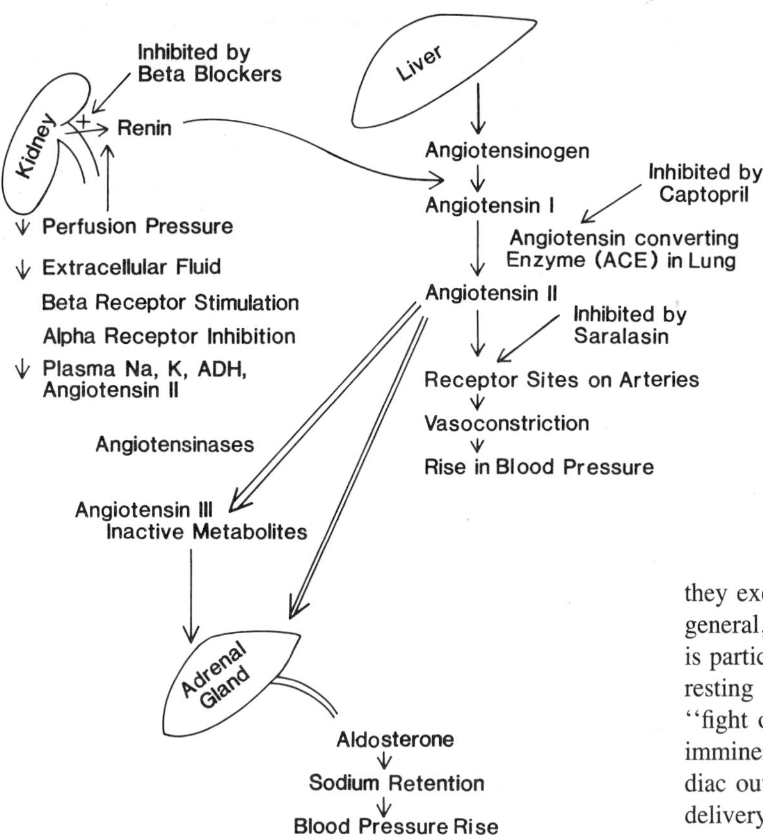

Fig. 28-10. Renin-angiotensin system.

specific inhibitors of components of this system, which offer the chance of treating hypertension in selected subjects with specific monotherapy.

ROLE OF CATECHOLAMINES IN ARTERIAL TENSION

The two major catecholamines are epinephrine and norepinephrine. Norepinephrine is the main catecholamine released at the terminals of the sympathetic nervous system, whereas epinephrine comes mainly from the adrenal medulla. Both hormones, particularly norepinephrine, produce an elevation in blood pressure through their effect on the heart and blood vessels, but there are significant differences in their individual effects. Thus epinephrine causes vasodilatation of the arterioles of the skeletal muscles while vasoconstricting those of the skin, mucous membranes, and the viscera. It is an important cardiac stimulant, increasing the rate and force of cardiac contraction and thus increasing cardiac output. It is also prone to induce cardiac rhythm abnormalities. Norepinephrine has less effect on cardiac output than epinephrine; its effect is that of general vasoconstriction. The catecholamines also have many other effects on the visceral smooth muscle and uterus and diverse metabolic effects. Both hormones may be regarded as general regulators of bodily functions, and they exert a significant effect on arterial tone. Arteries, in general, are never fully relaxed, and catecholamine input is particularly important in maintaining a certain degree of resting tone. The catecholamines are important in the "fight or flight" reaction when the circulation adjusts for imminent demands on its reserves with an increase in cardiac output and an elevation in blood pressure to enhance delivery of blood to essential organs, such as skeletal muscle. Emotional stress will similarly produce an increase in blood pressure, mediated largely through the sympathetic nervous system and possibly, to a lesser degree, through the adrenal medulla.

PATHOLOGICAL CONDITIONS[16,17]

Essential or primary hypertension

In the majority of cases of hypertension, so-called essential hypertension, the cause of the high blood pressure is not known. There are no diagnostic biochemical features associated with this disease. It is, however, a very common problem in developed societies, and the incidence tends to increase with age. Among suggested causal factors are stress and the intake of excessive salt during childhood, but the theory that currently has the most support suggests that hypertension is a mosaic of multiple factors, including genetic and environmental components. Idiopathic, or essential, hypertension has to be differentiated from secondary hypertension, in which an underlying pathological abnormality can be found to explain the hypertension.

Secondary hypertension

Although secondary causes of hypertension amount to only some 5% of cases of hypertension, they are important because they do offer the possibility of a surgical cure, thereby avoiding the necessity of medication for the rest of the person's life. Several diseases affecting the kidney may cause hypertension, but the one of particular significance is narrowing of the renal artery or arteries, since this produces an increase in renin production as the juxtaglomerular apparatus senses a reduction in blood flow. Excessive renin production eventually produces excess angiotensin II, which produces generalized systemic vasoconstriction to maintain blood perfusion to vital organs. Paradoxically, this renal-artery narrowing (stenosis) may protect the kidney from this elevated blood pressure, but it leaves other vital organs, such as the heart and brain, at risk of eventual failure.

Among other causes of secondary hypertension that have to be considered are those with an endocrine cause, such as Cushing's syndrome, which results from excess hormone production by the adrenal gland; Conn's syndrome, in which an adenoma of the zona glomerulosa of the adrenal cortex produces aldosterone; and pheochromocytoma, in which a tumor of the adrenal medulla or other chromaffin tissue of the body produces an excess of catecholamines. In general terms, hypertension is only one facet of the general clinical picture of excessive hyperactivity of a particular gland in question. In Conn's syndrome, patients often complain of muscle weakness and a renal profile can demonstrate a dramatically reduced potassium level. In Cushing's syndrome, the clinical presentation is often classic, with a patient developing abdominal striae caused by alterations in fat distribution, a buffalo hump because of deposition of fat at the nape of the neck, hypertension, hirsutism, and glycosuria.

Pheochromocytoma is unique in that hypertension is a major consequence of the condition, whereas in the previously noted syndromes it is often only one aspect. In pheochromocytoma there is an excess production of epinephrine and norepinephrine. The hypertension so produced is usually paroxysmal, initially with devastating headache, palpitation, and a feeling of dread, all well-known attributes of the fright reaction. Eventually, however, as the tumor grows, the hypertension tends to become persistent and blood levels of the catecholamines being produced by the tumor become persistently elevated. Blood levels in pheochromocytoma require fairly sensitive assays, and screening procedures rely much more on measurement of urinary production of free catecholamines or of their metabolites. In a few rare patients, an exclusively dopamine-producing pheochromocytoma has been described, and these patients in fact present with hypotension, since dopamine is a vasodilator. Most cases of pheochromocytoma, however, produce norepinephrine or epinephrine, and hypertension is the problem. The use of blood measurements is particularly useful in localizing the tumor, since one often finds a localized change in the plasma concentration of the catecholamines in the venous system draining the site of the tumor, thereby permitting its precise localization. These tumors are usually benign, but sometimes they are malignant (about 10%) and can metastasize to the liver or lung. One problem with this relatively rare condition (about 1 in a 1000 cases of secondary hypertension) is that pheochromocytomas can be multiple and recurrence is not unusual.

FUNCTION TEST: BLOOD PRESSURE

Blood pressure is defined by the ratio of the blood pressure measured when the heart is in the systolic phase over the blood pressure measured when the heart is in the diastolic phase.

In Western societies, blood pressure shows a tendency to rise with age. This rise is not found in primitive societies, and it may well be that the rise is abnormal. It should be emphasized that blood pressure is a continuous variable, and many normotensive subjects will stray into the hypertensive range if they are emotionally stressed or are exercising. However, if the elevation of the blood pressure persists even in the resting basal state, the patient is defined as hypertensive. The actual cutoff point for defining hypertension is strongly age related. Thus in a 2-year-old child, a blood pressure 120/80 is abnormal, whereas this is normal in young adults. For most young adults, an upper limit of 140/90 is generally accepted, though many would argue that even this is too high, and one could argue in favor of treatment to reduce this blood pressure still further into the ''normotensive'' range. In older subjects, a blood pressure level of 160/90 to 95 would be accepted by many as being normal.

DRUG THERAPY[16,17,22]

There are many drugs that may be useful for treating hypertensive persons. In general terms, assays of serum levels of these drugs have not played an important part in hypertension treatment, simply because the general approach has been to titrate the dose of drug against blood pressure level. By observation, it has been found that the best guidelines for drug therapy are reduction in blood pressure level and avoidance of side effects. Since one can assess both these aspects, laboratory procedures are for the most part unreliable. Thus the standard approach to a person with hypertension requiring medical treatment has consisted in using either a diuretic or a beta blocker as the drug of first choice.

Diuretics

It is usually found that the reduction in blood pressure with a diuretic is quickly limited, if the dose is increased, by the appearance of side effects, such as prerenal azotemia. Accordingly clinical practice generally limits the use of diuretics in hypertension to relatively modest doses. Monitoring the serum potassium level is particularly important in the many subjects who are treated with a diuretic for hypertension, since the majority of diuretics in current use will induce hypokalemia. Some potassium-sparing diuretics have the obverse capability of producing hyperkalemia. In the majority of cases, drastic alterations of potassium level can occur without clinical symptoms, and it becomes important to assess potassium routinely in subjects on diuretic therapy in case they run into the major complications of hypokalemia and hyperkalemia, that is, cardiac arrhythmias.

The kidney is an important target organ in hypertension, as well as being causal in a significant number of secondary hypertensives. It is of particular relevance to a laboratory assessment, since unlike the heart and brain, renal damage caused by hypertension is often manifest as the development of renal insufficiency, which may be occult. Consequently, many patients with hypertension have routine serial assessments of renal function, with particular attention being paid to blood urea nitrogen or creatinine levels. A rise in these parameters, even of modest size, is particularly disturbing because it may well imply inadequate control of the blood pressure. The alternative possibility is that the therapy being used can in fact reduce renal blood flow, and one may develop a degree of prerenal azotemia. This has to be guarded against, particularly with diuretic therapy.

Of the many diuretics available, the most commonly used to treat hypertension is a thiazide diuretic. Thiazides are gentler in inducing a diuresis than furosemide or ethacrynic acid; this benefit is important in maintaining good patient compliance. The fall in blood pressure is initially caused by sodium and water excretion and reduction in the vascular compartment, but a more prolonged effect appears to be the result of a direct relaxant effect on the arteriole.

Although thiazides may cause potassium loss in common with most diuretics (except the potassium-retaining ones), they have certain metabolic complications that have to be monitored regularly. These problems include hyperglycemia, which stresses the pancreas and may occasionally induce frank diabetes, and hyperuricemia, which can result in gout or urinary tract calculi. Thus regular checks on serum glucose and uric acid and a renal profile are in order.

Potassium-retaining diuretics (such as spironolactone, which antagonizes the action of aldosterone on the renal tubule) are useful in eliminating the need for potassium replacement therapy, but regular checks for hyperkalemia are essential. These diuretics should *not* be given with potassium supplements.

Beta blockers

Beta blockers (beta adrenergic receptor–blocking agents) are a group of drugs that act by blocking the stimulatory input (neural) of the beta-adrenergic receptor sites of muscle cells.

The use of a beta-blocker drug for hypertension is limited by the clinical observation that increasing increments of beta blocker produce progressively less satisfactory reductions in blood pressure as the dosage is pushed to high levels. It is, of course, possible to measure the serum level of propranolol (the most commonly used beta blocker) and to adjust the dose of propranolol accordingly, but in practice this is rarely done. Most clinicians gradually increase the dose of propranolol to a level with which they feel comfortable. Then, if adequate control of the blood pressure has not been obtained, they add a second medication, such as a diuretic.

Other drugs

The standard practice after use of a diuretic and beta blocker is to add a vasodilator drug, such as hydralazine or a centrally active drug such as methyldopa. The next step would be to use even more potent drugs, such as minoxidil or guanethidine. However, with these very potent drugs comes an increased incidence of severe side effects. The common poblems with antihypertensive medications include interference by methyldopa with cross-matching of blood and the production of inflammatory damage to body tissues by drugs such as hydralazine. The potential for hydralazine to induce this complication, termed the "lupus-like syndrome," is dose-related and related to duration of therapy and is particularly prone to occur in persons who do not break down hydralazine rapidly (slow acetylators). It is possible to detect slow acetylators fairly simply, but the common clinical practice is to give the person the medication and then to perform periodic antinuclear-antibody titers, since this titer appears to become positive and to rise before the frank clinical syndrome develops.

The development of the capability for assessing the renin-angiotensin system is important, not only for assessing the pathogenesis of hypertension in subjects but also in regard to therapy. Persons with hypertension resulting from the activity of the renin-angiotensin system can be treated with a specific antagonist to angiotensin-converting enzyme, such as captopril, or with a less specific but safer suppressor of renin production, such as a beta blocker. On the other hand, those persons with low plasma renin activities are much more likely to respond to a diuretic or vasodilator drug in the treatment of their hypertension. It is very likely that in the not too distant future, further inhibitors of this system that do not have the side effects of captopril (proteinuria, reduction in the number of circulating white cells, skin rash, and loss of taste) will become available. This will permit the treatment of hypertension in a more scientific fashion in comparison to the current stepped-care approach, which is designed primarily to avoid side effects of the various medications available.

CHANGE OF ANALYTE IN DISEASE

Renin-angiotensin system

Excess activity of this particular system is a common component in the production and maintenance of hypertension. Renin activity in serum is of little value on a casual basis because random samples from normotensive and hypertensive subjects show a wide overlap. The subject usually has to be prepared, by being ambulant for 2 hours and by taking a diuretic to stimulate renin production, to give a peripheral venous sample adequate discriminating power to accurately classify hypertensives. Some investigators find renal-vein renin assays more valuable for evaluating renal involvement in hypertension: a unilateral increase in secretion rate is highly suggestive of a renal cause (particularly renal artery stenosis) of the hypertension. These samples have to be obtained from the renal veins and require femoral vein catheterization. A ratio of 2:1 (some laboratories use a ratio of ≥ 1.5:1) is very suggestive of a renal cause of the hypertension and would indicate that a surgical approach (unilateral nephrectomy for unilateral parenchymal disease or renal artery surgery for renal artery stenosis) may well be curative. With the advent of the saralasin infusion test, it is easier to assess this particular system by competitive antagonism of angiotensin II uptake at arterial receptors. With the patient resting supine, a drop of ≥ 8 mm Hg in the systolic pressure is virtually diagnostic of significant involvement of this system in the maintenance of hypertension.

Catecholamines

Surprisingly, despite the commonly held belief that stress is a major factor in the induction and maintenance of blood pressure, catecholamine assays have little use in the assessment of most hypertensive persons. In labile hypertension (that is, blood pressure that fluctuates above and below a rather arbitrary cutoff level), there is some evidence that excess catecholamine production is characteristic and, if sustained, may result in permanent hypertension. Catecholamine assays are of particular value if pheochromocytoma is being considered, since in this condition hypertension is caused by excessive production of catecholamines.

Renal analytes

A renal profile is a common and regular procedure performed on hypertensive persons. Poorly controlled hypertension frequently damages the kidneys, and renal insufficiency may become manifest only as a progressive rise in blood urea nitrogen (BUN) or creatinine. A rise in BUN, however, occurs late in the disease, by which time a significant amount of renal damage may have occurred. Proteinuria is also a common manifestation of hypertension and is usually measured to monitor the disease. Potassium is an important assay not only because it may point to the underlying pathological condition (such as, severe, persistent hypokalemia in Conn's syndrome) but also because hypokalemia is a common problem with diuretic therapy and its development can be so insidious (mild fatigue) that it may eventually present as a life-threatening cardiac arrhythmia. The converse problem, hyperkalemia, may occur with potassium-conserving diuretics, especially if the patient continues to take potassium supplements or if renal insufficiency develops. A creatinine clearance is often useful in separating renal and prerenal azotemia.

REFERENCES

1. Braunwald, E., editor: Heart disease: a textbook of cardiovascular medicine, Philadelphia, 1980, W.B. Saunders Co.
2. Hearse, D.J.: Enzymes in cardiology: diagnosis and research, New York, 1979, John Wiley & Sons, Inc.
3. Chou, T.C.: Electrocardiography in clinical practice, New York, 1979, Grune & Stratton, Inc.
4. Borer, J.S., Brensike, J.F., Redwood, D.R., et al.: Limitations of the electrocardiographic response to exercise in predicting coronary-artery disease, N. Engl. J. Med. **293**:367, 1975.
5. Feigenbaum, H.: Electrocardiography, ed. 3, Philadelphia, 1981, Lea & Febiger.
6. Singh, B.N., Collett, J.T., and Chew, C.Y.B.: New perspectives in the pharmacologic therapy of cardiac arrhythmias, Prog. Cardiovasc. Dis. **224**:243-301, 1980.
7. Schwartz, J.B., Keefe, D., and Harrison, D.C.: Adverse effects of antiarrhythmic drugs, Drugs **21**:23-45, 1981.
8. Guzy, P.M.: Creatinine phosphokinase-MB (CPK-MB) and the diagnosis of myocardial infarction, West. J. Med. **127**:445-460, 1977.
9. Lott, J.A., and Stang, J.M.: Serum enzymes and isoenzymes in the diagnosis and differential diagnosis of myocardial ischemia and necrosis, Clin. Chem. **26**:1241-1250, 1980.
10. Blomberg, D.J., Kimber, W.D., and Burker, M.D.: Creatine kinase isoenzymes; predictive value in the early diagnosis of myocardial infarction, Am. J. Med. **59**:464-469, 1975.
11. Neufeld, H.N., Rabinowitz, B., Clejan, S., et al.: Isoenzymes of creatine phosphokinase in acute myocardial infarction, Angiology **28**:853-864, 1977.
12. Galen, R.S.: Isoenzymes: which method to use, Diagn. Med. **1**:42-63, 1978.
13. Irvin, R.G., Cobb, F.R., and Roe, C.R.: Acute myocardial infarction and creatine phosphokinase, Arch. Intern. Med. **140**:329-334, 1980.
14. Foxall, C.D.: Changes in creatine kinase and its isoenzymes in human fetal muscle during development, J. Neurol. Sci. **24**:483-492, 1975.
15. Galen, R.S., and Gambino, S.R.: Beyond normality, New York, 1975, John Wiley & Sons, Inc.
16. Kaplan, N.L.: Clinical hypertension, ed. 2, Baltimore, 1978, The Williams & Wilkins Co.
17. Marshall, A.J., and Bassett, D.W.: The hypertensive patient, London, 1980, Pitman Pub. (Pitman Medical).
18. Effects of treatment on morbidity in hypertension: I. Results in patients with diastolic blood pressures averaging 115 through 129 mmHg, Veterans Administration Cooperative Study Group on Antihypertensive Agents, J.A.M.A. **202**:1028-1034, 1967.
19. Effects of treatment on morbidity in hypertension. II. Results in patients with diastolic blood pressure averaging 90 through 114 mmHg, Veterans Administration Cooperative Study Group on Antihypertensive Agents, J.A.M.A. **213**:1143-1152, 1970.
20. Five year findings of the hypertension detection and follow-up program. II. Mortality by race, sex and age, Hypertension Detection and Follow-up Program Cooperative Group, J.A.M.A. **242**:2572-2577, 1979.
21. Vanhoutte, P.M.: Vasodilatation, New York, 1981, Raven Press.
22. Scriabine, A., editor: Pharmacology of antihypertensive drugs, New York, 1980, Raven Press.

Chapter 29 Muscle disease

Béla Nagy

acetylcholine (ACh) A quaternary ammonium compound, released at the motor end-plate, acetylcholine combines with a receptor on the muscle membrane, thereby transmitting the nerve impulse.

actin The protein component of thin filaments in the sarcomere.

action potential The potential difference between the interior and exterior of the living cell is the membrane potential; brief positive changes in the membrane potential initiating an action is termed ***action potential*** (also ***nerve impulse*** or ***spike***).

adenosine triphosphate (ATP) A nucleotide compound found in all living cells. It serves as energy storage (fuel) for cell activity.

adenosine deaminase An enzyme in muscle that converts AMP into inosine.

alanine aminotransferase (ALT, formerly GPT or SGPT) A transaminase found in muscle and other tissues.

aldolase An enzyme of the glycolytic pathway found in muscle and other tissue.

aspartate aminotransferase (AST, formerly GOT or SGOT) A transaminase found in muscle and other tissues.

cardiac muscle Found only in heart, it is cross-striated like skeletal muscle but the cells are branched and interconnected so that the heart contracts and relaxes as a whole. Cardiac muscle normally is not under voluntary control.

creatine kinase (CK) An enzyme in muscle that catalyzes the transfer of a phosphate group from creatine phosphate to ADP to yield ATP and creatine.

creatine phosphate A creatine–phosphoric acid compound in cells present in large amounts in muscle, another form of energy storage used to rephosphorylate ADP to ATP.

end plate An expansion of the motor nerve terminal, also called a ''neuromuscular junction,'' that contains the presynaptic terminal of the nerve, the synaptic cleft, and the postsynaptic element, which is the skeletal muscle fiber membrane.

glycogen A readily mobilized storage form of glucose. A branched polymer of glucose.

glycolysis Degradation of glucose into three-carbon compounds with a net production of ATP.

involuntary muscles Smooth muscles not under conscious control.

isoenzymes One of the multiple forms in which an enzyme may exist in a single organism, differing chemically and physically but catalyzing the same reaction.

lactate dehydrogenase (LD) An enzyme in muscle and other tissues producing lactate from pyruvate when the amount of oxygen is limiting.

lactic acidosis Lowering of the pH in muscle and serum caused by accumulation of lactic acid.

motor nerve endings Branched or plaquelike endings of axon terminals of voluntary muscles that convey motor impulses.

motor unit The motor nerve fiber together with the muscle fibers that it innervates.

muscular atrophy A motor unit dysfunction, the major cause of which is loss of efferent innervation.

muscular dystrophy A motor unit dysfunction; a hereditary myopathy involving the muscle fibers.

myofibrils The smallest long threads in the muscle fiber, composed of numerous myofilaments.

myogenic Pertaining to muscles, such as cardiac and smooth muscles, that do not require nerves to initiate and maintain contractions.

myoglobin A heme-containing protein of muscle that serves as a reserve supply of oxygen.

myokinase (MK) A muscle enzyme converting two ADP into an ATP and an AMP.

myosin The protein that constitutes the thick filaments of the sarcomere.

myopathy General motor unit disorder involving the muscle fibers.

neurogenic Muscles with well-defined end plates in which contraction is initiated by nerve impulses. If the nerves are cut, the muscles they supply are paralyzed and degenerate.

oxidative phosphorylation An ATP-generating process in which oxygen serves as an ultimate electron acceptor. It occurs in mitochondria and is the major source of ATP generation in aerobic organisms. Also called "respirations."

sarcomere The smallest functional unit of the myofibril, a region between two Z disks from the middle of one I band to the middle of the next I band.

sarcoplasmic reticulum (SR) Well-developed system of tubules around several myofibrils.

skeletal muscle Striated muscles under voluntary control that are attached to bones and usually cross one joint.

sliding filaments The interdigitated thick and thin filaments in the sarcomere. In muscle contraction they slide past each other, shortening the sarcomere without changing their length.

smooth muscle Muscle with narrow, long cells with no cross-banding, no end plates, and innervation. It contracts slowly.

tropomyosin Protein component of thin filaments.

troponin Protein component of thin filament, together with tropomyosin, regulates actin and myosin interactions.

ANATOMY OF MUSCLE

The recent advances in muscle research have greatly enhanced our understanding of the anatomy, molecular structure, physiology, and biochemistry of normal muscle function. There are three distinct groups of muscle: skeletal, cardiac, and smooth. The *skeletal muscle* is the most throughly studied and possesses a large degree of structural regularity. It consists of parallel bundles of unbranched, cylindrical muscle cells, or myotubes, fibers with multiple

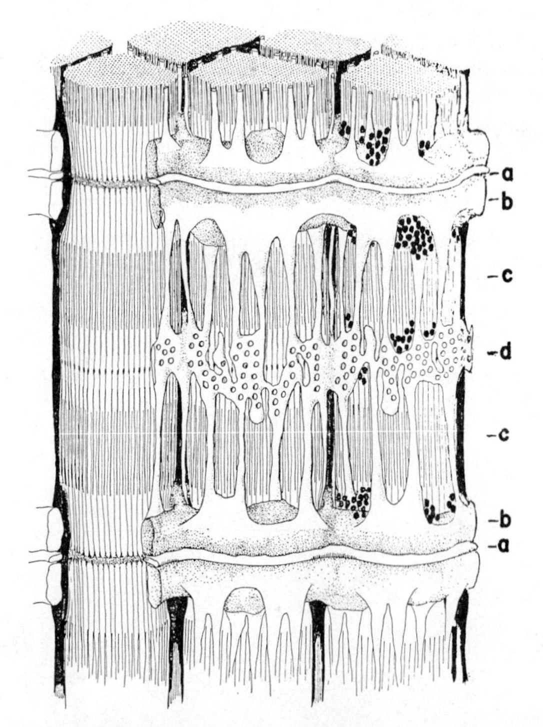

Fig. 29-1. Diagram of skeletal muscle filaments with sarcoplasmic reticulum. Major features of sarcoplasmic reticulum at different levels marked with letters on diagram: *a*, transverse tubules; *b*, terminal cysternae; *c*, longitudinal vesicles; *d*, fenestrated collar at center of A band. Dense granules are glycogen. Transverse tubules and terminal cysternae form the triads at so-called T system. (From Peachey, L.D.: J. Cell Biol. **25**:209-231, 1965.)

nuclei as a result of fusion of many single cells during myogenesis. The fibers run the whole length of the muscle and microscopically show characteristic crossbanding. They have well-defined nerve end plates and are under voluntary control. Contractions are initiated by nerve impulses and are termed *neurogenic*.

The vertebrate skeletal muscle is made up of many cells or fibers, which are extremely long, running parallel for the length of the muscle. The cells are multinucleated (Fig. 29-1). The diameter of the cells ranges from about 20 to 100 μm, and in each cell there is a bundle of cylindrical *myofibrils*, each about 1 to 2 μm across. The myofibrils are surrounded by an extensive network of tubular channels, the *sarcoplasmic reticulum*. *Nerve endings* are attached to the outer surface of the sarcolemma through the motor end plate of an axon. Within the sarcolemma-enclosed space the fibrils are bathed by the intracellular fluid of muscle, the *sarcoplasm*. The sarcoplasm contains *glycogen, glycolytic enzymes, ATP, ADP, AMP, creatine phosphate, creatine, inorganic electrolytes, amino acids, peptides*, and numerous other cell components. The fibers or cells of the muscle are linked together by collagenous *connective tissue*, which form, at the ends of the muscle, the *tendons* between the muscle and the skeleton. The muscle then exerts its action on the skeleton as a lever, as either a *flexor* or an *extensor*.

Within the myofibrils the contractile material is organized into repeating units, termed *sarcomeres*. The sarcomeres are seen by light micrographs giving a pattern of alternating light and dark bands, yielding the characteristic cross-striation (Fig. 29-2). The band pattern is caused by the organized contractile material that forms two sets of filaments, with alternating optical properties, named the "anisotropic band (A band)" and the "isotropic band (I band)," respectively. The organized contractile material within the fibrils is composed largely of the proteins *myosin* and *actin* in an alternating filamentous form in each sarcomere. The myosin is mainly in the A band, overlapping with actin filaments that are located mainly in the I band. The proteins found in the myofibrillar structure are listed in Table 29-1.

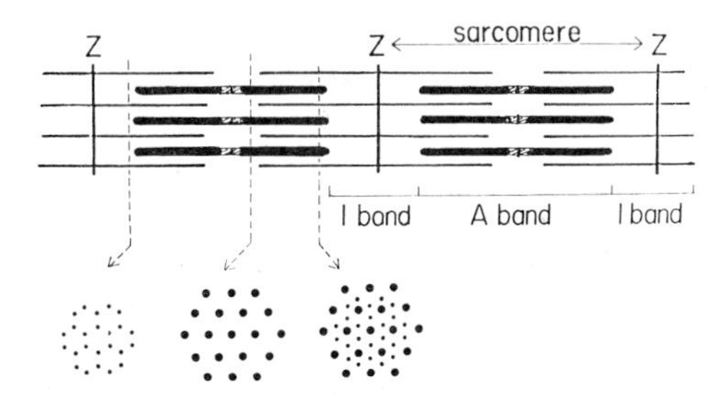

Fig. 29-2. Structure of sarcomer of striated muscle. The sarcomer is the repeating functional unit of muscle. In phase-contrast light microscopy the light I bands alternate with dark A bands. Z line contains material that anchors thin filaments. Transverse sections of sarcomere at different levels are shown in lower part of drawing. At left is a cross section through thin filaments (I band), middle is that of thick filaments (A band), and overlap region cross section is on far right. In overlap array each thick filament is surrounded by six thin filaments. In contraction there is interaction between thin and thick filaments, and they slide past each other to about half the distance of relaxed sarcomere length.

Cardiac muscle is found only in the heart. It has crossbanding, as skeletal muscle does, but the cells are branched and interconnected. There are no defined end plates in cardiac muscle, and in normal conditions it is not under voluntary control; that is, it does not require nerves to initiate and maintain its contraction and therefore it is termed *myogenic*.

Smooth muscle is composed of narrow, fusiform cells about 2 to 4 μm in diameter, with a single nucleus. It does not have crossbanding. The smooth muscle cell does not have a structurally defined end plate and is not under voluntary control; therefore it is also termed *involuntary* muscle. Smooth muscle is found usually in walls of tubes or sacs such as blood vessels, the wall of the uterus, the urinary bladder, the intestines, and the bronchioles. It is characterized by slow contraction, and it can maintain tension or a given length without fatigue at low energy cost.

FUNCTION

Mechanism

The general function of the muscles is a mechanical response to stimulation that produces shortening and force development, both usually occurring together. The skeletal muscle function is modified by leverage as a result of its attachment to the skeleton.

In cardiac muscle the force development is manifested by the development of pressure within the chambers of the heart during cardiac muscle shortening, which results in reduced chamber size.

Table 29-1. Myofibrillar proteins and their localization in vertebrate skeletal muscle

Proteins	Localization	Amount of protein as percentage of total cellular protein
Contractile proteins		
Myosin	A band	60
Actin	I band	20
Regulatory proteins		
Tropomyosin	I band	3
Troponin	I band	4.5
M protein	M line (M band)	Less than 1
C protein	A band	Less than 1
Actinin	Z line, I band	1
Other minor proteins		
Connectin	A and I band	Less than 1
Z protein	Z line	Less than 1
Desmin	Z line	Less than 1

Smooth muscle shortening is most demonstrative when smooth muscle sacs or cavities are emptied, as in the expulsion of urine from the bladder or of a child from the uterus. It is apparent that considerable force is developed during such smooth muscle activity.

Skeletal muscle can shorten to as much as a third of its original length. A model, termed the *sliding filament model,* is used to explain muscle contraction in this system. The two discontinuous muscle filaments found in the sarcomere interdigitate, and the thick and thin filaments slide past each other in contraction. During this sliding the length of the sarcomere decreases because the two types of filaments overlap more in the contracted state.

The hydrolysis of ATP supplies the immediate energy for muscle contraction. The amount of ATP hydrolyzed in contraction is not constant but depends on the duration of contraction and on the amount of work done by the muscle. The muscle uses up more ATP (''fuel'') when the work load made on it is greater than at resting. However, ATP is hydrolyzed to a noticeable extent even when there is no apparent external work; for example, the muscle is not allowed to shorten when one holds a heavy suitcase for a long time, and this leads to a tired feeling, just as lifting a heavy load does. The ATP-hydrolysis sites are on the cross bridges interacting between thick (myosin) and thin (actin) elements of the sarcomere, and the ATPase is highly active only when myosin and actin interact.

Relaxation of the muscle is a passive process. When there is no cross-bridge formation, the sarcomere acquires its resting length. The interdigitated filaments slide back to a less overlapped position, thereby increasing the length of the muscle. The same basic principle applies to skeletal and cardiac muscle and even to smooth muscle contraction; however, in smooth muscle there are no sarcomeres and the thick and thin filaments are less organized. In smooth muscles the actin filaments are anchored either to the cell membrane or within the cytoplasm to ''dense bodies,'' structures resembling developing Z lines in skeletal muscle.

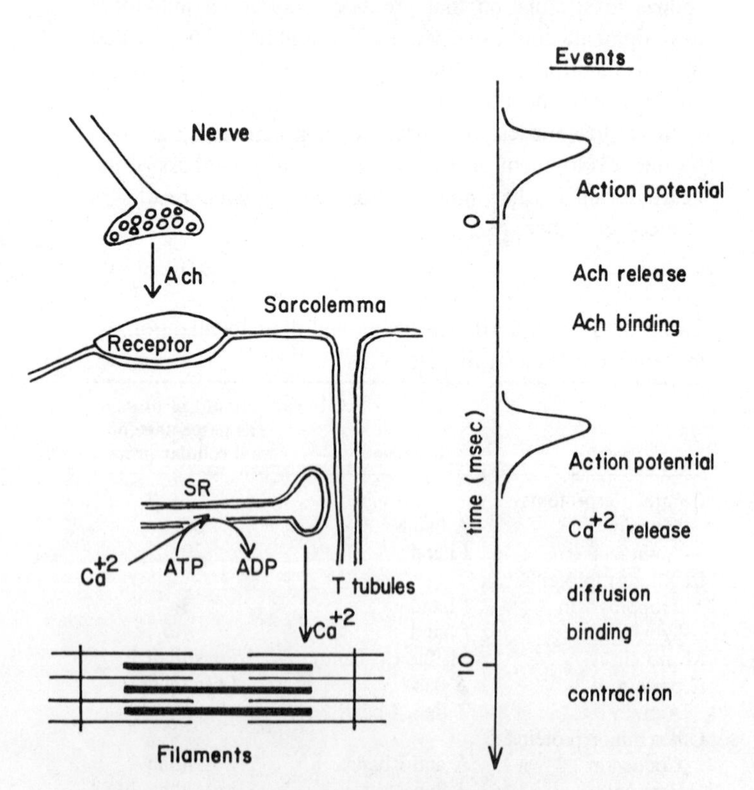

Fig. 29-3. Diagram of nerve impulse and muscle contraction. *Ach,* Acetylcholine; *SR,* sarcoplasmic reticulum.

Acetylcholine as neuromuscular transmitter

Motor nerve endings do differ, depending on the main fiber type. ''Slow-twitch'' fibers, which respond to nerve stimulus with prolonged contraction, are in general innervated by multiple nerve endings. ''Fast-twitch'' fibers are innervated usually by individual end plates. At both types of motor nerve terminals acetylcholine is synthesized and stored in vesicles. These vesicles contain approximately 10,000 acetylcholine molecules; one unit of such vesicle content is called a ''quantum.'' Conducted nerve impulses release large numbers of acetylcholine molecules, about 150 to 200 quanta per impulse or about 1 million molecules, giving rise to a large amplitude end-plate potential. Such end-plate potential triggers a complex sequence of events, starting with depolarization of the muscle membrane and leading to contraction of the muscle.

Excitation-contraction coupling

In striated muscle, at a single end plate of a motor nerve, the nerve impulse is transmitted to the muscle fiber membrane (sarcolemma) by the transmitter chemical from the nerve, acetylcholine (Ach) (Fig. 29-3). The plasma membrane depolarization is initiated by acetylcholine binding to the acetylcholine receptor. The depolarization spreads as action potential along the sarcolemma and down the T tubules around the myofilaments, in some way affecting the sarcoplasmic reticulum (SR). The response is a massive release of calcium from the sarcoplasmic reticulum. The free Ca^{2+} concentration rises near the myofilaments 100 to 200 times over the resting concentration. The calcium binds to the troponin complex in the thin filaments, and the interaction with thick filaments is initiated. The filaments slide by each other and contraction occurs.

On the left of Fig. 29-3 is the scheme of chronological events from nerve to myofilaments. On the right side of the figure is the time scale of the events.

In the meantime the released and receptor-bound acetylcholine is destroyed by a very fast acting enzyme, acetylcholinesterase. The calcium level is decreased by active accumulation into the sarcoplasmic reticulum. The calcium from the troponin complex is released and muscle activity ceases until a new train of events occurs.

NORMAL BIOCHEMISTRY

Contractile proteins of muscle (Table 29-1)

Myosin. Myosin is the major protein of skeletal muscle and composes 60% of the total protein in myofibrils, which are located in the A band in thick filaments. The individual myosin molecule has a molecular mass of about 480,000 daltons and a long, rodlike shape. It consists of two apparently identical polypeptide chains—the *heavy chains* (H), with a mass of about 200,000 daltons each, and four smaller peptides, termed *light chains* (L), in the mass range of 16,000 to 30,000 daltons.

There are three different isomeric forms of light chains, designated L_1, L_2, and L_3, respectively. Each myosin molecule contains two L_2 and a pair of either L_1 or L_3, depending on the type of muscle. There appears to be a slight variation between the light chains in the different muscle types; however, the L_2 chains appear to be related by their ability to undergo phosphorylation catalyzed by a *light-chain kinase*. The kinase is activated by the Ca^{2+}-binding protein *calmodulin*. The L_2 chain, which undergoes phosphorylation, has been implicated in the regulation of smooth muscle contraction.

In a single myosin molecule, the *functional part* is located mainly in the globular head part of each peptide, which contains the *ATPase site* and the *actin-binding site,* the two main biological activities of myosin.

Actin. Actin is the second most abundant protein in muscle, consisting of 20% of the total protein in myofibrils. Structurally, actin is a single polypeptide protein folded into a globular shape. Actin binds strongly with one nucleotide, ATP or ADP, and one divalent cation, Ca^{2+} or Mg^{2+}, per molecule. It is composed of 374 amino acids. The molecular mass of actin is 41,785 daltons, or 42,300 daltons when bound ATP and Ca^{2+} are taken as part of the molecule.

Actin has been found in every eukaryotic organism tested and is probably one of the most conserved proteins throughout biological evolution. There are some microheterogeneities between actins of different species, but in skeletal and cardiac muscle only a single type of actin, α-actin, is found. Two other species of actin, β-actin and γ-actin, are found in nonmuscle cells.

Actin interacts with a large number of proteins. Within the myofibrils some interactions are essential to its biological function. It interacts with *α-actinin,* which in striated muscle is located in the Z line of the sarcomere, and its role together with other Z-line proteins is probably to anchor the actin filaments. The other protein that interacts with actin is *tropomyosin,* which is an integral part of the thin filament in myofibrils and is discussed below. The major functional interaction of actin is with myosin, specifically with the globular part of myosin. The interaction stimulates the myosin ATPase activity by accelerating the release of ADP from myosin, which appears to be the rate-limiting step in the ATPase activity. The actomyosin complex then is dissociated by ATP, and the cycle of actin and myosin interaction starts again. This process leads to contraction, which was described on pp. 511 and 512.

Regulatory proteins of muscle

Tropomyosin and *troponin* are the two regulatory proteins involved with Ca^{2+}-mediated regulation of actin and myosin interactions. They are located in the thin filaments of the myofibrils with strong but noncovalent association with the double-stranded actin in the filament.

Tropomyosin. Tropomyosin is a fully α-helical, two-subunit, coiled-coil molecule with a molecular mass of about 65,000 daltons. It consists of about 3% of the total proteins in myofibrils. The long, coiled tropomyosin molecules form filamentous aggregates through overlapping amino and carboxyl ends. This long tropomyosin strand binds to actin and occupies the two grooved sides of the double-stranded actin filament. The dislocation of these tropomyosin strands on an actin filament from one site to another site is regulated through *troponin,* the regulator being the Ca^{2+} concentrations. The tropomyosin-troponin-Ca^{2+} regulation of contraction is proposed to be through tropomyosin dislocation of actin filament, which then is allowed to interact with myosin.

Troponin. Troponin is a three-subunit protein with a molecular mass of about 70,000 daltons and consists of about 4.5% of the total protein of the myofibrils. There are three subunits of troponin that are denoted as troponin C (the Ca^{2+}-binding component), troponin T (the tropomyosin-binding component), and troponin I (the inhibitory component, inhibiting the actin and myosin interactions in the presence of ATP but in the absence of Ca^{2+}).

The thin filament, consisting of actin, tropomyosin, and troponin and whose combined mass consists of about 27.5% of the total proteins in the myofibril, is localized in the I band of the sarcomere.

Other muscle proteins

Myoglobin. Although hemoglobin, which is found in the red cells, is the oxygen transporter in vertebrates, myoglobin, which is located in muscle, facilitates the movement of oxygen from blood to muscle and serves as a reserve supply of oxygen. It is a single-chain, compact molecule, with a molecular mass of 17,200 daltons, about one fourth the size of hemoglobin. Myoglobin carries only one heme group. However, the α and β chains of the hemoglobin subunits are strikingly similar to myoglobin, which is usually detected by immunological analyses.

Enzymes in muscle

Creatine kinase (CK). The enzymatic hydrolysis of creatine phosphate and transfer of the high-energy phosphate to adenosine diphosphate was shown in the early 1930s. The reversible reaction catalyzed is:

$$\text{ATP} + \text{Creatine} \rightleftharpoons \text{ADP} + \text{Creatine phosphate}$$

Table 29-2. Creatine kinase activity and isoenzyme patterns in fresh human tissues

Tissue	Creatine kinase activity (U/g wet weight)	Percent of total activity		
		MM	MB	BB
Skeletal muscle	2000-3200	>95	0-5	0
Heart	400	78	22	0
Brain	160	0	0	100
Prostate	10	4	4	92
Uterus	50	2	1	97
Kidney	15	12	0	88
Liver	4	90	6	4

From Tsung, S.H.: Clin. Chem. **22**:173-175, 1976.

The enzyme involved in this reaction is referred to as "creatine kinase." Previously the enzyme has been variously known as creatine phosphokinase, creatine phosphopherase, creatine phosphotransferase, ATP-creatine phosphopherase, ATP-creatine transphosphorylase, and also the Lohmann enzyme.

Creatine kinase is widely distributed in tissues. Its function is ATP regeneration especially in contractile and transport systems. In muscle it may represent 10% to 20% weight/volume of cytoplasmic protein. It has a molecular mass of 81,000 daltons and a dimeric composition. Three isoenzymes exist: MM (muscle), MB (hybrid, in heart), and BB (brain). See Table 29-2 for variation in composition of CK in various types of tissues.

Aldolase. Aldolase is an enzyme of the lyase group. Its effect is a reversible cleavage of substrate into two compounds without hydrolysis. The following reaction is catalyzed by adolase:

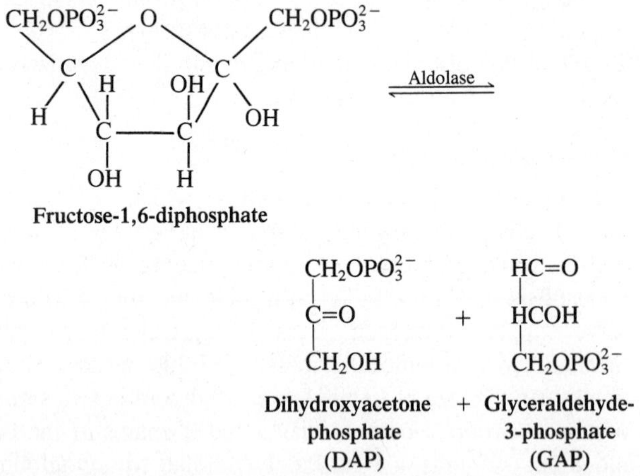

$$\text{Fructose-1,6-diphosphate (ring: } CH_2OPO_3^{2-}\text{, O, } CH_2OPO_3^{2-}\text{; C; H, H, OH, OH; C—C; OH, H)} \xrightleftharpoons{\text{Aldolase}}$$

$$\begin{array}{c} CH_2OPO_3^{2-} \\ | \\ C{=}O \\ | \\ CH_2OH \end{array} + \begin{array}{c} HC{=}O \\ | \\ HCOH \\ | \\ CH_2OPO_3^{2-} \end{array}$$

Dihydroxyacetone phosphate (DAP) + Glyceraldehyde-3-phosphate (GAP)

Aldolase is a significant enzyme in the glycolytic pathway and is found in every living cell. In tissues in which glycolysis supplies a large part of the energy need, high aldolase activity can be found. As an example, about 300 mg of skeletal muscle contains as much aldolase as is found in normal circulating blood volume. The fructose diphosphate aldolases are present in different isoenzyme forms. The three major forms are aldolase A, predominantly in muscle; aldolase B, in the liver; and aldolase C, in the brain. All three aldolases contain four polypeptide subunits, which differ in amino acid composition.

Aspartate aminotransferase (AST) or glutamic oxaloacetic transaminase (GOT). Transaminases catalyze the transfer of an α-amino group from an α-amino acid to an α-keto acid. Aspartate aminotransferase (glutamic oxaloacetic transaminase) catalyzes the following reaction:

$$\begin{array}{c} COOH \\ | \\ CH_2 \\ | \\ CHNH_2 \\ | \\ COOH \end{array} + \begin{array}{c} COOH \\ | \\ CH_2 \\ | \\ CH_2 \\ | \\ C{=}O \\ | \\ COOH \end{array} \xrightleftharpoons{\text{AST (GOT)}} \begin{array}{c} COOH \\ | \\ CH_2 \\ | \\ C{=}O \\ | \\ COOH \end{array} + \begin{array}{c} COOH \\ | \\ CH_2 \\ | \\ CH_2 \\ | \\ CHNH_2 \\ | \\ COOH \end{array}$$

Aspartic acid, α-Ketoglutamic acid, Oxaloacetic acid, Glutamic acid

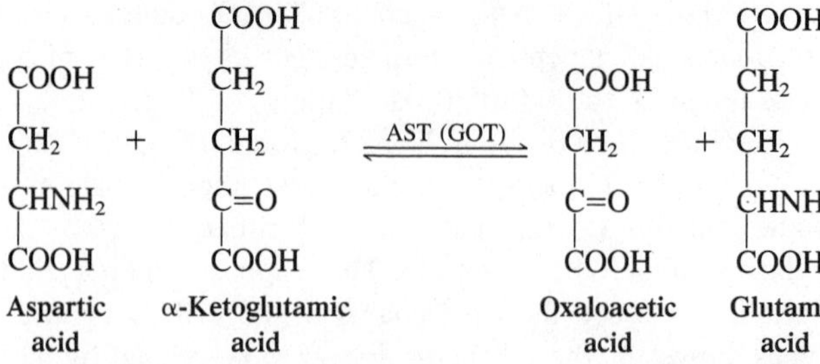

All tissues contain aspartate aminotransferase, particularly the heart, liver, and skeletal muscles. There are two isoenzymes, one mitochondrial (m-AST) and the other from the cytosol (c-AST). The two isoenzymes differ greatly in primary structure and chemical and physical properties; however, both catalyze the same reaction with only slight differences in the catalytic steps.

Alanine aminotransferase (ALT), or glutamic pyruvic transaminase (GPT). Alanine aminotransferase catalyzes the transfer of the amino group from alanine to α-ketoglutamic acid:

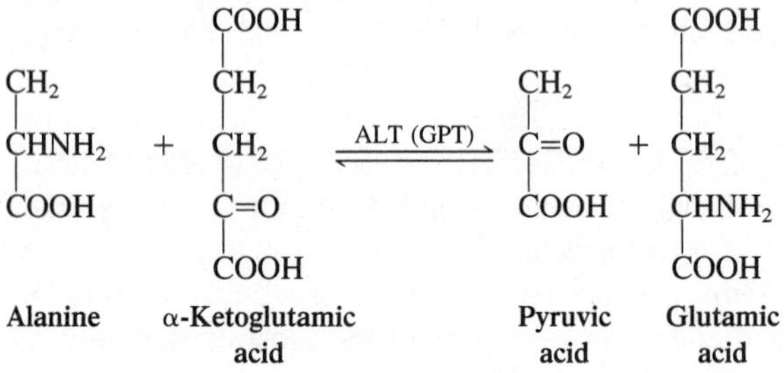

$$\begin{array}{c} CH_2 \\ | \\ CHNH_2 \\ | \\ COOH \end{array} + \begin{array}{c} COOH \\ | \\ CH_2 \\ | \\ CH_2 \\ | \\ C{=}O \\ | \\ COOH \end{array} \xrightleftharpoons{\text{ALT (GPT)}} \begin{array}{c} CH_2 \\ | \\ C{=}O \\ | \\ COOH \end{array} + \begin{array}{c} COOH \\ | \\ CH_2 \\ | \\ CH_2 \\ | \\ CHNH_2 \\ | \\ COOH \end{array}$$

Alanine, α-Ketoglutamic acid, Pyruvic acid, Glutamic acid

Alanine aminotransferase (ALT) is also present in all tissues and its appearance in the serum is a marker of tissue damage similar to AST. However, the serum values for ALT are more pronounced in liver damage than in myocardial damage and is a more specific indicator of liver damage.

Lactate dehydrogenase (LD). Lactate dehydrogenase catalyses the transfer to two electrons and one hydrogen ion from lactate to NAD:

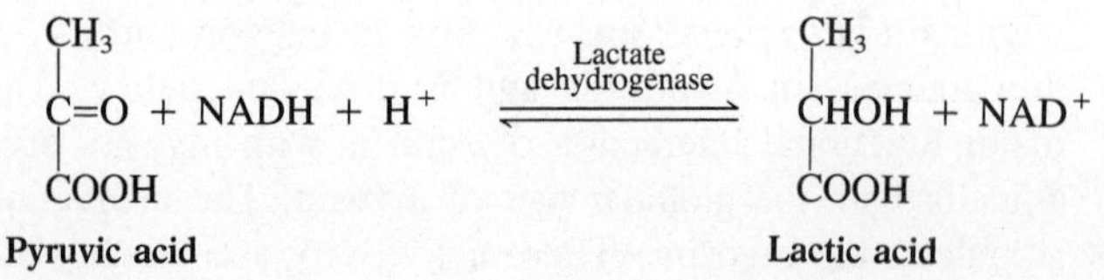

$$\begin{array}{c} CH_3 \\ | \\ C{=}O \\ | \\ COOH \end{array} + NADH + H^+ \xrightleftharpoons{\text{Lactate dehydrogenase}} \begin{array}{c} CH_3 \\ | \\ CHOH \\ | \\ COOH \end{array} + NAD^+$$

Pyruvic acid, Lactic acid

Lactate is formed from pyruvate produced by the glycolytic pathway when the amount of oxygen is limiting, as is the case in muscle during intense activity. In tissue damage caused by trauma or disease, lactate dehydrogenase appears in the serum; one can detect its presence by its ability to catalyze the above reaction.

Energetics of muscle contraction

ATP as energy source. The immediate source of energy for muscle contraction is the hydrolysis of ATP (P_i is inorganic phosphate):

$$ATP \rightarrow ADP + P_i + \text{Energy}$$

In resting muscle the [ATP]/[ADP] ratio is high. This balance is quickly changed during the contraction events. Since the contractile events are very rapid, there is a time lapse before appropriate metabolic signals can reach the targets and increase metabolic rate. To overcome this time discrepancy between immediate energy need and delayed metabolic response, there is an inbuilt reservoir in the muscle that responds first and reestablishes the high [ATP]/[ADP] ratio.

Energy reservoir in muscle

Creatine phosphate and creatine kinase. The energy reservoir in muscle is the high-energy bond of the phosphate group in creatine phosphate (CP). This high-energy phosphate reservoir uses the enzymes creatine kinase (CK) and myokinase (MK) to maintain an equilibrium concentration of ATP, ADP, and CP. The immediate effect of increased ADP concentrations caused by the hydrolysis of ATP during contraction is a disturbance of the equilibrium of the creatine kinease–catalyzed reaction:

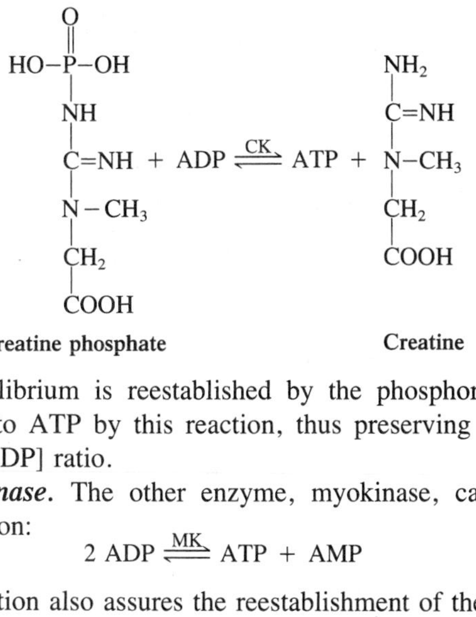

$$\text{HO-P(=O)(-OH)-NH-C(=NH)-N(CH_3)-CH_2-COOH} + ADP \underset{}{\overset{CK}{\rightleftharpoons}} ATP + \text{NH_2-C(=NH)-N(CH_3)-CH_2-COOH}$$

Creatine phosphate **Creatine**

The equilibrium is reestablished by the phosphorylation of ADP to ATP by this reaction, thus preserving a high [ATP]/[ADP] ratio.

Myokinase. The other enzyme, myokinase, catalyzes the reaction:

$$2\ ADP \overset{MK}{\rightleftharpoons} ATP + AMP$$

This reaction also assures the reestablishment of the original high [ATP]/[ADP] ratio.

Adenosine deaminase. A third enzyme present in the muscle, adenosine deaminase (AD), prevents the accumulation of AMP produced by the myokinase reaction by deamination:

$$AMP \overset{AD}{\rightleftharpoons} IMP + NH_3^+$$

Inosine monophosphate (IMP) then either returns to the nucleoside pool as inosine or is degraded further to uric acid. For a single contraction (twitch) or for short periods of muscle activity, the only measurable change in the high-energy phosphate pool is a small change in [CP].

Replenishment of ATP in skeletal muscle. The source of energy for rephosphorylation of ADP is oxidative phosphorylation and anaerobic glycolysis (see Chapter 30). Human skeletal muscle contains both red and white fibers, which differ in their metabolic properties. Red, or slow, fibers are rich in myoglobin and mitochondria. In these fibers the main metabolic pathway is oxidative phosphorylation. White, or fast, fibers contain little myoglobin and mitochondria and the main route for energy metabolism is glycolysis. The rate of glycolysis and respiration, and thus ATP production, are dependent on the rate of ATP consumption and are regulated by a series of feedback controls. When the [ATP]/[ADP] ratio is high, as in the resting state, both the rate of glycolysis and the rate of oxidation through the tricarboxylic acid cycle are low. During maximal muscle activity, when the [ATP]/[ADP] ratio is greatly reduced by the abrupt breakdown of ATP to ADP and P_i, both glycolysis and the tricarboxylic acid cycle are greatly increased.

At maximal activity of skeletal muscle the oxygen uptake can increase twentyfold or more in transition from rest to full activity in response to the oxygen need for the oxidative processes. However, at maximal activity the muscles still are relatively oxygen-poor (anoxic) and lactate, as the end product of anerobic metabolism, increases in blood.

Acidosis occurs when either metabolic or other abnormal processes result in a lower than normal pH of the arterial blood. Lactic acidosis results from an excessive production of lactic acid and can occur in normal muscles after excessive exercise. The localized acidosis in muscle contributes to fatigue and can result in muscle cramps and pain, especially when it is accompanied with excessive sodium chloride loss.

Fatty acids as energy source. Oxidative phosphorylation by means of the tricarboxylic acid cycle uses either pyruvate derived from glycolysis or fatty acids through oxidation. Both pyruvate and fatty acid oxidation products enter the cycle as acetyl CoA. There are three biochemical steps that occur before fatty acid utilization of the oxidative cycle in the mitochondria can begin: (1) ATP-driven activation of fatty acids, (2) acyl-group transfer to the carrier molecule carnitine, and (3) re-formation of activated fatty acyl CoA and free carnitine. Fatty acids are oxidized in the mitochondrial matrix. The long-chain fatty acids do not transverse the inner mitochondrial membrane. They are activated on the outer mitochondrial membrane and then carried across the inner mitochondrial membrane by a carrier molecule, carnitine.

The three stages of fatty acid utilization before the oxidative cycle in the mitochondria begins are as follows (Fig. 29-4):

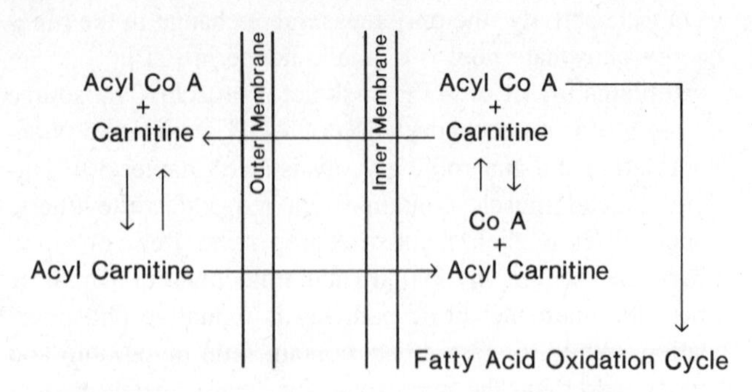

Fig. 29-4. Schema of fatty acid (acyl) transport into mitochondria.

1. Activation of fatty acids:

   $$R\text{-}COO^- + CoA + ATP \rightleftharpoons \text{Acyl CoA} + AMP + PP_i$$

   CoA is coenzyme A, PP_i is inorganic pyrophosphate, and $R\text{-}COO^-$ is the acyl fatty acid.
2. Acyl group transfer to the carrier molecule carnitine:

   Acyl CoA $R{-}C{-}S{-}CoA$

   +

   Carnitine $H_3C{-}N^+(CH_3)_2{-}CH_2{-}CH(OH){-}CH_2{-}C(=O)O^-$

   ⇅

   CoA $HS{-}CoA$

   +

   Acylcarnitine $H_3C{-}N^+(CH_3)_2{-}CH_2{-}CH(O{-}C(=O){-}R){-}CH_2{-}C(=O)O^-$

   Acylcarnitine diffuses across the inner mitochondrial membrane. On the matrix side step 3 occurs.
3. Transacylation to form acyl CoA and free carnitine in the intramitochondrial matrix

   $$\text{Acylcarnitine} + CoA \rightleftharpoons \text{Carnitine} + \text{Acyl CoA}$$

The transacylation reactions are catalyzed by the enzyme fatty acyl CoA:carnitine fatty acid transferase.

Acyl CoA in the mitochondrial matrix enters the fatty acid oxidation cycle. The carnitine diffuses to the outer membrane and steps 1 to 3 are repeated.

Defects in the transferase or a deficiency of carnitine impair the oxidation of fatty acids causing muscle fatigue and cramps on fasting, exercise, or high-fat diet. The fatty acyl CoA becomes the substrate for the fatty acid oxidation system in the inner matrix compartment of the mitochondria. Carnitine is particularly abundant in muscle and stimulates the long-chain fatty acid oxidation by mitochondria, especially in the red muscle fibers.

Glycogen as fuel supply for glycolysis. Glycogen is a readily mobilized, stored form of glucose present as a polymer in which glucose is linked by α-1,4-glycosidic bonds and with branches formed by α-1,6-glycosidic bonds. Glycogen is present in the cytosol in the form of granules containing the enzymes for synthesis and degradation of glycogen. The synthesis and degradation occur by different reaction pathways (see Chapter 30 for details).

Energy metabolism in different fiber types

It was recognized for over a hundred years that in skeletal muscles there was a correlation between speed of contraction and the color of the muscles. It was observed, in general, that red muscles contracted more slowly than white muscles. This simple classification as red-slow and white-fast became more complicated as more histological and chemical data accumulated. No muscle contains only one type of fiber; instead each muscle type is determined by the proportion of various fiber types it contains. Muscle fibers are classified into three general types: *slow-twitch oxidative, fast-twitch oxidative-glycolytic,* and *fast-twitch glycolytic*. The classification is based on physiology, histochemistry, and biochemistry (Table 29-3). There are differences in many isoenzymes that appear in different forms in different muscle types. There are differences of the light-chain components of different types of myosin. Fast-twitch myosin contains all three light chains, L_1, L_2, and L_3. Slow-twitch myosin contains only two types of light chains, L_1 and L_2. Recently there have also been found differences in tropomyosin and troponin subunits in different fiber types. The original differences based on the color of the muscle relates to the main energy source of the muscle. In fast-twitch muscle, the ATP supply comes primarily from glycolytic processes, which make a quick but rapidly exhaustible supply of energy. These muscles are white because of the paucity of myoglobin and mitochondrial cytochromes, which results in a pale appearance to the muscle. On the other hand, muscles made up mostly from slow-twitch oxidative or fast-twitch oxidative-glycolytic fibers gain energy through oxidative processes involving the Krebs cycle. These muscles are rich in mitochondria and contain significantly higher amounts of myoglobin. The cytochromes of mitochondria and myoglobin content give these muscles a red appearance used for early classification. The terminology of different muscle types is given in Table 29-3.

PATHOLOGICAL CONDITIONS OF MUSCLE

Muscular disorders

The diseases of muscle are all characterized by motor dysfunction, such as muscular weakness. The abnormality

Table 29-3. Terminology of fiber types of mammalian skeletal muscle

Source	Contractile muscle type and contraction velocity		
	I Slow-twitch	II Fast-twitch	III Fast-twitch
Energy supply	Oxidative	Oxidative-glycolytic	Glycolytic
Color	Red	Intermediate	White
Myoglobin content	High	Intermediate	Low
Creatine kinase activity	High	High	High
Lactate dehydrogenase	Low	Intermediate	High
Aldolase activity	Low	Intermediate	High
Myokinase activity (adenylate kinase)	Low	Intermediate	High
Mg^{2+}-stimulated ATPase (actinomyosin ATPase)	Low	Intermediate	High
Fumarase	High	Intermediate	Low
Citrate synthase	High	Intermediate	Low
Succinate dehydrogenase	High	Intermediate	Low
Myosin stability at pH 9	Low	Intermediate	High
Mitochondrial distribution	High	Intermediate	High
Cytochrome oxidase	High	Intermediate	Low

is in some part of the motor unit. The *motor unit* consists of a single lower motor neuron, the anterior horn cell, its efferent root, the peripheral nerve fiber, and its terminal arborization. It also includes the motor end plate at the neuromuscular junction together with the 10 to 600 muscle fibers innervated by the neuron.

The disorders are conveniently classified according to the part of the motor unit principally affected.

The *muscular atrophies* are caused by a loss of efferent innervation as a result of a degeneration of either an anterior horn cell or an axon at the level of an anterior efferent root or peripheral nerve cell.

The *myopathies* are characterized by major defects at the level of the muscle fibers. Certain hereditary progressive myopathies are called, by convention, *muscular dystrophies*. *Nonhereditary myopathies* can result from inflammation or from endocrine or metabolic abnormality.

The three major categories of muscle disorders, according to the part of the motor unit affected, are (1) neurogenic muscular atrophies, (2) muscle fiber disorders, and (3) disturbances of the neuromuscular junction. Within each major class there are further distinctions based on the loci or known origin of the defect. These categories are listed below.

Neurogenic disorders

Neurogenic muscular atrophies (motor neuron diseases). In this class of disorders the muscle weakness and wasting is caused by denervation. Two major categories are to be distinguished, depending on the loci of defect:

Amyotrophic lateral sclerosis and its variants *progressive spinal muscular atrophy,* or Aran-Duchenne muscular atrophy, and *progressive bulbar palsy,* or Duchenne's paralysis. These are late-onset diseases, generally occurring after 40 years of age.

Infantile spinal muscular atrophy, or Werdnig-Hoffmann disease, often is fatal before 2 years of age.

Juvenile spinal muscular atrophy, or Wohlfart-Kugelberg-Welander disease, has a longer life expectancy than infantile spinal muscular atrophy. Autosomal recessive inheritance is frequently noted in amyotrophic lateral sclerosis and juvenile spinal atrophy.

Anterior root and peripheral nerve involvements

Peroneal muscular atrophy, or Charcot-Marie-Tooth disease.

Hypertrophic interstitial neuropathy, or Déjerine-Sottas disease. The disease begins usually between childhood and 30 years of age and is slowly progressive.

Acute polyneuropathy, or Guillain-Barré syndrome. It is a parainfectious and postinfectious disease and presumed to be caused by immunological attack on peripheral nerves.

Metabolic neuropathies seen with metabolic diseases such as diabetes mellitus or malnutrition.

Disorders of muscle fibers

Muscular dystrophies. Muscular dystrophy is a general name for a group of chronic diseases of muscle. The general characteristics are a progressive weakness and degeneration of skeletal muscule with no evidence of neural degeneration. They are inherited diseases with different inheritance patterns. The age of onset, the course of the disease, and the effect on the different fiber types differ among the individual diseases.

Pseudohypotrophic muscular dystrophy, or Duchenne muscular dystrophy, is a sex-linked recessive disorder. It affects boys typically between 3 to 7 years of age with steady progression of proximal muscle weakness. First the pelvic girdle and then the shoulder girdle muscles are affected, and most patients are confined to wheelchairs by

10 to 12 years of age. Serum enzymes are greatly elevated in the disease even before symptoms develop; especially noted is the rise in creatine kinase. In heterozygous males the creatine kinase levels lie between those of affected homozygous males and normal persons. The creatine kinase values in heterozygous females and normal persons overlap; unfortunately only about 70% of heterozygous females are positively identifiable. Nevertheless, a screening program and genetic counseling has significantly changed the occurrence of the disease in the last 10 years, which is fortunate, given the severity of the disease and the lack of therapy.

Limb-girdle muscular dystrophy, an inherited autosomal recessive disease, begins later in childhood. This disease has two forms, the scapulohumeral or Erb type, which starts at the shoulder girdle, and the pelvifemoral or Leyden-Möbius type, which begins at the pelvic girdle. The involvement of muscles are similar to that in the Duchenne type of muscular dystrophy, but the progression is less predictable.

Facioscapulohumeral muscular dystrophy, or Landouzy-Déjerine muscular dystrophy, involves the facial and the shoulder girdle muscles. Its onset is in adolescence and progresses throughout life with normal life expectancy.

Myotonic muscular dystrophies or myotonic myopathies are characterized by abnormally slow relaxation of voluntary muscles after a contraction. One form is *Steinert's disease,* also known as *myotonia atrophica,* or *myotonic muscular dystrophy.* It is an autosomal dominant disorder, occurs at any age, and is variable in severity. The muscular weakness is combined with myotonia. Another form is called *myotonia congenita, Thomsen's disease, or ataxia muscularis.* It is a relatively rare autosomal dominant myotonia, and usually the onset is in early life. The muscles become hypertrophied, and muscle stiffness is the most important problem. Therapy can diminish stiffness and cramping, but the weakness is not affected. Exercise is of limited help.

Other, rare muscular dystrophies include *benign juvenile muscular dystrophy,* or *Becker's muscular dystrophy,* which resembles Duchenne's muscular dystrophy in heredity and also in clinical features but is more benign and does not present as early as Duchenne's dystrophy does. *Distal muscular dystrophy,* or *Gower's muscular dystrophy,* has its onset in later years, and the progressive weakness begins in the hands and feet and extends proximally. The weakness is only moderate. *Ocular myopathy,* another rare muscular dystrophy, is manifested in slowly progressive ptosis and ophthalmoplegia. It also begins in later life.

Glycogen-storage diseases of muscle. Glycogen-storage diseases are characterized by abnormal accumulation of glycogen in muscle. Inheritance is autosomal recessive. Diagnosis is by demonstrating absence of the specific enzyme in a biopsy of the muscle.

McArdle's disease, or *glycogen storage disease, type V,* is characterized by clinical symptoms of skeletal muscle weakness, cramps on exercise, and no lactate rise. The missing enzyme is myophosphorylase b. Clinically, this disease can be mild or severe. In normal patients the ischemic exercise results in an acid shift in muscle pH. In diseased patients no such shift in pH is present.

Tauri's disease or *glycogen storage disease, type VII,* results in muscle pain, contracture, and exercise intolerance after ischemic work, the same symptoms as in type V disease. It is caused by the absence of muscle phosphofructokinase. In both types V and VII there is occasionally myoglobinuria.

Pompe's disease, or *glycogen-storage disease, type II,* is caused by absence of α-1,4-glucosidase and is the most severe form of glycogen-storage diseases. Besides the symptoms listed above, it can also result in degeneration of anterior horn cells. It may affect the heart, causing cardiomegaly.

Familial periodic paralysis. There is a relatively rare group of disorders that are inherited. The pattern of inheritance is autosomal dominant. The attack is of flaccid paralysis, not affecting conciousness and lasting from an hour to several days. There are three types: hypokalemic, normokalemic, and hyperkalemic paralysis, with the severity in decreasing order. Documentation of serum potassium during an attack is important for the diagnosis.

Endocrine myopathies. Endocrine myopathies are observed in thyroid, parathyroid, and adrenal disorders, in hypercalcemia, and in hypophosphatemia.

Neuromuscular junction disturbances and diseases

Myasthenia gravis. Myasthenia gravis is a causative defect related to impairment of acetylcholine-mediated nerve-impulse transmission to muscle. Recent evidence points to an immunologically induced abnormality in the acetylcholine-receptor protein of the nerve-muscle junction. Antibodies to the receptor protein can often be demonstrated in this disease.

Drug-induced neuromuscular junction block. This block is caused by cholinergic agents, organophosphate insecticides, and most nerve gases, which are cholinesterase inhibitors, prevent breakdown of acetylcholine, and cause persisting depolarization at the neuromuscular junction. Symptoms are myasthenia-like weakness and muscular twitching, miosis, tightness in the chest, gastrointestinal hyperactivity, and increases in bronchial and salivary secretion.

Trauma

Trauma in muscles is produced in innumerable ways, such as a mechanical blow, heating, freezing, and electric shock.

Extensive necrosis of muscle fibers from any cause results in liberation of muscle enzymes (CK, LD, AST,

Table 29-4. Change of analyte in disease

	Aldolase	Myoglobin	CK	AST	ALT	LD	Lactic acid
Myocardial infarction	—, ↑	↑	↑	↑	↑	↑	—
Skeletal muscle damage	↑	↑	↑	↑	↑	↑	—
Muscular dystrophies	↑	—	↑	—	—	—	—
Fatigue	—	—	—	—	—	—	↑

ALT, Alanine aminotransferase; *AST*, aspartate aminotransferase; *CK*, creatine kinase; *LD*, lactate dehydrogenase.

ALT) and myoglobin into the bloodstream. After trauma a fetal form of myoglobin appears. It is to be found in the regenerative phase of all forms of muscle necrosis. The effectiveness of regeneration depends on how much of the basement cell membranes and supporting tissues remained intact and whether the basic pathological process has subsided. Only in cases of great trauma, hemorrhage, infection, and infarction are a complete disorientation of regenerating fibers and an excessive proliferation of fibroblasts and vessels observed. This so-called anisomorphis regeneration then may result in tumor formation. In all regenerative phases of muscle the creatine kinase MM isoenzyme activity decreases and BB and MB isoenzyme activities increase. In later phases of regeneration the MB and MM isoenzymes become more active, and finally the MM is firmly dominating. During the early phases of regeneration, a sufficiently high MB isoenzyme level appears in serum to interfere with the clinical picture of a myocardial infarction.

CHANGE OF ANALYTE IN DISEASE (Table 29-4)

In general there is a release of the cellular constituents into circulation when muscle tissue is damaged. Muscle contains the enzymes aldolase, aspartate transaminase, alanine transaminase, lactate dehydrogenase, and creatine kinase, and these are released into circulation. Myoglobin is also released. The major difference between muscle types is their isoenzyme composition. For example, CK-MM is virtually the exclusive form in skeletal muscle, whereas some of the MB form is present in heart tissue. Serum lactic acid elevations occur in cases of shock, muscle fatigue, and tissue hypoxia.

For specific diseases such as myasthenia gravis, specific analytes such as antibodies to the acetylcholine receptor may be found.

BIBLIOGRAPHY

Bais, R., and Edwards, J.B.: Creatine kinase, CRC Crit. Rev. Clin. Lab. Sci. **16**:291-335, 1982.

Dayton, W.R., Goll, D.E., Stromer, M.H., et al.: Some properties of a Ca^{2+}-activated protease that may be involved in myofibrillar protein turnover. In Reif, E., Rifkin, D.B., and Show, E., editors: Proteases and biological control, Cold Spring Harbor, N.Y., 1975, Cold Spring Harbor Lab., pp. 551-577.

Ebashi, S., Maruyama, K., and Endo, M., editors: Muscle contraction: its regulatory mechanism, New York, 1980, Springer-Verlag New York, Inc.

Flockhart, D.A., and Corbin, J.D.: Regulatory mechanisms in the control of protein kinases, CRC Crit. Rev. Biochem. **16**:133-186, 1982.

Foreback, C.C., and Chu, J.-W.: Creatine kinase isoenzymes: electrophoretic and quantitative measurements, CRC Crit. Rev. Clin. Lab. Sci. **15**:187-230, 1981.

Goldberg, A.L., and Odessey, R.: Regulation of protein and amino acid degradation in skeletal muscle. In Milhorat, A.T., editor: Exploratory concepts in muscular dystrophy, New York, 1974, American Elsevier Publishing Co., pp. 187-199.

Grob, D.: Myasthenia gravis: pathophysiology and management, retrospect and prospect, Ann. N.Y. Acad. Sci. **377**:xiii-xvi, 1981.

Guth, L.: Trophic influences of nerve on muscle, Physiol. Rev. **48**:645-687, 1968.

Huxley, A.F., and Simmons, R.M.: Proposed mechanism of force generation in striated muscle, Nature **233**:533-538, 1971.

Huxley, H.E.: Molecular basis of contraction in cross-striated muscles. In Bourne, G.H., editor: The structure and function of muscle, vol. I, pt. I, New York, 1972, Academic Press, Inc., pp. 302-384.

Iodice, A.A., Perker, S., and Weinstock, I.M.: Role of the cathepsins in muscular dystrophy. In Walton, J.N., Canal, N., and Scarlato, G., editors: Muscle diseases, Amsterdam, 1969, Excerpta Medica, pp. 313-318.

Korn, E.D.: Biochemistry of actomyosin-dependent cell motility (a review), Proc. Natl. Acad. Sci. USA **75**:588-589, 1978.

Korn, E.D.: Actin polymerization and its regulation by proteins from nonmuscle cells, Physiol. Rev. **62**:672-737, 1982.

Kretsinger, R.H.: Mechanism of selective signaling by calcium, Neurosci. Res. Prog. Bull. **19**:214-332, 1980.

Morgan, H.E., and Wildenthal, K., chairman: Symposium of the American Physiological Society: Protein turnover in heart and skeletal muscle, Fed. Proc. **39**: 7-52, 1980.

Nagy, B., and Samaha, F.J.: Physiology of normal and diseased muscle. In Frolich, E.D., editor: Pathophysiology, Philadelphia, 1983, J.B. Lippincott Co.

Peachey, L.D.: The Sarcoplasmic reticulum and transverse tubules of the frog's sartorius, J. Cell Biol. **25**:209-231, 1965.

Perry, S.V.: The regulation of contractile activity in muscle, Biochem. Soc. Trans. **7**:593-617, 1979.

Squire, J.: The structural basis of muscular contraction, New York, 1981, Plenum Press.

Stanbury, J.B., Wyngaarden, J.B., and Frederickson, D.S., editors: The metabolic bases of inherited disease, New York, 1972, McGraw-Hill Book Co.

Szent-Györgyi, A.: Chemistry of muscular contraction, ed. 2, New York, 1951, Academic Press, Inc.

Tsung, S.H.: Creatine kinase isoenzyme patterns in human tissue obtained at surgery, Clin. Chem. **22**:173-175, 1976.

Walton, J., editor: Disorders of voluntary muscle, Edinburgh, 1981, Churchill Livingstone.

Zak, R., Martin, A.F., and Blough, R.: Assessment of protein turnover by use of radioisotopic tracers, Physiol. Rev. **59**:407-447, 1979.

Chapter 30 Diabetes mellitus

Thomas Skillman

acetylcoenzyme A (acetyl CoA) An important intermediate in the citric acid cycle (Krebs cycle) and the chief precursor of lipids.

anaerobic glycolysis The breaking down of sugars into simple compounds without molecular oxygen.

autoimmunity A condition characterized by a specific humoral or cells-mediated immune response against the constituents of the body's own tissues; it can result in hypersensitivity or, if severe, in autoimmune disease.

autonomic neuropathy Self-controlling, functionally independent disturbances in the peripheral nervous system.

basement membrane The delicate layer of extracellular condensation of mucopolysaccharides and proteins underlying the epithelium of mucous membranes and secreting glands.

cellulitis Inflammation of cellular tissue, especially purulent inflammation of the loose subcutaneous tissue.

cortisol An adrenocortical steroid hormone, 17-hydroxycorticosterone or hydrocortisone (USP), as it is referred to pharmaceutically. It promotes gluconeogenesis, plays a role in fat and water metabolism, affects muscle tone and the excitation of nerve tissue, increases gastric secretion, alters the connective tissue response to injury, and impedes cartilage production.

Cushing's syndrome A condition most commonly seen in females and caused by hyperadrenocorticism resulting from neoplasms of the adrenal cortex or anterior lobe of the pituitary; associated with altered glucose tolerance test.

disaccharides Any of a class of sugars that yield two monosaccharides on hydrolysis and have the general formula $C_n(H_2O)_{n-1}$, or $C_{12}H_{22}O_{11}$. They include sucrose, lactose, and maltose.

epinephrine A hormone secreted by the adrenal medulla in response to splanchnic stimulation and stored in the chromaffin granules. It is released predominantly in response to hypoglycemia; also called "Adrenalin" or "adrenaline" (Great Britain).

glucagon A polypeptide hormone secreted by the alpha cells of the islets of Langerhans in response to hypoglycemia or to stimulation by growth hormone of the anterior pituitary; it stimulates glycogenolysis in the liver.

gluconeogenesis The formation of carbohydrates from molecules that are not themselves carbohydrates, such as amino acids, fatty acids, or related molecules.

glucose D-Glucose, or monosaccharide, $C_6H_{12}O_6$, occurring in certain foods, especially fruits, and in normal blood. Glucose is a primary source of energy for living organisms.

glycogen A polysaccharide $(C_6H_{10}O_5)_n$, the chief carbohydrate storage material in animals. It is formed by and largely stored in the liver and to a lesser extent, in muscles.

glycogenolysis The splitting up of glycogen in the body tissues yielding glucose.

glycolytic Pertaining to, characterized by, or promoting glycolysis.

growth hormone Any substance that stimulates growth, exerts a direct effect on protein, carbohydrate, and lipid metabolism, and controls the rate of skeletal and visceral growth.

hexose sugar A monosaccharide containing six carbon atoms in a molecule.

hyperglycemic, hyperosmolar nonketotic coma Diabetic coma in which the level of ketone bodies is normal; caused by hyperosmolarity of extracellular fluid and resulting in dehydration of intracellular fluid, often a consequence of overtreatment with hyperosmolar solutions.

hyperthyroidism Excessive functional activity of the thyroid gland characterized by increased basal metabolism and glucose catabolism.

insulin A protein hormone formed by the beta cells of the islets of Langerhans in the pancreas and secreted into the blood, in which it regulates carbohydrate, lipid, and amino acid metabolism. Also a preparation of the active principle of the pancreas; used therapeutically in diabetes mellitus.

islets of Langerhans Irregular microscopic structures scattered throughout the pancreas and composing its endocrine portion; composed of three types of cells: the alpha cells, which secrete the hyperglycemic factor, glucagon; the beta cells, which are the most abundant and secrete insulin; and the delta cells, which secrete gastrin. Degeneration of the beta cells, whose secretion (insulin) is important in carbohydrate metabolism, is one of the causes of diabetes mellitus.

malabsorption Impaired intestinal absorption of nutrients.

microangiopathy Disease of the small vessels in diabetes, wherein there is thickening of basement membrane of capillaries throughout many vascular beds.

mononeuropathy The disease affecting a single nerve.

oliguria Secretion of a diminished amount of urine in relation to the fluid intake.

pancreatitis Acute or chronic inflammation of the pancreas, which may be asymptomatic or symptomatic. Caused most often by alcoholism or biliary tract disease; less commonly it is associated with hyperlipidemia, hyperparathyroidism, abdominal trauma (accidental or operational injury), vasculitis, or uremia.

peripheral neuropathy A general term denoting functional disturbances or pathological changes in the peripheral nervous system.

phenformin Phenformin hydrochloride (USP). Chemical name: 1-phenethylbiguanide monohydrochloride; used as a hypoglycemic agent.

polydipsia Excessive thirst persisting for long periods of time.

polyphasia Excessive or voracious eating, with craving for all types of foods.

polyuria The passage of a large volume of urine in a given time.

postprandial Pertaining to time after a meal, especially dinner.

proinsulin A precursor of insulin with a molecular weight of 8000 to 10,000 daltons. Proinsulin has low biological activity but yields true insulin on tryptic digestion in vitro.

sulfonylurea An oral hypoglycemic agent used in the treatment of diabetes mellitus.

The syndrome of diabetes mellitus affects every facet of our health care delivery system.[1] It is probable that more than 10 million Americans are diabetic and that over 40,000 deaths each year result from this chronic, incurable metabolic disorder. The economic burden of diabetes exceeds $5 billion per year and the morbidity of this disease defies intelligent calculation. More than half a million new cases of diabetes are diagnosed yearly, and it has been predicted that by 1990 there will be at least 20 million diabetic persons.[2]

Although atherosclerosis is the chief cause of death in diabetic persons[3] (as it is in nondiabetic persons), there is good reason to associate certain metabolic derangements in diabetes with premature and severe vascular disease.[4] Hospital statistics indicate that a diabetic person is twice as likely to suffer myocardial infarction as a nondiabetic person with equal risk factors. Peripheral vascular occlusion is at least five times more common in the diabetic person. Chronic renal insufficiency is a major cause of morbidity and death in persons who develop insulin-dependent diabetes in youth.[5] Table 30-1 shows the remarkable difference in life expectancy in the diabetic person as compared to that of the nondiabetic person.[3]

Table 30-1. Life expectancy for diabetic and nondiabetic persons

Attained age of diabetic person	Expected years of additional life of diabetic person	Expected years of additional life of nondiabetic person	Years lost to diabetes
10	44.3	61.5	17.2
15	40.0	56.7	16.7
20	36.1	51.9	15.8
25	32.8	47.2	14.4
30	30.1	42.5	12.4
35	27.2	37.9	10.7
40	23.7	33.3	9.6

From Entmacher, P.S.: Long term prognosis in diabetes mellitus. In Sussman, K.E., and Metz, R.J., editors: Diabetes mellitus: diagnosis and treatment, ed. 4., New York, 1975, American Diabetes Association.

All major medicine textbooks allocate large sections to the epidemiology, diagnosis, clinical manifestations, complications and treatment of diabetes. The goal of this section is to interrelate the biochemical features of the diabetes syndromes with clinical expressions of the disease.

NORMAL BIOCHEMISTRY: GLUCOSE HOMEOSTASIS

Elements of carbohydrate metabolism

Carbohydrates are molecules that contain three or more carbons plus two hydrogens and one oxygen for each carbon. Dietary carbohydrates contain simple carbohydrates such as the hexose sugars (glucose, fructose, and galactose), the disaccharides (sucrose, lactose, and mannose), and complex carbohydrates such as starch. Intestinal activity digests starch to its hexose components and the disaccharides to hexoses. Glucose is the major hexose derivative.

There is very little glucose in the cells because glucose is converted to glucose-6-phosphate by intracellular enzymatic hexokinase activity. As demonstrated in Fig. 30-1, glucose has at least five possible fates after it enters the cell: (1) storage as glycogen, (2) anaerobic glycolysis to pyruvate and lactate, (3) oxidation to carbon dioxide and water to provide energy through the citric acid pathway (Krebs cycle), (4) conversion to fatty acids, and (5) release from the cell as glucose.

Fig. 30-2 shows the conversion of glucose to glucose-6-phosphate by hexokinase. This enzyme may have as many as four different forms and phosphorylates glucose, using ATP as a substrate and Mg^{++} ions as a cofactor.

Glycogen synthesis. Carbohydrate is stored within the cell as glycogen. This large molecule is a polysaccharide made up of glucose units linked in 1,4 bonds with branch points at 1,6 linkages as illustrated in Fig. 30-3. Glycogen is present in the cytosol in the form of granules with diameters of 10 to 40 nm, containing the enzymes for synthesis and degradation of glycogen. The rate of glycogen formation is controlled by the enzyme glycogen synthetase. This enzyme is activated from an inactive state when phosphate groups are lost through activity of the messenger, cyclic adenosine 3′,5′-monophosphate (cAMP). Increases in the concentration of glucose or insulin stimulate this process.

Glucose-1-phosphate is activated to UDP-glucose, and then glycogen synthetase transfers UDP-glucose to the C-4 terminus of glycogen to form 1,4-glycosidic linkage and UDP.

$$\text{UDP-glucose} + \underset{(n \text{ residues})}{\text{Glycogen}} \rightarrow \underset{(n+1 \text{ residues})}{\text{Glycogen}} + \text{UDP}$$

Synthetic pathways employ branching enzymes that create large numbers of nonreducing terminal residues, thus

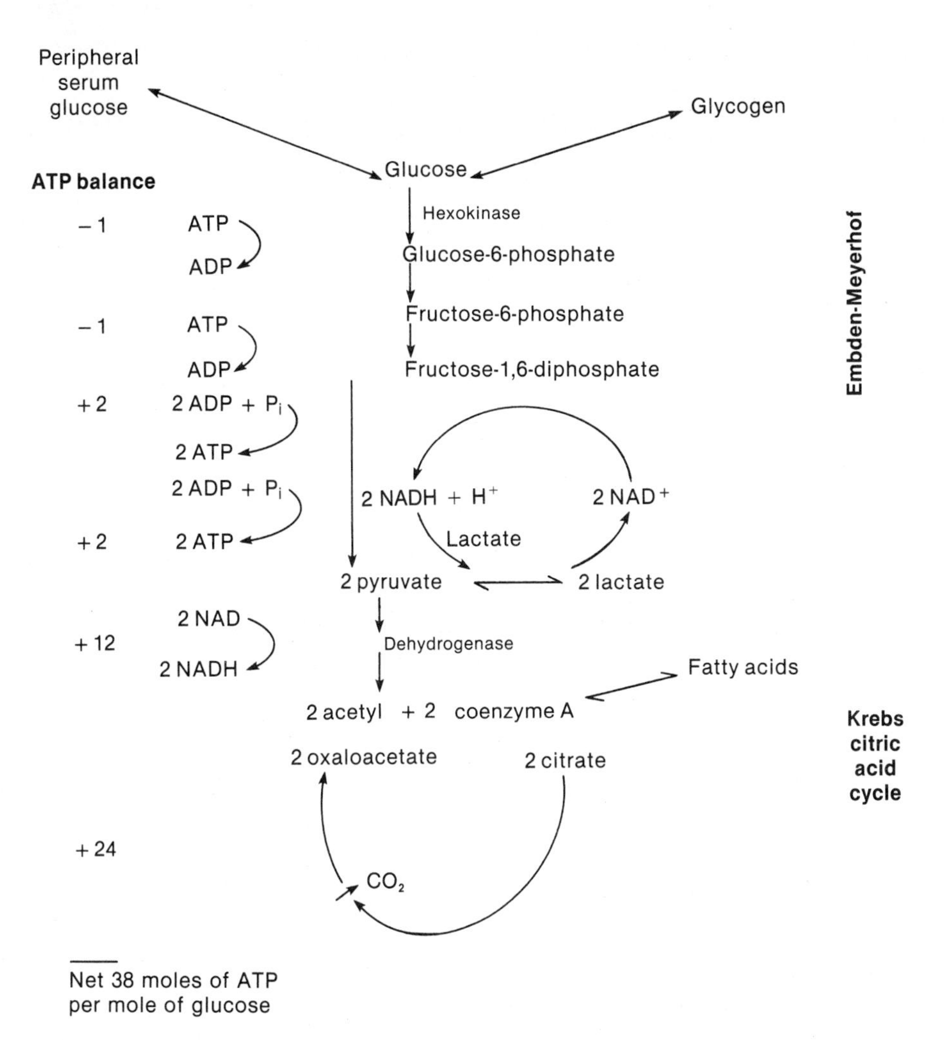

Fig. 30-1. Metabolism of glucose in cells requires phosphorylation to glucose-6-phosphate. This compound is then further phosphorylated and broken down through Embden-Meyerhof pathway, yielding two moles of ATP per mole of glucose in anaerobic glycolysis. One molecule of glucose (a 6-carbon compound) is converted to two molecules of pyruvate (a 3-carbon compound), which in aerobic glycolysis is further converted to two molecules of acetyl coenzyme A with production of 14 moles of ATP per mole of glucose. Acetyl CoA is then degraded in citric acid cycle with production of a further 24 moles of ATP. Thus aerobic glycosysis produces a net balance of 38 moles of ATP per mole of glucose. In anaerobic glycolysis, end product is lactate, which absorbs reducing equivalents (NADH) and produces nicotinamide adenine diphosphate (NAD^+) for reutilization in the Embden-Meyerhof pathway.

α-D-Glucose → (Hexokinase, Mg^{++}; ATP → ADP) → **α-D-Glucose-6-phosphate**

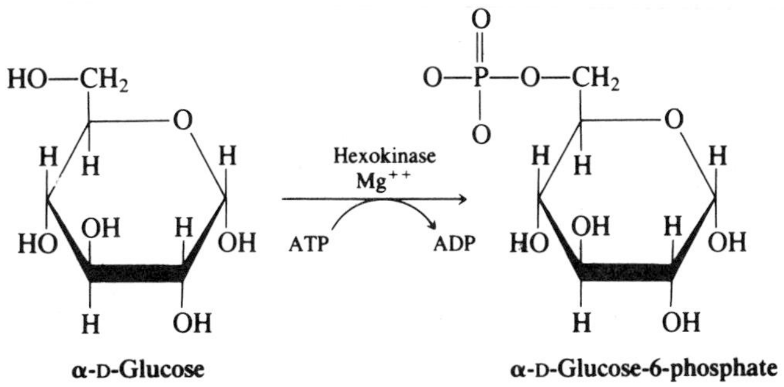

Fig. 30-2. Conversion of glucose to glucose-6-phosphate by hexokinase.

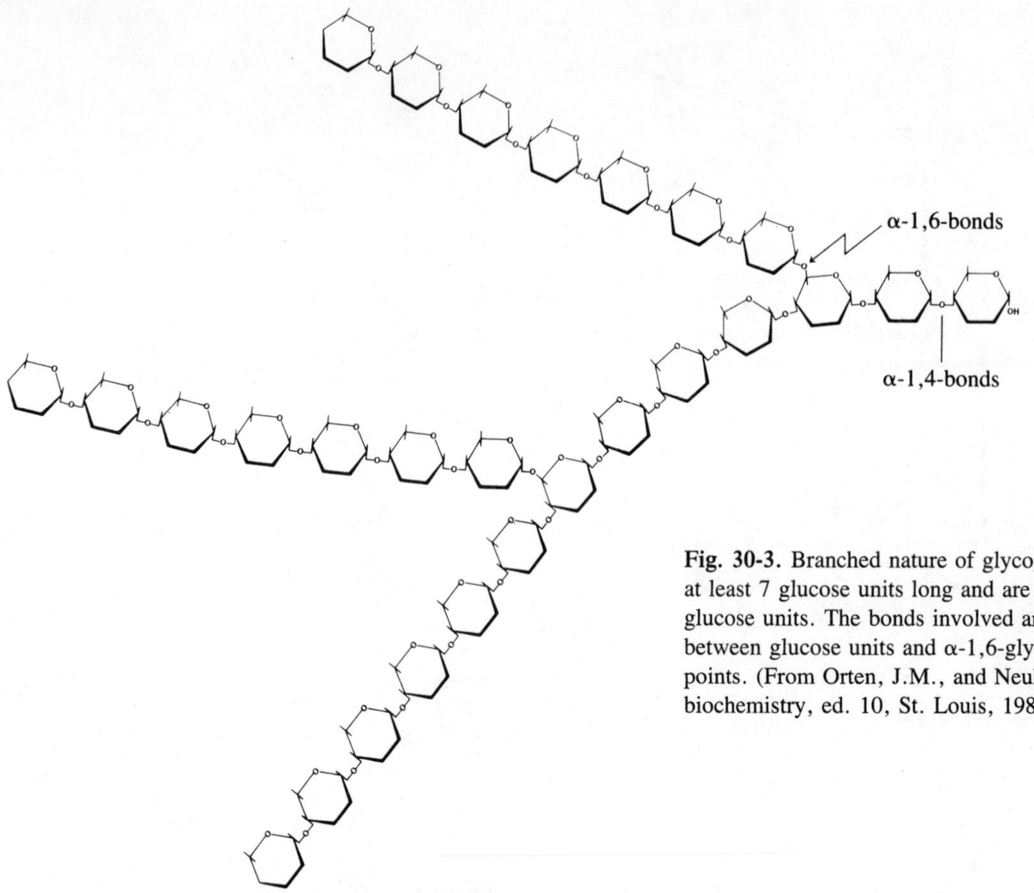

Fig. 30-3. Branched nature of glycogen. As seen, branches are at least 7 glucose units long and are separated by at least 3 glucose units. The bonds involved are α-1,4-glycosidic linkages between glucose units and α-1,6-glycosidic units at the branch points. (From Orten, J.M., and Neuhaus, O.W.: Human biochemistry, ed. 10, St. Louis, 1982, The C.V. Mosby Co.)

increasing the rate of glycogen synthesis and degradation.

Glycogenolysis. The breakdown of glycogen is controlled by another system. The breakdown is catalyzed by glycogen phosphorylase:

$$\underset{(n \text{ residues})}{\text{Glycogen}} + P_i \rightarrow \text{Glucose-1-phosphate} + \underset{(n - 1 \text{ residues})}{\text{Glycogen}}$$

The product glucose-1-phosphate is converted to glucose-6-phosphate and enters the glycolytic pathway. Glycogen phosphorylase exists in muscle in two forms: an active phosphorylase *a* and an inactive phosphorylase *b*. The inactive form is converted to the active form by phosphorylation by a specific enzyme. This activation is by a hormone–cyclic AMP system but can also be activated by Ca^{2+}; so in muscle the glycogen breakdown is activated by a transient increase in the Ca^{2+} level in the cytoplasm. In muscle the inactive form of phosphorylase *b* is activated also by cAMP through an allosteric mechanism: cAMP binds to the nucleotide-binding site and alters the conformation of the inactive phosphorylase *b* to an active phosphorylase *a*. ATP acts as a negative allosteric effector by competing with cAMP. Glucose-6-phosphate is also an inhibitor by binding to another site. In physiological conditions phosphorylase *b* is inactive because of inhibitions by ATP and glucose-6-phosphate. The other form, phosphorylase *a*, is fully active, irrespective of levels of cAMP, ATP, and glucose-6-phosphate. The enzyme phosphorylase becomes activated while glycogen synthetase becomes inactivated. These processes are stimulated by a decrease in concentration of glucose or insulin and by an increase in the intracellular concentrations of glucagon and epinephrine. The first step in glycogenolysis is removal of glucose from the terminal 1,4 linkages of glycogen by an enzyme that inserts a phosphate at the carbon-1 position, forming glucose-1-phosphate. The source of the phosphate is ATP.

There is a coordinated control of glycogen synthesis and breakdown by the following reaction sequence:

1. Epinephrine binds to muscle cell membrane and activates adenyl cyclase.
2. Adenyl cyclase catalyzes the reactions

$$\text{ATP} \rightarrow \text{Cyclic AMP} + PP_i$$

3. Cyclic AMP activates protein kinase.
4. Protein kinase phosphorylates both phosphorylase kinase and glycogen synthetase, activating the first and inactivating the second.

A scheme of glycogen breakdown and synthesis control is shown in Fig. 30-4.

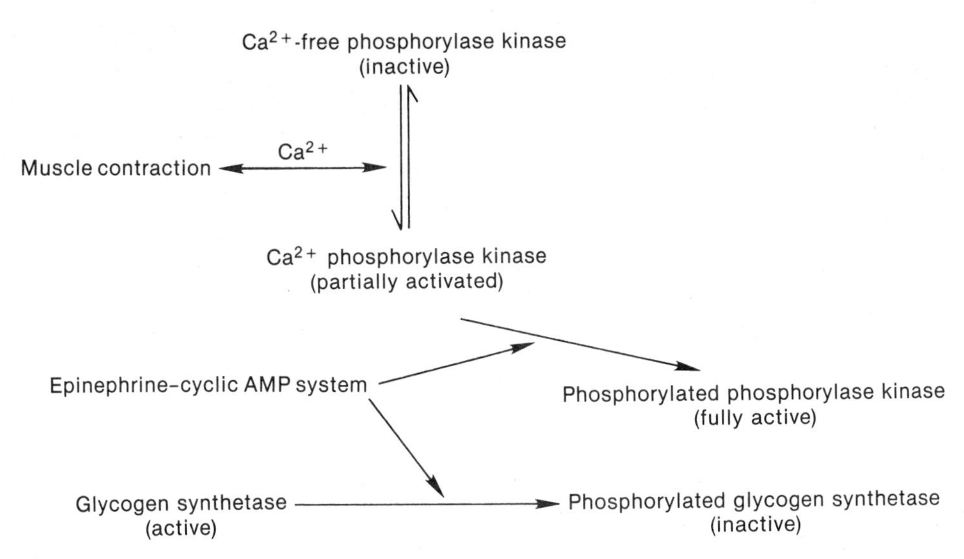

Fig. 30-4. Control of glycogen synthesis and degradation. Glycogen is a very efficient storage form of glucose. Storage consumes slightly more than one ATP per glucose-6-phosphate, the more than one ATP needed per glucose-6-phosphate is attributable to branching. The complete oxidation of glucose-6-phosphate yields 39 ATP. Comparing ATP used up for storage with yields from complete oxidation, one finds overall efficiency of storage is about 97%. Delicate balances of degradation and synthesis depicted in above scheme makes the two processes work according to immediate energy needs of muscle. Obviously a defect in glycogen metabolism leads to impairment of energy utilization.

Anaerobic glycolysis. Anaerobic glycolysis refers to the catabolism of glucose to pyruvate and lactate. The pathway employed is the Embden-Meyerhof cycle. It permits some chemical energy stored in glucose to be stored in ATP and may take place in the absence of oxygen. The controlling enzymes are located in the cytoplasm, and if there is a lack of oxygen, the end product is lactic acid. The use of this pathway in preference to other fuels occurs in red blood cells and exercising cardiac and skeletal muscle cells. When glucose, a six-carbon molecule is metabolized to lactic acid, two three-carbon molecules are formed. The enzyme phosphofructokinase serves as a major control point in the series of chemical steps employed to generate lactate. If there is a deficiency in cellular ATP and aerobic oxidation is depressed, phosphofructokinase is stimulated. The final step in anaerobic glycolysis, conversion of pyruvate to lactate, is controlled by the enzyme lactate dehydrogenase.

Aerobic glycolysis. When cellular enzymes use oxygen to produce energy, carbon dioxide, and water, the pathway most employed is the tricarboxylic acid (Krebs) cycle. This sequence is the final common pathway for carbohydrates, fatty acids, and amino acids. The enzymes are located with the mitochondria and are coupled so that energy

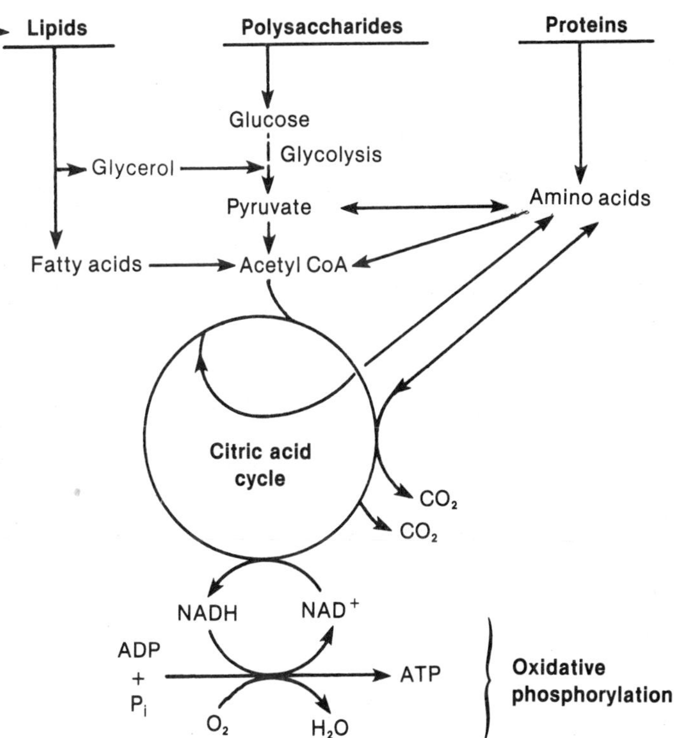

Fig. 30-5. Final common pathways for food substrate utilization.

is transferred for oxidative phosphorylation. Pyruvate dehydrogenase is the enzyme that controls the decarboxylation of pyruvic acid to generate the two-carbon fragment acetate. Next, acetate condenses with coenzyme A to form acetyl CoA. Fig. 30-5 shows major details. The cycle is established by the combination of oxaloacetate with acetyl CoA to form the tricarboxylic acid citrate. ATP availability controls the general activity of the cycle. As the amount of oxaloacetate decreases, ATP becomes less available, and substrate use swings toward gluconeogenesis.

Gluconeogenesis. Because there is a constant need for glucose when hexose is not available from food, gluconeogenesis, or formation of glucose-6-phosphate from noncarbohydrate sources, is necessary. Glucose may be derived from lactate, pyruvate, glycerol, amino acids, and to a minor extent certain fatty acids. The amino acid most used is alanine, but all amino acids derived from protein except leucine may be used. For the most part substrates are converted to pyruvate or oxaloacetate before glucose is generated. The liver is the site of most gluconeogenesis though the kidney contributes to this process after prolonged fasting. During fasting, insulin concentration diminishes and protein breakdown results in greater availability of glucose precursors. Also, fatty acids are mobilized and liver concentrations of acetyl CoA are increased. These changes stimulate the key enzyme for gluconeogenesis, pyruvate carboxylase.

Glucose-6-phosphate formed within a cell, either by glycogenolysis or gluconeogenesis, is usually used within the cell. Only a few organs contain the enzyme glucose-6-phosphatase, which produces glucose from its phosphorylated derivative. These organs, primarily the liver and kidney, thus are the only organs that can act as a source of serum glucose. In fact, the liver is the major nondietary source of serum glucose of the body.

Normal control of glucose during fasting and after food intake

When external sources of food are not available, the body's energy requirements must be met by selective delivery of stored calories from specific tissues. When food is available, it must be stored in a compact system that uses minimal energy and allows quick remobilization. The principle controller of both fuel storage and release is the polypeptide hormone insulin. Other humoral controls are

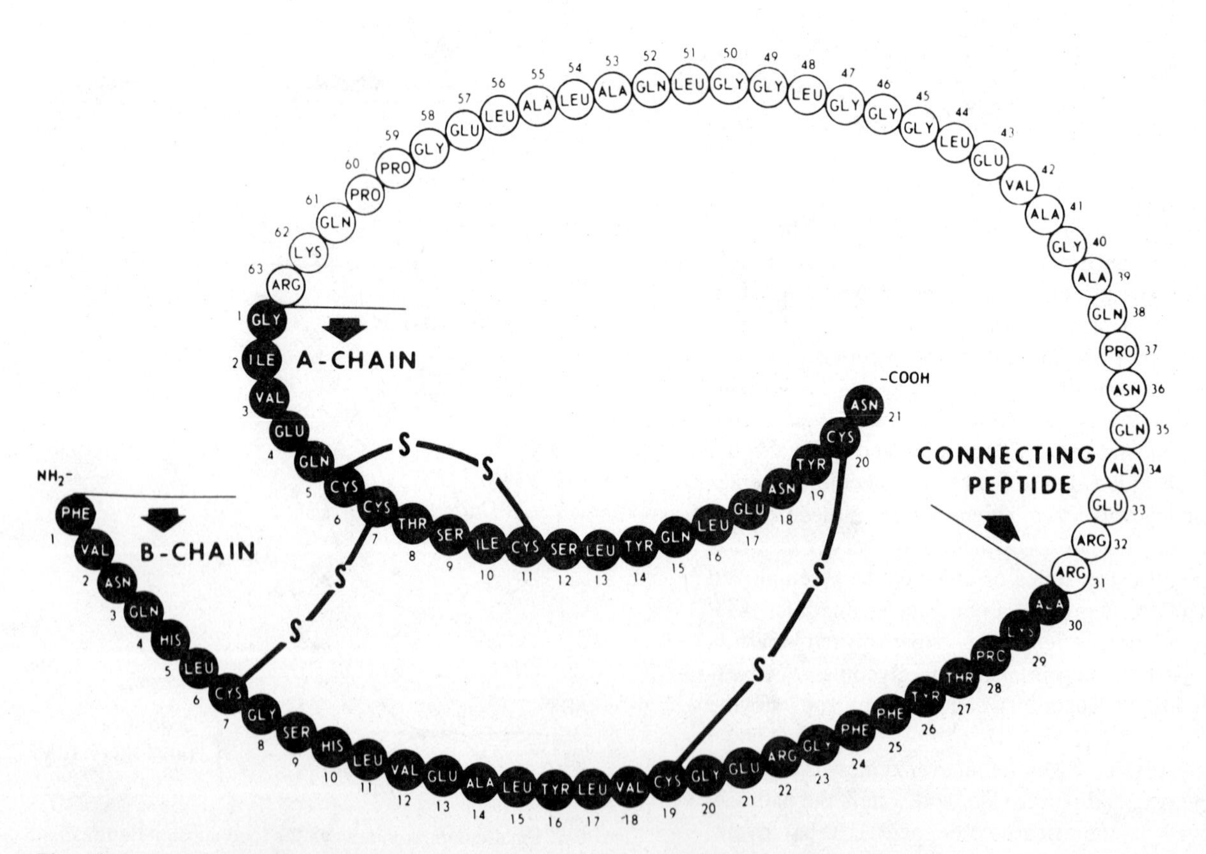

Fig. 30-6. Bovine proinsulin. Notice that A chain and B chain are formed when proteolysis occurs at arginine. Arginine is cleaved from glycine at A-chain junction with C chain connecting peptide and from alanine at carboxyl terminal of B chain.

less important, but modulation of fuel homeostasis by epinephrine, growth hormone, cortisol, the intestinal hormones, and especially glucagon occurs.

Insulin. Insulin is derived from a precursor substance named "proinsulin"[6] (Figs. 30-6 and 30-7). The latter is a single-chain polypeptide that may be described as composed of three parts: the 21-amino acid A chain, the 30-amino acid B chain, and a connecting C chain. Proinsulin is constantly synthesized by pancreatic beta cells of the islets of Langerhans. It is packaged by the Golgi apparatus so that a large number of molecules can be stored within a double-walled secretory vesicle. This sphere begins to migrate toward the interface of the beta cell and its capillary blood supply when a secretory effect is evoked by neural, food-substrate, or hormonal stimuli. During migration, the C chain is cleaved from the A and B chains by proteolytic activity at specific arginine positions. It has been accepted that the C chain of proinsulin ensures that the three-dimensional configuration of combined A and B chains and their important disulfide bridges is maintained.[7] When the granules are disrupted and discharge their content into the circulation, equal quantities of insulin and C chain and a small amount of proinsulin are released. No biological activity has been ascribed to the C chain, but proinsulin has a modest insulin effect[8] (Fig. 30-8).

The pattern of insulin release has been studied in great detail. Even before food is absorbed, there may be a brief period of release, which may account for about 2% of total secretion[9] (Fig. 30-9). Gastric inhibitory polypeptide (GIP) is a hormone that stimulates insulin secretion, which, when elaborated by small intestinal cells in concert with glucose absorption, effects the release of insulin. The major peak of insulin secretion occurs during later glucose absorption, and its chief stimulus is entry of glucose into beta cells.[10] Certain amino acids are potent stimuli for insulin release, and in some ruminants such as sheep, short-chain fatty acids exercise intense control of insulin secretion.

Fasting. Cahill[11] long ago stressed the similarity of the fasting state, wherein macromolecules are mobilized to readily oxidize substances, to the diabetic state, wherein the same process becomes intensely more catabolic. In a normal-weight adult the fasting concentration of plasma insulin ranges from 5 to 25 μU/mL, depending on the interval since food ingestion. This level is maintained by secretion of 0.2 to 1 unit of insulin per hour.[12] In an insulin-dependent diabetic person, insulin secretion approaches zero. Thus small concentrations of insulin exercise precise control of catabolism so that just enough fatty acids and glycerol are released to satisfy energy demands of tissues like muscle and organs like the heart. This limited amount of insulin is not sufficient to permit significant transport of glucose from the circulation to muscle; thus serum glucose use during fasting is limited to non–insulin dependent tissues such as the red blood cells, white blood cells, and the central nervous system (Fig. 30-10). There is abundant evidence that glucagon rises during fasting and that its action enhances hepatic gluconeogenesis and glycogenolysis.[13] Fig. 30-10 illustrates the major pathways used during the fasting state.

Feeding. The ingestion of food immediately transfers metabolic behavior from a catabolic to an anabolic mode. A minimal increment in insulin stimulates the onset of storage of small molecules derived from dietary fat, car-

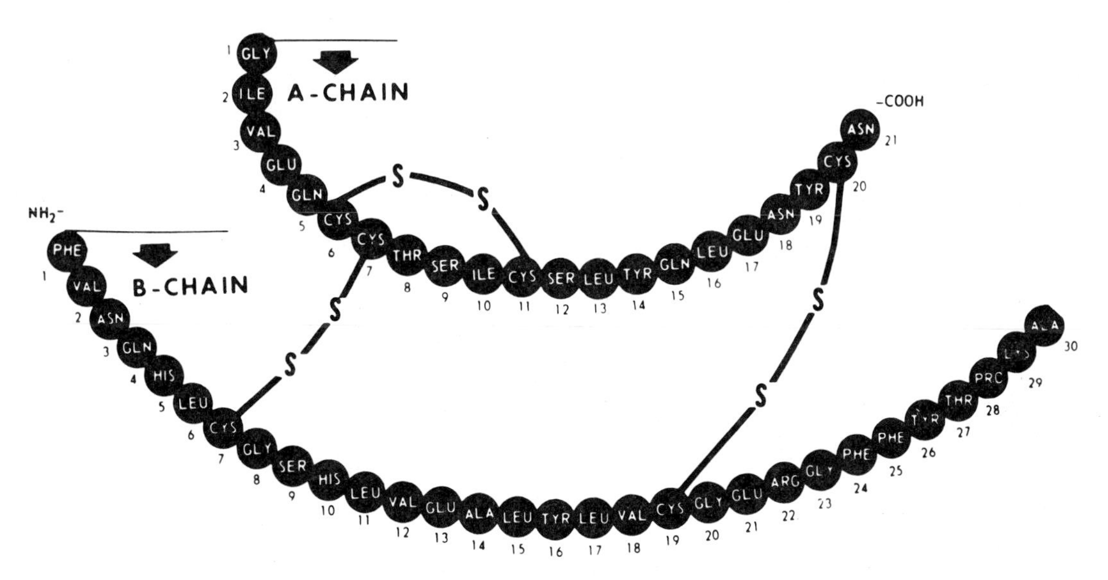

Fig. 30-7. Bovine insulin. Notice disulfide linkages between cystines connecting A and B chains, and disulfide "ring" of A chain connecting cystine 6 and 11. Major species differences occur in amino acids 8, 9, and 10.

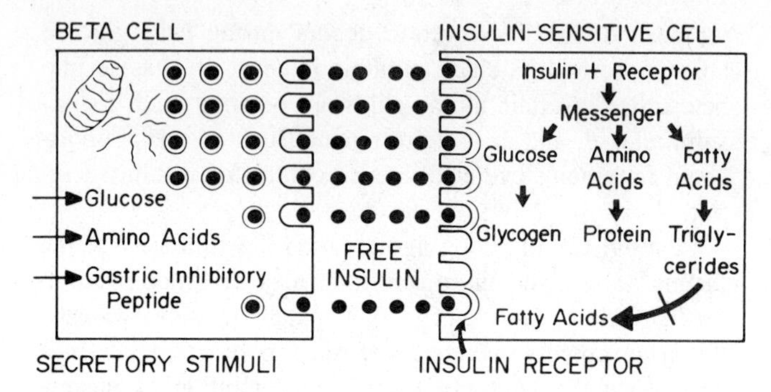

Fig. 30-8. Schematic concept for control and mechanisms of insulin secretion. Stimuli such as D-glucose, certain amino acids, and gastric inhibitory polypeptide direct beta cell to release packaged insulin and synthesize new insulin. Hormone then is transported in blood to specific receptor sites in insulin-sensitive tissues such as muscle, adipose tissue, and liver. Insulin and its receptors generate a messenger, which then stimulates synthesis of glycogen, protein, and fat macromolecules. Lipolysis and glycogenolysis are also inhibited.

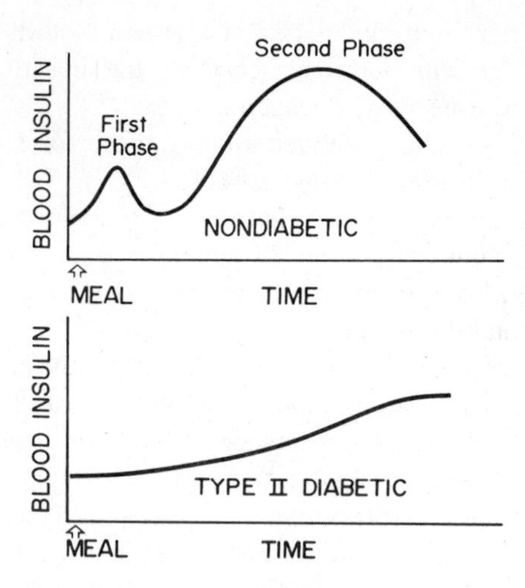

Fig. 30-9. Two phases of insulin secretion. *Vertical axis,* Relative amounts of insulin concentration in blood; *horizontal axis,* time after insulin-secretory stimulation. Notice that first phase is short and not intense and also that it may be absent in type 2 diabetics.

bohydrate, and protein to synthesize triglycerides from fatty acids, glycogen from glucose, and protein from amino acids. Felig has estimated that if 100 g of glucose were to be absorbed, 60 would be retained by the liver for glycogen synthesis, 25 would be stored in insulin-dependent tissues such as muscle and fat, and 15 would be taken up by non–insulin dependent tissues such as the central nervous system.[14] Important actions of insulin are stimulation of hepatic glycogen synthesis, stimulation of transfer of glucose and amino acids from the blood to insulin-dependent tissues for macromolecular storage and inhibition of catabolic processes such as glycogenolysis, lipolysis, and proteolysis. Modest increments in plasma glucose inhibit glucagon action, and lower glucagon levels complement higher insulin concentrations to ensure anabolic function.[15]

Glucose-elevating hormones

From the above description it is obvious that insulin acts as the body's only hypoglycemic agent. In concert with insulin receptors, the role of insulin is to shift extracellular glucose to intracellular storage sites in the form of macromolecules (such as glycogen, fats, and proteins). Thus glucose is stored away in times of plenty for times of need.

In periods of fasting, a series of hyperglycemic agents, in response to low blood glucose, act on intermediary metabolic pathways to form glucose from storage macromolecules. Thus proteins and glycogen are broken down to form glucose (gluconeogenesis). Glucose formed in the liver serves as the major source of nondietary blood glucose.

The most important hyperglycemic agents are glucagon, epinephrine, cortisol, thyroxine, growth hormone, and certain intestinal hormones. The behavior of each of these agents in regulating blood glucose is different; however, whereas insulin promotes anabolic metabolism (synthesis of macromolecules), these hormones in part induce catabolic metabolism to break down large molecules.

Glucagon is produced in the α-cells of the pancreas (see Chapter 26). It acts in the liver to accelerate the reactivation of phosphorylase, a key enzyme in glycogenolysis. The breakdown of stored liver glycogen is accelerated by glucagon and leads to a rapid rise of blood glucose. Glycogen also acts to increase the release of fatty acids from adipose cells.

Epinephrine also works to activate phosphorylase but stimulates phosphorylase in both liver and muscle.

Cortisol and other corticosteroids increase the rate of gluconeogenesis from proteins and amino acids. Gluconeogenesis in the liver results in increased blood glucose levels.

Thyroxine, growth hormone, and *other hormones* act in a manner similar to the corticosteroids.

CONCEPTS OF PATHOGENESIS OF DIABETES SYNDROMES

The concept that diabetes is an expression of several different types of pathogenic perturbations is not modern. The notion that diabetes represented a single disease was emphasized when Minkowski was able to make dogs diabetic by pancreatectomy and Banting was able to restore most metabolic aberrations by injecting insulin derived from pancreatic extracts. However, in 1936 Himsworth clearly demonstrated that there were two fundamental varieties of clinical diabetes.[16] One was a disease completely dependent on insulin for control of symptoms and preser-

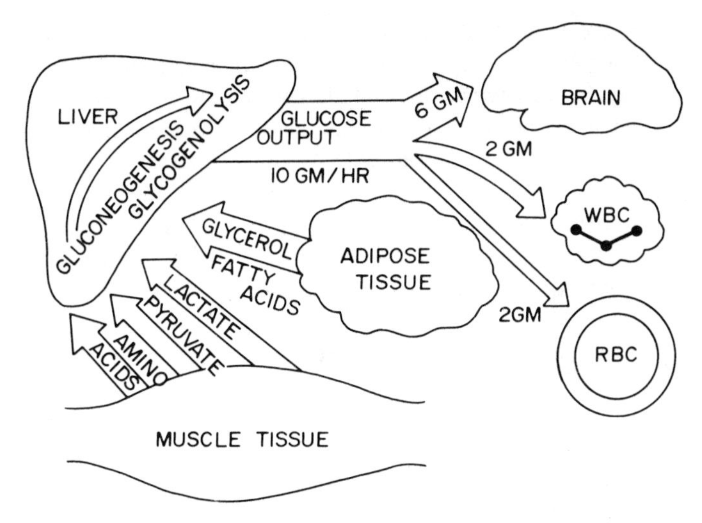

Fig. 30-10. Since brain requires glucose as its principle metabolic fuel, insulin levels fall so that very little glucose is transferred from blood to muscle and fat. About 6 grams per hour are passively transferred to brain, and moderate amounts are utilized by red and white blood cells. Hepatic gluconeogenesis and glycogenolysis are augmented by transfer of glycerol and fatty acids from adipose tissue.

Current classification of diabetes

Diagnoses associated with glucose intolerance

A. Diabetes mellitus (DM)
   1. Type I, insulin-dependent (IDDM)
   2. Type II, non–insulin dependent (NIDDM)
   3. Diabetes associated with certain conditions and syndromes
      a. Pancreatic diabetes
      b. Drug- or chemical-induced diabetes
      c. Endocrinopathies
      d. Insulin receptor disorders
   4. Certain genetic syndromes

B. Impaired glucose tolerance (IGT) (glucose levels between normal and diabetic)

C. Gestational diabetes (GDM) (glucose intolerance during pregnancy)

vation of life, whereas a second type was not dependent on insulin. It has been shown that certain diabetics are resistant to the effects of insulin[17,18] and such insulin resistance may be caused by excess adiposity. The development of a system to label insulin molecules with a single atom of radioactive tracer enabled Roth and colleagues to propose that insulin-sensitive cells contained specific receptors for insulin and that obese, insulin-resistant diabetic persons had a deficiency in receptors.[19] Contemporary research strongly indicates that the cellular response to insulin and the interaction between insulin and its receptor is synthesis of a messenger that governs enzymatic activity within the cells. It has been suggested that the messenger is a protein of small molecular weight; it is clear that the substance is generated by insulin and receptor interaction.[20] The concepts that diabetes mellitus may exist because there is insulin deficiency, insulin receptor deficiency, or messenger deficiency are thus all tenable.[21]

Classification

Until a few years ago, there was little agreement about criteria establishing the diagnosis of diabetes (see box). In 1979, the National Diabetes Data Group published a document entitled *Classification of Diabetes Mellitus and Other Categories of Glucose Intolerance*.[22] This paper accomplishes several goals. It gives criteria that clearly designate persons as nondiabetic or diabetic. It also accounts for those who display abnormalities in glucose tolerance that are not "normal" but still do not reach values that are clearly "diabetic." The classification defines the differences in type 1 and type 2 diabetes and also discusses the significance of glucose tolerance criteria that are abnormal during pregnancy or might have been abnormal in the past but are currently within normal limits. Care has been taken to set down specific directions for performing a glucose tolerance test (see p. 544). Much epidemiological information has been evaluated so that one can assign an actual risk of future diabetes in subjects with a family history of diabetes or a past aberration in glucose tolerance. Although the publication is thorough and thoughtful, it has failed to satisfy the special wishes of many expert authorities. Nevertheless, these criteria provide solid information that is currently being used by medical societies, insurance underwriters, and others who require precise numerical categorization.

Type 1 diabetics. Type 1 diabetics have essentially no insulin synthesis. As a result, they are identified as having insulin-deficient (or insulin-dependent) diabetes mellitus (IDDM). About 10% of all diabetic persons are classified as type 1.

Type 2 diabetics. Type 2 diabetics are not dependent on insulin to sustain life. Nearly 90% of all diabetic persons are of type 2. These major classes of diabetics are discussed at length in the subsequent text. A description of other types of diabetes is briefly discussed below.

Unusual types of diabetes. Recent advances in technology have shown that there are several rare but nevertheless interesting causes of diabetes. It is improbable that everyday medical practice is likely to identify such a case, but if one ritualistically measures both plasma glucose and insulin in the fasting state, causes of "insulin resistance" in which plasma insulin is extremely high or low may indicate that an unusual type of diabetes may be present. Specific listings include (1) severe reduction in receptor number,[23] (2) production of antibodies against insulin receptors,[24] (3) synthesis of an insulin with an abnormal sequence of amino acids and thus reduced biological activity,[25] (4) secretion of proinsulin rather than insulin,[26] and (5) severe deficiency after the receptor-binding step.[27]

Secondary diabetes. In a small proportion of cases of diabetes, disordered metabolism is secondary to a specific syndrome rather than caused by the genetic and environmental settings described as type 1 and 2 diabetes mellitus. There are at least 40 rare hereditary disorders, of which diabetes may be a component.[28] Usually the disease itself is obvious and often the diabetic state may be of less importance. Examples are the Willi-Prader syndrome and Friedreich's ataxia.

Disorders that produce insulin antagonists. This category of secondary diabetes includes acromegaly, Cushing's syndrome, and pheochromocytoma. In acromegaly, growth hormone excess often impairs the insulin effect, and about one fourth of all acromegalic persons display diabetes, sometimes in its insulin-resistant form. In Cushing's syndrome, hypercortisolism also antagonizes the insulin effect, and its mineralocorticoid effect can cause potassium depletion, which impairs insulin secretion. Pheochromocytoma associated with pronounced alpha-adrenergic stimulation is frequently associated with modest but reversible diabetes. A glucagonoma syndrome characterized by chronically elevated glucose is caused by pancreatic alpha-cell tumors that metastasize to the liver. Most patients with glucagonoma have diabetes.[29]

Drug-induced diabetes. A multitude of commonly used pharmaceutical agents are capable of modifying carbohydrate homeostasis. Fortunately, only a few drugs are directly implicated as actually causing diabetes, and often one can speculate that use of the drug has merely unmasked type 2 diabetes. The thiazide diuretics are good examples. These agents may seriously deplete total body potassium, but withdrawal of therapy plus potassium replacement may restore carbohydrate homeostasis. Phenytoin (Dilantin) exercises a direct effect on beta-cell insulin release and may cause glucose intolerance. Despite the widespread use of thiazides and phenytoin, it is remarkable that so few instances of disordered glucose metabolism occur. The same statement probably applies to adrenergic blocking agents such as propranolol, which are often given in enormous doses with chronic usage.

Direct beta-cell injury. A few diseases cause sufficient destruction of the pancreas to cause diabetes. The pancreas can become infiltrated with iron, and fibrosis may result. If more than 90% of the islets are destroyed, clinical diabetes may be found. In some instances there can be severe insulin resistance. Although acute and chronic pancreatitis are common, neither results in clinical diabetes very often. If a person develops chronic calcific pancreatitis, there may be sufficient islet cell destruction to cause insulin-deficiency diabetes. Often this disease is accompanied by malabsorption associated with exocrine pancreatic failure.

Prevalence

It has been estimated that only about half of persons living in the United States who have diabetes have been *diagnosed* as diabetic. This may be true because many patients diagnosed as "new" type 2 diabetics relate symptoms of this disease that are of several years' duration. Diabetologists accept prevalence data that describe about 2% of the population as diabetic.[2] This figure is derived from cross-sectional surveys that measure fasting or postprandial glucose in a very large proportion of all persons living in a circumscribed community. The criteria for diagnosis are therefore artificial but liberal enough to avoid overestimation. Type 1 diabetes accounts for less than 10% of known diabetes, and it is unlikely that cases of type 1 diabetes are excluded from surveys. Type 1 diabetes involves men and women with about equal frequency, and about 50% of type 1 patients are diagnosed during childhood. Most type 1 diabetes develops before 40 years of age, but many exceptions are found. Type 2 diabetes is much more common in adults than in children and increases in frequency with age. It is more common in women than in men and in blacks than in whites. Type 2 diabetes is more prevalent in people who are sedentary and especially develops in racial or ethnic populations who adopt a Western life-style.[30]

Etiology

The postulate that there are separate and genetically distinct varieties of diabetes has been endorsed by the report of the National Diabetes Data Group.[22] According to their deliberations, most diabetic persons are classified as having one of two major types of diabetes with correspondingly different causes.

Type 1 diabetes. The disorder previously known as "juvenile diabetes" is now identified as type 1, or insulin-deficient, diabetes mellitus (IDDM). Its development is seemingly multifactorial as genetic factors, infection, and

Factors that may possibly interact to cause beta-cell destruction in type 1 diabetes mellitus

1. Viral disease
2. Autoimmune reaction
3. Combination of 1 and 2
4. Genetic susceptibility plus 1, 2, and 3

perturbation of the immune defense system all appear to play a role (see box on p. 530). Its genetics and immune characteristics seem to be closely interrelated.

Genetic transmission. Genetic transmission of type 1 diabetes is important but difficult to define because environmental influences exercise at least as important a role. The long-term studies of twins by Pyke and Nelson have shown this.[31] The life course of identical twins was followed after one member of a set of twins developed type 1 diabetes. Surprisingly they discovered that the other twin developed IDDM only about 50% of the time, though his genetic constitution was precisely the same. One can postulate that the twin developing diabetes may have had an unmasking environmental experience. Similar behavior has been described by Fajans and Tattersall.[32] They made a prospective survey in first-degree relatives of persons with type 1 diabetes. The frequency of diabetes was only about 10% in the parents and 2% in the siblings of these subjects. Continuing observations now suggest that viral infections, HLA type, and immune response may represent controlling factors in the pathogenesis of type 1 diabetes.

HLA types. The relationship of certain human lymphocyte antigens (HLA) to specific diseases can be best described by appreciation of the fact that approximately 95% of patients with ankylosing spondylitis have the HLA subtype B15. Exhaustive analysis has disclosed that HLA-B8 and HLA-B15 are highly correlated with IDDM in whites and in natives of India. The HLA's are present on the surface of all nucleated cells, and their type is controlled by loci on the short arm of chromosome 6. It is believed that the HLA system directs both host defense and self-recognition. The fact that the presence of HLA-B7 is negatively correlated with the risk of type 1 diabetes emphasizes that the HLA system is more than a simple but incomplete marker.[33] HLA-B8 presence also correlates with circulating autoantibodies to islet cell antigens in persons with type 1 diabetes.

Viral infections. There has long been a notion that the onset of type 1 diabetes may bear a relationship to a recent infective disease. In 1977 Cudworth made a prospective analysis of 100 patients who developed IDDM.[34] It was found that nearly twice as many new cases of diabetes occurred during the spring and fall, when common viral infections were most prevalent. A trend toward clustering of cases in geographical locals was also noted. Serological examination indicated that most newly diagnosed diabetic patients had evidence of rising titers of antibodies to common viruses such as coxsackieviruses B1 and B4. Furthermore, the patients also had a significantly higher frequency of HLA patterns B8 and B15. There is other, less direct evidence of the association of IDDM with viruses.[35] Rubella, infectious hepatitis, and mononucleosis have been implicated. The literature is replete with virus-induced diabetes in susceptible mice. An islet cell lesion pathognomonic for diabetes has been described.[36]

Autoimmunity. Multiple types of autoimmune destruction of endocrine organs have been recognized for three decades. Chronic lymphocytic thyroiditis represents the best proved case. In addition, there is ample evidence that Graves' disease with or without hyperthyroidism, chronic adrenal insufficiency, and primary hypoparathyroidism may have an autoimmune origin. The concordance of IDDM in subjects with endocrine disorders postulated to be autoimmune led to an intensive search for an expression of abnormal defense system behavior in this type of diabetes.[37] Antibodies to islet cell antigens were found in high titer in subjects with thyroid disease and diabetes[23] and later in patients with IDDM alone.[38] Although as many as 95% of persons newly diagnosed as type 1 diabetes may have antibodies to islet cells,[39] the fact that 5% do not lends doubt to a hypothesis that declares autoimmunity a full qualification for development of IDDM. Circulating antibodies probably reflect a response of the islet cell to injury and could indicate that the injury might result from direct antibody attack. Early clinical studies to test this hypothesis by treatment of newly discovered diabetics with prednisone to subdue an immune response have been inconclusive.[40]

Type 2 diabetes. Unlike type 1 diabetes, there is no evidence of interactions of HLA subtypes, viral infections, or autoimmune responses (see following box). Genetic transmission, however, seems to exercise an even more important role in type 2 than type 1 diabetes. The Pyke-Nelson study of twins[31] showed that when one identical twin developed type 2 diabetes, the probability of the other twin becoming diabetic within just a few years exceeded 90%. A second but very important factor in the genesis of type 2 is obesity. More than two thirds of type 2 diabetics are overweight when diabetes is diagnosed. Obesity decreases insulin receptor number and thus insulin action. Although some obese diabetics secrete more insulin than nondiabetics do, they rarely secrete more insulin than a nondiabetic of similar weight.[21] Aging, physical inactivity, Western culture life-style, and the use of drugs that impair insulin or insulin receptor activity also are factors in unmasking type 2 diabetes. For example, if a type 2 diabetic capable of secreting a normal amount of insulin is treated with large doses of exogenous insulin, there is down regulation or reduction in the number of insulin receptors. On

Factors that may be responsible for type 2 diabetes

Genetic
Obesity
Aging
"Wearing out" of beta cells

the other hand, restriction of caloric intake or weight loss and exercise evoke an increase in insulin receptor number and improved control of diabetic hyperglycemia. There is emerging evidence that some type 2 diabetics may also have deficiency occurring after receptor-binding activity.[41] The box on p. 529 organizes contemporary analysis of the pathogenesis of the major diabetes syndromes.

Table 30-2. Characteristics of type 1 and type 2 diabetes

Variable	Type 1	Type 2
Insulin deficiency	Always present Often total	Variable Never absolute
Beta-cell destruction		
Viral cause	Proved in some	Never proved
Autoimmune factors	Characteristic	Unlikely
HLA type	Significant	Not significant
Patterns of genetic transmission	Not very well defined	Often very predictable
Onset of symptoms	Usually abrupt	Usually insidious
Ketoacidosis	Common	Rare
Need for exogenous insulin	Necessary for life	Not necessary for life
Obesity	Usually absent	Very common
Insulin receptor activity	Normal	Decreased
Microvascular complications	Develop in nearly all	Usually not a major problem
Macrovascular complications	Develop in many early in life	Major cause of death and morbidity

Clinical manifestations of diabetes syndromes

Although hyperglycemia is traditionally used as a marker for chemical analysis of the biochemical behavior of all types of diabetes, other indicators of this disorder of carbohydrate, fat, and protein perturbation could also be employed.

Type 1 diabetes. Type 1 diabetes is almost always symptomatic within a few days or weeks of its onset. Since there is invariable decreased pancreatic synthesis and release of insulin (insulinopenia), various degrees of the expected manifestation of hypoinsulemia become rapidly and progressively manifest. Many of these symptoms are caused by the high serum glucose levels (hyperglycemia). Polyuria (increased urine volume) caused by a glucose osmotic diuresis may be first recognized. When normally functioning kidneys are presented with a serum glucose concentration greater than 1800 mg/L (180 mg/dL), all the filtered glucose cannot be totally reabsorbed. This is defined as the renal threshold for glucose, that is the plasma glucose concentration above which glucose is excreted in urine. The excreted glucose must be accompanied by water for solubilization. Healthy persons can concentrate urine to about 1200 mOsm/kg. A mole of glucose constitutes 180 g and would require at least 1000 mL of extra water for excretion. The amounts of glucose excreted by hyperglycemic persons can be quite large, and urine outputs of 2000 mL or more per day are very common. Increased urine volume is recognized as a net loss of body water. There is associated increased thirst (polydipsia) and often an increase in appetite (polyphagia). Despite these symptoms, there is usually a loss of weight because not only are anticatabolic calories lost, but because of decreased anticatabolic effect of insulin there is also a heavy loss of nitrogen, potassium, sodium, and water. If insulin secretion falls to levels at which antilipolytic control fails, mobilization of free fatty acids from peripheral adipose tissue exceeds the liver's quantitative capacity to oxidize fatty acids or resynthesize them to triglycerides. At this point, overproduction of "keto acids" occurs. These substances are beta-hydroxybutyric acid (which is not a keto acid) and acetoacetic acid. Hydrogen-ion concentrations associated with generation of these substances accumulate and tax the ability of the kidney to neutralize and excrete them. If acid production becomes critical or water intake diminishes, hypovolemia occurs and compromises renal blood flow. The accumulation of hydrogen ions results in diabetic ketoacidosis. Ketoacidosis is defined as the presence of elevated levels of ketones in the blood (ketonemia) plus a condition of acidosis (metabolic) (see Chapter 22). Table 30-2 compares the major clinical and biochemical findings in types 1 and 2 diabetes.

In some persons with type 1 diabetes there is progression to total insulin deficiency in a short time. In others there may be a partial or near-complete (but always temporary) restitution of function. The latter is termed "honeymoon diabetes." Serial measurements of plasma or urine C-peptide can be quite helpful in the estimate of residual insulin secretory capacity.[8] It is generally believed that type 1 diabetic persons who retain modest insulin secretory reserve are more easily controlled.[42] Because of the absolute dependency of persons with type 1 diabetes on exogenous insulin, these persons are at greater risk for entering into a state of ketoacidosis. Thus patients presenting with ketoacidosis are most likely to have type 1 diabetes.

Type 2 diabetes. As noted, about half of all persons with type 2 diabetes living in the United States have not been diagnosed as diabetic. Thus type 2 diabetes can be asymptomatic, or its symptoms can be quite mild or overshadowed by those of other chronic diseases such as heart failure, arthritis, or peripheral vascular disease. On the other hand, some patients with type 2 diabetes display all the typical symptoms of hyperglycemia: polyuria, polydipsia, excessive hunger, and weight loss. Furthermore, although persons with type 2 diabetes do not usually have a risk of ketoacidosis, a stressful event may be associated with sufficient insulin resistance to permit development of hyperosmolar stupor or even ketoacidosis. It is very common to discover diabetes when patients are admitted to a

Table 30-3. Mean daily peripheral insulin delivery

Patient description	Units
Nondiabetic patients	
Lean	31
Massively obese	114
Diabetic patients	
Type 1 childhood onset	4
Type 1 adult onset	14
Type 2 obese	46

From Genuth, S.M.: Plasma insulin and glucose profiles in normal, obese, and diabetic persons, Ann. Intern. Med. **79**:812, 1973.

hospital with diseases such as acute cholecystitis, cellulitis, pneumonia, cataract, hip fracture, or myocardial infarction. Most often, type 2 diabetes is established when a routine office visit yields laboratory results that show fasting or postprandial hyperglycemia. There are a few symptoms that always suggest that a search for type 2 diabetes should be made.

Vaginitis. Hyperglycemia is particularly conducive to florid growth of *Candida* in the vagina. It has been estimated that about one in three women presenting with this form of vaginitis has diabetes mellitus. Vaginal itching and discharge of a thick lumpy whitish material is usually found. As a rule, vaginitis is not controlled until better control of hyperglycemia is established.

Myopia. When an adult complains of myopic vision, more often than not there is underlying diabetes. It is believed that the lens becomes thicker and the focal point of vision falls short of the macula. The deposition of polyhydroxy compounds (polyols) such as sorbitol, an alcohol derived from the action of aldose reductase on glucose, may serve as an osmotic factor, causing the lens to increase its hydration.

Furuncles. Persons with diabetes may display severe forms of boils with seemingly poor defense against staphylococcal infection. Patients who present with any variety of recurrent skin infections should be evaluated for possible diabetes.

Carpal tunnel syndrome. Compression of the median nerve is very common. Usually impairment of sensation of the middle three digits of the hand represents the earliest symptom. As a rule, no clear-cut neurological deficit may be found, but there may be tingling of the fingers when the median nerve is compressed at the wrist. When this entrapment syndrome is associated with diabetes, diagnosis and control of hyperglycemia should be instituted before surgical correction of the carpal tunnel syndrome is recommended.

Peripheral neuropathy. Quite frequently the undiagnosed diabetic may complain of bilateral, symmetric "burning" of the feet. Usually this symptom is the result of a sensory afferent neuropathy involving multiple peripheral nerves. There may be loss of the ankle and knee reflexes and often decrease in vibratory touch, pain, and position sense. Although there are many causes of peripheral neuropathy, symptoms of the variety described are more often associated with diabetes than with any other disease.

BIOCHEMICAL MANIFESTATIONS OF DIABETES

It is worthwhile to develop a concept of the total daily amount of insulin secreted by normal persons and those with various types of diabetes. Genuth[43] reported several years ago that IDDM patients secrete very small amounts of insulin, whereas many NIDDM patients secrete insulin as much as or more than nondiabetic persons do. His values for a large number of persons are given in Table 30-3.

Insulin secretion in type 1 diabetes

Information about insulin secretion complements estimates of insulin receptor activity. Although type 1 diabetic persons have very little circulating insulin and C-peptide, they probably have an increased number of insulin receptors on their cells. They are therefore very sensitive to exogenous insulin and nearly all are at risk to repeated insulin-induced hypoglycemia. On the other hand, obese type 2 diabetics may have approximately half as many insulin receptors as lean nondiabetics but often secrete more insulin.[44] It is probable that few obese diabetics secrete as much insulin as comparably obese nondiabetics do.[45]

Insulin secretion in type 2 diabetes

By definition, the non–insulin dependent diabetic (NIDDM) secretes sufficient insulin to protect against ketoacidosis. The fact that lesser concentrations of insulin are needed to restrain lipolysis than either hepatic glycogenolysis or transfer of glucose into muscle and fat partially explains this protection. However, under conditions of severe or prolonged stress, even the person with non–insulin-dependent diabetes can develop ketoacidosis and, if treated successfully, not require insulin after recovery. The fasting blood glucose (FBS) concentration of the NIDDM patient provides good evidence of his insulin secretory capacity.[9] If the FBS consistently exceeds 1400 mg/L, there is probably less total insulin secretion than if the FBS is normal. The pattern of insulin secretion may also be abnormal in NIDDM. Porte has described a loss of the first phase of insulin secretion. In addition, the second phase of insulin secretion may be delayed and sometimes prolonged[46] (see Fig. 30-9).

Hyperlipemias in diabetes

Almost every conceivable variety of hyperlipoproteinemia has been described in diabetes. For example, there may be increased production of VLDL, LDL, or HDL, or decreased removal of chylomicrons, VLDL (and its remnants), LDL, and even HDL.[47] It is probable that some insulin action is necessary for VLDL secretion by the liver. Lipoprotein lipase activity is low in patients with complete insulin deficiency and can be restored by insulin administration.[48] Patients presenting with diabetic ketoacidosis often have gross hyperchylomicronemia. Persons with type 2 diabetes often display high levels of VLDL.[49] There is a concept indicating that although these patients may have insulin resistance in regard to the ability to transfer glucose to muscle and fat, the liver is sensitive to insulin and responds by increased VLDL secretion. If VLDL removal rate in such patients is reduced, elevated VLDL may be found. Hypercholesterolemia has repeatedly been described in certain diabetics.[50] LDL cholesterol elevation has been documented in type 1 diabetics and in Pima Indians.[51] HDL has been described as elevated or normal in most groups of type 1 diabetic persons in whom it has been studied. This seems paradoxical, since diabetic persons are clearly prone to atherosclerosis and HDL is believed to protect against it. Lower than normal HDL levels have been reported in some persons with type 2 diabetes. Recently it has been proposed that HDL containing excess glucose may be cleared rapidly from the serum, causing low HDL in some diabetic persons.

It is probable that obesity, genetic hyperlipoproteinemias, variations in insulin sensitivity, dietary habits, exercise, and other factors interact with diabetes and its control to an extent that at this time precludes specific categorization of lipoprotein abnormalities in different types of diabetes. Certain generalizations, however, are permissible: (1) chylomicron hyperlipidemia may be present in states of severe insulin deficiency and may be improved with insulin therapy, (2) persons with type 2 diabetes often display elevated VLDL, which yields to a reduction in VLDL and triglyceride concentration when carbohydrate intake is reduced, (3) total cholesterol is higher than normal in many persons with diabetes and is improved with a low–saturated fat, low-cholesterol diet, and (4) HDL may be high in diabetes but subfraction analyses suggest that nascent HDL can be low and thus the protective effect of HDL can be lost.

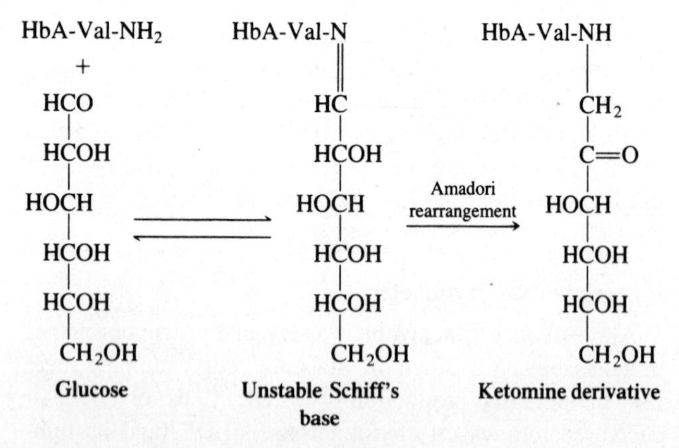

Fig. 30-11. Reaction of glucose with protein to form glycosylated protein; example of reaction with hemoglobin (Hb).

Glycosylation of proteins

The aldehyde moiety of glucose can spontaneously and nonenzymatically condense with accessible amino groups of proteins. The reaction initially forms a Schiff base. This product can spontaneously rearrange itself (Amadori rearrangement) to form a stable end product. The equilibrium between free glucose and the Schiff base favors the formation of free glucose.[52,53] However, in the presence of large concentrations of glucose, the equilibrium is shifted to the formation of the stable glycosylated product. The overall reaction is shown in Fig. 30-11. In conditions in which glucose concentration is elevated, more glycosylated proteins are formed. Thus the measurement of glycosylated proteins reflects the concentration of serum glucose and is useful as a long-term indicator of diabetic blood glucose control.

The most frequently measured glycosylated protein is hemoglobin.[54,55] The N-terminal amino acid and the side-chain amino group of lysine, of both the α and β chains of hemoglobin, can be glycosylated. There are several variants of glycosylated hemoglobin, and techniques are available to measure either the total glycosylated fraction (HbA_1) or the major glycosylated variant (HbA_{1c}). Since the average life-span of a red blood cell is 3 months, the percentage of glycohemoglobin present at any time reflects the average glucose level over the previous several weeks.

The other proteins can also form a glycosylated condensation product. The measurement of glycosylated albumin has been suggested as another useful indicator of glucose serum levels.

PATHOLOGICAL CONDITIONS: ACUTE COMPLICATIONS OF DIABETES MELLITUS

There are two acute life-threatening complications of diabetes: diabetic ketoacidosis (DKA) and hyperglycemic, hyperosmolar, nonketotic coma (HHNC). DKA occurs as a frequent complication in insulin-dependent diabetes and can be the first manifestation of the diabetic state.[56] HHNC usually occurs in elderly persons with type 2 diabetes but also may be encountered in type 1 patients. HHNC is often the first manifestation of diabetes. Most cases of DKA could probably be prevented by improved patient education and better communication with health delivery systems. In theory, the occurrence of HHNC might be reduced in frequency by identification of asymptomatic diabetic persons through detection drives. The complications differ enormously in prognosis. DKA mortality may be as high as 10%, but most medical centers report a gross mortality of under 2%. HHNC has a mortality of about 50%. Often it is an underlying and precipitating factor such as a major cardiovascular accident or infection that really causes death in both DKA and HHNC. Insulin-induced hypoglycemia must be regarded as a third acute complication of diabetes. Although it probably occurs in all IDDM patients, it is a rare cause of death.

Diabetic ketoacidosis (DKA)

Biochemical features. DKA is truly the end stage of uncontrolled diabetes. It expresses failure of the key actions of insulin. There is a mobilization of free fatty acids that are oxidized to the low molecular weight acids acetoacetate and β-hydroxybutyrate. Glucose use is decreased, resulting in hyperglycemia. Proteins are degraded to amino acids, which are used to augment gluconeogenesis (Figs. 30-12 and 30-13). Glucose output from the liver is augmented by diminished or absent insulin effect and by hyperglucagonemia as well. Amino acids derived from the periphery enter the oxaloacetate pathway and are converted to glucose-6-phosphate or glycogen. Glucose leaves the

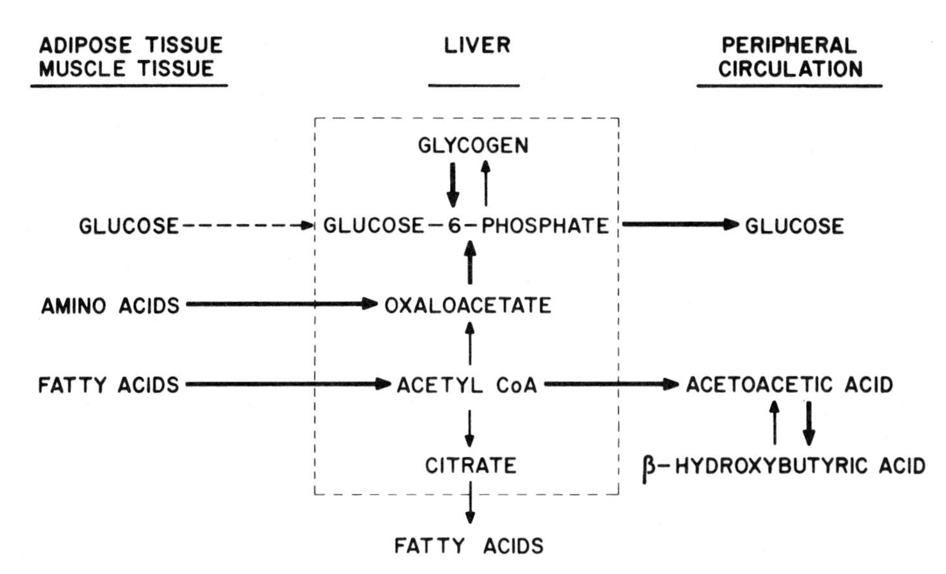

Fig. 30-12. Major pathways for ketogenesis and hyperglycemia in diabetic ketoacidosis. Metabolic pathways undergo remarkable alterations in liver and produce ketoacidosis in absence of insulin effect. *Heavy arrows,* Increased hepatic uptake of amino acids which are metabolized to glucose and glycogen by means of oxaloacetate causing hyperglycemia. Increased fatty acid uptake results in elevated acetoacetic and β-hydroxybutyric acids by means of acetyl CoA.

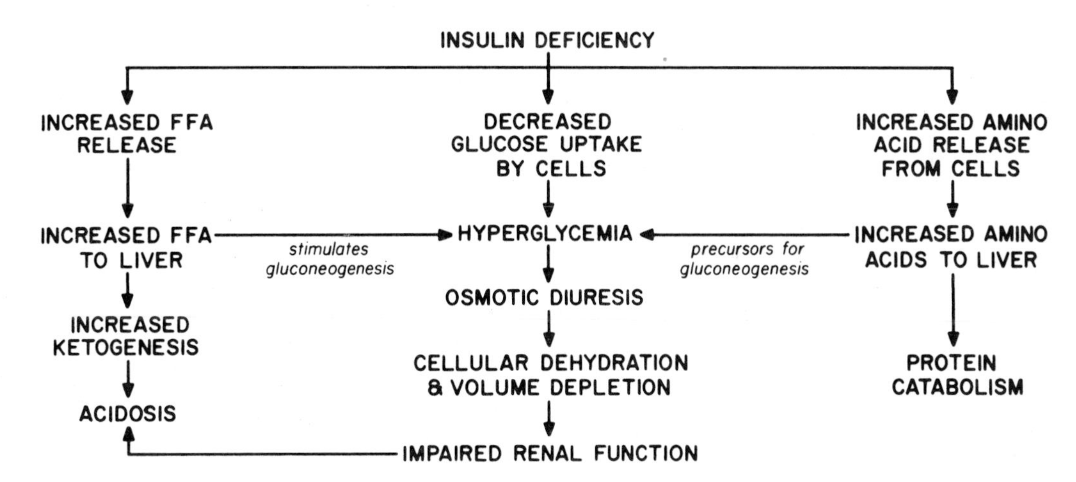

Fig. 30-13. Metabolic consequences of insulin deficiency. Deficiency of insulin (or its action) brings about a series of interrelated events. On the left, reduced effect of antilipolysis results in mobilization of free fatty acids, FFA, which both stimulate gluconeogenesis and become oxidized to acids. In the center, glucose uptake is reduced and adds to the hyperglycemia causing an osmotic diuresis, dehydration, and impaired renal function. On the right, reduced insulin activity mobilizes amino acids and promotes gluconeogenesis.

liver in large quantities but is not used effectively by muscle or adipose tissue because of impaired insulin effect. More than 150 g may be excreted in the urine within a few hours. When hypovolemia develops, renal concentrating ability becomes impaired and even more water is excreted, contributing to worse hypovolemia. Impaired tissue perfusion results, which can lead to an increased buildup of metabolic products such as lactic acid. In addition, there is increased movement of potassium from the intracellular to the extracellular space and eventually into urine, resulting in a total body deficit of potassium. Glucagon, epinephrine, and growth hormone excesses contribute to inhibition of peripheral glucose uptake and cause relative insulin resistance.

Clinical features. Some patients who develop DKA have been known to be clinically and chemically healthy and yet succumb within a few hours. More often the onset of DKA is insidious and progressive. In a majority of instances a precipitating factor is identified. Most often it is an infection within the respiratory, renal, and dermal systems. Unfortunately, many IDDM patients wake in the morning, feel nauseated, and omit their insulin. If metabolic perturbations have been the cause of their not feeling well, this identifiable omission of insulin may be the key factor that permits progression from poor control to DKA.

The early symptoms of DKA are usually not very specific. Polyuria, thirst, and polydipsia (increased fluid intake) may be common enough in certain IDDM patients that they do not signal events to come. Loss of body water causes weakness and orthostatic hypotension, and sometimes the patient will complain of headaches. The onset of gastrointestinal symptoms usually indicates progression from poor control to the onset of ketoacidosis. Nausea, vomiting, vague abdominal pain, and generalized abdominal tenderness are very common, but their basis is not well explained. Kussmaul-Kien (rapid, deep) respiration may be noted in a majority of patients with DKA. The respiratory rate may not be very rapid, but the respirations may be deep and labored. Sometimes the patient may complain of dyspnea. As blood pH falls further, the respiratory rate may increase but the depth of the ventilatory effort can decrease. Always, the odor of acetone on the breath provides the observer with a good clue to the diagnosis.

Acetoacetic acid is converted to acetone on contact with oxygen in the alveolus, and the fruity-sweet odor of the gas is detectable by any observer who has an ability to identify it.[57] It is not unusual to detect the odor of acetone as one enters a hospital room. Patients with modest acetonemia such as that generated by starvation do not have sufficient acetone concentration to exhibit this sign. Persons with alcoholic ketoacidosis pass in transition from a state in which the odor of alcohol changes to that of acetone as alcohol becomes metabolized.

Other clinical signs include dry skin, dry mouth, cool extremities, modest hypothermia, abdominal distension with few or no bowel sounds, and absent reflexes. Many patients exhibit a drop in systolic and diastolic blood pressure when moved from a recumbent to sitting position. About one in 10 patients is in clinical shock.

Diagnosis and laboratory findings. Although it is possible to diagnose DKA and treat it successfully without the use of a clinical laboratory, there are few instances in medicine in which the course and recovery from a life-threatening disease is better charted by successive laboratory maneuvers.

The diagnosis of DKA is established in about 10 minutes by study of a sample of arterial blood (Fig. 30-14). In this time the differential diagnosis in a mentally obtunded patient can be rather firmly established. One must consider the possible metabolic cause of the coma: (1) hypoglycemia, (2) DKA, (3) HHNC, (4) alcoholic ketoacidosis, or (5) lactic acidosis.[56] As soon as heparinized arterial blood is drawn, a small portion should be set aside and retained while the remainder is sent to the laboratory for pH and P_{CO_2} analysis. Plasma from the small specimen is placed on Dextrostix or Chemstrips. If the glucose is elevated, one immediately knows he is dealing with a diabetes-related problem. If it is low, a glucose infusion should be started because it is probable that the stupor or coma is caused by hypoglycemia.

The plasma sample should then be semiquantitatively assessed for acetoacetic acid (AcAc). Most laboratories use crushed Acetest tablets or Denver Chemical Powder. Both contain sodium nitroprusside, which reacts with acetoacetic acid to form a purple color. It is worthwhile to make serial dilutions with water at 1:1 through 1:32. If a moderate or strong reaction is found in a 1:2 dilution, there is significant ketonemia. If the patient has low pH, low P_{CO_2}, and positive AcAc, there is diabetic acidosis if the glucose is high. If the glucose is low, the diagnosis of alcoholic ketoacidosis is probable.[58] If pH and P_{CO_2} are low, ketones are trace or absent, and glucose is high, it is probable that a metabolic acidosis such as renal failure or lactic acidosis is present.

It is important to remember that the nitroprusside test measures mostly acetoacetate and does *not* detect beta-hydroxybutyrate (β-HBA) *at all.* In the early stages of acute DKA, the major metabolite of free fatty acids is in fact β-HBA and not AcAc. Thus, the test for serum or urinary ketones may be negative even though DKA is present. If the glucose is quite elevated (more than 10 g/L) and AcAc is weak or negative and pH and P_{CO_2} are near normal, the most probable diagnosis is HHNC.

Many laboratories have the capacity to measure blood lactic acid in emergency situations. If so, the diagnosis of lactic acidosis may be confirmed as the patient is receiving massive infusions of bicarbonate while an intense search for an occult cause of this state is made. Some laboratories have a facility for enzymatic measurement of AcAc and β-

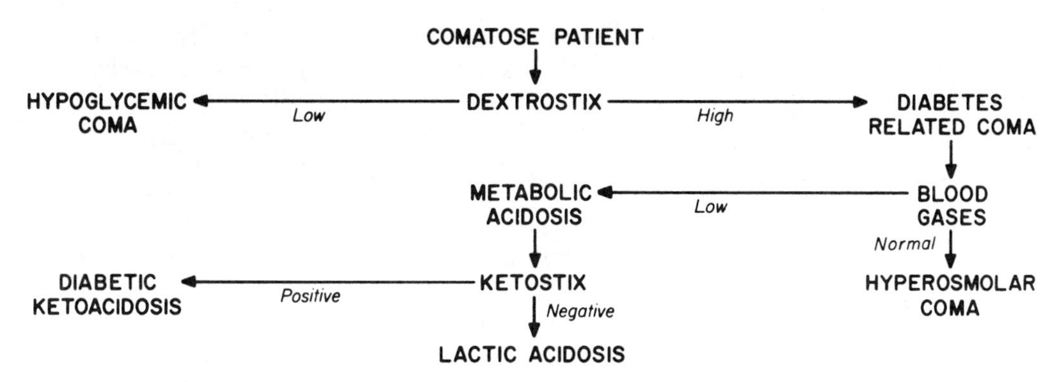

Fig. 30-14. Differential diagnosis of coma in a patient with possible diabetes. Three rapidly done tests almost always provide the proper diagnosis in diabetic persons encountered with impaired mentation. Blood applied to Dextrostix or Chemstrips bg separates hypoglycemia and diabetes-related stupor. Arterial pH identifies acidosis from hyperosmolar coma, and a nitroprusside test (Acetest or Ketostix) separates ketoacidosis from other causes of acidosis. Patients with alcoholic ketoacidosis are not hyperglycemic.

HBA, but most often these determinations are made in settings where clinical investigation is being done. I should point out that although estimation of AcAc by use of diluted plasma and the nitroprusside reaction frequently describes the true severity of ketonemia when the patient is first examined, changes in redox potential tend to shift from the usual 1 mol of AcAc to 3 mol of β-HBA to ratios that may be as high as 1:30. Thus sequential alternative chemistries such as blood glucose and blood gases should complement ketone estimates. The unwary might assume ketonemia is clearing as AcAc diminishes whereas β-HBA could be increasing.

As in all seriously ill patients, it is imperative that the laboratory be used to detect signs of inapparent associated metabolic problems. For example, it is not unusual for the DKA patient to present with a serum urea nitrogen of up to 1000 mg/L even though renal function might not be permanently impaired. The serum potassium might be elevated initially, though there could be depletion of 300 to 400 mEq of the body's potassium stores. A very low level of serum phosphate probably reflects phosphate depletion and could well indicate that 2,3-DPG is low and that rapid correction of pH by bicarbonate infusion could interfere with dissociation of oxygen from oxyhemoglobin.

Treatment. The first maneuvers in therapy are directed toward preservation of vital functions. An adequate airway should be assured, and if the patient is in a coma, a gastric tube should be placed to prevent aspiration of gastric contents. Blood volume restoration should be initiated by rapid infusion of isotonic saline at a rate of 1 liter within 30 minutes and up to 2 more liters if necessary. If acidosis is severe and pH is below 7.0, 50 to 100 mEq of sodium bicarbonate should be infused.

From the beginning, one should keep a flow chart describing time, laboratory values, clinical signs, water and electrolyte input, urine and other output, medication, and vital signs. The keys to correct therapy are appropriate infusions and monitoring of saline, insulin, potassium, phosphate, and, when indicated, glucose.

Insulin. Alberti showed that frequent "small" doses of insulin were effective in management of DKA.[59] Insulin may be administered intravenously, intramuscularly, or subcutaneously, but the study of Fisher et al.[60] suggests that the intravenous route is best. Insulin treatment should be started by infusion of a bolus of 10 units and then use of a pump or other infusion method to deliver about 8 units per hour. If an intravenous drip is to be employed, 100 units of crystalline insulin may be mixed with 250 mL of 0.9% saline solution and infused at a rate of 20 mL (8 units) per hour. Serial estimates of blood glucose must be made. Blood glucose should fall at a rate of about 20% per hour if an average response is attained. In very rare cases there may be insulin resistance, and very large doses of insulin may be necessary.

Fluids and electrolytes. The patient with DKA presents with hypertonic body fluids. The use of isotonic replacement therapy permits some water to reach the intracellular space and also allows for rapid expansion of intravascular volume. Urine flow is usually a good determinant of adequate replacement, though some patients may undergo a period of oliguria (low urine output) and care must be taken to assess blood volume to avoid risk of overhydration. If sodium bicarbonate is used, the pH should rise rapidly and out of proportion to the small amounts of bicarbonate administered. When the pH rises, there will be some impairment of dissociation of oxygen from oxyhemoglobin. For this reason some argue against early use of bicarbonate. With or without bicarbonate therapy, frequent estimations of acid-base status should be made.

Another determinant in oxygen delivery is the concentration of 2,3-DPG. In DKA it is low and often reflected by a serum inorganic phosphate that may be below 10 mg/L. Organic phosphate is not regenerated in a short time, and so depletion of 2,3-DPG may be prolonged. Nevertheless its regeneration is stimulated by the use of phosphate salts. Since nearly all patients with DKA have phosphate and potassium depletion,[61,62] it seems wise to give buffered potassium phosphate. Potassium begins to fall in most patients within an hour or two after saline solution and insulin are given. Serial measurements of serum potassium are necessary, and the effect of potassium infusion is estimated after one takes serial electrocardiograms and measures the serum potassium as well. It is rarely necessary to infuse potassium phosphate at a rate greater than 20 to 40 mEq per hour. In rare instances clinical emergencies of potassium or phosphate depletion may be identified. Potassium depletion is often heralded by muscle weakness that begins first in the hands and can involve the muscles of respiration, causing ventilatory distress. Phosphate depletion may not be easily identified at onset but may be associated with apnea, cerebral edema, hemolytic anemia, and even death.[63]

Hyperglycemic hyperosmolar nonketotic coma (HHNC)

Biochemical features. The complication hyperglycemic hyperosmolar nonketotic coma is very similar to DKA. In fact, it escaped specific description until 1957, and cases of HHNC were reported along with those of DKA until that time.[64] The precise reason why one patient develops HHNC and another develops DKA is not known. Studies reporting levels of insulin, glucagon, growth hormone, and cortisol show that concentrations of these hormones are about the same in the two acute disorders.[65] The major differences are that in HHNC (1) serum osmolality is somewhat higher (usually 350 to 450 mOsm/kg), (2) serum glucose is almost always above 8000 mg/L, (3) patients tend to be older (though HHNC has been described in youth), and (4) the duration of symptoms before presentation is longer. In addition, many HHNC patients are not known to have diabetes at presentation, and some of them are manageable without insulin after recovery.[66] Ninety-five percent of all HHNC patients have type 2 diabetes.

Clinical features. Dehydration is the cardinal feature of HHNC. Although electrolyte loss is severe, deficiency is extracellular and intracellular water is dominant. The eyeballs may appear sunken, the abdominal wall may feel doughy, and the mucous membranes become dry. Hypovolemia and orthostatic hypotension, cool skin, and an undetectable or thready pulse are found. The level of unconsciousness appears to be directly related to the increase in serum osmolality.[67] In most instances the underlying cause of HHNC is masked. Common precipitating diseases include myocardial infarction, pulmonary embolism, acute pancreatitis, sepsis, renal failure, and cerebral infarction. Very often recent use of a drug known to decrease insulin secretion such as propranolol, hydrochlorothiazide, or phenytoin is identified.[68]

Diagnosis and laboratory findings. The discipline of obtaining emergency blood analysis in all extremely ill patients has made the diagnosis of HHNC easy. In nearly all instances the blood glucose exceeds 8000 mg/L. It should be pointed out that semiquantitative test-paper assessment of whole blood glucose does not always permit accurate definition of this very high level. When an automated quantitative device is employed, the range may not cover this level and sometimes analysis must be repeated. In contrast to the pH in DKA, the blood pH is either normal or slightly below normal, though in rare instances lactic acidosis resulting from very poor perfusion may be present. Qualitative analysis for acetoacetic acid can indicate only a trace of the keto acid. The serum sodium may be high, normal, or low.[69] It has been stated that serum sodium falls 2 mEq/L for each 1000 mg/L of blood glucose. As in DKA, the plasma potassium is usually normal or elevated even though there is nearly always *total* body potassium depletion. When the diagnosis is suspect, it is helpful to measure serum osmolality. If osmolality cannot be measured, a calculation is reasonably accurate. The formula is:

$$\text{mOsm/L} = 2\,(\text{Na} + \text{K}) + \frac{\text{Glucose}}{180} + \frac{\text{BUN}}{28}$$

where Na and K are in mEq/L and glucose and BUN are in mg/L. Although urea freely permeates intracellular fluid, calculated osmolality using urea expresses true osmolality better than an equation that omits it. Many patients with HHNC have borderline or worse renal function, and the hypovolemia typical of the syndrome further reduces glomerular filtration so that blood urea nitrogen levels exceeding 1000 mg/L are common. Other indices of dehydration, such as a high hematocrit, can be found.

Treatment. There are only a few differences in the therapy of HHNC and DKA. Most experienced clinicians advocate the use of 0.45% saline solution rather than 0.9%. Most recommend that insulin be given intravenously rather than intramuscularly because they believe that absorption from a subcutaneous injection site in a poorly perfused patient is uncertain. The dosage of insulin is not critical. Regimens using a 20-unit intravenous bolus and an hourly infusion of 10 units are popular.[70] Because most HHNC patients are elderly and their cardiac and renal function is compromised, care must be used to ensure that volume overload does not occur. Central venous pressure monitoring is useful and if such monitoring is available, measurement of pulmonary capillary wedge pressure is even better. During therapy a search for occult, precipitating causes of HHNC must be made. Blood cultures, enzyme measurement for myocardial infarction and pancreatitis, and repeated neurological evaluations are routine. It should be stressed that HHNC alone can give rise to severe but usually transient neurological changes. Although it may be argued that the use of hypotonic saline solution might augment the risk of cerebral edema because brain osmolality is very high, there is little clinical proof that this is so.[71]

The best assessment of clinical response is improvement in mental activity, but sometimes the patient may remain in a stupor for a prolonged time even though hyperosmolality and electrolyte loss have been corrected. Hypokalemia and its attendant risk of cardiac arrhythmias and respiratory paralysis is as much a problem in HHNC as it is in DKA. The use of intravenous potassium is often necessary because oral electrolytes may not be useful because gastric atony and intestinal absorption deficiency occur frequently. It is wise to obtain the values for serial serum potassium levels. The best guide to chemical improvement is the rate of fall of plasma glucose. Hourly measurements are required.

Prognosis. Age, severity of mental obtundancy, duration of HHNC before diagnosis, severity of underlying disease or diseases, and quality of treatment are all factors in prognosis. Contemporary reports cite mortality as high as 70%, but most hospitals provide data showing a survival rate of about 70%.

Insulin-induced hypoglycemia

Biochemical features. It is almost certain that all persons with type 1 diabetes suffer frequent episodes of insulin-induced hypoglycemia. The majority experience at least one admission to an emergency room or hospital because of severe hypoglycemia, and a very few suffer permanent neurological damage or die. The latter morbidity is almost always the result of many hours of cerebral glucose deprivation, though injury caused by failure to operate an automobile or other mechanical equipment properly doubtless occurs in a few patients. Fortunately, the body's protective response to hypoglycemia is powerful and multifaceted. The primary initial reaction is release of epinephrine. Cerebral stimulation of the adrenal medullary results in epinephrine release within a brief time after glucose falls, and epinephrine secretion is greater when the rate of fall and depth of hypoglycemia are most severe.[72] Other counterregulatory responses include glucagon release and, over a longer time, continued cortisol and growth hormone hypersecretion. As discussed previously, these hyperglycemic hormones tend to mobilize both stores of glucose and increase gluconeogenesis to counter the fall in blood glucose.

Clinical features. The pattern of response to hypoglycemia is important. Most patients subconsciously sense hypoglycemia and if awake tend to search for food even before epinephrine-generated symptoms such as excessive sweating, palpitation, tachycardia, tremor, and weakness may be noted. Type 1 patients with autonomic neuropathy may escape the epinephrine-related warning and have been described as developing cerebral glucose deprivation as a primary event.

Neurological and behavorial aberrations are the most unpredictable manifestations of insulin-induced hypoglycemia. It is not uncommon for a physician to admit a patient to an emergency room with a tentative diagnosis of brainstem infarction only to find in a few minutes that the patient's blood glucose is below 100 mg/L and it has fully responded to a few hundred milliliters of 5% glucose in water. The same patient may escape all overt neurological findings and drive to another city, park his car, eat a meal in a restaurant, and recover but suffer retrograde amnesia. Permanent neurological damage is more likely in the diabetic person who is allowed to remain hypoglycemic for an extended number of hours. This is especially true if the hormonal responses are dulled by medication such as those causing adrenergic blockade and the circulation is compromised by hypotension, congestive heart failure, or atheromatous cerebral disease. In some instances hypoglycemia may result in motor or sensory abnormalities of peripheral nerves or plexuses.

Diagnosis. The diagnosis of insulin-induced hypoglycemia is easily made. It is probably a good investment for the insulin-taking, insulin-sensitive diabetic person to purchase a Dextrometer or have his close associates and family learn to use Dextrostix. Measurement of blood or plasma glucose by any rapid method is adequate. One should remember that blood glucose measured several minutes after the onset of clinical hypoglycemia may not reflect a much lower glucose than may have been present before the onset of symptoms. One must also appreciate that many persons with type 1 diabetes behave quite normally despite having blood glucose levels in the range of 250 to 300 mg/L for short periods of time.

Treatment. Therapy for hypoglycemia depends on a given situation. If there is some mental obtundation but the patient can swallow, very small amounts of oral glucose concentrate or orange juice will be all that is required. One should remember that rapid response should be followed by anticipation of a possible second episode. Glucagon, 1 mg administered subcutaneously, is a rapid and safe way to reverse most episodes of hypoglycemia. Glucagon powder must be mixed with diluent and injected with an insulin syringe. Many firemen and emergency squads handle this emergency with skill. Intravenous glucose given as a 50% infusion per 25 mL syringe is a time-honored maneuver. In some patients a lack of good visible veins causes difficulty.

Hypoglycemia secondary to oral sulfonylurea drug therapy may pose a serious and perhaps lethal threat.[72] Prolonged and profound hypoglycemia may develop in the persons with type 2 diabetes, especially if there is chronic debilitation, lack of food intake, heavy use of alcohol, or use of a drug that might inhibit epinephrine effect. Hospitalization with long-term intravenous glucose and electrolyte therapy may be required.

When episodes of hypoglycemia become frequent and

more severe, consideration of an added defect in glucose homeostasis should be made. Adrenal or thyroid failure, chronic renal or hepatocellular disease, hypopituitarism, or intentional self-overdose of insulin are possibilities.

CHRONIC COMPLICATIONS OF DIABETES

Were it not for the long-term complications of diabetes, the disease would not impose its enormous threat to the quality and length of life.[73] Well-defined but poorly understood vascular and nonvascular disorders are potential in all the diabetic syndromes but exercise their greatest morbidity in the insulin-dependent who develops diabetes before puberty.

Before the availability of insulin, diabetic ketoacidosis was identified as the cause of death in at least half of those diabetic persons seeking treatment at the Joslin Clinic in Boston. A contemporary report from that institution indicates that 77.9% of deaths are attributable to vascular disease, whereas cancer causes mortality in 9.5% and infections in 5.8%.[74] Persons with type 1 diabetes often develop coronary, carotid, cerebral, and peripheral occlusive disease before they reach 40 years of age. Furthermore, involvement is more general and manifestations are more severe in the diabetic than the nondiabetic. It is probable that microvascular disease, which seems almost inevitable in IDDM patients, is related to atherosclerosis. However, microvascular disease is specific for diabetes and seems related to both the duration of diabetes and its severity in terms of control of hyperglycemia.[75] It has been pointed out that there are as yet no prospective clinical studies that thoroughly convince the skeptical scientist that poor control results in earlier and more severe vascular disease.[76] One reason for this is that complete control of the metabolic manifestations of total diabetes is not possible even with devices such as open-loop insulin pumps.[77] Nevertheless, animal data increasingly support the axiom that the early microvascular lesions of diabetes are reversible when diabetes is curable.[78]

Microangiopathy

"Microangiopathy" is an anatomical term used to describe degenerative lesions in small blood vessels such as capillaries. It is typified by thickening of the basement membranes of vessel walls.[76] The changes occur in both type 1 and type 2 diabetes, but not all persons with diabetes have basement membranes that are thicker than those of nondiabetic persons of the same age.[79] The chemical composition of basement membrane material remains a controversial subject, but it is a collagen-like protein and probably contains more glycosylated material in diabetic compared to nondiabetic persons. In diabetes the two organs critically involved are the eye and kidney.

Retinopathy

Diabetic retinopathy is the chief cause of new cases of blindness in adult Americans. Persons with diabetes have a risk of blindness that exceeds that in nondiabetic persons 13 times.[80] Approximately 10 out of every 100,000 adults are blind because of diabetic retinopathy. It is important to know that retinopathy is subdivided into two major classes: disease that is confined to the retina itself and disease that extends from the retina into the vitreous.[81,82] Diabetes is also associated with cataracts, which cause loss of vision at a rate that is greater than that for nondiabetic persons.

Diabetic nephropathy

Chronic renal failure is a leading cause of death in patients with type 1 diabetes. Nearly one fourth of the patients participating in chronic renal dialysis programs have diabetes. Although many separate but related lesions contribute to kidney failure in the diabetic person, glomerulosclerosis is the most common, most progressive, and most characteristic.[83] If patients attending a large diabetic clinic are subjected to renal biopsy, the majority will have microscopic evidence of glomerulosclerosis. Morphological changes begin in persons with type 1 diabetes after 10 or more years of diabetes and are found in a majority of them after 20 years. Persons with type 2 diabetes often develop similar lesions, but dating their onset to development of diabetes is difficult because these persons often have undiagnosed diabetes for long periods of time. As renal disease progresses, proteinuria is a common laboratory finding.

Renal failure modifies the metabolic setting of the insulin-dependent diabetic person. The daily insulin dose may decrease because the rate of degradation of insulin by the diseased kidney is diminished. There may be some reduction in gluconeogenesis and a trend toward lower fasting blood glucose. Sensitivity to glucagon may also be decreased and postprandial glucose may be higher as the transfer of glucose from blood to the periphery lessens. Modifications of insulin dosage and timing, meal content and timing, and exercise pattern and protection against hypoglycemia may be required.[84]

Nerve diseases

Almost every abnormality theoretically possible can occur in the peripheral and autonomic nervous system of the person with diabetes.[85] Some lesions are probably secondary to diabetic vasculopathy, whereas others appear metabolic in pathogenesis. Some type of neuropathy is present in nearly all diabetic persons who have had the disease for more than 20 years.

Mechanisms of nerve damage include (1) increased uptake of glucose by Schwann cells in peripheral nerves, resulting in enzymatic conversion of glucose to fructose with intracellular osmolar damage secondary to "storage" of sorbitol[86]; (2) defective synthesis of myelin; (3) deficiency of myoinositol[87]; (4) excessive glycosylation of proteins; and (5) infarction of a nerve secondary to anatomical abnormalities in the arterioles.[88]

Diabetic nerve disease may be grossly classified into

three different categories: (1) peripheral neuropathy, (2) autonomic neuropathy, and (3) mononeuropathy and mononeuropathy multiplex.

Peripheral neuropathy. Peripheral neuropathy represents the most characteristic and predictable nerve syndrome. It usually involves the most distal segments of long nerves, is found in symmetric distribution, and manifests mostly sensory symptoms though both motor and sensory fibers are involved. Signs of peripheral neuropathy are present in about two thirds of all persons with type 2 diabetes.[89] The first symptom is often sensory: the patient complains of severe hypersensitivity of the toes and feet. Loss of pain sense coupled with loss of position sense results in destructive lesions in the skin, ligaments, and bones if motor activity is preserved.[90]

Autonomic neuropathy. Disturbances in function of the autonomic nervous system are very frequently found in patients with peripheral neuropathy.[91] Involvement of the pupils, cardiovascular system, gastrointestinal system, and genitourinary system generates symptoms that cause enormous discomfort. Excessive sweating of the face is actually rather common in diabetic-autonomic neuropathy.

Cardiac neuropathy is potentially a great danger.[92] Loss of nerve control of reflex cardiac responses can occur first. For example, there may be a resting tachycardia but no increase in heart rate after exercise. There is evidence that patients with cardiac denervation are at risk for sudden death.[93]

Orthostatic hypotension is a common manifestation of autonomic neuropathy. The initial symptom is light-headedness or sometimes fainting when the patient rises from bed in the morning or rises to urinate during the night. Typical of this sympathetic nervous system damage is failure of the heart rate to increase when blood pressure falls.

Impotence in the diabetic man may be the most disturbing manifestation of diabetic autonomic neuropathy.[94] Its earliest effect may be inability to sustain an erection. Later, erection may be impossible.

Mononeuropathy and mononeuropathy multiplex. Mononeuropathy is characterized by involvement of a single peripheral or cranial nerve. If multiple nerves are involved, the term ''multiplex'' is used. It seems clear that the cause of mononeuropathy is ischemia that results in infarction of the central part of the nerve with sparing of the periphery.[88]

TREATMENT OF DIABETES SYNDROMES

Just as there is a spectrum of varieties of diabetes, there are widely divergent approaches to therapy of diabetes. Since neither type 1 nor type 2 diabetes is curable, individualized but constantly modifiable general measures must be generated. It is very important to ensure long-term monitoring of diabetes, but it is imperative to not allow the treatment of diabetes to completely dominate and thereby greatly change the life-style of a given patient. Although some goals in therapy may be sharply individualized, a few goals should be sought in all patients, such as (1) making an intense, lifelong effort to maintain blood glucose as near to normal as possible, (2) striving to reach ideal body weight, and (3) participating in lifelong self-education about diabetes.

Control of hyperglycemia

There has long been a philosophical dispute as to whether control of hyperglycemia deters or prevents the microvascular complications of diabetes. Recently, evidence obtained from animals made diabetic with streptozotocin and then cured of their diabetes by pancreatic transplantation strongly supports the concept that control inhibits retinal and renovascular disease.[95] The American Diabetes Association has endorsed measures that offer improvements in diabetes control.[90] At present, it is appropriate to state that no proof of establishment of long-term normoglycemia has been achieved without the use of measures that are undesirable. For example, normal plasma glucose control and freedom from the need for insulin injections has been gained in a patient who underwent pancreatic transplantation and renal transplantation as well. However, to avoid organ rejection, this person must take immunosuppressive therapy.

Despite the debate, it seems sensible for the physician to attempt to establish the very best blood glucose control possible in each of his patients. Better diabetes control is achieved by programs that (1) improve the patient's knowledge of diabetes, (2) enable him to cook and ingest at appropriate times a diet that is designed especially for him, (3) allow him to understand and self-administer insulin or oral sulfonylurea drugs in a manner that best fits his specific needs, (4) use timed periods of physical exercise to both control hyperglycemia and avoid exercise-related hypoglycemia, and (5) enable him to monitor his own blood glucose variation with whatever blood or urine glucose measurement devices offer maximum ease, comfort, and reliability.[96]

Education

There is probably no other chronic disease comparable to diabetes in the hour-to-hour need for self-treatment over a lifetime. If the patient is to make intelligent and proper decisions, he must attain and maintain a knowledge of diabetes that in a specific sense (knowledge about *his own diabetes*) exceeds that of his physician.

Diet

In nonspecific terms, ''diet'' may simply designate the amount, proportion of carbohydrate, fat, and protein, and times that a *normal* person should eat.[97,98]

Calories. The number of calories ingested by a diabetic subject should be equal to the calories expended if he is of ideal weight, less if he is overweight, and more if he is underweight. Trial and error are part of establishing appropriate caloric intake, but such intake is usually balanced among the major food classes.

Carbohydrates. In most instances about 50% of the day's calorie intake should consist of carbohydrate.[99] At least 85% of this should be complex carbohydrate. Simple carbohydrates such as sucrose, glucose, fructose, and galactose are rapidly absorbed and quickly raise blood glucose. Complex carbohydrates, especially those contained in food with a high fiber content, tend to be absorbed more slowly and appear to require less insulin for storage as glycogen or conversion to triglycerides. A few diabetic persons develop significant increases in very low density-lipoprotein levels when given a 50% or greater carbohydrate diet.[100] In some persons with diabetes the hyperlipoproteinemia is temporary, but in others it may persist. It is therefore appropriate to obtain serum triglyceride levels after 4 to 6 weeks and, if they are elevated, reduce carbohydrate intake.

Fat. For many years a diet with relatively high fat and cholesterol content was recommended for the treatment of diabetes.[101] The concept of using a low fat, low-cholesterol diet has gained enthusiasm. The reasoning, of course, is that highly saturated fat, high-cholesterol diets are atherogenic and vascular disease is the leading cause of death in diabetes. It is usually recommended that 35% of the day's caloric intake be constituted of fat and that about one third of the fat be polyunsaturated. A recommendation that cholesterol intake be limited to less than 300 mg/day is also usually made.[100] Emerging clinical evidence possibly indicates that the newer diets significantly lower LDL cholesterol and improve the ratio of LDL cholesterol to alpha-lipoprotein cholesterol (HDL).

Protein. Protein intake is closely linked to fat intake because most foods rich in protein contain just about as many fat calories as protein calories. A diet consisting of about 15% protein calories is satisfactory in most patients. It is quite possible for a healthy person to stay in normal nitrogen balance with a protein intake of as little as 50 g per day. Many nutritionists recommend a minimum protein intake of 0.5 g per pound of ideal body weight. In the 158-pound man this would amount to 75 to 80 g/day.

Diet prescription. The frequency of meals and the number of calories taken at each meal must be carefully integrated with the type of diabetic pattern of insulin use. Both the type and amount of insulin administered must be taken into account.

The person with type 1 diabetes must carefully integrate the frequency of meals and the number of calories taken at each meal with the prescribed pattern of insulin use. Both the type and amount of insulin administered must be taken into account.[102]

The person with type 2 diabetes should time his meals so that they are as far apart as possible beause he does not wish to provide rapidly successive or maximal challenges to his less-than-ideal capacity to secrete insulin. Percentages of the daily diet need to be modified by trial and error so that best daily blood glucose may be established.[103]

Exercise

It has become evident that exercise rivals diet in importance in both type 1 and 2 diabetes.[104] In type 1 diabetes, exercise augments the rate of entry of glucose into insulin-dependent tissues, possibly by increasing the number of insulin receptors. A significant problem in insulin-treated patients is that insulin is released from an injection depot at a fairly constant rate. If blood glucose disposal is increased by muscular exercise, the effects of insulin action become accentuated and there is risk of insulin-dependent hypoglycemia. For this reason it is recommended that the insulin-dependent diabetic person plan his activities so that exercise is countered by delivery of glucose from gastrointestinal absorption of food: after either a meal or a snack. In type 2 diabetes exercise is not associated with risk of hypoglycemia but improved control is clearly demonstrated if exercise is carried out at regular intervals. It is theoretically possible that exercise done 30 to 120 minutes after meals might decrease the peak postprandial blood glucose. Certainly, since sedentary activity is a well-identified pathogenetic factor in type 2 diabetes, caloric expenditure brought about by exercise-related weight loss should increase insulin-receptor number and improve diabetes control.

Oral sulfonylurea drugs

It is probable that more patients are being treated with oral sulfonylurea compounds than with insulin. It is now clear that sulfonylureas are useful only in persons with type 2 diabetes who have moderate or better capacity to secrete endogenous insulin and who comply with their diet.[105]

Pharmacology. The sulfonylurea compounds presently used in the world market include the "first-generation" drugs tolbutamide, chlorpropamide, tolazamide and acetohexamide. When they are compared to the "second-generation" agents glyburide (glibenclamide), glipizide, and glibornuride, there is an enormous difference in the magnitude of dosage (Fig. 30-15). All compounds are rapidly absorbed, and 50% absorption is probably average. All bind to serum proteins but dissociate easily. Major differences among the drugs are related to their routes of metabolism and excretion. Tolbutamide is rapidly methylated by the liver, and its inactive metabolite is excreted in the urine. Its biological activity is believed to be only 6 to 8 hours. Tolazamide is more slowly metabolized to six compounds. Three of these are weakly active as hypoglycemic agents, and thus the bioactivity is estimated to be about 24 hours. Acetohexamide has an activity between that of tolbutamide and tolazamide because one of its metabolites, L-hydroxyhexamide, has a potency about 2.5 times that of the parent drug. Chlorpropamide is metabolized to several

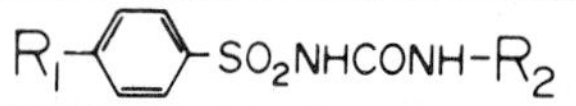

R_1–⟨benzene ring⟩–$SO_2NHCONH$–R_2				
FIRST-GENERATION COMPOUNDS				
NAME	R_1	R_2	DOSAGE RANGE (mg)	TABLET SIZE (mg)
TOLBUTAMIDE	CH_3-	$-(CH_2)_3CH_3$	500 – 3000	500
CHLORPROPAMIDE	Cl –	$(CH_2)_2CH_3$	100 – 500	100, 250
TOLAZAMIDE	CH_3-	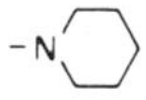 –N⟨ring⟩	100 – 750	100, 250
ACETOHEXAMIDE	CH_3CO-	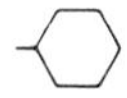 –⟨ring⟩	500 – 1,500	250, 500
SECOND-GENERATION COMPOUNDS				
NAME	R_1	R_2	DOSAGE RANGE (mg)	TABLET SIZE (mg)
GLIBENCLAMIDE	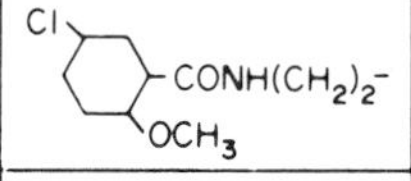 Cl, OCH_3–⟨ring⟩–$CONH(CH_2)_2-$	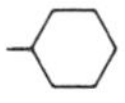 –⟨ring⟩	2.5 – 20	5
GLIBORNURIDE	CH_3-	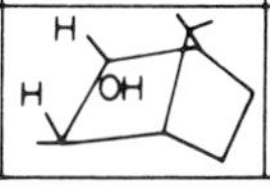 H, H, OH ⟨bicyclic ring⟩	12.5 – 100	12.5

Fig. 30-15. Sulfonylurea compounds given orally comprise two classes. The older drugs have dosage ranges of 100 to 1500 mg per day, and their R_1 side chains are simple structures. The second-generation drugs have similar composition but more complex R_1 side chains. Their average dose is about a hundredth that of the first-generation agents.

compounds, but the state of their activity is not known. Nevertheless, its biological activity is prolonged to over 36 hours. The dose of each of the presently available second-generation compounds is 1/50 or 1/100 that of the older agents. Each is reported to have a biological hypoglycemic effect that lasts about a day, and the structure of each is somewhat more complex than the parent sulfonylureas. Glyburide is converted to two virtually inert metabolites by the liver and about half is excreted by the kidneys, whereas the other half appears in biliary secretions and leaves the body in the feces. Glipizide is also converted to inactive metabolites by the liver, and nearly 90% is excreted through the gastrointestinal tract. Glibornuride is metabolized to six compounds, and one has modest hypoglycemic effects. It is excreted 65% by the kidneys and 35% through the intestine.

Action. There are abundant data in normal and diabetic humans and animals to show that each of the sulfonylureas is capable of stimulating the beta cells to secrete insulin. Sulfonylureas have been shown to bind to the beta-cell membrane but their insulin secretagogue role is apparently enhanced when a meal increases plasma glucose compared to the effect of the drugs during the fasting state. It seems probable that a major hypoglycemic effect of the sulfonylureas is manifested by an extrapancreatic effect.[106] There is clear evidence that diabetic persons rendered euglycemic by diet plus sulfonylurea treatment are more sensitive to intravenous exogenous insulin than persons with type 2 diabetes who are made euglycemic by dietary measures alone. Multiple studies with first-generation sulfonylureas indicate that these drugs evoke about a 50% increase in insulin-receptor number in patients who respond. Since a major defect in type 2 diabetes is decreased insulin-receptor number, there is logic to the notion that their major pharmacological effect may be on receptors.[107]

Side effects. Cardiovascular complications have been described by the controversial University Group Diabetes Program.[108] Phenformin has been withdrawn from the market because of its association with lactic acidosis.[109]

Antidiuretic hormone–like actions have been described with chlorpropamide but not with tolazamide, acetohexamide, or glyburide. It is believed that chlorpropamide can both stimulate ADH release and potentiate its action. An anti-ADH effect has been ascribed to agents with saturated cyclohexyl rings such as glyburide. Rare instances of hypoosmolality and hyperosmolar urine have been reported and chlorpropamide has been used to treat patients with incomplete diabetes insipidus.[105] Potent uricosuric activity has been ascribed to acetohexamide. Its molecular similarity to uricosuric drugs has been given as the most probable cause for this effect.

The disulfiram (Antabuse) syndrome, characterized by

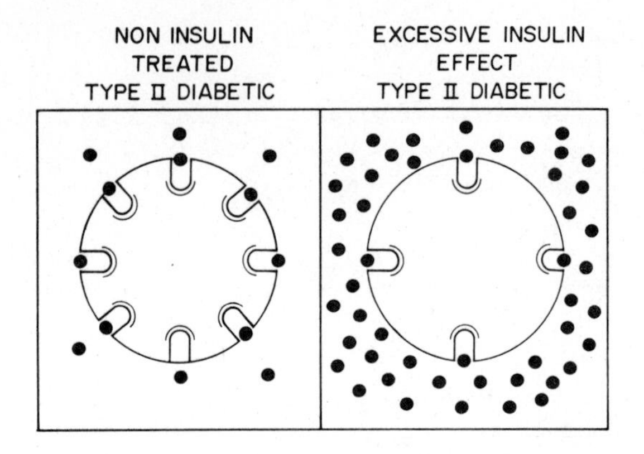

Fig. 30-16. *Left,* schematic depiction of eight insulin receptors upon surface of an insulin-sensitive cell surrounded by a near ideal concentration of free insulin. *Right,* reduction of receptors from eight to four on surface of cell engulfed in a milieu of excessive exogenous insulin. Here receptor number has become reduced because its intrinsic behavior has been influenced by too much insulin.

severe facial flushing in patients who drink alcohol, has been described in rare subjects taking first-generation sulfonylureas. It is possible that drugs like chlorpropamide may inhibit aldehyde dehydrogenase, a chief enzyme involved in the metabolism of alcohol.[110]

Efficacy. The sulfonylurea compounds have been generally abused by physicians who expect them to cause hypoglycemic activity in persons with type 2 diabetes who fail to diet, lose weight, exercise, or make an attempt to monitor their blood glucose. The agents are effective in most lean compliant patients and in some persons who are overweight but are able to diet. There is no acceptable rationale for use of a sulfonylurea in the patient who constantly displays hyperglycemia despite taking the drugs. The same lack of efficacy is usually found with insulin therapy in noncompliant patients. It is probable that most obese diabetic persons have a reduced number of insulin receptors. If exogenous insulin is administered to these patients, down regulation (further reduction in receptors) is likely and, even worse, poor control of hyperglycemia might occur (Fig. 30-16).

Insulin

Insulin use is by definition essential for treatment of type 1 diabetes and often is used in selected patients with type 2 diabetes. More than 1 million Americans inject insulin each day.[111]

The insulins are classified by time and related to the onset, offset, and peak ability to lower blood glucose (Table 30-4). Crystalline zinc insulin acts almost immediately. Its peak effect occurs from the second to fourth hour after subcutaneous injection. NPH insulin has an onset in hypoglycemic activity within 2 hours, peaks at 6 to 8 hours, and lasts for about 27 hours. Lente insulin has activities similar to NPH insulin. There are two very late onset and prolonged insulins: ultralente insulin and protamine zinc insulin.

In a few patients, successful control of total diabetes is achieved by use of an insulin pump.[112] This device consists of a power-driven syringe activated by a timing device and controls that may be adjusted so that very small amounts of regular insulin may be injected subcutaneously to establish basal insulin concentrations.[113] Although the presently available pumps are somewhat bulky and require intermittent electrical charging and need to be connected by a plastic tube to a needle placed subcutaneously, there is a rapid evolution of these devices, and there is hope that an artificial beta cell will be developed. Such a pump would contain a mechanism to measure blood glucose at frequent intervals, compute its significance in an individual patient, and then supply an appropriate dose of insulin.

FUNCTION TEST: ORAL GLUCOSE TOLERANCE TEST

In medical practice the diagnosis of diabetes mellitus is established by demonstration of the fact that the fasting plasma glucose exceeds 1400 mg/L in adults and children who are free from intercurrent illness, are ambulatory, and have been ingesting a normal diet. It is customary for one to confirm the first elevated value by doing another glucose test. In other persons the fasting glucose may be normal or borderline. In this case several measurements of plasma glucose that exceed 2000 mg/L done 2 hours after meals are sufficient to establish the diagnosis (Table 30-5).

Table 30-4. Varieties, modifying factors, onset, peak action, and hours of duration of action of contemporary insulin in the United States

Type of insulin	Action	Buffer	Protein	Peak (hours)	Duration (hours)
Regular (crystalline)	Rapid	None	None	1-2	5-6
Semilente	Rapid	Acetate	None	1-2	12-16
NPH (isophane)	Intermediate	Phosphate	Protamine	2-8	24-28
Lente	Intermediate	Acetate	None	2-8	24-28
PZI (protamine zinc)	Long	Phosphate	Protamine	8-12	36
Ultralente	Long	Acetate	None	8-12	36

Table 30-5. Criteria for diagnosis of diabetes using oral glucose tolerance test (glucose in plasma, mg/L, measured before and after a 75 g oral glucose load)

Time	Normal adults	Impaired glucose tolerance	Diabetic adults	Gestational diabetic patients	Diabetic children*
Fasting	< 1150	< 1400	> 1400	> 1050	> 1400
1 hour	< 2000	< 2000	>2000	> 1950	> 2000
2 hours	< 1400	> 1400 to 2000	> 2000	> 1650	> 2000
3 hours	< 1400	> 1400 to 2000	> 2000	> 1450	> 2000

*Glucose load: 1.75 g/kg.

When fasting and postmeal values are borderline, a glucose tolerance test is required.

Dose. It has been found that a dose of 75 g of glucose (or commercially prepared carbohydrate equivalent) is most discriminative in adults. For children the dose is 1.75 g of glucose per kilogram of ideal body weight.

Procedure. It should be established that the patient has fasted for at least 10 hours but not more than 16 hours. The patient should be seated, and smoking should not be permitted throughout the test. One should also determine that the person is not taking a drug that interferes with carbohydrate metabolism or with laboratory methods that measure serum glucose. A fasting blood sample should be collected and the glucose drunk in a solution of flavored water containing no more than 250 g/L. Zero time is the beginning of glucose ingestion, and the solution should be drunk within 5 minutes. Blood samples should be drawn at 1, 2, and 3 hours. Venous plasma glucose measurement is preferred. Accepted methods include glucose oxidase, hexokinase, and *o*-toluidine.

Interferences. A number of drugs have been reported to cause an impaired glucose tolerance test (see box in left column). Both the physician and laboratory personnel should be aware of such possible in vivo interference when they are attempting to interpret the test results.

Results. Table 30-5 and Fig. 30-17 indicate the temporal pattern of results obtained for normal, diabetic, and impaired glucose tolerance tests. It should again be emphasized that a single, abnormal glucose tolerance test does not confirm a diagnosis of diabetes mellitus. Furthermore, even a patient with a confirmed abnormal test will not necessarily become clinically diabetic.

Drugs reported as capable of causing impaired glucose tolerance

- Diuretics
  - Chlorthalidone
  - Clonidine
  - Diazoxide
  - Furosemide
  - Methalazone
  - Thiazides
- Hormones
  - ACTH
  - Glucagon
  - Glucocorticoids
  - Oral contraceptives
  - Somatotropin
  - Thyroxine
- Psychoactive agents
  - Chlorprothixene
  - Haloperidol
  - Lithium carbonate
  - Phenothiazines
  - Tricyclic antidepressants
- Catecholamines and neurologically active agents
  - Epinephrine
  - Isoproterenol
  - Levodopa
  - Norepinephrine
- Other
  - Phenytoin
  - Indomethacin
  - Nicotinic acid

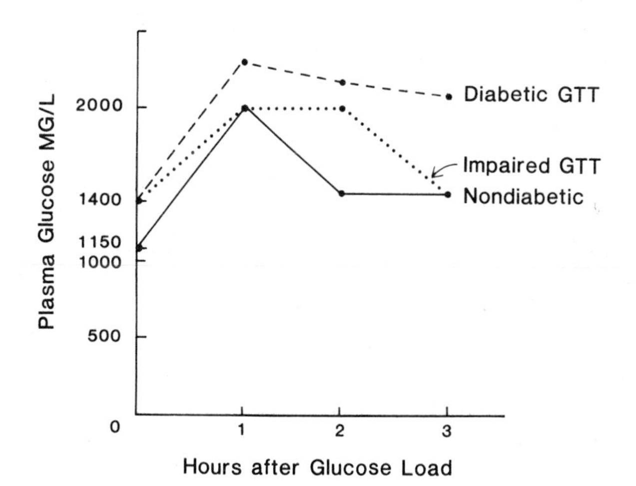

Fig. 30-17. Examples of glucose tolerance test curves (GTT). Plasma glucose (in milligrams per liter) versus hours after glucose load for diabetics, *dashes;* those with impaired glucose tolerance, *dots;* and normal subjects, *solid.*

Table 30-6. Changes of analyte in disease

	Acute presentations*			
Serum analyte	**Diabetic ketoacidosis**	**Hyperosmolar coma**	**Hypoglycemic coma**	**Insulinoma**
Glucose	↑	8->10 g/L	<0.5 g/L	N or ↑
Insulin	N or ↑	low N	N or ↑	↑
Triglycerides	↑	N	N	N
LDL	N	N	N	N
VLDL	↑	N	N	N
Na^+†	↑ or N†	↑ or N†	N	N
K^+	↑ or N†	↑ or N†	N	N
Urea	↑	↑	N	N
Hemoglobin A_1 (glycosylated hemoglobin)	↑	↑	N or ↑	↓ or N
pH	↓	N	N	N
HCO_3^-	↓	N	N	N
Ketones (acetoacetate β or β-OH)	↑	N	N	N

**N,* Normal; ↑, elevated; ↓, lowered.
†Serum levels of Na and K are dependent on the state of hydration of a patient. However, *total* body K is always in deficit in these conditions.

CHANGE OF ANALYTE IN DISEASE (Table 30-6)

Glucose. The degree of elevation of serum and urine glucose in diabetic persons will vary from person to person depending on the degree of effectiveness of treatment. Diabetics in relatively good control will have normal to slightly elevated serum glucose with no glucosuria. As control worsens, blood glucose levels rise until, above 1800 to 2000 mg/L, glucosuria will appear. Serum glucose values from 8 to 10 g/L are usually associated with nonketotic, hyperosmolar comas, whereas glucose values less than 0.5 g/L are associated with hypoglycemic coma.

Fasting blood glucose (FBG) is used as possible diagnostic evidence of diabetes. However, this test should be repeated along with a glucose tolerance test to ensure the proper diagnosis.

Ketones. Serum ketones are usually a hallmark of poorly controlled diabetes. The presence of ketones, especially acetoacetic acid, is an important laboratory finding in diabetic ketoacidosis. In some cases the screening test for ketones (nitroprusside test) can be negative for acetoacetic acid because the major ketone present is beta-hydroxybutyric acid. The latter analyte can be detected by more sophisticated procedures. Ketones are normally not detected in hypoglycemic or hyperosmolar comas.

pH. Decreased blood pH is a diagnostic finding of diabetic persons presenting in DKA. pH's as low as 7.0 to 7.3 are not uncommon, whereas lower pH's present greater risk for cardiovascular involvement. A pH less than 6.9 has a poor prognosis.

HCO_3^- or total CO_2. Serum HCO_3^- or total CO_2 are usually decreased in DKA but not in other acute presentations of diabetes unless renal acidosis is also present. Very low concentrations of HCO_3^- (high base deficit) can require bicarbonate therapy.

P_{CO_2}. Blood P_{CO_2} follows the HCO_3^- concentrations, usually being decreased in DKA.

Insulin and C-peptide. Insulin levels in persons with type 1 diabetes are quite low, reflecting the causal factors of this disease. Persons with type 2 diabetes show variable blood insulin levels presenting with slightly low to elevated concentrations. Elevated insulin in obese persons with type 2 diabetes with peripheral insulin resistance can often be seen.

Insulin and C-peptide levels are elevated in cases of hypoglycemia not caused by improper use of the sulfonylurea drugs. Either inappropriate use of insulin administration or the presence of an insulin-secreting tumor can be the source of insulin. In these cases, measurement of C-peptide levels can be useful, since in the former situation, although insulin levels are elevated, C-peptide concentrations are normal or low.

Electrolytes (Na, K). Abnormal serum electrolyte concentrations are usually associated with the acute presentations of hyperglycemia (DKA, HHNC). However, no rigorous description can be made, since the effects of hypovolemia and dehydration also present in these conditions tend to mask the loss of Na and K from the body. Thus acutely presenting diabetic persons have serum Na and K levels that are low, normal, and most frequently elevated. It must be realized that in DKA, total body K levels are always depressed because of the combined effects of hypoinsulinemia and acidosis. Potassium levels must therefore be closely monitored during therapy for DKA.

Glycohemoglobin (HbA_1 or HbA_{1c}). This analyte is elevated in proportion to the extent and duration of hyperglycemia. Since the average red blood cell lives approximately 3 months, the HbA_1 level at any one time reflects the average blood glucose levels over the previous 3 weeks. Elevations caused by acute hyperglycemia are seen but are minimized by pretreatment of the red blood cells (dialysis or chemical treatment). Glycohemoglobin measurements are mostly used to monitor therapy designed to maintain relatively "normal" blood glucose levels. HbA_1 levels near the normal range (6.5% to 8.5%) reflect good control, whereas greatly elevated values (over 12%) reflect poor control.

Very low HbA_1 levels have been reported in insulinomas.

Urea (BUN). Serum urea nitrogen will often be elevated in DKA or HHNC because of both the hemoconcentration effects of hypovolemia and the resulting prerenal failure. Thus blood urea measurements are measured during therapy of these disorders to monitor successful rehydration of the patient and the kidneys. Of course, since diabetics often have coexisting renal disease, chronic elevations of urea will reflect the nephrotic disorder.

Osmolality. Serum osmolality is measured or estimated by calculation to help make the diagnosis of hyperosmolar coma. Osmolalities over 350 mOsm/L would suggest such a disorder.

Triglyceride. Serum triglyceride levels are elevated as a result of increased synthesis of VLDL.

REFERENCES

1. Diabetes forecast, special edition, pp. 1-60, Dec. 1975.
2. Podolsky, S.: Symposium of diabetes (foreword), Med. Clin. North Am. **62:**625, 1978.
3. Entmacher, P.S.: Long term prognosis in diabetes mellitus. In Sussman, K.E., and Metz, R.J.S., editors: Diabetes mellitus: diagnosis and treatment, ed. 4, New York, 1975, American Diabetes Association.
4. Skyler, J.S., and Cahill, G.F.: Diabetes mellitus (foreword), New York, 1981, Yorke Medical Books, Dun and Bradstreet.
5. Melton, L.H., Machen, K.M., Palumbo, P.J., and Elveback, L.: Incidence and prevalence of clinical peripheral vascular disease in a population based cohort of diabetic patients, Diabetes Care **3:**650, 1980.
6. Steiner, D.F., and Oyer, P.E.: The biosynthesis of insulin and a probable precursor of insulin by a human islet cell adenoma, Proc. Natl. Acad. Sci. USA **57:**473, 1967.
7. Melani, F., and Rubenstein, A.H.: Proinsulin and its derivatives. In Fajans, S.S., and Sussman, K.E., editors: Diabetes mellitus: diagnosis and treatment, ed. 3, New York, 1971, American Diabetes Association.
8. Rubenstein, A.H., Kuzaya, H., and Horwitz, D.L.: Clinical significance of circulating C-peptide in diabetes mellitus and hypoglycemic disorders, Arch Intern. Med. **137:**625, 1977.
9. Porte, D., Jr,. and Bagdade, J.D.: Human insulin secretion: an integrated approach, Annu. Rev. Med. **21:**129, 1970.
10. Cataland, S., Crockett, S.E., Brown, J.C., and Mazzaferri, E.L.: Gastric inhibitory polypeptide (GIP) stimulation by oral glucose in man, J. Clin. Endocrinol. Metab. **39:**232, 1974.
11. Cahill, G.F.: Physiology of insulin in man, Diabetes **20:**785, 1971.
12. Sherwin, R., and Felig, P.: Pathophysiology of diabetes mellitus, Med. Clin. North Am. **62:**695, 1978.
13. Unger, R.H.: Diabetes and the alpha cell (the Banting Memorial Lecture), Diabetes **25:**136, 1975.
14. Felig, P.: The endocrine pancreas. In Felig, P., Baxter, J.D., Broadus, A.E., and Frohman, L., editors: Endocrinology and Metabolism, New York, 1981, McGraw-Hill Book Co.
15. Sherwin, R., Fisher, M., Hendler, R., and Felig, P.: Hyperglucagonemia and blood glucose regulation in normal, obese and diabetic subjects, N. Engl. J. Med. **94:**455, 1976.
16. Himsworth, H., and Kerr, R: Insulin-sensitive and insulin-insensitive types of diabetes, Clin. Sci. **4:**119, 1939.
17. Yalow, R.S., and Berson, S.A.: Plasma insulin in man (editorial), Am. J. Med. **29:**1, 1960.
18. Karam, J.H., Grodsky, G.M., Forsham, P.M., et al.: Excessive insulin response to glucose in obese subjects as measured by immunochemical assay, Diabetes **12:**197, 1963.
19. Bar, R.S., and Roth, J.: Insulin receptor status in disease states in man, Arch. Intern. Med. **137:**474, 1977.
20. Larner, J., Galasko, G., Cheng, K., et al.: Generation by insulin of a chemical mediator that controls protein phosphorylation and dephosphorylation, Science **206:**1408, 1979.
21. Rizza, R.A., Mandarino, L.J., and Gerich, J.E.: Mechanisms of insulin resistance in man, Am. J. Med. **70:**169, 1981.
22. National Diabetes Data Group: Classification and diagnosis of diabetes mellitus and other categories of glucose intolerance, Diabetes **28:**1039, 1979.
23. Botzao, G.F., Florin-Christensen, A., and Doniach, D.: Islet cell antibodies in diabetes mellitus with polyendocrine deficiencies, Lancet **2:**1279, 1974.
24. Flier, J.S., Kahn, C.R., Roth, J., and Bar, R.S.: Antibodies that impair insulin receptor binding in an unusual diabetic syndrome with severe insulin resistance, Science **190:**63, 1975.
25. Given, B.D., Mako, M.E., Tager, H., et al.: Circulating insulin with reduced biological activity in a patient with diabetes, N. Engl. J. Med. **302:**129, 1980.
26. Gabbay, K.H., DeLuca, K., Fisher, J.N., Jr., et al.: Familial hyperproinsulinemia: an autosomal-dominant defect, N. Engl. J. Med. **294:**911, 1976.
27. Kobayashi, M., Olefsky, J.M., Elders, J., et al.: Insulin resistance caused by a defect distal to the insulin receptor (demonstration in a patient with leprechaunism) Proc. Natl. Acad. Sci. USA **75:**3469, 1978.
28. Podolsky, S., and Viswanathan, M.: Secondary diabetes: spectrum of the diabetes syndromes, New York, 1980, Raven Press.
29. Mallinson, C.M., Bloom, S.R., Warin, A.B., et al.: A glucagonoma syndrome, Lancet **2:**1, 1974.
30. WHO Expert Committee on diabetes mellitus: Second report, WHO Tech. Rep. Ser. 646, 1980.
31. Pyke, D.A., and Nelson, P.G.: Diabetes mellitus in identical twins. In Creutsfeldt, W., and Kobberling, J.V., (editors): The genetics of diabetes mellitus, New York, 1976, Springer-Verlag New York, Inc.
32. Tattersall, R.B., and Fajans, S.S.: A difference between the inheritance of classical juvenile-onset type diabetes and maturity onset diabetes of young people, Diabetes **24:**44, 1975.
33. Goldstein, S., and Podolsky, S.: The genetics of diabetes mellitus, Med. Clin. North Am. **62:**639, 1978.
34. Cudworth, A.G., and Woodrow, J.C.: Genetic susceptibility in diabetes mellitus: analysis of HLA association, Br. Med. J. **2:**846, 1975.
35. Yoon, J.W., Austin, M., Onodera, T., and Notkins, A.: Isolation of a virus from the pancreas of a child with diabetic ketoacidosis, N. Engl. J. Med. **300:**1173, 1979.
36. Craighead, J.E.: Viral diabetes mellitus in man and experimental animals, Am. J. Med. **70:**127, 1981.
37. Nerup, J., and Lernmark, A.: Autoimmunity in insulin-dependent diabetes mellitus, Am. J. Med. **70:**135, 1981.
38. Lendrum, R., Walker, G., and Gamble, D.R.: Islet cell antibodies in juvenile diabetes of recent onset, Lancet **1:**1880, 1975.
39. Doberson, M.J., et al.: Cytotoxic autoantibodies to beta cells in the serum of patients with insulin-dependent diabetes mellitus, N. Engl. J. Med. **303:**1493, 1980.
40. Irvine, W.J., McCallum, C.J., Gray, R.S., et al.: Pancreatic islet cell antibodies in diabetes mellitus correlated with the duration and type of diabetes coexistent autoimmune disease and HLA-type, Diabetes **26:**138, 1977.
41. Olefsky, J.M., and Kolterman, O.G.: Mechanisms of insulin resistance in obesity and noninsulin-dependent diabetes (type II), Am. J. Med. **70:**151, 1981.
42. Masbad, S., Alberti, K.G.M.M., Binder, C., et al.: Role of resid-

ual insulin secretion in protecting against ketoacidosis in insulin dependent diabetes, Br. Med. J. **2**:1257, 1979.
43. Genuth, S.M.: Plasma insulin and glucose profiles in normal, obese and diabetic persons, Ann. Intern. Med. **79**:812, 1973.
44. Eaton, R.P., Allen, R.C., Schade, D.S., and Standifer, J.C.: "Normal" insulin secretion: the goal of artificial insulin delivery systems, Diabetes Care **3**:270, 1980.
45. Chiles, R., and Tzagournis, M.: Excessive insulin response to oral glucose in obesity in mild diabetes: study of 501 patients, Diabetes **19**:458, 1970.
46. Williams, R.H., and Porte, D. Jr.: The pancreas. In Williams, R.H., editor: Textbook of endocrinology, ed. 5, Philadelphia, 1974, W.B. Saunders Co.
47. Goldberg, R.B.: Lipid disorders in diabetes, Diabetes Care **4**:561, 1981.
48. Chen, Y.I., Risser, T.R., Cully, M., and Reaver, G.R.: Is the hypertriglyceridemia associated with insulin deficiency caused by decreased lipoprotein lipase activity? Diabetes **28**:893, 1979.
49. Brunzell, J.D., Porte, D., and Bierman, E.L.: Abnormal lipoprotein-lipase mediated plasma triglyceride removal in diabetes mellitus associated with hypertriglyceridemia, Metabolism **28**:901, 1979.
50. Kaufmann, R.L., Soeldner, J.S., Wilmhurst, E.G. et al.: Plasma lipid levels in diabetic children, Diabetes **24**:672, 1975.
51. Wilson, D.E., Schreibman, P.H., Day, V.C., and Arky, R.A.: Hyperlipidemia in an adult diabetic population, J. Chronic Dis. **23**:501, 1970.
52. Koenig, R.J., Blobstein, S.H., and Cerami, A.: Structure of carbohydrate of hemoglobin A_{1c}, J. Biol. Chem. **252**:2992-2997, 1977.
53. Rahbar, S.: Glycosylated hemoglobins: history, biochemistry and clinical implications, NY State J. Med. **80**:553-557, 1980.
54. Kaplan, L.A., Cline, D., Gertside, P., et al.: Hemoglobin A_1 in hemolysates from healthy and insulin-dependent diabetic children, as determined with a temperature-controlled mini-column assay, Clin. Chem. **28**:13-18, 1982.
55. Bunn, H.F.: Evaluation of glycosylated hemoglobin in diabetic patients, Diabetes **30**:613-617, 1981.
56. Skillman, T.G.: Diabetic ketoacidosis, Heart Lung **7**:594, 1978.
57. Sulway, M.J., and Malins, J.M.: Acetone in diabetic ketoacidosis, Lancet **2**:736, 1970.
58. Jenkins, D.W., Eckle, R.E., and Craig, J.W.: Alcoholic ketoacidosis, J.A.M.A. **217**:177, 1971.
59. Alberti, K.G.G.M., Hockaday, T.D.R., and Turner, R.C.: Small doses of intramuscular insulin in the treatment of diabetic "coma", Lancet **2**:215, 1973.
60. Fisher, J.N., Shahshahni, M.N., and Kitabchi, A.E.: Diabetic ketoacidosis: low dose insulin by various routes, N. Engl. J. Med. **297**:238, 1977.
61. Beigelman, P.M.: Potassium in severe diabetic ketoacidosis, Am. J. Med. **54**:419, 1973.
62. Newman, J.H., Neff, T.A., and Ziporin, P.: Acute respiratory failure associated with hypophosphatemia, N. Engl. J. Med. **296**:1101, 1977.
63. Kreisberg, R.A.: Phosphate deficiency and hypophosphatemia, Hosp. Pract. **12**:121, 1977.
64. Sament, S., and Schwartz, M.D.: Severe diabetic stupor without ketosis, S. Afr. Med. J. **31**:893, 1957.
65. Arieff, A.I., and Carroll, H.J.: Nonketotic hyperosmolar coma with hyperglycemia: clinical features, pathophysiology, renal function, acid base balance, plasma cerebrospinal equilibrium and effects of therapy in 37 cases, Medicine **51**:73, 1972.
66. Podolsky, S.: Hyperosmolar nonketotic coma: underdiagnosed and undertreated clinical diabetes: modern management, New York, 1980, Appleton-Century-Crofts, pp. 209-235.
67. Arieff, A.I., and Carroll, H.J.: Cerebral edema and depression of sensorium in nonketotic hyperosmolar coma, Diabetes **23**:525, 1974.
68. Podolsky, S.: Hyperosmolar nonketotic diabetic coma: a complication of propranolol therapy, Metabolism **22**:685, 1973.
69. DeGraef, J., and Lips, J.B.: Hypernatremia in diabetes mellitus, Acta Med. Scand. **157**:71, 1957.
70. Quibrera, R., Nava, M., DeLeon, E.D., et al.: Treatment of diabetic ketoacidosis, hyperosmolar coma and severe diabetes with low IV intermittent doses of insulin, Rev. Invest. Clin. **28**:1, 1976.
71. Guisado, R., and Arieff, A.I.: Neurological manifestations of diabetic comas: correlation with bichemical alterations in the brain, Metabolism **24**:665, 1975.
72. Seltzer, H.S.: Drug-induced hypoglycemia: a review based on 473 cases, Diabetes **21**:955, 1972.
73. Kessler, I.: Mortality experience in diabetic patients: a 26 year followup study, Am. J. Med. **51**:715, 1971.
74. Goetz, F.C.: Prognosis in diabetes mellitus. In Diabetes mellitus: diagnosis and treatment, ed. 2, New York, 1967, American Diabetes Association.
75. Pirart, J.: Diabète et complications dégénératives, Presentation d'une étude prospective portant sur 4400 cas observés entre 1947 et 1973, Diabète Métab. **3**:97-107, 173-182, 265, 1977.
76. Siperstein, M.D., Unger, R.H., and Madison, L.L.: Studies of muscle basement membranes in normal subjects, diabetic, and prediabetic patients, J. Clin. Invest. **47**:1973, 1968.
77. Peterson, C.M., Jovanovic, L.B., Brownlee, M., et al.: Closing the loop, practical and theoretical, Diabetes Care **3**:318, 1980.
78. Mauer, S.M., Steffes, M.W., Sutherland, D.E.R., et al.: Studies of the rate of regression of the glomerular lesions in the diabetic rats treated with pancreatic islet transplantation, Diabetes **24**:280, 1975.
79. Kilo, C., Vogler, N.J., and Williamson, J.R.: Basement membrane thickening in diabetes. In Fagans, S.S., and Sussman, K.E., editors: Diabetes mellitus: Diagnosis and treatment, ed. 3, New York, 1971, American Diabetes Association.
80. Kahn, H.A., and Miller, R.: Blindness caused by diabetic retinopathy, Am. J. Ophthalmol. **78**:58, 1974.
81. Beaumont, P., and Hollows, F.C.: Classification of diabetic retinopathy with therapeutic complications, Lancet **1**:419, 1972.
82. L'Esperance, F.A., Jr.: Diabetic retinopathy, Med. Clin. North Am. **62**:767, 1978.
83. Friedman, E.A.: Glomerular manifestations. In Podolsky, S., editor: Clinical diabetes: modern management, New York, 1980, Appleton-Century-Crofts.
84. Amico, J.A., and Klein I.: Diabetic management in patients with renal failure, Diabetes Care **4**:430, 1981.
85. Colby, A.D.: Neurologic disorders in diabetes mellitus, Diabetes **14**:424, 1965.
86. Gabbay, K.H.: The sorbitol pathway and the complication of diabetes, N. Engl. J. Med. **288**:831, 1973.
87. Clements, R.W., Jr., Vourganti, B., Kuba, T., et al.: Dietary myoinositol intake and peripheral nerve function in diabetic neuropathy, Metabolism **28**:477, 1979.
88. Asbury, A.K., Aldrege, H., Hershberg, R., et al.: Ischemic mononeuropathy in diabetes mellitus: a clinicopathological entity, Brain **93**:555, 1970.
89. Ellenberg, M.: Diabetic neuropathy: clinical aspects, Metabolism **25**:1627, 1976.
90. Levin, M.E., and O'Neal, L.W.: The diabetic foot, ed. 3, St. Louis, 1982, The C.V. Mosby Co.
91. Clements, R.S., Jr., and Bell, D.S.: Diabetic neuropathy: peripheral and autonomic syndromes, Postgrad. Med. **71**:50, 1982.
92. Page, M.M., and Watkins, P.J.: The heart in diabetes: autonomic neuropathy and cardiomyopathy, Clin. Endocrinol. Metab. **6**:377, 1977.
93. Page, M.M., and Watkins, P.J.: Cardiorespiratory arrest in diabetic autonomic neuropathy, Lancet **1**:14, 1978.
94. Furlow, W.L.: Diagnosis and treatment of male erectile failure, Diabetes Care **2**:218, 1979.
95. Mauer, S.M., Steffes, M.W., and Sutherland,D.E.R.: Studies of the rate of regression of glomerular lesions in diabetic rats treated with pancreatic transplantation, Diabetes **24**:280, 1975.
96. Cahill, G.F., Etzwiler, D., and Freinkel, N.: Blood glucose control in diabetes, Diabetes **25**:237, 1976.
97. Diabetes Forecast, p. 15, July-Aug. 1978.
98. Arky, R.A.: Current principles of dietary therapy of diabetes mellitus, Med. Clin. North Am. **62**:655, 1978.
99. American Diabetes Association and American Dietetic Association: A guide for professionals: the effective application of "exchange lists for meal planning," New York, 1977, the Association.

100. Bierman, E.L., and Porte, D.: Carbohydrate intolerance and lipemia, Ann. Intern. Med. **68:**926, 1968.
101. Joslin, A.P., and White, P.: The dietary management of diabetes, Med. Clin. North Am. **49:**905, 1965.
102. Phillips, M., Simpson, R.W., Holman, R.R., and Turner, R.C.: A simple and rational twice daily insulin regimen: distinction between basal and meal insulin requirements, Q. J. Med. **48:**493, 1979.
103. Streja, D., Boyko, E., and Rabkin, S.W.: Nutrition therapy in non–insulin dependent diabetes mellitus, Diabetes Care **4:**81, 1981.
104. Richter, E.A., Ruderman, N.B., and Schneider, S.H.: Diabetes and exercise, Am. J. Med. **70:**201, 1981.
105. Skillman, T.G., and Feldman, J.M.: The pharmacology of sulfonylureas, Am. J. Med. **70:**361, 1981.
106. Lebovitz, H.E., and Feinglos, M.N.: Sulfonylurea drugs: mechanism of action and therapeutic usefulness, Diabetes Care **1:**189, 1978.
107. Beck-Nielsen, H., Pederson, O., and Lindskov, H.O.: Increased insulin sensitivity and cellular insulin binding in obese diabetics following oral treatment with glibenclamide, Acta Endocrinol. **90:**451, 1979.
108. The University Group Diabetes Program: A study of the effects of hypoglycemic agents on vascular complications in patients with maturity onset diabetes, Diabetes: **19**(suppl.2):747, 1970.
109. ''HEW secretary suspends general marketing of phenformin,'' FDA Drug Bull. **7:**14, 1977.
110. Podgainey, H., and Bressler, R.: Biochemical basis of the sulfonylurea-induced Antabuse syndrome, Diabetes **17:**679, 1978.
111. Galloway, J.A.: Insulin treatment for the early 80s: facts and questions about old and new insulins and their usage, Diabetes Care **3:**615, 1980.
112. Soeldner, J.S.: Treatment of diabetes mellitus by devices, Am. J. Med. **70:**589, 1981.
113. Tattersall, R., and Gale, E.: Patient self-monitoring of blood glucose and refinements of conventional insulin treatment, Am. J. Med. **70:**101, 1981.

Chapter 31 Disorders of lipid metabolism

Herbert K. Naito

amphipathic (*amphi-*, 'on both sides' + *-pathic*, of feeling'); pertaining to a molecule having two sides with characteristically different properties, or a detergent, which has both a polar (hydrophilic) end and a nonpolar (hydrophobic) end but is long enough so that each end demonstrates its own solubility characteristics.

apolipoprotein The protein component of lipoprotein complexes.

chylomicron Large lipid-protein complexes that are made by the gut, and these molecules serve an important function in the transport of fats (mainly dietary triglycerides).

endogenous (*endo-*, 'within' + *-gen*, 'one that produces' + *-ous*). Originating or produced within the organism or one of its parts.

etiology (*aitia*, 'cause' + *-logy*, 'study of'). The study or theory of the factors that cause disease; all the factors that contribute to the occurrence of a disease.

gluconeogenesis (*glykys*, 'sweet' + *neos*, 'new' + *genesis*, 'birth, origin'). The formation of glucose from molecules that are not themselves carbohydrates, as from amino acids, lactate, and the glycerol portion of fats.

HDL This lipid-protein complex is also called alpha-lipoprotein and is the most dense of the lipoproteins; high-density lipoprotein.

hydrophilic (*hydro-*, 'water' + *-philic*, 'loving'). A term denoting the property of attracting water molecules, possessed by polar radicals or ions; the opposite of ''hydrophobic.''

hydrophobic (*hydro-*, 'water' + *-phobic*, 'fearing'). Repelling water, said of molecules or side chains that are more soluble in organic solvents.

IDL This lipid-protein complex has a density between VLDL and LDL (low-density lipoprotein) and is a product that has a relatively very short half-life and is in the blood in very low concentrations in a normal person. In a type III hyperlipoproteinemic person the IDL concentration in the blood is found to be elevated.

LDL This lipid-protein complex (low-density lipoprotein) is also called ''beta-lipoprotein'' and is the end product of VLDL catabolism. This lipoprotein complex is the major carrier of cholesterol.

lipoproteins Lipid (apoprotein)–protein complexes consisting of discrete families of macromolecules with known physical, chemical, and physiological properties.

polar Pertaining to a molecule or area on a molecule that builds up a net charge so that a ''dipole'' is created, that is, two equal and opposite charges separated in space. Each ''pole'' (positive or negative) is attracted by the opposite pole of another polar molecule but will repel a pole of the same charge. Water is a very polar molecule, and therefore polar substances are usually very hydrophilic.

pathogenesis (*pathos*, 'suffering' + *genesis*, 'origin, birth'). The development of morbid conditions or disease; more specifically the cellular events, reactions, and other pathological mechanisms occurring in the development of disease.

VLDL Also called ''pre-beta-lipoprotein'' (very-low-density lipoprotein). It is a relatively large lipid-protein complex that transports mainly endogenously synthesized triglycerides.

Part I: Lipids

NORMAL PHYSIOLOGY OF LIPIDS

Composition of foods

Since the quality and quantity of food we eat contribute to the concentration and distribution of lipids in the various lipoprotein fractions, it is appropriate to discuss briefly the composition of the food we eat. The "typical" American diet has an average caloric intake of about 3000 kcal, which generally consists of about 40% fat, 48% carbohydrate, and 12% of the total calories as proteins. The composition of fat intake is mostly animal fat, which is mainly saturated fat. Polyunsaturated fat (mainly vegetable fat) constitutes a small fraction (about 13%) of the fat intake.

When one examines the chemistry of food fats, they are composed mainly of triglycerides, about 98% to 99%, of which 92% to 95% is fatty acid and the remainder is glycerol. The remaining 1% to 2% of the lipids includes cholesterol, phospholipids, diglycerides, monoglycerides, fat-soluble vitamins, steroids, terpenes, and other fats. Most fats are mixtures of triglycerides containing four or five major fatty acids and many more minor or trace constituents. The individual glyceride molecules in most food fats contain both saturated and unsaturated fatty acids. Fully saturated glycerides are rare and appear in only a few natural fats, such as beef tallow and coconut oil. Thus "saturated" and "unsaturated" are not truly definitive terms when applied to food fats, but only indicate the proclivity for saturation levels.

In foods for human consumption, myristic, palmitic, and stearic acids are the most abundant of the saturated fatty acids. Of the unsaturated acids, oleic acid with one double bond is the most abundant, and of the polyunsaturated acids, linoleic acid is the most abundant and constitutes a high percentage of the commonly used vegetable oils. Several polyunsaturated acids (linoleic, linolenic, and arachidonic acids) cannot be synthesized in the animal body and must be provided in the diet. These have been termed "essential fatty acids" (EFA) and are mainly represented by linoleic acid.

In the naturally occurring polyunsaturated fatty acids, the double bonds are not continuous but are almost always separated by a methylene group. In addition, the configuration is usually of the *cis* type in which the hydrogens are on the same side of the chain, and a 120-degree bend is introduced into the chain itself. However, fatty acids containing *trans* bonds, in which the hydrogens are on opposite sides, are found in small amounts in natural fats and in greater amounts after processes involving catalytic hydrogenation.

The intake of essential fatty acids in the American diet is mostly from vegetable sources.

In addition to the triglycerides, glycerol esters of the fatty acids, in which one or two hydroxyl groups remain unesterified, may be encountered in food fats. Trace amounts of the monoglycerides and diglycerides and of free fatty acids are present in natural fats. These lipids are found during digestion and absorption and are present in the circulating lipids of the plasma.

The small amount of unsaponifiable matter in food fats consists of sterols, fatty alcohols, hydrocarbons, pigments, glycerol esters, and various other compounds. One of the principal differences between animal and vegetable fats is the nature of the sterols present. Cholesterol, found almost exclusively in animal tissue, differs from phytosterols (plant sterols), in that phytosterols (Fig. 31-1) have more highly branched side chains and may have a second double bond in the nucleus. Cholesterol occurs in all animal fats. It is a normal constituent of every animal tissue and a major component of brain, liver, and other tissues. Most sterols are cholesterol, but depending on the diet, phytosterols (beta-sitosterol, campesterol, and sigmasterol) can make up an appreciable percentage of the total sterols, particularly in people on vegetarian diets. The phytosterols are important because they compete with cholesterol for uptake by the mucosal cell. Thus the more phytosterols consumed, the less dietary cholesterol is absorbed by the mucosal cells of the gut.

Fat digestion, absorption, and metabolism of lipids

Fat absorption occurs in three phases: the *intraluminal,* or digestive, phase, during which the fats are modified both physically and chemically before absorption; the *cellular,* or absorptive, phase, in which the digested material enters the mucosal cell where it is reassembled into its preabsorptive form; and the *transport* phase, during which the absorbed lipids are shuttled from the mucosal cell to other tissues through the lymphatics and blood (Fig. 31-2).

Intraluminal phase. Although a small amount of fat digestion may take place in the stomach, most digestion of food fat is carried on in the intestine through the action of intestinal and pancreatic enzymes (lipases) and of bile ac-

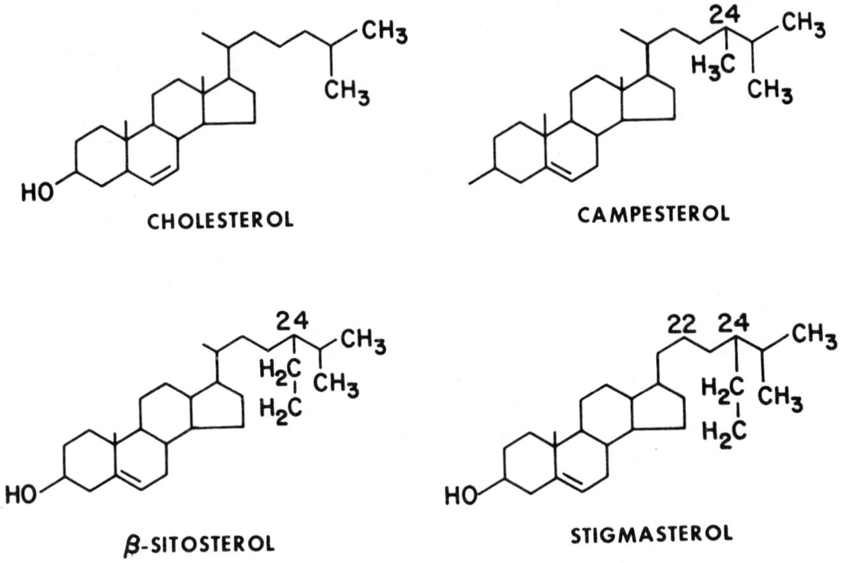

CHOLESTEROL

CAMPESTEROL

β-SITOSTEROL

STIGMASTEROL

Fig. 31-1. Structure of cholesterol and phytosterols or plant sterols (campesterol, β-sitosterol, stigmasterol).

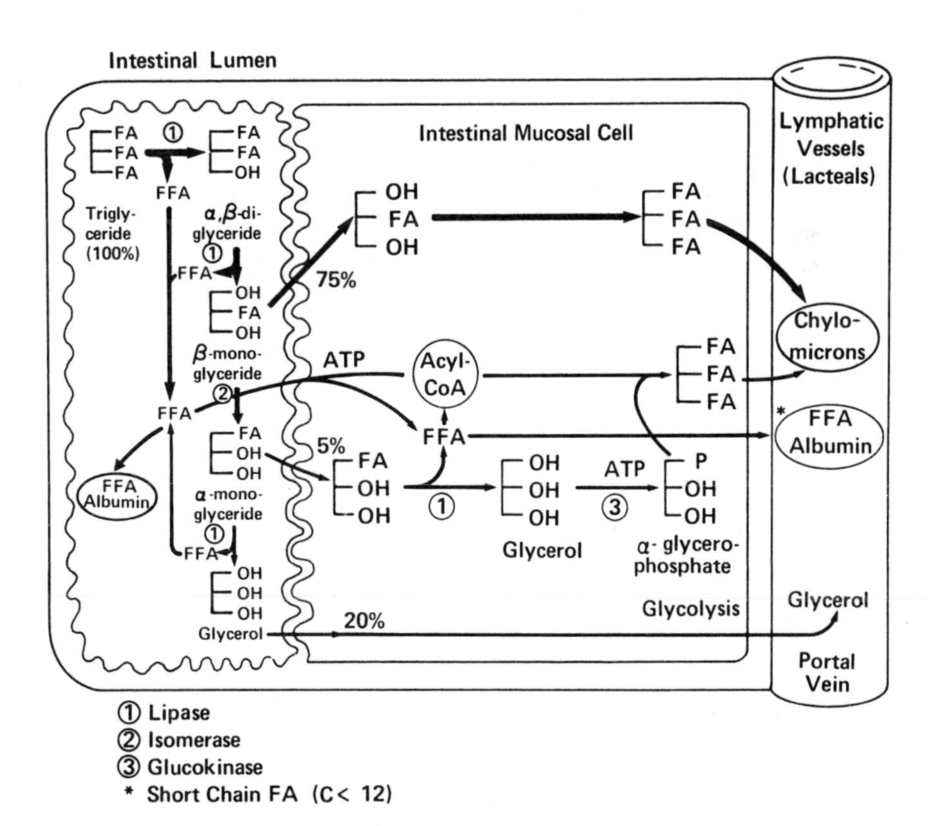

Fig. 31-2. Scheme of digestion of triglyceride in intestinal lumen, uptake of its breakdown products by mucosal cell, and resynthesis of newly formed lipid into chylomicrons. *1,* Lipase; *2,* isomerase; *3,* glucokinase; *, short-chain (and chain of less than 12 carbons) fatty acid; *FA,* fatty acid; *FFA,* free fatty acid.

ids. Bile salts, produced by the liver and stored in the gallbladder as bile, play an important role in this breakdown of food fats because of their *amphipathic* properties. The bile salt molecule is both strongly hydrophilic and strongly hydrophobic. Such molecules tend to arrange themselves on the surface of small triglyceride particles with their hydrophobic end turned inward and the hydrophilic end outward toward the water phase. Because of their surface-active properties, the bile salts emulsify the dietary triglyceride into very small particles with a diameter of approximately 1 μm. The emulsification process thus forms particles that can be readily acted upon by digestive enzymes.

The main path of fat digestion progresses from triglycerides to 1,2-diglycerides to 2-monoglycerides and fatty acids. Only a small percentage of the fat is hydrolyzed to free fatty acids and glyercol, perhaps after isomerization to the 1-monoglyceride (Fig. 31-2).

During digestion, an exchange of free fatty acids with glyceride fatty acids occurs. Furthermore, some synthesis of triglycerides from the monoglycerides and diglycerides takes place simultaneously with hydrolysis of the fats. The earlier stages of fat digestion are reversible processes and modify the chemical composition of ingested food fat.

In the intestinal lumen the action of pancreatic lipase on ingested fat results in a complex mixture of triglycerides, diglycerides, monoglycerides, and free fatty acids (FFA). Cholesterol esters are hydrolyzed to free cholesterol and free fatty acid, the reaction being catalyzed by the enzyme cholesterol esterase. In addition, the entry of bile into the duodenum contributes important amounts of bile salts and lecithin, the latter quickly undergoing hydrolysis to lysolecithin. Both of these classes of compounds are essential for solubilization of the lipids in the intestinal contents, which form a two-phase system—an oil phase containing almost all the triglycerides and diglycerides and a water-clear micellar solution of monoglycerides, bile salts, lyso-

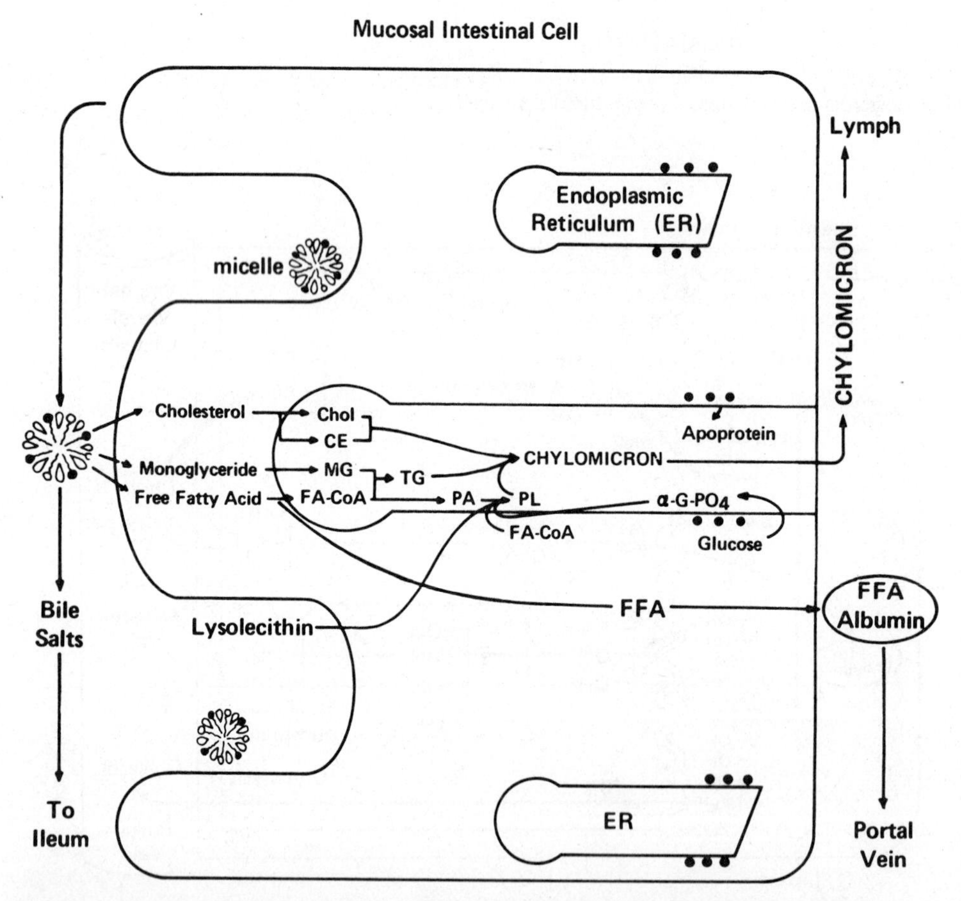

Fig. 31-3. Diagram of micelles along mucosal cell of gastrointestinal tract. After uptake of lipids, they are reassembled with apoproteins in endoplasmic reticulum *(ER)* to form triglyceride-rich particles, chylomicrons, which are carried away by lymph to systemic circulation. Short-chain fatty acids are attached to albumin and are transported through portal vein. *Chol,* Cholesterol; *CE,* cholesterol ester; *MG,* monoglyceride; *FA-CoA,* fatty acid coenzyme A; *PA,* phosphatidic acid; *PL,* phosphatidyl lecithin; *FFA,* free fatty acid; $\alpha G\text{-}PO_4$, α-glycerol phosphate; *TG,* triglyceride.

lecithin, and soaps. Bile salts combine with the resulting monoglycerides and fatty acids, forming negatively charged polymolecular aggregates, or micelles with a diameter in the range of 5 nm. In the conversion of fats from an emulsion phase to a micellar phase, the diameter of the fat-containing particles has been reduced further, approximately a hundredfold, and the surface area increased tenthousandfold. The 5 nm micelle now has access to the intramicrovillus spaces (50 to 100 nm) of the intestinal membrane where penetration into the intestinal mucosal cell can take place (Fig. 31-3).

Absorptive phase. How monoglycerides, fatty acids, and other lipids from the intestinal lumen micelles are taken into the mucosal cell is not yet clear. It is generally accepted, however, that the micelle disintegrates on contact with the brush border of microvilli of the mucosal cell membrane, allowing differential uptake of various micelle components (Fig. 31-3).

After monoglycerides and fatty acids enter the endoplasmic reticulum of the mucosal cell, presumably by diffusion, the monoglycerides and fatty acids are reesterified into triglycerides (Fig. 31-2). There are two main pathways for the resynthesis of triglycerides within the mucosal cell. The monoglyceride pathway is peculiar to the intestinal mucosa and involves the direct acylation of the absorbed monoglyceride from the lumen with activated free fatty acid. The alpha-glycerophosphate pathway present in most tissues involves the acylation of glycerophosphate to form phosphatidic acid, dephosphorylation of the phosphatidic acid to form diglyceride, and further acylation to form triglyceride. Fatty acid utilization for triglyceride resynthesis requires its activation by formation of a coenzyme A (acyl CoA) derivative of the fatty acid. This reaction, which requires ATP, is catalyzed by the enzyme fatty acid:CoA ligase (AMP). This enzyme has a pronounced specificity for longer-chain fatty acids. Thus long-chain fatty acids appear in thoracic duct lymph transported as triglycerides in the chylomicrons, whereas short- and medium-chain fatty acids are transported bound to albumin and transported in the portal circulation.

Transport phase. Once triglycerides have been resynthesized within the intestinal mucosal cell, they are assembled in the mucosal cell endoplasmic reticulum and the Golgi apparatus into water-soluble macromolecules—chylomicrons and, to a small extent, very-low-density lipoprotein (VLDL). The intestinal lipoproteins leave the mucosal cells presumably by reverse pinocytosis. They first appear in the lymphatic vessels of the abdominal region and later in the systemic circulation. These larger lipid-protein complexes are mixtures of triglycerides, some proteins (as apoproteins), small amounts of cholesterol, and phospholipids.

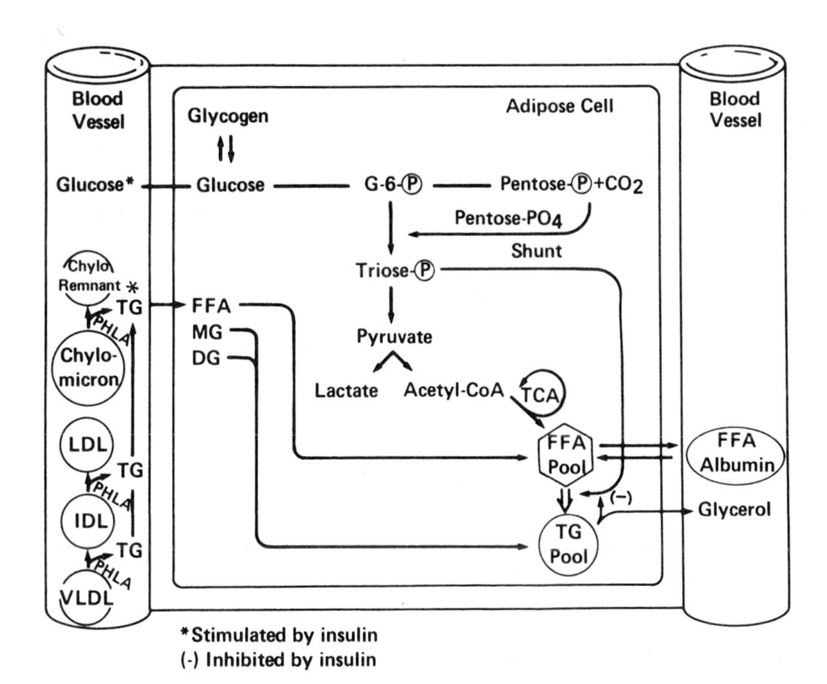

Fig. 31-4. Fat metabolism in adipose cell after meal. This metabolic pathway favors storage of energy (as triglycerides). Insulin promotes entry of glucose, necessary for triglyceride synthesis, and insulin also inhibits hormone-sensitive lipase, the enzyme that promotes triglyceride breakdown to free fatty acids and glycerol. The influx of free fatty acids from chylomicron and very-low-density lipoprotein is also an integral part of triglyceride synthesis. *, Stimulated by insulin; (−), inhibited by insulin; *PHLA,* post-heparin lipolytic activity, *G-6-P,* glucose-6-phosphate; *TCA,* tricarboxylic acid cycle; *DG,* diglyceride; remainder of abbreviations as in Fig. 31-3 and text.

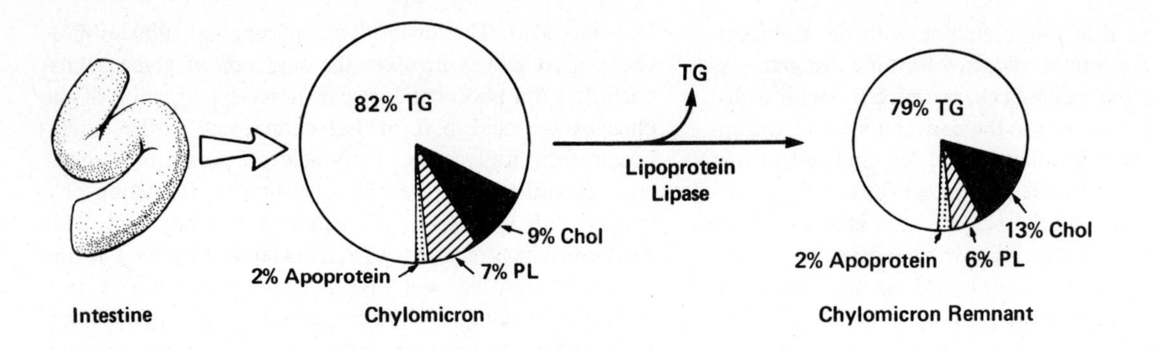

CATABOLIC PATHWAY FOR CHYLOMICRON AND VLDL

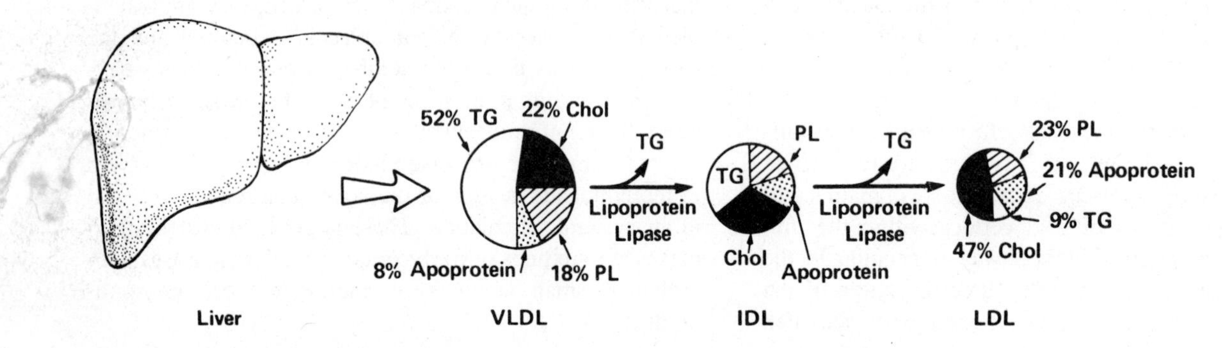

Fig. 31-5. Origin and catabolic pathway of chylomicron and very-low-density lipoprotein *(VLDL)*. End product of chylomicron is chylomicron remnant; end product of VLDL is LDL. *Chol,* Cholesterol; *PL,* phosphatidyl lecithin, *TG,* triglyceride.

The bloodstream transports chylomicrons and VLDL to all tissues in the body, including adipose tissue, which is probably their principal site of uptake (Fig. 31-4). The larger chylomicrons (more heavily laden with triglyceride) (Fig. 31-5) are removed rather rapidly (within minutes) as compared to the circulating VLDL of intestinal origin (within hours).

Chylomicron and VLDL triglycerides are hydrolyzed by an enzyme called "triglyceride (or lipoprotein) lipase." Chylomicrons and VLDL are cleared from the circulation in a matter of minutes and a few hours, respectively, and they are normally present only in trace or small amounts in fasting blood samples. The process apparently takes place at the luminal surface of the capillary endothelium and probably at the vascular endothelium in other sites. Morphological studies on mammary gland and adipose tissue have localized the site of lipoprotein lipase activity to the endothelial cell surface. In most organs, lipoprotein lipase is present in two compartments: the functional one located at the surface of cells that come into contact with circulating triglycerides; the other compartment is intracellular, the site of enzyme production and perhaps its storage.

Under normal conditions, chylomicron catabolism proceeds in two known phases. In the first, triglycerides are hydrolyzed at extrahepatic tissue sites under the influence of a triglyceride lipase localized in the capillary endothelium. This process results in a relatively triglyceride-poor, cholesterol-rich remnant particle. In the second phase, the remnant particle is removed by the liver.

In the first phase of chylomicron metabolism, unesterified fatty acids are released into the bloodstream while the diglycerides and monoglycerides are taken up in vacuoles and transported across the capillary wall for hydrolysis. Very little triglyceride lipase activity is present in the plasma of normal subjects. The enzyme is bound to the capillary endothelial cells in muscle and adipose tissue and can be released by intravenous administration of heparin. Triglyceride lipase activity is often referred to as "postheparin lipolytic activity" (PHLA). Whereas the adipose tissue enzyme is dependent on apoprotein C-II and insulin for activation, the hepatic enzyme is not.

The mechanism for the hydrolysis of chylomicron triglycerides by enzymes localized within the endothelial lining of the capillary wall is not known. It has been suggested that the chylomicron fuses with the outer phospholipid layer of the endothelial cell membrane. After fusion, there is hydrolysis of one triglyceride fatty acyl chain followed by uptake of the diglyceride. Partially degraded or remnant chylomicrons are then released into the circulation to be taken up and further metabolized by the liver. Chylomicron cholesterol and cholesteryl esters within the liver may then be (1) converted to bile acids and secreted in the bile, (2) incorporated into lipoproteins and released into the plasma, or (3) secreted into the bile as neutral sterol.

The fatty acids derived from the triglyceride (in chylomicron or very-low-density lipoprotein) enter the adipose tissue cell, where they can be stored while glycerol is released into the circulation (Fig. 31-4). The free fatty acids can also be utilized for energy when needed by all muscular tissue, especially the heart. Free fatty acids are used also for cellular phospholipid synthesis, which leads to membrane formation. Essential fatty acids, in addition to their function in membrane synthesis, are the only source for synthesis of prostaglandins, the ubiquitous local hormones.

The catabolism of VLDL and chylomicron triglyceride plays a prominent role in the supply of free fatty acids for milk formation and is mandatory for the entry of free fatty acids into storge organs, such as adipose tissue. It appears also that in the lactating mammary gland the catabolism of VLDL and chylomicrons provides free chlesterol, which is utilized for milk secretion. The uptake of the triglycerides leads to less triglyceride-rich particles: chylomicron remnant particles and VLDL remnant particles, the former being cleared by the liver. Further catabolism of the VLDL remnant occurs at an extracellular site and results in the formation of low density lipoproteins, LDL, a cholesterol-rich particle (Fig. 31-5).

Role of liver in metabolism of lipids

The liver plays an important role in the uptake and storage of nutrients arriving directly to it through the portal circulation. During fasting, the liver releases glucose and maintains its concentration in blood at the level necessary for central nervous system functioning. This function is possible because the liver cell is equipped with a dual system of enzymes that permits the flow of metabolites toward storage or toward glucose release. The direction depends on the composition of the blood that perfuses the organ, which in turn reflects the body's metabolic state. Hormonal factors and adaptive changes in the activity of key enzymes also play a role in regulating hepatic function. Newly synthesized fatty acids and those produced by elongation of short-chain fatty acids derived from diet are esterified to triglyceride.

Lipid metabolism in liver after meal. Nutrient metabolism in the liver after absorption is summarized in Fig. 31-6. Glucose, amino acids, and short-chain fatty acid concentrations rise in portal blood. Glucokinase phosphorylates glucose to glucose-6-phosphate. Glycogen is synthesized until the hepatic stores are repleted. If portal hyperglycemia persists, glucose is converted to fatty acid by acetyl CoA. Excess amino acids are deaminated and contribute to the acetyl CoA and pyruvate pool. Some acetyl CoA undergoes further oxidation to CO_2 by the tricarboxylic cycle, thus yielding the energy necessary for hepatic functioning. Newly synthesized fatty acids and those produced by elongation of short-chain fatty acids derived from diet are esterified to glycerol, monoglyceride, or diglyceride to form triglyceride.

The newly synthesized triglyceride is coupled with phospholipid, cholesterol, and proteins to form VLDL. These macromolecules are then released into the circulation and transported to adipose tissue. Fatty acids in plasma may also be utilized for energy by a variety of other tissues, such as heart, muscle, liver, and kidney.

Hepatic triglyceride synthesis is accelerated when the diet is rich in carbohydrate. This results in VLDL overproduction, which may explain the occasional transient hypertriglyceridemia observed in normal persons when they consume diets particularly rich in simple sugars.

Another important source of hepatic and VLDL triglyceride are the free fatty acids released by adipose tissue. Free fatty acids are removed by the liver in proportion to their blood concentration; thus the fluctuations in the blood levels of these metabolites are dampened. Free fatty acids of adipose tissue origin are esterified to triglycerides and incorporated into hepatic VLDL and then released into the bloodstream. During periods of stress and in certain metabolic conditions, like uncontrolled diabetes, free fatty acids are the principal precursors of hepatic VLDL. It is not yet established to what extent chylomicrons and VLDL of intestinal origin released through lymphatic channels into the general circulation are taken up by the liver. These particles could represent yet another source of hepatic triglyceride in the fed state.

In summary, the liver is a very important organ in the metabolism of lipids. The liver stores alimentary glucose in the form of glycogen and converts the excess to VLDL triglyceride. Free fatty acid from adipose tissue represents another important precursor of hepatic VLDL triglyceride. Possibly chylomicron triglycerides also contribute to hepatic VLDL triglyceride.

Lipid metabolism in liver during fasting. During fasting, blood glucose concentration falls and insulin levels diminish. As a consequence, amino acids are mobilized from muscle, and fatty acids are released from adipose tissue. These molecules are taken up by the liver where gluconeogenesis and ketogenesis predominate as diagrammed in Fig. 31-7. Alanine, the principal gluconeogenic precursor, and other amino acids are incorporated into glucose-6-phosphate by deamination and reverse glycolysis. Glucose-6-phosphate (G-6-P) is converted to glucose by the action of glucose-6-phosphatase and then released into the circulation. Free fatty acids derived from adipose tissue are taken up by the liver, and their oxidation to ketone bodies

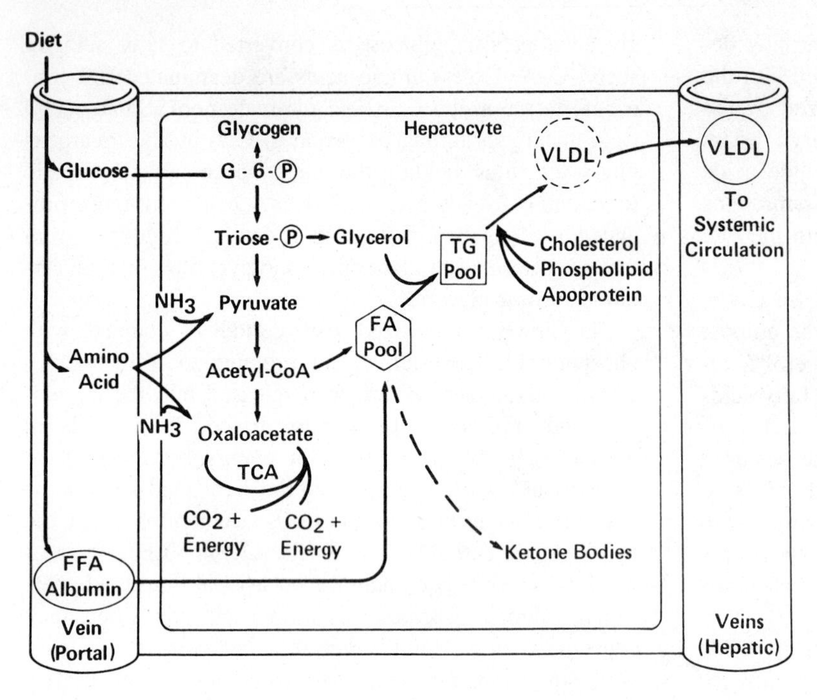

Fig. 31-6. Fat metabolism in liver after meal. Dietary components add to fatty acid pool in liver and ultimately to triglyceride pool. Lipids and apoprotein are packaged in endoplasmic reticulum and released as VLDL in systemic circulation. Abbreviations as in Figs. 31-3 and 31-4.

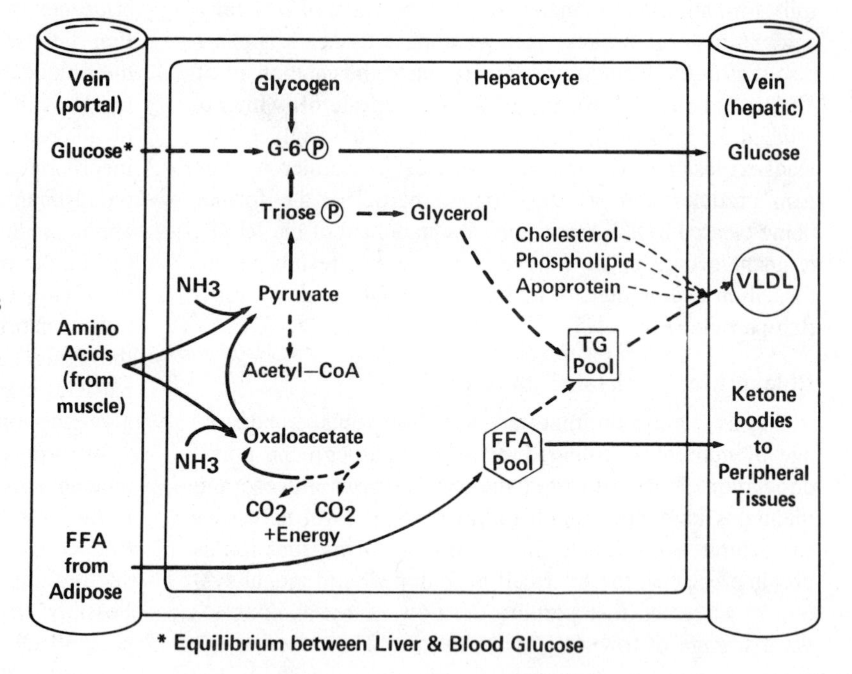

Fig. 31-7. Fat metabolism in liver during fasting conditions. VLDL synthesis and release by hepatocytes are decreased. Glucose becomes major fuel source from liver during fasting, whereas FFA is major fuel source from adipose tissue. Unused FFA goes back to liver. Abbreviations as in Figs. 31-3 and 31-4.

provides energy for gluconeogenesis. Hepatic VLDL triglyceride synthesis is diminished during fasting.

Role of adipose tissue in metabolism of lipids

If the free fatty acids are not immediately utilized for energy purposes by the peripheral tissues, this concentrated form of energy is stored in the fat cells (adipocytes).

After a meal containing fat and carbohydrate, chylomicron and VLDL fatty acids enter the intracellular free fatty acid pool. Glycerol, on the other hand, is released into the circulation, since adipose tissue lacks glycerokinase, the enzyme necessary to reesterify glycerol (Fig. 31-8). The absence of this enzyme means that fatty acid conversion to triglyceride depends on the availability of glucose to provide alpha-glycerophosphate needed for esterification. The enzyme glucokinase in the adipocyte is responsible for the phosphorylation of glucose, which then makes the glucose-6-phosphate available for esterification with the fatty acids.

Lipid metabolism in adipocyte after meal. After a meal, insulin is secreted by the pancreas into the portal

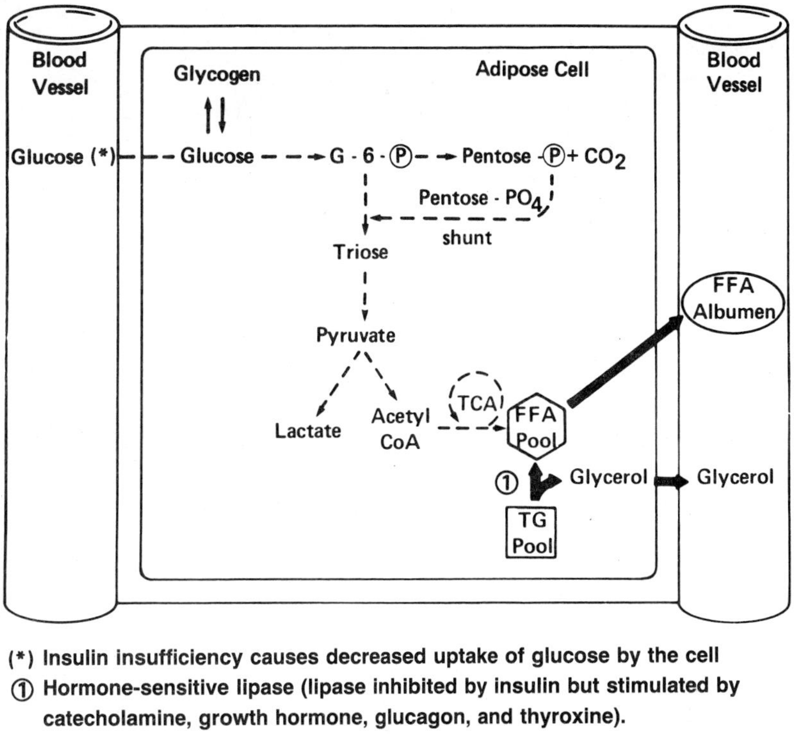

Fig. 31-8. Fat metabolism in adipose cell during fasting conditions. While synthetic TG synthesis rate decreases, rate of TG lipolysis to FFA and glycerol increases because of elevated activity of hormone-sensitive lipase. Abbreviations as in Figs. 31-3 and 31-4.

(*) Insulin insufficiency causes decreased uptake of glucose by the cell
① Hormone-sensitive lipase (lipase inhibited by insulin but stimulated by catecholamine, growth hormone, glucagon, and thyroxine).

bloodstream and reaches the liver in high concentration, and it is here that it exerts a major influence on carbohydrate metabolism. However, not to be neglected is its role in lipid metabolism.

Insulin accelerates glucose entry into adipose cells where it is metabolized to glycerophosphate and also to fatty acid chains. The uptake of lipoprotein triglycerides depends on the activity of the lipoprotein lipase and the availability of alpha-glycerophosphate for reesterification of fatty acids into triglycerides (Fig. 31-4). High levels of insulin accelerate both processes, and after a meal, adipose tissue operates as an effective storage system of alimentary lipids and carbohydrates. Triglycerides in adipose tissue are constantly subjected to the catabolic action of an intracellular enzyme (different from serum lipoprotein lipase) stimulated by hormones such as catecholamines, thyroxine, and glucagon, which accelerate the hydrolysis of triglycerides to fatty acid and glycerol.

Lipid metabolism in adipocyte during fasting. When glucose is in short supply (that is, during fasting), reesterification (or triglyceride synthesis) is inhibited, and at the same time hydrolysis of triglycerides is increased resulting in enlargement of the adipose cell fatty acid pool. Fatty acids are then released into the circulation where they are bound to albumin (Fig. 31-8).

In summary, adipose tissue is one of the major sites of chylomicron and VLDL triglyceride removal from the circulation and for storage. This process depends on the activity of lipoprotein lipase, the availability of glucose, and the presence of insulin. When glucose and insulin concentrations fall or when catecholamines and other hormones (thyroxine, growth hormone, glucagon) are secreted, lipolysis predominates and free fatty acids are released from the storage depot for utilization by liver or other cells.

BIOENERGETICS

The body requires energy to meet basal metabolic needs, support physical activity and growth, maintain body temperature, and cover the cost of digestion and absorption of food. When the energy intake meets these needs, all the involuntary and voluntary activities of the body are satisfied.

Energy that the body requires can be derived from foods or the body's own reserves (chiefly fat). Any excess of energy above immediate needs is frugally stored as fat. For complete utilization of the energy from fat, some carbohydrate (glucose) must be utilized simultaneously. If necessary, the body can synthesize glucose from certain amino acids. The body maintains an internal store of glucose in the form of glycogen in the liver and muscles. The carbohydrate reserve of hepatic glycogen is sufficient for about 16 hours. This reserve needs to be replenished by carbohydrate from foods (Fig. 31-9). It should be evident from Fig. 31-9 that an interrelationship exists among carbohydrates, fats, and proteins in the bioenergetic pathway.

The energy value of food is expressed in terms of a unit of heat, the kilocalorie (kcal).

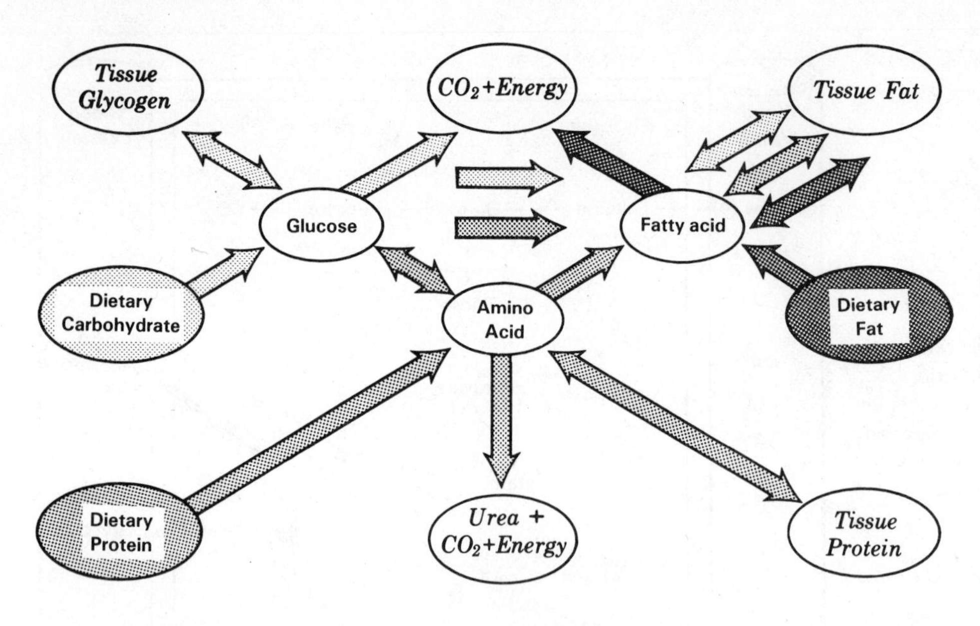

Fig. 31-9. Bioenergetics of fat, carbohydrate, and proteins. Notice interrelationship of these three energy compounds.

Table 31-1. Physiological fuel value of major nutrients

	Carbohydrate (Cal/g)	Fat (Cal/g)	Protein (Cal/g)	Ethanol (Cal/g)
Heat of combustion	4.1	9.45	5.65	7.1
Energy from combustion of nitrogen unavailable to the body	—	—	1.3	—
Net heat of combustion	4.1	9.45	4.35	7.1
Coefficient of digestibility	0.98	0.95	0.92	Small loss in urine and breath
Physiological fuel value	4.0	9.0	4.0	7.0

The first line of Table 31-1 presents the energy from heat of combustion obtained from burning or oxidation of 1 g of carbohydrate, fat, protein, or alcohol, respectively. The heat of combusion represents the total energy produced by the oxidation of the carbon of the food in the bomb calorimeter to carbon dioxide, the hydrogen to water, and the nitrogen of the protein to nitrous oxide. In the tissue cells the digested food products are also oxidized. However, unlike the bomb calorimeter, the heat lability of tissue cells prohibits as wasteful and destructive to the cell the direct oxidation of food. Rather, oxidation is accomplished by decarboxylations and by the gradual removal of the hydrogen and electrons of food through a respiratory cycle until the hydrogen and electrons can be united at the end of the cycle with molecular oxygen to form water. During this process the bond energies of the food molecule are released and captured in forming high-energy adenosine triphosphate (ATP) from adenosine diphosophate (ADP) present in the cell fluid. The combined process of the respiratory cycle and the capture of food bond energies as high-energy ATP is called "coupled oxidative phosphorylation." The energy of ATP can be transferred to creatine phosphate for temporary storage or utilized directly to drive the physiologic processes of the living cell. However, the animal cell cannot release or utilize the complete energy potential of nitrogen in protein. Therefore a reduction must be made in the case of proteins because these are not so completely oxidized in the body as in the bomb calorimeter. It is the nitrogen-containing product of protein, urea, that is not oxidized but excreted in the urine. The latent heat of this excreted nitrogen, which amounts to 1.3 Calorie for each gram of protein burned in the tissue cell, must be subtracted from the heat of combustion, which is 4.35 Cal/g (see line 3 of Table 31-1).

Since the human body is not 100% efficient in digesting and absorbing or metabolizing the major nutrients, one needs to determine the amount of energy available to the body from the ingested nutrient to calculate the coefficient of digestibility for each nutrient. This coefficient expresses the percentage of the nutrient ultimately available to the body as fuel.

CHOLESTEROL METABOLISM

Biological functions

Cholesterol is a member of a large class of biological compounds called "steroids" that have a similar ring structure, a cyclopentanoperhydrophenanthrene derivative (Fig. 31-10).

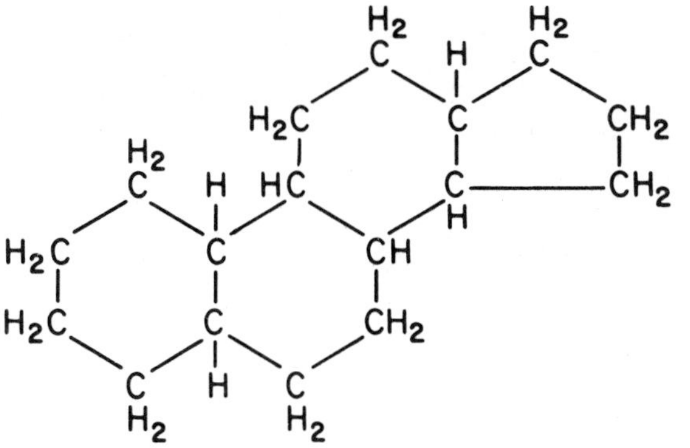

Cyclopentanoperhydrophenanthrene

Fig. 31-10. Chemical structure of cyclopentanoperhydrophenanthrene ring. This common four-ring structure is basic structure of all steroids.

Because of the well-established positive association between plasma cholesterol concentration and coronary heart disease (CHD), we are apt to think of cholesterol as a harmful substance. Contrary to that belief, cholesterol is essential for normal functioning of the organism because it is as follows:

1. Essential structural component of membranes of all animal cells and subcellular particles
2. Obligatory precursor of bile acids
3. Precursor of all steroid hormones, including sex and adrenal hormones.

Physiology of cholesterol metabolism

The normal human body contains about 2 grams of cholesterol per kilogram of body weight. Much of this is in constant exchange with plasma cholesterol, in which the turnover rates and the amount that is exchangeable with the plasma are variable from one tissue to another.

The fractional rate of turnover of the pool, because of loss and replacement, is such that about 2% is renewed each day. Since the main channel for outflow from the pool is the gastrointestinal tract, the absolute rate of turnover (in grams per day) can be estimated by measurement of the daily fecal output of bile acids plus that of neutral steroids (cholesterol). There may be small losses of cholesterol by excretion through skin and by conversion into steroid hormones, some of which may be lost in the urine, but these losses are very small compared with the main outflow through the intestine. Measurements of fecal steroid output indicate probably 1 to 2 g/day for turnover of cholesterol in man, with excretion of bile acids accounting for about half the total turnover. A schematic representation (Fig. 31-11) illustrates that the concentration of a given cholesterol pool is under the influence of cholesterol input, output, and turnover rates. It should be stressed that because of the continuous cycling of cholesterol into and out of the bloodstream, the plasma cholesterol concentration is not a simple additive function of dietary cholesterol intake and endogenous cholesterol synthesis. Rather, it reflects the rates of synthesis of the cholesterol-carrying lipoproteins and the efficiency of the receptor mechanisms that determine their catabolism. A detailed discussion of the dynamics of lipoprotein concentration can be found in the section on lipoprotein metabolism.

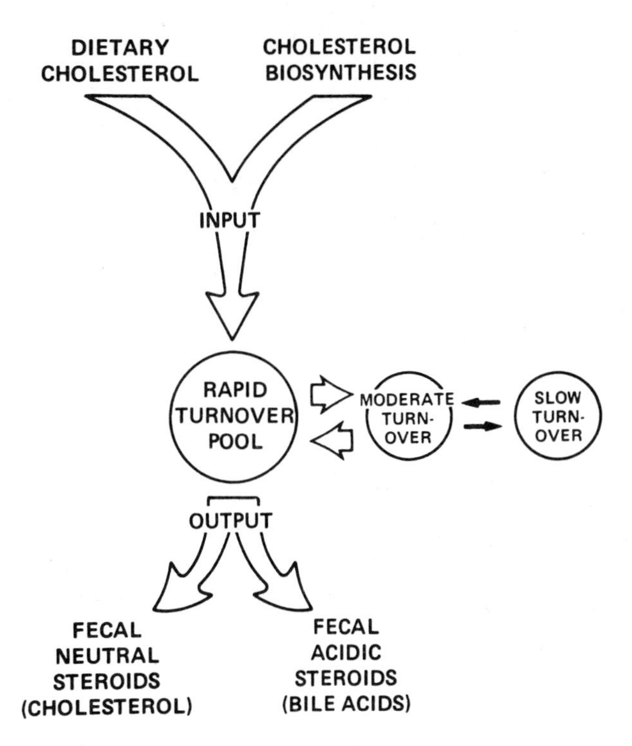

Fig. 31-11. Scheme of dynamics of cholesterol metabolism.

Cholesterol is present in all plasma lipoproteins, but about 60% of the total cholesterol in plasma from a fasting human subject is carried in the LDL. About two thirds of the plasma total cholesterol is esterified with long-chain fatty acids, with linoleic acid being the predominant fatty acid in man. The cholesteryl esters in the plasma are in a state of constant turnover because of their continual hydrolysis and resynthesis. Hydrolysis of cholesteryl esters takes place in the liver, but synthesis occurs mainly in the plasma by transfer of a fatty acid residue from lecithin to free cholesterol. This reaction is catalyzed by a plasma enyzme known as "lecithin:cholesterol acyltransferase," or LCAT (Fig. 31-12). The preferred lipoprotein for human LCAT is high-density lipoprotein (HDL), and it seems likely that the bulk of the esterified cholesterol in the plasma is formed on HDL. The cholesteryl ester then is transferred from HDL to LDL and VLDL, partly in exchange for triglyceride.

There is reason to believe that an LDL molecule consists of a micellar core of hydrophobic lipids including triglycerides and cholesteryl esters, surrounded by a liquid crystalline membrane of protein and amphipathic lipids (free cholesterol and phospholipids).

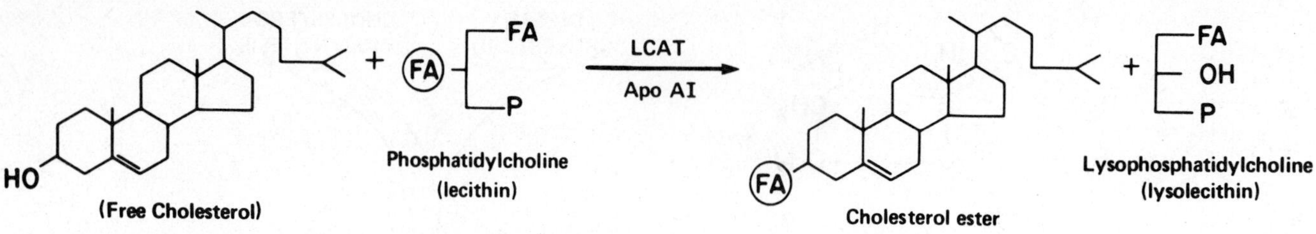

(Free Cholesterol) + Phosphatidylcholine (lecithin) —LCAT, Apo AI→ Cholesterol ester + Lysophosphatidylcholine (lysolecithin)

Fig. 31-12. Esterification reaction of free cholesterol; lecithin contributes a free fatty acid to form ester cholesterol. This reaction is dependent on enzyme lecithin:cholesterol acyltransferase, *LCAT,* and an apolipoprotein, *Apo AI*. *FA,* Fatty acid; *P,* choline.

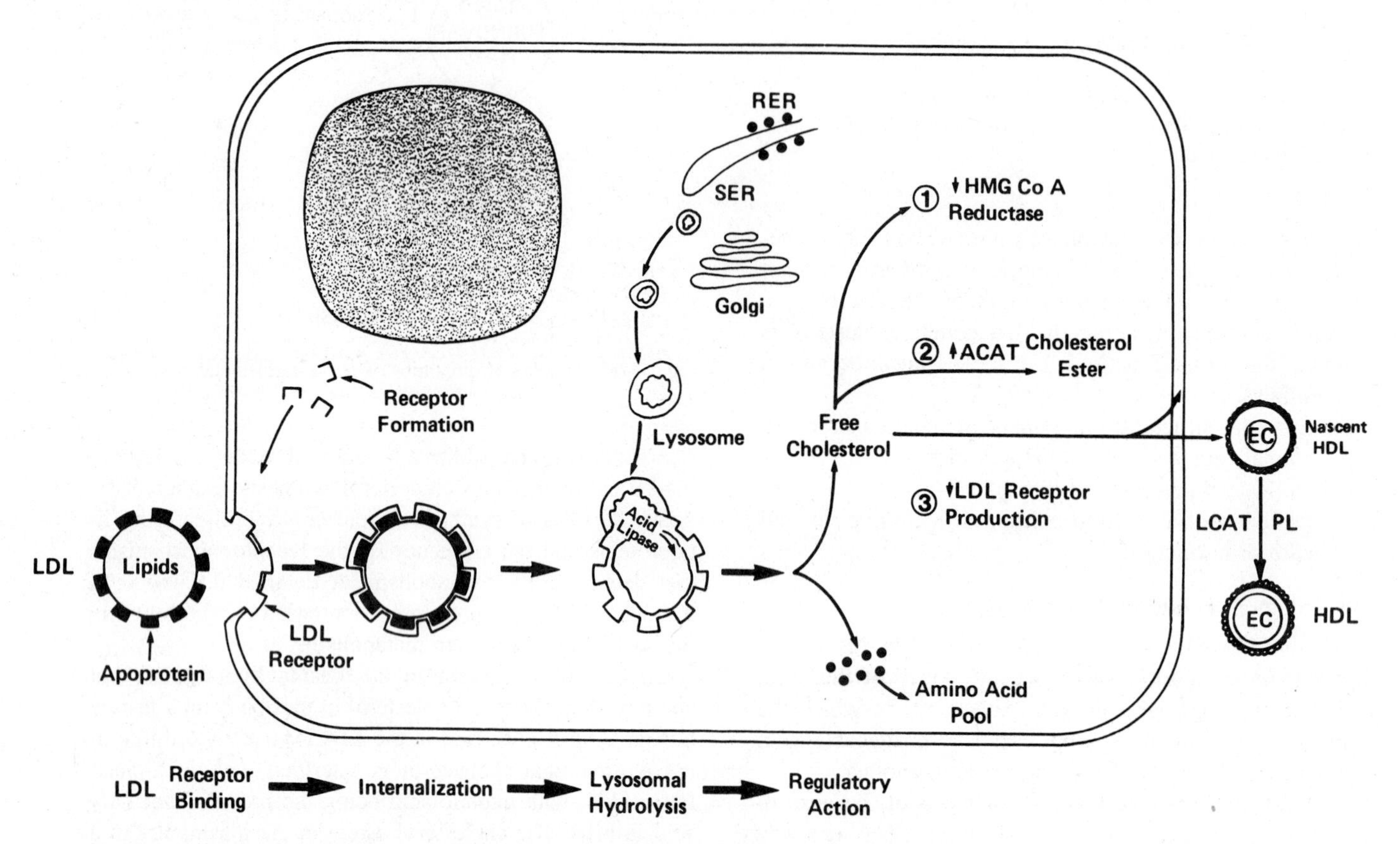

Fig. 31-13. Scheme of LDL uptake and catabolism by a cell. Mechanism not only clears LDL from circulation but also aids in regulation of cholesterol synthesis and storage. High density lipoprotein, *HDL,* plays an integral role in removing cellular cholesterol, esterifying free cholesterol in blood, and transporting cholesterol to liver for catabolism. *ACAT,* Acetylcholesterol acyltransferase; *HMG CoA reductase,* β-hydroxy-β-methylglutaryl-coenzyme A reductase; *EC,* cholesterol ester; *LCAT,* lecithin:cholesterol acyltransferase; *PL,* phospholipid; *RER,* rough endoplasmic reticulum, *SER,* smooth endoplasmic reticulum.

It has been suggested that one of the functions of LDL is to transport cholesterol, in esterified form, from the tissues to the liver. In this view, the function of LCAT is to generate esterified cholesterol from the free cholesterol formed in extrahepatic tissues. One might then envisage the sequence of events shown in Fig. 31-13. Tissue-free cholesterol from peripheral tissues is transferred to high-density lipoprotein; it is then esterified by LCAT, enabling HDL to take up more free cholesterol. The esterified cholesterol formed on high-density lipoprotein is transferred on to LDL and VLDL, where it is incorporated into the nonpolar core of the lipoprotein molecules. Low-density lipoprotein, carrying its load of cholesteryl ester, reaches the liver, where the cholesteryl esters are hydrolyzed. The free cholesterol so formed enters the pool of free cholesterol in the hepatocyte that is available for removal through

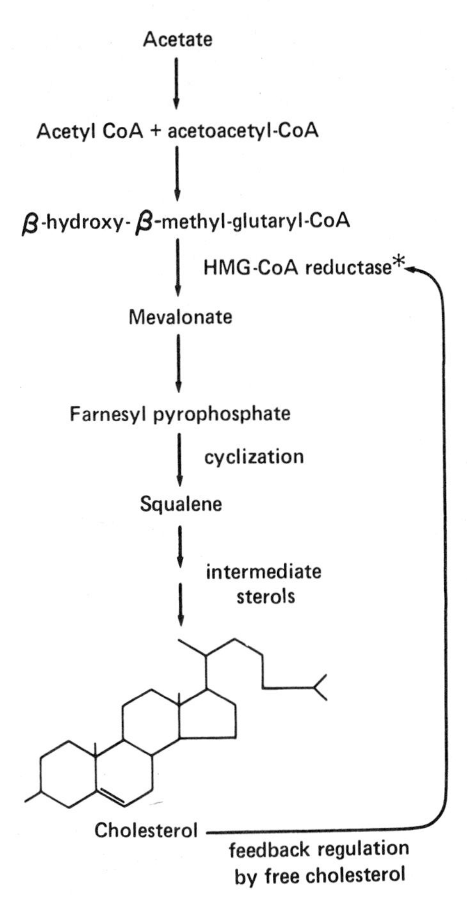

Fig. 31-14. Metabolic pathway of cholesterol synthesis, emphasizing negative feedback end-product inhibition step at β-hydroxy-β-methylglutaryl CoA step with the important enzyme HMG-CoA reductase.

the bile, conversion into bile acids, or reincorporation into plasma lipoprotein (VLDL).

Synthesis

Almost all animal tissues synthesize cholesterol from acetyl CoA. In adults the most actively synthesizing organs are liver and intestinal wall, these two tissues probably supplying over 90% of one's plasma cholesterol of endogenous origin. Hepatic cholesterogenesis, unlike intestinal cholesterol synthesis, is inhibited by dietary cholesterol. Cholesterol production rate (absorbed cholesterol plus endogenously synthesized cholesterol) amounts to about 1 g/day. As stated previously, cholesterol intake averages 600 mg/day, of which a maximum of 300 mg has been shown to be absorbed. The efficiency of cholesterol absorption (more properly called "coefficient of absorption") is somewhere between 25% and 40%, depending on the amount and type of other fats in the diet. The phenanthrene ring is not catabolized in the body; therefore, in the steady state, approximately 1 g of sterol is secreted into the feces as bile acids and neutral steroids. In most tissues the rate of synthesis of cholesterol is determined by the capacity of beta-hydroxy-beta-methylglutaryl CoA (HMG-CoA) reductase, the major enzyme catalyzing a rate-limiting step in the biosynthetic sequence from acetyl CoA to cholesterol. Although this appears to be the main rate-limiting reaction, there appear to be other sites of suppression in the biosynthetic cholesterol pathway. Hepatic HMG-CoA reductase is subject to induction and repression by several hormonal and dietary factors. A brief scheme on control of hepatic cholesterologenesis is shown below (Fig. 31-14).

Feedback control of hepatic cholesterologenesis is mediated by cholesterol itself and directly or indirectly by bile acids. Although virtually every cell utilizes cholesterol, the liver is one of the major organs that removes cholesterol from circulation. It does so by two routes: (1) by the excretion of free cholesterol into bile or (2) after conversion to primary bile acids (Fig. 31-15). The formation of bile acids is believed to regulate bile acid synthesis by negative product-inhibition feedback.

Absorption

Cholesterol is absorbed from the jejunum in the presence of bile salts. After entering the cells of the intestinal mucosa, it is incorporated into chylomicrons, which enter the blood circulation through the lymphatic system (see Fig. 31-3). The percentage of cholesterol that is absorbed from the diet is self-regulating. Increased levels of triglyceride in the diet tend to promote cholesterol absorption. The size of the bile acid pool appears to have a profound effect on the rate of cholesterol absorption. Diversion of bile acids from the intestinal lumen essentially stops cholesterol absorption, whereas the expansion of the bile acid pool by feeding of exogenous bile acids results in a greater than normal rate of cholesterol absorption. When the continuous intake of cholesterol is less than about 300 mg/day, more is absorbed (about 40% to 60%). If the intake is increased to 2 to 3 g/day, as little as 10% may be absorbed. In a typical American diet, about 600 mg of steroids are consumed per day in which one can anticipate about a 25% to 40% coefficient of absorption of dietary cholesterol. It should be stressed that there is a large individual variation in cholesterol absorption. This could account for, in part, the difference in individual responsiveness or lack of responsiveness to diet-induced hypercholesterolemia.

Catabolism

In man, increased absorption of cholesterol is followed by increased excretion of cholesterol from the exchangeable pool. Increased conversion of cholesterol into bile acids is also brought about by interruption of the enterohepatic circulation of bile salts. Bile salts returning to the liver from the intestine repress the formation of an enzyme catalyzing the rate-limiting step in the conversion of cho-

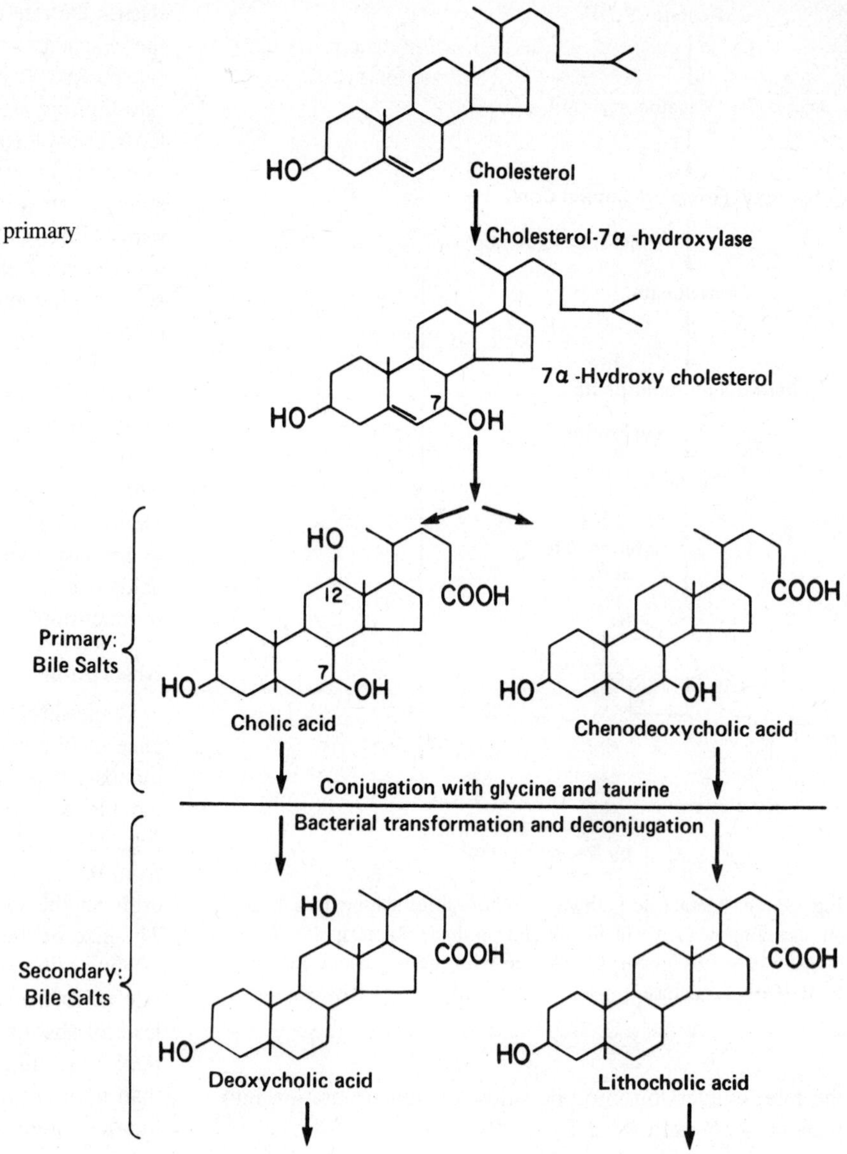

Fig. 31-15. Catabolic pathway of cholesterol to primary and secondary bile acids.

lesterol into bile acids. When bile salts are prevented from returning to the liver, the activity of this enzyme increases and degradation of cholesterol to bile acids is stimulated. This effect may be exploited therapeutically in the treatment of hypercholesterolemia by the use of unabsorbable resins, which bind bile acids in the lumen of the intestine and so prevent their return to the liver.

The percentage of neutral sterols and acidic sterols excreted in the bile is on the average 55% and 45% respectively. The main neutral sterol excreted in the bile is cholesterol.

The second mechanism that results in net loss of sterol from the body is the conversion of cholesterol to bile acid, which in turn is eventually excreted in the acidic sterol fraction of the feces. The simplified scheme in Fig. 31-15 describes what are assumed to be the principal pathways in the conversion of cholesterol to bile acids in man.

The mechanisms just stated for excretion of cholesterol via bile acids or cholesterol in the bile are dependent on the receptor-mediated activity in the hepatocytes. The hepatocytes have receptor sites that are specific for apoproteins B and E. The major function of the liver in lipoprotein clearance is to remove chylomicron remnants (which contain dietary cholesterol) from plasma. Lipoproteins containing apoproteins E (such as, chylomicron remnants and VLDL remnants) and B (such as LDL) are cleared by the same mechanism. However, the apo-E-containing lipoproteins are cleared with much greater efficiency than apo-B-containing lipoproteins. For this reason chylomicron remnants and VLDL remnants (IDL) are not measurable in normal individuals (see Fig. 31-5) because of their rapid clearance.

The uptake of LDL by the peripheral tissues is also receptor-site dependent. The binding of LDL to the receptor site followed by internalization and hydrolysis of the LDL leads to free cholesterol in the cell, which then functions (1) as a regulator for the rate of receptor synthesis, (2) as a regulator for cholesterol synthesis by the end-product negative-feedback mechanism, or (3) as a regulator for ACAT (acyl-CoA:cholesterol acyltransferase) activity, which determines how much cholesterol is stored in the cell as cholesteryl oleate, a cholesterol ester. It is believed that one of the factors that causes the efflux of cholesterol from the cell into the blood is the availability of HDL. By this process, the cholesteryl oleate in the cell is hydrolyzed to free cholesterol and fatty acid. The free cholesterol then crosses the cell membrane and binds to the nascent HDL, whereby it is transported back to the liver. During its transit in the blood to the liver, the free cholesterol is esterified to form cholesteryl ester by the enzyme LCAT. The liver, and to some extent the gastrointestinal tract and other organs, such as the adrenal glands and gonadal tissues, take up the HDL and catabolize the HDL to its protein and lipid constituents (including that of cholesterol).

Summary of cholesterol metabolism

In any attempt to summarize what is clearly a complicated interrelation in the intact animal, it is necessary to consider the interlocking nature of the various feedback systems involved.

Regulation of cholesterol biosynthesis in liver. Since cholesterol synthesis in the liver is controlled by the amount of dietary cholesterol absorbed, it is apparent that this system will be influenced by (1) the amount of dietary cholesterol, (2) the amount of sterol actually absorbed (which will in turn be influenced by the bile acids available to promote overall absorption), (3) the amount of lipoprotein available as its carrier, (4) species variation, and (5) the amount of cholesterol actually reaching the intracellular sites of cholesterol biosynthesis. Clearly, bile acids are major determinants of absorption and must therefore have a major role as a secondary mediator of hepatic cholesterol synthesis.

Regulation of cholesterol biosynthesis in intestine. Bile acids are also vital in the regulation of cholesterol biosynthesis in the intestinal wall, but it has not been possible up to the present to delineate whether this is a direct and primary action or the action is secondary to their essential role in promoting cholesterol uptake by the mucosa. It is clear, however, that any factors that cause a decrease in the steady-state concentration of bile acids in the gastrointestinal tract accelerate cholesterol synthesis in intestine and liver. In contrast, factors that produce an expanded bile acid pool cause an inhibition of cholesterol synthesis, either primarily or secondarily, in both liver and gastrointestinal tract.

Regulation of bile acid formation by cholesterol. Since the rate of bile acid formation may rise in response to or in compensation for enhanced cholesterol absorption at least in some species, it is clear that any absorbed cholesterol that exceeds the amount ordinarily synthesized by the liver must be disposed of through this mechanism or be deposited in the miscible cholesterol pools.

Regulation of bile acid formation by bile acids. Whether bile acids influence bile acid formation primarily or secondarily is likewise unsettled, but it is clear that the net effect of bile acid deprivation is to cause an acceleration of cholesterol and bile acid formation.

Normal expected cholesterol values

A problem arises in defining what levels of plasma lipids separate persons with elevated blood fats from the rest of the "normal" population. Cut-off values for plasma lipids are assumed to represent the 95th percentile of their distribution in a given population. Thus, before one can address the question of what is adequate dietary or drug therapy for the control of serum lipid and lipoprotein concentrations, one needs first to consider the degree to which blood lipids should be lowered. In other words, at what level should the clinician consider a sample "hyperlipidemic?" Many clinical laboratories and practicing physicians still use normal ranges in evaluating whether a sample can be considered normal or abnormal. As with many blood constituents, lipid nomograms are based on the sampling of "apparently normal" persons and arbitrarily defining hyperlipidemia as being present when the plasma cholesterol or triglycerides are above the 95th-percentile value for the population to which the persons belong. Unfortunately, because of the way we have defined "normal" in the past, many abnormal test results do not correlate very well with disease states or health-risk conditions on an individual basis. Thus the "normal" range for a given population may not be a "healthy" level. For example, a cholesterol value of 2500 to 2800 mg/L may be within the 95th percentile of the distribution of an apparently normal male population between 51 and 59 years of age in the United States, but about 40% to 50% of these persons eventually develop coronary heart disease. The frustrating question that frequently confronts the clinician is, "What is normal?"

Reference values for serum lipids and lipoproteins that are highly predictive of disease or disease risk, irrespective of the "normal distribution," should be established.

Cholesterol values depend on many factors. The main variables to consider are age and sex. Table 31-2 gives the distribution of total cholesterol by age and sex. These values were based on large-scale U.S. population studies (n = 42,653 subjects). The mean values are expressed as milligrams per liter with a range that represents values between the 5th and 95th percentiles. These are fasting serum cholesterol values. Plasma values are lower because of the dilution effect (about 3%) of anticoagulants, which draw water from the blood cells. Because of a physiological diurnal variation of 5% to 8%, two to three separate cho-

Table 31-2. Distribution of total cholesterol values in U.S. population*

	Total cholesterol in serum (mg/L)			
	Male (n = 24,485)		Female (n = 18,168)	
Age (years)	Mean	Range†	Mean	Range†
0-Cord blood	650	370-960	650	370-960
0-1	1510	1020-1900	1560	1130-2010
2-3	1590	1130-1990	1600	1170-2040
4-5	1620	1170-2180	1630	1200-2020
6-7	1630	1240-2040	1670	1300-2100
8-9	1660	1250-2110	1700	1280-2130
10-11	1650	1290-2100	1660	1300-2090
12-13	1610	1210-2050	1620	1260-2040
14-15	1550	1160-2010	1600	1230-2050
16-17	1540	1150-2040	1610	1230-2060
18-19	1570	1170-2040	1640	1230-2130
20-24	1720	1270-2240	1690	1250-2220
25-29	1870	1360-2510	1760	1310-2290
30-34	1980	1420-2620	1800	1330-2370
35-39	2070	1500-2780	1890	1440-2490
40-44	2130	1550-2760	2000	1510-2600
45-49	2180	1620-2850	2090	1560-2730
50-54	2190	1620-2860	2240	1660-2940
55-59	2200	1600-2850	2380	1770-3100
60-64	1690	1620-2850	2380	1770-3060
65-69	2190	1620-2820	2400	1760-3130
70+	2130	1550-2780	2350	1740-2980

*Cooper, G.R.: In Faulkner, R., and Meites, S., editors: Selected methods of clinical chemistry, vol. 9, Washington, D.C., 1982, American Association for Clinical Chemistry.
†Range is 95th and 5th percentiles.

lesterol values should be obtained for true base-line values.

Ideally, serum total cholesterol concentration should not exceed 2200 mg/L under 30 years of age, or 2500 mg/L under 60 years of age; children should have values less than 200 mg/L. Most institutions, however, just use the upper limit of "normal" for total cholesterol values (between 2500 to 2850), disregarding age, sex, and whether the sample is plasma or serum.

It should be emphasized that a person's serum or plasma cholesterol concentration is under the influence of several factors:

Genetics. This factor probably has the most important influence on a person's cholesterol concentration.

Age. Serum cholesterol concentration starts out around 650 mg/L at birth and steadily increases with age.

Sex. Cholesterol concentration in the blood of males is always higher than that in premenopausal females. After menopause, the cholesterol concentration is higher in females than that in males. Serum cholesterol levels in males seem to plateau by 50 to 60 years of age.

Diet. Saturated fat in the diet increases serum cholesterol levels, whereas polyunsaturated fat decreases cholesterol concentration. Monounsaturated fats have no apparent effect. Dietary cholesterol appears to elevate serum cholesterol levels. Plant sterols and certain types of fiber tend to decrease serum cholesterol concentration.

Physical activity. Physical activity tends to lower serum total cholesterol. Much of this effect depends on the type, intensity, duration, and frequency of physical activity. Exercise lowers LDL cholesterol but increases HDL cholesterol concentration.

Hormones. Growth hormone, thyroxine, and glucagon decrease serum cholesterol levels.

Primary disease states. Diabetes mellitus, thyroid dysfunction, obstructive liver disease, acute porphyria, dysgammaglobulinemias, and nephrotic syndrome have an effect on blood cholesterol concentrations.

TRIGLYCERIDE METABOLISM

Biological functions

Triglycerides represent the major form of fat found in nature, and their primary function is to provide energy for the cell. The cell burns fatty acids to CO_2 and H_2O at the expense of molecular oxygen. One gram of fatty acids liberates about 9 kcal. The human body stores large amounts of fatty acids in ester linkages with glycerol in the adipose tissue. This form of storage of reserve energy is highly efficient because of the magnitude of free-energy change that occurs when fatty acids undergo catabolism. The energy release of 9 kcal/g when fatty acid is burned to CO_2 and H_2O is about 2½ times higher than that of the other foodstuffs, that is, protein and carbohydrate, which yield about 4 kcal/g when catabolized. Moreover, fatty acids in the form of triglycerides are in an almost anhydrous form, whereas carbohydrates and proteins are stored in an aqueous environment. It is evident that, in terms of the energy to mass ratio, fat is a much more efficient means of storing energy than carbohydrates or protein.

Chemistry

Triglycerides are by far the most abundant subclass of neutral glycerides in nature. Mammalian tissues also contain some diglycerides and monoglycerides, but these occur in trace levels when compared to triglycerides.

Most triglyceride molecules in mammalian tissues are mixed glycerides. Although it had been postulated that the different kinds of fatty acids are distributed among the α, β, and α' positions of glyceride-glycerol in a random fashion, recent evidence clearly establishes that the distribution is nonrandom. The distribution between the β position and the α and α' positions seems to be governed by chain length and unsaturation. The shorter and more unsaturated fatty acids tend to occupy the β position. Fatty acids are also asymmetrically distributed between the α and α' positions, with the α position having an excess of palmitic acid and the α' position an excess of oleic acid.

Because of their water insolubility, triglycerides are transported in the plasma in combination with other more polar lipids (phospholipids) and proteins, as well as with cholesterol and cholesterol esters, in the complex lipoprotein macromolecules. The structure of the lipoproteins does not depend on covalent linkages between the various components but is determined rather by a combination of polar and nonpolar interactions between them. It appears that the essentially nonpolar triglyceride (and cholesterol ester) is largely in the center of the lipoprotein, with the more polar protein and phospholipid components at the surface, their polar groups being directed outwards so as to stabilize the whole structure in the aqueous plasma environment.

The concentration of triglyceride in the plasma at any given time is a balance between the rate of entry into the plasma and the rate of removal. A change in concentration may therefore be the result of a change in either or both of these factors. Moreover, a primary change in one may result in a secondary change in the other. Thus perhaps the main problem to be considered in any situation where the plasma triglyceride concentration is abnormally high is whether this is attributable to a rise in the rate of entry or to a fall in the rate of removal.

Entry of triglyceride into plasma

From intestine. Digestion in the intestinal lumen breaks down (hydrolyzes) triglyceride into free fatty acids and monoglycerides, and these substrates are absorbed by the intestinal cells and resynthesized into triglyceride, which is then released into the lymphatics as lipoproteins called "chylomicrons." These lipid-protein complexes contain about 82% triglyceride, 9% cholesterol (mainly as the ester), 7% phospholipid, and a very small amount (less than 2%) of protein (see Fig. 31-5). Although the amount of protein is small, there is a good deal of evidence that its presence is necessary for the release of the chylomicrons. For example, in abetalipoproteinemia (a genetically determined disease in which an apoprotein cannot be made in the body) triglyceride is not released from the intestinal cells.

The details of the processes involved in chylomicron formation and release are still obscure, and Fig. 31-3 merely provides a diagram of what takes place in general terms. The chylomicrons are probably formed in the cisternae of the rough and smooth endoplasmic reticulum and the Golgi apparatus of the intestinal cell, where they are transported to the lateral surfaces and released into the interstitial spaces. From the interstitial spaces they enter the lymphatics and are conveyed to the bloodstream through the thoracic duct.

An increased influx of chylomicron triglyceride into the plasma occurs after the ingestion of each meal, and this persists for the several hours during which fat is absorbed. It causes an increase in plasma triglyceride concentration, and because the chylomicrons are large enough to scatter light (up to 0.5 μm in diameter), the plasma becomes lactescent (turbid) so as to produce what is commonly called an "alimentary lipemic response." However, this rise in plasma triglyceride concentration is relatively small when related to the magnitude of the total triglycerides transported by the plasma because the rate of removal of the chylomicron triglyceride also rises rapidly to approximate itself to the increased rate of entry.

Even in starvation or in persons on diets containing little fat, some triglyceride is released from the intestinal cells into the lymphatics, though the amount is naturally much reduced. In such circumstances the lipoprotein complexes themselves are relatively small; on the other hand, when the amount released is large, as during the absorption of a meal containing fat, the complexes increase in size by acquiring a greater load of triglyceride until they become chylomicrons, as these are commonly defined. There may also be an increase in the number of chylomicrons released when a great deal of triglyceride has to be transported from the intestinal cells, but it nevertheless seems likely that much of the rise in output is achieved simply by an increase in the triglyceride loading of each chylomicron unit.

From liver. The liver is the second site of triglyceride release into the plasma. The source of the fatty acids present in the triglyceride entering the blood from this organ depends greatly on the nutritional state. Thus, in a fasting individual, fatty acids are mobilized from the adipose tissue stores and are transported in the plasma in unesterified form bound to the plasma albumin. Most are carried directly to tissues such as muscle and are utilized as a primary source of energy. However, some (about 30% to 40% in the resting state) enter the liver, and here a proportion is directly oxidized, either completely to carbon dioxide and water or partially to ketone bodies, while a proportion is reesterified within the liver and released again into the plasma as triglyceride. In this state, therefore, the fatty acids of the plasma triglyceride are derived indirectly from the fatty acids mobilized from adipose tissue.

Contrast this with the situation immediately after the ingestion of a meal by an individual on a low-fat, high-carbohydrate diet. Here part of the dietary carbohydrate is converted to fatty acid in the process known as "lipogenesis." This conversion occurs both in adipose tissue, in which case the fatty acid is directly esterified and stored in that tissue as triglyceride and in the liver. In the liver, fatty acids are also esterified to yield triglyceride, which is then released into the plasma as lipoproteins. In this situation the fatty acid moiety of the plasma triglyceride has come mainly from the dietary carbohydrate.

Despite these variations in the source of the triglyceride fatty acids released by the liver, it is important to realize that the release process itself is a continuous one and that, except during the absorption of dietary fat, the liver is the main contributor of triglyceride to the plasma.

Our knowledge of the process of triglyceride release from the liver indicates that it is probably very similar to triglyceride release from the intestine. Again, as in the intestine, the size of the lipoprotein complexes formed by the liver varies according to the amount of triglyceride that is being released. Thus high rates of release result in large complexes with a high triglyceride load and a correspondingly low density. In fact, the lipoprotein complexes released from the liver under such conditions may reach a size not much below that of the chylomicrons, even though they normally have a somewhat lower triglyceride content and therefore a higher density.

Transport of triglyceride by lipoproteins in plasma

Some lipoproteins present in the plasma are essentially those that have already been described as being produced by the intestine and the liver. These are known either as the VLDL because they are the plasma lipoprotein fraction most readily separated by flotation in the ultracentrifuge or as pre-beta-lipoproteins on the basis of their electrophoretic behavior. During the absorption of dietary fat there are also present chylomicrons, which, because they are of still lower density than the VLDL, are even more readily separated by flotation in the ultracentrifuge.

Although most plasma triglycerides are present in chylomicrons and VLDL, the plasma also contains other triglyceride-carrying lipoproteins. They are called either "low-density lipoproteins" (LDL) or "high-density lipoproteins" (HDL) on the basis of their relative densities and lipid/protein ratios, and they have, respectively, beta-globulin and alpha-globulin electrophoretic mobility. As more knowledge is acquired, it is clear that this relatively simple classification into four main groups of lipoproteins is far from adequate.

Removal of triglyceride from plasma

Since the rate of removal of triglyceride from the bloodstream is normally adjusted quickly to balance its rate of entry, whenever the latter varies, the former will also change. Indeed, in experimental animals, the circulating half-life of the chylomicron triglyceride fatty acid under such conditions is only a few minutes.

This rapid removal has to be reconciled with the fact that the lipoprotein complexes (of which the triglyceride is a part) are so large that their passage across the blood vessel walls in most body tissues could only be extremely slow. The rapid removal of the chylomicron (and VLDL) triglyceride fatty acid must therefore entail the operation of a process acting selectively on the triglyceride component.

It is now recognized that this selective process involves the hydrolysis of the triglyceride moiety by a lipase called "clearing-factor lipase," or "lipoprotein lipase." Present evidence suggests that the site of action of this enzyme on the VLDL and chylomicron triglyceride is at the luminal surface of the capillary endothelium of the extrahepatic tissues and, in order to be acted upon, the VLDL and chylomicrons have to be trapped or sequestered at this site as they pass through the capillary lumen. This sequestration is of particular interest because it requires an association to take place at the endothelial cell surface between the enzyme and the chylomicrons or VLDL in the bloodstream. In fact, it appears that the association of HDL apoprotein with VLDL and chylomicrons (which occurs mainly after these have been released into the plasma) is a prerequisite for the hydrolysis of the triglyceride fatty acids from the bloodstream.

The fatty acids released by hydrolysis at the endothelial cell surface can readily pass out of the bloodstream into the tissues. The uptake of triglyceride fatty acids by a particular tissue depends on the activity of clearing-factor lipase in that tissue's capillaries.

The action of clearing-factor lipase at the endothelial cell surface not only facilitates the removal of triglyceride fatty acid from the blood but also determines where it is utilized, and this has important consequences. For example, in a state of caloric excess, the proportion of the triglyceride fatty acid in the bloodstream that is in excess of the immediate caloric needs is taken up by adipose tissue. Most fatty acids are reconverted to triglyceride therein and stored, whereas the remainder of the uptake in such a situation occurs predominantly in muscle where most of the fatty acid is probably oxidized directly. In contrast, in a state of calorie deficit (as during fasting) the tissues derive their energy primarily from the oxidation of unesterified fatty acids carried to them in the blood after their mobilization from adipose tissue. There is still triglyceride in the blood in VLDL under these conditions, but instead of being taken up by adipose tissue for storage, it is now directed away from this tissue and toward muscle to supplement the supply of energy from the mobilized fatty acids. This switch in triglyceride fatty acid uptake is achieved through changes in the activity of clearing-factor lipase in the tissues concerned. Thus fasting results in a fall in the activity of the enzyme in adipose tissue and an increase in its activity in muscle.

Control of intracellular adipose cell lipase

If clearing factor lipase determines the pattern of triglyceride fatty acid removal from the blood, what regulates the mobilization of triglyceride fatty acids from adipose cells into the circulation? It appears likely that, at least with respect to the enzyme in adipose tissue, the control is humoral. This enzyme is concerned with the mobilization of the fatty acids that are stored in adipose tissue as triglyceride and is not active in the tissue when clearing factor lipase is least active.

It has been shown in various in vitro systems that insulin favors an increase in the activity of the enzyme in such tissue, whereas other hormones (such as the catecholamines, ACTH, and glucagon) inhibit the rise that is promoted by insulin. This intracellular adipose triglyceride enzyme is called "hormone-sensitive lipase" because it is converted from an inactive to an active form by epinephrine, norepinephrine, adrenocorticotropin (ACTH), thyroid-stimulating hormone (TSH), and glucagon. Moreover, its activity is promoted by growth hormone. On the other hand, insulin inhibits the activity of this lipase. It is believed that adenosine 3′,5′-monophosphate (cyclic 3′,5′-AMP) functions as an activator of the hormone-sensitive lipase and that the hormones that stimulate its activity do so by increasing the intracellular level of cyclic 3′,5′-AMP, whereas insulin is believed to decrease the level of this substance. Corticosteroids and thyroid hormone are believed to play a permissive role in the activation of this enzyme. Unlike the lipoprotein lipase of adipose tissue, hormone-sensitive lipase of other tissue exhibits increased activity during fasting, possibly because of falling insulin levels. It is believed that hormone-sensitive lipase plays an important role in fat mobilization from adipose tissue.

Adipose tissue contains a monoglyceride lipase that vigorously promotes the hydrolysis of beta-monoglycerides and, at a slower rate, alpha-monoglycerides to glycerol and free fatty acids. The rate of monoglyceride and diglyceride hydrolysis is much greater than that of triglyceride hydrolysis resulting from the action of hormone-sensitive lipase. However, the activity of the monoglyceride lipase is not influenced by hormones.

There is much more to be learned about the hormonal control of clearing factor lipase activity. For example, it is not yet clear how a particular hormone may alter the activity of the enzyme in adipose tissue in one direction and that of the enzyme in muscle in the opposite one. However, we are at least beginning to understand how changes in nutritional and hormonal balance may affect the activity of the enzyme in particular tissues and hence the pattern of triglyceride fatty acid removal from the blood. The activity of functional clearing-factor lipase in the tissues may be inhibited in some hypertriglyceridemic conditions, and this may well be the prime cause of one type of hypertriglyceridemia (type I hyperlipoproteinemia). However, such a deficiency may also be attributable either to a lack of total enzyme in the tissue or to a deficiency in the conversion of the inactive to the active form of the enzyme. Such consideration as the foregoing may have particular implications for clinical studies on hypertriglyceridemia in man. For example, at the present time the most convenient assay for clearing-factor lipase in man involves the measurement of the activity of the enzyme from the plasma after the intravenous injection of heparin (10 U/kg of body weight). This release of the enzyme into the bloodstream by heparin has not been discussed here because it tends to obscure a proper appreciation of the normal functioning of the enzyme in the tissues. However, although such post-heparin-plasma activity may give some measure of the total activity of the functional enzyme in a person's body tissues (that is, it probably represents the enzyme that is released by heparin from the endothelial cell surface), neither can it give any measure of the total activity in particular tissues, nor can it indicate the capacity to renew the supply of functional enzyme.

Normal expected triglyceride values

Like many cholesterol ranges that represent the "normal" U.S. population, most triglyceride data are also obtained from large-scale epidemiological studies. Table 31-3 provides "normal" expected values for both males and females according to age (Cooper, 1982).

Clinical chemistry laboratories usually set their upper limit of normal for fasting plasma triglyceride levels somewhere between 1350 to 2000 mg/L based on 95th-percentile determinations for given populations. However, the clinical significance of these statistical limits is uncertain.

PHOSPHOLIPID METABOLISM

Biological functions

Like cholesterol, phospholipids are important structural components for cell membrane formation and maintenance. In digestion, phospholipids serve as important amphipathic compounds, which combine with bile salts in specific combinations to solubilize cholesterol. For example, lithogenic bile is characterized by an altered ratio of cholesterol, phospholipid, and bile salts. Phospholipids are also important for the formation of lipoproteins for fat transport. Phospholipids serve in another vital structural

Table 31-3. Distribution of triglyceride values in population of U.S.A.*

	Total cholesterol in serum (mg/L)			
	Male (n = 24,485)		Female (n = 18,168)	
Age (years)	**Mean**	**Range†**	**Mean**	**Range†**
0-Cord blood	360	100-980	360	310-980
0-1	700	310-820	780	310-980
2-3	550	300-840	660	320-1030
4-5	550	280-910	590	320-970
6-7	560	310-1050	600	330-1070
8-9	580	300-1090	640	340-1100
10-11	590	300-1090	720	370-1240
12-13	720	360-1380	810	420-1380
14-15	760	360-1380	790	400-1370
16-17	800	380-1570	730	400-1200
18-19	850	430-1690	790	410-1350
20-24	1030	450-2070	740	370-1350
25-29	1190	470-2560	770	380-1490
30-34	1320	510-2740	810	400-1550
35-39	1490	550-3310	880	410-1810
40-44	1550	560-3300	1010	460-1970
45-49	1560	590-3370	1080	470-2200
50-54	1560	590-3300	1180	530-2400
55-59	1450	590-2940	1290	560-2700
60-64	1460	590-3000	1310	570-2460
65-69	1410	580-2750	1350	610-2500
70+	1340	590-2660	1350	610-2440

*Cooper, G.R.: In Faulkner, R., and Meites, S., editors: Selected methods of clinical chemistry, vol. 9, Washington, D.C., 1982, American Association for Clinical Chemistry.
†Range is 95th and 5th percentiles.

function, that is, in the esterification of free cholesterol. It seems the fatty acid moiety of the cholesterol ester is derived from the β position of the phospholipid molecule.

Another important biological function of phospholipid is the production of dipalmityl lecithin in the lung as an essential constituent of its surfactant. It has been shown that evaluation of the phospholipid patterns in amniotic fluid provides a reliable index of fetal lung maturity. (See Chapter 35 for details.) Table 31-4 summarizes the phospholipid subfractions and describes their origin and biological function.

Chemistry

Phospholipids are characterized by the presence of nonpolar hydrophobic side chains and polar hydrophilic head groups. These chemical groupings make them particularly suitable compounds to serve as major constituents for biological interfaces, such as membranes. The structure of hydrophobic paraffin chains varies from phospholipid to phospholipid and greatly influences their physiochemical behavior.

Lecithin appears to be quantitatively the most important glycerophosphatide, both in membrane systems and in lipid-transporting mechanisms. There are two characteristics of the fatty acid esterification pattern that have concerned investigators in the field: (1) the tendency for the fatty acids in the 1 position to be predominantly saturated and for those in the 2 position to be unsaturated or polyunsaturated and (2) the much higher proportion of polyunsaturated long-chain fatty acids, such as arachidonic acid (20:4ω6) in the 2 position of lecithin than in diglycerides or triglycerides. The biological significance of this pattern is not clear at this time.

Digestion and absorption of phospholipids

Phospholipids are hydrolyzed by phospholipases to form lysolecithin and fatty acids, which promote the absorption of these substrates by the mucosal cell whereby they are again resynthesized to phospholipids. Fig. 31-3 illustrates the digestion and absorption process of dietary phospholipids and their fate in lipoprotein transport.

Synthesis

The first reaction leading to synthesis of phospholipid molecules anew is the acylation of glycerol 3-phosphate giving rise to phosphatidic acid (see Fig. 31-16). After phosphatidic acid is formed, there are separate pathways for the synthesis of neutral and acidic glycerophosphatides. The neutral glycerophosphatides, phosphatidyl choline and phosphatidyl ethanolamine, are synthesized from diglycerides formed by the action of phosphatidate phosphatase (EC 3.1.3.4) on phosphatidic acid. However, the acidic

Table 31-4. Different phospholipids and their source and function

Name of phospholipid	Source	Remarks
Phosphatidic acid	Animals, higher plants, microorganisms	Only minute amounts found; main importance as a biosynthetic intermediate
Phosphatidyl choline (lecithin)	Animals; first isolated from egg yolks; higher plants; rare in microorganisms	Most abundant animal phospholipid
Phosphatidyl ethanolamine	Animals, higher plants, microorganisms	Widely distributed and abundant; major component of old "cephalin" fraction; *N*-acetyl derivatives in brain; fatty amides in wheat flour, peas
Phosphatidyl serine	Animals, higher plants, microorganisms	Widely distributed but in small amounts; minor component of old "cephalin" fraction
Phosphatidyl inositol	Animals, higher plants, microorganisms	Natural lipid is found as a derivative of *myo*-inositol 1-phosphate only
Phosphatidyl glycerol	Mainly higher plants and microorganisms	"Free" glycerol has opposite stereochemical configuration to the acylated glycerol, that is, 1,2-diacyl-*sn*-glycerol; probably most abundant phospholipid

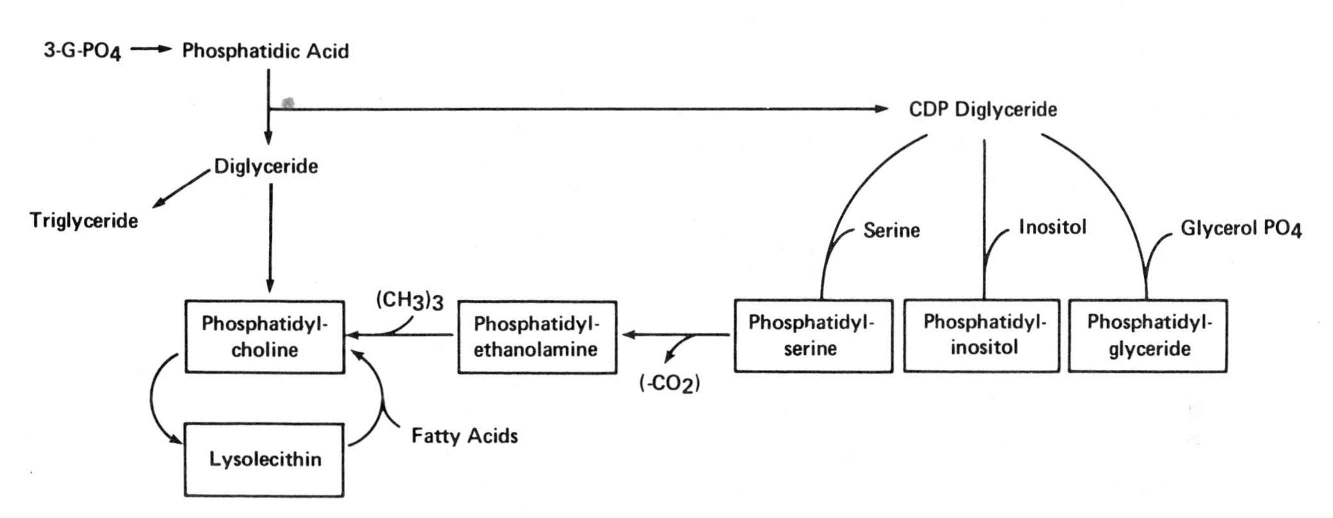

Fig. 31-16. Diagram of phospholipid synthesis. *CDP,* Cytidine diphosphate; *3-G-PO$_4$,* glycerol-3-phosphate.

glycerophosphatides are synthesized from cytidine diphosphate-diglyceride (CDP-diglyceride); this intermediate half-lipid, half-nucleotide can be regarded as a nucleotide-activated diglyceride. Thus synthesis of the neutral phosphatides is accomplished by activation of the bases choline and ethanolamine with cytidine triphosphate, whereas the acidic phosphatides are synthesized from an activated diglyceride without the further activation of the polar head group. The pathway of lecithin biosynthesis from glycerol 3-phosphate by means of diglyceride and CDP-choline was the first pathway of phospholipid biosynthesis to be established in detail and is referred to as the "Kennedy pathway."

Fig. 31-16 illustrates the major pathways for phospholipid biosynthesis. The first is the trimethylation of the ethanolamine residue of phosphatidyl ethanolamine with *S*-adenosylmethionine to produce phosphatidyl choline; the other is the decarboxylation of phosphatidyl serine to produce phosphatidyl ethanolamine. The former of these two reactions (the methylation pathway) appears to be quantitatively important in several mammalian tissues.

However, this outline of the biosynthetic pathways leading to the glycerophosphatides gives no indication of the mechanisms involved in establishing the 1-saturated, 2-unsaturated distribution of fatty acid esterification in lecithin and other phosphatides. The turnover of the fatty acids of lecithin is faster in relation to the glycerol backbone than the fatty acids of triglycerides. There are tissue enzymes that can remove fatty acids independently from the 1 and 2 positions of lecithin and other enzymes that can reesterify new ones back in their place. The enzyme system acylating a free hydroxyl at the 1 position was shown to be specific for saturated acyl CoA's, whereas the enzyme acylating a free hydroxyl at the 2 position was specific for unsaturated and polyunsaturated fatty acids. Soon after, the presence of tissue phospholipases (A_1 and A_2) was demonstrated, and this completed the Lands/Van Deenen cycle. This cycle will obviously take a lecithin molecule synthesized anew by the Kennedy pathway and remold it to form a lecithin with a 1-saturated, 2-unsaturated distribution.

The pathways that have been discussed so far indicate that there are three mechanisms for introducing fatty acids into lecithin: (1) the acylation of glycerol 3-phosphate, (2) methylation of phosphatidyl ethanolamine, and (3) fatty acid exchange reactions. There are also three mechanisms for introducing a choline residue onto the glycerol backbone: (1) choline phosphotransferase, (2) synthesis of choline by methylation of phosphatidyl ethanolamine, and (3) another pathway not yet mentioned, a lecithin-free choline exchange. It was originally believed that although these parts of the lecithin molecule could originate in several different ways the glycerol backbone was always derived from glycerol-3-phosphate. Recent work has shown that there are two other possible sources, namely, the triose phosphates of the glycolytic pathway.

Normal expected phospholipid values

The serum phospholipid concentration in normal healthy individuals is in about the same concentration range as total cholesterol. The ratio of phospholipid to cholesterol remains around unity. Any change in cholesterol concentration results in a corresponding change in phospholipids in a similar direction.

SPHINGOLIPID METABOLISM

Biological function

Sphingolipids are a heterogeneous group of compounds that have in common either the 18-carbon amino alcohol sphingosine or a closely related compound. In these sphingolipids, the amino group of sphingosine or its related compounds is an amide linkage with a fatty acid; that is, sphingosine is present as *N*-acylsphingosine compounds, which are called "ceramides." Ceramide is a fundamental unit found in all sphingolipids.

Sphingomyelin. Sphingomyelin is a phosphorus- and choline-containing sphingolipid. It is often grouped together with the phosphoglycerides under the classification of phospholipids. The phosphate in sphingomyelin is in diester linkage, one ester bond linking it with the primary alcohol of ceramide and the other with the hydroxyl of choline. Some sphingomyelin contains dihydrosphingosine rather than sphingosine. The fatty acid amide linkage in sphingomyelin is relatively stable to alkaline hydrolysis, and it was in this way that sphingomyelins were originally distinguished from choline-containing phosphoglycerides.

From a physical-chemical point of view, sphingomyelin is similar to phosphatidyl choline; that is, the ceramide is a hydrophobic region, as are the fatty acyl groups of phosphatidyl choline, and the highly polar region of both structures is the zwitterion phosphoryl choline. Because of the strongly basic character of the choline, sphingomyelin is soluble in 95% ethanol. It is also soluble in benzene but not in acetone, ether, and water.

Sphingomyelin is found in all tissues of the body. In the brain, which is particularly rich in phospholipids, about 10% of the lipid phosphorus is in the form of sphingomyelin. In kidney, spleen, lung, erythrocytes, and plasma, from 10 to 25% of the lipid phosphorus is in the form of sphingomyelin; in most other tissues, only 5% to 10% of the lipid phosphorus is in this form.

The fatty acids in amide linkage in sphingomyelin are mainly long-chain saturated and monounsaturated acids. Stearic, lignoceric, and nervonic acids comprise more than 75% of the fatty acids in the sphingomyelin of the brain. In other tissues, palmitic acid is also quantitatively important in this lipid.

Tissues of patients with Niemann-Pick disease contain greater than normal amounts of sphingomyelin. The spleen has been the subject of most study, but other tissues are involved, particularly certain regions of the brain. This rare disease, which nearly always affects young children, causes both physical and mental retardation. The cells of the liver and spleen acquire a foamy appearance, and the ganglion cells of the nervous system swell. Sphingomyelinase activity is greatly reduced or is absent from the spleen, liver, and kidneys of patients with the classic and the visceral forms of Niemann-Pick disease. Presumably, deficiency of this enzyme is the cause of the accumulation of excessive amounts of sphingomyelin in the tissues of these patients. However, in the late infantile form of Niemann-Pick disease there are significant amounts of sphingomyelinase in the tissues; thus a different biochemical lesion must be responsible for the sphingomyelin accumulation in these patients. The sphingomyelinase level of leukocytes can be used as an aid in the diagnosis of Niemann-Pick disease.

Cerebrosides and other glycosylceramides. Glycosylceramides are a family of sphingolipids in which the primary hydroxyl group of the ceramide is linked by a glycosidic bond to either a monosaccharide or an oligosaccharide chain. When linked to a monosaccharide, the compounds are called "cerebrosides." The glycosylceramides are also classified as glycolipids, as are the cerebroside sulfates and gangliosides.

The first member of the glycosylceramides to be discovered is the ceramide-galactoside of brain (commonly called "cerebroside"). The galactose is usually linked to the ceramide by a beta-glycosidic bond. Large amounts are found in the white matter of the central nervous system; 4% of the wet weight of brain is cerebroside. Kidney also contains ceramide-galactoside. Specific names are given to the ceramide-galactosides, based on the nature of the fatty acid group in the amide linkage (the major components of the brain cerebrosides are listed in Chapter 48 on the classification of lipids). About half of these fatty acids of brain cerebrosides are alpha-hydroxy fatty acids. Of these, cerebronic and oxynervonic are by far the most abundant. The brain apparently contains enzymes that catalyze the alpha-oxidation of fatty acids to yield alpha-hydroxy acids. In Refsum's disease, there is an enzymatic deficiency in the alpha-oxidation of phytanic acid.

The major nonhydroxy fatty acids in brain cerebrosides are 18 carbons or longer, even-carbon saturated, and monounsaturated acids, along with a small but significant amount of odd-carbon acids.

Although far less abundant than in the central nervous system, cerebrosides are found in most if not all mammalian tissues. In plasma, spleen, and liver, the cerebroside contains glucose rather than galactose. Glucose cerebrosides accumulate in liver, spleen, and lymph nodes in Gaucher's disease, an inherited lipidosis. Spleen also contains an enzyme that specifically catalyzes the hydrolysis of glucocerebrosides to yield ceramide and glucose; that is, this enzyme cannot utilize galactocerebroside as a substrate. It is the deficiency of this glucocerebroside-specific enzyme that is the primary enzyme deficiency in Gaucher's disease. On the basis of current knowledge, it seems likely that the following series of events cause glucocerebrosides to accumulate in the reticuloendothelial system of patients with Gaucher's disease. Globoside (see below), which is released from erythrocytes as they are phagocytized by the reticuloendothelial system, is hydrolyzed by the action of enzymes in that tissue. Glucocerebroside is one of the hydrolysis products; it normally is rather rapidly degraded by the glucocerebroside-specific enzyme, but since this enzyme is deficient in patients with Gaucher's disease, the glucocerebroside accumulates in the reticuloendothelial system.

Kidney also contains a ceramide-dihexoside in which the disaccharide is composed entirely of galactose units. Another ceramide-dihexoside has been found in kidney, plasma, liver, spleen, and erythrocytes in which the disaccharide is composed of glucose and galactose units.

A ceramide-trihexoside containing one glucose unit and two galactose units in the trisaccharide has been found in plasma, liver, spleen, and kidney. This lipid appears to be the main one accumulating in renal glomeruli and tubules and in blood vessels throughout the body in Fabry's disease, a sex-linked lipidosis.

Ceramide-tetrahexosides are also present in many tissues. The tetrasaccharide contains two galactose units, one glucose unit, and one acetylgalactosamine unit. Globoside of the red blood cell stroma is a ceramide-tetrahexoside and has been characterized as having the following structure: *N*-acetylgalactosaminyl (1→6) galactosyl (1→4) galactosyl (1→4) glycosylceramide. This ceramide-tetrahexoside, or a structure similar to it, is also present in plasma, kidney, liver, and spleen.

Sulfates (cerebroside sulfates). Besides cerebrosides, the brain contains sulfate esters of cerebroside with the sulfate group located on C-3 of the galactose moiety. These compounds are called "cerebroside sulfates," or "sulfatides." In the brain, the ratio of cerebroside to cerebroside sulfate is about 4 to 1. Although found most abundantly in brain, cerebroside sulfates are also present at low levels in liver, lung, kidney, spleen, skeletal muscle, and heart. The glycosidic linkage in cerebroside sulfate is of the beta configuration. The fatty acid composition of cerebroside sulfates is qualitatively similar to that of cerebrosides, but quantitatively there are differences; for example, brain cerebrosides have a much greater percentage of alpha-hydroxy fatty acids than brain cerebroside sulfates do.

Cerebroside sulfates accumulate in the central nervous system and the kidneys of patients afflicted with metachromatic leukodystrophy. Indeed, the metachromatic material is sulfatide. This disease is a diffuse demyelinating disorder, and its study has led to a consideration of the role of cerebroside sulfates in the formation of myelin and of plasma membranes in general.

Recently, a ceramide-dihexoside sulfate containing equal amounts of glucose and galactose has been isolated from the kidney. This sulfatide is clearly closely related to glycosylceramides containing oligosaccharide chains. Possibly a broad spectrum of sulfatides will be found, and they may correspond to the oligosaccharide-containing glycosylceramides, discussed in the preceding section.

Gangliosides. Gangliosides are a complex group of glycosphingolipids that are differentiated from other glycosylceramides by the presence of sialic acid in the carbohydrate chain. They are called "gangliosides" because of the high concentration of the compounds in the neurons of the central nervous system, but they have also been isolated from spleen and red blood cell stroma. They are probably widely distributed in mammalian tissues but are at low levels in most.

"Sialic acid" is the group term for acyl derivatives of neuraminic acid, the parent substance, with the acyl group being in amide linkage with the amino group of neuraminic acid. *N*-Acetylneuraminic acid (NANA) is the predominating species in brain gangliosides. The cyclic form predominates, and the compound is generally linked to other monosaccharides through the hemiketal grouping. Gangliosides found in tissues other than the nervous system have *N*-glycosylneuraminic as the predominant form of sialic acid. Because of their sialic acid moiety, gangliosides are often called "mucolipids."

Four major gangliosides have been isolated from human brain with the following basic structure in common: gal-

Table 31-5. Summary of biochemical defects of some lipid-storage diseases

Disease	Lipid that accumulates	Enzyme defect
Sphingomyelinosis (Niemann-Pick)	Sphingomyelin	Sphingomyelinase
Gangliosidoses		
Tay-Sachs	$G_{GNTrII1}$	Specific *N*-acetylgalactosidase
Neurovisceral gangliosidoses	G_{GNT1}	Specific β-galactosidase
Cerebrosides		
Gaucher	Cer-Glc	Glucocerebrosidase
Fabry	Cer-Glc-Gal-Gal	Specific galactosidase–ceramide-trihexosidase
Metachromatic leukodystrophy	Sulfatide	Sulfatase

Cer, Ceramide; G_{GNT}, ganglioside; *Gal,* galactose; *Glc,* glucose.

actosyl (1→3) *N*-acetylgalactosaminyl (1→4) galactosyl (1→4) glucosylceramide. The major gangliosides differ in the number and position of the sialic acid units attached to the basic structure. In one of the major disialogangliosides, both galactose units of the basic structure are substituted at C-3 with *N*-acetylneuraminic acid. In the other major disialoganglioside, the two sialic acids are bound to each other and to the middle galactose molecule. Besides these major ganglioside structures, the brain also contains several minor gangliosides. The gangliosides of other tissues are similar to those of the brain but differ in various specific features.

Sphingosine and its C-20 homolog are the major sphingosine-like moieties in gangliosides, but small quantities of their dihydro derivatives are also present. In brain gangliosides, about 90% of the fatty acid in amide linkage with the sphingosine bases is stearic acid, and small quantities of many other fatty acids are also present. Gangliosides from tissues other than the central nervous system contain longer-chain fatty acids, lignoceric being a major component.

There is a dramatic accumulation of gangliosides in brain, in both gray and white matter, in Tay-Sachs disease, and brain gangliosides are elevated in other amaurotic idiocies such as "late infantile systemic lipidoses." The ganglioside that accumulates most in Tay-Sachs disease is a monosialic species that lacks the terminal galactose of the major monosialic species of normal brain. In "late infantile systemic lipidosis" the accumulated ganglioside is the major monosialoganglioside type found in normal brain. In gargoylism, a modest elevation in total ganglioside occurs, but there are also rather distinctive changes in the chemical nature of the gangliosides present. Indeed, virtually all the sphingolipidoses involve changes in ganglioside of some kind, but in most of these diseases, the changes appear to be secondary to another primary lipid anomaly, such as sphingomylin accumulation in Niemann-Pick disease.

Table 31-5 provides a summary of the various forms of lipid-storage diseases associated with sphingolipids.

Part II: Lipoproteins

As discussed earlier, lipids are insoluble in aqueous media, including that of plasma. It is only when the hydrophobic lipids are bound to protein (such as lipid-protein complexes called "lipoproteins") that they become soluble in the bloodstream. Because lipoproteins are generally viewed as a class of macromolecules associated with lipid transport, recommendations were made in about 1967 to transfer the diagnostic emphasis from hyperlipidemia to hyperlipoproteinemia.

A lipoprotein can be visualized most simplistically as a globular structure with an outer solubilizing coat of protein and phospholipid and an inner hydrophobic, neutral core of triglyceride and cholesterol (Fig. 31-17). The protein and phospholipid impart solubility to the otherwise insoluble lipids. The binding of the inner lipid to the phospholipid and protein coat is noncovalent, occurring primarily through hydrogen bonding and van der Waals forces. The protein that is free of lipid is called "apolipoprotein." Note that the lipids, which are weakly bound to the protein and phospholipid, are bound loosely enough to allow the ready exchange of lipid betweeen the serum lipoproteins themselves, as well as between serum and tissue lipoproteins, yet strong enough to allow the lipid and protein moieties to be separated in the analytical systems that are used to isolate and classify the lipoproteins.

CLASSIFICATION OF LIPOPROTEINS

Over the years several analytical systems have been used to isolate, separate, and characterize lipoproteins. Most have been based on one or another physiochemical property of the lipid-protein complex. The four most frequently used systems are based on analytical ultracentrifugation, preparative ultracentrifugation, electrophoresis, and precipitation techniques. The most frequently used systems are those based on ultracentrifugation and electrophoresis (Fig. 31-18). With a paper or agarose support medium being used, electrophoretic patterns show that chylomicrons remain at the origin, while pre-beta-lipoproteins and beta-lipoproteins migrate in the $beta_1$- and $beta_2$-globulin area, respectively, and alpha-lipoproteins migrate in the $alpha_1$-globulin area.

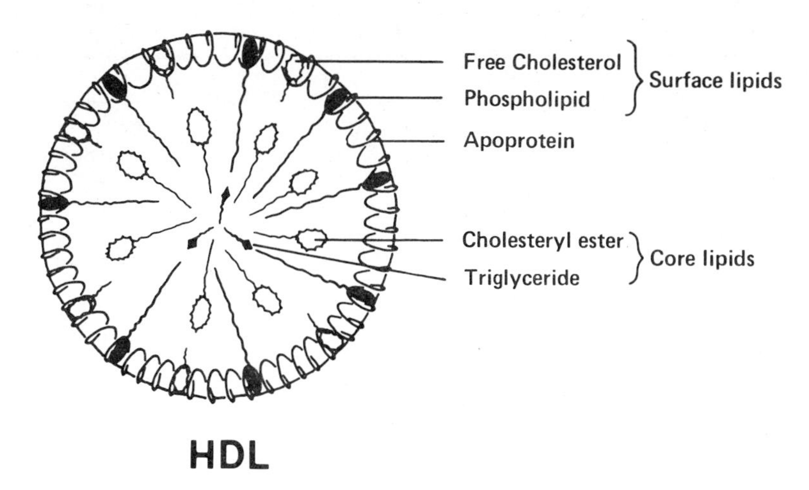

Fig. 31-17. Scheme of a lipoprotein complex showing polar outer surface and a core filled with neutral lipids. *HDL,* High-density lipoproteins.

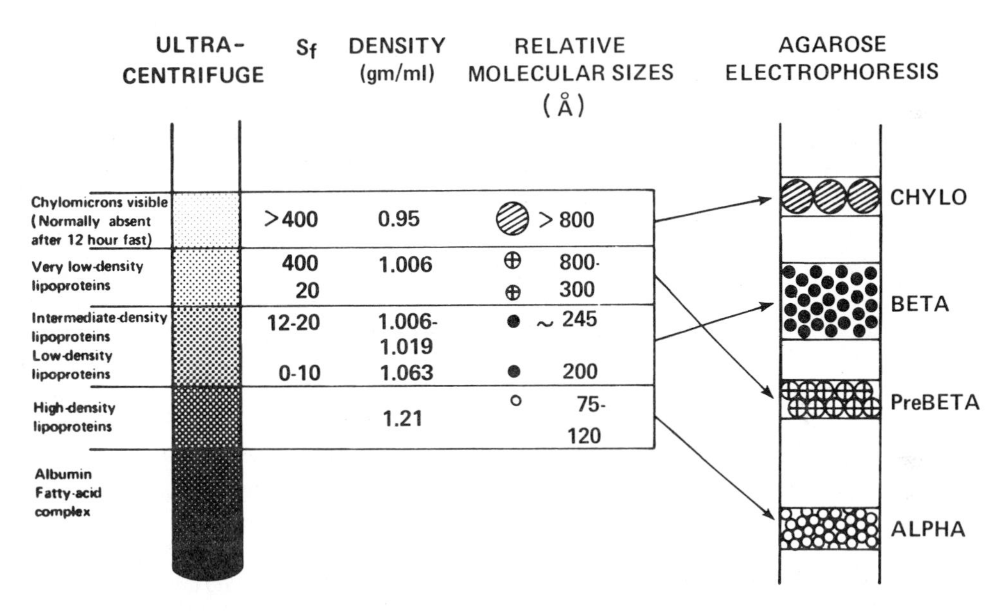

Fig. 31-18. Overview of major types of lipoproteins, showing some basic chemical and physical properties. S_f, Svedberg flotation rate; *chylo,* chylomicrons; *beta;* beta-lipoprotein; *prebeta,* a very-low-density lipoprotein; *alpha,* alpha-lipoprotein.

Using the ultracentrifuge and taking advantage of the fact that lipoproteins are lighter than the other serum proteins, one can separate the lipoproteins into chylomicrons (the lightest lipoproteins, of a density less than plasma), very-low-density lipoproteins (VLDL) at $d < 1.006$ g/mL (after chylomicron removal), low-density lipoproteins (LDL) separated between densities 1.006 and 1.063 g/mL, and high-density lipoproteins (HDL) of density 1.063 to 1.210 g/mL.

These lipoprotein classes correlate with electrophoretic patterns; for example, pre-beta-lipoprotein is generally synonymous with VLDL, beta-lipoprotein with LDL, and alpha-lipoprotein with HDL. Table 31-6 and Fig. 31-18 summarize the physical, chemical, and physiological description of the major plasma lipoproteins.

Chylomicrons

"Chylomicron" is the term originally used to describe the microscopically visible particle appearing in the plasma after a fatty meal. Subsequently, chylomicrons have been characterized as triglyceride-rich particles secreted by the intestine, the major transport form of dietary fat. During digestion, dietary fat is emulsified and hydrolyzed in the lumen of the duodenum by the combined actions of pancreatic lipase and biliary secretions. Fatty acids of fewer than 10 carbon units are transported through the portal cir-

Table 31-6. Physical and chemical description of plasma lipoproteins in man

Feature	Chylomicrons	VLDL	IDL	LDL	HDL
Density (g/mL)	<1.006	<1.006	1.006-1.019	1.019-1.063	1.063-1.21
Electrophoretic mobility	Origin	Pre-beta	Beta	Beta	Alpha
Flotation rate (S_f)	>400	20-400	12-20	0-10	—
Diameter (nm)	80-500	40-80	24.5	20	7.5-12
Lipids (% by weight)	98	92	85	79	50
Cholesterol	9	22	35	47	19
Triglyceride	82	52	20	9	3
Phospholipid	7	18	20	23	28
Apoproteins (% by weight)	2	8	15	21	50
Major	A-I, A-II				A-I, A-II
	B	B	B	B	
	C-I, C-II, C-III	C-I, C-II, C-III			
	E	E	E		
Minor		A-I, A-II		C-I, C-II, C-III	C-I, C-II, C-III
					D
					E

culation directly to the liver. Other degradation products, largely monoglycerides and free fatty acids of greater than 10 carbon units, enter the intestinal mucosal cell and serve as precursors in triglyceride synthesis. The resynthesized triglyceride is combined with cholesterol and small amounts of phospholipid and specific apoproteins (Apo B, Apo A, Apo C, and Apo E) to form the chylomicron molecule (Table 31-6). Over 80% of the chylomicron by weight is triglyceride. These water-insoluble lipids are maintained in a stable, emulsified form as they circulate in the blood stream. Most models for chylomicron structure have assumed that the neutral lipids (triglycerides and ester cholesterol) are partially surrounded by an outer shell of phospholipid, free cholesterol, and protein. Under fasting conditions (more than 12 hours after a meal), no chylomicrons are generally found in the blood.

Very-low-density lipoprotein (VLDL)

An average preparation of VLDL contains 52% triglyceride, 18% phospholipid, 22% cholesterol, and about 8% protein. Cholesterol and cholesteryl esters occur in a ratio of about 1:1 by weight. Sphingomyelin and phosphatidyl choline are the major phospholipids. The larger the size of a VLDL particle, the greater the proportion of triglycerides and apoprotein C and the smaller the proportion of phospholipid, apoprotein B, and other apoproteins. Apo B appears to be present in a constant absolute quantity in all VLDL fractions. Apo B accounts for approximately 30% to 35%, with Apo C comprising over 50% of the apoprotein content in VLDL. Apo E, and possible varying quantities of other apoproteins, may also be present. The relative quantity of each protein varies with the individual and with the degree of hyperlipidemia.

VLDL has a hydrated density of less than 1.006 g/mL. On paper and agarose support medium it migrates ahead of the beta-lipoprotein. On polyacrylamide gel it migrates behind the LDL because of the molecular-sieving effect of the gel.

Low-density lipoprotein (LDL)

LDL contains, by weight, 80% lipid and 20% protein. Consistent with this increased protein content, LDL is smaller (21 to 25nm) and is of higher hydrated density (1.006 to 1.063 g/mL) than that of VLDL and chylomicrons. About 60% of LDL lipid is cholesterol. LDL constitiutes 40% to 50% of the plasma lipoprotein mass in man. Its average concentration in normal adult American males is about 4000 mg/L, and in females it is 3400 mg/L. LDL is the major carrier of cholesterol. Apo B is the major apoprotein of normal LDL and represents 90% to 95% of the total plasma Apo B. Experimental evidence suggests that the LDL Apo B in normal man is derived almost entirely from VLDL Apo B in plasma. LDL is frequently separated into two classes, LDL_1 (IDL) and LDL_2, on the basis of flotation density. The lower density fraction, IDL (1.006 to 1.109 g/mL), is more lipid rich than LDL_2 (1.019 to 1.063 g/mL) and probably represents an intermediate in VLDL catabolism (see Fig. 31-5). Thus a comparison of IDL with LDL_2 demonstrates the gradual disappearance of triglyceride and of apoproteins more characteristic of VLDL (Apo C and Apo E) and an enrichment with Apo B and cholesterol ester.

High-density lipoprotein (HDL)

The HDL macromolecular complex (Fig. 31-17) contains approximately 50% protein and 50% lipid. HDL is the smallest of the lipoproteins (9 to 12 nm) and floats at the highest density (1.063 to 1.21 g/mL) of any of the lipoprotein molecules. The quantitatively most important HDL lipid is phospholipid, though HDL cholesterol is of particular interest. The major phospholipid species is phosphatidyl choline (also known as lecithin), accounting for

70% to 80% of the total phospholipid. It has an important functional role as a reactant in plasma cholesterol esterification, which is catalyzed by the enzyme lecithin: cholesterol acyltransferase (LCAT).

HDL may be further subfractionated by differential ultracentrifugation into HDL_2 (with a density of 1.063 to 1.110 g/mL) and HDL_3 (1.110 to 1.21 g/mL), the former being present in premenopausal women at about three times its concentration in men. Persons with lower HDL_2 levels are apparently more susceptible to premature coronary heart disease.

Other lipoproteins

Floating beta-lipoprotein, or beta-migrating VLDL. This lipoprotein fraction is found in persons with type III hyperlipoproteinemia, or "broad-beta disease" (derived from the broad smear from beta- to pre-beta-lipoprotein regions frequently present on whole plasma lipoprotein electrophoresis in these subjects). This fraction has a density of 1.006 g/mL, or a VLDL characteristic, but has a beta-lipoprotein migration pattern. The abnormal lipid composition of VLDL in type III hyperlipoproteinemic persons is attributable to a proportionately larger amount of cholesterol in that fraction.

Lp(a) or sinking pre-beta-lipoprotein. Similarities in lipid composition, concentration, and density (1.05 to 1.10 g/mL) between Lp(a) and LDL prevented clear discrimination of these two lipoproteins until immunological tests demonstrated the uniqueness of their protein moieties. Sixty-five percent of Lp(a) protein is Apo B, but another 15% is albumin, and the remainder is an apoprotein unique to Lp(a), called "Apo Lp(a)." Despite a high frequency in the population, the functional significance is uncertain. Recent studies suggest that this lipoprotein is very atherogenic.

Lipoprotein X. Although lipoprotein X has a flotation density similar to that of LDL, the lipid and protein compositions are quite different. This abnormal lipoprotein migrates and is characterized by an unusually high proportion of plasma phospholipid and unesterified cholesterol and by a low protein content consisting of Apo B, Apo C, and albumin. It is found most characteristically in plasma of patients with biliary obstruction.

LIPOPROTEIN METABOLISM

Chylomicrons

Synthesis. As discussed previously, chylomicrons are made exclusively by the intestine and traverse the lymphatic system to the thoracic duct where they then enter the systemic circulation. The major function of the chylomicrons is the transport of dietary or exogenous triglycerides. In the absorptive state, plasma triglyceride levels reflect mainly chylomicron (and to a lesser extent, intestinal VLDL) synthesis and secretion and respond accordingly to changes in both the amount and type of dietary fat. Normally, chylomicron triglycerides are rapidly metabolized (within 15 minutes of reaching the circulation), a process explaining why plasma triglyceride levels fluctuate with meal patterns. The measurement of triglyceride after a 12- to 14-hour overnight fast has the advantage of avoiding these fluctuations. Measurement of fasting triglyceride levels is recommended because of the difficulty in interpretation of a single random triglyceride determination as a result of the large variation in values obtained. Note that under fasting conditions (of more than 12 hours) no chylomicrons should be present in a normal healthy person's plasma.

In contrast to the diurnal fluctuation in plasma triglyceride levels, plasma cholesterol levels are relatively stable and do not reflect the chylomicron tide. This is consistent with the quantitatively small contribution made by chylomicrons and VLDL to total plasma cholesterol transport.

Catabolism. It is postulated that the newly synthesized and secreted chylomicron (80 to 500 nm) from the intestinal mucosal cells ultimately picks up Apo C-II from HDL as HDL and chylomicron circulate in the blood. Apo C-II then catalyzes lipoprotein triglyceride hydrolysis by the enzyme lipoprotein lipase. The hydrolysis results in the liberation of free fatty acids and monoglycerides. The free fatty acids enter adjacent muscle or adipocytes, where they are either metabolized for energy purposes or reesterified into triglycerides for storage.

As shown in Fig. 31-5, endothelial cell lipoprotein lipase-catalyzed hydrolysis results in progressive triglyceride depletion of the chylomicron molecule resulting in the chylomicron remnant particle. This transformation involves maintenance of lipoprotein structure by simultaneous removal of phospholipid, unesterified cholesterol, and Apo C peptides from the lipoprotein surface to plasma HDL. A reciprocal transfer of cholesterol ester from HDL may occur; Apo D may aid in this transfer process. The circulating chylomicron remnant particle is then released from the capillary wall and cleared from circulation through the liver, where it is metabolized. This particle, now smaller (30 to 80 nm), retains its ester cholesterol and Apo B and Apo E, which play an important role in the uptake of these particles by a high-affinity hepatic receptor uptake mechanism (see Fig. 31-13). When binding occurs, the remnants are immediately internalized by receptor-mediated endocytosis and degraded in hepatic lysozymes. The liver disposes of the dietary cholesterol through the bile by neutral and acidic sterol output.

Very-low-density lipoprotein

Synthesis. Similar to chylomicron formation, the major stimulus for hepatic VLDL (30 to 80 nm) synthesis is the demand for triglyceride transport. In the intestine this demand is created by influx of dietary fat, but in the liver the stimulus is the availability of precursors for endogenous triglyceride synthesis. These precursors are nonesterified or free fatty acids (FFA) and activated glycerol.

Free fatty acids are the principal stimuli for triglyceride synthesis and arise from several sources, the quantitative significance of which varies with the metabolic state. In the absorptive state, glucose and residual chylomicron triglyceride fatty acids serve as fatty acid precursors, whereas in the postabsorptive state free fatty acids mobilized from adipose tissue predominate.

When dietary cholesterol is sufficient, the liver uses that source of steroid derived from the receptor-mediated uptake of chylomicron remnants for VLDL synthesis. When dietary cholesterol is insufficient, the liver synthesizes its own cholesterol.

After the postprandial rise in chylomicron triglyceride, a secondary rise in triglyceride concentration occurs 4 to 6 hours after a meal. This represents predominantly hepatic VLDL triglyceride synthesized from glucose and chylomicron triglyceride not hydrolyzed in the peripheral tissue. The relative contributions of glucose and dietary fat vary with diet composition. Consumption of a high carbohydrate diet may lead to a phenomenon known as ''carbohydrate-induced hypertriglyceridemia.'' With high dietary carbohydrate, glucose influx into the hepatocyte is in excess both of energy demands and of glycogen-storage capacity. This results in shunting of acetyl CoA into fatty acid synthesis and of dihydroxyacetone phosphate into activated glycerol; thus the enzymatic machinery is tuned for triglyceride synthesis. This phenomenon may not persist in normal persons, but others may be unusually susceptible to carbohydrate induction of VLDL synthesis. This is the basis for reduction of dietary carbohydrate (simple sugars and alcohol) in the treatment of hypertriglyceridemia, but this approach is not successful if the hypertriglyceridemia has other causes, such as overproduction or a clearance defect.

Hepatic biosynthesis of triglyceride, cholesterol, and phospholipid occurs in the smooth endoplasmic reticulum. These lipid components contribute to the lipoprotein core and surface coat respectively, but, as in the intestine, apoprotein incorporation (especially Apo B) is essential before secretion.

Normally, VLDL's represent about 10% to 15% of the total circulating lipoproteins in a normal healthy individual.

Catabolism. VLDL triglycerides are believed to have a fate similar to that of the lipids from chylomicrons (see section on chylomicron catabolism, p. 577). During the catabolism by VLDL, more than 90% of Apo C is transferred to HDL, whereas essentially all the Apo B remains with the original lipoprotein particle (see Fig. 31-5). According to this postulated catabolic pathway, VLDL leads to the formation of the cholesterol-rich particle, LDL. LDL plays an important role in serving as an acceptor macromolecule of Apo C and unesterified cholesterol and phospholipids, the excess surface materials from a saturated VLDL. Apo C may recycle from HDL to newly synthesized chylomicrons or VLDL. The half-life of VLDL is 1 to 3 hours.

Intermediate-density lipoprotein (IDL)

Synthesis. This transient particle (22 to 28 nm) is usually present in very low concentrations in fasting human plasma. IDL, as discussed previously, is a lipoprotein derived from VLDL catabolism. The HDL particles interact with the plasma enyzme lecithin:cholesterol acyltransferase (LCAT), which esterifies the excess HDL free cholesterol with fatty acids derived from the 2 position of lecithin, the major phospholipid of plasma. The newly synthesized cholesteryl ester is transferred back to the IDL particles from HDL, apparently through the action of a plasma cholesteryl ester exchange protein (possibly Apo D). The net result of the coupled lipolysis and exchange reactions is the replacement of most of the original triglyceride core of VLDL with ester cholesterol.

Catabolism. After lipolysis, the IDL particles are released from the capillary wall into the circulation. They then undergo a further conversion in which most of the remaining triglycerides are removed and all the apoproteins except Apo B are lost. The resultant particle, which contains almost pure cholesteryl ester in the core and Apo B at the surface, is LDL. The site of the final conversion of IDL to LDL is unknown, but there is speculation that it occurs in the liver sinusoids. During this conversion a portion of the cholesteryl ester of IDL is removed, but the mechanism is unknown. In addition, some IDL particles are catabolized by the liver without being converted to LDL.

Low-density lipoprotein (LDL)

Synthesis. As discussed previously, LDL formation occurs primarily from the catabolism of VLDL. In normal healthy persons, LDL cholesterol constitutes about two thirds of the total plasma cholesterol, the amount in women being slightly less than that in men (except after menopause).

All tissues of the body are capable of cholesterol synthesis, but the liver and intestine are the major sources of endogenous cholesterol destined for transport. As discussed previously, the major steps in cholesterol biosynthesis are illustrated in Fig. 31-14; the rate-limiting enzyme is HMG-CoA reductase catalyzing the reduction of β-hydroxy-β-methyl-glutaryl-CoA to form mevalonic acid.

One might infer from this relation between dietary cholesterol and cholesterol synthesis that variations in dietary cholesterol can easily be monitored and cholesterol pools adjusted accordingly. Dietary cholesterol does, however, modify individual cholesterol levels, and this regulation possibly indicates a more complex relationship, perhaps relating to cholesterol synthesis by extrahepatic tissues that are not directly responsive to chylomicron cholesterol.

LDL delivers cholesterol to extrahepatic tissues (and to the liver), where it is utilized, deposited, or excreted.

Catabolism. The LDL particles are removed from the plasma with a fractional catabolic rate of about 45% of the plasma pool per day. Delivery of the LDL particles to peripheral tissue is accomplished when the LDL binds to high-affinity receptors located in regions of the plasma membrane called "coated pits." These pits invaginate into the cell and pinch off to form endocytic vesicles that carry the LDL to the lysosomes (Fig. 31-13). Fusion of the vesicle membrane with the lysosomal membrane exposes the LDL to a host of hydrolytic enzymes that degrade the Apo B to amino acids. The cholesteryl esters are hydrolyzed by an acid lipase, and liberated free cholesterol leaves the lysosomes for use in cellular reactions. As a result of this uptake mechanism, extrahepatic cells have low rates of cholesterol synthesis, relying instead on LDL-derived cholesterol. The free cholesterol thus released may cross the lysosomal membrane and gain access to cellular compartments, where it is used for membrane synthesis and serves to regulate, that is, depress cellular cholesterol synthesis by HMG-CoA reductase. This LDL internalization also regulates synthesis of the LDL receptor itself.

Excess cholesterol activates the enzyme acyl-CoA:cholesterol acyltransferase (ACAT), leading to intracellular cholesterol ester storage. Thus the net result of LDL binding and internalization is the reciprocal inhibition and activation of enzymes synthesizing and storing cellular cholesterol and a reduction in the number of receptors available to bind LDL.

The potential significance of this process for regulation of plasma cholesterol levels in man is illustrated by patients with the homozygous form of familial hypercholesterolemia. These patients are deficient in LDL receptors and have excessive LDL production and defective LDL catabolism, presumably because of an inability of tissues to bind, internalize, degrade, and thus regulate cholesterol synthesis. Except for the receptor-deficient state of familial hypercholesterolemia, however, the role of the LDL receptor in the final control of plasma cholesterol levels is uncertain and is probably only complementary to other regulatory processes. It has recently been recognized that the specificity of the LDL receptor extends to lipoproteins containing Apo E and Apo B as well. It appears that although extrahepatic receptors take up LDL readily, hepatic receptors take up chylomicron remnants with great efficiency (about 20 times greater) and LDL with much less efficiency. This difference is probably attributable to the Apo E content of chylomicron remnants, which has a higher affinity than that of Apo B.

In addition to its normal degradation mechanism, the high-affinity LDL receptor pathway, plasma LDL can be degraded by less efficient mechanisms that require high plasma levels to achieve significant rates by removal. One of these mechanisms occurs in scavenger cells (macrophages) of the reticuloendothelial system. When the plasma level of LDL rises, these scavenger cells degrade increasing amounts of LDL. When overloaded with cholesteryl esters, they are converted into "foam cells," which are classic components of atherosclerotic plaques. In man, estimates of the proportion of plasma LDL degraded by the LDL receptor system range from 33% to 66%. The remainder is degraded by the scavenger cell system and perhaps by other mechanisms not yet elucidated.

High-density lipoprotein

Synthesis. Nascent HDL molecules are synthesized in intestinal mucosal cells and in hepatocytes by a process analogous to that of VLDL and chylomicron synthesis. This involves microsomal lipid and protein synthesis followed by Golgi packaging and secretion. During the synthetic process, phospholipid and free cholesterol are combined with specific apoproteins to form disklike structures that undergo extensive compositional and structural modifications after secretion. The most important of these modifications is the esterification of free cholesterol to form ester cholesterol by an enzymatic reaction catalyzed by LCAT (Fig. 31-12). In man this is the major source of plasma ester cholesterol. Persons with LCAT deficiency have an accumulation of these ester cholesterol–deficient particles in plasma. This finding possibly indicates that the ester cholesterol formed in the LCAT reaction allows the expansion of the disklike structures to form spheres characteristic of normal plasma HDL (Fig. 31-13). Ester cholesterol thus formed may be transferred to VLDL during catabolism.

The apoprotein profile of nascent HDL is modified concomitantly with changes in lipid content. Apo E is a major component of newly secreted (nascent) HDL relative to Apo A and Apo C, unlike the plasma HDL, which is characterized by a predominance of Apo A with minor contributions by Apo C and Apo E. The functional significance of this modification is not completely understood, but Apo A-I is an activator of LCAT, and its acquisition must facilitate all LCAT reactions. In addition, HDL participates in the regulation of triglyceride catabolism and ester cholesterol formation by providing the respective cofactors, Apo C-II for activation and Apo C-III for inhibition of lipoprotein lipase activity. Also normal HDL may balance LDL transport by mediating cholesterol removal from peripheral sites to degradative and excretory sites. This role of HDL in reverse cholesterol transport may be the basis for the protection afforded by HDL against cardiovascular disease (Fig. 31-13).

Catabolism. The plasma half-life in normal subjects ranges from 3.3 to 5.8 days. The half-life of HDL apolipoproteins is similar if injected as Apo HDL or as HDL. The Apo A-I catabolism and Apo A-II catabolism within HDL are similar. HDL catabolism is enhanced in nephrotic patients but decreased in hypertriglyceridemic subjects, especially those with hyperchylomicronemia. It is also increased in subjects on high-carbohydrate diets and is greatly enhanced in patients with familial HDL deficiency (Tangier disease). It appears that changes in HDL catabolism may play a major role in regulating HDL levels in plasma.

RECAPITULATION OF LIPID AND LIPOPROTEIN METABOLISM

Plasma cholesterol and triglyceride levels are the result of a dynamic equilibrium between anabolic and catabolic aspects of lipoprotein metabolism. Triglyceride levels reflect the extent to which chylomicron and VLDL input from the intestine and liver is balanced by lipoprotein lipase–catalyzed triglyceride hydrolysis and subsequent utilization. Factors enhancing lipoprotein production or impairing removal may alter the balance in favor of hypertriglyceridemia.

Sources of cholesterol input are diet and endogenous synthesis; the latter predominates but is under modulation by dietary input. Opposing forces, including cholesterol degradation by hepatic and extrahepatic tissues, and excretion through the hepatobiliary system are equally important determinants of the steady-state level of plasma cholesterol.

LDL and HDL are the major cholesterol-transporting lipoproteins. Plasma levels of these lipoproteins are affected by subtle changes in VLDL lipid and apoprotein composition. Hypercholesterolemia consists of a heterogeneous group of disorders reflecting a primary alteration in either synthetic or degradative processes. The interdependence of these processes emphasizes the difficulty in identification of the initiating factors in hyperlipidemia.

HYPERLIPIDEMIA

By definition, hyperlipidemia is an elevated concentration of lipids in the blood. The major plasma lipids of interest are total cholesterol (free cholesterol + ester cholesterol) and the triglycerides. Rarely is measurement of plasma phospholipids helpful, except as in cases of obstructive liver disease. When one or more of these major classes of plasma lipids is elevated, a condition that is referred to as "hyperlipidemia" exists. Tables 31-2 and 31-3 provide some guidelines used to define hypercholesterolemia and hypertriglyceridemia.

The normal values and ranges for total cholesterol and triglycerides are presented in Tables 31-2 and 31-3. These data are based on epidemiological studies conducted in the United States, which include apparently healthy persons. It is usually necessary to set arbitrary statistical limits to normal concentrations on the basis of examination of a sufficient number of healthy-appearing subjects of different ages. The distribution of values is rarely gaussian. The limits of normal also have to be set high to minimize the effect of variation from environmental factors in presumed "genetically normal" subjects. This means that only the most obvious genetic variants will be recognized; in the study of these obvious variants there may be a tendency to oversimplify apparent modes of genetic determination that are actually complex or polygenic.

Cholesterol and triglyceride concentrations can be used to detect hyperlipoproteinemia. Over 95% of persons with hyperlipidemia, as defined previously, have hyperlipoproteinemia. The major exceptions are subjects with excessive amounts of low-density lipoprotein whose plasma cholesterol is kept within normal limits by a concomitant decrease in high-density lipoprotein. It should be noted that the concentration of the serum lipids is age and sex dependent.

As indicated earlier, these lipid concentrations do not indicate healthy values. In fact, the American Heart Association suggests that treatment in this country should be initiated when an adult has a serum cholesterol level above 2200 mg/L. The upper limit of serum triglyceride levels is less clear. Perhaps levels in excess of 1500 mg/L can be considered unhealthy. The rationale for initiating dietary therapy when the above limits are exceeded is to prevent or minimize the development of coronary heart disease, since atherosclerosis seldom occurs with a total cholesterol concentration of less then 2000 mg/L over the life-span of a person, unless other risk factors such as genetics, high blood pressure, smoking, and obesity are playing a dominant role.

Although serum lipids are not to be neglected, the influence of other risk factors associated with coronary heart disease should also be simultaneously considered. They include smoking, hypertension, obesity, stress, lack of physical activity, hormonal imbalances, and other primary disease states directly or indirectly associated with abnormal lipid and lipoprotein metabolism.

HYPERLIPOPROTEINEMIA

Hyperlipoproteinemia is an elevation of serum lipoprotein concentrations. With one exception (type III), hyperlipoproteinemia too is defined in quantitative terms. Arbitrary limits of lipoprotein concentrations have been established in a manner similar to that used for the lipid concentrations.

The quantitation of lipoprotein fractions is usually done by analytical ultracentrifugation. Simpler, but more indirect, procedures can be employed, such as lipoprotein electrophoresis and determining the cholesterol in the isolated VLDL, LDL, and HDL fractions. Both methods are routinely used in the clinical laboratory.

The classification of hyperlipoproteinemia begins with the determination of the type of abnormal lipoprotein pattern. However, other analyses are always necessary, for example:

1. Separation of hyperlipoproteinemia into primary and secondary forms (Table 31-7). The secondary form

Table 31-7. Causes of secondary hyperlipoproteinemia

Pattern	Causes
Type I	Insulinopenic diabetes mellitus Dysglobulinemia Lupus erythematosus Pancreatitis
Type II	Nephrotic syndrome Hypothyroidism Obstructive liver disease Porphyria Multiple myeloma Portal cirrhosis Viral hepatitis, acute phase Myxedema Stress Anorexia nervosa Idiopathic hypercalcemia
Type III	Hypothyroidism Dysgammaglobulinemia Myxedema Primary biliary cirrhosis Diabetic acidosis
Type IV	Diabetes mellitus Nephrotic syndrome Pregnancy Hormone use (oral contraceptives) Glycogen-storage disease Alcoholism Gaucher's disease Niemann-Pick disease Pancreatitis Hypothyroidism Dysglobulinemia
Type V	Insulinopenic diabetes mellitus Nephrotic syndrome Alcoholism Myeloma Idiopathic hypercalcemia Pancreatitis Macroglobulinemia Diabetes mellitus (insulin independent)

is caused by another known disease (such as insulinopenic diabetes mellitus, gout, hypothyroidism), which can result in secondary hyperlipoproteinemia manifesting itself in any of the six types of lipoprotein patterns.

2. Differentiation of primary hyperlipoproteinemia into heritable and nonheritable forms.
3. Determination of the relative concentration of the lipoprotein fractions, that is, VLDL cholesterol, LDL cholesterol, and HDL cholesterol.

Nearly all the patients with heritable hyperlipidemia have one of six abnormal lipoprotein patterns. These patterns are summarized in Fig. 31-19, where one can observe that only three of the four lipoprotein families serve as determinants. These three families are (1) chylomicrons, (2) VLDL, and (3) LDL. The original Fredrickson phenotyping system disregarded the importance of HDL or other lipoproteins discussed in this chapter. Because of the fairly recent worldwide epidemiological finding that there is a statistically significant inverse relationship between HDL cholesterol concentration and risk for coronary heart disease (CHD), laboratories now do LDL cholesterol and HDL cholesterol determinations as part of the overall lipid-lipoprotein profile.

CLASSIFICATION OF HYPERLIPOPROTEINEMIAS

Although the classification of hyperlipoproteinemia (Fig. 31-19) is based upon identification of elevated concentration of blood lipids and abnormal lipoprotein patterns, I should emphasize again that each form of dyslipoproteinemia is not a homogeneous entity from a genetic, clinical, or pathological point of view.

The hyperlipoproteinemias are now described in somewhat greater detail, with emphasis on distinctive clinical, diagnostic, genetic, biochemical-pathophysiological, and therapeutic aspects.

Type I hyperlipoproteinemia

This form of hyperlipoproteinemia is characterized by elevated plasma triglyceride concentration, chylomicronemia, with a slight increase in VLDL-triglyceride levels. LDL and HDL are often low, whereas VLDL may be slightly elevated. Chylomicron removal is reduced, and subjects have deficiencies in lipase activities, such as postheparin lipolytic activity (PHLA) (specifically extrahepatic, protamine-inactivated lipoprotein lipase). Primary type I hyperlipoproteinemia usually manifests itself early in childhood. Although type I hyperlipoproteinemia is usually primary and familial, the lipoprotein pattern may be produced by several other disease or metabolic states. Thus, secondary type I should be ruled out (Table 31-7).

Once the obvious secondary causes have been ruled out, the following should be checked: (1) presence of eruptive xanthomas, hepatosplenomegaly, lipemia retinalis, abdominal pain, and pancreatitis early in life, (2) intake of drugs that can cause secondary hypertriglyceridemia, (3) presence of reduced plasma levels of triglyceride lipases, (4) reduction in triglyceride levels and disappearance of chylomicronemia on a fat-free diet, and (5) confirmation by family screening of inheritance as an autosomal recessive trait.

Type IIa hyperlipoproteinemia

Type IIa is characterized by elevated concentration of plasma cholesterol with normal triglyceride levels and clear plasma. The lipoprotein pattern is characterized by an elevation of the LDL with normal VLDL. This lipoprotein disorder is recognized as familial hypercholesterolemia, which exhibits the following features: (1) a deficient number of functional LDL receptors in the fibroblast cul-

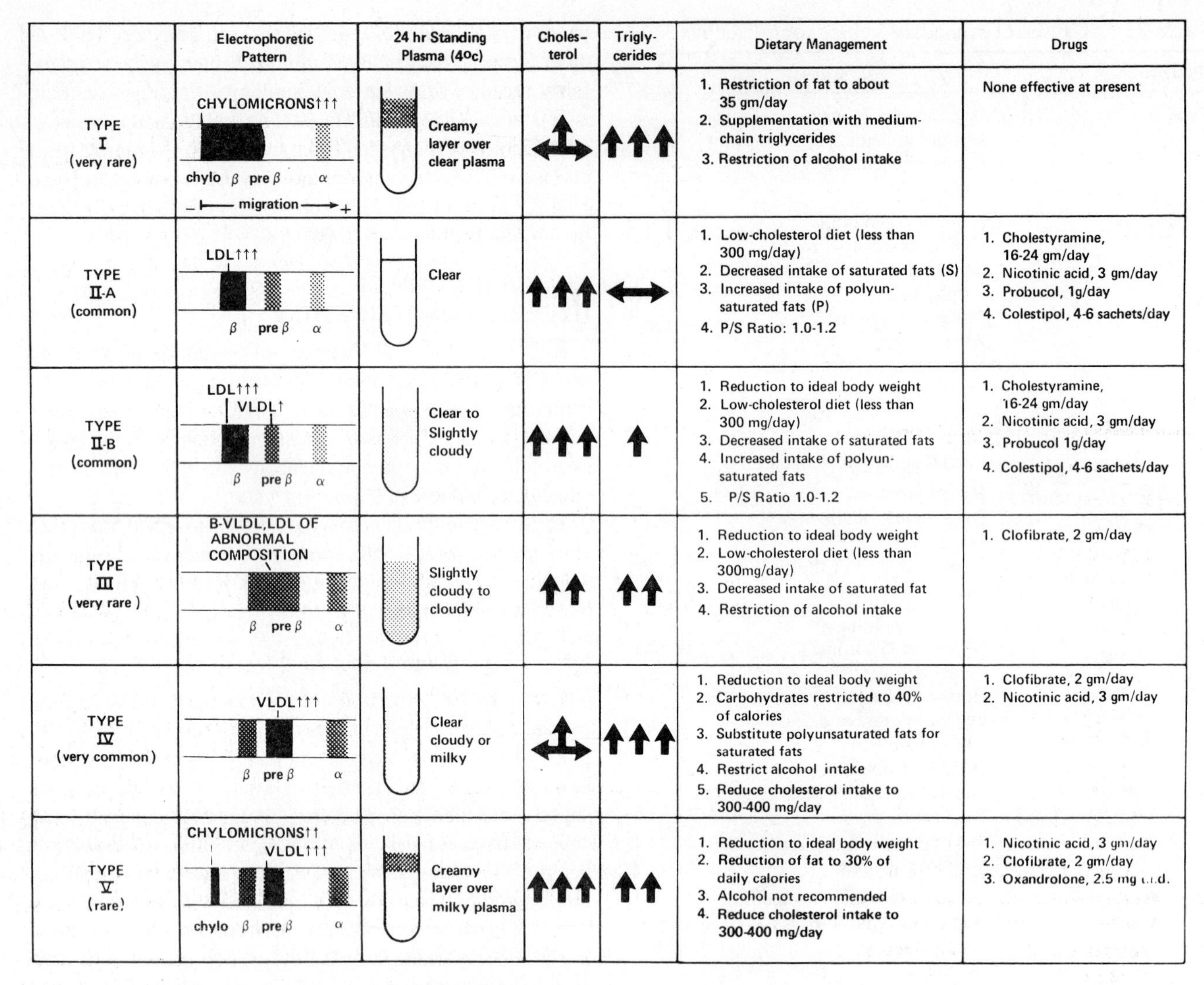

	Electrophoretic Pattern	24 hr Standing Plasma (4°c)	Cholesterol	Triglycerides	Dietary Management	Drugs
TYPE I (very rare)	CHYLOMICRONS↑↑↑; chylo β pre β α; – ⊢— migration —→ +	Creamy layer over clear plasma	↑↔	↑↑↑	1. Restriction of fat to about 35 gm/day 2. Supplementation with medium-chain triglycerides 3. Restriction of alcohol intake	None effective at present
TYPE II-A (common)	LDL↑↑↑; β pre β α	Clear	↑↑↑	↔	1. Low-cholesterol diet (less than 300 mg/day) 2. Decreased intake of saturated fats (S) 3. Increased intake of polyunsaturated fats (P) 4. P/S Ratio: 1.0-1.2	1. Cholestyramine, 16-24 gm/day 2. Nicotinic acid, 3 gm/day 3. Probucol, 1g/day 4. Colestipol, 4-6 sachets/day
TYPE II-B (common)	LDL↑↑↑; VLDL↑; β pre β α	Clear to Slightly cloudy	↑↑↑	↑	1. Reduction to ideal body weight 2. Low-cholesterol diet (less than 300 mg/day) 3. Decreased intake of saturated fats 4. Increased intake of polyunsaturated fats 5. P/S Ratio 1.0-1.2	1. Cholestyramine, 16-24 gm/day 2. Nicotinic acid, 3 gm/day 3. Probucol 1g/day 4. Colestipol, 4-6 sachets/day
TYPE III (very rare)	B-VLDL,LDL OF ABNORMAL COMPOSITION; β pre β α	Slightly cloudy to cloudy	↑↑	↑↑	1. Reduction to ideal body weight 2. Low-cholesterol diet (less than 300mg/day) 3. Decreased intake of saturated fat 4. Restriction of alcohol intake	1. Clofibrate, 2 gm/day
TYPE IV (very common)	VLDL↑↑↑; β pre β α	Clear cloudy or milky	↑↔	↑↑↑	1. Reduction to ideal body weight 2. Carbohydrates restricted to 40% of calories 3. Substitute polyunsaturated fats for saturated fats 4. Restrict alcohol intake 5. Reduce cholesterol intake to 300-400 mg/day	1. Clofibrate, 2 gm/day 2. Nicotinic acid, 3 gm/day
TYPE V (rare)	CHYLOMICRONS↑↑; VLDL↑↑↑; chylo β pre β α	Creamy layer over milky plasma	↑↑↑	↑↑	1. Reduction to ideal body weight 2. Reduction of fat to 30% of daily calories 3. Alcohol not recommended 4. Reduce cholesterol intake to 300-400 mg/day	1. Nicotinic acid, 3 gm/day 2. Clofibrate, 2 gm/day 3. Oxandrolone, 2.5 mg t.i.d.

Fig. 31-19. Summary of six types of hyperlipoproteinemias. Abbreviations as in Fig. 31-18.

tures (the pathognomonic feature), (2) an expression of type II pattern in infancy, (3) xanthomatosis in severely affected members, and (4) premature coronary heart disease seen by the third and fourth decade.

Secondary type II disorders, such as hypothyroidism, acute intermittent porphyria syndrome, dysgammaglobulinemia, obstructive liver disease, and highly saturated fat and cholesterol diets should be ruled out.

Once secondary hyperlipoproteinemia has been ruled out, the primary disorder can be confirmed by (1) family screening, including children, (2) persistent hypercholesterolemia even after 8 weeks of low cholesterol (less than 300 mg/day), highly polyunsaturated fat (polyunsaturated fat to saturated fat, or P/S, ratio 1 to 1.2) diet, and (3) presence of tendinous xanthomas, xanthelasma, and corneal arcus.

Type IIb hyperlipoproteinemia

Another form of familial hyperlipidemia is familial combined hyperlipidemia. The important features include (1) no abnormality in the number of functional LDL receptors in the fibroblast culture, (2) absence of the type II pattern in children, (3) early expression of hypertriglyceridemia, (4) multiple lipoprotein patterns in affected relatives in successive generations, and (5) hypercholesterolemia in the common type II pattern.

This is the most recently described familial hyperlipoproteinemic syndrome and also probably the most common of the primary hyperlipoproteinemias. The characteristic feature of this disorder is a scatter of lipoprotein phenotypes within a family. Most commonly patients will have an elevation in both LDL and VLDL; however, within a family the presence of hyperbetalipoproteinemia (type IIa) and hyperprebetalipoproteinemia (type IV) will also be found affecting different persons. In contrast to familial hyperbetalipoproteinemia, subjects with familial combined hyperlipoproteinemia generally do not manifest their disease until adulthood. Clinically, these patients have an increased incidence of coronary artery disease. They also are frequently diabetic, have a tendency for hyperuricemia, and show a low incidence of tendinitis and tuberous xanthomas. These features suggest a closer clinical relation to familial hyperprebetalipoproteinemia than to familial hyperbetalipoproteinemia. The mode of inheritance of familial combined hyperlipoproteinemia is still in doubt; however, that is clearly a familial disorder that presents most commonly with a type IIb pattern.

Remember that subjects presenting with hyperbetalipoproteinemia or hyperprebetalipoproteinemia may also belong within a kindred having familial combined hyperlipoproteinemia.

This lipoprotein disorder is characterized by elevated total cholesterol, LDL, triglycerides, and VLDL with the absence of floating beta-lipoprotein. Any secondary hypercholesterolemia and hypertriglyceridemia should be ruled out before confirmation of the primary type IIb. Family screening is mandatory for recognition of this lipid abnormality. Accurate diagnosis of the type IIb pattern also requires an appreciation of the factors that determine triglyceride levels. Studies in free-living populations in the United States have documented increases in triglyceride levels with age and have indicated that as many as one fourth of middle-aged men exceeded previously published cutoff values for triglyceride levels. Thus, although statistically valid, the critical limits for triglyceride concentrations may not represent physiological limits. Therefore one might expect a greater prevalence of type IIb patterns in older age groups.

The possible effects of dietary carbohydrates should not be overlooked when one is assessing the type IIb pattern. It has been shown that fasting hypertriglyceridemia in patients with type IIb could be attributable to an acute increase in dietary carbohydrates. It appears that triglyceride values greater than 4000 mg/L are rare in patients with type IIb patterns. The few reported cases in the medical literature occurred in postmenopausal women.

Lipoprotein electrophoresis is rarely necessary for the diagnosis of the type II pattern. If the type IIa pattern is present, the elevated cholesterol and the normal triglyceride levels leave little for the lipoprotein electrophoretic pattern to clarify. If the type IIb pattern is present, the decisive diagnostic procedure is to differentiate it from either the type III or type IV patterns. This task is not easily done with lipoprotein electrophoresis. Also note that the total plasma cholesterol may be normal despite an elevated LDL cholesterol value. This relationship also occurs in familial hypercholesterolemia. Although the type II pattern has been reported as a marker for preclinical hypothyroidism (when routine thyroid function tests are still within normal limits), a careful clinical and laboratory evaluation should preclude a firm diagnosis.

The "broad-beta" pattern is no longer considered unique for type III hyperlipoproteinemia and has been noted in cases homozygous for familial hypercholesterolemia. The presence of a discrete band attributable to high concentrations of Lp(a) or "sinking" pre-beta-lipoprotein can cause diagnostic confusion with the type IIb pattern. When the triglyceride value is normal, the appearance of the IIb pattern of electrophoresis suggests the presence of Lp(a). Thus the electrophoretic pattern is insensitive and is not a highly specific tool for diagnosing type IIb or type III patterns.

Type III hyperlipoproteinemia

Type III hyperlipoproteinemia is characterized by an elevation of both plasma cholesterol and triglyceride concentrations and an abnormal LDL, which floats in the fraction with a density less than 1.006 g/mL. This abnormal LDL often merges with the prebeta band on electrophoresis to produce a "broad-beta" band (Fig. 31-19). For accurate diagnosis of type III, an ultracentrifugal study with measurement of cholesterol and triglyceride in the $d<1.006$ g/mL fractions is required to document the presence of the floating beta-lipoprotein.

Measurements of the lipid composition of lipoproteins of a density less than 1.006 appear to offer a more reliable means of identifying type III than is afforded by the presence of floating beta-lipoproteins by means of electrophoresis alone. The ratio used clinically is that of the cholesterol content of the VLDL divided by the plasma triglyceride concentration. This ratio appears to be most useful for documenting type III hyperlipoproteinemia, when the triglyceride level is at least 1500 mg/L, but it may be subject to error when triglyceride exceeds 10,000 mg/L. It has been suggested that if the VLDL cholesterol: triglyceride ratio is 0.30 or more, the subject may have type III hyperlipoproteinemia. However, when the ratio is less than or equal to 0.25 and not more than 0.29, a diagnosis of "possible type III hyperlipoproteinemia" should be considered.

The clinical characteristics of subjects with type III vary widely as a function of age, sex, degree of adiposity, and presence of associated disorders such as hypothyroidism and alcoholism.

The most characteristic xanthoma in subjects with type III is called "xanthoma striatum palmare" (in the literature both "xanthoma striatum palmaris" and "xanthoma striata palmaris" are improper latin). In their most subtle form these lesions produce an orange or yellowish discoloration of the palmar creases (xanthochromia striata palmaris), a phenomenon most easily detected in subjects of fair complexion. When more advanced, these lesions may produce planar elevations and even the virtual obliteration of the palmar and digital creases. Raised lesions can occasionally affect the remaining palmar surfaces and in the severe form produce tuberous, incapacitating xanthomas.

Various forms of coronary heart disease have been reported in association with type III hyperlipoproteinemia. The prevalence of such disease among subjects with the type III pattern clearly relates to the reason for detection of the hyperlipoproteinemia, which is readily treated by diet and drugs.

The form of cardiovascular disease in type III differs significantly from that in familial type IIa in that peripheral, and even cerebrovascular, disease appears to be as common as coronary heart disease.

Secondary type III hyperlipoproteinemia has been associated with hypothyroidism, gout, and diabetes mellitus and is found in patients with acute renal failure on maintenance hemodialysis.

Type IV hyperlipoproteinemia

Type IV hyperlipoproteinemia has also been called "endogenous" or "carbohydrate-induced hyperlipemia." The latter term is no longer accepted by most workers, since carbohydrate induction of hypertriglyceridemia is also observed in normolipemic individuals. The term "endogenous" is preferred by some because it allows differentiation from exogenous or dietary fat-induced hyperlipemia. Endogenous hyperlipemia excludes the rare type I hyperlipoproteinemia but includes the uncommon type V as a mixed endogenous and exogenous hyperlipemia. The "pure" endogenous forms include types IIb, III, and IV.

By definition, type IV hyperlipoproteinemia is an elevation of VLDL (and triglyceride) levels above an arbitrary cutoff point in the absence of either chylomicrons or the abnormal VLDL of type III. LDL levels are normal, and LDL cholesterol measurement discriminates between types IV and IIb.

A tentative diagnosis of type IV may be made if the triglyceride concentration is increased, the total cholesterol is normal or moderately elevated, and the standing plasma reveals no chylomicrons. The biochemical diagnosis is confirmed if electrophoresis reveals a distinct pre-beta-lipoprotein band and the LDL cholesterol is within normal limits. Keep in mind that the presence of a pre-beta band with normal plasma triglyceride levels occurs with "sinking" pre-beta-lipoprotein, a triglyceride-poor, apparently normal lipoprotein variant observed in up to 35% of healthy subjects.

Once the biochemical pattern has been confirmed (that is, based on more than one sample under standard conditions), it should be classified according to cause as primary type IV—either familial or sporadic—or as secondary type IV (Table 31-7).

The diagnosis of primary familial type IV depends on the following criteria: (1) a type IV pattern, (2) one or more first-degree relatives with type IV, and (3) no close relative with primary type I, III, or V. Other common features are a normal cholesterol level if the triglyceride level is less than 4000 mg/L, a triglyceride level usually below 15,000 mg/L, and a family history of diabetes. The diagnosis of secondary type IV is sometimes obvious when it results from endocrine, gastrointestinal, or renal disease. Sometimes the conditions are not so obvious. The following are known to cause an elevation of VLDL: (1) alcoholism or excessive alcohol intake, (2) hypothyroidism, (3) nephrotic syndrome, (4) uremia, (5) estrogen-progesteronal contraceptives, (6) pregnancy (last trimester), (7) corticosteroids, (8) pancreatitis (usually alcoholic), (9) glycogen-storage diseases, (10) dysproteinemias, such as systemic lupus erythematosus, (11) lipid-storage diseases, such as Gaucher's, Niemann-Pick, and LCAT deficiency, (12) obesity, and (13) juvenile diabetes mellitus.

Since a large proportion of our society imbibes alcohol, a brief comment will be made. Although not regarded as a major cause of hyperlipidemia, ethanol is known to cause acute but transient hypertriglyceridemia with an elevation in primarily VLDL, causing mainly type IV hyperlipoproteinemia (sometimes type V hyperlipoproteinemia).

In the hypertriglyceridemia produced by ethanol intake, the following salient features stand out:

1. The hypertriglyceridemia is usually moderate in extent and limited in duration. The level of triglyceridemia rarely exceeds 10,000 mg/L, and the lipemia peaks in 12 to 14 hours and disappears after 25 to 40 hours. This apparently transient effect appears to hold true, especially for normolipemic individuals.
2. This hypertriglyceridemia is, for the most part, the result of increased VLDL and possibly chylomicrons. Some reports indicate that IDL is elevated.
3. The fatty liver associated with alcohol intake plays a vital role in the form and extent of the induced hyperlipoproteinemia, and the resultant changes may be related to the stage of the hepatic damage.
4. The triglycerides in alcoholic hyperlipidemia are intimately related to the quality and quantity of dietary fatty acid intake. It is well known that simultaneous ingestion of ethanol with fat (as in complex meals or singly as corn oil) produces a prolonged and augmented rise in serum triglyceride concentration.

Although it is obvious to conclude that hyperlipemia can be produced either by an excessive production and release of lipids (hence, lipoproteins) into circulation or by defective removal or clearance from the blood or by a combination of these physiological processes, the precise mechanism or mechanisms of ethanol-induced hyperlipemia is still not known.

Type V hyperlipoproteinemia

Another form of hyperchylomicronemia is the type V hyperlipoproteinemia. It is distinguished as having both elevated VLDL and chylomicrons in the plasma of fasting subjects on a regular diet. This disorder can present as an apparent primary genetic defect and thus represents a second form of familial hyperchylomicronemia. Triglyceride levels similar to those in type I may be observed. As in that disorder, the occurrence of abdominal syndromes, including pancreatitis, and the physical findings of eruptive xanthomas, lipemia retinalis, and hepatosplenomegaly are related to the level of plasma triglyceride. Although the pathophysiology of these manifestations of type V probably does not differ from that observed in type I, a variety of other differences exist in clinical, genetic, metabolic, and biochemical observations.

In sharp contrast to type I hyperlipoproteinemia, most subjects with type V present in adulthood. Full penetrance of the abnormality may not occur until the fifth or sixth decade of life, with the females presenting later than the males. Several children with familial type V have been described.

Extremely high triglycerides and hyperchylomicronemia are usually not attributable to primary type V but are found in the setting of several disorders that can lead to secondary type V. These disorders are particularly prone to produce type V if they occur in a patient with primary type IV. For example, pancreatitis may be associated with pronounced hyperchylomicronemia, but later only mild hypertriglyceridemia is found when the patient is reevaluated under stable conditions. It is also well known that type IV can present as type V in the poorly controlled insulin-dependent diabetic and in alcoholics.

Although type V hyperlipoproteinemia is associated with elevated triglycerides (hence, VLDL and chylomicrons), plasma cholesterol concentrations may be slightly to moderately increased. LDL and HDL cholesterols are usually normal to low. The plasma is usually opaque, and a floating cream layer above the turbid plasma may be observed.

The presence of chylomicrons in patients with type V may be difficult to discern separately in the presence of higher than normal elevations of VLDL. When total triglyceride is over 10,000 mg/L, visual appreciation of a discrete floating "creamy" supernate over turbid plasma becomes difficult, and the use of a combination of ultracentrifugation (to remove the chylomicrons) and electrophoresis provides a clear electrophoretic pattern for type V determination. The ultracentrifuge may be used to separate the chylomicrons from the VLDL for individual quantification if needed. Usually, a qualitative assessment of the electrophoresis strip is sufficient to document elevated VLDL and chylomicron lipid as well. Since in practice some overlap of VLDL levels may occur between types I and V, a postheparin lipoprotein lipase measurement should be made as a final characterization. This enzyme should be present in type V subjects. In the absence of a specific assay for the enzyme, a reasonably reliable clue to its presence may be sought by observation of the change in the lipoprotein electrophoresis pattern obtained with a plasma sample drawn 10 minutes after heparin injection (10 U/kg of body weight). The average patient with type V usually has considerably higher triglyceride levels than the patient with type IV. The plasma cholesterol to triglyceride ratio in type V (0.23 ± 0.02) is lower than in type IV (0.86 ± 0.03) because the chylomicrons incorporate less cholesterol than the VLDL do. Major qualitative difficulties can arise in distinguishing type V from type IV hyperlipoproteinemia (endogenous hypertriglyceridemia) because the type V pattern is often transient, with many type V subjects losing their chylomicron band with moderate reductions in triglyceride.

Type V hyperlipoproteinemia is often secondary to a wide variety of diseases, drugs, and dietary habits (see discussion on secondary effects of hypertriglyceridemia on p. 584). Since there are many ways of acquiring type V hyperlipoproteinemia, a careful distinction between primary and secondary causes must be made. A routine history of ethanol intake and estrogen or steroid administration, a urinalysis, and the measurement of fasting or 2-hour postprandial blood glucose, liver, thyroid, and renal function tests are all useful in this distinction. Superimposition of poorly controlled diabetes mellitus, alcoholic excess, or estrogens or estrogen-containing oral contraceptives in an individual with preexisting type IV hyperlipoproteinemia will often produce the type V pattern. In familial Type V, particularly with plasma triglycerides above 15,000 mg/L, lipemia retinalis, hepatosplenomegaly, and eruptive xanthomas may be present.

The biochemical defect in type V hyperlipoproteinemia is still not clear. The presence of high levels of chylomicrons in the fasting plasma of a subject whose diet contains a normal or low content of fat clearly indicates that clearance mechanisms are inadequate. A primary defect in removal of plasma triglyceride could also explain the elevated VLDL. That this defect is not simply a variant of type I is established by the easily detectable and often quantitatively normal heparin-releasable plasma PHLA from type V subjects. Thus a different type of problem must lead to this failure in lipoprotein uptake. One possibility is that the synthesis of endogenous triglyceride, and the resulting secretion of VLDL from the liver, may proceed at an abnormally high rate, sufficient to saturate pathways of removal that are shared by chylomicrons, thus leading to an elevation of both lipoproteins. Studies utilizing lipoproteins labeled with radioisotopes have indicated that many patients with endogenous hypertriglyceridemia have elevated synthesis of VLDL triglyceride.

Since types I and V hyperlipoproteinemia are similar in many respects and can at times be hard to differentiate, a summary is warranted.

Familial hyperchylomicronemia may be divided into types I and V hyperlipoproteinemia. Both disorders are manifested by very elevated triglyceride levels and frequently present with eruptive xanthomas, lipemia retinalis, hepatosplenomegaly, and abdominal pain. Type I is caused by a pronounced deficiency of lipoprotein lipase. The triglyceride elevation appears with ingestion of dietary fat and is thus manifested in very young children. Type V may rarely be detected in childhood, but the usual presentation is in the adult. It also differs from type I in that lipoprotein lipase is measurable and glucose intolerance and hyperuricemia are commonly associated findings. The only effective treatment for type I is a low-fat diet. Type V is most effectively treated by diet-induced weight loss and will frequently respond to one of the following drugs: nicotinic acid, norethindrone, oxandrolone, benzafibrate, or clofibrate.

TRANSFORMATION OF HYPERLIPIDEMIA TO HYPERLIPOPROTEINEMIA

Limitation of classification of types of hyperlipoproteinemia

The limitations and potentials of the lipoprotein-typing system of Frederickson, Levy, and Lees are well recognized. However, it should be stressed that plasma lipoprotein patterns are not a substitute for an etiological classification of the hyperlipoproteinemias. The approach of Fredrickson, Lee, and Levy is to be regarded as provisional, pending a more fundamental understanding of the causes of the hyperlipidemias.

Using quantitative measurements of cholesterol and triglyceride alone, one can divide patients into three major groups: those with hypercholesterolemia alone, those with hypertriglyceridemia alone, and those with a combination of the two. Subjects with pure hypercholesterolemia usually have the type IIa lipoprotein pattern and those with pure hypertriglyceridemia without chylomicrons, the type IV pattern. Subjects having relatively high cholesterol and triglyceride levels may include lipoprotein patterns of types IIb, III, or IV. In regard to the utility of the lipoprotein-typing system, it is known that there are four areas where the typing system retains general validity: (1) Lipoprotein patterns are useful in sharpening the focus upon the diverse metabolic abnormalities that underlie hyperlipidemia. (2) The types of hyperlipoproteinemias so identified are not disease states but represent disorders that similarly affect the concentrations of particular lipoproteins. (3) Each lipoprotein type is often associated with certain distinctive clinical features. (4) Each lipoprotein type, irrespective of cause, is generally more successfully handled by a specific diet and therapeutic approach.

The typing system has four major limitations: (1) For distinction of type III, the diagnosis requires substantiation by measurement of the ratio of cholesterol in VLDL divided by the plasma triglyceride as well as electrophoretic confirmation of beta-migrating VLDL. In addition, in subjects with mixed elevations of cholesterol and triglyceride, quantitation of LDL (LDL cholesterol) is important for accurate distinction. (2) The second major area where the typing system has limitations is genetics. No specific type of hyperlipoproteinemia should be considered genotypic; lipid and lipoprotein determinations cannot provide the diagnosis of a specific genetic disorder in a single patient; and there is increasing evidence of strong heterogeneity in lipoprotein patterns in first-degree relatives from families with monogenic familial hyperlipidemia. (3) The third area that the typing system did not cover was the evaluation of alpha-lipoprotein in the classification. It is now known that a low concentration of HDL cholesterol is an independent risk factor for coronary heart disease. (4) Finally, the present electrophoretic systems are limited in their ability to resolve other unusual lipoprotein fractions, such as HDL_c, beta-migrating VLDL, and IDL.

From lipids to lipoproteins: laboratory considerations

In the transformation of hyperlipidemia to hyperlipoproteinemia, lipid analyses alone can be used to determine the lipoprotein disorder with a fair degree of accuracy. If the plasma is clear, the triglyceride level is most likely to be either normal or near normal (less than 2000 mg/L). When triglyceride increases to about 3000 mg/L or higher, the plasma is usually hazy to turbid in appearance and is not translucent enough to allow for clear reading of newsprint through the tube. When plasma triglyceride is over 6000 mg/L, the plasma is usually opaque and "milky" (lipemic, lactescent). If chylomicrons are present, after several hours at 4° C a thick homogeneous "cream" layer may be observed floating at the plasma surface. As summarized in Fig. 31-19, a uniformly opaque plasma sample usually denotes a type IV pattern. An opaque plasma sample with a cream layer on top is usually consistent with the type V pattern. A thick chylomicron cream layer with generally clear plasma infranate is usually consistent with a type I pattern.

In patients with hypercholesterolemia without hypertriglyceridemia, most often with raised LDL levels, the plasma is clear but may have an orange-yellow tint since carotene is carried with LDL. After visual observation, which is simple and "free," the diagnosis of the lipid abnormality can be made in as many as 90% of subjects by quantitation of plasma cholesterol and triglyceride alone. Accurate and precise measurement of plasma cholesterol and triglyceride levels is within the capability of most laboratories. Wide variations in "normal" values are partly produced by differing techniques and analytical equipment.

By analyses of cholesterol and triglyceride, by use of

lipoprotein electrophoresis, immunoelectrophoresis, the preparative ultracentrifuge, and differential precipitation of lipoproteins, six lipoprotein phenotypes can be differentiated. The more expensive and cumbersome analytical techniques are often beyond the range of feasibility for small laboratories. Although quantitation of the different lipoprotein classes provides increased information about the clinical and lipoprotein status of the patient, an extensive array of methods is required to facilitate the diagnosis of type III hyperlipoproteinemia and to document the presence or absence of the floating beta-lipoprotein in this disorder. LDL cholesterol levels are conventionally measured by use of the preparative ultracentrifuge but can conveniently be estimated by the Friedewald formula:

$$\text{LDL cholesterol} = \text{Total cholesterol} - (\text{Triglyceride} \div 5 + \text{HDL cholesterol})$$

This estimation requires measurement of HDL cholesterol by various precipitation techniques and is accurate in patients whose triglycerides are less than 4000 mg/L; those over 4000 mg/L lead to an inconsistency in the VLDL triglyceride; that is, the cholesterol ratio does not permit division by a fixed number, and the formula cannot be used with great accuracy.

Use of electrophoresis alone, without quantitation of cholesterol and triglyceride, is not a logical first step for the following reasons: (1) Except under very controlled conditions, electrophoretic patterns are difficult to quantitate consistently so that they can provide relative or absolute concentrations for the lipoprotein classes. Most of these difficulties relate to differing intensities of dye, application problems, and other variables that are difficult to control routinely in electrophoresis. (2) The "sinking" pre-beta-lipoprotein, a ubiquitous lipoprotein, Lp(a), often appears as a "prebeta band" on electrophoresis but is not associated with any triglyceride elevation.

After measurement of cholesterol, triglyceride, and lipoprotein quantitation (if available), it is necessary to have extensive population data to determine the "normal." The definition of "normal" usually rests on some arbitrary cutoff point selected from the continuous distribution of lipid and lipoprotein levels. Age-adjusted cholesterol, triglyceride, VLDL, and LDL "suggested normal levels" of the National Heart and Lung Institute (NHLI) group to identify specific lipoprotein phenotypes, types I, IIa, IIb, III, IV, and V hyperlipoproteinemia, can be used.

Secondary hyperlipoproteinemia

Elevated levels of lipids and lipoprotein in the blood may occur under a variety of circumstances. Although attention is most often focused on the hyperlipoproteinemias whose origin is familial or dietary, many cases occur as a result of, or at least in association with, other pathological conditions (Table 31-7). The spectrum of disease that may be associated with hyperlipoproteinemia is wide, including such diverse disorders as diabetes, nephrosis, hepatic obstruction, and dysgammaglobulinemias. No single pathophysiological mechanism can be evoked to explain the hyperlipoproteinemia that occurs in these disorders. In many cases, the mechanisms involved remain obscure.

All the lipoprotein types that have been recognized to occur in a "primary" or familial form may also be secondary to or be aggravated by other diseases.

Some general comments about the differentiation between primary and secondary hyperlipoproteinemia are in order before discussion of the specific disorders with which hyperlipoproteinemia may be associated. In general, lipoprotein quantification and typing alone will not distinguish the primary from the secondary forms. Even the diagnosis of a disorder that is likely to cause secondary hyperlipoproteinemia does not necessarily establish it as the cause of a patient's hyperlipoproteinemia. Reversal of the lipid abnormality accompanying treatment of the suspected causative disorder, however, is compelling evidence of the secondary nature of the hyperlipoproteinemia. Failure of such reversal to occur implies that the hyperlipoproteinemia may be primary and indicates the need for family screening.

Some disorders that are associated with hyperlipoproteinemia will be obvious from the patient's history and physical examination. Others will require blood or urine tests for diagnosis. If such a screening reveals no abnormalities, it is reasonable to assume that the patient has a primary hyperlipoproteinemia. Whether the hyperlipoproteinemia is established as being familial in origin will depend on the results of family screening.

Other forms of dyslipoproteinemias

The familial lipoprotein disorders discussed in this chapter have thus far revolved around the premise that one or more of the four lipoprotein fractions (chylomicron, VLDL, beta-migrating VLDL, and LDL) are in elevated concentrations. These lipoprotein families are interrelated; the protein complexes are composed of the same lipids in differing proportions, and in some cases they give up some of these lipids to other complexes. The lipoproteins serve different metabolic functions, not all of which are known, and the plasma concentrations of each are subject to some independent controlling mechanisms. These in turn may be affected by one or more mutations.

There are other contrasting lipoprotein abnormalities to the hyperlipoproteinemias reviewed so far. There are three genetically determined disorders in which one or more of the lipoprotein families are absent from plasma, or occur in concentrations that are extremely low. The first of these to be discovered was abetalipoproteinemia, in which chylomicrons, VLDL, and LDL are missing. This is accompanied by malabsorption of fat and later by severe degenerative changes in the nervous system. Many circulating erythrocytes have a thorny appearance (acanthocytosis).

The probable inherited defect is one involving synthesis of the major protein moiety of LDL, Apo B. Another disease is hypobetalipoproteinemia (familial LDL deficiency), in which no lipoproteins are missing but LDL concentrations are far below normal. Nervous system dysfunction and possibly acanthocytes may be present, but usually the patients appear to be well. In the third disorder, familial hypoalphalipoproteinemia, HDL circulates but contains an abnormal proportion of the two major HDL apolipoproteins. A defect in the synthesis of one Apo A-I is the probable locus affected by this rare mutation. With this abnormal high density lipoprotein, patients with Tangier disease have low plasma HDL concentration, store cholesteryl esters in most parts of the body, and often have neuropathic changes for reasons that are not understood. Their large orange tonsils form an unforgettable part of the syndrome.

All these diseases are usually detected initially because of a common manifestation, hypocholesterolemia. They can be readily differentiated, however, if one has a basic understanding of the plasma lipoprotein system.

Last, although not classified as a dyslipoproteinemia, hyperalphalipoproteinemia is a condition in which the HDL is elevated in the blood beyond two standard deviations for a given age and sex. This condition is under genetic influence, and these persons appear to have a longer life-span than those with "normal" concentrations of HDL. There are no other clinical symptoms associated with this lipoprotein feature. A more detailed discussion of each of these lipoprotein abnormalities is provided below.

Abetalipoproteinemia. The rare disorder abetalipoproteinemia has five basic features: low plasma LDL concentration, malabsorption of fat, acanthocytosis, retinitis pigmentosa, and ataxic neuropathic disease. Several of these names under which it has been reported are not specific for this lipoprotein condition. Acanthocytosis may occur in other diseases in which lipoproteins are not deficient. In abetalipoproteinemia LDL is absent, not merely deficient. The plasma cholesterol concentration does not exceed 800 mg/L and is likely to be no higher than 300 mg/L. This is accompanied by concentrations of triglycerides lower than those seen in any other disease, usually less than 200 mg/L. The total phospholipid concentration is also low, that is, less than 1000 mg/L. Both the phospholipid partition and the fatty acid composition of the plasma lipids are abnormal and are frequently reflected in similar abnormalities in erythrocytes and adipose tissue. The most prominent alterations are a decrease in the concentration of linoleic acid and a decrease in the ratio of lecithin to sphingomyelin. The decrease in linoleic acid affects all plasma lipid fractions, including cholesteryl esters, lecithin, triglycerides, and cholesterol. There may be an increased proportion of palmitic, palmitoleic, and oleic acids. It is likely that these changes reflect a decrease in fat absorption and an increase in lipogenesis. The gastrointestinal problems of patients with abetalipoproteinemia are stereotypical. Fat malabsorption is present from birth, and the neonatal period is characterized by poor appetite, vomiting, loose voluminous stools, and little weight gain. The neuromuscular manifestations of abetalipoproteinemia are devastating. The cause of the neuromuscular abnormalities in abetalipoproteinemia is obscure. Attention has focused on the abnormal amount of ceroid pigment (lipofuscin) in the cerebellum and other tissues in this disorder. It is speculated that abnormal lipid peroxidation of the highly unsaturated phosphatides of myelin, secondary to prolonged vitamin E deficiency caused by fat malabsorption, underlie the neuronal degeneration.

In abetaliproteinemia there is a functional disturbance in fat transport. Chylomicrons never enter the plasma, and net transport of endogenous glycerides in VLDL appears to be either absent or sustained at some unchanging minimum level. Heavy feeding of carbohydrate for days, which promotes a brisk rise in level of plasma glycerides and VLDL in patients with Tangier disease and in nearly all other subjects, fails to do so in patients with abetalipoproteinemia. The defect in abetalipoproteinemia is not known. The most likely one is a failure to synthesize the apoprotein B. The intracellular assembly point of Apo B containing lipoprotein is another possible site of dysfunction, as is its secretion. Whichever of these may be primary, none of these defects allow one to explain the plural secondary manifestations of the disease.

Diagnosis of abetalipoproteinemia can be made in a patient with one of the following abnormalities: severe malabsorption of fat, retinitis pigmentosa or other macular degeneration, unexplained neurological abnormalities resembling Friedreich's ataxia and including ataxic neuropathy, or acanthocytosis. The possibility of dysglobulinemia producing antibodies to LDL should also be kept in mind. The single most important laboratory test for screening is the plasma cholesterol determination. The finding of a subnormal value, particularly any concentration below 1000 mg/L, should be followed by a triglyceride determination and lipoprotein electrophoresis. The definitive diagnosis depends on immunochemical demonstration of the absence of LDL apoprotein B in the plasma. In all patients in whom the diagnosis is made, it is also important to check, in all obligate heterozygotes, the concentration of LDL by immunochemical or ultracentrifugal analyses. In familial hypobetalipoproteinemia, the heterozygote has lower than normal concentrations of LDL and apoprotein B.

Hypobetalipoproteinemia. Apparently unrelated to abetalipoproteinemia is another genetic disorder, hypobetalipoproteinemia, in which plasma LDL concentrations are about one tenth of normal. This lipoprotein abnormality is inherited as an autosomal dominant trait. The plasma total cholesterol concentrations can be as low as those seen in abetalipoproteinemia. The percentage of cholesterol esterified is normal. The triglyceride levels may be well within the normal range but sometimes are in the lower limits of accurate measurement. The phospholipid concentrations may vary from 100 to 1800 mg/L and are usually in the low-normal borderline in most patients. Vitamin A and E concentrations are normal or low but, if low, are not decreased to the level seen in abetalipoproteinemia. A faint beta-lipoprotein band is visible on electrophoresis. HDL concentrations as measured by precipitation, preparative ultracentrifuge, or analytic ultracentrifuge are normal; VLDL is usually modestly reduced but is present.

LDL is present in serum as measured by immunoprecipitin tests. These measurements have suggested concentrations of LDL that are one eighth to one sixteenth of normal, levels that are consistent with chemical or optical measurements of LDL concentrations.

Hypoalphalipoproteinemia or analphalipoproteinemia. Tangier disease (familial HDL deficiency) is characterized by severe deficiency or absence of normal HDL in plasma and by the accumulation of cholesteryl esters in many tissues throughout the body, which include liver, spleen, lymph nodes, thymus, intestinal mucosa, skin, and cornea. A combination of two features is pathognomonic: a low plasma cholesterol concentration in combination with normal or elevated triglyceride levels and hyperplastic orange-yellow tonsils and adenoid tissue. Some persons may exhibit peripheral neuropathy. The small amounts of HDL in Tangier plasma differ qualitatively and quantitatively from normal HDL, particularly with respect to apolipoprotein content (Apo A-I). The disorder appears to be attributable to an autosomal recessive gene affecting HDL synthesis or catabolism. Heterozygotes in families with known homozygotes can usually be identified by low HDL concentrations (50% below normal) and do not develop neuropathy and cholesteryl ester accumulation. Among the lipoprotein-deficiency states and, indeed, among all known diseases, the combination of very low cholesterol and elevated triglyceride concentrations gives Tangier disease a unique signature. Some patients may have normal triglyceride levels in the postabsorptive state, however, and may superficially resemble those with LDL deficiency. The plasma total cholesterol level ranges from about 400 to 1250 mg/L, within the range also observed in abetalipoproteinemia and hypobetalipoproteinemia. Individual variation in the plasma triglyceride levels is considerable and is highly contingent on diet. Substitution of carbohydrate for fat often paradoxically lowers the plasma triglyceride concentration in this disorder. The plasma lipoprotein pattern is distinctive: (1) The alpha-lipoprotein band is absent, irrespective of the support medium used. Immunoelectrophoresis may occasionally generate a faint precipitin line of alpha-globulin mobility against anti-HDL antiserum. Most useful for detecting the small amounts of the A apoproteins is immunodiffusion of plasma with specific antiserums to the A-I and A-II apolipoproteins. The reactivity with anti-apo-A-II is generally stronger than with anti-apo-A-I. Estimation of the cholesterol content of the plasma lipoproteins after sequential preparative ultracentrifugation or after ultracentrifugation and selective heparin-manganese precipitation confirms the paucity of HDL.

In addition to HDL absence or deficiency, the following diseases must be excluded: (1) Familial deficiency of LCAT. Here HDL is very low, but the plasma cholesterol level is normal or high and most of the cholesterol is unesterified. (2) Obstructive liver disease, in which the plasma HDL and apoprotein A may be reduced to levels as low as those seen in Tangier disease. In this disorder the cholesterol level is not low, but high, and most of the cholesterol is not esterified. Appropriate tests of liver function should permit the correct diagnosis. (3) Severe malnutrition or hepatic parenchymal disease in which high-density lipoproteins are decreased; the decrease in cholesterol will also be associated with low triglyceride and LDL levels. (4) Acquired HDL deficiency attributable to dysglobulinemia, including possible development of antibodies to HDL. (5) Other storage diseases associated with foam cells and hepatosplenomegaly. In these conditions HDL levels are higher than those seen in Tangier disease, and typical tonsillar abnormalities are absent.

CLINICAL IMPLICATIONS OF HYPERLIPIDEMIA

Why should there be any concern for hyperlipidemia or hyperlipoproteinemia?

Hyperlipidemia is usually a symptomless biochemical state that, if present for a sufficiently long time, may be associated with the development of atherosclerosis and its complications. Occasionally hyperlipidemia may be associated with specific overt symptoms or signs directly attributable to the presence of hyperlipidemia. Examples are abdominal pain, pancreatitis, and the cutaneous manifestations of hyperlipidemia, such as xanthomas.

Risk factors associated with coronary artery disease

Coronary artery disease is almost always the result of atherosclerosis—hardening of the arteries. Coronary atherosclerosis primarily results from the accumulation of fatty deposits in the walls of coronary arteries, which leads to the formation of fibrous tissue in the vessel wall. Coronary artery disease is the most common type of heart disease and the leading cause of death in the United States and many other countries. In the United States, an estimated 50% of the adults die each year from coronary artery disease.

The exact sequence of events that ultimately leads to severe obstruction of coronary arteries by atheromatous plaques is unknown. Most authorities believe that the process is lifelong, beginning in childhood or early adulthood. Fatty materials, principally cholesterol, move from the bloodstream directly into the lining of the blood vessel and are usually deposited near the beginning of the artery, particularly at the point where it branches. In the coronary circulation, these lesions occur in the large vessels and are rarely seen once the branches of the coronary vessels dip into the muscle of the heart before branching into capillaries. The fatty deposits cause a reaction within the vessel wall, and scarlike fibrous tissue may build up around the deposit, forming a plaque, which can become calcified. Generally, the plaque does not cover the entire circumference of the vessel but only part of it, protruding into the lumen (the opening) and progressively narrowing it until in some cases total obstruction or occlusion takes place. Significant interference with the blood flow does not occur until well over half the vessel is occluded by a plaque. Coronary artery disease may be present therefore without evidence of coronary heart disease (see Chapter 30).

Coronary heart disease or symptoms of disturbed myocardial function resulting from interference in the blood flow occasionally take place when only one vessel is obstructed. More commonly, however, by the time a patient develops angina pectoris or has an acute heart attack or myocardial infarction, two or three major vessels are affected by one or more atherosclerotic plaques (atheromas). At times, an acute myocardial infarction may result when a coronary atheroma leads to the secondary formation of a blood clot (thrombus) that totally occludes or blocks the vessel. Blood platelets (cellular components involved in blood coagulation) tend to stick to atherosclerotic plaques, and the aggregation of these platelets may lead to the formation of a blood clot that completely occludes the vessel. When this happens, an acute heart attack may ensue. However, an acute heart attack may also occur without the presence of a recent thrombus if the plaque occludes a large segment of the vessel.

Coronary artery disease is much more common among males than among females, affecting 10 times more males than females under 45 years of age. The sex preference falls off rapidly between 45 and 60 years of age, though males in this group still have about twice as many heart attacks as females do. In the very elderly, the incidence is about the same for both sexes.

Although the basic cause of coronary artery disease is unknown, scientists have identified a number of factors that are associated with a distinct increase in the likelihood that a person will develop a heart attack later in life. These factors, which correlate with the presence of coronary heart disease, are spoken of as "risk factors." Some risk factors are unavoidable, such as racial and genetic susceptibility, prevalence in males, and increased likelihood of having a heart attack as aging occurs. A number of known risk factors are, however, susceptible to modification. Particularly important among these are high blood pressure, cigarette smoking, and elevated serum cholesterol. Approximately 50% of those who have heart attacks are persons who have one or more of these three risk factors. Recent studies show that modification of some of these risk factors may decrease, in part, the original susceptibility to a future heart attack or stroke. A prudent approach therefore appears to be elimination or attenuation of these risk factors whenever they are identified. The box above lists the primary and secondary risk factors associated with coronary heart disease.

Primary and secondary risk factors associated with coronary heart disease

Primary
- Genetic predisposition for coronary heart disease
- Hypertension
- Cigarette smoking
- Elevated total cholesterol (LDL cholesterol)
- Decreased HDL cholesterol

Secondary
- Lack of exercise
- Obesity
- Age
- Male sex
- Stress
- Diabetes mellitus
- Gout and hyperuricemia
- Renal failure patients on hemodialysis
- Subjects on oral contraceptives

Recently, reappraisal of the importance of HDL or HDL-cholesterol levels as predictors of coronary heart disease has created much interest. According to the Framingham data, there is a clear gradient of coronary heart disease incidence rates in relation to serum HDL-cholesterol concentrations. Persons with levels below 350 mg/L have eight times the coronary heart disease rate of persons with HDL-cholesterol levels of 650 mg/L or greater. In 1951 it was first pointed out that a high proportion of serum total cholesterol carried in HDL in babies was similar to that found in animals that appear to have a high resistance to atherosclerotic disease.

These findings tend to support the hypotheses on the atherogenicity of LDL and the seemingly protective effect of HDL. It has been demonstrated that LDL from hypercholesterolemic serum stimulates growth of smooth muscle cells in tissue culture and leads to intracellular lipid accumulation, whereas HDL has little effect. One possible mechanism for the protective role of HDL is that HDL competes with LDL for cell-surface receptors and may not be internalized as extensively as the latter lipoprotein, in effect reducing the cholesterol uptake and its accumulation in the smooth muscle cell. It is also possible that HDL, in concert with phospholipids, may remove cholesterol from cholesterol-laden tissues, such as the arterial intima. Although the mechanism of cholesterol removal from cells is still not known, it has been suggested that lecithin:cholesterol acyltransferase (LCAT) enzyme may play a role in the transport of cholesterol from peripheral tissues to the liver.

Etiology and pathogenesis

In reviewing the complex subject of the etiology and pathogenesis of atherosclerosis, I adopted a synthetic or holistic approach, rather than subscribing preferentially to one or more of the mutually exclusive hypotheses. One would hope that it will become apparent that the cause of atheroma is not the sole domain of the platelet, lipoproteins, high blood pressure, lysosome, or smooth cell, but rather that each of these factors plays a role in the sequential assembly of an exceedingly complex mosaic to which we ascribe the term ''atherogenesis.''

Initiating factors. It is not unreasonable to make the basic and probably correct assumption that the essential event or sine qua non of atherogenesis is enhanced transendothelial transport of macromolecules, resulting either from frank endothelial injury or, alternatively, from more subtle modifications in endothelial structure and function. The factors that enhance this transendothelial permeability to macromolecules, including low density lipoprotein, which results in a net influx or retention within the subendothelium and tunica media, may be conveniently considered as initiating factors. Retention may result not only from an enhanced influx but also from a decreased efflux. Decreased efflux may reflect binding of the macromolecules to components of the arterial wall or, alternatively, molecular modifications that occur within the arterial wall and render the molecules less able to transverse the endothelium to the lumen.

Focal hemodynamically induced endothelial injury with enhanced permeability is the probable determinant of the consistent and discrete localization of the atheromatous process. Hemodynamic effects may be mediated by shear, stretch, vibration, or pressure. Additional initiating factors, other than those associated with focal hemodynamics, include the release of platelet constituents, hypertension, carbon monoxide, antigen-antibody complexes, and hyperlipidemia. Cigarette smoking may prove to be an important initiating factor that exerts its effects either directly through an immune mechanism or indirectly through released platelet constituents or carbon monoxide.

Accelerating factors. Factors that influence the nature or rate of lesion development at all subsequent stages are conveniently classified under the generic term ''accelerating factors.'' Accelerating factors of note include hypercholesterolemia with an associated excess of low density lipoprotein and disturbances in platelet function, hemostasis, and thrombosis. These accelerating factors, namely LDL and platelet factors, may influence the nature and rate of plaque development by stimulating the proliferation of smooth muscle cells and by stimulating to varying degrees the synthesis by smooth muscle cells (SMC) of collagen, elastin, and the glycosaminoglycans. In addition, lipoproteins may significantly influence SMC lipid metabolism, including cholesterol synthesis, and the uptake and accumulation of lipids within smooth muscle cells.

In regard to interaction between initiating and accelerating factors, one might anticipate, there may be several levels of complex interaction, not only among differing accelerating factors, but also between accelerating and initiating factors. For example, platelets, through the release of their constituents, including 5-hydroxytryptamine, histamine, thromboxane A_2, and lysosomal enzymes, may directly cause endothelial injury or modify endothelial permeability. This modification of endothelial permeability will serve to enhance the entry of accelerating factors, such as LDL, and certain platelet proteins, which subsequently stimulate smooth muscle cell proliferation.

In addition, platelets may directly contribute to plaque growth by serving as components of mural thrombi. LDL may directly induce a spectrum of changes consistent with endothelial injury. In type II hyperlipoproteinemia, LDL has been shown as well to be associated with modified platelet function, which might influence the risk of thrombus formation.

Other pathogenic processes. Other processes not yet discussed that are of importance in pathogenesis include the concept of a monoclonal origin of the cells in atheroma and also the concept that atheroma is a lysosomal storage disease, with the large accumulation of cholesterol esters resulting from an overload of the lysosomal enzyme systems. Both the monoclonal explanation for the cellular monotypism, which in essence implies that atheromatous lesions are benign smooth muscle cell tumors, and the possible role of a lysosomal enzyme deficit in the accumulation of arterial lipids are compatible with the overview presented. Just how important each of these mechanisms is in the pathogenesis of atheroma has yet to be established.

There remain many unanswered questions and unresolved issues. For example, the precise relationship between fatty streaks and fibrous plaques in the evolution of atheroma has still to be clarified. Furthermore, how does the fibrous atherosclerotic plaque begin and progress, are the earliest lesions as characteristic of the disease as the atherosclerotic plaque is, and what extramural factors directly or indirectly influence the composition of the precursor lesions and the rate of their progression to the plaque? Other important problems relate to the time scale for the development or regression of clinically significant disease, the reasons for the striking sex-associated differences in disease prevalence and severity, and the mechanisms underlying the consistently greater frequency of clinical disease in patients with diabetes mellitus.

Finally, there may exist risk factors that so far have not been recognized and therefore are not measured. Identification of measurable characteristics associated with the probability of atheroma would make possible testing their relationship to coronary heart disease.

TREATMENT

Control of serum cholesterol concentrations: effect of diet and drugs

In the treatment of hypercholesterolemia, the first thing that needs to be determined is whether there are any primary diseases that may be contributing to the elevation of the serum cholesterol concentration. Before consideration of treatment, in order to determine whether the condition is sporadic or familial, the patient's first-degree relatives must be screened. Treatment begins with determining whether the subject is overweight. Reduction to ideal body weight will lower plasma lipids, particularly plasma triglycerides. Further reduction requires special dietary considerations.

Knowing that the quantity and quality of dietary fat influences blood cholesterol concentration, one should reduce saturated fat and cholesterol intake. Cholesterol intake of less than 300 mg/day is recommended. The polyunsaturated fat (P)/saturated fat (S) (or P/S) ratio is quantitatively more important than cholesterol consumption in lowering plasma cholesterol concentration.

In the average American diet, the P/S ratio is only about 0.3. For a therapeutic diet to lower plasma cholesterol in a patient with severe type II hyperlipidemia, it is recommended that the ratio be increased to a range of 1 to 1.2. To accomplish this, the patient must consume about 2 tablespoons of polyunsaturated fats or vegetable oil per day and drastically reduce the animal fat intake.

If diet fails to work or has minimal effect on cholesterol reduction in the blood, the next step is drugs. There are no precise criteria for drug therapy in the treatment of hyperlipidemia. Important considerations in prescribing medicaton are the age of the patient; the severity of the hyperlipidemia; the clinical manifestations, including those of arteriosclerosis; the family history; and the presence of coexisting diseases.

The combination of diet and drug therapy is more effective than drug therapy alone. In providing treatment for the patient, the physician should emphasize that diet and drugs are not curative; however, they can help control hyperlipidemia only as long as the patient complies. There are currently about a half dozen cholesterol-lowering drugs.

Cholestyramine. Cholestyramine is a nonabsorbable anion-exchange resin that is an effective cholesterol-lowering agent in the dose range of 16 to 24 g of base per day. Its main effect is to reduce serum cholesterol and LDL levels, but it has no capacity to reduce triglyceride levels. Occasionally cholestyramine will actually increase triglyceride levels while cholesterol levels are falling.

Cholestyramine interrupts the enterohepatic circulation by binding the bile salts and thus making them unavailable for cholesterol absorption. To compensate for this, the liver cells convert more cholesterol to bile salt, ultimately resulting in a fall in plasma cholesterol concentration. A compensatory, but usually lesser, increase in endogenous cholesterol synthesis also results.

Colestipol. Colestipol resin is a relatively new anion-exchange resin that is similar in most practical respects to cholestyramine. Individual patients who are unable to tolerate cholestyramine because of gastrointestinal side effects may be managed with colestipol.

Nicotinic acid (niacin). Nicotinic acid is a useful reserve drug in the treatment of hypercholesterolemia, hypertriglyceridemia, or mixed hyperlipidemia. It is a potent suppressor of LDL and VLDL levels in most hyperlipidemic states. It is the most effective cholesterol-lowering drug. The severity of side effects accompanying its rapid absorption restricts its wider use.

Nicotinic acid is a peripheral vasodilator and will produce noticeable skin flushing at the beginning of therapy. One can minimize this by using a small dose at first and gradually increasing the dose. Also, if it is consumed just before the meal with aspirin, the side effects should decrease.

Nicotinic acid reduces serum lipid levels, probably through multiple mechanisms, including reduction of lipid and lipoprotein synthesis in the liver, partly because of the reduction in the flux of substrate-free fatty acids through plasma from adipose tissue to the liver.

Probucol. Probucol is the newest lipid-lowering drug to come into routine use. It is poorly absorbed from the gut, becomes concentrated in various lipid-rich tissues, and persists in the body for a very long period of time. At an oral dosage of 500 mg twice daily, it will predictably lower serum cholesterol levels by approximately 15%,with negligible influence on triglyceride levels.

Betafibrate. This drug is a chemical derivative of clofibrate. It has been used in Europe but, as of 1983, not in the United States. It effectively lowers serum cholesterol but appears to cause side effects in some persons, that is, elevated ALT (alanine aminotransferase) and uric acid.

Summary. In considering the possible mode of action of any agent that leads to a change in plasma cholesterol concentration, remember that cholesterol circulates in the plasma in lipoproteins. Therefore any change in cholesterol concentration will reflect an underlying change in lipoprotein concentration if, as is likely, the metabolism of the lipid moiety of a lipoprotein molecule is intimately linked with that of a protein moiety. Hence the effect of a drug or dietary modification on plasma cholesterol metabolism may be secondary to the effect on the metabolism of the protein component of one or another of the plasma lipoproteins.

Control of serum triglyceride concentrations: effects of diet and drugs

In general, weight reduction and low-fat diets are the most effective means of controlling serum triglyceride levels. Serum triglyceride concentrations are very sensitive to weight changes. In both the obese and hypertriglyceridemic individual, a low-calorie diet is quite successful. In the nonobese hyperlipidemic subject, the choice of the dietary regimen should be guided by the nature of the plasma lipoprotein abnormality. Exogenous fats, whether saturated or not, contribute to the formation of chylomicrons and should be limited in the diet of patients who have an excess of these particles in their fasting plasma (types I and V).

This is the basis for the treatment of familial lipoprotein lipase deficiency (type I). Medium-chain triglycerides, which are absorbed and transported to the liver through the portal system, do not contribute to chylomicron formation and may be used as a substitute for vegetable oil in this disease. The total daily fat intake should be kept between 25 and 35 g/day, and no alcohol should be permitted, since it may promote or enhance pancreatitis.

An excess of simple carbohydrates in the diet usually enhances endogenous hypertriglyceridemia, which is found in types IIb, III, IV, and V. Affected patients should be encouraged to limit their intake of simple sugars. In some cases, a dietary intake reduced in simple sugar, without a reduction in total caloric intake, may completely correct an endogenous hypertriglyceridemia. Alcohol intake should also be limited because it causes acute, transient hypertriglyceridemia.

Drugs are of no value in familial hyperchylomicronemia. In treating familial hypertriglyceridemia (type IV), clofibrate is a first-line drug if the response to the diet has been inadequate. In case of resistance to this medication, one is left with nicotinic acid as a major hypotriglyceridemic agent.

SUGGESTED READINGS

Brewster, M.A., and Naito, H.K., editors: Nutritional elements and clinical biochemistry, New York, 1980, Plenum Pub. Corp.

Christie, W.W.: Lipid analysis, New York, 1980, Pergamon Press, Inc.

Cooper, G.R.: Introduction to analysis of cholesterol, triglyceride, and lipoprotein cholesterol: reference values and conditions for analyses. In Faulkner, W.R., and Meites, S., editors: Selected methods of clinical chemistry, vol. 9: Selected methods for the small clinical chemistry laboratory, Washington, D.C., 1982, American Association for Clinical Chemistry, pp. 157-164.

Fredrickson, D.S., and Lees, R.S.: Familial hyperlipoproteinemia. In Stanbury, J.B., Wyngaarden, J.B., and Fredrickson, D.S., editors: The metabolic basis of inherited disease, ed. 2, New York, 1966, McGraw-Hill Book Co., pp. 429-485.

Fredrickson, D.S., Levy, R.J., and Lees, R.S.: Fat transport in lipoproteins: an integrated approach to mechanisms and disorders, N. Engl. J. Med. **276:**32-44, 94-103, 148-156, 215-224, 273-281, 1967.

Gurr, M.I., and James, A.T.: Lipid biochemistry: an introduction, Cambridge, England, 1980, University Press.

Havel, R.J.: Approach to the patient with hyperlipidemia, Med. Clin. North Am. **66**(2):319-333, 1982.

Havel, R.J.: Familial dysbetalipoproteinemia, Med. Clin. North Am. **66**(2):441-454, 1982.

Levy, R.J., Rifkind, B.M., Dennis, B.H., and Ernst, N., editors: Nutrition, lipids, and coronary heart disease, New York, 1979, Raven Press.

Lewis, L.A., and Opplt, J.J., editors: Handbook of electrophoresis, vol.1: Lipoproteinemia: basic principles and concepts; vol. 2: Lipoproteins in disease, Boca Raton, Fla., 1980, CRC Press, Inc.

Lewis, L.A., editor: Handbook of electrophoresis, vol. 3: Lipoprotein methodology and human studies, Boca Raton, Fla., 1983, CRC Press, Inc.

Lewis, L.A., and Naito, H.K., editors: Handbook of electrophoresis, vol. 4: Lipoprotein studies of nonhuman species, Boca Raton, Fla., 1983, CRC Press, Inc.

Masoro, E.J.: Physiological chemistry of lipids in mammals, Philadelphia, 1968, W.B. Saunders Co.

Miller, M.E., and Lewis, B., editors: Lipoproteins, atherosclerosis and coronary heart disease, New York, 1981, American Elsevier Publishing Co., Inc.

Naito, H.K., editor: Nutrition and heart disease, New York, 1982, Spectrum Publications, Inc.

Nelson, G.J., editor: Blood lipids and lipoproteins: quantitation, composition, and metabolism, New York, 1972, John Wiley & Sons, Inc.

Perkins, E.G.: Analysis of lipids and lipoproteins, Champaign, Ill., 1975, American Oil Chemists' Society.

Rifkind, B.M., and Levy, R.I., editors: Hyperlipidemia: diagnosis and therapy, New York, 1977, Grune & Stratton.

Story, J.A., editor: Lipid research methodology, New York, Alan R. Liss, Inc. (In press.)

Chapter 32 Alcoholism

Charles L. Mendenhall
Robert E. Weesner

alcohol withdrawal syndrome The clinical symptoms associated with cessation of alcohol consumption. These may include tremor, hallucinations, autonomic nervous system dysfunction, and seizures.

alcoholic cirrhosis Cirrhosis resulting from chronic excess alcohol consumption. The cirrhotic process in the liver is a pathological process in which progressive injury produces fibrotic bands or scar tissue that entraps liver cells and results in loss of normal microscopic lobular architecture (nodular regeneration).

alcoholic hepatitis Acute toxic liver injury associated with excess ethanol consumption. This is characterized by necrosis, polymorphonuclear inflammation, and in many instances Mallory bodies.

alcoholic ketoacidosis The fall in blood pH (acidosis) sometimes seen in alcoholics and associated with a rise in serum ketone bodies (acetone, beta-hydroxybutyric acid, and acetoacetic acid).

fatty liver Excessive accumulation of fat in the liver parenchymal cells. These are primarily neutral lipids, triglycerides, and cholesterol. Fatty liver predictably develops after exposure to a variety of hepatotoxins of which ethyl alcohol is the most common.

fetal alcohol syndrome A group of fetal abnormalities resulting from maternal alcohol consumption during gestation.

hemochromatosis A disorder of iron metabolism characterized by excess iron deposits in tissues such as those composing the liver, pancreas, and heart leading to organ injury. The organ injury may manifest itself as cirrhosis, diabetes mellitus, or heart failure.

hepatic fibrosis The deposition of collagen and fibrous tissue in the liver before the development of nodular regeneration and cirrhosis.

kwashiorkor A nutritional disease resulting from protein deprivation. It is characterized by depleted visceral proteins and immunological dysfunction. Total calorie consumption may be deficient, adequate, or even excessive.

Mallory bodies (alcoholic hyalin) An eosinophilic cytoplasmic inclusion that accumulates in the liver cells. It is typically, but not always, associated with acute alcoholic liver injury (alcoholic hepatitis).

marasmus A nutritional disease resulting from calorie deprivation. It is characterized by weight loss and wasting of muscle mass and fat stores.

Wernicke-Korsakoff syndrome A disease of the central nervous system occurring in alcoholics and attributed to thiamin deficiency. Wernicke's encephalopathy consists of ocular disturbances, ataxia, and impaired mental functions. If untreated it may merge into Korsakoff's syndrome, which includes memory impairment, confabulations, and deranged perception of time.

Alcoholism represents one of the most serious worldwide socioeconomic and health problems. In the United States, alcohol-related liver disease represents the sixth leading cause of death.[1] An alcoholic is usually a person who consumes an amount of ethanol (ethyl alcohol) capable of producing pathological changes.[2] The amount of ethanol capable of producing disease is variable and depends on a variety of factors including genetic predisposition,[3] malnutrition,[4] and concomitant viral infection of the liver (viral hepatitis).[5] For a susceptible person this may be as low as 35 g/day,[6] which is equivalent to the daily consumption of three cocktails made with 100-proof whiskey. However, for most persons the grams of alcohol necessary to produce disease are in excess of 80 g/day for at least 10 to 15 years. When this amount is converted to the quantity of alcoholic beverages consumed, it represents eight 12-ounce 6% beers, a liter of 12% wine, or a half pint of 80-proof whiskey per day. It should be pointed out that susceptible persons may develop disease with much smaller amounts of ethanol. Rydberg[6] found that a pathological condition did not develop even in susceptible persons when the ethanol level was below 7 g/day.

DIAGNOSIS OF ALCOHOLISM

Characteristically the alcoholic may attempt to conceal his excessive drinking, and so the history of consumption may be unreliable. Verification from the spouse frequently is necessary to determine an accurate drinking history.

Several laboratory tests have been proposed as diagnostic tools to confirm the excess consumption of ethanol.[7-9] Increases in mean red cell volume (MCV), believed to result from ethanol-altered red blood cell membranes, have been shown to occur early and closely parallel the amount of ethanol consumed.[7] As few as two drinks a day has resulted in significant increases in MCV.[7] However, the changes are small and so base-line predrinking values may be necessary for recognition of the effect of ethanol.

Serum gamma-glutamyl transferase (GGT) has been shown to be readily inducible by a variety of compounds including ethanol.[10] This is an easy test to perform and is available in many hospital laboratories as a test of liver dysfunction. In the absence of other abnormal laboratory tests for liver disease, elevated levels of GGT may be useful for recognition of the heavy drinker in the predisease state.

BIOCHEMICAL AND METABOLIC ALTERATIONS

Ethanol metabolism

Although ethanol is usually considered an organic solvent in the laboratory, it may also be synthesized endogenously in trace amounts.[11] This is a more common problem in ruminant animals in which grain may be stored and fermented in the rumen, which results in intoxication and bizarre behavior.[12] In humans ethanol is primarily an exogenous compound consumed in alcoholic beverages and readily absorbed from the entire gastrointestinal tract. Most of the absorbed ethanol is then degraded by oxidative processes, primarily in the liver, first to acetaldehyde and then to acetate. There are at least three enzyme systems that are capable of ethanol oxidation: (1) alcohol dehydrogenase (ADH), (2) microsomal ethanol oxidizing system (MEOS), and (3) catalase. Only 2% to 10% is excreted unoxidized in the urine and lungs.

ADH appears to be the principal pathway for ethanol oxidation. This is especially true for acute intoxication. The involved reaction is as follows:

$$\text{Ethanol} \xrightarrow[\text{NAD}^+ \;\rightarrow\; \text{NADH} + \text{H}^+]{\text{ADH}} \text{Acetaldehyde}$$

The initial step with ADH appears to be rate limiting for the clearance of ethanol from the blood, but it is not specific for ethanol. Methanol, retinol (vitamin A), and a number of other alcohols are also oxidized by this enzyme. This fact is important clinically in the treatment of acute methanol poisoning.[13] In this instance, ethanol is given clinically to compete with methanol for binding sites on the ADH enzyme; thus oxidation of methanol to the toxic formaldehyde is delayed.

MEOS appears to be a secondary enzyme system for ethanol clearance. The reaction is as follows:

$$\text{Ethanol} \xrightarrow[\text{NADPH} + \text{H}^+ \;\rightarrow\; \text{NADP}^+ + \text{H}_2\text{O}]{\text{Oxygen} \quad \text{MEOS}} \text{Acetaldehyde}$$

This microsomal enzyme is in the smooth endoplasmic reticulum (SER) of the hepatocytes, and in the chronic alcoholic it is associated with enzyme induction[14] and hypertrophy of SER as seen by electron microscopy.[15] The induction of MEOS activity is further associated with increased activity of various other constituents of the SER involved with drug metabolism, such as cytochrome P-450 reductase and cytochrome P-450. These changes have clinical significance because they render the alcoholic more resistant to the effects of many common sedatives and barbiturates, resulting in the necessity for a larger-than-normal dose when sedation is required. If, however, ethanol and barbiturates are consumed simultaneously, competitive inhibition for these enzymes results in a reduction in clearance[16] and abnormally high blood levels. Indeed, overdosage and death have resulted from this interaction.

The role of catalase in the biological oxidation of ethanol is controversial[17,18] In vitro in the presence of a peroxide (H_2O_2)–generating system, catalase has been shown to be capable of oxidizing ethanol by the following reaction:

$$\text{Ethanol} + H_2O_2 \xrightarrow{\text{Catalase}} 2H_2O + \text{Acetaldehyde}$$

It appears that the slow rate at which peroxide can be generated from NADPH oxidase or xanthine oxidase prevents catalase from contributing to more than 2% of the in vivo ethanol oxidation.

The product of these varied oxidation pathways is acetaldehyde. It appears that various organ disorders and biochemical alterations are induced by either ethanol or acetaldehyde, but not necessarily both. For example, the fetal alcohol syndrome appears to be an ethanol effect independent of acetaldehyde,[19] whereas liver fibrosis and collagen formation are more closely associated with acetaldehyde than with ethanol.[20] The increase in NADH with a concomitant decrease in NAD^+ associated with ADH activity may also produce sequential metabolic changes with clinical consequences. The increased availability of NADH results in increased lactate production, which can result in hyperlactacidemia.[21] The hyperlactacidemia results in acidosis, which reduces the capacity of the kidney to excrete uric acid, leading to a secondary hyperuricemia.[22] In susceptible persons this may aggravate or precipitate attacks of gout.[23]

Altered lipid, protein, and carbohydrate biochemistry

Lipids accumulate in most tissues in which ethanol is metabolized, and such accumulation results in fatty liver, fatty myocardium, fatty renal tubules, and so on. The mechanism appears to be multifactorial, resulting from both increased lipid accumulation and decreased lipid removal. In the liver, which has been studied most extensively, the lipid accumulation is primarily of dietary origin.[24] The concomitant ingestion of ethanol with a fatty meal greatly increases intestinal uptake of fat into chyle[25] with an accompanying increase in both the hepatic and intestinal lymph flow.[26] Lipid synthesis is also accelerated. The increase in the $NADH/NAD^+$ ratio is associated with rises in alpha-glycerol phosphate production, which favors hepatic triglyceride formation.[27] Furthermore, enzymes associated with both fatty acid[28,29] and cholesterol synthesis[30] have increased activity in the alcoholic. Added to the increased synthesis are impaired fatty acid oxidation[31] and altered lipoprotein secretion.[32] Although the secretion of serum lipoproteins is low relative to the lipid load accumulating in the liver, the total amount secreted is increased above normal, and alcoholic hyperlipemia may result.[33] This is a type IV hyperlipemia with increases in both serum triglycerides and cholesterol esters and with a lipoprotein density of less than 1.006 (very-low-density lipoproteins and chylomicron-like particles of hepatic origin).

Changes in protein function and biochemistry are also seen. Enzyme activities related to ethanol oxidation and drug metabolism are frequently altered. These changes may be produced either directly by enzyme induction or suppression or indirectly by the shift in reducing equivalents ($NADH/NAD^+$). In addition, the availability of amino acids for protein synthesis may be altered. Amino acids are actively transported across the intestinal mucosal membrane. The (Na^+, K^+)- and Mg^{++}-ATPase activity necessary for this active transport has been shown to be decreased by ethanol[34] with a concomitant suppression of neutral amino acid transport.[35]

Once synthesized, the cellular release of proteins may be altered. This has been shown to be the case for glycoproteins[36] and alpha-fetoprotein[37] but not for albumin.[38] As a consequence of the altered release of protein from the liver cell, one of the earliest liver changes seen in the alcoholic is the accumulation of protein,[39] which occurs concurrently with fat and contributes to the development of the enlarged liver (hepatomegaly) that is seen in over 90% of alcoholics with liver disease.[40]

Changes in carbohydrate metabolism are also seen. As with amino acids, sugars are actively transported during the absorption process. In vitro, ethanol has been shown to inhibit the active transport of D-glucose[41] into intestinal cells. Furthermore, ethanol impairs galactose utilization by inhibiting its conversion to glucose by means of UDP-galactose 4-epimerase. This reaction is NAD^+ dependent and is impaired by the NADH generated from ethanol.[42]

Gluconeogenesis is similarly impaired by ethanol by a variety of mechanisms. Gluconeogenesis is the process whereby glucose is formed from noncarbohydrate sources, that is, glycerol, pyruvate, and a number of amino acids. Because ethanol promotes glycerol-lipid formation and impairs amino acid transport, the availability of these compounds for gluconeogenesis is decreased. The excess NADH from ethanol metabolism favors lactate formation and retards its oxidation for gluconeogenesis. Glutamic acid dehydrogenase activity is also diminished by NADH, decreasing the availability of alpha-ketoglutarate[43] necessary for amino acid transamination before its conversion into glucose.

Typically the storage of glycogen in the liver is also diminished. This results from both poor dietary intake and the liver disease so frequently associated with chronic alcoholism (fatty liver, alcoholic hepatitis, alcoholic cirrhosis). The clinical implications of these changes in liver glycogen and impaired gluconeogenesis are discussed later along with insulin and other hormonal involvement in the development of alcoholic hypoglycemia.

In the absence of liver disease, malnutrition, or other predisposing conditions, ethanol itself produces only a slight increase or no change in blood glucose. If it is present, the peak rise in glucose occurs 30 to 45 minutes after ingestion and represents usually no more than a 7% to 10% increase.

NUTRITIONAL ALTERATIONS ASSOCIATED WITH ALCOHOLISM AND ALCOHOLIC LIVER DISEASE

Protein-calorie malnutrition

Alcoholic beverages are high in calories. Each gram of ethanol yields 7 kcal. However, these are low in nutritive value.[44]

Chronic alcoholism has long been associated with both malnutrition and liver disease. Malnutrition in the form of protein-calorie deficiency results from a variety of causes: (1) poor dietary intake, (2) malabsorption of consumed nutrients as a result of alcohol-induced gastritis, pancreatitis, or diarrhea, and (3) altered biochemical and physiological processes.

Protein-calorie malnutrition has been classified as marasmus or a kwashiorkor-like disease, depending on whether a caloric deficit (marasmus) or a protein deficit (kwashiorkor-like) predominates.[45] Marasmus is associated with weight loss, loss of depot fat, and decreased skeletal-muscle mass. In the more advanced disease the immune response is also impaired. A kwashiorkor-like syndrome caused by protein depletion may not be associated with weight loss or the other findings of marasmus. However, visceral proteins such as albumin, transferrin, and retinol-binding protein are more noticeably depleted, and the immune system is impaired.

In a recent study of 284 alcoholics with alcoholic hepatitis, all patients had some degree of protein-calorie malnutrition, and 80% had features of both marasmus and kwashiorkor-like disease.[4]

The role of malnutrition in the development of liver injury is still controversial. Certainly liver injury can develop in the absence of malnutrition. However, the more severe forms are invariably associated with severe malnutrition. Recognition of nutritional deficits appears to be important for the alcoholic with liver disease because correction of these abnormalities by appropriate nutritional therapy has been observed to increase survival and accelerate improvement of the liver injury.[46,47]

Vitamin abnormalities

Excessive alcohol use commonly leads to vitamin deficiency.[48] Not only is the liver a major storage depot for

Table 32-1. Vitamins in the alcoholic person

Vitamin	Classification of vitamin	Effect of alcohol on intestinal absorption	Incidence (%) of low serum values in alcoholic cirrhosis[1]
A	Fat soluble	Decrease	12
B_1 (thiamin)	Water soluble	Decrease	58
B_2 (riboflavin)	Water soluble	None	6
Nicotinic acid	Water soluble	Not evaluated	33
B_6 (pyridoxine)	Water soluble	None	60
Folic acid	Water soluble	None	78
B_{12} (cyanocobalamin)	Water soluble	Decrease	25
C (ascorbic acid)	Water soluble	Not evaluated	25
D	Fat soluble	Not evaluated	?
E	Fat soluble	Not evaluated	15
K	Fat soluble	Not evaluated	?

vitamins,[49] but it also converts vitamins into metabolically useful forms.[50] Therefore injury to the liver alters vitamin metabolism. Vitamins can be released from necrotic liver cells, lost from the body, and not adequately replaced.[51] Alcohol can increase the body's need for vitamins because of increased nucleic acid synthesis and liver regeneration.[52,53] Fat-soluble vitamins (vitamins A, D, K, and E) may not be adequately absorbed by the intestine because of the increased fat loss in the stool seen in some patients with liver disease.[48] Alcohol can directly inhibit intestinal absorption of some vitamins,[54] and the malnutrition associated with alcohol abuse can also affect intestinal absorption.[55] The poor dietary intake of alcoholics can also lead to vitamin deficiency.[56] Because of the many interactions among the liver, alcohol, and vitamins, vitamin replacement plays a major role in treating the alcoholic patient. Table 32-1 gives the incidence of low serum values in persons with alcoholic cirrhosis. See Chapter 34 for a discussion of the biochemistry and physiology of vitamins.

Vitamin A. Vitamin A is absorbed as retinol, which must be oxidized to retinal with the help of alcohol dehydrogenase in order to be functional. Alcohol competitively inhibits the conversion of retinol to retinal in the liver.[57]

Vitamin B_1 (thiamin). Thiamin deficiency is encountered frequently in alcoholics.[48] Thiamin is absorbed from the intestine both passively and by active transport (requiring energy).[58] The requirement for thiamin is greatest when carbohydrate is the major source of energy,[58] a situation sometimes found in the alcoholic. The active transport of thiamin across the intestine is inhibited by alcohol.[54] Thiamin deficiency can produce a neuropsychiatric disorder called "Wernicke-Korsakoff syndrome" in a small number of alcoholics; this syndrome can be genetically determined.[59] There is also increasing evidence that thiamin deficiency may contribute to other forms of brain injury in alcoholics.[57]

Vitamin B_2 (riboflavin). Riboflavin deficiency is associated with cheilosis and corneal vascularization.[58] This vitamin is rarely a problem in the alcoholic.[48]

Nicotinic acid. Nicotinic acid is converted into important coenzymes involved in tissue respiration. A severe deficiency of nicotinic acid leads to pellagra and can be found in the alcoholic population.[58]

Vitamin B_6 (pyridoxine). Pyridoxine is converted to pyridoxal 5-phosphate, which is a coenzyme in the metabolic transformation of amino acids. In alcoholic hepatitis the serum aspartate aminotransferase (AST) (glutamic-oxaloacetic transaminase or SGOT) is greater than the serum alanine aminotransferase (ALT) (glutamic-pyruvic transaminase or SGPT) in 90% of the patients.[60] Animal studies possibly indicate that this decrease in ALT activity in patients with alcoholic hepatitis may be caused in part by pyridoxal phosphate deficiency.[61]

Folic acid. Folic acid deficiency is the most common vitamin deficiency among alcoholics.[48] Folic acids act as coenzymes in the synthesis of purine and pyrimidine nucleotides and in a number of amino acid interconversions.[58] Folate deficiency contributes to the megaloblastic anemia noted in alcoholics.[62] Alcohol does not inhibit intestinal absorption of food-associated folate.[63] Malnutrition is commonly found in alcoholics and may decrease folate absorption.[55] Poor dietary intake, increased requirements, and increased excretion are believed to be the major causes of folate deficiency in the alcoholic.[57] In addition, alcohol may directly block folate metabolism.[64]

Vitamin B_{12} (cyanocobalamin). Vitamin B_{12} is found only in foods of animal origin. Vitamin B_{12} is not broken down by the body but is excreted unchanged primarily in the bile.[58] Although alcohol in some way inhibits vitamin B_{12} absorption at the ileum,[57,65] vitamin B_{12} deficiency is rarely a problem.[48] Because vitamin B_{12} is stored in the liver, acute liver cell necrosis of alcoholic hepatitis may actually produce a noticeable increase in serum B_{12} levels that parallels the severity of the liver injury.

Vitamin C (ascorbic acid). Vitamin C deficiency is not common in alcoholics and does not produce any clear metabolic problems. Low levels in alcoholics have been attributed to poor dietary intake.[66]

Vitamin D. Vitamin D deficiency is not a major problem among alcoholics.[58]

Vitamin E. Chronic alcoholics are susceptible to vitamin E deficiency because of low dietary intake and general malnutrition. Pure vitamin E deficiency is known to produce a testicular lesion, but whether this plays a role in alcoholic hypogonadism remains unclear.[57]

Vitamin K. Vitamin K plays a key role in the synthesis of clotting factors II, VII, IX, and X. Because the stores of vitamin K are small, biliary obstruction or severe parenchymal liver disease can produce bleeding abnormalities.[58] Alcoholics often have prolonged prothrombin times.[40] When these patients are given vitamin K parenterally, some show corrected prothrombin time, indicating vitamin K deficiency. Those who do not improve with vitamin K have liver disease too severe to utilize the vitamin.[58]

Mineral abnormalities

Iron. Iron is essential for oxygen transport by the red blood cells.[67] The body has no physiological mechanism for excreting excess iron. Therefore iron balance is controlled by regulation of iron absorption from the duodenum and proximal jejunum.[67] When this regulatory mechanism breaks down, hemochromatosis may occur. *Hemochromatosis* refers to a group of disorders in which total body iron stores progressively increase with iron deposition in parenchymal cells of the liver, heart, pancreas, and other organs. This deposition of iron can lead to cellular damage with resulting cirrhosis, heart failure, and diabetes mellitus (bronze diabetes). Refer to Chapter 33 for a detailed review of iron metabolism.

The mechanism for mild to moderate iron accumulation in the liver of some patients with alcoholic liver disease is unclear and controversial. First, some alcoholic beverages, especially red wines, contain iron. Making beer in iron kettles has led to hemachromatosis among the South African Bantu.[68] Finally, the effect of alcohol on folate could influence iron absorption. Folate deficiency is associated with ineffective red blood cell formation (erythropoiesis) and iron overloading.[69] Alcoholics with significant iron overload may, in fact, carry the gene for primary idiopathic hemochromatosis.[70] Some alcoholics have low iron levels because of poor dietary intake or chronic blood loss from the intestinal tract (gastritis, peptic ulcer, esophageal and gastric varices, and poor clotting).[58]

Zinc. Zinc, an essential micronutrient element, is important for RNA and DNA synthesis and for a variety of zinc-dependent enzymes including alcohol dehydrogenase.[57] Zinc deficiency can lead to growth retardation, hypogonadism, anorexia with altered taste and smell, acrodermatitis, impaired wound healing, and impaired immunity.[71] Vitamin A metabolism can also be altered in zinc deficiency.[72] Alcoholics can have low zinc levels from both reduced oral intake and increased urine loss.[68] Although alcoholics suffer many symptoms seen in zinc deficiency, many factors must be considered. Further studies are needed to define the role of zinc in alcohol-related hypogonadism and other related problems.[57]

Lead. The lead content of some wines is high.[57] In addition, when old radiators or lead pipes are incorporated into homemade stills for the production of "moonshine" whiskey, lead poisoning can result.[68] For descriptive purposes lead poisoning can be divided into an alimentary form, a neuromuscular form, and an encephalopathic form. Anemia, weakness, insomnia, headache, dizziness, and irritability are signs and symptoms common to all these forms. On physical examination, lead sulfide may be deposited near the gingival margin, producing a black-blue "lead line" on the teeth. Stippling of the retina adjacent to the optic disk can also be seen. Basophilic stippling of red cells is regularly seen with anemia associated with lead poisoning. Lead toxicity interferes with the incorporation of iron into hemoglobin. Delta-aminolevulinic acid (ALA), a precursor of hemoglobin, is increased and excreted in the urine in excessive amounts.[68]

Magnesium. Magnesium activates and modifies many enzymes. The phosphatases, which transfer organic phosphate groups involving adenosine 5′-triphosphate (ATP) reactions, are activated by magnesium. ATP is required for carbohydrate, protein, nucleic acid, and fat utilization.[68] Magnesium deficiency enhances thiamin deficiency.[73] Alcoholics may not respond to thiamin replacement until the magnesium is also replaced. In addition, the body's response to parathyroid hormone (PTH) extract is abnormal when magnesium deficiency is present.[74] Chronic alcoholics commonly have magnesium deficiency. This is a result of poor dietary intake, increased urine loss secondary to a direct effect of alcohol, starvation ketosis, and vomiting. Magnesium is often low during the alcohol-withdrawal syndrome and may contribute to the signs and symptoms of this disorder.[75]

Calcium. Calcium metabolism is regulated primarily by vitamin D and PTH[58] (Chapters 25 and 42). Patients with alcoholic liver disease may have low levels of calcium. Alcohol has been found to increase calcium excretion.[76] Patients with cirrhosis have lower calcium absorption, which may be a result of decreased liver hydroxylation of vitamin D.[77] If the alcoholic has fat malabsorption, the calcium in the intestinal lumen may form insoluble soaps with the fats, thus preventing intestinal absorption. If the alcoholic has low magnesium, he may not respond appropriately to PTH.[74]

Phosphorus. Phosphorus is among the most abundant constituents of all tissues.[78] Hypophosphatemia occurs in about 50% of hospitalized alcoholics.[79] Serum phosphorus is depressed acutely by as much as 10 to 15 mg/L after ingestion of carbohydrates. Presumably this is caused by cellular uptake and formation of phosphate esters[78] and may explain why alcoholics reach their lowest phosphorus levels on the second to fourth day of hospitalization.[57] Hypophosphatemia results in decreased levels of 2,3-diphosphoglyceric acid and ATP in erythrocytes. When this occurs, oxygen is more tightly bound to hemoglobin, and less oxygen is delivered to peripheral tissue. Hypophosphatemia may be associated with anorexia, dizziness, bone pain, proximal muscle weakness, and waddling gait. In severe cases there are sharp rises in serum creatine kinase (CK), an indication of possible muscle injury superimposed on myopathy.[78] The low phosphorus levels in alcoholics may be a result of alcohol itself, poor food intake, diarrhea, vomiting, magnesium deficiency, or use of aluminum-containing antacids.

PATHOLOGY: SYSTEMIC DISEASES OF THE ALCOHOLIC

Liver

Ethyl alcohol is a direct systemic toxin that produces injury to all tissues, depending on the dose and duration of exposure. The degree of injury varies among organ systems. The liver is the organ where predominant exposure and metabolism occur and as such develops the highest incidence and severity of injury.

Three types of liver pathology develop, forming a continuum of disease from very mild reversible changes to life-threatening irreversible disease. The interrelationships among these changes are depicted in Fig. 32-1. The mild form is characterized by fatty infiltration (alcoholic fatty liver) and by some minimal degree of inflammation and necrosis.

In more severe cases of fatty liver, fibrosis may be present, especially around central venous channels. Clinically this is manifested by liver enlargement (hepatomegaly), tenderness over the liver, and anorexia. Laboratory changes are minimal with slight elevation of aspartate aminotransferase (AST), bilirubin, and alkaline phosphatase levels. When available, the serum bile acids and dye-clearance tests (indocyanine green and sulfobromophthalein) appear to be the most reliable parameters to detect early disease. Alcoholic fatty liver is usually considered a benign, reversible disease. However, sudden deaths have been observed in a small percent of alcoholics, presumably from fat emboli.

The more severe toxic form of injury, alcoholic hepatitis, is also characterized by fat and hepatomegaly. In addition, inflammation and necrosis are more extensive, and fibrosis is a prominent feature. In the recent large Veterans Administration Cooperative Study[40] of patients with clinical alcoholic hepatitis, the disease had progressed to cirrhosis in 54.7% of 97 patients in whom histological specimens were available. In a high percentage of patients (76.2%) an irregularly outlined homogeneous meshwork of eosin-staining material could be seen in the cytoplasm of the liver cells (Mallory bodies or alcoholic hyalin). Its presence is not diagnostic for alcoholic hepatitis because it has been seen in a variety of other liver conditions.[80-88] However, in an alcoholic with liver disease its presence most likely represents alcoholic hepatitis. It has been suggested that alcoholic hyalin results from degenerating microfilaments, which then act as irritants to stimulate fibrosis with chemotoxic properties to attract leukocytes.[89,90]

The clinical and laboratory manifestations of alcoholic hepatitis are shown in Tables 32-2 and 32-3. None of the symptoms or changes are pathognomonic for this condition.

As the disease progresses, all standard laboratory tests for liver injury become abnormal to varying degrees. Of special note are the serum enzyme changes. Although AST is elevated in over 75% of even the mild cases, the magnitude of the elevation rarely exceeds 10 times the upper limits of normal. The mean value observed was 84 ± 6.0 mU/mL. Similarly ALT was only mildly elevated, usually less than 100 mU/mL and almost invariably less than the AST. ALT values of 200 mU/mL or greater than the AST should indicate a chronic, persistent or chronic, aggressive

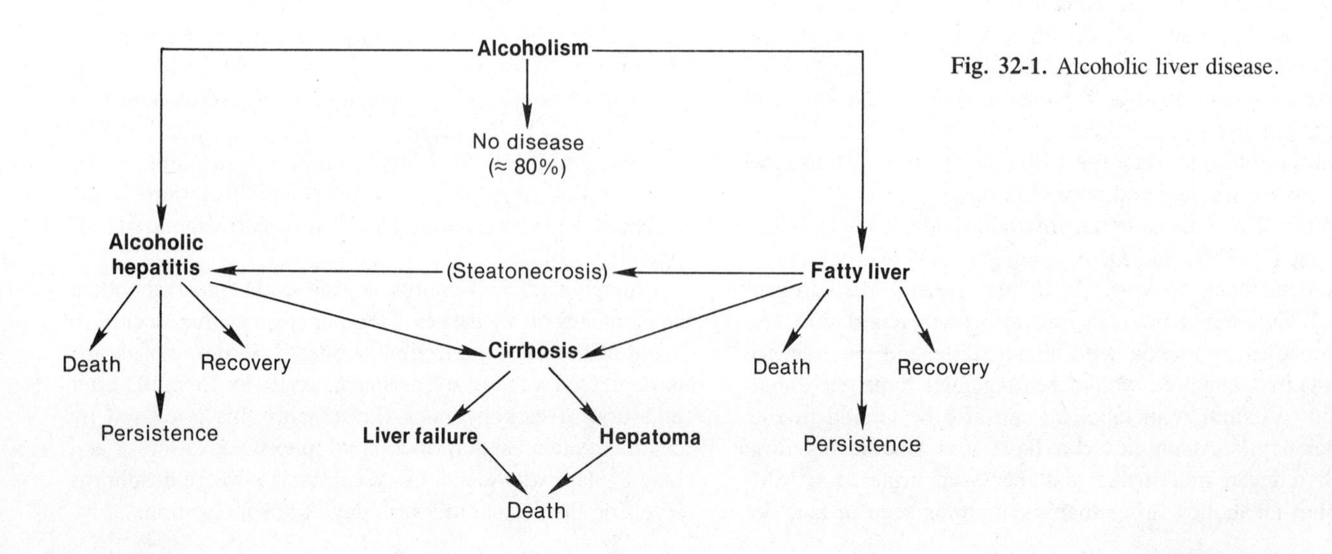

Fig. 32-1. Alcoholic liver disease.

hepatitis rather than uncomplicated alcoholic hepatitis. In these instances only a liver biopsy can differentiate the cause of liver pathosis.

Tests shown to be most reliable in predicting severity and survival are the bilirubin level and prothrombin time. These are not, however, sensitive tests to diagnose minimal disease. In very mild cases they may be normal or near normal. However, as the disease progresses, their usefulness increases. Patients are at high risk to die when the bilirubin level becomes greater than 200 mg/L and the prothrombin time becomes more than 4 seconds prolonged. In this group mortality exceeds 75%.

One application of the bilirubin test and prothrombin time has been for the classification of disease severity as to mild, moderate, or severe disease.[40] Mild disease is considered present when the bilirubin is less than 50 mg/L with a normal or only slightly increased prothrombin time. Moderate disease is present when the bilirubin is equal to or greater than 50 mg/L and the prothrombin time is normal to moderately elevated (less than 4 seconds prolonged). Severe disease is present when the bilirubin exceeds 50 mg/L and the prothrombin time is more than 4 seconds prolonged. As severity increases, the incidence of ascites and encephalopathy also increases, whereas 1-year survival decreases progressively from 91% to 46% (Table 32-2).

The end stage of chronic alcoholic liver disease is the development of cirrhosis with extensive liver fibrosis, nodular regeneration, distortion of the liver architecture, and ultimately all the clinical complications of chronic liver disease, liver failure, and death.

In some instances cirrhosis may be hard to diagnose clinically, especially when active alcoholic hepatitis with acute necrosis and inflammation are also present. Indeed, of 155 patients recently studied with alcoholic hepatitis without cirrhosis, 13% had grade 2 to 4 hepatic encephalopathy, and 26% had moderate to severe ascites.[91] Only the liver biopsy specimen provided an accurate diagnosis of cirrhosis.

Pancreas

The association between alcohol and pancreatitis is at least as strong as the relationship between alcohol and liver disease is. At postmortem examination 18% to 47% of alcoholics dying of causes other than pancreatitis have some histological evidence of pancreatitis.[92] Alcohol is responsible for at least 30% and in some series 60% to 90% of the cases of pancreatitis in the United States.[93] Although there is a wide range of individual variation, the average duration of alcohol consumption before the initial diagnosis of pancreatitis is made in males is 18 years and in females 11 years. This suggests that women have an in-

Table 32-2. Initial clinical features and complications of alcoholic hepatitis*[4]

Feature or complication	Severity of disease		
	Mild	Moderate	Severe
Anorexia†	46.2	63.0	65.7
Weight loss†‡	36.8	27.1	16.2
Fever	18.0	26.2	19.2
Hepatomegaly	85.9	97.1	88.9
Splenomegaly†	24.5	38.6	46.2
Infection	5.2	16.8	8.1
Pancreatitis	13.6	10.3	10.1
Gastrointestinal bleeding	10.4	7.5	14.1
Ascites			
Mild	18.7	19.4	17.2
Moderate	11.0	38.0	45.5
Severe	1.3	20.4	29.3
Combined†	31.0	77.8	92.0
Encephalopathy			
Grade 1	17.3	29.6	25.3
Grade 2	4.5	24.1	41.4
Grades 3 and 4	0.7	3.7	0
Combined†	22.5	57.4	66.7
Probability of surviving 1 year§	0.91 ± 0.027	0.75 ± 0.045	0.46 ± 0.055

*The diagnoses of the various complicating features and conditions were not defined in the protocol but were determined by the clinical judgment of the participating investigator. Because *n* varied among severity groups (mild, 156; moderate, 108; severe, 99), values are expressed as a percentage of total. Data analysis consisted of an overall chi-squared test, which, when significant at the 0.05 level, was followed by a Bartholomew's test for order. The hypothesized order was $p_{mild} < p_{moderate} < p_{severe}$ except for weight loss where the presence of ascites reversed the hypothesized order.

†$p < .005$ (Bartholomew's test).

‡The decreasing incidence of weight loss with increasing severity represented an increasing incidence of ascites.

§$p < .006$. All pairwise comparisons (mild:moderate:severe) determined by normal distribution test statistics.

Table 32-3. Laboratory values in alcoholic hepatitis*[40]

Laboratory tests†	Severity of disease		
	Mild	Moderate	Severe
Hemoglobin (140 − 180 g/L)	126 ± 2 (74% abnormal)	118 ± 2 (85% abnormal)	110 ± 3 (93% abnormal)
Hematocrit (47% ± 5%)	37.8 ± 0.6 (76% abnormal)	35.7 ± 0.7 (83% abnormal)	33.1 ± 0.9 (95% abnormal)
Mean corpuscular volume (MCV) (80-94 mm^3)	100.2 ± 0.9 (73% increased) (2% decreased)	102.2 ± 1.9 (83% increased) (8% decreased)	104.7 ± 1.4 (90% increased) (0% decreased)
White blood count (WBC) (5.0-10.0 × 1000/mm^3)	8 ± 0.3 (13% increased) (11% decreased)	10.7 ± 0.7 (47% increased) (9% decreased)	12.4 ± 1 (54% increased) (8% decreased)
Aspartate aminotransferase (AST) (10-40 mU/mL)	84 ± 6 (79% abnormal)	124 ± 10 (98% abnormal)	99 ± 9 (91% abnormal)
Alanine aminotransferase (ALT) (10-30 mU/mL)	56 ± 6 (62% abnormal)	56 ± 5 (73% abnormal)	57 ± 15 (56% abnormal)
Total bilirubin (1-10 mg/L)	16 ± 1 (53% abnormal)	135 ± 12 (100% abnormal)	187 ± 24 (100% abnormal)
Alkaline phosphatase (40-120 mU/mL)	165.8 ± 8.5 (67% abnormal)	276.3 ± 14.5 (100% abnormal)	224.9 ± 14.2 (88% abnormal)
Prothrombin time (seconds above control)	0.9 ± 0.1 (65% abnormal)	2.4 ± 0.2 (90% abnormal)	6.4 ± 0.4 (100% abnormal)
Blood urea nitrogen (BUN) (40-200 mg/L)	100 ± 10 (3% increased) (3% decreased)	140 ± 10 (17% increased) (0% decreased)	190 ± 50 (36% increased) (0% decreased)
Creatinine (6-17 mg/L)	10 (2% abnormal)	13 ± 1 (17% abnormal)	23 ± 4 (29% abnormal)
Albumin (35-50 g/L)	37 ± 1 (36% abnormal)	27 ± 1 (90% abnormal)	24 ± 1 (96% abnormal)
Cholylglycine (0-600 μg/L)	4590 ± 570 (85% abnormal)	1512 ± 1490 (100% abnormal)	9030 ± 930 (100% abnormal)
Immunoglobulins			
IgG (6.39-13.49 mg/mL)	15.50 ± 0.97 (49% abnormal)	17.04 ± 0.92 (72% abnormal)	21.40 ± 2.08 (83% abnormal)
IgA (0.7-3.12 mg/mL)	5.19 ± 0.34 (84% abnormal)	8.03 ± 0.53 (98% abnormal)	8.77 ± 1.07 (93% abnormal)
IgM (0.56-3.52 mg/mL)	2.32 ± 0.19 (17% abnormal)	25.8 ± 0.18 (26% abnormal)	3.35 ± 0.36 (43% abnormal)

*Values are expressed as the mean ± standard error of mean, and the incidence of abnormal occurrence is expressed as percentage of total; n = 89 with mild disease, 58 with moderate, 37 with severe.
†Numbers in parentheses represent normal range for each test.

creased susceptibility. With each approximately 50 g increase in daily alcohol consumption, the risk of developing pancreatitis doubles.[92,94]

Alcohol stimulates the secretion of at least some acid from the stomach. When acid comes into contact with duodenal mucosa, it stimulates the flow of pancreatic juice.[95] The pancreatic juice induced by alcohol early in the disease process has a high concentration of protein, which precipitates and forms protein plugs. These plugs form within the pancreatic ductules and subsequently obstruct the ductules. This obstruction leads to ductular dilatation, ductular proliferation, dilatation of acinar tissue, and ductal sclerosis. In time the protein plugs may calcify, giving the x-ray picture of chronic calcific pancreatitis. As scarring of the pancreas continues, the protein or enzyme concentration in the pancreatic juice decreases until not enough enzymes are present for food digestion, and malabsorption occurs.[95-98] For reasons that are not clear, acute painful attacks of pancreatitis characterized by elevations in amylase and lipase may be superimposed on this chronic process.

Because a long interval of steady alcohol consumption is required before the first clinical attack of pancreatitis, most patients develop the first attack of pancreatitis between 30 and 40 years of age.[95] Once pancreatitis has been established, a fairly high percentage of patients who continue to drink have chronic pain.[99] After 80% to 90% of the pancreas has been destroyed, pancreatic insufficiency develops, requiring oral enzyme replacement.[100] In addition to the enzyme-secreting cells being destroyed, insulin- and glucagon-producing cells may also be destroyed. When this occurs, the alcoholic develops diabetes mellitus and the need for insulin. Because the glucagon level is also decreased and is too low to adequately stimulate glucose synthesis, these patients are brittle diabetics and are prone to hypoglycemic attacks.[95] This makes treatment of these patients difficult, especially if drinking continues.[101]

Other less-common systemic involvement

Heart. Among alcoholics the incidence of clinically significant heart disease appears to be much less than that of liver or pancreatic disease. It is estimated that 20% to 30% of alcoholics develop liver disease. In two large series of alcoholics totaling 278 cases at autopsy, 11% to 15% had alcoholic cardiomyopathy.[102,103] Because these estimates pertain only to autopsied cases, the overall incidence among alcoholics is much less. Hartel et al.[104] studied 100 severely alcoholic males and failed to detect a single case of cardiomyopathy. The true incidence is unknown. When it occurs, ethanol produces an acute depressant effect on myocardial function[105,106] with a progressive cardiac dilatation and a low output type of cardiac failure. At autopsy the usual findings consist grossly of moderate hypertrophy of the left ventricle and dilatation of the right ventricle and both atria. Microscopically one finds scattered areas of muscle degeneration, patchy fibrosis, intracellular edema, patchy areas of lymphocytic infiltration, and large numbers of lipid droplets within the cytoplasm of the muscle cells. When death occurs, it is most often a result of gross cardiomegaly and chronic intractable cardiac failure complicated by embolic phenomena.

Central and peripheral nervous system. The most common manifestations of altered brain function in the alcholic are the symptoms of inebriation, followed in frequency by the withdrawal syndromes of tremor, confusion, hallucinations, "delirium tremens," and convulsions ("rum fits"). These are usually reversible functional changes. Much less common but frequently irreversible are the metabolic disorders associated with protracted steady drinking and in most instances nutritional deficiencies. These include Wernicke-Korsakoff syndrome, peripheral neuropathy, amblyopia (loss of vision), cerebellar degeneration, cerebral degeneration, central pontine myelinolysis, and progressive dementia with demyelination of the corpus callosum (Marchiafava-Bignami disease). Both clinical and animal studies on mnemonic brain function indicate an ethanol-induced impairment.[107,108] It is beyond the scope of this chapter to detail each of these syndromes. Refer to a neurology text. Because specific animal models are absent for most of these conditions, mechanisms for their development are poorly understood.

Cancer. Primary liver cancer, a rare tumor, is seen in 5% to 10% of alcoholics with alcoholic cirrhosis.[109] This is believed to be secondary to the cirrhotic process. In addition to the hepatomas associated with alcoholic cirrhosis, prospective and retrospective epidemiological studies indicate that chronic alcohol consumption by itself is a cancer hazard.[110] Numerous clinical studies of alcoholics have arrived at this conclusion. Heavy drinkers have an increased incidence of cancer of the mouth, pharynx, larynx, and esophagus[110-112] with a possible relationship to cancer of the pancreas,[113] the cardia of the stomach,[114] and the colon.[115] Such epidemiological evidence is most abundant in the case of cancers of the mouth and pharynx so that a heavy drinker and heavy smoker has a 15 times greater risk of developing oral malignancy than a nondrinker and nonsmoker. It appears that the risk from ethanol is both time and dose related,[115-117] which has led some to conclude that ethanol acts as either cocarcinogen or promoter.[118]

Endocrine effects. The endocrine effects of alcohol are multiple and vary depending on whether the alcohol intake is acute or chronic.[119] Moderate alcohol consumption has little effect on cortisol from the adrenal gland.[120] Alcohol blood levels in excess of 1 g/L, however, may produce an elevated plasma cortisol level. Occasionally chronic alcoholics will present with clinical features of Cushing's syndrome and high cortisol levels.[121] Abstention causes the cortisol level to return to normal in 2 to 3 weeks.[119]

Some alcoholics have evidence of pituitary insufficiency. Low blood glucose normally produces a rise in cortisol. This response is decreased or absent in 25% of chronic alcoholics because the pituitary gland does not secrete enough adrenocorticotropic hormone to stimulate the adrenal gland.[119] Low levels of growth hormone and prolactin after low blood glucose stimulation are further evidence for pituitary insufficiency.[122] The low testosterone levels seen in alcoholics are now related to pituitary insufficiency.[119]

Hematopoietic effects. The effects of alcohol on the blood and bone marrow are the result of both a direct toxic effect and associated nutritional deficiencies[119] (see the discussion of vitamins and minerals). Not only do alcoholics have low folate levels,[48] but also alcohol directly inhibits folate metabolism.[64] Vitamin B_{12} absorption is inhibited by alcohol.[57,65] Each of these factors alone or in combination can lead to abnormally large red blood cells and megaloblastic anemia.[58,62]

Iron metabolism is affected by alcohol in a number of ways. Alcohol can produce chronic blood loss (ulcers, gastritis, variceal bleeding) leading to iron-deficiency anemia.[58] An increase in serum and tissue iron can also be seen with excess alcohol consumption and causes abnormal iron utilization with the bone marrow (ring sideroblasts).[123] The latter effect may be caused by alcohol blocking the conversion of pyridoxine (vitamin B_6) to pyridoxal phosphate because the sideroblastic changes can be reversed by intravenous administration of pyridoxal phosphate.[124] Finally, alcohol excess can cause acute hemolytic anemia.[123]

Inhibition of the bone marrow by alcohol can cause a low number of circulating white blood cells[125] and platelets.[126] Alcohol also can impair the function of white blood cells,[127] and alcoholic cirrhosis impairs platelet function,[128] which may lead to increased susceptibility to infection and bleeding. Vitamin K deficiency, seen in alcoholics, causes a prolonged prothrombin time because of a lack of vitamin K–dependent clotting factors normally produced in the liver. When liver disease is severe, inadequate clotting factors are produced even when vitamin K is present.[123]

Immune system. Ethanol is known to strongly affect the immune system. Increased circulating B-cell antibodies have been identified in most patients with alcoholic liver disease[40] (see discussion on p. 605). Specific antibodies against alcoholic hyalin have been identified in patients with alcoholic hepatitis,[129] which may explain the frequent elevation in IgA.[130] This explanation is still controversial. The primary changes have been observed in the cell-mediated immune system. Total numbers of T cells are decreased[131]; their ability to synthesize DNA is impaired (lymphocyte transformation)[132]; their response to antigens and mitogens is suppressed[132]; and their cytotoxic properties are increased.[133] These are manifested clinically by a low lymphocyte count, anergy to skin tests,[4] altered response to vaccinations,[40,134] and increased susceptibility to infections.[135-138]

In addition to their role in defense against infections, these immunological alterations may be in part responsible for some of the liver cell injury and progression of the disease seen in patients with alcoholic hepatitis.[40,133]

CHANGE OF ANALYTE IN DISEASE (Table 32-4)

Enzymes

Increases in serum enzyme levels in liver disease may result from leakage of enzymes into the serum because of damaged cell membranes (AST, ALT, and LD), or they may result from increased enzyme production (alkaline phosphatase and gamma-glutamyl transpeptidase).

AST-ALT. Serum aspartate aminotransferase (AST) and serum alanine aminotransferase (ALT) are two serum enzymes that are found in the liver and readily leak from the cells during necrosis and cell injury. In the case of viral injury to the liver (viral hepatitis) the serum activity rises to high values, which may exceed several thousand units. The ALT level is frequently higher than the AST level.[60] Although alcoholic liver injury is characterized by liver cell necrosis and inflammation, the increase in these enzymes is minimal to moderate, with the rise in AST almost always exceeding that observed in ALT. The abnormalities in AST occur early and are one of the most frequently occurring, but the magnitude of the change may

Table 32-4. Prevalence of laboratory changes in alcoholic hepatitis[56]

Laboratory test	Direction of analyte change	Incidence of abnormal results, % (ULN × $\bar{X}$ of group)*		
		Group I	Group II	Group III
Hemoglobin	Decrease	74 (0.70)	85 (0.66)	93 (0.60)
Hematocrit	Decrease	76 (0.80)	83 (0.76)	95 (0.70)
MCV	Increase	73 (1.1)	83 (1.1)	90 (1.1)
	Decrease	2	8	0
WBC	Increase	13	7 (1.1)	54 (1.24)
	Decrease	11 (0.8)	9	8
AST (SGOT)	Increase	79 (2.1)	98 (3.1)	91 (2.5)
ALT (SGPT)	Increase	62 (1.9)	73 (1.9)	56 (1.9)
Total bilirubin	Increase	53 (1.6)	100 (13.5)	100 (18.7)
Alkaline phosphatase	Increase	67 (1.4)	100 (2.3)	88 (1.9)
Prothrombin time	Increase	65	90	100
BUN	Increase	3	17 (0.7)	36 (1.4)
	Decrease	3 (0.5)	0	0
Creatinine	Increase	2 (0.60)	17 (0.76)	29 (1.35)
	Decrease			
Albumin	Decrease	36 (0.74)	90 (0.54)	96 (0.48)
Cholylglycine	Increase	85 (8.2)	100 (25.2)	100 (15.1)
Immunoglobulins				
IgG	Increase	49 (1.1)	72 (1.3)	83 (1.6)
IgA	Increase	84 (1.7)	98 (2.6)	93 (2.8)
IgM	Decrease	17 (0.66)	26 (0.73)	43 (0.95)

*Data are expressed as the change in analyte, and the incidence of abnormal occurrence is expressed as the percentage of total and the magnitude of the change (times upper limit of normal [ULN] for mean); $n = 89$ in group I, 58 in group II, 37 in group III. See Table 32-3 for a description of three groups.

not parallel the clinical severity of the liver injury.[40] Explanations for this minimal response to injury are inadequate. Because pyridoxal phosphate is required for transamination reactions, deficiency in pyridoxine may in some alcoholics contribute to the diminished response.[61]

Alkaline phosphatase. In the presence of liver injury, especially that involving the bile ducts and excretory function, the concentrations of this enzyme increase both in the liver and serum.[139] In alcoholic liver disease the increase in activity tends to parallel the changes seen in bilirubin.

5′-Nucleotidase. 5′-Nucleotidase is not as sensitive to liver injury as alkaline phosphatase. Its main value is with alkaline phosphatase to determine whether a rise in the latter enzyme is the result of hepatobiliary disease or bone disease.

Gamma-glutamyl transferase (GGT). Increases in serum levels of gamma-glutamyl transferase have been reported in alcoholics with minimal or no liver disease. This has been attributed to microsomal enzyme induction rather than liver injury[140] (see discussion of diagnosis of alcoholism, p. 595).

Bilirubin

Although alcohol does not directly alter bilirubin metabolism,[141] the liver injury produced by alcohol does. Jaundice, the clinical sign of elevated bilirubin, is seen in about 60% of patients with alcoholic hepatitis.[40] Bilirubin elevation, along with a prolonged prothrombin time unresponsive to vitamin K, depressed serum albumin,[142] and hepatic encephalopathy[143] correlate well with the severity of alcoholic hepatitis. In some cases of alcoholic hepatitis the bilirubin will rise along with the alkaline phosphatase level so that common bile duct obstruction is suspected.[44] Because the patient with alcoholic hepatitis does not tolerate surgery well, a diagnostic procedure such as endoscopic or percutaneous cholangiography should be done to confirm the diagnosis of extrahepatic obstruction before surgery.

Proteins

Albumin. Many tests used to evaluate disease in the liver are based on alterations in biochemical functions of the liver. Because serum albumin is synthesized and secreted by the liver, its serum concentration has been used as a test of liver function. Serum levels of albumin are depressed in all forms of liver disease, including those associated with alcoholism. Depending on the laboratory, the normal albumin range is from 35 to 50 g/L. It is not uncommon to find values below 20 g/L in severe disease.[40] Of 111 patients with severe alcoholic hepatitis in the Veterans Administration Cooperative Study[40] 18% had values in this range, whereas only 1% had values within the normal range. In patients with alcoholic liver disease the decline in serum albumin may not be related only to decreased synthesis secondary to liver disease. Many of these patients have moderate to severe malnutrition; thus the decreased albumin level may reflect decreased visceral proteins resulting from poor dietary protein intake (kwashiorkor). Furthermore, many of these patients have ascites with expanded extravascular pools. The low serum albumin in these persons may represent a shift of albumin from the intravascular to the expanded extravascular space.

Globulins. Changes in serum globulins are not truly a test of liver function because the gamma globulins that change during liver disease are produced by lymphocytes and plasma cells in response to antigenic stimulation. This is more a test of immune changes in liver disease. Serum globulins usually rise as the albumin falls. The increase in many instances may be disproportionately high. Of the three types of gamma globulins (IgG, IgA, and IgM), IgA increases most readily in alcoholic liver disease. It is elevated in 90% of cases, with a mean increase of 118%.[40] IgM increases the least and is elevated in 25% of cases. IgG is intermediate at 64%, with a mean increase of 25%. The mechanism for the increased globulins is not well understood.

The IgG increase seems to result from failure of the damaged liver to remove intestinal bacterial antigens, particularly *Escherichia coli*.[145,146] Hence, increases tend to correlate with intrahepatic portosystemic shunting.[146].

The IgA elevation is more difficult to explain. The changes show no correlation with intrahepatic shunting and are independent of IgG. Some have suggested that they represent antibodies to alcoholic hyalin.[130]

Lipids

Triglycerides and cholesterol. Triglyceride and cholesterol changes in serum are frequently observed and are discussed on p. 596. The application of these changes is not commonly used in the diagnosis of alcoholic liver disease.

Bile acids. Bile acids represent the end products of cholesterol degradation. Because the liver is responsible for almost all phases of their metabolism, it is not surprising that they are very sensitive indicators of even minimal liver injury. In alcoholic liver disease, serum bile acids, especially cholic acid conjugates, have been correlated with biopsy findings.[147] Fatty liver alone was not associated with elevated levels, whereas alcoholic hepatitis and cirrhosis were accompanied by significant increases in serum levels. In the Veterans Administration Cooperative Study on Alcoholic Hepatitis, cholylglycine was abnormal with greater frequency than any other parameters in liver injury.[40] These observations indicate that cholylglycine can be most useful in detecting the presence of early injury. Differences in bile acid have also been reported in alcoholic liver disease with a reversal of the cholic acid to chenodeoxycholic acid ratio and a decrease in the levels of secondary bile acids.[147]

Although results of most laboratory tests used in the diagnosis of liver injury are abnormal to varying degrees in alcoholic liver disease, none of these tests is pathognomonic. In many instances only a liver biopsy can establish the diagnosis with certainty.

Carbohydrate

Hyperglycemia. Only rarely has alcohol been responsible for producing gross glucose intolerance or frank diabetes.[101] Hepatocellular damage, regardless of its cause, is an important cause of glucose intolerance and may play a major role in the hyperglycemia found in some alcoholics.[148] Alcohol-induced elevation in cortisol, especially when sufficient to produce pseudo-Cushing's syndrome, is another contributing factor in alcoholic hyperglycemia.[149]

Hypoglycemia. Although hypoglycemia may occur in severe liver disease with fulminant hepatic failure, it may also occur in association with alcohol use and a relatively normal liver.[119] Hypoglycemia may be the cause of sudden death among alcoholics, with an 11% mortality in adults and a 25% mortality in children.[150] This form of hypoglycemia (alcohol-induced fasting hypoglycemia) occurs in chronically malnourished alcoholics or when moderate-to-large amounts of alcohol are consumed after a 6- to 36-hour fast.[101] The patients are stuporous or comatose, smell of alcohol, and are often hypothermic. The diagnosis is made on the clinical findings of hypoglycemia and elevated blood alcohol. Lactic acidosis is not uncommon.[119] Plasma insulin levels are low and glucagon levels high.[151,152] Although growth hormone and cortisol levels are elevated, the elevation is still less than expected for the severity of the hypoglycemia.[151,152] The mechanism for the hypoglycemia is probably multifactorial, with alcohol inhibition of gluconeogenesis playing a major factor.[153] Other contributing factors include a relatively mild adrenocortical insufficiency and a defect in growth hormone secretion.[119]

Another form of hyoglycemia occurs when alcohol is consumed with carbohydrates (alcohol-induced reactive hypoglycemia).[101] Alcohol has the capacity to potentiate the insulin-stimulating properties of glucose.[154] Therefore alcohol consumed at a meal may potentiate insulin secretion and lead to nocturnal hypoglycemia. In addition, large amounts of sweetened coffee used to "sober up" may be ill advised because a rapid and severe reactive hypoglycemia could contribute to an accident during the drive home.

Alcohol can potentiate the action of hypoglycemic medications, especially insulin (alcohol potentiation of drug-induced hypoglycemia).[101] Although alcohol inhibits the liver's ability to mobilize glucose from glycogen (gluconeogenesis), the exact mechanism for hypoglycemia associated with hypoglycemic drugs is unknown. There is also slender evidence that reactive hypoglycemia may occur more frequently in chronic alcoholics than in controls ("essential" reactive hypoglycemia in alcoholics).[101]

Alcoholic ketoacidosis. Ketoacidosis is an uncommon condition occurring in nondiabetic alcoholics who usually have abnormal blood chemistry values.[119] Patients often present with abdominal pain, vomiting, and a history of no recent food intake. The patients are acidotic but conscious with high levels of serum ketones, salt and water depletion, and normal glucose levels.[155] The mechanism for alcoholic ketosis is uncertain.[119] Serum insulin concentrations are low with high cortisol levels and mild elevations of growth hormone. Cortisol may be ketogenic when insulin levels are low, and alcohol may be directly converted to ketone bodies through acetate production.[119]

TESTS OF LIVER FUNCTION

Bromsulphalein (BSP) dye excretion. BSP (sulfobromophthalein) dye excretion is useful for detecting early or minor changes in liver function. BSP is handled by the liver cell much as bilirubin. The dye is taken up by the liver cell, conjugated with glutathione (bilirubin is conjugated with glucuronide), and excreted in the bile.[156] The excretion test involves the infusion of a known amount of dye into an arm vein with venous blood samples collected at 30- and 45-minute intervals. Normally up to only 10% of the dye is retained at 30 minutes, and up to 3% is retained at 45 minutes. Because of cost, occasional side effects, and inconvenience, the test is rarely performed today.

Indocyanine green (ICG) dye excretion. Compared with BSP, ICG is free of side effects; as with BSP, it is rapidly excreted by the liver. Unlike BSP and bilirubin, ICG is not conjugated by the liver before excretion into the bile.[156] This dye has the advantage of being monitored accurately and continuously with a dichromatic ear densitometer, thus one can eliminate venous blood sampling.[157] Disappearance of the dye is believed to be related to hepatic blood flow. Although the test is not used routinely, it is probably the current dye excretion test of choice.[156]

Aminopyrine breath test. This test is based on the fact that ^{14}C-labeled aminopyrine given orally is demethylated by the microsomal enzyme system of the liver to produce $^{14}CO_2$, which is eliminated in the breath. The aminopyrine breath test is more sensitive than serum albumin, bilirubin, and prothrombin time for detection of liver disease. However, it is no more sensitive than AST, 2-hour postprandial bile acids, ICG excretion, or galactose elimination. The main value of the test appears to be in assessment of the status of the microsomal enzyme system rather than in detection of hepatocelluar injury.[156]

REFERENCES

1. U.S. Bureau of the Census: Statistical abstract of the United States, 1975, Washington, D.C., 1975, U.S. Government Printing Office.
2. Criteria Committee, National Council on Alcoholism: Criteria for the diagnosis of alcoholism, Ann. Intern. Med. **77**:249-258, 1972.
3. Bailey, R.J., Krasner, N., Feddleston, A.L.W., et al.: Histocompatibility antigens, autoantibodies, and immunoglobulins in alcoholic liver disease, Br. Med. J. **2**:727-729, 1976.
4. Mendenhall, C.L., Weesner, R.E., Crolic, K., et al.: A clinical study of nutritional alterations associated with alcoholic hepatitis, Am. J. Med. (In press.)
5. Pettigrew, N.M., Goudie, R.B., Russell, R.I., et al.: Evidence for a role of hepatitis B virus in chronic alcoholic liver disease, Lancet **2**:724-725, 1972.
6. Rydberg, U., and Skerfuing, S.: Toxicity of ethanol: a tentative risk evaluation. In Gross, E.M., editor: Alcohol intoxication and withdrawal, New York, 1977, Plenum Publishing Corp., pp. 403-418.
7. Kristensson, H., Trell, E., Eriksson, S., et al.: Serum-γ-glutamyltransferase in alcoholism, Lancet **1**:609, 1977.
8. Whitehead, T.P., Clarke, C.A., and Whitfield, A.G.W.: Biochemical and haematological markers of alcoholic intake, Lancet **1**:978-981, 1978.
9. Lieber, C.S.: Pathogenesis and early diagnosis of alcoholic liver injury, N. Engl. J. Med. **298**:888-893, 1978.
10. Dragosics, B., Ferenci, P., Pesendorfer, F., et al.: Gamma-glutamyltanspeptidase (GGTP): its relationship to other enzymes for diagnosis of liver disease. In Popper, H., and Schaffner, F., editors: Progress in liver disease, New York, 1976, Grune & Stratton, Inc., pp. 436-449.
11. Iwata, K.: A review of the literature on drunken symptoms due to yeasts in the gastrointestinal tract. In Iwata, K., editor: Yeasts and yeast-like micro-organisms in medical science, Tokyo, 1972, University of Tokyo Press, pp. 260-268.
12. White, R., Lindsey, D., and Ash, R.: Ethanol production from glucose by *Torulopsis glabrata* occurring naturally in the stomachs of newborn animals, J. Appl. Bacteriol. **35**:631-646, 1972.
13. Mendenhall, C.L., and Weesner, R.E.: Alcohols and glycols. In Hanenson, I.B., editor: Quick reference to clinical toxicology, Philadelphia, 1980, J.B. Lippincott Co., pp. 148-153.
14. Lieber, C.S., and DeCarli, L.M.: Effect of drug administration on the activity of the hepatic microsomal ethanol oxidizing system, Life Sci. **9**:267-276, 1970.
15. Iseri, O.A., Lieber, C.S., and Gottlieb, L.S.: The ultrastructure of fatty liver induced by prolonged ethanol ingestion, Am. J. Pathol. **48**:535-555, 1966.
16. Lieber, C.S., and DeCarli, L.M.: The role of the hepatic microsomal ethanol oxidizing system (MEOS) for ethanol metabolism in vivo, J. Pharmacol. Exp. Ther. **181**:279-287, 1972.
17. Freytmans, E., and Leighton, F.: Effects of pyrazole and 3-amino-1,2,4-triazole on methanol and ethanol metabolism by the rat, Biochem. Pharmacol. **22**:349-360, 1973.
18. Barlett, G.R.: Does catalase participate in the physiological oxidation of alcohols? Q. J. Stud. Alcohol **13**:583-589, 1952.
19. Mathinos, P.R.: Determination of the proximal teratogen of the fetal alcohol syndrome in CD/Mice, doctoral dissertation, Cincinnati, 1982, University of Cincinnati.
20. Mendenhall, C.L., and Kromme, C.: Acetaldehyde as a stimulus to liver collagen synthesis in the alcoholic, paper presented at the thirty-second annual meeting of the American Association for the Study of Liver Diseases, Chicago, Nov. 1981, **1**:531, 1981 (abstract).
21. Lieber, C.S., Jones, D.P., Losowsky, M.S., et al.: Interrelation of uric acid and ethanol metabolism in man, J. Clin. Invest. **41**:1863-1870, 1962.
22. Olin, J.S., Devenyi, P., and Weldon, K.L.: Uric acid in alcoholics, Q. J. Stud. Alcohol **34**:1202-1207, 1973.
23. Newcombe, D.S.: Ethanol metabolism and uric acid, Metabolism **21**:1193-1203, 1972.
24. Mendenhall, C.L.: The origin of hepatic triglyceride fatty acids: a quantitative estimation of the relative contributions of linoleic acid by diet and adipose tissue in controls and ethanol fed rats, J. Lipid, Res. **13**:177-183, 1972.
25. Mendenhall, C.L., Greenberger, P.A., Greenberger, J.C., et al.: Dietary lipid assimilation after acute ethanol ingestion in the rat, Am. J. Physiol. **22**:377-382, 1974.
26. Baraona, E., and Lieber, C.S.: Intestinal lymph formation and fat absorption: stimulation by acute ethanol administration and inhibition by chronic ethanol feeding, Gastroenterology **68**:495-502, 1975.
27. Nikkila, E.A., and Ojala, K.: Role of hepatic L-α-glycerophosphate and triglyceride synthesis in production of fatty liver by ethanol, Proc. Soc. Exp. Biol. Med. **113**:814-817, 1963.
28. Mendenhall, C.L., Muto, U., and Shinohara, N.: Changes in enzyme activities contributing to the production of ethanolic fatty liver, paper presented at American Society for the Study of Liver Disease; Gastroenterology **58**:307, 1970 (abstract).
29. Lieber, C.S., and Schmid, R.: The effect of ethanol on fatty acid metabolism: stimulation of hepatic fatty acid synthesis in vivo, J. Clin. Invest. **40**:394-399, 1960.
30. Lefevre, A.F., DeCarli, L.M., and Lieber, C.S.: Effect of ethanol on cholesterol and bile acid metabolism, J. Lipid Res. **13**:48-55, 1972.
31. Blomstrand, R., Kager, L., and Lantto, O.: Studies on the ethanol induced decrease of fatty acid oxidation in rat and human liver slices, Life Sci. **13**:1131-1141, 1973.
32. Baraona, E., and Lieber C.S.: Effects of chronic ethanol feeding on serum lipoprotein metabolism in the rat, J. Clin. Invest. **49**:769-778, 1970.
33. Losowsky, M.S., Jones, D.P., Davidson, C.S., et al.: Studies of alcoholic hyperlipemia and its mechanism, Am. J. Med. **35**:794-803, 1963.
34. Israel, Y., Kalant, H., and Laufer, I.: Effects of ethanol on Na, K, Mg-stimulated microsomal ATPase activity, Biochem. Pharmacol. **14**:1803-1814, 1965.
35. Israel, Y., Salazar, I., and Rosenman, E.: Inhibitory effects of alcohol on intestinal amino acid transport in vivo and in vitro, J. Nutr. **96**:499-504, 1968.
36. Sorrell, M.F., and Tuma, D.J.: Selective impairment of glycoprotein metabolism by ethanol and acetaldehyde in rat liver slices, Gastroenterology **75**:200-205, 1978.
37. Weesner, R.E., Mendenhall, C.L., and Morgan, D.D.: Serum α-fetoprotein: changes associated with acute and chronic ethanol ingestion in the resting and regenerating rat liver, J. Lab. Clin. Med. **95**:725-736, 1980.
38. Morland, J., Rothschild, M.A., Oratz, M., et al.: Protein secretion in suspensions of isolated rat hepatocytes: no influence of acute ethanol administration, Gastroenterology **80**:159-165, 1981.
39. Baraona, E., Leo, M., Borowsky, S.A., et al.: Alcoholic hepatomegaly: accumulation of protein in the liver, Science **190**:794-795, 1975.
40. Mendenhall, C.L., and the Cooperative Study Group on Alcoholic Hepatitis: Pathogenesis, diagnosis, and treatment of alcoholic hepatitis, Clin. Gastroenterol. **10**:417-441, 1981.
41. Chang, T., Lewis, J., and Glazko, A.J.: Effect of ethanol and other alcohols on the transport of amino acids and glucose by everted sacs of rat small intestine, Biochim. Biophys. Acta **135**:1000-1007, 1967.
42. Isselbacher, K.J., and Krane, S.M.: Studies on the mechanism of the inhibition of galactose oxidation by ethanol, J. Biol. Chem. **236**:2394-2398, 1961.
43. Friden, C.: Glutamic dehydrogenase. I. The effect of coenzyme on the sedimentation velocity and kinetic behavior, J. Biol. Chem. **234**:809, 1959.
44. Pirola, R.C., and Lieber, C.S.: The energy cost of the metabolism in drugs, including ethanol, Pharmacology **7**:185-196, 1972.
45. Blackburn, G.L., Bristrian, B.R., Maini, B.S., et al.: Nutritional and metabolic assessment of the hospitalized patient, J. Parent. Ent. Nutr. **1**:11-22, 1977.
46. Patek, A.J., Jr., and Post, J.: Treatment of cirrhosis of the liver by a nutritious diet and supplements rich in vitamin B complex, J. Clin. Invest. **20**:481-505, 1941.
47. Nosrallah, S.M., and Galambos, J.T.: Amino acid therapy of alcoholic hepatitis, Lancet **2**:1276-1277, 1980.

48. Leevy, C.M., Thompson, A., and Baker, H.: Vitamins and liver injury, Am. J. Clin. Nutr. **23:**493-498, 1970.
49. Cherrick, G.R., Baker, H., Frank, O., et al.: Observations on hepatic avidity for folate in Laennec's cirrhosis, J. Lab. Clin. Med. **66:**446-451, 1965.
50. Fennelly, J., Frank, O., Baker, H., et al.: Red blood cell transketolase activity in malnourished alcoholics with cirrhosis, Am. J. Clin. Nutr. **20:**946-949, 1967.
51. Frank, O., Baker, H., and Leevy, C.M.: Vitamin binding capacity of experimentally injured liver, Nature **203:**302-303, 1964.
52. Leevy, C.M.: In vitro studies of hepatic DNA synthesis in percutaneous liver biopsy specimens from man, J. Lab. Clin. Med. **61:**761-779, 1963.
53. Leevy, C.M., ten Hove, W., Frank, O., et al.: Folic acid deficiencies and hepatic DNA synthesis, Proc. Soc. Exp. Biol. Med. **117:**746-748, 1964.
54. Hoyumpa, A.M., Jr., Breen, K.I., Schenker, S., et al.: Thiamine transport across the rat intestine. II. Effect of ethanol, J. Lab. Clin. Med. **86:**803-816, 1975.
55. Halsted, C.H., Robles, E.A., and Mezey, E.: Decreased jejunal uptake of labeled folic acid (^{3}H-PGH) in alcoholic patients: roles of alcohol and nutrition, N. Engl. J. Med. **285:**701-706, 1971.
56. Leevy, C.M., Cardi, L., Frank, O., et al.: Incidence and significance of hypovitaminemia in a randomly selected municipal hospital population, Am. J. Clin. Nutr. **17:**259-271, 1965.
57. Thomson, A.D., and Majumdor, S.K.: The influence of ethanol on intestinal absorption and utilization of nutrients, Clin Gastroenterol. **10:**263-293, 1981.
58. McIntyre, N., and Morgan, M.Y.: Nutritional aspects of liver disease. In Wright, R., Alberti, K.G.M.M., Karran, S., et al., editors: Liver and biliary disease, W.B. Saunders Co., Philadelphia, 1979, pp. 108-133.
59. Blass, J.P., and Gibson, G.E.: Abnormality of a thiamine-requiring enzyme in patients with Wernicke-Korsakoff syndrome, N. Engl. J. Med. **297:**1367-1370, 1977.
60. Cohen, J.A., and Kaplan, M.M.: The SGOT/SGPT ratio—an indicator of alcoholic liver disease, Dig. Dis. Sci. **24:**835-838, 1979.
61. Ludwig, S., and Kaplowitz, N.: Effect of pyridoxine deficiency (B_6-D) on serum and liver transaminases (T) in experimental liver injury, Gastroenterology **76:**1290, 1979 (abstract).
62. Herbert, V., Zalusky, R., and Davidson, C.S.: Correlation of folate dificiency with alcoholism and associated macrocytosis, anemia, and liver disease, Ann. Intern. Med. **58:**977-988, 1963.
63. Baker, H., Frank, O., Zetterman, R., et al.: Inability of chronic alcoholics with liver disease to use food as a source of folates, thiamin and vitamin B_6, Am. J. Clin. Nutr. **28:**1377-1380, 1975.
64. Sullivan, L.W., and Herbert, V.: Suppression of hematopoiesis by ethanol, J. Clin. Invest. **43:**2048-2062, 1964.
65. Lindenbaum, F., Saha, J.R., Shea, N., et al.: Mechanism of alcohol-induced malabsorption of vitamin B^{12}, Gastroenterology **64:**762, 1973 (abstract).
66. Beattie, A.D., and Sherlock, S.: Ascorbic acid deficiency in liver disease, Gut **17:**571-575, 1976.
67. Powell, L.W., and Halliday, J.W.: Iron metabolism, iron absorption, and iron storage disorders, Viewpoints Dig. Dis. **14:**13-16, 1982.
68. Flink, E.B.: Mineral metabolism in alcoholism. In Kissin, F., and Begleiter, H., editors: The biology of alcoholism, vol. I, Biochemistry, New York, 1971, Plenum Publishing Corp., pp. 377-395.
69. Celada, A., Rudolph, H., and Donath, A.: Effect of experimental chronic alcohol ingestion and folic acid deficiencies on iron absorption, Blood **54:**906-915, 1979.
70. Powell, L.W.: The role of alcoholism in hepatic iron storage disease, Ann. N.Y. Acad. Sci. **252:**124-134, 1979.
71. McClain, C., Soutor, C., and Zieve, L.: Zinc deficiency: a complication of Crohn's disease, Gastroenterology **78:**272-279, 1980.
72. Smith, J.C., McDaniel, E.G., Fan, F.F., et al.: Zinc: a trace element essential in vitamin A metabolism, Science **181:**954-955, 1973.
73. Zieve, L.: Influence of magnesium deficiency on the utilization of thiamine, Ann. N.Y. Acad. Sci, **162:**732-743, 1969.
74. Estep, H., Shaw, W.A., Wathington, C., et al.: Hypocalcemia due to hypomagnesemia and reversible parathyroid hormone unresponsiveness, J. Clin. Endocrinol. **29:**842-848, 1969.
75. Wolfe, S.M., and Victor, M.: The relationship of hypomagnesemia to alcohol withdrawal seizures and delirium tremens, Ann. N.Y. Acad. Sci. **162:**973-984, 1969.
76. Kalbfleisch, J.M., Lindeman, R.D., Ginn, H.E., et al.: Effects of ethanol administration on urinary excretion of magnesium and other electrolytes in alcoholic and normal subjects, J. Clin. Invest. **42:**1471-1475, 1963.
77. Jung, R.T., Davie, M., Chalmers, J.O., et al.: Abnormal vitamin D metabolism in cirrhosis, Gut **19:**290-293, 1978.
78. Krane, S.M., and Potts, J.T.: Skeletal remodeling and factors influencing bone and bone mineral metabolism. In Isselbacker, K.J., Adams R.D., Braunwald, E., et al., editors: Principles of internal medicine, New York, 1980, McGraw-Hill Book Co., pp. 1821-1832.
79. Knochel, J.P.: The pathophysiology and clinical characteristics of severe hypophosphatemia, Arch. Intern. Med. **137:**203-220, 1977.
80. Smetana, K., Gyorkey, R., Gyorkey, P., et al.: Studies on nucleoli and cytoplasmic fibrillar bodies of human hepatocellular carcinomas, Cancer Res. **32:**925-932, 1972.
81. Smetana, H.F., Hadley, G.G., and Sirat, S.M.: Infantile cirrhosis and analytical review of the literature and a report of 50 cases, Pediatrics **28:**107-127, 1961.
82. Marubbio, A.R., Jr., Buchwald, H., Schwartz, M.Z., et al.: Hepatic lesions of central pericellular fibrosis in morbid obesity and after jejunoileal bypass, Am. J. Clin. Pathol. **66:**684-691, 1976.
83. Monroe, S., French, S.W., and Zamboni, L.: Mallory bodies in a case of primary biliary cirrhosis: an ultrastructural and morphogenetic study, Am. J. Clin. Pathol. **59:**254-262, 1973.
84. Norkin, S.A., and Campagna-Pinto, D.: Cytoplasmic hyaline inclusions in hepatoma, Arch. Pathol. **86:**25-32, 1968.
85. Nayak, N.C., Sagreiya, E., and Ramalingaswami, R.: Indian childhood cirrhosis: the nature and significance of cytoplasmic hyaline of hepatocytes, Arch. Pathol. **88:**631-637, 1969.
86. Gerber, M.A., Orr, W., Denk, H., et al.: Hepatocellular hyalin in cholestasis and cirrhosis: its diagnostic significance, Gastroenterology **64:**89-98, 1973.
87. Schaffner, R., Sternlieb, I., Borka, R., et al.: Hepatocellular changes in Wilson's disease, Am. J. Pathol. **41:**315-327, 1962.
88. Roy, S., Ramalingaswami, R., and Nayak, N.C.: An ultrastructural study of the liver in Indian childhood cirrhosis with particular reference to the structure of cytoplasmic hyaline, Gut **12:**693-701, 1971.
89. Popper, H.: The problem of hepatitis, Am. J. Gastroenterol. **55:**335-346, 1971.
90. Christoffersen, P., and Nielsen, K.: Histological changes in human liver biopsies from chronic alcoholics, Acta Pathol. Microbiol Scand. (A); **80:**557, 1972.
91. Mendenhall, C.L.: Unpublished data presented at the Fifth Study Group Meeting, Veterans Administration Cooperative Study No. 119 on Alcoholic Hepatitis, Chicago, 1982.
92. Durbec, J.P., and Sarles, H.: Multicenter survey of the etiology of pancreatic diseases: the relationship between the relative risk of developing chronic pancreatitis and alcohol, protein, and lipid consumption, Digestion **18:**337-350, 1970.
93. Camerson, J.L., Zuidema, G.D., and Margolis, S.: A pathogenesis for alcoholic pancreatitis, Surgery **77:**754-763, 1975.
94. Sarles, H.: Alcohol and the pancreas, Adv. Exp. Med. Biol. **85A:**429-448, 1977.
95. Banks, P.A., editor: Pancreatitis, ed.1, New York, 1979, Plenum Publishing Corp.
96. Sarles, H.: Chronic calcifying pancreatitis—chronic alcoholic pancreatitis, Gastroenterology **66:**604-616, 1974.
97. Nakamura, K., Sarles, H., and Pagan, H.: Three-dimensional reconstruction of the pancreatic ducts in chronic pancreatitis, Gastroenterology **62:**942-949, 1972.
98. Sarles, H., and Tiscornia, O.: Ethanol and chronic calcifying pancreatitis, Med. Clin. North. Am. **58:**1333-1346, 1974.
99. Amman, R.W., Largiades, F., and Akovbiantz, A.: Pain relief by

surgery in chronic pancreatitis? Relationship between pain relief, pancreatic dysfunction, and alcohol withdrawal, Scand. J. Gastroenterol. **14**:209-215, 1979.

100. DiMagno, E.P., Go, V.L.W., and Summerskill, W.H.J.: Relationship between pancreatic enzyme outputs and malabsorption in severe pancreatic insufficiency, N. Engl. J. Med. **288**:813-815, 1973.
101. Marks, V.: Alcohol and carbohydrate metabolism, Clin. Endocrinol. Metab. **7**:333-349, 1978.
102. Schnek, E.A., and Cohen, J.: The heart in chronic alcoholism: clinical and pathologic findings, Pathol. Microbiol. **35**:96-104, 1970.
103. Lumseth, J.H., Olmstead, E.G., and Abboud, R.: A study of heart disease in 108 hospitalized patients dying with portal cirrhosis, Arch. Intern. Med. **102**:405-413, 1958.
104. Hartel, G., Louhija, A., and Konttinen, A.: Cardiovascular study of 100 chronic alcoholics, Acta Med. Scand. **185**:507-513, 1969.
105. Fisher, V.J., and Kavaler, F.: The action of ethanol upon the contractility of normal ventricular myocardium. In Rothschild, M.A., Oratz, M., and Schreiber, S.S., editors: Alcohol and abnormal protein biosynthesis, Elmsford, N.Y., 1975, Pergamon Press, Inc., pp. 187-202.
106. Regan, T.J.: Metabolic adaptation to alcohol in the heart. In Rothschild, M.A., Oratz, M., and Schreiber, S.S., editors: Alcohol and abnormal protein biosynthesis, Elmsford, N.Y., 1975, Pergamon Press, Inc., pp. 247-272.
107. Tamarin, J.S., Weiner, S., Poppen, R., et al.: Alcohol and memory, Am. J. Psychol. **127**:1659-1667, 1971.
108. Goodwin, D.W., Powell, B., and Bremer, D.: Alcohol and recall state dependent effects in man, Science **163**:1358-1360, 1969.
109. Sherlock, S.: Alcohol and the liver: treatment, early recognition. In Sherlock, S., editor: Diseases of the liver and biliary system, ed. 6, London, 1981, Blackwell Scientific Publications, Ltd., p. 343.
110. Keller, A.Z., and Terris, M.: The association of alcohol and tobacco with cancer of the mouth and pharynx, Am. J. Public Health. **55**:1578-1585, 1965.
111. Vincent, R.G., and Marchetta, F.: The relationship of the use of tobacco and alcohol to cancer of the oral cavity, pharynx or larynx, Am. J. Surg. **106**:501-505, 1963.
112. Martínez, I.: Retrospective and prospective study of carcinoma of the esophagus, mouth, and pharynx in Puerto Rico, Bol. Asoc. Med. P. Rico **62**:170-178, 1970.
113. Burch, C.E., and Ansari, A.: Chronic alcoholism and carcinoma of the pancreas: a correlative hypothesis, Arch. Intern. Med. **122**:273-275, 1968.
114. MacDonald, E.C.: Clinical and pathological features of adenocarcinoma of the gastric cardia, Cancer **29**:724-732, 1972.
115. Williams, R.R., and Horm, J.W.: Association of cancer sites with tobacco and alcohol consumption and socioeconomic status of patients: interview study from the Third National Cancer Survey, J. Natl. Cancer Inst. **58**:525-547, 1977.
116. Wynder, E.L., Bross, I.J., and Feldman, R.M.: A study of the etiological factors in cancer of the mouth, Cancer **10**:1300-1323, 1957.
117. Wynder, E.L., and Stellman, S.D.: Comparative epidemiology of tobacco-related cancers, Cancer Res. **37**:4608-4622, 1977.
118. McCoy, G.D., and Wynder, E.L.: Etiological and preventive implications in alcohol carcinogenesis, Cancer Res. **39**:2844-2850, 1979.
119. Johnston, D.G., and Alberti, K.G.M.M.: The liver and the endocrine system. In Wright, R., Alberti, K.G.M.M., and Karran, S., et al., editors: Liver and biliary disease, Philadelphia, 1979, W.B. Saunders Co., pp. 134-158.
120. Jenkins, J.S., and Connolly, J.: Adrenocortical response to ethanol in man, Br. Med. J. **2**:804-805, 1968.
121. Merry, J., and Marks, V.: Hypothalamic-pituitary-adrenal function in chronic alcoholics. In Cross, M.M., editor: Alcohol intoxication and withdrawal: experimental studies, Advances in Experimental Medicine and Biology, New York, 1973, Plenum Publishing Corp., pp. 167-179.
122. Chalmers, R.J., Bennie, E.H., Johnson, R.H., et al.: Growth hormone, prolactin and corticosteroid responses to insulin hypoglycaemia in alcoholics, Br. Med. J. **2**:745-748, 1978.
123. Chisholm, M.: Haematological disorders in liver disease. In Wright, R., Alberti, K.G.M.M., Karran, S., et al. editors: Liver and biliary disease, Philadelphia, 1979, W.B. Saunders Co., pp. 159-181.
124. Hines, J.D., and Cowan, D.H.: Studies on the pathogenesis of alcohol-induced sideroblastic bone marrow abnormalities, N. Engl. J. Med. **283**:441-446, 1970.
125. Liu, Y.K.: Leukopenia in alcoholics, Am. J. Med. **54**:605-610, 1973.
126. Cowan, D.H., and Hines, J.D.: Thrombocytopaenia of severe alcoholism, Ann. Intern. Med. **74**:37-43, 1971.
127. McFarland, W., and Leibre, E.P.: Abnormal leukocyte response in alcoholism, Ann. Intern. Med. **59**:865-877, 1963.
128. Thomas, D.P., Ream, V.J., and Stuart, R.K.: Platelet aggregation in patients with Laennec's cirrhosis of the liver, N. Engl. J. Med. **276**:1344-1348, 1967.
129. Chen, T., Kanagasundaram, N., Kakumu, S., et al.: Serum autoantibodies to alcoholic hyalin in alcoholic hepatitis, (abstract), Gastroenterology **A13**:813, 1975.
130. Zinneman, H.H.: Autoimmune phenomena in alcoholic cirrhosis, Am. J. Dig. Dis. **20**:337-345, 1970.
131. Bernstein, I.M., Webster, K.H., Williams, R.C., Jr., et al.: Reduction in circulating T lymphocytes in alcoholic liver disease, Lancet **2**:488-490, 1974.
132. Hsu, C.C.S., and Leevy, C.M.: Inhibition of PHA-stimulated lymphocyte transformation by plasma from patients with advanced alcoholic cirrhosis, Clin. Exp. Immunol. **8**:749-760, 1971.
133. Kakumu, S., and Leevy, C.M.: Lymphocyte cytotoxicity in alcoholic hepatitis, Gastroenterology **72**:524-526, 1977.
134. Smith, W.I., Jr., VanThiel, D.H., Whiteside, T., et al.: Altered immunity in male patients with alcoholic liver disease: evidence for defective immune regulation, Alcohol Clin. Exp. Res. **4**:199-206, 1980.
135. Cherubin, C.E., Marr, J.S., Sierra, M.F., et al.: *Listeria* and gram-negative bacillary meningitis in New York City, Am. J. Med. **71**:199-209, 1981.
136. Conn, H., and Fessel, J.: Spontaneous bacterial peritonitis and cirrhosis: variations on the theme, Medicine **50**:161-197, 1971.
137. Feingold, A.O.: Association of tuberculosis and alcoholism, South. Med. J. **69**:1336-1337, 1976.
138. Nolan, J.: Alcohol as a factor in the illness of university service patients, Am. J. Med. Sci. **249**:135-142, 1965.
139. Kaplan, M.M.: Alkaline phosphatase, Gastroenterology **62**:452-468, 1972.
140. Rosalki, S.B.: Gamma-glutamyl transpeptidase, Adv. Clin. Chem. **17**:53-62, 1975.
141. McGill, D.B.: Steatosis, cholestasis, and alkaline phosphatase in alcoholic liver disease, Dig. Dis. Sci. **23**:1057-1060, 1978.
142. Maddrey, W.C., Boitnott, J.K., Bedine, M.S., et al.: Corticosteroid therapy of alcoholic hepatitis, Gastroenterology **75**:193-199, 1978.
143. Helman, R.A., Temko, M.H., Nye, S.W., et al.: Alcoholic hepatitis: natural history and evaluation of prednisolone therapy, Ann. Intern. Med. **74**:311-321, 1971.
144. Perrillo, R.P., Griffin, R., DeSchryver-Kecskemeti, K., et al.: Alcoholic liver disease presenting with marked elevation of serum alkaline phosphatase: a combined clinical and pathological study, Dig. Dis. Sci. **23**:1061-1066, 1978.
145. Triger, D.R., and Wright, R.: Hyperglobulinaemia in liver disease, Lancet **1**:1494-1496, 1973.
146. Pomier-Layrargues, G., Huet, P.M., Richer, G., et al.: Hyperglobulinemia in alcoholic cirrhosis: relationship with portal hypertension and intrahepatic portal–systemic shunting as assessed by Kupffer cell uptake. Dig. Dis. Sci. **25**:489-493, 1980.
147. Milstein, H.J., Bloomer, J.R., and Klatskin, G.: Serum bile acids in alcoholic liver disease, Am. J. Dig. Dis. **21**:281-285, 1976.
148. Lundquist, G.A.R.: Glucose tolerance in alcoholism, Br. J. Addict. **61**:51-55, 1965.
149. Rees, L.H., Besser, G.M., Joffcoate, W.J., et al.: Alcohol-induced pseudo-Cushing's syndrome, Lancet **1**:726-728, 1977.
150. Madison, L.L., Lochner, A., and Wulff, J.: Ethanol induced hy-

poglycemia. II. Mechanism of suppression of hepatic gluconeogenesis, Diabetes **16:**252-258, 1967.
151. Joffe, B.I., Seftel, H.C., and Van As, M.: Hormonal responses in ethanol-induced hypoglycaemia, J. Stud. Alcohol **36:**550-554, 1975.
152. Palmer, J.P., and Ensinck, J.W.: Stimulation of glucagon secretion by ethanol induced hypoglycaemia in man, Diabetes **24:**295-300, 1975.
153. Arky, R.A., and Freinkel, N.: Alcohol hypoglycemia. V. Alcohol infusion to test gluconeogenesis in starvation, with specific reference to obesity, N. Engl. J. Med. **274:**426-433, 1966.
154. O'Keefe, S.J., and Marks, V.: Lunchtime gin and tonic, a cause of reactive hypoglycaemia, Lancet **1:**1286-1288, 1977.
155. Cooperman, M.T., Davidoff, F., Spark, R., et al.: Clinical studies of alcoholic ketoacidosis, Diabetes **23:**433-439, 1974.
156. Price, C.P., and Alberti, K.G.M.M.: Biochemical assessment of liver function. In Wright, R., Alberti, K.G.M.M., Karran, S., et al., editors: Liver and biliary disease, Philadelphia, 1979, W.B. Saunders Co., pp. 381-416.
157. Leevy, C.M., Smith, F., Longueville, J., et al.: Indocyanine green clearance as a test for hepatic function: evaluation by dichromatic ear densitometry, J.A.M.A. **200:**236-240, 1967.

Chapter 33 Hemoglobin, porphyrin, and iron metabolism

John D. Bauer

anemia Reduction in hemoglobin or number of red cells.
anoxia Lack of oxygen in tissues.
chromatography Separation of substances by selective adsorption in mixture of solvents and adsorbents.
chromosome Nuclear DNA component, bearer of genetic information.
crisis Acute symptoms developing in sickle cell anemia because of exaggerated hemolysis and vascular occlusion.
cyanosis Bluish color caused by lack of oxygen.
dimer Combination of two like molecules.
dyad A bivalent element. Also *diad*.
electrophoresis Separation of particles in an electric field.
erythrocytosis Above-normal increase in number of red cells (polycythemia).
erythropoietin Red cell formation–stimulating hormone.
feedback Regulation of the output of a hormone or enzyme through the input of a controlling system: negative feedback leads to reduction of output; positive feedback leads to increase of output.
gene DNA unit of heredity of a chromosome, which reproduces itself exactly at each cell division and mediates protein synthesis and genetic information.
Heinz body Precipitated hemoglobin in red blood cells.
heme An iron-containing porphyrin (protoporphyrin), which with a protein (globin) composes hemoglobin.
hemoglobin (Hb) An iron-containing respiratory pigment of red blood cells.
hemolysis Destruction of red cells resulting in release of hemoglobin.
hemosiderosis Accumulation of hemosiderin in various organs.
hepatoslenomegaly Increased size of liver and spleen.
hypochromia Reduction of hemoglobin in red cells.
hypoxia Reduced oxygen level in tissues.
ligand Attachment of an organic molecule to a metal ion.
locus Specific site.
monomer Molecular unit of a large structure.
mutation Hereditary change in the structure of a gene.
myoglobin Oxygen-carrying protein in muscle.
normochromia Normal amount of hemoglobin in red cells.
ontogeny Development of an individual organism.
oxidation Addition of oxygen, removal of hydrogen, and loss of electrons.
oxygenation Combination with oxygen.
Plasmodium A blood parasite and causative organism of malaria.
polarity State of having opposite charges.
polypeptide A protein formed by the union of two or more amino acids.
radical Group of atoms that act as a unit in chemical reactions.
reduction Loss of oxygen, addition of hydrogen, and gain of electrons.
reticulocyte Young red cell containing a network of precipitated organelles visualized when treated with vital dyes.
reticuloendothelial system Collective term for phagocytic cells in tissues, blood vessels, spleen, liver, bone marrow, and lymph nodes.
synthase Enzyme catalyzing the synthesis of a compound without involving the breakdown of a pyrophosphate bond (lyase).
synthetase Enzyme catalyzing the synthesis of a specific compound by involving the breakdown of a pyrophosphate bond (ligase).
tetramer Composed of four parts.

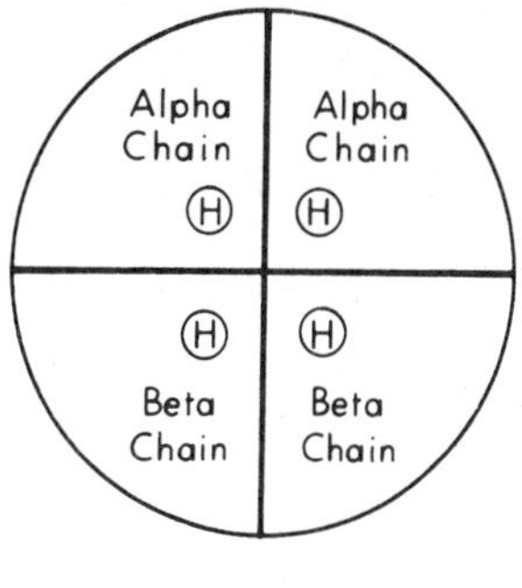

Alpha Chain
Alpha Chain
H
H
H
H
Beta Chain
Beta Chain

Fig. 33-1. Diagram of structure of Hb A molecule. Four heme groups are attached to one globin molecule, which consists of four polypeptide chains, two of which have an identical amino acid sequence of one type (α chain) and the other two an identical amino acid sequence of another type (β chain). Each polypeptide chain is conjugated to one heme moiety. *H,* Heme. (From Bauer, J.D.: Clinical laboratory methods, ed. 9, St. Louis, 1982, The C.V. Mosby Co.)

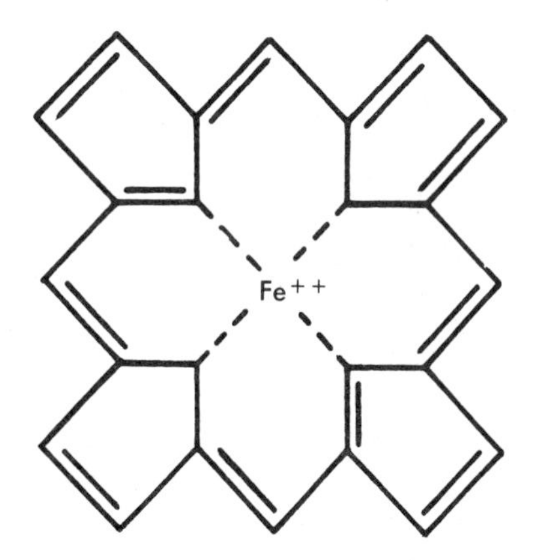

Fig. 33-2. Diagram of structure of heme. One iron atom (Fe^{++}) is related to four pyrrole rings, which are joined to each other through methylene bridges. (From Bauer, J.D.: Clinical laboratory methods, ed. 9, St. Louis, 1982, The C.V. Mosby Co.)

Part I: Hemoglobin

STRUCTURE AND FUNCTION OF HEMOGLOBIN

Structure

Hemoglobin (Hb) is the red-pigmented, oxygen carrying protein of the red blood cells of vertebrates. It is a globoid tetramer (molecular weight, 68,000 daltons) consisting of two pairs of unlike polypeptide chains (Fig. 33-1). Each chain carries an iron-containing porphyrin derivative called *heme,* a ferroprotoporphyrin IX, in which one iron atom is deposited in the center of the porphyrin ring (Fig. 33-2). The polypeptide chains (minus heme) are collectively called the *globin moiety* of hemoglobin. Each polypeptide chain is designated by a Greek letter: α, β, γ, δ, ϵ, ζ. These chains have quite similar three-dimensional structures. The *primary structure,* the amino acid sequence, has been known since the 1960s and demonstrates a closer relationship between β, γ, δ, and ϵ chains than between ζ and α chains. The α chain contains 141 amino acid residues (Fig. 33-3) and the δ, ϵ, and β chains 146 (Fig. 33-4). The amino acid sequence is genetically controlled and the substitution of one single amino acid can be responsible for at times fatal pathophysiological changes.[1-4,8] Three fourths of each polypeptide chain are arranged in coiled structures called α helices, the *secondary structure* (Fig. 33-5). Perutz[5,6,8] is credited with the elucidation of the further folding of the helices into a specific irregular pretzel configuration, the three-dimensional *tertiary structure* (Fig. 33-6). Alternating helical and nonhelical segments are coiled, a configuration that determines the orientation of polar (lysine and glutamic acid), dipolar (tyrosine and glutamine), and nonpolar (valine and leucine) groups. This arrangement is responsible for the close contact of interior amino acids with nonpolar hydrophobic side chains with neighboring residues. They are attracted to each other by weak electrical charges, which stabilize the hemoglobin molecule and keep its interior free of water, at the same time maintaining the heme molecule in its reduced form (Fe^{++}), a prerequisite of oxygenation. The surface of the hemoglobin molecule is covered primarily by polar hydrophilic side chains of the amino acid residues, which are able to react with solvents such as water but do not react with other side chains. The heme molecule lies in a pocket between the E and F helices, deeply embedded in the hydrophobic interior. Interatomic forces between the heme molecule and the globin chains (proximal and distal histidine) further stabilize the hemoglobin molecule. The hemoglobin tetramer, as mentioned previously, is composed of four interrelated subunits or monomers, each subunit consisting of a polypeptide chain and its heme moiety. The tetramer forms a globular, doughnutlike molecule that has a central water-filled cavity (Fig. 33-7) and a twofold dyad axis of symmetry. The relationship of the subunits to each other in the hemoglobin tetramer is the *quaternary structure* (Fig. 33-8). The molecule

```
 1   2   3   4   5   6   7   8   9   10  11  12  13  14  15
VAL-LEU-SER-PRO-ALA-ASP-LYS-THR-ASN-VAL-LYS-ALA-ALA-TRY-GLY

 16  17  18  19  20  21  22  23  24  25  26  27  28  29  30
LYS-VAL-GLY-ALA-HIS-ALA-GLY-GLU-TYR-GLY-ALA-GLU-ALA-LEU-GLU

 31  32  33  34  35  36  37  38  39  40  41  42  43  44  45
ARG-MET-PHE-LEU-SER-PHE-PRO-THR-THR-LYS-THR-TYR-PHE-PRO-HIS

 46  47  48  49  50  51  52  53  54  55  56  57  58  59  60
PHE-ASP-LEU-SER-HIS-GLY-SER-ALA-GLN-VAL-LYS-GLY-HIS-GLY-LYS

 61  62  63  64  65  66  67  68  69  70  71  72  73  74  75
LYS-VAL-ALA-ASP-ALA-LEU-THR-ASN-ALA-VAL-ALA-HIS-VAL-ASP-ASP

 76  77  78  79  80  81  82  83  84  85  86  87  88  89  90
MET-PRO-ASN-ALA-LEU-SER-ALA-LEU-SER-ASP-LEU-HIS-ALA-HIS-LYS

 91  92  93  94  95  96  97  98  99 100 101 102 103 104 105
LEU-ARG-VAL-ASP-PRO-VAL-ASN-PHE-LYS-LEU-LEU-SER-HIS-CYS-LEU

106 107 108 109 110 111 112 113 114 115 116 117 118 119 120
LEU-VAL-THR-LEU-ALA-ALA-HIS-LEU-PRO-ALA-GLU-PHE-THR-PRO-ALA

121 122 123 124 125 126 127 128 129 130 131 132 133 134 135
VAL-HIS-ALA-SER-LEU-ASP-LYS-PHE-LEU-ALA-SER-VAL-SER-THR-VAL

136 137 138 139 140 141
LEU-THR-SER-LYS-TYR-ARG
```

Fig. 33-3. Alpha chain of human hemoglobin, sequence of amino acids (primary structure). *ALA,* Alanine; *ARG,* arginine; *ASP,* aspartic acid; *ASN,* asparagine; *CYS,* cysteine; *GLU,* glutamic acid; *GLN,* glutamine; *GLY,* glycine; *HIS,* histidine; *ILU,* isoleucine; *LEU,* leucine; *LYS,* lysine; *MET,* methionine; *PHE,* phenylalanine; *PRO,* proline; *SER,* serine; *THR,* threonine; *TRY,* tryptophan; *TYR,* tyrosine; *VAL,* valine. (From Miale, J.B.: Laboratory medicine hematology, ed. 6, St. Louis, 1982, The C.V. Mosby Co.)

```
 1   2   3   4   5   6   7   8   9   10  11  12  13  14  15
VAL-HIS-LEU-THR-PRO-GLU-GLU-LYS-SER-ALA-VAL-THR-ALA-LEU-TRY

 16  17  18  19  20  21  22  23  24  25  26  27  28  29  30
GLY-LYS-VAL-ASN-VAL-ASP-GLU-VAL-GLY-GLY-GLU-ALA-LEU-GLY-ARG

 31  32  33  34  35  36  37  38  39  40  41  42  43  44  45
LEU-LEU-VAL-VAL-TYR-PRO-TRY-THR-GLN-ARG-PHE-PHE-GLU-SER-PHE

 46  47  48  49  50  51  52  53  54  55  56  57  58  59  60
GLY-ASP-LEU-SER-THR-PRO-ASP-ALA-VAL-MET-GLY-ASN-PRO-LYS-VAL

 61  62  63  64  65  66  67  68  69  70  71  72  73  74  75
LYS-ALA-HIS-GLY-LYS-LYS-VAL-LEU-GLY-ALA-PHE-SER-ASP-GLY-LEU

 76  77  78  79  80  81  82  83  84  85  86  87  88  89  90
ALA-HIS-LEU-ASP-ASN-LEU-LYS-GLY-THR-PHE-ALA-THR-LEU-SER-GLU

 91  92  93  94  95  96  97  98  99 100 101 102 103 104 105
LEU-HIS-CYS-ASP-LYS-LEU-HIS-VAL-ASP-PRO-GLU-ASN-PHE-ARG-LEU

106 107 108 109 110 111 112 113 114 115 116 117 118 119 120
LEU-GLY-ASN-VAL-LEU-VAL-CYS-VAL-LEU-ALA-HIS-HIS-PHE-GLY-LYS

121 122 123 124 125 126 127 128 129 130 131 132 133 134 135
GLU-PHE-THR-PRO-PRO-VAL-GLN-ALA-ALA-TYR-GLN-LYS-VAL-VAL-ALA

136 137 138 139 140 141 142 143 144 145 146
GLY-VAL-ALA-ASN-ALA-LEU-ALA-HIS-LYS-TYR-HIS
```

Fig. 33-4. Beta-polypeptide chain, sequence of amino acids (primary structure). Abbreviations as in Fig. 33-3. (From Miale, J.B.: Laboratory medicine hematology, ed. 6, St. Louis, 1982, The C.V. Mosby Co.)

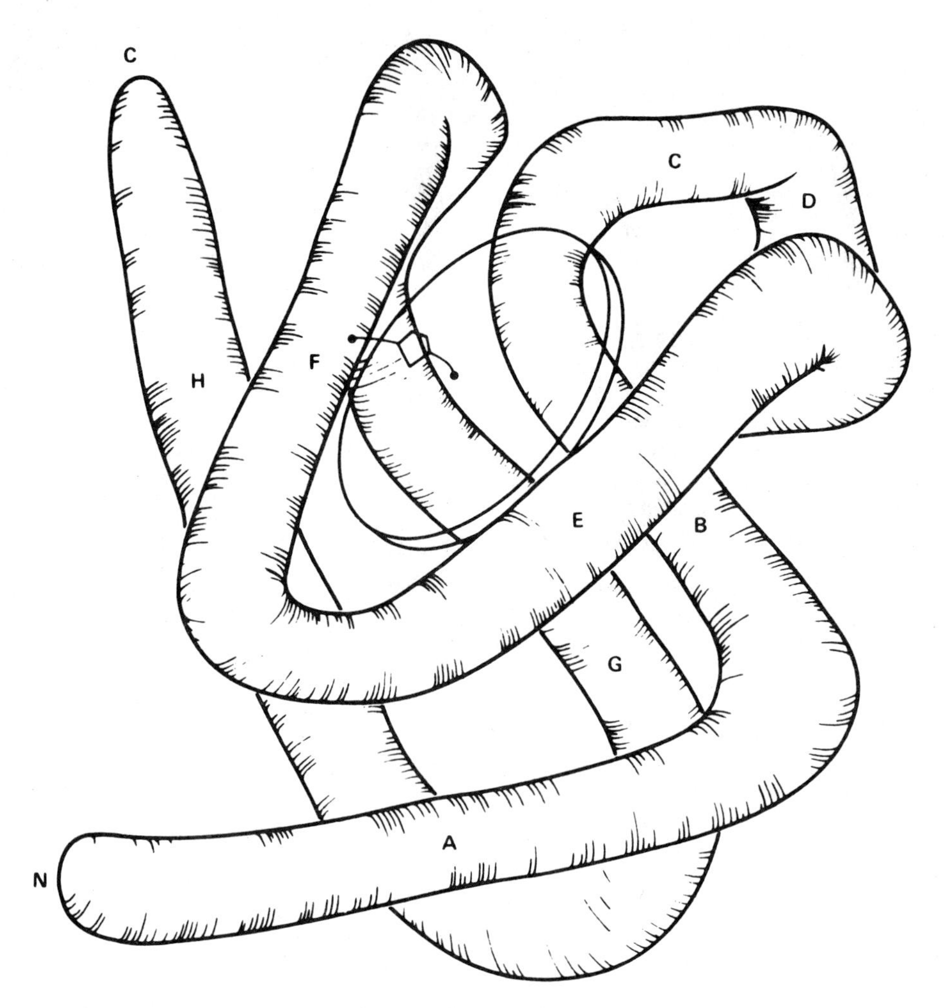

Fig. 33-5. Scheme of secondary structure of a globin chain. Any of globin chains (or myoglobin) could fit diagram with only minor alteration. The α-helical regions are lettered, and heme is shown as a disk. The only covalent link between heme and globin is between iron atom and His F8, the "proximal" histidine. (From Winslow, R.M., and Anderson, W.F.: In Stanbury, J.B., Wyngaarden, J.B., and Fredrickson, D.S., editors: The metabolic basis of inherited disease. New York, 1978, McGraw-Hill Book Co.)

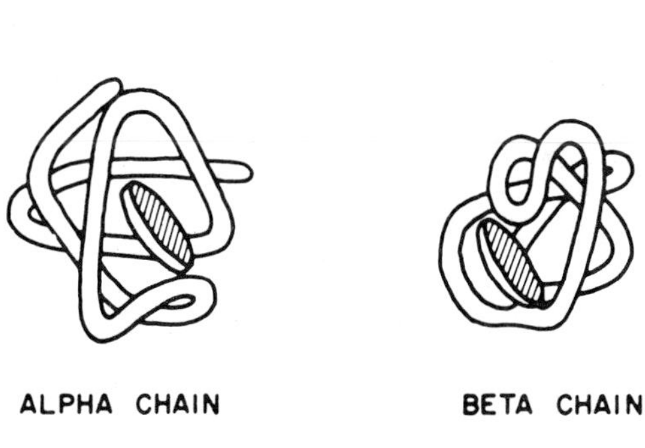

Fig. 33-6. Tertiary structure of polypeptide chains of normal hemoglobin (Hb A). Heme group is shown as a disk, which in actual size is much smaller. (From Miale, J.B.: Laboratory medicine hematology, St. Louis, 1982, The C.V. Mosby Co.)

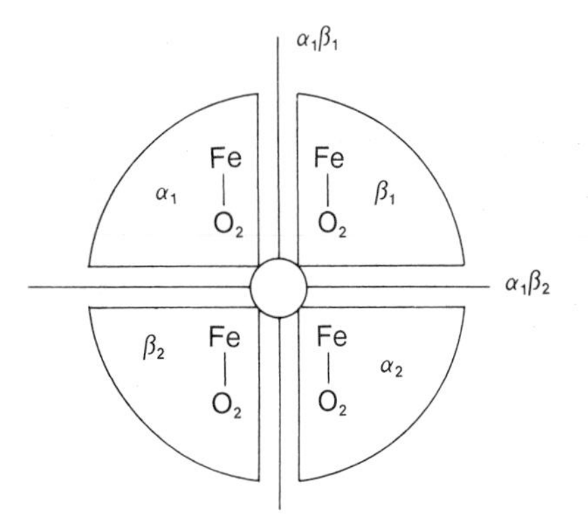

Fig. 33-7. Diagram of hemoglobin molecule (oxyhemoglobin) showing four subunits, with αβ dimers, contact planes, central cavity, and heme iron and oxygen.

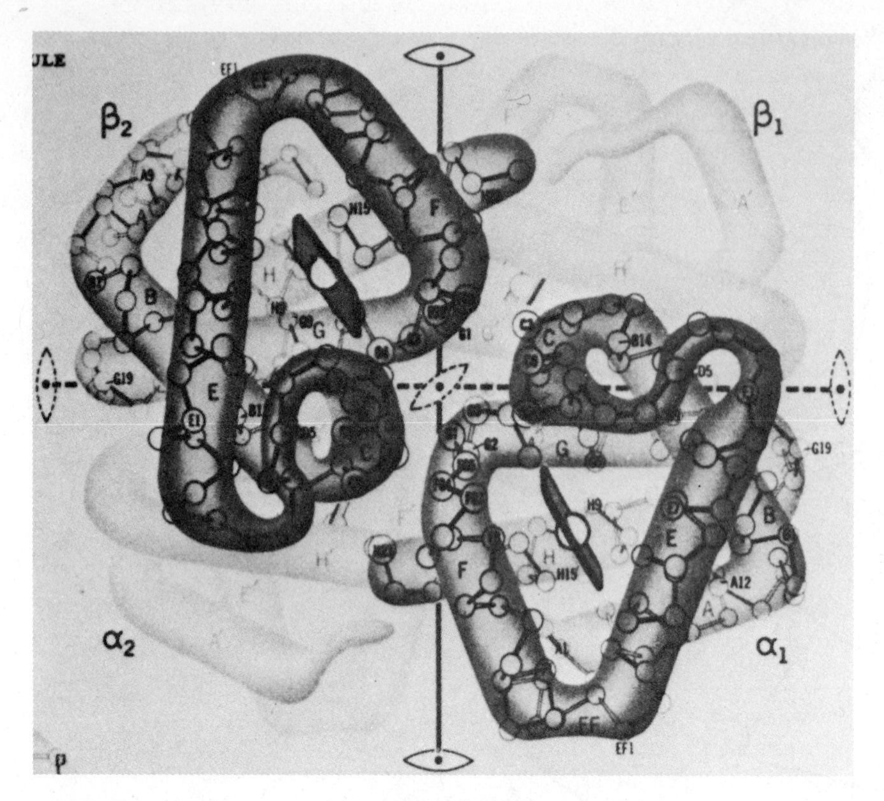

Fig. 33-8. Quaternary structure of hemoglobin. The α_1 and β_2 chains are in foreground, and $\alpha_1\beta_2$ contact is at center. (From Dickerson, R.E., and Geis, I.: The structure and action of proteins, Menlo Park, Calif., 1969, Benjamin/Cummings.

is symmetric about the axis of the cavity and also in a plane perpendicular to this axis. The four subunits are held in position by hydrophobic bonds at contact points. The central cavity allows the entrance of 2,3-diphosphoglycerate (2,3-DPG) and salts.

In summary, the amino acid sequences of hemoglobin represent the primary structure. The α helices form the secondary structure and the three-dimensional conformation forms the tertiary and quaternary structures. The allosteric configuration of hemoglobin allows it to fulfill its role as respiratory protein.

Globin chain synthesis and genetics

The differences of the various normal human hemoglobins are based on the variations in the amino acid sequences (the primary structure) of six different globin chains (α, β, γ, δ, ϵ, and ζ), the synthesis of which is controlled by at least six structural genes. The β, δ, and probably the ζ and ϵ chains are coded by one pair of genes, whereas the γ and α chains are coded by two pairs of genes.[7] The α genes are on chromosome 16 and the non-α genes (γ, δ, and β) are closely linked on chromosome 11 (Fig. 33-9). The positions of the ϵ and ζ gene loci are not known. The genetic code is translated into the final globin polypeptide chains by the usual protein synthetic pathway.[11-13] Mutation in any of the steps can cause decreased globin synthesis, as demonstrated in the various

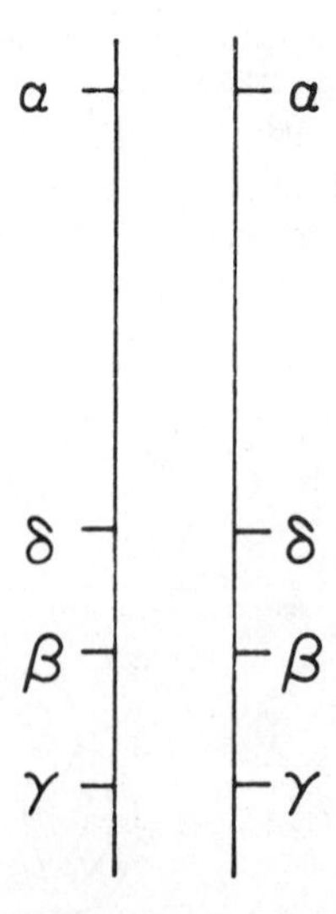

Fig. 33-9. "Working" diagram of arrangements of four of structural genes for hemoglobin synthesis. The α and non-α loci probably lie on different chromosomes, represented in diagram by distance between the two. The δ, β, and γ loci are closely linked. Both α and γ loci may be multiple. Not shown are loci for ϵ and ζ chains, since their position is not known. (From Miale, J.B.: Laboratory medicine hematology, St. Louis, 1982, The C.V. Mosby Co.)

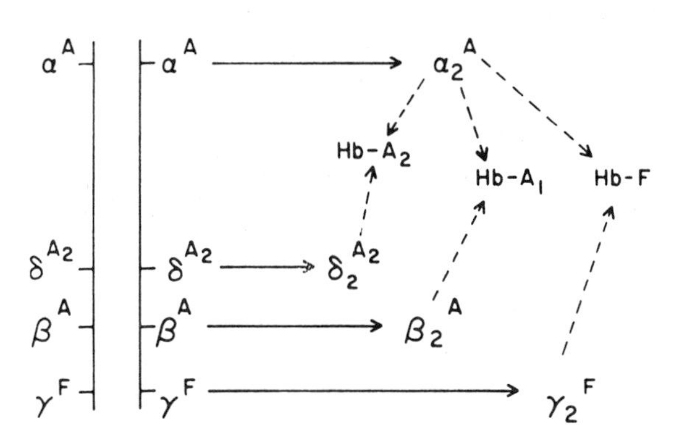

Fig. 33-10. Structural loci and combination of dimers to form the three normal hemoglobins. α_2, β_2, γ_2, and δ_2 dimers are synthesized and recombined to form Hb A_1, Hb A_2, and Hb F. Since normal adult blood contains about 3.5% Hb A_2, 1% to 2% Hb F, and the remainder Hb A, the different chains must be produced in different quantities if, as it is in fact, few or no excess chains are usually found. (From Miale, J.B.: Laboratory medicine hematology, St. Louis, 1982, The C.V. Mosby Co.)

forms of thalassemia. The ontogenetic sequence of hemoglobin production at different stages of development is also under genetic control, which requires suppression or activation by separate chromosomes or by the gene itself.[14] The α chains are formed early in embryonal life and once released from the ribosomes they spontaneously, quickly, and sequentially combine with the ε and γ, β, and δ chains to form the embryonic (ε), fetal (γ), A_2 (δ), and adult hemoglobins (β). Two single peptide chains of the same type combine to form dimers, and the dimers of different types then combine to form the various tetramers (Fig. 33-10).[15-17] The proportions and the rate of formation of the dimers conform to the various percentages of the normal hemoglobins, but the factors involved in the switching of intrauterine to adult hemoglobins are unknown.[18]

Inheritance patterns

The hemoglobin phenotype is the expression of paired genes on paired autosomal chromosomes. Each locus is therefore represented twice. If the effect of one member of a pair is greater than that of the other, the effect is said to

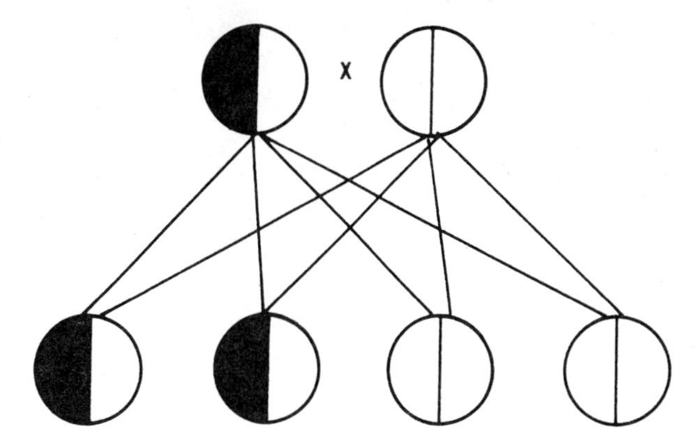

Fig. 33-11. Autosomal dominant inheritance. *Black,* Dominant gene for disease expression; *white,* recessive nondiseased gene. Result: one half of offspring express dominant gene phenotype and have the disease.

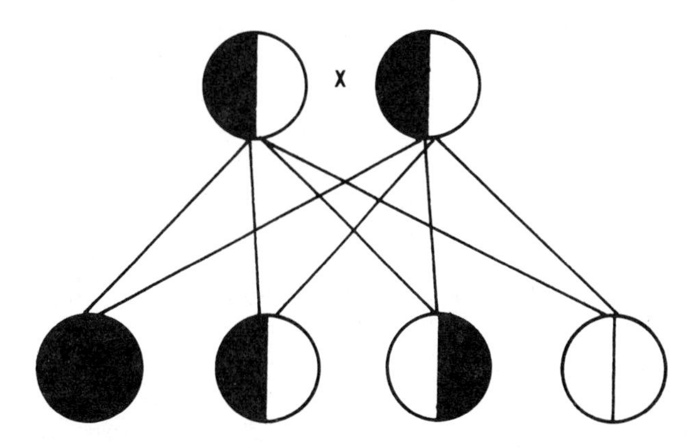

Fig. 33-12. Autosomal recessive inheritance. *Black,* Dominant nondiseased gene; *white,* recessive gene for disease expression. Result: one fourth of offspring express recessive gene and have the disease, and one half express the carrier state.

be dominant. It is incompletely dominant if the effect cannot be fully achieved. If a gene expresses itself in the homozygous or heterozygous state, it is dominant. Homozygosity implies the existence of identical genes on paired loci of homologous chromosomes. Heterozygosity implies the presence of different allelic genes on one or more

Table 33-1. Expected inheritance pattern with certain matings

		Expected proportion of children		
Parent A	Parent B	Normal (%)	Heterozygous (%)	Homozygous (%)
Heterozygous	Normal	50	50	0
Heterozygous	Heterozygous	25	50	25
Homozygous	Normal	0	100	0
Homozygous	Heterozygous	0	50	50
Homozygous	Homozygous	0	0	100

From Lee, G.R.: Disorders of globin synthesis: the hemoglobinopathies and thalassemia. In F. Tice's practice of medicine, ed. 6, Hagerstown, Md. 1970, Harper & Row, Publishers, Inc.

Table 33-2. Allelic and nonallelic inheritance patterns in doubly heterozygous persons

Patient	Spouse	Expected proportion of children		
		Normal	Heterozygous	Doubly heterozygous
Two genes affecting same chain (allelic), such as Hb S and C	Normal	0%	100% (50% S, 50% C)	0%
Two genes affecting different chains (nonallelic), such as Hb S, Hopkins II	Normal	25%	50% (25% S, 25% Hopkins II)	25%

From Wintrobe, M.M., et al.: The abnormal hemoglobins: general principles in clinical hematology, ed. 8, Philadelphia, 1981, Lea & Febiger.

paired loci of homologous chromosomes. If a heterozygous person with a hemoglobin abnormality that expresses itself in the heterozygous state (such as congenital Heinz body hemolytic anemia) mates with a normal person, 50% of the children will be normal and 50% heterozygous and clinically affected (Table 33-1). This inheritance pattern is called autosomal dominance or autosomal codominance (Fig. 33-11). If the abnormal gene is recognizable only in the homozygous state, an autosomal recessive inheritance (Fig. 33-12) pattern is observed. The heterozygous state is called a ''trait'' or ''carrier'' and is clinically normal and unaffected. A frequently seen example of a recessive condition is hemoglobin S. If both parents carry the trait they are clinically normal, but 25% of their children will be homozygous for Hb S, 50% heterozygous, and 25% normal. If a normal person mates with a heterozygous Hb S person, there will be 50% normal and 50% heterozygous offspring. If both parents are homozygous, all children will be homozygous, whereas if one parent is homozygous and the other heterozygous, 50% of the children will be heterozygous and 50% homozygous for Hb S. A double heterozygote is a person with two different hemoglobin genes, the inheritance pattern of which depends on whether they are alleles or nonalleles. Alleles occupy the same gene locus on a specific chromosome and therefore affect the same polypeptide chain. Nonalleles occupy positions on different chromosomes or different loci on the same chromosome. Mating of a normal person with a person doubly heterozygous for β-chain defects (Hb SC) results in 100% heterozygous offspring. Mating of a normal person with a person with a double nonallelic gene defect, such as Hb S (β-chain defect) and Hb Hopkins 2 (α-chain defect), produces normal, heterozygous, and doubly heterozygous children[22,23] (see Table 33-2).

NORMAL BIOCHEMISTRY

Heme synthesis

Heme synthesis[24-26] is the result of a series of enzyme-controlled steps that occur during erythroid maturation and are described in more detail later in this chapter. In summary, activated glycine is condensed with succinyl CoA and the resulting product is spontaneously decarboxylated to form δ-aminolevulinic acid, the condensation of two molecules of which forms porphobilinogen. The latter is then converted into uroporphyrinogen III, which is decarboxylated to coproporphyrinogen and finally converted into protophorphyrinogen and by oxidation into protoporphyrin. The final step is the chelation of Fe^{++} by protoporphyrin to form heme.

Iron metabolism (summary)

Iron metabolism is described in detail later in this chapter. The differentiation of the cells of the erythron[27] of the bone marrow requires the delivery of large amounts of iron to support the maturation of the primitive erythroblasts to the nonnucleated circulating red cells. Two pathways allow iron to gain access to the young red cells. The pathway that probably plays a minor role in the iron transfer but is demonstrable histologically involves a microphagocytic process similar to pinocytosis, called ''rhopheocytosis.''[28] The phagocytic cells of the reticuloendothelial system (reticulum cells) of the bone marrow obtain iron from dying aged red cells and from transferrin, storing it as hemosiderin or ferritin. Electron microscopic investigations[29] confirm the transfer of reticulum cell ferritin to the surfaces of normoblasts, probably by means of appoferritin receptors.[28,30] It is then taken up by micropinocytotic vesicles and carried into the interior of the cytoplasm.[31] The major pathway of iron transfer involves transferrin iron, which transiently adheres to transferrin attachment sites on the surface of erythroblasts.[32,33] After the release of iron from its carrier protein it migrates to the mitochondria, where it is incorporated into protoporphyrin by heme synthase[34] to create the heme moiety of hemoglobin. The interplay of heme and globin synthesis and of the available iron is regulated by a feedback-inhibitor system that coordinates all three factors with the demand for hemoglobin.[35,36]

Assembly of hemoglobin

Adult hemoglobin is a tetrameric protein consisting of two α and two β polypeptide chains. Both are synthesized in equal amounts, though there may be an excess of α chains[37] in the cytoplasm of young red cells. The assembly process starts with the release of α and β chains from the ribosomes into the cytoplasm. They immediately incorporate heme and form monomer combinations and dimer aggregates followed by the synthesis of tetramers.[38] In hemoglobinopathies the concentration of the two like chains (such as β^A and β^S) may differ, even though their rates of synthesis are the same. There is evidence that the relative rates of assembly (rather than synthesis) of the two hemoglobins (Hb A and Hb S) from their subunits may differ because of difference in affinities of β^A and β^S for α chains. The latter, if in short supply, prefer to combine with normal β chains rather than with the variant chains,[39,40] the excess of which is removed by proteolysis.

Relationship between function and structure of hemoglobin

Hemoglobin and oxygen: the oxygen-dissociation curve. Hemoglobin is a protein composed of four subunits, each of which contains a heme moiety deep in the pocket of the globin chains, leaving one edge of the heme exposed to receive the oxygen. Heme is a complex made up of protoporphyrin and one atom of ferrous iron. Each of the four iron atoms can bind reversibly one oxygen molecule. As the iron stays in the ferrous form, the reaction is an oxygenation, not an oxidation.[41]

To fulfill its function as respiratory pigment, hemoglobin must specifically bind (have high affinity) to large quantities of oxygen, transport them, and unload them at the relatively high oxygen tension of tissues. Approximately 1.34 mL of oxygen is bound by each gram of hemoglobin. The tetrameric structure of hemoglobin is responsible for its unique oxygen-binding capacity and renders it physiologically superior to single hemoglobin subunits or to myoglobin. The heme iron has six valences, four of which are occupied by the four pyrrole rings of heme. The fifth iron valency bond attaches heme to globin, leaving the sixth iron valency for a reversible combination with oxygen or other ligands.[42]

The affinity of hemoglobin for oxygen varies with different partial pressures of oxygen (Po_2). A plot of the oxygen content (percent oxygen saturation) against Po_2 in case of myoglobin or hemoglobin subunits results in a hyperbolic oxygen dissociation curve, but a similar plot using hemoglobin gives a sigmoid curve (Fig. 33-13). The hyperbolic curve indicates appreciable release of oxygen at very low partial pressures only, whereas the sigmoid curve indicates release of oxygen much earlier at relatively high oxygen tension to allow adequate oxygenation of tissues. In the lung, at a Po_2 of about 95 mm Hg, arterial blood becomes 97% saturated with oxygen and carries 20 volumes of O_2 per equivalent blood volume. In the capillary bed, venous blood at a Po_2 tension of about 40 mm Hg is still about 75% saturated with oxygen but is nevertheless able to give up 46 volumes of oxygen per 1000 mL of blood. The 75% of hemoglobin returned to the lung in oxygenated form establishes a large reservoir for improved oxygen delivery to tissues. This reservoir is tapped when the oxygen dissociation curve, discussed below, is shifted to the right; thus the oxygen affinity of hemoglobin is decreased without the tissue oxygen tension being lowered. At the venous Po_2 of 40 mm Hg the dissociation curve of normal hemoglobin is steep, an indication of the release of large amounts of oxygen in response to small changes in Po_2. Its sigmoid shape is related to the fact that oxygenation of one heme group increases the oxygen affinity of the others, a phenomenon called the *heme-heme interaction* (or subunit cooperativity), which is responsible for the physiologically efficient uptake and release of oxygen. There is a progressive change in oxygen affinity as each heme molecule becomes oxygenated, the affinity for oxygen being low at first but increased as each heme molecule takes up oxygen. In the case of myoglobin, the hyperbolic

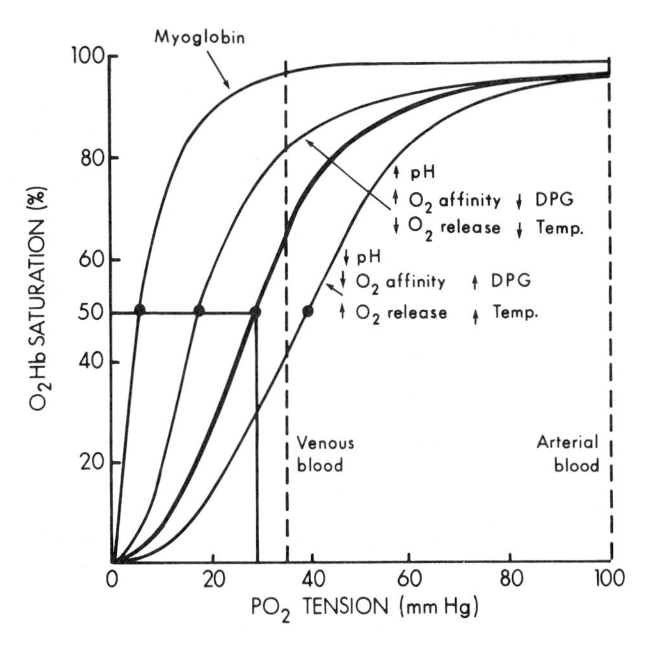

Fig. 33-13. Oxygen-dissociation curves of normal human hemoglobin. *Heavy middle line,* Dissociation curve of normal adult blood (temperature 37° C, pH 7.4, Pco_2 35 mm Hg). *Dots,* P_{50} values, partial pressure of oxygen (27 mm Hg) at which hemoglobin solution is 50% oxyhemoglobin and 50% deoxyhemoglobin. If temperature increases, pH decreases, or carbon dioxide tension (Pco_2) increases, the curve shifts to right. This shift increases release of oxygen from hemoglobin at given oxygen tension by decreasing its oxygen affinity. If temperature decreases, pH rises, or carbon dioxide tension decreases, the oxygen-dissociation curve moves to left. This shift increases oxygen-binding capacity of hemoglobin at given oxygen tension; thus there is a decrease in oxygen release. (From Bauer, J.D.: Clinical laboratory methods, ed. 9, St. Louis, 1982, The C.V. Mosby Co.)

dissociation curve indicates that each molecule is oxygenated independently.

The position of the oxygen dissociation curve is determined by a number of factors (see next column) affecting the affinity of hemoglobin for oxygen. It is conventionally indexed by the P_{50} value, the Po_2 at which the hemoglobin is 50% saturated with O_2, which normally occurs at a Po_2 of 27 mm Hg. The higher the P_{50}, the lower the affinity of hemoglobin for oxygen. The P_{50} value is derived from point-by-point measurement of fractional oxyhemoglobin levels over a range of partial oxygen pressures with the help of an oxygen dissociation analyzer. P_{50} serves as a measure of the oxygen affinity of hemoglobin and as an indicator of the position of the oxygen dissociation curve. A decreased P_{50} indicates a shift to the left and increased oxygen affinity of hemoglobin and an impaired oxygen release to tissues. It is seen (1) in high concentration of Hb F, the γ chain of which binds DPG (2,3-diphosphoglycerate) poorly, (2) in hemoglobin ligands, such as methemoglobin and carboxyhemoglobin, (3) in certain hemoglobin variants, such as Hb Ranier, and (4) after massive transfusions with DPG-depleted blood (Fig. 33-13). The shift to the right indicates a decreased oxygen affinity, which eases the delivery of oxygen to tissues. It is seen in various types of hypoxia, such as high altitude, severe anemia (Fig. 33-14), and heart and lung disease.[43,44]

Oxygen affinity and transport. Oxygen affinity and transport depend not only on Po_2 (see previous discussion of oxygen dissociation curve) but also on temperature,[41] pH (Bohr effect), and 2,3-diphosphoglycerate concentration.

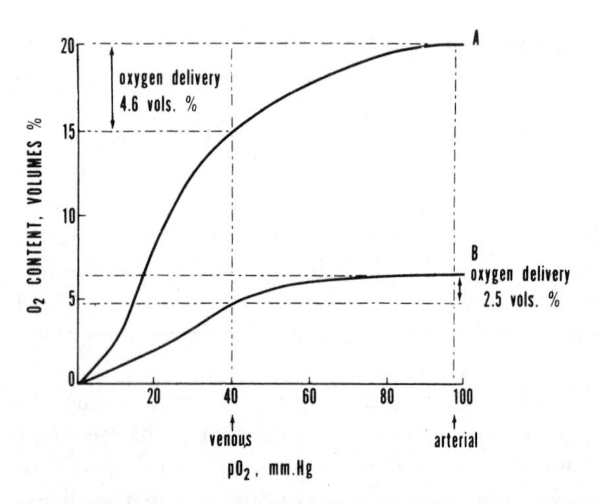

Fig. 33-14. Oxygen-dissociation curves of a normal subject (Hb 150 g/L), *A*, and of a subject with severe anemia (Hb 57 g/L), *B*, at pH 7.4, 37° C. The Po_2 of venous blood is taken as 40 mm Hg and that of arterial blood, as slightly less than 100 mm Hg. Difference between oxygen content of venous blood and arterial blood for each curve is amount of oxygen that can be delivered to tissues, roughly half of normal in anemia subject. (From Miale, J.B.: Laboratory medicine hematology, St. Louis, 1982, The C.V. Mosby Co.)

Bohr effect. The Bohr effect[4,5] expresses the fact that the oxygen affinity of hemoglobin varies with the pH. In the physiological pH range the affinity of hemoglobin for oxygen decreases as the acidity increases and the dissociation curve shifts to the right. The Bohr effect aids in the transfer of oxygen (and of CO_2) in the acid milieu of tissues in which carbon dioxide and acid metabolites accumulate. It is related to the fact that deoxyhemoglobin binds hydrogen (protons) more actively than oxyhemoglobin does because the latter is a stronger acid than deoxyhemoglobin. The preferential binding of protons to deoxyhemoglobin, the concomitant linkage to 2,3-DPG, and the increased CO_2 tension at tissue levels shift the dissociation curve to the right, decreasing the affinity of hemoglobin to oxygen. The uptake of hydrogen ions by hemoglobin aids in the buffering of the acid metabolites and in the stabilization of deoxyhemoglobin.

2,3-Diphosphoglycerate. Unlike other tissues in which only trace amounts are present, red cells contain large quantities of the organic ester 2,3-diphosphoglycerate (2,3-DPG), the concentration of which is equimolar with that of deoxyhemoglobin. Of the factors discussed that affect oxygen release (temperature, pH, Po_2, Pco_2), 2,3-DPG is the most important.[46] It is the most abundant glycolytic red cell intermediate, the result of the alternate metabolic pathway of glycolysis in red cells (the Rapoport-Leubering shunt, Fig. 33-15). Diphosphoglycerate mutase (DPGM) converts 1,3-diphosphoglycerate to 2,3-DPG, which in turn is converted to 3-phosphoglycerate by diphosphoglycerate phosphatase.[47] 2,3-DPG combines with deoxyhemoglobin, and by reducing the affinity of hemoglobin for oxygen it shifts the dissociation curve to the right. It acts as an allosteric modifier of oxygen dissociation; firmly binding to the β-chains bordering the central cavity of deoxyhemoglobin, thus stabilizing the deoxyhemoglobin. 2,3-DPG is ejected from oxyhemoglobin.[48] The rate of synthesis of 2,3-DPG depends on the concentration of 1,3-diphosphoglycerate and of phosphofructokinase. Its degradation to 3-phosphoglycerate is controlled by 2,3-phosphatase. Functionally, the 2,3-DPG concentration is closely related to the efficiency of oxygen transport and depends on the red cell pH and Po_2. When the pH drops, as in acidosis, the oxygen dissociation curve moves to the right, but the resulting inhibition of 2,3-DPG corrects the shift by an equal change to the left. When the red blood cell pH rises, DPG synthesis also rises. The elevated pH pushes the dissociation curve to the left, but the rising 2,3-DPG concentration shifts it to the right, returning it to the base position. The common denominator of the DPG–Bohr effect interaction is the rate of glycolysis,[49] which is stimulated by alkalosis and suppressed by acidosis[50] because the former stimulates phosphofructokinase activity and the latter suppresses it.[51]

The action of DPG on a molecular basis is explained as follows: the central cavity of hemoglobin flanked by the α and β chains in the transition from the oxy (R) to the deoxy (T) state (see discussion below) enlarges to accommodate one DPG molecule per hemoglobin tetramer. The binding of DPG to deoxyhemoglobin lowers the oxygen affinity of the tetramer, facilitating the release of oxygen.[52-54]

Molecular changes in oxygenation and deoxygenation

Oxygen transport. The sequence of molecular changes in oxygenation is as follows: oxygen is added to the first α chain, followed by rupture of salt bridges and changes in the tertiary configuration in the area of the heme pocket. Oxygen is then added to the second α chain and sequentially to the β chains. With each addition salt bridges disintegrate and tertiary changes occur in the subunits. After the third oxygenation the quaternary changes occur, the molecule is fixed in the R form, 2,3-DPG is expelled, and the salt and hydrophobic bonds at the $\alpha_1\beta_2$ plane are broken. The oxygen affinity progressively increases and oxygen is bound to the β chains[55] (Fig. 33-16).

Carbon dioxide transport. The solubility of carbon dioxide in blood is about 20 times that of oxygen. Only about one fourth of the CO_2 load combines with amino groups of deoxyhemoglobin (not with heme and not with oxyhemoglobin) to form carbaminohemoglobin (Hb-NH-COO^-). The remaining CO_2 diffuses into the red cells where, with the aid of carbonic anhydrase, it is hydrated to H_2CO_3. The latter dissociates to H^+ and HCO_3^-. The liberated hydrogen is accepted by the deoxyhemoglobin.[41] Increased CO_2 concentration lowers the oxygen affinity of oxyhemoglobin, encouraging it to release oxygen (Bohr effect).[56]

Hemoglobin and its derivatives

Oxyhemoglobin. The prime role of hemoglobin, that is, to carry oxygen, results in two physiological forms: oxyhemoglobin and deoxyhemoglobin. The oxygen-binding capacity of hemoglobin depends on (1) the type of hemoglobin (adult Hb P_{50} is 25 to 28 mm Hg, newborn and infant Hb [high Hb F] P_{50} is 18 to 24 mm Hg), (2) the partial pressure of oxygen, (3) the temperature, (4) the pH of blood (Bohr effect), (5) protons, (6) concentration of CO_2, and (7) concentration of 2,3-DPG (see preceding sections).

Deoxyhemoglobin. Tissue metabolism is responsible for the production of carbon dioxide (CO_2), which lowers the pH and shifts the oxygen dissociation curve to the right so that oxygen release is facilitated and oxyhemoglobin is converted into deoxyhemoglobin[65] (see the discussion of CO_2 transport).

Carboxyhemoglobin. Carbon monoxide (CO) is a ligand that, like oxygen, binds reversibly to the ferrous ion of hemoglobin but forms a toxic compound carboxyhemoglobin (Hb CO). It also binds other heme-containing proteins, such as myoglobin, cytochrome P-450, and cytochrome oxidase.[58] The affinity of hemoglobin for CO is 218 times greater than that for oxygen.[59] Because both CO and O_2 compete for the same heme-binding sites, CO displaces O_2, reduces the concentration of oxyhemoglobin,

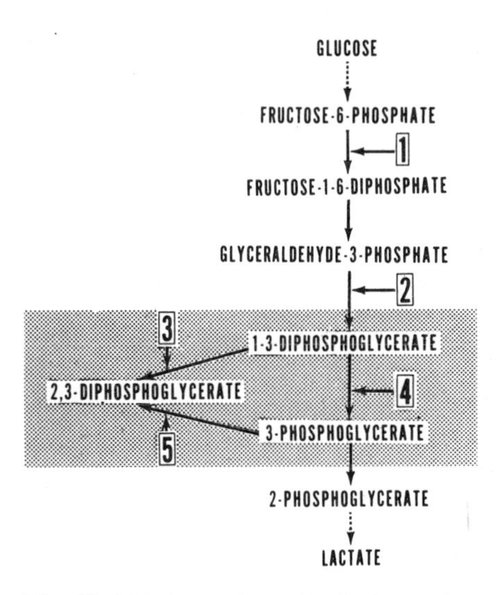

Fig. 33-15. The Rapoport-Luebering pathway for synthesis and degradation of 2,3-DPG *(reactions in shaded area)*. Pertinent enzymes are numbered: *1,* phosphofructokinase (PFK); *2,* glyceraldehyde-3-phosphate dehydrogenase (GAPDH); *3,* diphosphoglycerate mutase; *4,* phosphoglycerate kinase (PGK); *5,* 2,3-diphosphoglycerate phosphatase. (From Miale, J.B.: Laboratory medicine hematology, St. Louis, 1982, The C.V. Mosby Co.)

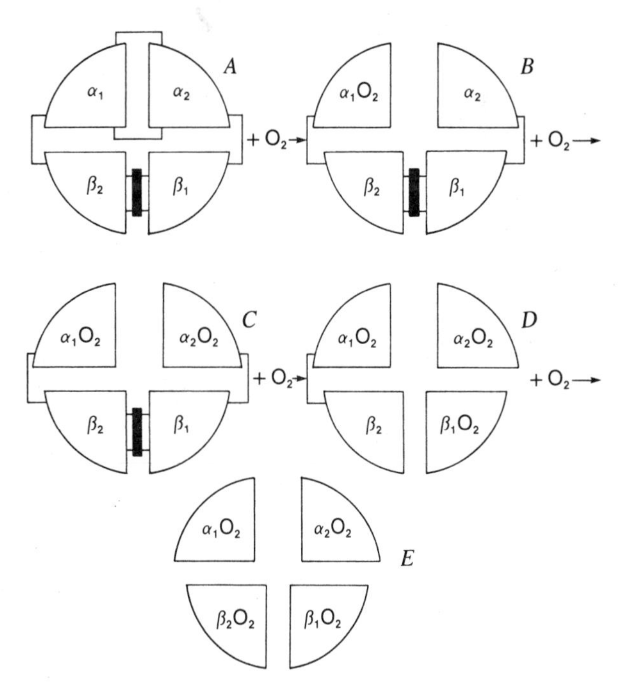

Fig. 33-16. Diagram of oxygenation of hemoglobin molecule. *A,* Deoxyhemoglobin; *B, C, D,* partially liganded hemoglobins; *E,* oxyhemoglobin; [, salt bonds; ▌, 2,3-DPG. Salt bonds are sequentially broken, and 2,3-DPG is expelled.

and prevents the formation of deoxyhemoglobin. At a CO concentration of 0.1% in the inhaled air more than 50% hemoglobin is not available for O_2 transport. CO combines with hemoglobin more slowly than oxygen does, but the union is much firmer and the release of CO is 10,000 times slower than the release of oxygen from O_2-Hb. In the presence of CO even the oxyhemoglobin dissociates more slowly because the iron atoms not bound to CO have a higher affinity for O_2, causing the oxygen dissociation curve to shift to the left. CO binds to heme in a manner similar to oxygen. It binds to the α chains first and then to the β chains (or other non-α chains).[58] Identical to oxygen, the degree of CO-heme binding depends on the partial pressure of CO.[58]

Methemoglobin. Methemoglobin (Met-Hb) is oxidized hemoglobin (oxy or deoxy forms) in which the ferrous ion of hemoglobin has been oxidized to the ferric state to form ferrihemoglobin. More correctly, methemoglobin is defined as an oxidation product of hemoglobin in which the sixth coordination position of ferric heme is in the acid form bound to a water molecule (brown color) or in the alkaline form to a hydroxyl group (dark red color).[59-61] Because of the loss of electrons from the heme ions, methemoglobin cannot bind oxygen reversibly and is unable to act as an oxygen carrier. If present in high enough concentrations (over 30% of total hemoglobin), it is responsible for hypoxia and cyanosis (methemoglobinemia).

Normally, after prolonged standing (auto-oxidation) oxyhemoglobin turns brown because of methemoglobin formation, which is also responsible for the brown color of blood in acid urine.

Physiologically, methemoglobin is continuously being formed within erythrocytes because of spontaneous oxidation of hemoglobin but is prevented from accumulating within the red cells by the reduction of oxidized heme by a number of enzyme systems that restrict its normal concentration to less than 1% of total hemoglobin.

Clinically, patients with hereditary methemoglobinemia have erythrocytosis and slate gray cyanosis since birth, not associated with cardiopulmonary disease.[62] Methemoglobin concentrations of 10% to 20% of total hemoglobin produce cyanosis but no other ill effects. Thirty to 40% methemoglobin may be responsible for headache and dyspnea, and concentrations of 70% and over may be fatal. Jaffe et al.[63] reported a low incidence of mental retardation and early death in cases of methemoglobinemia caused by NADH methemoglobin reductase deficiency, a disease that lends itself to antenatal diagnosis.[64]

The M hemoglobins are variant hemoglobins that are characterized by amino acid substitutions in the α or β chains at or near the heme pocket. These substitutions and structural globin alterations affect the hemoglobin bonds and the oxygen affinity of the M hemoglobins. In all M hemoglobins the amino acid substitutions firmly stabilize the heme groups, maintain the iron in the ferric state, have only two oxygen-binding sites per hemoglobin tetramer,[65] and resist the normal reductive systems of the red cells.[66] The β-chain substitutions have no effect on the oxygen affinity, but the α-chain substitutions lower it and eliminate the Bohr effect.[65]

The M hemoglobins—like all hemoglobin variants—show a recessive inheritance pattern and unlike many hemoglobinopathies do not produce hemolytic anemias. If only the heterozygous form is encountered—the homozygous form being incompatible with life—only one pair of globin chains (either α or β) is affected and the methemoglobin level does not exceed 25% to 30%. If α chains are involved, the cyanosis may be present at birth, whereas β-chain substitutions are responsible for cyanosis in the later months[64] because of the later appearance of these chains.

Hemichromes. When the oxidation of hemoglobin to methemoglobin continues, methemoglobin is converted to compounds called ''hemichromes,'' which ultimately precipitate to form *Heinz bodies,* which are responsible for the lysis of the affected red cells.[67,68] Hemichromes are greenish ferric compounds with a characteristic absorption spectrum. The first stage in the denaturation of methemoglobin and in the formation of Heinz bodies is an abnormal bond between the ferric iron and globin, a change that is followed by an alteration of the tertiary globin structure that leads to the formation of the reversible hemichrome 1. It is designated as reversible because under anaerobic conditions it converts to deoxyhemoglobin when treated with reducing agents.[64] Further distortion of the subunit results in the formation of the irreversible hemichrome 2, which is followed by the precipitation of hemoglobin as Heinz bodies, as seen in unstable hemoglobin disorders and in thalassemias.[70] These steps are as follows:

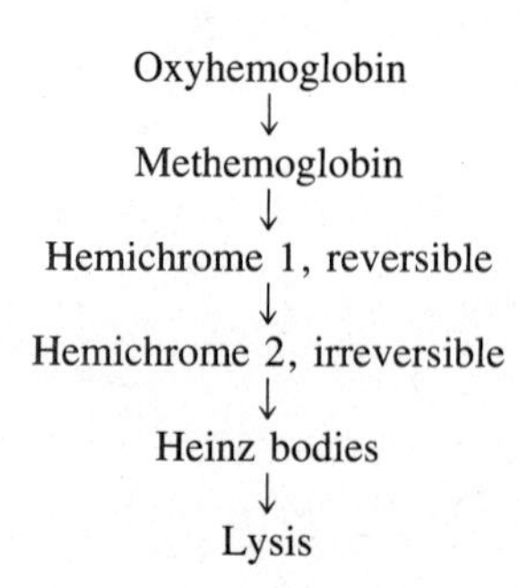

Sulfhemoglobin. Sulfhemoglobin (S-Hb) is a stable compound resulting from the linkage of sulfur to hemoglobin. The toxic effects of certain drugs on hemoglobin not only lead to the formation of methemoglobin but also to the concomitant S-Hb production.[71] Sulfhemoglobinemia appears in some persons after exposure to sulfonamides, phenacetin, acetanilid,[72] and trinitrotoluene (TNT). The structure of S-Hb is unknown, but sulfur is probably linked to heme.[73] The S-Hb complex is stable and irreversible (thus differing from the reversible Met-Hb) and does not

Table 33-3. Normal human hemoglobins

Designation	Tetrameric structure	Hemolysate (%) Adult	Newborn
Adult			
Hb A	$\alpha_2\beta_2$	95-98	20-30
Hb A_2	$\alpha_2\delta_2$	2-3	0.2
Fetal			
Hb F	$\alpha_2\gamma_2$	<1	80
Embryonic			
Gower 1	$\zeta_2\epsilon_2$	0	0
Gower 2	$\alpha_2\epsilon_2$	0	0
Hb Portland	$\zeta_2\gamma_2$	0	0

disappear from the circulation until the involved red cells complete their life cycle. Sulfhemoglobinemia produces anoxia and cyanosis, which clinically is indistinguishable from that of methemoglobinemia.

Normal human hemoglobins

Molecular structure. Adult and fetal hemoglobins contain α subunits linked to β, γ, or δ chains. In the embryo the ζ and ε chains are their precursors (Table 33-3). Hemoglobin is a mixture of variants, some genetically controlled and some acquired. The genetically controlled variants differ from each other in the structure of their globin chains (Table 33-3).

Hemoglobin A ($\alpha_2\beta_2$). Hemoglobin A makes up the major portion (95% to 98%) of the adult hemolysate. Small amounts of Hb A are produced in the last six weeks of fetal life (Fig. 33-17), the remainder being Hb F. During the ensuing 6 to 12 months of the infant's life Hb A reaches adult concentrations and is accompanied by only traces of Hb F and A_2. One pair of polypeptide chains of Hb A is composed of α chains and the other of β chains, so that the complete molecule has the $\alpha_2\beta_2$ formula.

Hemoglobin A_{1c}. Using cation-exchange resin chromatography, Allen et al.[74] demonstrated the heterogeneous nature of Hb A, 90% of which is made of slower-moving Hb A and the remainder, of faster-moving minor components, Hb A_{1a}, and A_{1c}, which are collectively referred to as the *fast hemoglobins*.[75] Postsynthetic nonenzymatic glycosylation of the terminal amino group (valine) of the β chain is responsible for the production of Hb A_{1c}.[76,77] Variations of this hemoglobin are the adduct of glucose-6-phosphate[78] or of fructose-1,6-diphosphate to the β chain.[76]

Hemoglobin A_3. Hb A_3 is demonstrated by starch block electrophoresis. It is probably a storage artifact caused by complexing of glutathione with hemoglobin.[79]

Hemoglobin A_2. Hb A_2 contains a pair of α chains and a pair of δ chains, so that its molecular structure is $\alpha_2\delta_2$. It is a minor component of hemoglobin that makes its first appearance close to term (0.2% of cord blood hemolysate)[119] and remains at low concentration (2.5%) throughout adult life. Its exact function is unknown but is probably similar to that of Hb A.[80] Its rise and fall in a number of hematological disorders gives its concentration differential diagnostic importance. Its concentration is increased in β-thalassemia and in β-chain unstable hemoglobins. Normal or decreased values are seen in α-thalassemias and in hemoglobin Lepore heterozygotes.[81] In hemoglobin Lepore homozygotes, Hb A_2 is absent because there is no δ chain synthesis.[82] Hb A_2 is decreased in iron deficiency and in lead poisoning.

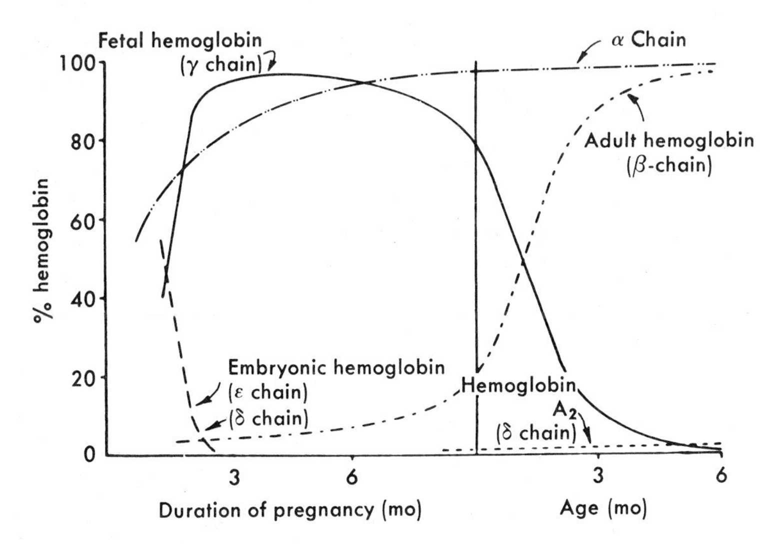

Fig. 33-17. Developmental changes in human hemoglobins. α-Chain synthesis begins well before third month of fetal life and continues into adult life. γ-Chain synthesis replaces embryonic ε-chain at about third month, and at about same time β-chain synthesis begins. (From Bauer, J.D.: Clinical laboratory methods, ed. 9, St. Louis, 1982, The C.V. Mosby Co.)

Fetal hemoglobin (Hb F). Hemoglobin F is the major hemoglobin of fetal life, preceded by the embryonic hemoglobins Gower I, Gower II, and Portland. Its molecular formula is $\alpha_2\gamma_2$. It is a mixture of two molecular species in which the γ chains have either glycine ($^G\gamma$) or alanine ($^A\gamma$) at position 136.[83] At birth the Hb F $^G\gamma/^A\gamma$ ratio is about 3:1, whereas in the normal adult hemoglobins the $^G\gamma/^A\gamma$ ratio of the small amount of Hb F (less than 1%) is 2:3. In the first months of fetal life Hb F competes with the Gower hemoglobins, which are replaced by Hb F at the end of the second month at which time the fetal hemoglobin concentration is about 90%. It remains at this level until just before term. At birth the red blood cells contain about 70% to 90% Hb F, though higher concentrations have been reported.[84] After birth, Hb F decreases rapidly to about 50% to 70% at the end of the first month, to 25% to 60% at the end of the second month, and to 10% to 30% at the third month. After 6 months and in the first year of life, the Hb F concentration falls from 8% to 2%, in the second year to 1.8%, in the third year to 1.0%, and it finally levels off to the adult level of less than 0.4% (see Fig. 33-17),[85] a level that is not detectable by routine laboratory methods.

Characteristics and function of Hb F

1. Electrophoretically it is slower than Hb A.
2. It resists alkali denaturation, a feature that is the basis of the Singer test for Hb F.[86]
3. It is twice as resistant to acid elution as Hb A, a characteristic that forms the basis of the Kleihauer elution technique.[87]
4. It is oxidized to Met-Hb twice as fast as Hb A, so that the newborn readily develops cyanosis as a result of methemoglobin concentration of 15 to 20 g/L blood in response to oxidant chemicals and drugs (see the discussion of methemoglobin).
5. It has a higher oxygen affinity than Hb A, since it binds 2,3-DPG to a lesser degree than adult hemoglobin because of its γ chain. This characteristic allows the oxygen transport across the placental villi, despite their low oxygen concentration (80%).[88]

Hb F shows three distribution patterns: (1) both Hb A and Hb F are present in strictly separated cell populations, (2) both hemoglobins are present in equal concentrations in all cells, and (3) both hemoglobins are irregularly distributed within the cells, some cells containing only Hb A, some only Hb F, and some varying amounts of both.

1. Hb A and F present in strictly *separate cell populations:* This distribution pattern is seen in fetal-maternal hemorrhage if the mother's blood is examined or in maternal-fetal hemorrhage if the infant's blood is examined.
2. *Even distribution* of Hb F and A within red cells: This distribution is seen in hereditary persistence of Hb F.
3. *Uneven distribution* of Hb A and Hb F in the red cells: This pattern is seen in thalassemia, S-S disease, Fanconi anemia, and hereditary spherocytosis.

Embryonal hemoglobins. The embryonal hemoglobin is a mixture of hemoglobins Gower I ($\zeta_2\epsilon_2$), Gower II ($\alpha_2\epsilon_2$), Portland ($\zeta_2\gamma_2$), and Hb F ($\alpha_2\gamma_2$). The first two hemoglobins are not detectable after the first months of gestation,[85] whereas small amounts of Hb Portland may persist into fetal life.[89] The embryonic hemoglobins are able to combine with oxygen at the low oxygen tension and low pH of interstitial fluid.[90] They are detectable in red cells by a modification of the Kleihauer method for Hb F.

Hemoglobin switching during ontogeny. The above discussion of the normal hemoglobins in humans points out the fact that the various globin chains (α, β, γ, δ, ϵ and ζ) appear at different stages of development. The changing of hemoglobin phenotypes during ontogeny is called "hemoglobin switching." In humans there is an early transition from embryonic to fetal hemoglobin followed by a later perinatal transition from fetal hemoglobin to adult hemoglobin. The understanding of hemoglobin switching and its timing aids in the elucidation of hemoglobinopathies and of the thalassemia syndromes. Defects in the α or γ chains will manifest themselves at birth, but β chain defects will remain hidden in the first 6 months of life unless very sensitive methods are used for their detection.

PATHOLOGY

Hemoglobinopathies

The inherited disorders of hemoglobin, the hemoglobinopathies, are genetic disorders of structure and synthesis of one or more of the globin polypeptide chains. Inherited disorders of the heme moiety are not known. One can divide hemoglobinopathies into a number of overlapping groups:[91,92] (1) the structural hemoglobin variants that involve substitution, addition, or deletion of one or more amino acids of the globin chain; (2) the thalassemias, a group of disorders in which there is a quantitative defect in globin chain production; (3) combinations of types 1 and 2 that result in complex hemoglobinopathies; and (4) hereditary persistence of fetal hemoglobin, an asymptomatic disorder.

Structural hemoglobin variants

Nomenclature. Hemoglobin variants are assigned letters of the alphabet (such as *S, C,* and *E*), places of discovery (such as Hb D-Los Angeles, Hb-Köln and Hb C-Harlem), and names of families in which they were first discovered (such as Hb Lepore). A more logical nomenclature is based on the position of the amino acid substitution, such as Hb S = Hb $\alpha_2\beta_2^{6Val}$. Recent recommendations indicate elimination of the superscript and the normal chain but designation of the affected chain, the helical position, and the substitution, such as Hb S =

Table 33-4. Clinical manifestations associated with some abnormal hemoglobins

Disorder	Abnormal Hb	Structural change	Comments
Hemolytic anemia	H	$alpha_2beta_2 \rightarrow beta_4$	Unstable hemoglobin occurring in some forms of alpha-thalassemia; precipitation of hemoglobin and hemolysis are accelerated by certain drugs
	S	beta 6 glu → val	Forms molecular aggregates when deoxygenated, producing sickle cell anemia in homozygotes
	C	beta 6 glu → lys	Low solubility lessens plasticity of red cells, causing hemolytic anemia in homozygotes.
	D_{Punjab}	beta 121 glu → gln	Mechanism unknown
	E	beta 26 glu → lys	
	Zurich	beta 63 his → arg	Unstable hemoglobin precipitated by certain drugs, producing hemolytic anemia in heterozygotes
	Köln	beta 98 val → met	
	Sydney	beta 67 val → ala	
	Santa Ana	beta 88 leu → pro	
	Philly	beta 35 tyr → phe	Unstable hemoglobin causes congenital nonspherocytic hemolytic anemia in heterozygotes; precipitated hemoglobin tends to form inclusion bodies within red cells, under certain conditions
	Gun Hill	beta deletion of 5 residues between 90 and 96	
	Wien	beta 130 try → asp	
	Torino	alpha 43 phe → val	
	Bibba	alpha 136 leu → pro	
	Hammersmith	beta 42 phe → ser	
	Genova	beta 28 leu → pro	
	Sabine	beta 91 leu → pro	
	Boras	beta 88 leu → arg	
Cyanosis due to methemoglobinemia	M_{Boston}	alpha 58 his → tyr	
	M_{Iwate}	alpha 87 his → tyr	Methemoglobin causes cyanosis in heterozygotes; some also have evidence of hemolytic anemia
	$M_{Milwaukee}$	beta 67 val → glu	
	$M_{Saskatoon}$	beta 63 his → tyr	
	$M_{Hyde\ Park}$	beta 92 his → tyr	
	Freiburg	beta 23 val deleted	
Cyanosis due to increased deoxyhemoglobin	Kansas	beta 102 asn → thr	Decreased oxygen affinity of hemoglobin causes cyanosis in heterozygotes
Polycythemia	$J_{Capetown}$	alpha 92 arg → gln	
	Chesapeake	alpha 92 arg → leu	Increased oxygen affinity of hemoglobin hinders release of oxygen to tissues, causing compensatory polycythemia in heterozygotes
	Yakima	beta 99 asp → his	
	Kempsey	beta 99 asp → asn	
	Rainier	beta 145 try → cys	
	Hiroshima	beta 146 his → asp	
Hydrops fetalis	Bart's	$alpha_2gamma_2 \rightarrow gamma_4$	Unstable hemoglobin with high oxygen affinity occurring in high concentration in stillborn fetuses with homozygous alpha-thalassemia

From Schmidt, R.M., and Brosious, E.M.: Basic laboratory methods of hemoglobinopathy detection, Atlanta, Ga., 1978, Centers for Disease Control.

Hb β6(A3) Glu-Val.[93] There are about 500 variant hemoglobins identified at present (Table 33-4).

Excellent tables dealing with various aspects of hemoglobin variants, such as nomenclature, molecular structure, clinical manifestations, and electrophoretic mobility are found in reference 93.

Classification. Hemoglobin variants are classified according to (1) the molecular mechanisms responsible for the abnormal hemoglobin (or thalassemia), (2) clinical and functional manifestations, and (3) their electrophoretic behavior.

Molecular mechanisms responsible for structural hemoglobin variants. Five basic types of structural changes are responsible for most hemoglobin variants. They are (1) amino acid substitution, (2) deletions and insertions, (3) unequal crossing over (fusion genes), (4) chain elongation, and (5) frame shift variance.[92]

Clinical consequences of structural alterations of hemoglobin molecule

Structural alterations of the hemoglobin molecule are responsible for a wide range of clinical manifestations. Most mutations are asymptomatic because they do not interfere with hemoglobin function. Others produce disease because they affect the stability, shape, or function of the hemoglobin molecule. A person homozygous for an abnormal hemoglobin may have striking clinical manifestations (such as sickle cell anemia—Hb S–Hb S), whereas a person heterozygous for the abnormal hemoglobin (Hb A–Hb S) may be asymptomatic. Even in the homozygous state, some hemoglobin variants (Hb C, Hb D, Hb E) produce only mild symptoms, whereas others are responsible for almost specific pathophysiological changes, such as cyanosis and erythrocytosis.

The clinical disorders can be grouped as follows[92] (Table 33-4):

1. *Hemolytic anemias*. Hb S and Hb C alter the shape and deformability of the red cells and may form intraerythrocytic crystals. Unstable hemoglobins are responsible for intraerythrocytic Heinz body inclusions. The affected cells are prematurely destroyed in the spleen; thus their life span is greatly shortened.
2. *Cyanosis*. Hemoglobins M are responsible for methemoglobinemia caused by amino acid substitution near the heme pocket.
3. *Erythrocytosis*. Amino acid substitution causes high oxygen affinity and tissue hypoxia, with the latter being responsible for erythropoietin stimulation. Hb Chesapeake, Rainier, and Ypsilanti fall in this group.
4. *Hypochromic anemias*. The mutation reduces the hemoglobin output. Examples are Hb Lepore and Constant Spring.

Sickling disorders: sickle hemoglobin

The sickling disorders embrace the homozygous form of the sickle cell gene, sickle cell anemia, the heterozygous form of the sickle cell gene, sickle cell trait, and the combination with other structural hemoglobin variants or with various types of thalassemia.

Sickle hemoglobin (Hb S) is the result of the substitution of valine for the normally occurring glutamine residue at the 6 position of the β chain, and so its structural formula is Hb $\beta^{6(A_3)Glu\text{-}Val}$. The result of this substitution is the formation of sickled cells by oxygenated Hb S.[94]

Hb S is not the only hemoglobin that sickles, but it is the most important one. In America and Africa, Hb S is the most common hemoglobin variant, with an incidence of approximately 8% in American blacks (heterozygous form) and 30% in African blacks. The mutation probably originated in Central Africa and spread to countries bordering the Mediterranean Sea so that it can also be found in nonblack inhabitants of these areas. In Africa the high frequency of the sickle cell gene has persisted because heterozygotes for Hb S are somewhat protected from malaria because the *Plasmodium* fails to grow in Hb S and also leads to premature destruction of the affected cell, the Hb S of which sickles readily because of the intraerythrocytic acidosis.[95]

Molecular mechanism of sickling. Deoxygenated Hb S tetramers aggregate to form rodlike fibers that align and are responsible for an intraerythrocytic phase change from solute to gel, which is designated as *gelation*.[95-97] The fibers align in bundles that force the sickle shape on the red cell and cause it to be less deformable, thus condemning it to premature destruction. The critical process, the gel formation, is dependent on (1) the concentration of deoxyhemoglobin S, favored by a high concentration, (2) the endothermic temperature (37° C is optimal),[98] (3) an optimal pH of 6.8, so that acidosis favors gelation,[98] (4) a high concentration of 2,3-DPG,[99] (5) the ionic strength (high salt concentrations inhibit gelation),[98] and (6) the state of oxygenation (oxygenation of Hb S prevents gel formation).[97] Sickling is reduced by a lowered concentration of Hb S, by a lowered mean corpuscular hemoglobin concentration (MCHC), by the presence of other hemoglobins (Hb A, F, and C), and by increased oxygen tension. All these points are of clinical significance (Table 33-5). The mixture of Hb S, A, F, and C produce a clinically milder form of sickle cell disease.[100] The interference of high concentrations of Hb F with sickling must be kept in mind if newborns and patients with persistence of Hb F are tested for Hb S by standard clinical methods. The admixture of Hb D and O is responsible for a clinically more severe disease because these hemoglobins strengthen the intermolecular binding sites of deoxyhemoglobin S.[101] Sickling depends on the reduction of the oxygen tension beyond a critical point. In homozygous sickle cell anemia, sickling follows a fall of oxygen saturation to 85% and is complete at 38% saturation. In the heterozygous sickle cell trait, sickling does not occur until the oxygen saturation falls to 40%.[91] Normal oxygen tension reverses the sickling process unless repeated sickling cycles generate permanently sickled cells.

Table 33-5. Varying clinical severity of the different sickle syndromes

Genotype	% of hemoglobin S	% of non-S hemoglobin	Clinical severity
SA	30-40	60-70 (A)	0
SF*	70	30 (F)	0
SS	80-90	5-15 (F)	+ +/+ + + +
S-thalassemia	80	20 (A + F)	+/+ + +
SC	50	50 (C)	+/+ + +
SO.SD	30-40	60-70 (O.D)	+ +/+ + + +

*Double heterozygous state for hemoglobin S and hereditary persistence of fetal hemoglobin.

From Bunn, H.F.: Sickle cell anemia and other hemoglobinopathies. In Beck, W.S., editor: Hematology, Cambridge, Mass., 1981, MIT Press.

Pathophysiology of sickle cell disease. Hb S is inherited as an autosomal codominant trait (see Fig. 33-11). With certain exceptions, the clinical course of the patient with Hb S depends on the concentration of Hb S, the biological behavior of which explains most clinical phenomena.

The sickle-shaped cells temporarily or permanently block the microcirculation and the resulting stasis leads to hypoxia and ischemic infarcts of various organs.[97] The vascular endothelial lining, damaged by lack of oxygen, attracts platelets, which initiate the process of diffuse intravascular coagulation (DIC).[102]

Sickled cells lack the deformability of normal cells because the polymerization of deoxyhemoglobin S produces a membrane defect that greatly shortens the life-span of the affected cells. The ensuing hemolytic anemia is augmented by the inability of the bone marrow to respond adequately to the anemia because of ineffective erythropoiesis.[65] The hemolysis is responsible for hyperbilirubinemia, reticulocytosis, bone marrow erythroid hyperplasia, gallbladder pigment stones, and osteoporosis as a result of the expanding bone marrow.

Sickle cells exhibit oxygen-transport abnormalities.[103] In sickled cells the oxygen dissociation curve is shifted to the right. The resulting decreased oxygen affinity favors the release of oxygen at higher oxygen tensions but also supports the formation of deoxyhemoglobin and sickling.[91] The shift to the right of the oxygen equilibrium is caused by an elevated 2,3-DPG concentration[104] and to a Hb S polymerization–mediated effect.[105] The fall in oxygen affinity is also related to the mean corpuscular hemoglobin (MCH)[106] and to the pH, with acidosis favoring an increase in deoxy–Hb S crystals.[98]

Sickle cell trait (Hb AS). About 8% of American blacks (30% of Central African blacks) are heterozygous for Hb S because they inherit Hb S gene (β^S) from one parent and normal Hb A (β^A) from the other.

Persons with sickle cell trait are usually asymptomatic and have a normal hemogram and red cell survival. The demonstration of Hb S is of no clinical significance but should suggest genetic counseling. There are rare reports of sickling complications in AS patients. They include (1) spontaneous hematuria in about 3% of patients and more frequently hyposthenuria because of the impairment of the concentrating power of the kidneys, both signs pointing to sickling within the vessels of the medulla,[107] (2) rupture of the infarcted spleen,[108] (3) sudden death after strenuous exercise at high altitude,[109] (4) sickling crisis in flight,[110] and (5) rarely, proliferative retinopathy.[111]

Sickle cell anemia (Hb SS). Sickle cell anemia is a chronic, moderate-to-severe hemolytic anemia in a person homozygous for Hb S, having inherited the Hb S gene from both parents, a fact that may have to be ascertained to differentiate sickle cell anemia from sickle cell–β thalassemia or Hb S–hereditary persistence of Hb F. The disease, a β chain variant, is not evident at birth and does not manifest itself until the α chains of the newborn are replaced by β^S chains after 3 to 6 months of life.

The clinical manifestations of Hb SS disease vary from patient to patient. There are a few fairly asymptomatic patients who contrast with those whose lives are dotted with constant painful crises. The pathophysiological consequences of Hb SS can be summarized under (1) the hemolytic anemia, and (2) the sickle cell crisis, both the result of the polymerization of Hb S within the red cells. Variations in the clinical presentation of Hb SS disease are the result of the influence of genes other than the β^S genes, which affect hemoglobin formation.[112]

The hemoglobin values hover around 70 to 80 g/L accompanied by a greatly elevated reticulocytosis (10%). The hemoglobin electrophoretogram shows absence of Hb A (no β^A chains), 80% to 95% Hb S, 2% to 4% Hb A_2, and 2% to 20% Hb F. Detailed hematological findings are beyond the scope of this presentation, but three outstanding biochemical findings should be mentioned: hyperuricemia in patients with altered tubular function,[113] reduced zinc levels in plasma, red cells, and hair,[114] and high LDH levels in patients in crisis.

Sickle cell–Hb C disease. Sickle cell–Hb C disease has a relatively high incidence (1:833 births among blacks in the United States) because Hb S and Hb C are the most commn hemoglobin variants.[115] The patient inherits one hemoglobinopathy from each parent and the resulting disease is a mild to moderate hemolytic anemia associated with the same vaso-occlusive complications as seen in Hb SS disease, but usually of lower frequency. Since the genes are allelic β-chain mutations, no normal β chains are formed and Hb A is absent.

The smear of the peripheral blood shows many target cells, rare sickle cells, and red cells with straight or curved, free and intracellular hemoglobin crystals. The hemoglobin values range from 100 to 130 g/L. The reticulocyte count varies from 3% to 10%. The alkaline cellulose acetate electrophoretogram (pH 8.6) shows only Hb C and Hb S bands. Hb A is absent. Hb A_2 is normal, but its band is hidden by the Hb C band. Hb F may be slightly elevated (1% to 2%). At alkaline pH, Hb O has the same mobility as Hb C but can be separated by citrate agar electrophoresis. Family studies confirm the diagnosis.

Hb C trait and disease

Hb C trait (Hb AC). Hb AC disease affects about 3% of American blacks. It is asymptomatic and the peripheral smear is normal, except for a few target cells. Electrophoresis patterns show about 30% to 40% Hb C, 50% to 60% Hb A, 3% to 4% Hb A_2, and 1% Hb F.

Hemoglobin C disease (Hb CC). The homozygous form is rare, occurring in 1 out of 10,000 American blacks[116] and is asymptomatic.

Unstable hemoglobin disorders

The unstable hemoglobins are structural variants of Hb A, in which the mutant hemoglobin is less stable than normal hemoglobin. There are about 8 unstable hemoglobins described, representing the largest single group of hemoglobin variants.[117] According to White,[118] the stability of the hemoglobin molecule depends on a number of structural factors: (1) the heme-contact areas, (2) the structure of the α helices,[119] (3) the nonpolar amino acids of the heme pocket, and (4) the stability of the $\alpha_1\beta_1$ contacts. If any of these structures is disturbed, hemoglobin instability results, which is responsible for a hemolytic anemia of varying intensity and for Heinz body formation. It must be emphasized that not all unstable hemoglobins give rise to hemolytic anemias and to Heinz bodies.[117] White's criteria exclude unstable hemoglobins such as Hb E and H. Hemoglobin instability may result from a number of mutations.[91,119,120]

The molecular distortion responsible for unstable hemoglobins produces a series of pathophysiological effects that one can evaluate by laboratory methods though they are not equally expressed by all unstable hemoglobins. They are (1) hemolytic anemia, (2) increased methemoglobin and sulfhemoglobin production, (3) hemichrome formation, (4) inclusion (Heinz) body formation, (5) altered oxygen dissociation, (6) drug sensitivity, (7) altered electrophoretic mobility (rare), (8) altered response to hemoglobin stability tests, and (9) passage of dark urine.

There are two levels of investigation. The initial phase includes electrophoresis at pH 8.6 and quantitation of hemoglobin F and A_2. The results of the latter two tests must be carefully interpreted because underlying diseases (cancer) and hematological disorders (leukemia, iron deficiency) may significantly influence their concentrations. The second phase might include tests for sickling hemoglobins and for unstable hemoglobins, chain separation, oxygen dissociation, peptide mapping, and quantitation of methemoglobin. Most unstable hemoglobins are not identifiable by hemoglobin electrophoresis but require tests that reveal their unstable bonding and are based on their abnormal response to thermal and oxidative stresses. Such tests are the heat denaturation and the isopropanol solubility test.[119]

The deeply pigmented urine is caused by mesobilifuscin, a dipyrrole, derived from the catabolism of Heinz bodies or free heme.[121]

Thalassemias

Definition. Thalassemias are inherited hemoglobinopathies resulting from a decreased rate of production of one or more of the globin chains of hemoglobin.[91,122-124] They are quantitative hemoglobinopathies that differ from the qualitative hemoglobinopathies by the fact that the structure of the affected globin chain (or chains) is normal but its synthesis is reduced or absent. The decreased hemoglobin synthesis results in decreased red cell hemoglobin, hypochromia, microcytosis, and a variable hemolytic component.

Defective synthesis of one set of globin chains results in excess production of the unaffected pair,[91] which precipitates in the red cells in form of inclusion bodies, resulting in hemolysis (imbalanced globin-chain synthesis).

Classification. The classification of thalassemias in thalassemia major, intermedia, minor, and minima describes the clinical severity of the disorder and disregards the genetic makeup. The preferred genetic classification is based on the particular polypeptide chain that is deficient. In α thalassemia the synthesis of α chains is diminished, in β thalassemia the synthesis of β chains is diminished, and so on.

Thalassemias involving γ, ϵ, or ζ genes may lead to fetal or embryonic death.[124] A thalassemia-like condition that is asymptomatic is the hereditary persistence of fetal hemoglobin (HPFH). The main forms of thalassemia are classified in the following box.

The inheritance of thalassemia is autosomal and is similar to that of Hb S. From the clinical point of view it is recessive because the heterozygous form is asymptomatic. Similar to Hb S (β^S) gene, the thalassemia gene may express itself in homozygous, heterozygous, and doubly heterozygous states.

The clinical spectrum varies from normalcy to a severe, life-threatening condition[125,126] and can include retardation of growth, hepatosplenomegaly, bone overgrowth, bone pain, and jaundice.

Alpha thalassemias. The alpha (α) thalassemias are a group of genetic disorders that result in defective α-chain synthesis.[127] Diminished α-chain synthesis depresses Hb A, F, and A_2 because they contain α chains, and it leads to excess β and γ chains, which polymerize to the tetrameric forms γ_4 (Hb Barts) and β_4 (Hb H). The presence of these hemoglobins is the hallmark of α thalassemia.

Main forms of thalassemias

- α Thalassemias
  - α Thalassemia 1 (α^o Thal)
  - α Thalassemia 2 (α^+ Thal)
  - Hb Constant Spring
- β Thalassemias
  - β^+ Thalassemia
  - β^+ Thalassemia, high F
  - β^o Thalassemia
  - $(\delta\beta)^o$ Thalassemia
  - $(\delta\beta)^+$ Thalassemia
- Hb Lepore
- Hereditary persistence of fetal hemoglobin (HPFH)

Table 33-6. Laboratory findings in α thalassemias

Genotype	Anemia (hypochromic)	Hb types	α-Chain deletion
α Thalassemia 1 trait	±	Birth: Hb Barts 5%-10% Hb CS 1%-2% Adult: Hb A,A_2,F	2
α Thalassemia 1/α thalassemia 1 (hydrops)	+ + +	Birth: Hb Barts 80% Traces of Hb H and Portland Adult: Not compatible with life	4
α Thalassemia 2/trait	±	Birth: Hb Barts 1%-2% Hb CS 1%-2% Adult: Hb A,A_2,F	1
α Thalassemia 1/α thalassemia 2 (Hb H disease)	± (Inclusions)	Birth: Hb Barts 1%-15% Hb H 4%-30% Adult: Hb A,A_2,F Hb H 8%-10%	3
α Thalassemia 1/Hb CS (Hb H/CS)	+ + (Inclusions)	Birth: Hb Barts Hb H, Hb, CS Adult: Hb H Hb A,A_2,F,CS	2 plus α-chain termination mutation
α Thalassemia 2/Hb CS	+	Birth: Hb Barts Adult: Hb A, CS	1 plus α-chain termination mutation
Hb CS/Hb CS	+	Birth: Hb Barts Adult: Hb A,A_2,F,CS	α-chain termination mutation

There are at least three α-thalassemia genes:[91] alpha thalassemia^0 (α thalassemia 1) in which no α chains are produced, alpha thalassemia$^+$ (α thalassemia 2) in which the α-chain production is reduced, and Hb Constant Spring (Hb CS), in which the α-chain production is thalassemia-like reduced.[128] For a summary of laboratory findings on α thalassemia see Table 33-6.

Beta thalassemia. The beta (β) thalassemias are a group of genetic disorders that result in diminished (β^+ thalassemia) or absent (β^0 thalassemia) β-chain synthesis. There are at least two forms of β^+ thalassemia, one with about one-half (β^+) and one with two-thirds (β^{++}) normal β-chain production. On a clinical basis the severe homozygous disorders of both types of β thalassemia have been called "thalassemia major," and the heterozygous carrier states "thalassemia minor" and "minima." "Thalassemia intermedia" describes the clinical manifestations of a form of β thalassemia more severe than the trait and milder than the homozygous form.

Beta thalassemias are widely distributed throughout the world but occur most frequently in the Mediterranean populations, in Southeast Asia, the Middle East, India, and Pakinstan. In Greeks and in American blacks the β^+ thalassemia is most common, whereas in Italy the β^0 thalassemia is predominant.

Like α thalassemia, β thalassemia is transmitted as a mendelian autosomal recessive characteristic. The output of β chains is reduced or absent because of a defect in transcription of the β-thalassemia genes. In β^0 thalassemia homozygotes no β-chain synthesis occurs, whereas in β^+ and β^{++} thalassemia homozygotes, β-chain synthesis occurs at a low rate.

Heterozygous β thalassemia, whether β^+ or β^0, is an asymptomatic disorder that may or may not have a mild degree of anemia.[129] It is the most commonly found thalassemic disease in North America.[124] Characteristically there is a slight-to-moderate erythrocytosis of poorly hemoglobinized cells. The mean corpuscular hemoglobin (MCH) and the mean corpuscular volume (MCV) are always strikingly decreased (MCH 20 pg and MCV between 55 and 56 fL). The mean corpuscular hemoglobin concentration (MCHC) is variable. MCV values greater than 75 fL rule out thalassemia.[130] A discriminant factor, a mathematical formula described by England and Frazer, is based on the MCV, the red blood cell count, and the hemoglobin level and discriminates between thalassemia trait and iron deficiency in 90% of cases,[131] but there are other features that aid in the exclusion of iron deficiency, such as slight reticulocytosis, elevated Hb A_2 and F, red cell morphology, and the normal plasma iron level. Free red cell protoporphyrin is elevated, a feature distinguishing β-thalassemia trait from iron deficiency.[132] Pregnancy may lead to severe anemia in patients with thalassemia trait.[133] The hemoglobin electrophoretogram shows a slightly elevated Hb F (1% to 7%) in 50% of cases and a diagnostic elevation of Hb A_2 (3.5% to 7.5%).[126] Distribution of Hb F within the red cells demonstrated by the acid elution technique reveals a heterogeneous pattern.

Homozygous β^0 thalassemia leads to complete suppression of β-chain synthesis and to complete absence of Hb A. It is the cause of a severe lethal transfusion-dependent hemolytic anemia accompanied by characteristic clinical and hematological findings. Homozygous β^+ thalassemia is a heterozygous disorder that on the basis of the amount of Hb A synthesized is divided into three main types: Type 1, which synthesizes 5% to 15% Hb A, is the Mediterranean and Oriental form characterized by a severe transfusion-dependent anemia. Type 2, of African background, has 20% to 30% Hb A and is responsible for a milder disease. Type 3 leads to a mild form of thalassemia intermedia.[91]

Hemoglobin electrophoresis.[134] The Hb F concentration can vary from 10% to 90%, and the distribution in the red cells as determined by the Kleihauer technique is heterogeneous in type. The Hb A_2 concentration may vary from 1.4% to 20%, embracing low, normal, elevated, and very high values. In homozygous β thalassemia, even if the Hb A_2 value is low or normal in the patient, it will be high in both parents. If the Hb A_2 level in the patients is expressed as a percentage of the total hemoglobin, it spans the previously mentioned spectrum from low to high; but if it is expressed in relation to the Hb A value only, the ratio is decreased in all cases of β-thalassemia, that is, an A/A_2 ratio of about 10:1 as compared to the normal A/A_2 ratio of about 40:1. If the Hb A_2 level is greatly increased, Hb F is normal or only slightly elevated, and vice versa. In β^0 thalassemia there is a total absence of Hb A, and so the patient's hemoglobin consists only of Hb F and A_2, whereas in β^+ thalassemia diminished amounts of Hb A are found (5% to 20%). In both thalassemias, free α-chains may be seen close to the application point of the electrophoretogram at alkaline pH.[135] The severe reduction in β-globin chains leads to a β/α ratio of less than 0.25 to 0.3.[129]

Iron metabolism. The serum iron and the bone marrow iron deposits are increased. The iron is seen not only in an increased number of sideroblasts and reticuloendothelial macrophages but also in the form of ring sideroblasts.[134] The total iron-binding capacity is completely saturated.

Other related biochemical findings. Because of the hemolytic component of the anemia, serum unconjugated bilirubin levels are elevated and haptoglobin is absent. Serum aspartate aminotransferase (AST), lactic dehydrogenase (LD), and erythropoietin concentrations are raised. The erythropoietin elevation is responsible for the 20% to 30% increase in erythropoietic marrow and is a result of the anemia and the high oxygen affinity of Hb F, which further increases the tissue anoxia. The liver involvement (transfusion hemosiderosis) can lead to a bleeding tendency. Gross examination of the urine may show the brown color of dipyrroles, caused by excessive intramedullary hemolysis of normoblasts.

Heterozygous HPFH. Hereditary persistence of fetal hemoglobin (HPFH) is a benign condition characterized by the continuous synthesis of Hb F into adult life. It is considered to be a form of δβ thalassemia[136] because the persistence of the γ-chain synthesis compensates for the deficient δ- and β-chain production. The HPFH mutation is allelic for the δ- and β-gene structural loci, which are close to each other on the same chromosome.[129] HPFH occurs in the heterozygous and homozygous forms and in association with β-chain hemoglobin variants. It can be broadly classified according to the distribution pattern in the red cells into pancellular and heterocellular groups.[137] In the first group, Hb F is uniformly distributed among the red cells, whereas in the latter it shows a heterogeneous pattern. According to geographical and other characteristics, it is divided into several types: Greek,[138] African,[139] British,[140] Swiss,[141] Georgian,[142] and others. The African type is also encountered in American blacks and is characterized by complete absence of the δ- and β-chain synthesis, and Hb F exhibits a pancellular distribution. In some of these cases the $^{A}\gamma$ locus has also been deleted, but in most the $^{G}\gamma$ and $^{A}\gamma$ chains are synthesized but in abnormal ratios.[129]

The hemoglobin electrophoretogram reveals about 25% Hb F, which by the Kleihauer method shows a pancellular distribution. Hb A_2 concentration is lower than normal (average value of 1.6%),[139] and the remaining hemoglobin mass is Hb A.

A mild erythrocytosis (6 million/mm^3) in response to the high oxygen affinity of Hb F is associated with a mild microcytosis (MCV 70 fL) and an equally mild reticulocytosis (1% to 2%).[91] Hemoglobin analysis reveals 100% Hb F in equal portions of $^{G}\gamma$ and $^{A}\gamma$ chains.[143] The α/non-α-globin chain synthesis ratio reveals the imbalance characteristic of thalassemia.[143]

Methemoglobinemia

Methemoglobinemia is classified into acquired and hereditary forms.

Acquired methemoglobinemia. The acquired form is caused by various drugs and chemicals (or their metabolites) that are able to oxidize hemoglobin at such a rate as to overwhelm the reductive capacity of the red cells.[59] These substances include nitrites, nitrates, aniline dyes, sulfonamides, and phenacetin. Exposure to toxins may be responsible for very high levels of acute methemoglobinemia.

Physiologically, heme is shielded from excessive oxidation by its position in the heme pocket and by a number of enzymes that can destroy certain oxidizing agents. The protection generated by the heme pocket is jeopardized in the unstable hemoglobins, which exhibit structural alterations in the heme pocket area and form methemoglobin spontaneously and at an accelerated rate. When superoxide (O_3) and hydrogen peroxide are involved in the production of methemoglobin, the normal reductive systems of red cells, superoxide dismutase, glutathione peroxidase, and catalase can counteract their oxidizing effects. The increased tendency of newborns to develop methemoglobinemia when brought into contact with oxidizing agents has been related to their decreased activity of glutathione peroxidase.[144]

Hereditary methemoglobinemia. The hereditary forms are further subdivided into two types. One is associated with Hb M, variant hemoglobins in which the altered globin moiety stabilizes methemoglobin and renders it resistant to reduction. The other is caused by deficiency in the methemoglobin-reducing ability of the red cells.

Four metabolic pathways for the reduction of the met-Hb to hemoglobin are available:[145] (1) the NADH methemoglobin reductase pathway, (2) the reverse (NADPH) methemoglobin reductase pathway, (3) the reduction by ascorbic acid, and (4) the reduction by reduced glutathione.

NADH reductase (or diaphorase or NADH cytochrome b_5 reductase) catalyzes the reduction of methemoglobin to hemoglobin, using NADH generated in the anaerobic glycolytic pathway of Embden-Meyerhoff. Methemoglobinemia caused by deficiency of NADH methemoglobin reductase is an inherited disease transmitted as an autosomal recessive trait.[61]

Table 33-7. Reference values for hemoglobin in "apparently healthy" subjects, white and black

Subjects	HGB (g/L)
Adult men	151 (139-163)
Adult women	135 (120-150)
Boys	
Birth	200 (185-215)
1 month	170 (155-185)
3 months	150 (135-165)
6 months	140 (130-160)
9 months	130 (120-140)
1 year	121 (100-140)
2 years	123 (105-142)
4 years	126 (112-143)
8 years	134 (120-148)
14 years	140 (125-150)
Girls	
Birth	195 (180-210)
1 month	170 (158-189)
3 months	148 (133-164)
6 months	138 (128-148)
9 months	128 (117-139)
1 year	122 (100-140)
2 years	122 (105-142)
4 years	127 (113-142)
8 years	130 (115-145)
14 years	132 (116-148)

From: Miale, J.B.: Laboratory medicine: hematology, ed. 6, St. Louis, 1982, The C.V. Mosby Co.

CHANGE OF ANALYTE IN DISEASE

Interpretations of hemoglobin values

The normal mean values for hemoglobin in adults are 151 g/L for men and 135 g/L for women. The values for newborns, infants, and children up to puberty differ (Table 33-7).

There are conflicting reports of normal,[146] increased,[147] or decreased[148] hemoglobin values in persons over 65 years of age. This controversy is probably caused by the difficulty in establishing criteria for perfect health in the older age group, but it also applies to younger persons. Miale points out that the 95% range equivalent to 2 standard deviations, if the distribution is normal, still excludes 5% of normal persons and that the probability of abnormality is increased if a given value falls between the second and third standard deviation.[65] Hemoglobin values are influenced by physiological variations and by pathological processes. The physiological variations include age, sex, physical exercise, posture, dehydration, and altitude. The influence exerted by age is readily apparent in Table 33-7. The high birth level falls rapidly during the first year of life and then gradually rises to the adult level attained at 14 years of age. Sex differences make their appearance at puberty and fade away in old age. Estrogen can influence the lower levels in women though they are not corrected by oral contraceptives.[149] Strenuous exercise raises the hemoglobin level, probably through fluid loss, and a transient increase is also experienced after one changes from the recumbent to the standing position.[150] Dehydration is responsible for a rise in hemoglobin concentration of such magnitude as to mask a significant anemia. High altitude is responsible for increased hemoglobin levels because of the erythropoietin-stimulating effect of hypoxia.[151] Early literature ascribes higher morning levels to diurnal variations in the hemoglobin concentration.[152]

Pathological levels are encountered in anemias and in polycythemias. The overall reduction in the hemoglobin level in the peripheral circulation is one of the main parameters used in the diagnosis of all types of anemia. Because of the physiological variations and the wide normal and usual ranges of hemoglobin concentrations, other factors must also be considered in the diagnosis of anemia. There are three main causes of anemia: impaired production of red cells by the bone marrow, excessive blood loss and impaired delivery of red cells to the peripheral blood, or a combination of these factors. Decreased erythropoiesis is seen in iron deficiency and in impaired synthesis of heme, globin, and DNA (B_{12}, folate deficiencies) and in impaired bone marrow production. Excessive blood loss may be caused by hemorrhage, hemolysis, abnormal splenic sequenstration, parasites, toxins, and trauma to red cells.

Increased hemoglobin values are encountered in polycythemia vera, erythrocytosis, dehydration, newborns, acquired or congenital cyanosis, chronic heart and lung disease, high altitude, renal cysts, and a number of erythropoietin-producing tumors.

Hb F

Shortly before birth, γ chains give way to β chains within the neonatal red cells.[153] The temporary suppression of erythropoiesis after birth is followed by an explosion of Hb A production that almost completely suppresses Hb F genesis, but minute amounts of Hb F may remain demonstrable for many years,[154] during which time the $^{G}\gamma/^{A}\gamma$ ratio changes.[155] Wood,[156] using fluorescent anti–Hb F antibodies, identified 0.2% to 7% of adult red cells as Hb F–containing cells, which represent a specific group of red blood cells called *F cells*.[157] In normal adult persons the concentration of F cells is fairly constant, but there are both genetic and acquired hematological conditions in which their concentration is increased, probably as the result of erythroid hyperplasia. The genetic disorders include thalassemias (β and δβ), hereditary persistence of Hb F, sickle cell anemia, and unstable β-chain variants. The acquired conditions include pregnancy at about midterm, recovery from bone marrow depression,[158] leukemias (highest values in Philadelphia chromosome negative juvenile chronic myelocytic leukemia),[159] thyrotoxicosis, and hepatoma.[160]

Hb A_{1c}

Hb A_{1c} is formed continuously during the life-span of the red cell by means of a stable ketoamine linkage[161] to glucose and is therefore a lasting indication of a period of high glucose levels.[162] Its normal concentration is 5% to 8.5% (average 6.5%) of total hemoglobin, but it may increase threefold in insulin-dependent diabetics.

In hemolytic anemias the Hb A_{1c} level is low because the red cell life-span is shortened by lysis.[163] If the latter is age related, one can use differential high-speed centrifugation or density gradients to separate the young cells and allow the measurement of Hb A_{1c} in the denser and heavier old cells.[164]

Hb CO

Some carboxyhemoglobin is produced endogenously as 1 mole of CO is generated by the degradation of 1 mole of heme to bilirubin.[165] Although this endogenous production of carboxyhemoglobin can present a hazard when exhaled air is concentrated in ill-ventilated small spaces, the exogenous generation of CO from combustion of organic material in confined spaces is more likely to cause intoxication. Exogenous CO is derived from the exhaust of automobiles[166] and from industrial pollutants such as coal gas, charcoal burning, and tobacco smoke. The affect of CO is insidious because it is a colorless and odorless gas, the toxic effect of which depends on the preexisting hemoglobin level (anemia), the duration of exposure, the CO concentration,[167] and the preexisting CO-Hb level. Under the most favorable conditions the CO-Hb concentration in blood is 0.2% to 0.8%. It is never zero because of its endogenous generation and can physiologically be elevated in hemolytic anemias.[168] In smokers the level may vary from 4% to 20%. In garage workers and others who have a greater exposure to CO, the average level may be 10%. Because of its firm liaison to hemoglobin, long exposure to even low CO concentrations can lead to toxic accumulations to which the most oxygen-dependent organs, such as brain and heart, are most susceptible. Mild symptoms such as slight headaches and slight dyspnea on exertion can occur at levels of 10% to 15% saturation. At levels of 20% to 30% the headaches will be more severe, and accompanied by fatigue and impaired vision and judgment. Levels of more than 50% cause increasingly severe symptoms, coma, and convulsions, and levels of 60% and over are usually fatal, though death has occurred at levels as low as 20%. The half-life of elimination of CO is about 4 hours[169] for a person breathing atmospheric air,[170] though in smokers the level may remain high.[171] Chronic exposure to CO may be responsible for a relative polycythemia.[172]

Oxygen saturation

Clinically the oxygen saturation is used as an indicator of hypoxia or hyperoxia. Tissue hypoxia is produced by decreased oxygen content of inspired air, as in high altitude, or by decreased alveolar-capillary oxygen exchange in the lungs, as in pulmonary fibrosis, emphysema, and chronic heart disease with right to left shunt. Tissue hypoxia is also produced (1) by a defect in the erythrocytic oxygen transport as in severe anemia (Fig. 33-14), (2) when hemoglobin ligands are present that prevent oxygen binding, such as carboxyhemoglobin, sulfhemoglobin, methemoglobin, (3) in hemoglobinopathies, (4) in inappropriate concentrations of erythrocytic 2,3-DPG, and (5) in intraerythrocytic enzyme deficiencies. Therapeutic hyperoxia must be carefully monitored because of the danger of oxygen toxicity. In newborns it can be responsible for retrolental fibroplasia and in adults for hyaline membrane disease of the lungs (adult respiratory distress syndrome).

2,3-Diphosphoglycerate

2,3-DPG plays an active role in the adaptation to hypoxia as seen in anemia,[52] exposure to high altitude,[53] and heart and lung disease. The DPG and hemoglobin interaction in hypoxia is related to the increased intraerythrocytic deoxyhemoglobin, which binds large amounts of DPG. This binding results in a feedback mechanism that stimulates glycolysis and DPG synthesis. Increased deoxyhemoglobin concentrations raise the pH, which in turn also stimulates the synthesis of DPG. Carrell et al.[46] emphasize that there is a reciprocal relationship between hemoglobin and DPG concentrations. Pyruvate kinase deficiency leads to a buildup of DPG, decreased oxygen affinity, and a low hemoglobin concentration. Hexokinase deficiency leads to a decrease in DPG and to a compensatory erythropoietic response with increased hemoglobin values.[46]

Part II: Iron metabolism

allele Pair of genes that occupy corresponding loci on homologous chromosomes.
chelate Bond formed between a metal and a molecule.
ferritin Iron-storage protein, apoferritin plus iron.
fibrogenic Producing fibrous tissue.
genotype Genetic constitution of an individual in contrast to that individual's appearance (phenotype), and gene combination at one locus or several loci.
haplotype The genetic composition of an individual that refers to one allele of a pair of allelic genes.
hemosiderin Iron-storage protein; denatured ferritin.
hepatocyte Liver cell.
homologous chromosomes Pair of chromosomes that carry identical genes (alleles) on identical loci.
human leukocyte antigens (HLA), histocompatibility antigens at locus A Antigens present in all nucleated cells of all tissues of the body.
MCH Mean corpuscular hemoglobin.
MCHC Mean corpuscular hemoglobin concentration.
MCV Mean corpuscular volume.
parenchyma The functional tissue of an organ (excluding the fibrous framework).
transferrin Carrier protein of iron.
WHO World Health Organization.

PHYSIOLOGICAL FUNCTION OF IRON

The prosthetic group of hemoglobin is the iron complex of protoporphyrin IX (heme) in which the centrally located iron atom acts as a stabilizer of oxyhemoglobin.[173] Although iron is essential for the existence of all plants and animals, only trace amounts are present in living cells. These small but biologically indispensable amounts are related to the relative insolubility of iron and to the specialized systems required for its transport and incorporation into the iron proteins: hemoglobin, myoglobin, heme enzymes (cytochromes), nonheme iron enzymes (flavoproteins), ferritin, transferrin, and hemosiderin. This list vividly demonstrates how critically iron is involved in cellular metabolism and oxidation systems of all living things.

Iron is distributed as follows: the total iron amounts to about 50 mg/kg of body weight in adult men and 35 mg/kg in adult women,[174] 66% to 70% of which is found in hemoglobin, 4% in myoglobin, 30% in storage pools (hemosiderin and ferritin) and trace amounts in iron-containing enzymes (catalases, peroxidases, and flavoprotein-cytochrome enzymes).[175]

The total body requirements vary from 1 to 4 mg/day, depending on the person's age and sex. For a normal adult 1.7 mg/day is usually adequate, but higher intake levels are required during pregnancy, lactation, and puberty.[175] In infants and children the total daily requirements for iron are relatively high: first year, 1.05 mg/day; seventh year, 0.80 mg/day; and thirteenth year, (women) 1.90 mg/day.[176,177]

IRON METABOLISM

Iron absorption

The body carefully controls its stores of iron by limiting iron absorption and by reusing iron derived from breakdown of hemoglobin and catabolism of nonheme iron proteins. Although there is no iron excretion in the physiological sense, there is a daily obligatory loss of 1 mg of iron that must be offset by the intestinal absorption of 5% to 10% of the dietary 10 to 30 mg/day.[178] The physiological iron loss includes iron lost through normal shedding of cutaneous and mucosal cells and in proteinaceous fluids such as sweat and bile and in a small number of red cells in feces and urine.[179] Increased iron loss is associated with menstruation, pregnancy, and lactation. The last is not compensated by the interruption of menses (Fig. 33-18). The iron needs must be satisfied by the absorption of heme and nonheme food iron. The first is in the divalent ferrous form (Fe^{++}), which is more readily absorbed, and the second is in the trivalent ferric state (Fe^{+++}), which requires separate processing. Both are absorbed from the intestinal stream by the duodenal and jejunal mucosa, but by different mechanisms.[180] The nonheme ferric food iron is ionized and reduced to the absorbable ferrous form by the hydrochloric acid and enzymes contained in gastric juice or made soluble by the formation of soluble low molecular weight chelates with sugars,[181] ascorbic acid,[182] and amino acids.[183] Compounds that form insoluble complexes with iron decrease its absorption. They include phosphates (eggs, cheese, milk), oxalates, phytates (vegetables), and tannins.[183-185] Heme iron (meat, fish) is processed differently. Gastric and intestinal proteases separate heme from globin, and the heme, still encased in the porphyrin ring, is absorbed intact and ionic iron is released within the mucosal cells by a heme-splitting enzyme.[183,186] The heme iron absorption is more efficient and greater than the nonheme iron assimilation and is not affected by dietary factors.[187-188]

Plasma iron turnover

The total plasma iron of a 70 kg man amounts to 3 to 4 g. Approximately 30 to 40 mg of iron leaves the plasma per day (the normal plasma iron turnover), 22 mg of which are used for daily hemoglobin synthesis and appear in red cells. One mg of iron is derived from intestinal absorption of food iron, the remaining, much larger portion originates from the breakdown of hemoglobin.

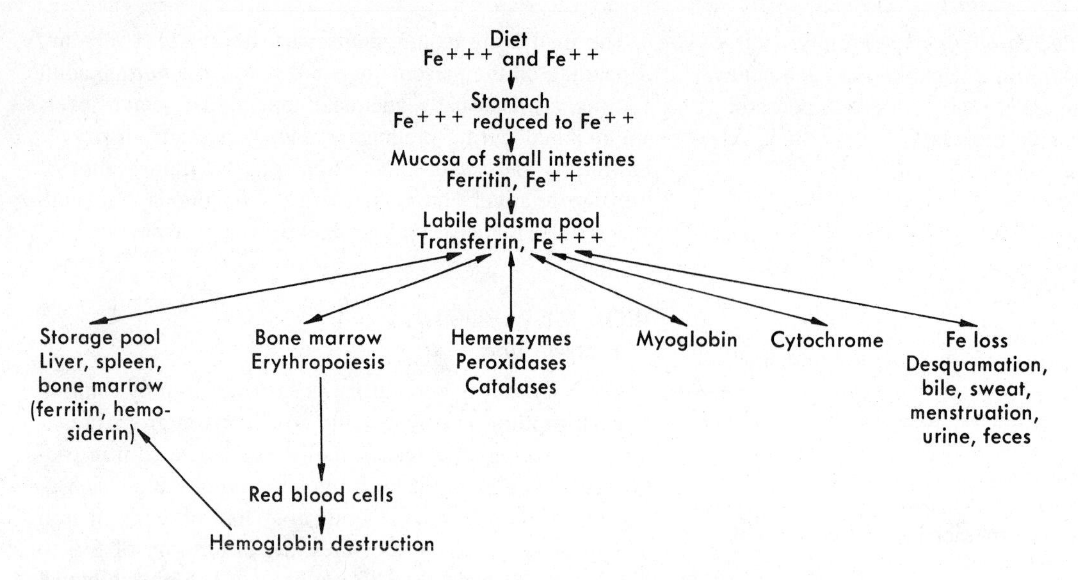

Fig. 33-18. Outline of iron metabolism. (From Bauer, J.J.: Clinical Laboratory methods, ed. 9, St. Louis, 1982, The C.V. Mosby Co.)

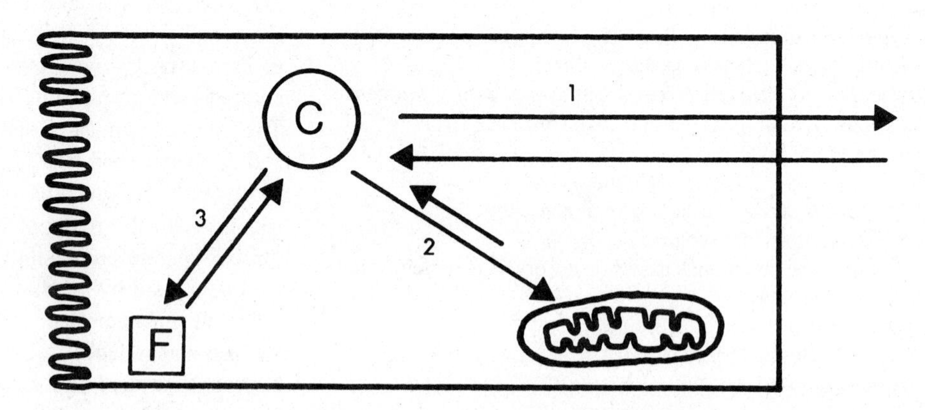

Fig. 33-19. Metabolic pathways in small intestine epithelial cell: *1,* iron transfer (from carrier pool C) to plasma; *2,* iron transfer to mitochondria for heme synthesis; *3,* iron incorporation in ferritin, *F.* (From Jacobs, A.: The mechanism of iron absorption, Clin. Haematol. **2:**323, 1973.)

Based on the plasma iron concentration, the plasma volume, and the plasma iron clearance, the amount of iron passing through the plasma in 24 hours can be calculated.[189] The plasma iron turnover is elevated in polycythemia, hemolytic states, and ineffective erythropoiesis; it is decreased in hypoplastic anemias, uremia, infections, and radiation damage.

Molecular mechanism of iron absorption

The molecular mechanism of iron absorption at the mucosal level can be broken down into two steps: (1) the mucosal uptake and (2) the transfer of iron to plasma. Initial mucosal uptake is probably a rate-limiting passive diffusion–controlled process,[190] with the mucosal cell acting as a temporary holding area for iron between intestinal contents and blood (plasma).[180] The mucosal absorption-regulating mechanism in humans is not known though it may be controlled by intracellular ferritin[191] and a transferrin-like protein. Iron is first bound to iron receptors of the brush borders of the mucosal cells and then transported into a hypothetical labile or transient iron pool of the cell interior (cell sap or cytosol) by means of a transferrin-like carrier protein (Fig. 33-19). The cytosol iron is available for transfer (1) to plasma transferrin contained in the mucosal blood vessels, (2) to mitochondria for iron enzyme synthesis, and (3) to apoferritin to form ferritin.[192] The control of the transfer of iron from the intestinal mucosa to the plasma is not well understood but is probably related to the total body iron stores, being part of an equilibrium between plasma iron and the exchangeable iron in all tissues.[192,193] This concept explains the observed inverse relationship between iron stores and absorption.[194] The synthesis of apoferritin is stimulated by the cytosol iron pool, which also influences (or controls) the iron uptake from the intestinal contents. The iron transferred to the mitochondria and to intracellular apoferritin is lost when the mucosal cells are sloughed.

Iron transport

On entering the bloodstream, ferrous iron combines with a carrier protein, a β_2 globulin called *transferrin*. It is responsible for the distribution of iron and its delivery to and from sites of absorption, storage, and utilization.[195] Transferrin and its iron represent the constantly changing plasma iron pool that is the temporary stopover for iron in transit and turnover. Iron enters the pool from the intestinal mucosa and from hemoglobin breakdown and leaves it to be complexed with membrane receptors to enter developing red cells for hemoglobin formation or liver cells for storage as ferritin. Transferrin is a single-chain polypeptide with a 4 day half-life and two specific iron-binding sites.[196] The sites allow it to accept and to donate to tissues increasing amounts of iron as plasma iron concentrations increase.[197] The iron binding involves only ferric iron and is reversible. The iron-donating property depends on specific cell surface receptors for transferrin iron of the various tissues that control the plasma iron turnover[198] and the mucosal regulation of iron absorption.[197] The reversible binding of iron depends on the concomitant binding of an anion (bicarbonate).[189] Specific cell surface receptors for transferrin iron have been identified in various tissues (liver, placenta, tumors),[189] but by sheer number the bone marrow reticulocytes and erythroid precursor cells offer the greatest receptor area and thus control plasma iron turnover.[197]

Transferrin is synthesized in the liver and probably in lymphocytes.[199] The hepatic synthesis is inversely proportional to the intracellular ferritin level of hepatocytes so that in iron deficiency its plasma concentration is increased.[200] There are about 20 molecular species of transferrin, with a wide range of electrophoretic mobilities, that are of genetic importance but do not differ in their avidity for iron. The most frequent transferrin is Tf. Other transferrins (Tf^b, Tf^d, and so forth) are quite rare, as is the absence of the carrier protein.[201,202]

Storage of iron

Ferritin. Ferritin is a heterogeneous group of water-soluble, relatively heat-stable, iron-storage proteins found in nearly all body cells but mainly in the liver, spleen, and bone marrow.[203] The ferritin molecule has a polynuclear iron core (hydrous ferric oxide–phosphate) coated by a protein shell.[204] Hemosiderin is similar to ferritin but lacks the solubilizing protein shell.[205] The protein shell (minus iron) is called *apoferritin* and has the qualities of an iron-storage protein: it takes up Fe^{++}, oxidizes it to Fe^{+++}, deposits it within the iron core, and releases it when confronted with reducing agents.[206] The synthesis of apoferritin is stimulated by iron. Because of differences in the amino acid sequence of the protein shell, ferritins from different species and from different tissues of the same species differ and are separable by immunoradiometric methods.[207,208] The different forms of the same molecule are called *isoferritins*.[204]

The iron-storage function of ferritin and its physiological significance have long been recognized. Ferritin packages and isolates large numbers of iron molecules by coating them with protein, thus preventing any toxic action free iron may exert on tissue cells. It also provides a readily available source of reserve iron.

Serum ferritin. A clinically important isoferritin (serum ferritin) is found in serum in low concentrations and is detectable by sensitive radioimmunoassay,[209] enzyme-linked immunoassays, and immunoradiometric techniques.[210] Ferritin in the serum may be the result of a nonspecific release of ferritin from damaged cells, a specific secretory activity of the endoplasmic reticulum, or both.[211] The adult serum ferritin concentration has a wide span of 2000 to 3000 ng/L. Despite the fact that it varies with age and sex,[206] it is less variable than serum iron concentration.[212] During infancy and childhood serum ferritin reflects the rapidly changing iron levels of the first months and years of life.[213] In adult men serum ferritin levels are about three times higher than in women of childbearing age, whose serum ferritin does not achieve the levels of men until after menopause.[214,215]

Hemosiderin. Hemosiderin is an insoluble iron-storage protein derived from ferritin that has lost some surface apoferritin, resulting in a higher iron concentration (up to 37%) than that found in the original ferritin (up to 24%).[175] Thus hemosiderin is considered a form of iron storage that develops only when ferritin molecules become saturated with iron.[216] One third of the body iron is stored in the liver, one third in the bone marrow, and the remaining one third in muscle and spleen.[175] Hemosiderin iron can be mobilized for hemoglobin formation, but the process is slower than the release of iron from ferritin.[175] Hemosiderin is visible by light microscopy as brown granular pigment, which stains blue with the Prussian blue stain.

PATHOLOGY

Iron deficiency

Iron deficiency is the most common cause of anemia and is significantly more frequent in women of childbearing age, in developing countries (men and women), and in children of low socioeconomic background.[217,218] The primary symptoms associated with anemia are those related to tissue hypoxia (lack of adequate tissue oxygen) secondary to decreased hemoglobin concentration. The classical symptoms of anemic hypoxia such as weakness, palpitations, pallor, and dizziness may be absent in even moderate to severe (less than 100 g/L hemoglobin) anemias, whereas persons with ''latent'' iron deficiency (low serum iron, depleted iron stores, but no anemia) may complain of several of these symptoms.[217] It is difficult to identify special features of iron deficiency that are not shared by all anemias. Reports of the effect of iron therapy on the symptoms of iron-deficient patients give conflicting results. In some groups iron therapy improved the hemoglobin level and the symptoms,[219] whereas others failed to show any relationship between severity of symptoms and hemoglobin levels.[220] More recent investigations demonstrate a significant relationship between iron deficiency and physical fitness and work capacity in adults[221] and between iron deficiency and mental performance and irritable behavior in children.[222] In iron deficiency, monoamine oxidase activity (MAO) is decreased, and because it is an important enzyme in norepinephrine catabolism, norepinephrine excretion in the urine is increased. Noreinephrine is believed to influence behavior in humans.[222,223]

Iron deficiency and iron loading affect iron absorption and iron handling by the mucosal cell (Fig. 33-20).

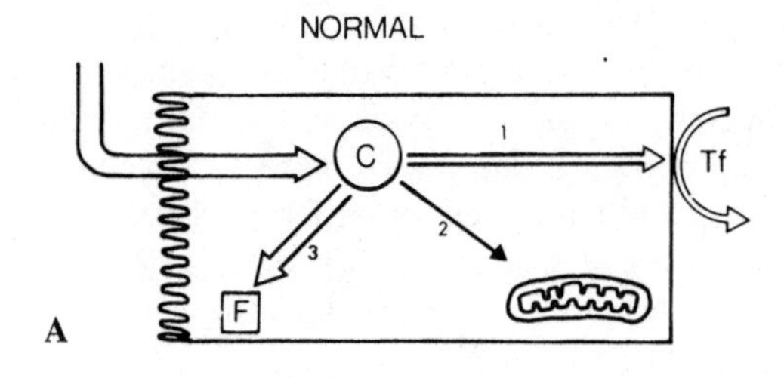

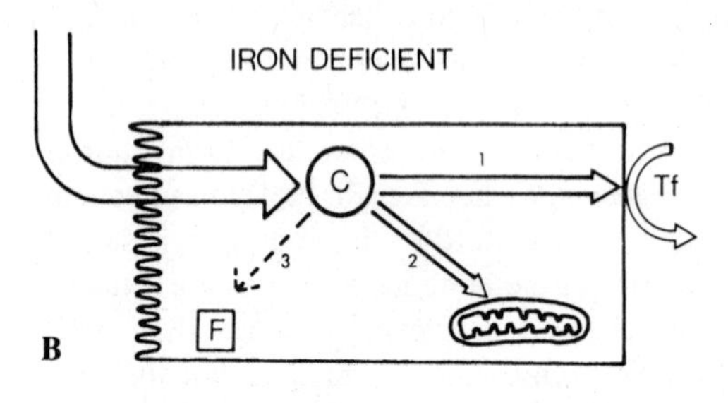

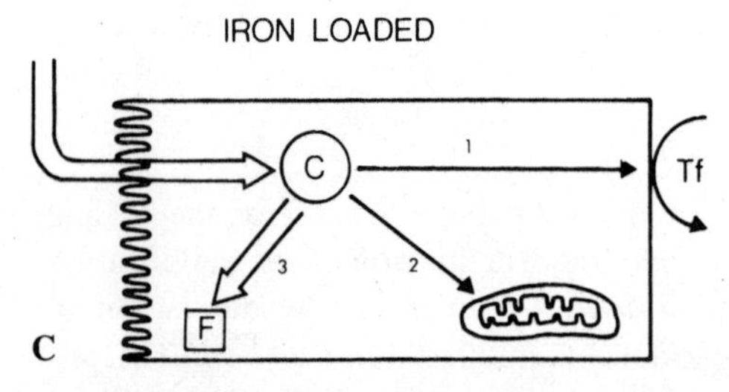

Fig. 33-20. Iron uptake and metabolic pathways within the small intestinal epithelial cell, *C*. **A,** Normal iron status. **B,** Iron deficiency: increased mucosal uptake caused by mitochondrial requirements; increased transfer along pathways 1 and 2 with no transfer along pathway 3. **C,** Iron overload: normal iron uptake with decreased transfer along pathway 1 and increased ferritin, *F*, formation by pathway 3. *Tf*, Transferrin. (From Jacobs, A.: Clin. Haematol. **2**:323, 1973.)

Iron deficiency results in increased iron absorption and accelerated transfer of iron to the unsaturated plasma transferrin. The incorporation into the ferritin iron pool is decreased to rapidly satisfy the increased body requirements and because the low iron pool level fails to stimulate apoferritin synthesis.

Iron overload

The term ''iron overload'' denotes the presence of excessive amounts of iron in the body. Iron overload may arise from the following[224]:

1. Excessive, inappropriate chronic absorption of normal dietary iron. This situation is seen in idiopathic hereditary *primary hemochromatosis*.
2. Conditions secondary to
   a. Abnormally high dietary iron intake
   b. Repeated blood transfusions
   c. Alcoholic cirrhosis
   d. Anemias with ineffective erythropoiesis (thalassemias, sideroblastic anemias)

Iron overload allows transfer of iron from the saturated transferrin back into the cell and restricts the transfer of iron from the cytosol pool to transferrin. Because the binding of luminal iron remains normal, the resultant high concentration of cytosol pool iron is responsible for increased ferritin formation. The iron transfer to mitochrondria remains normal.[179,225]

Idiopathic (primary hereditary) hemochromatosis. Idiopathic hemochromatosis (IH) is a hereditary disorder characterized by a progressive increase in body iron stores resulting in the deposition of iron in liver, heart, pancreas, and other organs.[224] Parenchymal iron overload eventually leads to fibrosis and cirrhosis.[224] The genetic transmission is autosomal recessive, and homozygosity is required for full development of the disease.[226] The hemochromatosis gene is located on chromosome 6 in tight linkage with the human leukocyte histocompatibility antigens (HLA).[227] The discovery of the mode of inheritance and of the HLA linkage allows the determination of the genotypes of the relatives of patients with clinically established IH. Based on the HLA haplotype, normal individuals and those heterozygous and homozygous carriers of IH can be differentiated.[228,229] The basic defect leading to the inappropriate absorption of iron in excess of the needs of erythropoiesis is unknown. The most important manifestation of the IH iron overload is the deposition of iron in the parenchymal cells, which is responsible for the clinical manifestations of the disease, such as hepatomegaly, skin pigmentation, heart failure, late onset diabetes, and rapid deterioration.

The most useful diagnostic laboratory tests are the serum iron and the percentage of iron saturation of transferrin, which serve as an indicator of the genotype in persons at risk.[227] Although HLA genotyping remains the most reliable method, serum ferritin concentration, chelation studies, and liver biopsy (stainable parenchymal iron) identify the degree of iron loading already present.[230]

Secondary hemochromatosis. Iron overload can develop in normal persons if there are high concentrations of iron in the diet as in the South African Bantu (home-brewed beverages),[231] or in persons with a long history of consumption of therapeutic iron preparations. It also develops in certain anemias as a consequence of ineffective erythropoiesis and repeated blood transfusions.[232] In secondary hemochromatosis the iron is deposited in hepatocytes and in the reticuloendothelial (RE) cells of liver, spleen, and bone marrow, a distribution that differs from that seen in IH, in which the parenchymal cells are primarily involved. The RE cell iron overload (siderosis) is relatively innocuous and leads to organ damage only in a minority of patients. Iron is only a low-grade fibrogenic agent but probably renders the liver more susceptible to other insults, such as alcohol.[233] There are a number of complications of RE iron retention that interfere with the release of iron from the RE stores. They are ascorbic acid deficiency,[233] inflammation, and hereditary RE siderosis. When massive secondary iron deposits eventually lead to cirrhosis, the iron-distribution pattern changes. The hemosiderin deposits, instead of being confined to liver and RE cells, spread to pancreas, heart, and endocrine glands so that the picture resembles IH.[233]

FUNCTION TEST

Iron-chelation test

Iron stores are assessed by measurement of the urinary excretion of iron after injection of an iron-chelating agent such as deferoxamine mesylate[234] or diethylenetriaminepentaacetic acid (DTPA).[235] These tests correlate well with the mobilizable storage iron in managing patients with hemochromatosis.[236]

CHANGE OF ANALYTE IN DISEASE (Table 33-8)

The clinical evaluation of the iron states requires a series of hematological and biochemical investigations that must be interpreted against the backdrop of the patient's age (newborn, child, childbearing woman, elderly), sex, physiological state (postoperative, pregnant) and history (drugs, diet, supplemental iron, birth control pills).

Definition

Iron deficiency leads to anemia because of the unavailability of iron for hemoglobin synthesis. The WHO criteria for anemia expressed in hemoglobin concentration are as follows: adult males, below 130 g/L; menstruating women, below 120 g/L; pregnant women, below 110 g/L.

The laboratory evaluation of the iron status can be grouped as follows[237]:

1. Iron deficiency: hemoglobin, red cell morphology.
2. Iron-deficient erythropoiesis: transferrin saturation, free erythrocyte protoporphyrin, red cell indices, sideroblast count.
3. Storage iron depletion: reticuloendothelial bone marrow iron, serum ferritin, serum transferrin concentration, and iron-chelation tests.

Hemoglobin as parameter of iron deficiency

The disadvantage of expressing iron deficiency in terms of hemoglobin concentration is the difficulty of identifying the stage that constitutes truly low hemoglobin concentra-

Table 33-8. Laboratory measurements of iron metabolism

Disease	Serum ferritin (ng/mL)	Serum iron (μg/L)	Iron-binding capacity (μg/L)	Transferrin saturation* (%)	Free erythrocyte protoporphyrin (μg/L of cells)	Tissue iron stores
Normal	20-300†	650-1850‡	2500-4400	30	120-790	N§
Iron depletion, early; no anemia	↓	N	N	N	N	↓
Iron depletion, intermediate; no anemia	↓	↓	↑	↓	↑	↓
Iron-deficiency anemia	↓	↓	↑	↓	↑	↓
Anemia of chronic infection	↑	↓	↓	N	↑	↑
Thalassemia	↑	↑	↑	↑	↑	↑
Hemochromatosis	↑	↑	↑	↑	N	↑
Lead poisoning	N	N	N	↑	↑	N

*Varies with age. Averages 38% at birth, 23% at 1 year.
†Varies with age and sex. Average for newborn, 100 ng/mL; average at age 1 month, 350 ng/mL; adult women, average 36.4, range 25 to 175 ng/mL; adult men, average 96, range 6 to 329 ng/mL.
‡Varies with age. Averages 1200 g/L at 1 month; 770 g/L at 6 months to 1 year.
§*N*, Normal; ↓, decreased; ↑, increased.
(From Miale, J.B.: Laboratory medicine: hematology, ed. 6, St. Louis, 1982, The C.V. Mosby Co.)

tion because there is great overlap in the frequency distribution curves of anemic and normal persons.[237] Despite the problem of the correct cutoff point, hemoglobin concentration appears to be the most precise discriminator and the best predictor of response to iron therapy.[217] However, there remains a significant uncertainty in the range of 100 g/L hemoglobin that requires further investigation to eliminate iron deficiency.

Red blood cell morphology as parameter of iron deficiency

Details of the hypochromic microerythrocytes of the peripheral blood smear are beyond the scope of this section. There are some pitfalls in the examination of the blood films. First, hypochromic microcytes are not diagnostic of iron deficiency because they are also found in thalassemias and sideroblastic anemias. Second, although their concentration usually reflects the severity of the anemia, they may fail to do so. Finally, the evaluation is subjective—How many abnormal cells need there be to be significant?

Red blood cell indices

Red blood cell indices are average values and in an early iron deficiency may be normal. In established iron deficiency the MCV and the MCH values are reduced, though the MCHC level is usually less abnormal even in severe anemia.[188] In the developing anemia of chronic disease and malignancy the reduction of the MCHC precedes the fall in the MCV.[238]

Serum iron

Apart from the diet, pathological and physiological states influence iron absorption.[188] Anemia (with the exception of the anemia of chronic disorders), pregnancy, iron deficiency, low transferrin saturation, erythroid hyperplasia, erythropoietin, and oral iron administration *enhance* iron absorption. Malabsorption, transfusion polycythemia, increased iron stores, chronic transferrin saturation, high gastric pH, postgastrectomy state, oral administration of alkali, and fever *suppress* iron absorption.

The iron present in the plasma bound to transferrin is largely derived from hemoglobin breakdown. It is part of the transport iron pool that is normally turned over 10 to 20 times each day, and so each iron atom does not remain in the plasma longer than 2 hours.[239,240]

Serum iron is iron in transit, and even if reduced, the amount delivered to the bone marrow is usually normal[239] as evidenced by a normal plasma iron turnover found in iron-deficient patients.[239,240] The diagnostic usefulness of the serum iron determination is impeded by its variability and fluctuation. There is a significant diurnal variation, with a 20 to 30% fall in concentration in the evening and there may be significant day-to-day variations ranging from levels suggesting iron deficiency to levels indicating iron overload.[239] The reference value for adult men is 420 to 1350 μg/L (20% less for women). Physiologically, serum iron is low in infants and in women during menstruation. Pathologically, serum iron is reduced in iron deficiency, in the anemia of chronic disorders and of infections, and in malignancies and Hodgkin's disease. Increased serum iron concentrations[239] are encountered in megaloblastic anemia, thalassemia, sideroblastic anemia,[241] bone marrow hypoplasia, hepatic necrosis, and iron overload and in some normal persons. The assay methods used are critical and should follow the recommendations of the International Committee for Standardization in Hematology.[235] Measurement of serum iron should always be combined with the assay of the total iron-binding capacity and the transferrin saturation.

Total iron-binding capacity (TIBC) and transferrin saturation

The serum iron-binding capacity measures the transferrin concentration and its iron-transport capabilities. One can quantitate it by immunological methods, but one usually assays it by adding excess iron, discarding what is not bound, and then measuring the iron concentration of the sample. This method ignores iron that binds to proteins other than transferrin.[243] The result is expressed as total iron-binding capacity or as the percentage of transferrin that is saturated with iron (usually 30%). The transferrin level is reciprocally related to the iron concentration in the liver.[244] In iron-deficiency anemia the TIBC is elevated and the transferrin saturation is lowered to 15% or less. The pattern of low serum iron and high TIBC is characteristic of iron deficiency but is also seen in pregnancy without iron deficiency and in idiopathic pulmonary siderosis.[188] Low serum iron associated with low TIBC is characteristic of the anemia of chronic disorders, malignant tumors, and infections. High transferrin saturation (up to 100%) is characteristic of thalassemia, sideroblastic anemia, and hemochromatosis.

The transferrin concentration in healthy adults is 2500 to 4400 μg/L of serum, 30% of which is saturated with iron, corresponding to an average plasma iron concentration of 1020 μg/L, leaving a latent iron-binding capacity of about 3400 μg/L. There are significant diurnal and day-to-day variations on the order of 30% to 40%.[245] The transferrin level is decreased in women and in persons with malignancies, nephrosis, inflammation, chronic diseases, starvation, and megaloblastic and hemolytic anemias[204] and before menses. The levels of transferrin are increased in iron-deficiency anemia, hepatitis, and pregnancy and after ingestion of oral contraceptives.

Free erythrocyte protoporphyrin (FEP)

In the process of heme synthesis protoporphyrin incorporates iron. If the iron supply is deficient, free protoporphyrin accumulates in the red blood cells above the normal level of 700 μg/L.[246] The assay of FEP is thus a measure of iron availability and is increased not only in iron deficiency anemia but also in the absence of iron deficiency in malignancies, chronic disease, and infections.[239] In defects of the porphyrin metabolism, as in lead poisoning and erythropoietic protoporphyria, the FEP is also increased.[188]

Red blood cell zinc protoporphyrin (RBC ZP) screen

Heme is synthesized from iron and protoporphyrin IX, but when iron stores are depleted, zinc replaces the iron in the protoporphyrin ring and the level of RBC ZP rises. One can rapidly measure the latter by hematofluorometry.[247-249]

Serum ferritin

Serum ferritin originates from and varies with the intracellular stored excess iron not used for hemoglobin synthesis. Reduction in the reticuloendothelial iron stores is responsible for a low serum ferritin level,[250,251] hence its use as a diagnostic parameter of iron deficiency. There is a significant inverse relationship between serum ferritin and iron absorption in normal persons and a high direct correlation between it and storage iron concentration.[206] Serum ferritin levels are low in persons with iron-deficiency anemia (120 ng/L), in infants, and after blood loss and therapeutic phlebotomy, all of which exhibit a reduction in the level of the reticuloendothelial iron stores.[206] Serum ferritin levels are increased independently of iron-storage levels in hepatic necrosis, tumors of pancreas, lung, breast and liver, acute inflammatory conditions,[239] folic acid deficiency, Hodgkin's disease,[203] and excess iron and transfusion therapy. They parallel increased iron storage in transfusion siderosis and in idiopathic hemochromatosis,[252] though the latter picture is clouded by the frequently encountered cirrhosis and hepatoma.

Part III: Heme biosynthesis and the porphyrias

cholestatic Stopping the flow of bile from liver cells into the common bile duct.
erythema Redness of skin.
hepatoma Tumor of liver.
hypertrichosis Excessive growth of hair.
isomer Compound existing in two or more forms, which have the same structural formula, but differ in spatial configuration.
Krebs cycle Series of mitochondrial reactions, which bring about the catabolism of acetyl CoA, a high-energy compound.
mechanical fragility of skin Tendency of skin to break in response to light trauma.
melanosis Brown pigmentation. *Hypomelanosis*—loss of pigmentation; *hypermelanosis*—abnormal dark brown pigmentation.
mitochondrion An organelle of the cytoplasm of cells containing the enzymes of the synthesis of fatty acids, cholesterol, and steroids.
organelle Specialized structure in cytoplasm of cell dedicated to a specific function.
PAS stain Periodic acid–Schiff stain used to demonstrate polysaccharides in tissues.
photosensitivity Sensitivity of skin to light.
ribosome RNA-rich organelle dedicated to protein synthesis.
sideroblast Erythroblast (nucleated red cell precursor) containing ferritin granules, which stain blue with the Prussian blue reaction and form a perinuclear ring in the pathological ring sideroblast.
Watson-Schwartz test Test for porphobilinogen using Ehrlich's *p*-dimethylaminobenzaldehyde reagent.

NORMAL BIOCHEMISTRY

The synthesis[253-255] of heme in all aerobic mammalian cells, including the red blood cell precursors, is the result of a sequence of enzyme-controlled steps (Fig. 33-21). The first step is the condensation of "activated" glycine of the general amino acid pool and succinyl coenzyme A of the Krebs cycle to produce δ-aminolevulinic acid (ALA). The enzyme that catalyzes this reaction is the mitochondrial ALA synthase[256] which uses pyridoxal phosphate derived from vitamin B_6 as cofactor to "activate" glycine. ALA synthase activity is the rate-limiting step of heme synthesis. This enzyme activity is regulated by a negative-feedback control of a pool of free heme,[257] and so a heme deficit results in an increased ALA synthase; in contrast, excess heme inhibits this enzyme activity.

In the next step, two molecules of ALA self-condense to form the basic monopyrrole porphobilinogen (PBG), the precursor of heme and porphyrins. This reaction is catalyzed by the cytoplasmic enzyme ALA dehydrase after the ALA diffuses out of the mitochondria into the cytoplasm. The ALA dehydrase is characterized by its sensitivity to toxic lead levels[258] and its unique requirement for zinc.[259] PBG forms a pink pigment with Ehrlich's aldehyde reagent, the basis of the widely used Watson-Schwartz test.

The next step involves two cytoplasmic enzymes, uroporphyrinogen I synthase and uroporphyrinogen cosynthase III, which act jointly to condense four molecules of PBG to form cyclic or ring tetrapyrrolic structures, the uroporphyrinogens. At the branch point, out of the four theoretically possible porphyrinogen isomers, only types I and III are formed in nature[260] and only the latter forms heme. The small quantities of type I isomer produced are excreted in the urine. Fig. 33-21 also demonstrates that the intermediate products between PBG and porphyrins are not porphyrins but their reduced forms, porphyrinogens. The latter differ from porphyrins by the lack of certain double bonds, by being in the reduced form, and by the fact that they are unstable, colorless, and nonfluorescent. The porphyrins, on the other hand, are irreversibly oxidized, sta-

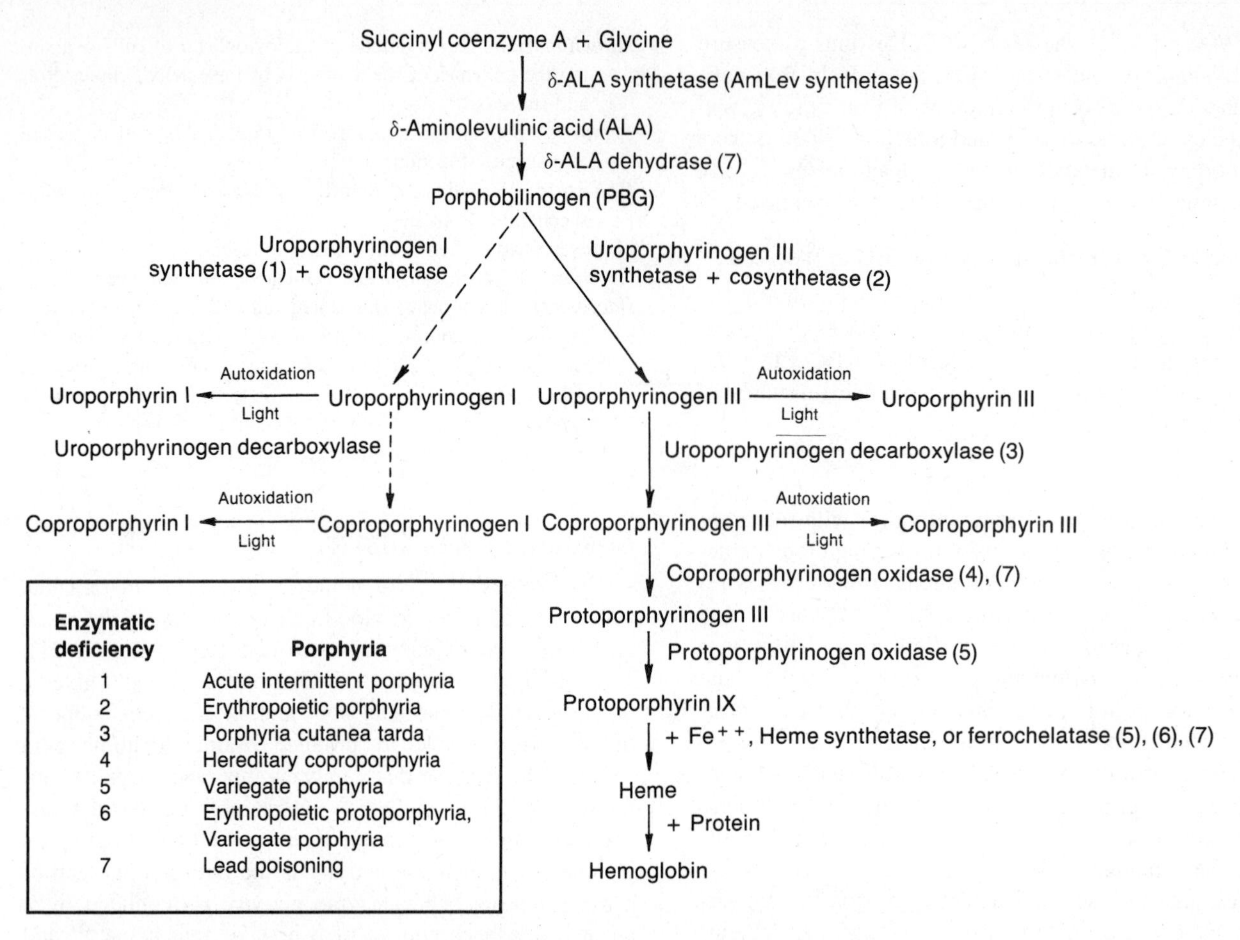

Enzymatic deficiency	Porphyria
1	Acute intermittent porphyria
2	Erythropoietic porphyria
3	Porphyria cutanea tarda
4	Hereditary coproporphyria
5	Variegate porphyria
6	Erythropoietic protoporphyria, Variegate porphyria
7	Lead poisoning

Fig. 33-21. Outline of biosynthesis of heme and of enzymatic deficiencies in porphyrias. (Modified from Bauer, J.D.: Clinical laboratory methods, ed. 9, St. Louis, 1982, The C.V. Mosby Co.)

ble, colored (red-violet at acid pH and red-brown at alkaline pH) compounds that fluoresce red when excited by light near 400 nm. These characteristics are important in clinical and laboratory diagnoses of the porphyrias and explain the darkening of the porphyrinogen-containing urine on standing. All porphyrins have a characteristic absorption band at or near 400 nm, the Soret band, named after its discoverer.

With the exception of protoporphyrins, the porphyrins are a by-product of heme synthesis. In the presence of light, oxygen, or oxidizing agents, porphyrinogens are readily oxidized to the corresponding porphyrins.[255] Once formed, they are stable but do not contribute to the biosynthesis of heme. The difference between type I and III isomers of porphyrinogen lies in the arrangement of some side groups (Fig. 33-22). The levels and the sequence of the action of the uroporphyrinogen I synthase and uroporphyrinogen III cosynthase are balanced so that the synthase produces a substrate on which the cosynthase acts[261,262] to preferentially produce the type III isomer of uroporphyrinogen. Under pathological conditions, one or the other may be deficient.[260] Decarboxylation of some of the side chains of both isomers of uroporphyrinogen leads to the stepwise formation of coproporphyrinogens I and III by the enzymatic action of uroporophyrinogen decarboxylase, a cytoplasmic enzyme. The fact that the type III isomer is decarboxylated faster than the type I isomer[263] leads to a preferential formation of the coproporphyrinogen III.[260] The pathway now returns to the anaerobic level of the mitochondria from which it began. Coproporphyrinogen III is converted to protoporphyrinogen III by the catalytic action of coproporphyrinogen oxidase, an enzyme sensitive to lead intoxication.[264] A further enzymatic step oxidizes protoporphyrinogen III to protoporphyrin. The specific enzyme is protoporphyrinogen oxidase.[265] As only the ninth protoporphyrin isomer is found in mammalian tissues, the protoporphyrin is labeled IX. The final step in the pathway is the insertion of Fe^{+++} into protoporphyrin with the help of the mitochondrial enzyme ferrochelatase or heme synthetase to form heme. The chelating enzyme is sensitive to lead intoxication. After its synthesis in the mitochondria, heme is transported into the cytoplasm for hemoglobin (or hemeprotein) generation by the cytoplasmic ribosomes. Under pathological conditions, heme may remain within the mitochondria and, when suitably stained, produces the picture of ring sideroblasts.

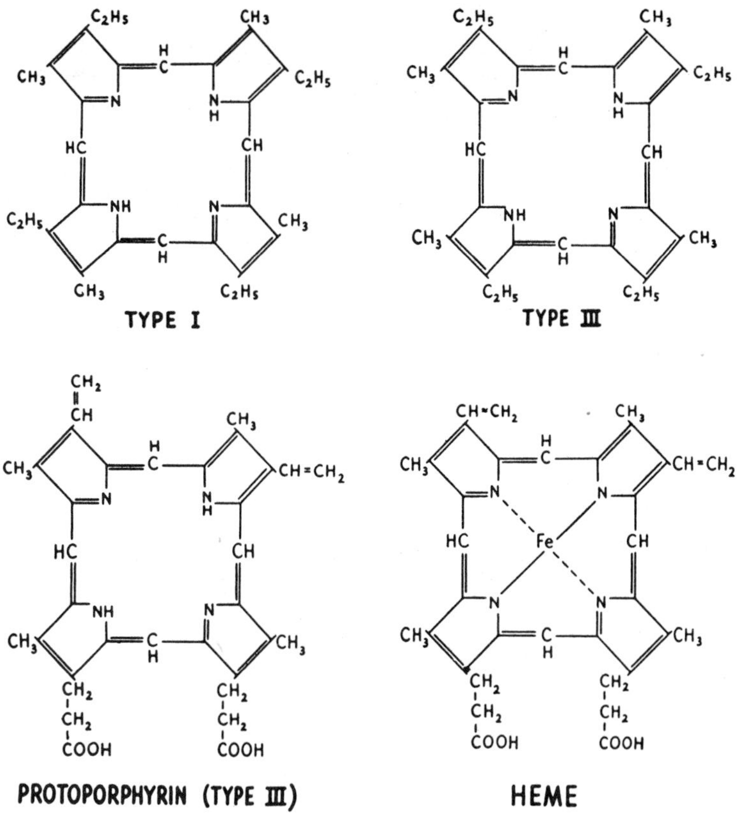

TYPE I

TYPE III

PROTOPORPHYRIN (TYPE III)

HEME

Fig. 33-22. Structure of porphyrin compound: types I and III isomers, protoporphyrin (type III), and heme. (From Miale, J.B.: Laboratory medicine hematology, ed. 6, 1982, The C.V. Mosby Co.)

PATHOLOGY: THE PORPHYRIAS

Definition

Porphyrias are a group of disorders that exhibit a defect in the conversion of glycine to heme, resulting in an abnormal porphyrin metabolism and the overproduction of heme precursors. Each of the porphyrias is the result of a deficiency, often congenital, of one of the enzymes of the porphyrinogen-heme pathway. The particular type of porphyria arising depends on which enzyme is reduced in activity.[266] The congenital forms must be distinguished from the acquired porphyrias, which are caused by alterations of the heme biosynthesis in response to drugs, liver disease, various anemias, and toxic metals.[267]

Congenital porphyrias

Classification. The classification of porphyrias is not based on the etiologically significant enzyme defects but rather on the organ primarily involved (liver or bone marrow) in the overproduction of porphyrins or their precursors, which accumulate in urine or feces, red cells, and other tissues (skin). *Hepatic porphyrias* include acute intermittent porphyria, coproporphyria, variegate porphyria, and porphyria cutanea tarda, because the effect seems to lie mainly in the liver cells. The bone marrow erythron is primarily involved in *erythropoietic porphyrias,* which include erythropoietic protoporphyria, erythropoietic porphyria, and erythropoietic coproporphyria. From the clinical point of view porphyrias are classified as (1) latent—despite the overproduction of heme precursors there are no clinical manifestations, (2) manifest—overt clinical expressions, and (3) carrier—the genetic defect is present but heme synthesis is not disturbed.[268] Although porphyrias are congenital disorders, persons with latent disease or in the carrier state of a genetic disposition may require the exogenous stimuli of drugs and foreign chemicals to precipitate overt manifestations of porphyrias.[269]

Alterations of the heme metabolism occur in a variety of pathological processes (lead poisoning, liver disease) other than congenital porphyrias but produce a porphyria-like picture called *acquired porphyria.* The latter must be differentiated from the congenital porphyrias because both groups (the acquired and the congenital) involve liver and bone marrow.

Acute hepatic porphyrias. The acute hepatic porphyrias include acute intermittent porphyria (AIP), hereditary coproporphyria (HC), and variegate porphyria (VP). They have many features in common, such as attacks of abdominal pain, neurological and psychiatric dysfunction, skin sensitivity to light (variegate porphyria and hereditary coproporphyria), accumulation of porphyrinogens and porphyrin in tissues, and excretion in urine of large quantities of δ-aminolevulinic acid and porphobilinogen. They are all uncommon and are transmitted by autosomal dominant modes of inheritance.[270] The site of the abnormal porphyrin synthesis is in the liver and expresses itself as a partial defect in hepatic heme synthesis. There is no involvement of the erythropoietic tissue. The latter feature is difficult to explain because the defect should involve all metalloporphyrin-carrying cells. It may have to do with the already mentioned feedback inhibition of ALA synthase by heme.

Acute intermittent porphyria. Acute intermittent porphyria (AIP) is frequently familial and is inherited as an autosomal dominant trait with variable pentrance, hence the variablity of symptoms. There is evidence of heterogeneity of the genetic defect as uroporphyrinogen I synthase forms four intermediates during the synthesis of uroporphyrinogen I.[271] AIP is the most common form of porphyria (1 in 100,000), occurring more commonly in women than in men. The age of onset is usually after puberty, suggesting a steroid-linked trigger mechanism.[272,273] The symptoms are intermittent, lasting for hours to days and include crampy abdominal pain, nausea, vomiting, peripheral neuropathy such as paresthesia, and sometimes psychiatric manifestations.[270] In patients with latent AIP, attacks are often precipitated by drugs such as barbiturates, hydantoins, and estrogens, but there may be no obvious precipitating agent.

The designation of the disease as acute intermittent porphyria is somewhat a misnomer because the porphyrins are not involved but rather the porphyrin precursors, aminolevulinic acid (ALA) and porphobilinogen (PBG). The enzymatic lesion is a deficiency of uroporphyrinogen I synthase in red cells and other tissues, and this synthase partially blocks the conversion of porphobilinogen into porphyrinogens (Fig. 33-21).[274,275] The deficiency of uroporphyrinogen I synthase leads to an increase in urinary PBG excretion, which is often coupled with an increased ALA excretion. The latter is the result of stimulated activity of ALA synthase secondary to diminished feedback inhibition by heme.[276]

During the attack large amounts of PBG and ALA are

Table 33-9. Classification and characteristics of congenital porphyrias

Disease	Enzymatic defect	Metabolites present in excess		
		Urine	Feces	Erythroid cells
Erythropoietic porphyria				
Congenital erythropoietic porphyria	Uroporphyrinogen I synthase and uroporphyrinogen III cosynthase	Uroporphyrin I ++++ Coproporphyrin I ++	Coproporphyrin I ++++ Uroporphyrin I ++	Uroporphyrin I ++++ Coproporphyrin I ++
Congenital erythropoietic protoporphyria	Ferrochelatase	Normal	Protoporphyrin ++++ Coproporphyrin +	Protoporphyrin ++++ Coproporphyrin +
Erythropoietic coproporphyria	?	Normal	Normal	Coproporphyrin III ++++
Hepatic porphyria				
Intermittent acute	Uroporphyrinogen I oxidase	δ-Aminolevulinic acid ++++ Porphobilinogen ++++	Normal	Uroporphyrinogen synthase ↓
Hereditary coproporphyria	Coproporphyrinogen synthase	Coproporphyrinogen III +++	Coproporphyrin III ++++ Protoporphyrin +	Normal
Variegate porphyria	Protoporphyrinogen oxidase	δ-Aminolevulinic acid ++ Porphobilinogen ++	X porphyrin ++++ Protoporphyrin ++++ Coproporphyrin ++++	Normal
Porphyria cutanea tarda	Uroporphyrinogen decarboxylase	Uroporphyrin I and III ++++	Protoporphyrin N to ++ Coproporphyrin N to ++	Normal

*+ to ++++, Degrees of increases; ↓, diminished.
Modified from Miale, J.B.: Laboratory medicine: hematology, ed. 6, St. Louis, 1981, The C.V. Mosby Co.

excreted in the urine.[277] During the latent periods the urinary PBG and ALA levels are still elevated though much lower than the acute attack concentrations and may even be normal.[277]

The elevated urinary PBG levels (fresh urine specimen) are not accompanied by increases in porphyrins, but on standing, nonenzymatic formation of porphyrins can take place in the urine specimens. The urinary excretion of PBG can amount to 150 to 200 mg/24 hr (normal 1 mg/24 hr) and that of ALA may reach 180 mg/24 hr (normal 2.5 mg/24 hr) (Table 33-9).[255] Carriers of AIP have normal ALA and PBG excretion.[255]

The diagnostic test is the measurement of the erythrocyte uroporphyrinogen I synthase, which is reduced in the latent and as well in the manifest forms of AIP.[278,279] The ALA synthase activity in leukocytes and liver is increased. Both assays lend themselves to the intrauterine diagnosis (amniotic cells) of AIP.[280]

Hereditary coproporphyria. Hereditary coproporphyria (HC) is the least common of the hepatic porphyrias,[270] and like all hepatic porphyrias it is inherited as an autosomal dominant trait. The clinical symptoms are similar to those encountered in AIP, but one third of the patients show a blistering, photosensitive skin.

The biochemical defect is a reduced activity of the coproporphyrinogen oxidase (Fig. 33-21), the reduction of which is demonstrable in leukocytes.[281,282] This defect results in the excretion of large amounts of coproporphyrin III in feces and urine (Table 33-9).

Variegate porphyria. Variegate porphyria (VP) is inherited in an autosomal dominant pattern. Clinically it resembles acute intermittent porphyria with its visceral and neurological manifestations, but in addition it shows blistering photosensitivity and mechanical cutaneous fragility. Until recent reports to the contrary, the disease, which is common in South Africa, has been considered rare in the United States.[283] Attacks are precipitated by exposure to drugs such as barbiturates, sulfonamides, alcohol, anesthetic agents, and estrogens (oral contraceptives). These agents can produce cholestatic hepatitis, and so porphyrin produced in the liver is not excreted by bile into the intestine but is shunted by the hepatic vein into the general circulation and into the skin.[284]

The enzymatic defect is a deficiency of protoporphyrinogen oxidase[285] (Fig. 33-21 and Table 33-9), possibly associated with a second defect, an inactive ferrochelatase.[286] There is a compensatory induction of ALA synthase activity leading to overproduction of ALA and PBG[287] when there is an increased demand for heme porphyrins, as in drug detoxification. Without this stimulus the reduced hepatic heme synthesis should lead to a depression of ALA synthase. High levels of coproporphyrin, protoporphyrin, and X porphyrins (hydrophilic porphyrin complexes)[287] are demonstrated in feces even when the patient is in clinical remission. During an acute attack the urine contains large amounts of ALA and PBG. Bothwell reported high serum-iron levels in 75% of patients suffering from variegate porphyria.[288]

Porphyria cutanea tarda. Porphyria cutanea tarda (PCT), a not uncommon disease, is associated with liver disease but, contrary to the other hepatic porphyrias, it lacks abdominal and neuropsychiatric symptoms. The major manifestations are limited to the skin, which exhibits excessive mechanical fragility, changes in pigmentation (hypermelanosis and hypomelanosis), photosensitivity, and diffuse thickening, erosions, and blisters.[286] The disease can be acquired or familial.[289] In the latter case it is inherited as an autosomal dominant trait. The defect can remain dormant until activated by liver disease,[290] alcoholism,[291] liver tumors,[292] siderosis,[292] therapeutic estrogens,[284] and diabetes mellitus.[291]

The enzymatic defect is a deficiency of uroporphyrinogen carboxylase[294] (Fig. 33-21 and Table 33-9), which results in the excretion of large amounts of uroporphyrin I and III in the urine. Excretions of ALA and PBG are usually normal. The fecal porphyrin contents vary from normal to high, consisting mainly of coproporphyrin and protoporphyrin.[295]

Erythropoietic protoporphyria. Congenital erythropoietic protoporphyria (EPP) is a relatively frequent disease, inherited as an autosomal dominant trait. Latent and incomplete forms are common.[284] The acute attack expresses itself as mild to severe photosensitivity of exposed skin usually of face and hands. Exposure to sunlight causes itching, burning, erythema, swelling, and superficial scarring.[296] Because liver disease and cholelithiasis are common complications of EPP[297,298] there is some doubt that this disease is purely erythropoietic, and in some cases the biochemical pattern suggests a hepatic origin of the excess protoporphyrin.[255,268,299]

The enzymatic defect is a 20% to 30% deficiency of ferrochelatase activity in all tissues including bone marrow[300] and liver[301] (Fig. 33-21 and Table 33-9), but it is not followed by the expected decreased heme synthesis or the overproduction of ALA, PBG, or ALA synthase. An essential but not diagnostic feature is the high concentration of protoporphyrin in red cells, which fluoresce when subjected to fluorescence microscopy, a phenomenon also seen in lead poisoning. There is no excess excretion of porphyrins in the urine, but there may be some increase in fecal protoporphyrin and coproporphyrin. Some patients show a variant biochemical pattern consisting in elevated fecal excretion of protoporphyrin without increased red cell protoporphyrin,[302] a possible indication of the hepatic origin of the excess protoporphyrin. Biopsy of exposed skin reveals a characteristic picture if the specimen is stained with PAS stain.[296]

Erythropoietic porphyria. Erythropoietic porphyria (EP) is a rare disease inherited as an autosomal recessive trait that manifests itself early in life, even before birth, and primarily affects the erythropoietic system. Clinically there is severe photosensitivity appearing early in life from the time of birth to a few years thereafter. It is responsible for an itching, bullous erythema that may involve the entire body. Porphyrin deposition colors the teeth brown and allows them to fluoresce bright red under ultraviolet light. Hypertrichosis, scar formation, and hyperpigmentation or hypopigmentation significantly affect the appearance of patients with EP.

The enzymatic defect probably involves uroporphyrinogen III cosynthase (Fig. 33-21 and Table 33-9) in red cells[303] and in skin fibroblasts.[304] The enzymatic deficiency leads to an increase in ALA synthase and ALA and to an increased production primarily of the I porphyrins because of the availability of excess substrate.[305] The biochemical characteristics of EP are large quantities of uroporphyrinogen I and coproporphyrinogen I, which are readily converted into uroporphyrin I and coproporphyrin I, the first exceeding the latter.[255]

The large amounts of porphyrin in the urine are responsible for its port-wine color, which, combined with the cutaneous photosensitivity, allows a rapid clinical diagnosis even in the first years of life. Greatly increased amounts of urinary and red cell uroporphyrins establish the diagnosis of EP. Some circulating red cells and most bone marrow normoblasts contain high concentrations of porphyrins, and so they fluoresce when examined under the fluorescence microscope (fluorocytes). They are responsible for a shortened red cell life-span and the resulting hemolytic anemia.

Erythropoietic coproporphyria. Erythropoietic coproporphyria is extremely rare. Only one case is described by Heilmeyer et al., who stress the photosensitivity and the significant increase in erythrocytic coproporphyrin III.[306]

Acquired (secondary) porphyrias

Alterations of heme synthesis occur in a wide variety of diseases other than the porphyrias,[267] leading to porphyria-like biochemical and clinical patterns. Most are the result of toxins or drugs producing pathological changes in specific red cell enzyme systems.

Lead poisoning. The toxicity of lead has been known for centuries, but lead poisoning is still with us. It is most common in young children who eat chips of lead-containing paint (pica)[307] but is also encountered in adults who are exposed to lead fumes (burning of batteries) or to ''moonshine'' whiskey distilled in lead-containing equipment.[308]

The most clearly defined effect of lead is defective heme synthesis resulting from inhibition of ALA dehydrase (Fig. 33-23), ferrochelatase, and coproporphyrinogen oxidase. The inhibition of ferrochelatase leads to an accumulation of free protoporphyrin in red cells, which fluoresce when examined under ultraviolet light. Increased activity of the

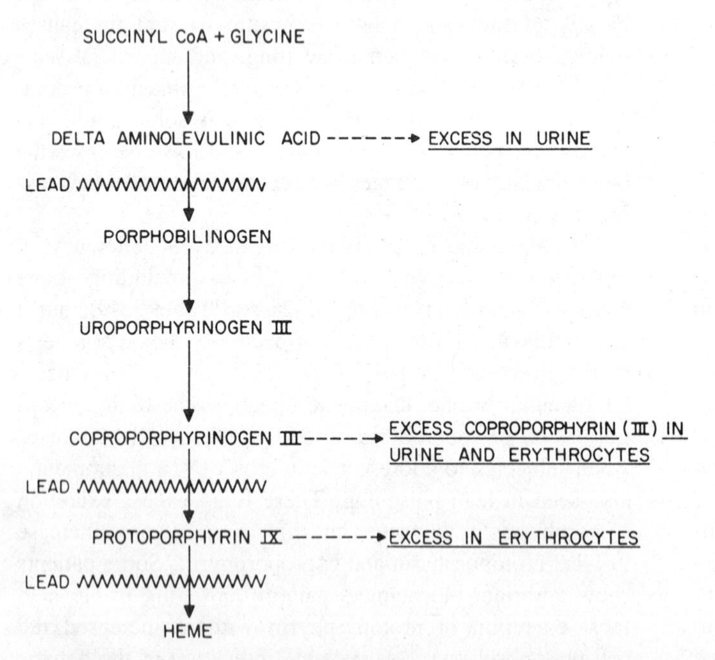

Fig. 33-23. Effect of lead on heme synthesis. *First block* in pathway of heme synthesis occurs at conversion of ALA (δ-aminolevulinic acid) to porphobilinogen, probably by inhibition of ALA dehydrase with accumulation and increased excretion of ALA in urine. *Second block* occurs at conversion of coproporphyrinogen to protoporphyrin, probably by inhibition of oxidase, resulting in excess coproporphyrin in urine and in erythrocytes. *Third block* is in synthesis of heme from protoporphyrin and iron, both of which accumulate in erythrocytes and precursors. (From Miale, J.B.: Laboratory medicine hematology, ed. 6, St. Louis, 1982, The C.V. Mosby Co.)

rate-controlling ALA synthase is the result of decreased heme production.[264] The consequence of these enzymatic inhibitions is an increase in urinary ALA, urinary porphyrins (mainly uroporphyrins), and free erythrocytic protoporphyrin (FEP). The assay for the latter lends itself to mass screening for lead poisoning, provided that a special spectrofluorometer is used.[309] The quantitation of erythrocyte ALA dehydrase is also useful as a screening test.[310] It must be stressed that the enzymatic tests are only screening tests and do not replace the more sensitive blood lead level assays.[314] The upper limit of normal lead levels is about 150 μg/L,[311] which corresponds to an FEP level of about 210 μg/L.

Most patients with lead poisoning have some anemia that is of multifactorial etiology and includes hemolysis,[312] depression of heme synthesis, and inhibition of pyrimidine 5′-nucleotidase,[313] the last contributing to the hemolysis and to basophilic stippling.

Other acquired porphyrias. Secondary porphyrinuria is common in all types of liver disease (hepatitis, cirrhosis, hepatoma) and has also been reported in excessive alcohol intake. In chronic alcoholism Brodie et al.[314] found increased concentrations of ALA synthase and uroporphyrinogen decarboxylase and decreased levels of ALA dehydrase and coproporphyrinogen oxidase. The urine contains increased concentrations of coproporphyrin[315] (Fig. 33-21).

CHANGE OF ANALYTE IN DISEASE

(Table 33-10)

Porphobilinogen (PBG)

Urinary PBG is elevated in a number of the porphyrias, including acute intermittent, variegate, and erythropoietic porphyrias. PBG levels are normal in porphyria cutanea tarda.

δ-Aminolevulinic acid (ALA)

δ-Aminolevulinic acid (ALA) tends to parallel PBG and is elevated in acute intermittent and variegate porphyrias but is normal in cutanea tarda. Urinary ALA levels are increased in lead poisoning.

ALA synthase in red blood cells

ALA synthase is increased in a number of diseases, including acute intermittent, variegate, and erythropoietic porphyrias. Increases in this enzyme activity are also seen in lead poisoning and in liver disease associated with chronic alcohol ingestion.

Coproporphyrin III

The fecal levels of coproporphyrin III are increased in hereditary coproporphyria and variegate porphyria. Normal to elevated levels are detected in porphyria cutanea tarda and erythropoietic protoporphyria. Red blood cell levels of this analyte are also increased in hereditary coproporphyria, whereas urinary levels are elevated in some secondary porphyrias associated with liver disease.

Uroporphyrin I and III

Urinary levels are elevated in porphyria cutanea tarda, erythropoietic porphyria, and lead poisoning.

Protoporphyrin

Fecal levels are elevated in variegate porphyria and normal to elevated in erythropoietic protoporphyria. RBC lev-

Table 33-10. Change of analyte with disease

	Analyte*						
Porphyria	Urinary PBG	Urinary ALA	RBC ALA synthase	Coproporphyrin III	Urinary uroporphyrin I and III	Protoporphyrin	RBC ALA dehydrase
Primary porphyrias							
Acute intermittent porphyria	↑	↑	↑	N	N	N	N
Hereditary coproporphyria	N	N	N	↑-feces ↑-RBC	N	N	N
Variegate porphyria	↑	↑	↑	↑-feces	N	↑-feces	N
Porphyria cutanea tarda	N	N	N	N, ↑-feces	↑	N	N
Erythropoietic protoporphyria	N	N	N	N, ↑-feces	N	↑-feces ↑-RBC	N
Erythropoietic porphyria	↑	N	↑	N	↑	N	N
Secondary porphyrias							
Lead poisoning	N	↑	↑	N	↑	↑-RBC	↑
Liver disease	N	N	↑	↑-urine	N	N	↓

*N = No change, normal; ↑, increased concentration or activity; ↓, decreased.

els are increased in lead poisoning and erythropoietic protoporphyria.

ALA dehydrase

RBC levels of this enzyme are elevated in lead poisoning, but decreased activity is seen in chronic liver disease associated with chronic alcohol abuse.

Part IV: Hemoglobin destruction (catabolism)

ampulla of Vater Opening of the common bile duct and the pancreatic duct in the duodenum.
bile Yellow-green secretion of liver.
bile canaliculi Network of ducts into which bile synthesized by liver cells is discharged.
colon Large bowel.
complement lysis Heat-labile plasma proteins that are "activated" by antigen-antibody complexes, causing hemolysis of red cells.
connective tissue Tissue of mesodermal origin.
connective tissue disorders Group of diseases involving connective tissue, blood vessels, and corpuscular blood components, probably caused by autoantibodies.
disseminated (diffuse) intravascular coagulation Consequence of diffuse intravascular deposition of fibrin.
duodenum First segment of small bowel extending from stomach to jejunum (second segment of small bowel).
endoplasmic reticulum Network of cytoplasmic vesicles with or without ribosomes.
erythrophagocytosis Engulfing of red cells by cells of the reticuloendothelial systems.
erythropoiesis The production of erythrocytes and their precursors.
fibrin The end product of coagulation.
glucose-6-phosphate dehydrogenase (G-6-PD) Enzyme of the glycolytic pathway.
ineffective erythropoiesis Erythropoiesis that fails to deliver cells to the blood.
leukemia Malignancy of white blood cells.
lymphoma Tumor composed of neoplastic lymphocytes or cells of lymphoreticular origin.
lymphoreticular tissues Tissues (or tumors) composed of lymphocytes (antibody producing) and reticulum cells (phagocytic).
membrane deformability Flexibility of the red cell membranous skeleton.
microsomal Pertaining to microsomes, segments of endoplasmic reticulum.
neoplastic Pertaining to a neoplasm.
osmotic lysis Loss of red cell hemoglobin in hypotonic solutions.
reticulum cell Phagocytic cell of liver, spleen, bone marrow, and lymph nodes (organs of the reticuloendothelial system).

On the average, 95% of erythrocytes live for 100 days; the remaining 5% survive 120 days. The death of normal red cells is primarily a function of senescence, but a small number of red cells are destroyed at random, irrespective of age. Red cell survival studies are used in the investigation of iron metabolism, hemolytic anemias, and survival of transfused red cells.

DESTRUCTION OF RED BLOOD CELLS

Aging of red blood cells begins as soon as they extrude their nuclei and lose their ribosomes. It results in (1) an irreversible decrease in the activity of several enzymes (G-6-PD, hexokinase, aldolase, pyruvate kinase, and so on)[316]; (2) loss of adenosine triphosphate (ATP)–dependent energy, adversely affecting cation transport, phospholipid exchange, and membrane deformability[317]; (3) reduction in membrane area, deformability, and enzymes[318]; and (4) irreversible oxidative damage to hemoglobin.[317] A number of mechanisms play a role in the destruction of nonviable aged cells: fragmentation, osmotic lysis, erythrophagocytosis, complement lysis, and hemoglobin denaturation. Approximately 90% of aged cells are destroyed by phagocytic reticuloendothelial cells in the spleen by extravascular destruction (hemolysis), the remaining 10% being destroyed intravascularly.[319] These two mechanisms of cell death are not always clinically separable but are distinguishable by contrasting routes of hemoglobin catabolism.

Extravascular hemolysis

Most extravascular erythrocyte destruction occurs in the reticuloendothelial components of the liver (Kupffer cells) and the spleen (connective-tissue macrophages, sinusoidal lining cells). Severely damaged red cells are preferentially destroyed in the liver[320] but also in all other reticuloen-

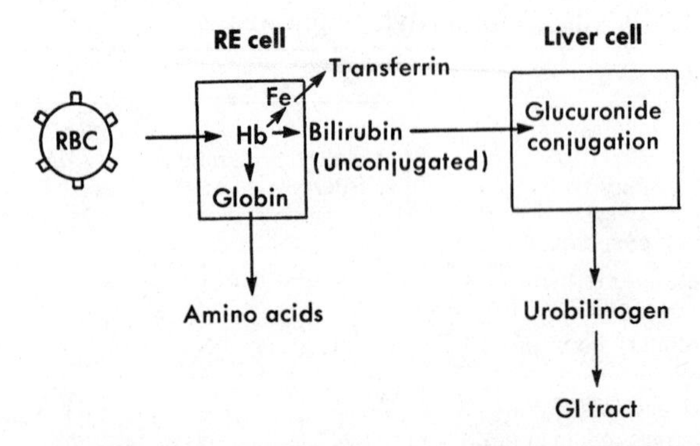

Fig. 33-24. Extravascular hemolysis. Damaged erythrocytes are not destroyed within circulation but are destroyed within macrophages lining sinusoids of spleen and liver. These reticuloendothelial *(RE)* macrophages release bilirubin pigment into blood, which is excreted by liver and kidney. (From Miller, W.V.: In Sonnenwirth, A.C., and Jarrett, L., editors: Gradwohl's clinical laboratory methods and diagnosis, ed. 8, St. Louis, 1980, The C.V. Mosby Co.)

Intrinsic red cell defects
1. Membrane defect Hereditary spherocytosis 2. Intracellular enzyme defects Glucose-6-phosphate dehydrogenase deficiency 3. Hemoglobinopathies Sickle cell anemia 4. Defect in rate of hemoglobin synthesis Thalassemia

dothelial organs, whereas minimally defective cells are destroyed in the spleen.[323] Degraded hemoglobin releases iron and bilirubin into the circulation (Fig. 33-24). The former is shunted to the bone marrow for hemoglobin synthesis,[322] and the latter starts its journey to the liver to be excreted in bile (see the discussion of formation and catabolism of bilirubin, p. 648).

Increased rates of extravascular hemolysis are found when red cells with intrinsic defects are destroyed (see box above).

Intravascular hemolysis

In intravascular hemolysis the red cells are destroyed within the circulation rather than within macrophages, and so the released hemoglobin is discharged directly into the plasma from which it is cleared by several mechanisms[321] (Fig. 33-25). In mild hemolysis all hemoglobin is bound to haptoglobin (Hp), a specific hemoglobin-binding plasma protein. One liter of plasma has enough haptoglobin to bind 1000 to 1500 mg of hemoglobin. The complex Hb-Hp is rapidly cleared by the liver, and so the haptoglobin reserves are quickly depleted. The plasma Hb-Hp complex is too big to be cleared by the renal glomeruli. At higher plasma hemoglobin levels the Hp clearance system is saturated, and the hemoglobin in excess of the haptoglobin-binding capacity circulates in the plasma as free protein. Some of this free hemoglobin is removed by the liver,[324] some is excreted in the urine (hemoglobinuria), and some is oxidized to methemoglobin. The free plasma hemoglobin dissociates into α and β dimers, which because of their small size pass through the glomeruli.[325] A large percentage of the hemoglobin in the glomerular filtrate is absorbed by the renal tubules, and so it does not appear in the urine until its threshold is exceeded[326,326a] (1.43 ± 0.96 mg/min or approximately 11 mg/L).[327] Small amounts of methemoglobin are normally present in the plasma (1% of total blood pigment).[328] Methemoglobin dissociates spontaneously into heme and globin.[329] The insoluble free heme is bound by two plasma proteins, *hemopexin* and *albumin,* which render it soluble. The heme-hemopexin complex is cleared by the liver.[330] Hemopexin is a glycoprotein synthesized in the liver. Its normal plasma range is 500 to 1000 μg/mL.[331] In intravascular hemolysis hemopexin depletion is preceded by haptoglobin absence.[332] Like haptoglobin, it is a member of the acute-phase reactant proteins, a group of proteins that are nonspecifically increased in the serum of patients with cancer, acute and chronic infections, and after major surgery and stress. *Methemalbumin* is formed from the combination of free heme with albumin. Because it is not normally present in the plasma, when detected it is diagnostic of intravascular hemolysis.[333] Its formation is preceded by haptoglobin exhaustion and is catabolized in the liver.[321]

Increased rates of intravascular hemolysis result from extracorpuscular factors that may rapidly destroy normal red cells (see following box on p. 648).

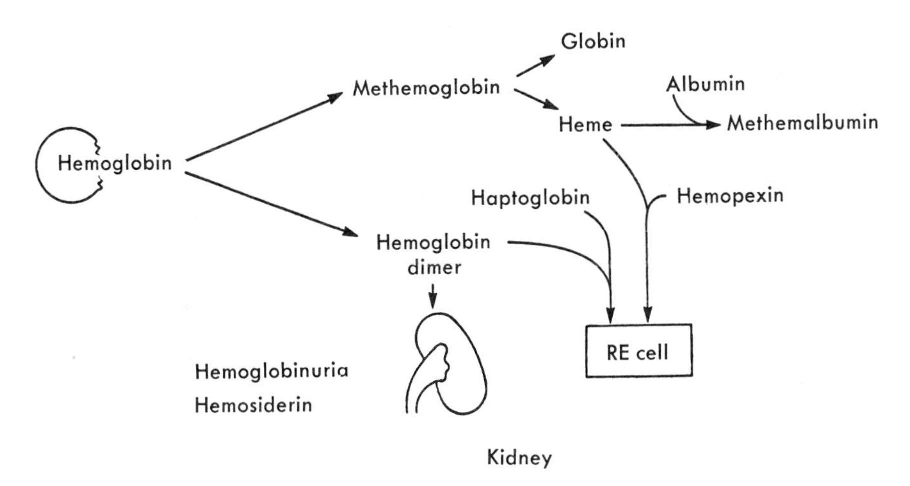

Fig. 33-25. Intravascular hemolysis. When erythrocytes are destroyed within vascular space, hemoglobin may either be cleared by reticuloendothelial cells *(RE)* or excreted directly by kidneys. (From Miller, W.V.: In Sonnenwirth, A.C., and Jarett, L., editors: Gradwohl's clinical laboratory methods and diagnosis, ed. 8, St. Louis, 1980, The C.V. Mosby Co.)

Extracorpuscular defects
Autoimmune processes caused by anti–red blood cell antibodies and complement
Primary (idiopathic)
Secondary: lymphoma, infection, connective tissue disease
Mechanical injury
Disseminated intravascular coagulation (DIC)
Animal agents
Snakebite

Haptoglobin

Haptoglobin (Hp) is an α_2 glycoprotein that belongs to the group of acute-phase reactant and transport proteins. It shares these two characteristics with transferrin (iron transport), hemopexin (heme transport), and ceruloplasmin (copper transport), which are binding proteins that nonspecifically increase in various neoplastic and inflammatory conditions. Haptoglobin specifically binds free hemoglobin, probably in the dimeric form in a molar ratio of 1:1,[334] each dimer first binding the α and then the β chains. The Hp molecule consists of two α and two β chains, reminiscent of the light and heavy chains of immunoglobulins. Hp is synthesized and destroyed by the liver. The half-life of unbound Hp is 5 days, but when complexed with Hb it is from 9 to 30 minutes.[321]

Three principal phenotypes, Hp 1-1, Hp 2-2, and Hp 2-1 are demonstrable by starch gel electrophoresis (Fig. 33-26). Originally it was believed that one single pair of genes, Hp^1 and Hp^2, was responsible for three genotypes, Hp^1/Hp^1, Hp^1/Hp^2, and Hp^2/Hp^2, but the discovery of α-haptoglobulinemia and of haptoglobin mutants (subtypes) expanded the list to five alleles and 10 phenotypes (Table 33-11).

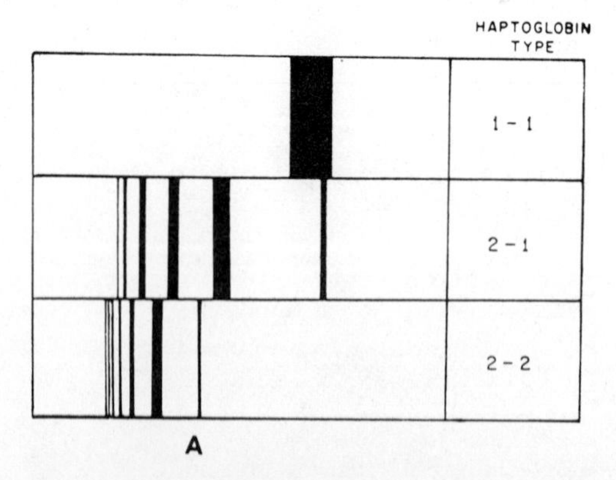

Fig. 33-26. Haptoglobin types (starch electrophoresis). (From Miale, J.B.: Laboratory medicine hematology, ed. 6, St. Louis, 1982, The C.V. Mosby Co.)

Table 33-11. Some haptoglobin types and subtypes

Haptoglobin type (phenotype)	Haptoglobin subtypes
Common phenotypes	
Hp 1-1	hp 1S-1S
	hp 1S-1F
	hp 1F-1F
Hp 2-1	hp 2-1S
	hp 2-1S
Hp 2-2	hp 2-2
	hp 2-2M
Rare phenotypes	
Hp 1 (Johnson)	hp 1 (J)F
	hp 1 (J)S
Hp 1B	None
Hp 1P	None

From Miale, J.B.: Laboratory medicine: hematology, ed. 6, St. Louis, 1982, The C.V. Mosby Co.

FORMATION AND CATABOLISM OF BILIRUBIN

Aged or damaged red cells are removed from the bloodstream by the reticuloendothelial system, which comprises the phagocytic cells of the spleen, the liver, and the bone marrow (Fig. 33-27). Within the phagocytic cells, hemoglobin and other heme proteins are catabolized to bilirubin by a series of enzymatic reactions. They first split the globin from the heme and then open the heme ring at the α-methane bridge. The cleavage is achieved by the microsomal heme oxygenase system, which converts heme into equimolar amounts of biliverdin, iron, and carbon monoxide (CO).[335-338] Globin is returned to the plasma protein and amino acid pool, heme iron to the plasma iron pool, and carbon monoxide is excreted by the lungs (Fig. 33-28). Biliverdin is rapidly reduced by biliverdin reductase to bilirubin.[338a] The activity of the oxygenase system is greatest in the spleen, followed by that in bone marrow, liver, brain, kidney, and lung.[8] Seventy to 80% of bilirubin is the metabolic end product of the continuing turnover of hemoglobin. The remaining 20% to 30% arise from ineffective erythropoiesis in the bone marrow and from heme-containing microsomal enzymes in the liver. The contributions from the catabolism of extrahepatic heme proteins such as myoglobin, cytochromes, and other heme protein enzymes have not been confirmed in humans.[338] After its formation within phagocytic cells, bilirubin is released into the circulation and is reversibly bound to albumin.[340] In this form the bilirubin is *unconjugated.* Bilirubin unbound to any protein is only sparingly soluble and toxic to cells.[341] Unconjugated bilirubin is cleared rapidly from the blood by the liver. Before entering the liver cells, bilirubin is dissociated from its carrier protein (albumin),[342] accepted by an as yet not proved membrane carrier protein of liver cells, and transported into the cell interior (cytosol) bound to two organic-anion binding proteins, *ligandin* and *Z protein,*[342] which carry it to the

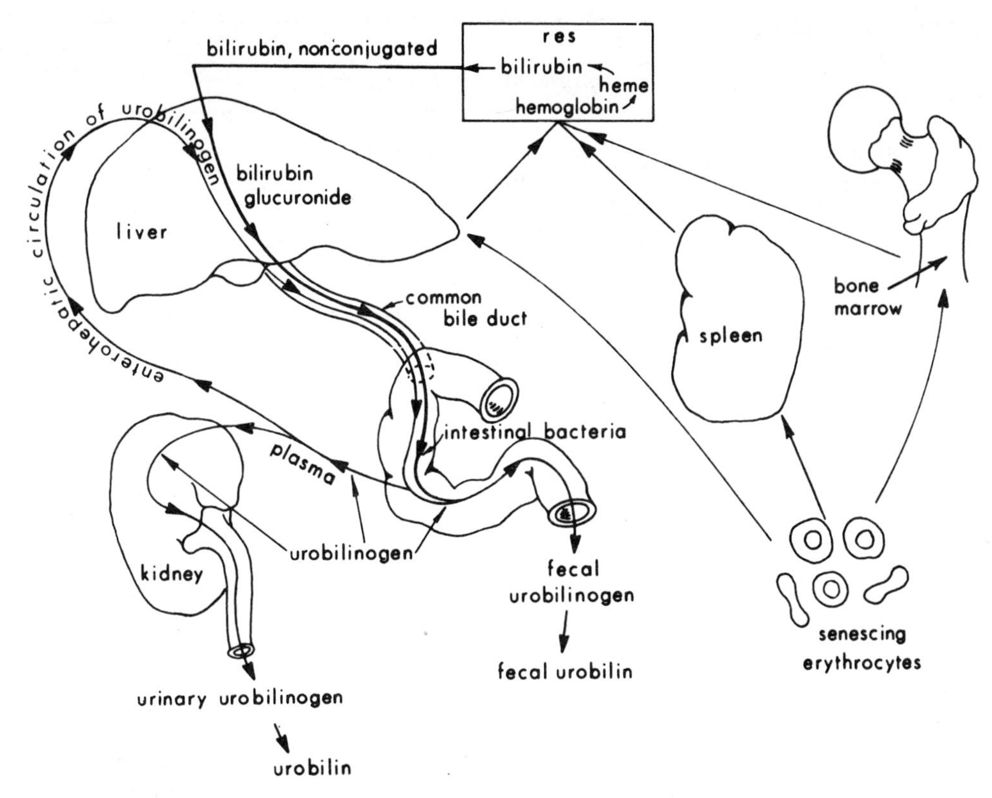

Fig. 33-27. Bilirubin metabolism. *res,* Reticuloendothelial system. (From Bauer, J.D.: Clinical laboratory methods, ed. 9, St. Louis, 1982, The C.V. Mosby Co.)

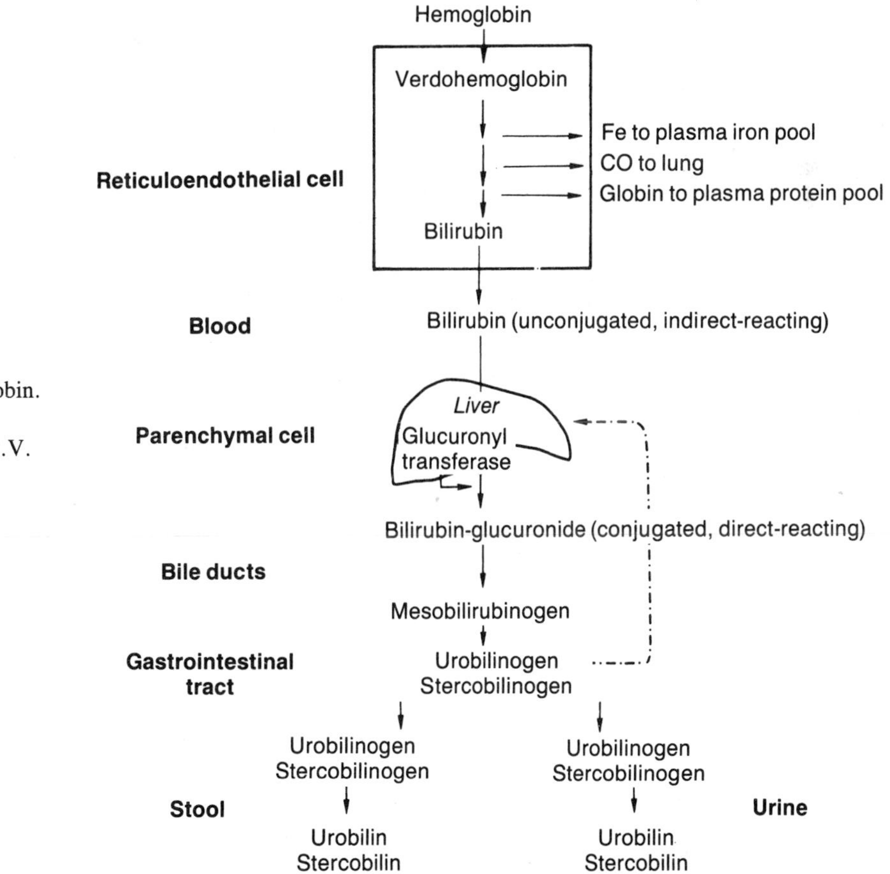

Fig. 33-28. Catabolic pathways of hemoglobin. (From Miale, J.B.: Laboratory medicine hematology, ed. 6, St. Louis, 1982, The C.V. Mosby Co.)

smooth endoplasmic reticulum for conjugation with glucuronic acid (the acid form of glucose). In this form the bilirubin is *conjugated*. The glucuronic acid donor is uridine diphosphoglucuronic acid, and the enzyme glucuronosyl transferase is required for conjugation. The conjugated bilirubin is water soluble and nontoxic. Most conjugated bilirubin is in the form of diglucuronide with a minor component as the monoglucuronide.[339] The conjugates are excreted into the bile canaliculi and through the common bile duct into the duodenum. Conjugated bilirubin does not spill into the bloodstream unless injury of liver cells interferes with the transport of bilirubin from hepatocytes to bile canaliculi, allowing conjugated bilirubin to accumulate in the plasma. The appearance of conjugated bilirubin in plasma is a sensitive marker of liver cell injury. When plasma concentration of conjugated bilirubin is increased, it also appears in the urine. In massive hemolysis, conjugated and unconjugated bilirubin enrich the plasma in the absence of liver injury. In the intestines bilirubin is reduced by bacteria to the colorless compounds called *urobilinogens*,[343,344,344a] which include mesobilirubinogen, stercobilinogen, and *d*-urobilinogen. They are excreted in feces and urine and react with Ehrlich's aldehyde reagent to form a red-violet compound. On standing they are easily oxidized to urobilins, which are colored compounds that fail to react with Ehrlich's reagent. Approximately 20% of urobilinogen is absorbed from the colon into the bloodstream, returned to the liver by the portal circulation, and reexcreted by the liver cells into bile (enterohepatic circulation of urobilinogen). A small fraction of the absorbed urobilinogen remains in the plasma to be excreted by the kidneys in the urine. In hepatocellular injury without cholestasis or when there is overproduction of bile pigment, as in hemolysis, urobilinogen excretion by the liver is overwhelmed and increased amounts of urobilinogen appear in the plasma and the urine. On the other hand, if the intestinal bacterial flora is sterilized by antibiotics, the conversion of bilirubin to urobilinogen is interfered with, resulting in low fecal and urinary urobilinogen excretion.[345] Because the urinary urobilinogen excretion is a function of glomerular filtration and tubular secretion and absorption, renal insufficiency affects its urine concentration.[346]

CHANGE OF ANALYTE IN DISEASE

Hyperbilirubinemia

Hyperbilirubinemia results from increased plasma concentrations of conjugated, unconjugated, or both bilirubins. Bilirubin is assayed by the methods that measure (1) total bilirubin and (2) conjugated, direct-reacting bilirubin. The unconjugated bilirubin fraction is calculated from the difference between the total and direct-reacting values.[347]

Hyperbilirubinemia is of two types: unconjugated or conjugated (see accompanying box). The differentiation is not always clear because liver injury may be coupled with obstruction or with hemolysis and renal failure.

Causes of hyperbilirubinemia

- Increase in plasma unconjugated bilirubin
  - Increased hemolysis
  - Genetic errors
  - Neonatal jaundice
  - Ineffective erythropoiesis
  - Drugs competing for glucuronide
- Increase in plasma conjugated bilirubin
  - Cholestasis
  - Genetic errors
  - Hepatocellular damage

Increase in plasma concentration of unconjugated bilirubin

1. Accelerated breakdown of hemoglobin leads to an increased plasma concentration of unconjugated bilirubin that exceeds the capacity of the liver cells to bind and conjugate it, so that it accumulates in the plasma. Increased hemolysis is seen in isoimmune hemolytic disease, congenital hemolytic anemia, and sepsis. In these conditions the total plasma bilirubin is almost 100% unconjugated pigment.
2. Diminished plasma bilirubin clearance is a feature of Crigler-Najjar[348] and Gilbert's syndromes,[349] the result of genetic errors (see Chapter 24 for details).
3. Transient neonatal hyperbilirubinemia is caused by a defect in hepatic intracellular binding and conjugation of bilirubin, which reflects the newborn's inability to synthesize adequate supplies of uridine diphosphoglucuronic acid and glucuronosyl transferase.[350]
4. Ineffective erythropoiesis leads to unconjugated hyperbilirubinemia, as seen in folic acid and B_{12} deficiencies.
5. Antibiotics and some tranquilizers compete with glucuronide and thus interfere with the hepatic clearance of unconjugated bilirubin.

Increased plasma concentration of conjugated bilirubin

1. Obstruction to the biliary flow anywhere in the biliary tract from bile canaliculi to the ampulla of Vater raises the total plasma bilirubin level, which is almost 100% conjugated pigment.
2. Genetic errors in bilirubin metabolism are responsible for hyperbilirubinemia in the Dubin-Johnson[351] and Rotor syndromes.[352] The essential feature of the Dubin-Johnson syndrome consists in an inherited defect in the transport of conjugated bilirubin from liver cell to bile canaliculi. Sixty percent of the plasma bilirubin is direct reacting. The Rotor syndrome has the biochemical features of cholestasis associated with coproporphyrinuria.
3. Hepatocellular damage. Severe liver disease is usu-

ally accompanied by cholestasis and renal impairment and leads to very elevated plasma bilirubin levels, 80% of which is conjugated.

Haptoglobin

Serum haptoglobin levels are high in inflammatory and neoplastic diseases, which include rheumatoid arthritis, lymphomas, leukemias, myeloma, and some forms of cancer. Levels are low or not demonstrable in newborns,[335] in hemolytic anemias, and hemolytic transfusion reactions characterized by intravascular hemolysis. Because of the rapid destruction of the Hp-Hb complex the low or absent level of Hp serves as a sensitive indicator of acute or chronic hemolysis. If the hemolysis ceases, it takes days to replenish the supply of Hp.

Free hemoglobin

Normal plasma contains small amounts of free hemoglobin (mean, 3 mg/L). Intravascular hemolysis can increase plasma levels of up to 2000 mg/L (hemoglobinemia). When the plasma levels exceed approximately 11 mg/L, free hemoglobin is found in urine.

REFERENCES

Hemoglobin

1. Lehmann, H., and Carrell, R.: Br. Med. Bull. **25:**14, 1969.
2. Winslow, R.M., and Anderson, W.F.: The hemoglobinopathies. In Stanbury, J.B., Wyngaarden, J.B., and Fredrickson, D.S., editors: The metabolic basis of inherited disease, ed. 4, New York, 1978, McGraw-Hill Book Co.
3. Braunitzer, G.: Hoppe Seylers Z. Physiol. Chem. **312:**72, 1958.
4. Braunitzer, G., Hilschmann, N., Rudloff, V., et al.: Nature **190:**480, 1961.
5. Rhinesmith, H.S., Schroeder, W.A., and Pauling, L.: J. Am. Chem. Soc. **79:**4682, 1957.
6. Perutz, M.F., Muirhead, H., Cox, J.M., et al.: Nature **219:**131, 1968.
7. Perutz, M.F.: Sci. Am. **239:**92, 1978.
8. Nienhuis, A.W., and Benz, E.J., Jr.: N. Engl. J. Med. **297:**1318, 1371, 1430, 1977.
9. Bradley, T.B., Boyer, S.H., and Allen, E.H.: Bull. Johns Hopkins Hosp. **108:**75, 1961.
10. Nienhuis, A.W., and Benz, E.J., Jr.: Ann. Intern. Med. **91:**883, 1979.
11. Safer, B., and Anderson, W.F.: CRC Crit. Rev. Biochem. **5:**261, 1978.
12. Benz, E.J., Jr., and Forget, B.G.P.: Semin. Hematol. **11:**463, 1974.
13. Deisseroth, A., Baker, J., Anderson, W.F., et al.: Proc. Natl. Acad. Sci. **72:**2682, 1975.
14. Gaul, A., and Axel, R.: Proc. Natl. Acad. Sci. USA **72:**3966, 1976.
15. Nienhuis, A.W., and Anderson, W.F.: Clin. Haematol. **3:**437, 1974.
16. Rabinowitz, M.: Ann. N.Y. Acad. Sci. **241:**322, 1974.
17. Lodish, H.F.: Annu. Rev. Biochem. **45:**39, 1976.
18. Weatherall, D.J.: J. Clin. Pathol. **8**(suppl.):1, 1974.
19. Engel, E.: The chromosome basis of human heredity. In Stanbury, J.B., Wyngaarden, J.B., and Fredrickson, D.S., editors: The metabolic basis of inherited disease, New York, 1978, McGraw-Hill Book Co.
20. Ranney, H.M.: J. Clin. Invest. **33:**1634, 1954.
21. Smith, E.W., and Torvert, J.F.: Johns Hopkins Med. J. **102:**38, 1958.
22. Wong, S.C., and Huisman, T.H.J.: Clin. Chim. Acta **38:**473, 1972.
23. Charache, S., Ostertag, W., and von Ehrenstein, G.: Nature (London) **234:**248, 1972.
24. Granick, S., and Levere, R.D.: Heme synthesis in erythroid cells, Prog. Hematol. **4:**1, 1964.
25. Meyer, V.A., and Schmid, R.: The porphyrins. In Stanbury, J.B., Wyngaarden, J.B., and Fredrickson, D.S., editors: The metabolic basis of inherited diseases, ed. 4, New York, 1978, McGraw-Hill Book Co.
26. Levere, R.D., and Granick, S.: J. Biol. Chem. **242:**1903, 1967.
27. Boycott, A.E.: Proc. R. Soc. Med. **23:**15, 1929.
28. Bessis, M.C., and Breton-Gorius, J.: Blood **19:**635, 1962.
29. Bessis, M.C.: Harvey Lectures **58:**125, 1963.
30. Tanaka, Y., Brecher, G., and Bull, B.: Blood **28:**758, 1966.
31. Tanaka, Y., and Brecher, G.: Blood **37:**121, 1971.
32. Jandl, J.H., Inman, J.K., Simmons, R.L., et al.: J. Clin. Invest. **38:**161, 1959.
33. Katz, J.H., and Jandl, J.H.: The role of transferrin in the transport of iron into the developing red cell. In Gross, F., editor: Iron metabolism, Berlin, 1964, Springer-Verlag.
34. Granick, S., and Levere, R.: Controls of hemoglobin synthesis. Proceedings of the Twelfth Congress of the International Society of Hematology, New York, 1968.
35. Ponka, P., and Neuwirt, J.: Blood **33:**690, 1969.
36. Neuwirt, J., and Ponka, P.: The regulation of heme synthesis in erythroid cells by the feedback of inhibition of cellular uptake of substrates. In Martin, H., and Nowicki, L., editors: Synthesis, structure and function of hemoglobin, Munich, 1972, J.F. Lehmanns Verlag, pp. 61-66.
37. Clegg, J.B., and Weatherall, D.J.: Nature **240:**190, 1972.
38. McDonald, M.J., Shaeffer, J.R., Turei, S.M., and Bunn, H.F.: Subunit assembly of hemoglobin A&S. In Brewer, G.J., editor: The red cell, Fifth Ann Arbor Conference, New York, 1981, Alan R. Liss, Inc.
39. Shaeffer, J.R.: J. Biol. Chem. **255:**2322, 1980.
40. Shaeffer, J.R., Kingston, R.E., McDonald, M.J., et al.: Nature **276:**631, 1978.
41. Ganong, W.F.: Review of medical physiology, Los Altos, Calif., 1979, Lange Medical Publications.
42. Roughton, F.J.W.: Transport of oxygen and carbon dioxide. In Senn, W.O., and Rahn, H., editors: Handbook of physiology, respiration, Washington, D.C., 1964, American Physiological Society.
43. Bunn, F.H.: Hemoglobin I. Structure and function. In Beck, W.S., editor: Hematology, Cambridge, Mass., 1981, MIT Press.
44. Bellingham, A.J.: Ninth symposium on abnormal medicine: Proceedings of a conference at the Royal College of Physicians, Turnbridge Wells, England, 1973, Pitman Medical Publishing Co., Ltd.
45. Riggs, A.: Physiol. Rev. **45:**619, 1965.
46. Carrell, R.W., and Lehmann, H.: Abnormal hemoglobins. In Brown, S.S., Mitchell, F.L., and Young, D.S., editors: Chemical diagnosis of disease, Amsterdam, 1979, Elsevier North-Holland Biomedical Press.
47. Rapoport, S., and Leubering, J.: J. Biol. Chem. **183:**507, 1950.
48. Benesch, R., Benesch, R.E., and Yu, C.I.: Proc. Natl. Acad. Sci. **59:**526, 1968.
49. Rorth, M.: Ser. Haematol. **5:**1, 1972.
50. Astrup, P.: N. Engl. J. Med. **283:**202, 1970.
51. Rabitzis, E.T., and Mills, G.C.: Biochem. Biophys. Acta **141:**439, 1967.
52. Torrance, J., et al.: N. Engl. J. Med. **283:**165, 1970.
53. Moore, L.G., Bunn, G.J.: Am. J. Phys. Anthrop. **53:**11, 1980.
54. Clench, J., Ferrell, R.E., Schull, W.J., et al.: Hematocrit and hemoglobin, ATP and DPG concentrations in Andean man; The interaction of altitude and trace metals with glycolytic and hematologic parameters in man. In Brewer, G.J., editor: The red cell: Fifth Ann Arbor Conference, New York, 1981, Alan R. Liss, Inc.
55. Perutz, M.: Nature **228:**734, 1970.
56. Kilmartin, J.V.: Br. Med. Bull. **32:**209, 1976.
57. Garby, L., and Robert, M.: Acta Physiol. Scand. **84:**482, 1972.
58. Urbanetti, J.S.: Carbon monoxide poisoning. In Wallach, D.F.H.,

editor: The function of the red blood cells: erythrocyte pathobiology, New York, 1981, Alan R. Liss, Inc.
59. Jaffe, E.R.: Methemoglobin pathophysiology. In Wallach, D.F.M., editor: The function of the red blood cells: erythrocyte pathobiology, New York, 1981, Alan R. Liss, Inc.
60. Hsieh, H.S., and Jaffe, E.R.: The metabolism of methemoglobin in human erythrocytes. In Surgenor, D.MacN., editor: The red blood cells, New York, 1975, Academic Press, Inc.
61. Schwartz, J.M., and Jaffe, E.R.: Hereditary methemoglobinemia with deficiency of NADH dehydrogenase. In Stanbury, J.B., Wyngaarden, J.B., and Fredrickson, D.S., editors: The metabolic basis of inherited diseases, New York, 1978, McGraw-Hill Book Co.
62. Jaffe, E.R.: Am. J. Med. **41:**786, 1966.
63. Jaffe, E.R., Neumann, G., Rothberg, H., et al.: Am. J. Med. **41:**42, 1966.
64. Jaffe, E.R.: Clin. Haematol. **12:**99, 1981.
65. Miale, J.B.: Laboratory medicine: hematology, ed. 6, St. Louis, 1982, The C.V. Mosby Co.
66. Jaffe, E.R., and Heller, P.: Methemoglobinemia in man. In Moore, C.V., and Brown, E.B., editors: Progress in hematology, New York, 1966, Grune & Stratton, Inc.
67. Winterbourn, C.C., and Carrell, R.W.: Br. J. Haematol. **25:**585, 1973.
68. Carrell, R.W., Winterbourn, C.C., and Rachmilewitz, E.A.: Br. J. Haematol. **30:**259, 1975.
69. Rachmilewitz, E.A., and Harari, E.: Br. J. Haematol. **22:**357, 1972.
70. Rachmilewitz, E.A.: Semin. Hematol. **11:**441, 1974.
71. Finch, C.A.: N. Engl. J. Med. **239:**470, 1948.
72. Reynolds, T.B., and Ware, A.: J.A.M.A. **149:**1538, 1952.
73. Berzofsky, J.A., Peisach, J., and Blumberg, W.F.: J. Biol. Chem. **246:**3367, 1971.
74. Allen, D.W., Schroeder, W.A., and Balog, J.: Am. Chem. Soc. **80:**1628, 1958.
75. Gambino, R., editor: Lab. Report Physicians **3:**83, 1981.
76. McDonald, M.J., Shapiro, R., Bleichman, M., et al.: J. Biol. Chem. **253:**2327, 1978.
77. Bunn, F.H.: Modification of hemoglobin and other proteins by nonenzymatic glycosylation. In Wallach, D.F.H., editor: The function of red blood cells: erythrocyte pathobiology, New York, 1981, Alan R. Liss, Inc.
78. Haney, E.N., and Bunn, H.F.: Proc. Natl. Acad. Sci. USA **73:**3534, 1976.
79. Huisman, T.H.J., and Dozy, A.M.: J. Lab. Clin. Med. **60:**302, 1962.
80. Bunn, H.F., and Biehl, R.W.: J. Clin. Invest. **49:**1088, 1970.
81. Efremov, G.D., Miadenovisky, B., Petkov, G., et al.: New Istanbul Contrib. Clin. Sci. **12:**211, 1978.
82. Roberts, A.V., Weatherall, D.J., and Clegg, J.B.: Biochem. Biophys. Res. Comm. **47:**81, 1972.
83. Schroeder, W.A., Huisman, T.H.J., Shelton, R., et al.: Proc. Natl. Acad. Sci. USA **60:**537, 1968.
84. Oski, F.A., and Naiman, J.L.: Hematologic problems of the newborn, Philadelphia, 1982, W.B. Saunders Co.
85. Huehns, E.R., and Beaven, G.H.: Developmental changes in human hemoglobin. In Benson, P., editor: The biochemistry of development, London, 1971, Spastics International Medical Publications, William Heinemann Medical Books.
86. Singer, K., Chernoff, A.J., and Singer, L.: Blood **6:**413, 1951.
87. Kleihauer, E.: Beihefte Arch. Kinderheilkd. **53** (Suppl.), 1966.
88. Perutz, M.F., and Lehmann, H.: Nature (London) **219:**90, 1968.
89. Pataryas, H.A., and Stamatoyannopoulos, G.: Blood **39:**688, 1972.
90. Wells, R.M.G.: J. Comp. Physiol. **129:**333, 1929.
91. Weatherall, D.J., and Clegg, J.B.: The thalassemia syndrome, Oxford, 1981, Blackwell Scientific Publications, Ltd.
92. Weatherall, D.J.: Abnormal hemoglobins and thalassemias. In Hoffbrand, A.V., Brain, M.C., and Hirsh, J., editors: Recent advances in hematology, London, 1977, Churchill Livingstone.
93. Fairbanks, V.F.: Nomenclature and taxonomy of hemoglobin variants. In Fairbanks, V.F., editor: Hemoglobinopathies and thalassemias, New York, 1980, Brian C. Decker, Publisher.
94. Ingram, V.M.: Biochem. Biophys. Acta **28:**539, 1958.
95. Briehl, R.W.: Physical chemical properties of sickle cell hemoglobin. In Wallach, D.F.H., editor: The function of red blood cells: erythrocyte pathobiology, New York, 1981, Alan R. Liss, Inc.
96. Harris, J.W.: Proc. Soc. Exp. Biol. Med. **75:**197, 1950.
97. Bunn, H.F.: Hemoglobin II, sickle cell anemia and other hemoglobinopathies. In Beck, W.S., editor: Hematology, Cambridge, Mass., 1981, MIT Press.
98. Briehl, R.W., and Ewert, S.: J. Mol. Biol. **80:**445, 1973.
99. Briehl, R.W.: J. Mol. Biol. **123:**521, 1978.
100. Huehns, E.R.: The structure and function of hemoglobin in clinical disorders due to abnormal hemoglobin structure. In Hardesty, R.M., and Weatherall, D.J., editors: Blood and its disorders, Oxford, 1974, Blackwell Scientific Publications, Ltd.
101. Milner, P.F., Miller, C., Gray, R., et al.: N. Engl. J. Med. **283:**1417, 1970.
102. Hibbel, R.P.: J. Clin. Invest. **65:**154, 1980.
103. Nagel, R.L., and Bookchin, R.M.: Oxygen transport and the sickle cell. In Wallach, D.F.H., editor: The function of red blood cells: erythrocyte pathobiology, New York, 1981, Alan R. Liss, Inc.
104. Charache, S., Grisolia, S., Fiedler, A.J., et al.: J. Clin. Invest. **49:**806, 1970.
105. Gill, S.J., Shold, R., Fall, L., et al.: Science **201:**362, 1978.
106. May, A., and Huehns, R.E.: Haemat. Bluttransfus. **10:**129, 1972.
107. Vega, R., Shanberg, A.M., and Malloy, T.R.: J. Urol. **105:**552, 1971.
108. Conn, H.O.: N. Engl. J. Med. **251:**417, 1954.
109. Jones, S.R.: N. Engl. J. Med. **282:**373, 1970.
110. Green, R.L.: Br. Med. J. **4:**593, 1971.
111. Nagpal, K., Asdourian, G.K., Patrianakos, D., et al.: Arch. Intern. Med. **137:**325, 1977.
112. Milner, P.F.: The clinical effects of Hb S, an overview. In Wallach, D.F.H., editor: The function of red blood cells: erythrocyte pathobiology, New York, 1981, Alan R. Liss, Inc.
113. Diamond, H.S., Meisel, A.D., and Holden, D.: Ann. Intern. Med. **90:**752, 1979.
114. Prasad, A.S., Schoomaker, E.B., Ortega, J., et al.: Clin. Chem. **21:**582, 1975.
115. Matulsky, A.G.: N. Engl. J. Med. **288:**31, 1973.
116. Schneider, R.G.: J. Lab. Clin. Med. **44:**133, 1954.
117. Milner, P.F., and Wrightstone, R.N.: The unstable hemoglobins: a review. In Wallach, D.F.H., editor: The function of red blood cells: erythrocyte pathobiology, New York, 1981, Alan R. Liss, Inc.
118. White, J.M.: Br. Med. Bull. **32:**219, 1976.
119. Jereb, J.A.: Hemoglobin stability tests. In Fairbanks, V.F., editor: Hemoglobinopathies and thalassemias, laboratory methods and case studies, New York, 1980, Brian C. Decker, Publisher.
120. White, J.M.: Clin. Haematol. **3:**333, 1974.
121. Kreimer-Birnbaum, M., Pinkerton, P.H., Bannerman, R.M., et al.: Br. Med. J. **2:**396, 1966.
122. Patrick, C.W.: Genetic hematology, American Society of Clinical Pathologists National Meeting, Washington, D.C., 1974.
123. Patrick, C.W., Steine, E.A., Doughtery, W.M., and Evans, V.S.: The erythrocytes, vol. 2, Hemolytic anemias, American Society of Clinical Pathologists National Meeting, Washington, D.C., 1974.
124. Fairbanks, V.F.: Thalassemias and hereditary persistence of fetal hemoglobins (HPFH). In Fairbanks, V.F., editor: Hemoglobinopathies and thalassemias, laboratory methods and case studies, New York, 1980, Brian C. Decker, Publisher.
125. Haight, V.: Thalassemia syndromes. In Red cell seminar II, acquired and genetic red cell defects, Milwaukee, 1979, School of Allied Health Professions, University of Wisconsin.
126. Wasi, P., Pootrakul, S., and Winichagon, P.: Screening for thalassemia in Thailand. In Schmidt, R.M., editor: Abnormal hemoglobins and thalassemias, New York, 1975, Academic Press, Inc.
127. Wasi, P., Na-Nahorn, S., and Pootrakul, S.: Clin. Haematol. **3:**383, 1974.
128. Worwood, M., Edwards, A., and Jacobs, A.: Nature **229:**409, 1971.
129. Okene-Frempong, K., and Schwartz, E.: Pediatr. Clin. North Am. **27:**403, 1980.
130. Hunt, J.A., and Ingram, V.M.: Nature **184:**640, 1959.
131. England, J.M., and Fraser, P.M.: Lancet **1:**145, 1979.
132. Loria, A., Konijn, A.M., and Hersko, C.: Isr. J. Med. Sci. **14:**1127, 1978.

133. Sicuranza, B.J., Tisdall, L.H., Saveck, R., et al.: N.Y. State J. Med. **78**:1691, 1978.
134. Fessas, P., and Loukopoulos, D.: Clin. Haematol. **3**:411, 1974.
135. Fessas, P., and Loukopoulos, D.: Science **143**:590, 1964.
136. Charache, S., Clegg, J.B., and Weatherall, D.J.: Br. J. Haematol. **34**:527, 1976.
137. Boyer, S.H., Margolet, L., Boyer, M.L., et al.: Am. J. Hum. Genet. **29**:256, 1977.
138. Fessas, P., and Stamatoyannopoulos, G.: Blood **24**:223, 1964.
139. Conley, C.L., Weatherall, D.J., Richardson, S.N., et al.: Blood **21**:261, 1963.
140. Weatherall, D.J., Cartner, R., Clegg, J.B., et al.: Br. J. Haematol. **29**:205, 1975.
141. Marti, H.R., and Bütler, R.: Acta Haematol. **26**:65, 1961.
142. Huisman, T.H.J., Schroeder, W.A., Adams, H.R., et al.: Blood **36**:1, 1970.
143. Ringlehann, B., Acquaye, C.T.A., Oldham, J.H., et al.: Biochem. Genet. **15**:1053, 1977.
144. Gross, R.T., Bracci, R., Schroeder, E., et al.: Blood **29**:481, 1967.
145. Jaffe, E.R.: Clin. Haematol. **10**:99, 1981.
146. Elwood, P.C., Shinton, N.K., Wilson, C.I.D., et al.: Br. J. Haematol. **21**:557, 1971.
147. Cruickshank, J.M.: Br. J. Haematol. **18**:523, 1970.
148. Vellar, O.D.: Acta Med. Scand. **182**:681, 1967.
149. Burton, J.L.: Lancet **1**:978, 1967.
150. Eklund, L.G., Eklund, B., and Kaijsur, L.: Acta Med. Scand. **190**:335, 1971.
151. Hurtado, A., Merino, C., and Delgado, E.: Arch. Intern. Med. **75**:284, 1964.
152. Stengle, J.M., and Schade, A.L.: Br. J. Haematol. **3**:117, 1957.
153. Shepard, M.K., Weatherall, D.J., and Conley, C.L.: Bull. Johns Hopkins Hosp. **110**:293, 1962.
154. Betke, K.: Proceedings of the Eighth International Congress of Hematology, Tokyo, 1960, Pan-Pacific Press.
155. Zargo, M.A., Wood, W.G., Clegg, J.B., et al.: Blood **53**:977, 1979.
156. Wood, W.G., Stamatoyannopoulos, G., Lim, G., et al.: Blood **46**:671, 1975.
157. Boyer, S.H., Belding, T.K., Margolet, L., et al.: Science **188**:361, 1975.
158. Dover, G.J., Boyer, S.H., and Zinkham, W.H.: J. Clin. Invest. **63**:173, 1979.
159. Hardisty, R.M., Speed, D.E., and Till, M.: Br. J. Haematol. **10**:551, 1964.
160. Weatherall, D.J., Pembrey, M.E., and Pritchard, J.: Clin. Haematol. **3**:467, 1974.
161. Bunn, H.F., Haney, D.N., Gabbey, K.H., et al.: Biochem. Biophys. Res. Commun. **67**:103, 1975.
162. Ambler, J.: Electrophoresis Today **3**:1, 1982.
163. Bunn, H.F., Haney, D.N., Kain, S., et al.: J. Clin. Invest. **57**:1652, 1976.
164. Bernstein, R.E.: Clin. Chem. **26**:174, 1980.
165. Coburn, R.F.: J. Clin. Invest. **46**:346, 1967.
166. Goldsmith, J.R., and Landau, S.A.: Science **162**:1352, 1968.
167. Davis, G.L., and Gartner, G.E.: J.A.M.A. **230**:996, 1974.
168. Landau, S.A., and Winchell, H.S.: Blood **36**:642, 1970.
169. Godin, G., and Shepard, R.J.: Respiration **29**:317, 1972.
170. Astrup, P.: J.A.M.A. **230**:1064, 1974.
171. Castledon, C.M., and Cole, P.V.: Br. Med. J. **4**:736, 1974.
172. Smith, J.R., and Landau, S.A.: N. Engl. J. Med. **298**:6, 1978.

Iron metabolism

173. Antonini, E., and Brunori, M.: Hemoglobin and methemoglobin. In Surgenor, D.M., editor: The red blood cell, New York, 1975, Academic Press, Inc.
174. Committee on iron deficiency of the AMA Council on Food and Nutrition, J.A.M.A. **203**:407, 1968.
175. Pollycove, M.: Hemochromatosis. In Stanbury, J.B., Wyngaarden, J.B., and Fredrickson, D.S., editors: The metabolic basis of inherited disease, New York, 1978, McGraw-Hill Book Co.
176. Oski, F.A.: Audio Digest Pediatrics, **27** 1981.
177. Burman, D.: Clin. Haematol. **2**:257, 1973.
178. Wintrobe, M.M., Lee, G.R., Boggs, D.R., et al.: Clinical Hematology, Philadelphia, 1981, Lea & Febiger.
179. Bothwell, T.H., Charlton, R.W., Cook, J.D., and Finch, C.A.: Iron metabolism in man, Oxford, 1979, Blackwell Scientific Publications, Ltd.
180. Dallman, P.R.: Iron deficiency and related nutritional anemias. In Nathan, D.G., and Oski, F.A., editors: Hematology of infancy and childhood, Philadelphia, 1981, W.B. Saunders Co.
181. Pollack, S., Kaufman, R.M., and Crosby, W.H.: Blood **24**:577, 1964.
182. Moore, C.V.: Iron metabolism and nutrition, Harvey Lect. **55**:1736, 1961.
183. Forth, W., and Rummel, W.: Physiol. Rev. **53**:724, 1973.
184. Disler, P.B., Lynch, S.R., Charlton, R.W., et al.: Gut **16**:193, 1975.
185. Peters, T.J., Apt, L., and Ross, J.F.: Gastroenterology **61**:315, 1971.
186. Weintraub, L.R., Weinstein, M.B., and Huser, H.J.: J. Clin. Invest. **47**:531, 1968.
187. Björn-Rasmussen, E., Hallberg, L., et al.: J. Clin. Invest. **53**:247, 1974.
188. Miale, J.B.: Laboratory Medicine: hematology, ed. 6. St. Louis, 1982, The C.V. Mosby Co.
189. Chanarin, L.: Hematologic biochemistry. In Brown, S.S., Mitchell, F.L., and Young, D.S., editors: Chemical diagnosis of disease, Amsterdam, 1979, Elsevier North-Holland Biomedical Press.
190. Aison, P.: Clin. Haematol. **11**:241, 1982.
191. Savin, M.A., and Cook, J.D.: Blood **56**:1029, 1980.
192. Jacobs, A.: Clin. Haematol. **2**:323, 1973.
193. Cavill, J., Worwood, M., and Jacobs, A.: Nature **256**:328, 1975.
194. Jacobs, A.: Disorders of iron metabolism. In Hoffbrand, A.V., Brian, M.C. and Hirsh, J., editors: Recent advances in hematology, New York, 1977, Churchill Livingstone, Inc.
195. Aison, P., and Brown, E.B.: Structure and function of transferrin. In Brown, E.B., editor: Progress in hematology, vol. 9, New York, 1975, Grune & Stratton, Inc.
196. Vanderle, C., Kroos, M.J., Vannort, W.L., et al.: Clin. Sci. Mol. Med. **60**:185, 1981.
197. Huebers, H., and Finch, C.A.: Semin. Hematol. **19**:3, 1982.
198. Finch, C.A., and Huebers, H.: Clin. Res. **29**:507a, 1981.
199. Soltys, H.D., and Brody, J.L.: J. Lab. Clin. Med. **75**:250, 1970.
200. Lane, R.S.: Br. J. Haematol. **12**:249, 1966.
201. Bearn, A.C., and Litwin, S.D.: Deficiencies of circulating enzymes and plasma proteins. In Stanbury, J.B., Wyngaarden, J.B., and Fredrickson, D.S., editors: The metabolic basis of inherited disease, New York, 1978, McGraw-Hill Book Co.
202. Riegel, C., and Thomas, D.: New Engl. J. Med. **255**:434, 1956.
203. Miller, R.T., and Ward, P.C.J.: Serum ferritin. Check sample no. H 81-10 (H-117), Chicago, 1981, American Society of Clinical Pathologists Commission on Continuing Education.
204. Harrison, P.M.: Semin. Hematol. **14**:55, 1977.
205. Fischbach, F.A., Gregory, D.W., Harrison, P.M., et al.: J. Ultrastruct. Res. **37**:495, 1971.
206. Worwood, M.: Clin. Haematol. **11**:275, 1982.
207. Jones, B.M., and Worwood, M.: Clin. Chim. Acta **85**:81, 1978.
208. Jones, B.M., and Worwood, M.: J. Clin. Pathol. **28**:540, 1975.
209. Miles, L.E., Lipschitz, D.A., Beaber, C.P., et al.: Anal. Biochem. **61**:209, 1974.
210. Addison, G.M., Beamish, M.R., Hales, C.N., et al.: J. Clin. Pathol. **25**:326, 1972.
211. Drysdale, J.W., Adelman, T.G., Arosio, P., et al.: Semin. Hematol. **14**:71, 1977.
212. Pilon, V.A., Howanitz, P.J., Howanitz, J.H., et al.: Clin. Chem. **27**:78, 1981.
213. Saarinen, V.M., and Siimes, M.A.: Acta Paediatr. Scand. **67**:745, 1978.
214. Sismes, M.A., Addiego, J.E., Jr., and Dallman, P.R.: Blood **43**:581, 1974.
215. Gvist, I., Norden, A., and Olofsson, T.: Scand. J. Clin. Lab. Invest. **40**:609, 1980.
216. Hoy, T.H., and Jacobs, A.: Br. J. Haematol. **49**:593, 1981.
217. Garby, L.: Clin. Haematol. **2**:245, 1973.
218. Woodruff, C.W.: Pediatr. Clin. North Am. **24**:85, 1977.

219. Morrow, J.J., Dagg, J.H., and Goldberg, A.: Scottish Med. J. **13**:78, 1968.
220. Elwood, P.C., Waters, W.E., Green, W.J.W., et al.: J. Chronic Dis. **21**:615, 1969.
221. Viteri, F.E., and Torun, B.: Clin. Haematol. **3**:609, 1974.
222. Dallman, P.R.: Semin. Hematol. **19**:19, 1982.
223. Voorhees, M.L., Stuart, M.J., Stockman, J.A., et al.: J. Pediatr. **86**:542, 1975.
224. Holliday, J.W., and Powell, L.W.: Semin. Hematol. **19**:42, 1982.
225. Jacobs, A.: Bibl. Nutr. Dieta **22**:61, 1975.
226. Edwards, C.G., Dadone, M.M., Skolnick, M.H., et al.: Clin. Haematol. **11**:411, 1982.
227. Dadone, M.M., Kushner, J.P., Edwards, C.G., et al.: Am. J. Clin. Pathol. **78**:196, 1982.
228. Simon, M., Bourel, M., Genetet, B., et al.: N. Engl. J. Med. **297**:1017, 1977.
229. Edwards, C.G., Skolnick, M.H., and Kushner, J.P.: Prog. Hematol. **12**:43, 1981.
230. Milder, M.S., Cook, J.D., Stray, S., et al.: Medicine **59**:34, 1980.
231. Charlton, R.W., Bothwell, T.H., and Seftel, H.C.: Clin. Haematol. **2**:383, 1973.
232. Bothwell, T.H., and Charlton, R.W.: Semin. Hematol. **19**:54, 1982.
233. Goldberg, L., and Smith, J.P.: Am. J. Pathol. **36**:125, 1960.
234. Balcerzak, S.P., Westerman, M.P., Heine, E.W., et al.: Ann. Intern. Med. **68**:518, 1968.
235. Barry, M., Cartei, G.C., and Sherlock, S.: Gut **11**:891, 1970.
236. Olsson, K.S.: Acta Med. Scand. **192**:401, 1972.
237. Cook, J.D.: Semin. Hematol. **19**:6, 1982.
238. Cartwright, G.E., and Lee, G.R.: Br. J. Haematol. **21**:147, 1971.
239. Cavill, I.: Clin. Haematol. **11**:259, 1982.
240. Cavill, I., Ricketts, C., Napier, J.A.F., et al.: Br. J. Haematol. **35**;33, 1977.
241. Finch, C.A., Deubelbeis, K., Cooks, J.D., et al.: Medicine **49**:17, 1970.
242. International Committee for Standardization in Haematology: Br. J. Haematol. **38**:291, 1978.
243. Vander Heul, C., Van Eijb, N.G., Wiltink, W.F, et al.: Clin. Chim. Acta **38**:347, 1972.
244. Morton, A.G., and Tavill, A.S.: Br. J. Haematol. **36**:383, 1977.
245. Statland, B.E., and Winkel, P.: Am. J. Clin. Pathol. **67**:84, 1977.
246. Langer, E.E.: Blood **40**:112, 1972.
247. Schifman, R.B., Rivers, S.L., Finley, P.R., et al.: J.A.M.A. **248**:2012, 1982.
248. Lamola, A.A., and Yamana, T.: Science **186**:936, 1974.
249. Labbe, R.F., Finch, C.A., Smith, N.J., et al.: Clin. Chem. **25**:87, 1979.
250. Worwood, M.: Serum ferritin. In Cook, J.D., editor: Methods in hematology, vol. 1, Iron, New York, 1980, Churchill Livingstone, Inc.
251. Konijn, A.M., and Hershko, C.: Br. J. Haematol. **37**:7, 1977.
252. Holliday, J.W., Russo, A.M., Cowlishau, J.L., et al.: Lancet **11**:621, 1977.

Heme biosynthesis and the porphyrias

253. Moore, M.R.: Clin. Haematol. **9**:227, 1980.
254. Meyer, V.A., and Schmid, R.: The porphyrias. In Stanbury, J.B., Wyngaarden, J.B., and Fredrickson, D.S., editors: The metabolic basis of inherited disease, New York, 1978, McGraw-Hill Book Co.
255. Tschudy, D.P.: Porphyrias. In Brown, S.S., Mitchell, F.L., and Young, D.S., editors: Chemical diagnosis of disease, Amsterdam, 1979, Elsevier/North Holland Biomedical Press.
256. Tschudy, D.P.: Porphyrin metabolism and the porphyrias. In Bandy, P.K., and Rosenberg, L., editors: Duncan's diseases of metabolism, ed. 7, Philadelphia, 1974, W.B. Saunders Co.
257. Granick, S., Sinclair, P., Sassa, S., et al.: J. Biol. Chem. **250**:9215, 1975.
258. Moore, M.R., Meredith, P.A., and Goldberg, A.: Lead and hemebiosynthesis. In Singhal, R.L., and Thomas, J., editors: Lead toxicity, Baltimore, 1980, Urban & Schwartzenberg, Inc., pp. 79-117.
259. Cheh, A., and Neilands, J.B.: Biochem Biophys. Res. Commun. **55**:1960, 1073.
260. Robinson, S.H., and Glass, J.: Disorders of heme metabolism, sideroblastic anemia and the porphyrias. In Nathan, D.G., and Oski, F.A., editors: Hematology of infancy and childhood, ed. 2, Philadelphia, 1981, W.B. Saunders Co.
261. Jordan, P.M., Burton, G., Nordlov, H., Schneider, M.M., et al.: J. Chem. Soc., Chem. Communications **5**:204, 1979.
262. Burton, G., Fagerness, P.E., Hosozawa, S., et al.: J. Chem. Soc., Chem. Communications **5**:202, 1979.
263. Jackson, A.H., Sancerich, H.A., Ferramola, A.M., et al.: Trans. R. Soc. London **273**:191, 1976.
264. Campbell, B.C., Brodie, M.J., Thompson, G.G., et al.: Clin. Sci. Mol. Med. **53**:335, 1977.
265. Jackson, A.H., Elder, G.H., and Smith, H.G.: Int. J. Biochem. **9**:877, 1978.
266. Ellepson, R.D.: Mayo Medical Laboratories Communique **7**(6), 1982.
267. McColl, K.E.L., and Goldberg, A.: Clin. Haematol. **9**:427, 1980.
268. Bonkowsky, H.L.: Porphyrin and heme metabolism and the porphyrias. In Zakim, D., and Boyer, T.D., editors: Hepatology, Philadelphia, 1982, W.B. Saunders Co.
269. Smith, A.G., and DeMatteis, F.: Clin. Haematol. **9**:399, 1980.
270. Brodie, M.J., and Goldberg, A.: Clin. Haematol. **9**:253, 1980.
271. Anderson, P.M., Reddy, R.M., Anderson, K.E., et al.: J. Clin. Invest. **68**:1, 1981.
272. Stein, J.A., and Tschudy, D.P.: Medicine **49**:1, 1970.
273. Tschudy, D.P., Valsamia, M., and Magnussen, C.R.: Ann. Intern. Med. **83**:851, 1975.
274. Miyagi, K., Cardinal, R., Bessenmeier, I., et al.: J. Lab. Clin. Med. **78**:683, 1971.
275. Gambino, R., editor: Lab Report for Physicians, **4**:9, 1982.
276. Strond, L.J., Fletcher, B.F., Redeker, A.B., et al.: Proc. Natl. Acad. Sci., USA **67**:1215, 1970.
277. Anderson, K.E., Sassa, S., and Kappas, H.: Ann. Intern. Med. **95**:784, 1981.
278. Meyer, V.A.: Enzyme **16**:334, 1973.
279. Strand, L.J., Meyer, V.A., Felcher, B.F., et al.: J. Clin. Invest. **51**:2530, 1972.
280. Sassa, S., Solish, G., Levere, R.D., et al.: J. Exp. Med. **142**:722, 1975.
281. Grandchamp, B., Phung, N., Grelier, M., et al.: Nouv. Presse Med. **6**:1537, 1978.
282. Brodie, M.J., Thompson, G.G., Moore, M.R., et al.: Q. J. Med. **46**:229, 1977.
283. Muhlbauer, J.E., Pothak, M.A., Tishler, P.V., et al.: J.A.M.A. **247**:3095, 1982.
284. Magnus, I.A.: Clin, Haematol. **9**:273, 1980.
285. Bremer, D.A., and Bloomer, J.R.: N. Engl. J. Med. **302**:765, 1980.
286. Kramer, S.: Clin. Haematol. **9**:303, 1980.
287. Kramer, S., Becker, D.M., and Viliden, J.D.: Br. J. Haematol. **37**:439, 1977.
288. Bothwell, T.H., Kramer, S., Jacobs, P., et al.: S. Afr. J. Med. Sci. **25**:141, 1960.
289. Waldenström, J.: Am. J. Med. **22**:758, 1957.
290. Cam, C., and Nigogosyan, G.: J.A.M.A. **183**:88, 1963.
291. Brunsting, L.A.: A.M.A. Arch. Dermatol. Syphilol. **70**:551, 1954.
292. Tio, T.H., Leijnse, B., Jarrett, A., et al.: Clin. Sci. **16**:517, 1957.
293. Turnbull, A.: Br. J. Dermatol. **84**:380, 1971.
294. Elder, G.H., Lee, G.B., and Touey, J.A.: N. Engl. J. Med. **299**:274, 1978.
295. Elder, G.H., Magnus, I.A., Handa, F., et al.: Enzyme **17**:29, 1974.
296. Rimington, C., Magnus, I., Ryan A., et al.: Q. J. Med. **36**:29, 1967.
297. DeLeo, V.A., Poh-Fitzpatrick, M.B., Matthews-Roh, M.M., et al.: Am. J. Med. **60**:8, 1976.
298. Schmidt, D., and Stich, W.: Blut **22**:202, 1971.
299. Miale, J.B.: Laboratory medicine: hematology, ed. 6, St. Louis, 1982, The C.V. Mosby Co.
300. Becker, D.M., Viljoen, J.D., Katz, J., et al.: Br. J. Haematol. **36**:171, 1977.

301. Bloomer, J.R.: Yale J. Biol. Med. **52:**39, 1979.
302. Gipps, D.J., and Mac Eachern, W.N.: Arch. Pathol. **91:**497, 1971.
303. Moore, M.R., Thompson, G.G., Goldberg, A., et al.: Int. J. Biochem. **9:**933, 1978.
304. Gidari, A.S., and Levere, R.D.: Semin. Hematol. **14:**145, 1977.
305. Eriksen, L., and Eriksen, N.: Scand. J. Clin. Lab. Invest. **33:**323, 1974.
306. Heilmeyer, L., and Clotten, R.: German Med. Month. **9:**353, 1964.
307. Chisholm, J.I., Jr.: Hosp. Pract. **8:**127, Nov. 1973.
308. Gutcher, J.C.: Ann. Intern. Med. **59:**707, 1963.
309. Blumberg, W.E., Eisinger, J., and Lamola, A.A.: Clin. Chem. **23:**270, 1977.
310. Berlin, A., and Schaller, K.H.: Z. Klin. Chem. Klin. Biochem. **12:**389, 1974.
311. Gambino, R., editor: Lab Report for Physicians **4:**69, 1982.
312. Waldron, H.A.: Br. J. Intern. Med. **23:**83, 1966.
313. Paglia, D.E., and Valentine, W.N.: Clin. Haematol. **10:**81, 1981.
314. Brodie, M.J., Thompson, G.G., Moore, M.R., et al.: Acta Hepatogastroenterol. **26:**122, 1979.
315. Waldenström, J., and Haeger-Aronson, B.: Prog. Med. Genet. **5:**58, 1967.

Hemoglobin destruction (catabolism)

316. Kahn, A.: Clin. Haematol. **10:**123, 1981.
317. Ganzoni, A.M., and Oakes, R.: J. Clin. Invest. **50:**1373, 1971.
318. Mohandas, N., and Shobet, S.B.: Clin. Haematol. **10:**223, 1981.
319. Keitt, A.S.: Red cell maturation and survival. In Nathan, D.G., and Oski, F.A., editors: Hematology of infancy and childhood, Philadelphia, 1981, W.B. Saunders Co.
320. Rosse, W.F., and Dacie, J.V.: J. Clin. Invest. **45:**736, 749, 1966.
321. Wintrobe, M.M.: Clinical hematology, Philadelphia, 1981 Lea & Febiger.
322. Gosby, W.H.: Blood **50:**643, 1977.
323. Cooper, R.A., and Shattil, S.J.: N. Engl. J. Med. **285:**1514, 1971.
324. Bissell, D.M.: Blood **40:**812, 1972.
325. Bunn, H.F.: J. Exp. Med. **129:**909, 925, 1969.
326. Lathem, W.: J. Clin. Invest. **38:**652, 1959.
326a. Lathem, W.: J. Clin. Invest. **39:**840, 1960.
327. Lowenstein, J.: J. Clin. Invest. **40:**1172, 1961.
328. Kravitz, H.: Am. J. Dis. Child **91:**1, 1956.
329. Bunn, H.F., and Jandl, J.H.: J. Biol. Chem. **243:**465, 1968.
330. Sears, D.A.: J. Clin. Invest. **49:**5, 1970.
331. Muller-Eberhard, U.: N. Engl. J. Med. **283:**1090, 1970.
332. Fertakis, A., Panitras, G., and Angelopoulos, B.: Acta Haematol. **50:**149, 1973.
333. Todd, D.: Clin. Haematol. **4:**63, 1975.
334. Waks, M., and Alfsen, A.: Arch. Biochem. Biophys. 113:304, 1966.
335. Miale, J.B.: Laborabory medicine: hematology, ed. 6, St. Louis, 1982, The C.V. Mosby Co.
336. Landow, S.A., Callahan, E.W., and Schmid, R.: J. Clin. Invest. **49:**914, 1970.
337. Blanchaett, N., and Schmid, R.: Physiology and pathophysiology of bilirubin metabolism. In Zakim, D., and Boyer, T.D., editors: Hepatology, 1982, Philadelphia, W.B. Saunders Co.
338. Jones, E.A., and Berk, P.D.: Liver functions. In Brown, S.S., Mitchell, F.L., and Young, D.S., editors: Chemical diagnosis of disease, Amsterdam, 1979, Elsevier/North-Holland Biomedical Press.
338a. Schmid, R.: Gastroenterology **74:**1307, 1978.
339. Billing, B.H.: Gut **19:**481, 1978.
340. Ostrow, J.D., and Schmid, R.: J. Clin. Invest. **42:**1286, 1963.
341. Gambino, R., editor: Lab Report for Physicians, **2:**21, 1980.
242. Arias, J.M., Fleischner, G., Ligtowsky, I., et al.: On the structure and function of ligandin and Z protein. In Taylor, W., editor: Hepatobiliary system, New York, 1976, Plenum Press.
343. Lester, R., and Texler, R.F.: Gastroenterology **56:**143, 1960.
344. Watson, C.J.: J. Lab. Clin. Med. **54:**1, 1959.
344a. Watson, C.J.: J. Clin. Pathol. **16:**1, 1963.
345. Dearing, W.H.: Mayo Clin. Proc. **33:**646, 1958.
346. Bourke, E., Milne, M.D., and Stokes, G.S.: Br. Med. J. **2:**1510, 1965.
347. Bloomer, J.R., Berre, P.D., Howe, R.B., et al.: J.A.M.A. **218:**216, 1971.
348. Crigler, J.F., and Najjar, V.A.: Pediatrics **10:**169, 1952.
349. Berke, P.D., Bloomer, J.R., Howe, R.B., et al.: Am. J. Med. **49:**296, 1970.
350. Odell, G.B.: N. Engl. J. Med. **277:**193, 1967.
351. Dubin, I.N., and Johnson, F.B.: Medicine **33:**155, 1954.
352. Rotor, A.B., Manahon, L., and Florentin, A.: Acta Med. Philippines **5:**37, 1948.

Chapter 34 Vitamins

Marge A. Brewster

ameloblasts Cells that form dental enamel.
angular stomatitis Inflammation at the corner of the mouth.
anorexia Appetite loss.
antioxidant A substance that protects against oxidation, usually by itself becoming oxidized.
aphonia Loss of voice; motions of crying but little sound.
avidin A glycoprotein in raw egg white with strong affinity for biotin.
Batten's disease A progressive childhood encephalopathy with disturbed metabolism of polyunsaturated fatty acids.
blepharospasm Spasm of eyelid.
carotenoids Compounds structurally similar to β-carotene (provitamin A) occurring naturally in vegetables and pigmented fruits.
cholelithiasis Presence of gallstones.
CNS Central nervous system.
coagulopathy A disorder of coagulation.
confabulation Lying.
creatinuria Excess excretion of creatine in the urine.
cyanosis Blue skin color caused by lack of oxygen.
dry beriberi Thiamin deficiency resulting in poor appetite, fatigue, and peripheral neuritis.
dyspnea Difficult or painful breathing.
ecchymoses Skin discolorations caused by oozing of blood into tissues.
ene-diol $-\underset{OH}{\underset{|}{C}}=\underset{OH}{\underset{|}{C}}-$
fibrinolysis The process of dissolution of fibrin clots.
flavins Riboflavin, flavin adenine dinucleotide (FAD), and flavin mononucleotide (FMN).
glossitis Smooth tongue.
hydrolases Enzymes that cleave ester bonds by addition of water.
hyperuricemia Increased concentration of serum uric acid.
hypogeusia Reduced taste.
ischemia Tissue deprivation of blood.
megaloblastic anemia Anemia in which marrow and blood cells are large and have multilobed nuclei.
osteoblasts Cells that form bone.
osteoclasts Cells that degrade bone.
ozone O_3; increased in air in areas of much automobile traffic.
pancytopenia Decrease in all blood cells and platelets.

paresthesias Abnormal sensations such as burning, prickling, or tingling.
pellagra Niacin deficiency resulting in diarrhea, dementia, and dermatitis.
pernicious anemia Anemia (no longer considered pernicious) caused by antibodies interfering with vitamin B_{12} absorption.
photophobia Abnormal sensitivity to light.
PUFA Polyunsaturated fatty acids.
pyrexia Fever.
pyridine nucleotides NAD, NADH, NADP, NADPH.
RDA Recommended Dietary Allowances, quantity of vitamin recommended to be ingested daily to meet essential needs of a healthy person.
rebound scurvy Symptoms of scurvy occurring on sudden withdrawal of megadosage of ascorbic acid.
rickets Muscle hypotonia and skeletal deformities in children.
scurvy Ascorbic acid deficiency characterized by swollen gums with loss of teeth, skin lesions, and pain and weakness in the lower extremities.
steatorrhea Excessive lipid in feces.
subclinical vitamin deficiency Chemical indices of vitamin status indicate deficiency, but there is no *apparent* clinical symptom.
thioester Ester bond involving −SH.
vitamins Small molecular weight components required for metabolic activity that must be supplied by dietary intake.
Wernicke-Korsakoff syndrome Neurological/behavioral alterations seen in some patients with thiamin deficiency.
wet beriberi Thiamin deficiency resulting in edema and cardiac failure.
xanthomatosis Presence of yellow skin deposits.
xerophthalmia Dry, thickened, lusterless eyeballs.

The vitamins comprise a group of small molecular weight compounds with diverse functions in biological tissues. Their primary role is to serve as cofactors of numerous enzymatic reactions. Without these cofactors a wide range of enzymes that play critical roles in cellular metabolism become inactive. The term *vitamin* is derived from early evidence that one such compound, thiamin, was an amine vital for health. These compounds or their biologically inactive precursors must be obtained, at least partially, from food sources or in some instances from intestinal bacterial synthesis. Inadequacies in supply (diet or absorption) are properly termed *vitamin deficiencies*. Abnormalities of metabolism requiring an abnormally high supply may be termed *vitamin insufficiencies* or *vitamin dependencies,* depending on the level of supply demanded for physiological function. Variabilities in clinical expression of vitamin abnormalities result from differences in specific etiology and degree and duration of vitamin inadequacy. Additional factors contributing to this variability include the genetic constitution of the individual, simultaneous presence of multiple nutritional insufficiencies, or increased metabolic demands such as infection, pregnancy, or tumor. We still have limited knowledge of human mechanisms of absorption, transport, storage, and metabolism and have much to learn with regard to the interactions of diseases, therapies, and exogenous agents upon these processes.

Because of their pervasive actions in metabolic processes, the clinical symptoms of vitamin deficiencies are usually nonspecific and vague, often delaying a definitive diagnosis. This is especially true in the early stages or in mild chronic deficiency states; also vitamin deficiency may be complicated by other simultaneous processes. A combination of dietary history, physical examination, and biochemical measurements is often required to diagnose a vitamin deficiency. In addition, therapeutic trial may also be required for a definitive diagnosis. Because vitamin metabolism is complex and interactive, vitamin supplementation without proper studies may lead to other nutrient deficien-

Table 34-1. 1980 Recommended Dietary Allowances[1]

Age (years)	Vitamin A (μg RE*)	Vitamin E (mg α-TE†)	Vitamin D (μg‡)	Vitamin C (μg)	Thiamin (mg)	Riboflavin (mg)	Niacin (mg NE§)
Infants							
0-6 months	420	3	10	35	0.3	0.4	6
6-12 months	400	4	10	35	0.5	0.6	8
Children							
1-3	400	5	10	45	0.7	0.8	9
4-6	500	6	10	45	0.9	1.0	11
7-10	700	7	10	45	1.2	1.4	16
Males							
11-14	1000	8	10	50	1.4	1.6	18
15-18	1000	10	10	60	1.4	1.7	18
19-22	1000	10	7.5	60	1.5	1.7	19
23-50	1000	10	5	60	1.4	1.6	18
50+	1000	10	5	60	1.2	1.4	16
Females							
11-14	800	8	10	50	1.1	1.3	15
15-18	800	8	10	60	1.1	1.3	14
19-22	800	8	7.5	60	1.1	1.3	14
23-50	800	8	5	60	1.0	1.2	13
50+	800	8	5	60	1.0	1.2	13
Pregnant	200‖	2‖	5‖	20‖	0.4‖	0.3‖	2‖
Lactating	400‖	3‖	5‖	40‖	0.5‖	0.5‖	5‖

*One retinol equivalent (RE) = 1 μg of retinol or 6 μg of β-carotene; 3 μg of retinol = 10 U.
†One mg of α-tocopherol equivalent (TE) = 1 mg of *d*-α-tocopherol.
‡As cholecalciferol; 10 μg of cholecalciferol = 400 U.
§One mg of niacin equivalent (NE) = 1 mg of niacin or 60 mg of dietary tryptophan.
‖RDA needed in addition to normal requirements for females.

cies, toxicities, or an irreversible state of untreated deficiency with symptoms masked by inappropriate therapy.

Suspicion of dietary deficiency of a vitamin arises primarily from knowledge of dietary sources and dietary practices likely to provide inadequate intake or absorption. Recommended Dietary Allowances (RDA) are defined by the Food and Nutrition Board[1] as the levels of intake of essential nutrients considered, "on the basis of available scientific knowledge, to be adequate to meet the nutritional needs of practically all healthy persons." These levels are defined from information on the daily intake requirements needed to avoid deficiency symptoms and to maintain specific functions. RDA values are generally set high enough to meet the needs of 97.5% of the population; they are sometimes even higher if the nutrient is poorly absorbed or insufficiently used. Naito[2] states:

> It is important to emphasize that RDA do not take into consideration the losses of nutrients during the processing and preparation of foods, nor are the RDA designed to cover increased needs resulting from severe stress, disease, or trauma above the usual minor infections and stresses of everyday life. The processing losses must be allowed for separately in planning a food supply, and the increased needs represent clinical problems that must be given consideration.

These RDA are summarized in Table 34-1.

Biochemical indices of vitamin status become abnormal before development of obvious clinical changes, thus allowing detection of a vitamin deficiency at a subclinical or nonclassical stage and helping to confirm the clinical diagnosis as well. Chemical determination of human vitamin status has been approached in the following ways:

1. Measurement of the active cofactor(s) or precursor(s) in biological fluids or blood cells
2. Measurement of urinary metabolite(s) of the vitamin
3. Measurement of a biochemical function requiring the vitamin (such as enzymatic activity) with and without in vitro addition of the cofactor form
4. Measurement of urinary excretion of vitamin or metabolite(s) after a test load of the vitamin
5. Measurement of urinary metabolites of a substance, the metabolism of which requires the vitamin, after administering a test load of the substance

Each of these approaches has merit in certain causes of vitamin deficiency or logistical advantages for specific circumstances such as population screening and sample trans-

Table 34-2. Chemical values associated with classical deficiency symptoms*†

Vitamin	Chemical value
A	<0.1 mg/L of plasma retinol
E	<5.0 mg/L of plasma α-tocopherol
K	Plasma prothrombin time > normal
D	See Chapter 25
Ascorbic acid (C)	<2.4 mg/L of serum ascorbate <3 mg/L of whole blood ascorbate <80 mg/L of leukocyte ascorbate
Riboflavin (B_2)	>1.4 AC§ of erythrocytic glutathione reductase <0.1 mg of riboflavin/L of erythrocytes <0.12 mg of urinary riboflavin per day ≤0.08 mg of urinary riboflavin per gram of creatinine
Pyridoxine (B_6)	≥1.5 AC of erythrocytic AST ≥1.25 AC of erythrocytic ALT <0.8 mg of urinary 4-pyridoxic acid per day >25 mg of urinary xanthurenic acid per day
Thiamin	>1.25 AC of erythrocyte transketolase <0.1 mg of urinary thiamin per day
Niacin	≥1 urinary ratio (α-pyridone/*N*′-methylnicotinamide)
Biotin‡	<0.7 μg/L of whole blood? <15 μg of urinary biotin per day?
Pantothenic acid‡	<1.0 mg/L of whole blood pantothenate <1.0 mg of urinary pantothenate per day
B_{12}	<150 ng/L of serum vitamin B_{12} ≥24 mg of urinary methylmalonic acid per day
Folate	<140 μg/L of erythrocyte folate <3.0 μg/L of serum folate ≥30 mg of urinary N^5-formiminoglutamic acid (FIGLU) per 8 hours

*These are general guidelines with normal values dependent on age and methodology employed.
†Compiled from references 4-6 and 25.
‡Deficient values for biotin and pantothenate are not well established.
§*AC*, Activity coefficient; ratio of activities with and without added cofactor.

port, and varies widely as to instrumention and labor requirements. Reduced serum concentrations of a vitamin do not always indicate a deficiency that interrupts cellular function; on the other hand, normal values do not always reflect adequate function. Interpretation of chemical values must be done with caution and with knowledge of physiological and methodological factors, which can mislead.

Table 34-2 lists representative biochemical data that are usually associated with classical deficiency symptoms. These values are, of course, somewhat variable with age and methodology. Major resources are listed for laboratory assessment methods[3,4] and their selection,[4-6] toxicological effects,[6] drug interactions,[4,6,7] environmental effects,[8] and metabolism[4,6-10] of the vitamins.

FAT-SOLUBLE VITAMINS

Because the fat-soluble vitamins (A, E, K, and D) are absorbed as part of the chylomicron complex, their absorption depends on the presence of adequate bile and pancreatic secretions and on healthy bowel mucosa as well. Therefore, in addition to dietary deprivation, chronic malabsorptive states are often associated with a deficiency of one or more of these vitamins. The malabsorptive states include biliary tract disease, pancreatic disease, fistula, small bowel obstruction, and alcoholic liver disease. Deficiency of this class of vitamins generally develops slowly because of stored supplies. Vitamin A can be stored in liver parenchymal cells for a year or longer, and vitamin E can be stored in body fat for several months. Paradoxically, although they are fat-soluble, vitamins K and D appear to be stored only for days or weeks.

Vitamin A

First described in 1909 and found to prevent night blindness in 1925, vitamin A is now known to be composed of three biologically active forms: retinol, retinal, and retinoic acid. These major vitamin A compounds all contain a trimethyl-cyclohexenyl group and an all-*trans* polyene chain with four double bonds (Fig. 34-1). These are derived directly from dietary sources, primarily as retinyl esters, or from metabolism of dietary carotenoids (provitamin A), primarily β-carotene. Major dietary sources of these compounds include animal products and pigmented fruits and vegetables (carotenoids). Each of these compounds is soluble in organic solvents, with retinoic acid being more polar than the others. Oxidation of retinol or retinal by peripheral cells is not reversible; thus neither retinoic acid nor retinal is metabolically converted to retinol.

Metabolism. Enzymes of the small intestinal mucosa convert dietary β-carotene and retinal to the predominant form of vitamin A, retinol (Fig. 34-2). The retinyl esters of dietary animal products are cleaved to retinol by pancreatic and mucosal hydrolases (vitamin A esterases). Once in the mucosal cell, retinol is reesterified, and these retinyl esters (primarily retinyl palmitate) are transported in lymph chylomicrons to the systemic circulation. After the chylomicrons release their triglycerides to adipose tissue, the retinyl esters are transported to the liver for storage associated with lipid droplets in hepatocytes. The more polar retinoic acid does not require this lipoprotein transport route but is directly absorbed into the portal circulation. However, this form is not stored in liver but is excreted through bile as a glucuronide conjugate.

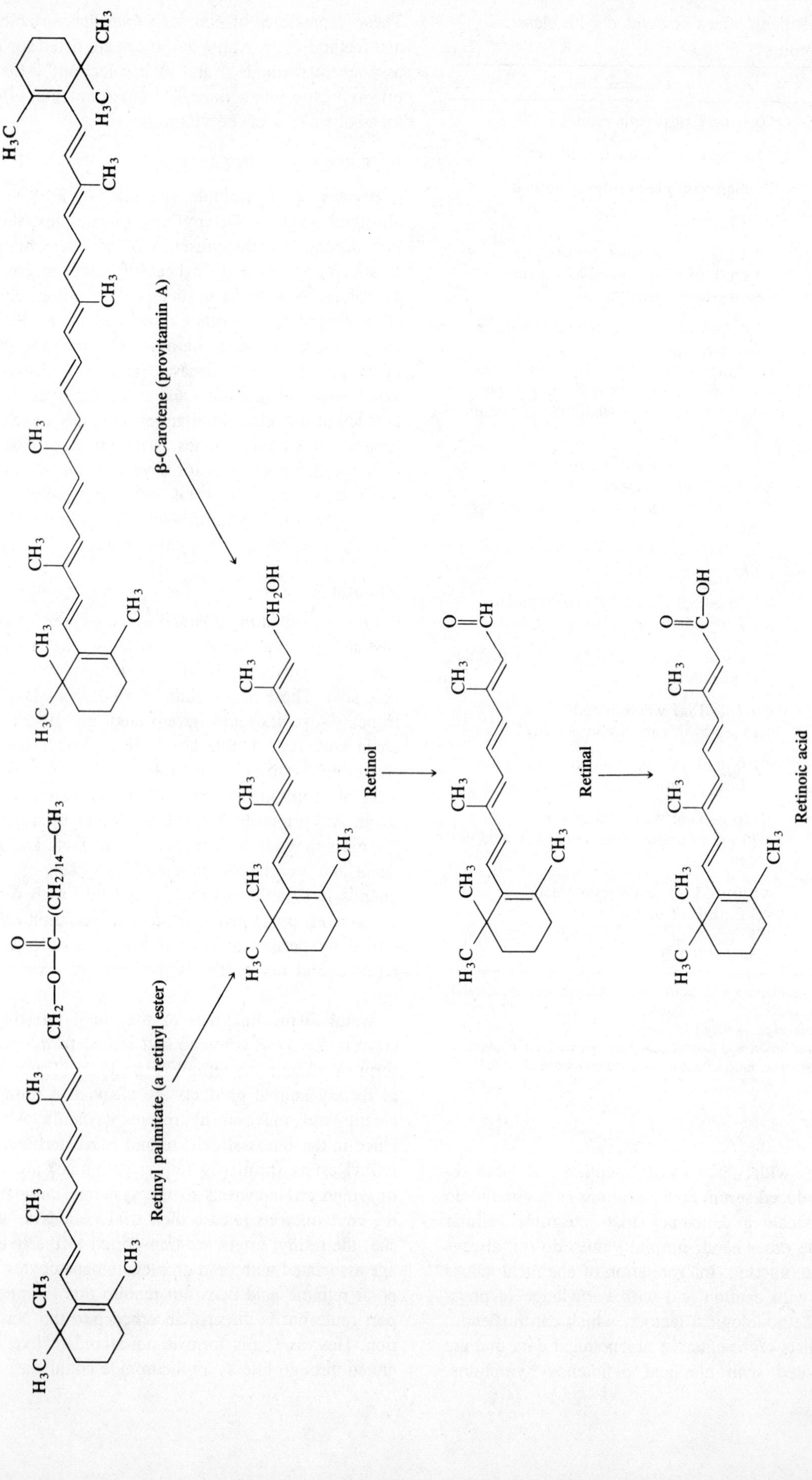

Retinyl palmitate (a retinyl ester)

β-Carotene (provitamin A)

Retinol

Retinal

Retinoic acid

Fig. 34-1. Structures of vitamin A (retinol) with its precursors and metabolites.

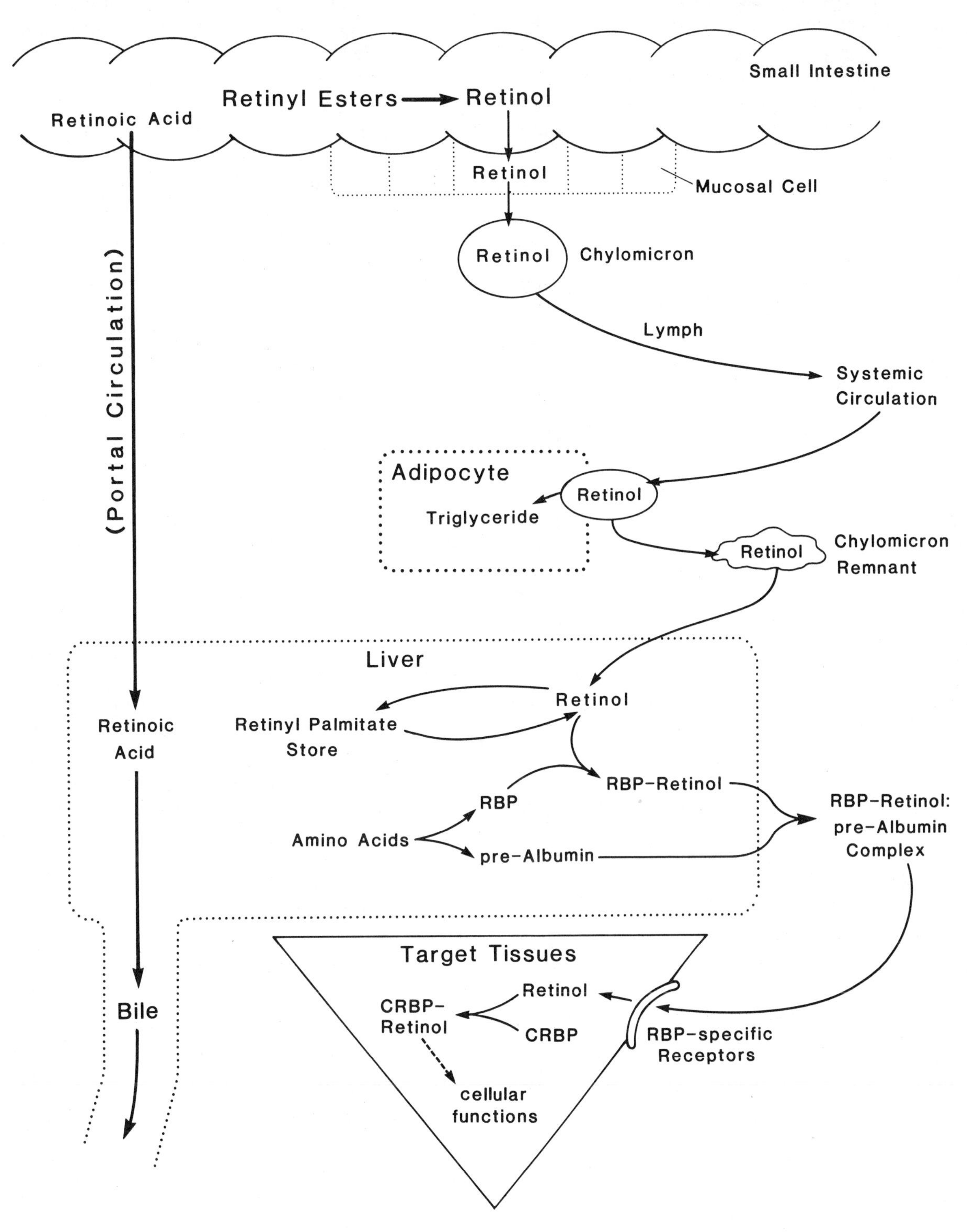

Fig. 34-2. Retinol metabolism. *RBP*, Retinol-binding protein; *CRBP*, cytoplasmic retinol-binding protein.

When body demands require mobilization of hepatic vitamin A, the stored retinyl palmitate is hydrolyzed and the free retinol combines with retinol-binding protein (RBP). RBP-retinol is then secreted into the circulation, where it complexes with prealbumin. This large complex then circulates to target tissues that have specific receptor sites for RBP, and retinol is transferred intracellularly to another specific binding protein termed "cytosol-retinol–binding protein" (CRBP). CRBP-retinol presumably transports the retinol to its functional site within the cell. Retinol metabolism is shown in Fig. 34-2.

Function. Vitamin A is known to function in vision, growth, reproduction, cellular differentiation, mucus secretion, and immune responses. Other than its role in vision, well-studied cellular mechanisms are unclear at this time. Retinol is oxidized in the rods of the eye to retinal, which, when complexed with opsin, forms rhodopsin, allowing dim-light vision. Vitamin A maintains epithelial cells in a soft, moist state. In deficiency states these epithelial cells (outer skin layers, lining of the gastrointestinal, respiratory, and urogenital tracts) become dry and keratinized. This process may relate to protection against infectious organisms. Vitamin A is required for osteoblast-to-osteoclast conversion in bone and for normal differentiation and function of ameloblasts (cells forming dental enamel). In vitamin A deficiency, several reproductive effects occur: epididymal fluid lacks spermatozoa; testicular germinal epithelium degenerates; sperm implantation into the egg is hindered; and estrous cycle changes are seen. Retinoic acid, a quantitatively minor component of vitamin A, is known to function in growth and epithelium maintenance but not in vision and reproduction.

Clinical deficiency signs. Clinical signs of deficiency include growth retardation, hypogeusia (reduced taste), anorexia (appetite loss), recurrent infections, dermatitis, dry mucous membranes, and night blindness. Later signs include bone growth failure, aspermatogenesis, and xerophthalmia (dry, thickened, lusterless eyeballs).

Chemical deficiency signs. The primary chemical sign of deficiency is reduction in plasma vitamin A. Garry[11] discussed the measurement approaches available and the strong dependence of "normal values" on methodology. Normal serum concentrations by fluorometric assay ranged from 0.34 to 1.33 mg/L, the highest values being obtained with retinyl acetate standard and no correction for interference of phytofluene (a tomato pigment). Generally retinol values below 0.1 mg/L are associated with clinical symptoms, and values above 0.2 mg/L are not. Vitamin A itself is not excreted in human urine. Although several metabolites are excreted, they do not seem to reflect the tissue status of vitamin A. Serum vitamin A measurement is currently performed by either fluorometry or high-performance liquid chromatography (HPLC).

Because retinol and RBP are secreted from liver as a 1:1 complex, low plasma concentrations of both are seen in deficiency. Adequate concentration of plasma retinol usually indicates dietary and tissue adequacy, but low concentrations *do not* always indicate dietary deficiency. Factors that reduce hepatic synthesis of RBP or secretion of the RBP-retinol complex lower plasma concentrations of retinol and RBP, even though dietary intake and the hepatic retinol store are adequate. These states are primarily recognized by no increase in plasma retinol after oral therapy with vitamin A.

Pathophysiology. In circumstances of reduced plasma retinol with adequate liver stores, target tissue deficiency, that is, clinical symptoms of deficiency, may or may not result. Protein-calorie malnutrition with reduced hepatic RBP synthesis is a prime example of adequate liver stores and reduced plasma retinol. Liver disease is a complex state associated with reduced plasma retinol not always increased by vitamin A therapy. Zinc deficiency also lowers plasma retinol by a RBP effect despite adequate liver retinol stores. Patients with cystic fibrosis have reduced concentrations of plasma retinol, RBP, and prealbumin that are not corrected by oral vitamin A therapy.

In dietary insufficiency, as liver stores decline, the plasma retinol also decreases gradually, an indication that secretion of RBP-retinol is controlled by the hepatic concentration of retinol. Ninety percent of the body's vitamin A reserves are in the liver; this liver reserve appears to rise with age and does not correlate well with serum vitamin A concentrations. A nonbiopsy approach to assessment of liver reserves has been developed (measuring the dilution of tritium- or deuterium-labeled vitamin A by the body stores by means of a blood sample) but has not yet been clinically applied. Because vitamin A in the liver can be inadequately mobilized and transported to target tissues, creating a true functional deficiency, serum vitamin A is currently believed to be the best available index of the functional state of health.

Because the kidneys are normally the main catabolic sites for RBP, it is not surprising that both RBP and vitamin A are greatly increased in the plasma of patients with chronic renal disease. Prealbumin is normal in this disease state, since its large size does not normally allow glomerular filtration. Free RBP, small enough to be filtered, is normally resorbed and degraded by renal tubules so that little appears in urine unless tubular function is abnormal.

Exogenous influences. Use of oral contraceptives increases plasma RBP and vitamin A by 30% to 70% (possibly requiring 3 months to normalize after discontinuation) and decreases plasma carotenoids. Stilbestrol, given to prevent lactation, decreases circulating vitamin A. In pregnancy there is a measurable drop in plasma vitamin A by 6 weeks after conception, further decreasing to about half the preconception level by 12 weeks. The significance of these hormonal influences is unclear.

Organophosphates inhibit vitamin A esterase and decrease plasma vitamin A. This vitamin effect occurs without any clinical symptoms of pesticide toxicity, a possible indication that persons employed in organophosphate-involved occupations may be at risk for subclinical vitamin A deficiency and possibly epithelial cancers.[8]

Toxicity and therapeutic uses. Vitamin A therapy has been shown to prevent noise damage to auditory function and sometimes improves inner ear deafness. Topical applications of retinoic acid (the form best absorbed by skin) and synthetic analogs are being studied in patients with a variety of skin disorders. The anticarcinogenic role of high-dosage vitamin A has been well established in animals, and there is active search for structural analogs with this same antitumor activity but without the toxicity of high-dosage retinol.[12] Tumor responsiveness to vitamin A treatment has been correlated to the presence within the tumor of cytosol-binding protein for retinoic acid.[13]

Because vitamin A can be stored for a year or longer, toxicity can occur with chronic excess intake. Toxicity probably involves loss of membrane integrity, and major clinical signs include skin and mucous membrane desquamation and hypercalcemia. Young children show abnormalities in bone. Less frequently reported signs include anemia, loss of hair, hemorrhage (secondary to vitamin K deficiency), increased intracranial pressure, blurred vision, and degenerative atrophy of various organs. Excess vitamin A intake during pregnancy can result in congenital malformation. Toxicity can occur after consumption of animal livers high in vitamin A (polar bear, shark, seal, halibut, tuna, cod, whale, and fox). More commonly it is the result of voluntary overdose.

Excess intake of vitamin A can exceed liver storage capacity and result in circulation of free retinol and retinyl esters in lipoproteins. In this instance retinyl ester concentration in plasma becomes quite high. A test of fat malabsorption involves an oral loading dose of vitamin A. With adequate absorption, plasma retinol is unchanged and retinyl palmitate is elevated within 4 hours. For toxicity detection and for the fat malabsorption test, high-performance liquid chromatography (HPLC) affords ready assessment of retinyl esters.

Vitamin E

A factor in vegetable oils that restored fertility to rats was isolated in the early 1920s as vitamin E; later it was given the generic name *tocopherol* and was shown to include several biologically active isomers (Fig. 34-3). The

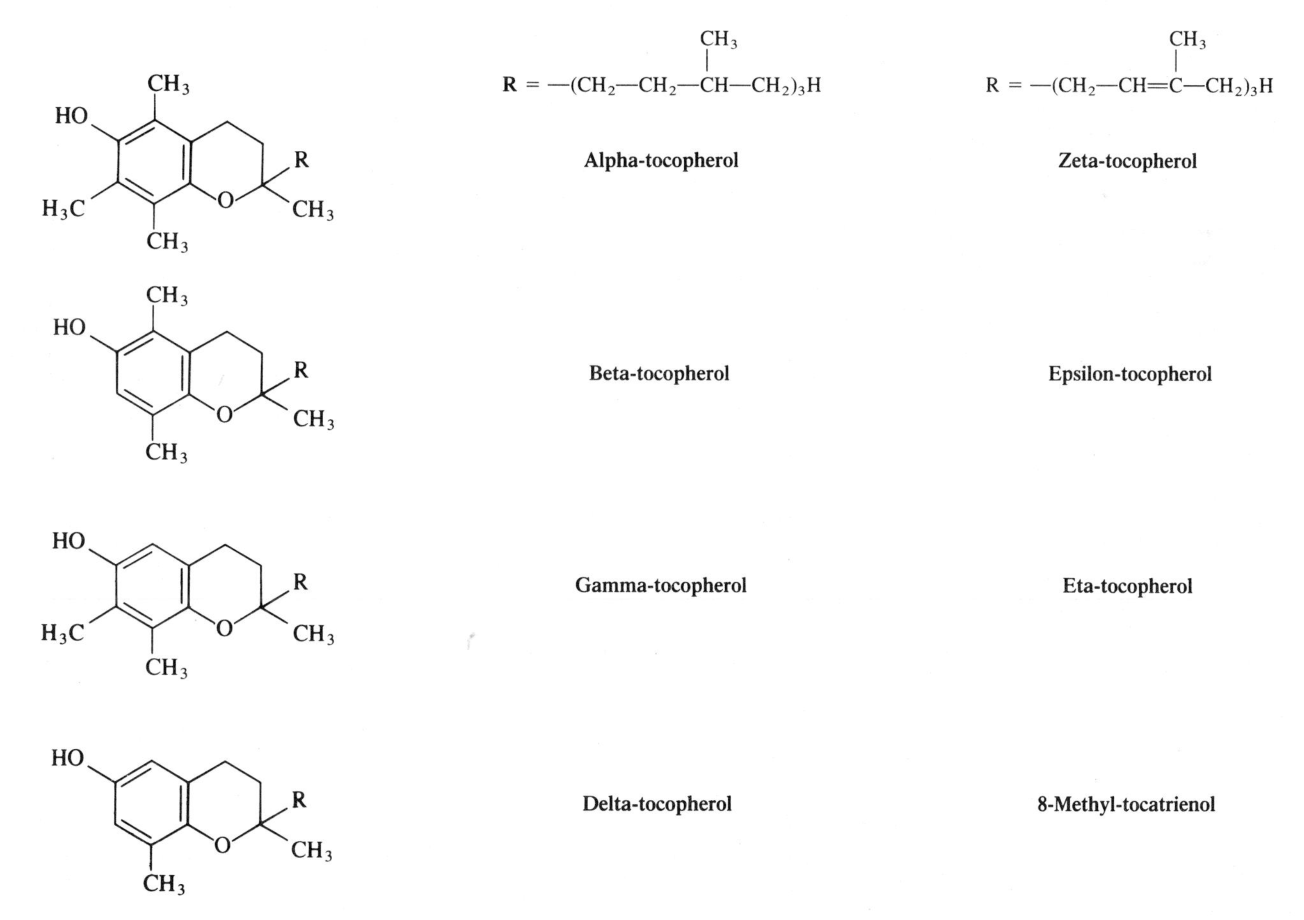

Fig. 34-3. Vitamin E isomers.

word *tocopherol* is of Greek derivation, meaning 'an oil that brings forth in childbirth,' but the fertility role of these compounds in humans is still questionable.

α-Tocopherol is the predominant isomer in plasma and is the most potent isomer by current biological assays. Whether the tocopherol isomers have separate physiological effects is unknown.

Dietary sources of tocopherols include vegetable oils, fresh leafy vegetables, egg yolk, legumes, peanuts, and margarine. Diets suspect for vitamin E deficiency are those low in vegetable oils or fresh green vegetables or those high in unsaturated fats.

Metabolism. Absorption, transport, storage, and metabolism of tocopherol is only partially understood. Absorption is believed to be associated with intestinal fat absorption and may also involve the stomach. Approximately 40% of ingested tocopherol is absorbed; the percentage is affected by the amount and unsaturation of dietary fat as well as by the isomer type. The physiological requirement for vitamin E increases with increasing polyunsaturated fatty acids (PUFA) in the diet. Absorbed vitamin E is first associated with circulating chylomicrons and very low density lipoproteins (VLDL) and later with other lipoproteins. Vitamin E is predominantly found in adipose tissue, though increased dietary α-tocopheryl acetate is reflected by increased concentrations in all animal tissues, including plasma, erythrocytes, and platelets.

Function. Vitamin E functions as an antioxidant, protecting unsaturated lipids from peroxidation (cleavage of fatty acids at unsaturated sites by oxygen addition across the double bond and formation of free radicals).[14] The metabolic effects of peroxidized lipids are not clear, but their presence in human erythrocytes is known to increase erythrocyte susceptibility to hydrogen peroxide–induced hemolysis. The role of vitamin E in protecting the erythrocyte membrane from oxidant stress is presently the major documented role in human physiology.

Clinical deficiency signs. Although such vitamin E usage is still controversial, premature newborns are commonly supplemented with vitamin E to prevent hemolytic anemia. Iron supplementation of potentially vitamin E–deficient newborns can exacerbate hemolysis by increasing free radical peroxidation. Protein-calorie malnutrition in infants has been associated with a vitamin E–responsive anemia. There is evidence for preventive roles of vitamin E in retrolental fibroplasia and bronchopulmonary dysplasia of neonates with respiratory distress syndrome. Premature infants suffering from vitamin E deficiency reportedly excrete large amounts of methylmalonic acid, which presumably reflects impaired utilization of vitamin B_{12}. A role of vitamin E in heme synthesis has been suggested, and its deficiency in animals does intensify the toxicity of lead. Animal studies also show that vitamin E reduces the mutagenicity of a variety of agents and has an effect on platelet aggregation and prostaglandins. Other suggested roles include slowing of the aging process, prevention of heart conditions, promotion of athletic ability, increased sexual endurance, and drug metabolism.[14]

Chemical deficiency signs. Plasma concentrations of α-tocopherol below 5 mg/L are associated with increased erythrocyte hemolysis by hydrogen peroxide and are thus quoted as being "deficient".[5] There is a strong correlation between plasma α-tocopherol and plasma lipids, suggesting that plasma concentrations should be interpreted relative to plasma lipid levels; 0.8 mg of tocopherol per gram of total plasma lipids appears to indicate adequate levels of vitamin E in infants. Elevation of plasma total lipids above 15 g/L can apparently shift erythrocyte tocopherol to plasma, potentially altering erythrocyte susceptibility despite "adequate" plasma concentrations of tocopherol in hyperlipidemic states.[15,16] The most current approach is HPLC assessment of the α-tocopherol isomer.

Pathophysiology. At the present time assessment of vitamin E status is primarily indicated in newborns, in those with potential fat malabsorption states, and in persons receiving synthetic diets. Bland[14] states:

> Increasing numbers of reports have demonstrated the direct correlation between the classic signs of vitamin E deficiency (including creatinuria, ceroid deposition, muscle weakness, increased serum creatine phosphokinase) and subnormal levels of α-tocopherol in blood or adipose tissue in cases of cystic fibrosis, biliary atresia, nontropical sprue, xanthomatosis, biliary cirrhosis, chronic pancreatitis, and a host of other conditions generally characterized by steatorrhea. Thus, there appears to be little question that malabsorption syndromes lead to vitamin E deficiencies in the human.

Elevated values of serum vitamin E have been reported during pregnancy and in patients with Batten's disease (a progressive childhood encephalopathy with disturbed PUFA metabolism).

Exogenous influences. Intestinal absorption of vitamin E may be reduced by organophosphate insecticides (nonsymptomatic doses). Although difficult to extrapolate to humans, there is impressive evidence that vitamin E has a protective effect on lung toxicity of the inhaled oxidants ozone and nitrogen dioxide. It has been suggested that persons with glucose 6-phosphate dehydrogenase (G6PD) deficiency who ingest inadequate vitamin E may be at increased risk of ozone-induced hemolytic crises. Other potential environmental oxidants have been inadequately studied regarding their interactions with vitamin E.[8]

Toxicity. Possible toxic effects suggested for humans and other animals include inhibition of prostaglandin synthesis, decreased platelet aggregation, potentiation of coagulopathy (resulting from vitamin K deficiency), inhibition of fibrinolysis, weakness and fatigue with creatinuria, and impaired wound healing.

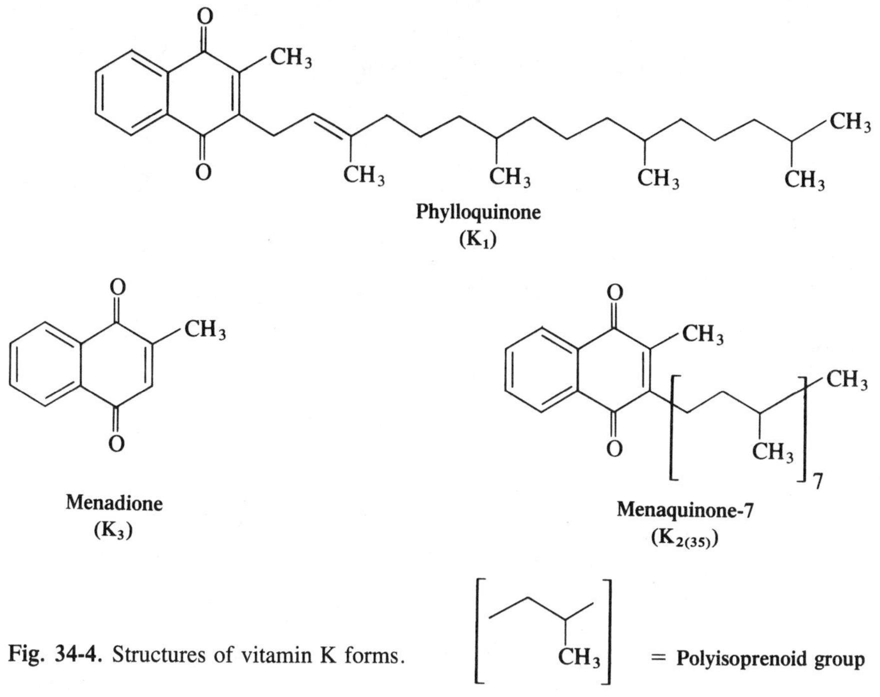

Phylloquinone
(K_1)

Menadione
(K_3)

Menaquinone-7
($K_{2(35)}$)

Fig. 34-4. Structures of vitamin K forms. = Polyisoprenoid group

Vitamin K

Experiments in the mid-1930s on the essentiality of cholesterol in chickens led to the discovery of the antihemorrhagic factor later called *vitamin K* (from German, *Koagulation*). Purification efforts revealed several quinone-containing compounds possessing this antihemorrhagic activity, and the term "vitamin K" is now used as a generic descriptor for menadione and derivatives exhibiting this activity. There are a large number of these compounds that are related to those shown in Fig. 34-4 by number and substituents of polyisoprenoid side chains and degree of unsaturation.[17]

The K vitamins are unstable in acidic or alkaline conditions and are readily oxidized. Major dietary sources are cabbage, cauliflower, spinach and other leafy vegetables, pork liver, soybeans, and vegetable oils. Uncomplicated dietary deficiency is considered rare in healthy children and adults, but a study of elderly persons revealed deficiencies in a large percentage correctable by oral administration of vitamin K.[18]

Metabolism. In infants vitamin K is absorbed in the colon, where bacterial synthesis is the major source of this vitamin. Older children and adults absorb vitamin K in the upper small intestine, where the contribution of intestinal bacteria is insignificant. Gastrointestinal infection with diarrhea, combined with prolonged antibiotic therapy and non–K supplemented milk-substitute formulas, has led to hemorrhagic symptoms in infants. Recognition of this combination of causes of vitamin K deficiency has led to supplementation of formulas and additional vitamin K administration when appropriate, making this circumstance rare.

Absorption of vitamin K in the intestines is chylomicron mediated, and vitamin K malabsorption states include cystic fibrosis, biliary atresia, cholelithiasis and obstructive jaundice secondary to hemolytic anemias, as well as other disorders leading to dysfunction of the upper small intestine. Depletion of body stores sufficient to manifest deficiency usually requires 3 weeks.

Function. First shown to be involved in prothrombin synthesis, vitamin K was later found to be required for other clotting factors (factors VII, IX, and X). This involvement is through vitamin K–dependent γ-carboxylation of glutamic acid residues of inactive precursor proteins in the liver (Fig. 34-5). These γ-carboxyglutamic acid residues in prothrombin are required for the calcium-dependent binding of prothrombin to phospholipid surfaces. Evidence suggests that this unique type of carboxylation might be involved in processes other than coagulation.[17]

Clinical deficiency signs. The clinical manifestation of a vitamin K deficiency is hemorrhage secondary to reduction in prothrombin and other clotting factors. Hemorrhagic symptoms are nonspecific, ranging from easy bruising to massive ecchymoses, mucous membrane hemorrhage, or posttraumatic bleeding; central nervous system hemorrhage sometimes occurs.

Chemical deficiency signs. Direct measurement of vitamin K in blood is not usually carried out. Determination of prothrombin time (velocity of clotting after addition of thromboplastin and calcium to citrated plasma) is an excellent index of prothrombin adequacy. This time is prolonged in deficiency of vitamin K and also in liver diseases characterized by decreased synthesis of prothrombin. De-

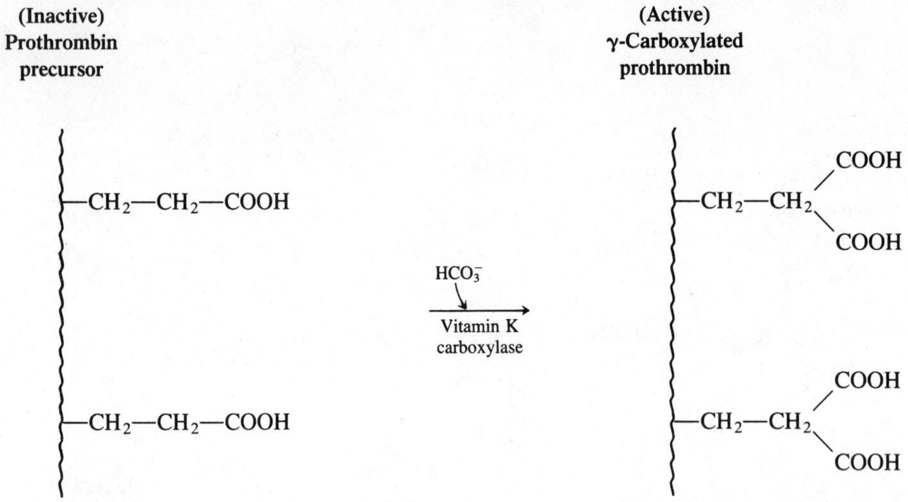

(Inactive)
Prothrombin
precursor

$-CH_2-CH_2-COOH$

$-CH_2-CH_2-COOH$

HCO_3^-
Vitamin K
carboxylase

(Active)
γ-Carboxylated
prothrombin

$-CH_2-CH_2(COOH)_2$

$-CH_2-CH_2(COOH)_2$

Fig. 34-5. Vitamin K–dependent γ-carboxylation of glutamyl residues.

ficiency of vitamin K also results in prolongation of the partial thromboplastin time, but thrombin time is normal. Clinical improvement of these clotting times is noted within a few hours of vitamin K administration.

Pathophysiology. The vitamin K malabsorptive states are described on p. 665. The prothrombin reduction seen in patients with alcoholic fatty liver is corrected by administration of vitamin K only if there is associated malnutrition or malabsorption. Paradoxically, administration of vitamin K to patients with preexisting subclinical hepatic dysfunction often results in prolongation of a previously normal prothrombin time.

Exogenous influences. Vitamin K action is antagonized by coumarin or indandione anticoagulants. Oral contraceptive use increases the levels of prothrombin and factors VII, IX, and X and apparently reduces the requirement for vitamin K.

Toxicity. High doses of the naturally occurring fat-soluble form of vitamin K (K_1) appear to be nontoxic, but the water-soluble forms of menadione (K_3) have produced serious side effects in high doses, especially in newborn infants. Large doses of menadione given to newborns or to their mothers during labor have resulted in hemolytic anemia. Premature infants have less tolerance to excess vitamin K than full-term infants do. Adult toxicity signs are primarily circulatory and involve a variety of cardiac and pulmonary signs.[19]

Vitamin D

Since 1822 it has been known that rickets (muscle hypotonia and skeletal deformities) could be cured with cod-liver oil. This antirachitic factor was identified as vitamin D a century later and soon was shown to have multiple forms. These antirachitic compounds are now collectively termed "vitamin D." The naturally occurring fish oil vitamin is cholecalciferol (vitamin D_3); it is produced in the skin from ultraviolet activation of 7-dehydrocholesterol. It is now known that vitamin D_3 is a prohormone that is converted in the liver to 25-hydroxycholecalciferol (calcidiol), which is further hydroxylated in the kidney to form the hormone 1,25-dihydroxycholecalciferol (calcitriol). Calciferol (vitamin D_2) is the primary dietary form (the form used in food fortification) and is similarly hydroxylated (see Chapter 25).

Major dietary sources include irradiated foods and commercially prepared milk. Small amounts occur in butter, egg yolk, liver, salmon, sardines, and tuna fish.

Physiological actions, regulation, and assessment of the hormone forms of vitamin D are discussed in Chapter 25.

Deficiency results in rickets (children) or osteomalacia (adults), both of which are forms of abnormal bone synthesis.

Acute toxicity does occur with excess intake; presenting signs and symptoms include anorexia, vomiting, headache, drowsiness, and diarrhea. The average person in the United States ingests six to seven times the recommended 400 IU/day of this fat-soluble compound. This has created concern for the effects of this chronic state of hypervitaminosis D on arterial metabolism, especially on development of arteriosclerosis.[20]

WATER-SOLUBLE VITAMINS

The nine water-soluble vitamins are absorbed without the involvement of fat absorption, and excess intakes are almost immediately excreted in the urine. Thiamin is stored for only a few weeks, and most other B-complex vitamins and ascorbic acid are stored for less than 2 months. Development of deficiency can therefore be quite rapid. Vitamin B_{12}, though a water-soluble vitamin, is stored in the liver for several years. The water-soluble vi-

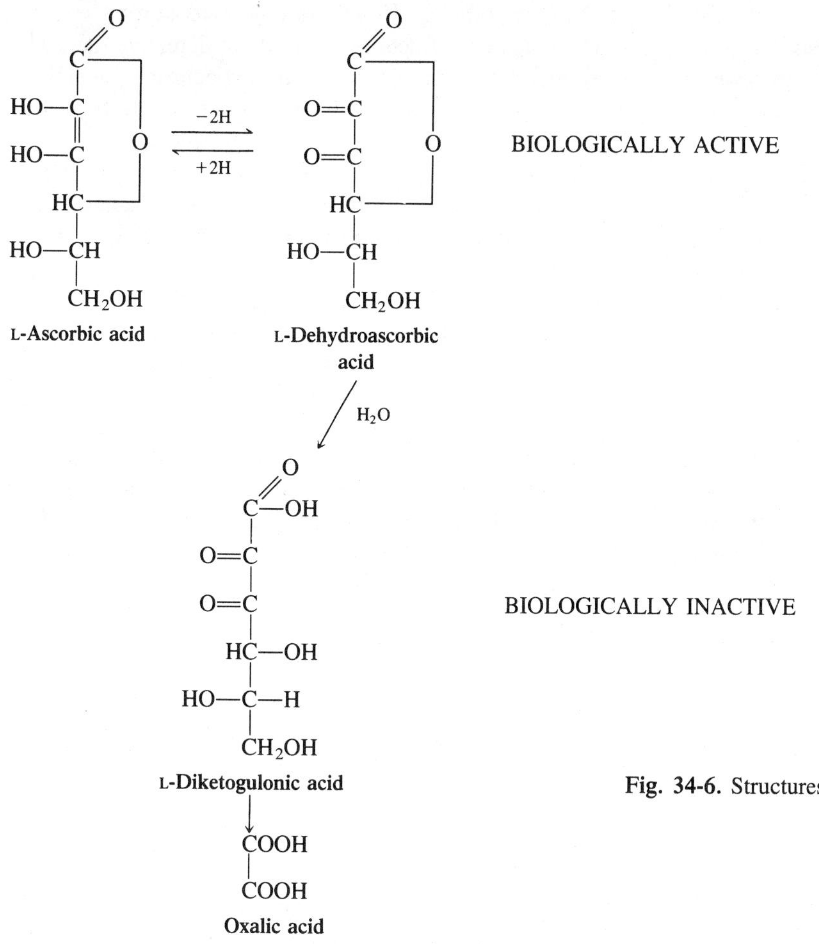

O
C
HO—C
HO—C O
HC
HO—CH
CH_2OH
L-Ascorbic acid

−2H ⇌ +2H

O
C
O=C
O=C O
HC
HO—CH
CH_2OH
L-Dehydroascorbic acid

BIOLOGICALLY ACTIVE

H_2O

O
C—OH
O=C
O=C
HC—OH
HO—C—H
CH_2OH
L-Diketogulonic acid

COOH
COOH
Oxalic acid

BIOLOGICALLY INACTIVE

Fig. 34-6. Structures of ascorbic acid and metabolites.

tamins function as coenzymes, and all except ascorbic acid and biotin must be metabolically converted to active forms.

Ascorbic acid (ascorbate)

The symptom cluster known as *scurvy* (swollen gums with loss of teeth, skin lesions, and pain and weakness in the lower extremities) was clearly described during the Crusades and became commonplace when long sea voyages began. A naval surgeon experimenting with diets to cure scurvy identified the efficacy of citrus fruits in 1747. This antiscurvy agent, vitamin C, was isolated in 1932 and later given the name "ascorbic acid" (from its antiscorbutic effect). The structures of this water-soluble vitamin and its oxidized form, dehydroascorbate, are shown in Fig. 34-6. By virtue of its ene-diol group, ascorbate is a very strong reducing compound. Although plants and most animals can synthesize this vitamin, humans cannot, and so dietary ingestion is essential.

Major dietary sources include fruits (especially citrus) and vegetables (tomatoes, green peppers, cabbage, leafy greens, potatoes). As ascorbate is labile to heat and oxygen, fresh and uncooked foods are highest in ascorbate content. Dietary deficiency is highly unlikely in a person who eats one daily serving of a fresh fruit or vegetable but is commonly seen in infants exclusively receiving cow's milk.

Metabolism. Absorption of ascorbate occurs in the small intestine. It is widely distributed in tissues (most concentrated in the adrenal cortex and pituitary) and passes the placenta readily. The normal body store requires several months to deplete before the appearance of symptoms of scurvy. Cerebrospinal fluid (CSF) concentrations are higher than those in plasma. Excess ascorbate and a metabolite, oxalate, are readily eliminated in urine (the ascorbate status determines the proportion being metabolized). Generally ascorbate metabolism accounts for about half the urinary oxalate (Fig. 34-6).

Function. The most important known role of ascorbic acid is in formation and stabilization of collagen, a major component of all connective tissues, by hydroxylation of proline and lysine for cross-linking. Deficiency, therefore, results in breakdown of connective tissues, teeth, and bones and in poor wound healing. Ascorbate also functions in the conversion of tyrosine to catecholamines (by dopamine β-hydroxylase) and is present in high concentrations in the brain. It enhances absorption of iron and is possibly involved in heme function. Shown to be required for conversion of cholesterol to bile acids, by a cytochrome P-450–mediated reaction, ascorbate's roles in reducing serum cholesterol and dissolving gallstones remain controversial. This same cytochrome mechanism is necessary for other hydroxylations, an indication that it also functions in steroid synthesis and in detoxification (or activation) of a variety of foreign organic compounds (xenobiotics). There is strong evidence that ascorbate protects against lipid peroxidation.[2,21]

Clinical deficiency signs. Early symptoms of ascorbate deficiency are weakness, lassitude, irritability, and vague aches and pains in joints and muscles. Children may also exhibit impaired growth. Clinical scurvy, which develops when the body pool falls to about 300 mg, includes hemorrhages into skin, alimentary and urinary tracts, conjunctiva, and possibly muscle and periosteum. Skin becomes rough and dry and develops hyperkeratotic (scaly) changes in the hair follicles. Bones become osteoporotic and exhibit spontaneous fractures in children; defective tooth formation is also seen. A microcytic hypochromic anemia is common, as is pyrexia (feverishness), delayed wound healing, and increased susceptibility to infection.

Chemical deficiency signs. Clinical signs of scurvy have been associated with serum ascorbate values below 2.4 mg/L or whole blood ascorbate levels below 3 mg/L. Transient low values, however, do not necessarily reflect a tissue deficiency. Increased ascorbate intake raises serum ascorbate levels up to a maximum of about 14 mg/L, at which time renal clearance rises sharply. Leukocyte ascorbate is considered to more closely reflect tissue stores but is technically more difficult to assay. Measurement of urinary ascorbate is not recommended for status assessment, since it reflects recent intake and has numerous analytical difficulties. Drugs known to increase the urinary excretion of ascorbate include aspirin, aminopyrine, barbiturates, hydantoins, and paraldehyde.

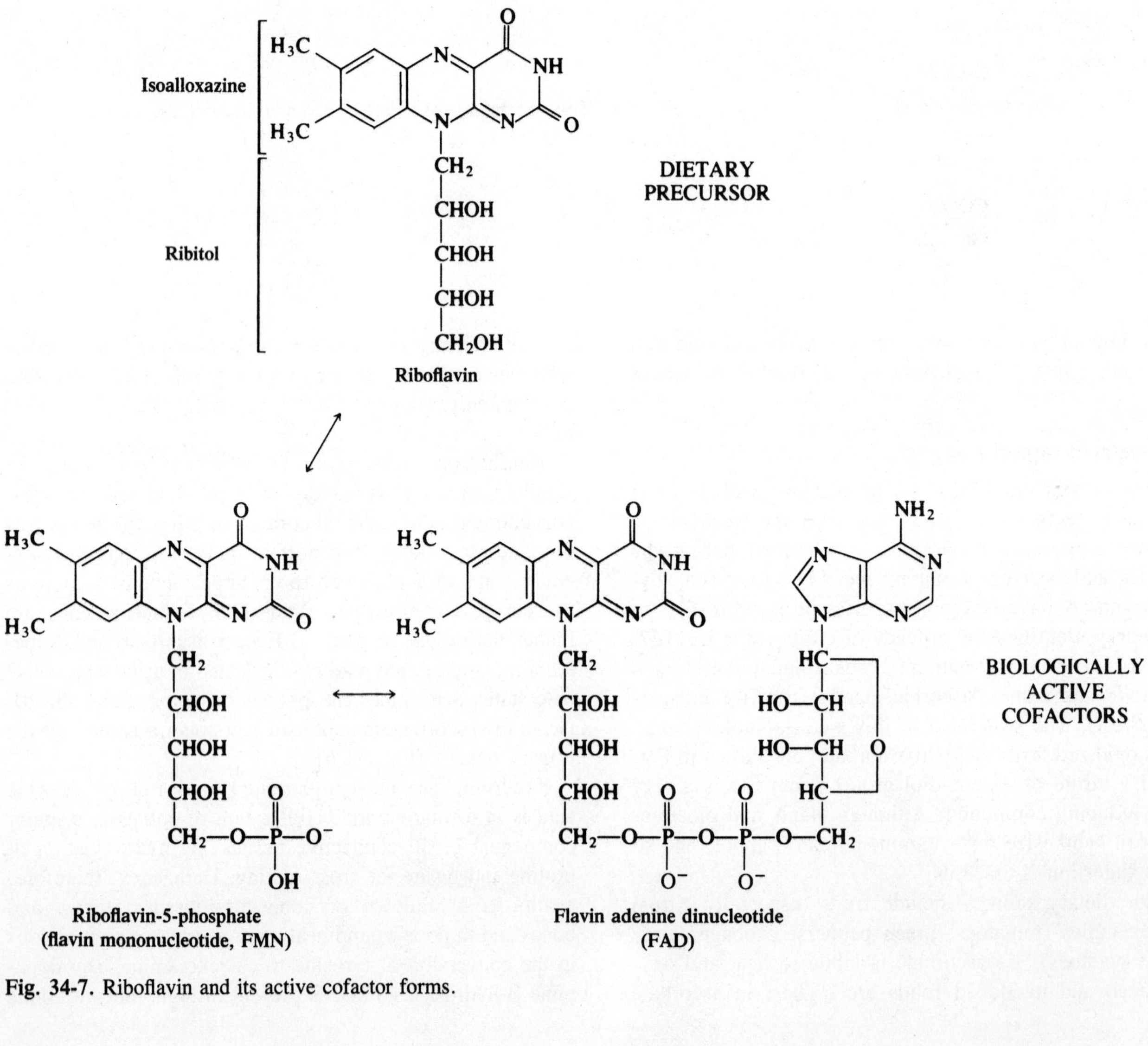

Isoalloxazine

Ribitol

Riboflavin

DIETARY PRECURSOR

Riboflavin-5-phosphate (flavin mononucleotide, FMN)

Flavin adenine dinucleotide (FAD)

BIOLOGICALLY ACTIVE COFACTORS

Fig. 34-7. Riboflavin and its active cofactor forms.

Pathophysiology. Requirements are increased in chronic illness, during pregnancy, or during oral contraceptive use. Animal studies of alterations in ascorbate status evoked by a variety of environmental intoxicants suggest that human ascorbic acid needs may be increased in exposed populations. Smoking is believed to reduce serum ascorbate levels.

Toxicity and therapeutic uses. Ingestion of pharmacological levels of ascorbate (0.5 to 1 g, rarely to 30 g/day) is commonly practiced to avoid or cure colds, though several studies have not supported the usefulness of this dosing. This controversial practice has stimulated investigation of ascorbate toxicity. Overall there appears to be little toxicity except in a few persons who may experience diarrhea, kidney stones (a result of increased levels of urinary oxalate), hemolysis, or allergic reactions. The popularity of megadosage of ascorbic acid has revealed a rebound scurvy on sudden discontinuation of the megadose intake. Excess intake may interfere with metabolism of vitamin B_{12} and with drug actions (aminosalicylic acid, tricyclic antidepressants, and anticoagulants).

Riboflavin

The yellowish fluorescent pigment in milk, eggs, yeast, and liver discovered in the nineteenth century (named *flavin* from the Latin for 'yellow'), the "yellow enzyme" found in yeast, and the yellow heat-stable animal growth factor crystallized from milk in the 1930s were finally realized to be the same substance—a vitamin consisting of a pigment, flavin, attached to the carbohydrate D-ribitol—riboflavin. Its structure and that of its two cofactor-active forms, riboflavin 5′-phosphate (flavin mononucleotide; FMN) and flavin adenine dinucleotide (FAD) are shown in Fig. 34-7. FAD is the most water-soluble of these three flavins and exhibits orange fluorescence, whereas the other two fluoresce greenish yellow. Aqueous solutions of flavins are stable to heat and oxidizing agents, but the riboflavin in milk is dramatically reduced upon exposure to sunlight.

Foods high in riboflavin include milk, liver, eggs, meat, and some green leafy vegetables. Intestinal organisms synthesize riboflavin, but the distal location of these bacteria may preclude absorption.

Metabolism. Consumed protein complexes of FAD and FMN release these flavins upon gastric acidification and proteolysis. After dephosphorylation they are absorbed in the proximal small intestine, rephosphorylated, and then appear in the circulation weakly bound to albumin and other plasma proteins, including specific riboflavin-binding proteins. Most flavins are quickly taken up by the liver and the kidneys, though measurable amounts are found in most tissues. The three flavin forms can be interconverted enzymatically, with FAD becoming the predominant tissue form. Minor amounts of flavin are excreted in bile, feces, sweat, and breast milk; urinary flavin excretion is greater, mostly in the free riboflavin form, and is dependent on tissue stores and the amount ingested. Only a small amount of the ingested riboflavin is metabolized by the body; hence metabolites are not detected in urine (except for those formed by bacteria). Riboflavin, but not FMN or FAD, is the moiety transferred from plasma into brain. The flavins in blood are mostly in the erythrocytes and function in the two FAD enzymes, glutathione reductase and methemoglobin reductase. The fetus obtains free riboflavin derived from maternal erythrocytic FAD in the placenta.

Function. Flavoprotein enzyme systems contain FAD or FMN as prosthetic groups. Most flavoproteins catalyze removal of either a hydride ion or a pair of hydrogen atoms from the substrate. These hydrogen atoms are first transferred to the FAD (or FMN) moiety of the enzyme and then transferred to the acceptor molecule:

$$\text{Glucose} + \text{Enzyme-FAD} \rightarrow \text{Gluconic acid} + \text{Enzyme-FADH}_2$$

$$O_2 \rightarrow H_2O_2$$

In this fashion these flavins act as cofactors in a variety of oxidative reactions, including fatty acid oxidation, the tricarboxylic acid cycle, and electron transport (see box below).

Clinical deficiency signs. Clinical symptoms associated with ariboflavinosis are angular stomatitis (mouth lesions), glossitis (smooth tongue), photophobia, blepharospasm (eyelid spasm), conjunctival congestion and other ocular changes, dermatological changes (seborrhea, blepharitis

Examples of flavoprotein enzyme reactions

D-Amino acid oxidase:

$$\text{Alanine} + H_2O + O_2 \xrightarrow{\text{enz-FAD}} \text{Pyruvate} + NH_3 + H_2O_2$$

Pyruvate dehydrogenase:

$$\text{Pyruvate} + \text{Oxidized lipoic acid} \xrightarrow[\text{TPP, Mg}]{\text{enz-FAD}} \text{6-}S\text{-Acetyldihydrolipoate} + CO_2$$

Xanthine oxidase:

$$\text{Xanthine} + O_2 \xrightarrow[\text{Fe,Mo}]{\text{enz-FAD}} \text{Uric acid} + H_2O_2$$

Glutathione reductase:

$$\text{Oxidized glutathione (GSSG)} + \text{2NADPH} \xrightarrow{\text{enz-FAD}} \text{Reduced glutathione (2GSH)} + \text{2NADP}$$

NADH dehydrogenase:

$$\text{NADH} + 2\ \text{oxidized cytochrome}\ c \xrightarrow{\text{enz-FMN}} \text{NAD} + 2\ \text{reduced cytochrome}\ c$$

angularis), neurological alterations (behavioral changes, decreased hand grip strength, burning feet in adults, retarded intellectual development and electroencephalogram changes in children), and hematological dyscrasia (anemia and reticulocytopenia).

Chemical deficiency signs. Erythrocytic concentrations of riboflavin, FMN, and FAD are more sensitive indices of riboflavin status than flavin measurements in urine or plasma are, but these blood indices are altered late in the progression of deficiency. A functional approach to assessment of riboflavin status involves measurement of the increase in erythrocytic glutathione reductase (EGR) activity on in vitro addition of FAD. Deficiency states are thus reflected by increased ratio of this enzyme's activity plus or minus FAD (EGR index). This EGR index plateaus rapidly and does not continue to increase as deficiency progresses; its validity is also questionable in states of altered erythrocyte enzymes.

Pathophysiology. Ariboflavinosis is most commonly encountered when intake is inadequate as a result of poverty. It often develops as a consequence of pregnancy, especially in the pregnant adolescent. American surveys have indicated inadequate riboflavin nutrition in up to 10% of children, up to 47% of teenagers, and up to 32% of geriatric subjects. Ariboflavinosis can occur secondary to disease states, including prolonged febrile illness, malignancy, hyperthyroidism, cardiac failure, diabetes mellitus, and gastrointestinal diseases. The malnutrition associated with alcoholism affects riboflavin and other nutrients as well. Its decomposition is accelerated by phototherapy for neonatal jaundice. Negative nitrogen balance, seen in all catabolic states (including stress, physical exertion, fasting, prolonged bed rest), results in increased urinary riboflavin excretion. Riboflavin deficiency often occurs concomitantly with deficiencies of other B vitamins. Additionally, riboflavin interacts in metabolic processes involving other vitamins. These interactions can therefore result in a mixed clinical picture and also may explain how therapy with one vitamin may improve symptoms of another vitamin deficiency.

Exogenous influences. Phenothiazines, oral contraceptives, and hormonal imbalance can impair riboflavin utilization; prednisolone reportedly interferes with conversion to coenzymes.

Toxicity. Riboflavin toxicity is low. Because its solubility is enhanced by administration as a suspension or dissolved in organic solvents, some reported toxicities may be caused by these agents rather than by riboflavin itself.

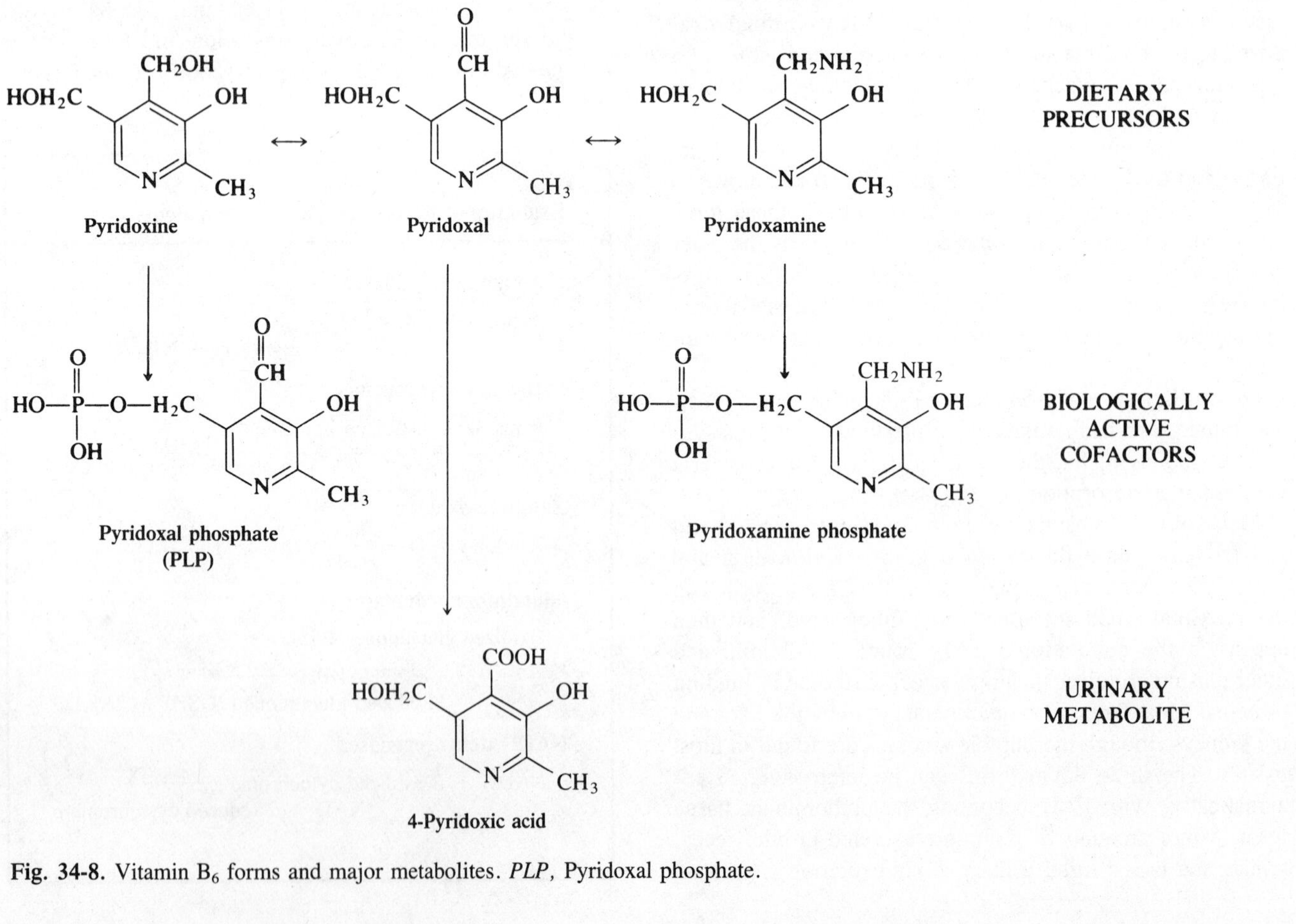

Pyridoxine ⟷ **Pyridoxal** ⟷ **Pyridoxamine** — DIETARY PRECURSORS

Pyridoxal phosphate (PLP), **Pyridoxamine phosphate** — BIOLOGICALLY ACTIVE COFACTORS

4-Pyridoxic acid — URINARY METABOLITE

Fig. 34-8. Vitamin B_6 forms and major metabolites. *PLP,* Pyridoxal phosphate.

Pyridoxine

Vitamin B_6, isolated in the early 1930s as the agent preventing skin lesions in rats, is known chemically as pyridoxine. Later research first identified pyridoxal and pyridoxamine with this same activity; then their phosphorylated forms were recognized as active cofactors. Thus the term "vitamin B_6" now refers to the family of compounds structurally related to pyridoxal phosphate (Fig. 34-8). Pyridoxine occurs mainly in plants, whereas pyridoxal and pyridoxamine are present mainly in animal products. These three pyridine derivatives are metabolically interconverted.

Major dietary sources of vitamin B_6 are meat, poultry, fish, potatoes, and vegetables; dairy products and grains contribute lesser amounts. The predominant food form is pyridoxal phosphate, which is readily lost in food processing.

Metabolism. Pyridoxine does not bind to plasma proteins; pyridoxal and pyridoxal phosphate bind mainly to albumin. Erythrocytes rapidly take up pyridoxine, convert it to pyridoxal phosphate and pyridoxal, and then release pyridoxal into plasma. Metabolism of pyridoxal appears to occur primarily in the liver with formation of 4-pyridoxic acid, which is excreted in urine (Fig. 34-8). Pyridoxal phosphate synthesis is dependent on a flavin enzyme, which interrelates riboflavin and pyridoxine. Vitamin B_6 concentration is high in the brain and the CSF; the nonphosphorylated forms best enter the CSF, choroid plexus, and brain.

Function. Almost all pyridoxal phosphate–catalyzed reactions are concerned with transformations of amino acids, the major exception being phosphorylases. These pyridoxine cofactor forms act in over 60 different enzyme systems (see examples in box below), catalyzing a variety of reaction types. All reactions involve initial formation of a Schiff base between pyridoxal phosphate and the substrate, facilitating bond-changing reactions. Perhaps best known are its roles in conversion of tryptophan to 5-hydroxytryptamine (serotonin) and the separate pathway of tryptophan to nicotinic acid ribonucleotide (the "niacin pathway"), both shown in Fig. 34-9. A high percentage of women receiving oral contraceptives exhibit abnormal tryptophan

Examples of pyridoxal phosphate–catalyzed reactions

Racemizations:

$$H_2N{-}CH(CH_3){-}COOH \rightleftharpoons CH_3{-}CH(NH_2){-}COOH$$

Transaminations:

$$R_1{-}CH_2{-}CH(NH_2){-}COOH + R_2{-}CH_2{-}C(=O){-}COOH \rightleftharpoons R_1{-}CH_2{-}C(=O){-}COOH + R_2{-}CH_2{-}CH(NH_2){-}COOH$$

Decarboxylations:

$$\text{(imidazole: N, NH)}{-}CH_2{-}CH(NH_2){-}COOH \longrightarrow \text{(imidazole: N, NH)}{-}CH_2{-}CH_2{-}NH_2$$

α,β-Eliminations:

$$HO{-}CH_2{-}CH(NH_2){-}COOH \xrightarrow{-H_2O} CH_2{=}C(NH_2){-}COOH \xrightarrow[NH_3]{H_2O} CH_3{-}C(=O){-}COOH$$

β,γ-Eliminations:

$$HS{-}CH_2{-}CH_2{-}CH(NH_2){-}COOH \xrightarrow{-H_2S} CH_3{-}CH{=}C(NH_2){-}COOH \xrightarrow[NH_3]{H_2O} CH_3{-}CH_2{-}C(=O){-}COOH$$

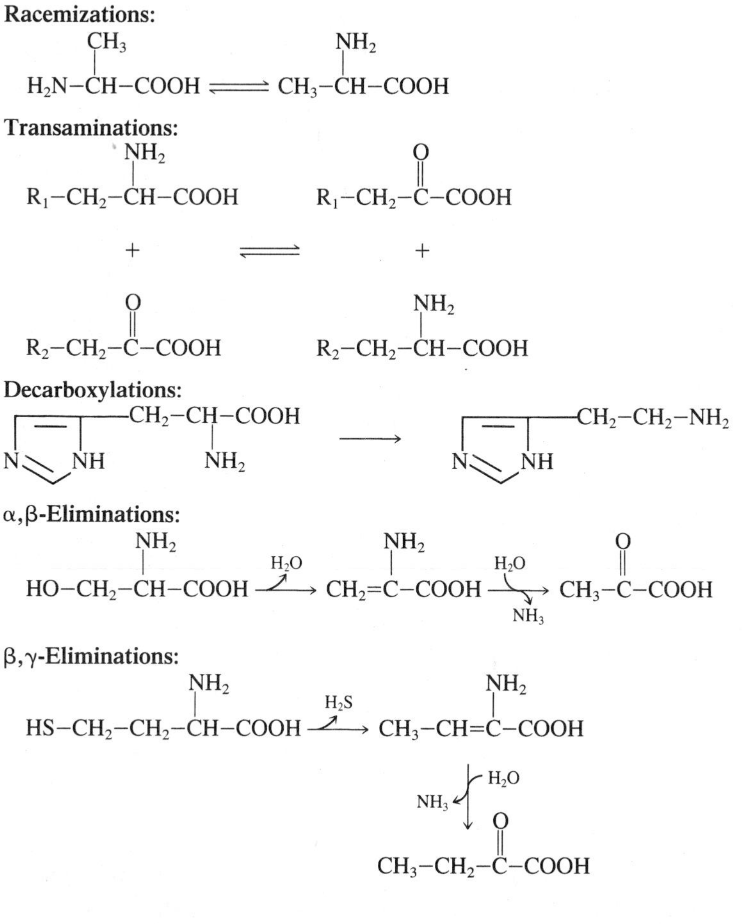

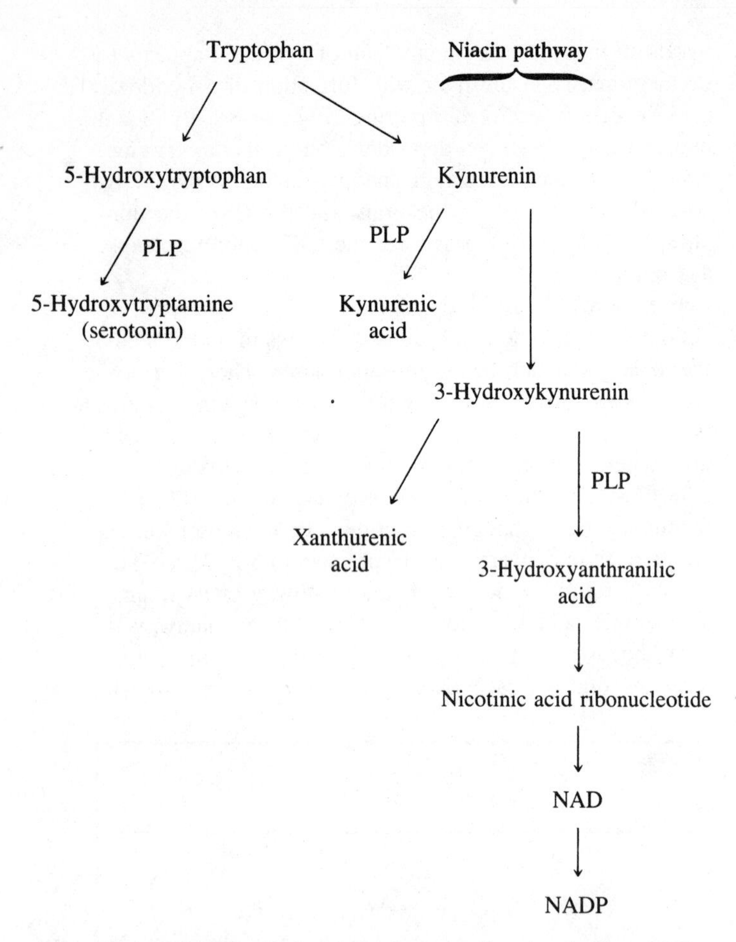

Fig. 34-9. Role of vitamin B_6 in tryptophan metabolism. *NAD*, Nicotinamide adenine dinucleotide; *NADP*, nicotinamide adenine dinucleotide phosphate; *PLP*, pyridoxal phosphate.

metabolism, and there is some evidence to support a link between depressive mood changes of some patients receiving oral contraceptives and the relative lack of 5-hydroxytryptamine created by the B_6-deficiency and the resulting diversion of tryptophan to the niacin pathway. There is controversy concerning the role of vitamin B_6 deficiency in gestational diabetes.

Clinical deficiency signs. Symptoms of pyridoxine deficiency described in infants given a commercial formula in which pyridoxine had been destroyed by milk processing include irritability, seizures, anemia, vomiting, weakness, ataxia, and abdominal pain. Adult volunteers deprived only of pyridoxine for 7 weeks showed no clinical symptoms (despite chemical evidence of deficiency) until a pyridoxine antagonist was fed; this caused facial seborrheic dermatitis (yellowish, oily scales). Pyridoxine inadequacy during pregnancy has been linked to suboptimal birth outcomes (infants with low Apgar scores and low birth weight).

Chemical-deficiency signs. Chemical indices of pyridoxine depletion include reduction in plasma and erythrocyte concentrations of pyridoxine or pyridoxal phosphate. Urinary pyridoxine (normally representing less than 10% of the pyridoxine intake) and pyridoxic acid, the major urinary metabolite, are also reduced. An oral tryptophan load to deficient persons results in excretion of several tryptophan metabolites in higher amounts than usual—xanthurenic acid being the one most commonly measured. The involvement of other metabolic and hormonal factors in this pathway necessitates cautious interpretation of this loading test. The tissue status of pyridoxal phosphate can be assessed by measurement of the increment of erythrocytic aspartate (or alanine) aminotransferase (AST or ALT, respectively, but GOT or GPT by older terminology) after in vitro addition of the pyridoxal phosphate cofactor. Elevation in the ratio of activity plus or minus pyridoxal phosphate (the EAST index) is suggestive of inadequate tissue stores. Pyridoxine-depleted subjects may require several months of repletion to increase enzymatic activity and reduce the stimulation index. It has been suggested that urinary pyridoxic acid and erythrocyte enzyme stimulation both be measured to evaluate short-term and long-term pyridoxine status, respectively. The newer methods of plasma pyridoxal phosphate determination may prove to be equal to (or better than) the widely accepted EAST index.

Pathophysiology. Conditions associated with low pyridoxine indices include celiac disease, acute alcoholism, psychoses such as paranoia and schizophrenia, epilepsy, ulcerative colitis, renal calculi, and lactation. Pyridoxine requirements increase during pregnancy as a result of fetal demand and hormonal induction of maternal enzymes, which increases maternal requirements. Certain inborn errors of metabolism can present in the first few days after birth with convulsions and can lead to mental retardation unless treated with high doses of pyridoxine. Vitamin B_6–responsive anemias have been described in adults.

Exogenous influences. Drugs known to antagonize pyridoxine include isonicotinic acid hydrazide (isoniazid, INH), steroids, and penicillamine. Oral contraceptives lower indices of pyridoxine status but are not associated with classical deficiency symptoms. Pyridoxine deficiency enhances susceptibility to carbon disulfide toxicity, suggesting that workers in the numerous occupations using this chemical may be at risk if pyridoxine intake is marginal.

Toxicity. The vitamin B_6 compounds are of low systemic toxicity, and no teratogenic effect has been detected. Very large doses may inhibit secretion of prolactin and consequently reduce milk production. Toxicity has been described in adult males ingesting 300 mg/day—a dose far in excess of any recommended therapy and not obtainable from diet alone.

Niacin

Over 200 years ago pellagra (from the Italian, meaning 'rough skin'), which is associated with diarrhea, dementia,

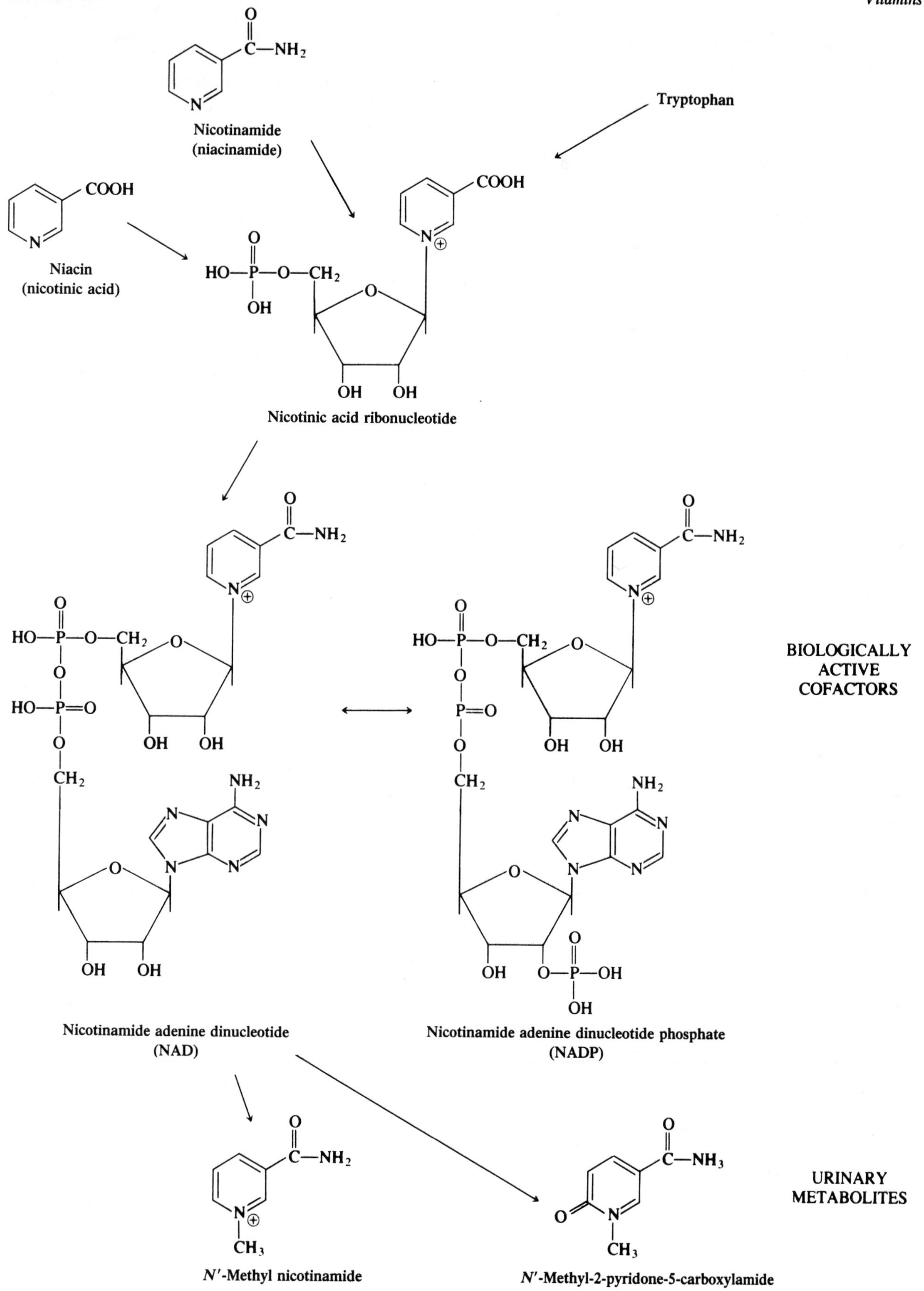

Nicotinamide
(niacinamide)

Tryptophan

Niacin
(nicotinic acid)

Nicotinic acid ribonucleotide

BIOLOGICALLY
ACTIVE
COFACTORS

Nicotinamide adenine dinucleotide
(NAD)

Nicotinamide adenine dinucleotide phosphate
(NADP)

URINARY
METABOLITES

N′-Methyl nicotinamide

N′-Methyl-2-pyridone-5-carboxylamide

Fig. 34-10. Cofactor forms and metabolites derived from niacin or tryptophan.

dermatitis, and death (the "four D's"), was attributed to poor diet. From 1910 to 1935 pellagra was the worst nutritional disease outbreak in U.S. history, with over 150,000 cases reported annually, most of whom were poor persons in the South. In 1912 nicotinic acid was extracted from rice polishings and was claimed to have vitamin-like effects, but it was not until 1935 that nicotinic acid (also called "niacin") was shown to cure black-tongue in dogs (a disease similar to pellagra in humans). Thus the responsible nutrient was identified as niacin, and its deficiency was associated with diets high in corn.

Niacin is a simple derivative of pyridine and is extremely stable. Moderately resistant to heat, acid, and alkali, it is related chemically to nicotine but has very different physiological properties. The active cofactor forms of NAD and NADP (Fig. 34-10) derived from niacin can also be synthesized from liver tryptophan (Fig. 34-9) so that sufficient dietary tryptophan can abolish the requirement for niacin. (Note that this conversion does not involve formation of niacin.)

Meats and grains are major sources of niacin, and many manufactured food products are supplemented with this vitamin, especially those made from cereals. Dietary deficiency is most likely in persons with diet histories reflecting mainly corn consumption (such as a diet of pork fat and hominy grits or sorghum). Corn is especially poor in tryptophan, and part of the niacin in corn is not absorbed. The high concentrations of leucine in cereals somehow interfere with the niacin pathway and with conversion of niacin to its cofactor forms. The niacin equivalent of tryptophan is approximately 60 mg of tryptophan equaling 1 mg of niacin. Most diets have ≥600 mg of tryptophan, but diets low in protein have much less.

Metabolism. Both niacin and nicotinamide are readily absorbed in the gut. Niacin is transported in blood mainly in association with erythrocytes. There is little storage in the body, and urine contains mostly the forms of nicotinamide and metabolites (Fig. 34-10). Plasma nicotinamide readily enters the CSF, whereas niacin does not. Brain tissue does not express the niacin pathway of tryptophan and so must use plasma nicotinamide from the diet or dephosphorylated forms of the cofactors.

Function. The active cofactor forms of niacin, collectively called the "pyridine nucleotides," were the first coenzymes to be recognized. NAD and NADP are involved in a large number of oxidation-reduction reactions catalyzed by dehydrogenases (see box in next column). Reduction yields the dihydronicotinamide that has a strong absorption at 340 nm—a feature widely used in assays of pyridine nucleotide–dependent enzymes (see Chapter 49).

Clinical deficiency signs. The clinical syndrome of niacin deficiency has early symptoms of lassitude, anorexia, weakness, digestive disturbances, anxiety, irritability, and depression. Progressive deficiency (classic pellagra) results in chronic dermatitis, mucous membrane inflammation (glossitis, stomatitis, esophagitis, diarrhea, urethritis, proctitis, vaginitis), and possibly severe weight loss, disorientation, delirium, dementia, and hallucinations. The triad of dementia, diarrhea, and dermatitis should suggest niacin deficiency.

Chemical deficiency signs. Chemical measures of niacin status primarily involve the two major urinary metabolites *N*′-methylnicotinamide and *N*′-methyl-2-pyridone-5-carboxylamide. The ratio of the 2-pyridone compound to the *N*′-methylnicotinamide is reduced in niacin deficiency; reduction in the individual metabolites is also seen. These metabolites are normally present in plasma also. Only a small percentage of administered niacin is excreted as niacin or as nicotinamide. Measurement of niacin or its nucleotides in blood is generally not considered a reliable index of niacin status, though erythrocytes of deficient persons have lower levels of NAD and NADP and higher amounts of nicotinamide ribonucleotide.

Exogenous influences. As for pyridoxine, niacin deficiency results in greater susceptibility to toxicity of carbon disulfide. Oral contraceptives stimulate the niacin pathway of tryptophan.

Toxicity and therapeutic uses. Niacin and nicotinamide are widely used in megadoses in the treatment of sprue, psychiatric conditions such as schizophrenia, and a wide range of circulatory disorders including hypertension, cerebral thrombosis, and intermittent claudication (limping). Niacin and its analogs are used in treatment of some hyperlipidemic states to reduce heart and vascular effects. Toxicities of both niacin and nicotinamide are low; niacin has a pronounced peripheral vasodilatory effect, and there are some reports of gastrointestinal disturbance, mild liver dysfunction, jaundice, altered glucose tolerance, hyperuricemia, and adverse mental changes.[19,22]

Examples of pyridine nucleotide–catalyzed reactions

Alcohol dehydrogenase:
Ethanol + NAD ⇌ Acetaldehyde + NADH

Glutamate dehydrogenase:
Glutamic acid + NAD ⇌
α-Ketoglutaric acid + NH_3 + NADH

Glycerolphosphate dehydrogenase:
L-Glycerol-3-phosphate + NAD ⇌
Dihydroxyacetone phosphate + NADH

Glucose 6-phosphate dehydrogenase:
Glucose 6-phosphate + NADP →
6-phospho-D-gluconolactone + NADPH

Nitrite reductase:
3NADPH + Nitrite → 3NADP + NH_4OH + H_2O

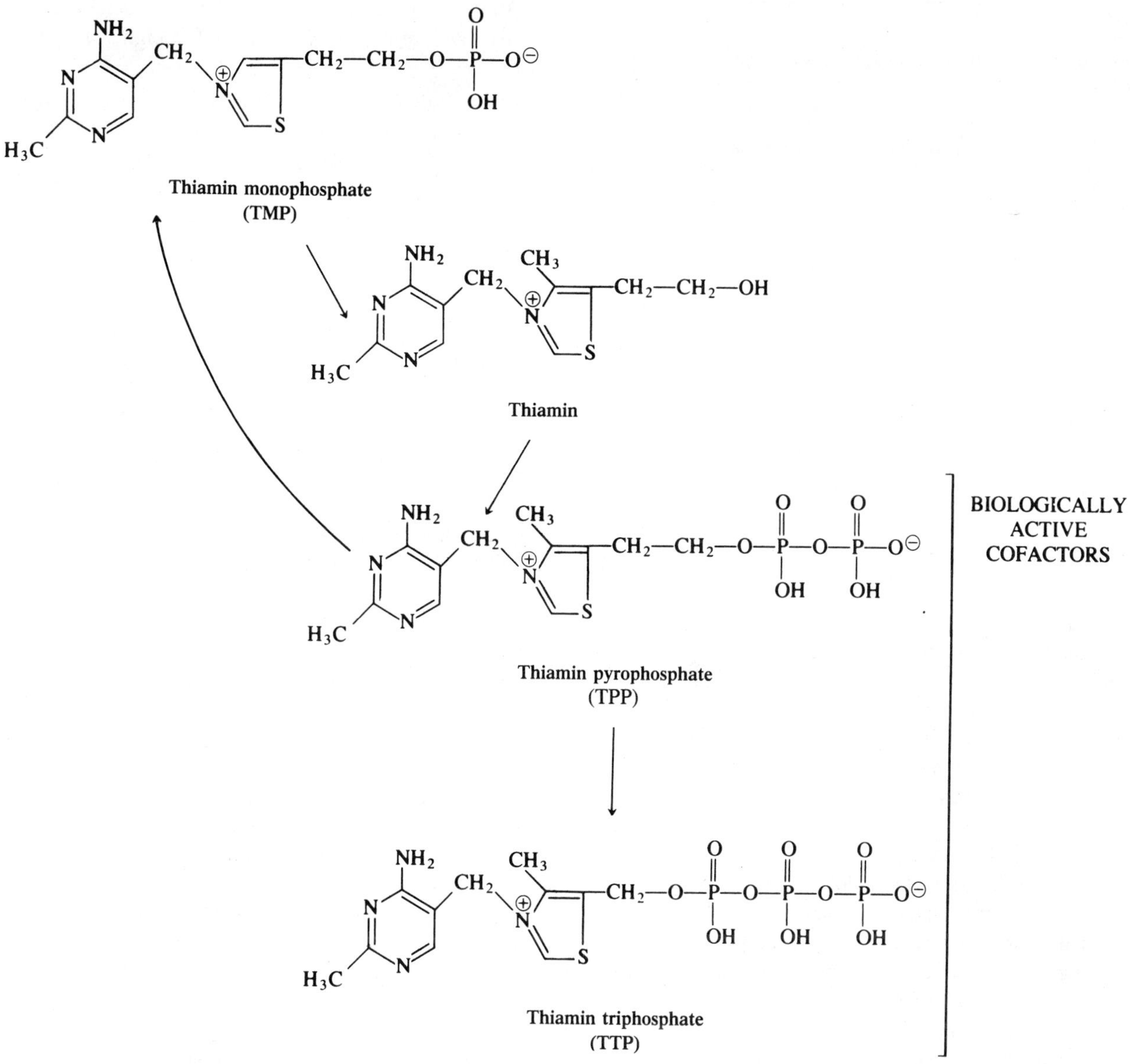

Thiamin monophosphate (TMP)

Thiamin

Thiamin pyrophosphate (TPP)

Thiamin triphosphate (TTP)

BIOLOGICALLY ACTIVE COFACTORS

Fig. 34-11. Thiamin and its cofactor forms.

Thiamin

In the early 1900s an observant physician noted that hens fed a diet of leftover cooked, polished rice developed a polyneuritis similar to the human disease of beriberi. The hen neuropathy was abolished with feeding of rice husks or whole grain. First believed to be caused by a toxin that was neutralized by rice husks, this state was later shown to be a nutrient deficiency caused by removal of an essential factor from rice as it was polished. This antiberiberi factor was finally isolated and crystallized in 1925; then its structure was shown to be a substituted pyrimidine linked by a methylene group to a substituted thiazole (Fig. 34-11). Thus the name reflects its components of amine and sulfur (thia-) groups.

Thiamin is easily destroyed in alkaline media. It resists temperatures up to 100°C but is destroyed at greater temperatures, a fact significant for fried foods or those cooked under pressure. Highly water soluble, it is easily leached out of foodstuffs being washed or boiled.

Sources of thiamin include yeast, wheat, whole grain, and enriched breads and cereals, nuts, peas, potatoes, and most vegetables. Dietary deficiency is suspect in persons ingesting polished rice, alcoholics, persons with anorexia, vomiting, or diarrhea, and postoperative patients. Deficiency of thiamin is common in the elderly.[23]

Metabolism. Dietary thiamin is absorbed in the intestine by a carrier-mediated process that is saturated at an oral intake of about 10 mg. Blood thiamin appears in the CSF and brain to a small degree; the phosphorylated forms are found even less commonly. Thiamin is excreted un-

Examples of thiamin pyrophosphate–catalyzed reactions

Nonoxidative decarboxylations of α-keto acids:

$$R-\overset{O}{\overset{\|}{C}}-COOH \rightarrow R-\overset{O}{\overset{\|}{C}}H + CO_2$$

Oxidative decarboxylations of α-keto acids:

$$R-\overset{O}{\overset{\|}{C}}-COOH + \tfrac{1}{2}O_2 \rightarrow R-COOH + CO_2$$

Formation of α-ketols:

$$R-\overset{O}{\overset{\|}{C}}-COOH + R'-CHO \rightarrow R-\overset{O}{\overset{\|}{C}}-\overset{OH}{\overset{|}{C}}H-R' + CO_2$$

changed or after cleavage between the ring systems by intestinal microorganisms. Thiamin pyrophosphate (TPP) is the predominant moiety in tissues, whereas the major form in plasma is thiamin. Erythrocytes contain TPP at concentrations about fivefold higher than in plasma. Liver, heart, and brain have higher concentrations than muscle or other organs.

Function. In its TPP cofactor form (Fig. 34-11), thiamin catalyzes the decarboxylation of α-keto acids (pyruvate and α-ketoglutarate), the oxidative decarboxylation by α-keto acid dehydrogenases, and the formation of ketols. Examples of these reaction types are shown in the box above. Transketolase (TK) activity in erythrocytes, commonly measured as an index of deficiency, is an example of α-ketol formation (see below). Through these reaction types, thiamin pyrophosphate functions in both major carbohydrate pathways. Thiamin triphosphate (TTP) may be the thiamin derivative released from nerves after electrical stimulation and as such may play a role in sodium-ion conductance. There is some evidence for decreased brain TTP and an inhibitor of its synthesis (demonstrable in tissues and urine) in patients with Leigh's encephalopathy.

Clinical deficiency signs. Infants breast-feeding from nonsymptomatic thiamin-deficient mothers display infantile dyspnea and cyanosis and may also develop diarrhea, vomiting, wasting, and aphonia (loss of voice, or motions of crying but little sound). Two other forms of thiamin deficiency are described: poor appetite, fatigue, and peripheral neuritis (termed "dry beriberi") and edema and cardiac failure ("wet beriberi"). Severe deficiency is often associated with encephalopathy. Neurologists in the United States generally use the term "Wernicke-Korsakoff syndrome" to refer to the central nervous system manifestations of thiamin deficiency, whereas Europeans refer to "Wernicke's encephalopathy" (intelligence disturbance, apathy, ataxia, double vision, nystagmus, drooping of eyelids) and "Korsakoff's psychosis" (deterioration of intelligence and loss of recent memory, often compensated by confabulation). Wernicke-Korsakoff syndrome responds to thiamin therapy, and there is evidence for an abnormality of neurotransmitter metabolism, perhaps involving TTP. These patients typically accumulate excessive amounts of pyruvate and lactate in physiological fluids.[24]

Genetic variations in TPP-dependent enzymes modify the effect of dietary thiamin deficiency. Although most patients with thiamin deficiency do not develop Wernicke-Korsakoff syndrome, mild deficiency leads to impairments in higher integrative functions (including memory). More severe deficiency leads either to "dry" or "wet" beriberi, but seldom do both occur together.

Chemical deficiency signs. Chemical indices of thiamin deficiency commonly employed are reduction in urinary thiamin, reduction in erythrocyte transketolase (ETK) activity, and stimulation of ETK by in vitro TPP. Prolonged deficiency lowers the ETK apoenzyme so that the ETK stimulation test may underestimate the magnitude of deficiency. There is also evidence of reduction of ETK in undernutrition, diabetes, and liver disease without a TPP stimulation effect. As is possible with all proteins, genetic heterogeneity of ETK has been demonstrated in humans. Rats made deficient in thiamin showed decreased TPP in

Xylulose 5-phosphate		Ribose 5-phosphate		Sedoheptulose 7-phosphate		Glyceraldehyde 3-phosphate
				CH_2OH		
CH_2OH		CHO		C=O		
C=O		HC–OH		HO–CH		CHO
HO–CH	+	HC–OH	$\xrightarrow[TPP]{TK}$	HC–OH	+	HC–OH
HC–OH		HC–OH		HC–OH		$CH_2OPO_3H_2$
$CH_2OPO_3H_2$		$CH_2OPO_3H_2$		HC–OH		
				$CH_2OPO_3H_2$		

red cells at 4 days, increased ETK stimulation at 7 days, and increased TPP stimulation of leukocyte pyruvate decarboxylase at 21 days. Correlation between ETK stimulation and dietary thiamin or clinical signs is not always seen. There are conflicting reports as to the usefulness of blood thiamin levels; this is possibly related to low concentrations and measurement difficulties. A newer approach to measurement of erythrocyte TPP using yeast apopyruvate decarboxylase shows greater sensitivity to thiamin status of animals than ETK does.

Pathophysiology. Populations of Southeast Asia who eat foods rich in antithiamin substances commonly develop beriberi. Milder forms of thiamin deficiency are common among pregnant women, elderly persons, and alcoholics. Magnesium deficiency (common in alcoholics) impairs thiamin utilization. Oral contraceptive use may induce deficiency.

Toxicity. The toxic dose of thiamin is 1000 times the usual therapeutic dose and leads to anxiety, headache, convulsions, weakness, trembling, and neuromuscular collapse.

Biotin

Growth factors separately discovered under the names of bios II, coenzyme R, and biotin were noted to be similar; other research demonstrated the symptom complex produced in rats by feeding them raw egg white was corrected by cooking the eggs or by addition of other foods presumably containing "vitamin H," or "protective factor X." Around 1940 the linkage of vitamin H and biotin was made, and the clinical role of biotin was established when human volunteers ingested large amounts of raw egg white and confirmed the "egg white injury" of animals that is correctable with biotin (structure in Fig. 34-12).

Numerous foods contain biotin, though none is especially rich (up to 2 mg/100 g). Dietary intake is low in the neonatal period despite the fact that concentrations increase as newborns switch from colostrum to mature breast milk. Enteric bacteria are known to synthesize biotin, though the relative contribution of this source is not clear.

Metabolism. Biotin is absorbed in the proximal half of the small intestine and circulates in blood largely bound to plasma proteins. Avidin, a glycoprotein of egg albumin, has very high affinity for biotin, thus explaining the biotin deficiency resulting from raw egg-white ingestion.

Function. Biotin is a coenzyme for carbon dioxide fixation (carboxylation) reactions and for carboxyl group exchange (examples are in box on next page). The biotin is covalently linked to these apoenzymes by a specific enzymatic reaction to form the holocarboxylase:

$$\underset{\text{(Inactive)}}{\text{Apocarboxylase}} + \text{Biotin} \rightarrow \underset{\text{(Active)}}{\text{Holocarboxylase}}$$

Carbon dioxide reacts with the holocarboxylase-bound biotin (Fig. 34-12) before transfer to the accepting molecule. Concentrations of the apocarboxylases do not change in biotin deficiency or in repletion after deficiency.

Clinical deficiency signs. Deficiency created by egg-white ingestion results in dermatitis (3 to 4 weeks) and progresses to mental and neurological changes, nausea, anorexia, and peripheral vasoconstriction or coronary ischemia in some. Hematocrit decreases are seen along with increases in plasma cholesterol and bile pigments. Reported dietary deficiency cases are rare, but each has a history of raw eggs as a large dietary component for months to years.

Chemical deficiency signs. Dietary deficiency is accompanied by decreased urinary biotin (to 10% of controls) and increased urinary organic acids, indicating functional deficiency of β-methylcrotonyl CoA carboxylase and propionyl CoA carboxylase; these enzyme deficiencies are confirmed in leukocytes. Competitive-binding assays for biotin capitalize on the very high affinity of avidin for biotin[25] and are replacing earlier biological assays.

Genetic alterations in these carboxylases may result in biotin-dependent states that cause metabolic acidosis and require pharmacological biotin doses. These defects should be recognized by analysis of organic acid patterns and the biotin dependency evaluated by clinical trial.

Pathophysiology. Long-term total parenteral nutrition would be a suspect circumstance for development of biotin deficiency. Recognition of disease and therapy impacts on biotin deficiency awaits wider availability of biotin or organic acid assays. No biotin toxicity has been described.[26]

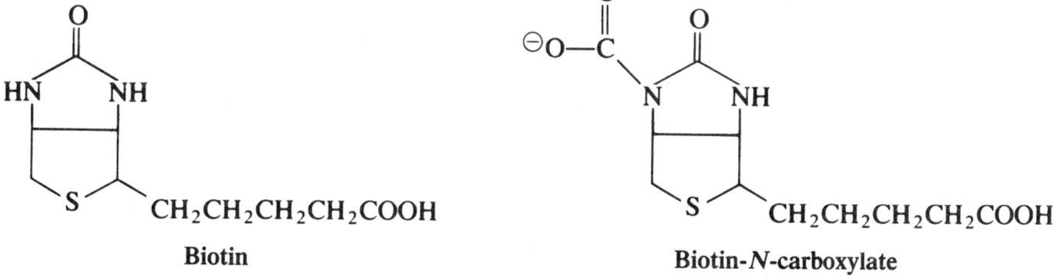

O
HN NH
S $CH_2CH_2CH_2CH_2COOH$

Biotin

O O
⊖O—C
N NH
S $CH_2CH_2CH_2CH_2COOH$

Biotin-*N*-carboxylate

Fig. 34-12. Biotin and its active form.

Examples of biotin-dependent reactions

Acetyl CoA carboxylase:

$$CH_3{-}\overset{O}{\overset{\|}{C}}{-}SCoA + CO_2 + ATP \rightarrow HOOC{-}CH_2{-}\overset{O}{\overset{\|}{C}}{-}SCoA + ADP + P_i$$

Propionyl CoA carboxylase:

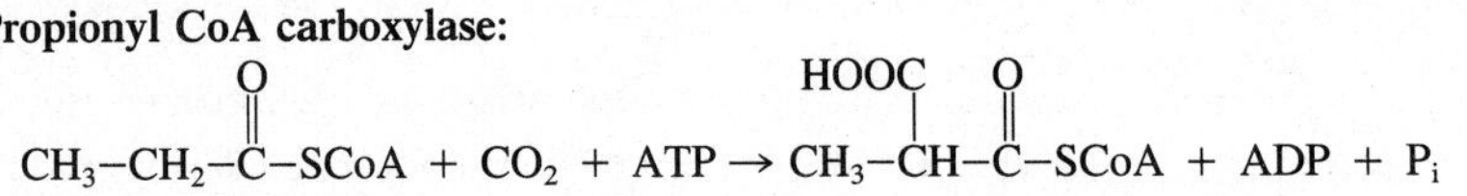

$$CH_3{-}CH_2{-}\overset{O}{\overset{\|}{C}}{-}SCoA + CO_2 + ATP \rightarrow CH_3{-}\overset{HOOC}{\overset{|}{C}H}{-}\overset{O}{\overset{\|}{C}}{-}SCoA + ADP + P_i$$

Pyruvate carboxylase:

$$CH_3{-}\overset{O}{\overset{\|}{C}}{-}COOH + CO_2 + ATP + H_2O \rightarrow \text{Oxaloacetate} + ADP + P_i$$

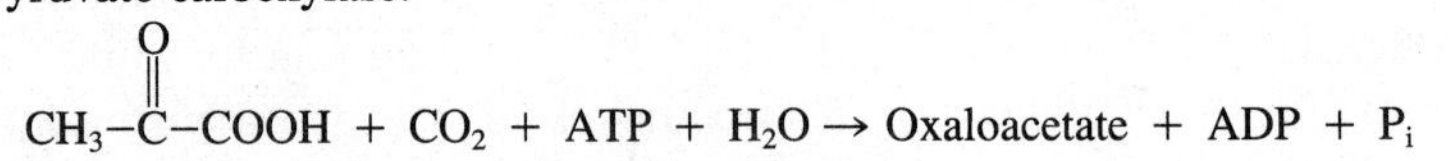

Methylmalonyl-oxaloacetic transcarboxylase:

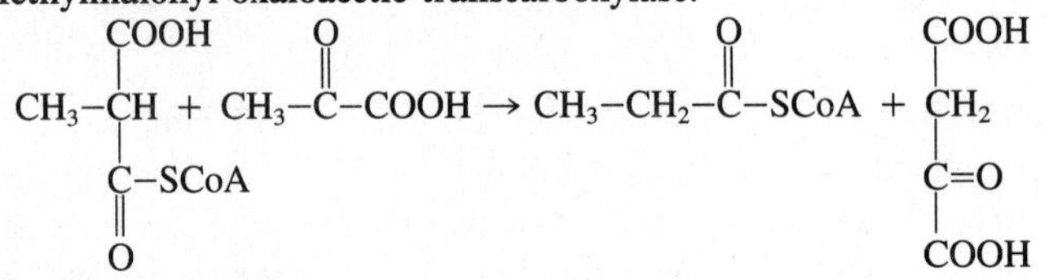

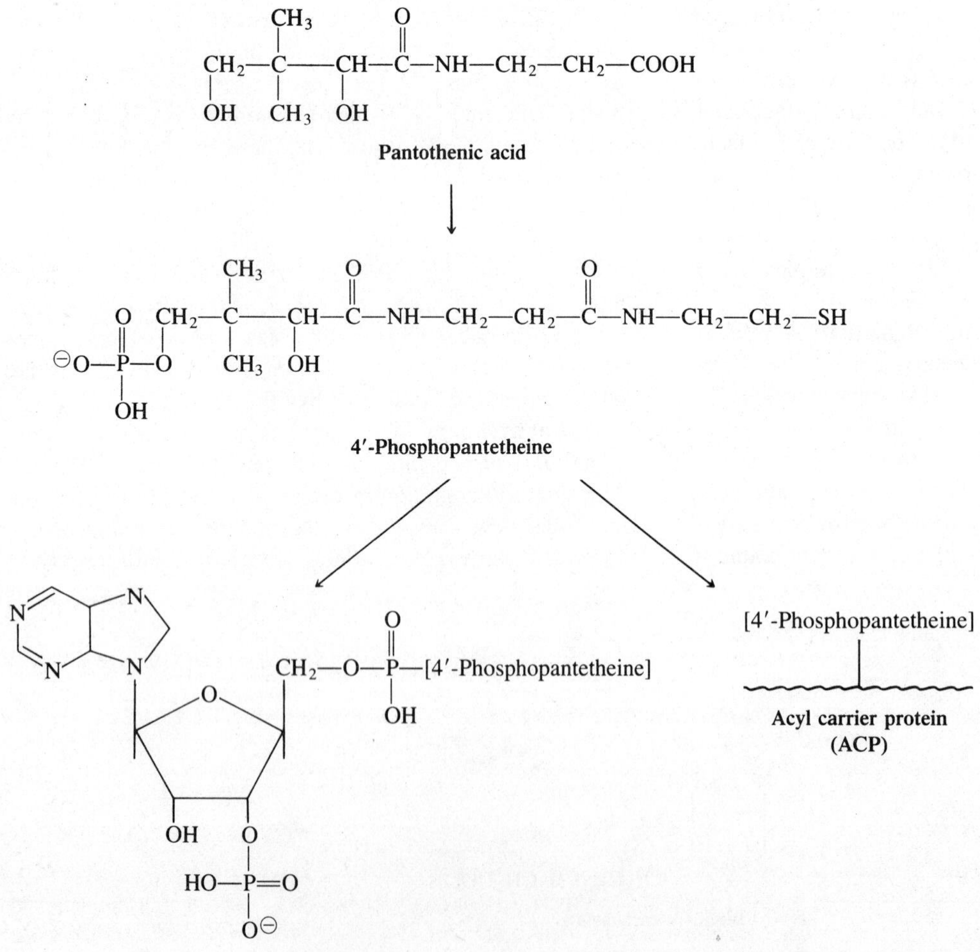

$$CH_3{-}\overset{COOH}{\overset{|}{\underset{\underset{\underset{O}{\|}}{\underset{|}{C{-}SCoA}}}{CH}}} + CH_3{-}\overset{O}{\overset{\|}{C}}{-}COOH \rightarrow CH_3{-}CH_2{-}\overset{O}{\overset{\|}{C}}{-}SCoA + \overset{COOH}{\overset{|}{\underset{\underset{\underset{COOH}{|}}{\underset{|}{C{=}O}}}{CH_2}}}$$

$$\underset{OH}{\underset{|}{CH_2}}{-}\overset{CH_3}{\overset{|}{\underset{\underset{CH_3}{|}}{C}}}{-}\underset{\underset{OH}{|}}{CH}{-}\overset{O}{\overset{\|}{C}}{-}NH{-}CH_2{-}CH_2{-}COOH$$

Pantothenic acid

↓

$$\ominus O{-}\overset{O}{\overset{\|}{\underset{\underset{OH}{|}}{P}}}{-}O{-}CH_2{-}\overset{CH_3}{\overset{|}{\underset{\underset{CH_3}{|}}{C}}}{-}\underset{\underset{OH}{|}}{CH}{-}\overset{O}{\overset{\|}{C}}{-}NH{-}CH_2{-}CH_2{-}\overset{O}{\overset{\|}{C}}{-}NH{-}CH_2{-}CH_2{-}SH$$

4′-Phosphopantetheine

↙ ↘

Adenine–ribose–$CH_2{-}O{-}\overset{O}{\overset{\|}{\underset{\underset{OH}{|}}{P}}}{-}$[4′-Phosphopantetheine], with ribose 2′-OH and 3′-O–$\overset{}{P}(=O)(OH)O^{\ominus}$ (HO—P=O, O⊖)

Coenzyme A

[4′-Phosphopantetheine]

Acyl carrier protein (ACP)

Fig. 34-13. Pantothenic acid and its active cofactors.

Pantothenic acid

A growth factor occurring in all types of animal and plant tissue was first designated "vitamin B_3" and later named "pantothenic acid" (from Greek, meaning 'from everywhere'). This factor was chemically identified in 1938, and its deficiency was linked to the "burning feet syndrome" in 1949.

Dietary sources include liver and other organ meats, milk, eggs, peanuts, legumes, mushrooms, salmon, and whole grains. Approximately 50% of pantothenate in food is available for absorption. Pantothenate is unstable to acid, alkali, heat, and some salts. Intestinal microorganisms are a source for animals, possibly including humans.

Metabolism. As shown in Fig. 34-13, pantothenate is metabolically converted to 4′-phosphopantetheine, which becomes covalently bound to either serum acyl carrier protein (ACP) or to coenzyme A. Little is known of pantothenate metabolism. Urinary excretion of pantothenate correlates well with intake; no saturation is seen with intakes of 10 mg/day. Free pantothenate is the major form in both urine and serum, whereas coenzyme A is the major erythrocytic form. In contrast to other B vitamins, pantothenate tissue repletion is very gradual. Tissues known to contain pantothenate are liver, adrenal glands, brain, kidneys, and heart.

Function. Coenzyme A is a highly important acyl-group transfer coenzyme that is involved in a large number of reactions of great variety of reaction types. Acyl derivatives of coenzyme A are first formed (by thioester linkage), followed by transfer of the acyl group to an acceptor molecule. Examples of these coenzyme A–dependent reactions are in the box below. 4′-Phosphopantetheine is also covalently bound to a serine residue in ACP, which similarly forms thioester linkages with acyl groups. This mode of acyl transfer occurs in synthesis of fatty acids:

$$CH_3-\overset{O}{\overset{\|}{C}}-S(ACP) + HOOC-CH_2-\overset{O}{\overset{\|}{C}}-S(ACP) \rightarrow CH_3\overset{O}{\overset{\|}{C}}CH_2-\overset{O}{\overset{\|}{C}}-S(ACP) + HS(ACP) + CO_2$$

$$\xrightarrow{2H^+} CH_3CH_2CH_2-\overset{O}{\overset{\|}{C}}-S(ACP) \xrightarrow{HSCoA} CH_2CH_2CH_2-\overset{O}{\overset{\|}{C}}-SCoA + HS(ASP)$$

Clinical deficiency signs. Clinical signs of deficiency (experimentally induced) include apathy, depression, increased infection, paresthesias (burning sensations), and muscle weakness.[27] No clear-cut case of dietary deficiency has been reported. Because it is seldom measured and the known symptoms are so nonspecific, the contribution of pantothenate deficiency to human disease is truly unknown.

Chemical deficiency signs. Whole blood pantothenate less than 1000 μg/L and urinary excretion of less than 1.0 mg/day are regarded as indicative of deficiency. Most past measurements have employed biological assays for free pantothenate, releasing pantothenate from coenzyme A by multiple enzymatic treatments. Functional tests based on acetylation of sulfanilamide by erythrocytes and urinary excretion of acetylated *p*-aminobenzoic acid (PABA) after an oral PABA load have been described, but their usefulness is not established. A newer competitive binding assay for pantothenate appears promising.[28]

Pathophysiology. Low urinary excretion and reduced blood levels of pantothenate have been reported in patients with chronic malnutrition, acute alcoholism, and acute rheumatism. Patients with circulatory and cardiovascular diseases and those with peptic ulcers have reduced circulating pantothenic acid; chronic alcoholics have increased excretion, an indication of impaired utilization.

Toxicity and therapeutic uses. Pantothenate has been given postoperatively to stimulate the gastrointestinal tract and also to treat streptomycin-induced neuropathy. No toxicity is known.

Vitamin B_{12} and folic acid

Pernicious anemia was described over 100 years ago and was recognized as a disease amenable to liver extract ther-

Examples of reactions requiring coenzyme A

Acetyl SCoA + D-2-Glucosamine → *N*-Acetyl-D-glucosamine + HSCoA
2 Fatty acyl SCoA + L-α-Glycerophosphate → Diglyceride phosphate + 2HSCoA
Acetyl SCoA + H_2O → Acetate + HSCoA
Acetyl SCoA + Oxaloacetate → Citrate + HSCoA
Acetoacetyl SCoA + Succinate → Acetoacetate + Succinyl SCoA

apy. The antipernicious anemia factor (extrinsic factor), finally isolated, crystallized, and characterized between 1948 and 1956, is now known as "vitamin B_{12}." In the early 1930s a similar entity, "pernicious anemia of pregnancy," was shown to be different in that it did not respond to liver extract therapy but did respond to an autolyzed yeast substance. Purification of this substance led to its identification as pteroylglutamic acid, more commonly known as "folic acid." Thus similar clinical symptoms were found to result from deficiencies of two totally different structures (Figs. 34-14 and 34-15), neither of which could replace the other. The clinical similarities and the metabolic interactions of vitamin B_{12} and folic acid usually dictate their simultaneous assessment.[29]

Vitamin B_{12}

Vitamin B_{12} bears a corrin ring (containing pyrroles similar to porphyrin) linked to a central cobalt atom. The pyrroles are almost saturated with side chains (methyl, acetamide, or propionamide groups). Different corrinoid compounds, or cobalamins, are distinguished by the substituent linked to the cobalt with methylcobalamin and 5′-deoxyadenosylcobalamin being the two known coenzyme forms.

Dietary sources of vitamin B_{12} are of animal origin (meat, eggs, milk) except for the plant comfrey. Total vegetarian diets are therefore likely settings for deficiency. (Animals derive their vitamin B_{12} from intestinal microbial synthesis.) The average daily diet contains 3 to 30 μg, of which 1 to 5 μg is absorbed. The frequency of dietary deficiency increases with age, occurring in over 0.5% of persons over age 60.

Metabolism. Most vitamin B_{12} absorption is by a complex with intrinsic factor (IF), a protein secreted by gastric parietal cells (Fig. 34-16). This IF-B_{12} complex binds with specific ileal receptors. Blocking IF-antibodies prevent binding of vitamin B_{12} to IF, and "binding" antibodies can combine with either free IF or the IF-B_{12} complex, thus preventing attachment of the complex to ileal receptors. Parietal cell antibodies have also been identified as a cause of pernicious anemia. Released from the IF complex within the mucosal cell, vitamin B_{12} circulates through specific transport proteins to liver, bone marrow, and other tissues. There is a significant enterohepatic circulation of vitamin B_{12}. As a result of this biliary reabsorption, 10 to 12 years are required for a strict vegetarian to become clinically deficient. A person with normal B_{12} stores but lacking IF requires less time (1 to 4 years) for deficiency to become evident.

Transcobalamin II (TC II) appears to be the major serum protein transporting exogenous vitamin B_{12} to tissues. Cobalophilin (previously R, or rapidly migrating, binding protein, or TC I) transports endogenous B_{12} and is the binder of food cobalamins. Saliva, breast milk, and granulocytes contain large amounts of this binding protein compared with relatively small amounts of cobalamin. Plasma contains both types of transport protein and the three forms of vitamin B_{12} (hydroxycobalamin, methylcobalamin, and deoxyadenosylcobalamin). Absorption and transport of this vitamin are illustrated in Fig. 34-16.

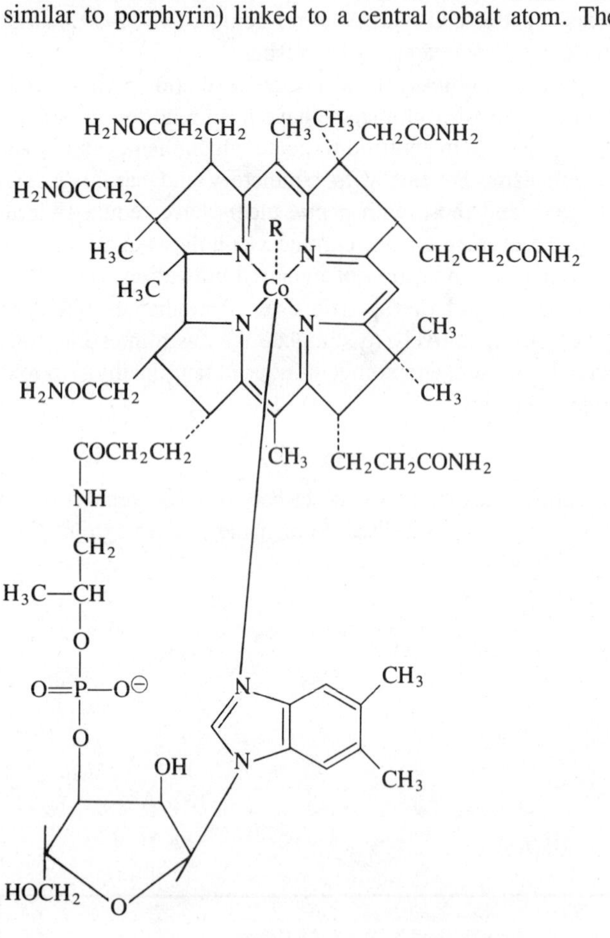

	R		
(Me-B_{12})	CH_3	**Methylcobalamin**	ACTIVE COFACTOR FORMS
(Ado-B_{12})	5′-Deoxy-adenosine	**Deoxyadenosyl-cobalamin**	
(OH-B_{12})	OH	**Hydroxocobalamin**	DIETARY FORMS AND THERAPY
(CN-B_{12})	CN	**Cyanocobalamin**	

Fig. 34-14. Vitamin B_{12} forms.

Functions. In its adenosyl coenzyme form (Ado-cbl), vitamin B_{12} functions in formation of myelin (the protective sheathing around the axons of nerve cells) through metabolism of odd-chain fatty acids. These fatty acids are metabolized through propionate to methylmalonyl CoA, which requires Ado-cbl for further conversion to succinyl CoA, a carbohydrate pathway intermediate. The other known coenzyme form, methylcobalamin (Me-cbl) functions in the synthesis of methionine from homocysteine and indirectly participates in synthesis of the folate forms required for synthesis of purines and pyrimidines and, ultimately, for synthesis of DNA. Both roles of vitamin B_{12} relate to group transfer reactions.

Clinical deficiency signs. Clinical symptoms of vitamin B_{12} deficiency include megaloblastic anemia (reflecting impaired DNA synthesis) and neurological abnormalities beginning with paresthesias and progressing to spastic ataxia secondary to myelin sheath degeneration. In the few case reports of inherited deficiency of TC II, symptoms of pancytopenia and failure to thrive develop within a few months of birth.

Chemical deficiency signs. Diagnostic tests for vitamin B_{12} deficiency include measurement of serum B_{12} by microbiological or radioligand assay methods, measurement of urinary methylmalonic acid, and the Schilling test. Definition of the cause of vitamin B_{12} deficiency may require

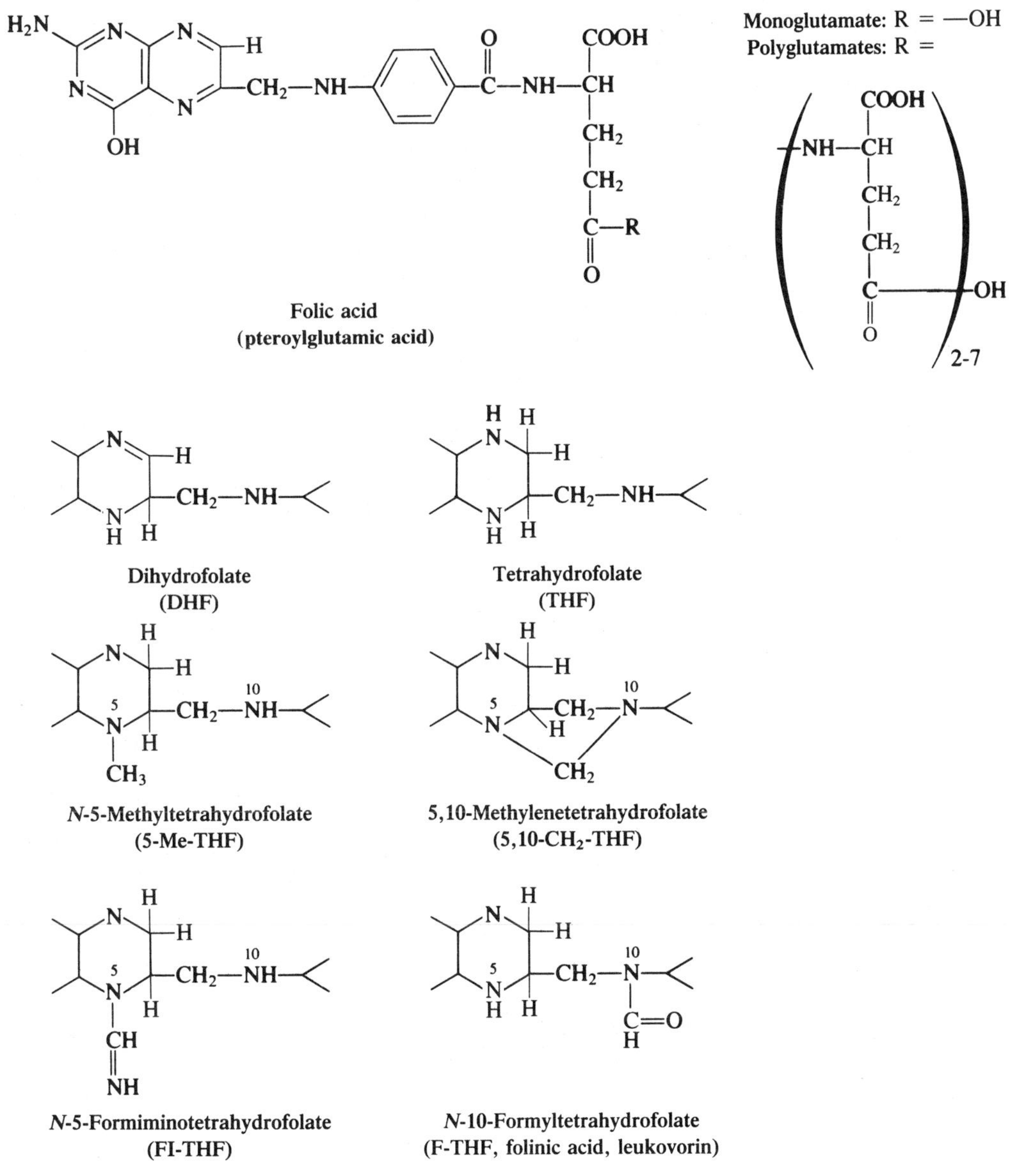

Folic acid
(pteroylglutamic acid)

Monoglutamate: R = —OH
Polyglutamates: R =

Dihydrofolate
(DHF)

Tetrahydrofolate
(THF)

N-5-Methyltetrahydrofolate
(5-Me-THF)

5,10-Methylenetetrahydrofolate
(5,10-CH_2-THF)

N-5-Formiminotetrahydrofolate
(FI-THF)

N-10-Formyltetrahydrofolate
(F-THF, folinic acid, leukovorin)

Fig. 34-15. Structures of folic acid forms.

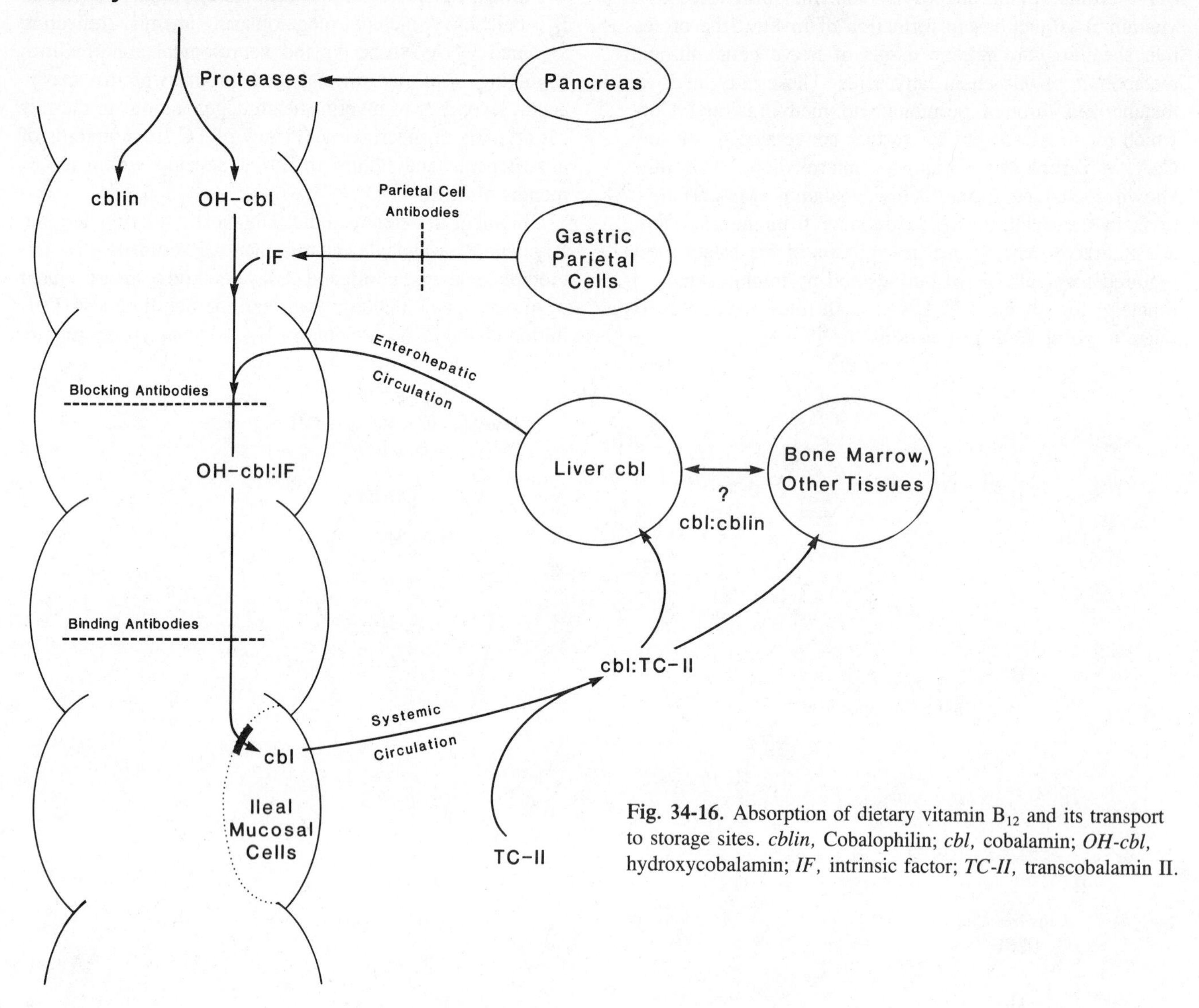

Fig. 34-16. Absorption of dietary vitamin B_{12} and its transport to storage sites. *cblin,* Cobalophilin; *cbl,* cobalamin; *OH-cbl,* hydroxycobalamin; *IF,* intrinsic factor; *TC-II,* transcobalamin II.

additional assays including tests for specific antibodies, the deoxyuridine suppression test, and assessment of the transport proteins. The Schilling test evaluates absorption of an oral dose of radioactive vitamin B_{12} by measurement of urinary excretion of the radiolabeled vitamin. If little of the radiolabel is excreted, the test can be repeated, including an oral dose of IF. Since the Schilling test employs a pharmacological dose of vitamin B_{12}, it should be performed only after other diagnostic tests are done. The deoxyuridine suppression test is based on the fact that preincubation of normal bone marrow with an appropriate concentration of deoxyuridine severely suppresses the subsequent incorporation of tritiated thymidine into DNA. This suppression is subnormal with bone marrow from patients deficient in either vitamin B_{12} or folate and is correctable in vitro by addition of the appropriate deficient vitamin. This test will also show abnormal results in patients with megaloblastic changes resulting from neoplastic, chemotherapeutic, or other agents interfering with DNA synthesis. Serum B_{12} competitive binding methods have employed B_{12} binders of variable purity and binding specificity, resulting in the charge that this approach is unreliable. This aspect of the methodology has been improved and should provide more correct clinical information than in the past.

Pathophysiology. Inadequate secretion of IF may accompany lesions of the gastric mucosa, gastric atrophy, gastrectomy, iron deficiency, and some endocrine disorders. The IF-B_{12} complex may be inadequately formed in pancreatic insufficiency as a result of lack of pancreatic protease activity splitting dietary vitamin B_{12} from cobalophilin in the duodenum. This IF-B_{12} complex may be inadequately absorbed in ileal malfunction (sprue, enteritis, ileal resection, neoplasias, granulomas, and so on). The term "pernicious anemia" is now most commonly applied to vitamin B_{12} deficiency secondary to lack of IF. Antibodies to IF and to parietal cells are very common in pernicious anemia patients, in their healthy relatives, and in patients with other autoimmune disorders. Blocking antibodies can result in normal or high serum B_{12} levels in patients with pernicious anemia. Numerous drugs induce vitamin B_{12} malabsorption, and excessive ascorbate intake may convert vitamin B_{12} to analog, blocking forms. Cobalophilin is elevated in patients with chronic myeloproliferative disorders, polycythemia vera, and various neoplasms.

Toxicity. Vitamin B_{12} has extremely low systemic toxicity. In the past cyanocobalamin was widely used as the therapeutic form, but it is being replaced with hydroxycobalamin largely on theoretical grounds of reducing cyanide exposure. The infrequent adverse reactions are mostly allergic reactions, some of which may be related to contaminants or preservatives (such as phenylcarbinol).

Folic acid

Structural relatives of pteroylglutamic acid (folic acid) are the metabolically active compounds usually referred to as "folates" (Fig. 34-15). Up to eight glutamate residues may be found in these naturally occurring compounds.

Food folates are primarily found in green and leafy vegetables, fruits, organ meats, and yeast. Excessive boiling of foods and use of large quantities of water result in folate destruction. The average American diet may be inadequate in folate for adolescents and for pregnant or lactating females.

Metabolism. The naturally occurring folate polyglutamates are hydrolyzed to monoglutamate forms before absorption (which occurs primarily in the proximal jejunum) by the intestinal mucosal cells. After this, folate enters the liver through the portal circulation. The liver converts some of these folate monoglutamates to polyglutamates, which are presumably then stored; another fraction of the folate is excreted in bile as N^5-methyltetrahydrofolate (MeTHF), which is reabsorbed and is the major circulating form of folate. Serum folate (MeTHF) is in the monoglutamate form and readily enters the choroid plexus and the CSF. Folic acid, on the other hand, is readily transported from CSF to plasma. A folate-binding protein has been identified in the choroid plexus, probably accounting for the high CSF/plasma ratio. The CSF form is mainly MeTHF; brain folates are predominately polyglutamate forms of dihydrofolate (DHF). Folate catabolism involves cleavage of the pterin ring, followed by acetylation to form the excreted product, *p*-acetamidobenzoylglutamic acid.

Function. Fig. 34-17 illustrates the various types of one-carbon transfers dependent on folate (such as methy-

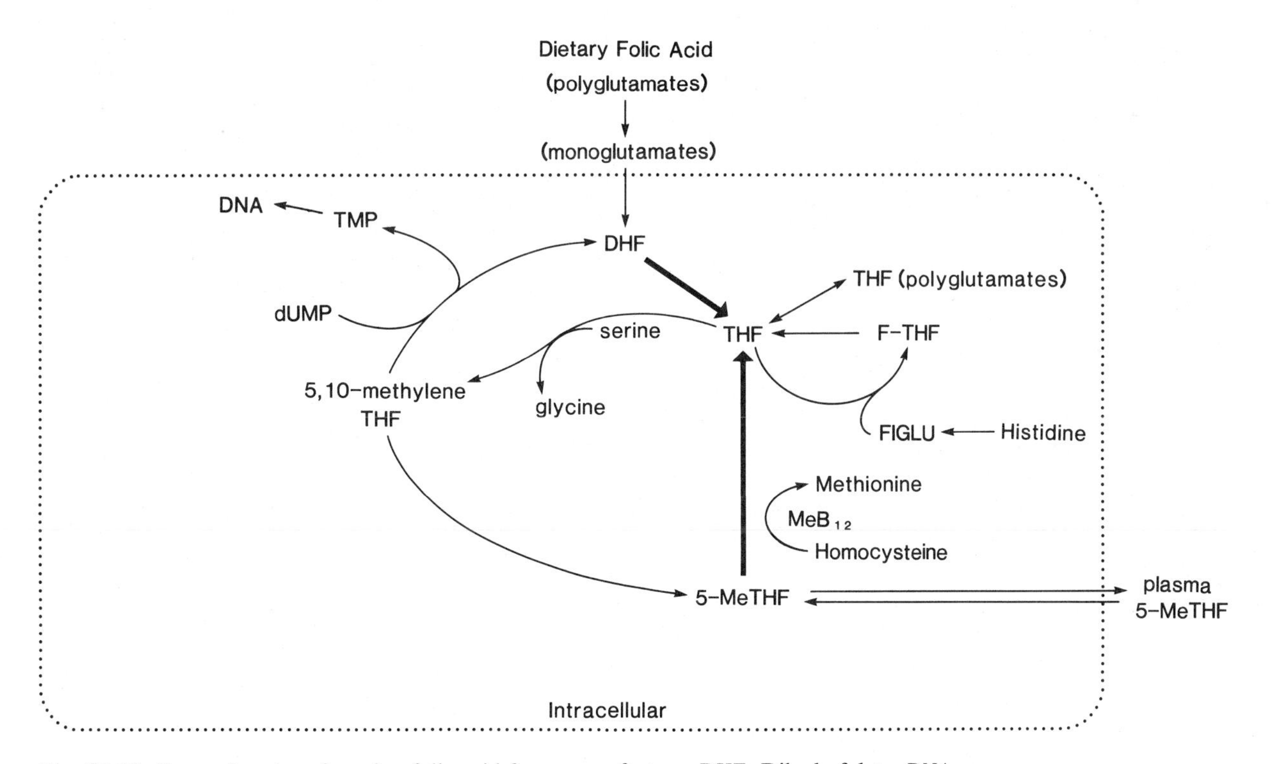

Fig. 34-17. One-carbon transfer using folic acid forms as cofactors. *DHF,* Dihydrofolate; *DNA,* deoxyribonucleic acid; *dUMP,* deoxyuridine monophosphate; *FIGLU,* formimino-L-glutaric acid; *F-THF,* folinic acid; *5-MeTHF,* *N*-5-methyltetrahydrofolate; *MeB_{12},* methyl cobalamin; *THF,* tetrahydrofolate; *TMP,* thymidine monophosphate.

lation and formylation). After cellular uptake, the MeTHF is converted to THF while transferring a carbon to homocysteine to yield methionine. As mentioned previously, this reaction requires vitamin B_{12}. In the absence of vitamin B_{12}, the folate is essentially trapped in the MeTHF form, making it unavailable for other reactions, including the synthesis of thymine for DNA (Fig. 34-17).

Clinical deficiency signs. The major clinical symptom of folate deficiency is megaloblastic anemia; there is continued debate over whether organic mental changes, especially affective disorders, result from this deficiency.

Chemical deficiency signs. Indices of deficiency named in order of occurrence include low serum folate, hypersegmentation of neutrophils, high urinary FIGLU (a histidine metabolite accumulating in absence of folate), low erythrocyte folate, macro-ovalocytosis, megaloblastic marrow, and finally anemia. The deoxyuridine suppression test, discussed with vitamin B_{12} is also an index of folate status. Serum folate, though an early index of deficiency, can frequently be low despite normal tissue stores. Hypersegmentation of neutrophils may not be seen in folate deficiency of pregnant women. The urinary FIGLU test requires ingestion of an oral load of histidine, followed by a timed urine collection, and can be abnormally high in deficiencies of either folate or vitamin B_{12}. Because most folate storage occurs after the vitamin B_{12}–dependent step, erythrocyte folate can also be reduced in deficiency of either B_{12} or folate. Despite this overlap, erythrocyte folate concentration is currently the best accepted laboratory index of folate deficiency.

Pathophysiology. Folate requirement is increased during pregnancy and especially during lactation, the latter partially resulting from the presence in milk of high-affinity folate binders. Other instances of increased folate requirement include hemolytic anemias, iron deficiency, prematurity, and multiple myeloma. Patients receiving dialysis treatment rapidly lose folate. Folate deficiency because of malabsorption can occur in sprue, celiac disease, inflammatory bowel diseases, cardiac failure, and systemic bacterial infections. Both acidic and alkaline conditions can interfere with intestinal transport; the organic anions that accumulate in uremia also impair folate utilization. Genetic alterations of most of the folate-interconverting enzymes have been reported. Several mentally retarded children have been described who are unable to transfer folates from blood into CSF; this is probably a result of abnormal folate transport-binding systems in the central nervous system.

Exogenous influences. Phenytoin (Dilantin) therapy accelerates folate excretion (as well as interferes with folate absorption and metabolism). Alcohol interferes with folate's enterohepatic circulation, whereas the chemotherapeutic agent, methotrexate, inhibits the enzyme dihydrofolate reductase. Although decreased serum folate can occur with use of oral contraceptives (cycle-day dependent), this is not believed to cause a functional deficit unless some other problem is also present. There has, however, been concern expressed that there may be a relationship between oral contraceptive–induced cervical folate deficiency and cervical carcinoma. A specific folate-binding protein, possibly of granulocytic source, has been reported in the serum of a number of women who were receiving oral contraceptives or who were pregnant.

Therapeutic uses. The therapeutic form of folate is 5-formyl-THF (also known as leucovorin, citrovorum factor, or folinic acid). This form of folate can bypass MeTHF and enter the cycles of folate's one-carbon transfer reactions (Fig. 34-17). This feature is useful in the "leucovorin rescue" of cancer patients given high-dose methotrexate therapy with toxic levels of methotrexate.

REFERENCES

General

1. Food and Nutrition Board: Recommended Dietary Allowances, ed. 9, Washington, D.C., 1980, National Academy of Sciences.
2. Naito, H.K.: Role of vitamin C in health and disease. In Brewster, M.A., and Naito, H.K., editors: Nutritional elements and clinical biochemistry, New York, 1980, Plenum Publishing Corp., pp. 69-116.
3. Association of Vitamin Chemists: Methods of vitamin assay, ed. 3, New York, 1966, Interscience.
4. Labbé, R.F., editor: Clinics in laboratory medicine, vol. 1, Laboratory assessment of nutritional status, Philadelphia, 1981, W.B. Saunders Co.
5. Sauberlich, H.E., Dowdy, R.P., and Skala, J.H.: Laboratory tests for the assessment of nutritional status, Boca Raton, Fla., 1974, CRC Press, Inc.
6. Briggs, M.H., editor: Vitamins in human biology and medicine, Boca Raton, Fla., 1981, CRC Press, Inc.
7. Garry, P.G., editor: Human nutrition: clinical and biochemical aspects, Washington, D.C., 1981, American Association for Clinical Chemistry.
8. Calabrese, E.J.: Nutrition and environmental health, vol. 1, The vitamins, New York, 1980, John Wiley & Sons, Inc.
9. Brewster, M.A., and Naito, H.K., editors: Nutritional elements and clinical biochemistry, New York, 1980, Plenum Publishing Corp.
10. Kreutler, P.A.: Nutrition in perspective, Englewood Cliffs, N.J., 1980, Prentice-Hall, Inc.

Vitamin A

11. Garry, P.J.: Vitamin A. In Labbé, R.F., editor: Clinics in laboratory medicine, vol. 1, Laboratory assessment of nutritional status, Philadelphia, 1981, W.B. Saunders Co.
12. Spron, M.B., Dunlop, W.M., Newton, D.L., et al.: Prevention of chemical carcinogenesis by vitamin A and its synthetic analogs (retinoids), Fed. Proc. **35:**1332, 1976.
13. Shamberger, R.J.: Vitamin A alterations in disease. In Brewster, M.A., and Naito, H.K., editors: Nutritional elements and clinical biochemistry, New York, 1980, Plenum Publishing Co., pp. 117-130.

Vitamin E

14. Bland, J.: Lipid antioxidant nutrition. In Brewster, M.A., and Naito, H.K., editors: Nutritional elements and clinical biochemistry, New York, 1980, Plenum Publishing Corp., pp. 139-167.
15. Farrell, P.M., and Bieri, J.G.: Megavitamin E supplementation in man, Am. J. Clin. Nutr. **28:**1381, 1975.
16. Bieri, J.G., Evarts, R.P., and Thorp, S.: Factors affecting the exchange of tocopherol between red cells and plasma, Am. J. Clin. Nutr. **30:**686, 1977.

Vitamin K

17. Suttie, J.W.: Role of vitamin K in the synthesis of clotting factors. In Draper, H.H., editor: Advances in nutritional research, vol. 1, New York, 1977, Plenum Publishing Corp.
18. Hazell, K., and Baloch, K.H.: Vitamin K deficiency in the elderly, Gerontol. Clin. **12**:10, 1970.
19. Cumming, F., Briggs, M., and Briggs, M.: Clinical toxicology of the vitamins. In Briggs, M.H., editor: Vitamins in human biology and medicine, Boca Raton, Fla., 1981, CRC Press, Inc., pp. 187-244.

Vitamin D

20. Taylor, C.B., and Peng, S.: Vitamin D—its excessive use in the U.S.A. In Brewster, M.A., and Naito, H.K., editors: Nutritional elements and clinical biochemistry, New York, 1980, Plenum Publishing Corp., pp. 132-138.

Ascorbic acid

21. Sauberlich, H.E.: Ascorbic acid. In Labbé, R.F., editor: Clinics in laboratory medicine, vol. 1, Laboratory assessment of nutritional status, Philadelphia, 1981, W.B. Saunders Co.

Niacin

22. Wahlqvist, M.L.: Effects on plasma cholesterol of nicotinic acid and its analogues. In Briggs, M.H., editor: Vitamins in human biology and medicine, Boca Raton, Fla., 1981, CRC Press, Inc., pp. 81-84.

Thiamin

23. Flint, D.M., and Prinsley, D.M.: Vitamin status of the elderly. In Briggs, M.H., editor: Vitamins in human biology and medicine, Boca Raton, Fla., 1981, CRC Press, Inc., pp. 65-80.
24. Blass, J.P.: Thiamin and the Wernicke-Korsakoff syndrome. In Briggs, M.H., editor: Vitamins in human biology and medicine, Boca Raton, Fla., 1981, CRC Press, Inc., pp. 107-136.

Biotin

25. Roth, K.S., Allen, L., Yang, W., et al.: Serum and urinary biotin levels during treatment of holocarboxylase synthetase deficiency, Clin. Chim. Acta **109**:337, 1981.
26. Roth, K.S.: Biotin in clinical medicine: a review, Am. J. Clin. Nutr. **34**:1967, 1981.

Pantothenic acid

27. Fry, P.C., Fox, H.M., and Tao, H.G.: Metabolic response to a pantothenic acid deficient diet in humans, J. Nutr. Sci. Vitaminol. **22**:399, 1976.
28. Wyse, B.W., Wittwer, C.W., and Hansen, R.G.: Radioimmunoassay for pantothenic acid in blood and other tissues, Clin. Chem. **25**:108, 1979.

Vitamin B_{12} and folic acid

29. Steinkamp, R.C.: Vitamin B_{12} and folic acid: clinical and pathophysiological considerations. In Brewster, M.A., and Naito, H.K., editors: Nutritional elements and clinical biochemistry, New York, 1980, Plenum Publishing Corp., pp. 169-240.

SUGGESTED READINGS

Colman, N.: Laboratory assessment of folate status. In Labbé, R.F., editor: Clinics in laboratory medicine, vol. 1, Laboratory assessment of nutritional status, Philadelphia, 1981, W.B. Saunders Co.

Komindr, S., and Nichoalds, G.E.: Clinical significance of riboflavin deficiency. In Brewster, M.A., and Naito, H.K., editors: Nutritional elements and clinical biochemistry, New York, 1980, Plenum Publishing Corp., pp. 685-698.

Schuster, K., Bailey, L.B., and Mahan, C.S.: Vitamin B_6 status of low-income adolescent and adult pregnant women and the condition of their infants at birth, Am. J. Clin. Nutr. **34**:1731, 1981.

Tonkin, S.Y.: Vitamins and oral contraceptives. In Briggs, M.H., editor: Vitamins in human biology and medicine, Boca Raton, Fla., 1981, CRC Press, Inc., pp. 30-64.

Wolf, B., et al.: Phenotypic variation in biotinidase deficiency, J. Pediatr. **103**:233-237, 1983.

Chapter 35 Pregnancy and fetal function

Paul T. Russell

anencephaly A defective development of the brain wherein the cerebral and cerebellar hemispheres are absent.

blastocyst An early stage in embryonic development characterized by a fluid-filled cavity within the cell mass and covered by trophoblast.

hemopoietic Related to the process of formation and development of the various types of blood cells.

hydramnios The presence of an excessive amount of amniotic fluid.

keratinization The process in skin development and differentiation whereby keratin, a proteinaceous substance, is produced.

meningomyelocele A protrusion of the membranes and spinal cord resulting from a defect in the vertebral column.

transudation The passage of a substance through a membrane as a result of a difference in hydrostatic pressure.

trophoblast The cell layer covering the blastocyst, which erodes the inner lining of the uterus during the process of implantation through which the embryo receives nourishment from the mother. Trophoblastic cells do not become part of the embryo itself but contribute to the formation of the placenta.

ANATOMICAL AND PHYSIOLOGICAL INTERACTION OF MOTHER AND FETUS

Placenta and fetal membranes

In the human the fertilized egg enters the uterus 3 days after ovulation and implants itself into the uterine lining 5 to 8 days later. After implantation a series of very complex and remarkable relationships unfold that establish a supportive environment for the developing embryo and later the fetus. The capsule-like wall of the blastocyst, the trophoblast, establishes an intimate nutritive relationship with the uterus (Fig. 35-1), and it provides hormonal functions. The invasion of the endometrium by the trophoblast is not a one-sided phenomenon, for the maternal tissues also adapt to the new relations and demands. Both sets of coordinated changes lead to the production of a specialized organ, the placenta.

As the demands for nourishment and oxygen increase by the rapidly growing embryo, the trophoblast increases its surface area by forming branching processes, or villi. The resulting surface of these villi is enormous, and from the villi the fetal circulation of the placenta is established. An innermost membrane, the amnion, immediately surrounds the embryo and is fluid filled. This fluid, bathing the fetus, preventing desiccation, and buffering the fetus against physical shocks, is referred to as the *liquor amnii* or *amniotic fluid*.

Fetal growth and nutrition

The fetus grows and develops according to the nutrition provided from maternal blood. The rate of this growth is shown in Table 35-1. Gestation is calculated from the last menstrual period. Unlike the *infant* whose diet is a complex mixture of carbohydrates, fats and proteins, the *fetus* has a diet consisting largely of glucose together with sufficient quantities of amino acids to satisfy the nitrogen requirements of protein synthesis. Also included are small amounts of materials, such as fatty acids, vitamins, and minerals, that are essential to normal growth and function. Glucose provides the energy needed for the synthesis of tissues.

The source of glucose for the fetus is maternal blood. How much glucose is available to the fetus depends on its concentration in maternal blood, which is regulated by the absorption of glucose from the gut and by the action of a number of hormones, including insulin and the placental hormone, human chorionic somatomammotropin (also called "placental lactogen, HPL").

The passage of glucose from the maternal to the fetal circulation occurs readily by means of an active-transport system. When glucose is rapidly infused into a pregnant woman, a rise in the glucose concentration in maternal blood is followed rapidly by a comparable increase in its concentration in fetal blood. The two levels do not become equal, however, for a concentration gradient from mother to fetus is always present, the concentration being higher in the maternal blood. In addition, a significant proportion of glucose is consumed by the placenta to meet its own energy requirements.

Since glucose levels in the fetus mirror those in the mother, there is neither need nor opportunity for the fetus to regulate its own blood glucose concentrations. Mechanisms for doing so do develop during the fetal period but are largely dormant until birth, when the supply of glucose, via the placenta, ends abruptly. Nevertheless, two important processes are active from an early stage. The first is the storage of glucose as glycogen or fat to provide for the metabolic needs of the newborn until feeding begins. The second is the control of the rate at which glucose is used by the growing tissues, and this is attributable primarily to the action of insulin secreted by the fetal pancreas. The highly regulated rate of glucose utilization under conditions of constant glucose availability is the important feature influencing growth under conditions of adequate nutrition and normal health.

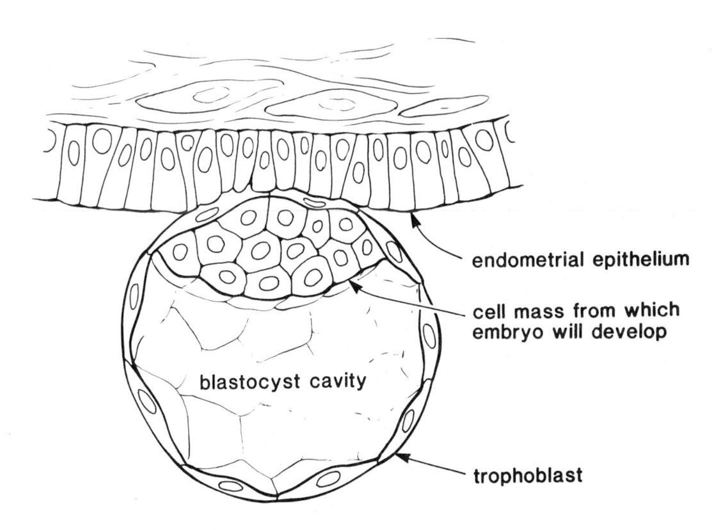

Fig. 35-1. Attachment of blastocyst to endometrial wall.

Table 35-1. Correlation between length of gestation and embryonic and fetal weight

	Weeks of gestation	Weight in grams	Percent of weight at 40 weeks
Embryo	7	0.07	0.002
	8	0.22	0.006
	10	3.5	0.1
	12	11	0.3
	14	33	1
	16	80	2
	18	170	5
	20	316	9
	22	460	14
	24	630	19
	26	823	24
Fetus	28	1045	31
	30	1323	39
	32	1680	49
	34	2074	61
	36	2478	73
	38	2914	86
	40	3405	100

Role of placenta in gas exchange

To meet its metabolic needs, the fetus is completely dependent on a continuous delivery of oxygen across the placenta. Transplacental exchanges, including those for gases, are accomplished as a function of both perfusion and permeability. Placental perfusion is a composite of uterine and umbilical blood flows, whereas permeability is a characteristic of the placental membrane. Under normal conditions with a well-oxygenated mother, oxygen transport across the placenta is primarily regulated by blood flow. With maternal hypoxia produced by disease or exposure to high altitude, oxygen transport across the placenta becomes limited primarily by the membrane.

The total oxygen uptake by the umbilical circulation increases as gestation progresses and the fetus grows. To support this demand, uterine blood flow increases manyfold from nonpregnant levels. There is still debate regarding the factors that produce the enormous increase in uterine blood flow in pregnancy, though growth of the vascular bed and vasodilatation are known responses to estrogen.

Permeability of the placental membrane also changes during gestation. By late gestation, there is a sharp increase in the total surface area of the trophoblast. This increase occurs at a time when the placenta decreases or stops its rate of growth but fetal growth continues at a rapid rate. The increase in the diffusing capacity of the placenta is a function of both increased folding of trophoblast, providing increased surface area, and of a reduction in the diffusing distance from the fetal capillaries to the maternal blood.[1] As might be expected, a cause-and-effect relationship exists between small placentas and small babies.

Table 35-2. Amniotic fluid volume in normal pregnancy

Gestational age (weeks)	Volume of fluid (mL)
12	5-200
14	50-220
16	150-300
18	200-400
20	225-775
22	300-500
24	500-675
26	500-700
28	500-875
30	400-1300
32	400-1375
34	500-1350
36	525-1500
38	300-1525
40	325-1450
42	600

The fetus excretes carbon dioxide also by means of a gradient across the placenta. The umbilical PCO_2 tends to equilibrate with uterine venous PCO_2, though the former remains slightly higher. Thus, when one compares arterial blood of the fetus (umbilical venous blood) with arterial blood of the mother, the PCO_2 is higher in the fetal circulation. Since bicarbonate of fetal blood is within the normal adult range, fetal pH is slightly lower than in the adult, reflecting a small but significant difference in PCO_2. The normal fetus does not, however, have evidence of metabolic acidosis. The PCO_2 in fetal blood changes rapidly in response to changes in maternal blood PCO_2 levels. Maternal metabolic acidosis or alkalosis with appropriate changes in maternal PCO_2 will affect the fetal pH.[2]

Formation of amniotic fluid

Amniotic fluid volume at any point in time is the result of a dynamic balance between production and removal.[3] Amniotic fluid originates from multiple sources, and the relative contribution from each source varies depending on the stage of fetal development. The multiple sources include the placenta, fetal kidneys, skin, membranes, lungs, and intestine. Much of our understanding of the pathways of flow of amniotic fluid has come from correlations of volume changes associated with certain fetal anomalies.

Amniotic fluid can be detected as early as the eighth week of pregnancy, its volume increasing throughout pregnancy until it reaches a maximum at about 36 weeks of gestation. Normal variations in volume throughout pregnancy are large (Table 35-2).

In the first half of pregnancy, the composition of amniotic fluid is similar to that of extracellular fluid and should be considered as an extension of the fetal extracellular fluid space.[4] The large fetal skin capillary bed may serve as a site for amniotic water and solute exchange in early pregnancy, but after keratinization of this layer at about 24 to 26 weeks of gestational age, this tissue becomes impermeable to water and most solutes.

Later in pregnancy, fetal kidneys and lungs assume the major role in amniotic fluid formation. The hourly production of nearly 30 mL by the fetal kidneys indicates the major contribution that these organs can make to amniotic fluid production. In pregnancies where fetal urination is impossible because of urinary tract malformations, amniotic fluid is absent or greatly diminished in volume.[5,6]

Fluid flowing out of the trachea and pharynx into the esophagus can enter the amniotic cavity. This source is important, since it provides a basis for our understanding about the appearance of pulmonary surfactant in amniotic fluid. Respiratory movements by the fetus readily mix fluid with surfactant because the movements produce a tidal volume exchange (in and out) of about 600 to 800 mL/day[7] through the fetal lungs throughout the third trimester. The association between hydramnios and congenital anomalies or tumors that interfere with fetal breathing movements is recognized. The fetal respiratory tract, besides being a site of fluid production, also appears to play a role in the net reabsorption of water from this cavity.

Amniotic fluid disappearance is in part effected by fetal swallowing. It is estimated that between 200 and 450 mL of amniotic fluid per day flow out from the amniotic cavity by this route accounting for about half of the daily urine production of the fetus. Since the amniotic cavity gains a fluid volume of no more than 10 mL/day in the third trimester (the total solute concentration always remains in the normal range), a sizable quantity of urine must be reabsorbed by other pathways.[8]

BIOCHEMISTRY OF AMNIOTIC FLUID

General comments

Excellent monographs on the subject of amniotic fluid are available and contain compendiums of the biochemical constituents of amniotic fluid.[9,10] Only a brief overview is presented at this time for orientation to the subsequent sections of this chapter.

As mentioned previously, there is an important shift in the source of amniotic fluid occurring about midway through pregnancy. Before the keratinization of the skin, amniotic fluid can result as a transudation from the surface of the fetus. After keratinization and with progressive development of the renal system, fetal urine makes a more prominent contribution to the amniotic fluid compartment. One can therefore expect that the biochemical composition of amniotic fluid would reflect the routes of formation of the fluid that relate to the developmental stage of the fetus.

Water, electrolytes, and urea

It has been estimated that approximately 4 liters of water (2800 mL in the fetus, 400 mL in the placenta, and 800 mL in amniotic fluid) accumulate within the human uterus during pregnancy. Exchange of water and solute between amniotic fluid and fetal compartments occurs throughout pregnancy. However, the sites of entry and removal differ for the many constituents because multiple mechanisms of transfer or transport may be involved. Thus detection of a transfer site for a single constituent, including water, does not define the sites of formation or removal of all components of amniotic fluid (see Liley[11] for a thorough discussion of this subject).

Two important generalizations can be made.[12] First, there is a dramatic effect of maturity on amniotic fluid dynamics so that in later gestation the fetus assumes a more important role as an intermediary in amniotic fluid water turnover. Second, although the exchanges between mother and fetus and amniotic fluid are of similar magnitude throughout pregnancy, there is sufficient imbalance to produce a net water circulation from mother to fetus to amniotic fluid and back to mother.

Amniotic fluid is isotonic during early pregnancy but by term becomes hypotonic compared with fetal and maternal plasma. This changing concentration of amniotic fluid is believed to reflect the maturation of fetal renal function.[13] Between 12 and 22 weeks of gestation, amniotic fluid sodium and urea concentrations are very slightly below fetal plasma levels, though closer to fetal than to maternal values.[4] Amniotic fluid very early in pregnancy is roughly equivalent to isotonic extracellular fluid derived from maternal or fetal compartments by dialysis across tissue layers largely impermeable to protein. The fact that the sodium concentration in amniotic fluid is very slightly lower than in fetal serum likewise possibly indicates that, even at this stage of development, small amounts of fetal urine may be influencing the final composition.

Amniotic fluid becomes moderately hypotonic (mean total solute concentration 255 to 260 mOsm/kg of water) near term and is believed to be the product of numerous exchanges with the fetus. These exchanges include (1) fetal swallowing; (2) fetal urination, and (3) exchanges within the fetal respiratory tract where the alveolar capillary bed is perfused by amniotic fluid transported in and out by respiratory movements. It is not clear to what extent this latter exchange contributes, since radiopaque material introduced into amniotic fluid does not appear to any significant extent in the lungs.[14]

Amniotic fluid at term has a fairly constant sodium concentration of about 120 mEq/L, and this concentration is maintained in a volume of about 800 mL, despite the fact that a normal fetus will remove some fluid by swallowing and contribute urine with a sodium concentration of 70 mEq/L or less. If the fetus swallows about 5 mL/kg/hr as suggested by Pritchard,[15] one can calculate that some source or sources must provide about 0.7 mEq of sodium per hour to the amniotic fluid. Thus it is necessary to postulate a site of sodium entry into amniotic fluid fast enough to keep the concentration at about 120 mEq/L but too slow to allow equilibration with maternal or fetal extracellular fluid. The fetal membranes may be that site.

Low molecular weight substances

Dissolved carbon dioxide appears to equilibrate quickly between mother, fetus and amniotic fluid while the bicarbonate anion crosses very slowly between the three compartments. Therefore changes in fetal pH secondary to P_{CO_2} fluctuations would be rapidly transmitted to the amniotic fluid in most circumstances, whereas fetal pH alterations accompanying changes in bicarbonate concentration would be poorly reflected in this space. Up to now, however, correlations between amniotic fluid pH and fetal status at birth in the human have been inconsistent.

Proteins and enzymes

A number of sources for proteins in amniotic fluid are possible. Proteins can cross from the maternal serum at the site of the amnion. Maternal extracellular fluid proteins can likewise enter amniotic fluid at sites away from the placental attachment by transudation across the amnion. Other sources of proteins include fetal skin, the membranes of pregnancy, fetal urine, meconium, and the nasopharyngeal, oral, and lachrymal secretions of the fetus. Fetal lungs also contribute proteins to the fluid. One should recognize that the relative contributions of these various sources change during the course of pregnancy.

Removal of protein from the amniotic fluid probably occurs primarily by three processes: permeation, fetal swallowing, and proteolysis. Very little is known about the relative contribution of these processes, but because about 500 mL of amniotic fluid per day are removed by fetal swallowing,[15,16] this route could account for much of the protein disappearance.

Over 50 enzymes have been identified in amniotic fluid[9,10] and fall into two categories, those whose activity is maximally expressed early in pregnancy (12 to 20 weeks) and those active at the latter times of pregnancy (35 to 40 weeks). The origin of many of these enzymes and their significance in the fluid are not understood. However, some enzymes of fetal origin assist in the diagnosis of particular inborn errors of metabolism.

Some hormones that have been identified in amniotic fluid are listed in the box in the next column. This list includes steroid, polypeptide, and amino acid–derived hormones such as the catecholamines. Although many of these hormones are products of urinary or biliary excretion from the fetus, a few have clinical usefulness and are discussed later. A more extensive list is available.[9,10]

Examples of hormones identified in amniotic fluid

Protein and polypeptide
- Insulin
- Growth hormone
- Somatomedin
- Somatostatin
- Prolactin
- Follicle-stimulating hormone
- Luteinizing hormone
- Human chorionic gonadotropin
- Adrenocorticotropic hormone
- Endorphin
- Human placental lactogen
- Oxytocin
- Relaxin
- Thyrotropin
- Thyroxine
- Renin
- Angiotensin

Steroids
- Estriol
- Estradiol
- Estrone

Prostaglandins
- E_2
- $F_{2\alpha}$

MATERNAL BIOCHEMICAL CHANGES DURING PREGNANCY

Human chorionic gonadotropin (HCG)

The urine and serum of pregnant women contain high concentrations of human chorionic gonadotropin (HCG or hCG), which provide the basis of tests for the diagnosis of pregnancy. Specific and sensitive analytical methods for the β-chain subunit of HCG permit the early detection (as early as 8 days after ovulation, 1 day after implantation) of pregnancy. HCG concentrations climb early in pregnancy, becoming maximum near 8 to 10 weeks (Fig. 35-2). HCG is produced by the trophoblast, which explains why its presence can be detected so soon after fertilization.

Human chorionic gonadotropin is one of a family of closely related glycoprotein hormones that regulate reproductive and metabolic functions. Besides HCG, the other hormones are follicle-stimulating hormone (FSH), luteinizing hormone (LH), and thyroid-stimulating hormone (TSH). All these hormones are composed of two polypeptide subunits referred to as alpha (α)- and beta (β)-chains and all contain carbohydrate, as the term ''glycoprotein'' indicates. The α-chains of HCG, LH, FSH, and TSH are nearly identical in their amnio acid sequence, a finding that accounts for the immunological similarity of these protein hormones.[18]

The β-chains of the glycoprotein hormones impart their biological specificity because they all differ in amino acid sequence except for HCG and LH. These latter two hormones are virtually identical, differing to the largest degree in carbohydrate content, with HCG containing about 30% carbohydrate by weight and LH approximately half

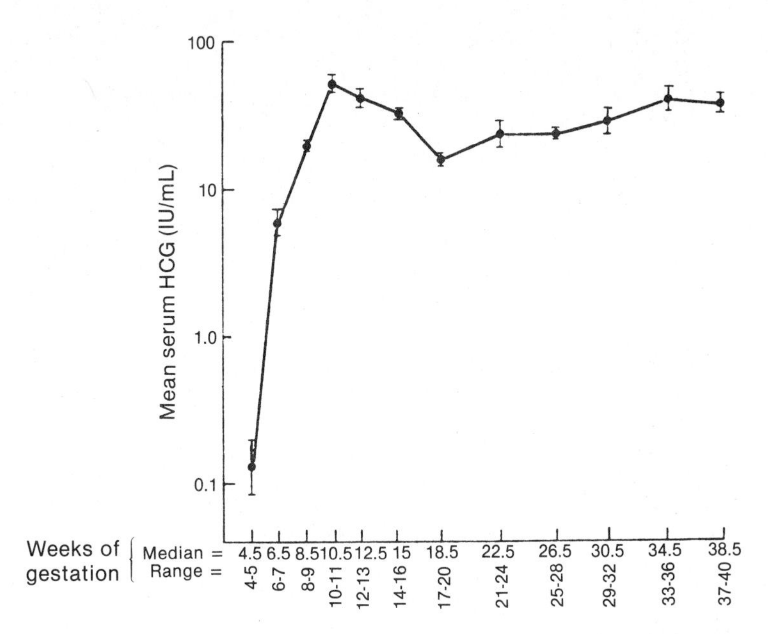

Fig. 35-2. Serum chorionic gonadotropin (HCG) concentrations during pregnancy. (From Goldstein, D.P., Aon, T., Taymor, M.T., et al.: Am. J. Obstet. Gynecol. **102:**110-114, 1968.)

that amount. The trophoblast of the placenta produces HCG, whereas LH is a product of the anterior pituitary. This carbohydrate-content difference perceptibly influences the metabolic patterns of these hormones, imparting a biological half-life to HCG of about 6 hours, whereas the biological half-life of plasma LH is 12 to 45 minutes.[19] These differences in metabolic clearance correlate well with our understanding of the functions of these hormones, with LH regulating the complicated process of ovulation and subsequent formation of the corpus luteum and HCG providing continued stimulation of the corpus luteum to ensure uninterrupted progesterone production until the placenta can provide sufficient progesterone to maintain the pregnancy.

Estrogens

Nearly a half century ago it was recognized that a substantial increase in estrogen excretion was associated with pregnancy. The predominant estrogen was identified as estriol, not the usual ovarian estrogens, estradiol or estrone (Fig. 35-3). It was not until many years later that the unique relationships were elucidated that described the pathway for estriol formation.[20] Estriol formation during pregnancy occurs only in the higher primates, and the obligatory interaction of fetus and placenta for the formation of this ''estrogen of pregnancy'' provides the basis for one of the few biochemical tests available to monitor fetal well-being.

Estrogen formation proceeds in an obligatory sequence of reactions that converts cholesterol to progestins, then to androgens, and then to estrogens (refer to Chapter 41). In the ovary, this sequence occurs solely within that organ. In pregnancy, for this sequence to occur, a complementary relationship between the placenta, the fetal adrenal cortex, and the fetal liver leads to estriol formation. This unique relationship is referred to as the ''fetoplacental unit''[21] (Fig. 35-4).

Fetal adrenal cortex converts acetate to cholesterol, then to pregnenolone, and then to the androgen dehydroepiandrosterone (DHEA). DHEA quantitatively is the major steroid produced by the fetal adrenal gland. The liver of the human fetus hydroxylates androgens (such as DHEA) to provide the 16α-hydroxylated precursor for estriol formation. The 16α-hydroxylated androgen precursor, 16α-hydroxy-DHEA, is eventually converted, in the placenta to

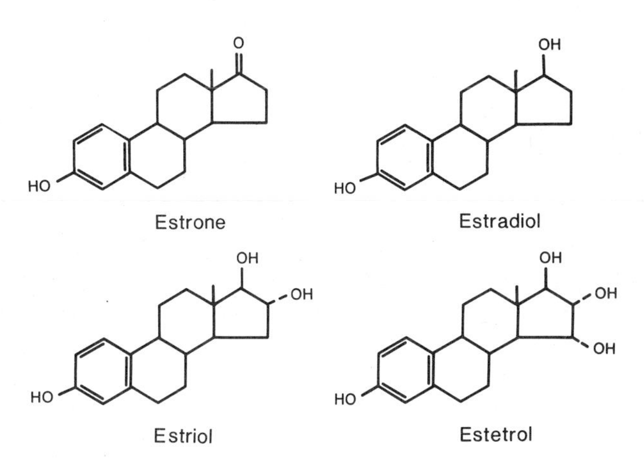

Fig. 35-3. Structures of estrogens. *Dashed lines,* α-Stereoconfigurations; *solid lines,* β-stereoconfigurations of hydroxyl groups.

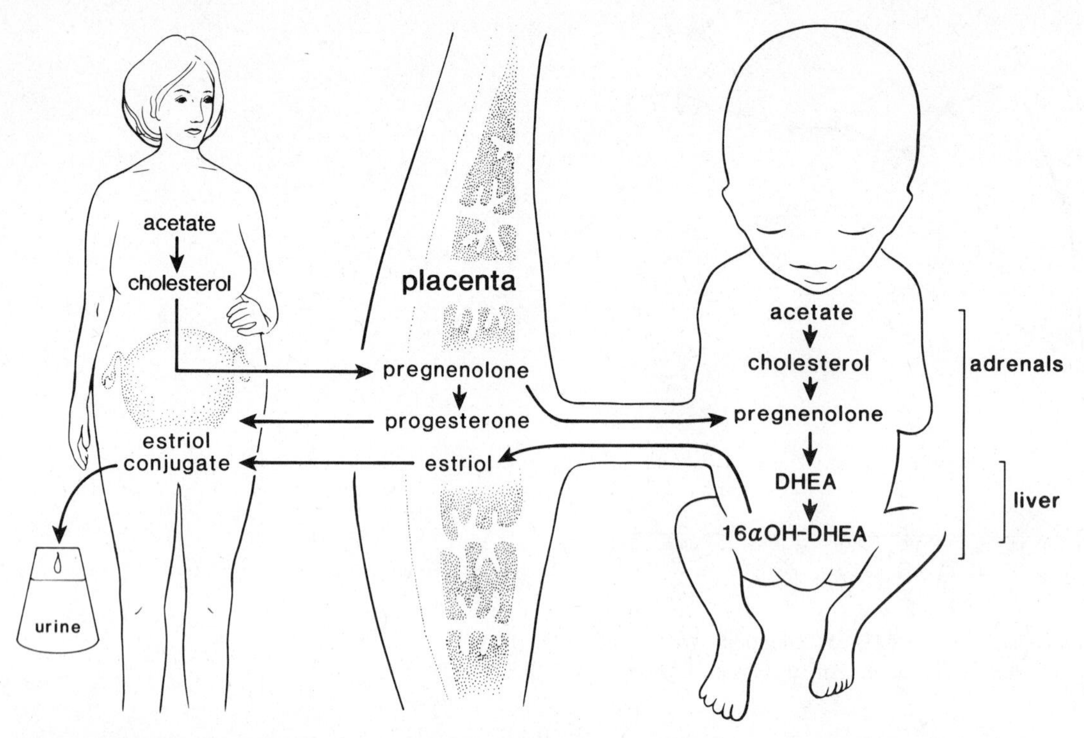

Fig. 35-4. Scheme of fetoplacental unit. *DHEA,* Dehydroepiandrosterone; *16α-OH-DHEA,* 16α-hydroxydehydroepiandrosterone.

estriol by the aromatizing enzyme complex located in that tissue.

The fetoplacental steroid relationship is further complicated by the fact that both fetal liver and fetal adrenals are active in the sulfation of steroids with hydroxyl functions at position 3; thus many circulating fetal steroids are sulfate conjugated. The fetus, however, lacks the ability to remove the sulfate, but the placenta possesses a very active sulfatase. Sulfation-desulfation appears to be integrally involved in the transfer of steroids from the placenta, since, in conditions where the placental sulfatase is absent, low estriol levels are characteristically observed.[22]

Estriol comprises over 90% of the known maternal estrogens of pregnancy. It is metabolized via conjugated forms by maternal liver, both as the sulfate and as the glucuronide, the primary excretory forms. Concentrations in maternal serum increase with advancing gestation and increase to nearly 40 ng/mL at term.[23] Fig. 35-5 indicates the normal increase in plasma estriol and a pattern seen with a diabetic patient, with fetal death, and with growth retardation. Estriol is also found in amniotic fluid.[24]

The functional role for estriol in pregnancy has prompted much speculation. In many biological test systems, estriol is a weak estrogen, demonstrating only one hundredth the potency of estradiol and one tenth the potency of estrone per unit weight. However, estriol can be demonstrated to be equipotent to estradiol in its ability to promote uteroplacental blood flow. For this reason, its role in pregnancy may be to ensure optimal blood flow in the gravid uterus.[25]

More recently, the estrogen estetrol (Fig. 35-3) is believed to offer more information than estriol about the status of the fetus in utero.[26] Estetrol (E_4) is not further metabolized as estriol (E_3) is in the maternal liver. In addition, E_4 specifically reflects fetal liver activity, whereas E_3 is dependent on fetal adrenals. Once E_4 is formed by fetal liver, it goes to maternal circulation and is excreted in urine as E_4-glucuronide. The presumed advantage of E_4 is based upon the belief that this estrogen reflects more directly fetal activity. Clinical evaluations of fetal well-being have not shown a clear advantage of estetrol over estriol, however.[27]

Thyroid

The thyroid gland enlarges during pregnancy. The reasons for this is not clear, though alterations in iodine metabolism and the production of placental thyroid stimulators may be involved. Human chorionic gonadotropin and TSH have nearly identical α-subunits, which may explain why HCG appears to have thyroid-stimulating activity, but whether this hormone plays a role in thyroid enlargement is a matter of speculation.[28]

During pregnancy there is an increase in total serum thyroxine (T_4) and triiodothyronine (T_3). This increase in serum thyroid hormones is stimulated by increases in the synthesis of thyroxine-binding globulin (TBG) by maternal

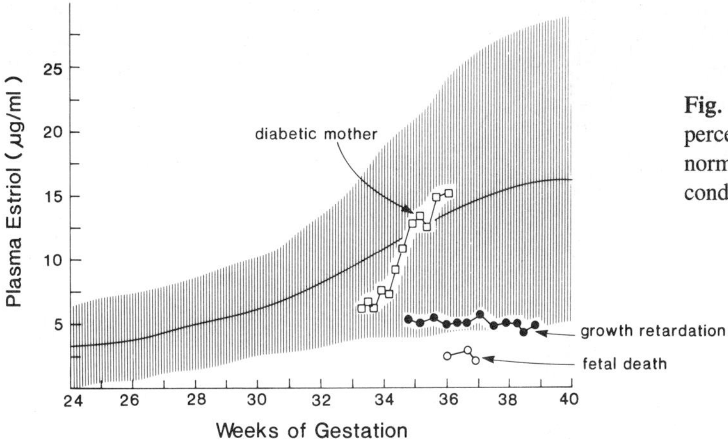

Fig. 35-5. Mean *(solid line)* and estimated 5th and 95th percentiles *(shaded area)* for plasma unconjugated estriol during normal pregnancy. Estriol patterns from three actual pregnancy conditions are shown.

liver, which in turn are attributable to the increase in estrogen production.[28,29] However, despite this increase in total serum thyroxine, the level of free hormone remains the same. The metabolism of thyroxine in pregnancy is altered because of the increase in TBG.[28]

Net thyroxine turnover is unchanged in normal pregnancy because turnovers of 90 μg/day in nonpregnant and 97 μg/day in pregnant women have been observed.[30] These two values become virtually identical when they are expressed as units per square meter of body surface.

About 60% of thyroxine is bound to TBG, 30% to thyroxine-binding prealbumin, and the rest to albumin. Little change occurs with the thyroid-binding prealbumin fraction with pregnancy. Serum TBG doubles between 8 and 12 weeks of pregnancy and rises progressively until term.[28] Levels return to normal by 6 weeks after delivery.

Whether plasma TSH levels are normal or elevated in pregnancy has not been resolved. TSH may rise early in pregnancy and return to normal in the second half of gestation.[28] Some reports of elevated serum TSH might represent cross-reactivity with HCG.

Passage of the thyroid hormones and of TSH across the placenta does not occur. An extensive discussion of all facets of thyroid and parathyroid function in pregnancy can be found elsewhere[28] and in Chapters 40 and 42.

Electrolytes

The concentrations of most electrolytes in serum are slightly lower in pregnancy than in the nonpregnant state.[31] Plasma osmolality also decreases from nonpregnant values around 290 to about 280 mOsm/L.[32] This decrease occurs in the first trimester because no significant change occurs from 12 weeks to term.[33]

Serum lipids

Hyperlipidemia develops with normal pregnancy. Total serum lipid concentration rises progressively from the end of the first trimester and is 40% to 50% above nonpregnant levels at term.[34] All components of the serum lipids are increased, but the triglyceride fraction shows the largest proportionate rise.[35]

Serum proteins and liver function

The total concentration of serum proteins decreases by about 1 g/L during pregnancy. Most of the decrease occurs during the first trimester. The decrease is mainly in serum albumin.[32,33,36] Alpha$_1$, alpha$_2$, and beta globulins rise slowly and progressively. Gamma globulin probably decreases slightly. The IgG component, which is the major immunoglobulin transferred to the fetus, falls progressively.[37] Colloid osmotic pressure decreases in parallel with serum albumin levels,[32,33] but a minimal influence is provided by the relatively inactive globulin fractions.

Throughout pregnancy fibrinogen increases progressively, with term values 30% to 50% above nonpregnant levels. Clotting factors VII, VIII, IX, and X are also increased,[38,39] whereas prothrombin and factors V and XII are reduced.[39,40] Alterations that occur in the levels of clotting factors and plasminogen are probably brought about by estrogen action on the liver. The placenta, too, is presumed to reduce fibrinolytic activity, for this activity increases promptly after delivery.[41]

Cortisol, a catabolic hormone, causes mobilization of amino acids from muscle protein and concomitantly stimulates the uptake of amino acids by the liver and the induction of enzymes required for gluconeogenesis. Thus cortisol effects a "translocation" of protein from muscle to the liver with an overall negative nitrogen balance. Although serum cortisol is doubled during late pregnancy, much increase is attributable to biologically inactive hormone bound to transcortin or corticoid-binding globulin (CBG), which is in turn elevated under the influence of estrogen.

There appears to be no distinct histological changes in the maternal liver as a consequence of normal pregnancy.[42] A number of laboratory function tests do change, however. Nonspecific alkaline phosphatase activity in serum nearly doubles during normal pregnancy and can reach levels that would be considered abnormal in the nonpregnant woman. Much increase is attributable to isozymes of this enzyme originating from the placenta.

Glycosuria

Pregnancy is potentially diabetogenic. Diabetes mellitus may be aggravated by pregnancy, and clinical diabetes may appear in some women only during pregnancy. The renal handling of glucose during pregnancy is of particular interest because of the frequent appearance of clinical glycosuria and the necessity to differentiate this "renal glycosuria" from that of pregnancy-aggravated diabetes mellitus. It seems that "renal glycosuria" in pregnancy results from a preexisting deficiency in tubular function that is aggravated during gestation.

FETAL BIOCHEMICAL CHANGES DURING PRENATAL DEVELOPMENT

Liver function

In fetal liver and in other fetal tissues examined, free amino acid levels are invariably higher than they are in the respective adult tissues. As pregnancy advances, amino acids show a tendency to decrease.[43] This decline is not uniform, but occurs at different rates for individual amino acids.[44,46]

The protein-to-DNA ratio shows a pronounced increase in fetal liver with advancing pregnancy. Besides synthesis there is a high rate of protein degradation as reflected by urea production. Plasma urea concentration of the human fetus is higher than that of maternal plasma, and urea production is estimated at 540 mg/kg per day, which exceeds by a significant margin that of adults, which range from 200 to 400 mg/kg per day.[47] A gradual increase of urea in the amniotic fluid is observed in the latter weeks of pregnancy (Fig. 35-6).

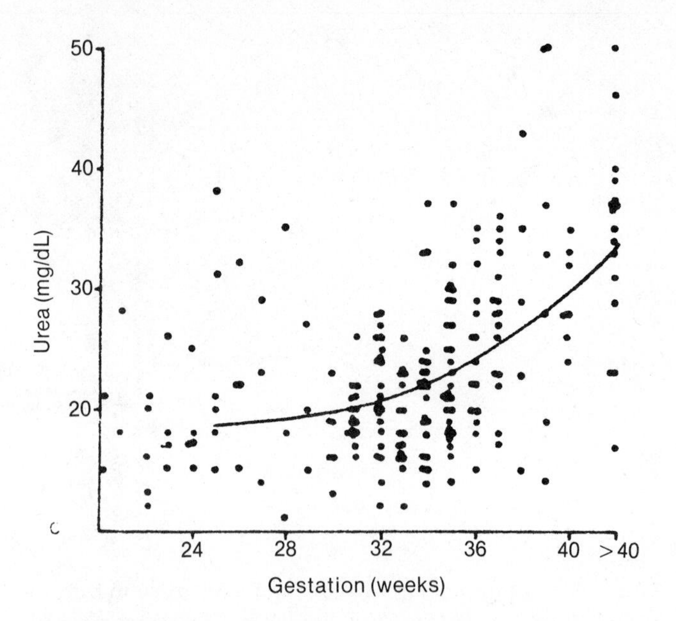

Fig. 35-6. Distribution and regression curve of amniotic fluid urea concentration in milligrams per deciliter. (From Lind, T.: In Fairweather, D.V.I., and Eskes, T.K.A.B., editors: Amniotic fluid: research and clinical application, ed. 2, Amsterdam, 1978, Excerpta Medica.)

The need for increased quantities of amino acids for protein synthesis is satisfied by placental transport. This is an active process against a concentration gradient that is dependent on placental blood flow and, to a lesser degree, on the concentration of amino acids in the maternal plasma.[43] Both total amino nitrogen and individual amino acids are elevated in umbilical cord blood,[43,45] and there is a tendency for these levels to decrease as pregnancy progresses.[47] Tyrosine, threonine, and taurine are exceptions, however, in that their levels remain fairly constant.

Fetal liver contains hematopoietic cells in abundance, and they gradually disappear after birth. In very early fetal life, the liver is the major blood-forming organ, but this role decreases in relative importance to the bone marrow by 22 to 24 weeks of gestation. Because of widely varying amounts of the different cell types in the newborn liver, enzyme-activity patterns are distorted.

Fetal liver along with the fetal yolk sac produces alpha-fetoprotein (AFP) which is released into the fetal circulation. It passes from the bloodstream by way of the urine to amniotic fluid. It is normally cleared from amniotic fluid by fetal swallowing. Alpha-fetoprotein appears in maternal serum throughout gestation (Fig. 35-7), and AFP from this source is more easily obtained than it would be from amniotic fluid in early pregnancy.[50]

Enzymes at birth and in the neonatal period are discussed by Stave.[51] A number of enzyme activities change drastically at birth or in the neonatal period. Gluconeogenesis, for example, is minimal during the developmental period and assumes its prominent role after birth. Suffice it to say, many changes can be induced by the events surrounding the process of birth, or through independent genetic control.

Human liver isozyme patterns for lactate dehydrogenase changes drastically during development.[52] Isozyme subunits necessary for early development favor a shift from subunit types reflecting aerobic conditions toward those more characteristic of an oxygen-deficient environment at birth. A detailed description of lactate dehydrogenase isozyme patterns during development has been published by Vesell and Philip.[53]

Because maturation of the fetus is not totally complete by the time of birth, some jaundice regularly occurs in virtually every newborn during the first week of life. This is known as "physiological jaundice." The yellow pigmentation, or jaundice, is caused by bilirubin pigments that come from the normal destruction of red blood cells,

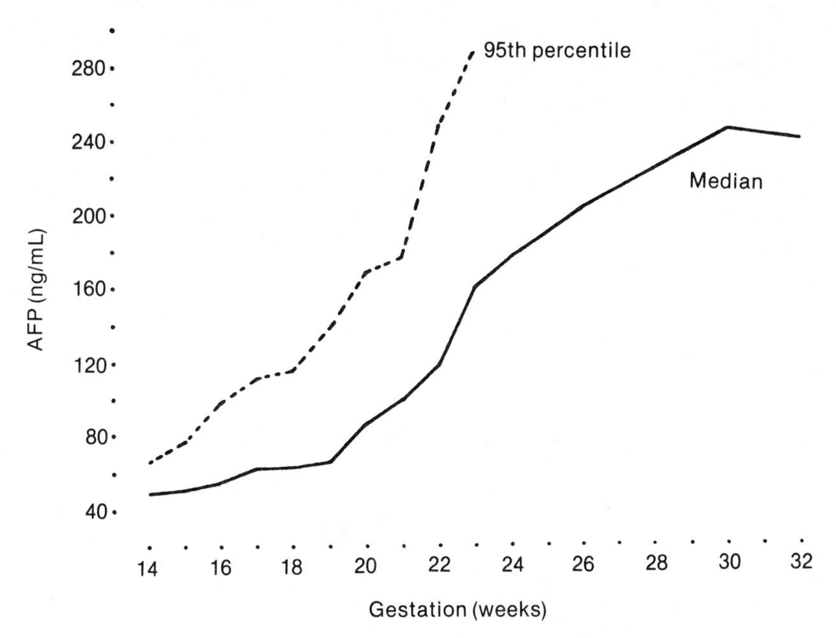

Fig. 35-7. Median and 95th percentile of alpha-fetoprotein, *AFP,* in maternal serum. (From Crandall, B.F.: In Kirkpatrick, A.M., Nakamura, R.M., editors: Alpha-fetoprotein: laboratory procedures and clinical applications, New York, 1981, Masson Publishing Co.)

and the immature fetal liver has not developed its full capability to clear bilirubin from the blood.

Renal function

The primary function of the mammalian kidney is to maintain water and electrolyte homeostasis. This is accomplished by selective excretion or retention of water and solutes as conditions dictate. In the fetus, body water and electrolyte balance are maintained largely by the placenta. For this reason fetuses without functional kidneys often show no water or electrolyte abnormalities. Thus the development of renal function coincides more with its extrauterine role.

In the human nephrogenesis ceases at about 36 weeks of gestation, though many nephrons are still immature. This anatomic immaturity disappears within several weeks of birth, however, and renal development from this stage consists solely in nephron growth. There is a complicated interplay of renal pressure and resistances during kidney maturation. This interplay provides blood to those nephrons that are most mature and presumably those that function best. As new nephrons mature, intrarenal vascular resistances are altered so that more blood can be distributed to these nephrons. This process results in a reasonable degree of nephronperfusion balance throughout the renal maturational period.

As mentioned previously, fetal urine is hypotonic to plasma as it passes directly into the amnion. The primary reason that the urine is not so concentrated as that of an adult is that the kidneys have smaller medullary solute gradients than those of the adult have. Only after the fifth month of gestation does the major growth of the cortical and outer medullary regions occur.

Organic nitrogenous compounds such as urea (Fig. 35-6), uric acid, and creatinine (Fig. 35-8) gradually increase in concentration in amniotic fluid as the renal system of the fetus matures. In early pregnancy these compounds are present in concentrations similar in amniotic fluid to concentrations in maternal and fetal blood and increase gradually to become significantly higher than levels in maternal or fetal blood. A sharp rise in creatinine at about the thirty-seventh week of gestation elevates the amniotic fluid concentrations to become two to three times higher than that of normal serum.[54,55]

For comprehensive discussions on kidney function and electrolyte and water metabolism, refer to Kleinman[56] and Kerpel-Fronius[57] and Assali, DeHaven, and Barrett.[58]

Lung development

At birth there is abrupt transition that requires the newborn infant to assume vital functions that were previously handled by the maternal circulation. The lung is shifted from a fluid-filled organ to a gas-exchange system in a few brief minutes. This functional transition is possible only if sufficient maturation of the fetal lung has occurred during development. The lungs should have developed alveoli with adequate surface area to sustain ventilatory movements for gas exchange; also a surfactant system must be

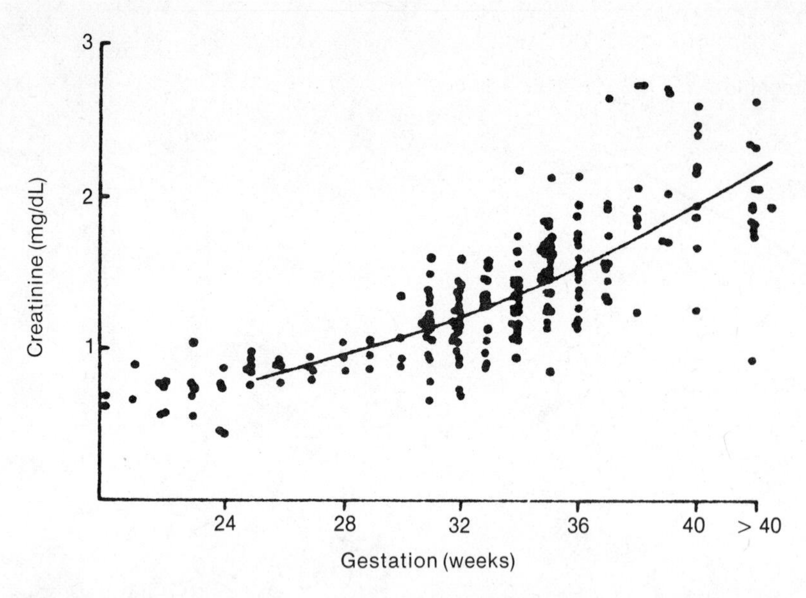

Fig. 35-8. Distribution and regression curve of amniotic fluid creatinine concentration in milligrams per deciliter. (From Lind, T.: In Fairweather, D.V.I., and Eskes, T.K.A.B., editors: Amniotic fluid: research and clinical application, ed. 2, Amsterdam, 1978, Excerpta Medica.)

in place to accommodate the physical chemical requirements for pulmonary function. The development of these processes are highly organized and are coordinated by the timing of anatomical and biochemical events.

The composition of surfactant is principally phospholipid, containing highly saturated fatty acid moieties.[59,60] Tables 35-3 and 35-4 list its composition. In addition to the highly unusual saturated lecithins, other important constituents of the surface-active system include phosphatidyl glycerol[60,63] and "surfactant apoprotein."[64] Gluck and Kulovich[65] were the first to correlate phospholipid concentrations in amniotic fluid to the functional status of the fetal lung. The lecithin-sphingomyelin ratio (L/S) (Fig. 35-9) is used widely for that purpose.

The site of formation for the surfactant is believed to be the large alveolar epithelial cells known as type II pneumonocytes, which compose about 10% of the cellular makeup of the lung. Synthesis of the phospholipid component takes place by the CDP-choline pathway[66] and the rate-limiting step, mediated by the enzyme cholinephosphotransferase, is under the control of cortisol.[67] Cholinephosphotransferase represents only one of several enzymes that are dependent on cortisol for their activity during development.

Surfactant facilitates pulmonary function in at least two ways: it maintains alveolar stability by preventing collapse of the terminal respiratory tree and it reduces the pressure that is needed to distend the lung in the initial phase of inspiration. Infants who develop the respiratory distress syndrome (RDS) have a higher surface tension at the alveolar air-liquid interface as a result of a pulmonary surfactant deficiency. Infants with RDS show increased res-

Table 35-3. Composition of rat lung phospholipids (%)

Phosphatidyl choline (lecithin)	54.3 ± 1.4*
Phosphatidyl ethanolamine	20.2 ± 1.1
Phosphatidyl serine and phosphatidyl inositol	5.5 ± 0.5
Sphingomyelin	12.6 ± 0.6
Phosphatidic acid and cardiolipin	3.9 ± 1.0
Phosphatidyl glycerol	3.9 ± 0.3

*Mean ± SEM.

From Godinez, R.I., Sanders, R.L., and Longmore, W.J., et al.: Biochemistry **14:**830–840, 1975.

Table 35-4. Composition of dog pulmonary surface-active material (%)

Phosphatidylcholine fatty acid components	
Palmitic acid (16:0)	71.0
Myristic acid (14:0)	6.1
Stearic acid (18:0)	3.6
Palmitoleic acid (16:1)	11.0
Oleic acid (18:1)	3.9
Unidentified	3.6
Chemical components	
Lipid	85
Protein	13
Hexose	<1.7
Nucleic acid	<0.7
Hexosamine	<0.5

Modified from King, R.J., and Clements, S.A.: Am. J. Physiol. **223:**715-726, 1972.

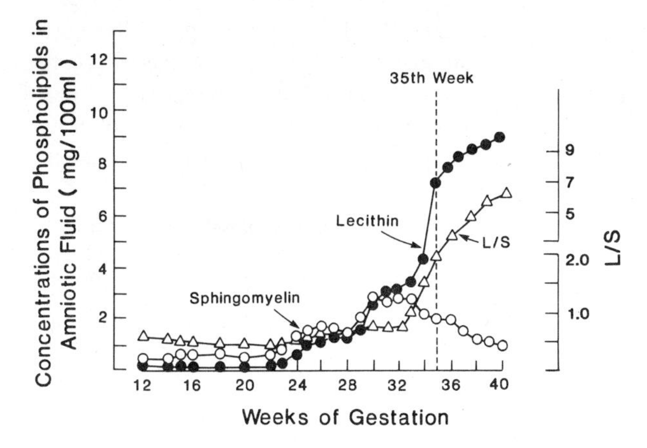

Fig. 35-9. Lecithin, sphingomyelin, and lecithin/sphingomyelin ratios in amniotic fluid during normal pregnancy. (Adapted from Gluck, L., and Kulovich, M.V.: Am. J. Obstet. Gynecol. **115**:539-546, 1973.)

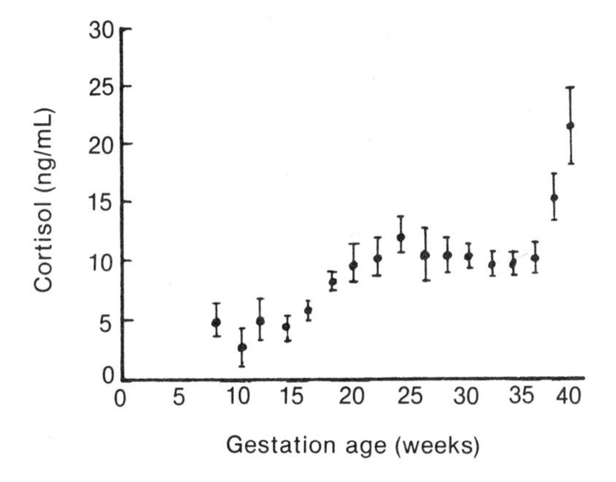

Fig. 35-10. Amniotic fluid cortisol during gestation (mean ± standard error of mean). (From Murphy, B.E., Patrick, J., and Denton, R.L.: J. Clin. Endocrinol. **40**:164-167, 1975.)

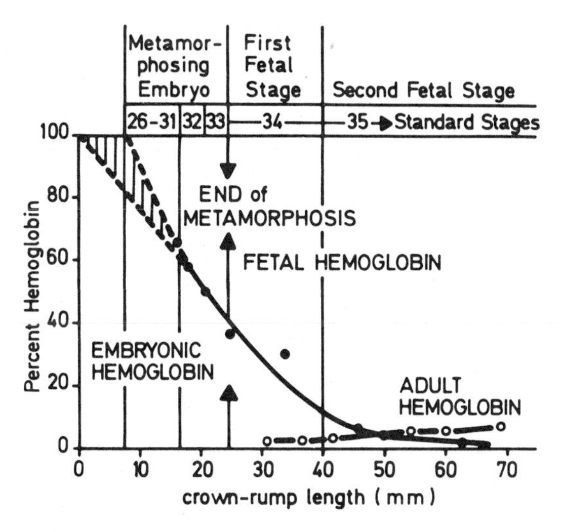

Fig. 35-11. Relationship between hemoglobin types and developmental stages in early human life. *Dashed lines and hatched area,* Expected development. (From Kleihauer, E.: In Stave, U., editor: Perinatal physiology, New York, 1978, Plenum Publishing Co.)

piratory effort. Clinical management is aimed at maintaining the infant in an oxygen-rich environment and keeping the alveoli open artificially, during spontaneous or mechanically assisted breathing, until such time as surfactant synthesis by the type II alveolar cells is adequate for unassisted respiration.[68]

A description of the histological development of the lung is available[66] and comprehensive reviews on the research observations that provide the background for our current understanding provide interesting reading.[68,70] More detailed accounts of the function and composition of pulmonary surfactant are published[68-70] as well as the accepted nomenclature.[71]

Cortisol production by the fetal adrenal increases dramatically late in pregnancy, and this is reflected in the cortisol seen in amniotic fluid (Fig. 35-10). Cortisol near term expedites the process of pulmonary maturation through control of cholinephosphotransferase. Exogenous administration of synthetic corticoids have been used clinically to hasten pulmonary maturation under circumstances where delivery of a preterm infant is imminent.[73]

Hemoglobin

Embryos have a hemoglobin that is unique to the embryonic stage of development and is replaced during fetal life by "fetal hemoglobin" (Hb F) and finally by adult hemoglobin (Hb A). The transition from one to the subsequent type is gradual because of different life-spans of the red cells and because of the increasing production of red cells during the periods of red cell production.[74]

Our current understanding about the pattern of hemoglobins formed during development is presented in Fig. 35-11. By approximately 10 weeks of gestation, the embryonic hemoglobins decrease to 10% of the total hemoglobin present. Fetal hemoglobin has been found to compose 34% of the total in an embryo just under 7 weeks of age.[75] Before the end of the first trimester (less than 12 weeks) the Hb F has increased to approximately 90% of the total, with the remaining percentage constituted by Hb A. From this point on Hb F remains constant until about the thirty-sixth week of gestation when there is a decline. The decline is primarily caused by an increase in Hb A synthesis rather than by a decrease in Hb F. Sharp increases in Hb A are seen in reticulocytes and erythrocytes by birth. The developmental changes in hematopoietic sites, the red cell morphology, and the hemoglobin types are shown in Fig. 35-12.

The structural differences between fetal and adult hemoglobins undoubtedly account for the functional differences between these two hemoglobins. However, the physiological significance of each functional difference is not well understood at this time. These differences are summarized by Kleihauer[74] and include differences in the affinities of the hemoglobins for oxygen, resistance to acid, base, and heat, electrophoretic and chromatographic

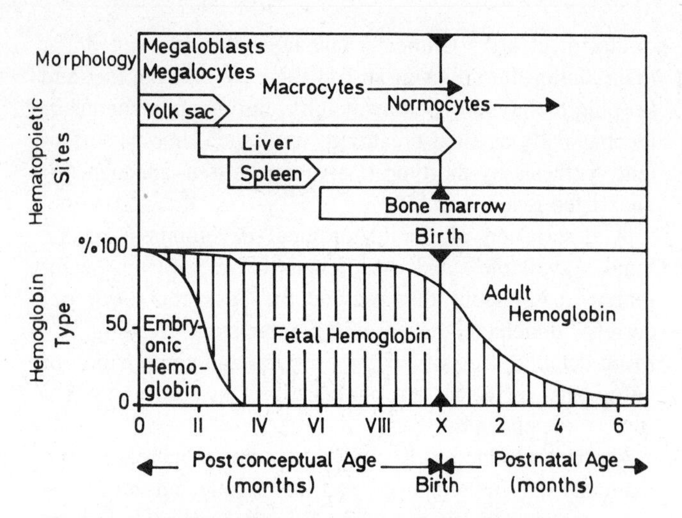

Fig. 35-12. Developmental changes in hematopoietic sites, red blood cell morphology, and hemoglobin types. (From Kleihauer, E.: In Stave, U., editor: Perinatal physiology, New York, 1978, Plenum Publishing Co.)

properties, and others. The difference in affinity that fetal hemoglobin has for oxygen is seen in the fetal oxyhemoglobin saturation curve, which lies to the left of the maternal curves under standard conditions.[76] In fact, when oxygen is diffusing from maternal blood with a Po_2 of about 34 mm Hg to fetal blood at a Po_2 of about 30 mm Hg, the oxygen is actually moving against a concentration gradient.[2]

Bilirubin

Erythrocyte destruction precedes the formation of bilirubin. This course is presented in simplifed terms:

$$\text{Hemoglobin} \xrightarrow[\text{Globin}]{} \text{Biliverdin} \rightarrow \text{Bilirubin}$$

It is generally believed that plasma concentrations of bilirubin are low in the fetal circulation except in unusual circumstances, as in severe erythroblastosis fetalis. Plasma concentrations rise sharply in the newborn infant shortly after birth if the production of bilirubin exceeds the capacity of the liver for uptake, conjugation, and excretion. Conjugation of bilirubin to the glucuronide and its prior uptake by the liver probably occurs in the fetus at a very low rate. The probable course of events then is for bilirubin to be transferred from the fetal circulation, across the placenta to the maternal circulation, and from there to be conjugated by the maternal liver and excreted in maternal bile.

Biliverdin is the principal and primary degradation product of hemoglobin and an important intermediate pigment in the formation of bilirubin. The conversion of biliverdin to bilirubin is by the enzyme biliverdin reductase.

Even in circumstances where the rapid breakdown of erythrocytes leads to accelerated bilirubin production, as in severe maternal-fetal blood group incompatibility, cord blood bilirubin rarely exceeds 50 mg/L. This fact attests to the rapid, efficient transfer of this pigment across the placenta and the equally efficient disposal of fetal bilirubin by the mother. A discussion of the transport and liver metabolism of bilirubin can be found in Chapter 24.

After birth, the newborn loses the placental mechanism for bilirubin removal. As a result, there is a modest accumulation of unconjugated bilirubin in the plasma. The jaundice resulting from this change in physiological circumstance is related to the limited uptake, conjugation, and excretion of bilirubin by the immature liver. The degree of neonatal jaundice occurring at birth is dependent on the maturity and health of the fetal liver at birth.

Fetal liver contains hematopoietic cells, which gradually disappear after birth. During the third trimester of pregnancy, the hematopoietic cells are so diffusely integrated into the parenchyma that the lobulation of the hepatic tissue is not visible. This widely varying amount of a profoundly different cell type in the newborn liver affects the analysis of enzyme activities and results in a particular distortion of the hepatic enzyme pattern.

PATHOLOGICAL CONDITIONS ASSOCIATED WITH PREGNANCY AND PERINATAL PERIOD

Placental disorders

Adequate exchange across the placenta between the maternal and fetal circulations is essential for normal fetal growth and metabolism.[2] Less than optimal quantities of nutrient results in small-for-gestational-age (SGA) fetuses, whereas in circumstances where nutrient is excessive, as with maternal diabetes, large-for-gestational-age (LGA) infants are common.

Few pathological conditions involving the placenta exist where the monitoring of chemicals by the laboratory can be applied. These instances, when they arise, usually take advantage of the unique hormone production by the placenta. One such example is with the hydatidiform mole. Molar tissue is a developmental anomaly of the placenta that has the potential for malignant growth. It is the most common lesion antecedent to choriocarcinoma and is characterized with a proliferation of chorionic epithelium. Since the mole is trophoblastic tissue, human chorionic gonadotropin is produced, resulting in a positive pregnancy test. If serum or urinary HCG levels exceed values typical of specific times in pregnancy, a presumptive diagnosis of mole may be made, though because of the variable gonadotropin values for normal pregnancy, no single value can be established as the borderline between normal and abnormal.

HCG is used to monitor trophoblastic tissue during its active life in the body. In women with hydatidiform moles, chorionic gonadotropin and thyrotropin levels are often elevated, whereas placental lactogen is lower than in normal pregnancies of the same gestational age.[77] After

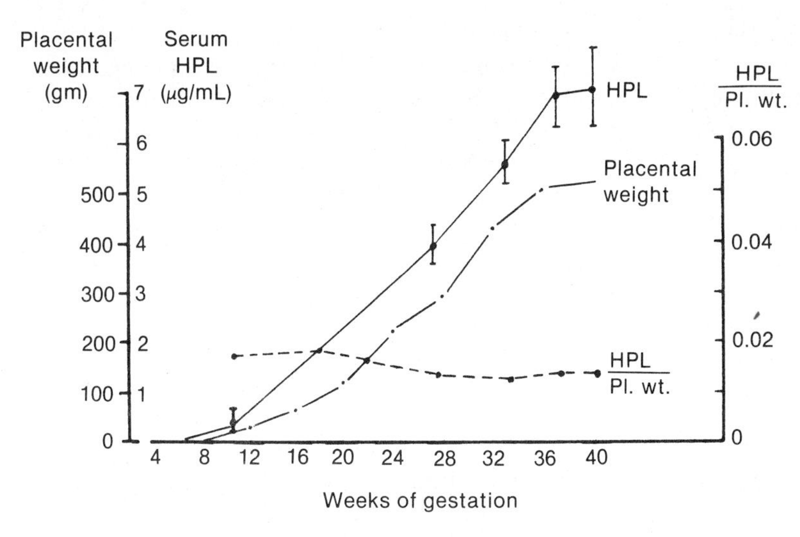

Fig. 35-13. Concentrations of human placental lactogen, *HPL,* during normal pregnancy. (From Selenkow, H.A., Saxena, S.M., Dana, C.L., and Emerson, K., Jr.: In Pecile, A., and Fenzi, P., editors: The foeto-placental unit, Amsterdam, 1969, Excerpta Medica.)

evacuation of the mole, the time required for the disappearance of these hormones is longest for HCG, shorter for chorionic thyrotropin, and much shorter for placental lactogen. Because highly sensitive and specific methods are available for monitoring serum HCG, this hormone is useful in monitoring the response to therapy.

Human placental lactogen (HPL) is produced by the placenta, and since its pattern of production is proportional to the size (mass) of the placenta (Fig. 35-13), this hormone can be used to gain insight into the growth of the placenta.

Fetal immaturity

The last of the organ systems to mature sufficiently to support extrauterine life is the lungs. Fetal "immaturity" generally implies a failure of the pulmonary system to permit gas exchange sufficient to sustain life. The respiratory distress syndrome (RDS), occurs most often when insufficient lung surfactant is present, an association between lung surfactant and the resiratory distress syndrome first discovered by Avery and Mead.[79]

A number of laboratory observations support the hypothesis that the respiratory distress syndrome is the result of pulmonary surfactant deficiency.[71] Very little evidence argues against the primacy of diminished surfactant in the respiratory distress syndrome. It is reasonable to assert therefore that in most cases the respiratory distress syndrome occurs as a consequence of the relative inability to synthesize or secrete surfactant in amounts sufficient for normal neonatal respiration.

Following methods generally of the type originally proposed by Gluck,[80] L/S values of 2 or greater have been found to correlate with fetal lung maturity. Before 34 weeks of gestation, lecithin and sphingomyelin are present in amniotic fluid in approximately equal amounts (Fig. 35-9), but at about 34 weeks, the concentration of lecithin begins to rise. It was possible to demonstrate (reference 81, for example) that when the concentration of lecithin in amniotic fluid became at least twice that of the sphingomyelin, the likelihood of respiratory distress after delivery was minimal. However, because of a greater risk of the respiratory distress syndrome associated with the diabetic pregnancy,[82,83] values of 2.5 or greater have often been used. In addition, because the L/S method is less successful in predicting fetal lung maturity in diabetic pregnancies, tests for other surface-active lipids such as phosphatidyl glycerol (PG) have been developed and used either with the L/S ratio, or solely as independent tests.

Therapy for newborns who have the respiratory distress syndrome is basically supportive.[68] Thermoregulation and assisted ventilation are two primary measures. Because of inadequate ventilation, acid-base disturbances are determined by pH, P_{CO_2}, bicarbonate, and changes in buffer base. It is also customary to maintain a venous hematocrit of 40% to 45% during the acute phase of the respiratory distress syndrome to support an adequate oxygen-carrying capacity. The administration of aerosolized phospholipids to infants with the respiratory distress syndrome has had only limited therapeutic success, probably because of the problems associated with duplicating the composition of natural surfactant.

Fetal hemolytic disorders (Rh problems)

Hepatic excretory capacity does not become fully mature until nearly 4 weeks post partum in full-term human infants. The hepatic processing of bilirubin therefore falls short of maximum before that time. When uptake and conjugation of bilirubin are forced to operate at rates exceeding the capacity of the liver to excrete the quantity formed, as in infants with severe hemolytic disease, conjugated bilirubin accumulates in liver and serum.

Erythroblastosis fetalis, caused by the incompatibility of an Rh-negative mother and an Rh-positive father, is a common cause of rapid red blood cell destruction. If the fetus is also Rh-positive, fetal cells entering the maternal circulation may elicit an antibody response to the Rh blood factor. The antibodies (IgG) cross the placenta into the fetus where they destroy fetal red cells.

The fetus normally generates approximately 35 mg of bilirubin from the catabolism of 1 g of hemoglobin. The high maternal to fetal plasma protein gradient facilitates rapid transplacental extraction of unconjugated fetal bilirubin and at the same time suppresses glucuronide conjugation by the fetal liver. Bilirubin so transferred is conjugated and excreted by the mother.[84] This mechanism is so efficient that it is uncommon for the neonate to have an elevated cord blood bilirubin level. However, in severe erythroblastosis, particularly if coupled with placental deterioration, unconjugated and blood bilirubin levels can run as high as 80 mg/L. Fetuses receiving intrauterine transfusions are often born with high levels of conjugated bilirubin, probably arising from stimulation of fetal glucuronide formation coupled with decreased placental permeability to the bilirubin glucuronide.

Also unique to the newborn and related to developmental immaturity is the tissue toxicity of unconjugated bilirubin, especially to the brain. In the adult, serum bilirubin elevations are viewed as an important clinical or laboratory sign of disease or altered physiological state. In the neonate, hyperbilirubinemia has a dual significance, as a clinical sign and also as a toxin.

Conjugated bilirubin is highly water soluble and is therefore readily excreted in fluids. Unconjugated or indirect bilirubin, on the other hand, is insoluble in aqueous solution but highly soluble in lipids. Under normal circumstances, unconjugated bilirubin is bound to plasma albumin, and this binding prevents the entrance of free or unbound indirect bilirubin into the lipid-rich central nervous system. It is when the albumin binding capacity is exceeded that unbound, unconjugated bilirubin readily passes into the central nervous system cells. Unconjugated bilirubin is toxic to the central nervous system and causes necrosis, a pathological process referred to as "kernicterus." Surviving infants may have mental retardation, hearing deficits, or cerebral palsy. Many affected infants, particularly those of low birth weight, may have no neonatal symptoms but later in childhood can develop hearing deficits, perceptual handicaps, and hyperkinesis.

Usually there is no detectable bilirubin in amniotic fluid when fetus and mother are normal.[85] However, a neonate who demonstrates significant elevations of unconjugated bilirubin in serum frequently also passes bilirubin into its amniotic fluid. The route by which bilirubin is transferred into amniotic fluid from the fetus is unclear.

Maternal diabetes

There is an increased association of intrauterine deaths, congenital malformations, and perinatal mortality and morbidity in fetuses of diabetic women. For this reason, the pregnant diabetic woman is monitored closely throughout the course of her pregnancy.

The fetal pancreas does not ameliorate maternal diabetes, since insulin does not cross the placenta. Exogenous insulin therapy and diet are necessary therefore for management of the mother's insulin-deficient state.

Maternal metabolism adapts to provide adequate nutrition for the mother and the growing fetoplacental unit. At term the human fetus requires approximately 30 g of glucose per day.[86] The fetus not only draws heavily on maternal glucose, but lipids also become an important metabolic fuel. Hyperlipidemia occurs with pregnancy, and human placental lactogen (HPL) stimulates lipolysis in adipose tissue. The fatty acids released lead to a reduction in glucose utilization acting to "spare" glucose for the fetus. HPL levels change with alterations in maternal food intake, rising when the mother is fasted but falling when adequate glucose is available.[87] When fasted, the pregnant woman rapidly develops hypoglycemia, hypoinsulinemia, and ketosis.[88] Maternal mechanisms to counter these effects include increased protein catabolism and accelerated renal gluconeogenesis.[89]

The type 2 diabetic state is characterized by an exaggerated rate and amount of insulin release associated with an apparent decrease in sensitivity to insulin at the cellular level. In addition, the placenta produces an enzyme capable of destroying insulin.[90,91] As placental size increases, larger amounts of HPL and other factors are present to modify or oppose the action of insulin. Also maternal pancreatic β-cell hyperplasia frequently occurs with pregnancy. Thus, normal insulin-glucose homeostasis is maintained in the face of increased glucose demand. When the homeostatic system goes awry, elevated maternal blood glucose results. Mild maternal diabetes usually disappears after cessation of pregnancy.

The placenta not only produces hormones that regulate maternal metabolism, but also controls the transport of nutrients and protein hormones to the fetal compartment. The placenta is essentially impermeable to protein hormones such as glucagon, growth hormone, and HPL and, as has been mentioned, insulin.[92,93] Maternal glucose reaches the fetus by "facilitated diffusion," crossing the placenta at a faster rate than would be expected on physiochemical grounds alone.[94] Fetal blood glucose levels usually remain 20% to 30% lower than those in the maternal compartment. Ketone bodies diffuse freely across the placenta and may be an important fetal fuel during maternal fasting.

Hyperglycemia of the diabetic mother increases the blood glucose of the fetus because of the facility with which glucose crosses the placenta. The increased levels of glucose in the fetus elicit the expected insulin response so that increased deposition of fat, protein, and glycogen lead to birthweights over 4000 g in 25% of the pregnancies complicated by diabetes. Another unfavorable complication of this fetal hyperglycemic-hyperinsulinemic state occurs at delivery when the glucose source is removed. The excess insulin rapidly decreases the glucose available, and so the newborn becomes hypoglycemic. Hypoglycemic crisis is frequently observed in untreated infants of diabetic mothers. Exogenous administration of glucose to the newborn is needed until proper glucose-insulin balance can be achieved.

A significant increase (five or six times) in incidence of the respiratory distress syndrome (RDS) is not unusual in pregnancies complicated by diabetes. In the not-too-distant past, before surfactant assessment was a common practice, early delivery was usually planned for diabetic women based primarily on gestational age. Since clinical indications for fetal size and gestational age are often misleading in the diabetic pregnancy, a large but premature infant who developed respiratory distress syndrome and died was not infrequent. Amniocentesis has greatly aided in the management of the pregnant diabetic, but some unanswered questions still exist. The reason why surfactant production is retarded in the fetus from a diabetic pregnancy is not fully known but may be related to the antagonistic effect of insulin on cortisol stimulation of surfactant production on the developing fetal lung.[95]

Toxemia (hypertensive disease of pregnancy)

Toxemia of pregnancy is characterized by a triad of symptoms including hypertension (blood pressure over 140/90), edema, and proteinuria. It is subdivided in classification to preeclampsia and eclampsia. Eclampsia is characterized by convulsions and is believed to be the sequela to the preeclamptic state. A large but as yet incomplete literature about this fascinating condition has developed over many years (for example, reference 96).

Much current thinking supports the contention that the precipitating cause for pregnancy hypertension is a compromised uteroplacental blood flow. The specific cause, however, remains in the realm of speculation but includes immunological processes, dietary considerations, and altered vascular reactivity. Management is directed toward bed rest, dietary restriction of salt, and under hospitalization the use of magnesium sulfate to blunt exaggerated neuronal reflex activities. The only satisfactory "cure" is delivery, and for this reason information about the pulmonary status of the fetus is of considerable importance. The frequent association of toxemia with diabetes and with vascular disease provides a basis for the correlation with SGA fetuses in this condition. Routine tests for creatinine and protein are the most common laboratory means for monitoring the toxemic pregnancy.

Spina bifida and anencephaly

Spina bifida and anencephaly are two of the common neural tube defects that constitute a large portion of the serious congenital malformations in man. In spina bifida there is a midline defect of the spine that results in a protrusion of the meninges or spinal cord or other neural elements. In anencephaly the brain is a disorganized mass of neural tissues and the cranial vault is absent.

Prenatal diagnosis of neural tube defects by alpha-fetoprotein from amniotic fluid has become routine in recent years. In addition, anencephaly specifically affects estriol formation. The absence of fetal pituitary function and hence fetal ACTH results in very low rates of production of DHEA (dehydroepiandrosterone) from the fetal adrenal. Since this androgen is a precursor to estriol, estriol levels are characteristically low.

A comprehensive discussion of the neural tube defects can be found in the monograph by Kirkpatrick and Nakamura.[97]

Tubal pregnancy

Fertilization occurs in the fallopian tubes. Under usual circumstances, the fertilized egg migrates down the tube, enters the uterus, and becomes implanted. Occasionally, this migration does not occur, and implantation takes place in the tube itself. This "out-of-place" pregnancy is an "ectopic pregnancy," and when implantation occurs in the tube, it is called a "tubal pregnancy." Reasons for this occurrence can include endocrine imbalances, since tubal motility is dramatically influenced by estrogen and progesterone, and can be from residual effects stemming from infections or other causes resulting in a narrowing of the tubal lumen.

Clinical symptoms include lower abdominal pain and vaginal bleeding. Amenorrhea is not a characteristic feature. Tubal pregnancies before rupture of the fallopian tube usually give a positive pregnancy test. However, since compromised placentas, that is, those in the stage of abruption, degeneration, or penetration of the tubal wall, cannot produce chorionic gonadotropin in usual quantities, the tests may turn negative and become misleading. The presence of fetal erythrocytes in the maternal circulation might assist in the diagnosis of cases where the usual tests fail to establish the presence of a pregnancy.[98]

CHANGE OF ANALYTE IN DISEASE

Over the past two decades significant advances elucidating the course of growth and development of the fetus have been made. Part of this new-found knowledge has been possible through the development and application of improved analytical techniques, such as radioimmunoassay, and the improved safety of amniocentesis through ultrasound. Application of much of this knowledge has dramatically altered the course of management of problem, or "high-risk," pregnancies (see references 99 and 110 for a

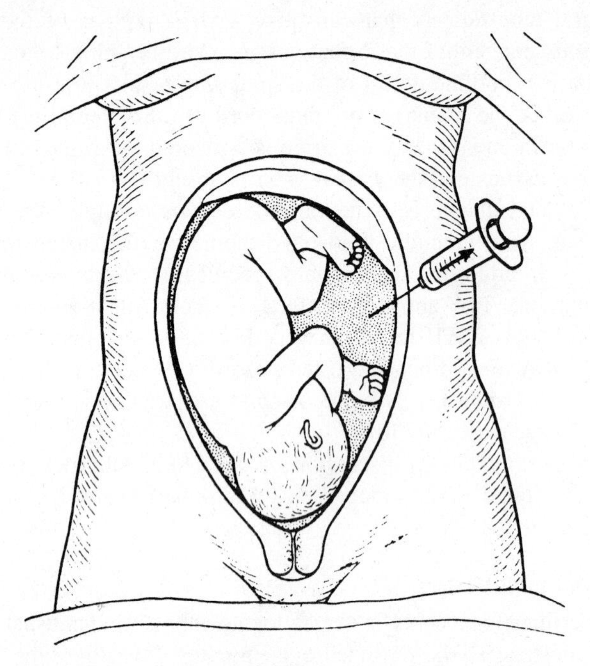

Fig. 35-14. Diagram demonstrating amniocentesis. (From Queenan, J.T.: Clin. Obstet. Gynecol. **9**:491-507, 1966.)

discussion of these techniques). Fig. 35-14 presents a representation of the technique of amniocentesis, and Fig. 35-15 indicates the multifaceted array of its applications.

Three clinical problem areas have been the primary beneficiaries of amniocentesis: (1) the management of Rh-antigen incompatibility of mother and fetus, (2) the identification of the earliest possible time in pregnancy that delivery can be performed with minimal risk of prematurity, and (3) the identification of potential genetic disorders. The last subject area is discussed in Chapter 44.

Human chorionic gonadotropin (HCG)

Chorionic gonadotropin is secreted by the trophoblast, and consequently is used to identify and follow the course of pregnancy and trophoblastic disease. In a normal pregnancy, urinary levels of HCG rise to a range of 20,000 to 100,000 U/day and then decrease to a range of 4,000 to 11,000 U/day later in the pregnancy. Pregnancy can be detected with urinary HCG by about a week after a missed period. By assays of plasma HCG particularly with the sensitive and specific methods such as those for the HCH β-chains, pregnancy can be detected as early as a few days after conception. In cases of hydatidiform mole, urinary HCG titers rise to over 300,000 U/day. After molar evacuation these values drop within 1 month, and in about 90% of cases HCG is not detectable by urinary assay after 3 months. In cases where trophoblastic tissue remains, such

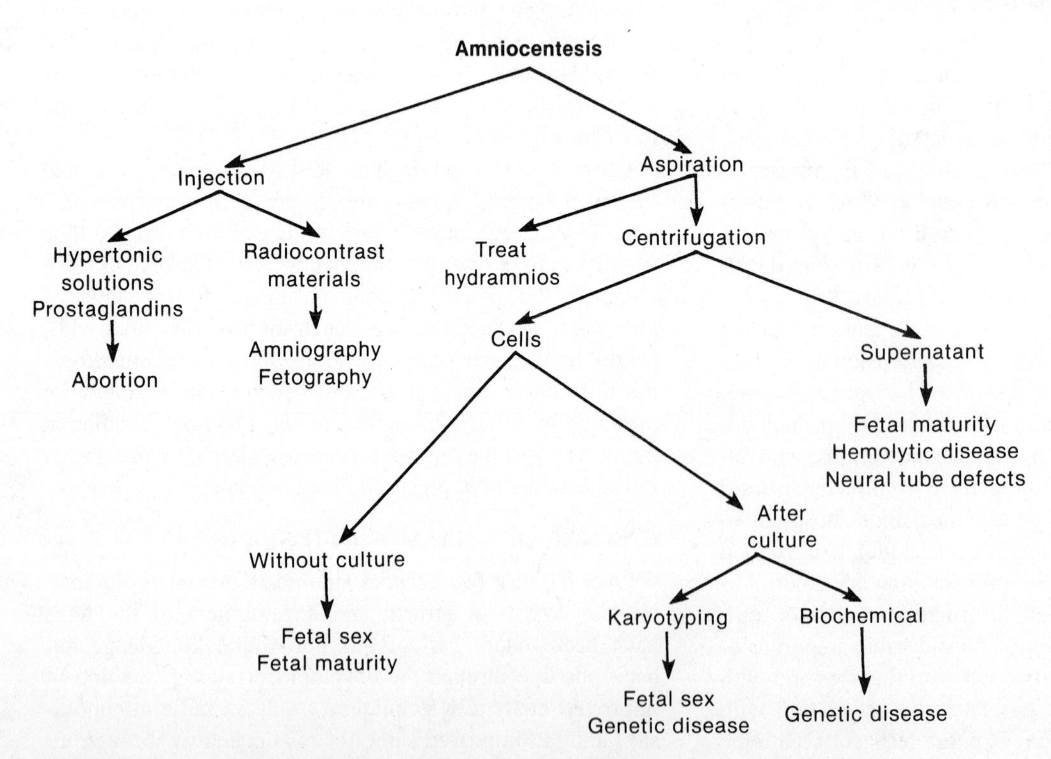

Fig. 35-15. Clinical applications of amniocentesis. (From Pritchard, J.A., and MacDonald, P.C.: In Pritchard, J.A., and MacDonald, P.C., editors: Williams obstetrics and gynecology, ed. 15, New York, 1976, Appleton-Century-Crofts.)

as retained choriocarcinoma, values remain elevated, and serial assays are of great value in determining the results of treatment, usually chemotherapy. One can also use blood serum assays by monitoring HCG or its β-subunit.

On a molecular basis, HCG shows about 1/4000 the thyrotropic activity of pituitary TSH.[103] If HCG levels are very high, thyroid-stimulating activity is possible. For this reason, the levels of HCG attained in molar pregnancies are believed to be the reason hyperthyroidism is often associated with molar pregnancies.[103-105] HCG values exceeding 100,000 U/day in urine or 300 U/mL in serum are in the range where the association of hyperthyroidism and hydatidiform mole should be suspected.[106]

Chorionic gonadotropin levels in amniotic fluid reflect those in sera, with a peak at 11 to 14 weeks and a sharp decline to term. At term, HCG levels are greater in mothers with female infants.[107] Women with a diagnosis of threatened abortion who have low HCG levels for the estimated time of gestation have been shown to go on to complete abortion. Human chorionic gonadotropin concentrations in pregnancies complicated with diabetes, toxemia and Rh-isoimmunization have been studied with variable findings. HCG assays in these conditions are not helpful clinically.

Chorionic gonadotropin levels are often usually low for the gestational age with ectopic, including tubal, pregnancies. Pregnancy tests have been found as positive in as few as 50% of ectopic pregnancies. A negative test therefore does not exclude ectopic pregnancy. Practical benefit may come from the use of the β-subunit radioimmunoassay for HCG because this assay is capable of detecting as little as 15 milliunits/mL or less of serum from women with ectopic pregnancies.[108]

Placental lactogen

Human placental lactogen (HPL) is used to measure placental (trophoblast) function (Fig. 35-13). It is not entirely satisfactory as a clinical measurement because there is a wide range of normal values as gestation advances. HPL is not used to follow metastatic trophoblastic disease because there is little HPL in comparison to the larger concentrations of HCG. There are, however, certain instances where HPL may contribute to diagnosis or status of the fetus. Spellacy and co-workers[109] have extensively investigated the use of HPL for identifying the fetus in jeopardy and found that values were less than 4 μg/mL after 39 weeks of gestation in pregnancies in which the fetus subsequently died. HPL values almost always exceeded 4 μg/mL after 30 weeks of gestation in normal pregnancies. The correlation beween low HPL and fetal death was especially strong with severe maternal hypertension.

Estriol

Low estriol values or, more importantly, declining trends carry unfavorable prognostic significance[110] (see Fig. 35-5, showing serial estriol values during various pregnancies). As a rough guideline, a decline of 30% to 50% from the mean of 3 previous days indicates probably impending danger to the fetus.[111] Because of the pronounced diurnal variation in estrogen formation (particularly excretion), time of day for sampling is an important consideration. Since the precursors of estriol are androgens produced by the fetal adrenal, drugs, such as synthetic corticoids, that can cross the placenta and suppress ACTH secretion depress estriol levels in both blood and urine.

Serum or urinary estriol levels that are greater than the 95th percentile should suggest the possibility of twins. Estriol values are good predictors of impending fetal death in hypertensive disease, renal disease, and diabetes, particularly when there is a vascular component. Conditions when chronically low values are seen include toxemia, anencephaly, and placental sulfatase deficiency. Estriols are not helpful in monitoring erythroblastosis fetalis.

Because the final step in estrogen formation is the aromatization of fetal androgens, estriol has been mentioned as a useful indicator of "placental function." Over the years, monitoring has shown pregnancies that were associated with very small placentas, progesterone values about 10% of normal, but estriol values in the normal range. I believe that estriol production is related more to fetal than to placental factors. Progesterone, pregnanediol, or human placental lactogen would be better indicators of placental function than estriol is.

Phospholipids

Lecithin (phosphatidyl choline) is the most abundant of the phospholipids identified in amniotic fluid. Even though fetal maturity can be assessed by determination of the concentration of phospholipids in amniotic fluid,[112,113] the use of the ratio of lecithin to sphingomyelin (L/S) has become the more popular method. This approach was originally proposed by Gluck et al.[65] and offers the advantages that the L/S ratio is independent of the volume of amniotic fluid (found to vary widely even within acceptable limits of what are considered to be "normal pregnancies") and the ratio accommodates less than quantitative extractions of the phospholipids from the fluid, as long as the losses are comparable for both lecithin and sphingomyelin.

In a composite compilation of experience with the L/S of 2 or greater, the risk of respiratory distress was slight unless the mother had diabetes. With diabetes, the risk of respiratory distress and in turn death from respiratory distress was much greater even though the L/S was greater than 2. If the L/S ratio was 1.5 to 2, respiratory distress was found in 40%, and, if below 1.5, in 73%. Although 73% of infants did develop respiratory distress when the L/S ratio was below 1.5, it proved fatal in only 14% (Table 35-5). At times, obviously the risk to the fetus from a hostile environment will be greater than the risk of death from respiratory distress.

Table 35-5. Relationship of lecithin-sphingomyelin ratio to development of respiratory distress

L/S ratio	Number of infants	Respiratory distress (%)	
		All cases	Deaths
<1.5	162	73	14
1.5-2.0	223	40	4
>2.0	1596	2.2	0.1
>2.5	543	0.9	0

From Harvey, D., Parkinson, C.E., and Campbell, S.: Lancet **1**:42, 1975.

Table 35-6. Fetal lung maturation in relation to disorders

Accelerated fetal lung maturation with:	Delayed fetal lung maturation with:
Maternal hypertension	Diabetes mellitus
Renal	Classes A, B, C
Cardiovascular	Hydrops fetalis
Severe, "chronic" toxemia	
Hemoglobinopathy—sickle cell disease	Infections
	Chorioamniotitis
Narcotics addiction (heroin, morphine)	Fetal infections, viral
	Placental conditions
Diabetes mellitus, classes D, E, F	Chronic abruptio placentae
	Prolonged rupture of membranes

From Gluck, L.: In Lung maturation and the prevention of hyaline membrane disease, Proceedings of the 7th Ross Conference on Pediatric Research. Columbus, Ohio, 1976, Ross Laboratories.

Table 35-7. Phosphatidyl glycerol (PG) with L/S ratio less than 2

	Respiratory distress syndrome	
	No RDS	RDS
PG present	91.6%	8.4%
PG absent	20.0%	80.0%

From Bent, A.E., Gray, J.H., Luther, E.R., et al.: Am. J. Obstet. Gynecol. **139**:259-263, 1981.

Gluck and Kulovich[65] and others have reported that in certain pathological pregnancy conditions pulmonary maturation appears to be accelerated, whereas in others it is delayed (Table 35-6). Diseases in which fetal lung maturation may be delayed include diabetes mellitus and hemolytic disease in the fetus. Maternal hypertension and premature rupture of the membranes with delayed delivery have been reported to hasten maturation through surfactant production by the fetal lung.[65,116,117] Observations of accelerated lung maturation have not been universally noted, however.[118]

False mature L/S values have been reported after intrauterine transfusions,[119] and blood-contaminated amniotic fluid may give falsely lowered values.[120] Also the respiratory distress syndrome can develop in an asphyxiated neonate despite a mature L/S value.[81]

Even though phosphatidyl glycerol (PG) represents only a minor fraction of the phospholipids of surfactant (Table 35-3), it contributes significantly to the functional properties of surfactant. Phosphatidyl glycerol, isolated from amniotic fluid, can carry prognostic value for fetal lung maturity. In Gluck's early report with PG,[121] PG becomes measurable in extracts of amniotic fluid when the L/S ratio exceeds 2. Functional lung maturity is associated with measurable quantities of PG,[122-124] but the absence of PG does not necessarily mean that the respiratory distress syndrome is inevitable (Table 35-7). Since PG appears to be but a very small constituent of blood,[122] the measurement of PG is especially valuable at times when fetal lung status must be predicted from blood-contaminated amniotic fluid samples. The measurement of PG is considered important also when one is evaluating the maturity of fetuses of diabetic mothers in whom the L/S ratio may be less reliable.

Surface tension

In an attempt to reduce the time and effort for obtaining the L/S ratio, the foam-stability test, or so-called shake test, was introduced by Clements and associates.[125] The foam-stability test depends upon the ability of surfactant to generate stable bubbles in the presence of ethanol. Compilation of experience reported from the use of this test shows that when the test was positive within 72 hours of parturition, less than 1% of infants developed respiratory distress syndrome. About 50% of infants with a negative test developed the respiratory distress syndrome.[126] Particular care must be taken to ensure that glassware for the test is clean and the reagents and amniotic fluid samples are not contaminated.

Cortisol

Amniotic fluid levels of cortisol are low in pregnancies with an anencephalic fetus and associated with diabetes. Amniotic fluid levels are high in infants with Rh incompatibility, infants in the postmaturity state or during premature labor, and stressed infants of toxemic mothers.[72,127] Since there is wide variation in amniotic fluid cortisol levels, in part because of dilutional effects, individual values do not necessarily reflect fetal gestational age, weight, or pulmonary maturation. Hence amniotic fluid cortisol is not useful alone as a reliable indicator of fetal maturity or well-being.[128-130]

Biliburin (ΔA_{450})

A maternal antibody titer (indirect Coombs' test) of 1 to 16 or more in most cases warrants amniocentesis and appropriately timed measurements of bilirubin pigment in

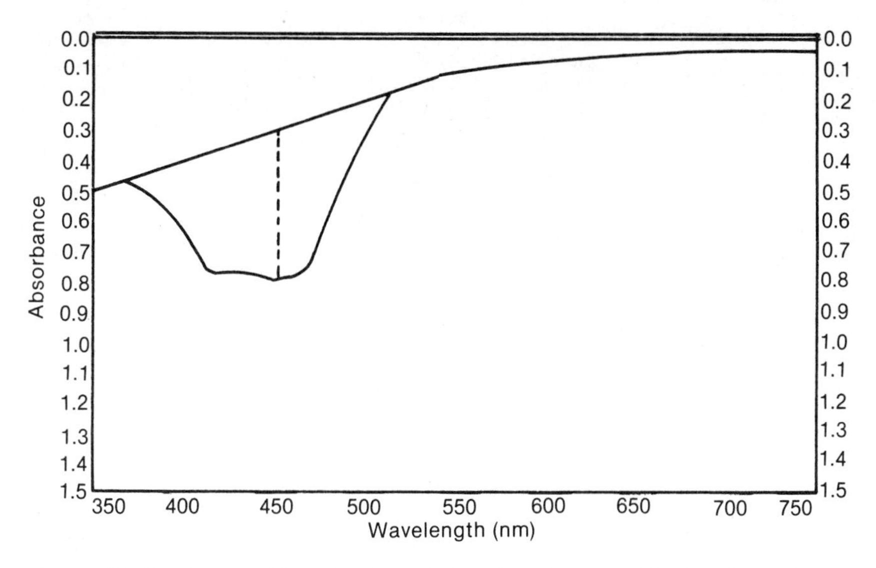

Fig. 35-16. Spectrum of bilirubin. *Dashed line,* Absorbance at 450 nm. (From Queenan, J.T.: Clin. Obstet. Gynecol. **14:**505-536, 1971.)

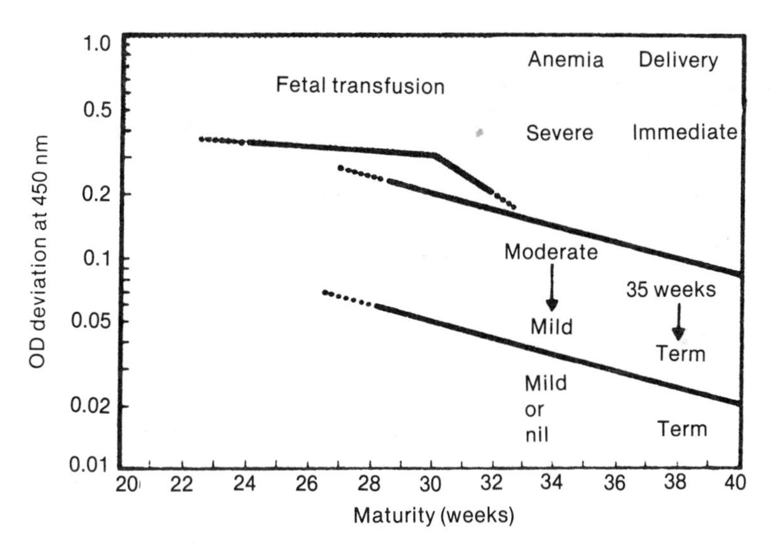

Fig. 35-17. Relationship of absorbance at 450 nm, gestational age of amniotic fluid associated with fetal anemia, and suggested clinical management. *O.D.*, Optical density (absorbance). (From Liley, A.W.: Am. J. Obstet. Gynecol. **86:**485-494, 1963.)

amniotic fluid. The absorbance of bilirubin, when measured in a continuously recording spectrophotometer, is demonstratable as a "hump" or inflection with maximum absorbance at 450 nm (Fig. 35-16). The magnitude of the increase in optical density above the base line value (ΔOD_{450}) usually but not always correlates well, for any gestational age, with the intensity of the hemolytic disease.

Liley[132] has constructed a graph that reasonably predicts the severity of the disease and recommends clinical management (Fig. 35-17). The higher the ΔOD_{450}, the more severe is the hemolytic disease, according to the gestational age of the pregnancy. In general, a decreasing amniotic fluid biluribin trend indicates a fetus that will survive, and a horizontal or rising biluribin indicates that there is severe erythroblastosis fetalis.

Amniotic fluid analyses, when properly used, are essentially totally effective in detecting the fetus that will die in utero. The main source of error is the fetus with polyhydramnios because it can cause a false low-bilirubin determination.

The majority of erythroblastotic neonates do not have significant cord blood bilirubin elevations; however, in severe disease, levels of 45 to 60 mg/L and occasionally as high as 80 mg/L are seen.[133] These levels are important in deciding whether to use immediate or delayed exchange transfusions. In neonates receiving intrauterine transfusions, it is not uncommon to see significantly elevated levels of direct bilirubin.[134]

Glucose

After delivery, the glucose concentration in offspring of diabetic mothers declines rapidly below that observed in normals. Approximately 60% of babies from insulin-dependent mothers have glucose concentrations below 300 mg/L in the first 6 hours of life.[135,136]

Renal function tests

During normal pregnancy, renal blood flow and the glomerular filtration rate are significantly increased above nonpregnant levels. With the development of pregnancy-induced hypertension, renal perfusion and glomerular filtration are reduced. Most often, therefore, the creatinine or urea concentration in plasma is not appreciably elevated. The plasma uric acid concentration is much more commonly elevated, especially in women with more severe disease. The elevation is a result primarily of decreased renal clearance of uric acid by the kidney, a decrease that exceeds the reduction in glomerular filtration rate and creatinine clearance.[137] In the experience of some, however, measurement of plasma uric acid levels are of little value for diagnosis or prognosis.[137]

Alpha-fetoprotein

When anencephaly or meningomyelocele has occurred in a preceding pregnancy, the risk of recurrence of either defect is approximately 5%. This is detected when one finds elevated alpha-fetoprotein in the amniotic fluid. If the alpha-fetoprotein is not elevated, anencephaly can be reliably excluded, but only about 90% of the pregnancies with meningomyelocele give positive results.[97]

REFERENCES

1. Meschia, G., Breathnach, C.S., Cotter, J.R., Hellegers, A., Barron, D.H.: The diffusibility of urea across the sheep placenta in the last two months of gestation, Q. J. Exp. Physiol. **50:**23-41, 1965.
2. Longo, L.D.: Disorders of placental transfer. In Assali, N.S., editor: Pathophysiology of gestation, vol. 2, New York, 1972, Academic Press, Inc., pp. 1-76.
3. Ryan, K.J.: Hormones and the placenta, Am. J. Obstet. Gynecol. **84**(Part 2):1695-1713, 1962.
4. Lind, T., Kendall, A., and Hytten, F.E.: The role of the fetus in the formation of amniotic fluid, J. Obstet. Gynecol. Br. Commonw. **79:**289-298, 1972.
5. Bain, A.D., and Scott, J.S.: Renal agenesis and severe urinary tract dysplasia: a review of 50 cases with particular reference to the associated anomalies, Br. Med. J. **1:**841-846, 1960.
6. Mauer, S.M., Dobrin, R.S., and Vernier, R.L.: Unilateral and bilateral renal agenesis in monoamniotic twins, J. Pediatr. **84:**236-238, 1974.
7. Duenhoelter, J.H., and Pritchard, J.A.: Fetal respiration: quantitative measurements of amniotic fluid inspired near term by human and rhesus fetuses, Am. J. Obstet. Gynecol. **125:**306-309, 1976.
8. Seeds, A.E.: Current concepts of amniotic fluid dynamics, Am. J. Obstet. Gynecol. **138:**575-586, 1980.
9. Sandler, M., editor: Amniotic fluid and its clinical significance, New York, 1981, Marcel Dekker, Inc.
10. Fairweather, D.V.I., and Eskes, T.K.A.B., editors: Amniotic fluid: Research and clinical application, Amsterdam, 1978, Excerpta Medica.
11. Liley, A.W.: Disorders of amniotic fluid. In Assali, N.S., editor: Pathophysiology of gestation, vol. 2, New York, 1972, Academic Press, Inc., pp. 157-206.
12. Hutchinson, D.L., Gray, M.J., Plentl, A.A., et al.: The role of the fetus in the water exchange of the amniotic fluid of normal and hydramniotic patients. J. Clin. Invest. **38:**971-980, 1959.
13. Lind, T., Billewicz, W.Z., and Cheyne, G.A.: Composition of amniotic fluid and maternal blood through pregnancy, J. Obstet. Gynecol. Br. Commonw. **78:**505-512, 1971.
14. McLain, C.R., and Russell, P.T.: Amniography studies of chronic fetal distress in hypertensive disorders of pregnancy, Am. J. Obstet. Gynecol. **107:**673-681, 1970.
15. Pritchard, J.A.: Deglutition by normal and anencephalic fetuses, Obstet. Gynecol. **25:**289-297, 1965.
16. Rosa, P.: Etude de la circulation du liquide amniotique humain, Gynecol. Obstet. **50:**463-476, 1951.
17. Goldstein, D.P., Aon, T., Taymor, M.T., et al.: Radioimmunoassay of serum chorionic gonadotropin activity in normal pregnancy, Am. J. Obstet. Gynecol. **102:**110-114, 1968.
18. Pierce, J.G., and Parsons, T.F.: Glycoprotein hormones: structure and function, Annu. Rev. Biochem. **50:**465-495, 1980.
19. Catt, K.J., and Pierce, J.G.: Gonadotropic hormones of the adenohypophysis (FSH, LH, and prolactin). In Yen, S.S.C., and Jaffe, R.B., editors: Reproductive endocrinology, Philadelphia, 1978, W.B. Saunders Co., pp. 34-62.
20. Diczfalusy, E.: Endocrine functions of the human fetus and placenta, Am. J. Obstet. Gynecol. **119:**419-433, 1974.
21. Eberlein, W.R.: The fetal adrenal cortex. In Christy, N.P., editor: The human adrenal cortex, New York, 1971, Harper & Row, Publishers, pp. 317-327.
22. France, J.T., Seddon, R.J., and Liggins, G.C.: A study of a pregnancy with low estrogen production due to placental sulfatase deficiency, J. Clin. Endocrinol. Metab. **36:**1-9, 1973.
23. Goebelsmann, U.: The uses of oestriol as a monitoring tool, Clin. Obstet. Gynecol. **6:**223-241, 1979.
24. Schindler, A.E., and Siiteri, P.K.: Isolation and quantitation of steroids from normal human amniotic fluid, J. Clin. Endocrinol. Metab. **28:**1189-1198, 1968.
25. Resnick, R., Killam, A.P., Battaglia, F.C., Makowski, E.L., and Meschia, G.: The stimulation of uterine blood flow by various estrogens, Endocrinology **94:**1192-1196, 1974.
26. Kundu, N., Carmody, P.J., Didolkar, S.M., and Petersen, L.P.: Sequential determination of serum human placental lactogen, estriol, and estetrol for assessment of fetal morbidity, Obstet. Gynecol. **52:**513-520, 1978.
27. Notation, A.D., and Tagatz, G.E.: Unconjugated estriol and 15α-hydroxyestriol in complicated pregnancies, Am. J. Obstet. Gynecol. **128:**747-756, 1977.
28. Mestman, J.H.: Thyroid and parathyroid diseases in pregnancy. In Quilligan, E.J., and Kretchmer, N., editors: Fetal and maternal medicine, New York, 1980, John Wiley & Sons, Inc., pp. 489-531.
29. Oppenheimer, J.H.: Role of plasma proteins in the binding, distribution and metabolism of the thyroid hormones, N. Engl. J. Med. **278:**1153-1162, 1968.
30. Dowling, J.T., Appleton, W.G., and Nicoloff, J.T.: Thyroxine turnover during human pregnancy, J. Clin. Endocrinol. Metab. **27:**1749-1750, 1967.
31. Newman, R.L.: Serum electrolytes in pregnancy, paturition and puerperium, Obstet. Gynecol. **10:**51-55, 1957.
32. Robertson, E.G.: Increased electrolyte fragility in association with osmotic changes in pregnancy serum, J. Reprod. Fertil. **16:**323-325, 1968.
33. Robertson, E.G., and Cheyne, G.A.: Plasma biochemistry in relation to oedema of pregnancy, J. Obstet. Gynecol. Br. Commonw. **79:**769-776, 1972.
34. DeAlvarez, R.R., Gaiser, D.F., Simkins, D.N., Smith, E.K., and Bratvold, G.E.: Serial studies of serum lipids in normal human pregnancy, Am. J. Obstet. Gynecol. **77:**743-759, 1959.

35. Peters, J.P., Heineman, M., and Man, E.B.: The lipids of the serum in pregnancy, J. Clin. Invest. **30**:388-394, 1951.
36. MacDonald, H.N., and Good, W.: Changes in plasma total protein, albumin, urea and α-amino nitrogen concentrations in pregnancy and the puerperium, J. Obstet. Gynecol. Br. Commonw. **78**:912-917, 1971.
37. Studd, J.W., and Wood, S.: Serum and urinary proteins in pregnancy. In Wynn, R.M., editor: Obstetrics and gynecology annual, New York, 1976, Appleton-Century-Crofts, pp. 103-111.
38. Todd, M.E., Thompson, J.H., Bowie, E.J.W., and Owen, C.A.: Changes in blood coagulation during pregnancy, Mayo Clin. Proc. **40**:370-383, 1965.
39. Nossel, H.L., Lanzkowsky, P., Levy, S., Mibashan, R.S., and Hansen, J.D.L.: A study of coagulation factor levels in women during labour and in their newborn infants, Thrombosis et Diathesis Haemorrhagica **16**:185-197, 1966.
40. Talbert, L.M., and Langdell, R.D.: Normal values of certain factors in the blood clotting mechanism in pregnancy, Am. J. Obstet. Gynecol. **90**:44-50, 1964.
41. Ratnoff, O.D., Colopy, J.E., and Pritchard, J.A.: The blood clotting mechanism during normal parturition, J. Lab. Clin. Med. **44**:408-415, 1954.
42. Ingerslev, M., and Teilum, G.: Biopsy studies on the liver in pregnancy. II. Liver biopsy on normal pregnant women, Acta Obstet. Gynecol. Scand. **25**:352-359, 1946.
43. Levy, H.L., and Montag, P.P.: Free amino acids in human amniotic fluid: a quantitative study by ion-exchange chromatography, Pediatr. Res. **3**:113-120, 1969.
44. Dallaire, L., Potier, M., Melancon, S.B., and Patrick, J.: Fetomaternal amino acid metabolism, J. Obstet. Gynecol. Br. Commonw. **81**:761-767, 1974.
45. Schulman, J.D., Queenan, J.T., and Doores, L.: Gas chromatographic analysis of concentrations of amino acids in amniotic fluid from early, middle and late periods of human gestation, Am. J. Obstet. Gynecol. **114**:243-249, 1972.
46. Scott, R.C., Teng, C.C., Sagerson, R.N., and Nelson, T.: Amino acids in amniotic fluid: changes in concentrations during the first half of pregnancy, Pediatr. Res. **6**:659-663, 1972.
47. Gresham, E.L., Simons, P.S., and Battaglia, F.C.: Maternal-fetal urea concentration differences in man, J. Pediatr. **79**:809-811, 1971.
48. Lind, T.: The biochemistry of amniotic fluid. In Fairweather, D.V.I., and Eskes, T.K.A.B., editors: Amniotic fluid: research and clinical application, ed. 2, Amsterdam, 1978, Excerpta Medica, pp. 59-80.
49. Crandall, B.F.: Second trimester maternal serum screening to identify neural tube defects. In Kirkpatrick, A.M., and Nakamura, R.M., editors: Alpha-fetoprotein: laboratory procedures and clinical applications, New York, 1981, Masson Publishing Co., p. 99.
50. Kirkpatrick, A.M., and Nakamura, R.M.: Isolation of alpha-fetoprotein and production of antisera to it. In Kirkpatrick, A.M., and Nakamura, R.M., editors: Alpha-fetoprotein: laboratory procedures and clinical application, New York, 1981, Masson Publishing Co., pp. 5-29.
51. Stave, U.: Liver enzymes. In Stave, U., editor: Perinatal physiology, New York, 1978, Plenum Publishing Co., pp. 499-521.
52. Fine, I.H., Kaplan, N.O., and Kuftinec, D.: Develomental changes of mammalian lactic dehydrogenase, Biochemistry **2**:116-121, 1963.
53. Vesell, E.S., and Philip, J.: Isozymes of lactic dehydrogenase: sequential alterations during development, Ann. N.Y. Acad. Sci. **111**:243-256, 1963.
54. Pitkin, R.M., and Zwirek, S.J.: Amniotic fluid creatinine, Am. J. Obstet. Gynecol. **98**:1135-1139, 1967.
55. Van Geuns, H.J., and Van Kessel, H.: Creatinine in amniotic fluid and fetal renal function. In Fairweather, D.V.I., and Eskes, T.K.A.B., editors: Amniotic fluid: research and clinical application, ed. 2, Amsterdam, 1978, Excerpta Medica, pp. 81-92.
56. Kleinman, L.I.: The kidney. In Stave, U., editor: Perinatal physiology, New York, 1978, Plenum Publishing Co., pp. 589-616.
57. Kerpel-Fronius, E.: Electrolyte and water metabolism. In Stave, U., editor: Perinatal physiology, New York, 1978, Plenum Publishing Co., pp. 565-587.
58. Assali, N.S., DeHaven, J.C., and Barrett, C.T.: Disorders of water electrolyte and acid-base balance. In Assali, N.S., editor: Pathophysiology of gestation, vol. 3, New York, 1972, Academic Press, Inc., pp. 154-231.
59. King, R.J., and Clements, J.A.: Surface active materials from dog lung. II. Composition and physiological correlations, Am. J. Physiol. **223**:715-726, 1972.
60. Body, G.B.: The phospholipid composition of pig lung surfactant, Lipids **6**:625-631, 1971.
61. King, R.J., and Clements, S.A.: Surface active materials from dog lung. I. Method of isolation, Am. J. Physiol. **223**:715-726, 1972.
62. Godinez, R.I., Sanders, R.L., and Longmore, W.J.: Phosphatidylglycerol in rat lung. I. Identification as a metabolically active phospholipid in isolated perfused rat lung, Biochemistry **14**:830-834, 1975.
63. Godinez, R.I., Sanders, R.L., and Longmore, W.J.: Phosphatidylglycerol in rat lung. II. Comparison of occurrence, composition, and metabolism in surfactant and residual fractions, Biochemistry **14**:835-840, 1975.
64. Gikas, E.G., King, R.J., Mescher, E.J., et al.: Radioimmunoassay of pulmonary surface-active material in the tracheal fluid of the fetal lamb, Am. Rev. Resp. Dis. **115**:587-593, 1977.
65. Gluck, L., and Kulovich, M.V.: Lecithin/sphingomyelin ratios in amniotic fluid in normal and abnormal pregnancies, Am. J. Obstet. Gynecol. **115**:539-546, 1973.
66. Blackburn, W.R., Travers, H., and Potter, D.M.: The role of the pituitary-adrenal axes in lung differentiation, Lab. Invest. **26**:306-318, 1972.
67. Farrell, P.M., and Zachman, R.D.: Induction of choline phosphotransferase and lecithin synthesis in the fetal lung by corticosteroids, Science **179**:297-298, 1973.
68. Martin, R.J., Fanaroff, A.A., and Skalina, M.E.L.: The respiratory system. In Fanaroff, A.A., and Martin, R.J., editors: Behrman's neonatal-perinatal medicine, St. Louis, 1983, The C.V. Mosby Co., pp. 432-437.
69. Farrell, P.M., and Avery, M.E.: Hyaline membrane disease, Am. Rev. Resp. Dis. **111**:657-688, 1975.
70. Farrell, P.M., and Kotas, R.V.: The prevention of hyaline membrane disease: new concepts and approaches to therapy, Adv. Pediatr. **23**:213-269, 1976.
71. Farrell, P.M., and Zachman, R.D.: Pulmonary surfactant and the respiratory distress syndrome. In Quilligan, E.J., and Kretchmer, N., editors: Fetal and maternal medicine, New York, 1980, John Wiley & Sons, Inc., pp. 221-242.
72. Murphy, B.E., Patrick, J., and Denton, R.L.: Cortisol in amniotic fluid during human gestation, J. Clin. Endocrol. Metab. **40**:164-167, 1975.
73. Liggins, G.C., and Howie, R.N.: A controlled trial of antepartum glucocorticoid treatment for prevention of the respiratory distress syndrome in premature infants, Pediatrics **50**:515-525, 1972.
74. Kleihauer, E.: The hemoglobins. In Stave, U., editor: Perinatal physiology, New York, 1978, Plenum Publishing Co., pp. 215-239.
75. Hecht, F., Jones, R.T., and Koler, R.D.: Newborn infants with Hb Portland 1, an indicator of α-chain deficiency, Ann. Hum. Genet. **31**:215-218, 1967.
76. Hellegers, A.E., and Schruefer, J.J.P.: Nomograms and empirical equations relating oxygen tension, percentage saturation, and pH in maternal and fetal blood, Am. J. Obstet. Gynecol. **81**:377-384, 1961.
77. Tojo, S., Mochizuki, M., and Kanazawa, S.: Comparative assay of HCG, HCT and HCS in molar pregnancy, Acta Obstet. Gynecol. Scand. **53**:369-373, 1974.
78. Selenkow, H.A., Saxena, S.M., Dana, C.L., and Emerson, K., Jr.: Measurements and pathologic significance of human placental lactogen. In Pecile, A., and Fenzi, P., editors: The foeto-placental unit, Amsterdam, 1969, Excerpta Medica.
79. Avery, M.E., and Mead, J.: Surface properties in relation to atelectasis and hyaline membrane disease, Am. J. Dis. Child **97**:517-523, 1959.
80. Gluck, L., Kulovich, M.V., and Borer, R.C.: Estimates of fetal lung maturity, Clin. Perinatol. **1**:125-139, 1974.
81. Donald, I.R., Freeman, R.K., Goebelsman, U., et al.: Clinical ex-

perience with the amniotic fluid lecithin/sphingomyelin ratio, Am. J. Obstet. Gynecol. **115**:547-552, 1973.
82. Robert, M.F., Neff, R.K., Hubbell, J.P., Taeusch, H.W., and Avery, M.E.: Association between maternal diabetes and the respiratory-distress syndrome in the newborn, N. Eng. J. Med. **294**:357-360, 1976.
83. Myers, J.L., Harrell, M.J.P., and Hill, F.L.: Fetal maturity: biochemical analyses of amniotic fluid, Am. J. Obstet. Gynecol. **121**:961-967, 1975.
84. Davies, J.: Feto-maternal interaction. In Avery, G.B., editor: Neonatology, Philadelphia, 1975, J.B. Lippincott Co., pp. 44-45.
85. Liley, A.W.: The administration of blood transfusions to the foetus in utero, Triangle **7**:184-189, 1966.
86. Crenshaw, C., Jr.: Fetal glucose metabolism, Clin. Obstet. Gynecol. **13**:579-585, 1970.
87. Gaspard, U., Sandront, H., and Luyckx, A.: Glucose-insulin interaction and the modulation of human placental lactogen (HPL) secretion during pregnancy, J. Obstet. Gynecol. Br. Commonw. **81**:201-209, 1974.
88. Felig, P.: Maternal and fetal fuel homeostasis in human pregnancy, Am. J. Clin. Nutr. **26**:998-1005, 1973.
89. Freinkel, N., Metzger, B.E., Nitzan, M., et al.: "Accelerated starvation" and mechanisms for the conservation of maternal nitrogen during pregnancy, Isr. J. Med. Sci. **8**:426-439, 1972.
90. Freinkel, N., and Goodner, C.J.: Carbohydrate metabolism in pregnancy. I. The metabolism of insulin by human placental tissue, J. Clin. Invest. **39**:116-131, 1960.
91. Posner, B.I.: Insulin metabolizing enzyme activities in human placental tissue, Diabetes **22**:552-563, 1973.
92. Adams, P.A.J., King, K.C., Schwartz, R., and Teramo, K.: Human placental barrier to ^{125}I-glucagen early in gestation, J. Clin. Endocrinol. Metab. **34**:772-782, 1972.
93. Kalhan, S.C., Schwartz, R., and Adam, P.A.J.: Placental barrier to human insulin-I^{125} in insulin-dependent diabetic mothers, J. Clin. Endocrinol. Metab. **40**:139-142, 1975.
94. Widdas, W.F.: The inability of diffusion to account for placental glucose transfer in the sheep and consideration of the kinetics of a possible carrier, J. Physiol. (London) **118**:23-39, 1952.
95. Smith, B.T., Giroud, C.J.P., Robert, M., and Avery, M.E.: Insulin antagonism of cortisol action on lecithin synthesis by cultured fetal lung cells, J. Pediatr. **87**:953-955, 1975.
96. Chesley, L.C., editor: Hypertensive disorders in pregnancy, New York, 1978, Appleton-Century-Crofts.
97. Kirkpatrick, A.M., and Nakamura, R.M.: Alpha-fetoprotein: laboratory proceedings and clinical application, New York, 1981, Masson Publishing Co.
98. Wingate, M.B., Iffy, L., Kelly, J.V., and Birnbaum, S.: Diseases specific to pregnancy. In Romney, S.L., Gray, M.J., Little, A.B., et al., editors: Gynecology and obstetrics: the health care of women, New York, 1975, McGraw-Hill Book Co., Inc., p. 721.
99. Bennett, M.J.: Amniocentesis. In Sandler, M., editor: Amniotic fluid and its clinical significance, New York, 1981, Marcel Dekker, Inc., pp. 27-36.
100. Mellows, H.J.: Ultrasound. In Sandler, M., editor: Amniotic fluid and its clinical significance, New York, 1981, Marcel Dekker, Inc., pp. 113-128.
101. Queenan, J.T.: Amniocentesis and transamniotic fetal transfusion for Rh disease, Clin. Obstet. Gynecol. **9**:491-507, 1966.
102. Pritchard, J.A., and MacDonald, P.C.: Technics to evaluate fetal health. In Pritchard, J.A., and MacDonald, P.C., editors: Williams obstetrics and gynecology, ed. 16, New York, 1980, Appleton-Century-Crofts, pp. 265-293.
103. Hershman, J.M.: Hyperthyroidism induced by trophoblastic thyrotropin, Mayo Clin. Proc. **47**:913-918, 1972.
104. Kenimer, J.G., Hershman, J.M., and Higgins, H.P.: The thyrotropin in hydatidiform mole is human chorionic gonadotropin, J. Clin. Endocrinol. Metab. **40**:482-491, 1975.
105. Nisula, B.C., and Ketelslegers, J.: Thyroid-stimulating activity and chorionic gonadotropin, J. Clin. Invest. **54**:494-499, 1974.
106. Higgins, H.P., Hershman, J.M., Kenimer, J.G., et al.: The thyrotoxicosis of hydatidiform mole, Ann. Intern. Med. **83**:307-311, 1975.
107. Root, A.W., and Reiter, E.O.: Human perinatal endocrinology. In Quilligan, E.J., and Kretchmer, N., editors: Fetal and maternal medicine, New York, 1980, John Wiley & Sons, Inc., pp. 15-58.
108. Kosasa, T.S., Taymor, M.L., Goldstein, D.P., and Levesque, L.A.: Use of a radioimmunoassay specific for human chorionic gonadotropin in the diagnosis of early ectopic pregnancy, Obstet. Gynecol. **42**:868-871, 1973.
109. Spellacy, W.N., Buki, W.C., and McCreary, S.A.: Measurement of human placental lactogen with a simple immunodiffusion kit, Obstet. Gynecol. **43**:306-309, 1974.
110. Little, B., and Billar, R.B.: Endocrine disorders. In Romney, S.L., Gray, M.J., Little, A.B., et al., editors: Gynecology and obstetrics: the health care of women, New York, 1975, McGraw-Hill Book Co., Inc., p. 398.
111. Dumont, M., Cohen, M., Cohen, H., and Bertrand, J.: Diagnosis of fetal distress in utero by determination of plasma levels of nonconjugated estrogens, Rev. Franc. Gynecol. **70**:171-178, 1975.
112. Arvidson, G., Ekeland, H., and Asted, B.: Phospholipid composition of human amniotic fluid during gestation and at term, Acta Obstet. Gynecol. Scand. **51**:71-75, 1972.
113. Falconer, G.F., Hodge, J.S., and Gadd, R.L.: Influence of amniotic fluid volume on lecithin estimation in prediction of respiratory distress, Br. Med. J. **2**:689-691, 1973.
114. Harvey, D., Parkinson, C.E., and Campbell, S.: Risk of respiratory-distress syndrome, Lancet **1**:42, 1975.
115. Gluck, L.: The evaluation of fetal lung maturity by analysis of phospholipid indicators in amniotic fluid. In Lung maturation and the prevention of hyaline membrane disease, Proceedings of the seventh Ross Conference on Pediatric Research, Columbus, Ohio, 1976, Ross Laboratories, pp. 47-49.
116. Bauer, C.R., Stern, L., and Colle, E.: Prolonged rupture of membranes associated with a decreased incidence of respiratory distress syndrome, Pediatrics **53**:7-12, 1974.
117. Richardson, C.J., Pomerance, J.J., Cunningham, M.D., and Gluck, L.: Acceleration of fetal lung maturation following prolonged rupture of the membranes, Am. J. Obstet. Gynecol. *118*:1115-1118, 1974.
118. Freeman, R.K., Bateman, B.G., Goebelsman, U., Arce, J.J., and James, J.: Clinical experience with amniotic fluid lecithin/sphingomyelin ratio, Am. J. Obstet. Gynecol. **119**:239-242, 1974.
119. Lemons, J.A., and Jaffe, R.B.: Amniotic fluid lecithin/sphingomyelin ratio in the diagnosis of hyaline membrane disease, Am. J. Obstet. Gynecol. **115**:233-237, 1973.
120. Gibbons, J.M., Huntley, T.E., Joachim, E., Ruperto, S., and Corral, A.G.: Amniotic fluid analysis for fetal maturity: effect of maternal blood contamination, Obstet. Gynecol. **39**:631, 1972.
121. Hallman, M., Kulovich, M., Kirkpatrick, E., Sugarman, R.G., and Gluck, L.: Phosphatidylinositol and phosphatidylglycerol in amniotic fluid: indices of lung maturity, Am. J. Obstet. Gynecol. **125**:613-617, 1976.
122. Bent, A.E., Gray, J.H., Luther, E.R., Oulton, M., and Peddle, L.J.: Phosphatidylglycerol determination on amniotic fluid 10,000 × g pellet in the prediction of fetal lung maturity, Am. J. Obstet. Gynecol. **139**:259-263, 1981.
123. Bent, A.E., Gray, J.H., Luther, E.R., Oulton, M., and Peddle, L.J.: Assessment of fetal lung maturity: relationship of gestational age and pregnancy complications to phosphatidylglycerol levels, Am. J. Obstet. Gynecol. **139**:664-669, 1981.
124. Painter, D.C.: Simultaneous measurement of lecithin, sphingomyelin, phosphatidylglycerol, phosphatidylinositol, phosphatidylethanolamine, and phosphatidylserine in amniotic fluid, Clin. Chem. **26**(Part 2):1147-1151, 1980.
125. Clements, J.A., Platzker, A.C.G., Tierney, D.F., et al.: Assessment of the risk of respiratory distress syndrome by a rapid test for surfactant in amniotic fluid, N. Engl. J. Med. **286**:1077-1081, 1972.
126. Harvey, D.R., and Parkinson, C.E.: In Barson, A.J., editor: Laboratory diagnosis of fetal disease, London, 1981, John Wright PSG, Inc., pp. 267-298.
127. Nwosu, U.C., Bolognese, R.J., Wallach, E.E., and Bongiovanni, A.M.: Amniotic fluid cortisol concentrations in normal labor, premature labor, and postmature pregnancy, Obstet. Gynecol. **49**:715-717, 1977.
128. Sharp-Cageorge, S.M., Blicher, B.M., Gordon, E.R., and Mur-

phy, B.E.P.: Amniotic fluid cortisol and human fetal lung maturation, N. Engl. J. Med. **296:**89-92, 1977.

129. Gewolb, I.H., Hobbins, J.C., and Tan, S.Y.: Amniotic fluid cortisol as an index of fetal lung maturity, Obstet. Gynecol. **49:**462-465, 1977.
130. Gewolb, I.H., Hobbins, J.C., and Tan, S.Y.: Amniotic fluid cortisol in high-risk human pregnancies, Obstet. Gynecol. **49:**466-470, 1977.
131. Queenan, J.T.: Amniotic fluid analysis, Clin. Obstet. Gynecol. **14:**505-536, 1971.
132. Liley, A.W.: Amniocentesis and amniography in hemolytic disease. In Greenhill, J.P., editor: Yearbook of obstetrics and gynecology, 1964-1965 series, Chicago, 1964, Year Book Medical Publishers, pp. 256.
133. Bellingham, A.J.: Hemoglobins with altered oxygen affinity, Br. Med. Bull. **32:**234-238, 1976.
134. Benz, E.J., and Forget, B.G.: The biosynthesis of hemoglobin, Semin. Hematol. **11:**463-523, 1974.
135. Pennoyer, M.M., and Hartmann, A.F., Sr.: Management of infants born of diabetic mothers, Postgrad. Med. **18:**199-206, 1955.
136. Cornblath, M., Nicolopoulos, D., Ganzon, A.F., et al.: Studies of carbohydrate metabolism in the newborn infant. IV. The effect of glucagon on the capillary blood sugar in infants of diabetic mothers, Pediatrics **28:**592-601, 1961.
137. Pritchard, J.A., and MacDonald, P.C.: Hypertensive disorders in pregnancy. In Williams obstetrics, ed. 15, New York, 1976, Appleton-Century-Crofts, p. 556.

Chapter 36 Extravascular biological fluids

Lewis Glasser

arthritis Inflammation of a joint.

ascites Serous fluid in the peritoneal cavity.

chyle Fatty lymph fluid originating from the intestinal lymphatics. It is milky white in appearance.

chylous effusion Chyle in a body cavity.

colloid osmotic pressure of plasma The difference in osmotic pressure between the plasma and interstitial fluid that drives water into the bloodstream from the interstitial spaces.

effusion Pathological accumulation of fluid in a body cavity.

empyema Usually pus in the pleural cavity.

epicardium The visceral layer of pericardium.

exudate The accumulation of a fluid having a high concentration of protein in a body cavity caused by increased capillary permeability usually secondary to inflammation.

gout An inflammatory arthritis of the joint secondary to crystallization of monosodium urate in the joint.

hemothorax Blood in the pleural cavity secondary to rupture of the blood vessels.

hyaluronic acid A high molecular weight polymer made up of repeating units of the disaccharide *N*-acetyl glucosamine and glucuronic acid.

hydrostatic pressure The lateral pressure of water within a vessel that tends to drive fluid out of the capillaries into the interstitial space.

joint An articulation between bones.

neuroarthropathy Disease of a joint secondary to a disease of the nervous system.

osmotic pressure The force with which a solvent passes through a semipermeable membrane.

osteoarthritis A degenerative form of arthritis that is primarily a disease of the bones with joint involvement.

osteochondromatosis. A joint disease characterized by the development of cartilaginous nodules in the synovial tissues.

paracentesis Aspiration of fluid from a body space.

parapneumonic effusion An accumulation of fluid in the pleural space secondary to pneumonia.

parietal The wall of a cavity.

pericardium The sac enclosing the heart.

peritoneum The serous membrane lining the abdominal cavity and the organs of the abdominal cavity.

permeable Allowing the passage of fluids through a membrane.

pigmented villonodular synovitis A disease of the joints of unknown cause characterized by finger-like proliferative growths of the synovial tissue with hemosiderin deposition within the synovial tissue.

pleura The serous membrane lining the inner surface of the thorax, the diaphragm, and the outer surface of the lungs.

pseudogout An inflammatory arthritis of the joint secondary to crystals of calcium pyrophosphate.

psoriatic arthritis A chronic destructive joint disease that occurs in some patients with the skin disease psoriasis.

Reiter's syndrome A syndrome of unknown cause characterized by inflammation of the joints, urethra, and conjunctivae.

rheumatoid arthritis A chronic progressive inflammatory disease of unknown cause involving multiple joints.

serous fluid Fluid having the characteristics of serum.

synovial fluid Joint fluid.

systemic lupus erythematosus A multisystem disease of an autoimmune cause involving mostly the skin, kidneys, joints, and serosal membranes.

thoracentesis Removal of fluid from the pleural cavity.

transudate The accumulation in a body cavity of a fluid having a low concentration of protein.

SEROUS FLUIDS

In this chapter the term *serous fluid* is restricted to pleural, pericardial, or peritoneal fluid. The word "serous" is derived from *serum* and accurately expresses the derivation of the body fluids from the plasma. Body fluids are designated by a variety of medical terms. *Pleural fluid* (thoracic or chest fluid) is obtained by surgical puncture of the chest wall (*thoracentesis* or *thoracocentesis*). *Empyema* refers to pus in the pleural cavity. Peritoneal fluid is frequently designated by the nonanatomical term *ascitic fluid*. *Ascites* is derived from the Latin word for 'bag' and describes the bloated abdomen of the patient afflicted with a massive accumulation of peritoneal fluid. *Paracentesis* means aspiration of fluid from a cavity, and *abdominal paracentesis fluid* is synonymous with *peritoneal fluid*. Whole blood in the body cavities is designated with the prefix "hemo," as in *hemothorax*. A *chylous effusion* refers to the accumulation of lymph (chyle) in the body cavity.

Formation

Normal formation. Each body cavity is lined by a thin glistening membrane. The lining of the body wall is the parietal membrane, and the lining of the organs is the visceral membrane. The two membranes are continuous, and the space between them is the body cavity (Fig. 36-1). Histologically, the membrane is composed of a thin layer of connective tissue containing numerous capillaries and lymphatics and a superficial layer of flattened mesothelial cells.

Serous fluid is an ultrafiltrate of plasma derived from the rich capillary network in the serosal membrane. Its formation is similar to the production of extravascular interstitial fluid anywhere in the body. Three factors are important: hydrostatic pressure, colloid osmotic pressure, and capillary permeability. Hydrostatic pressure drives fluid out of the capillaries and into the body cavities. Impermeable protein molecules in the plasma counteract the hydrostatic pressure and absorb fluid. This force is called the

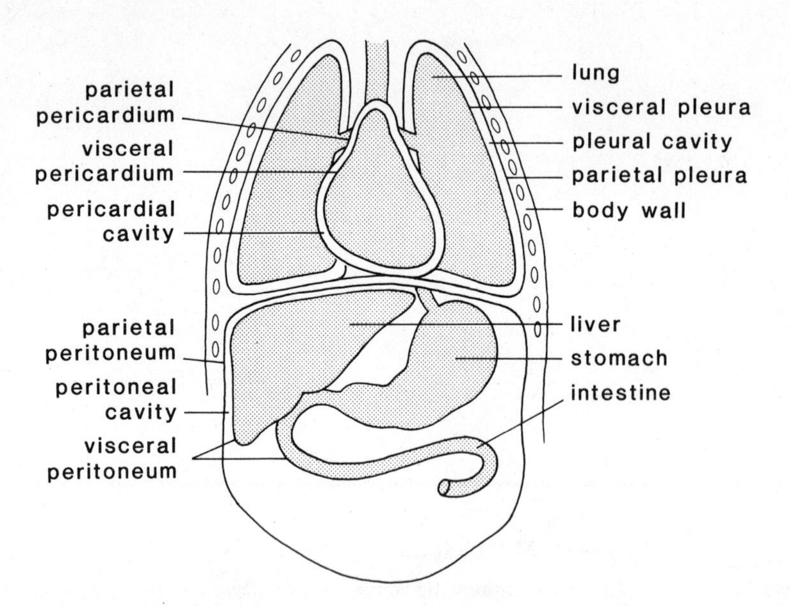

Fig. 36-1. Relationships of serous membranes, body cavities, and viscera. Heart is enclosed in pericardial sac. Outer layer of pericardium is called "parietal pericardium." Lining the exterior surface of heart is visceral pericardium, which is also called "epicardium." Parietal peritoneum lines wall of abdominal cavity. Visceral peritoneum invests stomach, liver, and intestines. Peritoneal cavity is space between two layers of peritoneum.

"colloid osmotic pressure" (COP) and is porportional to the molal concentration of protein. Albumin contributes more than globulins to the colloid osmotic pressure. Lymphatics also play an important role in the absorption of water, protein, and particulate matter from the extravascular space. In the thoracic (chest) cavity, fluid is formed at the parietal pleura because the high hydrostatic pressure of the systemic circulation exceeds the colloid osmotic pressure, and fluid is reabsorbed at the visceral pleura, where capillary colloid osmotic pressure exceeds the low hydrostatic pressure of the pulmonary circulation (Fig. 36-2). Some fluid is also reabsorbed by the lymphatics. In the abdomen both the parietal and visceral peritoneum are supplied by the systemic circulation and fluid is removed to a great extent by the subdiaphragmatic lymphatics. Normally there is less than 10 mL of fluid in each pleural cavity, 20 to 50 mL in the pericardial sac, and less than 100 mL in the peritoneal cavity.[1]

Abnormal formation. Effusions will form when the normal physiological mechanisms responsible for the formation or absorption of fluid are impaired. Thus fluid will accumulate if capillary permeability increases, hydrostatic pressure increases, colloid osmotic pressure decreases, or lymphatic drainage is obstructed. Hydrostatic pressure is increased in congestive heart failure. This is a frequent cause of effusions. Hypoproteinemia decreases the colloid osmotic pressure. A decreased plasma protein can be secondary to decreased synthesis or increased loss. Albumin synthesized in the liver is the most important protein in the maintenance of colloid osmotic pressure. Diseases of the liver may impair albumin synthesis; the liver disease most frequently associated with hypoproteinemia and effusions is cirrhosis. Hypoalbuminemia is also caused by an increased loss of protein, which occurs in the nephrotic syndrome. Capillary permeability increases if the pleural surfaces are inflamed. Not only is capillary permeability increased but also protein is lost from the vascular space and so the physical forces that lead to excess fluid formation are accentuated. Conditions causing an increase in capillary permeability include inflammatory diseases, infection, and metastatic tumors. If the lymphatics are obstructed, a protein-rich fluid will accumulate. Neoplasms of the lymph nodes frequently produce pleural effusions. The causes of effusions and the underlying pathogenesis are listed in Table 36-1.

Change of analyte in disease

Transudates and exudates. Serous effusions are designated as transudates or exudates, depending on the protein content of the fluid. The distinction is important because transudates are noninflammatory fluids caused by disturbances of hydrostatic or colloid osmotic pressure, whereas exudates are inflammatory fluids caused by increased capillary permeability secondary to diseases that directly involve the surfaces of the body cavities. The separation of transudates and exudates involve arbitrary medical decision levels that have been empirically determined. The higher the protein content, the more likely it is that the fluid is caused by a process that alters the capillary permeability and involves the surfaces of the body cavity.

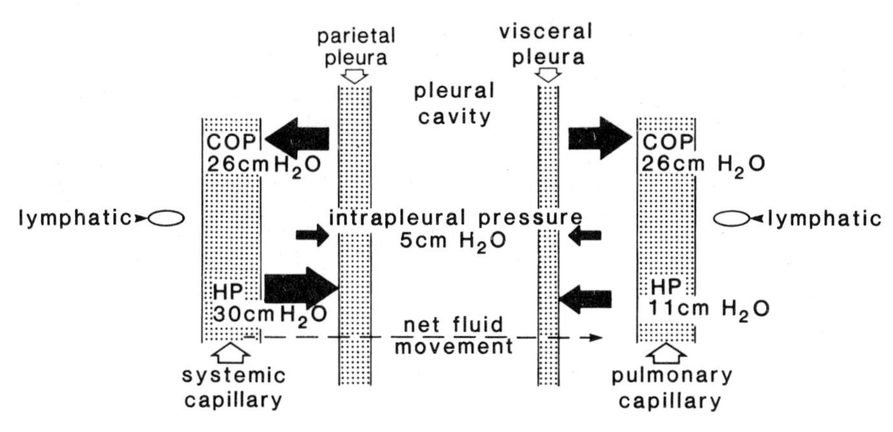

Fig 36-2. Pleural fluid is formed at parietal pleura as net forces for flow of fluid out of systemic capillaries exceed net colloid osmotic pressures. Movement of fluid is toward visceral pleura where net colloid osmotic pressure exceeds outward forces because of low hydrostatic pressure in pulmonary capillaries. Lymphatics play a role in absorption of water, protein, and particulate matter. *COP*, Colloid osmotic pressure; *HP*, hydrostatic pressure.

Table 36-1. Causes of effusions

Cause		Pathogenesis
Transudates		
Congestive heart failure	↑ HP	Systemic and pulmonary venous hypertension
Hepatic cirrhosis	↑ HP	Portal and inferior vena cava hypertension
	↓ COP	Hypoalbuminemia
Nephrotic syndrome	↓ COP	Hypoalbuminemia
Exudates		
Pancreatitis	↑ CP	Inflammation secondary to chemical injury
Bile peritonitis	↑ CP	Inflammation secondary to chemical injury
Rheumatoid disease	↑ CP	Inflammation of serosa
Systemic lupus erythematosus	↑ CP	Inflammation of serosa
Infections (bacterial, tuberculosis, fungal, viral)	↑ CP	Inflammation secondary to microorganisms
Infarction (myocardial, pulmonary)	↑ CP	Inflammation secondary to extension of process to serosal surface
Neoplasms	↑ CP	Increased permeability of capillaries supplying tumor implants; pleuritis secondary to obstructive pneumonitis
	↓ LyD	Lymphatic obstruction secondary to lymph node infiltration
Chyle		
Trauma	↓ LyD	Disruption of lymphatic ducts
Surgery		
Neoplasms		
Idiopathic		

HP, Hydrostatic pressure; *COP*, colloid osmotic pressure; *CP*, capillary permeability; *LyD*, lymphatic drainage.

Measuring the specific gravity will indirectly measure the protein. Pleural fluids are classified as exudates if the specific gravity is greater than 1.015 g/mL or the total protein is 30 g/L (3.0 g/dL) or greater. A total protein measurement is preferable to measurement of the specific gravity. The separation of exudates and transudates in pleural fluids is even more precise if the fluid protein is compared to the serum total protein. Dividing the concentration of the protein in the fluid by the concentration of the protein in the serum gives a ratio. A ratio of 0.5 or greater is indicative of an exudate.[2] The distinction is further improved if a large protein molecule such as lactate dehydrogenase (LD) is used as a marker of capillary permeability. Pleural fluid to serum LD ratios of 0.6 or greater are diagnostic of exudates.[2] The difference between transudates and exudates in pleural effusions are summarized in Table 36-2. Different cutoff values are used for peritoneal fluid. Protein levels greater than 25 g/L (2.5 g/dL) classify the fluid as an exudate.[1]

Glucose. Pleural fluid glucose concentrations are similar to plasma levels in normal fluids and transudates. The difference is less than 100 mg/L (10 mg/dL). Glucose is de-

Table 36-2. Diagnostic criteria of transudates and exudates in pleural fluid

Test	Transudate	Exudate
Appearance	Clear	Cloudy
Fibrinogen	No clot	Clots
Specific gravity	<1.015	≥1.015
Total protein	<30 g/L	≥30 g/L
Total protein (fluid/serum)	<0.5	≥0.5
Lactate dehydrogenase (fluid/serum)	<0.6	≥0.6
Glucose	~ serum	Often <600 mg/L

creased in exudative processes such as bacterial infection, tuberculosis, neoplasia, and rheumatoid disease.[3,4] A pleural fluid glucose less than 600 mg/L (60 mg/dL) or a difference between the plasma and fluid glucose of greater than 300 mg/L (30 mg/dL) is clinically significant. Only low glucose levels are diagnostically useful and the various diseases associated with low levels are also associated with normal values. Any etiological diagnosis on the basis of a low glucose level alone is unreliable. Two mechanisms are operative in producing low values. One is increased glucose use, and the second is a relative block in the transport of glucose from the blood to the fluid. The latter occurs in rheumatoid effusions.[5] Interpretation of low glucose concentrations in peritoneal and pericardial fluid is similar to pleural fluid.

pH. Pleural fluid pH is clinically useful in the management of patients with parapneumonic effusions. Patients with pneumonia develop effusions because the infectious process extends to the visceral pleura, causing exudation of fluid into the pleural space. Complications of parapneumonic effusions include loculation and pus in the pleural cavity. Fluids are divided into potentially benign or complicated effusions on the basis of the pH. Fluids with a pH >7.30 resolve spontaneously, whereas a pH <7.20 is an indication for tube drainage.[6,7] A cautionary note: the specimen must be collected anaerobically in a heparinized syringe, stored on ice, and measured electrometrically at

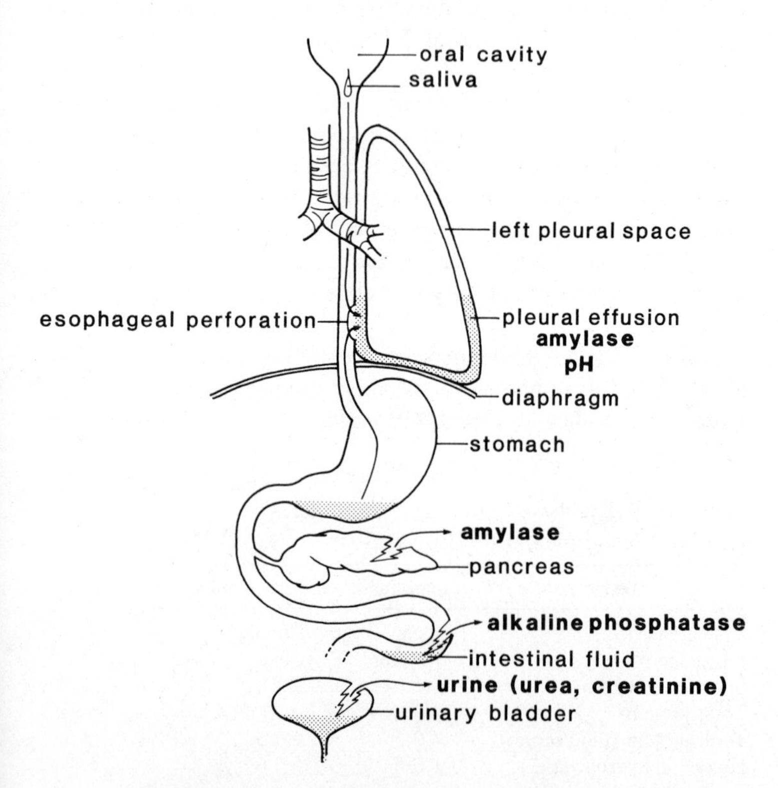

Fig. 36-3. Chemical determinations of body fluids as markers for specific organ involvement.

37° C. Measurement of peritoneal or pericardial fluid pH is of no clinical value.

Lipid. Chyle is a milky white emulsion of fatty lymph fluid originating from the intestinal lymphatics. The accumulation of chyle in the pleural space is rare. Even less frequent is chyle accumulation in the peritoneal or pericardial cavities. Chylous fluid accumulates because of the disruption of the thoracic duct. Chylomicrons on lipoprotein analysis are the best evidence for a chylous effusion. Triglyceride values above 1100 mg/L are highly suggestive of chylous effusion.[8] Cholesterol values do not distinguish between chylous and nonchylous effusions.

Analytes as markers for organs. Chemical substances can serve as markers for specific organ involvement. The rationale for these tests is easily understood from the anatomical location of the viscera and normal biochemistry (Fig. 36-3). Analytes that have been used as markers include amylase, pH, alkaline phosphatase, urea nitrogen, and creatinine. Pleural effusions accompany most cases of esophageal rupture. The perforation allows secretions from both the oral cavity and stomach to contaminate the effusion fluid. Pleural fluid amylase can be elevated and the levels are higher than the serum amylase. Electrophoretic studies indicate that the amylase is from the saliva.[9] Another indicator of esophageal perforation is the pH. Normal gastric juice has a pH below 3.5. Leakage of gastric contents through the esophageal tear will acidify the pleural fluid.[10,11] A pH below 6.0 is clinically significant. The measurement may be performed at the bedside using pH reagent paper.

Amylase is a well-accepted marker of pancreatic disease. In acute pancreatitis, amylase-rich fluid seeps into the peripancreatic tissue and causes a chemical peritonitis, with the formation of small amounts of peritoneal fluid in most cases. One study reports fluid amylase levels of 15,030 ± 4086 Somogyi units/dL.[12] The fluid levels are higher and persist longer than the corresponding blood amylase levels.[13] *Pancreatic ascites* is the chronic accumulation of massive amounts of fluid in association with pancreatitis. It is not certain whether the fluid represents leakage of pancreatic secretions from ruptured ducts or exudation of fluid from the serosal surfaces secondary to chemical irritation.[14] In pancreatic ascites, peritoneal fluid amylase ranges from 370 to 70,000 Somogyi units/dL.[15] Pleural effusions are present in 5% to 15% of cases of pancreatitis. Increased amylase levels are caused by transdiaphragmatic lymphatic drainage or seepage of the enzyme across the diaphragm.[15] In a rare case pleural fluid is present because of a direct communication between the pleural and peritoneal cavities.[16]

Alkaline phosphatase has been shown to be a marker enzyme for pathological processes of the small intestine. The source of the enzyme can be leakage of alkaline phosphatase–rich fluid from the intestinal contents or extravasation from the wall of the intestine. In dogs, intestinal fluid directly aspirated from the proximal duodenum to the distal ileum has values from 100 to 10,000 times the serum levels.[17] The enzyme is elevated in peritoneal serous effusions in association with intestinal perforation and infarctions of the small bowel and in peritoneal blood in patients with traumatic injuries of the small intestine.[18] The values are higher than corresponding peripheral blood levels.

Both urea nitrogen and creatinine are helpful in the differential diagnosis of a ruptured urinary bladder after abdominal trauma. Extravasated urine will have high levels of both urea nitrogen and creatinine. The former is freely diffusible and will also elevate the blood urea nitrogen; however, creatinine is relatively impermeable to the peritoneum and blood levels will not increase. In uncomplicated serous effusions the fluid urea nitrogen and creatinine are low. If the physician inadvertently aspirates urine from the bladder, both urea nitrogen and creatinine will be high but their concentrations in the blood will be normal.

Analytes as markers for tumors. Malignancy is a major cause of serous effusions. The diagnosis of a malignant effusion is made by a cytological test. A number of proteins have been described as tumor markers and are potentially useful in the diagnosis of malignant effusions. These include carcinoembryonic antigen (CEA), alpha-fetoprotein, beta$_2$-microglobulin, and orosomucoid. Not all studies are in agreement as to the clinical importance of these tests. One study suggests that the concentration of CEA in the fluid is related to the tumor burden.[19] The protein is produced by the tumor cells. Orosomucoid is an acute-phase protein that increases nonspecifically in cancer. Levels above 100 mg/L discriminated between benign and malignant effusions in one report.[20] Beta$_2$-microglobulin is the light chain moiety of HLA antigens, which are on the surface of all cells. The determination is of no value in the diagnosis of most malignant effusions[20] but may have a place in the diagnosis of hematological malignancies.[21] Alpha-fetoprotein is of limited value in the diagnosis of malignant effusions.[20]

SYNOVIAL FLUID

Joints, some immovable, others movable, are articulations between bones. The freely movable ones are composed of hyaline articular cartilage and a fibrous capsule lined on its inner surface by a membrane (Fig. 36-4). Synovial fluid fills the joint cavity and acts as a lubricant, keeping to a minimum the friction between bones during movement or weight bearing. The fluid also provides the sole nutrition for cartilage by diffusion and by a spongelike effect when it is compressed and relaxed. The term *synovium* is derived from the Greek *syn* ('with') and the Latin *ovum* ('egg'), suggesting the fluid's resemblance to raw egg white. The term was coined by Paracelsus in AD 1520.

Synovial fluid is a dialysate of plasma mixed with hyaluronic acid. Ultrafiltration in the rich vascular network in the synovial tissue produces this fluid, whereas hyaluronic acid, a mucoprotein, is secreted into the dialysate by type B synovial cells. These cells have a well-developed rough endoplasmic reticulum characteristic of other secreting cells. The synovial membrane is also lined with type A cells that function as phagocytes, sometimes containing vacuoles indicative of endocytosis.

Normal synovial fluid

The fluid volume in the normal joint depends on the size of the structure. The knee joint usually contains 0.1 to 3.5 mL of fluid.[22] Normal synovial fluid is rarely obtained. If it is obtained, the fluid is clear or pale yellow with a specific gravity close to plasma but a high viscosity relative to water because of the protein-polysaccharide complex, hyaluronic acid. This composes 99% of the mucoproteins present in the fluid and is a long-chain, high molecular weight polymer made up of repeating units of *N*-acetyl glucosamine and glucuronic acid. The molecule is destroyed in inflammatory states by hyaluronidase, an enzyme contained in the neutrophils. When this occurs, the fluid viscosity decreases significantly, giving the clinician a bedside test for the presence of inflammatory fluid.

Synovial fluid protein concentrations are related to their molecular weight because of the dialytic effect in the synovium. Thus albumin is present in relatively higher concentrations than the higher molecular weight globulins are. Fibrinogen is not present because of its high molecular

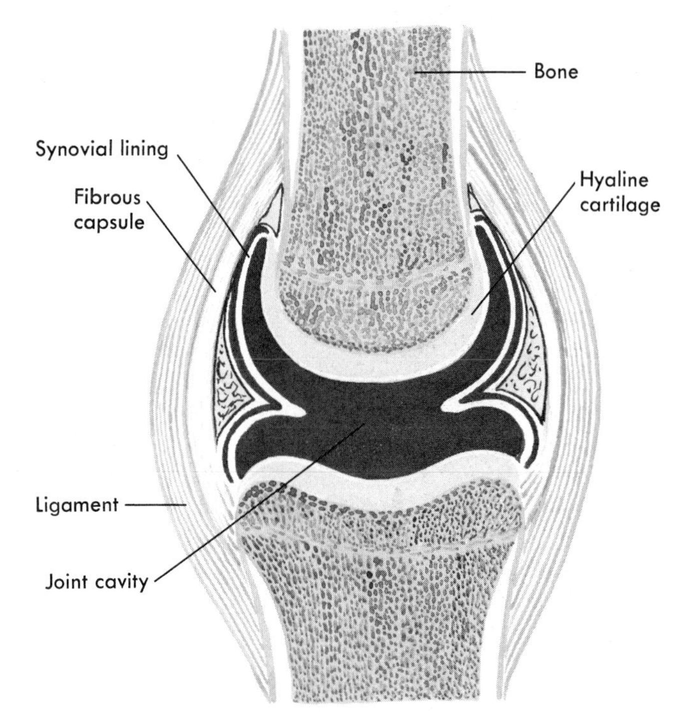

Fig. 36-4. Diagram of normal synovial joint. (From Beck, E.W.: Mosby's atlas of concise functional human anatomy, St. Louis, 1982, The C.V. Mosby Co.)

Table 36-3. Physical and chemical characteristics of normal synovial fluid

	Mean	Range
Volume (mL)*	1.1	0.13-3.5
Relative viscosity at 38° C	235	5.7-1160
Hyaluronic acid (g/L)	3600	1700-4050
Total protein (mg/L)	17.2	10.7-21.3
Immunoglobulins (mg/L)		
IgG	4530	330-8500
IgA	740	270-1770
IgM	370	0-840
Fibrinogen (mg/L)	0	0
Complement (CH_{50} U/mL)	20†	16-25
Glucose (mg/L)	‡	650-1200
Uric acid (mg/L)	‡	25-72

*Knee.
†Values are approximately 10% of plasma values.
‡Fasting values are similar to plasma values.

weight, and so normal synovial fluid does not coagulate. Glucose and uric acid diffuse freely into the synovial and in the fasting state are concentrated much as in plasma.

Normal synovial fluid values are summarized in Table 36-3.[23-26]

Change of analyte in disease

The physical and chemical changes that occur in the synovial fluid during disease are reflections of basic pathological processes occurring in the joint; therefore selected testing of chemical constituents is useful in predicting the underlying disease. A pathological classification of synovial fluids with the diseases associated with each category is summarized in Table 36-4. The laboratory tests that are discussed in this section include viscosity, fibrinogen, total protein, complement, glucose, and uric acid.

Clinically there is no need for a sophisticated measurement of viscosity. Instead, one measures it at the time of aspiration by placing a finger at the tip of the syringe and stringing out the fluid. Noninflammatory fluids will "string out" longer than 4 cm, simulating egg albumin in consistency. One can also drip fluid off the needle and syringe, observing it string out if it is a noninflammatory fluid and drip like water if it shows inflammation. The depolymerization of the hyaluronic acid moiety by neutrophil hyaluronidase decreases the viscosity in inflammatory disease.

The mucin clot test *(Ropes' test)* is seldom advocated but is frequently described in the literature. This test reflects the degree of hyaluronate polymerization. One performs it by dropping fluid into a dilute acid solution and determining the quality of the clot. Although the test is often mentioned without critical comment in discussions of synovial fluid, the quality of the information is inferior to that of other procedures and the test should be considered obsolete.

The practical importance of fibrinogen is not in its measurement but in its presence. Normal synovial fluid contains no fibrinogen, but because inflammatory synovitis permits the passage of large molecular weight proteins into the fluid, fibrinogen can be present and spontaneous clotting can occur. Thus anticoagulants are necessary when specimens are collected for microscopic and bacteriological examination.

Unlike serous fluids, the synovial fluid total protein concentration is not used to distinguish noninflammatory from inflammatory fluids, since the leukocyte count is used to make that distinction. Thus total protein is not included in the routine examination of synovial fluids; however its measurement can be helpful in interpreting complement levels.[27] Complement proteins normally have low values in synovial fluid compared to serum.[28 to 30] In systemic inflammatory conditions, complement behaves as an acute-phase reactant and there is hypercomplementemia. In some conditions, such as Reiter's disease, the joint fluid complement has been reported to be even higher than that in the serum.

Table 36-4. Pathological classification of synovial fluids

Test	Noninflammatory	Inflammatory	Septic	Hemorrhagic
Volume (mL)	>3.5	>3.5	>3.5	>3.5
Color	Yellow	Yellow-white	Yellow-green	Red-brown
Viscosity	High	Low	Low	Low
Leukocytes (cells/µL)	200-2000	2000-100,000	10,000->100,000	>5000
Neutrophils (%)	<25	>50	>75	>25
Glucose (mg/L)	~ serum	>250 mg/L lower than serum	>250 mg/L lower than serum	~ serum
Culture	Negative	Negative	Positive	Negative
Diseases	Osteoarthritis Osteochondritis dissecans Osteochondromatosis Traumatic arthritis Neuroarthropathy	Gout Pseudogout Psoriatic arthritis Reiter's syndrome Rheumatoid arthritis Systemic lupus erythematosus	Bacterial infection Fungal infection Tuberculous infection	Hemophilia Trauma Pigmented villonodular synovitis

In systemic immune complex diseases such as systemic lupus erythematosus (SLE), complement is consumed widely and can be low in both the serum and synovial fluid. In other diseases, such as rheumatoid arthritis and viral synovitis, complement is consumed locally in the synovia and serum levels are usually normal or high.

Assays for C_3, C_4, and CH_{50} are now available for routine use in many clinical laboratories. C_3 and C_4 are more stable than CH_{50}, which will be falsely low if the fluid sits out at room temperatures. A decreased C_4 level suggests classical pathway activation by immune complexes and is more sensitive than C_3 or CH_{50}. The latter two are low in both classical or alternative pathway activation of complement and, if low, suggest more profound systemic activity. The proper approach to interpreting complement levels in synovial fluid is controversial. For practical purposes compare synovial fluid and serum complement levels and consider synovial fluid complement low if it is less than 30% of serum levels.[27] However, in SLE and other severe immune complex diseases both levels may be low. In such a situation one can compare the synovial fluid and serum complement levels with total protein in each fluid.[28]

Interpretation of synovial fluid glucose levels requires knowledge of the patients' simultaneous serum glucose. This is best done in the fasting state, but such preparation is not always clinically feasible. In the ideal situation of an 8-hour fast, the serum to synovial fluid difference is less than 100 mg/L; levels 250 mg/L or more below the serum level suggest inflammation, and differences greater than 400 mg/L suggest sepsis. In the nonfasting state, synovial fluid glucose levels less than half of serum levels should definitely arouse suspicion of a septic process. Rarely, such findings are noted also in rheumatoid arthritis effusions.

Serum uric acid levels are important in the diagnosis of gout. The synovial fluid uric acid level is similar to serum. Its measurement in synovial fluid is of no diagnostic value[1]; however, formation of monosodium urate crystals and their identification by polarized light microscopy in synovial fluid is the cardinal diagnostic feature of gouty arthritis.

REFERENCES

1. Krieg. A.F.: Cerebrospinal fluid and other body fluids. In Henry, J.B., editor: Clinical diagnosis and management by laboratory methods, Philadelphia, 1979, W.B. Saunders Co., pp. 657-667.
2. Light, R.W., MacGregor, M.I., Luchsinger, P.C., and Ball, Jr., W.C.: Pleural effusions: the diagnostic separation of transudates and exudates, Ann. Intern. Med. **77**:507-513, 1972.
3. Light, R.W., and Ball, Jr., W.C.: Glucose and amylase in pleural effusions, J.A.M.A. **225**:257-260, 1973.
4. Carr, D.T., and Mayne, J.G.: Pleurisy with effusion in rheumatoid arthritis, with reference to the low concentration of glucose in pleural fluid, Am. Rev. Respir. Dis. **85**:345-350, 1962.
5. Dodson, W.H., and Hollingsworth, J.W.: Pleural effusion in rheumatoid arthritis: impaired transport of glucose, N. Engl. J. Med. **275**:1337-1342, 1966.
6. Potts, D.E., Levin, D.C., and Sahn, S.A.: Pleural fluid pH in parapneumonic effusions, Chest **70**:328-331, 1976.
7. Light, R.W.: Management of parapneumonic effusions, Chest **70**:325-326, 1976.
8. Staats, B.A., Ellefson, R.D., Budahn, L.L., et al.: The lipoprotein profile of chylous and nonchylous pleural effusions, Mayo Clin. Proc. **55**:700-704, 1980.
9. Sherr, H.P., Light, R.W., Merson, M.H., et al.: Origin of pleural fluid amylase in esophageal rupture, Ann. Intern. Med. **76**:985-986, 1972.
10. Dye, R.A., and Lafaret, E.G.: Esophageal rupture: diagnosis by pleural fluid pH, Chest **66**:454-456, 1974.
11. Abbott, O.A., Mansor, K.A., and Logan, W.D.: Atraumatic so-called "spontaneous" rupture of the esophagus, J. Thorac. Cardiovasc. Surg. **59**:67-82, 1970.
12. Geokas, M.C., Olsen, H., Carmack, C., and Rinderknecht, H.: Studies on the ascites and pleural effusion in acute pancreatitis, Gastroenterology **58**:950, 1970.
13. Keith, L.M., Zollinger, R.M., and McCleery, R.S.: Peritoneal fluid amylase determinations as an aid in diagnosis of acute pancreatitis, Arch. Surg. **61**:930-936, 1950.
14. Donowitz, M., Kerstein, M.D., and Spiro, H.M.: Pancreatic ascites, Medicine **53**:183-195, 1974.
15. Salt, W.B., and Schenker, S.: Amylase—its clinical significance: a review of the literature, Medicine **55**:269-289, 1976.
16. Goldman, M., Goldman, G., and Fleischner, F.G.: Pleural fluid amylase in acute pancratitis, N. Engl. J. Med. **266**:715-718, 1962.
17. Lee, Y.N.: Alkaline phosphatase in intestinal perforation, J.A.M.A. **208**:361, 1969.
18. Delany, H.M., Moss, C.M., and Carnevale, N.: The use of enzyme analysis of peritoneal blood in the clinical assessment of abdominal organ injury, Surg. Gynecol. Obstet. **142**:161-167, 1976.
19. DiStefano, A., Tashima, C.K., Fritsche, H.A., et al.: Carcinoembryonic antigens in pleural fluids obtained from patients with mammary cancer, Am. J. Clin. Pathol. **73**:386-389, 1980.
20. Vladutiu, A.O., Brason, F.W., and Adler, R.H.: Differential diagnosis of pleural effusions: clinical usefulness of cell marker quantitation, Chest **79**:297-301, 1981.
21. Vladutiu, A.O., Adler, R.H., and Brason, F.W.: Diagnostic value of biochemical analysis of pleural effusions. Carcinoembryonic antigen and $beta_2$ microglobulin, Am. J. Clin. Pathol. **71**:210-214, 1979.
22. Ropes, M.W., Rossmeisl, E.C., and Bauer, W.: The origin and nature of normal human synovial fluid, J. Clin. Invest. **19**:795-799, 1940.
23. Hamerman, D., and Schuster, H.: Hyaluronate in normal synovial fluid, J. Clin. Invest. **37**:57-64, 1958.
24. Hoeprich, P.D., and Ward, J.R.: The fluids of the parenteral body cavities, New York, 1959, Grune & Stratton, Inc.
25. Pekin, T.J., and Zvaifler, N.J.: Hemolytic complement in synovial fluid, J. Clin. Invest. **43**:1372-1382, 1964.
26. Pruzanski, W., Russell, M.L., Gordon, D.A., and Ofryzlo, M.A.: Serum and synovial fluid proteins in rheumatoid arthritis and degenerative joint diseases, Am. J. Med. Sci. **265**:483-490, 1973.
27. Bunch, T.W., Hunder, G.G., McDuffie, F.C., et al.: Synovial fluid complement determination as a diagnostic aid in inflammatory joint disease, Mayo Clin. Proc. **49**:715-720, 1974.
28. Fostiropoulos, G., Austen, K.F., and Block, K.J.: Total hemolytic complement (CH50) and second component of complement (C'2hu) activity in serum and synovial fluid, Arthritis Rheum. **8**:219-232, 1965.
29. Lundh, B., Hedberg, H., and Laurell, A.B.: Studies of the third component of complement in synovial fluid from arthritis patients. I. Immunochemical quantitation and relation to total complement, Clin. Exp. Immunol. **6**:407-411, 1970.
30. Ruddy, S., and Austen, K.F.: The complement system in rhuematoid synovitis. I. An analysis of complement component activities in rheumatoid synovial fluids, Arthritis Rheum. **6**:713-722, 1970.

Chapter 37 Central nervous system

Gerald Moriarty

anticonvulsant drugs Drugs given therapeutically to prevent seizures of various types.

arachnoid membrane Middle of the three membranes covering the brain.

blood-brain barrier The barrier between the brain and its membranes and extravascular fluid so that a specified composition of cerebrospinal fluid is maintained.

cerebrospinal fluid (CSF) Clear, colorless fluid contained within the four ventricles of the brain, the subarachnoid space, and the spinal cord.

choroid plexus Vascular folds in the pia membrane of the third, fourth, and lateral ventricles that synthesize cerebrospinal fluid.

coma State of unconsciousness from which patients cannot be aroused, even by the strongest stimuli.

dura Outermost of the three membranes covering the brain; it is the toughest and most fibrous of the three.

epilepsy Transient disturbances of brain function involving impairment of consciousness, motor function, and other neurological functions of the brain.

IgG index Ratio (CSF IgG × serum albumin)/(serum IgG × CSF albumin), used as an indicator of the source of elevated cerebrospinal fluid protein; that is, within the central nervous system versus leakage across the blood-brain barrier.

meninges The three membranes covering the brain and spinal cord; dura, arachnoid, and pia.

meningitis Inflammation of the meninges, often caused by viral or bacterial infections.

pia Innermost of the three membranes, directly covering the brain and spinal cord.

seizure Sudden attack of epilepsy.

stroke Sudden onset of symptoms caused by acute ischemia in the brain caused by hemorrhage, embolism, and so on; it is evidenced by loss of neurological functions.

subarachnoid space The space between the arachnoid and pia membranes.

ventricles Four cavities within the brain filled with cerebrospinal fluid and lined by the pia and the choroid plexus.

xanthochromia A yellow coloring to the cerebrospinal fluid caused by the presence of breakdown products of hemoglobin.

BASIC NEUROANATOMY

This chapter deals with diseases of the nervous system and the ways laboratory values are altered in those disease states. The topics covered are presented in greater depth in various reference texts, which one is strongly encouraged to consult when dealing with specific situations, including neurological disease.

The central nervous system consists of the brain and spinal cord. The brain includes the two cerebral hemispheres, which are roughly mirror images of one another; the brainstem, a narrow structure through which all the pathways entering and leaving the two hemispheres must pass and in which one also finds the centers that control breathing, heart rate, eye movements, and many other critical functions; and the cerebellum, a rounded structure about the size of a baseball that helps control movement and balance (Fig. 37-1). The cerebellum is attached to the back of the brainstem and located just beneath the cerebral hemispheres. The lower brainstem flows into the spinal cord. The spinal cord is the point of exit for nerves on their way out to the muscles they control and the point of entry for sensory fibers returning from the body's sensory organs. All the nerves outside the central nervous system are collectively called the "peripheral nervous system."

The two cerebral hemispheres are built around a connecting system of hollow spaces called the "ventricular system." The "ventricles" are filled with cerebrospinal fluid (CFS)(Fig. 37-2).

The brain and spinal cord are both covered by a double membrane called the "meninges" (Fig. 37-3). Its inner membrane, called the "pia," lies directly on the brain. Its outer membrane, the "arachnoid," lies next to the outermost covering of the brain and spinal cord, the "dura." The dura is a tough, nonelastic membrane that essentially wraps the brain and spinal cord in a nondistensible sac. The brain, blood, and CSF are thus sealed within a space whose volume is fixed. The space between the pia and the arachnoid is called the "subarachnoid space" and communicates directly with the ventricular system.

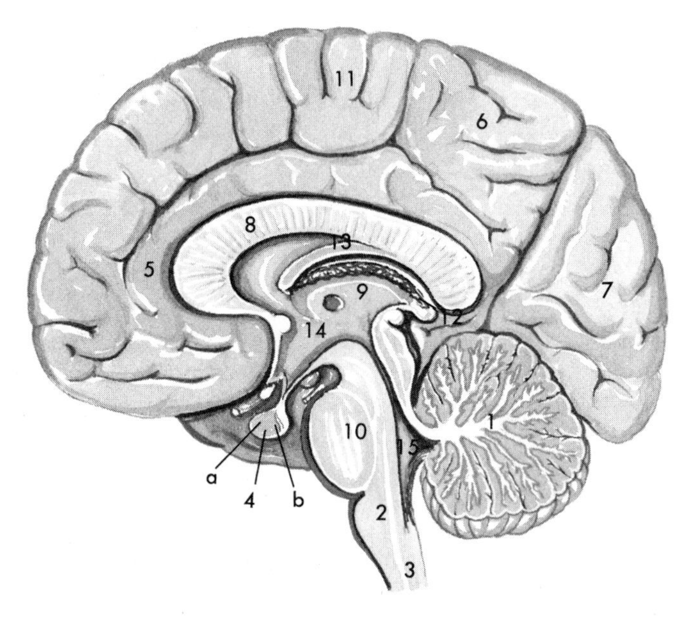

Fig. 37-1. Scheme of functional or motor-control areas of brain (right hemisphere, medial view). (From Beck, E.W.: Mosby's atlas of functional human anatomy, St. Louis, 1982, The C.V. Mosby Co.)

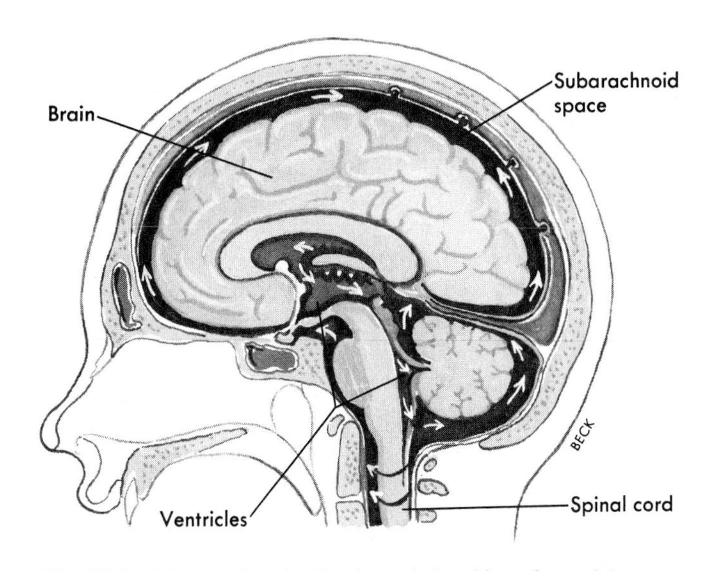

Fig. 37-2. Scheme of brain showing relationships of ventricles and subarachnoid space with rest of brain. (From Beck, E.W.: Mosby's atlas of functional human anatomy, St. Louis, 1982, The C.V. Mosby Co.)

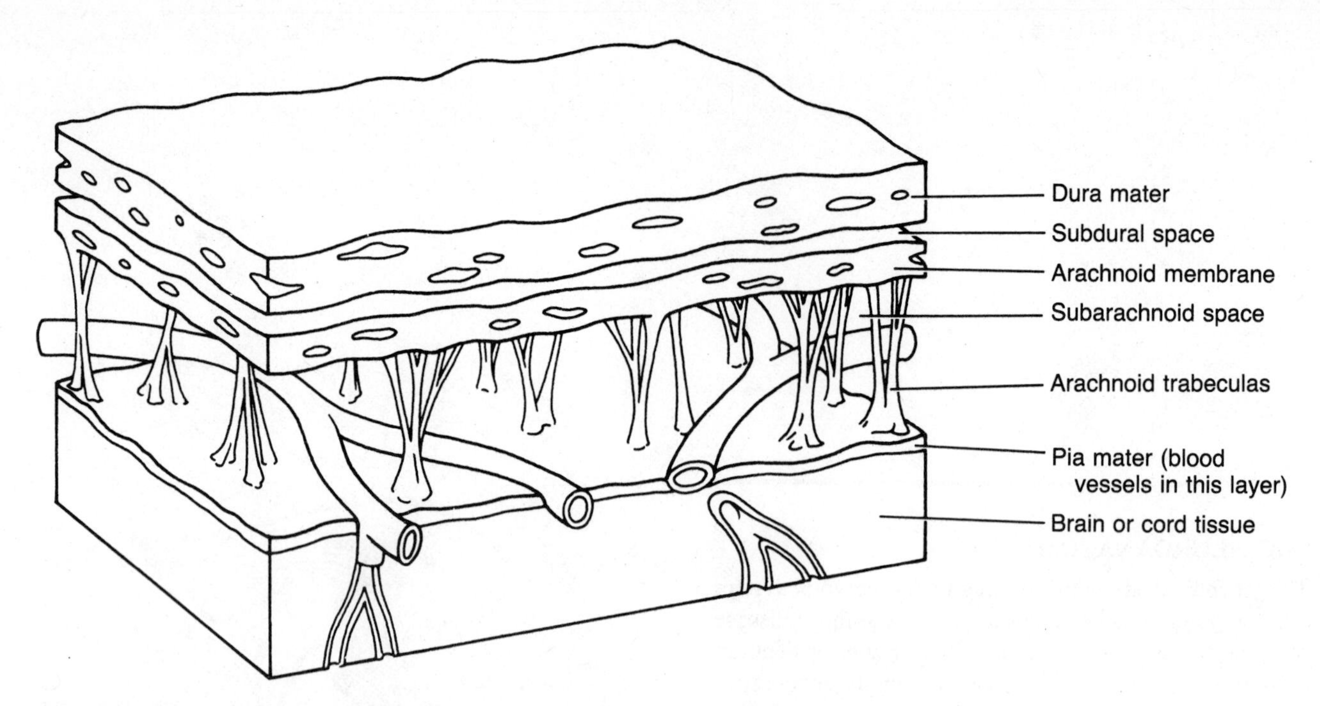

Fig. 37-3. Scheme of meninges. Arrangement may be compared to an underground parking garage. Dura and arachnoid form roof with pia membrane as floor. Cerebrospinal fluid flows in subarachnoid space. (From Prezbindowski, K.S.: Guide to learning anatomy and physiology, St. Louis, 1982, The C.V. Mosby Co.)

PHYSIOLOGY AND BIOCHEMISTRY

Formation of cerebrospinal fluid

The ventricular system and the subarachnoid space are filled with cerebrospinal fluid (CSF). The total volume of CSF in adults is about 150 mL, including 30 mL within the cerebral ventricles and 120 mL in the subarachnoid space (including its spinal segment). CSF is constantly produced and reabsorbed at a rate of approximately 500 mL/day (0.35 mL/minute). This means the total amount of CSF is replaced every 4 to 6 hours.

CSF is produced in the ventricles by a specialized spongelike structure called the "choroid plexus." Beginning in the lateral ventricles, where it is formed, it circulates into the third ventricle and then into the fourth ventricle. It leaves the fourth ventricle by three small openings, or foramina, to circulate through the intracranial and spinal subarachnoid spaces. Circulation may be blocked in any of the ventricles or at the foramina between them, leading to an "obstructive" hydrocephalus (accumulation of fluid in the brain).

If new CSF is constantly produced but its total volume remains constant, it has to be leaving the subarchnoid space in some fashion. In fact, it is *absorbed* at the "arachnoid villi and granulations." These are specialized outpouchings of the arachnoid membrane that penetrate the dura and protrude into the venous system around the brain. These small structures are actually labyrinths of folded tubes whose walls separate the subarachnoid space from the venous space and allow fluid to flow in only one direction by their valvelike action. These arachnoid villi and granulations are scattered along the entire inner table of the skull and down the spinal canal to the points at which the spinal nerves exit the dura. Thus CSF reabsorption can occur along the entire neuroaxis. If absorption is impaired (for example, after meningeal inflammation, bacterial meningitis, or subarachnoid hemorrhage) CNS pressure and CSF volume both rise and the situation is called a "communicating hydrocephalus."

Factors that determine the rate at which CSF is formed and absorbed are complex and not completely understood. It is known that if artificial CSF is infused rapidly up to rates of six times the normal rate of CSF production, the absorption will keep pace and CSF volume and pressure will not rise significantly. Above that rate of infusion, absorption capacity is overwhelmed. Any increase in the size of one component (that is, brain, CSF, or blood) leads to a sharp increase in pressure within the system unless there is a corresponding decrease in the volume of one of the other two components. For example, if the total volume of CSF increases (through increased production or decreased reabsorption), there is less room for blood and the pressure in the system and around the blood vessels increases overall. The same is true if brain volume increases through swelling: there is more pressure and less room for blood and CSF. If volume is added to the dural space, the brain may suffer from direct effects of the abnormally high pressure or perhaps from having its blood flow decreased or through a variety of other mechanisms.

Blood-brain barrier

In 1885, Paul Ehrlich discovered that if an aniline dye is injected intravenously it will stain the dura and many body tissues but fail to stain the brain or CSF. But if dye is injected directly into the subarachnoid space (by the intracisternal route) it will diffuse through the CSF and stain brain tissue. In 1921, Stern and Gautier used the term "blood-brain barrier" to refer to a physiological barrier separating brain and CSF from substances borne in the blood. As described later, the blood-brain barrier allows brain and CSF composition to be maintained at levels quite different from those of blood with respect to proteins, ions, and other molecular elements. The blood-brain barrier is extremely important in clinical practice. It determines the access of antibiotics to the brain and meninges and contributes to the exquisite control exercised over the brain's chemical milieu despite simultaneous changes occurring in the peripheral blood.

The composition of CSF is largely determined at the cell surfaces on which it is produced (the choroid plexus), where it is absorbed (the arachnoid villi and pacchionian granulations), at the arachnoid membrane, which lines the rest of the cerebral ventricles and subarachnoid space, and equally important, at the boundary between blood and brain that exists along the cerebral capillaries themselves.

The extracellular fluid (ECF) compartment of the brain is in relatively free communication with the CSF, whereas the barrier exists between capillary blood on the one hand and the ECF compartment and CSF on the other.

Factors that significantly influence the access of substances to brain and CSF include molecular weight, protein binding, and lipid solubility. With molecular weight, entry is inversely related to size, hence the 1:200 CSF-to-plasma ratio of albumin (molecular weight 69,000 daltons). Drugs that are highly protein bound enter the CSF much less readily than unbound smaller molecular weight substances. For example, phenytoin is 95% protein bound and 5% free in blood. Only 5% of the total measured blood phenytoin level is easily able to enter the CNS, and only that 5% is "active." In certain cases a clinician may need to know both the free and bound fractions to determine the level of active drug at a given dosage. Calcium, magnesium, and metabolites such as bilirubin are also highly protein bound and thus relatively restricted from CSF. Highly lipid-soluble substances such as carbon monoxide and neuroactive drugs and alcohol readily enter the CNS. Substances that are highly ionized at physiological pH are relatively excluded. Highly polar substances, such as some amino acids, enter slowly and require an active-transport mechanism.

Bicarbonate equilibrates slowly between CSF and blood, whereas the opposite is true of carbon dioxide. Thus, if bicarbonate is infused intravenously, serum pH rises but CSF pH drops for a period until equilibration occurs. The reason is that systemic alkylosis is accompanied by respiratory slowing and CO_2 retention. Increasing CO_2 immediately enters the CNS while CSF bicarbonate is still low, and since pH is determined by the ratio of bicarbonate:CO_2, CSF pH drops paradoxically. Thus in cases of established systemic acidosis, such as diabetic ketoacidosis, the rapid venous infusion of bicarbonate can lead to coma.

The blood-brain barrier is readily permeable to water but not to electrolytes. The major cations, sodium and potassium, require hours to reach equilibrium with CSF after changes in peripheral blood. Changes in blood osmolality are followed by parallel CSF changes after a lag time of a few hours.

In the case of drugs, their pK_a, or "ionic dissociation constant," is important in determining how readily they cross the blood-brain barrier. pK_a refers to the pH at which 50% of a compound is ionized. A nonionized drug is relatively lipid soluble and so more freely enters the CNS. The polar ionized fraction is relatively excluded. The degree of ionization of a drug at any pH (for example, physiological pH of 7.4 for blood, 7.32 for CSF) is given by the Henderson-Hasselbalch equation.

$$\text{Weak-acid pH} = pK_a + \log \frac{[\text{Salt}]}{[\text{Acid}]}$$

$$\text{Weak-base pH} = 14 - pK_a - \log \frac{[\text{Salt}]}{[\text{Base}]}$$

Thus phenytoin, with a pK_a equal to 8.3, is almost entirely unionized at pH 7.4, and so its steady-state concentration in brain and CSF is determined by the free or non–protein bound fraction in blood.

Stated somewhat differently, weak acids are relatively excluded from acidic compartments (those with low pH). Since phenobarbital is a weak acid, barbiturate intoxication can be more of a problem in a setting of systemic acidosis, since that condition favors its entry into CNS with consequently higher CNS concentrations.

Although these remarks illustrate the importance of the blood-brain barrier, they only begin to suggest the complexity of factors that determine the access of substances to the brain. In practice, blood levels of some compounds and even their CSF levels may give little indication of their concentration or activity within actual neural tissue. For example, it is not uncommon to see a delirium caused by lithium intoxication continue to worsen into coma and even death after serum levels of lithium have begun to drop. Thus one must consult available literature for details about individual substances and remain aware that frequently in clinical medicine a discrepancy remains between what theory predicts and what actually occurs.

Finally, the characteristics of the blood-brain barrier can be dramatically altered by disease states. Penicillin, an acidic substance, is normally excluded from the CNS after parenteral injection and yet it is an effective agent for treating meningitis. The reason is that meningeal inflammation alters (damages) the blood-brain barrier, allowing greater access of drugs, such as penicillin, that normally would not reach infected tissue.

Some specific factors that alter permeability of the blood-brain barrier are as follows:

1. In regard to inflammation, pneumococcal meningitis has been shown experimentally to increase the ease of entry into the nervous system of macromolecules such as albumin and penicillin.
2. Neovascularity, in association, for example, with tumor, trauma, or ischemia, alters the blood-brain barrier. This may be caused by defects in the new vessels or to their immaturity.
3. Toxins can change blood-brain barrier characteristics, and some agents used in radiographic studies (such as iodopyracet [Diodrast] and diatrizoate meglumine [Hypaque]) increase its permeability by direct toxic effects. When they are injected in hyperosmolar concentrations, the effect is greater.
4. Adrenal steroids and thyroid hormone actually help stabilize the integrity of the blood-brain barrier. Destabilization may contribute to the elevated CSF protein frequently seen in hypothyroid states.
5. Finally, the blood-brain barrier of the immature nervous system is more permeable to a variety of substances. For infants below the age of 6 months, CSF protein is normally as high as 1000 mg/L.

Functions of cerebrospinal fluid

Why should the brain be suspended in and bathed by this distinctive fluid? First, CSF provides mechanical support to the brain, and a "floating brain" weighs less than if it were simply resting on the bony table of the skull. Second, CSF probably functions to help remove metabolic products or waste from the brain, a function that is poorly understood but probably important in both normal and diseased states. Third, there is some evidence that CSF transports biologically active compounds that may function as chemical messengers. Finally, it plays an important role in maintaining the chemical environment of the brain. Although its communication with the plasma compartment is tightly regulated, CSF seems to be in relatively free communication with the brain's extracellular fluid compartment, which aids brain cells themselves. The following section examines the composition of CSF more closely.

Composition of cerebrospinal fluid (see box at right)

The ionic and molecular composition of CSF differs from that of plasma for some components and is the same for others. Changes in serum sodium are followed by corresponding changes in CSF sodium so that after a lag time of about 1 hour sodium values are nearly the same. However, CSF potassium is lower than in plasma and furthermore is maintained within a very narrow concentration range in CSF despite wide fluctuations in plasma values. Active transport in and out of the CSF space appears to be largely responsible for maintaining these differences. Chloride and magnesium are somewhat higher in CSF than in plasma, and bicarbonate is somewhat lower.

Characteristics of normal spinal fluid

Total volume: 150 mL
Color: colorless, like water
Transparency: clear, like water
Osmolality at 37° C: 281 mOsm/L
Specific gravity: 1.006 to 1.008
Acid-base balance:

pH	7.31
P_{CO_2}	47.9 mm Hg
HCO_3^-	22.9 mEq/L

Sodium: 138 to 150 mEq/L
Potassium: 2.7 to 3.9 mEq/L
Chloride: 116 to 127 mEq/L
Calcium: 2.0 to 2.5 mEq/L (40 to 50 mg/L)
Magnesium: 2.0 to 2.5 mEq/L (24.4 to 30.5 mg/L)
Lactic acid: 1.1 to 2.8 mmol/L
Lactic acid dehydrogenase: Absolute activity depends on method; approximately 10% of serum value
Glucose: 450 to 800 mg/L
Proteins: 200 to 400 mg/L
At different levels of spinal tap:

Lumbar	200 to 400 mg/L
Cisternal	150 to 250 mg/L
Ventricular	50 to 100 mg/L

Normal values in children:

Up to 6 days of age	700 mg/L
Up to 4 years of age	244 mg/L

Electrophoretic separation of spinal fluid proteins (% of total protein concentration):

Prealbumin	2% to 7%
Albumin	56% to 76%
α_1-globulin	2% to 7%
α_2-globulin	3.5% to 12%
β- and γ-globulin	8% to 18%
γ-globulin	7% to 12%
IgG	10 to 40 mg/L
IgA	0 to 0.2 mg/L
IgM	0 to 0.6 mg/L
κ/λ ratio	1

Erythrocyte count:

Newborn	0 to 675/mm^3
Adult	0 to 10/mm^3

Leukocyte count:

<1 year of age	0 to 30/mm^3
1-4 years of age	0 to 20/mm^3
5 years of age to puberty	0 to 10/mm^3
Adult	0 to 5/mm^3

CSF glucose normally ranges between 450 and 800 mg/L, that is, between 60% and 80% of the blood glucose concentration after equilibration. Blood and CSF glucose equilibrate only after a lag period of about 4 hours, so that CSF glucose at a given time reflects blood glucose levels during the past 4 hours. When a lumbar spinal puncture (LP) is performed and CSF glucose is to be determined, a simultaneous sample of peripheral blood must also be drawn. When glucose determination is critical (as it may be, for example, in the diagnosis of bacterial or carcinomatous meningitis), the LP and blood glucose should be obtained only after the patient has fasted for at least 4 hours. CSF glucose is altered by certain disease processes, as is discussed later. Equilibrated CSF glucose is definitely abnormal when it is less than 40% of the simultaneous blood glucose value; values less than between 400 and 450 mg/L are almost always abnormal.

One should also be aware that the expected ratio of CSF to blood glucose (60% to 80%) falls as blood glucose rises. That is, one would expect a CSF-to-blood ratio of 0.5 when blood glucose values reach 5000 mg/L and a ratio of 0.4 if blood glucose reaches 7000 mg/L.

Proteins found in the CSF ordinarily originate from serum and reach the CSF space by pinocytosis across the capillary endothelium. The normal ratio of serum to CSF protein is 200:1 (with serum equal to 70 g/L and CSF equal to 350 mg/L).

Brain metabolism

The rest of this chapter contains some fundamental ideas and information about what happens to the brain in common pathological states such as systemic hypoglycemia, anoxia, infection, and so on. A brief discussion of normal brain physiology will help explain the changes that occur with disease.

The brain's metabolic rate is one of the highest of any of the body's organs, whether one is awake or asleep. But unlike most other organs, which store and reserve some supplies of energy to sustain themselves, the brain has almost no energy reserve. It is entirely dependent on an uninterrupted supply of glucose and oxygen delivered by peripheral blood. The brain uses glucose almost exclusively to supply its energy needs. To get an idea just how hungry the brain is and how dependent on a constant, swift flow of blood, consider that under resting conditions total cerebral blood flow equals 15% to 20% of cardiac output (or about 500 mL per 100 g of brain per minute). The percentage of cardiac output flowing to the adult brain is much higher than that in infants and young children. Although *total* cerebral blood flow remains remarkably constant, discrete areas within the brain show striking variability. Gray matter (predominantly composed of cell bodies) enjoys a flow of blood three to four times greater than that of white matter (predominantly composed of the fiber connections between nerve cells). Gray matter lies on the surface of the brain (the "cortex") and is also arranged in some nuclear structures deep inside it. The rest of the brain is composed of white matter connections. Moreover, *regional* blood flow is known to vary during performance of certain tasks, with regional flow increasing in the appropriate areas during tasks such as hand movement, speaking, or mental problem solving. Blood flow is also altered in response to disease states, as in stroke.

The brain depends exclusively on respiration and glycolysis to supply its energy needs. It needs energy for several critical reasons: to maintain membrane potentials, to continue transmitting impulses, and continuously to synthesize new structural elements (since in any living organ, structural components are being constantly broken down and replaced). Available energy is packaged in high-energy phosphate bonds, and the compound that contains the bonds is ATP. Glycolysis and respiration produce energy or ATP according to the following reactions:

Glycolysis: $1 \text{ glucose} + 2 \text{ ADP} + 2 \text{ P}_i \rightarrow \text{lactate} + 2 \text{ ATP}$

Respiration: $1 \text{ glucose} + 6 \text{ O}_2 + 38 \text{ ADP} + 38 \text{ P}_i \rightarrow 6 \text{ CO}_2 + 2 \text{ H}_2\text{O} + 38 \text{ ATP}$

Respiration provides 38 mol of ATP per mol of glucose compared to only 2 mol of ATP derived from glycolysis. Respiration is clearly able to produce much more energy than glycolysis, but respiration requires oxygen and glycolysis does not. It is uncertain how much, if anything, glycolysis contributes to the brain's energy metabolism under normal conditions (very low levels of lactate are found in the brain under normal conditions and lactate is derived from glycolysis).

It is clear that glucose is the predominant substrate for cerebral metabolism, with 0.31 μmol (5.5 mg) of glucose used per minute per each 100 g of brain tissue. During starvation or in states of ketoacidosis, ketone bodies can be used as metabolic substrates and can provide up to 30% of the brain's oxidative metabolic needs. Under ordinary circumstances, however, potential substrates other than glucose, such as fatty acids, have their access to the brain narrowly limited by the blood-brain barrier. Glucose, on the other hand, moves readily across the blood-brain barrier by a process of "facilitated transport." Once available, about 35% of the entering glucose is rapidly metabolized to CO_2, and the remainder is used for synthetic and structural needs by being incorporated into amino acids, proteins, and lipids.

PATHOLOGICAL CONDITIONS

Damage to *discrete* (focal) areas of the brain or spinal cord produce predictable circumscribed signs and symptoms (for example, paralysis of an arm, leg, or side of the body, loss of ability to speak or comprehend spoken language, incoordination, and so on). *Diffuse* impairment of cerebral tissue, on the other hand, leads to its own characteristic clinical picture. Failure of various "intellectual" functions such as attention, concentration, judgment, memory, problem-solving ability, and insight are early findings with mild diffuse disease. Other symptoms include changes in alertness beginning with clouding of consciousness and proceeding to drowsiness, stupor, and coma. Seizures can accompany both diffuse and focal damage.

Various disease states tend to produce either focal or diffuse brain damage, so that the pattern of deficits is often helpful to the clinician in working backward toward a specific diagnosis. Some examples of conditions that cause focal damage are stroke caused by arterial occlusion or hemorrhage; trauma; cerebral abscess; and tumors. Many of these conditions also cause changes in the CSF by damaging the blood-brain barrier (elevating protein) and stimulating inflammatory changes (with leukocytosis), tissue necrosis (elevating CSF protein and cell count), or shedding of tumor cells in cytological specimens.

Examples of conditions associated with generalized cerebral dysfunction (encephalopathic states) are anoxia, generalized ischemia, hypoglycemia, sepsis, thyroid abnormalities, disseminated intravascular coagulation, and the entire group of toxic and metabolical derangements. The diagnosis of these states often rests on laboratory findings.

The clinical and pathological changes commonly found in a number of conditions are briefly discussed. Not all the neurological diseases known to produce changes in laboratory values are discussed, and none are presented in depth.

Coma

Coma is most simply defined as a state of unconsciousness from which the patient cannot be aroused. A coma is but one aspect of altered states of consciousness that can be present in patients. "Confusion" is the least altered state, in which there is disorientation with respect to time, associated drowsiness, and altered attention span. "Stupor" describes a state in which the patient is unresponsive but can be aroused back to a near-normal state with appropriate stimuli.

A patient with an altered mental state, as seen acutely in an emergency room, must first be given any life support necessary to maintain vital functions, such as ventilation. The next step is to determine the underlying cause of the altered mental status. Readily treatable causes can be treated if they exist, such as administration of dextrose to relieve coma caused by severe hypoglycemia. Table 37-1 lists the most important causes of coma and altered mental states, which include metabolic, structural, or infectious causes. Many of these causes are described in detail below.

Intracranial bleeding

Bleeding from a vessel on the surface of the brain, such as an arterial aneurysm, pours blood between the brain's surface and the pia-arachnoid and is called a "subarachnoid hemorrhage." Blood thus mingles with CSF and red blood cells appear in it. Furthermore, because blood is an extremely irritating substance when it escapes from its usual vascular channels, it may provoke an inflammatory response in the meninges called a "chemical meningitis." That is, leukocytes are shed into the CSF by the irritated meninges; and since meninges are pain sensitive, a subarachnoid hemorrhage is very painful (patients complain of the worst headache in their lives).

Infectious and inflammatory disease

Both the bacterial or viral organism and the intracranial structures they invade help determine CSF changes seen in the infectious process. The CSF parameters that reflect CNS invasion by an infectious agent are white cell count and differential, glucose, and to a lesser extent, protein concentration.

Meningitis is an inflammation of the meninges leading to several clinical patterns. Bacterial or "purulent" men-

Table 37-1. Causes of coma and altered mental states

Type	Cause	Laboratory findings
Metabolic	Alcoholism	Increased blood ethanol, metabolic acidosis, and ketosis.
	Hyperosmolar coma	Blood glucose $\geq$ 10,000 mg/L, no ketosis, dehydration
	Diabetic ketoacidosis	Increased blood glucose, ketosis, acidosis, dehydration
	Metabolic acidosis of other origin	Decreased pH, HCO_3^-; increased lactic acid
	Hypoglycemia	Decreased blood glucose (< 500 mg/L)
	Hypercalcemia or hypocalcemia	Changes in calcium levels; hypomagnesiumia can be found with hypocalcemia
Systemic metabolic diseases	Hepatic coma	Increase in blood ammonia, increased liver function tests
	Uremic coma	Increased serum urea, creatinine with metabolic acidosis
	Ischemic, cardiac, pulmonary	Tissue hypoxia, lactic acidosis
Encephalopathy	Epilepsy	Subtherapeutic levels of antiepileptic drugs
	Intracranial hemorrage	Bloody spinal tap
Trauma	—	None
Infectious	Bacterial, viral	Decreased CSF glucose, increased protein
Psychiatric	—	None

ingitis is associated with a CSF polymorphonuclear (PMN) leukocytosis, with counts ranging from a very few to many hundreds—even frank pus—containing countless PMN's. The glucose levels may be depressed to strikingly low values and protein may be elevated. Later in the course of illness lymphocytes can become prominent or dominant, especially if the infection has been partially treated with antibiotics. Partial treatment of bacterial meningitis obscures the CSF findings, making diagnosis difficult, but fails to eradicate the disease—a potentially treacherous circumstance.

Viral meningitides cause a predominantly or exclusively lymphocytic leukocytosis. Glucose usually remains within the range of normal, and protein is usually normal or only slightly elevated.

Fungi also invade the CNS and may cause no change in CSF other than a lymphocytosis and increased protein. Glucose usually remains normal.

Bacteria should be searched for on Gram stain and culture. Viruses can be sought with appropriate serological tests or even culture, and fungi can be found with culture of sensitivity or immunological procedures as well as with appropriate staining. (India ink, for example, may reveal *Cryptococcus* if carefully performed.)

The CSF findings in syphilis depend on the stage of illness, disease activity, and whether previous treatment was given in adequate amounts. The subject should be reviewed in a more detailed text.

Organisms can invade the brain substance, in which case the term "encephalitis" is used. CSF findings are comparable to findings in meningitis or may be quite minimal.

Abscess formation in the brain may produce no CSF changes even though a potentially deadly infection is rampant.

Finally, any of the above conditions can and frequently do lead to increased CSF pressure. This is especially true of meningitides, which obstruct the usual flow of CSF, causing an obstruction hydrocephalus. Meningitis can also impair CSF absorption, causing a communicating hydrocephalus.

Ischemia

Although immensely dependent on glucose, the brain's own reserves of that substance are tiny. It contains about 1 mmol of free glucose per kilogram and 3 mmol of glycogen per kilogram (70% of which can be immediately converted to glucose). *With oxygen available* and the brain's metabolical needs being supplied by respiration using only glucose stored in the brain itself, those stores could supply the brain's needs for only 2 to 3 minutes. If *both* glucose and oxygen are cut off—as in cardiac arrest—glycolysis would become the dominant source of energy. In that relatively inefficient state, glucose stores in the brain could support its energy metabolism for only about 14 seconds. When energy metabolism ceases, the integrity of cellular membranes essentially fails and potassium begins leaking from the cells, osmotic balance is lost and fluid rushes into the damaged cells, and within seconds cells begin to die.

The term "ischemia" refers to a situation in which the flow of blood to a tissue is inadequate. In the case of the brain, ischemia can be present for many reasons. It does not distinguish between decreased perfusions and total cessation of flow, situations that have different implications for tissue viability. For example, if the heart stops, total cerebral blood flow ceases; if the blood pressure drops low enough, the flow of blood becomes inadquate; if a single large vessel such as the carotid artery becomes narrowed, too little blood passes through. If a cerebral blood vessel becomes occluded by an embolus or an atherosclerotic plaque, the tissue that it irrigates becomes ischemic. Other blood vessels usually try to supply the area and make up the difference, but if the area of brain supplied by an occluded vessel cannot be supplied with blood from surrounding vessels, that area dies. The event is called a "cerebral infarction," or a "stroke."

Stroke. Consider what happens to an area of brain that becomes severely ischemic during a stroke. The drop in blood flow is accompanied by a rapid rise in tissue osmolality to a level in the range of 600 mOsm, a change that leads to severe osmotic cellular damage. As a result, the involved area of brain (whether the whole brain or a small area) begins to swell. Swelling leads to rising intracranial pressure, which in turn can further reduce cerebral perfusion. If total intracranial pressure exceeds the systemic arterial pressure, cerebral blood flow stops. In the presence of inadequate oxygen and glucose, tissue respiration ceases and glycolysis becomes the dominant source of energy for a short period. During that time, glycolysis produces high levels of lactic acid, which of course accumulates rapidly because, with blood flow stopped, it cannot be removed. High levels of tissue lactic acid and the accumulation of other possibly toxic products of failing energy metabolism cannot be removed and so add to the damage being done. Cellular membrane integrity is lost; the cells swell and rapidly die.

Clinically, the functions served by the infarcted region are damaged. For example, if the area controlling strength and movement in one extremity or one side of the body is infarcted, that extremity or side becomes weak or paralyzed. If the area subserving speech is damaged, the patient may lose the ability to talk or to comprehend what is heard. It is important to note that the dying or dead area of brain releases protein into the cerebrospinal fluid. Also, because blood vessels in the area are also damaged by ischemia, some bleeding may occur. Only a few cells may appear in the CSF, an indication that only a little blood has escaped from the area, or the CSF may become frankly bloody if an actual hemorrhage occurs in the damaged area. Finally, as the brain begins to clear away the damaged tissue, some white cells may appear in the CSF. In summary then, after a stroke that has not involved significant hemorrhage, one can expect to find the CSF either normal or more likely containing an elevated protein level, a few red cells, and possibly scattered white cells.

On occasion a clinician may wish to administer anticoagulant drugs to a patient to prevent the evolution of a stroke or prevent recurrent strokes. At those times, the clinician will perform a lumbar puncture to help rule out the possibility of intracranial bleeding by examining the fluid for red cells or xanthochromia. Evidence of recent bleeding would ordinarily constitute an absolute contraindication to the use of anticoagulant drugs.

Diffuse cerebral ischemia and hypoxia. If total cerebral circulation stops, as in cardiac arrest, consciousness is lost within 6 to 8 seconds. On the other hand, if oxygen supply becomes inadequate but circulation continues, the clinical result is usually a feeling of lightheadedness followed by mental confusion in mild cases, proceeding to loss of consciousness, seizures, and coma with moderate to severe hypoxia. Precipitating events include pulmonary edema, carbon monoxide poisoning, pulmonary embolism, strangulation, respiratory failure during mechanical ventilation, and exposure to ambient air at high altitudes.

Failure to restore cerebral circulation and oxygenation within 4 to 5 minutes after their total cessation may result in cell death and irreversible damage. The precise metabolic events occurring in brain tissue subjected to hypoxia, anoxia, ischemia, hypoglycemia, and so on in various combinations are incompletely understood.

Clinically, the onset of acute generalized cerebral ischemia, hypoxia, or severe hypoglycemia all lead to signs and symptoms of "diffuse cerebral impairment" as can occur in toxic, metabolic diseases.

Chronic hypoxia

Anemia, congestive heart failure, pulmonary disease, and changes in hemoglobin that interfere with oxygen binding (for example, carbon monoxide poisoning, methemoglobinemia) are examples of conditions associated with chronic insufficiency of cerebral perfusion or oxygenation. Lethargy, confusion, disorientation, and failing memory and judgment are early signs in such instances. Seizures, myoclonus (that is, a sudden single jerk of a large muscle group or even the whole extremity), focal neurological deficits, and finally stupor (unconsciousness from which a patient can be aroused) and coma (unconsciousness from which the patient cannot be aroused) may all occur when the defects become severe or other disease is superimposed on the state of chronic insufficiency. A blood Po_2 less than 40 mm Hg is necessary to produce significant cerebral symptoms if cerebral blood flow is adequate. Similarly, with oxygenation available, preserved cerebral blood flow must fall to 20 mL per 100 g per minute before loss of consciousness occurs. The brain's response to mild or moderate hypoxia is dilatation of medium-sized cerebral blood vessels (perhaps accounting for the common symptoms of headache in that condition) with a consequent increase of cerebral blood flow. This increase in cerebral profusion initially helps offset the decreased amount of oxygen in the blood. But in chronic states at least, cerebral blood flow can only increase to twice its usual amount, thus limiting the effectiveness of this compensatory adjustment. Arterial Pco_2 is the principal determinant of the total cerebral blood flow. This autoregulation of total cerebral blood flow protects the brain to a large extent from fluctuations in systemic blood pressure. Changes in regional cerebral blood flow are less well understood but are almost certainly of great importance.

Epilepsy and other seizure disorders

A seizure is a paroxysmal, unregulated burst of electrical firing at some point on the cerebral cortex. If the region of abnormal discharge is small and remains confined, it is called a "focal seizure." The signs and symptoms of such a focal seizure depend on the site of the electrically discharging tissue.

During such an event the patient's arms, legs, or face may twitch, or the patient may experience a period of confusion. The patient may smell an odor or find vision distorted. Some patients abruptly fall down with only a brief alteration in consciousness, others stare blankly into space and then resume talking in midsentence with no awareness that anything has happened (petit mal seizure of children). On the other hand, the region of abnormal discharging may not remain confined; it can become "generalized" to large areas of cerebral cortex and perhaps even involve structures deep inside the brain. This "generalized" seizure, if it is mild, can cause a lapse of consciousness or a "grand mal," or major motor seizure with loss of consciousness, violent shaking, and often loss of urinary continence. The patient may bruise himself as he falls down, bite his tongue or lips, or even break bones.

The cause of seizures in most instances is not known. Seizures occur in association with other diseases: birth injuries, brain tumors, strokes, metabolic abnormalities such as hypoglycemia or heavy alcohol use (especially around the time when the patient stops heavy drinking), head trauma, infections such as meningitis or encephalitis, ingestion of toxins such as illegal drugs, or even medications. Each person who comes to the doctor with seizures has to be evaluated carefully to see if there is some underlying disease or provocation. Often none is found.

Some persons continue to have seizures throughout their lives and are said to have a "seizure disorder," or epilepsy. Most people with seizure disorders are perfectly normal in every other way. Others have associated brain damage or disease.

Persons with a seizure disorder are placed on anticonvulsant drugs usually for months or even for the rest of their lives in an attempt to stop the seizures entirely or at least reduce their frequency. In other cases, in which some underlying problem such as an infection or metabolic encephalopathy can be found and corrected, anticonvulsant drugs may be needed only during the active phase of the illness.

Intoxication with drugs and poisons

Many drugs and poisons affect the nervous system directly, producing confusion, drowsiness, stupor, coma, seizures, or psychotic states. Drugs may also cause respiratory depression, alter the systemic metabolic balance or otherwise indirectly damage the nervous system. In many cases, the differential diagnosis of these states requires laboratory confirmation of the presence of an offending drug or toxin. When specific drugs are known or suspected to be available to the patient, the search is simplified. Often, however, a "toxic screen" for common substances is necessary because the circumstances surrounding an ingestion are unclear.

Physical findings may raise suspicion of a certain class of drugs; for example, small pupils suggest opiates and widely dilated pupils suggest drugs with atropine-like effects such as tricyclic antidepressants or others like the amphetamines with adrenergic actions. Unfortunately, many situations of intoxication involve multiple substances, whether "street" drugs or medications.

Effect of metabolic diseases

Hypoglycemia. The earlier discussion of cerebral metabolism made clear how important is an uninterrupted supply of glucose to the brain. At a blood glucose of 300 mg/L or lower, consciousness is impaired and then lost. At levels of 200 to 300 mg/L, oxygen use is reduced by 25%, and at 100 to 200 mg/L, oxygen use drops by half. Prolonged hypoglycemia eventually leads to energy failure, a drop in ATP levels, loss of membrane integrity and damage to various cellular processes, and finally irreversible neuron damage. The accumulation of hydrogen ions in tissue during these conditions can contribute in a major way to the damage done.

The most common causes of hypoglycemia are insulin overdose and iatrogenic hyperinsulinemia, but it also occurs in alcoholics and with various pancreatic and hepatic diseases and occasionally with insulin-secreting tumors.

Blood glucose levels approaching dangerous levels (300 to 400 mg/L) should be corrected quickly; since one cannot know, given a single sample, whether glucose levels are rising, falling, or remaining steady, dangerously low levels must be reported without delay. Prolonged hypoglycemia can lead to permanent and profound dementia if the patient survives.

Hyperglycemia. Diabetic ketoacidosis also causes impaired consciousness. The associated hyperosmolar state usually seen with blood glucose levels of more than 10,000 mg/L is almost certainly a primary cause, but the effects of dehydration, systemic acidosis, and perhaps the local accumulation of hydrogen ion in brain tissue may also contribute. There is little correlation, however, between acidosis and coma in this condition.

Liver failure. Signs of diffuse cerebral dysfunction ranging from mild encephalopathic features such as quiet apathetic delirium to seizures and coma can accompany liver disease. Examples are cirrhosis or the side effects of "portal systemic shunts."

Several possible toxic agents have been proposed as contributing to brain dysfunction in hepatic disease, but none fully explains the findings. Blood *ammonia* levels rise as the urea-producing system of the liver fails to detoxify nitrogenous products from the gut. Although it is clear that elevated ammonia levels do not fully explain hepatic encephalopathy, clinicians often follow blood ammonia levels to monitor the progress of liver-induced encephalopathy. When the diagnosis is in doubt, highly elevated CSF α-ketoglutamate levels are highly indicative of liver-induced encephalopathy.

In liver failure, routine CSF examination is usually normal, though some bilirubin may be found if the serum bilirubin level is high. Serum bilirubin levels of 40 to 60 mg/L are necessary before that substance appears in the spinal fluid. In practice, when bilirubin does appear in the CSF, serum bilirubin will usually be in the 100 to 150 mg/L range or greater.

Severe liver damage can also result in the loss of glycogen stores so that hypoglycemia occurs. Finally, hepatic encephalopathy is almost invariably accompanied by hyperventilation, with a resulting lowering of arterial Pco_2 (an accompanying metabolic alkalosis may mask the increased respiratory drive, however). Since an encephalopathy *not* accompanied by respiratory alkalosis (or metabolic alkalosis) is almost certainly not caused by liver failure, arterial gases may be important diagnostically.

Renal disease. Kidney failure leads to the uremic state and an associated encephalopathy characterized by confusion, delirium, and finally coma. Seizures and myoclonic jerks are common. A restless agitation is perhaps more characteristic of rapidly developing renal failure, and a quiet, drowsy confusioned state is more usual in chronic states. As with liver failure, the precise mechanisms of CNS impairment are incompletely understood. An altered blood-brain barrier, which allows entry of substances normally excluded from the CNS (such as certain dialyzable molecules) can be crucial. In any case, the magnitude of azotemia (elevation of BUN) is not well correlated with neurological symptoms, and to some extent the rapidity of onset of the uremic state rather than the absolute rise of BUN seems more likely to determine the severity of symptoms. Urea itself is not the offending agent, since urea infusions do not lead to the typical uremic encephalopathy.

The CSF may show changes indicative of an aseptic meningitis with lymphocytes, PMN's (generally less than $200/mm^3$), and moderately elevated protein (rarely greater than 1000 mg/L). Systemic acidosis is generally not reflected in the CSF. Finally, some element in water intoxication may be present when serum osmolality is less than 260 mOsm/L.

Dialysis may itself cause neurological abnormalities. Because the blood-brain barrier is only slowly permeable to urea and other "osmogens," as the serum osmolality falls during dialysis the brain becomes hyperosmolar relative to blood and water enters the CNS. If the dialysis is too rapid, the resultant CNS water intoxication can lead to its own encephalopathy. Although serious changes are uncommon and occur in less than 5% of patients, as many as half of all patients on dialysis experience at least mild dialysis disequilibrium symptoms.

Pulmonary disease. The lungs perform two critical functions of gas exchange: they allow oxygen from ambient air to diffuse rapidly into the blood and allow carbon dioxide to diffuse rapidly from the blood into the lungs and out into the ambient air. Some neurological consequences of hypoxia and anoxia have previously been considered in this chapter. But rising carbon dioxide levels (so-called hypercarbia) has its own important impact on cerebral function.

The rapidity of onset of respiratory insufficiency is a significant factor determining its impact on brain function. For example, patients with chronic lung disease who habitually run Pco_2 levels of 60 mm Hg may show no cerebral symptoms, or only the slightest dulling of intellectual processes. The same level of hypercarbia developing rapidly will likely render a patient stuporous. The degree of CSF acidosis accompanying elevated Pco_2 is probably an important determinant of impairment. Severe hypercarbia acts as an anesthetic; it has little effect on O_2 consumption but leads to decreases of 40% to 50% in glucose use.

There are many diseases that cause CO_2 narcosis, but they fall into three general categories: (1) those involving intrinsic damage to alveolar surfaces leading to impaired gas exchange, such as emphysema; (2) those that impair the rate or depth of respiration, such as neuromuscular disease, poisoning with respiratory depressants such as barbiturates, brainstem damage, and many others; and (3) those that impede or block air flow, such as strangulation, aspiration, or increased dead space.

The symptoms of CO_2 narcosis or encephalopathy usually begin with dull headache and progress to mental confusion, drowsiness, stupor, and finally coma.

Endocrine diseases

Diabetes. Diabetes mellitus can affect the nervous system by a number of paths. In this chapter some examples that were already discussed are hyperosmolar states, diabetic ketoacidosis, hypoglycemia usually caused by insulin overdose or relative fasting, complications of the treatment of ketoacidosis (that is, CNS acidosis resulting from too-rapid correction of systemic acidosis) together with CNS hyperosmolarity leading to an influx of fluid into the CNS, and consequent cerebral edema. The latter also results from a too-rapid correction of systemic hyperosmolarity.

Adrenal disease. Inadequate release of cortisol affects the brain in ways that are complex and not well understood. In chronic untreated states, apathy, depression, fatigue, and even mild delirium are not uncommon. Stupor and coma usually occur only in abrupt severe worsening of chronic illness, the so-called addisonian crisis. Several metabolic derangements—hyponatremia, hyperkalemia, hypoglycemia, and hypotension—occur and add their own insult to the brain.

Excess glucocorticoid products and the administration of steroid medications are associated in some patients with disturbances of mood (depression, elevation, or hypomania), mild confusion, delusions, hallucinations, impaired insight, and grossly inappropriate behavior.

Thyroid disease. The thyroid hormones thyroxine (T_4) and triiodothyronine (T_3) have important effects on brain function. Both hyperthyroidism (thyrotoxicosis) and hypothyroidism (myxedema) predictably affect the brain.

Hypothyroidism. In the fetus and during infancy, hypothyroidism can cause irreversible brain damage and profound mental retardation (cretinism) unless the condition is corrected without delay. Chronic thyroid insufficiency is associated with depression or lability of mood, listlessness, confusion, and sometimes psychosis (that is, delusions and hallucinations usually of a depressive or persecutory nature). Peripheral neuropathy and unsteady gait related to impaired cerebellar functions also occur together with abnormal deep tendon reflexes. For obscure reasons, elevated CSF protein is a common finding in hypothyroidism.

So-called myxedema coma accompanied by decreased body temperature, slowed respirations, and hypometabolism usually occurs in a setting of chronic hypothyroidism on which some acute event is superimposed, such as infection, surgery, trauma, or congestive heart failure. Because myxedema coma is rapidly fatal, correct diagnosis and prompt treatment are essential.

Thyrotoxicosis. Signs of hypermetabolism distinguish the state of thyroid excess that is associated with disturbances in thinking and emotion.

CHANGE OF ANALYTE IN DISEASE (Table 37-2)

Appearance of cerebrospinal fluid

CSF is normally crystal-clear and free from all pigmentation, that is, "clear and colorless." One should examine it in a glass tube while comparing it with a tube of water. Both are held in white light against a pure white back-

Table 37-2. Change of analyte in central nervous system diseases

Disease	Analyte in central nervous system					
	Glucose	Total protein	IgG	IgG index	Xanthochromia	Lactic acid
Stroke (cerebral infarction)	N	↑	N	↓	N, ↑	N, ↑
Hemorrhage	N	N, ↑ ↑	N	N	↑ ↑	N
Epilepsy	N	N	N	N	N	N
CNS tumor	N, ↓	↑	N, ↑	↓	N, ↑	N, ↑
Infection						
Fungal, bacterial	↓	↑	↑	↑	N	↑
Viral	N	N	↑	↑	N	N
Coma	↑ ↑ hyperosmolar ↓ hypoglycemia	↑ (trauma)	N	N	N, ↑ (trauma) N	N
Meningitis, viral	N	N, ↑	N, ↑	↑	N	N

N, Little or no change; ↑ , Increase; ↑ ↑ , Large increase; ↓ , Decrease.

ground. It is best to look down the long axis of the tube, and at least 1 ml of fluid should be observed.

A red cell count of $500/mm^3$ gives a pink or yellow tinge to the fluid. White counts of $200/mm^3$ will give a slightly cloudy appearance. Xanthochromia (a yellow tinge) will appear when blood has been mixed with CSF. This yellowing does not occur immediately but requires from 2 to 4 hours. Since blood in the CSF may represent either subarachnoid bleeding or be simply the result of a "traumatic tap" (a common occurrence), the CSF sample should be centrifuged immediately. If this is done promptly, bleeding from a traumatic tap should produce no xanthochromia. Xanthochromia suggests that the bleeding occurred at least 2 to 4 hours before observation of the sample. It may be helpful to know that even after a subarachnoid hemorrhage, CSF may not become xanthochromic until 2 to 4 hours have passed. As many as 10% of patients with subarachnoid hemorrhage actually have clear CSF at 12 hours, but beyond that time 100% will show xanthochromia if the sample is examined carefully.

Protein will also give a slightly xanthochromic appearance when present in concentrations greater than 1500 mg/L. Hemoglobin from hemolyzed red cells appear in CSF after only about 10 hours. When a patient is jaundiced (that is, when serum bilirubin is elevated, as it may be in liver failure), bilirubin may enter the CSF. However, this requires serum bilirubin levels of at least 100 to 150 mg/L before CSF xanthochromia is found.

Proteins of cerebrospinal fluid

In diseased states the local production or modification of proteins within the central nervous system may lead to diagnostically useful changes in CSF protein patterns. In general, diseases that interrupt the integrity of the capillary endothelial barrier lead to an increase in total CSF protein. Examples are brain tumor, purulent (bacterial) meningitis, cerebral infarction, and trauma. The elevation in total CSF protein that often occurs in hypothyroid states is poorly understood but probably reflects changes in the blood-brain barrier.

Immunoelectrophoresis allows further fractionation of CSF protein constituents. The major "immunoglobulins" in CSF are IgG, IgA, and IgM (with only trace amounts of IgD and IgE). Of all these, IgG is quantitatively the most important. It is often useful to know whether an elevated IgG value is caused by local production of that immunoglobulin within the central nervous system (as may be the case in some demyelinating diseases such as multiple sclerosis) or by its leaking across a damaged blood-brain barrier (as in some infections). Since the normal serum IgG is 15% to 18% of total serum protein and normal CSF IgG is 5% to 12%, the ratio of IgG to total protein is sometimes used to estimate the source of IgG elevation. That is, if the ratio in a sample more nearly approximates the ratio ordinarily found in serum, one tends to suspect that the IgG has been somehow transferred into the CSF from serum. But this is a crude and not especially reliable estimate. Another measure currently popular is the IgG-albumin index. The formula for determining it is as follows:

$$\text{IgG index} = \frac{\text{IgG (CSF)} \times \text{Albumin (serum)}}{\text{IgG (serum)} \times \text{Albumin (CSF)}}$$

The upper limit of normal for this index must be determined for each laboratory, but generally it ranges between 0.25 and 0.85. The IgG index is elevated in diseases in which there is increased CNS IgG production and an intact blood-brain barrier (as in infections) and decreased CNS IgG production when the blood-brain barrier is compromised, allowing serum proteins to cross into the CSF (as in strokes, some tumors, and some forms of meningitis).

Myelin basic protein has aroused interest as a potential indicator of demyelination. Myelin is a complex substance that surrounds many CNS axons like the insulation on a wire cable and is necessary for normal conduction of nerve impulses down the axon. "Demyelinating diseases" are a group of disorders in which the primary insult is some form of damage to the myelin coating of CNS axons. The resulting abnormal propagation of impulses in the brain and spinal cord leads to a variety of clinical signs and symptoms. Myelin basic protein (MBP) is one constituent of normal myelin and has been found to be elevated in a variety of conditions involving myelin damage: multiple sclerosis is an important example but not the only disease associated with elevated myelin basic protein. In other words, MBP is a very nonspecific indicator of disease; because it lacks specificity, it is being used less commonly in clinical neurology.

Gamma globulin synthesis within central nervous system

Although elevations of CSF gamma globulin may be secondary to changes in serum proteins (for example, small molecular weight Bence Jones proteins in multiple myeloma cross the blood-brain barrier and appear in the gamma fraction of CSF proteins), there is evidence that local CNS immunoglobulin production occurs in many diseases. Examples include multiple sclerosis, subacute sclerosing panencephalitis (a devastating process of myelin damage that occurs in children and young adults in association with greatly elevated CSF measles titers), many chronic and acute infections (neurosyphilis, tuberculous meningitis, abscess, viral meningoencephalitis, and sarcoidosis), and some brain tumors. As a practical point, in those settings in which CSF total protein rises as a result of increased permeability of the blood-brain barrier, the addition of serum protein (which normally contains 15% to 18% gamma globulin) raises the CSF gamma fraction. It thus becomes difficult to estimate the upper limit of normal for gamma globulin as a percentage of total protein when total protein is significantly elevated. In any case, when CSF gamma globulin is elevated, a clinician may order a simultaneous serum protein electrophoresis to help determine the source of the increased CSF gamma fraction.

Oligoclonal bands

The gamma fraction is composed of a variety of proteins. Agarose gel electrophoresis performed on concentrated CSF demonstrates elevation of a population of proteins within the gamma range that share the same electrophoretic mobility. The population of gamma proteins called "oligoclonal bands" are believed to derive from a few groups of immunocompetent cells or "clone." The appearance of oligoclonal bands has been reported in 79% to 90% of patients with multiple sclerosis and in the variety of CNS inflammatory conditions. Interestingly, this change in the composition of the gamma fraction may occur without any increase in the total gamma globulin concentration. The significance of oligoclonal immunoglobulins in the CSF is under active investigation. They seem to signify that some antigenic stimulus (perhaps a virus or altered protein) has triggered the replication of a family of immunocompetent cells, all responsive to the same antigen. How these immune cells relate to CNS damage is also under active investigation. They may be either a cause or effect of damage to the nervous system, but at the present time it is not known which. Whatever their relation to demyelination, inflammation, and some degenerative conditions (disorders in which parts of the nervous system degenerate for reasons as yet not understood), CSF total protein, gamma globulin, and oligoclonal bands are currently of considerable diagnostic importance.

A final practical point about protein determinations in CSF. When there is blood in the CSF from a traumatic tap or bleeding within the nervous system, one expects that the blood will elevate the CSF protein value. One can still determine the CSF protein level by correcting for the amount of blood present. Simply allow 10 mg/L of protein for every 1000 red blood cells per cubic millimeter. For example, if the red cell count is 10,000 in the CSF sample and its total protein is 1000 mg/L, the corrected total protein equals 900 mg/L. The cell count should be performed as rapidly as possible after lumbar puncture, preferably in the first half hour and certainly not later than 2 hours, since hemolysis will be occurring after that time.

Glucose in cerebrospinal fluid

Determination of CSF glucose helps distinguish bacterial from viral meningitis—the glucose value often being quite *low* (less than 40% to 45% of simultaneously analyzed, equilibrated serum glucose) in bacterial meningitis and tuberculous meningitis and generally normal in viral disease. Carcinomatous meningitis (widespread infiltration of the meninges by tumor cells) also drives CSF glucose values below the normal range.

Anticonvulsant drugs

Fortunately, a number of effective anticonvulsant drugs are available: phenobarbital, phenytoin, valproic acid, primidone, and carbamazepine, which are excellent for a wide range of seizure types; diazepam (Valium) and lorazepam, which are used acutely only to interrupt a seizure

Table 37-3. Commonly used anticonvulsant drugs

Drug	Optimal therapeutic range (μg/mL)
Phenytoin	10-20
Phenobarbital	15-40
Carbamazepine	4-8
Ethosuximide	40-100
Valproic acid	50-100
Primidone	5-12

or flurry of seizures; ethosuximide used for petit mal epilepsy; and clonazepam used primarily for myoclonic seizures. Table 37-3 lists some optimal therapeutic ranges of commonly used anticonvulsant drugs.

In general, a physician begins therapy with one anticonvulsant drug believed likely to be effective and increases the dosage until either adequate control of the seizures is achieved or a toxic side effect that cannot be overcome occurs. The dosage is then lowered, a second drug is added, and the process is repeated. (A hypersensitivity reaction such as rash, great suppression of white count, and evidence of actual liver damage requires that the drug be stopped.)

Although the patient's clinical response—control of seizures, failure to control, or development of side effects—is the principal concern, monitoring serum levels of anticonvulsant drugs is quite useful. For example, failure to control seizures is often traced to inadequate serum levels, and that inadequacy is in turn traced to poor compliance, poor absorption of the drug, or too-rapid metabolism of it. One can adjust dosage accordingly. Symptoms can be related to drug toxicity or toxicity may be ruled out as the explanation. And finally, monitoring of anticonvulsant levels gives important information about drug kinetics and interactions. Periodic measurements of anticonvulsant serum levels is a routine part of the management of seizures.

BIBLIOGRAHY

Adams, R.D., and Victor, M.: Principles of neurology, New York, 1977, Mc-Graw-Hill Book Co.

Fishman, R.A.: Cerebrospinal fluid in diseases of the nervous system, Philadelphia, 1980, W.B. Saunders Co.,

Plum, F., and Posner, J.B.: The diagnosis of stupor and coma, ed. 3, Philadelphia, 1980, F.A. Davis Co.

Glasser, L.: Tapping the wealth of information in CSF, Diagn. Med., pp. 23-33, Jan.-Feb. 1981.

Killingsworth, L.M., Cooney, S.K., Tyllia, M.M., and Killingsworth, C.E. Deciphering cerebrospinal fluid patterns, Diagn. Med., pp. 23-29, March-April 1980.

Chapter 38 Endocrine function

acromegaly Pathological condition caused by the hypersecretion of growth hormone in adults.

activation Change in the tertiary structure of a cytoplasmic, steroid hormone receptor induced by the binding of a specific hormone to the receptor; allows translocation of the complex to the nucleus.

adenohypophysis Anterior lobe of the pituitary gland that secretes trophic hormones.

amenorrhea Absence or abnormal stoppage of menstrual flow.

cyclic AMP Cyclic 3′,5′-adenosine monophosphate; acts as a second messenger, relaying the message of a protein hormone to the nucleus where specific gene stimulation occurs, leading to the final biochemical change or changes specific for that hormone.

diabetes insipidus A disorder caused by a deficient quantity of antidiuretic hormone manifested by excessive urine secretion.

endocrine gland A ductless gland that secretes hormones directly into the bloodstream.

feedback loop A loop modulating an interaction between two endocrine organs.

galactorrhea Persistent secretion of milk, irrespective of nursing.

gigantism A pathological condition caused by the hypersecretion of growth hormone in children.

hormone Chemical used to transmit messages to a distant target organ to elicit a specific response or action.

hormone transport The manner in which a hormone is carried in the blood, that is, protein bound or free.

hypothalamic nucleus Anatomical structure composed of neurons with a similar function.

hypothalamus Portion of the brain involved in the endocrine and autonomic functions.

immunocytochemical stains Cytological method for detection of hormones based on labeled antibodies.

neurohypophysis Part of the pituitary gland that is an extension of the central nervous system.

panhypopituitarism Deficient secretion of all the pituitary hormones.

pituicyte Cell of the neurohypophysis.

pituitary pericapillary space Intermediate space between the secretory cell and the capillary sinusoid.

pituitary portal system Vascular system connecting the hypothalamus with the adenohypophysis.

pituitary stalk Structure connecting the pituitary gland with the hypothalamus.

receptor Specific cellular protein that must first bind a hormone before a cellular response can be elicited; located in either the cell membrane or the cytoplasm.

release-inhibiting factors Peptides secreted by the hypothalamus that inhibit secretions by the adenohypophysis.

releasing factors Peptides secreted by the hypothalamus that stimulate secretions by the adenohypophysis.

sella turcica A depression in the base of the skull that accommodates the pituitary gland.

sexual precocity Unusually early development of primary and secondary sex characteristics.

target endocrine organ Endocrine gland whose function is controlled by the pituitary gland.

translocation Process by which a steroid hormone–cytoplasmic receptor complex moves from the cytoplasm into the nucleus, where it can activate specific genes.

trophic hormones Hormones secreted by the adenohypophysis that stimulate the target organs. The suffix "-tropic" means a 'turning' or, 'changing'; "trophic" means a 'nurturing' or 'causing growth'; both are used in regard to hormones.

Part I: General aspects of hormone physiology

BIDY KULKARNI

INTRODUCTION AND DESCRIPTION OF ENDOCRINE FUNCTION

Transmission of information in the body is brought about by the nervous and the endocrine systems. In the nervous system the message is carried between the peripheral tissue and the brain by means of a specially built nerve network, whereas in the endocrine system the message is carried from one cell to another by the bloodstream through chemical substances called "hormones." These hormones are chemical messengers produced by specialized cells in endocrine or ductless glands such as the adrenals, gonads, and pituitary. The hormones are released directly into the blood for transport to a distant site, called the "target tissue," where they bring about their specific action. The hormones possess a high degree of structural specificity, and any alteration in their molecular composition results in a dramatic change in their physiological activity.

There are three main classes of hormones, each with different characteristics with regard to structure, chemical composition, transport, metabolism, and mechanisms of action. These hormones, listed in Table 38-1, fall into three chemical classes: (1) the protein hormones such as adrenocorticotropic hormone (ACTH), follicle-stimulating hormone (FSH), luteinizing hormone (LH), and thyroid-stimulating hormone (TSH); (2) the aromatic amines such as thyroxine and triiodothyronine, and (3) the steroid hormones such as testosterone, estradiol, and cortisol. The names and locations of the various endocrine glands in the body are shown in Fig. 38-1.

Among the endocrine glands, the most important are the anterior pituitary and the hypothalamus. These two glands function as the primary regulators of the entire endocrine system by synthesizing and releasing a series of protein and polypeptide hormones into the bloodstream. In re-

Table 38-1. Summary of endocrine hormones*

Hormones	Target tissues	Chemical nature	Primary actions
Steroid hormones			
Androgens (male sex hormones)			
Testosterone (and other androgens)	Peripheral tissues, testis	Steroid	Stimulates development of secondary sex characteristics in men, protein anabolic effect
Esterogens (female sex hormones)			
Estrone, estradiol	Peripheral tissues, ovary	Steroid	Stimulates development of secondary sex characteristics in women, protein anabolic effect
Progesterone	Uterus	Steroid	Prepares uterus for implantation of ovum
Adrenal cortex hormones			
11-Deoxycortisol	Renal distal tubule, large intestine	Steroid	Maintains electrolyte and water balance
Cortisone, or hydrocortisone (cortisol)	Muscle, liver, adipose cells	Steroids	Stimulates gluconeogenesis from amino acids, anti-insulin effects on glucose and fat metabolism
Aldosterone	Renal distal tubule, large intestines	Steroid	Regulates retention of sodium ions, excretion of potassium ions
Amino acid–derived hormones			
Adrenal medulla hormones			
Epinephrine Norepinephrine	Liver, adipose cells	Phenolic amines	Stimulates glycogenolysis, hypertensive effect
Thyroid hormones	Most tissues	Amino acid (iodinated): thyroxine or derivatives	Regulates rate of metabolism, increases serum glucose
Protein and polypeptide hormones			
Anterior pituitary			
Thyroid-stimulating hormone (TSH), thyrotropin	Thyroid	Protein	Stimulates development and secretion of thyroid gland
Adrenocorticotropic hormone (ACTH)	Adrenal cortex	Protein	Stimulates growth and secretion of adrenal cortex
Follicle-stimulating hormone (FSH)	Ovarian follicle (women) Seminiferous tubules (men)	Protein	Stimulates growth of follicles and production of estrogen in women, formation of spermatozoa in men
Luteinizing hormone (LH)	Corpus luteum (women) Testes (men)	Protein	Triggers formation of corpus luteum and production of progesterone in women, production of androgens by interstitial cells in men
Prolactin (lactogenic hormone, LTH)	Milk glands	Protein	Initiates lactation
Growth hormone (GH)	Most tissues	Protein	Stimulates growth (also affects fat and carbohydrate metabolism)
Posterior pituitary hormones			
Vasopressin, (antidiuretic hormone, ADH)	Renal collecting ducts, bladder	Peptide	Stimulates reabsorption of water in kidney tubule
Oxytocin (Pitocin)	Uterus, mammary glands	Peptide	Contracts uterus
Calcitonin (thyroid gland)	Bone	Peptide	Lowers serum calcium
Parathyroid hormone	Bone, small intestines, kidney	Protein	Regulates blood calcium
Pancreatic hormones			
Insulin	Most cells	Peptide	Facilitates carbohydrate catabolism
Glucagon	Liver	Peptide	Raises blood glucose by hepatic glycogenolysis
Gastrointestinal hormones			
Secretin	Pancreas, gallbladder	Peptide	Stimulates flow of pancreatic juice and bile to a much smaller extent
Cholecystokinin-pancreozymin	Gallbladder, pancreas	Peptide	Contraction of gallbladder; stimulates secretion of pancreatic enzymes
Gastrin	Stomach	Peptide	Stimulates secretion of gastric juice (HCl)
Renal erythropoietin	Bone	Peptide	Stimulates red cell formation
Hypothalamic factors	Anterior pituitary	Peptides	Stimulates or inhibits release of corresponding trophic hormones

*From Toporek, M.: Basic chemistry of life, St. Louis, 1980, The C.V. Mosby Co.

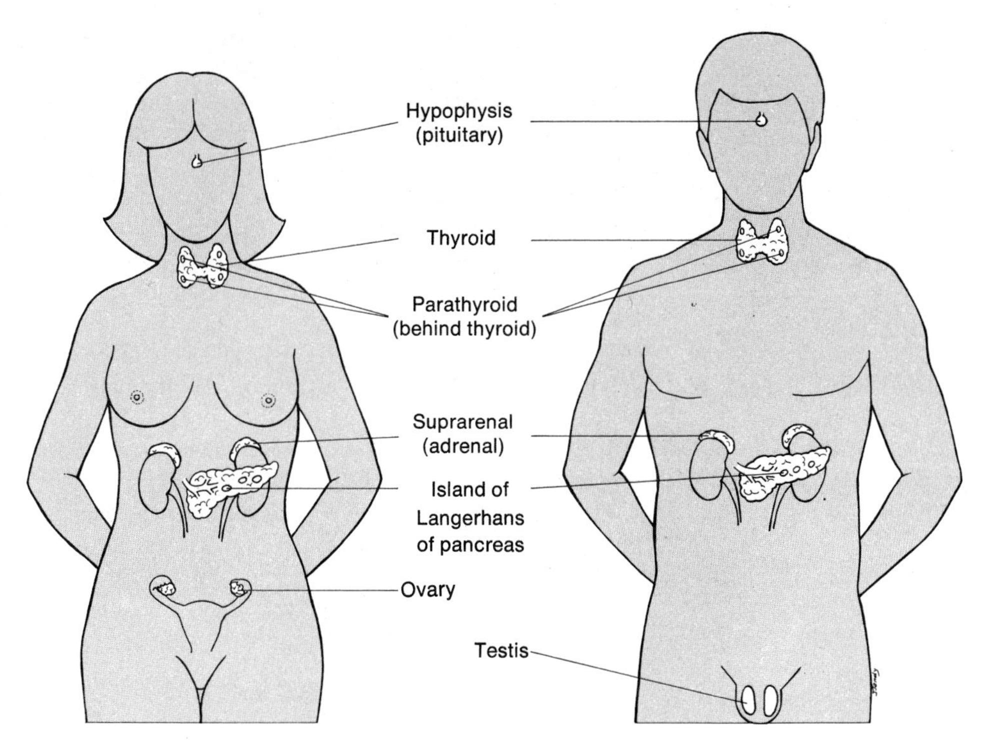

Fig. 38-1. Location of endocrine glands. (From Toporek, M.: Basic chemistry of life, St. Louis, 1980, The C.V. Mosby Co.)

sponse to external and internal stimuli, the hypothalamus secretes a number of neuropeptides, called "releasing hormones," into the portal circulation of the pituitary gland. These hormones, formerly called "releasing factors," include thyrotropin-releasing hormone (TRH), gonadotropin-releasing hormone (GnRH), somatostatin releasing-inhibiting factor (SRIF) and prolactin releasing-inhibiting factor (PIF or PRIF). These small polypeptide hormones are characterized by a rapid onset of action and a brief biological half-life. They regulate production and release of the anterior pituitary hormones.

Under control of the hypothalamic hormones, the anterior pituitary secretes a group of protein hormones, called "trophic hormones," which are named for the target organ they affect. These hormones include thyroid-stimulating hormones (TSH), luteinizing hormones (LH), follicle-stimulating hormone (FSH), adrenocorticotropic-stimulating hormone (ACTH), growth hormone (GH), melanin-stimulating hormone (MSH), and prolactin. These pituitary hormones regulate hormonal release from their respective target endocrine glands such as the gonads, the thyroid, and adrenals. FSH and LH together are referred to as "pituitary gonadotropins." The third gonadotropin, human chorionic gonadotropin (HCG), is secreted by the placenta during pregnancy (see Chapter 35).

Table 38-2. Serum carrier proteins for hormones

Hormone	Protein (in order of importance)
Testosterone	Sex hormone–binding protein (SHBG) Albumin
Estradiol	Albumin Sex hormone–binding protein
Progesterone	Corticosteroid-binding globulin (CBG)
Cortisol	Corticosteroid-binding globulin
Thyroxine (T_4)	Thyroxine-binding globulin (TBG) Thyroxine-binding prealbumin (TBPA) Albumin

HORMONE TRANSPORT

Once the hormone is released by the endocrine gland, it is transported to the site of action by the bloodstream. The protein hormones, being water-soluble substances, move freely in the bloodstream. The steroid and thyroid hormones, being insoluble in water, are rendered soluble by binding to carrier proteins such as albumin, sex hormone–binding protein, and transcortin or cortisol-binding globulin (Table 38-2). The aromatic amines are also transported in the blood attached to specific serum proteins, namely, thyroxine-binding globulin and thyroxine-binding prealbumin.

MECHANISMS OF REGULATORY CONTROL OF HORMONAL LEVELS

Secretion and release of a hormone by an endocrine gland is not a steady and continuous process but is a dy-

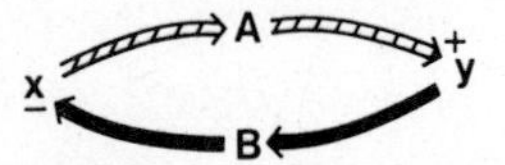

Fig. 38-2. Scheme of feedback loop system in which gland *X* releases product *A*, which has a positive feedback on gland *Y* to produce substance *B*. Increased concentrations of *B* feedback on gland *X* to inhibit the release of product *A*.

namic event varying in response to endogenous and exogenous stimuli. There are two regulatory mechanisms that govern the activity of an endocrine gland. One is the feedback mechanism, which directly controls the secretory activity of the endocrine gland, and the other is a series of interacting regulatory mechanisms of the hypothalamus–pituitary–target gland, which maintains the integrity of the entire endocrine system.

The primary mechanism with which the body maintains the levels of circulating hormones within set physiological limits is by the feedback mechanism. A feedback mechanism is defined as one in which two variables are interdependent in regard to their degree of function. There are two general types of feedback systems operative in biological systems. These are the positive feedback system, which is rarely seen, and the more common negative feedback system. In a typical endocrine system an increase in one variable, for example, A, causes an *increase* in the other variable, B. In a negative-feedback system, an increase in B causes a *decrease* in A. These two systems can combine into one continuous loop. For example, compound A causes an increase in the synthesis of B, which causes a decrease in the synthesis of A, thus tending to oppose the primary change (A) (Fig. 38-2). This combined system is used for the hypothalamic–pituitary–target gland control system.

Control of endocrine gland function is based on the use of the feedback principle. This control is exerted either directly on the gland or by the hypothalamic pituitary axis. An example of feedback control directly on the endocrine gland is the regulation of extracellular fluid (ECF) calcium by parathyroid hormone (PTH). Rising levels of PTH increase the concentration of calcium in ECF through its effect on bone, kidney, and gut (positive feedback). The rising levels of calcium in ECF cause the parathyroid to suppress secretion of PTH (negative feedback).

The functional relationship between the hypothalamus, pituitary, and an endocrine gland is discussed in greater detail in Part II of this chapter. Fig. 38-3 illustrates the relationships between the pituitary gland hormones and their target organs.

MECHANISMS OF HORMONE ACTION

Steroid and protein hormones act as messengers, transporting information to all parts of the body through the bloodstream. The hormones, however, circulate in extremely low concentrations. Blood levels of these hormones range from 10^{-10} to 10^{-12} M, whereas glucose circulates in the blood at the level of 10^{-2} M, and sodium at the level of 10^{-1} M. It is therefore imperative that these chemical messengers be selectively picked up only by the

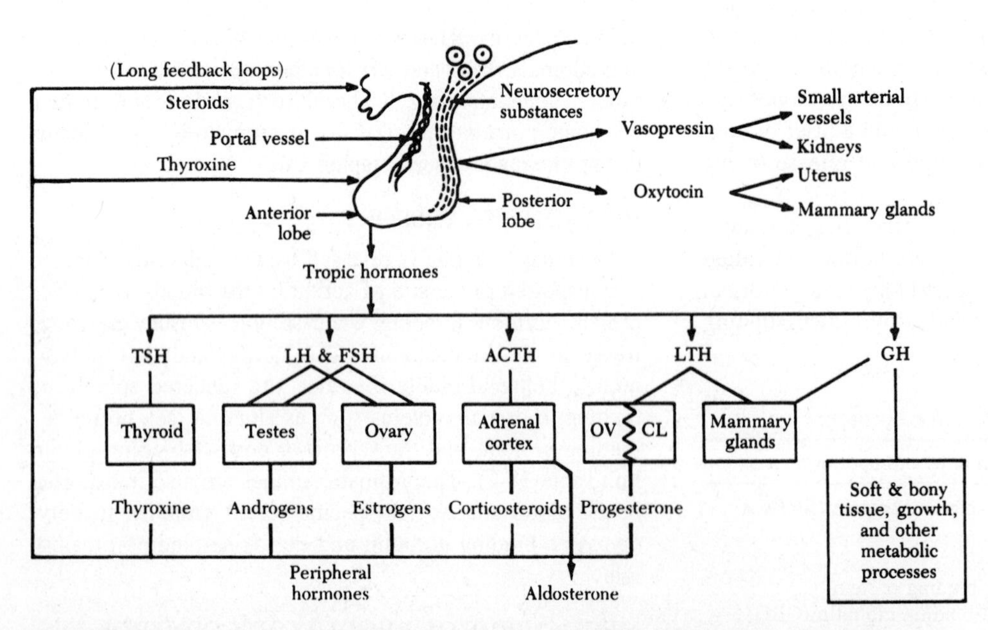

Fig. 38-3. Scheme of relationships between hormones of anterior and posterior lobes of pituitary gland and their target structures. *CL,* Corpus luteum; *OV,* ovary. (From Orten, J.M., and Neuhaus, O.W.: Human biochemistry, ed. 10, St. Louis, 1982, the C.V. Mosby Co.)

target cells and that the message they carry be accepted, processed, and translated into the appropriate response. This highly selective function is carried out by means of hormone receptors located only in the target cells, which thus determine end-organ specificity.

Receptors are cellular proteins with the unique ability to bind to a single type of hormone molecule. For example, estrogen receptors will bind to estrogens and not to androgens or progesterones. The receptors of the ovary will specifically bind only to LH. This selectivity of receptor binding is caused by their specific primary, secondary, and tertiary molecular structures. Alterations in receptor structures can compromise the specificity, resulting in a physiological disorder. For example, the lysine-lysine-arginine-arginine sequence of a receptor protein is extremely important in the binding of ACTH to the adrenal cell. Receptors of adrenal tumor–like cells apparently lose this amino acid sequence and in turn the receptor specificity, thereby permitting binding by TSH, FSH, or LH to this site.

The mechanism of action of protein hormones differs considerably from that of steroid hormones with respect to intracellular location of the receptors and their mode of action. Protein and thyroid hormone receptors are present in the plasma membrane of the target cell, whereas the steroid hormone receptors are exclusively within the cytoplasm. In addition, most of the protein hormones initiate their actions by activating the membrane-located enzyme adenyl cyclase without entering the cell, whereas the steroid hormones' actions are mediated by the cytoplasmic and nuclear receptor proteins after the entry of the steroid hormone into the cytoplasm of the target cell.

Mechanism of action of protein hormones

Protein hormones, being unable to pass through the plasma membrane because of their molecular composition, attach themselves to receptors on the external surface membrane of the target cell (Fig. 38-4). After binding of the hormone to the receptor, the entire complex is brought into the cell membrane, thus protecting it from degradation by extracellular enzymes. (This is called "internalization.") Once within the cell membrane, the receptor-hormone complex stimulates the membrane-bound enzyme adenyl cyclase, which activates the conversion of adenosine triphosphate (ATP) to cyclic 3′,5′-adenosine monophosphate (cAMP). The cAMP functions as a second messenger, relaying the message of the protein hormone (first messenger) to initiate biochemical processes inside the cell:

$$\text{Endocrine gland} \xrightarrow[\text{(hormone)}]{\text{First messenger}} \text{Target cell} \xrightarrow[\text{(cAMP)}]{\text{Second messenger}} \text{Metabolic effects}$$

It is the cAMP that directly transmits the message, probably by activation of the enzyme called protein kinase (PK), which in turn activates other regulatory intracellular enzymes. The cAMP firmly and specifically binds to a receptor protein localized in the cytosol and endoplasmic reticulum of the cell to form a cAMP-receptor complex, which activates the release of PK. The cAMP-receptor complex renders the cAMP resistant to the action of phosphodiesterase, which readily degrades the unbound cAMP to 5′-AMP. Protein kinase, which is present in both the

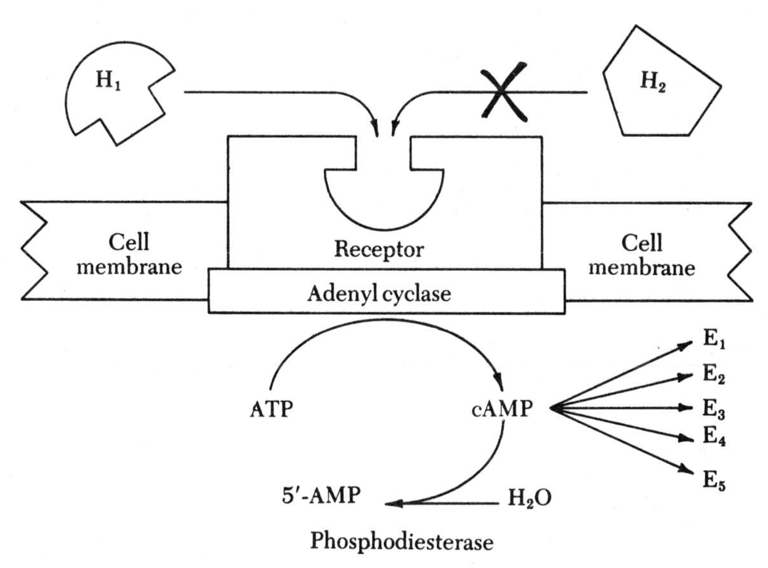

Fig. 38-4. Concept of hormone action by way of second messenger, cyclic AMP. H_1 is hormone recognized by specific membrane receptor site, whereas H_2 is a hormone that is not sensed in the environment. *E_1 to E_5*, Various metabolic effects of increased cAMP. (From Orten, J.M., and Neuhaus, O.W.: Human biochemistry, ed. 10, St. Louis, 1982, The C.V. Mosby Co.)

cytoplasm and the nucleus, acts as a link between cAMP and the activation of the biochemical pathway by bringing about an array of complex enzymatic reactions involving phosphorylase, esterase, and desmolase, leading to the final regulatory effect of the hormone.

An example of this process is the effect of luteinizing hormone (LH) (first messenger) on the ovarian cell. LH, interacting with the cell membrane reactor, activates adenyl cyclase to synthesize cAMP (second messenger) to accelerate steroidogenic processes in the ovarian cell. These actions of enzymes activated by cyclic AMP may be summarized as follows: (1) increased cholesterol formation from cholesterol ester by activated esterase, (2) transport of cholesterol into the mitochondria, where the side-chain cleavage enzyme is located, (3) activation of side-chain cleavage enzyme system, (4) increase in concentration of the cofactor nicotinamide adenine dinucleotide phosphate (NADPH) for conversion of cholesterol to pregnenolone, (5) transport of an end product such as pregnenolone out of the mitochondria.

Mechanism of action of steroid hormones

The iodinated thyroid hormones and all steroid hormones act on a wide variety of target cells to alter their function, using the same general mechanism to effect changes in cellular activity through participation of cell receptors.

Fig. 38-5 summarizes the most generally accepted mechanism of steroid hormone action, though significant differences do exist between hormones. The various steps involved in the mechanism of the steroid hormone action are (1) passage through the target cell membrane, (2) binding to a cytoplasmic receptor, (3) activation of the cytoplasmic receptor, (4) transfer of the activated cytoplasmic receptor (ACR) complex across the nuclear membrane to the nucleus, (5) binding of the ACR complex to specific sites in nuclear chromatin DNA, (6) synthesis of new RNA, and (7) synthesis of proteins resulting in cellular growth or normal function.

The unbound or free steroid hormone, being a low-molecular-weight (less than 500 daltons) and fat-soluble compound, diffuses freely across the plasma membrane into the cytoplasm of any tissue cell. The responsiveness of a cell to the hormone depends on the presence of an appropriate receptor protein. All the target tissue cells are bestowed with the unique feature of having specific receptors in their cytoplasm. Steroid hormone receptors are proteins with molecular weights of about 200,000 daltons and are divided into the three functional classes of steroid hormones that they bind: estrogens, androgens, and progester-

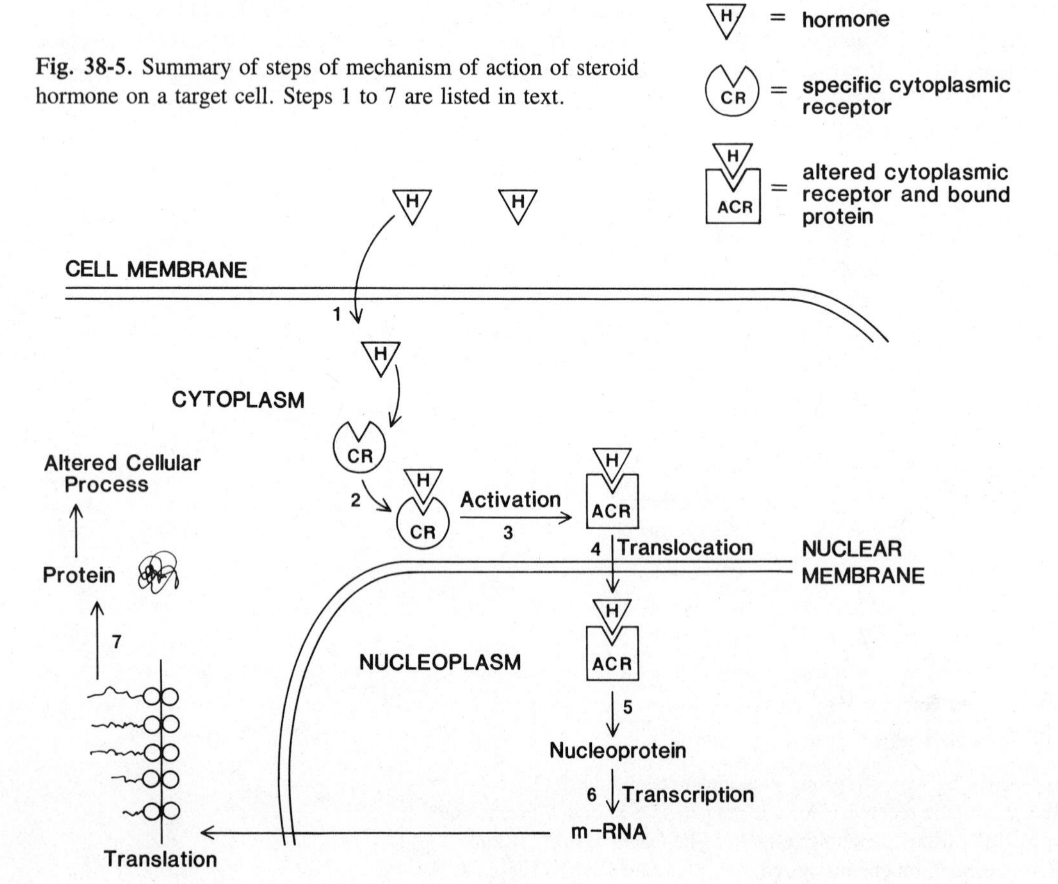

Fig. 38-5. Summary of steps of mechanism of action of steroid hormone on a target cell. Steps 1 to 7 are listed in text.

ones. These receptors have been identified in target cells for estrogens, progesterone, androgens, cortisol, aldosterone, and even the sterol 1,25-dihydroxyvitamin D_3.

The concentration of the receptors in the tissue determines its responsiveness. Many tissues that have few or no receptors are not responsive to that hormone. Testicular feminization, a syndrome of congenital androgen insensitivity, is an example of absence of androgen receptors in the gonad. Variation in receptor concentration rather than total absence may explain some problems of inadequate hormone response.

The specific role of the cytoplasmic receptors is to form a complex with the steroid. Each receptor binds to its respective hormone with great specificity and is present in a significant amount only in the target tissue cells. This complex is then transported, or translocated, to the nucleus. Like estradiol, progesterone and testosterone stimulate response in target tissues by their specific cytoplasmic receptors, namely, progesterone receptors and testosterone receptors. Unlike estradiol and progesterone, however, testosterone is not a potent hormone but rather a prehormone. After its entry within the cytoplasm it does not directly bind with the androgen-receptor sites. Instead, it is first metabolized by the membrane-bound enzyme 5′Δ-reductase, which converts it to a more potent androgen, dihydrotestosterone (DHT). It is the DHT that binds to the cytoplasmic receptor to form the receptor complex for subsequent transmission to the nucleus.

Once the steroid hormone enters the target cell by diffusion through the cell membrane, it is quickly bound by the specific receptor in the cytoplasm. After hormone binding the receptors are believed to undergo conformational change called *transformation,* a steroid-induced effect that allows the passage of the steroid-receptor (SR) complex through the nuclear membrane. *Translocation* is the passage of SR complex from the cytoplasm through the nuclear membrane into the nucleus after transformation.

Within the nucleus the SR complex binds to nonhistone protein of chromatin. This interaction leads to stimulation of a specific RNA polymerase. The formation of new messenger RNA (mRNA) from a segment of DNA is called "transcription" and is catalyzed by the enzyme RNA-polymerase. The mRNA then migrates into the cytoplasm, where it attaches to ribosomes. The ribosomes "read" the genetic code, a process called "translation." Each code word specifies an amino acid that is brought to the ribosome by transfer RNA and oriented for protein formation by ribosomal RNA. The ribosomes sequentially join the amino acids together to form proteins.

Thus steroid hormones bring about their effect by stimulating protein synthesis, which begins with the transfer of genetic information from DNA to mRNA in the target cell after translocation of the steroid receptor complex.

Catabolism of hormones

The protein hormones are degraded by proteases present in serum and cells. The small polypeptides are filtered by the glomerulus and degraded by tubular cells. The aromatic amino acids are catabolized by deiodination or oxidation to inactive forms.

The protein-bound steroid hormones are converted to more water-soluble, inactive forms by hydroxylation, oxidation, or reaction with glucuronic acid or sulfate to form conjugates of these compounds, which are excreted by the kidney. Peripherally, the active steroids are oxidized to less-active hormones, which are then converted by the liver to inactive forms.

Part II: The hypothalamopituitary system

BOLESLAW H. LIWNICZ
REGINA G. LIWNICZ

ANATOMY AND HISTOLOGY

Pituitary gland

Gross appearance. The pituitary gland (hypophysis cerebri) lies at the base of the brain in a depression of the sphenoid bone called the "sella turcica" (Turkish saddle). The pituitary gland is connected to the hypothalamus (a portion of the brain) by a stalk that exits the sella turcica through an opening in a membranous diaphragm (diaphragma sella). It weighs approximately 0.6 g in an adult man and 1 g in an adult woman during her reproductive age.

The pituitary gland is composed of two major parts: the anterior lobe (adenohypophysis) and the posterior lobe (neurohypophysis). These two parts are of different embryological origin, which is reflected in their function. The adenohypophysis, contributing almost three fourths of the weight of the gland, is composed of three parts: pars dis-

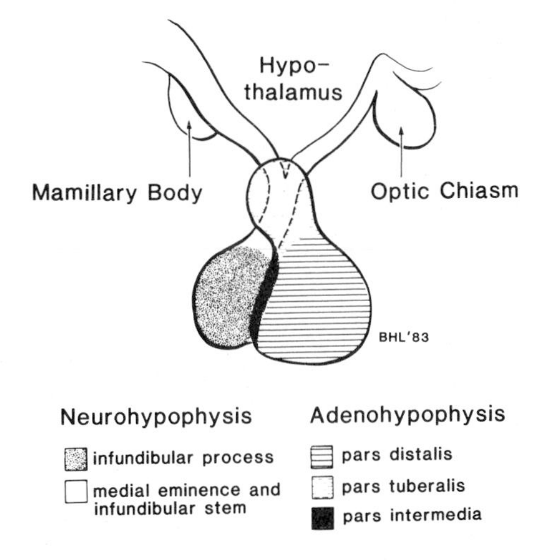

Fig. 38-6. Major parts of pituitary gland.

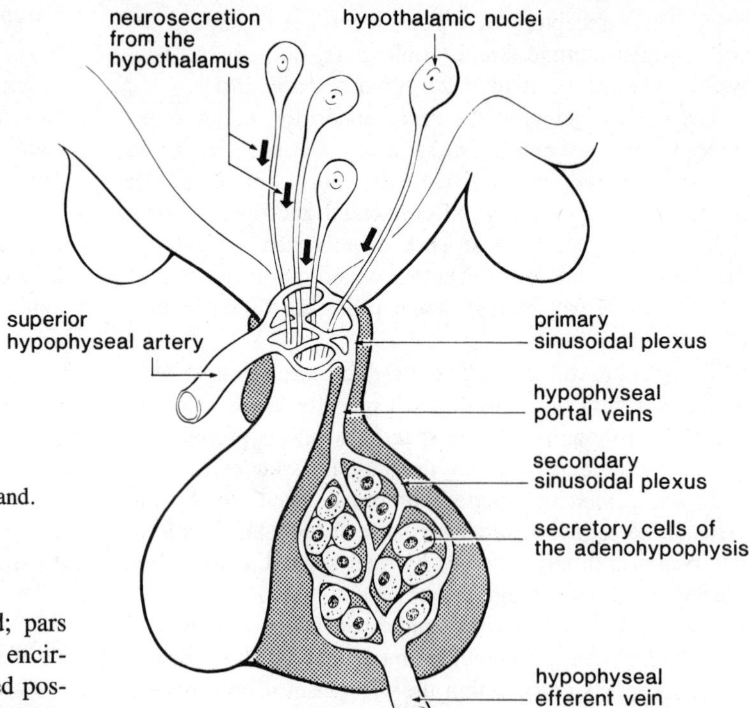

Fig. 38-7. Sinusoid portal system of pituitary gland.

talis (the largest), the frontal portion of the gland; pars tuberalis, located superiorly to the pars distalis and encircling the pituitary stalk; and pars intermedia, situated posteriorly on the border with the neurohypophysis. The neurohypophysis is composed of two parts—the neural lobe and the infundibulum, which communicates with the hypothalamus (Fig. 38-6).

There is no direct arterial supply to the pars distalis of the adenohypophysis. The superior hypophyseal arteries form a capillary bed that extends upward into the medial eminence of the hypothalamus and drains into portal veins of the pars tuberalis, which in turn give rise to a second capillary bed within the pars distalis (Fig. 38-7). This sinusoid portal system has a significant function in connecting the hypothalamus with the adenohypophysis. The hypothalamic hormones, which modulate the secretion of trophic hormones of the adenohypophysis, are released into this system.

Cells of adenohypophysis. On routine hematoxylin-and-eosin stain the adenohypophysis shows three cell populations: acidophils, staining red; basophils, staining blue; and chromophobes, not staining. Experimental studies and clinicopathological correlation show that acidophils secrete growth hormone (GH) and prolactin (PRL), basophils secrete gonadotropins (follicle-stimulating hormone [FSH] and luteinizing hormone [LH]) and thyroid-stimulating hormone (TSH), and the chromophobes secrete adrenocorticotropic hormone (ACTH). The cells secreting the same hormone, however, may differ significantly in appearance by light microscopy, depending on the amount of secretory granules and endoplasmic reticulum. For example, prolactin-secreting cells are acidophilic when containing abundant secretory granules and chromophobic when depleted. Only immunocytochemical stains based on labeled antibody against pituitary hormones allow a definite determination of the type of secretory cells (Table 38-3).

The basic functional unit of the adenohypophysis is a secretory cell within the surrounding sinusoidal capillaries. The secretory cells release the hormone to the pericapillary space, from which it freely diffuses through the endothelial pores into the lumen of the sinusoidal capillaries (bloodstream). The same route, but in the opposite direction, serves for transport of the releasing factors into the secretory cells (Fig. 38-8).

Cells of neurohypophysis. Cells of the neurohypophysis, the pituicytes, resemble astrocytes and have no known secretory function. The neurohypophysis is an endocrine storehouse. The two hormones found in it, antidiuretic

Table 38-3. Cell types of adenohypophysis

Hematoxylin and eosin	Trophic hormones
Acidophilic cells	Prolactin (PRL)
	Growth hormone (GH)
Basophilic cells	Follicle-stimulating hormone (FSH)
	Luteinizing hormone (LH)
	Thyroid-stimulating hormone (TSH)
Chromophobic cells	Adrenocorticotropic hormone (ACTH)
	Prolactin (PRL)
	Growth hormone (GH)
	Follicle-stimulating hormone (FSH)
	Luteinizing hormone (LH)
	Thyroid-stimulating hormone (TSH)
	No hormone

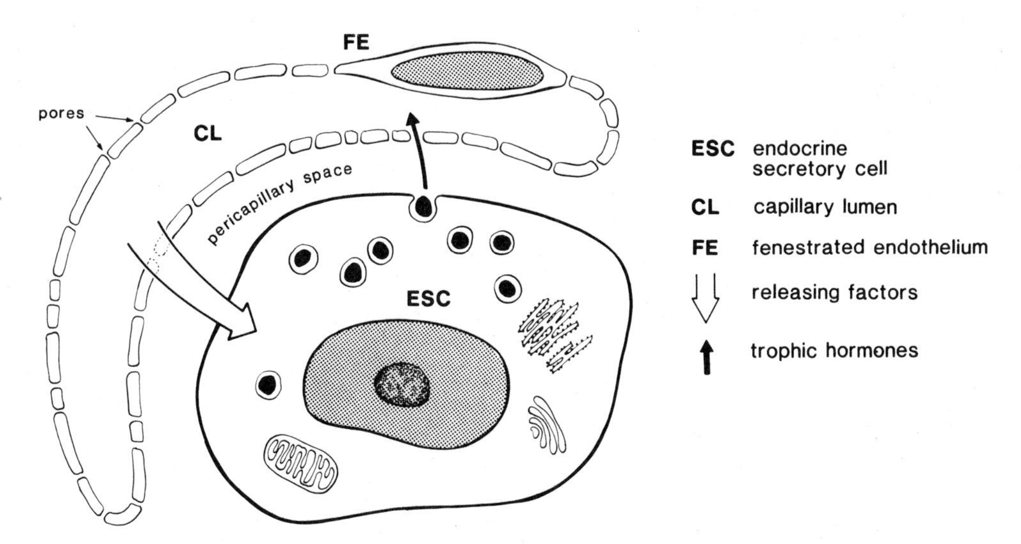

Fig. 38-8. Basic functional unit of adenohypophysis (secretory cell, pericapillary space, and sinusoid capillary).

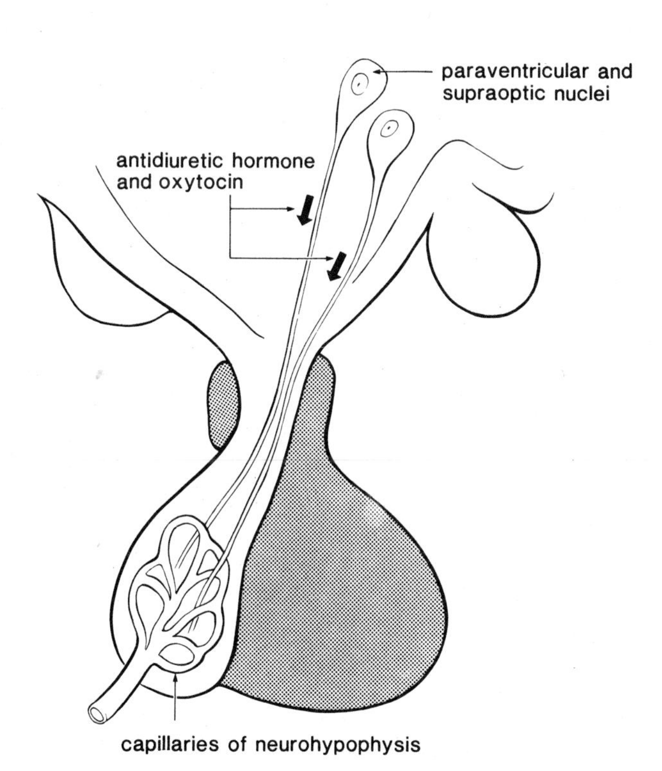

Fig. 38-9. Secretory pathway of neurohypophysis.

hormone (ADH, vasopressin) and oxytocin, are synthesized in the hypothalamic nuclei and transported through the pituitary stalk into the neurohypophysis, where they are secreted directly into the bloodstream or stored in extracellular globules (Herring bodies) (Fig. 38-9).

Hypothalamus

The hypothalamus is a portion of the central nervous system. It lies at the base of the brain beneath the thalamus and is connected to the pituitary gland below by the pituitary stalk. The hypothalamic neurons form many nuclear groups that either have neuroendocrine function or control the autonomic nervous system. The nuclei of the autonomic system control appetite, sexual drive, body temperature, and water and calorie balance. Small neurosecretory cells forming parvicellular nuclei (arcuate, ventromedial, and suprachiasmatic nuclei and nuclei of the medial eminence) secrete releasing or release-inhibiting neuropeptides, which are transported to the adenohypophysis by the pituitary portal system. Another type of neurosecretory cell is large, with long processes. These cells form paraventricular and supraoptic nuclei. The processes of these cells transverse the pituitary stalk and enter the neurohypophysis, where they release their hormones.

FUNCTION OF HYPOTHALAMOPITUITARY SYSTEM

The adenohypophysis secretes trophic hormones that stimulate the secretion of target endocrine glands, or di-

Table 38-4. Action of hormones of pituitary gland

Trophic hormone	Structure (molecular weight in daltons)	Target tissue	Hormone of target endocrine organ
GH	Protein, MW 27,000	Liver, kidney, soft tissue, bone	Somatomedin
PRL	Protein, MW 25,000	Breast	None
FSH	Glycoprotein, MW 30,000 (α and β subunits)	Ovary, testis	Estradiol
LH	Glycoprotein, MW 26,000 (α and β subunits)	Ovary, testis	Progesterone, testosterone
TSH	Glycoprotein, MW 25,000 (α and β subunits)	Thyroid	Thyroxine, triiodothyronine
ACTH	Peptide, MW 45,000	Adrenal cortex	Cortisol
ADH	Cyclic octapeptide, MW 1000	Collecting ducts of renal nephrons	None
Oxytocin	Cyclic octapeptide, MW 1000	Uterus, breast	None

Table 38-5. Hormones or factors of hypothalamus acting on adenohypophysis

Hormone or factor	Abbreviation	Effect on adenohypophysis: Stimulation	Effect on adenohypophysis: Inhibition
Thyrotropin-releasing hormone	TRH	TSH PRL (may not be physiological) GH FSH (in men may not be physiological)	
Luteinizing hormone-releasing hormone or gonadotropin-releasing hormone	LHRH or GnRH	FSH LH	
Prolactin-releasing factor	PRF	PRL	
Growth hormone-releasing factor	GHRF	GH	
Corticotropin-releasing factor	CRF	ACTH	
Growth hormone release–inhibiting hormone (somatostatin)	GHRIH (SRIF)		GH TSH
Prolactin release–inhibiting factor	PIF		PRL

rectly, the metabolism and growth of tissues. The main trophic hormones are growth hormone (GH), prolactin (PRL), follicle-stimulating hormone (FSH), luteinizing hormone (LH), thyroid-stimulating hormone (TSH), and adrenocorticotropic hormone (ACTH) (Table 38-4). Their secretion depends on (1) the blood level of the hormone secreted by the target endocrine organ, (2) the metabolic state of the patient, and (3) the control of the central nervous system through secretion of releasing factors and factors inhibiting the release of pituitary trophic hormones (Table 38-5).

The interaction of these components is complex and the action of the hypothalamic–pituitary–target endocrine gland–end cell system is explained by a system of feedbacks. The feedback loop is composed of two endocrine organs, the pituitary and the target endocrine gland. The pituitary secretes the trophic hormone, which stimulates the target endocrine gland to secrete its hormone. The hormone of the target endocrine gland in turn affects the secretion of the pituitary gland. The hormone secreted by the target endocrine gland can inhibit the secretory function of the pituitary, lowering the level of trophic hormones (negative feedback), or it can stimulate the pituitary to secrete more of its trophic hormone (positive feedback). For example, negative feedback exists between the pituitary and the thyroid gland. The pituitary secretes TSH, which stimulates the secretion of thyroxin by the thyroid. The elevation of the thyroxin serum level has an inhibitory effect on the secretion of TSH by the pituitary (and the hypothalamus) (Fig. 38-10). This mechanism protects the organism from excessive thyroxin secretion, which could elevate the metabolism to a damaging level. An example of positive feedback is the interaction between the secretion of FSH and LH by the pituitary and the secretion of estradiol by the graafian follicle of the ovary. At a threshold plasma level, estradiol stimulates the secretion of FSH and LH. FSH in turn further stimulates the secretion of estradiol. Potentially this could raise the level of estradiol dangerously high. Fortunately, when the graafian follicle reaches maturity, a synergistic action of FSH and LH results in its rupture and the destruction of the cells secreting estradiol (Fig. 38-11). The negative feedback preserves the homeostasis of the organism, whereas the positive feedback occurs only in systems in which a transient exacerbation of growth is needed.

The above examples describe only the interaction between the pituitary and endocrine target organ. One can further expand the feedback model to include the hypo-

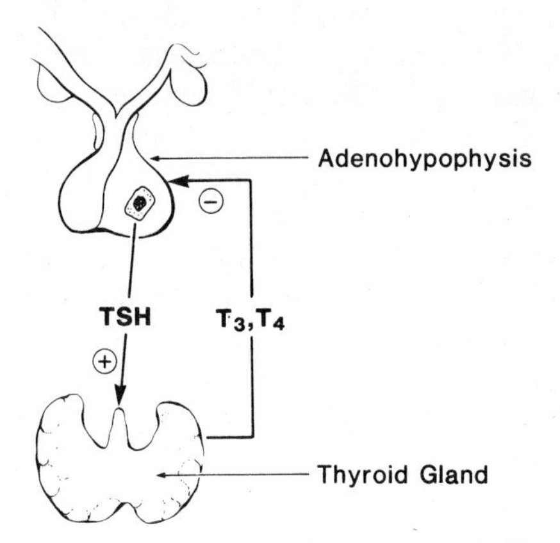

Fig. 38-10. Negative-feedback loop between pituitary gland and thyroid gland. T_3, Triiodothyronine; T_4, thyroxine; ⊕, positive effect; ⊖, negative feedback.

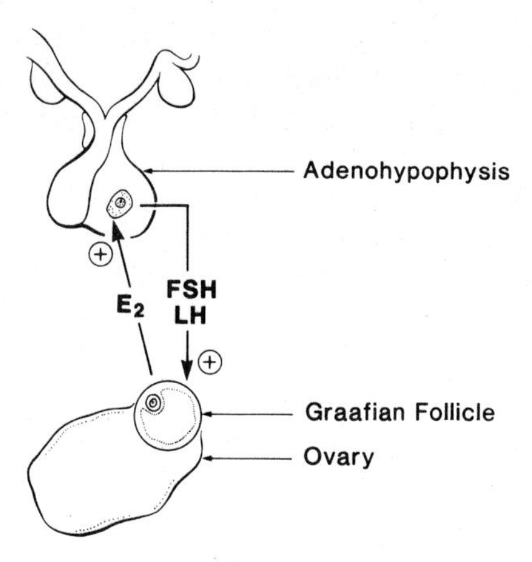

Fig. 38-11. Positive-feedback loop between pituitary gland and graafian follicle of ovary. This functional unit is unique for a living organism. E_2, Estradiol; *FSH,* follicle-stimulating hormone; *LH,* luteinizing hormone; ⊕, positive feedback.

thalamus by adding a feedback loop between the hypothalamus and the target endocrine organ. This feedback loop incorporates a releasing factor secreted by the hypothalamus, which stimulates the release of a corresponding trophic hormone from the pituitary, which in turn stimulates the hormone of the endocrine target organ. This hormone stimulates the hypothalamus to secrete an inhibitory releasing factor. This type of feedback is called the indirect inhibitory feedback because it incorporates an intermediate organ, the pituitary. It also differs from the direct negative feedback in that it not only inhibits the secretion of the releasing factor but can also stimulate the secretion of a release-inhibiting factor. An example of this system is the hypothalamic-pituitary-thyroid axis (Fig. 38-12). Another version of the negative feedback is prolactin secretion, in which there is no known target endocrine organ; therefore a short feedback loop involves only the hypothalamus and the pituitary. In the positive feedback there is probably also an indirect loop involving the hypothalamus, pituitary, and ovaries.

Two trophic hormones, GH and PRL, function by direct stimulation of the target cells without the intermediate involvement of the target endocrine organ (Table 38-4). PRL directly influences secretions of the breast, and GH promotes growth in soft tissue, cartilage, and bones. The four other adenohypophyseal hormones act through the stimulation of intermediary target endocrine organs. TSH stimulates the thyroid gland to synthesize and secrete its hormones: triiodothyronine (T_3) and thyroxine (T_4) (Table 38-4). FSH in women stimulates the development of follicles of the ovary and in men the spermatogenesis in the testes. LH in women precipitates ovulation by acting synergistically with FSH and later maintains the secretory function of corpus luteum. In men, this hormone is called "interstitial cell–stimulating hormone" (ICSH) and acts primar-

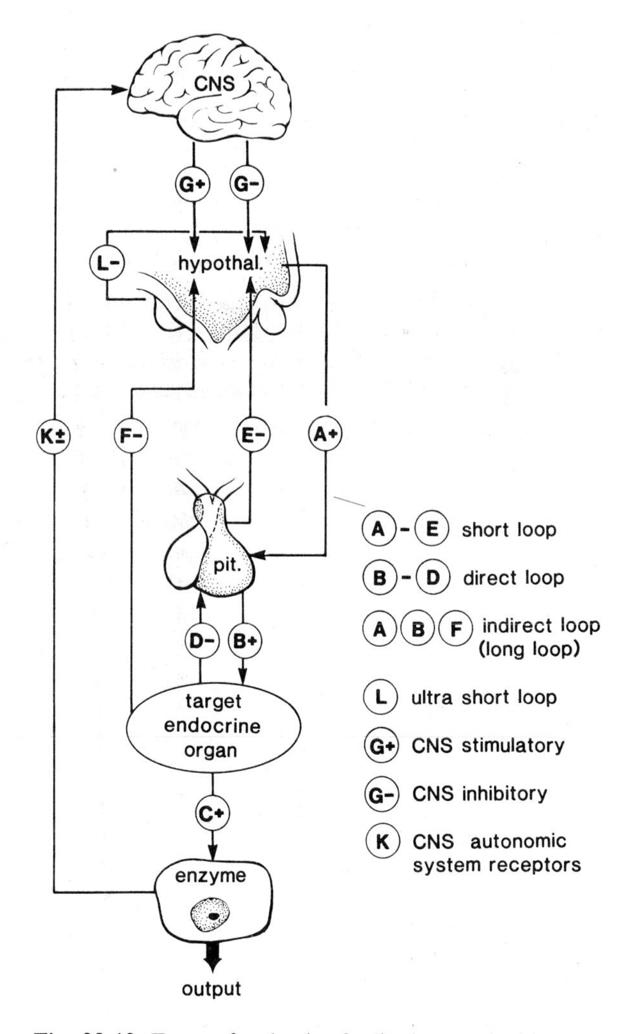

Fig. 38-12. Types of endocrine feedbacks. +, Positive effect; −, negative feedback.

ily by stimulating the interstitial Leydig cells in the testes to secrete testosterone. ACTH stimulates the two inner zones of the cortex of the adrenal glands, the zonae fasciculata and reticularis, which secrete the glucocorticoid hormones and a small quantity of sex hormones.

The neurohypophysis stores and releases two hormones made by the hypothalamic neurons: antidiuretic hormone (ADH), which acts directly on the collecting ducts of the renal nephrons regulating the reabsorption of water, and oxytocin, which in women stimulates the contractions of the muscles of the uterus and the lactating mammary glands.

PATHOLOGICAL CONDITIONS

Adenohypophysis

The clinical manifestations of disorders of adenohypophysis can be the result of hormone hypersecretion or hyposecretion. The hypersecretion usually involves only one hormone but is not uncommonly associated with hyposecretion of other trophic hormones. For example, hypersecretion of PRL can be associated with hyposecretion of FSH. The hyposecretion usually involves more than one trophic hormone.

Hypersecretion. The following factors can cause pituitary hypersecretion:

- Primary factors
  - Pituitary adenomas, benign tumors of the pituitary gland.
  - Pituitary hyperplasia.
- Secondary factors
  - A lack of release-inhibiting factors caused by damage in the hypothalamus or the hypothalamopituitary tract. This is known to cause the hypersecretion of PRL.
  - Extrapituitary (ectopic) secretion of trophic hormones by carcinomas (Table 38-6).
  - Hyposecretion of endocrine target organs.

Growth hormone. The effect of hypersecretion of GH depends on the age of the patient. In adults, it leads to a progressive enlargement of the acral (distal) parts of the extremities, called "acromegaly." The patient's feet, hands, and jaw enlarge, appearing strong and massive. This appearance is deceiving because the thickening of the bone is not associated with remodeling; thus it leads to loss of bone mass called "osteoporosis" (friable bone). A similar process occurring in the vertebrae can lead to scoliosis, pathological curving of the spinal column. The soft tissues expand, which is best seen on the face, giving coarse features. There is also a hypertrophy of the internal organs. The hypertrophy of the heart muscle can lead to cardiac insufficiency. The symptoms usually progress slowly and initially can escape detection. In children, the hypersecretion of GH causes gigantism (characterized by excessive increase in size), which can lead to a growth of over 8 feet. Unfortunately, the accelerated growth is associated with osteoporosis and muscle weakness. Acromegaly and gigantism are usually caused by a pituitary adenoma secreting GH and compressing the adjacent tissues of the pituitary gland, causing hyposecretion of other trophic hormones.

Prolactin. Hypersecretion of PRL causes galactorrhea (lactation). In women this can be associated with infertility and amenorrhea (Forbes-Albright syndrome). PRL hypersecretion is usually caused by pituitary adenoma.

Adrenocorticotropic hormone. The symptoms are caused by enhanced quantities of cortisol and androgens secreted by the adrenal cortex. The resulting clinical picture (Cushing's syndrome) is described in Chapter 43. Cushing's syndrome in about 75% of cases is caused by excessive production of ACTH. The remaining cases are caused by the primary diseases of the adrenal gland. Hypersecretion of ACTH by the pituitary is usually caused by pituitary microadenomas or hyperplasia and is called "Cushing's disease."

Thyroid-stimulating hormone. In the past, hyperthyroidism (Chapter 39) was believed to be always caused by hypersecretion of TSH, but this view is no longer held. In fact, TSH secretion is greatly reduced in thyrotoxicosis. Only very rarely is the thyrotoxicosis caused by oversecretion of TSH, and when it occurs, it usually is associated with acromegaly.

Gonadotropins. Sexual precocity associated with increased gonadotropin secretion is usually induced by cerebral tumors in the region of the hypothalamus.

Hyposecretion. Pituitary hyposecretion occurs as hyposecretion of one hormone, a group of hormones, or a total hyposecretion of all hormones, otherwise called "panhypopituitarism." A single hormone hyposecretion is commonly the result of lesions within the hypothalamus. They are related to an uncontrollable elevation of release-inhibiting factors or a lack of releasing factors. In pituitary adenomas, not uncommonly there is a hypersecretion of one hormone, which is made by the tumor, accompanied by hyposecretion of the remaining pituitary hormones be-

Table 38-6. "Ectopic" pituitary hormones produced by nonendocrine tumors

Hormone	Tumor
GH	Poorly differentiated lung carcinoma
PRL	Oat cell carcinoma of lung
	Renal carcinomas
ACTH	Oat cell carcinoma of lung
	Carcinoids of lung, thymus, gut, and pancreas
	Medullary carcinoma of thyroid
	Pheochromocytoma
	Neuroblastoma
ADH	Oat cell carcinoma of lung
Oxytocin	Oat cell carcinoma of lung

cause of a destruction of the pituitary gland by the compressing tumor. Adenomas can destroy up to three fourths of the pituitary mass without any evident clinical symptoms, and a patient can survive a destruction of up to 90% of the pituitary mass. However, the pituitary does not regenerate to a significant degree; therefore a massive loss of cells of the pituitary has to be treated with administration of the pituitary hormones. There are many factors that can destroy the pituitary gland (Table 38-7). The clinical symptoms do not depend on the etiological factor but merely reflect the lack of trophic hormones.

Depending on the location of the destruction, a different set of clinical symptoms will occur. In intrasellar lesions the symptoms are usually secondary to deficiency of all the pituitary hormones. Hypothalamic lesions can result in the hyposecretion of a single hormone and also involve other autonomic functions, such as centers of appetite, thirst, or sexual drive. Therefore a patient can have hypogonadism caused by the lack of gonadotropins associated with obesity as a result of excessive appetite. An increase in size of an intrasellar tumor can lead to a compression of the optic chiasm, which manifests itself in a lack of peripheral vision.

Growth hormone. Deficiency of GH in children leads to pituitary dwarfism. The child is small but proportionally built. This condition is usually not recognized until after the first year of life.

Prolactin. The only known clinical effect of the deficiency of PRL is the lack of lactation in postpartal women.

Adrenocorticotropic hormone. The clinical picture is essentially similar to changes seen in Addison's disease (Chapter 43), but there is no hyperpigmentation and only slight impairment in aldosterone secretion.

Thyroid-stimulating hormone. This effect of the hormone may be difficult to differentiate from primary defect of the thyroid gland. Usually there is evidence of other trophic hormone deficiencies, for example, gonadal deficiency.

Gonadotropins. These are usually the first trophic hormones to be affected in lesions of adenohypophysis. The child is often tall for his age, and at adolescence secondary sex characteristics fail to develop. In adults, secondary sex characteristics show signs of atrophy, with scanty or absent pubic and axillary hair, infertility, and possible loss of libido (Chapter 41).

Neurohypophysis

Hypersecretion of antidiuretic hormone (vasopressin). An excessive secretion can occur in a wide variety of clinical conditions, such as meningitis, brain abscesses, head injury, cerebrovascular thrombosis, tuberculosis, hypoadrenalism, hypothyroidism, and hepatic serosis. This syndrome is associated with hyponatremia and urine osmolality in excess of that of plasma with normal renal and adrenal function. The clinical symptoms include weakness, malaise, and poor mental status.

Hyposecretion of antidiuretic hormone (vasopressin). The clinical entity associated with the hyposecretion of ADH is diabetes insipidus, also known as "inappropriate secretion of vasopressin." It can be caused by lesions directly involving the neurohypophysis, hypothalamus, or hypothalamopituitary tract. The clinical symptoms are insatiable thirst and intense polyuria (large urine volume).

Table 38-7. Etiological factors in hypopituitarism

Factors within sella	Factors outside sella (hypothalamus)
Tumors (pituitary adenomas and craniopharyngiomas, gliomas, meningiomas or metastatic tumors)	Tumors (craniopharyngioma, gliomas, germinomas, teratoma or metastatic tumors)
Infections (tuberculosis, other basal meningitides, syphilis, encephalitis)	Infections
Granulomas (sarcoidosis, Hand-Schüller-Christian disease)	Granulomas
Vascular (postpartum necrosis, hypertensive hemorrhages, infarcts, diabetes mellitus, cranial arteritis, vascular malformations)	Vascular
Iatrogenic (surgical hypophysectomy, irradiation, prolonged therapy with hormones)	Trauma
Trauma	
Amyloidosis	
Hemosiderosis	

Table 38-8. Range of normal levels of pituitary hormones in blood

Hormone	Conditions	Value
ACTH	8:00 AM	40-120 ng/L
	8:00 PM	< 5-50 ng/L
FSH	Children 1 year of age to puberty	> 2.5 U/L
	Adult men	1-7 U/L
	Adult women	
	Premenopausal	1-10 U/L
	Midcycle peak	6-25 U/L
	Postmenopausal	30-120 U/L
LH	Men	1.5-10 U/L
	Women	
	Early follicular	2.5-15 U/L
	Midfollicular	Up to 20 U/L
	Midcycle	5-70 U/L
	Luteal	Up to 13 U/L
PRL		Up to 800 mU/L
TSH		Up to 5 mU/L
GH	Hypoglycemia	> 0.5 mU/L

The greatest danger in a rapid onset of diabetes insipidus is dehydration. The urine concentration is usually of the order of 100 mOsm/kg. Plasma urea and electrolytes and full blood count show the expected changes of dehydration.

FUNCTION TESTS

To assess the pituitary-hypothalamic function, it is essential to separate the function of the hypothalamopituitary system from the secretory function of the target endocrine organ. This is done by radioimmunoassays for specific trophic hormones and the use of challenge tests. The challenge tests also allow for the separation of endocrine abnormalities caused by pituitary lesions from those caused by hypothalamic lesions. Table 38-8 gives the range of normal values of pituitary hormones in blood, and Table 38-9 lists examples of challenge tests.

Table 38-9. Tests of hypothalamopituitary function

Hormones	Test	Procedure	Normal response	Pituitary lesion	Hypothalamic lesion
GH/ACTH/PRL	Insulin tolerance	0.1-0.3 U/kg of weight intravenously	Increase in GH (above 40 mU/L), ACTH, PRL, corticosteroids (above 550 nmol/L).	Panhypopituitarism: no increase in GH, ACTH, PRL, and corticosteroids Partial hypopituitarism: no increase in GH	No response of ACTH and corticosteroids with normal response to vasopressin
GH/ACTH/PRL	Glucagon (used when insulin is contraindicated)	1 mg subcutaneously	Increase in GH (men at least 14 mU/L, women 20 mU/L), corticosteroids (above 550 nmol/L)	As above	As above
GH	Arginine	30 g intravenously (children 1.5 g/kg)	Increase (at least 14 mU/L)	Hypopituitarism: no increase	
GH in children	Bovril	20 g/1.5 m^2 body surface orally	Increase (above 20 mU/L)	As above	
GH	Oral glucose tolerance	50 g orally	Decrease (4 mU/L)	Hypersecretion (acromegaly and gigantism): no decrease	
LH/FSH	Clomiphene stimulation	50 mg daily for 5 days	Increase in LH and FSH above laboratory norms	Hypopituitarism: no increase in LH and FSH	No increase in LH and FSH
ADH	Water deprivation	No fluids for 8 hours	Urine 600 mOsm/kg or above, plasma not above 300 mOsm/kg, urine flow rate less than 0.5 mL/min		Hyposecretion of ADH (diabetes insipidus): urine less than 270 mOsm/kg, plasma above 300 mOsm/kg, urine volume not reduced
GH/ACTH/PRL/TSH/LH/FSH	Combined pituitary function test	Insulin 0.1-0.3 U/kg of body weight TRH 200 μg LHRH 100 μg intravenous	To insulin—increase in GH (above 40 mU/L), PRL (usually masked by response to TRH) and corticosteroids (above 250 nmol/L) To TRH—increase in TSH (3-15 mU/L) and PRL (1800 mU/L) To LHRH—increase in LH (15-42 U/L) and FSH	Panhypopituitarism: no increase in GH, ACTH, PRL, corticosteroids, TSH, LH, and FSF Hypopituitarism: no increase in individual hormones	Increase in PRL, TSH, LH, and FSH

BIBLIOGRAPHY

General aspects of hormone physiology

Orten, J.M., and Neuhaus, O.W.: Human biochemistry, ed. 10, St. Louis, 1982, The C.V. Mosby Co.
Litwack, G., editor: Biochemical actions of hormones, 2 vols., New York, 1970-1972, Academic Press, Inc.
Chan, L., and O'Malley, B.W.: Mechanism of action of the sex steroid hormones, N. Engl. J. Med. **294:**1322-1326, 1372-1379, 1976.
Sutherland, E.W.: Studies on the mechanism of hormone action, Science **177:**401, 1972.
Robison, G.A., Butcher, R.W., and Sutherland, E.W.: Cyclic AMP, New York, 1971, Academic Press, Inc.

Hypothalamopituitary system

Lee, T., and Laycock, J.L.: Essential endocrinology, Oxford, 1978, Oxford University Press.
Hall, R., Anderson, J., Smart, G.A., and Besser, M.: Fundamentals of clinical endocrinology, Tunbridge Wells, England, 1980, Pitman Medical Publishers, p. 1-74.
Jeffcoate, S.L.: Efficiency and effectiveness in the endocrine laboratory, New York, 1981, Academic Press, Inc.
Alsever, R.N., and Gotlin, R.W.: Handbook of endocrine tests in adults and children, Chicago, 1978, Year Book Medical Publishers.
Beardwell, C., and Robertson, G.L., editors: The pituitary, London, 1981, Butterworth & Co., Publishers, Ltd., pp. 47-75, 175-210, 238-264.
McCain, S.M., editor: Endocrine physiology II, Int. Rev. Physio. **24:**97-156, 1981.
Tolis, G., Martin, J.B., Labrie, F., and Naftolin, F., editors: Clinical neuroendocrinology: a pathophysiological approach, New York, 1979, Raven Press, pp. 1-46, 89-128, 366-384, 437-445.

Chapter 39 Thyroid

Mariano Fernandez-Ulloa
Harry R. Maxon

acromegaly Enlargement of the extremities (especially hands and feet) caused by an increased secretion of pituitary growth hormone.

adrenergic activity Metabolic effects caused or mimicked by the hormone epinephrine (adrenaline).

anabolic agents Compounds (usually androgens) that promote increase or construction of new tissue.

androgen Natural or synthetic substance (usually hormones) that produces masculinizing effects.

aplasia Lack of development of any organ.

C cells Calcitonin-secreting cells of the thyroid.

calorigenesis Production of heat and energy.

choriocarcinoma Highly malignant tumor originating from trophoblastic cells of the placental epithelium.

competitive protein binding (CPB) Property of a protein to bind to another molecule, enabling a labeled molecule to compete with an unlabeled one for the same binding site.

corticosteroid Any natural or synthetic steroid compound with metabolic actions similar to those produced in the cortex of the suprarenal glands.

cretinism Hypothyroid condition caused by congenital lack of thyroid hormone secretion and characterized by impaired physical and mental development.

Cushing's syndrome Group of metabolic alterations and clinical manifestations caused by increased production or administration of corticosteroid hormones.

DIT Diiodotyrosine.

Down's syndrome (mongolism) Condition characterized by mental retardation and distinctive physical characteristics associated with a chromosomal abnormality.

dysgenesis Abnormal or defective development.

dysplasia Abnormal development of an organ resulting in alteration of configuration, size, or cell organization.

embryonal carcinoma Rare and highly malignant tumor of the testis originating from primitive germinal cells.

endemic Present in a particular geographical location at all times.

endocytosis The uptake by a cell of material from the environment by invagination of its plasma membrane.

factitial Produced by artificial means; unintentionally produced.

follicular cells Thyroid hormone–producing cells arranged in spherical vesicle-like units.

gluconeogenesis Production of sugars from non–carbohydrate related substances.

goiter Enlargement of the thyroid gland; *multinodular*—enlarged thyroid containing numerous superficial and deep indurations.

Graves' disease Immune disorder caused by antibodies binding to thyroid-stimulating hormone (TSH) receptors, resulting in an unregulated increase in thyroid hormone production and release.

Hashimoto's thyroiditis Inflammatory process of the thyroid caused by a derangement of the immune system, which may or may not lead to abnormal thyroid function.

hydatidiform mole An intrauterine tumor derived from abnormal epithelial proliferation and cavitation of the ovum and presenting as a mass with cystlike appearance.

hyperfunction Increased function of any organ or system.

hyperlipidemia Increased level of fats in blood.

hyperthyroidism Metabolic and clinical state caused by an increase in circulating active thyroid hormone.

hypofunction Decreased function of any organ or system.

hypoproteinemia Diminished level of proteins in blood.

hypothyroidism Metabolic and clinical state caused by decreased levels of circulating active thyroid hormone or increased tissue resistance; *primary*—decreased thyroid function caused by disease of the thyroid gland; *secondary*—decreased thyroid function caused by disease of the pituitary gland; *tertiary*—decreased thyroid function caused by disease of the hypothalamus.

iatrogenic Any adverse condition resulting from the actions of a physician.

interstitium Pertaining to the intercellular spaces of a tissue.

iodine trapping The ability of the thyroid gland to sequester the iodine against a concentration gradient.

Klinefelter's syndrome Characterized by small atrophic testes, varying degrees of masculinization, infertility, and chromosomal abnormalities.

medullary carcinoma (thyroid) Cancer of the C cells.

MIT Monoiodotyrosine.

monodeiodination Loss or removal of a single iodine atom.

myxedema Advanced hypothyroid state characterized clinically by distinctive external appearance: pallor, skin edema (swelling) of face and hands, apathy, and so on.

neonatal Period of time comprising the first 4 weeks after birth.

nephrotic syndrome Condition characterized by augmented loss of proteins in the urine, decreased albumin in blood, and increased tissue accumulation of fluid (edema) caused by various diseases affecting the kidneys.

organogenesis Incorporation of ionic form of iodine into the molecular structure of tyrosine.

parafollicular C cells Calcitonin-secreting cells located between follicles.

parenchyma A group of basic morphological and functional cellular units that constitute any organ.

porphyria Disease characterized by an alteration of the porphyrin metabolism.

prohormone A compound requiring chemical transformation to become an active hormone.

protein-bound iodine (PBI) The total amount of iodine bound to proteins in the plasma.

resin uptake test Measurement of the number of available binding sites of plasma thyroid hormone–transporting proteins.

solitary nodule Localized enlargement of a portion of the thyroid gland.

struma ovarii Rare teratoid tumor of the ovary composed almost entirely of thyroid tissue.

T_3 Thyroid hormone with iodine atoms in positions 3, 5, and 3′ (triiodotyrosine).

T_4 Thyroid hormone with iodine atoms in positions 3, 5, 3′, and 5′ (tetraiodotyrosine).

rT_3 Triiodotyrosine with iodine in positions 3, 3′, and 5′.

thyroglobulin A glycoprotein of molecular weight 660,000 daltons produced by the follicular cells and containing the precursors of T_3 and T_4.

thyroid colloid The material found within the follicles of the thyroid and containing thyroglobulin and thyroid hormone.

thyroiditis A general term for inflammation of the thyroid gland.

thyrotoxicosis Condition caused by excess thyroid hormone secretion, often used as synonym for hyperthyroidism.

thyroxine-binding globulin (TBG) A glycoprotein of alpha-mobility that transports thyroid hormone in the blood.

TRH (thyrotropin-releasing hormone) A tripeptide that promotes release of TSH.

trophic action Stimulation of cell reproduction and enlargement.

trophoblastic tumor Tumor originating from extraembryonal cells of ectodermic nature located in the blastocyst.

TSH (thyroid-stimulating hormone, thyrotropin) The glycoprotein composed of alpha and beta subunits released from the pituitary that promotes thyroid hormone production and release.

TSI Thyroid-stimulating immunoglobulin.

Wolff-Chaikoff effect Decreased formation and release of thyroid hormone in the presence of excess iodine.

EMBRYOLOGY

The thyroid gland originates from the floor of the pharynx approximately 4½ weeks after conception. Initially an invagination of the epithelium appears and continues to grow and migrate caudally, forming the thyroglossal duct. Eventually the connection with the pharynx disappears, and the thyroid gland evolves into an isolated bilobulated structure located in the anterior part of the neck. The fetal thyroid gland begins to function at about 10 to 12 weeks of gestation.

ANATOMY AND HISTOLOGY

The thyroid is formed of two lobes, one on either side of the neck, which are connected by a narrow band of thyroid tissue called the "isthmus."

A portion of accessory thyroid tissue, called the "pyramidal lobe," is present in most people. It is a remnant of the thyroglossal duct and is located in a position superior to the isthmus.

The thyroid gland consists of two types of cells, follicular and parafollicular (Fig. 39-1). Follicular cells are arranged spherically in a single layer. Follicular cells have an apical end facing the center of the follicle and a basal end facing the interstitium, which contains the blood supply and parafollicular cells. Follicular cells produce thyroid hormone, which is then stored in the central portion of the spherical follicle in a material called "colloid." Parafollicular cells are located in the interstitium between follicles and secrete the hormone calcitonin. For this reason they are called "C cells."

THYROID PHYSIOLOGY

The thyroid gland has as its main function the production and secretion of metabolically active compounds or hormones that are essential for the regulation of various metabolic functions. Thyroid hormones are produced within the cells of the follicles from proteins, the amino acid tyrosine, and the halogen element iodine. The two

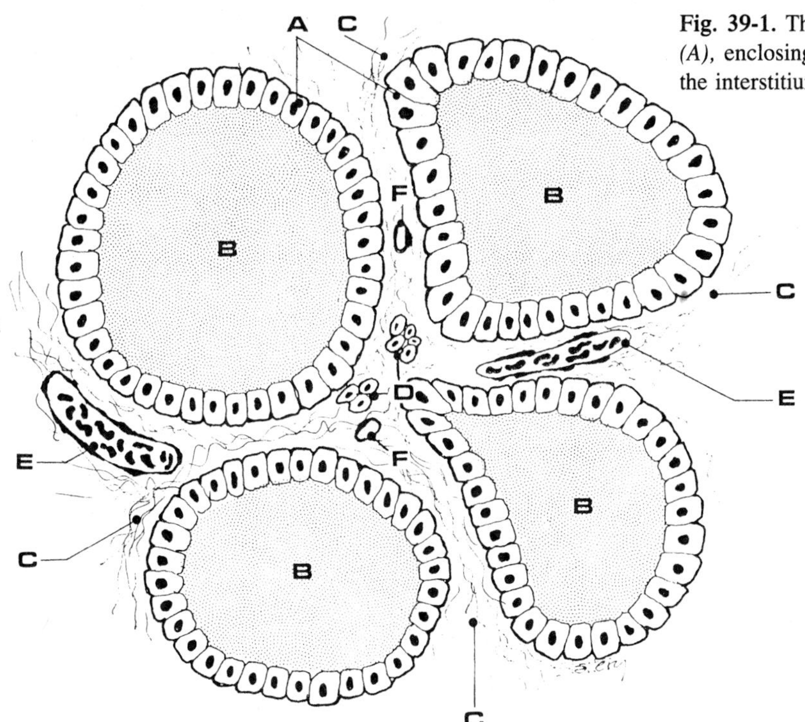

Fig. 39-1. Thyroid-gland structure consists of follicular cells *(A)*, enclosing colloid *(B)*, and parafollicular "C" cells *(D)*, in the interstitium *(C)*. *E*, Venule; *F*, capillary.

most important thyroid hormones are thyroxine (T_4), which contains four iodine atoms, and triiodothyronine (T_3), which contains three iodine atoms.

Thyroid hormones are released from the colloid and secreted into circulation in response to stimulation of the thyroid gland by the pituitary hormone called "thyroid-stimulating hormone" (TSH).

Metabolism of iodine and thyroid hormone synthesis

Iodine is a natural component of many foods and in the United States is provided in adequate amounts by a well-balanced diet. Extra amounts of iodine are currently provided by the ingestion of iodine-enriched foods[1] and numerous iodine-containing drugs such as "vitamin pills."

The daily intake of iodine varies widely in different parts of the world. In the United States iodine intake ranges from 250 to 700 μg or more daily. In countries such as Japan, intake may reach several milligrams per day, whereas in some areas of Africa, South America, Asia, and Europe daily intake may be as low as 50 μg.

Under physiological conditions, iodine is absorbed in the small bowel and then enters either the excretory or metabolic pathways (Fig. 39-2). Between 60% and 80% of the ingested iodine is excreted by the kidneys. Small amounts are excreted through the intestinal route. Fecal excretion is derived for the most part from hormones degraded by the liver and excreted into the bowel by the biliary tract.[2] The remainder of the iodine is distributed into the extracellular and thyroid compartments. The intrathyroid iodine compartment contains about 90% of the total body iodine and can amount to as much as 6000 to 12,000 μg. The extracellular compartment contains most other iodine, except for a small but important amount within cells. The metabolism of iodine is closely related to the process of thyroid hormonogenesis.

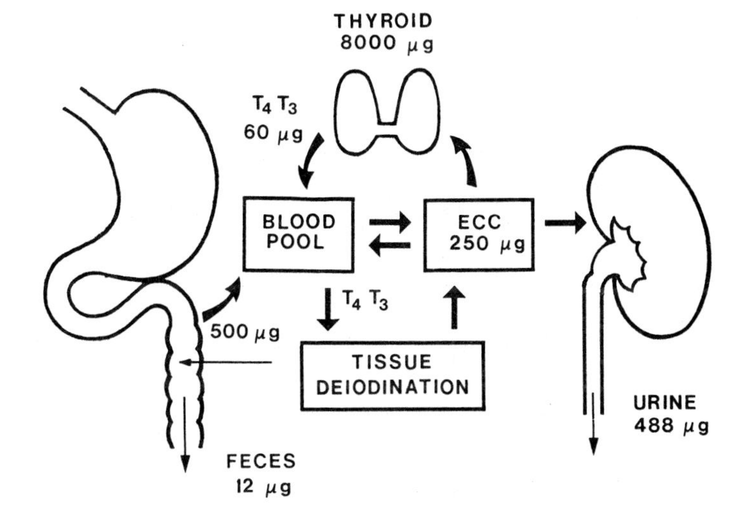

Fig. 39-2. Iodine metabolic pathway in a 24-hour period. *ECC*, Extracellular compartment.

Classically, intrathyroid iodine metabolism has been divided into the following stages: (1) iodine trapping or uptake of iodine by the follicular cells, (2) organification, (3) coupling, (4) storage, and (5) secretion (Fig. 39-3). During

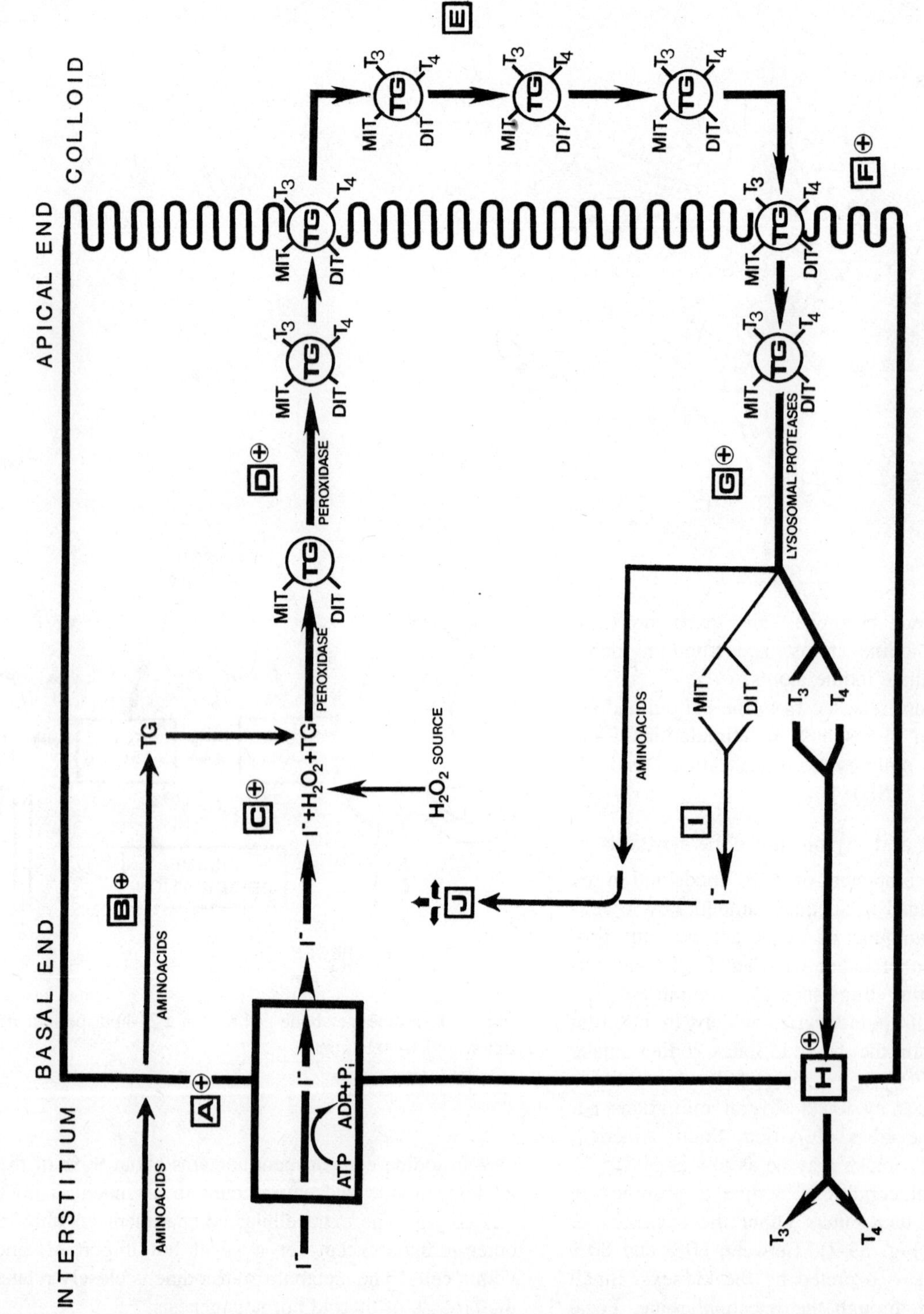

Fig. 39-3. Thyroid cell. Schema depicting stages of thyroid hormonogenesis and intrathyroidal iodine metabolism. *A*, Iodine transport; *B*, thyroglobulin (TG) synthesis; *C*, iodine organification; *D*, intrathyroglobulin coupling; *E*, storage; *F*, endocytosis; *G*, hydrolysis; *H*, hormone secretion; *I*, intrathyroidal deiodination; *J*, recycling; steps influenced by the thyroid-stimulating hormone (TSH) are indicated by the symbol ⊕.

trapping, thyroid cells concentrate iodine against high chemical and electrical gradients that require an active energy-dependent mechanism at the level of the cell membrane. An adenosine triphosphate (ATP)–dependent system is postulated to be responsible for this transport of iodine across the cell membrane.[3]

This trapping mechanism for iodine has been exploited for many years in clinical tests to assess thyroid function. Radioactive iodine is given orally to patients, and its degree of concentration in the thyroid is subsequently measured. Various physiological and pharmacological factors influence trapping. The most important factor is thyroid-stimulating hormone (TSH), which stimulates trapping of iodine.[4] Iodine excess inhibits the transport of iodine; iodine deficiency stimulates it.[5] Ions such as perchlorate and thiocyanate inhibit the transport of iodine.[6] These inhibitory effects of perchlorate have been used in clinical practice to test the function of the thyroid gland, specifically organification defects.[7]

The second step of intrathyroid iodine metabolism is organification by means of which iodine is incorporated rapidly into thyroid hormone.[8] The iodine used comes from trapping and from the intrathyroid deiodination of stored thyroid hormone precursors. Iodine in the thyroid is oxidized in the presence of a peroxidase enzyme into a reactive form that combines with the protein thyroglobulin.[8]

Thyroglobulin is a glycoprotein with a molecular weight of about 660,000 daltons.[9] It contains 10% carbohydrate and approximately 5000 amino acids arranged in four chains of polypeptides. The entire thyroglobulin molecule has 115 tyrosyl groups, which account for 2% of its molecular weight.[10] Thyroglobulin serves as a preformed matrix containing tyrosyl groups to which reactive iodine is attached to form residues of monoiodotyrosine (MIT), diiodotyrosine (DIT), triiodothyronine (T_3), and thyroxine (T_4). After their formation, coupling of MIT and DIT takes place to form intrathyroglobulin T_3 and T_4. This process involves the intramolecular transference and rearrangement of iodotyrosyls (MIT and DIT) within the thyroglobulin molecule and coupling with other iodotyrosyl groups.[8] Peroxidase is also responsible for this coupling function of iodotyrosyls. The storage function of thyroglobulin provides a constant pool of thyroid hormone. The thyroglobulin is then released from the cells into the colloid of the follicle where it remains stored.

When the TSH stimulus to secrete thyroid hormone affects the gland, a series of cellular and biochemical changes takes place, all directed toward the release of thyroid hormone.[11] Initially, the apical portion of the cells form pseudopodia, which surround and eventually engulf droplets of colloid. These droplets fuse with cytoplasmic lysosomes, which carry proteases. The proteolytic enzymes digest the thyroglobulin to its basic constituent amino acids. In this process, MIT, DIT, T_3, and T_4 are released from the basic thyroglobulin matrix. Intracellular MIT and DIT are immediately deiodinated, and their iodines are reused in subsequent thyroid hormone synthesis. The released T_3 and T_4 are resistant to intrathyroid deiodination and are secreted as active hormones. The daily secretion of thyroid hormone includes about 100 nanomoles (72 μg) of T_4 and 5 nmol (4 μg) of T_3.[12,13] Small amounts of rT_3 (less than 5 nmol daily) are also secreted by the thyroid.

Transport of thyroid hormones

On their release into the bloodstream, circulating free T_3 and T_4 hormones enter body cells, where they become part of an intracellular pool of hormone and exert their metabolic effects.[14] The T_3 and T_4 also are transported to portions of the brain and pituitary gland, where they participate in the regulation of pituitary TSH secretion, and to the organs of degradation and excretion. In the bloodstream the thyroid hormones are transported in two forms, protein bound and free. The free form is metabolically active but represents only a small fraction (less than 1%) of the total plasma thyroid hormone content. The bound hormone is metabolically inactive and serves as a stable reservoir of hormone that, by virtue of its size and state of equilibrium with the free hormone compartment, maintains constant amounts of hormone available to tissues.

The binding of thyroid hormones to proteins in the blood is accomplished by three plasma proteins.[15] The most important is the thyroxine-binding globulin (TBG), which is a glycoprotein synthesized in the liver with an electrophoretic mobility between that of α_1 and α_2 globulin. The second most important is the thyroxine-binding protein prealbumin (TBPA) with a molecular weight of 50,000 daltons. The third transporting protein is albumin.

The role of each of these binding proteins in the transport of T_3 and T_4 depends on their relative affinities for each of the thyroid hormones and on their relative concentrations in plasma. Almost all the circulating T_4 (99.95%) is bound to these plasma proteins. Under physiological circumstances, TBG transports 70% to 75% of total T_4, and TBPA and albumin transport 15% to 20% and 10% of T_4, respectively.[16,17] The binding of T_3 to plasma proteins occurs to a lesser extent, with 99.7% of T_3 circulating in bound form. The affinity of TBG for T_3 is lower than its affinity for T_4, and binding of T_3 to TBPA is negligible. T_3 is bound to albumin. Abnormalities of the binding proteins may result in abnormal total (bound) hormone concentrations in the blood with normal amounts of free hormone.

Changes in concentration or binding capacity of plasma proteins affect the total serum concentration of thyroid hormones. Since both T_4 and T_3 are bound mainly to TBG, changes of TBG levels are more likely to affect T_4 and T_3 levels. Changes in TBG concentrations in blood do not have any direct effect on the negative feedback mechanisms of thyroid hormone regulation. Rather, these effects on TSH and thyroid hormone secretion are indirect[18]

Table 39-1. Physiological relationship between thyroxine-binding globulin (TBG) and serum thyroid hormone concentrations

Biological modulators	Initial biochemical changes	Intermediate biological response	Final equilibrium conditions
Increased TBG Pregnancy Oral contraceptives	Increased TBG levels Augmented binding of hormone (T_4, T_3) Decreased free T_4, T_3	Decreased negative feedback mechanism Increased serum TSH Increased T_4 and T_3 production	Increased TBG Elevated serum T_4 and T_3 levels Normal TBG saturation Normal free T_4 and T_3
Decreased TBG Androgens Malnutrition Liver disease	Decreased TBG Diminished binding of hormone (T_4, T_3) Increased free T_4 and T_3	Increased negative feedback mechanism Decreased serum TSH Decreased T_4 and T_3 production	Decreased TBG Decreased T_4 and T_3 Normal TBG saturation Normal free T_4 and T_3 serum

(Table 39-1). The relationship between the thyroid hormones in blood and TBG is governed by the law of mass action:

$$TBG + FTH \overset{k}{\rightleftarrows} TTBG$$

where FTH is free thyroid hormone, TBG is thyroid-binding globulin, TTBG is TBG-bound hormone, and k is a reaction-rate constant. An increase in TBG will cause a relative decrease of TBG saturation. Correction of this new biological imbalance will result in an increased binding of free thyroid hormone with subsequent decrease of circulating free hormone. This immediately triggers the secretion of TSH, which results in increased thyroid hormone production. All these changes are transient, and a new equilibrium with preservation of the normal thyroid status is reached. Decreases of TBG will cause the opposite biological changes.

Metabolism of thyroid hormones

Circulating T_3 and T_4 are either incorporated into the intracellular pool, where they undergo partial transformations and exert their metabolic effects, or are degraded and eliminated by excretory organs.

All circulating T_4 originates in the thyroid gland, which secretes 80 to 100 μg of T_4 per day. The more metabolically important T_3 originates from both direct thyroid secretion (about 20%) and peripheral conversion of T_4 to T_3 by monodeiodination (about 80%).[12,13,19] The total daily production of T_3 ranges from 22 to 47 μg.[13] Although T_3 has been established as the main active hormone, the role of T_4 as a hormone with direct biological activity has been questioned. For this reason, many workers consider T_4 to be a prohormone, that is, a compound that requires chemical transformation to become an active hormone. Limited evidence does suggest, however, that T_4 indeed has some direct biological activity,[20] albeit less than T_3.

The thyroid hormones are metabolized through deiodinative and nondeiodinative mechanisms. The following are some of the most important metabolic steps (Fig. 39-4):

1. Both T_4 and T_3 exert their biological effects by binding to specific cellular receptors and are subsequently degraded through successive deiodinations.
2. Deiodination accounts for 80% to 85% of the metabolism of T_4 and T_3.[21-24]
3. About 35% to 50% of the T_4 undergoing deiodination is converted to T_3.[22,23,25]
4. About 50% to 65% of the deiodinated T_4 is converted into rT_3.[22,23]
5. Most of the T_4, T_3, and rT_3 are metabolized through a chain of successive deiodinations, resulting in the formation of iodinated intermediary metabolites and ultimately thyronine (Fig. 39-5).
6. Both T_4 and T_3 undergo oxidative deamination and decarboxylation of the alanine side chains to form the acetic acid analogs tetrac and triac.[13,27,28]
7. Small amounts of free T_4 are eliminated in the bile and urine.[29]
8. Small amounts of T_3, rT_3, and indirectly T_4 are metabolized through processes of conjugation with glucuronic acid and sulfate.[26]

Conversion of T_4 to T_3 is one of the most important metabolic pathways of T_4 and takes place in many tissues, particularly the liver and the kidney.[13,25] In T_3, iodine atoms are located at positions 3 and 5 of the inner (nonphenolic) ring and 3′ of the outer (phenolic) ring (Fig. 39-5). In the case of rT_3, iodine atoms are found in position 3 of the inner ring and positions 3′ and 5′ of the outer ring. Secretion of rT_3 by the thyroid gland is negligible, accounting for only 2.5%,[24] and almost all rT_3 derives from T_4 monodeiodination.[23,24] In contrast to T_3, rT_3 is essentially inert without metabolic activity.[30,31]

Mechanisms of action of the thyroid hormones at the cellular level have been the focus of intensive research. One major mechanism for the initiation of thyroid hormone effects is through the binding to specific nuclear receptors, which stimulates the activities of certain genes.[32-35] Subsequently, messenger ribonucleic acid (mRNA) is formed, and it directs the synthesis of proteins and enzymes responsible for metabolic functions.[36,37] Other postulated mechanisms of action at the cellular level

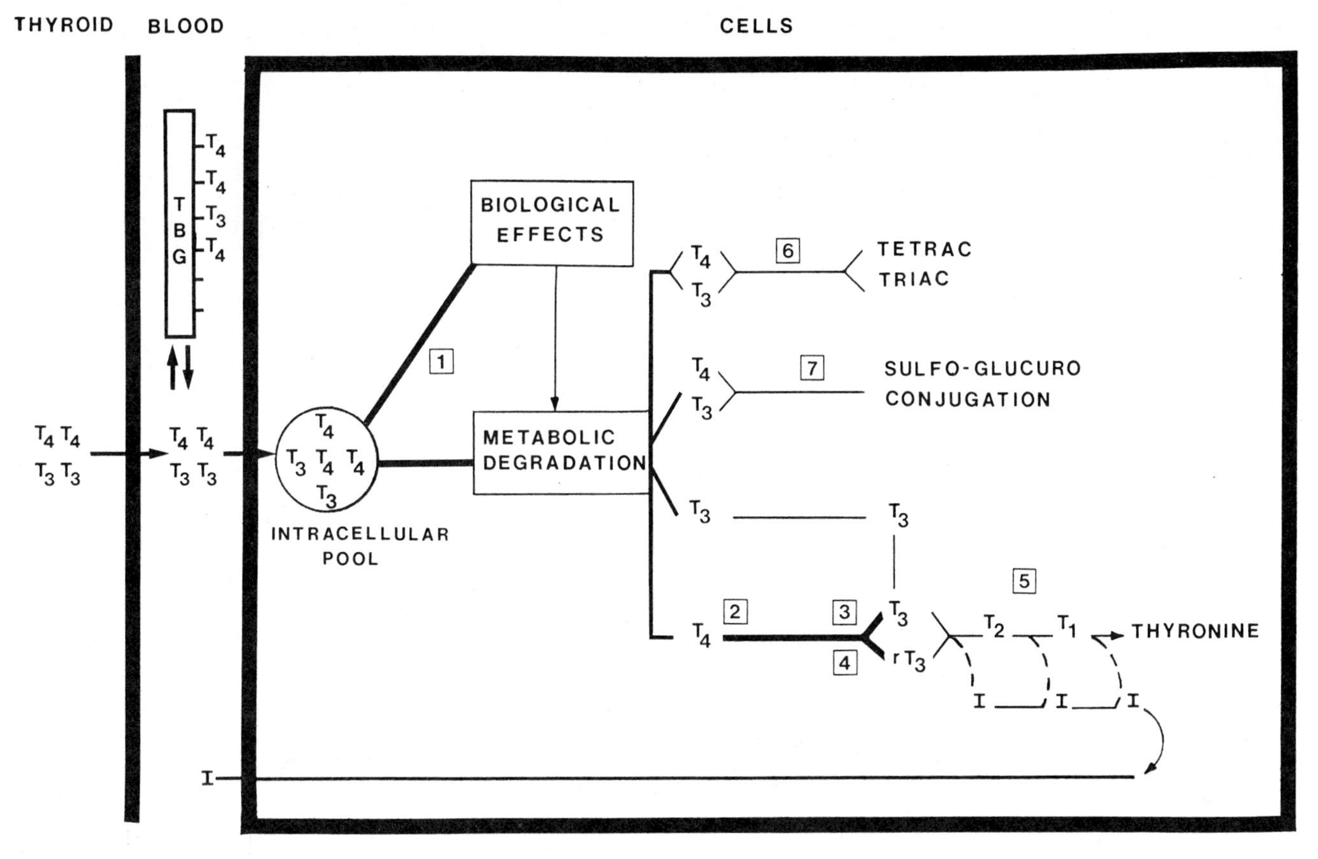

Fig. 39-4. Metabolic pathways of thyroid hormone. *1,* Biological effects through binding to intracellular receptors; *2,* main deiodinative pathway for T_4; *3,* conversion of T_4 into T_3; *4,* conversion of T_4 into rT_3; *5,* serial deiodinations of T_3 and rT_3; *6,* deamination and decarboxylation pathway; *7,* conjugative pathway.

are (1) mitochondrial activation,[38] (2) stimulation of Na^+-K^+ adenosinetriphosphatase (ATPase) activity,[39-42] (3) stimulation of cell membrane functions probably through a specific receptor,[41] and (4) interaction with the adrenergic system.

Control and regulation of thyroid function

Regulation of synthesis and release of thyroid hormones

Hypothalamic-pituitary-thyroid axis (HPTA). The HPTA comprises a group of physiologically interrelated neuroendocrine and endocrine organs that regulate and control the secretion of thyroid hormone (Fig. 39-6). The ultimate effector in this axis is the thyroid gland, which is directly in charge of producing, storing, and secreting the hormones thyroxine and triiodothyronine. The hypothalamus, which is part of the diencephalic portion of the brain, acts as an important regulating organ.

Thyrotropin-releasing hormone (TRH) is a tripeptide[44] produced in the hypothalamus and secreted into the venous system, connecting this portion of the brain with the anterior hypophysis. The TRH attaches to receptor sites in the pituitary, where it causes increased production and secretion of thyroid-stimulating hormone (TSH).[45-47] TSH is a glycopeptide structurally composed of two subunits, alpha and beta.[48,49] The beta subunit confers on TSH the specific physiological properties that differentiate it from other pituitary glycopeptides. TSH is released from the pituitary into the bloodstream by which it reaches the thyroid gland. At the thyroid, TSH attaches to specific cell receptors and exerts two main actions. The first is a trophic one, which stimulates cell reproduction and hypertrophy. The second effect is to stimulate production and secretion of thyroid hormone by the thyroid cell. The TSH action on thyroid cells is mediated through the cyclic adenosine monophosphate (cAMP) system.[50]

In normal persons an increase in blood TRH levels will affect the blood levels of TSH and thyroid hormone. After the intravenous administration of synthetic TRH, blood levels of TSH begin to increase within 10 minutes, reach a maximum at 15 to 45 minutes, and return to normal base levels in 1 to 4 hours.[51]

Elevations of TSH that take place several hours after the administration of TRH result in increases in serum T_4 and

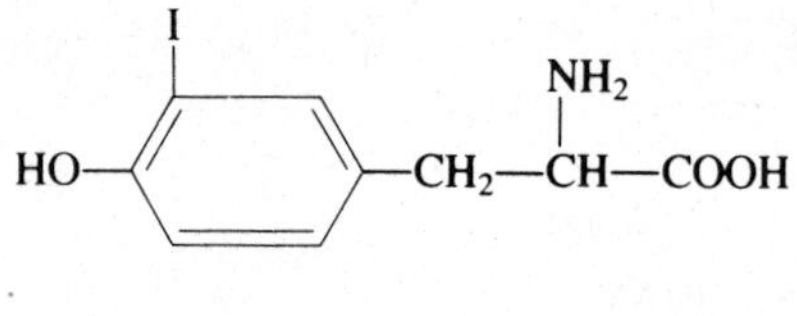

MIT
(Monoiodotyrosine)

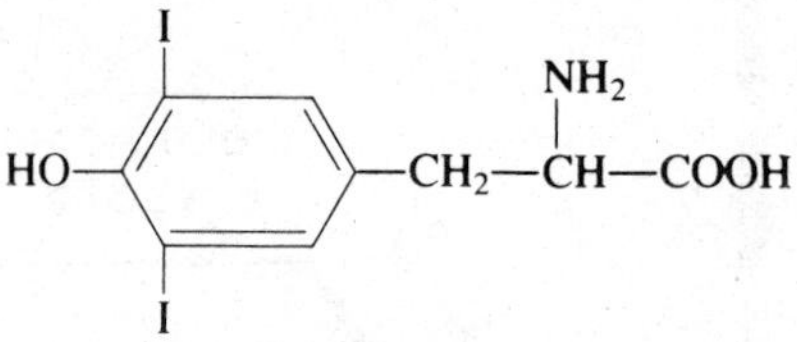

DIT
(Diiodotyrosine)

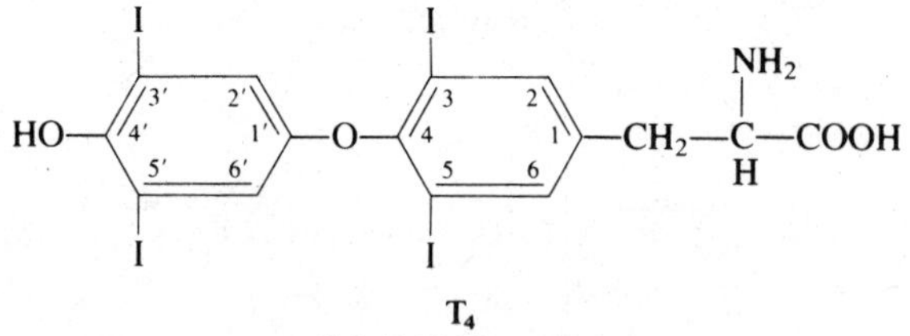

T_4
(3,5,3′,5′-Thyroxine)

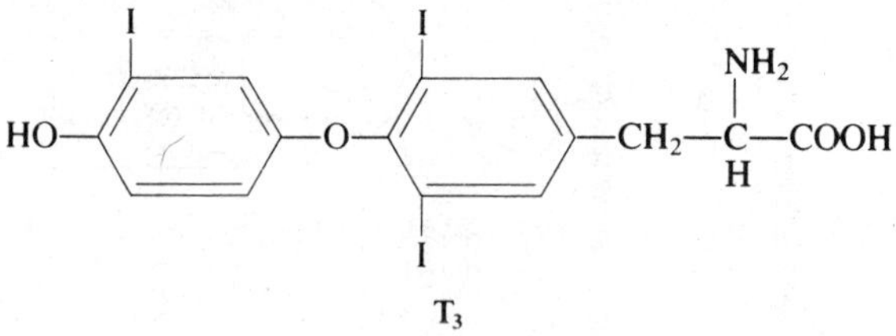

T_3
(3,5,3′-Triiodothyronine)

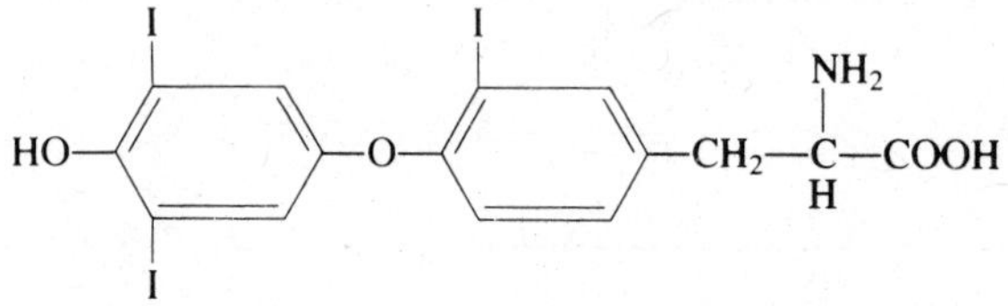

rT_3
(3,3′,5′-Triiodothyronine)

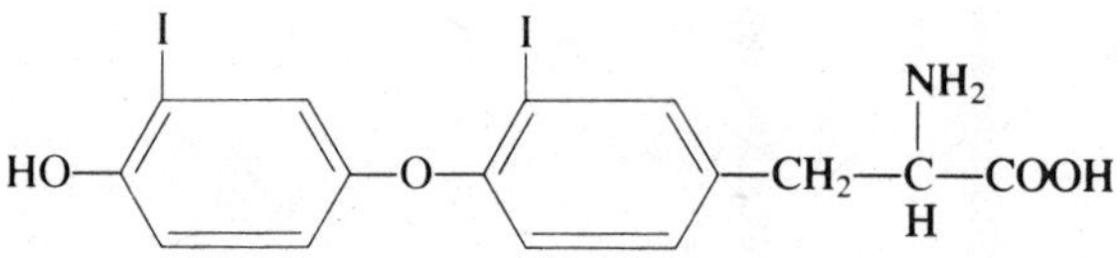

3,3′-T_2
(3,3′-Diiodothyronine)

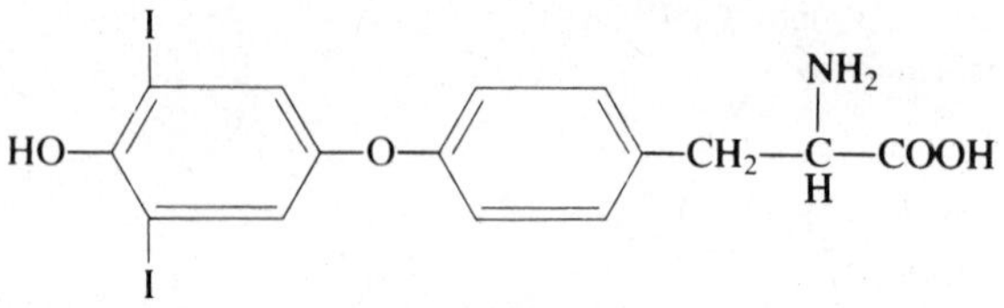

3′,5′-T_2
(3′,5′-Diiodothyronine)

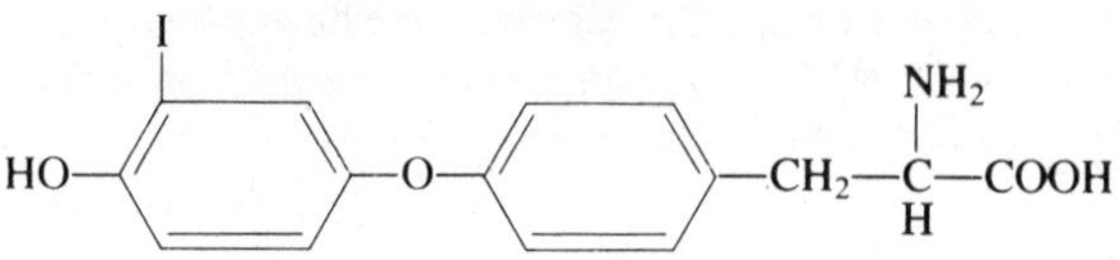

3′-T_1
(3′-Monoiodothyronine)

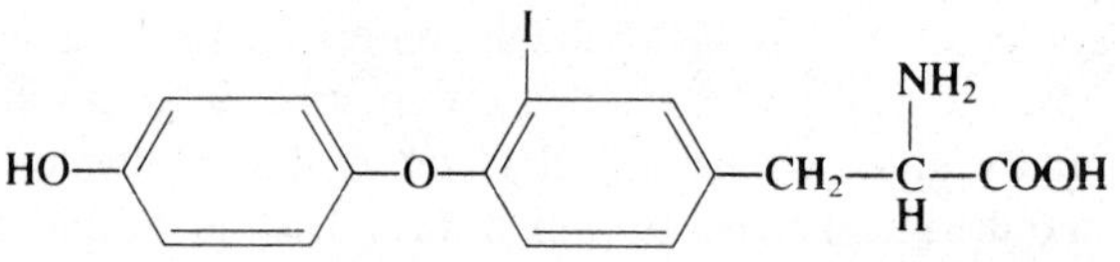

3-T_1
(3-Monoiodothyronine)

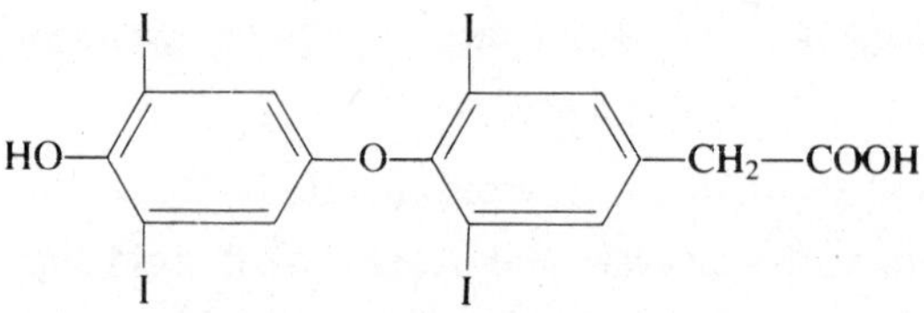

Tetrac
(3,5,3′,5′-Tetraiodothyroacetic acid)

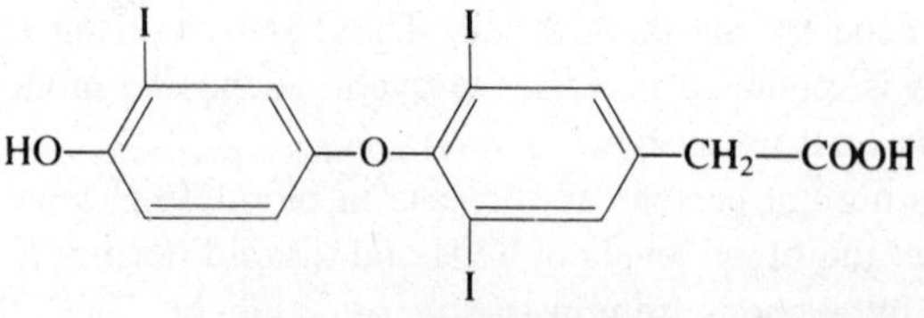

Triac
(3,5,3′-Triiodothyroacetic acid)

Fig. 39-5. Chemical structure of thyroid hormones and iodinated precursors and metabolites.

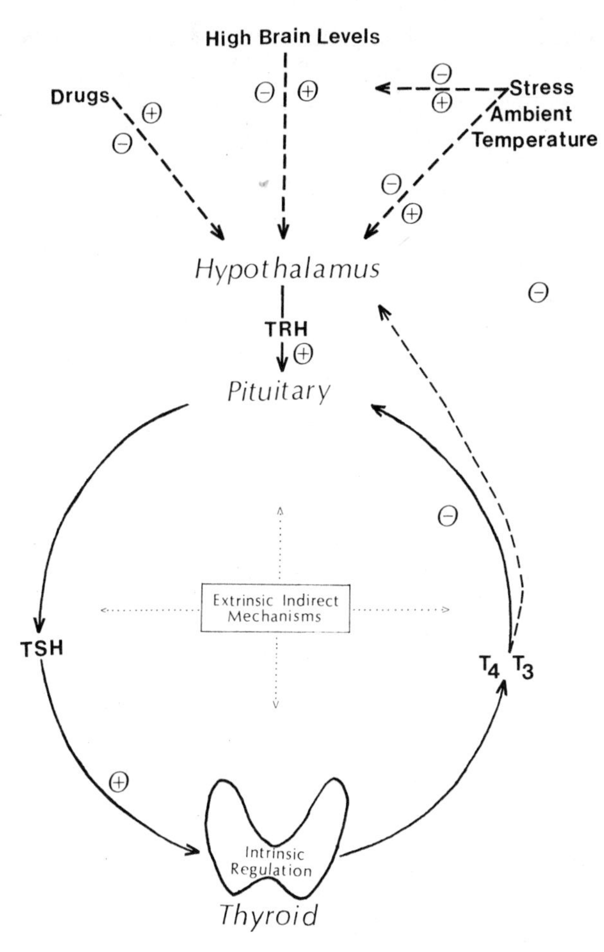

Fig. 39-6. Hypothalamic-pituitary-thyroid axis (HPTA). Stimulatory, ⊕, or inhibitory, ⊖, effect of agent.

T_3. The initial increases of T_3 are higher (75% increase over basal levels) than those of T_4 (15% to 50% over basal levels). The T_3 and T_4 levels subseqently drop slowly. The TSH, T_3, and T_4 responses to intravenous administrations of synthetic TRH are currently used for diagnostic purposes.

Negative-feedback system on pituitary secretion of TSH. Increased levels of thyroid hormones in blood inhibit TSH secretion by the pituitary (so-called negative feedback). The mechanism of this inhibitory effect appears to involve circulating free T_3 and T_4. The pituitary gland also is able to convert intrapituitary T_4 directly to T_3. This locally generated T_3 then acts with circulating T_3 on the pituitary gland cells to inhibit the secretion of TSH in response to TRH. Conversely, decreased levels of thyroid hormone result in increased secretion of TRH and TSH.[52]

Other factors affecting thyroid function. Thyroid function is finely regulated by extrinsic and intrinsic mechanisms. The extrinsic direct mechanism is represented by the HPTA (hypothalamic-pituitary-thyroid axis). The extrinsic indirect mechanisms encompass a host of collateral factors of neurogenic, metabolic, and pharmacological nature that exert inhibitory and stimulating effects centrally at all levels of the HPTA and peripherally on the metabolism of the thyroid hormones. Some of these mechanisms are discussed below. The intrinsic mechanisms are those taking place within the thyroid cells and are concerned with the maintenance of adequate amounts of intrathyroid hormone. These are closely dependent on the availability and effects of iodine.

For instance, an increase of serum iodine is not accompanied by an increase of thyroid hormone secretion.[53] When the thyroid is exposed to rapid and strong increases of iodine, there is decreased formation and release of thyroid hormone. This acute inhibitory action of excess iodine is known as the "Wolff-Chaikoff effect."[54,55] If the thyroid continues to be exposed to increased concentrations of iodine, the inhibitory effects of excess iodine on hormone formation and release decrease and eventually cease.[54,56] This protective mechanism of adaptation by the thyroid appears to be mediated by an inhibition of transport of iodine across the cell membrane,[56] resulting in decreased intrathyroid iodine.

Effects of drugs and other compounds on thyroid function

Various drugs and substances may interfere with thyroid function. The pharmacological effects of these may take place at one or several levels. Drugs acting on the hypothalamus and pituitary include dopamine, levodopa, and metoclopramide. Dopamine, an agent used for treatment of shock, inhibits the secretion of TSH. Continuous administration of dopamine causes decreases of serum T_4 and TSH and blunts TSH response to TRH stimulation.[57] Blunting of TSH response to TRH, without changes of T_4 and TSH, is seen during administration of levodopa,[58] which metabolizes into dopamine in the central nervous system.

Amphetamine has been found to increase serum levels of T_4, free T_4, and T_3,[59] apparently by acting on the hypothalamus or pituitary gland. Metoclopramide, used for gastrointestinal disorders, inhibits dopamine and may increase serum TSH.[60]

Numerous drugs alter thyroid function by interfering with hormonogenesis. Perchlorate blocks the iodine-trapping mechanism of the thyroid cell. Thiocyanates, found in certain foods, have actions similar to perchlorate and have been believed to be contributing factors to goiter in areas with low iodine intake. Nitroprusside, used as a hypotensive agent, is metabolized into thiocyanate and therefore indirectly inhibits iodine trapping, resulting in decreased serum levels of T_4 and free T_4 and increased TSH. Propylthiouracil (PTU) and methimazole are pharmaceuticals used for the therapy of hyperthyroidism because they interfere with the organification process.[61] In addition to the organification block, PTU decreases peripheral conversion of T_4 to T_3.[62,63] Other agents blocking organification include hypoglycemic sulfonylurea agents, antibacterial sulfonamide agents, and the chemotherapeutic drug 6-mercaptopurine, among others. These agents cause decreased total and free T_4 and increased TSH serum levels.

Lithium, currently used in carbonate form for the treatment of psychiatric disorders, inhibits the release of hormone from the thyroid.[64] Although hypothyroidism rarely develops, enlargement of the thyroid gland and elevated serum TSH levels may be found in patients treated with lithium.[65,66]

Various other drugs affect TBG levels and thus total T_3 and T_4 determinations (see accompanying box).

Causes of abnormalities in thyroxine-binding globulin (TBG)

- Quantitative
  - Increased TBG serum levels
    - Pregnancy
    - Estrogen therapy
    - Oral contraceptives
    - Perphenazine
    - Acute hepatitis
    - Hypothyroidism
    - Neonatal period
    - Acute intermittent porphyria
    - Genetic TBG excess
  - Decreased TBG serum levels
    - Androgens
    - Anabolic agents
    - Cirrhosis
    - Acute illness
    - Surgical stress
    - Severe chronic illness
    - Severe hypoproteinemia
    - Nephrotic syndrome
    - Hyperthyroidism
    - Corticosteroid therapy
    - Active acromegaly
    - Klinefelter's syndrome
    - Cushing's syndrome
    - Down's syndrome
    - Type III hyperlipidemia
    - Chronic metabolic acidosis
    - Genetic deficiency
- Qualitative
  - Genetic
    - Genetic increase in binding affinity
    - Genetic decrease in binding affinity
  - Drugs competing with T_4 and T_3 for TBG-binding sites
    - Phenytoin (diphenylhydantoin, Dilantin)
    - Dicumarol
    - Heparin
    - Atromid S
    - Aspirin
    - Phenylbutazone

Degradation of thyroid hormone is enhanced by phenytoin (Dilantin) and phenobarbital,[67,68] both of which are used for treatment of seizure disorders. This increased degradation is caused by an induction of liver enzymes that metabolize thyroid hormones. As a consequence, T_4 levels in serum may decrease slightly and TSH levels may increase.

Glucocorticoid agents affect thyroid function in several ways. They decrease secretion of TSH and diminish TSH response to TRH, probably through a direct negative-feedback effect on the pituitary. Glucocorticoids also decrease conversion of T_4 to T_3.

Finally, the effects of thyroid hormone on the cardiovascular system may be decreased by beta-blocking agents such as propranolol. Numerous beta blockers are currently used for hypertension and coronary artery disease. These drugs are also used to ameliorate the symptoms of hyperthyroidism, and they do not alter thyroidal hormonogenesis itself, although they may decrease the peripheral monodeiodination of T_4 to T_3.

Interrelationships between thyroid gland and other functions

In addition to the close interrelationships the thyroid has with the hypothalamus and hypophysis, there are other physiological factors that indirectly influence thyroid function and the effects of thyroid hormones.

The effects of gender and sex hormones on thyroid function are variable. Thyroid disease is known to predominate in females. Estrogens cause an increase in TBG, and therefore total serum T_3 and T_4 content, whereas androgenic hormones have the opposite effect. On the other hand, states of decreased (hypothyroidism) and increased (hyperthyroidism) thyroid function are associated with alterations of the reproductive system, abnormalities of the menstrual cycle, delayed or precocious puberty, infertility, and growth retardation. These result from alterations of the transport and metabolism of nonthyroid hormones secondary to the thyroid disease.[69,70]

Age also affects thyroid function. The newborn has lower levels of T_3 and higher levels of T_4 and rT_3 than adults have.[71] Immediately after birth there is an increase of TSH, with corresponding elevations of T_3 and T_4 that reach their zenith the second day after birth. These elevated levels of T_3 and T_4 gradually decrease toward normal adult levels by the end of the first year of life. There also is a slow progressive decrease of T_3 levels in adults during the aging process.[72]

Striking changes of thyroid function occur during pregnancy. The normal placenta produces significant amounts of nonthyroid hormones, which stimulate the formation of increased TBG levels. Both total T_4 and, to a lesser extent, total T_3 levels increase during pregnancy because of these increases in TBG concentration.[73]

Factors such as stress and starvation also affect thyroid

function. In fasting states, changes in levels of thyroid hormone in serum occur within 24 to 48 hours. Initially serum T_3 decreases rapidly and subsequently, if fasting is maintained, more slowly. The decrease of T_3 levels is caused by decreased peripheral conversion of T_4 to T_3. Concomitantly there is an increase in rT_3 levels. Decreased caloric intake also results in a decline of thyroxine-binding prealbumin (TBPA), probably caused by decreased hepatic production.

General metabolic and physiological effects of thyroid hormone

Abnormal function in thyroid disease results from alterations of the rate and quantity of hormone produced and secreted by the thyroid gland. An increase and decrease of hormone production will result in states of hyperfunction (hyperthyroidism) and hypofunction (hypothyroidism), respectively. Clinical presentation of hyperthyroidism and hypothyroidism is understandable when one knows the basic effects of thyroid hormone on various organs and metabolic systems.

Table 39-2 depicts some general effects that thyroid hormone has on various metabolic functions and systems, some of which are still poorly understood. One can divide the effects of thyroid hormone according to their clinical expression into (1) general metabolic effects, (2) growth and maturation effects, and (3) organ-specific effects. The general metabolic effects are those the thyroid hormone exerts on the basic metabolisms of the carbohydrates, proteins, and lipids. Generally speaking, both intermediate and specific metabolism are stimulated by the thyroid hormone. Two of the most conspicuous results of these effects are increased oxygen consumption and calorigenesis. Thyroid hormone has important actions as a promotor of cell differentiation, growth, and maturation. Deficiency of thyroid hormone in early life results in severe impairment of physical growth, maturation, and brain development. The third class represents a number of direct effects exerted on specific organs and systems.

Finally, increased thyroid function is associated with a hypersympathetic (hyperadrenergic) state important in the genesis of some symptoms of hyperthyroidism. This thyroid-catecholamine (adrenergic) interrelationship has been found at two levels. The first relates to the increase of adrenergic beta receptors induced by the thyroid hormone in the cardiovascular system and other tissues.[74] The second refers to the chemical similarity of thyroid hormones and catecholamines, since both are tyrosine analogs.[75]

Decreased thyroid hormone results in deposition of mucoproteins and mucopolysaccharides in various tissues such as the skin, muscles, and heart. This phenomenon partly explains the thick skin, muscle weakness, enlarged heart, and symptoms of heart failure seen in hypothyroidism.

Thyroid function in nonthyroid disease (NTD)

There is a group of conditions that do not directly alter the output of hormone by the thyroid but rather interfere

Table 39-2. Basic physiological effects of thyroid hormone and their relationship with syndromes of thyroid dysfunction

System	Thyroid hormone effects	Usual symptoms	
		Hyperthyroidism	**Hypothyroidism**
Metabolic	Increased calorigenesis and O_2 consumption Increased heat dissipation Increased protein catabolism Increased glucose absorption and production (gluconeogenesis) Increased glucose use	Heat intolerance Flushed skin Increased perspiration Increased appetite and food ingestion Muscle wasting and weakness Weight loss	Cold intolerance Dry and pale skin Decreased appetite and food ingestion Generalized weakness Weight gain
Cardiovascular	Increased adrenergic activity and sensitivity Increased heart rate Increased myocardial contractility Increased cardiac output Increased blood volume	Palpitations Fast heart rate (tachycardia) Increased blood pressure, mainly systolic Bouncy, hyperdynamic arterial pulses Shortness of breath	Slow heart rate (bradycardia) Low blood pressure Heart failure Heart enlargement
Central nervous	Increased adrenergic activity and sensitivity	Restlessness, hypermotility Nervousness Emotional lability Fatigue Exaggerated reflexes	Apathy Mental sluggishness Depressed reflexes Mental retardation
Gastrointestinal (GI)	Increased motility	Hyperdefecation	Constipation

with the normal transport and metabolism of thyroid hormones. These conditions produce changes in standard blood tests of thyroid function but are generally not accompanied by true thyroid dysfunction. These disorders are of clinical importance because they may mimic true thyroid disease or confound the diagnosis of concomitant thyroid dysfunction.

Nutrition, renal function, and hepatic function are the three most important nonthyroid factors leading to abnormalities of thyroid function tests. The effect that decreased caloric intake has on thyroid hormone metabolism has been previously mentioned. States of malnutrition are often seen in chronic and debilitating diseases such as congestive heart failure, diabetes mellitus, chronic obstructive lung disease, cancer, and postsurgical states. The most conspicuous physiological abnormality found in all these entities is the impaired peripheral conversion of T_4 to T_3.[76-78]

The liver has important actions in the metabolism of thyroid hormones. These include production of thyroxine-binding proteins,[79] conversion of T_4 to T_3 or rT_3, and clearing and metabolism of thyroid hormones.[80] Alterations of these functions can occur in liver disease, resulting in abnormal test results in the absence of thyroid disease.

The kidneys have two main functions in relation to the thyroid. The first one concerns iodine metabolism, since the kidneys represent a major elimination pathway. The second is to prevent excessive losses of thyroxine-binding globulins in the urine. Kidney disease can result in either decreased iodine clearance and excretion with secondary increases in blood and interstitial pools of iodine as seen in renal failure, or augmented loss of thyroxine-binding proteins as seen in the nephrotic syndrome. Finally, although less important than that of the liver, the kidney has a metabolic function of converting T_4 to T_3.

PATHOLOGICAL CONDITIONS

Hyperthyroidism

Causes. Hyperthyroidism refers to the clinical syndrome caused by an excess of circulating active thyroid hormone. Hyperthyroidism is caused by multiple and heterogeneous pathological conditions that have as a common denominator an increase of circulating thyroid hormone (see box in right-hand column).

Graves' disease is caused by an immunological disorder in which serum autoantibodies bind to TSH receptors in the thyroid cell and stimulate the production and release of thyroid hormone. These antibodies compose a heterogeneous group of serum immunoglobulins that belong to the IgG fraction and are generically termed *thyroid-stimulating immunoglobulins* (TSI)[81] because of their ability to stimulate function in thyroid cells. The long-acting thyroid stimulator (LATS) antibody was the first TSI described. LATS is present in about 60% of patients with Graves' disease. Other TSI's have been described and given different names according to the assays employed for their detection. These include human thyroid stimulator (HTS), thyroid-stimulating antibody (TSAB), and thyrotropin binding–inhibiting immunoglobulin (TBII).

Causes of hyperthyroidism

Primary hyperthyroidism
- Primary thyroid abnormalities
  - Toxic multinodular goiter
  - Thyroid adenoma
  - Thyroid carcinoma
  - Struma ovarii

Secondary hyperthyroidism
- Endogenous: increased serum levels of TSH or thyroid-stimulating substances, resulting in thyroid hyperactivity:
  - Graves' disease
  - Neonatal hyperthyroidism
  - Pituitary tumors
  - Trophoblastic tumors
    - Hydatidiform mole
    - Choriocarcinoma
    - Embryonal carcinoma of testes
- Exogenous
  - Iatrogenic
  - Factitious hyperthyroidism

Thyroiditis
- Subacute thyroiditis
- Lymphocytic (Hashimoto's) thyroiditis
- Radiation

Graves' disease is characterized clinically by the presence of diffuse goiter (enlarged thyroid gland), symptoms and signs of hyperthyroidism (Table 39-2), ophthalmopathy, and, occasionally, pretibial edema. Thyroid hyperfunction can be clinically obvious or only minimally apparent. Incomplete manifestations occur in patients who present with isolated thyroid hyperfunction or ophthalmopathy. Graves' ophthalmopathy is characterized by myxedematous infiltration of the tissues and muscles of the orbit, resulting in protrusion of the eyes and ocular muscle dysfunction. The ophthalmopathy may exist without accompanying thyroid hyperfunction (euthyroid Graves' disease).[82]

Toxic multinodular goiter is a frequent cause of hyperthyroidism and usually appears in patients with preexisting nodular goiters. The term "multinodular" refers to the presence in the thyroid gland of areas of normal tissue alternating with multiple areas of thyroid cell hyperplasia (increased number of thyroid cells). The exact pathogenesis of this condition is unknown, but the result is that portions of the thyroid gland become autonomous and secrete excess amounts of thyroid hormone. This condition occurs more frequently in elderly patients and is not accompanied by ophthalmopathy or pretibial edema.

Thyroid adenomas are benign tumors that can occasionally produce excess thyroid hormone. Unlike normal thyroid tissue, adenomas may not be influenced by changes in serum TSH levels and may behave as autonomous tissue. In addition to signs and laboratory tests of thyroid hyperfunction, patients with adenomas have thyroid nodules that concentrate radioactive iodine avidly (hot nodules). It should be emphasized that most adenomas do not cause thyroid hyperfunction.

Thyroid cancer as a cause of hyperthyroidism is a rare condition. The excess of thyroid hormone is produced by tumor cells. Struma ovarii is a very rare cause of hyperthyroidism. In this condition, abnormal ectopic thyroid tissue harbored in certain tumors of the ovary is the source of the increased hormone production. Typically, no functioning thyroid tissue is found in the neck in radioisotopic studies.

''Thyroiditis'' is a general term used to describe an inflammation of the thyroid gland. All forms of thyroiditis can potentially cause hyperthyroidism by releasing large quantities of hormone from the inflamed and disrupted follicles. Subacute thyroiditis (SAT) is considered to be caused by a viral infection of the gland and follows two phases. The early phase is characterized by active inflammation of the thyroid, which results in enlargement and tenderness of the gland and clinical and laboratory findings of thyroid hyperfunction caused by release of hormones.[83-87] During the late phase, recuperation takes place and function usually returns to normal. Some patients may evolve through an intermediary state of hypothyroidism.[84]

Painless thyroiditis, or so-called silent thyroiditis, is a variation of SAT. This form is pathophysiologically similar to classic subacute thyroiditis but differs from it by the lack of neck pain.[88,89] Several authors have suggested that painless thyroiditis is actually a form of chronic lymphocytic thyroiditis.[86,90]

Chronic lymphocytic thyroiditis (Hashimoto's thyroiditis) is occasionally associated with an overactive thyroid state but more often results in hypothyroidism.

A sudden release of hormone may be seen after irradiation of the thyroid gland (radiation thyroiditis). For instance, patients with hyperthyroidism treated with ^{131}I may experience an exacerbation of symptoms 7 to 10 days after treatment. External radiation to the neck for the treatment of extrathyroid disease and ^{131}I ablative therapy as part of the treatment of thyroid cancer can also result in a sudden release of thyroid hormone from damaged thyroid cells. Thyroid hyperfunction caused by this radiation thyroiditis as suggested by elevations of serum T_3 and T_4[91] is usually mild and self-limited.

Exogenous hyperthyroidism is caused by the administration of excessive thyroid hormone. This situation appears as a complication of thyroid hormone replacement administered by the physician (iatrogenic) or as a result of surreptitious intake of thyroid hormone by patients (factitious).

Tumors originating from the trophoblast, or outer cellular layer of the forming embryo, can secrete large amounts of human chorionic gonadotropin (HCG). This

Table 39-3. Hyperthyroidism: laboratory findings in various clinical conditions

Clinical entity	T_4	T_3	FT_4	T_3RU	FT_4I	TSH	TRH stimulation	TSI	Thyroid ^{123}I uptake
Graves' disease	↑	↑	↑	↑	↑	↓,U	Blunted	+	↑
Euthyroid Graves' disease	N	N	N	N	N	N	Blunted, N	+	N
Toxic multinodular goiter	↑	↑	↑	↑	↑	↓,U	Blunted	−	↑,N
Toxic adenoma	↑	↑	↑	↑	↑	↓,U	Blunted	−	↑,N
T_3 toxicosis	N	↑	N	N, ↑	N	↓,U	Blunted	+, −	N, ↑
Hyperthyroidism in pregnancy	↑	↑	↑	N, ↓	↑	↓,U	Blunted	+, −	*
Neonatal hyperthyroidism	↑	↑	↑	↑	↑	↓,N	Blunted	+	*
Subacute thyroiditis	↑,N	↑,N	↑,N	↑,N	↑,N	N, ↓,U	Blunted, N	−	↓,N
Exogenous hyperthyroidism with T_4	↑	↑	↑	↑	↑	↓,U	Blunted	−	↓
Trophoblastic tumors	↑	↑	↑	↑	↑	↑,N, ↓	Blunted	−	↓,N, ↑
Pituitary TSH-secreting tumors	↑	↑	↑	↑	↑	↑	N, ↑	−	↑
Pseudohyperthyroidism	↑	↑,N	↑,N	↑,N	↑,N	↑,N	N	−	N or ↑

FT_4, Free thyroxine; *FT_4I*, free thyroxine index; *N*, normal; *TRH*, thyrotropin-releasing hormone; *T_3RU*, triiodothyronine resin-uptake test; *TSH*, thyroid-stimulating hormone; *TSI*, thyroid-stimulating immunoglobulins; *U*, undetectable.
↑, Elevated; ↓, decreased; +, present; −, absent; *, test contraindicated or not recommended.

hormone has been found to have a weak thyroid-stimulating action.[92,93] As a result, such tumors may cause an increased secretion of thyroid hormones. The incidence of clinically apparent hyperthyroidism is low, though mildly elevated levels of T_3 and T_4 are often found in such patients.[94]

Hyperthyroidism caused by pituitary tumors that secrete high levels of TSH is a rare occurrence.[95] In these instances overstimulation of the thyroid gland caused by increased levels of TSH results in increased hormone secretion and thyroid hyperfunction.

Laboratory findings. Graves' disease is the classic example of hyperthyroidism. Its laboratory abnormalities are displayed in Table 39-3. As one can see, abnormal laboratory tests include (1) elevation of thyroid hormones in serum, (2) decreased serum levels of TSH, and (3) blunted responses to TRH. In addition, patients with Graves' disease may have elevated serum levels of thyroglobulin[96] and thyroid-stimulating immunoglobulin.

Although most patients with Graves' disease have increased levels of both T_4 and T_3 in serum, in general there is a disproportionately higher rate of production of T_3.[97] This is reflected by greater increases of T_3 relative to those of T_4. Occasionally a condition known as T_3 toxicosis is found. A syndrome has been described with symptoms of hyperthyroidism and normal total T_4 and free T_4 concentrations, normal or mildly elevated thyroidal radioactive iodine uptake, and elevated T_3 levels in blood (T_3 thyrotoxicosis). This entity may present as part of Graves' disease but also has been observed in patients with toxic nodular goiter, after iodine ingestion in patients with previous iodine deficiency, and with recurrent hyperthyroidism after treatment with radioactive iodine, thyroid blocking agents, or surgery.[98-102]

A less frequently found entity is T_4 toxicosis, which refers to a condition in which the T_3 is normal or only mildly elevated and T_4 is quite elevated. This is caused by an alteration in conversion of T_4 to T_3 frequently found in patients with chronic debilitating diseases. Disproportionate elevations of T_4 relative to those of T_3 have also been described in patients with hyperthyroidism after receiving iodinated contrast material used in radiographic procedures.[103]

Neonatal hyperthyroidism refers to a state of increased thyroid function seen in newborn infants of mothers whose serum may contain thyroid-stimulating immunoglobulins (TSI). This condition is postulated to be caused by transplacental transfer of maternal TSI.[104] Its course is benign with spontaneous remission. Treatment is rarely required.

Treatment. The treatment of hyperthyroidism is aimed at either eliminating excess functioning thyroid tissue (surgery or radioiodine) or blocking the production of hormone (thyroid-blocking drugs). All these methods are currently used, and the choice depends on the cause of the hyperthyroidism, special clinical situations, and the physician's personal practices and preferences.

Hypothyroidism

Causes. A clinical state of hypothyroidism develops whenever insufficient amounts of thyroid hormone are available to tissues. By and large the most common group of entities causing hypothyroidism are those that involve the thyroid gland itself (see following box).

Hashimoto's thyroiditis is probably the single most common cause of hypothyroidism. It is believed to result from a derangement of the immune system. Alterations of both cell-mediated immunity and humoral immunity (that is, antibody formation) have been proposed.[105,106]

The most important characteristics of Hashimoto's thyroiditis are the presence of an organification defect and lymphocytic infiltration of the gland with concomitant loss of thyroid tissue. These elements of the disease are reflected clinically by the presence of an enlarged thyroid gland and in some cases, symptoms, signs, and laboratory findings of thyroid hypofunction. The alteration of the immune system is reflected by the presence in serum of antithyroid antibodies.

Subacute thyroiditis usually resolves with a return of thyroid function to normal. Mild hypothyroidism can occur during the course of the disease, but this is usually tran-

Causes of hypothyroidism

Primary thyroid dysfunction
- Parenchymal damage
  - Thyroiditis
    - Chronic lymphocytic (Hashimoto's) thyroiditis
    - Subacute thyroiditis
  - Therapeutic ablation
    - After ^{131}I therapy
    - After surgery
  - Thyroid dysgenesis
    - Aplasia
    - Dysplasia
  - Thyroid infiltration
    - Tumors
- Abnormal hormonogenesis
  - Iodine deficiency
  - Iodine excess
  - Thyroid-blocking drugs
  - Congenital and acquired defects of hormone synthesis and thyroglobulin metabolism defects

Pituitary hypothyroidism (TSH deficiency)

Hypothalamic hypothyroidism (TRH deficiency)

Reduced peripheral response to thyroid hormone

Table 39-4. Laboratory findings in hypothyroidism

Type	T_4	T_3	T_3RU	FT_4I	TSH	TRH stimulation*
Primary	↓	↓, N	↓, N	↓	↑	↑
Secondary	↓	↓	↓	↓	↓, N	↓
Tertiary	↓	↓	↓	↓	↓, N	N
Peripheral unresponsiveness	↑	↑	N	↑ or N	↑ or N	N or ↑

*Assessed by response of serum TSH to TRH administration.
N, Normal; ↑, elevated; ↓, decreased. See Table 39-3 for abbreviations.

sient. On rare occasions hypothyroidism may not be resolved.

Hypothyroidism often follows surgical thyroidectomy or therapy with radioiodine. In the case of Graves' disease, the natural history of the disease is such that many patients eventually become hypothyroid regardless of the type of treatment used.[107,108]

Developmental abnormalities can result in hypothyroidism caused by a total (aplasia) or partial (dysgenesis) absence of functioning thyroid tissue. Tumors and other infiltrative disorders displacing and destroying thyroid tissue can occasionally cause hypothyroidism.

Hypothyroidism can result from various other conditions, including congenital defects in hormonogenesis and the effect of drugs. Pituitary and hypothalamic disease are rare conditions that occasionally lead to hypothyroidism caused by inadequate TRH or TSH secretion.

Laboratory findings. Regardless of the cause, the laboratory findings in hypothyroidism are characterized by decreased total serum T_4, T_3, FT_4, FT_4I, and T_3RU. The degree of abnormality of these tests varies widely, depending on the cause of hypothyroidism and stage of the disease.

Some distinctive laboratory findings can be found in Table 39-4. For instance, hypothyroidism caused by intrinsic thyroid disease (primary hypothyroidism) will display elevated TSH levels and exaggerated TSH responses to TRH stimulation. Hypothyroidism caused by pituitary (secondary) and hypothalamic (tertiary) disease exhibits low or borderline normal levels of TSH and TRH test responses as noted above.

Hypothyroidism presenting in the neonatal period represents a special problem. Hypothyroidism can lead to severe growth and maturation retardation of the central nervous system (cretinism). Therefore it is extremely important to detect this abnormality early in the neonatal period. Routine screening of neonates for hypothyroidism has provided an important tool in the diagnosis of hypothyroidism and a significant step in preventive pediatric medicine.[109]

Treatment. The treatment of hypothyroidism consists in thyroid hormone replacement given orally, which reverses the abnormal laboratory findings and clinical symptoms and signs, provided that there are no abnormalities of transport and peripheral use of thyroid hormones.

Goiter

Goiter refers to an enlargement of the thyroid gland. Anatomically goiters can be diffuse or multinodular. Diffuse goiter is characterized by uniform hypertrophy and enlargement of the thyroid gland. In multinodular goiter the thyroid gland is enlarged in a nonuniform fashion, resulting in nodules located both superficially and deep within the gland.

Depending on the mode of appearance, goiters have been classified as endemic, familial, or sporadic. Endemic goiter is found clustered in certain geographical areas and is caused by iodine deficiency or excess, or ingestion of substances that impair the production of thyroid hormone (goitrogens). Familial goiters are usually produced by inheritable defects of thyroid hormonogenesis. Iodine deficiency as a cause of goiter in the United States has become rare because of the enrichment of foods with iodine.

Sporadic goiter occurs in areas not associated with endemic goiters. It may be caused by congenital errors of thyroid hormone synthesis or by ingestion of goitrogens. Patients with sporadic goiter have varying degrees of thyroid enlargement and hypofunction. Defects of thyroid hormone synthesis can occur at any step in the formation of thyroid hormone. In many instances the goiter is of unknown cause and termed "idiopathic."

Solitary nodule

A solitary nodule refers to the presence of a localized enlargement of a portion of the thyroid gland. Although most of these nodules represent benign conditions such as cysts, localized hemorrhages, focal thyroiditis, and adenomas, they may also represent malignant tumors of the thyroid. External radiation therapy used in the past for acne and enlarged tonsils and other conditions of the head, neck, and chest has been associated with the subsequent development of thyroid cancer.[110,111]

Thyroid cancer

The most common primary malignant thyroid tumors originate from the epithelial cells of the follicles and include papillary carcinoma, follicular carcinoma, and undifferentiated anaplastic carcinoma in decreasing order of frequency. Patients with thyroid cancer usually do not display significant abnormalities of thyroid function. These tumors often release thyroglobulin into the circulation, where it may be followed as a tumor marker. Since other thyroid diseases can also result in increased levels of thyroglobulin, this parameter is most useful for following activity of the disease rather than for specific diagnostic purposes.[112]

A different type of thyroid cancer originates from the parafollicular C cells of the thyroid and is termed "medullary carcinoma." These tumors secrete calcitonin, a hormone that lowers calcium in the blood, which is a useful tumor marker for diagnosing and following these patients. In patients with borderline calcitonin levels, stimulation tests with infusions of calcium or pentagastrin may elicit excess calcitonin secretion.

Pseudohyperthyroidism

Increased levels of thyroid hormone in the serum are caused by true hyperthyroidism, abnormalitites of the thyroxine-binding proteins, or abnormalities in peripheral thyroid hormone metabolism. In the absence of quantitative abnormalities of the thyroxine-binding proteins, elevated levels of thyroid hormone in the serum are usually suggestive of hyperthyroidism. However, cases of abnormally high T_4 and T_3 hormone levels and normal levels of TBG have been described in patients with otherwise normal thyroid function.[113-115] A defect characterized by decreased affinity and responsiveness of tissue receptors has been postulated.[116] A different familial defect with impairment of T_3 production from T_4 also has been recently described.[117] Laboratory findings in these cases demonstrate elevated T_4 and free T_4 levels with normal levels of T_3.

TESTS OF THYROID FUNCTION

The iodine-concentrating property of the thyroid is used to estimate thyroid function and to obtain functional anatomical images of the gland. For these studies tracer amounts of a radioactive form of iodine are administered to the patient. Radioactivity emitted in the form of gamma rays by the radioiodine concentrated in the thyroid is detected by specially designed imaging and counting devices and transformed into thyroid radioiodine uptakes and images. The two most commonly used radioiodines are ^{131}I and ^{123}I. The more desirable radioisotope is ^{123}I, since it produces better images and delivers lower radiation doses to the patient's thyroid. Serum samples for hormone radioassays that are obtained after the administration of radioactive materials to the patient must be checked for residual radioactivity before the assay is started or spurious results may be obtained.

Thyroid iodine uptake

This parameter measures the percentage of an administered dose of radioiodine that concentrates in the gland by the trapping and organification mechanisms. This measurement may be obtained at various intervals after radioiodine administration. Theoretically, the degree of thyroid uptake in the thyroid at a given time reflects thyroid hormone synthesis and excretion and therefore reflects the functional status of the thyroid. In general, it is believed that increased iodine ingestion has caused the values of iodine uptake to decrease in the United States. Regional variations in iodine uptake make interpretation more difficult. Low thyroid uptakes are found not only in hypothyroidism but also in certain conditions actually associated with thyroid hyperfunction, including (1) thyrotoxicosis factitia caused by exogenous thyroid hormone, (2) subacute thyroiditis, (3) iodine-induced hyperthyroidism, and (4) certain forms of chronic thyroiditis.

TRH stimulation test

The TRH stimulation test takes advantages of the interrelationship between the TRH and TSH secretions. Normally, after the intravenous administration of TRH, there is an increase of TSH levels in blood that in turn elicit an elevation of T_3 and T_4 serum levels.[118,119] Various factors influence TSH responses to TRH stimulation. For instance, the degree of TSH increase appears to be directly proportional to the base-line levels of TSH.[120] Other factors influencing TSH responses to TRH administration include age and gender of the patient and exposure to various medications. Abnormal responses to TRH stimulation are of two types. The first type is characterized by lower-than-normal responses of serum TSH to TRH stimulation. In these cases the TSH response to TRH is said to be blunted. The second type of abnormal response consists in higher-than-normal increments of serum TSH after TRH stimulation.

The main clinical applications of the TRH stimulation test are (1) diagnosis of subclinical hyperthyroidism or confirmation that exogenous thyroid hormone doses are sufficient to fully suppress TSH secretion in patients with thyroid cancer, (2) evaluation of patients with ophthalmopathy without hyperthyroidism, and (3) diagnosis of hypothalamic and pituitary hypothyroidism.

Patients with overt or subclinical hyperthyroidism[118,121,122] and some patients with Graves' ophthalmopathy[123,124] have blunted responses of TSH to TRH stimulation. In patients with thyroid cancer the suppression of TSH secretion is often desirable therapeutically to decrease rates of tumor growth. Adequate doses of thyroid hormone in such patients will result in blunted TSH response to TRH.

Patients with hypothyroidism caused by hypothalamic lesions and intact pituitary glands show delayed but quantitatively normal responses to TRH stimulation, whereas patients with hypothyroidism secondary to pituitary tumors have subnormal increments of TSH after administration of TRH.[125] Patients with primary hypothyroidism have an exaggerated TSH response to TRH.

TSH stimulation test

Administration of exogenous TSH will stimulate all phases of thyroid function, which will be reflected by increases in the radioiodine thyroid uptake and blood levels of thyroid hormone. The TSH stimulation test is performed by monitoring of the radioiodine uptake before (base line) and after the administration of bovine TSH for 3 consecutive days. An abbreviated TSH stimulation test is sometimes performed. In this case bovine TSH is administered for 1 day only. Normally there is an increase of serum T_4 and radioiodine uptake to more than 1.5 times the base-line value in response to TSH administration. The chief use of this test has been to differentiate primary hypothyroidism from secondary or tertiary hypothyroidism, and it has been replaced largely by TSH determinations in serum and by the TRH stimulation test.

Triiodothyronine (T_3) or Cytomel suppression test

This test consists of a determination of 24-hour thyroid radioiodine uptake before (base line) and then after the oral administration of T_3 for 7 to 10 days.[126] Images and thyroid uptake determinations are obtained simultaneously. The purpose of this test is to establish the presence of thyroid tissue that has become autonomous and unresponsive to TSH changes. Normally the administration of thyroid hormone (T_3 or T_4) will shut off the secretion of TSH by the pituitary gland. The subsequent absence of TSH results in diminished thyroid concentration of radioiodine. A thyroid gland that is overactive is often autonomous, and so suppression of TSH production by administration of exogenous thyroid hormone will not be followed by corresponding declines of thyroid radioiodine uptake. Normally a drop in thyroid uptake (suppression) to 30% to 50% of the base-line value occurs after T_3 administration.

Perchlorate discharge test

The perchlorate discharge test detects defects of iodine organification present in conditions such as Hashimoto's thyroiditis and congenital goiters. It is also used to determine the degree of organification defect caused by certain thyroid-blocking drugs used for treatment of hyperthyroidism to assess their therapeutic effects.

CHANGE OF ANALYTE IN DISEASE

Thyroid function and its alterations caused by disease can be assessed by determination of various analytes in blood. Tables 39-3 to 39-5 summarize the changes of some analytes in various disease states.

Protein-bound iodine (PBI)

Historically PBI was used to indirectly assess the concentrations of thyroid hormone in the blood. This test measured iodine contained in thyroid hormones, which in turn were bound to plasma proteins. However, iodine of nonhormonal origin also binds to proteins and alters PBI values in the absence of true T_4 elevations.[127] For this reason, the PBI measurement has been abandoned as an indirect indicator of T_4 levels in blood. The PBI may be useful to detect increases of endogenous non-T_4 iodoproteins that may be released into the serum in conditions such as Hashimoto's and subacute thyroiditis. In these situations the difference between PBI and T_4-iodine ($0.65 \times$ total T_4) values provides an estimation of the non-T_4 iodoproteins in plasma.

Serum T_4

Elevated total serum T_4 can occur as a result of increased hormone synthesis, increased hormone release from the thyroid cells, and increased binding capacity of plasma proteins, especially TBG. Increased hormone secretion is most frequently seen in states of hyperthyroidism such as Graves' disease, toxic multinodular goiter, and toxic thyroid adenoma. Other less frequent causes of hyperthyroidism are listed in the box on p. 760.

Increased T_4 release occurs in subacute thyroiditis and Hashimoto's thyroiditis and after radiation. Increased levels of serum TBG of various causes (see box, p. 758) produce elevations of total serum T_4. These conditions are not accompanied by hyperthyroidism.

Serum T_3

In general, elevations of serum T_3 parallel those of T_4 and are found in most states of hyperthyroidism. In addition, isolated elevations of T_3 are found in T_3 thyrotoxicosis. This entity is found in Graves' disease, toxic multinodular goiter, and toxic adenoma and in patients with recurrent hyperthyroidism after treatment with antithyroid drugs, radioactive iodine, and surgery.[99,100] Increased serum concentration or binding capacity of TBG also results in elevations of T_3.

Resin-uptake test (T_3RU or T_4RU)

It is important to remember that thyroid hormones in the blood are distributed in two compartments: protein-bound and unbound or free compartments (Fig. 39-7). Variations of measurable total thyroid hormone in blood can result from changes in the binding-protein levels. Hypothyroidism and hyperthyroidism will only occur if a net persistent decrease or increase of free unbound thyroid hormone exists in the blood. Blood level determinations of T_4 and T_3 are clinically more meaningful if the functional levels of thyroid hormone–binding protein in blood are known. Thus the resin-uptake test does not measure a specific analyte per se but the functional state (ability to bind hormone) of an analyte (such as TBG).

The resin-uptake test reveals the number of free available binding sites in the plasma thyroid hormone–transporting proteins, especially TBG, since this is the most important transporter of thyroid hormones. Assessment of the degree of TBG concentration is most commonly accomplished by the resin-uptake test. Samples of patient serum are mixed with radiolabeled thyroid hormone (T_3 or

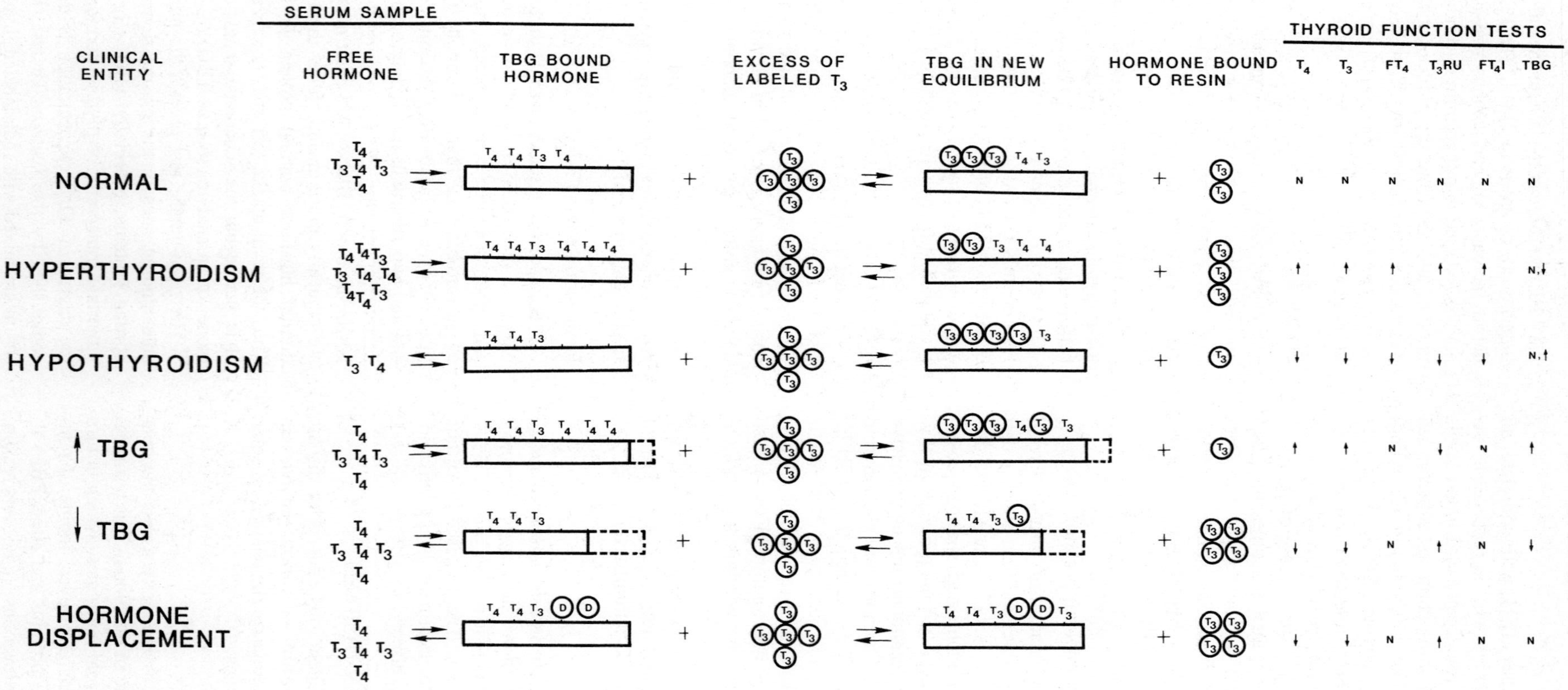

CLINICAL ENTITY	T_4	T_3	FT_4	T_3RU	FT_4I	TBG
NORMAL	N	N	N	N	N	N
HYPERTHYROIDISM	↑	↑	↑	↑	↑	N,↓
HYPOTHYROIDISM	↓	↓	↓	↓	↓	N,↑
↑ TBG	↑	↑	N	↓	N	↑
↓ TBG	↓	↓	N	↑	N	↓
HORMONE DISPLACEMENT	↓	↓	N	↑	N	N

Fig. 39-7. Interrelationships between serum TBG (thyroxine-binding globulin), the T_3RU, and other thyroid function tests. *D*, Drugs occupying binding sites on TBG; *FT_4I*, free thyroxine hormone index.

T_4). The amount of thyroid hormone that remains unbound in the mixture is inversely proportional to the number of available binding sites (Fig. 39-7). This unbound hormone is separated from the mixture when one adds relatively low affinity binding resin to the system (Fig. 39-7). The resin is then separated from the serum, and the amount of uptake of radioactive hormone by the resin is determined. In cases of increased TBG in plasma, although free thyroid hormone levels are normal, the total T_4 is increased, since the total number of binding sites in TBG is increased. Labeled T_3 will find an increased pool of TBG (available binding sites), and less labeled T_3 will remain free to bind resin. Therefore the T_3 resin uptake (T_3RU) is low. The opposite changes occur in states of decreased TBG. Conditions that cause abnormalities of the binding sites (TBG) are listed in Table 39-3. Certain drugs compete with thyroid hormone for the TBG-binding sites. This phenomenon is reflected by normal free T_4, low total T_4, and high T_3RU values.

In hyperthyroidism, free thyroid hormone in plasma increases. This results in complete or almost complete saturation of binding sites in TBG, which in turn will result in fewer molecules of labeled T_3 bound to TBG and increased amounts of free-labeled T_3 (increased T_3RU).

Free thyroxine hormone index (FT_4I)

It is the free hormone that induces metabolic and biological effects in target cells. The free T_4 index (FT_4I) indirectly assesses the level of free T_4 in blood and adjusts for most interferences caused by binding-protein abnormalities.[128] The FT_4I is determined from total T_4 and resin-uptake values obtained on the same sample. The FT_4I is calculated as follows:

$$\frac{\text{Total serum } T_4 \times \text{Patient's resin uptake}}{100}$$

As expected, most alterations of binding proteins produce reciprocal changes in resin uptake and total serum T_4 or T_3, resulting in normal values of FT_4I, whereas true alterations of free thyroid hormone content cause concordant unidirectional changes of FT_4I. Total T_4 and FT_4I are still considered the most practical and valuable tests in the initial evaluation of thyroid status. The FT_4I is elevated in hyperthyroidism (Table 39-3). It is decreased in states of hypothyroidism (Table 39-4).

The free T_3 index is calculated in a similar fashion as the FT_4I from the total serum T_3.[129] It has similar applications and significance as the FT_4I. The FT_3I may be helpful to exclude T_3 toxicosis in some patients taking oral contraceptives in whom isolated serum T_3 increases without corresponding elevations of T_4. In general, the FT_3I offers no advantages to the FT_4I and is used less frequently in clinical practice.

Free T_4 and free T_3

FT_4 and FT_3 tend to parallel changes in total T_4 and total T_3 concentrations. Measurement of the free hormone levels is useful in instances in which other test results are borderline or conflicting. Elevations and decreases of free T_4 and free T_3 theoretically are true reflections of hyperthyroidism and hypothyroidism. However, the results are very dependent on the analytical method used to measure "free" hormone.

Thyroxine-binding globulin (TBG)

Many diseases produce alterations of TBG (see box, p. 758) and other thyroid hormone–transporting proteins in plasma. Changes in concentrations of such proteins affect the total plasma concentrations of T_4 and T_3. Since most of the T_4 and T_3 are bound to TBG, changes in this protein level are the most important clinically.

Abnormalities of TBG may be of two types, quantitative and qualitative. Quantitative abnormalities are those characterized by absolute increases or decreases of TBG levels in plasma. Consequently, the total amount of thyroid hormone transported in plasma will increase (increased TBG) or decrease (decreased TBG).

Qualitative abnormalities of TBG refer to those stemming from alterations of the binding affinity rather than from absolute changes of TBG amounts in plasma. The binding affinity of the TBG may be increased or decreased in intrinsic TBG defects. Various compounds and drugs may strongly bind to TBG and result in displacement of thyroid hormone from binding sites. The result is decreased total serum T_4 and T_3 (Fig. 39-7).

Serum thyroid-stimulating hormone (TSH)

The principal use of thyroid-stimulating hormone (TSH) determinations in serum is the diagnosis of hypothyroidism. Patients with untreated hypothyroidism stemming from intrinsic thyroid defects (primary hypothyroidism), regardless of the cause, have elevated serum levels of TSH. Patients with hypothyroidism caused by pituitary lesions (secondary) or hypothalamic (tertiary) lesions have normal or low TSH levels. Differentiation between secondary and tertiary hypothyroidism is accomplished by TRH stimulation testing.

Most patients with hyperthyroidism have low or undetectable TSH levels in serum, reflecting the inhibitory effects of high levels of circulating thyroid hormone on the hypothalamic-pituitary axis. Upper normal or elevated TSH levels in such patients indicate a possibly different cause from primary hyperthyroidism.

Immunoglobulins

It has been suggested that immunological abnormalities play an important role in thyroid pathology.[81] Antibodies against components of thyroid cells are found in many patients with thyroid disease. High levels of antibodies against thyroglobulin or thyroid microsomes may be found in patients with Hashimoto's thyroiditis. High titers of thyroid-stimulating immunoglobulins may be found in Graves' disease.

Laboratory findings in nonthyroid disease (NTD)

As previously mentioned, patients with NTD may develop abnormalities of thyroid tests. These changes can mimic thyroid disease or modify and confound laboratory findings in patients with thyroid disease. Chronic diseases that alter laboratory tests of thyroid function are more frequent than thyroid disease. NTD includes patients with general chronic debilitating conditions and patients with specific conditions, namely, alcoholic cirrhosis, hepatitis, and renal failure. Table 39-5 depicts the most important laboratory findings in several nonthyroid diseases.

As a group, euthyroid patients with NTD have either normal or low total T_4, normal or low FT_4I, normal or low total T_3 and free T_3, normal or high rT_3, normal or high T_3RU, and normal or slightly high TSH levels in serum.[76,130-132] The free T_4 index (FT_4I), which is often a reliable indicator of thyroid function,[133] is usually normal but may also be low in patients with NTD.[76,134]

Low total T_4, low FT_4I, and low or subnormal T_3 levels may suggest the presence of hypothyroidism in patients with NTD. The presence of high serum levels of rT_3 is useful in excluding hypothyroidism in this circumstance.[130] TSH concentrations are usually normal.

Liver disease is characterized by abnormalities of protein synthesis, including the thyroid-binding proteins, and by alterations of T_4 metabolism. Alcoholic liver cirrhosis is accompanied by decreased binding capacity of thyroid hormones, resulting in high T_3RU[135] (Table 39-5). Total T_4 is usually normal or low,[80,135] and free T_4 is slightly elevated. There is depressed conversion of T_4 to T_3[13] with decreased serum levels of T_3 and occasionally also the free T_3.

Acute hepatitis is characterized by an increase of the serum TBG[136] (Table 39-5), which results in increased levels of T_4 and decreased T_3RU. In general, chronic active hepatitis also produces the same changes in blood as those seen in acute hepatitis. Some patients with chronic active hepatitis have an associated autoimmune involvement of the thyroid gland. In these cases concomitant changes caused by slightly decreased thyroid function also appear. These include (1) lower than expected T_4, (2) low free T_4 index, (3) decreased free T_4, (4) normal serum TSH, and (5) abnormally increased TSH response to TRH.[137]

In patients with renal failure the most common findings are decreased total and free T_3[138] in the serum, caused by diminished peripheral conversion of T_4 to T_3. Serum thyroxine-binding protein, T_3RU, and total T_4 usually are normal. However, the total T_4 and free T_4 may be slightly low.[139,140] Diminished total T_4 can result from renal failure with severe catabolic states, probably caused by decreased thyroxine-binding proteins in serum. Concentration of TSH in serum is normal.

Association of NTD with hyperthyroidism is found occasionally. In these situations there is a disproportionate increase of T_4 in serum with only modestly elevated or normal T_3 levels. This is caused by a decreased peripheral conversion of T_4 to T_3 concomitant with the increased T_4 production.[141]

Laboratory findings after therapeutic interventions for thyroid disease

The two most frequently found clinical situations that require closely monitored treatment are hyperthyroidism and hypothyroidism.

Graves' disease is a good example of hyperthyroidism. When successfully treated surgically, usually with subtotal thyroidectomy, regression of symptoms occurs in 4 to 10 weeks. From the laboratory point of view, normalization of T_4, T_3, T_3RU, RT_4I, and TSH occur concomitantly. In patients treated with antithyroid drugs such as propylthiouracil (PTU) and methimazole (MM), improvement of thyroid function tests may be seen after 2 weeks of treatment but usually takes longer. Because of the blocking effect of PTU on conversion of T_4 to T_3, a disproportionately rapid decrease in T_3 may be seen in some cases.[71]

Table 39-5. Thyroid-function tests in nonthyroid diseases (NTD)*

	T_4	T_3RU	FT_4I	Free T_4	Free T_3	T_3	rT_3	TSH	TBG
General NTD†	N, ↓	↑, N	↓, N	N, ↑	↓, N	↓, N	↑, N	N	↓, N
Acute hepatitis	N, ↑	↓	N, ↓	N, ↑	N, ↓	↓, N	↑	N	↑
Chronic active hepatitis	N, ↑	↓	N, ↓	N, ↑	N	N	↑	N	↑
Alcoholic cirrhosis	N, ↓	↑, N	N, ↑	N, ↑	↓	↓	↑	N	↓, N
Renal failure	N, ↓	N	N, ↓	N, ↓	↓	↓	N	N	N

*Normal thyroid function in patients with very complex problems; findings in individual patients may vary widely.
†Includes all patients with NTD.
N, Normal; ↑, augmented, ↓, diminished.

Therapy with ^{131}I is now used more frequently. Clinical improvement is seen at about 6 to 12 weeks after treatment begins. Laboratory tests show a decrease of T_4, T_3, T_3RU, and FT_4I approximately 6 to 10 weeks after treatment. Residual radioactivity in the serum must be considered when one is using radioassay methods within 90 days of therapy.

Patients treated for thyrotoxicosis by any of the methods may follow one of three courses. Ideally normal thyroid function ensues and is reflected by normal laboratory values. Occasionally patients remain hyperthyroid and abnormal laboratory tests persist. More often patients become hypothyroid with low T_4, T_3RU, and FT_4I values. The earliest indication of the ensuing hypothyroidism is a depressed T_4 followed by an elevation of TSH levels as the hypothalamic pituitary axis recovers. The TSH may remain elevated with normal T_4 levels, a condition known as biochemical hypothyroidism.

Patients with treated Graves' disease may develop an ongoing hyperthyroid status with normal or subnormal T_4 values but elevated T_3 levels. This situation occurs most frequently in patients treated with thyroid-blocking drugs and ^{131}I.

The treatment of hypothyroidism consists in thyroid hormone replacement. During the monitoring of thyroid hormone replacement it is very important to keep in mind the effect that changes in TBG levels produce on thyroid hormone levels. Laboratory tests should therefore include those that indirectly or directly assess TBG levels, such as the T_3RU, FT_4I, and TBG determinations. When preparations containing T_4 and T_3 are used, normalization of both T_3 and T_4 values is expected.[142] When preparations containing only T_3, such as Cytomel, are used, T_3 normalizes or becomes elevated but T_4 remains low. With pure T_4 preparations the T_4 may be mildly elevated but T_3 values are usually normal. Normalization of T_4 and T_3 values is accompanied by a corresponding decline in TSH levels.

ACKNOWLEDGMENT

We wish to acknowledge Lisa A. Fooks, Ruth M. McDevitt, R.N., B.S.N., and Patrick Slattery for their assistance in the compilation of this chapter and preparation of illustrations.

REFERENCES

1. Oddie, T.H., Fisher, D.A., McConahey, W.M., and Thompson, C.S.: Iodine intake in the United States: a reassessment, J. Clin. Endocrinol. Metab. **30:**659-665, 1970.
2. Harrison, M.T., Harden, R., McG., Alexander, W.D., et al.: Iodine balance studies in patients with normal and abnormal thyroid function, J. Clin. Endocrinol. **25:**1077-1084, 1965.
3. Wolff, J.: Transport of iodide and other anions in the thyroid gland, Physiol. Rev. **44:**45-79, 1964.
4. Tong, W.: Thyrotropin stimulation of tyrosine synthesis in isolated thyroid cells treated with perchlorate, Endocrinology **75:**968-969, 1964.
5. Sherwin, J.R., and Tong, W.: The actions of iodide and TSH on thyroid cells showing a dual control system for the iodide pump, Endocrinology **94:**1465-1474, 1974.
6. Wyngaarden, J.B., Wright, B.M., and Ways, P.: The effect of certain anions upon the accumulation and retention of iodide by the thyroid gland, Endocrinology **50:**537-549, 1952.
7. Takeuchi, K., Suzuki, H., Horiuchi, Y., and Mashimo, K.: Significance of iodine-perchlorate discharge test for detection of iodine organification defect of the thyroid, J. Clin. Endocrinol. Metab. **31:**144-146, 1970.
8. DeGroot, L.J., and Niepomniszcze, H.: Biosynthesis of thyroid hormone: basic and clinical aspects, Metabolism **26:**665-718, 1977.
9. Edelhoch, H.: The structure of thyroglobulin and its role in iodination, Recent Prog. Horm. Res. **21:**1-31, 1965.
10. Spiro, M.J.: Studies on the protein portion of thyroglobulin: aminoacid compositions and terminal amino acids of several thyroglobulins, J. Biol. Chem. **245:**5820-5826, 1970.
11. Dumont, J.E.: The action of thyrotropin on thyroid metabolism, Vitam. Horm. **29:**287-412, 1971.
12. Cavalieri, R.R., and Rapoport, B.: Impaired peripheral conversion of thyroxine to triiodothyronine, Annu. Rev. Med. **28:**57-65, 1977.
13. Chopra, I.J., Solomon, D.H., Chopra, U., et al.: Pathways of metabolism of thyroid hormones, Recent Prog. Horm. Res. **34:**521-567, 1978.
14. Oppenheimer, J.H.: Role of plasma proteins in the binding distribution and metabolism of the thyroid hormones, N. Engl. J. Med. **278:**1153-1162, 1968.
15. Woeber, K.A., and Ingbar, S.H.: The interactions of the thyroid hormones with binding protein. In Greer, M.A., and Solomon, D.H., editors: Thyroid, American handbook of physiology, vol. 3, Washington, D.C., 1973, American Physiological Society, pp. 187-196.
16. Woeber, K.A., and Ingbar, S.H.: The contribution of thyroxine-binding prealbumin to the binding of thyroxine in human serum, as assessed by immunoabsorption, J. Clin. Invest. **47:**1710-1721, 1968.
17. Lutz, J.H., and Gregerman, R.I.: Dependence of the binding of thyroxine to prealbumin in human serum, J. Clin. Endocrinol. Metab. **29:**487-496, 1969.
18. Stahl, T.J.: Radioimmunoassay and the hormones of thyroid function, Semin. Nucl. Med. **5:**221-246, 1975.
19. Braverman, L.E., Ingbar, S.H., and Sterling, K.: Conversion of thyroxine (T_4) to triiodothyronine (T_3) in athyrotic subjects, J. clin. Invest. **49:**855-864, 1970.
20. Ingbar, S.H., and Braverman, L.E.: Active form of the thyroid hormone, Annu. Rev. Med. **26:**443-449, 1975.
21. Ingbar, S.H., and Freinkel, N.: Simultaneous estimation of rates of thyroxine degradation and thyroid hormone synthesis, J. Clin. Invest. **34:**808-819, 1955.
22. Schimmel, M., and Utiger, R.D.: Thyroidal and peripheral production of thyroid hormones, Ann. Intern. Med. **87:**760-768, 1977.
23. Gavin, L., Castel, J., McMahon, F., et al.: Extrathyroidal conversion of thyroxine to 3,3′,5′-triiodothyronine (reverse T_3) and to 3,5,3′-triiodothyronine (T_3) in humans, J. Clin. Endocrinol. Metab. **44:**733-742, 1977.
24. Chopra, I.J.: An assessment of daily production and significance of thyroidal 3,3′,5′-triiodothyronine (reverse T_3) in man, J. Clin. Invest. **58:**32-40, 1976.
25. Surks, M.I., Schadlow, A.R., Stock, J.M., et al.: Determination of iodothyronine absorption and conversion of L-thyroxine (T_4) to L-triiodothyronine (T_3) using turnover rate techniques, J. Clin. Invest. **52:**805-811, 1973.
26. Sakurada, T., Rudolph, M., Fang, S.-L.L., et al.: Evidence that triiodothyronine and reverse triiodothyronine are sequentially deiodinated in man, J. Clin. Endocrinol. Metab. **46:**916-922, 1978.
27. Nakamura, Y., Chopra, I.J., and Solomon, D.H.: An assessment of the concentration of acetic acid and propionic acid derivatives of 3,5,3′-triiodothyronine in human serum, J. Clin. Endocrinol. Metab. **46:**91-97, 1978.
28. Galton, V.A., and Pitt-Rivers, R: A quantitative method for the separation of thyroid hormones and related compounds from serum and tissues with an anion-exchange resin, Biochem. J. **72:**310-313, 1959.

29. Oddie, T.H., Fisher, D.A., and Rogers, C.: Whole body counting of ^{131}I-labeled thyroxine, J. Clin. Endocrinol. Metab. **24:**628-637, 1964.
30. Pittman, J.A., Brown, R.W., and Register, H.B.: Biological activity of 3,3′5′-triiodo-DL-thyronine, Endocrinology **70:**79-83, 1962.
31. Chopra, I.J.: A radioimmunoassay for measurement of 3,3′5′-triiodothyronine (reverse T_3), J. Clin. Invest. **54:**583-592, 1974.
32. Oppenheimer, J.H., Koerner, D., Schwartz, H.L., et al.: Specific nuclear triiodothyronine binding sites in rat liver and kidney, J. Clin. Endocrinol. Metab. **35:**330-333, 1972.
33. Surks, M.I., Koerner, D., Dillman, W., et al.: Limited capacity binding sites for L-triiodothyronine in rat liver nuclei: localization to the chromatin and partial characterization of the L-triiodothyronine-chromatin complex, J. Biol. Chem. **248:**7066-7072, 1973.
34. Oppenheimer, J.H., Schwartz, H.L., and Surks M.I.: Tissue differences in the concentration of triiodothyronine nuclear binding sites in the rat: liver, kidney, pituitary, heart, brain, spleen and testis, Endocrinology **95:**897-903, 1974.
35. Surks, M.I., Koerner, D.H., and Oppenheimer, J.H.: In vitro binding of L-triiodothyronine to receptors in rat liver nuclei: kinetics of binding extraction properties, and lack of requirement for cytosol proteins, J. Clin. Invest. **55:**50-60, 1975.
36. Oppenheimer, J.H.: Initiation of thyroid-hormone action, N. Engl. J. Med. **292:**1063-1068, 1975.
37. Sterling, K.: Thyroid hormone action at the cell level, N. Engl. J. Med. **300:**117-123, 1979.
38. Sterling, K., Milch, P.O., Brenner, M.A., et al.: Thyroid action: the mitochondrial pathway, Science **197:**996-999, 1977.
39. Ismail-Beigi, F., and Edelman, I.S.: Mechanism of thyroid calorigenesis: role of active sodium transport, Proc. Natl. Acad. Sci. USA **67:**1071-1078, 1970.
40. Edelman, I.S., and Ismail-Beigi, F.: Thyroid thermogenesis and active sodium transport, Recent Prog. Horm. Res. **30:**235-257, 1974.
41. Edelman, I.S.: Thyroid thermogenesis, N. Engl. J. Med. **290:**1303-1308, 1974.
42. Ismail-Beigi, F., Bissell, D.M., and Edelman, I.S.: Thyroid thermogenesis in primary hepatocyte cultures, J. Clin. Invest. **25:**464, 1977 (Abstract).
43. Segal, J., Gordon, A., and Gross, J.: Evidence that L-triiodothyronine (T_3) exerts its biological action not only through its effects on nuclear activity. In Robbins, J., and Braverman, L.E., editors: Thyroid research: Proceedings of the 7th International Thyroid Conference, Amsterdam, 1976, Excerpta Medica, pp. 331-333.
44. Burgus, R., Dunn, T.F., Desiderio, D., et al.: Characterization of ovine hypothalamic hypophysiotropic TSH-releasing factor, Nature **226:**321-325, 1970.
45. Anderson, M.S., Bowers, C.Y., et al.: Synthetic thyrotropin-releasing hormone: a potent stimulator of thyrotropin secretion in man, N. Engl. J. Med. **285:**1279-1283, 1971.
46. Labrie, F., Barden, N., Poirier, G., et al.: Binding of thyrotropin-releasing hormone to plasma membranes of bovine anterior pituitary gland, Proc. Natl. Acad. Sci. USA **69:**283-287, 1972.
47. Poirier, G., Borden, N., Labrie, F., et al.: Partial purification and some properties of adenyl cyclase and receptor for TRH from anterior pituitary gland. In Proceedings of the IV International Congress on Endocrinology, Excerpta Medica Int. Congr. Ser. No. 256, Amsterdam, 1972, Excerpta Medica, p. 85.
48. Pierce, J.G.: Eli Lilly lecture: the subunits of pituitary thyrotropin—their relationship to other glycoprotein hormones, Endocrinology **89:**1331-1344, 1971a.
49. Pierce, J.G., Liao, T., et al.: Studies on the structure of thyrotropin: its relationship to luteinizing hormone, Recent Prog. Horm. Res. **27:**165-212, 1971b.
50. Amir, S.M., Carraway, T.F., and Kohn, L.D.: The binding of thyrotropin to isolated bovine thyroid plasma membranes, J. Biol. Chem. **248:**4092-4100, 1973.
51. Haigler, E.D., Jr., Pittman, J.A., Jr., Hershman, J.M., et al.: Direct evaluation of pituitary thyrotropin reserve utilizing synthetic thyrotropin releasing hormone, J. Clin. Endocrinol. Metab. **33:**573-581, 1971.
52. Silva, J.E., and Larsen, P.R.: Contributions of plasma triiodothyronine and local thyroxine monodeiodination to triiodothyronine to nuclear triiodothyronine receptor saturation in pituitary, liver, and kidney of hypothyroid rats, J. Clin. Invest. **61:**1247-1259, 1978.
53. Greer, M.A., and DeGroot, L.J.: The effect of stable iodine on thyroid secretion in man, Metabolism **5:**682-696, 1956.
54. Wolff, J.: Iodide goiter and pharmacologic effects of excess iodide, Am. J. Med. **47:**101-124, 1969.
55. Wolff, J., and Chaikoff, I.L.: Plasma inorganic iodide as a homeostatic regulator of thyroid function, J. Biol. Chem. **174:**555-564, 1948.
56. Braverman, L.E., and Ingbar, S.H.: Changes in thyroidal function during adaptation to large doses of iodide, J. Clin. Invest. **42:**1216-1231, 1963.
57. Kaptein, E.M., Spencer, D.A., Kamiel, M.D., et al.: Prolonged dopamine administration and thyroid hormone economy in normal and critically ill subjects, J. Clin. Endocrinol. Metab. **51:**387-393, 1980.
58. Spaulding, S.W., Burrow, G.N., Donabedian, R., et al.: L-Dopa suppression of thyrotropin-releasing hormone response in man, J. Clin. Endocrinol Metab. **35:**182-185, 1972.
59. Morley, J.E., Schafer, R.B., Elson, M.K., et al.: Amphetamine-induced hyperthyroxinemia, Ann. Intern. Med. **93:**707-709, 1980.
60. Healy, D.L., and Burger, H.G.: Increased prolactin and thyrotrophin secretion following oral metoclopramide: dose-response relationships, Clin. Endocrinol. **7:**195-201, 1977.
61. Taurog, A.: The mechanism of action of thioureylene antithyroid drugs, Endocrinology **98:**1031-1046, 1976.
62. Kaplan, M.M., Schimmel, M., and Utiger, R.D.: Changes in serum 3,3′,5′-triiodothyronine (reverse T_3) concentrations with altered thyroid hormone secretion and metabolism, J. Clin. Endocrinol. Metab. **45:**447-456, 1977.
63. Oppenheimer, J.H., Schwartz, H.L., and Surks, M.I.: Propylthiouracil inhibits the conversion of L-thyroxine to L-triiodothyronine: an explanation of the antithyroxine effects of propylthiouracil and evidence supporting the concept that triiodothyronine is the active thyroid hormone, J. Clin. Invest. **51:**2493-2497, 1972.
64. Spaulding, S.W., Burrow, G.N., Bermudez, F., and Himmelhoch, J.M.: The inhibitory effect of lithium on thyroid hormone release in both euthyroid and thyrotoxic patients, J. Clin. Endocrinol. Metab. **35:**905-911, 1972.
65. Emerson, C.H., Dyson, W.L., and Utiger, R.D.: Serum TSH and T_4 concentrations in patients receiving lithium carbonate, J. Clin. Endocrinol. Metab. **36:**338-346, 1973.
66. Singer, I., and Rotenberg, D.: Mechanisms of lithium action, N. Engl. J. Med. **289:**254-260, 1973.
67. Cavalieri, R.R., Gavin, L.A., Wallace, A., et al.: Serum T_4, free T_4, T_3 and rT_3 in diphenylhydantoin-treated patients, Metabolism **29:**1161-1165, 1979.
68. Cavalieri, R.R., Sung L.C., and Becker, C.E.: Effects of phenobarbital on T_4 and T_3 kinetics on Graves' disease, J. Clin. Endocrinol. Metab. **37:**308-316, 1973.
69. Akande, E.D., and Anderson, D.C.: Role of sex-hormone–binding globulin in hormonal changes and amenorrhea in thyrotoxic women, Br. J. Obstet. Gynecol. **82:**557-561, 1975.
70. Chopra, I.J., and Tulchinsky, D.: Status of estrogen-androgen balance in hyperthyroid men with Graves' disease, J. Clin. Endocrinol. Metab. **38:**269-277, 1973.
71. Abuid, J., Stinson, D.A., and Larsen, P.R.: Serum triiodothyronine and thyroxine in the neonate and the acute increases in these hormones following delivery, J. Clin. Invest. **52:**1195-1199, 1973.
72. Brunelle, P.H., and Bohuon, C.: Baisse de la triiodothyronine sérique avec l'âge, Clin. Chim. Acta **42:**201-203, 1972.
73. Selenkow, H.A., Birnbaum M.D., and Hollander, C.S.: Thyroid function and dysfunction during pregnancy, Clin. Obstet. Gynecol. **16:**66-68, 1973.
74. Williams, L.T., Lefkowitz, R.J., Watanabe, A.M., et al.: Thyroid hormone regulation of β adrenergic receptor number, J. Biol. Chem. **252:**2787-2789, 1977.
75. Dratman, M.B.: On the mechanism of action of thyroxine, an amino acid analogue of tyrosine, J. Theor. Biol. **46:**255-270, 1974.
76. Bermudez, F., Surks, M.I., and Oppenheimer, J.H.: High incidence of decreased serum triiodothyronine concentration with nonthyroidal disease, J. Clin. Endocrinol. Metab. **41:**27-40, 1975.
77. Reichlin, S., Bollinger, J., Nejad, I., et al.: Tissue thyroid hor-

mone concentration of rat and man determined by radioimmunoassay: biologic significance, Mt. Sinai J. Med. **40**:502-510, 1973.
78. Portnay, G.I., O'Brian, J.T., Bush J., et al.: The effect of starvation on the concentration and binding of thyroxine and triiodothyronine in serum and on the response to TRH, J. Clin. Endocrinol. Metab. **39**:191-194, 1974.
79. Glinoer, D., Gershengorn, M.C., and Robbins, J.: Thyroxine-binding globulin biosynthesis in isolated monkey hepatocytes, Biochim. Biophys. Acta **418**:232-244, 1976.
80. Chopra, I.J., Chopra, U., Smith, S.R., et al.: Reciprocal changes in serum concentrations of rT_3 and T_3 in systemic illnesses, J. Clin. Endocrinol. Metab. **41**:1043-1049, 1975.
81. McKenzie, J.M., Zakarija, M., and Sato, A.: Humoral immunity in Graves' disease, Clin. Endocrinol. Metab. **7**:31-45, 1978.
82. Solomon, D.H., Chopra, I.J., Chopra, U., and Smith, F.J.: Identification of subgroups of euthyroid Graves' ophthalmopathy, N. Engl. J. Med. **296**:181-186, 1977.
83. Volpe, R., and Johnston, M.W.: Subacute thyroiditis: a disease commonly mistaken for a pharyngitis, Can. Med. Assoc. J. **77**:297-307, 1957.
84. Volpe, R., Johnston, M.W., and Huber, N.: Thyroid function in subacute thyroiditis, J. Clin. Endocrinol. **18**:65-78, 1958.
85. Christiansen, N.J.B., Sierboek-Nielson, K., Hansen, J.E.M., and Christiansen, L.K.: Serum thyroxine in the early phase of subacute thyroiditis, Acta Endocrinol. **64**:359-363, 1970.
86. Woolf, P.D., and Daly, R.: Thyrotoxicosis with painless thyroiditis, Am. J. Med. **60**:73-79, 1976.
87. Dorfman, S.G., Cooperman, M.T., Nelson R.L., et al.: Painless thyroiditis and transient hyperthyroidism without goiter, Ann. Intern. Med. **86**:24-28, 1977.
88. Hamburger, J.L.: Subactue thyroiditis: diagnostic difficulties and simple treatment, J. Nucl. Med. **15**:81-89, 1974.
89. Blonde, L., Witkin, M., and Harris, R.: Painless subacute thyroiditis simulating Graves' disease, West. J. Med. **125**:75-78, 1976.
90. Gorman, C.A., and Duick, D.S.: Transient hyperthyroidism in patients with lymphocytic thyroiditis, Mayo Clin. Proc. **53**:359-365, 1978.
91. Shafer, R.B., and Nuttall, F.Q.: Acute changes in thyroid function in patients treated with radioactive iodine, Lancet **2**:635-637, 1975.
92. Kenimer, J.G., Hershman, J.M., and Higgins, H.P.: The thyrotropin in hydatidiform moles is human chorionic gonadotropin, J. Clin. Endocrinol. Metab. **40**:480-489, 1975.
93. Nisula, B.C., Morgan, F.J., and Canfield, R.E.: Evidence that chorionic gonadotropin has intrinsic thyrotropic activity, Biochem. Biophys. Res. Comun. **59**:86-91, 1974.
94. Nagataki, S., Mizuno, M., Sakamoto, S., et al.: Thyroid function in molar pregnancy, J. Clin. Endocrinol Metab. **44**:254-263, 1977.
95. Tolis, G., Bird, C., Bertrand G., et al.: Pituitary hyperthyroidism: case report and review of the literature, Am. J. Med. **64**:177-181, 1978.
96. Uller, R.P., and Van Herle, A.J.: Effect of therapy on serum thyroglobulin levels in patients with Graves' disease, J. Clin. Endocrinol. Metab. **46**:747-755, 1978.
97. Kaplan, M.M., and Utiger, R.D.: Diagnosis of hyperthyroidism, Clin. Endocrinol. Metab. **7**:97-113, 1978.
98. Hollander, C.S., Mitsuma, T., Nihei, N., et al.: Clinical and laboratory observations in cases of triiodothyronine toxicosis confirmed by radioimmunoassay, Lancet **1**:609-611, 1972.
99. Shenkman, L., Mitsuma, T., Blum, M., and Hollander, C.S.: Recurrent hyperthyroidism presenting as triiodothyronine toxicosis, Ann. Intern. Med. **77**:410-413, 1972.
100. Hollander, C.S., Mitsuma, T., Shenkman, L., et al.: T_3 toxicosis in iodine deficient area, Lancet **2**:1276-1278, 1972.
101. Hollander, C.S., Shenkman, L., Mitsuma, T., et al.: Hypertriiodothyronemia as a premonitory manifestation of thyrotoxicosis, Lancet **2**:731-733, 1971.
102. Ahmed, M., Doe, R.P., and Nuttall, F.Q.: Triiodothyronine thyrotoxicosis following iodine ingestion: a case report, J. Clin. Endocrinol. Metab. **38**:574-576, 1974.
103. Birkhauser, M., Burer, L., and Burger, A.: Diagnosis of hyperthyroidism when serum-thyroxine alone is raised, Lancet **2**:53-56, 1977.
104. McKenzie, J.M., and Zakarija, M.: Pathogenesis of neonatal Graves' disease, J. Endocrinol. Invest. **2**:183-189, 1978.
105. Volpe, R., Edmonds, M., Lamki, L., et al.: The pathogenesis of Graves' disease: a disorder of delayed hypersensitivity? Mayo Clin. Proc. **47**:824-834, 1972.
106. Brown, J., Solomon, D.H., Beall, G.N., et al.: Autoimmune thyroid disease—Graves' and Hashimoto's, Ann. Intern. Med. **88**:379-391, 1978.
107. Nofal, M.M., Beierwaltes, W.H., and Patno, M.E.: Treatment of hyperthyroidism with ^{131}I: a 16-year experience, J.A.M.A. **197**:605-610, 1966.
108. Wood, L.C., and Ingbar, S.H.: Hypothyroidism as a late sequela in patients with Graves' disease treated with antithyroid drugs, J. Clin. Invest. **64**:1429-1436, 1979.
109. Dussault, J.H., Coulombe, P., Laberge, C., et al.: Preliminary report on a mass screening program for neonatal hypothyroidism, J. Pediatr. **86**:670-674, 1975.
110. Maxon, H.R., III, Saenger, E.L., Thomas, S.R., et al.: Clinically important radiation-associated thyroid disease: a controlled study, J.A.M.A. **244**:1802-1805, 1980.
111. Maxon, H.R., III, Thomas, S.R., Saenger, E.L., et al.: Ionizing irradiation and the induction of clinically significant disease in the human thyroid gland, Am. J. Med. **63**:967-978, 1977.
112. LoGerfo, P., Stillman, T., Colacchio, D., and Feind, C.: Serum thyroglobulin and recurrent thyroid cancer, Lancet **1**:881-882, 1977.
113. Agerbaek, H.: Congenital goiter presumably resulting from tissue resistance to thyroid hormones, Isr. J. Med. Sci. **8**:1859-1860, 1972.
114. Bode, H.H., Danon, M., Weintraub, B.D., et al.: Partial target organ resistance to thyroid hormone, J. Clin. Invest. **52**:776-782, 1973.
115. Refetoff, S., DeWind, L.T., and DeGroot, L.J.: Familial syndrome combining deaf mutism, stippled epiphyses, goiter, and abnormally high PBI: possible target organ refractoriness to thyroid hormone, J. Clin. Endocrinol. Metab. **27**:279-294, 1967.
116. Bernal, J., Refetoff, S., and DeGroot, L.J.: Abnormalities of triiodothyronine binding to lymphocyte and fibroblast nuclei from a patient with peripheral tissue resistance to thyroid hormone action, J. Clin. Endocrinol. Metab. **47**:1266-1272, 1978.
117. Maxon, H.R., Burman, K.D., Premachandra, B.N., et al.: Familial elevations of total and free thyroxine in healthy, euthyroid subjects without detectable binding protein abnormalities, Acta Endocrinologica **100**:224-230, 1982.
118. Azizi, F., Vagenakis, A.G., Portnay, G.I., et al.: Pituitary-thyroid responsiveness to intramuscular thyrotropin-releasing hormone based on analyses of serum thyroxine, triiodothyronine, and thyrotropin concentrations, N. Engl. J. Med. **292**:273-277, 1975.
119. Snyder, P.J., and Utiger, R.D.: Response to thyrotropin-releasing hormone (TRH) in normal man, J. Clin. Endocrinol. Metab. **34**:380-385, 1972.
120. Sawin, C.T., and Hershman, J.M.: The TSH response to thyrotropin-releasing hormone (TRH) in young adult men: intra-individual variation and relation to basal serum TSH and thyroid hormones, J. Clin. Endocrinol. Metab. **42**:809-816, 1976.
121. Hershman, J.M., and Pittman, J.A., Jr.: Utility of the radioimmunoassay of serum thyrotrophin in man, Ann. Intern. Med. **74**:481-490, 1971.
122. Franco, P.S., Hershman, J.M., Haigler, E.D., Jr., et al.: Response to thyrotropin-releasing hormone compared with thyroid suppression tests in euthyroid Grave's disease, Metabolism **22**:1357-1365, 1973.
123. Chopra, I.J., Chopra, U., and Orgiazzi, J.: Abnormalities of hypothalamohypophyseal thyroid axis in patients with Graves' ophthalmopathy, J. Clin. Endocrinol. Metab. **37**:955-967, 1973.
124. Ormston, B.J., Alexander, L., Evered, D.C., et al.: Thyrotrophin response to thyrotrophin-releasing hormone in ophthalmic Graves' disease: correlation with other aspects of thyroid function, thyroid suppressibility, and activity of eye signs, Clin. Endocrinol. **2**:369-376, 1973.
125. Faglia, G., Beck-Peccoz, P., Ferrari, C., et al.: Plasma thyrotropin response to thyrotropin-releasing hormone in patients with pituitary

and hypothalamic disorders, J. Clin. Endocrinol. Metab. **37:**595-601, 1973.

126. Werner, S.C., and Spooner, M.: A new and simple test for hyperthyroidism employing L-triiodothyronine and the 24 hour ^{131}I uptake method, Bull. NY Acad. Med. **31:**137-145, 1955.
127. Davis, P.J.: Factors affecting the determination of the serum protein-bound iodine, Am. J. Med. **40:**918-940, 1966.
128. Stein, R.B., and Price, L.: Evaluation of adjusted total thyroxine (free thyroxine index) as a measure of thyroid function, J. Clin. Endocrinol. Metab. **34:**225-228, 1972.
129. Sawin, C.T., Chopra, D., and Albano, J.: The free triiodothyronine (T_3) index, Ann. Intern. Med. **88:**474-477, 1978.
130. Chopra, I.J., Solomon, D.H., Hepner, G.W., et al.: Misleadingly low free thyroxine index and usefulness of reverse triiodothyronine measurement in nonthyroidal illnesses, Ann. Intern. Med. **90:**905-912, 1979.
131. Chopra, I.J., Solomon, D.H., Chopra, U., et al.: Alterations in circulating thyroid hormones and thyrotropin in hepatic cirrhosis: evidence for euthyroidism despite subnormal serum triiodothyronine, J. Clin. Endocrinol. Metab. **39:**501-511, 1974.
132. Grenn, J.R.B.: Thyroid function and thyroid regulation in euthyroid men with chronic liver disease evidence of multiple abnormalities, Clin. Endocrinol. **7:**453-461, 1977.
133. Rosenfeld, L.: "Free thyroxine index": a reliable substitute for "free" thyroxine concentration, Am. J. Clin. Pathol. **61:**118-121, 1974.
134. Howorth, P.J.N., and Ward, R.I.: The (T_4) free thyroxine index as a test of thyroid function of first choice, J. Clin. Pathol. **25:**259-262, 1972.
135. Inada, M., and Sterling, K.: Thyroxine turnover and transport in Laennec's cirrhosis of the liver, J. Clin. Invest. **46:**1275-1282, 1967.
136. Tabei, A., and Shimoda, S.: Increased TBG-T_4-binding capacity in acute hepatitis, Folia Endocrinol. Jpn. **49:**1025-1033, 1973.
137. Schussler, G.C., Schaffner, F., and Korn, F.: Increased serum thyroid hormone binding and decreased free hormone in chronic active liver disease, N. Engl. J. Med. **299:**510-515, 1978.
138. Spector, D.A., Davis, P.J., Helderman, J.H., et al.: Thyroid function and metabolic state in chronic renal failure, Ann. Intern. Med. **85:**724-730, 1976.
139. Felicetta, J.V., Green, W.C., Haas, L.B., et al.: Decreased serum thyrotropin induced by fasting, J. Clin. Endocrinol. Metab. **45:**560-568, 1977.
140. Lim, V.S., Fang, V.S., Katz, A., et al.: Thyroid dysfunction in chronic renal failure, J. Clin. Invest. **60:**522-534, 1977.
141. Engler, D., Donaldson, E.B., Stockigt, J.R., and Taft, P.: Hyperthyroidism without triiodothyronine excess: an effect of severe nonthyroidal illness, J. Clin. Endocrinol. Metab. **46:**77-82, 1978.
142. Stock, J.M., Surks, M.I., and Oppenheimer, J.H.: Replacement dosage of L-thyroxine in hypothyroidism: a re-evaluation, N. Engl. J. Med. **290:**529-533, 1974.

Chapter 40 Gastrointestinal hormones

Stephen N. Joffe

APUD cells Acronym for cytochemical properties of an endocrine cell that has an *a*mine content, an amine *p*recursor *u*ptake, and *d*ecarboxylation.

brain-gut axis Similar peptides found in the gut, nerves, and central nervous system.

GEP hormones Gastroenteropancreatic hormones.

neurocrine hormones Peptides released from nerve synapses of the autonomous system.

paracrine hormones Peptides released into extracellular space where they can exert their effect on adjacent cells.

triple hormone control Endocrine, paracrine (locally acting), and neurocrine, including neurotransmission.

Gastrointestinal hormones compose a heterogeneous group of biologically active substances that are involved in the regulation of gastrointestinal function. They are secreted by endocrine cells found widely distributed through the gut mucosa in such large amounts that the gut has been labeled the largest endocrine organ in the body[1]. In addition, a significant number of these biologically active peptides are present in the nerves of the gastrointestinal tract[2] and in the central nervous system[3] (Table 40-1). Thus it is reasonable to assume that, although some of these peptides function as classical hormones, others are paracrine or neurocrine in their modes of action (Fig. 40-1).

Many gut peptides appear unlikely to act through the circulation. They have extremely short biological half-lives after injection and have multiple actions on many different organs. Peptides may be released from nerve synapses analogous to the release of catecholamines or acetylcholine, and these peptidergic nerves form part of the autonomic nervous system. Peptides, instead of acting locally as neurotransmitters, can be released into the general circulation and have neuroendocrine functions.

Endocrine cells may either release the contents of their storage granule (hormones) into the bloodstream where they become classical hormones acting at distant sites, or they may act locally in paracrine roles. The paracrine method of transmission involves the release of peptides into the extracellular space where they can interact with adjacent cells. As a result, peptides may spill over into the lumen of the gastrointestinal tract (''lumones'').

Circulating hormones from the gastrointestinal tract are only one specialized branch of the tripartite control system. Initially it seemed enough to investigate the physiology of single hormones. It soon became apparent that organs such as the pancreas are controlled by several different hormones and measurement of only one gives an incomplete picture. It now appears that the story is even more complicated because the measurement of the complete circulating gut hormone profile is inadequate to describe the physiological control of digestive function. Even with the available knowledge, unfortunately, technology still lags behind. It is extremely difficult to develop accurate quantitation of the influence of neural and paracrine systems. However, even the role played by circulating hormones of the gut is, in most cases, the subject of considerable controversy. We are passing through an era of rapid advance where, in the midst of general confusion, the fundamental cornerstones are being laid for the future. Comprehensive understanding of neuroendocrine control of the gut is an important subject. There are numerous troublesome disorders of the alimentary tract, the pathophysiology of which is unknown. Furthermore, the gut forms a model system whose mechanisms may be directly applicable to the understanding of the control of other vital organs. The realization that many apparently unrelated endocrine tumors can be classed together as APUDomas provides a convenient, unified concept that increases understanding and may in time lead to more effective methods of treatment. Many of these tumors are rare; however, with increased understanding, better methods of investi-

Table 40-1. Human gut hormones

Classical hormones	Candidate hormones (paracrine)	Biologically active	Immunoreactive
Gastrin*	Motilin*	Enterogastrone	Bombesin-like*
Cholecystokinin*	Enkephalin*	Antral chalone	Eledoisin-like
Secretin	Enteroglucagon	Bulbogastrone	Thyrotropin releasing hormone–like*
Gastric inhibitory polypeptide (GIP)	Vasoactive intestinal polypeptide (VIP)*	Gastrone	
		Villikinin	
Insulin	Urogastrone	Vagogastrone	
Glucagon	Neurotensin*	Entero-oxyntin	
	Substance P*	Coherin	
	Endorphin	Enterocrinin	
	Somatostatin		
	Pancreatic peptide (PP)		
	Chymodenin		

*Present in both brain and gut.

gation, and earlier diagnosis, more effective treatment can be planned, and the overall prognosis should improve.

ANATOMY

In 1902 Bayliss and Starling provided evidence of the existence of a chemical messenger in the duodenal mucosa. They used the term "hormone" ('to arouse an activity') to describe a substance released into the blood with an action on a distant organ. In 1953 Feyrter[4] described a system of clear cells that was diffusely scattered throughout the gut. He proposed that substances secreted by these cells acted only in a local environment and suggested that the agents be called "mediators." A large number of cells that secreted biologically active substances were thus scattered throughout the gut and other endocrine organs. In 1968 Pearse[5] suggested that these cells were derived from a neuroectodermal origin and could be classified under their common cytochemical characteristics, which he designated "APUD." This implies that the cells have an *a*mine content, amine *p*recurser *u*ptake, and *d*ecarboxylation. Although this theory was attractive and provided a central framework to clarify the heterogeneous gut endocrine system, recent work by Le Douarin[6] suggests that not all endocrine cells are APUD and that many may be of endodermal origin rather than ectodermal origin.

Gastrointestinal hormones are produced by cells belonging to the APUD series[5]. These cells are scattered throughout the gut and nervous system (Fig. 40-2). APUD cells may be considered in two groups: those that secrete polypeptides and those whose polypeptide product (if there is one) has not been identified. The first group includes most endocrine cells of the alimentary tract, including the islets of Langerhans and cells of the anterior pituitary, pineal, and thyroid glands (Table 40-2). The second group (Table 40-3) includes cells in many parts of the body, the most important probably being in the adrenal medulla.

Brain-gut axis

In recent times it has become clear that considerable numbers of hormones originally described in the gut are also found in nerves and the central nervous system. Conversely, certain brain peptides have also been identified in the gut.[7] Thus gastrin, cholecystokinin, vasoactive intestinal polypeptide, and motilin are found in the peripheral and central nervous systems, whereas the brain peptides, such as neurotensin, substance P, enkephalin, corticotropin, somatostatin, and thyrotropin-releasing hormones, have been identified in endocrine cells of the gut.[8] A morphological explanation of this dual neuroendocrine localization has been provided by the APUD theory, which postulates that endocrine peptide-producing cells and nervous system cells originated in common from the neuroectoderm. It has been suggested that the nervous system exists in three functional units: autonomic, somatic, and neuroendocrine[9] The knowledge that gut hormones are distributed not only in endocrine cells but also in peripheral and central nerves has established the fact that these peptides function not only as hormones but also as neurotrans-

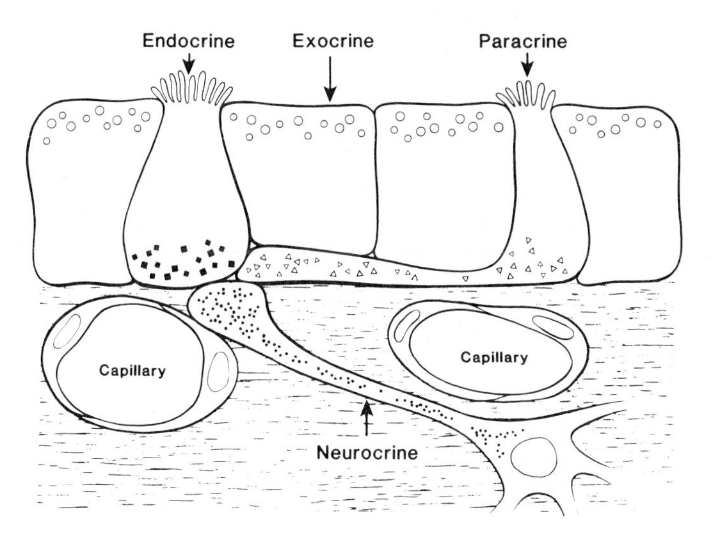

Fig. 40-1. Schematic representation of endocrine, paracrine, and neurocrine systems.

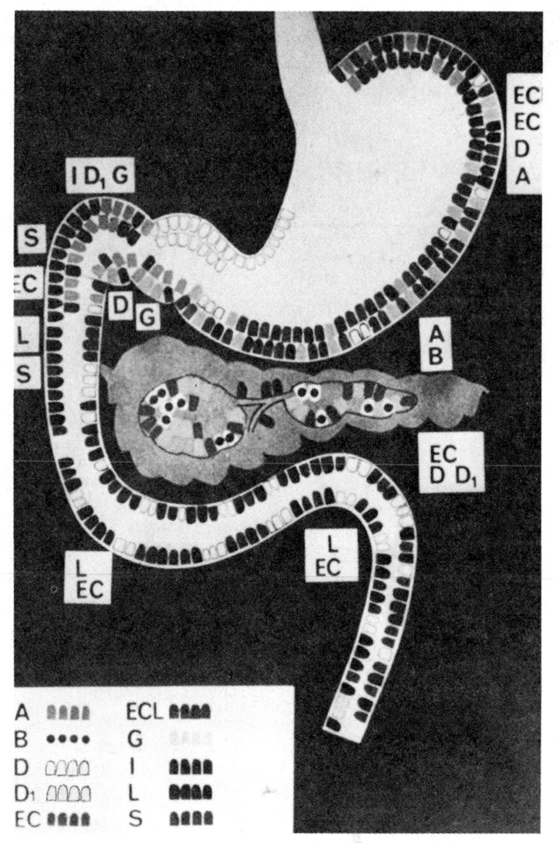

Fig. 40-2. Distribution of APUD cells in stomach, small bowel, and pancreas.

Table 40-2. APUD cells with known polypeptide products with related APUDomas and syndromes

Organ	Cell	Polypeptide	Amine	APUDoma	Syndrome
Alimentary tract					
Islets of Langerhans	B (β)	Insulin	Dopamine or 5-HT*	Hyperplasia	Hypoglycemia
	A (α)	Glucagon		Adenoma	Diabetes, dermatitis
	D ($\beta_1\alpha_1$)	Somatostatin		Carcinoma	—
		? Gastrin			Zollinger-Ellison
Stomach	G	Gastrin	—	Hyperplasia Carcinoma	Zollinger-Ellison
	AL (A-like)	Enteroglucagon	—	—	—
	D	Somatostatin	—	—	—
Duodenum and small intestine	S	Secretin	—	—	—
	D_1	Gastric inhibitory polypeptide (GIP)	—	—	—
	EC	Motilin ? Substance P	5-HT	Carcinoid	Malignant carcinoid
	D	Somatostatin	—	—	—
Large intestine	D	Vasoactive intestinal polypeptide (VIP)	—	—	—
	EC	Enteroglucagon	—	—	—
Other sites					
Anterior pituitary	c (corticotroph)	ACTH†	(Tryptamine)	Hyperplasia	Cushing's
	m (melanotroph)	Melanocyte-stimulating hormone (MSH)	(Tryptamine)	Adenoma	Pigmentation
	s (somatotroph)	Growth hormone (GH)		Carcinoma	Acromegaly or gigantism
	l (lactotroph)	Prolactin			Forbes-Albright
Pineal	P	Melatonin	5-HT	Pinealoma	Hypogonadism
Thyroid	C	Calcitonin	5-HT	Medullary carcinoma	Medullary carcinoma

*5-Hydroxytryptamine.
†Adrenocorticotropic hormone.

Table 40-3. APUD cells without known but *possible* polypeptide products and related APUDomas and syndromes

Organ	Cell	Amine	Possible polypeptide	APUDoma	Syndrome
Alimentary tract					
Islets of Langerhans	D_1	—	—	—	—
Stomach	EC	?5-HT	Substance P	Carcinoid	Atypical carcinoid
Duodenum and small intestine	I	—	VIP	—	—
	K	—	GIP	—	—
	D_1	—	—	—	—
	G	—	Gastrin	—	—
Large intestine	EC	5-HT	Substance P	Carcinoid	—
	H	—	VIP	—	—
Other sites					
Carotid body	Type 1 (glomus)	Catecholamines and 5-HT	—	Carotid body tumor (chemodectoma)	—
Skin	Melanocyte	—	—	—	—
Adrenal	A (E)	Epinephrine (adrenaline)	—		
	NA (NE)	Norepinephrine (noradrenaline)	—	Pheochromocytoma	Hypertension
Lung	P (Feyrter)	—	VLP*	—	—
	EC	?5-HT	—	Carcinoid	Atypical carcinoid
Urogenital tract	U	—	Urogastrone	—	—

*Vasoactive lung peptide.

mitters. Implications of this concept are far reaching and include the interpretation of all the currently proposed theories of gut physiology.[10]

Gut hormone structure

The two main families of gut hormones are the gastrin and the secretin families. Their basic structures are outlined in Table 40-4 and their molecular weights in Table 40-5. The *gastrin* family consists primarily of gastrin and cholecystokinin, but, in addition, motilin and enkephalin share a number of structural identities. The *secretin* group includes secretin, gastric inhibitory polypeptide (GIP), vasoactive intestinal polypeptide (VIP), glucagon, and bombesin.[11]

A further feature of gut hormones is that they exhibit molecular heterogeneity. This may be either macroheterogeneity, whereby there are a number of different hormone forms that differ considerably, or microheterogeneity, in which the differences involve one amino acid residue or only a side chain. Thus gastrin has been found to circulate as four components—I, II, III, and IV—and each of these components has various forms. This molecular heterogeneity has also been described for cholecystokinin and glucagon, and it presumably exists in many of the gut hormones. One molecular form probably represents a principal hormone, and the other forms either biosynthetic precursors or postsecretory degradation products.[13]

The particular amino acid sequences are important because they confer a specific biological activity to the peptide and also determine the potency of action on a particular organ (Table 40-4). In the gastrin-cholecystokinin (CCK) family, the COOH terminal pentapeptide amide sequence (Gly-Trp-Met-Asp-Phe-NH_2) is the main site of similarity and is a biologically active site for the two molecules. It is therefore not surprising that the two hormones share similar biological activity. The NH_2-terminal part of the molecule, however, influences potency for different targets. Thus CCK acts mainly on the gallbladder and pancreas, whereas gastrin is a more potent stimulus for gastric parietal cells. In addition, the localization of the sulfated tyrosyl residue is important in determining both the potency of the peptide and in influencing whether its action will be gastrinlike or CCK-like. If the sulfated residue is moved only one position, the hormone activity can be

Table 40-4. Basic structures of gut hormones

	Gastrin-cholecystokinin family					Secretin family				
	Gastrin (17)	Cholecystokinin (17-33)	Cerulein	Enkephalin	Motilin (14-21)	Secretin	Glucagon	VIP	GIP (1-29)	Bombesin
	Glp	Asp				His	His	His	Tyr	
	Gly	Pro				Ser	Ser	Ser	Ala	
	Pro	Ser				Asp	Gln	Asp	Glu	
	Trp	His				Gly	Gly	Ala	Gly	
	Leu	Arg				Thr	Thr	Val	Thr	
	Glu	Ile			Gln	Phe	Phe	Phe	Phe	
	Glu	Ser			Glu	Thr	Thr	Thr	Ile	
	Glu	Asp	Glp		Lys	Ser	Ser	Asp	Ser	
	Glu	Arg	Gln		Gul	Glu	Asp	Asn	Asp	
	Glu	Asp	Asp		Arg	Leu	Tyr	Tyr	Tyr	Glp
	Ala	Tyr(SO_3H)	Tyr(SO_3H)	Tyr	Asn	Ser	Ser	Thr	Ser	Glu
	Tyr(SO_3H)	Met	Thr	Gly	Lys	Arg	Lys	Arg	Ile	Arg
Pentagastrin	Gly	Gly	Gly	Gly	Gly	Leu	Tyr	Leu	Ala	Leu
Pentagastrin	Trp	Trp	Trp	Phe		Arg	Leu	Arg	Met	Gly
Pentagastrin	Met	Met	Met	Met		Asp	Asp	Lys	Asp	Asn
Pentagastrin	Asp	Asp	Asp			Ser	Ser	Gln	Lys	Gln
Pentagastrin	Phe	Phe	Phe			Ala	Arg	Met	Ile	Trp
Pentagastrin	NH_2	NH_2	NH_2			Arg	Arg	Ala	Arg	Ala
						Leu	Ala	Val	Gln	Val
						Gln	Gln	Lys	Gln	Gly
						Arg	Asp	Lys	Asp	His
						Leu	Phe	Tyr	Phe	Leu
						Leu	Val	Leu	Val	Met
						Glu	Gln	Asn	Asn	
						Gly	Trp	Ser	Trp	
						Leu	Leu	Ile	Leu	
						Val	Met	Leu	Leu	
						NH_2	Asp	Asn	Ala	
							Thr	NH_2	Gln	

Table 40-5. Size and molecular weight of gut hormones

Hormone	Number of amino acids	Molecular weight	Successful synthesis
Pancreatic polypeptide	36	4226	Yes
Motilin (porcine)	22	2700	Yes
GIP (porcine)	43	5105	Yes
Neurotensin (bovine)	13	1673	Yes
VIP (porcine)	28	3326	Yes
Somatostatin (ovine)	14	1640	Yes
Bombesin (amphibian)	14	1620	Yes
Substance P (bovine)	11	1348	Yes

changed from gastrinlike to CCK-like, or if the sulfate radical is removed, it becomes far less potent.[8] The effect of amino acid changes on activity are not so clear in the secretin family, but all share common actions such as inhibition of acid secretion, stimulation of insulin release, and hepatic glycogenolysis. The main sites of origin and action of the gut hormones are given in Table 40-6.

PHYSIOLOGY

An in-depth discussion of the digestive process is found in Chapter 27. Gut hormones are regarded as regulators of digestion and absorption. They are released in response to the presence of nutrients in the lumen of the gastrointestinal tract and provide acid, bicarbonate, and enzymes for digestion of food. Once the nutrients enter the blood, the pancreatic metabolic hormones are then released. This link between gut factors and metabolism led to the term "enteroinsular axis." Different steps of digestion, absorption, and storage can be both stimulated and inhibited by different gastroenteropancreatic (GEP) peptides. This interplay prevents an overreaction and has a self-limiting negative feedback control.

Somatostatin is a unique peptide that inhibits most of the gastrointestinal secretory and motor functions. It slows down the entry of nutrients and thus prevents an overflow of metabolic pathways. It is released by nutrients, acid, and several gut hormones and probably acts as a circulatory hormone and a local paracrine substance as well.

For each gastrointestinal function there are several agonists and antagonists. Final control thus depends on a fine balance of numerous influences. In the case of gastric acid secretion, at least 21 different factors appear important in its normal control. When summarizing some of the major functions of the gastrointestinal tract, one finds that motilin, gastrin, VIP, and glucagon are the major hormones involved in the control of secretion, absorption, motility, and growth in the stomach and intestine. Secretin, CCK, VIP, and pancreatic polypeptide control exocrine pancreatic function, whereas insulin and GIP are involved in the enteroinsular axis. Insulin, glucagon, and somatostatin are primarily involved in the metabolism of carbohydrate, fats, and protein. Substance P, VIP, and the enkephalins

Table 40-6. Main sites of origin and action of gut hormones

Tissue	Hormone	Action
Stomach	Gastrin	Stimulation of gastric acid; gastric mucosal growth
Pancreas	Pancreatic polypeptide	Inhibition of gallbladder contraction; inhibition of pancreatic exocrine secretion
Upper small intestine	Secretin	Stimulation of pancreatic bicarbonate
	Cholecystokinin	Gallbladder contraction; stimulation of pancreatic enzyme secretion; stimulation of pancreatic growth
	Motilin	Stimulation of upper GI motor activity
	GIP	Enhancement of insulin release
Lower small intestine	Enteroglucagon	Inhibition of intestinal transit; enhancement of mucosal growth
Colon	Neurotensin	?
	Enteroglucagon	Inhibition of intestinal transit; enhancement of mucosal growth

have a major neurotransmitter involvement in the central, peripheral, and autonomic nervous systems.

Progress in gut hormone research has considerably increased our knowledge of gastrointestinal physiology. The pathophysiological role of many of these GEP hormones awaits elucidation. Morphological and biochemical methods rarely can be applied to gastrointestinal endocrinology because of the special design of the gut hormone system.

The following reviews some characteristics of the GEP hormones.

CHARACTERISTICS OF GASTROENTEROPANCREATIC (GEP) HORMONES AND POLYPEPTIDES

Gastrin

Chemistry. Gastrin is known to exist in multiple molecular forms, the relationship of which is complex. The four main components, gastrin I, II, III, and IV, differ in the number of amino acids in the polypeptide chain. Component II is gastrin 34 (G 34) or big gastrin: component III is gastrin 17 (G 17); and component IV is gastrin 14 (G 14). The numbers identify the number of amino acids present. Gastrin tetradecapeptide probably occurs naturally. Gastrin I has not yet been characterized. "Big, big gastrin" probably does not exist except as an artifact in radioimmunoassay or in certain tumors that may produce macromolecules.[13]

Distribution and biosynthesis. The cellular origin of gastrin is the G cell of the gut. These cells extend into the gut lumen, where they terminate in a tuft of microvilli. Because luminal stimuli regulate gastrin release, it is possible that the microvilli bear receptors for these active luminal factors. The secretory granules that contain gastrin are located in the basal parts of the cells, which allows direct secretion of the hormone into the bloodstream. Gastrin is mainly produced by the antral G cells and to a lesser extent is produced by G cells of the proximal duodenum.[14]

The biosynthesis of many secretory proteins and peptides, including hormones and enzymes, proceeds by way of translation from mRNA of large precursor polypeptides that are packaged into granules and cleaved by intracellular proteases to yield the final secretory product. By analogy the biosynthesis of gastrin probably proceeds through a large precursor that is converted first to G 34 and then to G 17. The relatively high concentrations of G 34 in human duodenal mucosa might suggest that at this site the conversion of G 34 to G 17 proceeds less readily than in antral G cells. There is evidence to support the concept that G 34 is a precursor of G 17. Cleavage of G 34 in vitro with trypsin liberates G 17, and the site of cleavage is two consecutive lysine residues.[15]

Biological effects and control. The availability of pure gastrin and its active analog, pentagastrin, has allowed studies with regard to its spectrum of actions. Gastrin affects almost all the smooth muscle and secretory cells of the gut. This creates problems in deciding which actions are physiological and which pharmacological. Progress has been made by relating the concentration of the hormone in the blood to biological responses. Gastrin stimulates acid secretion, gastric motility, and the growth of fundic small bowel and mucosa.[16]

Secretion of gastrin is mediated by luminal, nervous, and bloodborne stimuli. The principal luminal stimuli are the products of protein digestion. Most amino acids release gastrin, but there is no simple relationship between gastrin release and the rate of acid secretion evoked by intragastric amino acids.[17] Other luminal stimuli include calcium, alcohol, and pH. In normal subjects gastrin release is dependent on intragastric acidity. At pH 5.5 the rate of gastrin release is three times as high as at pH 2.5. The inhibition effect of acid provides a feedback mechanism for autoregulation of gastrin release. Carbohydrates and fats have little effect on gastrin release. Distension of the antral pouch has long been recognized to release gastrin, and both cholinergic nerve stimulation and noncholinergic mechanisms also play an important role. The feedback regulation of gastrin release is mediated by antral acidification and by neural and hormonal regulatory mechanisms generally characterized by the concept of an enterogastric axis.[18] Several peptides known to inhibit the release of gastrin include secretin, GIP, VIP, glucagon, calcitonin, and somatostatin. Peptides may also modulate gastrin release by paracrine or local action.

The kidney and small intestine are important sites of gastrin degradation. Degradation of G 17 by the liver appears to be no greater than that by other tissues, though the smaller COOH terminal fragments of gastrin such as the pentapeptide may be almost completely cleared by the liver on a single pass.[18] The normal fasting serum gastrin concentration varies among laboratories, but the usual normal range is less than 25 pmol/L or 50 pg/mL (human G 17 equivalents). Variations in antibody specificity and in the potency of the different standard preparations may account for some of the interlaboratory variation. In addition, serum proteins cause a nonspecific inhibition in the gastrin radioimmunoassay which varies among antisera. In normal subjects a meal such as eggs and toast stimulates an increase in total serum gastrin, which reaches a peak in about 30 to 60 minutes and then declines to basal over the next 1 to 2 hours. Gel filtration analysis indicates that the postprandial serum increases are almost entirely a result of G 17 and G 34. Maximal concentrations of G 17 are achieved 20 minutes after the meal, whereas a peak concentration of G 34 is reached at 45 to 60 minutes.[16]

Circulating G 17 and G 34 are almost equally potent on molar basis to stimulate acid secretion, but G 17 has a half-life of 6 minutes, whereas that of G 34 is 40 minutes. These potencies are compared on the basis of an exogenous dose. However, if potency is described in terms of concentration in the blood required for a given rate of acid secretion, that is, endogenous potency, then in both humans and dogs G 17 is about five times more potent than G 34. It follows that total serum gastrin concentration is not a reliable indicator of circulating biological activity, and an interpretation of the physiological significance of serum gastrin concentration depends on a knowledge of the relative concentration of the two forms. It seems likely that G 17 accounts for nearly all the circulating gastrin that is biologically active after a normal meal.[18]

Pathology

Gastrinomas or Zollinger-Ellison syndrome. Plasma gastrin is particularly important in the diagnosis of pancreatic endocrine tumors (gastrinomas). The Zollinger-Ellison syndrome is characterized by peptic ulceration, gastric acid hypersecretion, and a non–beta cell islet cell tumor of the pancreas.[19] These tumors produce large quantities of gastrin, hence the descriptive name "gastrinoma." Basal serum gastrin concentrations above 500 pmol/L (1000 pg/mL) accompanied by gastric acid hypersecretion are virtually diagnostic of gastrinoma. Patients with achlorhydria (decreased HCl output) may have serum gastrin in the gastrinoma range[20] but can be readily distinguished on the basis of acid secretory tests. A very small number of patients may have hypergastrinemia and hyperchlorhydria resulting from an isolated antrum from previous surgery. The majority of gastrinomas are believed to be malignant and, at the time of discovery, have metastasized to other parts of the body.

It is now recognized that five types of APUDomas (tumors of the APUD system) may produce excessive gastrin and cause a variant of the Zollinger-Ellison syndrome. Those recognized include G cell hyperplasia or carcinoma in the stomach; islet cell tumor (adenoma or carcinoma) or D-cell hyperplasia in the pancreas; and carcinoma or adenoma in the duodenum or splenic hilum.[21]

In patients with renal failure there are elevated serum gastrin concentrations that return to normal after renal transplantation.[22] Serum gastrin concentrations are also raised in patients after massive small bowel resection, which may be a result of either decreased clearance of gastrin or removal of an intestinal factor that normally suppresses gastrin release.

Peptic ulcer. The role of gastrin in peptic ulceration unrelated to gastrinoma has been the subject of intensive study. There is now evidence to indicate that patients with duodenal ulcers secrete more gastrin in response to feeding and are also more sensitive than normal to gastrin than other persons. However, the role of gastrin in the pathogenesis of peptic ulceration remains uncertain. Basal serum gastrin concentrations in duodenal ulcer patients are similar to those in normal subjects, but several groups have shown that the gastrin response to feeding is higher than normal.[23] As in normal subjects, G 17 and G 34 account for most of the increase in gastrin after a meal, with G 34 accounting for the greater increase. Because G 34 is a weak stimulant of acid secretion relative to G 17, it is not yet certain whether there is a significant elevation of biologically active gastrin in duodenal ulcer patients. It has been established, however, that patients with duodenal ulcers tend to have higher-than-normal basal and maximal acid secretion; they are also more sensitive than normal to exogenous gastrin.[24] From the available work it appears that there is a defect in the mechanisms of acid inhibition of gastrin release in duodenal ulcer patients.[23]

Patients with gastric ulcer have both elevated basal and postprandial serum gastrin concentrations. These patients have lower basal and maximal acid output than normal, and it seems likely that the higher serum gastrin is the result of a decreased inhibition. After a meal, the peak increase in G 14 is similar to normal, but G 34 increases about four times higher than normal, which is similar to levels in duodenal ulcer patients.[16]

Function tests. In recent years it has become clear that many gastrinoma patients have serum gastrin levels that overlap with the upper end of the range for normal subjects and patients with ordinary peptic ulceration. In 15 patients with gastrinoma, 6 were found to have serum gastrin concentrations less than 500 pg/mL.[25] One or more provocation tests are therefore needed to establish the presence of gastrinoma in such patients. The most widely used tests are the *standard test meal,* the *secretin test,* and the *calcium challenge.*[26] The secretion of gastrin in a patient with a gastrinoma caused by a pancreatic tumor is usually unaltered by food but does increase in a patient with antral G-cell hyperplasia. Rapid intravenous injection of secretin (2 to 4 clinical units per kilogram) stimulates a pronounced increase of gastrin in about 80% of all gastrinoma patients with peaks at 2½ to 10 minutes that reach 100% above the basal level.

At peak concentrations there is relatively more G 17 in the serum than in the basal state, suggesting that secretin acts by releasing tumor stores of gastrin. In normal subjects and in patients with ordinary peptic ulceration, achlorhydria, or isolated retained antrum, secretin tends to depress the serum gastrin. The reason for the paradoxical effect of secretin on the release of gastrin from gastrinomas is unknown. The structurally related hormone, glucagon, has an effect similar to that of secretin.

After a calcium infusion, the serum gastrin level noticeably increases in patients with gastrinoma.[27] In one study all patients responded with a two- to fiftyfold increase in plasma gastrin. Calcium, however, may provoke a modest rise in serum gastrin levels in normal subjects and in patients with duodenal ulcers, but these responses usually fall short of those seen in gastrinoma patients. There is an association between gastrinoma and hyperparathyroidism, and the hypercalcemia may exacerbate through the release of gastrin from tumors.

Cholecystokinin-pancreozymin (CCK-PZ)

Chemistry. It is now well recognized that cholecystokinin (CCK) and pancreozymin (PZ) are a single substance.[28] Cholecystokinin was first discovered and named by Ivy and Oldberg.[29] The molecule isolated from hog duodenum by Jorpes and Mutt is a basic peptide of 33 amino acid residues. The final stage of purification includes a step to remove what was first considered a contaminating material but is now known to include a peptide of 39 residues, which is the larger form known as CCK.[30] By analogy with abbreviated designations based on the polypeptide length of the different forms of gastrin, it is convenient to refer to the cholecystokinins as CCK 33, CCK 39, and CCK 8. These amino acid residue forms may also be sulfated or desulfated. Sulfation is particularly important for conferring biological activity, though it is possible that unsulfated forms may have an alternative physiological role as neurotransmitters. The carboxyl-terminal octapeptide portion of CCK is at least tenfold more potent than the intact hormone.[30]

Site and mode of synthesis. The distribution of CCK has been studied by bioassay, immunoassay, and immunocytochemistry. CCK is found in the brain and in the K cells of the upper small intestinal mucosa, which correspond to the I cell previously identified by electron microscopy.

The first indication of heterogeneity of CCK was the isolation from the hog intestine of the peptides CCK 33 and 39. These remain the only forms of CCK to be isolated from the intestine and fully characterized.[31] However, CCK 8 has now been isolated from the sheep brain; in addition, there is accumulating evidence to suggest that several other immunoreactive forms of CCK occur in the brain and gut.[32] The molecular forms of CCK identified in immunochemical studies of tissue extracts depend on both the antibody specificity and the methods of extraction used. In acid extraction, primarily CCK 33 and 39 are obtained. In contrast, after extraction in boiling water, CCK 33 and CCK 8 are obtained in relatively large amounts. It is probable that CCK 33 and 39 are biosynthetic precursors of the smaller CCK-like peptides such as CCK 8 but may themselves be the product of cleavage of larger precursors.

Biological effects. The finding of CCK-like peptides in the central nervous system and in the gut suggests that these peptides function both as neurotransmitters and hormones. The physiological role is related to the regulation of motility of the gallbladder and intestine and secretion by the pancreas. Physiological actions of CCK in the pancreas include the stimulation of enzymes and the potentiation of the action of secretin and a trophic effect in stimulating the growth of the pancreas. CCK also affects smooth muscle other than the gallbladder. It appears to inhibit gastric emptying, relaxes the sphincter of Oddi, and stimulates the small intestine to contract. In addition, it has been reported to stimulate the secretion of pepsin, Brunner's glands, and hepatic biocarbonate and water. Other activities probably relate to its gastrin homologs and include the weak stimulation of gastric acid secretion and antagonism of secretin-stimulated bicarbonate secretion.[33] Its function in the central nervous system is not yet clear, but it has been suggested to play a role in the satiety (state of satisfied appetite) mechanisms.[34] Structural features modify the potency for particular actions. In the dog CCK 8 is about three times more potent than CCK 33 in stimulating gallbladder contraction and enzyme secretion.

Release of CCK 33 has been studied for the most part by biological methods that involve monitoring gallbladder contraction and pancreatic enzyme secretion in response to the instillation of food into the small intestine. This approach has several limitations, but the balance of evidence indicates that CCK is released by the products of protein and fat digestion. Like gastrin, pure undigested protein does not stimulate hormone release, but after partial proteolysis the mixture of polypeptides and amino acids is a strong stimulus.[35] Fatty acids with chains longer than nine carbons also stimulate CCK release, with the activity being improved by the formation of micelles.

Fasting values of CCK concentration in normal humans vary from 26 pg/mL (equivalent to porcine CCK 33) to much higher values. Estimates of CCK release after stimulation by feeding are also highly variable. Standard meals can increase CCK from 200 to 300 pg/mL.[36] The half-life of exogenous CCK is reported to be 2½ minutes, and the expected increase of plasma CCK during an infusion of 250 pg/kg/hour would be about 100 pmol/L / (400 pg/mL).

Pathology. Plasma CCK is reported to increase in patients with celiac disease and pancreatic exocrine insufficiency.[37]

Secretin

Chemistry. In 1902 Bayliss and Starling discovered secretin,[33] but it took 65 years before it was finally obtained in a homogeneous state and its structure elucidated.[31] Secretin isolated from hog duodenum was characterized as a basic peptide of 27 amino acid residues with strong similarities in sequence to glucagon. Synthesis of porcine secretin was first achieved by Bodansky in 1966 and has been achieved by several other groups.[39]

Early studies of structure and activity relationships indicate that the intact peptide is needed for full activity and that there is no minimum active fragment comparable to that of gastrin or CCK. Several experiments suggested that the NH_2-terminal sequence of secretin is required for biological activity and that affinity for the receptor is increased by the COOH-terminal portion of the molecule.[40]

Site and mode of synthesis. Secretin is predominantly located in the S cells of the mucosa of the duodenum and jejunum.[41] The frequency of S cells in the jejunum is about one third of that in the duodenum, but the greater length of the jejunum relative to the duodenum means that the bulk of intestinal secretin is jejunal in origin.

Biological activity. Secretin has numerous pharmacological actions. These include inhibition of smooth muscle contraction and decrease in gastric acid secretion, lowering of the lower esophageal sphincter pressure, and a positive trophic effect on pancreatic growth. In addition to these effects, it stimulates water and bicarbonate secretion from the liver and Brunner's glands and augments gallbladder contraction. It has a synergistic action with CCK in the stimulation of gallbladder contraction and pancreatic enzyme secretion.[36] The primary physiological role of secretin appears to be the modulation of pancreatic bicarbonate secretion. This is accomplished by the stimulation of pancreatic ductal cyclic AMP (cAMP).

A principal stimulus for secretin release is acid, but in the adult jejunum there is seldom if ever likely to be sufficient acid to liberate secretion.[42] This suggests that other factors might also affect the release of the hormone. Also, fatty acids with 10 or more carbons in the chain weakly stimulate its release, but undigested fat has no activity. The threshold for release of secretin is a pH of 4.5, and below this the amount of secretin released is proportionate to the load of acid entering the small intestine.[36] The optimum conditions for secretin release, that is, pH less than 4.5, are normally obtained only in the first few centimeters of duodenum. The rise in plasma secretin therefore is very modest after a normal meal. The secretion of bicarbonate from the pancreas closely matches the entry of acid into the duodenum. There is little doubt therefore that gastric acid drives pancreatic bicarbonate output. It seems likely that the increase in plasma secretin after a normal meal is by itself insufficient to stimulate the pancreas, but its action is strongly potentiated by CCK.

The concentrations of secretin in human fasted plasma vary from 0.6 to 50 pg/mL (0.2 to 20 pmol/L).[43] Plasma proteins can cause nonspecific inhibition of binding of labeled secretin to antiserum and thus give misleading high estimates of secretin concentrations. After a meal, there is a small but measurable increase in normal subjects, and increases of about 2 pmol/L to a peak of 5 to 6 pmol/L have been shown.[43] These small increases are likely to be effective in stimulating the pancreas because of the accompanying presence of CCK and a strong potentiation between the two hormones. Plasma concentrations rise rapidly after stimulation by acid and reach a peak about 3 minutes before declining to basal levels in about 60 minutes, suggesting a half-life of about 3 minutes.

Pathology. Because acid is the main stimulus for secretin release, it is not surprising that secretin concentrations are elevated in patients with gastric acid hypersecretion (such as gastrinomas).[44] Other conditions characterized by elevated plasma secretin have not yet been described. A possible role for secretin in the pathogenesis of duodenal ulcers has been postulated but not confirmed. In patients with celiac disease an increased number of S cells was found by immunocytochemistry, and the granule content appeared higher than normal, possibly as a result of failure of hormone release. Evidence has been suggestive of decreased secretion of secretin in patients with celiac disease.[43]

Secretin is clinically useful in confirming the diagnosis of gastrinoma. As mentioned before, for reasons as yet unclear, the administration of secretin as a bolus intravenously (2 clinical units/kilogram) provokes a rapid rise in plasma gastrin in patients with gastrinoma. It has little or no effect on the plasma gastrin levels when hypergastrinemia results from other causes.[45]

Vasoactive intestinal polypeptide (VIP)

Chemistry and site of synthesis. VIP has 28 amino acids and a molecular weight of 3320 daltons. It was originally isolated from porcine intestinal mucosa and identified by using bioassay measurements of its powerful vasodilatory activity.[46] VIP has subsequently been shown to be present in both endocrine cells and nerves of the gut and central nervous system[47]; there are a number of molecular forms of VIP.

Biological effects. VIP has a wide spectrum of activities including inhibition of gastric acid secretion, stimulation of insulin release, stimulation of pancreatic water and bicarbonate secretion, stimulation of hepatic glycogenolysis, stimulation of intestinal fluid and electrolyte secretion, stimulation of mucosal cAMP, relaxation of smooth muscle, and a positive inotropic action on the heart.[48] Its exact mode of action is uncertain, but it probably functions primarily as a paracrine neurotransmitter substance rather than as a circulating hormone.[49,50] There is no significant release of VIP into the blood after a meal. Through use of sensitive radioimmunoassay, the serum concentration mean fasting plasma VIP is 1.5 pmol/L, but there is a skewed distribution so that the upper end of the normal range is 21 pmol/L, with a mean normal of 2.1 pmol/L.[48,49]

Pathology. Abnormally high tissue levels of VIP have been found in Crohn's disease, whereas in the aganglionic colon segment of patients with Hirschsprung's disease VIP levels have been reported to be extremely low.[11] The most important application in the measurement of VIP is that elevated levels are found in the plasma in some patients with watery diarrhea syndrome (Verner-Morrison syndrome). Verner and Morrison in 1958 reported on two patients with pancreatic non–beta islet cell tumor causing watery diarrhea without peptic ulceration.[51] Subsequently it was named "WDHA" after the initial letters of its main characteristics: *w*atery *d*iarrhea, *h*ypokalemia, and hypo- or *a*chlorhydria. The syndrome is rare, being about one tenth as common as Zollinger-Ellison syndrome, and it is sometimes part of multiple endocrine neoplasia.

A non–beta islet cell tumor (vipoma) of the pancreas is usually present (D_1 cells); about half of these tumors are malignant. Tumors occur elsewhere, such as bronchial (probably oat cell) carcinoma or retroperitoneal neuroblastoma, and secrete VIP, a polypeptide humoral agent normally produced by the gut. In large doses VIP causes vasodilatation with facial flushing, increases intestinal blood flow, induces watery diarrhea, and inhibits gastric secretion, which causes the clinical features. The diarrhea is explosive, consists of up to 30 stools per day, and does not respond to simple measures. This causes a profound hypokalemia (1 to 3 mmol/L), but the associated hypercalcemia is unexplained.[52]

The diagnosis is made by elimination of the common cause of watery diarrhea and hypokalemia. Hypochlorhydria is found on gastric secretion tests, and the diagnosis is confirmed by measurements of elevated blood levels of immunocreactive VIP.[53] A characteristic tumor blush of dilated vessels caused by the local action of VIP is found on selective arteriography. The pseudo–Verner-Morrison syndrome refers to patients with watery diarrhea, hypokalemia, and achlorhydria who have normal serum VIP levels and no obvious tumor.

Preoperatively the electrolytes and dehydration need to be corrected, and if the diarrhea is severe and continuous, steroids (prednisone 60 mg/day) or somatostatin may be given. Treatment requires early and aggressive surgery to remove both the primary and as much metastatic tumor as possible.[54]

Plasma VIP measurement is a very useful screening test

for the detection of vipomas and is an effective tumor marker to detect occult metastases.

Pancreatic polypeptide (PP)

Chemistry. PP was first discovered as a contaminant of chicken insulin,[55] and subsequently a very similar peptide was found to contaminate mammalian insulins.[56] Both contain 36 amino acids arranged in a linear sequence. The molecular weight of bovine PP is 4226 and shows only 4 amino acid differences from human PP. The entire biological activity is displayed by the last 6 amino acids,[57] which do not differ among pig, bovine, and human PP. PP is fairly freely available in the by-products of insulin manufacture and appears to be quite stable both in plasma and on the shelf.

Site of synthesis. PP is found almost entirely in the pancreas, where it is distributed in the D_2 cells of the islets, but it is also scattered throughout the acinar cells of the exocrine pancreas and the walls of the pancreatic ducts. Most of the PP cells are in the head of the pancreas.

Biological effects. Pharmacological studies show that it opposes the effects of cholecystokinin.[57] It therefore inhibits the pancreatic output of enzymes and gallbladder contraction. At somewhat higher doses it has a biphasic effect, initially stimulating and then later inhibiting pancreatic bicarbonate output. Gastric acid output is stimulated, whereas motor activity of the upper intestine is inhibited.[58] PP may play a role in the modulation of insulin and glucagon secretion by the islets. Plasma PP levels increase with age, and although the exact reason for this is not yet clear, it has been suggested that it reflects alteration in vagal tone.[59]

Fasting human PP concentrations have been variously reported as 20 to 80 pmol/L.[60] There is a significant tenfold biphasic rise occurring after a meal, with peak values being achieved at 30 minutes. The main constituents of the meal that are responsible for the PP release appear to be fat and protein, and the PP levels are undetectable in patients after pancreatectomy. Postprandial rise reflects signals from the gut to the pancreas. Changes in plasma concentration of amino acids, fat, or glucose do not affect PP release directly. During insulin hypoglycemia a large rise in PP concentration is seen, which is abolished by vagotomy, suggesting an important vagal influence. PP is liberated into the blood in large quantities after a meal by mechanisms that appear to be dependent on vagal nerve stimulation.[61]

Pathology. Plasma PP levels have been documented to be considerably elevated in a high percentage of patients with vipomas, insulinomas, gastrinomas, and glucagonomas.[62,63] Although some of these tumors contain PP cells, in others only PP cell hyperplasia in the rest of the pancreas is evident. This suggests that elevated PP levels might serve as biochemical markers for pancreatic endocrine tumors. A number of patients with diarrhea have been noted to have pancreatic endocrine tumors composed of only PP cells (PPoma). Pure pancreatic polypeptide APUDomas (PPomas) are rare, though elevated plasma PP levels are found in one third of pancreatic APUDomas.

PP levels may be altered in chronic pancreatitis, with both basal and meal-stimulated levels being decreased.[64] The peptide has been implicated in the pathogenesis of diabetes mellitus, but only one study has demonstrated elevated PP levels in such patients.[60] Diabetic patients treated with unicomponent insulin have normal fasting PP plasma levels.

Gastric inhibitory polypeptide (GIP)

Chemistry and site of synthesis. Gastric inhibitory peptide (also called *g*lucose-dependent *i*nsulin-releasing *p*eptide) was first isolated as an intestinal factor capable of inhibiting stimulated gastric secretion. It is a straight-chain polypeptide of 43 amino acids with a molecular weight of 5105 daltons and shows considerable sequence homology with the other gastrointestinal hormones.[65] In mammals there is no evidence of any interspecies difference in amino acid sequence. Considerable difficulties have been encountered in an attempt to synthesize porcine GIP, and even apparently pure preparations seem to have biological activity considerably below that of natural GIP. GIP is found in the K cells of the jejunal mucosa and to a lesser extent in the duodenum and ileum.[65]

Biological effects. Reported fasting human plasma levels of GIP range from 15 to 100 pmol/L. There is a rise in GIP after a meal that is relatively prolonged.[66] Infusion of GIP in human volunteers has shown that at a blood level near the physiological range there is considerable enhancement of glucose-induced insulin release.[67] Little effect on insulin is seen, however, when subjects are studied in the fasted state. Thus GIP has been renamed "*g*lucose-dependent *i*nsulin-releasing *p*eptide" and has been proposed as the main hormone enhancing insulin release when nutrients are taken orally and thus may be the major component of the enteroinsular axis.[63]

Pharmacological studies have shown that GIP has the capacity to inhibit gastric acid and pepsin secretion as well as intestinal motility. The physiological role of GIP in the control of acid secretion is at present controversial. GIP is also a potent stimulant of small intestinal fluid and electrolyte secretion.

Pathology. Although it is tempting to speculate that the abnormalities of GIP release might be linked to the pathogenesis of diabetes, studies in this area have produced conflicting results.[69] High levels of GIP are found on prolonged fasting and in diabetic patients. In mature-onset diabetes, where insulin is available but not released appropriately, GIP release has been shown to be either normal or decreased. Further investigation of GIP release in diabetes secondary to alimentary disease may be more important.

Motilin

Chemistry and site of synthesis. Motilin was first discovered by Brown in 1973[70] as a result of the observation that duodenal stimulation causes gastric contraction independent of the vagus. Motilin is a straight-chain peptide of 22 amino acids with a molecular weight of 2700 daltons. The porcine form appears to be different from canine motilin. A synthetic motilin has all the actions of the natural peptide.

Biological activity. Motilin is found in the enterochromaffin (EC) cells of the small intestine, predominantly in the upper portion, though immunohistochemical studies have possibly indicated that it is also present in the central nervous system.[71] Motilin is released into the plasma after a meal, particularly if the fat content is high. In humans duodenal acidification causes an increase in plasma motilin. Because it is released by physiological stimulus (a meal), it has been suggested that motilin plays a role in the modulation of gastrointestinal function.

The pharmacological actions of motilin include stimulation of small intestine and gastric motility, increase in lower esophageal sphincter pressure, and stimulation of pepsin output. Recent studies have indicated that motilin accelerates gastric emptying in humans and that it is also probably responsible for emptying the small intestine between meals.

Fasting plasma motilin concentrations in humans show a skewed distribution, with the upper limits of normal at about 300 pmol/L.[72] Motilin circulates in only one molecular form, and no disease has yet been recognized in which plasma motilin levels are abnormal.

Glucagon-like immunoreactive peptides

Chemistry and site of synthesis. An intestinal material that cross-reacts with pancreatic glucagon assays was first detected by Unger et al.[73] in 1961. The material was found to have physicochemical properties different from those of glucagon. In particular, there was a larger molecular weight component, and so the term ''glucagon-like immunoreactivity'' was introduced. Further antibodies that reacted with the C-terminal portion of pancreatic glucagon do not detect this enteric material, whereas N-terminal bodies react with it to a variable extent. It has been found that there are several molecular forms of this family of hormones, each probably having separate physiological roles, though only one cell type containing glucagon-like immunoreactivities is found in the intestine.[74] Nomenclature is thus still confusing, with names such as GLI, N-GLI, gut glucagon, or enteroglucagon being applied. It has been suggested that enteroglucagon contains the entire amino acid sequence of pancreatic glucagon with an additional N-terminal sequence.[75] The full amino acid sequence has not yet been determined. The pancreatic alpha cells produce a polypeptide of 29 amino acids with a molecular weight of 3485 daltons called ''pancreatic glucagon,'' which shows numerous amino acid homologies with secretin; consequently many of the biological functions are similar to those of VIP, GIP, and secretin.[8] This peptide is therefore called ''pancreatic glucagon.''

Biological activity. Glucagon and the glucagon-like immunoactive peptides have a number of biological actions that include relaxation of smooth muscle, inhibition of pancreatic enzyme secretion, inhibition of gastric acid secretion, stimulation of intestinal fluid and electrolyte secretion, and stimulation of cardiac output. Pancreatic glucagon is secreted primarily in response to hypoglycemia and is important in the mobilization of hepatic glycogen stores and carbohydrate homeostasis.[76] The inhibitory action of glucagon on gastrointestinal motility has enabled it to be of clinical use to endoscopists and radiologists. Its positive effect in shocklike states has not been confirmed clinically. Fasting human plasma enteroglucagon levels have been reported to be about 20 pmol/L and rising after a normal meal to 40 pmol/L. Enteroglucagon is considered a trophic hormone, or the growth hormone of the gut. In favor of this speculation is the finding of low enteroglucagon levels after starvation and high enteroglucagon levels associated with hyperphagia (overeating). After partial gut resection, enteroglucagon rises rather than falls, and very high enteroglucagon levels are found in patients after jejunoileal bypass. Similarly, in states of mucosal damage often associated with malabsorption (such as celiac disease), enteroglucagon levels are found to be elevated. Enteroglucagon is probably an important hormone in the physiology of the alimentary tract, but wider recognition of this is at present prevented by the controversy over nomenclature and failure to obtain sufficient quantities of pure material to develop direct assays or to be able to study its pharmacology.

Pathology. McGarvan in 1966 reported the first patient with diabetes, skin rash, and a high concentration of plasma tumor pancreatic glucagon.[77] The characteristic necrolytic migratory erythematous rash led Mallinson in 1974 to diagnose nine additional cases.[78] The diagnosis is easily confirmed by the radioimmunoassay measurements of elevated plasma glucagon levels, and the site of the tumor is confirmed by selective angiography and pancreatic venography.[79] An arginine provocation test has been reported to be of some use in the diagnosis of cases with equivocal plasma glucagon levels.

Somatostatin

Chemistry and site of synthesis. Somatostatin was isolated from the bovine hypothalamus as an inhibitor of growth hormone secretion by the pituitary.[80] It is a peptide of 14 amino acids with a molecular weight of 1640 daltons, bridged across at the third and fourteenth positions by a cystine link. The original molecule and several active analogs are freely available in the synthetic form. Two forms of biologically active somatostatin are somatostatin 14 and somatostatin 28.

Certain of the analogs are relatively selective in their action, and the analog DES^{1,2,4,5,12,13} D Tryp^{8} is almost as potent as the parent molecule, despite missing 6 of the 14 amino acids. (The superscripts represent carbon positions.)

Somatostatin is found throughout the brain, but high concentrations are present in the hypothalamus, particularly in the median eminence. In the gut the greatest amounts are found in the antrum of the stomach and the upper small intestine as well as in the pancreas. It is mainly localized in the mucosa of the alimentary tract, in the D cells of the islets of Langerhans, and in a small number of fine nerve fibers.[81]

Biological activity. With somatostatin having such diverse effects as an inhibitory hormone throughout the body, a number of biological actions have been found. These include inhibition of gastric emptying and pepsin secretion, delay of gastric emptying, inhibition of gallbladder contraction, inhibition of bile and pancreatic enzyme secretion, and hypotension.[82] Because somatostatin has so many effects, it is probable that it functions as a local hormone and part of the paracrine system. There is evidence that somatostatin acts beyond the cAMP system either by interfering with exocytosis of peptide granules or by altering the handling of bivalent cations.[83]

At present there is considerable interest in this hormone. Pharmacological studies have demonstrated that it is one of the most potent inhibitors of endocrine secretions known.[84] It has now been shown that it not only completely inhibits the release of growth hormone but also inhibits the release of thyroid-stimulating hormone, insulin, glucagon, pancreatic polypeptide, gastrin, secretin, motilin, enteroglucagon, and neurotensin.[85] In addition, it can inhibit the effect of these hormones on their target tissues and thus will prevent acid secretion during pentagastrin infusion and pancreatic bicarbonate secretion during secretin infusion.[82]

Pathology. A number of somatostatin-producing tumors of the pancreas have recently been described and are known as somatostatinomas.[86] These tumors produce an ill-defined clinical syndrome that includes diarrhea, gallbladder disease, and diabetes. In patients with other pancreatic endocrine tumors, somatostatin cell hyperplasia is often present in the surrounding pancreatic tissue. The reason for this is not clear, but it has been suggested that this represents an attempt by the pancreas to decrease tumor hormone production. Because somatostatin is such a potent inhibitor of peptide release in both physiological and pathological situations, its use as a therapeutic agent in the management of tumors would be appropriate. The administration of long-acting somatostatin analogs has resulted in inhibition of peptide secretion in a number of patients with gastrinomas, glucagonomas, and VIPomas with amelioration of the clinical symptoms.[87] Somatostatin has also recently been implicated in the etiology of peptic ulceration.

Miscellaneous peptides

With the advent of improved methodology and increased interest in the gut as an endocrine organ, a large number of substances have been postulated and, in some cases, identified (see Table 40-1). The exact role and function in the gut of many of these substances are still speculative. Their presence, however, in gut endocrine cells or nervous system tissue requires explanation; they are therefore considered either as candidate hormones or in terms of their described pharmacological actions.[10] Terms such as "incretin" or "enterogastrone" have been used to show the probable existence of a gut peptide that respectively shares functions in either modulating insulin release or inhibiting gastric acid secretion. In many cases such peptides are known only in crude biological extracts, and in other cases the term is used merely to describe the function rather than a known agent.

Peptides found in both gut and nervous system (brain-gut axis). Several of these peptides have been dealt with previously; the list includes gastrin, cholecystokinin, VIP, motilin, somatostatin, neurotensin, bombesin, substance P, and enkephalin. The most recently identified peptides in this group are neurotensin, substance P, and the endogenous opiate group.

Neurotensin. Neurotensin is a peptide of 13 amino acids originally isolated from the bovine hypothalamus.[88] In humans neurotensin is localized mainly in the mucosa of the ileum in specific endocrine cells called "N cells." A number of molecular forms of neurotensin exist in both plasma and tissue. Neurotensin appears to be released independently of the vagus, but its exact physiological function is not yet known.

Substance P. This decapeptide has a molecular weight of 1348 daltons and was initially identified in extracts of brain and intestine.[89] More recently it has been demonstrated in the brain, intestinal nerve plexus, and enterochromaffin cells of the gut mucosa. The pharmacological actions of substance P include stimulation of intestinal motility, salivary secretion, and increase in esophageal sphincter pressure.[90] Elevated plasma levels have been found in patients with carcinoid syndrome. In Hirschsprung's disease substance P levels are extremely low in the aganglionic segments.

Endogenous opiates. This group consists of two main subgroups: the pentapeptide enkephalins and the endorphins.[91] The two enkephalins differ in structure by only one amino acid. They are found in high concentrations both in the brain and in the G cells and in the myenteric plexus of the antrum and duodenum, where they probably have a paracrine function. Morphine and methionine enkephalin are partial angonists of gastric acid secretion, and their action can be blocked by naloxone, atropine, or metiamide.[82] Evidence suggests that this group of substances may have a role in the regulation of gastric acid secretion and gut motility.

Peptides related to acid secretion. Human *urogastrone* was isolated and synthesized by Gregory[93] in 1979. It is a polypeptide of 53 amino acids with a molecular weight of about 6000 daltons, though larger molecular forms occur. Urogastrone has a structure and function similar to those of epidermal growth factor. Both peptides are powerful inhibitors of gastric acid secretion and stimulate epithelial cell proliferation.

Other peptides, the biological effects of which would regulate acid secretion, have been postulated but have not yet been isolated. These include antral chalone, bulbogastrone, gastrone, enterogastrone, vagogastrone, and entero-oxyntin.

Peptides related to pancreatointestinal function. A number of these peptides have been identified and occasionally isolated. They include chymodenin, coherin, duocrinin, enterocrinin, and villikinin. Unfortunately, further work is required to isolate and confirm both their existence and action.

Other tumors

Mixed pancreatic islet cell tumors. Mixed tumors are of clinical importance because the spectrum of symptoms may change with time and metastases may contain only some of the primary tumor cell types. Because raised plasma concentration of hormones may originate either from the primary or metastatic tumor tissue or from hyperplastic extratumoral cells, immunocytochemical studies are necessary. Ideally the plasma concentration of as many hormones and polypeptides as possible should be measured, tumor tissue investigated by immunocytochemistry, and hormones measured postoperatively to confirm removal of all tumor tissue. APUDomas that secrete a combination of several polypeptides, such as ACTH, MSH, and gastrin, have been described.[94]

Multiple endocrine neoplasia. Multiple endocrine neoplasia describes a group of syndromes, often familial, in which two or more endocrine glands undergo hyperplasia or tumor formation in the same individual, either at the same time or consecutively. The hyperfunctioning glands secrete their normal major hormones (orthoendocrine syndromes) and abnormal hormones (paraendocrine syndromes). There are two main varieties.

Multiple endocrine neoplasia type 1 (MEN I or MEA I), or Werner's syndrome. This manifests from the second decade into old age with an equal sex distribution. The areas involved in order of frequency are parathyroids (88%), pancreatic islets (81%), anterior pituitary (65%), adrenal cortex (38%), and thyroid follicular cells (19%).

The parathyroids usually undergo chief cell hyperplasia or less commonly multiple adenoma formation. The islet cells may be involved by an adenoma (multiple or single), carcinoma, and very rarely a generalized hyperplasia. In the pituitary an adenoma or rarely a carcinoma or hyperplasia may involve any of the different cell types. The adrenal cortices usually show bilateral diffuse or nodular hyperplasia, but adenomas have been reported. The changes in the thyroid are variable, and occasionally carcinoid tumors are present in the lungs, pancreas or intestines. Peptic ulceration, especially duodenal, is a common feature in some families and may affect more than half the patients. These may form part of the Zollinger-Ellison syndrome as a result of an associated pancreatic gastrinoma or hyperparathyroidism. Many of these lesions are APUDomas, and the syndrome may result from a widespread dysplasia of the APUD cells.

Multiple endocrine neoplasia type II (MEN II or MEA II), or Sipple's syndrome. This syndrome is usually inherited, affects sexes equally, and may manifest itself from the first decade onward, especially between 20 to 40 years. Three forms of the syndrome are recognized. The most common tumor is medullary carcinoma of the thyroid with a pheochromocytoma in one or both adrenal glands. Hyperparathyroidism resulting from hyperplasia or adenoma may be present and is probably caused by hypercalcemia. Another variant is a medullary carcinoma of thyroid and pheochromocytoma with multiple small subcutaneous and submucous neuromas of the eyelids, tongue, and buccal mucosa with a diffuse hypertrophy of the lips. These lesions are present from birth. In the last type, all features are present together with autonomic ganglioneuromatosis and various other congenital abnormalities.

REFERENCES

1. Pearse, A.C.E., Polak, J.M., and Bloom, S.R.: The newer gut hormones, Gastroenterology **72:**746, 1977.
2. Polak, J.M., and Bloom, S.R.: Neuropeptides of the gut, World J. Surg. **3:**393, 1979.
3. Rehfeld, J.F.: Gastrointestinal hormones. In Crane, R.K., editor: International review of physiology: gastrointestinal physiology III, vol. 19, Baltimore, 1979, University Park Press, p. 291.
4. Feyrter, F.: Über die peripheren endokrinen (parakrinen) Drusen des Menschen, Vienna/Dusseldorf, 1953, Verlag Maudinch, p. 27.
5. Pearse, A.G.E.: The cytochemistry and ultrastructure of polypeptide hormone-producing cells of the APUD series and the embryologic, physiologic, and pathologic implications of the concept, J.Histochem. Cytochem.**17:**303, 1969.
6. Le Douarin, N.: The migration of neural crest to the wall of the digestive tract in avian embryo, J. Embryol. Exp. Morphol. **30**(1):31, 1973.
7. Rehfeld, J.E.: Immunochemical studies in cholecystokinin. II. Distribution and molecular heterogeneity in the central nervous system and small intestine of man and hog, J.Biol. Chem. **253:**4022, 1978.
8. Walsh, J.H.: Gastrointestinal peptide hormones and other biologically active peptides. In Sleisenger M.H., and Fordtran, J.S., editors: Gastrointestinal disease pathophysiology, diagnosis, management, Philadelphia, 1978, W.B. Saunders Co., p. 109.
9. Pearse, A.G.E., and Takor, T.: Neuroendocrine embryology and the APUD concept, Clin. Endocrinol. **5:**(Suppl.) 229S, 1976.
10. Grossman, M.I.: Neural and hormonal regulation of gastrointestinal function: an overview, Annu. Rev. Physiol. **41:**27, 1979.
11. Bloom, S.R., and Polak, J.M.: In Glass, B. editor: Progress in gastroenterology, New York, 1977, Grune & Stratton, Inc., pp. 109-151.
12. Rehfeld, J.F.: Heterogeneity of gastrointestinal hormones, World J. Surg. **3:**415, 1979.

13. Rehfeld, J.F., Schwartz, T.W., and Stadil, F.: Immunochemical studies on macromolecular gastrins: evidence that "big big gastrin" in blood and mucosa is artifactual—but truly present in some large gastrinomas, Gastroenterology **72:**469, 1977.
14. Larsson, L.I.: Peptides of the gastrin cell. In Rehfeld, J.H., and Amdrup, E., editors: Gastrin and the vagus, New York, 1979, Academic Press, Inc., p. 5.
15. Gregory, R.A.: A review of some recent developments in the chemistry of the gastrins, Bioorgan. Chem. **8:**497, 1979.
16. Dockray, G.J.: Gastrin overview. In Bloom, S.R., editor: Gut hormones, Edinburgh, 1978, Churchill Livingston, p. 129.
17. Strunz, U.T., Walsh, J.H., and Grossman, M.I.: Stimulation of gastrin release in dogs by individual amino acids (40072), Proc. Soc. Exp. Biol. Med. **157:**440, 1978.
18. McGuigan, J.E.: Gastrointestinal hormones, Annu. Rev. Med. **29:**307, 1978.
19. Zollinger, R.M., and Ellison, E.H.: Primary peptic ulceration of the jejunum associated with islet cell tumors of the pancreas, Ann. Surg. **142:**709, 1955.
20. Yalow, R.S., and Berson, S.A.: Radioimmunoassay of gastrin, Gastroenterology **58:**1, 1970.
21. Welbourn, R.B., and Joffe, S.N.: The apudomas. In Taylow, S., editor: Recent advances in surgery, Endinburgh, 1977, No. 9, Churchill Livingstone.
22. Hamsky, J., King, R.W., and Holdsworth, S.: In Thompson, J.C., editor: Gastrointestinal hormones, Austin, 1978, University of Texas Press, p. 115.
23. Walsh, J.H., and Grossman, M.I.: Gastrin, N. Engl. J. Med. **292:**1324, 1975.
24. Isenberg, J.I., Grossman, M.I., Maxwell, V., et al.: Increased sensitivity to stimulation of acid secretion by pentagastrin in duodenal ulcer, J.Clin. Invest. **55:**330, 1975.
25. Creutzfeldt, R.A., and Frerichs, H.: Insulinomas and gastrinomas. In Bloom, S.R., editor: Gut hormones, Edinburgh, 1978, Churchill Livingstone, pp. 589-598.
26. Primrose, J.N., Ratcliffe, J.G., and Joffe, S.N.: Assessment of secretion provocation test in the diagnosis of gastrinoma, Br. J. Surg. **67:**744, 1980.
27. Passaro, E., Basso, N., and Walsh, J.H.: Calcium challenge in the Zollinger-Ellison syndrome, Surgery **72:**60, 1972.
28. Jorpes, J.E., and Mutt, V.: Secretin, cholecystokinin, pancreozymin, and gastrin, New York, 1973, Springer-Verlag New York, Inc.
29. Ivy, A.C., and Oldberg, E.: Hormone mechanism for gallbladder contraction, Am. J. Physiol. **85:**381, 1928.
30. Lin, T.M.: Actions of gastrointestinal hormones and related peptides on the motor function of the biliary tract, Gastroenterology **69:**1006, 1975.
31. Mutt, V.: Gastrointestinal hormones: a field of increasing complexity, Scand. J. Gastroenterol. **77:**133, 1982.
32. Dockray, G.J.: Evolutionary relationships of the gut hormones, Fed. Proc. **38:**9, 1979.
33. Brooks, A.M., and Grossman, M.I.: Effect of secretin and cholecystokinin on pentagastrin stimulated gastric secretion in man, Gastroenterology **59:**114, 1970.
34. Sturdevant, R.A.L., and Goetz, H.: Cholecystokinin both stimulates and inhibits human food intake, Nature (Lond.) **261:**713, 1976.
35. Meyer, J.H., and Kelly, G.A.: Canine pancreatic responses to intestinally perfused proteins and protein digests, Am. J. Physiol. **231:**682, 1976.
36. Rayford, P.L., Miller, T.A., and Thompson, J.C.: Secretin, cholecystokinin, and newer gastrointestinal hormones, N. Engl. J. Med. **294:**1093, 1976.
37. Harvey, R.F., Dowsett, L., Hartog, M., et al.: A radioimmunoassay for cholecystokinin-pancreozymin, Lancet **2:**826, 1973.
38. Bayliss, W.M., and Starling, E.H.: On the causation of the so-called "peripheral reflex secretion" and the pancreas, J. Physiol. (Lond.)**28:**352, 1902.
39. Bodansky, M., Ondetti, M.A., Levine, S.D., et al.: Synthesis of a heptacosapeptide amide with the hormonal activity of secretin, Chem. Industr. **42:** 1957, 1966.
40. Robberecht, P., Conlon, T.P., and Gardner, J.D.: Interaction of porcine vasoactive intestinal peptide with dispersed pancreatic acinar cells from the guinea pig. J.Biol. Chem. **251:**4653, 1976.
41. Polak, J.M., Pearse, A.G.E., Joffe, S.N., et al: Quantification of secretin release by acid using immunocytochemical and radioimmunoassay, Experientia **31:**462, 1975.
42. Schaffalitzky de Muckadell, O.B., Fahrenkrug, J., Watt-Boolsen, S., et al: Pancreatic response and plasma secretin concentration during infusion of low dose secretin in man, Scand. J. Gastroenterol. **13**(3):305, 1978.
43. Hacki, W.H., Greenberg, G.R., and Bloom, S.R.: In Bloom, S.R., editor: Gut hormones, Edinburgh, 1978, Churchill Livingstone, p. 182.
44. Straus, E., and Yallow, R.S.: Hypersecretinemia associated with marked basal hyperchlorhydria in man and dog, Gastroenterology **72:**992, 1977.
45. Deveney, C.W., Deveney, K.S., Jaffe, B.M., et al.: Use of calcium and secretin in the diagnosis of gastrinoma, Ann. Intern. Med. **87**(6):680, 1977.
46. Said, S.I., and Mutt, V.: Isolation from porcine intestinal wall of vasoactive octacosapeptide related to secretin and glucagon, Eur. J. Biochem. **28:**199, 1972.
47. Bryant, M.G., Bloom, S.R., Polak, J.M., et al.: Possible dual role for vasoactive intestinal peptide as gastrointestinal hormone and neurotransmitter substance, Lancet **1:**991, 1976.
48. Said, S.I.: Vasoactive intestinal peptide (VIP) current status. In Thompson, J.C., editor: Gastrointestinal hormones, Austin, 1975, University of Texas Press, p. 591.
49. Fahrenkrug, T.V.: Vasoactive intestinal polypeptide: measurement, distribution and putative neurotransmitter function, Digestion **19**(3):149, 1979.
50. Fahrenkrug, J., Schaffalitzky de Muckadell, O.B., and Holst, J.J.: Plasma secretin concentration in anaesthetized pigs after intraduodenal glucose, fat, amino acids or meals with various pH, Scand. J. Gastroenterol. **12:**273, 1977.
51. Verner, J.V., Jr., and Morrison, A.B.: Islet cell tumor and a syndrome of refractory watery diarrhea and hypokalemia, Am. J. Med. **25:**374, 1958.
52. Bloom, S.R., Polak, J.M., and Pearse, A.G.E.: Vasoactive intestinal peptide and watery diarrhea syndrome, Lancet **2:**14, 1973.
53. Ebeid, A.M., Murray, P.D., and Fischer, J.E.: Vasoactive intestinal peptide and the watery diarrhea syndrome, Ann. Surg.,**187:**411, 1978.
54. Devine, B.L., Carmichael, A., Russell, R.I., et al.: Cyclical release of vasoactive intestinal polypeptide (VIP) from a pancreatic islet cell apudoma, Postgrad. Med. **54:**566, 1978.
55. Kimmel, J.R., Hayden, L.J., and Pollack, H.G.: Isolation and characterization of a new pancreatic polypeptide hormone, J. Biol. Chem. **250:**9369, 1975.
56. Lin, T.M., and Chance, R.E.: Spectrum gastrointestinal actions of a new bovine pancreatic polypeptide (BPP), Gastroenterology **62:**852, 1972.
57. Lin, T.M., Evans, D.C., Chance, R.E., et al.: Bovine pancreatic polypeptide: action on gastric and pancreatic secretion in dogs, Am. J. Physiol. **232:**E311, 1977.
58. Parks, D.L., Gingerich, R.L., Jaffe, B.M., and Akande, B.: Role of pancreatic polypeptide in canine and gastric acid secretion, Am. J. Physiol. **236**(4):E488, 1979.
59. Schwartz, T.W.: Pancreatic polypeptide as indicator of vagal activity. In Rehfeld, J.H., and Amdrup, E., editors: Gastrin and the vagus, New York, 1979, Academic Press, Inc., p. 175.
60. Floyd, J.C.: Human pancreatic polypeptide, Clin. Endocrinol. Metab. **8**(2):379, 1979.
61. Modlin, I.M., Lamers, C.B., and Jaffe, B.M.: Evidence for cholinergic dependence of pancreatic polypeptide (PP) release by bombesin—a possible application, Surgery **88:**75, 1980.
62. Polak, J.M., Bloom, S.R., Adrian, I.E., et al.: Pancreatic polypeptide in insulinomas, gastrinomas, VIPomas, and glucagonomas, Lancet **1:**328, 1976.
63. Schwartz, T.W.: Pancreatic polypeptide and endocrine tumors of the pancreas, Scand. J. Gastroenterol. **14**(Suppl 53):93, 1979.
64. Stern, I., Robet-Thomson, I.C., and Hansky, J.: Correlation between pancreatic polypeptide response to secetin v. ERCP., chronic pancreatitis, Gut **23:**235, 1982.
65. Brown, J.C., Pederson, R.A., Jorpes, E., et al.: Preparation of a

highly active enterogastrone, Can. J. Physiol. Pharmacol. **47:**113, 1969.
66. Anderson, D., Elahi, D., Brown, J.C., et al.: Oral glucose augmentation of insulin secretion, J. Clin. Invest. **62**(1):152, 1978.
67. Brown, J.C.: Physiology and pathophysiology of GIP, Adv. Exp. Med. Biol. **106:**169, 1978.
68. Brown, J.C., and Otto, S.C. GIP and the entero-insular axis, Clin. Endocrinol. Metab. **8**(2):365, 1979.
69. Ross, S.A., Brown, J.C., Dryburgh, J.R., et al.: Hypersecretion of gastric inhibitory polypeptide in diabetes mellitus, Clin. Res. **21:**1029, 1973.
70. Brown, J.C., and Dryburgh, J.R.: Discovery of motilin, Scand. J. Gastroenterol. **11**(Suppl. 39):15, 1976.
71. Polak, J.M., Heitz, P., and Pearse, A.G.E.: Differential localization of substance P and motilin, Scand. J. Gastroenterol. **11**(Suppl. 39):39, 1976.
72. Bloom, S.R., Mitznegg, P., and Bryant, M.G.: Measurement of human plasma motilin, Scand. J. Gastroenterol. **11**(Suppl. 39):47, 1976.
73. Unger, R.H., Eisentraut, A., Sims, K., et al.: Sites of origin of glucagon in dogs and humans, Clin. Res. **9:**53, 1961.
74. Grimelius, L., Polar, J.M., Solcia, E., et al.: Gut hormones, Edinburgh, 1978, Churchill Livingstone, p. 365.
75. Moody, A.J., Sundby, F., Jacobsen, H., et al.: The tissue distribution and plasma levels of glicentin (gut GLI-1), Scand. J. Gastroenterol. **13**(Suppl. 49):127, 1978.
76. Moody, A.J., Frandsen, E.K., Jacobsen, H., et al.: Heterogeneity of gut glucagon-like immunoreactivity (GLI). In Foa, P.P., Bajaj, J.S., and Foa, N.L., editors: Glucagon: its role in physiology and clinical medicine, Amsterdam, 1978, Excerpta Medica, p. 129.
77. McGavran, M.H., Under, R.H., Recant, L., et al.: A glucagon-secreting alpha-cell carinoma of the pancreas, N. Engl. J. Med. **274:**1408, 1966.
78. Mallinson, C.N., Bloom, S.R., Warin, A.P., et al.: A glucagonoma syndrome, Lancet **2:**1, 1974.
79. Hardy, J.D., and Doolittle, R.D.: Zollinger-Ellison syndrome, Ann. Surg. **185:**661, 1977.
80. Schally, A., Dupont, A., Arimura, A., et al.: Isolation and structure of somatostatin from porcine hypothalami, Biochemistry **15:**509, 1976.
81. Polak, J.M., Pearse, A.G.E., Grimelius, L., et al.: Growth hormone release—inhibiting hormone in gastrointestinal and pancreatic D cells, Lancet **1:**1220, 1975.
82. Creutzfeldt, W., and Arnold, R.: Somatostatin and the stomach: exocrine and endocrine aspects: First International Somatostatin Symposium, Freibarg, Germany, 1977, Metabolism **27**(9 Suppl. 1):1309, 1978.
83. Schofield, J.C., Mira, F., and Orei, L.: Somatostatin and growth hormone secretion in vitro: a biochemical and morphological study, Diabetology **10:**385, 1974.
84. Raptis, S., and Gerich, J.E.: Metabolism, vol. 27 (Suppl. 1), 1978.
85. Vale, W., Rivier, C., and Brown, M.: Regulatory peptides of the hypothalamus, Annu. Rev. Physiol. **38:**473, 1977.
86. Krejs, G.J., Orci, L., Conlon, J.M., et al.: Somatostatinoma syndrome: biochemical morphology and clinical features, N. Engl. J. Med. **301**(6):285, 1979.
87. Bloom, S.R.: New specific long-acting somatostatin analogues in the treatment of pancreatic endocrine tumors, Gut **19:**446, 1978.
88. Carraway, R., and Leeman, S.: The isolation of a new hypotensive peptide, neurotensin, from bovine hypothalami, J. Biol. Chem. **24**(19):6854, 1973.
89. von Euler, U.S., and Pernow, B., editors: Substance P: proceedings of Nobel Symposium, New York, 1978, Raven Press.
90. Skrabanek, P., and Powell, D.: Substance P, vol. 1, Edinburgh, 1977, Eden Press.
91. Ambinder, R.F., and Schuster, M.M.: Endorphins: new gut peptides with a familiar face, Gastroenterology **77**(5):1132, 1979.
92. Konturek, S.J., Tasler, J., Cieszkowski, M., et al.: Comparison of methionine—enkephalin and morphine in the stimulation of gastric acid secretion in the dog, Gastroenterology **78:**294, 1980.
93. Gregory, H.: The identification of urogastrone in serum, saliva, and gastric juice, Gastroenterology **77**(2):313, 1979.
94. Joffe, S.M., Elias, E., Rehfeld, J.F., et al.: Clinically silent gross hypergastrinaemia from a multiple hormone-secreting pancreatic apudoma, Br. J. Surg. **65:**277, 1978.

Chapter 41 The gonads

Elizabeth J. Kicklighter
Bidy Kulkarni

amenorrhea Absence or abnormal cessation of menstruation.

androgens Sex steroid hormones responsible for the development of the male secondary sex characteristics. The major androgens are testosterone, dihydrotestosterone, and androstenedione.

anorchia Congenital absence of the testis.

endometrium Inner layer of the uterus, the thickness and structure of which vary with the phase of the menstrual cycle.

epididymis Elongated cordlike structure along the posterior portion of the testis that contains ducts capable of storing spermatozoa.

estrogens Sex steroid hormones responsible for the development of the female secondary sex characteristics. The major estrogens include estradiol, estrone, and estriol.

fallopian tube Long, slender, hollow tube that extends from the upper lateral portion of the uterus to the ovary on the same side.

Frommel-Chiari syndrome Postpartum condition possibly caused by hypothalamic dysfunction; characterized by atrophy of the uterus, persistent lactation, galactorrhea, prolonged amenorrhea, and low levels of estrogens and gonadotropins.

gametogenesis The development of male and female sex cells, or gametes.

germinal aplasia Absence of the primordial germ cells within the gonads.

gonadotropins Hormonal substances having affinity for or a stimulating effect on the gonads.

graafian follicle Ovarian structure consisting of the ovum and its encasing cells.

granulosa cells Modified epithelial cells that surround the ovum and secrete the follicular fluid. After ovulation, they are transformed into glandular cells of the corpus luteum.

gynecomastia Benign glandular enlargement of the mammary glands in males.

hirsutism More hair than is cosmetically acceptable for women in a given culture.

Kallmann's syndrome Condition characterized by the absence of the sense of smell because of agenesis of the olfactory bulbs and by secondary hypogonadism because of the lack of LHRH.

Klinefelter's syndrome Condition characterized by small testes with fibrosis of seminiferous tubules, impairment of function and clumping of Leydig cells, and by increased gonadotropins. It is believed to be caused by the presence of an extra sex (X) chromosome.

Lawrence-Moon-Bardel-Biedl syndrome Condition characterized by obesity, hypogenitalism, and mental deficiency.

Leydig cells Interstitial cells of the testes that produce the male sex steroids.

menopause (climacteric) Syndrome of endocrine and somatic changes occurring at the termination of the reproductive period in the adult woman.

menstruation Cyclic, physiological uterine bleeding that normally occurs, usually at approximately 28-day intervals, during the reproductive period of the female in the absence of pregnancy.

myometrium Smooth muscle portion of the uterus that forms the main portion of the uterus.

ovum Female reproductive cell produced in the ovary.

polycystic ovary syndrome (Stein-Leventhal syndrome) Condition associated with bilateral polycystic ovaries and characterized by secondary amenorrhea and anovulation.

pseudohermaphroditism Condition in which the gonads are of one sex, but contradictions exist in the morphological expression of the sex, for example, a genetic and gonadal female with partial masculinization.

Sertoli cells Elongated cells in the tubules of the testes that provide support, protection, and apparent nutrition of spermatids until they become transformed into mature spermatozoa.

Sheehan's syndrome Postpartum pituitary necrosis.

spermatogonia Undifferentiated germ cell of a male originating in the seminal tubule.

Turner's syndrome Condition characterized by a female phenotype, short stature, undifferentiated (streak) gonads, and variable abnormalities including webbing of the neck and cardiac defects. It is associated with an absence of the second (X) chromosome.

virilism Development of male secondary sexual characteristics in a female, including enlargement of the clitoris, growth of facial and body hair, stimulation of sebaceous glands (often with acne), and deepening of the voice.

Hormones are regulators of a variety of physiological processes. In particular, all reproductive processes are regulated closely by a complicated series of interrelated systems involving the hypothalamus, the pituitary gland, and the male and female gonads. Among the most complex of the hormonal actions are those involving the reproductive system.

Contrary to popular belief, both men and women secrete the same sex steroid hormones, that is, androgens, estrogens, and progesterone. The difference in steroid hormone secretion between the sexes is not of an absolute or all-or-none type, but rather a matter of degree. Testosterone is the major sex steroid in men, whereas estradiol is the primary sex hormone in women.

SEX DIFFERENTIATION AND ANATOMY OF REPRODUCTIVE SYSTEM

Sex differentiation

Sex is basically determined by the type of sex chromosomes present in the cells, two X chromosomes, resulting in a genetic female, and one X and one Y, resulting in a genetic male. The presence of the Y chromosome is a prerequisite for the development of the male gonad. This in turn gives rise to testosterone secretion during fetal life. Lack of fetal testosterone secretion will always result in the establishment of the female gonad. If testosterone production fails, the embryonic gonads will develop as female, regardless of the chromosomal sex. In both sexes the gonads have a dual function—the production of germ cells (gametogenesis) and the secretion of sex hormones.

Anatomy of female reproductive system

The female organs of reproduction are classified as external and internal structures. The external organs and the vagina serve for copulation, whereas the internal organs located within the pelvis provide for development and birth of the fetus. The external organs are the vulva, mons pubis, the labia (majora and minora), the clitoris, the ves-

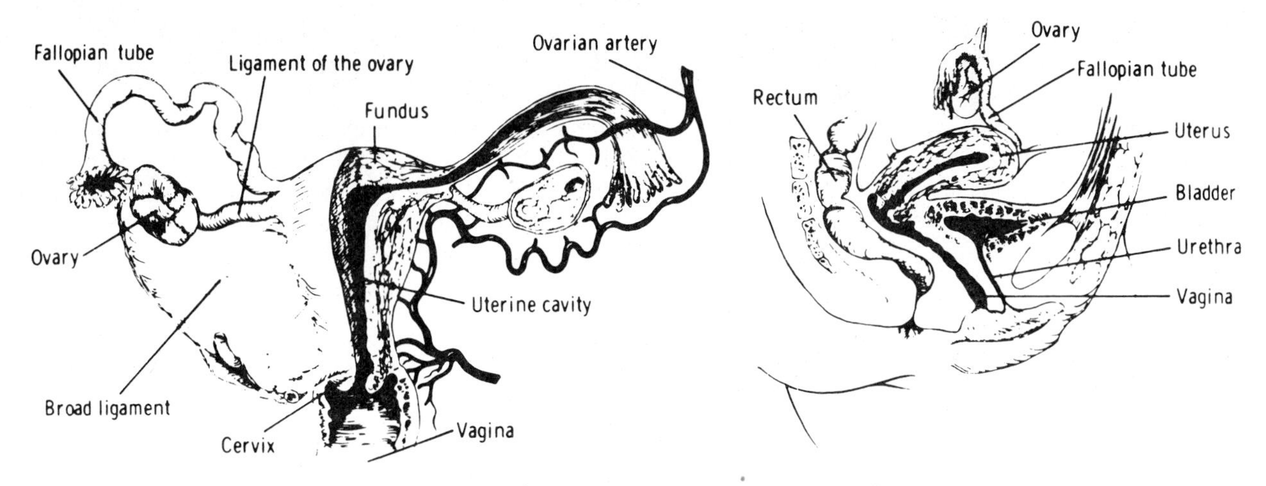

Fig. 41-1. Female reproductive system. (From Ganong W.F.: Review of medical physiology, ed. 11, Los Altos, Calif., 1983, Lange Medical Publications.)

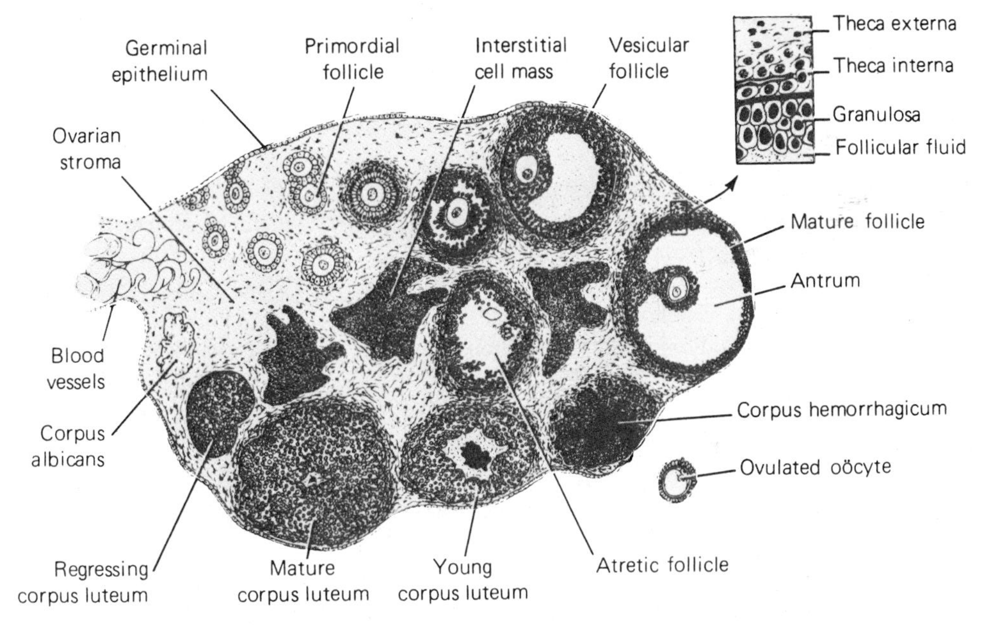

Fig. 41-2. Diagram of ovary, showing sequential development of a follicle and formation of corpus luteum. Section of wall of a mature follicle is enlarged at upper right. (From Gorbman, A., and Bern, H.A.: A textbook of comparative endocrinology, New York, 1962, John Wiley & Sons.)

tibule of the vagina, and mammary glands as accessories to the reproductive system.

The female reproductive tract's internal organs consists of two ovaries, two fallopian tubes, the uterus, and the vagina (Fig. 41-1). Unlike the male reproductive system, the internal female reproductive organs show regular cyclic changes that may be regarded as periodic preparations for fertilization and pregnancy.

The ovaries are located on either side of the uterus and serve the dual purpose of production of the ova and secretion of the female sex hormones—the estrogens and progesterone (Fig. 41-2). Every female at birth has about 400,000 microscopically small, immature follicles in the ovaries, each containing an immature ovum. During the reproductive life, however, only 300 to 400 follicles reach maturity. The mature follicle is composed of three layers of cells—the theca externa, the theca interna, and the granulosa cells. The cells of the theca interna are the primary source of estrogens.

The follicle that ruptures at the time of ovulation promptly fills with blood, forming what is sometimes called a "corpus hemorrhagicum." The granulosa and theca cells of the follicle lining promptly begin to proliferate, and the clotted blood is rapidly replaced with yellowish, lipid-rich luteal cells, forming the corpus luteum. The luteal cells secrete estrogens and progesterone. If pregnancy occurs, the corpus luteum persists and there are usually no more menstrual periods until after delivery. If there is no pregnancy, the corpus luteum begins to degenerate about 4 days before the next menses and is eventually replaced by scar tissue to form a corpus albicans.

The fallopian tubes, also known as "oviducts" or "uterine tubes," lead from the vicinity of the ovaries to the uterus. They serve to transport the sperm from the uterine cavity and the fertilized egg to the uterus as well and serve as the location for fertilization to occur.

The uterus is a thick-walled muscular chamber connecting the fallopian tubes and the vagina in which the fetus develops until the time of delivery. The muscular layer of the uterus is called the "myometrium." The innermost layer is the "endometrium." The endometrium undergoes cyclic changes, and during menstruation the endometrium is shed.

Anatomy of male reproductive system

The male reproductive system (Fig. 41-3) consists of two testes producing sperm and androgens, a penis, and a system of exocrine glands whose secretions constitute the bulk of seminal fluid in which sperm are conveyed. These glands are the two bulbourethral glands (Cowper's gland), two seminal vesicles and one prostate, and a system of bilateral ducts through which sperm pass outside of the body, namely, the epididymis, the vas deferens, and the ejaculatory duct.

The normal mature testis (Fig. 41-4) is enclosed by a thick white capsule of three layers—the tunica vaginalis, the tunica albuginea, and the inner tunica vasculosa. Smooth muscle fibers are present and can aid in sperm transport from the testis to the epididymis. The capsule maintains pressure on approximately 250 pyramidal lobules of seminiferous tubules, which are separated by the fibrous septa. The tubules compose over 85% of the volume of the testis.

Surrounding the central lumen of the seminiferous tubule is a structured epithelium containing Sertoli cells and spermatogenic cells. Around the tubules is a thin basement membrane, a fibrous tunica propria, and myoid cells. The interstitial tissue between the tubules contains Leydig cells, which are responsible for the production of testicular androgens.

PHYSIOLOGY OF SEX STEROID HORMONES

General physiology

The interrelationships between the hypothalamus, pituitary, and gonads are extremely complex. The hypothalamic decapeptide, luteinizing hormone–releasing hormone (LHRH) (also known as gonadotropin-releasing hormone [GnRH]), enhances the release of both pituitary gonadotropins, luteinizing hormone (LH) and follicle-stimulating hormone (FSH), affecting LH more than FSH.

The pituitary gonadotropins, FSH and LH, play an important role by influencing different functions of both the male and female internal sex organs. In women, FSH induces the growth of the follicles of the ovary and causes them to secrete estrogens, whereas LH acts in the final stages of follicular growth to bring about ovulation (or release of the ovum from the follicle) and acts as a luteal cell stimulator inducing progesterone production. On the other hand, FSH in men brings about spermatogenesis, whereas LH stimulates testosterone production. In other words, the same hormones can act on different target tissues to produce different biological effects. This is also true for the effects the primary sex hormones, testosterone and estradiol, have on their target tissues.

Levels of both gonadotropins (LH and FSH) vary with the age and sex and in the female, with the phase of the

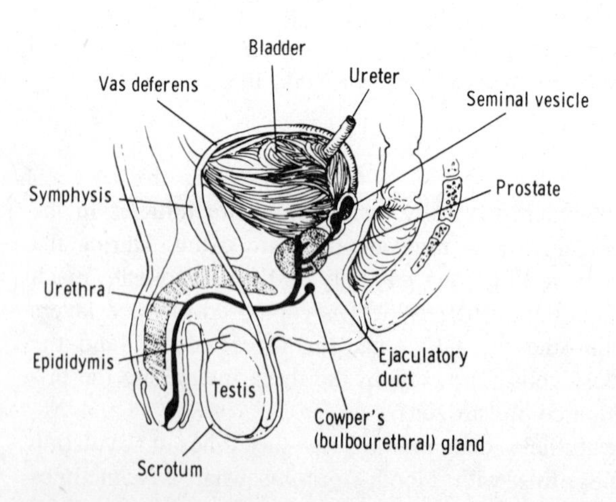

Fig. 41-3. Male reproductive organs. (From Ganong, W.F.: Review of medical physiology, ed. 11, Los Altos, Calif., 1983, Lange Medical Publications.)

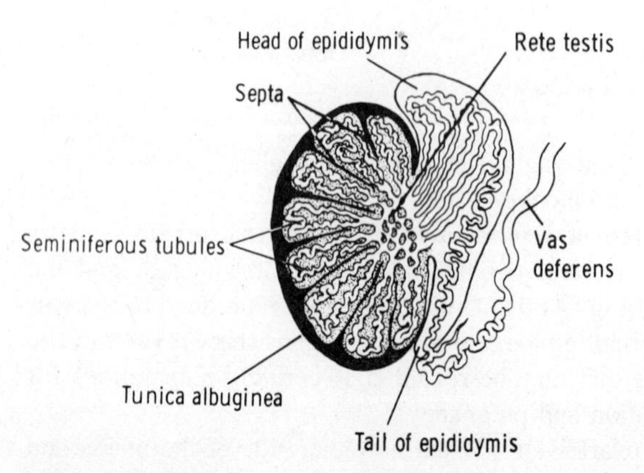

Fig. 41-4. Anatomy of testis. (From Ganong, W.F.: Review of medical physiology, ed. 11, Los Altos, Calif., 1983, Lange Medical Publications.)

menstrual cycle. In addition, both hormones are secreted in rhythmic pulses several times each hour.

Feedback of the various sex steroids at the pituitary and hypothalamic levels is complicated. Several hormones may exert additive effects. The effects of various sex steroids may be positive under one set of circumstances and negative with another; for example, estrogen in low levels has a negative feedback on the pituitary, whereas estrogen in high levels has a positive effect. These effects vary at different times during the menstrual cycle (for example, pituitary responsiveness to LHRH depends on the phase of the menstrual cycle).

Estradiol is the primary secretory product of the ovary. Estrone, another important estrogen, is the product of ovarian secretion and the peripheral conversion of prohormones of adrenal or ovarian origin as well. Progesterone is produced by the corpus luteum and is present in significant amounts only after ovulation.

Testosterone is the primary secretory hormone of the testis. In females, testosterone is derived from peripheral conversion of prohormones, mainly androstenedione.

Sex steroid hormones in both sexes during adolescence stimulate characteristic spurts of growth and union of epiphyses with bone shafts, which terminate future growth. In addition, they promote protein synthesis. Although androgens affect the skeletal muscle, giving a muscular appearance in men, estrogens more strongly affect smooth muscle. The distribution of subcutaneous fat and smooth character of the skin in women are estrogen dependent. Androgens induce secondary sex charcteristics typical of the male, whereas estrogens stimulate typical female secondary sex characteristics and as well growth and development of the female reproductive organs. Estrogens, together with progesterone and other pituitary factors, stimulate the growth and development of the mammary glands. Progesterone may be regarded as the hormone of the mother-to-be, and both estrogen and progesterone bring about characteristic developments during the menstrual cycle.

Menstrual cycle

The physiology of reproduction in the female is dominated by the events in the menstrual cycle. Menstruation is a regular cyclic shedding of the surface layer of the endometrium along with some blood that passes out of the body in the menstrual flow. These periodic uterine events are the result of a cyclic supply and withdrawal of hormonal support and are well coordinated with the morphologic and hormonal changes taking place in the ovary. The hormonal control of the menstrual cycle involves a complex interplay between the hypothalamus, the anterior pituitary hormones FSH and LH, and the ovarian hormones, estradiol and progesterone (Fig. 41-5). The menstrual cycle, an interval between successive menstruations, is 28 days, starting from the first days of menstruation. The menstruation period, the period of active uterine bleeding, in an idealized menstrual cycle is 4 days. The length of the menstrual cycle can vary among individual women from 21 to 42 days and the menstrual period from 3 to 7 days.

Ovarian events during menstrual cycle. The ovarian events during the normal menstrual cycle include both morphological and hormonal changes. Under the influence of FSH, several follicles start to develop at the beginning of the cycle, and only one will mature in about 10 to 12 days. The one developing follicle that matures and ruptures, releasing the ovum, is called the "graafian follicle," and the process of rupture to release the ovum is called "ovulation." The ovulation presumably takes place on day 14 of the 28-day cycle and continues to occur until the ovarian function ceases 30 to 35 years after puberty (menopause). The ruptured follicle undergoes morphological changes to become a wrinkled, yellow structure called the "corpus luteum," which becomes a new endocrine organ. At about day 25 the corpus luteum begins to shrink and degenerate if conception has not occurred. This degeneration is soon followed (within 2 to 3 days) by menstruation, which heralds the beginning of the next menstrual cycle. When the follicles are the dominant ovarian structures, the ovary is said to be in the follicular phase, and this occurs during the first half of the cycle. When the corpus luteum is the dominant ovarian structure, the ovary is said to be in the luteal phase of the cycle, and this occurs during the second half of the cycle.

Changes in ovarian morphology such as the graafian follicle development and corpus luteum formation correlate well with the changes in ovarian steroid hormone production. With the above uterine morphological changes occurring during the menstrual cycle, simultaneous hormonal changes take place in the ovary. The most important ovarian hormones (those that effect the uterine changes during the menstrual cycle) are estradiol and progesterone. Estradiol alone dominates the events during the follicular phase, and progesterone, combined with estradiol, dominates the events during the luteal phase. Thus the ovary plays a key endocrine role, acting as the pivotal unit of the menstrual cycle. The morphological and hormonal function of the ovary in a rhythmic fashion depends solely on the appropriately timed release of both FSH and LH by the pituitary in response to hypothalamic gonadotropin-releasing hormone of the hypothalamus.

Hormonal changes during menstrual cycle. Early in the cycle when the levels of estrogen and progesterone are relatively constant and low, the FSH levels are rising and high and the LH levels are low (Fig. 41-5). These high levels of FSH stimulate the follicular growth and its output of estrogens, particularly estradiol. By days 7 and 8, the rise of estradiol is at a rapid rate, and it reaches its first peak before ovulation. The rising levels of estradiol result in a negative feedback to the hypothalamus and pituitary

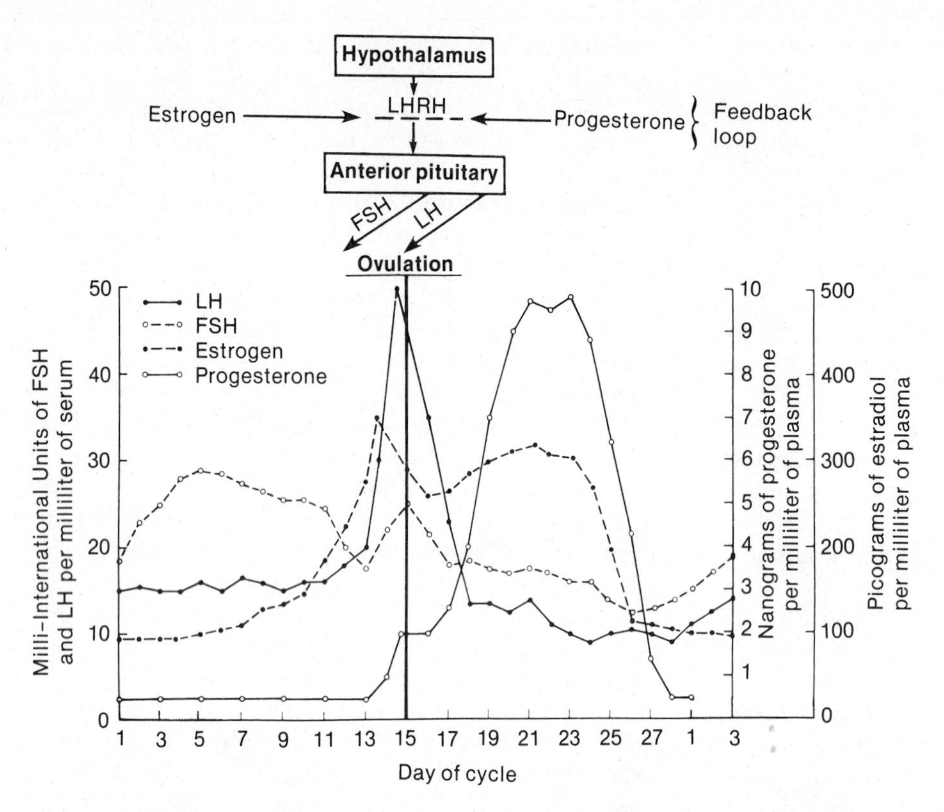

Fig. 41-5. Changes in blood levels of various hormones throughout menstrual cycle. Also shown is hypothalamic-pituitary axis and feedback exerted by estrogens and progesterone. (Pituitary actually produces only one hormone, gonadotropin-releasing hormone also known as LHRH, because it stimulates secretion of LH more than FSH. A separate FSHRH has never been identified.) (Modified from Taymor, M.L., Berger, M.J., Thompson, I.E., and Karamo, K.S.: Am. J. Obstet. Gynecol. **114**:445, 1972.)

glands and cause a fall in FSH levels because of the inhibitory action of estradiol on FSH release. Concurrently, the rise in estradiol triggers a rapid rise in LH (positive-feedback effect). Estradiol reaches a maximum on the day before LH peak. During the midcycle there is a peak rise of LH, which leads to maturation of the graafian follicle and its rupture, releasing the ovum (ovulation) 16 to 24 hours after LH peak. Before the LH surge and before ovulation, the estradiol level drops considerably and then rises again after ovulation. The ruptured follicle becomes the corpus luteum. The progesterone levels released by the corpus luteum begin to rise, causing an inhibition of the secretion of LH. A sharp increase in progesterone follows, reaching a maximum in about 8 or 9 days after the LH peak (days 23 to 25 of the cycle). As estradiol and progesterone increase, FSH and LH decline throughout the luteal phase. As the corpus luteum regresses, the levels of both estradiol and progesterone begin to diminish. The removal of the inhibitory effect of these two compounds results in the increase of FSH, which stimulates the growth of a new crop of follicles in the ovary. During the menstruation phase, estradiol, progesterone, and LH are at a relatively constant but low level, whereas FSH is the only hormone present in elevated and rising levels.

Effect of ovarian hormones on uterus (endometrium). As a physiological result of the ovarian hormones, estradiol and progesterone, definite changes occur in the endometrium in preparation to receive and implant a fertilized ovum. The rising levels of estradiol stimulate the reconstruction of the endometrium, blood vessels, and secretory glands of the uterus. This change in the endometrial growth is called the "proliferative phase" and corresponds with the follicular phase of the ovary. The newly formed glands begin to release glandular substances. Hence this phase is called the "secretory phase" of the endometrium, and it corresponds to the luteal phase of the ovary. During days 1 to 5 of the cycle, enough estrogen is secreted by the follicles to start the deep layer of the endometrium growing, but it is not enough to support the thickened secretory endometrium of the previous cycle. As a result, this well-developed surface of the endometrium sloughs off as the estradiol levels decrease. This is menstruation. Thus menstruation occurs during the later stages of the proliferative phase of the endometrium because of with-

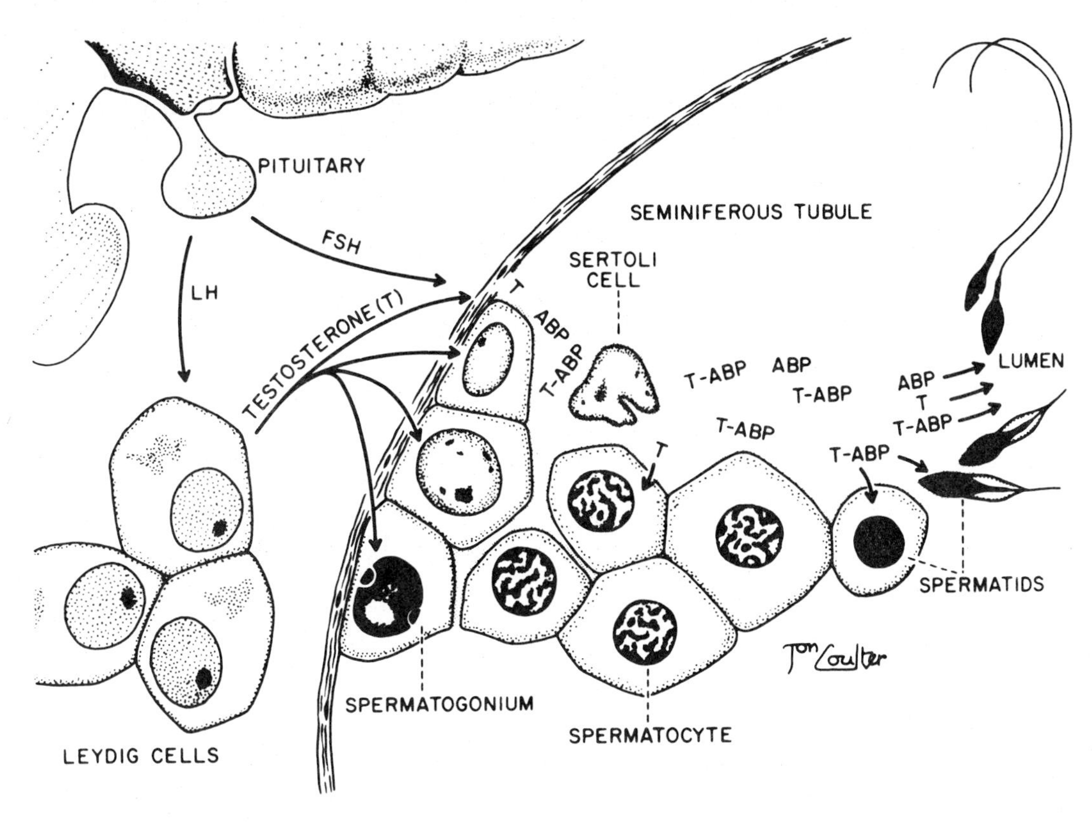

Fig. 41-6. Relationship of gonadotropins and testosterone to tubular function. (From Felig, P., et al.: Endocrinology and metabolism, New York, 1981, McGraw-Hill Book Co.)

drawal of hormonal support. If a fertilized egg is successfully implanted in the enriched endometrium, the implantation causes additional estrogens to be produced; thus menstruation is prevented and the development of a pregnancy is begun (see Chapter 35),

Hormonal regulation in the male

The testes, like the ovaries produce both steroid hormones and gametes. These two activities, androgen synthesis and sperm production (spermatogenesis), are carried out by two anatomically different structures. Androgen synthesis occurs in the Leydig cells and spermatogenesis takes place in the seminiferous tubules (Fig. 41-6). The anterior pituitary hormones control both of these functions.

FSH stimulates spermatogenesis, and LH (called "interstitial cell–stimulating hormone," or ICSH, in the male) stimulates testosterone production. LH binding in the testis occurs on Leydig cell membranes located in the interstitial tissue of the testis. LH stimulates Leydig cell synthesis of testosterone through cAMP-mediated processes.

FSH receptors in the testis are located in the Sertoli cells of the seminiferous tubules. The primary effect of FSH is the stimulation of protein synthesis, especially androgen-binding protein (ABP) by cAMP. Testosterone (T) binds to ABP as testosterone or as dihydrotestosterone (DHT) after Δ^4-5α-reductase action. Because of high androgen affinity, ABP concentrates testosterone and dihydroxytestosterone as they diffuse into the tubule, thus transporting androgens to the site of spermatogenesis. Testosterone acts first on spermatogonia and then stimulates primary spermatocytes to complete meiotic division and to form secondary spermatocytes and young spermatids. FSH then induces maturation of late spermatids to spermatozoa.

Increased levels of serum testosterone feed back upon the hypothalamic pituitary axis to modulate the amounts of LH being released. These events proceed normally at a rather fixed rate during the adult life.

BIOSYNTHESIS OF STEROID HORMONES

The two trophic hormones produced by the anterior pituitary that have a direct effect on gonadal function are FSH and LH. FSH and LH are glycoproteins composed of two polypeptide subunits: the alpha-subunit is common to FSH, LH, HCG and TSH; the beta-subunit confers biologic specificity. The pituitary hormones are transported by the systemic circulation to the target organs.

Under control of the anterior pituitary hormones, the ovaries, the testes, and the adrenals secrete a group of steroid hormones. The steroids are characterized by the presence of a basic ring structure—the cyclopentan-

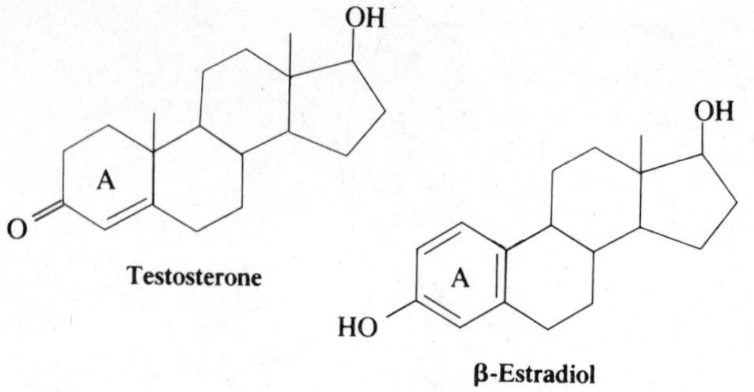

Testosterone

β-Estradiol

Fig. 41-7. Basic structures of testosterone and estradiol. (Modified from Orten, J.M., and Neuhaus, O.W.: Human biochemistry, ed. 10, St. Louis, 1982, The C.V. Mosby Co.)

operhydrophenanthrene nucleus. The steroid hormones are classified as estrogens, androgens, progestogens, and corticosteroids, depending on the number of carbon atoms in the molecule and their substituents. C-21 steroids include progesterone (and the glucocorticoids), C-19 steroids include testosterone, dihydrotestosterone, and the adrenal androgens, and estrogens are in the C-18 group. The basic structures of testosterone and estradiol are depicted in Fig. 41-7.

Biologically active steroid hormones are synthesized through a series of enzymatic reactions starting with a common precursor, cholesterol, which serves as the basic skeleton for all steroid hormones. The pertinent enzymes necessary to bring about the specific changes in the cholesterol molecule are synthesized after stimulation by the pituitary hormones.

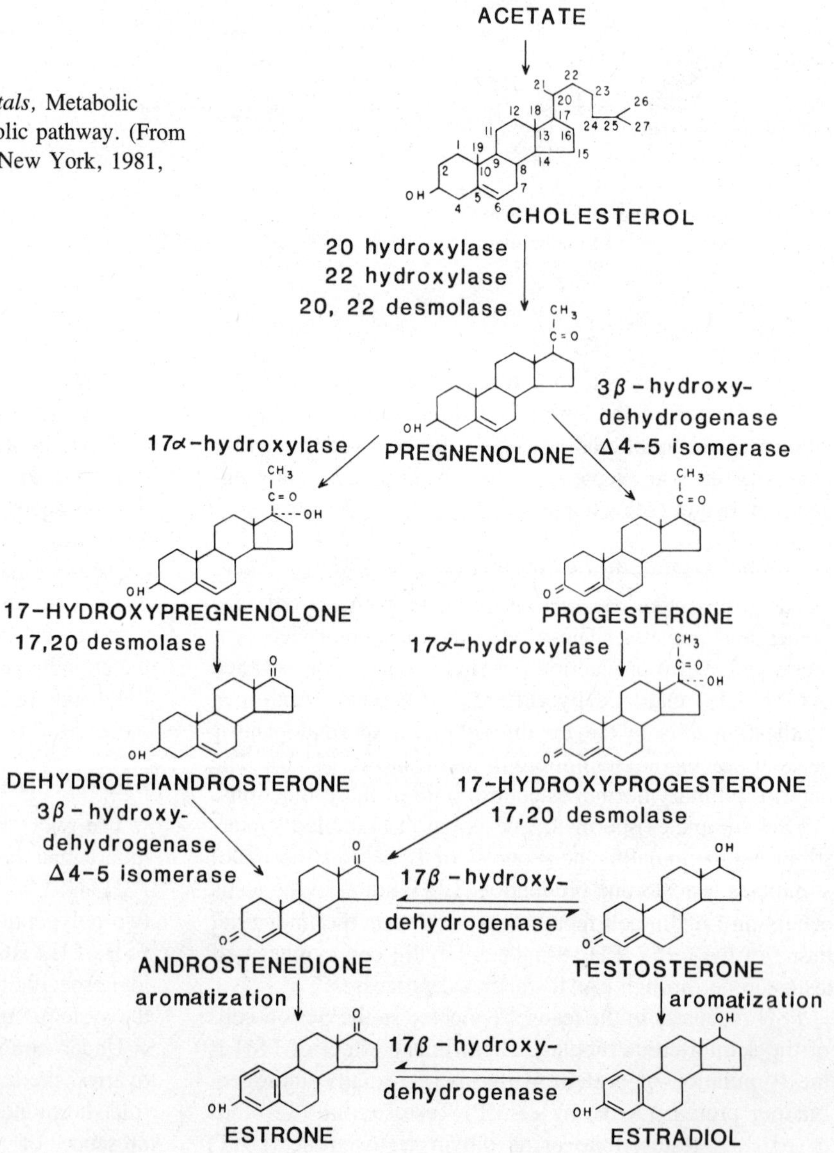

Fig. 41-8. Pathways of sex steroidogenesis. *Capitals,* Metabolic products; *lower-case lettering,* enzymes in metabolic pathway. (From Felig, P., et al.: Endocrinology and metabolism, New York, 1981, McGraw-Hill Book Co.)

Pathways of steroid hormone biosynthesis are shown in Fig. 41-8. The initial step in steroid hormone synthesis is the conversion of cholesterol (C-27 compound) to pregnenolene (C-21 compound) by hydroxylation at carbon 20 and 22 and subsequent cleavage between these two carbon atoms. These reactions are catalyzed by the enzymes 20-hydroxylase, 22-hydroxylase, and 20,22-desmolase. This constitutes a key step in the steroid hormone biosynthesis regulated by the trophic hormone LH in the ovary and testis, and ACTH in the adrenals.

From pregnenolone, steroid hormone biosynthesis proceeds along either of the two pathways: (1) the Δ^5 pathway via 17α-hydroxypregnenolone or (2) the Δ^4 pathway via progesterone. The conversion of Δ^5-steroids (pregnenolone) to Δ^4-steroids is brought about by two enzymes, 3β-hydroxydehydrogenase and $\Delta^{4,5}$-isomerase. The choice of the pathway is governed by the cell type. The Δ^5 pathway predominates in the follicle of the ovary for estrogen production and the Δ^4 pathway predominates in the alternative pathway for progesterone synthesis. The 17α-hydroxylation of pregnenolone and progesterone gives 17α-hydroxypregnenolone and 17α-hydroxyprogesterone respectively. Conversion of C-21 compounds to C-19 compounds (androgens) is effected by an enzyme 17,20-desmolase to yield androgens, namely, dehydroepiandrosterone (DHEA) and androstenedione by the respective Δ^5 and Δ^4 pathways. DHEA may be converted to androstenedione by the enzyme 3β-hydroxydehydrogenase, and androstenedione to testosterone by the enzyme 17β-hydroxydehydrogenase. The conversion of C-19, androgens, to C-18, estrogens (estrone and estradiol), is brought about by a group of enzymes called the "aromatizing system" involving C-19 hydroxylation, loss of the carbon at position 19 and reduction, resulting in a phenolic or aromatic ring; hence the process is called "aromatization." This process is important for the production of ovarian hormones.

Thus all classical steroid hormones are synthesized from the same precursor, cholesterol, and basically utilize the same biosynthetic pathway. The specific hormone synthesized will depend on which genes of the endocrine gland are activated. The ovaries and testes have the enzymatic pattern to synthesize estrogens and androgens respectively, but they lack the 21-hydroxylase and 11β-hydroxylase system, characteristic of the adrenocortical cell. The testes differ from the ovaries enzymatically by the functional absence of the aromatizing enzyme system. Under normal circumstances, the adrenals do not appear to contain desmolase and the aromatizing system and hence do not synthesize biologically active androgens and estrogens. Some adrenal tumors do produce testosterone or estrogens, or both, indicating the activation and presence of the desmolase and aromatizing systems. All the steroid-producing glands have the ability to synthesize other hormones in addition to their own; however, the synthesis of these enzymes are repressed during normal cell activity.

Ovarian steroid hormones

The ovary produces estrogens, androgens, and progesterone, following the pathways of steroid biosynthesis as outlined in Fig. 41-8. Estradiol is the chief secretory product of the maturing follicles, and progesterone is the chief secretory product of the corpus luteum. Estradiol, a C-18 steroid, is characterized by the presence of a phenolic ring and hydroxy group in the 17β position, whereas progesterone, a C-21 steroid, is characterized by a keto group in ring A and a 2-carbon side chain at C-17 with a keto group at the C-20 position.

The predominance of each of these steroids at certain points during the menstrual cycle has been explained by two theories: the "two-pathway" theory and more recently the "two-cell" hypothesis. According to the two-pathway theory, the Δ^4-3-ketone (progesterone) pathway seems to be predominant in luteal tissue, whereas the Δ^5-3-hydroxy (pregnenolone) pathway is characteristic of nonluteinized tissue. The predominance of estrogen before ovulation is explained on a morphological basis. Limitation of vascularization within the follicular layers ensures that estrogen is the major secretory product. The implication is that, because of the absence of vascularization in the granulosa layer until after ovulation, progesterone synthesized by the granulosa cells during the preovulatory period must diffuse toward the theca, where it will be utilized for estrogen production. In the corpus luteum where the vascularization of the granulosa layer has been achieved, both estrogen and progesterone are major secretory products by the Δ^4-3-ketone pathway.

The two-cell hypothesis presents a more logical explanation of ovarian steroidogenesis (Fig. 41-9). It proposes

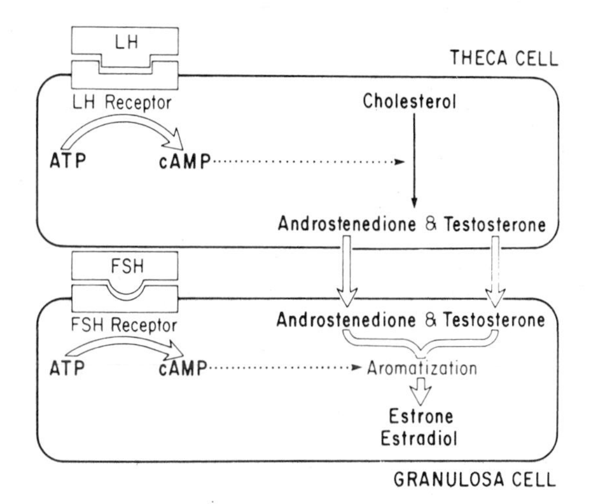

Fig. 41-9. "Two-cell hypothesis" of ovarian estrogen production. Luteinizing hormone stimulates androgen production in thecal cells. Androgens then diffuse into granulosa layer, where they are aromatized into estrogens, in an action specifically stimulated by FSH. (From Felig, P., et al.: Endocrinology and metabolism, New York, 1981, McGraw-Hill Book Co.)

that LH acts on the theca cells of the ovary to produce androgens. These androgens are transferred from theca cells into granulosa cells where, before ovulation, under the influence of FSH, the androgens are converted to estrogens by the enzyme aromatase. After ovulation, the granulosa cells secrete progesterone and estrogen directly into the bloodstream under the influence of LH.

Estrone, the other major biologically important estrogen, is derived from peripheral conversion from prohormones, particularly androstenedione, from interconversion of estradiol or from secretion by the ovary. In the postmenopausal woman, estrone is the dominant plasma estrogen.

Very little estriol is produced by nonpregnant females. Among the androgens secreted by the ovaries, androstenedione is the chief secretory androgen.

Testicular steroid hormones

The testis in adult men produces large amounts of testosterone in the Leydig cells (about 5 to 12 mg/day) plus a very small amount of other steroids, including estradiol, androstenedione, and dehydroepiandrosterone (Fig. 41-8). Testosterone, a C-19 steroid, is characterized by the presence of a Δ^4-3-keto group in ring A, a hydroxyl group in position 17, and no side chain. Secretion of testosterone by the testes is episodic; a circadian pattern can be demonstrated, with a maximum in the early morning (about 7:00 A.M.) and minimum about 13 hours later.

Peripheral conversion by adrenal androstenedione accounts for approximately 5% of testosterone. The major adrenal androgen dehydroepiandrosterone has weak biological potency.

TRANSPORT AND METABOLISM

Transport

Steroid hormones are water-insoluble compounds that need to be transported in the bloodstream to their site of action. This is achieved by their binding to albumin and other binding proteins in the plasma. Circulating testosterone and estradiol are largely bound to sex hormone–binding globulin (SHBG), though there is some binding to albumin. Progesterone is preferentially bound to corticosteroid-binding globulin (CBG). The binding proteins are synthesized in the liver.

The binding of a hormone to its carrier protein is a reversible process resulting in an equilibrium between bound and free fractions. It is important to understand that, as for all nonprotein hormones, the *free* fraction is the only biologically active hormone. The percentage of total serum estradiol that is found in the free fraction is 3%, whereas 2% of serum testosterone is not bound to serum protein. The significance of this equilibrium is that the plasma level of active, free testosterone becomes a variable fraction of total amount of testosterone, depending on the level of SHBG. The free testosterone is therefore a better index of androgenicity than the total serum testosterone.

Testosterone and dihydrotestosterone bind more strongly to SHBG than estradiol does. Albumin, on the other hand, binds estrogen more strongly than testosterone does. The concentration of SHBG is decreased by administration of testosterone or growth hormone and is increased by estrogen or thyroid hormone.

When the binding capacity of the serum increases by increased levels of binding protein, removal of steroid from plasma by peripheral tissue and subsequent metabolism are reduced. For example, in a patient with hyperthyroidism, the SHBG is increased. This increased binding capacity reduces the level of free testosterone, which in turn stimulates the production of additional testosterone. As a result, the total plasma level of testosterone increases, however the newly formed testosterone also binds in large part to SHBG. At a certain point the new levels of SHBG and testosterone reach concentrations where the free testosterone level is in the desired range.

Steroid catabolism

The catabolic reactions are mostly reductive and occur mainly, though not exclusively, in the liver. The steroid hormones are converted by the liver to inactive compounds by a group of hepatic enzymes such as hydrogenases, dehydrogenases, and hydroxylases. The Δ^4-5-hydrogenases reduce the double bond between carbon 4 and 5 to give 5α- and 5β-isomers. Not only are the Δ^4-5β-hydrogenases distinct from Δ^4-5α-hydrogenases, but they are also differently distributed within the liver cell, the former being present in the soluble, cytoplasmic fraction whereas the latter is localized in the microsomes.

The hydroxysteroid dehydrogenases bring about the reduction of the 3-ketone group to a secondary alcohol of the 3-hydroxy configuration. 17β-Hydroxysteroid dehydrogenase, a soluble liver enzyme, rapidly converts 17-hydroxysteroids to 17-ketone and vice versa, whereas the 11-hydroxysteroid dehydrogenase catalyses similar reversible reaction at the C-11 position.

Steroid hydroxylases of the liver bring about hydroxylations at various positions. They do not play as important a role as the dehydrogenases and hydrogenases do in steroid catabolism.

The metabolic products of these reactions are a series of more polar steroids; however, because the products are still not very hydrophilic, they remain bound to serum proteins and hence cannot be readily filtered by the kidneys. To aid their renal excretion, these metabolites are converted to still more polar metabolites—largely glucuronide and sulfate conjugates. The conjugation is carried out in the liver by specific enzymes, sulfokinase catalyzing the conversion of hydroxysteroids to corresponding sulfates and glucuronyl transferase catalyzing the conversion of hydroxysteroids to corresponding glucuronides. The sulfokinases are cytoplasmic enzymes present in a number of other organs such as the adrenals, the kidneys, and the placenta. Most steroids are excreted largely as glucuro-

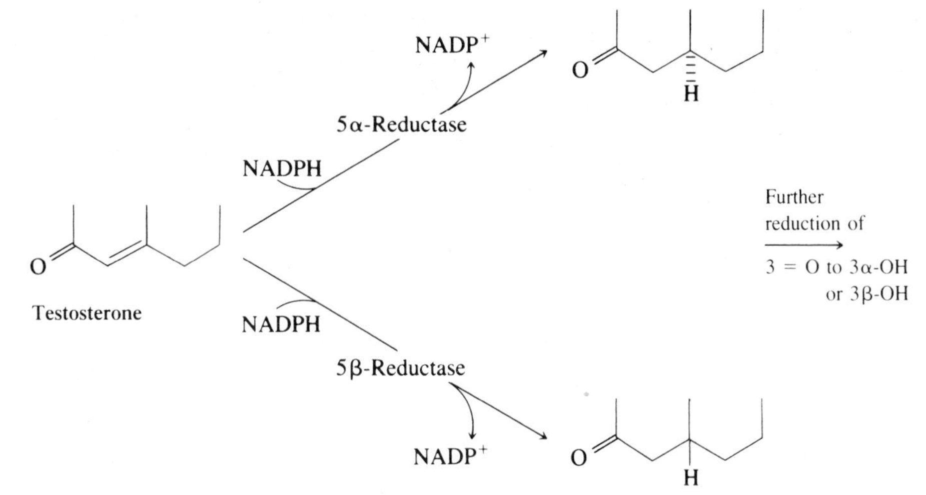

NADP$^+$
5α-Reductase
NADPH
Testosterone
NADPH
5β-Reductase
NADP$^+$
Further reduction of 3 = O to 3α-OH or 3β-OH
Some metabolites of testosterone
Androsterone
Etiocholanolone
Epiandrosterone

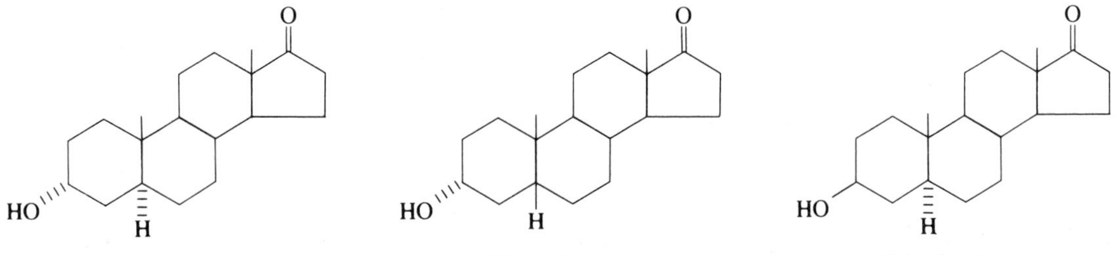

Fig. 41-10. Catabolism of testosterone, and the major metabolites. (From Felig, P., et al.: Endocrinology and metabolism, New York, 1981, McGraw-Hill Book Co.)

nides and filtered readily by the kidney, except for dehydroepiandrosterone (DHEA), which is largely excreted as a sulfate. The sulfate conjugates are cleared much more slowly by the kidneys.

Catabolism of androgens. Basically there are four main steps by which the androgens are metabolized and deactivated in the liver by liver enzymes, namely, (1) reduction of Δ^4-3-keto grouping in ring A, (2) interconversion of secondary alcohol and a ketone group, (3) additional hydroxylation at different carbon atoms (not in all cases), and (4) conjugation as glucuronides or sulfates or both to render the metabolites soluble (Fig. 41-10).

By means of reaction (1) testosterone and androstenedione are converted chiefly to 5α- and 5β-androstene-3,17-diones, and by reaction (2) the 17-OH group is reduced to a 17-keto group yielding (3α,5α-)androsterone, (3α,5β-)-etiocholanolone, and (3β,5α-)epiandrosterone. These last three metabolites, along with dehydroepiandrosterone, constitute the major metabolites of androgens, referred to as 17-ketosteroids, which can be measured in urine.

These metabolites are rendered water soluble by conjugation as sulfates and glucuronides. Androsterone, epiandrosterone, and etiocholanolone are found predominantly as glucuronides. DHEA is found largely as a sulfate.

It should be pointed out that about two thirds of the urinary 17-ketosteroids are of adrenal origin, and one third is of testicular origin. Testosterone itself is also excreted directly into the urine conjugated with glucuronic acid, where its level correlates with the production level of testosterone.

Catabolism of progesterone. Progesterone is metabolized along pathways similar to those for androgen metabolism (Fig. 41-11). Progesterone is metabolized by reduction at C-20 by a variety of tissues to give 20α- and 20β-hydroxyprogesterone, both of which still retain progesterone activity. Reduction occurring at the 4,5 double bond alone gives rise to two pregnanediones, the 5α- and the 5β-pregnane-3,20-dione. Reduction occurring in ring A at the 4,5 double bond and ketone at carbon 3 simultaneously give rise to three pregnenolones, of which 3α-hydroxy-5β-pregnan-20-one is most important. Finally, the reduction occurring in ring A at the 4,5 double bond and ketone at carbon 3 and carbon 20 give rise to three pregnanediols, of which 5β-pregnane-3α,20α-diol is the predominant metabolite and accounts for 10% to 30% of progesterone secreted in the urine. This metabolite also arises from catabolism of deoxycorticosterone and from pregnenolone.

Hydroxylation of progesterone at the 17 position gives

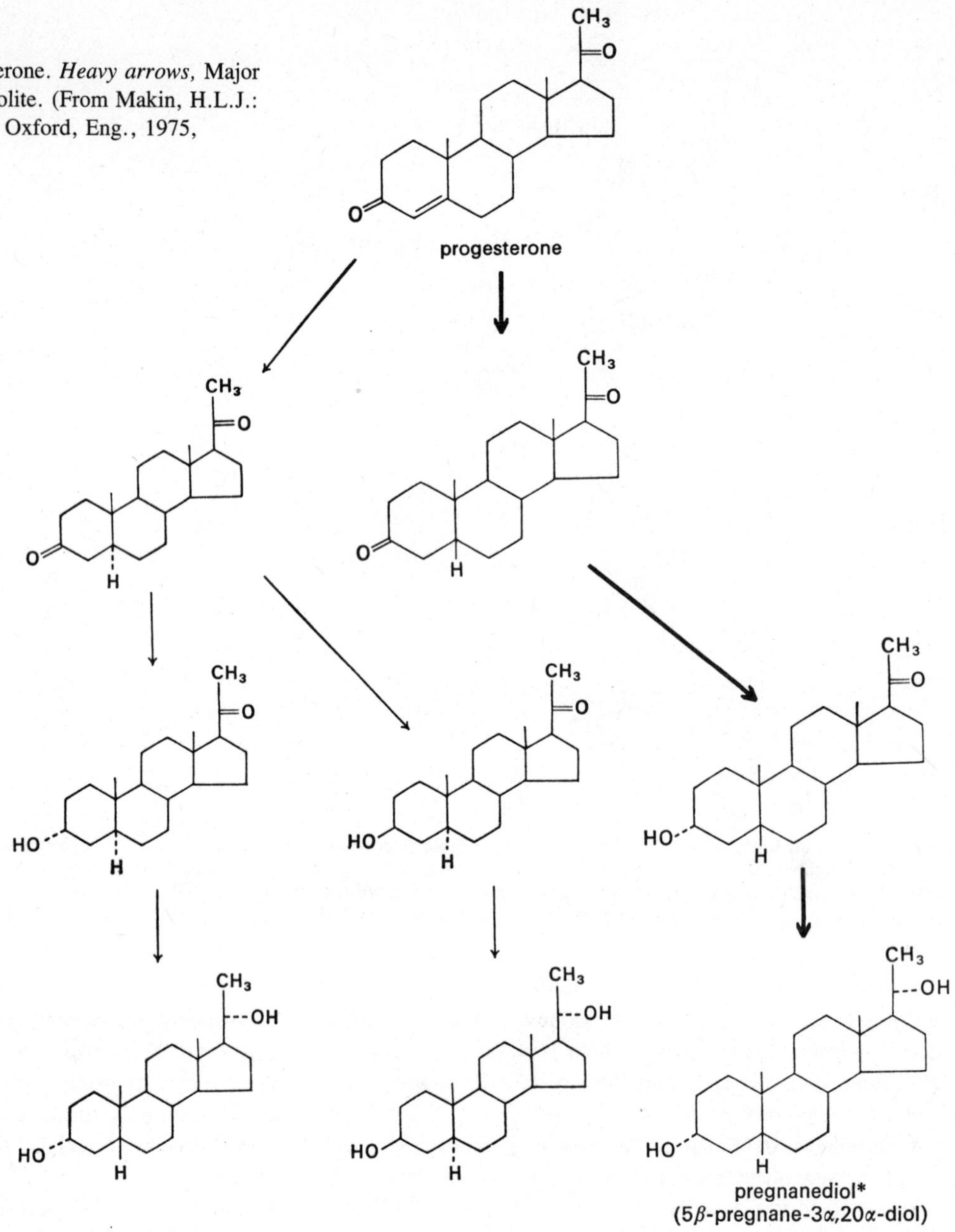

Fig. 41-11. Catabolism of progesterone. *Heavy arrows,* Major pathway; (*), major urinary metabolite. (From Makin, H.L.J.: Biochemistry of steroid hormones, Oxford, Eng., 1975, Blackwell Scientific Publications.)

rise to 17-hydroxyprogesterone. When fully reduced, it gives rise to three pregnanetriols. 5β-Pregnane-3α,17α,20α-triol is the major and unique urinary metabolite of 17-hydroxyprogesterone and is present in diseased states such as congenital adrenal hyperplasia.

Catabolism of estrogens. The metabolism of estrogens differs from that of androgens in that the aromatic ring A is not reduced.

Basically, the estrogens undergo metabolism by hydroxylation at different positions (Fig. 41-12). Hydroxylation at the C-2 position in the phenolic ring A is followed by methylation to give 2-methoxyestradiol or 2-methoxyestrone. Hydroxylation of estradiol at position 16 gives estriol, an abundant metabolite in pregnancy. These and other estrogen metabolites are conjugated with glucuronide and sulfate for excretion in the urine. Paradoxically, estrogen sulfates are known to circulate in the blood as biologically active compounds. Although urine represents the major (50% to 70%) route of elimination of estrogen metabolites, considerable amounts of their metabolites do appear in the feces.

Route of excretion of steroid metabolites. A great proportion of steroid metabolites can enter the intestinal tract but are reabsorbed through the portal circulation and are recirculated through the liver. After conjugation by the liver, the metabolites finally appear in the urine. Some me-

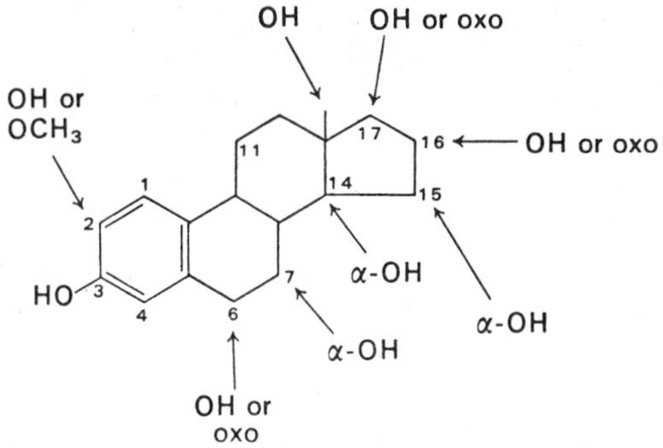

Fig. 41-12. Catabolism of estrogen. (From Makin, H.L.J.: Biochemistry of steroid hormones, Oxford, Eng., 1975, Blackwell Scientific Publications.)

tabolites, especially the less polar steroids, are excreted in the bile. The phenolic steroids, such as estrogen, are secreted to a considerable extent into the bile and appear in the feces. The metabolites of androgens and corticosteroids, on the other hand, largely appear in the urine.

PATHOLOGICAL CONDITIONS

Abnormalities of testicular function

Abnormalities in testicular function can be divided into two categories: those resulting in decreased androgen production, or hypogonadism, and those resulting in excessive androgen production, or hypergonadism. Either disorder may be attributable to a primary dysfunction of the testes, or may be secondary to a derangement of the pituitary-hypothalamic axis.

Hypogonadism. The clinical picture of hypogonadism is directly related to the time of development of androgen deficiency. In prepubertal hypogonadism, absence of androgen production by the testis is associated with persistent infantile genitalia, a barely palpable prostate, female distribution of pubic and axillary hair, a partial to total lack of facial and body hair, and the absence of a deepening voice. Although the growth rate is decreased, it continues with the absence of epiphyseal closure, eventually resulting in tall stature with long arms and legs. Prepubertal hypogonadism is usually inapparent until the adolescent period, when normal adolescent development including genital and secondary sexual changes do not occur.

Postpubertal hypogonadism results in minimal changes. In young males, there is usually diminished beard growth and thinning axillary and other body hair. The prostate atrophies, and sexual desire and performance decrease. The genitalia may decrease somewhat in size. In older men, none of these changes may be noted. Beard and body hair growth usually persist, and there may be no noticeable change in libido or sexual function.

Primary hypogonadism. Because of the lack of androgenic feedback on the pituitary-hypothalamic axis, primary hypogonadism is manifested by increased serum and urine gonadotropins and by decreased serum androgen levels and decreased urinary 17-ketosteroid levels. The testicular abnormality may be secondary to a developmental abnormality, such as a genetic or embryological defect, or it may occur at any time later in life as a result of either testicular infections or after trauma, irradiation, a neoplasm, or castration. Developmental abnormalities account for most prepubertal cases of hypogonadism. Many syndromes are included in these disorders. The majority are very rare, including Reifenstein's syndrome (male pseudohermaphroditism), male Turner's syndrome, germinal aplasia, and anorchia (congenital absence of testes). Probably the most common syndrome resulting in a developmental abnormality is Klinefelter's syndrome. These patients possess an extra sex chromosome (X) resulting in a karyotype consisting of 44 autosomes plus two X chromosomes and one Y chromosome. Klinefelter's syndrome has an incidence of approximately 1 in 400 males. The spectrum of clinical manifestations is broad, including obviously feminized males on one end and normally virilized males with only microscopic and biochemically detected disease on the other end.

The majority of cases of primary hypogonadism manifesting themselves after puberty are the result of testicular infections, trauma, irradiation, a tumor that has replaced the testicular parenchyma, or surgical or accidental castration. Other rare congenital disorders may manifest themselves as a primary hypogonadal state after puberty. These include such conditions as myotonia dystrophica and cystic fibrosis.

Another condition that may be considered primary hypogonadism is that of Leydig cell failure occurring as a result of male climacteric. Population studies in males indicate total serum testosterone levels remain stable until about 70 years of age, after which they begin to decline. LH and FSH levels begin a slight but definite rise at this time.

Secondary hypogonadism. Secondary hypogonadism results from failure of the pituitary to produce LH (ICHS) and FSH. This is usually the result of primary hypopituitarism, though rarely this condition may be secondary to a defect within the hypothalamus resulting in a failure to release LHRH. This results in an absence of positive feedback on the pituitary gland and a lack in the synthesis of FSH and LH. Kallman's syndrome is a rare genetic condition in which a lack of LHRH leads to secondary hypogonadism.

In rare cases of primary hypopituitarism, isolated deficiencies of gonadotropic hormones have been described. In most cases, however, there is an associated loss of other pituitary hormones resulting in decreased thyroid, adrenal, and gonadal function at all ages. This panhypopituitarism may be idiopathic, part of a congenital disorder or secondary to a neurohypophysial lesion such as a neoplasm, cyst, or granulomatous process. When there is a progressive loss of the pituitary function because of a neurohypophysial lesion, a decrease in gonadotropins is at times the first deficiency observed and the patient may present with isolated hypogonadism.

The absence of serum and urinary gonadotropins after the age of adolescence in patients with diminished gonadal function is diagnostic of secondary hypogonadism. Studies of growth, thyroid, adrenal, and antidiuretic hormones may reveal clinically unsuspected deficiencies in other pituitary hormones.

Hypergonadism. Hypergonadism may occur as a primary process because of excessive androgen production from a testicular tumor (Leydig cell or interstitial cell carcinoma). Hypergonadism may also occur secondary to altered pituitary-hypothalamic axis function with increased LH/FSH secretion. Primary hypergonadism is noted for high serum androgen levels, high urinary 17-ketosteroids, and low serum gonadotropins. Secondary hypergonadism is differentiated by elevated androgens and their urinary metabolites, as well as the gonadotropins.

The production of excessive quantities of androgenic hormones in adult males results in little if any morphological change. However, this excessive production in children results in precocious puberty. This usually manifests itself by early onset of puberty with enlargement of genitalia, appearance of pubic or axillary hair, deepening of the voice, and the beginning of acne. Sexual precocity secondary to increased gonadotropins in males, though uncommon, is often familial in contrast to secondary precocious puberty occurring in females. When precocious puberty occurs in males without a family history, it is almost always associated with a space-occupying lesion in the region of the third ventricle of the brain. The continued production of androgens excessive for the age of these patients results in accelerated skeletal growth followed by early closure of the epiphysis, resulting in an adult who is frequently shorter than his contemporaries.

Premature puberty in the male may also result from excessive adrenal androgens. The pattern of growth and development is similar to that produced by increased testicular steroids; however, the testes remain prepuberal in size.

Abnormalities in ovarian function

Endocrine disorders of the ovary may also be classified as ''hypo-'' or ''hyper-''; primary ovarian in origin or secondary to disturbances of hypothalamic-pituitary secretion; and congenital or acquired. Clinical manifestations may be very subtle because the ovary is mainly an organ of reproduction. Its disturbances may be picked up only in a workup for infertility. Other ovarian disorders are manifest clinically by precocious or delayed puberty.

Ovarian hypofunction. As in the male, the symptoms of gonadal hypofunction are dependent on whether the condition manifests itself before puberty or after puberty. Ovarian hypofunction that develops in the prepubertal period will manifest itself clinically as delayed or absent menarche or primary amenorrhea. Ovarian hypofunction developing after puberty may manifest itself as secondary amenorrhea.

Primary ovarian hypofunction. Because of the lack of estrogenic feedback on the hypothalamic-pituitary axis, primary ovarian hypofunction is characterized by increased levels of gonadotropins in association with decreased estrogen levels. The two most common causes of primary ovarian failure are Turner's syndrome and the female climacteric, or menopause.

Turner's syndrome is a genetic disorder characterized by 45 chromosomes, consisting of 44 normal autosomal chromosomes with only one X chromosome. In this condition, the primordial germ cells appear to degenerate for obscure reasons without undergoing transformation into premordial follicles, resulting in primitive or ''streak'' gonads incapable of ovulation or estrogen secretion. The presence of other congenital abnormalities such as short stature, webbing of the neck, short metacarpals, and coarctation of the aorta often suggest the diagnosis during the prepubertal years. Cyclic replacement hormonal therapy is indicated to prevent osteoporosis, premature aging of the skin, and other consequences of estrogen deficiency.

Menopause is a form of primary ovarian hypofunction that is experienced by all women who survive to middle life. The current median age of onset of menopause is 50 years. In the majority of instances, no explanation can be found. It appears to reflect the progressive depletion of oocytes and the loss of their inductive effect upon the granulosa and thecal cells. With a diminution of the steroid feedback influence upon the hypothalamus or pituitary, the gonadotropin titer rises and the release of FSH and LH is no longer coordinated. Termination of menstrual function before 40 years of age may be considered premature. Symptoms associated with menopause can be variable. The most common is the occurrence of ''hot flashes,'' which occurs in 85% of the patients. In patients in whom prior cancer of the breast or uterus has been excluded, estrogen therapy is usually given to relieve hot flashes and delay the onset of osteoporosis.

Secondary ovarian hypofunction. Secondary ovarian hypofunction is characterized by decreased estrogen and progesterone levels in association with decreased gonadotropin levels. It may be attributable to hypothalamic, pituitary, or constitutional disturbances. Hypothalamic disorders that result in hypofunction of the ovary include the Lawrence-Moon-Bardet-Biedl syndrome, which is a congenital disorder, and the Frommel-Chiari syndrome, which occurs after childbirth. Emotional strain appears to alter the afferent neural input into the hypothalamic centers controlling the secretion of LHRH and is perhaps the commonest cause of secondary amenorrhea, apart from pregnancy.

Tumors of the pituitary and necrosis resulting from postpartum hemorrhages (Sheehan's syndrome) are the most frequent pituitary lesions resulting in ovarian insufficiency. Granulomatous diseases may also result in diminished pituitary function.

Severe constitutional illnesses such as congenital heart disease, chronic renal disease, or rheumatoid arthritis may cause amenorrhea and sterility. Dieting with too rapid weight loss, anorexia nervosa, poorly controlled diabetes mellitus, and hyperthyroidism may also impair reproductive function. The exact mechanism by which these different disorders affect the hypothalamic-pituitary-ovarian axis is not known.

Ovarian hyperfunction

Primary ovarian hyperfunction. The main cause of primary ovarian hyperfunction is estrogen-secreting tumors. Granulosa and thecal cell tumors are the most common of the estrogen-producing tumors. Approximately 5% arise before puberty, 55% during the period of reproductive life, and 40% after menopause. The symptoms depend on the age of the patient. Precocious puberty and intermittent uterine bleeding result from function of such tumors during the premenarchal years. Irregular uterine bleeding frequently alternating with periods of amenorrhea is common during the active reproductive life. Uterine bleeding is the characteristic manifestation of these tumors during the postmenopausal years. Primary ovarian hyperfunction results in decreased levels of FSH and LH because of increased negative feedback on the hypothalamic-pituitary axis.

Secondary ovarian hyperfunction. Secondary ovarian hyperfunction is characterized by increased levels of gonadotropins resulting in increased estrogen secretion. Sexual precocity from pituitary stimulation of ovarian function is generally idiopathic in origin. Gonadotropin secretion can usually be reduced, with regression of the secondary sexual characteristics, by intramuscular injections of progesterone. An unusual form of precocious puberty is associated with hypothyroidism, where the ovary has increased sensitivity to endogenous gonadotropins. The precocity can be reversed by treatment of the hypothyroidism. These observations contrast with the effects of hypothyroidism on adult women who commonly experience a failure of ovulation.

Miscellaneous conditions

Hirsutism. Hirsutism is usually defined as more hair than is cosmetically acceptable to a woman living in a certain culture. Many women with increased hair growth are not found to have a gross hormonal abnormality. Increased body hair in women is commonly attributable to familial traits. Idiopathic hirsutism and polycystic ovary syndrome make up the largest group of the remaining women. Simple increase in hair may be caused by androgen production from any source, or by an increase in the sensitivity of the hair follicles to androgens, or by other abnormalities.

The hormonal abnormalities leading to hirsutism are heterogeneous. Excess androgens may be produced by the ovaries, by the adrenal glands, or by peripheral conversion of prohormone, or by all these sources. There is considerable controversy at present whether ovarian, adrenal, or mixed disorders predominate.

On the other hand, virilism (frontal balding, hirsutism, deepening of the voice, and acne) is almost always associated with excess testosterone, usually from an ovarian source, such as the rare testosterone-secreting tumors of the ovary. An arrhenoblastoma is the most common masculinizing tumor of the ovary.

Polycystic ovary syndrome (Stein-Leventhal syndrome). The Stein-Leventhal syndrome is characterized by bilateral polycystic ovaries in association with infertility and anovulatory menstrual irregularities. It may present initially with oligomenorrhea, hirsutism, or both, in varying degrees. This disorder is one of the most confusing of all endocrine diseases and thus far has lacked a concrete definition. According to some, the diagnosis requires the presence of bilaterally enlarged ovaries characteristically with multiple bluish follicular cysts. There have been described numerous patients who cannot be differentiated from this group on clinical grounds in whom the ovaries are normal or only slightly enlarged. The anatomic findings, the laboratory abnormalities, and the clinical presentation of these women span a wide spectrum from near normal to extremes. Obesity is common but is not a criteria for diagnosis. Plasma testosterone levels (total) and urinary 17-ketosteroid excretion are usually at or slightly above the upper limits of normal. LH levels may be elevated, with normal or low levels of FSH. A ratio of LH to FSH over 1.5 would support the diagnosis of polycystic ovary syndrome.

FUNCTION TESTS

The assessment of the functional status of an endocrine gland is an important step in the diagnosis of an endocrine disorder. The endocrine gland is not an autonomous entity in regard to both its secretory and functional activities. The gonad is a part of the hypothalamic-pituitary-gonadal axis (control system), which produces a hormone, which in turn interacts with the target tissue (compliance system). As a result of this complex system, the cause of an endocrine disorder may rest within any one of the components of the control system or the compliance system. Differential diagnosis involving a series of stimulation and suppression tests may be necessary to establish the proper diagnosis of these endocrine disorders. The following are provocative tests of hormonal function.

Gonadotropin-releasing hormone (GnRH) test or LH-releasing hormone (LHRH) test

This test is performed to assess the status of gonadotropin (FSH-LH) secretion by the pituitary gland. A dose of synthetic LHRH is given intravenously. Venous blood is obtained 30 minutes and 15 minutes before the drug is given to serve as control and every 15 to 30 minutes thereafter for 3 hours. The blood samples are analyzed for FSH and LH concentrations. A rise in gonadotropins normally occurs promptly, with a peak in serum LH occurring in about 30 minutes and in serum FSH about 45 minutes after the administration of LHRH. This rise in disease-free persons suggests the presence of a normally responsive pituitary gland. (There should, however, be little response in children before puberty.) In postpubertal patients with hypogonadotropic hypogonadism, a depressed or absent response to the LHRH test is strong evidence that the condition is secondary to hypopituitarism. A normal LHRH test in a postpubertal hypogonadotropic hypogonadal patient excludes hypopituitarism as the cause of the hypogonadism. The possibility that abnormality may be the result of a defect in release of LHRH by the hypothalamus (such as seen in Kallman's syndrome, Lawrence-Moon-Bardet-Biedl syndrome and Frommel-Chiari syndrome) (see p. 802) may be confirmed with the clomiphene citrate test (see below). Note that patients with hypothalamic lesions may have impaired response to a single injection of LHRH, whereas repetitive administration of LHRH may result in a normal pituitary response.

Clomiphene citrate test

The clomiphene citrate test is another means of assessing the hypothalamic-pituitary-gonadal axis. Clomiphene citrate (Clomid), a nonsteroidal agent, probably acts by competitive inhibition of estradiol at hypothalamic receptor sites. As a result of the decreased binding of estradiol to the receptor sites, the hypothalamus ''senses'' less gonadal steroids to be present. This activates the negative-feedback relationship, with release of LHRH, resulting in an increased secretion of FSH and LH by the pituitary.

The rise in gonadotropins thus depends on an intact hypothalamus and intact pituitary as well. (Before puberty there should be no response to clomiphene administration.) Blood is obtained before administration of the drug. The drug is administered for 5 to 10 days. Blood is obtained on the first and last day of the treatment and analyzed for FSH and LH. A positive response of the test is indicated if there is a twofold rise in serum LH. A diminished response to clomiphene with a good response to the LHRH test possibly indicates that a hypothalamic lesion is the basis for secondary (hypogonadotropic) hypogonadism.

CHANGE OF ANALYTE IN DISEASE

(Table 41-1)

Gonadotropins (LH and FSH)

Measurement of serum LH and FSH concentrations provide important indices of hypothalamic-pituitary axis function and helps distinguish primary from secondary gonadal dysfunction syndromes. Remember that normal levels in the female vary widely throughout the reproductive cycle (refer to Fig. 41-4). In the male, pulsatile fluctuations of LH are commonly noticed, but plasma FSH levels generally remain fairly constant throughout the day. It is therefore advisable to obtain duplicate or triplicate sampling of the blood at approximately 30-minute intervals. Assays of LH and FSH are performed by radioimmunoassay and are

Table 41-1. Change of analyte in disease

Disease	Analyte				
	FSH	LH (ICSH)	Estradiol	Progesterone	Testosterone
Female					
Primary ovarian hypofunction (including menopause)	↑	↑	↓	↓	—
Secondary ovarian hypofunction	↓	↓	↓	↓	—
Primary ovarian hyperfunction—feminizing tumors	↓	↓	↑	—	—
Masculinizing tumors	—	—	—	—	↑
Secondary ovarian hyperfunction	↑	↑	↑	↑	—
Polycystic ovary syndrome	nl, ↓	↑	—	—	sl ↑
Male					
Primary hypogonadism	↑	↑	—	—	↓
Secondary hypogonadism	↓	↓	—	—	↓
Primary hypergonadism	↓	↓	—	—	↑
Secondary hypergonadism	↑	↑	—	—	↑

↑, Increased; ↓ decreased; *nl,* normal; *sl,* slightly.

expressed in terms of international units of a reference preparation of human menopausal gonadotropin or in nanograms of a human pituitary FSH and LH standard.

Increased levels of gonadotropins. Increased levels of FSH and LH in the female are seen in primary ovarian hypofunction such as Turner's syndrome or menopause. Increased levels are also seen with secondary ovarian hyperfunction. In polycystic ovary syndrome, LH levels may be elevated with normal or low levels of FSH. A ratio of LH to FSH over 1.5 would support the diagnosis of polycystic ovary syndrome.

Increased levels of FSH and LH in the male may be attributable to either primary hypogonadism or to secondary hypergonadism.

Decreased levels of gonadotropins. Low circulating levels of LH or FSH, or both, in a male or female with gonadal failure indicate that the primary defect is either with the hypothalamus or in the pituitary.

Androgens

Increased androgens. Increased androgens may be seen in a female with masculinizing tumors or with idiopathic hirsutism. Slightly elevated levels may be part of the polycystic ovarian syndrome. Elevated androgen levels in the male may be seen in primary or secondary hypergonadism.

Decreased androgens. In males, decreased levels of androgens signify testicular hypofunction, either primary or secondary.

Estrogens

The primary secretory product of the ovary is estradiol. Estrone is the other major biologically important estrogen. For women in the reproductive age, the ratio of estrone to estradiol should be less than 1 except at the time of the midmenstrual cycle peak. A ratio over 2 indicates that ovarian function is probably abnormal. In the postmenopausal woman, estrone is the major plasma estrogen. Elevated estriol levels are seen in women with pregnancy (see Chapter 35).

Increased levels of estrogen. Increased levels of estrogen in the female are seen in all cases of ovarian hyperfunction, whether they are secreted from a feminizing tumor or secondary to hypothalamic-pituitary-ovarian axis dysfunction. In the male, increased levels of circulating estradiol and estrone may reflect not only aromatization and secretion by the testis and adrenal but also altered aromatization activity elsewhere, as in liver disease, which may in turn affect normal testicular function. Increased levels in the male are usually manifest as gynecomastia.

Decreased levels of estrogen. In the female, decreased levels of estrogen may be seen with ovarian hypofunction, either primary or secondary. Decreased levels of estrogen in the male are not known to be of any clinical significance.

Progesterone

Progesterone levels in postpubertal premenopausal females vary with the phase of the reproductive cycle. Progesterone levels in males and prepubertal or postmenopausal females should be relatively stable. Progesterone levels are determined by radioimmunoassay. A single progesterone level of about 3 ng/mL, obtained between 4 and 10 days before the onset of the next menstruation, is presumptive evidence of ovulation.

Increased progesterone. Increased progesterone levels may be seen in pregnancy or in secondary ovarian hyperfunction.

Decreased progesterone. Decreased progesterone levels are seen in primary and secondary ovarian hypofunction.

BIBLIOGRAPHY

Crigler, J., Rose, L., and Rosenburg, E.: Diseases of the testes. In Thorn, G.W., et al.: Harrison's principles of internal medicine, New York, 1977, McGraw-Hill Book Co.

Erikson, G.: Normal ovarian function, Clin. Obstet. Gynecol. **21**:31, 1978.

Forest, M.G., De Peretti, E., and Bertrand, J.: Hypothalamic-pituitary-gonadal relationships in man from birth to puberty, Clin. Endocrinol. **5**:551, 1976.

Ganong, W.: The gonads: development and function of the reproductive system. In Review of medical physiology, Los Altos, Calif., 1977, Lange Medical Publications.

Kase, N.: Steroid synthesis in abnormal ovaries: polycystic ovaries, Am. J. Obstet. Gynecol. **90**:1268, 1964.

McArthur, J.: Disease of the ovary. In Thorn, G.W., et al.: Harrison's principles of internal medicine, New York, 1977, McGraw-Hill Book Co.

Makin, H., editor: Biochemistry of steroid hormones, Oxford, England, 1975, Blackwell Scientific Publications.

Novak, E., Jones, G., and Jones, H.: Gynecology, Baltimore, 1975, The Williams & Wilkins Co.

Orten, J.M., and Newhaus, O.W.: Human biochemistry, St. Louis, 1982, The C.V. Mosby Co.

Sherman, B.M., West, J.H., and Korenman, S.G.: The menopausal transition, J. Clin. Endocrinol. Metab. **42**:629, 1976.

Snell, R.S.: Clinical anatomy for medical students, Boston, 1973, Little, Brown & Co.

Speroff, L.: The ovary. In Felig, P., Baxter, J., Broadus, A.E., and Frohman, L.A., editors: Endocrinology and metabolism, New York, 1981, McGraw-Hill Book Co., Chapter 17.

Stryer, L.: Biochemistry, San Francisco, Calif., 1975, W.H. Freeman & Co.

Troen, P., and Oshima, H.: The testis. In Felig, P., Baxter, J., Broadus, A.E., and Frohman, L.A.: Endocrinology and metabolism, New York, 1981, McGraw-Hill Book Co., Chapter 16.

Watts, N., and Keffer, J.: Reproductive endocrinology. In Practical endocrine diagnosis, Philadelphia, 1982, Lea & Febiger.

Wentz, A.C.: Clinical applications of LHRL, Fertil. Steril. **28**:901, 1977.

Yen, S., and Lein, A.: The apparent paradox of the negative and positive feedback control system on gonadotropin secretion, Am. J. Obstet. Gynecol. **126**:942, 1976.

Chapter 42 Parathyroid gland

Richard J. Kozera

chief cells Portion of the parathyroid gland actively synthesizing and secreting parathyroid hormone (PTH).

hyperparathyroidism Condition defined by elevated serum PTH levels caused by hypersecretion of the parathyroid gland (primary hyperparathyroidism) or secondarily in response to decreased serum calcium levels.

hypoparathyroidism Condition defined by serum parathyroid hormone levels that are decreased from established reference ranges. Can be caused by primary parathyroid dysfunction or secondarily by elevated serum calcium levels.

ionized calcium The ionized, unbound, noncomplexed fraction of serum calcium that is biologically active.

multiple endocrine neoplasia Simultaneously occurring tumors of different endocrine glands.

osteoblasts Bone cells responsible for laying down new bone material.

osteoclast activating factor Factor released from certain tumors that enhances osteoclastic activity and results in increased levels of serum calcium. The factor is not parathyroid hormone.

osteoclasts Bone cells responsible for bone resorption.

pre-proparathyroid hormone (pre–pro-PTH) The initial translation product of the parathyroid hormone gene; this 13,000 molecular weight compound is degraded to a second precursor, pro-PTH.

pro-PTH An immediate precursor to active parathyroid hormone formed by cleavage of pre–pro-PTH.

pseudohypoparathyroidism Clinical conditions of hypocalcemia caused by cellular resistance to PTH. This includes pseudo-pseudohypoparathyroidism.

PTH resistance Cellular dysfunction in which PTH molecules do not elicit a normal response from target cells.

tetany A muscular spasm that is characteristic of hypocalcemic conditions.

ANATOMY

Human parathyroid glands are small, ovoid, yellowish brown structures. They usually number four but may be as few as two and as many as 12. Each gland is about 6 mm by 4 mm by 2 mm and weighs about 30 mg. They are paired structures usually embedded in the posterior capsule of the thyroid gland but may be located in other parts of the neck or upper chest.

The glands are populated with cells of epithelial origin. They are divided into two basic groups. In the first group are two subpopulations of chief cells, one that is lighter staining and considered to be inactive or resting and one that stains more darkly because it is densely packed with secretory granules and considered to be actively producing hormone. The second group of cells are the oxyphils, which are acidophilic, abundant in granules, and loaded with mitochondria. Their function is unknown.

PHYSIOLOGY

Biosynthesis of parathyroid hormone

The biosynthesis of parathyroid hormone provides another example of the intracellular process of protein synthesis. It encompasses a large number of complex reactions in which information originally encoded in the language of the gene (DNA) is finally expressed in the polyamino acid language of the biologically active PTH. This takes place in four sequences. The first of these is the process of *transcription* in which RNA is synthesized from the DNA template. The second step is *posttranscriptional modification,* with messenger RNA formed from precursor RNA by excision and rejoining of RNA segments. The third is *translation,* with assembly of the amino acids by appropriate pairings with amino acids on messenger RNA followed by polymerization of the newly assembled amino acids into a polypeptide chain. The last sequence is *posttranslational modification,* with rearrangement of molecular configuration. The end product of this process is the 115 amino molecule pre-proparathyroid hormone, which has a molecular weight of 13,000 daltons.

As translation proceeds, the first 25 amino acids are cleaved off, leaving a 90 amino acid product, pro-PTH, with a molecular weight of 10,200 daltons. This is transported through the endoplasmic reticulum to the Golgi apparatus. The first six amino acids of pro-PTH are cleaved, yielding the 84 amino acid parathyroid hormone (Fig. 42-1), which is packaged in granules and either stored for subsequent secretion or degraded. Found with PTH in the storage granules is an acidic glycoprotein, parathyroid secretory protein (PSP), which is also synthesized by the chief cells. It exists as a dimer of molecular weight 140,000 daltons and is associated with six molecules of parathyroid hormone. Its function is unknown but may relate to transport within the parathyroid cell. The entire process of synthesis to storage of PTH takes about 15 minutes.

Secretion of parathyroid hormone

Biological activity of the parathormone molecule resides in about the first 30 amino acids of its amino terminal end. PTH seems to be secreted intact but is quickly cleaved into an amino-terminal fragment with a half-life of only a few

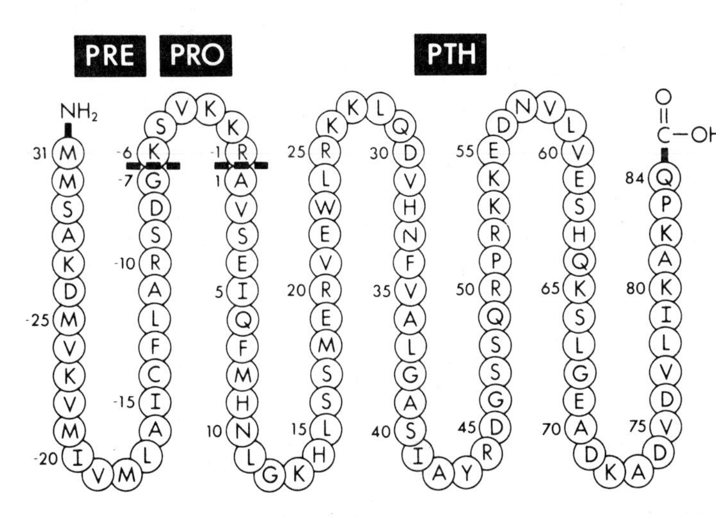

Fig. 42-1. Amino acid sequence of bovine pre–pro-PTH. Dashed lines separate "pre" and "pro" amino acid segments and native hormone (PTH, parathormone). (From Cohn, D.V., and MacGregor, R.R.: Endocrine Rev. 2:1, 1981; copyright The Endocrine Society.)

minutes and a carboxy-terminal fragment with a half-life of an hour or more. Most carboxy-terminal fragments are biologically inactive but constitute the largest pool of parathyroid hormone fragments in the circulation. Other biologically inactive fragments of parathyroid hormone may be found in the circulation in varying but small quantities. These fragments are mostly produced during the peripheral degradation of PTH and are not believed to be direct secretory products of the chief cell. There is some evidence that other precursors, intermediaries, and associates of PTH metabolism within the cell may be secreted, but the magnitude and importance of this is poorly understood. Unlike many other proteins that circulate in a free and bound (carrier) state, PTH exists as freely circulating intact molecules and fragments.

Regulation of parathyroid hormone secretion

The single most important determinant of the release of PTH from the parathyroid glands is the serum concentration of ionized calcium, which is usually adequately reflected in the concentration of total serum calcium. Parathyroid hormone secretion is stimulated as the calcium concentration falls and is suppressed as the calcium concentration rises. There is a sharp increase in the secretion rate of PTH when the total serum concentration falls below 90 mg/L. The rate is maximal after an acute fall of serum calcium to 70 or 80 mg/L. The change in PTH secretion rate is linear, with changes of calcium concentrations between 90 and 105 mg/L. When calcium concentrations are greater than 105 mg/L, the PTH secretion rate is greatly diminished but not completely suppressed. Studies of subjects with hyperplastic or neoplastic parathyroid glands have shown similar relationships between PTH secretion or serum concentration and serum calcium concentration, but higher calcium concentrations were required to suppress parathyroid hormone secretion in these abnormal glands.

Other substances are probably best regarded as modifiers of the calcium-PTH interaction. The most important of these is magnesium, which has an effect on PTH that both parallels and modulates the effect of calcium. In healthy persons and in those with hyperparathyroidism who have normal total body magnesium stores, an acute increase in the serum magnesium concentration will, if anything, decrease the amount of circulating parathyroid hormone. In subjects who are both hypocalcemic and hypomagnesemic the PTH concentration increases with administration of magnesium regardless of whether the PTH was low, normal, or high to begin with. This suggests that magnesium at very low concentrations probably suppresses PTH secretion. The likely explanation for this is that the generation of cyclic AMP by adenylate cyclase, a magnesium-dependent process, is inhibited, blocking the cyclic AMP–dependent release of parathyroid hormone. In hypomagnesemic states PTH is quickly released with replenishment of magnesium. It is probable that this is a release from prepackaged stores and that the synthetic processes through the point of storage are not magnesium dependent.

Persons with high circulating levels of PTH who develop hypomagnesemia show a decreased effect of PTH at the target cell. In this circumstance the low magnesium level inhibits generation of cyclic AMP by the target cell, thereby inducing a state of tissue resistance.

The role of biogenic amines in PTH secretion is somewhat unclear. Generally, substances that raise the level of cyclic AMP in the parathyroid cells tend to enhance PTH secretion and those that lower cyclic AMP tend to inhibit PTH secretion. Cyclic AMP levels may be raised by prostaglandin E_1, prostaglandin E_2, dopamine, cholera toxin, aminophylline, dibutyryl cyclic AMP, and the beta-agonists epinephrine, nonepinephrine, and isoproterenol. Cyclic AMP levels may be lowered by beta antagonists (propranolol), alpha agonists (methoxamine), prostaglandin F_2, and nitroprusside. There may be different responses to varying concentrations of the same drug. For example, isoproterenol at low concentrations has been shown to stimulate PTH release but at higher concentrations has had no effect on PTH release. Although there are several explanations for this phenomenon, including decreased blood flow and some masking by simultaneous alpha-agonist effects, it indicates the need to separate physiological from pharmacological effects of biologically active agents.

Although vitamin D is discussed extensively in Chapter 25, it is important to recognize here that there is probably direct interaction between vitamin D metabolites and parathyroid gland function. Parathyroid cell receptors for 1,25-dihydroxycholecaliciferol (calcitriol) have been demonstrated in several species. Some investigators have shown effects of vitamin D metabolites on serum calcium concentration, parathyroid gland size, and modulation of synthesis or secretion of parathyroid hormone. The role of these interactions in humans still remains to be clearly defined.

Physiological actions of parathyroid hormone

Mineral homeostasis is governed in large part through the action of parathyroid hormone on its two major target organs, bone and the kidney. The exact nature of the interaction between target cell and PTH is unknown, but biological activity is initiated by the interaction between PTH and cellular receptors and is almost certainly mediated through the generation of cyclic AMP within the target cell. This has been most clearly demonstrated in renal tissue. Administration of exogenous PTH to humans or enrichment of culture media results in an increase in nephrogenous cyclic AMP production 10 to 70 times the normal level. Dibutyryl cyclic AMP in vivo reproduces the effects of PTH on calcium and phosphorus handling by the kidney. Phosphodiesterase hydrolyzes cAMP. Phosphodiesterase inhibitors enhance, and phosphodiesterase stimu-

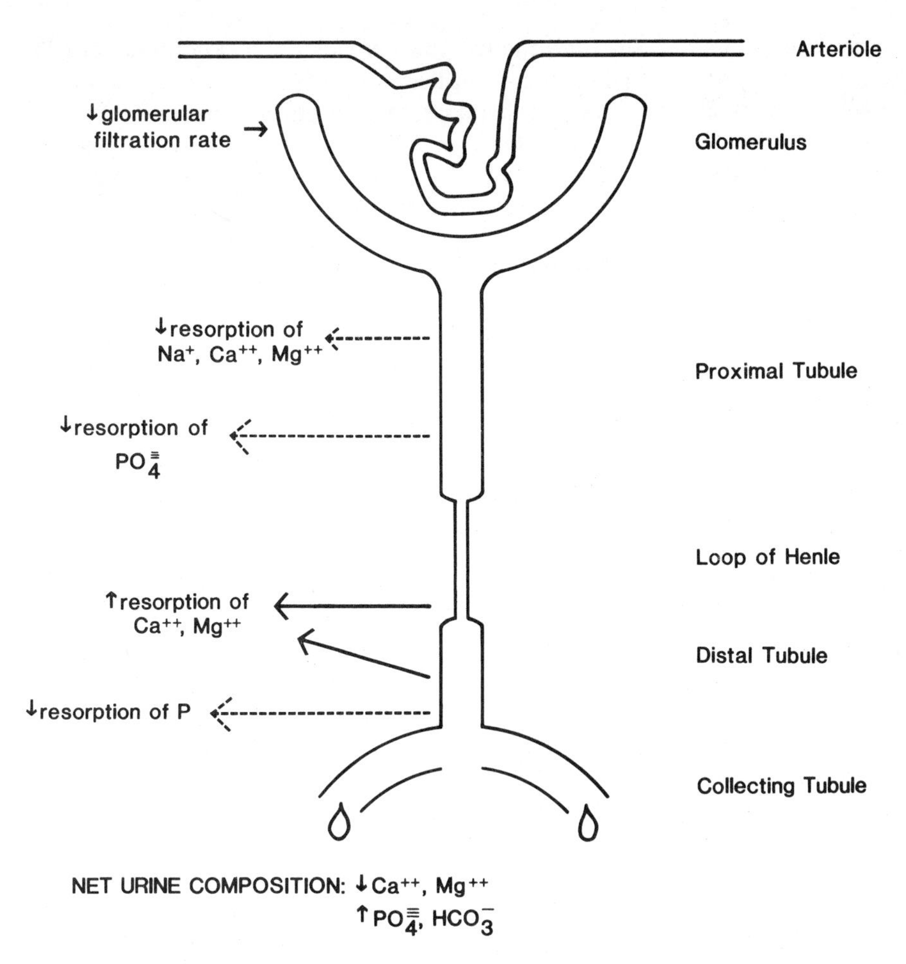

Fig. 42-2. Effect of PTH on renal tubules. *Arrows,* Effect of PTH on each analyte at each segment of tubule.

lators decrease, PTH effects. There are less data on bone, but in vitro studies of bone cells suggest similar PTH effects.

Despite the ignorance of molecular mechanisms of PTH action, the physiological effects have been well described. Although the control of calcium and phosphorus in the kidney depends on a complex interrelationship between PTH, vitamin D, calcium, and phosphorus, the net effect of PTH on the renal tubule is calcium and magnesium retention and phosphate and bicarbonate loss (Fig. 42–2). Specifically, PTH causes increased resorption of both calcium and magnesium in the ascending limb of the loop of Henle and in the distal tubule. It may cause a slight decrease in the proximal tubular resorption of isotonic fluid, which contains sodium, calcium, and magnesium. It causes decreased phosphate reabsorption in the proximal tubule and possibly in the distal tubule. It also causes a slight decrease in the glomerular filtration rate.

The effect of PTH on bone is equally complex and again requires the presence of the hormone, an appropriate receptor system, vitamin D metabolites, calcium, and phosphorus. As mentioned in Chapter 25, there are three major types of cells found in bone: osteoblasts, responsible for the provision of new bone; osteoclasts, primarily responsible for bone resorption; and osteocytes, which have many complex functions. There are two primary effects of PTH. The first is mediated by the osteoclasts and over a period of time results in bone remodeling. The second is probably mediated through the osteocytes, and an important result is the rapid efflux of calcium from bone into the extracellular fluid. These processes are interdependent and help serve both the acute minute-to-minute maintenance of mineral homeostasis and the long-term demands of proper bone structure.

PATHOLOGICAL CONDITIONS

Calcium

States of abnormal parathyroid function most often present with signs and symptoms of hypocalcemia or hypercalcemia. The measurement of serum calcium therefore assumes major importance. Remember that calcium circulates in both an ionized (''free'') fraction that is biologically active and a fraction that is chiefly ''bound'' to serum albumin. Each fraction composes about half the total serum calcium. For most clinical purposes the total serum calcium varies in proportion to the ionized fraction and is considered a good indicator of the level of ionized calcium. In the following sections, the terms "hypocalcemia" and "hypercalcemia" are used to indicate a decrease or increase in both total and ionized serum calcium concentrations.

Hypocalcemia

Tetany, or muscle spasm, is the most characteristic of the major signs and symptoms of acute or chronic hypocalcemia. It is the direct result of enhanced neuromuscular excitability. One may demonstrate latent tetany by (1) lightly tapping the seventh cranial (facial) nerve just in front of the ear and observing the muscle contractions around the eye, nose, and mouth (Chvostek's sign) or (2) inflating a blood pressure cuff applied around the upper arm to levels above the systolic pressure for 3 minutes, causing local ischemia to the nerve trunk and resulting in severe flexion at the wrists (carpal spasm, Trousseau's sign).

Other clinical findings are largely related to chronic hypocalcemia, which may arise from a variety of conditions including hypoparathyroidism. Emotional or behavioral symptoms can range from mild irritability to frank psychosis. Skin changes can include mild dryness, brittle nails, severe dermatitis, and psoriasis. The scalp hair may be coarse and brittle or fall out in scattered patches (alopecia areata). The lenses of the eyes may develop cataracts. If hypocalcemia is present early in childhood, the teeth may fail to erupt, be underdeveloped, or show poor enamel formation. The heart may develop both contractile and electrical abnormalities. Hypotension or heart failure may occur, probably the result of defective actin and myosin coupling, which is a calcium-dependent process. The electrocardiogram may show a prolonged S-T segment, representing a longer interval between the end of depolarization and the beginning of repolarization. Gastrointestinal manifestations include steatorrhea, gastric achlorhydria, and impaired absorption of vitamin B_{12}. Mucocutaneous infection by *Candida albicans* and the bony changes of osteomalacia are more related to PTH deficiency and vitamin D deficiency than they are to hypocalcemia itself.

Hypocalcemic disorders

Clinical disorders resulting in hypocalcemia may be grouped into those associated with deficient PTH concentration or activity and those that have a primary effect on calcium deposition or excretion that is largely independent of PTH activity. Vitamin D–deficiency states are of major importance in both the classification and pathogenesis of hypocalcemia but are discussed elsewhere in this book. A bewildering terminology has developed for the classification of hypoparathyroid disorders. It is reasonable to consider them in two large groups: (1) hyposecretion of PTH and (2) tissue resistance to PTH.

Primary hypoparathyroidism. Hyposecretion of PTH or primary hypoparathyroidism is most commonly seen after neck surgery. This may be treatment for thyroid disorders such as goiter, malignancy, or hyperthyroidism, for hyperparathyroidism, or for nonendocrine malignancies with or without primary involvement of endocrine tissues. The clinical manifestations are those of hypocalcemia. The severity and duration of the disorder depends on the extent of surgery and the amount of parathyroid tissue injured or removed. The condition may be transient, usually the result of trauma, or permanent, caused either by extensive trauma or removal of excessive tissue. If transient, it may be asymptomatic and recognized only by a drop in the serum calcium concentration, or it may be more severe and cause latent or obvious tetany.

Radiation therapy has occasionally been implicated as a cause of hypoparathyroidism. The source of irradiation may be an external beam usually used for treatment of malignancy or radioactive iodine (^{131}I) commonly used for treatment of hyperthyroidism or thyroid malignancy.

Idiopathic hypoparathyroidism is an unusual condition that can be either sporadic or familial and occurs either independently or linked to other endocrine deficiency states. The serum calcium concentration tends to be quite low, averaging around 55 mg/L and the serum phosphorus high, averaging around 75 mg/L. Patients are affected early in life and present primarily with symptoms of hypocalcemia, which may be quite severe, even to the point of generalized seizures or asphyxia caused by laryngeal obstruction. There is often associated mental retardation. Most bases of polyendocrine failure (that is, hypofunctioning of several endocrine glands simultaneously) are familial and are believed to show an autosomal recessive inheritance. Associated disorders include failure of the adrenals, thyroid glands, or ovaries, mucocutaneous candidosis, steatorrhea, myasthenia gravis, cirrhosis, alopecia, and vitiligo.

Tissue resistance to PTH. Syndromes of PTH resistance are divided into two groups. Both show normal or increased circulating levels of PTH, and both involve multiple organ systems. In the first group, pseudohypoparathyroidism (PHP), there is lack of the expected renal responsiveness to PTH. The associated hypocalcemia is not caused by a decrease in PTH synthesis but by a lack of cellular response to PTH. The group is further subdivided into two types. PTH administered to subjects with PHP type I does not cause an increase in nephrogenous cAMP. PTH administered to subjects with type II PHP does increase nephrogenous cAMP, but this does not result in the expected increase in calcium reabsorption or phosphorus excretion. In both types the clinical presentation can vary considerably depending on the organ systems involved but most commonly includes a round face, short stature, stocky habitus, mental retardation, thickening of the calvarium, short metacarpals or metatarsals, ectopic calcification or ossification, and the signs and symptoms of hypocalcemia.

The second group, pseudo-pseudohypoparathyroidism (PPHP), demonstrates normal renal responsiveness to PTH, including stimulation of nephrogenous cAMP, increased calcium resorption, and increased phosphorus excretion. The clinical presentation excludes the signs and symptoms of hypocalcemia, but the physical habitus is that of PHP as described above. This is probably caused by resistance of selected tissues other than the kidney to the effects of PTH.

There is at least one report in the literature of a patient with manifestations of hypoparathyroidism that were caused by the release of PTH that was immunologically but not biologically active. The major signs and symptoms that developed were those of hypocalcemia and not parathyroid hormone resistance. The disorder has been termed "pseudoidiopathic" because of these features.

As noted previously, hypomagnesemic states will lead to ineffective generation of AMP by PTH at the target cell level and may also result in defective PTH production or release from the parathyroid gland. This can be seen in malabsorption states, renal losses of magnesium (use of diuretics), chronic alcoholism, diabetic ketoacidosis, and, occasionally, hyperthyroidism. The tissue resistance is reversible on replenishment of magnesium to serum levels above 10 mg/L.

Other causes of hypocalcemia. Hypocalcemia is a common finding in acute pancreatitis. The reason for this is unclear, but suggested mechanisms have included the formation of calcium soaps within the pancreas or in other soft tissues, inappropriate release of PTH, abnormal metabolism of PTH, or activation of other hormones including gastrin, glucagon, and calcitonin. The hypocalcemic effect of these hormones has been caused only by their administration in pharmacological rather than physiological doses. These hormones are therefore less likely causes of the hypocalcemia.

Persons with hyperparathyroidism resulting in bone and mineral loss who undergo curative surgery may have severe postoperative hypocalcemia not related to PTH insufficiency but rather to have an increased rate of deposition of calcium within the bone. This is usually a temporary problem that resolves in a matter of days or weeks with administration of adequate amounts of calcium.

Conditions resulting in hyperphosphatemia, including major surgery, cytotoxic therapy, or renal failure may cause hypocalcemia either by local effects on bone, suppression of formation of 1,25-dihydroxycholecalciferol, or precipitation of calcium phosphate complexes within soft tissues.

Drugs may cause hypocalcemia in a variety of ways: inhibiting resorption of calcium and bone matrix, inhibiting calcium resorption from bone but leaving the matrix undisturbed, promoting calcium deposition in bone and soft tissues, or chelating calcium within the blood. Some of the more important hypocalcemia-causing agents are estrogens, mithramycin, calcitonin, fluoride, CPDA1 (a citrate-phosphate-dextrose–containing preservative used in storing blood), and ethylene glycol. Diuretics that are active in the loop of Henle may cause sufficient calciuresis, resulting in hypocalcemia. States of vitamin D deficiency, rickets, or osteomalacia are commonly associated with hypocalcemia.

Hypocalcemia is commonly seen in the perinatal period. In the first few days of life it is most likely caused by transient developmental hypoparathyroidism. If it occurs 2 to 3 weeks after birth, it is usually caused by excess ingestion of phosphorus or immaturity of the renal tubule.

Hypercalcemia

Generally, a hypercalcemic disorder is suspected if the total serum calcium concentration is above 102 mg/L on repeated occasions but may also be suspected if on repeated determinations the serum calcium fails to fall below the level of 100 mg/L. The obtaining of proper specimens for serum calcium determination is perhaps most important when a hypercalcemic disorder is suspected and the serum calcium is in the range of 100 to 110 mg/L. The sample should be collected from a subject, in a relaxed state for at least 30 minutes, who has been fasting for 4 hours. No tourniquet should be used. Collection under other conditions can increase the serum calcium level by as much as 5 to 10 mg/L.

The signs and symptoms of subjects with hypercalcemia may be related to the hypercalcemia itself or to the underlying disease process causing the hypercalcemia, or both. Mild hypercalcemia may be entirely asymptomatic and an unexpected finding on routine serum screening. The prevalence of asymptomatic hypercalcemia is unknown but has been estimated as high as 1 in 1000. More commonly hypercalcemia is seen in association with a chronic disease process and may present in several ways: as polyuria and polydipsia caused by a defect in renal concentrating ability; as progressive renal insufficiency; as nephrolithiasis; in association with hypertension, ulcers, or pancreatitis; or as an acid-base disorder. Subjects with hypercalcemia but without hyperparathyroidism may have a metabolic alkalosis caused by direct stimulation of hydrogen-ion secretion by the distal renal tubule or may show a distal renal tubular acidosis. If hypercalcemia develops acutely, the subject may present with anorexia, nausea, vomiting, confusion, stupor, and coma.

There are no specific physical signs. Laboratory studies may provide clues. The electrocardiogram may show a shortening of the Q-Tc interval as a result of shortened S-T segment. At greatly elevated levels of serum calcium, however, there is prolongation of the T-wave and the Q-T interval may return toward normal or actually become prolonged. The S-T segment still remains short. There may be a metabolic acidosis or alkalosis as just discussed. The urine may show a lack of concentrating ability and may contain an excess of triple phosphate crystals. If renal stones are present, there may be hematuria.

A frequently difficult problem in clinical medicine is establishing the diagnosis of hyperparathyroidism in subjects with or without underlying disorders that may cause hypercalcemia. Determination of the serum PTH level assumes major importance in these cases.

Hypercalcemic disorders

Hypercalcemic disorders are broadly classified into five categories (see box below).

Hypercalcemia can be either a primary or secondary feature in the diseases listed below. The clinical presentation may be as subtle as an abnormal laboratory test in an asymptomatic subject or as severe as coma in a person with advanced malignancy. The disorders in the list that are responsible for most cases of hypercalcemia are malignancy and hyperparathyroidism.

Hypercalcemic disorders

- Malignancy
  - Multiple myeloma
  - Leukemia
  - Lymphoma
  - Carcinoma
- Endocrine disorders
  - Hyperparathyroidism
  - Hyperthyroidism
  - Adrenal insufficiency
  - Acromegaly
  - Paget's disease
- Drug-induced disorders
  - Vitamin D
  - Vitamin A
  - Thiazide diuretics
  - Lithium
  - Milk and alkali
- Granulomatous disease
  - Sarcoidosis
  - Tuberculosis
  - Berylliosis
  - Systemic mycosis
- Miscellaneous causes
  - Familial hypocalciuria
  - After transplantation
  - Recovery from acute renal failure

Primary hyperparathyroidism. Primary forms may be subclassified as adenomas or hyperplasia. Adenomas tend to involve single glands, whereas hyperplasia tends to involve multiple glands. Many published series in the past indicated that the frequency of adenomas in subjects with hyperparathyroidism was approximately 80%. More recent experience suggests that this may be an overestimate and that hyperplasia accounts for an increasing proportion of this population.

Hyperparathyroidism can exist as an independent condition or can be associated with other endocrine disorders in a category termed ''multiple endocrine neoplasia (MEN).'' Although MEN have been divided into specific groups, more overlap is being recognized among the groups. Type I MEN consists of lesions of the pituitary, pancreas, and parathyroid glands. Type II-A (Sipple's syndrome) consists of hyperparathyroidism, medullary carcinoma of the thyroid, and pheochromocytoma. Type II-B (also referred to as type III) consists of medullary carcinoma of the thyroid, pheochromocytoma, and mucosal neuromas. The histological appearance of the parathyroid glands in types I and II-A shows hyperplasia. The clinical manifestations are those of the underlying disease and hypercalcemia.

Symptomatic hyperparathyroidism presents predominantly as bone disease or renal stones. The renal stones may cause pain and hematuria or obstruction, hydronephrosis, and renal failure. The bone changes include osteoporosis and subperiosteal resorption. In the past a severe disorder termed osteitis fibrosa cystica was seen and was characterized by pain, fractures, and skeletal deformities. This has largely vanished from the clinical scene, presumably because of the earlier recognition and greater ease of diagnosis of parathyroid disorders. Nonspecific features of hyperparathyroidism include vague neuromuscular symptoms, mild lethargy, altered behavior, probably peptic ulcer disease, acute or chronic pancreatitis, nonspecific abdominal pain, gallstones, and hypertension. Central nervous system abnormalities are almost always seen at serum calcium levels of greater than 130 mg/L, with stupor and coma becoming prevalent as the serum calcium exceeds 170 mg/L.

Secondary hyperparathyroidism. Secondary hyperparathyroidism is regarded as a compensatory biological response to a hypocalcemic state. The most common example is seen in patients with renal failure whose kidneys become unable to excrete phosphorus. The serum phosphorus is increased leading to a secondary decrease in serum calcium. This stimulates the secretion of PTH in an attempt to restore the serum calcium concentration to normal, but this is done largely at the expense of demineralization of bone. This process can usually be kept in control with adequate treatment of renal failure, including dialysis and the use of substances that bind phosphorus within the gastrointestinal tract so that it cannot be absorbed. If the serum phosphorus can be maintained within the normal range, the serum calcium concentration will stay within the normal range. Some poorly controlled subjects may have a return of calcium to or above the normal range though the serum phosphorus is still elevated. This condition has been called ''tertiary hyperparathyroidism'' to signify an increased degree of autonomy of parathyroid gland function. It is suggested that the previously hyperplastic parathyroid tissue has now become autonomous, but such a pathogenetic mechanism is difficult to verify.

Other forms of secondary hyperparathyroidism are commonly seen in osteomalacic states, which are caused by lack of absorption of calcium through the gastrointestinal tract or excessive renal loss of calcium.

Hypercalcemia of malignancy. Hypercalcemia related to malignancy is most commonly seen in squamous cell carcinoma of the lung or of the head and neck but may also be seen in a variety of other solid tumors, including those of the breast, kidney, uterine cervix, or prostate. Of the ''liquid'' tumors, leukemia and lymphoma are infrequently associated with hypercalcemia, but multiple myeloma also has a frequent association. Factors responsible for tumor hypercalcemia include production of PTH or a PTH-like substance by the tumors, particularly those of squamous cell origin, production of prostaglandins, especially by breast tumors, and elaboration of a factor stimulating osteoclastic activity (OAF), which is seen in multiple myeloma. Demineralization of bone is caused by localized invasion by many tumors, both solid and liquid. Measurement of PTH in subjects with suspected ectopic PTH production may well not show elevated levels, since the PTH fragments may be immunologically different from native PTH. This is different from tumors that produce other polypeptide hormones because those polypeptides seem to be both biologically and immunologically similar to the naturally produced hormone.

The other conditions causing hypercalcemia listed in the box on p. 812 are generally diagnosed rather easily, and the hypercalcemia can be corrected by appropriate treatment of the underlying disorder. An interesting phenomenon is seen in some patients with sarcoidosis, a systemic granulomatous disease frequently accompanied by hypercalcemia. There is an increased rate of conversion of 25-hydroxycholecalciferol to the 1,25-dihydroxy form, which increases the transport of calcium from the gut to the extracellular fluid. It has been suggested but not proved that the granulomatous tissue itself is responsible for this biochemical alteration. Additionally, primary hyperparathyroidism and sarcoidosis can coexist, and measurement of serum PTH assumes major diagnostic importance, being normal or suppressed in sarcoidosis alone and elevated in the combination of sarcoidosis and hyperparathyroidism.

CHANGE OF ANALYTE IN DISEASE (Table 42-1)

Calcium

In ideal circumstances the normal daily human diet contains about 1000 mg of calcium. Each day about 825 mg is excreted in the feces and 175 mg in the urine. This is a net result, for there is an ongoing flux of serum calcium among the gut, extracellular fluid, bone, and kidney. Under normal circumstances there is considerable variation in urinary calcium excretion, which is primarily dependent on the amount of calcium ingested. It is usually less than 300 mg/day. More critical values may be obtained when the calcium excretion is expressed in terms of the amount of creatinine excreted or in terms of the glomerular filtration rate. Critical analyses must be made with the patient on a ''metabolic diet'' of known calcium content for a period of several days. Such determinations are not very practical in the diagnosis or management of hypercalcemic states but may be of some help in the diagnosis and management of hypercalciuric states unaccompanied by hypercalcemia.

In conditions resulting in decreased serum albumin concentration, the total serum calcium is not a good indicator of the metabolically important ionized fraction, which usually remains at normal levels. Albumin is the protein to which serum calcium is predominantly bound. Decreases in the serum albumin concentration result in proportional

Table 42-1. Change of analyte with disease

Analyte	Primary hyperparathyroidism	Secondary hyperparathyroidism (osteomalacia)	Primary hypoparathyroidism	Pseudohypoparathyroidism type I	Pseudohypoparathyroidism type II	Sarcoidosis
PTH-N terminal	Frequently normal	Normal to ↑	↓	↑	↑	↓
PTH-C terminal	↑	↑	↓	↑	↑	↓
Calcium	↑	Low to normal	↓	↓	↓	↑
Inorganic phosphorus	Normal to ↓	Normal to high	↑	↑	↑	Normal to ↓
Nephrogenous cAMP (urine)	↑	↑	↓	↓	?	↓
Nephrogenous cAMP response to PTH	—	—	↑	↓	↑	—
Urinary calcium excretion	↑	↓	↓	—	—	↑

↑, Increased; ↓ decreased.

decreases in total serum calcium concentration. A variety of formulas have been suggested to estimate the ionized calcium from the total calcium in states of altered albumin concentrations. At physiological pH, each gram of albumin binds about 0.8 mg of calcium. The normal range of serum albumin is about 35 to 55 g/L. The measured albumin concentration may be subtracted from 46 (''normal'' concentration), and the difference is multiplied by 0.8. The product is then added to the measured calcium concentration. This provides the ''corrected'' serum calcium concentration. In another method of calculation, the measured serum albumin (in grams per liter) may be subtracted from the measured serum calcium (in milligrams per liter) and 40 is added to the difference. A corrected value in the normal range for total serum calcium would indicate a normal circulating ionized calcium concentration. In states of acute acidosis or alkalosis the change in ionized calcium concentration is inversely proportional to the change in pH. In chronic states of acidosis or alkalosis, compensatory mechanisms usually allow for return of the ionized calcium to the normal range.

Phosphorus

Approximately 1400 mg of phosphorus is ingested each day, with a net excretion of 500 mg in the feces and 900 mg in urine. As with calcium there is a continuing flux among gut, extracellular fluid, bone, and kidney. Measurement of serum phosphorus is helpful in the diagnosis of hypoparathyroidism and may be helpful in discriminating among some hypercalcemic states. Measurement of urinary phosphorus is now rarely used in the diagnosis or management of hypercalcemic states.

Parathyroid hormone

The serum PTH concentration should be interpreted in light of the serum calcium concentration. Subjects with hypercalcemia caused by hyperparathyroidism usually but not always have elevated levels of PTH. Subjects with hypercalcemia from other causes may have PTH levels that are measurable and within the ''normal'' range or levels that are below the limits of detectability of the assay. Subjects with secondary hyperparathyroidism usually have low serum calcium values and elevated circulating PTH. Subjects with primary hypoparathyroidism rarely, if ever, have detectable PTH.

These phenomena demand that caution be taken in the interpretation of ''normal'' PTH levels, especially in the evaluation of hypercalcemic states. Thus the PTH values cannot be properly interpreted unless the serum calcium concentration is also known.

Efforts to clarify ambiguous or confusing PTH levels by stimulation or suppression of PTH secretion through lowering or raising of serum calcium levels have generally not provided diagnostic information that is more helpful than measurement of basal levels. Such studies have demonstrated, however, that either hyperplastic or adenomatous parathyroid tissue is responsive to changes in extracellular calcium concentrations but at a higher set point.

Nephrogenous cyclic AMP

About 40% of the cAMP that appears in the urine is produced by the kidney in response to PTH. The remainder is derived from the plasma as it is filtered by the glomerulus. It is primarily of adjunctive value in states of altered PTH production. In cases of suspected hyperparathyroidism when serum PTH levels are in the normal range, elevated urinary cAMP may more clearly define the diagnosis. In states of PTH resistance (pseudohypoparathyroidism), response of urinary cAMP to injected PTH can establish the diagnosis and type of pseudohypoparathyroidism.

Calcitonin

Calcitonin is secreted from the C cells of the thyroid gland. In animal species its behavior generally contrasts with that of PTH in response to changes in ionized calcium. A similar role in humans has not been clearly defined. The major diagnostic role of serum levels of calcitonin is in medullary carcinoma of the thyroid and not in states of abnormal parathyroid function.

BIBLIOGRAPHY

Physiology

Agus, Z.S., Wasserstein, A., and Goldfarb, S.: Disorders of calcium and magnesium homeostasis, Am. J. Med. **72**:473-488, 1982.

Cohn, D.V., and MacGregor, R.R.: The biosynthesis, intracellular processing, and secretion of parathormone, Endocr. Rev. **2**:1-26, 1981.

Epstein, F.H.: Calcium and the kidney, Am. J. Med. **45**:700-714, 1968.

Habener, J.F.: Responsiveness of neoplastic and hyperplastic parathyroid tissue to calcium in vitro, J. Clin. Invest. **62**:436-450, 1978.

Heath, H., III: Biogenic amines and the secretion of parathyroid hormone and calcitonin, Endocr. Rev. **1**:319-338, 1980.

Lemann, J., Jr., Adams, N., and Gray, R.W.: Urinary calcium excretion in human beings, N. Engl. J. Med. **301**:535-541, 1979.

Robinson, C.J.: The physiology of parathyroid hormone, Clin. Endocrinol. Metab. **3**:389-417, 1974.

Stoff, J.S., Phosphate homeostasis and hypophosphatemia, Am. J. Med. **72**:489-495, 1982.

Clinical pathology

Benson, R.C., Jr., Riggs, B.L., and Pickard, B.M.: Immunoreactive forms of circulating PTH in primary and ectopic hyperparathyroidism, J. Clin. Invest. **54**:175-181, 1974.

Berson, S.A., and Yalow, R.S.: Immunochemical heterogeneity of parathyroid hormone in plasma, J. Clin. Endocrinol. Metab. **28**:1037-1047, 1968.

Broadus, A.E.: Nephrogenous cyclic AMP as a parathyroid function test, Nephron **23**:136-141, 1979.

Chen, I.W., Park, H.M., King, L.R., et al.: Radioimmunoassay of parathyroid hormone: peripheral plasma immunoreactive parathyroid hormone response to ethylenediaminetetraacetate, J. Nucl. Med. **15**:763-769, 1974.

Flueck, J.A., DiBella, F.P., Edis, A.J., et al.: Immunoheterogeneity of parathyroid hormone in venous effluent serum from hyperfunctioning parathyroid glands, J. Clin. Invest. **60**:1367-1375, 1977.

Habener, J.F., and Segre, G.V.: Parathyroid hormone radioimmunoassay, Ann. Intern. Med. **91**:782-784, 1979.

Clinical disorders

Bone, H.G., II, Snyder, W.H., III, and Pak, C.Y.C.: Diagnosis of hyperparathyroidism, Annu. Rev. Med. **28:**111-117, 1977.

Goldsmith, R.E., Sizemore, G.W., and Chen, I.W.: Familial hyperparathyroidism: description of a large kindred with physiologic observations and a review of the literature, Ann. Intern. Med. **84:**36-43, 1976.

Mallette, L.E., Bilezikian, J.P., Heath, D.A., and Aarback, D.G.: Primary hyperparathyroidism: clinical and biochemical features, Medicine **53:**127-146, 1974.

Nusynowitz, M.L., Frame, B., and Kolb, F.O.: The spectrum of the hypoparathyroid states, Medicine **55:**105-119, 1976.

Scholz, D.A., Purnell, D.C., Edin, A.J., et al.: Primary hyperparathyroidism with multiple parathyroid gland enlargement, Mayo Clin. Proc. **53:**792-797, 1978.

Skrabanek, P., McPartlin, J., and Powell, D.: Tumor hypercalcemia and "ectopic hyperparathyroidism," Medicine **59:**262-282, 1980.

Chapter 43 Adrenal hormones

Robert Hoff

ACTH stimulation test A function test to evaluate adrenocortical function. It is primarily used to differentiate between primary and secondary hypocortical diseases.

Addison's disease Also known as primary hypoadrenalism it is decreased adrenocortical hormone secretion, usually without known cause.

adrenal cortex The outer portion of the adrenal gland, which produces various steroid hormones.

adrenal medulla The inner portion of both adrenal glands, which produces catecholamines.

adrenocorticotropic hormone (ACTH) This hormone is produced in the pituitary gland and is the primary stimulator of glucocorticoid secretion and to a much lesser extent of mineralocorticoid secretion.

adrenocorticotropic hormone (ACTH) stimulation test After administration of ACTH, plasma cortisol or urinary products are measured. It is an excellent screening test for hypoadrenalism.

adrenogenital syndrome A group of diseases that result from various enzyme deficiencies and lead to various increases in sex steroid hormone.

catecholamines Epinephrine and norpinephrine, which are produced in the adrenal medulla and are responsible for maintenance of blood pressure.

chromaffin cells These cells are found in the adrenal medulla and other sites throughout the body and produce catecholamines.

clonidine-suppression test A newly described function test to diagnose pheochromocytoma.

Conn's syndrome Another name used to denote primary hyperaldosteronism.

corticosteroid-binding globulin See *transcortin*.

corticotropin-releasing factor (CRF) A polypeptide that is produced in the hypothalamus and stimulates ACTH secretion.

Cushing's disease Hypersecretion of glucocorticoids caused by elevated ACTH secretion.

Cushing's syndrome Elevation of glucocorticoids secretion because of either increased ACTH or autonomous adrenal secretion.

dexamethasone suppression test A function test that is used in the diagnosis and differentiation of various causes of Cushing's syndrome.

diurnal release Characteristic daily cyclic release of cortisol in response to ACTH.

glucocorticoids A group of steroid hormones, most notably cortisol, that have multiple functions throughout the body.

hypoadrenalism It results from a decrease or lack of adrenocortical hormone production. There are two subclassifications: primary hypoaldosteronism (Addison's disease) and secondary hypoaldosteronism.

hyperaldosteronism Also known as Conn's syndrome and results in increased aldosterone secretion, either because of elevated renin or autonomous adrenocortical secretion.

metyrapone test This adrenocortical function test previously enjoyed greater use in diagnosis of Cushing's syndrome by testing the ability of the pituitary to secrete ACTH. It has largely been superseded by other function tests.

mineralocorticoids A group of steroid hormones that are produced in the adrenal cortex, most notably aldosterone, and stimulate reabsorption of sodium in the distal tubules of the kidneys.

pheochromocytoma A tumor that is usually benign but can be malignant and results in hypersecretion of epinephrine and norepinephrine.

renin-angiotensin system This system is responsible for stimulation of aldosterone secretion from the adrenal cortex.

transcortin This diglobulin protein binds a majority of cortisol in the plasma. It is also known as corticosteroid-binding globulin.

zona fasciculata The middle portion of the adrenal cortex in which glucocorticoids and various sex hormones are produced.

zona glomerulosa The outer portion of the adrenal cortex in which the mineralocorticoids are produced.

zona reticularis The innermost portion of the adrenal cortex, next to the adrenal medulla, which acts in consort with the zona fasciculata.

ANATOMY

The human body contains a pair of adrenal glands, each weighing approximately 5 g. Both adrenal glands are located on the upper pole of each kidney, each gland being roughly pyramidal in shape (Fig. 43-1).

Each adrenal gland consists of two functionally distinct entities, an outer cortex and inner medulla (Fig. 43-1). The cortex is bright yellow and is composed of three distinct zones: the zona glomerulosa, zona fasciculata, and zona reticularis. These zones produce hormones that regulate salt and water metabolism, carbohydrate metabolism, and a small number of sex hormones. The medulla is pearly gray and consists of catecholamine-secreting cells known as chromaffin cells.[1]

PHYSIOLOGY OF ADRENAL HORMONES

The zona glomerulosa produces mineralocorticoids, which regulate salt and water balance. The zona fasciculata and zona reticularis both are involved in the production of various sex hormones and as well the glucocorticoids, which have multiple effects, principally regulation of carbohydrate metabolism. All adrenocortical hormones have the same basic structure, consisting of four rings composed of 17 carbon atoms (Fig. 43-2). They differ only in degree of carbon saturation and in spatial orientation of the side chains. These relatively minor differences in chemical structure result in the major biological differences as discussed below.

Glucocorticoids

Cortisol and corticosteroid are the major glucocorticoids produced by the adrenal gland. The term "glucocorticoid" is a misnomer because a number of important effects of these hormones occur outside of their effect on carbohydrate metabolism. Some of the more important glucocorticoid effects are summarized in the box on p. 819. These hormones are necessary for life, especially when the human body is stressed. In general, they are antagonistic to insulin and act to increase plasma glucose (see Chapter

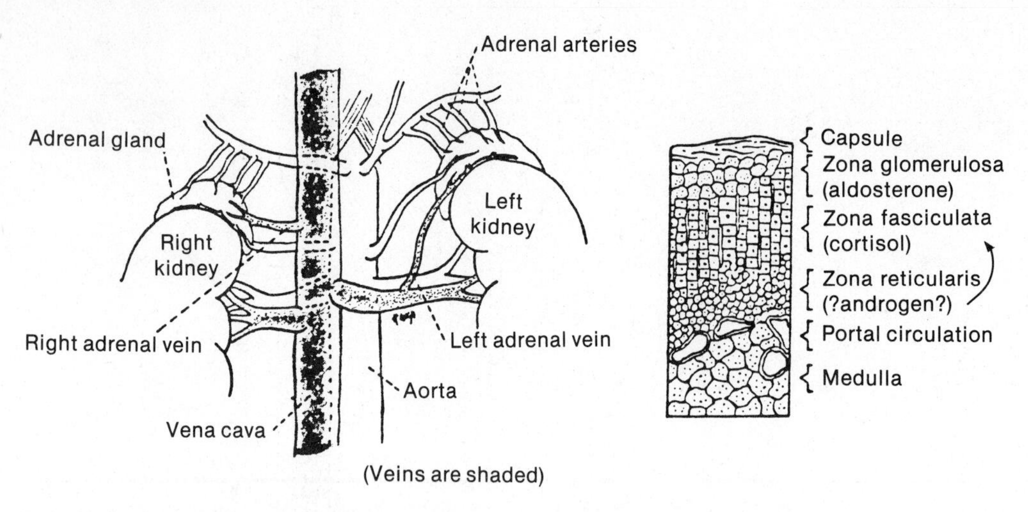

Fig. 43-1. Adrenal gland anatomy and histology. (From Ryan, W.: Endocrine disorders, Chicago, 1980, Year Book Medical Publishing Co.)

30), catabolize proteins to amino acids, and result in a net increase in serum fatty acid levels for use as an energy source.[2] Other functions take on importance in relative deficiency or excess states.[3,4] The glucocorticoids are capable of suppressing immunological and inflammatory responses. This occurs by selective depletion of T lymphocytes relative to B lymphocytes, lymphocytopenia, neutrophilia with increased release of neutrophils from the bone marrow, and increased accumulation of neutrophils at inflammatory sites.[5] These effects are often used pharmacologically to decrease excessive inflammatory or immunological activity in a number of diseases. However, elevated glucocorticoid levels from any cause can also lead to increase in frequency and severity of infections, as seen in Cushing's syndrome (see p. 824), and cause increased erythropoiesis (that is, production of red blood cells) with a mild polycythemia. The exact mechanism for this latter phenomenon is not known. Other effects include thinning of bone, osteoporosis, and inhibition of fibroblasts, which leads to easy bruising and poor wound healing, whereas excesses or deficiencies of glucocorticoids may result in changes of moods and behavior.[6,7] In addition, glucocorticoids are necessary for regulation of blood pressure.[8] The exact mechanism is unknown; however, it is believed that it is unrelated to the mineralocorticoid action. One of their most essential functions in response to stress is maintenance of blood pressure. Without this increase in cortisol production during stress, life ceases to exist.[9]

From 10 to 25 mg of cortisol are produced per day. In normal circumstances there is great variability in release from the highest levels, achieved in early morning, to the lowest levels, achieved in the late evening. This *diurnal* variation parallels the rhythmic character of the hypothalamus in secreting ACTH, which is the regulatory center for cortisol production.[10]

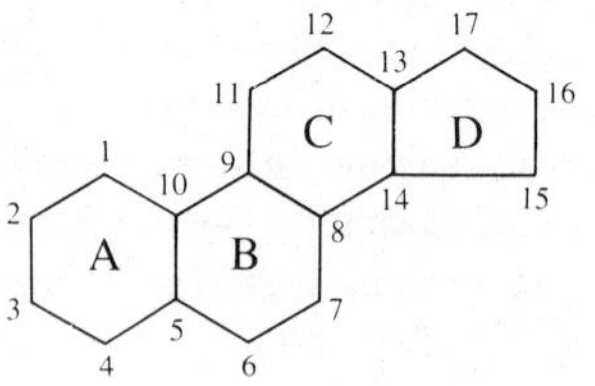

Basic steroid nucleus

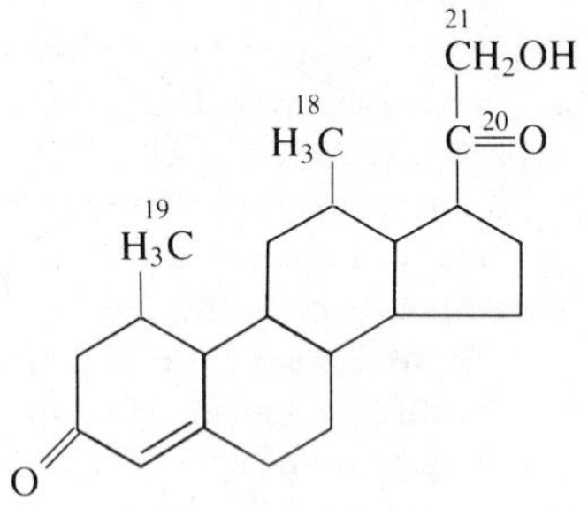

Corticosteroid nucleus

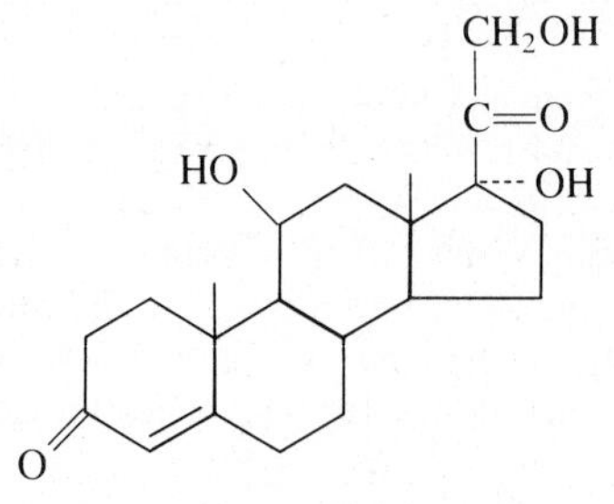

Cortisol

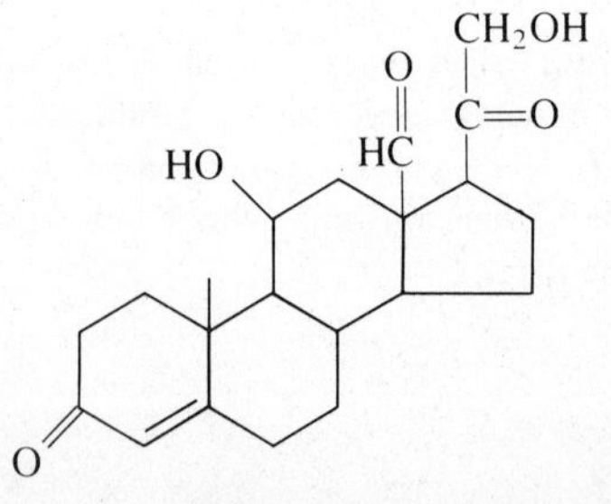

Aldosterone

Fig. 43-2. Structures of adrenocortical hormones.

Metabolic and antiinflammatory actions of glucocorticoids

Carbohydrate metabolism
- Increase in blood glucose
- Decrease in glucose use
- Increase in gluconeogenesis from body protein
- Increase in liver glycogen formation

Protein metabolism
- Increase in protein catabolism
- Decrease in protein synthesis

Inhibition of allergic and inflammatory reactions
- Decrease in antibody formation
- Decrease in number of circulating lymphocytes and eosinophils
- Decrease in mass of lymphatic tissue
- Inhibition of activity of granulocytes and monocytes

From Feldkamp, C.S., Powsner, E.R., Zak, B., and Epstein, E.: Adrenal cortex. In Sonnenwirth, A.C., and Jarett, L., editors: Gradwohl's clinical laboratory methods and diagnosis, ed. 8, St. Louis, 1980, The C.V. Mosby Co.

Glucocorticosteroid enters a target cell by unknown mechanisms and then binds to cytoplasmic receptors. This induces a change in the receptors, known as "activation," which results in binding of the hormone-receptor complexes to nuclear chromatin. This activity induces increased synthesis of specific messenger RNA and a subsequent increase in the synthesis of proteins coded by them. The newly synthesized proteins (that is, enzymes) are directly responsible for steroid-induced action.[11]

Mineralocorticoids

The most potent mineralocorticoid synthesized in the adrenal gland is aldosterone. It regulates electrolyte metabolism by modifying sodium and potassium transport across membranes throughout the body. The principal result is maintenance of blood pressure and conservation of body sodium at the expense of potassium. This is achieved principally in the distal convoluted tubules of the kidney through promotion of sodium reabsorption with concomitant secretion of hydrogen or potassium. Aldosterone also regulates electrolyte balance in the intestine, salivary glands, and sweat glands by decreasing sodium excretion and increasing potassium excretion in those organs. However, these actions are not known to have physiological importance. Cortisol, and other corticosteroids such as corticosterone and deoxycorticosterone (DOC) have some mineralocorticoid activity that can become clinically significant when serum levels are high. Table 43-1 compares relative glucocorticoid and mineralocorticoid activities. As shown in Table 43-1, it is evident that physiological levels of cortisol would not contribute significantly to mineralocorticoid activity. However, increased serum cortisol levels such as that found in Cushing's syndrome could contribute to the body's mineralocorticoid activity. Physiological levels of aldosterone do not exert effects on carbohydrate metabolism and do not exert any feedback on ACTH levels.

Table 43-1. Relative glucocorticoid and mineralocorticoid activity

	Glucocorticoid activity	Mineralocorticoid activity
Cortisol	1	1
Corticosterone	0.5	1.5
Deoxycorticosterone	0	20
Dexamethasone	30	0
Aldosterone	0.3	300

Aldosterone binds to cytosol receptors, which are present in mineralocorticoid-responsive tissues. This results in an increase of messenger RNAs whose protein products presumably mediate the mineralocorticoid hormone responses.[12]

Medullary hormones

The main secretory products of the adrenal medulla consist of the catecholamines, epinephrine and norepinephrine. These hormones were previously known as adrenaline and noradrenaline, respectively. The hormones are produced in episodic bursts. Production of these hormones is not limited to the adrenal medulla but also occurs in the neurons of the sympathetic nervous system and in the central nervous system and scattered groups of chromaffin cells found in multiple areas of the body, including retroperitoneum, mediastinum, and neck. Norepinephrine is the principal product synthesized in the central nervous system, and epinephrine is the principle catecholamine produced by the adrenal glands.

The catecholamines are responsible for the physiological responses that are summarized by the well-known phrase, "fright, flight, or fight." The various effects of the catecholamines are provided by two main groups of cell membrane receptors in the various target organs. The receptors that primarily bind norepinephrine are known as alpha receptors, whereas the beta receptors bind epinephrine. In general, the alpha receptors are responsible for such functions as vasoconstriction and beta receptors' functions include vasodilatation and bronchodilatation. Both epinephrine and norepinephrine increase the respiratory rate and increase somewhat the depth of respiration. Although both result in bronchodilatation, epinephrine is significantly more active than norepinephrine in this regard.

Both catecholamines have extremely important metabolic functions, though epinephrine is about eight times more active in this regard than norepinephrine. They both stimulate glycogenolysis in the liver by stimulating activation of phosphorylated kinase, leading to the formation of glucose-1-phosphate and eventually free glucose, which is released into the blood. Thus these hormones act as antagonists to insulin. Muscle glycogenolysis is also stimulated, but since there is an absence of glucose-6-phosphatase in muscle, the glucose remains in the muscle cells and is available for muscular work. Phosphorylation of glycogen synthase leads to decreased activity and thus reduced glycogenesis. Both epinephrine and norepinephrine increase free fatty acid liberation from triglyceride stores in mature adipose tissue. Other effects include erection of body hair through activation of the arrector pili muscles of the skin (''hair standing on end''), increasing wakefulness through its effect on the higher centers of the central nervous system, and increasing the blood supply to the skeletal muscle and increasing the force of muscular contraction as well.

BIOSYNTHESIS

Glucocorticoids and mineralocorticoids

All adrenocortical hormones are steroids derived from either acetate or cholesterol.[13] The major source of cholesterol appears to be low density lipoproteins found in the plasma (Chapter 31). The synthesis of the various hormones occurs by enzymes located either in mitochondria, microsomes, or the cytoplasm. The locations of the various enzymes dictate the region of the adrenal cortex in which the various hormones are produced. Both the mineralocorticoids and glucocorticoids have 21 carbon atoms in their chemical structures and are thus known as the C-21 steroids. It should be emphasized that all steroids that

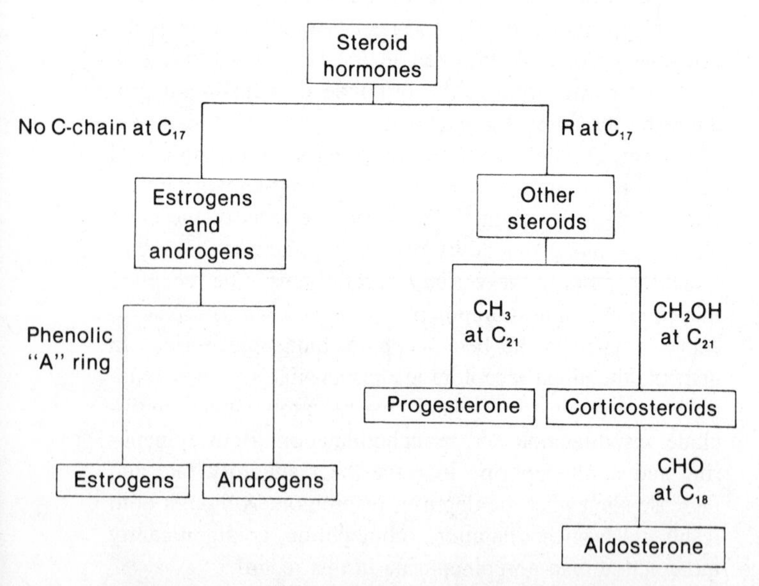

Fig. 43-3. Major subclasses of steroid hormones. (From Montgomery, R., Dryer, R.L., Conway, T.W., and Spector, A.A.: Biochemistry: a case-oriented approach, St. Louis, 1980, The C.V. Mosby Co.)

are C-21 from the adrenal gland have some mineralocorticoid and glucocorticoid activities. The sex hormones that are produced in the adrenal cortex are usually produced in relatively small amounts in comparison with the levels produced by the ovaries and testes. These sex hormones contain 19 carbon atoms and are thus known as C-19 steroids (Fig. 43-3). Regulation of these reactions as discussed in Chapter 41 is achieved by stimulation or inhibition of certain of these enzymes, and the so-called adrenogenital syndromes are a collection of diseases resulting from specific enzyme deficiencies.

As shown in Fig. 43-4, the conversion of cholesterol to Δ^5-pregnenolone and of Δ^5-pregnenolone to progesterone occupies central positions in the synthetic scheme of the hormones. Until this point the pathway is identical for all steroids produced, and the proportions of the final hormones produced are based on enzyme activities of each subsequent pathway. The major regulation of adrenocortical hormone production occurs at the reaction of cholesterol to Δ^5-pregnenolone. ACTH produced in the pituitary gland and angiotensin produced in the liver stimulate this reaction.

Aldosterone, the primary mineralocorticoid in humans, is synthesized from cholesterol by the conversion of progesterone to 11-deoxycorticosterone, by the enzyme 21β-hydroxylase, which is then converted to corticosterone by the enzyme 11β-hydroxylase. Further hydroxylation occurs to produce 18-hydroxycorticosterone, which is converted by the enzyme 18-hydroxysteroid dehydrogenase to aldosterone. The last enzyme, 18-hydroxysteroid dehydrogenase, is located only within the zona glomerulosa and thus aldosterone is synthesized only in this region of the adrenal cortex.

The principal glucocorticoid, cortisol, is produced in a similar manner to the production of aldosterone. From progesterone, it is converted to 17α-hydroxyprogesterone by the enzyme 17α-hydroxylase. This is converted by 21β-hydroxylase to 11-deoxycortisol, which is converted to cortisol under the action of the enzyme 11-hydroxylase.

Medullary hormones

The synthesis of these hormones begins with the hydroxylation of the amino acid tyrosine (Fig. 43-5). Tyrosine is believed to be derived primarily from dietary sources. Tyrosine is converted to dihydroxyphenylalanine (dopa) by the enzyme tyrosine hydroxylase. This enzyme is the rate-limiting enzyme in catecholamine synthesis and is inhibited by both epinephrine and norepinephrine (end-product inhibition). Dopa is decarboxylated to form dopamine. This reaction is catalyzed by the enzyme aromatic L-amino acid decarboxylase. Dopamine is the major product in some neurons, especially in the central nervous system. In these particular neurons the subsequent enzymes are absent. Dopamine β-hydroxylase catalyzes the hydroxylation of dopamine to norepinephrine. Finally, norepi-

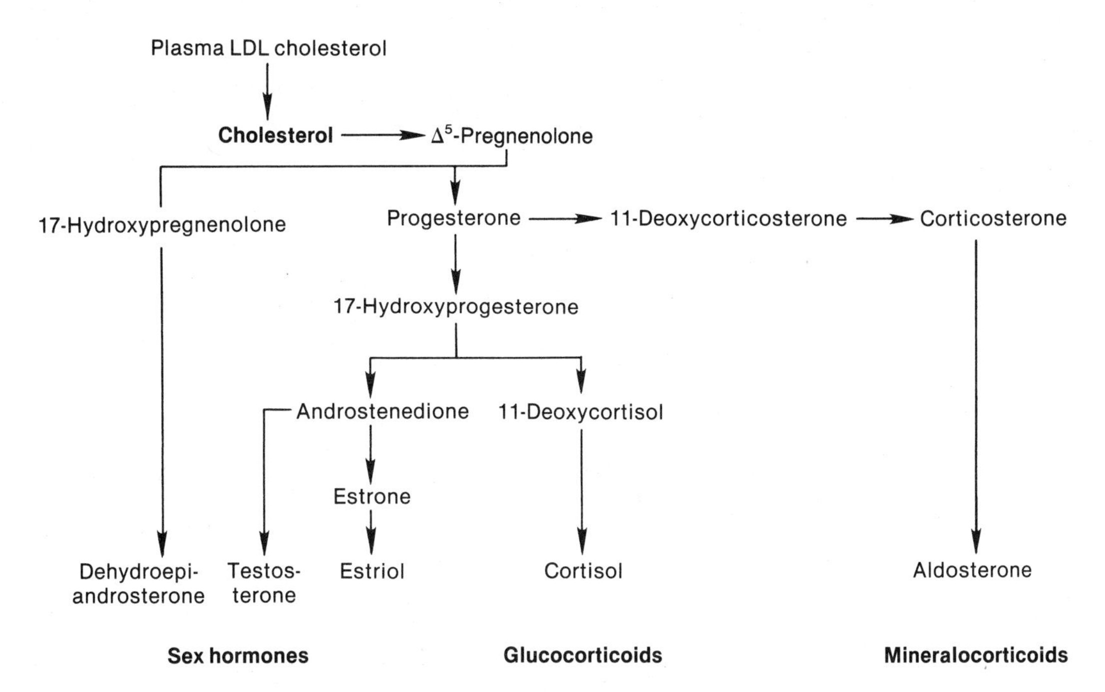

Fig. 43-4. Scheme of biosynthetic pathways of cortical hormones.

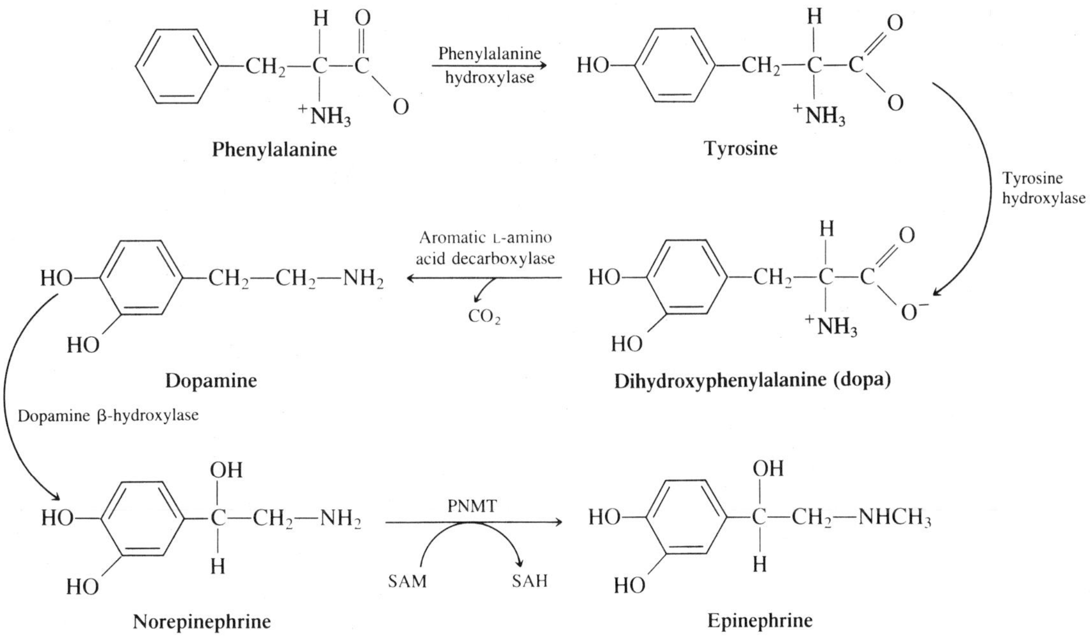

Phenylalanine → (Phenylalanine hydroxylase) → Tyrosine → (Tyrosine hydroxylase) → Dihydroxyphenylalanine (dopa) → (Aromatic L-amino acid decarboxylase; CO_2) → Dopamine → (Dopamine β-hydroxylase) → Norepinephrine → (PNMT; SAM → SAH) → Epinephrine

Phenylalanine: $C_6H_5-CH_2-CH(^+NH_3)-COO^-$

Tyrosine: $HO-C_6H_4-CH_2-CH(^+NH_3)-COO^-$

Dihydroxyphenylalanine (dopa): $(HO)_2C_6H_3-CH_2-CH(^+NH_3)-COO^-$

Dopamine: $(HO)_2C_6H_3-CH_2-CH_2-NH_2$

Norepinephrine: $(HO)_2C_6H_3-CH(OH)-CH_2-NH_2$

Epinephrine: $(HO)_2C_6H_3-CH(OH)-CH_2-NHCH_3$

Fig. 43-5. Synthesis of medullary hormones. *PNMT,* Phenylethanolamine *N*-methyl transferase; *SAM, S*-adenosylmethionine; *SAH, S*-adenosyl homocysteine. (From Orten, J.M., and Neuhaus, O.W.: Human biochemistry, St. Louis, 1982, The C.V. Mosby Co.)

nephrine is methylated by the action of phenylethanol amine *N*-methyl transferase to form epinephrine. The final step of converting norepinephrine to epinephrine normally occurs only in the adrenal medullary cells. After synthesis, both epinephrine and norepinephrine are stored in chromaffin granules in the medullary cells until they are released into the blood.

Sex hormones

Adrenal sex hormones have similar functions to those produced in the ovary and testes (see Chapter 41) but are much weaker in biological activity. They assume great importance in states in which their levels are greatly increased, as occurs in the adrenogenital syndromes.

TRANSPORT AND CATABOLISM

Glucocorticoids

After release of cortisol by the adrenal gland, approximately 90% to 93% is bound by plasma proteins. Of this, 80% is bound to corticosteroid-binding globulin (CBG), also known as "transcortin." CBG binds cortisol with high affinity and specifity. Albumin also binds a small amount of cortisol.[14] Bound cortisol is biologically *inactive*.

Cortisol is metabolized and inactivated by the liver to compounds that are later excreted by the kidney. The most important catabolic product is dihydrocortisol, which is converted to tetrahydrocortisol by 3-hydroxysteroid dehydrogenase.[15] These products are excreted by the kidney as urinary 17-hydroxysteroids (17-OHS). Detection of 17-OHS in the urine provides a reliable and convenient method for assessment of adrenocortical function.

Mineralocorticoids

Although most steroids are bound in plasma in significant amounts to proteins, there is no protein that binds aldosterone with a high affinity. However, it is weakly bound to plasma proteins, including albumin and cortisol-binding globulin (CBG). Approximately 55% of total aldosterone in the plasma is bound to protein. Of this, 10% is bound to CBG. Therefore high concentrations of cortisol can decrease total aldosterone binding. In any case the percentage of free aldosterone is relatively stable in contrast to the wide fluctuations seen with other steroids.[16]

Aldosterone is metabolized in the liver by reaction and conjugation with glucuronic acid. These water-soluble metabolites, known as tetrahydroaldosterone glucosiduronate and aldosterone 18-glucosiduronate, are eliminated by the kidney into urine.[17]

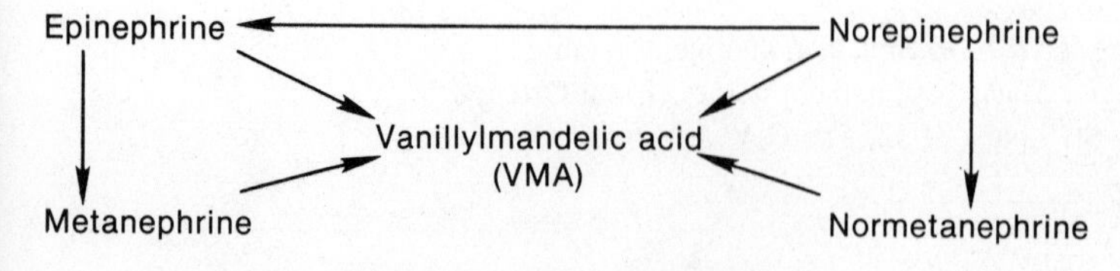

Fig. 43-6. Metabolism of medullary hormones.

Medullary hormones

The catecholamines are found tightly bound to ATP in the storage chromaffin granules of adrenal medullary cells, which protects them from degradation. Generally, both epinephrine and norepinephrine are inactivated rapidly once released from cells. As noted in Fig. 43-6, catecholamines are degraded by two distinct enzyme systems, the catechol-*O*-methyltransferase (COMT) and the monoamine oxidase (MAO) enzymes. The majority of the inactivation occurs in the liver, though both enzymes are present in many tissues of the body, including chromaffin cells. The common end metabolic product of both enzyme systems is vanillylmandelic acid (VMA), which is excreted in the urine. However, metanephrine and normetanephrine represent other major urinary metabolites. All three metabolites are used in the diagnosis of adrenal medullary disease. In addition, small amounts of free and conjugated epinephrine and norepinephrine are excreted in the urine.

CONTROL AND REGULATION

Glucocorticoids

Adrenoglucocorticoid and androgen production is regulated by the hypothalamic-pituitary axis. As shown in Fig. 43-7, corticotropin-releasing factor (CRF) is released by the hypothalamus in response to certain factors and transported by the blood to the anterior pituitary gland, where it stimulates the synthesis and release of adrenocorticotropic hormone (ACTH). ACTH then acts on adrenocortical cells to stimulate production and release of cortisol. Within 2 or 3 minutes of ACTH production, there is increased release of cortisol through increased synthesis.

CRF is released in response to endogenous rhythms in the brain, which result in the characteristic circadian (diurnal) pattern of ACTH and cortisol release. Various emotional and physical stimuli result in increase of CRF. ACTH release is inhibited by increased blood cortisol levels.

Mineralocorticoids

Aldosterone secretion is regulated by many factors, including serum levels of potassium, sodium, ACTH, serotonin, and components of the renin-angiotensin system. Of these factors, the renin-angiotensin system and potassium are of primary importance. Aldosterone secretion is increased by a fall in serum sodium, an elevation of serum potassium, and a reduction in blood volume. The net result of increased aldosterone production is a reversal of these situations by an increase in the renal reabsorption of sodium and water into the extracellular fluids. The dominant stimulus, as previously stated, is provided by production of angiotensin by the renin-angiotensin system (Fig. 43-8).

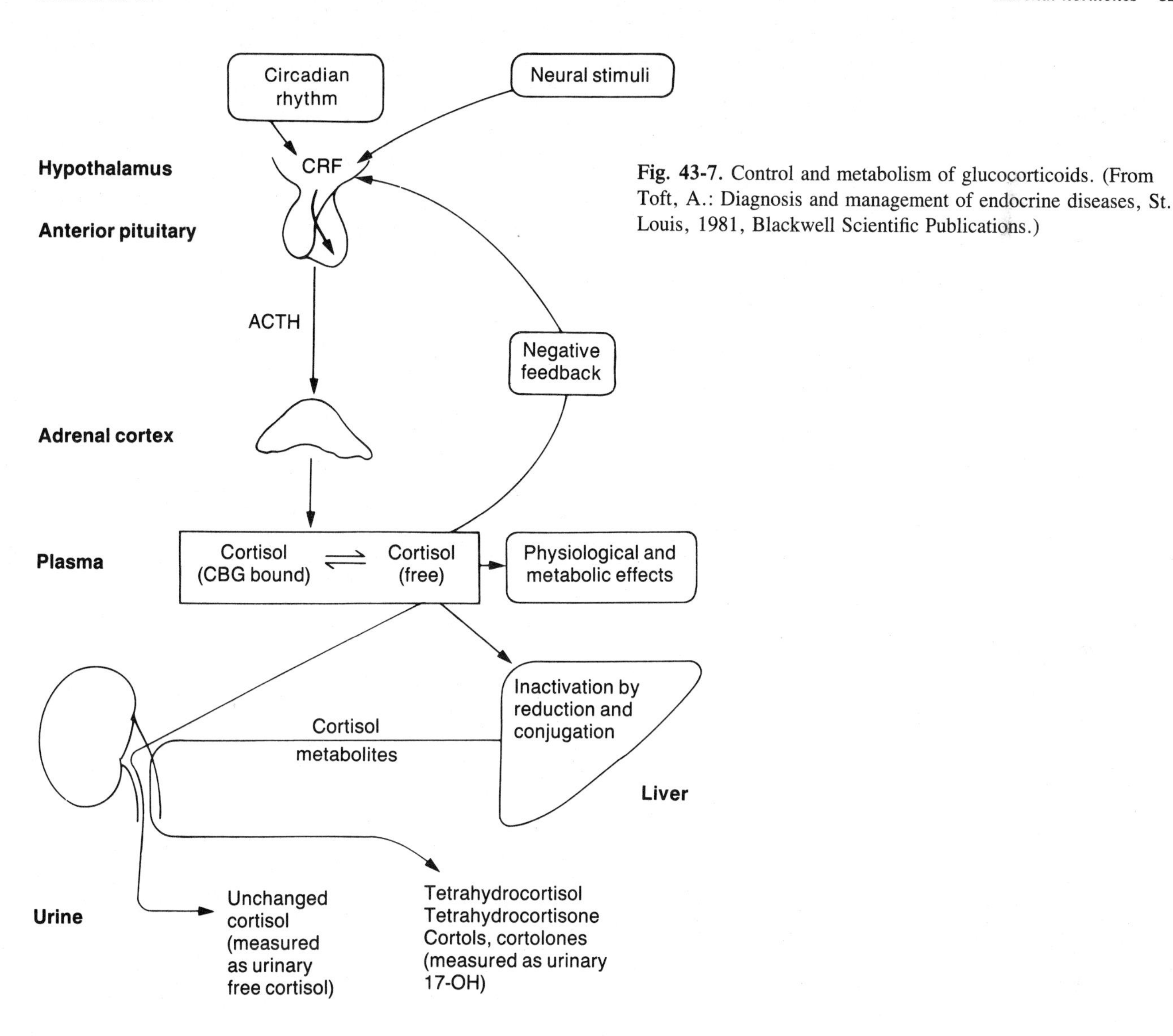

Fig. 43-7. Control and metabolism of glucocorticoids. (From Toft, A.: Diagnosis and management of endocrine diseases, St. Louis, 1981, Blackwell Scientific Publications.)

Renin is synthesized in the kidney in response to decreased blood pressure and low sodium concentration. The renin acts to convert angiotensinogen, which is produced in the liver, to angiotensin I. Angiotensin I is then converted by the angiotensin I–converting enzyme (ACE), primarily produced in the lungs, to angiotensin II. Angiotensin II then stimulates a rapid increase of aldosterone production in the zona glomerulosa and, additionally, has a powerful vasoconstrictive effect, which further augments blood pressure maintenance. There is no associated increase in cortisol production. Angiotensin II stimulates conversion of cholesterol to pregnenolone and corticosterone to 18-hydroxycorticosterone and finally to aldosterone. Angiotensin II is rapidly degraded in plasma by tissue angiotensinases so that the circulating half-life is only 1 to 2 minutes.

As noted above, an increase in plasma potassium also stimulates aldosterone production.[18] This stimulation is independent of sodium and angiotensin II and also has no effect on cortisol production. The mechanism by which potassium acts is not known. Sodium deficiency increases aldosterone production, usually by increasing production of renin in the kidney. However, there is some evidence that sodium can act directly to increase aldosterone production. ACTH is not a major regulator of aldosterone production but may increase aldosterone levels to a minor degree by stimulating conversion of cholesterol to pregnenolone.

Medullary hormones

As stated previously, the synthesis of epinephrine and norepinephrine is inhibited by both epinephrine and norepinephrine (end-product inhibition). The catecholamines are released in response to hypotension, hypoxia, exposure to cold, muscular exertion, pain, and emotional disturbances.

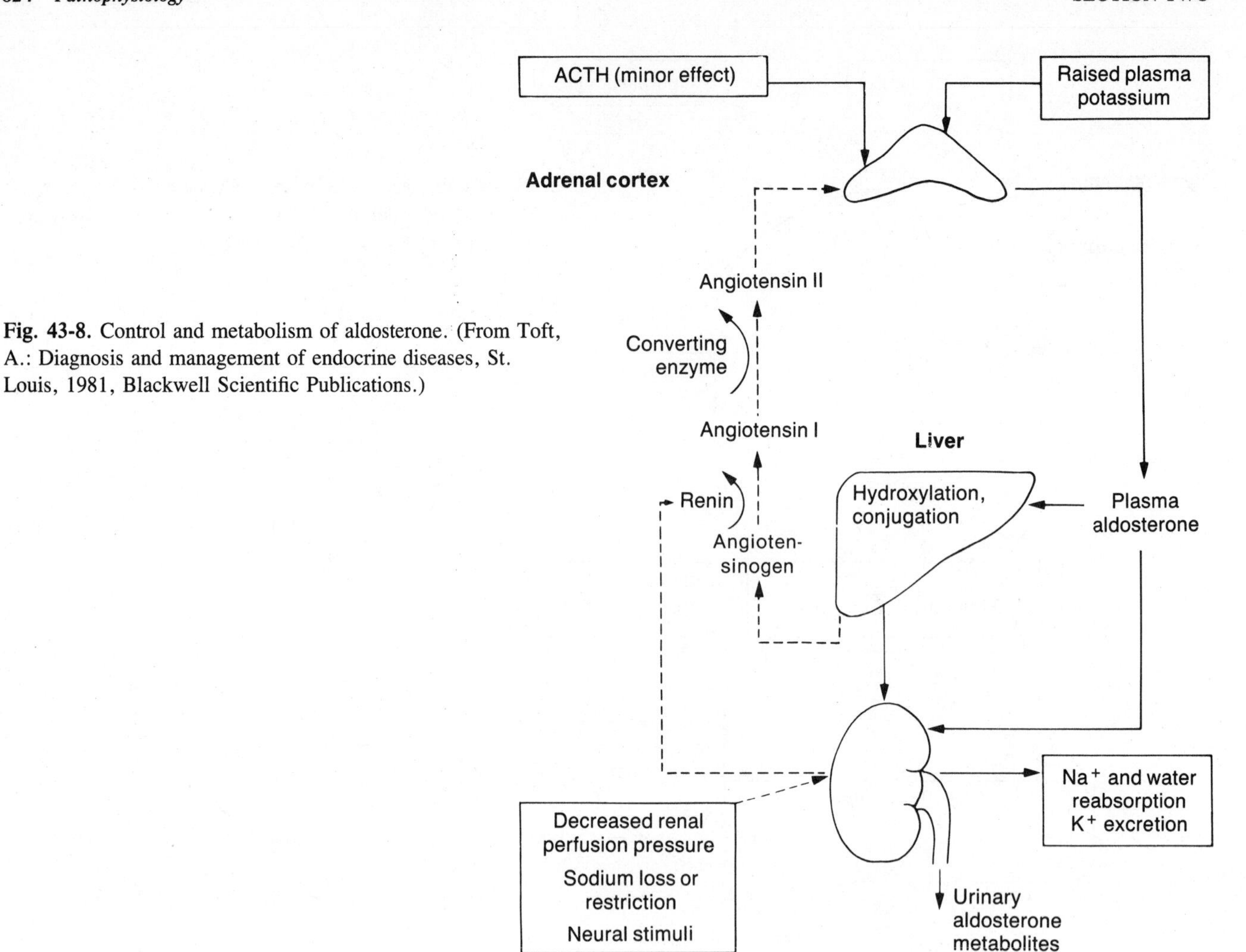

Fig. 43-8. Control and metabolism of aldosterone. (From Toft, A.: Diagnosis and management of endocrine diseases, St. Louis, 1981, Blackwell Scientific Publications.)

PATHOLOGICAL CONDITIONS

Functionally, the disorders of the adrenal cortex are classified into those that result in an increase, decrease, or lack of steroid release and those that do not have an effect on steroid release.

Hyperadrenalism, or hypercorticalism, is defined as a process that results in increased release of adrenocortical hormones. It can result from neoplastic or hyperplastic conditions. Hypoadrenalism or hypocorticalism is defined as the process that results in decreased release of adrenocortical hormones. It has many causes, including inflammatory, infectious, neoplastic, or idiopathic conditions. Additionally, certain congenital enzyme deficiencies result in a particular type of hyperadrenalism known as the "adrenogenital syndrome." There is only one disease of importance affecting the adrenal medulla, pheochromocytoma. It results in hypersecretion of epinephrine or norepinephrine, or both.

Hyperadrenalism

Conditions that result in increased synthesis can affect each of the three basic adrenocortical hormone groups. Therefore an increase in cortisol production is known as Cushing's syndrome, an increase in aldosterone is known as "aldosteronism," and an increase in adrenal androgens is known as "congenital adrenal hyperplasia."

Cushing's syndrome. "Cushing's syndrome" has become the generic term for elevated cortisol. This increase in circulating plasma cortisol results in a distinctive grouping of symptoms that consist in excessive fat deposition resulting in central truncal obesity, moon face, and buffalo hump, edema, hyperglycemia, hypertension, osteoporosis, skin striae, thinning of the skin, hirsutism, menstrual disturbances, acne, weakness, and emotional lability. The most common cause is iatrogenic, that is, from chronic treatment or use of potent glucocorticoids.[19] Other causes include excess ACTH production caused by a benign pituitary tumor (a form of secondary hyperadrenalism known as "Cushing's disease"); excess cortisol production from either a benign or a malignant (rare) tumor; an adrenal gland tumor; or production of ACTH by a nonpituitary (ectopic) tumor.[20]

The diagnosis is suspected when the above grouping of signs and symptoms is found. Confirmation of the diag-

Major features of Cushing's syndrome

Clinical findings

Obesity	Osteoporosis
Hypertension	Personality change
Glucose intolerance	Acne
Hirsutism	Edema
Amenorrhea	Poor wound healing
Striae	Headache

Laboratory findings
- Elevated 17-OHS
- Elevated urinary free cortisol
- Loss of diurnal variation in serum cortisol

Function tests
- Lack of response to low-dose dexamethasone
- High-dose dexamethasone test: used to differentiate betweeen various causes
- Elevated ACTH; depends on cause (see p. 830)

nosis is made using specific chemical determinations. Characteristic laboratory findings include elevation of plasma cortisol with resultant loss of diurnal rhythm, elevation of urinary 17-hydroxycorticosteroids, and elevation of urinary free cortisol. Routine laboratory determinations are of little value in the diagnosis of Cushing's syndrome. A summary of clinical and laboratory findings is found in the box above.

Specific causes are more difficult to ascertain. Characteristic laboratory findings in secondary hyperadrenalism caused by excess ACTH production include elevated plasma ACTH with concomitant loss of diurnal rhythm and loss of ACTH suppressibility by dexamethasone with suppression of cortisol secretion when a larger dosage of dexamethasone is given. Hyperpigmentation is a common finding in Cushing's disease because of amino acid sequence homology between ACTH and melanocyte-stimulating hormone (MSH), which causes skin pigmentation. Adrenal tumors account for about only one fourth of Cushing's syndrome cases, with malignant adrenal carcinomas being slightly more common than benign adrenal adenomas. Diagnostic features include wide variability in secretory activity and nonsuppressibility by even large doses of dexamethasone.[21]

The effect of ectopic ACTH production on the adrenal glands results in excess mineralocorticoid activity secondary to the tremendous increase in cortisol, corticosterone, and DOC. This produces hypokalemia, alkalosis, weakness, and weight loss. This usually results from malignant tumors, of which oat cell carcinoma of the lung is most common.[22]

Hyperaldosteronism. Hyperaldosteronism or aldosteronism results in sodium retention and potassium excretion caused by excess aldosterone activity in the distal tubules of the kidney. The sodium retention in turn leads to an increase in plasma water volume, resulting in hypertension. Hyperaldosteronism is separated into two categories, primary and secondary.

Major features of aldosteronism

Clinical findings
- Hypertension (may be episodic)
- Increased urine volume
- Weakness
- Tetany
- Paresthesias

Laboratory findings
- Low serum potassium
- Elevated urinary potassium
- Elevated serum sodium
- Alkalosis
- Elevated urine aldosterone: best when done with prior salt loading
- Elevated plasma aldosterone: best when done with prior salt loading
- Serum renin: of use in differentiating primary aldosteronism from secondary aldosteronism

Function tests
- Salt loading results in accentuated hypokalemia
- Nonaldosterone mineralocorticoid suppression: used in differentiating primary aldosteronism from secondary aldosteronism

Primary aldosteronism, also known as "Conn's syndrome," is usually caused by an adrenal adenoma but can also result from adrenal carcinoma or hyperplasia. Therefore plasma renin is decreased in response to the autonomous elevation of aldosterone. The usual symptoms consist of weakness, tetany (caused by decreased serum K) and paresthesias (prickling). Chemical abnormalities include elevated serum sodium, decreased serum potassiuim with alkalosis, elevated urine potassium, increased serum aldosterone, and decreased plasma renin. Screening is done in patients with hypertension, low serum potassium, and high urine potassium. Diagnosis is based on an elevated plasma aldosterone level with prior salt loading. The tumor is localized with computed tomography (CT) scanning.[23] A summary of clinical and laboratory findings is found in the box above.

Secondary aldosteronism represents increased aldosterone that is caused by nonadrenal stimuli. Some causes include cardiac failure, nephrotic syndrome, and obstructive renal disease. In contrast to primary aldosteronism, plasma renin is elevated in secondary aldosteronism.

Adrenogenital syndrome. This disorder consists in a group of diseases manifested by excessive secretion of adrenal androgens. The various causes revolve around inborn errors of metabolism in which specific enzyme deficiencies result in elevation of intermediate products proximal to the defect and decreased production of biochemical products distal to the defect. The virilizing effects are dependent on the sex and age of onset. Signs and symptoms include baldness, acne, hirsutism, deepening of the voice, and menstrual irregularities. Causes include congenital virilizing adrenal hyperplasia and virilizing adenomas or carcinomas. Urinary 17-ketosteroids and plasma ACTH are increased with decreased urinary 17-hydroxycorticosteroids and cortisol.[24]

Adrenal hypofunction

Primary adrenal hypofunction or insufficiency, a rare disease, is also known as "Addison's disease."[25] There are two distinct etiological groups responsible for primary adrenal hypofunction. Approximately 80% of cases are caused by idiopathic atrophy of the adrenal cortex, which is believed to have an autoimmune basis. In this variety, there is a high incidence of associated endocrine disorders such as ovarian failure, thyroid disorders, diabetes mellitus, and hypoparathyroidism. Vitiligo and pernicious anemia are often found. The remaining cases are caused by granulomatous diseases such as tuberculosis or histoplasmosis, hemorrhagic disorders, neoplasms, amyloidosis, or infection. In the last case, fulminant meningococcal sepsis is a well-recognized cause.

Symptoms begin to appear after about 90% destruction of both adrenal cortices. ACTH levels increase because there is progressive loss of cortisol, resulting in hyperpigmentation because of its melanin-stimulating hormone effect. The metabolic effects such as muscular weakness, fatigue, weight loss, and orthostatic hypotension are rather vague, and their mechanisms are poorly understood. There is sodium loss and potassium retention because of decreased mineralocorticoid biosynthesis, and there may be hypoglycemia from loss of glucocorticoid activity. The severity of the clinical picture depends on the acuteness of insufficiency. Acute adrenocortical insufficiency is life threatening primarily because of hypotension and may occur as the initial presentation or in a patient with known insufficiency who is extremely stressed (as in infection). Without replacement therapy the patient may die.

Characteristic laboratory findings include low serum sodium, high serum potassium, low or low normal plasma glucose, and an elevated BUN (caused by dehydration). Specific diagnosis is made by the failure of adrenals to increase plasma cortisol in response to ACTH administration. Alternatively, elevated plasma ACTH with a low plasma cortisol level is diagnostic. A summary of clinical and laboratory findings is listed at right. The prognosis is excellent with chronic hydrocortisone and fluorocortisone therapy.

Major features of hypoadrenalism

Clinical findings
- Hyperpigmented skin: caused by elevated ACTH as response to decreased cortisol (only in primary hypoadrenalism)
- Salt craving: caused by sodium loss
- Weight loss: caused by disordered glucose metabolism
- Hypotension: caused by fluid and electrolyte disturbance
- Muscular weakness

Laboratory findings
- Low plasma cortisol
- Low urine 17-OHS
- Elevated serum potassium and decreased serum sodium: caused by decreased aldosterone
- Hypoglycemia: caused by decreased cortisol
- Autoantibodies
- Elevated BUN

Function tests
- ACTH stimulation test:
  - Primary: no response
  - Secondary: partial response

Adrenal hypofunction also can result from lack of ACTH either because of panhypopituitarism, or in sudden cessation of corticosteroid therapy to which the body cannot immediately respond with cortisol production. This is known as "secondary adrenocortical hypofunction" or "insufficiency." The presentation is similar to primary adrenocortical insufficiency. There are two important differences, however. These patients do not have hyperpigmentation, since ACTH is depressed, and usually they lack the signs or symptoms of mineralocorticoid deficiency. Therefore serum sodium and potassium are normal.[26]

Disease of adrenal medulla: pheochromocytoma

Pheochromocytoma, the only significant medullary disease, is a tumor of chromaffin cells that results in hypersecretion of catecholamines—epinephrine and norepinephrine. Although it accounts for approximately 0.5% of cases of hypertension, detection of pheochromocytoma is important because it is often surgically curable and, even more significant, can cause death by an acute hypertensive attack. Approximately 90% of pheochromocytomas occur in the adrenals, with the remainder occurring in extra-adrenal chromaffin cells. Most are benign and approximately 20% consist of multiple primary tumors. Pheochromocytomas are found as part of a group of diseases in various inherited family syndromes, such as neurofibromatosis (von Recklinghausen's disease) and multiple endocrine neoplasia type II syndrome (MEN II), which is composed of pheochromocytoma, hyperparathyroidism, and medullary carcinoma of the thyroid. These patients are more

Major features of pheochromocytoma

- Clinical findings
  - Hypertension: paroxysmal or sustained
  - Tachycardia
  - Excessive perspiration
  - Tachypnea
  - Severe headaches
  - May have associated family syndromes
  - Sense of doom
  - Nausea
  - Weight loss
  - Orthostatic hypotension
- Laboratory findings
  - Elevated metanephrines
  - Elevated vanillylmandelic acid
  - Elevated plasma norepinephrine or epinephrine, or both
- Function tests
  - Clonidine suppression tests
  - Provocative (challenge) tests: dangerous

likely to have multiple pheochromocytomas. The tumors may release their hormones in a sustained or intermittent (episodic) fashion.

Clinically, the most significant finding is either paroxysmal or sustained hypertension, which is found in almost all cases. Other features include tachycardia, excessive perspiration, tachypnea, severe headaches, palpitations, nausea, and a sense of impending doom. Signs of hypermetabolism such as orthostatic hypotension, glucose intolerance, and weight loss may also be present. Diagnosis is confirmed by laboratory testing in the proper clinical setting.[27,28]

The principal metabolic products of epinephrine and norepinephrine, the metanephrines and vanillylmandelic acid, are increased in urine. Provocative tests are extremely dangerous and are to be used only with the utmost caution in the diagnosis of pheochromocytoma. Laboratory abnormalities can also include elevated blood glucose because of the hyperglycemic effects of the catecholamines, a rise in the hematocrit, and leukocytosis with an increase in neutrophilic bands. Two separate elevations of metanephrine or vanillylmandelic acid in a 24-hour urine collection or elevated plasma catecholamine levels are the most reliable means of diagnosing pheochromocytoma.[29-31] A summary of typical clinical and laboratory findings is listed at left. Localization of the tumor is done by CT scanning. Surgical removal is the treatment of choice.

FUNCTION TESTS (see Table 43-2)

Adrenocortical function tests

Dexamethasone suppression test. Dexamethasone is a potent synthetic glucocorticoid that suppresses ACTH production by the previously discussed feedback inhibition mechanism *without* initiating a significant increase in steroid urine excretion. Its biological activity is approximately 25 to 40 times greater than cortisol; therefore only small amounts are necessary for the tests. It can be given in two doses: the low-dose dexamethasone suppression test and the high-dose dexamethasone suppression test. To administer the low dose test, 1 mg of dexamethasone is given to the patient at 11 P.M. The next morning at 8 A.M. the plasma cortisol is measured. A normal response is a de-

Table 43-2. Summary of adrenocortical function tests

	Hypercortical disease					Hypocortical disease	
Test	Primary	Cushing's disease	Ectopic ACTH	Primary hyperaldosteronism	Secondary hyperaldosteronism	Primary	Secondary
Dexamethasone suppression							
Low dose	No response	No response	No response				
High dose	No response	17-OH shows >50% suppression	No response				
ACTH stimulation						No response	Partial response
Metyrapone ACTH stimulation	No response	↑	Unpredictable				
Salt loading				↑	Variable response		
Nonaldosterone mineralocorticoid suppression				No response	↓		

↑, Increased; ↓, decreased.

creased plasma cortisol level of less than 50 μg/L of plasma (alternatively measured by a decrease in urinary 17-hydroxysteroid output), and this would tend to rule out the presence of Cushing's syndrome, especially with a normal urinary free cortisol (see below). If the low-dose dexamethasone is abnormal (that is, no suppression of corticoid synthesis), the high-dose dexamethasone suppression test is performed to help delineate the cause of Cushing's syndrome.

One can perform the high-dose test as a 2-day test or as an overnight 8 mg dose test. A lack of cortisol suppression accompanied by elevated ACTH is consistent with ectopic ACTH syndrome, whereas a lack of cortisol suppression with low levels of ACTH is consistent with an adrenal tumor. Normal to elevated serum ACTH levels with partial cortisol suppression, that is, more than 50% of base line by the high-dose dexamethasone suppression test, is consistent with Cushing's disease (pituitary hyperfunction). In summary, pituitary hypersecretion of ACTH is partially suppressible, and ectopic ACTH is not suppressible.

False-negative results from the dexamethasone suppression tests can occur when patients have delayed clearance and thus higher than usual levels of dexamethasone. In these cases, urinary free cortisol would be elevated and thus would be indicative of Cushing's syndrome. A second albeit unusual cause of false-negative results is an episodic production of cortisol in Cushing's syndrome. In these cases, repeated evaluation is warranted to establish a diagnosis of Cushing's syndrome. False-positive results are more common and include patients who have illnesses that result in stress and increased release of cortisol and who fail to take dexamethasone or are taking phenytoin (Dilantin) or phenobarbital, both of which enhance microsomal enzymes in the liver and can increase dexamethasone degradation.[32] This problem is alleviated when the patient is tested during an unstressed period.

ACTH stimulation test. The basis of this test is to administer ACTH intravascularly and subsequently measure plasma cortisol or urinary products. An example of a normal response would be elevation of plasma cortisol to 100 μg/L at 1 hour after ACTH administration. It is an *excellent* screening test for hypoadrenalism. In primary hypoadrenalism (Addison's disease), there is no response, whereas secondary hypoadrenalism shows a partial response. A possible dangerous side effect of adrenal crisis is avoided when one gives dexamethasone or fluorocortisone during the test. These steroids do not interfere with measurements.[33,34]

Metyrapone-ACTH stimulation test. This test measures the ability of the pituitary gland to respond to decreased levels of circulating cortisol. Metyrapone acts by inhibiting the synthesis of cortisol in the adrenal cortex by reducing the activity of the enzyme 11β-hydroxylase. After administration of metyrapone, serum cortisol levels should fall and stimulate ACTH production and subsequent stimulation of the steroid synthetic pathways of the adrenal. Although cortisol production is blocked by metyrapone, precursor compounds are synthesized and are detectable in the urine as 17-hydroxysteroids. Since it detects the response by measurement of adrenal products, it should be performed only after the adrenals respond normally to an ACTH challenge test.[35]

Metyrapone, 500 to 750 mg, is given orally every 4 hours for 24 hours. Urine is collected, started at the first dose for 2 consecutive 24-hour periods. In the second 24-hour period the urine should have greater than two times the normal 17-OHS levels. An elevation in 17-OHS occurs in pituitary-induced Cushing's syndrome (Cushing's disease). The failure of a rise in 17-OHS possibly indicates an impaired pituitary production of ACTH. This test has been used in the past to differentiate between the various causes of Cushing's syndrome. However, it is of little diagnostic value, and the plasma ACTH level test is much more reliable. Chlorpromazine interferes with the response to metyrapone and drugs that induce the hepatic microsomal enzymes, such as phenobarbital and phenytoin (Dilantin), increase degradation of metyrapone.[36]

Salt-loading test. The basis of this test is that sodium loading intensifies potassium depletion in primary aldosteronism. In this test, the patient ingests greater than 200 mEq of sodium chloride per day for 4 days. In patients who have aldosteronism the serum potassium falls to usually 3.5 mEq or less per liter; in those who do not have aldosteronism there is no change in the serum potassium. This test provides strong albeit indirect evidence of aldosteronism.

Nonaldosterone mineralocorticoid suppression test. This test is used if there ambiguity with the various aldosterone and renin testing. In this test, deoxycorticosterone acetate or lurocortisol is given and then plasma aldosterone is measured. There is no effect in primary aldosteronism, whereas plasma aldosterone is inhibited in secondary aldosteronism.

Adrenal medullary function tests

Provocative tests. Provocative, or challenge, testing was formerly used to diagnose pheochromocytoma. However, precise and accurate measurement of both plasma catecholamines has rendered these tests obsolete for the most part. In addition, these tests are extremely dangerous and can precipitate hypertensive or hypotensive crises. Nonetheless, some physicians advocate the use of these tests in difficult diagnostic cases when administered by a physician who is well-versed in the various side effects.

The tests are based on either a hypotensive or hypertensive response to a catecholamine-releasing agent or inhibitory agent, respectively. Examples of the former include histamine, glucagon, and tyramine, and an example of the latter is phentolamine (Regitine). Patients who have pheochromocytoma respond with a greater than 60 mm Hg elevation of blood pressure with the catecholamine-releasing agents or greater than a 35 mm Hg decrease of blood pressure in response to phentolamine. Both tests suffer from numerous false-positive and false-negative results.

Clonidine-suppression test. This test has just been recently described and its value is not well established.[37] Clonidine inhibits sympathetic hormone release in a normal person. Patients who have pheochromocytoma and are given 0.3 mg of clonidine do not show a decrease in the elevated plasma norepinephrine after 3 hours. Patients with essential hypertension show suppression of plasma norepinephrine levels.

CHANGE OF ANALYTE IN DISEASE (Table 43-3)

Adrenocortical analytes

Twenty-four-hour urinary-free cortisol. Although only 1% of total adrenocortical secretion is found in the urine as cortisols, it is quite useful as an index of adrenal hyperfunctioning because it is not affected by an increase in cortisol-binding globulin. In addition, it is filtered by the glomerulus in the kidney and only a portion is reabsorbed. Thus elevated serum cortisol levels result in amplified elevations in the urine of free cortisol. It is measured in a 24-hour total urine specimen. The reference range is 20 to 100 μg excreted per 24-hour period. The dexamethasone-suppression test has been found to be of great value in diagnosis or exclusion of Cushing's syndrome.[38,39] Because of the overlap of the reference range, it is *not* of value in the diagnosis of hypoadrenalism. The disadvantages include difficulty in accurate timing of urine collections, spuriously low values in patients with renal diseases, and high values in those using topical steroids. This test is not affected by liver disease, and obesity does not appear to elevate urinary cortisol levels.

Urinary 17-hydroxysteroids (17-OHS). The rationale for this determination is that measurement of the degradation products reflects adrenal gland function. A 24-hour urine collection reflects the quantitative output. Elevated levels are characteristic of Cushing's syndrome. Although the test is noninvasive and inexpensive to perform, it suffers from false-positive results in patients who are obese, stressed, or hyperthyroid. This is the major reason for the limited use of this particular test. Normal values rule out adrenal insufficiency. However, low values may occur in patients with cachexia, liver disease, severe depression, or hypothyroidism. In these conditions, free urine cortisol levels are usually normal. The reference range is 2 to 10 mg excreted per 24-hour period. This test has largely been superseded by measurements of urinary free cortisol or plasma cortisol.

Plasma cortisol. It should be noted that the serum cortisol levels include free and bound cortisol. Reference values for serum cortisol are 100 to 250 μg/L in the morning and 20 to 100 μg/L in the early evening. Measurement of serum cortisol may be used as part of the ACTH stimula-

Table 43-3. Change of analyte with disease

Disease	24-hour urinary free cortisol	17-OHS	Plasma ACTH	Urinary aldosterone	Plasma aldosterone	Plasma cortisol	Serum renin	Plasma catecholamine	Vanillylmandelic acid and metanephrine
Hypercortical disease									
Primary Cushing's syndrome	↑	↑	↓			±			
Cushing's disease (secondary)	↑	↑	±			±			
Ectopic ACTH	↑	↑	↑			±			
Primary hyperaldosteronism				↑	↑		↓		
Secondary hyperaldosteronism				↑	↑		↑		
Hypocortical disease									
Primary	±	↓	↑			↓			
Secondary	±	↓	↓			↓			
Pheochromocytoma								↑	↑

↑, Elevated; ±, variable response; ↓, diminished.

tion test. It is used in the diagnosis of Cushing's syndrome; however, some patients who truly have the disease may have values in the reference range and people who have increased cortisol-binding globulin, as seen in pregnancy and estrogen therapy, have increased serum cortisol.[42] In both of the above situations, urinary free cortisol is usually more helpful.

Plasma ACTH. Plasma ACTH levels are used to distinguish various causes of Cushing's syndrome. Drawbacks include expense and rapid in vitro degradation of ACTH. Patients who have Cushing's disease have normal to slightly elevated plasma ACTH levels (greater than 200 pg/mL), whereas patients who have ectopic ACTH syndrome have plasma ACTH levels that are frequently quite elevated. Low levels (that is, less than 75 pg/mL) are indicative of autonomous adrenal hyperactivity. It should be noted that this test is also used to distinguish between primary and secondary adrenal hypofunction. Those patients who have the signs and symptoms of adrenal hypofunction and have high ACTH levels are placed in the primary group and those who have low ACTH are placed in the secondary group. Because of the overlap of ACTH levels in the various causes of Cushing's syndrome, it is best used with the dexamethasone suppression test, especially the high-dose test. The disadvantage of the plasma ACTH test is that it is quite insensitive in low ranges and thus of little value in detection of hypopituitary function.[40]

Urinary aldosterone excretion. In this test, a 24-hour sample of urine is collected at approximately pH 5, with preservatives used to prevent bacterial growth. The urine is then treated with acid to convert the 18-glucosiduronate excretion product of aldosterone to free aldosterone, which is measurable by radioimmunoassay methods. The test is performed when the patient is on a high salt intake and lying down. It is important that the patient maintain a high salt intake before the test because in essential hypertension (the most common type of hypertension) any decrease in salt intake will normally lead to an increase in aldosterone production and false-positive results. The normal urinary aldosterone levels are 4 to 17 μg/24 hours; in patients who have adenomas resulting in aldosteronism the mean values are 37.6 ± 3.1 μg/24 hours. In patients with hyperplasia responsible for the aldosterone the values were 22.5 ± 1.5 μg/24 hours.[41]

Plasma aldosterone levels. Because aldosterone is present in extremely low quantities in the blood, very sensitive tests are needed to measure it. Plasma aldosterone levels are measured by chromatography and radioimmunoassay methods. Reference range is 10 to 50 ng/L in the supine patient. There are a number of different approaches to this test. However, elevated aldosterone in the presence of low plasma renin is diagnostic of primary aldosteronism. It is important that the patient discontinue diuretics for at least 3 weeks and other antihypertensives for at least 1 week before this test and maintain a high salt intake as for the urinary aldosterone excretion test (see above).

Plasma renin. There are many technical difficulties in the measurement of renin, and it has not found great use. One can determine renin by measuring its activity or indirectly measuring its concentration by quantitating the rate of angiotensin I formation by radioimmunoassay.

Adrenal medullary analytes

Plasma catecholamines. Both epinephrine and norepinephrine are measurable by high-performance liquid chromatography or radioenzymic methods. Blood is collected from patients in the morning after an overnight fast. It is important that the patient rest for approximately 30 minutes in the supine position, with a heparinized indwelling catheter inserted 20 to 30 minutes before sampling to avoid any increase in catecholamines secondary to stress. The use of plasma catecholamine measurements provides a quick albeit expensive method of diagnosing pheochromocytoma. To avoid breakdown of the catecholamine, one should keep the blood specimen on ice and centrifuge it at 4° C within 1 hour of collection. If the specimen is not used immediately, it should be stored at −20° C for up to several months. It is believed that abnormal levels of either norepinephrine or epinephrine persist in patients with pheochromocytoma, even in those who are normotensive at the time of the study.

Twenty-four-hour urinary vanillylmandelic acid (VMA) and metanephrines. It is desirable to measure both VMA and metanephrines in a 24-hour urine specimen that has been collected in acid. Although urine tests are usually fraught with multiple problems such as compliance, it is believed that high levels of both analytes on two separate occasions are diagnostic of pheochromocytoma. One should note, however, that there are two schools of thought. Some clinicians advocate use of 24-hour VMA and metanephrine levels, but others advocate measurement of plasma catecholamine levels to diagnose pheochromocytoma.[31] Total urinary metanephrines can increase in times of severe stress.

REFERENCES

1. CIBA collection of medical illustrations. In Netter, F.H.: Endocrine system and selected metabolic diseases, vol. 4, Summit, N.J., 1965, CIBA Pharmaceutical Co.
2. Baxter, J.D.: Glucocorticoid hormone action. In Gill, G.N., editor: Pharmacology of adrenal cortical hormones, Oxford, 1976, Pergamon Press, Ltd.
3. Glaman, N.H.: How corticosteroids work, J. Allergy Clin. Immunol. **55**:145-151, 1975.
4. Fauci, A.S., Dale D.C., and Balow, J.E.: Glucocorticoid therapy: mechanisms of action and clinical considerations, Ann. Intern. Med. **84**:304-315, 1976.
5. Craddock, C.G.: Corticosteroid-induced lymphopenia, immunosuppression, and body defense, Ann. Intern. Med. **88**:564-566, 1978.
6. Aronow, L.: Effects of glucocorticoids on fibroblasts. In Baxter, J.D., and Rousseau, G.G., editors: Glucocorticoid hormone action, New York, 1979, Springer-Verlag New York, Inc., pp. 327-340.
7. Glaser, G.H.: Psychotic reactions induced by corticotropin (ACTH) and cortisone, Psychosom. Med. **15**:280-291, 1953.
8. Sambhi, M.P., Weil, M.H., and Udhoji, V.N.: Acute pharmacologic effects of glucocorticoids: cardiac output and related hemody-

namic changes in normal subjects and patients with shock, Circulation **31**:523-530, 1965.
9. Schumer, W.: Steroids in the treatment of clinical septic shock, Ann. Surg. **184**:333-341, 1976.
10. Krieger, D.T.: Rhythms in CRF, ACTH, and corticosteroids. In Krieger, D.T., editor: Endocrine rhythms, New York, 1979, Raven Press, pp. 123-142.
11. Johnson, L.K., Baxter, J.D., and Rousseau, G.G.: Mechanisms of glucocorticoid receptor function. In Baxter, J.D., and Rousseau, G.G., editors: Glucocorticoid hormone action, New York, 1979, Springer-Verlag New York, Inc., pp. 135-160.
12. Ludens, J.H., and Fanestil, D.D.: The mechanism of aldosterone function. In Gill, G.N., editor: Pharmacology of adrenal cortical hormones, New York, 1979, Pergamon Press, Inc., pp. 143-184.
13. Brooks, R.V.: Biosynthesis and metabolism of adrenocortical steroids. In James, V.H.T., editor: The adrenal gland, New York, 1979, Raven Press, pp. 67-92.
14. Daughday, W.H.: The binding of corticosteroids by plasma protein. In Eisenstein, A.B., editor: The adrenal cortex, Boston, 1967, Little, Brown & Co., pp. 385-403.
15. Peterson, R.E.: Metabolism of adrenal cortisol steroids. In Christy, N.P., editor: The human adrenal cortex, New York, 1971, Harper & Row, Publishers, Inc., pp. 87-189.
16. Zipser, R.D., Meidor, V., and Horton, R.: Characteristics of aldosterone binding in human plasma, J. Clin. Endocrinol. Metab. **50**:158-162, 1979.
17. Hellman, L., Kream, J., and Rosenfeld, R.S.: The metabolism and 24-hour plasma concentrations of androsterone in man, J. Clin. Endocrinol. Metab. **45**:35-39, 1977.
18. McKenna, T.J., Island, D.P., Nicholson, W.E., and Liddle, G.W.: The effects of potassium on early and late steps in aldosterone biosynthesis in cells of the zona glomerulosa, Endocrinology **103**:1411-1416, 1978.
19. Christy, N.P.: Iatrogenic Cushing's syndrome. In Christy, N.P., editor: The human adrenal cortex, New York, 1971, Harper & Row, Publishers, Inc., pp. 395-425.
20. Huff, T.A.: Clinical syndromes related to disorders of adrenocorticotrophic hormone. In Allen, M.B., Jr., editor: The pituitary: a current review, New York, 1977, Academic Press, Inc., pp. 153-168.
21. Crapo, L.: Cushing's syndrome: a review of diagnostic tests, Metabolism **28**:955, 1979.
22. Singer, W., Kovacs, K., Ryan, N., and Horvath, E.: Ectopic ACTH syndrome: clinicopathological correlations, J. Clin. Pathol. **31**:591-598, 1978.
23. Weinberger, M.H., Grim, C.E., Hollifield, J.W., et al.: Primary aldosteronism: diagnosis, localization, and treatment, Ann. Intern. Med. **90**:386, 1979.
24. Brook, C.G.D.: Congenital adrenal hyperplasia: pathology, diagnosis, and treatment. In James, V.H.T., editor: The adrenal gland, New York, 1979, Raven Press, pp. 243-257.
25. Knowlton, A.I.: Addison's disease: a review of its clinical course and management. In Christy, N.P., editor: The human adrenal cortex, New York, 1971, Harper & Row, Publishers, Inc., pp. 329-358.
26. Nerup, J.: Addison's disease—a review of some clinical, pathological and immunological features, Dan. Med. Bull. **21**:201-217, 1974.
27. Gifford, R.W., Jr., Kvale, W.F., Mahler, R.T., et al.: Clinical features, diagnosis and treatment of pheochromocytoma: a review of 76 cases, Mayo Clin. Proc. **39**:281, 1964.
28. Melicow, M.M.: One hundred cases of pheochromocytoma (107 tumors) at the Columbia-Presbyterian Medical Center, 1926-1976: a clinicopathological analysis, Cancer **40**:1987, 1977.
29. Gitlow, S.E., Mendlowitz, M., and Bertani, L.M.: The biochemical techniques for detecting and establishing the presence of a pheochromocytoma: a review of ten years' experience, Am. J. Cardiol. **26**:270, 1970.
30. Kaplan, N.M., Kramer, N.J., Holland, O.B., et al.: Single-voided urine metanephrine assays in screening for pheochromocytoma, Arch. Intern. Med. **137**:190, 1977.
31. Bravo, E.L., Tarazi, R.C., Gifford, R.W., et al.: Circulatory and urinary catecholamines in pheochromocytoma: diagnosis and pathophysiologic implications, N. Engl. J. Med. **301**:682, 1979.
32. Jubiz, W., Meikle, A.W., Levinson, R.A., et al.: Effect of diphenylhydantoin on the metabolism of dexamethasone: mechanism of the abnormal dexamethasone suppression in humans, N. Engl. J. Med. **283**:11-14, 1970.
33. Grieg, W.R., Jasani, M.K., Boyle, J.A., and Maxwell, J.D.: Corticotrophin stimulation tests, Mem. Soc. Endocrinol. **17**:175-192, 1968.
34. Lindholm, J., Kehlet, H., Blichert-Toft, M., et al.: Reliability of the 30-minute ACTH test in assessing hypothalamic-pituitary-adrenal function, J. Clin. Endocrinol. Metab. **47**;272-274, 1978.
35. Spiger, M., Jubiz, W., Meikle, A.W., et al.: Single-dose metyrapone test: review of a four-year experience, Arch. Intern. Med. **135**:698-700, 1975.
36. Spiger, M., Jubiz, W., Meikle, A.W., et al.: Single-dose metyrapone test: review of a four-year experience, Arch. Intern. Med. **135**:698, 1975.
37. Bravo, E.L., Tarazi, R.C., Fouad, F.M., et al.: Clonidine-suppression test: a useful aid in the diagnosis of pheochromocytoma, N. Engl. J. Med. **305**:623, 1981.
38. Burke, C.W., and Beardwell, C.G.: Cushing's syndrome: an evaluation of the clinical usefulness of urinary free cortisol and other urinary steroid measurements in diagnosis, Q. J. Med. **42**:175-204, 1973.
39. Crapo, L.: Cushing's syndrome: a review of diagnostic tests, Metabolism **28**:955-977, 1979.
40. Kehlet, H., Blichert-Toft, M., Lindholm, J., et al.: Short ACTH test in assessing hypothalamic-pituitary-adrenocortical function, Br. Med. J. **1**:249, 1976.
41. Biglier, E.G.: Effect of posture on plasma concentrations of aldosterone in hypertension and primary hyperaldosteronism, Nephron **23**:112, 1976.
42. Aron, D.C., Tyrrell, J.B., Fitzgerald, P.C., et al.: Cushing's syndrome: problems in diagnosis, Medicine **60**:25-35, 1981.

BIBLIOGRAPHY

Felig, P., Baxter, J., Broadus, A., and Frohman, L.: Endocrinology and metabolism, New York, 1981, McGraw-Hill Book Co., pp. 385-626.

Ryan, W.: Endocrine disorders, Chicago, 1980, Year Book Medical Publishers, Inc.

Toft, A.D., Campbell, M.B., and Seth, J.: Diagnosis and management of endocrine diseases including diabetes mellitus, Oxford, 1981, Blackwell Scientific Publishing Co., Ltd.

Chapter 44 Diseases of genetic origin

Thaddeus E. Kelly

allele Various forms of a gene that may appear at a specific locus.

amniocentesis A transabdominal aspiration of the uterus by syringe in order to obtain amniotic fluid.

aneuploidy A chromosomal abnormality caused by the addition or absence of an entire chromosome.

autosomal Pertaining to any of the 22 chromosomes except the X and Y chromosomes.

Barr body The condensation of nuclear (genetic) material of the inactivated X chromosome.

chromosome Nuclear structure containing a linear array of genes. Humans have 23 pairs of chromosomes.

diploid The duplicate representation of each gene.

dominant trait A trait that is expressed or determined by the heterozygous presence of an allele at the locus on the chromosome.

Down's syndrome A condition noted by mental retardation and physical abnormalities caused by trisomy of chromosome 21 (formerly called "mongolism").

dysmorphogenesis Physical defects caused by intrinsically altered embryonic development.

galactosemia A toxicity syndrome associated with intolerance to dietary galactose and characterized by deficiency of the enzyme galactose-1-phosphate uridyl transferase.

gene The smallest biological unit of heredity, located on specific sites of specific chromosomes. A gene contains the encoding for one specific protein.

Guthrie test A microbiological assay for the amount of serum phenylalanine; used to screen for PKU.

heterozygous Pertaining to a state in which a pair of alleles are dissimilar at both positions of the same loci.

homozygous Pertaining to a state in which a pair of alleles are the same at both positions of the same loci.

karyotype The chromosomal makeup of a nucleated cell.

Lesch-Nyhan syndrome A rare sex-linked error of purine metabolism characterized by mental retardation and self-mutilation. There is a deficiency of the enzyme hypoxanthine-guanine phosphoribosyl transferase.

locus The particular location on a given chromosome occupied by a structural gene.

lyonization A process of random inactivation of the X chromosome to compensate for the double-gene dosage of the X chromosome in females.

meiosis Segregation of genes and chromosomes with reduction from a diploid to a haploid number in sperm and egg formation.

mitogen A chemical substance that induces cells in culture to divide.

monosomy Absence of one chromosome from an otherwise diploid cell ($2n-1$). Only known disorder is absence of Y or X chromosome in Turner's syndrome.

phenylketonuria (PKU) An autosomal, recessively inherited disorder resulting from a defect in conversion of phenylalanine to tyrosine because of a phenylalanine hydroxylase deficiency.

recessive trait The expression of a trait that requires that both genes consist of the same allele.

Tay-Sachs disease Infantile form of a recessive hereditary disorder caused by a defect in lipid metabolism in which sphingolipids accumulate in brain, resulting in progressive mental and physical degeneration. The disease primarily occurs in children of Jewish ancestry.

trait An observable feature of an organism that is visually apparent or laboratory derived.

translocation A term used in genetics to describe the movement of a portion of one chromosome into the structure of another.

trisomy Presence of one additional chromosome of a specific pair in an otherwise diploid cell ($2n+1$ chromosomes). See *Down's syndrome*.

Turner's syndrome Monosomal condition caused by absence of the Y or X chromosome (XO phenotype) and characterized by physical defects.

X linked Pertaining to a trait carried on the X chromosome.

GENETIC BASIS OF INHERITANCE

In the 1860s, after a series of simple experiments, Gregor Mendel proposed two laws regarding the behavior of the transmission of traits from generation to generation. The terms *trait, character,* and *phenotype* are synonymous and refer to an observable feature of an organism, whether that observation be a visually apparent or laboratory-derived feature. These two laws formed the basis of our understanding of the segregation of single gene–determined traits. In the first law Mendel stated that the segregation of a trait followed predictable patterns in the pure-bred and hybrid states. That is, one could anticipate the proportion of offspring expressing various forms of a given trait. In the second law he recognized that the transmission of two or more traits occurred as a random process. That is, the segregation of one trait was unrelated to the proportions observed of the second trait. Mendel described the inheritance of traits as being determined by "factors." Approximately 40 years later, with the capability of observing cellular division under the microscope, it was recognized that the behavior of chromosomes during cellular division was consistent with the predictions made by Mendel regarding the segregation of traits. This led to a new appreciation of Mendel's work, as now a biological basis for his factors and their segregation was appreciated. Thus we now recognize the factors described by Mendel as *genes* and that these were carried by *chromosomes* whose segregation could be observed in cellular division.

Chemical basis

Genes must be described in functional terms. Genes are composed of DNA and occasionally RNA, which, through their nucleotide sequence, encode a specific function. These genes occupy specific locations on chromosomes and their chemical structure is varied and complicated. Structural genes code for the amino acid sequence of a polypeptide that has a specific physiological function. That polypeptide may function as a single protein or combine with

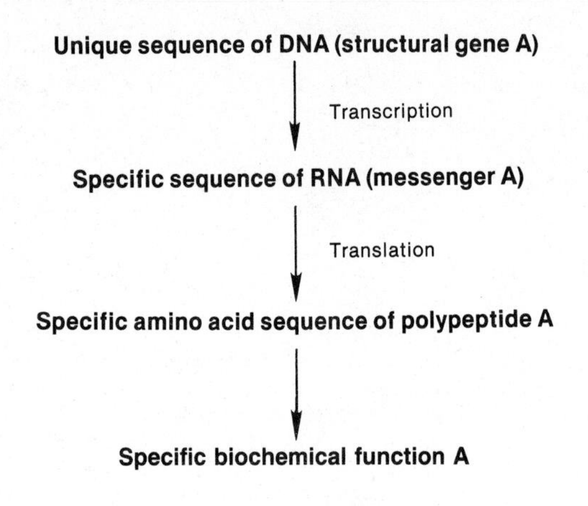

Fig. 44-1. Sequence of events from a structural gene leading to a specific, normal biochemical function.

Table 44-1. Diseases caused by various genetic mutations

Protein type	Consequence of mutation	Example
Enzyme	Loss of enzyme activity	Phenylketonuria
Hemoglobin	Altered oxygen affinity	Sickle cell disease
Structural protein (cartilage collagen)	Defective bone matrix	Osteogenesis imperfecta
Receptor	Altered metabolic regulation	Familial hypercholesterolemia
Membrane protein	Altered membrane transport	Cystinuria
Coagulation protein	Defective coagulation	Hemophilia
Carrier protein	Inability to transport compound	Hemochromatosis

other polypeptides to form a heteropolymeric functional unit. The newly formed or nascent polypeptides often undergo structural changes, or posttranslational modification, to attain their ultimate functional structure. However, the basic premise remains that one structural gene encodes a unique polypeptide with a specific function (Fig. 44-1). Changes in the DNA sequence of a structural gene, a mutation, will result in a structurally altered polypeptide. This in turn, depending on the particular alteration in the polypeptide, will produce a change in the phenotype through a functional change in the gene product. The kinds of diseases that result from such mutations depend on the function of the polypeptide. Examples are illustrated in Table 44-1.

The genetic material of human cells is *diploid* in that each gene is represented twice, one of each occurring at a specific location on a pair of similar or homologous chromosomes. The segregation of genes and chromosomes through *meiosis* occurs with reduction of the chromosome number from a diploid to haploid number. Thus humans have the chromosomal or modal number of 46 in somatic cells and 23 in gametes. The particular location on a given chromosome that is occupied by a structural gene is called a *locus*. Thus on the long arm of chromosome 1 is the locus for the Duffy blood group. At that locus, various alternative forms of the gene, or alleles, may occur. Numerous changes may occur in the DNA sequence of a structural gene; some of these include deletion of all or part of the gene, substitution of a different nucleotide for any of the nucleotides in the DNA sequence, and failure to initiate or terminate proper reading of the DNA code. Thus many different kinds of mutations may affect a specific gene; these mutations constitute the series of alleles that can exist at any one gene locus. A series of allelic mutations may impart a range of functional consequences on the polypeptide gene product. Many different mutations occur in the structural gene for the beta chain of hemoglobin. This series of alleles result in functional changes in beta-hemoglobin chains that range from insignificant, no disease, to lack of the gene product and severe disease, beta-thalassemia.

Virtually every structural gene locus has a series of various alleles that may occur. In most instances these mutations represent a single nucleotide change in the structure of the gene and thus a single amino acid substitution in the gene product. Many of these gene products are equally functional but may be recognized by different electrophoretic or immunological properties of the gene product. The two most common alleles at the Duffy blood group locus are designated *a* and *b*. The gene products for the *a* and *b* alleles of the Duffy blood group are recognized by standard blood-typing techniques. When the pair of alleles at a given locus are the same, the person is said to be *homozygous* at that locus. Thus a person with the *a* allele at both positions on the pair of 1 chromosomes for the Duffy blood group is homozygous *a*. When the pair of alleles are dissimilar, such as, the Duffy allele *a* and *b*, the person is a heterozygote.

Single gene patterns

The pairs of chromosomes numbered 1 to 22 are called *autosomes*, and the remaining pair are called the *sex chromosomes* (X and Y). When a *trait* is determined by the heterozygous presence of an allele at the locus on an autosome, that trait is inherited as an *autosomal dominant trait*. When the expression of the trait requires that both genes consist of the same allele, the trait is said to be *autosomal recessively inherited*. The terms *dominant* and *recessive* refer to the trait in question and not the genes determining that trait. For example, the presence of the S and A alleles for the beta chain of hemoglobin gives rise to the presence of normal hemoglobin A and sickle hemoglobin S in a person. That condition being present in the heterozygous state is dominantly inherited. Thus the presence of sickle hemoglobin by electrophoresis and a

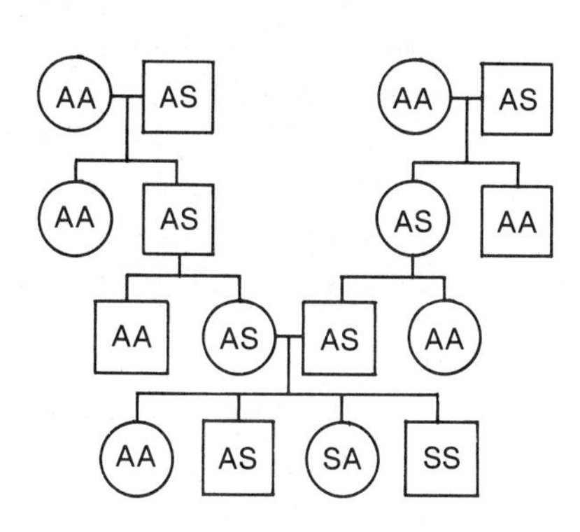

AA, Homozygous for normal hemoglobin A
AS, Heterozygous for hemoglobins A and S (sickle cell carrier)
SS, Homozygous for mutant hemoglobin S (sickle cell disease)

Fig. 44-2. Pedigree that demonstrates segregation of single-gene mutations. Heterozygous state (carrier for sickle-cell disease in this example) is inherited from a single parent, whereas homozygous state (a person affected with sickle cell disease in this example) requires inheritance of a mutant gene from each parent.

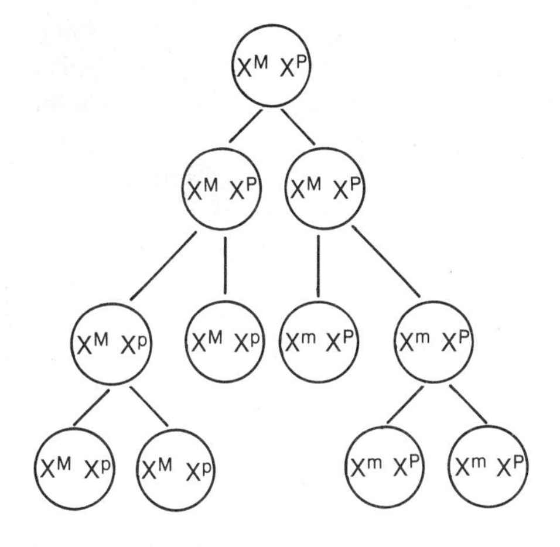

$\mathbf{X^M}$, Active maternally derived X chromosome
$\mathbf{X^P}$, Active paternally derived X chromosome
$\mathbf{X^m}$, Inactive maternally derived X chromosome
$\mathbf{X^p}$, Inactive paternally derived X chromosome

Fig. 44-3. Random inactivation of the X chromosome in the 46,XX female. Once an X chromosome is inactivated, that particular X is inactivated in all subsequent progeny of that cell.

tendency for sickling of red blood cells to occur under reduced oxygen tension is inherited as an autosomal dominant trait. However, the disease *sickle cell disease* requires that both alleles code for S hemoglobin. The person with sickle cell disease is homozygous for the S allele and sickle cell disease is inherited as an autosomal recessive disorder. With this definition it becomes apparent that one parent may transmit an autosomal dominant disorder to their offspring as it is determined by the presence of a single allele, but that an autosomal recessively inherited disorder requires the inheritance of mutant alleles from both parents. It can be seen from Fig. 44-2 that one may inherit the carrier state for sickle cell disease from a single parent, but sickle cell disease itself must be inherited from both parents. In general, the disorders that are characterized by altered biochemical metabolism and are diagnosable through biochemical means are conditions that are autosomal recessively inherited.

Of the 46 chromosomes of the human male, the two sex chromosomes consist of an X chromosome and a Y chromosome. The human female has two X chromosomes. The enzyme glucose-6-phosphate dehydrogenase (G-6-PD) is coded for by a gene on the X chromosome. Because a female has two X chromosomes and a male has one X chromosome, it might be expected that a female would make twice as much G-6-PD as a male does. On the average, however, females and males make equal amounts of G-6-PD. Thus there is gene dosage compensation that occurs by a mechanism called *lyonization,* or random inactivation of the X chromosome. During early embryonic development of the 46,XX female, one of the two X chromosomes is randomly inactivated; it no longer produces gene products. Because this is a random process, each female is a mosaic, or mixture, of cells in which one or the other, but not both, X chromosomes is functioning (Fig. 44-3). The Y chromosome determines maleness by a mechanism that currently is not clearly understood. Associated with the Y chromosomes of males is a factor called the H-Y antigen. It is believed that the Y chromosome does not code for the H-Y antigen but does regulate its expression. The H-Y antigen stimulates the embryonic gonad to develop into a testis, which in turn directs the development of the embryo into a male. In the absence of the H-Y antigen, the gonad becomes an ovary and the embryo a female. The Y chromosome apparently does not carry any structural genes other than those related to male determination. The X chromosome, however, carries a similar complement of structural genes, as seen on the medium-sized autosomes.

This difference in sex chromosome constitution of males and females is the basis for the particular pattern of inheritance that is known as X linked. If there is a mutation on the single X chromosome of a male, that male will always transmit that mutant gene to all of his daughters and to none of his sons. The female, however, may give either of her X chromosomes, thus the normal or mutant gene, to a son or daughter. If the mutation on the X chromosome results in a disease when present in the heterozygous state of the female, the trait is known as an *X-linked dominant disorder* and will be observed in both males and females. Because the single X of the male carries a mutation, the disease is usually more severe than in the female who has a mutant gene and a normal gene. If the mutation is not expressed as a recognizable trait in the heterozygous state of the female, the disorder is *X-linked recessively inherited*. Thus males will manifest the disorder, and females will carry but not manifest the disorder. Females are carriers for hemophilia and Duchenne muscular dystrophy, and the disease occurs among males who receive the mutant gene. There are tests that will identify the carrier female of most X-linked recessive diseases. In general these tests demonstrate a partial expression of the defect seen in affected males.

A number of traits such as height and intelligence and birth defects such as cleft lip and congenital heart disease are determined by the joint actions of a number of genes and environment. This mode of inheritance is called *multifactorial causation*. Such traits are familial and have a major genetic input but do not segregate in families, similar to that seen for autosomal dominant and recessive traits. It is the consequence of a number of genes received from each parent and a variable environmental influence that determines the expression of such a trait. In addition to isolated birth defects seen in infants, multifactorial causation is responsible for a number of common diseases of adults. Examples include insulin-independent diabetes mellitus, osteoarthritis, gout, and certain forms of hypertension and coronary artery disease.

PATHOLOGICAL DISORDERS ASSOCIATED WITH ABNORMAL CHROMOSOMES

Aneuploidy

In addition to genetic disorders being determined by the action of single gene mutations and the combination of a number of genes from both parents, abnormalities are associated with alterations of chromosomes that involve large blocks of genes. Disorders that occur as the result of a chromosome abnormality involve a change in the gene dosage rather than a change in the gene structure. This produces a change in the reading of the blueprint for the structural development of the embryo. Although it is sometimes possible to show a decrease or increase in the amount of gene product made, the diagnosis is made by an analysis of the chromosomes rather than a change of any analyte. Chromosomal abnormalities that result in phenotypic abnormalities are characterized by a quantitative change in the amount of chromosomal material present. Such an abnormality comes about in two ways. First, there is the presence or absence of an entire chromosome, an *aneuploidy*. This is exemplified by the presence of an extra 21 chromosome in *trisomy 21*, which causes *Down's syndrome*, and by *monosomy* for an X chromosome, resulting in *Turner's syndrome*. *Trisomy* means the presence of three copies of a chromosome rather than the normal two and *monosomy* means a single copy. Second, there are structural alterations in chromosomes that result in the loss or addition of part but not all of a chromosome. A variety of structural abnormalities of chromosomes can lead to such a deviation from the normal amount of chromosomal material.

Chromosomal abnormalities produce phenotypic abnormalities not through the presence of abnormal genes but rather through an alteration in the total amount of genetic material present. This quantitative change in chromosomal material in essence represents an alteration of the blueprint for the structural development of the embryo. Thus chromosomal abnormalities involving the autosomes, whether it be a partial or complete loss or addition of a chromosome, always result in altered morphogenesis expressed as major and minor birth defects plus altered mental development in the form of mental retardation. The altered physical development in the structure of a chromosomally abnormal embryo is called *dysmorphogenesis;* the individual physical abnormalities are called *dysmorphic features*. In Down's syndrome this consists of such features as small ears, unusual creases on the palms, upward slant of the eyes, small head size, congenital heart disease, short stature, and others.

The analysis of Down's syndrome provides an excellent example of the principles involving abnormalities in autosomes. Down's syndrome is an easily recognized combination of major and minor physical abnormalities associated with mental retardation. It occurs as the specific result of the presence of a triple dose of chromosome 21 material. This occurs most commonly as trisomy 21, the presence of three separate 21 chromosomes. Down's syndrome most often is the result of *meiotic nondysjunction* in the second cell division of meiosis. *Meiosis* is the special form of cell division that occurs in germ cells that produce ova or sperm. This division reduces the chromosome number from 46 to 23, with one member of each pair of chromosomes being represented. *Nondysjunction* is the failure of one of a pair of chromosomes to go to each daughter cell during division; instead both go to one daughter cell (Fig. 44-4). The frequency of such nondysjunction increases with maternal age. For that reason, prenatal diagnosis through *amniocentesis* with cytogenetic study is recommended for women 35 years of age or older. This form of Down's syndrome is not recognized to occur with an increased familial incidence.

Structural abnormalities

A familial form of Down's syndrome does occur because of a structural abnormality known as a *robertsonian translocation*. This translocation is formed by the fusion of the centromeres, most commonly, of chromosomes 14

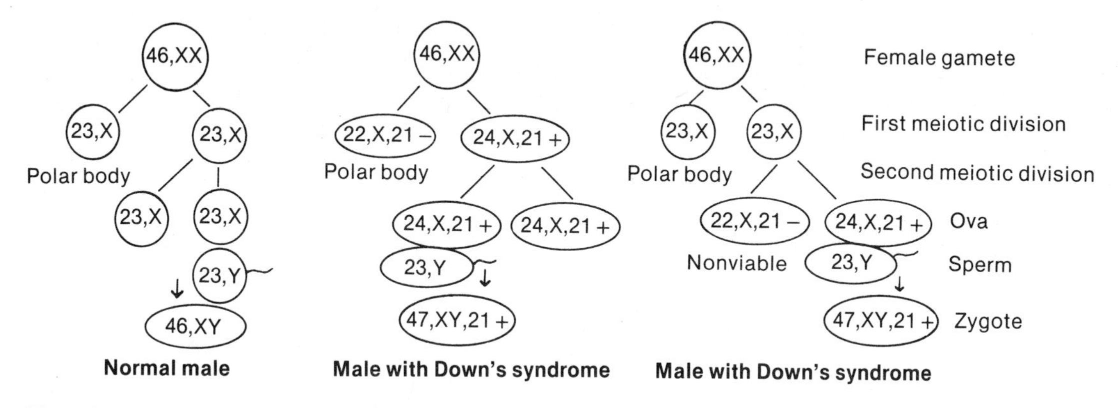

Fig. 44-4. Meiotic nondisjunction. *Left,* Normal meiosis of an ovum fertilized by a normal sperm yielding a normal zygote. *Center,* Error in first cell division of meiosis results in an aneuploid ovum that when fertilized yields a 47,XY,21+ male with Down's syndrome. *Right,* Error has occurred in second cell division of meiosis with same result.

	14, 21 t14/21	45,XX,t14/21	Carrier
Gametes		**Zygotes**	**Phenotype**
Ovum Sperm	14, 21 14, 21	46,XX	Normal
Ovum Sperm	t14/21 14, 21	45,XX,t14/21	Normal (carrier)
Ovum Sperm	t14/21,21 14, 21	46,XX,t14/21	Down's syndrome

Fig. 44-5. Translocation 14/21. Carrier has 45 chromosomes with one normal 14, one 21, and genetic content of a 14 and 21 contained in translocation. A person with Down's syndrome secondary to such a translocation has two separate 21's and a third 21 as part of translocation. Such a person has a total of 46 chromosomes but genetic content of three number 21's, that is, trisomy 21.

and 21. Thus a carrier will have a total chromosomal number of 45. However, the translocation chromosome contains the normal amount of chromosomal material of two separate chromosomes. The carrier, though having only 45 chromosomes, is phenotypically normal in that he or she has the normal genetic make up of 46 chromosomes. Meiosis in a carrier of such a translocation can result in a variety of different gametes. Thus among the offspring of carriers of a 14:21 translocation, one may observe normal infants with a normal 46 chromosomal constitution, phenotypically normal infants with 45 chromosomes and a translocation similar to the parent, or infants with 46 chromosomes of which there are two normal 21 chromosomes and the material of a third 21 present in the translocation (Fig. 44-5). Such children have typical Down's syndrome. Thus the translocation may occur in multiple members of a family, each of whom is at increased risk of having a child with Down's syndrome. One can differentiate the familial and sporadic forms of Down's syndrome by chromosomal studies. For that reason it is advisable that each person with Down's syndrome undergo chromosomal analysis to determine if there is an increased risk for other family members of having children with Down's syndrome.

Sex chromosome abnormalities

Alterations in the number of sex chromosomes are common and occur as several types of aneuploidy. These conditions were originally recognized through the use of *buccal smear* for *Barr body analysis*. Each X chromosome in excess of 1 in the cell nucleus is inactivated and is seen

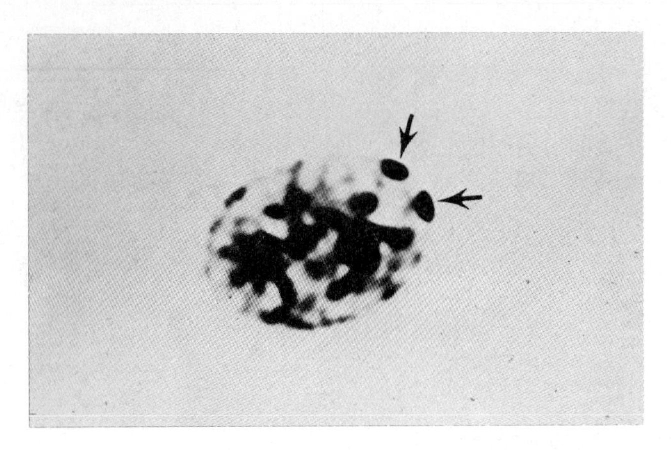

Fig. 44-6. Barr body. This cell has two Barr bodies seen as nuclear condensations just inside nuclear membrane at 3-o'clock position *(arrow)*. Two Barr bodies would be seen in a male with 48,XXXY and a female with 47,XXX.

Table 44-2. Buccal smear findings in sex chromosome abnormalities

Karyotype	Sex	Barr body	Y body
46,XX	F	+	0
46,XY	M	0	+
45,X	F	0	0
47,XXX	F	++	0
48,XXXX	F	+++	0
49,XXXXX	F	++++	0
47,XXY	M	+	+
47,XYY	M	0	++
48,XXYY	M	+	++
49,XXXXY	M	+++	+

microscopically as the Barr body. The Barr body, named for its discoverer, is the nuclear condensation of the inactivated X chromosome and appears as a clump inside and adjacent to the nuclear membrane (Fig. 44-6). The cells of a normal male contain only one X chromosome; thus no Barr body is present. Cells of a normal female contain two X chromosomes and demonstrate a single Barr body.

Abnormalities of sex chromosomes in females occur most commonly as a 45,X and a 47,XXX karyotype. The former is the most common chromosomal finding in Turner's syndrome. This disorder has three major features: short stature, minor dysmorphic abnormalities and congenital malformations, and sexual infantilism. The sexual infantilism occurs as a result of fibrosis of the infantile ovary so that at the time of puberty there are neither follicles nor hormone-producing cells present. In its classic form this syndrome is diagnosed by buccal smear, which will show the absence of Barr bodies. However, a variety of X chromosome abnormalities may result in Turner's syndrome; for that reason a full lymphocyte karyotype should always be done regardless of the findings on buccal smear. The 47,XXX karyotype is not associated with significant clinical abnormalities. A large series of such persons has disclosed a mild decrease in intellectual function and height, but in general such persons should be considered clinically normal. Such females have normal ovarian function and are fertile. There has been no recognized increase in risk of chromosomal anomalies occurring in the offspring of such women.

The most common abnormalities of sex chromosomes in males include the 47,XXY and 47,XYY karyotypes. The Klinefelter syndrome (47,XXY) has three major clinical findings: minor to no dysmorphic abnormalities, normal to mild retardation in cognitive function, and failure of testicular development with secondary sexual infantilism and a eunuchoid body habitus. Because of the lack of testosterone production at the age of puberty, such males may respond to ovarian estrogens with breast enlargement or gynecomastia. Surgical correction of gynecomastia and testosterone replacement therapy will correct most of the problems associated with this disorder. There is considerable controversy regarding the type and frequency of clinical manifestations associated with the 47,XYY karyotype. These are physically and sexually normal men, but there are data that suggest that a significant proportion of such men have problems with sociopathic behavior. As shown in Table 44-2 there are other aneuploidies of males involving the X and Y chromosomes. With each additional X chromosome present the degree of mental and physical disability increases and all such persons have primary hypogonadism.

Cytogenetic methods

The most common method of chromosome analysis is the culture of peripheral lymphocytes. These are mature cells circulating in the peripheral blood that are not actively undergoing cell division. Such cells can in culture, however, be stimulated to divide by the addition of a *mitogen* to the culture medium. The most commonly used mitogen in human cytogenetics is phytohemagglutinin (PHA). Lymphocytes in the presence of PHA are commonly cultured for 72 hours. Toward the end of the culture period, colchicine is added to the culture. This agent acts as a microtubular poison and thereby disrupts the mitotic spindle during cell division. Thus cells can be arrested in metaphase, the only time during cell division when chromosomes can be easily visualized and studied (Fig. 44-7). At the end of the culture period the cells are harvested, spread on a slide, and stained with a number of fluorogens or chromagens. By special fixative techniques the chromosomes take on a banded appearance. These bands are specific for each chromosome and allow detailed analysis of the structure of each chromosome.

With the use of fluorescent dyes to obtain banding of chromosomes, it was recognized that the normal Y chromosome contains a large fluorescent segment in the long

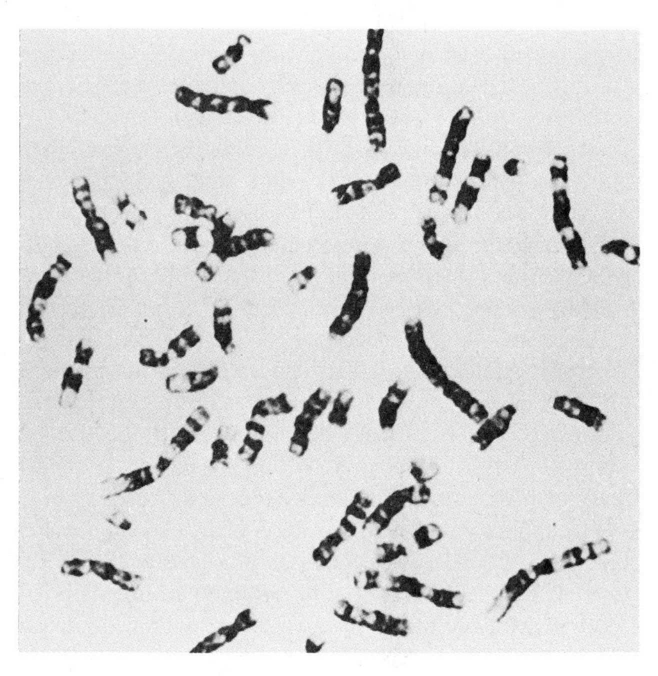

Fig. 44-7. Metaphase spread. Chromosomes as seen through microscope in a spread stained by G-banding technique.

arm. Staining of a buccal smear with a fluorescent dye will reveal a bright fluorescent spot called the *F body,* or *Y body*. Thus, through the use of Giemsa staining for Barr bodies and fluorescent staining for Y bodies, it is possible to determine the sex chromosome constitution from a buccal smear. The use of buccal smears for diagnosis in infants with ambigous genitalia or older children or adolescents with suspected abnormalities of sex chromosomes requires considerable experience and the use of controls in such an analysis. The recognition of both Barr bodies and Y bodies is sufficiently subjective that an experienced technician with control cells from a normal male and female is necessary for an accurate determination. The expected findings of Barr body–Y body determination on buccal smear for normal and abnormal sex chromosomal constitutions are shown in Table 44-2.

Chromosomal markers of disease

The analysis of the chromosomes in dividing cells may reveal abnormalities of three types. First, a change in the amount of chromosomal material present as an aneuploidy or partial duplication or deletion. Such a finding is the direct cause of clinical abnormalities in persons with an unbalanced karyotype. Second, there may be no alteration in the amount of chromosomal material but specific chromosomal changes that are diagnostic of specific disease states. Such changes may take a variety of forms and require special culture and staining techniques for their demonstration. Third, there may be nonspecific changes in chromo-

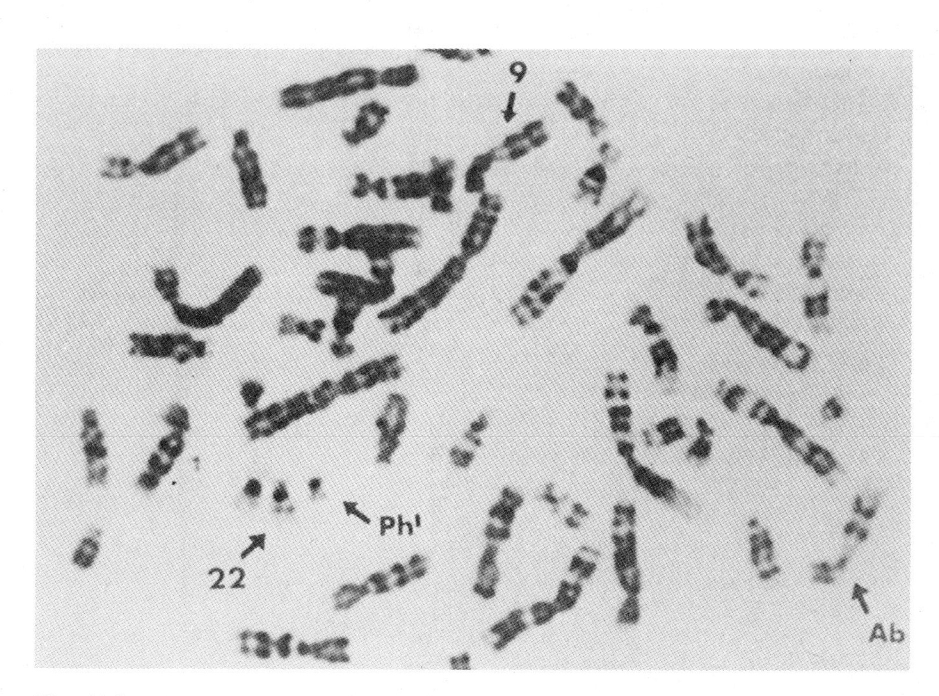

Fig. 44-8. Philadelphia chromosome. A metaphase spread with banded chromosomes shows small number 22, or the Ph^1 chromosome, and a normal 22 plus a normal chromosome 9 and the chromosome 9(Ab) involved in the 9/22 translocation producing the Ph^1 chromosome.

somal structure or number that reflect the consequences of environmental exposures such as drugs or radiation or abnormal chromosomal behavior in malignant cells. These changes are the consequence of external factors and do not reflect a primary chromosomal abnormality or specific marker.

Improvements in culture techniques and staining of chromosomes have revealed that malignant cells often demonstrate a consistent and specific chromosomal abnormality that is of major diagnostic assistance. Such an abnormality is best exemplified by the Philadelphia chromosome (Ph^1) found in the majority of persons with chronic myelogenous leukemia (CML). This Philadelphia chromosome represents a deletion of chromosome 22, which renders it a small acrocentric chromosome (Fig. 44-8). The deleted material is most often translocated to a 9 chromosome. Thus there is no loss or gain of chromosomal material, but there is a structural rearrangement. The Philadelphia chromosome is confined to the malignant cells of the bone marrow, and analysis for the Philadelphia chromosome is done on bone marrow aspirate. As these blast cells are actively dividing, no mitogen need be added to the culture. Thus the culture interval of 4 hours with the addition of colchinine will allow for analysis of malignant cells in the bone marrow. The Philadelphia chromosome is found in the malignant cells of myeloid origin. The finding of the Philadelphia chromosome in a bone marrow aspirate in a person suspected of having chronic myelogenous leukemia is highly diagnostic and suggests a better prognosis for chemotherapy than chronic myelogenous leukemia in which the Philadelphia chromosome is not present. It is being increasingly recognized that malignant cells not only have a high frequency of chromosomal abnormalities but that specific chromosomal changes characterize specific types of malignancy.

A second type of chromosomal marker used diagnostically is that of chromosomal breakage. This is most dramatically illustrated in the Fanconi anemia. Persons with this form of autosomal recessively inherited aplastic anemia demonstrate a pronounced spontaneous frequency of chromosomal breaks and gaps in cultured peripheral lymphocytes. Aplastic anemia occurs secondary to failure of the bone marrow to produce blood cells. Fanconi anemia cells are quite sensitive to agents such as mitomycin C, which further increases the frequency of chromosomal breakage. Chromosomal breaks and gaps can be seen in normal cells, but in Fanconi anemia the frequency is on an order of tenfold greater. This abnormality is not restricted to lymphocytes but is also demonstrable in cultured skin fibroblasts. Although these findings are of diagnostic significance, it is not known how they are related to the primary gene abnormality in this disorder.

The semiconservative replication of DNA is demonstrable cytogenetically by the addition of a base analog to cultured cells. The base analog, bromodeoxyuridine, (BrDU), when added to cultured cells, will be incorporated into one DNA strand of a chromatid rather than thymidine. Continued culture will incorporate BrDU in the chromatid. The presence of BrDU in one chromatid will diminish the fluorescent staining property of that chromatid when a flu-

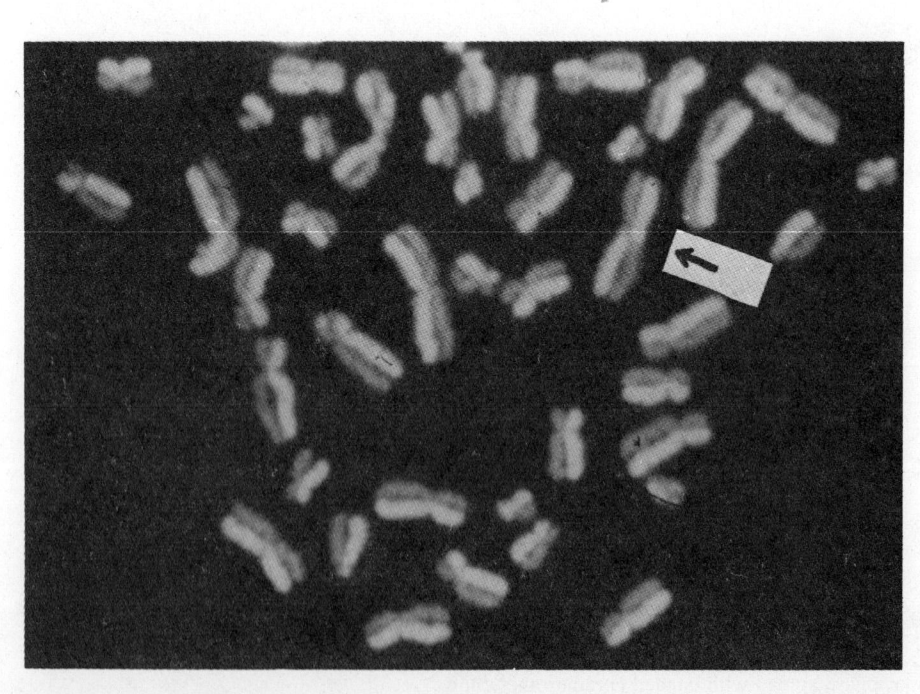

Fig. 44-9. Sister chromatid exchange. Two chromatids of each chromosome are differentially stained; one chromatid is fluorescent and the other is dull. A sister chromatid exchange, *arrow*.

orescent dye is used. Thus, after the incorporation of BrDU and staining with the fluorogen, a chromosome will show one chromatid brightly fluorescent and one with diminished fluorescence. Thus if exchanges occur between the chromatids during cell division, this is easily recognizable as an alternating pattern of fluorescence and nonfluorescence of the chromatid. The autosomal recessively inherited disorder Bloom's syndrome is characterized by a sharp increase in the frequency of sister chromatid exchanges in cultured lymphocytes (Fig. 44-9). As with Fanconi's anemia, the mechanism for the sister chromatid exchange and how it relates to the disease is unknown, but it is well recognized that this represents a highly diagnostic finding in patients suspected of having this disorder. Bloom's syndrome is characterized by intrauterine growth retardation and short stature thereafter. Such patients have increased sensitivity of their skin to ultraviolet light and a significant risk of malignancies including leukemia during late childhood and early adult years. Carriers for Bloom's syndrome who are heterozygous for this mutation demonstrate an increased number of sister chromatid exchanges, but less than that demonstrated in the homozygously affected person.

A fourth type of specific chromosomal marker is the fragile X chromosome recognized in one X-linked form of mental retardation (Fig. 44-10). The characteristic findings in this disorder include postpubertal enlargement of the testes, large, simply formed ears, a jovial personality, and moderate to severe mental retardation. Such affected males have generally normal health and physical findings otherwise. Approximately 10% of carrier females for this form of X-linked mental retardation show a mild form of mental retardation. The culture of lymphocytes with standard growth media used for cytogenetic studies will not demonstrate the fragile X chromosome. However, the use of media that are deficient in folic acid will show a fragile X chromosome in a significant percentage of the cells from an affected male. The designation *fragile X* is used because the X chromosome in these males appears to have a break at the end of the long arm. This is a common form of mental retardation and probably is the major disorder that contributes to the greater frequency of mental retardation in males than females. Because of its X-linked inheritance and therefore the potential for a large family with many affected males, this disorder should be specifically looked for by use of special culturing techniques in mentally retarded males for whom no explanation is otherwise found.

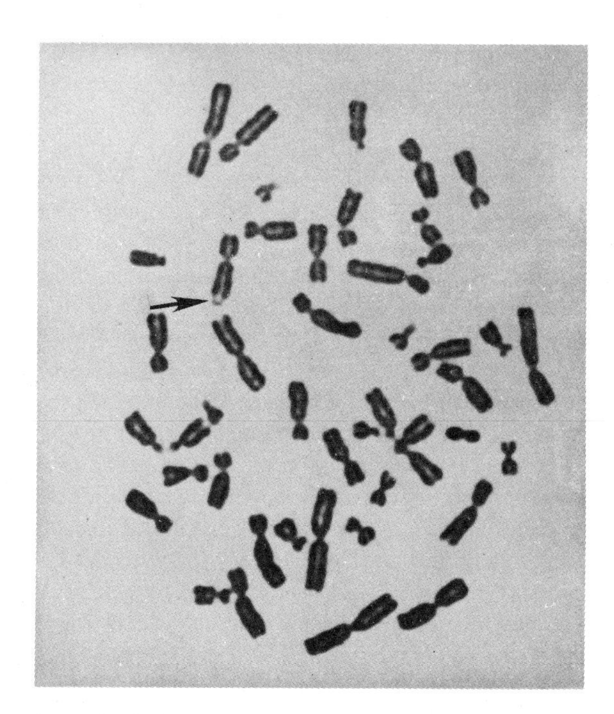

Fig. 44-10. Fragile X chromosome. Fragile X has a small area on ends of long arms, which appear to be attached by a thread *(arrow)*.

It should be pointed out that chromosomal abnormalities that are visible under the microscope result in differences in gene dosage and are the molecular basis for the clinical syndrome. The actual gene products, specific proteins that are involved, are largely unknown as is the way in which the clinical syndrome is produced. These abnormalities do not reflect an alteration in the gene structure or in the structure of the gene product as occurs with mutations in single genes that produce inherited genetic diseases. The chromosomal changes observed do serve as specific markers for the clinical syndromes involved. Most genetic diseases occur as a result of changes in individual genes and thus are not associated with any visible change in the chromosomes.

PATHOLOGICAL DISORDERS ASSOCIATED WITH BIOCHEMICAL CHANGES

Lysosomal storage diseases

Lysosomal storage diseases are recessively inherited disorders that are the result of individual deficiencies of an acid hydrolase. These acid hydrolases are located within membrane bound cytoplasmic organelles called *lysosomes*. Within the lysosome a number of these hydrolases active at acidic pH degrade macromolecular compounds. The necessity for breakdown of these compounds is the normal turnover in cell metabolism. If there is a complete or nearly complete deficiency of a specific hydrolase, the macromolecular compound on which it operates will accumulate within tissue, and this will eventually result in clinical manifestations. The primary organs affected are dependent on the tissue distribution of that macromolecular compound. Lysosomal storage diseases are characterized by the nature of the macromolecular compounds that accumulate. Additionally, they generally result in three types of organ system manifestations. These include (1) the reticuloendothelial system, (2) the central nervous system, and (3) the skeleton and connective tissues plus other so-

Classification of lysosomal storage disease

- Mucopolysaccharidoses
- Mucolipidoses (ML)
  - (ML I—now called "sialidosis")
  - ML II or I-cell disease
  - ML III or pseudo-Hurler polydystrophy
  - ML IV
- Gangliosidoses
  - GM_1 gangliosidosis or generalized gangliosidosis
  - GM_2 gangliosidosis
    - Tay-Sachs disease
    - Sandhoff disease
- Leukodystrophies
  - Metachromatic leukodystrophy
  - Krabbe leukodystrophy
  - Adrenoleukodystrophy
- Glycoproteinosis
  - Mannosidosis
  - Fucosidosis
  - Sialidosis
  - Aspartylglucosaminuria
- Others
  - Ceramidosis
  - Cholesterol ester storage disease
  - Pompe disease or glycogen-storage disease II

matic tissues. In a given disorder the clinical manifestations may be apparent in only one or any combination of these three system involvements. The box above lists a classification scheme for lysosomal storage diseases. These diseases are the result of abnormalities of metabolism of mucopolysaccharides, gangliosides, glycoproteins, ceramide, cholesterol esters, and glycogen. Varying degrees of deficiency or altered activity of a hydrolase may result in several different clinical syndromes occuring as a result of the same enzyme deficiency. This latter feature is most ably demonstrated by deficiency of the hydrolase α-L-iduronidase. Deficiency of this enzyme results in the accumulation of the glycosaminoglycans dermatan sulfate and heparan sulfate. In its most severe form with the greatest degree of enzyme deficiency Hurler's syndrome occurs, with severe skeletal, somatic, and central nervous system manifestations and death by 5 to 8 years of age. In its mildest form, deficiency of this enzyme results in the Scheie syndrome with normal height, life expectancy, and intelligence but with skeletal, ocular, and cardiac abnormalities. An intermediate form, known as the Hurler-Scheie syndrome, is intermediate in its clinical manifestations. The diagnosis of a lysosomal storage disease is most accurately made by the concomitant demonstration of substrate accumulation and specific enzyme deficiency. However, a number of less specific diagnostic tests are routinely used in screening for these disorders. These

$$\text{Ceramide}\overset{\beta}{—}\text{Glucose} \xrightarrow[+H_2O]{\beta\text{-Glucosidase}} \text{Ceramide} + \text{Glucose}$$

Fig. 44-11. Gaucher's disease. Deficiency of β-glucosidase results in intralysosomal accumulation of glucosylceramide.

diagnostic approaches are illustrated by the selective disorders described in detail.

Reticuloendothelial system involvement. The two disorders in which involvement of the liver and spleen is most prominent and may represent the major manifestations are *Gaucher's* and *Niemann-Pick diseases*. Gaucher's disease occurs as a result of the deficiency of the acid hydrolase β-glucosidase (Fig. 44-11). Deficiency of this enzyme results in the accumulation of phospholipids. This disorder occurs in several forms. The most severe form is known as neuronopathic Gaucher's disease in which, although liver and spleen enlargement occur, massive accumulation of phospholipids within the central nervous system predominates and affected infants die within the first year of life. The other extreme is illustrated by the adult form of Gaucher's disease recognized by hepatosplenomegaly, but in which good general health and normal life expectancy occur. Before the use of enzyme assays for specific diagnoses, Gaucher's disease was most often recognized by the demonstration of Gaucher cells in a bone marrow aspirate. These are large macrophages that stain with characteristics that allow identification of the Gaucher cell. There are no simple urine or blood screening tests for this disorder. The second disorder in this group, Niemann-Pick disease, also occurs in several forms ranging from a severe infantile to a mild adult disease. This disorder occurs secondary to a deficiency of sphingomyelinase (Fig. 44-12). It too is associated with enlargement of the liver and spleen and demonstration of storage cells on bone marrow aspirate that represent macrophages with engorged storage material.

Central nervous system predominance. The lysosomal storage diseases that affect primarily the central nervous system are reflected clinically principally by early involvement in grey matter or white matter. When the initial accumulation is within white matter the disorder is known as a *leukodystrophy* and motor abnormalities are seen early in the course of the disease. Severe mental deficiency occurs during the course of the disorder. The most commonly recognized leukodystrophy is metachromatic leukodystrophy. This designation is derived from the fact that neural tissue demonstrates a particular metachromatic staining property that for many years was the primary diagnostic method. Several screening tests are available for this disorder and include the examination of urine for metachromatic granules. The deficient enzyme in this disorder is called arylsulfatase A, or cerebroside sulfate sulfatase (Fig. 44-13). In normal urine adequate amounts of this enzyme are present for a simple colorimetric assay of arylsulfatase activity. Measurement of arylsulfatase A activity in urine is a useful

$$\text{Sphingomyelin} \xrightarrow[+H_2O]{\text{Sphingomyelinase}} \text{Ceramide} + \text{Phosphorylcholine}$$

Fig. 44-12. Niemann-Pick disease. Deficiency of sphingomyelinase results in intralysosomal accumulation of sphingomyelin.

$$\text{Ceramide} \overset{\beta}{—} \text{Galactose sulfate} \xrightarrow[+H_2O]{\text{Arylsulfatase A}} \text{Ceramide} \overset{\beta}{—} \text{Galactose} + \text{Sulfate}$$

Fig. 44-13. Metachromatic leukodystrophy. Deficiency of arylsulfatase A results in storage of sulfated galactosylceramide in white matter of brain.

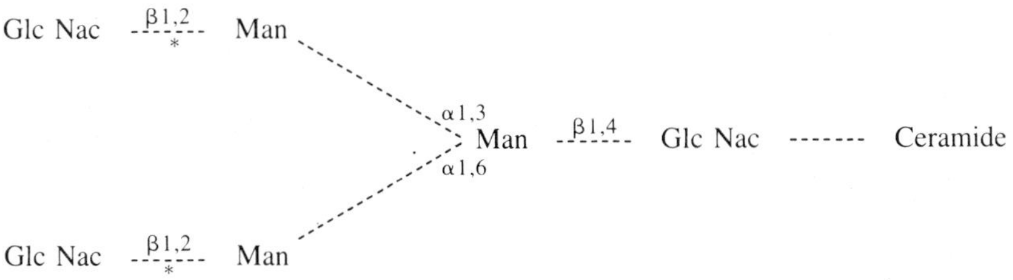

Glc Nac -- β1,2 (*) -- Man -- α1,3 --

Man -- β1,4 -- Glc Nac ------- Ceramide

Glc Nac -- β1,2 (*) -- Man -- α1,6 --

G_{M2} ganglioside

*Cleaved by hexosaminidase A (*N*-acetyl-β-1,2-glucosaminidase)

Fig. 44-14. Tay-Sachs disease. G_{M2} ganglioside accumulates in the gray matter of brain because of deficiency of *N*-acetyl-β-1,2-glucosaminidase (hexosaminidase A). *Glc Nac,* *N*-Acetylglucosamine; *Man,* mannose.

screening test for metachromatic leukodystrophy. The presence of an adequate amount of enzyme activity will exclude this diagnosis, but lack of enzyme activity is not diagnostic and suggests that more specific assay systems should be used. The specific diagnosis of metachromatic leukodystrophy requires the demonstration of arylsulfatase A deficiency in homogenized peripheral leukocytes or cultured skin fibroblasts. Arylsulfatase occurs in two lysosomal forms known as A and B. By quantitative enzyme assay these are distinguished by the use of inhibitors added to the enzyme assay system. Thus incomplete inhibition is seen, but quantitative analysis can be made. The use of cellulose acetate electrophoresis is a second diagnostic method. In metachromatic leukodystrophy a normal arylsulfatase B band will be observed and no A band will be recognized. There is a variant of metachromatic leukodystrophy caused by the deficiency of multiple sulfatases. Clinically this disorder is quite similar to metachromatic leukodystrophy and is best recognized by the use of cellulose acetate electrophoresis for diagnosis. In this latter disorder both arylsulfatase A and B band will be absent. Carrier detection for metachromatic leukodystrophy is made difficult by genetic variation in normal levels of enzyme activity in families.

The lysosomal storage disease that results in macromolecular compound accumulation within gray matter that is best known is Tay-Sachs disease. This disorder occurs as a result of the accumulation of G_{M2} ganglioside within the central nervous system. Clinically it results in early regression of CNS function so that by 1 year of age such children have delayed cognitive development, impaired hearing and sight, and a characteristic cherry-red spot found on fundoscopic examination. The accumulation of G_{M2} ganglioside within the central nervous system occurs as a result of deficiency of the acid hydrolase *N*-acetyl-β-D-galactosaminidase (Fig. 44-14). Assay of this enzyme activity with artificial substrates is done using either a glucosamine or galactosamine. Therefore the enzyme activity so measured is referred to as *hexosaminidase*. Hexosaminidase occurs as a heat-labile form, hexosaminidase A, and a heat-stabile form, hexosaminidase B. Tay-Sachs disease is characterized by the specific deficiency of hexosaminidase A, whereas a similar disorder, Sandhoff's disease, is characterized by the deficiency of hexosaminidase A and B. Levels of hexosaminidase A in body fluids are such that this enzyme determination can be done on tears, sweat, plasma, urine, leukocytes, cultured skin fibroblasts, tissue extracts, and cultured amniotic fluid cells. Measurement of enzyme activity with and without heat inactivation gives levels of hexosaminidase B and total hexosaminidase, respectively. The difference between the two determinations, or the heat labile form is thus the calculated level of hexosaminidase A. One can also separate and visualize hexosaminidase A and B by several types of electrophoresis.

Because Tay-Sachs disease occurs most commonly among Ashkenazic Jewish infants, a target population is available for carrier detection. Additionally, serum assay of this enzyme represents an excellent screening test for carrier detection. Carriers for Tay-Sachs disease have a partial deficiency in hexosaminidase A. This is, however, not readily apparent by the simple determination of the total activity of hexosaminidase A in serum. Rather, on empirical grounds, it is recognized that the percentage of hexosaminidase A present as a function of total hexosaminidase activity represents the most accurate screening method for detecting carriers of Tay-Sachs disease. The method of carrier detection involves measurement of hexosaminidase activity in serum with and without heat inactivation and a calculation of total hexosaminidase A activity. This, in turn, is expressed as the percentage of total hexosaminidase represented by hexosaminidase A. In establishing an assay system for Tay-Sachs disease carriers, it is necessary that each laboratory determine its reference values and values for Tay-Sachs disease carriers from obligate heterozygotes. However, in general, most laboratories use a normal hexosaminidase A percentage of 45% or greater. Given a lower percentage of hexosaminidase A, confirmation of the carrier state requires study of hexosaminidase A activity in peripheral leukocytes and additional family studies if necessary to resolve questionable results. Given a single person with borderline values of hexosaminidase A by leukocyte assay, study of other family members should resolve the carrier status of that person. The use of such carrier tests among Ashkenazic Jews and subsequent use of prenatal diagnosis has significantly reduced the incidence of Tay-Sachs disease in the United States.

Connective tissue and skeletal predominance. Lysosomal storage diseases in which involvement of connective tissue predominates, especially the skeletal system, are referred to as *Hurler-like disorders*. With skeletal system involvement there may be central nervous system or reticuloendothelial involvement. The common feature to this group of disorders, however, is the skeletal involvement. Table 44-3 lists a classification of lysosomal disorders that present with this phenotype. Clinical diagnostic procedures used for this group of disorders are illustrated through three representative diseases: Hurler syndrome, fucosidosis, and mucolipidosis II, or I-cell disease.

Hurler's syndrome represents the prototype of diseases involving glycosaminoglycan metabolism and resulting in a connective tissue disorder. This form of α-L-iduronidase deficiency is characterized by multisystem involvement that is usually apparent by 6 months of age (Fig. 44-15). It follows a rapidly progressive course thereafter with all organ systems involved and mental deterioration predominating late in the disorder with death ultimately occurring between 5 and 8 years of age. These diseases were classified initially on the basis of the urinary pattern of mucopolysaccharide or glycosaminoglycan excretion. In Hurler's syndrome massive amounts of dermatan sulfate and heparan sulfate are excreted in the urine. Mucopolysacchariduria is detected by a variety of screening tests including a urine spot test, the toluidine blue test. Urine is dropped on filter paper and stained with toluidine blue. A positive test is a bluish discoloration to the urinary spot with an increase in urinary mucopolysaccharides being noted. A second urine screening test involves the use of CPC (cetylpyridinium chloride), which forms an insoluble precipitate with mucopolysaccharides and results in turbidity. One can measure this quantitatively on a spectrophotometer and express it as turbidity units.

It is now uncommon to require the use of 24-hour urine assays for mucopolysaccharides in the diagnosis of these disorders. Rather, the clinical phenotype and urine screening tests are followed now by study of mucopolysaccharide metabolism in cultured fibroblasts and assay of specific hydrolases resulting in the mucopolysaccharidoses. The use of cultured skin fibroblasts greatly facilitates the specific diagnosis of all lysosomal storage diseases and especially the mucopolysaccharidoses. Inorganic sulfate added to tissue culture media is exclusively incorporated into cultured fibroblasts as sulfated glycosaminoglycans. Thus the use of radioactive sulfate allows the recognition of accumulation of these sulfated compounds within cultured fibroblasts and therefore provides direct evidence for tissue accumulation of glycosaminoglycans. Given a clinical

Table 44-3. Hurler-like disorders

Disorder	Screening test
Mucopolysaccharidoses	Urinary screening for mucopolysacchariduria
Mucolipidoses	
ML II and ML III	Serum levels of acid hydrolases Urinary bound sialic acid
Glycoproteinosis	Urinary oligosaccharide TLC Urinary bound sialic acid

IA—*Glc Nac—GA—Glc Nac—IA—*Glc Nac

Dermatan sulfate

IA—*Gal Nac—GA—Gal Nac—IA—*Gal Nac

Heparan sulfate

*Site of hydrolysis by α-L-iduronidase in oligosaccharide chain of dermatan and heparan sulfate

Fig. 44-15. Hurler's syndrome. Deficiency of lysosomal hydrolase α-L-iduronidase results in widespread accumulation of both heparan and dermatan sulfate. *GA,* Glucosamine; *Gal Nac, N*-acetylgalactose; *Glc Nac, N*-acetylglucosamine; *IA,* iduronic acid.

Table 44-4. Mucopolysaccharidoses

Eponym	Number	Mucopolysacchariduria*	Enzyme deficiency
Hurler	MPS I-H	DS, HS	α-L-Iduronidase
Scheie	MPS I-S	DS, HS	α-L-Iduronidase
Hurler-Scheie	MPS I-H/S	DS, HS	α-L-Iduronidase
Hunter	MPS II	DS, HS	Iduronide sulfate sulfatase
Sanfilippo	MPS III-A	HS	Heparan-*N*-sulfate sulfatase
Sanfilippo	MPS III-B	HS	α-Glucosaminidase
Sanfilippo	MPS III-C	HS	Acetyl-CoA:α-glucosaminide *N*-acetyltransferase
Morquio A	MPS IV-A	KS	Galactosaminyl-6-sulfate sulfatase
Morquio B	MPS IV-B	KS	β-Galactosidase
Maroteaux-Lamy	MPS VI	DS	Galactosaminyl-4-sulfate sulfatase (arylsulfatase B)
Sly	MPS VII	DS or HS or DS, HS	β-Glucuronidase

**DS,* Dermatan sulfate; *HS,* heparan sulfate; *KS,* keratan sulfate.*

suspicion of a Hurler-like disorder and abnormal accumulation of radioactive sulfate in cultured skin fibroblasts, a specific diagnosis can then be further pursued by the assay of the individual hydrolases that, when deficient, result in a mucopolysaccharidosis. The pattern of mucopolysacchariduria and the specific acid hydrolase deficient in the mucopolysaccharidoses are shown in Table 44-4.

Fucosidosis was originally described as a variant form of Hurler's syndrome because of a similarity in clinical and radiographic findings. Affected persons, however, do not excrete elevated amounts of mucopolysaccharides in the urine. Rather, by thin-layer chromatography one can demonstrate an increased urinary excretion of fucose-containing oligosaccharides. The pattern of oligosaccharide excretion is characteristic for fucosidosis. This disorder, like the other lysosomal storage diseases, results in several clinical forms that breed true within individual families. The specific diagnosis of fucosidosis is made by demonstration of a specific deficiency of the acid hydrolase α-L-fucosidase, which can be best accomplished in the study of cultured skin fibroblasts (Fig. 44-16).

A third type of disorder that results in a Hurler-like clinical picture is represented by *mucolipidosis II,* or I-cell disease. Children with this disease are similar clinically to those with Hurler's syndrome but present at an earlier age and follow a more rapid downhill course with death by 3 to 5 years of age. There is no excess mucopolysaccharide in the urine. When cultured skin fibroblasts from affected children were first analyzed under phase microscopy, a significant increase in cytoplasmic inclusions were noted;

$$\text{Fucose-}\alpha\,1,2\text{—}(*) \xrightarrow[+H_2O]{\alpha\text{-Fucosidase}} \text{Fucose} + (*)$$

(*) = Number of different sugars in oligosaccharide chains of glycoproteins and glycolipids

Fig. 44-16. Fucosidosis. Lysosomal enzyme, α-L-fucosidase, cleaves terminal fucosyl residues from a number of oligosaccharide chains of glycoprotein and glycolipid origin.

hence the name *I-cell,* or *inclusion cell, disease.* Lysosomal hydrolases are glycoproteins in structure. Each lysosomal enzyme has a unique structural gene that codes for the protein component of the glycoprotein. There is, however, commonality in the posttranslational modification of these proteins in the oligosaccharide chain. Thus genetic abnormalities in this posttranslational modification can result in alterations in numerous acid hydrolases. In I-cell disease a variety of abnormalities were observed before the basic mechanism was recently recognized. Cultured skin fibroblasts from affected children show a deficient level of numerous acid hydrolases in cultured cells. However, the media in which such cells were grown contain elevated amounts of these enzymes. From this it was concluded that the primary defect resided in abnormal localization of these hydrolases intracellularly. Body fluids including tears, urine, and serum from affected children contain significant elevations in the levels of these acid hydrolases similar to those found in tissue culture media. Additionally, cellulose acetate electrophoresis of these hydrolases showed altered banding patterns for many of these enzymes. The study of radioactive-sulfate turnover in cultured fibroblasts from I-cell disease patients showed accumulation of mucopolysaccharides similar to the mucopolysaccharidoses. Further, the urine of such patients showed not only increased levels of lysosomal hydrolases but also an increase in the amount of sialic acid oligosaccharides. The measurement of bound sialic acid in urine is a relatively simple procedure and represents an excellent screening test for a number of lysosomal storage diseases in which excessive urinary exretion of oligosaccharides occurs. It is now recognized that the abnormal radioactive sulfate turnover in fibroblasts and the elevated urinary excretion of sialic acid–oligosaccharides are the result of the numerous intracellular acid hydrolase deficiencies that occur in this disease. It has recently been recognized that mannose-6-phosphate sugars in the oligosaccharide chain of the glycoprotein structure of acid hydrolases are required for their appropriate intracellular localization. This

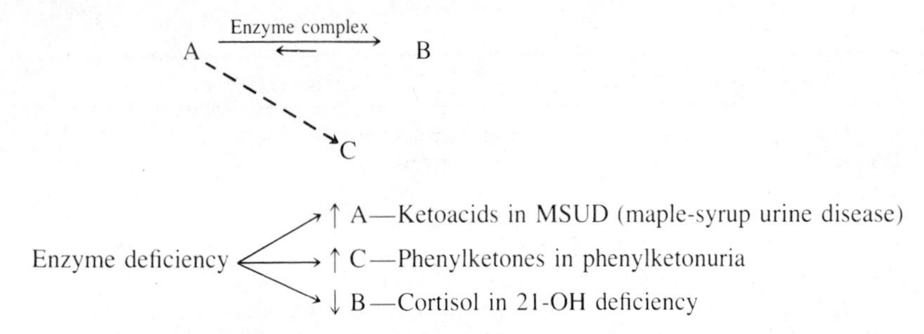

Fig. 44-17. Enzyme deficiency. Deficiency of enzyme in metabolic pathway may produce clinical symptoms because of accumulation of substrate, *A;* lack of production of product, *B;* or shunting of substrate to alternate pathway with production of a toxic compound, *C*.

is accomplished through a number of posttranslational enzymatic steps in the modification of the oligosaccharide chain. Deficiency of a transferase involved in such posttranslational modifications has been shown to occur in I-cell disease and is now presumed to represent the primary molecular defect. There are therefore a number of screening tests involving urine, cultured cells, and serum that will indicate the laboratory diagnosis, given the clinical phenotype of I-cell disease. Specific diagnosis requires the use of a number of fibroblast assays and specific demonstration of the transferase deficiency.

Disorders of intermediary metabolism

Disorders of intermediary metabolism occur as a result of three basic mechanisms: an enzyme deficiency with defective substrate conversion, a membrane transport defect resulting in failure of absorption or excessive excretion of a compound, and defects in receptors involved in mediating metabolism. Given such a defect in a metabolic pathway, the biochemical basis for the resulting symptom occurs as a result of (1) accumulation of a substrate to levels that become toxic, (2) deficiency through lack of production or excessive loss of a needed compound, and (3) conversion of an elevated compound to an altered metabolite that itself represents a toxic material. Fig. 44-17 shows only one example of these modes of abnormalities. In general, defects in intermediary metabolism result in alterations of molecular weight compounds smaller than those seen in lysosomal storage diseases. As a result they are more likely to result in more acute manifestations and are less likely to be associated with the striking physical features that dominate the lysosomal storage disorders. Defects in intermediary metabolism result in a disruption of a normal metabolic process and produce an elevated urinary excretion of a normal metabolite or urinary excretion of an abnormal metabolite. Therefore urine screening tests are the principal means of recognizing the presence of such a disorder, prompting more specific diagnostic assays. Although a large number of defects in intermediary metabolism exist, their mode of clinical presentation is limited to a smaller number of manifestations. Thus these disorders will present, because of their altered metabolism, as ketosis, acidosis, hypoglycemia, toxic effects of an abnormal metabolite, or alterations in mineral metabolism. The clinical effects of such derangements will present as an acute illness in a child, altered mental development, or retarded growth.

Amino acidopathies. Table 44-5 is a list of the more common amino acid disorders and their clinical chemical characterizations. *Phenylketonuria* (PKU) is the best known and best understood of the amino acid disorders. This autosomal recessively inherited disorder occurs as a defect in phenylalanine conversion to tyrosine because of a deficiency of the enzyme phenylalanine hydrolase (Fig. 44-18). This disease occurs in approximately one in 15,000 live-born infants. This enzymatic activity is restricted to liver, and therefore diagnosis is accomplished through demonstration of alterations in phenylalanine metabolism. At one time this disorder was called ''Fölling's disease'' based on the pioneering biochemical studies done by this Norwegian chemist. He used ferric chloride in the isolation of phenylpyruvic acid in the urine of mentally retarded siblings. It was later recognized that the deficiency of phenylalanine hydrolase results in an increase level of plasma phenylalanine. This level results in a shunting of phenylalanine metabolism to several ketoacids, the principal one being phenylpyruvic acid. Phenylketonuria does not result in acute symptoms, but rather an alteration in protein metabolism within the central nervous system. Affected children often demonstrate less pigmentation

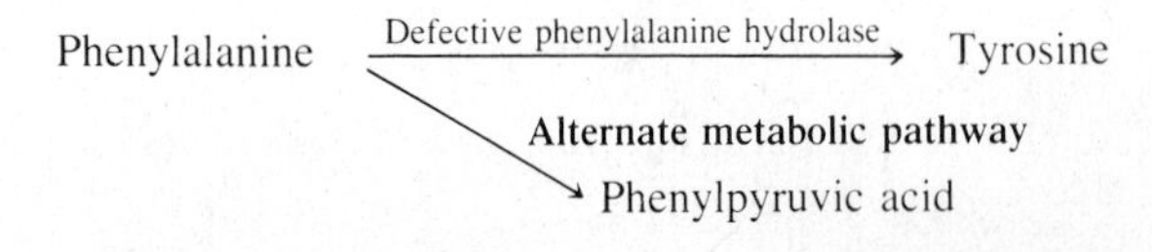

Fig. 44-18. Phenylketonuria. Deficiency of phenylalanine hydrolase results in shunting of elevated phenylalanine to form phenylketones, one of which is phenylpyruvic acid.

Table 44-5. Inborn errors of intermediary metabolism of amino aids

Condition	Defective enzyme	Biochemical features	Clinical features	Treatment
Alkaptonuria*	Homogentisate oxygenase	Urinary excretion of homogentisic acid	Urine darkens; ochronosis; arthritis in later life	None known
Phenylketonuria**	Phenylalanine 4-hydroxylase	Phenylalanine accumulates in blood, CSF, etc.; urinary excretion of phenylpyruvic acid and related compounds	Severe mental deficiency, epilepsy, abnormal EEG, eczema, behavioural disorders	Diet low in phenylalanine beginning at early age
Albinism***	*o*-Diphenol oxidase (tyrosinase)	Lack of melanin in skin, hair and eyes	Photophobia, nystagmus, carcinomata of the skin	None known
Goitrous cretinism (several types)	(1) Tyrosine iodinase (2) Coupling enzyme (3) Deiodinase	Lack of thyroid hormone	Cretinism, goitre	Thyroid, thyroxine or triiodothyronine
Maple syrup urine disease (leucinosis)	Enzyme responsible for oxidative decarboxylation of α-ketoisocaproic, α-keto-β-methyl-*n*-valeric and α-ketoisovaleric acids.	Leucine, isoleucine and valine accumulate in blood, CSF, etc.; urinary excretion of the 3 keto acids and related compounds	Cerebral degeneration; usually early death. Milder form with partial enzyme deficiency, symptomless except during infections, etc.	Diet low in leucine, isoleucine and valine
Cystinosis	Cystine reductase (?)	Cystine is deposited in reticulo-endothelial system; amino-aciduria, glucosuria, proteinuria, phosphaturia, dilute urine	Dwarfism, photophobia, renal acidosis, hypokalaemia, vitamin-resistant rickets; death before puberty. A benign (non renal?) variant occurs in adults	Palliative: potassium salts, alkalis, vitamin D. Diet low in cystine and methionine (efficacy doubtful)
Homocystinuria	L-Serine dehydratase	Urinary excretion of homocystine	Mental retardation, retinal defects, dislocated lenses, malar flush, thromboses	Diet low in methionine, high in cystine. Pyridoxine
Hyperglycinaemia (several types)	(Uncertain, depends of type)	Glycine accumulates in blood, etc.; urinary excretion of glycine and, in one type, methylmalonic acid	Neonatal lethargy and ketosis, neutropenia, hypo-γ-globulinaemia; mental retardation	Diet low in protein
Oxalosis	Excessive conversion of glycine to oxalic acid	Calcium oxalate accumulates in kidneys, heart, bone marrow and cartilages	Nephrocalcinosis leading to progressive renal failure	None known
Histidinaemia	Histidine ammonialyase	Urinary excretion of β-imidazolylpyruvic acid and related compounds	Speech defects; mental retardation in some	Diet low in histidine
Familial tyrosinaemia	(Uncertain)	Tyrosine level in blood and urine raised; urinary excretion of phenolic acids related to tyrosine; generalized amino-aciduria; glucosuria; fructosuria	Rapidly enlarging liver; jaundice; hypoprothrombinaemia; death common in infancy; survivors may have vitamin D-resistant rickets and acidosis	Diet low in tyrosine and phenylalanine (efficacy doubtful)
Hyperprolinaemia Type I Type II	 Pyrroline-5-carboxylate reductase Pyrroline-5-carboxylate dehydrogenase	Hyperprolinaemia; urinary excretion of proline, glycine and hydroxyproline	Mental retardation, convulsions, renal disease, deafness	None known

From Geigy scientific tables, ed. 7, 1970, CIBA-Geigy Corp., Summit, N.J.
*Incidence 1 in 100,000. **Incidence varies from 1 in 3200 to 1 in 10^7 according to locality. ***Incidence 1 in 13,000.

Continued.

Table 44-5. Inborn errors of intermediary metabolism of amino aids—cont'd

Condition	Defective enzyme	Biochemical features	Clinical features	Treatment
Hydroxyprolinaemia	3-Hydroxypyrroline-5-carboxylate reductase (?)	High levels of hydroxyproline in blood and urine	Mental retardation (?)	None known
Citrullinaemia	Argininosuccinate synthetase	High blood and urinary levels of citrulline; blood ammonia increased; urea excretion normal	Mental retardation, epilepsy, vomiting, ammonia intoxication	Diet low in protein
Argininosuccinic aciduria	Argininosuccinate lyase	Urinary excretion of argininosuccinic acid; high blood and CSF ammonia levels; urea excretion normal	Mental retardation, convulsions, hair abnormalities, ammonia intoxication	Diet low in protein
Hyperammonaemia				
Type I	Ornithine carbamoyltransferase	Blood ammonia about 10 mg/L; urea excretion normal	Mental retardation, ammonia intoxication	Diet low in protein (?)
Type II	Carbamoyl-phosphate synthase			

of the eyes, hair, and skin than their relatives but are first recognized clinically because of irreversible mental retardation. Once the disease had been recognized in a child, it was appreciated that a 25% recurrence risk existed for subsequently born children in that family. Such children could be diagnosed through the use of ferric chloride screening of the urine in the newborn. Subsequently born children in the family, when treated with diets low in phenylalanine, had better cognitive development than their older affected siblings. This then was the basis of an effort for screening of all newborn infants to allow for dietary management from early infancy. Such a screening technique was devised by Dr. Robert Guthrie and is now widely used and referred to as the Guthrie test. This test involves a bacterial inhibition assay. Blood samples are collected from newborn infants and placed on filter paper. They are stable for several years. The samples are usually analyzed in a central state facility using automated equipment. A circular punch removes a portion of the blood-stained filter paper, and this is placed on a sheet of agar. The agar contains thienylalanine, an isomer of phenylalanine. The plate is then inoculated with *Bacillus subtilis,* which is inhibited in its growth by the presence of thienylalanine. However, if the blood disk contains an elevated level of phenylalanine, this will support the growth of bacteria around the disk, and one can get a semiquantitative determination of the blood level of phenylalanine. In newborn screening a Guthrie test is considered positive when the levels exceed a range of 20 to 40 mg/L. Given a positive result, approximately 1 in 10 to 1 in 20 such infants will in fact have phenylketonuria. The others represent a transient newborn elevation in tyrosine and secondarily phenylalanine or a genetically determined elevation of phenylalanine, but one that is not at a level characteristic of phenylketonuria. The specific diagnosis of phenylketonuria after a positive screening test requires the demonstration of plasma levels of phenylalanine greater than 200 mg/L on 2 consecutive days while normal feedings are given. Such children will also have a positive ferric chloride test indicating urinary excretion of ketoacid metabolites of phenylalanine, principally phenylpyruvic acid. Electroencephalograms at this stage are mildly abnormal. After institution of dietary restriction in phenylalanine intake, urine screening tests will revert to normal, as will the electroencephalograms.

Subsequent monitoring of children on dietary management requires quantitative plasma determinations of phenylalanine levels. Periodically, all amino acids need to be quantitated to ensure proper protein metabolism. The plasma level of phenylalanine most often chosen as indicative of appropriate dietary control is 80 to 100 mgL. With newborn screening and appropriate dietary treatment several infants were recognized who, despite optimal dietary management, underwent mental deterioration and seizures. These children were then recognized to have a different form of phenylketonuria in which a reductase deficiency

$$\mathrm{R{-}\overset{\overset{\displaystyle O}{\|}}{C}{-}COOH + NAD^+ + CoA{-}SH \xrightarrow{BCKADH} R{-}\overset{\overset{\displaystyle O}{\|}}{C}{-}S{-}CoA + CO_2 + NADH + H^+}$$

$\mathrm{R{-}\overset{\overset{\displaystyle O}{\|}}{C}{-}COOH}$ = Branched-chain keto acid

BCKADH = Branched-chain keto acid dehydrogenase complex, a multisubunit enzyme complex involved in a five-step reaction

Fig. 44-19. Maple-syrup urine disease. Branched-chain keto acid decarboxylase enzyme complex when deficient results in elevation of branched-chain amino acids valine, isoleucine, and leucine as well as their keto acid analogs.

produced not only elevated phenylalanine and phenylketonuria but also alterations in neurotransmitters. The dietary management controlled levels of phenylalanine but was unable to correct the alterations in neurotransmitter metabolism.

Maple-syrup urine disease was so named because of the obvious odor of urine in affected infants. This disease occurs in roughly one in 250,000 live-born infants. Many disorders of amino acid metabolism are characterized by unusual odors that usually are the result of the excessive urinary excretion of organic acids that impart the peculiar odor. The three branched-chain amino acids, leucine, isoleucine, and valine, undergo transamination and transketolation and decarboxylation through a common pathway (Fig. 44-19). Maple-syrup urine disease occurs as a result of a deficiency in the multicomponent enzyme, branch-chain ketoacid decarboxylase. Deficiency of this enzyme results in the elevated urinary excretion of the ketoacid analogs of these three branch-chained amino acids and gives positive ferric chloride and dinitrophenol hydrazine screening tests. The plasma is characterized by elevations in the levels of leucine, isoleucine, and valine. This disorder is screened for in the newborn using a bacterial inhibition assay system similar to the Guthrie test, which detects elevated levels of plasma leucine in the newborn. After the diagnosis, special diets that restrict the intake of isoleucine, leucine, and valine with careful monitoring can provide adequate control of this disease. Affected children who are not detected by newborn screening within the first month of life generally become acutely ill with hypoglycemia and ketoacidosis. Without early recognition and treatment death occurs frequently, and in those in whom diet is instituted late, significant mental retardation results.

Although phenylketonuria is recognized in older infants because of delayed development and microcephaly (small head), and maple-syrup urine disease presents as an acute illness in the neonatal period, homocystinuria is generally recognized in late childhood because of physical abnormalities with or without associated mental retardation. Homocystine is an intermediate amino acid in the metabolism of methionine to cystine (Fig. 44-20). Deficiency of the enzyme cystathionine synthetase results in elevated plasma and urinary levels of homocystine. The elevation in plasma levels of homocystine results in conversion of most of this substance to methionine. Thus one can screen for homocystinuria by newborn blood testing analyzing plasma levels of methionine. This condition results in clinical manifestations through two mechanisms: (1) the elevated levels of homocystine have been shown to be toxic to vascular endothelium and account for the thromboembolic phenomenon (plugging of brain or lung blood vessels by blood clots) of this disease, and (2) the failure of production of cystine results in a deficiency of this essential amino acid in connective tissue, specifically collagen, metabolism. Thus these patients present with skeletal and ocular abnormalities much like those seen in Marfan's syndrome (a dominantly inherited connective tissue disease affecting the eyes, skeleton, and heart). The enzymatic activity of cystathionine synthetase has as its cofactor the B vitamin, pyridoxine. Different mutations at the structural

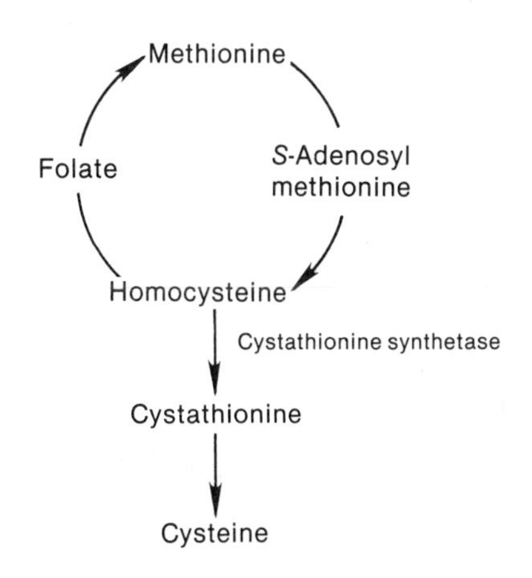

Fig. 44-20. Homocystinuria. Most common form of homocystinuria results from deficiency of cystathionine synthetase. Residual enzyme activity in some patients can be enhanced by administration of pyridoxine.

locus for cystathionine synthetase result in two distinct forms of homocystinuria. With a complete lack of enzymatic activity, the use of pyridoxine will not enhance any residual enzyme activity. These patients, in addition to the thromboembolic phenomenon and connective tissue manifestions of this disease, also demonstrate mild to moderate mental retardation. Approximately half the patients with cystathionine synthetase deficiency will have sufficient enhancement of residual activity by the dietary use of therapeutic doses of pyridoxine that their disease can be essentially cured. The diagnosis of homocystinuria is most often suspected because of the connective tissue complication of dislocated lenses. Screening for this disease is accomplished through the use of the urine-screening test, cyanide-nitroprusside. Nitroprusside combines with sulfur-containing amino acids to produce a red color. The cyanide reduces disulfide bonds and gives a positive screening result. Both homocystinuria and cystinuria give a positive cyanide-nitroprusside reaction, whereas silver nitroprusside, by not reducing the disulfide bond of homocystine, gives only a positive result with cystinuria present. The analysis of plasma amino acids in these patients demonstrates an elevated level of blood methionine. This and the resulting cystathionine synthetase deficiency tend to distinguish homocystinuria from two more rare forms of this disease that involve defects in folate and cyanocobalamin metabolism. These two more rare forms of homocystinuria are less severe and generally easier to manage.

Organic acidurias. The metabolism of amino acids involves a number of intermediate steps producing organic acids. The defect in subsequent metabolism of these compounds results in disorders that are characterized by severe metabolic acidosis. Among this group of disorders, the ketotic hyperglycinemia syndrome is composed of several defects in propionic acid and methylmalonic acid metabolism. The initial reports of this group of disorders described infants with recurrence of severe ketoacidosis and mental retardation who were intolerant to protein and were found to have striking hyperglycinemia. Subsequently, it has been recognized that this syndrome involves defects in propionic acid metabolism and methymalonic acid metabolism, each of which occurs in several forms. The amino acids leucine, isoleucine, valine, threonine, and methionine and as well odd-numbered fatty acids lead to the production of propionyl CoA, which is converted to methylmalonyl CoA. Methylmalonyl CoA is converted into succinyl CoA, which then can enter the tricarboxylic acid cycle (Fig. 44-21). Infants with these disorders usually present within the first few weeks of life with severe ketoacidosis, which may be accompanied by hypoglycemia. The urine yields a strongly positive dinitrophenyl hydrazine reaction, and by amino acid analysis shows striking elevations in glycine. Given such an infant with these laboratory findings, one can screen the urine by a simple semiquantitative measure of methylmalonic acid in the urine. If methylmalonic acid is not found in the urine, one can make a presumptive diagnosis of propionic aciduria. A definitive diagnosis of this disorder can then be further pursued through the demonstration of the presence of methylcitrate in the urine, the gas-liquid chromatographic analysis of organic acids in the urine, or the assay of propionyl CoA carboxylase in cultured cells. The conversion of propionyl CoA to methymalonic CoA uses biotin as a cofactor, which may in therapeutic doses enhance residual enzyme activity. The conversion of methylmalonic CoA to succinyl CoA involves the metabolism of B_{12}, which in therapeutic doses can also result in significant biochemical improvement. In those instances in which vitamin therapy does not result in a significant therapeutic response, a low-protein diet is used. The difficulty in dietary management of such infants is based on the fact that a diet sufficiently low in protein to prevent the recurring episodes of ketoacidosis may not provide adequate protein for proper growth and mental development. Because the enzyme defects in both of these disorders can be assayed in cultured cells, prenatal diagnosis exists for both conditions.

Defects in carbohydrate metabolism. The term *galactosemia* has been used for a toxicity syndrome associated with an intolerance to dietary galactose. This recessively inherited disorder occurs as a result of a deficiency of the enzyme galactose-1-phosphate uridyl transferase (Fig. 44-22). Lactose, composed of galactose and glucose, is the major disaccharide in mammalian milk. Hydrolysis of lactose by the intestine results in the release of the monosaccharides glucose and galactose. The main pathway of galactose metabolism in humans involves the conversion of

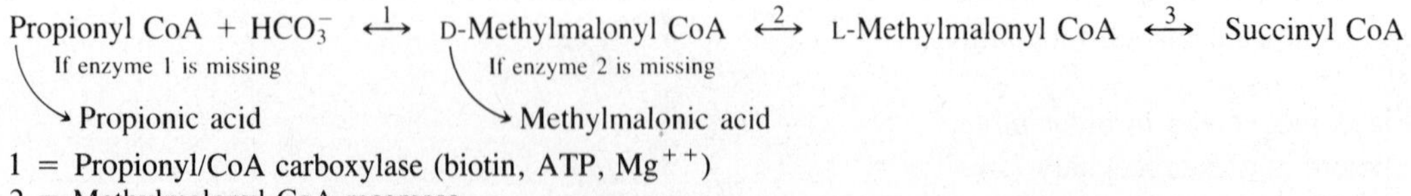

Propionyl CoA + HCO_3^- $\xleftrightarrow{1}$ D-Methylmalonyl CoA $\xleftrightarrow{2}$ L-Methylmalonyl CoA $\xleftrightarrow{3}$ Succinyl CoA

If enzyme 1 is missing → Propionic acid

If enzyme 2 is missing → Methylmalonic acid

1 = Propionyl/CoA carboxylase (biotin, ATP, Mg^{++})
2 = Methylmalonyl-CoA racemase
3 = Methylmalonyl-CoA mutase

Fig. 44-21. Propionic aciduria and methylmalonic aciduria. This sequence of metabolism of organic acids can be interrupted at several points, each of which results in a specific organic aciduria.

galactose to glucose by epimerization of the hydroxyl group at the carbon-4 position. The reaction catalyzed by galactose-1-phosphate uridyl transferase involves galactose-1-phosphate plus UDP glucose, yielding UDP galactose plus glucose-1-phosphate. The UDP galactose can by further conversion yield UDP plus glucose-1-phosphate. Humans are thus capable of metabolizing large amounts of galactose. However, with deficiency of the transferase, galactose is reduced to galactitol and oxidized to galactonate. It is the presence of these two intermediate products of galactose metabolism that has direct toxic effects and results in the clinical manifestations of galactosemia. The classic clinical presentation of galactosemia is one of failure to thrive in early infancy complicated by vomiting and diarrhea. In addition, these infants show deranged hepatic function with jaundice and hepatomegaly. Severe hemolysis can also occur, and cataracts may be noted shortly after birth. Without dietary therapy, retarded mental development can be apparent after only a few months of age. There are variants of galactosemia that do not present so strikingly in the newborn period but may be later recognized in late infancy or early childhood by poor growth, deranged hepatic function, retarded mental development, and cataracts. With early diagnosis, removal of dietary sources of galactose can accomplish excellent metabolic control of this disorder with prevention of cataracts, and normal growth and mental development. For this reason a number of states have instituted new-born screening for galactosemia using the disk of blood collected for the Guthrie test as a means of assaying for galactose-1-phosphate uridyl transferase. Metabolic screening tests of the urine from acutely ill infants should include a test for reducing substances in the urine. Urine dip sticks for glucose are specific for glucose in that they are impregnated with glucose oxidase. Thus such a urine screening would be negative in most infants with galactosemia. An infant with the clinical manifestations described above who has a negative dip-stick urine test for glucose but a positive copper sulfate test (such as Clinitest) for reducing substances in the urine has strong presumptive evidence for galactosemia. Galactosemia will result in Fanconi's syndrome of abnormal renal tubular function. This includes the renal loss of amino acids and glucose, which may further compound urine metabolic screening. Dietary control of the disease is important in the first few years of life; with advancing age some tolerance to galactose develops.

A second defect in galactose metabolism occurs secondary to deficiency of the enzyme *galactokinase* (Fig. 44-23). This disorder is recognized most commonly based on urine metabolic screening of patients with cataracts. Deficiency of galactokinase results in the accumulation of galactitol as an end product in galactose metabolism with this defect. This compound accumulates within the lenses and is responsible for the cataracts. There are no hepatic or other systemic manifestations of this defect. One can assay the enzyme in red blood cells, and the excretion of galactose and galactitol in urine will give a positive reducing substance reaction. These two tests are the primary means of diagnosis of galactokinase deficiency. Recognition of this disorder and removal of galactose from the diet will prevent the development of cataracts.

Lactic acidosis is a commonly observed complication of many diseases that result in increased anaerobic metabolism. Primary abnormalities in carbohydrate metabolism leading to lactic acidosis are rare. Anaerobic metabolism and defective carbohydrate metabolism may give rise to elevations in both pyruvic and lactic acid. However, as pyruvate is shunted rapidly to lactate and most clinical laboratories perform lactate assays but not pyruvate, lactic acidosis rather than pyruvic acidosis is more commonly recognized and referred to. A primary abnormality in pyruvate metabolism or in muscle mitochondrial metabolism of glucose will result in an increased pyruvate and lactate level after a glucose load. Additionally, the derangement in carbohydrate metabolism will also lead to a decreased rate of gluconeogenesis reflected as an elevation in plasma alanine after a glucose load. In patients with lactic acidosis in whom no abnormality in tissue blood profusion or oxygenation is apparent as a cause for a secondary lactic acidosis caused by increased anaerobic metabolism, a glucose

$$\text{Galactose-1-phosphate} + \text{UDP glucose} \xrightarrow[\text{uridyltransferase}]{\text{Galactose-1-phosphate}} \text{UDP galactose} + \text{Glucose-1-phosphate}$$

Fig. 44-22. Galactosemia. Utilization of galactose-1-phosphate requires that it be converted into UDP-galactose by reaction shown.

$$\text{Galactose} + \text{ATP} \xrightarrow{\text{Galactokinase}} \text{Galactose-1-phosphate} + \text{ADP}$$

Fig. 44-23. Galactokinase deficiency. Utilization of galactose requires that it be first converted into galactose-1-phosphate.

Table 44-6. Inborn errors of glycogen deposition or utilization

Condition	CORI type	Biochemical features	Clinical features
Glucose-6-phosphatase deficiency (GIERKE's disease)	1	Normal glycogen accumulates in liver and kidney	Hepatomegaly, hypoglycaemia; stunted growth with retarded bone age, etc.
Idiopathic generalized glycogenosis (POMPE's disease)	2	Normal glycogen accumulates in all organs	Cardiac failure, muscle hypotonia, neurological disorders, death in infancy
Dextrin-1,6-glucosidase (debrancher) deficiency (limit dextrinosis; FORBES' disease)	3	Abnormal glycogen with short banches deposited in liver and, sometimes, skeletal and cardiac muscle	Hepatomegaly, hypoglycaemia; less severe than GIERKE's disease
α-Glucan-branching glycosyltransferase (brancher) deficiency (amylopectinosis; ANDERSEN's disease)	4	Abnormal cabohydrate with long inner and outer branches deposited in liver, spleen and lymph nodes	Hepatic cirrhosis; death within two years of birth
Glycogen phosphorylase (glycogen phosphorylase of the muscle) deficiency (MCARDLE's syndrome)	5	Moderate accumulation of normal glycogen in skeletal muscles; lactate and pyruvate levels in blood fall during exercise	Generalized muscular fatiguability and pain
Glycogen phosphorylase (hepatic glycogen phosphorylase) deficiency (HERS' disease)	6	Normal glycogen accumulates in liver; phosphorylase content of liver and leucocytes reduced	Hepatomegaly; relatively benign
Deficiency of UDP glucose glycogen glucosyltransferase (glycogen synthetase)	7	Liver glycogen almost completely absent	Severe fasting hypoglycaemia

From Geigy's scientific tables, ed. 7., 1970, CIBA-Geigy Corp., Summit, N.J. From FIELD, R.A., in STANBURY et al. (Eds.), *The Metabolic Basis of Inherited Disease,* 2nd ed., McGraw-Hill, New York, 1966, page 141; HERS, H.G., *Advanc. Metab. Disord.*, **1**,1 (1964); *Control of Glycogen Metabolism,* Ciba Foundation Symposium, Churchill, London, 1964.

tolerance test with determinations of plasma alanine, pyruvate, and lactate may suggest a primary abnormality in carbohydrate or pyruvate metabolism.

Glycogen-storage diseases. The glycogen-storage diseases combine two different types of defects in metabolism. First, because of the alterations in glycogen metabolism, inadequate glucose stores are available, and therefore acute hypoglycemic symptoms occur as a result of the lack of available glucose for metabolic needs. Second, the accumulation of glycogen results in the long-term chronic effects of the storage disease whose consequences are dependent on the glycogen distribution in the storage process. Table 44-6 lists a classification scheme of the glycogen-storage diseases.

The classic form of glycogen-storage disease is von Gierke's disease, which occurs secondary to a deficiency of glucose-6-phosphatase. All metabolic sources of blood glucose are channeled through the intrahepatic formation of glucose-6-phosphate. Glucose in this form cannot be transported outside the liver cell. Thus the formation of glucose from amino acids through gluconeogenesis or the conversion of other carbohydrates into glucose utilizes the intermediate of glucose-6-phosphate. With deficiency of glucose-6-phosphatase, the only carbohydrate available to maintain blood glucose is the glucose metabolite glucose-1-phosphate. A simplified scheme of glycogen metabolism and the consequences of glucose-6-phosphatase deficiency are shown in Fig. 44-24. Infants with type 1 glycogen-storage disease, glucose-6-phosphatase deficiency, usually present within the first few days of life with recurring episodes of hypoglycemia that are often accompanied by ketoacidosis and lactic acidosis. At the time of presentation, in addition to hypoglycemia, these infants show an elevated level of blood lactate, uric acid, elevated platelet count, and hepatomegaly. These infants are unable to respond to glucagon by glycogenolysis. After the administration of glucagon, if there is no elevation in blood glucose, a defect in glycogen metabolism must be assumed. A number of the glycogen-storage diseases will fail to respond to glucagon by elevating blood glucose, and therefore a specific diagnosis of type 1 glycogen-storage disease requires a liver biopsy for the demonstration of glucose-6-phosphatase deficiency. Because this enzyme is restricted to hepatic tissue, the diagnosis cannot be made in cultured skin cell fibroblasts, and therefore prenatal diagnosis is not available. Children with type 1 glycogen-storage disease are unable to maintain an adequate blood glucose for more than 2 to 2½ hours after a normal feeding. The management of infants with this disorder cannot be satisfactorily accomplished through the use of frequent oral feedings. For that reason, nasogastric tubes and feeding gastrostomies have been used to greatly improve the management of this disease during the first few years of life. The dietary treatment of glycogen-storage disease is best accomplished through the use of complex carbohydrates, which will result in a slower rise and fall in blood glucose after a feeding. The use of simple sugars will immediately elevate blood glucose but will contribute to continued intra-

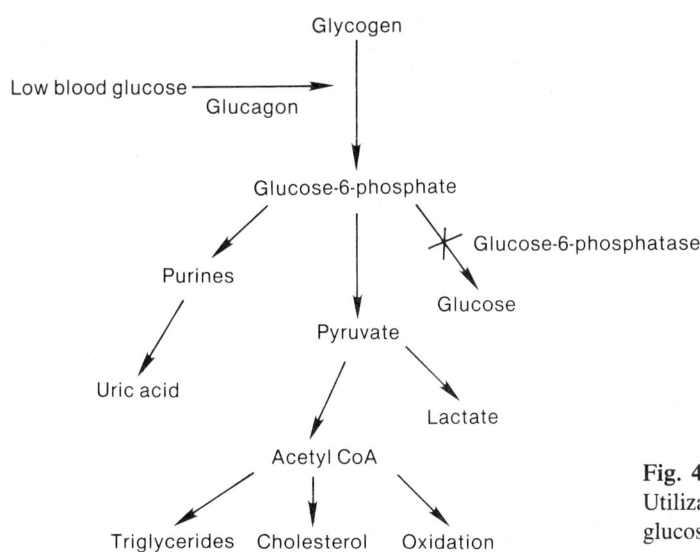

Fig. 44-24. Glycogen-storage disease type 1 (von Gierke's disease). Utilization of glycogen to form free glucose requires hepatic action of glucose-6-phosphatase to transport glucose extracellularly.

hepatic storage of glycogen and will not maintain adequate blood levels of glucose for a satisfactory period of time. With such an aggressive approach toward the management of a stable and adequate blood glucose over a 24-hour period by nasogastric or gastrostomy feedings, the management of this disease has been significantly improved in the past few years.

Pompe's disease or type 2 glycogen-storage disease represents a condition entirely different from the remainder of the glycogen-storage diseases. The reason is that this disorder is the result of a defect in lysosomal degradation of glycogen. The enzyme deficiency is α-glucosidase, an acid hydrolase similar to that of other lysosomal storage diseases. Deficiency of this enzyme results in widespread, systematic lysosomal storage of glycogen and presents in infancy with hepatomegaly, cardiomegaly, and central hypotonia. The disease follows a rapidly downhill course, with death by 1 to 2 years of age. This alteration of glycogen metabolism does not affect normal glucose homeostasis. Because the defect is one of lysosomal degradation, dietary approaches to modify glucose metabolism have no effect on the course of this disease. Diagnosis of this form of glycogen-storage disease is indicated by lysosomal accumulation of glycogen demonstrated on muscle or liver biopsy. One can assay the enzyme α-glucosidase in peripheral lymphocytes or in cultured skin fibroblasts to arrive at a specific diagnosis. Because this enzyme is present in cultured cells, prenatal diagnosis exists for this form of glycogen-storage disease.

Transport defects. There are a number of conditions that lead to altered metabolic states as a result of a defect in membrane transport of metabolites rather than an enzyme deficiency in a metabolic pathway. Pentosuria and renal glucosuria are two such abnormalities that are clinically benign. They result from a renal tubular defect in the reabsorption of these two carbohydrates.

After glomerular filtration there is the selective reabsorption through the renal tubules of plasma contents such that urine ultimately contains waste products with the conservation of nonwaste products. Single gene–determined components of the plasma membrane of the renal tubule are responsible for the selective reabsorption of individual compounds. Single-gene abnormalities therefore can result in the selective renal tubular loss of a variety of plasma components. One such selective renal tubular reabsorption mechanism exists for the four amino acids cystine, lysine, ornithine, and arginine. Homozygotes for a mutation involving this reabsorptive system have a pronounced renal loss of these four amino acids. The condition is known as *cystinuria* because it is the renal loss of cystine that contributes to the clinical disorder. Cystine is relatively more insoluble in an acid urine than an alkaline urine. With an increased elevation in renal loss of cystine this amino acid precipitates out to form stones in renal papillae and the collecting system. Thus cystinuria presents in early childhood as a form of chronic renal insufficiency usually first manifest as growth retardation. Metabolic screening of urine because of failure to thrive or short stature will detect cystinuria through a positive cyanide-nitroprusside reaction. Given such a positive test, quantitative amino acid analysis of urine will show significant increases in renal losses of cystine, lysine, ornithine, and arginine.

Management of this disease through a limited dietary intake of cystine is impractical. The primary means of treatment is the alkalinization of urine so as to increase the solubility of cystine and prevent crystal formation with stones. If a large fluid intake and alkalinization of urine are not successful in controlling this disease, the chelating

Table 44-7. Errors in renal amino acid transport

Disorder	Urine analysis	Findings
Cystinuria	Urine cyanide-nitroprusside	Renal failure secondary to cystine stones
Hyperdibasic aminoaciduria	Elevated urinary lysine, ornithine, arginine	Autosomal dominant asymptomatic
Fanconi's syndrome	Elevated urinary amino acids, bicarbonate, phosphate, glucose, potassium, uric acid	Growth failure in primary form, occurs secondarily in Wilson's disease and galactosemia
Glucoglycinuria	Glucosuria, glycinuria	Autosomal dominant asymptomatic
Hartnup's disease	Elevated neutral and ring amino acids	Asymptomatic with good nutrition, especially nicotinic acid
Familial renal iminoglycinuria	Elevated urine proline, hydroxylproline, glycine	Autosomal recessive asymptomatic
Lowe's (oculocerebrorenal) syndrome	Renal aminoaciduria, proteinuria, aciduria, phosphaturia	X-linked recessive cataracts, mental retardation, and hypotonia

agent penicillamine is useful. Penicillamine will bind to cystine and further decrease the crystallization of cystine in the collecting system. Heterozygotes for cystinuria will have a partial defect in renal tubular reabsorption of these amino acids, and this is demonstrable by quantitative amino acid analysis of a 24-hour urine specimen. The difference in the heterozygote and homozygote is the absolute amount of amino acids excreted. Thus the heterozygote may on occasion demonstrate a positive cyanide-nitroprusside screening test and quantitative analysis will be necessary to determine whether this is a heterozygote manifestation or actual cystinuria in the homozygous state. The heterozygote for cystinuria does not require therapy because inadequate cystine is present to result in significant crystalization. Children with short stature are often investigated for growth hormone deficiency through the use of an arginine-insulin test of growth hormone response. After an arginine-insulin tolerance test, the urine will be cyanide-nitroprusside positive, and amino acid analysis will demonstrate a pattern similar to cystinuria. This occurs because the infused arginine overloads the renal tubular reabsorption of these four amino acids and the resulting pattern of cystinuria is observed.

A number of conditions exist in which there is defective renal transport of amino acids; these are shown in Table 44-7.

Vitamin D–resistant rickets represents a defect in renal tubular reabsorption of phosphate. This condition is inherited as an X-linked dominant disorder. As a result it is quite variable in heterozygous females who may show no clinical manifestations to moderate short stature and evidence of rickets. In affected males there is hypophosphatemia and hyperphosphaturia with short stature and bowing of the legs accompanying the roentgenographic evidence of rickets. This disorder is most commonly recognized by demonstration of hypophosphatemia in the evaluation of a young male with short stature or bowing of the legs. Definitive diagnosis is then made by demonstration of a high renal loss of phosphate with a greatly elevated 24-hour urine phosphate excretion. This disorder is called vitamin D–resistant rickets because massive doses of vitamin D will not correct the altered plasma and urine levels of calcium and phosphate. The disorder is resistant to vitamin D because the defect does not involve vitamin D metabolism. One can satisfactorily accomplish treatment of this condition, however, through a greatly increased dietary intake in phosphate. Phosphate is a common component of many laxatives, and therefore the limiting factor in oral phosphate intake is the complication of diarrhea. Phosphate is rapidly cleared from plasma when taken orally and therefore replacement therapy requires phosphate intake at a rate of every 4 hours. With regular intake of the maximal gastrointestinal tolerance of phosphate, males with this condition can have significant improvement of rickets and enhancement of their adult height.

Disorders of mineral metabolism. The most common and best known disorder involving mineral metabolism is *Wilson's disease,* or *hepatolenticular degeneration*. This autosomal recessively inherited disorder involving copper metabolism may present initially as cirrhosis in adolescents or teen-agers or as a psychiatric disorder in older teenagers or young adults. Although the exact genetic basis of the specific abnormality in copper metabolism remains unknown, derangements in tissue and urinary levels of copper are diagnostic for the disease. Clinically this disorder is characterized by a triad of findings: a peculiar neurological syndrome, cirrhosis of the liver, and Kayser-Fleischer rings of the cornea. The neurological abnormalities take two forms: The first is lenticular degeneration (loss of cerebellar function), which is also known as the dystonic form (abnormal movement and tone of muscles) of the disease. This results in spasticity, rigidity, drooling of saliva, dysarthria, and dysphasia as predominant early features. The progression of the disease is unrelenting, and unexplained febrile episodes accompany later stages of the disorder. The second neurological form is pseudosclerosis. This involves flapping tremors of the wrist and shoulders associated with rigidity and spasticity. Psychiatric involve-

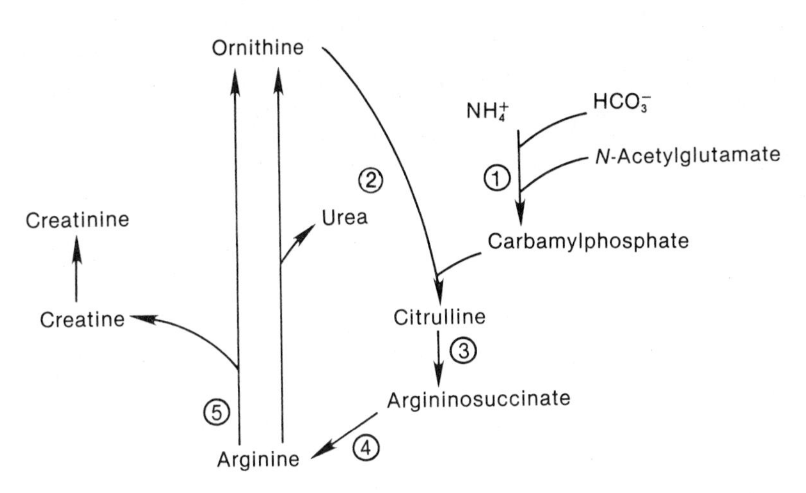

Fig. 44-25. Urea cycle defects. Five primary hyperammonemias include type 1, or carbamyl phosphate synthetase deficiency; type 2, or ornithine transcarbamyl transferase deficiency; type 3, or citrullinemia caused by argininosuccinate synthetase deficiency; type 4, or argininosuccinic aciduria caused by argininosuccinase deficiency; type 5, or hyperargininemia caused by arginase deficiency.

ment includes personality disturbances that are usually of an aggressive nature and childish behavior. Intellectual function remains intact despite these neurological manifestations. The hepatic involvement of this disease is first manifest as an enlarged liver with associated splenomegaly. Thereafter the course is one similar to that of chronic hepatitis. Ultimately the disease progresses to the full-blown picture of cirrhosis. The Kayser-Fleischer ring is still considered as the single most important clinical diagnostic finding of this disease. It is a ring of a golden-brown or greenish discoloration that appears at the margin of the cornea near the limbus. On slit-lamp examination it is recognized as moderately coarse granules deposited on the inner surface of the cornea in Descemet's membrane.

There are a number of laboratory approaches to the diagnosis of Wilson's disease. Most but not all patients with this disorder will have a depressed plasma level of ceruloplasmin. Ceruloplasmin is a metalloprotein containing copper and has a glycoprotein structure. It acts as an oxidase in the oxidation of iron from the ferrous to the ferric state. Early in the course of the disease the 24-hour urine level of copper may be normal to slightly increased but is uniformly strikingly elevated in advanced stages of the disorder. Given a clinical suspicion of this disorder, one diagnostic approach is the 24-hour urine measurement of copper followed by a 24-hour urine measurement of copper during administration of penicillamine. With increased tissue levels of copper the chelating agent penicillamine will result in a striking increase in copper excretion. The most definitive and in some cases the only assured way of making a diagnosis is by liver biopsy and tissue analysis of copper content. Given the diagnosis of Wilson's disease in one member of a family, measurement of urinary copper and plasma ceruloplasmin levels will often assist in the detection of the disease in younger presymptomatic siblings. Wilson's disease can be essentially controlled by the long-term administration of penicillamine.

A second disorder of copper metabolism is known as *Menkes' syndrome,* or *kinky-hair disease.* This disorder is an X-linked recessive neurological disease that presents in early infancy with failure to thrive, lethargy, hypothermia, and myoclonic seizure activity. Affected male infants have pallid skin and a similar facial appearance. The hair may appear normal at birth, but thereafter a characteristic pattern develops. The hair lacks luster and is somewhat depigmented; it also has a steely feel. Microscopic examination of the hair reveals pili torti. Children with Menkes' syndrome have low serum levels of copper and ceruloplasmin. However, attempts at treatment through copper administration have been unsuccessful. A defect in copper transport is demonstrated through the study of radioactive copper metabolism in cultured skin fibroblasts. This approach has also been succesfully used in the prenatal diagnosis of this disease.

Disorders of urea cycle and hyperammonemias. The principal end product of nitrogen metabolism in humans is urea. Protein metabolism results in the production of ammonia, which enters the urea cycle (Fig. 44-25) in the synthesis of carbamyl phosphate and exits through urea in the conversion of arginine to ornithine. Within this cycle there have been recognized five distinct enzymatic deficiencies, each of which results in hyperammonemia. Disruption of the urea cycle does not produce acidosis, ketosis, or hypoglycemia but manifests itself clinically as the direct toxic effect of ammonia on the central nervous system. The urine metabolic screening test employed for the detection of amino acidurias and organic acidurias will generally not give a clue as to the presence of a urea cycle defect. Affected infants thus are not acidotic or hypoglycemic, but rather experience central nervous system depression because of the direct effect of ammonia. Therefore the clinical diagnosis should be suspected in infants with central nervous system depression after protein intake, a condition that prompts an analysis of blood ammonia level. The degree of hyperammonemia is massive and in itself does not indicate a particular disorder but rather an interruption in the urea cycle. Quantitative determination of plasma and urine amino acids may show elevations in amino acids suggestive of defects in earlier steps in the urea cycle or demonstrate the specific compound elevated in the latter steps of the urea cycle. Carbamyl phosphate synthetase deficiency and ornithine transcarbamylase may be suspected clinically but require liver biopsy for specific diagnosis using an enzymatic analysis. Citrullinemia, argininosuccinicaciduria, and argininemia are recognizable clinically by elevation of these substrates through two-dimensional paper chromatography of urine or column chromatography for amino acid analysis. The enzyme defect in these latter three conditions is demonstrated in cultured skin fibroblasts. Thus prenatal diagnosis through enzyme assay in cultured amniotic fluid cells can be accomplished for the latter three disorders but not for carbamyl phosphate synthetase or ornithine transcarbamylase deficiency.

A variety of therapeutic approaches have been used in the treatment of these disorders with varying degrees of success. Carbamyl phosphate synthetase and ornithine transcarbamylase deficiencies are the more difficult to treat and generally cannot be adequately controlled by a restricted protein diet. A novel approach has been the dietary use of the ketoacid analogs of the essential amino acids. Intake of these ketoacid analogs allows for transamination of them into the essential amino acids, thereby using body nitrogen without increasing the requirement for nitrogen excretion.

Receptor defects. Receptors are discrete gene products that function as membrane-bound or cytoplasmic agents required for the transport of specific compounds across membranes or as the intracellular regulators of metabolic activity. Knowledge about receptor function has been greatly enhanced through the study of single gene disorders that result in specific receptor defects. Such genetic disorders are illustrated by the conditions familial hypercholesterolemia and testosterone-resistant syndromes (testicular feminization syndrome).

Familial hypercholesterolemia refers to a dominantly inherited disorder that is the most common single gene cause of coronary artery disease and myocardial infarction in young adults. In humans, low-density lipoprotein (LDL) is the carrier for most of the cholesterol in plasma. After cholesterol is synthesized within the liver, it is transported bound to LDL to cells, where it is internalized for the production of hormones and other lipid-containing macromolecules. The internalization of cholesterol is accomplished through the binding of LDL molecules to a specific cell surface receptor. After binding of LDL molecules to the cell surface receptors an endocytotic vesicle is formed, which fuses with lysosomes. Within the lysosome the LDL molecules are degraded to free amino acids and the cholesterol ester, which is hydrolyzed to yield unesterified free cholesterol. As adequate amounts of cholesterol are internalized, there is suppression of 3-hydroxy-3-methylglutaryl coenzyme A reductase (HMG CoA reductase), which causes a reduction in cholesterol synthesis. Conversely, an intracellular requirement for cholesterol stimulates HMG CoA reductase activity. The disorder familial hypercholesterolemia results from a mutation affecting the LDL receptor. With a reduced number of cell surface receptors for LDL cholesterol and thus reduced uptake of cholesterol, there is inadequate suppression of HMG CoA reductase activity. This results in stimulation of further cholesterol synthesis. The net result is familial or type IIa hypercholesterolemia. The elucidation of the molecular defect in this disorder serves as a prime example of approaches in molecular genetics. There are rare persons who are homozygous for hypercholesterolemia. During early childhood such persons have greatly elevated levels of plasma cholesterol and severe coronary artery disease, usually leading to death by adolescence. Study of cultured skin fibroblasts from the homozygously affected persons allowed elucidation of the defect in LDL receptors and the concomitant effect on the regulation of HMG CoA reductase activity. After this demonstration it was possible to show that the more common heterozygous form of autosomal dominant hypercholesterolemia represented a partial defect in LDL receptors and regulation of HMG CoA reductase activity.

The term *testicular feminization syndrome* is an older designation for a disorder involving abnormality in testosterone metabolism. Affected individuals are phenotypic females who develop female secondary sexual characteristics at puberty but fail to undergo menarche. A laparotomy in such a person will show absence of a uterus and the presence of intra-abdominal testes. Chromosome analysis reveals a normal male 46,XY karyotype. These chromosomal males embryologically develop the external genitalia of a female because of a resistance to the action of testosterone. There is normal testosterone generation by the testes, but an unresponsiveness of target tissues to testosterone. The action of testosterone (androgens) on target cells is mediated by specific cytoplasmic receptors. Dihydrotestosterone combines with a cytosol-binding protein (the receptor) to form a hormone-protein complex that is transported into the cell nucleus, where it exerts its action on chromatin.

This cytosol protein receptor for androgens is the gene product encoded by a single locus on the X chromosome. Affected persons who have a deficiency of receptors are infertile. The pedigree analysis of affected families is consistent with either an autosomal recessive sex-limited disorder or an X-linked recessively inherited disorder. The most convincing evidence for X-linked inheritance of the testicular resistance syndromes was accomplished through experiments that demonstrated lyonization of this defect in cultured cells from heterozygous females. Skin biopsy specimens were taken from heterozygous females and fibroblast cultures established. Clones were grown from individual fibroblast cells, and these clones behaved in two distinct fashions. Some clones demonstrated normal receptor activity for testosterone, and other clones were deficient in the cytosol receptor for androgens. Such an observation could only occur if the cytosol receptor were encoded by the X chromosome and heterozygous females demonstrated lyonization of this defect.

Disorders of purine metabolism. The best known disorder involving a defect in purine metabolism is the *Lesch-Nyhan syndrome*. The elucidation of this disorder is an interesting story, and the use of this mutation has been a powerful tool in the study of molecular genetics. As a medical student, Lesch evaluated a number of mentally retarded persons with simple urine screening tests. He saw two brothers with a disorder of self-mutilation, mental retardation, and cerebellar dysfunction. Their urine was noted to contain excessive amounts of uric acid crystals. Working with his mentor, Nyhan, it was recognized that hyperuricemia characterized this X-linked recessively inherited disorder. The enzymic defect was later found to be hypoxanthine-guanine phosphoribosyltransferase (HGPRT). This enzyme is involved in the metabolism of the purine nucleotides guanylic acid (GMP) and adenylic acid (AMP). These two purines exert feedback inhibition on the activity of a number of enzymes in purine metabolism including HGPRT. The deficiency of this enzyme results in an accelerated rate of purine biosynthesis, and because of the block, increased amounts of hypoxanthine and xanthine, which are readily converted in the liver by xanthine oxidase to the massive amounts of uric acid that are formed. Allopurinol is used to lower the serum and urinary levels of uric acid in these patients, as is done with gout, but this does not alter the central nervous system abnormalities that are part of the disorder.

Affected males with the severe form of the disease appear in the first 6 months of life to have cerebral palsy. This motor disability combined with delayed speech, in part caused by dysarthria of cerebellar origin, suggests that additionally these children are mentally retarded. Gross ataxia becomes apparent, and by several years of age they are easily agitated and react by self-mutilation. This usually takes the form of biting off the tips of fingers or the lips. While earlier in the disease the mental abilities are better than the initial impression suggested, cognitive function deteriorates with time. The bizarre behavior and mental retardation often lead to institutionalization by 6 to 8 years of age. Death occurs in late teens or the early twenties. The mechanism by which the deranged purine metabolism produces the central nervous system abnormalities in this condition has not been elucidated.

When cells deficient in HGPRT are cultured, they are resistant to the effects of purine analogs such as 8-azaguanine and 6-mercaptopurine, which inhibit the growth of normal cells. Because the female carrier for this X-linked condition by lyonization has cells with HGPRT and cells without HGPRT, fibroblasts will grow in both normal media and media with one of these purine analogs. Cells from a noncarrier female will grow in the normal media but not in media with the analog. This selective medium system has proved to be a powerful tool in the detection of carriers for the Lesch-Nyhan syndrome. As prenatal diagnosis is available, study of females in a family in which this condition has occurred is important for genetic counseling.

GENETIC SCREENING

Role of mass screening

Screening tests play a major role in the diagnosis and population study of genetic disease. A screening test is not designed as a primary diagnostic tool. Rather, its purpose is to identify a subpopulation of persons on whom definitive diagnostic testing would be cost beneficial. In general, the more specific a particular screening test, the greater is its cost; thus its utility in screening large populations is reduced. A definitive diagnostic test for a given entity is one that contains close to 100% sensitivity and specificity. The sensitivity of a test is a measure of how frequently a test will detect a disorder under question. The specificity of a test is a measure of how well a test discriminates one disorder from several disorders. The sensitivity and specificity that are required of a screening test are dependent on the population to be studied and the purpose for which the screening test is undertaken. This is best illustrated when one considers the purposes for screening tests: (1) screening tests for the early detection of disease in presymptomatic state when effective therapy exists (phenylketonuria), (2) carrier detection for inherited disease (Tay-Sachs disease), and (3) epidemiological studies of the frequency of a given disorder within the population (47,XXY or 47,XYY karyotype). Given a disease that can be effectively treated when detected early and the consequences of nondetection that are unacceptable, it is desirable to have a test with maximum sensitivity in order to detect all possibly affected persons. Those persons ascertained through a positive screening test can undergo additional testing with more specific assays that will establish a specific diagnosis. For a disease in which treatment is not readily available, successful, less-sensitive screening tests may be used. By lowering the sensitivity one may improve the specificity. The few persons with the disease who are missed by the screening test will not be unduly compromised.

The use of screening tests for the early detection of treatable disease is best exemplified by the Guthrie test for phenylketonuria. Before 1960 it was recognized that phenylketonuria is a recessively inherited disorder that results in mental retardation. With detection of phenylketonuria in early infancy, use of diets low in phenylalanine would largely ameliorate the mental retardation seen in untreated children. However, one could only anticipate the likelihood of the presence of phenylketonuria in an infant through their having had an older mentally retarded sibling with phenylketonuria. Thus it was obvious that a populational screening test for phenylketonuria would be able to significantly reduce mental retardation from this disorder. In 1960 it was estimated that approximately 1% of the institutionalized persons with mental retardation in the United States had phenylketonuria. The test for newborn screening of phenylketonuria developed by Dr. Robert Guthrie involves a bacterial inhibition assay.

Through the use of mass newborn screening employing the Guthrie test, the subsequent specific diagnosis of phenylketonuria and the institution of diets low in phenylalanine, mental retardation secondary to this amino acid disorder has been essentially eliminated. However, a second major problem has been created through this program. Screening throughout the United States for phenylketonuria was instituted in the early 1960s. In general, dietary restrictions of phenylalanine were lifted after children reached 6 years of age. Thus there are at the present time a significant number of young women entering childbearing age with treated phenylketonuria and normal intelligence. However, pregnancies undertaken by these women without dietary therapy usually result in infants with mental retardation because of their in utero exposure to an environment of phenylketonuria. A major effort therefore is now being undertaken to identify these young women, advise them of the risk of pregnancies without treatment, and institute low-phenylalanine dietary therapy so that they might successfully undertake pregnancies. After the development of the Guthrie test a number of other newborn-screening tests have been developed that use a similar bacterial inhibition assay. Maple-syrup urine disease is screened for with assays for blood levels of leucine, and homocystinuria is screened for by measurement of the levels of methionine. This same disk is also used for the screening of galactosemia by measurement of uridyl transferase and newborn screening for congenital hypothyroidism by assays of T_4 and TSH.

Screening for normal persons who are carriers of recessively inherited disease and therefore capable of having affected children is exemplified by screening tests for sickle cell carriers and carriers of Tay-Sachs disease. The effective use of such screening tests has several requirements. First, there must be an identified population with a high incidence of the disorder on whom large-scale screening is practical and feasible. For sickle cell disease this is accomplished through the testing of teen-aged and young adult blacks. For Tay-Sachs disease, the screening is targeted for the Ashkenazic Jewish population. This is the result of ethnicity of disease, that is, certain recessively inherited disorders that occur with a disproportionately high frequency among certain ethnic populations. Second, a simple inexpensive screening test with high sensitivity and specificity must be available. For sickle cell disease this is accomplished through the Sickledex test or hemoglobin electrophoresis. For Tay-Sachs disease this is accomplished through serum assay of the lysosomal enzyme hexosaminidase A. Third, a reliable definitive diagnostic test must be available for persons with a positive screening test. This exists for both sickle cell carriers and Tay-Sachs disease carriers. Finally, the detection of the carrier state must provide useful information to those persons recognized to be carriers. Because both sickle cell disease and Tay-Sachs disease are recessively inherited disorders, the purpose of screening tests is to identify couples, both of whom are carriers and therefore capable of having an affected child. Once such couples have been recognized, a number of alternatives are available to them in future family planning. These include (1) undertaking a pregnancy with a known 25% risk, (2) use of adoption as an alternative means of raising a family, (3) artificial insemination using sperm from a noncarrier donor, (4) and prenatal diagnosis after a pregnancy by the couple. Accurate prenatal diagnosis now exists for sickle cell disease and Tay-Sachs disease.

The third use of screening tests is for the epidemiological survey of disorders. This was well demonstrated through the use of buccal smears among males in penal institutions, mental health institutions, and institutions for the mentally retarded for the detection of Klinefelter's syndrome. Screening for this purpose is not intended to provide information that will be directly useful to the persons screened but rather to provide information about the population frequency of the condition. Males with Klinefelter's syndrome have a Barr body present in cells on a buccal smear because of the presence of two X chromosomes. This screening test can lead to a confirmatory diagnosis through a karyotype analysis of cultured peripheral lymphocytes. Through the use of fluorescent dyes and analysis for fluorescent Y-bodies in buccal smears, this technique can also be expanded to include screening for the 47,XYY karyotype. The initial suggestion that the 47,XYY karyotype may be associated with antisocial behavior came from such a survey in penal institutions in England.

Detection of heterozygote carriers

During the past 20 years there has amassed considerable information regarding the molecular abnormality in many genetically determined disorders. In recessively inherited disorders, application of this knowledge of the molecular defect can be used for the detection of carriers of these diseases through demonstration of a partial expression of this molecular defect in heterozygous individuals. The accuracy of these carrier detection tests generally is directly proportional to how nearly the assay system measures the specific molecular defect of the disorder under question. A number of biochemical abnormalities have been noted in cystic fibrosis, most notably an elevation in the chloride content of sweat. However, the primary abnormality in cystic fibrosis remains unknown, and the application of numerous biochemical tests to known carriers of cystic fibrosis has to date failed to yield a means of carrier detection. It is therefore unlikely that an accurate carrier test for cystic fibrosis will become available until the specific molecular defect in cystic fibrosis is known. A number of assay systems exists for the detection of carriers of recessively inherited disorders.

For recessively inherited disorders that occur as a result of deficiency of a specific enzyme, assay of that enzyme in phenotypically normal persons is often useful for the detection of heterozygous carriers of that disease. To establish the accuracy of such an assay system for carrier detection, it is necessary to carry out assays on a large number of controls and obligate heterozygotes. Obligate heterozygotes are recognized as such by being the parent of an affected child. In general, enzyme assay in these two groups of persons will show that the heterozygous carriers have levels of enzyme activity that on the average are approximately half that of the control group. A large number of assays are required, however, to determine the accuracy of the assay in a person as to his or her carrier status. Such assay studies will vary in the accuracy of carrier detection among different disorders and usually requires considerable experience and expertise with the assay system. For example, assay of enzymatic activity of α-L-iduronidase in peripheral leukocytes of a control population and of parents of children with Hurler's syndrome showed that the parents as a group had levels of enzyme activity approximately one half that of the control group. However, within the group of parents there were certain persons whose levels of enzyme activity overlapped with the lower levels detected among the control group. Therefore the assay may not provide a definitive answer on each person tested. A sensitive method of carrier detection for X-linked recessively inherited disorders is the demonstration of lyonization for that trait in females. As discussed above for testosterone-resistance syndromes, cloning of cultured fibroblasts from females who are potential carriers has been successfully used in the carrier detection of disorders such as Hurler's syndrome, Menkes' syndrome, and Lesch-Nyhan syndrome. When the enzyme defect is known but assays on potential carriers are impractical, such as phenylalanine hydrolase determinations, tolerance tests may be helpful in a demonstration of the carrier state. With reduced levels of enzyme activity, administration of large amounts of substrate to control and carrier persons will often show tolerance curves that allow separation. Among disorders with defects in renal tubular transport, the carrier state can often be recognized through a partial expression of the primary abnormality. Thus heterozygotes for cystinuria will often demonstrate a modest elevation in the 24-hour urine level of cystine, indicative of the carrier state.

Prenatal diagnosis

With the rapid advances in genetic technology over the past few years the capabilities for prenatal diagnosis of genetic diseases have increased at a rapid pace. There are several criteria that a diagnostic method must meet for its successful use in prenatal diagnosis. These include the following. First, there must be a means of identifying couples at sufficient risk for a particular condition to warrant the risks and costs of amniocentesis and other diagnostic modalities. This ascertainment is most often accomplished by the previous birth of a child with a condition carrying a significant recurrence risk, determination that the parents are carriers for an autosomal recessively inherited disorder by previous carrier detection, or risks based on the age of mother.

Second, the disorder under question should by its nature warrant prenatal diagnosis. At present this is generally true in that practically all conditions for which prenatal diagnosis is available are severe disorders for which there is no effective therapy for affected infants.

Third, the accuracy of the diagnostic method employed should be well established, as there is little margin for error with these studies. For diagnoses of many cytogenetic disorders and biochemical disorders using cultured amniotic fluid cells, the laboratory receives a 20 to 30 mL sample for a one-time opportunity of establishing a culture and performing a limited number of assays. The results of these assays will be used to determine whether the pregnancy will be continued. One wants supreme confidence in the systems used under these circumstances.

On p. 860 is a list of the various diagnostic modalities used in prenatal diagnoses, with an example of each.

Studies in cultured amniotic fluid cells are useful for the diagnosis of any condition for which an assay system has been devised in cultured skin fibroblasts. There are several important differences, however. In doing a cytogenetic study on peripheral lymphocytes the sample has been taken because of an abnormal phenotype in the patient and there is clinical information to assist in the interpretation of the karyotype results. With prenatal diagnosis this is not true because the only information of use in arriving at a conclusion is the karyotype itself, established from the cultured amniotic fluid cells. When an unusual result is found, it is possible to analyze the chromosomes of the parents; such information may be necessary to make an informed decision regarding the changes found in the amniotic fluid cells. The recent advent of the use of coverslip

Methods of prenatal diagnosis

Cultured amniotic fluid cells
- Cytogenetic
  - Advanced maternal age for Down's syndrome
- Biochemical
  - Hexosaminidase assays for Tay-Sachs disease
- Morphological analysis
  - Electron microscopy for mucolipidosis IV
- Substrate analysis
  - Radioactive copper in Menkes' syndrome

Uncultured amniotic fluid cells
- Restrictive endonuclease analysis
  - Beta-hemoglobin gene study in sickle cell disease
- Genetic polymorphisms for linkage diagnosis
  - HLA determination and 21-hydroxylase deficiency

Amniotic fluid analysis
- Quantification of marker components
  - Alpha-fetoprotein for open neural tube defects
- Analysis of abnormal metabolites
  - Methylmalonate in methylmalonicaciduria
- Genetic polymorphisms for linkage diagnosis
  - Secretory status and myotonic dystrophy

Imaging techniques
- Ultrasonography
  - Autosomal recessive polycystic kidneys
- Roentgenography
  - Skeletal dysplasia—achondrogenesis
- Fetoscopy
  - Polydactyly in Ellis–van Creveld syndrome

Fetal sampling
- Fetal blood sampling
  - Factor VIII antigen assay in hemophilia A
- Skin biopsy
  - Histological appearance in epidermolysis bullosa dystrophica

culture techniques with clonal analysis has greatly assisted in reducing two sources of error in the interpretation of amniotic fluid cell cultures. With this method, clones are established on coverslips from single cells; by analyzing the cells on the coverslip, one can keep the clones intact. Amniotic fluid cells have a high rate of cultural artifacts, which may be mistakenly diagnosed as mosaicism for a chromosomal abnormality in the fetus. If cultured cells are harvested from a flask into a cell pellet and a smear is made, a finding of 10 abnormal cells of 25 cells analyzed would suggest possible mosaicism. By the coverslip method, however, if all the abnormal cells were derived from a single clone, this would argue for a clonal origin in culture of these abnormal cells. The other use of the coverslip method is to assist in the recognition of maternal contamination of the amniotic fluid cell culture. If the fetus is a female and there is maternal contamination, all the cells will be 46,XX. However, when analyzing clones, one can recognize a different appearance of some clones or more easily compare clones with maternal lymphocyte karyotypes to recognize polymorphic differences in the appearance of the fetal and maternal chromosomes.

With biochemical assays in cultured amniotic fluid cells the levels of enzyme activity are generally lower than they are in cultured fibroblasts. This may create a problem in distinguishing between a homozygously affected fetus and a heterozygously normal fetus with a partial enzyme deficiency. For this reason such assays should be done with an adequate series of positive and negative controls.

There are several diagnostic techniques used in prenatal diagnosis that have not been previously mentioned but are important because they are or will be used increasingly in other clinical settings. The imaging techniques of prenatal diagnosis are currently in wide use in clinical medicine; these include roentgenography, ultrasonography, and fetoscopy. These approaches are used for structural abnormalities for which there are no biochemical or cytogenetic means of diagnosis. Examples would include skeletal dysplasias with obvious changes in the fetal skeleton, renal agenesis or polycystic kidneys, omphalocele or gastroschisis, and other structural anomalies.

It is interesting that the use of genetic linkage for diagnosis has been applied more for prenatal diagnosis than any other situation. This approach is based on an exception to Mendel's second law that two different traits are assorted randomly. When the loci for these two traits are on the same chromosome, the two alleles on the same chromosome will not be assorted but will be passed together, depending on how close physically they are to one another on that chromosome. An example of this diagnostic method is illustrated in Fig. 44-26, in which determination of the HLA haplotypes are used for the prenatal diagnosis of 21-hydroxylase deficiency or congenital virilizing adrenal hyperplasia. This is possible because the locus for 21-hydroxylase is close to the HLA loci on chromosome number 6. Because of the great genetic variation that exists at the HLA loci, it has been possible to determine several close linkage relationships with HLA. Two other examples of HLA linkage useful diagnostically include autosomal recessively inherited hemochromatosis and one form of dominantly inherited olivopontocerebellar atrophy.

The recombinant DNA techniques that have proved so useful in the study of gene structure and function have provided means for accurate prenatal diagnosis of several conditions. The best known example is the test for the prenatal diagnosis of sickle cell disease. Restrictive endonucleases, as the name suggests, are enzymes, generally purified from bacteria, that make cuts in DNA at specific locations. These locations are generally determined by a 5-nucleotide sequence in the DNA. At any point in the DNA of a cell that the particular sequence recognized by a restrictive endonuclease occurs, a cut will be made in the DNA. In human cells use of one of these enzymes will usually result in the production of 1 million or so frag-

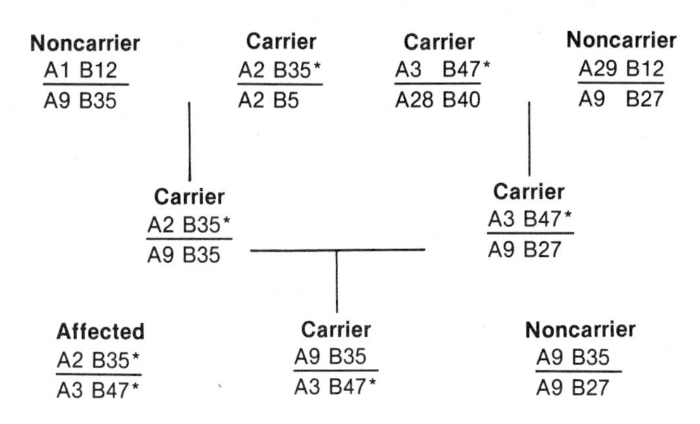

Fig. 44-26. Prenatal diagnosis by linkage analysis of HLA genotype and congenital adrenal hyperplasia. In this example mutant genes for 21-hydroxylase are linked to the HLA haplotypes A2, B35, and A3 B47, whereas normal 21-hydroxylase genes are linked to the other HLA haplotypes.

ments of DNA of varying lengths. The selection of a particular fragment is made possible by probes that are usually radioactively labeled, synthesized copies of DNA, cDNA, which by their homology bind the fragment of DNA carrying the gene of interest. With sickle cell disease it has been possible to combine the use of restrictive endonucleases that recognize the specific nucleotide change of the mutant sickle cell gene with probes that bind to that portion of the beta chain of hemoglobin gene. This technique does require neither the culturing of amniotic fluid cells nor special studies on the parents provided that they are known carriers for sickle cell disease. It is possible to imagine a time when probes and restrictive endonucleases will exist for the diagnosis of all genetic disorders.

COMMENT

The diagnosis and management of genetic diseases offer an important lesson for all aspects of clinical medicine. The closer the assay system comes to measuring the specific cause of a condition, the more accurate that diagnostic system will be. At the same time, the closer therapy of disease comes to correcting the specific molecular abnormalities of that condition, the more likely it will be that effective treatment for that condition is offered. At the present time, the ability to provide accurate, specific diagnoses for genetic conditions far outreaches the ability to effectively correct those abnormalities by some therapeutic measure. However, if significant progress in the treatment of many chronic diseases that confront adults and children is to be made, it can only come about by strategies that are based on a molecular understanding of the disease process.

CHANGE OF ANALYTE IN DISEASE

Each disease caused by an inborn genetic error results in a unique defect in a protein or enzyme function, or both. The consequence of this defect is a singular pattern of analyte change in biological fluids. Tables 44-4 to 44-9 list the pathognomonic analyte changes for a number of these genetic defects. In addition, the major clinical findings for each disease are also given.

Table 44-8. Inborn errors of purine and pyrimidine metabolism

Condition	Defective enzyme or system	Biochemical features	Clinical features	Incidence and genetics
Gout (hyperuricaemia)	Excessive synthesis of uric acid from precursors	Concentration of uric acid increased in serum and often in urine	Acute arthritic attacks, chronic arthritis with urate deposition in tissues; urinary urate calculi causing kidney damage; asymptomatic in 80% of cases	Hyperuricaemia in 1-2%, clinical gout in 2-4 per 1000; probably autosomal dominant with variable and sex modified expression
Xanthinuria	Deficiency of xanthine oxidase and defective renal tubular reabsorption of xanthine	Xanthine excreted in large amounts	Xanthine calculi in urinary tract	Rare
Orotic-aciduria	Absence of orotidine-5′-phosphate pyrophosphorylase and/or decarboxylase	Orotic acid accumulates and is excreted in urine	Severe megaloblastic anaemia, orotic-acid crystalluria	Very rare; recessive
β-Aminoisobutyric-aciduria	Deficiency of a catabolic enzyme	High urinary excretion of β-aminoisobutyric acid	Harmless	0-46% depending on ethnic group; recessive

From Scientific tables, ed. 7, 1970, CIBA-Geigy Corp., Summit, N.J.

Table 44-9. Lipidoses

Condition	Lipid accumulating	Site	Clinical features	Age at which symptoms appear	Genetics
GAUCHER'S disease (a) 'Adult' (b) Acute infantile (c) Juvenile and adult neurological	Glucocerebroside	Spleen, liver, bone marrow, leucocytes. Brain in (b) and (c); lung in (b)	Splenomegaly, often gross; hepatomegaly; anaemia; bone disorder; purpura; cerebral degeneration in (b) and (c)	(a) 1-60 years (b) 1st or 2nd half-year of life (c) 6-20 years	(a), (b) and (c) in different families; all recessive
TAY-SACHS disease (infantile amaurotic familial idiocy)	Ganglioside G_{M2} (G_0), amino glycolipid	White and grey matter of the brain	Cherry red spot; progressive cerebral degeneration; death at age 1-5 years	Usually 4-6 months, some times earlier	Recessive
Juvenile and adult amaurotic familial idiocy	Ganglioside G_{M1} (G_1)	Brain (moderate increase)	Progressive loss of vision and cerebral degeneration	From 5 years onwards	Probably recessive
NIEMANN-PICK disease (a) Acute infantile (b) Cerebral juvenile (c) Noncerebral	Mainly sphingomyelin	Spleen, bone marrow, liver; usually also brain and retina	Often cherry-red spot; hepatosplenomegaly; hepatic cirrhosis; usually cerebral degeneration and death in first 2½ years. Some adult cases are without neurological involvement	(a) From birth (b) Childhood (c) Up to 30 years or later	(a) Recessive (b) Recessive (c) Uncertain
Metachromatic leucodystrophy (a) Infantile (b) Adult	Sulphatides	Brain, kidney, urine, gallbladder	(a) Cerebral and cerebellar degeneration; spasticity; dementia; death after 1-6 years (b) Psychotic changes; blindness; aphasia; tetraplegia. Death after 3-12 years	(a) 1-2 years (b) Late childhood or adulthood	(a) Recessive (b) Uncertain
Essential familial hyperlipaemia	Triglycerides, lipoproteins	Blood plasma (chylomicrons)	Hepatosplenomegaly; sometimes xanthomata. Relatively benign	Usually early childhood	Complex
Hypercholesterolaemia	Cholesterol (free and esterified), phosphatides, sometimes triglycerides	Blood plasma (lipoproteins), tendons, skin, blood vessels	Cutaneous and tendinous xanthomata; atheroma of endocardium, coronary arteries or great vessels	From childhood onwards	Usually dominant

*From Scientific tables, ed. 7, 1970, CIBA-Geigy Corp., Summit N.J.

BIBLIOGRAPHY

Kelly, T.E.: Clinical genetics and genetic counseling, Chicago, 1980, Yearbook Medical Publishers, Inc.

Scriver, C.R., and Rosenberg, L.B.: Amino acid metabolism and its disorders, Philadelphia, 1973, W.B. Saunders Co.

Stanbury, J.B., Wyngaarden, J.B., Fredrickson, D.S., et al., editors: Metabolic basis of inherited disease, New York, 1983, McGraw-Hill Book Co.

McKusick, V.A.: Mendelian inheritance in man, Baltimore, 1983, John Hopkins University Press.

Simpson, J.L.: Disorders of sexual differentiation, New York, 1976, Academic Press, Inc.

Thomas, G.H., and Howell, R.R.: Selected screening tests for genetic metabolic diseases, Chicago, 1973, Yearbook Medical Publishers, Inc.

Grouchy, J. de, and Turleau, C.: Clinical atlas of human chromosomes, New York, 1977, John Wiley & Sons, Inc.

Chapter 45 Psychiatric disorders

David L. Garver

affective disorder A disorder of mood regulation manifest clinically by episodes or sustained periods of depression or mania, or both.

antipsychotic drugs Drugs that are utilized for the reversal or attenuation of psychotic symptoms (hallucinations, delusions, and disorders of cognition). All current antipsychotic drugs act by a single mechanism: blockade of postsynaptic dopamine receptors by stereospecific attachment to the receptor.

bimodal distribution Distribution of quantitative data around two separate modes, suggestive of two separate normally distributed populations from which the data are drawn.

bipolar affective disease Affective disorder in which both poles of mood disturbance are episodically present in the same patient: episodes of both depression and of mania.

cognitive disorder (schizophrenic) Thinking disorder characterized by speech in which ideas shift from one subject to another that is completely unrelated or only obliquely related without the speaker's showing any awareness that the topics are unrelated; frank incoherence results from total lack of logical or meaningful connections between words, phrases, or sentences.

computerized axial tomography (CAT or CT) A computer-assisted roentgenological procedure that constructs slices of brain or other body tissues from multipositioned x-ray projections useful for detection of deviations from normal caused by tumor, malformations, degeneration, and so forth. Cerebral and ventricular size can be estimated; cortical and cerebellar atrophy can be detected.

delusion False belief based on incorrect inference about external reality that is not held by other members of an individual's sociocultural milieu and is so firmly held that evidence to the contrary will not shake the false belief.

depression A mood disturbance often described as being sad, blue, hopeless, low, "down in the dumps," or irritable, accompanied by pervasive loss of interest or pleasure in almost all usual activities or pastimes.

genetic patterns of transmission Familial biological patterns of illness occurrence as manifested by increased density of particular or related illnesses both within families and in the adopted-away offspring (who share common genes with their biological as opposed to adopting families).

hallucinations A sensory perception (such as voices, visions, tactile sensations) that has the immediate sense of reality of a true perception but without external stimulation of the relevant sensory organ.

limbic system A loosely defined group of brain structures, including the hippocampus and dentate gyrus with their archicortex, the cingulate gyrus, septal areas, amygdala, and parts of the hypothalamus, that are believed to regulate not only emotion, but also certain aspects of cognition (as well as olfaction and autonomic function).

mania A periodic disturbance of mood in which the mood is elevated, expansive, or irritable and is accompanied by hyperactivity, pressure of speech, flight of ideas, inflated self-esteem, decreased need for sleep, distractibility and excessive involvement in activities that have a high potential for unrecognized painful consequences.

neurotransmitter A chemical substance released by one neuron onto a specific receptor or an adjacent cell, producing information transfer in the form of regulating or modulating the resting potential and discharge rates of the postsynaptic cell.

precursor loading A process whereby neurotransmitter levels are increased by an increase in the availability of precursors for enzymatic synthesis of the neurotransmitter.

receptor A protein complex embedded in the cell membrane that stereospecifically identifies a particular neurotransmitter and responds to its attachment by initiating a series of events that alter the membrane's resting potential and rate of firing.

reuptake An energy-dependent process by which neurons conserve their own neurotransmitter by recovering it from the synaptic cleft for storage and subsequent re-release.

schizophrenia A deteriorating psychotic illness with characteristic delusions, hallucinations, and disorders of cognition that has lasted more than 6 months and from which full recovery is not expected.

schizophreniform disorder An episodic psychotic illness with symptoms similar to those of schizophrenia, but the psychotic episodes of which last less than 6 months; recovery of function is full between illness episodes.

stereospecific binding Specific attachment of a neurotransmitter or drug to a receptor; the attachment is dependent on complimentary configurations of the compound and its receptor; stereospecificity of the attachment indicates failure of a stereoisomer (a structual mirror image of the compound) to attach to the receptor.

storage granules Membranous vesicles within the cell cytoplasm that both store neurotransmitter for deplorization-induced release and protect the neurotransmitter from deamination by intracellular enzymes.

synapse The structural junction of two neurons where chemical messages are carried from the presynaptic neuron to the postsynaptic receptor by neurotransmitters.

unipolar affective disease Affective disorders in which episodes of depression alone occur without episodes of mania.

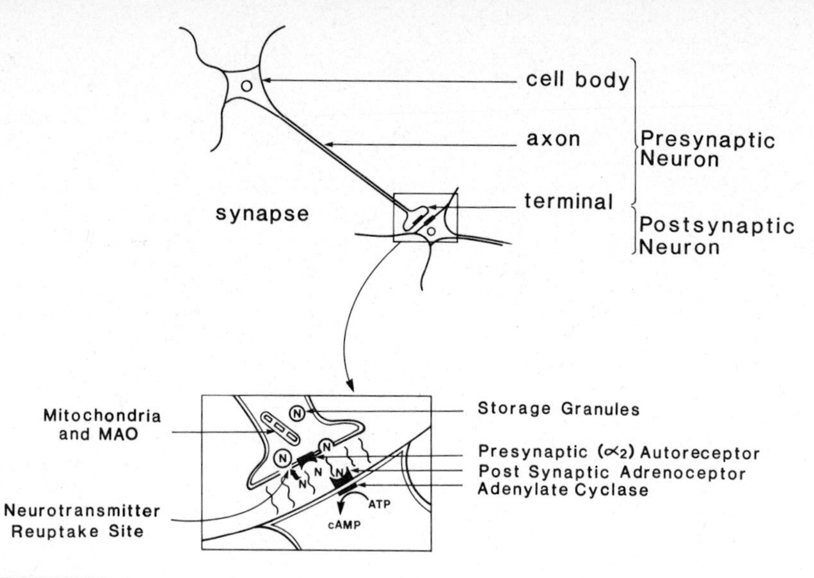

Fig. 45-1. Norepinephrine neuron, synapse, and postsynaptic connections.

A new alliance between psychiatrists and clinical chemists is developing as a result of new interest in biological causes for many of the major psychiatric disorders. Such an alliance has been hampered in the past by a diagnostic approach to mental illness, which relied almost exclusively on either clinicodescriptive or psychological emphases in defining psychiatric disturbances. Retrospectively, such syndrome or psychological emphases have resulted in the blurring of potentially useful biological information, which could aid in discriminating several different biopathological processes that are relevant to valid diagnosis and improved treatment of some forms of mental illnesses.

The syndrome diagnoses of psychiatric syndromes (based on signs and symptoms) have often been like the diagnosis of pneumonia—a syndrome that may have many diverse etiological origins. Progress in the diagnosis and treatment of the pneumonias was made when investigators found that some pneumonias were the result of infection with the pneumococcus, other pneumonias were the result of autoimmune disorders, others from toxic inhalents, and others from still more diverse etiological agents. On the other hand, it was found that the pneumococcus, autoimmune disorders, and so on do not invariably present with a pneumonia-like syndrome; they may present with a meningitis, an arthritis, renal failure, and so on. The major psychiatric (syndrome) diagnoses, like the diagnosis of "pneumonia," imply little concerning the diverse underlying pathobiological condition, illness course, or specific treatment approaches for a particular illness.

The future of psychiatric research into the major psychiatric disorders is therefore the search for valid diagnoses—diagnoses that reflect the underlying biopathological conditions, diagnoses that are predictive of illness course, diagnoses that imply specific intervention to reverse or compensate for the underlying biopathological condition. Such diagnosis may be aided by the discovery of biological markers or laboratory tests that directly reflect the underlying biopathological condition. Enduring "trait markers" of such a condition may be present throughout the life cycle and can serve for identification of persons at risk for the illness. If a particular disposition to illness is transmitted genetically, valid diagnoses also imply that relevant laboratory tests related to trait and risk for illness should also be found in some family members who share critical gene pools with the identified patient. Finally, such valid diagnoses, based on relevant laboratory parameters should permit the identification of infants and children at special risk for subsequent development of psychiatric disorders, and permit early, preventive intervention. Psychiatry of the 1980s often remains distant from such valid diagnoses; yet the process of identifying each underlying biopathological condition of the psychiatric illnesses is currently underway and will undoubtedly be aided by the skill of the clinical and research chemists in decades to come.

CENTRAL NERVOUS SYSTEM: ANATOMY, BIOCHEMISTRY, AND PHYSIOLOGY

The focus of biological psychiatry is clearly on abnormalities of the basic units of central nervous system function: neuronal systems. A brief review of the anatomy has been presented in Chapter 37.

Neurons within the central nervous system process information arising from multiple internal and external sources. In maintaining physiological and psychobiological homeostasis, central nervous system neurons communicate both with one another and eventually with effectors outside the central nervous system by means of neurochemical packets (neurotransmitters) released by each neuron onto specific receptors. Neurons are often characterized not only by their anatomical distribution (location of their cell bodies) and path of projection to their terminal areas, but also by the nature of the neurochemical hormone/transmitter that they synthesize and release.

Central norepinephrine systems

A norepinephrine (NE) neuron and its synaptic links are shown in Fig. 45-1. The presynaptic neuron contains the machinery necessary for the synthesis, storage, release, reuptake, and inactivation of NE. In the NE neuron, the precursor amino acid tyrosine is taken up and stepwise enzymatically converted to levodopa, dopamine, and norepinephrine (Fig. 45-2). Uptake of dopamine and NE into storage granules within the neuron protect the neurotransmitter from enzymatic degradation (by mitochondrial MAO) to inactive aldehydes and acids. With each electrical depolarization of the neuron, the granular packets migrate and fuse to the synaptic membrane, releasing their neurochemical contents into the synaptic cleft.

Once released into the synaptic cleft, NE acts upon the postsynaptic NE receptor to alter the electrical resting potential of the postsynaptic cell membrane, either inhibiting or facilitating the rate of electrical and neurochemical discharge by the postsynaptic cell. In many cases, the changes in resting potential triggered by NE receptor activation are mediated by a second intracellular messenger, cyclic adenosine monophosphate (cAMP). NE released in the cleft also interacts with presynaptic ($alpha_2$) adrenoceptors, whose activation reduces the quantity of NE released with each electrical depolarization that arrives at the terminal area of the NE neuron.

The presynaptic NE neuron does not abandon the majority of neurotransmitters once used. Rather, the neuron has a reuptake mechanism that actively recovers NE from the synaptic cleft and again stores the recovered NE for

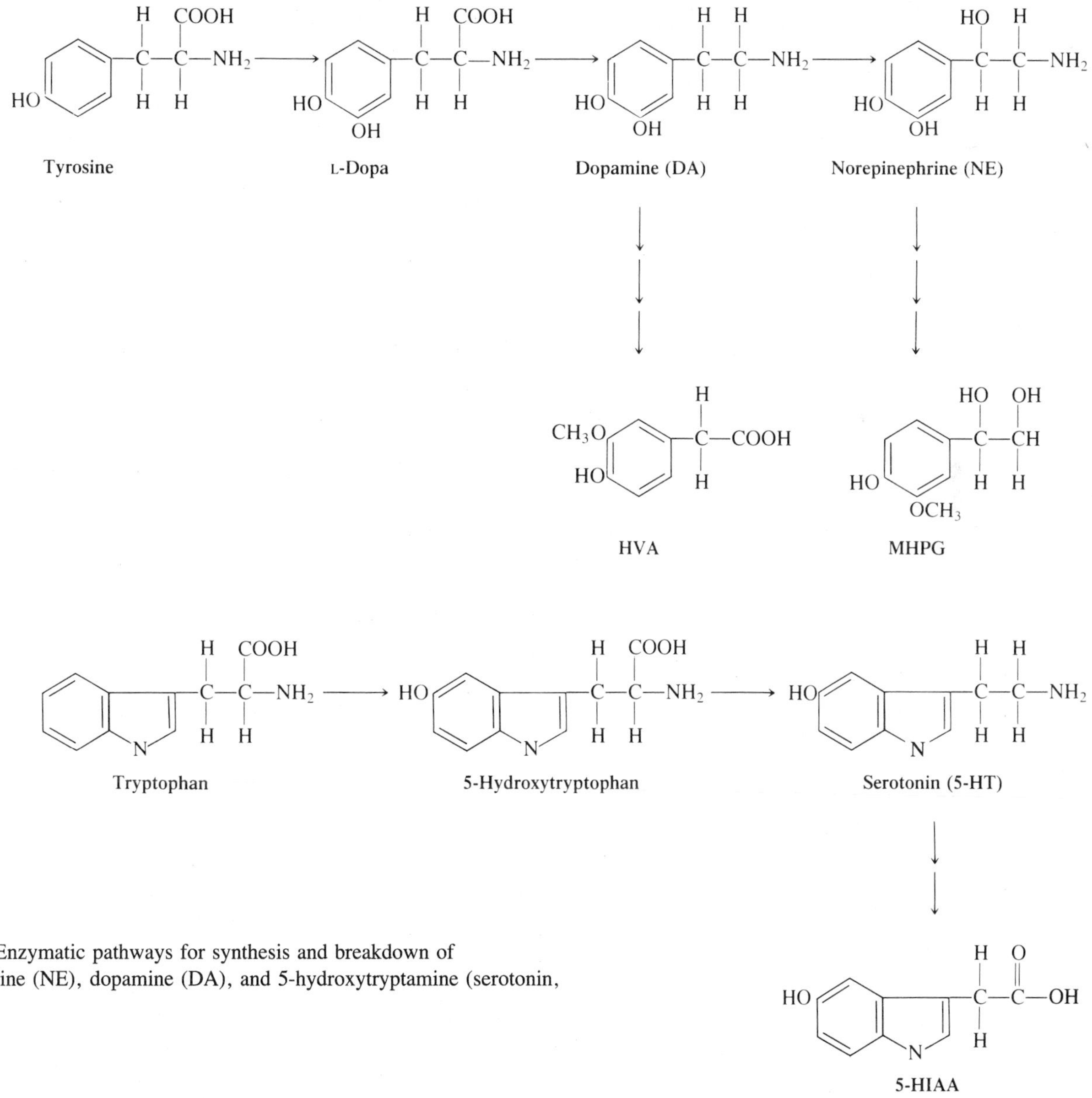

Tyrosine → L-Dopa → Dopamine (DA) → Norepinephrine (NE)

Dopamine (DA) → HVA

Norepinephrine (NE) → MHPG

Tryptophan → 5-Hydroxytryptophan → Serotonin (5-HT) → 5-HIAA

Fig. 45-2. Enzymatic pathways for synthesis and breakdown of norepinephrine (NE), dopamine (DA), and 5-hydroxytryptamine (serotonin, 5-HT).

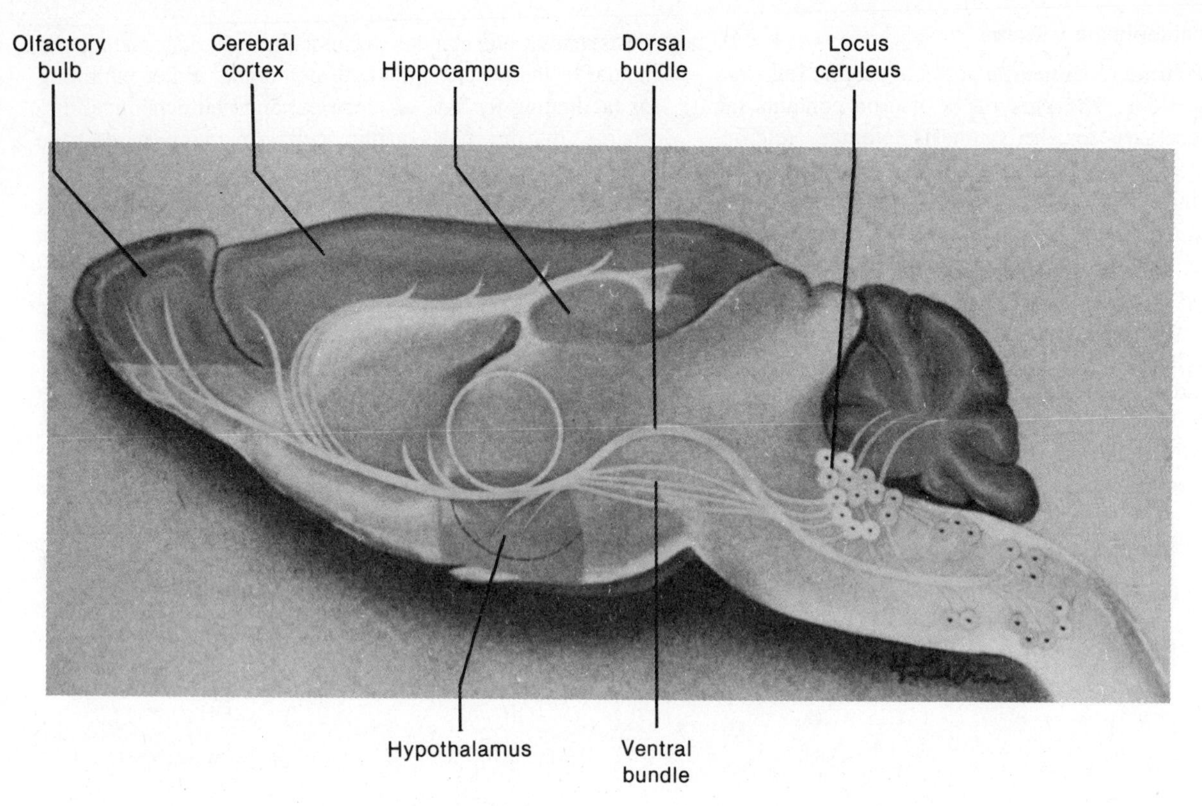

Fig. 45-3. Sagittal representation of major norepinephrine (NE) pathways of mammalian brain. (From Lader, M.: Introduction to psychopharmacology, Kalamazoo, Mich., 1980, The Upjohn Co.)

subsequent release. Some central NE, however, is metabolized to 3-methoxy-4-hydroxyphenylglycol (MHPG) (Fig. 45-2), which can be measured in the plasma or urine. The quantity of MHPG in a 24-hour urine sample can be used as an estimate of synthesis and breakdown of central NE: NE turn-over.

There are two major clusterings of NE cell bodies from which axons arise to innervate targets throughout the neuronal axis. The largest collection of noradrenergic cells is located in the nucleus locus ceruleus in the pons (Fig. 45-3). From the locus ceruleus axonal projections rise to the thalamic and hypothalamic nuclei, olfactory bulb, hypocampus, and cerebral cortices, as well as to the cerebellum, and descend to the spinal cord. Axons from a group of lateral tegmental NE nuclei ascend in the ventricle bundle to innervate the amygdala and septum and also descend into the spinal cord.

Central dopamine systems

Dopamine (DA) neurons also utilize the precursor amino acid tyrosine from which they synthesize their neurotransmitter, but unlike NE neurons they lack the enzyme dopamine β-hydroxylase, which converts DA to NE. DA is stored in the granules, which are released into the DA synapse after electrical depolarization of the DA neuron. DA is released not only onto the postsynaptic receptor, but also onto a presynaptic DA autoreceptor (D_3), which, when activated, partially inhibits the activity of the DA neuron. Like the NE neuron, the DA neuron recovers synaptically released neurotransmitter by an energy-dependent process and stores the neurotransmitter again for subsequent rerelease. Some DA is metabolized to homovanillic acid (HVA) (Fig. 45-2), which can be quantified in spinal fluid. Three major tracks or bundles of such dopamine neurons innervate the striatum (caudate, putamen), the limbic-cortical areas (nucleus accumbens, septal nuclei, amygdala, cingulate gyrus, and frontal cortex), and the tuberoinfundibular areas (hypothalamus and median eminence) (Fig. 45-4).

Central serotonin systems

Serotonin neurons within the CNS utilize the amino acid precursor tryptophan, from which they synthesize serotonin (5-hydroxytryptamine, or 5-HT) (Fig. 45-2). Release of 5-HT onto stereospecific 5-HT receptors, followed by reuptake of the neurotransmitter is similar to that found with NE and DA neurons. 5-HT is metabolized to 5-OH-indole acetic acid (5-HIAA) which like HVA, can be measured in the cerebrospinal fluid.

The cell bodies of serotonergic neurons, for the most

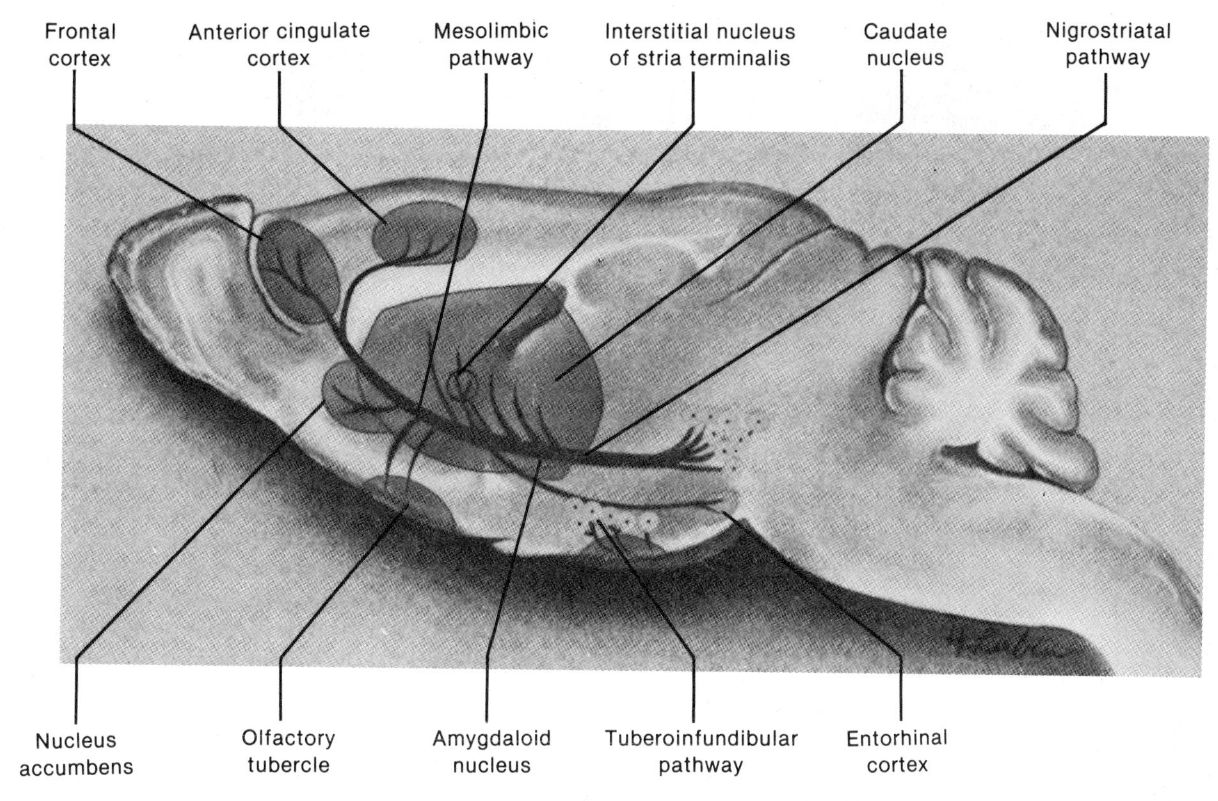

Fig. 45-4. Sagittal representation of major dopamine pathways of mammalian brain. (From Lader, M.: Introduction to psychopharmacology, Kalamazoo, Mich., 1980, The Upjohn Co.)

part, are located in the raphe of the mesencephalon. Serotonergic axons project from the raphe nuclei to the limbic septum (hippocampus, amygdala, septum, and hypothalamus), in additioh to projecting to the cortex (Fig. 45-5).

Other central neurotransmitter systems

NE, DA, and 5-HT neurons constitute a very small proportion of the neuronal systems within the central nervous system. Gamma-amino-butyric acid (GABA), glutamic acid, and acetyl choline systems are distributed through the central nervous system, as are neurons, which use as their neurotransmitter complex polypeptides such as β-endorphin, the enkephalins, cholecystokinin, substance P, adrenocorticotropic hormone (ACTH), and somatostatin. The distribution of such neurotransmitter systems and their biochemistry and physiology are currently being studied. Their significance for psychiatric disorders is unknown.

PSYCHIATRIC DISORDERS

Schizophrenias

Description. The schizophrenias (syndrome diagnosis) are primarily disorders that manifest characteristic delusions, hallucinations, and disordered cognition.[1] Specifically, they are characterized by the following:

1. At least one of the following during a phase of the illness:
   a. Bizarre delusions, such as delusions of being controlled, thought broadcasting, thought insertion, or thought withdrawal
   b. Somatic, grandiose, religious, nihilistic, or other delusions without persecutory or jealous content
   c. Delusions with persecutory or jealous content, if accompanied by hallucinations of any type
   d. Auditory hallucinations in which either a voice keeps up a running commentary on the individual's behavior or thoughts, or two or more voices conversing with each other
   e. Auditory hallucinations on several occasions, with content of more than one or two words, having no apparent relation to depression or elation
   f. Incoherence, noticeable loosening of associations, noticeable illogical thinking, or noticeable poverty of content of speech if associated with at least one of the following:
      (1) Blunted, flat, or inappropriate affect
      (2) Delusions or hallucinations
      (3) Catatonic or other grossly disorganized behavior

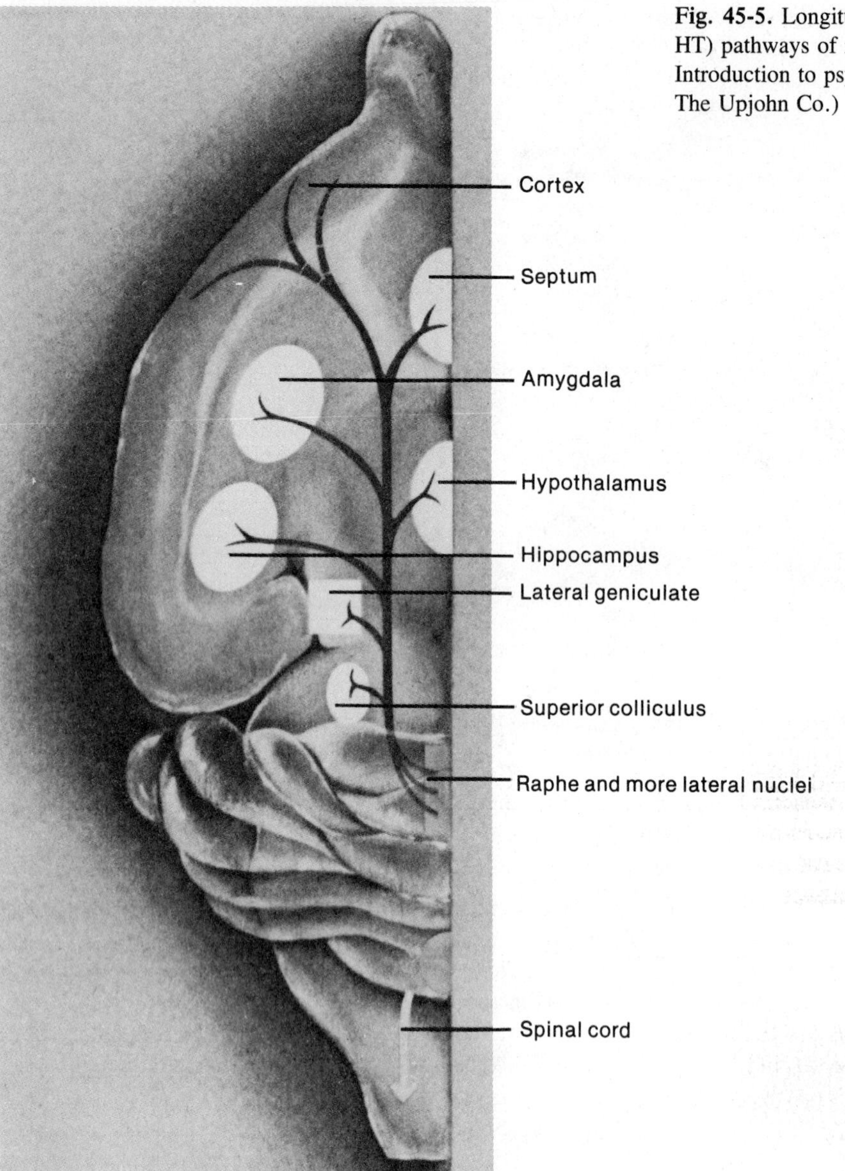

Fig. 45-5. Longitudinal representation of major serotonin (5-HT) pathways of mammalian brain. (From Lader, M.: Introduction to psychopharmacology, Kalamazoo, Mich., 1980, The Upjohn Co.)

2. Deterioration from a previous level of functioning in such areas as work, social relations, and self-care
3. Duration: continuous signs of the illness for at least 6 months at some time during the person's life, with some signs of the illness at present. The 6-month period must include an active phase during which there were symptoms from point 1, with or without prodromal or residual phases, as defined below.

Prodromal phase. There is a clear deterioration in functioning *before* the active phase of the illness not attributable to a disturbance in mood or to a substance-use disorder and involving at least two of the symptoms noted below.

Residual phase. There is persistence, after the active phase of illness, of at least two of the symptoms noted below, not attributable to a disturbance in mood or to a substance-use disorder.

Prodromal or residual symptoms:

1. Social isolation or withdrawal
2. Great impairment in role functioning as a wage earner, student, or homemaker
3. Peculiar behavior (collecting garbage, talking to self in public, or hoarding food)
4. Great impairment in personal hygiene and grooming
5. Blunted, flat, or inappropriate affect
6. Digressive, vague, overelaborate, circumstantial, or metaphorical speech
7. Odd or bizarre ideation, or magical thinking, such as superstitiousness, clairvoyance, telepathy, ''sixth

sense," "others can feel my feelings," overvalued ideas, ideas of reference

8. Unusual perceptual experiences, such as recurrent illusions, sensing the presence of a force or person not actually present.

A preliminary attempt is being made in the current diagnostic nomenclature[1] to separate with some degree of validity the larger disease group into distinct subentities. This is accomplished by isolation of a group of similarly symptomatic disorders whose episodic course differs in both duration and clinical response. One group's episode duration is shorter (less than 6 months), and such patients exhibit essentially full recovery between psychotic episodes. Additional findings support differentiation of at least three separate disorders that have traditionally been grouped as "schizophrenia." Such clinical differentiation is important, since it separates those patients who can be treated by certain drugs from those who cannot. The three subtypes to be discussed below include classical schizophrenias, organic psychoses, and atypical psychoses.

Types of psychotic schizophrenia-like psychoses

Classical schizophrenia. It has been observed that drugs that can prevent dopamine-mediated neurotransmission clinically have an antipsychotic effect (reduce the quantity or quality of delusions, hallucinations, and disordered cognition of many psychotic patients). Such drugs bind stereospecifically to a form of the postsynaptic dopamine receptor, each with an affinity that is inversely related to the average dose of the particular antipsychotic drug required to reduce psychotic symptoms clinically.

At postmortum examination, substantial numbers of chronic psychotic (schizophrenic) patients have been found to have an increased number of dopamine-related postsynaptic receptors. Such receptors can be identified and quantitated by measurement of the quantity of stereospecific binding of a radiolabeled antipsychotic drug to the specific receptor. Stereospecific ^{3}H-haloperidol and ^{3}H-spiroperidol binding per milligram of membrane protein has been found to be significantly greater in such chronic psychotic patients than in age and sex matched controls in both the caudate (striatum) and the nucleus accumbens (limbic system).[2] Although chronic treatment with antipsychotic drugs is known to increase artificially such binding, increased numbers of such binding sites have also been observed in chronic psychotics who have never received antipsychotic drug for the treatment of their illness. Such findings suggest that the apparent dopamine-related functional hyperactivity underlying many chronic psychotic states may be attributable not to excess of central dopamine but to a proliferation of dopamine-related postsynaptic receptors: a sensitization of the receiver (receptor) rather than excess transmission of the signal.

Attempts to monitor functional dopamine system hyperactivity directly within the central nervous system of psychotic patients have been largely limited by the obvious invasiveness of such procedures as implantation of electrodes in postsynaptic cells. Some investigators have suggested that functional dopamine system activity can be monitored by measurement of the frequency of eye blinks. Rates of eye blinks higher than in controls have been found in some groups of psychotic patients; antipsychotic drugs, consistent with their antidopamine effects, have been shown to reduce the frequency of eye blinks in such psychotics.[3]

Monitoring neuroendocrine hormone release has been suggested as another indirect means of assessing the functional activity of central dopamine systems. The tuberoinfundibular dopamine system is concerned primarily with such neuroendocrine regulation. In particular, the *suppression* of prolactin release from the anterior pituitary is mediated by dopamine released from the median eminence. In contrast, the *facilitation* of growth hormone release from the pituitary is indirectly controlled in part by dopamine acting upon releasing-factor neurons in the hypothalamus. Evidence has been accumulated possibly indicating that the regulation of these releasing pathways are impaired in some subsets of psychotic patients.[4]

Organic psychoses. If one group of chronic psychotic patients shows evidence of altered dopamine receptor number, such a group must be discriminated from another group of psychotic patients who present with similar-appearing psychotic illness, but who fail to have evidence of dopamine hyperactivity. Many of this latter category of chronic psychotic patients fail to show increased frequency of eye blinks and fail to respond to antipsychotic drugs that block dopamine receptors. Many such psychotic patients do have signs of at least mild neuropsychological deficits. Chronic psychotics of such a separate group have evidence of enlarged cerebral ventricles by computerized axial tomography (CAT scan)[5] and perhaps of diminished size of the anterior cerebellar vermis[6] as well. It is presently not clear whether such chronic "organic" psychoses are the result of a chemical insult or trauma to brain substance with brain atrophy, or whether such a group of chronic psychotic illnesses represent a developmental disorder affecting the differentiation of CNS tissues. The clinical history of poor premorbid adjustment (poor level of functioning before the first frankly psychotic episode) possibly indicates that abnormal developmental processes may be the most relevant in this subgroup of chronic psychotic illnesses. Genetic patterns of transmission of such "organic" psychoses have not been explored thoroughly as yet for contrast with the bulk of the chronic patients who do demonstrate genetic patterns of transmission and who do benefit from drug antipsychotic treatment (dopaminergic blockade).

Atypical psychoses (schizophreniform disorders). A third group of psychoses are those that appear to present with symptoms similar to the other groups but have a course that is intermittent (as opposed to chronic) and with essentially full remission of the psychotic and dysfunctional process between episodes (DSM-III schizophreniform disorder).[1] Supersensitivity of the tuberoinfundibular dopamine system, which can be demonstrated by excessive growth hormone release after apomorphine is characteristic of this group of psychotics. Deficiencies in membrane countertransport of lithium and altered in vitro lithium ratios (ratios of red blood cell lithium to medium lithium)[7] are also found frequently in such patients, as is a possible deficiency in the methylation of membrane phospholipids in the formation of phosphatidylcholine.[8] Such patients appear to respond to a variety of therapeutic interventions, including lithium alone[9] as well as classical antipsychotic drugs, and frequently have family histories of affective disorders, but not of the chronic deteriorating schizophrenias and the schizophrenic spectrum diseases.

The early diagnosis of such good-prognosis, schizophreniform, psychotic illnesses is important, and the discrimination of such patients from the bulk of the chronic progressively deteriorating schizophrenias is vital because of the necessity to anticipate the illness course in treatment planning. Most such good-prognosis patients with proper treatment can be expected to return to their homes and to employment at a level of functioning comparable to what was present before the psychotic episode. Although further episodes of psychosis may occur, such episodes may be either reduced in frequency or prevented altogether with proper maintenance medication, which appears to compensate for the underlying biological deficit. The use of lithium for both acute treatment and maintenance in this group of psychotics[9] emphasizes the group's similarity in response and prevention, as well as its similarity in illness course with the affective disorders, though, unlike the affective disorders, it presents with classical delusions, hallucinations, and disorders of cognition, which characterize the schizophrenic, psychotic syndrome.

Affective disorders

Description. The affective disorders are diseases of mood regulation and are accompanied by altered rates of cognition (slowed or speeded) and by physical signs and symptoms. Numerous investigators have observed that affective disorders may appear in either a unipolar or a bipolar form. In *unipolar disease,* only periods of depression occur. Specifically, such major depressive episodes are characterized by the following:

1. Dysphoric mood or loss of interest or of pleasure in all or almost all usual activities.
2. At least four of the following symptoms for at least 2 weeks:
   a. Poor appetite or weight loss or increased appetite or weight gain
   b. Sleep difficulty or sleeping too much
   c. Psychomotor agitation or retardation
   d. Loss of interest or pleasure in usual activities, decrease in sexual drive
   e. Loss of energy, fatigue
   f. Feelings of worthlessness, self-reproach, or excessive or inappropriate guilt
   g. Complaints or evidence of diminished ability to think or concentrate
   h. Recurrent thoughts of death or suicidal behavior
3. Neither of the following dominate the clinical picture when the affective syndrome is absent:
   a. Preoccupation with a mood-incongruent delusion or hallucination
   b. Bizarre behavior
4. Not superimposed on schizophrenia, schizophreniform disorders, or a paranoid disorder
5. Not attributable to any organic mental disorder or uncomplicated bereavement.

Bipolar affective disease is characterized by episodes of mania and by episodes of depression. Specifically, manic episodes are characterized by the following:

1. One or more distinct periods with a predominantly elevated, expansive, or irritable mood
2. Duration of such a period of at least 1 week during which time at least three of the following symptoms have persisted and have been present to a significant degree (four, if the mood is only irritable):
   a. Increase in activity or physical restlessness
   b. More talkative than usual or pressured to keep talking
   c. Flight of ideas or subjective experience that thoughts are racing
   d. Inflated self-esteem
   e. Need for sleep
   f. Distractibility
   g. Excessive involvement in activities that have a high potential for painful consequences that is not recognized by the patient
3. Neither of the following dominates the clinical picture when an affective syndrome is absent:
   a. Preoccupation with a mood-incongruent delusion or hallucination
   b. Bizarre behavior
4. Not superimposed upon either a schizophrenia, schizophreniform disorder, or a paranoid disorder
5. Not attributable to any organic mental disorder such as substance intoxication.

The effective treatment of the affective disorders frequently requires careful attention to determine the underlying biological process that needs to be corrected or compensated for by pharmacological intervention. Therapeutic response to one affective type of medication but not to another has been observed in affective disordered patients. The correct choice of drug for an affective disordered patient requires choosing a drug that compensates for specific underlying biological abnormalities. The identification of these abnormalities, or the clustering of the series of clinicobiological findings associated with the abnormalities is already beginning to guide selection of psychopharmacological agents.

Several groups of investigators have explored the incidence of affective disorders in first-degree relatives of disordered probands. Most investigators have found that bipolar illness breeds relatively true in families. In a study by Perris et al.[10] only 0.5% of first-degree relatives of patients with unipolar illness had bipolar disease, whereas 10.6% of such relatives had previous episodes of only depression. In contrast, 16.3% of first-degree relatives of patients with bipolar illness had at some time manifested both depression and mania, whereas only 0.8% of such relatives had experienced depression alone. Such evidence clearly suggests that unipolar and bipolar affective illnesses are distinct entities. It follows that biological studies might reveal different biological abnormalities in bipolar patients as opposed to unipolar depressed patients. Similarly, drugs that specifically correct or compensate for a primary biological abnormality in bipolar diseased patients may be ineffective in unipolar populations.

Types of affective disorders

Unipolar affective disorders. After studying the occurrence of unipolar disease in families, Winokur et al.[11] suggested that unipolar disease may itself have at least two subtypes. He distinguished *depressive spectrum disease* from *pure depressive disease*. Depressive spectrum disease has an early onset (before 40 years of age); it is associated with a history of greater incidence of frankly affective disorder in first-degree female relatives and a prominent history of alcoholism or antisocial personality, or both, in first-degree male relatives. In contrast, pure depressive disease is characterized by a later onset, essentially equal incidence of affective disorder in male and female first-degree relatives, and little or no family history of alcoholism or antisocial personality.

Although such subtypes of unipolar disorders are indistinguishable symptomatically, these family history findings suggest that the primarily underlying biological abnormality of the two unipolar disorders might be different. One might expect that pharmacological agents with a high degree of specificity of action might be effective in one, but probably not in both of the unipolar disorders. Indeed, there is growing evidence that this may be the case with commonly used, relatively specific antidepressant drugs.

Serotonin-related depressions. Evidence for the role of serotonin (5-HT) in depressive disease comes from numerous, often conflicting reports of diminished quantities of its metabolite 5-hydroxyindoleacetic acid (5-HIAA) in the cerebrospinal fluid (CSF) of depressed patients and from findings that 5-hydroxytryptophan, a precursor of 5-HT, is an effective antidepressant only in depressed patients with decreased CSF 5-HIAA.

The routine evaluation of cerebrospinal fluid 5-HIAA in depressed patients might therefore identify a subgroup of depressive patients who have deficiencies in serotonin turnover, who might therefore benefit from precursor loading with tryptophan or 5-hydroxytryptophan.[12] Such 5-HT-deficient patients may also be candidates for treatment with more conventional antidepressant drugs that increase the 5-HT in the serotoninoergic synapse by partially inhibiting the reuptake of 5-HT into the presynaptic neuron. Pharmacological agents such as amitriptyline and trazodone have been shown to inhibit selectively the neuronal reuptake of the 5-HT from the synapse.

Norepinephrine-related depressions. Much of our understanding of the biological basis of affective disorders comes from work that has pursued the hypothesis that a norepinephrine (NE) deficiency at critical CNS synapses underlies the cyclic emergence of depression in many affective disordered patients.

The presence of large NE-containing cells in the pontine locus ceruleus and in the more ventral reticular system, with axons projecting forward to supply NE to the hypothalamus, thalamus, preoptic and septal areas, and cortices (Fig. 45-3), suggests the possibility of a critical regulating role of NE in a variety of activities that are impaired in depression: appetite, sexual function, sleep, cognition, and so forth.

MHPG IN NOREPINEPHRINE DEPRESSIONS. The early studies of depressive disease examined whether the turnover of central NE was abnormal in depressed patients. Urinary 3-methoxy-4-hydroxyphenylglycol (MHPG), half of which has its origin in central NE turnover, has been shown to be deficient in many depressed patients.[13] However, concentrations of the peripheral NE metabolites vanillylmandelic acid and normetanephrine were both found to be similar in depressed patients in age- and sex-matched controls. Such findings suggest that in some forms of depression, though the rates of synthesis and breakdown of NE in the periphery appear to be normal, turnover of NE in CNS is diminished. Several investigators have demonstrated that patients who are deficient in urinary MHPG, and therefore presumably have a deficit in central NE, have a favorable therapeutic response to antidepressants that selectively inhibit the reuptake of NE from the synaptic cleft into the presynaptic neuron.[14,15] The consequence of such inhibition of reuptake is that more NE is available in the synaptic cleft, acting upon the postsynaptic receptor. Such augmented activity of central NE may underlie the antidepressant response clinically. Antidepressant drugs that are relatively selective in inhibition of neuronal reuptake of NE include desipramine and maprotiline. Imipramine has also been shown to produce good antidepressant response in low-MHPG depressions, presumably because it is itself metabolized rapidly to desipramine, which has selective effects on NE reuptake.

ABNORMAL CORTISOL SECRETION. Considerable numbers of depressed patients have been shown to have an abnormality of cortisol secretion.[16] In some depressed patients, the usual diurnal rhythm of plasma cortisol is frequently obscured by continuous elevation of cortisol throughout the day. Such depressed patients are found to have a pronounced elevation of urinary free cortisol as well. The dexamethasone suppresion test (DST) (in which normals suppress plasma cortisol for 36 hours after dexamethasone) is also abnormal in many depressed patients, showing either total nonsuppression or early escape of cortisol from dexamethasone suppression.[17] As is discussed below, nonsuppression of cortisol by dexamethasone in depression is also consistent with a deficiency of hypothalamic NE, which indirectly suppresses cortisol release.

RECEPTOR SENSITIVITY. One of the newer, intriguing markers of depression, which may be of some relevance to abnormalities of the central NE system, is the recent finding that receptors that ordinarily aid in the homeostatic feedback regulation of NE neurons may be abnormal in some depressive illnesses.

Direct evidence of abnormalities of such $alpha_2$-adrenergic autoreceptors in some major depressive disordered patients has come from studies of a peripheral tissue: the platelet. Like the NE neurons of the CNS, platelets have $alpha_2$-adrenergic receptors in the outer phospholipid bilayer of the cell membrane. Increased numbers of these receptors, identified by ^{3}H-clonidine binding, are found in depressed patients than in control subjects. Moreover, treatment with antidepressant drugs led to a significant decrease (normalization) in the number of platelet $alpha_2$-adrenergic receptors.[18]

$Alpha_2$-adrenergic receptors within the CNS are part of the feedback loop that regulates the synthesis of NE and the firing rate of NE neurons. Activation of $alpha_2$-adrenergic receptors, which are located on NE neuron membranes, decreases NE synthesis and release, resulting in a diminished plasma and urinary MHPG.[19] If increased numbers of $alpha_2$-adrenergic receptors occur on NE neurons of the CNS in certain depressed patients—similar to the increase of $alpha_2$-adrenergic receptors found in the peripheral platelet—the increased central $alpha_2$-sensitivity would be expected to cause both reduction of the synthesis of central NE (as demonstrated by diminished plasma and urinary MHPG) and a functional NE deficit at central NE synapses. The striking finding that chronic but not acute treatment with antidepressant drugs decreases the number of $alpha_2$ receptors in platelets[20] indicates that similar events may be occurring at central $alpha_2$-adrenergic sites after chronic administration of antidepressant drugs.

Bipolar disorders

Clinical presentation and history. As noted previously, classical bipolar disordered patients have a history of major depressive episodes and episodes of mania with affectively normal intervals interspersed. The diagnosis of bipolar illness is not always obvious from clinical presentation and clinical history. Many bipolar illnesses can present with a first or even second episode of depression without a previous episode of diagnosable mania. In addition, many bipolar patients can present with psychotic symptoms suggestive of schizophrenia. Clinically, the history of mania in a first-degree relative should suggest the presence of a bipolar disorder in a patient who is in a major depressive episode or undergoing a brief psychotic decompensation. The treatment of a bipolar disorder and that of unipolar disease and schizophrenia differ; the correct diagnosis early in the episode is of importance for therapeutic management.

Neurotransmitter dysregulation. The biology of bipolar affective disorder indicates possible dysregulation of several neurotransmitter systems. During the depressed phase of bipolar illness, decreases in urinary MHPG have been clearly documented[21] and indicate a possible abnormality of central NE function. Manic episodes can be precipitated by dopamine-like drugs such as amphetamine. The episodes are at least partially resolved acutely by the use of dopamine receptor–blocking agents, such as an antipsychotic drug.[22] However, it appears that central cholinergic systems may also have a role in bipolar illness. Mania can be attenuated not only by a blockade of dopamine receptors, but also by an increase in central cholinergic activity, such as after the infusion of a cholinesterase inhibitor such as physostigmine.[23] The increased functional activity of cholinergic systems promptly relieves mania. As the effect of physostigmine wears off, mania recurs. Bipolar illness then appears to be a dysregulation of balance between several central neurotransmitter systems. The restoration of such balance by pharmacological intervention appears to be associated with cessation of the affective episode.

Membrane dysfunction. The lithium ratio test examines the ability of the red blood cell membrane to transport lithium against a gradient. The proportion of internal to external Li is the Li ratio. Although the lithium ratio may be a reasonably satisfactory marker for bipolar illness, the relationship of the marker to the underlying biopathological condition in the patient is more obscure. Lithium treatment of such patients does not result in normalization of the Li ratio. More recent work has suggested that membrane-compositional abnormalities may be present in such patients. These deficiencies involve low membrane phosphatidylcholine levels and an abnormality of the critical choline pathway that converts phosphatidylethanolamine to phosphatidylcholine.[24] Lithium may increase the activity of methyltransferase I and methyltransferase II required for the formation of phosphatidylcholine and thereby may normalize membrane composition within critical microdomains of the membrane. Considerable further work needs to be done to clarify the relevant mechanism of action of lithium.

FUNCTION TESTS

Dexamethasone suppression test

In some depressed patients, plasma cortisol levels do not vary during the course of the day but tend to remain elevated; the neuroregulatory mechanism is altered. Ordinarily, the drug dexamethasone suppresses the secretion of

Table 45-1. Challenge or function test used in diagnosis of psychiatric disorders

Test	Sample	Procedure	Interpretation
DST (dexamethasone suppression)	Plasma	1 mg of dexamethasone at 11 P.M. followed by blood drawing for plasma cortisol at 4 P.M. and 11 P.M. the following day	4 P.M. or 11 P.M. cortisol greater than 50 μg/L suggestive of depression responsive to antidepressant drugs
Lithium ratio in vitro	RBC	Incubation of red blood cells with 1.5 mmol of Li for 24 hours	High ratio (greater than 39) suggestive of lithium-treatable bipolar or schizophreniform illness
Protirelin	Plasma	500 μg of protirelin infused over 30 min; basal, 15, 30, 60, and 90 min plasma samples assayed for thyroid-stimulating hormone (TSH)	Blunted increase from base line of TSH (less than 7 μU/mL) suggestive of affective disorder rather than schizophrenia
Apomorphine–growth hormone	Plasma	0.75 mg of apomorphine with 40, 60, and 80 min plasma samples assayed for growth hormone (GH)	Elevated GH peak response (greater than 20 ng/mL) suggestive of lithium-responsive psychotic illness

cortisol into plasma. Some depressed patients do not suppress cortisol secretion after dexamethasone or fail to suppress it completely. The details of the dexamethasone suppression test (DST) are given in Table 45-1.

Experiments in animals have shown that NE tonically inhibits the release of the peptide corticotropin-releasing factor (CRF) from the hypothalamus,[25] which itself causes release of adrenocorticotropic hormone (ACTH) from the anterior pituitary. Continuing the cascade, ACTH causes the release of cortisol from the adrenals. Diminished functional central NE activity might therefore increase CRF release, resulting in increased ACTH and subsequently, increased cortisol release into the vascular system.

The dexamethasone suppression test is commonly used to detect a melancholic (unipolar) depression, which responds characteristically to NE drugs such as imipramine, desipramine, and maprotiline. Although early studies suggested that tricyclic drugs such as imipramine and desipramine were more effective in depressed dexamethasone nonsuppressors than drugs that block serotonin reuptake, such as amitriptyline, further evaluations of this important issue are required before one draws any firm conclusions. Several investigators have since suggested that dexamethasone nonsuppression predicts the response to all antidepressant agents, without discriminating well between NE and serotonin types of antidepressants[26] (Table 45-1).

Lithium ratio

In the absence of the history of mania either in the patient or in first-degree relatives, only three laboratory tests appear relevant for a choice of drug treatment, and these may have limited value. The in vitro lithium ratio monitors the ability of the patient's red blood cell membrane to extrude intracellularly accumulated lithium against a concentration gradient extracellularly. One can measure the in vitro lithium ratio by incubating patients' red blood cells with lithium for 24 hours and determining the intracellular lithium concentration in relation to that of the incubation medium[27] (Table 45-1). Elevated lithium ratios have been documented in at least one third of patients with biopolar affective illness and, when present, are associated with beneficial response to lithium both acutely and prophylactically.[28] Note, however, that many classical manic-depressive patients do not have the lithium ratio abnormality, and many individual patients without the abnormality still appear to gain therapeutic effects and protection from future episodes by the use of lithium. The test is then marginally sensitive in detecting lithium responders and is marginally specific for lithium response when the abnormality is present. The elevated lithium ratio abnormality appears to be under genetic control and is present in first-degree family members of bipolar patients.[29]

Protirelin test

Another test that may have some usefulness in discriminating manics who present with psychotic symptoms from schizophrenics is the protirelin test (Table 45-1). Many manic (as well as depressed) affective-disorder patients show a blunted thyroid-stimulating hormone (TSH) response to protirelin (thyrotropin-releasing hormone). Schizophrenics, on the other hand, tend to show a more brisk release of TSH after protirelin infusion, similar to that of nonpsychiatric controls.[30] The mechanism for such a blunted TSH response in atypical, psychotic manics is not clear, but it may be related to excessive, chronic adrenergic-receptor stimulation causing a chronic release of TRH, with subsequent homeostatic down regulation of pituitary TRH receptors on TSH-secreting cells. Like an abnormal lithium ratio, a blunted TSH response to protirelin (TRH) may be an indication that the psychotic patient is really an atypical manic (bipolar), requiring lithium as a primary treatment modality.

Apomorphine–growth hormone test

Some psychotic patients demonstrate hypersensitivity of the tuberoinfundibular dopamine system, such that administration of apomorphine results in an excessive release of growth hormone (GH) into the plasma (Table 45-1).

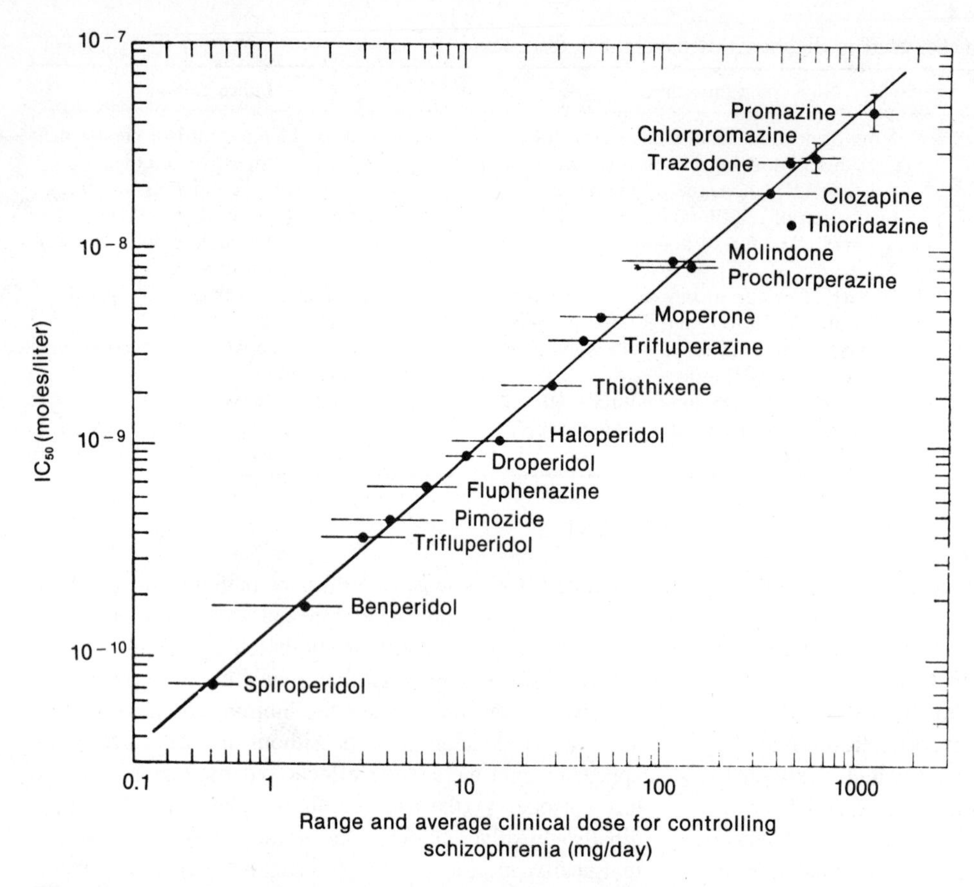

Fig. 45-6. Correlation between moles per liter required to displace ^{3}H-spiroperidol from its binding site in calf caudate (IC_{50}) and average daily dose required clinically for acute antipsychotic effects. (From Seeman, P.: Neuroleptics (major tranquilizers). In Seeman, P., Sellers, E., and Roschlau, W.H.E., editors: Principles of medical pharmacology, Toronto, 1980, The University of Toronto Press.)

Schizophrenic-like, psychotic patients who have such a hypersensitive GH response to apomorphine frequently can be successfully treated with lithium.[31]

TREATMENT

Treatment of schizophrenia: antipsychotic drugs

Mechanism of action. The classical antipsychotic drugs, those that successfully block postsynaptic dopamine receptors have been shown since the mid-1950s to be moderately successful in the reversal of acute psychotic symptoms and the prevention of subsequent episodes in the majority of schizophrenic patients. It is well known, however, that differing doses of the various antipsychotic drugs are required for treatment of the schizophrenic symptoms (Table 45-2). The dose of each antipsychotic drug required for treatment has been found to be inversely related to the potency (affinity) with which each of the drugs binds to a specific form of the central dopamine receptor (Fig. 45-6). Fluphenazine, which has a high affinity for the relevant receptor, can therefore be given at low doses; chlorpromazine, which has comparatively lower affinity for the receptor, reverses schizophrenic symptoms clinically only at higher doses in the average patient.

Dose and blood level relationships for antipsychotic response. Although most clinical studies have suggested a sigmoid dose-response curve for antipsychotic drugs, with the optimal dose for the average schizophrenic patient above a range of 500 to 1000 mg of chlorpromazine equivalents per day,[32] recent studies relating blood levels of such drugs to therapeutic effects have indicated that for at least some antipsychotic drugs patients may appear to have an optimal therapeutic range, both below which and above which the antipsychotic effects of the drug are less. One can therefore anticipate that in the not-too-distant future antipsychotic drug levels will be adjusted in accordance with the concept of optimal drug levels for response.

There is wide variation in drug blood levels among patients receiving the same dose of the same antipsychotic drug. Hundredfold between-patient variations have been reported for plasma chlorpromazine; fortyfold variations have been reported for butaperazine and fluphenazine. At average antipsychotic drug doses, substantial numbers of

Table 45-2. Comparable doses and relative potencies of antipsychotic drugs defined from dose-ranging studies clinically

Generic name	Empirically defined relative potency	Average daily dose for acute treatment (mg)
Fluphenazine decanoate	163.9	—
Fluphenazine	83.3	9
Haloperidol	62.5	12
Trifluoperazine	35.7	21
Thiothixene	22.7	32
Molindone	16.7	44
Lexapine	11.4	64
Perphenazine	11.1	66
Butaperazine	11.1	66
Piperacetazine	9.1	81
Prochlorperazine	7.1	103
Acetophenazine	4.4	169
Carphenazine	4.0	184
Triflupromazine	3.6	206
Chlorprothixene	2.2	322
Mesoridazine	1.8	411
Thioridazine	1.3	712
Chlorpromazine	1.0	734

From Garver, D.: In Hippus, H., and Winokur, G., editor: Psychopharmacology, 1. Part 2: Clinical psychopharmacology, Amsterdam, 1982, Excerpta Medica, pp. 250-270.

patients do appear to have extremely low or extremely high blood-and-tissue drug levels because of the pronounced between-patient differences in drug absorption or metabolism.[33] Some chronic antipsychotic drug–resistant patients have such poor absorption or rapid metabolism that they fail to respond at conventional doses by virtue of very low antipsychotic drug levels. On the other hand, patients who, by reason of slow metabolism, develop very high levels of antipsychotic drug on conventional dosage frequently fail to respond; such patients frequently show therapeutic response after lowering of the dose and the consequent reduction of the drug blood level into a more therapeutic range.

A "therapeutic window" for plasma fluphenazine levels in acutely admitted patients has recently been described by Dysken et al.[34] Failure of response was observed both in patients with plasma levels less than 0.2 ng of fluphenazine per milliliter of plasma or greater than 2.8 ng/mL. A similar nonresponse at both high and low plasma levels of haloperidol has been indicated in a recent study of Magliozzi[35] and an older study with chlorpromazine by Rivera-Calimlim.[36]

Not all investigators have found clear-cut relationships between plasma antipsychotic drug levels and therapeutic response, perhaps because of the fact that many antipsychotic drugs also have active metabolites, which generally are not measured by the usual gas chromatographic (GC) or liquid chromatographic (LC) or high-performance liquid chromatographic (HPLC) assays. Ignoring such active metabolites of the drug results in only a portion of the functional drug level actually being assayed. Relationships between total active antipsychotic drug and active metabolites and antipsychotic effects may therefore be obscured.

Receptor-binding assay, a sophisticated bioassay in which the quantity of functional antipsychotic drug together with its functional, active metabolites are monitored, has recently been shown to have considerable potential in monitoring of the quantity of functionally active drug and metabolites. Such a receptor-binding procedure measures the total dopamine receptor–blocking potency in a patient sample (plasma, cerebrospinal fluid, tissue). This is accomplished by determination of the ability of the patient sample to displace a reference radiolabeled antipsychotic drug from especially prepared striatal dopamine receptors.[37] Preliminary studies of the relationship between receptor binding assay–determined antipsychotic drug plasma levels and clinical response in schizophrenic populations have been undertaken by Calil for haloperidol[38] and by Tune[39] and Cohen[40] for a variety of antipsychotic drugs. Using such techniques, virtually all investigators have found that it was possible to define a lower range of drug levels at which therapeutic response was impaired and an increasing therapeutic response at higher receptor binding assay–determined antipsychotic levels. Therapeutic ranges for drug plasma levels, as determined by the receptor-binding assay, and those of GC or HPLC differ significantly for some antipsychotic drugs. Reasons for such differences are not immediately apparent but probably include the successful monitoring of active metabolites, which are generally ignored in GC or HPLC assays.

As soon as further studies are completed (which can be anticipated to define the optimal therapeutic range of antipsychotic drug and metabolites for the treatment of some forms of the schizophrenias), both assay techniques will likely be useful in the routine monitoring of antipsychotic drug concentrations. Relevant drug doses can then be individualized for each patient. At the time of this writing, only therapeutic ranges for plasma haloperidol, as monitored by GC methods, have been reproduced by sufficient numbers of laboratories to warrant a definition of haloperidol's therapeutic range: 8 to 15 ng/mL of plasma.

Treatment of affective disorders

The mainstays of the pharmacological treatment of the depressive disorders are the tricyclic antidepressant drugs and the monoamine oxidase inhibitors (MAOI). Lithium is used in the treatment of bipolar (manic-depressive) disease.

Reuptake inhibitors: tricyclic antidepressants

Choice of antidepressant drug. The most widely used drugs for treating depression are the tricyclic antidepressants. Tricyclic antidepressants are generally well toler-

Table 45-3. Commonly used antidepressant drugs and their approximate effective dose

Drug	Effective dose range (mg/day)
Amine-reuptake inhibitors	
Amitriptyline	150-300
Desipramine	150-300
Doxepin	75-300
Imipramine	150-300
Maprotiline	150-300
Nortriptyline	50-100
Trazodone	200-600
Monoamine oxidase inhibitors	
Phenelzine	45-75
Tranylcypromine	20-40

ated, cause minimum side effects, and are relatively safe (though drug overdose may cause severe cardiac toxicity). As yet, no clinical signs or epidemiological aspects of the patient's history have been proved useful in the choice of a tricyclic antidepressant to be used for a particular patient. Previous response of the patient or of family members to a specific tricyclic may be the most reliable method to use in the selection of a drug. The future use of urinary MHPG, of cerebrospinal fluid 5-HIAA concentrations, or of certain neuroendocrine tests such as the dexamethazone suppression test, and receptor abnormalities as well do show some promise in aiding the clinician in the selection of an appropriate tricyclic drug for a particular patient. For now, these must be considered research tools that provide evidence for a theoretical framework or rationale for changing to other types of tricyclates when the first drug tried is not effective.

In the physically healthy patient, the dosage of the tricyclic should be advanced to the average mean dose as shown in Table 45-3 within 1 week. If the patient has not shown improvement in depressive symptoms after 2 weeks at this average dose, the first question is whether the plasma levels of drug are too high or too low.

Use of antidepressant drug levels in adjustment of antidepressant drug dosage. The recognition of major differences in the absorption in metabolism of tricyclic drugs in individual patients has resulted in increased interest in relating plasma levels (rather than simply dosage) of antidepressant drugs to therapeutic response. For example, daily doses of 150 mg of amitriptyline given to a series of different patients results commonly in threefold differences in the quantity of plasma amitriptyline plus its active metabolite nortriptyline. Such differences in steady state plasma levels after comparable oral doses may account for the different doses of drug required by individual patients. Patients who absorb an antidepressant poorly or metabolize it rapidly need a higher oral dose in order to reach adequate therapeutic levels. On the other hand, patients who absorb the drug rapidly and metabolize it slowly require less drug to reach therapeutic drug levels.

The definition of therapeutic plasma and serum levels of antidepressant drugs has been the center of considerable effort and some controversy over the past years. Relationships between plasma levels of the tricyclic antidepressant nortriptyline and therapeutic response have been most fully studied. The available evidence suggests an inverted U type of relationship between nortriptyline plasma levels and antidepressant effects in tricyclic responsive, endogenously depressed inpatients. Maximal therapeutic efficacy is achieved with plasma nortriptyline levels between 50 and 175 ng/mL. Patients whose blood levels are below 50 ng of nortriptyline per milliliter generally fail to respond to the drug; importantly, patients whose plasma nortriptyline levels are in excess of 175 ng/mL also fail to respond.[41]

For patients receiving the antidepressant drug imipramine, the relationship between plasma levels of imipramine plus its active metabolite desipramine and clinical response appears to be linear in nondelusional, endogenously depressed, tricylic drug–responsive patients.[42] That is, antidepressant response to imipramine occurs when a threshold plasma concentration of imipramine and desipramine is reached (approximately 200 ng/mL of plasma). Patients whose blood levels are significantly in excess of this threshold level continue to have an antidepressant response yet may be burdened by excessive side effects of the drug.

There are less clear relationships betwen plasma levels of other antidepressant drugs and therapeutic response. The available evidence for amitriptyline suggests that a blood level in excess of 120 ng/mL of plasma amitriptyline plus nortriptyline produces good antidepressant response in tricyclic-responsive, endogenously depressed patients.[41] Some reports suggest that even better response occurs at higher blood concentrations of amitriptyline plus its metabolite, but other reports suggest that blood concentrations over 180 ng/mL reduce the therapeutic efficacy of the drug[41]; it is not clear whether the relationship of amitriptyline plus nortriptyline and their response is a linear or curvilinear one.

For other antidepressants such as protriptyline, desipramine, doxapine, maprotiline, and amoxapine, significant relationships, if any, await further elucidation by systematic studies.

The most important indications for obtaining antidepressant drug levels are as follows:

1. A patient who has received what should have been an adequate dosage of an antidepressant drug for 2 to 3 weeks and failed to respond. If a low plasma concentration is present and the patient has been compliant with the dose schedule, the dosage should be increased to a generally accepted therapeutic range; antidepressant response may then occur. With some drugs such as nortriptyline, and perhaps ami-

triptyline and desipramine, excessive concentrations of drug are an indication for dosage reduction, which sometimes is accompanied by antidepressant response.
2. Higher than usual doses of antidepressant drug merit measurements of plasma concentrations. For example, doses of amitriptyline or imipramine of more than 300 mg a day may result in altered clearance with a nonlinear increase in plasma concentrations. Monitoring could help avoid problems with toxicity.
3. When treating very old or very young patients, plasma concentrations should be kept at about 50% of what one might wish in robust patients in the middle years of life. Protein binding of these drugs is less at extremes of age, and more drug is present in pharmacologically active free form; the usual interpretations of plasma concentration may not hold under such circumstances.

Monoamine oxidase inhibitors

Clinical use. Monoamine oxidase inhibitors have been generally found to be less effective than other antidepressant drugs, except in chronically depressed persons who are nonresponsive to other antidepressants. This generalization is somewhat surprising because monoamine oxidase inhibitors raise the levels of both serotonin and NE and appear to have effects in altering receptor sensitivity similar to that of the more conventional antidepressant drugs. It is likely that previous trials of monoamine oxidase inhibitors have not utilized adequate drug doses to reach optimal degrees of monoamine oxidase inhibition for therapeutic response.

Platelet MAO inhibition and antidepressant response. During treatment with monoamine oxidase inhibitors such as phenelzine, the degree of monoamine oxidase inhibition can be related to therapeutic response. The inhibition of monoamine oxidase activity in peripheral tissues, such as the platelet, is assumed to be proportional to the degree of inhibition produced by the monoamine oxidase inhibitor in central nervous system tissue. Several laboratories have found that therapeutic response to phenelzine occurs only when monoamine oxidase activity of the platelet can be reduced to between 10% and 20% of base-line activity.[43] Utilizing a platelet assay for monoamine oxidase activity, one can titrate the oral dose of phenelzine to achieve optimal therapeutic response with minimal toxic side effects; such side effects generally appear when platelet monoamine oxidase inhibition exceeds 95%. Laboratory monitoring of monoamine oxidase activity at base line and during inhibitory treatment may offer antidepressant response to patients who are otherwise resistant to conventional antidepressant drug therapies.

Lithium

Clinical use. Lithium is clearly the treatment of choice for bipolar illness. It appears to be a specific agent for the treatment of mania, dramatically converting an acute manic episode into a normal mood state without sedation. The major drawback with lithium therapy is the lag period between the beginning of lithium administration and its clinical effect. For this reason, acutely manic patients are frequently treated initially with antipsychotic drugs such as haloperidol, in addition to lithium, to achieve a rapid antimanic effect. After approximately 1 to 2 weeks, the neuroleptic drug can generally be gradually withdrawn, with lithium alone maintaining significant protection from both manic and depressive future episodes.

Lithium is also effective acutely in the treatment of depression, especially the bipolar type, but again 3 to 4 weeks are generally required for lithium to produce an acute antidepressant effect. Generally, bipolar depressed patients are initially treated with a combination of a norepinephrine type of antidepressant, such as desipramine, and lithium. The antidepressant can usually be withdrawn after 3 to 4 weeks with continued maintenance on lithium. Addition of an antidepressant early in the course of treatment of bipolar depression generally shortens the acute depressive episode by about 2 weeks as compared to treatment with lithium alone.

Fifteen percent of depressed, bipolar patients treated with antidepressant drugs alone (rather than with lithium) switch into mania. In general, bipolar depressed patients are therefore not maintained on conventional antidepressants alone. Neither do bipolar patients presenting with mania receive full therapeutic benefit from an antipsychotic alone. Such patients treated with antipsychotic feel constricted in physical movements (extrapyramidal effects) and often retain an antipsychotic agent–induced dysphoria. Maintenance of such patients on lithium is therefore clearly indicated.

The diagnosis of a lithium responsive disorder, when not apparent on clinical grounds (episodes of mania in the patient or in first-degree relatives), can sometimes be aided by the use of the in vitro ratio or the protirelin test.

Use of lithium plasma levels in adjustment of lithium dose. During both lithium treatment and lithium maintenance (prophylaxis) it is important for the clinician to adjust the lithium dose so as to achieve therapeutic plasma levels of lithium. Therapeutic plasma levels of lithium appear to vary widely from patient to patient, some patients requiring as little as 0.5 mEq of lithium per liter of plasma with others requiring up to 1.5 mEq for acute antimanic or antidepressant effects. In general, for acute treatment lithium levels are maintained between 1 and 1.2 mEq/L. Dosage of lithium to achieve such levels varies between 600 and 2700 mg of lithium carbonate per day. Lithium toxicity including tremor and confusional states occurs at lithium levels above 1.5 mEq/L. Some patients appear more sensitive to toxic side effects and will have similar side effects at much lower levels. A combination of laboratory lithium values and clinical judgment is necessary to determine the optimal plasma concentration of lithium for acute therapeutic effects and for prophylaxis. Many elderly patients develop an energy-deficit syndrome on lithium at concentrations above 0.8 mEq/L; yet they receive therapeutic benefit at concentrations of lithium between 0.5 and 0.7 mEq/L.

Since lithium prevents antidiuretic hormone from having full functional effect the kidney, polyuria may occur in some lithium-treated patients. The response of the thyroid to thyroid-stimulating hormone is also less in the presence of lithium; hypothroidism or goiter may occur. Basal and follow-up tests of thyroid function are often necessary during chronic lithium prophylaxis. A rare but possible effect of lithium on the kidney (sclerosis of both glomeruli and tubules)[44] generally requires periodic evaluation of kidney function (BUN, creatinine). Base-line creatinine clearance and urinary concentration abilities should be undertaken in patients for whom chronic lithium treatment is warranted. In case of possible abnormality of kidney function during lithium treatment, creatinine clearance and urinary concentration abilities should be repeated as necessary and compared to base-line values.

CHANGE OF ANALYTE IN DISEASE

3-Methoxy-4-hydroxyphenolglycol (MHPG)

In 24-hour *urine* samples decreased MHPG (less than 1400 μg/24 hours in males or 1200 μg/24 hours in females) is suggestive of NE-related depressive disease.[15]

Although many clinicians routinely collect 24-hour urines for the evaluation of quantities of urinary MHPG in depressed patients, there are considerable problems determining the depressive subtype from such means. Most studies demonstrating clear relationships between low MHPG and response to NE type of antidepressants are based on studies in which patients were drug free for 2 or more weeks before MHPG determinations. Many drugs have prolonged effects on NE turnover and urinary MHPG. In particular, antidepressants themselves generally cause a suppression of NE turnover (lower MHPG), particularly in nonresponsive patients. After withdrawal of antidepressant drug therapy, MHPG precipitously rises (escapes) and remains elevated for up to 2 weeks. The use of MHPG as a means of identifying NE type of depressions is impractical because of the necessity of having patients 2 or more weeks drug free before one makes a decision (based on MHPG) concerning the choice of antidepressant drug. Urinary MHPG, though being an intriguing research tool, may therefore have limited clinical practicality for patient subtype identification and the initiation of proper drug treatment.

5-Hydroxyindoleacetic acid (5-HIAA)

5-HIAA (5-hydroxyindoleacetic acid) is collected from a single lumbar-puncture sample of cerebrospinal fluid. When 5-HIAA is decreased (less than 15 ng/mL), that level is suggestive of 5-hydroxytryptophan (5-HT)–related depressive disease.

Not all unipolar depressed patients have evidence of decreased 5-HIAA in the cerebrospinal fluid. Several investigators have found that CSF 5-HIAA is bimodally distributed in depressed patients.[45] A precursor of serotonin, 5-hydroxytryptophan has been shown to be an effective antidepressant only in patients with the lower distribution of 5-HIAA.[12] This observation supports the idea that loading with 5-HT precursors may be effective only in a subgroup of depressive patients who have diminished central serotonin turnover.

Alpha$_2$-adrenoceptors

Blood is collected and made into a platelet-rich fraction for assessment of the number of alpha$_2$-receptors per milligrams of platelets. Stereospecific ^{3}H-clonidine binding to platelet alpha$_2$-adrenergic receptors has been found to be significantly greater in a group of depressed patients than was found for similar ^{3}H-clonidine binding to platelets from a control population.[18] Although such increased numbers of alpha$_2$-adrenoceptors may be related to NE type of depressions, clinical studies have not yet demonstrated preferential antidepressant response after any particular antidepressant drug in depressed patients having the receptor abnormality.

REFERENCES

1. American Psychiatric Association: Diagnostic and statistical manual of mental disorders, ed. 3, (DSM-111), Washington, D.C., 1980, the Association.
2. Lee, T., and Seeman, P.: Brain dopamine receptors in schizophrenia. In Usdin, E., and Hanin, I., editors: Biological markers in psychiatry and neurology, New York, 1982, Pergamon Press, Inc., pp. 219-226.
3. Karson, C.N., Kleinman, J.E., Freed, W.J., et al.: Blink rates in schizophrenia. In Usdin, E., and Hanin, I., editors: Biological markers in psychiatry and neurology, New York, 1982, Pergamon Press, Inc., pp. 339-345.
4. Hirschowitz, J., Zemlan, R.P., and Garver, D.L.: Growth hormone levels and lithium ratios as predictors of success of lithium therapy in schizophrenia, Am. J Psychiatry **139**:646-649, 1982.
5. Golden, C.J., Moses, J.A., Zelazowski, R., et al.: Cerebral ventricular size and neuropsychological impairment in young chronic schizophrenics, Arch Gen. Psychiatry **37**:619-626, 1980.
6. Winberger, D.R., and Wyatt, R.J.: Structural pathology of the cerebellum in schizophrenia: CT and postmortum studies. In Jansson, B., Perris, C., and Struve, G., editors: Biological psychiatry 1981, New York, 1982, Elsevier/North Holland Inc., pp. 272-275.
7. Garver, D.L., Hitzemann, R.J., and Hirschowitz, J.: Lithium ratio, lithium response and the schizophrenias. In Usdin, E., and Hanin, I., editors: Biological markers in psychiatry and neurology, New York, 1982, Pergamon Press, Inc., pp. 169-175.
8. Hitzemann, R.J., and Garver, D.L.: Abnormalities in membrane lipids associated with deficiencies in lithium counterflow. In Usdin, E., and Hanin, I., editors: Biological markers in psychiatry and neurology, New York, 1982, Pergamon Press, Inc., pp. 177-182.
9. Hirschowitz, J., Casper, R., Garver, D.L., and Chang, S.: Lithium response in good prognosis schizophrenia, Am. J. Psychiatry **137**:916-920, 1980.
10. Perris, C.: A study of bipolar (manic-depressive) and unipolar recurrent depressive psychoses: 1. Genetic investigation, Acta Psychiatr. Scand. **194**(suppl.):15-44, 1966.

11. Winokur, G., Cadoret, K., Dorzab, J., and Baker, M.: Depressive disease: a genetic study, Arch. Gen. Psychiatry **24:**135-144, 1971.
12. Van Praag, H.M., and Korf, J.: Endogenous depression with and without disturbances in the 5-hydroxytryptamine metabolism: a biochemical classification? Psychopharmacology **19:**148-152, 1971.
13. Maas, J.W., Fawcett, J., and Dekirmenjian, H.: 3-Methoxy-4-hydroxyphenylglycol (MHPG) excretion in depressive states, Arch. Gen. Psychiatry **19:**129-134, 1968.
14. Maas, J.W., Fawcett, J.A., and Dekirmenjian, H.: Catecholamine metabolism, depressive illness and drug response, Arch. Gen. Psychiatry **26:**252-262, 1972.
15. Beckmann, H., and Goodwin, F.K.: Antidepressant response to tricyclics and urinary MHPG in unipolar patients, Arch. Gen. Psychiatry **32:**17-21, 1975.
16. Carrol, B.J., Curtis, C.G., and Mendels, J.: Neuroendocrine regulation in depression. I. Limbic system-adrenocortical dysfunction, Arch. Gen. Psychiatry **33:**1039-1058, 1976.
17. Carrol, B.J., Feinberg, M., Greden, J.F., et al.: Specific laboratory test for diagnosis of melancholia, Arch. Gen. Psychiatry **38:**15-23, 1981.
18. Garcia-Sevilla, J.A., Zis, A.P., Hollingsworth, P.J., et al.: Platelet $alpha_2$-adrenergic receptors in major depressive disorder, Arch. Gen. Psychiatry **38:**1327-1333, 1981.
19. Charney, D.S., Heninger, G.R., Sternberg, E.D., et al.: Presynaptic adrenergic receptor sensitivity in depression, Arch. Gen. Psychiatry **38:**1334-1343, 1981.
20. Siever, L.J., Cohen, R.M., and Murphy, D.L.: Antidepressants and $alpha_2$-adrenergic autoreceptor desensitization, Am. J. Psychiatry **138:**681-682, 1981.
21. Jones, F.D., Maas, J.W., Dekirmenjian, H., and Fawcett, J.A.: Urinary catecholamine metabolites during behavioral changes in a patient with manic-depressive cycles, Science **179:**300-302, 1973.
22. Gerner, R.H., Post, R.M., and Bunney, W.E.: A dopaminergic mechanism in mania, Am. J. Psychiatry **133:**1177-1180, 1976.
23. Janowsky, D.W., El-Yousef, M.K., Davis, J.M., and Sekerke, J.H.: Parasympathetic suppression of manic symptoms by physostigmine, Arch. Gen. Psychiatry **28:**542-547, 1973.
24. Hitzemann, R.J., and Garver, D.L.: Membranes, methylation and lithium responsive psychoses, Proc. Am. Psychiat. Assoc., New Research Abstracts NR 11, 1982.
25. Jones, M.T., Hillhouse, E., and Burden, J.,: Secretion of corticotropin-releasing hormone in vitro. In Martini, L., and Ganong, W.F., editors. Frontiers in neuroendocrinology, vol. 4, New York, 1976, Raven Press, pp. 129-168.
26. Greden, F.J.: The dexamethasone suppression test: an established biological marker of melancholia. In Usdin, E., and Hanin, I., editors: Biological markers in psychiatry and neurology, New York, 1982, Pergamon Press, Inc., pp. 229-240.
27. Pandey, G.N., Baker, J., Chang, S., and Davis, J.M.: Prediction of *in vivo* red cell/plasma lithium ratios by *in vitro* methods, Clin. Pharmacol. Ther. **24:**343-349, 1978.
28. Flemenbaum, A., Weddige, R., and Miller, J.: Lithium erythrocyte/plasma ratio as a predictor of response, Am. J. Psychiatry **135:**336-338, 1978.
29. Dorus, E., Pandey, G.N., Shaughnessy, R., et al.: Lithium transport across red cell membrane: a cell membrane abnormality in manic-depressive illness, Science **205:**932-934, 1979.
30. Extein, I., Pottash, A.L.C., Gold, M.S., and Cowdry, R.W.: Using the protirelin test to distinguish mania from schizophrenia, Arch. Gen. Psychiatry **39:**77-81, 1982.
31. Hirschowitz, J., Zemlan, F.P., and Garver, D.L.: Growth hormone levels and lithium ratios as predictors of success of lithium therapy in schizophrenia, Am. J. Psychiatry **139:**646-649, 1982.
32. Davis, J.M., Schaffer, C.B., Killian, G.A., Kinard, C., and Chang, C.: Important issues in the drug treatment of schizophrenia. In Keith, S.J., and Mosher, L.R., editor: Special report: Schizophrenia 1981, Washington, D.C., 1980, U.S. Government Printing Office, pp. 109-126.
33. Garver, D.L.: Drug therapy of psychiatric disease: schizophrenias and related psychoses. In Graham-Smith, D.G., Hippus, H., and Winokur, G., editors: Psychopharmacology. 1. Clinical psychopharmacology, Amsterdam, 1982, Excerpta Medica, pp. 250-270.
34. Dysken, M.W., Javaid, J.L., Chang, S.S., et al.: Fluphenazine pharmacokinetics and therapeutic response, Psychopharmacology **73:**205-210, 1981.
35. Magliozzi, J.R., Hollister, L.E., Arnold, K.V., and Earle, G.M.: Relationship of serum haloperidol levels to clinical response in schizophrenic patients, Am. J. Psychiatry **138:**365-367, 1981.
36. Rivera-Calimlim, L., Nasrallah, J., Strauss, J., and Lasagna, L.: Clinical response and plasma levels: effect of dose, dose schedules and drug interactions on plasma chlorpromazine levels, Am. J. Psychiatry **133:**646-652, 1972.
37. Creese, I., and Snyder, S.H.: A novel, simple and sensitive radioreceptor assay for antischizophrenic drugs in blood, Nature **270:**1980-182, 1977.
38. Calil, H.M., Avery, D.H., Hollister, L.E., Creese, I., and Snyder, S.H.: Serum levels of neuroleptics measured by dopamine radioreceptor assay and some clinical observations, Psychiatry Res. **1:**39-41, 1979.
39. Tune, L.E., Creese, I., DePaulo, J.R., et al.: Clinical state and serum neuroleptic levels measured by radioreceptor assay in schizophrenia, Am. J. Psychiatry **137:**187-190, 1980.
40. Cohen, B.M., Lipinski, J.F., Harris, P.Q., et al.: Clinical use of radioreceptor assay for neuroleptics, Psychiatry Res. **1:**173-177, 1980.
41. Risch, S.C., Huey, L.Y., and Janowsky, D.S.: Plasma levels of tricyclic antidepressants and clinical efficacy: review of the literature—Part I, J. Clin. Psychiatry **40:**4-16, 1979.
42. Risch, S.C., Huey, L.Y., and Janowsky, D.S.: Plasma levels of tricyclic antidepressants and clinical efficacy: Review of the literature—Part II, J. Clin. Psychiatry **40:**58-69, 1979.
43. Robinson, D.W., Nies, A., Ravaris, L., Ives, J.O., and Bartlett, D.: Clinial pharmacology of phenelzine, Arch. Gen. Psychiatry **35:**629-635, 1978.
44. Ramsey, A.T., and Cox, M.: Lithium and the kidney: a review, Am. J. Psychiatry **139:**443-449, 1982.
45. Asberg, M., Thoren, P., Traskman, L., Bertilsson, L., and Ringberger, V.: ''Serotonin depression''—a biochemical subgroup within affective disorders? Science **191:**478-480, 1976.

Chapter 46 Neoplasia

Bernard E. Statland
Per Winkel

carcinoembryonic antigen A molecule produced by or associated with cancer cells, which is also expressed by fetal cells. Small levels are detected in normal circulation. Detection is by immunochemical analysis.

carcinogen An agent, usually a chemical, that transforms a cell from a normal to a cancerous state.

cocarcinogen An agent that, by itself, does not transform a normal cell into a cancerous state but in concert with another agent can effect the transformation.

complementary test A second test usually based on a different type of biological principle or observation that is used to confirm a diagnosis; for example, the sputum cytology test is a complementary test for chest x-ray examination.

confirmation Use of a second test with 100% specificity to verify the observation of a less specific test, for example, biopsy to verify a mass as a tumor.

dedifferentiation The process by which cells go from the more specific to the more general in nature. Usually such cells lose their morphological architecture and ability to synthesize specific cell components, for example estrogen receptors.

dissemination The phase of cancer in which the cells spread to various parts of the body distant from the site of origin.

ectopic tumors Cancers that produce hormones.

estriol receptor The specific tissue membrane receptor in breast that binds the hormone estriol. Its presence signifies a differentiated tumor.

heterogeneity A variation in gene expression among cancer cells. They are not uniform clones because differences between cells exist. Not all cells within a tumor are positive for the same antigen or respond to the same drug.

induction phase The period of time during which a normal cell becomes transformed into a cancerous cell.

in situ A term to show that cancer cells are localized at the place of origin.

invasion Process by which malignant cells move into deeper tissue and through the basement membrane and gain access to blood vessels and lymphatic channels.

metastasis Cancer cells that have spread to other organs and have formed colonies that are growing and often invading the organ.

monitoring Measurement of a biochemical marker of cancer after a confirmed diagnosis, for example, carcinoembryonal antigen (CEA).

oncofetal protein A protein produced by, or associated with, cancer cells, but also one that is made by the fetus; however, it is normally produced in the child or adult at very low levels.

Pap (Papanicolaou) smear A common screening test for cancer in which cells from the cervix are examined for cytological abnormalities consistent with cancer.

staging A process of diagnosis in which the pathologist determines the position of cancer in the cycle of induction, in situ, invasion, or dissemination.

tumor marker A misused term applied to molecules that can be used to diagnose or monitor the presence or growth of a cancer. Usually such markers are not specific for cancer.

SCOPE OF PROBLEM

Incidence

As a cause of death, cancer and its complications rank second in the United States and in most civilized countries; it is second only to cardiovascular disease. Approximately one in five persons in the United States will die of cancer. In addition to the high incidence of this disease, it accounts for a large proportion of the total money spent in health care in the United States. The various complications of cancer, the type of therapy used, and the long rehabilitation play a role in the costs. Finally, the psychological fear and dread of this disease, as well as the long-term anxiety when a patient knows he has cancer, represent an added burden caused by this disease.

Distribution of deaths by site

Table 46-1 presents the percent of cancer deaths by site and sex as tabulated in 1981 in the United States. As noted in Table 46-1, the leading cancer sites associated with

Table 46-1. Estimated cancer deaths in 1981 in the United States by site and sex

Site	Sex	
	Male (%)	Female (%)
Breast	—	19
Colon and rectum	12	15
Blood and lymphoid tissue (leukemia and lymphomas)	9	9
Lung	34	15
Oral cavity	3	1
Ovary	—	6
Pancreas	5	5
Prostate gland	10	—
Skin	2	1
Urinary tract	5	3
Uterus	—	5
Other	20	21

mortality in males of all ages are, in descending order, lung, colorectal area, prostate, pancreas, and stomach. In females of all ages the sites are breast, colorectal area, lung, uterus, and ovary. Obviously, if the survival of patients with cancer in general is to improve, it is imperative to attempt to find cures for the commonly seen cancers. It should be mentioned that the distribution of deaths by site is dependent on geographical considerations as well. For example, in Japan esophageal and stomach cancers represent the leading types of malignancy, whereas this pattern is not true in the United States.

CANCER: NATURE OF THE DISEASE

Description of cancer

Clinical manifestations. The clinical manifestations of cancer vary widely, depending on the tissue affected. For example, cancer of the gastrointestinal tract is manifested by obstruction, hemoptysis, and bloody stools. Cancer of the lung is manifested by hypoxia, chest pain, and often various neurological symptoms. The clinical manifestations are related to the primary origin of the cancer and to other organs involved as well. For example, cancer of an endocrine gland can result in production of excess hormone with many systemic hormonal effects. An important aspect relates to the spread (metastasis) of the cancer cells to other organs, with invasion and subsequent destruction. Cancer spreads through the lymphatic system and the bloodstream as well, resulting in commonly seen liver, bone, and pulmonary metastases.

Time as a factor

Cancer as a long-term process. Cancer is a long-term process and progresses through four obligatory phases: an induction phase, an in situ phase, an invasion phase, and a dissemination phase. During the *induction phase,* which can last up to 30 years, the cells are exposed to one or more carcinogens. These environmental carcinogens may include radiation or various toxins. It has been estimated that approximately three fourths of all human cancers may be caused by these environmental factors.

It is now believed that many years of exposure may be necessary before a carcinogen is able to have its effect on the host. The histological changes begin with severe dysplasia and eventually become a definite cancer. It should be obvious that not everyone who is exposed to the same carcinogen will develop cancer. Additionally, other factors play a role in deciding which individual may get cancer. These factors include individual or tissue susceptibility, the presence of other carcinogens or cocarcinogens, the site at which the carcinogen may act, the duration of exposure, and obviously the nature, amount, and concentration of the carcinogen under question. Often the time between the induction phase and the clinically apparent cancer can be as long as 20 years.

After induction there is the *in situ phase*. The in situ phase represents that time during which the cell actually becomes truly cancerous but remains localized in the original site and does not invade other tissues.

The third phase is called the *invasion phase*. During the invasion phase the malignant cells multiply and invade into the deeper tissues through the basement membrane, thereby gaining access to blood vessels and lymphatic channels.

The fourth stage is that of dissemination. During the *dissemination phase,* which lasts 1 to 5 years, the invading cancer spreads to various parts of the body distant from the site of origin.

It is critical to detect cancer early, before metastatic spread. Ideally it should be detected during the induction phase. This is impossible because before the in situ phase one is *not* certain if cancer will actually develop in the individual. The next approach is to detect the cancer in the in situ phase. This has been done with great success in patients with cancer of the cervix. Here the Pap (Papanicolaou) smear technique has been of great benefit. When in situ cancer of the cervix is detected, the prognosis is excellent. Most cancers are detected during the invasion phase. If dissemination has not yet occurred, the prognosis is reasonable. Detection of local spreading with or without involvement of the lymph nodes often leads to a cure. However, if dissemination has already occurred, the prognosis is very poor.

Invasion by cancer cells of surrounding tissue. Several factors play a role in determining the cancer's ability to invade the surrounding tissue. Such factors include increased motility of the cells, increased pressure within the tumor caused by active multiplication of the cells, elaboration by the cancer of lytic substances, lack of intercellular bridges found between all normal cells, decreased cohesiveness between cells, and eventual spread of the tumor cells to the regional lymph nodes. Much work is now being done to assess these critical factors that play a role in the invasiveness of cancer cells. However, when the metastases are still microscopic (micrometastases), the clinician's ability to detect them is very poor. It has been estimated that approximately half the patients who appear to be clinically free of metastases do in fact have unrecognized distant micrometastases at the time of initial diagnosis and treatment.

Change in cell division. Cancer is manifested often by change in cellular division rate. Although most cancers are associated with an increased rate of division, there are examples where this is not always the case, such as nephroma.

Dedifferentiation of cells. A common phenomenon of cancer is dedifferentiation, in which cells go from the more specific to the more general. Thus it is not uncommon for cancer cells to elaborate various proteins that are normally present only in the embryonic or fetal stage. On the other hand, as cells dedifferentiate, they may lose certain specific cellular properties such as receptor activity or an enzyme activity. These changes can be used as prognostic indicators.

Chromosomal changes in cancer. Chromosomal changes in cancer have been studied mainly in patients with leukemia. In fact, various types of leukemia can be confirmed on the basis of these chromosomal changes.

Etiology

Recently Sager[1] presented an intriguing approach to the origin of cancer. She discusses it as a multistage genetic process. The stages are of three types:

1. Initial DNA damage
2. Chromosome breakdown and rearrangement
3. Selection of successfully growing mutant cells

The initial changes in cellular DNA can be caused by radiation, chemicals, viruses, or unknown agents. This leads to faulty growth control and to loss of chromosome stability.

The chromosome breakage and rearrangement occurs in several continuous phases. There is first an initiation of cell division. This is later manifested in terms of aberrant chromosomal transpositions, which lead to genomic rearrangements. Genetic and phenotypic changes cascade as the aberrations continue to arise.

The selection of successfully growing mutant cells relates to the following issues. The genomic rearrangements generate new phenotypes. Selection favors those phenotypes of proliferating and well-adapted cells. Finally, specific phenotypes will succeed in different tissues.

Carcinogens are present in the environment in increasing amounts. Although various carcinogens are known, there probably are many that still are unknown. Viruses also can cause certain cancers. Some of these, such as Burkitt's lymphoma, are seen in humans. There is experimental and epidemiological support for the presence of hepatic viruses and herpesviruses that may cause hepatic and cervical carcinoma in humans. However, viruses as general cancer-causing agents are not very well substantiated at this point.

Diversity of cancer cells

Variation of gene expression. There is a broad spectrum of possible combinations of gene expression in the human cell. The spectrum goes from normal cells to the most atypical cancer cells. The phenotypic variation occurs not only from cancer cells to normal cells, or from cancer type to cancer type, but also within particular cancer types. For example, in patients with cancer of the breast there is a heterogeneity of genes expressed by various cells; that is, not all cells express the same genes.

Variable gene expression and its manifestations. Variable gene expression leads to biological and biochemical diversity of cancer cells; consequently various tumor-specific markers are not necessarily elaborated by all cancer cells of the same type or even of a single cancer over a period of time. This is very important clinically in trying to determine which protein to follow in monitoring patients with known malignancy.

OVERVIEW OF ROLES OF LABORATORY TESTS

Detection (screening)

There are four major functions that laboratory tests can serve in the field of neoplasia. They include detection or screening, confirmation, classification (staging), and monitoring. In the detection role the laboratory test is used to determine whether an otherwise healthy individual might have cancer. The population tested may either be a general adult group in a particular area or a high-risk group that has a greater incidence of cancer. An example of the latter is smokers, who are at higher risk for developing cancer of the lung. A laboratory test that is used to detect cancer should have a very low rate of false-negative results; that is, the test should approximate 100% diagnostic sensitivity. Ideally all individuals with cancer should give positive test results. Thus, if a patient has a "negative" result, he should be able to assume that he is free of the disease.

Examples of tests that are used to detect cancer are stool hemoglobin for cancer of the colon, Pap smear for cancer of the cervix, and mammography for breast cancer. Tests for detection of cancer may suffer from two obvious problems. The first problem is that very early cancers (in situ) will not always give a positive test result. Furthermore, there may be a high percentage of "false positives," which can lead to unnecessary testing and unnecessary worry on the part of the patient.

The evaluation of mass screening for cancer appears to be extremely difficult, and there is controversy about the merits of nearly every screening program. In assessment of any proposed mass screening program the following four basic questions should be answered[1a]

1. What is the effect of the current screening program?
2. How will this effect change when certain conditions change with screening elsewhere or in the future (such as improved screening tests, and improved treatment)?
3. What is the best age range in which to perform screening tests, and what should be the intervals between subsequent screens?
4. Is this screening strategy sufficiently cost effective to justify mass screening?

An essential feature of any screening program is the screening test. Table 46-2 lists a number of screening tests for early detection of cancer.[2] The quality of a screening test is usually expressed by its sensitivity and specificity. The observations from the screening tests are divided into negative and positive results. Each person examined is classified as either a diseased or nondiseased person.

A rigid classification of test results into positive and negative results may sometimes be too simplistic. Outcomes of screening tests can usually be ordered from very negative to very positive. The latter approach allows for a

Table 46-2. Screening tests for early detection of cancer

Site	Test
Bladder	Cytologic analysis of urine
Breast	Mammography, physical examination, self-examination
Cervix	Papanicolaou smear, pelvic examination
Colon and rectum	Testing stool for occult blood, sigmoidoscopy
Hodgkin's disease	Physical examination and roentgenography
Lung	X-ray, cytologic analysis of sputum
Oral cavity	Visual examination
Prostate	Digital palpation per rectum, prostatic massage and cytologic examination, determination of serum acid phosphate concentration
Skin	Visual inspection
Stomach	Photofluorography, saline wash and cytologic examination of gastric contents, examination of stool for occult blood

From Habbema, J.D.F., van Oortmarssen, G.J., and van der Maas, P.J.: In Statland, B.E., and Winkel, P., editors: Laboratory measurements in malignant disease, vol. 2, Philadelphia, 1982, W.B. Saunders Co.

more sophisticated test interpretation in actual screening programs. One example is early recall for a repeat screen for patients whose results are not negative but also are not alarming enough to justify immediate diagnostic action.[3] Another example is a stepwise screening policy in which only individuals with positive results at the first screening test are subject to further testing.[4]

Sometimes the use of more than one screening test may seem advantageous. However, assessment of the sensitivity and specificity of a combination of screening tests based on data available for the individual tests is complicated by the fact that usually the tests are not independent in a statistical sense. In general, it is more effective to combine two tests that are complementary, that is, directed at different anatomical and other features of the tumor, than to combine tests directed at the same types of features.

Complementary tests include sputum cytology and chest x-ray examination for lung cancer screening.[5] Palpation and mammography in breast cancer screening are an example of two related tests. They both detect tumors largely on the basis of size. A recent study[6] showed that when mammography was performed, the physical examination proved to be almost completely redundant.

Confirmation

Additional tests are used to confirm the suspicion of cancer based on clinical symptoms or signs. Tests that tend to confirm the presence of a cancer include, for example, bone marrow examination for leukemia, urinary catecholamines for pheochromocytoma, and alpha-fetoprotein for testicular cancer. The confirmatory results must be above a certain decision level.[7] For a laboratory test result to be confirmatory, it should possess 100% diagnostic specificity, that is, contain no false-positive results. For example, all cases where the catecholamine level is above a certain value should be associated with pheochromocytoma.

Classification and staging

Surgical pathologists have developed various staging approaches based on the size and extent of invasion of surrounding tissues by the tumor, the number of positive lymph nodes, and the presence or absence of metastases. This has been called the TNM (tumor, nodes, metastases) system.[8] The purpose of such staging is to give reasonable estimates of prognosis, that is, recurrence of cancer, appropriate response to therapy, or likely course of the disease. In addition to staging based on gross or microscopic pathological data, it would be of great value to have biochemical tests that also would classify cancers appropriately.

One important type of classification is to differentiate cancer with metastases from cancer without metastases. A further subdivision might be cancer with bone metastases versus cancer with liver metastases. In the patient known to have cancer, such classification would serve an important role, from both a prognostic and a therapeutic perspective.

Monitoring

The fourth function of laboratory tests is that of monitoring the course of the disease or response to therapy. Recently Winkel et al.[9] have developed various strategies to monitor patients known to have breast cancer. The problem addressed was that of predicting on the basis of sequential values whether or not a patient would have recurrence of this disease. Other approaches have been used to monitor patients with colon cancer on the basis of carcinoembryonal antigen (CEA) in colon cancer.[10] An increased CEA value is a signal to explore the patient surgically again to remove additional cancer. It is assumed that the CEA-producing tumor has recurred when the serum CEA values reach a certain threshold.

• • •

All four major functions, screening, confirming, classifying, and monitoring, are possible roles for laboratory tests for neoplasia.

DEFINITION OF IDEAL TUMOR MARKER

The ideal tumor marker is positive in all cases of cancer and negative in the absence of cancer. It is difficult to know whether a tumor marker is as good as its early proponents claim. Most tumor markers have the following three phases. First, the tumor marker is claimed to be per-

Table 46-3. Classes of biochemicals used as tumor markers

Class of biochemical	Examples	Use
Increased production of endogenous biochemicals	Hormones, enzymes, polyamines, and so on	Confirmation, diagnosis, monitoring
Synthesis of biochemicals of previously quiescent genes	Oncofetal proteins, cell surface antigens, enzymes	Monitoring, prognosis
Receptors	Estriol receptor (breast cancer), androgen receptor (prostate cancer)	Prognosis, treatment
Modification of usual cell or organ function	Gamma-glutamyl transferase (GGT) or 5′-nucleotidase	Diagnosis

fect. Its proponents claim it has 100% sensitivity and 100% specificity. The second phase consists of disbelief. Additional populations are tested, and the tumor marker is seen to fail in many cases of patients who do not have cancer (that is, false positives) and with patients who do have cancer (that is, false negatives). Investigators studying the marker will claim that it is worthless in many cases. The third phase is that of compromise. Here it is agreed that the tumor marker is not as good as originally claimed, but it still has merit in selected cases or for certain functions.

A good example of such a marker is carcinoembryonic antigen (CEA). Initially CEA was considered the ideal tumor marker. After many years of investigation critics claimed that CEA did not have much value because it was positive in patients who were smokers but did not have cancer and it was negative in patients with various types of cancers. Furthermore, critics claimed it could not be used to detect (screen) populations for cancer. Laboratories have now entered the third phase of evaluation of CEA, and it is now realized that CEA cannot be used as a general cancer marker. However, CEA does have an important role in monitoring the patient known to have a CEA-producing tumor. An example of such a cancer is carcinoma of the colon.

Coombes and Neville[11] have suggested that the *ideal* tumor marker should fulfill the following criteria:

1. Be easy and inexpensive to measure
2. Be specific to the tumor studied and commonly associated with it
3. Have a stoichiometric relationship between plasma level of the marker and tumor cell number
4. Have an abnormal plasma level, urine level, or both in the presence of micrometastases, that is, at a stage at which no clinical or presently available diagnostic methods reveal their presence
5. Have plasma levels, urine levels, or both that are stable—not subject to wild fluctuations
6. If present in normal plasma, exist at a much lower concentration than that found in association with all stages of cancer

Obviously much additional research must be done before such ideal tumor markers will be found. However, it is important to recognize that the evaluation of an ideal tumor marker should relate to the clinical setting. In order to do so, it has been suggested that all tumor markers should also comply with the folowing major criteria[12]:

1. They should prognosticate a higher or lower risk for eventual development of recurrence.
2. They should change as the current status of the tumor changes over time.
3. They should precede and predict recurrences before they are clinically detectable.

All tumor markers should be analyzed both according to the criteria that Coombes and Neville have presented and according to the considerations just mentioned. Often it is seen that a tumor marker may have some value; however, when examined in a clinical setting, it is obvious that a history of the patient's disease or a routine physical examination would also yield equivalent information. For a tumor marker to be of some value, it must give information beyond that readily seen on the basis of physical examination or history, and it must give this information with a reasonably long lead time so as to give appropriate therapy in a timely manner. Lead time means the time elapsed between a test being positive and the time the disease is clinically evident or advanced.

TYPES OF ANALYTES

Classes of biochemicals used as tumor markers

(Table 46-3)

This section will review a number of biochemical tests that have been used either as primary tumor markers or as secondary tests to note invasion or dissemination of cancer. The types of analytes are listed in the following box and are discussed in terms of their clinical usefulness and applications. This chapter cannot deal with all tests that have been suggested as tumor markers, but rather it discusses those assays that are commonly in use or seem to have potential value. Furthermore, the analytical procedures are not dealt with in great detail. They are mentioned only if such procedures are critical in the interpretation of results. A recent review[13] that covers a number of these assays in greater depth is recommended for further reading.

Types of analytes

Oncofetal proteins
- Carcinoembryonal antigen (CEA)
- Alpha-fetoprotein (AFP) and human chorionic gonadotropin (HCG)
- Tissue polypeptide antigen (TPA)
- Tennessee antigen (TENAGEN)
- Fetal sulfoglycoprotein (FSA)
- Pancreatic oncofetal antigen (POA)
- Breast cancer–associated markers

Various proteins
- Glycoproteins
- Ferritin
- Casein

Enzymes
Collagen-breakdown products
Polyamines
Nucleosides
Cellular markers
Hemostasis-related factors

Oncofetal antigens

Carcinoembryonal antigen (CEA). CEA is a glycoprotein present in colonic adenocarcinoma and fetal gut; it was first described by Gold and Freedman.[14] Very sensitive radioimmunoassays have since revealed CEA-like substances in other malignant and nonmalignant tissues. The detection of CEA in various tissues or serum is complicated by the occurrence of CEA–cross-reacting antigens. Studies of such antigens indicate that classes of CEA-like molecules exist.[15-17] In examining the clinical literature, one should bear in mind that a number of unresolved problems regarding purification and characterization of CEA and CEA-related antigens still remain; therefore the clinical conclusions should be tempered accordingly.

In general, CEA plasma levels increase with increasing age and smoking. This has limited the use of CEA levels for the purpose of screening.[18,19] In the Busselton population study[19] 2372 analyses resulted in the identification of two new cancer cases. Screening programs confined to subpopulations with higher-than-average risk of developing cancer have been equally discouraging.[20,21]

Usually a CEA measurement is not a useful adjunct to cancer diagnosis. Neither the sensitivity nor the specificity of CEA justifies its use to support by itself a diagnosis of cancer.[22,23] In particular situations, however, CEA has proved of diagnostic value, for example, for the detection of primary colorectal cancer[24] when used in combination with a barium enema and when used in combination with radioiodide imaging for the detection of carcinoma metastatic to the liver. According to the consensus statement of the National Cancer Institute,[18] only values five to ten times the upper normal reference limit in patients with symptoms should be considered strongly suggestive of the presence of cancer. In a number of cancers, including colorectal and breast cancer, the plasma level of CEA and the frequency of elevated values are positively correlated with the severity of the disease as assessed by clinical staging.* In general, CEA-producing tumors tend to be more aggressive after initial treatment than nonsecreting tumors are.[33,34] Alteration in CEA levels correlated well with change in disease status in patients with metastatic breast carcinoma and in patients with metastatic colorectal cancer.[10,35] However, because the relationship between clinical course and CEA plasma level is not completely consistent, it is difficult to assess the potential clinical usefulness of these observations.

Postoperative monitoring of plasma CEA levels for the detection of recurrence or metastases has proved valuable in colorectal cancer. Mackey et al.[30] found that of 53 patients who developed metastases, 36 had elevated CEA levels 3 to 18 months before the metastases were clinically evident. Such findings have stimulated interest in second-look operations in patients with surgically treated colorectal cancer. Balz et al.[36] found that 19 of 22 patients with postoperatively rising CEA levels had otherwise undetected recurrent disease. Among these, the recurrent tumor was localized and could be resected in six patients. Five of these were asymptomatic with stable CEA values over the next 37 months.

Postoperative CEA levels are less frequently elevated in breast cancer patients who eventually develop overt metastatic disease than in corresponding patients operated for colorectal cancer. Recent reviews[37,38] indicate that in only 10% to 15% of such breast cancer patients does the plasma CEA level rise to values above 10 μg/L. On the average this happens 4 to 6 months before the appearance of overt metastatic disease.

Alpha-fetoprotein (AFP) and human chorionic gonadotropin (HCG). AFP is an oncofetal glycoprotein. In early embryonic life it is a predominant component of the serum proteins. It is first synthesized by the yolk sac and later by the fetal liver. Later in life it is mainly produced in the liver. AFP was found by Bergstrand and Czar[39] in 1956 and first recognized as a tumor marker by Abele in 1963.[40]

Serum AFP values should be less than 10 μg/L in healthy subjects. In benign hepatic disorders, moderate (40 μg/L) elevations may be seen. Values above 400 μg/L are almost always associated with hepatocellular carcinoma, germ cell carcinoma (such as testicular carcinoma), chronic aggressive hepatitis, or subacute hepatic necrosis.

It is now possible to determine if the AFP produced is of yolk sac or liver cell origin; in some cases this may be

*See references 8, 13-19, 21, and 25-32.

Table 46-4. WHO classification of germ cell tumors and associated tumor markers

WHO classification	Immunohistochemistry		Serology		Comments
	AFP*	HCG†	AFP	HCG	
Seminoma (S)	−	±	No	± Yes	HCG in giant cells
Embryonal carcinoma (EC)	+	+	± Yes	± Yes	HCG in giant cells AFP controversial, may occur in undiagnosed yolk sac elements
Yolk sac tumor (YST)	+	−	± Yes	No	
Choriocarcinoma (CC)	−	+	No		
Teratoma (TT)	−	−	No?	No?	

From Norgaard-Pedersen, B., and Hangaard, J.: In Statland, B.E., and Winkel, P., editors: Laboratory measurements in malignant disease, Philadelphia, 1982, W.B. Saunders Co.
*AFP = alpha-fetoprotein.
†HCG = human chorionic gonadotropin.

necessary to distinguish liver metastases from AFP-producing tumors.[41]

HCG is a glycoprotein hormone that shares indistinguishable biological activity and extensive structural homology with its pituitary counterpart, human luteinizing hormone (HLH).[42] Both hormones have two subunits, designated alpha and beta. Although the alpha subunits of HLH and HCG are essentially identical, the beta subunits can be differentiated on the basis of specific immunoassay techniques.[43]

Tumors of the placenta and the testes that contain trophoblastic tissue secrete excessive amounts of HCG. Specific and sensitive assays have revealed that many cancers secrete HCG. However, available data[44] clearly show that HCG determinations are of no value in screening for cancer.

The main clinical use of AFP and HCG is related to the diagnosis, therapy, and follow-up of germ cell tumors.[45] Table 46-4 presents the World Health Organization (WHO) classification of germ cell tumors and associated markers in tissue and serum. In general, AFP and HCG provide the most information about tumor status when they are persistently elevated. The absence of a marker does not preclude the presence of germ cell tumors, in that these tumors may be composed of several cell lines with different sensitivities to therapy.

Tissue polypeptide antigen (TPA). TPA is a single-chain polypeptide that has been isolated from a large number of malignant tumors and from placental tissue as well. This protein has been studied primarily by the Swedish group headed by Björklund.[46,47] TPA has a molecular weight of approximately 22,000. The decision level for a serum TPA value is 0.09 units/mL; when there is a value greater than this concentration, there is a high likelihood of malignancy. The Swedish group evaluated 15,000 patients who have had malignant disease or benign disease and a normal (healthy) group. Their results are as follows: 3.5% of the healthy subjects had values above 0.09 units/mL; 16% of the subjects with benign disease had increased TPA values; 55% of patients with primary malignancy but no metastases had elevated values; 81% of patients with metastatic disease had increased TPA values. The percentage of patients having elevated TPA values ranged from 53% of patients who had prostatic cancer to 82% with colon cancer. Intermediate values included 80% with lung cancer, pancreatic cancer, and cervical cancer and 75% with stomach cancer or bladder cancer.

Tennessee antigen (TENAGEN). TENAGEN is a glycoprotein with a molecular weight of 50,000. The antigen was isolated first from primary and metastatic adenocarcinomas. Early results by Lovins[48] revealed an overall diagnostic sensitivity of 75% to 85% for various cancers, including lung, stomach, pancreas, and colorectal; 92% of a normal population was found to be negative for TENAGEN. As compared to CEA, there was no smoker bias. In the past few years little has been written about this antigen. The coming years will indicate whether it will have clinical use as a tumor marker.

Fetal sulfoglycoprotein (FSA). FSA is found in the fetal gastrointestinal tract and in gastric carcinoma. It is noteworthy that this antigen is not present in either normal or diseased adult tissues. In a study of patients with gastric cancer,[49] 97% were found to have FSA in the gastric juice. Consequently this assay has been suggested as an excellent screening test for gastric cancer. It has been noted by Hakkinen[49] that 91% of persons without gastric cancer are FSA negative. Furthermore, in only 10% of patients with gastric carcinomas was the FSA test falsely negative.[49]

Pancreatic oncofetal antigen (POA). Pancreatic cancer has become the fourth leading cause of cancer deaths in American males. Unfortunately the prognosis is poor. A number of workers[50,51] have suggested that a POA, which is a glycoprotein with a molecular weight of approximately 900,000 daltons, would be of some value in helping to diagnose this entity. Goldman[52] noted that 63% of patients with pancreatic cancer have elevated POA values. It is noteworthy that POA elevations are seen in benign conditions (8% to 16%) affecting the pancreas or the biliary tract. In healthy subjects the elevations are seen very rarely and amount to only approximately 1%. The major disadvantage of POA testing is that the assay is not a good screening test but rather one of confirmation of pancreatic cancer.

Breast cancer–associated markers. The candidate tumor markers for breast carcinoma have recently been reviewed.[38] Either most of them have proved to be useless from a clinical point of view, or they have not yet been subjected to a sufficiently careful clinical assessment so that their clinical utility can be defined.

The role of CEA in the postoperative prediction of recurrence and the monitoring of the treatment of metastatic disease has been discussed elsewhere in this chapter. Only few breast cancers secrete CEA, and so there is a need for additional breast cancer markers.

A gross cystic disease fluid glycoprotein (GCDFP-15) is a specific plasma marker of some breast carcinomas, mainly those with apocrine features.[53] Haagensen et al.[54] developed a radioimmunoassay for this protein. The assay cannot distinguish women with early breast carcinoma from those with benign breast gross cystic disease,[38] but it may be useful for monitoring purposes.[55]

Other proteins

Glycoproteins. Among the various serum glycoproteins are the acute-phase reaction proteins synthesized by the liver. These proteins tend to be elevated in patients with malignancy. Although not very specific, the acute phase reactants (haptoglobin, $alpha_1$ acid glycoprotein and $alpha_1$ acid antitrypsin) may be very sensitive indicators of cancer.

Silverman et al.[56] found a glycoprotein with serum electrophoresis methodology. This glycoprotein appears between the $alpha_1$ and beta-globulin regions. Only one of 18 patients with benign breast disease had this glycoprotein. However, 84% of 31 patients with breast cancer had this glycoprotein present.

Harvey et al.,[57] after isolating the glycoproteins associated with sialic acid (*N*-acetylneuraminic acid, NANA), studied 28 patients with widespread malignancy whose responses to chemotherapy were graded as progressive, stable, or responsive. Serum specimens were analyzed for NANA levels before and after therapy. Eleven of the 28 patients demonstrated tumor regression or stable disease, and 10 of them had decreased NANA levels. The remaining 17 showed tumor progression with widespread progression of metastatic disease. Serum NANA rose in 11 of the latter 12. In 8 of the 11 the increase preceded clinical relapse. In 4 of 5 patients with local site relapse, there was no elevation.[58]

Ferritin. Marcus and Zinberg[58] measured serum ferritin in various populations. The serum ferritin levels exceeded the upper reference limit in preoperative sera of 40% of women with breast cancer and in 67% of women with locally recurrent or metastatic breast cancer. It should be pointed out, however, that hepatic inflammatory disease also results in elevated serum ferritin values; for example, 431 of patients with hepatitis or cirrhosis had values above the upper reference limit. Furthermore, 13% of patients with ulcerative colitis or gastric duodenal ulcers had elevated values.

Casein. Hendrick and Franchimont[59] found that patients with breast cancer had elevated values of casein. However, these elevations were mainly present in women with advanced breast cancer. Lung cancer may also be associated with elevated values of serum casein.

Enzymes

Recently Schwartz[60] reviewed the use of enzyme tests in the management of patients with cancer. The following box presents various enzymes that have been used for this purpose. As with many putative tumor markers discussed in this chapter, the use of enzyme markers is fraught with difficulties. Not all patients with a particular cancer type have elevations in an enzyme (poor sensitivity), and many noncancer diseases are associated with elevations of many of these enzymes. Thus the most frequent uses of these enzymes are as objective markers to give semiquantitative estimates of response to therapy or as prognostic indicators.

Acid phosphatase (ACP). There are two isoenzymes of the 13 known for ACP that appear to have important roles in managing patients with cancer. The first one is the prostatic isoenzyme, and the second is the bone isoenzyme. The former is inhibited by L-tartrate, whereas the bone isoenzyme is tartrate resistant.

The success in interpreting elevations of the prostatic acid phosphatase (PAP) isoenzyme for diagnostic purposes has been poor.[61-64] Most assays investigated have low sen-

Enzymes useful in cancer detection

- Acid phosphatase (ACP)
- Alkaline phosphatase (ALP)
- Creatine kinase BB (CK-BB)
- Gamma-glutamyl transferase (GGT)
- Glycosyl transferases
- Lactate dehydrogenase (LD)
- Lysozyme (muramidase)
- 5′-Nucleotidase
- Pancreatic enzymes (amylase, lipase, ribonuclease, trypsin)
- Phosphohexose isomerase (PHI)
- Terminal deoxynucleotidyl transferase (TDT)

sitivities in the early, more treatable stages of prostatic carcinoma (A and B) (Table 46-5). The use of many of these assays to confirm the presence of prostatic carcinoma in men with urological complaints has also been limited.[65]

The bone isoenzyme of ACP has been used as a diagnostic aid for finding patients with osteolytic metastases. Tavassoli et al.[66] studied this isoenzyme in 65 patients with breast cancer; 36 had bone metastases, and 29 did not. The mean value for the patients with bone metastases was 6.5 U/L; patients without metastases had a mean value of 1.9 U/L. These workers concluded that the bone isoenzyme of ACP may be a useful aid in monitoring cancer patients in whom osteolytic metastases would be very likely.

Alkaline phosphatase (ALP). There are three major uses of ALP measurements in following up patients with cancer, as follows: ALP derived from patients with osteoblastic metastases, ALP elevated from liver metastases, and ALP derived from the tumor itself. This last isoenzyme has been called the Regan isoenzyme. Fishman described the Regan isoenzyme in a patient with lung cancer.[67] This isoenzyme was resistant to inhibition by L-phenylalanine.

Kahan et al.[68] reported a homoarginine-sensitive isoenzyme of ALP in the sera of 60% of patients with lung cancer, 77% of patients with pancreatic cancer, 36% of patients with benign pancreatic disease, 65% of patients with uterine cancer, and 9% of patients with benign gynecological disease.

Creatine kinase (CK). Recently Griffiths[69] reviewed CK-BB in relationship to its usefulness as a marker for neoplastic disease. It should be pointed out that there are pathological conditions other than those of neoplasia in which CK-BB would be elevated. They include aortal coronary artery bypass surgery, severe myocardial infarction with hypoxic brain damage, condition after cardiopulmonary resuscitation, malignant hyperthermia, hypothermia, acute cerebral disorders (cerebral contusion, cerebral hemorrhage, and ischemic brain infarction), and pulmonary disease or damage. CK-BB has been studied in patients with lung cancer, where 54% of patients with small cell carcinoma, 27% of patients with squamous cell cancer, and 12% of patients with lung adenocarcinoma had elevated CK-BB values. It has also been studied in patients with prostatic cancer; 52% of patients with stage D prostatic cancer had elevated CK-BB.

Thompson et al.[70] reported CK-BB in serum of 4 to 6 patients with primary breast cancer and in 27 of 35 women with metastatic disease.[70].

Gamma-glutamyl transferase (GGT). GGT is an extremely sensitive indicator of liver disease. Consequently it has been used as an early marker of hepatic metastases. Recently Griffiths[71] has suggested that a particular isoenzyme of GGT may be a constant finding in patients with pancreatic cancer. Unfortunately this finding is present as well in patients with pancreatic disease other than pancreatic cancer.

Glycosyl and other carbohydrate transferases. Numerous workers have studied sialyl transferase in the sera (also plasma) of patients with cancer. The response of patients to surgery has been monitored by plasma sialyl transferase levels. Galactosyl transferase isoenzymes have been the object of much interest. The isoenzyme galactosyl transferase II (GT II) has been examined by many workers.[72] In a study of over 200 patients with cancer, elevated values of GT II were observed in 71% of the patients, including 73% of the patients with colorectal cancer, 83% with pancreatic cancer, 75% with gastric cancer, 78% of women with breast cancer, and 65% with lung cancer. It is noteworthy that this enzyme was not found in the sera of 58 control patients who were healthy or in patients with ulcerative colitis, Crohn's disease, pancreatitis, viral hepatitis, or biliary tract disease. However, in 20% of patients with severe alcoholic hepatitis and in 90% of patients with celiac disease, the concentration of GT II was elevated. In addition to the sialyl transferases and the galactosyl transferases, fucosyl transferase has been elevated.[73] This enzyme seems to be of some value in monitoring patients with colon cancer, breast cancer, or both. The fucosyl transferase is also elevated in patients with acute myelogenous leukemia and in patients with nonresponding non-Hodgkin's lymphoma.

Lactate dehydrogenase (LD). Total LD values are elevated in many patients with cancer. Usually these elevated levels are most highly correlated in the case of liver metastases. LD isoenzyme 5 is the predominant form in most cancer tissue, and as such the LD_5-to-LD_1 ratio is generally elevated in cancer tissue as compared to the ratio in benign or normal tissue fom the same organ.[74]

Changes in LD closely parallel changes in HCG and alpha-fetoprotein in patients with germ cell testicular tumors.[75] LD activity seems to reflect the tumor mass and may be useful in monitoring the course of the disease. Ini-

Table 46-5. Diagnostic sensitivities (percent positive) of several Pap assays for prostatic carcinoma*

Stage	Chemical†	Radioimmunoassay			
		†	‡	§	‖
A	12	33	13	0	8
B	15	79	26	20	21
C	29	71	30	33	40
D	60	92	94	79	86

*Sensitivities were calculated from data presented in the reference listed for each assay.
†Foti, A.G., Cooper, J.F., and Herschman, H., et al.: N. Engl. J. Med. **297:**1357-1361, 1977.
‡Mahan, D.E., and Doctor, B.P.: Clin. Biochem. **12:**10-17, 1979.
§Chu, T.M., Wang, M.C., Scott, W.W., et al.: Invest. Urol. **15:**319-323, 1978.
‖New England Nuclear, April 1979, Study Report, Boston.

Table 46-6. Liver metastasis and serum enzyme activity in 95 patients

Enzyme	Patients with metastasis		Patients without metastasis		Number of patients*
	Normal	Abnormal	Normal	Abnormal	
Alkaline phosphatase (ALP)	9	31	36	18	94
5′-Nucleotidase (5′-NT)	13	24	50	4	91
Gamma-glutamyl transferase (GGT)	1	35	30	25	91
Glutamate dehydrogenase (GD)	11	26	36	19	92

From Kim, N.K., Yasmineh, W.G., Freier, E.F., et al.: Clin. Chem. **23:**2034-2038, 1977.
*The total is less than 95 because some enzyme determinations were omitted.

tial LD values may be useful in prognosis and in monitoring patients, particularly those with seminoma where neither alpha-fetoprotein nor HCG is elevated.

Lysozyme (muramidase). Lysozyme is a useful enzyme in classifying patients with leukemia. It was noted that serum lysozyme was elevated in 45% of patients with acute granulocytic leukemia, in 77% of patients with acute myelomonocytic leukemia, and in 38% of patients with acute myeloblastic leukemia.[76] In only 1% of patients with acute lymphoblastic leukemia was there an elevation of lysozyme. Sequential determinations of lysozyme are also useful in following up patients with the disease.[77]

5′-Nucleotidase. Kim et al.[77] compared four enzymes in terms of their ability to identify patients with liver metastases. Of the 95 patients who were examined, 40 had liver metastases and 55 did not. The presence or absence of metastases was made on the basis of liver scan, liver echogram, liver biopsy, or some combination of these. Table 46-6 presents the four enzymes studied. They include ALP, 5′-nucleotidase, GGT, and glutamate dehydrogenase. The relationship of abnormality (elevation) and normality in the patients with metastases and in patients without metastases as well can be seen. These workers concluded that the best test for the detection of liver metastases is 5′-nucleotidase.

Pancreatic enzymes (amylase, lipase, ribonuclease, and trypsin). Serum amylase elevations have been found in 8% to 40% of patients with pancreatic cancer. Unfortunately lipase assays are not more useful than are amylase assays. Various workers have examined ribonuclease as a marker of pancreatic cancer. Reddi and Holland[78] observed that this enzyme was elevated in 90% of patients with pancreatic cancer and in only 10% of patients with pancreatitis. Fitzgerald repeated these studies and found that 50% of patients with pancreatic cancer but, unfortunately, 35% of patients with noncancerous pancreatic disease had elevated ribonuclease values.[79]

Immunoreactive trypsin has been used to study patients with pancreatic cancer. In one study all of 17 patients with pancreatic cancer had elevated renal clearance of immunoreactive trypsin, whereas 67% of patients with acute pancreatitis had elevated values. This should be compared with the fact that all patients with chronic pancreatitis had values within the control range.[80]

Phosphohexose isomerase (PHI). PHI had been suggested to be a sensitive indicator of liver metastases. Cuniette and Greco compared serum PHI to stage of disease. Regardless of the type of cancer, they found that patients with advanced cancer had higher PHI values. A significant reduction of the mean PHI values was found in patients who showed an objective response to therapy as compared to those who did not.[74]

Terminal deoxynucleotidyl transferase (TDT). TDT is a polymerase that is found in high concentrations in normal thymus and in T and non-T, non-B acute lymphoblastic leukemia cells. High TDT activity in peripheral blood lymphocytes and in bone marrow is observed in most patients at initial diagnosis of acute lymphoblastic leukemia.

Collagen-breakdown products

Collagen contains unique derivatives of amino acids including hydroxyproline (HYP) and hydroxylysine (HYL). Once released during collagen breakdown, these compounds are not incorporated into other proteins but are either further catabolized or excreted into the urine.[81] HYP excretion rates have been expressed as HYP/day, as HYP/m^2 body surface area, and as the urinary concentration of HYP over that of creatinine (HYP/Cr). The latter two correct for the effect of differences in body size. The ratio HYP/Cr measured in a morning urine after a 12-hour overnight fast is the easiest and probably also the most clinically relevant quantity to measure.[82,83]

A review of the literature shows that 80% of patients with x ray–verified breast carcinoma metastases in the bones and 26% of breast cancer patients without x ray–verified bone metastases had increased HYP excretion. The corresponding percentages were 70% and 11% in patients with carcinoma of the prostate, 68% and 46% in patients with carcinoma of the lung, and 69% and 0% in patients with multiple myeloma. It has been speculated that the high percentage of x ray–negative patients with increased HYP excretion seen among patients with pulmonary carcinoma could be attributed to an increased collagen turnover caused by ectopic production of parathyroid hormone by the tumor cell.[84] When bone metastases were detected by bone scan followed by x-ray studies to confirm the malignant nature of the processes, 69% of untreated patients with breast carcinoma metastases in the bones and 16% of breast cancer patients without bone metastases had increased HYP excretion. In patients with prostatic carcinoma the percentages were 84% and 15% respectively.

It has been shown that many breast cancer patients who have a normal bone survey but who excrete increased amounts of HYP will subsequently develop other evidence of bone metastases.[82,83] In general, for the detection of bone metastases, bone scan and HYP excretion measurements probably provide complementary rather than redundant information.[81] Guzzo et al.[82] found that 5 of 14 breast cancer patients who had normal radiographs but increased HYP/Cr developed bone lesions, whereas 7 patients with normal excretion rates remained free of metastases. Cuschieri et al.[83,85,86] and Basu et al.[87] confirmed these findings and showed that increased HYP excretion was associated with shorter survival. Other investigators have been unable to demonstrate that HYP excretion is predictive for the development of bone metastases.[88-90] In the study by Guzzo et al.[82] it was found that in 5 of 10 patients with primary carcinoma of the lung who had increased HYP excretion without bone lesions, bone metastases appeared within 12 months. There are some indications that in patients with lymphomas increased HYP excretion rate is an indication of presence of active disease rather than indication of bone lesions.[91]

Bone healing may result in short-term increases in HYP excretion.[82,92,93] In the long-term follow-up of breast cancer patients with bone metastases, excretion becomes normal if remission is maintained.[94] The treatment response in patients whose pretreatment HYP excretion is increased can be monitored by serial measurements of HYP. Results obtained from patients with carcinoma of the prostate suggest that HYP excretion is of value in monitoring response to treatment in patients with bone metastases.[96,97] Often the HYP excretion indicates changes before these become evident by clinical or radiological criteria. There are some indications that HYP excretion reflects response to treatment in patients with multiple myeloma.[98]

Polyamines

The polyamines spermidine, spermine, and their diamine precursor putrescine are highly basic compounds that are involved in a variety of cellular processes. During the past decade the polyamines have been studied intensively as potential tumor markers.[99] In 1971 Russell et al.[100] first demonstrated that polyamines are elevated in the urine of patients with cancer. The best correlation between polyamine concentration in the urine and in tumors is found in patients with Burkitt's lymphoma, colorectal carcinoma, lung carcinoma, medulloblastoma, acute leukemia, and multiple myeloma. Poor correlations have been found in patients with breast cancer, gynecological cancer, or testicular cancer.

Nucleosides

Increased urinary excretion of nucleosides represents increased turnover of transfer RNA. Among the various nucleosides examined are 2-*N*-dimethylguanosine, 1-methylinosine, and pseudouridine. Tormey and co-workers[90] found that the excretion of 2-*N*-dimethylguanosine was increased in as many as 58% of patients with metastatic breast cancer and in 44% of patients with local lymph node involvement with the disease. Cimino[101] and co-workers have studied pseudouridine and have found it to be an excellent indicator of neoplastic disease. It is especially useful in following up patients with cancer. Levine et al.[102] have found increased concentrations of 2-*N*-dimethylguanosine and 1-methylinosine in plasma of patients with acute leukemia.

Cellular markers

A number of markers associated with the plasma membrane, cytoplasm, or nuclei of the lymphoid cell have been identified. Various techniques have been used. These techniques, which tend to be immunological in nature, include rosetting, immunofluorescence, and immunoenzymatic testing. The rosetting technique is based on a reaction between an indicator cell (usually an erythrocyte) and the lymphoid cell to form rosettes in cases in which the lymphoid cell carries a particular membrane marker. By such techniques the cells may be mixed directly, or the indicator cell may first be coded with antibody or complement to demonstrate receptors for the Fc part of immunoglobulin or complement components.

It appears that the various antigens demonstrated by these techniques are not tumor-specific antigens but rather are tumor-associated differentiation antigens that represent the expression of oncofetal antigens not normally expressed by differentiated cells.

Lymphocytic leukemias and non-Hodgkin lymphomas have been subdivided into clinically usable subgroups on the basis of biochemical cell markers. The most striking evidence of the value of typing lymphocytes with a panel of markers comes from studies of acute lymphocytic leukemia (ALL). Table 46-7 represents the five prognostically distinct groups of ALL and as well the relevant markers for each. The groups are ordered according to prognosis. The cells of the B-cell type are characterized by the presence of surface membrane immunoglobulin (SmIg), as are normal mature B cells. The cells of the T-cell type are characterized by the presence of sheep erythrocyte receptors and human thymocyte antigen, as are mature T cells. The cells of the pre–B cell type are characterized by a cytoplasmic IgM heavy chain but no SmIg, which corresponds to the characteristics of an early stage during the B-cell differentiation.

Table 46-7. Phenotypic heterogeneity of acute lymphocytic leukemia*

	E	HTA	SmIg	Cyμ	CALLA	HLA-DR	TdT
Common ALL	−	−	−	−	+	+	+
Pre–B ALL	−	−	−	+	+	+	+
Null-ALL	−	−	−	−	−	+	+
T-ALL	+	+	−	−	−	−	+
B-ALL	−	−	+	−	−	+	−

From Plesner, T., Wilken, M., and Avenstrøm, S.: In Statland, B.E., and Winkel, P., editors: Laboratory measurements in malignant disease, Philadelphia, 1982, W.B. Saunders Co.

*Abbreviations: E = sheep erythrocyte receptor; HTA = human thymocyte antigen(s); Sm = surface membrane; Cy = cytoplasmic; μ = IgM heavy chain; CALLA = common ALL antigen; HLA-DR = human Ia-like antigen; TdT = terminal deoxynucleotidyltransferase; B-ALL = B-cell type of acute lymphoblastic leukemia; T-ALL = T-cell type of acute lymphoblastic leukemia.

The terminal deoxynucleotidyl transferase and the common ALL antigen are of value not only for the classification of ALL (see Table 46-7) but also because they are very useful in distinguishing between acute lymphoblastic and myeloblastic leukemia.

ALL may be classified into B-cell leukemia (95%), which is characterized by low-density monoclonal surface membrane immunoglobulins, usually IgM or IgM and IgD with one light chain, and the more rare, but also more aggressive, T-ALL (5%). The cells of the last type form E rosettes and have T-antigens but lack SmIg.

Until recently lymphomas have been classified in a variety of ways, mainly based on histological patterns, that is, cytological features and cell size. Lukes and Collins[105] proposed an immunologically oriented histopathological classification that characterizes lymphomas into T-cell, B-cell, histiocytic, or undefined. The median survival of patients with B-cell lymphomas is longer than that of persons with T-cell and null-cell lymphomas. The latter category, however, seems to include a subgroup of patients who are curable with currently available therapy.

Refer to reviews by Plesner et al.[103] and Parker[106] to gain more information concerning the use of cell markers with leukemia and lymphoma.

Hemostasis-related factors

Ito and Statland recently reviewed selected hemostatic abnormalities in patients with neoplastic disease.[106] Plasma fibrinogen levels are generally elevated in patients with cancer. However, in patients with disseminated intravascular coagulation (DIC) hypofibrinogenemia has also been noted. As expected, DIC is also associated with decreased values for antithrombin III (AT III). Consequently, in patients with DIC associated with cancer, AT III values will be decreased.

Increased fibrinolytic or fibrinogenolytic activity has been reported in patients with cancer. Consequently the fibrinogen-degradation products (FDPs) are often elevated in the plasma or urine in patients with cancer.

It is interesting that plasminogen activators are often elevated in patients with cancer. Unfortunately, techniques presently available to measure the plasminogen activators are still under development and are not readily available for routine use.

SUMMARY

The scope of the problem is too large to be covered in one chapter in a textbook of clinical chemistry. More important than merely enumerating all the assays available is the fact that there is a challenge that must be presented for any candidate tumor marker. The challenge is that it be useful clinically in the management of patients with the disease or suspected of having the disease. Unfortunately the present scene is a very pessimistic one. It is hoped that with further research and clear thinking additional strategies and new markers will be made available.

REFERENCES

1. Sager, R.: Explorations on the origin of cancer, Focus, Harvard U. **2/3:**1-3, 1983.

1a. Habbema, J.D.F., van Oortmarssen, G.J., and van der Maas, P.J.: Mass screening for cancer: the interpretation of findings and the prediction of effects on morbidity and mortality. In Statland, B.E., and Winkel, P., editors: Laboratory measurements in malignant disease, vol. 2, Philadelphia, 1982, W.B. Saunders Co., pp. 627-638.

2. Habbema, J.D.F., and van Oortmarssen, G.J.: Performance characteristics of screening tests. In Statland, B.E., and Winkel, P., editors: Laboratory measurements in malignant disease, vol. 2, Philadelphia 1982, W.B. Saunders Co., pp. 639-656.

3. EVAC: Rapport Eerste Screeningsronde, Leidschendam, 1980, Ministerie van Volksgezondheid en milieuhygiene.

4. Tabar, L., and Gad, A.: Screening for breast cancer: the Swedish trial, Radiology **138:**219-222, 1981.

5. Woolner, L.B., Fontant, R.S., Sanderson, D.R., et al.: Mayo Lung Project: evaluation of lung cancer screening through December 1979, Mayo Clin. Proc. **56:**544-555, 1981.

6. Shapiro, S.: Evidence on screening for breast cancer from a randomized trial, Cancer **39:**2772-2782, 1977.

7. Statland, B.E.: Clinical decision levels for lab tests, Oradell, N.J., 1983, Medical Economics Co.

8. Rubin, P.: Clinical oncology for medical students and physicians, ed. 5, New York, 1978, American Cancer Society.

9. Winkel, P., Bentzon, M.W., Statland, B.E., et al.: Predicting recurrence in patients with breast cancer from cumulative laboratory results: a new technique for the application of time series analysis, Clin. Chem. **28:**2057-2067, 1982.

10. Ravry, M., Moertel, C.G., Schutt, A.J., et al.: Usefulness of serial serum carcinoembryonic antigen (CEA) determinations during anticancer therapy or long-term follow-up of gastrointestinal carcinoma, Cancer **34:**1230-1234, 1974.

11. Coombes, R.C., and Neville, A.M.: Significance of tumor-index-substances in management. In Stoll, B.A., editor: Secondary spread in breast cancer, Chicago, 1978, William Heinemann Medical Books.
12. Statland, B.E.: The challenge of cancer testing, Diagnostic Med. **4:**2-8, 1981.
13. Statland, B.E., and Winkel, P., editors: Clinics in laboratory medicine: laboratory measurements in malignant disease, vol. 2, Philadelphia, 1982, W.B. Saunders Co.
14. Gold, P., and Freedman, S.O.: Demonstration of tumor-specific antigens in human colonic carcinomatta by immunological tolerance and absorption techniques, J. Exp. Med. **121:**439-462, 1965.
15. Gold, P., Shuster, J., and Freedman, S.O.: Carcinoembryonic antigen (CEA) in clinical medicine, Cancer **42:**1399-1402, 1978.
16. Hammarström, S., Svenberg, T., Hedin, A., et al.: Antigens related to carcinoembryonic antigen, Scand. J. Immunol. **7:**35-40, 1968.
17. Shively, J.E., and Todd, C.W.: Carcinoembryonic antigen, Scand. J. Immunol. **7:**19-25, 1978.
18. CEA as a cancer marker, National Institutes of Health, Consensus Development Conference Summary, Bethesda, Md., 1981, vol. 3, no. 7.
19. Mackey, I.R.: Carcinoembryonic antigen use of screening. In Heberman, R.B., and McIntire, K.R., editors: Immunodiagnosis of cancer, New York, 1979, Marcel Dekker, Inc., pp. 255-260.
20. Doos, W.G., Wolff, W.I., Shinya, H., et al.: CEA levels in patients with colorectal polyps, Cancer **33:**583-590, 1974.
21. Holyoke, E.D., Chu, T.M., and Murphy, G.P.: CEA as a monitor of gastrointestinal malignancy, Cancer **35:**830-836, 1975.
22. Reynoso, G.: Tumor associated phase specific antigens in cancer diagnosis, Proceedings of the First Invitational Symposium on the Serodiagnosis of Cancer, AFRRI SP 74-1, Bethesda, Md., Armed Forces Radiobiological Research Institute, Defense Nuclear Agency, March 1974.
23. Costanza, M.E., Das, S., Nathanson, et al.: Carcinoembryonic antigen: report of a screening study, Cancer **33:**583-590, 1974.
24. McCartney, W.H., and Hoffer, P.B.: The value of carcinoembryonic antigen (CEA) as an adjunct to the radiological colon examination in the diagnosis of malignancy, Radiology **110:**325-328, 1974.
25. McCartney, W.H., and Hoffer, P.B.: Carcinoembryonic antigen assay in hepatic metastases detection, J.A.M.A. **236:**1023-1027, 1976.
26. Meeker, W.R., Kashmiri, R., Hunter, L., et al.: Clinical evaluation of carcinoembryonic antigen test, Arch. Surg. **107:**266-274, 1973.
27. Khoo, S.K., and Mackay, E.V.: Carcinoembryonic antigen by radioimmunoassay in the detection of recurrence during long-term follow-up of female genital cancer, Cancer **34:**542-548, 1974.
28. Livingston, A.S., Hampoon, L.G., Shuster, J., et al.: Carcinoembryonic antigen in the diagnosis and management of colorectal carcinoma, Arch. Surg. **109:**259-264, 1974.
29. LoGerfo, P., and Herter, F.P.: Carcinoembryonic antigen and prognosis in patients with colon cancer, Ann. Surg. **181:**81-84, 1975.
30. Mackey, A.M., Patek, S., Carter, S., et al.: Role of serial plasma CEA levels in detection of recurrent and metastatic colorectal carcinomas, Br. Med. J. **4:**382-385, 1974.
31. Wanebo, H.J., et al.: Preoperative carcinoembryonic antigen level as a prognostic indicator in colorectal cancer, N. Engl. J. Med. **200:**448-452, 1978.
32. Myers, R.E., Sutherland, D.J., Meakin, J.W., et al.: Carcinoembryonic antigen in breast cancer, Cancer **42:**1520-1525, 1978.
33. Zamcheck, N., Doos, W.B., Prodente, R., et al.: Prognostic factors in colon carcinoma: correlation of serum carcinoembryonic antigen level and tumor histopathology, Hum. Pathol. **6:**31-45, 1975.
34. Chu, T.M., Holyoke, E.D., Cedermark, R., et al.: A long-term follow-up of CEA in resective colorectal cancer. In Fishman, W.H., and Sell, S., editors: Oncodevelopmental gene expression, New York 1976, Academic Press, Inc., pp. 427-431.
35. Skarin, A.T., Delwiche, R., Zamcheck, N., et al.: Carcinoembryonic antigen: clinical correlation with chemotherapy for metastatic gastrointestinal cancer, Cancer **33:**1239-1245, 1974.
36. Balz, J.B., Martin, E.W., and Minton, J.P.: CEA as an early indicator for second-look procedure in colorectal carcinoma, Rev. Surg. **34:**1-5, 1977.
37. Statland, B.E., and Winkel, P.: Usefulness of clinical chemistry measurements in classifying patients with breast cancer, CRC Crit. Rev. Clin. Lab. Sci. **26:**255-290, 1982.
38. Haagensen, D.E.: Tumor markers for breast carcinoma. In Statland, B.E., and Winkel, P., editors: Laboratory measurements in malignant disease, Philadelphia, 1982, W.B. Saunders Co., pp. 543-565.
39. Bergstrand, C.G., and Czar, B.: Demonstration of a new protein fraction in serum from the human fetus, Scand, J. Clin. Lab. Invest. **8:**174, 1956.
40. Sell, S., and Becker, F.F.: Alpha-fetoprotein, J. Natl. Cancer Inst. **60:**19-26, 1978.
41. Norgaard-Pedersen, B., and Hangaard, J.: Germ cell tumors and biochemical markers in clinical and experimental research. In Statland, B.E., and Winkel, P., editors: Laboratory measurements in malignant disease, Philadelphia, 1982, W.B. Saunders Co., pp. 431-458.
42. Vaitukaitis, J.L.: Secretion of human chorionic gonadotrophin by tumors. In Carcino-embryonic proteins, vol. I., New York, 1979, Elsevier/North-Holland, Inc., pp. 447-455.
43. Vaitukaitis, J.L., Braunstein, G.D., and Ross, G.T.: A radioimmunoassay, which specifically measures human chorionic gonadotropin in the presence of human luteinizing hormone, Am. J. Obstet. Gynecol. **112:**751-758, 1972.
44. Braunstein, G.D.: Human chorionic gonadotropin in nontrophoblastic tumors and tissues. In Talwar, G.P., editor: Recent advances in reproduction and regulation of fertility, Amsterdam, 1979, Elsevier/North Holland Biomedical Press.
45. Anderson, C.K., Jones, W.G., and Ward, A.: Germ cell tumors, London, 1981, Taylor and Francis, Ltd.
46. Björklund, B.: Tissue polypeptide antigen (TPA): biology, biochemistry, improved assay methodology, clinical significance in cancer and other conditions, and future outlook, Antibiot. Chemother. **22:**16, 1978.
47. Nemoto, T., Constantine, R., and Chu, T.M.: Human tissue polypeptide antigen in breast cancer, J. Natl. Cancer Inst. **63:**1347-1350, 1979.
48. Lovins, R.: Tennessee antigen: a colorectal marker for patient management. In Griffiths, J.C., and Kessler, A., editors: Cancer and enzymes, Berkeley, 1980, University of California Press.
49. Hakkinen, I.: Fetal sulfoglycoprotein in gastric cancer. In Heberman, R.B., editor: Compendium of assays for immunodiagnosis of human cancer, New York, 1979 Elsevier/North-Holland, Inc., pp. 241-246.
50. Banwo, O., Versey, T., and Hobbs, J.R.: New oncofetal antigen from human pancreas, Lancet **1:**643-645, 1974.
51. Gelder, F.B., Reese, C.J., Moossa, A.R., et al.: Purification, partial characterization, and clinical evaluation of a pancreatic oncofetal antigen, Cancer Res. **38:**313-324, 1978.
52. Goldman, L.: Pancreatic oncofetal antigen: the diagnostic applications to pancreatic cancer. In Griffiths, J.C., and Kessler, A., editors: Cancer and enzymes, Berkeley, 1980, University of California Press.
53. Mazoujian, G., Pinkus, G., Davis, S., et al.: Immunoperoxidase localization of a breast gross cystic disease fluid protein, Lab. Invest. **46:**52A-53A, 1982.
54. Haagensen, D.E., Jr., Mazoujian, G., Dilley, W.G., et al.: Breast gross cystic disease fluid analysis. I. Isolation and radioimmunoassay for a major component protein, J. Natl. Cancer Inst. **62:**239-244, 1979.
55. Haagensen, C.D., Bodian, C., and Haagensen, D.E., Jr.: Breast carcinoma: risk and detection, Philadelphia, 1981, W.B. Saunders Co.
56. Silverman, L.M., Dermer, G., and Tokes, Z.: Electrophoretic patterns for serum glycoproteins reflect the presence of human breast cancer, Clin. Chem. **23:**2055-2058, 1977.
57. Harvey, H.A., Lipton, A., et al.: Glycoproteins and human cancer: correlation between circulating level and disease status, Cancer **47:**324-327, 1981.
58. Marcus, D., and Zinberg, N.: Measurement of serum ferritin by radioimmunoassay: results in normal individuals and patients with breast cancer, J. Natl. Cancer Inst. **55:**791-795, 1975.
59. Hendrick, J.C., and Franchimont, P.: Radioimmunoassay of casein in the serum of normal subjects and of patients with various malignancies, Eur. J. Cancer **10:**725-730, 1974.

60. Schwartz, M.K.: Enzyme tests in cancer. In Statland, B.E., and Winkel, P., editors: Laboratory measurements in malignant disease, Philadelphia, 1982, W.B. Saunders Co. pp. 479-491.
61. Foti, A.G., Cooper, J.F., Herschman, H., et al.: Detection of prostatic cancer by solid phase radioimmunoassay of serum prostatic acid phosphatase, N. Engl. J. Med. **297**:1357-1361, 1977.
62. Mahan, D.E., and Doctor, B.P.: A radioimmunoassay for human prostatic acid phosphatase levels in prostatic disease, Clin. Biochem. **12**:10-17, 1979.
63. Chu, T.M., Wang, M.C., Scott, W.W., et al.: Immunological detection of serum prostatic acid phosphatase: methodology and clinical evaluation, Invest. Urol. **15**:319-323, 1978.
64. New England Nuclear, April 1979, Study Report, Boston.
65. Watson, R.A., and Tang, D.B.: The predictive value of prostatic acid phosphatase as a screening test for prostatic cancer, N. Engl. J. Med. **303**:497-499, 1980.
66. Tavassoli, M., Rizo, M., and Yam, L.T.: Elevation of serum acid phosphatase in cancers with bone metastases, Cancer **45**:2400-2403, 1980.
67. Fishman, W.H., Inglis, N.R., Green, S., et al.: Immunology and biochemistry of Regan isoenzyme of alkaline phosphatase in human cancer, Nature **219**:697-699, 1968.
68. Kahan, L., Go, V.W., and Larsen, F.C.: Increased activity in serum of an alkaline phosphatase isoenzyme in cancer: analytical method and preliminary clinical studies, Clin. Chem. **27**:104-107, 1981.
69. Griffiths, J.C.: Creatine kinase isoenzyme 1. In Statland, B.E., and Winkel, P., editors: Laboratory measurements in malignant disease, Philadelphia, 1982, W.B. Saunders Co., pp. 493-506.
70. Thompson, R.J., Kynoch, P.A.M., and Sarjant, J.: Immunohistochemical localization of creatine kinase BB isoenzyme to astrocytes in human brain, Brain Res. **201**:423-426, 1980.
71. Griffiths, J.C.: Personal communication.
72. Podolsky, D., McPhee, M.S., Alpert, E., et al.: Galactosyltransferase isoenzyme II in the detection of pancreatic cancer: comparison with radiologic, endoscopic, and serologic tests, N. Engl. J. Med. **304**:1313-1318, 1981.
73. Bauer, C.J., Reutter, W.G., Erhart, K.P., et al.: Decrease of human serum fucosyltransferase as an indicator of successful tumor therapy, Science **201**:1232-1234, 1978.
74. Schwartz, M.K.: Measurement of CEA, Clin. Chem. **19**:1214-1220, 1973.
75. Bosl, G.J., Lange, P.H., Nochomovitz, L.E., et al.: Tumor markers in advanced nonseminomatous testicular cancer, Cancer **47**:572-576, 1981.
76. Perille, P.E., and Finch, S.L.: In Osserman, E.F., Canfield, R.E., and Beychock, S., editors: Lysozyme, New York, 1974, Academic Press, Inc., pp. 359-365.
77. Kim, N.K., Yasmineh, W.G., Freier, E.F., et al.: Value of alkaline phosphatase, 5′-nucleotidase, γ-glutamyltransferase, and glutamate dehydrogenase activity measurements (single and combined), in serum in diagnosis of metastasis to the liver, Clin. Chem. **23**:2034-2038, 1977.
78. Reddi, K., and Holland, J.F.: Elevated serum ribonuclease in patients with pancreatic cancer, Proc. Natl. Acad. Sci. USA **73**:2308, 1976.
79. Fitzgerald, P.J., Fortner, J.G., Watson, R.C., et al.: The value of diagnostic aids in detecting pancreas cancer, Cancer **41**:868-879, 1978.
80. Lake-Bakaar, G., McKavanaugh, S., and Summerfield, J.A.: Urinary immunoreactive trypsin excretion: a non-invasive screening test for pancreatic cancer, Lancet **2**:878-880, 1979.
81. Kelleher, P.C., and Smith, C.J.P.: Collagen metabolities in the urine and serum of patients with cancer. In Statland, B.E., and Winkel, P., editors: Laboratory measurements in malignant disease, Philadelphia, 1982, W.B. Saunders Co., pp. 519-542.
82. Guzzo, C.E., Pachas, W.N., Pinals, R.S., et al.: Urinary hydroxyproline excretion in patients with cancer, Cancer **24**:382-387, 1969.
83. Cuschieri, A.: Urinary hydroxyproline excretion and survival in cancer of the breast, Clin. Oncol **1**:127-130, 1975.
84. Klein, L., Lafferty, F.W., Pearson, O.H., et al.: Correlation of urinary hydroxyproline, serum alkaline phosphatase, and skeletal calcium turnover, Metabolism **13**:272-284, 1964.
85. Cuschieri, A.L.: Urinary hydroxyproline excretion in early and advanced breast cancer—a sequential study, Br. J. Surg. **60**:800-803, 1973.
86. Cuschieri, A., and Felgate, R.A.: Urinary hydroxyproline excretion in carcinoma of the breast, Br. J. Exp. Pathol. **53**:237-241, 1972.
87. Basu, T.K., Donaldson, D., and Williams, D.C.: Urinary hydroxyproline and plasma mucoprotein as an indication of the presence of bone metastases, Oncology **30**:197-200, 1974.
88. Gielen, F., Dequeker, J., Drochmans, A., et al.: Relevance of hydroxyproline excretion to bone metastasis in breast cancer, Br. J. Cancer **34**:279-285, 1976.
89. Black, R.B., Roberts, M.M., Stewart, H.J., et al.: The search for occult metastases in breast cancer: does it add to established staging methods? Aust. N.Z. J. Surg. **50**:574-578, 1980.
90. Coombes, R.C., Powles, T.J., Gazet, J.C., et al.: Assessment of biochemical tests to screen for metastases in patients with breast cancer, Lancet **1**:296-297, 1980.
91. Nehlawi, M.J., Shaw, D., Mitchell, P.E.G., et al.: Urinary hydroxyproline excretion in patients with Hodgkin's disease and non-Hodgkin's lymphoma, Clin. Oncol. **5**:109-112, 1979.
92. Hosley, H.F., Taft, E.G., Olson, K.B., et al.: Hydroxyproline excretion in malignant neoplastic disease, Arch. Intern. Med. **118**:565-571, 1966.
93. Roberts, J.G., Williams, M., Henk, J.M., et al.: The hypronosticon test in breast cancer, Clin. Oncol. **1**:33-35, 1975.
94. Powles, T.J., Leese, C.L., and Bondy, P.K.: Hydroxyproline excretion in patients with breast cancer and response to treatment, Br. Med. J. **2**:164-166, 1975.
95. Bishop, M.C., and Fellows, G.J.: Urine hydroxyproline excretion—a marker of bone metastases in prostatic carcinoma, Br. J. Urol. **49**:711-716, 1977.
96. Mundy, A.R.: Urinary hydroxyproline excretion in carcinoma of the prostrate: a comparison of four different modes of assessment and its role as a marker, Br. J. Urol. **51**:570-574, 1979.
97. Katz, F.H., and Kappas, A.: Influence of estradiol and estriol on urinary excretion of hydroxyproline in man, J. Lab. Clin. Med. **71**:65-74, 1968.
98. Niell, H.B., Neely, C.L., and Palmieri, G.M.: The postabsorptive urinary hydroxyproline (Spot-HYPRO) in patients with multiple myeloma, Cancer **48**:783-787, 1981.
99. Oredsson, S.M., and Marton, L.J.: Polyamines—the elusive cancer markers. In Statland, B.E., and Winkel, P., editors: Laboratory measurements in malignant disease, Philadelphia, 1982, W.B. Saunders Co., pp. 507-518.
100. Russell, D.H., Durie, B.G.M., and Salmon, S.E.: Polyamines as predictors of success and failure in cancer chemotherapy, Lancet **2**:797-799, 1975.
101. Cimino, F.: Personal communication.
102. Levine, L., Waalkes, T., and Stolbach, L.: Serum levels of N^2,N^2-dimethylguanosine and pseudouridine as determined by radioimmunoassay for patients with malignancy, J. Natl. Cancer Inst. **54**:341-343, 1975.
103. Plesner, T., Wilken, M., and Avnstrøm, S.: The contribution of immunologic methods to the classification of leukemias and malignant lymphomas. In Statland, B.E., and Winkel, P., editors: Laboratory measurements in malignant disease, Philadelphia, 1982, W.B. Saunders Co., pp. 579-597.
104. Fu, S.M., Chiorazzi, N., Kunkel, H.C., et al.: Induction of in vitro differentiation and immunoglobulin synthesis of human leukemic B lymphocytes, J. Exp. Med. **148**:1570-1578, 1978.
105. Lukes, R.J., and Collins, R.D.: Functional approach to the classification of malignant lymphoma, Recent Results Cancer Res. **46**:18-30, 1974.
106. Parker, J.W.: A new look at malignant lymphomas, Diagnost. Med. **4**(4):77-85, 1981.

Chapter 47 Toxicology

Alphonse Poklis
Amadeo J. Pesce

acute toxicity Usually refers to the harmful effect of a toxic agent that manifests itself in seconds, minutes, hours, or days after entering the patient.

analgesics A class of drugs that reduce pain.

antidote Any agent that counteracts the effects of a poison.

benzodiazepines A group of antianxiety sedative drugs.

biotransformation The chemical modifications of chemicals as they pass through the body.

chronic toxicity Usually refers to the long-term harmful effects of an agent weeks, months, and years after entering the patient.

drug interaction The observation that when two drugs are given, their actions or effects are not independent.

drug screen A test that qualitatively identifies the presence of any one of a class of drugs.

forensic toxicology A branch of the discipline of toxicology concerned with the medical and legal aspects of the harmful effects of chemicals or poisons.

hypnotics A class of drugs often used as sedatives.

LD_{50} The amount of a substance that will cause death in half the test animal population.

lethal dose The amount of a substance that, after being ingested, causes death.

mixed function oxidases A group of enzymes present in the microsomes of the liver that adds oxygen to a drug.

phenothiazines. Drugs used to treat psychoses.

therapeutic dose The amount of a substance that will produce a desired pharmacological effect.

toxicant or poison A substance or toxicant that, when taken in sufficient quantity, will cause sickness or death.

toxicokinetics The passage of a toxicant with time through an affected person. Similar to pharmacokinetics.

toxicology The study of poisons.

tricyclic antidepressants Mood-elevating drugs.

DEFINITIONS

Toxicology is the study of poisons. More specifically, toxicology is concerned with the chemical and physical properties of poisons, their physiological or behavioral effects on living organisms, qualitative and quantitative methods for their analysis in biological and nonbiological materials, and the development of procedures for the treatment of poisoning. A poison (or toxicant) is regarded as any substance that when taken in sufficient quantity will cause sickness or death. The key phrase in this definition is ''sufficient quantity.'' As the sixteenth century physician Paracelsus observed, ''All substances are poisons; there is none which is not a poison. The right dose differentiates a poison from a remedy.'' For example, minute quantities of cyanide, arsenic, lead, and dichlorodiphenyltrichlorethane (DDT) are regularly ingested from food sources or inhaled as environmental contaminates and retained by the human body. However, the amounts of toxicants are insufficient to cause obvious deleterious effects. On the other hand, a substance as apparently innocuous as pure water will, if ingested in sufficient quantity, cause incapacitating electrolyte imbalance or even death.

There is often no difference between the mechanism of action of a drug and a poison. A drug is administered in doses that alter physiological function in order to produce a desired therapeutic effect. If administered in greater than therapeutic quantities, a drug may produce toxic (harmful) effects. Thus toxicology is a quantitative discipline that seeks to identify the amount of a substance that, in particular exposure situations, will cause deleterious effects in a particular animal or patient.

Toxicology is a multidisciplinary field that freely borrows from the basic sciences. Because of a diversity of concerns and applications, modern toxicology has developed into three specialized branches: environmental, clinical, and forensic. Environmental toxicology is concerned primarily with the harmful effects of chemicals that are encountered incidentally because they are in the atmosphere, in the food chain, or present in occupational or recreational environments. Clinical toxicology is concerned with the harmful effects of chemicals that are intentionally administered to living organisms for the purpose of achieving a specific effect. The desired effect achieved may be beneficial to the organ (therapeutic), in which case the toxicologist is interested in the adverse or side effects of the agent in question. Forensic toxicology is the branch of toxicology that is concerned with the medical and legal aspects of the harmful effects of chemicals or poisons.

DOSAGE RELATIONSHIPS

Toxicity is a relative term used to compare one substance with another. To state that one substance is more toxic than another is to make a quantitative comparison of dose. Such a comparison is only valid in the source organism under identical exposure situations. A quantitative toxicity rating of the lethal dose of a substance for a ''normal man'' (body weight 150 pounds or 70 kg) is presented in Table 47-1. In the common vernacular, ''toxic substance'' refers to substances with a toxicity defined in Table 47-1 as ''extremely'' or ''super'' toxic (arsenic, cyanide, strychnine).

Important quantitative parameters in toxicology include the amount of agent introduced into a person (that is, a dose), the route of administration, the number of doses, and the time period over which the agent is administered. Time is an essential quantity, since the effect of many agents is time dependent. For convenience, terms such as *acute toxicity* and *chronic toxicity* are used. To some extent, these terms overlap. In general, the toxic reactions observed in the emergency room are caused by the acute ingestion, inhalation, or injection of some agent, whereas those caused by environmentally toxic agents are more often observed in the chronic-care setting. Thus toxic agents can have an acute effect when the action of the chemical is observed from seconds to days, and chronic when the time scheme is in weeks, years, or decades.

The dose-response relationship occurs when an in-

Table 47-1. Toxicity rating chart

Toxicity rating or class	Probable lethal dose (human)	
	mg/kg	for 70 kg man (150 pounds)
6 Super toxic	Less than 5	A taste (less than 7 drops)
5 Extremely toxic	5-50	Between 7 drops and 1 teaspoonful
4 Very toxic	50-500	Between 1 teaspoonful and 1 ounce
3 Moderately toxic	500-5 g/kg	Between 1 ounce and 1 pint (or 1 pound)
2 Slightly toxic	5-15 g/kg	Between 1 pint and 1 quart
1 Practically nontoxic	Above 15 g/kg	More than 1 quart

creased toxic effect is observed as the dose is increased. However, the increased toxic effect may not be observed in all persons at the same dose. Only a portion of the population may be affected at a given dose, and as increasing amounts of the agent are given, the dose becomes great enough so the entire population is affected. This dose-response curve is often semilogarithmic in form, and many toxic agents affect a population in this manner. Thus 10% of a population may be affected at one dose, and only at 100 times this level may 90% of the population be affected. This sounds extreme, but the important fact to remember is that the effect of a toxic agent will usually follow that of a semilogarithmic distribution. The reasons behind such variability in response are many, including age, sex, race, health status, and genetic factors. All contribute to the distribution of the effect of an agent on a population.

One important quantitative parameter is the dose necessary to cause the death of an animal. If increasing doses are given, an increasing percentage of animals will die. A semilogarithmic plot of such a curve is presented in Fig. 47-1. The lethal dosage for 50% of the population, termed the LD_{50}, is an important value and used for many drugs or agents. For humans, such values are extrapolated from survivors or deaths reported in the literature. The agent does not have to affect the entire population. If only 1% of the animals died at a given dosage, the term would become LD_1, the dosage lethal for 1% of the animals. If 99% died, the term would then be LD_{99}.

MECHANISMS OF TOXICITY

There are many ways in which toxic agents can cause their effects. They can be subtle, such as "promoter" molecules, which affect cells only if a carcinogen has been given, or dramatic, such as that of a caustic agent, which destroys tissue on contact. The examples given in the subsequent text illustrate only a few of the ways that toxic agents act.

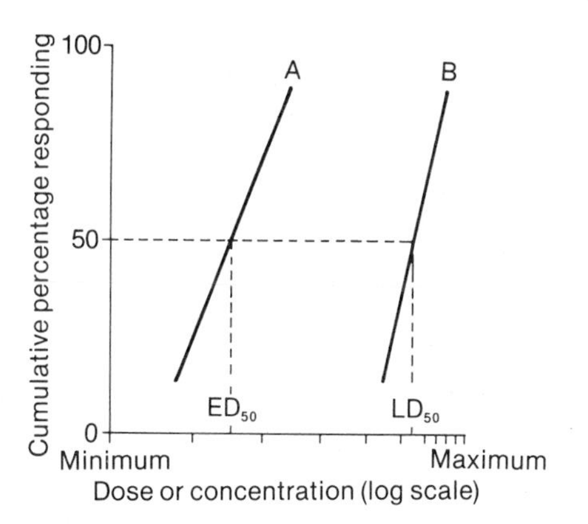

Fig. 47-1. Hypothetical dose-response curves for a drug administered to a uniform population of animals. *Curve A,* Therapeutic effect (such as anesthesia); *curve B,* lethal effect. (From Loomis, T.A.: Essentials of toxicology, ed. 3, Philadelphia, 1978, Lea & Febiger Publishers.)

Interference with action of enzyme systems

Some agents are effective because they interfere with the normal catalytic activity of enzymes.

Irreversible enzyme inhibition. This occurs when the agent attaches itself irreversibly to the "active" site of an enzyme. Enzyme inhibitors can be specific or nonspecific. Specific enzyme inhibitors act only on selected enzymes. Usually the agent will have a structure very similar to that of the normal substrate. One example is that of the organophosphates and their action on the enzyme acetylcholinesterase. Acetylcholinesterase catalyzes the reaction

$$\text{Acetylcholine} \rightarrow \text{Acetate} + \text{Choline}$$

The organophosphates are a group of insecticides that have a structure of the form

R_1O O
P
R_2O OR_3

which, when acted on by the enzyme acetylcholinesterase, forms a complex of the type

R_1O O
P
R_2O O-Serine

This covalent linkage is irreversible and the enzyme becomes inactivated. If given to a human, the person can no longer hydrolyze acetylcholine at the nerve endings and death may ensue because of the loss of control at the neuromuscular junctions.

In contrast, heavy metals, such as arsenic, lead, and mercury, have a high affinity for the SH groups of proteins. Such groups are often an integral part of the active site of enzymes. Toxic agents react with many different enzymes, inhibiting them to varying degrees. When enzymes that catalyze a critical reactions are inhibited, toxic effects result.

Reversible enzyme inhibition. These chemicals are usually specific for one enzyme or a group of enzymes having the same cofactors. These agents closely resemble the normal substrate or cofactor and bind reversibly to the enzyme's active site. Such a compound is methotrexate, which is used in cancer therapy. It functions by competing for the active site of enzymes, such as thymidylate synthetase, thus blocking the normal synthesis of DNA.

Uncoupling of biochemical reactions. Agents, such as dinitrophenol, interact with the oxidative phosphorylation system of the mitochondria in such a manner that, although oxidation occurs, phosphorylation of ATP does not take place. This uncoupling of the reaction can result in the death of a cell if adequate amounts of ATP are not synthesized.

Lethal synthesis. The toxicant usually resembles a normal substrate, and an enzyme or series of enzymes will convert it to a subsequent metabolite resembling another substrate in the metabolic sequence. Somewhere in the sequence a metabolite of the toxicant either inhibits or inactivates an enzyme. One example of this is fluoroacetic acid. This molecule is readily incorporated into citric acid, forming fluorocitric acid. Fluorocitric acid cannot be isomerized by the enzyme, aconitase, but is recognized as an inhibitor. Thus the enzyme is inhibited by this agent synthesized from fluoroacetic acid.

Removal of metals essential for enzymatic activity. Many enzymes have metals as part of their catalytic site. Even though they have a high affinity for these metals, other compounds (such as chelators) can have a greater affinity. These agents will then remove significant amounts of the metal, leaving the body deficient and the enzymes inactive. One group of such agents is the dithiocarbamates, which cause the enzyme acetaldehyde dehydrogenase to be inhibited.

Inhibition of oxygen transfer. Cyanide is one of a number of agents, including azide, that act on the molecule cytochrome oxidase, preventing the binding of oxygen to the sequence of molecules necessary for oxidative phosphorylation. Failure to produce ATP in the accumulation of intermediary products results in cell death.

Blockage of hemoglobin oxygen transport

One common agent that blocks oxygen transport by hemoglobin is carbon monoxide. This molecule has a greater affinity for the hemoglobin molecule than molecular oxygen. When present in large enough amounts, a very significant portion of the hemoglobin is converted to the carbomonoxy form (HbCO). Levels of 40% HbCO are considered potentially lethal. The interaction is reversible, and the CO is removed from the hemoglobin if the patient is removed from the CO environment and given oxygen.

Several agents, such as amyl nitrate, cause the conversion of hemoglobin to methemoglobin. That is, the iron of normal hemoglobin that is in the ferrous state is converted by oxidation to the ferric form. In this form, hemoglobin cannot carry oxygen, resulting in oxygen starvation, and if the conversion is great enough, death. The body has some mechanisms to convert small amounts of methemoglobin to hemoglobin, but when this mechanism is overwhelmed, the methemoglobin is catabolized along the pathway given in Chapter 33.

A number of agents, particularly surfactants (detergent-like compounds), hydrazine derivatives, and arsine gas affect the surface of red cells, causing lysis. This hemolytic process can cause renal failure when it occurs in modest amounts and oxygen starvation if great quantities of red blood cells are hemolyzed.

Interference with general functions of cell

Although the process is not completely elucidated, chemicals, such as anesthetics, seem to exert their effect by interacting with the lipid phase of the cell, particularly the cell membrane. Such incorporation affects the normal cell membrane potential or its reaction to a depolarizing pulse. In particular, anesthetics interfere with the neurotransmission between synapsis or neuroeffector junctions.

Interference with DNA and RNA synthesis

Agents that affect DNA and RNA prevent cell division, cause lethal changes, or alter the gene material of the cell. Agents that have a cytostatic action inhibit cell division. One such group of compounds is the alkylating agents, the nitrogen mustards. These compounds form covalent bonds between and within the DNA chains, preventing duplication of the DNA and stopping protein synthesis.

Agents that have a mutagenic action change the genetic properties of the cell, such as the alteration of a protein or an enzyme so that it has a different property or loses its original property (catalysis, if there is an enzyme). Other changes include the expression or lack of expression of a gene at an inappropriate time or sequence during the development of a cell or group of cells.

Carcinogens are similar to mutagens in that they appear to act on the DNA of a cell. It is not completely clear how these agents function and whether cell division or interaction with other molecules is needed. It is clear that the result is an altered cell that has lost the capacity to control its ability to divide (see Chapter 46).

Hypersensitivity reactions

Hypersensitivity reaction is a term usually restricted to toxic reactions with an immunological basis. The reaction is disproportionate to the amount of agent given; in fact, a few molecules inappropriately inhaled or placed on the skin of a sensitized person can cause extreme allergic or hypersensitivity reactions. The presumed mechanism is that when persons initially are exposed to the agent, the toxicant or its metabolic product is processed to give an immunological response. When a subsequent exposure occurs, the primed person can react to very few molecules. It is important to note that the response to the toxicant requires one or many previous exposures. In addition, keep in mind that usually only a few persons will have a hypersensitivity reaction. Many will have an immune response that will not lead to a hypersensitivity reaction.

FACTORS THAT INFLUENCE TOXICITY

Nature of toxicant

Chemical properties. One of the most important properties of a toxicant is its solubility in various tissues of the body. Toxicants that are water soluble tend to be less likely to penetrate the skin. Those that are lipid soluble tend to be more readily absorbed; however, the pH of the skin, the ionic form of the chemical, and its molecular size are all important chemical properties.

Physical properties. One important property of a toxicant is the physical state of the compound, whether gas, liquid, or solid. Gases are readily inhaled and pass immediately into circulation. In addition, they pass through skin. In contrast, liquids are not inhaled unless in a vaporous state or aerosol form. Liquids can pass through the epidermal barrier. Solids are not inhaled and are not absorbed quickly through the skin. They usually must be ingested or dissolved before they have an effect. Compounds with a liquid form that have a significant gas phase under exposure conditions can have an adequate amount of the compound in the gas phase to induce toxicity by vapor inhalation. Sometimes such toxicity may take years. One example of this is metallic mercury, a liquid at room temperature with a very low vapor pressure. The inhaled metallic mercury vapor present in such low amounts leads to pathological neurological behavior if the exposure occurs over a period of years.

Exposure situations

Dose dependency. For common pharmacological agents or drugs, increased dosage results in an increase in the number or percentage of a population that will be affected and as well an increase in the severity of the toxicological reaction. In contrast, continued exposure in very low concentrations can be noncumulative and relatively innocuous. Such an effect of concentration is observed with the drug acetaminophen. At low concentrations this analgesic is commonly used to treat pain and is metabolized in the liver by a pathway involving glutathione. At high concentrations of acetaminophen, this pathway is depleted of glutathione and free radical damage occurs. This can irreversibly damage liver cells causing cell death. Thus concentration is important because at low levels some agents are handled by usual biochemical transformation mechanisms or excretory systems, but at higher concentrations these normal detoxification mechanisms are exceeded, resulting in a harmful effect.

Also keep in mind that the symptoms of toxicological response may not be obviously related to the clinical response of the drug. In the case of the tricyclic antidepressants, the obvious effect is mood elevation by action on the central nervous system (CNS) receptors. In the overdose situation the drug may affect similar receptors in the heart, causing arrhythmia. As the amount of the tricyclic given increases, so do the arrhythmias that can cause heart failure and death.

Route, rate, and site of administration. The route of administration can be percutaneous, inhalation, oral, rectal, and parenteral. Of these routes, the percutaneous is usually not considered in a hospital environment because most agents are given by the other routes. However, some new systems that deliver medications using transcutaneous membranes are increasing the availability of percutaneous absorption. The skin has a very large surface and is permeable to many chemicals. Molecules, such as phenols, readily pass through the skin into the circulation.

Variation in the amounts of an agent causing death based on the route of administration can be quite large. For example, a twenty-fold difference between the effect of procainamide given intravenously versus subcutaneously has been demonstrated. For other compounds, such as pentobarbital, there is less than a twofold difference in dosage between these two routes to achieve the same lethal effect. Thus route of administration can lead to considerable differences in the dose-response curve to a toxic agent.

Duration and frequency of exposure. Duration of exposure is the time in seconds, minutes, hours, days, months, or years that a person is exposed to a toxicant. For some agents, such as carbon monoxide, the duration can be a few seconds or minutes, such as an exposure in a room with a high concentration of carbon monoxide, or chronic as in cigarette smoking. One calculation often used is exposure time × dosage. The derived value is used to compare persons with different levels of toxicant, such as the blood lead levels of workers, and their possible toxicological symptoms, such as renal disease. Frequent exposure in many cases leads to an accumulation of toxic effects, but this will vary, depending on the agent. The unusual case is that of delayed hypersensitivity reactions, in which minute amounts can result in a toxic anaphylactic reaction even though the amounts may be subtoxic for the same persons under presensitizing conditions.

Biological variables

There can be considerable difference in the rate of metabolism of compounds with age. In general, younger subjects have a more rapid metabolism than older ones. This is particularly true in the case of drug metabolism. Older patients are often given the same dosage as younger ones; however, because their metabolism may be one half to one third as rapid as younger ones, the drug may build up to toxic levels because of the inability to excrete and metabolize the drug. Part of the age difference is attributable to changes in organ function. For example, renal clearance is 20% to 50% less in an aged population as compared to young medical technology students. Age is of considerable importance when weighing the toxic effect of drugs, such as tricyclic antidepressants, and the aminoglycosides, such as gentamicin.

Sex differences do exist in metabolism of drugs and thus their possible toxic effect. Usually differences in the metabolism of these drugs are under direct endocrine control. In humans, it has been shown that the daughters of women given diethylstilbesterol during pregnancy have an increased risk of endometrial cancer. The hormonal status of women varies considerably with age. For example, there are prepuberty, adult, postmenopausal, and pregnancy states. The metabolism of compounds that are under endocrine control can vary widely with these different conditions.

The state of nutritional well-being also plays a role in toxic effects. In starvation there is a change in body metabolism, particularly catabolic pathways. There are no glycogen stores in starvation and as a result protein and fat are broken down in order to achieve gluconeogenesis. Other changes also occur; thus the metabolism of drugs and the effect of toxic agents can be quite different as compared to the normal state.

The difference in genetic makeup can be one of the most dramatic factors influencing the toxic effect of an agent. Succinyl choline, a muscle relaxant, is hydrolyzed to succinic acid and choline. However, a group of persons (1 in 3000) exists who do not have the enzyme that will hydrolyze this compound or have a defective form of the enzyme. Thus they have prolonged muscle relaxation and possible apnea (stopping of breathing) in contrast to normal persons and therefore are at risk when given this drug. It is possible to screen a population for this deficiency by measuring acetyl cholinesterase levels in the presence the inhibitor dibucaine and thus find those persons with this enzyme deficiency.

Another example of genetic differences in drug metabolism is the ability to *N*-acetylate certain compounds. One drug, procainamide, is *N*-acetylated as part of the normal metabolic detoxification pathway. It has been found that there are two groups of persons, those who are "fast" acetylators and those who are "slow" acetylators. This difference in acetylation has profound implications in the toxic reactions to the drug. The slow acetylators who form *N*-acetyl procainamide at a less rapid rate are more likely to have an immunological reaction to the drug. In addition, the *N*-acetylated drug is cleared more slowly, thus leading to higher blood levels at a given dosage. Thus drug levels are dependent on the *N*-acetylation rate. Since both forms of the drug are active, both must be monitored for toxic and pharmacological effects.

Toxicokinetics

Principles of toxicokinetics are similar to those of pharmacokinetics as discussed in Chapter 51 on therapeutic drug monitoring. The various phases of a toxicant's passage through the body must be divided into absorption, distribution, storage, biotransformation, and excretion. Some toxicants, such as caustic agents (lye, acid, corrosives), have a direct tissue-damaging effect and thus are not studied in this manner. Others, such as gases or drugs, must pass through membranes or be absorbed and then pass through the circulation before they exert their effect on a particular target tissue. In addition, many toxicants must be transformed by metabolic processes before they can exert action on receptor or cell. Finally, most drugs or compounds must be metabolized before they are excreted. In general one can use the same equations presented for therapeutic monitoring to follow these agents.

Biotransformation

Most chemicals or agents that enter the body are not passively received passengers but undergo changes in their structure. This change is termed *biotransformation*. There are numerous pathways for this to occur. A list of such reactions is presented in Fig. 47-2. Biotransformations result in molecules that are different from the parent or starting compound. Biotransformed molecules can have properties that make them biologically less active than the parent, more reactive, or equireactive. Atropine is hydrolyzed to the inactive moieties tropic acid and tropine. On the other hand, cyclophosphamide, an anticancer alkylating agent, must be activated by liver microsomes to be effective. Finally, the drug amitriptyline is demethylated to form nortriptyline, which is equipotent as a tricyclic antidepressant.

The microsomes of the liver contain a group of protein molecules that oxidize various chemicals such as steroids, fatty acids, drugs, pesticides, and carcinogens. These microsomal particles are termed *mixed-function oxidases*. The last oxidase in this chain of oxidases is cytochrome P-450, so termed because of its spectral absorption at 450 nm. Oxidation of a drug is by a reaction that involves oxygen, the drug or substrate, and the appropriate cytochrome P-450.

$$\text{Substrate} + O_2 + \text{NADPH} + H^+ \xrightarrow{\text{Cytochrome P-450}} \text{Oxidized substrate} + H_2O + \text{NADP}$$

Cytochrome P-450 exists in multiple forms, and the variability in structure has been estimated to be on the same order as the immunoglobulins. It has been shown that presentation of different substrates for oxidation results in the stimulation of the production of different forms of the cytochrome P-450 oxidase. The oxidations catalyzed by the microsomal cytochrome P-450 system include epoxidations, hydroxylations of aryl and alkyl hydrocarbons, oxidation of alcohols, removal of the alkyl group from ether-type linkages, *N*-hydroxylation, nitroxidation, *N*-dealkylations, oxidative deamination, sulfoxication, and dehalogenation. These enzymes are induced by some drugs, such as phenobarbital or alcohol, and inhibited by others, such as cimetidine, thus changing the rate of metabolism and effectively altering the steady-state level of these drugs when administered.

The hepatic P-450 system is recognized as one of the most important systems for the oxidation of drugs and other xenobiotic agents. In general, the drugs oxidized by the microsomal enzyme system are lipid soluble. After oxidation, they are more water soluble and thus more likely to be excreted.

DRUGS AND NONTHERAPEUTIC AGENTS ENCOUNTERED IN CLINICAL LABORATORY

Agents used in suicide

Table 47-2 summarizes deaths from suicide in the United States for the year 1978. The first item that should be noted is the large number of deaths caused by gases, particularly motor vehicle and exhaust gases. These data reveal that carbon monoxide must be considered one of the most lethal agents in our society. The three categories, barbituric acids and its derivatives, psychotherapeutic agents, and other drugs, account for the majority of the remaining other successful suicides.

Table 47-3 presents a classification of the drugs present at time of deaths from suicide in St. Louis, Missouri, in the period 1977 to 1979. Such a table will vary for different areas of the country. It must be noted that the number of drugs present in these cases exceeds 100% (the last column); that is, more than one drug was found at the time of death. Tables 47-2 and 47-3 illustrate that, except for carbon monoxide, most suicides are related to medications available to the person.

Table 47-2. Poisoning deaths from suicide during 1978

Agent	Number of deaths
Barbituric acid and derivatives	661
Salicylates and congeners	84
Psychotherapeutic agents	571
Other and unspecified drugs	1527
Corrosive aromatics	14
Strychnine	10
Lye and potash (caustic)	35
Arsenic and its compounds	8
Fluorides	3
Other and unspecified solids and liquids	594
Gases, domestic use	41
Other gases	2469
Motor-vehicle exhaust gas	2017
Other carbon monoxide	435
Other and unspecified gases and vapors	17
Total	6017

Table 47-3. Incidence of most frequently encountered drugs or class of drugs at autopsy in drug-related suicides in St. Louis, Missouri, 1977 to 1979

Drug or class of drug	Number of cases present	% of 147 deaths
Barbiturates	39	27
Benzodiazepines	39	27
Ethanol	38	25
Over-the-counter analgesics	28	20
Propoxyphene	20	14
Tricyclic antidepressants	27	25

Drugs in emergency situations

In a hospital setting, particularly that involved in emergency care, the pattern of toxic drug ingestion or abuse is somewhat different from that seen in suicide. Table 47-4 lists the most frequently ordered tests and the percentage positive from a survey conducted at a large hospital. Such tests are ordered on a variety of patients presenting clinically as hyperactive, hallucinating, or in coma. A description of the various grades of coma in toxic cases is given by Hanenson as follows:

Coma gradation (from Hanenson; see bibliography)

Grade 1—drowsy but responds to verbal commands

Grade 2—unconscious but responds to mild painful stimulus

Grade 3—unconscious and responds only to maximum painful stimulus

Grade 4—unconscious without response to painful stimulus

In general, patient history is not a reliable index of what drugs may have been taken by a patient (less than 50% reliability).

Physician judgment as to possible intoxicating or suicidal drugs is based on a number of observations and judgments regarding the patient's condition. Each drug has general and specific effects and possible consequences.

General effects of toxic agents can include symptoms such as respiratory depression, shock, convulsions, hyperthermia, or hypothermia. Some clinical effects and symp-

Table 47-4. Summarized ordering pattern from the Los Angeles County University of Southern California Medical Center (USCMC) for the ten most commonly ordered drugs or drug groups

Test	Requests	Positives (%)
Volatiles*	12,409	6,681 (53.8)
Barbiturates	6,778	3,095 (45.7)
Hypnotic screen†	7,518	1,227 (16.3)
Salicylates	2,754	667 (24.2)
Opiates	1,895	470 (24.8)
Amphetamine	1,426	112 (7.8)
Phenytoin	841	508 (60.4)
Phenothiazines	510	134 (26.3)
Phencyclidine	509	207 (40.7)
Tricyclic antidepressants‡	377	125 (33.2)
Totals	35,017	13,226
% of annual totals	94.7	93.0

*Included in this assay is the evaluation for ethanol, methanol, and isopropanol.

†Included in this total are both blood and urine requests for which the following drugs were analyzed: carisoprodol, chlordiazepoxide, ethchlorvynol, methaqualone, methyprylon, and oxazepam. The urine is also tested for the presence of ethchlorvynol and phenothiazines by colorimetric methods.

‡Includes data for amitriptyline, doxepin, imipramine, desipramine, nortriptyline, and protriptyline.

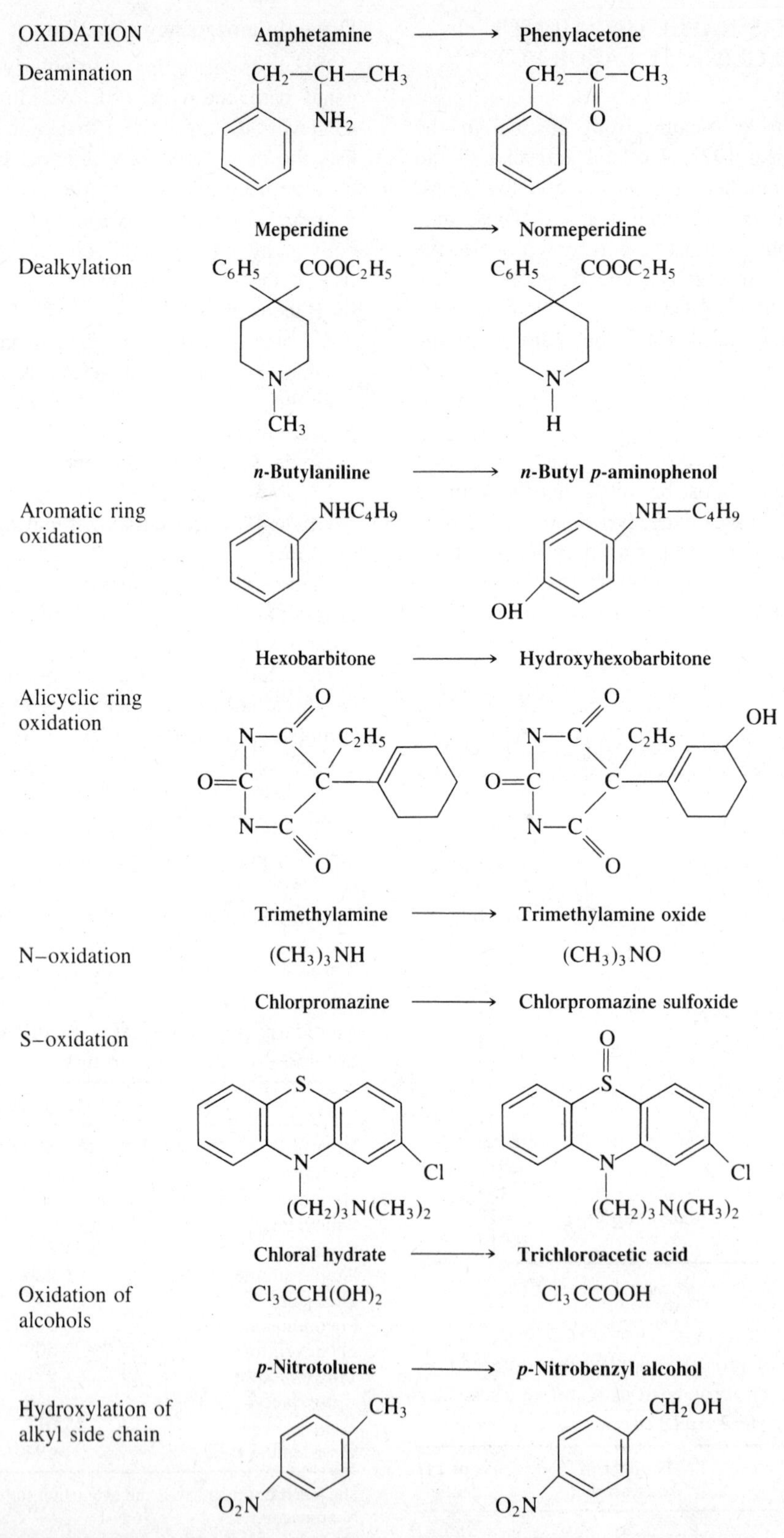

Fig. 47-2. Some types and examples of biotransformation mechanisms in animals. Oxidation and reduction reactions are catalyzed by liver microsomal enzyme systems. Hydrolysis, acetylation, and conjugation reactions may involve enzyme systems from other tissues. (From Loomis, T.A.: Essentials of toxicology, ed. 3, Philadelphia, 1978, Lea & Febiger Publishers.)

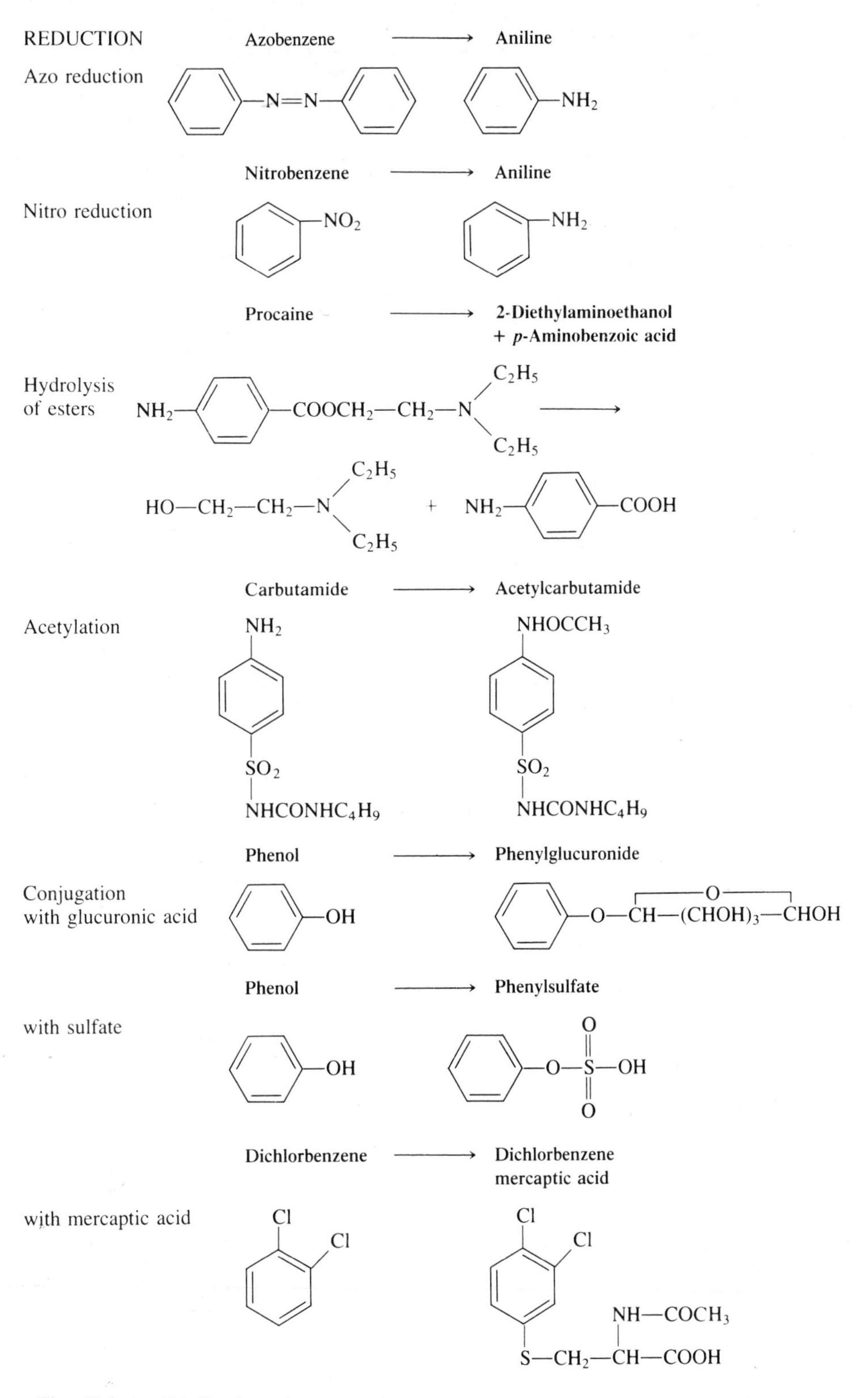

Fig. 47-2, cont'd. For legend see opposite page.

Table 47-5. Some clinical features of severe drug intoxication

Agent	Pupils	Respiration	Other
Acetaminophen	—	—	Liver toxicity
Barbiturates	Dilated, light reactive	↓	Flaccid paralysis, hypothermia, hypotension
Benzodiazepines	—	↓	CNS depression, extrapyramidal system changes, autonomic nervous system changes, hypotension
Ethanol	—	↓	CNS depression (in some grand mal seizures)
Narcotics and propoxyphene	Pinpoint	↓	CNS depression
Salicylates	—	↓	Metabolic acidosis (later)
Tricyclics	Constricted	—	Tachycardia, arrhythmia

toms of patients at very high doses of these agents are summarized in Table 47-5. The effect on pupils, respiration, cardiac status, and other systems are all important vital signs and can also give information as to the nature of the toxic substance. In contrast, an overdose of acetaminophen is only established initially by laboratory measurement because the poisonous effect of the agent may take several days to appear.

When a comatose or bizarrely acting patient is seen in an emergency room situation, the possibility of a drug overdose is one of the diagnoses that must be considered by a physician. Electrolyte, BUN, and glucose analysis can often rule out renal or diabetic disease states.

For the most efficient use of laboratory facilities, it is important for the physician to try and ascertain the nature or identity of drug or drugs used. Physician judgment as to possible intoxicating or suicidal drugs is based on a number of clinical observations (Table 47-5), and judgment regarding the patient's condition. A history obtained from an alert patient can be helpful, though in general, patient history is not a reliable index of what drugs may have been taken (less than 50% reliability).

If by observations or history the physician is highly suspicious of one or more drugs, laboratory tests may be ordered to confirm the presence of the drug or drugs and to determine the amount present. These data can establish the therapy to be used. If the clinician is not certain if drugs are present, a drug screen may be ordered. The drug screen will rapidly test for the presence of some of the major drug groups, and if it is positive, the presence will be confirmed by an alternative procedure (see p. 1355).

Drugs frequently encountered in overdose situations

Hypnotics. Hypnotics include the barbiturates chloral hydrate, ethchlorvynol, flurazepam, glutethimide, methaqualone, and methyprylon. They function as general depressants; that is, they decrease cellular function in many vital tissues, especially those of the central nervous system (CNS). As a result they are very often prescribed as sedatives.

Analgesics. The analgesics are compounds that reduce pain without causing a loss of consciousness. These compounds often have antipyretic properties; that is, they lower body temperature in addition to their antiinflammatory properties. This group of compounds includes the following narcotics: morphine, heroin, meperidine, codeine, dihydrocodeine, oxycodeine, hydromorphone, pentazocine, propoxyphene, butyrophenone, and methadone. It includes the salicylates acetaminophen, phenacetin, phenylbutazone, and oxyphenbutazone. The opiate-derived narcotics act as agonists with receptors in the brain and other tissues. These receptors also bind naturally occurring molecules such as endorphens and enkephalins. Aspirin and its derivatives function by inhibiting the release of prostaglandins from damaged or inflamed cells.

Tricyclics. Tricyclics are mood-elevating drugs used in the treatment of affective psychiatric disorders (such as depression). They are believed to function by blocking the receptors of the biogenic amines, 5-hydroxytryptamine and norepinephrine, at the nerve endings. They are also anticholinergic and in high doses give an anticholinergic response such as mydriasis (dilation of pupils), flushed dry skin, decreased bowel sounds, urinary retention, and tachycardia. These drugs include amitriptyline, nortriptyline, imiprimine, desipramine, and doxepin.

Benzodiazepines. Benzodiazepines are antianxiety-sedative drugs, which are used in the therapy of anxiety states. Although the mechanism of action is not completely known, there is central nervous system depression when taken. The most popular of these drugs are chlordiazepoxide (Librium) and diazepam (Valium).

Phenothiazines. Phenothiazines are drugs used to treat psychoses; that is, conditions that include delusions and hallucinations. Schizophrenia includes paranoid delusions and auditory hallucinations, disordered thinking, and emotional withdrawal. The mechanism of action of these drugs is believed to be similar to that of the benzodiazepines in that they act as antagonists of dopamine, a neurotransmitter in the brain. Some phenothiazines are chlorpromazine, fluphenazine, mesoridazine, perphenazine, and thioridazine.

ANTIDOTES AND TREATMENT

The history of poisons is intertwined with that of antidotes. Currently there are only a few antidotes specific for

Table 47-6. Specific antagonists and antidotes

Drug	Antagonist or antidote
Acetaminophen	Methionine, *N*-acetyl cysteine
Amphetamines	Chlorpromazine
Anticholinergic drugs (tricyclic antidepressants, atropine, scopolamine)	Physostigmine
Carbon monoxide	Oxygen
Cyanide	Sodium nitrite
Nitrates and nitrites	Methylene blue
Opiates	Naloxane
Organophosphates	Atropine and pralidoxime

those agents commonly observed in a hospital setting. These are presented in Table 47-6. Usually most therapy is supportive, involving treatment for respiratory depression, aspiration, shock, convulsions, delerium, methemoglobinemia, hyperthermia, and hypothermia.

Removal of the toxic agent can be effected by emesis (vomiting), gastric lavage, or skin decontamination. Other forms of treatment include agents that help remove the toxicant from the body. Dimercaprol is used to treat arsenic and mercury, and the salts of the divalent metals (lead, mercury, copper, nickel, zinc, cadmium, cobalt, beryllium, and manganese) are chelated by ethylenediaminotetraacetate (EDTA). The chemical deferoxamine is used to remove excess amounts of iron. Charcoal may be swallowed or placed in the stomach. This agent enhances the rate of removal of drug from the body.

A number of agents are removed more rapidly if urine flow is increased (forced diuresis). This is often accomplished by the intravenous administration of a diuretic such as mannitol. In addition, because of a lower concentration in the kidney under diuretic conditions, the potential for nephrotoxicity of some agents (such as *cis*-platinum) is less. The pH of urine is also important. For basic drugs such as the amphetamines, excretion is enhanced by an acid urine, in which the drugs are ionized. Barbiturate excretion is increased by alkaline urine, in which the molecule is ionized.

DRUG INTERACTIONS

Definitions and examples

Drug interaction is a term used to denote the observations that when two drugs are given to a patient their actions may not be independent. That is, one drug will affect the pharmacological action of the other. Such effects are not obvious, and much information about these phenomena has been gathered from case reports. Many types of interactions have been noted, and it is not possible to provide a comprehensive list here. Stockly has listed the types of interactions, though he notes that "interactions which occur when drugs are given concurrently are often the result of not a single mechanism, but of two or more mechanisms acting in concert. . . ." Table 47-7 gives this listing with an example of each type of interaction.

Effect of ethanol on drug metabolism

The combined use of ethanol and drugs can lead to a variety of drug interactions. A most important group of drug interactions are those between ethanol and other agents with central nervous system effects. Ethanol has a wide range of depressant effects on the central nervous system. Combined with narcotic analgesics, sedative-hyp-

Table 47-7. Examples of classes of drug interactions

Action	Example
Drugs with similar effects	Multiple nephrotoxic drugs such as gentamicin and cephalosporin yield increased nephrotoxicity.
Drugs with opposing effects	Hypnotics and caffeine result in antagonism to the hypnotic effect.
Absorption interactions	Tetracycline and iron (Fe^{++}) supplement result in decreased oral uptake of drug.
Drug displacement interactions	Theoretical displacement of bound drug from albumin yields increased free-drug level.
Drug metabolism interactions	
Enzyme induction	Barbiturates stimulate microsomal oxidation of drugs such as dicumarol, resulting in lower plasma levels.
Enzyme inhibition	Cimetidine blocks the P-450 oxidation pathway, slowing theophylline metabolism and resulting in a longer half-life and higher plasma levels.
Altered excretion interactions	
Changes in urine pH	Acid urine enhances basic drug excretion. The reverse is true for acidic drugs.
Competition for active tubular secretion	Probenecid decreases the active secretion of drugs such as penicillin, thus increasing serum levels.
Interactions at adrenergic neurons	Tricyclic antidepressants prevent the uptake of guanethidine into neurons, thus blocking its antihypertensive effect.

notics, anxiolytics, antidepressants, or even antihistamines, ethanol has a synergistic or additive effect (see Chapter 32).

Anxiolytics. Benzodiazepines appear to produce varying interactions with ethanol: diazepam (Valium) may be synergistic or additive with alcohol, whereas chlordiazepoxide (Librium) may be antagonistic to alcohol. The interactions between alcohol and diazepam can cause an increase in the degree of sedation and thus can significantly impair psychomotor performance. Moreover, this combination may be fatal.

Antidepressants. Amitriptyline and doxepin, which possess strong anticholinergic properties, have short-term additive effects with alcohol. With long-term use, tricyclic antidepressants antagonize the effects of ethanol because of their action on norepinephrine uptake in the brain.

Antihistamines. Many antihistamines produce drowsiness; when combined with ethanol, even more pronounced sedative effects may ensure.

Antipsychotics. Phenothiazines with strong anticholinergic potency (such as thioridazine) have an additive interaction with alcohol, which can cause severe respiratory depression. Also, the hypotension associated with phenothiazines may be potentiated by concomitant ingestion of alcohol.

Narcotic analgesics. When combined with ethanol, narcotics, such as morphine, meperidine, hydromorphone, and propoxyphene, can cause impairment of motor activities. This interaction is especially pronounced in alcoholic patients with cirrhosis, in whom metabolism of the narcotics may be impaired.

Sedative-hypnotics. The concomitant use of the barbiturates and alcohol can have deleterious effects. The lethal dose of barbiturates is almost 50% lower when combined with alcohol. This same effect is true with concomitant use of alcohol and agents such as chloral hydrate, paraldehyde, glutethimide, meprobamate, and other tranquilizers.

Central nervous system stimulants and caffeine. The use of amphetamines or caffeine together with ethanol can have unpredictable results. There is a mistaken notion that coffee is an antidote to the depressant effects of ethanol on the central nervous system. Studies suggest that stimulants such as caffeine may antagonize the effects of alcohol, leaving the intoxicated person with a false sense of security without improving muscle control or alertness.

BIBLIOGRAPHY

Bowman, W.C., and Rand, M.J.: Textbook of pharmacology, ed. 2, Oxford, 1980, Blackwell Scientific Publications, Ltd.

Casarett, L.J., and Doull, J., editors: Toxicology: the basic science of poisons, New York, 1975, Macmillan Publishing Co., Inc.

Dreisbach, R.H.: Handbook of poisoning, ed. 8, Los Altos, Calif., Lange Medical Publications.

Gilman, A.G., Goodman, L.S., and Gilman, A., editors: Goodman and Gilman's pharmacological basis of therapeutics, New York, 1980, Macmillan Publishing Co., Inc.

Hanenson, I.B., editor: Quick reference to clinical toxicology, Philadelphia, 1980, J.B. Lippincott Co.

Loomis, T.A.: Essentials of toxicology, ed. 3, Philadelphia, 1978, Lea & Febiger.

Stockley, I.: Drug interactions: a source book of adverse interactions, their clinical importance, mechanisms and management, Oxford, 1981, Blackwell Scientific Publications, Ltd.

Section three

Methods of analysis

Chapter 48 Classifications and description of proteins, lipids, and carbohydrates

aldose The chemical form of monosaccharides in which the carbonyl group is an aldehyde.

apoprotein Polypeptide chain not yet complexed to its specific prosthetic group.

carbohydrate Chemicals with the general formula of hydrated carbon, $(CH_2O)_n$, that are an aldehyde or ketone derivative of polyhydric alcohols. Commonly called "sugars."

compound (conjugated) proteins Polypeptide chain complex with other chemical classes such as lipids (lipoproteins), carbohydrates (glycoprotein), or nucleic acids (nucleoproteins).

conjugated lipids Esters of fatty acids and alcohols containing additional chemical moieties. Group includes phospholipids, sphingolipids, sterols, bile acids, and so on.

denaturation Unfolding of tertiary structure of a protein that renders it insoluble, causing it to precipitate out of solution.

derived lipids Lipids derived from the hydrolysis of simple and conjugated fats and include the fatty acids.

disaccharide Two monosaccharides linked together by a $1\rightarrow4$ or $1\rightarrow6$ glycosidic linkage, such as sucrose, maltose, or lactose.

furanose Five-membered rings of monosaccharides formed by intramolecular reaction between the carbonyl group and a hydroxyl group; present in alpha or beta stereoisomeric forms.

isoelectric point pH at which a molecule containing many ionizable groups is electrically neutral; that is, the number of positively charged groups equals the number of negatively charged groups.

ketose The chemical form of a monosaccharide in which the carbonyl group is a ketone.

monosaccharide Basic monomeric carbohydrate unit in which n in the formula $(CH_2O)_n$ ranges from 3 to 8.

polypeptide bond The covalent amide bond between a primary amino group of one amino acid and the carboxylic acid group of a second amino acid.

polysaccharide Polymer usually containing greater than 10 monosaccharides linked by glycosidic bonds; branched and unbranched chains up to many millions of molecular weight can be formed, as in cellulose, starch, or glycogen.

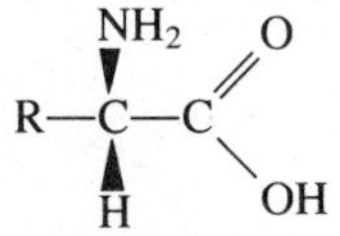

Fig. 48-1. General structure of amino acid of L-stereoisomeric form. *Heavy lines,* Bonds coming out of plane of page; *dotted lines,* bonds extending behind plane of paper.

primary structure The linear sequence of amino acids in a protein, defined by the genetic code in cellular DNA.

prosthetic group A nonprotein chemical group that is bound to a protein and is responsible for the biological activity of the protein. The functional complex between protein and a prosthetic group is called a ''holoprotein,'' and the protein without the prosthetic group is called an ''apoprotein.''

pyranose Six-membered rings of monosaccharides formed by intramolecular reaction between a carbonyl group and a hydroxylin group, present in an alpha or beta stereoisomeric form.

quaternary structure The three-dimensional spatial arrangement of polypeptide chains resulting from the combining of more than one polypeptide chain into a larger, stable complex.

Schiff's base Covalent complex between a primary amine and carbonyl function of an aldose.

secondary structure The spatial arrangement of a linear chain of amino acid in a polypeptide; the three groups are the beta-plated sheet, alpha-helix, and random coil.

sialic acids *N*-acetyl derivatives of neuraminic acid that are covalently linked to many proteins.

simple lipids Esters of fatty acids with various alcohols, including the triglycerides and some steroids.

simple proteins Polypeptide chain consisting only of amino acid groups.

steroids Lipids containing four six-membered rings and including many hormones, vitamins, and drugs.

tertiary structure The intramolecular folding of a polypeptide chain upon itself resulting from interactions between side-chain groups of individual amino acids.

zwitterion Molecule containing two ionized groups of opposite charge. (Pronounced tsvit'er-i'on.)

The analytes in this section are grouped as much as possible by common chemical class. This grouping is important because the chemical nature of an analyte is often the single most important factor in determining which type of method will be used for analysis. Other important factors for consideration are the physiological concentration of the analyte, biological variation in concentration, and the clinical situations for which the analytes are measured. These chemical classes are now reviewed, as well as the type of methods used for their analysis. This chapter is not intended to provide a complete biochemical review of the analytes measured in the chemistry laboratory. For this, readers are encouraged to use the excellent biochemistry texts listed in the bibliography. Instead, an emphasis has been placed on those properties of proteins, lipids, and carbohydrates that affect how the analytes may be measured.

Part I: Proteins

LAWRENCE A. KAPLAN

DEFINITION AND CLASSIFICATION

Proteins are linear polymers of alpha-amino acids. There are 20 natural amino acids with the general structure as seen in Fig. 48-1. These exist as the L-stereoisomeric form with the amino group placed on the alpha-carbon atom next to the carboxylic acid group. The pK_a of the carboxylic acid group is approximately 1.8 to 2.4, whereas the pK_a of the alpha-amino group is approximately 8.5 to

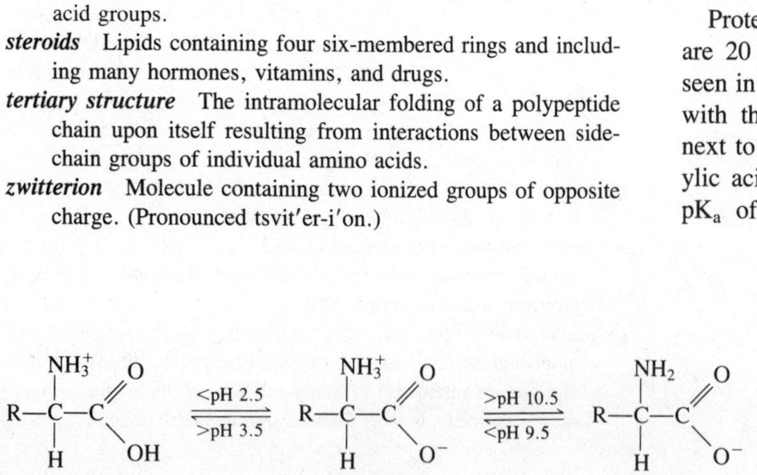

Zwitterion

Fig. 48-2. Various ionized and nonionized forms of amino acids present at various pH's. When two opposite charges are present on same molecule, molecule is called a ''zwitterion.''

Table 48-1. Classification and properties of side chain (R groups) for naturally occurring amino acids

R group (R–CH(NH₂)COOH)	L-Amino acid (symbol)	Amino acid molecular weight	pKₐ* (25° C) Primary –COOH	Primary $-NH_2$	Secondary groups
Nonpolar (hydrophobic)					
H–	Glycine (gly)	75.07	2.34	9.60	—
CH_3–	Alanine (ala)	89.09	2.34	9.69	—
$(CH_3)_2CH$–	Valine (val)	117.15	2.32	9.62	—
$(H_3C)_2CH-CH_2$–	Leucine (leu)	131.18	2.36	9.60	—
$CH_3CH_2-CH(CH_3)$–	Isoleucine (ile)	131.18	2.36	9.68	—
C_6H_5–CH_2	Phenylalanine (phe)	165.19	1.83	9.13	—
H_2C-CH_2, H_2C CH–, N, H (pyrrolidine ring)	Proline (pro)	115.13	1.99	10.60	—
$CH_3-S-CH_2CH_2$–	Methionine (met)	149.21	2.28	9.21	—
Neutral polar (hydrophilic)					
$OHCH_2$–	Serine (ser)	105.09	2.21	9.15	—
$CH_3CH(OH)$–	Threonine (thr)	119.12	—	—	—
$NH_2-C(=O)CH_2$	Asparagine (asp)	132.12	2.02	8.80	—
$NH_2-C(=O)CH_2CH_2$–	Glutamine (gln)	146.15	2.17	9.13	—
$HSCH_2$–	Cysteine (cys)	121.16	1.96 (30°)	10.28	8.18 (SH)
HO–C_6H_4–CH_2–	Tyrosine (tyr)	181.19	2.20	9.11	10.07 (OH)
Indole ring –C–CH_2–, CH, N, H	Tryptophan (trp)	204.23	2.38	9.39	—

From Cohn, E.J., and Edsall, J.T.: Proteins, amino acids and peptides, New York, 1943, Reinhold Co.
*The pKₐ values will be slightly different in a protein molecule.

Table 48-1. Classification and properties of side chain (R groups) for naturally occurring amino acids—cont'd

R group ($R-CH(NH_2)COOH$)	L-Amino acid (symbol)	Amino acid molecular weight	pK$_a$* (25° C) Primary –COOH	Primary $-NH_2$	Secondary groups
Acidic polar (hydrophilic)					
$HOOCCH_2-$	Aspartic acid (asp)	133.10	1.88	9.60	3.65 (COOH)
$HOOCCH_2CH_2-$	Glutamic acid (glu)	147.13	2.19	9.67	4.25 (COOH)
Basic polar (hydrophilic)					
$H_2NCH_2CH_2CH_2CH_2-$	Lysine (lys)	146.19	2.18	8.95	10.53 ($E-NH_3$)
$H_2N-C(=NH)-NH-CH_2CH_2CH_2-$	Arginine (arg)	174.20	2.17	9.04	12.48 (guanidinium)
$HC=C(-CH_2-)-NH-CH=N-$ (imidazole ring)	Histidine (his)	155.16	1.82	9.17	6.00 (imidazolium)

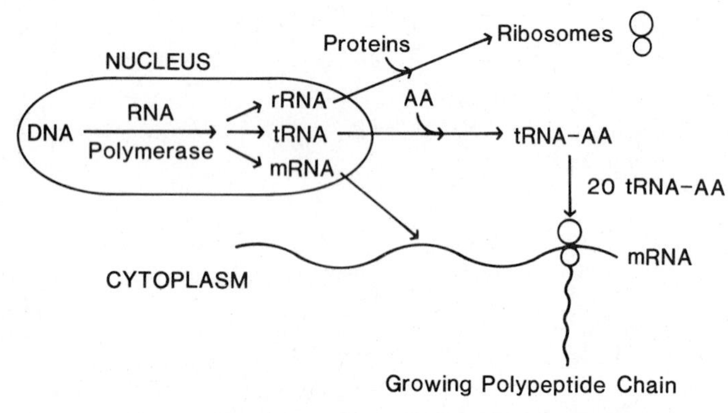

Fig. 48-3. Scheme of synthesis of proteins. *AA,* Amino acid; *DNA,* deoxyribonucleic acid; *tRNA,* transfer ribonucleic acid; *mRNA,* messenger ribonucleic acid; *rRNA,* ribosomal ribonucleic acid (18S and 28S forms); *tRNA-AA,* activated amino acid covalently bound to amino acid–specific tRNA.

groups are ionized. A compound, such as an amino acid, having two opposite charges is called a *zwitterion* (''hybrid ion'' or ''hermaphrodite ion'').

10.5. This means that at a pH less than 2, the carboxylic acid will be in the un-ionized form (COOH), whereas the alpha-amino group will remain ionized at pH's less than 9 (Fig. 48-2). At physiological pH (approximately 7.4), both groups are ionized. A compound, such as an amino acid, having two opposite charges is called a *zwitterion* (''hybrid ion'' or ''hermaphrodite ion'').

The side chain groups of the 20 amino acids are listed in Table 48-1, along with the pK$_a$'s of all ionizable groups. These side-chain groups can interact with one another to determine the overall chemical, physical, and biological properties of the polypeptide chain.

The amino acids are covalently linked together by the protein-synthesizing machinery of cells. The actual *order* or sequence of amino acids in a protein chain is predetermined by the genetic code within the cell. The sequence of genetic information in DNA is transcribed into messenger RNA, which is translated in the cytoplasm into protein (Fig. 48-3). The specific sequence of amino acids for protein is called its *primary structure*.

The amino acids are linked together by the peptide bond. As shown in Fig. 48-4, this bond has a specific spatial arrangement in three-dimensional space. The linear polypeptide chain can exist in three possible conformations, alpha-helix, beta-plated sheet, and random coil (Fig. 48-5). These conformations are called the *secondary structure* of the protein.

When a polypeptide chain is in solution, it is flexible enough for the molecule to bend, allowing the side chain groups to interact with one another. The types of interactions are listed in Table 48-2. Although the interaction energy of each side group is small, the net energy of all these interactions is great enough to stabilize proteins in a folded, convoluted, three-dimensional spatial arrangement called the *tertiary structure* (Fig. 48-6). Each protein's unique tertiary structure confers on it specific biological properties.

The folded polypeptide chains are often organized as aggregates with identical or different polypeptides. The specific number and type of these polypeptide chains deter-

Table 48-2. Types of intramolecular side-chain interactions of protein R-groups

Type of bond	Schematic*
Covalent	
Disulfide (cystine)	—S—S—
Lysinonorleucine (in collagen)	$-(CH_2)_3-CH_2-\overset{H}{N}-(CH_2)_4-$
Noncovalent	
Electrostatic	$C(=O)O^{\ominus}$ ··· $^{\oplus}NH_3$
Hydrogen	C_6H_4—O—H···O=C—O^-
Hydrophobic	(stacked aromatic rings)
Van der Waals	CH_2OH ··· CH_2OH

*Wavy line, Polypeptide chain.

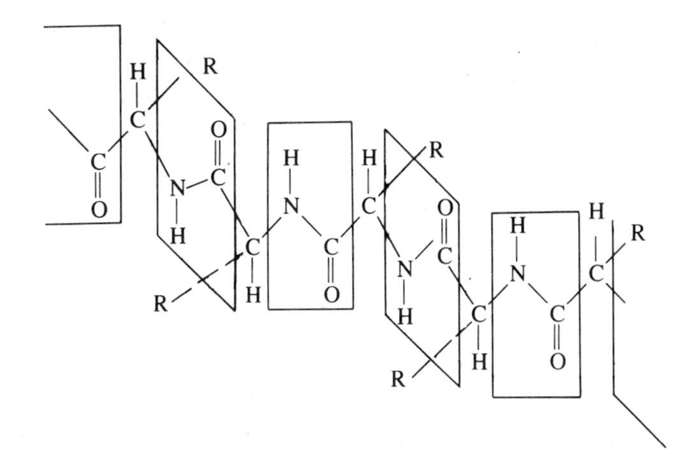

Fig. 48-4. Spatial relationships of a polypeptide bond. *C,* Carbon atom; *N,* nitrogen atom; *H,* hydrogen atom; *O,* oxygen atom. (From Orten, J.M., and Neuhaus, O.W.: Human biochemistry, ed. 10, St. Louis, 1982, The C.V. Mosby Co.)

Fig. 48-5. Scheme of three possible polypeptide chain conformations defining secondary structure of proteins.

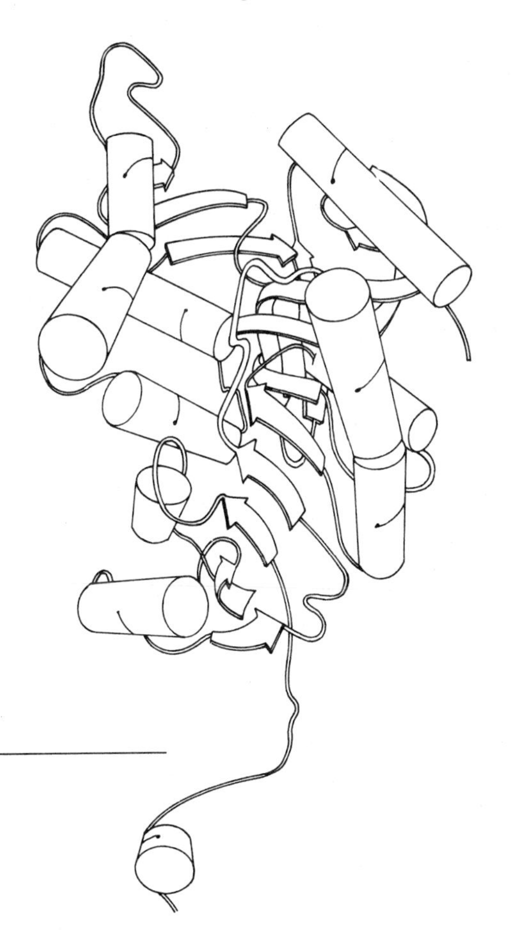

Fig. 48-6. Scheme of tertiary structure of a protein. *Cylinders,* α-helix; *flat arrows,* β-plated sheet; *lines,* random coil secondary structure of lactate dehydrogenase. (From Orten, J.M., and Neuhaus, O.W.: Human biochemistry, ed. 10, St. Louis, 1982, The C.V. Mosby Co.)

mines the specific properties of the entire complex. The spatial arrangement of these multichain proteins is called the *quaternary structure* of the protein. Usually the biological properties of such quaternary proteins consisting of subunit chains is the sum of each individual chain.

Proteins are generally classified into two major groups, simple and conjugated, with several subdivisions within each group. This classification scheme is based on the physical properties of the protein and on their chemical composition as well.

Simple proteins—generally not associated with other major chemical classes
- Globular proteins—relatively symmetric, water-soluble or saline-soluble proteins
  - Albumin—major serum protein
  - Globulins—most other serum proteins
  - Histones—basic proteins, found associated with nucleic acids
  - Protamines—strongly basic proteins found associated with nucleic acids
- Fibrous proteins—asymmetric proteins that are insoluble in water or dilute salts. Highly resistant to most proteolytic enzymes.
  - Collagens—major proteins of connective tissue, high in hydroxyproline content
  - Elastins—found in elastic tissue such as tendon and arteries
  - Keratins—major proteins in animal hair, nails, hooves, and so on

Conjugated proteins—these proteins are combined with other non–amino acid biochemicals. The conjugated protein is considered to consist of two components—the protein, called the *apoprotein,* and the nonprotein *prosthetic* group. The ability of the prosthetic group and apoprotein to dissociate varies from group to group.
- Nucleoproteins—the prosthetic groups are the nucleic acids (DNA or RNA)
- Mucoproteins—in which large amounts (more than 4% by weight) of complex carbohydrates are covalently linked to the protein
- Glycoproteins—these proteins also contain covalently linked carbohydrate residues but usually less than 4% by weight
- Lipoproteins—these contain cholesterol, triglycerides, and phospholipids associated with highly water-insoluble proteins
- Metalloproteins—these include proteins that contain metals strongly bound to the protein, either as the ion, or as complex metals such as the flavoproteins and hemoproteins
- Phosphoproteins—contain high concentrations of phosphate groups covalently linked to protein

The biological functions of the proteins are extraordinarily varied, but many functions (see below) are specific for only one or two of these classes of proteins.

CHEMICAL PROPERTIES

The chemical properties of proteins are based on the sum of their parts, that is, the constituent amino acids and prosthetic groups. The peptide bond is chemically reactive and is the basis of the most popular, specific method for quantifying total protein in serum. This is the biuret reaction. The amino acid side-chain groups are also chemically reactive, though only a few of these reactions are used in the chemistry laboratory. The amino groups at the N-terminal end of the polypeptide chain and those of lysine and the guanidino groups of arginine can react with several compounds to produce an intense fluorescence. These same groups can react with ninhydrin to give a blue color. Both of these reactions have been used to quantify total protein.

The phenolic group of tyrosine and the indole group of tryptophan react with the oxidizing reagent of the Folin-Wu or Lowry reactions to form a blue color. This method is employed with dilute solutions or microanalysis.

PHYSICAL PROPERTIES

The aromatic amino acids (tryptophan, phenylalanine, and tyrosine) give most proteins an absorption spectrum with the unique absorption maximum at 278 to 280 nm. The absorption at 280 nm is used to estimate the concentration of proteins in solution. In addition, complex proteins such as hemoglobin, which have prosthetic groups with unique absorption properties, can be individually quantitated without extensive purification on the basis of their specific absorption spectra. The Soret absorption band of hemoglobin at 415 nm is used extensively to quantitate the concentration of hemoglobins.

Polypeptide chains vary widely in molecular size. The smallest polypeptides, such as the endorphins or the hypothalamic hormones, contain 5 to 25 amino acids, whereas the largest proteins, containing several subunits, can have molecular weights in the millions of daltons. Although separation of protein on the basis of different molecular weights can be done, it is a method rarely used in the chemistry laboratory.

The density of most proteins falls within a fairly narrow range of about 1.33 g/mL. Lipoproteins represent an important exception because the lipid content of these proteins gives them an unusually low density, allowing the various classes of lipoproteins to be separated from each other and from other proteins on the basis of their density. This technique is used primarily in specialized laboratories.

An important physical property of proteins is the net charge on the protein molecule. The net charge on a protein is the sum of all ionic charges of the amino acids, carbohydrate, and prosthetic groups of the protein. Since the various chemical groups are ionized at different pH's, the net charge of a protein varies with pH. The pH at which a protein carries no net charge is called the "pI," or "isoelectric point"; the isoelectric point of a protein is the point at which the number of positively charged groups equals the number of negatively charged groups. At a pH greater than the pI, the protein will be negatively charged, whereas at a pH less than the pI, it will be more positively charged. At physiological pH, most serum proteins are negatively charged.

Since proteins differ in the number and type of constituent amino acids, they also differ in their pI's. Therefore at different pH's, proteins will carry different net charges. This difference in net charge is the basis of many procedures for separating and quantifying classes of proteins or individual proteins. The most common procedures are electrophoresis, ion-exchange chromatography, and isoelectric focusing. After separation, individual proteins are detected by use of specific stains.

Proteins found in body fluids are readily water soluble but can become insoluble in the presence of a wide range of denaturing or precipitating agents. These include organic solvents (such as acetone and acetonitrile), heavy metals (such as tungstic acid), certain salts (such as zinc hydroxide and ammonium sulfate), and strong acids (such as sulfosalicylic acid, trichloroacetic acid, and mineral acids). The ability to precipitate proteins from solution by one or more of these chemicals is the basis for a few routine clinical analysis. The turbidometric analysis of cerebrospinal fluid and urine protein is probably the most common. In addition, protein precipitation steps are often included as part of purification schemes for analytes before analysis.

BIOLOGICAL PROPERTIES

All proteins fulfill some physiological or biological function. The known functions of proteins cover a wide range of activities and are listed in Table 48-3. Often the known biological property of a protein is the basis of a method for its detection and quantification.

Of the important physiological functions of proteins, the most important to the clinical chemistry laboratory are the transport receptor and catalytic functions. Many serum proteins function as specific transporters of small molecules. Most transport proteins are globular proteins. Examples are thyroid-binding globulin (TBG), which binds thyroxine; transcortin, which binds cortisol; and albumin, which transports free fatty acids, unconjugated bilirubin, calcium, and many other endogenous and exogenous compounds. The specific binding properties of these transport proteins have been used as the basis for procedures to measure the serum concentrations of cortisol, TBG saturation, and other analytes. The lipoproteins function as transporters of lipids in serum (see Chapter 31).

Table 48-3. Biological functions of proteins

Function	Example
Transport of small molecules	Transcortin (cortisol), thyroxin (TBG)
Receptors	Estriol receptors (cytoplasmic), insulin receptors (surface)
Catalytic	All enzymes
Structural	Collagen
Nutritional (source of calories and amino acids)	Albumin
Oncotic pressure	Albumin
Host defense versus foreign antigens	Antibodies (all classes)
Hormonal	Thyroid-stimulating hormone (TSH)

Many cellular proteins act as intermediary information processors for hormone molecules. Each protein, called a receptor, binds a specific hormone and then acts to transmit the hormonal message to the cell. Receptor proteins are usually glycoproteins. Assays for specific receptors, such as estriol and progesterone receptors, are valuable for the management of certain types of cancers.

An important biological property of some proteins is their ability to catalyze the rates of biochemical reactions. The serum concentrations of these proteins, called "enzymes," are often important in determining the nature of a disease process. Most proteins exhibiting catalytic properties are globular or metalloproteins. An enzyme is most often measured by monitoring of the specific biochemical reaction it catalyzes. The conditions of the enzyme assay are carefully defined so as to give maximal enzymatic activity and therefore sensitivity of analysis (see Chapter 49).

One of the most important biological properties of proteins is their ability to act as antigens (see Chapter 9). An antigen inserted into an immunologically competent host will cause the synthesis of antibodies by the immune system of the host. Antibodies are also globular proteins. An antibody raised against a specific antigen will be able to bind specifically to that antigen. This antibody-antigen interaction is the basis of many assays for the sensitive and specific measurement of proteins that are not able to be detected by other means, and it is used to increase both the senitivity and specificity of the detection of older assays.

Proteins play a major structural role both intracellularly and in tissues. For example, connective tissue is composed primarily of collagen and mucoproteins. The proteins forming the cytoplasmic endoskeleton also fall into this group.

Part II: Lipids

HERBERT K. NAITO

DEFINITION AND CLASSIFICATION

Lipids (fats) comprise a wide spectrum of organic compounds that differ greatly in their chemical and physical properties and in their physiological roles. They include a variety of substances, such as fatty acids, sterols, triacylglycerides (more commonly called "triglycerides"), phosphorus-containing compounds (phospholipids), fat-soluble vitamins, bile acids, waxes, and other complex fats. As a consequence, it is difficult to provide a uniform and clear-cut definition of lipids that is broad enough to encompass all these diverse compounds. In general, however, one can say that lipids are substances that are insoluble in water but soluble in organic solvents such as alcohol, chloroform, ether, acetone, hexane, and benzene. Even with this general definition, there are some exceptions, such as phospholipids, which are rather insoluble in acetone, and some have a limited but significant ability to dissolve in water, such as phosphatidyl serine, phosphatidyl inositol, and phosphatidyl ethanolamine.

There is no generally agreed on system for the classification of lipids, but for simplicity, the following commonly used classification of lipids can be used.

Simple lipids

Simple lipids are esters of fatty acids with various alcohols.

Neutral fats. Neutral fats are esters of fatty acids and glycerol (triglycerides). Because they are uncharged, cholesterol and cholesterol esters, in addition to triglycerides, are also termed "neutral lipids." However, structurally they are steroids and not neutral fats.

The neutral fats contain mixtures of triglycerides—esters of glycerol and fatty acids (such as stearic, palmitic, or oleic acid). The general formula for such a fat is:

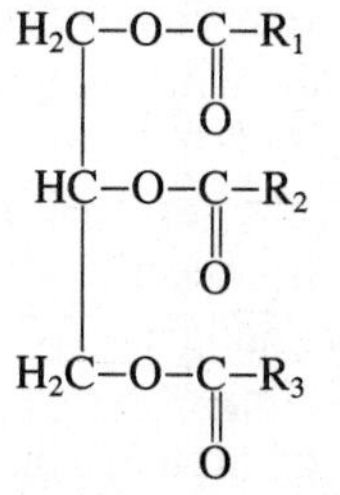

$$\begin{array}{l} H_2C-O-\underset{\displaystyle \|\atop O}{C}-R_1 \\ | \\ HC-O-\underset{\displaystyle \|\atop O}{C}-R_2 \\ | \\ H_2C-O-\underset{\displaystyle \|\atop O}{C}-R_3 \end{array}$$

If $R_1 = R_2 = R_3$ (where R = fatty acid), one has a simple triglyceride. If the R's are not the same, one has a mixed triglyceride. Naturally occurring fats usually exist as mixtures of mixed triglycerides.

The fats then are triesters of the trihydric alcohol (glycerol) and of certain, but not all, organic acids. Since all three glycerol alcohol radicals are esterified, they are termed "triacylglycerides" or more commonly called "triglycerides." A simple ester would be formed by the combination of an acid and an alcohol:

$$CH_3COOH + C_2H_5OH \rightarrow CH_3COOC_2H_5 + H_2O$$

A fat is formed by the combination of an acid (usually of relatively high molecular weight) with the alcohol glycerol.

Being esters, the fats are readily hydrolyzed:

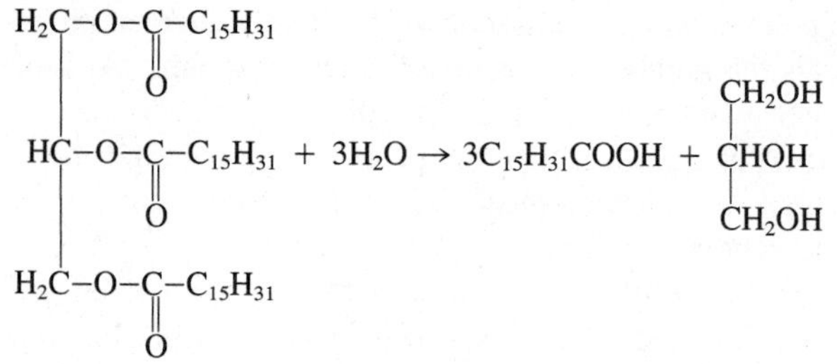

$$\begin{array}{l} H_2C-O-\underset{\displaystyle \|\atop O}{C}-C_{15}H_{31} \\ | \\ HC-O-\underset{\displaystyle \|\atop O}{C}-C_{15}H_{31} \\ | \\ H_2C-O-\underset{\displaystyle \|\atop O}{C}-C_{15}H_{31} \end{array} + 3H_2O \rightarrow 3C_{15}H_{31}COOH + \begin{array}{l} CH_2OH \\ | \\ CHOH \\ | \\ CH_2OH \end{array}$$

Tripalmitin Palmitic acid Glycerol

This hydrolysis is accomplished by use of acid, alkali, superheated steam, or an appropriate enzyme (such as pancreatic lipase). In acid hydrolysis, the free fatty acid is liberated. When alkali is used, a soap is formed, and the process is called *saponification:*

$$\underset{\text{Stearin}}{C_3H_5(O-CO-C_{17}H_{35})_3} + 3NaOH \rightarrow \underset{\text{Sodium stearate (a soap)}}{3C_{17}H_{35}COONa} + \underset{\text{Glycerol}}{C_3H_5(OH)_3}$$

The fats we eat are mostly triglycerides, which contain even-numbered fatty acids because of their mode of biosynthesis, ranging from butyric (C_4) to lignoceric (C_{24}) and probably higher fatty acids (see Tables 48-5 to 48-7). It should be noted that odd-numbered fatty acids do occur naturally.

Waxes. Waxes are esters of fatty acids with higher molecular weight alcohols than glycerol. Examples are carnauba wax, wool wax, beeswax, and sperm oil. Industrially, they are used in the manufacture of lubricants (sperm oil), polishes (carnauba wax), ointments (lanolin, which contains wool wax), candles (spermaceti), and so on.

Aside from cholesterol, the common alcohols found in waxes are cetyl alcohol ($C_{16}H_{33}OH$), ceryl alcohol ($C_{26}H_{53}OH$), and myricyl alcohol ($C_{30}H_{61}OH$).

Conjugated lipids

Conjugated lipids are esters of fatty acids containing groups in addition to an alcohol and a fatty acid (Table 48-4).

Phospholipids. Phospholipids are lipids having, in addition to fatty acids and glycerol, a phosphoric acid residue, nitrogen-containing bases, and other constituents. These lipids include phosphatidyl choline (lecithin), phosphatidyl ethanolamine, phosphatidyl inositol, phosphatidyl serine, sphingomyelins, and plasmalogens. Phosphatidyl ethanolamine, phosphatidyl serine, and phosphatidyl inositol (lipositol) are also known as cephalins.

Table 48-4. Classification of phosphatides and glycolipids

Name	Main alcohol component	Other alcohol components
Glycerophosphatides		
Phosphatidic acid	Diglyceride (= glycerol diester)	
Lecithin	Diglyceride (= glycerol diester)	Choline
Cephalin	Diglyceride (= glycerol diester)	Ethanolamine, serine
Inositide	Diglyceride (= glycerol diester)	Inositol
Plasmalogens (acetal phosphatides)	Glycerol ester and enol ether	Ethanolamine, choline
Sphingolipids		
Sphingomyelins	*N*-Acylsphingosine	Choline
Cerebrosides	*N*-Acylsphingosine	Galactose,* glucose*
Sulfatides	*N*-Acylsphingosine	Galactose*
Gangliosides	*N*-Acylsphingosine	Hexoses,*hexosamine,* neuraminic acid*

*These components are not present as phosphoric esters, but rather in glycosidic linkage; for this reason, cerebrosides, sulfatides, and gangliosides are called "glycolipids."

This class of complex lipids is also called glycerophosphatides, phosphoglycerides, glycerol phosphatides, or more commonly phospholipids. Note that not all phosphorus-containing lipids are phosphoglycerides; that is, sphingomyelin is a phospholipid because it contains phosphorus, but it is better classified as a sphingolipid because of the nature of the backbone structure to which the fatty acid is attached. In phospholipids, one of the primary OH groups of glycerol is esterified to phosphoric acid; the other OH groups are esterified to fatty acids. The parent compound of the phospholipids is phosphatidic acid, which contains no polar alcohol head group. The phospholipids are constituents of all animal and vegetable cells. They are present in abundance in brain, heart, kidney, eggs, soybeans, and so on. The phospholipids have been commonly separated into five distinct groups of compounds: phosphatidic acid, lecithin, cephalin, plasmalogens, and sphingolipids. In addition to carbon, hydrogen, and oxygen, the compounds contain the elements nitrogen and phosphorus. In lecithin and cephalin, the nitrogen-phosphorus ratio is 1:1; in sphingomyelin it is 2:1.

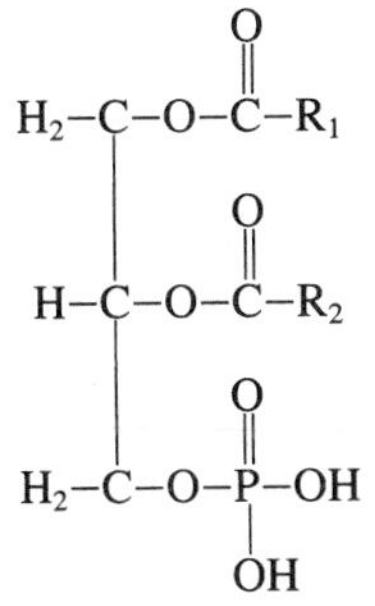

$$\begin{array}{l} H_2{-}C{-}O{-}C({=}O){-}R_1 \\ \quad | \\ H{-}C{-}O{-}C({=}O){-}R_2 \\ \quad | \\ H_2{-}C{-}O{-}P({=}O)(OH){-}OH \end{array}$$

Phosphatidic acid

Phosphatidic acid. Phosphatidic acid is important as an intermediate in the synthesis of triglycerides and phospholipids, but it is not found in any quantity in tissues. Phosphatidic acid is the simplest type of phospholipid. Phosphatidic acid is derived from glycerophosphoric acid by esterification of the two remaining OH groups with fatty acids.

Lecithins. On hydrolysis, a typical lecithin forms glycerol, 2 moles of fatty acids, phosphoric acid, and the nitrogenous base, choline. Most lecithins have a saturated fatty acid in the C-1 position and an unsaturated fatty acid in the C-2 position. The structural formula may be written as follows:

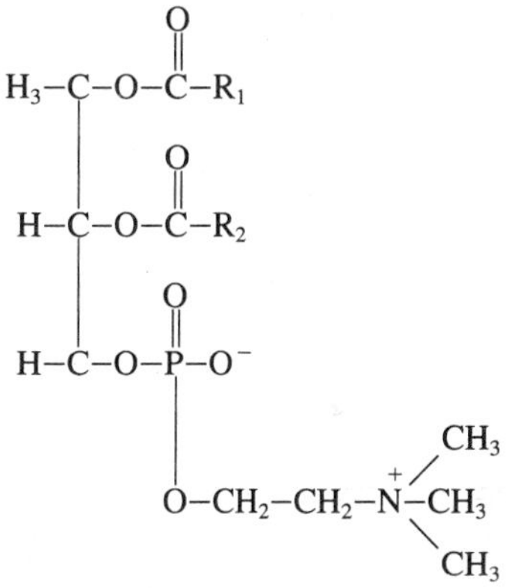

$$\begin{array}{l} H_3{-}C{-}O{-}C({=}O){-}R_1 \\ \quad | \\ H{-}C{-}O{-}C({=}O){-}R_2 \\ \quad | \\ H{-}C{-}O{-}P({=}O){-}O^- \\ \qquad\qquad | \\ \qquad\qquad O{-}CH_2{-}CH_2{-}\overset{+}{N}(CH_3)_3 \end{array}$$

Lecithin
(phosphatidyl choline)

The lecithins, like cholesterol, are common cell constituents that occur principally in animal tissue, having both structural (as part of cell membranes) and metabolic functions. Although not found in depot fat, they make up a considerable proportion of the liver and brain lipids. They also occur in the plasma as part of the lipid-protein complexes called lipoproteins; thus they are important for the formation of these macromolecules, which play an important role in fat transport. Lecithins play an important role in the esterification of free cholesterol to form ester cholesterol. Lecithins are an important precursor for the formation of lung surfactant. Lecithins form colloidal solutions with water, but one can prepare an aqueous solution of lecithin by the addition of bile salts. It is believed that a bile salt–lecithin compound is formed that is water soluble. Lecithins are not soluble in acetone; this property is used to separate them from other phospholipids and lipids.

Cephalins. The cephalins resemble the lecithins in structure except for the component corresponding to choline. There are three main fractions: the ethanolamine cephalins, serine cephalins, and inositol cephalins:

$$
\begin{array}{l}
H_2-C-O-\overset{O}{\overset{\|}{C}}-R_1 \\
\quad\ \ | \\
H-C-O-\overset{O}{\overset{\|}{C}}-R_2 \\
\quad\ \ | \\
H_2-C-O-\overset{O}{\overset{\|}{P}}-OH \\
\qquad\qquad\ \ | \\
\qquad\qquad\ O-\boxed{\text{Ethanolamine}} \text{ or } \boxed{\text{Serine}} \text{ or } \boxed{\text{Inositol}}
\end{array}
$$

The cephalins differ from lecithins in their insolubility in ethyl or methyl alcohol.

Phosphatidyl ethanolamine. This cephalin differs from lecithins in that ethanolamine replaces choline. Both alpha and beta cephalins are known. This is one of the more abundant cephalins found in higher plants and animals.

Phosphatidyl serine. This cephalin, which contains the amino acid serine rather than ethanolamine, has been found in tissues, like the brain.

Phosphatidyl inositol. This cephalin is found in phospholipids of brain tissue and of soybeans and in other plant phospholipids as well. The inositol is present as the stereoisomer myoinositol.

Plasmalogens. Plasmalogens constitute as much as 10% of the phospholipids of the membranes of nerves and muscles. They are also found in the liver and other organs. Their function is not known at present. Structurally, the plasmalogens resemble lecithins and cephalins but give a positive reaction when tested for aldehydes with Schiff's reagent (fuchsin–sulfurous acid) after pretreatment of the phospholipid with mercuric chloride. These phospholipids contain higher fatty aldehydes in place of fatty acids. Thus the basic units of this class of compounds include glycerol, phosphorus, fatty aldehyde, and ethanolamine.

Sphingolipids. The amino dialcohol group sphingosine characterizes all sphingolipids. It serves as a structural unit for substitution, just as the trihydroxyalcohol glycerol does in glycerides. Sphingosine is a long-chain C_{18} compound that contains a *trans*-double bond, an NH_2 group on C-2, and two OH groups (on C-1 and C-3). Sphingolipids are especially abundant in the brain. Some storage diseases are characterized biochemically by the accumulation of certain sphingolipids. There are four major categories (Table 48-2): sphingomyelins, cerebrosides, sulfatides, and gangliosides.

Sphingomyelins. These are found in the brain and in other organs. Stearic, lignoceric, and nervonic acid are the sole fatty acids present in brain sphingomyelins, whereas palmitic and lignoceric acids are the fatty acids in lung and spleen sphingomyelins. A typical formula is:

Sphingomyelin

$$
\underbrace{CH_3-(CH_2)_{12}-CH=CH-\underset{H}{\overset{HO}{C}}-\underset{\underset{\underset{\underset{CH_3}{|}}{(CH)_{22}}}{C=O}}{\overset{H}{C}}\!\!\!\!\!\!\underset{NH}{}-CH_2}_{\text{Sphingosine}}-\underbrace{O-\underset{O^-}{\overset{O}{\overset{\|}{P}}}-O-CH_2-CH_2-\overset{+}{N}(CH_3)_3}_{\text{Phosphoryl choline}}
$$

(Fatty acid: $C=O$, $(CH)_{22}$, CH_3 attached to NH)

and its two important constituents are:

$$
CH_3-(CH_2)_{12}-CH=CH-\underset{H}{\overset{HO}{C}}-\underset{NH_2}{\overset{H}{C}}-CH_2OH \qquad CH_3(CH_2)_{22}-COOH
$$

Sphingosine **Lignoceric acid**

Cerebrosides. Cerebrosides contain galactose or glucose, a high molecular weight fatty acid, and sphingosine. Thus cerebrosides have the following basic structure:

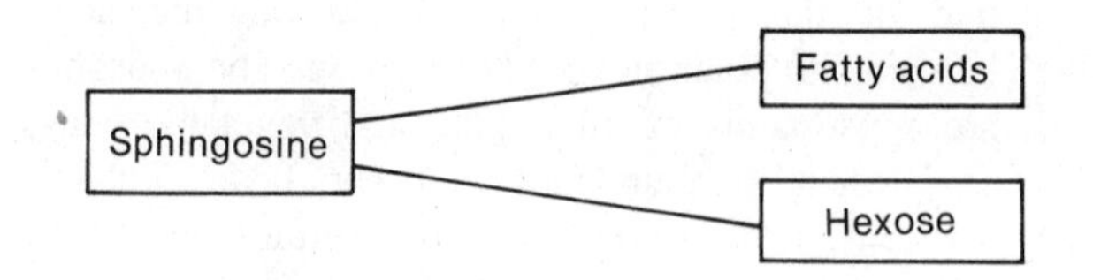

They are structurally similar to sphingomyelins. Cerebrosides may also be classified with the sphingomyelins as sphingolipids. Individual cerebrosides are differentiated by the type of fatty acid in the molecule: *kersins* contain lignoceric acid; *cerebrons* contain a hydroxylignoceric acid (cerebronic acid); *nervons* contain an unsaturated homolog of lignoceric acid called nervonic acid; and *oxynervons* apparently contain the hydroxyl derivative of nervonic acid as a constituent fatty acid.

$$CH_3-(CH_2)_{22}-COOH$$
Lignoceric acid

$$CH_3-(CH_2)_{21}-CH(OH)-COOH$$
Cerebronic acid

$$CH_3-(CH_2)_7-CH=CH-(CH_2)_{13}-COOH$$
Nervonic acid

$$CH_3-(CH_2)_7-CH=CH-(CH_2)_{12}-CH(OH)-COOH$$
Oxynervonic acid

Cerebrosides are found in many tissues besides the brain. In Gaucher's disease, the cerebroside content of the reticuloendothelial cells (as in the spleen) is very high. The cerebrosides are in much higher concentration in myelinated than in nonmyelinated nerve fibers.

Sulfatides. Sulfatides are sulfate derivatives of the galactosyl residue in cerebrosides.

Gangliosides. These substances are glycolipids occurring in the brain (in ganglionic cells). The main components are sphingosine, fatty acids, and branched-chain carbohydrates with as many as seven sugar residues. The construction of gangliosides is similar to that of cerebrosides, but the carbohydrate moiety is far more complex. The various gangliosides differ primarily in the number of sugar residues.

Derived lipids

Derived lipids are compounds derived from the hydrolysis of simple and conjugated fats. These include the compounds described below.

Fatty acids. Fatty acids are straight-chain carboxylic acids (both saturated and unsaturated). Over 100 different kinds of fatty acids have been isolated from various lipids of animals, plants, and microorganisms. All possess a long hydrocarbon chain and a terminal carboxyl group. Fatty acids are obtained from the hydrolysis of fats or can be synthesized from two carbon units (acetyl radicals). Fatty acids that occur in naturally occurring fats usually contain an even number of carbon atoms (because they are synthesized from two carbon units) and are straight-chain derivatives. The chain may be saturated (containing no double bonds) or unsaturated (containing one or more double bonds). Tables 48-5 and 48-6 list some fatty acids found in fats that occur commonly in human diets.

Some generalizations may be made on the fatty acids present in lipids of higher plants and animals. Nearly all have an even number of carbon atoms and have chains that are between 14 and 22 carbon atoms long; those having 16 or 18 carbons are by far the most abundant. In general, unsaturated fatty acids predominate over the saturated type, particularly in the neutral fats and in cells of poikilothermic (cold-blooded) organisms living at lower temperatures. Unsaturated fatty acids have lower melting points than saturated fatty acids. Most neutral fats rich in unsaturated fatty acids are liquid down to 5° C or lower. In most unsaturated fatty acids in higher organisms, there is a double bond between carbon atoms 9 and 10; additional double bonds usually occur between C-10 and the methyl end of the chain. In fatty acids containing two or more double

Table 48-5. Some naturally occurring straight-chain saturated acids

Shorthand notation C_n	Formula	Systematic name	Common name	Melting point (°C)	Typical occurrence
1:0	$HCOOH$	Methanoic	Formic	8.4	Ruminants
2:0	CH_3COOH	Ethanoic	Acetic	16.7	As alcohol acetates in many plants, and in some plant triacylglycerols. At low levels, widespread as salt or thioester; at higher levels in the rumen as salt
3:0	CH_3CH_2COOH	Propionic	Propionic	−22.0	At high levels in the rumen
4:0	$CH_3(CH_2)_2COOH$	Butanoic	Butyric	−7.9	At high levels in the rumen, also in milk fat of ruminants
6:0	$CH_3(CH_2)_4COOH$	Hexanoic	Caproic	−2.0	Found in milk fat
8:0	$CH_3(CH_2)_6COOH$	Octanoic	Caprylic	16.7	Very minor component of animal and plant fats
10:0	$CH_3(CH_2)_8COOH$	Decanoic	Capric	31.6	Widespread as a minor component
12:0	$CH_3(CH_2)_{10}COOH$	Dodecanoic	Lauric	44.2	Widely distributed, a major component of some seed fats
14:0	$CH_3(CH_2)_{12}COOH$	Tetradecanoic	Myristic	54.1	Widespread, occasionally as a major component
16:0	$CH_3(CH_2)_{14}COOH$	Hexadecanoic	Palmitic	62.7	Widespread usually as a major component; one of the most common fatty acids in both plants and animals
18:0	$CH_3(CH_2)_{16}COOH$	Octadecanoic	Stearic	69.6	Widespread, usually as a major component
20:0	$CH_3(CH_2)_{18}COOH$	Eicosanoic	Arachidic	75.4	Widespread minor component, occasionally a major component
24:0	$CH_3(CH_2)_{22}COOH$	Tetracosanoic	Lignoceric	84.2	Fairly widespread as minor component in seed fat triacylglycerols
26:0	$CH_3(CH_2)_{24}COOH$	Hexacosanoic	Cerotic	90.9	Major component of some plant waxes

Table 48-6. Some common unsaturated fatty acids

Shorthand notation C_n	Formula	Systematic name	Common name	Melting point (°C)	Typical occurrence
16:1	$CH_3(CH_2)_5CH{=}CH(CH_2)_7COOH$	9-Hexadecenoic	Palmitoleic	−0.5	Animal and vegetable fats, seed oils
18:1	$CH_3(CH_2)_7CH{=}CH(CH_2)_7COOH$	9-Octadecenoic	Oleic	13.4	Animal and vegetable fats
18:2ω6	$CH_3(CH_2)_4CH{=}CHCH_2CH{=}CH(CH_2)_7$-COOH	9,12-Octadecadienoic	Linoleic*	05.0	Linseed oil, cottonseed oil, and others
18:3ω3	$CH_3CH_2CH{=}CHCH_2CH{=}CHCH_2CH{=}CH(CH_2)_7COOH$	9,12,15-Octadecatrienoic	Linolenic*	−1.0	Linseed oil
18:3ω6	$CH_3(CH_2)_4CH{=}CHCH_2CH{=}CHCH_2CH{=}CH(CH_2)_3COOH$	6,9,12-Octadecatrienoic	γ-Linolenic		
20:3ω6	$CH_3(CH_2)_7CH{=}CHCH_2CH{=}CHCH_2CH{=}CH(CH_2)_3COOH$	8,11,14-Eicosatrienoic	Homo-γ-linolenic		Animal fats
20:3ω9	$CH_3(CH_2)_7CH{=}CHCH_2CH{=}CHCH_2CH{=}CH(CH_2)_3COOH$	5,8,11-Eicosatrienoic			
20:4ω6	$CH_3(CH_2)_4CH{=}CHCH_2CH{=}CHCH_2CH{=}CHCH_2CH{=}CH(CH_2)_3COOH$	5,8,11,14-Eicosatetraenoic	Arachidonic*	−49.5	Animal fats

*Essential fatty acids (cannot be synthesized by animal cells).

Table 48-7. Common unsaturated fatty acids, number of double bonds, and length of carbon chain

Fatty acid	Number of double bonds	Number of carbons
Palmitoleic	1	16
Oleic	1	18
Linoleic	2	18
Linolenic	3	18
Arachidonic	4	20

bonds, the double bonds are never found in conjugation but are separated by one methylene group. The double bonds of nearly all the naturally occurring unsaturated fatty acids are in the *cis* configuration. The most abundant unsaturated fatty acids in higher organisms are oleic, linoleic, linolenic, and arachidonic acids (Table 48-7).

Alcohols. Straight-chain alcohols and cyclic alcohols (such as the sterols) are a subclass of derived lipids.

The steroids may be classifed into the following groups:

Sterols
Bile acids
Substances obtained from cardiac glycosides
Substances obtained from saponins
Sex hormones
Adrenocorticosteroids
Vitamin D

These compounds are widely distributed in plant and animal tissues, either in the free state or in the form of esters (a combination with higher fatty acids). Chemically, they are known to be phenanthrene derivatives, or more correctly cyclopentanoperhydrophenanthrene derivatives.

Steroids. The best known steroid is cholesterol. It is present in all animal cells and is particularly abundant in nervous tissue and liver. Varying quantities of this steroid are found admixed in animal fats but not in vegetable fats. The structure of cholesterol is illustrated in Fig. 48-7.

Cholesterol is the precursor of many other steroids in animal tissues, including the bile acids, detergent-like compounds that aid in emulsification and absorption of lipids in the intestine; the androgens, or male sex hormones; the estrogens, or female sex hormones; the progestational hormones; and the adrenocortical hormones.

Cholesterol is a member of a large subgroup of steroids called the *sterols*. It is a steroid alcohol containing a hydroxyl group at carbon 3 of ring A and a branched aliphatic chain of eight or more carbon atoms at carbon 17. Sterols occur either as free alcohols or as long-chain fatty acid esters of the hydroxyl group at carbon 3; all are solids at room temperature. Cholesterol melts at 150°C and is insoluble in water but readily extracted from tissues with

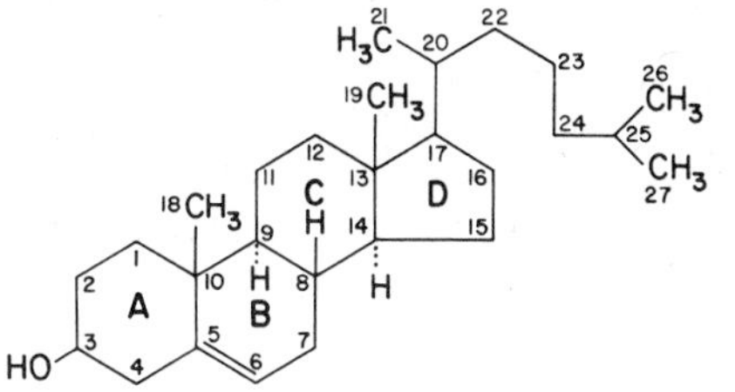

Fig. 48-7. Structure of cholesterol molecule, a C_{27} hydrocarbon sterol.

chloroform, ether, or hot alcohol. Cholesterol occurs in the plasma membranes of animal cells and in the lipoproteins of blood. Cholesterol is found only in animal tissues and fluids, never in plants.

Other similar steroids are phytosterols, which are steroids from plants. Among these are stigmasterol, campesterol, and sitosterol.

Fungi and yeasts contain still other types of sterols, the mycosterols. Among these is ergosterol, which is converted to vitamin D.

Bile acids. Bile acids (a C_{24} steroid) are digestion-promoting constituents of bile. They are surface-active agents. This means that they lower surface tension and thus can emulsify fats, an important step in the formation of micelles. Bile acids also activate gastrointestinal lipases. For these reasons, bile acids play an important physiological role in the digestion and absorption of fats.

The major primary bile acids are cholic acid and chenodeoxycholic acid (Fig. 48-2), which are made in the liver by the enzymatic cleavage of the terminal three carbons on the cholesterol molecule (a C_{27} hydrocarbon). Thus the bile acids are one of the end products of the metabolism of cholesterol; however, it should be noted that bile acid constitutes the acidic sterol fraction of the bile, which is about 50% to 60% of the total steroid excreted. The remainder of the steroid output in the bile is in the form of neutral steroids, such as cholesterol.

Hydrocarbons. The hydrocarbons are both aliphatic and cyclic compounds.

Vitamins. Since vitamins and hormones are not discussed in this chapter, their structural formulas are not presented. See Chapters 34 and 63.

Other compound lipids. Sulfolipids, aminolipids, and lipoproteins may also be placed in this category.

CHEMICAL AND PHYSICAL PROPERTIES

Melting point. The melting point of fatty acids is influenced by the chain length and degree of chain unsaturation. Increasing the chain length and decreasing the number of unsaturated double bonds will increase the melting point of fatty acids. The melting points of fatty acids and other lipids can be used to identify the compound, but this property is not routinely used in analysis.

Solubility. The relative insolubility of lipids in aqueous solutions is an important property of lipids. The major consequence of this insolubility is that analyses of lipids often require a prior treatment of the sample so as to extract the lipid into a more lipid-soluble medium, such as methanol, chloroform, or ether.

The insolubility of unesterified cholesterol has been used in the past for certain cholesterol assays. Cholesterol can be quantitatively separated from cholesterol esters by precipitation with digitonin as a 1:1 molecular complex of cholesterol digitonide. Measurement, gravimetrically, of the amount of digitonide complex formed has been the basis of cholesterol quantitation in the past.

Specific gravity. The specific gravity of all fat is less than 1.0 g/mL. Consequently, all fats float in water. This characteristic has enabled lipoproteins to be selectively separated from more dense proteins, and individual lipoproteins to be separated from each other on the basis of varying proportions of lipid content.

Alcohol groups of steroids. The chemically reactive alcohol group of steroids is the basis of many assays for quantitating cholesterol. The group can react with strong mineral acids, such as sulfuric acid and salts, to form a chromogen. The Burchard-Liebermann reaction is the most frequently used example of such a chemical reaction.

This hydroxyl group can also be specifically oxidized by the enzyme cholesterol oxidase. Monitoring of this reaction is the basis for the increasingly used enzyme assays for cholesterol.

Triglyceride composition. The chemical composition of triglycerides, that is, the esterification of glycerol by three fatty acids, is the basis of all methods for quantitating triglycerides. These techniques are based on the quantitation of glycerol released from triglycerides after chemical or enzymatic hydrolysis of the fatty acid esters. The glycerol can be chemically or enzymatically oxidized to form measurable chromogens.

Chemical composition. The phospholipids are detectable by specific reactions for phosphorus after chemical reaction.

BIOLOGICAL PROPERTIES

The most important biological properties of lipids are structural, nutritional, and hormonal. Almost all classes of lipids are used as structural components of membranes. The triglycerides are essential components in the formation of the bimolecular protein-lipid–lipid-protein membranes. Cell membranes also contain varying amounts of steroids, phospholipids, and other complex lipids.

Triglycerides also function as an important source of calories and as a source of carbon atoms for the synthesis of other macromolecules.

The role of steroids as hormones has a large impact on a laboratory because the measurement of many of these hormones is required. These hormones are often measured by some of the chemical techniques discussed above. For example, one can measure cortisol and estriol by the chromophores formed after reaction with hot sulfuric acid.

Part III: Carbohydrates

LAWRENCE A. KAPLAN

DEFINITION AND CLASSIFICATION

The earliest carbohydrates were found to have the empirical formula of $(CH_2O)_n$. Thus these chemicals were simply defined as compounds consisting of hydrated (H_2O) carbon; hence the name "carbohydrate." Subsequently the presence of complex carbohydrates containing other chemical moieties was noted to exist. Thus carbohydrates can be covalently linked to proteins, lipids, and nucleic acids. The various classes of carbohydrates are discussed below.

Simple monomeric carbohydrates (saccharides). Saccharides are also known as "sugars," and their common names all end with the suffix *-ose,* meaning "sugar." The smallest sugar units are called "monosaccharides," in which n in the above formula is from 3 to 8. If $n = 3$, the sugar is triose, if $n = 4$, a tetrose, and so on. The monosaccharides are straight carbon chains in which each carbon atom except one carries a hydroxyl group (−OH); the one remaining carbon atom has a carbonyl group. If the carbonyl group is on the first or last carbon atom, the carbonyl group is an aldehyde and the monosaccharide is called an "aldose." If the carbonyl group is on an internal carbon atom, it is a "ketone," and the monosaccharide is called a "ketose" (Fig. 48-8). Thus a 4-carbon aldose is an "aldotetrose," a 6-carbon ketose is a "ketohexose," and so on.

The monosaccharides found in nature are all stereoisomers. Stereoisomerism is defined by the ability of a molecule to rotate the plane of incident polarized light. The physical and chemical properties of, for example, all the eight aldohexoses (6-carbon chain) are exactly the same except for their different actions on polarized light. All the monosaccharides in human biochemistry are of the dextroisomeric (D) form. Examples of some monosaccharides are given in Fig. 48-9.

The pentose and hexose monosaccharides also have the ability to form ring structures by intramolecular reaction of the terminal hydroxyl group with the carbonyl function. The six-membered ring forms of the sugars are called "pyranoses," whereas the five-membered rings are called "furanoses." The aldohexoses, such as D-glucose, form six-membered rings, whereas an aldoketose, such as D-fructose, forms a five-membered ring (Fig. 48-9).

Glucose can form two types of six-membered rings. The rings differ in how the hydroxyl group at the number 1 carbon atom is positioned with respect to the plane of the ring. If the hydroxyl group is on the same side of the molecule as the ring oxygen (Fig. 48-9), the isomer is known as the α-D-glucose isomer, whereas if the hydroxyl group is on the opposite side of the ring oxygen, then this isomer is known as the β-D-glucose. Enzymes acting on carbohydrates usually have a specificity directed toward one of the isomers, usually the most common one found, such as β-D-fructose.

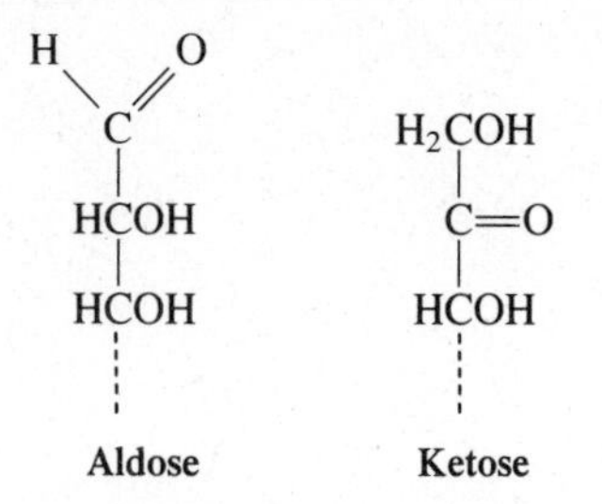

Fig. 48-8. Structural differences between aldoses and ketoses, which are aldehydes and ketones, respectively.

Derived monosaccharides. These sugars are formed by reduction or oxidation of the carbonyl groups. The products of reductive reactions are polyols (polyalcohols), such as D-sorbitol or D-mannitol, whereas the products of oxidation are acids, such as D-glucuronic acid (from D-glucose). Many acid forms of monosaccharides are important constituents of more complex carbohydrates, such as mucopolysaccharides.

An important group of derived monosaccharides is the result of the replacement of a hydroxyl group by an amino group. The term "sialic acid" is used to describe the important *N*-acetyl derivatives of neuraminic acid, which are often found covalently linked to proteins (Fig. 48-10).

Complex carbohydrates. These molecules are formed by linking two or more monosaccharides by a glycosidic linkage (Fig. 48-11). The simplest disaccharides, important nutritionally, are maltose (two glucoses), lactose (milk sugar, one galactose and one glucose), and sucrose (one fructose and one glucose). Oligosaccharides are often defined as carbohydrates containing two to 10 monosaccharide subunits. Polysaccharides are larger polymers of up to 100 million daltons. All three of the most important polysaccharides contain glucose as the monomeric subunit. Cellulose, a structural component of plant walls, consists of glucose units linked by a β-(1→4) glycosidic bond to form long, unbranched chains. Starch, a storage form of glucose in plants, consists of glucose residues connected by α-(1→4) glycocytic linkages, which, unlike the β-(1→4) linkages of cellulose, are amenable to degradation by human hydrolytic enzymes (such as amylase). Starch also differs from cellulose in that it is a branched molecule. Branching points are scattered throughout the molecule formed by α-(1→6) bonds. The two forms of starch are therefore called "amylose" (the straight-chain fraction) and "amylopectin" (the highly branched fraction). Glycogen is the glucose-storage molecule found in animal cells. Glycogen more closely resembles amylopectin than

α-D-Glucose (α-D-glucopyranose) D-Glucose (aldehyde form) β-D-Glucose (β-D-glucopyranose)

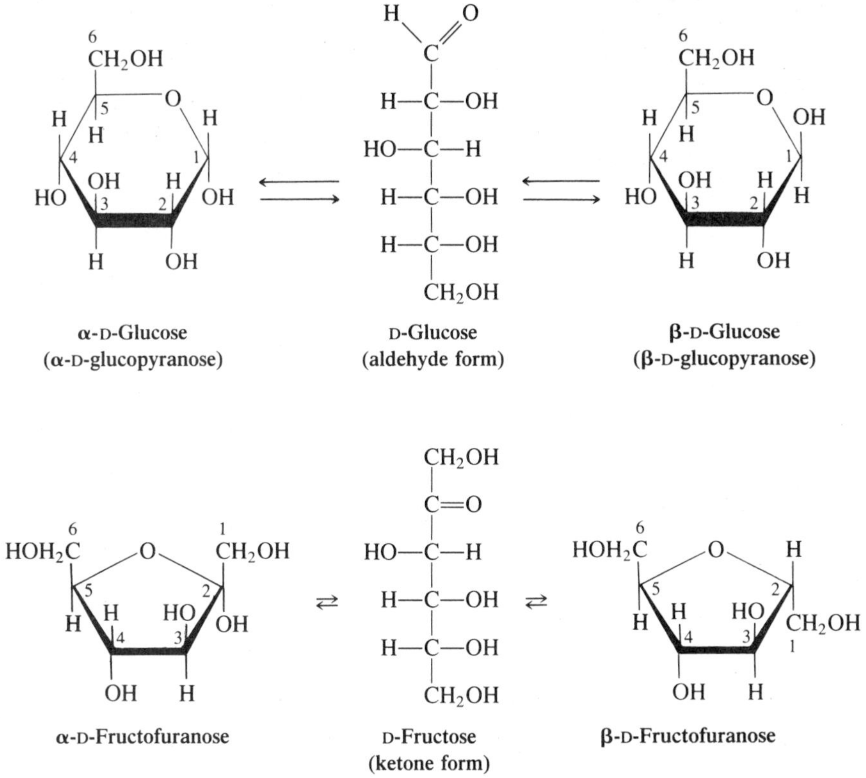

α-D-Fructofuranose D-Fructose (ketone form) β-D-Fructofuranose

Fig. 48-9. Interrelationships between straight-chain and ring forms of D-glucose and D-fructose, which form pyranose and furanose rings. (From Orten, J.M., and Neuhaus, O.W.: Human biochemistry, ed. 10, St. Louis, 1982, The C.V. Mosby Co.)

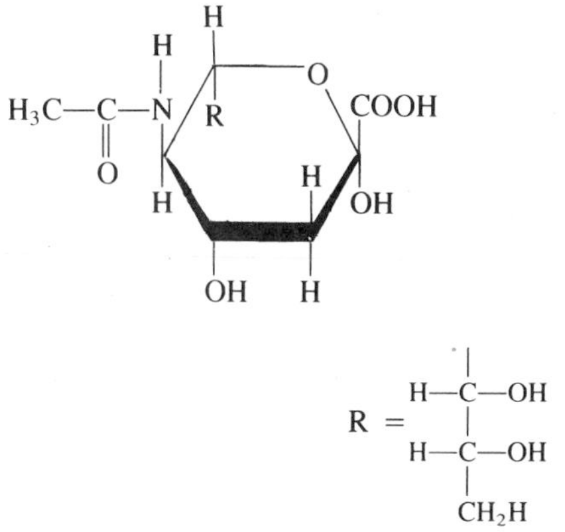

N-Acetylneuraminic acid

Fig. 48-10. Structure of *N*-acetylneuraminic acid (''sialic acid''). (From Orten, J.M., and Neuhaus, O.W.: Human biochemistry, ed. 10, St. Louis, 1982, The C.V. Mosby Co.)

amylose because of its highly branched nature (see Fig. 30-3).

Complex polysaccharides containing hyaluronic acid, chondroitin-4-sulfate, and keratin sulfates as the repeating subunits are important constituents of synovial fluid and connective tissue. Heparin is a complex polysaccharide containing *N*-acetyl-D-glucosamine-6-sulfate as the repeating subunit.

CHEMICAL PROPERTIES

The monosaccharides (pentoses and larger) can undergo dehydration in the presence of hot mineral acids to form the cyclic furfural derivatives. One can dehydrate glucose in this manner to form 3-hydroxymethylfurfural, a reaction which is the basis for a colorimetric assay for glycosylated proteins.

An important chemical property of the monosaccharides is the ability of these compounds to be oxidized or reduced and in turn reduce or oxidize some other compounds. The ability of reducing aldoses, such as glucose, to be oxidized to the acid form has been the historic basis for chemical assays for glucose. The glucose in turn reduced such compounds as Cu^{++} or $Fe(CN_6)^-$ with the formation of colored complexes of the reduced forms of these compounds (such as Cu^+ and Cu_2O).

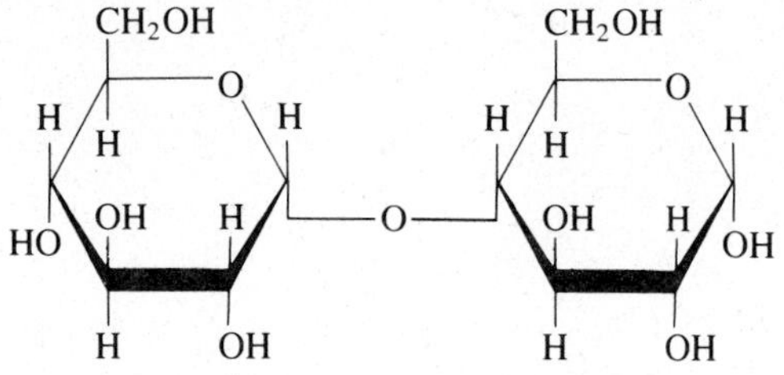

(D-glucose) (D-glucose)
(4-*O*-α-D-glucopyranosyl-D-glucopyranose)
Maltose (β-form)

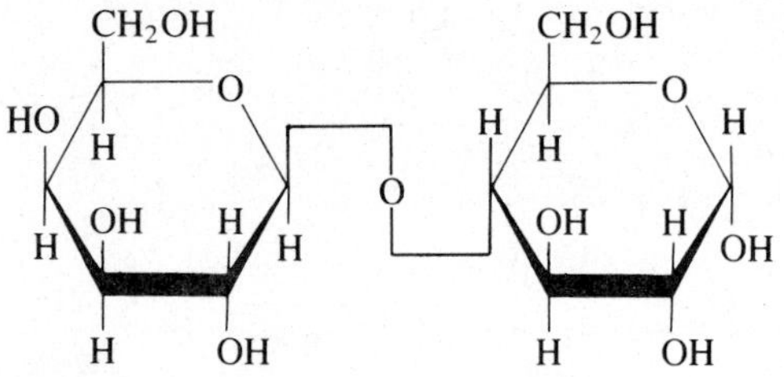

(D-galactose portion) (D-glucose portion)
(4-*O*-β-D-galactopyranosyl-α-D-glucopyranose)
Lactose (α-form)

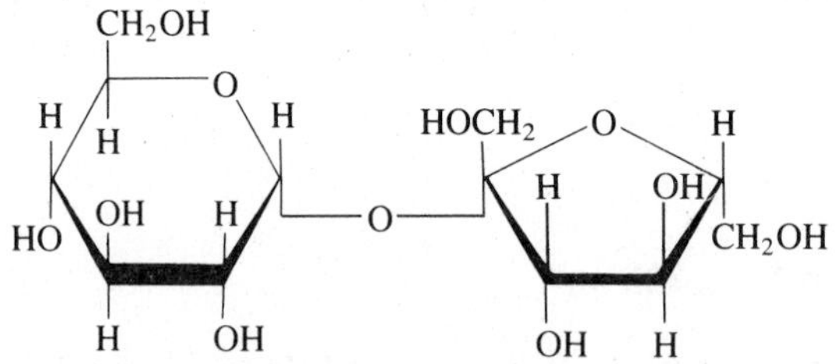

(D-glucose portion) (D-frutose portion)
(α-D-glucopyranosyl-β-D-fructofuranoside)
Sucrose

Fig. 48-11. Common disaccharides linked by β-glycosidic bonds. (From Orten, J.M., and Neuhaus, O.W.: Human biochemistry, ed. 10, St. Louis, 1982, The C.V. Mosby Co.)

The enzymatic oxidation of glucose by glucose oxidase is the basis of many of the current glucose assay procedures, whereas the oxidation of glucose-6-phosphate is the basis of the hexokinase assay for glucose.

Aldoses, such as glucose, can react with primary amines to form a Schiff base. This nonenzymatic condensation is the mechanism for the formation of glycoproteins in blood. In addition, the reaction of glucose with aromatic primary amines, such as *o*-toluidine, is the basis of an important method for measuring blood glucose.

PHYSICAL PROPERTIES

The commonly measured monosaccharides and disaccharides are highly water-soluble compounds. Assays for these analytes thus do not require prior extraction or purification. Separation of the monosaccharides by adsorption chromatography is possible, though this is usually performed by specialized metabolic laboratories. The simple monosaccharides, disaccharides, or polysaccharides are not readily distinguished by their spectral or electrophoretic properties.

BIOLOGICAL PROPERTIES

The monosaccharides and disaccharides are the major source of calories for the human body and as such serve as a primary form of nutrition. Polymeric forms of glucose, such as glycogen, serve as a storage for glucose in liver and muscle cells. Complex polysaccharides are found in body fluids and connective tissue.

BIBLIOGRAPHY

General

Frisell, W.R.: Human biochemistry, New York, 1982, Macmillan, Inc.

Orten, J.M., and Neuhaus, O.W.: Human biochemistry, ed. 10, St. Louis, 1982, The C.V. Mosby Co.

Proteins and amino acids

Anfinsen, C.B., Jr., Edsall, J.T., and Richards, F.M.: Advances in protein chemistry, vol. 1-26, New York, 1944-1972, Academic Press, Inc.

Dickerson, R.E., and Gris, I.: The structure and actions of proteins, New York, 1969, Harper & Row, Publishers, Inc.

Lipids

Gurr, M.I., and James, A.T.: Lipid biochemistry, ed. 2, New York, 1975, John Wiley & Sons, Inc.

Johnson, A.R., and Davenport, J.B.: Biochemistry and methodology of lipids, New York, 1971, John Wiley & Sons, Inc.

Carbohydrates

Whelan, W.J., editor: Biochemistry of carbohydrates, vol. 5. In MTP International Review of Science, Baltimore, 1975, University Park Press.

Pigman, W., and Horton, D., editors: The carbohydrates: chemistry and biochemistry, vol. 1A and 1B, New York, 1972, Academic Press, Inc.

Chapter 49 Enzymes

David C. Hohnadel

activation energy The energy required to convert reactants to activated or transition-state species that will spontaneously proceed to products.

activators Inorganic ions that are cofactors for an enzyme reaction.

active sites The specific areas on an enzyme where a substrate binds and catalysis takes place.

activity The activity of an enzyme is the amount of substrate for a particular reaction that is converted to product per unit time under defined conditions.

allosteric sites, or regulatory sites The sites, other than the active site or sites, of an enzyme that bind regulatory molecules.

apoenzyme An enzyme without any associated cofactors or with less than the entire amount of cofactors or prosthetic groups.

auxiliary enzyme In a coupled assay system, an enzyme that links the enzyme being measured with an indicator enzyme.

binding sites The sites on the surface of the enzyme that serve to bind the substrate or product of the reaction.

bond specificity The nature of enzyme action that causes the disruption of only certain bonds between atoms.

catalyst A substance that increases the rate of a reaction without being changed by the reaction.

catalytic site Another name for active site.

coenzymes Organic cofactor compounds, such as nicotinamide adenine dinucleotide phosphate ($NADP^+$) and pyridoxyl-5-phosphate.

cofactors Nonprotein substances associated with an enzyme that are needed for catalytic activity.

competitive inhibitor An inhibitor of an enzyme reaction that competes with the substrate by binding at the active site.

constitutive enzymes Enzymes that are present at a constant concentration during the life of a cell.

coupled assays Assays with several enzyme reactions leading to a reaction that has an easily measured substance.

denaturation The loss of the biological properties or a protein usually as a result of changes in tertiary or quaternary structure.

EC code The four-digit Enzyme Commission code for the systematic classification of enzyme reactions.

ELISA Enzyme-linked immunosorbent assay.

EMIT Enzyme-multiplied immunoassay technique.

end-point assays Assays whereby a single measurement is made at a fixed time.

endopeptidases Protein-hydrolyzing enzymes that break bonds in the interior of a protein substrate.

enzyme kinetics The study of enzyme reaction rates and the factors that affect them.

enzyme specificity The degree to which an enzyme will catalyze one or more reactions.

enzyme substrate complex An intermediate active complex formed between the substrate and the enzyme during the reaction.

enzymes Biological materials (proteins) with catalytic properties.

equilibrium constant The ratio of the concentration of product to the concentration of substrate when the reaction is at equilibrium.

exopeptidases Protein-hydrolyzing enzymes that break bonds proceeding from one end of the protein substrate toward the center of the substrate.

first-order kinetics State occurring when the rate of an enzyme reaction is proportional to the concentration of the substrate.

holoenzymes The complete enzyme-cofactor complex that gives full catalytic activity.

hydrophilic amino acids Polar, water-loving amino acids.

hydrophobic amino acids Nonpolar, water-hating amino acids.

inactivation A reversible denaturation of a protein.

indicator enzymes Enzymes that produce (or consume) an easily measured substance.

inducible enzymes Enzymes whose cellular concentrations can be made to increase.

inhibitors Materials that reduce the catalytic activity of an enzyme.

initial rates Enzyme measurements made at the start of a reaction just after the lag phase.

international unit of enzyme activity The amount of enzyme that catalyzes the conversion of one micromole of substrate per minute under defined conditions, $1\ U = 16.7 \times 10^{-8}$ katal.

in vitro systems Those outside of a living organism, that is, in a test tube.

isoenzymes Different forms of an enzyme that catalyze the same reaction.

katal (kat, K) An enzyme unit in moles per second defined by the SI system: $1\ K = 6.6 \times 10^{9}\ U$

K_m The symbol for the Michaelis-Menten constant.

kinetic assays Assays that form increasing amounts of product with time, usually monitored by multiple datum points.

labile enzymes Unstable or easily denatured proteins.

lag phase The early time in an assay when mixing occurs and temperature and kinetic equilibrium are becoming established.

linear phase Time when an assay is following zero-order kinetics producing a constant absorbance change per unit of time.

metalloenzymes Enzymes that contain very tightly bound metal ions.

Michaelis-Menten constant A constant related to the rate constants of an enzyme reaction and equal to the concentration of substrate that gives one half the maximal velocity.

noncompetitive inhibitor An inhibitor that binds to an allosteric site of an enzyme and does not compete with the substrate by binding at the active site.

optimal assay conditions Conditions for reaction concentrations of substrates, cofactors, activators, and buffer that produce the maximum rate of enzyme catalysis.

primary structure The sequence of amino acids of a protein.

prosthetic groups Cofactors that are so tightly bound that they are considered to be part of the enzyme structure.

quaternary structure The structural relationship of various enzyme subunits to one another.

reactivation is a restoration of biological properties of a protein after a temporary loss.

regulatory sites See *allosteric sites*.

secondary structure of an enzyme The twisting of amino acids into a semifixed steric relationship in two dimensions.

specific activity The enzyme activity expressed as units per milligram of protein.

stereoisomeric specificity The specificity of an enzyme for one form of a DL pair of compounds with an asymmetric carbon atom.
substrate-depletion phase The time late in an enzyme assay when the substrate concentration is falling and the assay is not following zero-order kinetics.
substrates The materials enzymes act upon.
subunits Single protein chains from enzymes composed of two or more peptide chains in an active form.
Système International d'Unités An international system of rational and internally consistent units for all types of scientific quantities; SI units.
tertiary structure of an enzyme The folding of amino acid chains into a three-dimensional structure.
turnover number The number of moles of substrate converted to product per minute per mole of enzyme per active site.
uncompetitive inhibitor An inhibitor that appears to bind only to the enzyme-substrate complex and not to free enzyme.
V_{max} The maximum rate of catalysis.
zero-order kinetics State occurring when the rate of an enzyme reaction is independent of the concentration of the substrate.

BIOCHEMICAL NATURE OF ENZYMES

Enzymes are biological materials with catalytic properties; that is, they increase the rate of chemical reactions in cells and in vitro systems that otherwise proceed very slowly. They are large naturally occuring compounds with molecular weights usually between 13,000 and 500,000 daltons. The study of these molecules and of the changes in enzyme activity that occur in body fluids over time has become a valuable diagnostic tool for the elucidation of various disease entities and for testing organ function.

Different tissues or cell types do not contain the same amounts or types of enzymes. The hundreds of different enzymes in each cell are attached to the cell walls and membranes and are also found in the cytoplasm, the nucleus, and many other specialized subcellular organelles (that is, microsomes, mitochondria, and lysosomes). Often the determination of one or several enzymes in plasma gives a pattern of results that is indicative of the tissue or cell type from which the enzyme or enzymes have been derived. Different cells or compartments within a single cell can even contain different forms of an enzyme (that is, isoenzymes) that catalyse the same chemical reaction. Assays for these different isoenzyme forms can sometimes be performed for specific determination of the tissue or compartment from which an enzyme has come. A few enzymes are found in plasma or other extracellular fluids where they seem to perform a physiological function, but most enzymes catalyze reactions inside cells or in the lumen of various organs.

Composition and structure

All enzymes are proteins: that is, they are compounds of high molecular weight; they contain amounts of carbon, hydrogen, oxygen, nitrogen, and sulfur that are similar to amounts found in other proteins; and hydrolysis with strong acid yields a mixture of amino acids and small peptides. Enzymes are distinguished from other proteins that are not enzymes by their catalytic action, which is usually quite specific for the materials they act upon. The structure of proteins (and enzymes) is discussed in Chapter 48.

The catalytic behavior of an enzyme is dependent on the primary, secondary, tertiary, and quaternary structures of the protein molecule. Changes in any one of these structures can affect the enzymatic activity of the protein. Some causes of such changes are discussed below.

Apoenzymes and cofactors

An enzyme may have nonprotein substances associated with it that are needed for optimal activity. These other materials called *cofactors* may be either loosely or tightly bound to the protein portion of the enzyme. Those that are loosely bound can often be removed by dialysis. These materials may be organic compounds such as nicotinamide adenine dinucleotide phosphate ($NADP^+$) and pyridoxyl-5-phosphate, which are called *coenzymes,* or inorganic ions like chloride (Cl^-) and magnesium (Mg^{+2}), which are called *activators*. Cofactors like the heme portion of peroxidase that are so tightly bound that they are considered to be part of the enzyme structure are termed *prosthetic groups*. Enzymes that have metal ions (that is, activators) bound very tightly are called *metalloenzymes*. Ferroxidase (ceruloplasmin, EC 1.16.3.1), an enzyme containing a relatively large amount of tightly bound copper, and carbonate dehydratase (carbonic anhydrase, EC 4.2.1.1) an enzyme with a large amount of zinc, are two examples of metalloenzymes.

The term "coenzyme" is often loosely used when referring to NADH (or NADPH) for a reaction like the lactate dehydrogenase reaction.

$$\text{Pyruvate} + \text{NADH} + H^+ \underset{}{\overset{\text{LD}}{\rightleftharpoons}} \text{L-Lactate} + \text{NAD}^+$$

In a formal kinetic sense, both pyruvate and NADH are substrates for the enzyme reaction, and lactate and NAD^+ are products. In this case, pyruvate and NADH react stoichiometrically with one another on a one-to-one basis, and the enzyme reaction does not recycle the NAD^+ that is produced back to NADH. The NADH is still called a "coenzyme," that is, a nonprotein organic material needed for maximal activity, perhaps for historic reasons, even though it should be more correctly called a "second substrate," or "cosubstrate."

Since it is possible to dialyze away loosely held cofactors from some enzymes and still retain some activity, an enzyme without the associated cofactors is referred to as an *apoenzyme,* and the complete enzyme-cofactor complex is termed a *holoenzyme*. In the clinical use of enzyme assays, the enzyme assay mixture must contain an excess of all the activators and coenzymes to ensure that the holoenzyme is the enzyme form being measured, rather than a mixture of apoenzyme and holoenzyme forms.

Catalysts

Enzymes function as biological catalysts. They are proteins that have the property of accelerating specific chemical reactions toward equilibrium without being consumed in the process. The material the enzyme reacts with is termed the *substrate,* and a simple enzymatic reaction for one substrate and one product is listed below:

$$E + S \underset{k_{-1}}{\overset{k_{+1}}{\rightleftharpoons}} \{ES\} \underset{k_{-2}}{\overset{k_{+2}}{\rightleftharpoons}} P + E \qquad \textit{Eq. 49-1}$$

In this case the enzyme is represented by *E,* the substrate on which the enzyme acts by *S,* a postulated enzyme-substrate complex by $\{ES\}$, and the product of the reaction by *P*. The forward reaction rate constants are represented by k_{+1} and k_{+2}, whereas the reverse reaction rate constants are represented by k_{-1} and k_{-2}. An example of a single substrate enzyme reaction is the action of the enzyme urease (EC 3.5.1.5) on the substrate urea, though in this case two products are produced:

$$\underset{\textbf{Urea}}{H_2N{-}\overset{\overset{\displaystyle O}{\|}}{C}{-}NH_2} + E \underset{k_{-1}}{\overset{k_{+1}}{\rightleftharpoons}} \{\text{Urea-E}\} \underset{k_{-2}}{\overset{k_{+2}}{\rightleftharpoons}} \underset{\textbf{Ammonia}}{2NH_3} + \underset{\textbf{Carbon dioxide}}{CO_2} + E$$

Water also participates in the reaction, but for these purposes it is not a substrate. These biological catalysts are similar to other chemical catalysts in many respects, except that they function in biological systems. Enzyme catalysts, though they are unstable and easily destroyed, have similar catalytic properties to those of other chemical catalysts. These include the following: they are effective in small amounts; they are unchanged by the reaction; they affect the speed of attaining equilibrium but do *not* change the final concentrations of the substrates and products of the *equilibrium* state; and they demonstrate a much greater degree of specificity than the usual chemical catalysts for the reactions they accelerate.

It is the first property that makes enzymes such a valuable diagnostic tool. Since they are effective in such small amounts, measurement of changes in enzyme concentrations is a very sensitive way to follow changes that have occurred in various types of tissues.

In conventional assays for other materials in serum, like glucose, the amount of material in a measured volume is determined and has usually been expressed as milligrams of glucose per deciliter, or in this book as milligrams of glucose per liter, or millimoles of glucose per liter.

In plasma enzyme assays, the amount of enzyme involved in the enzyme assay is very much smaller than the amount of glucose present in an assay for glucose, which makes conventional chemical assays for enzyme materials very difficult. Of the several thousand proteins in plasma, the measurement of the concentration of a single enzyme (protein), even if it is present at a very elevated value, is below the limit of detection for most chemical protein assays. What is easier to measure and is biologically related to many clinical conditions is the amount of catalytic activity of the enzyme and how it changes with time.

The activity of an enzyme is the amount of substrate for a particular enzyme reaction that is converted to product per unit time under defined conditions (see p. 943). The assumption that is made in the use of activity as a concentration, is that a given amount (that is, weight) of enzyme has a given number of units of activity. That is, the specific enzyme activity (in units per milligram of protein) remains constant even when the increase in enzyme activity observed during a particular disorder may come from a different tissue. The increased enzyme activity is assumed to occur because of the presence of more enzyme with the same specific activity rather than the presence of another form of the enzyme with a different and perhaps higher specific activity. In practice, the analyst uses activity measurements of enzymes as if they were enzyme concentrations. As long as the linearity (or high activity limit), timing, and sample volume constraints for a specific method are followed, the assumption seems to be generally valid. Additional information on conditions that affect the activity are considered later.

As shown in equation 49-1, if the enzyme were acting as a catalyst, it would be unchanged by the reaction. Because of the unstable nature of most enzymes this function is difficult to demonstrate. Many assays of 25 years ago were under the assumption that this catalytic property existed and used extremely long assay times (that is, 1 to 6 hours) for analysis. The instrumentation and chemical methods employed were not sufficiently sensitive to use the shorter assay times of more modern methods (that is, seconds to minutes). It is possible with many current assays to use very short analysis times with sufficient precision to calculate enzyme activity early in an assay period and again after 10 to 15 minutes without showing a decrease in enzyme activity. The amount of substrate converted during this time might be 5% to 10% of the initial amount present, whereas the enzyme activity determined at both times would be essentially unchanged, demonstrating that the enzyme did not participate in the reaction on a molar basis with the substrate. Thus the catalytic properties of many enzymes can now be confirmed using modern assays.

Another aspect of the biological catalysts is that they accelerate the attainment of equilibrium. One way of considering this process is to examine the effect of lactate dehydrogenase (LD, EC 1.1.1.27) on the conversion of pyruvate to lactate. In the presence of the enzyme LD and the coenzyme called reduced nicotinamide adenine dinucleotide (NADH) the conversion of pyruvate to lactate occurs rapidly, but without the enzyme the process is so slow that it can hardly be demonstrated.

$$\text{Pyruvate} + \text{NADH} + H^+ \xrightleftharpoons{\text{LD}} \text{L-Lactate} + \text{NAD}^+$$

$$\text{Pyruvate} + \text{NADH} + H^+ \xrightarrow[]{\text{No enzyme}} \text{No detectable reaction}$$

In addition, this is not a one-way process but an approach to the equilibrium concentrations of pyruvate and lactate, since the same enzyme converts lactate to pyruvate with the coenzyme nicotinamide adenine dinucleotide (NAD^+). In fact, the enzyme was named for this direction of the conversion process. The speed of the reaction and the conditions employed are not the same in both directions, since they are related to the equilibrium constant. It is possible to measure the conversion from either direction, and both methods are widely used to determine LD activity in the clinical laboratory.

Active sites and catalysis

The Gibbs free-energy change (ΔG) is the measure of the amount of work a chemical reaction can produce. All reactions that proceed from initial reactants to products have a net negative free energy ($-\Delta G$). However, the reactants do not become products directly but must absorb energy to pass through an activated or transition state. An energy diagram is given in Fig. 49-1 showing the effect of an enzyme on this process.

Enzymes lower the energy required for activation to the transition state. Without the enzyme present, even with a favorable negative free energy, the reaction may not proceed to any appreciable extent. The reactants must gain the energy to overcome this activation-energy barrier to enter the transition state (active state) and then pass on to products. Without a catalyst present, the reaction will occur only if enough heat (energy) can be added to the reaction system. With an enzyme catalyst, the reaction may easily proceed at normal physiological temperatures. Rewriting

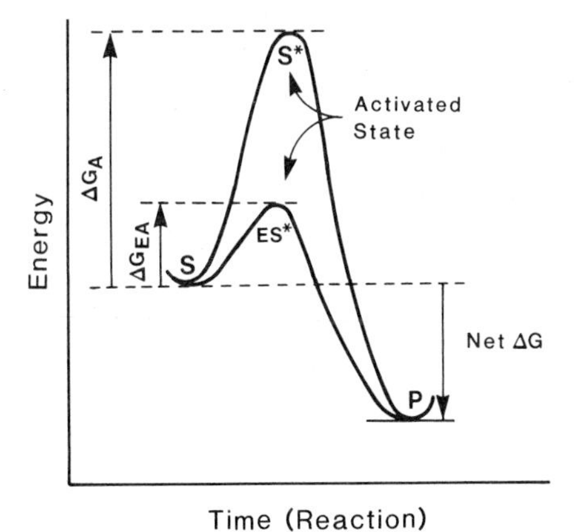

Fig. 49-1. Energy diagram showing reduction in activation energy $\Delta G_{EA} << \Delta G_A$ that occurs for same reaction with and without enzyme catalyst. *A,* Activated state; (also *); *EA,* enzyme activation; *ES,* enzyme substrate; *G,* energy; *P,* product; *S,* substrate.

equation 49-1 to account for this transition state in an enzyme-catalyzed reaction:

$$E + S \underset{k_{-1}}{\overset{k_{+1}}{\rightleftharpoons}} \{ES \rightarrow ES^* \rightarrow EP\} \underset{k_{-2}}{\overset{k_{+2}}{\rightleftharpoons}} P + E \qquad \textit{Eq. 49-2}$$

where ES* is the transition state form of the reactant (substrate) and ES and EP are enzyme-substrate and enzyme-product intermediate forms in which substrate and product are bound but not activated. Substantial reductions in the activation energy requirements are often found when enzymes are used to catalyse the process. For example, the activation energy necessary for the decomposition of hydrogen peroxide is 18,000 cal/mol, but in the presence of the enzyme catalase less than 2000 cal/mol are required.

One of the most difficult problems facing enzymologists is to explain how an enzyme can reduce the activation energy and at the same time remain unchanged by the reaction. Equation 49-2 shows one general possibility but is not detailed enough to understand what happens on a molecular level. A further examination of some of the details of enzyme structure will serve to clarify the mechanism of enzyme catalysis.

A wide variety of nonpolar *hydrophobic* (water-hating) and polar *hydrophilic* (water-loving) amino acids are present in enzyme proteins. The external surface of the enzyme is believed to be composed of mostly polar but generally unreactive side chains of amino acids. The unreactive amino acid side chains may contain structures like methyl and isopropyl groups (that is, $R{-}CH_3$ and $R{-}CH(CH_3){-}CH_3$), found in alanine and leucine.

Some areas of the enzyme surface are believed to contain amino acids with reactive side chains as a part of their structure. The reactive amino acid side chains may contain charged groups like carboxyl and amino groups (that is, $RCOO^-$ and RNH_3^+) found in aspartic and glutamic acids or lysine and arginine. Noncharged portions like the hydroxyl and sulfhydryl groups (that is, ROH and RSH) found in serine, tyrosine, and cysteine are also reactive. There are other types of reactive groups present in amino acids, such as histidine, which has an active nitrogen in a ring structure. These active areas of the enzyme may be on the surface or can exist in more hidden clefts or folds in the enzyme surface. These areas can be involved in the binding of substrates, products, activators, and inhibitors or be involved in the catalytic process itself. There is only a limited number of places on the enzyme where catalysis can take place. These specific areas are called *active sites* or *active centers* and may involve only five to 10 amino acids out of a total of 200 to 300 in the entire enzyme. The active site that has catalytic properties also serves to bind the substrate in a specific way to facilitate the breaking and forming of new bonds. It is believed that certain areas of the catalytic site contain reactive amino acids that bind portions of the substrate through ionic and hydrogen bonds. The substrate is positioned so that other reactive amino acids at the active site cause conversion to product.

The sites on the surface of the enzyme that serve to bind the substrate or product of the reaction are termed *binding sites*. Enzymes, particularly those of complex structure with several subunits, often have other sites that are far removed from the active site but seem to affect enzyme activity. These sites are called *allosteric sites* or *regulatory sites* and much information about enzyme mechanisms has been derived from studies with inhibitors that seem to affect these sites and consequently the enzyme activity.

A few schematic examples of different substrates with an enzyme are given in Fig. 49-2. Only the specific substrate S1 (in Fig. 49-2, *A*) properly fits (binds) or induces the fit at the active site and therefore enters into the reaction. Although one would expect great diversity among the types of catalytic sites to occur for the many kinds of reactions that enzymes catalyze, there are a number of common features that have been observed.

The substrate of the reaction binds to the active site and is oriented so that a particular bond is subject to attack. The reactive side-chain moieties of the enzyme interact with the group on the substrate so that the covalent bond to be altered becomes weakened. This bond weakening decreases the activation energy needed for chemical reaction. The weakened bond now undergoes a chemical reaction that breaks the covalent bond and allows new ones to

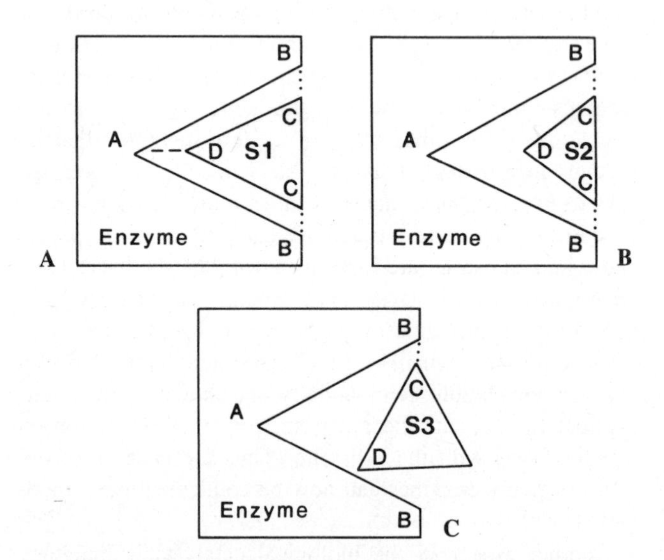

Fig. 49-2. Active site on enzyme is at point *A*, and binding sites are at points *B*. **A,** Correct substrate, *S1*, has complementary binding sites at *C*, and active site can react at point *D* on substrate. **B,** Substrate *S2* has complementary binding sites, but point *D* is too far away for catalysis to take place. If *S2* were present with *S1*, it could act as an inhibitor depending on relative binding constants by preventing *S1* from binding. **C,** Substrate *S3* has only one complementary binding site, *C*, and point *D* is not aligned correctly for catalysis.

form. The modified substrate, that is, product, no longer has the same affinity for the active site as the original substrate and will be released from the enzyme.

Changes in the amino-acid sequence of a protein could produce different enzymes presumably with different active and binding sites, or even similar proteins without catalytic activity. Such changes, caused by genetic mutations, are often the cause of *inborn errors of metabolism* and other diseases of genetic origin (see Chapter 45).

The chemical reactions in which these reactive amino acids take part not only define the enzyme's specific catalytic activity but also determine the sensitivity of the enzyme to losses of activity by such factors as heavy metals, detergents, or even other reactive parts of the same protein molecule. Metals or detergents may bind to active groups and inactivate them. Changes in surface tension, that is, vigorous shaking, may cause unfolding of the protein, or denaturation. As a result, the spatial relationships of these reactive amino acids with each other are disrupted; thus the usual reaction is prevented from taking place.

Specificity of reaction

Differences in enzyme specificity are believed to be related to physical differences at the active site. Some enzymes will react with many related compounds and are said to have a broad specificity. Acid phosphatase (EC 3.1.3.2) is one of these enzymes that exhibits a broad *bond specificity* by hydrolysis (that is, addition of water to a bond) of many types of organic phosphate ester bonds, such as β-glycerol phosphate, thymolphthalein phosphate, *para*-nitrophenyl phosphate, and α-naphthyl phosphate. At an acid pH, the enzyme-catalyzed reaction:

$$\text{R–O–P} + H_2O \xrightarrow{\text{Acid phosphatase}} \text{R–O–H} + P_i$$

produces an organic alcohol and inorganic phosphate (P_i).

Many proteases also exhibit a broad bond specificity where a large number and variety of peptide bonds within a protein substrate are hydrolyzed. A further distinction in the case of enzyme peptidases is that these enzymes can hydrolyze peptide bonds between amino acids that are either near the end of the protein or in the middle of the protein. If the peptide bonds of the substrate that are hydrolyzed are located on the inside of the protein, the enzyme is called an *endopeptidase,* like pepsin A (EC 3.4.23.1). The specificity of hydrolysis is also determined by one group of amino acids, with hydrolysis occuring next to a particular type of amino acid. For example, with the enzyme chymotrypsin (EC 3.4.21.1), another endopeptidase, cleavage of the peptide bond occurs on the carboxyl side of the amino acids tyrosine, tryptophan, phenylalanine, and leucine, no matter which amino acids are positioned next to these four amino acids. Alternatively, carboxypeptidases are enzymes that act on protein substrates cleaving peptide bonds from the carboxyl terminus of the substrate toward the middle of the protein. These enzymes are termed *exopeptidases* because they hydrolyze proteins from one end of the peptide substrate, and they also demonstrate a broad substrate specificity.

In contrast to the broad specifications of many peptidases, many enzymes are more specific in their action, in that they will catalyze only a definite reaction with a few substrates. In extreme cases, an almost absolute specificity is demonstrated where only a single compound will serve as a substrate for an enzyme reaction, such as phospho*enol*pyruvate, for the pyruvate kinase (EC 2.7.1.40) reaction.

$$\left.\begin{matrix}\text{Phospho}\textit{enol}\text{pyruvate}\\ \text{(PEP)}\\ +\\ \text{Adenosine diphosphate}\\ \text{(ADP)}\end{matrix}\right\} \underset{}{\overset{\text{PK}}{\rightleftharpoons}} \left\{\begin{matrix}\text{Pyruvate}\\ \text{(PYR)}\\ +\\ \text{Adenosine triphosphate}\\ \text{(ATP)}\end{matrix}\right.$$

Enzyme specificity should be described for each substrate involved in a reaction. In contrast to the absolute specificity shown for phospho*enol*pyruvate in the pyruvate kinase reaction, several natural and synthetic nucleoside diphosphates, such as UDP, IDP, GDP, and CDP will also serve as phosphate acceptors in the reaction in place of adenosine diphosphate (ADP).[1] Thus, although an absolute specificity is shown for one substrate, an intermediate degree of specificity is shown for the other substrate.

An intermediate degree of specificity for each substrate is shown by the hexokinase (HK, EC 2.7.1.1) reaction, in which D-glucose and several other sugars may be phosphorylated, that is, D-mannose, 2-deoxy-D-glucose, and D-glucosamine. However, D-galactose and 5-carbon sugars like D-xylose are not substrates. The enzyme can also use a variety of nucleoside triphosphates as phosphate donors, such as inosine triphosphate (ITP) and guanosine triphosphate (GTP) as well as adenosine triphosphate (ATP).[2]

$$\left.\begin{matrix}\text{D-Glucose}\\ +\\ \text{Adenosine triphosphate}\\ \text{(ATP)}\end{matrix}\right\} \overset{\text{HK}}{\rightleftharpoons} \left\{\begin{matrix}\text{Glucose-6-phosphate}\\ +\\ \text{Adenosine diphosphate}\\ \text{(ADP)}\end{matrix}\right.$$

Many enzymes demonstrate a *stereoisomeric specificity* for either the L-form or the D-form of a pair of compounds. Hexokinase is absolutely specific for the D-form of glucose; the L-form is not a substrate. Malate dehydrogenase (EC 1.1.1.37) acts only on the L-form of malate, not the D-form. Lactate dehydrogenase (EC 1.1.1.27) acts only on L-lactate not D-lactate. However, stereoisomeric specificity does not necessarily mean that the enzyme is absolutely substrate specific, since some forms of lactate dehydrogenase act on hydroxybutyrate as well as on lactate, and, as mentioned above, hexokinase works with several D substrates.

Subunits

Some commonly encountered enzymes in the clinical laboratory have been shown to be composed of subunits, but only the multiple subunit form has been found to have significant enzyme activity. The enzymes whose subunit structures have been most widely studied have been creatine kinase, CK, and lactate dehydrogenase, LD. The enzyme creatine kinase has been found to have two different subunits. These were historically designated "M" for muscle and "B" for brain, after tissues that were rich sources of the enzyme that had these subunits as their predominant form. Both the M and B subunit forms are inactive, and only enzymes containing the two subunits together as a dimer have enzyme activity. LD also has been found to have two different subunits. These are the "H" (heart) and "M" (muscle) forms. The active LD enzyme is a tetramer composed of four subunits.

In some unusual cases, an enzyme subunit may have catalytic activity by itself, such as glutamate dehydrogenase (EC 1.4.1.2.). In these cases, the natural enzyme form made up of several subunits has a greater activity than the sum of the activities of the separate subunits. In addition, the multiple subunit form of the enzyme often has activators and inhibitors that may more closely control the enzyme activity. Thus a more complex biological structure may result in more exact control of enzyme activity.

Anabolism and catabolism

The synthesis of all enzymes is assumed to occur by the usual intracellular protein synthetic pathways within the tissues that contain the enzymes. Extracellular enzymes like those involved in the coagulation process are synthesized in the liver and elaborated into the plasma. In some cases other organs, that is, kidney, lung, and pancreas, also contribute to the extracellular enzyme pool.

The large size of enzymes and complexity of structure results in molecular forms that are somewhat unstable and are therefore said to be *labile*. Many enzymes in vitro lose their catalytic activity with relatively slight changes in pH, temperature, or even salt concentration of the surrounding medium. It is presumed that similar processes occur intracellularly and that constant, though slight, synthesis of enzymes occurs in a steady-state fashion to maintain the required amounts of intracellular enzymes needed for intermediary metabolism.

A loss of enzyme activity can be either reversible and temporary or irreversible and permanent. *Denaturation* is a process where biological properties are lost by a protein: that is, enzyme activity is lost. It has been suggested that the denaturation process is an unfolding or "melting" of tightly coiled peptide chains leading to a more disorganized structure. There is much experimental support for this concept, including increased reactivity of side chains, changes in viscosity, and changes in the sedimentation behavior, of the protein solutions. *Irreversible denaturation* can occur when the enzyme protein chains unfold and are unable to refold to their biologically active form, or when a heavy metal ion (such as mercury or lead) binds tightly at or near the active site. Many other factors and events can lead to denaturation and loss of activity including changes in temperature (heating or cooling), the addition of strong acids or bases, exposure to high pressure, treatment with ultraviolet light, repeated freezing, and the addition of detergents, or organic solvents or the presence of high concentrations of urea or guanidine.

A *reversible denaturation* or loss of enzyme activity is called *inactivation*. For example, inactivation can occur if an enzyme solution is allowed to remain for some time at room temperature and the enzyme partially loses activity. This temporary activity loss can have several causes including heat instability with the breaking of hydrogen bonds or oxidation of sulfhydryl groups. In both of these cases, there is some loss of the natural structural form. With some enzymes, reducing the temperature of the solution or the addition of a sulfhydryl reducing agent like dithiothreitol may allow the enzyme to refold to the original active form, with re-formation of hydrogen bonds or reduction of oxidized sulfhydryl groups, thus producing a *reactivation* of the enzyme and a restoration of the lost activity.

Little is known about the mechanism of removal of enzyme proteins from the extracellular fluid compartment. Presumably, extracellular proteases inactivate enzymes that are lost from cells to the extracellular compartment, and the inactive proteins are then removed by a mechanism utilizing one of several excretory routes, that is, excretion in bile, intestine, liver, kidney, or reticuloendothelial system. It is not clear whether this is a general mechanism that applies to all proteins or whether it is related to those proteins that are recognized as intracellular. In addition, it is known that different enzymes have different half-lives, a finding that indicates that several mechanisms of removal may be present.

ENZYME NOMENCLATURE AND CLASSIFICATION

Many enzymes were first named for their function (such as lactate dehydrogenase), but some have also been named for the type of substrate on which they act: urease hydrolyses urea, lipase hydrolyzes lipids, and phosphatases act on organic phosphates. Many clinically important enzymes are still known by these trivial names that arose from historic circumstances and will continue to pervade the literature because of their simplicity. A more organized approach was needed because of the increasing complexity of names found in the literature and with the increasing number of enzymes being discovered and studied. The Enzyme Commission (EC) of the International Union of Biochemistry (IUB) developed and proposed a systematic convention for the naming of enzymes.[3]

IUB names and codes

The IUB systematic name describes the reaction catalyzed. The IUB also recognized that trivial names were important and assigned practical names to many enzymes

Table 49-1. Examples of enzyme nomenclature

EC code	Recommended name (trivial)	Abbreviation*	Systematic name	Other name or abbreviation
Oxidoreductases				
1.1.1.27	Lactate dehydrogenase	LD	L-Lactate: NAD^+ oxidoreductase	LDH
1.1.1.37	Malate dehydrogenase	MD	L-Malate: NAD^+ oxidoreductase	MDH
1.1.1.42	Isocitrate dehydrogenase ($NADP^+$)	ICD	*threo*-Ds-Isocitrate: $NADP^+$ oxido-reductase (decarboxylating)	
1.1.1.49	Glucose-6-phosphate dehydro-genase	GPD	D-Glucose-6-phosphate: $NADP^+$ 1-oxidoreductase	G6PDH
1.4.1.2	Glutamate dehydrogenase	GMD	L-Glutamate: NAD^+ oxidoreductase (deaminating)	—
1.16.3.1	Ferroxidase	—	Iron(II): oxygen oxidoreductase	Ceruloplasmin
Transferases				
2.1.3.3	Ornithine carbamoyltransferase	OCT	Carbamoylphosphate:L-ornithine car-bamoyltransferase	Ornithine carbamyl-transferase
2.3.2.2	γ-Glutamyl transferase	GGT	(5-Glutamyl)-peptide: amino acid 5-glutamyl transferase	—
2.6.1.1	Asparate aminotransferase	AST	L-Aspartate: 2-oxoglutarate amino-transferase	Glutamic oxaloacetic transaminase/SGOT
2.6.1.2	Alanine aminotransferase	ALT	L-Alanine: 2-oxoglutarate aminotrans-ferase	Glutamic pyruvic transaminase/SGPT
2.7.1.1	Hexokinase	HK†	ATP: D-hexose-6-phosphotransferase	—
2.7.1.40	Pyruvate kinase	PK	ATP: pyruvate 2-*O*-phosphotransfer-ase	—
2.7.3.2	Creatine kinase	CK	ATP: creatine *N*-phosphotransferase	CPK
Hydrolases				
3.1.1.3	Triacylglycerol lipase	LPS	Triacylglycerol acyl hydrolase	Lipase
3.1.1.8	Cholinesterase	CHS	Acylcholine acyl hydrolase	Pseudocholinesterase
3.1.3.1	Alkaline phosphatase	ALP	Orthophosphoric-monoester phospho-hydrolase (alkaline optimum)	—
3.1.3.2	Acid phosphatase	ACP	Orthophosphoric-monoester phospho-hydrolase (acid optimum)	—
3.1.3.5	5′-Nucleotidase	NT	5′-Ribonucleotide phosphohydrolase	—
3.2.1.1	α-Amylase	AMS	1,4-α-D-Glucan glucanohydrolase	Diastase
3.4.11.1	Aminopeptidase (cytosol)	LAS‡	α-Aminoacyl-peptide hydrolase (cy-tosol)	Arylaminadase/LAP; leucine aminopepti-dase
3.4.21.1	Chymotrypsin	—	None (preferred cleavage: Tyr, Trp, Phe, Leu)	Chymotrypsin A and B
3.4.21.4	Trypsin	TPS	None (preferred cleavage: Arg, Lys)	α- and β-trypsin
Lyase				
4.1.2.13	Fructose-bisphosphate aldolase	ALS	D-Fructose-1,6,-bisphosphate: D-glyc-eraldehyde-3-phosphatelyase	Aldolase
4.2.1.24	Porphobilinogen synthase	—	5-Aminolevulinate hydrolyase	—
Isomerases				
5.3.1.1	Triosephosphate isomerase	TPI	D-Glyceraldehyde-3-phosphate: ketol-isomerase	Triosephosphate mu-tase
5.3.1.9	Glucosephosphate isomerase	GPI	D-Glucose-6-phosphate: ketol-isomer-ase	Phosphohexose iso-merase
Ligases				
6.3.1.2	Glutamine synthetase	—	L-Glutamate: ammonia ligase (ADP-forming)	—

*Baron, D.N., et al.: J. Clin. Pathol. **24:**656-657, 1971 (ref. 4) and Baron, D.N., et al.: J. Clin. Pathol. **28:**592-593, 1975 (ref. 5) are *not* recommended by the International Union of Biochemistry but are in common use.

†Not listed in references 4 and 5 but in common use in biochemistry laboratories.

‡Reference 5 incorrectly lists (EC 3.4.11.2) the microsomal form of this enzyme as "leucine aminopeptidase."

but no abbreviations. For each enzyme the system provides a numeric code designation consisting of four numbers separated by periods. The first number assigns the enzyme to one of six categories of reaction. The next two numbers are the subclass and the sub-subclass of reaction, and the last is a serial number unique to the enzyme for the sub-subclass. A more complete description is given below.

EC classification

Since the EC classification of the enzymes will be used increasingly in the future, a description of its basis is given. All enzymes are placed in one of six classes depending on the type of reaction they catalyze. A few clinically important enzymes are listed in Table 49-1 along with the EC code and systematic names.

The first class includes the *oxidoreductases,* those enzymes that catalyze electron transfer or oxidation-reduction reactions, which can be illustrated schematically as follows:

$$A_{red} + B_{ox} \rightleftharpoons A_{ox} + B_{red}$$

Some common names of enzymes in this category include dehydrogenases, reductases, oxidases, and peroxidases. If the reaction involves a direct participation of oxygen, the enzymes are termed "oxidases." If the reaction involves the transfer of hydrogen, as in the change from L-malate to oxaloacetate, the enzymes are known as "dehydrogenases." This reaction is shown below for the enzyme malate dehydrogenase (EC 1.1.1.37):

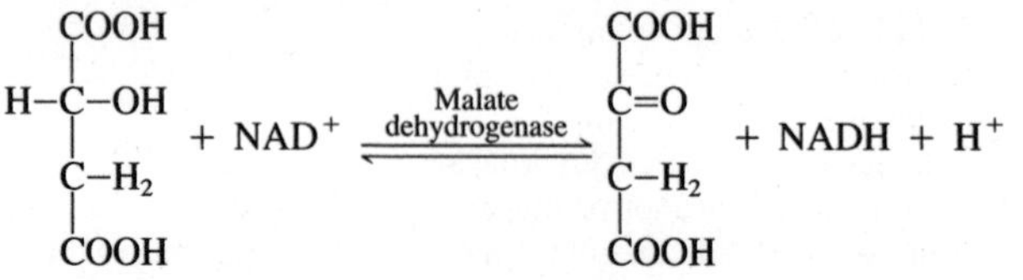

$$\text{COOH-CH(OH)-CH}_2\text{-COOH} + NAD^+ \xrightleftharpoons{\text{Malate dehydrogenase}} \text{COOH-C(=O)-CH}_2\text{-COOH} + NADH + H^+$$

L-Malate Oxaloacetate

The second group of enzymes contains the *transferases,* those enzymes that catalyze the transfer of a group, such as an amino, carboxyl, glucosyl, methyl, or phosphoryl group, from one molecule to another. These reactions can be listed schematically as follows:

$$A{-}X + B \rightleftharpoons A + B{-}X$$

Alanine aminotransferase (EC 2.6.1.2) is an example of this group; this enzyme catalyzes the transfer of an amino group stereospecifically from L-alanine to alpha-ketoglutarate, producing pyruvate and L-glutamate.

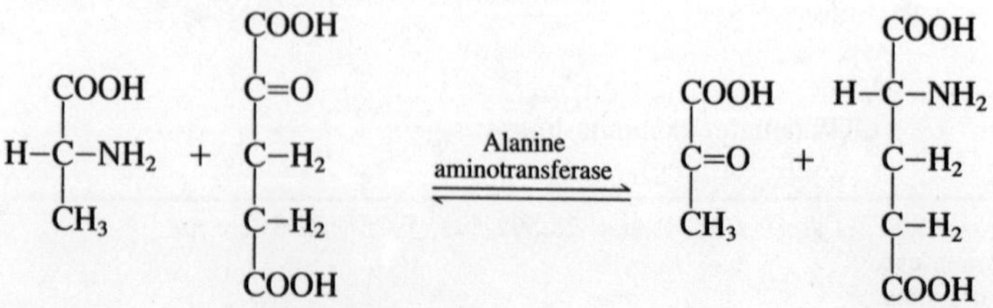

$$\text{COOH-CH(NH}_2\text{)-CH}_3 + \text{COOH-C(=O)-CH}_2\text{-CH}_2\text{-COOH} \xrightleftharpoons{\text{Alanine aminotransferase}} \text{COOH-C(=O)-CH}_3 + \text{COOH-CH(NH}_2\text{)-CH}_2\text{-CH}_2\text{-COOH}$$

L-Alanine α-Ketoglutarate Pyruvate L-Glutamate

The term "aminotransferase" may be replaced by "transaminase." Other common enzyme names in this category include kinases, transferases, and transcarboxylases.

A third group includes the *hydrolases,* which catalyze the cleavage of C–O, C–N, C–C and some other bonds with the addition of water. These hydrolysis reactions can be illustrated as follows:

$$A{-}B + H_2O \rightleftharpoons A{-}OH + B{-}H$$

An example of this group is acid phosphatase (EC 3.1.3.2), which catalyzes the hydrolysis of a number of organic phosphate monoesters, under acid conditions. For example, with *para*-nitrophenyl phosphate as a substrate:

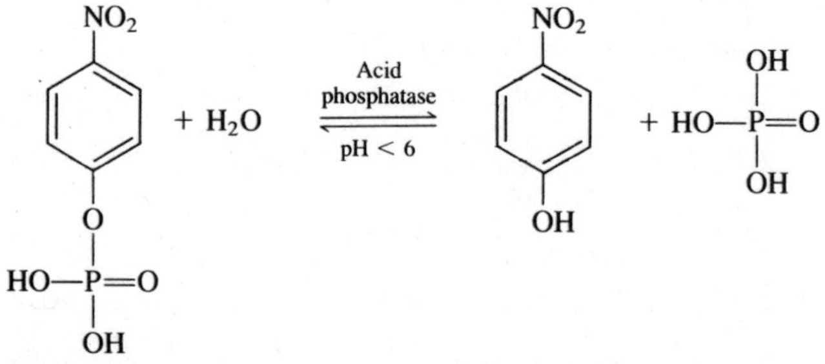

$$O_2N\text{-}C_6H_4\text{-}O\text{-}P(=O)(OH)_2 + H_2O \xrightleftharpoons[\text{pH} < 6]{\text{Acid phosphatase}} O_2N\text{-}C_6H_4\text{-}OH + HO\text{-}P(=O)(OH)_2$$

p-Nitrophenyl phosphate *p*-Nitrophenol Phosphate

Other common enzyme names in this category are amylase, urease, pepsin, trypsin, chymotrypsin, and various peptidases and esterases.

A fourth group contains the *lyases,* which hydrolyze C–C, C–O, and C–N bonds by elimination, with the formation of a double bond, or the reverse reaction, the addition of a group to a double bond. In cases where the reverse reaction is important the term "synthase" (not the group EC 6 synthetase) is used in the name. This type of reaction is illustrated as follows:

$$A + B \rightleftharpoons AB \text{ (synthase)}$$

or

$$AB \rightleftharpoons A + B \text{ (lyase)}$$

An examination of the EC listing shows that this and the subsequent groups contain relatively few enzymes that are used in clinical diagnosis. One example of an enzyme from this group that is clinically useful is porphobilinogen synthase (EC 4.2.1.24), which is also known by the older name δ-aminolevulinate dehydratase. The enzyme is strongly inhibited by heavy metals and has been used as an estimate of the severity of lead (Pb^{+2}) poisoning. Other enzymes in this group include aldolase and several decarboxylases.

```
COOH
|
C–H2           COOH
|              |
C–H2           C–H2          Porphobilinogen
|        +     |                 synthase
C=O            H2C          ===================>
|              |
H2–C           O=C–CH2–NH2
|
NH2
2(5-Aminolevulinate)

                    COOH
                    |
                    C–H2  COOH
                    |     |
                    C–H2  C–H2 + 2H2O
                    |     |
                    C=====C
                    |     |
                    H2C   C–CH2–NH2
                      \  //
                        N
                   Porphobilinogen
```

The fifth group includes the *isomerases,* which catalyze structural or geometric changes in a molecule. They may be called epimerases, isomerases, and mutases depending on the type of isomerism involved. This reaction can be illustrated as follows:

$$ABC \rightleftharpoons CAB$$

An example of this group is the enzyme glucose phosphate isomerase (EC 5.3.1.9), which catalyzes the change of glucose-6-phosphate into fructose-6-phosphate; one can see that the two compounds mentioned have exactly the same number of atoms ($C_6H_{13}O_9P$) except that they are arranged into a slightly different structure in each compound.

```
 O=C–H                                     CH2OH
   |                                         |
 H–C–OH                                     C=O
   |                                         |
HO–C–H                                   HO–C–H
   |            Glucose phosphate            |
 H–C–OH             isomerase             H–C–OH
   |           <==================>          |
 H–C–OH   OH                              H–C–OH   OH
   |      |                                  |     |
  CH2O–––P=O                               CH2O–––P=O
          |                                        |
          OH                                       OH
Glucose-6-phosphate                     Fructose-6-phosphate
   (C6H13O9P)                               (C6H13O9P)
```

A sixth and last group consists of the *ligases* (synthetases). In this case two molecules are joined, coupled with the hydrolysis of the pyrophosphate in ATP. Many of these enzymes are involved in DNA, RNA, and protein synthesis; none are currently used in clinical diagnosis. The synthetic reaction type is illustrated as follows:

$$A + B + ATP \rightleftharpoons AB + ADP + P_i$$

An example of this reaction is the enzyme glutamine synthetase (EC 6.3.1.2), which is an enzyme involved in an alternate pathway for the removal of ammonia.

```
  COOH                                          CONH2
  |                                             |
  C–H2                                          C–H2
  |                          Glutamine          |
  C–H2 + NH3 + ATP           synthetase         C–H2 + ADP + Pi
  |                     <==================>    |
H–C–NH2                                       H–C–NH2
  |                                             |
  COOH                                          COOH
L-Glutamate                                   L-Glutamine
```

In the EC code number, the second number denotes the subclass, which is often based on the type of group, such as amino group or hydroxyl group, that takes part in the reaction. The third number of the EC code indicates the different sub-subclass of reaction, often the acceptor group, and the last number is merely the serial number of the particular enzyme in this sub-subgroup. Thus for the enzyme lactate dehydrogenase (EC 1.1.1.27) the first number, 1, indicates that the enzyme is an oxidoreductase; the second number, 1, indicates that the enzyme acts on the CH–OH group of donors; the third number, 1, indicates that the acceptor is NAD^+ or $NADP^+$; and the fourth number, 27, is merely the serial number of the enzyme in the EC 1.1.1.*x* group. The actual reaction with structural formulas is as follows:

```
  COOH                                   COOH
  |                 Lactate              |
H–C–OH + NAD+     dehydrogenase          C=O + NADH + H+
  |             <=================>      |
  CH3                                    CH3
L-Lactate                              Pyruvate
```

Isoenzymes

Multiple natural forms of an enzyme in a single species are known as "isozymes," or "isoenzymes."[6] The IUB has designated that this term is to apply only to those forms of enzymes arising from genetically determined differences in primary (that is, amino acid) structure though there is not complete agreement on this designation. Isoenzymes are to be distinguished on the basis of electrophoretic mobility and subscripted with the first form having the mobility closest to the anode (+). Although there exist many reports to the contrary in the literature, isoenzymes are not to be labeled on the basis of tissue distribution (that is, heart type, brain type), since some confusion may arise because of differences in the predominant form found in various species. There are three groups of multiple enzyme forms that have been defined as isoenzymes by the IUB. These are grouped as follows: genetically independent proteins, such as mitochondrial and cytosol forms of malate dehydrogenase (EC 1.1.1.37); heteropolymers of two or more different subunits, such as CK (EC 2.7.3.2) and LD (EC 1.1.1.27); genetic variants in protein structure, such as glucose-6-phosphate dehydrogenases (EC 1.1.1.49), with more than 50 varieties known in man.

The polymeric forms of glutamate dehydrogenase (EC 1.4.1.2) are not isoenzymes by this definition, since they are polymers of a single subunit. Another similar example is the polymeric forms of phosphorylase (EC 2.4.1.1). Some additional forms of isoenzymes that do not fit the strict definition are variations that occur in the cleavage of a terminal segment of a protein chain producing various isoenzymes. Hexokinase (EC 2.7.1.1) and carbonate dehydratase (EC 4.2.1.1) are examples of this type of isoenzyme. A more complete description of isoenzymes is given in Chapter 50.

Nonstandard abbreviations

A variety of simple abbreviations containing four or fewer capital letters are also used to represent the enzymes in common usage. These abbreviations, suggested by Baron et al.[4,5] as a useful addition, are widely used in practice but are *not* part of the IUB system.

There is even some ambiguity from the IUB, since the names of enzymes are not to be abbreviated *except* in terms of substrates for which accepted abbreviations exist; for example, ATPase and RNAse are allowed but CK and LD are not allowed.[7] Although strictly correct, this policy is not followed by most of producers and users of enzyme data in the clinical area. There is widespread use of these unofficial abbreviations perhaps because the systematic names are so complex and awkward to use. These abbreviations have become so convenient and commonly used that it would be difficult to discard them completely. These nonstandard abbreviations are included in Table 49-1 along with the IUB systematic names.

MEASUREMENT OF ENZYMES

For most enzymatic procedures the reaction rates are found not to be constant with time, particularly if they are examined in detail. By observing the rate of change of absorbance at some wavelength, such as 340 nm, the amount of substrate converted to product can be followed. Initially, there is a *lag phase* with little change of absorbance per unit time when the reactants are mixed and reach thermal and kinetic equilibrium, then a *linear phase* of constant absorbance change per unit time, and finally a *substrate depletion phase* with little change of absorbance per unit time.

Enzyme assays must be performed during the linear phase of absorbance change where a constant amount of activity can be determined for a period of time (Fig. 49-3, *A*). Thus measurements do not start at zero time but begin after the lag phase has occurred. Measurements can be made at any time during the linear phase and can continue up to the substrate depletion phase. The time course for the same assay with increasing amounts of enzyme present is shown in Fig. 49-3, *B*. There is a time period when all three enzyme activities can be measured with the same assay conditions, but the length of time of the linear phase is different for each amount of enzyme.

If the lag phase is very short and the linear phase is relatively long, the activity can be measured from zero time up to the substrate depletion phase with only a small error. Many older manual assays included the lag phase in

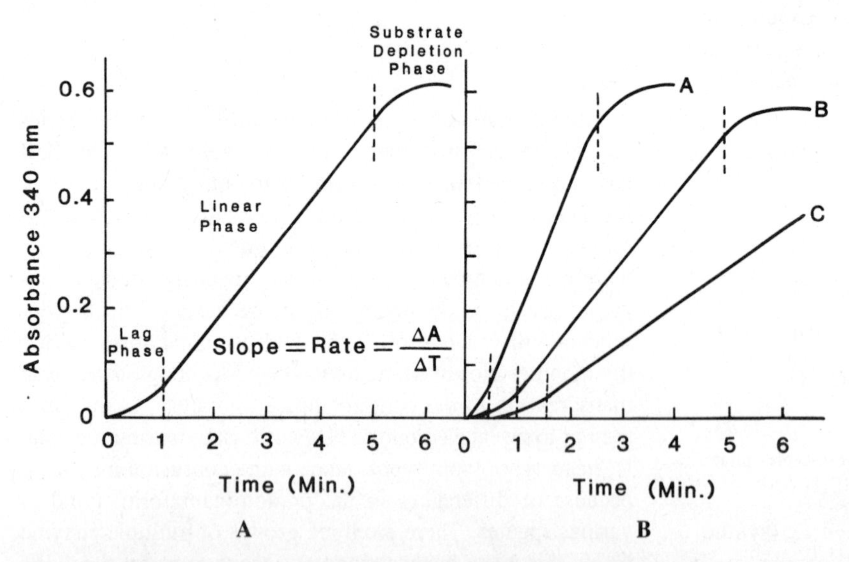

Fig. 49-3. **A,** Typical enzyme reaction with initial lag phase, linear change of absorbance, and final phase of substrate depletion. Enzyme activity is slope of linear phase. **B,** Time course of an enzyme reaction with three different amounts of enzyme present. Curve *A* has a high activity, *B* has a medium activity, and *C* has a low activity. As enzyme activity is increased in an assay system, lag phase decreases, linear phase decreases, and substrate depletion occurs sooner. ΔA, Change of absorbance; ΔT, change of time.

the determination of enzyme activity because the timing was started from zero rather than after the lag phase was completed. The result of timing assays this way was that halving or doubling the time of assay only *sometimes* halved or doubled the amount of product produced. The activity thus measured will be either half or double that originally measured only if the lag is relatively insignificant. In an absolute sense, if the rate is taken from the linear phase, one can usually assay samples with high activity by halving the measurement time and calculating a new $\Delta A/\Delta T$. An alternate way to handle samples with high activity is to dilute them two- or threefold with saline or water.[8] However, not all enzymes demonstrate linearity on dilution, particularly if the enzymes are active at a lipid-water interface, such as lipase (EC 3.1.1.3), or if there are inhibitors present in the sample, such as LD (EC 1.1.1.27) when measured in urine.

One of the more convenient methods of assaying enzyme activity is based on measurement of the absorbance of either the substrates or the products. There is a number of enzyme systems that involve the conversion of nicotinamide adenine dinucleotide (NAD^+) to its reduced form (NADH), or vice versa. The reduced form, NADH, has a much greater absorption at 340 nm than the oxidized form does, and consequently, reactions that convert one form to the other may be conveniently followed by measurement of the change in absorption at this wavelength. The difference in the absorption spectrum of the reduced and oxidized compounds is shown in Fig. 49-4.

A good example of an enzyme reaction that involves the conversion of NAD^+ to NADH is the reaction catalyzed by the enzyme malate dehydrogenase (EC 1.1.1.37):

$$\text{L-Malate} + NAD^+ \overset{MD}{\rightleftharpoons} \text{Oxaloacetate} + NADH + H^+$$

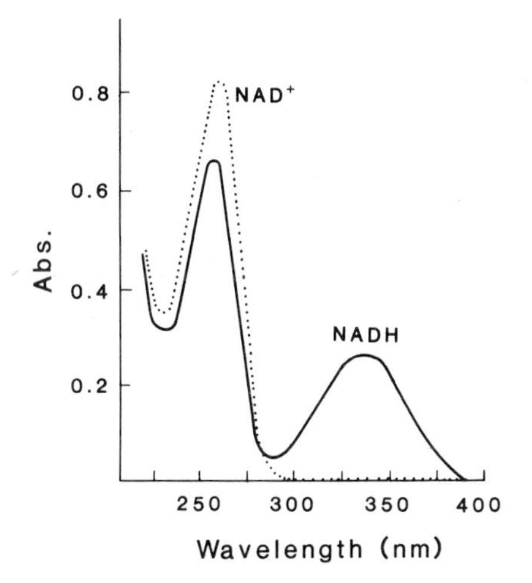

Fig. 49-4. Absorption spectrum, *Abs.*, of 5×10^{-5} M NAD^+ in 0.1 M Tris buffer, pH 7.5, and absorption spectrum of 4×10^{-5} M NADH in 0.1 M Tris buffer, pH 9.5.

In this case, the NAD^+, often termed a "coenzyme," is essential to the reaction as a second substrate; the enzyme activity is therefore easily followed directly. In some instances the NAD^+ may be replaced by $NADP^+$ (nicotinamide adenine dinucleotide phosphate), producing NADPH, which has properties similar to those of NADH. In addition, many other enzyme reactions that do not involve the NAD^+/NADH change directly can be coupled with another reaction involving NAD^+, so that the change in the NAD^+ becomes a measure of the enzyme reaction. Thus a number of enzyme reactions can be followed by measurement of the absorbance change at 340 nm.

Enzyme assays

Enzymes have been measured by several different techniques. The two most commonly employed methods are the one-point method at a fixed time, sometimes called *end-point methods,* and multipoint fixed-time assays, called *kinetic methods*.

In both the one-point and multiple-point cases, fixed time is a somewhat arbitrary term, since most current enzyme assays are performed for a fixed time of a few minutes, which is often less than the entire time of linearity during which the assay potentially could be performed. This larger time frame may be used by the analyst in a flexible way such as doubling the normal assay time for a sample of low activity to produce a more accurate result, or shortening the normal assay time for a sample of high activity as an alternative to the method of dilution mentioned earlier.

End-point methods are also used in the literature to mean assays that have come to an equilibrium or steady-state point, that is, assays that measure the amount of an analyte after no further reaction has occurred.

Most older methods employed an end-point, fixed-time approach, often for 30 or 60 minutes and in some cases for several hours. It is now known that many of these time periods were too long and that some enzyme degradation and loss of activity occurred. The assumption that the reaction rate was constant for the entire assay period would not be valid, since the enzyme activity would be reduced as the reaction progressed and as the enzyme was degraded. The calculated enzyme activity would therefore be underestimated.

End-point assays are still used in some cases, but in general shorter time periods are employed. In these assays a reaction is started and allowed to incubate at a constant temperature for a fixed time period, such as 30 minutes. The reaction is then stopped, perhaps by the addition of another reagent, and the amount of product is measured. The assumption for this type of assay is that a constant amount of product is produced throughout the entire assay period. This analytical approach is valid as long as the analyst checks at several points in time, when setting up the assay, to see that the amount of product produced is linearly related to the time of assay.

If the rate of reaction is followed continuously or with many points as a function of time, the assay is termed a *kinetic assay*. Usually the reaction time is short, that is, some seconds to a few minutes, and there is little danger of enzyme degradation. Pardue[9] has pointed out that the term *kinetic assays* describes those assays that form increasing amounts of product with time, whether they are one-point *end-point* assays or *multiple-point* assays. A *kinetic method* is a term often used to mean *continuous* monitoring of the progress of the reaction. Although this terminology is not strictly correct, the continuous or multiple-point assays are superior to the single-point fixed-time assays, since it is easier to demonstrate approximate linearity of the reaction over the entire measurement period. The term "fixed time" is not limited to the single-point assays, since almost all enzyme assays are performed for a fixed time whether a single point or many datum points are taken.

Current instrumentation often permits multiple absorbance readings of the reaction to be made automatically for the determination of enzyme activity. With some instruments, one can take hundreds of readings and average them to determine the rate or to calculate the rate several times during the assay period. In some cases, the enzyme activity will be calculated only if several rate measurements agree within certain limits, thereby demonstrating lack of substrate depletion. The use of multiple readings allows the analyst to determine when the lag phase is over and also if the enzyme activity in the sample is too high so that substrate depletion occurs during the usual measurement period. Such refinements are not possible with end-point fixed-time assays.

Principles of kinetic analysis

Enzyme kinetics is the study of enzyme reaction rates and the factors that affect them. Initially, many experiments are performed to examine the effects of different assay conditions on measurements of enzyme activity. Eventually, a series of specific conditions are established that give rise to the maximum rate of enzyme activity. In general, for measurements of enzyme activity, it is desirable to measure the increase in the amount of product from zero rather than measure the slight decrease in a high concentration of substrate, but there is no difference from a kinetic point of view.

The general enzyme reaction given previously for a *single substrate reaction* may be rewritten slightly for *initial rates,* as equation 49-3. In this case, the amount of product is very small and the reverse reaction of P combining with E and going back to form ES is ignored since initial rate measurements are to be made. The initial rate is the rate at the start of the reaction, after the lag phase and during the linear phase in Fig. 49-3, *A,* but before substantial product formation.

$$E + S \underset{k_{-1}}{\overset{k_{+1}}{\rightleftharpoons}} \{ES\} \xrightarrow{k_{+2}} P + E \qquad \textit{Eq. 49-3}$$

Keep in mind that k_{+1} and k_{+2} denote the forward reaction rate constants for this reaction whereas k_{-1} denotes the reverse reaction rate constant. For a given quantity of enzyme, the rate of activity that is observed increases with increasing amounts of substrate as shown in Fig. 49-5, *A*. At low substrate concentrations the rate is linearly dependent on the amount of substrate, this is, *first order,* but at high substrate concentrations the rate is essentially independent of substrate concentration, that is, *zero order*. A mathematical description of the reaction must explain how

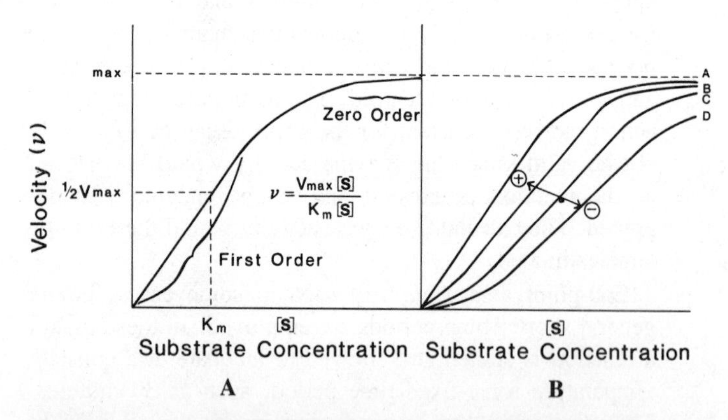

Fig. 49-5. A, Relationship of substrate, *S,* to velocity of reaction. At low substrate concentrations, the rate is first order (linearly dependent) with respect to substrate concentration. At high substrate concentrations the rate becomes zero order (independent) with respect to substrate concentration. K_m, Michaelis-Menten constant; V_{max}, maximal rate of reaction. **B,** Relationship between velocity and substrate concentration for an allosteric enzyme. Presence of positive or negative effectors shifts curve toward the + or − side respectively.

the reaction can be first order at low substrate concentrations and zero order at high substrate concentrations.

If the enzyme has limited number of active sites, at a low substrate concentration the rate will be dependent on the amount of substrate present, since there will be a large effective concentration of unfilled active sites per substrate. However, since the total number of sites, that is, enzyme, is limited, then as the amount of substrate is increased, the sites will become increasingly saturated with substrate until the reaction will appear to be independent of the substrate concentration. At these high substrate concentrations all the enzyme active sites are filled and the reaction proceeds at maximal velocity. Small changes in the substrate concentration will not affect the enzyme rate.

The second step, product formation, is assumed to be the rate-limiting step or the one that determines the overall activity. Another way of stating this is to say that the enzyme and substrate are in equilibrium. The equilibrium for the formation of the ES complex can be written as follows with the molar concentrations of all the reacting species expressed in brackets:

$$K_{eq} = \frac{k_{+1}}{k_{-1}} = \frac{[ES]}{[E][S]}$$

The equilibrium constant, K_{eq}, is equal to the ratio of the forward over the reverse rate constants. From equation 49-3, the rate of formation of the product, P, is the amount of [ES] times the rate, k_{+2}, at which the enzyme complex is converted to P. Thus the rate of formation of product is expressed as follows:

$$\text{Velocity or rate} = [ES] \times k_{+2}$$

Since the rate is the amount of product formed for some period of time:

$$\text{Rate} = \frac{\Delta P}{\Delta T} = [ES] \times k_{+2}$$

substituting K_{eq} [E] [S] for [ES] and rearranging gives:

$$\Delta P = K_{eq} \times [S] \times [E] \times k_{+2} \times \Delta T$$

The amount of product that is formed is proportional to the amount of enzyme present, the time of the assay, and the amount of substrate present. When a proportionality constant is substituted for the rate constants, the equation becomes:

$$\Delta P = K1 \times [S] \times [E] \times \Delta T$$

where ΔP is the amount of product formed during the assay time, $[E]$ is the amount of enzyme, $[S]$ is the amount of substrate, ΔT is the assay time, and $K1$ is a proportionality constant. The enzyme activity or rate of product formation over time is then given by the following:

$$\text{Rate} = \frac{\Delta P}{\Delta T} = K1 \times [S] \times [E]$$

Usually enzyme assays are performed at a high substrate concentration for a short enough time period so that the substrate concentration can be assumed to be constant. The value of this constant substrate concentration can be combined with $K1$ to produce a second proportionality constant $K2$, which is the product of $K1$ times the substrate concentration. The rate can then be expressed so that it is dependent only on the amount of enzyme present, that is, a zero-order reaction, independent of substrate concentration.

$$\text{Rate} = \frac{\Delta P}{\Delta T} = K2 \times [E]$$

This rate of reaction or velocity is often listed as v, or V_i, or V_o, in the enzyme kinetic literature.

K_m and V_{max}. The enzyme activity (that is, velocity) is dependent on the substrate concentration when the amount of substrate is low relative to the amount of enzyme present in an assay. This relationship for a single substrate reaction is shown graphically in Fig. 49-5, *A*, with the same enzyme concentration assayed at many different substrate concentrations.

At steady state, before much product is present, the rate of formation of the ES complex will equal the rate of breakdown. This can be described using the following rate equation:

Formation | Breakdown

$$k_{+1}\,[E]\,[S] = k_{-1}\,[ES] + k_{+2}\,[ES]$$

By collecting terms and rearranging, we can remove the rate constants and define a constant, K_m.

$$\frac{[E]\,[S]}{[ES]} = \frac{k_{-1} + k_{+2}}{k_{+1}} = K_m \qquad \textit{Eq. 49-4}$$

The rate or velocity of product formation, v, at any time, and free enzyme concentration, [E], are related by the following:

$$v = k_{+2}\,[ES]$$

and

$$[E] = [Et] - [ES]$$

where [Et] is the total amount of enzyme and [ES] is the amount complexed with substrate. When all the enzyme is present in the form of ES (that is, at a very high [S] in a zero-order reaction), the maximum rate, V_{max}, is as follows:

$$V_{max} = k_{+2}\,[Et]$$

Combining the above three equations gives:

$$[E] = \frac{V_{max}}{k_{+2}} - \frac{v}{k_{+2}} = \frac{V_{max} - v}{k_{+2}}$$

Since from equation 49-4:

$$[E] = \frac{K_m\,[ES]}{[S]}$$

and

$$[ES] = \frac{v}{k_{+2}}$$

then

$$\frac{K_m\,[ES]}{[S]} = \frac{V_{max} - v}{k_{+2}}$$

or

$$\frac{K_m \times v}{[S] \times k_{+2}} = \frac{V_{max} - v}{k_{+2}}$$

Rearranging gives:

$$K_m \times v = (V_{max} - v)\,[S]$$

or

$$v(K_m + [S]) = V_{max}\,[S]$$

When this equation is solved for v, it gives the *Michaelis-Menten equation*, which is the equation for the rectangular hyperbola shown in Fig. 49-5, *A*.

$$v = \frac{V_{max}\,[S]}{K_m + [S]} \qquad \textit{Eq. 49-5}$$

[S] is the concentration of substrate, v is the velocity or rate (that is, enzyme activity), V_{max} is the maximal rate of reaction when the enzyme is saturated with substrate (that is, when [S] is approximately a constant), and K_m, the Michaelis-Menten constant, is the substrate concentration that produces one half the maximal velocity (Fig. 49-5, *A*). Thus when K_m = [S]:

$$v = \frac{V_{max}\,[S]}{[S] + [S]} = \frac{1\cancel{[S]} \times V_{max}}{2\cancel{[S]}} = \tfrac{1}{2}V_{max}$$

At the fixed high substrate concentration found in the usual clinical laboratory assays, the velocity, v, approaches V_{max} and is proportional to the amount of enzyme present, since all other factors are constant. The reaction is said to be zero order with respect to substrate, that is, independent of the concentration of substrate. The common condition used for assaying enzyme activity is a high substrate concentration where [S] $\cong$ 10 K_m or higher. The rate at a substrate concentration of 10 K_m is given by the following equation:

$$v = \frac{V_{max}\,(10\,K_m)}{K_m + (10\,K_m)} = V_{max}\,\frac{10\,K_m}{11\,K_m} = 0.91\,V_{max}$$

Thus at [S] = 10 K_m the rate produced is greater than 90% of V_{max}. The result of various substrate concentrations on the enzyme activity is shown in Table 49-2.

If all other cofactors and activators are present in similar excess, the rate is said to follow zero-order kinetics; that is, it is independent of all other factors and depends only on the amount of enzyme present. When this is so, the rate approaches V_{max}, and changes in the enzyme activity are attributable to changes in the amount of enzyme present in a sample rather than to variations in the assay conditions.

Table 49-2. Enzyme activity as a function of substrate concentration (expressed as multiples of K_m)

Substrate × K_m	Activity (V_{max})
1	0.50
5	0.83
10	0.91
20	0.95
50	0.98
100	0.99

Another way to examine the Michaelis-Menten equation is to see if it is consistent with first-order kinetics at low substrate concentrations and zero-order kinetics at high substrate concentrations.

At low substrate concentrations where [S] << K_m, then:

$$v = \frac{V_{max}\,[S]}{K_m + [S]} \cong \frac{V_{max}\,[S]}{K_m}$$

and since K_m and V_{max} are constants:

$$v = K1[S]$$

showing that the rate is dependent only on the first power of the substrate concentration.

At high substrate concentrations where [S] >> K_m, then:

$$v = \frac{V_{max}\,[S]}{K_m + [S]} \cong \frac{V_{max}\,\cancel{[S]}}{\cancel{[S]}} = V_{max}$$

showing that the rate (v) does not depend on substrate concentration.

As was shown in Fig. 49-5, *A*, the relationship of substrate concentration to enzyme activity is a curve that is often similar to a rectangular hyperbola. For multisubstrate enzyme reactions the kinetics are more complex and the reaction curves are similar to those shown in Fig. 49-5, *B*. The presence of activators and inhibitors acting at allosteric or regulatory sites tends to make the curves sigmoid because of the complex kinetics. A high concentration of activators, which may sometimes be the substrates themselves, and a low concentration of inhibitors (or products) tend to make these curves approach the case shown for a single substrate in Fig. 49-5, *A*. The presence of inhibitors will tend to make the curve more sigmoid, and thus they are to be avoided if possible in setting up enzyme assays. The reactions with two substrates must have the concentrations of both substrates optimized for maximal activity and as well any other activators needed for catalysis. For a more advanced treatment of kinetics see the text by Orten and Neuhaus.[10]

The accurate determination of K_m and V_{max} for each substrate or activator from Michaelis-Menten curves as are found in Fig. 49-5, *A*, is very difficult, even if the curves

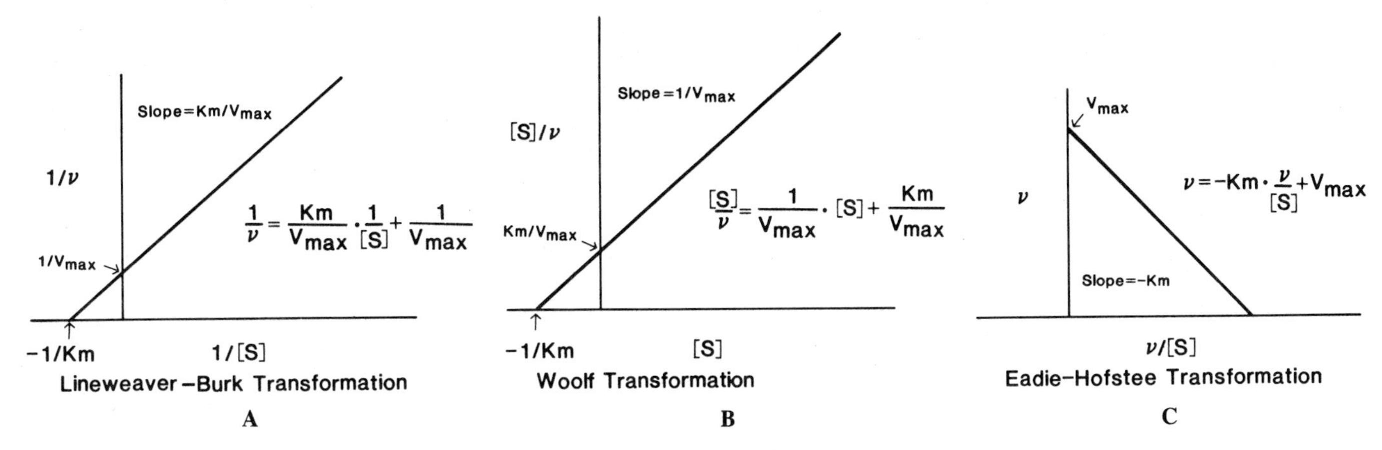

Fig. 49-6. Graphic representations of linear forms of Michaelis-Menten equation.

are not sigmoid. However, it is necessary to determine these constants so that assays may be established using optimal conditions to correctly measure enzyme activity. If the curve is transformed into a straight line, the K_m and V_{max} can be determined with greater accuracy. One may transform the Michaelis-Menten equation mathematically and obtain the equation of a straight line in several ways. The K_m and V_{max} can then be graphically determined from the line slopes and intercepts by use of these transformed equations. Several common graphic presentations are shown in Fig. 49-6.

Calculation of enzyme activity

The results of an enzyme determination are expressed as an activity unit in terms of the amount of product formed per unit of time, under specified conditions, for a given amount (volume) of sample (usually serum). Thus one unit of enzyme activity might be the amount that would, under certain specified conditions, cause the formation of 1 mg of the product, P, per minute when 1 mL of the sample was used. In older procedures arbitrary units like these were often employed.

Enzyme activity has been historically reported in the measuring units used by the authors who developed the assay; for example, for alkaline phosphatase using β-glycerol phosphate as a substrate, 1 Bodansky unit equals 1 mg of P formed per deciliter of serum per hour; and using phenolphosphate as a substrate, 1 King-Armstrong unit equals 1 mg of phenol formed per deciliter of serum per 30 minutes. The units were expressed as so many King-Armstrong units or Bodansky units after those who developed the assay. Each time an assay was modified a new unit would be created, which took on the name or names of the authors who proposed the modified assay. Such name units are no longer used and all assays are now expressed in international units (U) or in Système International units (K, katal), which are described below.

In 1961 the Enzyme Commission recommended the adoption of an international unit of enzyme activity. This unit, U, was defined as the amount of enzyme that would convert 1 micromole of substrate per minute under standard conditions.

$$1 \text{ U} = 1 \text{ micromole/minute}$$

In those instances where one molecule of substrate is transformed into a number of molecules of a product, the definition is per molecule (micromole) of product formed.

This recommendation of an international unit was an attempt to standardize the units of assay and to reduce the large number of units in use. This unit has been widely adopted, and in some respects it has standardized assay units. It has not reduced the number of reference ranges because if the standard conditions change, the apparent enzyme activity changes. For example, if a new buffer were used in the assay, it may affect the enzyme rate and produce a different number of international units of activity, that is, a larger absorbance change per minute, and thus would also change the reference range. In some cases this has added to the confusion since there may be several international units in common use for the same enzyme because of different standard conditions and the assay conditions are not usually included with the results. A clinician practicing at two different institutions where different international units were being used might receive results from these institutions expressed in terms of two levels of international units for the same enzyme assay, and he or she must remember that the reference ranges from the first institution are different from those from the second in order to interpret them properly.

The Système International d'Unités (SI), as originally adopted by the World Health Organization established the unit of enzyme activity as the katal, K. This is defined as 1 mol/sec of substrate changed. This unit is too large to be useful clinically, and so it has met with little acceptance in the United States though it was recommended by the EC in 1972.

For conversion of international units to katals:

$$1\ \text{U} = \frac{\text{Micromole}}{\text{Minute}} \times \frac{10^{-6}\ \text{mole}}{\text{Micromole}} \times \frac{1\ \text{min}}{60\ \text{sec}} = 1.67 \times 10^{-8}\ \text{K}$$

Thus 1.0 U = 16.7 nK (nanokatals).

These international units have been adopted by most workers in the field of clinical enzymology, and professional journals such as *Clinical Chemistry* require the use of international units in papers presented for publication. However, the use of the newer units in some clinical laboratories has been slower in gaining acceptance, since most clinicians are more familiar with the older name units.

Pure human enzyme materials are not available. Thus enzyme assays cannot be standardized in each laboratory by calibration with pure materials. Other methods of standardization for these assays must be used. The alternative method that is used most widely depends on having an accurately calibrated spectrophotometer. As was mentioned earlier, many enzyme assays are followed by making spectrophotometric measurements at a specific wavelength. With the spectrophotometric method usually one assumes that at 340 nm, NADH has a molar absorption coefficient of

$$A/(b \times c) = 6.22 \times 10^3\ \text{L·mol}^{-1}\text{·cm}^{-1}$$

where A is the actual absorbance of a solution, b is the light path in centimeters through the solution, and c is the concentration in moles per liter of the absorbing substance. Thus for a 1 cm light path and rearrangement for c:

$$c = A \times 10^{-3}/6.22$$

When the concentration is expressed in micromoles per liter (the international unit per liter) instead of moles per liter the expression will be:

$$c = A \times 10^3/6.22$$

Thus from the absorbance change that was measured and the volume of solution used, one can readily calculate the number of micromoles of NADH formed or used up during the enzyme measurement period.

$$\Delta c = \Delta A \times 10^3/6.22$$

For example, in the malate dehydrogenase reaction on p. 939, if a change in absorbance of 0.04 per minute was observed at 340 nm, and a 0.1 mL sample was used with a total assay volume of 3.0 mL, the calculation of activity would be as follows:

$$\text{International units/L} = \frac{0.04 \times 1000 \times 3.0}{6.22 \times 0.1} = 191\ \text{U/L}$$

Both enzyme units that have been described express the activity in terms of units per volume of sample. This is a particularly convenient unit of measure in the clinical laboratory when one is assaying enzymes in biological fluids like serum and plasma. If one is measuring an enzyme found in red blood cells or in white blood cells, another unit of measure is needed. In the case of RBC and WBC enzymes, the enzyme activity can be expressed as units per 10^{10} cells.

In biochemistry laboratories where enzyme purification is important the activity might be expressed per milligram of protein or per dry weight of cells or per microgram of DNA, but these are not convenient units for the clinical laboratory. Another way that enzyme activity is expressed in many research laboratories, particularly when comparing similar enzymes from different species, is to use the turnover number. This is a calculated number that can be derived from the usual activity measurements in in vitro systems. The *turnover number* is defined as the number of moles of substrate converted to product per minute per mole of enzyme. If the enzyme has more than one active site per molecule, the turnover number per active site is calculated for comparison. The calculation of the turnover number does require that an acurate measure of the molecular weight of the enzyme be known and that a relatively pure enzyme be used in an in vitro assay if it is to be expressed on the basis of units per milligram of protein.

Factors affecting enzyme measurements

The rate of reactions involving enzymes is greatly influenced by temperature, pH, concentration of substrate, and a number of other factors. Accordingly, all the details of a given procedure must be followed exactly so that precise and accurate results are produced.

Assays of enzyme activity should be performed under conditions of zero-order kinetics (that is, with the rate dependent only on the amount of enzyme present as shown at the top of the curve in Fig. 49-5, A) and optimal conditions. If experiments are not performed to determine the substrate concentration that will result in maximal activity, it may be that an assay will give different results if the substrate concentration is increased. It has been shown that one of the substrates used in the original assay[11] for alanine aminotransferase (ALT, EC 2.6.1.2) was insufficient for maximal activity.[12] If the determination of ALT activity is performed with the optimal substrate concentration, a considerably higher activity is obtained.

For optimization of an assay, such as the lactate dehydrogenase reaction given earlier, a series of reaction assays would be set up with increasing concentrations of lactate but at a high fixed NAD^+ concentration and a fixed amount of enzyme. The enzyme rates would then be measured, and a graph similar to Fig. 49-4 A, would be constructed. A second series of assays would then be performed with increasing concentrations of NAD^+, but at the fixed high concentration of lactate determined from the first experiment, that is, $[S] \cong 10\ K_m$ for lactate, and the same amount of enzyme present. The enzyme rates would again be determined and another figure created to determine the K_m for NAD^+. This same type of experiment would be performed for each item of the assay mixture (that is, metal ions, pH, buffer) until all the variables had been evaluated for maximal enzyme activity. The final conditions determined from this set of experiments would be the *optimal assay* conditions. These are the reaction concentrations of substrates, cofactors, activators, and buffer that produce the maximum rate of enzyme catalysis. These optimal conditions have been examined for LD (EC 1.1.1.27) by Howell[13] with highly purified NADH. Experiments to determine optimal assay conditions have been performed for the current clinically important enzymes, and diagnostic kits are commercially available with all the materials at optimal concentrations. It is important to check these concentrations because not all commercially available materials that are labeled "optimal" actually are.

At times optimal conditions cannot be used; that is, the substrate might have a limited solubility or might inhibit a secondary enzyme used in coupled assay, and then compromises in the optimal assay conditions must be made.

pH. Changes in pH will greatly affect the enzyme reaction rate. For most enzymes there is a definite pH range where the enzyme is most active. A pH near the center of this range is usually specified for the measurement of that particular enzyme. The optimal pH is different for different enzymes. Reduced activity is observed at pH values greater or less than the optimal.

At pH values other than the optimal pH, the enzyme activity may be affected because of changes in the structure of the enzyme. These changes may occur at the active site, or be attributable to conformational changes affecting the three-dimensional structure. There may also be changes in the charge of the enzyme or of the substrate. Since the active site of an enzyme often contains ionizable side chains of amino acids, such as $RCOO^-$ or RNH_3^+, a significant change in the pH can lead to the gain or loss of a proton. The result will be a substantial change in charge at the active site. The active site might therefore lose its ability to attract a substrate with a significant opposing charge. A similar loss of activity would occur if the change in charge were on the substrate molecule rather than on the enzyme. A change of pH might bring about an unfolding of the enzyme and loss of activity if the effect of pH change was to disrupt hydrogen-bonds and other intramolecular forces holding the enzyme in an active conformation.

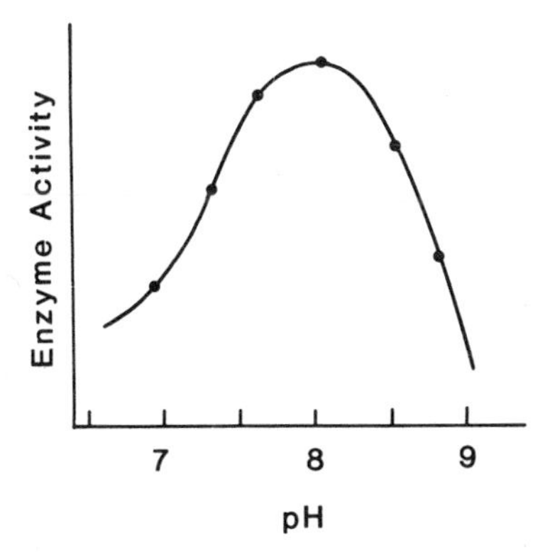

Fig. 49-7. Enzyme activity as a function of pH. Optimal pH range is 7.8 to 8.2; lower activities are observed at pH < 7.8 and pH > 8.2.

Buffer. In many cases, as the enzyme reaction proceeds, the products tend to alter the pH. Most assays include a buffer to maintain the assay pH within the optimal pH range. The buffer chosen should have a pK_a within 1 pH unit of the optimal pH of the enzyme in order to exert effective pH control.

A typical pH curve of enzyme activity is given in Fig. 49-7. This bell-shaped curve showing maximal enzyme activity versus factor (in this case pH) concentration is seen for other assay factors such as concentration of buffer, substrate, cofactors, and activators.

Buffers not only serve to regulate the pH of an assay, but they may also take part in the reaction as well. Alkaline phosphatase (ALP, EC 3.1.3.1) assays with *p*-nitrophenol phosphate as a substrate, as in the Bowers and McComb procedure,[14] use the buffer 2-amino-2-methyl-1-propanol (AMP) to maintain the pH at 10.2. The enzyme hydrolyzes the substrate into *p*-nitrophenol and inorganic phosphate in a multistep process, part of which involves a temporary phosphorylation of the enzyme. The final and rate-limiting step includes hydrolysis of the enzyme-phosphate bond to regenerate free enzyme. At similar pH values, buffers that are phosphate acceptors in a transphosphorylation process with the enzyme will produce rates of alkaline phosphatase activity higher than those of buffers that do not act as phosphate acceptors.[15] Thus AMP buffer produces higher rates of alkaline phosphatase activity at pH 10.2 than glycylglycine buffer does at pH 10.2 because glycine is not a phosphate acceptor. In the case of buffers that do not participate in the reaction, the concentration of buffer that gives maximal enzyme activity at the optimal pH must also be experimentally determined.

It has been found that the buffer and certain salts may have an unusual effect on the K_m. When the buffer-to-substrate ratio is very large, the buffer may compete with the substrate for the enzyme and make the enzyme activity appear to be related to substrate concentration in a nonlinear way. This has been observed with NADH in the LD reaction.[16] Here the buffer-to-substrate molar ratio is 10^4:1 and the rate of reaction is affected by Tris, phosphate, and NH_4HCO_3 buffers and certain salts, such as NaCl and $(NH_4)_2SO_4$, which are often found in the materials used to prepare enzyme assays. There seems to be no effect at buffer concentrations below 0.05 mol/L, a finding that is consistent with several recommendations for optimal assay conditions.[13,17,18] It would seem prudent to maintain as low a concentration of buffer as possible without compromising pH stability or enzyme rate.

Cofactors. Many enzymes require a nonprotein, often dialyzable material for maximal activity. Some of these materials are related to vitamin structures. For example, thiamin, or vitamin B_1, can be converted to thiamin pyrophosphate, a cofactor in many decarboxylation reactions. Niacin can be converted to nicotinamide adenine dinucleotide, and vitamin B_2, riboflavin, can be converted to flavin adenine dinucleotide. Both niacin and riboflavin are involved in many dehydrogenation reactions. Pyridoxine, vitamin B_6, is modified to pyridoxal phosphate, which is used in many transamination reactions.

In the lactate dehydrogenase reaction, NAD^+ is considered to be such a cofactor, but since it reacts on an equal molar basis with lactate, it is a substrate, though it is discussed as if it were a cofactor.

In analytical assays of transaminase activity pyridoxyl-5-phosphate is an example of a cofactor that is not a substrate. The optimal concentration of a cofactor is determined in the same way as a substrate so that assay conditions can be established with a cofactor concentration of approximately 10 K_m or higher.

Activators and inhibitors. Many enzymes require specific ions for maximal activity. All phosphate-transferring enzymes, such as hexokinase, require magnesium ions (Mg^{+2}). Other common metal ion activators are manganese (Mn^{+2}), calcium (Ca^{+2}), zinc (Zn^{+2}), iron (Fe^{+2}), and potassium (K^+). Amylase requires chloride (Cl^-) for maximal activity, and there are enzymes that require several ions for maximal activity; for example, pyruvate kinase requires magnesium (Mg^{+2}) and potassium (K^+). In each case, the optimal concentration of the activator must be determined just as the optimal concentration of substrate is determined.

Inhibitors are materials that reduce the catalytic activity of an enzyme. There are many types of inhibitors and several classes of inhibition. Inhibitors may act by removing an activator by chelation; for example, Ca^{+2} and Mg^{+2} are removed by EDTA or oxalate in the inhibition of hexokinase. They may also act by binding to the active site to compete with the substrate, as shown in Fig. 49-2, or by forming a complex at a different site, that is, an allosteric site, which may affect the enzyme activity.

Inhibitors are classed into three main groups. *Competitive inhibitors* bind at the active site and compete with the substrate, for binding sites. These materials demonstrate a reversible inhibition that can be reduced by use of a higher substrate concentration.

$$E \begin{array}{l} + \ S \rightleftharpoons \{ES\} \rightarrow P + E \\ + \ I \rightleftharpoons \{EI\} \end{array}$$

The maximum rate of reaction is not affected if enough substrate is present because of the reversibility of the reactions. The binding of the substrate is affected, and thus the apparent K_m will be higher while the V_{max} remains the same.

Noncompetitive inhibitors are a second group of materials that bind at an allosteric or regulatory site, which may be far removed from the active site. These inhibitors cannot be reversed by the addition of more substrate, since they are believed to bind at a different location on the enzyme surface.

$$\begin{array}{ccccc} E & + S \rightleftharpoons & \{ES\} & \rightarrow P + E \\ + & & + & \\ I & & I & \\ \upharpoonleft\downharpoonright & & \upharpoonleft\downharpoonright & \\ \{EI\} & + S \rightleftharpoons & \{ESI\} & \end{array}$$

Since the inhibitor does not compete with the substrate, the K_m will be unaffected but the amount of E or ES that

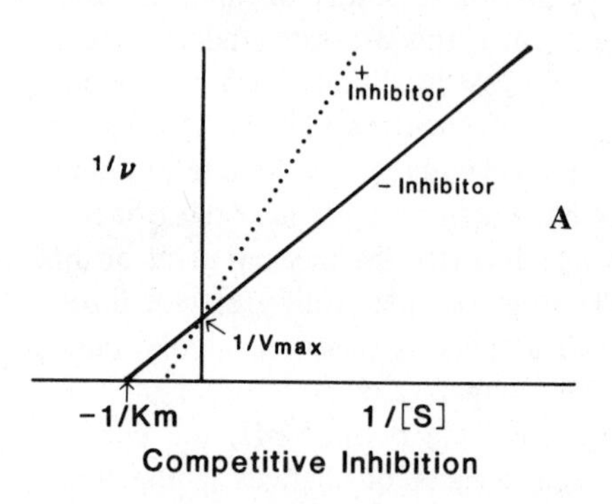

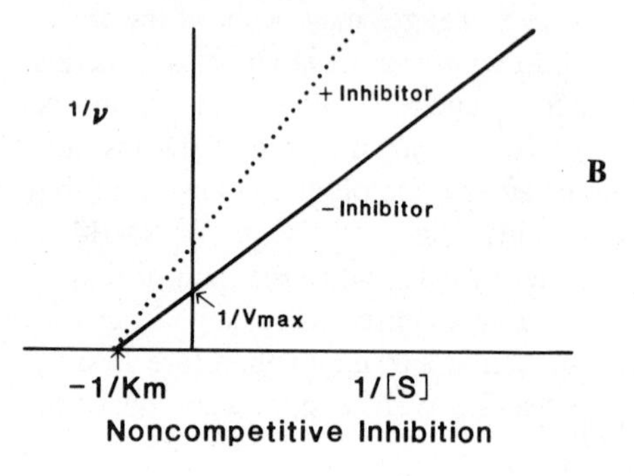

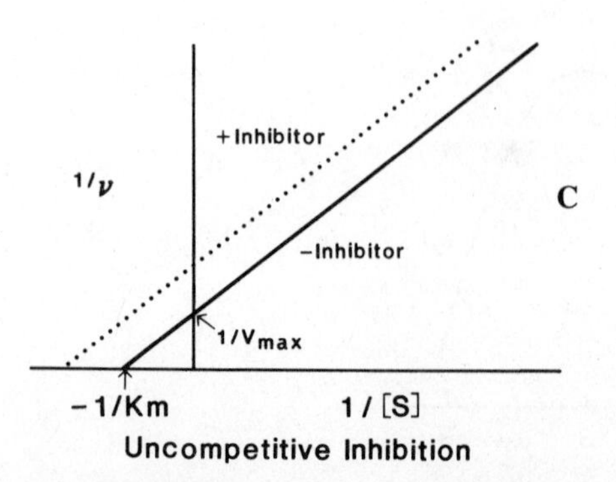

Fig. 49-8. The three types of inhibition are shown by use of Lineweaver-Burk graphic method to demonstrate effect of type of inhibition on K_m and V_{max}.

converts substrate to product will be reduced and the V_{max} will be lessened.

Uncompetitive inhibitors are a third group of inhibitors and are believed to bind with the enzyme substrate complex and not to the free enzyme. In this case, at low substrate concentrations, the addition of more substrate increases the inhibition, since it produces more enzyme-substrate complex to react with the inhibitor. The result of this type of inhibition is that the V_{max} is reduced and the K_m is decreased.

$$\begin{array}{c} E + S \rightleftharpoons \{ES\} \rightarrow P + E \\ + \\ I \\ \upharpoonleft\downharpoonright \\ \{ESI\} \end{array}$$

The type of inhibition a substance exerts can be determined when one examines the results of kinetic studies using a linear graph of enzyme activity with and without inhibitors, as shown in Fig. 49-8.

A brief summary of the effects of the types of inhibition is given in Table 49-3. The simple types of inhibition may be classified by examination of the kinetic effect on the K_m and V_{max}.

Coupling enzymes. Some enzyme reactions of interest, such as alanine aminotransferase (ALT) and aspartate aminotransferase (AST), do not form products that can be monitored directly. One may couple the initial enzyme reaction to a second *indicating* enzyme reaction that, for example, does contain the $NAD^+/NADH$ conversion to make a convenient assay. The AST enzyme reaction can be coupled to the malate dehydrogenase reaction (MD, EC 1.1.1.37):

$$\text{L-Aspartate} + \alpha\text{-Ketoglutarate} \xrightleftharpoons{AST} \text{L-Glutamate} + \text{Oxaloacetate}$$

$$\text{Oxaloacetate} + \text{NADH} + H^+ \xrightleftharpoons{MD} \text{L-Malate} + \text{NAD}^+$$

This gives the following net reaction:

$$\text{L-Asparate} + \alpha\text{-Ketoglutarate} + \text{NADH} + H^+ \leftrightarrows \text{L-Glutamate} + \text{L-Malate} + \text{NAD}^+$$

In this case, the substrate for the second reaction, oxaloacetate, is supplied as the product of the first reaction. Oxaloacetate from the AST reaction serves as the substrate, with the confactor NADH for the malate dehydrogenase reaction. This assay would have L-aspartate, α-ketoglutarate, NADH, and the enzyme malate dehydrogenase (MD) present at large excesses so that the rate-limiting item in the assay would be the amount of AST in the added sample.

There are other enzymes, such as creatine kinase (CK, EC 2.7.3.2) where the measurement of the first enzyme requires an intermediate *auxiliary* enzyme reaction and then an *indicator* enzyme. In the measurement of CK, hexokinase (EC 2.7.1.1) is used as an *auxiliary* enzyme and glucose-6-phosphate dehydrogenase (EC 1.1.1.49) is used as an *indicating* enzyme. Both of these additional enzymes would have to be present in large excesses for correct measurement of CK. It is difficult to correctly produce optimum assays that have more than two coupled reactions because of the large number of components in the assay system and the problems with maximizing all the components without causing inhibition of the limiting reaction.

Temperature. There is no optimal temperature for enzyme assays. Most enzymes show an increasing amount of activity as the temperature is raised over a limited temperature range, such as, 10° to 40° C, an example of which is shown in Fig. 49-9.

To minimize any losses of activity if the enzyme cannot be assayed immediately after collection, one should store samples at refrigerator temperatures, 2° to 6° C, or frozen (Table 49-4). In a few cases, some forms of enzymes, such as LD_4 and LD_5 have been found to be more stable at room temperature than at refrigerator temperatures. The repeated freezing and thawing of a specimen will often cause denaturation and loss of activity. Above 40° C most enzymes are rapidly denatured and lose almost all activity after a short time. An exception to this general rule is amylase, which seems to be stable up to about 60° C before significant losses of activity occur.

There is an approximate doubling of the enzyme activity for a 10 Celsius degree increase in temperature within the limits given above. To describe this phenomena, a quantity

Table 49-3. Kinetic effects of inhibition

Type of inhibition	Change in K_m	Change in V_{max}
Competitive	Increased	No change
Noncompetitive	No change	Decreased
Uncompetitive	Decreased	Decreased

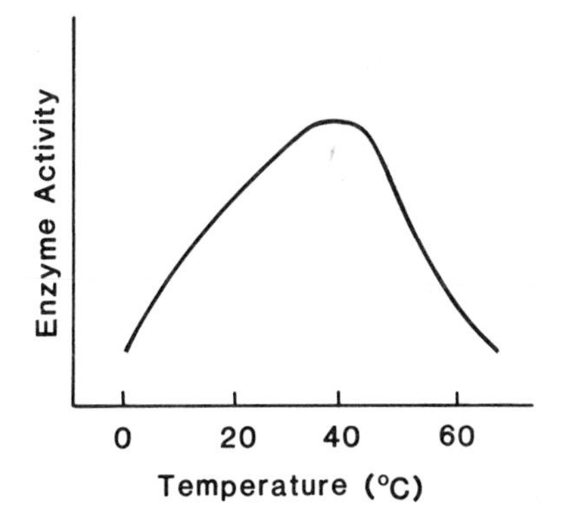

Fig. 49-9. Enzyme activity as a function of temperature of assay. At low temperatures activity decreases. As temperature is raised, activity increases until rate of denaturation is greater than increase in activity.

Table 49-4. Enzyme stability under various storage conditions (less than 10% change in activity)

Enzyme	Room temperature (about 25° C)	Refrigeration (0° to about 4° C)	Frozen (−25° C)
Aldolase (ALS)	2 days	2 days	Unstable*
Alanine aminotransferase (ALT, GPT)	2 days	5 days	Unstable*
α-Amylase (AMS)	1 month	7 months	2 months
Aspartate aminotransferase (AST, GOT)	3 days	1 week	1 month
Ferroxidase I (ceruloplasmin)	1 day	2 weeks	2 weeks
Cholinesterase (CHS)	1 week	1 week	1 week
Creatine kinase (CK)	1 week	1 week	1 month
γ-Glutamyl transferase (GGT)	2 days	1 week	1 month
Isocitrate dehydrogenase (ICD)	1 day	2 days	1 day
Lactate dehydrogenase (LD)	1 week	1 to 3 days†	1 to 3 days†
Leucine aminopeptidase (LAS)	1 week	1 week	1 week
Lipase (LPS)	1 week	3 weeks	3 weeks
Phosphatase, acid (ACP)	4 hours‡	3 days§	3 days§
Phosphatase, alkaline (ALP)	2 to 3 days‖	2 to 3 days	1 month

*Enzyme does not tolerate thawing well.
†Depending on isoenzyme pattern in the serum.
‡Unacidified.
§With added citrate or acetate to pH~5.
‖Activity may increase.

was defined, Q10, which is the ratio of activity at the higher temperature over the activity at the lower temperature for a 10° difference in temperature. For many enzymes a 1° change in temperature would produce about a 10% change in activity. A tolerance of ±0.1° C for temperature control of an enzyme analyzer is recommended, since this would produce approximately a ±1% change in the activity that was measured. This amount of variation would be small enough to be ignored for most clinical work as an insignificant source of error. A recommendation of ±0.05° C for temperature control has also been suggested,[19] and it would reduce the change in activity to ±0.5%.

Different enzymes show variable increases with temperature, and so it is not always the case that the activity will be doubled if the temperature is increased by 10 Celsius degrees; that is, the Q10 for many different enzymes ranges from about 1.5 to 2.5. As more work has been performed on temperature variablity, it has been determined that the Q10 for an enzyme in not fixed but varies depending on the 10° temperature range that is picked to compare enzyme activities. However, because almost all clinical measurements of enzyme activity are performed in the 12° range of 25° to 37° C, this is the only range of practical concern. The apparent increase in activity with increasing temperature means that assays that are performed at higher temperatures, such as 37° C, will be more sensitive to slight changes in the amount of enzyme in a sample. The common enzymes employed for clinical diagnosis are less stable at this temperature than at 25° to 30° C, and therefore assays to 37° C must be performed with relatively short assay times so that any enzyme denaturation that would occur would be minimized.

It would be expected that instrument manufacturers would tend to prefer an assay temperature of 37° C, since it would provide the most sensitivity. Reagent manufacturers, on the other hand, would tend to favor a lower temperature, since less substrate would be required in the assay mixture and the assay material would be less costly to produce. Arguments for both higher, that is, 37° C, and lower temperatures, that is, 25° C, have appeared in the literature[19,20] based primarily on scientific and technical reasoning. A reasonable compromise seems to be measurement at 30° C, which is the recommendation of the International Federation on Clinical Chemistry (IFCC).[21] There is a very accurate gallium standard melting point cell that is now available to all laboratories from the National Bureau of Standards.[22] This material has a melting temperature plateau of 29.772° C and can be used to calibrate or check the assay temperature of a wide variety of instruments.

Enzymes as reagents

It is possible to determine substrate concentrations using many principles of enzyme kinetics applied in a slightly different way. The enzyme activity at low substrate concentrations is first order, that is, linear, with respect to substrate concentration (Fig. 49-5,A). In order to measure the concentration of pyruvate in a sample, a special assay mixture would be prepared. The samples containing the unknown amounts of pyruvate are diluted (or only a small amount used) so that pyruvate concentration in the assay is low, that is, less than the K_m. An assay mixture that contains an excess of lactate dehydrogenase (LD), an excess of coenzyme NADH and a buffer would be used. The reduction in the amount of NADH in this assay could be

Table 49-5. Loss of enzyme activity on storage at 4° C (%)

	1 day	3 days	5 days	7 days
LD_1	0	0	3	5
LD, total	0	8	9	12
CK*	0	0	0	0
AST (GOT)	2	8	10	12
ALT (GPT)	2	10	14	20
AMS	0	0	0	0
LPS	0	0	0	0
ALP†	2	2	4	—
ACP‡	0	0	0	0
GGT	0	0	0	0

*Assayed with reagents containing thiol (−SH) groups.
†Values for ALP are increases in activity on storage. (From Massion, C.G., and Frankenfeld, J.D.: Clin. Chem. **18:**366-373, 1972.)
‡After acidification of serum to pH 5.5 to 6.0.

Table 49-6. Loss of enzyme activity on storage at 25° C (%)

	1 day	3 days	5 days	7 days
LD_1	0	0	0	5
LD, total	0	2	10	15
CK*	0	1	4	6
AST (GOT)	2	10	11	13
ALT (GPT)	8	17	19	39
AMS	0	0	0	0
LPS	0	0	0	0
ALP†	3	4	6	—
ACP‡	0	0	0	0
GGT	0	0	0	0

*Assayed with reagents containing thiol (−SH) groups.
†Values for ALP are increases in activity on storage. (From Massion, C.G., and Frankenfeld, J.D.: Clin. Chem. **18:**366-373, 1972.)
‡After acidification of serum to pH 5.5 to 6.0.

related to the amount of pyruvate present from the millimolar absorption coefficient of NADH at 340 nm. Alternatively, a series of pyruvate standards could be used to calibrate the assay. Other enzymatic assays of a similar nature that are commonly used in the clinical laboratory include the determination of glucose, urea, ethanol, cholesterol, triglycerides, and uric acid. There are many other uses of enzymes as reagents. They are often used as indicators in EMIT or in ELISA assays for drugs.

Storage of enzymes

Most enzymes that are utilized clinically are stable at refrigerator temperatures from 2 to 3 days to about a week and at room temperature for a shorter time. Table 49-4 summarizes data for three temperatures.[23,24] Additional information on enzyme stability can be found in Tables 49-5 to 49-7. Several enzymes deserve particular comment. Acid phosphatase is particularly unstable at all temperatures unless the pH of the serum is reduced to about 5 or 6 with citrate or acetate. Alkaline phosphatase in human serum demonstrates a linear increase in activity dependent on temperature and time.[25] At 96 hours (4 days) there is a 6% increase at room temperature, a 4% increase at refrigerator temperature, and a 1% increase at −20° C. Enzymes in control materials are usually of nonhuman origin and are much more varied; some are more stable and some are less stable than enzymes in human serum.

CLINICAL RATIONALE FOR ENZYME MEASUREMENTS

In 1954 after LaDue, Wróblewski, and Karmen[26] found a temporary increase in serum aspartate aminotransferase (EC 2.6.1.1) activity after an acute myocardial infarction, the measurement of changes in plasma enzyme activity gained importance as a means of following the course of a disease or of improving clinical diagnosis. Many investigators began to look for changes in enzyme activity that were specific for a disease state or organ, using newer and more sensitive methods.

Table 49-7. Plasma half-lives for clinically important enzymes

Enzyme	Half-life (hours) (mean ± 2 SD)
LD_1	53-173
LD_5	8-12
CK	15
AST (GOT)	12-22
ALT (GPT)	37-57
AMS	3-6
LPS	3-6
ALP	3-7 days
GGT	3-7 days

It was reasonable to hypothesize that variation in enzyme concentrations in tissue cells would reflect changes in states of health and disease of the tissue. However, cellular enzyme concentrations in intact tissues cannot be assayed on a routine basis. Changes in enzyme activity in the plasma (or serum) are followed, since it is known that enzymes are primarily intracellular constituents that are released after cell damage or cell death has taken place in a specific organ or tissue. In a few specialized cases, enzyme activities are measured in lysates of red or white cells because these kinds of tissues can be sampled relatively easily and the contents can be assayed without too much difficulty.

Although many disease-related changes have been found, the original hope of finding a simple marker for a specific organ or disease has not been realized. The changes that occur with many diseases or in a particular organ can now be understood, but only by examining the pattern of several enzyme or isoenzyme changes over a period of hours or days.

Extracellular versus cellular enzymes

The enzymes that are found in plasma can be categorized into two major groups. These major subdivisions are the plasma-specific enzymes and the non–plasma specific enzymes.[27]

The plasma-specific enzymes are those enzymes that have a definite and specific function in plasma. Plasma is their normal site of action, and they are present in plasma at higher concentrations than in most tissues. Among these are the enzymes involved in blood coagulation, and as well ferroxidase, pseudocholinesterase, and lipoprotein lipase. These enzymes are synthesized in the liver and are constantly liberated into the plasma to maintain a steady-state concentration. These enzymes are clinically of interest when their concentration decreases in plasma, and some have historically been used as estimates of liver function.

The non–plasma specific enzymes are those enzymes with no known function in plasma. Their concentrations in plasma are usually found to be lower than in most tissues, and there may be a deficiency in activators or cofactors that are necessary for enzyme activity. These enzymes can be further divided into two groups, the enzymes of secretion and the enzymes of intermediary metabolism.

The enzymes of secretion are those known to be from exocrine glands, that is, the pancreas and prostate, and some enzymes from the gastric mucosa and the bones. Enzymes in this group are clinically important when their concentrations are either higher or lower than normal. Elevated values are found when the normal mode of excretion is blocked or when the amount of enzyme produced is increased. Decreases in the amount of enzyme are found when the tissue that normally produces the enzyme is damaged or necrotic. Common examples of this group are amylase, lipase, and acid and alkaline phosphatases.

The other major group of non–plasma specific enzymes are the enzymes of metabolism. The concentrations of these enzymes in tissues are very high, sometimes thousands of times higher than in the plasma. Cellular damage resulting in leakage or necrosis allows a fraction of these proteins to escape into the plasma and causes a sharp rise in the concentration normally found. Some common examples are creatine kinase (CK), lactate dehydrogenase (LD), alanine aminotransferase (ALT), and aspartate aminotransferase (AST).

The enzyme content that is observed for a tissue lysate or body fluid is a steady-state result of the synthetic and degradative processes. In a cell or tissue, the enzyme content is controlled by many factors. Some enzymes are produced at a fixed rate and do not seem to be affected by other metabolic events. These enzymes, which remain at a constant concentration, are called *constitutive enzymes*. Creatine kinase, an enzyme involved in energy storage in muscle, is a constitutive enzyme. Changes in the amount of this enzyme in serum are related to cell necrosis or muscle damage, but the amount found in muscle is constant and related to the total mass of the cells.

Many other enzymes are produced in greater quantities when the environment (plasma) contains the substrates for these enzymes. Enzymes whose concentration can be increased or decreased because of environmental stimuli are called *inducible enzymes*. This is particularly true for those enzymes, found in the liver in mammals, that help to process the products of digestion for the body. Some enzymes in the metabolism of carbohydrates and amino acids can also be shown to increase in concentration because of metabolic loads of their substrates. For example, in the liver, glucose is phosphorylated by hexokinase, a constitutive enzyme, and glucokinase, an inducible enzyme. In response to a high-carbohydrate diet, the glucokinase that is usually found in low concentrations in the liver is induced to higher concentrations to help metabolize the additional glucose. High glucose concentrations will cause the pancreas to produce insulin. Liver cells do not need insulin to allow the entry of glucose into these cells, in contrast to muscle and adipose tissue, which do require insulin. The rise in blood insulin has several effects. It allows glucose to enter muscle and adipose tissue, thus reducing serum glucose directly, and it induces the production of glucokinase in liver cells. With the additional glucokinase, the extra carbohydrate load can be metabolized by this secondary inducible enzyme. The mixed-function cytochrome P-450 oxidase (EC 1.6.2.4) is increased when many drugs or toxic materials are ingested, presumably to increase the rate of detoxification of these compounds.

Other aspects of enzyme control also deserve mention. Product inhibition can result in a temporary reduction in enzyme activity. Hexokinase is subject to this inhibition, and a large glucose load will produce enough glucose-6-phosphate to inhibit the enzyme.

Negative feedback also plays a role in reducing enzyme activity. The product of a series of reactions may affect an early reaction in a synthetic pathway in such a way that the enzyme reaction is reduced, resulting in the reduction of the amount of end product produced. Alanine is an amino acid produced from pyruvate. The pyruvate kinase reaction in liver, which produces pyruvate from phospho*enol*pyruvate, can be inhibited by L-alanine. In the presence of high concentrations of L-alanine the pyruvate kinase reaction is inhibited and phospho*enol*pyruvate is diverted to the production of glucose instead of amino acids.

SIGNIFICANT FACTORS AFFECTING ENZYME-REFERENCE VALUES

There are a number of important factors that affect the reference ranges for enzyme determinations. If these factors are not accounted for in the interpretation of the results, a misdiagnosis is possible. In the following items a brief comment on the problem and an example will be given. Further details are covered in Section Three, Methods of Analysis (Chapters 49, 50, and 56) or the section on clinical interpretation (Section Two) and Chapter 18.

Age

There appear to be variations in the amounts of enzymes normally present in serum that are the result of differences in age between various subgroups in the population. There are perhaps three principle times to consider whether age will be important to an assay. These are during the first year of life as various organs (such as the liver) are coming to maturity, during puberty, and in late middle age when hormonal changes are occurring.

Perhaps some of the most dramatic changes are seen with the enzyme alkaline phosphatase.[28] Using the Bowers and McComb procedure at 30° C the following values are found; 135 to 270 U/L for children 6 months to 10 years of age, to 90 to 320 U/L for children 10 to 18 years of age, and 40 to 100 U/L for adults.

Sex

Differences may occur with some enzymes that are either related to muscle mass, exercise, or hormone concentration. The assignment of the ultimate cause is difficult, but these assays are usually said to have a difference related to sex.

An example of these effects is seen with the enzyme creatine kinase, where males are reported to have higher reference ranges than females, an effect that is most likely attributable to increased muscle mass.[29]

Race

Race may also be a factor for a limited number of enzymes, but data are sparse. Black populations are reported to have higher reference ranges than comparable white populations for creatine kinase,[30] but the effect may be an indirect result of several factors other than race in the two populations.

Exercise

Exercise and ambulation are important variables in the consideration of reference ranges for several enzymes. Patients who have been at complete bed rest for several days are found to have 20% to 30% lower values for creatine kinase than ambulatory patients.[31] Normal amounts of exercise also elevate creatine kinase.[32] The additional creatine kinase attributable to normal exercise is of the MM type, CK_3. Thus the distinction between these elevations and those that are caused by an acute myocardial infarction, which is the MB type, CK_2, is easily accomplished by determination of the isoenzyme pattern. The increases seen after exercise usually disappear after 2 to 24 hours. This distinction is more difficult for the extremes of exercise. In ultralong-distance runners the normal CK-MB was found to be up to threefold higher than usual and the total CK was found to be up to fortyfold higher than usual.[33]

Sampling time

Since enzymes do not undergo any significant circadian rhythmn, sampling time is unimportant for the determination of enzyme reference ranges. On the other hand, the sampling time may be important for detecting a variety of acute and chronic conditions if the changes observed are sufficiently rapid. The classic mean time for maximum elevation for a series of enzymes in patients with a myocardial infarction was reported to be: CK-MB, 6 hours; CK, 18 hours; AST, 24 hours; and LD, 48 hours.[34] Since not all patients follow the classical pattern, see Chapter 28.

Assay type

Although there is an optimal set of conditions for the assay of an enzyme, it is clear that not all assays are optimal. At times the differences between assays may not appear to be significant and yet the results obtained will be substantially divergent. The effect of various components of an assay upon one another is even more significant when one considers a coupled assay. These assays not only have the concentrations of substrates and activators of the primary reaction to consider but they must also have excesses of the auxiliary and indicating enzymes with their associated additional activators.

Alanine aminotransferase (ALT) is often measured by addition of an excess of lactate dehydrogenase and NADH to a mixture containing L-alanine and α-ketoglutarate and buffer. The usual commercially available kits often specify that about 500 U/L of LD are present as an indicating enzyme and perhaps that the source of this enzyme is ''animal.'' This is not sufficient to completely define the assay since the K_m of pyruvate varies with the isoenzyme type. About four times as many units of M_4-LD_5 would be required as H_4-LD_1, to achieve an equivalent reaction rate. A crude mixture of isoenzymes would be somewhere in between these extremes. Even if the units of LD added were the same, the measured enzyme rates might vary with each lot of a kit if the indicating enzyme were added without regard to the isoenzyme content. This same kind of variability would occur between manufacturers with the ''same'' concentrations of substrates, activators, and units of LD if a different source, such as bacterial, of the indicating enzyme was used. It is for this reason that a reference range should be checked by each laboratory, particularly when changing reagent manufacturers.

BIBLIOGRAPHY

Brown, S.S., Mitchell, F.L., and Young, D.S., editors: Chemical diagnosis of disease, New York, 1979, Elsevier/North Holland Biomedical Press, Chapter 7.

Richterich, R., and Colombo, J.P.: Clinical chemistry: theory, practice and interpretation, New York, 1981, John Wiley & Sons, Chapters 3 and 4.

Zilva, J.F., and Pannall, P.R.: Clinical chemistry in diagnosis and treatment, ed. 3, Chicago, 1979, Year Book Medical Publishers, Inc., Chapter 15.

REFERENCES

1. Hohnadel, D.C., and Cooper, C.: The effect of structural alterations on the reactivity of the nucleotide substrate of rabbit muscle pyruvate kinase, FEBS Lett. **30:**18-20, 1973.
2. Hohnadel, D.C., and Cooper, C.: The effect of structural modifica-

tions of ATP on the yeast-hexokinase reaction, Eur. J. Biochem. **31:**180-185, 1972.
3. Nomenclature committee of the International Union of Biochemistry: Enzyme nomenclature 1978, Recommendations of the nomenclature committee of the International Union of Biochemistry on the nomenclature and classification of enzymes, New York, 1979, Academic Press, Inc.
4. Baron, D.N., Moss, D.W., Walker, P.G. and Wilkinson, J.H.: Abbreviations for names of enzymes of diagnostic importance: J. Clin. Pathol. **24:**656-657, 1971.
5. Baron, D.N., Moss, D.W., Walker, P.G., and Wilkinson, J.H.: Revised list of abbreviations for names of enzymes of diagnostic importance: J. Clin. Pathol. **28:**592-593, 1975.
6. Commission of Editors of Biochemical Journals—International Union of Biochemistry: Biochemical Nomenclature and Related Documents, London, England, 1978, William Clowes & Sons, Limited.
7. International Union of Pure and Applied Chemistry—International Union of Biochemistry Commission on Biochemical Nomenclature: Abbreviations and symbols, Clin. Chem. **23:**2344-2349, 1977.
8. Fendley, T.W., Hochholzer, J.M., and Frings, C.S.: Effect of various diluents on the activity of several enzymes present in serum, Clin. Chem. **19:**1079-1080, 1973.
9. Pardue, H.L.: A comprehensive classification of kinetic methods of analysis used in clinical chemistry, Clin. Chem. **23:**2189-2201, 1977.
10. Orten, J.M., and Neuhaus, O.W.: Human biochemistry, St. Louis, 1982, The C.V. Mosby Co., Chapter 5.
11. Wróblewski, F., and LaDue, J.S.: Serum glutamic pyruvic transaminase in cardiac and hepatic disease, Proc. Soc. Exp. Biol. Med. **91:**569-571, 1956.
12. Henry, R.J., Chiamori, N., Golub, O.J., and Berkman, S.: Revised spectrophotometric methods for the determination of glutamic-oxalacetic transaminase, glutamic-pyruvic transaminase, and lactic acid dehydrogenase, Am. J. Clin. Pathol. **34:**381-398, 1960.
13. Howell, B.F., McCune, S., and Schaffer, R.: Lactate-to-pyruvate or pyruvate-to-lactate assay for lactate dehydrogenase: a re-examination, Clin. Chem. **25:**269-272, 1979.
14. Bowers, G.N., Jr., and McComb, R.B.: A continuous spectrophotometric method for measuring the activity of serum alkaline phosphatase, Clin. Chem. **12:**70-89, 1966.
15. McComb, R.B., and Bowers, G.N., Jr.: Study of optimum buffer conditions for measuring alkaline phosphatase activity in human serum, Clin. Chem. **18:**97-104, 1972.
16. Howell, B.F., McCune, S., and Schaffer, R.: Influence of salts on Michaelis-constant values for NADH, Clin. Chem. **23:**2231-2237, 1977.
17. Bergmeyer, H.U., Büttner, H., Hillman, G., et al.: Standardization of methods for the estimation of enzyme activity in biological fluids, Z. Klin. Chem. Klin. Biochem. **8:**659-660, 1970.
18. Keiding, R., Hörder, M., Gerhardt, W., et al.: Recommended methods for the determination of four enzymes in blood, Scand. J. Clin. Lab. Invest. **33:**291-306, 1974.
19. Bergmeyer, H.U.: Standardization of the reaction temperature for the determination of enzyme activity, Z. Klin. Chem. Klin. Biochem. **11:**39-45, 1973.
20. Duggan, P.F.: Activities of enzymes in plasma should be measured at 37° C, Clin. Chem. **25:**348-352, 1979.
21. Committee on Standards—Expert Panel on Enzymes: Provisional recommendation (1974) of IFCC methods for the measurement of catalytic concentrations of enzymes, Clin. Chem. **22:**384-391, 1976.
22. Bowers, G.N., Jr., and Inman, S.R.: The gallium melting-point standard: its evaluation for temperature measurements in the clinical laboratory, Clin. Chem. **23:**733-737, 1977.
23. Bergmeyer, H.U.: Standardization of enzyme assays, Clin. Chem. **18:**1305-1311, 1972.
24. Labrosse, K.R., Bixby, E.K., and Lakatua, D.J.: Stability of cardiac enzymes at different storage temperatures, Clin. Chem. **26:**1026, 1980.
25. Massion, C.G., and Frankenfeld, J.D.: Alkaline phosphatase: Lability in fresh and frozen human serum and in lyophilized control material, Clin. Chem. **18:**366-373, 1972.
26. LaDue, J.S., Wróblewski, F., and Karmen, A.: Serum glutamic oxaloacetic transaminase activity in human acute transmural myocardial infarction, Science **120:**497-499, 1954.
27. Hess, B.: Enzymes in blood plasma, New York, 1963, Academic Press, Inc.
28. McComb, R.B., Bowers, G.N., Jr., and Posen, S.: Alkaline phosphatase, New York, 1979, Plenum Press, Chapter 9.
29. Garcia, W.: Elevated creatine phosphokinase levels associated with large muscle mass, J.A.M.A. **228:**1395-1396, 1974.
30. Blight, M., Wagman, E., Shastri, S., and Nevins, M.: Race-related differences in reference intervals for creatine kinase, Clin. Chem. **26:**1928-1929, 1980.
31. Tietz, N.W., editor: Fundamentals of clinical chemistry, Philadelphia, 1976, W.B. Saunders Co., p. 688.
32. LaPorta, M.A., Linde, H.W., Bruce, D.L., and Fitzsimons, E.J.: Elevations of creatine phosphokinase in young men after recreational exercise, J.A.M.A. **239:**2685-2686, 1978.
33. Kielblock, A.J., Manjoo, M., Booyens, J., and Katzeff, I.E.: Creatine phosphokinase and lactate dehydrogenase levels after ultra long-distance running, S. Afr. Med. J. **55:**1061-1064, 1979.
34. Blomberg, D.J., Kimber, W.D., and Burke, M.D.: Creatine kinase isoenzymes: predictive value in the early diagnosis of acute myocardial infarction, Am. J. Med. **59:**464-469, 1975.

Chapter 50 Isoenzymes

Lawrence M. Silverman
Vicki N. Daasch
John F. Chapman

artifactual modification Change in protein structure caused by in vitro manipulation.

dimer A protein, such as creatine kinase, composed of two polypeptide subunits.

heteropolymer Subunits of an isoenzyme are different.

homopolymer All subunits of an enzyme are the same.

immunoinhibition Reduction of enzyme activity of a specific polypeptide subunit by its reaction with antibody.

isoenzymes Multiple forms of an enzyme family catalyzing the same biochemical reaction.

macroenvironment Distribution of the enzyme in organs or tissues.

microenvironment Distribution of the enzyme within a cell.

Regan isoenzyme Name given to an alkaline phosphatase isoenzyme "specific" for some cancers.

subunit A polypeptide chain constituting a protein or enzyme.

tetramer A protein, such as lactate dehydrogenase, composed of four polypeptide subunits.

DEFINITIONS

Isoenzyme properties

Isoenzymes (isozymes) are multiple forms (isomers) of an enzyme family that catalyze the same biochemical reaction. Each isoenzyme of a family has a different affinity for substrates and cofactors. Thus the Michaelis constant (K_m) and specificity for different substrates may vary.[1] The various isoenzymes of a family may also differ in the ability to be inhibited by specific agents.[1] They may also differ in their physical properties, such as heat stability, charge, and amino acid composition. However, isoenzymes of a particular family usually do not differ in molecular size. Isoenzymes of a family have different immunological reactivities.

Most or all of these properties have been used to differentiate the various forms of an isoenzyme family. Assays based on these differences are discussed on p. 960.

Structural basis

The most commonly encountered isoenzyme families in clinical chemistry are proteins composed of two or more polypeptide chains or subunits. If the subunits are identical in primary, secondary, and tertiary structure, the resultant enzyme is a homopolymer. If the subunits are different, the enzyme is a heteropolymer. The final composition of an isoenzyme is dependent on the number of polypeptide subunits that compose the complete molecule. For example, if there are two subunits and two different subunit types (A and B), the three possible combinations are AA, AB, and BB. If there are three subunits with two different types, the four possible subunits are AAA, AAB, ABB, and BBB. With four subunits, five different combinations would result: AAAA, AAAB, AABB, ABBB, and BBBB. Therefore with these combinations the enzyme may be present as a homopolymer or a heteropolymer. An example of a dimeric isoenzyme family is creatine kinase (CK), which is formed by variable combinations of the subunits M and B. The subunits differ from each other in primary, secondary, and tertiary structure. The resulting isomeric forms are CK-MM, CK-MB, and CK-BB. The lactate dehydrogenase (LD) isoenzyme family is an example of different enzyme forms composed of combinations of four of the H and M subunits. The resulting tetrameric structures are HHHH, HHHM, HHMM, HMMM, and MMMM (H_4, H_3M_1, H_2M_2, H_1M_3, and M_4).

ORIGIN

Genetic basis

Since the early studies establishing the existence of isoenzymes, much progress has been made in elucidating the chemical bases for structural diversification. For example, tryptic digestion of the H and M subunit chains of the LD isoenzymes suggests that these subunits are the products of two related but separate genes.[2] Some amino acid sequences within the chains are identical, including the enzymatically active site, but there are differences in the amino acid composition and the antigenic properties of the H and M chains. These differences suggest that the diversity may have arisen through gene duplication, followed by independent mutations of the two genes, resulting in different amino acid sequences. One could surmise that evolutionary selection influences the retention of genes for isoenzyme forms that impart some biological advantage to the organism.

Posttranslational and artifactual modifications

Isomeric forms of enzymes may occur by posttranslational modification of the parent enzyme structure.[3] Varying aggregation of different monomer units of the enzyme can cause variable enzymatic properties.[3] Oxidation or reduction of functional groups in the enzyme molecule may also result in identifiable isoenzyme forms.[3] Variability in terminal modifications, such as addition of sialic acid, could result in variable net charge, which would allow separation and identification of the isoenzyme activities.[3] Addition of neutral sugars such as glucose and mannose is not likely to alter the overall electrical charge of the enzyme; therefore these alterations cannot be detected by electrophoretic methods. Binding of a low molecular substance such as NAD may influence the electrophoretic pattern of an enzyme.

PURPOSE OR FUNCTION

Microenvironmental factors

The functional significance of isoenzymes remains an intriguing biological question. This question has been approached from the level of individual cells, organization of cells into tissues, and the developmental process of the organism as a whole. Observations of the compartmentalization of isoenzymes within organelles of individual cells have led to theories related to subcellular interactions and metabolic requirements. Net charge of the isoenzyme may influence its interaction with other charged molecules within the cell. This may result in differential location of specific isoenzymes with respect to specialized parts of the cell. Charge probably plays a role in the preferential location of LD_1 in the mitochondrion.[4] This may result in improved metabolic activity, since the mitochondrion has a high aerobic metabolism, and it has been suggested that the kinetic properties of the LD_1 isoenzyme are optimal under aerobic conditions. Aggregation of monomer enzymes may play a role in intracellular compartmentalization by hindering diffusion of the enzyme through the cell membrane.

Macroenvironmental factors

Differential location of isoenzymes has also been observed on a larger scale, that is, within different tissues. For example, tissues such as heart and brain show a predominance of LD_1, whereas other tissues, such as muscle, have a high content of LD_5. The tetramer of H chains that forms LD_1 has an affinity for pyruvate that is 10 times greater than the affinity of LD_5, a tetramer of M chains. Cahn et al.[5] suggested that because of its kinetic properties LD_1 isoenzyme predominates in tissues rich in oxygen supply that undergo oxidative metabolism and do not accumulate lactate or pyruvate. The LD_5 isoenzyme is the major form in skeletal muscle, which undergoes anaerobic glycolysis with the accumulation of pyruvate and lactate. Thus it has been proposed that isoenzyme patterns vary according to the particular needs and metabolic environment of various tissues.

In muscle cells, which frequently must undergo anaerobic metabolism, biological need demands a predominance of LD activity, which forces the conversion of pyruvate to lactate in order to regenerate NADH for aerobic glycolysis. Pyruvate cannot be successfully metabolized under anaerobic conditions, and an inability to regenerate NADH prevents the available energy-producing reactions from occurring.

In oxygen-rich tissues such as the heart or brain the predominant LD favors the catalysis of lactate (picked up from blood and derived from muscles) to pyruvate in order to fully metabolize it to CO_2, H_2O, and energy (adenosine 5′-triphosphate, ATP).

Because the actual tissue distribution does show this pattern, it must be assumed that the isoenzymes exist to fulfill different metabolic needs for the varying conditions that exist within the body.

Developmental factors

If one considers the development of the organism itself, the pattern of isoenzymes in various tissues that evolves during ontogeny may reflect the tremendous changes in the interaction of the organism with its environment. As the environment changes during fetal development, the organism must adapt in order to optimize enzymatic reaction. Dramatic changes in intracellular aerobic metabolism require major shifts in the isoenzyme composition. Although not a traditional enzyme, the switch from the fetal form to the adult form of hemoglobin illustrates adaptation to the environmental conditions (true respiration) that begin at birth. The oxygen affinities of purified fetal and adult hemoglobin structures are about the same, but organic phosphate in the blood lowers the oxygen affinity of adult hemoglobin more than fetal hemoglobin. Maternal hemoglobin provides transport of oxygen to the fetal circulation, where oxygen exchange can occur as a result of the increased oxygen affinity of fetal hemoglobin. Thus different hemoglobins, as with certain isoenzymes, exist in response to different developmental biological needs.

Changes in isoenzyme patterns may also occur during early embryological development. For example, CK is initially present as the BB isoenzyme in all tissues. In developing myocardium MB and MM isoenzymes gradually appear as the M gene is expressed in association with myofibrillar contractile elements.[6] CK-MB and CK-MM are the predominant isoenzyme forms in adult heart tissue, with CK-MB composing 14% to 42% of the enzyme activity (reports vary concerning the percentage of activity). In a similar way, varying expression of the M and B genes in skeletal muscle tissue results in appearance of CK-MB and CK-MM during development. However, the shift to greater expression of the M gene proceeds to the point where almost 100% of the CK enzyme activity in skeletal muscle is of the CK-MM isoenzyme form. Changing patterns of isoenzyme expression, like that of hemoglobin, occur as a result of differential gene activation and repression in response to changing biological needs.

The LD isoenzymes of undifferentiated embryonal tissues have maximal activity in the hybrid isoenzymes LD_2, LD_3, and LD_4.[7] As tissues differentiate, the LD pattern appears to change to adapt to the type of energy production required for the tissues' specific function. Tissues such as heart, brain, and renal cortex have predominantly aerobic energy metabolism and show maximal enzyme activity in the LD_1 and LD_2 isoenzymes. Anaerobic energy production such as occurs in skeletal muscle, liver, and renal medulla is associated with increased activity of LD_4 and LD_5 isoenzymes.

TISSUE DISTRIBUTION OF MAJOR ISOENZYME FAMILIES

CK-MM is the predominant isoenzyme in skeletal muscle tissue, with only a small quantity of CK-MB present. Unlike skeletal muscle, myocardium contains 14% to 42% CK-MB (reports vary), with the remainder of activity being CK-MM. Brain tissue primarily expresses the CK-BB isoenzyme (Table 50-1).

The heart is the organ richest in LD_1 isoenzyme (H_4), whereas LD_5 (M_4) is found predominately in liver and skeletal muscle. Red blood cells are also enriched in LD_1 activity. The lung has large amounts of LD_2 and LD_3 (Table 50-2).

Table 50-1. Creatine kinase activity in various human tissues

Tissue	Isoenzyme distribution in U/g of wet tissue (% total activity)		
	MM	MB	BB
Skeletal muscle	3281 (100)	0-623 (0-19)	0 (0)
Heart	313 (78)	56-169 (14-42)	0 (0)
Brain	0 (0)	0 (0)	157 (100)
Colon	4 (3)	1 (1)	143 (96)
Stomach	4 (3)	2 (2)	114 (95)
Uterus	1 (2)	1 (3)	45 (95)
Thyroid	7 (26)	0.3 (1)	21 (73)
Kidney	2 (8)	0 (0)	19 (92)
Lung	5 (35)	0.1 (1)	9 (64)
Prostate	0.3 (3)	0.4 (4)	9.3 (93)
Spleen	5 (74)	0 (0)	2 (26)
Liver	3.6 (90)	0.2 (6)	0.2 (4)
Pancreas	0.4 (14)	0 (1)	2.6 (85)
Placenta	1.4 (48)	0.2 (6)	1.4 (46)

From Chapman, J., and Silverman, L.: Bull. Lab. Med. (NCMH), no. 60, pp. 1-7, Jan. 1982.

Table 50-2. Lactate dehydrogenase activity—percentage activity distribution

Organ	Isoenzyme distribution				
	H_4	H_3M_1	H_2M_2	H_1M_3	M_4
Heart	60	30	5	3	2
Kidney	28	34	21	11	6
Cerebrum	28	32	19	16	5
Liver	0.2	0.8	1	4	94
Skeletal muscle	3	4	8	9	76
Skin	0	0	4	17	79
Lung	10	18	28	23	21
Spleen	5	15	31	31	18

From Pfleiderer, G., et al.: In Schmidt, E., et al., editors: Advances in clinical enzymology, Hanover, West Germany, 1979, S. Karger A.G.

Table 50-3. Aldolase activity in various human tissues

Tissue	Isoenzyme distribution in U/g of wet tissue (% total activity)		
	A	B	C
Skeletal muscle	19,600 (97)	—	—
Heart	2068 (90-96)	0-224 (0-10)	0-264 (0-12)
Lung	254 (96)	—	—
Spleen	470 (96)	—	—
Kidney	224 (40)	336 (60)	—
Liver	114 (14)	700 (86)	—
Small intestine	255 (50)	255 (50)	—
Cerebrum	2125 (70)	—	875 (30)
Cerebellum	1200 (33)	—	2400 (66)

From Pfleiderer, G., et al.: In Schmidt, E., et al., editors: Advances in clinical enzymology, Hanover, West Germany, 1979, S. Karger A.G.

Skeletal muscle has the largest amount of aldolase per gram of wet tissue (Table 50-3). Whereas aldolase A isoenzyme activity is detected in all tissue, aldolase B activity is the predominant form in liver and kidney, and aldolase C isoenzyme is observed primarily in brain tissue.

Two thirds of alkaline phosphatase activity normally found in serum is of liver origin. The remainder is derived from bone. Intestinal alkaline phosphatase activity rarely is seen in serum; it is most often found in persons of AB blood group, especially after a meal. The tissue distribution of alkaline phosphatase activity is shown in Table 50-4.

Acid phosphatase is a ubiquitous enzyme; located in the lysosomes of all cells. Red blood cells, white blood cells, platelets, and the prostate gland have particularly enriched levels of this enzyme. The molecules from the last three tissues listed are closely related immunologically. The majority of the enzyme activity found in serum is most likely derived from blood cells, whereas only a small fraction is from the prostate gland.

It is important to realize that many enzymes exist in isomeric or isoenzyme forms. However, except for a few of the isoenzyme families, such as the ones described here, there has been little demonstrable value for doing isoenzyme analyses. The clinical utility of isoenzyme analysis is described later.

CHANGE IN ISOENZYME PATTERNS SECONDARY TO PATHOLOGICAL PROCESSES

Adaptation of isoenzyme patterns may occur in response to pathological processes such as ischemia, atherosclerosis, or cancer. Wilhelm[7] showed that LD activity in nor-

Table 50-4. Alkaline phosphatase activity in human tissues*

Tissue	Activity (U/g of wet tissue)	
	MAP†	DEA‡
Adrenal	30	66
Placenta	36	—
Liver	12.6	27
Bone	7.5	18
Spleen	7.5	18
Lung	6.6	15
Intestine	4.8	9
Kidney	4.2	11
Prostate	3.3	6.6
Thyroid	2.1	5.1
Heart	1.8	3.6
Erythrocytes	0.02	—

Data calculated from Bowers, G.N., et al.: Clin. Chem. **21:**1988-1995, 1975.
*Mean activity in two buffer systems of tissue specimens from human autopsies. Note the greater than twofold activity between the buffer systems.
†2-Methyl-2-amino-1-propanol buffer.
‡Diethylamine buffer.

mal adult aortic tissue is found primarily in the LD_3 fraction. In atherosclerotic aortic tissue, maximal LD activity is present in the LD_5 fraction. Likewise, myocardial LD activity shifts from predominantly LD_1 to LD_3 during the progression of ischemic heart disease.[7] CK activity in ischemic myocardial tissue also shows greater expression of CK-MB activity compared with normal adult myocardium.[7]

Pathological conditions that result in regeneration of damaged tissue may result in changes in isoenzyme composition that resemble the patterns observed during embryological development. For example, many types of muscle disease, such as Duchenne's muscular dystrophy, are characterized by muscle fiber regeneration; that is, new muscle fibers are formed as a result of destruction of adult muscle fibers. Immature muscle fibers may contain isoenzymes that are normally expressed only during early development. Thus appearance of CK-MB isoenzyme in skeletal muscle may indicate an abnormality in that tissue. However one must use caution in the interpretation of isoenzymes, as the fetal isoenzyme form in one tissue may be the adult isoenzyme form in another tissue. For example, the CK-MB isoenzyme represents a significant proportion of the CK activity in both fetal and adult myocardium. However, increased amounts of CK-MB isoenzyme in the sera of normal adults would probably represent damage to the heart. For a more complete discussion of the interpretational problems associated with the presence of CK-MB in patient sera, refer to Chapter 28.

In addition to the processes of destruction and regeneration, the processes of transformation and dedifferentiation associated with the development of malignancy may change isoenzyme patterns. The isoenzymes associated with tumors are often referred to as *oncofetal tumor markers* because of the similarities with the isoenzyme expression observed during early embryological development. For example, although CK-BB is the predominant isoenzyme in all early embryonic tissue, its expression is more restricted in the adult and is associated primarily with the brain and some gut-associated tissues. In patients without malignancy, detection of CK-BB in the serum is often associated with a pathological condition affecting the nervous, pulmonary, or gastrointestinal systems.[10] However, during the process of cell transformation, some cells may express significant amounts of CK-BB, which may be detected in the serum.[10]

In lymphoid malignancies, the LD isoenzymes in serum are predominantly LD_2, LD_3, and LD_4.[11] This pattern reflects the presence of increased numbers of lymphoid cells resulting from malignant proliferation. The LD isoenzyme pattern in the malignant cells reflects the pattern in normal lymphoid cells.

A shift toward LD_5 is observed in many solid tumors, especially in carcinomas of the genitalia or the digestive tract.[11] However, this shift does not always occur; in some tumors there is a shift toward LD_1 expression. The inconsistency of the shift toward LD_5 expression argues against the relationship of isoenzyme expression and increased anaerobic glycolysis.[11] What seems clear is that the development of most malignancies is accompanied by an isoenzyme expression that is different from the normal adult tissue of origin.

Isoenzymes may provide flexibility within the organism to interact with its environment most efficiently. Changes in the environment of the individual cells may occur during the process of ontogenesis, as a result of pathological destruction and the process of regeneration, or even with the development of the malignant state.

CLINICAL SIGNIFICANCE OF SPECIFIC ISOENZYMES

For all practical purposes, serum has been the only clinical specimen examined for isoenzyme markers of specific tissue abnormalities. The release of intracellular isoenzymes after damage or disease can be demonstrated in tissue culture or other in vitro situations; however, the release and significance of intracellular isoenzymes in vivo do not necessarily follow the same rules that govern the in vitro situations. For instance, various mechanisms have been proposed to explain how isoenzymes, with molecular weights usually exceeding 15,000 daltons, can be released from living cells without irreversible damage to the cellular membrane. The fact that living cells apparently do release these molecules has been frequently demonstrated; this constitutes the base-line levels of activities of isoenzymes and enzymes that are found in serum. These levels define laboratory reference intervals (often referred to as "normal" values). Because these base-line levels exist for most enzymes and isoenzymes released from the cytoplasm of cells, reference intervals represent normal cell leakage and cellular turnover. Increases in the levels of enzymes and isoenzymes in excess of these reference ranges are associated with a variety of pathological abnormalities and are the basis of the clinical utility of enzyme and isoenzyme determinations. The purpose of the next section is to discuss the major abnormalities associated with increased levels of isoenzymes and to provide the basis for interpretation of abnormal serum isoenzyme values. Determination of isoenzymes in other body fluids is rare and is discussed briefly at the conclusion of this section.

Creatine kinase (CK)

As previously mentioned, CK is found primarily in skeletal muscle, cardiac tissue, and brain. Although the majority of the isoenzymes are cytoplasmic, there are reports describing CK in other subcellular locations, particularly in the mitochondrion. Significant increases in serum CK levels usually reflect either skeletal muscle or cardiac release. By fractioning CK into its isoenzymes, skeletal muscle release usually can be discriminated from cardiac tissue release. However, the complexities of enzyme release and the various clearance mechanisms by which the body recycles proteins require the interpretation of serum enzyme or isoenzyme levels to be made in context of the clinical situation. Generally, multiple serum markers yield information that makes interpretation of enzyme values more practical. Thus the combined use of LD isoenzymes with CK isoenzymes yields the necessary information in most cases to evaluate patients for myocardial infarction. Nevertheless, multiple specimen analyses appropriately spaced in time provide even more useful information (see Chapter 28).

A more common interpretative problem exists when skeletal muscle abnormalities result in significant elevation of CK activity in serum. Because skeletal muscle contains 5 to 10 times more CK than cardiac tissue does per gram, small areas of muscle damage or disease can result in serum levels of CK consistent with substantial damage to the heart. The use of isoenzyme fractionation can usually differentiate the source of the elevated CK serum activity, since skeletal muscle usually consists of more than 90% of the CK-MM isoenzyme. However, certain diseases of skeletal muscle result in an increased amount of CK-MB content frequently associated with muscle fiber regeneration. Thus diseases such as Duchenne's muscular dystrophy or polymyositis often result in serum elevations of total CK and an abnormal increase in serum CK-MB activity (usually elevated to 5% to 15% of the total CK activity). Because the majority of patients with these muscle diseases are not being evaluated for myocardial infarction, misinterpretation of these CK-MB elevations is infrequent. However, clinical information must be available in order for proper interpretation of isoenzyme values.

Further difficulty in CK isoenzyme interpretation may be encountered in patients undergoing thoracic surgery, particularly coronary artery bypass graft surgery. Surgical procedures involving the heart release myocardial enzymes and isoenzymes that reach levels consistent with myocardial infarction. In such patients the clinician is frequently concerned about infarction either during surgery or during recovery. Therefore isoenzyme levels are extremely difficult to interpret. Experience has led many laboratories to require multiple serum samples for CK isoenzymes during the recovery period (4, 12, 24, and 48 hours and 3, 5, and 7 days postoperatively). In an uncomplicated recovery the CK-MB levels return to normal within 24 hours or decrease to levels approaching normal. With serious complications involving extension of myocardial necrosis or reinfarction, CK-MB levels continue to rise during the recovery period or rise after an initial diminution. Additional information on this subject is found in Chapter 28.

Occasionally CK isoenzymes can be used to evaluate tissues other than the heart or muscle. For example, CK-BB activity can be elevated in the sera of patients with various conditions, including malignancy and prostatic, pulmonary, and neurological disorders. One cannot make the assumption that abnormal isoenzymes are associated only with a particular tissue, as evidenced here for CK or other isoenzymes used in the clinical laboratory.

Acid phosphatase

Although acid phosphatase isoenzymes are found in a variety of cells, including red blood cells, white blood cells, and platelets, the most common isoenzyme measured in the clinical laboratory is the so-called prostatic acid phosphatase. The method of measurement of prostatic acid phosphatase affects the clinical usefulness of the results and is discussed on p. 1084.

For nearly 40 years prostatic acid phosphatase has been shown to be frequently elevated in patients with advanced stages of prostatic cancer. The levels of this isoenzyme are also of some use when one monitors patients receiving treatment as a means of assessing tumor burden (the relative size of the tumor that remains during or after treatment). Thus this isoenzyme activity can be used to assess the success of various types of treatment, but this method is rarely successful for diagnosis of early stages of prostate cancer.

Prostatic acid phosphatase levels may be normal even when a patient has prostate cancer. Therefore false-negative rates are high. Interpretative problems also exist with false-positive elevations. The most common source of false-positive results is benign prostatic hypertrophy in which there is increased growth of the prostate leading to various urological problems. Unfortunately persons with this condition are frequently the very patients in whom prostate cancer is a strong possibility. Other false-positive elevations have been observed in patients with various malignancies other than prostate cancer, such as leukemia. Again, interpretation of isoenzyme levels depends on adequate clinical information.

Other clinical situations in which acid phosphate isoenzyme studies may be ordered include the diagnosis of Gaucher's disease (an inborn error of metabolism) and various leukemias in which leukocyte acid phosphatase can be measured (also leukocyte alkaline phosphatase). Total acid phosphatase and isoenzymes associated with prostatic secretions have been used in rape cases as legal evidence of sexual assault.

Alkaline phosphatase

Alkaline phosphatase isoenzyme studies are frequently ordered for a variety of clinical situations. As in the case of acid phosphatase, the method of measurement plays a significant role in the clinical usefulness of the results. Although there are at least five different isoenzymes of alkaline phosphatase reported in serum, many methods are able to discriminate only the bone isoenzyme from the liver form. In laboratories in which these methods are used, the only useful clinical situations for fractionation of alkaline phosphatase isoenzymes are those in which liver or bone is involved. With the advent of other tests to monitor liver involvement and dysfunction, for example, gamma-glutamyl transferase, the necessity to fractionate alkaline phosphatase isoenzymes has diminished. Even in those laboratories in which methods exist to fractionate alkaline phosphatase into five or more components, there is controversy over the clinical usefulness of these additional fractions. For example, one isoenzyme (Regan isoenzyme) has been reported to be an effective tumor marker in certain malignancies (such as bronchogenic carcinoma) but is certainly no more useful for diagnosis than other tumor markers.

By far the most frequent uses of alkaline phosphatase isoenzymes involve diseases affecting the bone, liver, and intestine. In particular, alkaline phosphatase isoenzyme evaluations are often ordered in patients with cancer as a means of monitoring metastases. Because bone and liver are two of the more common metastatic sites, the ability to diagnose early metastases to either of these tissues is important. Unfortunately not all tumors that have metastasized to bone or liver are associated with increased alkaline phosphatase activity or an abnormal isoenzyme pattern. Again, the use of other markers for liver and bone involvement has decreased the usefulness of alkaline phosphatase isoenzyme determinations. In addition, the methodological variations in isoenzyme fractionation may result in errors in identification or the inability to totally resolve isoenzymes, leading to a high degree of uncertainty in interpretation.

Lactate dehydrogenase (LD)

Much discussion of LD has been covered in the previous section on CK and in Chapter 28, since LD isoenzyme levels are most frequently ordered in combination with CK isoenzymes for the diagnosis of myocardial infarction. Methodological considerations can also determine clinical usefulness of LD isoenzymes, particularly with the advent of newer, more sensitive immunological techniques. This discussion is confined to results obtained by use of conventional electrophoresis.

LD isoenzymes have been more widely investigated than most other isoenzyme families on which tests are currently performed routinely in clinical laboratories. Still the clinical usefulness of this isoenzyme family is somewhat limited to myocardial infarction, partly because of its tremendous variety of tissue sources. LD is found in virtually every tissue, though there is some tissue specificity for the various isoenzymes. Nevertheless, with only five isoenzyme forms commonly found in serum, there is considerable overlap in isoenzyme-tissue specificity. For example, though the heart-specific isoenzyme (referred to as LD_1) is considered the most clinically useful isoenzyme form, elevations of this isoenzyme are also observed in red blood cells and in kidney cells. Thus any process that results in hemolysis can increase the LD_1 fraction. Also, although LD_5 isoenzyme is frequently used to ascertain damage to skeletal muscle, it is also the predominant isoenzyme in liver. A similar situation exists for each of the other three isoenzymes, which leads most clinicians to depend on other more specific tests as their primary liver, muscle, or cardiac assessment tools. As mentioned, much of this lack of specificity may be method-dependent, a situation that may be resolved with newer techniques.

Nevertheless, LD isoenzyme fractionation is most useful for ruling out myocardial damage and muscle injury and occasionally for monitoring progression of certain malignancies. For several decades an increase in LD_5 isoenzyme has been observed in the sera of patients with various types of cancer. Again, the use of isoenzymes as tumor markers is a nonspecific finding fraught with possible misinterpretation.

Other isoenzymes

Although various laboratories perform isoenzyme fractionation for other families besides those just described, the clinical usefulness of these other procedures is questionable. For completeness, isoenzymes of amylase, aspartate aminotransferase, and aldolase should be mentioned because they are reported to have some clinical applications. Fractionation of these and other isoenzyme families is performed by some laboratories, though the clinical usefulness of these data is not so well established as for CK and others previously discussed in this chapter.

Fluids other than serum

Various reports indicate isoenzyme fractionation of several enzymes has significance in cerebrospinal fluid, pleural effusions, urine, and so on. These fluids are rarely examined, and the conditions of isoenzyme fractionation are occasionally very different from serum. Because so little data exist, laboratories that are involved in analyzing these fluids generally determine their own guidelines for clinical interpretation.

MODES OF ISOENZYME ANALYSIS

Most of the varying physical and catalytic differences between individual isoenzymes of a family have been exploited, at one time or another, to determine the isoenzyme concentrations in serum (Table 50-5). All these methods depend on the differences of individual subunit polypeptide chains that impart differences to the complete isoenzyme molecule.

Many older methods, based on the catalytic and inhibitory properties of the isoenzymes, were in widespread use before the advent of more sophisticated methods. These latter methods differentiated between the isoenzymes based on physical or immunological differences of the subunit chains. Thus the use of substrate affinity and catalytic rates to differentiate between CK isoenzymes is rarely used today. Similarly, the use of β-hydroxybutyrate as a "specific" substrate for the measurement of LD_1 activity is now viewed as an inadequate and unsatisfactory method. Heat stability and catalytic inhibition methods for the differentiation of isoenzymes with alkaline phosphatase activity are also rarely used for routine clinical analysis. In contrast, methods using L-tartrate to inhibit prostatic acid phosphatase and dibucaine to inhibit cholinesterase isoenzymes are still frequently used.

The older catalytic and inhibitory isoenzyme procedures have been largely replaced by methods that physically separate the individual isoenzymes before isoenzyme measurement. These methods are based on differences in the net charge of each isoenzymatic member of a family. Thus electrophoretic techniques to separate CK and LD isoenzymes have come into widespread use, whereas the reference method for CK-MB analysis is the anion-exchange method of Mercer.[12] Electrophoretic methods for the separation and quantitation of amylase, gamma-glutamyl transferase, alkaline and acid phosphatases, and many other isoenzyme families have been developed. The popularity of electrophoretic methods is based on their relative ease of use and sample throughput.

The development of isoenzyme analysis based on immunological differences between isoenzymes has been a relatively recent phenomenon. These methods are based on the different amino acid sequences and thus the different immunological characteristics of subunit polypeptides. The immunological methods include radioimmunoassay (RIA) and immunoinhibition techniques. Because of the ease and sensitivity of most enzymatic assays for isoenzymes, radioimmunoassays have not come into widespread usage. The immunoinhibition assays for CK and LD_1 are used with more frequency. The immunoprecipitation assays for acid phosphatase may be viable replacements for the older tartrate-inhibition assays.

For additional information on isoenzyme analysis of the more commonly analyzed isoenzymes, refer to the individual methods listed on pp. 1098, 1116, and 1127.

Table 50-5. Modes of isoenzyme analysis

Technique	Principles of analysis	Isoenzyme family
Electrophoresis	Subunits have different charges; isoenzymes separated in an electrical field	All
Ion-exchange chromatography	Subunits have different charges; isoenzymes separated by differential affinity for ion-exchange resin	CK, LD
Immunoinhibition	Antibody reacts specifically with one subunit type; this property can be used to render an isoenzyme or isoenzymes catalytically inactive or be used to physically remove an isoenzyme or isoenzymes from solution	CK, LD, acid phosphatase
Radioimmunoassay	Antibody reacts specifically with one subunit type; extent of reaction monitored by use of radioisotopes	CK, LD, acid phosphatase
Heat stability	Individual isoenzyme subunits are rendered catalytically inactive at different temperatures	Alkaline phosphatase
Catalytic inhibition	Individual isoenzyme subunits bind low molecular weight inhibitors with different affinities: such binding results in different inhibition of each isoenzyme	Acid phosphatase (L-tartrate), alkaline phosphatase (urea and L-phenylalanine), cholinesterase (dibucaine)
Substrate specificity	Each isoenzyme subunit binds a substrate with different affinities (K_m), giving each isoenzyme various rates of activity	CK
	Also each isoenzyme subunit may bind various substrates with different affinities; different isoenzymes have increased catalytic rates with certain substrates, whereas others have very low activities.	LD_1

BIBLIOGRAPHY

Foreback, C.C., and Chu, J.W.: Creatine kinase isoenzymes: electrophoretic and quantitative measurements, CRC Crit. Rev. Clin. Lab. Sci. **15**(3):187-230, 1981.

Vesell, E.S.: Multiple molecular forms of enzymes: introduction, Ann. N.Y. Acad. Sci. **151:**5-13, 1968.

REFERENCES

1. Wilkinson, J.H.: Principles and practice of diagnostic enzymology, London, 1976, E.J. Arnold & Son, Ltd.
2. Markert, C.L.: The molecular basis for isoenzymes, Ann. N.Y. Acad. Sci. **151:**15-39, 1968.
3. Rothe, G.M.: A survey of the formation and localization of secondary isoenzymes in mammalia, Hum. Genet. **56:**129-155, 1980.
4. Agostoni, A., Vergani, C., and Villa, L.: Intracellular distribution of the different forms of lactic dehydrogenase, Nature **209:**1024-1025, 1966.
5. Cahn, R.D., Kaplan, N.O., Levine, L., and Zwilling, E.: Nature and development of lactic dehydrogenases, Science **136:**962-969, 1962.
6. Chapman, J., and Silverman, L.: Creatine kinase in the diagnosis of myocardial disease, Bull. Lab. Med. (NCMH), no. 60, pp. 1-7, 1982.
7. Wilhelm, A.: Topochemical variation of LDH and CK isoenzyme patterns in aorta, Artery **8:**362-367, 1980.
8. Pfleiderer, G., et al.: Tissue enzymes in phylogenetic and ontogenetic development. In Schmidt, E., et al., editors: Advances in clinical enzymology, Hanover, West Germany, 1979, S. Karger A.G., pp. 17-29.
9. Bowers, G.N., McComb, R.B., Statland, B.E., et al.: Measurement of total alkaline phosphatase activity in human serum (selected method), Clin. Chem. **21:**1988-1995, 1975.
10. Lang, H., and Wurzburg, U.: Creatine kinase, an enzyme of many forms, Clin. Chem. **28:**1439-1447, 1982.
11. Schapira, F.: Isoenzymes and cancer, Adv. Cancer Res. **18:**77-153, 1973.
12. Mercer, D.W., and Varat, M.A.: Detection of cardiac-specific creatine kinase isoenzyme in sera with normal or slightly increased total creatine kinase activity, Clin. Chem. **21:**1088-1092, 1975.

Chapter 51 Therapeutic drug monitoring (TDM)

Wolfgang Ritschel

absorption Uptake of unchanged drug into circulation.

absorption rate constant Value describing how much drug is absorbed per unit of time.

active transport Movement of drug across a membrane by binding to a carrier molecule and delivery to the opposite side with expenditure of energy.

bioavailability The amount of drug in the formulation that the system of the patient can absorb.

biophase The site of interaction between the drug molecule and its receptor.

bound drug A pharmacological agent that exists in blood complexed with another molecule (usually protein or lipid).

C_{max} Maximum plasma level of drug.

C^{ss}_{av} Average steady-state concentration.

C^{ss}_{max} Maximum steady-state concentration (peak concentration).

C^{ss}_{min} Minimum steady-state concentration (trough concentration).

compartment A pharmacokinetic term consisting of the drug concentration, C, and a volume of distribution.

distribution Proportional division of drug into different compartments of the body, such as blood and extracellular fluid.

elimination Final excretion of an agent.

first-order kinetics The rate of change of plasma drug concentration that is dependent on the concentration itself; that is, a constant proportion of drug is removed with time, or $dC/dt = -k \cdot C$

free drug A pharmacological agent that exists in biological fluids unbound by other molecules.

half-life ($t_{1/2}$) The amount of time required to reduce a drug level to one half its initial value. Usually it refers to time necessary to reduce the plasma value to one half of its initial value. The term is also applied to the disappearance of the total amount of drug from the body.

LADME An acronym for the time course of drug distribution: *l*iberation, *a*bsorption, *d*istribution, *m*etabolism, and *e*limination.

liberation The process of drug release from the dosage form.

limited fluctuation method of dosing A method of dosing in which the drug given is not to exceed or go below specified limits.

maintenance dose The amount of drug required to keep a desired mean steady-state concentration.

MEC, MIC The minimum effective concentration or the minimum inhibitory concentration for a drug to be active. Drug is effective at any level above this value.

metabolism The biotransformation of the parent drug into metabolites.

Michaelis-Menten kinetics The method of transforming drug plasma levels into a linear relationship using the parameters of drug concentration and a constant, K_m.

passive diffusion The transport of drug by a concentration gradient across the membrane.

peak concentration The highest concentration reached after a dosage (usually soon after the dose is given).

peak method of dosing A method whereby the drug must reach a specified maximum level to be effective.

pharmacokinetics The quantitative study of drug disposition in the body.

pharmacological effect The influence of a drug on a patient's biochemical or physiological state (such as lowering of blood pressure and bacteriostasis).

prodrug A parent compound that is usually not active and must be metabolized to the active form.

receptor The structure in the body with which the drug interacts, yielding its pharmacological effect. Most often it is located on a cell membrane or other cellular component.

slow release A dosage form of drug that allows the drug to be slowly placed into solution.

steady state A condition in which drug input and drug output are equal. This is obtained when, after multiple dosing, the peak concentration and the trough concentration after each dose oscillate within a certain range.

subtherapeutic A level of drug less than that necessary to have the desired clinical effect.

t_{max} The time of maximum drug concentration.

τ Dosing interval.

terminal disposition rate constant The overall elimination of drug from the body per unit time.

therapeutic index The ratio between the plasma concentrations yielding the desired and undesired effects of a drug.

therapeutic range The relationship between the desired clinical effect of a drug and the concentration of the drug in the plasma.

therapeutic window A term describing a bell-shaped response curve of drug level versus pharmacological response.

total clearance (Cl_{tot}) A term that describes how much of the volume of distribution is cleared per unit of time.

toxic Implies poisonous or deleterious, sometimes fatal, side effects from too-high levels of a therapeutic agent.

trough concentration The lowest drug concentration reached, usually before the next dose is given.

zero-order kinetics The rate of change of plasma concentration, independent of the plasma concentration. A constant amount is eliminated per unit of time, or $dC/dt = -k_0$

zero-time blood level A hypothetical blood concentration obtained by extrapolating back to the initial or zero time period of administration. Usually this yields a maximal value.

FATE OF DRUG AND NEED FOR THERAPEUTIC DRUG MONITORING

Concept of therapeutic range

For many drugs a relationship has been established between the clinical effects and the drug concentration in plasma. In general, in order to achieve the pharmacological effect necessary (such as lowering of blood pressure, pain relief, and bacteriostasis) certain concentrations must be reached in the biophase, the site of interaction between the drug molecule and the receptor (cell membrane, cell component) to elicit the clinical effect. One can describe the effect of most drugs by a log dose-response, or log concentration-response curve, which usually is sigmoid. Any dose or concentration that does not result in any measurable or quantifiable effect is subtherapeutic. Any dose or concentration larger than the minimum dose or concentration that gives 100% effectiveness is unwarranted and may be toxic. Indeed, a dose or concentration in the upper region of the log dose-response curve may already show toxicity.

Similar to the log dose-response curve, there usually is a log dose-toxicity curve. And often one finds an overlap between the upper portion of the log dose-response and the lower portion of the log dose-toxicity curve. For most drugs the therapeutic range is a concentration range somewhere in the lower third to middle portion of the log concentration-response curve. The steepness of the log concentration-response curve indicates the magnitude of the therapeutic range. Absolute toxicity is less important than the ratio between the toxic and therapeutic dose or concentration. This ratio, called the "therapeutic index," may be narrow for some drugs (digoxin, lithium) or wide for others. Hence the therapeutic range may also be narrow or wide. It is particularly desirable to monitor drug concentrations for drugs having a narrow therapeutic range and low therapeutic index (digoxin, lithium, gentamicin) or for drugs having dose-dependent elimination kinetics (phenytoin) or showing great individual variability in metabolism (tricyclic antidepressants). It is well known that drugs may not be equally effective for all patients having similarly diagnosed disorders. In addition to individual variability in metabolism and elimination, the rate and extent of absorption and distribution can also influence the drug concentration in plasma. Thus it is important for the physician to know whether the drug is present in a concentration within the therapeutic range.

The purpose of this chapter is to describe the fate of drugs once administered, to provide some insight as to how the type of dosage regimen is related to achieving the therapeutic range, and to describe some basic principles of pharmacokinetics.

LADME system to describe drug disposition

It is generally accepted that changes of drug concentrations in the body, which occur with time, are related to the course of the pharmacological effects. The change of drug concentration with time is described by the LADME system, in which the *l*iberation, *a*bsorption, *d*istribution, *m*etabolism, and *e*limination of a drug is considered in sequence. Liberation examines the process of drug release from the dosage form. Absorption is the uptake of the unchanged drug into the systemic circulation. Distribution is the proportional division of the drug into different body compartments such as blood, other body fluids, and tissues. Metabolism is the biotransformation of the parent drug molecule to usually less effective or ineffective metabolites, though in some cases the metabolites may be more effective than the parent agent or equally effective. Elimination is the process of final excretion of an agent independent of the pathway, as by kidney, liver, lung, skin, saliva, or milk.

Liberation or drug release from a dosage form. A prerequisite for a drug to be absorbed is its presence in the form of a true solution at the site of absorption. Hence the active ingredient of any dosage form except those that are true solutions (such as intravenous injection, peroral elixir, peroral syrup, rectal enema, eye drops, and nose drops) has to be released from the dosage form before the drug can be absorbed. The release or liberation is the process of the drug passing into solution. When given orally by tablets, capsules, or suspensions, the drug dissolves in gastric fluid. After intramuscular or subcutaneous injection of suspensions, the drug dissolves in tissue fluid. After rectal administration, suppositories melt in the rectum and the drug dissolves in rectal fluid. After application of ointments, the drug dissolves in the water of perspiration at the interface between the skin and the ointment. These are a few cases in which liberation is necessary for the drug to be absorbed.

Sustained or controlled-release dosage forms are preparations with slow release rates. These are designed for those drugs that do not remain long in the body. Since the drug cannot be absorbed faster than it is released, the apparent absorption rate becomes a function of the release rate and the entire absorption process takes longer, resulting in a prolonged duration of clinical effect. Controlled-release dosage forms are comparable to infusions.

Absorption. Absorption is the process of the drug molecule being taken up into systemic circulation. Systemic circulation is defined as the blood stream, except for the blood in the portal vein and in the mesenteric veins. The process of absorption is applicable to all *extravascular* applications; that is, perorally, orally, intramuscularly, subcutaneously, rectally, topically, and so on. Whenever a drug is given *intravascularly,* no absorption takes place because the drug is directly introduced into the bloodstream, that is, intravenously, intra-arterially, intracardiacly.

There are various mechanisms of absorption, including passive diffusion, active transport, facilitated transport, convective transport, and pinocytosis. *Passive diffusion* is the most prominent one, applicable to about 95% of all drugs. Passive diffusion depends on the drug concentration of the nonionized moiety on both sides of the absorbing membrane, with the concentration of drug being higher on one side of the membrane than the other. As long as there is a concentration gradient across the membrane, the drug will be absorbed into the region of lower concentration. For weak electrolytes, the drug's pK_a and the pH at the absorption site (such as stomach pH 1.5 to 3, intestines pH 5 to 7, rectum pH 7.8, or skin pH 5) influence the degree of ionization. The pH of blood is 7.4 and rather constant. As a general rule, ionized drug species are passively absorbed much *less* readily than the nonionized form. At two pH units below an acid drug's pK_a and 2 pH units above a basic drug's pK_a, the drugs will be 99% nonionized and have maximal rate of absorption.

The next important absorption mechanism is *active transport,* which requires binding of the drug molecule to a carrier (protein) of the membrane. The carrier delivers the drug to the opposite site of the membrane by an expenditure of energy. The process moves the drug against a concentration gradient (such as cardiac glycosides, hexoses, monosaccharides, amino acids, riboflavin). *Facilitated transport* is a similar mechanism, with the only difference being that it follows the concentration gradient (such as vitamin B_{12}). *Convective transport* is the mechanism of absorption for small molecules to enter systemic circulation through water-filled pores in the membrane (such as urea). For all these mechanisms it is a prerequisite that the drug is in true aqueous solution at the absorption site.

A unique absorption mechanism is *pinocytosis* for fats and solid particles by formation of engulfing vesicles in the membrane that open at the opposite site, releasing the fat droplets or particles (such as vitamins A, K, D, and E and parasite eggs, fats, and starch).

Distribution. Once drug molecules are absorbed, they distribute within the bloodstream and can (1) be confined to the blood space (such as heparin), (2) leave the bloodstream and enter other extravascular fluids (such as interstitial fluid), or (3) migrate into various tissues and organs. The entire process of transfer of drug from the blood stream to other compartments is called "distribution." This process may be completed within a few minutes but usually takes between 30 minutes and 2 hours; however, it may take longer for some drugs (distribution time for methotrexate is 15 hours).

Metabolism. Metabolism is the process of biotransformation of the parent drug molecule to one or more metabolites. The metabolites are usually more polar, that is, more water soluble, and can thus be more easily excreted by the kidney. Metabolism occurs primarily in the liver and the kidney but also takes place in plasma and muscle tissue. Usually metabolites are less active and less toxic than their parent compounds, though exceptions exist. However, at this point a group of drugs, known as *prodrugs,* should be mentioned. The prodrug as parent compound is usually not active and must be metabolized to the active form (for example, the inactive cancer drug cyclophosphamide is biotransformed to the active compound 4-hydroxycyclophosphamide). The active form of prodrugs is either unstable, not readily soluble, or poorly absorbed.

Some drugs form metabolites that are also active. For example, the active drug procainamide is biotransformed to the equipotent metabolite acetylprocainamide. The knowledge of active metabolites is particularly important for therapeutic drug monitoring to correlate all active forms with pharmacological effects.

Elimination. The final excretion of the drug from the body either as unchanged parent compound or in the form of metabolites is called "elimination." The major routes are through the kidney into urine and through the liver into bile and consequently into feces. Other pathways of elimination are through skin (sweat), lungs (expired air), mammary glands (milk), and salivary glands (saliva).

The elimination half-life is the time required to reduce the blood level concentration to one half after equilibrium is obtained. Once the drug is absorbed and distributed, it takes one half-life to eliminate 50% of the drug, seven

Table 51-1. Elimination half-lives

Number of half-lives	Percentage of drug remaining in body	Percentage of drug eliminated
1	50	50
2	25	75
3	12.5	87.5
4	6.25	≅94
5	3.125	≅97
6	1.5625	≅98.5
7	0.78125	≅99.2
8	0.3906	≅99.6
9	0.195	≅99.8
10	0.097	≅99.9

half-lives to eliminate 99% of the drug, and 10 half-lives to eliminate 99.9% of the drug (Table 51-1).

Effect of biological variation on LADME. If a drug is given in identical amounts by the same route of administration at the same time of day to identical twins, the pharmacokinetic parameters will differ only very slightly. In fraternal twins there will be larger differences. Greater differences will occur within a population group, even if this group is homogeneous with regard to sex, age, body weight, and health. These differences are genetically based variations in drug handling. These variations may influence absorption, distribution, metabolism, and elimination and also drug-receptor interactions. Hence pharmacokinetic parameters for healthy subjects reported in the literature are means with ranges. They are actually valid only for the group studied.

Physiological and pathological factors influencing drug disposition. Apart from biological variations caused by genetic differences, there are a number of physiological and pathological factors that may considerably alter a drug's disposition.[1]

The most prominent physiological factors are body weight and composition, neonatal period, infancy, age, temperature (hyperthermia and hypothermia), gastric emptying time and gastrointestinal motility, bloodflow rates (during rest and exercise), environment (high altitude, mountain sickness), nutrition, pregnancy, and circadian rhythm.

Among the pathological factors of most importance are renal impairment, liver impairment, acute congestive heart failure, burns, shock, trauma, and gastrointestinal diseases.

Blood levels as indicators of clinical response

The rationale for the use of blood levels as an indicator of clinical response is based on the concept that for reversibly acting drugs, the drug concentration at the site of action will determine the intensity and duration of the pharmacological effect. The term "reversibly acting drugs" applies to those that interact at a receptor site without being changed. When a drug is present, the normal activity of the receptor is altered by a blocking or enhancing effect of the drug. When removed, the receptor reverts to its original state. Since it is usually not possible to sample at the site of action or biophase (such as the cell membrane), the next alternative is to sample whole blood, plasma, or serum, which is the biological fluid in closest equilibrium with the receptor site that can be easily sampled. The course of the drug concentration versus the time data can be evaluated by pharmacological models, and the shape of the curve can be used to differentiate between one- and two-compartment models. Once the distribution phase is complete, the drug concentration in the central and peripheral compartments will decline in parallel. At this point a pseudoequilibrium of distribution is obtained regardless of whether the site of action is in the central compartment or in any peripheral compartment. Although the total drug concentration may differ considerably between central and peripheral compartments, the concentration of *free* (nonbound) drug will be the same. Hence, once the pseudoequilibrium of distribution is reached, a correlation should exist between pharmacological effect and drug concentration in blood. Usually only the total drug concentration is measured in plasma. This is quite acceptable under normal conditions because individual differences in plasma-protein binding seem to be small.[2] However, if two acidic drugs are given simultaneously and both are extensively bound,

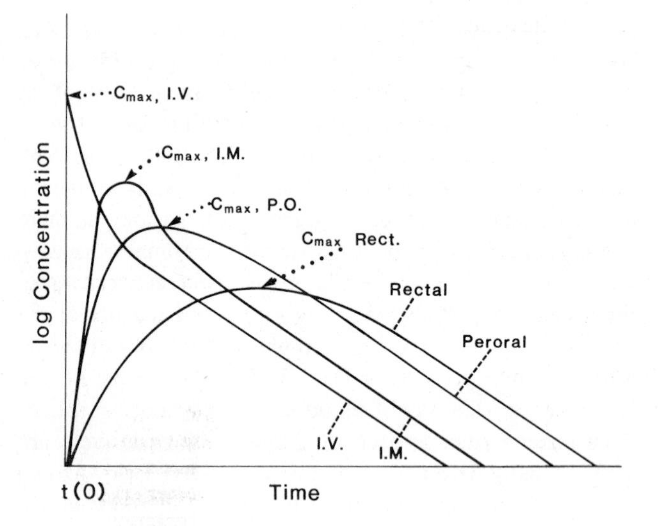

Fig. 51-1. Blood level–time curves of a hypothetical drug upon different routes of administration.

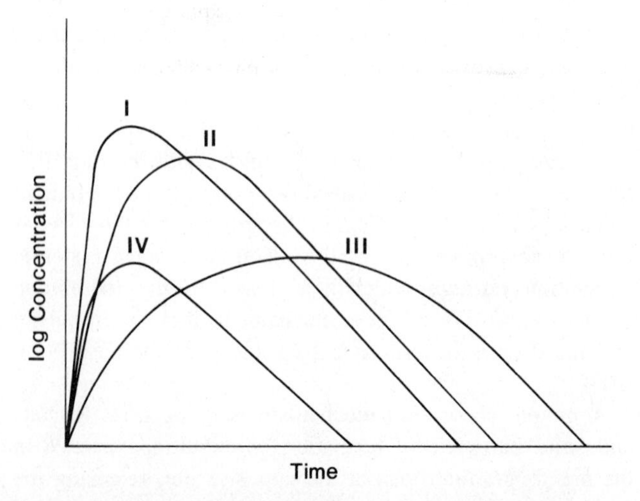

Fig. 51-2. Influence of liberation process on course of blood level–time curves. *I,* Fast-dissolving tablet; *II,* tablet with slower dissolution rate; *III,* sustained-release tablet; *IV,* tablet with poor bioavailability.

one drug may displace the other one from its binding site. The pharmacological action of the displaced drug will therefore increase. If the displaced drug is extensively metabolized, the increased portion of the drug in the free form available for metabolism will increase even though the total drug concentration may decrease. Hence an increased pharmacological or even toxic effect may occur at lower total drug concentration.[3] Models for correlation between pharmacological response and drug disposition have been described in the literature.[4-8]

Blood levels after single dose of drug

The shape and course of a blood level–time curve depend on the route of administration and the LADME system.

On rapid intravenous administration, all the drug is instantly in the systemic circulation. If the drug is given by extravascular administration, none will be in systemic circulation at the moment of administration, that is, at time zero. After the drug is released from the dosage form, the blood level–time curve rises with continuous absorption. Once absorbed, a molecule is exposed to distribution, metabolism, and elimination. Since initially a greater proportion is absorbed than is distributed, metabolized, and eliminated, the blood level–time curve rises until input and output are equal. At this time (t_{max}) the peak concentration (C_{max}) is reached and the blood level–time curve declines (Fig. 51-1). Each LADME process may have a profound influence on the course of the blood level–time curve.

Regarding liberation, there are two factors that may change the shape of the curve: the rate and the extent of liberation. Drug products from different manufacturers may release the drug at various rates. A slow release may also be intentional, as in the case of slow-release (sustained-release) dosage forms. However, if all the drug is released, the areas under the blood level–time curves of different formulations will be the same. If the drug is not fully released, a so-called bioavailability problem might be present and the area under the curve will be reduced (Fig. 51-2). *Bioavailability* refers to the amount of drug systemically absorbed.

The absorption process can be influenced by many factors. Food (when the drug is given orally before, during, or after meals) may have no effect on the absorption, can accelerate or prolong the absorption rate, and may also influence the extent of absorption. For instance, the blood level of griseofulvin is greatly enhanced when the drug is given with fat, whereas a tetracycline blood level decreases when the drug is ingested with milk (Fig. 51-3).

The volume of distribution may change in various pathological conditions. If the volume of distribution increases, the blood level decreases and vice versa. In congestive heart failure the volume of distribution for certain drugs is reduced (digoxin, quinidine). The same dose will therefore result in a higher concentration (Fig. 51-4).

For drugs that are extensively metabolized, changes in the course of blood levels may result from impaired metabolism (liver damage) or other drugs given concomitantly that either compete for metabolic pathways (enzyme inhibition) or accelerate metabolism (enzyme induction) (Fig. 51-5).

Elimination of drugs, particularly of those predominantly eliminated through the kidney, may be tremendously prolonged in case of renal failure and in aged persons. The reduced elimination can result in a manyfold prolonged elimination half-life. Classic examples are the

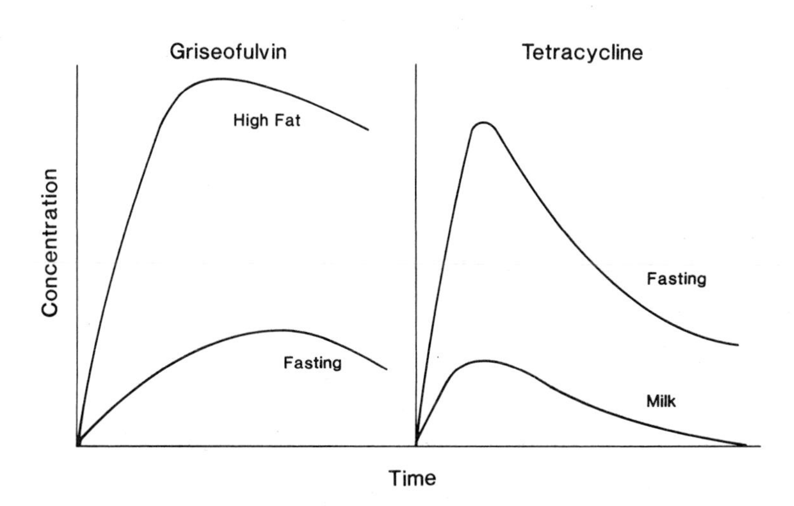

Fig. 51-3. Influence of absorption process on course of blood level–time curves of griseofulvin and tetracycline. High-fat meal results in increased absorption of griseofulvin. Milk causes decrease in blood level of tetracycline.

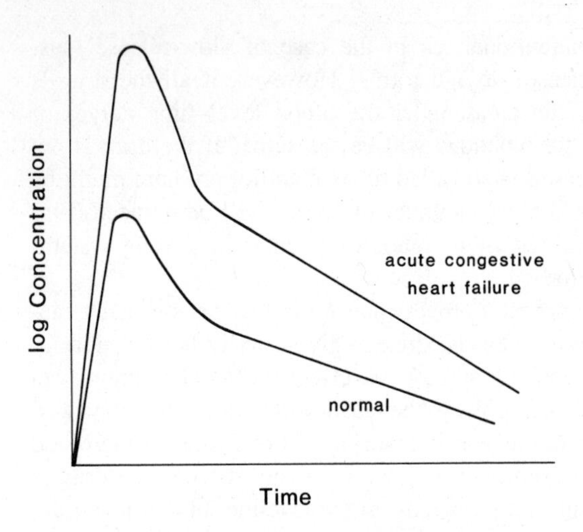

Fig. 51-4. Influence of distribution process on course of blood level–time curves of digoxin. In acute congestive heart failure a higher blood level is observed because of decreased volume of distribution.

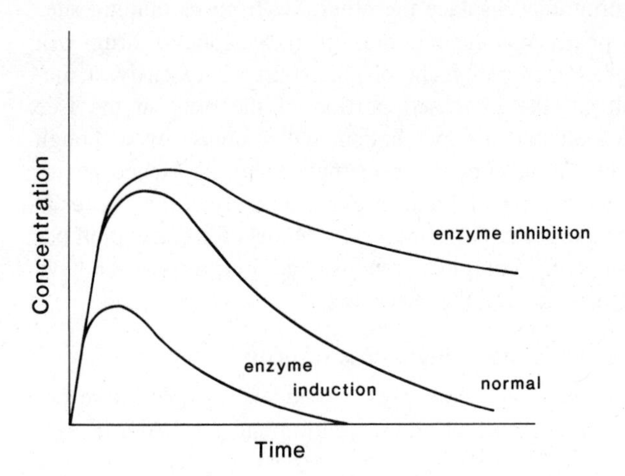

Fig. 51-5. Influence of metabolism processes on course of blood level–time curves. Enzyme inhibition and liver damage may greatly increase blood level, whereas enzyme induction may decrease it.

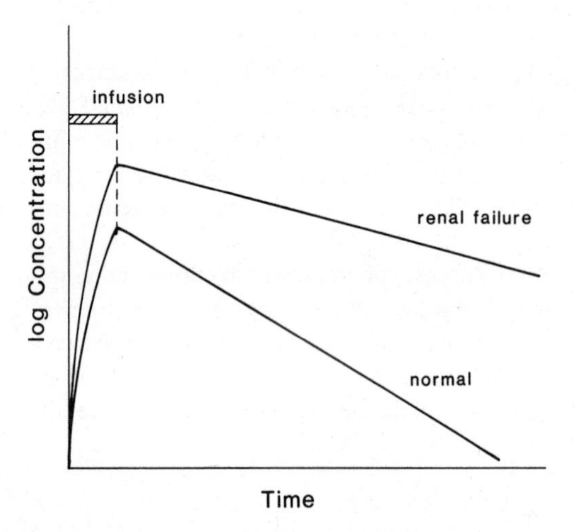

Fig. 51-6. Influence of elimination processes on course of blood level–time curve of gentamicin. In presence of renal failure, peak concentration after short-term infusion is higher and blood level remains elevated with a longer elimination half-life.

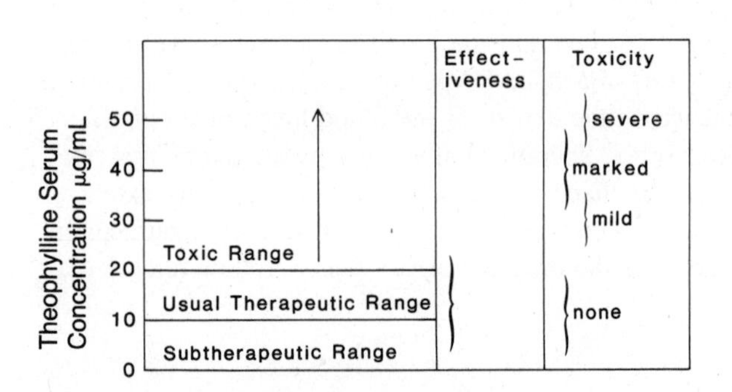

Fig. 51-7. Relationship between serum theophylline concentration and effectiveness and toxicity.

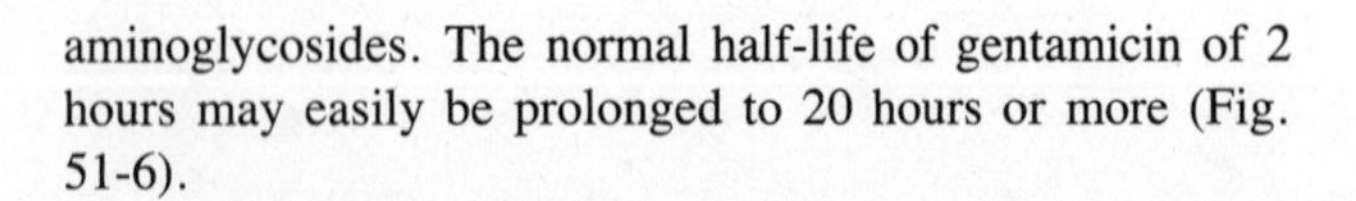
aminoglycosides. The normal half-life of gentamicin of 2 hours may easily be prolonged to 20 hours or more (Fig. 51-6).

Effects of high levels of drugs

As stated before, for most drugs there is a relationship between drug concentration in the blood and the pharmacological and toxic response. Hence an increase in the blood level is usually associated with an increase not only in intensity of clinical effectiveness but also of toxicity. For most drugs it is desirable to either reach a therapeutic range with the peak concentration or maintain the blood level throughout the dosage interval within the therapeutic range. A concentration below the therapeutic range is subtherapeutic or ineffective, and a concentration above the therapeutic range is likely to be toxic and cause side effects. However, the therapeutic and subtherapeutic range and the therapeutic and toxic range often overlap. Additionally, an established therapeutic range may be applicable for the majority of patients but may be too low or too high for an individual patient. One such example is theophylline, for which the usual therapeutic range is between 10 and 20 µg/mL. However, some patients are perfectly controlled with levels as low as 5 µg/mL. On the other hand, whereas most patients do not experience theophylline side effects with levels

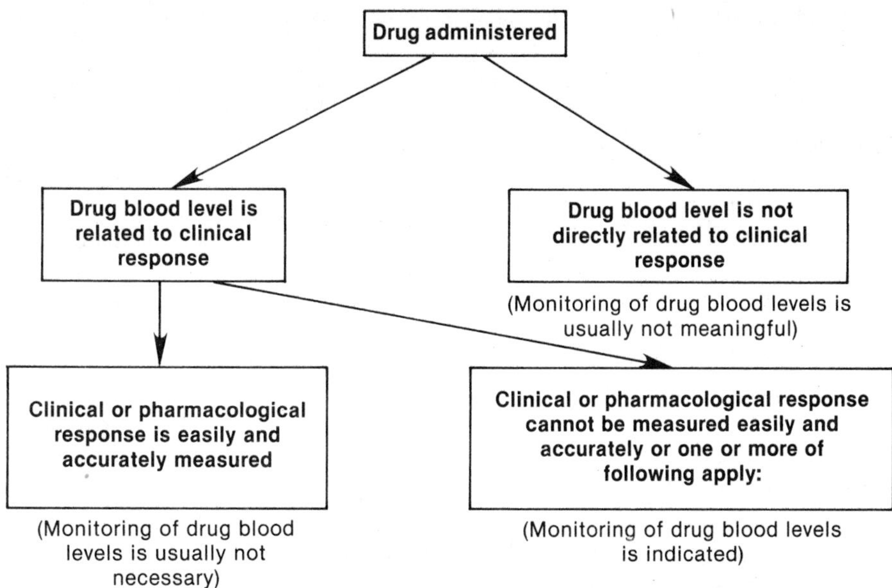

1. Steady-state blood levels show wide individual differences
2. Low therapeutic index
3. Narrow therapeutic range
4. Poor relationship between dose size and blood level
5. Suspected bioavailability problem
6. Suspected malabsorption
7. Toxic effects not easily differentiated from disease symptoms
8. Drug used as a prophylactic
9. Suspected poor compliance
10. Suspected tolerance
11. Suspected drug interaction
12. Multipharmacon therapy
13. Renal impairment for drugs predominantly eliminated in unchanged form by the kidneys
14. Liver impairment for drugs that are predominantly eliminated by hepatic metabolism

Fig. 51-8. Scheme to identify cases and situations when drug monitoring is indicated. (Modified from Pippenger, C.E.: Ther. Drug Monit. **1**:3-9, 1979.)

of 23 μg/mL, some already show toxic signs at this level (Fig. 51-7).

Need for monitoring of drug therapy

Many patients, regardless of whether they are hospitalized or ambulatory, receive more than one drug during any given day,[9,10] which increases the probability of drug-induced diseases, drug interactions, and side effects. One of the most widely used drugs, cimetidine, has been reported to interact with 21 different drugs.[11] A definite need for drug monitoring is also based on the finding that 30% to 50% of dosage administrations in hospitals and nursing homes were in error.[12]

Another reason for drug monitoring is noncompliance with a prescribed dosage regimen. One report states that the percentage of patients failing to take their medication as directed ranges between 20% and 82%.[13]

Drug monitoring for all drugs and all patients is neither possible nor feasible because of problems with logistics, manpower, and time and for financial reasons. Furthermore, total drug monitoring is, at least at present, not relevant for all drugs. For the so-called hit-and-run drugs, that is, those drugs for which, at this time, there is no known correlation between pharmacological response and pharmacokinetic drug disposition (monoamine oxidase inhibitors, reserpine, nitroglycerin, and so on), therapeutic drug monitoring is not feasible.

For some drugs for which a pharmacological response is easily, quickly, and accurately measured, it is clinically more relevant to directly monitor the clinical response (such as blood pressure, blood glucose, electrolyte excretion) instead of the blood level.

If a patient is responding well to the drug therapy without any signs of toxicity, this dosage regimen should be maintained even though the blood level might be outside the usual therapeutic range. One can answer the question

For which drugs is therapeutic drug monitoring indicated? as follows. Monitoring is indicated for a number of drug groups such as antiepileptic drugs, antiarrhythmic agents, peroral anticoagulants, theophylline, tricyclic antidepressants, lithium, and aminoglycosides that either show large individual variation or are toxic above the therapeutic range. A flow chart showing the factors underlying the need for monitoring is given in Fig. 51-8.[14]

DOSAGE REGIMENS USED IN ACHIEVING THERAPEUTIC TARGET CONCENTRATION

Prediction of dosage for steady-state therapeutic levels

Most drugs are not administered as a single dose, but by a regimen of a given dose size in specified dosing intervals throughout the entire course of drug therapy. If the drug is administered repeatedly in dosing intervals shorter than the time required to eliminate the drug remaining in the body from the preceding dose, the drug will *accumulate* until a steady state is reached, that is, one in which input and output are equal. Steady state is obtained when, on multiple dosing, the peak concentration (C^{ss}_{max}, or maximum steady-state concentration) and trough concentration (C^{ss}_{min}, or minimum steady-state concentration) after each dose oscillate within a certain range. Of course, the goal is to achieve a C^{ss}_{max} or the C^{ss}_{max} and C^{ss}_{min} within the therapeutic range. In other words, the therapeutic range is the target concentration, and a dose safe at a given dosing interval has to be calculated to achieve a drug concentration within the desired therapeutic range. By obtaining a blood level–time curve after a single dose, one can derive from such a curve the necessary parameters to predict the steady state and in turn the dose required to achieve a desired steady state.

The *maintenance dose,* which is required to maintain a desired mean steady-state concentration, C^{ss}_{av}, at a given dosing interval, τ, depends on the magnitude of C^{ss}_{av} (the required drug concentration in blood to elicit the pharmacological response), the pharmacokinetic parameters of drug disposition, and the patient's body weight. The generalized equation for maintenance dose is given in the box below.[15]

The drug's disposition in this equation is characterized by the total clearance, Cl_{tot}, which is the product of the apparent volume of distribution, V_d, and the terminal disposition rate constant, β. The V_d in turn is influenced by several factors, such as the drug's apparent lipid/water partition coefficient, the total body fluid, the fat content of the body, the extent of protein binding, the protein structure, and the tissue blood flow rates. The overall terminal disposition rate constant, k_e or β, is mainly influenced by the two major pathways of elimination, metabolism and urinary excretion. Regarding metabolism, the liver blood flow rate, enzyme activity (enzyme induction, enzyme inhibition), and subcellular changes (liver diseases) are of importance. For urinary excretion the effective renal plasma flow rate, the glomerular filtration rate, active tubular transport, and morphological changes of the kidney (renal diseases) are the major determining factors.

The desired C^{ss}_{av} is a concentration within the therapeutic range. This range may not be constant but may be influenced by receptor sensitivity, number of receptors present, age, or disease.

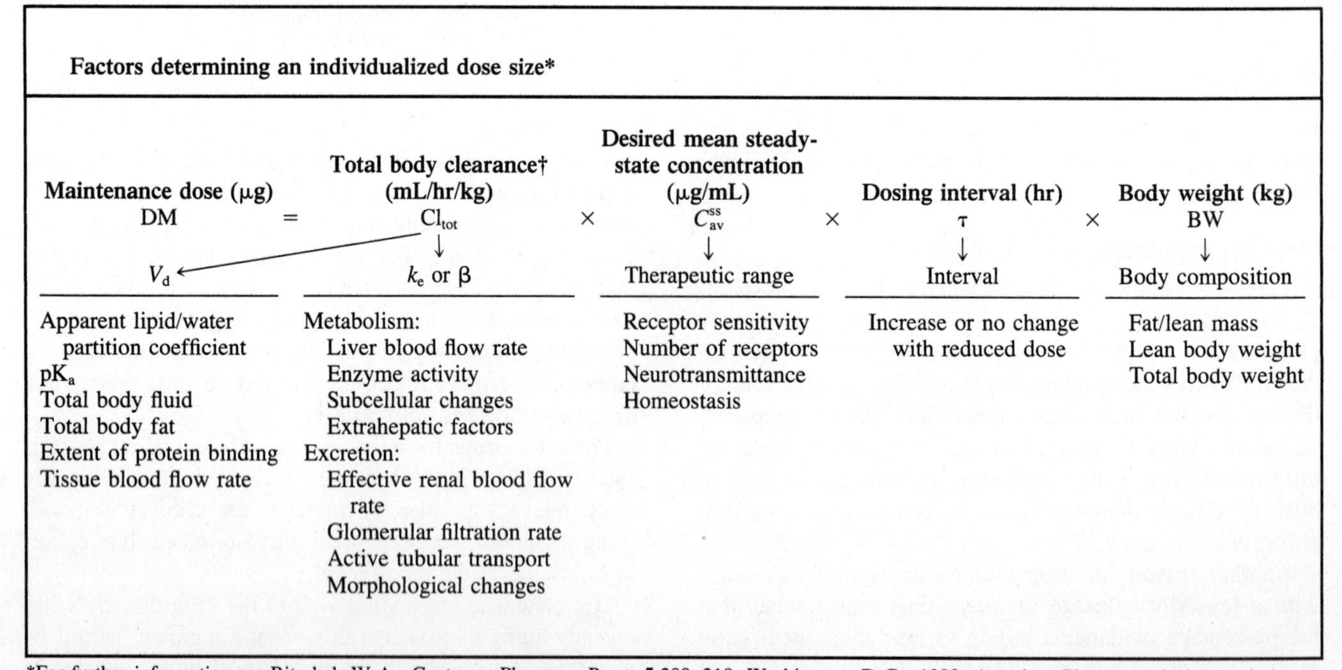

Factors determining an individualized dose size*

Maintenance dose (μg) DM	=	Total body clearance† (mL/hr/kg) Cl_{tot}	×	Desired mean steady-state concentration (μg/mL) C^{ss}_{av}	×	Dosing interval (hr) τ	×	Body weight (kg) BW
V_d ←		↓ k_e or β		↓ Therapeutic range		↓ Interval		↓ Body composition
Apparent lipid/water partition coefficient pK_a Total body fluid Total body fat Extent of protein binding Tissue blood flow rate		Metabolism: Liver blood flow rate Enzyme activity Subcellular changes Extrahepatic factors Excretion: Effective renal blood flow rate Glomerular filtration rate Active tubular transport Morphological changes		Receptor sensitivity Number of receptors Neurotransmittance Homeostasis		Increase or no change with reduced dose		Fat/lean mass Lean body weight Total body weight

*For further information see Ritschel, W.A.: Contemp. Pharmacy Pract. **5**:209–218, Washington, D.C., 1982, American Pharmaceutical Association.
†V_d, Apparent volume of distribution (mL/kg); k_e or β, overall terminal disposition rate constant (hr^{-1}); $Cl_{tot} = V_d \cdot \beta$.

The dosing interval, τ, is freely chosen within a wide range, most often at times less than $t_{1/2}$ (half-life). It may have to be increased in renal or hepatic diseases since $t_{1/2}$ is often greatly extended in these cases. In general, at the end of four half-lives (if a dosing interval less than the half-life is chosen) a steady-state level is reached with multiple dosing.

The body weight is not necessarily the actual body weight. In geriatric or obese patients it is more appropriate to use the lean or ideal body weight rather than the actual one for some drugs.

It is easily recognized that a changed body composition, as in obesity, infancy, old age, and severe edemas, can in turn alter the apparent volume of distribution. Since these pharmacokinetic parameters are derived from studies with healthy, young adult volunteers and may not apply to an individual patient, drug monitoring seems to be a logical consequence.

Dosing regimens

One can design the dosage regimen for multiple dosing maintenance therapy according to five different methods, depending on the desired target concentration to be achieved or maintained throughout each dosing interval. For monitoring purposes it is necessary to know which method will be used because the optimal blood sampling for laboratory analysis depends on the method in question. The five methods for dosage regimen design[16] are as follows:

- Minimum effective concentration (MEC) or minimum inhibitory concentration (MIC) method
- C^{ss}_{max} or peak method
- C^{ss}_{max}-C^{ss}_{min} or limited fluctuation method
- C^{ss}_{av} or log dose-response method
- TW or therapeutic window method

All methods refer to steady-state concentrations.

MEC or MIC method. For some drugs to be effective it is necessary to reach and maintain a minimum inhibitory concentration (MIC), or a minimum effective concentration (MEC), at steady state. Above the MIC or MEC the drug will be effective regardless of how high a peak level is reached as long as the entire steady-state blood level–time curve is above the required MIC or MEC. If the blood level–time curve falls below the MIC or MEC level, the drug will be ineffective as long as the concentration stays below this level. The dosage regimen for such cases is usually based on the C^{ss}_{min} method, which is shown schematically in Fig. 51-9. Drugs that are often based on this method are bacteriostatic antibiotics and other antimicrobial agents (sulfonamides) that have a relatively large *therapeutic index*.

However, one has to realize that each drug has a different MIC for each type of microorganism, and even then there are sometimes wide ranges for different strains of the same microorganism. For instance, the MIC of erythro-

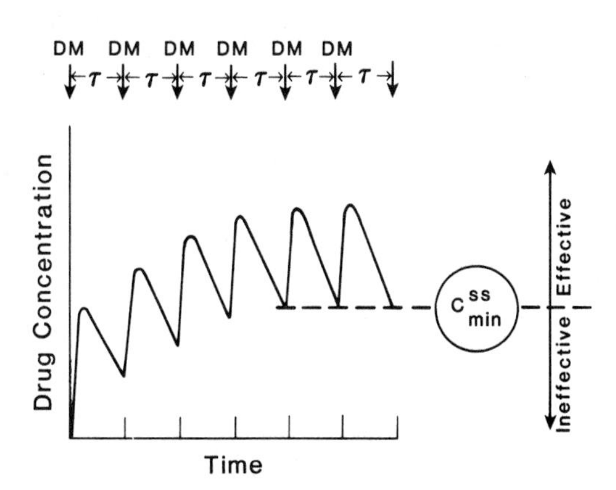

Fig. 51-9. MIC (minimum inhibitory concentration) or MEC (minimum effective concentration) method for calculation of a dosage regimen.

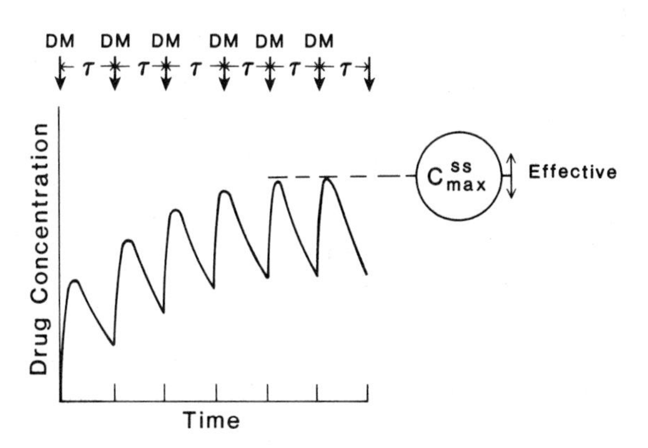

Fig. 51-10. C^{ss}_{max} (maximum steady-state concentration) or peak method for calculation of a dosage regimen

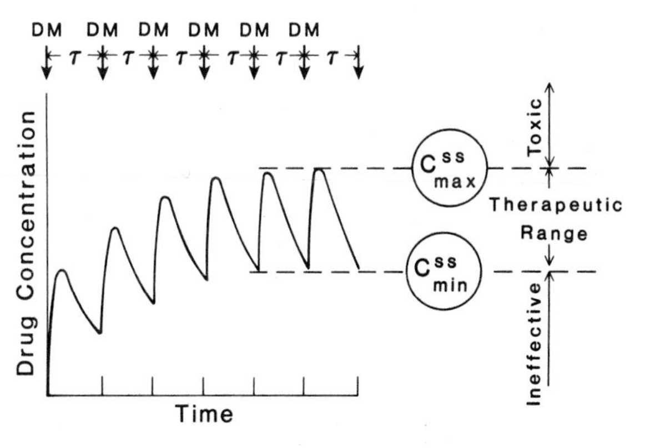

Fig. 51-11. C^{ss}_{max}-C^{ss}_{min} (maximum to minimum steady-state concentration) or limited fluctuation method for calculation of a dosage regimen.

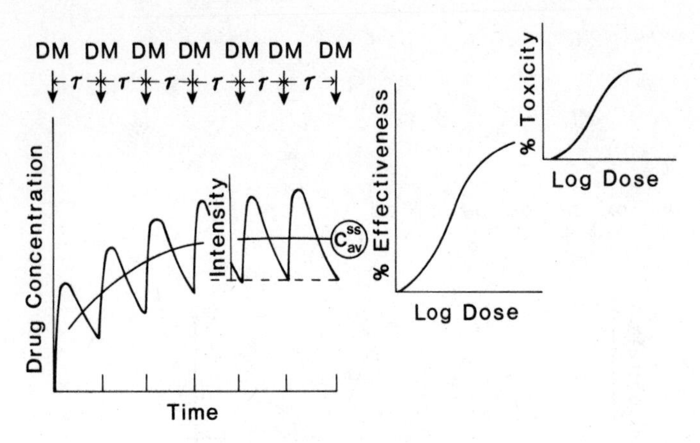

Fig. 51-12. C^{ss}_{av} (mean or average steady-state concentration) or log dose-response method for calculation of a dosage regimen.

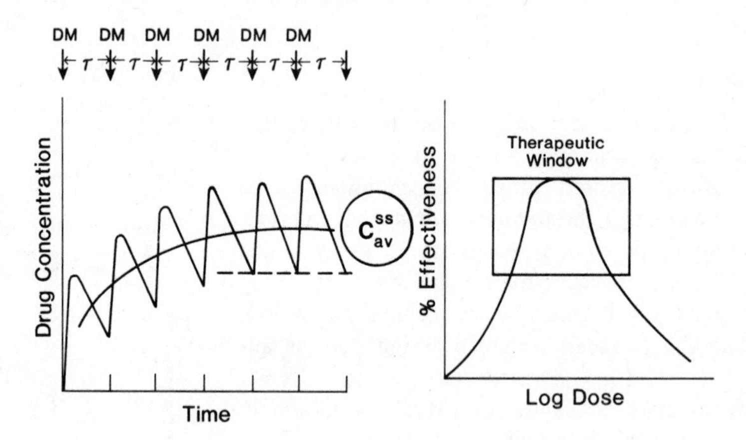

Fig. 51-13. TW, or therapeutic window, method for calculation of a dosage regimen.

mycin against *Staphylococcus aureus* may vary between 0.01 and 2.5 μg/mL. For the proper selection of the MIC for the dosage regimen calculation, a sensitivity test is indicated.

C^{ss}_{max} or peak method. For some drugs it is desirable to reach a certain steady-state peak concentration at each dosing interval. However, for the remainder of the dosing interval it is not required that the drug concentration remain above a minimum level. On the contrary, sometimes it is believed to be desirable that the drug blood level drops below an MIC or MEC value. This is particularly the case with bactericidal drugs, which act only on the proliferating microorganisms. In these cases one does not want to inhibit growth of those microorganisms that have not been killed by the previous dose. The peak pattern is shown in Fig. 51-10.

Drugs that are often based on the C^{ss}_{max} or peak method are penicillins, cephalosporins, gentamicin, kanamycin, and so on.

C^{ss}_{max}-C^{ss}_{min} or limited-fluctuation method. For some drugs it might be desirable to maintain at steady state an MIC or MEC throughout the dosing interval but never exceed a certain peak value. This is particularly the case if the drug has a narrow therapeutic range. The limited fluctuation method is shown in Fig. 51-11.

Drugs that might be dosed by this method are gentamicin, kanamycin, streptomycin, isoniazid, theophylline, and so one. This pattern is particularly useful in treatment of patients with renal or hepatic impairment.

C^{ss}_{av} or log dose-response method. For drugs whose clinical effect follows a log dose-response curve, a mean or average steady-state concentration is often required that is within the therapeutic range (Fig. 51-12). The lower level of the therapeutic range coincides approximately with the dose size at which the log dose-response curve starts. In most cases, drug doses are selected to be in the lower portion of the log dose-response curve, so that the desired steady-state concentration is usually in the lower third of the log dose-response curve. The log dose-toxicity curve may start already in the upper region of the log dose-response curve. For drugs following a log dose response the intensity of effect (and of toxicity) increases with increasing peak size. The dosage regimen for such drugs is often based on the mean steady-state C_{av}^{ss}. A schematic diagram is shown in Fig. 51-12.

Drugs that are often based on this pattern are digoxin, lidocaine, procainamide, theophylline, quinidine, bactericidal antibiotics, analgesics, antipyretics, and hypoglycemic agents.

TW or therapeutic window method. With some drugs, such as antidepressants and antipsychotics, the clinical effect increases with dose size only up to a certain point and then actually diminishes as the dose size is further increased. Instead of a therapeutic range there exists a therapeutic window, showing a more or less bell-shaped log dose-response curve (Fig. 51-13).

PHARMACOKINETICS

Pharmacokinetics is the quantitative study of drug disposition in the body. For most drugs one can demonstrate a relationship between the drug concentration in blood, plasma, or serum and the pharmacological or clinical response. Pharmacokinetics permits one (1) to mathematically describe the fate of a drug on administration in a given dosage form by a given route of administration, (2) to compare one drug with others or other dosage forms, and (3) to predict blood levels in the same or other persons on single or multiple dosing under various conditions of either changed dosage regimens or disease states.

Basically, in pharmacokinetics three types of kinetic processes are used to characterize the fate of drugs in the body: first-order or linear kinetics; zero-order or nonlinear kinetics; and Michaelis-Menten or saturation kinetics.

First-order kinetics

Most processes of drug uptake (absorption), diffusion and permeation in the body (distribution), and excretion (urinary elimination) can be described by first-order, or linear, kinetics. This means that the rate of change of concentration of drug is dependent on the drug concentration. When the concentration *versus* time data are plotted on numeric or cartesian graph paper, a concave curve is obtained; when plotted on semilog paper, a straight line is obtained as seen in Fig. 51-14. The relationship is expressed by equation 51-1:

$$dC/dt = -k \cdot C \qquad \textit{Eq. 51-1}$$

where C is the concentration of the drug, k is the first-order rate constant, and t is time.

The rate of change of drug concentration per unit of time is equal to the drug concentration multiplied by the rate constant for that particular first-order process. The minus sign indicates that the drug concentration decreases with time. Elimination of drugs by first-order kinetics is by a constant percentage of drug being eliminated per unit of time. Examples of drugs exhibiting first-order elimination kinetics are all antibiotics and sulfonamides, digoxin, lidocaine, procainamide, and theophylline. First-order kinetics describes the elimination of most drugs.

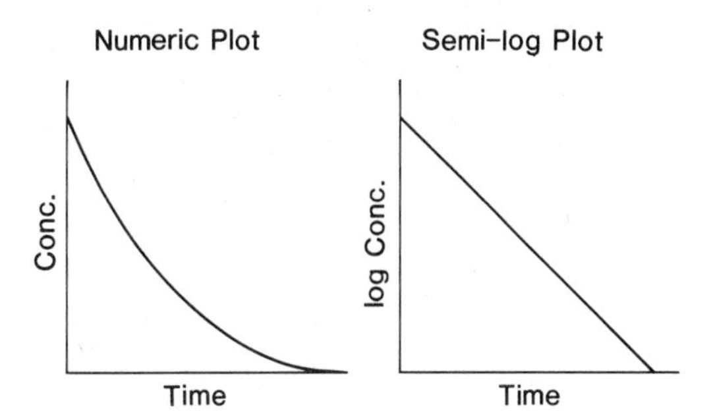

Fig. 51-14. Scheme of first-order kinetics.

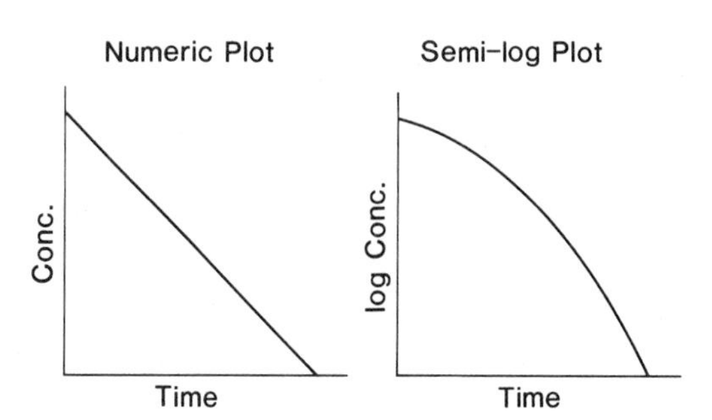

Fig. 51-15. Scheme of zero-order kinetics.

Zero-order kinetics

If the rate of elimination of few compounds from the body is not proportional to the concentration of the drug taken, the elimination usually follows zero-order, or nonlinear, kinetics. This means that the rate of change of concentration is independent of the concentration of the particular drug. In other words, a constant amount of drug, rather than a constant proportion, is eliminated per unit of time (elimination depends on the amount per unit of time). When the concentration versus time data are plotted on numeric or cartesian graph paper, a straight line is obtained whereas on semilog paper a convex curve is obtained (Fig. 51-15). The classic example for zero-order kinetics is the disposition of alcohol (ethanol).

The relationship can be expressed by equation 51-2:

$$dC/dt = -k_0 \qquad \textit{Eq. 51-2}$$

where the rate of change of concentration, dC/dt, is equal to the zero-order rate constant k_0, which has the units of amount per unit of time.

Michaelis-Menten kinetics

In metabolism nearly all biotransformation processes are catalyzed by specific enzyme systems with a limited capacity for the drug. Also in active transport of drugs across membranes the carriers have a limited capacity. Whenever

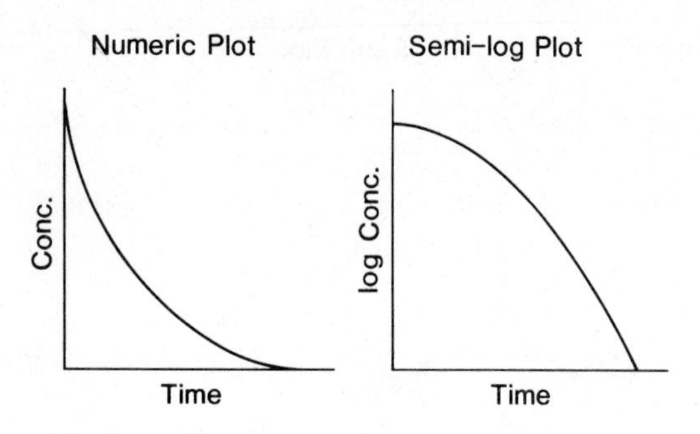

Fig. 51-16. Scheme of Michaelis-Menten kinetics.

the drug concentration present in a given system exceeds the capacity of the system, the rate of change of concentration is most precisely described by the Michaelis-Menten equation.

When the drug concentration versus time is plotted on numeric or cartesian graph paper, one obtains a curve that in the upper portion is only slightly concave but becomes more concave in its lower portion. On semilog paper the upper portion of the curve decays in a convex fashion, whereas the lower portion ends in a straight line as shown in Fig. 51-16.

The rate of change of concentration dC/dt for the Michaelis-Menten process is expressed by equation 51-3:

$$dC/dt = -(V_m \cdot C)/(K_m + C) \qquad \textit{Eq. 51-3}$$

where C is the drug concentration, t is the time, V_m is a constant representing the maximum rate of the process, and K_m is the Michaelis constant, the drug concentration at which the process proceeds at exactly one half its maximal rate.

For simplification of this presentation, one can often approximate a Michaelis-Menten process by either one of the following modifications: (1) if $C << K_m$, then C can be removed from the denominator of equation 51-3, and V_m and K_m, both being constants, can be combined to a new constant, k. Hence equation 51-3 is simplified to a first-order process as given in equation 51-1; (2) if $C >> K_m$, then K_m can be removed from the denominator of equation 51-3, and C cancels out. Hence the equation is simplified to:

$$dC/dt = -V_m \qquad \textit{Eq. 51-4}$$

indicating that the rate of change of concentration is equal to a constant, that is, follows zero-order kinetics.

Examples of saturation elimination kinetics are phenytoin, high doses of barbiturates, and glutethimide.

It is important to note that first-order processes of absorption and distribution, but particularly of metabolism and elimination, may change from first-order to Michaelis-Menten kinetics or pseudo–zero order kinetics with increasing dose size (usually beyond the therapeutic range), in certain disease states, or because of concomitant administration of other drugs.

Compartment models

To describe the quantitative processes of a drug in the organism, pharmacokinetics uses the concept of compartments. A compartment is a unit characterized by two parameters: the drug concentration, C, and the volume, V_d. By multiplying the drug concentration by the apparent volume of distribution, one obtains the amount, A, of the drug in that compartment:

$$C \cdot V_d = A \qquad \textit{Eq. 51-5}$$

A given compartment model is not necessarily specific for a given drug. For instance, a drug given intravenously is often described by a two-compartment open model, whereas the same drug given orally or by any other extravascular route may be described by a one-compartment open model. "Open" means that there is input to and output from the compartment.

In reality the human body is a multimillion compartment system. However, usually one has in the intact organism easy access to only two kinds of biological fluids, blood (serum, plasma) and urine. Being restricted to blood or urine specimens, the drug has a fate in the body usually described by either a one-compartment or two-compartment open model. Clinically speaking, the concept of the one-compartment and two-compartment models is usually satisfactory for therapeutic use. The main application of pharmacokinetics in a clinical setting is to predict, monitor, and adjust dosage regimens to gain therapeutically effective and safe blood levels of the drug. Most mathematical procedures for these purposes are actually based on the one-compartment open model and are applicable for the two or higher compartment models under certain conditions and assumptions. The difference between a one compartment and a two-compartment model is that in the former the distribution occurs instantly whereas in the latter the distribution process needs a measurable time before pseudoequilibrium is obtained.

Compartment models are usually visualized by block diagrams. Each compartment is characterized by volume and concentration terms. The compartments are connected by arrows indicating the rate of drug transfer in units of time. In Fig. 51-17 the most important compartment models are shown, listing the block diagrams with their corresponding numeric and semilog concentration versus time curves.

Absorption rate constant

The differences between the actual blood level data during the absorption phase and the back-extrapolated C_0 (theoretical concentration at zero time) from the k_e-slope in the one-compartment model (Fig. 51-18) are used to calculate the absorption rate constant, k_a. The k_e slope is extrapolated back to time zero. This yields C_0, a theoretical concentration roughly equivalent to that obtained from an intravenous injection of the same amount of drug. By

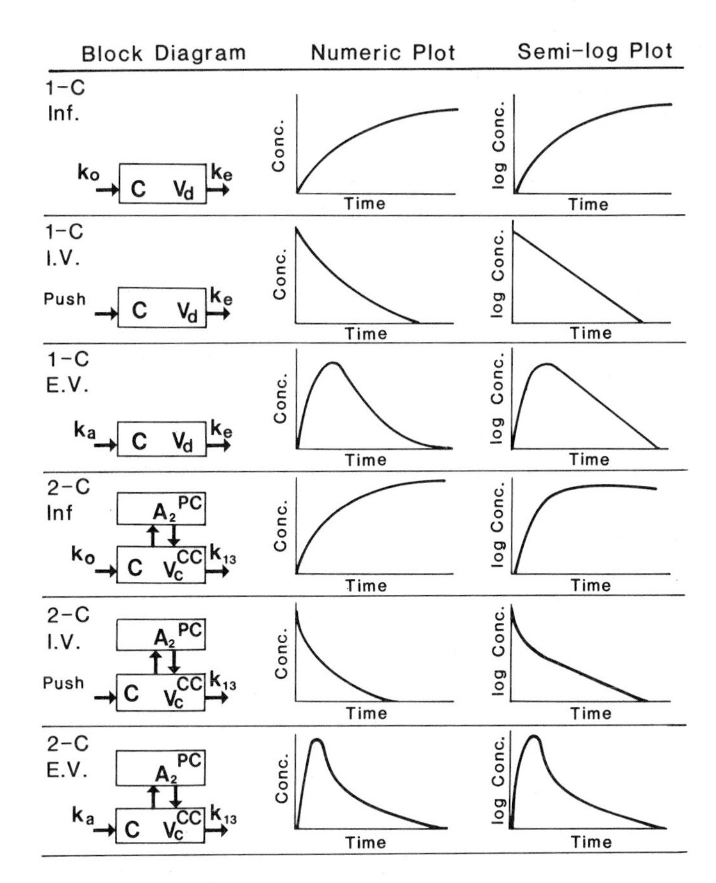

Fig. 51-17. Scheme of pharmacokinetic compartment models in form of block diagrams *(left column),* and blood level–time curves as numeric *(middle column)* and semilog plots *(right column)* for intravenous infusion, intravenous administration, and extravascular administration according to the open one-compartment model and open two-compartment model. A_2, Drug amount in peripheral compartment; *C,* drug concentration in blood; *1-C,* open one-compartment model; *2-C,* open two-compartment model; *CC,* central compartment (blood and any organ or tissue that is in immediate equilibrium with drug concentration in blood); *E.V.,* extravascular; *I.V.,* intravascular; k_0, zero-order infusion rate; k_a, absorption rate constant; k_e, overall elimination rate constant; k_{12}, distribution rate constant from central to peripheral compartment; k_{13}, elimination rate constant for loss of drug from central compartment; k_{21}, distribution rate constant from peripheral to central compartment; *PC,* peripheral compartment (any organ or tissue that has at any time a drug concentration different from that in blood); V_d, apparent volume of distribution. (From Ritschel, W.A.: Graphic approach to clinical pharmacokinetics, Barcelona, 1983, J.R. Prous.)

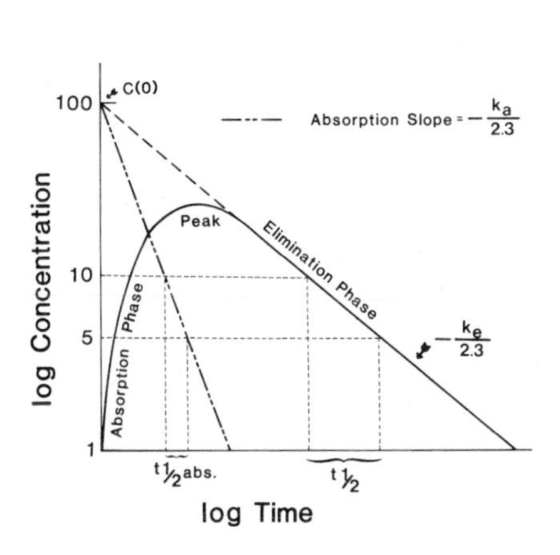

Fig. 51-18. One-compartment model blood level–time curve after extravascular administration, with monoexponential slopes for elimination, k_e, and absorption, k_a. (From Ritschel, W.A.: Graphic approach to clinical pharmacokinetics, Barcelona, 1983, J.R. Prous.)

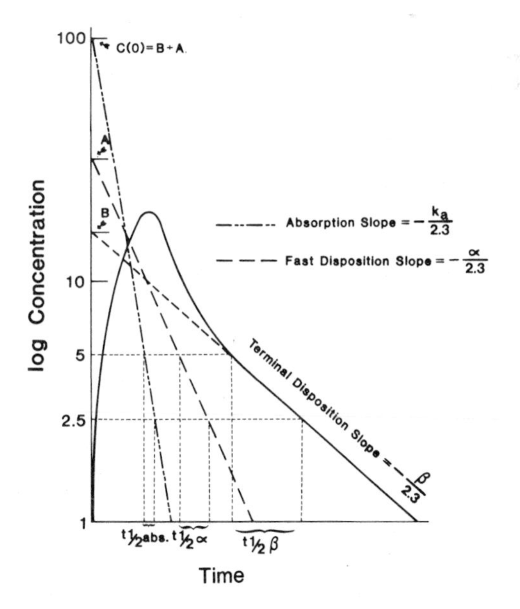

Fig. 51-19. Two-compartment model blood level–time curve after extravascular administration, with monoexponential slopes for slow disposition, β; fast disposition, α; and absorption, k_a. (From Ritschel, W.A.: Graphic approach to clinical pharmacokinetics, Barcelona, 1983, J.R. Prous.)

subtraction of the observed drug concentration during the absorption phase from the concentrations read from the back-extrapolated k_e slope, "residual" points are obtained. When plotted on semilog paper, they are described by a straight line, the slope of which is the absorption rate constant k_a (Fig. 51-18) in the one-compartment model. In the open two-compartment model the "residuals" have to be taken first for the α-phase and then for the absorption phase.

Terminal disposition rate constant

The last or terminal portion of the straight (monoexponential) slope of a semilog blood level–time curve gives, in the one-compartment open model, the overall elimination rate constant, k_e (metabolism, renal excretion, and other pathways of elimination). In the two-compartment model it is the rate constant, β, the slow-disposition rate constant (Figs. 51-18 and 51-19).

Zero-time blood level

Back-extrapolation of the blood level–time curve after intravenous administration results in the zero-time blood level C_0. On extravascular administration, the "fictitious" zero-time blood level, C_0, is the intercept of the k_e slope with the ordinate on a semilog plot in the one-compartment model, and the sum of the intercepts $A + B$ of the α and β slopes in the two-compartment model (Figs. 51-18 and 51-19).

Elimination half-life

Whenever a monoexponential straight line is obtained, one can calculate a half-life. Note the line describing the terms k_e and β in Figs. 51-18 and 51-19. Other half-lives frequently used are the absorption half-life ($t_{1/2}$ abs) and distribution half-life ($t_{1/2}$).

The terms *half-time, half-life, plasma half-life, elimination half-life,* and *biological half-life* are often used interchangeably. Half-life refers to the total dose or amount of drug in the *body* and is equal to the time required for elimination of one half the total dose of drug from the body. The elimination half-life, or plasma half-life, refers to the half-time of the terminal elimination phase (k_e slope or β slope) for the drug in the blood (plasma or serum). In those instances in which the decline of drug concentrations in all tissues does not parallel the decline of drug concentration in plasma, blood, or serum, half-life and elimination half-life will be different. Most statements on drug disposition refer to the elimination half-life.

Volume of distribution

The volume of distribution is not a real volume and usually has no relationship to any physiological space or body fluid volume. It is simply a term to make the mass-balance equation valid. On intravenous administration the amount of drug in the body is known; however, only the blood can be sampled. Because an amount of drug, A, equals the product of concentration and volume (μg/mL × mL), the volume of distribution is the hypothetical volume that would be required to dissolve the total amount of drug at the same concentration as found in blood.

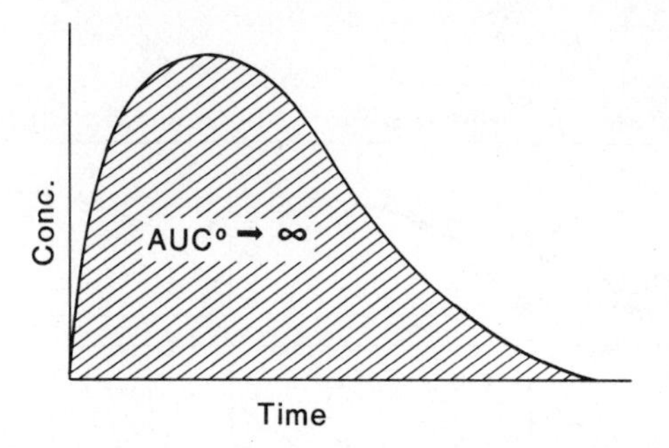

Fig. 51-20. Scheme of total area under the blood level–time curve. $AUC^{0\rightarrow\infty}$, Area under curve from time = 0 to time = ∞. (From Ritschel, W.A.: Graphic approach to clinical pharmacokinetics, Barcelona, 1983, J.R. Prous.)

The volume of distribution is expressed in milliliters. If this value is divided by the patient's body weight, the distribution coefficient Δ′ is obtained in mL/g (or L/kg).

Area under blood level–time curve

The integral under a blood level–time curve is a measure of the total amount of drug in the body. One can calculate the area under the blood level–time curve from time zero to infinity, AUC, or one may approximate it by cutting out the area and weighing it. The AUC is shown in Fig. 51-20.

Total clearance

The total clearance in pharmacokinetics describes how much of the volume of distribution is cleared of the drug per unit of time, regardless of the pathway for the loss of drug from the body. In effect it is the sum of all clearances by different pathways. The major pathways are through the liver (metabolism) and the kidney (urinary excretion). Other normal and nonhepatic pathways are expired air (lung), skin (sweat), and mammae (milk). The total clearance is the product of the apparent volume of distribution and the terminal disposition rate constant.

Steady state

Steady state refers to the accumulation of drug in the body in multiple dosing when input and output are equal within a dosing interval. The magnitude of accumulation depends on the drug's elimination half-life and the dosing interval. The smaller the dosing interval for a given dosage, the greater the accumulation and the smaller the fluctuation around the mean serum value. At steady state the drug concentration oscillates around a mean steady-state concentration, C_{av}^{ss}, with a definite maximum steady-state concentration, C_{max}^{ss}, and a minimum steady-state concentration, C_{min}^{ss}. Only in the case of an intravenous constant rate infusion are C_{max}^{ss}, C_{min}^{ss}, and C_{av}^{ss} identical.

APPLICATION OF PHARMACOKINETICS TO TDM

Clinical assessment

Clinical (physician) estimation of patient response is the first and most important task in therapeutic monitoring. One should not forget that therapy requires an approach that considers all aspects of a patient's condition, including the disease symptoms, the disease itself, other diseases present, and the patient's physical condition, age, nutritional status, and psychological aspects. Furthermore, one should not forget that clinical pharmacokinetics is only a tool to assist with but never substitute for clinical evaluation.

Application. The clinical evaluation of patient response is made in the hospital at the bedside during routine rounds or by the practitioner during the office visit or house call. It comprises the evaluation of vital signs and change of symptoms in response to the drug therapy. This is, if indicated, supported by measurement of pharmacological responses such as blood pressure, pulse rate, electrocardiogram, measurement of edema, and urinary output. Furthermore, supportive laboratory analyses may be required, such as serum glucose and electrolytes. In all these cases the pharmacological response is evaluated clinically, either as direct measurement of pharmacological effect or as a measurement of body constituents, but *not* measurement of drug concentration in biological fluid.

Limitations. Sometimes the clinical evaluation might be difficult because of the presence of more diseases with similar or overlapping symptoms, polypharmacy (many drugs are given simultaneously), or *unexpected* results. Unexpected results that might occur are that (1) the patient is not responding as expected, showing either no effectiveness or limited effectiveness of therapy, or (2) the patient may exhibit unexpected toxic or side effects. In the first case the cause of the reduced effect might reside with the drug product having a lower bioavailability or with the patient because of noncompliance (not taking the drug in the amount and times as prescribed), or with the presence of the malabsorption syndrome, in which less drug is absorbed because of certain disease states. In the second case one has to consider possible drug interactions, enzyme induction, enzyme inhibition, renal or hepatic impairment, edema, dehydration, and so on. In these cases it is advisable, when possible, to request drug monitoring in biological samples (Fig. 51-8).

Assessment by drug analysis

When the therapeutic target concentration, therapeutic range, or toxic concentration is known, therapeutic drug monitoring might be indicated to support the clinical evaluation and is recommended when the patient is treated with drugs of low therapeutic index, such as aminoglycosides, digoxin, and lithium, or when unexpected side effects or toxicity are occurring (Fig. 51-8).

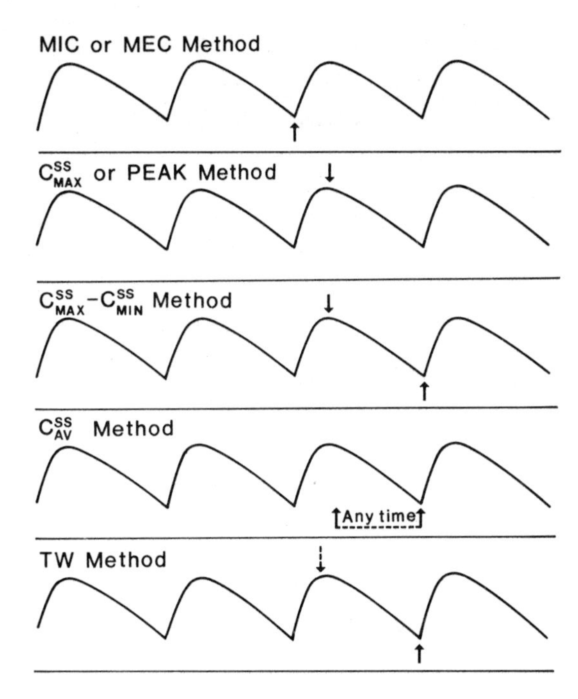

Fig. 51-21. Scheme showing optimal sampling times for monitoring for different methods used for dosage regimens.

Basis for monitoring. For monitoring, it is essential to observe several "musts." Otherwise, any evaluation will be in error[17]:

1. The dose size, dosage form, and route of administration must be known.
2. The dosage regimen must be followed.
3. The time of the last dose given relative to the blood sample must be known.
4. The blood sampling time or times must be recorded exactly.
5. The sampling times must be appropriate.

Most of the "musts" listed above are self-explanatory.

Regarding the optimal sampling times, one should remember that any samples taken during the absorption or distribution phase are useless for monitoring. Samples taken at peak time allow only an approximation of pharmacokinetic data. Samples taken at peak time and during the terminal elimination phase will result in an overestimate of the elimination rate constant and an underestimate of the elimination half-life.

The sampling time should be chosen according to the appropriate type of dosage regimen (p. 971). The optimal sampling times for the various dosage regimen methods are shown in Fig. 51-21. Sampling times of commonly monitored drugs are listed in Table 51-2.[18,19]

A review of the most important pharmacokinetic char-

Table 51-2. Recommended sampling times for commonly monitored drugs[18,19]

Drug	Recommended sampling time
Amikacin	0.5 to 1 hour after dose and end of dosing interval
Carbamazepine	End of dosing interval
Clonazepam	3 to 5 hours after dose
Diazepam	3 to 5 hours after dose
Desmethyldiazepam	3 to 5 hours after dose
Digitoxin	Any time at least 6 hours after dose
Digoxin	Any time at least 6 hours after dose
Dipropylacetate	1 to 2 hours and 8 hours after dose
Ethosuximide	2 to 5 hours after dose
Gentamicin	0.5 to 1 hour after dose and end of dosing interval
Lidocaine	24 hours after start of infusion
Lithium	End of dosing interval
Oxazepam	1 to 3 hours after dose
Phenobarbital	End of dosing interval
Phenytoin	End of dosing interval
Primidone	End of dosing interval
Procainamide	End of dosing interval
Salicylate	1 to 3 hours after dose
Theophylline	Intravenous infusion: 24, 48, and 72 hours after start of infusion
	Oral or intravenous injection: 2 hours after dose or end of dosing interval, or both
	Oral sustained release: 6 hours after dose and end of dosing interval
Tobramycin	0.5 to 1 hour after dose and end of dosing interval

acteristics of some drugs is given in Table 51-3, listing the therapeutic ranges.[20,21]

Limitations. The limitations to drug monitoring by blood samples are primarily that (1) not all information required by the ''musts'' discussed previously are available or accurate, (2) a reliable assay method is not available, and (3) laboratory analysis time is not reasonable. Most difficulties experienced in monitoring reside in limited or inaccurate information. One also has to consider the precision of the assay. The coefficient of variation of the assay may be important when the found drug concentration is either at the lower or upper end of the therapeutic range. Furthermore, the therapeutic ranges reported in Table 51-3 are mean values applicable to the majority of patients. Nevertheless, the therapeutic or toxic concentration may be different for a particular patient (see Fig. 51-7).

Assessment by pharmacokinetic calculations

Winter et al.[22] presented an algorithm for evaluating and interpreting plasma drug levels. To evaluate blood samples pharmacokinetically, one must know whether the drug concentration is at steady state. To decide if a steady state has been achieved, one needs to know the dosage regimen (dose size and dosing interval) and how long this dosage regimen has been in effect. Usually it is assumed that a steady state is reached when the dosage regimen has been implemented for a time greater than four times the drug's elimination half-life. If it is less than $4t_{1/2}$, one should

Table 51-3. Pharmacokinetic characteristics of some important drugs

Drug	Therapeutic range (μg/mL)	Normal Δ′ (L/kg)	Normal $t_{1/2}$ (hours)	Important metabolites	Major mode of clearance
Analgesics					
Acetaminophen	10-20	1.1	2.5	Conjugates	M
Aminopyrine	5		2.7	4-Aminopyrine,* methylrubazoic acid	M
Meperidine	0.6	4.7	3.5	Normeperidine,* meperidinic acid, normeperidinic acid	M
Morphine	0.07	1.0	2.3	Normorphine,* glucuronide	M
Salicylate	20-100	0.2	2.4	Glycine-glucuronic conjugates	M
Antibiotics					
Amikacin	10-25	0.25	2.5		R
Amoxicillin	7-8	0.47	1.2		R
Ampicillin	2-6	0.52	0.9		R
Carbenicillin	10-125	0.15	0.8		R
Cefazolin	0.1-63	0.17	1.6		R
Cephalexin	6-50	0.33	0.9		R
Cephaloridine	0.1-16	0.23	1.2		R
Chloramphenicol	5-40	0.57	2.7		M
Chlortetracycline	0.5-6	1.3	6.0		M
Cloxacillin	7-14	0.15	0.5		R
Doxycycline	1-2	1.0	20.0		R
Erythromycin	0.5-2.5	0.6	1.4		M
Gentamicin	0.5-8	0.25	2.0		R

*Pharmacologically active. *M*, Metabolized by liver; *R*, renal elimination.
From Ritschel, W.A.: Handbook of basic pharmacokinetics, ed. 2, Hamilton, Ill., 1980, Drug Intelligence Publications, and Sadee, W., and Beelen, G.C.M.: Drug level monitoring, New York, 1980, John Wiley & Sons, Inc.

Table 51-3. Pharmacokinetic characteristics of some important drugs—cont'd

Drug	Therapeutic range (μg/mL)	Normal Δ′ (L/kg)	Normal $t_{1/2}$ (hours)	Important metabolites	Major mode of clearance
Griseofulvin	0.3-1.3	1.5	14.0		M
Kanamycin	2-8	0.25	2.0		R
Methicillin	1-6	0.3	1.0		R
Nafcillin	0.03-1	0.3	0.5		M
Oxacillin	5-6	0.3	1.0		R+M
Oxytetracycline	0.05-3	1.9	9.2		R
Penicillin G	1.5-3	0.5	0.7		R
Penicillin V	3-5	0.4	0.6		M+R
Rifampin	0.5-10	0.6	3.0		M
Streptomycin	20-25	0.26	2.4		R
Tetracycline	0.5-2	1.3	6.8		R
Tobramycin	2-10	0.25	2.0		R
Anticonvulsants					
Carbamazepine	5-12	1.2	10-60	Epoxide*	M
Ethosuximide	40-100		24-72		M
Phenobarbital	10-40	0.7	84-108		M+R
Phenytoin	10-20	0.65	12-36	4-OH-glucuronide	M
Primidone	5-12	0.8	6-18	Phenobarbital*	
Barbiturates					
Amobarbital	1-8	1.1	21	4-OH-amobarbital	M
Hexobarbital	2-4	1.3	4.1	Desmethylhexobarbital*	M
Pentobarbital	1-4	1.0	22.3	Hydroxylated metabolites	M
Phenobarbital	10-40	0.7	84-108		M
Thiopental			8		M
Antiarrhythmics					
Digoxin	0.5-2	6.3	40		R
Digitoxin	0.01-0.035	0.6	164		M
Lidocaine	1.5-7	1.5	1.8	Monoethylglycine xylidide*	M
Procainamide	4-10	2.0	2.5-5	Acetylprocainamide*	M+R
Quinidine	1.5-4	3.0	6.3	Hydroxyquinidine*	M
Antirheumatics					
Phenylbutazone	50-150	0.25	72	Oxyphenylbutazone*	M
Salicylate	300-400	0.2	2-4	Glycine–glucuronic acid conjugates	M
Antidepressants and antipsychotics					
Amitriptyline	0.3-0.9	8.8	17.1		M
Chlordiazepoxide	1-3	0.3	15	Desmethylchlordiazepoxide,* oxazepam glucuronide	M
Chlorpromazine	0.05-0.30			Monodesmethylchlorpromazine,* 7-OH-chlorpromazine	M
Desipramine	0.15-0.3	41.9	17		
Diazepam	0.1-1	2.0	33	Desmethyldiazepam,* oxazepam glucuronide	M
Imipramine	0.15-0.5	30	7	Desimipramine	M
Lithium	0.6-1.4 mmol/L	0.8	19		R
Lorazepam	0.02-0.05	0.8	13		M
Meprobamate	5-15	0.7	12		M
Nitrazepam	0.03-0.06	2.1	31		M
Nortriptyline	0.1-0.8	20	27		M
Oxazepam	1-2	1.2	12	Glucuronide	M
Miscellaneous					
Amphetamine	0.02-0.03	4.0	12.2		M+R
Cimetidine	0.25-1	1.8	2.0		R
Dicumarol	5-10	0.13	8.2		M
Ethambutol	1-5	2.3	3.5		R
Isoniazid	5-10	0.6	2.3	*N*-acetylisoniazid	M
Sulfadiazine	100-150	0.9	17	Acetylsulfadiazine	M+R
Sulfisoxazole	90-150	0.16	6	Acetylsulfisoxazole*	M+R
Tolbutamide	50-100	0.12	7	Carboxytolbutamide	M
Valproate	20-100	0.14	12		M
Warfarin	1-10	0.1	40	Warfarin alcohols*	M

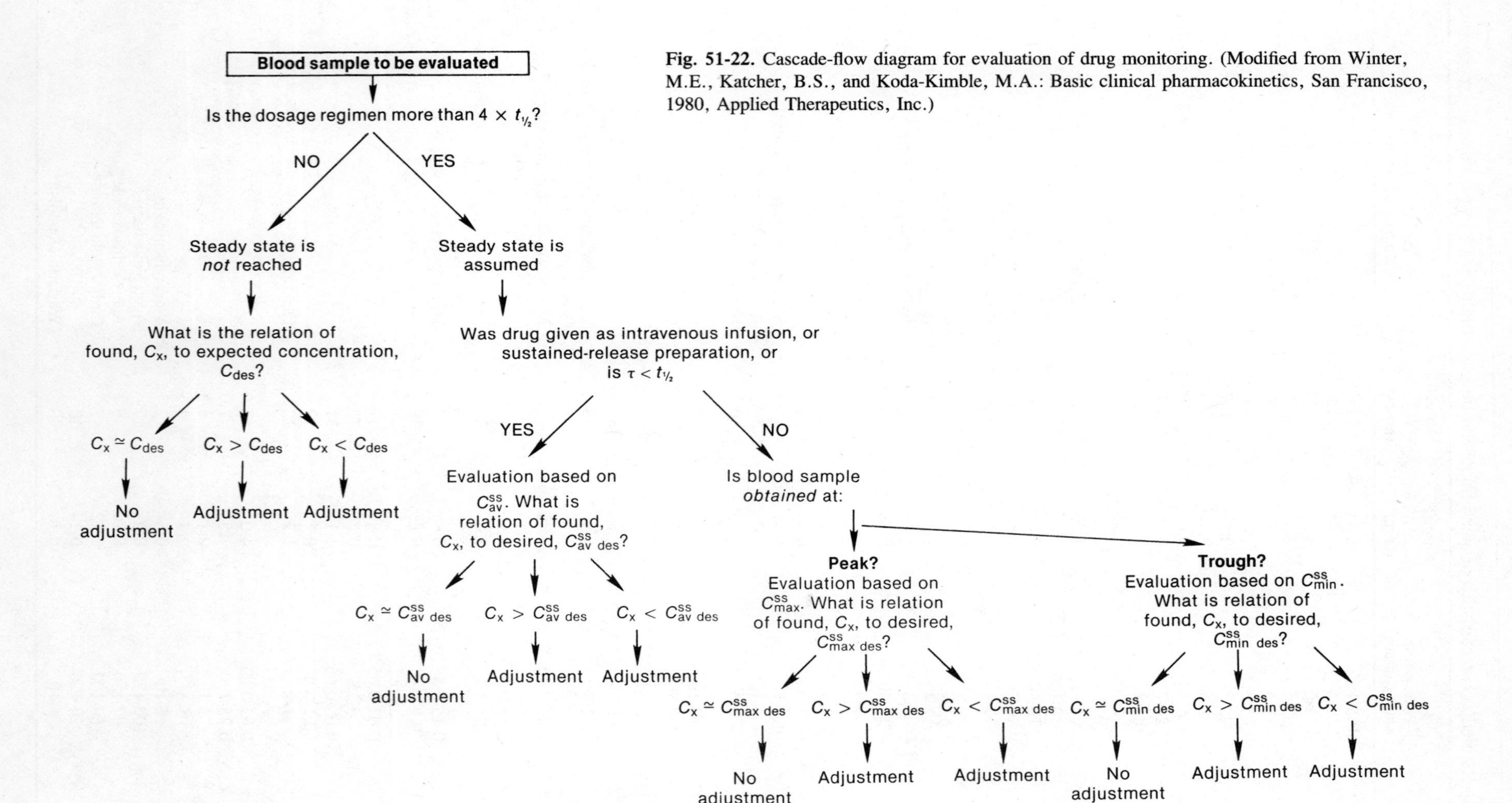

Fig. 51-22. Cascade-flow diagram for evaluation of drug monitoring. (Modified from Winter, M.E., Katcher, B.S., and Koda-Kimble, M.A.: Basic clinical pharmacokinetics, San Francisco, 1980, Applied Therapeutics, Inc.)

know if the sample was obtained after the first dose (that is, a single dose) or how many doses have been given. Using classic pharmacokinetic equations one can calculate what the drug concentration should be, based on mean pharmacokinetic parameters from the literature and patient information (age, body weight, sex, height, renal status, and so on). The purpose of drug monitoring is to evaluate the observed drug concentration with respect to the expected or desired one. The observed or found drug concentration may be equal to, smaller than, or greater than the desired one. In the latter two cases an adjustment of the dosage regimen may be indicated and recommended. A cascade flow diagram for evaluation of drug monitoring is given in Fig. 51-22.

Not only might an adjustment of the dosage regimen be indicated if C_x (the found drug concentration) differs from the desired concentration (such as $C^{ss}_{av\ des}$ or $C^{ss}_{max\ des}$), but one should also try to find the reason for the deviation. Some general and probable causes are stated below.

First, if the found concentration is higher than the expected one, the reasons could be associated with an increased bioavailability of a drug of generally low bioavailability (for example, cimetidine increases bioavailability of propranolol), or the patient used more drug (or more often) than prescribed (noncompliance), or the total clearance is decreased (renal or liver failure, acute congestive heart failure), or the protein binding might be increased.

Second, if the found concentration is less than the expected one, the reasons could reside with decreased bioavailability (possibly drug interaction), or insufficient drug dosing (or in longer dosing intervals, missed doses or noncompliance), increase in the total clearance (drug interaction or enzyme induction), or decreased protein binding.

Adjustment of maintenance dose (DM) using pharmacokinetic data

If a designed dosage regimen results in a steady-state concentration that deviates from the desired one, adjustments can be made as follows:

$$DM_{new} = \frac{C^{ss}_{av\ desired}}{C^{ss}_{av\ found}} \cdot DM_{used} \qquad Eq.51\text{-}6$$

$$DM_{new} = \frac{C^{ss}_{max\ desired}}{C^{ss}_{max\ found}} \cdot DM_{used} \qquad Eq.51\text{-}7$$

$$DM_{new} = \frac{C^{ss}_{min\ desired}}{C^{ss}_{min\ found}} \cdot DM_{used} \qquad Eq.51\text{-}8$$

where DM_{new} is the calculated *new* maintenance dose to be given at the same dosing intervals as before so that the desired therapeutic level is achieved, DM_{used} is the dose size given when the blood sample or samples were drawn, $C^{ss}_{desired}$ is the desired therapeutic concentration as average, maximum, or minimum steady-state concentration, and C^{ss}_{found} is the drug concentration found by analysis for the average, maximum, or minimum steady-state concentration.

More complex pharmacokinetic analyses

The equations presented here are basic ones. Space and background information do not permit more detail. The purpose of this chapter is to present a general review and to transmit a general understanding of the pharmacokinetic principles involved, including dosage regimens.

The equations for unequal dosing intervals, or when the blood samples are not taken at the moment before the next dose is given, or calculation of a new loading dose (DL) after change of the dosage regimen are much more complex.

A proper understanding of basic pharmacokinetics and its application is essential for reliable, safe, and accurate clinical pharmacokinetics. Although numerous standard equations and programs for hand-held calculators and computers are available, their use requires a thorough understanding of the principles and limitations involved.

The flow chart (Fig. 51-22) demonstrates that drug monitoring is not simply measurement of a drug concentration in a biological fluid and plugging of the result into a standard equation to obtain a clear-cut answer regarding dosage regimen adjustment. It also demonstrates that a proper evaluation of a found drug concentration can be made only with pertinent information intrinsic to the case and patient.

A number of articles and books, published on the topic of pharmacokinetic assessment of drug monitoring can provide more information.[23-30]

REFERENCES

1. Ritschel, W.A.: Handbook of basic pharmacokinetics, ed. 2, Hamilton, Ill., 1980, Drug Intelligence Publications. pp. 345-355.
2. Borga, O., Azarnoff, D.L., Plym Forshell, G., and Sjöqvist, F.: Plasma protein binding of tricyclic antidepressants in man, Biochem. Pharmacol. **18:**2135-2143, 1969.
3. Reidenberg, M.M., Odar-Cederlof, I., von Bahr, C., et al.: Protein binding of diphenylhydantoin and desmethylimipramine in plasma from patients with poor renal function, N. Engl. J. Med. **285:**264-267, 1971.
4. Levy, G.: Kinetics of pharmacologic effects, Clin. Pharmacol, Ther. **7:**362-372, 1966.
5. Gibaldi, M., Levy, G., and Weintraub, H.: Drug distribution and pharmacologic effects, Clin. Pharmacol. Ther. **12:**734-742, 1971.
6. Levy, G., and Gibaldi, M.: Pharmacokinetics of drug action, Annu. Rev. Pharmacol. **12:**85-98, 1972.
7. Levy, G.: Relationship between pharmacological effects and plasma or tissue concentration of drugs in man. In Davies, D.S., and Pritchard, B.N.C., editors: Biological effects of drugs in relation to their plasma concentrations, Baltimore, 1973, University Park Press, pp. 83-95.
8. Sjöqvist, F., and Bertilsson, L.: Plasma concentrations of drugs and pharmacological response in man. In Davies, D.S., and Pritchard, B.N.C., editors: Biological effects in drugs in relation to their plasma concentrations, Baltimore, 1973, University Park Press, pp. 25-40.
9. Jick, H., Miettinen, O.S., Shapiro, S., et al.: Comprehensive drug surveillance, J.A.M.A. **213:**1455-1460, 1970.
10. Stewart, R.B., Forgnone, M., and Cluff, L.E.: Drug utilization and reported adverse drug reactions in outpatients, Drugs in Health Care **2:**231-243, 1975.

11. Ritschel, W.A.: Pharmacokinetics of H_2-receptor antagonists, Sci. Pharm. **50:**250-259, 1982.
12. Barker, K.N., and McConnell, W.E.: How to detect medication errors, Mod. Hosp. **99:**95-106, 1962.
13. Stewart, R.B., and Cluff, L.E.: A review of medication errors and compliance in ambulant patients, Clin. Pharmacol. Ther. **13:**463-468, 1971.
14. Bochner, F., Carruthers, G., Kampmann, J., and Stiner, J.: Handbook of clinical pharmacology, Boston, 1978, Little, Brown & Co. pp. 22-25.
15. Ritschel, W.A.: The effect of aging on pharmacokinetics: a scientists' view of the future, Contemp. Pharmacy Pract. **5:**209-218, Washington, D.C., 1982, American Pharmaceutical Association.
16. Ritschel, W.A.: Simulation of single dose and multiple dose pharmacokinetics and clinical monitoring, Paper presented at Pharmacokinetics 79 Seminar, Los Angeles, March 2-4, 1979.
17. Ritschel, W.A.: Pitfalls and errors in blood level studies, Lecture presented at University of Cincinnati Medical Center, Cincinnati, May 14, 1982.
18. Slaughter, R.L., and Koup, J.R.: Clinical pharmacokinetic service. In McLeod, D.D., and Miller, W.A., editors: The practice of pharmacy, Cincinnati, 1981, Harvey Whitney Books, pp. 114-128.
19. van der Kleijn, E.: Clinical pharmacokinetics as component of the clinical pharmacy service at the Sint Radboud Hospital of the Catholic University of Nijmegen, The Netherlands. In Levy, G., editor: Clinical pharmacokinetics—a symposium, Washington, D.C., 1974, American Pharmaceutical Associations, Academy of Pharmacokinetics Sciences, pp. 17-30.
20. Ritschel, W.A.: Handbook of basic pharmacokinetics, ed. 2, Hamilton, Ill., 1980, Drug Intelligence Publications, pp. 412-427.
21. Sadee, W., and Beelen, G.C.M.: Drug level monitoring New York, 1980, John Wiley & Sons, Inc. pp. 4-11.
22. Winter, M.E., Katcher, B.S., and Koda-Kimble, M.A.: Basic clinical pharmacokinetics, San Francisco, 1980, Applied Therapeutics, Inc. pp. 63-67.
23. Levy, G.: An orientation to clinical pharmacokinetics. In Levy G., editor: Clinical pharmacokinetics—a symposium, Washington, D.C., 1974, American Pharmaceutical Association, Academy of Pharmaceutical Sciences, pp. 1-9.
24. Stewart, R.B.: Drug therapy monitoring In McLeod, D.L., and Miller, W.A., editors: The practice of pharmacy, Cincinnati, 1981, Harvey Whitney Books, p. 70-82.
25. Pippenger, C.E.: Therapeutic drug monitoring: an overview, Ther. Drug Monit. **1:**3-9, 1979.
26. Elenbaas, R.M., Payne, W.W., and Baumann, J.L.: Influence of clinical pharmacist consultations on the use of drug blood level tests, Am. J. Hosp. Pharm. **37:**61-64, 1980.
27. Ritschel, W.A.: Dosierungsschema und Monitoring als Aufgabe klinischer Pharmakokinetik, Dtsch. Apoth. Zeitung. **116:**1612-1616, 1976.
28. Sadee, W.: Perspectives in therapeutic drug level monitoring, Pharm. Intern. **1:**32-35, 1980.
29. Sadee, W., and Beelen, G.C.M.: Drug level monitoring New York, 1980, John Wiley & Sons, Inc., pp. 17-26.
30. Ritschel, W.A.: How clinical is clinical pharmacology? Special considerations of clinical pharmacokinetics, Methods Find. Exp. Clin. Pharmacol. **3**(suppl.):95-165, 1981.

Chapter 52 Interferences in spectral analysis

Lawrence A. Kaplan
Amadeo J. Pesce

Allen correction Multichromatic analysis of reaction to correct for background absorbance. Two wavelengths, in addition to the A_{max} of the chromophore, are monitored to subtract average background absorbance.

biochromatic analysis Spectrophotometric monitoring of a reaction at two wavelengths. Used to correct for background color.

chemical interferent A compound that either produces an endogenous color or interferes directly in the reaction or process being monitored.

end-point analysis Monitoring of a reaction after the reaction has been essentially completed.

hemolysis Breakage of red blood cells, either in vitro or in vivo. If in vivo, hemolysis will give plasma specimen a red color.

icteric Pertaining to orange color imparted to a sample because of the presence of bilirubin.

interferent Any chemical or physical phenomena that can interfere or disrupt a reaction or process.

in vitro interferent An interferent that is not caused by any in situ physiological process.

in vivo interferent An interfering process resulting from physiological processes within the body.

kinetic analysis Analysis in which the *change* of the monitored parameter with time is related to concentration, such as change of absorbance per minute.

lipemic Presence of lipid particles (usually very low density lipoprotein) in a sample, which gives the sample a turbid appearance.

reagent blank Reaction mixture *minus* the sample; used to subtract endogenous reagent color from the absorbance of the complete reaction (plus sample).

sample blank Sample plus diluent; used to correct absorbance of complete reaction mixture for endogenous sample color.

turbidity Scatter of light in a liquid containing suspended particles.

window Term used to denote a specific time during which reactions are monitored, a phenomenon can be observed, or a procedure is allowed to be initiated.

A chemistry laboratory uses a number of techniques, discussed below, for measuring the concentration of specific biochemicals. All these techniques are subject to interferences from a variety of sources. Specific interferences of one technique may not be important for another. There are, however, general concepts that, when understood, help to control and minimize their effects on method accuracy.

There are four basic types of interferences in laboratory analysis: (1) those arising from limitations of detectors, (2) in vitro substances that directly interfere with the analytical method, (3) disease states or exogenous agents that modify certain physiological processes, thus changing the concentrations of an analyte in vivo and (4) those occurring as a result of sample (blood) processing.

LIMITATIONS OF DETECTORS

Methods yielding quantitative answers usually employ a detector, such as a spectrophotometer, from which it is possible to obtain a linear relationship between the detector response and the concentrations of analytes in various samples. In the case of absorption spectrophotometry, there is a complex logarithmic relationship between concentration and detector response. When fluorescence is used, there is a linear response between concentration and the fluorescence signal. Similarly, when other optical properties, such as refractive index, or electrochemical properties, such as ion current from oxidation, are used as detectors, the response is also linear. This knowledge of the type of relationship between concentration and detector response is important.

The first section describes errors in absorption spectrophotometry because the clinical chemistry laboratory quantifies virtually all of its analytes by this technique. It is most important therefore to understand the interference problems associated with this mode of measurement. As the nature and sophistication of the labortary changes, other types of interferences will become important to consider when one is performing a laboratory analysis.

Absorption spectrophotometer

There are two types of error in spectrophotometric measurement that are interrelated. The first is caused by the nature of the mathematical relationship between absorbancy and percent transmission, and the second is related to limitations of the instrument.

Absorbance variance. As discussed in Chapter 3, there is a logarithmic relationship between the percentage of transmittance, which is the quantity actually measured, and the absorbance, which is calculated. Fig. 52-1 shows the relationship between the linear percent transmittance and the log absorbance scales. At very low percent transmittances, small changes in percent T result in proportionately large changes in absorbance. For example, a change in percent transmittance of 60% to 50% T represents an absorbance change of approximately 0.08 A. A change in percent transmittance from 15% to 5%, however, results in the change of absorbance of 0.65 A. Thus small changes of percent T at very low transmittances will result in disproportionately large changes in the calculated absorbance at this part of the scale.

One can consider the error of a spectrophotometric measurement as a function of the total or full-scale deflection of the meter or electronics. When the absorbance scale is set at 0.000 or the transmittance scale is adjusted to 100%, the maximum electronic signal is obtained. For error analysis, the variation in this measurement is presumed to be constant throughout all readings on the scale. Since percent T directly reflects the electrical signal, some simple calculations using percent T can be done. At full scale deflection (100% T), a 1% variation in percent T means an error of $\pm 1\%$ T. At half scale, a 1% variation of the 50% T value means a 2% absolute error ($1/50$), and at 10% T, this becomes a 10% error ($1/10$). However, it is not percent T that is directly proportional to concentration, it is absorbance. One can convert these percent T values into absorbance and calculate the error (Table 52-1).

Because the conversion of percent T to absorbance is a logarithmic function, both ends of the scale, 0.000 absorbance and high absorbance (more than 1.0), have the greatest error. In simple terms, when the solution has little color, it is difficult to tell the difference between no color

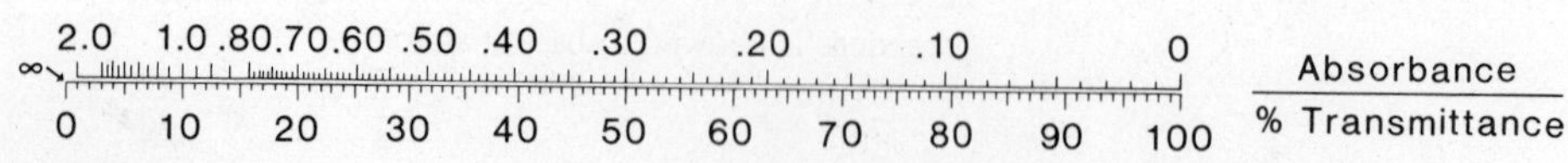

Fig. 52-1. Absorbance and percentage of transmittance scales juxtaposed.

Table 52-1. Absorbance error as a function of percent T

% T	Absorbance	Variation in absorbance	Percent error of absorbance measurement
4 ± 1	1.398	0.22	15.8
10 ± 1	1.000	0.041	8.60
25 ± 1	0.602	0.035	5.79
35 ± 1	0.456	0.025	5.44
50 ± 1	0.301	0.017	5.78
70 ± 1	0.155	0.012	8.03
90 ± 1	0.046	0.0097	21.2

and some color. In terms of relative error this can be huge, since this difference is so small. At high absorbance it is not easy to record accurately a small amount of light passing through the solution. Thus differences between the two values are minute compared to the total incident light used to calibrate 100% T, and it is difficult to measure these small changes.

The relative error of spectrophotometric measurements versus percent transmittance and absorbance is shown in Fig. 52-2. One should make most spectrophotometric measurements at absorbances between 0.1 and 1.1 to minimize this type of error.

Instrument limitations. Consider how a spectrophotometer functions. At 100% T or zero absorbance, all the light signal is converted to an electronic signal. Presume that this signal measures 1000 nanoamps (nA). If the absorbance changes to 0.010, there is a decrease to 990 nA, and the instrument must measure accurately 10/1000 nA or a 1% change in signal. To do this accurately (1%), it must measure the signal to $\pm$ 0.10 nA (1% of 10). Thus at zero absorbance there is the difficulty of accurate measurement of one part in 10^4 of signal. On the other hand, if the absorbance is 2.0, the signal to the photomultiplier is only 10 nA because only 1% of the light reaches the photodetector. To achieve the same degree of accuracy between a value of absorbance of 2 and 2.02, the instrument must measure the difference between 10 and 9.9 nA, or 0.1 nA. To do this accurately, it must measure to within 0.001 nA (1% of 0.1). Thus at high absorbance the limitations are caused by the inability of the detection system to accurately measure small differences between high levels of absorbance.

Thus in analyses with relatively high levels of absorbing compound or interferent there will be a large spectrophotometric error. An initial dilution to lower total absorption *plus* a sample blank may be needed to eliminate this problem (see the discussion of turbidity on p. 987 for an example).

Fluorescence spectrophotometer

Fluorescence measurements are different from those of absorption in that the intensity of the fluorescence signal is

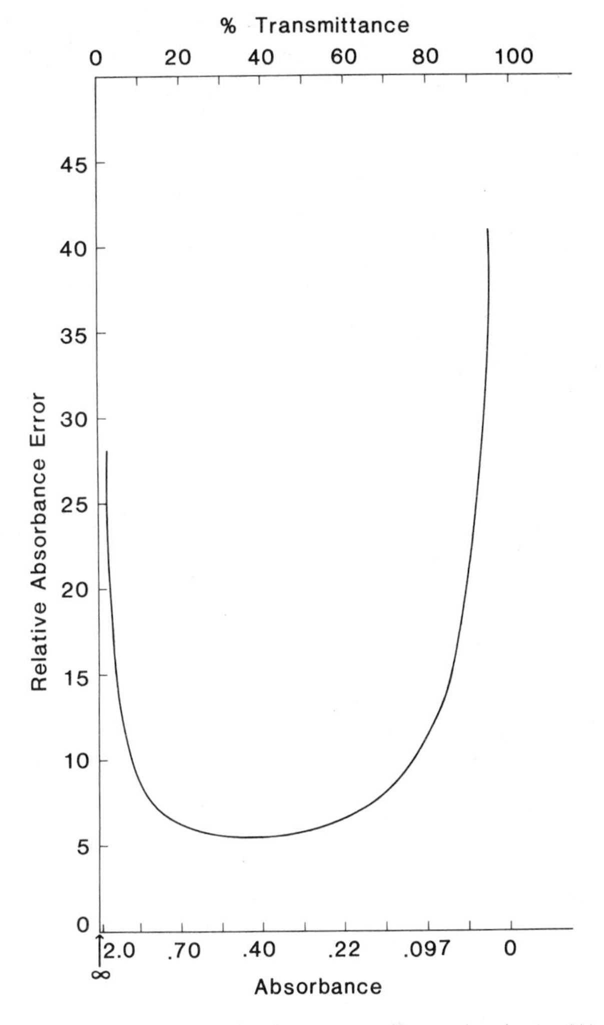

Fig. 52-2. Relative absorbance error versus absorbance *(A)* and percentage of transmittance (% T) for a $\pm 1\%$ error in measurement of transmitted light. Relative error is minimal at 36.8% T, and A is 0.434.

linearly related to concentration. A doubling of the fluorescence signal is indicative of a twofold increase in concentration. This is true only if there is very little light absorbed by the sample. If a significant portion of the light passing through a fluorescence sample is absorbed, the relationship is no longer linear; it becomes a more complex mathematical function. Thus fluorescence analysis should be performed with relatively dilute solutions the absorbance of which is less than 0.1

Scattering of light or the presence of stray light has pronounced effects on fluorescence measurements. If the sample scatters light, some may be observed by the detector as fluorescence. Similarly, if there is *stray light* (a term meaning that the light used to excite the sample was not pure, that is, not of a very narrow color band), this also may be recorded by the detector. Because only a small fraction of the incident light (less than 1% and often less than one part in a million of the total light input into the instrument) is detected as fluorescence, these extraneous signals have a disproportionate effect on the reading and thus on the error.

Fluorescence measurements, like absorption measurements, can be inaccurate or invalid because of high signals. Unlike absorption, this usually occurs because of the blanking system. If the fluorometer is adjusted to read zero for the blank, the entire detector sensitivity is adjusted for this zero reading. Presume that there is a blank solution that the instrument records as 10 units on its most sensitive scale. The instrument is now set to read zero fluorescence units. If the instrument can record accurately this zero units to ±one unit and if it can also measure full scale of 100 units in the same way, it can accurately measure a signal of 100 ± 1 units. If a different blank solution is used and this records as 100 units, this value can be set at zero by use of an electronic manipulation (subtraction of the signal). At this new full-scale deflection of 100, the detector is really recording 200, of which 100 is subtracted as the blank. Since the instrument is accurate to 1%, it is now accurate to 1% of the new full scale of 200 units. The accuracy of measurement is now ± 2 units. The accuracy is therefore twice as poor as that for the first example. Similarly, if a blank records as 1000, the accuracy is 1/10 as great. Therefore, the blank limits the accuracy of fluorescence measurements.

This same line of argument also applies to other linear measurements. The background can be blanked, but this must be considered in relation to the total signal.

IN VITRO INTERFERENCES

In vitro interferences arise from the fact that biochemical analyses are performed in the complex matrices that make up biological fluids (serum, plasma, urine, cerebrospinal fluid, and so on). These fluids contain hundreds of compounds that either have chemical groups that can react to some extent with the test reagents or can mimic the physical, chromatographic, or spectral properties of the desired analyte. This situation is further complicated because the chemical composition of body fluids can vary with the nature and extent of disease processes. This variability is increased by the possible presence of a large number of drugs. Each of these factors, alone or in combination, can result in a possible interference.

The in vitro interferences can be subclassified into those of a spectral nature and those caused by competing chemical reactions.

Spectral interferences

Absorbance. Spectral interferences are observed when a compound causes a response in the spectrophotometer similar to that of the analyte of interest, though the interferents do not themselves necessarily undergo any chemical change during the analytical reaction.

The simplest and most common example is the effect of hemoglobin (Hb) on many analytical procedures. A partial spectrum of HbO_2 (Fig. 52-3) shows significant absorption in the 500 to 600 nm portion of the visible spectrum. If one were monitoring the reaction of a colorimetric procedure in this region of the visible spectrum, there would be significant positive interference whenever Hb were contaminating the specimen. Other molecules, such as bilirubin, also cause a similar interference.

Hemoglobin interference is observed when a serum total protein (TP) concentration is determined by measurement of the A_{540} of the biuret reaction. A standard curve for this

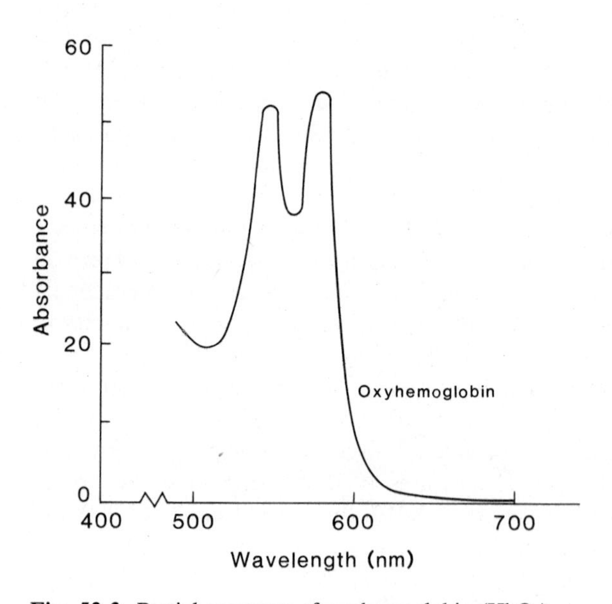

Fig. 52-3. Partial spectrum of oxyhemoglobin (HbO_2).

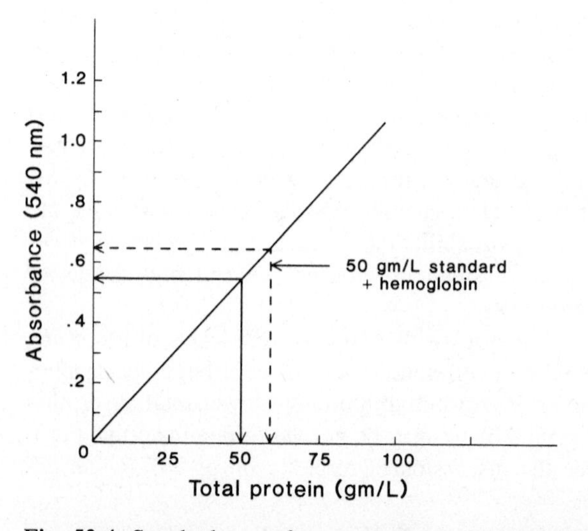

Fig. 52-4. Standard curve for measurement of total protein by the biuret reaction: A_{540} versus concentrations. *Solid arrow,* A_{540} for 50 g/L standard; *dotted arrow,* A_{540} for same standard containing hemoglobin.

reaction is depicted in Fig. 52-4. If a significant concentration of hemoglobin is added to a sample, the absorption at 540 nm is increased, thus giving a falsely high total protein reading. For example, the A_{540} of a 50 g/L standard of Fig. 52-4 is 0.550. If small amounts of hemoglobin are now added to this standard, the A_{540} is now 0.650. When this solution is read off the standard curve, a higher apparent concentration of protein is calculated (*dotted line*). Most spectral interferences give falsely elevated results in this manner.

Turbidity. A common type of spectral interference is caused by the turbidity of the sample. Turbidity is caused by large lipoprotein molecules called very low density lipoproteins (VLDL), which are suspended in serum. When a turbid specimen is analyzed in a colorimetric reaction, the lipoproteins cause the incident light to scatter, much as in nephelometry (see Chapter 3).

Since spectrophotometric analysis normally measures transmitted light at 180 degrees to the incident light, any light scattering tends to decrease the transmitted light and therefore increase the apparent optical absorbance of the specimen. This, of course, results in falsely elevated results. Sample blanks normally work poorly here, just as two-point kinetic analysis does (see below) because of the very high absorbances often encountered. The best method for eliminating interference caused by turbidity is to dilute the specimen. The extent that the sample can be diluted to minimize turbidimetric interference is limited by the ability of the analytic procedure to measure the diluted analyte. If possible, several dilutions should be analyzed simultaneously to determine the best response. An example of the effect and elimination of turbidometric interference is presented in the analysis of equal amounts of lactate dehydrogenase (LD) activity in a turbid and a nonturbid specimen (Table 52-2). When the nonturbid specimen is diluted out, all the corrected LD activities calculate out to the same approximate value. This indicates linearity of dilution. In contrast, when the turbid specimen is diluted, the calculated LD activity changes with dilution. Only at higher dilutions containing minimal turbidity do the calculated LD activities converge with the true values of the nonturbid specimen.

Fluorescence. Turbidity affects fluorescence measurements in a similar fashion. Here some scattered light will reach the detector set at 90 degrees to the incident light, thus giving an apparent increase in fluorescence and falsely elevated concentrations. Reducing problems of turbidity in fluorescence measurements is more difficult than for absorption spectroscopy. The best approach is to remove the source of light scattering by filtration or centrifugation.

Correction of spectral interferences

Sample blank. One can minimize spectral interferences by measuring the absorbance of the assay against a sample blank. The simplest sample blank is obtained by a mixture of the sample and diluent (instead of reagent). The correction for spectral interference is made by subtraction of the absorbance value of the blank from the absorbance of the complete reaction mixture. Any significant color inherent to the sample is eliminated by this calculation. In the example of the biuret reaction discussed above, the absorbance of the sample plus hemoglobin diluted with saline is 0.100. If this is subtracted from the absorbance of the complete reaction mixture (0.650), the true absorbance of the standard (0.550) is obtained. Such a sample blank can usually work unless there are *gross* amounts of the interferent present. In these cases, the very large total absorption ($A_{interferent} + A_{reaction}$) results in large spectrophotometric and calculation errors.

Reagent blanks (reagent plus diluent) are used in a similar fashion to correct for high absorbance of the reagent.

In the case of fluorescence, the sample blank allows for the correction of nonspecific fluorescence; however, this blank cannot be a great portion of the total fluorescence signal (see above).

Kinetic measurements. One increasingly used method to correct for spectral interference is to measure a typical end-point reaction as a two-point kinetic reaction. If the absorbance of a noninstantaneous, colorimetric reaction is monitored versus time, a reaction curve as shown in Fig. 52-5 will be observed. An *end-point* reaction is monitored at a single time point when the reaction is mostly completed (Fig. 52-5, *arrow 3*). If there are no spectral interferences present, the reaction curve should go through the

Table 52-2. Effect of turbidity on measurement of LD activity (U/L)

Dilution (with saline)	LD Activity (U/L)			
	Nonturbid sample		Turbid sample	
	Uncorrected	Corrected	Uncorrected	Corrected
Undiluted	440	—	28	—
1:2	245	450	32	64
1:4	136	444	30	120
1:8	62	496	26	208
1:16	30	480	25	400
1:32	14	450	13	416

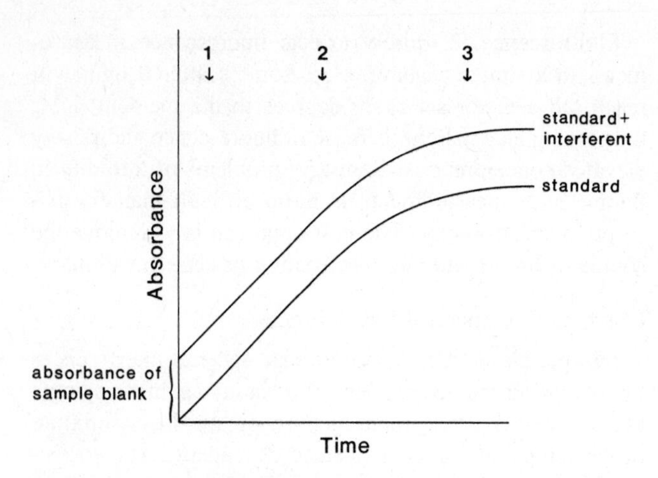

Fig. 52-5. Absorbance changes versus time for colorimetric reaction, with and without interferent present. *Arrows 1 and 2,* Time frame for kinetic analysis; *arrow 3,* end-point reading.

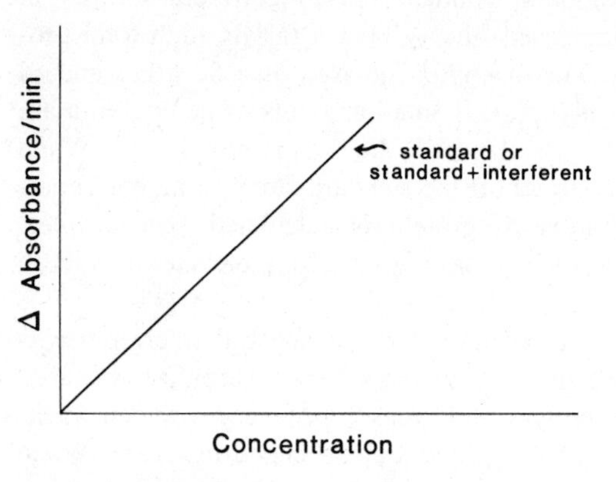

Fig. 52-6. Kinetic analysis of both reactions shown in Fig. 52-5. Change in absorbance (Δ absorbance) per minute versus concentration during linear portion of absorbance versus time curve of Fig. 52-5, between arrows 1 and 2.

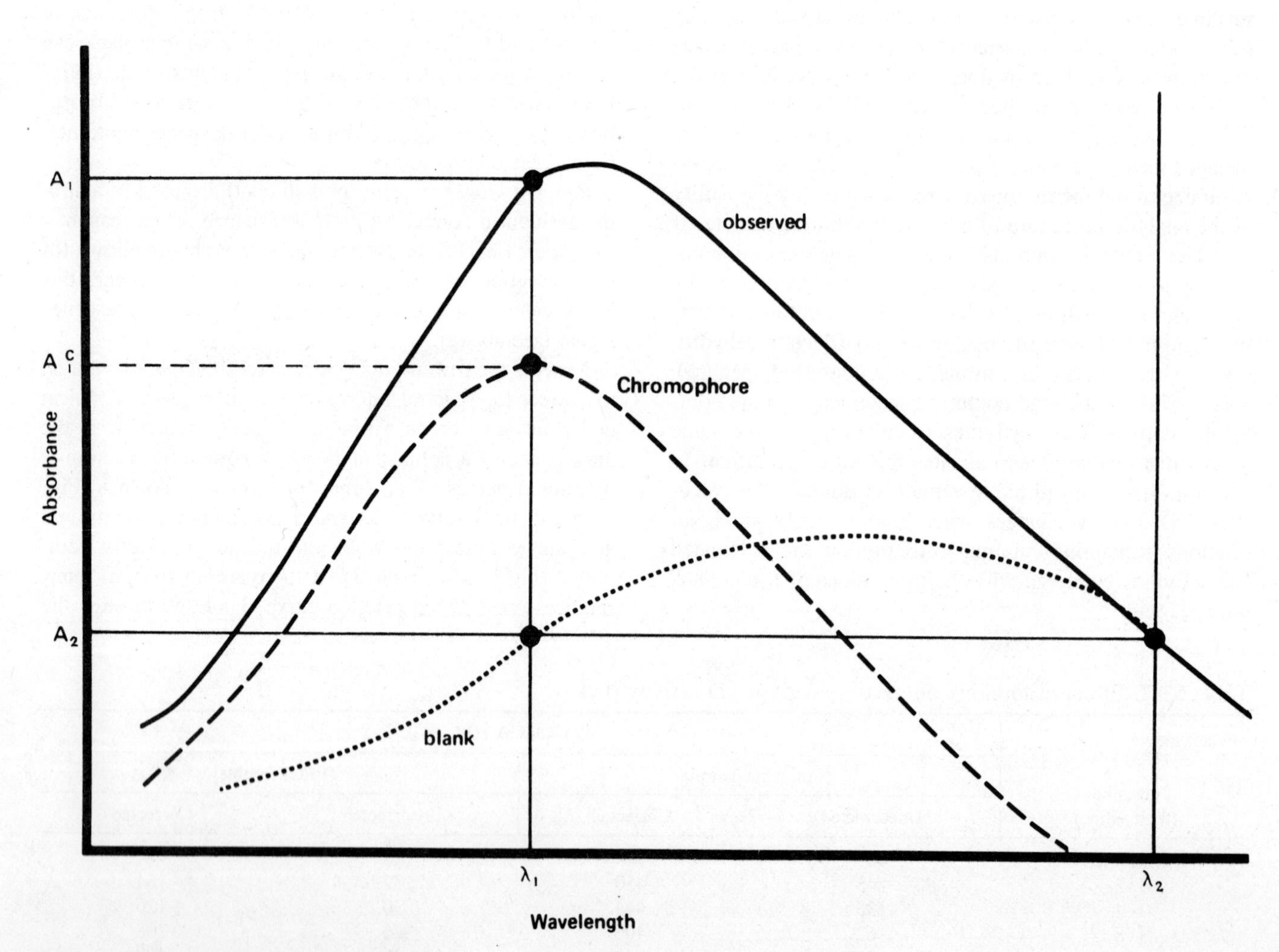

Fig. 52-7. Spectral curves for chromophore and nonreactive blank, where blank absorbance is equal at λ_1 and λ_2.

origin. If there are such interferences present, the curve will be *parallel* to the original curve but biased high because of the endogenous color present in the sample. If a sample blank was used to subtract endogenous color, a line identical to the sample containing no interferences would be obtained.

In a *two-point kinetic* assay, the absorbance is measured at *two* different time points (Fig. 52-5, *arrows 1* and *2*) when (1) the final color development has not occurred and in fact may be small and (2) the absorbance versus time response is still linear.

The initial absorbance reading (Fig. 52-5, *arrow 1*) is actually taken when almost no color formation has occurred. Thus any absorbance at this time is primarily caused by endogenous spectral interferents. A second reading is taken a short time later when only a small amount of color has formed and the response of absorbance versus time is still linear (Fig. 52-5, *arrow 2*). This absorbance therefore includes both the original endogenous color plus color formed because of the specific analytic reaction. By subtraction of the first reading from the second, the calculated *delta absorbance* (ΔA) is caused only by the specific color of the analytical reaction. Standard curves based on kinetic analysis have the change in absorbance (ΔA) plotted versus concentration (Fig. 52-6). In this standard curve the presence of a nonreacting, endogenous, colored interferent has no effect. Thus no separate sample blank reaction needs to be set up; a two-point kinetic reaction is self-blanking when there is no change in the nature of the interferent during the reaction. This is an important technique when one is performing automated chemical analysis on large numbers of specimens.

Bichromatic analysis.[1,2] Some instruments (ABA-100, du Pont ACA, CentrifiChem 500) use a different technique for correction of spectral interferences. This technique involves measurement of the absorbance of the reaction mixture simultaneously at two different wavelengths. These are the primary wavelength (λ_1) and one other wavelength (λ_2) close by. As shown in Fig. 52-7, λ_1 is the wavelength at which the chromagen maximally absorbs. At λ_2 there is minimal absorbance of the chromagen. Since the reaction is monitored simultaneously at two wavelengths, this is known as "bichromatic analysis." This technique is based on the premise that although a compound may give a spectral interference, the absorbance maxima of the interferent will differ from that of the actual analytical reaction. In addition, this procedure assumes that the absorption caused by the interfering compound is approximately the same at λ_1 as at λ_2. Although the measured absorbance at λ_1 will be caused by both the analytical reaction and the interferent, the absorbance at the second wavelength (λ_2) will be caused by only the interferent. This technique can also correct for instrument problems such as dirt on the cell, which causes light scattering or reflectance. Standard curves are then based on either $A_1 - A_2$ or the ratio of the two absorbances (A_1/A_2). Use of this procedure also allows each sample to serve as its own blank for endogenous color. Fig. 52-8 shows a dual filter used on the ABA-100 for automated, bichromatic analysis.

Another similar method to correct for background interference is to measure absorbance at the primary wavelength (A_{max}) and at two additional wavelengths, usually equidistant from the peak (A_1 and A_2). The absorbance readings at these last two wavelengths are averaged to give the average background absorbance in the specimen. This technique for correcting background absorbance by interfering substances is known as the "Allen correction."[3]

The Allen correction is only valid if the background absorbance is approximately linear with wavelength over the region in which the measurements are being taken. Thus the shape of the absorption curve for both the analyte plus interferent *(solid line)* and the interferent or interferents *(dotted line)* must be obtained as shown in Fig. 52-9. Use

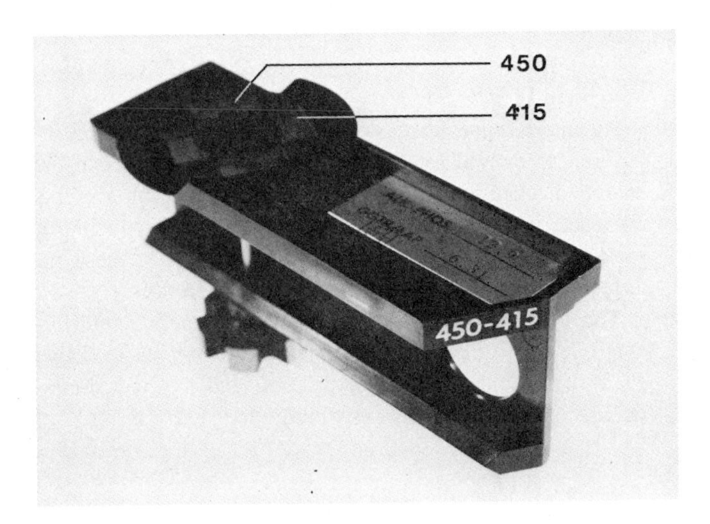

Fig. 52-8. Dual filter for bichromatic analysis on ABA-100. Two sets of filters at each wavelength are opposite each other.

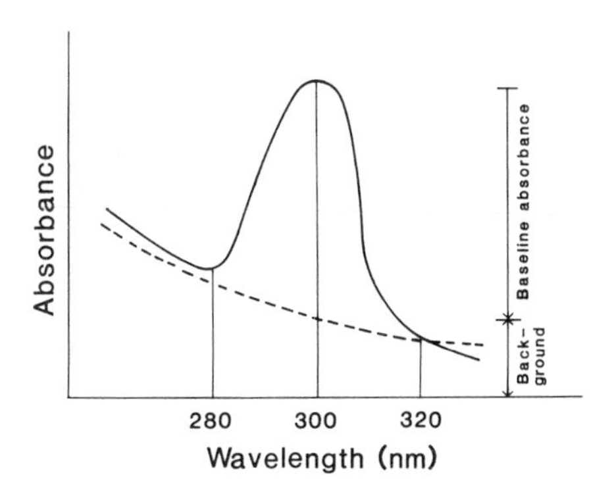

Fig. 52-9. Spectral curves of chromophore and interferent, *solid line,* and background interferents, *dotted line*. Average of A_{280} and A_{320} represents background absorbance at A_{max} for chromophore (300 nm).

of the Allen correction in this example, when the wavelengths used to correct for background are equidistant from the absorbance maxima, would give the following equation:

$$A_{300} \text{ corrected} = A_{300} - \frac{(A_{320} + A_{280})}{2}$$

The "A_{300} corrected" has the *average* background absorbance subtracted from the absorbance maximum to give the actual absorbance above base line. The Allen correction is widely used, but failure to use it properly, that is, with nonlinear background interference, can lead to even larger errors. Since the final corrected absorbance is based on three measurements, there is a decrease in the precision of the assay.

Dilution. As discussed for turbidity, dilution of a sample containing a spectral interferent can sometimes reduce the problem. The goal is not to overdilute the desired analyte or chromagen to a concentration below the minimum detectable level for a given assay. Several dilutions should be assayed simultaneously to determine the most effective dilution.

Chemical interferences

All interferences discussed above have been spectral interferences caused by compounds that do not react in the analytical chemical reaction. However, many interferents do react, in a nonspecific manner, with the chemicals of the analytical reaction. The reaction products of these interferences usually result in positive interferences.

The types of nonspecific, chemically reacting interferents can vary greatly, as seen in the examples in Table 52-3. Uric acid, creatinine, protein, and sulfhydryl-containing compounds (such as glutathione) react in most of the older reducing methods used for glucose measurement. The alkaline picrate reaction for the measurement of creatinine is known to have both positive (ketones, protein) and negative (bilirubin) interferences.

Table 52-3. Examples of chemically interfering biochemicals

Analyte	Method	Interferences*
Glucose	Reducing sugar	Uric acid (+), creatinine (+), protein (+), glutathione (+)
	Glucose oxidase–horseradish peroxidase	Uric acid (+), ascorbic acid (+), bilirubin (−)
	Glucose oxidase consumption–O_2	Hemoglobin (−), ascorbic acid (+)
	Hexokinase	None
Creatinine	Alkaline picrate	Ascorbic acid (+), glucose (+), protein (+), ketone (+)
Vanillylmandelic acid	Pisano	Certain foods (such as bananas), vanilla, aspirin (+)
	High-performance liquid chromatography	None

*(+), Positive interference; (−), negative interference.

Correction of chemical interferences

Elimination of many nonspecific chemical interferents is often achieved by one or more of the following techniques:

1. Diluting the interferent
2. Increasing the specificity of the reaction
3. Removing the interferent
4. Monitoring an assay by kinetic measurement
5. Monitoring an assay by bichromatic measurement

Dilution of the sample is an effective method in the case of interferents that do not react at the same rate or produce the same color intensity as the analyte. Interference by protein is minimized in many automated analyzers by a large specimen dilution.

Increased specificity of an analytic reaction is often achieved by use of specific enzymes as reagents. Examples of this approach include the measurement of glucose by hexokinase or glucose oxidase, uric acid by uricase, and urea by urease. Immunochemical-based reactions are also used to increase the specificity of the analysis. An example would be the measurement of theophylline by enzyme immunoassay versus the older methods, which employed ultraviolet absorbance.

Separation of an interferent from the analyte is achieved by the use of (1) a protein-free sample, (2) liquid-liquid extraction, and (3) adsorption or partition chromatography. Protein-free samples were originally prepared by precipitation of serum proteins and separation of the protein-free sample by filtration or centrifugation. Agents used to precipitate protein include tungstic acid (Folin-Wu procedure) and heavy-metal salts (such as barium and zinc; Somogyi-Nelson procedure). Protein-free specimens obtained by the dialysis technique are the basis of many Technicon AutoAnalyzer procedures.

Liquid-liquid extractions are used when the analyte and interferent or interferents can be separated into different liquid phases. Similarly, in adsorption and partition chromatography, the analyte and interferent are separated by their differential affinity for the stationary phase.

The basis for the elimination of nonspecific chemical reactants by use of a two-point kinetic reaction is that many interferents react at a different *rate* than the specific analyte of interest does. This is observed in the example of the Jaffé reaction with creatinine.[4]

Creatinine reacts with alkaline picrate at a finite rate (curve TC—true creatinine) (Fig. 52-10). Many of the nonspecific interferents (such as acetone) react at a faster kinetic rate (FR), whereas some (such as protein) react at a slower rate (SR), giving a complex change of absorbance with time for the reaction of a mixture of all three species (Fig. 52-11). Therefore, by properly choosing an optimal

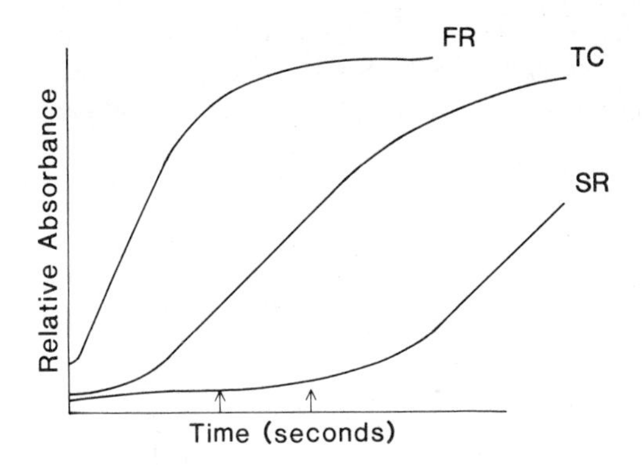

Fig. 52-10. Relative absorbance versus time curves for alkaline picrate reaction, for creatinine (TC), slow-reacting interferents (SR), and fast-reacting interferents (FR). *Arrows,* Time frame during which absorbance over time primarily reflects change attributable to the TC reaction.

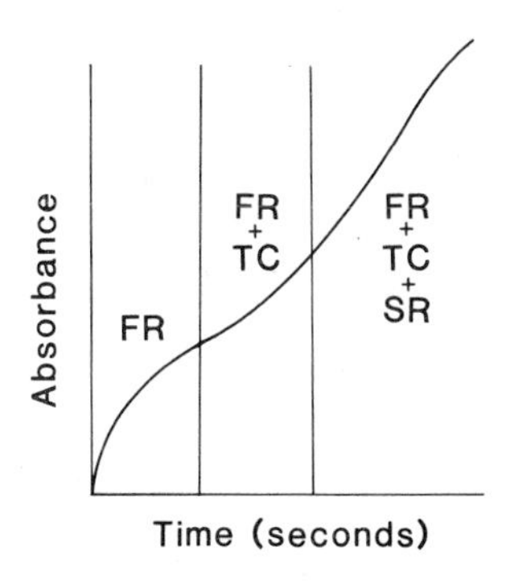

Fig. 52-11. Complex reaction of mixture containing fast-reacting (FR) and slow-reacting (SR) interferents plus creatinine (TC). Only by measurement of the change in absorbance over time can TC reaction be isolated and SR and FR interferences minimized.

window of time for the two absorbance readings for the kinetic analysis *(arrows),* one can minimize the effect of the fast-reacting and slow-reacting nonspecific interferents and isolate the absorbance change caused primarily by creatinine. During the window time period, the reaction of the FR interferents is essentially complete, whereas that of the SR interferents is not yet occurring (Fig. 52-11). The only ΔA during this time period is caused by the analyte, that is, creatinine. This concept has also been used for the glucose oxidase reaction and is, in fact, becoming a popular technique for increasing the specificity of reactions. Many newer instruments employ kinetic methods in place of traditional end-point methods (IL-508, Beckman ASTRA).

Chromatographic interferences

The third type of common in vitro, or methodological, interference is one of chromatographic interference. This often occurs when an interfering compound cochromatographs with the compound of interest to give falsely elevated results.

Currently, many analytical procedures use chromatography to separate the analyte to be measured from possible interfering compounds. A chromatographic method assumes that the desired analyte is completely isolated from other compounds that may be recorded by the detection system. However, no single set of chromatography conditions can possibly prevent interferences from cochromatographing or closely chromatographing compounds, especially when the patient may be receiving a number of potentially interfering drugs.

An example of this type of interference and its correction is the high-performance liquid chromatographic (HPLC) separation of catecholamines and a drug. Methyldopa (Aldomet), which is used in the treatment of hypertension, is eluted from the column just before norepinephrine. Since the pharmacological dose of Aldomet is much greater than the physiological concentrations of norepinephrine, it will obliterate or be confused with the norepinephrine peak. The only way to eliminate this type of interference is to remove the source of exogenous interferent. In the case of methyldopa, removal of the drug from the patient for 2 to 10 days is required. Drug or diet restrictions before biochemical analysis are often a necessity for many compounds.

There are two primary modes of minimizing chromatographic interferents: (1) increasing the specificity of the detector and (2) removing the interferent from the analyte. Detectors can measure compounds based on a variety of different principles. If the interferent has physical or chemical properties different from those of the analyte, it is possible to select a detector that will not respond significantly to a potential interferent. In liquid chromatography, refractive index detectors can detect almost any compound in the eluant and thus are very nonspecific. Fluorescence detectors have much higher specificities, since not all compounds fluoresce. By appropriate setting of excitation and emission wavelengths, the detector can be made even more specific. Electrochemical detectors also have high specificities because not all compounds are electrochemically active. The specificity can be further increased by selection of an electrode voltage at which the interferent will not react.

As discussed earlier, separation of the analytes from potentially interfering compounds is achieved by several techniques. These techniques are based on differences in solubility or chromatographic behavior between the analyte and interferents. The techniques commonly used are single liquid-liquid extractions, multiple extractions including back-extractions, and adsorption and ion-exchange chromatography. The complexity of the procedure used depends on the nature of the interference and required sensitivity. HPLC methods for serum theophylline present in relatively large amounts usually employ only a simple liquid-liquid extraction. On the other hand, the HPLC analysis of the tricyclic antidepressants present in nanogram quantities requires several extraction steps, including back-extractions. Chapter 4 discusses these types of procedures in more detail.

A technique for detecting contaminating or cochromatographing compounds is dual-detection analysis. This technique uses two different types of detectors to monitor the column eluate. The response of the two different detectors (D_1 and D_2) is determined for a standard, and the ratio of the responses is calculated (D_1 standard/D_2 standard). The probability that another compound would have a similar characteristic ratio is quite small. Thus significant deviations of the ratio for patient analyses would strongly suggest that the analyte peak contains a contaminating, coeluting compound.

Often the presence of a cochromatographing interferent is also recognized by an abnormal shape to the peak. The presence of a contaminant often causes a normally symmetric peak to be skewed. In these cases, rerunning the chromatogram at lower flow rates will sometimes separate or partially separate the analyte from the interfering compound to allow quantitation.

IN VIVO INTERFERENCES

NATHAN PICKARD

There are many in vivo factors that may play a significant role in the proper interpretation of results obtained by clinicians. This is of special concern for those tests in which the reference range interval is narrow and a result may be improperly considered abnormal when in fact it may be normal for a specific person.

Age and sex

Some tests, such as alkaline phosphatase, may be both age and sex dependent.[5] Reference intervals for all laboratory tests should reflect both variables, if they occur. Consideration should also be given to the patient's state of health. For example, interpretation of glucose tolerance tests is under the assumption that the patient is otherwise healthy.[6]

Time of day

The time of day at which a sample is drawn may affect the interpretation of numerous tests. Cortisol and estriol levels in blood vary significantly during the course of the day, as do catecholamine and iron levels.[7-9]

Drugs

Virtually every drug affects some laboratory procedure, and any laboratory procedure may be affected by at least one or more drugs. The interference may be either in vivo or in vitro. Examples of commonly encountered factors include alcohol ingestion, which may affect glucose, lactate, urate, bicarbonate, gamma-glutaryl transferase, and creatine phosphokinase levels. Smoking may alter catecholamine, cortisol, and blood-gas results.[10] A detailed compilation of the effects of drugs on laboratory tests is discussed on p. 994.

Diet

The patient's diet may considerably alter laboratory results. Fasting for 48 hours significantly elevates serum bilirubin, triglycerides, glycerol, nonesterified fatty acids, various amino acids, glucagon, secretin, insulin, and ketone bodies.[11] High-fat diets can result in elevated alkaline phosphatase levels of intestinal origin. Caffeine ingestion will increase plasma catecholamines and fatty acids and may interfere with certain theophylline procedures. Highly saturated fatty acid diets affect the cholesterol levels, and high-protein, meat diets can increase the urea, ammonia, and uric acid levels.[12]

Recent food ingestion produces alterations in bile acids, glucose, insulin, triglyceride, cortisol, and catecholamines. Therefore it is best to recommend for most procedures that samples be collected first thing in the morning after an overnight fast.[13]

Pregnancy and menses

Pregnancy has profound effects on the interpretation of laboratory tests of the mother because of, for example, hormonal changes and altered metabolism.[14] Many of these changes are difficult to quantitate, and probably unessential testing should be delayed if possible until at least a couple of months after the birth of the child. Test results, such as iron, may be affected by the presence or absence of menses, and one should consider this when evaluating the results.[15]

Other conditions

Physical conditioning, jogging, and swimming have resulted in elevated enzyme results, including CK-MB, lactate, and glucose levels, and thus affect test interpretation.[16] Physical and psychological stress affects laboratory results just as blood transfusions and other diagnostic manipulations such as surgery, physical examination, intravenous infusions, and intramuscular injections do.[17]

It would be an oversimplification to conclude that all the variables will always produce their effect. It often depends on the person, the duration of the effect, the time between initial stress and sample collection, and the degree of exposure. More important is a heightened awareness of the many factors occurring outside the laboratory in and around the patient that may affect the test result before the sample reaches the laboratory or even before the sample is collected.

Such problems of in vivo interferences are best eliminated when the clinician takes a good history and there is good communication of such information between the clinician and laboratory.

SAMPLE-PROCESSING ERRORS

The final category of interferences is the result of improper phlebotomy or sample processing.

Hemolysis

Release of intracellular constituents, especially Hb, can drastically affect test results.[18] The effect of hemolysis, discussed in Chapter 2, shows dramatic effects on K^+, acid phosphatase, lactate dehydrogenase, and other analytes. These effects are caused by spectrophotometric, methodological, and in vivo factors. One can best eliminate the effects of sample hemolysis by not performing those tests for which there is a known effect, such as acid phosphatase, K^+, lactate dehydrogenase, and Mg^{++}

Phlebotomy tube

The improper selection of the phlebotomy tube can lead to large laboratory errors. These errors most often arise from the use of anticoagulants, which can inhibit the test procedure. For example, EDTA or oxalate severely inhibits enzymes requiring divalent cations, such as alkaline

BARBITURATES
U AICAR **inc** **V** Occurs if megaloblastic anemia, *2138*
S ALANINE AMINOTRANSFERASE **inc** **V** Occurs with poisoning, probable muscle origin, *2206*
S ALKALINE PHOSPHATASE **inc** **V** Rare case of hepatotoxicity, *0217*
U AMINOLEVULINIC ACID **inc** **V** May precipitate acute porphyria, *0781*
S AMITRIPTYLINE **dec** **V** Stimulates metabolism of tricyclic antidepressants, *0878*
P AMMONIA **inc** **V** Impaired metabolism in dogs, *2045*
S ASPARTATE AMINOTRANSFERASE **inc** **V** Occurs with poisoning, probable muscle origin, *2206*
O BASAL METABOLIC RATE **dec** **V** Decreases rate by about 10%, *0153*
S BILIRUBIN **dec** **V** Induces glucuronyl transferase in newborn infants, *0879*
S BILIRUBIN **inc** **V** Rare cases of jaundice following use, *0217*
S BSP RETENTION **dec** **V** Increases conjugation with glutathione, *0798*
S BSP RETENTION **inc** **V** May increase retention if given within 24 hours, *0797*
S CALCIUM **Z** **M** No effect on fluorescence of calcein, *1384*
S CHOLINESTERASE **dec** **V** May inhibit activity, *0985*
B COPROPORPHYRIN **inc** **V** May precipitate acute cutaneous porphyria, *0781*
F COPROPORPHYRIN **inc** **V** May precipitate acute porphyria, *0781*
U COPROPORPHYRIN **inc** **V** May precipitate acute porphyria, *0781*
P CORTISOL **dec** **V** If used preoperatively lower concentration, *2081*
S CREATINE KINASE **inc** **V** Occurs with poisoning, probable muscle origin, *2206*
S CREATININE **inc** **V** Shock and renal failure in intoxication, *1400*
U DOPAMINE **Z** **V** No effect with addiction or withdrawal, *2067*
U EPINEPHRINE **Z** **V** No effect with addiction or withdrawal, *2067*
U ESTRIOL **inc** **V** Theoretical due to increased hydroxylation, *0014*
U FIGLU **inc** **V** Occurs if megaloblastic anemia, *2138*
S FOLATE **dec** **V** May cause megaloblastic anemia (impairs absorp), *2138*

Fig. 52-12. Example of format used in special issue of *Clinical Chemistry:* "Effects of drugs on clinical laboratory tests." (From O'Kell, R.T., and Elliott, J.R.: Clin. Chem. **16:**161-165, 1970.)

phosphatase. Even filling an EDTA phlebotomy tube before a serum tube can lead to carry-over of the anticoagulant into the serum tube when multiple collections are obtained from the same phlebotomy site.

Stability and storage

Improper storage and stabilization of the specimen also can result in invalid laboratory results. Analytes such as lactic acid and ammonia will rapidly increase in concentration unless either the sample is chilled or metabolism is inhibited until the analysis is performed. Storage of samples at room temperature before analysis can lead to thermal inactivation of enzymes, evaporation and concentration of the sample, and bacterial growth. All these factors can potentially result in significant changes in the true analyte concentration.

The laboratory personnel can prevent these problems of sample processing from occurring by fully understanding the precautionary steps required in general, and for each analyte in particular. The laboratory supervisory staff should educate both the technologists and clinical staff and then implement the necessary precautions. A more complete discussion on the collection and handling of patient samples is found in Chapter 2.

SOURCE-REFERENCE MATERIAL

This chapter is only a brief description of the wide variety and types of interferences in chemistry laboratory testing. Many common interfering substances are well documented, and an alert laboratory staff can often eliminate these as a source of interference.

However, an ongoing problem for the laboratory is the rapid proliferation of drugs produced by pharmaceutical companies and consumed by the public. The difficulty facing the laboratory is that these drugs can have both in vivo and in vitro effects on clinical laboratory analysis. The problem remaining for the laboratory is to know which drug is affecting a specific assay.

One of the few attempts to list the known effects of drugs and other interferences on chemical analysis was done by Donald Young in 1975. This compendium was published as an issue of *Clinical Chemistry* in April 1975.[19] Although it was outdated when published, it still remains the best and only single listing of drug effects on laboratory tests. One of the two major sections has the possible interferents (not only drugs) listed in alphabetical order. Listed under each interferent are those laboratory tests that may be affected by that interferent.

One section of Young's study is shown in Fig. 52-12. Barbiturates are the tested interferent. The initial capital letter indicates the type of body fluid in which the laboratory test was analyzed (U = urine, S = serum, and so on); the laboratory tests are then given in alphabetical order. After each test is the indication of how the tested interferent affects the lab test: *inc* = increases, *dec* = decreases, and z = no effect. The next capital letter indicates the type of interference: V = in vivo, M = methodological. A comment plus a reference finishes the test listing.

This section is cross-indexed by another section that lists, in alphabetical order, laboratory tests. Each laboratory test is listed by subsections according to decreased *(dec),* increased *(inc),* or no effects *(z)* of the listed interferents on that test. The laboratory test is also listed as either a urinary *(U)* or serum *(S)* analyte. Under each laboratory test is listed the interferent for which the effect *(inc, dec,* or *z)* has been shown.

The complexity and extent of Young's work highlights the awareness laboratory personnel must have concerning interferences with clinical laboratory tests. A complementary issue to the 1975 listing was published in 1980.[20] This volume lists the effects of disease on clinical laboratory tests. The format for this volume is similar to the one described above. The first section lists each analyte and those disease states in which changes in the concentration of that analyte have been noted. The second section lists diseases and those analytes that change during the course of the disease.

REFERENCES

1. Cowles, J.C.: Theory of dual-wavelength spectrometry for turbid samples, J. Opt. Soc. Am. **55**:690-693, 1969.
2. Hahn, B., Vlastelica, D.L., Synder, L.R., et al.: Polychromatic analysis: new applications of an old technique, Clin. Chem. **25**:951-959, 1979.
3. Allen, E., and Rieman, W.: Determining only one compound in a mixture, short spectrophotometric method, Anal. Chem. **25**:1325-1331, 1953.
4. Soldin, S.J., Henderson, L., and Hill, J.G.: The effect of bilirubin and ketones on reaction rate methods for the measurement of creatinine, Clin. Biochem. **11**:82-86, 1978.
5. Leonard, P.J.: The effect of age and sex on biochemical parameters in blood of healthy human subjects. In Proceedings of the Second International Colloquium, Automation and Prospective Biology, Pont-à-Mousson, France 1972, Basel, 1973, S. Karger AG, Medical and Scientific Publishers, pp. 134-137.
6. Klimt, C.R., Prout, J.E., Bradley, R.F., et al.: Standardization of the oral glucose tolerance test: report of the Committee on Statistics of the American Diabetes Association, June 14, 1968, Diabetes **18**:299-310, 1969.
7. Crane, J.P., Sauvage, J.D., and Arias, F.: A high risk pregnancy management protocol, Am. J. Obstet. Gynecol. **125**:227-235, 1976.
8. Jacobs, S.L., Sobel, C., and Henry, R.J.: Specificity of the trihydroxyindole method for the determination of urinary catecholamines, J. Clin. Endocrinol. Metabol. **21**:305-314, 1961.
9. Bowie, E.J.W., Tauxe, W.N., Sjoberg, W.E. Jr., and Yamaguchi, M.: Daily variation in the concentration of iron in serum, Am. J. Clin. Pathol. **40**:491-494, 1963.
10. Freer, D.E., and Statland, B.E.: The effect of ethanol (.75 g/kg body weight) on the activities of selected enzymes in sera of healthy young adults. I. Intermediate-term effect, Clin. Chem. **23**:830-834, 1977.
11. Stout, R.W., Henry, R.W., and Buchanan, K.D.: Triglyceride metabolism in acute starvation: the role of secretin and glucagon, European J. Clin. Invest. **6**:179-185, 1976.
12. Statland, B.E., and Winkel, P.: Effects of non-analytical factors on the intra-individual variation of analytes in the blood of healthy subjects: consideration of preparation of the subject and time of venipuncture, CRC Crit. Rev. Clin. Lab. Sci. **8**:105-110, 1977.
13. Schwartz, M.K.: Interferences in diagnostic biochemical procedures, Adv. Clin. Chem. **16**:1-45, 1973. See pp. 17-20.
14. O'Kell, R.T., and Elliott, J.R.: Development of normal values for

use in multitest biochemical screening of sera, Clin. Chem. **16:**161-165, 1970.
15. Latner, A.L.: Clinical biochemistry, Philadelphia, 1975, W.B. Saunders Co., p. 320.
16. King, S., Statland, B.E., and Savory, J.: The effects of a short burst of exercise on activity values of enzymes in sera of healthy subjects, Clin. Chem. Acta **72:**211-218, 1977.
17. Schwartz, M.K.: Interferences in diagnostic biochemical procedures, Adv. Clin. Chem. **16:**1-45, 1973. See pp. 21-26.
18. Frank, J.J., Bermes, E.W., Bickel, M.J., and Watkins, B.F.: Effects of in vitro hemolysis on chemical values for serum, Clin. Chem. **24:**1966-1970, 1978.
19. Young, D.S., Pestaner, L.C., and Gibberman, V.: Effects of drugs on laboratory tests, Clin. Chem. **21:**1D-432D, 1975.
20. Friedman, R.B., Anderson, R.E., Entine, S.M., and Hirshberg, S.B.: Effect of diseases on clinical laboratory tests, Clin. Chem. **26:**1D-476D, 1980.

Chapter 53 Examination of urine

G. Berry Schumann

acidosis A pathological condition resulting from accumulation of acid in, or loss of base from, the body.
aciduria The presence of acid in the urine.
acute tubular necrosis A disease that involves the destruction of renal tubular epithelial cells and is most commonly associated with reduced blood supply to the renal tubules (ischemic) or toxic exposures.
Addis count A quantitative urine sediment test in which the number of erythrocytes, leukocytes, and casts are quantified in a timed urine specimen.
albuminuria Increased albumin in urine.
alkalosis A pathological condition resulting from accumulation of base in, or loss of acid from, the body.
aminoaciduria Excess of one or more amino acids in urine.
amorphous crystals Shapeless, ill-defined crystals, usually phosphates.
ascorbic acid Vitamin C.
bacteriuria The presence of bacteria in urine.
bilirubinuria The presence of bilirubin in urine.
biliverdin Oxidized bilirubin.
calculus An abnormal concretion, usually composed of mineral salts, present in the urinary system or other tissues.
cast Molded, cylindrical structure that is formed as a result of cell conglutination and protein precipitation in the lumen of the distal convoluted tubule or collecting duct of the nephron and is extruded into the urinary sediment.
catheterization Passage of a thin, flexible, tubular instrument into the bladder or ureter for the withdrawal of urine.
crystalluria The presence of crystals in urine.
cylindruria The presence of casts in urine.
cystitis Inflammation of the bladder.
cytology The study of cells, their origin, structure, function, and pathological processes.
diuretic An agent that promotes the secretion of urine.
fixation A principle by which cells detached from tissue are preserved and rendered morphologically stable for subsequent staining and microscopic examination.
fixatives Chemical agents that structurally preserve urine constituents.
glitter cells Pale staining, swollen, and degenerated neutrophils found in dilute urine with cytoplasmic granules that exhibit a characteristic brownian movement.

glomeruli Coils of blood vessels projecting into the expanded end of the capsule of each of the uriniferous tubules of the kidney.

glycosuria The presence of glucose in urine.

hematuria The abnormal presence of blood or hemorrhage into urine.

hemoglobinuria The presence of free hemoglobin in urine.

hyaline cast Transparent cast composed of mucoprotein.

hydrometer An instrument used for determining the specific gravity of a fluid.

ketone Any compound containing the carbonyl group,–CO–, and having hydrocarbon groups attached to the carbonyl carbon.

ketonuria The presence of ketone bodies, which are intermediary products of fat metabolism, in urine, as in diabetes mellitus.

melanin The dark amorphous pigment of the skin, hair, and various tumors. It is produced by polymerization of oxidation products of tyrosine and dihydroxyphenol compounds and contains carbon, hydrogen, nitrogen, oxygen, and often sulfur.

myoglobinuria The presence of myoglobin, a pigment of muscle cells, in urine.

nephritis Inflammation of the kidney.

nephrosis Disease of the kidney.

osmolality The number of solute particles per unit mass of solvent (mOsm/kg).

osmolarity The number of osmotically active particles per unit volume of a solvent (mOsm/L).

Papanicolaou stain A differential stain that aids in the identification of nuclear chromatin, cytoplasmic properties (such as keratinization), noncellular entities (such as crystals and casts), and hematopoietic elements.

porphyrins A group of iron-free or magnesium-free pyrrole derivatives that occur universally in protoplasm. They constitute the basis of the respiratory pigments in animals and plants.

proteinuria Increased protein in urine.

pyuria An abnormal number of leukocytes in urine.

reagent-strip testing The use of a chemical test strip to determine whether pathological concentrations of various substances are present in the urine.

specific gravity The weight of a substance compared with that of an equal volume of another substance taken as a standard.

Sternheimer-Malbin stain A crystal violet and safranin stain used in urinalysis. This stain provides additional contrast for some cells and casts.

urobilinogen A group of colorless compounds formed from the reduction of conjugated bulirubin by intestinal bacteria; about 1% of the total urobilinogen produced reaches the urine.

virus A self-replicating agent that consists of a core of nucleic acid enclosed by a protein coating. This microorganism can multiply only within living host cells.

yeast Unicellular nucleated microorganism that reproduces by budding.

INTRODUCTION AND CLINICAL UTILITY OF URINALYSIS

Urinalysis can provide a wide variety of clinical information regarding an individual's kidneys and the systemic diseases that may affect this excretory organ. Both structural (anatomical) and functional (physiological) disorders of the kidney and lower urinary tract may be elucidated, as well as sequential information about the disease and its cause, and prognosis. Careful laboratory examination of urine often narrows the clinical differential diagnosis of numerous urinary system diseases. Usually these laboratory data may be obtained without pain, danger, or distress. Therefore properly performed and interpreted urinalysis will always remain an essential test in clinical practice.

Currently urinalysis procedures consist of two major components: (1) *macroscopic urinalysis,* or physicochemical testing (appearance, specific gravity, and reagent-strip measurements of several chemical constituents), and (2) *microscopic urinalysis,* or microscopic examination of urinary sediment.[1] Because of the rapidity with which urinalysis information can be generated, it is essential that this type of laboratory evaluation be carefully monitored.[2] The clinician's education regarding specimen requirements, test availability, the advantages and disadvantages of each procedure, and the pathophysiological basis and significance of each determination should enhance the effectiveness of the urine examination.[3]

The purpose of this chapter is to describe briefly common methodologies employed by most conventional urinalysis laboratories. Emphasis will be placed on the responsibilities of the urinalysis laboratory in the following areas: (1) common procedures and equipment; (2) quality reagents; (3) sensitivity, specificity, and limitations of each procedure; (4) confirmatory tests; (5) accurate identification of urine sediment entities; (6) quality control; and (7) continuing education of technical staffs to ensure competency in performing duties. The technical and diagnostic significance of urinalysis results and the mechanism of diseases that produce urinalysis abnormalities are discussed. Readers are encouraged to supplement their knowledge by using the reference list.

Urinalysis laboratories perform front-line tests for the detection of chemical and morphological abnormalities present in urine. Routine urinalysis is a standard part of almost every physical examination or hospital admission. Diabetic patients often monitor their own disease by the daily testing of urine specimens.[4] The purpose of examining urine is to screen asymptomatic, low-risk individuals, diagnose the symptomatic patient, and assist in the therapeutic monitoring of conditions that affect the urinary system.

As a screening procedure the value of urinalysis lies in the ease of obtaining specimens and the rapid determination of chemical constituents with reagent-strip technology. Urine tests can often provide rapid diagnostic information for hematuria, proteinuria, pathological casts, and so on that may confirm or possibly exclude a clinical diagnosis of renal or lower urinary tract disease.[5]

Urinalysis information is important in monitoring the treatment and convalescence of certain diseases. Clinical examples of therapeutic monitoring of urine are reduction of proteinuria in patients with lupus nephritis after the administration of steroids, clearing of hypercellular sediment that occurs in transplanted kidneys after antirejection therapy, and controlling of the cellular alterations of inflammation and urinary tract infection. Urine bilirubin tests are particularly useful in monitoring the recovery phases of hepatitis. Bilirubin will reappear in the urine if there is a recurrence of hepatitis. For certain individuals prone to bacteriuria, frequent monitoring of nitrites, leukocyte esterase tests, or urine microscopy may identify urinary tract infections.

SPECIMEN COLLECTION

Careful attention to collection of the urine specimen and prompt delivery to the laboratory are crucial for optimal information to be derived from urinalysis. Urine should be collected in a clean, sterile container that has a tightly fitting lid to prevent spillage, evaporation, and contamination. Specimen containers should be marked with patient's name, date, and time of collection. Complete urine collection systems are commercially available to assist the laboratory in quickly handling numerous specimens and standardizing results (Fig. 53-1).

The first voiding in the morning is usually the most desirable for urinalysis testing, since it provides the most concentrated urine. A clean-catch, midstream collection avoids contamination from the distal urethra. The sample must be free of vaginal secretions and other extraneous debris. Kunin[6] thoroughly describes a variety of collection techniques for urinary tract diseases.

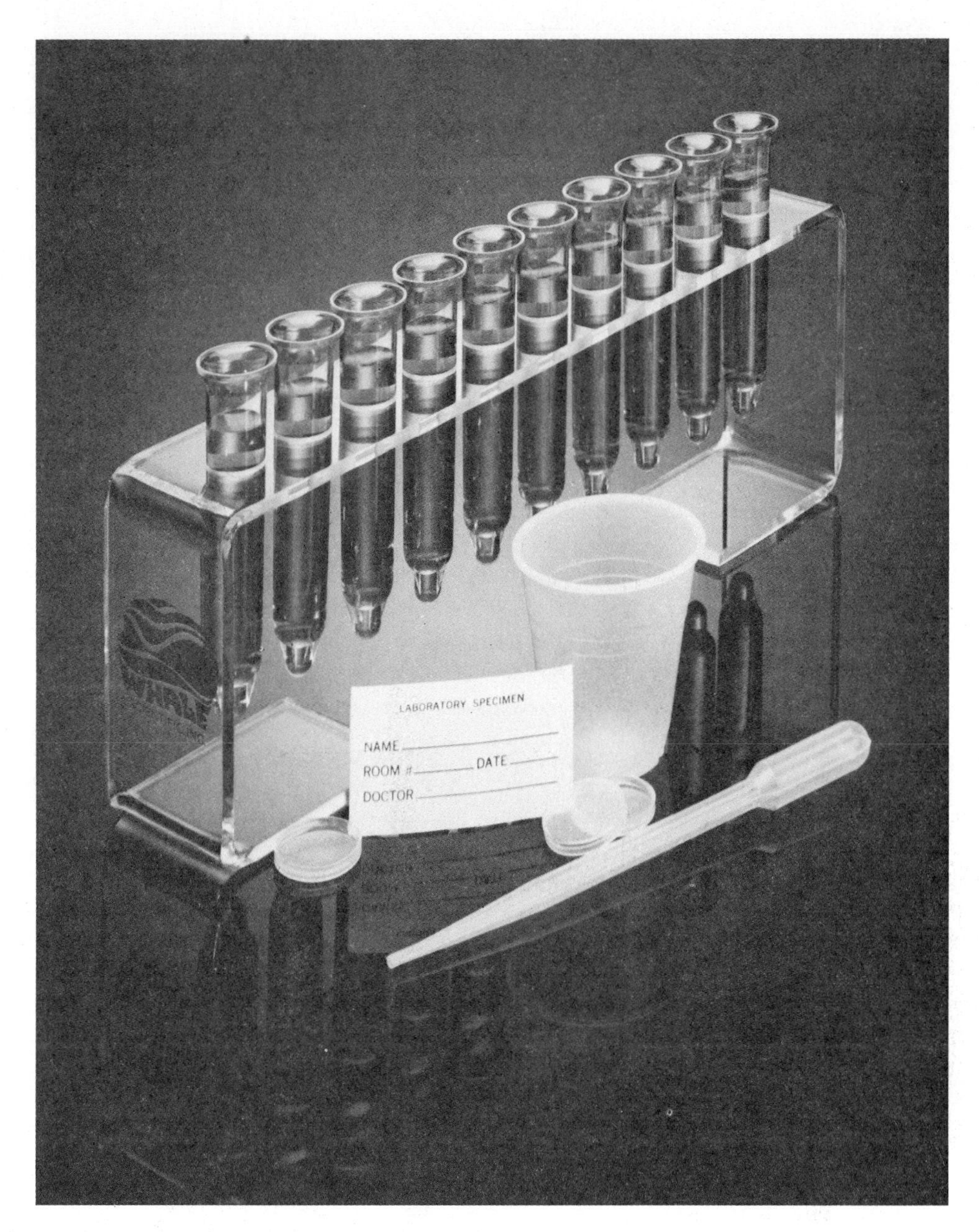

Fig. 53-1. Urine collection system. (Courtesy Whale Scientific, Inc., Commerce City, Colo.)

It is highly recommended that the urine be examined within 2 hours after it has been voided. The longer the urine is allowed to stand at room temperature, the less valuable it is. If a delay in analysis is unavoidable, the urine should be refrigerated (2° to 8° C), or an appropriate fixative or preservative should be added. Recommended types of specimens for various urine tests are listed in Table 53-1. Ethanol (95%) or commercially available fixatives, such as Mucolexx and Saccomanno, may be used to preserve cell structure. Laboratories should be responsible for the correct type and amount of urine preservatives used for cellular preservation. Toluene, phenol, thymol, and acid preservatives are commonly used for urine chemistry determinations.

Timed urine specimens are frequently utilized to quantitate various aspects of renal function. The urine must reflect excretion over a precisely measured duration of time. Urine that has been produced by the kidneys during a specific period should be collected. Timed urine specimens must not include urine in the bladder before the timed test. One would therefore obtain a 24-hour urine specimen by discarding the first-voided morning urine on the first day and collecting all subsequent urine up to and including the first-voided morning urine on the second day.

Table 53-1. Recommended types of urine specimens

Tests	Types of collection					Types of storage		
	Random	First morning	2-hour	12-hour	24-hour	Fresh	Refrigerated	Preserved
Addis count				×			×	
Albumin								
Qualitative analysis	×					×		
Quantitative analysis					×		×	
Aldosterone					×		×	
Amino acid–nitrogen					×		×	×
Bence Jones		×				×		
Bilirubin	×					×		
Blood	×					×		
Calcium					×		×	
Catecholamine					×			×
Coproporphyrin								
Qualitative analysis	×					×		
Quantitative analysis					×			×
Creatinine				×	×		×	
Cytological		×				×	×	
Estrogen	×				×		×	
Glucose	×					×		
17-Hydroxycorticosteroid					×		×	
17-Ketosteroid					×		×	
Ketone	×					×		
Lead		×					×	
Microbiological		×				×		
Nitrite		×				×		
Osmolality	×					×		
pH	×					×		
Phosphorus					×		×	
Porphobilinogen	×					×		
Potassium					×		×	
Pregnanediol					×		×	
Pregnanetriol					×		×	
Protein						×		
Qualitative analysis	×						×	
Quantitative analysis					×	×		
Sediment		×					×	
Sodium					×	×		
Specific gravity	×					×		
Sugar	×						×	
Urea nitrogen					×		×	
Uric acid					×	×		
Urobilinogen			×					×
			(afternoon)					×
Uroporphyrin					×		×	
Vanillylmandelic acid					×			×
Volume					×			

PHYSICAL EXAMINATION OF URINE

The physical examination of urine is the initial part of a routine urinalysis examination. This examination includes assessment of volume, odor, appearance (color and turbidity), specific gravity, and osmolality.

Volume

Urinary volume is influenced by the fluid intake; solutes to be excreted, primarily sodium and urea; loss of fluid in perspiration and respiration; and cardiovascular and renal status. Normally adults excrete 750 to 2000 mL/24 hours. Conditions that produce an increased or decreased amount of urine are discussed in Chapter 23. Although the volume of a random specimen is clinically insignificant, the volume of specimen received should be recorded for purposes of documentation and standardization.

Odor

Normal fresh urine is odorless. Although it is not diagnostically significant, odor can provide clues to certain

Table 53-2. Common colors of urine

Colors	Potential causes				Clinical conditions	Commonly associated diseases
	Foodstuffs	Metabolites	Drugs	Organisms		
Yellow to colorless					Polyuria	Diabetes insipidus Diabetes mellitus Chronic renal failure
Yellow	Food color	Urochrome	Quinacrine (Atabrine) Sulfasalazine (Azulfidine) Phenacetin Nitrofurantoin Riboflavin		Normal	
Yellow-orange	Food color Carotene Rhubarb	Urobilin	Sulfisoxazole and phenazopyridine (Azo Gantrisin) Riboflavin Furazolidone (Furoxone)		Dehydration Jaundice	Liver disease
Yellow-green		Bilirubin-biliverdin	Methylene blue Indican Amitriptyline (Elavil)		Jaundice	Liver disease Biliary obstruction
Yellow-red-brown	Beets Food color Rhubarb	Hemoglobin Myoglobin Porphyrin	Sulfisoxazole and phenazopyridine (Azo Gantrisin) Phenytoin (Dilantin) Pyrvinium pamoate (Povan) Phenazopyridine (Pyridium) Phenolsulfonphthalein Phenindione Amidopyrine	*Serratia marcescens*	Hematuria Hemoglobinuria Myoglobinuria Porphyrinuria Menstrual contamination	Hemolysis Transfusion Burns Renal disease Urological disease
Brown-black		Porphyrin Melanin Methemoglobin Homogentisic acid	Chloroquine (Aralen) Levodopa		Alkaptonuria	Melanoma Ochronosis Phenol poisoning
Blue-green		Indican	Methylene blue	*Pseudomonas*	Dysuria	Urinary tract infection

urine abnormalities. An ammonia-like odor suggests urea-splitting bacteria, fruity odors indicate the presence of acetone (ketone), and a sweet or foul odor suggests pus or inflammation. Odor is important in the clinical detection of maple-syrup urine disease (a congenital metabolic defect).

Appearance (color and turbidity)

The color of urine is determined to a large degree by the specific gravity. Normal urine varies widely from colorless to deep yellow. Interpretation of color is subjective and varies with each laboratory examiner. It may be helpful to the technologist to use standardized colored objects in the laboratory as reference points.

Red urine is perhaps the most clinically important discoloration; it may be a result of urinary hemoglobin or myoglobin. Intact erythrocytes, hemolyzed erythrocytes, or free hemoglobin (hemolysis) can be responsible for the red color. Hemolysis may occur intravascularly, as in a transfusion reaction, with malaria (blackwater fever), or after the urine has been voided. Myoglobin results from crushing damage to muscle. Urinary hemoglobin and myoglobin may not be differentiated by gross inspection or simple tests. Ingestion of beets may redden the urine, as may certain drugs, and thus such chromogens must be excluded as a source of urine coloration.

Urine characteristic of acute glomerulonephritis is brownish red. It is acidic and contains blood, a combination producing brownish colored, acidic hematin. Blood in the urine also may assume a smokey appearance. One can observe this only by shaking the urine while holding it up to the light. The "smoke," or turbidity, is actually erythrocytes suspended in solution. Centrifugation will yield clear urine with a red plug in the bottom of the tube. Do not confuse the smoky appearance of erythrocytes with the faint turbidity of other suspensions (such as crystals and cells.) Common colors of urine are listed in Table 53-2.

Table 53-3. Common causes of cloudy or turbid urine

Causes	Methods of clearing	Comments
Chemical		
Urates	Soluble at 6° C or alkali	Pink sediment
Phosphates and carbonates	Soluble at 6° C or acid	
Calcium oxalates	Soluble in dilute hydrochloric acid	
Mucus	—	Sticky
X-ray contrast media	Soluble in 10% sodium hydroxide	
Lipids	Soluble in ether	Opalescent
Chyle	Soluble in ether	Milky
Cells		
Bacteria		Foul-smelling odor
Fungi		Sweet-smelling odor
Erythrocyte		Red, smoky
Leukocyte		
Epithelium		
Spermatozoa		

Normally freshly voided urine is clear. When urine is allowed to stand, amorphous crystals, usually phosphates, may precipitate and cause urine to be cloudy. The turbidity of urine should always be recorded and microscopically explained. Table 53-3 lists common causes of cloudy urine.

Specific gravity

Urinary specific gravity serves as a partial assessment of the ability of the kidneys to concentrate urine. The normal range is 1.005 to 1.030 g/mL. A value of 1.020 or greater indicates good renal function and increased amounts of dissolved solutes excreted by the kidneys. Increased specific gravity greater than 1.030 is found in dehydration, diabetes mellitus, congestive heart failure, proteinuria, and adrenal insufficiency. Decreased specific gravity is found in patients with hypothermia and those using diuretics. In a patient with severe kidney disease, urine of a fixed specific gravity would correspond to that of the glomerular filtrate, approximately 1.010. X-ray contrast media and preservatives may cause an increase in specific gravity to greater than 1.040. Sediment constituents are often poorly preserved in dilute urine. Therefore a low specific gravity indicates that microscopic examinations of urine may not yield optimally accurate results.

A hydrometer and a suitable container may also be used to determine specific gravity. The hydrometer must float freely in the urine and not touch the sides or bottom of the vessel.

The calibrated paper inside the hydrometer may slide down and give erroneous readings. It is therefore important to float the hydrometer in distilled water regularly to ensure that it reads an accurate 1.000. If it does not, tap the hydrometer a few times until it reads accurately. To make a determination, one merely reads the bottom of the meniscus against the calibrated scale. One must realize that specific gravity is related to temperature. Refrigerated urine will give high values that must be corrected to 25° C (0.001 for each 3 Celsius degrees). Therefore it is desirable to allow the urine to reach room temperature before measuring specific gravity.

Most laboratories are now equipped with refractometers. The refractive index of urine is the ratio of the velocity of light in urine to that in a vacuum. It increases with the amount of dissolved solutes. This instrument requires only a single drop of urine for specific gravity determination as opposed to 15 to 20 mL for hydrometers.

1. Urinometer method
   a. Fill the urinometer vessel three fourths full with urine (15 mL minimum volume).
   b. Insert the urinometer with a slight spinning motion to make sure that it is floating freely.
   c. When the reading is taken, be sure that the urinometer is not touching the sides or bottom of the container.
   d. The specific gravity is affected by temperature and the protein and sugar content of urine. Subtract 0.004 for every 10 g/L of glucose, 0.003 for every 10 g/L of protein, and 0.001 for every 3° C below 20° C.
2. Refractometer method
   a. First clean the surfaces of the cover with a damp cloth and then dry.
   b. Using two applicator sticks or a capillary pipet, apply a drop of urine at the notched bottom of the cover so that it flows over the prism surface by capillary action.
   c. Point the refractometer toward a light source and read directly on the specific gravity scale, the sharp dividing line between light and dark contrast.
   d. For specific gravity over 1.030, dilute the urine with equal parts of water and read. Multiply the last three digits by three.[3]
   e. This specific gravity reading is temperature compensated. Distilled water should read 1.000, or the refractometer needs recalibration.

Osmolality

The normal kidney is capable of producing urine with a range of 50 to 1200 mOsm/kg. Urine ranges from one sixth to four times the osmolality of normal serum (280 to 290 mOsm/kg). Osmotic pressure is measured by an osmometer, though this instrument is sometimes not available in routine urinalysis laboratories.

Osmotic pressure is determined by the number of particles per unit volume, whereas the specific gravity is a re-

Table 53-4. Composition of urine from healthy subjects

Constituent	Value
Creatinine	1.2-1.8 g/24 hr
Glucose	<300 mg/L
Osmolality	>600 mOsm/L
Potassium	30-100 mEq/24 hr
pH	4.7-7.8
Sodium	85-250 mEq/24 hr
Specific gravity	1.005-1.030
Total bilirubin	(Not detected)
Total protein	<150 mg/24 hr
Urea nitrogen	7-16 g/24 hr
Uric acid	300-800 mg/24 hr
Urobilinogen	<1 mg/L

flection of the density (size or weight) of the suspended particles. Normal ranges are 300 to 1200 mOsm/kg for adults and 200 to 220 mOsm/kg for infants. Generally, specific gravity and osmolality are directly related in a linear fashion, though there are important exceptions. For example, after heavy, iodinated dyes are administered to a patient for intravenous pyelography, the specific gravity may reach as high as 1.070 or 1.080, though the osmolality will remain within normal limits. The dye particles have a sufficiently high mass to raise the specific gravity, but their numbers are few and are incapable of notably increasing the osmolality.

CHEMICAL EXAMINATION OF URINE

Several chemical constituents are routinely analyzed by urinalysis laboratories. Both qualitative reagent-strip or tablet tests are used, as well as quantitative methods for protein, electrolytes, and porphyrins. It is recommended that all positive reagent-strip tests be confirmed by more quantitative methods, if clinically indicated.

Normal constituents

Urine is composed of numerous chemical substances. Table 53-4 lists common chemical constituents measured by urinalysis laboratories.

Reagent-strip testing

Reagent-strip tests have enabled urinalysis laboratories to generate valuable qualitative chemical results in a rapid, accurate, and efficient manner. In general, properly performed urine test strips are sensitive, specific, and cost effective.

It is the responsibility of the urinalysis laboratory to select the most suitable type of reagent strip for its hospital or clinical setting. Table 53-5 lists various types of commercially available reagent strips. The following guidelines will ensure the best results:

1. Test urine promptly; use properly timed test readings only.
2. Beware of interfering substances.
3. Understand the advantages and limitations of the test.
4. Employ controls.

Strip tests should be performed on well-mixed urine equilibrated at room temperature. Each chemical parameter must be evaluated within a specific time interval, as suggested by the manufacturer's instructions. The correct number of strips needed for immediate analysis should be removed from the container and the lid tightly replaced. Reagent strips should be stored in a cool, moisture-free atmosphere, and special attention should be given to avoid using outdated strips.

After dipping the reagent strip into the urine, remove excess urine by gently tapping the strip on the edge of the specimen container. Compare individual reagent-pad reactions with the correct color chart in a properly lighted area. Automated reagent-strip instruments (reflectance photometers) are available to standardize results. Table 53-6 lists sensitivities of two commercially available multiple test strips. Reagent-strip results that are positive require confirmation with chemical and microscopic methods. Manufacturers' inserts should be reviewed to identify sources of inhibitors and false-positive and false-negative results. Quality control is essential to verify reagent-strip results.

Confirmatory testing

When addressing the topic of confirmatory tests, a definition or explanation of the term is essential. Many labo-

Table 53-5. Commercially available reagent strips used for urine testing

Sticks*	Strip†	pH	Protein	Glucose	Ketones	Bilirubin	Hemoglobin	Urobilinogen	Nitrite	Esterase	Specific gravity
Labstix	Chemstrip 5	×	×	×	×		×				
Bili-Labstix	Chemstrip 6	×	×	×	×	×	×				
Multistix	Chemstrip 7	×	×	×	×	×	×	×			
N-Multistix	Chemstrip 8	×	×	×	×	×	×	×	×		
—	Chemstrip 9	×	×	×	×	×	×	×	×	×	
N-Multistix SG‡	—	×	×	×	×	×	×	×	×		×

*Ames, Inc., Division of Miles Laboratories, Elkhart, Ind.
†BMC/Biodynamics, Indianapolis, Ind.
‡*SG*, "Specific gravity."

Table 53-6. Practical sensitivities of commercial reagent strips

Urine parameters	Chemstrip*	N-Multistix†
pH	±0.5 pH unit	±1.0 pH unit
Protein	60 mg/L	50-200 mg/L
Glucose	400 mg/L	800 mg/L
Ketones	50 mg/L acetoacetic acid	100-200 mg/L acetoacetic acid
	400-700 mg/L acetone	800-1400 mg/L acetone
Bilirubin	5 mg/L	2-4 mg/L
Blood	5 intact erythrocytes/μL or hemoglobin from 10 erythrocytes/μL	12 intact erythrocytes/μL or hemoglobin from 5 erythrocytes/μL
Urobilinogen	4 mg/L	1 mg/L
Nitrite	0.5 mg/L	0.75 mg/L
Esterase (neutrophils)	—	10 leukocytes/μL

*Ames, Inc., Division of Miles Laboratories, Elkhart, Ind.
†BMC/Biodynamics, Indianapolis, Ind.

ratory workers equate a confirmatory test with "rechecking" for a given parameter. This confirms nothing; it only establishes the precision for that parameter. A confirmatory test is one that will establish the accuracy or correctness of another procedure. Examples of confirmatory tests include quantitative protein analysis, protein electrophoresis, bacterial cultures and cytological appearance. A confirmatory test should have the same or better specificity, be based on a different principle, and have a sensitivity equal to or better than that of the original test.

Urinary pH

Urinary pH tests measure the concentration of free hydrogen ions. Although the standard of pH measurements uses glass electrodes, the urinary pH level is usually measured by indicator paper, since small changes in pH are of little clinical significance. Most urinalysis laboratories use multitest reagent strips that utilize two indicators. They are methyl red and bromthymol blue, providing a pH range from 5.0 to 9.0 that is demonstrated by a color change from orange (acid) to green to blue (alkaline). The normal pH range is 4.7 to 7.8. Extremely acidic or alkaline urine usually indicates a poorly collected specimen.

Table 53-7 lists common clinical causes of acidic and alkaline urine. The urinary pH is used to monitor the adequacy of treatment in several conditions. The average American ingests an acid-residue diet high in protein that results in acidic urine (5.0 to 6.5). In the case of alkaline urine (8.0 to 8.5) one should always suspect the presence of urease-producing ammonia bacteria, such as *Proteus* species. Patients with renal tubular acidosis, a clinical syndrome characterized by an inability to excrete acidic urine, may produce urine with a much higher pH, than would be expected on the basis of the acidosis.

The urinary pH is important in the management of renal

Table 53-7. Common clinical conditions causing acidic and alkaline urine

Acidic urine	Alkaline urine
Protein diet	Vegetable diet
Starvation	Vomiting
Dehydration	Renal tubular acidosis
Diarrhea	Respiratory and metabolic alkalosis
Diabetic acidosis	Ammonia-producing, urea-splitting bacteria
Metabolic and respiratory acidosis	Acetazolamide therapy
Metabolism of fats	Low-carbohydrate diets
Sleep	Chronic renal failure
Acid-producing bacteria	

stones or crystals. Uric acid stones precipitate in acidic urine and are more soluble in basic urines. Alkaline urine will precipitate calcium or calcium phosphate stones, whereas acidic urine will tend to dissolve them. Alkaline urine is desirable during sulfonamide and streptomycin therapy to prevent precipitation of the drugs in the kidneys and the formation of uric acid, cystine, and oxalate stones. The alkaline pH is also maintained during treatment of transfusion reactions and salicylate intoxication. The pH is kept acidic to combat bacteriuria and to prevent formation of "alkaline stones," such as calcium carbonate or calcium phosphate stones.[7] Technologists should be aware that alkaline urine interferes with the determination of proteins and may obscure the urine sediment examination.

Proteins

Albumin is not the only protein found in urine. Proteinuria may be caused by several types of proteins, such as Bence Jones proteins and globulins. In general, notably elevated protein levels occur when there is glomerular damage, and mildly elevated protein levels occur when there are renal tubular abnormalities. The protein determinations on reagent strips are not very accurate and cannot be used in a quantitative way. Reagent strips are pH sensitive and depend on the presence of protein for color generation (Sörensen's protein error). The presence of protein on the strip changes the pH environment of the dye embedded in the pad resulting in a change in color. Highly buffered, alkaline urine can result in a false-positive test. It is highly recommended that the laboratory simultaneously perform the reagent-strip test and a turbidity, acid precipitation test for the detection of proteins.

$$\text{Tetrabromphenol blue} \xrightarrow[\text{Protein}]{\text{pH 3}} \text{Positive results } \textit{(green-blue)}$$

$$\text{Tetrabromphenol blue} \xrightarrow[\text{No protein}]{\text{pH 3}} \text{Negative results } \textit{(yellow)}$$

False-positive results are not infrequent; therefore the reagent-strip test is best used as a screening procedure. A positive or faintly positive result should be confirmed with a more specific test such the sulfosalicylic acid (Kingsbury-Clark) test. The dye-binding properties of proteins make the reagent-strip procedures the most sensitive to the presence of albumin. A mildly positive test-strip result and a grossly positive turbidity test result may indicate the presence of drugs or Bence Jones proteins.

Sulfosalicylic acid test and semiquantitative turbidity method

Principle. Protein, denatured by acid, precipitates and renders the urine specimens progressively more turbid as protein concentration increases.

Materials

Sulfosalicylic acid, 3% (30 g/L) aqueous solution.

Test tubes, 13 × 100 mm.

Procedure

1. To 1 mL of centrifuged urine add 3 mL of reagent.
2. Mix and allow to stand for 5 minutes.
3. Observe the degree of turbidity and compare the test sample with the original urine specimen in a 13 × 100 mm test tube.

Reporting

Grade	Turbidity*	Protein range (mg/L)
None		≤75
Trace	Print can be easily seen and read when viewed through test tube	200
1+	Print can be easily seen and read with some difficulty when viewed through test tube	300-1000
2+	Print can be seen but not easily read when viewed through test tube	1000-2500
3+	Print cannot be seen when viewed through test tube	2500-4500
4+	Precipitate formed	4500

*Urine clarity when light is passed through specimen in a clear test tube.

In normal urine, turbidity should be absent and protein should represent less than 75 mg/L. Table 53-8 lists the sensitivities, types of proteins detected, and sources of false-negative and false-positive results for both the reagent-strip and the sulfosalicylic acid tests.

Sugars

Enzymatic Clinistix test. The reagent-strip test is an excellent test that is specific for glucose. It detects the oxidation of glucose to gluconic acid.

$$\text{Glucose} + \text{Oxygen in room air} \xrightarrow{\text{Glucose oxidase}} \text{Gluconic acid} + \text{Hydrogen peroxide}$$

$$\text{Hydrogen peroxide} + o\text{-Tolidine} \xrightarrow{\text{Peroxidase}} \text{Oxidized } o\text{-tolidine } (blue) + H_2O$$

Other reagent strip products use *o*-tolidine as the substrate for the indicator reaction.

Urine should be warmed to room temperature before the test is performed because refrigeration will interfere with this test. Sugar will appear in the urine in diabetes mellitus, renal glucosuria, and other clinical situations. The presence of ascorbic acid may mask the glucose and lead to erroneous results.

Copper reduction (Clinitest, Benedict's test)

$$\text{Cupric ions} + \text{Glucose (or reducing substances)} \xrightarrow[\text{Alkali}]{\text{Heat}} \text{Cuprous oxide } (red) + \text{Cuprous hydroxide } (yellow)$$

The Clinitest tablet (Ames, Inc.) provides another test for sugar, in addition to the reagent-strip test. It is a copper reduction test that measures other reducing sugars, and

Table 53-8. Tests for proteinuria

	Reagent strip	Sulfosalicylic acid	Results
Sensitivity	50-200 mg/L (albumin)	200 mg/L (all proteins	
Combined use	Negative result	Negative result	No protein
	Positive result	Negative result	Clinically insignificant protein level
	Negative result	Positive result	Bence Jones proteins
			Heavy-chain proteins (electrophoresis confirmation required)
Common urine constituents causing false-positive or false-negative results:			
Urine turbidity	—	False-positive and false-negative result	
X-ray contrast media	—	False-positive result	
Tolbutamide (Orinase)	—	False-positive result	
Penicillin (massive dose)	—	False-positive result	
Sulfisoxazole (Gantrisin)	—	False-positive result	
Highly buffered alkaline urine	False-positive result	False-negative result	
Quaternary ammonium salts	False-positive result	—	
Tolmetin sodium (Tolectin)	—	False positive result	

glucose as well. The Clinitest will identify sugars such as lactose in pregnant women and pentoses in pentosuria, as well as ascorbic acid and certain drugs (that is, nalidixic acid [NegGram], used to treat urinary tract infections; probenecid, used to treat gout; and cephalosporin, an antibiotic). Therefore subtraction of the reagent-strip test result from the Clinitest result may yield some interesting findings. Clinitest is also an important test in pediatric screening. A negative test-strip result (specific for glucose) but a positive Clinitest result (sensitive for any reducing sugar) may indicate the presence of an inherited metabolic disorder in the newborn.

The presence of sugar in the urine increases the specific gravity significantly. Each 1% (10 g/L) of sugar increases the specific gravity by 0.004 units. Because 4% (40 g/L) sugar will increase the specific gravity by 0.016, a patient with poor renal function may excrete urine with a specific gravity of 1.020 or 1.025 if it contains sufficient sugar.

Ketones

Ketone bodies appear in the serum and then in the urine of persons in starvation, with diabetes, or with increased fat utilization, because of incomplete fatty acid combustion. Reagent-strip testing uses a sodium nitroprusside reaction that measures acetone and acetoacetic acid, but not β-hydroxybutyric acid, a key ketone body. It is important to realize that the sodium nitroprusside reagent reacts primarily with acetoacetic acids; acetone has only a 20% reactivity compared to acetoacetate.

$$\text{Acetoacetic acid} + \text{Sodium nitroprusside} + \text{Glycine} \xrightarrow{\text{Alkaline pH}} \textit{Purple}$$

Ketone determinations are important in diabetes and ketoacidosis. Ketones should be determined whenever sugar determinations are made. If the ketone levels are extremely high, the ferric chloride test for acetoacetic acid will be positive. Urine (5 mL) is placed in a test tube, and 10% ferric chloride is added drop by drop, with thorough agitation until a precipitate forms and redissolves. Urine containing large quantities of acetoacetic acid will turn wine red. When present in the urine, coal-tar derivatives such as salicylates give false-positive results. Therefore urine from a patient taking aspirin would yield a false-positive result. These two situations can be differentiated by retesting with another aliquot of urine; however, this time the urine is boiled for 15 minutes and the evaporated volume replaced with water before the ferric chloride is added. Boiling will convert acetoacetic acid to acetone, which is boiled off. A positive result will persist for salicylates after boiling, whereas the positive result for ketone bodies will be negated. This test would establish a presumptive diagnosis for aspirin poisoning. Also false-positive results occur with L-dopa administration, phenylketonuria, sulfobromophthalein (Bromsulphalein) excretion tests for liver function, and phenolphthalein injection in renal function tests.

Bilirubin

Dark, yellow to brown, foamy urine probably contains conjugated bilirubin, a pigmentary constituent of bile. Normal urine does not contain bilirubin. Jaundiced patients with hepatocellular disease, such as hepatitis, or obstructive disease, such as biliary cirrhosis, may have conjugated bilirubin in the urine. The reagent-strip method for determining bilirubin involves a diazotization reaction.

$$\text{Bilirubin glucuronide} + \text{Diazonium salt} \xrightarrow{\text{Acid}} \text{Azobilirubin } \textit{(brown)}$$

Amber-colored urine or urine with yellow foam should be tested for the presence of bilirubin. Because of the difficulty in assessing the color reaction on reagent strips, one should confirm the positive results by using Ictotest tablets. False-negative results may occur if the urine is not fresh because urinary bilirubin may hydrolyze or oxidize when exposed to light.

Blood and myoglobin

As previously discussed, red urine usually means the presence of erythrocytes, hemoglobin, or myoglobin in the urine. Hematuria most often represents a combination of intact erythrocytes, degenerated erythrocytes, and free hemoglobin. Normal urine contains only a few erythrocytes (less than 5 erythrocytes per high-power field). Gross hematuria implies hemorrhage or fresh bleeding, and its presence in acidic urine is red to brown and turbid or "smoky." The reagent-strip method for hemoglobin and myoglobin utilizes the peroxidase activity of these proteins.

$$\text{Hydrogen peroxide } (H_2O_2) + o\text{-Tolidine} \xrightarrow{\text{Myoglobin or hemoglobin}} \text{Oxidized } o\text{-tolidine } \textit{(blue)} + H_2O$$

This extremely sensitive test indicates the presence of hematuria, hemoglobinuria, or myoglobinuria. A positive reagent-strip result indicates the need to confirm microscopically the presence of erythrocytes. Oxidizing agents such as iodides and bromides in the urine may cause false-positive results. As with glucose, large quantities of ascorbic acid (used in some antibiotics) in the urine may produce false-negative results.

Myoglobin is a ferrous porphyrin similar to hemoglobin. It is a product of muscle degeneration and is commonly seen after crush injuries and muscle trauma. When myoglobin is released in the circulatory system, it is not bound to proteins, and it is rapidly excreted by the kidney. Like that of hemoglobin its presence will produce pink to red urine, but it will not discolor plasma. Myoglobinuria should always be confirmed with rapid immunodiffusion or radioimmunoassay procedures.

Nitrite

The nitrite test is used in urinalysis laboratories to detect asymptomatic bacteriuria. The reagent-strip nitrite test depends on the reduction of nitrate to nitrite by the enzymatic action of certain bacteria in the urine. Under conditions of acid pH, nitrite reacts with *p*-arsanilic acid to form a diazonium compound, which in turn couples with *N*-1-naphthyl ethylenediamine to form a pink color.[8] The nitrite test should be performed on a first morning specimen or on a urine sample that has been collected at least 4 hours or more after the last voiding to allow the organisms in the bladder time to metabolize the nitrate. Stale urine may have bacterial contamination instead of a true chemical infection. The nitrite test is specific for gram-negative organisms; however, some false-negative results will occur if organims such as enterococci, streptococci, or staphylococci, which do not form nitrite, are present. The sensitivity of the nitrite test is about 60% when compared to microbiological procedures.[8] There are very few cases of false-positive nitrite results.

Nitrite testing is of dubious value in hospital clinical urinalysis because there is limited effective control over how and when the urine sample is collected. The nitrite test may have use in a clinic or physician's office because proper control over sampling can be better achieved, provided that the physician recognizes the possibility of false-positive and false-negative results and the importance of follow-up with urine cultures when appropriate.

Urobilinogen

Urobilinogen, a colorless compound, is formed in the intestine by the bacterial reduction of bilirubin. Some is excreted in the feces (urobilin), and some is reabsorbed in the intestine into the circulation and excreted by the liver in the formation of bile. Normal urine contains small amounts of urobilinogen. Decreased urobilinogen is found in infants, who lack reducing intestinal bacteria; in patients after administration of antibiotics that suppress intestinal flora; and in patients with obstructive liver disease. Complete obstruction of the common bile duct prevents bile or bilirubin from being reduced in the intestines. Increased urobilinogen is present in hemolytic anemia (increased bilirubin formation) and liver dysfunction.

Urinalysis laboratories should perform both the bilirubin and urobilinogen tests to detect liver disorders. These tests are important in the clinical differential diagnosis of biliary obstruction from hepatocellular disorders in jaundiced patients. The reagent-strip test for urobilinogen is a simple color reaction using *p*-dimethylaminobenzaldehyde.

$$p\text{-Dimethylaminobenzaldehyde} \xrightarrow[\text{Urobilinogen}]{\text{Acid pH}} \textit{Red}$$

Bilirubin may interfere with the grading of the color reaction.

Leukocyte esterasuria

The evaluation of urine for the presence of leukocytes (pyuria) is clinically important. Since accurate quantification of pyuria is often unreliable and time consuming, a reagent-strip test for pyuria has been developed. The Chemstrip Leukocyte Test (BMC/Biodynamics, Indianapolis, Ind.) detects both lysed and intact leukocytes. The principle of the 15-minute test is that neutrophilic leukocytes contain enzymes, called *esterases,* that catalyze the hydrolysis of a carboxylic acid ester of indoxyl to indigo. The indoxyl is oxidized to form a blue color. The 1-minute test strip has the indoxyl reacting with a diazonium salt to produce a purple color. The intensity of the color is proportional to the number of leukocytes present in the specimen. Kusumi, Grover, and Kunin[9] found the chamber count and esterase test more sensitive than conventional bright-field microscopy in detecting pyuria. Gillenwater[10] evaluated 300 patients with the strip test and found its sensitivity to be approximately 95% and its specificity to be 98%. In two separate trials, Avent, Schumann, and Vars[11] compared neutrophil counts in a standardized, Papanicolaou-stained urine sediment examination with the reagent-strip test. Their data showed a sensitivity of 81% and a specificity of 88%. Leukocyte esterasuria detected by reagent-strip testing is a valuable screening test for pyuria. Many laboratory workers believe a microscopic examination for leukocytes need be performed only on the urine specimens that are reagent-strip positive.

Porphyrins

Porphyrins are groups of intermediary products in the biosynthesis of heme and cytochromes, which are produced in the liver and bone marrow. They serve as ion chelates and are involved in numerous cytoplasmic biochemical reactions. Porphyrins are initially formed from glycine and succinyl coenzyme A, which combine to form δ-aminolevulinic acid and then porphobilinogen. Identification of various porphyrins and porphyrin precursors is important in the clinical diagnosis of porphyrias, a group of genetic or metabolic disorders. Normal urinary excretion of porphyrins is approximately 2 mg/day. Increased quantities of excreted porphyrins (porphyrinuria) give urine a red or wine color. Screening urinalysis tests for porphyrins are based on the red fluorescence produced after ultraviolet light is shined on the sample (Watson-Schwartz test).[12] Confirmatory and quantitative tests are available.[13]

Melanin

Normal urine does not contain melanin. Melanin is a dark pigment that may be excreted in the urine of patients with malignant melanoma. Patients with this malignancy will excrete a colorless precursor of melanin. When exposed to air, the precursor 5,6-dihydroxyindole melanogen will polymerize to form the dark pigment, melanin. In the past, urinalysis laboratories were involved in screening patients for melanoma. Screening tests, using ferric chloride, oxidize melanogen to melanin, which turns urine brown-black. Cytological procedures should be considered in identifying metastatic malignant cells of melanoma.

Fats

Fats are found in the urine of patients who have fat emboli after bone-crush injuries, fatty degeneration of the kidney, or nephrotic syndrome. Fat will appear on top of urine and is the last part of voided urine. Vacuolated epithelial cells may be found in the urinary sediment. Fat stains should be used for accurate identification of fat droplets in urine.

Inborn errors of metabolism

Urinalysis laboratories may be responsible for the qualitative or quantitative tests employed in the clinical diagnosis of inherited metabolic disorders. Bradley, Schumann, and Ward[5] discuss in detail common screening and confirmatory tests used in evaluating inborn errors of metabolism.

MICROSCOPIC EXAMINATION OF URINE

Methods

Accurate microscopic identification of urine sediment is important in the early recognition of infectious, inflammatory, and neoplastic conditions affecting the urinary system.[1] It is debatable whether all routine urine specimens require the more time-consuming urine microscopy. Instead, most laboratory workers agree that a microscopic examination should be performed when the patient is symptomatic, when the physician specifically requests this examination, and when the macroscopic urinalysis is abnormal, that is, hematuria, proteinuria, or a positive nitrite result.[14]

Several microscopic procedures are available for the sediment examination. Standardized bright-field microscopy is still the most common technique employed by urinalysis laboratories.[15] Supravital staining should be combined with bright-field microscopy to enhance cellular detail. Phase-contrast microscopy is probably the best method for rapid urine sediment evaluations without the use of stains. Commercially available standardized slide methods are far superior to conventional bright-field microscopy and often a practical alternative to hemocytometer chamber counts.[16,17] As an alternative to the once commonly used Addis quantitative cell counts, a combined cytocentrifugation-Papanicolaou method has been recommended.[18,19] This is the procedure of choice for patients with known urinary tract disease.

Bright-field microscopy of unstained urine. Unstained bright-field microscopy utilizes reduced light to delineate the more translucent formed elements of the urine, such as hyaline casts, crystals, and mucus threads.

Accurate identification of leukocytes, macrophages, renal tubular epithelial cells, and viral inclusion-bearing cells may be very difficult in unstained preparations. Cytological techniques, stained preparations, or both, should be used to confirm results, especially when cellular casts need to be distinguished and malignant cells are suspected.[19,20]

Procedure. The urine specimen must be examined while fresh, since cells and casts begin to disintegrate within 1 to 3 hours. Refrigeration (2° to 8° C) usually prevents the lysis of cells and pathological entities. Each specimen is concentrated tenfold or twentyfold for the purpose of standardization. The examination proceeds as follows:

1. Mix the specimen well.
2. Pour exactly 10 mL of urine into a graduated centrifuge tube.
3. Centrifuge at 1500 rpm or approximately 80 *g* for 5 minutes.
4. Remove the supernatant fluid by careful decantation or aspiration to the 1 mL mark; 0.5 or 0.25 mL may be used to increase the sediment concentration. Resuspend the sediment by gently tapping the bottom of the tube.
5. Place a drop of resuspended sediment on one area of a slide and cover with a 22 mm square cover slip, avoiding bubbles. (Too much fluid will cause the cover slip to float.)
6. Examine with both low- and high-power fields. Examine with low power (100×) and subdued light. Vary the fine focus continuously while randomly scanning the area under the cover slip.

Casts are often found along the edge of the cover slip. Count the number of casts per low-power field in 10 fields. Switch to high power (440×) and identify casts present.

Erythrocytes, leukocytes, and renal epithelial cells are identified with the high-power objective and counted in 10 fields. Squamous and transitional (urothelial) epithelial cells are noted if large numbers are present. Bacteria and yeast also should be noted. Crystals (such as cysteine or tyrosine crystals) are reported if their number is unusually large or they are abnormal (as being associated with certain disease states); these are counted under low power. The identity of certain abnormal crystals should be confirmed chemically. The urinalysis report should be numerical when possible. Averaging the contents of 10 representative fields is commonly used by routine laboratories.

Bright-field microscopy with supravital staining. Cellular detail is enhanced with stained sediments. A crystal violet–safranin O stain is often used in the rapid assessment of certain cellular elements.[21]

REAGENT (Sternheimer-Malbin stain)

SOLUTION 1	Crystal violet	3.0 g
	Ethyl alcohol (95%)	20.0 mL
	Ammonium oxalate	0.8 g
	Distilled water	80.0 mL

SOLUTION 2	Safranin O	1 g
	Ethyl alcohol (95%)	40 mL
	Distilled water	400 mL

Three parts of solution 1 and 97 parts of solution 2 are mixed and filtered. The mixture should be clarified by filtration every 2 weeks; discard the mixture after 3 months. Separately, solutions 1 and 2 keep indefinitely at room temperature. In highly alkaline urine, the stain will precipitate.

Procedure

1. Add one or two drops of crystal violet–safranin O stain to approximately 1 mL of precentrifuged, concentrated urine sediment.
2. Mix with a pipet and place a drop of this suspension on a slide under a cover slip.

Methylene blue and toluidine blue are also useful supravital stains. Improved supravital stains that facilitate identification of cells, casts, and their inclusions have recently been recommended.[21,22]

Phase and interference microscopy. Many urinalysis laboratories recommend phase microscopy for the detection of more translucent formed elements of the urinary sediment. Such sediment, notably hyaline casts, mucus, and bacteria, may escape detection using conventional unstained bright-field microscopy. Phase microscopy has the advantage of hardening the outlines of even the most ephemeral formed elements, making detection simple.[23] Even greater morphological detail of formed elements (notably casts and cells) is afforded by interference contrast microscopy.[24,25]

Standardized slide methods. The KOVA system (ICL Scientific, Fountain Valley, Calif.) and the Whale T-System (Whale Scientific, Inc., Commerce City, Colo.) offer complete standardized procedures that are technically more precise, reproducible, and reliable than conventional bright-field microscopy. Both systems include a clear, accurately graduated centrifuge tube, a transfer pipet, and supravital stains. The KOVA system uses a 1 × 3 inch, patented, optically clear plastic microscope slide with four individual, covered examination chambers. The Whale T-System contains a patent-pending chamber with a cover slip that must be properly aligned (Fig. 53-2). Both systems offer a measure of technical standardization that is apparently superior to conventional methods and should gain increasing popularity in urinalysis laboratories.

In some laboratories, the hemocytometer continues to be used for quantifying urine sediment entities. Kesson, Gregory, and Gyory[26] provide evidence that chamber counts are more reliable in detecting sediment abnormalities than conventional methods of counting cells under high-power fields.

Combined cytocentrifugation and Papanicolaou stain method. A combined cytocentrifugation (Cytospin, Shandon Southern Instruments, Sewickley, Pa.) and Papanicolaou staining method has been recommended to assess the urine sediment in renal allograft recipients during acute rejection, acute tubular necrosis, and other renal parenchymal diseases.[27] This more specialized method provides a simple, reproducible, and semiquantitative method for quantifying urine sediment entities. Cellular casts (leuko-

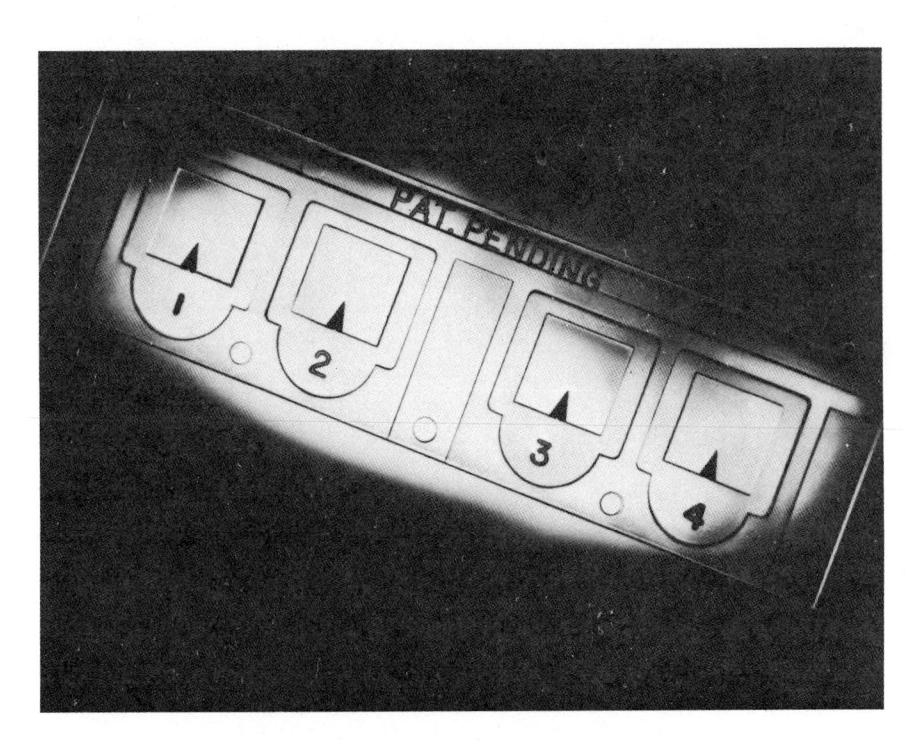

Fig. 53-2. Sediment chamber for standardized urine microscopy.

cyte versus renal tubular epithelial casts), mononuclear cells (such as plasma cells, lymphocytes, and macrophages), tissue fragments, and neoplastic cells may be clearly demonstrated with this method.

Procedure

1. Obtain urine (volumes ranging from 10 to 50 mL), preferably an early morning specimen.
2. On delivery to the laboratory the container is immediately refrigerated (2 to 8° C)
3. Provide an accompanying requisition noting the clinical history, such as transplantation, systemic disease or symptoms, radiation, and chemotherapy, with each specimen.
4. Spin a 10 mL sample of urine immediately in a standard tabletop centrifuge at 1500 rpm for 10 minutes.
5. Discard the supernatant solution using a pipet until 1 mL remains and resuspend the sediment.
6. Using a cytocentrifuge, prepare four slides using 0.25 mL of resuspended specimen per chamber and spin them at 900 rpm for 6 minutes.
7. After discarding the filter, apply one or two drops of Parlodion* to the cellular area of a horizontally held slide.
8. Fix the slide for 15 minutes in 95% EtOH and stain it using the Papanicolaou technique.*

Screen all four slides, noting the background pattern, cellularity, neutrophils, lymphocytes, renal tubular cells, cellular casts and fragments, erythrocytes, viral inclusions, and abnormal cells. Ten high-power fields are counted on the most cellular slide, for differentiation between neutrophils, lymphocytes, renal tubular cells, and casts. Various modifications of the Papanicolaou stain may be used, including a rapid 5-minute method.[1]

*Mixture of 200 mL of 95% ethanol, 200 mL of anhydrous ether, and 1 g of Parlodion (purified pyroxylin) (Mallinckrodt, Inc., St. Louis, Mo.).

*Aqueous alum-hematoxylin, OG-6, EA-36.

Crystals

Urinary crystals (Figs. 53-3 to 53-7) are commonly seen in urinalysis laboratories. In general, the formation of crystals should be regarded as an artifact of the system of collection.[8] Usually crystals are not present when urine is freshly voided. Various chemical constituents become saturated or undergo altered solubilities when urine is stored at cooler temperatures. Certain chemical substances such as albumin prevent crystallization. When heated to 37° C, most crystals disappear.[7] Those still present might have some diagnostic significance when correlated with clinical symptoms.[5]

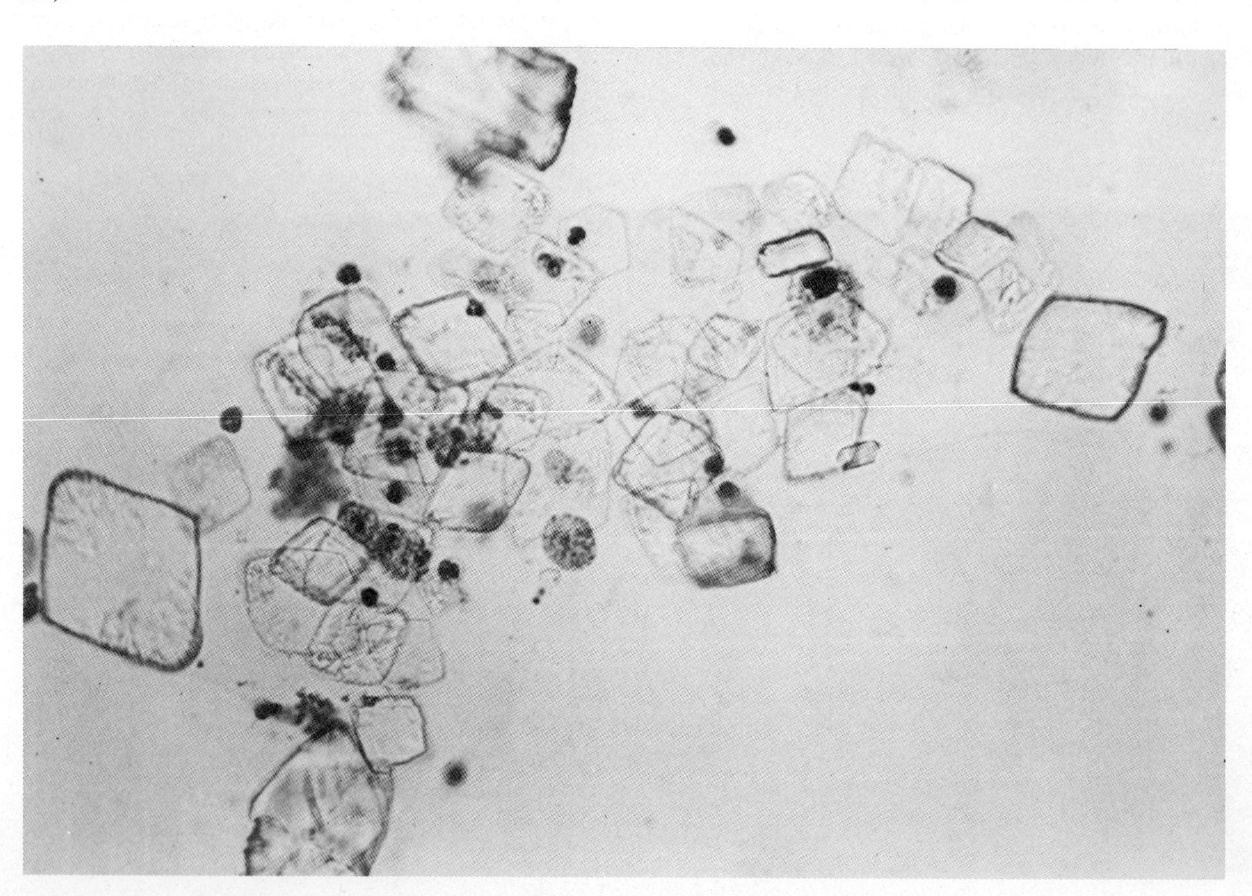

Fig. 53-3. Uric acid crystals. (Papanicolaou stain, 400×.)

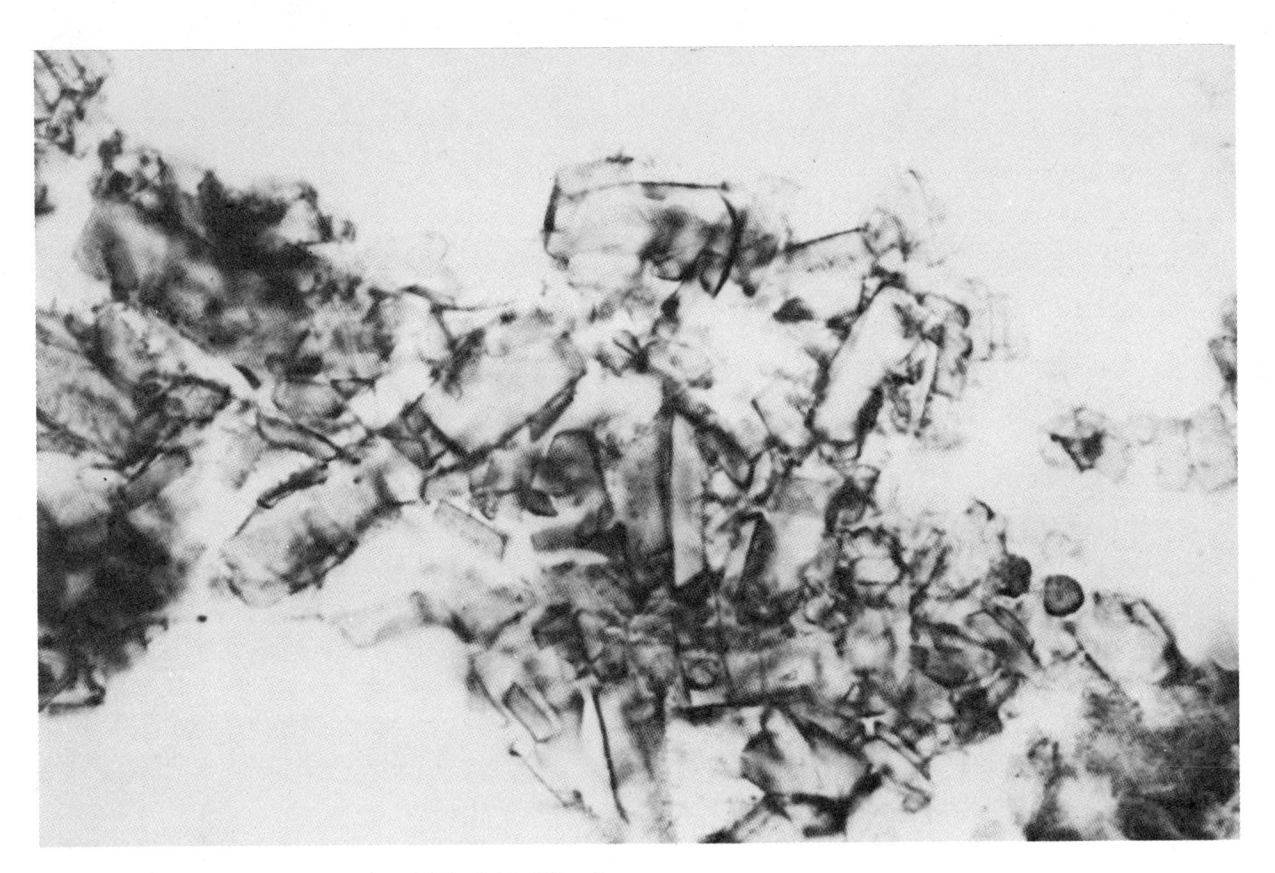

Fig. 53-4. Triple phosphate crystals. (Bright-field, 400×.)

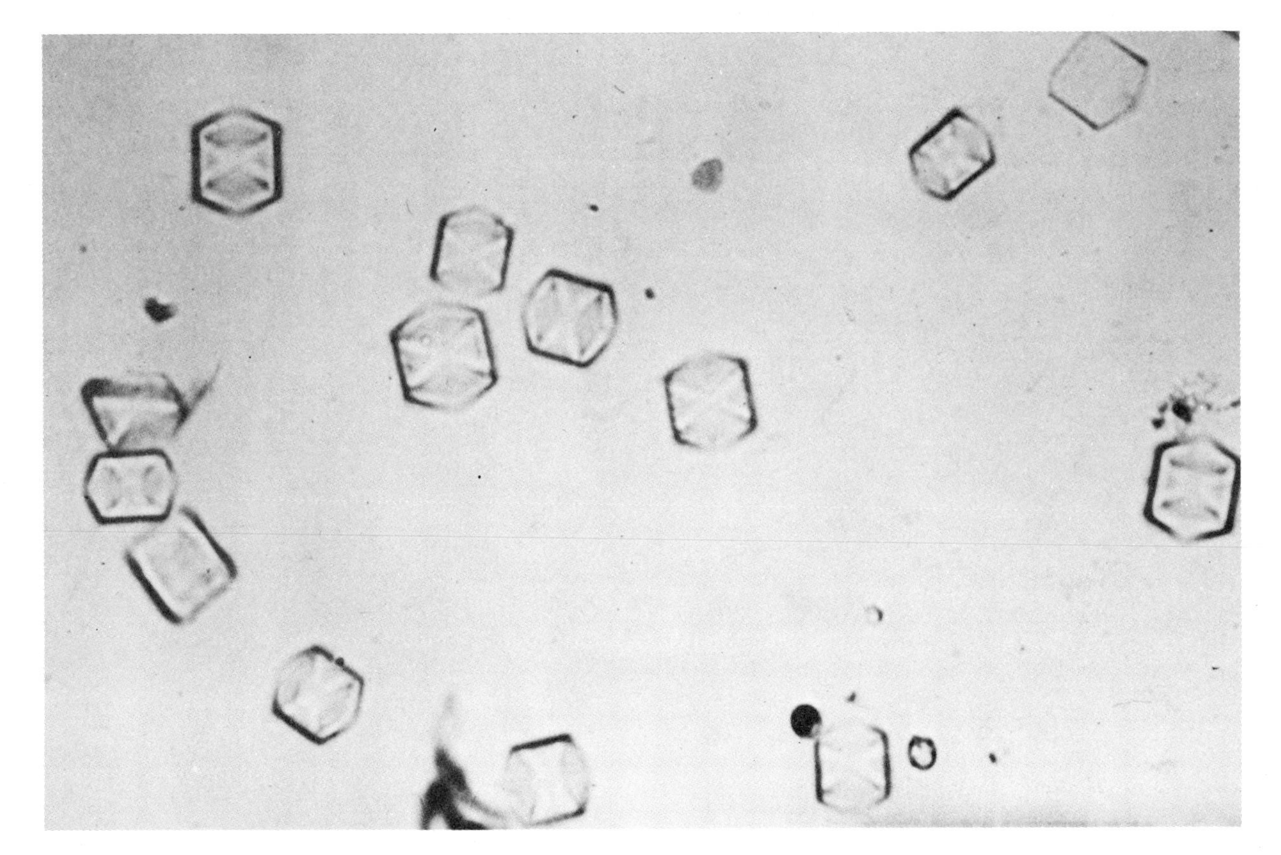

Fig. 53-5. Calcium oxalate crystals. (Bright-field, 1000×.)

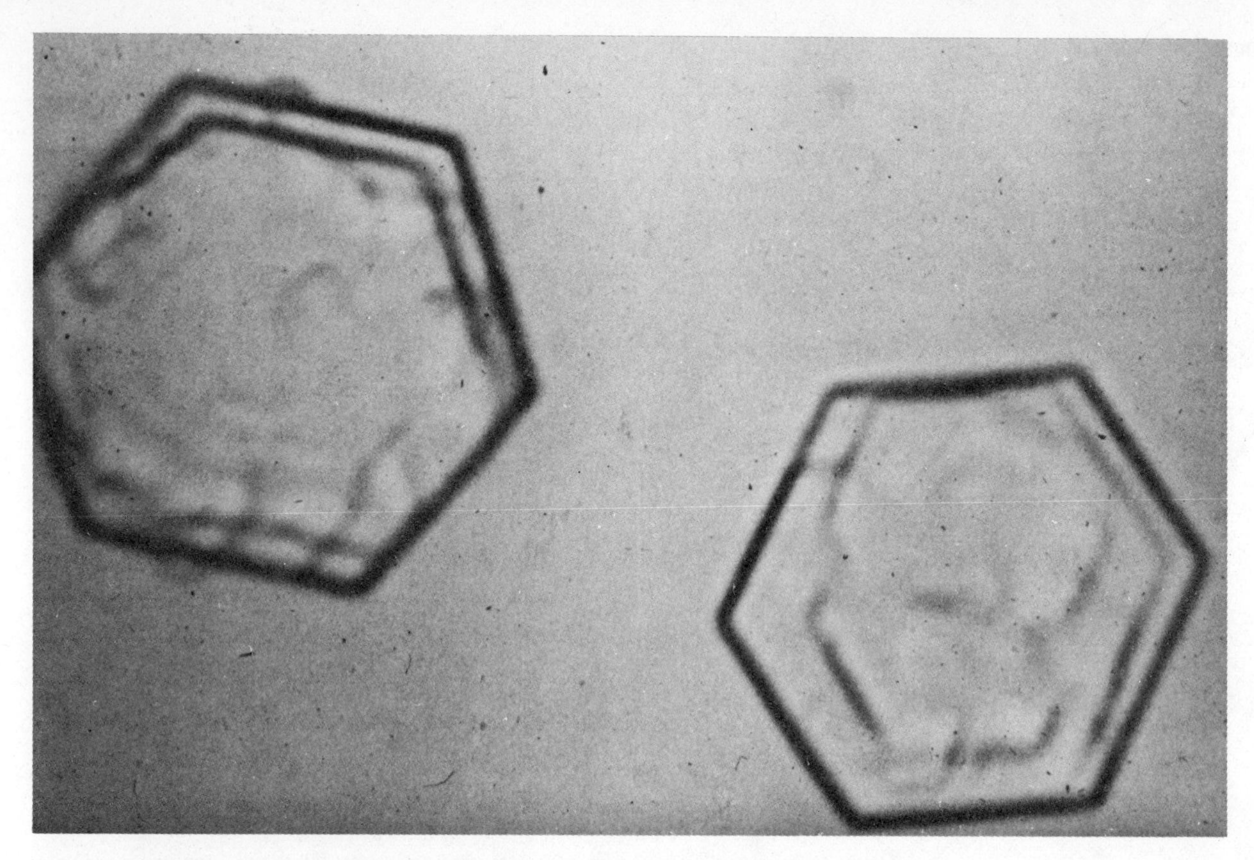

Fig. 53-6. Cystine crystals. (Bright-field, 2500 ×.)

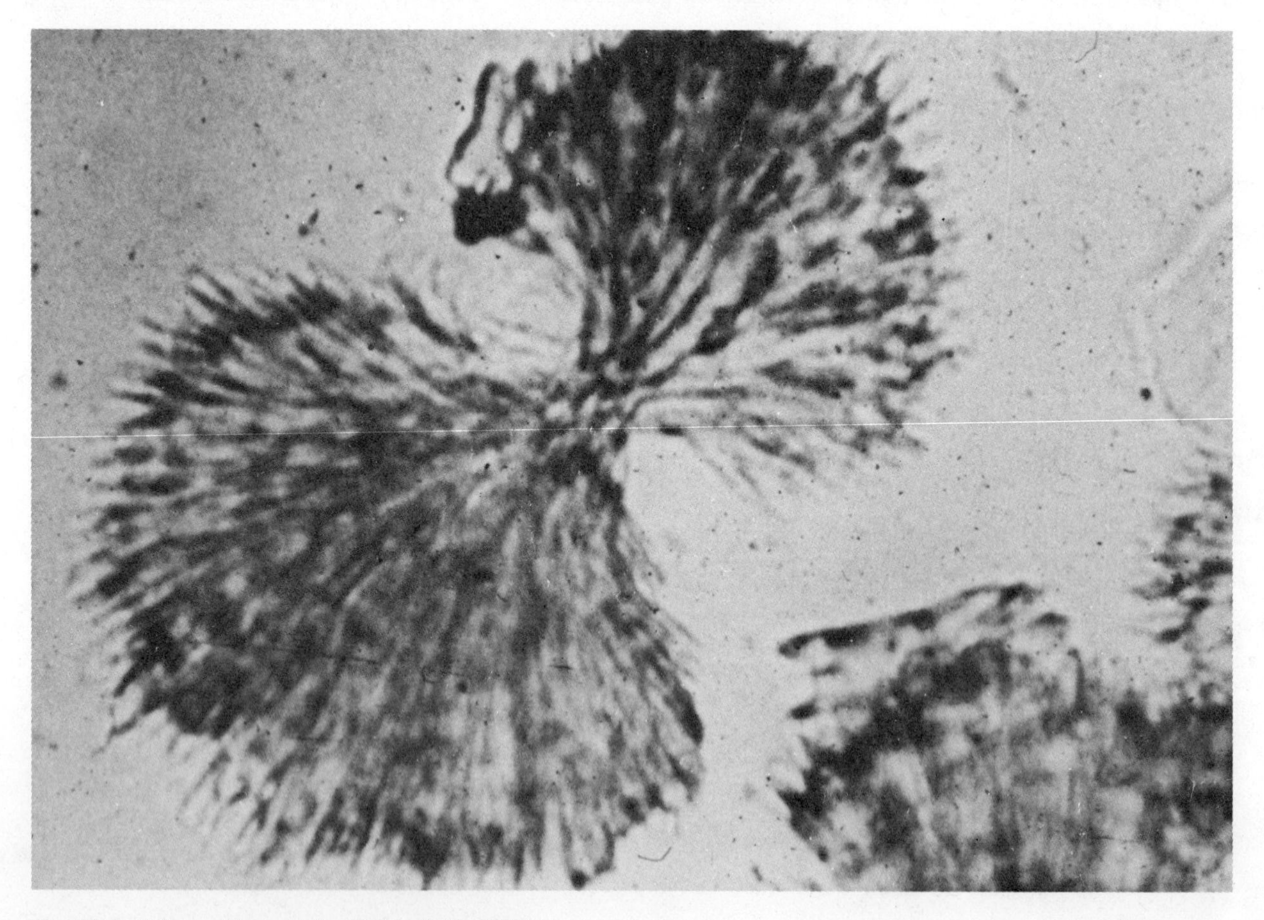

Fig. 53-7. Sulfonamide crystals. (Bright-field, 2500 ×.)

Table 53-9. Common urinary crystals

Types	Urine pH				Diagnostic morphological appearance	Solubility characteristics	Clinical significance
	Acid	Alkaline	Neutral	Variable			
Urates	×				Colorless, amorphous, spherical, or needle shaped	Soluble in alkali or at 60° C	Normal
Uric acid	×				Colorless to yellow brown, pleomorphic, rhombic, four-sided plates or rosettes	Soluble in alkali	Normal, chemotherapy, gout
Calcium carbonate				×	Colorless dumbbells or spheres	Soluble in acetic acid	Normal
Phosphates (triple phosphates)		×			Colorless, three- to six-sided prisms; "coffin lids"	Soluble in acetic acid	Normal
Ammonium urates		×			Brown, "thorn apple"	Soluble at 60° C with acetic acid	Normal
Calcium oxalate				×	Colorless, octahedron, dumbbell, or envelope or internal "X" forms	Soluble in dilute hydrochloric acid	Normal, glycol poisoning
Cystine				×	Colorless, hexagonal, flat	Soluble in alkali	Cystinosis
Cholesterol				×	Colorless, flat plates with corner notched	Soluble in chloroform and ether	Renal damage
Tyrosine				×	Yellow to colorless, fine needles, in sheaves or rosettes	Soluble in alkali	Liver damage, aminoaciduria
Leucine				×	Yellow spheroids with striations	Soluble in hot alcohol	Liver damage, aminoaciduria
Bilirubin				×	Reddish brown, amorphous, needles, rhombic plates, or cubes	Soluble in alkali	Bilirubinuria
Sulfonamides				×	Cubes, globules, or sheaves	Soluble in acetone	Antibiotic therapy

The types of urinary crystals formed depend on the pH of freshly voided urine. Table 53-9 lists common types, properties, and clinical significance of various urinary crystals. Cystine, uric acid, leucine, and tyrosine crystals are the most diagnostically important crystals to recognize. Because of the limited clinical significance of the urinary crystals, most laboratorians agree that time should not be wasted on their specific identification. Bradley, Schumann, and Ward[5] illustrate and describe the clinical significance of various types of crystals.

Organisms

In a properly collected and processed urine specimen, the presence of organisms is clinically significant. Microbiological organisms are important abnormalities found in urinalysis. Bacteria, fungi, parasites, and virally infected cells are frequently reported. Organisms seen in urine sediment are microscopically recognized as extracellular or intracellular structures.[19] Bright-field microscopy readily detects extracellular bacteria, fungi, and parasites. Detection of intracellular phagocytized bacteria and fungi, *Toxoplasma* organisms, and viral inclusion bodies usually require cytological procedures.

Accurate identification of organisms aids in the clinical differential diagnosis of urinary system infections. Stained preparations are important in the evaluation of organisms, identification of associated inflammatory cells, and assessment of epithelial exfoliation and renal cast formation for purposes of localization.[1] The lack of an inflammatory response may be misleading in immunosuppressed patients. Microbiological techniques should be used to confirm and fully classify various urinary organisms.

Bacteria (Fig. 53-8). Normal urine is sterile and does not contain bacteria. Some bacteria may be present because of contamination during collection or prolonged storage. If bacteria are seen in centrifuged specimens but not in unspun urine specimens, less than 10^5 bacteria/mL are present. Bacteria in an unspun specimen indicate greater than 10^5 bacteria are present. The presence of 10^5 bacteria or greater is suggestive of urinary tract infection. This number corresponds to 10 or more bacteria per high-power field.

Identification of bacteria, cocci, or rods can readily be accomplished by bright-field or phase-contrast microscopy. Occasionally, there is difficulty in distinguishing bacteria from amorphous crystals. Significant amounts of bacteria

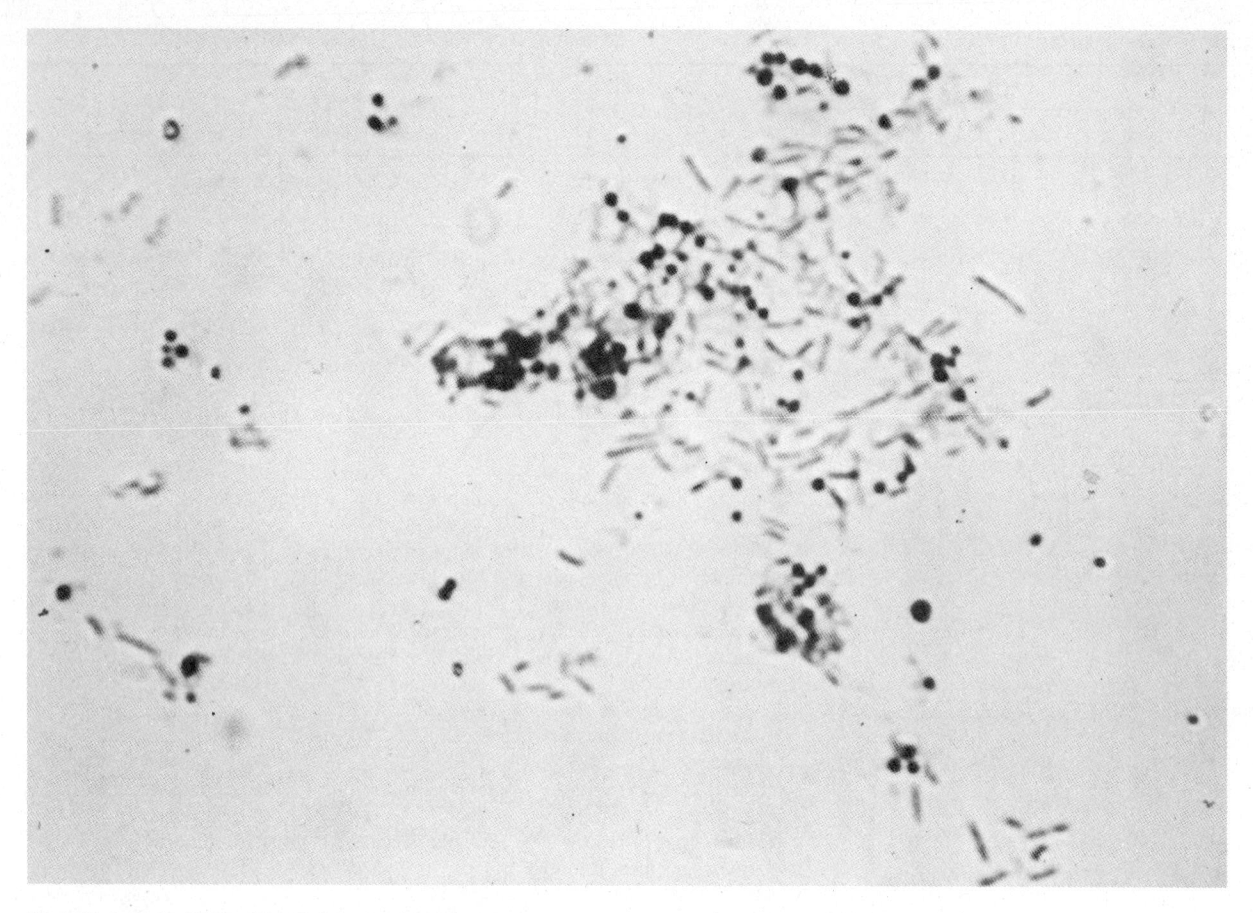

Fig. 53-8. Bacteria (rods). (Bright-field, 1000×.)

usually are accompanied by pyuria. For the purpose of accurate classification, microbiological techniques such as Gram's stain, acid-fast stain, and plating procedures should be used to identify and quantify the organisms.

Fungus. Urinary tract fungal infections are common in patients with diabetes, those taking birth control pills, or those undergoing intensive antibiotic or immunosuppressive therapy. *Cryptococcus* and *Histoplasma* organisms are rarely identified in urine sediment.

Candida albicans is the most common fungus and is identified as budding yeast or mycelia (Fig. 53-9). In general the budding yeast appearance indicates that the fungi are coexisting with the host, whereas the mycelial forms appear during tissue invasion. Yeasts of *Candida albicans* are oval and highly refractile and measure 3 to 5 μm. Often they can be misinterpreted as erythrocytes (7 μm). Unlike erythrocytes, yeasts are not lysed by acids.

Parasites. The presence of parasites in urine usually indicates vaginal or fecal contamination. *Trichomonas vaginalis,* a flagellate, is the most common parasite seen in urine. The incidence of this type of parasitism is very high in women and may be the cause of intense vaginitis. In men the parasite causes an asymptomatic urethritis. Because of the motion of this oval organism, bright-field microscopy is used for simple and rapid identification. Nonmotile trichomonads can easily be mistaken for leukocytes or epithelial cells.

Pinworm ova *(Enterobius vermicularis)* have been found in urine in children because of fecal contamination. Morphologically a pinworm ovum is surrounded by a thick two-layered transparent capsule, and a coiled embryo may be visible inside.

Trematode ova, *Schistosoma haematobium* (found in North Africa) and *Schistosoma mansoni* (found in Central America), may be found in urine. In countries with endemic schistosomiasis, there is an associated increased incidence of squamous cell carcinoma of the bladder.

Virally infected cells. Virus-induced cellular changes have been recognized with increased frequency in urine sediment, especially in immunosuppressed patients. Cytological techniques should be used for the accurate identification of cytomegalovirus (CMV) (Fig. 53-10), herpes simplex, and *Polyomavirus,* which produce diagnostic intranuclear inclusion cells and are the most common types of urinary system viral infections. Viral inclusion cells must be distinguished from nonviral sources of inclusion

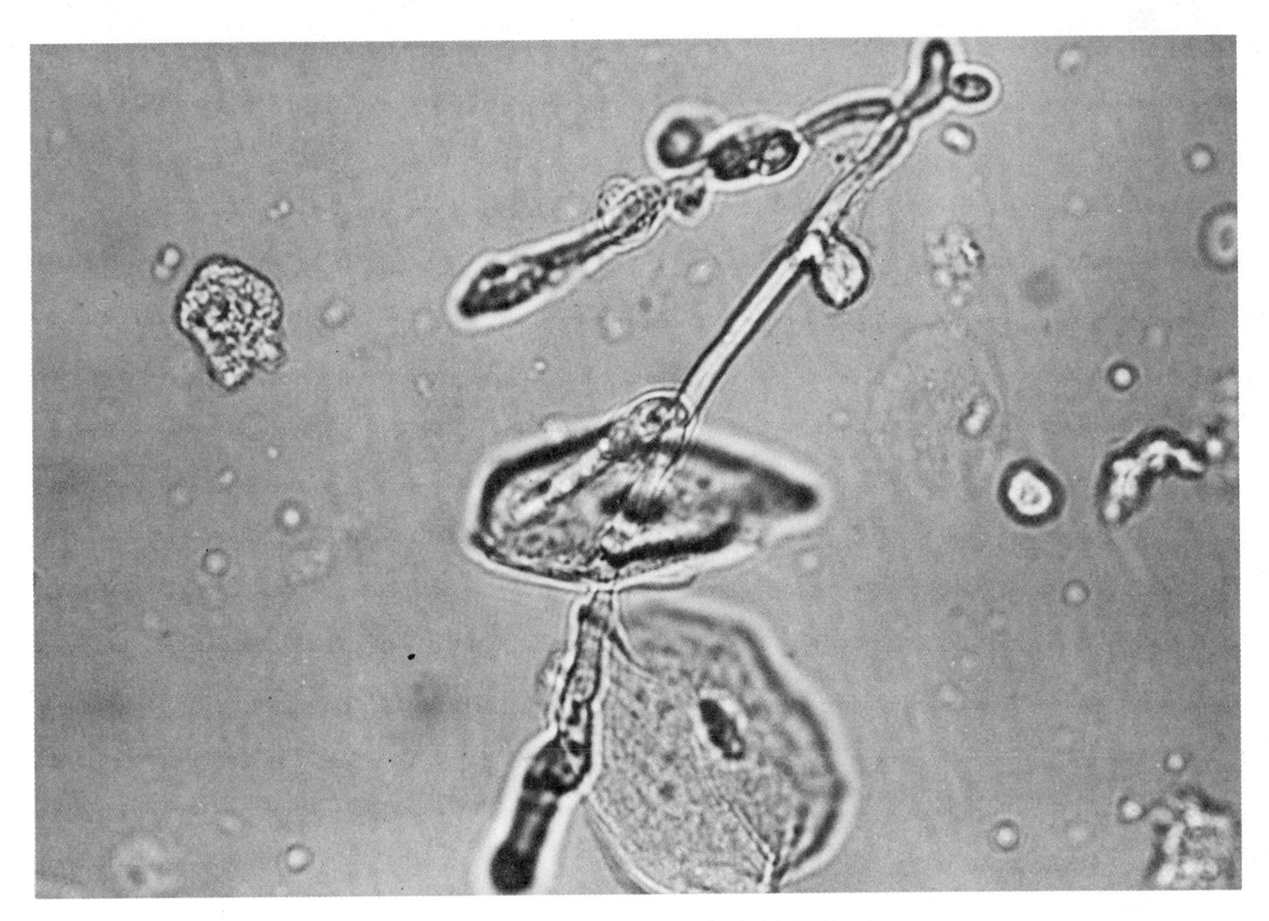

Fig. 53-9. Fungal yeast and mycelia, probable *Candida* species. (Bright-field, 2500×.)

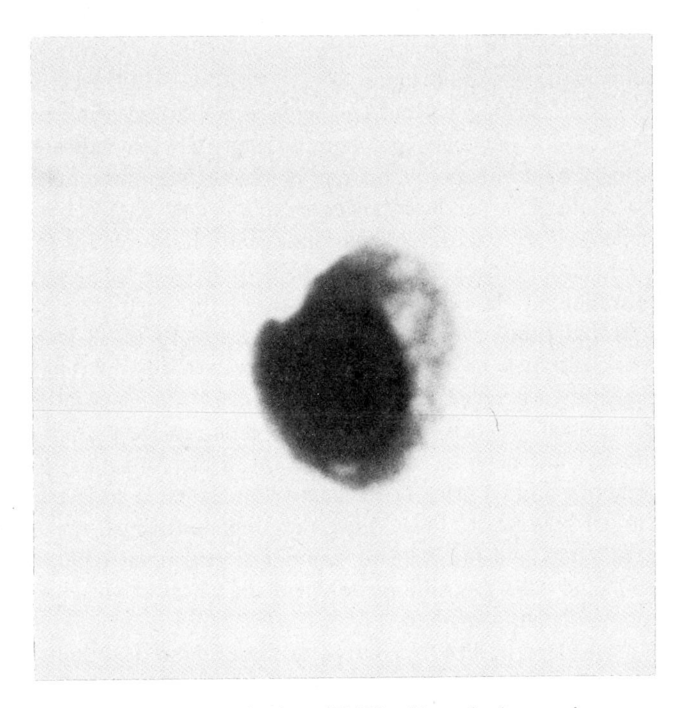

Fig. 53-10. Cytomegalovirus (CMV). (Papanicolaou stain, 1000×.)

cells such as heavy metal exposure (lead and cadmium) and nonspecific degenerative cellular changes. A detailed description of viral inclusion cells can be found in standard textbooks on urine cytology.

Cells

Microscopic identification and evaluation of cells is an important part of urinalysis. Common types of cells normally found in urine include a few erythrocytes, leukocytes, and epithelial cells of renal or lower urinary tract origin. All cells are usually quantified by high-power field examination. Standardization techniques allow assessment of increased or decreased numbers of cells. Special techniques such as cytocentrifugation followed by Papanicolaou stain are important for recognizing the various types of renal tubular epithelial cells, plasma cells, lymphocytes, and eosinophils. Atypical cells suggestive of malignancy require cytological evaluation.

Erythrocytes (Fig. 53-11). Normal urine should never contain more than a few erythrocytes per high-power field. They appear in the urine stream after vascular injury or disorders in the kidney or lower urinary tract. Erythrocytes accompanied by blood casts or dysmorphic erythrocytes

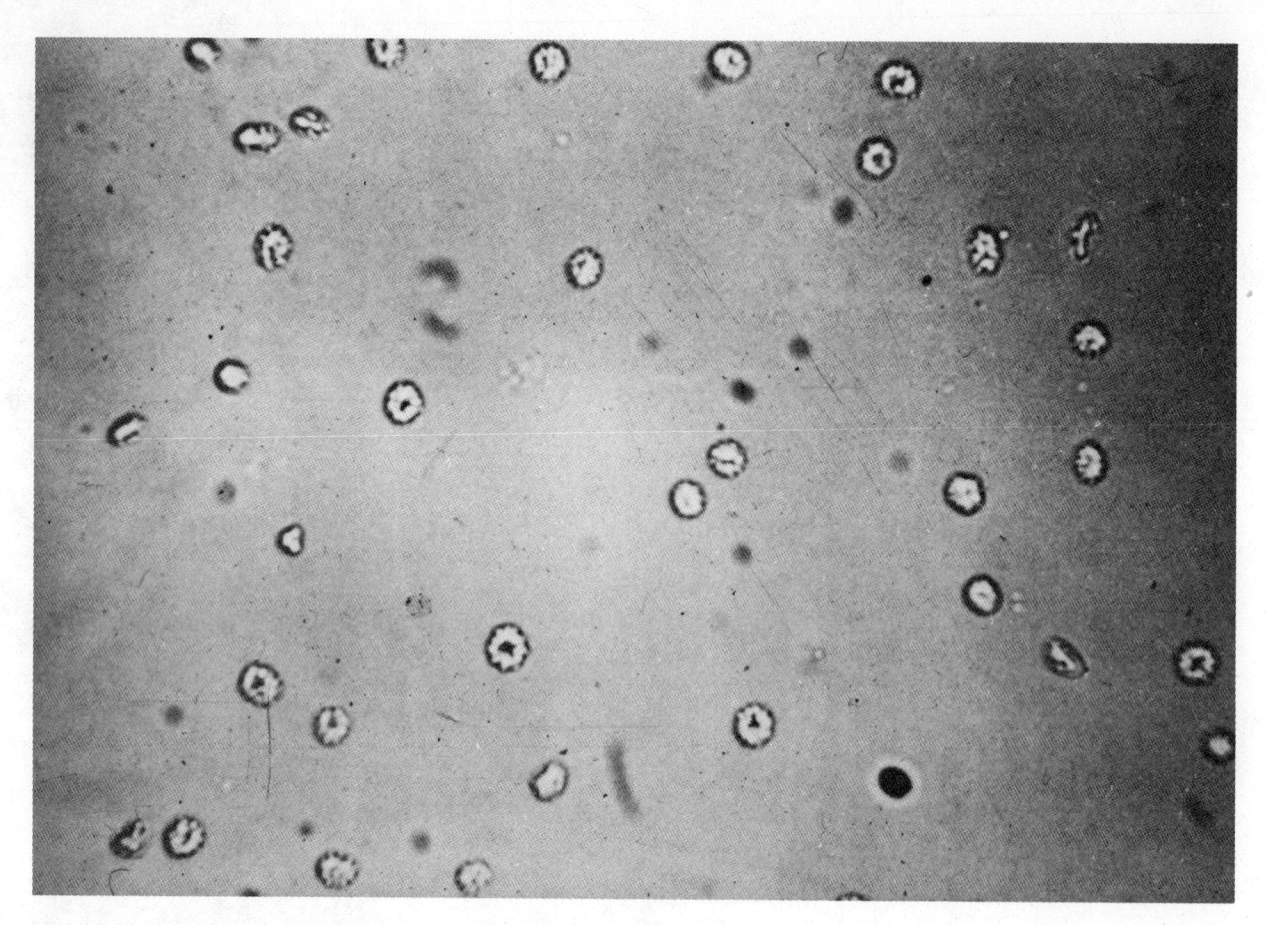

Fig. 53-11. Crenated erythrocytes. (Bright-field, 400×.)

suggest renal parenchymal bleeding. Quantification aids in the diagnosis and patient management. It is important that contamination from menstrual flow be avoided when urine from females is to be examined.

Erythrocytes measure approximately 7 μm in diameter, have a biconcave disk shape, and often appear pale yellow when bright-field microscopy is used. In hypertonic urine they are smaller and crenated, whereas in hypotonic urine they are larger and swollen. When erythrocytes have been in urine for a considerable time, the hemoglobin may have dissolved.

On occasion hemoglobin is detected in urine in the absence of microscopic erythrocytes in the sediment. It is strongly suggested that the pigment that appears in the urine (which may be hemoglobin or myoglobin) originates from filtration of these products from the blood.

Leukocytes (Fig. 53-12). The normal excretion rate for leukocytes in the urine is up to one leukocyte per three high-power fields, 3000 cells/mL, or up to about 200,000 cells per hour. Elevated numbers of leukocytes (pyuria) are associated with numerous urinary tract inflammatory and infectious conditions. Most leukocytes recognized by bright-field microscopy are segmented neutrophils. The identification of lymphocytes, plasma cells, and eosinophils requires special stains.

Little[32] has shown that leukocyte excretion rates in excess of 400,000 cells per hour virtually always indicate urinary tract infection. This rate corresponds to more than 10 neutrophils per high-power field. Patients with active upper urinary tract infections infrequently have more than 50 neutrophils per high-power field or have a leukocyte excretion rate in excess of 2 or even 3 million per hour.

Renal tubular epithelial cells (Figs. 53-13 to 53-15). Various types of renal tubular epithelial cells line the nephron and senescent or diseased cells are constantly being shed. As new renal epithelial cells replace the old, those of the nephron are exfoliated with the formation of urine. In normal urine fewer than two renal cells are present per high-power field. Since they represent actual renal exfoliation, the presence of more than two renal tubular epithelial cells per high-power field indicates active renal tubular damage or injury.

Considerable difficulty exists in the accurate identification of renal tubular cells, particularly in distinguishing them from other mononuclear cells commonly found in urine. By bright-field microscopy they are polygonal and

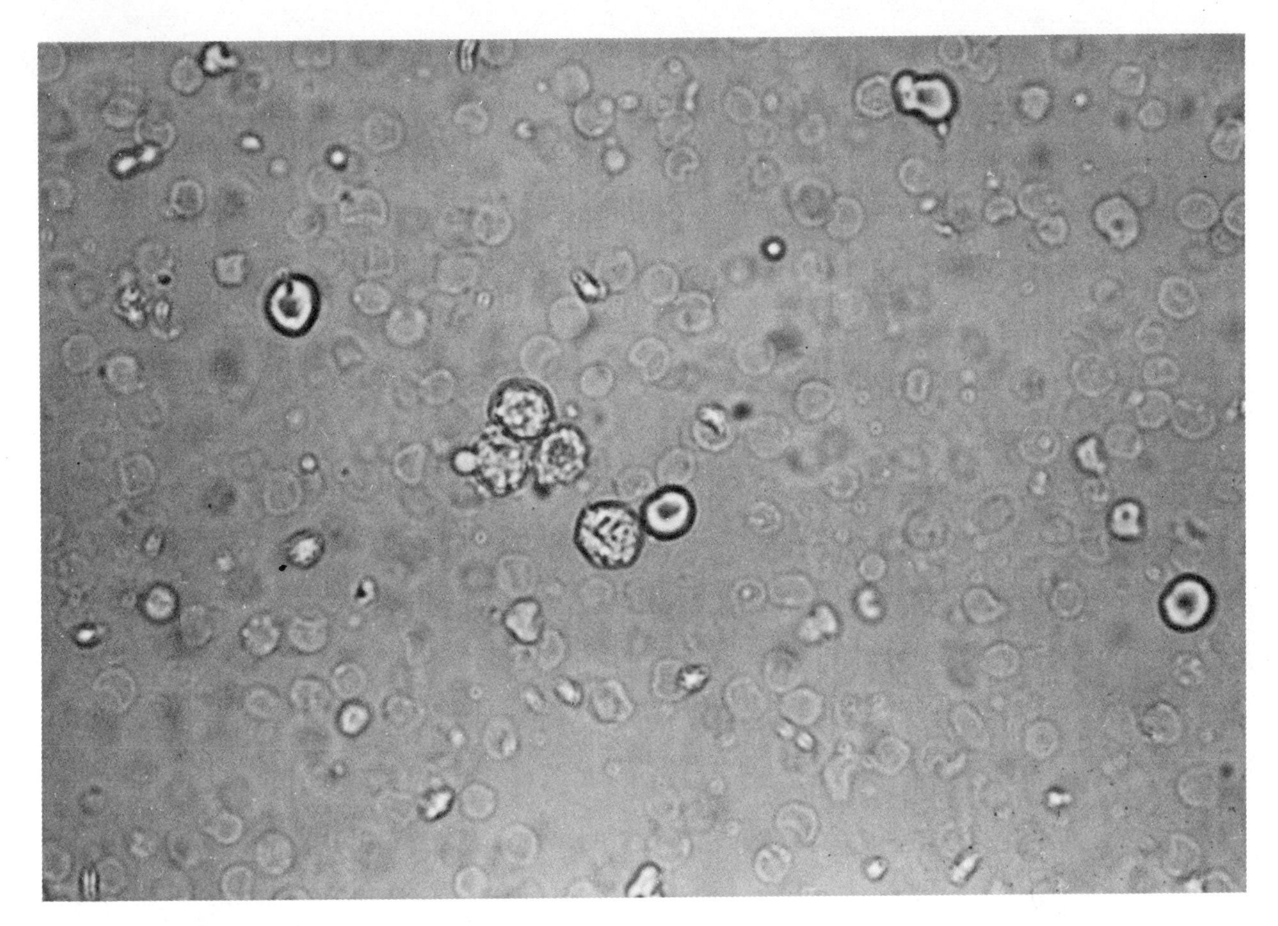

Fig. 53-12. Erythrocytes and leukocytes. (Bright-field, 400×.)

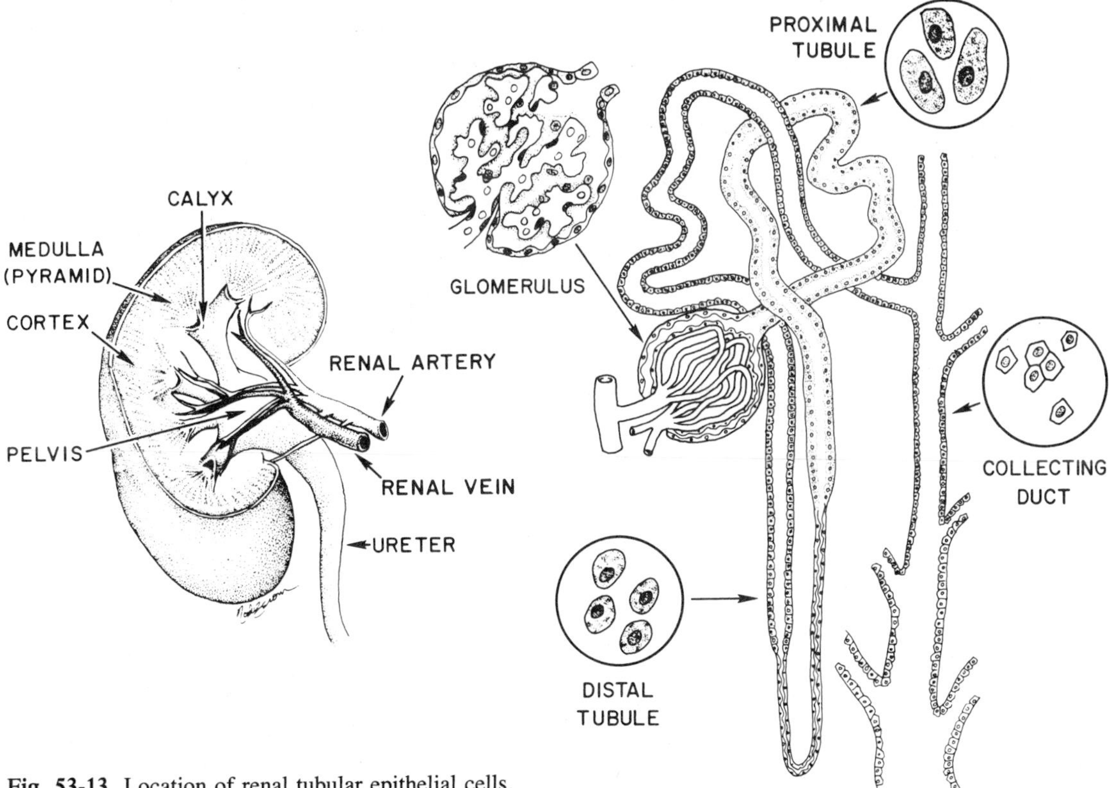

Fig. 53-13. Location of renal tubular epithelial cells.

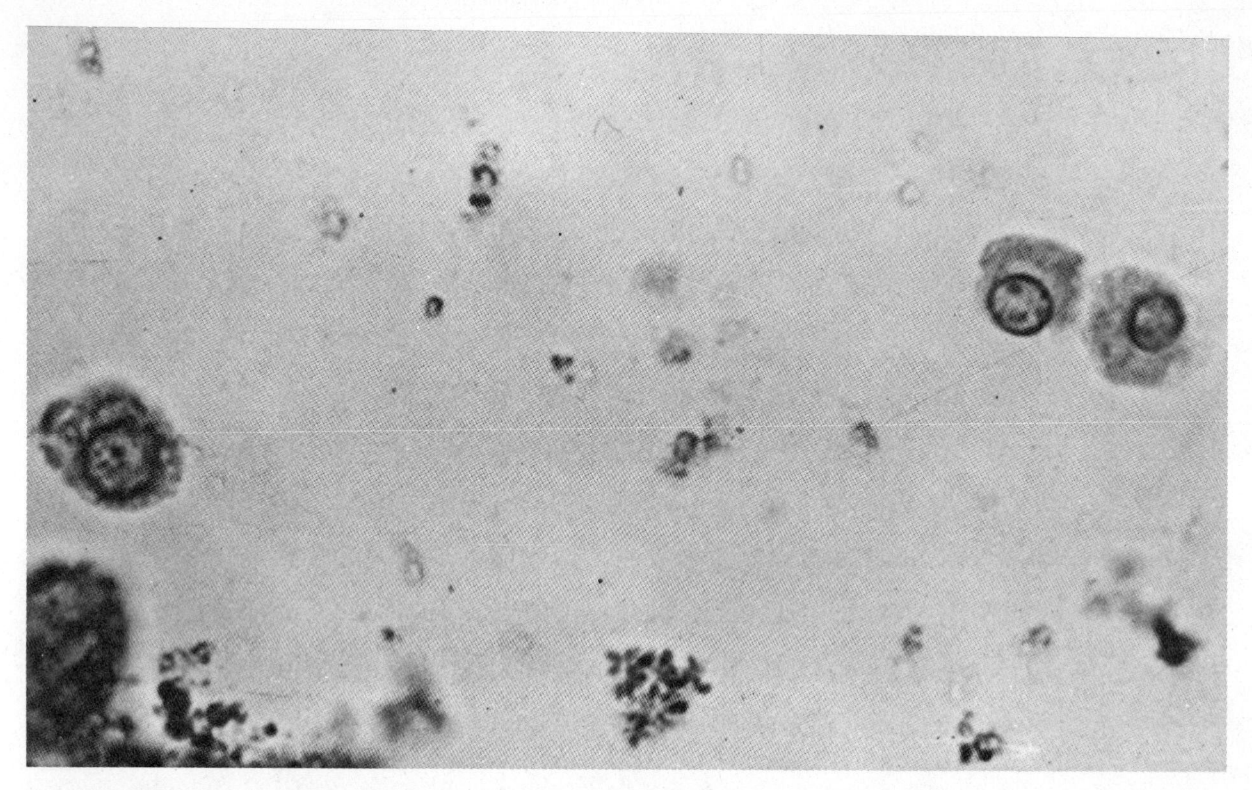

Fig. 53-14. Renal tubular epithelial cells. (Bright-field, 1000×.)

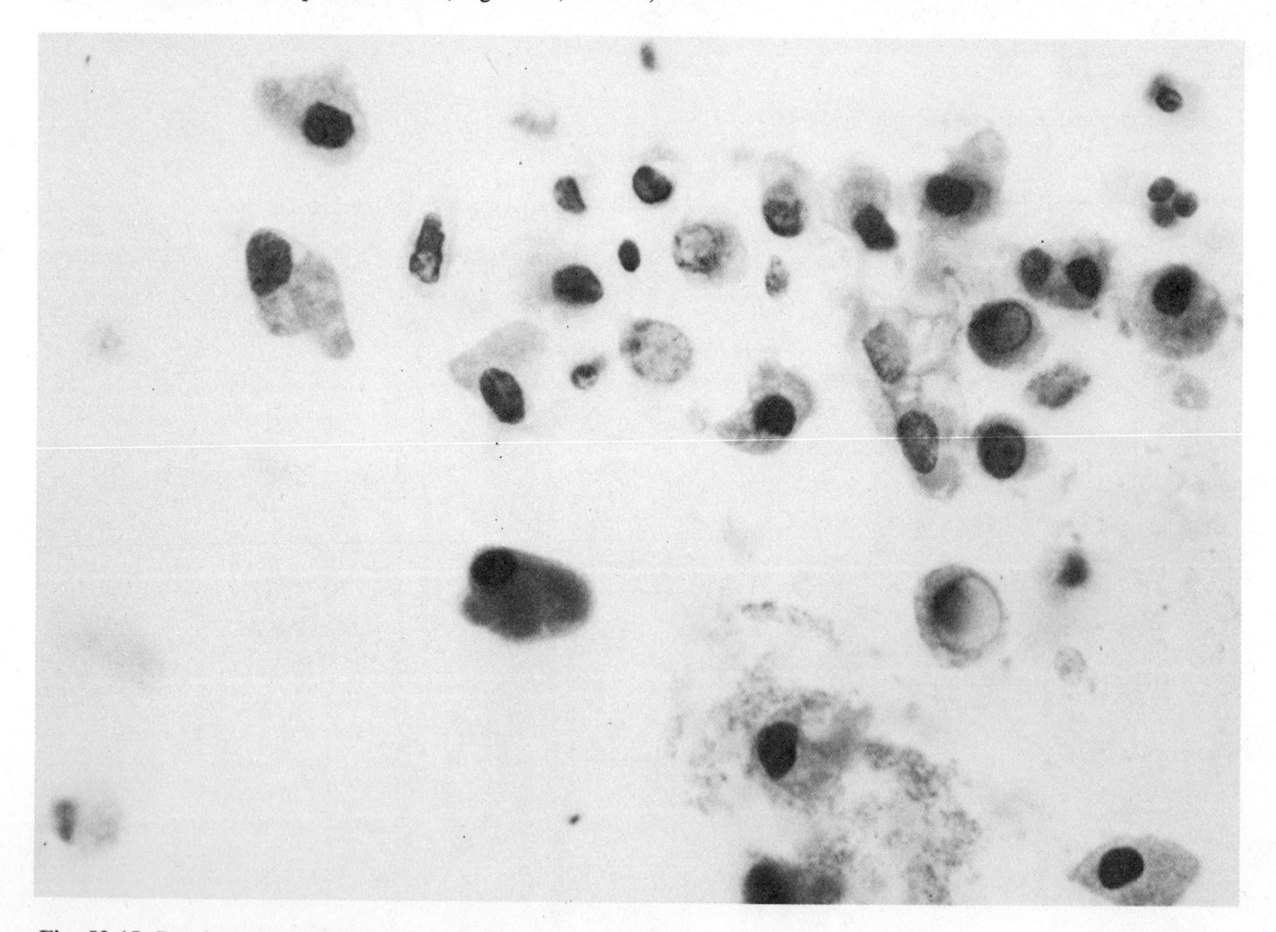

Fig. 53-15. Renal tubular epithelial cells. (Papanicolaou stain, 1000×.)

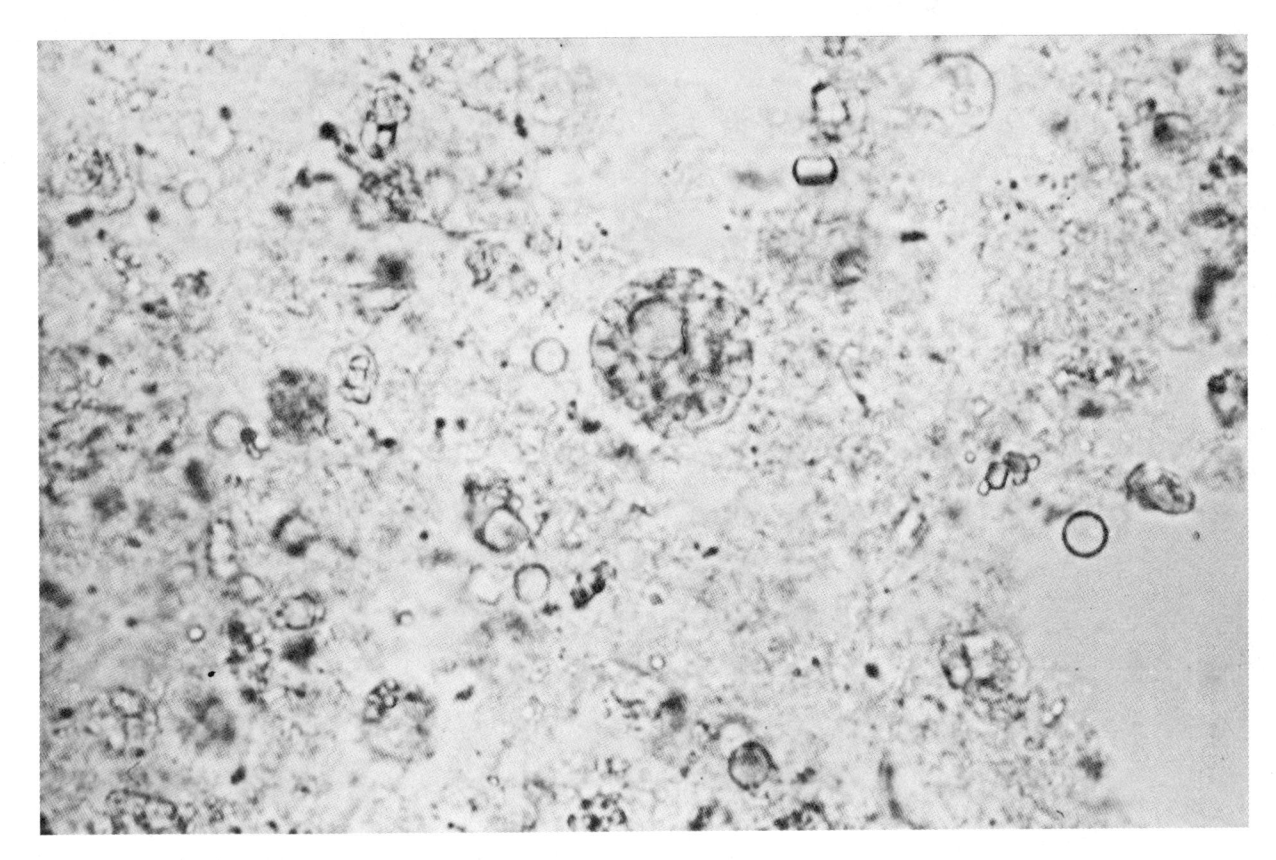

Fig. 53-16. Oval fat bodies. (Bright-field, 1250×.)

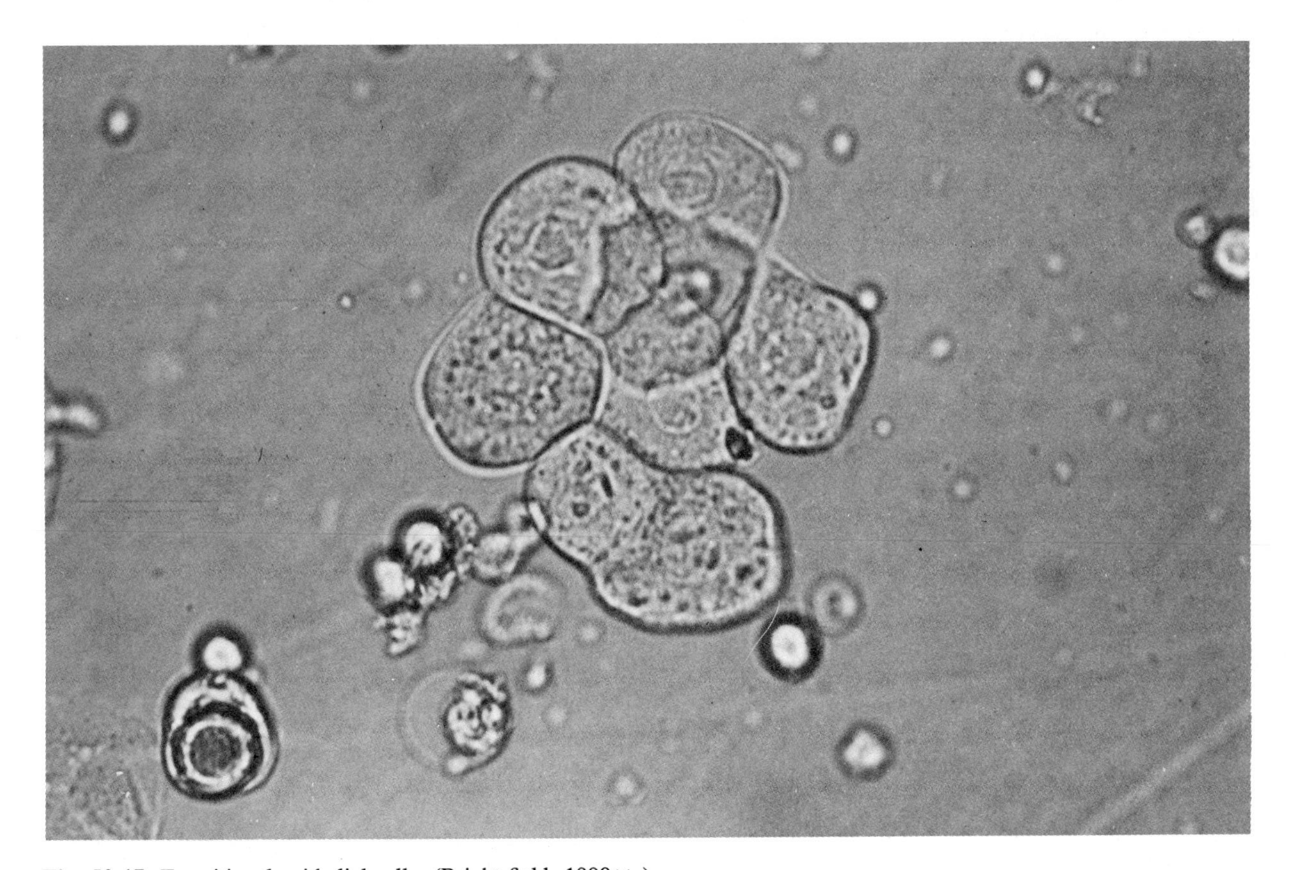

Fig. 53-17. Transitional epithelial cells. (Bright-field, 1000×.)

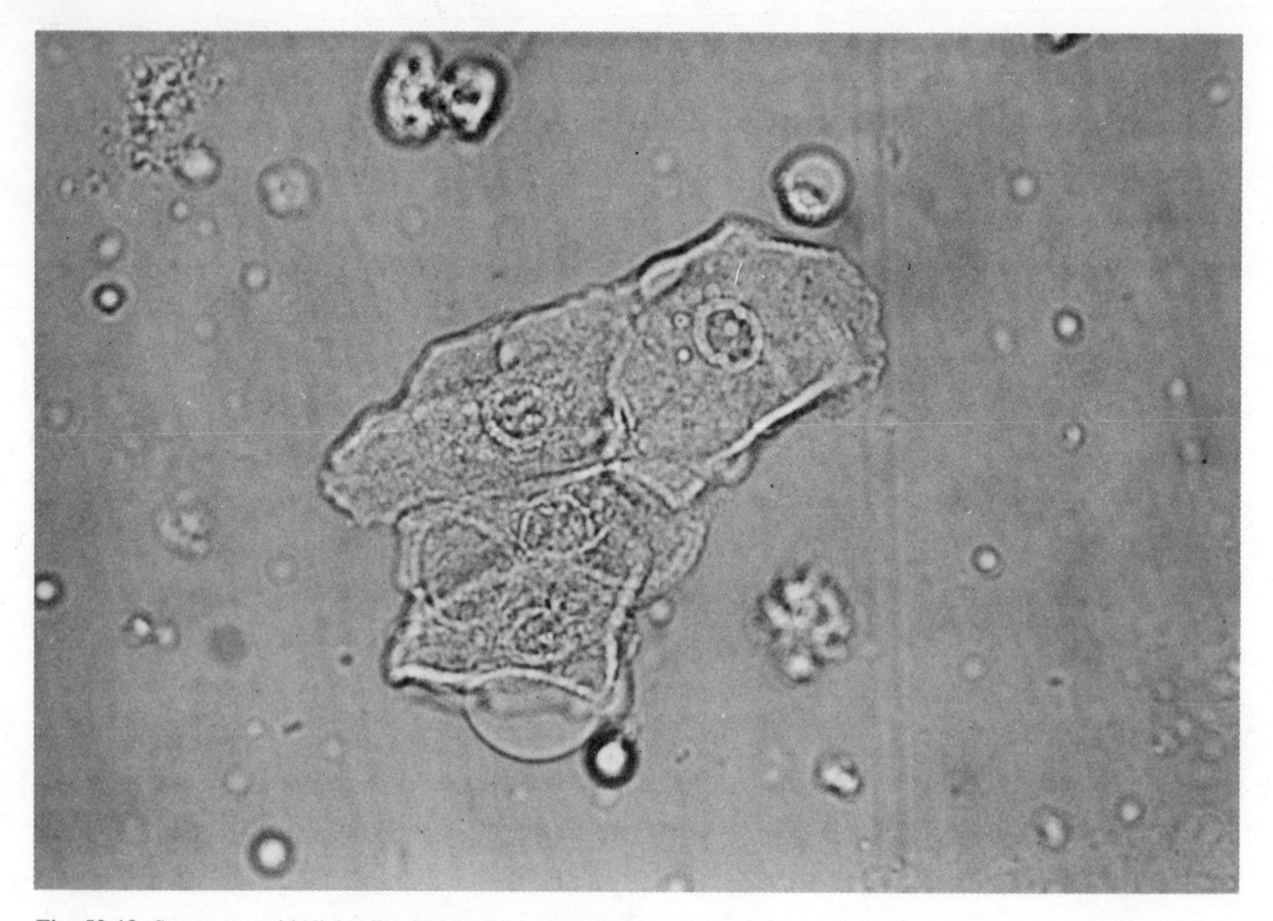

Fig. 53-18. Squamous epithelial cells. (Bright-field, 1000 ×.)

slightly larger than leukocytes. For accurate identification of various types of renal tubular cells (convoluted versus collecting duct) and sheets or fragments, cytological techniques are required. Quantifying renal epithelial cells is important in the documenting of progressive and regressive changes in several renal parenchymal diseases such as acute tubular necrosis, tubular interstitial inflammation, and renal allograft rejection.

Oval fat bodies (Fig. 53-16). Oval fat bodies are renal tubular epithelial cells that are filled with absorbed lipids or have undergone degenerative cellular changes. Oval fat bodies are often associated with lipiduria and are characteristically seen in the nephrotic syndrome and diabetes mellitus.

Transitional epithelial cells (Fig. 53-17). A few transitional (urothelial) cells can be found in normal urine. Inflammatory conditions, catheterization, or a pathological process such as malignancy may be indicated by large numbers of transitional cells.

By bright-field microscopy, transitional cells are round to oval, measure 40 to 60 μm, and have a centrally located nucleus. The cytoplasm borders of these cells appear thickened and crisp. When the nuclei of transitional cells become enlarged or irregular, cytological techniques should be suggested for the purpose of detecting a urinary system malignancy.

Squamous epithelial cells (Fig. 53-18). Squamous epithelial cells line the distal portion of the lower urinary tract and the female genital tract. Squamous cells are the largest of the epithelial cells found in urine. They have abundant flat cytoplasm with small nuclei. Frequently one or more corners of these cells may be folded. Squamous cells in urine usually indicate vaginal contamination or squamous metaplasia of the bladder and represent the least significant type of epithelial cell found in urine.

Tissue fragments in urine (Fig. 53-19). Any clumps or chunks of solid-appearing material in urine specimens should be noted. This material is identified during the initial inspection of the specimen because of its large size. It is often white or tan. Precise identification of such material is important for an accurate diagnosis. This involves transfer of the observed material to an appropriate fixative in order to preserve it for a cytological or histological evaluation. Renal papillary necrosis and bladder tumors are most frequently responsible for shedding large tissue fragments into the urine.

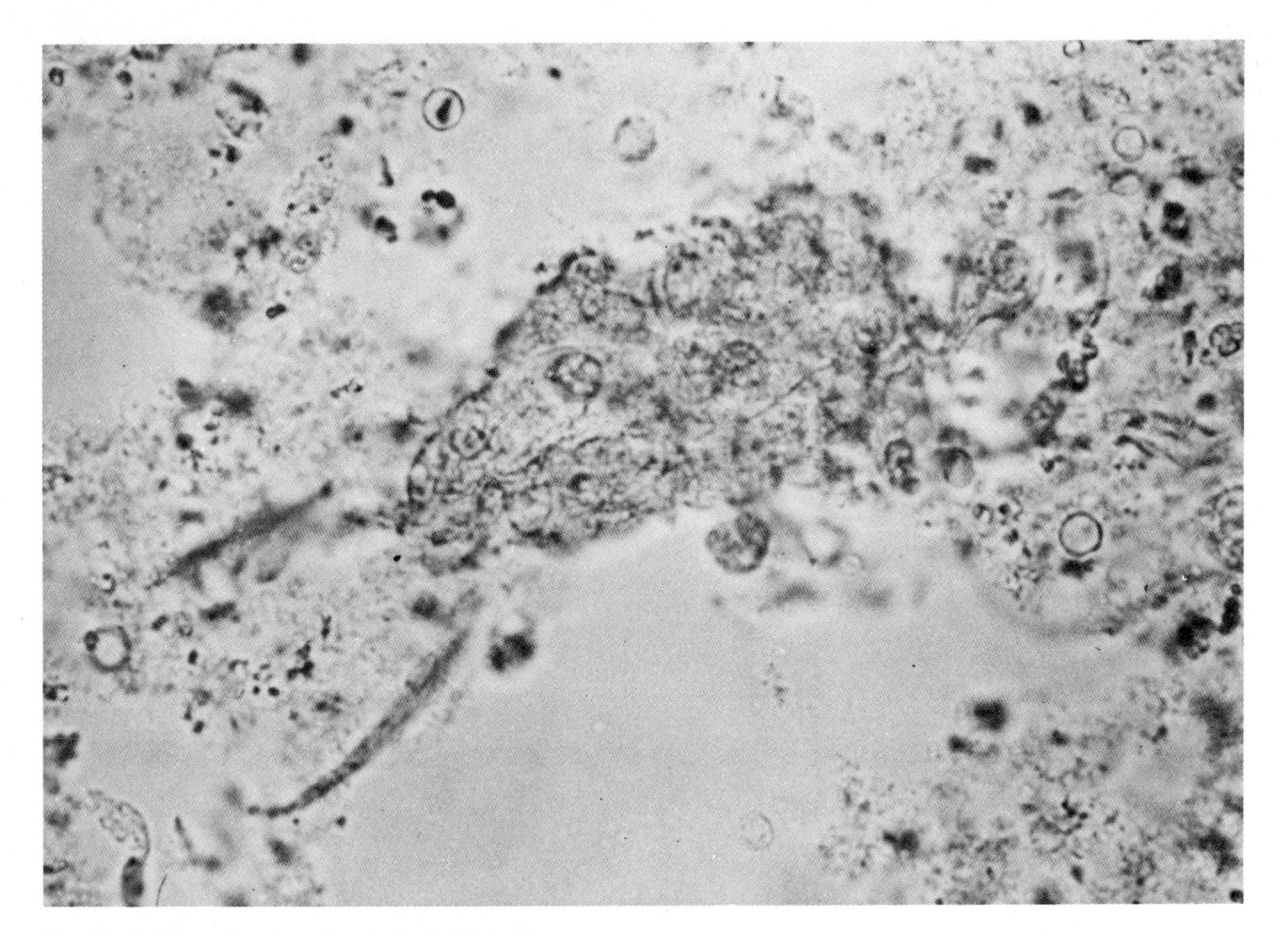

Fig. 53-19. Tissue in urine. (Bright-field, 1000×.)

Spermatozoa. Spermatozoa may be easily recognized in the urine of a man after ejaculation or in the urine of a woman as a vaginal contaminant after coitus. Their identification is of little clinical significance and the presence of spermatozoa is usually not reported.

Renal casts (Figs. 53-20 to 53-27)

Renal (urinary) casts are formed cylindrical structures that are organized in the nephron. Casts are significant because of their localizing value. Immunofluorescent studies have shown that casts are composed of uromucoid (Tamm-Horsfall mucoprotein), which is always present in urine, usually in solution, and they originate predominantly from the renal tubular epithelial cells of the ascending limb of the loop of Henle. Casts are formed as a result of the stasis of precipitated uromucoid. These cylindrical structures also form when the concentration of salts increases and when the pH value of urine declines. Since the precipitation of this protein depends on the concentration and composition of urine, casts are more likely to be formed at the distal portion of the nephron and in the collecting ducts of the kidney where the urine is more concentrated. Casts may form in the proximal convoluted tubules in patients with Bence Jones proteins (multiple myeloma).

The appearance of these cylindrical formed elements in urine reflects the shapes (long versus short, straight versus convoluted) and diameter (thin versus broad) of the original renal tubular lumens. Their number and measurable dimensions in size, form, and compositions are valuable clues of intrinsic renal parenchymal disease.

Microscopically casts are characterized by the appearance of the matrix (hyaline, granular, or waxy), by the cellular constituents (erythrocytes, leukocytes, or renal tubular epithelial cells), or by particulate matter embedded in the matrix (fine or coarse granules or fibrin).

Accurate identification of casts, especially the cellular types, is often difficult when unstained bright-field microscopy (wet preparation) is used. A skilled microscopist is essential to avoid misinterpretations. Phase contrast and interference contrast microscopy, as well as special stains, can be used to improve visualization. The cytocentrifugation-Papanicolaou technique is a superior method for the accurate identification of casts in urine specimens of patients with renal disease.

For diagnostic purposes, renal casts are classified as *physiological* or *pathological*. Table 53-10 lists the common types, properties, and clinical significance of casts.

Text continued on p. 1027.

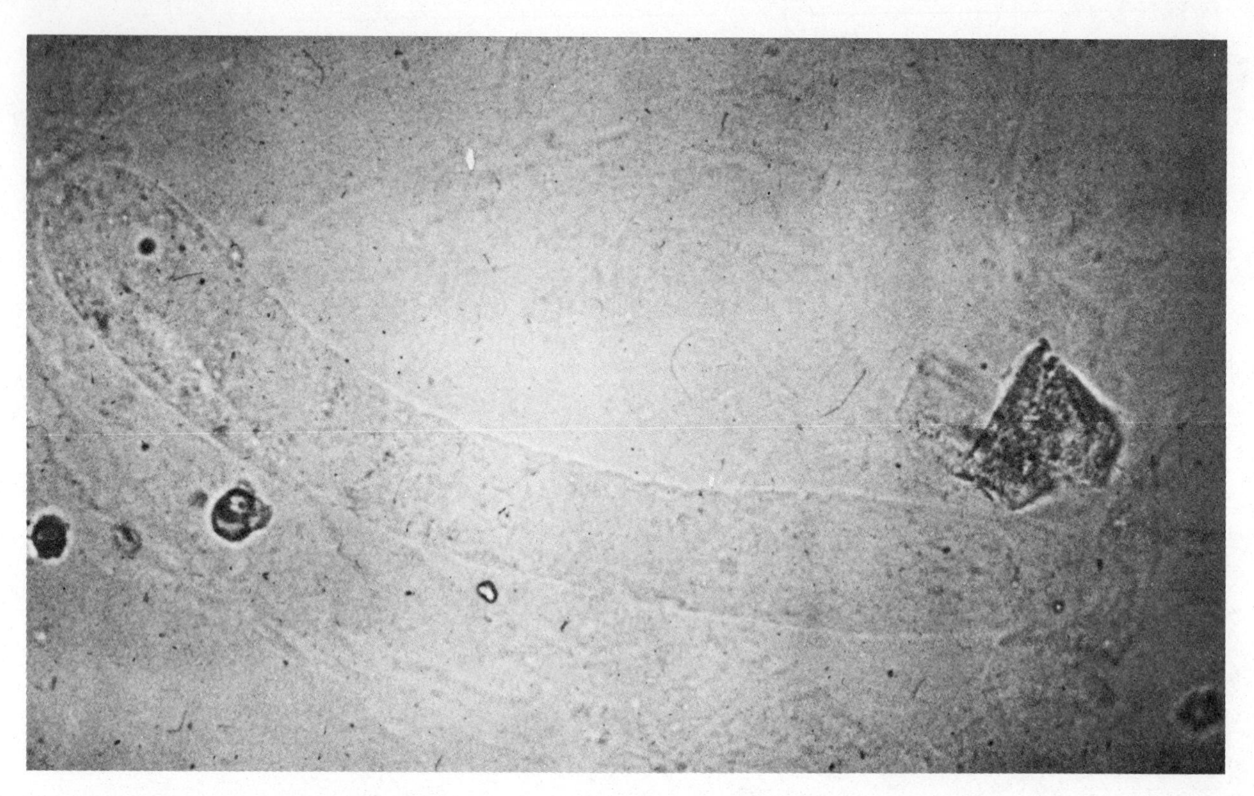

Fig. 53-20. Hyaline cast. (Bright-field, 1000×.)

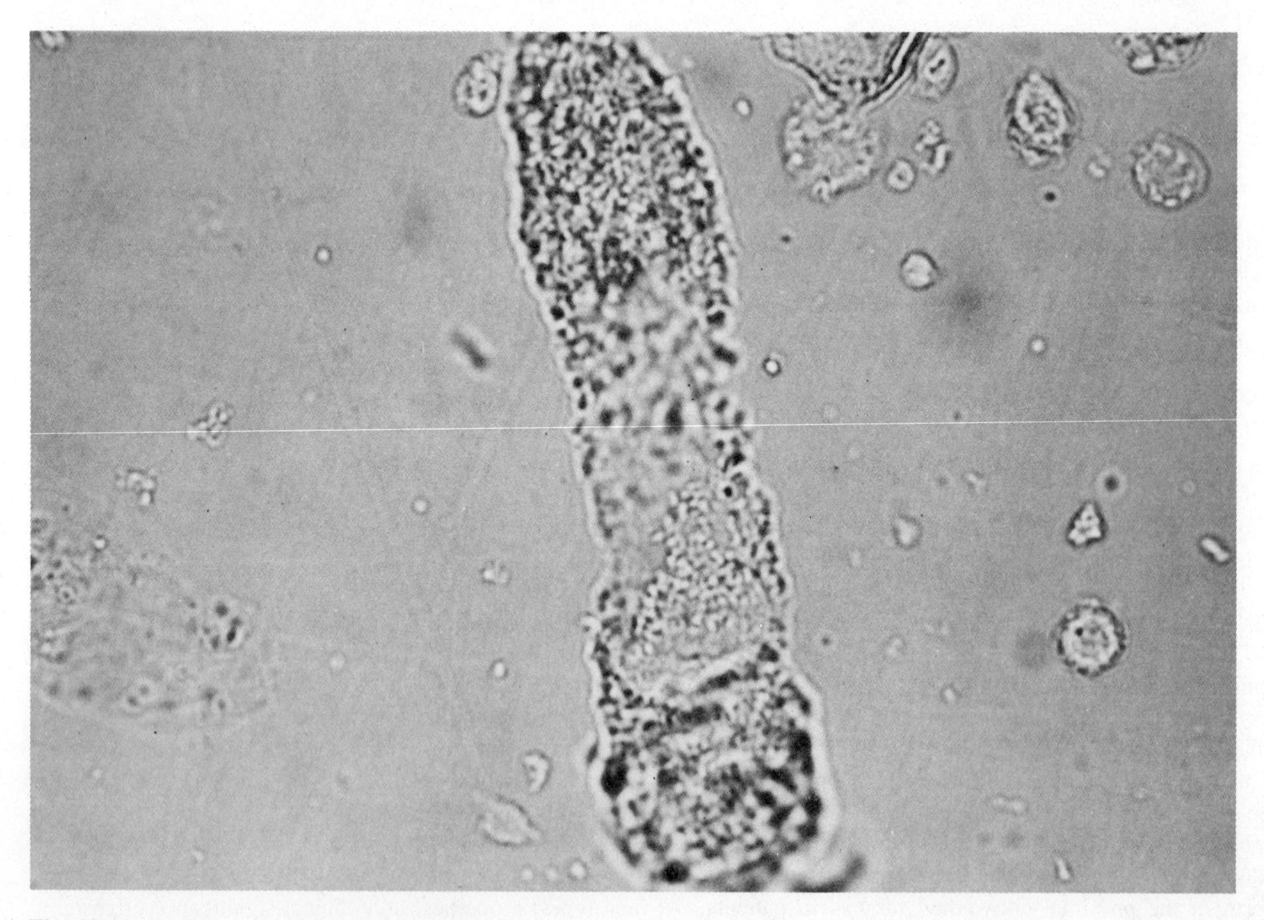

Fig. 53-21. Granular cast. (Bright-field, 1000×.)

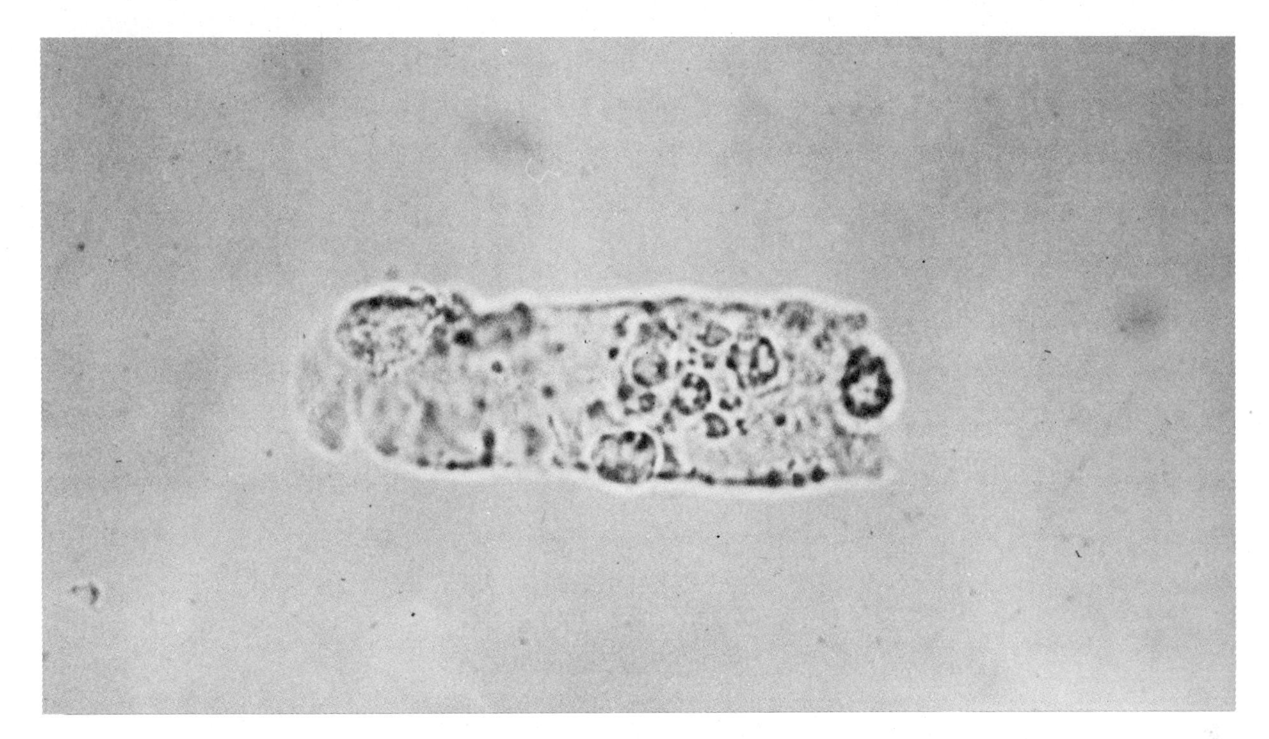

Fig. 53-22. Cellular cast. (Bright-field, 1000×.)

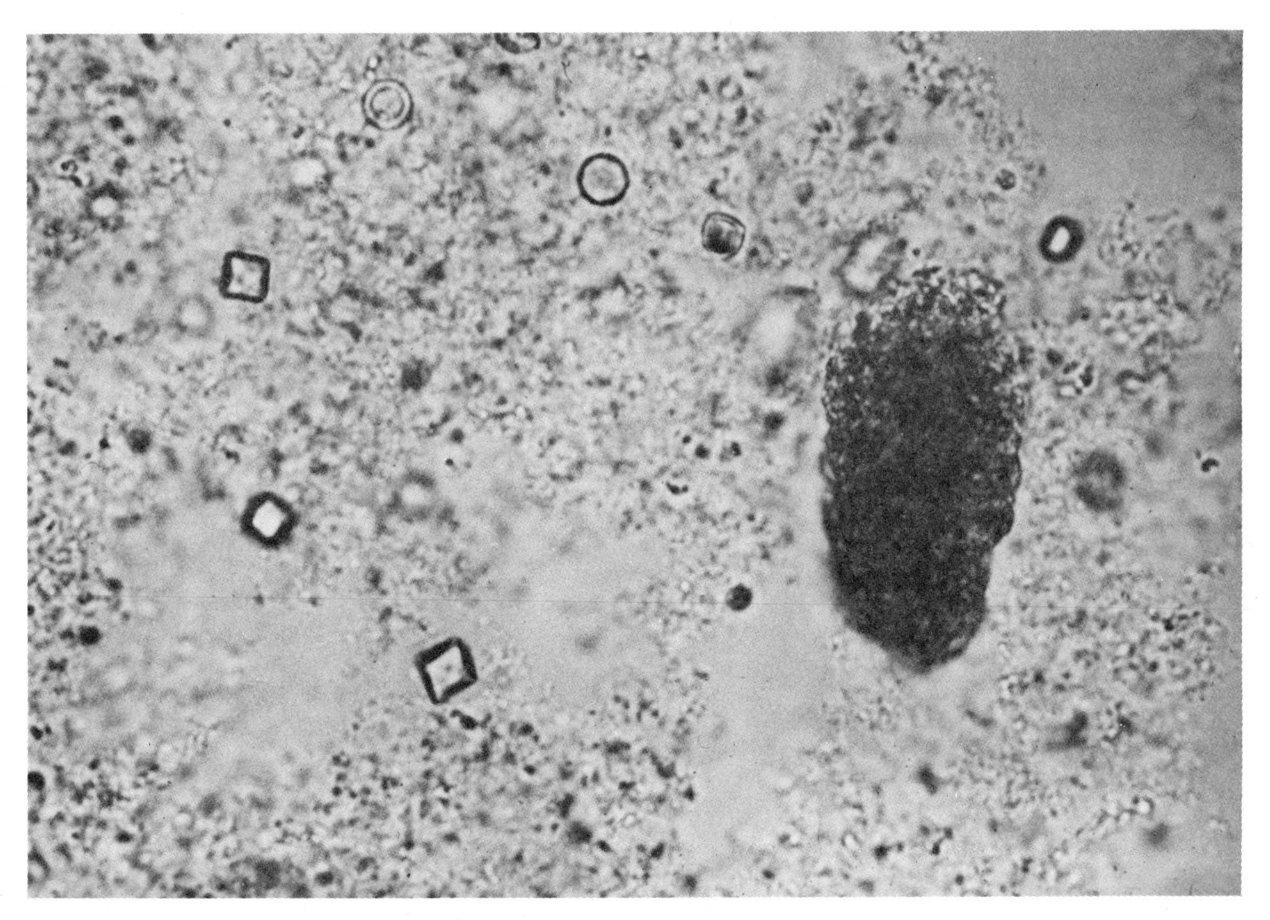

Fig. 53-23. Blood cast. (Bright-field, 1000×.)

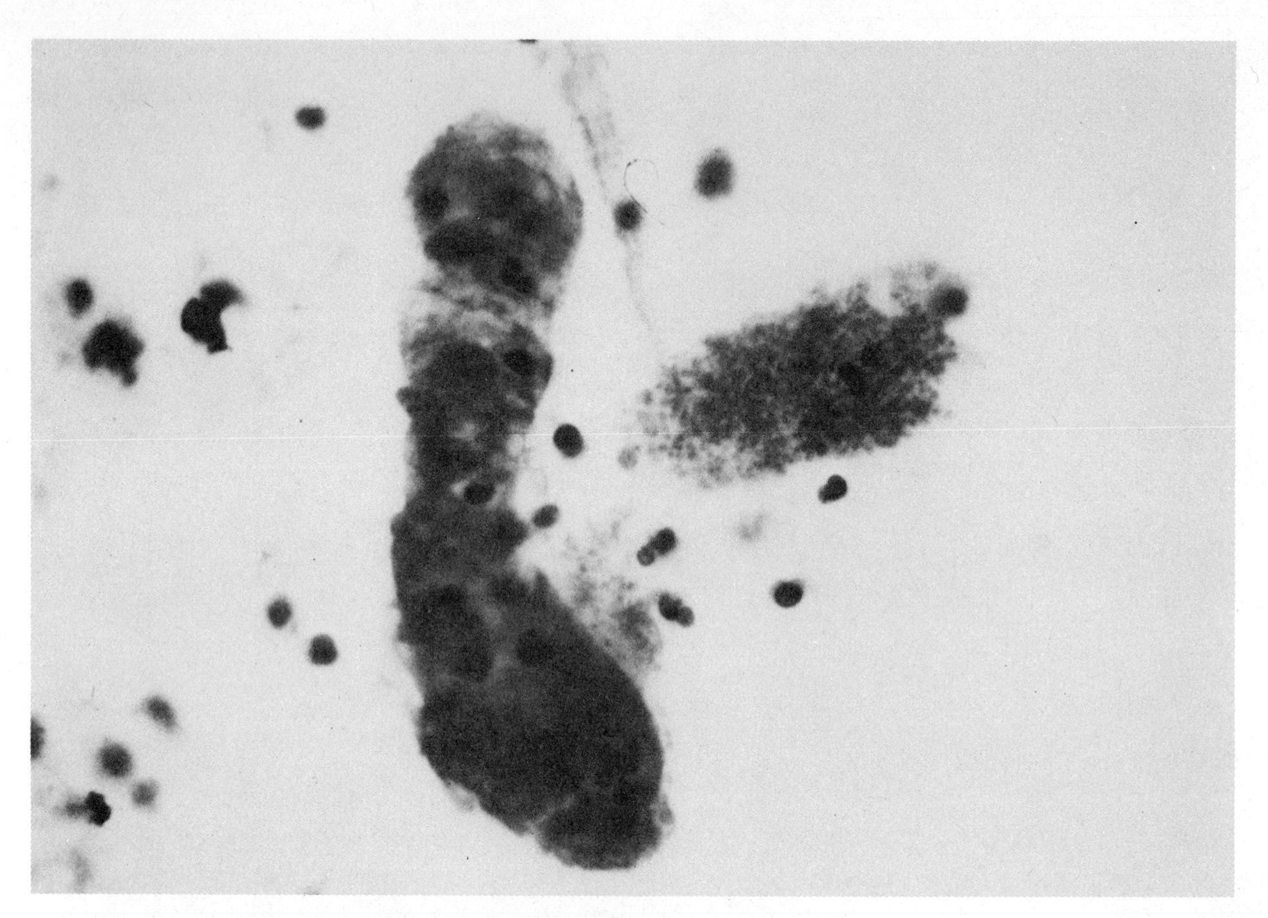

Fig. 53-24. Leukocyte cast. (Papanicolaou stain, 1000×.)

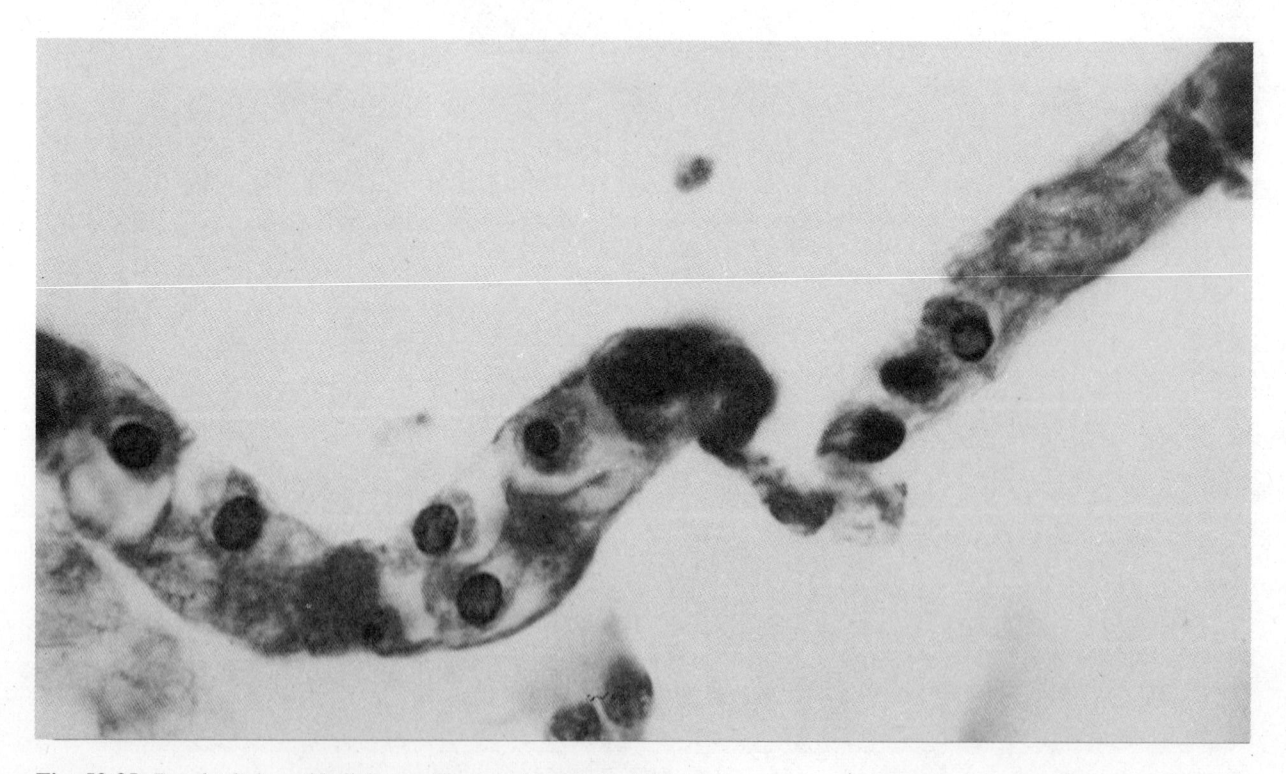

Fig. 53-25. Renal tubular epithelial cast. (Papanicolaou stain, 1000×.)

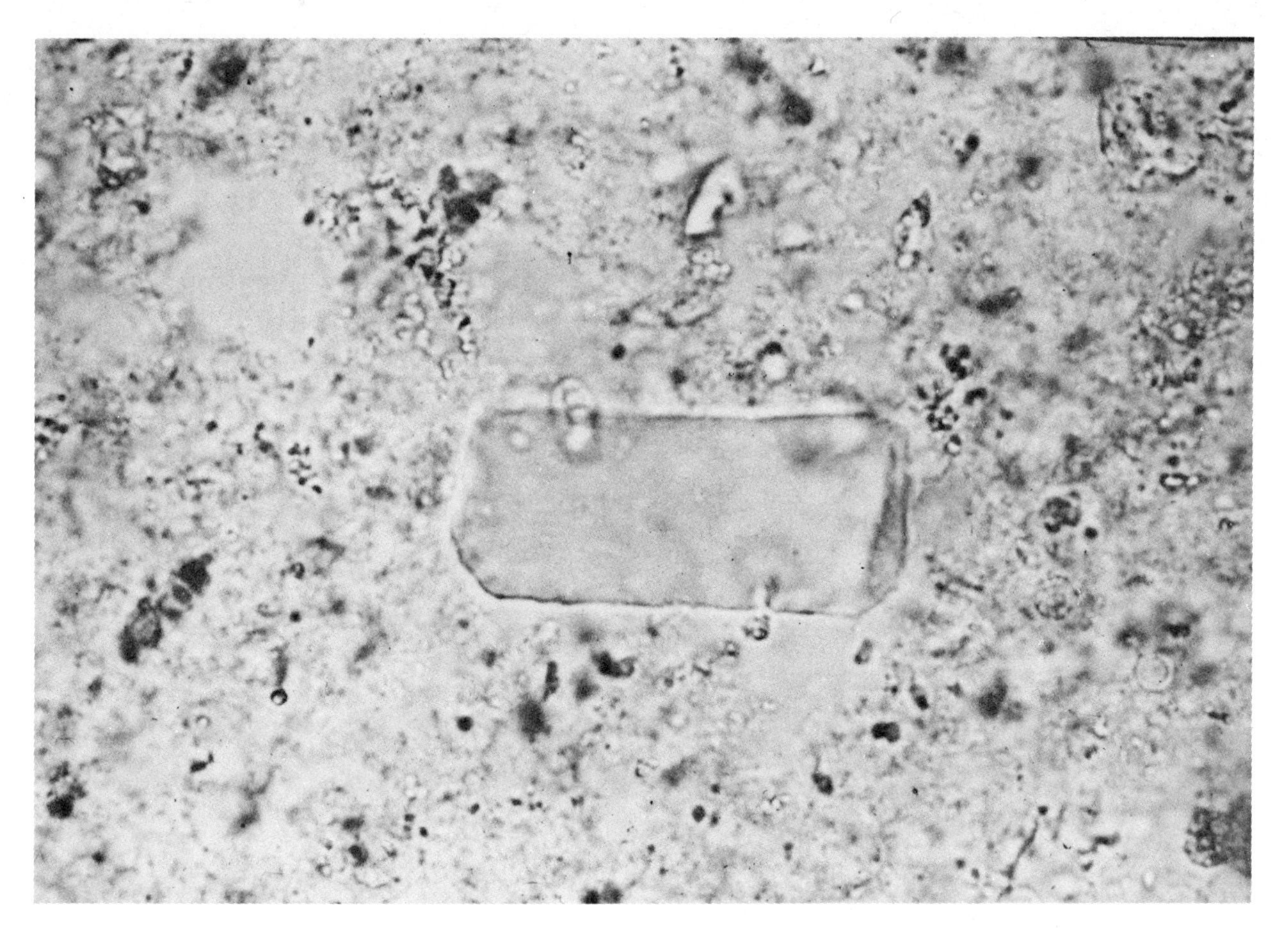

Fig. 53-26. Waxy cast. (Bright-field, 1000×.)

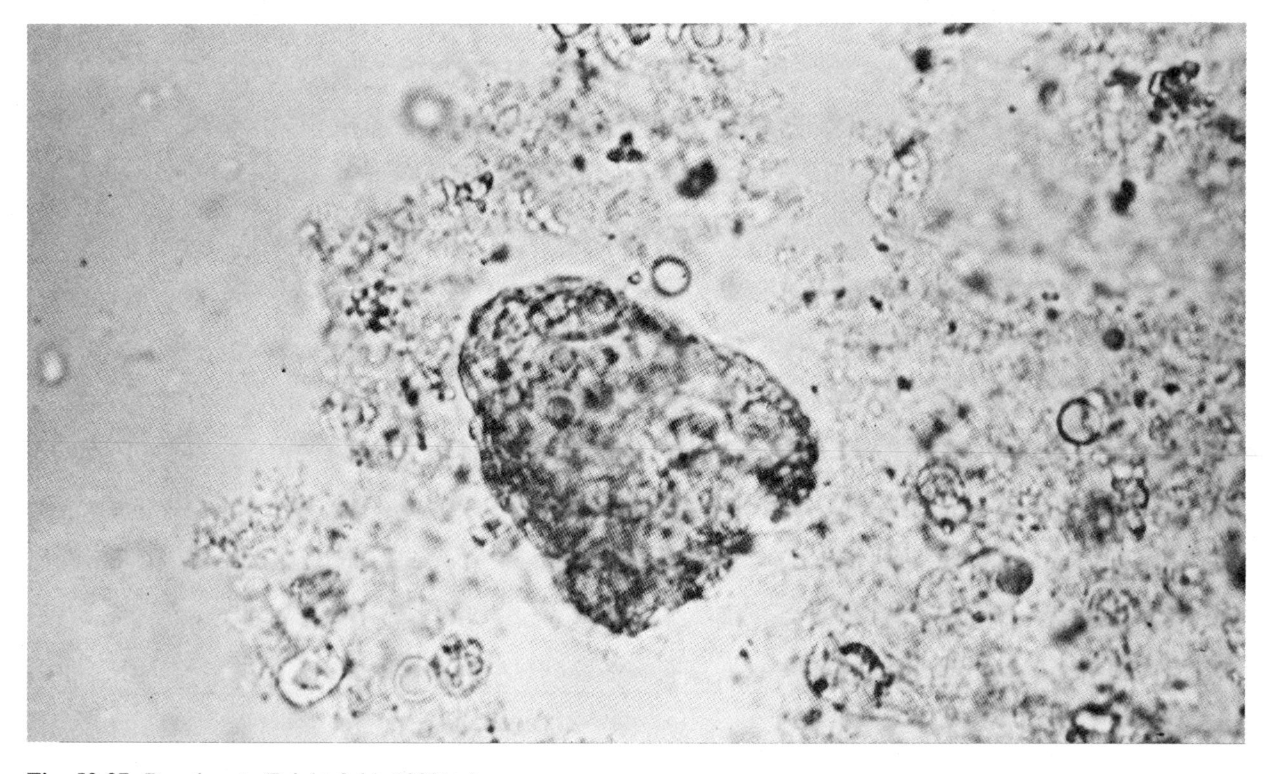

Fig. 53-27. Broad cast. (Bright-field, 1000×.)

Table 53-10. Common renal (urinary) casts

Types	Characteristic morphological appearance	Significance	Associated diseases
Physiological			
Hyaline	Transparent cylinder	Exercise, dehydration, and fevers	Nonspecific
Granular (fine)	Semitransparent cylinder containing fine refractile granules	Exercise, dehydration, and fever; accumulation of plasma proteins	Nonspecific
Pathological			
Cellular			
Erythrocytic	Semitransparent or granular cylinder containing distinct erythrocyte stroma	Renal parenchymal bleeding, glomerular leakage	Glomerular disease, interstitial hemorrhage (infarction)
Blood	Red-brown granular cylinder, but intact erythrocyte stroma is *not* seen	Same as above	Same as above
Leukocytic	Transparent granular or waxy cylinder containing segmented neutrophils	Renal inflammation	Tubulointerstitial inflammation (pyelonephritis), glomerular disease
Renal tubular epithelial	Semitransparent granular or waxy cylinder containing intact or necrotic renal tubular epithelial cells	Renal tubular damage	Renal tubular injury, acute tubular necrosis, acute allograft rejections, tubulointerstitial disease
Bacterial	Semitransparent or granular cylinder containing bacteria	Renal infection	Acute pyelonephritis
Fungal	Semitransparent or granular cylinder containing fungi	Renal infection, probably sepsis	Acute pyelonephritis, papillary necrosis
Noncellular			
Granular (coarse)	Semitransparent cylinder containing coarse refractile granules	Cellular degeneration, accumulation of plasma proteins	Nonspecific
Waxy	Sharply defined, highly refractile, homogeneous cylinder with broken-off borders and indentations	Cellular degeneration	Nonspecific
Fibrin	Transparent or granular cylinder containing thin long fibrils	Leakage of coagulation products	Glomerular disease, thrombosis
Fatty	Semitransparent or granular cylinder containing large highly refractile vacuoles or droplets	Lipiduria	Nephrotic syndrome
Bile	Deep yellow, transparent, granular waxy cylinder	Leakage of bile salts	Liver dysfunction, tubulointerstitial disease
Crystal	Crystalline inclusion in a semitransparent or granular cylinder	Cellular degeneration and malabsorption or excretion	Nonspecific
Other			
Broad	Width of cylinder is two to six times that of other casts; waxy and granular are most common types	Tubular dilatation and stasis	Advanced renal disease

URINE CYTOLOGY

Urine cytology is the study of exfoliated cells present in the urine sediment. It is used for the early detection of cancer of the bladder, particularly in persons exposed to certain aromatic amines. Other uses include detection of carcinoma in situ and recurrent carcinoma, renal transplant follow-up, and detection of opportunistic infections.[28,29] It is less valuable in patients with kidney tumors, except when the tumors communicate directly with the urinary stream. Fresh or refrigerated, spontaneously voided, or catheterized specimens are used for the differentiation of normal and abnormal cells. Various cellular fixatives such as 50% to 70% ethanol, Muccolexx, or Saccomanno are recommended to minimize cellular degeneration if examination cannot be performed within 48 hours. Papanicolaou or Wright-Giemsa staining methods are used for enhanced cellular detail.[19,29]

Members of the urinalysis laboratories should be aware of various procedures and entities recognized in urine cytology laboratories. Abnormal mononuclear cells suggestive of malignancy should be referred for urine cytology (Fig. 53-28). Holmquist[22] has advocated use of supravital stains for malignant cell detection and suggests a role for the urinalysis laboratory in cancer screening (Fig. 53-29).

CALCULI (LITHIASIS)

Urinary calculi are precipitates, concretions, or crystalloids embedded in a binding substance of mucus and protein. Bacteria and epithelial cells also may be included in the calculi.[7] Although the cause of calculus formation remains controversial, the detection and identification of calculi are important for diagnosis of urinary system conditions. Calculi formation in the kidney or lower urinary tract can cause serious anatomical damage and can be the source of excruciating pain for the patient. Knowledge of the specific composition of a passed or surgically removed calculus may aid the physician in effectively treating lithiasis and preventing future stone formation.[7]

Calculus (stone) analysis for chemical constituents is a complete laboratory procedure. Urinary system calculi are usually composed of calcium oxalate, calcium oxalate mixed with calcium phosphate, ammonium magnesium phosphate, uric acid, or cystine. Most urinalysis laboratories refer specimens for calculi analysis to more sophisticated laboratories. Bauer[7] describes in greater detail a variety of chemical procedures for identifying the elemental components of calculi. X-ray diffraction and infrared spectroscopy are used to identify the crystalline components in minute calculi.

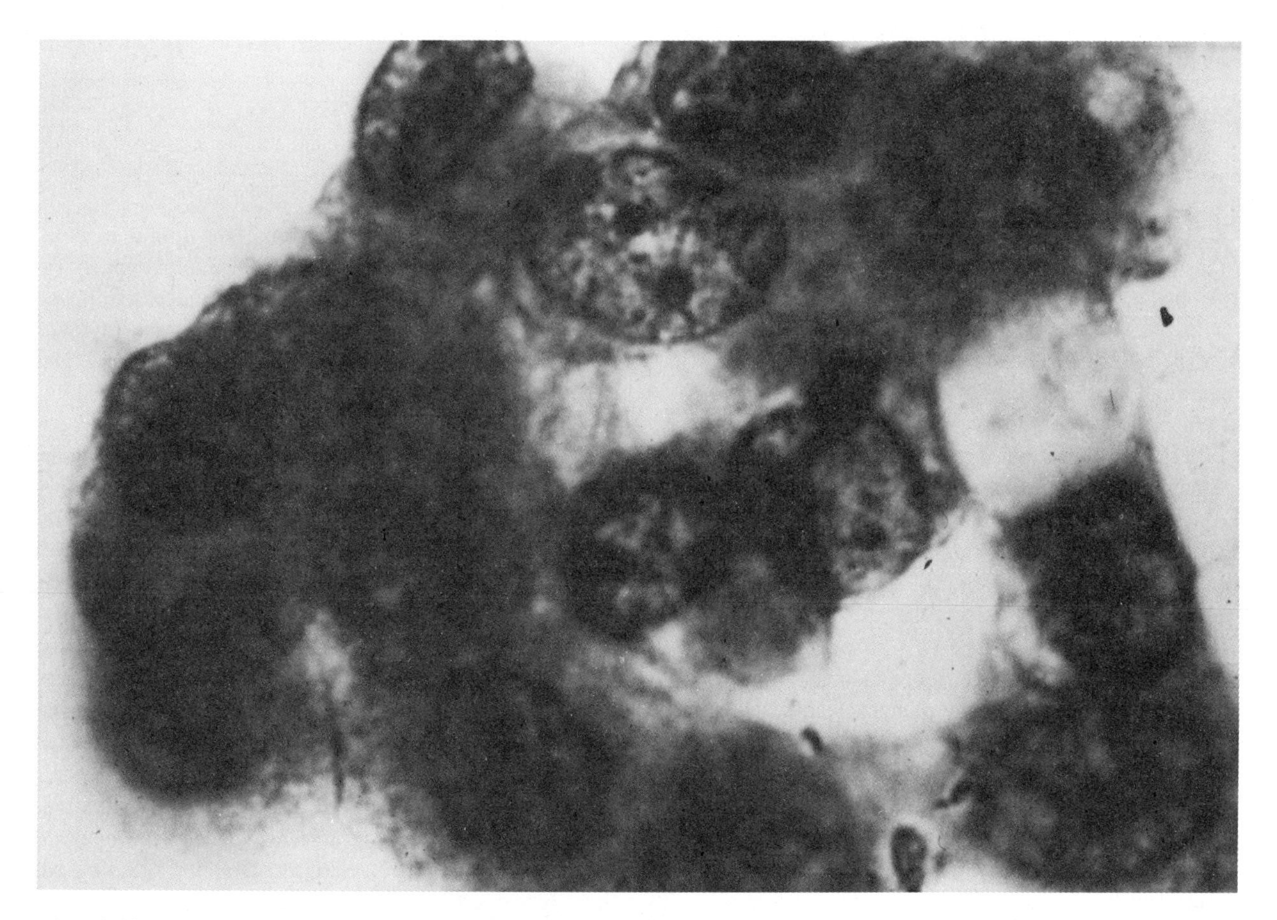

Fig. 53-28. Transitional cell carcinoma. (Papanicolaou stain, 1000×.)

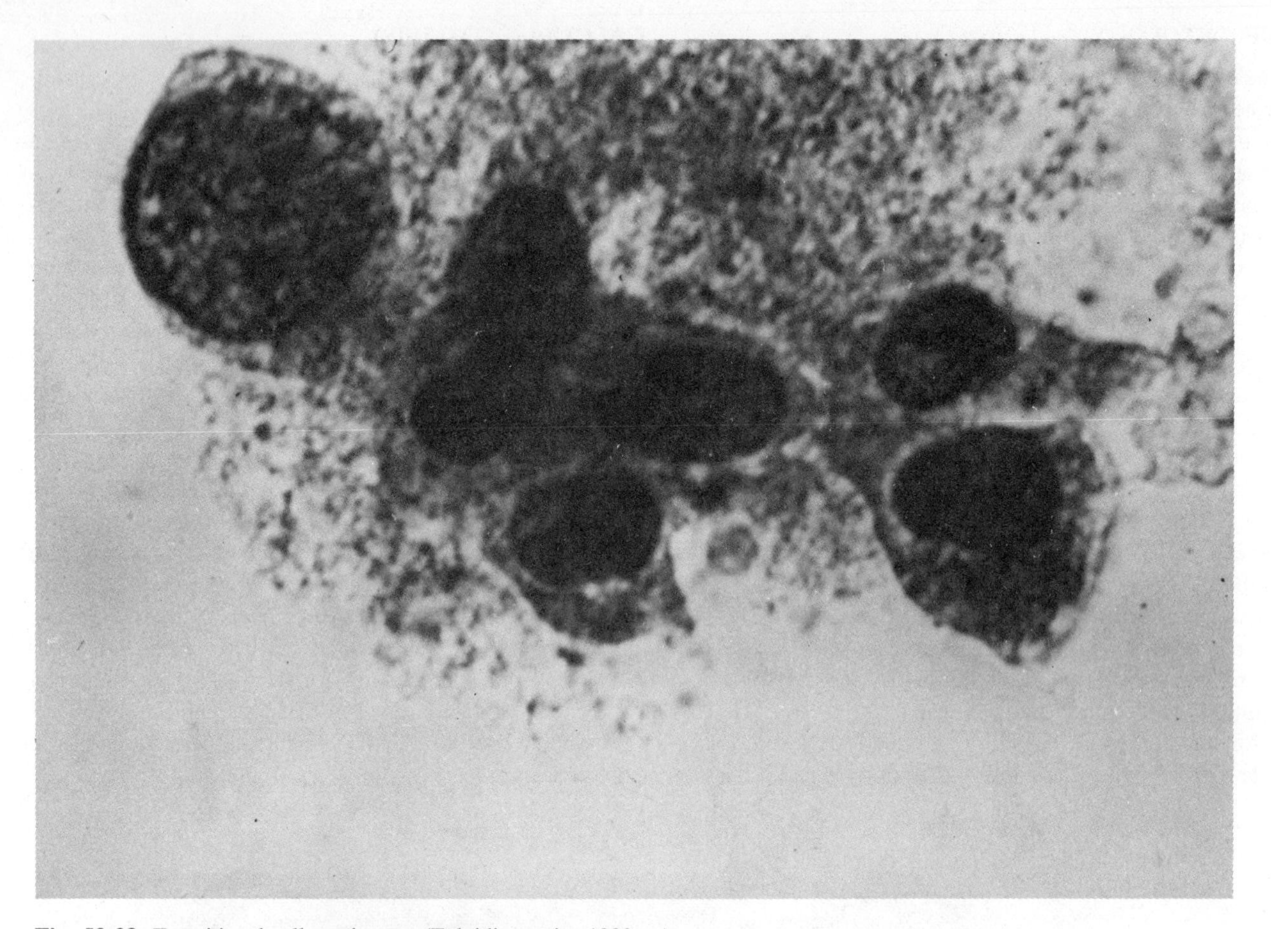

Fig. 53-29. Transitional cell carcinoma. (Toluidine stain, 1000×.)

URINE FINDINGS IN COMMON RENAL AND LOWER URINARY TRACT DISEASES

Numerous primary and secondary conditions and diseases occur in the kidney and lower urinary tract. An accurate diagnosis of urinary system disease requires correlation of abnormal urine findings with the patient history, physical examination, symptoms, signs, renal function tests, and other laboratory data. To be classified as abnormal, urine sediment must contain at least one of the following:

1. More than five erythrocytes or leukocytes per high-power field (400×)
2. More than two renal tubular cells per high-power field (400×)
3. More than three hyaline casts, more than one granular cast, or the presence of any pathological cast per low-power field (100×)
4. More than one bacterium per high-power field (400×)
5. Presence of fungus, parasites, or viral inclusion cells
6. Presence of pathological crystals (such as cystine) or presence of a large number of nonpathological crystals (such as uric acid)

A summary of urinalysis abnormalities found in common renal and lower urinary tract disease is shown in Table 53-11. Diagnostic findings of both the macroscopic and microscopic examination are also listed for quick review.

QUALITY CONTROL

An effective quality-control program is essential to ensure accuracy in urinalysis. A program of quality assurance that is similar to other areas of the clinical laboratory must be developed to achieve more reliable urinalysis results.[30]

Several quality-control preparations are commercially available in tablet or liquid form with varying concentrations of chemical constituents. Hoeltge and Ersts describe a 3-year experience using a synthetic-urine control prepared in their laboratory.[31] Currently there is no ideal commercial preparation for urine sediment elements. A product for the qualitative identification of casts, cells, and crystals can be prepared using formalin or Mucolexx as a fixative.

Most quality-control programs are developed to maintain reagents and supplies, equipment, and continuing education. A suggested urinalysis quality-control schedule is found in Table 53-12. It is important that reagents are

Table 53-11. Urinalysis abnormalities found in common urinary system diseases

Conditions	Diagnostic physicochemical findings (macroscopic urinalysis)	Diagnostic urine sediment findings (microscopic urinalysis)
Renal		
Acute glomerulonephritis	Decreased urine volume, increased turbidity (smoky), proteinuria (<2 g/24 hr), hematuria (often gross)	Erythrocytic and blood casts, erythrocytes (often dysmorphic), neutrophils, mixed cellular casts, renal epithelial cells, occasional leukocytic or renal tubuloepithelial casts
Nephrotic syndrome	Lipiduria, significant proteinuria (>4.5 g/24 hr)	Fatty and waxy casts, doubly refractile oval fat bodies (Maltese cross with polarized light), lipid-laden renal tubuloepithelial cells
Chronic glomerulonephritis	Occasional lipiduria, decreased and fixed specific gravity, proteinuria (>2 g/24 hr), hematuria	Pathological casts, especially broad types
Acute tubular necrosis	Decreased urine volume, decreased specific gravity, minimal proteinuria, hematuria	Intact and necrotic renal epithelial cells, renal epithelial fragments, pathological casts
Acute pyelonephritis (tubulointerstitial inflammation)	Occasional odor, increased turbidity, minimal proteinuria, positive nitrite reaction	Leukocyte casts, neutrophils, especially in clumps; bacterial, granular, and waxy casts; renal tubuloepithelial cells
Diabetes mellitus	Proteinuria, glycosuria, ketonuria	Fatty and waxy casts, oval fat bodies, renal epithelial cells, leukocytes
Systemic lupus erythematosus	Proteinuria	Pathological casts, renal epithelial cells, neutrophils, erythrocytes
Cystinosis	Minimal proteinuria or hematuria	Cystine crystals
Acute allograft rejection	Decreased urine volume, minimal proteinuria, hematuria	Renal epithelial cells, renal epithelial casts, lymphocytes, pathological casts, especially renal epithelial casts
Viral nephropathy (cytomegalic inclusion disease)	Minimal proteinuria, hematuria	Mononuclear cells (occasional giant cell forms) with prominent intranuclear or cytoplasmic inclusions
Lower urinary tract		
Bacterial urinary tract infection	Occasional odor, increased turbidity, positive nitrite reaction, occasional hematuria	Bacteria, neutrophils, reactive transitional epithelial cells, absence of cast formation
Fungal urinary tract infection	Increased turbidity, occasional hematuria	Fungi, neutrophils and lymphocytes, reactive transitional epithelial cells
Viral urinary tract infection	Occasional hematuria	Viral inclusion bodies, neutrophils, transitional cells
Eosinophilic cystitis	Hematuria	Eosinophils (numerous), reactive transitional epithelial cells, absence of cast formation
Transitional cell carcinoma	Hematuria	Increased numbers of malignant transitional epithelial cells with high nuclear/cytoplasmic ratio, hyperchromasia, and chromatin clumping; cells occur singly and as tissue fragments

Modified from Schumann, G.B.: Urine sediment examination, Baltimore, 1980, The Williams & Wilkins Co.

Table 53-12. Suggested urinalysis quality-control schedule

	Time				
Areas checked	Daily	Weekly	Monthly	Semiannually or annually	As needed
Reagents and supplies					
Reagent strip	×				
Reagent tablets	×				
Remaking protein standard				×	
Equipment					
Refrigerator temperature	×				
Freezer temperature	×				
Refractometer calibration		×			
Urinometer calibration		×			
Spectrophotometer calibration		×			
Microscope maintenance				×	
Thermometers		×			
Glassware			×		
Centrifuge maintenance				×	
Education					
Revise laboratory manual				×	
Technologist's proficiency testing			×		
Update library				×	
Clinicopathological correlations					×

dated on entering the laboratory and used before expiration. Urine control solutions should retain their utility if they are stored in a tightly stoppered container, refrigerated, and protected from light.[8] The new lot number should always be recorded. A laboratory manual containing operating instructions and documentation of equipment maintenance should be maintained and reviewed yearly. The importance of continuing education of technologists and utilization of current references cannot be overemphasized.

COORDINATED APPROACH TO URINALYSIS

The time-honored urinalysis continues to be one of the most commonly requested and demanding clinical laboratory procedures. Laboratories involved in the examination of urine must define new responsibilities for both the rapid, "routine" assessment of urine and the more time-consuming, "specialized" interpretive tests.

Current, rapid dipstick technology and standardization of bright-field microscopy provide a quality program for the analysis of urine specimens from asymptomatic individuals. If symptomatic patients are to receive a more comprehensive urine examination, there is a need to detect these specimens and process them for a more definitive evaluation and confirmation.

A coordinated approach to the examination of urine is described in Fig. 53-30. Proper utilization of this approach requires that the clinician and urine technologist understand the roles of both the routine laboratory, which offers basic urinalysis, and the specialized laboratory, which offers a more comprehensive sediment examination. After the collection of urine and macroscopic analysis, the physician or laboratory must differentiate between results from symptomatic and asymptomatic patients. Emphasis must be focused on the urinalysis technologist's ability to recognize the results that require additional follow-up testing. This coordinated approach represents a flexible system in which technical responsibilities can be shared, and additional laboratory confirmation procedures can be integrated into the system (Fig. 53-30). Communication between physicians, the laboratory technologist, and the medical director is essential to resolve inconclusive results and reestablish credibility of the urine examination.

REFERENCES

1. Schumann, C.B.: Urine sediment examination, Baltimore, 1980, The Williams & Wilkins Co.
2. Free, A.H., and Free, M.A.: Urinalysis in clinical laboratory practice, Boca Raton, Fla., 1975, CRC Press, Inc.
3. Schumann, G.B., Schumann, J.L., and Schweitzer, S.: Coordinated approach to the urine sediment examination, Lab. Management **1:**45-48, 1983.
4. Free, A.H., and Free, M.A.: Rapid convenience urine tests: their use and misuse, Lab. Med. **9:**9-17, 1978.
5. Bradley, M., Schumann, G.B., and Ward, P.C.J.: Examination of urine. In Henry, J.B., editor: Todd-Sanford clinical diagnosis by laboratory methods, ed. 16, Philadelphia, 1979, W.B. Saunders Co.
6. Kunin, C.M.: Detection, prevention and management of urinary tract infections, ed. 2, Philadelphia, 1974, Lea & Febiger.
7. Bauer, J.D.: Clinical laboratory methods, ed. 9, St. Louis, 1982, The C.V. Mosby Co.
8. Monte-Verde, D., and Nosanchuk, J.S.: The sensitivity and specificity of nitrite testing for bacteriuria, Lab. Med. **12:**755-757, 1981.
9. Kusumi, R.K., Grover, P.J., and Kunin, C.M.: Rapid detection of pyuria by leukocyte esterase activity, J.A.M.A. **245:**1653-1655, 1981.

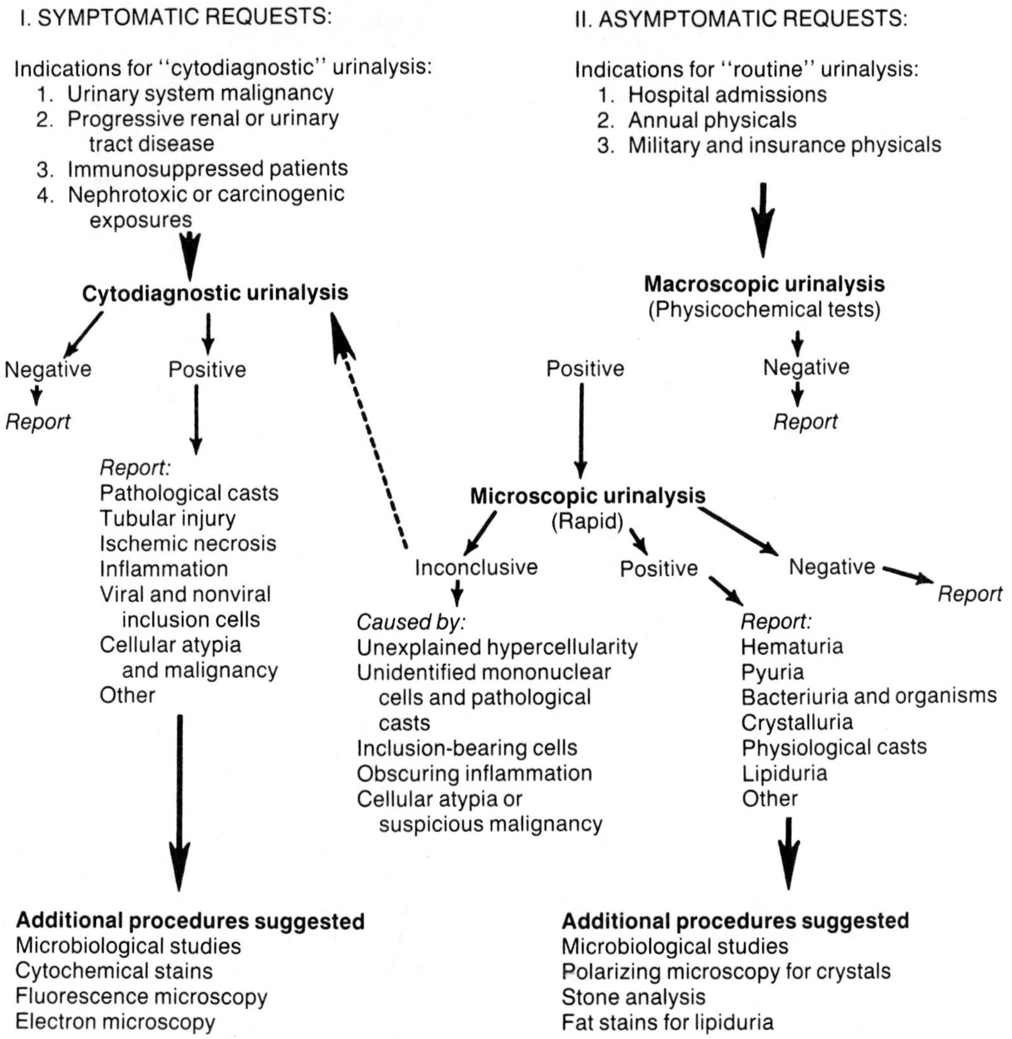

I. SYMPTOMATIC REQUESTS:

Indications for "cytodiagnostic" urinalysis:
1. Urinary system malignancy
2. Progressive renal or urinary tract disease
3. Immunosuppressed patients
4. Nephrotoxic or carcinogenic exposures

Cytodiagnostic urinalysis

Negative → *Report*

Positive → *Report:* Pathological casts; Tubular injury; Ischemic necrosis; Inflammation; Viral and nonviral inclusion cells; Cellular atypia and malignancy; Other

Additional procedures suggested
Microbiological studies
Cytochemical stains
Fluorescence microscopy
Electron microscopy

II. ASYMPTOMATIC REQUESTS:

Indications for "routine" urinalysis:
1. Hospital admissions
2. Annual physicals
3. Military and insurance physicals

Macroscopic urinalysis
(Physicochemical tests)

Negative → *Report*

Positive → **Microscopic urinalysis** (Rapid)

Inconclusive → *Caused by:* Unexplained hypercellularity; Unidentified mononuclear cells and pathological casts; Inclusion-bearing cells; Obscuring inflammation; Cellular atypia or suspicious malignancy

Positive → *Report:* Hematuria; Pyuria; Bacteriuria and organisms; Crystalluria; Physiological casts; Lipiduria; Other

Negative → *Report*

Additional procedures suggested
Microbiological studies
Polarizing microscopy for crystals
Stone analysis
Fat stains for lipiduria

Fig. 53-30. Coordinated approach to urine sediment examination. (From Schumann, G.B., Schumann, J.L., and Schweitzer, S.: Lab. Management **1**:47, 1983.)

10. Gillenwater, N.Y.: Detection of urinary leukocytes by Chemstrip-L, J. Urol. **125**:383-384, 1981.
11. Avent, J., Schumann, G.B., and Vars, L.: Comparison of the Chemstrip leukocyte test with a standardized Papanicolaou-stained urine sediment evaluation, Lab. Med. **14**:163-166, 1983.
12. Race, G.J., and White, M.G.: Basic urinalysis, Hagerstown, Md., 1979, Harper and Row, Publishers, Inc.
13. Ross, D.L., and Neely, A.E.: Textbook of urinalysis and body fluids, East Norwalk, Conn., 1983, Appleton-Century-Crofts.
14. Schumann, G.B., and Greenberg, N.F.: Usefulness of macroscopic urinalysis as a screening procedure: a preliminary report, Am. J. Clin. Pathol. **71**:452-456, 1979.
15. Winkel, P., Statland, B., and Jorgenson, K.: Urine microscopy: an ill-defined method examined by a multifactorial technique, Clin. Chem. **20**:436-439, 1974.
16. KOVA Technical bulletin, Fountain Valley, Calif., 1982, ICL Scientific.
17. Whale-T-System: Technical bulletin, Commerce City, Col., 1982, Whale Scientific, Inc.
18. Schumann, G.B., and Henry, J.B.: An improved technique for the evaluation of urine sediment, Lab. Management **1**:19-24, 1977.
19. Schumann, G.B., and Weiss, M.A: Atlas of renal and urinary tract cytology and its histopathologic bases, Philadelphia, 1981, J.B. Lippincott Co.
20. Koss, L.G.: Diagnostic cytology and its histopathologic bases, Philadelphia, 1979, J.B. Lippincott Co.
21. Sternheimer, R.: A supravital cytodiagnostic stain for urinary sediment, J.A.M.A. **231**:826-832, 1975.
22. Holmquist, N.: Detection of cancer with urinary sediment, J. Urol. **123**:188-189, 1980.
23. Brody, L.H., Webster, M.C., and Kark, R.M.: Identification of elements of urinary sediment with phase contrast, J.A.M.A. **206**:1777-1781, 1969.
24. Haber, M.H.: Interference contrast microscopy for identification of urinary sediment, Am. J. Clin. Pathol. **57**:316-319, 1972.
25. Haber, M.H.: Urinary sediment: a textbook atlas, Chicago, 1981, American Society of Clinical Pathologists.
26. Kesson, A.M., Talbott, J.M., and Gyory, A.Z.: Microscopic examination of urine, Lancet **2**:809-812, 1978.
27. Schumann, G.B., Burleson, R.L., Henry, J.B., et al: Urinary cytodiagnosis of acute renal allograft rejection using the cytocentrifuge, Am. J. Clin. Pathol. **67**:134-140, 1977.
28. Tweeddale, D.N.: Urinary cytology, Boston, 1977, Little, Brown & Co.
29. Voogt, H.J., Rathert, P., and Beyer-Boon, M.E.: Urinary cytology, New York, 1977, Springer-Verlag, New York, Inc.
30. Dudas, H.C.: Quality in urinalysis, Lab. Med. **12**:765-767, 1981.
31. Hoeltge, G.A., and Ersts, A.: A quality-control system for the general urinalysis laboratory, Am. J. Clin. Pathol. **73**:403-408, 1980.
32. Little, P.J.: A comparison of the urinary white cell concentration with the white cell excretion rate, Br. J. Urol. **36**:360-363, 1964.

Chapter 54 Carbohydrates and metabolites

Glucose

LAWRENCE A. KAPLAN

Glucose, dextrose
Clinical significance: p. 520
Molecular formula: $C_6H_{12}O_6$
Molecular weight: 180.16 daltons
Merck Index: 4290
Chemical class: carbohydrate

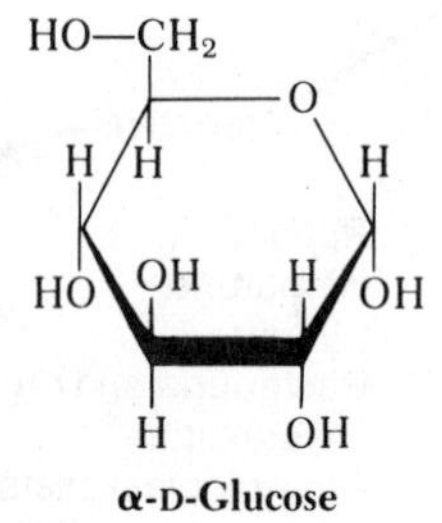

α-D-Glucose

PRINCIPLES OF ANALYSIS AND CURRENT USAGE

Chemically, glucose is an aldohexose in which the aldehyde form is in equilibrium with the endiol form (shown above). The latter is the favored structure at physiological pH. The aldehyde/endiol equilibrium allows glucose to be reduced and oxidized with facility.

Most older established methods for measurement of serum glucose were based on the ability of glucose to directly reduce cupric ions (Cu^{++}) to monovalent cuprous ions (Cu^{+}). In the presence of heat, the reduced cuprous ions can form cuprous oxide (Cu_2O), which can be detected by a variety of methods. The most popular method has been the reduction of phosphomolybdate (Folin-Wu) or arsenomolybdate (Somogyi-Nelson) to form blue molybdenum compounds (Table 54-1, methods 6 to 8). Neocuproine (2,9-dimethyl-1,10-phenanthroline) can also be reduced by Cu^{+} ions to form a highly colored complex (Table 54-1, method 5). Although it is a more sensitive procedure than the other copper-reduction methods, this method also lacks specificity. According to recent College of American Pathologists surveys, few laboratories use these procedures, and they are listed in Table 54-1 only for historic interest.

Benedict's modification of the copper-reduction methods (Table 54-1, method 8) is still used today, but only as a semiquantitative method for estimating urine glucose. This is marketed by the Ames Company (Division of Miles Laboratories, Elkhart, Indiana) as Clinitest. This procedure, sensitive to *total* reducing compounds present in urine, yields red Cu_2O and yellow CuOH precipitates. The greater the concentration of glucose, the redder the final color. The use of the Benedict reaction for measuring urinary glucose is restricted by its poor specificity. In combination with a more specific enzymatic glucose assay (see below), the Benedict reaction can be used to screen for genetic diseases of carbohydrate metabolism in newborns (see Chapter 5). A negative enzyme test and a positive Benedict reaction (Clinitest) is suggestive of such diseases.

The alkaline ferricyanide reaction (Table 54-1, method 9) method involves the reduction of yellow ferricyanide, Fe $(CN)_6^{-3}$, by glucose in alkaline conditions to colorless ferrocyanide, $Fe(CN)_6^{-4}$. Initial automation of this method on the Technicon AutoAnalyzer correlated the decrease in yellow color to glucose concentration. Later, adaptations of this method in the 1960s and 1970s to AutoAnalyzer equipment utilized direct measurement of ferrocyanide. The ferrocyanide was reacted with excess ferric ions to form ferric ferrocyanide (Prussian blue). However, this reaction is fairly nonspecific because of significant positive interference with such compounds as creatinine and uric acid. The alkaline ferrocyanide reactions have largely been replaced by more specific methods.

The *o*-toluidine reaction is based on the ability of many aromatic amines in acid solutions to condense with the aldehyde group of glucose to form glycosamines. The initial reaction product is most likely an unstable *N*-glycoside that is in equilibrium with the stable Schiff base (Table 54-1, method 4). The aromatic amine used currently, *o*-toluidine, is believed to be a carcinogen. The positive features of the reaction are its relatively good accuracy, sensitivity, and precision. The primary disadvantages are the noxious and carcinogenic character of the reagents and positive interferences by urea and other sugars, most notably mannose and galactose. The method is not routinely automated today.

The most commonly used procedures for glucose analysis employ enzymes as reagents to increase reaction specificity. These are the glucose oxidase and hexokinase re-

Table 54-1. Methods of glucose analysis

Method	Type of analysis	Principle	Usage	Comments
A. Enzymatic				
1. Hexosekinase (HK)	Quantitative, increased absorbance at 340 nm, K or EP	$\text{Glucose} + \text{ATP} \xrightleftharpoons{\text{HK}} \text{Glucose-6-phosphate} + \text{ADP}$ $\text{Glucose-6-phosphate} + \text{NADP}^+ \xrightleftharpoons{\text{G-6-PD}} \text{6-Phosphogluconate} + \text{NADPH} + \text{H}^+$	Serum, CSF, and urine Automated	Has been proposed as the basis of the reference method; very good accuracy and precision
Glucose oxidase (GO) 2. Oxygen consumption	Quantitative, polarigraphic measurement using O_2 electrode	$\text{Glucose} + O_2 \xrightarrow{\text{GO}} \text{Gluconic acid} + H_2O_2$ H_2O_2 consumed in side reactions	Serum and CSF Semiautomated and fully automated systems	Correlates best with the proposed reference method; very good accuracy and precision
3. Coupled reaction ("Trinder")	Quantitative, using various types of dyes as final O_2 acceptor, increased absorbance, K or EP	$\text{Glucose} + O_2 \xrightarrow{\text{GO}} \text{Gluconic acid} + H_2O_2$ $H_2O_2 + \text{Reduced dye} \xrightleftharpoons[\text{peroxidase}]{\text{Horseradish}} \text{Oxidized dye (colored)} + H_2O$ *(Peroxidase indicator reaction)*	Serum, urine, and CSF Easily and usually adapted to automated analysis	Second indicator reaction susceptible to false-positive interferences from a variety of compounds; good accuracy and precision
	Semiquantitative in dipstick screen, visual, or reflectance photometry		Serum, urine Used in all dipstick screens	
B.				
4. *o*-Toluidine	Quantitative, increased absorbance at 630 nm, usually EP	*o*-Toluidine + Glucose ⟶ Glycosamine (colored)	Serum or urine Rarely used in automated analysis	Suspected carcinogen; positive interferences from other sugars, especially mannose and galactose; positive bias from turbidity
C. Copper reduction 5. (Neocuproine)	Reactions 5 to 7 Quantitative EP	$Cu^{++} + \text{Glucose} \xrightarrow[OH^-]{\text{Heat}} Cu^+$ $Cu^+ + \text{Neocuproine} \longrightarrow$ Colored complex (2,9-dimethyl-1,10-phenanthroline HCl)	Reactions 5 to 7 are rarely used now; of historical interest	Large positive bias because of chemical interference by other sugars, creatinine, ascorbic acid, and other compounds
6. Phosphomolybdate (Folin-Wu) 7. Arsenomolybdate (Somogyi-Nelson)		$Cu^{++} + \text{Glucose} \xrightarrow[OH^-]{\text{Heat}} Cu_2O$ (colored) $Cu^+ + \text{Molybdate } (Mo^{++}) \longrightarrow$ Blue molybdenum complexes		
8. Benedict's modification	Qualitative, semiquantitative	$Cu^{++} + \text{Glucose} \xrightarrow[OH^-]{\text{Heat}} Cu_2O\downarrow$ (red) $+ CuOH\downarrow$ (yellow)	Formulation used in reaction 8 basis of semiquantitative tests for total reducing sugars in *urine*	Used in combination with more specific glucose oxidase urine screen to differentiate glucosuria from other sugars in urine, especially in neonates

K, Kinetic analysis mode; *EP*, end-point analysis mode.

Continued.

Table 54-1. Methods of glucose analysis—cont'd

Method	Type of analysis	Principle	Usage	Comments
D.				
9. Alkaline ferricyanide	Quantitative, decreased absorbance at 420 nm because of consumption of ferricyanide	$Fe(CN)_6^{-3}$ (ferricyanide) (yellow) $\xrightarrow[\text{Glucose-e}^-]{\text{Heat, OH}^-}$ $Fe(CN)_6^{-4}$ (ferrocyanide)	Rarely used, sometimes seen in Technicon systems; of historical interest	1 mg of creatinine = 1 mg of glucose; 0.5 mg of uric acid = 1 mg of glucose; very poor specificity
E.				
10. Mass fragmentography	Ratio of specific fragments for unlabeled and deuterated labeled glucose Quantitative	Fragmentation of glucose into specific detectable fragments	Rare	Proposed reference standards method

actions. Both procedures have been automated with resulting high specificity and precision.

The hexokinase method involves two coupled reactions. The hexokinase reaction (Table 54-1, method 1) phosphorylates hexoses with ATP. Glucose-6-phosphate dehydrogenase reacts specifically with glucose-6-phosphate produced by the hexokinase reaction to yield 1 mole of NADPH for each mole of glucose that is reduced. Although other hexoses can enter into the hexokinase reaction, normal serum concentrations of these sugars do not cause significant interference. The hexokinase procedure, using a protein-free filtrate as the sample, has been proposed as the product class standard for the measurement of glucose.[2]

One of the most frequently used glucose oxidase methods uses two coupled enzyme reactions (Table 54-1, method 3). In this case, the initial reaction is the specific one, and the indicator reaction is nonspecific. The first reaction employs glucose oxidase to oxidize glucose to glucuronic acid and hydrogen peroxide. The hydrogen peroxide is consumed by a peroxidase-dye indicator reaction in which the oxidized dye is colored, allowing the reaction to be monitored photometrically. The dyes employed as final oxygen acceptors can vary.[1] The two most widely used compounds are 3-methyl-2-benzothiazolinone hydrazone/*N,N*-dimethylaniline (MBTA/DMA) and phenylaminophenazone (PAP).

Automated and semiautomated procedures have also taken advantage of the unique specificity of the glucose oxidase reaction (Table 54-1, method 1) as a basis of highly specific and precise methods. These procedures follow only the specific glucose oxidase reaction by polarigraphic monitoring of the kinetic consumption of oxygen with an oxygen electrode. Hydrogen peroxide produced by the glucose oxidase reaction is removed by reaction either with iodides or with ethanol in the presence of catalase. Monitoring only the glucose oxidase reaction instead of the less specific peroxidase step greatly reduces the number of potential interferences.

REFERENCE AND PREFERRED METHODS

Both the copper reduction and ferricyanide reactions have been shown to be nonspecific, reacting with compounds such as creatinine and uric acid even after use of a protein-free filtrate as the sample. Although the *o*-toluidine procedure has acceptable accuracy and precision, it has disadvantages because of the noxious nature of its reagents and the nonspecific reaction with urea and other hexoses, most notably mannose and galactose. It offers only an economic advantage over enzymatic methods, and its use has rapidly decreased over the past few years.

The College of American Pathologists (CAP), in an extensive analysis of their survey data for serum glucose, suggested that the results reported from users of the hexokinase method showed the lowest overall bias from the group mean.[3] Glucose oxidase methods using phenylaminophenazone as the final indicator showed the next smallest overall bias. Oxygen rate–glucose oxidase assays showed a larger and consistently low bias, just as *o*-toluidine did. Ferricyanide and neocuproine had unacceptably large biases. The results of the same CAP analysis for urine glucose showed a similar pattern of results. The CAP survey showed the oxygen rate–glucose oxidase assays to have the best precision of all assays reported, with the other enzyme-based assays having the next best precision.

In 1977, the product evaluation subcommittee for the standards committee of the College of American Pathologists conducted a study comparing the proposed product class standard method for glucose analysis (hexokinase) with eight automated and two manual glucose procedures performed in a large reference laboratory.[4] The authors of the study concluded that the automated oxygen rate–glucose oxidase method compared best with the proposed reference method in terms of precision, accuracy, freedom from interferences, and sample carryover.

In 1976, Björkhem et al. suggested the use of a mass fragmentation procedure as a reference method for serum glucose measurements.[5] In their study they compared their method with two hexokinase and two glucose oxidase assays. These assays were performed by manual analysis except for the oxygen rate–glucose oxidase procedure. Although all the reviewed methods had nonsignificant error from a clinical view, the oxygen rate–glucose oxidase method compared best with the mass spectrometer method from an analytical view (accuracy and precision).

Thus the best methods available for the routine automated measurement of glucose in body fluids appear to be the enzyme methods used in the kinetic analysis mode. A comparison of these methods is found in Table 54-2. High throughput automated instruments that employ the oxygen rate–glucose oxidase methods are available. Use of coupled glucose oxidase hexokinase methods in the kinetic mode will give results as clinically acceptable as those given by the oxygen-rate procedure.

The disadvantage of the coupled glucose oxidase assays is that many compounds present in serum or urine (such as bilirubin, ascorbic acid, and uric acid) can be oxidized by the hydrogen peroxide produced by the glucose oxidase reaction, resulting in a negative bias. In addition, there may be compounds that can react by oxidizing the indicator dye, resulting in positive biases. The small positive biases observed in these procedures indicate that the latter type of interferences predominate. The specificity and accuracy of the hexokinase and coupled glucose oxidase procedures have been increased by a choice of reaction conditions that minimize this bias. The most common approach is to employ kinetic analysis of the reaction. By selection of the most appropriate initial and subsequent times for spectrophotometric reading, a reduction in both direct photometric and chemical interferences can be made. Kinetic analysis–based methods will most likely increase in frequency for the analysis of glucose.

The reference hexokinase and glucose oxidase/peroxidase methods minimize interference by protein and other serum constituents by means of a Somogyi protein precipitation step. Adaptations of these procedures for routine automated analysis employ large dilutions of the sample to minimize the effects of these interferences. Thus the fraction of sample volume noted for the last two methods listed in Table 54-2 are tenfold lower than for the reference methods. In addition, one can further minimize errors in accuracy resulting from the direct analysis of serum by performing the analysis as a kinetic reaction or by bichromatic analysis (see pp. 987 to 990).

The manual method that follows is an adaptation of an automated procedure based on Trinder's method and is listed in Table 54-2.

SPECIMEN

Serum or plasma, free of hemolysis, is the specimen of choice. Other body fluids, such as cerebrospinal fluid and urine, can also be used. Common anticoagulants (oxalate, fluoride, ethylenediaminetetraacetic acid [EDTA], citrate, or heparin) do not cause interference. Since glucose in whole blood at room temperature can undergo glycolysis at a rate of approximately 5% per hour, the sample should be centrifuged and removed from the clot or cells as soon as possible. Glucose in serum or plasma separated from blood cells is stable for up to 3 days at 2° to 8° C. Fluoride and iodoacetate have been used as inhibitors of glycolysis to preserve blood that cannot be separated rapidly. Since iodoacetate-preserved blood yields a better serum specimen, this might be the preferred preservative.

Urine can be analyzed by enzymatic methods if a large dilution step and kinetic analysis are employed.

REFERENCE RANGE

The range of serum glucose in healthy adults is listed below for several methods. There does not appear to be any significant difference in glucose levels between males and females or between races. There are some age-related differences. Below 5 years of age, normal glucose levels may be 10% to 15% below adult levels. Newborns can have blood glucose concentrations ranging from 200 to 800 mg/L, with blood glucose in premature infants even lower.[6] Glucose levels in cerebrospinal fluid are approximately 40% to 80% of serum or plasma levels.

Reference ranges for methods can vary significantly and should be evaluated by each laboratory.

Method	Serum glucose, adult (mg/L)
1. Glucose oxidase, phenylaminophenazone, end point (Technicon)	700 to 1050
2. Glucose oxidase, phenylaminophenazone, kinetic reaction (IL919)	600 to 950
3. Oxygen rate–glucose oxidase (Beckman ASTRA)	650 to 1100
4. Hexokinase, end point (ACA)	700 to 1100

Table 54-2. Comparison of reaction conditions for glucose analysis

	Hexokinase (proposed reference)	Glucose oxidase/peroxidase (proposed reference)	Glucose oxidase oxygen rate consumption*	Glucose oxidase/peroxidase (modified Trinder)†
Temperature	25° C	37° C	37° C	37° C
pH	7.5	7.0	7.0	7.0
Final concentration of reagent components	Hexokinase (yeast): 83 U/L Glucose-6-phosphate dehydrogenase: 83 U/L *(Leuconostoc mesenteroides)* NADP: 1.0 mmol/L ATP: 1.1 mmol/L Tris buffer: 80 mmol/L Magnesium acetate: 3.3 mmol/L	Glucose oxidase *(Aspergillus niger)*: 2600 U/L Peroxidase (horseradish): 5000 U/L Chromogen (*o*-dianisidine): 0.324 mmol/L 0.036 mol/L phosphate buffer in 40% glycerol	Glucose oxidase *(A. niger)*: 140 U/L Methanol: 5% Potassium iodide: 10 mmol/L	Glucose oxidase *(Aspergillus)*: ≥ 12,000 U/L Peroxidase (horseradish): ≥ 1200 U/L Chromogen (4-aminophenazone): 1.48 mmol/L Phenol: 4.25 mmol/L Phosphate buffer: 0.01 mol/L
Fraction of sample volume	0.16	0.20	0.01	0.006
Sample	Protein-free supernatant according to Somogyi	Protein-free supernatant according to Somogyi	Serum, CSF, and urine	Serum, plasma, CSF, and urine
Linearity	6000 mg/L	10,000 mg/L	6000 mg/L	7500 mg/L
Time of reaction	End point at 10 minutes	End point at 30 minutes	Kinetic, 9.6 seconds	Kinetic, 16 seconds
Major interferences (concentration at which compound interfered at glucose concentration within normal range)	None	Ascorbic acid (250 mg/L) L-Cysteine (1.5 g/L) Citric acid (15 g/L) Uric acid (150 mg/L) L-Dopa (100 mg/L)	Ascorbic acid: 250 mg/L L-Cysteine:1.5 g/L Citric acid: 15 g/L Hemoglobin: 10 g/L	Bilirubin: >100 mg/L Acetoacetic acid
Precision $\bar{X}$‡ (% CV)	468 (5%-10%)§ 921 (3%-7.9%) 3020 (2%-6.5%)	443 (6%-9.7%)‖ 911 (3.5%-6.1%) 4033 (2%-6.5%)	428 (4.4%-6.3%) 879 (3.4%) 2983 (2.8%)	396 (6.4%) 848 (3.6% 2897 (1.7%)

*Beckman Instruments, Inc., Fullerton, Calif.
†IL919, Instrumentation Laboratory, Lexington, Mass.
‡Data from 1980 College of American Pathologists comprehensive chemistry surveys; $\bar{X}$, mean in mg/L; *CV*, coefficient of variation.
§Group mean for all hexokinase methods (without blank).
‖Group mean for all glucose oxidase–phenylaminophenazone methods.

Ketones (ketone bodies)

LAWRENCE A. KAPLAN

Clinical significance: p. 520

Common name:	Acetoacetic acid	Acetone	β-Hydroxybutyric acid
Structure:	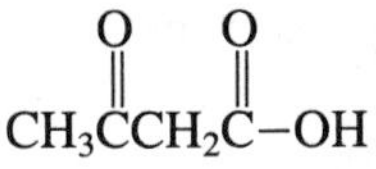 $CH_3C(=O)CH_2C(=O){-}OH$	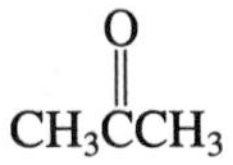 $CH_3C(=O)CH_3$	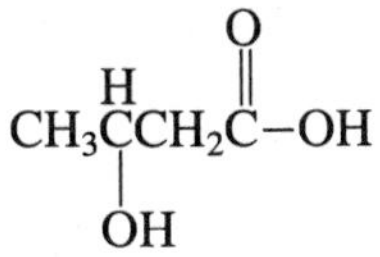 $CH_3CH(OH)CH_2C(=O){-}OH$
Molecular formula:	$C_4H_6O_3$	C_3H_6O	$C_4H_8O_3$
Molecular weight:	102.09	58.08	104.10
Merck Index:	46	52	4716
Chemical class:	Ketocarboxylic acid	Ketone	Hydroxy acid

PRINCIPLES OF ANALYSIS AND CURRENT USAGE

Ketone bodies are biochemicals that have either structural ketone moieties ($R_1{-}C(=O){-}R_2$) or are directly derived from ketones (that is, reduced ketones). Commonly found serum ketone bodies include the ketones, pyruvate, acetoacetic acid, and acetone, and the reduced chemical forms, lactic acid and β-hydroxybutyric acid. Acetoacetic acid, acetone, and β-hydroxybutyric acid are metabolites of fatty acids that accumulate in large amounts when the liver's capacity to metabolize the final oxidative catabolite of fatty acids, acetyl coenzyme A, is overwhelmed. This occurs in states of starvation, impaired carbohydrate metabolism (such as diabetes), or acute ethanol abuse. In these states, the oxidative capacity of liver cells is lessened and the ratio of NAD/NADH is greatly decreased, resulting in an inability to oxidize or transfer acetyl CoA. To allow continued metabolism and recycling of the coenzyme A cofactor, the hepatocytes begin active ketogenesis by the following series of reactions:

Thus a series of ketones or "ketone bodies" (that is, β-hydroxybutyric acid) are formed. The type and amount of the ketone bodies formed depend on the oxidative state of the cells. In early, severe diabetic ketoacidosis or in acute alcohol ketoacidosis, the cells are in a more reduced state (that is, decreased NAD/NADH ratio) and β-hydroxybutyric acid is usually the predominant ketone body form. If these states are associated with a concomitant lacticacidemia, there can be a greater tendency toward production of β-hydroxybutyric acid from acetoacetic acid. As the biochemical status of the patient improves and the NAD/NADH ratio increases, the β-hydroxybutyric acid levels decrease while the levels of acetoacetic acid increase. Generally the reported proportions of total serum ketone bodies are 2% acetone, 20% acetoacetic acid, and 78% β-hydroxybutyric acid.[36]

The analysis for ketones has thus been limited by the fact that the ketone bodies may be present in varied proportions during an illness. An assay that specifically mea-

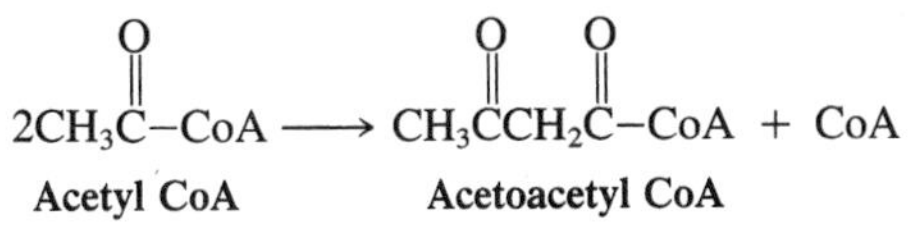

$$2CH_3C(=O){-}CoA \longrightarrow CH_3C(=O)CH_2C(=O){-}CoA + CoA \quad \text{Eq. 54-1}$$

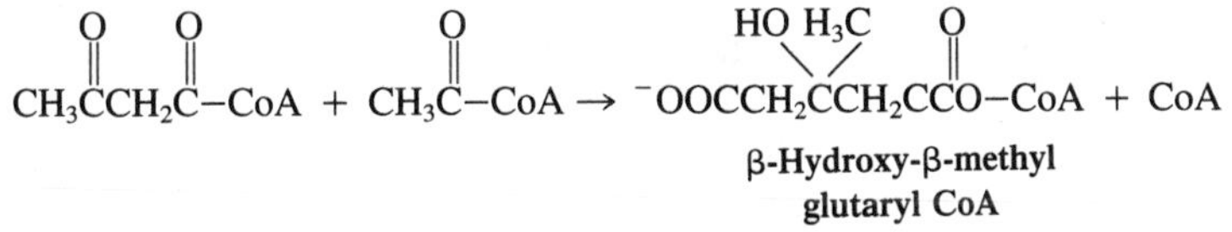

Acetyl CoA → Acetoacetyl CoA

$$CH_3C(=O)CH_2C(=O){-}CoA + CH_3C(=O){-}CoA \rightarrow {}^{-}OOCCH_2C(OH)(CH_3)CH_2C(=O){-}CoA + CoA \quad \text{Eq. 54-2}$$

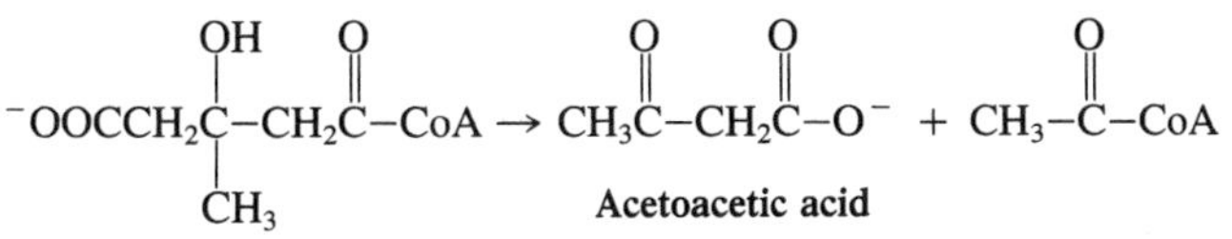

β-Hydroxy-β-methyl glutaryl CoA

$$^{-}OOCCH_2C(OH)(CH_3){-}CH_2C(=O){-}CoA \rightarrow CH_3C(=O){-}CH_2C(=O){-}O^{-} + CH_3{-}C(=O){-}CoA \quad \text{Eq. 54-3}$$

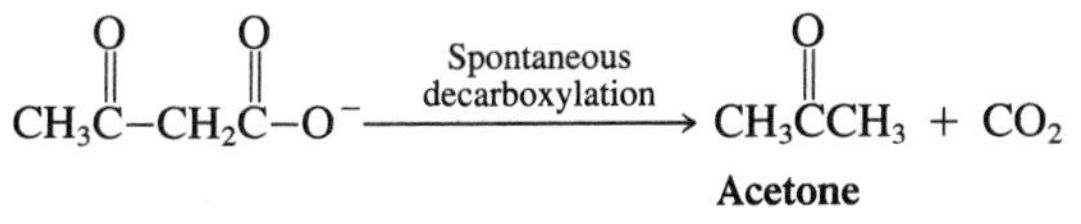

Acetoacetic acid

$$CH_3C(=O){-}CH_2C(=O){-}O^{-} \xrightarrow[\text{decarboxylation}]{\text{Spontaneous}} CH_3C(=O)CH_3 + CO_2 \quad \text{Eq. 54-4}$$

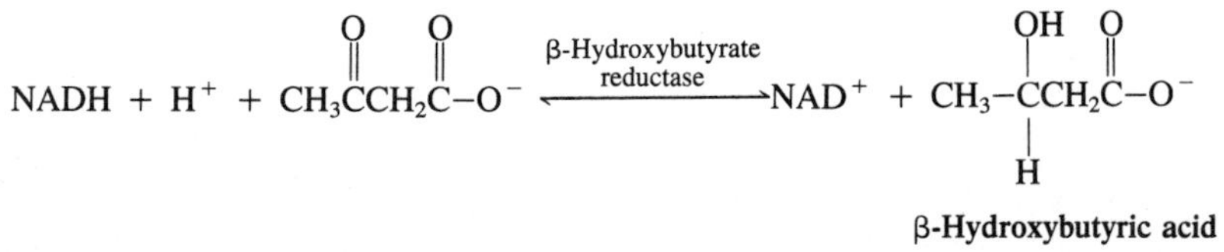

Acetone

$$NADH + H^+ + CH_3C(=O)CH_2C(=O){-}O^{-} \xrightleftharpoons[\text{reductase}]{\beta\text{-Hydroxybutyrate}} NAD^+ + CH_3{-}C(OH)(H)CH_2C(=O){-}O^{-} \quad \text{Eq. 54-5}$$

β-Hydroxybutyric acid

sures one of these biochemicals will almost certainly lack the sensitivity to detect total ketone bodies produced. In addition, any clinically useful test should be able to be performed rapidly, as a 24-hour stat. test.

The oldest quantitative methods for the analysis of ketones employed precipitation of acetone with mercury salts followed by gravimetric or titrimetric analysis (method 1, Table 54-3).[37,38] Colorimetric analyses that have been used include the reaction of acetone with salicylaldehyde to form a red color[39,40] and the formation of hydrazones between the ketones and dinitrophenylhydrazine.[41] More recently suggested colorimetric assays include the reaction between acetone and vanillin at 50° to 55° C to form divanillalacetone (method 2, Table 54-3), which is red (A_{max} = 415 nm) under the alkaline conditions of the reaction.[42] Nitroferricyanide, ferric chloride, and 2,5-dichlorobenzene diazonium chloride have also been used for colorimetric assays.[36]

The reaction between ketones and nitroprusside (sodium nitroferricyanide) under alkaline conditions is a widely used procedure (method 3, Table 54-3).[43,44] Acetoacetic acid and acetone both form a purple color, which is most frequently used as a semiquantitative measure of ketones in serum and urine. Ames (Ames, Inc., Division of Miles Laboratories, Elkhart, Indiana) markets Ketostix in the form of a paper impregnated with nitroprusside, glycine, and sodium phosphate. The phosphate serves to provide a suitably buffered pH for the reaction to occur. A similar reagent (including lactose as a color enhancer) is marketed as a spot test by Ames as Acetest. Both reagents can be used for serum or urine.

Recently developed have been several techniques that measure the individual ketone bodies with high specificity. Acetoacetic acid and β-hydroxybutyric acid can be quantitated by an enzyme assay (method 4, Table 54-3), which makes use of the following reversible reaction:

$$\text{NADH} + \text{H}^+ + \text{Acetoacetic acid} \xrightleftharpoons[\text{dehydrogenase}]{\beta\text{-Hydroxybutyrate}} \beta\text{-Hydroxybutyrate} + \text{NAD} \qquad \textit{Eq. 54-6}$$

The reaction can be monitored by following the absorption of NADH at 340 nm.[45,46] At pH 8.5 to 9.5, the reaction proceeds to the left as written and the concentration of β-hydroxybutyric acid is quantitated by monitoring of the increase in absorbance at 340 nm. When performed at pH 7.0, the reaction proceeds to the right and the amount of acetoacetic acid present is proportional to the decrease in absorbance at 340 nm.

Table 54-3. Methods for ketone analysis

Method	Type of analysis	Principle	Usage	Comments
Urine				
1. Precipitation	Gravimetric or titrimetric, quantitative	Acetone is precipitated by mercury salts	Historical	Measures only acetone
2. Colorimetric	Quantitative	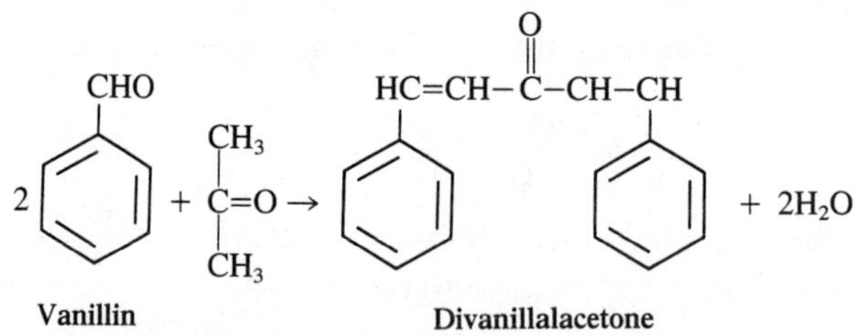 2 Vanillin (benzene ring–CHO) + CH_3–C(=O)–CH_3 → Divanillalacetone (ring–HC=CH–C(=O)–CH–CH–ring) + $2H_2O$	Rare	Measures only acetone
3. Colorimetric	Semiquantitative	$Na_2Fe(CN)_5NO$ + Acetone/acetoacetate → Purple color	Most common; frequently used as a stat. procedure	Sensitivity for acetoacetate five times greater than for acetone
4. Enzymatic	Quantitative	$\text{NADH} + \text{H}^+ + \text{AcAc} \xrightleftharpoons[\text{dehydrogenase, pH 8.5-9.5}]{\text{pH 7.0, }\beta\text{-Hydroxybutyrate}} \beta\text{-HB} + \text{NAD}$	Rare, not useful as a stat. procedure	By adjustment of pH and cofactor concentration (that is, NADH or NAD), reaction used to quantitate AcAc and β-HB
5. Gas chromatography	Quantitative	Acetone is detected by flame ionization detector; AcAc is converted to acetone by heating.	Rare, not useful as stat. test	Acetoacetate quantitated by subtraction of nonheated acetone from heated acetone

AcAc, Acetoacetic acid; *β-HB*, β-hydroxybutyric acid.

Acetone and acetoacetic acid have been quantitated by gas chromatography (method 5, Table 54-3). Acetone is quantified by a flame ionization detector, and acetoacetic acid is estimated by measurement of acetone before and after heating to convert the acetoacetic acid to acetone.[47] The difference represents the amount of acetoacetic acid present in the sample.

REFERENCE AND PREFERRED METHOD

There are drawbacks to most of the older, quantitative methods for the assay of ketones, the most important being that they detect only acetone. Since this ketone represents only a very small proportion of the total ketone bodies present in serum or urine, it can estimate only the total ketone load. Similarly, although the nitroprusside test for acetoacetic acid more accurately reflects total ketones, this test also can be misleading. As mentioned earlier, in diabetic ketoacidosis β-hydroxybutyric acid is the predominant ketone body formed, and tests for acetoacetic acid may be negative or weakly positive; therefore the presence of ketones is underestimated. In fact, as the diabetic ketoacidosis is resolved, more acetoacetic acid can be found than in the initial stages of the crisis.

The older methods based on precipitation were time and labor consuming and are no longer in use. The dinitrophenolhydrazone procedure requires extraction and is interfered with by ethanol and acetaldehyde.[36,48] The salicylate method is not quantitative because of color instability of the reaction product. The Gerhardt ferric chloride test for acetoacetate is very nonspecific, with false-positive results occurring with phenols and salicylates, and has poor sensitivity (250 to 500 mg/L).[36] The vanillan method is quantitative and fairly specific but only for acetone. Its precision is reported to be 15% for serum acetone levels within the normal range.[36] This method is not suitable as a rapid, stat. procedure.

Gas chromatography is very specific for acetone (and, by heating and subtraction, acetoacetic acid). However, this method requires specialized, expensive equipment and may not be suited for routine clinical needs.

The enzymatic procedures are very specific for acetoacetate or β-hydroxybuterate and have been readily adapted to automated instruments such as the ABA-100 and centrifugal analyzers with coefficients of variation of less than 8% for levels within the normal range.[45,46,49] The reagents for these assays can be easily prepared by most laboratories. Thus these methods are recommended for those laboratories interested in specifically quantitating the individual serum and urine ketone bodies.

Most analyses for ketones in serum and urine appear to be best determined by the semiquantitative nitroprusside test. The major drawback to this procedure is that it cannot detect β-hydroxybutyric acid at all and is five- to tenfold less sensitive for acetone than for acetoacetic acid. Acetone is rarely positive at serum concentrations less than 5 mmol/L. The primary advantage to the nitroprusside screen is its relative ease of use and reagent stability, making it well suited as a stat. screen for ketones. As with any of the dry chemical processes, however, the reagent can deteriorate unless properly protected from air.

The reported sensitivity of the Ketostix is 50 to 100 mg/L of acetoacetate and 25 to 50 mg/L for Acetest. The Ketostix can give false-positive result in the presence of large amounts of levodopa.[50]

SPECIMEN

Well-centrifuged urine, serum, or plasma can be used for the semiquantitative nitroprusside test. Serum or plasma can be used for the enzymatic assays as well.

Bacterial contamination of urine can hasten the disappearance of the ketones from urine. In addition, acetone is a volatile substance and can be lost by evaporation. Thus the sample should be refrigerated within 20 to 30 minutes of phlebotomy if not analyzed immediately.

PROCEDURE

Principle

Acetoacetic acid (and acetone to a lesser extent) reacts with nitroprusside (sodium nitroferricyanide) under alkaline conditions to form a pink to purplish color. The test can be most easily performed with commercially available reagent in a paper-impregnated form or as a solid tablet.

Assay

Follow manufacturer's instructions for the storage, stability, and use of reagents. The reagent strips are reported to give most accurate results for urines whose specific gravities are between 1.010 and 1.020 g/mL.

Linearity

The reagent strips appear to give satisfactory results up to 1600 mg/L of acetoacetate.

Interferences

The reagent strips will give false-positive results in the presence of levodopa or its metabolites. Highly colored urine or hemolyzed samples may yield false-positive results.

REFERENCE RANGE

Henry reports a reference range for serum acetoacetate of 5 to 30 mg/L,[36] which compares well with results of 3 to 23 mg/L obtained by our laboratory using an enzymatic procedure.[49] There does not appear to be any age-related differences in reference ranges for ketone bodies.[39] At these levels, serum and urines will be negative by the nitroprusside semiquantitative screening tests.

Lactic acid

NANCY GAU

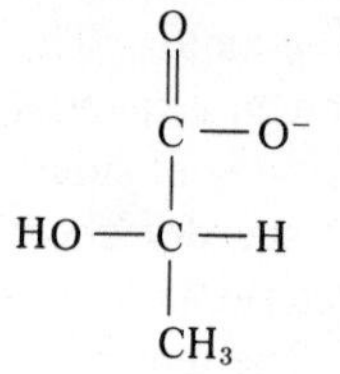

$$HO-\underset{CH_3}{\overset{\overset{O}{\|}\,C-O^-}{C}}-H$$

L-Lactate

L-Lactic acid, L-Lactate
Clinical significance: p. 387 and 520
Molecular formula: $C_3H_6O_3$
Molecular weight: 90.08 daltons
Merck Index: 5186
Chemical class: α-hydroxycarboxylic acid (glucose metabolite)

PRINCIPLES OF ANALYSIS AND CURRENT USAGE

Usually lactic acid is measured either by chemical or by enzymatic oxidation.

Chemical oxidation methods use either permanganate or manganese dioxide to degrade lactate to acetaldehyde, CO_2, or CO. The amount of lactate is proportional to the amount of product formed. Acetaldehyde can be measured colorimetrically,[7-9] by titration,[10,11] or by gas chromatography.[12,13] The CO formed by method 1a (Table 54-4) can be determined by titration[14] or gasometric analysis.[15] The CO_2 formed by method 1b can be quantified gasometrically by a Van Slyke manometer.[16,17]

Enzymatic methods[18-21] employ lactate dehydrogenase (LD) to oxidize lactic acid to pyruvic acid with the formation of the reduced form (NADH) of nicotinamide adenine dinucleotide (NAD) (method 2 in Table 54-4). Hydrazine or semicarbazide is added to form a complex with the pyruvate to remove it as a reaction product and therefore to drive the reaction to completion. The reaction is quantitated by measurement of the formation of NADH by either absorption spectroscopy or fluorometric methods. The analysis has been monitored as either an end point or a two-point kinetic reaction.

Recently, there have been introduced discrete lactate analyzers that employ electrochemical detection. The principle of measurement is derived from the older methods of lactate oxidation by ferricyanide, catalyzed by yeast L(+)-lactate dehydrogenase[22-27] (see method 3 listed in Table 54-4). The hexacyanoferrate II produced by the reaction is then oxidized at a platinum electrode against a silver–silver chloride electrode according to method 3 in Table 54-4. The amount of current generated at the platinum electrode is a measure of the concentration of hexacyanoferrate and thus of lactic acid.

REFERENCE AND PREFERRED METHODS

The chemical oxidation methods appear to be accurate and sensitive but time consuming and laborious. Excessive oxidation conditions can lead to further reaction of acetaldehyde and inaccurate results. Thus the reaction conditions of the methods measuring acetaldehyde must be carefully controlled. A gas chromatographic method for measuring the acetaldehyde formed is suitable if the proper equipment is available. Precision on duplicate analyses shows a 2.8% coefficient of variation; recovery, at levels of lactate ranging from 1.61 to 5.33 mmol/L, is from 97% to 104%.[13] The gas-liquid chromatographic columns, however, are easily contaminated, and care must be taken to avoid interfering substances (such as propyleneglycol and ethanol).[13]

Table 54-4. Methods of lactate analysis

Method	Type of analysis	Principle	Usage	Comments
1. Chemical oxidation with formation of:		Permanganate or MnO_2 oxidizes lactate to one of several products in one of the following reactions:	Historical	Accurate, precise, and sensitive, but time consuming
a. Acetaldehyde (CH_3CHO)	Quantitative, EP	$\text{Lactate} \xrightarrow[\text{Heat}]{H_2SO_4} \text{Acetaldehyde} + H_2O + CO$		
b. CO (carbon monoxide)	Quantitative, EP	As above		
c. CO_2 (carbon dioxide)	Quantitative, EP	$2\ \text{lactate} + O_2 \rightarrow 2\ \text{acetaldehyde} + 2CO_2 + H_2O$		
2. Enzymatic	Quantitative, EP or K	$\text{Lactate} + NAD^+ \xrightleftharpoons{LDH} \text{Pyruvate} + NADH + H^+$	Automated	Accurate and precise; automatable and rapid
3. Electrochemical	Quantitative, EP	$\text{Lactate} + 2Fe(CN)_6^{-3} \rightarrow \text{Pyruvate} + 2H^+ + 2Fe(CN)_6^{-4}$ $2Fe(CN)_6^{-4} \xrightarrow[\text{Pt electrode}]{} 2Fe(CN)_6^{-3} + 2e^-$	Automated	Accurate, precise, and sensitive, rapid analysis but equipment not readily available

K, Kinetic analysis mode; *EP*, end-point analysis mode.

The gasometric procedures measuring CO_2 produced by the oxidation of lactic acid have an advantage over the measurement of acetaldehyde in that excess oxidant, which can cause further oxidation of the acetaldehyde, does not affect the formation of CO_2. The other gasometric procedure, based on the analysis of CO, is less accurate and shows a positive bias for both blood and urine.[14,28] Both gasometric procedures require gasometers that are not routinely available in most laboratories.

Electrochemical devices for lactate analysis seem potentially quite useful in that they are very rapid (less than 3 minutes) and can be used with whole blood. However, only limited evaluation and comparison studies have been published on this method. Since the method requires the availability of a discrete specialized analyzer for a single, rarely ordered test, it cannot be recommended at this time.

Enzymatic methods provide direct measurement of lactic acid along with ease of operation and adaptation to available equipment. For these reasons, the enzymatic procedure is the method of choice. The Hohorst manual method is an end-point measurement in the presence of semicarbazide.[29] A modification of the method has a recovery of 100% with a precision (coefficient of variation) of 6%.[30] These reactions generally require a long incubation period and are very dependent on the concentration of various reactants.[20]

Recently a two-point kinetic method was introduced using an alkaline pH and adding hydrazine to remove the pyruvate. The method is easily adapted to centrifugal analyzers, which use microamounts of sample, and is both precise and accurate, with recovery of 101% to 104% and a coefficient of variation of less than 5%.[31,32] One limitation of the method is the low upper limit of linearity (5 mmol/L), which can easily be overcome by appropriate dilutions of the sample. Those dilutions that fall within the linear range are used for the calculations. The fluorescence method, which detects NADH by its emission at 450 nm, is an acceptable alternative to measurements using absorption spectroscopy. These methods have acceptable performance, with good recovery (98% to 101%) and precision (1.0% CV) and correlate well with chemical methods.[19,33]

SPECIMEN

One major problem with accurate lactic acid analysis is the need for rigorous sample processing. The reason is that blood cells will metabolize glucose to lactic acid, thereby giving a positive bias.[30,34] This metabolism is time and temperature dependent.

Preservation of lactate concentrations by deproteinization of whole blood or serum is neither convenient nor reliable.[32] Studies on the effect of sodium fluoride and sodium iodoacete showed an inhibitory effect on glycolysis with either compound and complete inhibition with a mixture of 1% sodium fluoride and 1% iodoacetate.[34] Since sodium fluoride can partially inhibit coagulation, the serum specimen from sodium fluoride phlebotomy tubes tends to be contaminated with protein clots.

A recent study showed that lactic acid levels in samples collected with sodium iodoacetate (final concentration, 0.5 g/L of blood) are stable for up to 2 hours at room temperature.[35] The iodoacetate does not affect coagulation and thus yields a clear serum sample. Heparinized plasma can be used for lactic acid analysis, but the specimen *must* be placed on ice and the plasma separated within an hour of collection. Oxaloacetate cannot be used as an anticoagulant, since it will inhibit lactate dehydrogenase used in the assay.

In terms of convenience and accuracy, the recommended sample for lactic acid analysis is a serum sample drawn in a sodium iodoacetate phlebotomy tube. The sample should be drawn with minimal stasis, preferably without a tourniquet. If a tourniquet must be used, it should be withdrawn after venipuncture. The phlebotomist should then allow several minutes to pass before actually drawing blood into the phlebotomy tube.

PROCEDURE

Principle

This method is based on the enzymatic oxidation of lactate to pyruvate by lactate dehydrogenase in the presence of NAD. The reaction is as follows:

$$\text{L-Lactate} + NAD^+ \underset{\text{pH 9.8}}{\overset{\text{LD}}{\rightleftharpoons}} \text{Pyruvate} + NADH + H^+$$

At this pH, the equilibrium favors the oxidation of lactate to pyruvate. However, if hydrazine is added, it reacts with pyruvate and removes the pyruvate as it is formed, further shifting the equilibrium to the right. The rate of increase in absorption of the NADH formed, measured at 340 nm, is related to the concentration of lactic acid in the sample.

Reagents

Buffer. Dissolve 60.5 g of hydroxymethylaminomethane (Tris), 4 g of disodium ethylenediaminetetraacetate, in 800 mL of distilled water. Add 11 mL of hydrazine hydrate. Adjust the pH to 9.8 and dilute to 1 L with distilled water. Stable 6 months at 4° C.

Beta-NAD (product no. 7004, Sigma Chemical Co.). Preweigh 65.4 mg of beta-NAD into test tubes. Cover and store at −20° C. Stable for at least 1 month. Dissolve preweighed NAD into 3 mL of distilled water before preparation of working reagent.

Lactate dehydrogenase (BMC Co., catalog no. 127230). An ammonium sulfate suspension of purified rabbit muscle LD, approximately 550 U/mg.

Working reagent. Mix 27 mL of buffer, 3 mL of dissolved NAD, and 40 μL of LD solution. Mix. Stable 24 hours at 4° C.

Stock lactate standard (20 mmol/L). Dissolve 192 mg of lithium L(+)-lactate in 100 mL of distilled water. Stable 6 months at 4° C.

Working standard

2 mmol/L: 1 mL of 20 mmol/L standard diluted volumetrically to 10 mL with distilled water.

8 mmol/L: 4 mL of 20 mmol/L standard diluted volumetrically to 10 mL with distilled water.

Both standards are stable for 2 months at 4° C.

Assay (Table 54-5)

1. Equipment: ABA-100 (Abbott Diagnostics, Chicago, Ill.)
2. Test parameters:

Temperature:	37° C
Mode:	End point
Normal/FRR:	Normal
Direction:	Up
Time:	5 minutes
Revolutions:	3
Filter:	340/380
Syringe plate:	1:51
Decimal:	0.000
Calibration factor:	0.500 (Set after initial dispensing)
Zero setting:	0 (Set after initial dispensing)

3. Follow normal operational procedures according to Abbott ABA-100 procedure manual.
4. Run both the 2 and 8 mmol/L lactate standards in each run. Analyze patient samples undiluted and diluted fivefold with distilled water.
5. After the last sample is aspirated and dispensed, push STOP button. Go to test mode.
6. Set 0 absorbance on position 1 with zero knob.
7. Push calibrate button in and set to 0.500 with scale knobs.
8. Return carousel to starting position and push RUN.
9. When carousel begins to move, spin it back to the starting position. It will now print first absorbance readings.
10. After second print (5 minutes later) perform calculations.

Calculation:

1. Calculate delta absorbance for each cuvette. Subtract first print from second print.
2. Use the following formula:

$$\frac{\Delta\ \text{Absorbance}_{unk}}{\Delta\ \text{Absorbance}_{std}} \times \text{Concentration}_{std} = \text{Concentration}_{unk}\ (\text{mmoles/L})$$

3. For unknowns less than 5 mmol/L, use the 2 mmol/L standard for the calculation. For unknowns greater than 5 but less than 10 mmol/L, use the 8 mmol/L standard for calculation. For unknowns greater than 10 mmol/L, use the diluted sample for calculation, again using either the 2 or 8 mmol/L standard as described above.

Table 54-5. Reaction conditions for lactate analysis on ABA-100

Temperature	37° C
pH	9.8
Final concentration of reagent components	Tris buffer: 0.44 mol/L NAD: 3 mmol/L LDH: 3600 U/L
Fraction of sample volume	0.02
Sample	Serum (iodoacetate) Plasma (heparin on ice) Cerebrospinal fluid
Linearity	5 mmol/L
Time of reaction	5 minutes
Major interferences	Intravenous injection of epinephrine, glucose, bicarbonate, or other infusions that modify the acid-base balance, causing an elevation in lactic acid. Avoid using hemolyzed samples.
Precision (between day)	10.7% (value: 3.5 mmol/L)

Notes

This method has also been adapted to a centrifugal analyzer (CentrifiChem 400, Baker Instruments, Pleasantville, N.Y.). Use the following parameters on this instrument:

Pipettor	Analyzer
10 μl of sample	30° C
50 μl of diluent	340 nm
350 μl of reagent (place reagent in 0 cuvette)	T_0: 3 seconds
	ΔT: 4 minutes
	Auto
	Terminal
	Concentration

Follow normal pipetting and operational guidelines for the CentrifiChem as outlined in the CentrifiChem procedure manual. Set print to 2 or 8 mmol/L standard according to the approximate concentration of the unknown.

REFERENCE RANGE

Less than 2 mmol/L

REFERENCES

1. College of American Pathologists: Comprehensive chemistry survey, Skokie, Ill., 1983.
2. Food and Drug Administration: In vitro diagnostic products for human use, proposed establishment of product class standards for detection or measurement of glucose, Federal Register 39, No. 126, 24136-24147, Washington, D.C., 1974, Department of Health, Education and Welfare.
3. Sheiko, M.C., Burkhardt, R.T., and Batsakis, J.G.: Glucose measurements: A 1977 CAP survey analysis, Am. J. Clin. Pathol. **72**(2):337-340, 1979.
4. Passey, R.B., Gillum, R.L., Fuller, J.B., Urry, F.M., and Giles, M.L.: Evaluation and comparison of ten glucose methods and the reference method recommended in the proposed product class standard, Clin. Chem. **23**:131-139, 1977.
5. Björkhem, I., Blomstrand, R., Falk, O., et al.: The use of mass

fragmentography in the evaluation of routine methods for glucose determination, Clin. Chem. Acta **72:**353-362, 1976.
6. Meites, S., editor-in-chief: Pediatric clinical chemistry, Washington, D.C., 1976, American Association for Clinical Chemistry, pp. 103-107.
7. Hochella, N.J., and Weinhouse, S.: Automated lactic acid determination in serum and tissue extracts, Anal. Biochem. **10:**304, 1965.
8. Barker, S.B., and Sumerson, W.H.: Colorimetric determination of lactic acid in biological material, J. Biol. Chem. **138:**535-554, 1941.
9. Barker, S.B.: Lactic acid. In Seligson, J.D., editor: In standard methods of clinical chemistry, Vol. 3, New York, Academic Press, 1961, p. 167.
10. Friedman, T.E., and Graeser, J.B.: The determination of lactic acid, J. Biol. Chem. **100:**291-308, 1933.
11. Long, C.: The stabilization and estimation of lactic acid in blood samples, Biochem. J. **40:**27, 1946.
12. Hoffman, N.E., Barboriak, J.J., and Hardman, H.F.: A sensitive gas chromatographic method for the determination of lactic acid, Anal. Biochem. **9:**175-179, 1964.
13. Savory, J., and Kaplan, A.: A gas chromatographic method for the determination of lactic acid in blood, Clin. Chem. **12**(9):559-569, 1966.
14. Ronzoni, E., and Wallen-Lawrence, Z.: Determination of lactic acid in blood, J. Biol. Chem. **74:**363-377, 1927.
15. Schneyer, J.: Eine Methode zur quantitativen Milchsäurebestimmung im Harne, Biochem. Z. **70:**294, 1915.
16. Avery, B.F., and Hastings, A.B.: A gasometric method for the determination of lactic acid in the blood, J. Biol. Chem. **94:**273-280, 1931.
17. Baumberger, J.P., and Field, J., Manometric method for quantitative determination of lactic acid, Proc. Soc. Exp. Biol. Med. **25:**87, 1927.
18. Pfleiderer, G., and Dose, K., Eine enzymatische Bestimmung der L(+)-Milchsäure mit Milchsäuredehydrase, Biochem. Z. **326:**436, 1955.
19. Loomis, M.E., An enzymatic fluorometric method for the determination of lactic acid in serum, J. Lab. Clin. Med. **57:**966, 1961.
20. Olson, G.F.: Optimal conditions for the enzymatic determination of L-lactic acid, Clin. Chem. **8:**1-10, 1962.
21. Parijs, J., and Barbier, F.: The enzymatic L(+)-lactate determination in blood, Z. Klin. Chem. **3:**74, 1965.
22. Williams, D.L., Doig, A.R., and Korosi, A.: Electrochemical analysis of blood glucose and lactate, Anal. Chem. **42:**118-121, 1970.
23. Durliat, H., Comtat, M., and Baudras, A., Spectrophotometric and electrochemical determinations of L(+)-lactate in blood by use of lactate dehydrogenase from yeast, Clin. Chem. **22**(11):1802-1805, 1976.
24. Racine, P., Engelhardt, R., Higelin, J.C., and Mindt, W.: An instrument for the rapid determination of L-lactate in biological fluids, Med. Instrum. **9:**11-14, 1975.
25. Racine, P., Klenk, H.O., and Kochsiek, K.: Rapid lactate determination with an electrochemical enzymatic sensor: clinical usability and comparative measurements, J. Clin. Chem. Clin. Biochem. **13:**533-539, 1975.
26. Soutler, W.P., Sharp, F., and Clark, D.M.: Bedside estimation of whole blood lactate, Br. J. Anaesth. **50:**445-450, 1978.
27. Piquard, F., Schaefer, A., and Haberey, P.: Bedside estimation of plasma lactate, Clin. Chem. **26**(3):532-533, 1980.
28. Maver, M.C.: The Schneyer method for the determination of lactic acid in urine, J. Biol. Chem. **32:**71-76, 1917.
29. Hohorst, H.J.: In Bergmeyer, H.U., editor: Methods of enzymatic analysis, New York, 1963, Academic Press, p. 266.
30. Drewes, P.A.: In Henry, R.J., Cannon, D.C., and Winkleman, J.W., editors: Clinical chemistry: principles and techniques, ed. 2, Hagerstown, Md., 1974, Harper & Row, Publishers, Inc., pp. 1330-1334.
31. Hadjiioannou, T.P., Hadjiioannou, S.I., Brunk, S.D., and Malmstadt, H.V.: Automated enzymatic determination of L-lactate in serum, with use of a miniature centrifugal analyzer, Clin. Chem. **22**(12):2038-2041, 1976.
32. Pesce, M.A., Bodourian, S.H., and Nicholson, J.F.: Rapid kinetic measurement of lactate in plasma with a centrifugal analyzer, Clin. Chem. **21**(13):1932-1934, 1975.
33. Antonis, A., Clark, M., and Pilkington, T.R.E.: A semiautomated fluorometric method for the enzymatic determination of pyruvate, lactate, acetoacetate, and β-hydroxybutyrate levels in plasma, J. Lab. Clin. Med. **68:**340-355, 1966.
34. Bueding, E., and Goldfarb, W.: The effect of sodium fluoride and sodium iodoacetate on glycolysis in human blood, J. Biol. Chem. **141:**539-544, 1941.
35. Kaplan, L.A., Gau, N., and Stein, E.A.: Collection and storage of serum lactic acid samples at room temperature without deproteinization, Clin. Chem. **26**(1):175-176, 1980.
36. Drews, P.A.: Carbohydrate derivatives and metabolites. In Henry, R.J., Cannon, D.C., and Winkleman, J.W., editors: Clinical chemistry: principles and techniques, ed. 2, Hagerstown, Md., 1974, Harper & Row, Publishers, Inc., Chapter 26.
37. Scott-Wilson, H.: A method for estimating acetone in animal fluids, J. Physiol. **42:**444-470, 1911.
38. Van Slyke, D.D.: Studies of acidosis VII: the determination of β-hydroxybutyric acid, acetoacetic acid, and acetone in urine, J. Biol. Chem. **32:**455-493, 1917.
39. Peden, V.H.: Determination of individual serum "ketone bodies," with normal values in infants and children, J. Lab. Clin. Med. **63:**332-343, 1964.
40. Procos, J.: Modification of the spectrophotometric determination of ketone bodies in blood enabling the total recovery of β-hydroxybutyric acid, Clin. Chem. **7:**97-106, 1961.
41. Tsao, M.U., Lowrey, G.H., and Graham, E.J.: Microdetermination of acetone in biological fluids, Anal. Chem. **31:**311-314, 1959.
42. Nadeau, G.: The interference of acetone in blood alcohol determinations: a simple method for the determination of blood acetone (Ravin modified), Can. Med. Assoc. J. **67:**158-159, 1952.
43. Rothera, A.C.H.: Note on the sodium nitroprusside reaction for acetone, J. Physiol. **37:**491-494, 1908.
44. Free, A.H., and Free, H.M.: Nature of nitroprusside reactive material in urine in ketosis, Am. J. Clin. Pathol. **30:**7-10, 1958.
45. Li, P.L., Lee, J.T., MacGilliray, M.H., et al.: Direct fixed-time kinetic assays for beta-hydroxybutyrate and acetoacetate with a centrifugal analyzer or a computer-backed spectrophotometer, Clin. Chem. **26:**1713-1717, 1980.
46. Hansen, J.L., and Frier, E.F.: Direct assays of lactate, pyruvate, beta-hydroxybutyrate and acetoacetate with a centrifugal analyzer, Clin. Chem. **24,** 475-479, 1978.
47. Eriksson, C.J.P.: Micromethod of determination of ketone bodies by dead space gas chromatography, Anal. Biochem. **47:**235-243, 1972.
48. Mayes, P.A., and Robson, W.: The determination of ketone bodies, Biochem. J. **67:**11-15, 1957.
49. Sacoolidge, J., Donohoo, R., and Kaplan, L.A.: Unpublished results.
50. Ketostix package insert, 1983, Ames, Inc., Elkhart, Ind.

Chapter 55 Electrolytes

Anion gap

W. GREGORY MILLER

Clinical significance: pp. 363, 389, and 520

Computation formula: Na-Cl-CO_2

PRINCIPLES OF ANALYSIS AND CURRENT USAGE

The anion gap, Na^+-(Cl^- + CO_2), is an estimate of the net amount of anions not directly measured in serum. This parameter is usually calculated by subtraction of the *measured* anions from the *measured* cations to give a positive gap:

$$\underset{\text{cations}}{\text{Measured}}\ (\text{Na}) - \underset{\text{anions}}{\text{Measured}}\ (\text{Cl} + \text{CO}_2) = +\text{Anion gap}$$

The above equation is equivalent to subtraction of the unmeasured anions from the unmeasured cations, which results in the same absolute value as the above equation, but with a negative value, as in the following equation:

$$\text{Unmeasured cations} - \text{Unmeasured anions} = -\text{Anion gap}$$

The major unmeasured cations are calcium (5 mEq/L) and magnesium (2 mEq/L), and the major unmeasured anions are protein (15 mEq/L), phosphate (2 mEq/L), sulfate (1 mEq/L), and organic acids (5 mEq/L). Since potassium (4 mEq/L) is usually not included in the computation of anion gap, the net "unmeasured anions," or anion gap, is approximately

$$(5 + 2 + 4) - (15 + 5 + 2 + 1) = -12 \text{ mEq/L}$$

The computation of the anion gap is useful clinically to alert the physician to the potential presence of metabolic disorders that alter electrolyte balance.[1] The box on p. 1045 lists some causes of both increased and decreased anion gaps. A normal anion gap is observed in hyperchloremic metabolic acidosis such as that produced in renal tubular acidosis or diarrhea. In these cases the bicarbonate loss is ionically compensated by increased chloride, with no net change in the calculated anion gap.

A recent study of electrolyte measurements on normal persons observed a range of anion gaps from 8 to 18 mEq/L, whereas a group of hospitalized patients had a range from 1 to 30 mEq/L, with 90% of the patients between 6 and 20 mEq/L.[2] Because the expected result for an anion gap calculation falls within a well-defined range, one can use it as a quality control criterion for each patient to detect spurious electrolyte results. If the anion gap falls outside the range 6 to 20 mEq/L, it *may* indicate the occurrence of an analytical error. In this case it is necessary to repeat the electrolyte measurements to confirm the results. The imprecision of the anion gap (typical coefficient of variation, CV, 9%) is much larger than for the individual electrolyte results (typical CV for sodium, 1%; for chloride, 1.5%; and for CO_2, 4%), since it includes the combined imprecision from each term. Consequently, the individual electrolyte results (not the anion gap) should agree within ± 2 mEq/L of the original results for confirmation. If the repeated results do not agree with the original, it is necessary to repeat the analysis again to determine which values are correct. If over a period of time all anion gaps are on the high side or on the low side, this trend may indicate the presence of a consistent analytical error and should be investigated.

Repeat analysis may include the use of alternative, confirmatory methods such as nondilutional, ion-selective electrodes or a chloridometer to obviate the presence of lipemia or bromism. Inspection of other laboratory findings such as protein, presence of ketones, or hyperglycemia may indicate a physiological reason for an abnormal anion gap.

It is more common to observe anion gaps greater than 20 than it is to observe gaps less than 6. A moderately increased anion gap (that is, 21 to 25) in the presence of greatly elevated BUN (more than 500 mg/L) or creatinine (more than 40 mg/L), or both, is consistent with renal failure, and electrolyte results do not need to be confirmed.

Anion gap changes

Causes of increased anion gap
- Decreased unmeasured cations
  - Hypocalcemia
  - Hypomagnesemia
- Increased unmeasured anions
  - Associated with metabolic acidosis
    - Uremia (renal failure)
    - Ketoacidosis
    - Lactic acidosis
    - Salicylate poisoning
  - Not necessarily associated with metabolic acidosis
    - Hyperphosphatemia
    - Hypersulfatemia
    - Large doses of antibiotics (such as penicillin and carbenicillin)
    - Treatment with lactate, citrate, or acetate
- Increase in net protein charge as in alkalosis

Causes of decreased anion gap
- Decreased unmeasured anions
  - Hypoalbuminemia
  - Hypophosphatemia
- Increased unmeasured cations
  - Hypercalcemia
  - Hypermagnesemia
  - Paraproteins
  - Polyclonal gamma globulins
  - Drugs such as polymyxin B or lithium
- Underestimation of serum sodium
  - Hyperproteinemia
  - Hypertriglyceridemia
- Overestimation of serum chloride
  - Bromism

Similarly, a moderately increased anion gap in the presence of greatly increased glucose (more than 300 mg/dL) is consistent with diabetic ketoacidosis, and electrolyte results do not need to be confirmed.

REFERENCE RANGE

Ambulatory	7 to 18 mEq/L
Hospitalized	6 to 20 mEq/L

The ambulatory reference range is based on measurement of electrolytes in 935 healthy adults; the reference range for the hospitalized patients is from Witte et al.[2]

REFERENCES

1. Garg, A.K., and Nanji, A.A.: The anion gap and other critical calculations, Diagn. Med., pp. 32-43, March/April 1982.
2. Witte, D.L., Rodgers, J.L., and Barrett, D.A.: The anion gap: its use in quality control, Clin. Chem. **22:**643-646, 1976.

Blood gases

JOHN E. SHERWIN
BERNDT B. BRUEGGER

PARTIAL PRESSURE OF OXYGEN (Po_2), CARBON DIOXIDE (Pco_2) AND THE pH OF WHOLE BLOOD

Clinical significance: pp. 363 and 387

PRINCIPLES OF ANALYSIS AND CURRENT USAGE

Blood gas analyses are usually performed to assess either the acid-base status or the respiratory oxygenation status of the patient. A blood gas analysis includes measurement of the blood pH, Pco_2, and Po_2 and may include any or all of the calculated parameters of oxygen content, oxygen saturation, total CO_2, bicarbonate, and base excess. These calculated parameters are discussed in Chapter 22 and are not considered further in this chapter. Measurement of blood pH is routinely performed using a glass electrode.[1]

One measures the Pco_2 by isolating a glass electrode in a weak bicarbonate buffer and separating this buffer from the blood sample by a gas-permeable membrane. The measurement of Po_2 relies on the electrochemical measurement of the oxygen that diffuses across a gas-permeable membrane into an electrolyte buffer that surrounds a Clark electrode.[2]

The transcutaneous measurement of Pco_2 and Po_2 currently relies on miniaturized electrodes of the same basic design as the electrodes of blood gas analyzers. These transcutaneous, noninvasive measurements are complicated by the presence of the skin, which acts as a second gas-permeable membrane. A more complete discussion of the electrodes is found in Chapter 14.

The majority of currently available blood gas analyzers require a whole blood sample of less than 250 μL and measure the pH, Pco_2, and Po_2 simultaneously. Many are now microprocessor-controlled so that calibration occurs automatically at preset intervals and both the measured parameters and the calculated parameters are displayed when analysis is complete. Maintenance has been greatly simplified by use of the microprocessor to monitor all system parameters and indicate potential malfunctions. The basic principles of electrode operation are unchanged from the original equipment, and all available instruments use equivalent electrodes. Nonetheless, the electrode configuration within the sample compartment can significantly affect both the sample volume and the required analysis time. Thus there can be statistical differences in measured parameters between instruments produced by different manufacturers as well as between different models produced by the same manufacturer.

In the past, instrument design has led to a variety of

required sample volumes for blood gas analysis. The modern, commonly used blood gas analyzers all have the capability of analyzing microsamples of less than 100 μL and conventional 1 mL samples as well. Most instruments currently available automatically aspirate the sample, thus reducing errors introduced by manual introduction of the sample. Therefore sample volume is less of a consideration than it once was and attention has now focused on techniques that eliminate the need to collect a blood specimen. Noninvasive instruments that measure Po_2 and Pco_2 have been developed by Huch et al.[3] The transcutaneous Po_2 electrode has been used routinely for more than 5 years and provides an excellent correlation with Po_2 of whole blood when the proper technique is used. Noninvasive transcutaneous Pco_2 electrodes have been available for several years. The correlation between the data obtained using these electrodes and the Pco_2 from blood gas analysis is acceptable. The difficulty with the routine use of transcutaneous Pco_2 monitor is that it does not provide a pH measurement for complete acid-base assessment. On the other hand, the transcutaneous Po_2 monitor provides data that are directly applicable to patient care.

REFERENCE AND PREFERRED METHODS

The methods described below use the AVL Model 940 blood gas analyzer* and the Radiometer transcutaneous Po_2 monitor† and are provided only as illustrative examples. The instrumentation mentioned reflects our experience and does not imply that other instruments are of less value. Table 55-1 lists a comparison of some features of several available blood gas analyzers.

SPECIMEN

Whole blood is collected in heparinized capillary tubes containing a small metal mixing wire (flea), which is added to facilitate mixing with the magnet provided. These are then sealed at both ends with clay. The capillaries are placed into a labeled test tube or plastic bag and delivered to the laboratory in crushed ice to prevent deterioration of the sample.

The time necessary for completion of the test should not exceed 15 minutes. This includes drawing of sample, de-

*AVL Scientific Corp., Schaffhausen, Switzerland, and Pine Brook, N.J.
†Radiometer, Inc., Westlake, Ohio.

Table 55-1. Comparison of several blood gas analyzers

	Instrument			
Characteristic	**ABL3***	**AVL 947†**	**IL 1303‡**	**Corning 178§**
Dynamic range (resolution)				
pH	6.3-8.0 (0.002)	6-8 (0.003)	6-8 (not specified)	6-8 (0.001)
Pco_2	5-250 mm Hg (0.5)	4-200 mm Hg (CV 1%)	1-800 mm Hg (not specified)	5-250 mm Hg (0.1)
Po_2	0-800 mm Hg (0.5)	0-800 mm Hg (CV 0.5% at 94 mm Hg)	0-800 mm Hg (not specified)	0-999 mm Hg (0.1)
Calibration of Po_2 and Pco_2	Liquid	Gas	Gas	Gas
Minimum sample volume	125 μL (automatic)	25 μL (manual)	65 μL (automatic)	40 μL (manual)
Cycle time	105 seconds	40 seconds	90-120 seconds	Not specified
Microprocessor controlled	Yes	Yes	Yes	Yes
Temperature correction	Yes	Yes	Yes	Yes
Calculation of O_2 saturation	Yes	Yes	Yes	Yes
Calculation of HCO_3^-, total CO_2, and base excess	Yes	Yes	Yes	Yes

CV, Coefficient of variation.
*Radiometer, Inc., West Lake, Ohio.
†AVL Scientific Corp., Schaffhausen, Switzerland, and Pine Brook, N.J.
‡Instrumentation Laboratory, Inc., Lexington, Mass.
§Corning Scientific, Medfield, Mass.

livery to laboratory, log-in, analysis, and reporting of results. If this maximum time limit is adhered to, no difference is observed whether plastic or glass syringes are used. A problem causing potentially greater error in Pco_2 measurement is the heparinization of syringes with liquid heparin. If more than 50 μL of solution remains in a 1 mL syringe and a minimal specimen (0.2 mL) is collected, a negative error of about 5 mm Hg of Pco_2 is observed. However, with proper care this is not a problem.

Glass versus plastic syringes

Early studies suggested that plastic substances absorb so much oxygen that the use of plastic syringes for blood gas samples would be ill advised. More recent studies have not substantiated these assumptions.[4] In essence, pH and Pco_2 values are not affected, whereas Po_2 values in excess of 400 mm Hg drop more rapidly for samples in plastic than in glass syringes. The rate of O_2 diffusion is also dependent on the temperature of the sample and the time required before analysis is completed, with diffusion increasing as both variables increase. It is doubtful that this circumstance is clinically significant. However, glass syringes are preferred for the following reasons:

1. The barrel has minimal friction with the syringe wall, and the pulsating arterial pressure is clearly visible as the blood fills the syringe; thus the nature of the sample is clearly identified to the clinician.
2. There is seldom a need to "pull back" on the barrel, a maneuver that can cause air bubbles to enter the sample around the barrel.
3. Small air bubbles adhere tenaciously to the sides of a plastic syringe, making it difficult to expel the air from the samples.

Thus, although glass is preferred, there is no *strong* valid objection to appropriately used plastic syringes.

Anticoagulants

Blood gas measurements must be performed on "whole blood," that is, unclotted and unseparated blood. An anticoagulant is used to inactivate the clotting mechanisms so that the sample remains unclotted in the syringe. Oxalates, EDTA, and citrates are not acceptable for blood gas samples because they significantly alter the blood sample.[5]

Heparin is the anticoagulant of choice; however, too much heparin may affect several parameters. The pH of sodium heparin is approximately 7.0; Pco_2 and Po_2 approach room-air values.

It has been shown that 0.05 mL of sodium heparin (1000 units/mL or 10 mg/mL) will adequately anticoagulate 1 mL of blood; whereas up to 0.1 mL per 1 mL of blood will not affect pH, Pco_2, or Po_2 values. When a 5 mL syringe is washed with sodium heparin and then ejected, the dead space of the syringe will contain approximately 0.15 to 0.25 mL of sodium heparin. Thus 2 to 4 mL of blood will theoretically contain at least 0.05 mL of heparin per 1 mL of blood.

Thus it is recommended that the syringe be flushed with sodium heparin (10 mg/mL) and then emptied; doing so will allow adequate anticoagulation of a 2 to 4 mL blood sample with assurance that the results will not be altered by the anticoagulant.

Anaerobic conditions

Room air contains a Pco_2 of essentially zero and a Po_2 of approximately 150 mm Hg. Air bubbles that mix with a blood sample will result in gas equilibration between the air and the blood. Air bubbles may thus *significantly* lower the Pco_2 values of the blood sample and cause the Po_2 to approach 150 mm Hg. The greater the amount of air mixed with a blood sample, the greater the error.

It is recommended that any sample obtained with *more than minor air bubbles be discarded.* The syringe must be *immediately* sealed with a cork or a cap after the sample is obtained.

Temperature during transport and storage

Blood is living tissue in which oxygen continues to be consumed and carbon dioxide continues to be produced, even after the blood is drawn into a syringe. Table 55-2 shows the approximate rate of change of a sample that is held in a syringe at 37° C. If the sample is immediately placed in an ice slush, the temperature rapidly falls below 4° C and the Pco_2 changes are insignificant over several minutes. Acceptable samples are those that have been properly iced immediately after being drawn and rapidly (less than 15 minutes) transported to the laboratory at 4° C.

Although a majority of pediatric blood gas specimens are capillary in nature, a variable percentage will be arterial. Arterial samples may be drawn into heparinized syringes either by a laboratory phlebotomist (from a catheter line only) or from the femoral or radial artery (by a physician only). The arterial sample is preserved in crushed ice and delivered to the laboratory.

PROCEDURE

Principle

The Automatic Gas Check AVL 940 System allows for direct determination in blood of actual pH, actual carbon

Table 55-2. In vitro blood gas changes*

	37° C	4° C
pH	0.01/10 min	0.001/10 min
Pco_2	1 mm Hg/10 min	0.1 mm Hg/10 min
Po_2	0.5 mm Hg/10 min	0.05 mm Hg/10 min

*Approximate changes with time and temperature after the sample is drawn into the syringe.
A temperature of 37° C implies that the blood remains at body temperature in the syringe.

dioxide tension (Pco_2), and actual oxygen tension (Po_2) by a set of microelectrodes. The base excess, bicarbonate concentration, and percent oxygen saturation are calculated by the AVL 940 from the measured parameters. Only 40 μL of blood are required for a fully automated analysis, whereas in the manual step mode a 25 μL sample can be analyzed.

Calibration

This section is included to illustrate the type of automatic calibration systems available in currently available blood gas instruments. The principal calibration of the entire measuring system occurs every 12 hours and requires a maximum of 20 minutes to complete. The pH electrode is calibrated with two precision buffers, one at pH 6.832 and another at pH 7.373. The Po_2 and Pco_2 electrodes are calibrated with two analyzed gases, one containing 5% CO_2 and 0% O_2 and the other containing 12% CO_2 and 21% O_2.

Further automatic calibrations occur at regular intervals so that the operational accuracy of the instrument is maintained. Three hours after the first calibration after one switches on the instrument, an automatic 1-point calibration of the pH electrode takes place with buffer 1 (pH 7.383).

If during this first recalibration of the pH system it is established that the deviations of the measured calibration values exceed the value limits stored in the program, the next recalibration should be after 2 hours. If the deviation is even greater, recalibration again takes place after 1 hour. If this drifting of the electrodes is rectified, the AVL 940 automatically returns to longer calibration intervals (2 or 3 hours).

During standby operation, gas 1 (5% CO_2 in air) continuously flows through the measuring capillary.[1] The gas 1 calibration values of the Pco_2 and Po_2 electrodes are corrected at intervals of 10 seconds.

Whenever one or more of the electrodes are changed, they are subject to a specific two-point calibration. This begins automatically when the cover of the instrument is closed after the electrodes have been installed.

The automatic initial calibration and recalibrations are announced 5 minutes in advance.

By pressing the start key one can interrupt the main calibration or recalibration or the 5-minute prewarning period in order to carry out one or more measurements. It is *not* possible to interrupt the first main calibration (after one switches on the instrument and presses the calibration key) or the specific calibrations initiated by outside influences (such as example exchange of electrodes).

Comments

Two different techniques are used for calibration of the Po_2 and Pco_2 electrodes. The majority of available blood gas analyzers use certified gases saturated with water for calibration. This technique has the advantage of providing a constant and highly reproducible calibration mixture. Nonetheless, it suffers from the fact that it is a gas rather than a liquid and therefore is not equivalent to the patient samples being analyzed. It is essential that the gas flow rate through the measuring chamber be strictly controlled. Variations will result in electrode cooling and in incomplete humidification, both of which result in significant calibration errors. The second calibration technique uses a gas-tonometered, buffered liquid for calibration. The advantages of this technique are that it is similar to the unknown samples being analyzed, and the calibration is independent of the flow rate of the gas and humidification is not a consideration. This technique results in a positive bias of about 5% to 8% for both Pco_2 and Po_2 as compared to the gas calibration technique. Therefore it is important that the results from instruments using different calibration techniques be identified. If multiple instruments are used, the calibrations can be cross-correlated to minimize this difference.

Assay

When the instrument is in the standby condition and neither the red service lights nor the yellow calibration lights are on, blood samples can be introduced into the AVL 940 as follows:

For a specimen in a capillary tube

1. Mix the sample by moving the magnet gently back and forth along the capillary tube to cause the metal rod in the capillary tube to move through the sample.
2. Use a file to score the tube just above the sealing wax plug and break the tube. Expel one or two drops of blood onto a gauze pad to be sure blood is flowing freely. If it is not, take the needle point at the wider end of the AVL Capillary Adapter and insert into cut end of capillary to remove clotted blood, which may impede blood flow.
3. To introduce the blood from the capillary into the instrument, push the capillary into the wider end of the capillary adapter. Then push the thin end of the capillary adapter in turn into the sample fill port until it stops.
4. By pressing the start key, the measuring process is initiated:
   a. The blood sample is drawn into the measuring capillary by the peristaltic pump. One can observe this process through the window that covers and protects the measuring capillary. The peristaltic pump is turned off when all four electrode seatings are filled with blood.

   IMPORTANT: There must be no air bubbles visible at the contact areas of the Po_2, Pco_2, and pH electrodes (the results of the measurements would be inaccurate and not reproducible). Small air bubbles on the pH reference electrode do not change the results so long as the electrical connection between the pH reference and pH electrode exists.

b. After the filling of the measuring capillary the analysis commences:

   Measurement of the blood values in the normal mode is carried out at 37° C and an assumed hemoglobin value of 150 g/L. If the patient's actual hemoglobin value is known, it can be programmed into the instrument.

c. After the measurement values appear on the displays (about 40 seconds), the measured values and the calculated values together with the general data are printed in the form of a blood gas report.

d. The measured and displayed values are now available for approximately 3 minutes. As long as these data are visible on the digital displays, one can repeat the printout as often as required by pressing the reprint key. After each printout, the values are retained for an additional 3 minutes.

e. After every measurement, a wash and dry process also takes place automatically before the AVL 940 returns to standby.

For a specimen in a syringe

1. Rotate the syringe between your palms for about 30 seconds to mix and warm the specimen.
2. Remove the needle and expel one or two drops into a gauze pad. This is to ensure that there are no blood clots present in the syringe.
3. A syringe adapter is first put into the sample fill port and then the other end of the syringe adapter (the thin tube) into the syringe.
4. When the green standby signal lights up, the start key can be pressed. The end of the syringe should point slightly upwards while the sample is being drawn in. When the sample has been drawn in the syringe, the adaptor can be withdrawn.
5. The measuring process is the same as that described for the capillary specimen.
6. The adapter is of no more use and may be discarded.
7. If the sample volume is insufficient, a warning lamp and audible signal are activated. If no blood enters the measuring capillary (for example, if the dried end of a capillary is not sufficiently opened), the red service signal flashes and the warning signal ''no sample'' apears.

Operation with extremely small samples. The minimum volume of a blood sample for fully automated analysis is approximately 40 μL. If there is less than 40 μL of blood but at least 25 μL, one can still perform an analysis using the procedure described by the manufacturer.

Temperature corrections

1. If the blood sample is taken from a patient whose body temperature deviates from 37° C, the measured and calculated values should be referred to the current patient temperature.
2. A temperature selected is valid for only one analysis, since the instrument returns to a value of 37° C when a new analysis is initiated.
3. If the selected temperature is changed after the first printing, and the reprint key is then hit, a report will be printed out with the parameters calculated on the basis of the new temperature.

Comments. Temperature correction is useful only for pH, P_{CO_2}, and for P_{O_2} values less than 50 mm Hg. Correction of calculated parameters is of little value because the calculation algorithms have not been shown to be applicable at temperatures other than 37° C. Temperature correction of pH, P_{CO_2}, and P_{O_2} may be of use for hypothermic patients undergoing cardiac surgery.

Quality control

1. At the beginning of each shift, after maintenance is performed, all three levels of a blood gas quality assurance system should be analyzed. Results should be recorded on a quality control log. If any value is out of control, appropriate corrective action should be taken.

   In addition, one additional level of a blood control should be analyzed every 2 to 3 hours on each instrument used. If an out-of-control value is encountered either in the morning quality control or during the shift, the instrument in question should be taken out of service until corrective action brings the value or values back into control.
2. All results on adults and neonates are logged on a flow chart. Log results should include color of the sample. Obviously inconsistent results should be rechecked.

Comments. There is general agreement that the best quality control material for blood gas analyzers is tonometered human whole blood. A tonometered solution is one that has been equilibrated with a gas. However, the technical difficulties associated with tonometry of whole blood in the laboratory have prevented its general use for preparation of quality control material. Most laboratories presently use one or more of the tonometered buffer, blood, or hemoglobin solutions prepared commercially and distributed in ampules. When handled according to the manufacturer's instructions and maintained at correct temperature, these solutions provide an excellent basis for blood gas quality control. However, several available instruments require transfer of the solution from the ampule to a syringe or capillary tube before analysis. The recovery of the stated values for P_{O_2} and P_{CO_2} is exceedingly difficult if buffered solutions are used. Blood solutions are more stable and are transferred more easily. This stability can potentially mask changes in electrode function, but use of the aqueous-buffer type of control potentially causes a search for problems where none exist. Either control is effective with appropriate care in handling.

REFERENCE RANGE (at 37°C)

	Arterial	Venous
pH	7.35-7.45	7.32-7.42
P_{CO_2} (mm Hg)	35-45	40-50
P_{O_2} (mm Hg)	80-105	25-47

TRANSCUTANEOUS PO_2 ($TcPO_2$) MEASUREMENT[6,7]

Specimen

Since this is a noninvasive technique, no specimen is required.

Principle

The measurement of $TcPO_2$ relies upon the electrochemical measurement of oxygen using a miniaturized Clark electrode.[8] The Clark electrode is based on the reduction of dissolved oxygen to water. The accuracy of these $TcPO_2$ measurements is a function of the vascularity of the dermis and the metabolic activity of the dermal and epidermal layers. Since the epidermis functions as a secondary membrane, the epidermal thickness relative to the thickness of the electrode membrane is an important consideration in controlling instrument response time. These two parameters are related by the equation

$$i = A \cdot a^2 \cdot F \frac{PO_2}{d/P_m + D/P_t}$$

where i is current; a, electrode radius; F, Faraday constant; d, membrane thickness; Pm, membrane permeability coefficient; P_t, tissue permeability coefficient; and D, thickness of the oxygen depletion tissue layer.

Therefore, if $d/P_m >> D/P_t$, the tissue permeability to oxygen is not a factor in the measured oxygen. This equation suggests that the electrode should be applied to skin areas where the epidermis is thin.

In order to obviate the effect of local blood flow on $TcPO_2$, vasodilatation is essential. Although one can use chemicals such as nicotine and caffeine, the variable response to these drugs compromises their routine use.[9] Thermal dilatation is now used almost exclusively. The mechanism for this local response is not fully understood, but temperatures of 42° to 45° C are commonly used. Temperatures as high as 45° C are occasionally accompanied by second-degree burns. At a temperature of 42° C burns are rare, but $TcPO_2$ values tend to be somewhat lower than corresponding arterial oxygen pressure (PaO_2) measurements.[10]

Instrumentation

A procedure using a Radiometer TCK-1 oxygen monitor is recommended, though other instruments function with equal effectiveness. A Radiometer TCM101 calibration unit was routinely used to calibrate the oxygen monitor (The London Company, Cleveland, OH 44145). Continuous data display was achieved when a strip-chart recorder was connected to the oxygen monitor (Linear Instruments Corporation, Irvine, CA 92717). All equipment was used according to the manufacturers' directions.

Assay

The manufacturer's recommended procedures were followed.

1. *Calibrations.* Using the TCM101 calibration tonometer, the electrode was set initally to zero $TcPO_2$. The high $TcPO_2$ calibration is generally achieved by use of tonometered room air (approximately 150 mm Hg). If the electrode has just been re-membraned, it must be polarized for at least 20 minutes before calibration if reproducible results are to be obtained. Calibration should be done at the same temperature at which patient analyses are performed if accurate measurements are to be made.
2. *Electrode application.* The electrode application site must be cleaned with soap and water and dried. The protective cover is removed from the electrode adhesive ring and the ring is gently secured to the chosen measuring site. Apply a small drop of water or contact gel to the measurement site. It is essential that no air be trapped between the skin and the electrode. One can easily confirm this by observing an abrupt decrease in the $TcPO_2$ on application of the electrode. It is good practice to further secure the electrode to the site with clinical adhesive tape.
3. *Measurement.* After electrode application the heater must be activated so that hyperthermization of the skin is achieved. The skin temperature must be 42° to 44° C so that reproducible data are achieved and one can avoid burning the patient's skin. The temperature of choice is 43° C because this temperature is well tolerated by nearly all patients. Monitoring the power output of the electrode heating unit is essential. This can be done by an alarm system or by actual recording of output. A relationship between blood flow and heater output has been reported.[12] However, the blood flow is more adequately assessed in these patients by a laser Doppler velocimetry technique.[11] The oxygen electrode can be left in place for up to 8 hours if the temperature is maintained at 43° C. However, at 44° C the electrode should be moved at least every 4 hours.
4. *Quality control.* It is essential to confirm the correlation between the blood oxygen and the $TcPO_2$ at the beginning of the measurements and at least every 4 hours if the electrode is to be left in place longer than 4 hours. The $TcPO_2$ and the PaO_2 should not differ by more than 5%, and $TcPO_2$ and the capillary

Po_2 should not differ by more than 8%. Frequent calibrations (as at 4-hour intervals) will correct for instrument drift, which is approximately 5% per hour. This should be accompanied by comparison with the patient's Pao_2.

Comments. There are some instances in which the transcutaneous monitor does not accurately reflect arterial Po_2.[12,13] Fortunately, these instances are relatively rare, being restricted to cases in which the blood pressure falls 2 standard deviations below normal and to infants given tolazoline, a pulmonary vasodilator, or anesthetics such as nitrous oxide and halothane.[14]

REFERENCES

1. Clark, L.C.: Monitor and control of blood and tissue oxygen tensions, Trans. Am. Soc. Artif. Intern. Organs **2**:41-48, 1956.
2. Evans, N.T.S., and Naylor, P.F.D.: The systemic oxygen supply to the surface of human skin, Respir. Physiol. **3**:21-27, 1967.
3. Huch, R., Lubbers, D.W., and Huch, A.: Reliability of arterial Po_2 in newborn infants, Arch. Dis. Child. **49**:213-218, 1974.
4. Evers, W., et al.: A comparative study of plastic (polypropylene) and gas syringes in blood gas analysis, Anesth. Analg. **50**:92, 1972.
5. Shapiro, B.A., Harrison R.A., and Walton, J.R.: In Clinical applications of blood gases, ed. 2, Chicago, Ill., 1977, Year Book Medical Publishers, Inc.
6. Enkema, L., Holloway, G.A., Piraino, D.W., et al.: Laser Doppler volocimetry vs heater power as indicators of skin perfusion during transcutaneous O_2 monitoring, Clin. Chem. **27**:391-396, 1981.
7. Oda, Y.: Clinical experience of transcutaneous Po_2 monitoring in high-risk newborn infants, Birth Defects **15**:585-588, 1979.
8. Graham, G., and Kenny, M.A.: Changes in transcutaneous oxygen tension during capillary blood-gas sampling, Clin. Chem. **26**:1860-1863, 1980.
9. Marshall, T.A., Kathwinkel, J., Berry, F.A., and Shaw, A.: Transcutaneous oxygen monitoring of neonates during surgery, J. Pediatr. Surg. **16**:794-804, 1980.
10. Peabody, J.L., Willis, M.M., Gregory, G.A., and Severinghaus, J.W.: Reliability of skin (tc) Po_2 electrode heating power as a continuous noninvasive monitor of mean arterial pressure in sick newborns, Birth Defects **15**:127-133, 1979.
11. Graham, G., and Kenny, M.A.: Performance of a radiometer transcutaneous oxygen monitor in a neonatal-intensive care unit, Clin. Chem. **26**:629-633, 1980.
12. Huch, R., Lubbers, D.W., and Hugh, A.: The transcutaneous measurement of oxygen and carbon dioxide tensions for the determination of arterial blood gas values with control of local perfusion and peripheral perfusion pressure, In Payne, J.P., and Hill, D.W., editors: Oxygen measurements in biology and medicine, London, 1975, Butterworth & Co., Publishers, Ltd., pp. 121-138.
13. Peabody, J.L., Willis, M.M., Gregory, G.A., et al.: Clinical limitations and advantages of transcutaneous oxygen electrodes, Acta Anaesth. Scand. **68**:76-82, 1978.
14. Gothgen, I., and Jacobsen, E.: Transcutaneous oxygen tension measurement. II. The influence of halothane and hypotension, Acta Anesth. Scand. **67**:71-75, 1978.

Calcium

E. CHRISTIS FARRELL

Clinical significance: pp. 439 and 809
Atomic symbol: Ca
Atomic weight: 40.08 daltons
Merck index: 1634
Chemical class: Alkaline earth element
Forms: Free Ca^{++}, protein bound, or in inorganic complexes

PRINCIPLES OF ANALYSIS AND CURRENT USAGE

Calcium offers a constant challenge to analytical principles because of problems with interfering compounds and because it exists in several physical forms. Calcium is found 46% free, 32% bound to albumin, 8% bound to globulins, and 14% in association with freely diffusible calcium complexes. All calcium is ionized no matter what it is bound to, but it may not be dialyzable or completely reactive with a chosen chromogen when complexed to another compound. Ion-specific electrode techniques for measuring the activity of the unbound fraction are available.

The oldest principles used for clinical determination of total calcium involved the quantitative precipitation of calcium with an excess of anions from oxalic, chloranilic, or naphthylhydroxamic acids.[1] In the case of oxalate, the precipitate was quantitated by a redox titration with potassium permanganate or ceric ions (method 1, Table 55-3). The most widely used oxalate methods were based on the modifications introduced by Kramer-Tisdall and Clark-Collip and were for years considered the reference method for calcium analysis.[2]

Methods using chloranilate or naphthylhydroxamate ions to precipitate calcium were based on quantification of the color of the precipitating ion. Chloranilate was added in an amount known to be in excess of the serum calcium. The calcium-chloranilate complex was precipitated and redissolved in alkaline solutions of ethylenediaminetetraacetic acid (EDTA). The intense red-purple color of the liberated chloranilic acid (A_{max} 520 nm) was proportional to the concentration of calcium in the specimen (method 2, Table 55-3). Similarly, naphthylhydroxamate has been used to precipitate calcium. The precipitate was dissolved in alkaline EDTA and, when acid ferric nitrate was added, a red-orange color was produced. These precipitation methods are rarely used today because they are time-consuming and insensitive.

A sensitive fluorescent method that is valuable for microanalysis of serum or urine is available.[3] In one method (Calcette) the calcium is added to an alkaline solution containing calcein (method 3, Table 55-3). The calcium combines with the calcein, which at high pH fluoresces at 520 nm (excitation at 490 nm). The fluorescent complex is ti-

Table 55-3. Methods of calcium analysis

Method	Principle	Usage	Comments
1. Precipitation by oxalate and redox titration	Ca^{++} + Oxalate → Ca oxalate (ppt) Ca oxalate (ppt) + H_2SO_4 → Oxalate + $CaSO_4$ $KMn^{7+}O_4$ + Oxalate $\xrightarrow{70° C}$ $Mn^{2+}SO_4$	Historical	Initial reference method
2. Precipitation by colored anions; spectrophotometric	Ca^{++} + Chloranilate → Ca-chloranilate (ppt) Ca-chloranilate (ppt) + EDTA $\xrightarrow{OH^-}$ Ca EDTA + Chloranilic acid (purple)	Historical	Labor intensive, imprecise, many other dyes available
3. Titration of fluorescent Ca^{++} complex	Ca^{++} + Calcein → Ca-calcein (fluorescent) EGTA + Ca-calcein → Ca^{+}-EGTA + Calcein (decreased fluorescence)	Stat. or small labs	Small sample size, dedicated instrument
4. Spectrophotometric measurement of Ca^{++} complexes			
a. Direct	Ca^{++} + *o*-Cresolphthalein $\xrightarrow{OH^+}$ Red complex (520 nm)	Most common	Early adapted to a variety of automated instruments; positive bias compared to atomic absorption
b. Dialysis	Ca^{++} complex + H^+ $\xrightarrow{Dialysis}$ Ca^{++} in recipient stream Ca^{++} detected as in (a)	On Technicon AutoAnalyzer	
5. Flame emission*	Ca^{++} $\xrightarrow[2e^-]{Heat}$ Ca^0 $\xrightarrow{Heat}$ Ca* → Ca^0 + Photon	Historical	Poor sensitivity
6. Atomic absorption	Ca^{++} $\xrightarrow{2e^-}$ Ca^0 Photon + Ca^0 → Ca*	Reference method	Excellent accuracy and sensitivity
7. Isotope-dilution mass spectrometry	Ca and known amount of Ca isotope; isolate Ca^{++} and record ratio of two isotopes on mass spectrometer	Definitive method	Available in reference centers only

Ca, Calcium atom in excited state; *Ca^0*, Calcium atom in ground state; *EGTA*, ethylene glycol-bis(β-aminoethylether)*N,N'*-tetraacetic acid; *ppt*, precipitate.

trated with ethylene glycol-bis(β-aminoethylether)-*N,N'*-tetraacetic acid (EGTA), which binds the calcium from the complex with a resulting decrease in fluorescence. When the base-line fluorescence is achieved, the volume of EGTA required to titrate the complex is directly proportional to the concentration of calcium. Magnesium and phosphate are reported not to interfere with this method.

Direct spectrophotometric measurements of calcium in serum or urine are also based on formation of color complexes between calcium and organic molecules. Examples of compounds that give colored reaction products with calcium are (1) glyoxal-bis(2-hydroxyanil), (2) alizarin, (3) chlorophosphonazo III, (4) methylthymol blue, and (5) *o*-cresolphthalein complexone. Only the last four of these methods are reported to be used at all by laboratories participating in the 1984 College of American Pathologists Comprehensive Chemistry Survey. Of these calcium-complexing, colorimetric reagents, the *o*-cresolphthalein is certainly the most commonly used for routine calcium analysis. Approximately 86% of the participants in the 1984 CAP survey used the *o*-cresolphthalein method.

The reaction of calcium with *o*-cresolphthalein produces a red complex (quantitated at 570 to 575 nm) at pH 10 to 12 (method 4, Table 55-3). The reaction product is stabilized by the addition of KCN (potassium cyanide), which also acts to eliminate interference from heavy metals. Interference by magnesium ions is eliminated by the addition of 8-hydroxyquinoline.[4] To reduce interference from proteins, one can dialyze the sample with an acid solution to release bound calcium (AutoAnalyzers, Technicon, Inc.) or make a large dilution of the sample (approximately forty to a hundredfold) with an acid solution. Most methods use diethylamine to achieve a high buffered pH. This buffer, however, is both volatile and irritating. The method of Moorehead and Biggs[5] uses the nonirritating, less volatile 2-amino-2-methyl-1-propanol (AMP) buffer. A commercially available kit (Dow Diagnostics, Inc.) uses 2-ethylaminoethanol as the buffer, which is claimed to increase sensitivity and base-line stability and reduce blanking problems. Another commercially available reagent eliminates the potentially hazardous cyanide and uses a sodium carbonate/bicarbonate buffer.

Methods for measurement of calcium in serum and urine using atomic emission (method 5, Table 55-3) or atomic absorption (method 6, Table 55-3) have been suggested, though only atomic absorption has proved to be practical in terms of specificity and sensitivity of the method (see Chapter 3). Atomic absorption relies on the principle that free calcium atoms in their ground state will absorb light at characteristic wavelengths (such as 422.7 nm). In the flame, which is used for dispersal of the atoms, only about 0.1% of the atoms are raised to a higher energy level, leaving 99.9% to be "viewed" by the beam of a hollow cathode lamp (see Chapter 3).

Calcium present in protein or inorganic complexes is detectable by flame atomic absorption only when steps are taken to dissociate calcium from these complexes. Acid is used for the dissociation of protein-bound calcium, and lanthanum (La^{++}) or strontium (Sr^{++}) ions are added to displace Ca^{++} from phosphate, oxalate, citrate, and other complexes. Dissociation of these complexes is accomplished by the addition of La^{++} or Sr^{++} to the sample diluent. Serum is diluted 1:50 with 1% La^{++} and urine is diluted 1:50 with 5% La^{++} to overcome the higher phosphate concentration found in urine. Protein precipitation may be accomplished by acid treatment of the sample. Precipitation results in a 2% to 3% contraction in sample volume when the precipitate is removed. One can mathematically correct for the contracted volume. Acid treatment without protein precipitation requires increased maintenance for the burner head to prevent accumulation of debris.

Interference by magnesium and other elements in atomic absorption is reduced by the use of narrow-band-pass, diffraction-grating spectrophotometers to specifically isolate the atomic absorption line of calcium (422.7 nm). Interference by sodium is eliminated by the addition of physiological concentrations of sodium ions to calcium standards. Atomic absorption is not frequently used for routine analysis (approximately 2% of CAP participants), probably because it has rarely been automated for high sample throughput. However, procedures for automating the method are available, making this an attractive routine procedure.

The definitive method for calcium measurements is isotope-dilution mass spectroscopy (method 7, Table 55-3). This method, available in only a few institutions, is the accuracy standard against which all methods must be compared.

REFERENCE AND PREFERRED METHODS

The older oxalate precipitation and titration methods are no longer in use for routine calcium analysis because they were laborious and relatively inprecise.

The most commonly used reference method for accurate calcium measurement is atomic absorption. Although it is not the definitive method, atomic absorption serves as a more available reference method for comparison of new methods. The precision of atomic absorption over the range of physiological concentrations is on the order of 2% to 3%. Atomic absorption does not show any important interferences by common compounds such as hemoglobin, bilirubin, or lipemia. The only class of interferent is calcium-chelating compounds. Atomic absorption methods

Table 55-4. Reaction conditions for calcium analysis

Condition	*o*-Cresolphthalein spectrophotometric*	Atomic absorption
Temperature	25°-37° C	2300° C
Sample volume	20 µL	50 µL
Volume fraction	0.01	0.02
Final concentration	*o*-Cresolphthalein: 0.06 mmol/L 8-Hydroxyquinoline: 9.6 mmol/L Diethylamine: 193.4 mmol/L Potassium cyanide: 3.8 mmol/L	Lanthanum diluent 1 g/L (serum) 5 g/L (urine)
Wavelength	575 nm	422.7 nm
Reaction mode	End point	End point
Time of reaction	30 seconds	12 seconds
Precision† (mean, percent coefficient of variation)	Serum 78.0 mg/L, 3.7% CV 147.7 mg/L, 3.3% CV Urine 159 mg/L, 7.9% CV 286 mg/L, 7.7% CV	Serum 78.2 mg/L, 2.3% CV 146.1 mg/L, 3.6% CV Urine 158 mg/L, 5.5% CV 290 mg/L, 7.6% CV
Linearity	150 mg/L	Highest standard, usually 120 mg/L
Interferences	Gross lipemia	Calcium chelators

*Baginski, E.S., Marie, S.S., Alcock, N.W., et al.: Calcium in biological fluids. In Faulkner, W.R., and Meites, S., editors: Selected methods of clinical chemistry, vol. 9, Washington, D.C., 1982, American Association for Clinical Chemistry Press.
†From College of American Pathologists Quality Assurance Survey data. Cresolphthalein data is from all multianalyzers. Individual means and coefficients of variation vary considerably from method to method.

are very sensitive, requiring 10 to 20 μL of sample for analysis. The method has been semiautomated with automatic sample aspiration (Perkin-Elmer, Inc.) after dilution with semiautomatic diluters. The analysis has been fully automated by the use of Technicon AutoAnalyzer equipment. These procedures show high sample throughput and precision.

The *o*-cresolphthalein methods have been adapted to a wide variety of automated analyzers with a resulting high precision. A significant problem with this technique is the requirement for good temperature control during the analysis, since the chromogen's molar absorptivity is temperature dependent. Most methods using direct analysis after dilution appear to be sensitive to interferences by lipemia, hemoglobin, and myeloma protein.[6-10] The *o*-cresolphthalein methods show a slight positive bias when compared to atomic absorption (10 to 15 mg/L).[6] The *o*-cresolphthalein methods can be fairly sensitive, requiring 25 to 50 μL of samples for analysis. The precision of the *o*-cresolphthalein method varies with the type of automated instrument used for analysis but ranges from 2% to 7%.[6,7]

The preferred clinical method is certainly atomic absorption on the basis of small sample size, precision, and degree of accuracy. The major drawbacks of atomic absorption are the high level of maintenance and care the equipment requires. However, it is a versatile instrument capable of measuring several other important analytes (magnesium, lithium, and, with flameless atomic absorption, heavy metals). Instrument problems may present an impediment for smaller laboratories. For these laboratories the *o*-cresolphthalein method will suffice for most routine needs. The reaction conditions used for these two methods are summarized in Table 55-4.

SPECIMEN

Serum or heparinized plasma is separated from cells as rapidly as possible. Blood anticoagulated with oxalate or EDTA is not acceptable because these chemicals will strongly chelate calcium. Venous stasis and erect posture can elevate calcium by 4 to 6 mg/L.[14] Stasis changes protein-bound Ca^{++}, and concentrations of free Ca^{++} will be changed by pH shifts.

Urine calcium can be kept dissolved by addition of 10 mL of 6 N HCl to the collection container before collection of a 24-hour specimen. Keep urine well mixed during collection period.

PROCEDURE

Principle

The calcium determinations by atomic absorption spectroscopy are based on the fact that atoms of an element in the "ground," or unexcited, state absorb light of the same wavelength as that emitted by the element in the excited state. Each element has its own characteristic absorption or resonance lines, and no two elements are known to have an identical resonance line.

Calcium in serum or urine is determined when diluted sufficiently with lanthanum oxide solution to avoid interference by chelating substances, especially phosphate ions.

Reagents

Use specially cleaned atomic absorption glassware and water (that is, free of calcium).

1. *Stock lanthanum (50 g/L):* Weigh 58.64 g of lanthanum oxide (La_2O_3); place in a 1-liter flask and add about 50 mL of deionized H_2O to wet powder. Add 250 mL of concentrated HCl *slowly* and *cautiously* under hood while swirling the flask. Dilute to 1 liter with deionized H_2O. Store in dark at room temperature. Stable for 6 months.
2. *Working lanthanum diluent for serum (1 g/L):* Dilute 20 mL of stock lanthanum to 1 liter with deionized H_2O. Store in dark bottle at room temperature. Stable for 6 months.
3. *Working lanthanum diluent for urine (5 g/L):* Dilute 100 mL of stock lanthanum to 1 liter with deionized H_2O. Store in dark bottle at room temperature. Stable for 6 months.
4. *Stock calcium standard (1000 mg/mL, 1000 ppm):* Obtain from Fisher Scientific (50-M-51) or dissolve 124.9 mg of reagent grade $CaCO_3$ in 3 mL of 1 mol/L HCl. Dilute to 100 mL with H_2O. Stable for 6 months at room temperature in a closed container.
5. *Calcium blank solution (140 mmol/L Na, 5 mmol/L K, 4 mg/L P):* Place 8.2 g of NaCl, 0.373 g of KCl, and 0.072 g of Na_2HPO_4 into a 1-liter volumetric flask. Dissolve salts in about 250 mL of distilled water and dilute to mark with water. Stable for 1 month at refrigerator temperatures.

Working standards

Ca (mg/L)	Stock Ca (mL)	Dilute to volume with calcium blank solution (mL)
50	5	100
100	10	100
120	12	100
150	15	100
200	20	100

Store at room temperature in glass flasks. Do not pipet from flasks. Pour aliquots daily into separate containers.

Assay

Equipment. Perkin-Elmer Atomic Absorption Spectrophotometer Model 460 (or other similar instrument) with a single-slot burner head and a calcium-magnesium hollow cathode lamp. Follow individual manufacturer's directions for maintenance of burner, optimization of lamp output ("peaking" flame), and air-acetylene fuel mixture, as well as the actual analytical procedure.

Notes

1. The dilution (1:50) of serum or urine prevents protein interference. The dilution can be best performed with a semiautomatic diluter (such as Micromedic Systems, Inc., Philadelphia, Pa.).
2. Sera left overnight in plastic sample cups may give low calcium values.
3. There is no appreciable variation in the patient's calcium levels throughout the day if exercise is avoided, and there is no difference between fasting and nonfasting calcium levels.
4. There is no difference between venous and capillary serum calcium levels.
5. Serum or plasma should not be allowed to remain in prolonged contact with the erythrocytes, since with time the cells become permeable to calcium.
6. Cerebrospinal fluid (CSF) can be analyzed for calcium when one uses the procedure for serum analysis. Since CSF concentrations may be lower than serum, use of a slightly lower standard is advisable.

Atomic absorbance maintenance

Daily

1. Soak burner head and rotate usage of two burner heads daily (see below).
2. Run copper wire through nebulizer.
3. Run blade or emery cloth *carefully* through burner slit to clean off accumulated debris.

Weekly

1. Clean waste line (remove line at instrument and clean up into outlet with pipe cleaner).
2. Perform linearity (50, 100, 120, 150 mg/L standards) check.
3. Peak nebulizer.

Daily maintenance to be performed first thing in the morning by first-shift technician

1. Turn flame on and aspirate 5% FL-70 (a detergent, supplied by Fisher Scientific Co., Fairlawn, N.J., #50-F-105) from probe for 20 minutes.
2. Turn flame off and let cool.
3. Remove burner head and pour 100 mL of distilled water down neck of chamber (where head was; water will come out waste line).
4. Position clean burner head, securing pin in place and slipping wires over hooks on either side of chamber; tighten burner head collar.
5. Set instrument for maximum absorbance (''peak'' flame), following manufacturer's instruction.
6. Place instrument back in concentration mode.
7. Place dirty burner head in 5% FL-70 to soak overnight.

REFERENCE RANGE

Many sources report serum calcium levels of 90 to 110 mg/L for disease-free persons (''normals''). Newer methods show lower ranges of about 85 to 104 mg/L.

Meites[11] lists premature infants as having serum calcium levels of 60 to 100 mg/L, with full-term infants having serum calcium concentrations of 73 to 120 mg/L. In the first 24 hours after birth, calcium drops from 100 to 80 mg/L. This is followed by a slight rise, a drop to 75 mg/L, and a return to 85 mg/L at 72 hours.[12]

There is no clinically significant difference in serum calcium for men and women. Urine calcium values vary considerably and are only meaningul if the patient is kept on a low-calcium, neutral-ash diet for 3 days before collection. Low output is 50 to 100 mg/day, average is 100 to 300 mg/day. Values are lower in persons above 70 years old.

Reference	Serum Ca	Urine Ca (mg/24 hours)
Atomic absorption—adults	85-105 mg/L (21-26 mmol/L)	Men, < 275 Women, < 250 Hypercalcemia, > 300 Average diet, 100-250
Cresolphthalein complexone[13]—adults	85-105 mg/L (21-26 mmol/L)	
Meites		
Premature infants[11]	60-100 (15-25 mmol/L)	
Full-term infants	73-120 (18.3-40 mmol/L)	
1-2 years (by atomic absorption)	100-120 (25-30 mmol/L)	

REFERENCES

1. Weissman, N., and Pileggi, V.J.: Inorganic ions. In Henry, R.J., Cannon, D.C., and Winkelman, J.W., editors: Clinical chemistry principles and techniques, ed. 2, New York, 1974, Harper & Row, Publishers, Inc., p. 646-653.
2. Clark, E.P., and Collips, J.B.: A study of the Tisdall method for the determination of blood serum calcium with suggested modification, J. Biol. Chem. **63**:461-464, 1925.
3. Jackson, J.E., Breem, M., and Cheng, C.: Fluorometric titration of calcium, J. Lab. Clin. Med. **60**:700-708, 1962.
4. Walmsley, T.A., and Fowler, R.T.: Optimum use of 8-hydroxyquinoline in plasma calcium determinations, Clin. Chem. **27**:1782, 1981.
5. Moorehead, W.R., and Briggs, H.G.: 2-amino-2-methyl-1-propanol as the alkalinizing agent in an improved continuous flow cresolphthalein complexone procedure for calcium in serum, Clin. Chem. **20**:1458-1460, 1974.
6. Brett, E.M., and Hicks, J.M.: Total calcium measurement in serum from neonates: limitations of current methds, Clin. Chem. **27**:1733-1737, 1981.
7. Leidtke, R.J., Kroon, G., and Batjer, J.D.: Centrifugal analysis with automated sequential reagent addition: measurement of serum calcium, Clin. Chem. **27**:2025-2028, 1981.
8. Porter, W.H., Carrol, J.R., and Roberts, R.E.: Hemoglobin interference with the DuPont automatic clinical analyzer procedure for calcium, Clin. Chem. **23**:2145-2147, 1977.
9. Hass, R.G., and Mushel, S.: Hemoglobin interference compensation using a two pack calcium method on the du Pont ACA, Clin. Chem. **25**:1126, 1979.
10. Ladenson, J.H., McDonald, J.M., and Goren, M.: Multiple myeloma and hypercalcemia, Clin. Chem. **25**:1821-1825, 1979. (Case conference.)
11. Meites, S.: Normal values for pediatric clinical chemistry, Washington, D.C., 1974, American Association for Clinical Chemistry.

12. Meites, S.: Serum calcium values in premature infants and newborns; spring meeting, Ohio Valley Section of American Association for Clinical Chemistry, 1974.
13. Baginski, E.S., Marie, S.S., Alcock, N.W., et al.: Calcium in biological fluids. In Faulkner, W.R., and Meites, S., editors: Selected methods of clinical chemistry, vol. 9, Washington, D.C., 1982, American Association for Clinical Chemistry Press.
14. Stewart, A.F., Adler, M., Byers, C.M., et al.: Calcium homeostasis in immobilization: an example of resorptive hypercalcuria, N. Engl. J. Med. **306:**1136-1140, 1982.

Carbon dioxide

W. GREGORY MILLER

CO_2, total CO_2 content
Clinical significance: pp. 363 and 387
Molecular formula: CO_2 or O=C=O
Molecular weight: 44.01 daltons
Merck Index: 1815
Chemical class: gas, organic acid in soluble phase

PRINCIPLES OF ANALYSIS AND CURRENT USAGE

Total carbon dioxide in serum or plasma exists in two major chemical forms: dissolved CO_2 (5%), and bicarbonate (HCO_3^-) anion (95%). Other quantitatively minor forms are carbonic acid (H_2CO_3), carbonate ions ($CO_3^=$) and carbamino derivatives of plasma proteins. The majority of procedures to quantitate total carbon dioxide content in serum or plasma involve acidification of the sample to convert all carbon dioxide forms to CO_2 gas and measurement of the amount of gas formed. The established reference method uses the Natelson Microgasometer [2,3] to measure the liberated CO_2 gas manometrically (method 1, Table 55-5). The pressure is corrected for other gases dissolved in serum by addition of alkali to totally absorb the CO_2 and measurement of the residual pressure.

A commonly used method for carbon dioxide determination by continuous flow analysis diffuses the CO_2 gas across a silicon-rubber membrane into an alkaline bicarbonate buffer containing a pH-indicating dye (phenolphthalein or cresol red) (method 2, Table 55-5). The CO_2 gas is quantitatively converted to bicarbonate and hydrogen ions, which produces a pH change and therefore a change in indicator color intensity, which is detected spectrophotometrically. Since other gases dissolved in the serum do not produce a pH change, they do not interfere.

A second commonly used method employs a P_{CO_2} electrode to quantitate the CO_2 gas produced by acidification of the sample. Here again the CO_2 gas diffuses across a silicon-rubber membrane and changes the pH of the bicar-

Table 55-5. Methods of total CO_2 analysis

Method	Analysis	Principle	Usage	Comment
1. Gas pressure	Manometric	CO_2 is released from samples after addition of acid, and the total gas pressure is measured manometrically. The CO_2 gas is absorbed by alkali and the residual gas pressure (other than CO_2) is measured. The difference is the P_{CO_2}. P_{CO_2} = Total gas pressure − Residual gas pressure	Rare	The reference method, used for calibration purposes
2. pH indicator	Spectrophotometric	$CO_2 + \text{Acid} \rightarrow CO_2\ \text{gas}$; $CO_2\ \text{gas} \xrightarrow{\text{Silicone membrane}} CO_2\ \text{dissolved}, \downarrow \text{pH}$; $H^+ + \text{pH indicator} \rightarrow \text{Color change}$	Common	Used in continuous flow analyzers
3. CO_2 electrode	Ion-selective (pH) electrode	After acidification of sample, CO_2 diffuses across a silicone membrane into dilute bicarbonate buffer. The resulting pH change, as measured by a pH electrode, is related to total CO_2 content.	Common	Same method by which P_{CO_2} is measured in blood gas analyzers
4. Enzymatic	Spectrophotometric	$HCO_3^- + PEP \xrightarrow{PEPC} Ox + P_i$; $Ox + NADH \xrightarrow{MDH} MAL + NAD\ (\downarrow A, 340\ nm)$	Common	Used on discrete analyzers, such as du Pont ACA

MAL, Malate; *MDH*, Malate dehydrogenase; *Ox*, Oxaloacetate; P_i, Phosphate; *PEP*, phospho*enol*pyruvate; *PEPC*, phospho*enol*pyruvate carboxylase.

bonate electrode buffer. The pH change is detected by a glass pH electrode within the P_{CO_2} electrode assembly (method 3, Table 55-6).

A final method used to measure total carbon dioxide by some discrete analyzers, such as the du Pont ACA, quantitatively converts all CO_2 forms to HCO_3^- by adding alkali to the serum. The bicarbonate is then enzymatically linked to NADH consumption and quantitated spectrophotometrically as follows (method 4, Table 55-5):

$$HCO_3^- + \text{Phospho}\textit{enol}\text{pyruvate} \xrightarrow{\text{PEPC}} \text{Oxaloacetate} + P_i$$

$$\text{Oxaloacetate} + \text{NADH} \xrightarrow{\text{MDH}} \text{Malate} + \text{NAD}^+$$

PEPC is phospho*enol*pyruvate carboxylase, and MDH is malate dehydrogenase. The decrease in absorption at 340 nm is related to the serum concentration of CO_2.

REFERENCE AND PREFERRED METHODS

The cresol red method on the Technicon SMA 6/60 uses Tris buffer, whose pH response is temperature sensitive. In addition, the method has a nonlinear response of absorbance versus concentration, which is compensated for by use of nonlinear recorder paper. Phenolphthalein-based reagents in bicarbonate buffer do not suffer from these deficiencies and are preferred.

The methods based on gas diffusion and pH change and the enzymatic procedure must be calibrated either with primary aqueous standards of sodium bicarbonate or serum-based secondary standards that have had carbon dioxide calibration values assigned to them.

The manometric technique requires no standards, since the result is calculated from physical properties of gases. Each microgasometer is supplied with a temperature correction chart that corrects the gas pressure and the glass expansion for the temperature of measurement. The technique is very accurate and specific and requires only 10 to 30 μL of sample. A drawback to the manometric procedure is that the instrument uses metallic mercury, which is toxic. Extra care therefore should be exercised when one is using this instrument. In addition, the analysis is time consuming and requires special care and attention.

All these methods are capable of producing accurate and precise results (Table 55-6). For smaller laboratories, the manometric reference method might be suitable. For laboratories with larger numbers of specimens, automated instruments using gas-diffusion methods or the enzymatic procedure are clinically acceptable.

SPECIMEN

One can use serum or heparinized plasma. Other anticoagulants cannot be used because they disturb the equilibrium between erythrocyte and plasma CO_2. Whole blood cannot be used because heme-bound hemoglobin and carbamino-bound CO_2 vary with the hematocrit and oxygen saturation and produce variable results.

The major errors in total CO_2 measurement are associated with sample handling. Samples should ideally be handled anaerobically, centrifuged at 37° C, and stored tightly stoppered before analysis. Once separated from erythrocytes and kept stoppered, serum or plasma total CO_2 content is stable several days at 4° C. Centrifugation at room temperature produces a decrease of approximately 0.5 mmol/L[4] because of a shift of CO_2 into the erythrocyte as a result of the pH drop, which occurs on changing blood from 25° to 37° C. Loss of CO_2 from uncapped tubes is approximately 4 mmol/L after 1 hour.[5] One can prevent CO_2 loss by adding NH_4OH to a pH of 9.0 to 9.3. Layering with oil is unreliable because of CO_2 solubility in the oil.

Table 55-6. Comparison of reaction conditions for total CO_2 analysis

Parameter	Microgasometer*	pH indicator†	CO_2 electrode‡	Enzymatic§
Temperature	Ambient	37.5° C	~40° C	37° C
Sample volume	30 μL	~90 μL	~175 μL	20 μL
Fraction of sample volume	0.25	0.05	0.05	0.004
Final concentration of reagents	Lactic acid: 833 mmol/L	H_2SO_4: 249 mmol/L Cresol red: 105 μmol/L Tris: 2.7 mmol/L	H_2SO_4: 80 mmol/L Surfactant	NADH: 320 μmol/L PEP-carboxylase: 160 U/L PEP: 16 mmol/L MDH: 3000 U/L
Wavelength	—	420 nm	—	340 nm
Reaction mode	End point	End point	End point	Kinetic
Time of reaction	2-3 minutes	~9 minutes	45 seconds	56 seconds
Linearity		10-50 mmol/L	10-50 mmol/L	5-40 mmol/L
Interferences	None	None	None	None

*Natelson Microgasometer, Scientific Industries, Bohemia, N.Y.
†Technicon, AutoAnalyzer Method No. SE4-0008FJ4, Technicon Instruments Corp., Tarrytown, N.Y.
‡IL 446, Instrumentation Laboratory, Inc., Lexington, Mass.
§Du Pont ACA, E.I. du Pont de Nemours & Co., Wilmington, Del.

REFERENCE PROCEDURE (NATELSON MICROGASOMETER)

Principle

A. $\underset{\text{(liquid)}}{O_2, N_2, CO_2\text{ (gas)}, HCO_3^-} + H^+ \longrightarrow \underset{\text{(liquid)}}{H_2O} + \underset{\text{(gas)}}{O_2, N_2, CO_2\text{ (gas)}}$

B. $\underset{\text{(gas)}}{CO_2\text{ (gas)}, O_2, N_2} + \underset{\text{(liquid)}}{OH^-} \longrightarrow \underset{\text{(liquid)}}{HCO_3^-} + \underset{\text{(gas)}}{O_2, N_2}$

Reagents

Lactic acid (1 mol/L). Lactic acid working reagent is prepared by mixture of 100 mL of 85% DL-lactic acid with water to 1 liter total volume. Stable 2 years at room temperature.

Sodium hydroxide (3 mol/L). NaOH working reagent is prepared by dissolving 120 g of NaOH in a total volume of 1 liter of water. Stable 2 years at room temperature in a plastic container.

Mercury. Liquid mercury (highly poisonous—handle with extreme caution).

Antifoam solution. Available from Scientific Industries (Bohemia, N.Y.), or 2-octanol may be used.

Working reagents. Working reagents are prepared by placing approximately 2 mL of mercury and 10 mL of reagent together in small vials.

Assay

Equipment. The Natelson Microgasometer is manufactured by Scientific Industries (Bohemia, N.Y.). The device (Fig. 55-1) is a manometer connected to a calibrated volume chamber and has a calibrated inlet tip to pipet serum and reagent volumes. Liquid is transported hydraulically by a manually operated plunger in the mercury reservoir. Detailed instructions are provided with each instrument and should be followed carefully. An outline of the procedure is presented here.

1. Serum or plasma (30 μL), lactic acid, antifoam solution, and water (or a special low-foam detergent from Scientific Industries) are pipetted into the instrument. Mercury segments are pipetted from the

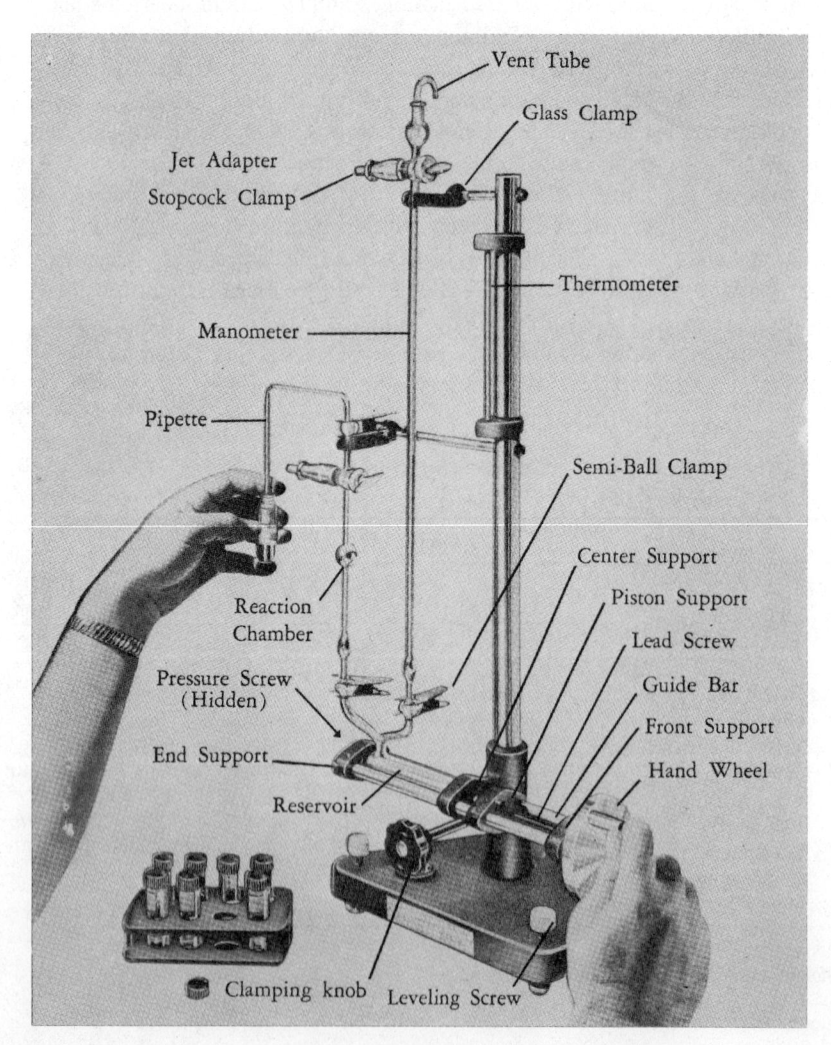

Fig. 55-1. Model no. 600 Natelson Microgasometer.

bottom of the vial between each reagent to prevent the introduction of air bubbles.

2. The stopcock is closed and the mercury withdrawn to produce a vacuum chamber over a liquid reaction chamber. The device is shaken thoroughly for 1 minute to extract CO_2 gas into the vacuum chamber.
3. The mercury is then advanced to compress the vacuum/gas chamber to a fixed constant volume and the temperature and pressure of gas are recorded.
4. Sodium hydroxide is then pipetted into the reaction chamber and the apparatus shaken again to reextract the CO_2 into the liquid phase.
5. The gas volume is then compressed to the same constant volume, and the residual pressure caused by oxygen and nitrogen is measured.

Calculation

The difference in pressure is proportional to total CO_2 content of the sample. The pressure (mm Hg) is multiplied by a temperature-dependent factor supplied with the instrument to obtain millimoles of CO_2 per liter of serum.

Notes

1. Complete and rigorous agitation is necessary to ensure quantitative release of CO_2 from the sample.
2. Great care must be exercised not to introduce air in between the sample and reagents, since this will falsely increase the results.
3. Mercury is toxic with a vapor pressure of 2×10^{-3} mm. The entire apparatus should be used inside a ventilation hood and any spills cleaned promptly.
4. One can use 10 μL of sample for pediatric specimens.

Interferences

The only interference is caused by anesthetics, such as ether, which are partially volatilized during the acid extraction into a vacuum and partially reabsorbed by the alkali. Consequently, the pressure is undercorrected and yields a falsely high result.

REFERENCE RANGE

Children	18-27 mmol/L
Adults	21-31 mmol/L

The reference ranges listed above are based on the manometric procedure. The ranges obtained by the other available methods may vary by 1 to 2 mmol/L, depending on the instrument used.

REFERENCES

1. Davenport, H.W.: The ABC of acid-base chemistry, ed. 5, Chicago, 1969, University of Chicago Press, p. 34.
2. Natelson, S.: Routine use of ultramicro methods in the clinical laboratory, Am. J. Clin. Pathol. **21:**1153-1172, 1951.
3. Meites, S., and Faulkner, W.R.: Manual of practical micro and general procedures in clinical chemistry, Springfield, Ill., 1962, Charles C Thomas, Publisher.
4. Astrup, P.: A simple electrometric technique for the determination of carbon dioxide tension in blood and plasma, total content of carbon dioxide in plasma, and bicarbonate content in "separated" plasma at a fixed carbon dioxide tension, Scand. J. Clin. Lab. Invest. **8:**33-43, 1956.
5. Gambino, R.S., and Schreiber, H.: The measurement of CO_2 content with the autoanalyser, Am. J. Clin. Pathol. **45:**406-411, 1966.

Chloride

W. GREGORY MILLER

Clinical significance: pp. 363, 387, and 403
Atomic formula: Cl^-
Atomic weight: 35.453 daltons
Merck Index: 2065
Chemical class: inorganic anion

PRINCIPLES OF ANALYSIS AND CURRENT USAGE

Chloride is the major anion in extracellular body fluids and is measured in serum, plasma, urine, sweat, and occasionally other body fluids. Early methods to quantitate chloride depended on the very low solubility of silver chloride and mercury chloride salts. Quantitation of the precipitate, a method originally described by Schales and Schales in 1953 and still used in a few laboratories, is by direct manual titration of chloride with mercuric nitrate (method 4, Table 55-7).[1] The end point is detected visually as the faint blue color of a mercuric diphenylcarbazone complex. A principal difficulty with this method is detection of the end point, which varies in shade and intensity in the presence of protein and is obscured by bilirubin, hemoglobin, and lipemia. Precision is also limited by the ability to read the volume on a buret and it will vary significantly among technologists. A spectrophotometric procedure involved the displacement of chloranilate from its mercuric salt by Cl^- ions. The color of the released chloranilate was used to determine the chloride concentration directly.

The most commonly used method today employs the quantitative displacement of thiocyanate by chloride from mercuric thiocyanate and subsequent formation of a red ferric thiocyanate complex, which is measured colorimetrically at 525 nm (method 1, Table 55-7).[2] The sensitivity and linear range of the reaction are adjusted by addition of an excess of mercuric ions as mercuric nitrate. The chloride will combine first with the free mercury ions (colorless) before displacing any thiocyanate from mercuric thiocyanate to produce a colored product measured at 525 nm. This approach will limit the minimum detectable chloride to 80 mmol/L but greatly improve the sensitivity (absorbance change per millimole) in the clinically important serum range of 80 to 125 mmol/L. One can optimize the concentrations and ratios of reagents and sample size for assays of various clinical specimens with differing concentration ranges. The ferric thiocyanate reaction is very tem-

Table 55-7. Methods of chloride measurement

Method	Type of analysis	Principle	Usage	Comments
1. Mercuric/ferric thiocyanate	Quantitative, end point	$2\ Cl^- + Hg(SCN)_2 \rightarrow HgCl_2 + 2\ (SCN)^-$ $3\ (SCN)^- + Fe^{+++} \rightarrow Fe(SCN)_3$ (red) A_{max}, 525 nm	Serum, plasma, urine; manual, automated	Most frequently used, good accuracy and precision
2. Coulometric titration	Quantitative titration, end point	$Ag^+ + Cl^- \rightarrow AgCl(\downarrow)$	Serum, plasma, urine, fluids, sweat; manual, automated	Reference method, highly accurate
3. Ion-selective electrode	Quantitative, potentiometric end point, or kinetic	Ag^+,AgCl(s) | Cl^- + AgCl, AgS Test solution | Reference electrode	Serum, plasma, urine, fluids, sweat; manual, automated	Increasingly used, good accuracy and precision
4. Mercuric nitrate	Quantitative titration	$2\ Cl^- + Hg(NO_3)_2 \rightarrow HgCl_2 + 2\ (NO_3)^-$ Excess Hg^{++} + Diphenylcarbazone → Mercuric diphenylcarbazone *(blue)*	Serum, plasma, urine; manual	Relatively uncommon, poor precision
5. Isotope dilution	Mass spectrometer	Dilution of stable isotope ^{36}Cl with ^{35}Cl	Research or reference laboratories	Definitive method

perature sensitive, and a constant temperature must be maintained to obtain accurate results. Failure to do so can result in a significant drift in some automated methods. Procedures that do not include a protein separation step, such as dialysis, are subject to interference from bilirubin, hemoglobin, and lipemia. However, procedures using small sample volumes (0.01 mL of serum to 1 mL of reagent) show no interference from bilirubin to 220 mg/L or from triglycerides to 6 g/L.[2]

A second very common method to quantitate chloride uses coulometric generation of silver ions with formation of insoluble silver chloride (method 2, Table 55-7).[3,4] The end point is detected amperometrically by a second pair of electrodes as a sudden increase in conductivity caused by free silver ions when all chloride ions are consumed. This procedure is accepted as the reference method for chloride. In principle one can determine the absolute amount of silver ions generated from the number of coulombs (current × time) produced and Faraday's constant (96,487 coulomb/equivalent). In practice, the time required to titrate a chloride standard solution or unknown sample with a constant current is measured. The unknown concentration is then calculated by direct proportion to the time taken to titrate a standard.

Methodological precision in coulometric titration is limited by the rate of mass transfer of silver ions from the generating electrode to the bulk solution. The time needed to detect the end point once all the Cl^- is consumed is called the "blank time" and depends on the rate of mixing in the vessel and the rate of Ag^+ formation. The blank time should be a small fraction of the total titration time for maximum precision. At very high chloride concentrations (400 mmol/L under typical test conditions) the physical bulk of precipitate interferes with mass transfer and obscures detection of the end point. Dirty electrodes will cause poor reproducibility of results, and some workers recommend cleaning with ammonium hydroxide and nitric acid solutions between each titration.[4] Gelatin in the original procedure and polyvinyl alcohol, which is more stable, in current procedures, prevent reduction of silver chloride at the indicating electrodes and promote uniform deposition of excess silver ions on the indicator cathode. This results in a smooth amperometric current and a reproducible detection of the end point.[1]

A fourth method for chloride measurement uses an ion-selective electrode, usually with a silver chloride, silver sulfide sensing element (method 3, Table 55-7). Electrodes and measuring equipment are increasingly available from several manufacturers.

All the methods for chloride will show positive interference from other halides. The only clinically important interfering halide is bromide, which is administered in some drug preparations. Bromide and chloride do not react equivalently in all analytical systems. Bromide ion showed a reactivity equivalent to 2.3 chloride ions with one ion-selective electrode instrument, an equivalency to 1.6 chloride ions with the continuous-flow mercuric thiocyanate method, but an equivalency to 1 chloride ion with the coulometric titration method.[5]

The definitive method for chloride analysis is isotope dilution—mass spectrometry using ^{36}Cl as the internal standard.[6]

Table 55-8. Reference and preferred methods of chloride measurement

Condition	Ferric thiocyanate*	Coulometric titration†
Temperature	~40° C	Ambient
pH	1.5-2.0	< 1
Final concentration of reagents	$Fe(NO_3)_3$, 50 mmol/L $Hg(SCN)_2$, 2 mmol/L $Hg(NO_3)_2$, 0.22 mmol/L	Acetic acid, 1.7 mol/L HNO_3, 0.1 mol/L Gelatin, 300 ng/L‡
Sample volume	175 µL	50 µL
Fraction sample volume	0.0025	0.012
Time of reaction	51 seconds	Variable
Precision (CAP survey data)	125 mmol/L, 2.2% CV 111 mmol/L, 1.9% CV 101 mmol/L, 1.6% CV	125 mmol/L, 2.5% CV 113 mmol/L, 2.4% CV 102 mmol/L, 2.0% CV
Interferences	Br^-, I^- Bilirubin (210 mg/L = 1 mmol/L of Cl^-) Lipemia	Br^-, I^-
Wavelength	525 nm	
Linearity	70-125 mmol/L	

CV, Coefficient of variation.
*IL463 (Instrumentation Laboratory, Inc., Lexington, Mass.)
†Cotlove chloridometer (Buchler Instruments, Fort Lee, N.J.)
‡Can be replaced by polyvinyl alcohol at a concentration of 1.8 g/L of acetic acid–nitric acid reagent.

REFERENCE AND PREFERRED METHODS

(Table 55-8)

Coulometric titration is the accepted reference method because of the precision with which silver ions are generated and its freedom from optical interferences.

All the methods discussed are capable of accurate chloride quantitation when calibrated with primary aqueous standards. Care should be exercised in the manual coulometric techniques to use pipets in the "to-contain" mode with rinsing of the contents to avoid viscosity errors between aqueous standards and serum samples. Coulometric titration at normal serum concentrations of chloride gives between-day coefficients of variation of 1% to 1.5% for automated methods and 2% to 2.5% for manual methods. Analysis with an automated ion-selective electrode method is reported to give a coefficient of variation less than 1%.[7] The manual coulometric titration method and any of the automated methods are recommended for routine analysis of serum. The mercuric/ferric thiocyanate method must be performed at constant temperature and may require a protein-removal step (such as dialysis) to eliminate optical interferences from serum components. The manual titration method is not recommended for routine serum analysis because of its notably poor precision and low sample throughput. Because of the wide range of chloride concentrations possible in urine and sweat, chloride analyses are best done with coulometric methods, though ion-selective electrodes can also be satisfactory for these applications.

SPECIMEN

Serum, heparinized plasma, urine, and other body fluids are acceptable. Serum, plasma, and other fluids should be promptly separated from cells to avoid shifts in ionic equilibriums with metabolism and pH changes. Chloride in serum, plasma, urine, and other fluids is stable at least 1 week at room, refrigerator, or frozen temperatures. Some ion-selective analyzers can measure chloride activity in whole blood, and such measurements should be made within 2 hours to avoid ionic shifts in electrolytes.

Sweat for chloride measurement is collected after iontophoretic delivery of pilocarpine to skin sweat glands to stimulate sweating. The area of skin that must be stimulated depends on the volume of sweat needed for analysis. Ion-selective electrode methods require less volume than coulometric titration methods. In any event, the rate of sweating must be greater than 1 $g/m^2/min$ to obtain clinically reliable results.[8] Lower sweating rates will produce unreliable results because of physiological variability in chloride concentration with sweating rate and because of insufficient quantity of chloride for accurate analytical measurement. Children under 2 to 3 weeks of age may have unusually high sweat electrolyte concentrations. It is therefore advisable to delay the sweat test until after this time.[9] Sweat specimens should always be collected in duplicate. This procedure is important in the diagnosis of cystic fibrosis.

PROCEDURE: SWEAT CHLORIDE

The following procedure for sweat collection is based on the method of Gibson and Cooke[9,10] intended for the coulometric titration of chloride. A procedure intended for ion-selective electrode measurement of chloride is described by Warwick and Hansen.[11]

Reagents

1. Pilocarpine nitrate, 5 g/L water.
2. Potassium sulfate, 10 g/L water.

Equipment. Equipment for iontophoretic delivery of pilocarpine is available from several supply houses, and specialized ion-selective electrode equipment is available from Orion Research Inc. (Cambridge, Mass.).

1. Battery-operated power supply (available from several supply houses).
2. Copper electrodes, 2.5 to 5 cm^2.
3. Gauze squares, 2 to 3 inches.
4. Plastic vials (15 to 25 mL) with caps.
5. Parafilm, 6.5 × 6.5 cm squares.
6. Waterproof plastic tape.
7. Schleicher and Schuell No. 589 green ribbon filter paper (2.5 cm) or equivalent.

Assay

1. Using forceps, place a filter paper into a plastic vial and weigh to 0.1 mg.

2. Wash the forearm with distilled water and dry with gauze square.
3. Thoroughly moisten a gauze square with pilocarpine reagent and position on the volar surface of the forearm. Fasten the positive electrode over the gauze and secure with a rubber strap.
4. Thoroughly moisten a gauze square with potassium sulfate solution and position on the posterior surface of the forearm. Fasten the negative electrode over the gauze and secure with the rubber strap.
5. Connect the electrodes to the power supply, observing the correct polarity, and slowly increase the current to 2 mA.
   NOTE: Discomfort during iontophoresis is caused by inadequate volume of electrolyte solutions or uneven contact of the electrode with the skin surface. Either add more pilocarpine or potassium sulfate to the appropriate gauze square or increase the pressure holding the electrode to the skin surface. If allergic symptoms are experienced by the patient, discontinue the iontophoresis immediately.
6. After 5 minutes turn off the current and disconnect the electrodes.
7. Using distilled water and gauze squares, thoroughly clean and dry the pilocarpine and potassium sulfate–treated areas.
8. Using forceps, remove the preweighed (tare weight) filter paper from the vial and place it over the pilocarpine-treated area. Cover with the Parafilm square and seal all four sides completely with waterproof plastic tape. Evaporation must be avoided.
9. After 30 minutes of sweat collection, press the Parafilm down to collect all sweat droplets in the filter paper. Then remove the Parafilm and with forceps immediately transfer the filter paper back into the preweighed vial.
10. Weigh the vial and substract the tare weight to obtain the amount of sweat collected.
11. For coulometric titration with Cotlove or similar titrating equipment, the appropriate reagent is added directly to the vial containing the gauze to extract sweat electrolytes and the supernatant is analyzed for chloride. Since the actual quantity of chloride ions is lower than that normally found in serum, a slower titration rate is usually necessary to obtain accurate results.

REFERENCE RANGE

1. *Serum or plasma,* 101 to 111 mmol/L
   These values were determined as the central 95% range obtained from analysis of 1205 (596 males, 609 females) healthy ambulatory persons 1 to 78 years old. No variation with sex or age was observed. The method used was automated mercuric/ferric thiocyanate, which was calibrated against the coulometric titration method using primary aqueous standards.
2. *Urine,* 110 to 250 mmol/24 hours
   Urine chloride levels are strongly dependent on dietary intake.
3. *Sweat*[7-10]

Normal	0 to 35 mmol/L
Ambiguous	35 to 60 mmol/L
Cystic fibrosis	60 mmol/L

Reliable sweat chloride measurements depend on achieving an adequate rate of sweating, as discussed earlier. Repeat determination of sweat chloride (new collection and analysis) should always be made to confirm a positive finding. Intermediate values do not indicate heterozygote status. Adults have a more variable sweat composition than children have and can have sweat chloride values up to 70 mmol/L.

REFERENCES

1. Mather, A.: Chloride, direct mercurimetric titration (Provisional). In Faulkner, W.R., and Meites, S., editors: Selected methods of clinical chemistry, vol. 9, Washington, D.C., 1982, American Association for Clinical Chemistry, pp. 153-155.
2. Levinson, S.S.: Chloride, colorimetric method. In Faulkner, W.R., and Meites, S., editors: Selected methods of clinical chemistry, vol. 9, Washington, D.C., 1982, American Association for Clinical Chemistry, pp. 143-148.
3. Cotlove, E.: Chloride: standard methods of clinical chemistry, vol. 3, New York, 1961, Academic Press, Inc., pp. 81-92.
4. Dietz, A.A., and Bond, E.E.: Chloride, coulometric-amperometric methods. In Faulkner, W.R., and Meites, S., editors: Selected methods of clinical chemistry, vol. 9, Washington, D.C., 1982, American Association for Clinical Chemistry, pp. 149-152.
5. Elin, E.J., et al.: Bromide interferes with determination of chloride by each of four methods, Clin. Chem. **27:**778-779, 1981.
6. Cotlove, E.: Determination of the true chloride content of biological fluids and tissues. II. Analysis by simple, nonisotopic methods, Anal. Chem. **35:**101-105, 1963.
7. Lustgarten, J.A., et al.: Evaluation of an automated selective-ion electrolyte analyzer for measuring NA^+, K^+, and Cl^- in serum, Clin. Chem. **20:**1217-1221, 1974.
8. Gibson, L.E.: The decline of the sweat test, Clin. Pediatrics **12:**450-453, 1973.
9. Hammond, K.B., and Johnston, E.J.: Sweat test for cystic fibrosis. In Faulkner, W.R., and Meites, S., editors: Selected methods of clinical chemistry, vol. 9, Washington, D.C., 1982, American Association for Clinical Chemistry, pp. 347-351.
10. Gibson, L.E., and Cooke, R.E.: A test for concentration of electrolytes in sweat in cystic fibrosis of the pancreas utilizing pilocarpine by iontophoresis, J. Pediatr. **55:**545-549, 1959.
11. Warwick, W.J., and Hansen, L.: Measurement of chloride in sweat with the chloride-selective electrode, Clin. Chem. **24:**2050-2053, 1978.

Iron and iron-binding capacity

GERARDO PERROTTA

Clinical significance: pp. 633
Atomic formula: Fe^{+2}(ferrous), Fe^{+3}(ferric)
Atomic weight: 56 daltons
Merck Index: 4942
Chemical class: element, metal

PRINCIPLES OF ANALYSIS AND CURRENT USAGE

There are many methods available for determining serum iron. All colorimetric procedures have in common a reaction in which ferric iron (Fe^{+3}) is reduced to the ferrous state (Fe^{+2}) by the addition of a strong reducing agent such as hydrazine, ascorbic acid, thioglycolic acid, or hydroxylamine.[1] Once the reduction has taken place, colorimetric reactions use an iron-complexing chromogenic agent to bind the ferrous iron. The more commonly used complexing agents are bathophenanthroline, 3-(2-pyridyl)-5,6-bis(4-phenylsulfonic acid)-1,2,4-triazine (FerroZine Iron Reagent), and tripyridyltriazine (TPZ).[1-3] In summary, all colorimetric procedures employ the following reaction (methods 1 and 3, Table 55-9).

More recently, the Environmental Sciences Associates Company (ESA) (Bedford, Mass.) has introduced a new instrument, the FerroChem 3050, in which the automated measurement of serum iron is based on principles of coulometry (method 2, Table 55-9). The method first removes iron from transferrin by the addition of an alcoholic HCl solution. The freed iron is exposed to different specific potentials and electrodes in a multielectrode sensor. The transfer of electrons between the Fe^{+2} and Fe^{+3} generates a current. The Ferrochem instrument detects the electrons in transition and relates the total electron flow to iron concentration[4] (equation 55-2, Table 55-9).

$$Fe^{+3} + e^- \rightleftharpoons Fe^{+2}$$

Other methods, such as radiometry and atomic absorption spectrophotometry, complete the list of methods available for assaying serum iron. Most recent CAP surveys show that less than 2% of the total participants use the radiometric methods, and none are listed as using the atomic absorption method.

Serum iron is almost always accompanied by a measurement of the total iron-binding capacity (TIBC). As indi-

$$\text{Transferrin }(Fe^{+3})_2 \xrightleftharpoons[\text{Reducing agents}]{\text{Buffer}} \text{Transferrin} + 2Fe^{+2}$$

$$Fe^{+2} + \text{Complexing chromogenic agent} \rightarrow \text{Colored complex (magenta, red, and so on) measured at 530 to 560 nm}$$

Table 55-9. Methods of analysis for iron

Method	Type of analysis	Principle	Usage	Comments
Manual				
1. FerroZine Iron Reagent	Colorimetric 560 nm; magenta color complex; no prior protein removal is necessary	***Eq. 55-1*** $\text{Transferrin }(Fe^{+3})_2 \xrightarrow[\text{Reducing agent}]{\text{Buffer}} 2\,Fe^{+2} + \text{Transferrin}$ $Fe^{+2} + 3\text{ FerroZine} \rightarrow Fe(\text{FerroZine})_3^{2+}$	Serum	Good coefficient of variation
Automated				
2. Electrochemical FerroChem 3050	Coulometry; sums up electron in transition between two possible valence states of Fe	***Eq. 55-2*** $Fe^{+3} + e^- \rightleftharpoons Fe^{+2}$ Sum of electrons in this transition	Serum	Correlates well with FerroZine method. Has very good accuracy and precision. Excellent for microsamples and stat. tests
Automated				
3. Du Pont ACA	Colorimetric at 540 nm; surfactant prevents protein precipitation	$\text{Transferrin }(Fe^{+3})_2 \xrightarrow{\text{pH 4.2}} 2\,Fe^{+3} + \text{Transferrin}$ $F^{+3} + \text{Hydroxylamine} \rightarrow Fe^{+2}$ $Fe^{+2} + \text{Bathophenanthroline} \rightarrow \text{Red complex}$ $Fe^{+2} + \text{Bathophenanthroline} \rightarrow \text{Red complex}$	Serum	Good for stat. tests
4. Spectroscopy	Atomic absorption; iron absorption line at 248.3 nm	Iron is concentrated by chelation with bathophenanthroline and is extracted into methyl isobutyl ketone (MIBK).	Serum	Impractical and low sensitivity; mainly for research

cated in Table 55-9, serum iron is bound to transferrin, but only a portion of the transferrin molecule is saturated with iron. The unsaturated iron-binding capacity of transferrin (UIBC) denotes the available iron-binding sites of serum. On the other hand, the amount of iron that serum transferrin can bind when completely saturated is the total iron-binding capacity (TIBC).

In most colorimetric methods, one measures the TIBC by first saturating the transferrin with excess ferric iron. The remaining iron that cannot be bound to transferrin is removed by a chelator, such as magnesium carbonate, which forms an insoluble complex with the excess ferric iron. The total amount of iron saturating the transferrin is then measured as described for total serum iron analysis.

The FerroChem instrument, on the other hand, uses an ion-exchange technique. A sample of serum is added to an aliquot of ion-exchange resin preloaded with iron. After a short period of equilibration, the serum sample (transferrin) is saturated with iron, since the affinity of iron for the transferrin is greater than that for the resin. An aliquot of this iron-saturated serum is then analyzed on the Ferrochem instrument the same way as for total iron analysis.

It is also possible to measure TIBC by an additive procedure. A known iron standard (5000 μg/L) is incubated with the serum sample at a pH of 8.5, and so the transferrin becomes saturated. The remaining unbound (free) iron is measured after the regular iron procedure. The 5000 μg/L standard minus the excess Fe not bound to transferrin is the UIBC, or the unbound iron-binding capacity.

$$\text{UIBC} + \text{Serum iron} = \text{TIBC}$$

REFERENCE AND PREFERRED METHODS

In the selection of the most accurate and precise serum iron procedure one must consider (1) the optimal chromogenic complexing agent for the colorimetric method, (2) whether to remove protein before analysis, and (3) the advantages and accuracy of the colorimetric procedures versus the coulombic (FerroChem), radiometric, and other assays. The comparison is complicated by the lack of a definitive method for the determination of iron. Thus only relative comparisons can be made at this time.

The methods using FerroZine or TPZ as iron-complexing reagents yield consistently higher values than the bathophenanthroline method. The most common comparison is between colorimetric methods that use prior protein removal and those without prior protein removal. Most laboratories reporting in a recent CAP survey used methods that fall into the category of analysis without prior protein removal. Overall, the summary of methods for serum iron determinations based on CAP surveys indicates that an automated bathophenanthroline method without prior protein removal had the lowest coefficient of variation (3%). Radiometric procedures, on the other hand, show the highest coefficient of variation of any method.

Itano[5] reviewed the results of a questionnaire that was used to determine the cause of the differences reported in the CAP survey between the various methods listed. His findings suggested that significantly lower values were common for the FerroZine and TPZ procedures not using

Table 55-10. Comparison of methods for serum-iron procedure

Condition	FerroZine (Harleco)	Du Pont ACA	FerroChem*	ICSH†
Temperature	37° C	37° C	Room temperature	Room temperature
pH				
Fe	4.0	4.2	Acid	Acid
TIBC (total iron-binding capacity)	8.5			
Sample volume				
Fe	0.5 mL	260 μL	50 μL	2 mL
TIBC	0.5 mL	260 μL	100 μL	
Linearity	up to 8800 μg/L	10,000 μg/L	10,000 μg/L	—
Time of reaction (min)	10	4.5	1	10
Wavelength (nm)	560	540	—	535
Interferences	Hemolysis Copper ≥ 1000 μg/L: Anticoagulants (EDTA, citrate)	Same as under FerroZine	Same as others, though copper interferences can be nulled out	Same as under FerroZine
Precision (coefficient of variation)‡	6%-7%	3%	Not available	—
Normal range				
Fe	400-1600 μg/L	420-1350 μg/L	400-1600 μg/L	Not available
TIBC	2650-4330 μg/L	2800-4000 μg/L	2650-4330 μg/L	
% saturated	20%-50%	20%-50%	20%-50%	

*Environmental Sciences Associates Co., Bedford, Mass.
†The International Committee for Standardization in Hematology.
‡For Fe values within the normal range (from CAP Comprehensive Quality Control Survey).

prior protein removal. However, there appeared to be no significant bias between these two chromogenic reagents used for either category. His study reinforced the need for some reference method. The International Committee for Standardization in Hematology[6] has proposed such a method, but very little information is available on any studies conducted in a clinical setting (Table 55-10).

Good comparisons between methods based on different principles are not yet possible since some methods, like the Ferrochem, are not included in CAP surveys. In general, radiometric methods show the lowest mean $(\bar{X})$ values when compared to all other methods.

Based on my experience, the FerroChem procedure has a recovery of 99% and a correlation coefficient of 0.98 for iron and 0.96 for TIBC compared to the Harleco kit procedure. In my laboratory the between-run coefficient of variation of the FerroChem method is approximately 5%. The small sample size, rapid analysis time, and acceptable precision make the FerroChem procedure an attractive method.

SPECIMEN

Traditionally, serum iron specimens have been collected in amber-top tubes. However, it has been found that one can use regular serum or serum separator tubes with no loss of accuracy. Phlebotomy should be accomplished with minimal stasis to allow the free flow of blood into the phlebotomy container. A fasting, early morning sample is desired for analysis because a diurnal variation may decrease the iron value by 30% during the course of a day. The specimen should be centrifuged as soon as possible after collection. Phlebotomy tubes without silicone barriers should be spun twice for 5 minutes with the angular velocity set at 5 if a Sorvall centrifuge is used. This is to make sure that all red blood cells have been removed from the serum sample to be analyzed. Serum separator tubes, on the other hand, need only to be centrifuged once. Serum iron is stable up to 1 week at 2° to 8° C. Transferrin is stable for at least a week at refrigerated temperatures and at least a month when frozen.

Hemolyzed samples are rejected, since erythrocytes contain iron and therefore falsely elevate the serum result.[7]

REFERENCE RANGE

The range for serum iron in healthy persons as established in my laboratory using the FerroChem system is 400 to 1600 μg/L for adults. Children less than 2 years old usually have iron consistently less than 1000 μg/L. However, they reach adult levels after that age.

There is also a diurnal variation for serum iron, with the highest value in the morning and lowest in the evening.

Women during childbearing years lose about 600 to 800 mg of iron during each pregnancy. This loss must be balanced by the absorption of an equivalent amount of dietary iron.[8]

REFERENCES

1. Peters, T., Giovanniello, T., et al.: A simple improved method for the determination of serum iron II, J. Lab. Clin. Med. **48:**280-288, 1956.
2. Levy, A., and Vitacce, P.: Direct determination and binding capacity of serum iron, Clin. Chem. **7:**241-248, 1961.
3. Goodwin, J., et al.: Direct measurement of serum iron and binding capacity, Clin. Chem. **12:**47-57, 1966.
4. Ferrochem Model 3050, Serum Iron and Total Iron Binding Capacity Analyzer instruction manual, ed. 2, Bedford, Mass., March 1979, rev. 2/1981, Environmental Sciences Associates, Inc.
5. Itano, M.: CAP comprehensive chemistry–serum iron survey, Am. J. Clin. Pathol. **70:**516-522, 1978.
6. Lewis, S.M.: International Committee for Standardization in Hematology and proposed recommendation for measurement of serum iron in human blood, Am. J. Clin. Pathol. **56:**543-545, 1971.
7. Hematology Laboratory, Department of Pathology and Laboratory Medicine, University of Cincinnati, unpublished data.
8. Sonnenwirth, A.C., and Jarett, L.: Gradwohl's clinical laboratory methods and diagnosis, ed. 8, St. Louis, 1980, The C.V. Mosby Co.

Magnesium

E. CHRISTIS FARRELL

Clinical significance: pp. 439
Atomic formula: Mg
Atomic weight: 24.31 daltons
Merck Index: 5474
Chemical class: alkaline earth element
Forms: free Mg^{++}, protein bound, or in inorganic complexes

PRINCIPLES OF ANALYSIS AND CURRENT USAGE

Magnesium, like calcium, has offered a challenge to analysts because of the ease of contamination and problems with binding to serum proteins. Seventy-one percent of magnesium is free, 22% is bound to albumin, and 7% is bound to globulins.[1]

The first methods used for magnesium analysis were precipitation techniques[2-4] in which the magnesium was precipitated from solution in the form of salts such as $MgNH_4PO_4$ (method 1, Table 55-11). The amount of precipitate, measured gravimetrically or by analysis of phosphorus in the precipitate, was directly related to the magnesium concentration. 8-Hydroxyquinoline was also used as a precipitant; the precipitate formed was quantified by a variety of techniques (method 2, Table 55-11). The manually performed precipitation methods are no longer in current use.

Compleximetric techniques using EDTA[5] and other chelators[1,6] have been developed but are rarely used. Sensitive fluorometric assays employing calcein,[1] *o,o'*-dihydroxyazobenzene,[7] and 8-hydroxyquinoline are also available.[8] In the latter method (method 3, Table 55-11), a chelated complex forms that has an excitation maximum at 420 nm and fluoresces at 530 nm. This technique has also been automated. Many fluorescence procedures employ chelating agents such as EGTA to inhibit interference from Ca^{++}.

Table 55-11. Methods for analysis of magnesium

Method	Principle	Usage	Comments
1. Precipitation by ammonium phosphate with gravimetric analysis	Mg^{++} + Ammonium phosphate → Mg-ammonium phosphate (precipitate)	Historical	One of first methods of analysis
With subsequent analysis for phosphate	Fisk-Subbarow phosphorus analysis of precipitate	Historical	
2. Reaction with 8-hydroxyquinoline	Mg^{++} + 2 HO— (N) + 2 NH_4OH → (N) —O—Mg—O— (N) + 2 NH_4^+ + 2 H_2O (precipitate) Alternate analyses a. Quantitation with Folin's phenol reagent b. Blue-green color with ferric iron c. Bromination and titration of excess bromate	Historical	Labor intensive; supplanted by better methods
3. Titration of fluorescent complex	a. Mg^{++} + Calcein → Mg-calcein (fluorescent)	Stat. or small laboratories	Difficulty with fluorescence background noise as well as quenching
	b. Mg^{++} + *o,o'*-Dihydroxyazobenzene → Fluorescent complex	Rarely used	
4. Flame emission	$Mg^{++} \xrightarrow{2e^-} Mg^0 \xrightarrow{Heat} Mg^* \rightarrow$ Photon + Mg^0 (370 nm, 383 nm)	Historical	Poor sensitivity
5. Calmagite	Mg^{++} + (OH, CH_3) —N=N— (OH) —SO_3H $\xrightarrow{PVP}$ Complex at 532 nm (violet)	Labs without ACA, or atomic absorption widely used; adapted to many automated analyzers	Slight negative bias; positive shift caused by lipemia.
6. Methylthymol blue	Mg^{++} + Methylthymol blue → Complex at 510 and 600 nm	Du Pont ACA, most commonly reported method	Good correlation with atomic absorption
7. Titan yellow	Mg^{++} + 2NaOH + Titan yellow $\xrightarrow{PVP}$ $Mg(OH)_2$–Titan yellow (red lake) (colloidal ppt)	Infrequently used	Only one eighth sensitivity of Calmagite
8. Atomic absorption	$Mg^{++} \xrightarrow[+2e^-]{Heat} Mg^0 \xrightarrow[Photon]{} Mg^*$	Used in a number of routine laboratories	Excellent accuracy, reference method
9. Neutron activation isotope dilution	Uses magnesium isotope ^{27}Mg	Definitive method	

One can measure magnesium by flame photometry at its band emission peaks at 370 or 383 nm, but this is also not a widely used procedure[1] (method 4, Table 55-11).

Several spectrophotometric methods that one can use directly on biological fluids with deproteinization have been developed. Several of these methods have been automated and are in widespread use today.

Gindler and Heth[9] introduced the use of Calmagite (1-[1-hydroxy-4-methyl-2-phenylazo]-2-naphthol-4-sulfonic acid), a metallochromic dye, for direct determination of magnesium without deproteinization (method 5, Table 55-11). The alkaline, blue-colored working reagent forms a pink magnesium-Calmagite complex with a resulting shift of the reagent color to reddish violet. This color is quantitated at 532 nm. EGTA is used to prevent Ca^{++} interference, and KCN is used to inhibit heavy metal complexes. Polyvinylpyrrolidone and its 9-ethyleneoxide adduct (Bion PVP, Bion Ne-9) have been employed to prevent serum proteins from shifting the absorbance maxima of the Mg-Calmagite complex.[10] The reaction is quite rapid, being completed in 60 seconds. This method has been adapted to a large number of automated analyzers, especially centrifugal analyzers, and is a widely used procedure.

Methylthymol blue, which has been used for calcium analysis,[1] has also been employed for the analysis of magnesium. Again, specific calcium chelators are used to increase the specificity of the analysis. The methylthymol blue (MTB) reaction has been automated on the du Pont ACA, on which the color is quantitated bichromatically at 510 and 600 nm (method 6, Table 55-11).

Another direct spectrophotometric method is based on the reaction between magnesium and the dye Titan yellow (Clayton yellow, Thiazol yellow, 2,2′-[(diazoamino) di-*p*-phenylene]bis[6-methyl-7-benzothiazolesulfonic acid]disodium salt). The reaction proceeds under alkaline conditions in which the colloid $Mg(OH)_2$ forms.[1,11] The colloidal particles adsorb a reactive fraction of the Titan yellow dye to form a red complex (lake) measured at 540 nm (method 7, Table 55-11). To prevent precipitation of the dye-colloid complex, stabilizers such as polyvinyl alcohol have been added to the reaction mixture. The stabilizers also increase the intensity of the color.[1] This method has been adapted for automated analysis but is a rarely used procedure.

A widely used method for magnesium analysis is atomic absorption (method 8, Table 55-11), which is also the reference method. Magnesium has a strong spectral emission or absorption line at 285.2 nm, which can be readily isolated and used to specifically measure magnesium concentration in biological fluids. Although atomic absorption is most frequently employed as a manual procedure, several steps have been semiautomated.

The definitive method for magnesium analysis has been neutron activation with ^{27}Mg (method 9, Table 55-11).

REFERENCE AND PREFERRED METHODS

When comparing methods and results of proficiency studies, one must be aware of the use of four systems of units of reporting magnesium concentrations:

$$0.5 \text{ mmol/L} = 1.0 \text{ mEq/L} = 1.22 \text{ mg/dL} = 12.2 \text{ mg/L}$$

The older methods employing manual precipitation techniques were relatively imprecise and difficult to automate. The fluorescent techniques, though sensitive, suffer from problems of both background fluorescence and fluorescence quenching. Quenching effects in urine samples renders fluorescence methods unsuitable for analysis of this fluid.

The direct spectrophotometric analyses, as adapted to automated techniques, are fairly accurate and precise (Table 55-12). A review of results of quality-control surveys and the linear regression analysis of several adaptations of the Calmagite reaction reveal a rather small, negative proportional bias ($r = 0.96$) for the Calmagite method as compared with atomic absorption.[12,13] The du Pont ACA, which uses a methylthymol blue color reaction, usually shows a closer correlation with atomic absorption. The precision of the methylthymol blue (MTB) method for serum magnesium concentrations within the normal range is approximately 7% to 10% CV.

A survey conducted in Canada by the Canadian government suggested that manual colorimetric and fluorometric procedures were inferior to atomic absorption analysis and the methylthymol blue method, but that atomic absorption had a higher variation for normal level specimens than was expected, about 9%. Overall, the laboratories that had low CV's on a daily basis were more accurate than those that did not, the major difficulty being the correct measurement of elevated-level specimens.[14] However, CAP quality assurance surveys suggest a coefficient of variation (CV) for atomic absorption analysis of magnesium of 5.5% to 8.5% for values within the normal serum range.

A problem with the Titan yellow dye method is the purity of the dye. Only a portion of the dye is reactive with magnesium,[1,15] and the degree of purity can affect the sensitivity and accuracy of this method. Titan yellow methods appear to show a 10% to 15% positive bias when compared to precipitation methods.[16] The precision of these assays for serum magnesium values in the normal range is 8% to 14% CV.

The Calmagite method has been adapted to many automated analyzers, including the Abbott ABA bichromatic analyzer[17] and the fast centrifugal analyzers.[12] Studies in my laboratory reveal that the absorbance change at 500 nm (used on the ABA) is 70% of the change at 532 nm, whereas the "blank" at 600 nm can have an absorbance of greater than 1.0. Although the reaction has also been monitored at 520 nm, I prefer to monitor the peak wave length at 532 nm. The reagents for the Calmagite reaction

Table 55-12. Reaction conditions for magnesium analysis

Condition	Calmagite*	Atomic absorption†	Methylthymol blue‡
Temperature	25°-37° C	2000° C	37° C
Sample volume	Variable	50 μL	20 μL
Fraction of sample volume	0.01	0.02	0.004
Final concentration of reagents	Calmagite: 0.06 g/L (10 parts dye plus 1 part base reagent) KCl: 0.34 mol/L Bion NE-9: 0.98 g/L Bion PVP: 0.91 g/L KCN: 0.280 mmol/L KOH: 25.6 mmol/L EGTA: 0.041 g/L	Lanthanum: 1 g/L for serum and CSF; 5 g/L for urine	EGTA‖: 500 μmol/L Methylthymol blue: 60 μmol/L Sodium metaborate: pH > 11
Wavelength	532 nm	285.2 nm	600 nm and 510 nm (primary and secondary)
Reaction	End point	End point	End point
Time of reaction	30 seconds	30 seconds	261.5 seconds
Linearity	40 mg/L	50 mg/L	200 mg/L
Interferences (negative)	Chelators (EDTA, oxalate)	Chelators	Chelators
Precision ($\bar{X}$, *CV*)§	39.4 mg/L, 11.7% CV 19.2 mg/L, 12.1% CV	40.9 mg/L, 7.0% CV 20.0 mg/L, 7.4% CV	40.2 mg/L, 4.8% CV 19.5 mg/L, 9.4% CV

*Performed on reagent obtained from Pierce Chemical Co. (Rockford, Ill.), adaptable to any spectrophotometer.
†Performed as described in text.
‡As used on the du Pont ACA.
§From CAP Quality Assurance Survey data; $\bar{X}$, mean; *CV*, coefficient of variation.
‖Ethylene glycol-bis(β-aminoethylether)*N,N'*-tetraacetic acid

are more stable separately and so should be added in two separate steps or the reagent should be mixed immediately before use. Using the Roche COBAS centrifugal analyzer, I have found it useful to take a dye plus serum blank and then add the alkaline cyanide reagent to develop the color. Subtracting the blank color from the final reaction color gives a change of absorbance proportional to magnesium concentrations up to 40 mg/L. In addition, it has been reported that the Calmagite will precipitate if left in contact with plastic labware for more than 15 to 20 minutes.[12] The coefficient of variation for the Calmagite methods for serum magnesium concentrations within normal reference limits is approximately 11% to 12%.

On the basis of precision and accuracy, atomic absorption must be the preferred method. Since the methylthymol blue method correlates well with atomic absorption, I recommend it as a second choice. For those laboratories without either atomic absorption or ACA's, the automated Calmagite method will provide adequate clinical data.

SPECIMEN

Serum must be separated from the clot as soon as possible or the level of magnesium will increase because of its elution from the red blood cells.[1] The erythrocytes contain about threefold more Mg^{++} than plasma, and the Mg^{++} is free within the interior of the cell, not bound to the membrane like Ca^{++}.[1] Thus, in general, hemolyzed samples are unacceptable for analysis. The presence of Ca^{++} gluconate in serum because of intravenous administration can cause results of the Titan yellow method to be 35% low.[1] Blood anticoagulated with oxalate or EDTA is unacceptable because these compounds can chelate the magnesium ions.

Urine should be acidified to pH 1 with concentrated HCl. If a precipitate forms, shake, mix, acidify, and warm to 60° C to redissolve it.[1]

PROCEDURE

Principle

The magnesium determinations by atomic absorption spectroscopy are based on the fact that atoms of an element in the "ground," or unexcited, state absorb light of the same wavelength as that emitted by the element in the excited state. Each element has its own characteristic absorption or resonance lines, and no two elements are known to have an identical resonance line.

Magnesium in serum or urine is determined when diluted sufficiently with lanthanum oxide solution to avoid interference by phosphate ions.

Reagents

Use special atomic absorption glassware and water that are free of calcium and magnesium contamination.

1. *Stock lanthanum, 50 g/L in 4 mol/L of HCl.* Weigh 58.64 g of lanthanum oxide (La_2O_3) and wet with about 50 mL of deionized H_2O. Add 250 mL of concentrated HCl *slowly* and *cautiously* under hood. Mix to dissolve powder. Dilute to 1 liter with deionized H_2O. Store in dark at room temperature. Stable 6 months.
2. *Working lanthanum diluent for serum, 1 g/L:* Dilute 20 mL of stock lanthanum to 1 liter with deionized H_2O. Store in dark bottle at room temperature. Stable 3 months.
3. *Working lanthanum diluent for urine, 5 g/L:* Dilute 100 mL of stock lanthanum to 1 liter with deionized H_2O. Store in dark bottle at room temperature. Stable 3 months.
4. *Stock magnesium standard (1000 mg/L, 1 mg/mL, 1215 mEq/L):* Commercially available from Fisher Scientific (50-M-51). Alternatively, dissolve 100 mg of magnesium metal (commercially available in powder or ribbon form) in 2 mL of 1 mol/L of HCl in a 100 mL volumetric flask. After metal has completely dissolved, dilute to 100 mL with distilled, deionized water. Store in a polyethylene bottle at 4° C. Stable for 12 months.

Working standards

Mg mg/L (mEq/L)	Stock Mg (mL)	Dilute to volume with working lanthanum, 1 g/L (mL)
10 (12.15)	1	100
20 (24.3)	2	100
30 (36.5)	3	100
40 (48.6)	4	100
50 (60.7)	5	100

Assay

Equipment. Perkin-Elmer Atomic Absorption Spectrophotometer Model 460 with a single-slot burner head and calcium-magnesium cathode lamp. Follow manufacturer's instructions for proper analytical procedure and for establishing a proper air-acetylene fuel ratio. See calcium analysis for preventive maintenance schedule, p. 1055.

Micro-Medic dilutor, Model 25000, is used to prepare 1:50 dilutions of samples with the appropriate lanthanum solution.

Notes

1. The dilution (1:50) of serum or urine prevents protein interference.
2. There is no appreciable variation in the patient's magnesium levels throughout the day if exercise is avoided, and there is no difference between fasting and nonfasting magnesium levels.
3. Newborns have essentially the same serum concentrations of magnesium as adults.
4. Table 55-13 indicates changes in serum magnesium concentrations in various disease states.

REFERENCE RANGE

Urinary excretion of magnesium is diet dependent but normally equals about one third of daily intake.[1]

Method	Mg^{++}
Serum and CSF	
Titan yellow[1] for serum	18.3-29.3 mg/L (1.5-2.4 mEq/L)
Average value[1] for CSF (no reported sex difference)	24.4-32.0 mg/L (2.0-2.7 mEq/L)
Atomic absorption	15.8-25.5 mg/L (1.3-2.6 mEq/L)
Urine	24-255 mg/24 hours (2-21 mEq/24 hours)

Table 55-13. Changes in disease

High magnesium	Low magnesium
Uremia	Mg^{++}-deficiency tetanus
Acute and chronic renal failure	Loss of gastrointestinal secretions
Chronic glomerulonephritis	Malabsorption syndrome
Addison's disease	Diuretic therapy
Intensive antiacid therapy	Diabetic acidosis
Oxalate poisoning	Treatment of diabetic coma
Lupus erythematosus	Parenteral fluids (prolonged administration)
Multiple myeloma	Hyperparathyroidism
	Rickets

Adapted from Diagnotes, Lufkin Medical Laboratories, Minneapolis, Minnesota 55440.

REFERENCES

1. Henry, R.J., Cannon, D.C., and Winkelman, J.W.: Clinical chemistry principles and techniques, ed. 2, New York, 1974, Harper & Row, Publishers, Inc.
2. Kramer, B., and Tisdall, F.F.: A single technique for the determination of calcium and magnesium in small amounts of serum, J. Biol. Chem. **47:**475-481, 1921.
3. Briggs, A.P.: A colorimetric method for the determination of small amounts of magnesium, J. Biol. Chem. **52:**349-355, 1922.
4. Denis, W.: The determination of magnesium in blood, plasma, and serum, J. Biol. Chem. **52:**411-415, 1922.
5. Schwartzenbach, G., Bidermann, W., and Banserter, F.: Komplexone VI: Neue einfache Titriermethoden zur Bestimmung der Wasserhärte, Helv. Chim. Acta **29:**811-818, 1946.
6. Thiers, R.E.: Magnesium (fluorometric). In Meites, S., editor: Standard methods in clinical chemistry, vol. 5, New York, 1965, Academic Press, Inc.
7. Brien, M., and Marshall, R.T.: An automated fluorometric method for the direct determination of magnesium in serum and urine using *o,o'*-dihydroxyazobenzene: studies on normal and uremic subjects, J. Lab. Clin. Med. **68:**701-712, 1966.
8. Schachter, D.: The fluorometric estimation of magnesium in serum and urine, J. Lab. Clin. Med. **54:**763-768, 1959.
9. Gindler, E.M., and Heth, D.A.: Colorimetric determination with bound "Calmagite" of magnesium in human blood serum, Clin. Chem. **17:**663 1971 (abstract).
10. Pierce Magnesium Rapid Stat Kit, Pierce Chemical Co., Rockford, Ill.; insert revised Feb. 1979.

11. Basinski, D.H.: Magnesium (Titan yellow). In Meites, S., editor: Standard methods in clinical chemistry, vol. 5, New York, 1965, Academic Press, Inc., p. 137.
12. Khayam-Bashi, K., Liu, T.Z., and Walter, V.: Measurement of serum magnesium with a centrifugal analyzer, Clin. Chem. **23**:289, 1977.
13. Cohen, S.A., and Daza, I.E.: Calmagite method for determination of serum magnesium modified (letter), Clin. Chem. **26**:783, 1980.
14. Weatherburn, M.W., and Trotman, R.B.B.: Serum magnesium methodology—an assessment based on application of a reference method in Canada, Clin. Chem. **26**:(abstract 322)1022, 1980.
15. King, H.G.C., and Pruden, G.: Component of commercial Titan yellow most reactive towards magnesium—its isolation and use in determining magnesium in silicate minerals, Analyst **92**:83, 1967.
16. Simonsen, D.G., Westover, L.M., and Wertman, M.: The determination of serum magnesium by the molybdivanadate method for phosphate, J. Biol. Chem. **169**:39-47, 1947.
17. Wong, H.K.: Fully automated procedure for serum magnesium (letter), Clin. Chem. **21**:169, 1975.

Osmolality

LAWRENCE A. KAPLAN

Clinical significance: pp. 232

PRINCIPLES OF ANALYSIS AND CURRENT USAGE

Osmolality is a colligative property of solutions that depends on the number of dissolved particles present in the solution. As the number of dissolved particles increases, the freezing point and vapor pressure are decreased, and the osmotic pressure and boiling point are increased.[1] Chapter 12 discusses in detail the theory of measurement of these colligative properties.

There are three types of solutes most often encountered in biological fluids: electrolytes, organic molecules, and colloids. In dilute solutions there is a linear change in the colligative properties as the solute concentration increases. It is important to realize that it is not the mass concentration but the molal concentration (that is, moles/kg of solvent) that is the basis of colligative properties. For example, in biological fluids the concentration of salts and small molecular weight organic compounds (such as glucose and urea) affect osmolality much more than albumin, which is present in a large mass amount but because of its large molecular weight is present at low molal concentrations.

The most frequently measured colligative property used to estimate fluid osmolality is that of freezing-point depression. When 1 mole of solute is added to 1 kilogram of water, the freezing point is depressed by 1.858° C. The laboratory osmometer actually measures the heat released by the freezing fluid and relates it to the freezing point of the fluid.

A less frequently measured parameter is that of vapor-pressure depression. The vapor-pressure osmometer actually measures the dew-point depression of the vapor, that is, the vapor in equilibrium with the solution being measured. The more dissolved particles present (increased osmolality), the lower the vapor pressure of the aqueous component of the solution. An important exception to this is when the solute is itself a volatile substance.

A frequently used "stat" procedure to estimate the osmolality of a fluid is the calculated osmolarity. The calculated osmolarity sums up the molar concentrations of the principal active osmolar solutes present in serum.[2] Although there are many equations used to calculate the osmolarity, the simplest and the most frequently used is:

$$\text{mOsm/L of } H_2O = 1.86\,[Na^+] + \frac{\text{Glucose (mg/L)}}{180 \text{ mg/mmol}} + \frac{\text{BUN (mg/L)}}{28 \text{ mg/mmol}}$$

where sodium concentration is expressed in milliequivalents per liter (mEq/L), and the concentrations of glucose and blood urea nitrogen (BUN) are divided by 180 and 28, respectively, so that one can convert these mass concentrations into millimoles per liter (mmol/L). It is important to note that although osmolality should be expressed as milliosmoles per kilogram (mOsm/kg) of water it is most frequently reported as osmolarity, that is, as milliosmoles per liter of water. The slight error introduced has little clinical significance.

REFERENCE AND PREFERRED METHODS

Serum osmolality, as measured by freezing-point depression osmometers, has a coefficient of variation of 1.1% to 1.8% for osmolalities within the normal reference range (Table 55-14). The vapor-pressure osmometers tend to have worse precision at all levels of osmolality, as es-

Table 55-14. Comparison of freezing-point and vapor-pressure osmometers

Conditions	Freezing point*	Vapor pressure†
Sample volume	250-2000 µL	8 µL
Time of analysis	< 60-70 sec	90 sec
Interferences	None	Volatile solutes (negative)
Precision‡ ($\bar{X}$, % *CV*)		
Serum	277 mOsm/kg of H_2O, 1.8% CV	274 mOsm/kg of H_2O, 3.0% CV
	386 mOsm/kg of H_2O, 1.5% CV	377 mOsm/kg of H_2O, 2.8% CV
Urine	643 mOsm/kg of H_2O, 1.3% CV	628 mOsm/kg of H_2O, 2.3% CV
	882 mOsm/kg of H_2O, 1.4% CV	872 mOsm/kg of H_2O, 2.0% CV

$\bar{X}$, Mean; *CV*, coefficient of variation.
*Osmette A, Precision Systems, Inc., Sudbury, Mass.
†Wescor model 5100C, Wescor, Inc., Logan, Utah.
‡College of American Pathologists Quality Assurance Survey data.

timated by CAP quality assurance surveys, that is, 2.6% to 2.9% CV within the normal range.

It has been pointed out that the two types of osmometers differ in their ability to respond to volatile solutes that may be clinically encountered.[3,4] The contribution to serum osmolality of several commonly seen alcohols (such as ethanol, methanol, and isopropranol) is directly proportional to their molal concentrations when measured by freezing-point depression osmometry. However, vapor-pressure osmometers do not detect these volatile substances. The reason is that volatile solutes themselves contribute to the total vapor pressure present above the solution, though they decrease the vapor pressure of the water portion of the solution.[5] Therefore the decreased vapor pressure of water is counterbalanced by the increase in vapor pressure caused by the presence of a volatile substance such as ethanol. As the concentration of ethanol increases in aqueous solutions, the vapor pressure above the solution actually increases, giving falsely low osmolality values.

Aside from the problem encountered with volatile solutes, the osmolality of most specimens as measured by freezing-point depression correlates well with the osmolality measured by vapor-pressure depression. The vapor-pressure osmometers have the advantage of requiring only 8 μL of sample as compared to 75 to 250 μL for most freezing-point osmometers. This can be an important consideration for use in pediatric laboratories.

In practice the vapor-pressure osmometers appear to be more difficult to maintain in working order than freezing-point depression osmometers. For smaller, especially pediatric, laboratories with fewer technologists using the osmometer, the vapor-pressure instrument may work quite well. For most laboratories, for its greater sturdiness and precision, the freezing-point osmometer is recommended.

The calculated osmolarity is not meant to provide a definitive result but only one sufficient for most acute emergency service work. Used as such, the calculated osmolality can provide clinicians with valuable information with no extra analysis time. The calculated osmolality will give falsely low results in the presence of volatile solutes.

SPECIMEN

Blood drawn without the use of any anticoagulant is the preferred sample. The serum should be removed from the clotted blood cells as rapidly as possible. Well-centrifuged urine or other body fluids are also acceptable samples. Random urine samples can vary considerably in the concentration of analytes and are rarely useful clinically. Therefore 24-hour urine collections are the preferred specimens for the measurement of urine osmolality.

Any fluid left uncovered will have an increased osmolality because of evaporation and concentration of sample solutes. Serum osmolality has been reported by one worker[6] to be stable for 3 hours at room temperature and 10 hours at 4° C and as long as 3 days at 4° C by another source.[7] Urine osmolality may be stable at 4° C for up to 24 hours.[7] Frozen samples may be stable for several weeks.

PROCEDURE FOR FREEZING POINT

Principle

The sample is supercooled below its freezing point and then agitated by a rapidly vibrating probe. This induces rapid crystallization of the sample, and it begins to freeze. Heat of fusion is released by the crystallization and is accurately measured by a thermistor probe present in the solution. The released heat is related to the freezing-point depression (1 milliosmole of solute depresses the freezing point by 0.001858° C), which is converted to osmolality.

Reagent

Osmolality standards can be purchased from a variety of commercial sources or can be provided by the instrument manufacturers. Calibrators are used to standardize the instrument for either serum or urine osmolalities.

A relatively simple procedure for preparing osmolality standards has been described.[6] Reagent-grade NaCl is heated to 200° C overnight to drive off any water present in the crystals. After it cools, weigh the desired amount of NaCl (see below) and add it to 1 kg of class I distilled water. The most convenient way to do this is to fill a one-liter class A volumetric flask with water at 20° C to the mark. Add exactly 1.8 mL of additional water and then add the salt to dissolve.

Desired osmolality (mOsm/kg)	Grams of NaCl/kg of H_2O
100	3.094
300	9.476
500	15.93
1000	32.12

The standard should be stored in clean polyethylene or borosilicate glass bottles. Always pour the standards into clean test tubes before use. When properly closed, the standards will be stable for 3 to 6 months at room temperature.

Assay

Follow manufacturers' instructions for the calibration, use, and maintenance of osmometers.

REFERENCE RANGE

Serum. Most reported ranges for serum osmolality fall between 282 and 300 mOsm/kg. There are no sex or age differences for serum osmolality.

Urine. Urine osmolality will vary considerably with diet. Twenty-four-hour osmolalities will vary between 50 and 1200 mOsm/24 hours.

Cerebrospinal fluid. CSF and other true body fluids will have an osmolality that is essentially equal to that of serum drawn at the same time.

REFERENCES

1. Stevens, S.C., Neumayer, F., and Gutch, C.F.: Serum osmolality as a routine test, Nebr. Med. J. **45:**447, 1960.
2. Weisberg, H.F.: Osmolality—calculated "delta" and more formulas, Clin. Chem. **21:**1182-1184, 1975.
3. Rocco, R.M.: Volatiles and osmometry, I., Clin. Chem. **22:**399, 1976.
4. Barlow, W.K.: Volatile and osmometry II., Clin. Chem. **22:**1230-1232, 1976.
5. Mercier, D.E., Feld, R.D., and Witte, D.L.: Comparison of dewpoint and freezing point osmolality, Am. J. Med. Technol. **44:**1066-1069, 1978.
6. Johnson, R.B., and Hoch, H.: Osmolality. In Meites, S., editor: Standard methods of clinical chemistry, vol. 5, New York, 1965, Academic Press, Inc.
7. Weissman, N., and Pilegg, V.J.: In Henry, R.J., Cannon, D.C., and Winkelman, J.W., editors: Clinical chemistry principles and techniques, New York, 1974, Harper & Row, Publishers, Inc.

Phosphorus

E. CHRISTIS FARRELL

Phosphorus, inorganic phosphorus, phosphate
Clinical significance: pp. 403, 439, and 806
Molecular weight: 97.00, 96.00 daltons (P is 30.98)
Chemical class: inorganic anions, (P is nonmetallic group 5 element)
Forms: $H_2PO_4^{-1}$, HPO_4^{-2}

PRINCIPLES OF ANALYSIS AND CURRENT USAGE

Elemental phosphorus does not exist to any appreciable extent in the body, and so methods have been directed toward analysis of the two phosphate anions, which interchange rapidly depending on pH. The monovalent and divalent anion forms are present in serum at about equal concentrations in acidosis, in a ratio of 1:9 in alkalosis, in a ratio of 1:4 at pH 7.4, and in a ratio of 100:1 in a pH 4.5 urine, making it impossible to say with any certainty what the molecular weight of inorganic "phosphate" is. Therefore the units traditionally chosen have been milligrams or millimoles of phosphorus (in a volume) but never milliequivalents of phosphate, since that would change rapidly with the charge.

The oldest and still most commonly used methods are based on the reaction of phosphate ions with molybdate to form complex structures such as ammonium phosphomolybdate, $(NH_4)_3[P(Mo_3O_{12})_4]$. One can measure phosphomolybdate complexes directly or convert them to molybdenum blue by a wide variety of reducing agents (methods 1 to 7, Table 55-15). Molybdenum blue is a complex heteropolymer of unknown structure whose absorbance is usually measured at 660 nm. Approximately one half of laboratories reporting in CAP proficiency surveys use a reduction assay.

The classical method of Bell and Doisy for phosphate was based on the reduction of phosphomolybdic acid by hydroquinone to molybdenum blue[1] (method 1, Table 55-15). It was limited by irregular and rapid fading of the color in alkaline solution. Fiske and Subbarow demonstrated that 1-amino-2-naphthol-4-sulfonic acid (ANS) was a superior reducing reagent that could be used at room temperature[2] (method 2, Table 55-15). Bartlett intensified the color by heating the reaction to 100° C for 5 minutes[3] (method 3, Table 55-15).

Kutter and Cohen showed that stannous chloride stock reagent was more stable than ANS and that it produced a more intense color with phosphomolybic acid[4] (method 4, Table 55-15). Hurst stabilized the dilute stannous chloride

Table 55-15. Methods of phosphorus analysis*

Method	Principle	Usage	Comment
1. Hydroquinone	Reduction of phosphomolybdate to molybdenum blue	Historical	Irregular, rapid fading of color
2. ANS (1-amino-2-naphthol-4-sulfonic acid)		Historical	Better than hydroquinone
3. ANS + 100° C, 5 min		Historical	Intensified color
4. $SnCl_2$		Early continuous flow	Replaced by improved methods because of reagent instability, poor linearity, and lack of precision
5. $SnCl_2$ + Hydrazine		Continuous flow	Good accuracy, limited linearity, better reagent stability than method 4
6. NH_4FeSO_4		Discrete analyzers	Not so sensitive as ultraviolet methods
7. *p*-Semidine		Lipid phosphorus	Research applications
8. No reduction	340 nm, monitoring	Discrete and continuous flow	Good accuracy, precision, sensitivity
9. No reduction	Bichromatic	Some discrete analyzers	Bichromatic (340 and 380 nm) about one-half sensitivity of 340 nm; blanking problems with hemolysis and lipemia

*General principle: $PO_4 + H_2SO_4 + (NH_4)_6Mo_7O_{24} \cdot 4\ H_2O \rightarrow Mo/PO_4$ complex
Mo/PO_4 complex + Reducing agent → Heteropolymeric blue complex

working reagent by combining it with hydrazine sulfate[5] (method 5, Table 55-15). This also permitted rapid development of the color at room temperature. Technicon[6] produced a method for the AutoAnalyzer and the Multichannel analyzers based on the work of Hurst and Kraml,[7] in which the stannous chloride concentration is doubled to ensure linearity.

Taussky and Shorr[8] proposed the use of ferrous ammonium sulfate as a reducing agent, which is still employed by several kit manufacturers (method 6, Table 55-15). The merits of several methods of phosphate analysis by reduction methods are discussed by Martinek in a 1970 review.[9]

Simonsen et al.[10] reported that phosphate could be quantitated using the 340 nm absorbance of unreduced phosphorus-molybdate (methods 8 and 9, Table 55-15). This made the reaction faster and simpler and improved reagent stability. It also produces three to four times more absorbance change than the molybdenum blue reductions. Daly and Ertingshausen[11] adapted this technique to the CentrifiChem (Baker Instruments), studied the reaction kinetics, and adjusted reagent concentrations so that only 2% of the reaction occurred before the 3-second initial read time on the CentrifiChem. Sera had half-reaction times of about 70 seconds and aqueous standards at 100 seconds. During the 12-minute total reaction time, 99% of the final color was produced. One interesting observation, which has been confirmed, is that the formation of the molybdate complex is slowed when the acidity is increased.

Monitoring of the unreduced phosphorus-molybdate complex has been adapted to a wide variety of automated and semiautomated analyzers, including those produced by Gilford, Abbott, Technicon, and du Pont. Over half the laboratories reporting proficiency test results to the CAP are now using some variety of this technique. Bichromatic instruments use 340 and 380 nm filters to monitor the unreduced complex.

Direct measurement of inorganic phosphate without reduction has also been done by use of a mixed vanadium-molybdenum heteropoly acid.[12] At low acidity a yellow-orange complex causes high blanks; at high acidity the rate is slowed and phosphate values are elevated.[13,14] A recent publication uses nitric acid and an unusual detergent, triethanolamine lauryl sulfate (Montosol RL-10), to shift the absorbance spectral line slightly to 350 nm with a slight improvement in sensitivity.[15-17] The detergent is not widely available in the United States.

REFERENCE AND PREFERRED METHODS

Any analyzer capable of good 340 nm photometry should be able to accomplish accurate, sensitive phosphate analysis employing direct analysis of the phosphorus molybdate complex. There are two factors that complicate what seems to be a simple, straightforward adaptation of this reaction to any automated system. The first factor is that aqueous standards react considerably slower than protein-based standards and may not attain the end point at the same time as serum unknowns.[14] One can add albumin to the standards to speed the reaction rate, but it may not be free of phosphate contamination. The second factor is the unusual influence of pH; acidity is necessary, but as the pH decreases to 1.0, the reaction is substantially slowed; as it is increased beyond 2.0, spontaneous reduction of the molybdate takes place.[11,14]

Gilford Diagnostics has adjusted concentrations for a kinetic variation of direct, nonreduction reaction. It is difficult to imagine linearity over a wide range for a kinetic modification without use of many datum points and complicated rate expressions, since the concentrations of molybdate, phosphate, acid, and protein are all factors affecting the rate. In the modification I use, which is performed on a Roche COBAS centrifugal analyzer, a small amount of diluent washes the serum from the sample probe and reduces the concentration of acid in order to speed the reaction (Table 55-16). The molybdate concentration is high enough in the reagent so that its dilution does not substan-

Table 55-16. Reaction conditions for serum phosphate analysis

Condition	Nonreduction at 340 nm*	Reduction method†
Sample volume	3 μL	10 μL
Fraction of sample volume	0.013	0.014
Final concentration of reagents	Ammonium molybdate: 2.25 mmol/L Sulfuric acid: 750 mmol/L, pH 1.55 Tween 80: 0.45 mL/L	Sodium acetate: 55.7 mmol/L Acetic acid: 328 mmol/L Sodium metabisulfate: 37.2 mmol/L *p*-Methyammonium phenol sulfate: 11.0 mmol/L Ammonium paramolybdate · 7 H_2O: 587 μmol/L H_2SO_4: 16.2 mmol/L Cu^{++}: as catalyst
Temperature	37° C	30° C
Wavelength	340 nm	620 nm
Reaction	End point with serum blank	Two-point kinetic
Time	200 seconds	300 sec
Precision‡ ($\bar{X}$, % CV)	25 mg/L, 9.2% CV 78.7 mg/L, 5.3% CV	24.9 mg/L, 7.2% CV 77.3 mg/L, 4.8% CV
Linearity	80 mg/L	150 mg/L

$\bar{X}$, Mean; *CV*, coefficient of variation.
*As analyzed on COBAS analyzer (Roche Inc., Nutley, N.J.), using reagent obtained from Gilford Diagnostics, Oberlin, Ohio.
†As analyzed on CentrifiChem Analyzer (Baker Instruments, Dallas, Texas), using reagent from Smith-Kline-Beckman (Sunnyvale, Calif.).
‡From College of American Pathologists quality assurance survey data for nonreduction and reduction method groups (overall).

tially slow the reaction. Depending on the type of pipetting available, the reagent may be prediluted if a serum "washout" is not needed.

In the centrifugal analyzer method (such as Roche COBAS), a true serum blank is needed because of the presence of a great number of 340 nm–absorbing compounds and drugs. The blank must also contain the surfactant Tween 80 for clearing of serum turbidity and sulfuric acid as well to correct for the effect of pH on ultraviolet absorption of many compounds. On the COBAS this is simple to achieve because the horizontal light path reads all the absorbance at all times regardless of the total volume. The Coulter DACOS uses traditional optics and will read the blank at one volume, add molybdate, read the final absorbance, and mathematically correct the absorbance for the differences in the two volumes.

Because of the similar slope of the reaction rate of the color complex produced, icteric sera should be properly blanked. This is readily achieved by the use of bichromatic analysis with the combination of 340 and 380 nm wavelengths. Lipemic sera, whose absorbance is higher at the shorter wavelengths, should produce a positive bias. Hemoglobin will produce a strong negative photometric influence, and losses from the erythrocytes will produce a positive influence. For these reasons, the best phosphate analysis requires correct blanking of the individual sera. Use of 340 and 380 nm filters or 340 nm monitoring with blanking should be superior.

Additional problems encountered with the phosphomolybdate-reduction type of assay include instability of the reducing reagent. This is especially true for the stannous chloride reagent, which also shows poor adherence to Beer's law.[13] Stabilization of $SnCl_2$ with hydrazine sulfate has been used to overcome this difficulty. I have found that many laboratories using the Technicon AutoAnalyzer procedure continue to find nonlinearity above 50 mg/L and that the introduction of the stannous chloride–hydrazine sulfate reducing reagent earlier (between coils 1 and 2 instead of between coils 3 and 4) will result in better linearity.

Crough and Malmstadt studied the kinetics of the overall molybdenum blue reaction and found that the rate was first order with respect to phosphate but that the velocity was highly sensitive to changes in pH and molybdate concentration.[16] It has also been noted using the ANS procedure that at higher pH's, such as 2 or 3, nonspecific reduction of the molybdate reagent takes place.[17]

The reaction conditions of these reduction assays are vigorous enough to hydrolyze glucose phosphate, creatine phosphate, and other organic phosphates, creating a positive error.

Table 15-17. Changes in disease

High phosphate	Low phosphate
Diet	Nonfasting specimen
Improper collection	Hypervitaminosis D
Alcohol ingestion	Primary hyperparathyroidism
Acromegaly (>50 mg/L)	Osteomalacia
Hypoparathyroidism (>60 mg/L)	Renal tubular disorders (Fanconi syndrome)
Osteomalacia	
Sarcoidosis	Antacids, which bind phosphorus
Milk alkali syndrome	
Phosphorus in enemas	Steatorrhea
Liver disease	Malabsorption
Bone metastases	
Diarrhea	
Vomiting	
Catastrophic events such as cardiac resuscitation, pulmonary embolism, overwhelming sepsis, or acidosis.	

Modified from Diagnotes, Lufkin Medical Laboratories, Minneapolis, Minn., circa 1977.

SPECIMEN

Serum values will be lower after meals, and so it is important to have the patient in the fasting state. Phosphate will also be lower during the menstrual period. Intravenous glucose and fructose also physiologically lower phosphate. Since phosphate is the major intracellular anion, hemolyzed samples should not be analyzed and the red cells and platelets should be removed promptly before phosphate loss to the serum can occur.

Urine may contain larger quantities of organic phosphates that can decompose on exposure to elevated temperatures. When acidified with HCl, urine phosphate is stable for more than 6 months.[13] Urine samples are usually diluted 1:10 with normal saline before analysis.

REFERENCE RANGES

Method	Males and females reported as mg of phosphorus/L (mEq P/L)	Urine (g/24 hr)
Molybdenum blue methods		
Henry TCA filtrate[13]		
in children	40-70 (2.32-4.06)	0.5-0.8
in adults	25-48 (1.45-2.76)	0.3-1.3
Unreduced phosphomolybdate methods		
Gilford 340 nm rate		
adults	25-48 (1.45-2.76)	
Abbott 340 nm kinetic		
adults	20-46 (1.16-2.67)	

NOTE: There is a diurnal variation in urine phosphate excretion, with the highest output occurring in the afternoon. There is no significant sex difference reported, but serum and urine levels do vary with age.

REFERENCES

1. Bell, R.D., and Doisy, E.A.: Rapid colorimetric methods for the determination of phosphorus in urine and blood, J. Biol. Chem. **44**:55, 1920.
2. Fiske, C.H., and Subbarow, Y.: The colorimetric determination of phosphorus, J. Biol. Chem. **66**:375, 1925.

3. Bartlett, G.R.: Phosphorus assay in column chromatography, J. Biol. Chem. **234**:466, 1959.
4. Kuttner, T., and Cohen, H.R.: Micro colorimetric studies. I. A molybdic acid stannous chloride reagent: the micro estimation of phosphate and calcium in pus, plasma, and spinal fluid, J. Biol. Chem. **75**:517, 1927.
5. Hurst, R.O.: The determination of nucleotide phosphorus with stannous chloride–hydrazine sulphate reagent, Can. J. Biochem. **42**:287, 1964.
6. Method No. SF4-0004FF5, Technicon Instruments Corporation, Tarrytown, N.Y. 10591, released June 1975.
7. Kraml, M.: A semi-automated determination of phospholipids, Clin. Chim. Acta **13**:422, 1966.
8. Taussky, H.H., and Shorr, J.: Microcolorimetric method for determination of inorganic phosphorus, J. Biol. Chem. **202**:675, 1953.
9. Martinek, R.G.: Review of methods for determining inorganic phosphorus in biological fluid, J. Am. Med. Technol. **32**:337, 1970.
10. Simonsen, D.G., Wertman, M., Westover, L.M., and Mehl, J.W.: The determination of serum phosphate by the molybdivanadate method, J. Biol. Chem. **166**:747, 1946.
11. Daly, J.A., and Ertingshausen, G.: Direct method for determining inorganic phosphorus in serum with the "CentrifiChem", Clin. Chem. **18**:263, 1972.
12. Quinlan, K.P., and DeSesa, M.A.: Spectrophotometric determination of phosphorus as molybdovanadophosphoric acid, Anal. Chem. **27**:1626, 1955.
13. Henry, R.J.: Clinical chemistry: pinciples and technics, ed. 2, New York, 1974, Harper & Row, Publishers, Inc., p. 726.
14. Michelsen, O.B.: Photometric determination of phosphorus as molybdovanado-phosphoric acid, Anal. Chem. **29**:60, 1957.
15. Munoz, M.A., Balon, M., and Fernandez, C.: Direct determination of inorganic phosphorus in serum with a single reagent, Clin. Chem. **29**:372, 1983.
16. Crouch, S.R., and Malmstadt, H.V.: A mechanistic investigation of molybdenum blue method for the determination of phosphate, Anal. Chem. **39**:1084, 1967.
17. Tietz, N.: Fundamentals of clinical chemistry, ed. 2, Philadelphia, 1976, W.B. Saunders Co., p. 915.
18. Lubell, D.: Spontaneous diurnal variation of urinary phosphate excretion and its relation to the parathyroid hormone test, Helv. Paediatr. Acta **12**:179, 1957.

Sodium and potassium

W. GREGORY MILLER

Clinical significance: pp. 363 and 403
Atomic formula: Na, K
Atomic weight: Na, 22.990; K, 39.098
Merck Index: Na, 8308; K, 7377
Chemical class: inorganic cation

PRINCIPLES OF ANALYSIS AND CURRENT USAGE

Sodium and potassium are usually quantitated simultaneously either by flame atomic emission spectroscopy (FAES) (Chapter 3) or by ion-selective electrode potentiometry (ISE) methods (Chapter 14) (Table 55-18). Equipment for these methods is available from several manufacturers.

FAES methods typically employ a 1:100 or 1:200 dilution of sample with a diluent containing an internal standard. Lithium at 1500 mEq/L has been the most widely used internal standard, but cesium at 1.5 mEq/L has been used with increased frequency by one manufacturer (Instrumentation Laboratory Inc., Lexington, Mass.).

The heat of the air-propane flame (about 1925° C) vaporizes the salts, which gain electrons from the reducing gases to form the ground-state atoms Na^0 and K^0. These atoms are heated in the flame, resulting in the formation of excited atoms, Na* and K*. The excited atoms instantly decay to the original ground state with the emission of light. Sodium emission is usually monitored at 589 nm, potassium at 766 nm, lithium internal standard at 671 nm, and cesium internal standard at 852 nm. The amount of transmitted light is directly proportional to the concentration of Na and K. Although only 1% to 5% of the atoms

Table 55-18. Methods of sodium and potassium analysis

Method	Type of analysis	Principle	Usage	Comments
1. Flame atomic emission spectroscopy (FAES)	Quantitation of mass concentration	Excited atom emits photon	Serum, urine, CSF, other body fluids	Reference method as well as most commonly used; dilution error possible
2. Ion-selective electrode potentiometry (ISE)	Quantitation of chemical activity	Ion-selective electrode measures potentiometric change as function of ion concentration	Serum, urine, CSF	Dilution error possible by indirect procedure; urine analysis may have limited linear range
3. Atomic absorption	Quantitation of mass concentration	Ground state atoms absorb incident light from hollow cathode lamp	All biological fluids	Highest sensitivity, but not useful for routine analysis

in the flame are excited to emission, this is a sufficient amount for accurate and precise quantitation.

The FAES method measures the concentration of ions in the total volume of specimen sampled. In the case of a serum or plasma sample for a normal person, the solvent volume is composed of 93.1% water, 5.4% protein, 0.6% lipid, and 0.9% other molecules.[1] The physiological effect of ions is a function of their concentration (activity) in the volume of water in which they are dissolved. Consequently, in diseases that alter the volume fraction of water in serum, that is,

$$\frac{\text{Water volume}}{\text{Total serum volume}}$$

the physiological control mechanisms will strive to keep ionic activity per water volume at a constant level. However, if the water volume fraction is reduced, as occurs in hyperproteinemia or hyperlipidemia, the actual concentration of ions in the *total* volume of sample will be reduced though the effective activity of ions in the *water* volume is approximately normal. This physiological interference in FAES methods is responsible for the *pseudohyponatremia* (or pseudohypochloremia) observed in such cases. When the interference is caused by lipemia, one can use an ultracentrifuge to remove the excess lipid volume from the serum and then analyze the sample by FAES to get a physiologically meaningful result. If the interference is caused by protein, the FAES method is inappropriate and a "direct" ISE method as described below must be used to obtain a clinically reliable result. This volume fraction error, when present, does not usually create a problem of clinical interpretation for potassium because the normal range is fairly large with respect to the magnitude of concentration of potassium in serum. For potassium, the size of the reference range (3.6 to 5 mmol/L) represents a 33% change in concentration, whereas for sodium the reference range (135 to 145 mmol/L) extends over only a 7% change in concentration. Hyperproteinemia or hypertriglyceridemia also produces an increased serum viscosity that does not affect aspiration into the flame at the dilutions usually employed but can introduce error into the dilution step itself, depending on the design of the diluter.

Ion-selective elctrode (ISE) methods use a glass ion-exchange membrane for sodium and a valinomycin neutral-carrier membrane for potassium measurement. Typical sodium electrodes have a thousandfold greater selectivity for Na^+ than K^+ and are insensitive to H^+ above pH 6. Typical potassium electrodes have a ten-thousandfold greater selectivity for K^+ than Na^+ and are insensitive to pH above 1. Although the Na^+/K^+ ratio in serum is approximately 30, the potassium electrode has adequate selectivity so that sodium does not interfere.

There are two general types of ISE measurements made on clinical samples. "Direct" potentiometric systems measure the ion activity in an undiluted sample. "Indirect" ISE systems measure the ion activity in a prediluted sample. Because ISE measurements determine the activity of an ion in the water volume fraction in which it is dissolved, "direct" measurements are unaffected by conditions such as hyperproteinemia or hyperlipidemia, which alter the volume fraction of water in serum. However, "indirect" methods are usually sensitive to this physiological effect because the dilution step itself is based on total volumes, and after dilution the volume occupied by soluble serum molecules becomes insignificant with respect to the total diluent volume.

Both sodium and potassium can be individually quantitated by atomic absorption spectrophotometry (Chapter 3). Diluted sample is introduced into the flame and the metals are vaporized and their atoms are reduced to the ground state. Monochromatic light of the specific resonance lines (766.5 nm for K and 589.0 nm for Na) produced by a hollow cathode lamp is passed through the flame and absorbed by the ground-state atoms present in the flame. The amount of absorbed light is directly proportional to the concentration of the metal being measured.

Historically, Na and K have routinely been measured only by flame emission spectroscopy. However, in recent years ISE's have achieved an increased importance in Na and K analysis. A recent College of American Pathologists (CAP) chemistry survey shows that ISE's and FAES's are each used by approximately 54% and 46% of the survey participants respectively. Approximately 61% of ISE's being used employ dilution of the sample.

REFERENCE AND PREFERRED METHODS

Reference methods based on FAES have been published.[5-7] The only significant error for FAES and indirect ISE is that attributable to alteration of the water volume fraction of serum in some diseases. Direct ISE methods are not sensitive to this problem but suffer from potential calibration bias because of electrochemical mismatch between standard solutions and serum. Direct ISE methods may be advantageous in those settings in which one can perform analyses on whole blood. Indirect ISE methods have the advantage of generally requiring a smaller sample volume.

There are no significant spectral interferences in the measurement of sodium and potassium in biological fluids by FAES. There is a small effect of sodium on potassium results caused by the influence of the relatively high sodium concentration on the degree of ionization of potassium atoms in the flame. The degree of interference depends on the degree of mismatch between sodium to potassium ratios in the standards used for calibration and in the patient samples. For the range of values found in serum or plasma the maximum potassium error is 0.2 mmol/L.[2] There is no clinical correlation between sodium and potassium levels in serum, and therefore calibration with a standard solution (140 mmol of Na/L, 5.0 mmol of

Table 55-19. Reaction condition for sodium and potassium analysis

Condition	Flame emission	Ion-selective electrode (indirect)
Temperature	1925° C	Ambient
pH	—	8.0
Final reagent composition	Li: 1500 mmol/L or Cs: 1.5 mmol/L	$MgSO_4$: 40 mmol/L Buffer: pH 8.0
Fraction of sample volume	0.01	0.029
Sample	Serum, urine, CSF	Serum, urine, CSF
Linearity		
Na	100-160 mmol/L (serum) 0-200 mmol/L (urine)	100-180 mmol/L (serum) 25-150 mmol/L (urine)
K	2-8 mmol/L (serum) 0-200 mmol/L (urine)	1-9 mmol/L (serum) 5-105 mmol/L (urine)
Time of reaction	30 seconds	1-2 seconds
Major interferences	Dilution effect by lipids or proteins Hemolysis (for K)	Hemolysis; ammonium heparin and sodium anticoagulants
Precision		
Na	0.7%-1.4% CV in normal range	0.7%-1.4% CV in normal range
K	1.5%-2% CV in normal range	1.5%-2% CV in normal range

CV, Coefficient of variation.

K/L) reflecting normal serum composition will minimize this error. The ionization interference can be larger in urine; however, the clinical impact is negligible because of the large reference range for potassium in urine. One can minimize the ionization interference by adding an excess of an easily ionized element, such as cesium, as an ionization suppressant. This has not been considered necessary in the clinical laboratory because of the small size of the erorr.

A significant source of error exists in the calibration of ISE systems because of the nonideal electrochemical behavior of ions in the high ionic strength (0.151 mol/L) and complex serum matrix. If a simple aqueous standard of NaCl and KCl is used to calibrate an ISE system, a typical ISE result will be 3% lower than an FAES result.[4] This is a result of the combined effect of a decreased activity coefficient for Na^+ or K^+ in the serum caused by interaction with the various anions and cations found in serum and to a difference in mobility of the serum ions participating in the salt bridge at the liquid junction. The ionic mobility effect is primarily caused by replacement of part of the chloride, the counterion in the aqueous standard, with bicarbonate and protein anions in the serum matrix. One can correct the problem by replacing chloride in the aqueous standard with other anions, such as acetate, whose concentrations are empirically adjusted to match the electrochemical behavior of serum in a particular ISE system. The magnitude of this calibration error will vary with the design details of an ISE system. The error is clinically more noticeable for sodium than for potassium because of the relatively small percent concentration interval of the reference range discussed earlier. This calibration bias is not usually observed in ''indirect'' ISE systems because the dilution produces an essentially constant ionic environment.

Measurement of urine sodium and potassium by ISE methods has been reported to have a limited linear range[3] and to be subject to interference by the ionic composition of the urine.[4] These effects will vary with design details of an ISE system.

Either FAES or ISE methods are satisfactory for clinical use with typical coefficients of variation of 1.0% for sodium and 1.5% for potassium at normal serum levels (Table 55-19). Atomic absorption has the advantage of higher sensitivity but is less precise and cannot be recommended for a routine work load.

SPECIMEN

Serum, heparinized plasma, heparinized whole blood, urine, and other body fluids are acceptable specimens. Heparinized plasma in which the heparin is present as the sodium salt is obviously unacceptable for Na analysis. Therefore, heparinized phlebotomy tubes with Li^+ or NH_4^+ heparin are preferred. Serum, plasma, and other fluids should be separated from cells within 3 hours to avoid shifts in ionic equilibrium with metabolism and pH changes. Similarly, whole blood should be analyzed within 3 hours. Sodium and potassium are stable for at least 1 week at room or refrigerator temperatures and for at least 1 year frozen. Whole blood analysis for Na and K is only possible for certain direct ISE methods. Hemolysis will falsely elevate potassium values, and moderately to grossly hemolyzed specimens should be rejected or indicated on the report form.

REFERENCE RANGE

Serum or plasma using FAES or indirect ISE methods

Sodium	135 to 145 mmol/L
Potassium	3.6 to 5.0 mmol/L

These values were determined as the central 95% range obtained from analysis of 1205 (596 males, 609 females) serum samples from healthy ambulatory persons from 1 to 78 years of age. No variation with sex or age was observed. Plasma potassium values may be 0.1 to 0.2 mmol/L lower than serum values because serum may contain a small amount of potassium released from platelets during clotting.

Serum, plasma, or whole blood using direct ISE methods

Sodium	145 to 155 mmol/L
Potassium	3.9 to 5.3 mmol/L

Direct ISE methods that show no aqueous standards calibration bias will have values 7% greater than ''total volume'' methods because of the 93% water-volume fraction in serum, plasma, or whole blood. The magnitude of calibration bias will vary among instruments, which may result in some instruments having a reference range closer to that found for FAES or indirect ISE methods.

Urine

Sodium	40 to 220 mmol/24 hours
Potassium	25 to 150 mmol/24 hours

Urine sodium and potassium levels are strongly dependent on dietary intake. Urine sodium by ISE methods cannot be performed on urine collected in acid. Urine potassium by ISE methods on samples collected in acid may be subject to a significant calibration bias because of liquid-junction potential errors.

REFERENCES

1. Czaban, J.D., Cormier, A.D., and Legg, K.D.: Establishing the direct-potentiometric ''normal'' range for Na/K: residual liquid junction potential and activity coefficient effects, Clin. Chem. **128:**1936-1945, 1982.
2. Anand, V.D., Lott, J.A., Grannis, G.F., and Mercier, J.E.: Effect of sodium on the autoanalyzer flame photometric determination of potassium in serum, Am. J. Clin. Pathol. **59:**717-730, 1973.
3. Ladenson, J.H.: Evaluation of an instrument (Nova-1) for direct potentiometric analysis of sodium and potassium in blood and their indirect potentiometric determination in urine, Clin. Chem. **25:**757-763, 1979.
4. Harff, G.A., and reply by Flores, O., and Buzza, E.: Sodium results for urine with the Beckman Electrolyte 2 Ion-selective Electrolyte Analyzer are dependent on potassium concentration, Clin. Chem. **28:**1232-1233, 1982.
5. Schaffer, R., et al.: A multilaboratory-evaluated reference method for the determination of serum sodium, Clin. Chem. **27:**1824-1828, 1981.
6. Velapoldi, R.A., et al.: A reference method for the determination of sodium in serum, NBS Special Publication 260-60, U.S. Dept. of Commerce, National Bureau of Standards, Washington, D.C., 1978, U.S. Government Printing Office.
7. Velapoldi, R.A., et al.: A reference method for the determination of potassium in serum, NBS Special Publication 260-63, U.S. Dept. of Commerce, National Bureau of Standards, Washington, D.C., 1978, U.S. Government Printing Office.

Chapter 56 Enzymes

Acid phosphatase

LENOX B. ABBOTT
WILLIAM C. WENGER
JOHN A. LOTT

Acid phosphatase, ACP, orthophosphoric monoester phosphohydrolase (acid optimum)
Clinical significance: p. 882
EC 3.1.3.2
Molecular weight: approximately 100,000 daltons; varies with isoenzyme
Chemical class: enzyme, protein
Known isoenzymes: prostate, erythrocyte, platelet, liver, spleen, kidney, bone marrow
Biochemical reaction:

$$\underset{\text{Organic phosphate ester}}{R_1{-}O{-}PO_3H_2} + \underset{\text{Alcohol}}{H{-}OH} \xrightarrow[\text{pH 5.0}]{\text{ACP}} \underset{\text{New alcohol}}{R_1{-}OH} + \underset{\text{New phosphate ester}}{H{-}O{-}PO_3H_2}$$

PRINCIPLES OF ANALYSIS AND CURRENT USAGE

The term "acid phosphatase" (ACP) refers to a group of phosphatases that show maximal activity near pH 5.0 and catalyze the hydrolysis of an orthophosphoric monoester to yield an alcohol and a phosphate group, as shown in the biochemical reaction. ACP has various isoenzymes and is present in most tissue.[1-4] The prostate gland has the highest ACP activity on the basis of international units per gram of tissue.

Numerous methods have been developed for the measurement of ACP activity in biological fluids. The methods vary with the type of substrate and buffer used, as well as temperature and time of incubation. Each modification represents an attempt to maximize the sensitivity for total ACP or specificity for a selected isoenzyme, such as prostatic acid phosphatase (PAP). A representative summary of ACP methods, including those of historical interest and those in current usage, is found in Table 56-1.

The natural substrate or substrates for ACP are not known, though the enzyme is capable of hydrolyzing many orthophosphoric monoesters. The selection of a particular substrate is influenced by the differing reactivity of various substrates with the isoenzymes of ACP.[5,6] For example, PAP reacts more readily with thymolphthalein monophosphate than does ACP derived from platelets or erthyrocytes,[5] but no substrate exhibits absolute specificity for a particular isoenzyme of ACP.[7,8]

The isoenzyme PAP has long been established as a confirmatory test for carcinoma of the prostate gland.[9] A common approach in measuring total ACP and PAP activity in the same serum specimen involves the use of inhibitors specific for PAP. Total ACP activity is measured, L-tartrate or formaldehyde is added to inhibit the PAP fraction, and the ACP is measured again. The difference in activity before and after the addition of inhibitor is assumed to be the PAP activity.

Many early assays for ACP were adaptions of methods already in use for alkaline phosphatase. Identical substrates were used; the adaptation in many cases simply was to buffer the reaction medium at an acid pH.

The first method to estimate ACP was described by Gutman and Gutman.[9] Monophenyl phosphate was used as substrate in a citrate buffer (pH 4.8 at 37° C). After 1 to 3 hours of incubation, the reaction was stopped by the addition of alkali. The amount of phenol liberated by the enzyme was measured colorimetrically with the Folin-Ciocalteu reagent (phosphotungstic-phosphomolybdic acid), which forms a blue complex with phenol in an alkaline medium. It has an absorption maximum at 750 nm.

Jaffe and Bodansky[10] used β-glycerophosphate as substrate in a barbiturate buffer (pH 5.0). After 3 hours of incubation at 37° C the reaction was terminated by the addition of trichloroacetic acid. Liberated phosphate was reacted with phosphomolybdic acid reagent, and the resulting chromogen was measured colorimetrically at 600 nm.

Table 56-1. Summary of methods for acid phosphatase (ACP) measurement in body fluids

Method	Type of analysis	Principle	Usage	Comments
Gutman and Gutman[9]	End-point spectrophotometric	Substrate: monophenyl phosphate Liberated phenol measured with Folin-Ciocalteu reagent	Serum, manual	Historical interest only Insensitive substrate
Jaffe and Bodansky[10]	End-point spectrophotometric	Substrate: β-glycerophosphate Measure liberated phosphate with color reaction	Serum, manual	Historical interest only Long incubation times High background absorbance
Hudson et al.[11]	End-point or kinetic spectrophotometric	Substrate: *p*-nitrophenyl phosphate Measure release of *p*-nitrophenol	Serum, manual/automated	Sensitive, self-indicating substrate; not specific for PAP
Huggins and Talalay[12]	End-point or kinetic spectrophotometric	Substrate: phenolphthalein phosphate Released phenolphthalein measured colorimetrically after addition of base	Serum, manual/automated	Self-indicating but relatively insensitive substrate
Seligman et al.[13]	End-point or kinetic spectrophotometric	Substrate: β-naphthyl phosphate Liberated β-naphthol complexed with *o*-dianisidine to form dye	Serum, manual/automated	Multistep procedure Colored product requires extraction before reading
Babson and Read[14]	End-point or kinetic spectrophotometric	Substrate: α-naphthyl phosphate Liberated α-naphthol complexed with *o*-dianisidine to form dye	Serum, manual/automated	Multistep procedure Precise timing required No extraction involved
Roy et al.[7]	Kinetic spectrophotometric	Substrate: thymolphthalein monophosphate Measure liberated thymolphthalein after addition of base	All body fluids and tissues, manual/automated	Simple procedure Sensitive, self-indicating substrate; high specificity for PAP
Ewen[8]	End-point or kinetic spectrophotometric	Substrate: thymolphthalein monophosphate Measure liberated thymolphthalein after addition of base	All body fluids and tissues, manual/automated	Modification of Roy et al.[7] Enhanced sensitivity of substrate Preferred method
Rietz and Guilbault[15]	Fluorometric kinetic	Substrate: 4-methylumbelliferone phosphate Measure rate of appearance of fluorescent 4-methylumbelliferone	All body fluids and semisolids, manual/automated	Highly sensitive substrate Equipment expense may be limitation
Foti et al.[16]	Radioimmunoassay (RIA) (solid phase)	Tubes coated with antibody directed to PAP; used for RIA with ^{125}I-labeled PAP	All body fluids and tissues, manual	Substrate specific for PAP with good analytical sensitivity Requires 48 hours of incubation
Chu et al.[17]	Counterimmunoelectrophoresis (CIEP)	Precipitin line formation between PAP and anti-PAP antiserum resulting from countermigration in electric field	Serum, manual	Sensitive substrate; specific for PAP Semiquantitative
Lee et al.[18]	Fluorescent immunoassay (FIA) (solid phase)	ACP in serum binds to IgG-Sepharose-coated tubes and is quantitated fluorometrically after reaction with α-naphthyl phosphatase	Serum, manual	Sensitive substrate; specific for PAP Quantitative Requires overnight incubation
Lee et al.[19]	Immunoassay spectrophotometric	Similar to Lee et al.[18] Bound PAP quantitated colorimetrically after dye reaction	Serum, manual	Modification of Lee et al.[18] to give colorimetric procedure

Hudson et al.[11] introduced the use of the self-indicating substrate *p*-nitrophenyl phosphate. The enzyme hydrolyzed the colorless substrate in an acetic acid–acetate buffer at pH 5.4 for 30 minutes at 38° C. The reaction was stopped by the addition of NaOH, and the liberated yellow *p*-nitrophenol was read at 400 nm.

Another self-indicating substrate, phenolphthalein phosphate, was used by Huggins and Talalay.[12] Serum was incubated with the substrate for 1 hour at 37° C in acetic acid–acetate buffer (pH 5.4). The reaction was terminated by addition of an alkaline glycine buffer (pH 11.2). At alkaline pH, the phenolphthalein liberated by the enzyme was intensely pink and was measured colorimetrically at 550 nm.

In the Seligman procedure,[13] serum was added to β-naphthyl phosphate in acetate buffer (pH 4.8) and incubated for 2 hours at 37.5° C. The liberated β-naphthol was complexed with tetrazotized orthodianisidine (*o*-dianisidine) to form a dye, which was extracted with ethyl acetate and then measured at 540 nm.

Babson and Read[14] introduced the use of α-napthyl phosphate as substrate, suggesting that it was relatively specific for the isoenzyme PAP. Serum was added to the substrate in citrate buffer (pH 5.2) and incubated at 37° C for precisely 30 minutes. The reaction mixture was cooled to 15° to 20° C, and tetrazotized *o*-dianisidine was added. The absorbance of the resulting colored solution was measured at 530 nm exactly 3 minutes after the addition of the color developer.

All these methods were manual end-point assays as originally reported. As such, they are of historical interest only. Some procedures have been adapted to automated methods of analysis and find limited current usage in the clinical laboratory. For the most part, however, they have been replaced by the following methods, which employ the highly sensitive substrate thymolphthalein monophosphate or take advantage of recent advances in technology, such as radioimmunoassay (RIA) for prostatic ACP.

Thymolphthalein monophosphate is a self-indicating substrate that exhibits a high degree of specificity for PAP. It was introduced for the quantitation of ACP by Roy et al.[7] Serum was incubated for 30 minutes at 37° C with the substrate in citrate buffer at pH 5.95. The reaction was stopped by the addition of NaOH. The liberated thymolphthalein produced a color in alkaline medium, which was measured at 590 nm. The method was subsequently modified to increase its sensitivity by Ewen,[8] who used an acetate buffer (pH 5.4), reduced the substrate concentration and reaction volume, and changed conditions for sample preservation. A summary of reaction conditions for the Ewen method is presented in Table 56-2.

Table 56-2. Reaction conditions for ACP analysis by preferred method*

Temperature	37° C
pH	5.4
Final concentrations of reagents	
Acetate buffer	0.15 mol/L
Thymolphthalein monophosphate	1.1 mmol/L
Brij-35	1.6 g/L
Fraction of sample volume	0.0303
Expected precision (within run)	2% at 1.7 U/L 1% at 5.0 U/L

*From Ewen, L.M.: In Faulkner, W.H., and Meites, S., editors: Selected methods for the small clinical laboratory, Philadelphia, 1982, W.B. Saunders Co.

A highly sensitive fluorometric assay for ACP was reported by Rietz and Guilbault.[15] ACP activity was determined from the rate of appearance of the fluorescent 4-methylumbelliferone liberated by the enzyme from 4-methylumbelliferone phosphate. They added 1 to 10 μL of serum to citrate buffer (pH 4.9) and measured the reaction rate in a fluorometer at 37° C (λ_{ex} = 365 nm; λ_{em} = 455 nm).

Foti et al.[16] introduced a solid-phase RIA method for PAP. Disposable polypropylene tubes were coated with monospecific antiserum directed against PAP. Sample (0.1 mL) was added to the tubes, along with ^{125}I-labeled PAP. The tubes are incubated for 48 hours at 4° C, decanted, washed, and counted. The percentage of ^{125}I bound to the antibody was calculated. PAP activities in the specimens were quantitated by use of an appropriate standard curve.

Another assay for acid phosphatase is based on counterimmunoelectrophoresis (CIEP) and was developed by Chu et al.[17] The principle of CIEP for detection of PAP is based on the reaction, in an electric field, between serum containing PAP and anti-PAP antiserum. At pH 6.5 PAP migrates anodically, whereas anti-PAP migrates cathodically. After placement of serum specimens in a well at the cathode and anti-PAP near the anode, a precipitin line will form at the point of contact. Electrophoresis is carried out in a 0.75% agarose solution at 4° C for 2 hours in a 50 mmol/L phosphate buffer. The precipitin line is visualized by incubation of the gels for 1 hour at 37° C in an ammonium acetate buffer (pH 5.0) containing 100 mg of α-naphthyl phosphate.

The solid-phase fluorescent immunoassay (FIA) of Lee et al.[18] combines an immunological and biochemical assay for determination of prostatic acid phosphatase. Serum is incubated with IgG-Sepharose 4B in tubes containing phosphate-buffered saline. Initial incubation is for 2 hours at ambient temperature, followed by overnight incubation at 4° C. After centrifugation and washing, the amount of enzyme bound to the Sepharose can be determined by incubation with α-naphthyl phosphate for 1 hour at 37° C. The reaction is stopped with NaOH, and the liberated α-naphthol is determined fluorometrically with excitation at 340 nm and emission at 465 nm. A standard curve is constructed by use of purified PAP. From this standard curve the PAP activity can be determined. A modification of the solid-phase FIA was suggested by Lee et al.[19] Quantitation is by a solid-phase immunoadsorbent assay. The principle is essentially the same as in the FIA, except that enzyme quantitation can be carried out colorimetrically. The PAP activity bound to the Sepharose is determined after incubation with α-naphthyl phosphate. The α-naphthol formed is determined by reaction with a 200 mg/L Fast Red B salt solution. The absorbance of the colored complex at 588 nm is proportional to the original activity of PAP in the sample. A standard curve is prepared with α-naphthol acting as the primary standard.

REFERENCE AND PREFERRED METHODS

The method of Roy et al.[7] as modified by Ewen[8] is the preferred method and is described in the following paragraphs. It measures all isoenzymes of ACP, but it reacts preferentially with the prostatic form. The procedure, presented as a manual method, is simple and can easily be automated.[20] It has been adapted to several commercially available instruments, and it has been used by the National Bureau of Standards and Cooperating Laboratories Enzyme Study Group to assay human serum calibrators.[21]

SPECIMEN

Serum or heparinized plasma give similar results with thymolphthalein monophosphate as substrate. EDTA does not interfere, whereas oxalate is reported to inhibit the enzyme. Trace hemolysis is tolerable, but gross hemolysis results in increases in ACP activity ranging from 6% to 315%, depending on the particular method used.[22] Any drug causing hemolysis will increase ACP activity, and detergent left on glassware can inhibit the enzyme. ACP is very unstable in separated serum at room temperature; the rise in pH caused by the loss of CO_2 from the serum results in the rapid and irreversible inactivation of the enzyme. Maintain an acid pH by adding citric acid tablets to the serum or by storing the serum in tubes containing an acetate buffer.[8] As soon as the serum or plasma is separated, add an acid stabilizer. The stabilizer tablets from Sigma Chemical Co. (St. Louis, MO 63178; catalog 104-9) are very convenient. Add 1 tablet per milliliter of serum. Control sera should be treated in the same way after reconstitution. The specimens from vaginal washings in suspected rape cases must also be acidified if storage is necessary.

PROCEDURE[7,8]

Principle

ACP hydrolyzes thymolphthalein monophosphate to thymolphthalein and phosphate. The reaction is shown below. The products are colorless at an acid pH, but thymolphthalein becomes a self-indicating chromogen when the reaction is terminated by the addition of NaOH. The increase in absorbance at 595 nm with time is proportional to both the thymolphthalein liberated and to the activity of ACP present in the specimen.

Reagents

All reagents should be at least ACS (American Chemical Society) grade.

Acetate buffer, 0.30 mol/L, pH 5.4. Add 3.84 g of sodium acetate trihydrate to 80 mL of distilled water and adjust the pH of the solution with a pH meter to 5.4 at 25° C with 0.1 mol/L of HCl. Dilute to 100 mL with distilled water. The solution is stable for 2 months at 4° C.

Brij-35 solution, 3.2 g/L. Dilute 10.8 mL Brij-35 (300 g/L) to 1 L with distilled water and mix well. This is stable at 4° C for several years.

Buffered substrate reagent, 1.1 mmol/L of thymolphthalein monophosphate. Dissolve 74.4 mg of thymolphthalein monophosphate (formula weight, 676) in 50 mL of the 3.2 g/L of Brij-35 solution. Dilute to 100 mL with the acetate buffer, and check that the pH is 5.4 at 25° C. Add 0.1 mol/L of HCl or 0.1 mol/L of NaOH if needed. Prepare daily.

NOTE. Commercial preparations of thymolphthalein monophosphate will vary in their degree of hydration. Carefully note the formula weight of each preparation to ensure that the appropriate amount of the substrate is added to give a final concentration of 1.1 mmol/L of substrate in the buffered substrate reagent.

Alkaline color development solution. Dissolve 10.6 g

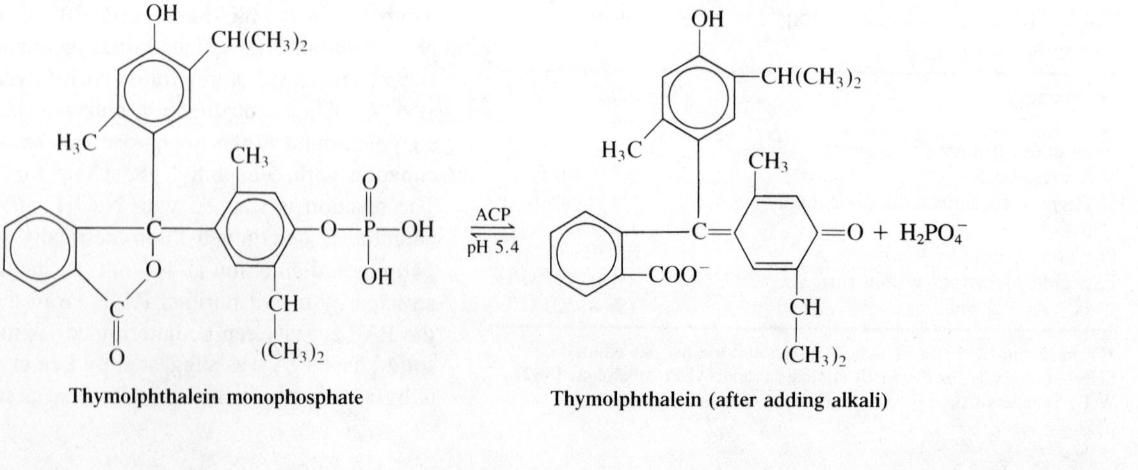

Thymolphthalein monophosphate $\xrightleftharpoons[\text{pH 5.4}]{\text{ACP}}$ Thymolphthalein (after adding alkali) $+ H_2PO_4^-$

Thymolphthalein monophosphate **Thymolphthalein (after adding alkali)**

of anhydrous Na_2CO_3 and 4 g of NaOH in 1 L of distilled water. Store in a polyethylene container. This solution is stable for at least 3 months at room temperature.

Thymolphthalein standard solution, 3 mmol/L. Dissolve 129.2 mg of thymolphthalein (formula weight, 430.52) in 100 mL of an *n*-propanol/water mixture (70:30 by volume). This is stable for 3 months at room temperature.

Assay

Equipment: a spectrophotometer with a band pass ≤10 nm and a 37° C incubator.

1. For each specimen, label two tubes, one "test" and the other "blank."
2. Into each tube pipet 550 μL of the buffered substrate reagent.
3. Warm all the tubes labeled "test" to 37° C in a water bath. At 30 second intervals, add 50 μL of specimen to its corresponding tube. Mix gently and incubate each assay tube at 37° C for exactly 30 minutes.
4. Terminate the assay by adding 1 mL of alkaline color development solution to each of the tubes. Mix well.
5. Blank samples are prepared during the incubation period in step 3 by the addition of 1 mL of alkaline color development solution to each of the blank tubes, followed by the addition of 50 μL of the corresponding patient specimen.
6. Read the absorbances of all samples at 595 nm. Subtract the blank from the test absorbance and determine ACP activity by one of the two following methods of calculation.

Calculations

One can determine ACP activity from the specimen's net increase in absorbance by using the molar absorptivity of thymolphthalein or a standard curve.

Method 1:

Use of molar absorptivity with a 1 cm light path

$$\text{ACP activity (U/L)} = (\Delta A/\Delta t) \times (\text{TV/SV}) \times 10^6 \times (1/\epsilon)$$

where $\Delta A/\Delta t$ = change in absorbance for t minutes
TV = total volume of assay
SV = sample volume
ϵ = molar absorptivity of thymolphthalein = 4.015×10^4

(It is recommended that each laboratory determine ϵ with thymolphthalein solutions of known concentration, in their spectrophotometer. See reference 9 for details of determining ϵ.)

$$\text{U/L} = (\Delta A/30\text{ min}) \times (1.6/0.05) \times 10^6 \times 2.4 \times 10^{-5}$$
$$\text{U/L} = \Delta A \times 26.57$$

Method 2:

Preparation of standard curve

A. *Dilution of standards.* Prepare dilutions of the thymolphthalein standard solution previously described. Pipet 0.0, 0.1, 0.2, 0.3, 0.4, and 0.5 mL of the standard solution into six different tubes, and bring the total volume of each tube to 1.0 mL with the *n*-propanol/water mixture (70:30 by volume). Mix well.

B. *Working standards.* Label a second set of tubes as 0, 10, 20, 30, 40, and 50 U/L. Add 550 μL of buffered substrate solution and 1 mL of alkaline color development solution to each tube. Pipet 50 μL of the corresponding diluted standard into each tube. Mix well and read the absorbances at 595 nm. Plot absorbances versus international units per liter (0 to 50 U/L) on linear graph paper.

Reference range[8]

The serum reference range for healthy individuals determined by the above method at 37° C is 0.5 to 1.9 U/L. Activities were similar for both sexes.

Serum acid phosphatase in disease

A partial listing of disease states causing increased serum ACP levels includes the following:

1. Paget's disease (advanced)
2. Gaucher's disease
3. Myelocytic leukemia
4. Niemann-Pick disease
5. Idiopathic thrombocytopenia purpura
6. Sickle cell disease in crisis
7. Prostatic carcinoma
8. Infarction of prostate
9. Trauma or manipulation of prostate
10. Metastatic carcinoma of bone
11. Multiple myeloma (some patients)
12. Hyperparathyroidism (some patients)

NOTE. Decreased serum ACP has no clinical significance.

The detection of high activities of ACP in vaginal washings indicates the presence of seminal fluid.

REFERENCES

1. Bodansky, O.: Acid phosphatase, Adv. Clin. Chem. **15:**43-147, 1972.
2. Yam, L.T.: Clinical significance of the human acid phosphatases: a review, Am. J. Med. **56:**604-616, 1974.
3. Kahn, R., Turner, B., Edson, M., et al.: Bone marrow acid phosphatase: another look, J. Urol. **117:**79-80, 1977.
4. Schumann, G.B., Badany, S., Peglow, A., et al.: Prostatic acid phosphatase—current assessment in vaginal fluid of alleged rape victims, Am. J. Clin. Pathol. **66:**944-952, 1976.
5. Ladenson, J.H., and McDonald, J.M.: Acid phosphatase and prostatic carcinoma, Clin. Chem. **24:**120-134, 1978.
6. Henderson, A.R., and Nealon, D.A.: Enzyme measurements by mass: an interim review of the clinical efficacy of some mass measurements of prostatic acid phosphatase and isoenzymes of creatine kinase, Clin. Chim. Acta **115:**9-32, 1981.
7. Roy, A.V., Brower, M.E., and Hayden, J.E.: Sodium thymolphthalein monophosphate: a new acid phosphatase substrate with greater specificity for the prostatic enzyme in serum, Clin. Chem. **17:**1093-1102, 1971.
8. Ewen, L.M.: Acid phosphatase activity (thymophthalein monophosphate substrate). In Faulkner, W.H., and Meites, S., editors: Selected methods for the small clinical laboratory, Philadelphia, 1982, W.B. Saunders Co.

9. Gutman, A.B., and Gutman E.B.: An "acid" phosphatase occurring in the serum of patients with metastasizing carcinoma of the prostate gland, J. Clin. Invest. **17**:473-478, 1938.
10. Jaffee, H.L., and Bodansky, O.: Diagnostic significance of serum alkaline and acid phosphatase values in relation to bone disease, Bull. N.Y. Acad. Med. **19**:831-848, 1943.
11. Hudson, P.B., Brendler, H., and Scott, W.W.: Simple method for the determination of serum acid phosphatase, J. Urol. **58**:89-92, 1947.
12. Huggins, C., and Talalay, P.: Sodium phenolphthalein phosphate as a substrate for phosphatase test, J. Biol. Chem. **159**:399-410, 1945.
13. Seligman, A.M., Chauncey, H.H., Nachlas, M.M., et al.: The colorimetric determination of phosphatases in human serum, J. Biol. Chem. **190**:7-15, 1951.
14. Babson, A.L., and Read, P.A.: A new assay for prostatic acid phosphatase in serum, Am. J. Clin. Pathol. **32**:88-91, 1959.
15. Rietz, B., and Guilbault, G.G.: Fluorometric assay of serum acid or alkaline phosphatase in solution or on a semisolid surface, Clin. Chem. **21**:1791-1794, 1975.
16. Foti, A.G., Herschman, H., and Cooper, J.F.: A solid phase radioimmunoassay for human prostatic acid phosphatase, Cancer Res. **35**:2446-2452, 1975.
17. Chu, T.M., Wang, M.C., Scott, W.W., et al.: Immunochemical detection of serum prostatic acid phosphatase, methodology and clinical evaluation, Invest. Urol. **15**:319-322, 1978.
18. Lee, C.I., Wang, M.C., Murphy, G.P., and Chu, T.M.: A solid-phase fluorescent immunoassay for human prostatic acid phosphatase, Cancer Res. **38**:2871-2878, 1978.
19. Lee, C., Killian, C.S., Murphy, G.P., et al.: A solid-phase immunoadsorbent assay for serum prostatic acid phosphatase, Clin. Chim. Acta **101**:209-216, 1980.
20. Chu, S.Y., and Turkington, V.E.: A discrete analyzer (ABA-100) method for prostatic acid phosphatase with thymolphthalein monophosphate as substrate, Clin. Chem. **28**:389-390, 1982.
21. Bowers, G.N., Jr., Cali, J.P., Elser, R., et al.: Activity measurements for seven enzymes in lyophilized human serum SRM 909, Clin. Chem. **26**:969, 1980 (Abstract).
22. Frank, J.J., Bermes, E.W., Bickel, M.J., et al.: Effect of in vitro hemolysis on chemical values for serum, Clin. Chem. **24**:1966-1970, 1978.

Acid phosphatase isoenzymes

JOHN F. CHAPMAN
LINDA L. WOODARD
LAWRENCE M. SILVERMAN

Acid phosphatase (ACP, orthophosphoric monoester phosphohydrolase [acid optimum])

Clinical significance: pp. 882 and 953

EC 3.1.3.2

Molecular weight: 100,000 daltons (varies with tissue source)

Chemical class: enzyme, glycoprotein

Known isoenzyme forms: ACP_0, ACP_1, ACP_2, ACP_3, ACP_{3b}, ACP_4, and ACP_5 are commonly recognized; 20 different isoenzymes have been reported[1,2]

Biochemical reaction: pH 5.0

$$\underset{\text{Organic phosphate ester}}{R_1{-}O{-}PO_3H_2} + \underset{\text{Alcohol}}{H{-}OH} \xrightarrow[\text{pH5.0}]{\text{ACP}} \underset{\text{New alcohol}}{R_1{-}OH} + \underset{\text{New phosphate ester}}{H{-}O{-}PO_3H_2}$$

Net reaction: "hydrolysis" if alcohol is H_2O

PRINCIPLES OF ANALYSIS AND CURRENT USAGE

Acid phosphatase (ACP) is the name given a group of phosphohydrolases that hydrolyze phosphoric monoesters at an acidic pH. Several ACP isoenzymes are found in human tissues and cells, including liver, spleen, kidney, prostate, erythrocytes, platelets, osteoclast, and hairy cell leukemias.[1,2] Heterogeneity in the carbohydrate portion of the molecule may result in 20 or more isoenzyme forms in human tissues, though the clinical relevance of most of these remains to be established.[3] Of the ACP isoenzymes currently identified, only prostatic acid phosphatase (PAP, ACP isoenzyme 2 by electrophoresis) and isoenzymes 1 and 5 from human spleen in Gaucher's disease are associated with demonstrated clinical utility. Prostatic tissue is unusually rich in PAP, which does not normally enter the circulation. Thus elevations in PAP are seldom associated with benign prostatic hypertrophy, whereas elevations are common in association with prostatic carcinoma.[4]

Historically, PAP enzymatic activity was first measured by use of substrates such as α-naphthyl phosphate, which were erroneously believed to be specific for this isoenzyme. After the discovery that ACP isoenzymes originating in leukocytes could also utilize this substrate,[5] the inhibitor L-tartrate was shown to be useful as a specific inhibitor of PAP (method 1, Table 56-3). By this approach, the test is run in the absence and presence of L-tartrate, and the difference in activity between the assays is attributed to the prostatic isoenzyme. Although this technique is simple to perform and widely used, its sensitivity and specificity have been criticized.[7]

More recently, a variety of immunoassay techniques have been developed for the quantitation of PAP (Table 56-3). Examples of these techniques include radioimmunoassay (RIA), [8] fluorescence immunoassay (FIA),[9] counterimmunoelectrophoresis (CIEP),[10] and immunoenzyme assays.[11] By several techniques specific antibodies to PAP are used to separate this isoenzyme from serum. This can be accomplished by an antibody-coated support (beads or coated tubes) or double-antibody precipitation. The RIA, FIA, and CIEP assays all quantify the mass amount of immunologically reactive PAP. Both RIA and FIA are competitive binding assays in which the label used for measurement is ^{125}I or a fluorescent ligand, respectively. The immunoenzyme techniques employ antibody-coated supports or double-antibody precipitation. By the former approach, anti–human PAP antibodies bound to polypropylene tubes selectively sequester the PAP during the initial incubation. The tubes are then emptied, washed, and incubated with substrate, after which enzyme activity is measured as an ACP end-point reaction. In the double-antibody precipitation technique an anti-PAP antibody (Ab_1) first binds to PAP molecules, and this complex is precipitated by a second antibody (anti-Ab_1). The precipi-

Table 56-3. Methods for measurement of prostatic acid phosphatase (PAP)

Method	Type of analysis	Principle	Usage	Comments
1. Enzyme reactions and tartrate inhibition	Colorimetric, kinetic, or end point	Total ACP is measured by usual method* with and without L-tartrate, which inhibits PAP; subtraction of two results gives PAP activity	Most frequently used	Poor sensitivity for prostatic cancer
2. Radioimmunoassay (RIA)	Competitive binding assay	^{125}I-labeled PAP competes with sample PAP for binding to anti-PAP antibodies	Some usage	Higher sensitivity than method 1
3. Fluorescence immunoassay (FIA)	Competitive binding assay	Fluorescent ligand–labeled PAP competes with sample PAP for binding to anti-PAP antibodies	Not frequently used	—
4. Counterimmunoelectrophoresis (CIEP)	Immunoelectrophoresis	Electrophoresis of PAP into countermigrating anti-PAP; measurement of height of precipitation line	Not frequently used	—
5. Enzyme immunoassay	Immunoprecipitation of PAP followed by end-point colorimetric assay	Antibody-coated support or double-antibody precipitation with first Ab anti-PAP specific; enzyme assay by any ACP assay	Some usage	Sensitivity for prostatic cancer between methods 1 and 2
6. Electrophoresis	Electrophoresis	Separation of all isoenzymes on basis of charge	Rarely used	Quantitates many ACP isoenzymes

*See previous section on total acid phosphatase for discussion on optimal assay.

tate is resuspended in buffer and substrate, and the enzyme activity is measured as in the coated-tube approach. Antibody binding of PAP thus appears to have no inhibitory effect on catalytic activity and may actually impart enzymatic stability, since these methods have been performed successfully by use of specimens stored at −20° C without acidification for up to 30 days with no significant loss of activity. ACP isoenzymes have been separated by electrophoretic techniques, but these methods have not been used for routine clinical analysis.

All these techniques possess the advantage of increased sensitivity over the so-called functional techniques such as tartrate inhibition, though specificity is dependent on the antiserum used, and elevated serum values have been reported in patients without apparent prostatic cancer.[12]

REFERENCE AND PREFERRED METHODS

As just described, a variety of techniques have been developed for separation and quantitation of ACP isoenzymes, particularly PAP, though no established reference method exists. The L-tartrate inhibition method for PAP is described in this section, since it is the established method against which all new methods are compared.

SPECIMEN

Nonhemolyzed serum is preferred. *Grossly hemolyzed specimens are not acceptable,* since blood-cell acid phosphatases can result in falsely increased PAP results. Icteric specimens may cause a depression in the acid phosphatase activity. Samples collected in tubes containing anticoagulants that inhibit enzyme activity (oxalate, sodium fluoride) *cannot* be used. If plasma is used, heparin is the anticoagulant of choice. All specimens must be separated from cells immediately. One must then preserve the sample by adding 20 μL of 5 M acetate buffer or a citrate tablet to 1 mL of serum to adjust the pH between 5.0 and 6.0.[13] Specimens not preserved in this manner are unsuitable for analysis. ACP activity may decrease by up to 50% in 1 hour if unbuffered. ACP activity in preserved serum is stable for 2 days at 4° C.

PROCEDURE[14-16]

Principle

This method is based on the original method of Babson et al.[14] as modified by Hillmann.[15] This procedure is commercially available through Smith-Kline Instruments, Inc.[16] In the presence of the enzyme ACP, α-naphthyl phosphate is hydrolyzed to α-naphthol and inorganic phosphate at pH 5.0; the rate of hydrolysis is proportional to the enzyme present (Table 56-4):

$$\alpha\text{-Naphthyl phosphate} + H_2O \xrightarrow{ACP} \alpha\text{-Naphthol} + \text{Inorganic phosphate}$$

To quantitate the reaction photometrically, the α-naphthol produced is coupled with a diazonium compound, Fast Red TR (diazotized 2-amino-5-chlorotoluene) to produce a colored complex that absorbs at 405 nm. Since the cou-

Table 56-4. Comparison of several assays for prostatic acid phosphatase (PAP)

Condition	Enzymatic assay with L-tartrate inhibition*	Radioimmunoassay†	Enzyme immunoassay‡
Temperature	30° or 37° C	Ambient	Ambient—primary incubation and precipitation enzyme 37° C—reaction
Sample volume	40 μL each for total and L-tartrate inhibition	50 μL	100 μL each for reaction and blank
Fraction of sample volume	0.0625	0.01	Assay performed on precipitate or in antibody-coated tube
Final concentration of reagents	α-Naphthylphosphate · NaH_2O: 740.6 mg/L Fast Red TR: 412.5 mg/L Citric acid: 8.5 mmol/L Sodium citrate: 53.4 mmol/L pH 5.0 ± 0.2 L-Tartrate (only in inhibition reaction): 18.2 mmol/L	—	*p*-Nitrophenyl phosphate: 10 mmol/L Citric acid: 30 mmol/L Sodium citrate: 57 mmol/L pH: 5.0 ± 0.2 NaOH: 0.32 mol/L (for color development)
Time of reaction	10 minutes	18 hours with first antibody 3 hours with radioligand	Total procedure time approximately 3 to 4 hours Colorimetric reaction 60 min
Wavelength	405	^{125}I counting by gamma-ray counter	405 nm
Linearity	45 U/mL	25 ng/mL	Approximately 20 U/L
Precision‡ $\bar{X}$ (mean), % coefficient of variation	1.0 U/L (69%) 21.3 U/L (8.4%)	3 ng/mL (4.7%-5.1%) 18 ng/mL (5.2%-5.9%)	1.1 U/L (14.5%) 19.4 U/L (8.2%)
Interferences	Observable hemolysis and lipemia pH <4.0 or >6.0	pH <4.0 or >6.0	Gross lipemia and hemolysis pH <4.0 or >6.0

*Smith-Kline, Inc., Division of Smith-Kline-Beckman, Inc., Sunnyvale, Calif. Assay described here.
†New England Nuclear Corp., Boston. Modified method of Hillmann.[14]
‡Unpublished data from University Hospital, University of Cincinnati, Departments of Pathology and Laboratory Medicine (chemistry laboratory) and Radiobiology, using reagents supplied by Smith-Kline-Beckman, Inc., Sunnyvale, Calif.

pling reaction is instantaneous, the rate of color appearance is limited only by the rate of α-naphthol production in the enzyme-mediated reaction.

As a specific inhibitor of PAP, L-tartrate is used with this method to establish differentially the amount of prostatic enzyme present. The test is run both in the absence and presence of L-tartrate, and the difference in activity between the assays is attributed to PAP.

Reagents

Substrate

α-Naphthyl phosphate, monosodium salt monohydrate, 79 mg

Fast Red TR (diazotized 2-amino-5-chlorotoluene), 44 mg

Citric acid 0.58 g

Sodium citrate 1.47 g

Dissolve in 100 mL of distilled H_2O. Test pH; it should be pH 5.0 ± 0.2. This reagent is stable for 1 week at 4° C if protected from light. Discard if turbidity or red color develops.

L-Tartrate, disodium salt (2 M), 4.6 g. Dilute to 10 mL in distilled H_2O. This reagent is stable for 3 months at 4° C. Discard if turbidity appears.

Acetate buffer (5 M)

Sodium acetate dihydrate, 20.3 g

Glacial acetic acid, 2.9 mL

Dilute to 50 mL in distilled water. Test pH; it should be pH 5.0 ± 0.2.

Assay

Equipment: a spectrophotometer (≤10 nm band pass) capable of reading at 405 nm and a 37° C incubator.

1. Follow manufacturer's directions for reconstitution of reagents.
2. Pipet 3 mL of reagent into each cuvette to be used. If tartrate-labile, prostatic acid phosphatase is to be determined, label one cuvette for total and one cuvette for nonprostatic activity for each specimen.
3. Add 25 μL of tartrate solution to cuvettes for the nonprostatic enzyme assay and mix.
4. Add 0.2 mL (200 μL) of sample.
5. Mix sample and reagent gently but thoroughly and place cuvette in incubator set at 37° ± 0.1° C.
6. Approximately 5 minutes after mixing, check that instrument is at zero absorbance and start time.

7. Read and record a first absorbance value for each cuvette. When working with multiple samples, space them at convenient intervals (30 seconds) so that each may be properly timed and there will be no confusion when taking absorbance readings.
8. Readings may now be taken and recorded at exact 1-minute intervals to establish the absorbance change per minute (ΔA/min).
   Alternatively, as when it is necessary to incubate outside the spectrophotometer, the second reading may be deferred for precisely 5 minutes after the first (step 8). Calculation is now made from the 5-minute absorbance change (ΔA/5 min).

Calculation

For each sample run, calculate either the average absorbance change per minute (ΔA/min) or the 5-minute absorbance change (ΔA/5 min). Multiply ΔA/min, or ΔA/5 min, by the appropriate factor *(F)* in the following table relative to the sample volume employed.

Use the same calculation factor for all cuvettes with and without tartrate. The volume difference caused by introduction of tartrate changes the calculation factor and final answer by less than 1%.

Sample volume (μL)	International unit (U) factors for 3 mL of reagent volume	
	F (1 min)	*F* (5 min)
50	4765	953
100	2420	484
200	1250	250

Prostatic ACP (U) = Total ACP (U) − Nonprostatic ACP (U)

EXAMPLE: Using a sample volume of 200 μL with 3.0 mL of reagent volume, two absorbance readings taken exactly 1 minute apart were 0.415 and 0.438 in the absence of tartrate. With tartrate present, readings of 0.361 and 0.367 were obtained.

First, calculate ΔA/min for total ACP: 0.438 − 0.415 = 0.023; for nonprostatic ACP: 0.367 − 0.361 = 0.006. Refer to table for appropriate factor, 1250. Multiply $F \times \Delta A$/min.

$$\begin{aligned} \text{Total ACP} &= 1250 \times 0.023 = 29 \text{ U/L} \\ \text{Nonprostatic ACP} &= 1250 \times 0.006 = 8 \text{ U/L} \\ \text{Prostatic ACP U} &= \text{U total} - \text{U nonprostatic} \\ &= 29 - 8 = 21 \text{ U/L} \end{aligned}$$

Reference range

Male, total ACP	2.4 to 5.0 U/L
Male, PAP	0 to 1.2 U/L

Females have approximately the same levels of prostatic acid phosphatase activity as males have.

Clinical significance

The greatest concentrations of ACP are found in the liver, spleen, erythrocytes, platelets, and prostate gland.

There are no disease states associated with decreased ACP activity in serum. Causes of *increased* ACP activity include the following:

Prostatic fraction

1. Carcinoma of prostate
2. Infarction of prostate
3. Operative trauma to prostate

Total acid phosphatase

1. Excessive destruction of platelets (as in idiopathic thrombocytopenic purpura)
2. Metastatic breast cancer
3. Diseases of bone (Paget's disease, metastatic carcinoma of bone)
4. Various liver diseases (hepatitis, obstructive jaundice)
5. Acute renal impairment
6. Niemann-Pick disease (occasionally)
7. Gaucher's disease

REFERENCES

1. Robinson, D.G., and Glew, R.H.: Acid phosphatase in Gaucher's disease, Clin. Chem. **26:**371-382, 1980.
2. Smith, J.K., and Whitby, I.G. The heterogeneity of prostatic acid phosphatase, Biochim. Biophys. Acta **151:**607-613, 1968.
3. Lam, K.W., Lee, P., Eastlund, P., and Yam, L.T.: Antigenic and molecular relationship of human acid phosphatase isoenzymes, Invest. Urol. **18:**209-211, 1980.
4. Bodansky, O.: Acid phosphatase, Adv. Clin. Chem. **15:**43-147, 1972.
5. Amador, E., Price, J.W., and Marshall, G.: Serum acid α-naphthyl phosphatase activity, Am. J. Clin. Pathol. **51:**202-206, 1969.
6. Fishman, W.H., and Lerner, F.A.: A method for estimating serum acid phosphatase of prostatic origin, J. Biol. Chem. **200:**89-97, 1953.
7. Townsend, R.M.: Enzyme tests in diseases of the prostate, Ann. Clin. Lab. Sci. **7:**254-261, 1977.
8. Vihko, P., Sanjanti, E., Janne, et al.: Serum prostate-specific acid phosphatase: development and validation of a specific radioimmunoassay, Clin. Chem. **24:**1915-1919, 1978.
9. Lee, C.I., Wang, W.C., Murphy, G.P., and Chu, T.M.: A solid-phase fluorescent immunoassay for human prostatic acid phosphatase, Cancer Res. **38:**2871-2877, 1978.
10. Quinones, G.R., Rohner, T.J., Jr., Drago, J.R., and Demers, L.M.: Will prostatic acid phosphatase determination by radioimmunoassay increase the diagnosis of early prostatic cancer? J. Urol. **125:**361-364, 1981.
11. Gericke, K., Kohse, K.P., Pfleiderer, G., et al.: Development and evaluation of a new solid-phase direct immunoenzyme assay for prostatic acid phosphatase, Clin. Chem. **28:**596-602, 1982.
12. Foti, A.G., Cooper, J.F., Herschman, H., and Malvarz, R.R.: Detection of prostatic cancer by solid phase radioimmunoassay of serum prostatic acid phosphatase. N. Engl. J. Med. **297:**1257-1261, 1977.
13. Chen, I.-W., Sperling, M.I., Maxon, H.R., and Kaplan, L.A.: Stability of immunological activity of human prostatic acid phosphatase in serum, Clin. Chem. **28:**1163-1166, 1982.
14. Babson, A.L., Read, P.A., and Phillips, G.E.: The importance of the substrate in assays of acid phosphatase in serum, Am. J. Clin. Pathol. **32:**83-87, 1959.
15. Hillmann, G.: Continuous photometric measurement of prostate acid phosphatase activity, Z. Klin. Chem. Klin. Biochem. **9:**273-274, 1971.
16. Smith-Kline Instruments, Inc.: Product no. 89507/89509, Sunnyvale, Calif., Nov. 1981.

Alanine aminotransferase

ROBERT L. MURRAY

Alanine aminotransferase, ALT, L-alanine:2-oxoglutarate aminotransferase, serum glutamate pyruvate transaminase, SGPT

Clinical significance: pp. 420, 509, and 594

EC 2.6.1.2

Molecular weight: 180,000 daltons

Chemical class: enzyme, protein

PRINCIPLES OF ANALYSIS AND CURRENT USAGE

Alanine aminotransferase (ALT) catalyzes the transfer of an amino group between L-alanine and L-glutamate; the corresponding keto acids in this process are α-ketoglutarate and pyruvate (Fig. 56-1). In vivo this reaction goes to the right (as in Fig. 56-1) to provide a source of nitrogen for the urea cycle. The pyruvate thus generated is available for entry into the citric acid cycle, whereas the glutamate is deaminated (catalyzed by glutamate dehydrogenase), yielding ammonia and α-ketoglutarate.

The reaction is reversible; the chemical equilibrium favors the formation of alanine and α-ketoglutarate. Because of the relative difficulty in the assay of those products, however, analytical techniques typically force the reverse reaction, allowing quantitation of pyruvate. Two methods of ALT analysis have enjoyed wide popularity for routine clinical use: the Reitman-Frankel method[2] (method 3, Table 56-5), involving measurement of activity by conversion of the reaction product, pyruvate, to its hydrazone[1]; and the Wróblewski method,[1] involving coupling of the ALT reaction to a lactate dehydrogenase (LD) reaction, with measurement of that reaction's products (method 1, Table 56-5).[2]

In the DNPH procedure,[2] the serum is incubated with L-alanine and α-ketoglutarate; after a measured length of time the reaction is stopped, and the newly formed pyruvate is reacted with dinitrophenylhydrazine (DPNH), producing the corresponding hydrazone (Fig. 56-2). This condensation is relatively rapid, even at room temperature. After condensation, the reaction mixture is alkalinized, producing a blue color caused by the anion form of the hydrazone. The absorbance is measured at 505 nm and compared to a standard curve. The dinitrophenylhydrazone of α-ketoglutarate produces negligible color, and little error is introduced by this source.[3]

In the other approach (methods 1 and 2) NADH is the reaction product that is quantitated. Lactate dehydrogenase (LD) and its required cofactors are added, with allowance for the enzymatic conversion of pyruvate to lactate and with simultaneous oxidation of reduced nicotinamide adenine dinucleotide (NADH) (Fig. 56-3). The disappearance of NADH is followed spectrophotometrically (at 340 nm) or fluorometrically. Theoretically, one could follow this reaction by either continuous monitoring or by an end-point determination; in practice the continuous monitoring technique is more frequently encountered. Only under very unusual circumstances would the increased sensitivity afforded by fluorescence be needed; for routine clinical use, the absorbance technique is adequate.[4-7]

REFERENCE AND PREFERRED METHOD

The coupled enzyme technique, with continuous ultraviolet monitoring of NADH disappearance, is recommended as the preferred method for clinical analysis of ALT. When performed as an end-point assay, the reaction suffers from feedback inhibition by pyruvate, which de-

Table 56-5. Methods of alanine aminotransferase (ALT) analysis

Method	Type of analysis	Principle*	Usage
1. Enzymatic (ultraviolet monitoring) (Wróblewski and LaDue[1])	Quantitative	*UV monitoring of NADH disappearance at 340 nm:* Ala + α-KG → Gl + Pyr $\text{Pyr} + \text{NADH} + \text{H}^+ \xrightarrow{\text{LD}} \text{Lac} + \text{NAD}^+$	Serum, most frequently employed procedure
2. Enzymatic (fluorescence)	Quantitative	*Fluorescence monitoring of NADH disappearance:* Same reactions as above	Serum, rarely used (high sensitivity)
3. Dinitrophenylhydrazine (DNPH) coupling (colorimetric) (Reitman and Frankel[2])	Quantitative	*End-point absorbance measurement at 505 nm:* Ala + α-KG → Gl + Pyr Pyr + DNPH → Pyr-DNP-hydrazone	Serum, rarely performed assay

**Ala*, Alanine; *α-KG*, α-ketoglutarate; *DNP*, dinitrophenyl; *DNPH*, dinitrophenylhydrazine; *Gl*, glutamate; *Lac*, lactate; NAD^+, nicotinamide adenine dinucleotide; *NADH*, reduced nicotinamide adenine dinucleotide; *Pyr*, pyruvate.

creases linearity. The kinetic approach is superior in terms of specificity, freedom from interference, linearity, reproducibility, and speed of analysis. Since most laboratories purchase reagents for enzyme assays as preformulated kits, recommendations as to specifications are of value only when a choice of reagent source is being made; it is not practical to modify a manufacturer's formulation after purchase.

The colorimetric procedure suffers from a limited linearity resulting from feedback inhibition of the ALT by pyruvate. Other activated carbonyl groups can interfere by coupling with DNPH. Similarly, the presence of endogenous pyruvate in high concentrations results in an elevated blank.

Table 56-6 gives various published recommendations for performance of the ALT assay using coupled enzyme methodology; the International Federation of Clinical Chemistry (IFCC) method is recommended. Substrate for the human enzyme is exclusively L-alanine; unlike the corresponding substrate for asparate aminotransferase (AST), where the D-isomer inhibits human AST, D-alanine has no inhibitory effect on ALT. However, the limited solubility of alanine makes the use of the racemic mixture unwise. Tris buffer is preferred to phosphate for two reasons: NADH is more stable in Tris than in phosphate buffer, and pyridoxal-5′-phosphate (PP) is a more effective activator of ALT in Tris buffer than it is in phosphate buffer. A lag time, before initiation of the reaction with α-ketoglutarate, must be provided to allow for consumption of endogenous or accidentally provided α-ketoglutarate and for any other NADH-consuming processes. Unless the absorbance change is monitored to ensure linearity, this lag phase should not be less than 90 seconds.

A 10-minute preincubation period of serum with the cofactor PP is recommended. The effect of adding this co-

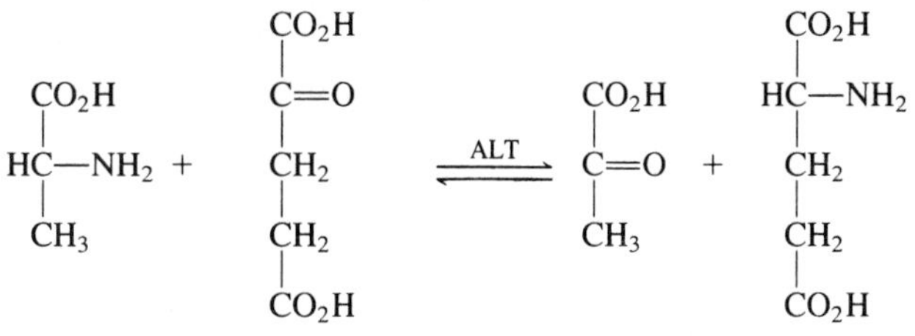

L-Alanine + α-Ketoglutarate $\xrightleftharpoons{ALT}$ Pyruvate + L-Glutamate

Fig. 56-1. Amino transfer catalyzed by ALT.

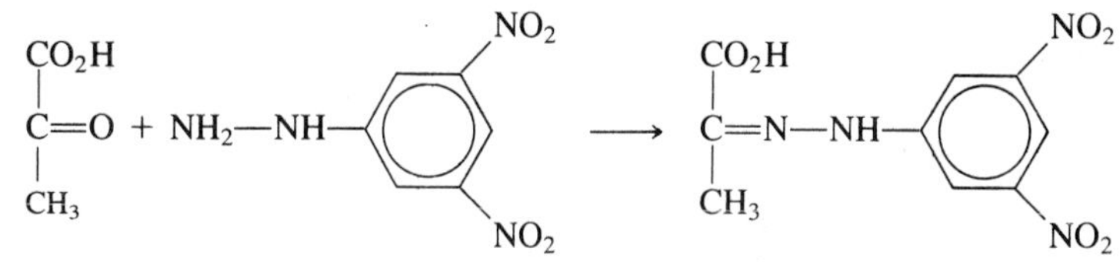

Pyruvate + DNPH → Pyruvate dinitrophenylhydrazone

Fig. 56-2. Condensation of pyruvate with DNPH.

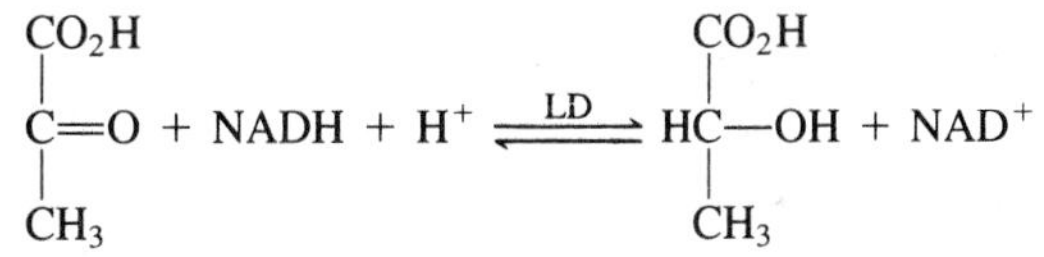

Pyruvate + NADH + H^+ $\xrightleftharpoons{LD}$ Lactate + NAD^+

Fig. 56-3. Enzymatic conversion of pyruvate to lactate.

Table 56-6. Comparison of enzymatic alanine aminotransferase (ALT) methods

Condition	AACC[3]*	British[4]	IFCC[5]*	Scandinavian[6]	German[7]
Temperature	30° C	25° C	30° C	37° C	35° C
Fraction of serum volume	0.082	0.067	0.083	0.120	0.135
Final concentration of reagent					
L-Alanine	500 mmol/L	250 mmol/L	500 mmol/L	400 mmol/L	800 mmol/L
α-Ketoglutarate	15 mmol/L	6.7 mmol/L	15 mmol/L	12 mmol/L	18 mmol/L
pH	7.3	7.4	7.5	7.4	7.4
Tris buffer	76 mmol/L	—	100 mmol/L	20 mmol/L	—
Phosphate	—	90 mmol/L	—	—	80 mmol/L
NADH	0.16 mmol/L	0.25 mmol/L	0.18 mmol/L	0.15 mmol/L	0.18 mmol/L
Pyridoxal phosphate	0.1 mmol/L	—	0.1 mmol/L	—	—
Lactate dehydrogenase	1.2 U/mL	0.15 U/mL	1.2 U/mL	2.0 U/mL	1.2 U/mL
EDTA	—	—	—	5 mmol/L	—

*AACC, American Association for Clinical Chemistry; IFCC, International Federation of Clinical Chemistry.

factor has been well documented,[8-11] with increases of 7% to 55% in the measured ALT activity in human serum reported. The wide variation is a result of differences in the samples studied. In a healthy population (with serum levels of PP that are likely to be adequate), little change would be expected from the addition of excess amounts of PP. In a sample of patients with renal disease (with a greater likelihood of subnormal serum levels of PP), addition of the required PP would be expected to show a significant increase in ALT activity. It appears that the degree of activation is less for ALT than it is for AST.

SPECIMEN

Serum is the preferred specimen. Oxalate, heparin, and citrate do not inhibit the enzymatic activity but may introduce slight turbidity. Hemolyzed specimens should be avoided, since erythrocytes contain three to five times more ALT than serum does. ALT is stable in serum for 3 days at room temperature and up to 1 week at 4° C. Urine has little or no activity and is not recommended for analysis.

REFERENCE RANGE

When analyzed at 37° C by methods employing activation with PP, the normal adult reference range is 0 to 55 U/L. Men have been reported to show slightly higher values than women. Because the levels of this enzyme are unusually sensitive to liver damage, increases in ALT may be a result of excessive use of alcohol or to exposure to a greater variety of hepatotoxic agents.

Normal newborns have been reported to show a reference range of up to double the adult upper level. These values decline to adult levels by approximately 3 months of age. This increased activity has been attributed to seepage from the neonate's hepatocytes, which, being immature, have more permeable membranes.

REFERENCES

1. Wróblewski, F., and LaDue, J.S.: Serum glutamic-pyruvic transaminase in cardiac and hepatic disease, Proc. Soc. Exp. Biol. Med. **91**:569-571, 1956.
2. Reitman, S., and Frankel, S.: A colorimetric method for the determination of serum glutamic oxalacetic and glutamic pyruvic transaminases, Am. J. Clin. Pathol. **28**:56-63, 1957.
3. Brétaudière, J.P., Burtis, C., Pasching, J., et al.: Study of the alanine aminotransferase kinetic assay by response surface methodology, Clin. Chem. **26**:1023, 1980 (Abstract).
4. Wilkinson, J.H., Baron, D.N., Moss, D.W., and Walker, P.G.: Standardization of clinical enzyme assays: a reference method for aspartate and alanine transaminases, J. Clin. Pathol. **25**:940-944, 1972.
5. Bergmeyer, H.U., and Hørder, M.: IFCC methods for the measurement of catalytic concentrations of enzymes. Part 3. IFCC method for alanine aminotransferase, J. Clin. Chem. Clin. Biochem. **18**:521-534, 1980.
6. Committee on Enzymes of the Scandinavian Society for Clinical Chemistry and Clinical Physiology: Recommended methods for the determination of four enzymes in blood, Scand. J. Clin. Lab. Invest. **33**:291-305, 1974.
7. Enzyme Commission of the German Society for Clinical Chemistry: Recommendations of the German Society for Clinical Chemistry, Z. Klin. Chem. Klin. Biochem. **10**:281-291, 1972.
8. Lustig, V.: Activation of alanine aminotransferase in serum by pyridoxal phosphate, Clin. Chem. **23**:175-177, 1977.
9. Bergmeyer, H.U., Scheibe, P., and Wahlefeld, A.W.: Optimization of methods for aspartate aminotransferase, Clin. Chem. **24**: 58-73, 1978.
10. Siest, G., Schiele, F., Galteau, M.-M., et al.: Aspartate aminotransferase and alanine aminotransferase activities in plasma: statistical distributions, individual variations, and reference values, Clin. Chem. **21**:1077-1087, 1975.
11. Miller, D.A., Glick, M.R., and Oei, T.O.: Results compared for plasma aminotransferase activity with and without pyridoxal phosphate activation in infants and children, Clin. Chem. **27**:1035, 1981 (Abstract).

Aldolase

STEPHEN GENDLER

Aldolase (ALS, fructose-1,6-diphosphate: D-glyceraldehyde-3-phosphate lyase)
Clinical significance: p. 509
EC 4.1.2.13
Molecular weight: approximately 160,000 daltons
Chemical class: enzyme, protein
Known isoenzymes: subunits A, B, and C in tetrameric form
Biochemical reaction:

$CH_2OPO_3H_2$–$C{=}O$–$HOCH$–$HCOH$–$HCOH$–$CH_2OPO_3H_2$ $\xrightarrow{\text{EC 4.1.2.13}}$ $CH_2OPO_3H_2$–$C{=}O$–CH_2OH + $HC{=}O$–$HCOH$–$CH_2OPO_3H_2$

Fructose-1,6-diphosphate (FDP) → **Dihydroxyacetone phosphate (DAP)** + **Glyceraldehyde-3-phosphate (GAP)**

PRINCIPLES OF ANALYSIS AND CURRENT USAGE

Aldolase (ALS) catalyzes the reversible biochemical reaction shown, a reaction that has been studied extensively. Aldolase forms a Schiff base intermediate with its ketonic substrate, utilizing the lysine at position 227.[1] It is present as a tetramer composed of two of three known subunits designated A, B, and C. When analyzed as homotetramers, aldolase A, B, and C have similar pH-activity profiles, Michaelis constants (K_m) for fructose diphosphate (FDP), and molecular weights. Using FDP as substrate, the pH optima in Tris-malonate buffer[2] and in collidine buffer[3] range from pH 7.0 to 8.0. The Michaelis constants with FDP as substrate range from 10^{-6} to 3×10^{-6} M. The molecular weights for the homomeric enzymes range from about 145,000 to 175,000 daltons.[2] The A homomer is found in high concentration in skeletal muscle. The B

Table 56-7. Comparison of methods of serum aldolase (ALS) analysis

Method	Type of analysis	Principle*	Usage	Comments
1. Dinitrophenylhydrazine (DNPH)	End-point spectrophotometric	FDP $\rightleftharpoons$ trioses; alkaline conditions cause hydrolysis/rearrangement and formation of purple hydrazone	Manual or automated	Suggested reference method (AACC)
2. NADH formation	End-point spectrophotometric	FDP $\rightleftharpoons$ DAP + GAP GAP $\xrightleftharpoons{TPI}$ DAP GAP + NAD$^+$ $\xrightleftharpoons[AsO_4^{-2}]{GAPDH}$ 3-Phosphoglycerate + NADH + H$^+$	Manual	Subject to large interferences
3. NADH degradation	End-point or kinetic spectrophotometric	FDP $\rightleftharpoons$ DAP + GAP GAP $\xrightleftharpoons{TPI}$ DAP 2 DAP + 2 NADH + 2 H$^+$ $\xrightleftharpoons{GDH}$ 2 Glycerol-1-phosphate + 2 NAD$^+$	Manual or automated	Accurate, precise, flexible, and easy to use

**DAP*, Dihydroxyacetone phosphate; *FDP*, fructose diphosphate; *GAP*, glyceraldehyde phosphate; *GAPDH*, glyceraldehyde phosphate dehydrogenase; *GDH*, glycerol-3-phosphate:NAD 2-oxidoreductase; *NAD*$^+$, nicotinamide adenine dinucleotide; *NADH*, reduced NAD; *TPI*, triosephosphate isomerase.

homomer predominates in liver, and the C homomer in brain and other tissues. The primary isoenzyme in normal serum is the A homomer. The hybrid isoenzyme composed of three A subunits and one C subunit is present in somewhat less concentration.[4]

Aldolase analyses in current usage allow estimation of enzymatic activity by employing FDP as the substrate, since aldolase cleaves FDP more rapidly than other substrates.[2] For analytic purposes the aldolase reaction is driven to the formation of trioses by using hydrazine, a trapping agent for the triosephosphates.[3,5,6] The aldolase reaction can also be shifted to trioses by removal of the glyceraldehyde phosphate (GAP) with the enzyme triosephosphate isomerase (TPI, D-glyceraldehyde-3-phosphate: ketol-isomerase, EC 5.3.1.1). This enzyme catalyzes the conversion of GAP to dihydroxyacetone phosphate (DAP).[7]

Aldolase activity cannot be measured by direct monitoring of the substrate or the products of the reaction. Instead, the reaction must be coupled to a second, indicator reaction. Indicator reactions currently either utilize the reaction of 2,4-dinitrophenylhydrazine (DNPH) with the triosephosphates to produce a chromophore or use coupled enzyme reactions and follow changes in the concentration of reduced nicotinamide adenine dinucleotide (NADH) by absorbance at 340 nm.

The colorimetric analyses use DNPH in an indicator reaction to form a colored hydrazone.[5,6,8] The colorimetric reaction requires carefully controlled conditions with alkali to cause hydrolysis and rearrangement of the triosephosphates to methylglyoxal. This compound is converted to its dinitrophenylhydrazone, which yields a deep purple color in an excess of alkali (method 1, Table 56-7). The manual colorimetric technique requires an ice bath (to slow the indicator reaction to a manageable time) and a 38° C water bath, and it involves rather intensive work. A colorimetric procedure has been adapted to the AutoAnalyzer (Technicon, Tarrytown, NY 10951). Correlation with the manual method is good, but precision and recovery values are not published.[9]

The ultraviolet spectrophotometric analyses use two coupled enzyme indicator reactions, with the concomitant formation or degradation of NADH (methods 2 and 3, Table 56-7). Coupled enzyme systems that result in the appearance of NADH have been shown to be subject to large errors caused by the presence of endogenous interfering enzymes such as glycerol dehydrogenase (GDH, glycerol-3-phosphate:NAD 2-oxidoreductase, EC 1.1.1.8) or TPI. Thus the most satisfactory design for a coupled enzymatic system is as follows:

$$FDP \xrightleftharpoons{ALS} DAP + GAP$$

$$GAP \xrightleftharpoons{TPI} DAP$$

$$2\ DAP + 2\ NADH + 2\ H^+ \xrightleftharpoons{GDH} 2\ glycerol\text{-}1\text{-}phosphate + 2\ NAD^+$$

In this scheme interferences can be eliminated in the preincubation phase. Interferences can be demonstrated by a decrease in A_{340} after the addition of the indicator enzymes and sample and before the addition of FDP. These interferences have not been characterized but may be endogenous substrates and dehydrogenases in serum that utilize NADH.[10] DAP or GAP contamination of the FDP preparation can also be demonstrated by a decrease in A_{340} after the addition of indicator enzymes and FDP and before the addition of serum sample.[11] Proposed by Beisenherz et al.[7] in 1953, this coupled reaction has been optimized by Ludvigsen,[12] Pinto et al.,[10] and Bergmeyer and Bernt,[13] adapted to kit form (Boehringer-Mannheim, Calbiochem, and other companies), and adapted to the kinetic measurement of enzyme activity.[14]

Radioimmunoassay (RIA) for the homotetramer of aldolase A[15] and aldolase C[16,17] have also been reported. RIA is not discussed further because the assay is not widely used. In addition, the clinical utility of the assay is as yet unproved.

REFERENCE AND PREFERRED METHODS

A colorimetric method was published as a "selected method" by the American Association for Clinical Chemistry (AACC) in 1961. The method requires a 38° C water bath for the 30-minute incubation with substrate, a trichloroacetic acid precipitation and centrifugation, 1 hour of incubation in an ice bath for methylglyoxal formation, 25 minutes at 38° C for hydrazone formation, and 10 minutes for chromophore formation in excess alkali. Incubation times alone for these steps total 140 minutes.[6] Because of the many manipulations and time required, this method is not recommended.

The various modifications of the coupled enzymatic techniques are difficult to compare, since the methods have not been compared by one laboratory in a controlled situation. Pinto et al.,[8] in the evaluation of their method, analyzed 10 normal and 10 abnormal specimens in two laboratories, resulting in a mean difference of 0.4 mU/mL and a standard deviation of the differences equal to 1.8 mU/mL. A modification of the Pinto procedure shows a coefficient of variation (CV) to be 20% in the normal range and 10% in the abnormal range.[18] Ludvigsen's method gives a CV of 3.0% to 3.3%, but these are apparently within-run experiments.[12] A method proposed by Bergmeyer and Bernt has a CV of 5%[13] (Table 56-8).

Table 56-8. Reaction conditions for adolase (ALS) analysis*†

Temperature	37° C
pH	7.4
Final concentration of reagent components‡	Collidine: 55 mmol/L Iodoacetate: 0.3 mmol/L FDP: 4 mmol/L NADH: 15 mmol/L GDH: >75 U/mL TPI: >500 U/mL LD: >233 U/mL
Sample volume	200 μL
Fraction of sample volume	0.07
Linearity	27.5 U/L
Precision (% CV)	2.7%-12.4%

*From Bergmeyer, H.-V., and Bernt, E.: In Bergmeyer, H.-V., editor: Methods of enzymatic analysis, ed. 2, New York, 1974, Academic Press, Inc.

†Sources of analytical error: any specimen with visible hemolysis will give elevated results because of release of aldolase from cells. Serum stored on the clot gives similar results.

‡For abbreviations see Table 56-7. *LD*, Lactate dehydrogenase.

Boehringer-Mannheim has adapted the Bergmeyer and Bernt method to a kit (catalog number 123838). The company claims day-to-day imprecision to be 2.7% in the normal range and 1.9% in the abnormal range, whereas in my laboratory, CV's are 12.4% in the normal range and 5.0% in the abnormal range. This kit was adapted to a semiautomated kinetic enzyme analyzer. Day-to-day precision was reported as 6.0% in the normal range.[14] For reasons of ease of use, good reproducibility, and adaptability to semiautomation, the method of Bergmeyer and Bernt is the method of choice.

SPECIMEN

The anticoagulants oxalate, citrate, fluoride, heparin, and EDTA have no effect on aldolase activity.[19] Whether serum or plasma is used, it is essential to remove the specimen from contact with the red cells and platelets. Refrigerated specimens not removed from the clot for 24 hours result in the aldolase level being 12% to 46% greater than a properly treated specimen.[20] The stability of the enzyme at room temperature is reported as 5 hours to 2 days, depending on the study. Boric acid powder, added directly to serum to achieve a concentration of 25 g/L, extends the stability of the enzyme at room temperature to 6 days.[21] The activity of aldolase remains unchanged for at least 6 months if the specimen is kept at −20° C.[9] Given that the concentration of aldolase in red cells is 10 times that of the serum, hemolysis is a cause for rejection of the specimen.[18]

PROCEDURE

Principle

$$\text{FDP} \xrightleftharpoons{\text{ALS}} \text{GAP} + \text{DAP}$$

$$\text{GAP} \xrightleftharpoons{\text{TPI}} \text{DAP}$$

$$2\ \text{DAP} + 2\ \text{NADH} + 2\ \text{H}^+ \xrightleftharpoons{\text{GDH}} 2\ \text{glycerol-1-phosphate} + 2\ \text{NAD}^+$$

Lactate dehydrogenase (LD) is present in the reagent to remove the interference of endogenous pyruvate. The oxidation of NADH is followed by absorbance measurement at 340 nm. The method is that proposed by Bergmeyer and Bernt.[13]

Reagents

Buffer/substrate solution. Dissolve 220 mg of $Na_3FDP \cdot 8\ H_2O$ or 370 mg of tricyclohexylammonium $FDP \cdot 10\ H_2O$ and 6.2 mg of iodoacetic acid in 90 mL of distilled water. Add 0.75 mL of collidine, mix, and adjust to pH 7.4 with approximately 0.6 mL of 5 M HCl. Dilute with distilled water to 100 mL. Final concentrations are 55 mmol/L of collidine buffer, pH 7.4; 0.3 mmol/L of iodoacetate; and 4 mmol/L of FDP. This solution is stable for 4 weeks at 2° to 8° C.

NADH solution. Dissolve 25 mg of Na_2-NADH and 20 mg of $NaHCO_3$ in 2 mL of distilled water. Final concentration is 15 mmol/L. This is stable for 4 weeks at 2° to 8° C.

GDH/TPI/LD suspension. Using 3.2 mol/L of ammonium sulfate, mix enzyme solutions so that final concentrations are >75 U/mL of GDH (25° C) (source: rabbit muscle); >500 U/mL of TPI (25° C) (source: rabbit muscle); >233 U/mL of LD (25° C) (source: rabbit muscle). This is stable at 2° to 8° C for about 1 year.

Procedure

Equipment: a 37° C-constant-temperature incubator and a spectrophotometer (band pass ≤10 nm) capable of reading at 340 nm

1. Place 2.50 mL of buffer/substrate solution, 0.05 mL of NADH solution, 0.01 mL of GDH/TPI/LD suspension, and 0.20 mL of serum or plasma in a 3.0 mL cuvette.
2. Mix well and incubate at 37° C for about 5 minutes.
3. Read the initial A_{340} (A_1).
4. After exactly 20 minutes at 37° C, read the final A_{340} (A_2).
5. If the ΔA $(A_1 - A_2)$ is greater than 0.500, dilute specimens five- to tenfold with isotonic saline and repeat the analysis using the same volumes of reagents and specimen.

Calculations

With use of the molar absorptivity (ϵ) of NADH as 6300 $\frac{L \cdot cm}{mole}$ the formula for calculation of enzyme activity is as follows:

$$U/L = \frac{\Delta A \text{ for } T \text{ min}}{T \text{ min}} \times \frac{\text{Total volume}}{\text{Sample volume}} \times \frac{1}{\epsilon} \times \frac{10^6 \ \mu\text{mol}}{\text{mol}}$$

where A = difference between A_1 and A_2 over analysis time, T minutes

ϵ = molar absorptivity of NADH[22]

10^6 = factor to convert concentration to μmoles per liter

For the assay described, the values are:

$$U/L = \frac{\Delta A}{20 \text{ min}} \times \frac{2.76 \text{ mL}}{0.2 \text{ mL}} \times \frac{1}{6300} \times \frac{10^6}{2} = \Delta A \times 55$$

The factor of 2 indicates that for every molecule of FDP catalyzed, two molecules of NADH are converted to NAD^+.

Interferences

Hemolysis is cause for rejection.[18] Also, refrigerated serum not removed from the clot results in the aldolase level being 12% to 46% greater than that of a properly treated specimen.[20]

REFERENCE RANGE[20]

Males	2.61 to 5.71 U/L
Females	1.98 to 5.54 U/L

REFERENCES

1. Hartman, F.C., and Norton, I.L.: Active-site-directed reagents of glycolytic enzymes, Methods Enzymol. **47:**494-496, 1977.
2. Penhoet, E.E., Kochman, M., and Rutter, W.J.: Molecular and catalytic properties of aldolase C, Biochemistry **8:**4396-4402, 1969.
3. Bruns, F.H.: Bestimmung und Eigenschaften der Serumaldolase, Biochem. Z. **325:**156-162, 1954.
4. Tzvetanova, E.: Aldolase isoenzymes in patients with progressive muscle dystrophy and in human fetuses, Clin. Chem. **17:**926-930, 1971.
5. Sibley, J.A., and Lehninger, A.L.: Determination of aldolase in animal tissues, J. Biol. Chem. **177:**859-872, 1949.
6. Fleisher, G.A.: Aldolase. In American Association of Clinical Chemists: Standard Methods of Clinical Chemistry, vol. 3, New York, 1961, Academic Press, Inc., pp. 14-22.
7. Beisenherz, G., Boltze, H.J., Bucher, T., et al.: Diphosphofructose-Aldolase, Phosphoglyceraldehyd-Dehydrogenase, Milchsäure-Dehydrogenase, Glycerophosphat-Dehydrogenase und Pyruvat-Kinase aus Kaninchenmuskulatur in einem Arbeitsgang, Z. Naturforsch. **8b:**555-577, 1953.
8. Pinto, P.V.C., Van Dreal, P.A., and Kaplan, A.: Aldolase. I. Colorimetric determination, Clin. Chem. **15:**339-348, 1969.
9. Kaldor, J., and Schiavone, D.J.: Automated procedure for serum aldolase estimation, Clin. Chem. **14:**735-739, 1968.
10. Pinto, P.V.C., Kaplan, A., and Van Dreal, P.A.: Aldolase. II. Spectrophotometric determination using an ultraviolet procedure, Clin. Chem. **15:**349-360, 1969.
11. Bruns, F.H., and Bergmeyer, H.-U.: Fructose-1,6-diphosphate aldolase. In Bergmeyer, H.-U., editor: Methods of enzymatic analysis, ed. 2, New York, 1974, Academic Press, Inc., vol. 2, pp. 724-731.
12. Ludvigsen, B.: DPNH method for the estimation of serum aldolase activity, J. Lab. Clin. Med. **61:**329-337, 1963.
13. Bergmeyer, H.-U., and Bernt, E.: Fructose-1,6-diphosphate aldolase, UV assay, manual method. In Bergmeyer, H.-U., editor: Methods of enzymatic analysis, ed. 2, New York, 1974, Academic Press, Inc., pp. 1100-1105.
14. Harjanne, A.: The kinetic measurement of serum aldolase, Clin. Chim. Acta **92:**311-313, 1979.
15. Asaka, M., Nagase, K., Miyozaki, T., et al.: Radioimmunoassay for human aldolase A, Clin. Chim. Acta **117:**289-296, 1981.
16. Willson, V.J.C., Graham, J.G., McQueen, I.N.F., and Thompson, R.J.: Immunoreactive aldolase C in cerebrospinal fluid of patients with neurological disorders, Ann. Clin. Biochem. **17:**110-113, 1980.
17. Willson, V.J.C., and Thompson, R.J.: Human brain aldolase C_4 isoenzyme: purification, radioimmunoassay, and distribution in human tissues, Ann. Clin. Biochem. **17:**114-121, 1980.
18. Demetriou, J.A., Drewes, P.A., and Gin, J.B.: Enzymes. In Henry, R.J., Cannon, D.C., and Winkelman, J.W., editors: Clinical chemistry: principles and techniques, Hagerstown, Md., 1974, Harper & Row, Publishers, Inc., pp. 971-974.
19. Ladenson, J.H.: Nonanalytical sources of variation in clinical chemistry results. In Sonnenwirth, A.C., and Jarett, L., editors: Gradwohl's clinical laboratory methods and diagnosis, ed. 8, vol. 1, St. Louis, 1980, The C.V. Mosby Co., pp. 153-154.
20. Sweetin, J.C., and Thomson, W.H.S.: Revised normal ranges for six serum enzymes: further statistical analysis and the effects of different treatments of blood specimens, Clin. Chim. Acta **48:**49-63, 1973.
21. Beardslee, R., and Owers, P.: Stabilization by boric acid of aldolase activity at room temperature, Clin. Chem. **22:**1543-1545, 1976 (Letter).
22. Bergmeyer, H.-U.: Neue Werte für die molaren Extinktion-Koeffizienten von NADH und NADPH zum Gebrauch im Routine-Laboratorium, Z. Klin. Chem. Klin. Biochem. **13:**507-508, 1975.

Alkaline phosphatase

WILLIAM C. WENGER
JOHN A. LOTT

Alkaline phosphatase (ALP, orthophosphoric monoester phosphohydrolase [alkaline optimum])
Clinical significance: pp. 420, 439, 806, and 953
EC 3.1.3.1
Molecular weight: varies with tissue source of enzyme
Chemical class: enzyme, protein
Known isoenzymes: bone, liver, placenta, intestine, kidney, Regan (fetal)
Biochemical reaction:

$$H_2O + R{-}O{-}\overset{\overset{O}{\|}}{\underset{\underset{OH}{|}}{P}}{-}O^- \xrightleftharpoons[pH>9]{ALP} R{-}OH + H_2PO_4^-$$

PRINCIPLES OF ANALYSIS AND CURRENT USAGE

The group of nonspecific phosphatases that catalyze the reaction shown is known collectively as alkaline phosphatase (ALP). Phosphatases transfer a phosphate moiety from one group to a second, forming an alcohol and a second phosphate compound. When water is the phosphate acceptor, inorganic phosphate is formed. Optimal activity of these enzymes is exhibited at a pH of approximately 10.0. ALP requires Mg^{2+} for activation and is inhibited by Ca^{2+} and inorganic phosphate.

Reaction rates are notably dependent on such variables as the tissue source of the enzyme, the type of substrate and buffer used, and the incubation temperature. ALP is denatured slowly at 37° C; thus a reaction temperature of 25° to 30° C is often recommended for clinical assays. The type of buffer used may enhance the enzyme rate by acting as a phosphate group acceptor in a process called "transphosphorylation." Examples of transphosphorylating buffers are 2-methyl-2-amino-1-propanol (MAP), diethanolamine (DEA), and Tris. Since the natural substrates of these enzymes are not known, widely used assays for ALP activity currently use *p*-nitrophenyl phosphate (*p*NPP) as substrate. At alkaline pH, *p*NPP is colorless, whereas the reaction product, *p*-nitrophenol (*p*NP), is intensely yellow.

Table 56-9. Methods of alkaline phosphatase (ALP) analysis

Method source	Type of analysis	Principle	Usage	Comments
Shinowara et al.[1]	Two-point spectrophotometric	Substrate: β-glycerophosphate Measure rate of release of inorganic phosphate; 1 hour incubation	All body fluids Manual	Suffers from long incubation time; high phosphate background in samples Considered obsolete
King and Armstrong[2]	End-point spectrophotometric	Substrate: phenylphosphate Measure rate of release of phenol with Folin-Ciocalteu reagent; 30 min incubation	All body fluids Manual	Samples require deproteinization Considered obsolete
Kind and King[3]	End-point spectrophotometric	Substrate: Phenylphosphate Measure rate of release of phenol with 4-amino-antipyrine as chromogenic reagent; 15 min incubation	All body fluids Manual/automated	Faster rate than King and Armstrong[2] Requires no deproteinization
Bessey et al.[4]	End-point or kinetic spectrophotometric	Substrate: *p*-nitrophenyl phosphate (*p*NPP) Measure rate of formation of yellow *p*-nitrophenoxide ion	All body fluids Manual/automated	Rapid; linear change in absorbance with time
Moss[5]	End-point or kinetic spectrophotometric	Substrate: α-naphthol monophosphate Measure rate of formation of α-naphthol at 340 nm	All body fluids Manual/automated	Rapid; convenience of measuring at 340 nm
Cornish et al.[6]	Fluorescent	Substrate: 4-methylumbelliferyl phosphate Measure fluorescent product	All body fluids Automated	Highly sensitive
Bowers and McComb[7,8]	Kinetic or end-point spectrophotometric	Substrate: *p*NPP Measure rate of release of *p*-nitrophenoxide in transphosphorylating buffer	Manual or automated	Proposed reference method; more sensitive than Bessey et al.[4]

The earliest assays for the quantitation of total ALP activity in serum measured the rate of release of inorganic phosphate. Subsequently, various ALP methods have been introduced in which refinements in the use of chromogenic substrates and rate-enhancing buffers have led to significant improvements in analytical sensitivity and precision. A summary of these methods is presented in Table 56-9.

The early assays based on the measurement of liberated inorganic phosphate are now only of historical interest. The substrates and reaction conditions used were relatively insensitive, and long incubation times (1 to 1½ hours) were required to produce acceptable results. Representative of these assays is the classic method of Shinowara et al.[1] Serum is incubated for 1 hour at 37° C in diethylbarbiturate buffer (pH 9.3) containing β-glycerophosphate as substrate. Liberated phosphate ion is reacted with phosphomolybdic acid reagent, and the resulting chromogen is measured colorimetrically at 600 nm.

Another classic procedure of historical interest is that of King and Armstrong.[2] Phenyl phosphate is used as substrate, and the liberated phenol is measured colorimetrically after the addition of Folin-Ciocalteu reagent (phosphotungstic-phosphomolybdic acid), which forms a blue color with phenol in an alkaline solution and has an absorption maximum at 750 nm.

Kind and King[3] reported a similar but improved method that also used phenyl phosphate as substrate but measured the liberated phenol with 4-amino-antipyrine, which reacts with phenol to yield a red-colored quinone. The quinone can be measured colorimetrically at 500 nm; it does not react with plasma proteins; thus protein precipitation and separation before analysis is not required, as is the case when the Folin-Ciocalteu reagent is used.

Bessey et al.[4] introduced the use of *p*NPP as substrate in ALP assays. The phosphatase hydrolyzes the colorless substrate to yield the yellow salt of *p*NP, which has an absorption maximum at 400 nm. The indicator *p*NP eliminates the need for additional color-producing reactions. The assay can be carried out on as little as 5 μL of serum. In the original procedure, serum is added to 50 μL of 0.1 mol/L of glycine buffer (pH 10.3) containing 2 g/L of *p*NPP. The samples are incubated at 38° C for 30 minutes, and the rection is stopped by the addition of 0.5 mL of 20 mmol/L of NaOH. The color is then read at 400 to 420 nm. Correction for background is performed by the addition of acid to the samples, which converts the yellow sodium salt into colorless free nitrophenol, and a second colorimetric reading is taken.

A rapid, continuous spectrophotometric assay that offers the convenience of measurement at 340 nm was reported by Moss.[5] Assays are carried out in 0.1 mol/L of sodium carbonate-bicarbonate buffer (pH 10.0 at 37° C) containing 5 mmol/L of $MgCl_2$ and α-naphthyl phosphate as substrate. The hydrolysis product, α-naphthol, absorbs at 340 nm. Reaction curves are recorded for 5 to 10 minutes, and reaction velocities are calculated from the initial, linear portions of the curves.

Cornish et al.[6] introduced a highly sensitive, automated fluorometric ALP assay using 4-methylumbelliferyl phosphate as substrate. The substrate is hydrolyzed to 4-methylumbelliferone, a highly fluorescent compound. The excitation peak is at 360 nm, and the fluorescent light is isolated by a secondary cutoff filter at 465 nm. The procedure requires that 10 μL of serum be added to 0.1 mol/L of carbonate-bicarbonate buffer (pH 9.2 at 37° C) containing 1.0 mmol/L of 4-methylumbelliferyl phosphate. Incubation time is approximately 8 minutes.

The method of Bowers and McComb[7,8] is widely used today because of its simplicity and sensitivity. The assay is accepted by many as the reference method for ALP quantitation; it uses *p*NPP as the substrate. The method is considered an advance over prior colorimetric assays because of its use of a transphosphorylating buffer (MAP) and because of carefully selected conditions of pH, tem-

Table 56-10. Reaction conditions for alkaline phosphatase (ALP) analysis*

Condition	Manual reference (Bowers and McComb[7])	German[9]	Scandinavian[10]
Temperature	30° C	25° C	37° C
pH	10.5	9.8	9.8
Final concentration of reagents	*p*NPP: 16 mmol/L	*p*NPP: 10 mmol/L	*p*NPP: 10 mmol/L
	MAP: 1.0 mol/L	DEA†: 1.0 mol/L	DEA†: 1.0 mol/L
	Mg^{+2}: 1.0 mmol/L	Mg^{+2}: 0.5 mmol/L	Mg^{+2}: 0.5 mmol/L
Fraction of sample volume	0.0164	0.009	0.009
Linearity (approximate)	To 500 U/L	To 500 U/L	Linear for 10 minutes or up to 1500 U/L
Precision (in reference range)	3%-6%	5%	5%

*Major interferences: EDTA, citrate, oxalate, inorganic phosphate, calcium, and ammonium sulfate.
†It has been reported that a contaminant in some lots of DEA causes significant loss of ALP activity.[7]

perature, and reagent concentrations. Specific parameters of the assay are summarized in Table 56-10.

REFERENCE AND PREFERRED METHODS

Although all the methods just described have at one time or another been used in the clinical laboratory, recent advances in our knowledge about the nature and reactivity of ALP have led to modifications in assay conditions that render most earlier methods obsolete. Controversy over conditions for optimal ALP activity continues, and up to now a universally accepted reference method has not been established. The relative merits and drawbacks of the methods listed in Table 56-9 are now discussed.

A major drawback of the early assays based on the measurement of liberated phosphate was their inability to be adapted to continuous monitoring systems. The harsh acidic conditions required for the colorimetric determination of inorganic phosphate denatured the enzyme, and so the course of the reaction could not be followed by multipoint analyses.

The method of King and Armstrong provides superior sensitivity when compared to methods that measure the release of inorganic phosphate.[2] For a given incubation period, up to three times as much phenol is liberated from the phenyl phosphate substrate than the inorganic phosphate released from previously used substrates. This increased sensitivity also permits a reduction of the incubation time to 30 minutes. Phenyl phosphate is self-indicating in the ultraviolet range, with an absorption maximum at 288 nm. The sensitivity of the method is poor when compared to more recent assays that use *p*NPP. The method also requires the precipitation and removal of serum proteins, which react with the Folin-Ciocalteu reagent, before the measurement of liberated phenol.

The use of 4-amino-antipyrine by Kind and King obviates deproteinization because the reagent does not react with serum proteins.[3] However, with phenyl phosphate as substrate, the sensitivity is still much lower than when *p*NPP is used.

The method of Bessey et al. has the advantage of using *p*NPP as substrate.[4] The substrate is self-indicating, and the hydrolysis product possesses a high molar absorptivity, which permits the use of a small sample size and short incubation periods. The method does not utilize a transphosphorylating buffer and therefore suffers from a lack of sensitivity when compared to more current procedures.

The use of α-naphthylphosphate introduced by Moss allows for measurement at 340 nm, the same wavelength used to measure enzyme reactions involving the NAD/NADH-coupled dehydrogenase systems.[5] The method uses a carbonate-bicarbonate buffer, which does not have transphosphorylating properties, and hence the method is relatively insensitive.

The automated fluorescent method of Cornish provides the advantage of high sensitivity.[6] Because of this sensitivity, it is superior to conventional methods when only small sample volumes or samples with low enzymatic activity are available, as in the neonatal and pediatric patient populations. Drawbacks of the method include equipment cost and the quenching effects inherent in all fluorometric assays. The substrate 4-methylumbelliferyl phosphate is also expensive.

The method of Bowers and McComb appears to be the most likely candidate for a reference method.[7,8] It offers the convenience and sensitivity of a self-indicating, highly reactive substrate in a transphosphorylating buffer. Conditions of temperature, pH, and reagent concentrations reflect an optimization of those factors known to influence ALP activity. The method and slight modifications of it are under active consideration by various national and international organizations interested in establishing a reference method for ALP. The German Society for Clinical Chemistry[9] and the Scandinavian Society for Clinical Chemistry and Clinical Physiology[10] have proposed reference methods similar to that of Bowers and McComb. A summary of the three proposed reference methods is given in Table 56-10.

All the methods listed in Table 56-9 do not differ greatly in terms of within-day precision. When carefully performed, the within-day coefficients of variation (CV's) for the manual methods are approximately 5%. Better within-day precision can be achieved with automated methods. Within-day precision of the automated methods listed in Table 56-9 ranges from 2% to 6%. Long-term precision within the laboratory is also acceptable. Over a 3-year period, a CV of about 5% is typical for procedures using *p*NPP. Interlaboratory precision of the methods is much poorer. In interlaboratory proficiency surveys, ALP is among the least precise of the common enzyme tests.[11] Reported interlaboratory CV's range from 17%[10] to 28%.[4] Factors that could contribute to poor interlaboratory precision are impure reagents, slight variations in reaction temperatures, subtle enzyme activation and inactivation phenomena, and variations in conditions of sample storage.

SPECIMEN

ALP can be found in all body fluids and tissues. Blood samples should be drawn after a fast of at least 8 hours. Serum and heparinized plasma give similar results. Other anticoagulants such as EDTA, oxalate, and citrate inhibit the enzyme by complexing Mg^{+2}. Slight hemolysis is tolerable, but gross hemolysis should be avoided. Bilirubin does not interfere with kinetic methods using *p*NPP as substrate. Any drug that causes hepatotoxicity can be expected to increase serum ALP activities. Most reports indicate that serum ALP activity increases slowly with storage; increases of 5% to 10% after storage at 4° C for 24 hours are typical. Stability can also be affected by the isoenzyme content of the specimen and by the binding of the enzyme to immunoglobulins. As a general rule, it is best to analyze ALP specimens the same day they are drawn.

PROCEDURE

The following procedure is adapted from the manual reference method used by the National Bureau of Standards.[8] It utilizes *p*NPP as substrate, MAP as the rate-enhancing buffer, and rigorously controlled conditions of pH (10.5) and temperature (30° C). With careful adherence to the procedure, correct standardization, and adequate temperature control, within-day CV's of 3% to 6% have been reported.

The German[9] and Scandinavian[10] societies have proposed reference procedures that also use *p*NPP as substrate, but they buffer the reaction at pH 9.8 with DEA. Reaction conditions for the three preferred methods are listed in Table 56-10.

Principle

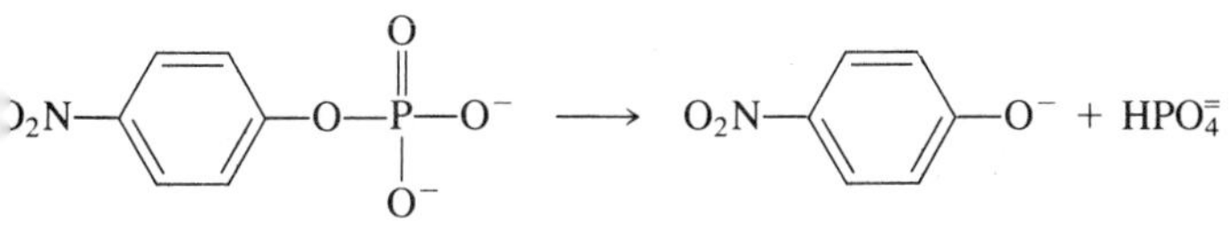

$$O_2N\text{—}C_6H_4\text{—}O\text{—}PO_3^{2-} \longrightarrow O_2N\text{—}C_6H_4\text{—}O^- + HPO_4^=$$

$$p\text{-Nitrophenylphosphate} + H_2O \xrightleftharpoons{ALP} p\text{-Nitrophenoxide} + \text{Phosphate}$$

The colorless substrate is converted at alkaline pH to the yellow *p*-nitrophenoxide ion. The reaction is followed by measurement of the increase in absorbance at 403 nm.

Reagents

MAP buffer (1.5 mol/L). Liquefy 2-methyl-2-amino-1-propanol by warming to 30° to 35° C. Weigh 135 g of the liquid directly into a one-liter volumetric flask, add 500 mL of distilled water, and mix. Carefully add 190 mL of 1.0 mol/L of HCl to the flask. When the solution has cooled to room temperature, dilute to 1 L with water. Confirm that the pH is 10.5 at 30° C. This buffer is stable for 1 month when stored in an airtight container at 25° C.

Magnesium acetate solution (3 mmol/L). Dissolve 650 mg of magnesium acetate · 4 H_2O in 1 L of water. This is stable indefinitely at 4° C.

***p*-Nitrophenylphosphate solution (24.5 mmol/L).** Dissolve 9.1 mg of disodium 2-nitrophenylphosphate · 6 H_2O per milliliter of MAP buffer. Prepare fresh daily.

***p*-Nitrophenol standard solutions (1 mmol/L).** Dissolve 139.1 mg of *p*-NPP in 1 L of water. This stock solution is stable for several months when stored in the dark. To prepare the working standard solution, add 25 mL of the stock solution to 900 mL of MAP buffer. Dilute to 1 L with water. The working standard solution is stable for at least 2 months.

Assay

Equipment. The procedure described is suitable for any recording spectrophotometer equipped with a temperature-controlled cell compartment. The temperature should be held constant at 30° ±0.1° C.

1. Add 50 μL of specimen to 1.0 mL of magnesium acetate solution in a test tube. Thoroughly mix and incubate for 5 minutes at 30° C.
2. Prewarm buffered *p*NPP solution to 30° C and add 2.0 mL to the incubation mixture from step 1. Agitate thoroughly.
3. Transfer the reaction mixture to a cuvette with a 1 cm light path and read the absorbance change versus time at 403 nm for 2 minutes. Readings can be taken immediately after mixing.

Calculations

International units (U) of activity are expressed as micromoles of *p*-nitrophenoxide formed per minute. Enzyme concentrations are expressed as international units per liter (U/L). U/L can be calculated from the change in absorbance by the following equation for a 1 cm light path:

$$U/L = [(\Delta A \text{ for } t \text{ min})/(t \text{ min})] \times (\text{Total volume/Sample volume}) \times 10^6/\epsilon$$

where ΔA = change in absorbance for time, t

ϵ = molar absorptivity for *p*-nitrophenoxide ($18.8 \times 10^3\ L \cdot mol^{-1} \cdot cm^{-1}$)

10^6 = factor to convert the concentration to micromoles per liter (10^6 μmol/mol)

The values are as follows:

$$U/L = \Delta A/\text{min} \times (3050\ \mu L/50 \mu L) \times 10^6/18{,}800$$
$$U/L = \Delta A/\text{min} \times 3245$$

The method is linear to approximately 500 U/L.

REFERENCE RANGE

The serum reference ranges for healthy individuals, determined by the method just outlined at 30° C, are[8,12]:

Group (age in years)	ALP (up to U/L)	Group (age in years)	ALP (up to U/L)
Newborns	250		
Females:		Males:	
1 to 12	350	1 to 12	350
10 to 14	280	12 to 15	500
15 to 19	150	10 to 14	275
20 to 24	85	15 to 19	155
25 to 34	85	20 to 24	90
35 to 44	95	25 to 34	95
45 to 54	100	35 to 44	105
55 to 64	110	45 to 54	120
65 to 74	145	55 to 64	135
75+	165	65 to 74	140
		75+	190

Factors affecting ALP activities in a normal population include exercise, periods of rapid growth in children, and pregnancy.

SERUM ALKALINE PHOSPHATASE IN DISEASE

Both increases and decreases of plasma ALP are of importance clinically. A partial listing of disease states in which abnormal serum activities of ALP are encountered follows.

Causes of increased plasma ALP

1. Paget's disease of bone
2. Obstructive liver diseases, including cholestasis
3. Hepatitis
4. Hepatotoxicity caused by drugs
5. Osteomalacia
6. Malignancy with bone or liver involvement

Causes of decreased plasma ALP

1. Cretinism
2. Vitamin C deficiency

ALP is often interpreted as "abnormal" because of the use of inappropriate reference ranges by a laboratory.

REFERENCES

1. Shinowara, G., Jones, L.M., and Reinhart, H.L.: Estimation of serum inorganic phosphate and "acid" and "alkaline" phosphatase activity, J. Biol. Chem. **142:**921-933, 1942.
2. King, E.J., and Armstrong, A.R.: A convenient method for determining serum and bile phosphatase activity, Can. Med. Assoc. J. **31:**376-381, 1934.
3. Kind, P.R.N., and King, E.J.: Estimation of plasma phosphatase by determination of hydrolysed phenol with amino-antipyrine, J. Clin. Pathol. **7:**322-326, 1954.
4. Bessey, O., Lowry, O.H., and Brock, M.J.: Method for the determination of alkaline phosphatase with five cubic millimeters of serum, J. Biol. Chem. **164:**321-329, 1946.
5. Moss, D.W.: A note on the spectrophotometric estimation of alkaline phosphatase activity, Enzymologia **31:**193-202, 1966.
6. Cornish, C.J., Neale, F.C., and Posen, S.: Automated fluorometric alkaline phosphatase microassay with 4-methylumbelliferyl-phosphate as a substrate, Am. J. Clin. Pathol. **53:**68-76, 1970.
7. Bowers, G.N., Jr., and McComb, R.B.: Measurement of total alkaline phosphatase activity in human serum, Clin. Chem. **21:**1988-1995, 1975.
8. Bowers, G.N., Jr., and McComb, R.B.: Alkaline phosphatase, total activity in human serum. In Faulkner, W.H., and Meites, S., editors: Selected methods for the small clinical laboratory, Philadelphia, 1982, W.B. Saunders Co.
9. German Society for Clinical Chemistry: Recommendations of the Enzyme Commission, Z. Klin. Chem. Klin. Biochem. **10:**281-291, 1972.
10. Scandinavian Society for Clinical Chemistry and Clinical Physiology: Recommended methods for the determination of four enzymes in blood, Scand. J. Clin. Lab. Invest. **33:**291-306, 1974.
11. Lott, J.A., and Massion, C.G.: Interlaboratory quality control of enzyme analyses: the CAP experience. In Hamburger, H.A., editor: Clinical and analytical concepts in enzymology, Skokie, Ill., 1983, College of American pathologists, pp. 233-256.
12. Munan, L., Kelly, A., Petitclerc, C., and Billon, B.: Atlas of blood data. Prepared by the Epidemiology Laboratory and the Laboratory of Clinical Biochemistry, University of Sherbrooke, Sherbrooke, Quebec, 1980.

Alkaline phosphatase isoenzymes

JOHN F. CHAPMAN
LINDA L. WOODARD
LAWRENCE M. SILVERMAN

Alkaline phosphatase, ALP, orthophosphoric monoester phosphohydrolase (alkaline optimum)
Clinical significance: pp. 420, 439, and 806
EC 3.1.3.1
Molecular weight: 120,000 daltons (varies considerably)
Chemical class: enzyme, protein
Known isoenzyme forms: bone, liver, fast liver, intestine, placenta, kidney, Regan, and Nagao
Biochemical reaction:

$$R_1{-}O{-}PO_3 \; + \; H{-}OH \xrightarrow[\text{pH 9.8}]{\text{ALP}} R_1{-}OH \; + \; H{-}O{-}PO_3H_2$$

Organic phosphate ester + Alcohol → New alcohol + New organic phosphate ester

Net reaction: "hydrolysis" if alcohol is H_2O

PRINCIPLES OF ANALYSIS AND CURRENT USAGE

Alkaline phosphatase (ALP) is an enzyme found in practically all body tissues, catalyzing the reaction shown above at an alkaline pH. Although "optimal" reaction conditions for the assay of total ALP have been reported,[1] it is important to remember that these conditions were established using serum samples containing almost exclusively liver and bone isoenzymes. Thus optimal conditions of analysis for ALP isoenzymes from other tissue sources are known to vary widely with respect to substrate, buffer, and pH optima.

Likewise, ALP isoenzymes from various tissue sources display considerable heterogeneity with respect to net molecular charge[2]; inhibition by L-phenylalanine,[3] other amino acids,[4] and urea[5]; and differential sensitivity to heat.[6] These unique characteristics have provided the basis for the development of a variety of techniques for the separation and identification of ALP isoenzymes, with each technique possessing respective advantages and disadvantages, leading to uniformly recognized "preferred" methods. Electrophoresis, differential heat sensitivity, and differential chemical inhibition are currently the methods of analysis most often used.

Zone electrophoresis followed by staining for enzymatic activity is a commonly used technique for the qualitative analysis of ALP isoenzymes in serum (method 1, Table 56-11). Various supporting media, including agarose,[7] cellulose acetate, [8] paper,[9] polyacrylamide gel,[10] and starch gel,[9] have been used with varying degrees of success.[11] In general, electrophoretic methods suffer from incomplete separation of many ALP isoenzymes of interest.

Polyacrylamide gel provides the best separation of ALP isoenzymes, whereas cellulose acetate provides the least. In polyacrylamide gels at a pH of approximately 9.0, the

Table 56-11. Methods for measurement of alkaline phosphatase (ALS) isoenzymes

Method	Principle	Usage	Comment
1. Electrophoresis	Total charge on various isoenzymes differ, allowing differential migration in electric field	Most frequently used technique	Incomplete separation; bone and liver often not completely resolved
a. Polyacrylamide gel	Separation by both charge and molecular weight	Limited usage	Best separation, difficulty in densitometric scanning; qualitative estimate of isoenzymes
b. Cellulose	Separation by charge	Some usage	Bone and liver not completely resolved
2. Heat inactivation	Heat inactivation rates of isoenzymes differ. Rate of inactivation after incubation at 56° C is suggestive of presence of bone or liver isoenzymes	Frequently used	Calculation is made more difficult by presence of intestinal or other more heat-stable isoenzyme forms
3. Chemical inactivation	L-Phenylalanine and urea inactivate different isoenzymes at different rates	Some usage; can be automated on a centrifugal analyzer	Heat-stable isoenzyme activities differentiated from bone and liver

charge of the ALP isoenzymes results in the normal liver isoenzyme (liver I) migrating anodically fastest, with bone and intestinal isoenzymes migrating progressively slower. Fig. 56-4 demonstrates a typical polyacrylamide gel electrophoretic separation of ALP isoenzymes. Typically, only liver I and bone isoenzymes are detected by any of the isoenzyme procedures. With the polyacrylamide method shown, these two activities are clearly distinguishable (Fig. 56-4, *b* and *c*). The intestinal isoenzyme (Fig. 56-4, *a*) is not normally detected, though some individuals of blood types O or B normally have a small amount of intestinal isoenzyme in serum, particularly after meals. The intestinal isoenzyme may also be elevated with intestinal disease. The liver II isoenzyme (Fig. 56-4, *d* and *e*) is an intracellular enzyme that is not usually observed in serum. Its presence is often associated with more severe parenchymal cell damage. The biliary isoenzyme is derived from the biliary tree, and its presence may suggest active cholestatic disease.

Liver and bone isoenzymes can generally be separated adequately by many electrophoretic methods to allow qualitative, but not quantitative, assessment of their relative activities in serum. Our and others' experience[12] seems to indicate that qualitative data alone may not be sufficient to provide clinically relevant information in cases when multiple ALP isoenzyme forms contribute to elevated ALP levels in serum or when it is desirable to follow changes in the activity of one or more serum ALP isoenzymes over the course of disease or treatment. In addition, certain ALP isoenzymes of placental or tumor origin exhibit

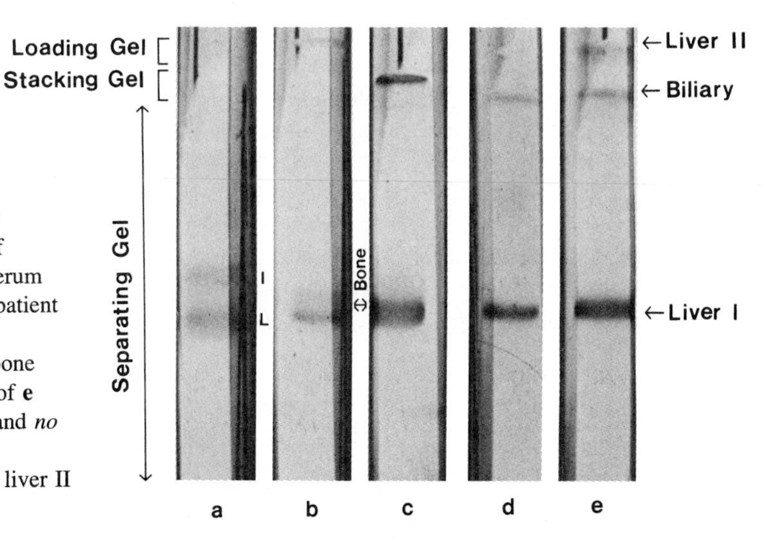

Fig. 56-4. Alkaline phosphatase isoenzymes separated by polyacrylamide disk electrophoresis. Normal blue stain of enzyme activity shown here as black bands. **a,** Control serum showing intestinal (I) and liver I isoenzymes. **b,** Heated patient sample of **c** showing residual bone and liver isoenzyme activities. **c,** Unheated patient sample showing elevated bone and liver isoenzyme activities. **d,** Heated patient sample of **e** showing residual liver I and biliary isoenzyme activities and *no* bone activity. **e,** Unheated patient sample showing large increase in liver I isoenzyme activity and the presence of liver II and biliary isoenzyme activities.

widely different electrophoretic migration patterns and often comigrate with the more common bone and liver forms, leading to potential misinterpretation of electrophoretic patterns.

Differentiation of ALP isoenzymes on the basis of selective inactivation at 56° C remains one of the most widely used techniques (method 2, Table 56-11). Whereas the use of selective inactivation by heat as a method of identifying ALP isoenzymes is a relatively straightforward concept, the technical considerations and pitfalls are numerous. Thus meticulous attention must be paid to details such as temperature control, exposure time, thickness of glass tubes used, and so on. In addition, the pH of the sample affects heat stability,[13] and either it must be tightly controlled through adequate buffering or the effects of variable pH must be negated through the construction of heat-inactivation curves by use of multiple estimates of residual activity.[14,15] Sample buffering generally involves unavoidable sample dilution and potential error, whereas construction of heat-inactivation curves requires excessive consumption of time or sophisticated instrumentation.

To the extent that the variables discussed can be controlled on a routine basis, selective inactivation at 56° C can provide qualitative and semiquantitative estimates of liver and bone ALP activity when these two isoenzymes forms predominate in serum. The presence of more than two ALP isoenzymes in serum, however, complicates the interpretation of results when one is using this method. The presence of as little as 2% placental or "Regan-like" isoenzymes in serum can have a significant effect on the "apparent" liver isoenzyme activity.[13]

The combined use of L-phenylalanine and urea as differential inhibitors of ALP isoenzymes (method 3, Table 56-11) was first introduced by Statland et al.[16] in 1972 for use in association with a centrifugal fast analyzer. The authors have employed this technique for a number of years in their own laboratory with a high degree of success. Whereas many analytical variables that affect heat-inactivation methods apply to chemical inhibition methods as well, performance of these assays by instruments such as centrifugal fast analyzers or other sophisticated enzyme analyzers is usually sufficient to control many of these variables. The combined use of L-phenylalanine and urea allows the calculation of bone and liver isoenzyme activities, as well as a determination of the relative activity of placental, intestinal, and Regan isoenzymes in serum. Although this technique does not differentiate between placental, intestinal, and Regan isoenzymes, it does allow the calculation of bone and liver activity in the presence of these other isoenzyme forms.

REFERENCE AND PREFERRED METHODS

Although there is currently no recognized reference method for ALP isoenzyme determination, both electrophoretic and heat-inactivation techniques are popular, relatively simple to perform, and yield clinically useful results under most circumstances. Chemical inhibition methods, though not as widely utilized, may offer certain advantages over other techniques and should probably be considered by those laboratories possessing adequate instrumentation. An example of a recommended electrophoretic method follows.

SPECIMEN

Serum from clotted venous blood is recommended. Plasma containing EDTA, fluoride, or oxalate should not be used. Samples are stable for 1 year at −20° C or for 1 week at 4° C.

PROCEDURE

Principle

Polyacrylamide gel electrophoresis provides separation of charged molecules in a region of electric potential gradient. Ionized molecules in solution migrate toward the region of opposite charge. In polyacrylamide, because of its molecular-sieving characteristics, the rate of migration of charged species depends on both net charge and molecular size. The most efficient separation of charged species is facilitated by use of three separate polymerized layers of polyacrylamide gel contained within small glass tubes:

- Loading gel—sample added before polymerization; acts as anticonvection medium; large pores
- Stacking gel—concentrates sample and causes alignment of charged species into contiguous fractions according to electrophoretic mobility; moderately sized pores
- Separating gel—pore size and pH adjusted for maximum resolving power; smaller pore size and higher pH than loading or stacking gels cause realignment of molecular species

This procedure is available from Ames Co. (division of Miles, Inc., Elkart, Ind.); the manufacturer's instructions should be followed.

Reagents and assay (See manufacturer's directions for details.)

Alkaline phosphatase enzyme stain is 5-bromo-4-chloro-3-indolyl-phosphate (*p*-toluidine salt) in *N,N′*-dimethylformamide. Electrophoresis takes approximately 1 hour; staining, 2 hours; and the entire procedure, approximately 4 to 5 hours. Each sample should be run in duplicate, with the second aliquot being heated at 56° C for 10 minutes before polymerization into the sample gel.

Interpretation of results

Stained gels may be photographed, evaluated qualitatively, or scanned in a densitometer at 600 nm. Representative locations of ALP isoenzymes are shown in Fig. 56-4.

Specimens with elevated total ALP are characterized by

a proportional elevation of one or more isoenzymes. Although liver and bone isoenzymes are clearly distinguishable when present at moderate strength, a strong elevation of one may overlap the other, making accurate interpretation difficult. This is rarely a problem if ALP activity is diluted to approximately 100 U/L before electrophoresis. The bone and liver isoenzymes can also be easily differentiated by comparison of the heat-treated aliquot to the unheated sample. Since bone is 85% inactivated after 10 minutes of heat inactivation at 56° C, a large decrease in activity will indicate the presence of bone isoenzyme. Although abnormal isoenzyme patterns are usually associated with elevated total ALP values, they occasionally occur along with a normal total ALP value.

REFERENCE RANGE

The ranges given below are based on the L-phenylalanine and urea inhibition method[15,17] but reflect the results observed by polyacrylamide electrophoresis.

	ALP isoenzyme		
Age group	Bone	Liver	Other (intestinal, and so on)
Child and adolescent	85%	5%	<10%
Adult	45%	45%	<10%
Elderly	30%	60%	<10%

Clinical significance

Bone ALP will increase with increased osteoblastic activity and is thus seen in growing children, secondary hyperparathyroidism, and osteoblastic metastasis. Liver ALP isoenzymes are noted in cases of extrahepatic obstruction and intrahepatic cholestasis. Common causes of increased liver ALP include cirrhosis, metastatic disease to the liver, and a variety of "space-occupying lesions" in the liver. Placental ALP is noted during pregnancy. The tumor (Regan) isoenzyme can be seen in malignancy but is relatively rare. The intestinal ALP isoenzyme is seldom encountered in large amounts.

REFERENCES

1. Bowers, G.N., and McComb, R.D.: Measurement of total alkaline phosphatase activity in human serum, Clin. Chem. **21**(13):1988-1995, 1975.
2. Moss, D.W.: Scientific foundations of the estimation of isoenzymes in diagnosis. In Schmidt, E., and Schmidt, F.W., editors: Multiple forms of enzymes, New York, 1982, S. Karger Pubs., Inc., pp. 23-30.
3. Fernley, H.N., and Walker, P.G.: Inhibition of alkaline phosphatase by L-phenylalanine, Biochem. J. **116**:543, 1970.
4. O'Carroll, D., Statland, B.E., Steele, B.W., and Burke, M.D.: Chemical inhibition method for alkaline phosphatase isoenzymes in human serum, Am. J. Clin. Pathol. **63**:564-572, 1975.
5. Bahr, M., and Wilkinson, J.H.: Urea as a selective inhibitor of human tissue alkaline phosphatases, Clin. Chim. Acta **17**:367, 1967.
6. Fishman, W.H., and Ghosh, N.K.: Isoenzymes of human alkaline phosphatase, Adv. Clin. Chem. **10**:255-370, 1967.
7. Sundblad, L., Wallin-Nilsson, M., and Brohult, J.: Characterization of alkaline phosphatase isoenzymes in serum by agar gel electrophoresis, Clin. Chim. Acta **45**:219, 1973.
8. Viot, M., Joulin, C., Cambon, P., et al.: The value of serum alkaline phosphatase isoenzyme in the diagnosis of liver metastases, preliminary results, Biomedicine **31**:74-77, 1979.
9. Kieding, N.R.: Differentiation into three fractions of the serum alkaline phosphatase and the behavior of the fractions in diseases of bone and liver, Scand. J. Clin. Lab. Invest. **11**:106-112, 1959.
10. Fishman, L.: Acrylamide disc gel electrophoresis of alkaline phosphatase of human tissues, serum and ascites fluid using Triton X-100 in the sample and gel matrix, Biochem. Med. **9**:309-315, 1974.
11. Moss, D.W.: Isoenzyme analysis, London, 1979, The Chemical Society, pp. 38-57.
12. Moss, D.W.: Alkaline phosphatase isoenzymes, Clin. Chem. **28**(10):2007-2016, 1982.
13. Moss, D.W., Shakespeare, M.J., and Thomas, D.M.: Observations on the heat stability of alkaline phosphatase isoenzyme in serum, Clin. Chim. Acta **40**:35-41, 1972.
14. Whitby, L.G., and Moss, D.W.: Analysis of heat inactivation curves of alkaline phosphatase isoenzymes activity in serum, Clin. Chim. Acta **59**:361-367, 1975.
15. Moss, D.W., and Whitby, L.G.: A simplified heat inactivation method for investigating alkaline phosphatase isoenzymes in serum, Clin. Chim. Acta **61**:63-71, 1975.
16. Statland, B.E., Nishi, N.H., and Young, D.S.: Serum alkaline phosphatase: total activity and isoenzyme determinations made by use of the centrifugal fast analyzer, Clin. Chem. **18**:1468-1474, 1972.
17. Gorman, L., and Statland, B.E.: Clinical usefulness of alkaline phosphatase isoenzyme determinations, Clin. Biochem. **10**:171-174, 1977.

Amylase

MICHAEL D.D. McNEELY

Amylase, AMS, α-1,4-glucan 4-glucanhydrolase
Clinical significance: pp. 460, 468, and 594
EC 3.2.1.1
Molecular weight: 40,000 to 50,000 daltons
Chemical class: enzyme, protein
Isoenzymes: at least seven in human tissue; salivary and pancreatic forms in serum

PRINCIPLES OF ANALYSIS AND CURRENT USAGE

Amylases are enzymes that degrade complex carbohydrate molecules into smaller components. Amylase (AMS) is produced by the exocrine pancreas and the salivary glands to aid in the digestion of starch. It is also found in the liver and in the lining of the fallopian tubes. Human amylase is termed "α-amylase" (or endoamylase) because of its ability to split polysaccharide α-1,4 linkages in a random manner. The α-1,6 linkages at the branch points remain untouched. The end product of α-amylase action on polysaccharide is the formation of dextrans, maltose, and some glucose molecules.

Amylase activity is increased in the serum and urine of patients suffering from pancreatitis. It is in the detection of this condition that amylase measurements are used. Pancreatitis often presents as a medical emergency; thus clinically appropriate amylase methods must be procedures suitable for stat use.

Some 200 methods for amylase measurement have been devised. All begin by combining the patient's sample with a buffered solution of polysaccharide. The assay requirements must be rigidly adjusted to suit the specific conditions of amylase. The pH optimum is 6.9 to 7.0, and calcium and chloride are absolutely required.

After incubation with the polysaccharide substrate, a variety of different detection techniques are available to measure the activity of amylase. The different approaches are the basis of the following classification consisting of five techniques.

Viscosimetric techniques (method 1, Table 56-12)

Viscosimetric techniques depend on the decrease in viscosity of the substrate that occurs after the action of amylase. Initially, the large polysaccharide substrate molecules form a highly viscid solution. After the action of amylase many molecules have been broken into smaller fragments, and therefore the solution is less viscid. Viscosity measurements before and after a timed incubation period are proportional to the amylase activity. This technique has been abandoned because of the inconvenience of performing viscosity measurements.

Turbidimetric and nephelometric techniques
(method 2, Table 56-12)

In turbidimetric techniques the unreacted substrate is designed to produce a turbid solution with a spectrophotometric absorbance of about 1.000. Amylase reduces the turbidity by fracturing the substrate. Turbidimetric and light-scattering measurements are carried out either kinetically or after a fixed time interval, and the change in turbidity or light scattering is proportional to amylase activity. These methods are difficult to standardize because of substrate variation, and in general they lack precision.[1]

Nephelometry has been used in a commercial instrument for amylase and lipase measurements (Model 91, Coleman Instruments Division, Perkin-Elmer Corp., Norwalk, Conn.).[2] This instrument uses a standardized substrate and carries out rate measurements over a short time. In this configuration the precision is acceptable and the method rapidly completed.

Iodometric (amyloclastic) techniques
(method 3, Table 56-12)

The iodometric methods are based on the ability of iodine to form a vivid blue color in combination with starch. The by-product of amylase action may also form colored substances with iodine but at different wavelengths from the characteristic starch-iodine complex. The methods are carried out by adding iodine color reagent to the substrate-sample mixture after an incubation period. The greater the amount of amylase activity, the lighter will be the color of the final solution.

An early version of the iodometric approach incubated a series of sample-starch mixtures on a reaction plate. The iodine reagent was added to the reaction mixtures at various time intervals. When no iodine color was formed, the time was recorded and the amylase activity was semiquantitatively estimated. The longer the time required to accomplish discoloration, the lower the sample activity.[3]

The other approach employs a fixed time interval.[4] The substrate and sample are mixed together and incubated for a fixed period. One then adds the iodine color solution and makes a spectrophotometric measurement. The lower the final absorbance (the greater the difference between the

Table 56-12. Methods of amylase measurement

Method	Principle	Usage	Comments
1. Viscosimetric	Amylase degradation of starch decreases viscosity of solution; viscosity inversely related to amylase concentration.	Historical	Neither precise nor accurate
2. Turbidimetric, nephelometric	Decrease in turbidity or scattered light (nephelometry) of a starch solution is directly related to amylase concentration	Rare	With constant substrate, automated technique is acceptable
3. Iodometric	Degradation of starch by amylase reduces the reaction of iodine with starch; reduction in iodine-starch product (A_{max} = 660 nm) inversely related to amylase activity	Common	Can be easily adapted by most laboratories
4. Saccharogenic	Glucose released from substrate is quantitated, usually by an enzymatic procedure, such as hexokinase or glucose oxidase	Very common	Can be easily adapted to many automated, discrete analyzers
5. Chromolytic	Liberation of a dye coupled to an insoluble polysaccharide	Common	Labor intensive, not suited for automation

blank and color), the higher the activity. Amyloclastic methods are widely employed, but because substrates vary, these methods are difficult to standardize.[5]

Saccharogenic techniques (method 4, Table 56-12)

Saccharogenic techniques depend upon the measurement of monosaccharides or disaccharides liberated by amylase's action on the substrate. The classical amylase method of Somogyi is a saccharogenic technique.[6] Sample and substrate are incubated for 30 minutes; a serum blank and an incubated sample reaction tube then undergo measurement for reducing substances. Since the by-products of amylase action are reducing substances, the enzyme's activity is directly proportional to the amount of reducing substance produced.

More recently, the saccharogenic approach has been modified for automation. In these techniques the maltose split from the substrate is converted into glucose by maltase, which is added to the reagent mixture. One measures the glucose liberated by this step by using an enzymatic glucose technique. The methods must be designed to account for glucose present in the patient's sample.

One can use a hexokinase assay to measure the glucose spectrophotometrically or fluorometrically.[7] In another approach one uses glucose oxidase and measures glucose by using electrometric detection of oxygen consumption.[8] Another similar approach employs a coupled color reagent instead of the O_2 electrode.[9]

The Beckman method[9] employs a maltotetraose substrate, which allows the maltose to be split from a shorter substrate. Since two glucose molecules are produced from 1 mole of substrate, a twofold amplification occurs.

The du Pont ACA method[10] employs a maltopentaose substrate, which generates five molecules of glucose measured by a hexokinase reaction. Glucose interference is eliminated by gel filtration of the sample before analysis.

The enzyme techniques are useful because the entire reaction can be carried out in a single tube. However, the amount of glucose originally in the patient sample must be taken into consideration. One can do this by hydrolyzing the glucose in the patient's sample before performing the complete analysis or by performing the analysis in two steps to allow the patient's glucose to react first, or one may use a serum blank. Such blanks may be very high, rendering spectrophotometric measurements difficult. In general, the reagents are very complex and therefore rather expensive. The reagents may also lose their effectiveness through a compromise among the many coupled enzyme reactions, which must go on simultaneously. They rarely have true, zero-order kinetics.

Chromolytic techniques
(method 5, Table 56-12)

In recent years commercial manufacturers have produced a variety of intriguing and convenient amylase methods that depend on the liberation of a dye coupled to a complex, insoluble polysaccharide. In some of these methods one incubates the sample with the synthetic substrate and after a fixed time measures the amount of color liberated as the dye is split from the original complex. The Pharmacia (Pharmacia, Inc. Piscataway, N.J.) method has been one of the most popular of this type.[11] These methods often require blanks, centrifugation steps, and decanting and are therefore difficult to automate.

An interesting version of this approach employs a substrate of reactone red 2 β-amylopectin in an agarose gel.[12] Amylase activity is indicated by the size of the cleared diffusion ring around a central well containing the sample. The method has a large measurement range and is suitable for multiple samples.

A somewhat different tactic depends on the release of small, water-soluble fragments in such a way that the color can be measured continuously.[13] This can be done with *p*-nitrophenol glycosides, which produce *p*-nitrophenol by direct hydrolysis or by a coupled reaction involving α-glucosidase and β-glucosidase. These methods are readily automated.

It is not practical for laboratories to synthesize their own artificial substrates. Therefore the selection of such a technique must be made carefully.

REFERENCE AND PREFERRED METHODS

The Amylase Conference of the German Society for Clinical Chemistry has proposed several characteristics of a good amylase method[12]:

1. A defined substrate with constant quality, economic cost, and well-defined reaction products.
2. A continuous monitoring method that obeys zero-order kinetics and has no lag phase.
3. Sensitive enough to read at 30° C.
4. Lack of endogenous glucose interference.

Several additional criteria are also obvious:

1. Assays must be able to be performed "stat." (by itself, minimal steps, in less than 30 minutes).
2. Methods should be suitable for serum and urine.
3. Dilutions should be avoided or must be easily performed.
4. Viscosimetric techniques are not practical.

In general, turbidimetric methods lack precision, sensitivity, and consistency. The Perkin-Elmer nephelometric approach has circumvented these deficiencies to produce a very rapid method requiring a specialized machine. It has been recommended for routine and emergency use by Lehane et al.[14]

The iodometric (amyloclastic) techniques are cheap, simple, quick, and sensitive but are limited by protein-iodine interaction, require blanks, and may lack consistency because of substrate variability. However, they employ readily available reagents and are the most suitable "in-house" method. See p. 1104 for an example procedure.

The traditional saccharogenic methods are the most reliable but are time consuming and have high sample blanks and variable substrates. They are recommended as reference procedures but not for routine use. The multicoupled, continuous saccharogenic methods are conveniently automated but may be hindered by glucose interference, high blanks, and complex, expensive reagents.

Chromolytic techniques can be fast, precise, and easy to perform but often have manual steps that defy convenient automation. In general, the synthetic substrates used by these methods are very consistent. The continuous chromolytic techniques satisfy all the criteria just enumerated above and will probably become the standard approach in the future.

Brétaudière et al. have examined sources for calibration and control materials.[15] Human salivary amylase was shown to be suitable for this purpose.

PROCEDURE

Principle

Starch is hydrolyzed by amylase in the sample to liberate smaller molecules. An iodine reagent is added, which forms a vivid blue color with the remaining starch. The amylase activity is inversely proportional to the amount of color in the final solution.

Reagents (Table 56-13)

Buffered starch substrate, pH 7.0. Dissolve 13.3 g of anhydrous disodium phosphate (Na_2HPO_4) and 4.3 g of benzoic acid (C_6H_5COOH) in approximately 250 mL of distilled water. Heat to boiling. Add 0.200 g of soluble starch in 5 mL of cold water. Add this starch mixture to the boiling buffer, stirring and rinsing to ensure complete transfer. Allow the mixture to cool to room temperature and dilute to 500 mL with water. Store at 4° C. The solution should remain clear. One should assess its stability by measuring the absorbance of mixed reagents (observing for a significant decline in the reagent blank value from run to run).

Stock iodine solution (0.1 mol/L). Dissolve 3.567 g of potassium iodate (KIO_3) and 45 g of potassium iodide (KI) in 800 mL of water. Add 9 mL of 12 M HCl slowly to the mixture, using ample mixing. Dilute to 1 L. Store in an amber bottle at 4° C. The solution is stable for 12 months.

Working iodine solution (0.01 mol/L). Dilute 50 mL of stock iodine solution to 500 mL. Store in an amber bottle at 4° C. The solution is stable for 2 months.

Assay

Equipment. For each test, two 25 mL volumetric flasks (or graduated cylinders) are used. One is labeled "test" and the other "serum blank." One tube is used as a reagent blank. Use a spectrophotometer with a ≤10 nm band pass capable of reading at 660 nm and a 37° C water bath.

1. Pipet 2.5 mL of starch substrate into each container and place in a 37° C water bath for 15 minutes.
2. Pipet 50 μL of sample into the test container. A series of tubes should be pipetted at timed intervals.
3. Incubate at 37° C for exactly 7½ minutes.
4. At timed intervals remove each test container from the water bath and pipet 1.5 mL of water and 2.5 mL of working iodide solution into each tube. Dilute to 25 mL with water. Mix each tube well, and read the absorbance against water at 660 nm.
5. Remove the serum blank containers from the water bath, and add 50 μL of serum, approximately 15 mL of water, and 2.5 mL of starch solution; dilute to 25 mL. This should be done in rapid sequence, and the absorbance read as in step 4.
6. If the amylase activity is more than 400 units, the

Table 56-13. Comparison of methods for amylase

Condition	Iodometric*	Saccharogenic†
Temperature	37° C	37° C
pH	7.0	6.7
Final concentration of reagents	Starch: 392 mg/L Na_2HPO_4: 188 mmol/L I^-: 1 mmol/L	Maltotetraose: 5 g/L NAD: 2.5 mmol/L Maltose phosphorylase: ≥3 U/mL β-Phosphoglucomutase: ≥1 U/mL Glucose-6-phosphate dehydrogenase: ≥6 U/mL
Sample volume	50 μL	10 μL
Fraction of sample volume	0.0196	0.042
Time of reaction	7.5 min	11 min
Wavelength	660 nm	340 and 380 nm
Linearity	8000 U/L	350 U/L
Interference	Turbidity	Turbidity

*Method listed in this text.
†Boehringer-Mannheim Diagnostics, Inc. (Indianapolis, Ind.): reagent employed on the ABA-100 (Abbott, North Chicago, Ill.).

test should be repeated with a 25 μL of sample, since the reaction does not proceed linearly when more than half the substrate has been hydrolyzed.

Calculation

$$\frac{\text{Absorbance (reagent blank)} - \text{Absorbance (test)}}{\text{Absorbance (reagent blank)}} \times \text{Dilution factor} \times 8000 = \text{Amylase U/L}$$

$$\text{Dilution factor} = \frac{2.5 \text{ mL}}{0.05 \text{ mL}} = 50$$

REFERENCE RANGE

Serum: less than 1800 U/L
Urine: less than 5000 U/24 hours

REFERENCES

1. Zinterhofer, L., Wardlaw, S., Jatlow, P., and Seligson, D.: Nephelometric determination of pancreatic enzymes. I. Amylase, Clin. Chim. Acta **43:**5-12, 1973.
2. Smeaton, J.R., and Marquardt, H.F.: A reaction rate nephelometer for amylase determinations, Clin. Chem. **20:**896, 1974 (Abstract).
3. Wohlgemuth, J.: A new method for the quantitative determination of amylolytic ferments, Biochemistry **9:**1-9, 1908. (Title translated into English.)
4. Caraway, W.T.: A stable starch substrate for the determination of amylase in serum and other body fluids, Am. J. Clin. Pathol. **32:**97-99, 1959.
5. Alpha-amylase methodology survey I. Center for Disease Control, U.S. Public Health Service, Atlanta, Ga., Nov. 1975.
6. Somogyi, M.: Modifications of two methods for the assay of amylase, Clin. Chem. **6:**23-35, 1960.
7. Guilbault, G.G., and Rietz, E.B.: Enzymatic, fluorometric assay of α-amylase in serum, Clin. Chem. **22:**1702-1704, 1976.
8. Tietz, N.W., Miranda, E.F., and Weinstock, A.: A kinetic assay for the measurement of amylase activity in serum, Clin. Chem. **19:**645, 1973 (Abstract).
9. Kaufman, R.A., and Tietz, N.W.: Recent advances in measurement of amylase activity—a comparative study, Clin. Chem. **26:**846-853, 1980.
10. Balkcom, R.M., O'Donnell, C.M., and Amano, E.: Evaluation of the DuPont ACA amylase method, Clin. Chem. **25:**1831-1835, 1979.
11. Ceska, M., Birath, K., and Brown, B.: A new and rapid method for the clinical determination of α-amylase activities in human serum and urine, optimal conditions, Clin. Chim. Acta **26:**437-444, 1969.
12. Tauschel, H.-D., and Rudolph, C.: A new sensitive radial diffusion method for microdetermination of α-amylase, Anal. Biochem. **120:**262-266, 1982.
13. Lorentz, K.: α-Amylase assay: current state and future development, J. Clin. Chem. Clin. Biochem. **17:**499-504, 1979.
14. Lehane, D.P., Wissert, P.J., Lum, G., and Levy, A.L.: Amylase activity in serum and urine: comparison of results by the amyloclastic, dyed-starch, and nephelometric techniques, Clin. Chem. **23:**1061-1065, 1977.
15. Brétaudière, J.-P., Rej, R., Drake, P., et al.: Suitability of control materials for determination of α-amylase activity, Clin. Chem. **27:**806-815, 1981.

Aspartate aminotransferase

ROBERT L. MURRAY

Aspartate aminotransferase, AST, L-aspartate:2-oxoglutarate aminotransferase, serum glutamate oxaloacetate transaminase, SGOT
Clinical significance: pp. 420, 509, and 594
EC 2.6.1.1
Molecular weight: 110,000 daltons
Chemical class: enzyme, protein
Biochemical reaction: (See Fig. 56-5 below.)

PRINCIPLES OF ANALYSIS AND CURRENT USAGE

Aspartate aminotransferase (AST) catalyzes the transfer of an amino group from specific amino acids (L-glutamate or L-aspartate) to specific keto acids (α-ketoglutarate or oxaloacetate) (Fig. 56-5). Although at physiological pH the reaction is energetically favored toward the formation of L-aspartate and α-ketoglutarate (to the left, as in Fig. 56-5), in vivo the reaction goes to the right to provide a source of nitrogen for the urea cycle. The glutamate thus produced is deaminated by glutamate dehydrogenase, resulting in ammonia and regeneration of α-ketoglutarate.

Two methods of AST analysis have enjoyed the widest popularity. The coupling of oxaloacetate with 2,4-dinitrophenylhydrazine (DNPH) to produce the blue hydrazone (method 1, Table 56-14) was introduced in 1957[1] and gained popularity because of its simplicity and relative accuracy. In this method, as in all others, it is necessary to force the AST reaction toward the energetically less-favored products, of which only oxaloacetate is easily quantitated. Thus L-aspartate and α-ketoglutarate are added in excess. After incubation the reaction is stopped with the addition of DNPH, which reacts with both α-ketoglutarate and oxaloacetate. After condensation the pH is made alkaline, and the absorbance at 505 nm is measured. Because the absorptivity of the oxaloacetate hydrazone is much greater than the corresponding hydrazone of α-ketoglutarate (present in great excess when the

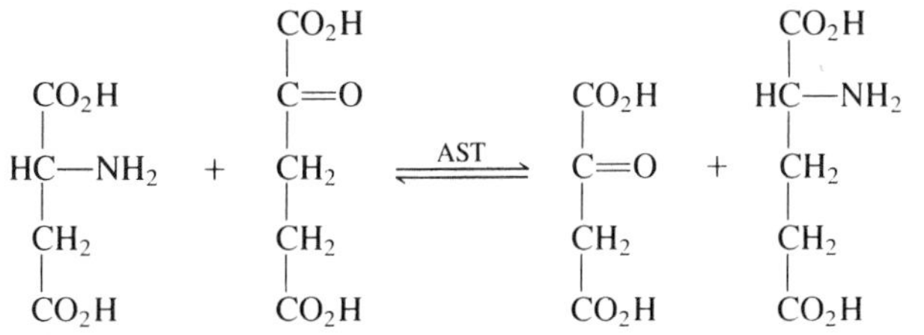

$$\underset{\text{L-Aspartate}}{\begin{array}{c}CO_2H\\|\\HC{-}NH_2\\|\\CH_2\\|\\CO_2H\end{array}} + \underset{\alpha\text{-Ketoglutarate}}{\begin{array}{c}CO_2H\\|\\C{=}O\\|\\CH_2\\|\\CH_2\\|\\CO_2H\end{array}} \xrightleftharpoons{AST} \underset{\text{Oxaloacetate}}{\begin{array}{c}CO_2H\\|\\C{=}O\\|\\CH_2\\|\\CO_2H\end{array}} + \underset{\text{L-Glutamate}}{\begin{array}{c}CO_2H\\|\\HC{-}NH_2\\|\\CH_2\\|\\CH_2\\|\\CO_2H\end{array}}$$

Fig. 56-5. Amino transfer catalyzed by AST.

Table 56-14. Methods of aspartate aminotransferase (AST) analysis

Method	Type of analysis	Principle*	Usage
1. Dinitrophenylhydrazone coupling (colorimetric) (Reitman and Frankel[1])	Quantitative	**Asp + α-KG $\xrightarrow{AST}$ Gl + Oa** 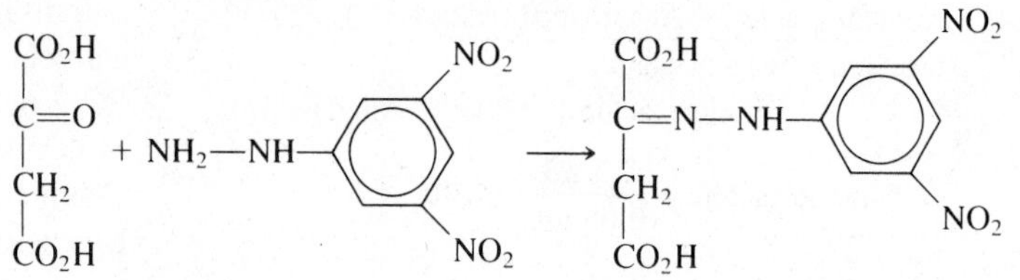 Oa (CO_2H–C(=O)–CH_2–CO_2H) + NH_2—NH–(2,4-dinitrophenyl, NO_2, NO_2) → CO_2H–C(=N—NH–(dinitrophenyl, NO_2, NO_2))–CH_2–CO_2H **Oa** **Dinitrophenylhydrazine** **Oa-dinitrophenylhydrazone (absorbance at 505 nm)**	Serum
2. Enzymatic (ultraviolet monitoring) (Karmen[2])	Quantitative	**Asp + α-KG $\xrightarrow{AST}$ Gl + Oa** 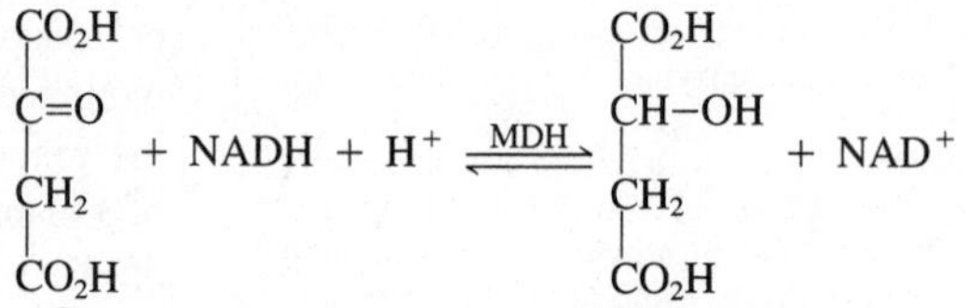 CO_2H–C(=O)–CH_2–CO_2H + NADH + H^+ $\overset{MDH}{\rightleftharpoons}$ CO_2H–CH–OH–CH_2–CO_2H + NAD^+ **Oa** **Malic acid**	Serum
3. Diazonium dye coupling (colorimetric)	Quantitative	**Asp + α-KG $\xrightarrow{AST}$ Gl + Oa** 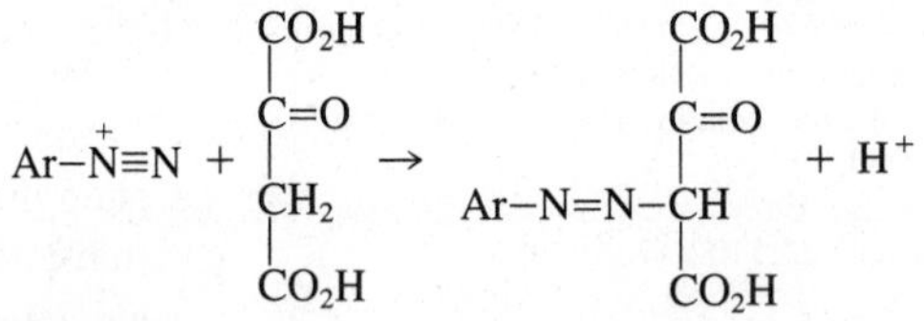 Ar–$\overset{+}{N}$≡N + CO_2H–C(=O)–CH_2–CO_2H → CO_2H–C(=O)–CH(–N=N–Ar)–CO_2H + H^+ **Diazonium dye** **Oa** **Oa diazonium dye**	Serum

**Asp,* Aspartic acid; *α-KG,* α-ketoglutaric acid; *Gl,* glutamic acid; *Oa,* oxaloacetic acid.

patient specimen has low AST activity), little error is introduced.

In the second approach, first described by Karmen,[2] oxaloacetate is again the product quantitated. Malate dehydrogenase (MDH) is used to convert the oxaloacetate to malate (method 2, Table 56-14). The decrease in absorbance at 340 nm caused by NADH (reduced nicotinamide adenine dinucleotide) consumption by the second reaction is used to follow the course of the AST reaction.

A number of photometric techniques have been described, all of which follow a similar principle—coupling an aryldiazonium salt to the active methylene of oxaloacetate[3] (method 3, Table 56-14). In all cases the aryl substituent (Ar) was a dye with absorbance characteristics that changed with formation of the azo bridge to oxaloacetate. The reaction was thus monitored by an increase in absorbance of the diazonium reaction product.

Identification and quantitation of AST isoenzymes has recently been reported.[4,5] The clinical utility and technical practicality of these have yet to be determined.

REFERENCE AND PREFERRED METHODS

When choosing a method for routine clinical use, one is frequently forced into a compromise between accuracy and practicality. With AST measurement this is fortuitously not the case. The coupled enzymatic method (method 2, Table 56-14) and the reference method have the characteristics of simplicity and speed, which make it the preferred method for routine operations as well.

The drawbacks of the dinitrophenylhydrazine method are that (1) the oxaloacetate produced by the reaction results in feedback inhibition of AST, thus specimens exhibiting high activity are spuriously lowered; and (2) any ketone in serum can react, though most do not result in an absorbance change in the region measured. However, acetoacetic acid and hydroxybutyric acid, both components of ketosis, do cause false elevations. The diazonium method similarly is limited by nonspecificity; any activated methylene group will react. This is seen in patients on certain drugs and those with renal failure.

There have been many minor modifications in the enzy-

Table 56-15. Comparison of methods for aspartate aminotransferase (AST) measurement

Conditions	Karmen[2] (1955)	AACC[7]* (1961)	Henry et al.[8] (1960)	British[8] (1972)	German[10] (1972)	Scandinavian[11] (1974)	IFCC[6]* (1977)
Reaction temperature (°C)	23-28	25	32	25	25	37	30
Sample volume (mL)	0.2	0.5	0.2	0.2	0.5	0.15-0.3	0.2
Total reaction volume (mL)	3.0	3.0	3.0	3.0	3.70	1.25-2.50	2.40
Fraction of sample volume	0.066	0.167	0.066	0.066	0.135	0.120	0.083
Buffer type	Phosphate	Phosphate	Phosphate	Phosphate	Phosphate	Tris/edetate	Tris
Final concentration of reagents							
Buffer concentration (mmol/L)	33	50	100	80	80	20	80
Final pH (at reaction temperature)	7.4	7.4	7.4	7.4	7.4	7.7	7.8
L-Aspartate (mmol/L)	33	33	125	125	200	200	200
α-Ketoglutarate (mmol/L)	6.7	6.7	6.7	6.7	12	12	12
Reduced NAD^+ (mmol/L)	0.16	0.16	0.18	0.24	0.18	0.15	0.18
Pyridoxal PO_4 (mmol/L)	—	—	—	—	—	(Optional)	0.10
Malate dehydrogenase (U/L)	16.6×10^3	16.6×10^3	83×10^3	46.5×10^3	166×10^3	166×10^3	166×10^3
Lactate dehydrogenase (U/L)	—	—	—	—	333×10^3	50×10^3	250×10^3
Preincubation time (min)	5	—	10	10	10	5-15	10

**AACC*, American Association for Clinical Chemistry; *IFCC*, International Federation of Clinical Chemistry.

matic technique since its introduction by Karmen.[2] The specifications listed by the International Federation of Clinical Chemistry (IFCC) [6] are recommended, primarily in evaluation of preformulated reagent mixtures offered by various manufacturers. Although it may be impossible or impractical to modify such purchased mixtures, the product most closely matching the stated conditions can be identified. Other groups have also published recommended conditions for measurement of AST (Table 56-15). Some differences among these groups reflect the span of almost 20 years during which these recommendations were developed.

In the IFCC procedure 0.20 mL of serum is preincubated for 10 minutes in 2.00 mL of a buffer-reagent that contains all reactants except α-ketoglutarate. During this preincubation period, the added lactate dehydrogenase (LD) rapidly converts the endogenous pyruvate in the serum; the pyridoxal phosphate cofactor joins with any inactive apoenzyme to form an increased amount of active AST. With the addition of 0.20 mL of α-ketoglutarate, the primary reaction is initiated, and the concentrations shown in Table 56-15 are reached, exclusive of the small increases caused by the presence of endogenous material in the serum. After steady state is reached, the rate of NADH oxidation is monitored repeatedly at 340 nm.

A striking difference between the IFCC method and the others is the inclusion of pyridoxal phosphate. The need for addition of this cofactor has been widely debated. Although the essential nature of the cofactor has long been recognized, its usual presence in human serum in adequate amounts has led many to omit it as an added component of the reaction mixture. In the occasional patient with severe vitamin B deficiency, this could lead to a serious underestimation of that patient's AST level. Addition of pyridoxal phosphate results in an increased activity, reportedly ranging up to 50%.[12] The higher values might be expected in severely ill patients, especially if elevated AST levels are present, and in those suffering from pyridoxal phosphate deficiency. As expected, the increased sensitivity afforded by the pyridoxal phosphate is accompanied by a decreased range of linearity.

Potential sources of error with the enzymatic method include glutamate dehydrogenase, aminotransferases (in the reagent preparation), very high pyruvate concentration, or the use of phosphate buffer. Glutamate dehydrogenase is present in some pathological sera, particularly in serum from patients with liver disease. This enzyme introduces error by catalyzing the formation of glutamate with NH_4^+ to form α-ketoglutarate, with reduction of NADH to

NAD^+. Since this source of error is possible only when ammonium ions are present, the IFCC method eliminates all deliberate addition of these. (Older methods sometimes included ammonium sulfate in the formulation.)

The appearance of aminotransferases in the reagents is possible if the LD is not carefully prepared. Good-quality enzymes will not pose a problem; in any event, a blank determination will identify the problem. The presence of pyruvate is a potential source of error, since in the presence of endogenous LD, pyruvate will be converted to lactate with simultaneous consumption of NADH. This is circumvented by the addition of a large excess of LD, so that endogenous pyruvate is converted during the preincubation period, eliminating interference during the measurement period.

The use of phosphate buffer retards the recombination of added pyridoxal phosphate with the apoenzyme; if a large amount of the inactive enzyme is present, a falsely low activity will be observed. However, NADH is somewhat less stable in Tris buffer than it is in phosphate buffer.[13] For this reason, the Tris concentration is kept relatively low, at 80 mmol/L.

SPECIMEN

Serum or plasma may be used; heparin, oxalate, EDTA, and citrate have been found to cause no enzyme inhibition.[14] (Anticoagulants with ammonium as cation should be avoided to reduce the possibility of glutamate dehydrogenase activity as a source of error.) Urine has little or no activity and is not recommended for analysis.

REFERENCE RANGE

Adults[12]	5 to 34 U/L (37°C)
	8 to 22 U/L (30° C)

Levels approximately twice the adult level are seen in neonates and in infants; these decline to adult levels by approximately 6 months of age.[15]

REFERENCES

1. Reitman, S., and Frankel, S.: A colorimetric method for the determination of serum glutamic oxalacetic and glutamic pyruvic transaminases, Am. J. Clin. Pathol. **28**:56-63, 1957.
2. Karmen, A.: A note on the spectrophotometric assay of glutamic oxalacetic transaminase in human blood serum, J. Clin. Invest. **34**:131-135, 1955.
3. Sax, S., and Moore, J.: Determination of glutamic oxalacetic transaminase activity by coupling of oxalacetate with diazonium salts, Clin. Chem. **13**:175-185, 1967.
4. Sampson, E., Hannon, W., McKneally, S., et al.: Column chromatography and immunoassay compared for measuring the isoenzymes of aspartate aminotransferase in serum, Clin. Chem. **25**:1691-1696, 1979.
5. Rej, R., Brétaudière, J., and Graffunder, B.: Measurement of aspartate aminotransferase isoenzymes: six procedures compared, Clin. Chem. **27**:535-542, 1981.
6. Bergmeyer, H., Bowers, G., Jr., Hørder, M., and Moss, D.W.: Provisional recommendations on IFCC methods for the measurement of catalytic concentrations of enzymes. Part 2. IFCC method for aspartate aminotransferase, Clin. Chem. **23**:887-899, 1977.
7. Frieman, M., and Taylor, T.: Transaminase. In American Association of Clinical Chemists: Standard methods of clinical chemistry, vol. 3, New York, 1961, Academic Press, Inc., pp. 207-217.
8. Henry, R., Chiamori, N., Golu, O., and Berkman, S.: Revised spectrophotometric methods for the determination of glutamic-oxalacetic transaminase, glutamic-pyruvic transaminase, and lactic acid dehydrogenase, Am. J. Clin. Pathol. **34**:381-391, 1960.
9. Wilkinson, J.H., Baron, D.N., Moss, D.W., and Walter, P.G.: Standardization of clinical enzyme assays: a reference method for aspartate and alanine aminotransferases, J. Clin. Pathol. **25**:940-948, 1972.
10. Recommendations of the German Society for Clinical Chemistry: Z. Klin. Chem. Klin. Biochem. **6**:37-45, 1972.
11. Committee on Enzymes of the Scandinavian Society for Clinical Chemistry and Clinical Physiology: Recommended methods for the determination of four enzymes in blood, Scand. J. Clin. Lab. Invest. **33**:291-296, 1974.
12. Bruns, D., Savory, J., Titheradge, A., et al.: Evaluation of the IFCC-recommended procedure for serum aspartate aminotransferase as modified for use with the centrifugal analyzer, Clin. Chem. **27**:156-159, 1981.
13. Rodgerson, D.O.: Development of a method for aspartate aminotransferase activity measurement by the AACC Subcommittee on Enzymes: a progress report. In Tietz, N.W., Weinstock, A., and Rodgerson, D.O., editors: Proceedings of the Second International Symposium on Clinical Enzymology, Washington, D.C., 1976, American Association for Clinical Chemistry, pp. 119-134.
14. Demetriou, J.A., Drewes, P.A., and Gin, J.B.: Enzymes. In Henry, R.J., Cannon, D.C., and Winkelman, J.W., editors: Clinical chemistry, principles and technics, ed. 2, Hagerstown, Md., 1974, Harper & Row, Publishers, Inc., pp. 815-1001.
15. Meites, S., editor: Pediatric clinical chemistry, Washington, D.C., 1981, The American Association for Clinical Chemistry, pp. 117-120.

Cholinesterase

MARY ELLEN KING

Cholinesterase, CHS, acylcholine acylhydrolase, pseudocholinesterase
Clinical significance: pp. 718 and 897
EC 3.1.1.8
Molecular weight: 360,000 daltons
Merck Index: 2205
Chemical class: enzyme, protein
Biochemical reaction:

$$\underset{\text{Acetylcholine}}{CH_3{-}C({=}O){-}O{-}CH_2{-}CH_2{-}\overset{+}{N}(CH_3)_3} + H_2O \xrightleftharpoons{\text{EC 3.1.1.8}} \underset{\text{Acetate}}{CH_3{-}C({=}O){-}O^-} + H^+ + \underset{\text{Choline}}{HO{-}CH_2{-}CH_2{-}\overset{+}{N}(CH_3)_3}$$

PRINCIPLES OF ANALYSIS AND CURRENT USAGE

Cholinesterase (CHS, choline esterase), which has also been referred to as plasma or serum cholinesterase, pseu-

Table 56-16. Methods of cholinesterase analysis

Method	Type of analysis	Principle	Usage	Comments
1. Colorimetric	pH dye color change, end point	$CH_3C(=O)O{-}CH_2CH_2\overset{+}{N}(CH_3)_3 + H_2O \xrightarrow{\text{Cholinesterase (CHS)}}$ **Acetylcholine** $CH_3{-}C(=O){-}O^- + HO{-}CH_2CH_2{-}\overset{+}{N}(CH)_3 + H^+$ **Acetate** **Choline** Dye + $H^+ \rightarrow$ Color change	Serum or plasma	Wide interlaboratory variation of results
2. pH change	pH measurement	Acetylcholine $\xrightarrow{CHS}$ Choline + Acetate + H^+	Serum or plasma	Wide interlaboratory variation of results
3. Electrometric	Titrate acid released, kinetic or end point	Acetylcholine $\xrightarrow{CHS}$ Choline + Acetate + H^+ H^+ + Added base $\rightarrow H_2O$ and constant pH	Serum or plasma	Requires automated pH titration
4. Colorimetric	Spectrophotometric, kinetic or end point	$C_2H_5{-}C(=O){-}S{-}CH_2CH_2\overset{+}{N}(CH_3)_3 + H_2O \xrightarrow{CHS}$ **Propionylthiocholine** $C_2H_5{-}COO^- + HS{-}CH_2CH_2{-}\overset{+}{N}(CH_3)_3$ **Propionate** **Thiocholine** Thiocholine + DNTB (O_2N, HO_2C-substituted phenyl–S–S–phenyl with CO_2H, NO_2) $\rightarrow$ 2 HS–C_6H_3(NO_2)–CO_2H + $[(CH_3)_3\overset{+}{N}CH_2CH_2{-}S]_2$ **DNTB** **5-Thio-2-nitrobenzoic acid** (yellow, A_{max} = 410 nm)	Serum or plasma	Recommended method

docholinesterase, and butyrylcholinesterase, is synthesized by the liver and present in plasma.[1-3] Its true physiological function is unknown. Although it can hydrolyze acetylcholine, cholinesterase is less specific for this substrate than is the red blood cell enzyme acetylcholinesterase (''true'' cholinesterase, EC 3.1.1.7), and so its function may be to hydrolyze other choline esters. Cholinesterase activity is usually measured for one of two reasons: (1) the most common is the use of a decrease in enzymatic activity as an indicator of exposure to organophosphorus compounds, including many pesticides, and (2) inherited abnormal variants of cholinesterase may be identified by an assay of both total activity and the extent of inhibition by either dibucaine or fluoride; some of these variants lead to prolonged apnea in patients receiving the anesthetic succinylcholine.

Most current assays for cholinesterase are spectrophotometric methods which use propionylthiocholine or acetylthiocholine as substrate (method 4, Table 56-16). 5,5′-Dithio-bis(2-nitrobenzoic acid) (DTNB) is added to react with the released thiocholine to form the yellow compound 5-thio-2-nitrobenzoic acid (absorption maximum, 410 nm.)[4,5] The reaction may be followed as a rate or end-point procedure.

Alternate procedures include titrimetric[6-10] and electrometric[11-13] methods that follow the release of hydrogen ions from a choline ester. One titrimetric method directly measures hydrogen-ion release by monitoring pH changes over a 1- to 1½-hour period (method 2, Table 56-16).[8,9] A second titrimetric colorimetric method quantifies the amount of hydrogen ion released by the hydrolysis of acetylcholine by following the color change of a phenol red indicator over a 30-minute incubation time (method 1, Table 56-16). The color change is related to the moles of substrate converted.[6,7]

A special type of titrimetric method usually referred to as electrometric[10-12] procedure uses highly automated titration instrumentation (''pH Stat'') to continuously add base to neutralize the hydrogen ions released from acetylcholine by cholinesterase (method 3, Table 56-16). This instrument maintains the enzyme reaction at constant pH, and the amount of base added is directly related to the number of moles of acetylcholine hydrolyzed on a mole-per-mole basis. The reaction rate is monitored automatically.

REFERENCE AND PREFERRED METHODS

(Table 56-17)

Described is a simple colorimetric method based on the procedure of Ellman requiring only a spectrophotometer and 10 minutes of time that has been proposed as a selected method by The American Association for Clinical Chemistry.[4] The substrate propionylthiocholine iodide is used, since it gives 27% greater activity than butyrylthiocholine iodide at 37° C and lower blank readings. Easier discrimination among some common cholinesterase phenotypes is observed with use of this substrate. Interrun coefficients of variation of 3% are reported.[4,5]

The procedures that simply monitor changes in the pH of the solution suffer from the problem that changes in pH are nonlinear, since pH is a logarithmic function. Also, approximated factors must be applied to the result to correct for both the change in cholinesterase activity as the reaction mixture pH changes and for nonenzymatic hydrolysis of substrate during the lengthy incubation.[8,9] Temperature control is also critical.[10] Because of these problems, interlaboratory variations of over 30% are reported for plasma titration assays, despite intralaboratory interrun coefficients of variation of 6% to 8%.[10]

The electrometric (pH Stat) procedure, in which the neutralization of the reaction is monitored by an automatic titrator, requires a specialized instrument dedicated to the assay; however, it offers the advantage of being a rapid, rate method.[11,12] Coefficients of variation of 4% to 5% are reported.[13] The instrument can also be used to measure red blood cell acetylcholinesterase.

Because of its ease of adaptation to most laboratory spectrophotometers, the Ellman method is recommended and is described below.

SPECIMEN

Either serum or heparinized plasma may be used. Moderate hemolysis will not interfere if the serum is well centrifuged to remove red blood cell ghosts.[4] Cholinesterase is stable at room temperature for up to 80 days and, if frozen at −20° C, for 3 years.[1] Samples should be stored and shipped cold to avoid high extremes of temperature.

Table 56-17. Conditions for cholinesterase analysis

Temperature	37° C
Wavelength	410 nm
pH	7.6
Final concentration of reagent	PCTI: 2 mmol/L DTNB: 0.253 mmol/L Phosphate: 25 mmol/L Dibucaine: 0.03 mmol/L Fluoride: 4 mmol/L
Sample volume	10 μL
Fraction of sample volume	0.002 (dilution corrected)
Reaction time	10 min
Coefficient of variation	3%[4]
Linearity	0 through normal range

PCTI, Propionylthiocholine; *DNTB,* dithionitrobenzene.

PROCEDURE

Principle

1. Propionylthiocholine + $H_2O \xrightarrow{\text{EC 3.1.1.8}}$ Propionic acid + Thiocholine
2. Thiocholine + DTNB $\longrightarrow$ Oxidized thiocholine + 5-Thio-2-nitrobenzoic acid

Phosphate buffer, pH 7.6, is used and the rate of reaction followed at 410 nm, an absorption maximum of 5-thio-2-nitrobenzoic acid. Dibucaine and fluoride inhibitions (measured separately) are determined in addition to total activity when cholinesterase phenotyping is requested. Use glass containers only. Inhibitors may be extracted from some plastics.

Reagents

1. *Phosphate buffer, pH 7.6.* Make a solution of 0.033 M Na_2HPO_4 by dissolving 4.73 g in 1 liter of distilled H_2O (final volume). Make a 0.1 M solution of KH_2PO_4 by dissolving 13.61 g in a final volume of 1 liter. Mix 50 mL of KH_2PO_4 with 950 mL of Na_2HPO_4 to achieve a final pH of 7.6. Stable for 3 months at 4° to 8° C. Check weekly for growth.
2. *Propionylthiocholine iodide solution (PCTI), 20 mmol/L.* Dissolve 606 mg in 100 mL final volume of distilled H_2O. Stable for 1 day at ambient temperature.
3. *5,5′-Dithio-bis(2-nitrobenzoic acid), (DTNB), 0.421 mmol/L.* Weigh out 167 mg of DTNB and dissolve in a final volume of 1 liter of buffer. Stable for 12 months when stored in a brown glass bottle at 4° C.
4. *Dibucaine HCl (Nupercaine Hydrochloride, CIBA), 0.3 mol/L.* Weigh out 57 mg and dissolve in distilled water to make a final volume of 500 mL. Stable for 1 year at room temperature.
5. *Sodium fluoride solution, 40 mmol/L.* Weigh out 84 mg of NaF and dissolve in 50 mL final volume of distilled water. Stable for 1 day at room temperature.
6. *Quinidine sulfate solution, 14 mmol/L (5 g/L).* Weigh out 500 mg and dissolve in a final volume of 100 mL of distilled H_2O. Stable for 1 year at room temperature.
7. *Working substrate reagents.* Prepare 3 mL of each solution per test sample.
   a. Dilute PCTI with an equal volume of water.
   b. Dilute PCTI with an equal volume of dibucaine solution.
   c. Dilute PCTI with an equal volume of NaF solution.

Assay

This procedure measures the activity of cholinesterase and the effect of two inhibitors, dibucaine and fluoride.

Equipment: spectrophotometer or photometer set at 410 nm, with band pass ≤ 10 nm, and 37° C water bath.

Assay for total activity

1. Add 3 mL of DTNB buffer to a test tube and allow to equilibrate 5 minutes in a 37° C water bath.
2. Add 1 mL of substrate (reagent 7a).
3. Add 1 mL of 1:100 dilution of serum. (Samples with low activity should be repeated on a smaller dilution, such as ten- or fiftyfold). Dilute the serum shortly before the assay, since the cholinesterase may slowly lose activity after dilution.
4. After exactly 3 minutes add 1 mL of quinidine reagent to stop the reaction.
5. Prepare blank readings by reversing steps 3 and 4.
6. Read absorbance within 5 minutes against the corresponding blank.

Assay for dibucaine or fluoride inhibition

1. Add 3 mL of DTNB buffer to a test tube and allow to incubate 5 minutes in a 37° C water bath.
2. Add 1 mL of substrate-inhibitor (reagent 7b or 7c).
3. Follow steps 3 to 6 of the assay for total activity.

Calculations

The cholinesterase activity is expressed in international units per milliliter (U/mL = µmol/min/mL) at 37° C and is calculated from the following:

$$\text{U/mL} = \frac{x \cdot \Delta A_{\text{unknown}}}{y \cdot 13.6 \cdot z} = 14.71 \cdot \Delta A_{\text{unknown}}$$

where x is the total volume in milliliters at which the absorbance is read, y is the volume of sample used, $\Delta A_{\text{unknown}}$ is the increase in absorbance corrected for the blank, *13.6* is the millimolar absorptivity of the 5-thio-2-nitrobenzoate for a 1 cm cuvette, and z is the number of minutes of incubation.

Inhibition by dibucaine or fluoride is calculated as follows:

$$\text{Percent inhibition} = 100 - \left(\frac{\text{U/mL}_{\text{with inhibitor}} \times 100}{\text{U/mL}_{\text{without inhibitor}}}\right)$$

REFERENCE RANGES (Table 56-18)

The most common phenotype (U, "usual") is not associated with prolonged response to succinylcholine and is inhibited approximately 84% by dibucaine and 80% by fluoride.[1-4] Other phenotypes that can occur include dibucaine-resistant A ("atypical") and fluoride-resistant F ("fluoride"). The so-called silent phenotypes S_1 and S_2 display little or no cholinesterase activity in the homozygous state.

Table 56-18. Reference values for genetic variants of cholinesterase*†

Phenotype	Activity at 37° C, µmol/min/mL	Percent inhibition by	
		Dibucaine	Fluoride
U	8.44 ± 1.78	83.6 ± 1.3	79.7 ± 1.2
A‡	1.90 ± 0.61	19.9 ± 2.7	84.0 ± 1.8
AS‡	1.30 ± 0.37	20.7 ± 4.1	82.3 ± 3.4
S_1‡	0.03 ± 0.01	5.3 ± 4.3	35.7 ± 6.1
S_2‡	0.18 ± 0.05	67.6 ± 4.3	67.6 ± 1.7
F‡	3.57	71.8	53.6
AF‡	3.65 ± 0.47	60.2 ± 3.1	68.3 ± 1.6
FS‡	3.47	76.7	64.9
UA	5.84 ± 1.76	72.7 ± 3.1	80.0 ± 1.6
UF	5.99 ± 1.26	79.8 ± 1.2	73.0 ± 1.7
US	4.61 ± 0.57	84.4 ± 0.8	79.3 ± 1.4

*Derived from Dietz, A.A., Rubenstein, H.M., and Lubrano, T.: In Cooper, G.R., and King, J.S., editors: Selected methods for the small clinical chemistry laboratory, vol.8, Washington, D.C., 1977, American Association for Clinical Chemistry.
†Values are mean ± standard deviation.
‡Phenotypes associated with prolonged apnea in response to succinylcholine administration.

REFERENCES

1. Brown, S.S., Kalow, W., Pilz, W., et al.: The plasma cholinesterases: a new perspective, Adv. Clin. Chem. **22:**1-123, 1981.
2. Willis, J.H.: Blood cholinesterase: assay methods and considerations Lab. Management **20:**53-64, 1982.
3. Lehmann, H., and Liddell, J.: The cholinesterase variants. In Wyngaarden, J.B., and Fredrickson, D.S., editors: Stanbury, J.B., The metabolic basis of inherited disease, ed. 3, New York, 1972, McGraw-Hill Book Co., pp. 1730-1736.
4. Dietz, A.A., Rubenstein, H.M., and Lubrano, T.: Colorimetric determination of serum cholinesterase and its genetic variants by the propionylthiocholine-dithio-bis(nitrobenzoic acid) procedure. In Cooper, G.R., and King, J.S., editors: Selected methods for the small clinical chemistry laboratory, vol. 8, Washington, D.C., 1977, American Association for Clinical Chemistry. pp. 41-46.
5. Price, E.M., and Brown, S.S., Scope and limitations of propionylthio-cholinesterase in the characterization of cholinesterase variants, Clin. Biochem. **8:**384-390, 1975.
6. Caraway, W.T.: Photometric determination of serum cholinesterase activity Am. J. Clin. Pathol. **26:**945, 1956.
7. Crane, C.R., Sanders, D.C., and Abbot, J.N.: Cholinesterase use and interpretation of cholinesterase measurements. In Sunshine, I., editor: Methodology for analytical toxicology, Cleveland, Ohio, 1975, CRC Press Inc., pp. 86-104.
8. Michel, H.O.: An electrometric method for the determination of red blood cell and plasma cholinesterase activity, J. Lab. Clin. Med. **34:**1564-1568, 1949.
9. Witter, R.F., Grubbs, L.M., and Farrior, W.L.: A simplified version of the Michel method for plasma or red cell cholinesterase, Clin. Chim. Acta **13:**76-78, 1966.
10. Ellin, R.I., and Vicario, P.: A pH method for measuring blood cholinesterase, Arch. Environ. Health **30:**263-265, 1975.
11. Nabb, D.P., and Whitfield, F.: Determination of cholinesterase by an automated pH stat method, Arch. Environ. Health **15:**147-152, 1967.
12. Aldrich, F.D., Walker, G.F., and Patnoe, C.A.: A micromodification of the pH-stat assay for human blood cholinesterase, Arch. Environ. Health **19:**617-620, 1969.
13. Serat, W.F., Van Loon, A.J., and Lee, M.K.: Microsampling techniques for human blood cholinesterase analysis with the pH stat, Bull. Environ. Contam. Toxicol. **17:**542-550, 1977.

CH3 N NH HOOC NH2 Creatine + HO—P—O—P—O—P—O—CH2 (OH OH OH / O O O) NH2 N N N N O H H H H OH OH Adenosine triphosphate ⇌ CK

CH3 N =NH HOOC NH HO—P=O O H Creatine phosphate + OH OH HO—P—O—P—O—CH2 O O NH2 N N N N O H H H H OH OH Adenosine diphosphate

Creatine kinase

LENOX B. ABBOTT
JOHN A. LOTT

Creatine kinase (CK, CPK, adenosine triphosphate: creatine *N*-phosphotransferase)
Clinical significance: pp. 488, 509, and 953
EC 2.7.3.2
Molecular weight: approximately 80,000 daltons
Chemical class: enzyme, protein
Known isoenzymes: CK_1 (CK-BB), CK_2 (CK-MB), CK_3 (CK-MM), CKm (mitochondrial—rare)
Biochemical reaction: (See at right)

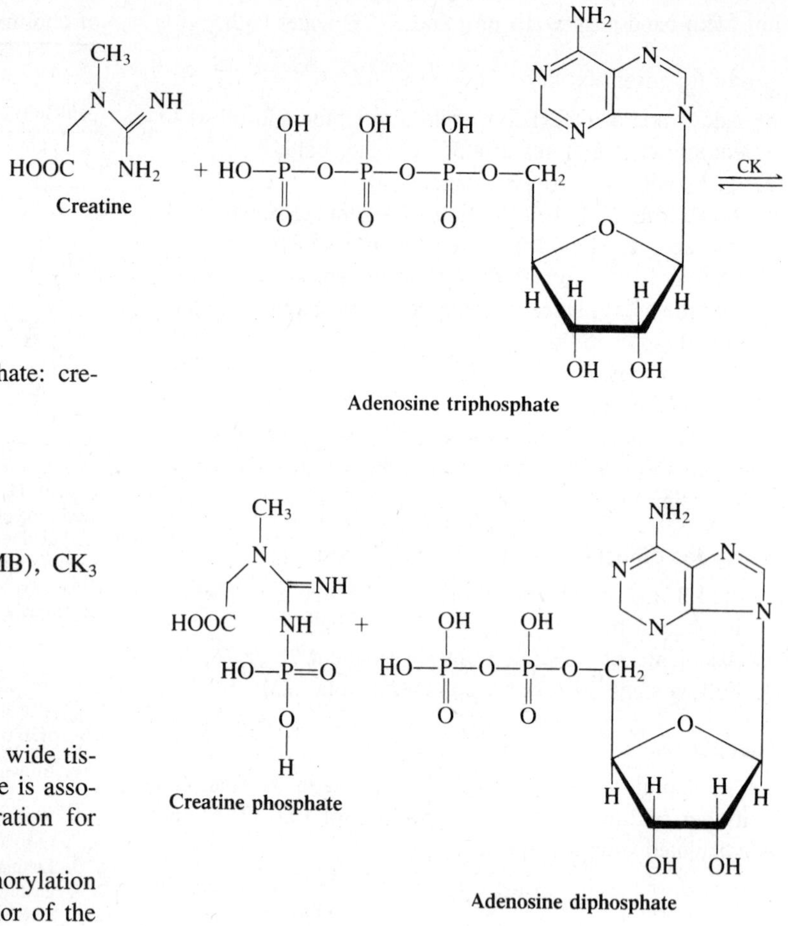

PRINCIPLES OF ANALYSIS AND CURRENT USAGE

Creatine kinase (CK) is a cellular enzyme with wide tissue distribution in the body. Its physiological role is associated with adenosine triphosphate (ATP) generation for contractile or transport systems.[1]

Creatine kinase catalyzes the reversible phosphorylation of creatine, a reaction in which ATP is the donor of the phosphate group. The reaction is as follows:

$$\text{Creatine + ATP} \underset{\text{pH 6.8}}{\overset{\text{pH 9.0}}{\underset{\text{CK}}{\rightleftharpoons}}} \text{Creatine phosphate + ADP}$$

(The "forward" reaction)

As indicated, the reaction is pH dependent, and at neutral pH the formation of ATP is favored. A pH of 9.0 is optimal for the formation of creatine phosphate, another high-energy compound.

Methods for serum creatine kinase determinations utilized either the products formed in the forward or reverse reaction. Creatine phosphate is more labile to acid hydrolysis than either ATP or ADP; thus under mild acid conditions, creatine phosphate is hydrolyzed to creatine and phosphate. The phosphate can be determined by the blue complex formed with molybdate (method 1, Table 56-19).

The ADP formed has been determined in a coupled reaction sequence described by Tanzer and Gilvarg.[2] The reaction sequence is outlined in method 2, Table 56-19. This assay is carried out at an alkaline pH so that both ATPase and alkaline phosphatase interfere, and a serum blank is required to correct for endogenous pyruvate. The Tanzer and Gilvarg method is still being used and has been automated; however its sensitivity is low compared to newer methods. In a further modification of this method, the pyruvate produced in the second reaction is reacted with 2,4-dinitrophenylhydrazine to form a colored hydrazone derivative (method 3, Table 56-19). Because it is an end-point method and subject to interferences by keto acids in serum, it is no longer used.

The reverse reaction proceeds six times faster than the forward reaction when each is carried out under optimum conditions. There are several methods using the reverse reaction. Hughes[3] complexed the creatine that is produced with diacetyl and α-naphthol to form a chromophore (method 4, Table 56-19). Because compounds other than creatine also react, the method is obsolete. Sax and Moore[4] reacted the creatine with ninhydrin in the presence of KOH to form a fluorophor, which can be measured fluorometrically (method 5, Table 56-19). Both these methods have been adapted to continuous flow analyzers[5-7] and require fixed incubation times, after which the fluorescent derivative of creatine is prepared.

Bioluminescence has also been applied to measure CK activity (method 6, Table 56-19). This approach couples the reverse reaction with the luciferin/luciferase reaction, as shown below:

$$\text{Creatine phosphate} + \text{ADP} \underset{Mg^{++}}{\overset{CK}{\rightleftharpoons}} \text{Creatine} + \text{ATP}$$

$$\text{ATP} + \text{Luciferin} + O_2 \underset{Mg^{++}}{\overset{\text{Luciferase}}{\rightleftharpoons}} \text{Adenosine monophosphate (AMP)} + \text{Oxyluciferin} + \text{Pyrophosphate} + CO_2 + \text{Light}$$

In the presence of ATP, luciferin, and oxygen, light is emitted. The intensity of the emitted light is proportional to the concentration of ATP and thus the CK activity. This assay is by far the most sensitive of any of the currently available CK methods. It is not widely used, since it requires expensive reagents and specialized instrumentation. The currently used methods for CK and the preferred method described here provide enough sensitivity, and so the extreme sensitivity of bioluminescence methods is not needed.

Oliver described a coupled reaction sequence (method 7, Table 56-19) for the determination of CK as follows[8]:

$$\text{Creatine phosphate} + \text{ADP} \underset{\text{pH } 6.8}{\overset{CK}{\rightleftharpoons}} \text{Creatine} + \text{ATP}$$

$$\text{ATP} + \text{Glucose} \overset{\text{Hexokinase}}{\rightleftharpoons} \text{Glucose-6-phosphate} + \text{ADP}$$

$$\text{Glucose-6-phosphate} + \text{NADP}^+ \overset{\text{G-6-PD}}{\rightleftharpoons} \text{6-Phosphogluconate} + \text{NADPH} + H^+$$

(Abbreviations: *NADP*, nicotinamide adenine dinucleotide phosphate; *G-6-PD*, glucose-6-phosphate dehydrogenase.)

Table 56-19. Creatine kinase (CK) methods

Method	Direction of reaction	Type of analysis	Principle	Usage	Comments
1. Kuby et al.[19]	Forward	End point, manual, modified for AutoAnalyzer	Acid hydrolysis of creatine phosphate to creatine and phosphate; phosphate then determined by Fiske-Subbarow method	All body fluids	Requires serum blank because of endogenous phosphate Interference from alkaline phosphatase Very laborious
2. Tanzer and Gilvarg[2]	Forward	Kinetic	Adenosine diphosphate (ADP) produced* Follow decrease in absorbance at 340 nm	All body fluids	Interference from ATPase and alkaline phosphatase Slow reaction Serum blank required to correct for endogenous pyruvate
3. Nuttal and Weddin[20]	Forward	End point	Pyruvate produced in method 2 reacts with 2,4-dinitrophenylhydrazine to form 2,4-dinitrophenylhydrazone, which is measured at 440 nm	All body fluids	Serum blank required to correct for endogenous pyruvate and other metabolites Not suited for clinical laboratory because of poor accuracy
4. Hughes[3]	Reverse	End point	Creatine produced complexes with diacetyl and α-naphthol by method of Voges and Proskauer Measured at 520 nm	All body fluids	Poor sensitivity Serum blank needed Reagent unstable; both diacetyl and α-naphthol must be prepared daily
5. Sax and Moore[4]	Reverse	Fluorometric	Creatine produced reacts with ninhydryin to produce a fluorophor	All body fluids	Needs serum blank Requires fixed incubation period
6. Witteveen et al.[21]	Reverse	Bioluminescence	ATP in reverse reaction is determined with luciferin-luciferase method	All body fluids	Highly sensitive Expensive reagents Needs specialized instrumentation
7. Oliver[8]; Rosalki[9]	Reverse	Kinetic	(See text)	All body fluids	No serum blank needed Preferred method

*ADP produced reacts as follows:

$$\text{ADP} + \text{Phospho}\textit{enol}\text{pyruvate} \overset{\text{Pyruvate kinase}}{\rightleftharpoons} \text{ATP} + \text{Pyruvate}$$

$$\text{Pyruvate} + \text{NADH} + H^+ \overset{LD}{\rightleftharpoons} \text{Lactate} + \text{NAD}^+$$

ATP, Adenosine triphosphate; *LD*, lactate dehydrogenase; *NADH*, reduced nicotinamide adenine dinucleotide (NAD^+).

The increase in absorbance at 340 nm is a measure of CK activity present in serum or other body fluids. This method was later modified by Rosalki to include a thiol compound in the reagent mixture to increase the CK activity by maintaining the sulfhydryl groups of CK in a reduced form.[9] The use of a thiol reducing agent in the reagent or serum permits storage of serum without irreversible loss of activity.

Many modifications of Rosalki's approach have appeared. Most have been designed to optimize the reaction conditions.[10-12] The conditions of the method described by Szasz et al. appear to be optimal,[10] and it has been further modified to include 10 μmol/L of diadenosine-5-pentaphosphate and 5 mmol/L adenosine monophosphate (AMP), which act as adenylate kinase inhibitors (see below).

REFERENCE AND PREFERRED METHOD

A preferred method should provide sufficient sensitivity and analytical precision to facilitate clinical decisions, and it should be easy to perform both manually and in automated procedures. The reverse reaction provides greater sensitivity, since it proceeds approximately six times faster than the forward reaction. The original method of Oliver[8] and Rosalki[9] appears to be the method of choice; the reagent conditions are suboptimal, but modifications have been described, and it appears that the conditions described here for the preferred method are optimal.

The variations described by Morin[12] and Szasz et al.[10] and the method commercially available from BMD (Boehringer-Mannheim Diagnostics, Indianapolis, Ind.) as "CK-NAC UV" (Table 56-20) are very similar. Since the recipe is complex, and most laboratories will prefer to buy rather than prepare the reagent, the BMD CK-NAC procedure is recommended here as the preferred method. It is sensitive, has adequate precision, and is relatively free from interference from adenylate kinase (AK).

The major differences among the commerically available "optimized" methods include the choice of buffer and source of G-6-PD, which also determines the type of pyridine nucleotide that must be used. Imidazole acetate appears to be the best buffer system, but Bis-Tris with a pK_a of 6.5 at 25° C has been shown to give the highest CK activity among all other buffers tested.[12] The choice of pyridine nucleotide depends on the source of G-6-PD. This enzyme is usually obtained from baker's yeast and from the bacterium *Leuconostoc mesenteroides*. The former requires NADP, whereas the latter can use either

Table 56-20. Final reaction conditions for creatine kinase*

Condition	Oliver[8]	Rosalki[9]	Morin[12]	Szasz[10]	CK-NAC†
Temperature	30°, 37° C	37° C	30°, 37° C	25°, 30°, 37° C	30°, 37° C
Buffer	50 Barbital	50 Tris	200 Bis-Tris	100 Imidazole	100 Imidazole
pH (at temperature)	8.6(?)	6.8 (25° C)	6.7 (30° C)	6.7 (30° C)	6.7 (30° C)
Final concentration of reagents					
Creatine phosphate	10	10	30	30	30
Adenosine diphosphate	1	1	2	2	2
Mg^{+2}	10	30	20	10	10
Salt of Mg^{+2}	Chloride	Chloride	Acetate	Acetate	Acetate
D-Glucose	20	20	15	20	20
Pyridine nucleotide	0.1 NADP	0.8 NADP	2 NAD	2 NADP	2 NADP
Hexokinase, U/L	—	600	3000	2500	2500
Glucose-6-phosphate dehydrogenase, U/L	—	300	2500	1500	1500
Thiol	None	5 Cysteine	20 Monothioglycerol	20 *N*-Acetyl cysteine	20 *N*-Acetyl cysteine
Adenylate kinase inhibitor	— AMP	10 AMP	25 Fluoride	5 AMP	5 AMP plus 10 μmol/L of diadenosine-5-pentaphosphate
Linearity to units per liter	—	300	—	1500 (37° C)	1500 (37° C)
Volume fraction	—	0.033	0.02	0.02	0.02
Precision (approximate % coefficient of variation)	—	3%	—	—	5%

*Concentrations in millimoles per liter in final reaction mixture.
†From Boehringer-Mannheim Diagnostics (Indianapolis, Indiana); the preferred method.

NADP or the less expensive NAD. The linearity of the system is limited by the activity of G-6-PD and the concentrations of pyridine nucleotide and ADP.

SPECIMEN

Blood should be collected by venipuncture into plain evacuated phlebotomy tubes or syringes and the serum promptly taken off the blood clot. Since serum CK is rapidly inactivated in vitro at room temperature, storage should be at 4° C if analysis is delayed beyond 4 hours. Long-term storage of samples at −20° C has resulted in minimal loss of activity over 2 months.[13]

Hemolyzed serum specimens with approximately 320 mg/L hemoglobin (trace hemolysis) do not appreciably change the measured CK activity.[14] Greater hemolysis produces an erroneously high CK activity because of the presence of adenylate kinase (AK, myokinase, EC 2.7.4.3), which is abundant in erythrocytes.

INTERFERENCES

The major interferent in CK assays is AK. Human erythrocytes contain high activities of AK and very little CK. AK catalyzes the direct conversion of ADP to ATP as follows:

$$2\ \text{ADP} \xrightleftharpoons{\text{AK}} \text{ATP} + \text{AMP}$$

The production of ATP leads to false increases in CK activity, since ATP is also a product of the reverse reactions used for CK assays.

Several AK inhibitors have been suggested: AMP, fluoride, and a combination of AMP and diadenosine-5-pentaphosphate. AMP is an effective inhibitor of AK; however, it also inhibits CK activity.[15] Sodium fluoride inhibits AK without having an effect on CK.[16] On storage, however, reagents with fluoride form insoluble magnesium fluoride.[17] The most effective inhibitor of AK appears to be a combination of 5 mmol/L of AMP and 10 μmol/L of diadenosine-5-pentaphosphate.[16,17]

PROCEDURE

Principle

Sample is added to CK-NAC reagent as listed in Table 56-20. Absorbance measurements at 340 nm are made over a 5-minute reaction period. The increase in absorbance is directly related to the CK activity per liter of sample.

Reagents

Purchase reagents from kit manufacturer.

Assay

Equipment: a spectrophotometer that can make measurements at 340 nm and is fitted with a thermostatting device to keep the reaction cuvette at 37 ± 0.1° C

1. Reconstitute the reagent as per the manufacturer's instructions. Invert the bottle until the dry reagent is completely dissolved.
2. Pipet 2.5 mL of the reconstituted reagent, warm to 37° C, and add 50 μL of serum or control into a cuvette or suitable container that can be incubated at 37° C.
3. Incubate the reagent-specimen mixture for 5 minutes. This allows the reaction rate to reach maximum velocity.
4. During the incubation time set the spectrophotometer to 340 nm and adjust it to zero absorbance with a water blank.
5. After the 5 minutes of incubation, place the reaction mixture into a cuvette with a 1 cm light path. Take absorbance readings immediately and at 1 minute intervals for 5 minutes.
6. If the change in absorbance (ΔA) per minute exceeds 0.180, dilute the specimen 1 + 1 with 154 mmol/L of NaCl. Repeat the assay on the diluted specimen and multiply the final result by 2.
7. Ensure that the control results are within the acceptable limits before reporting results of patient specimens.

Calculations

Average the five ΔA per minute values. If the ΔA values are decreasing or appear inconsistent, repeat the assay with a diluted specimen.

$$\text{CK activity: U/L} = \frac{\Delta A\ \text{(average)}}{\text{Minute}} \times \frac{10^6\ \mu\text{mol}}{\text{Mole}} \times \frac{1}{\text{Molar absorptivity}} \times \frac{\text{Total volume}}{\text{Sample volume}}$$

$$\text{U/L} = \frac{\Delta A\ \text{(average)}}{\text{Minute}} \times \frac{10^6}{6220} \times \frac{2.55}{0.05}$$

$$\text{U/L} = \frac{\Delta A\ \text{(average)}}{\text{Minute}} \times 8200$$

(The molar absorptivity of NADPH is 6220 $\text{L} \cdot \text{mol}^{-1} \cdot \text{cm}^{-1}$ at 340 nm.)

The method should be linear to 1500 U/L.

REFERENCE RANGE (SERUM)

Males:	up to 160 U/L
Females:	up to 130 U/L

Note that serum CK can be increased by exercise. Blacks tend to have higher values than whites, and large, well-muscled individuals have normal serum CK activities that are much greater than those of small, sedentary persons.

CREATINE KINASE CHANGES

Serum CK is increased in nearly all patients when there is injury, inflammation, or necrosis of skeletal or heart muscle. Decreased serum values have no meaning.[18] The following list includes some of the more important disease entities in which serum CK has been found to be abnormal.

1. Muscular dystrophy, especially Duchenne's dystrophy
2. Myocardial infarction
3. Polymyositis
4. Myopathies, except denervation myopathies
5. Trauma and after surgical trauma
6. Extreme physical exertion
7. Malignant hyperpyrexia
8. Hypothyroidism
9. Intramuscular injections
10. Myocarditis and pericarditis
11. Rhabdomyolysis
12. Drug overdose and poisonings
13. Coma
14. Certain infectious diseases that affect muscle

REFERENCES

1. Watts, D.C.: Creatine kinase (adenosine-5′-triphosphate—creatine phosphotransferase). In Boyer, P.D., editor: The enzymes, vol. 8, Part A, ed. 3, New York, 1973, Academic Press, Inc.
2. Tanzer, M.L., and Gilvarg, C.: Creatine and creatine kinase measurement, J. Biol. Chem. **234:**3201-3204, 1959.
3. Hughes, B.P.: A method for estimation of serum creatine kinase and its use in comparing creatine kinase and aldolase activity in normal and pathological sera, Clin. Chim. Acta **7:**597-603, 1962.
4. Sax, S.M., and Moore, J.S.: Fluorometric measurement of creatine kinase activity, Clin. Chem. **11:**951-958, 1965.
5. Fleischer, G.A.: Fluorometric measurement of creatine kinase activity, Clin. Chem. **11:**951-958, 1965.
6. Rokos, J.A.S., Rosalki, S.B., and Tarlow, D.: Automated fluorometric procedure for measurement of creatine phosphokinase activity, Clin. Chem. **18:**193-198, 1972.
7. Armstrong, J.B., Lowden, J.A., and Sherwin, A.L.: Automated fluorometric creatine kinase assay: measurement of 100-fold normal activity without serum dilution, Clin. Chem. **20:**560-565, 1974.
8. Oliver, I.T.: A spectrophotometric method for the determination of creatine phosphokinase and myokinase, Biochem. J. **61:**116-122, 1955.
9. Rosalki, S.B.: An improved procedure for serum creatine phosphokinase determination, J. Lab. Clin. Med. **69:**696-705, 1967.
10. Szasz, G., Gruber, W., and Bernt, E.: Creatine kinase in serum. 1. Determination of optimum reaction conditions, Clin. Chem. **22:**650-656, 1976.
11. Committee on Enzymes, Scandinavian Society for Clinical Chemistry and Clinical Physiology: Recommended method for the determination of creatine kinase in blood, Scand. J. Clin. Lab. Invest. **36:**711-723, 1976.
12. Morin, L.G.: Creatine kinase: re-examination of optimum reaction conditions, Clin. Chem. **23:**1569-1575, 1977.
13. Bowie, L.J., Griffiths, J.C., and Gochman, N. : The preferred method for creatine kinase. In Griffiths, J.C., editor: Clinical enzymology, New York, 1979, Masson Publishing USA, Inc.
14. Frank, J.J., Bermes, E.W., Bickel, M.J., and Watkins, B.F.: Effect of *in-vitro* hemolysis on chemical values for serum, Clin. Chem. **24:**1966-1070, 1978.
15. Szasz, G.: Laboratory measurement of creatine kinase activity. In Tietz, N.W., Weinstock, A., and Rodgerson, D.O., editors: Proceedings of the Second International Symposium on Clinical Enzymology, 1976, Washington, D.C., American Association for Clinical Chemistry, pp. 168-170.
16. Szasz, G., Gerhardt, W., and Gruber, W.: Creatine kinase in serum. 3. Further study of adenylate kinase inhibitors, Clin. Chem. **23:**1888-1892, 1977.
17. Rosano, T.G., Clayson, K.J., and Strandjord, P.E.: Evaluation of adenosine 5′-monophosphate and fluoride as adenylate kinase inhibitors in the creatine kinase assay, Clin. Chem. **22:**1078-1083, 1976.
18. Lott, J.A., and Stang, J.M.: Serum enzymes and isoenzymes in the diagnosis and differential diagnosis of myocardial ischemia and necrosis, Clin. Chem. **26:**1241-1250, 1980.
19. Kuby, S.A., Noda, L., and Lardy, H.A.: Adenosine triphosphate-creatine transphorylase J.Biol. Chem. **209:**191-201, 1954.
20. Nuttal, F., and Weddin, D.J.: A simple rapid colorimetric method for determination of creatine kinase activity, J. Lab. Clin. Med. **68:**324-332, 1966.
21. Witteven, S.A.G.J., Sobel, B.E., and DeLuca, M.: Kinetic properties of the isoenzymes of human creatine phosphokinase, Proc. Natl. Acad. Sci. **71:**1384-1387, 1974.

Creatine kinase isoenzymes

JOHN F. CHAPMAN
LINDA L. WOODARD
LAWRENCE M. SILVERMAN

Creatine kinase, CK, CPK, creatine phosphotransferase
Clinical significance: pp. 488 and 509
EC 2.7.3.2
Molecular weight: approximately 80,000 daltons
Chemical class: enzyme, protein
Known isoenzymes 4: CK_1 (CK-BB), CK_2 (CK-MB), CK_3 (CK-MM), CKm (mitochondrial—rare)
Biochemical reaction:

$$\text{Creatine} + \text{Adenosine triphosphate (ATP)} \underset{\text{pH } 6.8}{\overset{\text{EC 2.7.3.2}}{\underset{\text{pH } 9.0}{\rightleftharpoons}}} \text{Adenosine diphosphate (ADP)} + \text{Creatine phosphate}$$

PRINCIPLES OF ANALYSIS AND CURRENT USAGE

Creatine kinase (CK) is an enzyme present in several human tissues (skeletal muscle, myocardium, brain, prostate, uterus, and others). It catalyzes the reversible reaction shown. The molecule is a dimer composed of two nonidentical monomeric subunits, M and B. Three isoenzymes are routinely measured; designated as CK_1 (BB), CK_2 (MB), and CK_3 (MM). Methods for separation and quantitation of the CK isoenzymes include electrophoresis, ion-exchange chromatography, immunoinhibition, and radioimmunoassay (RIA).

One can perform electrophoretic separation by using agarose, cellulose acetate, or polyacrylamide as the support medium. Electrophoretic methods on agarose gels (method 1, Table 56-21) are currently the most frequently used technique for separation and quantitation of CK isoenzymes. The net charge on the M and B subunits differ, imparting a different total charge on each of the isoenzymes. At a pH of 8.6, the mobilities towards the anode are BB > MB > MM. At this pH, the MM remains at or near the point of application. Fig. 56-6 shows a CK separation in a 1% agarose gel at pH 8.6, employing the colorimetric procedure described on p. 1117. On agarose

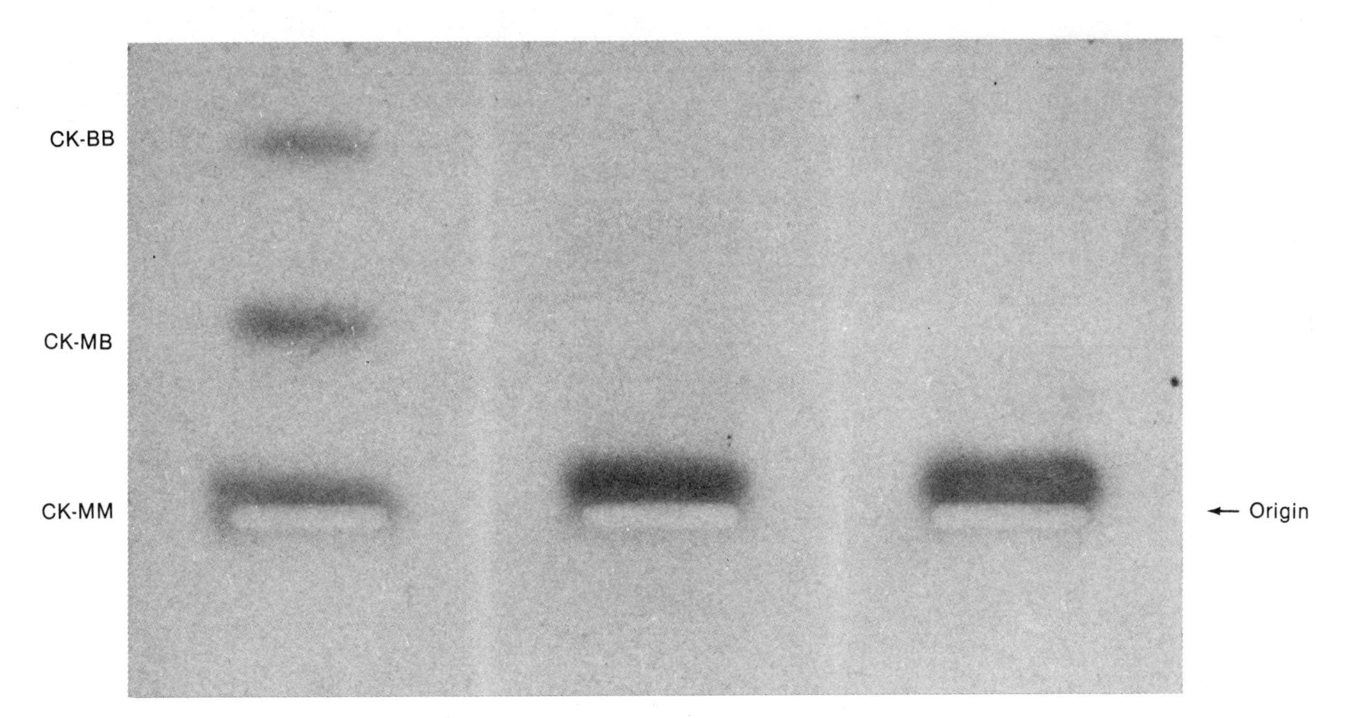

Fig. 56-6. Separation of creatine kinase isoenzymes on 1% agarose gel, with tetrazolium dye staining. Specimen on left is control; other two, showing only CK-MM activity, are from a patient.

gels, fluorometric determination of the relative amounts of each isoenzyme present is accomplished through incubation of the gels with substrate mixtures containing creatine phosphate, ADP, magnesium chloride, glucose, hexokinase, glucose-6-phosphate dehydrogenase (G-6-PD), and nicotinamide adenine dinucleotide phosphate (NADP). The following series of reactions occur in the presence of CK:

$$\text{Creatine phosphate} + \text{ADP} \xrightarrow[\text{MgCl}_2]{\text{CK}} \text{Creatine} + \text{ATP}$$

$$\text{ATP} + \text{Glucose} \xrightarrow{\text{Hexokinase}} \text{Glucose-6-phosphate} + \text{ADP}$$

$$\text{Glucose-6-phosphate} + \text{NADP} \xrightarrow{\text{G-6-PD}} \text{NADPH} + \text{6-Phosphogluconate}$$

The fluorescent NADPH produced in the agarose is considered to be directly proportional to the amount of CK present in the isoenzyme bands. Colorimetric determination of CK isoenzymes is also possible by the addition of phenazine methosulfate and nitroblue tetrazolium to the substrate, so that the following reaction can occur:

$$\text{NADPH} + \text{Phenazine methosulfate} + \text{Nitroblue tetrazolium} \rightarrow \text{Purple formazan}$$

Similar visualization of the isoenzymes is possible with other support media. Densitometric or fluorometric scanning of the resultant patterns yields relative percentages of the isoenzymes present.

Ion-exchange chromatography (method 2, Table 56-21) relies on separation of the isoenzymes through the use of minicolumns containing an anion-exchange material. These are either diethylaminoethyl-Sephadex A-50 or DEAE-cellulose.[1-3] The isoenzymes are absorbed onto the ion-exchange resin at pH approximately 6.7. At this pH the MM fraction is not charged and therefore is not absorbed and immediately elutes from the column. The MB and BB isoenzymes are eluted from the columns by stepwise increases of salt concentration in a Tris or imidazole-eluting buffer. The aliquots collected from the column may then be analyzed for CK activity by the usual procedures for total CK.

Immunoinhibition methods (method 3, Table 56-21) involve the use of specific antisera to selectively inhibit the B or, more frequently, the M subunits of CK; one then measures the remaining CK activity.[4] When anti-M antibodies are used, the residual B subunit activity is multiplied by 2 so that MB activity is calculated. This method assumes that there is no CK-BB activity present in the sample, unless a blank is employed.

Radioimmunoassay methods (method 4, Table 56-21) for CK isoenzymes have been reported[5,6] and may offer superior sensitivity in the diagnosis of myocardial infarction. Recently a radioimmunoassay (radiometric) specific for CK-MB has been developed,[7] and preliminary evaluation suggests that the high sensitivity and specificity of this

Table 56-21. Methods for creatinine kinase (CK) isoenzyme analysis

Method	Principle	Usage	Comments
1. Agarose gel electrophoresis	At pH 6.3, major isoenzymes have different mobilities in 1% agarose; CK activities detected by coupled enzymatic assay (fluorescent or visible)	Most frequently used technique	Separates all three major isoenzymes Rapid, relatively simple Very small sample volume Unusual isoenzyme forms such as macro CK easily detectable
2. Ion-exchange chromatography	At pH 6.7, CK-MM isoenzyme does not absorb onto column and elutes; CK-MB is eluted from column with buffer of different ionic strength	Used to some extent; most frequently used method for CK-MB quantitation	Reference method for quantitation of CK-MB activity Requires rigorous control of assay parameters, for accurate performance
3. Immunoinhibition	M subunits of CK-MM and CK-MB are inactivated by reaction with anti-M antibody; measure remaining B subunit activity	Recently introduced; used by a few laboratories	Can be subject to interference by CK-BB
4. Radioimmunoassay	Competitive binding assay for anti-B or anti-MB subunits	Rarely used	Clinical utility of this assay has not been proved

test may make it helpful in cases where the diagnosis of myocardial infarction is equivocal.[8]

REFERENCE AND PREFERRED METHODS

Electrophoretic methods are usually considered to be simple to perform, though considerable time and skill may be required for proper interpretation of unusual results. Some studies have indicated that electrophoresis is less sensitive than column chromatography,[9] though this conclusion appears to be variable, depending on which specific chromatographic and electrophoretic methods are compared. Although activity is measured indirectly, electrophoretic systems using agarose gels can consistently detect as little as 2 to 5 U/L of CK-MB in serum specimens. These methods also allow one to visualize the isoenzymes as discrete bands and to note unsatisfactory or unusual separations. In addition, CK-MM, CK-MB, and CK-BB are separated completely, provided that care is taken not to overload the system. Chromatographic methods offer no such absolute check on separation efficiency, and CK-MM activity can sometimes coelute with CK-MB or CK-BB with CK-MB, causing erroneous and potentially misleading results. The inability to ensure the separation characteristics of each column therefore represents a major disadvantage of this technique. This variability and the resulting imprecision has limited the use of the column technique. (See Table 56-22.)

Although the fluorescent visualization of the separated CK isoenzyme activities might be more sensitive than the colorimetric procedure, the latter technique has the advantage of providing a more permanent, easier-to-visualize record than does the fluorescent stain. In addition, the artifactual presence of CK-BB activity can occur as a result of the interaction of the fluorescent stain with proteins migrating in the CK-BB region of the electrophoretogram.

The newer immunoinhibition techniques are technically simple and can be performed rapidly. The methods also have the advantage of requiring no additional, specialized equipment beyond that used for total CK analysis. The major drawback of such methods is their lack of appropriate sensitivity and specificity.[9] The lack of specificity mainly arises from the interference from trace amounts of CK-BB activity in the specimens. The use of a ''blanking'' technique (assay sold by Roche Diagnostics, Inc. Nutley, N.J.) does much to overcome this drawback.

Radioimmunoassays are expensive, time consuming, and in some cases less specific than electrophoresis. It should be noted that most of the currently available immunoinhibition and radioimmunoassay methods fail to distinguish between CK-MB and CK-BB, a rather serious deficiency because elevations of the latter isoenzyme in the serum of seriously ill patients are frequently encountered.

There is no generally accepted reference method for CK isoenzyme fractionation. The Mercer ion-exchange column procedure under rigorously controlled conditions may be accepted as a reference procedure for CK-MB quantitation. However, this is not widely accepted as a routine laboratory procedure. For ease of use and the clinical usefulness of the resulting information, the electrophoretic procedures in agarose gels are the recommended, preferred methods.

Table 56-22. Comparison of conditions of several methods for creatine kinase (CK) isoenzyme analysis

Condition	Agarose gel electrophoresis	Ion-exchange chromatography	Immunoinhibition*
Temperature (for separation)	Ambient	Ambient	Ambient
Sample volume	1-2μL	500 μL	200 μL each for sample and blank
Assay time	20 min (separation) 30 min (analysis)	30 min (separation) 5 min (quantitation)	25 min (separation) 5 min (quantitation)

*Roche Diagnostics, Nutley, N.J.

SPECIMEN

Serum is the preferred specimen because CK activity is easily inhibited by EDTA, citrate, and fluoride. Because of the instability of CK at room temperature, all specimens should be separated from the cells as rapidly as possible and stored at 4° C for no longer than 24 hours. Specimens that are not to be assayed within 24 hours should be frozen at −20° C. As a result of the presence of adenylate kinase in red blood cells and its possible interference with electrophoretic methods, hemolyzed specimens are less than satisfactory. However, this interference may be reduced by the addition of an adenylate kinase inhibitor such as adenosine monophosphate (AMP), fluoride, or diadenosine-5-pentaphosphate[9,10] to the substrate reaction mixture.

PROCEDURE

An agarose electrophoresis method is described here[11] and is commercially available from Corning Medical and Scientific Inc. (Palo Alto, Calif.). The gels are 1% agarose.

Principle

$$\text{Creatine phosphate} + \text{ADP} \xrightarrow[\text{MgCl}_2]{\text{EC 2.7.3.2}} \text{Creatine} + \text{ATP}$$

$$\text{ATP} + \text{Glucose} \xrightarrow{\text{Hexokinase}} \text{Glucose-6-phosphate} + \text{ADP}$$

$$\text{Glucose-6-phosphate} + \text{NADP} \xrightarrow{\text{G-6-PD}} \text{NADPH} + \text{6-Phosphogluconate}$$

Reaction monitored by following production of NADPH

Reagents

MOPSO [3-(*N*-morpholino)-2-hydroxypropane sulfonic acid] buffer. Dissolve 22.18 g of sodium MOPSO and 2.32 g of MOPSO in 2 L of deionized water. The reconstituted buffer is stable for 6 months at room temperature.

Working CK substrate. Dry substrate includes the following ingredients at the final concentration in the reagent shown:

Creatine phosphate	189 mmol/L
Adenosine diphosphate	6 mmol/L
Mg^{+2}	33 mmol/L
Adenosine monophosphate	18 mmol/L
NADP	6 mmol/L
Glucose	63 mmol/L
Glutathione	84 mmol/L
Hexokinase	9×10^3 U/L
G-6-PD	3×10^3 U/L

Approximately 180 mg of the dry substrate should be mixed with 1 mL of MES [2-(*N*-morpholino)ethane sulfonic acid] buffer (pH 6.2). Working substrate should be prepared 20 minutes before completion of electrophoresis and is stable for 4 hours at 25° C and for 8 hours at 4° C.

The alternative colorimetric reagent can be obtained from P.L. Biochemicals, Inc. (Milwaukee, Wis.).

Marker. Add approximately 25 mg of bromphenol blue to 25 mL of MOPSO buffer and swirl to dissolve. The dye is stable at room temperature for 6 months.

Assay

Equipment: electrophoresis apparatus, incubator, drying oven, and densitometer capable of measuring absorbance or fluorescence.

1. Turn on the incubator (39° C) and drying oven (65° C), allowing a warm-up time of 30 minutes.
2. Prepare serum samples by adding 1.5 μL of bromphenol blue marker to 50 μL of serum. Mix thoroughly. Follow manufacturer's instructions for preparation of the electrophoresis system and agarose gels and for performing the electrophoresis procedure on the serum-dye mixtures prepared.
3. Allow 20 minutes for electrophoresis at 90 volts. Timing is critical.
4. After electrophoresis, remove the cell cover from the base and drain excess buffer from the cover without inverting it. Grasp the film by its edges and remove it from the cover.
5. Place the film, agarose side up, on a flat surface, having the cathode (−) edge toward you.
6. Place a 2 mL serological pipet lengthwise along the cathode edge of the agarose. Evenly dispense the working substrate onto the agarose along the edge of the pipet.
7. Slowly and smoothly push the pipet toward the anode edge of the film. Without lifting the pipet, pull it slowly back toward the cathode edge. This should uniformly coat the surface of the agarose with substrate.

8. Place the film, agarose side up, in the prewarmed incubator tray, on top of moistened filter paper. Return the tray to the incubator (39° C) for 20 minutes.
9. Place the film on a drying shelf in the oven. Dry at 65° C for 20 minutes.
10. Inspect the dry film under ultraviolet light (365 nm). Sites of CK activity will appear as bands of blue-white fluorescence. CK_3 (CK-MM) is at the point of application, CK_1 (CK-BB) is the most anodic fraction, and CK_2 (CK-MB) migrates in the α-2 region, between the other two isoenzymes. The bromphenol blue binds with albumin and may exhibit a reddish fluorescence slightly to the cathodic side of BB. Compare mobilities of unknown with those of the control.
11. Scan the gel patterns that show the presence of CK-MB or CK-BB, using appropriate fluorometric densitometer equipment.

Notes

1. Samples with total CK activity greater than 1000 U/L should be rerun after dilution with deionized water to bring the total CK activity into a range of 800 to 1000 U/L, if CK-MB is present.
2. Atypically migrating isoenzymes can occur, and their presence should be noted. Adenylate kinase migrates just slightly to the cathodic side of CK-MM and is seen frequently in hemolyzed specimens.[9] Two forms of macromolecular CK have been described. Macro-CK, type 1, a complex of CK-BB and IgG, migrates between MM and MB. Macro-CK, type 2, migrates to the cathodic side of MM and may be a polymerization product of mitochondrial CK.[12-14]

REFERENCE INTERVALS

	MM	MB	BB
Adult—1 month	96% to 100%	<4%	0%
		(or <5 U/L of MB)	

Varying levels of CK-BB have been demonstrated in neonatal serum samples. Reference intervals have not been established. In general, however, the CK-BB should disappear within the first month of life.

REFERENCES

1. Mercer, D.W.: Separation of tissue and serum creatine kinase isoenzymes by ion-exchange column chromatography, Clin. Chem. **20**:36, 1974.
2. Nealon, D.A., and Henderson, A.R.: Separation of creatine kinase isoenzymes by ion-exchange column chromatography (Mercer's method, modified to increase sensitivity), Clin. Chem. **21**:393, 1975.
3. Mercer, D.W., and Varat, M.A.: Detection of cardiac specific creatine kinase isoenzyme in sera with normal or slightly increased total creatine kinase activity, Clin. Chem. **21**:1088, 1975.
4. Wicks, R., Usategui-Gomez, M., Miller, M., and Warshaw; M.: Immunochemical determination of CK-MB isoenzyme in human serum. II. An enzymic approach, Clin. Chem. **28**:54-58, 1982.
5. Roberts, R., Sobel, B.E., and Parker, C.W.: Radioimmunoassay for creatine kinase isoenzymes, Science **194**:855-857, 1976.
6. Zweig, M.H., Van Steirteghem, A.C., and Schechter, A.N.: Radioimmunoassay for creatine kinase isoenzyme in human serum: isoenzyme BB, Clin. Chem. **24**:422-428, 1978.
7. Usategui-Gomez, M., Wicks, R.W., Farrenkopf, B., et al.: Immunochemical determination of CK-MB isoenzyme in human serum: a radiometric approach, Clin. Chem. **27**:823-827, 1981.
8. Kwong, T.C., Rothbard, R.L., and Biddle, T.L.: Clinical evaluation of a radiometric assay specific for creatine kinase isoenzyme MB, Clin. Chem. **27**:828-831, 1981.
9. Galen, R.S.: Isoenzymes: which method to use, Diag. Med. **1**:42-63, 1978.
10. Welch, S.L., and Swanson, J.R.: Adenylate kinase interference in creatine kinase isoenzyme electrophoresis, Clin. Chem. **27**:1026, 1981. (Abstract.)
11. Desjarlais, F., Morin, L.G., and Daigneault, R.: In search of optimum conditions for the measurement of creatine kinase activity: a critical review of nineteen formulations, Clin. Biochem. **13**:116-121, 1980.
12. Roe, C.R., Limbird, L.E., Wagner; G.S., and Nerenberg, S.T.: Combined isoenzyme analysis in the diagnosis of myocardial injury: application of electrophoretic methods for the detection and quantitation of the creatine phosphokinase MB isoenzyme, J. Lab. Clin. Med. **80**:4, 1972.
13. Wu, A.H.B., and Bowers, G.N., Jr.: Clinical correlation of macromolecular CK, type 2, Clin. Chem. **28**:1565, 1982.
14. Stein, W., Bohner, J., Steinhart, R., and Eggstein, M.: Macro creatine kinase: determination of two types by their activation energies, Clin. Chem. **28**:19, 1982.

Gamma-glutamyl transferase

STEPHEN GENDLER

Gamma-glutamyl transferase, GGT, γ-glutamyl transpeptidase, GGTP
Clinical significance: pp. 420 and 594
EC 2.3.2.2
Molecular weight: 90,000 daltons
Chemical class: enzyme, protein
Known isoenzymes: multiple forms are known, but their identity is as yet in question
Biochemical reaction:

$$\gamma\text{-Glutamyl cysteinylglycine (glutathione)} + \text{Amino acid} \xrightleftharpoons{\text{EC 2.3.2.2}} \gamma\text{-Glutamyl amino acid} + \text{Cysteinylglycine}$$

PRINCIPLES OF ANALYSIS AND CURRENT USAGE

γ-Glutamyl transferase (GGT) catalyzes the transfer of a γ-glutamyl group to amino acids and peptides (external transpeptidation), to another substrate molecule (internal transpeptidation), and to water (hydrolysis). It is an enzyme present primarily in the kidney, pancreas, liver, and prostate.[1] GGT in serum has a molecular weight of 90,000 daltons, as measured by electrophoresis.[2] This molecular weight is similar to papain-solubilized liver GGT. Serum GGT shows an electrophoretic mobility and lectin affinity reaction identical to the liver enzyme and different from GGT when kidney, urine, and pancreas are the sources.

Table 56-23. Methods for γ-glutamyl transferase (GGT) determination

Method	Type of analysis	Principle	Usage	Comments
1. Szasz[7]	Quantitative, kinetic	(γ-L-Glutamyl)-4-nitroanilide → 2-Nitroaniline (405 nm)	Most frequently used	Selected method of American Association for Clinical Chemistry Donor-substrate at supersaturating concentration Poor reagent stability
2. Scandinavian Society for Clinical Chemistry (SSCC),[9] modified Szasz	Quantitative, kinetic	As for method 1	—	Similar to method 1; different pH, buffer, and substrate concentrations
3. Substrate, modified SSCC[17]	Quantitative, kinetic	(γ-L-Glutamyl)-3-carboxy-4-nitroanilide → 2-Nitro-5-aminobenzoic acid (410 nm)	Not currently in use	Recommended method because of more soluble substrate donor
4. Fluorometric[14,15]	Quantitative	—	Not in current use	Most rapid, sensitive technique; requires specialized instrumentation

However, an antibody raised against liver GGT has the same reactivity against liver GGT as it does against kidney and pancreatic GGT. The data are consistent with (1) the isoenzymes having similar structures but differing primarily in their carbohydrate content, similar to the alkaline phosphatase isoenzymes, and (2) serum GGT originating from the liver.[3] Larger forms of GGT are reported with molecular weights greater than 200,000 daltons. This varies according to the detergent used for solubilization and molecular weight determination, since the enzyme is known to aggregate with lipoproteins and membrane fragments. Given the confusion in GGT characterization, it is prudent to wait for more evidence before one draws a conclusion.[4]

Glutathione was used as the donor substrate in the earlier methods of GGT analysis. Detection of substrate disappearance or product formation used manometric, chromatographic, or ultraviolet techniques. The first synthetic substrates were α-(*N*-γ-DL-glutamyl) aminoproprionitrile and (*N*-γ-DL-glutamyl) aniline. Also, (γ-L-glutamyl)-α-naphthylamide and (γ-L-glutamyl)-β-naphthylamide were substrates used in the 1960s. Many procedural steps, long incubations, drug interferences, and the carcinogenicity of these chemicals have caused these methods to be discarded.[1]

Modern methods use either (γ-L-glutamyl)-4-nitroanilide or (γ-L-glutamyl)-3-carboxy-4-nitroanilide because these substrates allow a direct reaction rate measurement without deproteinization or any chemical treatment of the cleavage products (methods 1 and 3, respectively, Table 56-23). Orlowski and Meister introduced the use of (γ-L-glutamyl)-4-nitroanilide in 1963.[5] Szasz used this substrate (4.0 mmol/L) in a kinetic method,[6] which was published as a "selected method" by the American Association for Clinical Chemistry.[7] Szasz preferred 100 mmol/L of 2-amino-2-methyl-propane-1,3-diol (ammediol) buffer, pH 8.6, because enzyme activity is not affected over a broad range of buffer concentration and the donor substrate has a higher solubility in this buffer.[6] The glutamyl group acceptor in this selected method is glycylglycine, 50 mmol/L. Although other substances, such as methionine, glutamine, and cystine, can act as acceptor substrates, glycylglycine is the acceptor of choice.[8]

The temperature used in the selected method is 25° or 30° C, with the reaction followed at 405 nm, where the hydrolysis product, 4-nitroaniline, absorbs but the substrate does not.[6] The day-to-day imprecision of this method in the manual mode measured as the coefficient of variation (CV) is 5.5% at a level 23.6 U/L for 22 days.

A modification of the Szasz procedure was also published by the Committee on Enzymes of the Scandinavian Society for Clinical Chemistry and Clinical Physiology (method 2, Table 56-23).[9] The (γ-L-glutamyl)-4-nitroanilide concentration for this method is 4 mmol/L. The buffer used is Tris-HCl, 100 mmol/L, pH 7.6. $MgCl_2$ is added to a concentration of 10 mmol/L to increase the solubility of the substrate. Glycylglycine is used at 75 mmol/L, with the reaction carried out at 37° C.

Fluorometric methods for GGT analysis have also been reported[14,15] (method 4, Table 56-23). Imprecision is similar to spectrophotometric GGT analyses, but sensitivity is 20 times greater. This allows smaller sample volumes and

Table 56-24. Comparison of methods for GGT determination

Reaction condition	AACC-Szasz*	SSCC-Szasz†	Recommended method‡
Temperature	25° or 30° C	37° C	30° C
Final concentration of reagents			
Substrate donor	(γ-L-Glutamyl)-4-nitroanilide: 4 mmol/L	(γ-L-Glutamyl)-4-nitroanilide: 4.0 mmol/L	(γ-Glutamyl)-5-carboxy-4-nitroanilide: 8.4 mmol/L
Substrate acceptor	Glycylglycine: 50 mmol/L	Glycylglycine: 75 mmol/L	Glycylglycine: 190 mmol/L
Buffer	2-Amino-2-methyl-propane-1,3-diol: 100 mmol/L	Tris-HCl: 100 mmol/L	Tris-HCl: 100 mmol/L
pH	8.6	7.6	8.16
Other		$MgCl_2$: 10 mmol/L	
Fraction of sample volume	0.0909 (1:11)	0.0909	0.0909
Wavelength	405 nm	405 nm	410 nm
Interferences	Gross hemolysis	Gross hemolysis	Gross hemolysis
Precision (at upper limit of normal) (% CV)	5.5%	—	—
Reference range			
Men	5-30 U/L	—	—
Women	3-20 U/L	—	—

*Selected method, American Association for Clinical Chemistry.[7]
†Committee on Enzymes of the Scandinavian Society for Clinical Chemistry.[9]
‡This text and reference 17.

shorter incubation times. These substrates have not come into common usage, probably because of the equipment requirements.

Table 56-24 compares the various methods for GGT determination.

REFERENCE AND PREFERRED METHODS

The two proposed methods do not solve the major problems encountered with GGT analysis. First, some investigators report an impurity when the (γ-L-glutamyl)-4-nitroanilide is synthesized by certain manufacturers.[10] This requires a pyridine extraction to remove the impurity, which acts as a competitive inhibitor.[11] Second, the low solubility of (γ-L-glutamyl)-4-nitroanilide requires a supersaturated solution to achieve a maximum reaction rate. The stability of this supersaturated solution is only a few hours.[9] Third, optimization of the assay is difficult because of the inhibitory effect of excess donor or acceptor substrates. This double inhibition is caused by the kinetic mechanism of GGT.[12] Donor substrate inhibition with both 4-nitroanilide compounds is pH dependent: maximal at acid pH, slight at pH 8.5, and absent at more alkaline pH.[13] Thus assay conditions must strike a balance that takes into account maximal substrate inhibition, which is both pH and substrate concentration dependent, and maximum reaction rate, which is also both pH and substrate concentration dependent.

The selected methods (methods 1 and 2, Table 56-23) do not use the currently preferred substrate, (γ-L-glutamyl)-3-carboxy-4-nitroanilide. This chemical is a preferred substrate since it is much more soluble than (γ-L-glutamyl)-4-nitroanilide. Spontaneous hydrolysis of the carboxy derivative is not detectable. Further, kinetic constants and serum GGT activities with the new substrate are comparable to the old substrate. However, controls supplemented with beef kidney GGT are on the average 19% lower with the carboxylated substrate. Thus the (γ-L-glutamyl)-4-nitroanilide is more frequently used, since higher GGT activation and greater sensitivities are achieved with this substrate.

Several studies have been done to optimize the GGT reaction by use of the carboxylated substrate.[13,16,17] Given the complicated kinetics of the enzyme, it is not surprising that several "optimized" methods have been proposed. The inhibition of GGT by donor substrate is present with (γ-L-glutamyl)-3-carboxy-4-nitroanilide. The extent of the reported inhibition varies with the investigator and the instrument used. Significant inhibition by the carboxylated donor substrate was reported as low as 3.5 mmol/L,[18] but most data show no inhibition until a concentration of 10 to 12 mmol/L is achieved.[13]

None of the supposed optimal assays have been thoroughly evaluated; however, equations that model the GGT assay have been developed. Equations derived using kinetic theory[16] and equations derived using a computerized optimization scheme (called "response surface methodology") have been shown to accurately model the enzyme assay over a broad range of GGT concentrations. Response surface methodology has shown that maximal GGT activity is achieved within a range of assay conditions: pH 7.8 to 8.5; glycylglycine, 129 to 250 mmol/L; and (γ-L-glutamyl)-3-carboxy-4-nitroanilide, 6.6 to 10.2 mmol/L; given a serum-to-reagent ratio of 1:11, 100 mmol/L Tris-HCl buffer, at a temperature of 30° C. None of the proposed assays using the carboxy derivative has conditions that fall in these ranges. Response surface methodology has shown that maximal activity is achieved using the given buffer, temperature, and fraction of sample volume with a pH of 8.16, a glycylglycine concentration of 190 mmol/L, and a substrate concentration of 8.4 mmol/L. These conditions are proposed in the recommended method that follows.[17]

The absorbance of the hydrolysis product, 2-nitro-5-aminobenzoic acid, is followed at 410 nm. This is a slightly longer wavelength than is used for the noncarboxylated substrate, since the carboxy derivative has a higher absorbance than the noncarboxylated substrate.[18] The longer wavelength reduces the blank absorbance.

SPECIMEN

Serum or lithium heparinized plasma is the recommended specimen in the procedure for GGT analysis.[7,9] These specimens give the same results. EDTA and citrate do not interfere with analysis, and hemolysis and prolonged contact with cells do not influence GGT values, according to one review.[19] Some investigators have shown fluoride, oxalate, and citrate, at a concentration of 1 g/L, to inhibit GGT by approximately 15%.[20]

Serum GGT values are stable at room temperature or 4° C for at least 7 days and are stable for at least 2 months when frozen.[19]

PROCEDURE

Principle

$$\text{(γ-L-Glutamyl)-3-carboxy-4-nitroanilide} + \text{Glycylglycine} \xrightleftharpoons{\text{EC 2.3.2.2}} \text{(γ-L-Glutamyl) glycylglycide} + \text{2-Nitro-5-aminobenzoic acid}$$

or

$$\text{(γ-L-Glutamyl)-3-carboxy-4-nitroanilide} + H_2O \xrightleftharpoons{\text{EC 2.3.2.2}} \text{Glutamic acid} + \text{2-Nitro-5-aminobenzoic acid}$$

This reaction is monitored at 410 nm.

Reagents

Buffer/substrate solution. Tris (100 mmol/L), 12.11 g; glycylglycine (190 mmol/L), 25.10 g; (γ-L-glutamyl)-3-carboxy-4-nitroanilide (8.4 mmol/L), 2.91 g are added to about 800 mL of water. Adjust pH to 8.16 with 1 mol/L of HCl and dilute to 1000 mL. This solution is stable 2 days at room temperature or 5 days at 2° to 8° C.

Assay

Equipment: spectrophotometer (≤10 nm band pass) equipped with cuvette heated at 30° C and 410 nm filter.

1. Place 2 mL of buffer/substrate solution and 0.20 mL of serum or plasma in a 3 mL cuvette.
2. After 1 minute of preincubation at 30° C, follow absorbance at 410 nm for 5 minutes.
3. The serum sample volume fraction is 0.091 (1:11).

Calculations

Using the absorptivity, ϵ, of 2-nitro-5-aminobenzoic acid as $9500 \frac{L \cdot cm}{mole}$, the formula for calculation of enzyme activity for a 1 cm path length is as follows:

$$U/L = \frac{\Delta A \text{ for } t \text{ min}}{t \text{ min}} \times \frac{\text{Total volume}}{\text{Sample volume}} \times \frac{1}{\epsilon} \times \frac{10^6 \ \mu mol}{mol}$$

where ΔA = difference between the initial and final absorbance over analysis time, t, in minutes
ϵ = molar absorptivity of 2-nitro-5-aminobenzoic acid
10^6 = factor to convert concentration to μmoles/L

For the assay, the values are:

$$U/L = \frac{\Delta A}{5 \text{ min}} \times \frac{2.20 \text{ mL}}{0.20 \text{ mL}} \times \frac{1}{9500 \frac{L \cdot cm}{mole}} \times 10^6 = \Delta A \times 231.6$$

Interferences

Gross hemolysis.

REFERENCE RANGE

Women, 5 to 24 U/L; men, 8 to 37 U/L

REFERENCES

1. Rosalki, S.B.: Gamma-glutamyl transferase, Adv. Clin. Chem. **17:**53-107, 1975.
2. Tsuji, A., Matsuda, Y., and Katunuma, N.: Characterization of human serum γ-glutamyltransferase, Clin. Chim. Acta **104:**361-366, 1980.
3. Huseby, N.-E.: Separation and characterization of human γ-glutamyltransferases, Clin. Chim. Acta **111:**39-45, 1981.
4. Henderson, A.R.: Clinical enzymology, Annu. Rev. Clin. Biochem. **1:**233-270, 1980.
5. Orlowski, M., and Meister, A.: γ-Glutamyl-*p*-nitroanilide: a new convenient substrate for determination and study of L- and D-γ-glutamyl transpeptidase activities, Biochim. Biophys. Acta **73:**679-681, 1963.
6. Szasz, G.: A kinetic photometric method for serum γ-glutamyl transpeptidase, Clin. Chem. **15:**124-136, 1969.
7. Szasz, G.: Reaction-rate method for γ-glutamyltransferase activity in serum, Clin. Chem. **22:**2051-2055, 1976.
8. Shaw, L.M., London, J.W., and Petersen, L.E.: Isolation of γ-glutamyltransferase from human liver, and comparison with the enzyme from human kidney, Clin. Chem. **24:**905-915, 1978.
9. Stromme, J.H., Chairman of The Committee on Enzymes of the Scandinavian Society for Clinical Chemistry and Clinical Physiology: Recommended method for the determination of γ-glutamyltransferase in blood, Scand. J. Clin. Lab. Invest. **36:**119-125, 1976.

10. Rowe, J.A., Tarlow, D., and Rosalki, S.B.: Impurity in γ-glutamyl-*p*-nitroanilide used as a substrate for D-glutamyl transferase, Clin. Chem. **19**:435-436, 1973 (Letter).
11. Huseby, N.E., and Stromme, J.H.: Practical points regarding routine determination of γ-glutamyl transferase (γ-GT) to serum with a kinetic method at 37° C, Scand. J. Clin. Lab. Invest. **34**:357-363, 1974.
12. Stromme, J.H., and Theodorsen, L.: γ-Glutamyl-transferase: substrate inhibition, kinetic mechanism, and assay conditions, Clin. Chem. **22**:417-421, 1976.
13. Schiele, F., Artur, Y., Bagrel, D., et al.: Measurement of plasma gamma glutamyltransferase in clinical chemistry: kinetic basis and standardization propositions, Clin. Chim. Acta **112**:187-195, 1981.
14. MacQueen, J.D., Driscoll, R.C., and Gargiulo, R.J.: New substrate for fluorometric determination of γ-glutamyltransferase activity in serum, Clin. Chem. **23**:879-881, 1977.
15. Prusak, E., Siewinski, M., and Szewczuk, A.: A new fluorometric method for the determination of γ-glutamyltransferase activity in blood serum, Clin. Chim. Acta **107**:21-26, 1980.
16. Shaw, L.M., London, J.W., Fetterolf, D., and Garfinkel, D.: γ-Glutamyltransferase: kinetic properties and assay conditions when γ-glutamyl-4-nitroanilide and its 3-carboxy derivative are used as donor substrates, Clin. Chem. **23**:79-85, 1977.
17. London, J.W., Shaw, L.M., Theodorsen, L., and Stromme, J.H.: Application of response surface methodology to the assay of gamma-glutamyltransferase, Clin. Chem. **28**:1140-1143, 1982.
18. Theodorsen, L., and Stromme, J.H., γ-Glutamyl-3-carboxy-4-nitroanilide: the substrate of choice for routine determinations of γ-glutamyl-transferase activity in serum? Clin. Chim. Acta **72**:205-210, 1976.
19. Ladenson, J.H.: Nonanalytical sources of variation in clinical chemistry results. In Sonnenwirth, A.C., and Jarett, L., editors: Gradwohl's clinical laboratory methods and diagnosis, vol. 1, ed. 8, St. Louis, 1980, The C.V. Mosby Co.

Lactate dehydrogenase

AMADEO J. PESCE

Lactate dehydrogenase, LD, LDH, L-lactate:NAD oxidoreductase
Clinical significance: p. 488
EC 1.1.1.27
Molecular weight: 140,000 daltons
Chemical class: enzyme, protein
Known isoenzymes: LD_1, LD_2, LD_3, LD_4, LD_5
Biochemical reaction:

$$\underset{\text{Pyruvate}}{\begin{array}{c}CH_3\\ |\\ C{=}O\\ |\\ COOH\end{array}} + H^+ + NADH \xrightleftharpoons{\text{EC 1.1.1.27}} \underset{\text{L-Lactate}}{\begin{array}{c}CH_3\\ |\\ CHOH\\ |\\ COOH\end{array}} + NAD^+$$

PRINCIPLES OF ANALYSIS AND CURRENT USAGE

Lactate dehydrogenase (LD) is an enzyme present in all cells of the body, catalyzing the reaction shown. The equilibrium of the reaction is pH dependent, with alkaline pH favoring the conversion of lactate to pyruvate and neutral pH favoring the reverse reaction.

Most current methods for LD quantitation use spectrophotometers to measure the interconversion of the coenzyme NAD (nicotinamide adenine dinucleotide) and NADH (reduced NAD) at 340 nm[1-4] (methods 1 and 2, Table 56-25). In contrast, most older reactions used an indicator dye after a set time to determine the amount of reactants consumed or formed.[5,6] The dinitrophenylhydrazine of method 3 (Table 56-25) forms a colored adduct with pyruvate. For the reaction to be measured in the pyruvate-to-lactate (P-L) direction, a significant amount of pyruvate must be converted. The tetrazolium reaction oxidizes NADH by a dye system to form a colored tetrazolium dye (method 4, Table 56-25).

It has been shown that the very *initial* rate of the lactate-to-pyruvate (L-P) reaction is nonlinear.[7] After 20 seconds the reaction becomes linear, provided that not enough product is formed to inhibit the reaction. In both L-P and P-L directions product inhibition can be observed. Therefore there is a limit to the maximum rate that can be followed before the reaction becomes nonlinear (usually about 0.5 absorbance units/min). The reaction is linear with dilution (zero-order kinetics), and high levels can be accurately quantified by the addition of smaller amounts of enzyme (dilution). In general, the best method of following the reaction is to measure the rate kinetically rather than by end-point analysis.

The reaction rate is temperature dependent, with significantly faster rates occurring at higher temperatures. Conversion factors have been established for comparing reaction rates at 25°, 30°, and 37° C (for example, U/L at 25° C × factor = U/L at 37° C). See Table 56-26.

REFERENCE AND PREFERRED METHODS

The problem of selecting the best method revolves around the heterogeneity of the enzyme itself, inability of various standardizing groups to agree on temperature, and disagreement on whether to follow the reaction in the L-P or P-L direction.

The most widely used procedures for LD determination employ the L-P reaction, since it is claimed that there is less dependence on the NAD and the lactate concentrations, as well as less contamination of NAD with inhibiting products.[3] In its favor, the P-L reaction has a faster rate and lower reagent cost.[1,2] The higher activities seen with the P-L reaction can result in slightly better precision. However, there is considerable difficulty in standardizing the reaction between laboratories. It has been shown that the maximal rate is strongly dependent on the type or source of the NADH used in the reaction. The National Bureau of Standards has been attempting to standardize these preparations.

In 1973 the Scandinavian Society for Clinical Chemistry and Clinical Physiology accepted a subcommittee report for a recommended method for LD in blood.[2] This method uses the P-L reaction. In contrast, the method most used in the United States is the L-P reaction, as described by Henry.[3] Since there is no generally accepted LD method, assays based on either of these two procedures are adequate for routine use.

Table 56-25. Methods of lactate dehydrogenase (LD) analysis

Method	Type of analysis	Principle	Usage	Comments
1. Lactate to pyruvate (L → P)	End-point or kinetic spectrophotometric	Reaction buffered at alkaline pH to favor equilibrium to pyruvate	All body fluids Automated	Kinetic Suggested reference method
2. Pyruvate to lactate (P → L)	End-point or kinetic spectrophotometric	Reaction buffered at physiological pH	All body fluids Automated	Claimed to be less linear than 1; however, a faster rate than 1
3. Dinitrophenylhydrazine	End-point spectrophotometric	Pyruvate → Lactate Unreacted pyruvate forms colored product with reagent	Manual	For laboratories without ultraviolet spectrophotometer; imprecise
4. Tetrazolium	End-point spectrophotometric	Lactate → Pyruvate + NADH Reduced NADH forms colored product by reduction of tetrazolium dye	Manual	For laboratories without ultraviolet spectrophotometer Most often used for electrophoretic isoenzyme detection

Table 56-26. Comparison of reaction conditions for LD analysis*

Condition	Scandinavian[2]	Henry[3]	Optimal concentration[7]‡	
			L-P	P-L
Temperature†	37° C	32° C	37° C	37° C
pH	7.4	9.0	8.7	7.0
Final concentration of reagent components	Tris: 50 mmol/L EDTA: 5 mmol/L Pyruvate: 1.2 mmol/L NADH: 0.15 mmol/L	2-Amino-2-methyl-1-propanol: 0.89 mol/L Lactic acid: 72 mmol/L NAD: 6 mmol/L	2-Amino-2-methyl-1,3-propane diol: 200 mmol/L Lithium lactate: 70 mmol/L NAD: 7 mmol/L	Imidazole: 100 mmol/L Pyruvate: 1.5 mmol/L NADH: 0.22 mmol/L
Fraction of sample volume	0.12	0.33-0.17	0.05	0.05
Linearity	20% decrease at 1000 U/L	Better than P → L at same concentration	—	—
Precision	3%	3%	—	—

*Sources of analytical error: Any specimen with visible hemolysis will give elevated results because of release of LD from red cells. *Interferences* include salicylate and inhibitors that may be present in uremic sera.
†CAP surveys indicate that the most common temperatures of running LD assays are at 30° or 37° C.
‡The optimal concentrations for NAD + lactate and NADH + pyruvate are temperature dependent. The maximal reaction rate can vary depending on the buffer used. These conditions have been worked out by Buhl and Jackson.[7]

SPECIMEN

LD can be found in all body fluids and tissues. The most accurate blood levels are obtained from serum rather than plasma, since some anticoagulants such as oxalate interfere with the reaction. Hemolysis results in the release of LD from red cells into the body fluid, giving possible falsely high results. Enzyme stability is temperature dependent. Since different isoenzymes are stable at different temperatures, there is no absolute method of storing samples without some loss of activity. Consistency in storage technique will yield the most precise results. Freezing results in loss of activity of the enzyme.

PROCEDURE

SCANDINAVIAN COMMITTEE ON ENZYMES[2]

Principle

$$\text{Pyruvate} + H^+ + \text{NADH} \xrightarrow{\text{EC 1.1.1.27}} \text{Lactate} + \text{NAD}^+$$

The reaction is determined with pyruvate and NADH at

alkaline pH. The initial reaction rate is followed by recording of the rate of consumption of NADH at 340 nm.

Reagents

Working Tris buffer. Tris(hydroxymethyl)aminomethane; ethylenediaminetetraacetate disodium salt, EDTA. To approximately 800 mL of distilled water add 6.8 g of Tris and 2.1 g of EDTA. Allow to dissolve, and bring to pH 7.4 with 1 M HCl. Dilute to 1 L. This solution is stable for at least 6 weeks at 4° C.

Tris-NADH reagent. Dissolve 13 mg of NADH in 90 mL of working Tris buffer. Measure the absorption, and bring to an absorbance of 1.0 by dilution with the same buffer. This is stable for 72 hours at 4° C.

Working pyruvate solution. Dissolve 149 mg of sodium pyruvate in 100 mL of distilled water. This solution is stable for 20 days at 4° C.

Assay

Equipment: spectrophotometer with a ≤10 nm band pass and a 37° C cuvette.

The procedure is described for any recording spectrophotometer, and the volume is calculated for a 3 mL cuvette. The temperature should be held constant at 37° C.

1. Place 2 mL of Tris-NADH reagent in a 3 mL cuvette, add 50 μL of serum, mix, and incubate 5 to 15 minutes at 37° C.
2. Initiate the reaction by adding 200 μL of the pyruvate working solution. Mix and rapidly insert the cuvette into the spectrophotometer. Immediately start the recording of the change in absorbance at 340 nm.
3. If the resulting curve is linear (ΔA per minute is constant over a period of analysis), proceed with the calculations described as follows. If the rate is too rapid to give a linear curve, dilute the sample with saline and repeat the assay.

Calculation

The international units of activity (U) are expressed as micromoles of NADH per minute, whereas enzyme concentrations are expressed as U/L. The change in absorbance is related to units by the equation for a 1 cm light path:

$$U/L = \frac{\Delta A \text{ for } t \text{ min}}{t \text{ min}} \times \frac{\text{Total volume}}{\text{Sample volume}} \times \frac{1}{\epsilon} \times \frac{10^6\ \mu\text{mol}}{\text{mol}}$$

where ΔA = measured absorbance for time t
ϵ = molar extinction coefficient for NADH
10^6 = factor to convert concentration to μmol/L. For the assay the values are as follows:

$$U/L = \Delta A/\text{min} \times \frac{2250\ \mu\text{L}}{50\ \mu\text{L}} \times \frac{1}{6220} \times 10^6 = 7235 \times \Delta A/\text{min}$$

METHOD OF HENRY[3]

Principle

$$\text{Lactate} + \text{NAD}^+ \xrightarrow{\text{EC 1.1.1.27}} \text{Pyruvate} + \text{H}^+ + \text{NADH}$$

The initial reaction rate is determined with lactate and NAD. The reaction is followed by recording the rate of formation of NADH at 340 nm.

Reagents

2-Amino-2-methyl-1-propanol. This is a liquid and is stable for at least 1 year.

Lactic acid, 85%. This is a liquid and is stable for at least 1 year.

Working lactic acid solution. To approximately 800 mL of water, add 63.9 mL of 2-amino-2-methyl-1-propanol plus 10 mL of 85% lactic acid. Mix and adjust to pH 9.0 ± 0.05 with 5 N NaOH. Store in refrigerator. This solution is stable 1 month.

Lactic acid–NAD reagent. For every 100 mL of working lactic acid solution to be used (enough for 30 samples), add 400 mg of NAD. Prepare fresh each day a sufficient volume for the number of specimens to be assayed. Keep solution in refrigerator or on ice at all times.

Assay

Equipment. The original procedure is described for a Gilford 2400 recording spectrophotometer equipped with a temperature-controlled cuvette compartment held at 32° C by a circulating constant-temperature bath. I recommend either 30° or 37° C, since 32° C is no longer used in the United States.

Place 2.9 mL of lactic acid–NAD reagent into a test tube and incubate in a water bath at 37° ± 0.5° C for 4 to 5 minutes. During this period the cuvettes are kept in the instrument's cuvette compartment so that they reach temperature equilibrium (spectrophotometer and circulating water bath are turned on at least 2 hours before use).

Add 0.1 mL of serum by to-contain pipet, mix, and transfer as quickly as possible to the prewarmed cuvette, which is rapidly reinserted into the cuvette compartment. The cuvette compartment lid is left open for as short a time as possible. Immediately start automatic recording of the change in absorbance at 340 nm.

If the resulting recorded curve is linear for at least 6 minutes, proceed with calculations as described next. If the rate is too rapid to give a linear curve, rerun using the sample diluted with 0.15 M NaCl (dilution required if result is more than 300 units).

Calculation

(See the previous Calculation section for method 1.)

For the assay in method 2 the values are as follows:

$$U/L = \Delta A/\text{min} \times \frac{3.0\ \text{mL}}{0.1\ \text{mL}} \times \frac{1}{6220} \times 10^6 = 4830 \times \Delta A/\text{min}$$

REFERENCE RANGE

	Method 1 (L-P)	Method 2 (P-L)
Temperature	37° C	37° C
Adult males	63 to 155 U/L	90 to 320 U/L
Adult females	62 to 131 U/L	90 to 320 U/L
Children	Approximately 10% to 15% higher	
Coefficient for 25° C		2.44
Coefficient for 30° C		1.68

Lactate dehydrogenase changes

The most significant change in concentration is an increase in serum caused by release by tissue. Decreases are not important clinically. The types of disease in which LD is elevated follow, but the list is not complete:

1. Hepatitis
2. Infectious mononucleosis
3. Cancers of pancreas, prostate, and so on; lymphomas
4. Anemias
5. Muscular dystrophy
6. Acute myocardial infarctions
7. Liver cirrhosis and necrosis
8. Pancreatitis

REFERENCES

1. Howell, B.F., McClure, S., and Schaffer, R.: Lactate-to-pyruvate or pyruvate-to-lactate assay for lactate dehydrogenase: re-examination, Clin. Chem. **25:**269-272, 1979.
2. Keiding, R., Hørder, M., Gerhardt, W., et al.: Recommended methods for determination of four enzymes in blood, Scand. J. Clin. Lab. Invest. **33:**291-306, 1974.
3. Demetriou, J.A., Drewes, P.A., and Gin, J.B.: Enzymes. In Henry, R.J., Cannon, D.C., and Winkelman, J.W., editors: Clinical chemistry: principles and techniques. New York, 1974, Harper & Row, Publishers, Inc., pp. 824-831.
4. Freer, D.E., Statland, B.E., Johnson, M., and Felton, H.: Reference values for selected enzyme activities and protein concentrations in serum and plasma derived from cord-blood specimens, Clin. Chem. **25:**565-569, 1979.
5. Natelson, S.: Techniques of clinical chemistry, ed. 3, Springfield, Ill., 1971, Charles C Thomas, Publisher, pp. 449-452.
6. Spiegel, H.E., Symington, J.A., Hordynsky, W.E., and Babson, A.L.: Colorimetric determination of lactate dehydrogenase (L-lactate:NAD oxidoreductase) activity, Stand. Methods Clin. Chem. **7:**43-46, New York, 1972, Academic Press, Inc.
7. Buhl, S.N., and Jackson, K.Y.: Optimal conditions and comparison of lactate dehydrogenase catalysis of the lactate-to-pyruvate and pyruvate-to-lactate reactions in human serum at 25°, 30° and 37° C, Clin. Chem. **24:**828-831, 1978.

Lactate dehydrogenase isoenzymes

JOHN F. CHAPMAN
LINDA L. WOODARD
LAWRENCE M. SILVERMAN

Lactate dehydrogenase, LD, LDH, L-lactate:NAD oxidoreductase
Clinical significance: pp. 488 and 953
EC 1.1.1.27
Molecular weight: 140,000 daltons
Chemical class: enzyme, protein
Known isoenzymes: LD_1, LD_2, LD_3, LD_4, LD_5
Biochemical reaction:

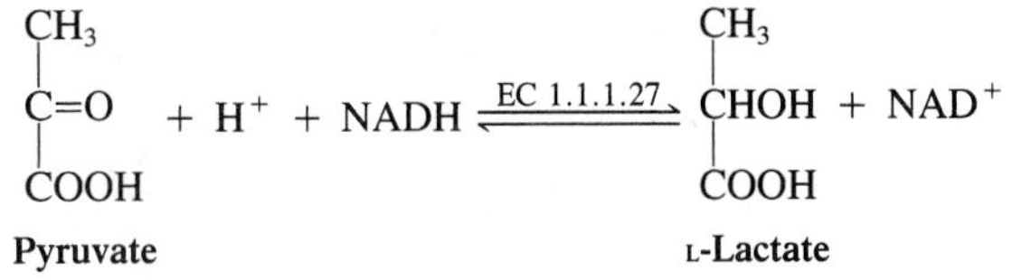

$$\underset{\text{Pyruvate}}{\begin{array}{c}CH_3\\|\\C{=}O\\|\\COOH\end{array}} + H^+ + NADH \underset{}{\overset{\text{EC 1.1.1.27}}{\rightleftharpoons}} \underset{\text{L-Lactate}}{\begin{array}{c}CH_3\\|\\CHOH\\|\\COOH\end{array}} + NAD^+$$

PRINCIPLES OF ANALYSIS AND CURRENT USAGE

Lactate dehydrogenase (LD) is an enzyme present in many tissues. Consequently, elevations in serum have to be considered nonspecific foı any single tissue or organ. Additional information can be obtained through the separation and quantitation of LD into the five known isoenzymes. There are two subunits designated H (heart) and M (muscle), and each isoenzyme is composed of four of these subunits. They are LD_1 (HHHH, or H_4), LD_2 (HHHM, or H_3M), LD_3 (HHMM, or H_2M_2), LD_4 (HMMM, or HM_3), and LD_5 (MMMM, or M_4). Methods used for separation and quantitation of these isoenzymes include electrophoresis, chromatography, and immunoinhibition (Table 56-27).

Support media for electrophoretic separation include agarose and cellulose acetate. Electrophoretic methods are currently the most frequently used procedures for LD isoenzyme separation. The charge on each subunit differs, thus imparting a different, total charge on each isoenzyme. At a pH of about 8.6, LD_1 (H_4) has the highest negative charge and so moves farthest toward the anode. LD migrates slightly toward the cathode, whereas LD_2, LD_3, and LD_4 migrate in order of decreasing anodic mobilities. Fig. 56-7 shows a LD isoenzyme separation in 1% agarose gel at pH 8.6. Colorimetric or fluorometric determination of the relative amounts of each isoenzyme present may be accomplished by the addition of substrate containing lactate, nicotinamide adenine dinucleotide (NAD), iodonitrotetrazolium chloride (INT), and phenazine methosulfate (PMS). LD catalyzes the following reaction:

$$\text{L-Lactate} + NAD \xrightarrow{LD} \text{Pyruvate} + \underset{\text{(fluorescent)}}{NADH}$$

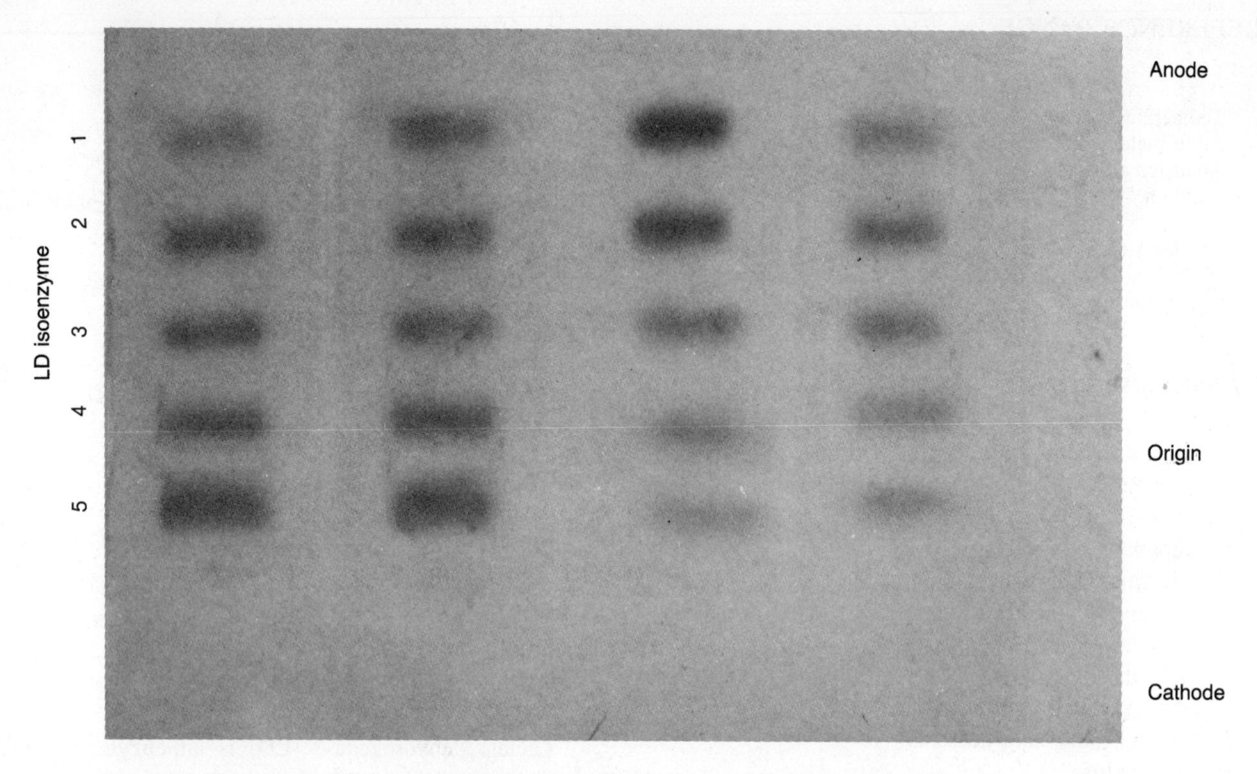

Fig. 56-7. Separation of five lactate dehydrogenase isoenzymes on 1% agarose gel, with tetrazolium dye staining. Specimen on right is a control; others are three specimens from a patient. Note LD "flip" on second specimen from right.

The NADH (reduced NAD) generated is then used to reduce the tetrazolium salt to a colored formazan compound (INF):

$$\text{NADH} + \text{INT} + \text{PMS} \longrightarrow \text{NAD} + \text{INF}$$

Densitometric scanning of the resultant pattern yields relative percentages of the isoenzymes present.

Ion-exchange methods for LD isoenzymes have been described[1,2] but are more tedious to perform and inaccurate if special care is not taken to ensure proper separation.[3] Immunochemical methods for the determination of LD_1 utilize antibodies against the M subunit of LD. The anti–M antibody/M isoenzyme complex then reacts with a second antibody directed against the first to remove all the M-bearing isoenzymes leaving LD_1 remaining in solution. Alternatively, the anti-M immunoglobulin can be covalently attached to an insoluble particle, and the bound M subunits—LD_2, LD_3, LD_4, and LD_5 isoenzymes—are removed from solution by centrifugation. The remaining LD activity is measured by the usual LD procedure. This method is generally considered to be a more sensitive indicator of early elevations in LD_1 after acute myocardial infarction.

REFERENCE AND PREFERRED METHODS

There has been no agreement on a reference method for LD isoenzyme analysis. Electrophoretic methods are generally simple to perform but may be time consuming for the analysis of many samples. In addition, unusual specimens may present interpretive problems. Electrophoretic isoenzyme analysis is most often used in the clinical context of the diagnosis of acute myocardial infarction. Most frequently, a "positive" LD isoenzyme result occurs when the ratio of LD_1/LD_2 is greater than 1. Since the usual ratio is approximately 0.72 to 0.79, a ratio > 1 is called an LD "flip." The presence of an LD flip, in the absence of hemolysis, has a very high specificity (approximately 100%) and can be considered very positive evidence for a myocardial infarction. The question of the effect of hemolysis on the LD flip has been controversial, depending on the electrophoresis system used. In agarose gel electrophoresis, the presence of observable hemolysis *can* cause a spurious flip but shows great variability from sample to sample.[4]

It has been suggested that the sensitivity of the LD isoenzyme analysis can be raised if one simply looks for an *elevation* in the LD_1/LD_2 ratio other than a large enough elevation to result in a flip.[5] This technique would of course require a "normal-range" cutoff value, with all the associated problems. The use of a normal-range cutoff value for the LD_1/LD_2 ratio will increase the sensitivity of the procedure for myocardial infarction, but at a cost of lowering the specificity.

Immunoinhibition methods claim greater sensitivity than

Table 56-27. Methods of lactate dehydrogenase (LD) isoenzyme analysis

Method	Principle	Usage	Comments
1. Electrophoresis	At pH 8.6, major isoenzymes have different mobilities in 1% agarose; LD activity quantitatively determined by coupled dye reaction	Most frequently used	Separates all five major isoenzymes Rapid and relatively simple
2. Ion-exchange chromatography	At pH 8.0, LD_1 and LD_2 are absorbed to a diethylaminoethyl (DEAE) type of ion-exchange resin column LD_5 is eluted initially, with LD_1 being eluted at the highest salt concentrations Either stepwise reactions or gradients elute the isoenzyme fractions Fractions are quantitated by standard assay	None	Separation of all isoenzymes not readily accomplished
3. Immunoinhibition	M subunits of H_3M, H_2M_2, HM_3, and M_4 (LD_{2-5}) removed by double-antibody solid-phase precipitation reaction	Recently introduced	Gives only LD_1 activity; rapid and relatively simple

either electrophoresis or chromatography. They measure only LD_1, which is useful primarily in the diagnosis of acute myocardial infarction. However, since the most frequent use of LD isoenzyme analysis is for the diagnosis of myocardial infarction, this limitation is not unreasonable. The Roche immunoinhibition assay is simple to perform and rapid. However, it may be subject to false-positive results in cases of trauma, surgery, and hemolytic disease (Table 56-28).

Electrophoresis on occasion can provide more information about elevations of total LD, especially in cases of muscle trauma or liver damage (with increases of LD_5) and a variety of malignancies. For this reason and because of our own experience with this specific method, an electrophoretic method is described.

SPECIMEN

The preferred specimen is serum, which should be separated from the clot as quickly as possible, since red cells contain 100 times the amount of LD as normal serum. If serum is allowed to stand on the clot, an elevation in LD_1 and LD_2 will be seen. For the same reason, hemolyzed specimens are unacceptable.

PROCEDURE

An agarose electrophoresis method is described here[6,7] and is essentially the system commercially available from Corning Medical and Scientific Inc. (Palo Alto, Calif.).

Principle

$$\text{L}(+)\text{-Lactate} + \text{NAD}^+ \xrightarrow[\text{pH 9.1}]{\text{LD}} \text{Pyruvate} + \text{NADH} + \text{H}^+$$

$$\text{H}^+ + \text{NADH} + \text{INT} + \text{PMS} \xrightarrow[\text{pH 9.1}]{} \text{NAD}^+ + \text{INF}$$

Table 56-28. Comparison of conditions for LD isoenzyme analysis

Condition	Electrophoresis	Immunoinhibition*
Temperature	Ambient	Ambient
Sample volume	1-2 μL	200 μL
Assay time	20 min (separation) 30 min (analysis)	25 min (separation) 5 min (quantitation)

*Roche Diagnostics, Nutley, N.J.

Reagents

Universal barbital buffer (0.05 M, 0.035% EDTA, pH 8.6). Dissolve 17.9 g of sodium barbital, 2.6 g of barbital, 1.0 g of sodium chloride, and 0.7 g of disodium EDTA in 2 L of deionized water. This buffer should be stored at 4° C and is stable for 6 months if bacterial contamination does not occur.

Working substrate. One vial of commercially supplied reagent per gel should be prepared before electrophoresis. Add 1.0 mL of adenosine monophosphate (AMP)–lactate solution (L(+)-lactate at 412 nmol/L concentration, adjusted with 2-amino-2-methylpropanol to pH 9.1) to one vial of color reagent (containing 2 mg of NAD, 4 mg of iodonitrotetrazolium chloride [INT], and 0.05 mg of phenazine methosulfate [PMS]). The reconstituted reagent is stable for 2 hours at room temperature.

Assay

Equipment: electrophoresis systems available from several commercial vendors, oven, incubator, and densitometer.

1. Turn on the incubator (39° C) and drying oven (65° C), allowing a warm-up time of 30 minutes.

2. Follow manufacturer's directions for preparation of the electrophoresis system and agarose gel plates.
3. Carefully fill the sample wells with 1.0 μL of serum. If the total activity of the sample is less than 150 U/L, it is advisable to use 2.0 μL of sample, applied in two separate applications of 1.0 μL each, allowing time for adsorption between applications. If the total activity of the sample is greater than 1000 U/L, the activity in one or more of the isoenzyme bands may exceed the range of linearity of the system. Appropriate dilution of the sample, in normal saline or deionized water, to lower the total activity is advised.
4. Allow 35 minutes for electrophoresis at 90 volts per plate (10.5 cm × 12 cm).
5. After electrophoresis, remove the cell cover from the cell base. Drain excess buffer without inverting the cell cover. Grasp the agarose film by its edges and remove it from the cover.
6. Place the film, agarose side up, on a flat countertop with the anode (+) edge toward you.
7. Place a 2 mL serological pipet lengthwise along the anode (+) edge of the film. Evenly dispense the working substrate onto the agarose along the edge of the pipet. With a single, smooth motion, slowly push the serological pipet across the agarose film. This should uniformly coat the surface of the agarose with substrate.
8. Insert the film, agarose side up, on moistened filter paper in a prewarmed incubator tray. Replace the tray in the incubator.
9. Incubate at 39° C for 20 minutes. Inspect the gels; if further color development is desired, an additional 10 minutes of incubation time is allowed.
10. After incubation, wash the film in the deionized water for 20 minutes. Cover the tray to protect the gel from light and to prevent buildup of background color.
11. Place gel on a drying shelf in the oven. Dry at 65° C for 20 to 30 minutes, or until dry. Overdrying can also result in development of excess background color.
12. Scan the gels using appropriate densitometric equipment.

REFERENCE RANGE

Reference intervals should be determined for the population of patients being tested. The following reference intervals are given only as a guide:

Isoenzyme fractions	Reference intervals (relative %)
LD_1	18 to 33
LD_2	28 to 40
LD_3	18 to 30
LD_4	6 to 16
LD_5	2 to 13

Clinical significance

The following chart lists some of the clinical conditions associated with elevations of different LD isoenzymes:

LD isoenzymes	Clinical condition
1,2	Hemolysis Myocardial infarction Renal cortex infarction Some tumors
4,5	Hepatocellular damage (necrosis, inflammation, congestion) Skeletal muscle trauma Crush injuries
2,3,4	Malignancies, especially leukemias and lymphomas Lung disease and congestion

REFERENCES

1. Mercer, D.W.: Simultaneous separation of serum creatine kinase and lactate dehydrogenase isoenzymes by ion-exchange column chromatography, Clin. Chem. **21**:1102, 1975.
2. Mercer, D.W.: Improved column method for separating lactate dehydrogenase isoenzymes 1 and 2, Clin. Chem. **24**:480, 1980.
3. Usategui-Gomez, M., and Wicks, R.W., and Warshaw, M.: Immunochemical determination of the heart isoenzyme of lactate dehydrogenase (LDH 1) in human serum, Clin. Chem. **25**:729, 1979.
4. Lott, J.A., and Stang, J.M.: Serum enzymes and isoenzymes in the diagnosis and differential diagnosis of myocardial ischemia and necrosis, Clin. Chem. **26**:1241-1250, 1980.
5. Leung, F.Y., and Henderson, A.R.: Thin layer electrophoresis of lactate dehydrogenase isoenzymes in serum: a note on the method of reporting and on the lactate dehydrogenase isoenzyme-1/isoenzyme-2 ratio in acute myocardial infarction, Clin. Chem. **25**:209-211, 1979.
6. Wilkson, J.H.: Isoenzymes, ed. 2, Philadelphia, 1970, J.B. Lippincott Co.
7. Data on file, Corning Medical, Medfield, Mass.

Lipase

MICHAEL D.D. McNEELY

Lipase, LPS, triacylglycerol acylhydrolase
Clinical significance: pp. 460 and 550
Molecular weight: approximately 38,000 daltons
EC 3.1.1.3
Chemical class: protein, enzyme
Biochemical reaction:

$$\begin{array}{l} \alpha{-}CH_2{-}O{-}\overset{O}{\overset{\|}{C}}{-}R_1 \\ \beta{-}CH{-}O{-}\overset{O}{\overset{\|}{C}}{-}R_2 \\ \alpha_1{-}CH_2{-}O{-}\overset{O}{\overset{\|}{C}}{-}R_3 \end{array} \xrightarrow{\text{Lipase}} \begin{array}{l} CH_2{-}OH \\ CH{-}O{-}\overset{O}{\overset{\|}{C}}{-}R_2 \\ CH_2{-}OH \end{array} \quad \begin{array}{c} R_1{-}\overset{O}{\overset{\|}{C}}{-}OH \\ + \\ R_3{-}\overset{O}{\overset{\|}{C}}{-}OH \end{array}$$

Triglyceride **β-Monoglyceride + 2 fatty acids**

PRINCIPLES OF ANALYSIS AND CURRENT USAGE

The measurement of serum lipase (LPS) activity has long been recognized as a useful method for diagnosing

Table 56-29. Methods of lipase (LPS) analysis

Method	Principle	Usage	Comment
1. Titration of released fatty acids	Reaction *A:* Triglyceride $\xrightarrow{LPS}$ Monoglyceride + 2 fatty acids		
	(a) Reaction *A*: fatty acid + NaOH titration to neutrality using phenolphthalein indicator	Used	Cherry-Crandall method[1]
	(b) Reaction *A*: titration of released acids using a pH meter to assess end point	Rarely used	Requires a pH meter accurate to ± 0.01 units
2. Colorimetric	(a) Reaction *A*, with released fatty acids reacting with Spectru Cationic dye to form a blue complex	Rarely used	
	(b) Reaction *A* with released fatty acids reacting with Cu^{++} ions	Rarely used	
	(c) Reaction *A* with released fatty acids changing the color of a pH indicator, methyl red	Rarely used	
3. Coupled enzymatic	Trilinolein $\xrightarrow{LPS}$ 1,2-Dilinolein + Linoleic acid Linoleic acid $+ O_2 \xrightarrow{Lipoxygenase}$ Linoleic acid–hydrogen peroxide (Reactions monitored polarographically by rate consumption of O_2)	Rarely used	
4. Emulsion clearing	Turbidimetric or nephelometric monitoring of decrease in size of emulsion of subtrate after action of lipase	Widely used	Very dependent on availability of stable, reproducible substitute

pancreatitis. Unfortunately, the enzyme has been very difficult to measure accurately and precisely and shortcut techniques have proved to be clinically unreliable. These factors have cast the diagnostic use of serum lipase into doubt.

Lipases are enzymes that hydrolyze the long-chain fatty acid (LCFA) triglyceride (triacylglycerol). They act on the glycerol to (LCFA) bonds (ester linkage) to liberate two molecules of fatty acid and one molecule of β-monoglyceride.

Essential to the understanding of lipase methodology is that the enzyme acts only at an ester-water interface. Thus lipase assay substrates must be established as emulsions. The reaction rate will increase with the dispersion of the emulsion (surface area). Failure to achieve a suitable ester-water interface will permit the action of nonlipase enzymes (carboxylic esterhydrolase, aryl-ester hydrolase, and lipoprotein lipase). The use of short-chain fatty acid triglyceride substrates, which are water soluble, will permit this false lipase reaction to occur.

The classical technique is that of Cherry and Crandall, [1] who allowed serum to incubate for 24 hours at 37° C, with a 50% emulsion of olive oil and 5% gum acacia in a pH 7.0 phosphate buffer. The liberated fatty acid was then quantified by titration against 0.05 M NaOH by use of a phenolphthalein indicator (method 1a, Table 56-29).

Subsequent methods have attempted to improve on ease, analytical sensitivity, and precision of the assay. Some approaches have refined the assay conditions or employed more sensitive indicators to allow incubation periods as short as 1 hour.[2-5]

Alternative detection methods to titration have been developed. For example, Tietz and Fiereck[6] have done considerable work using pH meter end-point measurements (method 1b, Table 56-29) and have published the application of a pH-stat apparatus.[7]

Gindler[8] introduced the Spectru Cationic Blue dye, which reacts directly with liberated fatty acids to form a blue complex that can be measured spectrophotometrically (method 2a, Table 56-29). A similar approach was used by Yang and Biggs[9] who added Cu^{++} to form colored complexes with the fatty acids, which could be extracted into chloroform and measured in a photometer (method 2b, Table 56-29).

The method of Massion and Seligson[10] quantitates the liberated fatty acids after 1 hour of incubation by extracting them and allowing them to change the color of an appropriately buffered methyl red indicator (method 2c, Table 56-29).

Substrates other than long-chain fatty acid triglycerides have been used to allow direct color production and a more rapid assay[11-13] or a highly sensitive fluorometric end point.[14] Unfortunately, these substrates react with nonlipase esterases and give confusing clinical results. It is for this reason that natural oil substrates must be purified by alumina absorption.

An interesting approach has been published by Griebel et al.[15] who employed an oxygen electrode (Beckman Glucose Analyzer) to measure the oxygen consumed when the substrate trilinolein is hydrolyzed to 1,2-dilinolein and linoleic acid, and the linoleic acid is oxidized to linoleic acid–hydrogen peroxide by the enzyme lipoxygenase (method 3, Table 56-29).

A completely different analytical approach depends on emulsion clearing and was introduced by Vogel and Zieve[16] and has been modified successfully by others.[17,18] In this method, monitored at 340 or 400 nm, the turbidity of the oil-water emulsion is reduced as lipase hydrolyzes the triglyceride molecules. The technique requires a highly stable, reproducible substrate with an appropriate initial absorbance. The sample is added to the substrate, and kinetic measurements are made turbidimetrically or nephelometrically (method 4, Table 56-29). A special purpose instrument has been developed for this method (Perkin-Elmer Corp., Norwalk, Conn.).

Three percent to 5% of specimens analyzed by emulsion clearing methods will display an anomalous absorbance increase caused by molecular aggregation. This phenomenon occurs more frequently when specimens are frozen and thawed or when they contain increased concentrations of IgM (such as rheumatoid factor). By avoidance of freezing or by the use of polyethylene glycol (PEG) to precipitate protein (mostly IgM) these problems will be reduced in frequency.[19]

An interesting variant of the emulsion clearing method uses immobilized olive oil in an agarose gel. The sample is placed in a well, and after an incubation period one measures a concentric ring of clearing whose diameter is proportional to the lipase activity.[20]

REFERENCE AND PREFERRED METHOD

Methods using substrates other than long-chain fatty acid (LCFA) triglycerides must be viewed with great suspicion.

The use of purified olive oil substrate, extraction of LCFA, and quantitation against pure standards is recommended as a standard technique. The modified method of Massion and Seligson[10] (detailed in following paragraphs) has these characteristics and can be carried out with microsamples in under 2 hours.

The emulsion-clearing methods are fastest and can be automated,[21] but the stability and reproducibility of the substrate is extremely critical. Commercial products employing this principle must prove the quality of their substrate over time. I have had success with the method of Ziegenhorn[22] (Boehringer-Mannheim Diagnostics, Inc., Indianapolis, Indiana).

All new methods for lipase must be compared to an olive oil substrate with acid measurement technique of known clinical utility.

SPECIMEN

Serum should be stored at 4° C until analyzed. Repeated freezing and thawing should be avoided. Urine lipase does not exist.

PROCEDURE: Spectrophotometric method[23]

Principle

A purified olive oil emulsion is used as the substrate. After 30 minutes of incubation, the fatty acids are extracted into petroleum ether. A carefully balanced methyl red indicator and buffer mixture is added. The color change is proportional to the fatty acid present and hence to the lipase activity.

Reagents

Absolute ethanol, reagent grade

Sulfuric acid (0.28 N). Add 4 mL of concentrated sulfuric acid to 500 mL of water. Stable for 1 year at room temperature.

Acid-alcohol mixture. Add 3 volumes of absolute ethanol to 2 volumes of 0.28 N H_2SO_4. Prepare fresh daily.

Petroleum ether (boiling range: 30° to 60° C), redistilled

Ethanol, 95%, reagent grade

Methyl red, 2 g/L (M-129, Fisher Scientific Co., Pittsburgh, Penn.). Dissolve powdered methyl red, 200 mg, in 100 mL of 95% ethanol. This process requires at least 12 hours using a magnetic stirrer. Stable for 6 months at room temperature.

Olive oil, reagent grade (no. 0-111, Fisher Scientific Co.). Separate the olive oil from the fatty acids by passing it through alumina (80-200 mesh, cat. no. A-540, Fisher Scientific Co.) contained in a column or separatory funnel plugged with glass wool. One volume of oil is purified by an equal volume of alumina. Check the fatty acid content of the oil by pipetting a 0.050 mL aliquot into 3 mL of methyl red reagent, shake, and centrifuge. If the absorbance increase is more than 0.020 in a 10 mm cuvette, the oil must be passed through more alumina. The oil is stable for at least 1 week at 4° to 8° C.

Substrate/buffer (triglyceride emulsion, 5%). Homogenize 5 mL of the purified oil in a high-speed household blender with 95 mL of buffer solution. Mix using repeated short bursts of blending (30 to 60 seconds) followed by short intervals of cooling to avoid overheating and hydrolysis of the oil. Continue blending until no film of oil is present after the foam settles. Adjust the pH to 8.50 at 25° C by the addition of 1 M HCl or NaOH. Store at 4° to 6° C. The doubling of the blank value is an indication that a fresh substrate buffer reagent must be prepared.

Buffer. Sequentially dissolve 2.42 g of tris(hydroxymethyl)aminomethane, 3.5 g of deoxycholic acid, and 0.2 g of ascorbic acid in 500 mL of water in a 1 L volumetric flask and dilute to volume with distilled water. Stable for 2 months at 4° to 8° C.

Methyl red reagent (buffered indicator solution). Prepare this reagent using photometric measurements to avoid variations in dye lots. Add 10 mL of 1 M NaOH to 1 L of 95% ethanol. Add enough 0.2% methyl red solution (usually 10 to 13 mL) to bring the absorbance to a range of 0.095 to 0.100 in a 10 mm cuvette at 502 nm. Add 1 mL of 1 M sodium acetate, mixing with a magnetic stirrer. Add 1 N HCl dropwise until a faint orange-red tinge persists. Add more acid carefully until the solution develops an absorbance of 0.200 ± 0.005 in a 10 mm cuvette. If this point is passed, the solution must be back-titrated with dilute NaOH. Avoid excessive dilution of this reagent. The reagent is stable for 1 month if stored at room temperature in a dark brown glass bottle.

Standard solution: stearic acid (no. A-293, Fisher Scientific Co.) (1 μEq/mL). Dissolve 28.5 mg of stearic acid in 100 mL of heptane or petroleum ether. Measure 0.25, 0.50, and 0.75 mL into screw-topped tubes and dry at 50° to 60° C in a water bath with a gentle stream of air. These standards represent 0.25, 0.50, and 0.75 μEq of fatty acid. Several sets of standards can be prepared at once, since the dry residue is stable for weeks in tightly capped tubes placed at 4° to 8° C.

Drying apparatus. A device to accelerate evaporation may be prepared when one cements 10 cm lengths of thin metal or rigid plastic tubing into holes drilled into the bottom of a small plastic box. The box is then filled with loosely packed glass wool. Compressed air is injected into one side of the box to deliver air into the petroleum ether tubes. Evaporation can be enhanced by placement of the tubes into a 50° to 60° C water bath.

Assay

Equipment: spectrophotometer with band pass ≤ 10 nm capable of reading at 502 nm and a 37° C water bath.

1. Prewarm aliquots of serum and of substrate for 5 minutes in a 37° C water bath.
2. Measure 1 mL of substrate and 50 μL of serum into two screw-capped glass culture tubes, mix, and cover.
3. Incubate for 30 minutes.
4. Measure 3.3 mL of acid-alcohol mixture into each tube.
5. To 1 mL of substrate in a separate tube, measure 3.3 mL of acid-alcohol mixture and 50 μL of serum to produce a blank.
6. Measure 3.3 mL of acid-alcohol mixture and 1 mL of substrate to another tube to serve as a reagent blank.
7. Measure 4 mL of petroleum ether into all tubes.
8. Tightly cap and shake all tubes vigorously for 2 minutes and then centrifuge at 2000 rpm for 5 minutes.
9. Remove 2.0 mL of petroleum ether phase (top layer) from each tube and transfer to a separate glass, screw-top tube.
10. Evaporate off the petroleum ether.
11. Add 3 mL of methyl red reagent to the oil residue of sample as well as the predried standards. Cap the tubes with Teflon-lined caps, and shake vigorously for 1 minute. Centrifuge at 2000 rpm to remove suspended oil.
12. Decant the supernatant and read in a 10 mm cuvette in a spectrophotometer at 502 nm.
13. If the absorbance is greater than the highest standard, the sample should be diluted with an equal volume of methyl red reagent.

Calculations

Plot absorbance values of standards and methyl red reagent blank versus the microequivalents (μEq) of fatty acids. Determine the amount of fatty acid in the samples and blanks by referring the measured absorbances to the standard curve and then applying the formula:

$$(\mu\text{Eq of sample} - \mu\text{Eq of serum blank}) \times \text{Dilution} \times \frac{\text{Extraction volume}}{\text{Volume evaporated}} \times \frac{1}{\text{Serum volume}} = \text{Fatty acid released (in } \mu\text{Eq/mL of serum)}$$

or

$$\text{Net fatty acid released} \times \text{Dilution} \times 2 \times 20 = \text{Units of lipase}$$

where 1 μEq = 1 unit of lipase activity per milliliter of serum.

REFERENCE RANGE

Serum from healthy persons will have 2 to 7.5 U of lipase activity per milliliter.

TURBIDIMETRIC METHOD (MODIFIED)[24]

Principle

The clearing of an emulsion of olive oil is measured turbidimetrically.

Reagents

Olive oil. Purify with an equal volume of alumina as previously described.

Tris buffer. Dissolve 3.0 g of tris(hydroxymethyl)aminomethane and 6.0 g of sodium deoxycholate in 1 L water. Adjust the pH to 8.8 with concentrated HCl.

Emulsion. Dissolve 1.0 g of purified olive oil in 100 mL of absolute ethanol. Add 4 mL of this solution slowly with stirring to 100 mL of Tris buffer. The reagent should be adjusted at 340 nm by addition of Tris buffer until the absorbance is within the measurable range of the photometer.

Standard. A lyophilized serum with known lipase activity is used.

Assay

Equipment: a ≤10 nm band-pass spectrophotometer with temperature-controlled cuvette.

1. To 1 mL of emulsion, add 50 μL of sample, standard or control. Mix immediately.
2. Read the absorbance of each tube at 340 nm at 1 minute (A1′) and again at 5 minutes (A5′).

Calculation

$$\frac{(A5' - A1')_{unknown} \times \text{Value}_{std}}{(A5' - A1')_{std}} = \text{Value}_{unknown}$$

Value_{std} is the lipase activity of a known standard.

REFERENCES

1. Cherry, I.S., and Crandall, I.A., Jr.: The specificity of pancreatic lipase: its appearance in the blood after pancreatic injury, Am. J. Physiol. **100:**266-273, 1932.
2. MacDonald, R.P., and LeFave, R.O.: Serum lipase determination with an olive oil substrate using a three-hour incubation period, Clin. Chem. **8:**509-519, 1962.
3. Roe, J.H., and Byler, R.E.: Serum lipase determination using a one-hour period of hydrolysis, Anal. Biochem. **6:**451-460, 1963.
4. Tietz, N.W., Borden, T., and Stepleton, J.D.: An improved method for the determination of lipase in serum, Am. J. Clin. Pathol. **31:**148-154, 1959.
5. Vogel, W.C., and Zieve, L.: A rapid and sensitive turbidimetric method for serum lipase based on differences of normal and pancreatitis serum, Clin. Chem. **9:**168-181, 1963.
6. Tietz, N.W., and Fiereck, E.A.: Measurement of lipase in serum. In Cooper, G.R., editor: Standard methods of clinical chemistry, vol. 7, New York, 1972, Academic Press, Inc., pp. 19-31.
7. Tietz, N.W., and Repique, E.V.: Proposed standard method for measuring lipase activity in serum by a continuous sampling technique, Clin. Chem. **19:**1268-1275, 1973.
8. Gindler, E.M.: Colorimetric estimation of lipase activity in blood serum with use of olive oil as substrate, Clin. Chem. **17:**633, 1971.
9. Yang, J.S., and Biggs, H.G.: Rapid, reliable method for measuring serum lipase activity, Clin. Chem. **17:**512, 1971.
10. Massion, C.G., and Seligson, D.: Serum lipase: a rapid photometric method, Am. J. Clin. Pathol. **48:**307-313, 1967.
11. Kramer, S.P., Batatos, M., Karpa, J.N., et al.: Development of a clinically useful colorimetric method for serum lipase, J. Surg. Res. **4:**23-35, 1964.
12. Patt, H.H., Kramer, S.P., Woel, G., et al.: Serum lipase determination in acute pancreatitis, Arch. Surg. **92:**718-723, 1966.
13. Whitaker, J.F.: A rapid and specific method for the determination of pancreatic lipase in serum and urine, Clin. Chim. Acta **44:**133-138, 1973.
14. Fleisher, M., and Schwartz, M.K.: An automated, fluorometric procedure for determining serum lipase, Clin. Chem. **17:**417-422, 1971.
15. Griebel, R.J., Knoblock, E.D., and Koch, T.R.: Measurement of serum lipase activity with the oxygen electrode, Clin. Chem. **27:**153-165, 1981.
16. Vogel, W.C., and Zieve, L.: A rapid and sensitive turbidimetric method for serum lipase based upon differences between the lipases of normal and pancreatitis serum, Clin. Chem. **9:**168-181, 1963.
17. Shipe, J.R., and Savory, J.: The simultaneous kinetic measurement of amylase and lipase using an automatic sampling fluoronephelometer, Clin. Chem. **19:**645, 1973.
18. Zinterhofer, L., Wardlaw, S., Jatlow, P., and Seligson, D.: Nephelometric determination of pancreatic enzymes. II. Lipase, Clin. Chim. Acta **44:**173-178, 1973.
19. Kannisto, H., Lalla, M., and Lukkari, E.: Characterization and elimination of a factor in serum that interferes with turbidimetry and nephelometry of lipase, Clin. Chem. **29:**96-99, 1983.
20. Goldberg, J.M., and Pagast, P.: Evaluation of lipase activity in serum by radial enzyme diffusion, Clin. Chem. **22:**633-637, 1976.
21. Feld, R.D., Witte, D.L., and Barrett, D.A.: Kinetic determination of serum lipase activity with the Abbott ABA-100, Clin. Chem. **22:**607-610, 1976.
22. Ziegenhorn, J., Neumann, V., Knitsch, D.W., and Zwez, W.: Determination of serum lipase, Clin. Chem. **23:**1067, 1979.
23. Massion, C.G., and McNeely, M.D.D.: An improved photometric method for lipase activity suitable for both routine and reference work, Ann. Clin. Lab. Sci. **2:**444-452, 1972.
24. Shihabi, Z.K., and Bishop, C.: Simplified turbidimetric assay for lipase activity, Clin. Chem. **17:**1150-1153, 1971.

Chapter 57 Hormones and their metabolites

Cortisol

SUMAN PATEL
JOHN A. LOTT

Cortisol, hydrocortisone, compound F, glucocorticoid
Clinical significance: p. 816
Molecular formula: $C_{21}H_{30}O_5$
Molecular weight: 362.47 daltons
Merck Index: 4674
Chemical class: steroid

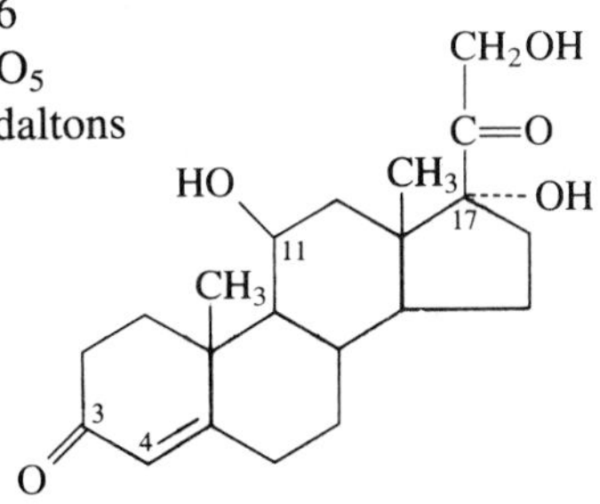

PRINCIPLES OF ANALYSIS AND CURRENT USAGE

Cortisol is a major hormonal product of the adrenal gland and is also known as compound F, hydrocortisone, and glucocorticoid. More than 90% of circulating cortisol is bound to proteins, 85% is bound to transcortin (corticosterone-binding globulin), and about 10% is bound to albumin.

Less than 1% of the total cortisol synthesized daily is excreted as such in urine, the rest is excreted as soluble metabolites and glucuronide conjugates. The excretory products of cortisol that contain the dihydroxyacetone group are known as 17-hydroxycorticosteroids (17-OH-CS). The measurement of urinary 17-OH-CS gives an indirect measure of the rate of cortisol secretion. Free cortisol in urine is derived from free-circulating cortisol.

17-OH-CS, the major metabolites of cortisol, are steroids characterized by a hydroxyl group at the C-17 position with or without oxygen at C-11. Before the availability of radioimmunoassay (RIA) methods for cortisol, urinary 17-OH-CS estimations by colorimetric or fluorometric methods were used routinely to assess the functional status of the adrenal cortex. With the development of more specific RIA procedures for cortisol, the colorimetric and fluorometric methods have become obsolete.

The colorimetric Porter and Silber procedure[4] (method 1, Table 57-1) for the analysis of cortisol and other 17-OH-CS was one of the popular methods for the assessment of adrenocortical function. In acid solution, 17,21-dihydroxy-20-ketosteroids form a 21-aldehyde that yields a yellow hydrazone when reacted with phenylhydrazine. This method, when used for urinary 17-OH-CS, requires hydrolysis of corticosteroid conjugates before the extraction. The method described by Silber[5] involves hydrolysis of corticosteroid conjugates, followed by extractions with chloroform and dichloromethane, an alkaline wash, and the color reaction with the Porter-Silber reagent at 60° C for 30 minutes. One measures the final reaction products at three different wavelengths and calculates results using an Allen correction.

Another colorimetric method, the Zimmerman reaction (method 2, Table 57-1), measures 17-ketosteroids. However, the method can be used to measure cortisol, its metabolites, and its precursors by reducing these compounds to their 17,20-dihydroxy derivatives using sodium borohydride, followed by oxidation to 17-ketosteroids by sodium *meta*-periodate. The conjugated steroids are hydrolyzed with mild alkali after the oxidation step. The 17-ketosteroids are extracted and estimated by their reaction with *m*-dinitrobenzene at alkaline conditions.[6] The compounds that react are called ''ketogenic'' steroids.

A fluorometric method originally developed by Mattingly[7] and its modifications[8] have been used to measure plasma cortisol (method 3, Table 57-1). Various interfering substances are removed with organic solvents.[9] In this method plasma cortisol is extracted with dichloromethane, and the organic phase is washed with 0.1 mol/L of NaOH. After an extraction with $CHCl_3$, powdered Na_2SO_4 is added to remove traces of water. Finally, an ethanolic sulfuric acid reagent is added to the dichloromethane phase and mixed. The acid phase is transferred to cuvettes, the solution is excited at 420 nm, and the fluorescence is re-

Table 57-1. Methods of cortisol analysis

Method	Type of analysis	Principle	Usage	Comments
1. Porter-Silber[4]	Colorimetric	Reaction of cortisol and related compounds with phenylhydrazine in alcohol–sulfuric acid solution 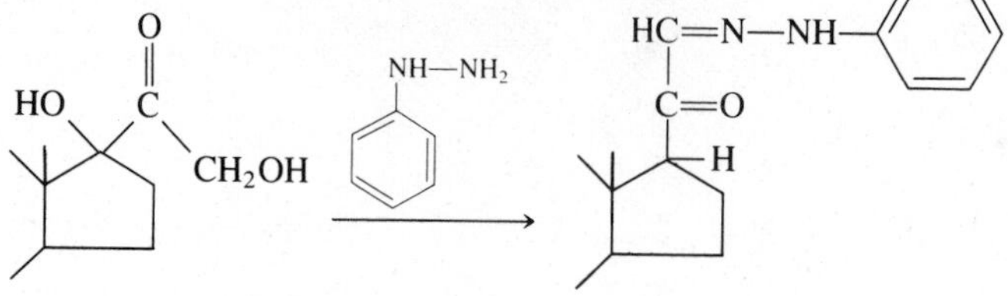 Yellow chromogen A_{max} = 410 nm	Serum/plasma	Requires organic extraction Interference from certain drugs, ketones, and glucose Rarely used
2. Zimmerman	Colorimetric	Measurement of 17-ketogenic steroids by reduction of 17-ketosteroids to C_{17},C_{20}-dihydroxy compounds with $NaBH_4$ followed by oxidation to 17-ketosteroids with sodium *meta*-periodate 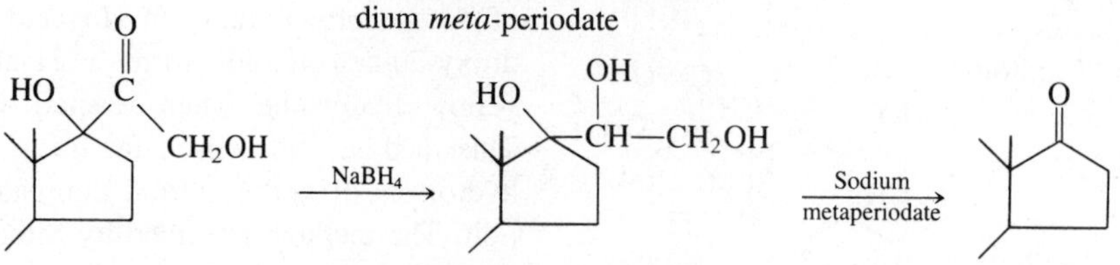 $NaBH_4$; Sodium metaperiodate Reaction of 17-keto compounds with *m*-dinitrobenzene to give colored product 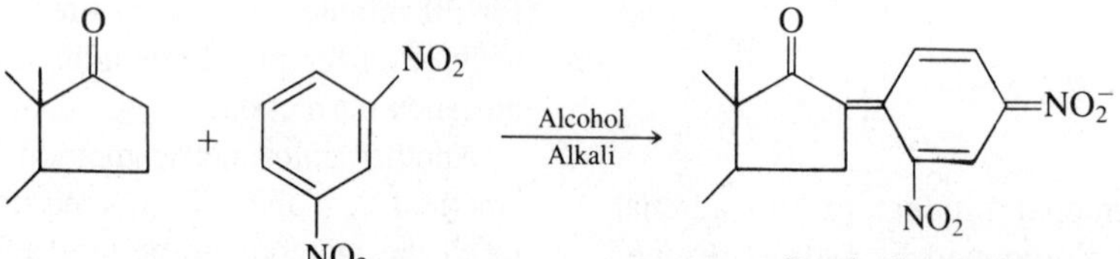 Alcohol, Alkali; *m*-Dinitrobenzene Purple chromogen A_{max} = 520 nm	Serum, plasma	Requires organic extraction Interference from drugs, ketones, and glucose Rarely used
3. Fluorometry[7-10]	Chemical derivatization	After extraction, ethanolic sulfuric acid converts cortisol to fluorometric chromogen $\text{Cortisol} \xrightarrow[H_2SO_4]{\text{Ethanol}}$ Fluorescent chromogen (λ_{exc}, 420 nm; λ_{em}, 520 nm)	Serum, plasma	Subject to interferences and quenching by reagents Fluorescence is time dependent
4. Competitive protein-binding[11-15]	Radioassay	Specific cortisol-binding protein titrated with ^{125}I-labeled cortisol Serum cortisol competes with radiolabeled cortisol Heterogeneous assay Reaction phase is liquid	Serum	Requires extraction of serum and phase separation

Table 57-1. Methods of cortisol analysis—cont'd

Method	Type of analysis	Principle	Usage	Comments
5. Radioimmunoassay	Competitive radioisotopic	Antibody specific for cortisol reacted with ^{125}I-labeled cortisol and serum Serum cortisol competes with radiolabeled cortisol Heterogeneous assay Reaction phase can be liquid or solid	Serum	Requires extraction of serum and phase separation Most sensitive method Widely used
6. Fluorescence-polarization[18] immunoassay	Competitive nonisotopic	Antibody specific for cortisol reacted with cortisol that is labeled with fluorescent dye Polarization of emitted fluorescent light indicates extent of binding Serum cortisol competes with labeled cortisol for antibody binding Homogeneous assay Reaction phase is liquid	Serum	Requires no extraction of serum and no phase separation
7. Enzyme immunoassay[20-22]				
a. Heterogeneous	Competitive binding	Enzyme labeled with cortisol allowed to compete with free cortisol for antibody Antibody complex precipitated with second antibody Heterogeneous assay in liquid phase	Serum	Not yet adopted for use in clinical laboratories
b. Homogeneous	Competitive binding	Enzyme labeled with cortisol has less activity when reacted with cortisol antibody Serum cortisol competes for antibody, changing enzyme activity Homogeneous assay in liquid phase	Serum	Requires pretreatment of serum and addition of enzyme substrate Assay available commercially
8. Chromatography[23,24]				
a. High-performance liquid (HPLC)	Separation	Cortisol separated from interfering compounds by chromatography detected at 254 nm	Serum, plasma	Requires extraction and separation
b. Gas (GC)	Separation	Cortisol derivatized and chromatographed; detected by flame ionization		Requires extraction and separation Derivatization can be difficult Yields vary
9. Gas chromatography–mass spectrometry[24]	Separation	Cortisol derivatized and chromatographed Quantitation by mass spectrometry		Same as for method 8 Candidate reference method

corded at 520 nm. Stahl et al.[10] reported that using the increase of fluorescence between 2 and 5 minutes considerably improved the specificity of the fluorometric assay and minimized interferences from nonsteroid substances.

Several competitive protein-binding assays (method 4, Table 57-1) have been developed for the measurement of serum cortisol.[11-14] Prior isolation of cortisol, using Sephadex LH-20 chromatography[15] or preheating, prewashing, and extraction followed by binding to horse transcortin in the presence of gelatin,[14] increases the sensitivity and specificity of the competitive protein-binding assay of cortisol.

RIA methods (method 5, Table 57-1) do not require

multiple extraction procedures, and they give better specificity and reproducibility. The specificity of RIA procedures depends largely on the properties of the antibody. Most RIA procedures use heat denaturation of native binding proteins or a single-step extraction of the steroids with methylene chloride or methanol before reaction with the antibody. Many commercial kits are available for the determination of plasma and urinary cortisol. Most recommend the use of 8-anilinonaphthalenesulfonic acid to aid in the extraction of plasma cortisol and methylene chloride extraction for urinary cortisol measurements.

Most antibodies to cortisol show significant cross-reactivity with 21-deoxycortisol, and thus RIA kits may give higher values for plasma cortisol in children with adrenal hyperplasia. Assays specific for plasma cortisol in patients with adrenal hyperplasia and in pregnant women require extraction and isolation of cortisol using chromatographic techniques.[16] Separation of bound and free antigen is an important aspect of RIA's. Dextran-coated charcoal techniques have been widely used; however, commercially available RIA kits using solid-phase techniques for separation of bound and free ligands are preferred.

A fluorescence-polarization immunoassay (method 6, Table 57-1) for serum cortisol has been described by Kobayashi et al.[18] The homogeneous assay uses fluorescein as a label and eliminates the antibody-bound/free ligand separation step. The fluorescence-polarization immunoassay results agree well with RIA methods.

A direct fluoroimmunoassay for cortisol described recently uses fluorescein-labeled cortisol and antibodies to cortisol coupled to magnetizable cellulose/iron oxide particles. Sodium salicylate was used as a blocking agent to prevent interference from endogeneous binding proteins in serum.[19] After incubation of the reaction mixture, the solid phase was sedimented on a multipolar ferrite magnet, and the supernatant aspirated to separate bound and free ligands. The antibody-bound fraction was eluted into an alkaline methanolic solution, and its fluorescence was measured. The assay was sufficiently sensitive and specific, and the results correlated well with an established RIA procedure.

Enzyme immunoassays (methods 7a and 7b, Table 57-1) have also been used for the determination of plasma cortisol.[20-22] Comoglio and Celada[20] used β-galactosidase conjugated to cortisol as the tracer in a heterogeneous, competitive binding assay. The antibody complex was separated from free tracer by a double-antibody (solid-phase) method.

A homogenous, competitive binding assay[23] is commercially available (Syva, Palo Alto, Calif.). Cortisol conjugated to glucose-6-phosphate dehydrogenase is used as the tracer in a two-reagent step reaction. Serum is pretreated with an acid solution to effect the release of cortisol from serum-binding proteins, which can then react with antibodies. Any cortisol-enzyme conjugate that binds to the remaining sites on the antibodies will have reduced enzymatic activity. Residual enzymatic activity, measured at 340 nm, is directly related to cortisol concentration in the sample.

Several high-performance liquid chromatographic (HPLC)[23] (method 8a, Table 57-1) and gas chromatographic–mass spectrometric[24] methods (methods 8b and 9 Table 57-1) have been described, but these methods are time consuming and not suitable for routine analysis. The HPLC method of Kabra et al.[23] was carried out by use of a 30 cm × 4 cm μBondpak C18 reversed-phase column (Waters Associates, Milford, Mass.). The column was eluted with acetonitrile/phosphate buffer, pH 3.2 (30:70 by volume), at a rate of 2 mL/min at 50° C and monitored at 254 nm. The authors suggested that cortisone, prednisone, and prednisolone may interfere with the analysis of cortisol.

REFERENCE AND PREFERRED METHODS

The chemical methods mentioned (Porter-Silber, Zimmerman, and fluorescence) are rarely used now because they are time consuming and lack specificity. A number of drugs, glucose, bilirubin, estrogens, progestins, and so on interfere with the measurement of cortisol and 17-OH-CS by the Zimmerman reaction, the Porter-Silber reaction, or fluorometric assays.[25] The Porter-Silber reaction also yields hydrazones with other aldehydes and ketones. As a result, prior purification by chromatography[26] or solvent extraction[27] is essential for accurate analysis. Also, hydrolysis of the glucuronide conjugates of corticosteroids with β-glucuronidase is required before extraction and measurement steps can be taken. A recent modification of the acid reagent increases the sensitivity of the assay.[28] The Porter-Silber method measures cortisol, 11-deoxycortisol, and cortisone, but because concentrations of steroids other than cortisol are low in blood, results obtained with the colorimetric method agree with those of competitive protein-binding assay.

The fluorescence assay has the advantages of low cost, ease of use, and high throughput. However, the fluorometric method gives consistently higher values than RIA because of acid-induced fluorescence of many steroids and other substances in urine. This nonspecific background fluorescence is equivalent to 20 to 30 μg/L of cortisol in urine. In addition, many compounds that can be present in serum, such as niacin, spironolactone, and quinidine, will also give false-positive fluorescence.

Transcortin, or cortisol-binding protein, used in the competitive protein-binding assay can also bind other steroids, such as 21-deoxycortisol, cortisone, and corticosteroid, and thus the method requires time-consuming purification steps in patients with congenital adrenal hyperplasia and in pregnant women.

RIA is the method of choice and is widely used because of its specificity, sensitivity, and precision and because results are available fairly rapidly. There are several com-

Table 57-2. Reaction conditions for serum cortisol analysis by radioimmunoassay (RIA) and enzyme-multiplied immunoassay (EMIT)

Condition	Manual RIA	EMIT (Syva), automated
Temperature	37° C	37° C
pH	Varies	8.1
Sample volume	10-25 μL	10 μL
Sample pretreatment	Not required	Pretreatment with acid and detergent for 15 min
Blocking agent	ANS*	Not required
Tracer	^{125}I-cortisol	Glucose-6-phosphate dehydrogenase with cortisol
Antibody	Rabbit or sheep; bound to glass beads[31] or tubes[32], or use solid-phase double antibody[33]	Sheep
Enzyme substrate	Not required	Glucose-6-phosphate + NAD†
Reaction time	45-90 min	5 min
Sensitivity	1-7 μg/L	5 μg/L
Precision, % coefficient of variation	5%-12%	10%
Interferences	Prednisolone, 21-deoxycortisol	Cortisone

*8-Anilino-naphthalenesulfonic acid.
†Nicotinamide adenine dinucleotide.

mercially available kits for RIA of plasma and urinary cortisol. Kits are also available for the measurement of urinary 17-OH-CS and 17-ketosteroids. Table 57-2 compares reaction conditions of the manual RIA, method and the enzyme-multiplied immunoassay technique (EMIT).

The gas chromatographic–mass spectrometric method of Björkem et al.[24] (method 9, Table 57-1) may be useful as a reference method. In this method methoxylamine and trimethylsilylimidazole are used to derivatize cortisol extracted from plasma, and 4-^{13}C-hydrocortisone is used as an internal standard.

SPECIMEN

Serum or plasma

Both serum and plasma (EDTA, heparin) are suitable. Specimens can be stored at 2° to 8° C for 2 days. For longer storage, specimens must be frozen.

Noted that abnormally high results may be obtained if the patient has taken contraceptive drugs; this results from the estrogenic compound inducing an increase in the transcortin concentrations. Similarly, concentrations of plasma cortisol may also be increased in pregnancy. Both prednisone and prednisolone interfere with the assay because of cross-reactivity with the antibody.

Urine

Free cortisol in urine can be determined by RIA. This procedure requires an extraction step using methylene chloride, and it is customary to monitor the efficiency of the extraction technique by extraction of a urine control with known cortisol concentration. A 24-hour urine sample is preferred. Ten grams of boric acid per liter of urine is used as a preservative. The sample must be frozen at −20° C if storage is necessary.

PROCEDURE

Follow the manufacturer's instructions for the assay of serum or plasma and urinary cortisol.

REFERENCE RANGE

Reference values vary to some extent, depending on the methods used for cortisol, 17-OH-CS, or 17-ketosteroids. Other factors affecting the reference range are age, sex, and diurnal variations. Values presented in Table 57-3 are

Table 57-3. Reference range values for cortisol and 17-hydroxycorticosteroids (17-OH-CS)

Test	Specimen	Reference range
Cortisol	Serum or plasma	Children[29] 7-8 A.M.: 50-70 μg/L to 210-303 μg/L (0.14-0.19 μmol/L to 0.58-0.84 μmol/L) 2-4 P.M.: 50% decrease from base-line values Adults[30] 8 A.M.: 50-230 μg/L (0.14-0.63 μmol/L) 4 P.M.: 30-150 μg/L (0.08-0.41 μmol/L) 8 P.M.: 50% of 8 A.M. values
17-OH-CS	24-hour urine	Children[30] 0.5-5.6 mg/day (1.4-15.4 μmol/day) Adults[30] Males: 4-12 mg/day (11-33 μmol/day) Females: 4-8 mg/day (11-22 μmol/day)

compiled from the literature. Plasma cortisol concentrations exhibit a circadian rhythm, reaching a peak level of about 200 μg/L between 4 and 6 A.M. and decreasing by midnight to about 60 μg/L.

REFERENCES

1. Walker, R.F., Riad-Fahmy, D., and Read, G.F.: Adrenal status assessed by radioimmunoassay of cortisol in whole saliva or parotid saliva, Clin. Chem. **14**:1460-1463, 1978.
2. Hiramatsu, R.: Direct assay of cortisol in human saliva by solid phase radioimmunoassay and its clinical applications, Clin. Chim. Acta **117**:239-242, 1981.
3. Walker, R.F., Read, G.F., and Fahmy, D.R.: The radioimmunoassay of for cortisol in parotid fluid and saliva, J. Endocrinol. **77**:26-27, 1978.
4. Porter, C.C., and Silber, R.H.: A quantitative color reaction for cortisone and related 17,21-dihydroxy-20 ketosteroids, J. Biol. Chem. **185**:201-207, 1950.
5. Silber, R.H.: Free and conjugated 17-hydroxycorticosteroids in urine, Stand. Methods Clin. Chem. **4**:113-120, 1963.
6. Rutherford, E.R., and Nelson, D.H.: Determination of urinary 17-ketogenic steroids by means of sodium metaperiodate oxidation, J. Clin. Endocrinol. Metab. **23**:533-538, 1963.
7. Mattingly, D.: A simple fluorometric method for the estimation of free 11-hydroxycorticoids in human plasma, J. Clin. Pathol. **15**:374-379, 1962.
8. Scriba, T.C. and Miller, O.A.: Determination of cortisol in serum by fluorometry. In Brewer, H., Hamel, D., and Kruskemper, H.L., editors: Methods of hormone analysis, New York, 1976, John Wiley & Sons, Inc., pp. 203-210.
9. Smith, E.K., and Muehlbaecher, C.A.: A fluorometric method for plasma cortisol and transcortin, Clin. Chem. **15**:961-978, 1969.
10. Stahl, F., Hubl, W., Schnore, D., and Dorner, G.: Evaluation of a competitive binding assay for cortisol using horse transcortin, Endocrinologie **72**:214-222, 1978.
11. Pokoly, T.B.: The role of cortisol in human parturition, Am. J. Obstet. Gynecol. **117**:549-553, 1973.
12. Talbert, L.M., Esterling, W.E., Jr., and Potter, H.D.: Maternal and fetal plasma levels of adrenal corticoids in spontaneous vaginal delivery and cesarean section, A. J. Obstet. Gynecol. **117**:554-559, 1973.
13. Turner, A.K., Carrol, C.J., Pinkus, J.L., et al.: Simultaneous competitive protein binding assay for cortisol, cortisone, and prednisolone in plasma, and its clinical application, Clin. Chem. **19**:731-736, 1973.
14. Murphy, B.E.P.: Non-chromatographic radiotransinassay for cortisol: application to human adult serum, umbilical cord serum, and amniotic fluid, J. Clin. Endocrinol. Metab. **41**:1050-1057, 1975.
15. Murphy, B.E.P.: "Sephadex" column chromatography as an adjunct to competitive protein binding assays of steroids, Nature (New Biol.) **232**:21-24, 1971.
16. Apter, D., Janne, O., and Vihko, R.: Lipidex chromatography in the radioimmunoassay of serum and urinary cortisol, Clin. Chim. Acta **63**:139-148, 1975.
17. Rash, J.M., Jerkunica, I., and Sgoutas, D.S.: Lipid interference in steroid radioimmunoassay, Clin. Chem. **26**:84-88, 1980.
18. Kobayashi, Y., Amitani, K., Watanabe, F., and Miyai, F.: Fluorescence polarization immunoassay for cortisol, Clin. Chim. Acta **92**:241-247, 1979.
19. Pourfarzaneh, M., White, G.W., Landon, J., and Smith, D.S.: Cortisol directly determined in serum by fluoroimmunoassay with magnetizable solid phase, Clin. Chem. **26**:730-733, 1980.
20. Comoglio, S., and Celada, F.: An immuno-enzymatic assay of cortisol using *E. coli* beta-galactosidase as label, J. Immunol. Methods **10**:161-170, 1976.
21. Ogihara, T., Miyai, K., Nishi, K., et al.: Enzyme-labelled immunoassay for plasma cortisol, J. Clin. Endocrinol. Metab. **44**:91-95, 1977.
22. Winfrey, L.J., Johns-Stevens, L., and Greenwood, H.M.: Development of an automated homogeneous enzyme immunoassay for serum cortisol on the ABA-100 Bichromatic Analyzer, Clin. Chem. **25**:1151, 1979, (Abstract).
23. Kabra, P.M., Tsai, L.L., and Marton, L.J.: Improved liquid chromatographic method for determination of serum cortisol, Clin. Chem. **25**:1293-1300, 1979.
24. Björkem, I., Blomstrand, R., Lantto, O., et al.: Plasma cortisol determination by mass fragmentography, Clin. Chim. Acta **56**:241-248, 1974.
25. Smith, E.K., and Tippit, D.F.: Evaluation of steroid hormone metabolism. In Kelley, V.C., editor: Metabolic, endocrine and genetic disorders of children, Hagerstown, Md., 1974, Harper & Row, Publishers, Inc., pp. 309-336.
26. Levy, S., and Schwartz, T.: A simple rapid colorimetric method for extraction and determination of urinary 17-ketosteroids using styrene divinylbenzene copolymer XAD-2 resin column, Clin. Chem. **19**:679, 1973.
27. Hall, F.F., Adams, H.R., Eddy, R.L., et al.: Simultaneous determination of urinary tetrahydro compounds and 17-hydroxycorticosteroids, Clin. Chem. **19**:678, 1973.
28. Hall, H.E.: Measurement of blood plasma cortisol by *p*-hydrazinobenzene sulfonic acid/H_3PO_4 reagent: a modified Porter-Silbert reagent, of increased sensitivity, Biochem. Med. **13**:353-358, 1975.
29. Meites, S., editor: Pediatric clinical chemistry, ed. 2, Washington D.C., 1981, American Association for Clinical Chemistry, p. 163.
30. Tietz, N.W., editor: Clinical guide to laboratory tests, Philadelphia, 1983, W.B. Saunders Co., p. 146.
31. Cortisol [^{125}I] radioimmunoassay, Corning Medical and Scientific, Corning glass works, Medfield, MA 01051.
32. GammaCoat™ [^{125}I] Cortisol radioimmunoassay kit, Clinical assays, Division of Travenol Laboratories, Inc., Cambridge, MA 02139.
33. Cortisol solid phase, Beckman Instruments, Inc., Immunosystems Operation, Brea, CA 92621.

Estriol

LAWRENCE A. KAPLAN

Clinical significance: pp. 686 and 789
Molecular formula: $C_{18}H_{24}O_3$
Molecular weight: 288.37 daltons
Merck Index: 3635
Chemical class: estrogenic (C_{18}) steroid

OH
OH
HO

PRINCIPLES OF ANALYSIS AND CURRENT USAGE

As described in Chapter 35, estriol is the major estrogen steroid of pregnancy. The concentration of plasma and urinary estriol rises a thousandfold during pregnancy and by the third trimester represents 85% to 90% of the C_{18} estrogenic steroids present in plasma and urine. The relative levels of the estriol derivatives are shown in Table 57-4. Urinary estriol is present only in the conjugated forms, since the protein-bound, unconjugated estriol is not filtered at the glomerulus.

Table 57-4. Estriol metabolites present in plasma and urine[1]

Estrogen	Urine (%)	Plasma (%)
Unconjugated estriol	0	9
Estriol-3-sulfate	4	15
Estriol-3-glucuronide	13	15
Estriol-16-glucuronide	73	21
Estriol-3-sulfate-16-glucuronide	10	40

Estriol is most frequently measured as part of the biochemical monitoring of high-risk pregnancies, and thus the sample turnaround time before a physician can obtain a result is an important consideration. The type of method used will depend most importantly on the decision to monitor estriol in either serum or urine. In either case, the monitoring is usually initiated in the beginning of the third trimester, when estriol levels rise to measurable amounts. The absolute measurement at one time is less important than changes over time.

The urinary procedures were traditionally measured on a 24-hour sample. This is a difficult specimen to accurately and repeatedly collect. When reported out as milligrams of estrogen per 24 hours, the collection error is added to the already significant biological and analytical errors. The introduction of reporting estrogen output normalized to creatinine output, the estrogen/creatinine (E/C) ratio,[3,4] improved the assay by eliminating the need for an accurate collection. The premise of the E/C ratio is that the kidneys excrete both the conjugated extriol and the creatinine by similar mechanisms. It has also been suggested that a random urine measured for an E/C ratio yields results as valid as for the longer collection.[5]

Urinary estriol output is dependent on fetal and placental synthesis as well as on maternal hepatic conjugation and renal clearance. It has thus been suggested that urinary estrogen monitoring is less valid than measurement of serum estriol for the assessment of the status of the fetoplacental unit (fetus plus placenta) because of the dependence on both maternal and placental function.[6,7] The measurement of plasma estriol has been recommended as being more specific for monitoring only the fetoplacental unit. In this regard, unconjugated estriol would have the highest specificity,[6,8] since with a half-life of only 20 minutes, it must reflect the output of the fetoplacental unit at the time of phlebotomy.

Few studies have demonstrated an overwhelming utility of one measurement (urine versus plasma) over the other. Many studies even suggest a good correlation between plasma estriol and urinary estriol measurements, both on entire populations and on individual, high-risk pregnancies.[9,10] There is a large variation in estriol output by the fetoplacental unit, and circadian rhythms, and hour-to-hours fluctuations have both been noted.[11,12] These variations appear to affect plasma estriol values more than urine values. Urinary estrogens, the E/C ratio, and total serum estriol seem to have maximal values at around midday with minimum concentrations noted about midnight, whereas the daytime concentrations for serum free estriol are greater than the nighttime values.[11] The daily fluctuations in total serum estriol appear greater than for free serum estriol, though the percentage of fluctuations for both appears to be similar.[10] The individual diurnal variations of plasma estriol can vary from 15% to 45%, with total urinary estrogens varying from 30% to 40% during the day.[12] However, when the estrogen output is reported as a E/C ratio, the variation is reduced to 12% to 15%.

Urine

The estrogen composition of urine of a pregnant woman in the third trimester is approximately 80% to 95% estriol conjugates (sulfate and glucuronate) with the remainder consisting of conjugates of estrone and estridiol.[1,2] The oldest and still most widely used procedure is the colorimetric reaction developed by Kober in 1931.[13] This method is based on the reaction of estrogens with hydroquinone in sulfuric acid to form a pinkish red, Kober chromogen of still unknown form[14] (method 1, Table 57-5). Ferrous sulfate is often added as a catalyst to the color reaction and to minimize certain interferences.[14,15] Since most urinary estrogens in the third trimester are estriol, this assay is effective in monitoring changes in excretion of estriol during pregnancy, though it is not specific for this compound.

The Kober reaction is usually performed after a hydrolysis step (acid or enzymatic) to convert conjugated estriol into the free form. Performing the hydrolysis step before extraction has been shown to increase the final estrogen yield by about 20%, but for a screening procedure this may not be a necessary step.[16] Hydrolysis of the estriols before extraction can also lead to interferences. The Kober reaction can be performed directly on the hydrolyzed urine, or the hydrolyzed estrogens can be extracted into an organic phase (such as ethyl acetate), which is removed by evaporation. Similarly, the unhydrolyzed estrogens can be extracted from the urine into a solvent. In either case, the Kober reaction is then performed on the residue. A pinkish red color is formed with estrogens, and a brownish color results from nonestrogenic compounds that are also extracted. Since the interfering compounds produce a linear background absorbance over the spectral region about the absorbance maximum of the Kober chromogen,[16] the Allen correction has been used to correct the absorbance for background interference. The Kober chromogen in aqueous solutions has an absorbance maximum at 514 nm, and the Allen correction measurements are made at 472 and 556 nm.[16]

A important modification of the Kober reaction was the extraction of the Kober chromophore into chloroform (containing 2% *p*-nitrophenol + 1% ethanol), the so-

Table 57-5. Methods of estriol analysis

Method	Type of analysis	Principle of analysis	Usage	Comment
Urine				
1. Kober reaction[13]—direct or indirect (prior extraction)	Spectrophotometric	Estriol heated in the presence of sulfuric acid and hydroquinone forms a pinkish red chromogen (absorbance maximum, 514 nm) Reaction can be performed with (indirect) or without (direct) prior extraction of estrogen (absorbance maximum, 350 nm)	Historically most popular method Frequently used today	Measures total estrogen Direct assay can have negative interference by glucose
2. Ittrich[17]	Fluorometric	Chromophore extracted and measured fluorometrically (excitation, 530 nm; emission, 550 nm) or colorimetrically	Rare	Measures total estrogen Automated on continuous-flow analyzers
3. Gas-liquid chromatography[24,25]	Flame ionization detector	Volatile derivatives chromatographed on OV-1 with N_2 mobile phase and flame ionization detector	Rare	Requires hydrolysis to free estriol Very specific
4. High-performance liquid chromatography (HPLC)[26-28]	Ultraviolet or fluorescence detection	Estriol chromatographed by reversed-phase (C_{18}) chromatography, detected at 280 nm (ultraviolet); fluorescence (excitation, 220 nm; emission, 608 nm) Dansyl derivatives detected by fluorescence	Rare	Requires hydrolysis to unconjugated estriol Very specific
5. Radioimmunoassay (RIA)[29-31]	Liquid scintillation or gamma-ray counting	Competitive binding assay, employing ^3H- or ^{125}I-labeled estriol* Free from bound label separated by second antibody, $(NH_4)_2SO_4$ or dextran-charcoal	Frequent	Requires hydrolysis to free estriol Specific
6. Enzyme immunoassay (EIA)[32]	Spectrophotometric, end point	Competitive binding assay; estriol-enzyme conjugate separated from free enzyme and enzyme reaction (peroxidase) monitored at 463 nm	Rare	Measures total estriol hydrolysis
Plasma				
7. RIA[10,12,33-39]—free or total estriol	See method 5	See method 5 Extraction required for free estriol, hydrolysis for total	Most frequent	Very specific Automated
8. EIA[40,41]	See method 6	See method 6	Rare	See method 6
9. HPLC[42]	Electrochemical	Extracted unconjugated estriol is chromatographed by reversed-phase chromatography and detected by electrochemical detection	Rare	Very specific

*^3H, tritium; ^{125}I, radioactive iodine.

called Ittrich extraction.[17] This extraction step was designed to remove the estrogen reaction product from other interfering chromogens. The Ittrich chromogen can be measured colorimetrically or fluorometrically. In chloroform, the chromophore has an absorbance maximum between 530 and 540 nm[16,17] (method 2, Table 57-5). The Allen correction has also been applied to the chloroform extract, with the secondary absorbance measurements being made at 505 and 565 nm.[18] When the extracted chromophore is excited at 530 nm, an intense, yellow-green fluorescence is produced, which can be monitored at 550 nm.[17,19] The fluorescent reaction has been adapted to continuous-flow, automated analysis.[15,20,21]

Brown and his co-workers employed several extraction steps to separate estriol from estrone and estradiol.[22,23] The estrogens were further purified and separated on alumina columns, where the fractions were monitored by the Kober reaction. This complex procedure allowed for quantitation of each individual estrogen (estriol, estrone, and estradiol).

Urinary estriol can be specifically quantitated by a variety of chromatographic techniques, all of which require a hydrolysis step to convert estriol to the unconjugated form. Gas-liquid chromatographic techniques usually require an extraction of the hydrolyzed estrogens before forming volatile derivatives, such as acetate, methyl, and tetramethylsialyl conjugates. Most separations have been on 3% OV-1 columns, with nitrogen gas as the mobile phase, with or without temperature programming, and employing flame ionization detection[24,25] (method 3, Table 57-5). High-performance liquid chromatography (HPLC) analyses of urinary estriol using reversed-phase columns have been reported. These methods employ several different types of detection methods (method 4, Table 57-5) including absorbance at 280 nm,[26] native fluorescence at 608 nm after excitation at 220 nm,[27] and derivatization with dansyl chloride and measurement of the fluorescence compound.[28]

Immunoassays developed for measurement of urine estriol include radioimmunoassay (RIA) and enzyme immunoassays (EIA) (methods 5 and 6, Table 57-5). These assays all require a hydrolysis step to convert estriol conjugates to the free form. Several RIA's have been reported using tritium (^{3}H) or radioactive iodine (^{125}I) labeling and employing a variety of techniques for separating bound from free label, including dextran-coated charcoal,[29] double-antibody precipitation,[30] and ammonium sulfate precipitation.[31] The EIA is a heterogeneous assay requiring polyethylene glycol to separate the antibody-bound estriol-enzyme conjugate from the free label.[32] The enzyme label used in these assays is horseradish peroxidase with 5-aminosalicylic acid as a substrate. The reaction is monitored as a 60-minute end-point analysis at 463 nm.

Plasma estriol

The thousandfold lower estriol concentration in serum as compared to that of urine has for the most part precluded the use of colorimetric or fluorometric assays[29] for analysis of blood estriol. Most assays for serum estriol, whether measuring total or only the unconjugated fraction, have been by RIA's (method 7, Table 57-5). The RIA's use either ^{3}H- or ^{125}I-estriol as the tracer and employ a variety of procedures to separate bound from free label,[33] including polyethylene glycol,[34] ammonium sulfate,[35] a

Table 57-6. Representative assays for urine and serum estriol

	Urine			Plasma		
Parameter	**Kober[16]**	**HPLC[26]**	**RIA***	**RIA (total)†**	**RIA (free)‡**	**HPLC[42]**
Reaction temperature	100° C	60° C (hydrolysis)	27° C (hydrolysis)	37° C (hydrolysis)	—	Ambient
Sample volume	2 mL	5 mL	20 μL	25 μL	250 μL	2 mL
Label or detection system	Spectrophotometric 514 nm	Spectrophotometric 280 nm	^{125}I	^{125}I	^{125}I	Chromatographic, electrochemical detection
Linearity	100 mg/L	100 mg/L	400 mg/L	80 μg/L	800 ng/L	0.4-8 ng
Assay time						
Incubation/extraction	20 min	60 min	15 min	75 min	2.5 hr/batch	25 min
Operational	40 min	15 min	90 min	30 min	—	25 min
Sensitivity	—	2-3 mg/L	3 mg/L	1000 ng/L	100 ng/L	1000 ng/L
Final concentration of reagents	Hydroquinone: 180 mmol/L H_2SO_4: 24.5 mol/L	—	90 × 10^3 dpm/tube§	50-70 × 10^3 dpm/tube	76 × 10^3 dpm/tube	—
Reported precision (between run): $\bar{X}$, % CV	—	15.9 mg/L, 3.7%	100 mg/L, 4.3%	150 μg/L, 8.4%	3900 ng/L, 8.8%	14.9 μg/L, 9.9%
Interferences	Ampicillin	Ampicillin	Ampicillin	Ampicillin	Estriol-3-sulfate, ampicillin	Ampicillin

*Amersham Corp., Arlington Heights, Ill.
†Nuclear Medical Systems, Inc., Newport Beach, Calif.
‡Becton-Dickinson, Automated RIA, Orangeburg, N.Y.
§*dpm*, Disintegrations per minute.

second precipitating antibody,[12,36] and solid-phase bound systems.[10,37] The use of the last technique has led to the automation of the assay.[38] Measurement of total serum estriol requires the hydrolysis of estriol conjugates to the free form before the analysis; either acid or enzyme hydrolysis is used.

Since there can be significant antibody cross-reactivity between unconjugated estriol and its conjugated forms (present in five to 10 times higher concentration), the analysis for free estriol requires separation of the free estriol from its conjugates. This is most frequently accomplished by a liquid-liquid extraction step, in which the less water-soluble, free unconjugated estriol is extracted into the organic phase with high efficiency. Alternatively, gel exclusion chromatography (such as Sephadex) has been used for this purpose.[39]

The measurement of total serum estrogens by EIA has been reported by several groups.[40,41] These assays are not specific for estriol but measure total serum estrogens, including conjugates of estriol. These assays also utilize a horseradish peroxidase antibody label, with the reaction being monitored at 463 nm (method 8, Table 57-5).

An HPLC assay for unconjugated serum estriol has also been reported.[42] The extracted estriol is separated from interfering compounds by reversed-phase chromatography (method 9, Table 57-5). Detection and quantitation of the estriol peak is by electrochemical analysis at +0.75 volts using a glassy carbon electrode.

Recent College of American Pathologists (CAP) quality assurance surveys indicate that most participating laboratories are using RIA to measure serum free estriol rather than total estriol. Approximately two thirds of those laboratories that report measuring serum estriol are measuring the unconjugated estriol form, whereas the remainder are measuring total estriol. For those laboratories measuring urinary estriol, approximately half are using either the Kober or Ittrich modifications, with the remainder using an RIA method.

REFERENCE AND PREFERRED METHODS

There is no reference method for the measurement of either urine or serum estriol. The lack of a reference method may be partly caused by the lack of a consensus as to how to best monitor estrogens during pregnancy. (See Table 57-6.)

Urine

The original colorimetric method for the measurement of urinary estriol, the Kober reaction, has sufficient sensitivity (0.2 μg) to detect only total pregnancy estrogens in urine. When the product of the Kober reaction is measured by fluorometry, the sensitivity can be increased to the low-nanogram range needed to measure serum estriol.[29] However, the poor specificity and variability of the endogenous fluorescence method reduce the usefulness of this assay unless the chromogen is further purified. Both the colorimetric and fluorometric procedures have been adapted to automated, continuous-flow analysis to increase the throughput and precision. Manual assays have reported between-day coefficient of variation (CV) of about 7.5%,[16] whereas the automated assays have reported CV's of 4% to 5% ($\bar{X}$: 12 to 29 ng/L).[21]

The direct Kober assay, performed on prehydrolyzed, unextracted urine, is subject to interferences by a wide number of compounds, especially those that form acetaldehyde during the acid hydrolysis step. The acetaldehyde reacts with the hydroquinone to give false-negative results. One of the most important of these interfering compounds is glucose, which, in the monitoring of the high-risk pregnancies of diabetic women, can cause significant negative interferences. Methenamine mandelate, as with glucose, is also known to interfere with the acid hydrolysis step and to cause false-negative results. One can reduce interference by glucose by performing the acid hydrolysis on diluted urine or extracting estriol before performing the Kober reaction.[16] A prior reduction step using sodium borohydride ($NaBH_4$) to eliminate glucose interference during a direct Kober reaction has been shown to give accurate results without the necessity of multiple readings for an Allen correction.[21] The use of enzymatic hydrolysis will certainly prevent the interference from glucose but requires a much longer time for the incubation step. General corrections for nonspecific color have been accomplished by the use of the Allen correction.

The use of chromatographic procedures for specific urinary estriol analysis appears to be both accurate and precise. Between-day CV's of 3.7% ($\bar{X}$: 15.9 mg/L)[26] and 8.5% ($\bar{X}$: 16.9 mg/L)[27] have been reported for HPLC assays. The HPLC analysis of urinary estriol with fluorescence detection may have the advantage of increased sensitivity and fewer possible interfering chromatographic peaks. Gas chromatographic analysis is probably not suited for clinical use because of the labor and time necessary for such assays. Measurement of urinary estriol by RIA is frequently performed, and good precision and excellent specificity are obtained. The intralaboratory precision for urinary RIA, as reported in a recent CAP quality assurance survey, was 15% to 20% ($\bar{X}$: 4 to 6 mg/L).

Plasma

The RIA's can be used for either urine or serum analysis, although they are most frequently employed to measure estriol levels of serum. The antibody used to measure total estriols after a hydrolysis step usually has a cross-reactivity of less than 1% to 2% with estradiol or estrone. Similarly, when one is measuring free serum estriol, the antibody should have low cross-reactivity with the conjugated forms of estriol, unless an extraction step is used to separate the conjugated from the unconjugated forms.

^{125}I seems to be the label of choice in most RIA's because of its high specific activity (higher counts) and because of its lesser environmental impact resulting from its shorter half-life. The shorter shelf-life of ^{125}I RIA's does not appear to be a major problem for most laboratories. The use of a second, precipitating antibody to separate bound from free radioactivity may have a small advantage over other techniques (polyethylene glycol, ammonium sulfate) in terms of reproducibility of technique.[10] The use of solid-phase assays has the important advantage of adaptation to automated analysis with a concomitant increase in precision. Automated, solid-phase free estriol assays are available from Becton Dickinson (Salt Lake City, Utah). The intralaboratory precision for estriol analysis, according to a recent CAP quality assurance survey, was 23% to 27% CV ($\bar{X}$: 1 to 2 ng/mL) and 10% to 13% CV ($\bar{X}$: 25 to 28 ng/mL) for free serum estriol, and 20% to 30% ($\bar{X}$: 11 to 12 ng/mL), and 10% to 17% ($\bar{X}$: 56 to 62 ng/mL) for total serum estriol.

The EIA's are quite sensitive, being able to detect approximately 0.2 nmol of estriol. However, these assays suffer from a lack of specificity, having significant cross-reactivity with several estrogenic compounds, and are primarily designed to measure total plasma estrogens.

The method of choice for the monitoring of pregnancy estrogens will depend on factors to be evaluated by each laboratory: cost, ease and reproducibility of analysis, and the clinical value of the information obtained. The lower biological day-to-day variability of the random urine E/C ratio, the low cost, and rapid turnaround time still makes this test an important method for consideration. The rapid turnaround time for a random E/C analysis is approximately the same as for free serum estriol. A rapid manual method for urinary estrogens is presented next. In addition, for those laboratories with large numbers of samples, an automated Kober method on continuous-flow analysis would be appropriate.

For those laboratories asked to measure plasma estriol, the free estriol analysis provides physicians with a slightly more rapid turnaround time than total serum estriol assays and gives results with less daily fluctuation. However, because of the cost of reagents, most laboratories batch their estriol analyses and cannot provide clinicians with estriol measurements throughout the day. Perhaps low-cost, rapid-automated HPLC analysis may provide a future alternative for results available 24 hours a day.

SPECIMEN

Urine specimens are either 24-hour collections or single voidings. Serum or plasma specimens are both valid for blood analysis. When one is monitoring patients with consecutive samples, the random urine or serum specimens should be obtained at approximately the same time of the day to minimize any error caused by biological variation. It has been shown that bacterial contamination of urine left unpreserved leads to an apparent increase in estrogens measured by an automated fluorometric assay.[43] Preservation of urine during collection with thimerosal (about 25 mg/L) or storage at 4° C effectively presented changes in measured estrogen.

Certain antibiotics, most notably ampicillin, have been shown to cause a decrease in the urinary excretion of estriol conjugates by reducing the intestinal bacteria that hydrolyze estriol conjugates.[5] This reduces the recirculation of estriol into serum and its eventual urinary excretion.

TOTAL URINE ESTROGEN DETERMINATION[16]

Principle

Urine is acidified and saturated with sodium chloride to reduce the water solubility of estriol. Estrogens are then extracted into ethyl acetate. An aliquot of this extract is evaporated, treated with hydroquinone in 64% sulfuric acid, and heated in a boiling water bath to develop color. Readings are taken at 472, 514, and 556 nm, and the Allen correction is used in the calculation of estrogen. This method has no interference by glucose because of the initial extraction.

Reagents

5 M HCl (5 mol/L). Dilute 486 mL of concentrated HCl to 1 L with distilled water. This is stable for 1 year at room temperature.

Crystalline NaC1, reagent grade

Ethyl acetate, reagent grade

Nitrogen gas

Hydroquinone in ethanol, 20 g/L (180 mmol/L). Dissolve 09.2 g of hydroquinone (Fisher, purified) in 10 mL of absolute ethanol. Prepare fresh daily.

Hydroquinone, 20 g/L (180 mmol/L) in 64% H_2SO_4 (24.5 mol/L). Add 664 mL of concentrated H_2SO_4 (98%) very slowly to 336 mL of distilled water in an ice bath. Mix after each addition and allow to cool slightly before adding more acid. Add 20 g of hydroquinone to final solution and heat to 50° C to dissolve. Store in refrigerator. This solution is stable for 2 months.

Estriol standard (1,3,5,10-estratrien-3,16α,17β-triol) (Mann Research Laboratories, New York, N.Y.)

Stock standard solution, 500 mg/L (1.73 mmol/L). Dissolve 50 mg of estriol in 100 mL of absolute ethanol. This is stable for 6 months when capped tightly at 6° to 8° C.

Working standard, 20 mg/L (6.94 μmol/L). Dilute 2 mL of stock solution to 50 mL with absolute ethanol. This is stable for 2 to 3 months when capped tightly at 6° to 8° C.

Assay

Equipment: a spectrophotometer with a band pass ≤10 nm; ground-glass stoppered, 50 mL extraction tubes.

1. Prepare the urine sample by mixing thoroughly and centrifuge to remove solid matter. The specific gravity

of each single-voided specimen should be measured with a refractometer and adjusted to 1.015 or less. Prepare a 1:20 dilution of urine for creatinine determination.

2. To each extraction tube add in the following order:
   a. 2 mL of urine
   b. 0.4 mL of 5 M HCl
   c. 2 g of NaCl
   d. 2 mL of distilled water
   e. 3 mL of ethyl acetate

   Cap the tube securely and shake vigorously for 30 seconds. Centrifuge for 5 minutes at 200 *g* at room temperature. The top layer is used for subsequent analysis.
3. To each color tube add:

Tube	2% hydroquinone in ethanol	Working standard	Ethyl acetate extract (top layer)
Blank	0.2 mL	—	—
5 μg standard	0.2 mL	0.25 mL	—
10 μg standard	0.2 mL	0.50 mL	—
Unknown	0.2 mL	—	0.5 mL

4. Place all tubes in a heating block set for no more than 50° C, and evaporate all solvent from the tubes under a stream of nitrogen.
5. Add 2 mL of hydroquinone in 64% H_2SO_4 and place in a boiling water bath. Swirl the tubes immediate to dissolve the dried extract. Stopper the tubes and heat for 40 minutes. Cool the tubes in tap water.
6. Add 1.7 mL of distilled water to each tube, mix thoroughly, and leave in the cold tap water until cool.
7. Read the absorbance of each standard and unknown tube against the reagent blank at 472, 514, and 556 nm on the spectrophotometer.

Calculations

1. A corrected absorbance at 514 nm, $A_{514\ (corr)}$, calculation for each tube by the Allen equation:

$$A_{514\ (corr)} = A_{514} - \frac{(A_{556} + A_{472})}{2} \qquad \textit{Eq. 57-1}$$

   where A_{472}, A_{514}, and A_{556} are the absorbances observed at designated wavelengths.
2. The concentration of estrogen (as estriol) in urine is calculated by the equation:

$$\mu g \text{ of estrogen/mL of urine} = A_{514\ (corr)} \times 3.6\ F \qquad \textit{Eq. 57-2}$$

$$\text{where } F = \mu g \text{ of standard}/A_{514\ (corr)} \text{ of standard}$$

3. Milligrams of estrogen excreted per 24 hours:

$$\text{mg of estrogen/24 hr} = \mu g \text{ of estrogen/mL of urine} \times \text{Urine volume (mL)} \qquad \textit{Eq. 57-3}$$

4. The estrogen/creatinine ratio (E/C):

$$E/C = \frac{\mu g \text{ of estrogen/mL of urine}}{\text{mg of creatinine/mL of urine}} \qquad \textit{Eq. 57-4}$$

Reference range

The reference range for both urinary and serum estriol varies with gestational age (see Chapter 35). The biological variability for each fetoplacental unit is shown by the last month of pregnancy, when the 95th percentile range is very broad. Even with estriol concentrations below the 95th percentile, good outcomes can be obtained, though very low values, less than 4 μg/L (13.9 nmol/L) of free estriol, are a cause for concern. Most often, rather than being concerned with the absolute concentrations, physicians follow consecutive measurements. A decrease in serum free estriol of 40% to 45% from the mean of the three previous determinations has been considered an early indicator of fetoplacental distress.[8]

REFERENCES

1. Goebelsmann, U.: The uses of estriol as a monitoring tool, Clin. Obstet. Gynecol. **6**:223, 1979.
2. Hobkirk, R., Anuman-Rajadhon, Y., and Nilsen, M.: Contribution of estriol to total urinary estrogens during pregnancy, Clin. Chem. **16**:235-238, 1970.
3. Dickey, R.P., Besch, P.K., Borys, N., and Ullery, J.C.: Am. J. Obstet. Gynecol. **94**:591-594, 1966.
4. Orsofsky, H.J., Long, R.E., O'Connell, E.J., Jr., and Marshall, L.D.: Extrogen excretion during pregnancy in a high-risk population, Am. J. Obstet. Gynecol. **109**:1-7, 1971.
5. Dickey, R.P., Grannis, G.F., and Hanson, F.W.: Use of the estrogen/creatinine ratio and the "estrogen index" for screening of normal and "high-risk" pregnancy, Am. J. Obstet. Gynecol. **113**:880-886, 1972.
6. Klopper, A.: Criteria for the selection of steroid assays in the assessment of fetoplacental function. In Klopper, A., editor: Plasma hormone assays in the evaluation of fetal well-being, New York, 1976, Churchill Division of Longmens Press, pp. 20-35.
7. Trolle, D., Bock, J.E., and Gaede, P.: The prognosis and diagnostic value of total estriol in urine and serum and of human placental lactogen hormone in the last part of pregnancy, Am. J. Obstet. Gynecol. **126**:834-842, 1976.
8. Distler, W., Gabbe, S.G., Freeman, R.K., et al.: Estriol in pregnancy. V. Unconjugated and total plasma estriol in the management of pregnant diabetic patients, Am. J. Obstet. Gynecol. **130**:424-431, 1978.
9. Dubin, N.H., Crystle, C.D., Grannis, G.F., and Townsley, J.D.: Comparison of maternal serum estriol and urinary estriol determinations as indices of fetal health, Am. J. Obstet. Gynecol. **115**:835-841, 1973.
10. Kirkish, L.S., Barclay, M.L., Parra, J.B., et al.: Plasma estriol vs. estrogen assays in 24-hour urines as an index to fetal status, Clin. Chem. **24**:1830-1832, 1978.
11. Townsley, J.D., Dubin, N.H., Grannis, G.F., et al.: Circadian rhythms of serum and urinary estrogens in pregnancy, J. Clin. Endocrinol. Metab. **36**:289-295, 1973.

12. Katagiri, H., Distler, W., Freeman, R.K., et al.: Estriol in pregnancy. IV. Normal concentrations, diurnal and/or episodic variations, and day-to-day changes of unconjugated and total estriol in late pregnancy plasma, Am. J. Obstet. Gynecol. **124**:272-280, 1976.
13. Kober, S.: Ein kolorimetrische Bestimmung des Brunsthormons (Menformon), Biochem. Z. **239**:209-212, 1931.
14. Brown, J.B.: Some observations on the Kober color and fluorescence reactions of the natural oestrogens, J. Endocrinol. **8**:196-210, 1952.
15. Lever, M., Powell, J.C., and Peace, S.M.: Improved estriol determinations in a continous flow system, Biochem. Med. **8**:188-198, 1973.
16. Grannis, G.F., and Dickey, R.P.: Simplified procedure for determination of estrogen in pregnancy urine. Clin. Chem. **16**:97-102, 1970.
17. Ittrich, G.: Eine neue Methode zur chemischen Bestimmung der oestrogenen Hormone in Harn, Z. Physiol. Chem. **312**:1-14, 1958.
18. Epstein, E., Zak, B., Powsner, E.R., and Feldkamp, C.S.: In Sonnenwirth, A.C. and Jarett, L., editors: Gradwohl's clinical laboratory methods and diagnosis, St. Louis, 1980, The C.V. Mosby Co., pp. 560-561.
19. Adessi, G., and Jale, M.F.: Rapid fluorometric estimation of estrogens in urine after twenty weeks of pregnancy, Ann. Biol. Clin. **30**:127, 1972.
20. Little, A.J., Aulton, K., and Payne, R.B.: Comparison of an all-aqueous automated urine pregnancy oestrogen method with one ring organic solvent extraction, Clin. Chim. Acta **65**:167-173, 1975.
21. Hammond, J.E., Phillips, J.C., and Savory, J.: Evaluation of an aqueous fluorometric continuous flow method for measurement of total urinary estrogens, Clin. Chem. **24**:631-634, 1978.
22. Brown, J.B.: The determination and significance of the natural estrogens. In Sobotka, H., and Stewart, C.P., editors: Advances in clinical chemistry, New York, 1960, Academic Press, Inc., vol. 3.
23. Beisher, N.A., Brown, J.B., and Wood, C.: The value of urinary estriol measurements during pregnancy, Obstet. Gynecol. Digest **10**:37, 1968.
24. Pinelli, A., and Formento, M.L.: A precise and sensitive method for the analysis of steriods in small urine samples by thin-layer chromatography and gas-liquid chromatography, J. Chromatogr. **68**:67-75, 1972.
25. Gotelli, G.R., Kabra, P.M., and Marton, L.J.: Determination of placental estriol in urine by gas-liquid chromatography with equilenin as internal standard, Clin. Chem. **23**:165-168, 1977.
26. Gotelli, G.R., Wall, J.H., Kabra, P.M., and Marton, L.J.: Improved liquid chromatographic determination of placental estriol in urine, Clin. Chem. **24**:2132-2134, 1978.
27. Taylor, J.T., Krotts, J.G., and Schmidt, G.J.: Determination of urinary placental estriol by reversed-phase liquid chromatography with fluorescence detection, Clin. Chem. **26**:130-132, 1980.
28. Schmidt, G.J., Vandemark, F.L., and Slavin, W.: Estrogen determination using liquid chromatography with precolumn fluorescent labeling, Anal. Biochem. **91**:636-645, 1978.
29. Ertel, N.H., Moskovitz, M., and Schiffer, M.A.: A modification of the rapid method for the assay of plasma estriol in pregnancy: use of unconjugated ^3H to correct for losses, J. Clin. Endocrinol. **29**:1266-1268, 1969.
30. Jawad, J.J., Wilson, E.A., and Kincaid, H.L.: Improved radioimmunoassay for total urinary estriol, Clin. Chem. **25**:99-102, 1979.
31. Anderson, D.W., and Goebelsmann, U.: Rapid radioimmunoassay for total urinary estriol, Clin. Chem. **22**:611-615, 1976.
32. Korhonen, M.K., Juntunen, K.O., and Stenman, U.H.: Enzyme immunoassay of estriol in pregnancy urine, Clin. Chem. **26**:1829-1831, 1980.
33. Product guide for radioassays and non-isotopic ligand assays, ed. 3, Clin. Chem. **29**:891-896, 1983.
34. Buster, J., and Abraham, G.E.: Radioimmunoassay of plasma dehydroepiandrosterone sulfate, Anal. Lett. **5**:543, 1972.
35. Brooks, C.T., Copas, J.B., and Oliver, R.W.A.: Total estriol in serum and plasma and determined by radioimmunoassay, Clin. Chem. **28**:499-502, 1982.
36. Goebelsmann, U., et al.: Estriol assays in obstetrics, J. Steroid Biochem. **6**:703-709, 1975.
37. France, J.T., Knox, B.S., and Fisher, P.R.: Evaluation of a new commercial solid-phase direct radioimmunoassay for unconjugated estriol in pregnancy plasma, Clin. Chem. **28**:2103-2105, 1982.
38. Astill, M.E., Larriva, M.L., Greno, J.V., et al.: Automated cortisol and free estriol methods using ^{125}I and the ARIA system, Clin. Chem. **27**:1055, 1981 (Abstract).
39. Haigh, W.G., Hoffman, L.F., and Barron, E.J.: Two-hour Sephadex column method for assay of unconjugated estriol in serum, Clin. Chem. **26**:309-312, 1980.
40. Busch, A.M., Dijkhuizen, D.M., Schuwrs, A.H., and van Weemen, B.K.: Enzyme immunoassay for total estrogens in pregnancy plasma or serum, Clin. Chim. Acta **89**:59-70, 1978.
41. Osterman, T.M., Juntunen, K.O., and Gothoni, G.D.: Enzyme immunoassay of estrogen-like substances in plasma with polyethylene glycol as precipitant, Clin. Chem. **25**:716-718, 1979.
42. Kaplan, L.A., and Hohnadel, D.C.: Measurement of unconjugated estriol in serum by liquid chromatography with electrochemical detection compared with radioimmunoassay, Clin. Chem. **29**:1463-1466, 1983.
43. Simkins, A., and Crawley, M.: Some chemical and bacterial contributions to analytical variations in urinary oestrogen quantitations during pregnancy, Med. Lab. Sci. **35**:325-334, 1978.

Human chorionic gonadotropin

LAWRENCE A. KAPLAN

Human chorionic gonadotropin (HCG, hCG, β-HCG)
Clinical significance: p. 686
Molecular weight: 47,000 daltons
Chemical class: glycoprotein, hormone

PRINCIPLES OF ANALYSIS AND CURRENT USAGE

Human chorionic gonadotropin (HCG) is a glycoprotein composed of two polypeptide subunits: the alpha and beta subunits. The polypeptide chains contain various amounts of carbohydrate moieties, including D-galactose, D-mannose, and *N*-acetylneuraminic acid, the last of which is essential for the biological activity of the hormone.[1] The α-HCG subunit is essentially identical to the alpha chain of several other pituitary polypeptide hormones, such as thyroid-stimulating hormone (TSH), follicle-stimulating hormone (FSH), and luteinizing hormone (LH). It is the varying beta subunit that gives each of these hormones their specific biological characteristics. Since it is the beta chain that specifies the biological activity of HCG, many current assays for HCG are designed to specifically detect this portion of the HCG molecule, and thus these assays are known as β-HCG assays.

HCG is synthesized by the placenta, and its appearance in urine and serum occurs relatively soon after conception. Thus the presence of this hormone serves as the basis for pregnancy testing. HCG is often elevated in nonuteral pregancies (ectopic, molar) and by the presence of certain cancers. In an acute care setting the clinician must quickly know whether a woman is pregnant. This knowledge is used in deciding whether to submit such a patient for abdominal roentgenograms or possible surgery or to rule out an ectopic or molar pregnancy. Pregnancy tests have thus been developed for use in acute care situations to provide this information accurately within 20 to 60 minutes of sample receipt.

Quantitative HCG measurements are also useful for determining the presence of trophoblastic tissue or for following threatened abortions. Rapid decrease in serum HCG can indicate a possible miscarriage.

The varying clinical needs of β-HCG analyses have produced a wide variety of methods. These procedures differ in their sensitivity and specificity, speed and ease of analysis, and cost. The most important considerations for use of any stat. assay for β-HCG are the sensitivity of the assay and time necessary to report a negative result.

The methods developed for measurement of urinary β-HCG are commonly referred to as either "slide tests" or "tube tests" (methods 1 and 2, Table 57-7). These tests use either agglutination or agglutination inhibition as visi-

Table 57-7. Methods for HCG β-HCG measurements

Method	Type of analysis	Principle	Usage	Comments
Urine				
1. "Slide tests"	a. Agglutination	Colored latex or other visible particles coated with antibodies to β-HCG Negative urines remain homogeneous; presence of HCG results in visible agglutination of particles ("clumpiness")	Infrequently used as stat. urinary pregnancy test	Least sensitive of all HCG methods Most rapid (2 to 3 min)
	b. Agglutination inhibition	Colored latex or other visible particles (red blood cells) coated with HCG; antibodies to HCG and urine are mixed with particles Negative urine results in visible agglutination; presence of HCG in urine inhibits agglutination (or protein flocculation)	Most frequently used as stat. urinary pregnancy test	
2. "Tube tests"	Same as method 1	Same as method 1; reaction occurs in tube	Also frequently used for stat. urine pregnancy tests	More sensitive than slide, some approach upper limit of sensitivity of RIA methods 45 to 120 min/assay
Serum and urine				
3. Radioimmunoassay (RIA)	Competitive inhibition	Radiolabeled (radioactive iodine, ^{125}I) HCG competes with sample analyte for binding to anti-HCG. Increased HCG in sample, decreased bound radioactivity	Infrequently used as stat. procedure Serum or urine	Most sensitive HCG assay available 40 to 60 min/assay
4. Enzyme-linked immunosorbent assay (ELISA)	Sandwich assay	a. Enzyme-labeled HCG reacts with sample HCG and binds to solid-phase anti-HCG Amount of bound enzyme activity directly proportional to amount of HCG in sample	Recently introduced serum assay, Not widely used	Reported sensitivity of 5 to 10 U/mL Assay time 1 to 3 hours
		b. Solid-phase, double-antibody sandwich ELISA in which HCG binds to antibody Enzyme-labeled antibody added, and residual activity directly related to HCG concentration	Recently introduced for urine and serum testing	Reported sensitivity 25-50 mU/mL 60-90 min assay Qualitative or quantitative assay
5. Radioreceptor assay (RRA)	Competitive inhibition	Radiolabeled HCG competes with sample analyte for binding to tissue receptor sites Increased HCG in sample, decreased bound radioactivity	Infrequently used as stat. procedure	Not as sensitive as RIA's, more sensitive than tube tests

ble indicators for a positive test. The agglutination techniques employ colored latex or some other highly visible particle that has an antibody to the β-HCG subunit bound to it. The tube tests often used red blood cells coated with antibody (hemagglutination). The urine specimen is mixed with the particle suspension. In the absence of urinary β-HCG, the particles remain in a homogeneous suspension. When β-HCG is present in the urine, it will react with the antibody-particle complex and cause agglutination of the particles. Thus, with a positive urine test, the particles and suspension will become clumpy or speckled, whereas for a negative one there will be a homogeneous cloudy suspension. Most slide tests are quite rapid, producing results within 2 to 3 minutes, whereas the tube tests require 60 to 70 minutes for a negative result to be ascertained. The agglutination tests are no longer commonly used.

Current tube and slide tests use the agglutination- (or flocculation-) inhibition method. The agglutination-inhibition procedures employ either latex particles or sheep red blood cells coated with the β-HCG molecules. These particles are mixed with the urine sample and then with a solution containing antibodies to the β-HCG subunit. In the absence of urinary β-HCG, the antibody is allowed to react with the β-HCG-coated particles and cause agglutination. When β-HCG is present in the urine sample, it will react with and neutralize the antibody, thus inhibiting particle agglutination. In these tube tests a positive urine test thus results in the formation of a visible ring of particles at the bottom of the reaction tube. A negative urine test will not have the visible ring, but a button of agglutinated particles will settle at the bottom of the tube (Fig. 57-1). In the slide tests, urine positive for HCG will remain a homogeneous suspension, whereas negative urine shows a clumped, heterogeneous suspension of agglutinated particles (Fig. 57-2). The agglutination-inhibition tube tests typically require approximately 60 to 90 minutes for a negative result, though a positive result can be detected sooner. A few methods take as long as 120 minutes for a negative result. The agglutination-inhibition slide tests are complete within 2 to 3 minutes.

There are several radioimmunoassays (RIA's) available for measurement of serum and urine β-HCG levels (method 3, Table 57-7). The RIA's are typical competitive binding assays in which the β-HCG in the sample competes with radioactive iodine labeled β-HCG for binding to a β-HCG antibody. Both solid-phase and double-antibody procedures are available. Many quantitative RIA's have been modified for shorter reaction times to act as qualitative stat. pregnancy tests.

A second type of immunoassay, only recently introduced, utilizes solid-phase, double-antibody, sandwich enzyme-linked immunosorbent assay (ELISA) technology (method 4a, Table 57-7). One method (Abbott Laboratories, Inc., Dallas, Tex.) uses horseradish peroxidase coupled to anti–β-HCG, which reacts with the β-HCG in the serum sample, and this complex binds to anti–β-HCG coated polystyrene beads. After the reaction is completed, the unbound anti-β-HCG–peroxidase complexes are removed from the tube, and the horseradish peroxidase bound to the tube is allowed to react with added substrate.

Fig. 57-1. Hemagglutination "tube" test for urinary HCG. Tube on left shows results of a urine positive for HCG. Tube on right shows results of a urine negative for HCG.

Fig. 57-2. Agglutination "slide" test for urinary HCG. *Left,* Test 1 shows results for a positive urine. *Right,* Test 2 shows results for a urine negative for HCG.

The chromogen (*o*-phenylenediamine) is oxidized and measured at 492 nm. As the levels of β-HCG increase in the sample, the amount of chromogen formed increases.

In addition, several ELISA's that employ monoclonal antibodies for the measurement of urinary HCG have been recently introduced (method 4b, Table 57-7). Hybridtech (San Diego, CA 92121) usees a bead coated with a monoclonal antibody that binds HCG present in the sample. After washing, a monoclonal antibody-alkaline phosphatase conjugate is added, and the amount of enzyme activity left after washing is related to the amount of HCG present in the sample. Monoclonal Antibodies (Mountain View, CA 94043) has a similar assay in which the initial antibody (monoclonal, anti–α-chain) is coated onto a tube. The second reagent is a monoclonal, anti-β-chain–alkaline phosphatase conjugate. Using monoclonal antibodies with different subchain specificity ensures that only intact HCG molecules will be measured. Both assays claim a sensitivity of approximately 50 mU/mL. Although they are sold as screening assays, it is also possible to quantitate the color maximum absorbance about (650 nm) developed and, by use of appropriate standards, give a more quantitative estimate of the amount of HCG present. Although none of the ELISA's is widely used at present, an increased usage of such procedures is probable.

Another type of assay system available for serum β-HCG measurements is a radioreceptor assay (RRA, method 5, Table 57-7). This method (Biocept-G, Wampole Laboratories, Cranbury, N.J.) is based on the competition between the β-HCG in the serum sample and a ^{125}I-labeled β-HCG for binding to β-HCG receptors present in tissue (such as bovine corpus luteum cell membranes). The RRA technique is not widely used.

Bioassays are available for β-HCG measurements but are only in specialized research centers.

REFERENCE AND PREFERRED METHODS

The method of choice must of necessity be a balance between the need for a rapid and yet accurate result. A false-negative result can result in a delay in the recognition of an ectopic or molar pregnancy or having the fetus subjected to unnecessary x rays or surgical procedures. On the other hand, a false-positive pregnancy test can result in a delay in the proper usage of such procedures, in an improper diagnosis, or in not providing the correct medical care in cases of acute lower abdominal pain in women.

The many tests currently available are intrinsically limited by the antibody used in the assay. Although the β-HCG and β-LH subunits are different, there is still extensive homology between the two polypeptides with resulting immunological cross-reactivity. Thus high levels of LH can be a source of false-positive results, depending on the nature of the antibody provided in each procedure. Elevated LH levels encountered in menopausal women can be a cause of false-positive results. The frequency of false-positive results in young women increases in midmenstrual cycle because of the high levels of LH found at that time. Drugs such as phenothiazines and promethazine, which may cause increased excretion of LH, may also cause false-positive results. A monoclonal antibody that reportedly has no detectable cross-reactivity to LH has been produced.[2] The use of monoclonal antibodies with absolute β-

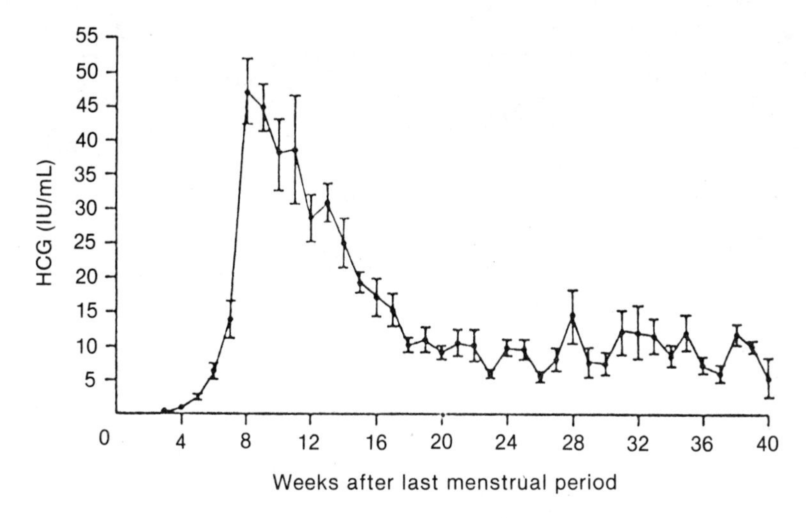

Fig. 57-3. Mean serum HCG levels throughout normal pregnancy. Arithmetical scale used on ordinate. Bars represent ±1 standard error of the mean. (From Braunstein, G.D., et al.: Am. J. Obstet. Gynecol. **126:**680, Nov. 1976.)

chain specificity will increase the accuracy of all β-HCG immunoassays in the future.

Another impediment to reproducibility within and between methods is the reported heterogeneity of HCG molecules. Precursor molecules ("big" HCG) and free β-HCG chains as well have been reported to be detected during pregnancy.[3] In addition, immunoreactive fragments of HCG have been found in the urine, but not serum, of normal pregnant women.[4] A large proportion (70%) of the immunoreactive HCG detected by RIA's in this study were HCG fragments. Variability in the sialic acid content of the molecules will also cause variation in HCG reactivity to an antibody and therefore in the accuracy of the various assays. The problem of securing acceptable, accurate HCG standards is still being resolved and will require further effort.

A source of possible false-positive results is the presence of protein in the urine sample.[5] The slide tests appear to be more sensitive to the presence of proteinuria than the tube tests, whereas the RIA's do not seem to be affected at all. The heterogeneous ELISA assays are also reported not to be affected by proteinuria. Since the response to proteinuria can vary from method to method, each laboratory must determine the rate of false positives for the slide or tube method in use. An increase in false-positive results with the tube tests may be seen in women with urinary tract infection or pelvic inflammatory disease.

The desired clinical sensitivity of the β-HCG assays is that needed to detect a pregnancy as soon after conception as possible. Thus the clinical sensitivity of the various methods depends on the levels of HCG present in early pregnancy. Serum HCG levels are detectable (approximately over 0.02 U/mL) by the third through fourth week after the last menstrual period (LMP) by RIA (Fig. 57-3).[3] This is approximately after 1 to 2½ weeks of gestation. The serum and urine β-HCG levels increase rapidly, peaking at approximately 8 weeks after LMP, and then fall off for the remaining 32 weeks of pregnancy.

However, the serum levels present in an ectopic (tubal) pregnancy are often considerably lower than for a uterine pregnancy.[6] Although by 4 weeks of gestation the β-HCG levels are approximately the same for both types of pregnancies, after 4 weeks the β-HCG levels fall off for the ectopic pregnancy, especially after a fallopian tube ruptures. Thus, after 4 to 8 weeks of ectopic gestation, serum HCG levels can range from less than 0.02 to 20 U/mL, whereas for normal pregnancies the serum β-HCG levels range from two to two-and-one-half times higher.

Urine levels of HCG can range from 0.05 to 5 U/mL during the 4 to 8 weeks after LMP. Urine HCG levels are normally not positive until the fourth to fifth weeks after LMP.

Against this background, the reported sensitivities of various commercially available β-HCG "pregnancy" tests are quite varied. In one report[7] the urine concentration needed to attain 90% to 100% sensitivity for the slide test was about 2.7 U/mL. This level is equivalent to approximately 3 weeks of intrauterine gestation or 4 to 5 weeks after LMP. On the other hand, the tube tests were reported to have 100% sensitivities at quite varied levels of β-HCG. Sensitex (Roche Diagnostics) had a 100% sensitivity at 0.10 U/mL, which approaches the sensitivity of some RIA's (less than 0.05 U/mL). The ELISA's that are now commercially available have reported sensitivities of 0.05 U/mL.

Recent surveys of the College of American Pathologists concerning slide and tube urine HCG methods showed a positivity rate greater than or equal to 98% at a level of

Table 57-8. Performance characteristics of representative urine and serum HCG methods*

Characteristic	Urine slide tests		Urine tube tests			Serum-radioassay		
	Gravindex†	UCG‡	Neocept§	Sensi-Tex‖	UCG-Beta-Stat‡	Beta-Tec‡	Beta-CG¶	Biocept‡
Temperature	Ambient	Ambient	Ambient	37° C	Ambient	Ambient	Ambient	Ambient
Sample volume	1 drop	1 drop	0.05 mL	1.0 mL	0.1 mL	100 μL	100 μL	100 μL
Time of reaction (for negative result)	2 min	2 min	60 min	90 min	60 min	10 min	70 min	50 min
Analytical sensitivity (U/mL)#	1.4	1.4	0.1-0.15	0.05-0.1	0.2-0.25	0.005	0.03	0.16
Indicator particle	Latex	Red blood cell	Red blood cell	Protein flocculation	Red blood cell	RIA	RIA	RRA
Interferences	Protein	Protein	—	—	—	—	—	—

*Data partially taken from reference 6. The urine tests listed are all agglutination- (or flocculation-) inhibition assays.
†Ortho Diagnostic, Ranton, N.J.
‡Wampole Laboratories, Cranbury, N.J.
§Organon Diagnostics, West Orange, N.J.
‖Roche Diagnostics, Nutley, N.J.
¶Monitor Science, Newport Beach, Calif.
#HCG concentration at which 50% of results were negative.

3.7 U/mL, a level at which all reported slide and tube tests should be positive. At a lower concentration, 0.6 U/mL, the rate of positivity was 96% to 100% for those tube procedures, with a stated sensitivity less than 0.6 U/mL (Gravindex, Neocept, Seni-Tex), except for UCG-Beta-Stat (tube test), which had a 78% positivity. The low rate of positivity for the UCG-Beta-Stat was reported because of possible spurious specimen-reagent interactions. Several methods with stated sensitivities less than the target value also had acceptable rates of positivity (92% to 96%). Table 57-8 reviews the lower limits of detection of several urine β-HCG assays.

Serum HCG assays are usually RIA's. Table 57-8 lists the sensitivities of several of these assays. The RIA's range from 0.005 to 0.1 U/mL sensitivity, which is adequate to detect normal pregnancies at 3 to 4 weeks after LMP and to be positive in greater than 80% of ectopic pregnancies. The RRA has a reported sensitivity less than that of RIA's but greater than that of the most sensitive tube tests for urine β-HCG.

Information on the enzyme immunoassays is still scarce, but these tests certainly have the potential to replace the agglutination (slide and tube) tests. Their reported sensitivities and short assay times make them attractive alternatives for review.

The choice of which method to have for "stat. pregnancy" testing is a difficult one. The slide and tube tests have the advantage of relatively low cost and stable reagents (7 to 18 months). Also, the specimen for these tests (urine) is readily available. The RIA kits are outdated 4 to 8 weeks after receipt and are very expensive for stat. testing. The slide and tube tests can be easily performed in a clinical chemistry laboratory 24 hours a day.

Although the quantitative RIA serum assays can take several hours to perform, they give the best sensitivity of all the available procedures. A compromise of performing a shortened RIA as a qualitative screening procedure may be attractive for some laboratories. The primary disadvantage of the RIA's is the need for specialized equipment and trained personnel; the latter may not be available 24 hours a day.

The recent introduction of ELISA assays, some using monoclonal antibodies, do provide the necessary features for an optimal stat. pregnancy test. They are sensitive (approximately 50 mU/mL), rapid (less than 60 minutes), and require no equipment or technique that is not available in most laboratories. Their primary disadvantage is the need for multiple washing steps—not a very burdensome handicap.

When choosing between the slide and tube tests for stat. pregnancy testing, it appears that the extra time needed to perform the tube test is more than warranted. The extra sensitivity and accuracy obtained from the tube tests are needed clinically, whereas the additional 60 minutes required is rarely critical for most clinical situations.

Thus the recommended procedure for stat. pregnancy tests would be one of the recently introduced serum or urine ELISA procedures. An RIA procedure should be available for serum HCG measurements for those conditions that require continued monitoring to ensure the absence of an ectopic pregnancy and to confirm all negative tube tests as well.

SPECIMEN

Urine. Centrifuge random urines at 900 *g* (approximately 4500 to 2000 rpm with a tabletop centrifuge) for 10 minutes. Urinary HCG levels are highest in an early morning specimen, approximating serum levels.

Serum. Obtain blood without the use of anticoagulants. Remove the serum from the clot as soon as possible.

Analyses for both urine and serum should be performed as soon as possible. For any undue delay, especially for serum analysis, the sample should be frozen at −20° C.

REFERENCE RANGE

The level varies widely with gestational age. See Chapter 35.

REFERENCES

1. Orten, J.M., and Neuhaus, O.W.: Human biochemistry, ed. 10, 1982, The C.V. Mosby Co., p. 608.
2. Wehmann, R.E., Harman, S.M., Birken, S., et al.: Convenient radioimmunoassay for urinary human choriogonadotropin without interference by urinary lutropin, Clin. Chem. **27:**1997-2001, 1981.
3. Braunstein, G.D., Rasor, J., Adler, D., et al.: Serum human chorionic gonadotropin levels throughout normal pregnancy, Am. J. Obstet. Gynecol. **126:**678-681, 1976.
4. Schroeder, H.R., and Halter, C.M.: Specificity of human β-choriogonadotropin assays for the hormone and for an immunoreactive fragment present in urine during normal pregnancy, Clin. Chem. **29:**667-671, 1983.
5. Hogan, W.J., and Price, J.W.: Proteinuria as a cause of false positive results in pregnancy tests, Obstet. Gynecol. **29:**585-589, 1967.
6. Dhont, M., Serreyn, R., Vandekerckhove, D., and Thiery, D.: Serum-chorionic-gonadotropin assay and ectopic pregnancy, Lancet **1:**559, 1978 (Letter).
7. Wold, L.E., Mangan, R., Bill, N., and Homburger, H.A.: An evaluation of four types of commercial immunoassays for human chorionic gonadotropin, Lab. Med. **12:**418-421, 1981.

5-Hydroxyindoleacetic acid

ARNOLD L. SCHULTZ

5-Hydroxyindoleacetic acid (5-HIAA)
Clinical significance: p. 864
Molecular formula: $C_{10}H_9NO_3$
Molecular weight: 191.19 daltons
Chemical class: acid

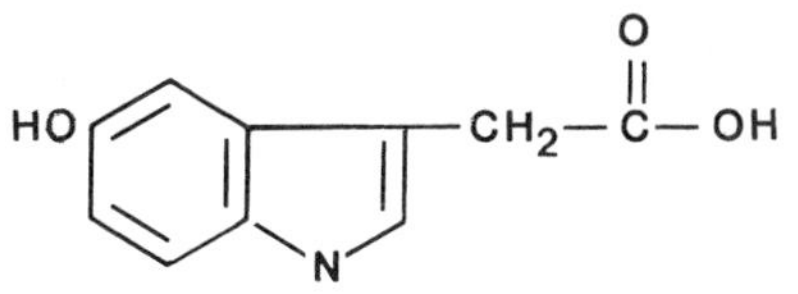

HO—(indole ring)—CH_2—C(=O)—OH

PRINCIPLES OF ANALYSIS AND CURRENT USAGE

The qualitative and quantitative analysis of the serotonin metabolite 5-hydroxyindoleacetic acid (5-HIAA) in urine was first described in 1955. The quantitative assay of 5-HIAA, based on the absorbance measurement at 540 nm of the violet chromophore formed when 5-HIAA reacts with 1-nitroso-2-naphthol in nitrous acid medium, was the first method reported[1] and is still widely used today (method 1, Table 57-9). A modification of this original determination adds 2-mercaptoethanol to the reaction mixture to improve the specificity of the assay (method 2, Table 57-9).[2] In this modification the absorbance is measured at 590 nm.

The qualitative test for 5-HIAA in urine[3] is an adaptation of the first quantitative assay. It is based on the reaction of 5-hydroxyindoles with 1-nitroso-2-naphthol and nitrous acid and is performed today as it was originally described (method 3, Table 57-9).

In the same year that the 1-nitroso-2-naphthol methodology was described, qualitative[4,5] and quantitative[4] determinations of 5-HIAA in urine with use of Ehrlich's aldehyde reagent (*p*-dimethylaminobenzaldehyde in dilute hydrochloric acid) were reported (methods 4 and 5, Table 57-9). The blue color produced in the quantitative test is measured with a spectrophotometer at 590 nm.

The quantitation of 5-HIAA in urine has been performed by gas-liquid chromatography on a silicone column utilizing a thermal conductivity detector after methylation by diazomethane (method 6, Table 57-9).[6] Spectrophotofluorometry (method 7, Table 57-9) has been used for the quantitation of 5-HIAA in both urine[7,8] and cerebrospinal fluid (CSF).[8-10] The spectrophotofluorometric procedures involve isolation of the 5-HIAA by column chromatography on either Sephadex G-10 or AG 1-X8 anion-exchange resin. The fluorescence intensity is then measured directly or after derivatization with *o*-phthalaldehyde. The direct measurement of the fluorescence is performed with excitation at either 295 or 300 nm and emission at 535 or 540 nm. If derivatization is performed before the fluorescence measurements, excitation is performed at either 360 or 355 nm, with emission measured at 480 or 475 nm. Radioimmunoassay (RIA) techniques (method 8, Table 57-9) have been applied to the measurement of 5-HIAA in tissue, blood, CSF, and perfusate.[11,12] Many high-performance liquid chromatography (HPLC) procedures (method 9, Table 57-9) for the measurement of 5-HIAA in urine and CSF using reverse-phase chromatography have been published. Some of these methods use fluorometric detection.[13,14] Electrochemical detection has been applied to the measurement of 5-HIAA in urine[15] and CSF[16] by HPLC.

REFERENCE AND PREFERRED METHODS

Two methods for the qualitative screening test for 5-HIAA acid in urine have been used. The 1-nitroso-2-naphthol procedure is more specific than the Ehrlich's aldehyde procedure and is used almost exclusively. The lower limit of detection of the 1-nitroso-2-naphthol screening test is 30 mg/L.

The quantitative test for 5-HIAA in urine using Ehrlich's aldehyde reagent has a sensitivity of 12.5 mg/L.[4] However, this method is not specific for 5-HIAA. Several other indole derivatives, such as 5-hydroxytryptamine and tryptamine, give a similar reaction.

The RIA technique has not been widely used for the determination of 5-HIAA because the antibody and labeled antigen are not readily available.

Table 57-9. Methods of 5-hydroxyindoleacetic acid (5-HIAA) analysis

Method	Type of analysis	Principle	Usage	Comments
1. 1-Nitroso-2-naphthol and nitrous acid	Spectrophotometric, quantitative	Formation of violet chromophore after extraction	Urine	Widely used Phenolic substances and metabolites of several drugs produce positive interference
2. 1-Nitroso-2-naphthol, nitrous acid, and mercaptoethanol	Spectrophotometric, quantitative	Formation of blue chromophore after extraction	Urine	Recommended method Increased sensitivity and specificity and decreased analysis time
3. 1-Nitroso-2-naphthol and nitrous acid	Qualitative	Formation of purple color Interfering chromogens extracted into ethylene dichloride	Urine	Recommended screening test Verify positive results with quantitative method
4. Ehrlich's aldehyde reagent	Spectrophotometric, quantitative	Formation of blue chromophore by reaction of reagent with indole derivatives after extraction	Urine	Lacks specificity
5. Ehrlich's aldehyde reagent	Qualitative	Direct formation of blue color	Urine	Lacks sensitivity and specificity
6. Gas chromatography	Quantitative	Gas chromatography after extraction	Urine	Tedious
7. Spectrophotofluorometry	Quantitative	Adsorption of 5-HIAA on Sephadex G-10 or AG 1-X8, elution of interfering fluorophores, followed by elution of 5-HIAA and fluorometry, with or without derivatization with *o*-phthalaldehyde	Urine, cerebrospinal fluid (CSF)	Increased sensitivity and specificity; 90% recovery
8. Radioimmunoassay	Quantitative	Radioimmunoassay	Tissue, blood, CSF, perfusate	Sensitive and specific Reagents not readily available
9. High-performance liquid chromatography	Quantitative	Fluorometric detection with and without derivatization or electrochemical detection	Urine, CSF	Sensitive and specific Still under development

The use of gas chromatography for the determination of 5-HIAA in urine has not been popular, since it involves time-consuming extractions, methylation, and a relatively long retention time.

The estimation of 5-HIAA in urine and CSF by spectrophotofluorometry after isolation on Sephadex G-10 is reported to be specific for 5-HIAA and to have a sensitivity of about 50 ng and a coefficient of variation (CV) of about 4%.[7] The procedure has a recovery of 90% to 95% and is not too difficult to perform. However, even with its apparent advantages, it does not appear that this procedure is often used.

Many HPLC methods for the determination of 5-HIAA in both urine and CSF have been published. These procedures have sensitivities in the range of 25 to 50 pg, making them the most sensitive of the methods. They also have the advantage of improved specificity; serotonin, indole, tryptamine, tryptophan, homovanillic acid, *p*-hydroxyacetanilide, and acetanilide do not interfere in the analysis.[15] However, the instrumentation utilized in these assays is not so readily available as that required for the spectrophotometric quantitation of 5-HIAA. There are many variations of the HPLC technique, and it is not possible to indicate which of the many procedures or detection methods is best.

The original spectrophotometric analysis of 5-HIAA in urine, with formation of a violet chromophore after extraction and reaction with 1-nitroso-2-naphthol and nitrous acid, is still widely used today. Drawbacks of this method include preliminary treatment of the urine with dinitrophenylhydrazine to react with any keto acids that may interfere, a double extraction with chloroform to remove indoleacetic acid from the urine, and interference from a number of phenolic subtances in the urine and from several drugs as well. *p*-Hydroxyacetanilide, derived from acetanilide or related drugs, reacts similarly. Acetaminophen, glyceryl guaiacolate, phenacetin, mephenesin, and methocarbamol all give rise to urinary metabolites that produce a positive reaction with 1-nitroso-2-naphthol. On the other hand, methenamin, phenothiazines, prochlorperazine, and promethazine inhibit color development.

Many problems with the original spectrophotometric method can be overcome by the addition of 2-mercaptoethanol to the 1-nitroso-2-naphthol and nitrous acid in the reaction mixture. The mercaptoethanol eliminates interference from phenols, indoleacetic acid, mephenesin, and glyceryl guaiacolate. It also converts the violet chromophore to a more intense blue chromophore.

A comparison of both the unmodified and the 2-mercaptoethanol-modified spectrophotometric methods with HPLC method using fluorometric detection supports the contention that the unmodified assay suffers from considerable interference, whereas the modified assay does not.[17] The correlation between the modified spectrophotometric assay and the HPLC assay is good (coefficient of correlation, $r = 0.9303$; number of determinations, $n = 44$). The linear regression line is represented by the equation:

$$\text{Modified spectrophotometric assay} = 0.85 \text{ HPLC assay} + 0.11$$

Given the availability of equipment, the mercaptoethanol-modified spectrophotometric method probably is the best for laboratories performing routine procedures. When HPLC equipment becomes available to the laboratory, this method might be the method of choice.

SPECIMEN

A random urine specimen is adequate for the screening test for 5-HIAA.

For quantitative analysis of 5-HIAA, a complete 24-hour urine collection in a bottle containing 12 g of boric acid is required. The specimen is stable for 1 week at room temperature and for periods over 1 month at 2° to 6° C.

Avocados, bananas, eggplants, pineapples, plums, and walnuts are rich in serotonin and will produce falsely elevated results. Such foods should be withheld for 3 to 4 days before the urine collection.

PROCEDURE

Principle

A. Qualitative test. With the addition of 1-nitroso-2-naphthol and nitrous acid, a purple color specific for 5-hydroxyindoles develops. Interfering chromogens are extracted into ethylene dichloride.

B. Quantitative test. 5-HIAA is extracted into ether from acidified urine. Salt is added to the urine to promote the transfer of 5-HIAA into the organic solvent. The 5-HIAA is recovered from the ether phase by reextraction into phosphate buffer at pH 7.0. Urinary phenols are not ionized at pH 7.0 and remain in the ether fraction. Treatment with 1-nitroso-2-naphthol and nitrous acid results in formation of a violet chromophore, which is converted to a blue chromophore on addition of 2-mercaptoethanol. Mercaptoethanol discharges the colors formed by indoleacetic acid and phenols other than the 5-hydroxyindole derivatives. Treatment with mercaptoethanol also eliminates interference by mephenesin and glyceryl guaiacolate. The blue chromophore produces a twofold increase in extinction relative to the violet chromophore. The absorbance maximum for the blue reaction product is 645 nm. However, Beer's law is not obeyed through a suitable range of 5-HIAA concentrations at this wavelength. A more useful linear range is obtained at 590 nm.

Reagents

A. Qualitative test

1-Nitroso-2-naphthol, 1 g/L in 95% ethanol (5.75 mmol/L). Dissolve 0.5 g of 1-nitroso-2-naphthol in 95% ethanol and dilute to 500 mL with 95% ethanol. Store in an amber-colored bottle. Use *caution,* since this solution is possibly carcinogenic.

Sulfuric acid, 1 mol/L. To approximately 800 mL of deionized water in a 1000 mL volumetric flask, add 56 mL of 17.8 mol/L (95%) sulfuric acid. Mix and dilute to volume with deionized water.

Sodium nitrite, 25 g/L (0.36 mol/L). Prepare fresh monthly. Store at 2° to 6° C.

Nitrous acid reagent (60 mmol/L). Prepare fresh before use by mixing 1 mL of 25 g/L of sodium nitrite with 5 mL of 1 mol/L sulfuric acid.

Ethylene dichloride. Redistill reagent-grade ethylene dichloride or use "Distilled in Glass" grade (Burdick & Jackson Laboratories, Inc., Muskegon, Mich.).

B. Quantitative test

Hydrochloric acid, 1 mol/L. To approximately 800 mL of deionized water in a 1000 mL volumetric flask, add 83 mL of concentrated hydrochloric acid. Mix and dilute to volume with deionized water.

Sodium chloride, reagent grade

Ether. The ether is freed of peroxide by shaking in a separatory funnel with one-half its volume consisting of aqueous ferrous sulfate (50 g/L), followed by two washes with deionized water.

Phosphate buffer, 0.1 mol/L, pH 7.0. Prepare using 2.59 g of potassium phosphate (monobasic) and 8.31 g of sodium phosphate (dibasic) heptahydrate or 4.40 g of sodium phosphate (dibasic), anhydrous, per 500 mL of solution. Check the pH with a pH meter. This buffer is stable for 2 months when stored at 2° to 6° C.

1-Nitroso-2-naphthol, 2 g/L in 95% ethanol (11.5 mmol/L). Store in an amber-colored bottle. Use caution, since this solution is possibly carcinogenic.

The commercial grade of 1-nitroso-2-naphthol (Eastman Practical, Rochester, N.Y.) is suitable for analytical use, though it may yield a light blank color. Obtain an improved reagent by dissolving 20 g of the compound in 100 mL of hot ethanol, adding decolorizing charcoal, and filtering through two sheets of Whatman No. 1 filter paper. Refrigerate the mixture overnight and filter off the brown crystalline deposit on a Büchner funnel. It can be dried under atmospheric conditions in subdued light or in a vacuum desiccator. This yields 10 to 12 g.

Sodium nitrite, 25 g/L (0.36 mol/L). Prepare fresh monthly. Store at 2° to 6° C.

Nitrous acid reagent (0.013 mol/L). Prepare fresh before use by mixing 1 mL of 25 g/L sodium nitrite with 25 mL of 1 mol/L hydrochloric acid.

Mercaptoethanol, 250 mL/L (3.57 mol/L). Prepare in a fume hood. Store at 2° to 6° C.

Ethyl acetate. Redistill reagent-grade ethyl acetate or use ''Distilled in Glass'' grade (Burdick & Jackson, Laboratories, Inc., Muskegon, Mich.).

Stock 5-HIAA standard, 250 mg/L (1.3 mmol/L). Prepare in 1 mmol/L of hydrochloric acid (1 mol/L of hydrochloric acid diluted 1:1000). This is stable when stored at 2° to 6° C in an amber-colored bottle.

Working 5-HIAA standard, 10 mg/L (50 μmol/L). Dilute 4.0 mL of the stock standard to 100 mL with aqueous thiourea (1 g/L). This is stable for 2 weeks when stored at 2° to 6° C in an amber-colored bottle.

Assay

A. Qualitative test (Table 57-10)

1. Run a normal urine control for comparison.
2. Pipet into a test tube 0.2 mL of urine, 0.8 mL of deionized water, and 0.5 mL of 1-nitroso-2-naphthol and mix.
3. Add 0.5 mL of nitrous acid reagent and mix again.
4. Let the reaction mixture stand at room temperature for 5 minutes.
5. Add 5 mL of ethylene dichloride and shake.
6. If turbidity results, the tube should be centrifuged.
7. A positive test is indicated by purple color in the top layer. No purple color will be seen with normal urine, though a slight yellow may be noticed.

B. Quantitative test (Table 57-11)

Equipment: spectrophotometer with ≤10 nm band pass, 37° C and 100° C incubators.

1. Measure the volume of the 24-hour urine collection in milliliters.
2. Place 5 mL of urine in a 50 mL glass-stoppered centrifuge tube.
3. Place 5 mL of deionized water in a 50 mL glass-stoppered centrifuge tube to be used for the blank.
4. Place 5 mL of the 10 mg/L working standard in a 50 mL glass-stoppered centrifuge tube to be used as the standard.
5. To each tube add 5 mL of 1 mol/L hydrochloric acid, a saturating amount of sodium chloride (approximately 4 g), and 25 mL of ether.
6. Shake the tubes for 5 minutes and centrifuge.
7. Transfer 20 mL of the ether extracts to clean 50 mL glass-stoppered centrifuge tubes.
8. Add 4 mL of phosphate buffer. Shake for 5 minutes and centrifuge.
9. Aspirate off the upper ether layer.
10. Transfer 2 mL aliquots of the lower aqueous layers to clean test tubes. Be careful not to carry over any ether. Do not discard the remaining aqueous solution until the analysis is completed.
11. Add 0.5 mL of 1-nitroso-2-naphthol to each tube with mixing.

Table 57-10. Reaction conditions for 5-HIAA, qualitative test

Condition	Qualitative 5-HIAA
Temperature	Room temperature
Reaction time	5 min
Final concentration of reagents	1-Nitroso-2-naphthol: 1.44 mmol/L HNO_3: 15.0 mmol/L H_2SO_4: 0.20 mol/L
Sample volume	0.2 mL
Fraction of sample volume	0.1
Sensitivity	20 mg/L

Table 57-11. Reaction conditions for 5-HIAA, quantitative test

Condition	Quantitative 5-HIAA
Temperature	37° and 85° to 100° C
pH	7.0
Reaction time	5 min and 5 min
Sample volume	5 mL (2 ml of extract)
Fraction of sample volume	0.54 (for extract)
Final concentration of reagents	1-Nitroso-2-naphthol: 1.64 mmol/L HNO_3: 3.7 mmol/L H_2SO_4: 0.28 mol/L 2-Mercaptoethanol: 193 mmol/L
Linearity	30 mg/L
Precision (CV)	4 %

12. Add 1 mL of nitrous acid reagent to each tube and mix.
13. Incubate all the tubes at 37° C for 5 minutes.
14. Add 0.2 mL of mercaptoethanol to each tube and mix.
15. Incubate all the tubes in a hot water bath (85° to 100° C) for 5 minutes.
16. Cool the tubes to room temperature.
17. Add 5.0 mL of ethyl acetate into each of the tubes. Stopper with a rubber finger cot and shake vigorously for about ½ minute. Let the layers separate.
18. Aspirate off the upper organic layer.
19. Read the absorbances *(A)* of the lower aqueous layers against the blank at 590 nm.

Calculation

$$\text{5-HIAA (mg)/24 hr} = \frac{A_{\text{unknown}}}{A_{\text{standard}}} \times 10 \times \frac{\text{24 hr urine volume (mL)}}{1000}$$

Notes

1. For the qualitative test a purple color will be seen at levels of 5-HIAA as low as 40 mg/L. At higher levels the color is more intense and is almost black at levels above 300 mg/L.

2. Color formation in the screening test may be inhibited in conditions that result in the excretion of large amounts of keto acids.
3. *p*-Hydroxyacetanilide reacts similarly to 5-HIAA in the qualitative test but is found in urine only after administration of acetanilide or related drugs.
4. Formation of a deep red color in step 11 of the quantitative test procedure may be indicative of the presence of metabolites of glyceryl guaiacolate (present in cough medicine) or related drugs.
5. If the absorbance *(A)* of the unknown is more than three times that of the standard, repeat the color development using 0.2 mL of phosphate extract from step 10. Add 1.8 mL of phosphate buffer to the sample and proceed with the analysis. For the calculation multiply the value obtained from the formula by 10 to correct for the dilution.
6. Negative interference from the presence of phenothiazine drugs and homogentisic acid has been reported for the unmodified 1-nitroso-2-naphthol and nitrous acid method. These compounds have not been examined under the conditions of the analysis presented here.

REFERENCE RANGES

A. Qualitative test. Negative
B. Quantitative test. 1.8 to 6.0 mg/24 hr

REFERENCES

1. Udenfriend, S., Titus, E., and Weissbach, H.: The identification of 3-hydroxy-3-indoleacetic acid in normal urine and a method for its assay, J. Biol. Chem. **216**:499-505, 1955.
2. Goldenberg, H.: Specific determination of 5-hydroxyindoleacetic acid in urine, Clin. Chem. **19**:38-44, 1973.
3. Sjoerdsma, A., Weissbach, H., and Udenfriend, S.: Simple test for diagnosis of metastatic carcinoid (argentaffinoma), J.A.M.A. **159**:397, 1955.
4. Hanson, A., and Serin, F.: Determination of 5-hydroxy-indole-acetic acid in urine and its excretion in patients with malignant carcinoids, Lancet **ii**:1359-1361, 1955.
5. Curzon, G.: A rapid chromatographic test for high urinary excretion of 5-hydroxy-indole-acetic acid and 5-hydroxytryptamine, Lancet **ii**:1361-1362, 1955.
6. Williams, C.M., and Sweeley, C.C.: A new method for the determination of urinary aromatic acids by gas chromatography, J. Clin. Endocrinol. Metab. **21**:1500-1504, 1961.
7. Contractor, S.F.: A rapid quantitative method for the estimation of 5-hydroxyindoleacetic acid in human urine, Biochem. Pharmacol. **15**:1701-1706, 1966.
8. Dombro, R.S., and Hutson, D.G.: A modified procedure for the determination of 5-hydroxyindoleacetic acid in the urine and cerebrospinal fluid of patients with hepatic cirrhosis, Clin. Chim. Acta **100:** 231-237, 1980.
9. Jonsson, J., and Lewander, T.: A method for the simultaneous determination of 5-hydroxy-3-indole-acetic acid (5-HIAA) and 5-hydroxytryptamine (5-HT) in brain tissue and cerebrospinal fluid, Acta Physiol. Scand. **78**:43-51, 1970.
10. Kemerer, V.F., Lichtenfeld, K.M., and Koch, T.R.: A column chromatographic method for the determination of 5-hydroxytryptamine (serotonin) and 5-hydroxyindole acetic acid in cerebrospinal fluid, Clin. Chim. Acta **92**:81-85, 1979.
11. Delaage, M.A., and Puizillout, J.J.: Radioimmunoassays of serotonin and 5-hydroxyindole acetic acid, J. Physiol. (Paris) **77**:339-347, 1981.
12. Puizillout, J.J., and Delaage, M.A.: Radioimmunoassay of 5-hydroxyindole acetic acid using an iodinated derivative, J. Pharmacol. Exp. Ther. **217**:791-797, 1981.
13. Beck, O., Palmskog, G., and Hultman, E.: Quantitative determination of 5-hydroxyindole-3-acetic acid in body fluids by high-performance liquid chromatography, Clin. Chim. Acta **79**:149-154, 1977.
14. Wahlund, K.-G., and Edlen, B.: Simple and rapid determination of 5-hydroxyindole-3-acetic acid in urine by direct injection on a liquid chromatographic column, Clin. Chim. Acta **110**:71-76, 1981.
15. Shihabi, Z.K., and Scaro, J.: Liquid-chromatographic assay of urinary 5-hydroxy-3-indoleacetic acid, with electrochemical detection, Clin. Chem. **26**:907-909, 1980.
16. Koch, D.D., and Kissinger, P.T.: Current concepts in analysis of biogenic amines and their metabolites by high pressure liquid chromatography and electrochemical detection. I. Liquid chromatography with pre-column sample enrichment and electrochemical detection. Regional determination of serotonin and 5-hydroxyindoleacetic acid in brain tissue, Life Sci. **26**:1109-1114, 1980.
17. Tracy, R.P., Wold, L.E., Jones, J.D., and Burritt, M.F.: Colorimetric vs liquid-chromatographic determination of urinary 5-hydroxyindole-3-acetic acid, Clin. Chem. **27**:160-162, 1981.

Insulin

MICHAEL D.D. McNEELY

Clinical significance: p. 520
Molecular weight: insulin, 6000 daltons
C-peptide, 3000 daltons
proinsulin, 9000 daltons
Merck Index: 4859
Chemical class: protein (hormone)

PRINCIPLES OF ANALYSIS AND CURRENT USAGE

Insulin is released as a 9000-dalton precursor molecule known as "proinsulin." Proinsulin is a union of the insulin molecule with a single polypeptide chain of 3000 daltons known as "C-peptide."

Active insulin is a 31–amino acid protein of 6000 daltons. It is composed of two chains (A and B) that are joined by disulfide bridges. The insulin molecules of humans, horses, cattle, sheep, pigs, and rabbits are biologically and to a lesser extent immunologically similar.

The analysis of insulin holds a position of singular importance in clinical chemistry, since it is this substance that was measured in the first radioimmunoassay (RIA, method 1, Table 57-12).[1] RIA remains the most widely used technique. Other assays include insulin radioreceptor methods using cultured leukocytes, bioassays that measure the stimulation of glucose oxidation, and most recently enzyme immunoassay (EIA, method 2, Table 57-12).

Many persons who have received diabetic therapy have circulating endogenous antibodies to insulin. These antibodies may increase or decrease the apparent insulin concentration because of an effect on the binding antibody or on the separation step.[2] The endogenous antibodies can be measured and can be removed by precipitation (see following discussion).

For assay purposes antibodies to insulin are usually

Table 57-12. Methods for insulin analysis

Method	Principle	Usage	Comments
1. Radioimmunoassay (RIA)	Radioactive iodine (^{125}I)–labeled insulin competes with sample insulin for binding sites on anti-insulin antibodies Separation of bound from free ligand can be accomplished by double-antibody precipitation, solid phase, and so on	Most frequently used	Can be affected by endogenous anti-insulin antibodies
2. Enzyme immunoassay (EIA)	Employs "sandwich" technique, in which second antibody is coupled to horseradish peroxidase Reaction products can be measured spectrophotometrically (500 nm) or fluorometrically ($\lambda_{excitation}$, 317 nm; $\lambda_{emission}$, 414 nm)	Research	As sensitive as RIA

raised in guinea pigs, using purified porcine insulin. Because human and rabbit insulin are very similar, antisera developed in rabbits generally do not have a high affinity for human insulin. Insulin is sufficiently immunogenic to initiate an antibody response without being covalently linked to a carrier. However, the hypoglycemia caused by the injection of pure insulin is a limiting factor when one is designing an immunization schedule. A good insulin assay will allow the detection of 10 to 30 pg (24 to 72 μU) of insulin per assay tube. Antibody cross-reactivity with human, porcine, and bovine insulin; human growth hormone; C-peptide; and proinsulin should be determined for all assays. Cross-reactivity with various animal insulins will be nearly 100%, with proinsulin up to 60%, but less than 1% with C-peptide and HGH.

For standardization purified insulin preparations are calibrated bioactively. One unit of bioactive insulin is defined as the activity contained in 0.04167 mg of the Fourth International Standard preparation, which is composed of approximately equal amounts of bovine and porcine insulin.

Insulin is readily labeled with radioactive iodine (^{125}I) using the chloramine-T procedure. Separation of bound and free labeled insulin has been carried out by use of most available techniques. These include dextran-coated charcoal, double-antibody precipitation, and polyethylene glycol. Endogenous insulin antibodies influence the separation step.

Enzyme immunoassays (EIA's) for insulin show good correlation with RIA procedures[3,4] but are not so easily performed. The EIA's are heterogeneous assays employing the sandwich technique. This technique has antibodies to insulin adsorbed onto a solid phase (such as disks or beads). Insulin from added sample remains bound to the solid-phase antibodies after a washing step. Anti-insulin antibodies conjugated to an enzyme are added to the solid phase. After washing, the amount of bound-antibody enzyme is determined by incubation of the solid phase with enzyme substrate. The enzyme most often used is horseradish peroxidase, reacting with such substrates as 5-aminosalicylic acid[4] and *p*-hydroxyphenylacetic acid.[5] The products of the enzyme reaction are measured spectrophotometrically (at 500 nm[4]) or fluorometrically[5] ($\lambda_{excitation}$, 317 or 325 nm; $\lambda_{emission}$, 414 nm). The amount of product formed is directly proportional to the amount of insulin present in the sample. A homogeneous EIA for insulin has not been developed.

Assays for C-peptide are used for detecting self-induced hyperinsulinism, since pharmaceutical preparations of insulin do not contain C-peptide.

C-peptide RIA development is difficult, since the molecule does not contain tyrosine, and synthetic or tyrosylated molecules must be used. Standardization is a source of interassay variability. It is generally agreed[5,6] that prior precipitation of endogenous antibodies is mandatory before one carries out a C-peptide assay.

REFERENCE AND PREFERRED METHODS

The reference method for insulin is a bioassay. Such methods are used to standardize insulin preparations and are only available in specialized laboratories.

For insulin assay RIA is the approach of choice. The selected method should not employ an antibody that cross-reacts (that is, at less than 1%) with C-peptide or growth hormone. With dextran charcoal, endogenous antibodies bind tracer and cause false low values. Double-antibody techniques may give falsely increased values in the presence of endogenous antibodies.[7] Therefore an appropriate method for removal of endogenous antibodies should be available. The sensitivity of the assay should be equivalent to at least 10 μU/mL (416.7 pg/mL) of sample. For C-peptide an RIA technique with a preliminary antibody precipitation is the method of choice. Before selecting a method for routine use, consult van Rijn et al.[6] Table 57-13 lists reaction conditions for insulin RIA.

A method for the detection of insulin antibodies is sometimes required. For this purpose the method of Gerbitz and Kemmier[8] is recommended.

SPECIMEN

Insulin concentrations are identical in serum and plasma. The hormone is stable for at least 12 hours at

Table 57-13. Assay conditions for insulin radioimmunoassay (RIA)

Condition	RIA
Reaction temperature	4° C
Sample volume	100 μL
Fraction of sample volume	0.2
Assay time	Approximately 20 hours
Final concentration of reagents	^{125}I-insulin: 0.2 ng/mL (40,000 cpm)
	Phosphate buffer (pH 7.5): 40 μmol/mL
	Albumin: 2 mg/mL
Sensitivity	Approximately 10 μU/mL
Linearity	10-288 μU/mL
Interferences	Anti-insulin antibodies

room temperature, for a week at 4° C, and for a month at −10° C. Random insulin assays are not useful. Specimens for insulin analysis should always be collected as part of a planned clinical protocol.

PROCEDURE

Antibody precipitation

Principle. Polyethylene glycol is added to serum or plasma in a concentration that causes precipitation of IgG (and therefore endogenous antibodies).

Reagents

Polyethylene glycol (250 g/L). Dissolve 2.5 g of PEG 8000 (7000 to 9000 daltons) (Fisher Scientific Carbowax PEG 8000) and bring to 10 mL with deionized water.

Procedure

1. Combine 1.0 mL of serum (or plasma) with 1.0 mL of PEG solution.
2. Mix on a vortex mixer.
3. Centrifuge at 3000 rpm at 4° C for 15 minutes.
4. Remove supernatant, which may now be used as a sample in the insulin assay. The result obtained on the supernatant must be multiplied by 2.

Insulin assay[9-11]

Principle. A guinea pig antiserum to porcine insulin is incubated with the test serum and ^{125}I-labeled human insulin. Separation is achieved with dextran-coated charcoal.

Reagents

Phosphate-BSA buffer, 0.05 mol/L, pH 7.5. Dissolve 8.90 g of $Na_2HPO_4 \cdot 2\,H_2O$ in about 500 mL of deionized water using mild heating. Add 2.5 g of bovine serum albumin. Adjust pH to 7.5 using solution described next. Bring to volume to 1000 mL with deionized water.

Dissolve 1195 g of $NaH_2PO_4 \cdot 2\,H_2O$ in about 200 mL of deionized water and bring to a volume of 250 mL. Use this solution to adjust the pH of the phosphate-BSA solution to exactly 7.50.

Insulin antiserum. Antiserum to porcine insulin, produced in guinea pigs, is available from New England Nuclear (549 Albany St., Boston, MA 02118).

For use, dilute the antiserum in phosphate-BSA buffer to produce 40% to 60% binding of tracer.

Stock standard (10 μg/mL: 0.24 U/mL, 1.67 nmol/mL). Dissolve 10.0 mg of crystalline porcine insulin (Sigma I3505, 24 U/mg) in 0.01 mol/L of HCl and dilute to 10 mL with phosphate-albumin buffer. Remove 1.00 mL of this solution and bring to 100 mL with phosphate-albumin buffer. This solution is stable for 1 month at −70° C in 1 mL aliquots and for 7 days at 4° C after thawing.

Working standards. Each day prepare a series of standards by diluting 250 μL of stock standard to 50 mL with phosphate-BSA to produce a 1200 μU/mL (50 ng/mL) standard.

The 1200 μU/mL standard is then diluted with phosphate-BSA according to the following values:

Concentration (μU/mL)	mL of 1200 μU/mL standard	Final volume (mL)
9.6	0.200	25
19.2	0.400	25
28.8	0.600	25
48.0	0.400	10
72.0	0.600	10
96.0	0.800	10
144.0	1.200	10
192.0	1.600	10
288.0	2.400	10

Tracer (^{125}I-labeled insulin). Obtain iodinated insulin from New England Nuclear (Boston, Mass.) (NEX-104), or label porcine insulin using the method of Greenwood et al.[12] Dilute the tracer in phosphate-BSA to produce a minimum of 20,000 cpm (counts per minute) in 200 μL with approximately 0.1 ng of insulin. This is stable for 48 hours at 4° C when diluted.

Assay

Equipment: a gamma-ray counter that will record ^{125}I.

1. Combine 200 μL of antiserum and 200 μL of labeled insulin tracer in a series of 12 × 100 mm tubes.
2. Add 100 μL of standards, controls, or unknown sera.
3. Prepare total-count (TC) and nonspecific binding (NSB) tubes by combining 200 μL of tracer and 300 μL of phosphate-BSA. Save TC tube of measurement of total counts.
4. A blank tube may be prepared for any sample by substitution of 200 μL of phosphate-BSA for antiserum.
5. Incubate overnight at 4° C (18 hours).
6. Add 2.0 mL of dextran-coated charcoal. Mix and centrifuge at 1000 *g* for 15 minutes at 4° C. Decant and count the supernatant.

Calculations

Subtract the NSB counts from all tubes. Compute for each specimen and standard:

$$\frac{\text{Counts} - \text{NSB}}{\text{Total count} - \text{NSB}} \times 100 = \%\ \text{B/T}$$

Plot % B/T versus concentration.

NOTE: If the value of the blank tubes is significantly greater than the NSB tubes, an endogenous insulin antibody should be suspected.

REFERENCE RANGE

There is no single range for insulin in serum. All values must be considered in relation to the clinical state of the patient, particularly the dynamics of the plasma glucose.

Fasting insulin values are usually less than 25 μU/mL (1042 pg/mL, 0.17 pmol/mL). During a glucose tolerance test the insulin value usually peaks between 50 and 100 μU/mL (2083 to 4167 pg/mL, 0.35 to 0.7 pmol/L) and follows the glucose curve by about 15 minutes.

In hypoglycemic attacks, in which the glucose level goes below 500 mg/L, the insulin should be suppressed below 20 μU/mL (833.4 pg/mL); during hypoglycemia the glucose/insulin ratio should be greater than 25. Otherwise, an insulin-secreting tumor should be suspected.

REFERENCES

1. Yalow, R.S., and Berson, S.A.: Immunoassay of endogenous plasma insulin in man, J. Clin. Invest. **39:**1157-1175, 1960.
2. Nisselbaum, J.S., Fleisher, M., and Schwartz, M.K.: Discrepancies between serum insulin RIA methods in the presence of endogenous antibodies, Clin. Chem. **23:**1167, 1977 (Abstract).
3. Hinsberg, W.D., III, Milby, K.H., and Zare, R.N.: Determination of insulin in serum by enzyme immunoassay with fluorimetric detection, Anal. Chem. **53:**1509-1512, 1981.
4. Yoshioka, M., Taniguchi, H., Kawaguchi, A., et al.: Evaluation of a commercial enzyme immunoassay for insulin in human serum, and its clinical application, Clin. Chem. **25:**35-38, 1979.
5. Meistas, M.T., Kumar, M.S., and Schumacher, O.P.: Diagnosis of self-induced hyperinsulinism in an insulin-dependent diabetic patient by radioimmunoassay of free C-peptide, Clin. Chem. **27:**184–186, 1981.
6. Van Rijn, H.J.M., Hoekstra, J.B.L., and Thijssen, J.H.H.: Evaluation of three commercially available C-peptide kits, Ann. Clin. Biochem. **19:**368-373, 1982.
7. Feldkamp, C.S., Chapin, E., and Shearer, G.: Blanking in insulin radioimmunoassay, Clin. Chem. **23:**1167, 1977 (Abstract).
8. Gerbitz, K.D., and Kemmier, W.: Method for rapid quantitation and characterization of insulin antibodies, Clin. Chem. **24:**890-894, 1978.
9. Albano, J.D., Ekins, R.P., Maritz, G., et al.: A sensitive, precise, radioimmunoassay of serum insulin relying on charcoal separation of bound and free hormone moieties, Acta Endocrinol. (Kbh.) **70:**487-509, 1972.
10. Velasco, C.A., Cole, H.S., and Camerini-Davalos, P.A.: Radioimmunoassay of insulin, Clin. Chem. **20:**700-702, 1974.
11. Jorensen, K.R.: Evaluation of the double antibody radioimmunoassay of insulin and the determination of insulin in plasma and urine in normal subjects, Acta Endocrinol. **60:**327-351, 1969.
12. Greenwood, F.C., Hunter, W.M., and Glover, J.S.: The preparation of I-131-labelled human growth hormone of high specific radioactivity, Biochem. J. **89:**114-123, 1963.

Thyroid-stimulating hormone

I-WEN CHEN
LINDA A. HEMINGER

Thyroid-stimulating hormone (TSH, thyrotropin)
Clinical significance: p. 748
Molecular weight: 28,300 daltons
Merck Index: 9154
Chemical class: glycoprotein, hormone

PRINCIPLES OF ANALYSIS AND CURRENT USAGE

Thyroid-stimulating hormone (TSH, thyrotropin) is a glycoprotein secreted by the anterior lobe of the pituitary gland that is capable of stimulating the normal thyroid gland to synthesize and secrete thyroxine (T_4) and triiodothyronine (T_3). TSH is structurally similar to two other pituitary glycoprotein hormones, follicle-stimulating hormone (FSH) and luteinizing hormone (LH), as well as to the placental glycoprotein hormone, chorionic gonadotropin (CG). They are all composed of two dissimilar noncovalently bound subunits; the α-subunit is hormone-nonspecific, whereas the β-subunit is distinct for each hormone and confers biological and immunological specificity. The primary amino acid structure of both subunits has been established.[1] The α-subunit of human TSH (molecular weight, 13,600 daltons) is a single-chain polypeptide consisting of 89 amino acids and two carbohydrate side chains; it is identical to the α-subunit of human LH and differs from that of human FSH and CG only by a tripeptide at the NH_2 terminal. The β-subunit of human TSH is also a single-chain polypeptide consisting of 112 amino acids and one carbohydrate side chain (molecular weight, 14,700 daltons); its primary amino acid structure is significantly different from those of FSH, LH, and CG. Available evidence indicates that the α- and β-subunits may be synthesized independently and then combined to form TSH before it is secreted from the pituitary gland.

Past measurement of TSH was dependent solely on bioassays, which were time consuming, tedious, insensitive, and nonspecific. The measurement of serum TSH did not become a routine laboratory test for clinical evaluation of the thyrometabolic state until purified human pituitary TSH became available for development of antibodies and radiolabeling and for use as a standard in a radioimmunoassay (RIA) specific for human TSH.[2,3] The RIA for TSH (method 1, Table 57-14) is based on competition between a constant trace amount of radiolabeled TSH and TSH from patient serum samples for a fixed and limited number of TSH-antibody binding sites. The amount of labeled TSH bound to the antibody is inversely related to the amount of unlabeled TSH present in the sample.

The antisera used in TSH RIA's frequently show significant cross-reaction with other glycoprotein hormones, es-

Table 57-14. Methods of TSH analysis

Methods	Principle	Marker	Comments
1. Radioimmunoassay (RIA)	Competitive binding of radiolabeled TSH and nonlabeled TSH to limited binding sites on antibody	^{125}I	Most widely used
2. Immunoradiometric assay	Binding of TSH to radiolabeled antibody	^{125}I	May be more sensitive and specific than RIA's Fully automated
3. Immunoenzymometric assay (IEMA or ELISA)	Binding of TSH to enzyme-labeled antibody	Enzyme	May become an excellent alternative to radioassay as sensitivity is improved
4. Immunoluminometric assay (ILMA)	Binding of TSH to luminescent molecule–labeled antibody	Luminescent labels such as luminol and luciferase	Has potential of being very sensitive assay Not yet available for routine use
5. Cytochemical bioassay	Detection of TSH-induced increase in lysosomal membrane permeability in thyroid follicular epithelial cells to a chromogenic substrate	Leucyl-β-naphthylamide	Most sensitive, but cumbersome and time consuming Not suitable for routine clinical laboratories

pecially LH and CG. This is partly a result of the presence of anitbodies directed to the α-subunit, which is shared by all the glycoprotein hormones mentioned. Thus the TSH RIA can be made more specific by adding sufficient quantities of CG, which contains the α-determinant, for absorption of the nonspecific cross-reacting antibodies. However, it has been reported that structural similarities among the β-subunits of the glycoprotein hormones are more important in determining the cross-reactivity of LH and CG in the TSH RIA.[4] Therefore, it is extremely important to evaluate the cross-reactivity with glycoprotein hormones of the antiserum used in the TSH assay.

The other complicating factor involved in the RIA of TSH is the variability of TSH preparations used as the standard. The standard preparations of human TSH are calibrated by bioassay and are expressed in international units (U). Such preparations are available from several agencies, such as the World Health Organization, the Medical Research Council (Great Britain), and the National Institutes of Health (United States). The standard TSH used in RIA's should therefore be calibrated against one of these standard preparations. However, it is important to remember, as with any immunoassay, that the TSH RIA measures the immunological activity, not the biological activity, of circulating TSH, though the results of TSH RIA's are customarily expressed as milli–international units (mU) of biological activity per liter of serum. It is possible that a TSH molecule loses its biological activity without losing its immunological activity. However, all available data indicate that TSH concentrations measured by RIA correlate well with the biological activity and are excellent biological indicators for the diagnosis of disorders of the hypothalamic-pituitary-thyroid axis.

In addition to the conventional RIA, a solid-phase two-site immunoradiometric assay (IRMA) developed by Miles and Hales[5] has also been used for the measurement of circulating TSH (method 2, Table 57-14). In this method TSH is assayed directly by reaction with excess radiolabeled specific antibodies rather than by competition with a radiolabeled TSH molecule for a fixed number of binding sites on a limited amount of antibody. In general, samples containing TSH are reacted with a radiolabeled antibody directed toward a unique site on the TSH molecule and with an immobilized solid-phase antibody directed against a different antigenic site on the same TSH molecule. The radiolabeled antibody/TSH/solid-phase antibody complex formed is separated from the reaction mixture, washed, and counted. The radioactivity measured is directly proportional to the concentration of TSH present in the test sample. An assay of this type is also known as a ''sandwich'' assay, since the antigen TSH becomes sandwiched between labeled and immobilized antibodies. Recently, Hybritech Inc. (San Diego, Calif.) has marketed a TSH IRMA kit using monoclonal antibodies with high affinity to TSH and virtually no cross-reactivity with LH, FSH, or CG.[6] The minimum detectable concentration of TSH by this assay is claimed by the manufacturers to be 0.2 μU/mL, which is five to 10 times more sensitive than most commercial RIA kits.

Immunoassays of TSH using enzymes as a label have also been recently developed. The major disadvantage of enzyme immunoassays (EIA's) has been the relatively low sensitivity, believed to be attributed primarily to the steric hindrance introduced into the antigen-antibody reaction by the presence of the enzyme macromolecule. The problem of the steric hindrance appears to be partly eliminated by the sandwich technique, in which enzyme-labeled antibodies instead of antigens are used. In the sandwich, solid-phase, two-site enzyme-linked immunosorbent assay (ELISA) developed by Ventrex Laboratories (method 3, Table 57-14), the test serum is allowed to react with two different anti-TSH monoclonal antibodies, one solid phase and the other labeled with horseradish peroxidase (HRPO). The HRPO-labeled antibody/TSH/solid-phase antibody complex formed is separated, washed, and reacted with a chromogenic substrate, *o*-phenylenediamine. The yellow color developed is measured spectrophotometrically at 492 nm. The concentration of TSH is directly proportional to the color intensity (enzyme activity). A somewhat different technique for labeling antibodies is used in the two-site ELISA for TSH recently developed by Abbott Laboratories (North Chicago).[7] In this method biotin (a growth factor, vitamin H) is used to label antibodies (biotinylation). The biotinylated antibody/TSH/solid-phase antibody complex formed is separated, washed, and incubated with avidin-HRPO conjugates. Avidin is a glycoprotein found in large amounts in raw egg white and is capable of avidly binding biotin. The resulting avidin-HRPO conjugate/biotinylated antibody/TSH/solid-phase antibody complex is separated, washed, and reacted with a chromogenic substrate for the determination of HRPO activity. The biotinylated antibody has several advantages over the HRPO-labeled antibody: (1) since biotin is a small molecule (244 daltons) compared with HRPO (40,000 daltons), and since HRPO is not incorporated into the sandwiched complex until after the antigen-antibody complex formation, the steric hindrance introduced into the antigen-antibody reaction by the label is minimal; (2) up to four biotin molecules have been incorporated into a molecule of antibody, resulting in an increase in assay sensitivity through signal enhancement; (3) biotinylation of antibodies is technically easier than labeling of antibodies with enzymes; and (4) the avidin-HRPO conjugate may be employed in any ELISA techniques using biotin as a label. Recently, Woodhead of the Welsh National School of Medicine reported development of a chemiluminescent immunoassay of TSH capable of detecting as low as 0.04 μU/L (method 4, Table 57-14).[8] The extreme sensitivity is said to be attributable to the high quantum efficiency of the luminescent compound used for labeling anti-TSH antibody.

The most sensitive method by far for detection of TSH is the cytochemical bioassay (method 5, Table 57-14).[9] This assay will detect as little as 5×10^{-5} mU/L of TSH—about 10^4 times more sensitive than current immunoassay techniques. Unfortunately, the method requires facilities for tissue culture, cryostat sectioning, and scanning microdensitometry, and only a very limited number of samples can be analyzed per run. Therefore it is not suitable for use in routine clinical laboratories. The technique depends on the detection of TSH-induced increase in the permeability to a chromogenic substrate of the lysosomal membrane in thyroid follicular epithelial cells. Segments of guinea pig thyroid tissue maintained in nonproliferative culture are exposed to test material for various time intervals, chilled in *n*-hexane at $-70°$ C, and subjected to cryostat sectioning. The cryostat sections are treated with the chromogenic substrate, leucyl-β-naphthylamide, which penetrates the lysosomal membrane to interact with the intralysosomal naphthylamidase. The colored reaction product formed in each follicle cell is measured at 550 nm by scanning integrating microdensitometry. The TSH concentration in the test material is directly proportional to the absorbance.

At present, radioassay, which includes the conventional RIA's and the two-site IRMA, are used almost exclusively in clinical laboratories for the routine measurement of serum TSH levels because of the availability of a wide variety of commerical kits. According to the 1982 College of American Pathologists (CAP) basic ligand assay survey, 23 commercial assay kits manufactured by 23 different companies were used in 1768 participating laboratories. About 63% of the participating laboratories reported using conventional RIA.

REFERENCE AND PREFERRED METHODS

The cytochemical bioassay may be considered as the reference method, since it measures the biological activity of TSH. However, because of the specialized techniques and equipment required, it is only available at research centers. In most commercial RIA kits the antibody-bound TSH is separated from unbound TSH by precipitation of the primary antibody with a second antibody directed against the primary antibody. The immunoprecipitation is frequently enhanced by the addition of polyethylene glycol, or the second antibody is bound to a solid supporting material such as cellulose or polymer particles. Polyethylene glycol alone has been used as the sole precipitating agent for the antibody-antigen complex in TSH RIA's but is seldom used now because of variations in nonspecific binding caused by varying concentrations of globulins in serum samples.[10]

Most commercial TSH RIA's are designed as a confirmatory test for primary hypothyroidism, that is, for the detection of an elevated serum TSH level, and thus usually place more emphasis on a shorter assay time than the sensitivity sufficient to discern between normal circulating TSH levels and levels in patients with TSH deficiency (Table 57-15). However, it is necessary for a TSH assay to be able to reliably distinguish normal from low concentrations of TSH in order for the assay to be useful in diagnosing hypothyroidism and hyperthyroidism as well. Such sensitivity can be achieved by proper selection of assay conditions[4]: (1) use of radioiodinated TSH with low specific activity (50 to 80 microcuries per microgram, (2) use of TSH-free human serum as a matrix in TSH standards, and (3) a delay of 48 to 72 hours in the addition of radiolabeled TSH in a nonequilibrium assay. Nisula and Louvet[11] were able to improve their TSH assay sensitivity

Table 57-15. Reaction conditions for TSH assays

Conditions	Radioimmunoassay (Nuclear Medical)	Manual immunoradiometric assay (IRMA) (Abbott)	Automated IRMA (ARIA-HT)	Immunoenzymometric assay (IEMA) (Ventrex)
Temperature	37° C	15°-30° C	37° C	37° C
Reaction time	2 hr without tracer 3 hr with tracer	2 hr	1 hr	1.5 hr
pH	7.8	8.0	8.0	8.0
Sample volume	200 μL	100 μL	240 μL	200 μL
Tracer	^{125}I-labeled TSH	^{125}I-labeled antibody	^{125}I-labeled antibody	Peroxidase
Antibodies	Rabbit primary Goat secondary	Goat: ^{125}I-labeled Rabbit: bead support	Rabbit antibodies: ^{125}I-labeled, polysaccharide support	Monoclonal antibodies: peroxidase labeled, bead support
Enzyme substrate	Not required	Not required	Not required	*o*-Phenylenediamine
Sensitivity*	1.0 mU/L	0.5 mU/L	1.2mU/L	0.5 mU/L
Precision†	1.9%-13.3% (3.0-25.8 mU/L)	2.0%-5.8% (7.1-27.7 mU/L)	2.4%-23.4% (1.5-16.4 mU/L)	3.2%-8.9% (3.3-36.7 mU/L)
Interferences	Lipemic sera, radioactive drugs	Radioactive drugs	Hemolyzed or turbid sera, radioactive drugs	Enzyme inhibitors

*Minimum detectable concentration.
†Coefficient of variation at concentration ranges indicated in parentheses, as reported by the manufacturers.

to 0.56 mU/L by concentrating human TSH in serum using concanavalin A covalently linked to Sepharose; they were then able to detect TSH in 20 of 21 normal adults. In our laboratory[12] we were able to improve the sensitivity of a commercial TSH RIA kit (Diagnostic Products Corp., Los Angeles) to 0.4 mU/L by reducing the incubation temperature, increasing the incubation time, and decreasing the incubation volume, and we were then able to detect TSH on all 50 healthy subjects studied. Efforts to develop more sensitive assays for TSH continue unabatedly.

Although the conventional RIA technique is used most frequently for routine clinical assays of TSH, the two-site IRMA technique seems to be gaining popularity in recent years. The main advantage of the sandwich method is that the radiolabeled antibody is usually more stable and can have a higher specific activity than the labeled antigen (TSH) used in the conventional RIA. In addition, the sandwich method theoretically can be more specific, since only those molecules possessing the two immunoreactive sites will be determined by the assay. It can also be more sensitive, since sensitivity depends on both the specific activity of the radiolabeled antibodies used and the signal-to-noise ratio, as well as the avidity of the antiserum, which is the most important limiting factor in setting the sensitivity of the conventional RIA. The major disadvantage of the sandwich method is the requirement of two purified antibodies capable of recognizing two different antigenic sites on the same antigen molecule, which requires tedious and time-consuming antibody selection and purification processes. Such problems can be simplified through the use of monoclonal antibody techology, which can provide large quantities of pure, homogeneous antibody with precisely defined immunochemical properties.

Two automated systems (ARIA/HT by Becton Dickinson Immunodiagnostics, Salt Lake City, Utah, and Concept 4 by Micromedic Systems, Inc., Horsham, Penn.) are capable of assaying TSH in a fully automated fashion. The two-site IRMA technique is used in both systems. In ARIA-HT, the test sample is incubated with radiolabeled antibody at 37° C for 60 minutes. The labeled antibody–TSH complex formed is separated from unreacted labeled antibody when the reaction mixture is passed through a chamber containing an excess of immobilized antibody. The immune complex is retained in the chamber, and the unreacted labeled antibody is washed away. The immune complex is then eluted from the chamber and counted, and the chamber is regenerated when it is washed with buffer for the next cycle. In Concept 4 the test sample and radiolabeled antibody are incubated in an assay tube coated with an antibody to TSH. After incubation the reaction mixture is removed by aspiration, and the labeled antibody–TSH complex bound to the tube wall is washed and counted. Since incubation time is 5 to 18 hours at room, temperature, it is more logical to set up this system for an overnight run. However, this system is also designed to perform off-line incubation; thus it can be used for other assays during the 5 to 18 hours of incubation. One drawback of the automated system for TSH assays is the requirement for a relatively large amount of serum sample: 240 μL for ARIA-HT and 200 μL for Concept 4 for a single assay.

TSH assays with nonradioactive labels as the tracer have not been widely used in clinical laboratories, primarily because of the limited availability of commercial kits. Sensitive immunoenzymometric assays using an enzyme-labeled antibody and immunoluminometric assays using a luminescent compound as a tracer have been developed and are expected to become excellent alternatives to radioassays for TSH, especially in laboratories without dedicated radioactivity counters.

SPECIMEN

Since TSH secretion is relatively constant throughout the day, useful clinical information can be obtained from a blood specimen drawn at any convenient time. Serum or plasma samples can be used, but some kit manufacturers recommend use of serum samples only. TSH in serum is stable for at least 5 days at 4° C, but if the test is not to be run within 24 hours, the serum should be kept frozen at −20° C. Frozen sample should be completely thawed and mixed to ensure homogeneity before testing. Repeated freezing and thawing of the sample should be avoided. The use of grossly hemolyzed or lipemic samples is not recommended.

In the TRH stimulation test of pituitary TSH reserve, a base-line TSH sample is drawn before the test. Five hundred micrograms of TRH is given intravenously to the patient, and blood samples are drawn again 20, 30, and 40 minutes after TRH injection. Minor side-effects caused by TRH injection include headaches, dizziness, nausea, and a momentary urge to urinate.

Since the immunological activity of TSH shows pronounced genus specificity, assay kits for human TSH cannot be used to measure TSH concentrations in the blood of other animal species.

REFERENCE RANGE

Most commercial kits for TSH are not sensitive enough to detect TSH in all healthy subjects. In our studies with a commercial kit with a minimum detectable limit of 1 mU/L, TSH was not measurable in 13 of 93 or about 14% of healthy normal subjects, but we were able to estimate the reference range of serum TSH to be 0.51 to 5.75 mU/L, with a mean value of 1.71 mU/L, by the use of the maximum-likelihood estimation method.[13] However, as with all diagnostic tests, it is recommended that reference ranges be determined by each laboratory to conform with the characteristics of the population being tested.

Lipson et al.[14] found no difference in TSH values between men and women ranging from 20 to 89 years old. Males have stable TSH levels throughout all decades, but females over age 60 show a significantly higher mean TSH level (3.4 ± 1.6 mU/L) than younger females (2.3 ± 1.3 mU/L).

In neonates serum TSH levels rise sharply within 10 minutes after delivery, reach a peak (ten- to twenty-four-fold increase) after 30 minutes, and then decline gradually and reach adult levels about 5 days after delivery.[15]

For the TRH stimulation test, the maximum increment of serum TSH from base line (ΔTSH) after TRH injection is 2 to 20 mU/L in euthyroid subjects. Elderly males are less responsive to TRH stimulation than females and young males. ΔTSH is also normal in hypothalamic (tertiary) hypothyroidism, but the response is usually delayed (60 to 180 minutes versus 30 minutes for normal subjects). ΔTSH is less than 2 mU/L in pituitary (secondary) hypothyroidism and also in hyperthyroidism, but it is greater than 20 mU/L in thyroidal (primary) hypothyroidism.

REFERENCES

1. Rathnam, P.: Structure-function relationship of pituitary hormones HFSH, HLH, and HTSH. In Abraham, G.E., editor: Radioassay systems in clinical endocrinology, New York, 1981, Marcel Dekker, Inc., pp. 21-34.
2. Utiger, R.D.: Radioimmunoassay of human plasma thyrotropin, J. Clin. Invest. **44:**1277-1286, 1965.
3. Odell, W.D., Wilber, J.F., and Paul, W.E.: Radioimmunoassay of thyrotropin in human serum, J. Clin. Endocrinol. Metab. **15:**1179-1188, 1965.
4. Kourides, I.A.: Clinical application of the radioimmunoassay for human TSH. In Abraham, G.E., editor: Radioassay systems in clinical endocrinology, New York, 1981, Marcel Dekker, Inc., pp. 57-71.
5. Miles, L.E.M., and Hales, C.N.: Labelled antibodies and immunological assay systems, Nature **219:**186-189, 1968.
6. Tandem™ R TSH Immunoradiometric Assay Publication 701081-013B, Jan. 1983, Hybritech Inc. San Diego, Calif.
7. Doss, R.C., Sun, M., Green, J.B., et. al.: Enzyme linked immunoassay for thyroid stimulating hormone. Syllabus of 9th Annual Meeting and Exhibit for the Clinical Ligand Assay Society, March 1983, p. 214.
8. Woodhead, S.: Data presented at the 9th Annual Meeting and Exhibit for the Clinical Ligand Assay Society, March 1983. (Manuscript in preparation.)
9. Bitensky, L., Alaghband-Zadeh, J., and Chayen, J.: Studies on thyroid stimulating hormone and the long-acting thyroid stimulating hormone, Clin. Endocrinol. **3:**363-374, 1974.
10. Chen, I.W., Heminger, L., Maxon, H.R., et al.: Nonspecific binding as a source of error in thyrotropin radioimmunoassay with polyethylene glycol as separating agent, Clin. Chem. **26:**487-490, 1980.
11. Nisula, B.C., and Louvet, J.P.: Radioimmunoassay of thyrotropin concentrated from serum, J. Clin. Endocrinol. Metab. **46:**729-733, 1978.
12. Chen, I.W., Heminger, L.A., Barnes, E.L., et al.: A sensitive radioimmunoassay (RIA) for detection of serum thyrotropin (TSH) in healthy subjects and patients with suppressed pituitary function, J. Nucl. Med. **24:**114, 1983.
13. Tsay, J.Y., Chen, I.W., Maxon, H.R., et al.: A statistical method for determining normal ranges from laboratory data including values below the minimum detectable value, Clin. Chem. **25:**2011-2014, 1979.
14. Lipson, A., Nickoloff, E.L., Hsu, T.H., et al.: A study of age-dependent changes in thyroid function tests in adults, J. Nucl. Med. **20:**1124-1130, 1979.
15. Fisher, D.A.: Thyroid physiology and function tests in infancy and childhood. In Werner, S.C., and Ingbard, S.H., editors: The thyroid, New York, 1978, Harper & Row, Publishers, Inc., p. 382.

Thyroxine

I-WEN CHEN
MATTHEW I. SPERLING

Thyroxine (T_4, 3,5,3′,5′-tetraiodothyronine)
Clinical significance: p. 748
Molecular formula: $C_{15}H_{11}I_4NO_4$
Molecular weight: 776.93 daltons
Merck Index: 9155
Chemical class: amino acid (thyroid hormone)

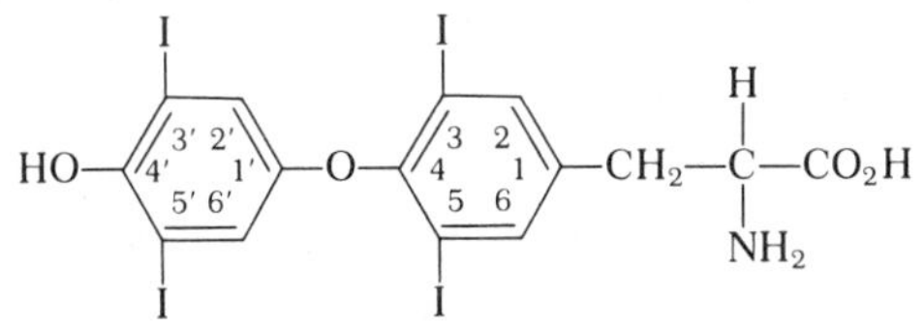

PRINCIPLES OF ANALYSIS AND CURRENT USAGE

Thyroxine (T_4) is an amino acid synthesized in and secreted from the thyroid gland. It plays an important role in the regulation of developmental and metabolic processes. The naturally occurring T_4 is the L-isomer. Although the D-isomer of T_4 has the affinity for nuclear binding proteins equal to that of the L-isomer, its biological activity is only 5% to 20% of the L-isomer. The differences in transport and clearance of the L- and D-isomers are probably responsible for differences in their metabolic activity.

Historically, serum T_4 concentration was estimated by measurement of the amount of iodine present in the partially purified T_4 fractions of serum samples. Such methods involved the measurement of the iodine content in a protein precipitate of serum (protein-bound iodine, PBI), in an alkali-washed butanol extract of a protein precipitate of serum (butanol-extractable iodine, BEI), and in an anion-exchange column–purified fraction of serum (T_4 by column). All these methods are relatively nonspecific and are subject to contamination by iodine-containing drugs and nonhormonal iodine. They are therefore used rarely in routine clinical chemistry laboratories at present. However, the test can be useful in a few clinical situations, as for detecting the presence of iodoprotein abnormalities.

A more specific and sensitive method for measuring total T_4 concentration in serum called "competitive protein-binding assay" (CPBA) was developed in the early 1960s by Murphy and Pattee (method 1, Table 57-16).[1] This method is based on the competition between serum T_4 and added radioactive T_4 for the limited binding sites on specific T_4 binding proteins. The radioactivity (of T_4) bound to the specific binding protein is counted after the unbound radioactivity has been removed from the assay mixture. The bound radioactivity is inversely proportional to the amount of T_4 present in serum samples. (A detailed description of the principle of assays of this type is given in Chapter 11.) The specific binding protein used in this assay is thyroxine-binding globulin (TBG) and is usually obtained by proper dilution of a human serum pool with a barbital buffer. The barbital buffer selectively inhibits any binding of T_4 to the other T_4 binding protein present in serum called "thyroxine-binding prealbumin," whereas binding of T_4 to albumin is a low-affinity binding and can be eliminated merely by dilution of the serum pool with a

Table 57-16. Methods of thyroxine (T_4) analysis

Methods	Tracer	Binder	Reaction phase	Usage	Comments
1. Competitive protein-binding analysis (CPBA)	^{125}I-labeled T_4	Thyroxine-binding globulin (TBG)	Liquid	Any biological specimens	Requires extraction of serum and phase separation
2. Radioimmunoassay (RIA) heterogeneous	^{125}I-labeled T_4	Antibody	Liquid or solid	Serum only, fully automated	Requires phase separation Most sensitive method
3. Enzyme immunoassay (EIA)					
a. Heterogeneous	Enzyme-labeled T_4	Antibody	Solid	Serum only, semiautomated	Requires phase separation and addition of enzyme substrate
b. Homogeneous	Enzyme-labeled T_4	Antibody	Liquid	Serum only, fully automated	Requires pretreatment of serum and addition of enzyme substrate
4. Fluorescence immunoassay (FIA)					
a. Heterogeneous	Fluorescein-labeled T_4	Antibody	Solid	Serum only, semiautomated	Requires phase separation
b. Homogeneous	Fluorescein-labeled T_4	Quencher-antibody	Liquid	Serum only, fully automated	Requires pretreatment of serum

barbital buffer. Since more than 99.9% of T_4 in serum is normally bound to TBG and the other thyroxine-binding proteins and since CPBA requires a constant quantity of TBG in each assay tube, it is necessary to extract T_4 and remove TBG from serum samples by an alcohol, usually ethanol. The dried T_4 extract is redissolved in the TBG solution and used in CPBA for T_4.

Although CPBA for T_4 is a direct analysis of T_4 and thus is free from interference by nonhormonal iodine in the blood, it has now largely been replaced by a more sensitive and specific radioimmunoassay (RIA) (method 2, Table 57-16). The higher sensitivity and specificity of RIA are achieved by the use of a high-affinity antibody. The affinity constant of the T_4 antibody could be as high as 10^{14} L/mol compared to 2×10^{10} L/mol for TBG. Cross-reactivity of other structurally similar thyroid hormone metabolites with the T_4 antibody is negligible. The other advantage of the T_4 RIA is avoidance of the time-consuming extraction step. This is achieved by the use of blocking agents, such as 8-anilinonaphthalenesulfonic acid, sodium salicylate, and thimerosal (Merthiolate), that selectively block binding of T_4 to TBG without significantly altering the interaction of T_4 to the antibody. Otherwise, the basic principles of RIA and CPBA are the same.

Many T_4 RIA kits are available commercially. The 1982 College of American Pathologists (CAP) survey listed 25 commercial kits manufactured by 20 different companies. These numbers could be doubled easily if small manufacturers of complete kits or kit components were included. Although all kits are based on the principle of RIA, many different techniques are used in the separation of antibody-bound and free radioligands. The most commonly used technique is the solid-phase separation procedure utilizing the T_4 antibody chemically or physically bonded to a solid support such as glass beads, plastic tubes, cellulose, and magnetic particles. Other separation methods involve the use of second antibodies or polyethylene glycol to precipitate the antibody-bound ligand. Charcoal and ion-exchange resins have been used as the separating agent, but they are largely abandoned in T_4 RIA because they are relatively nonspecific.

In RIA a radionuclide (radioactive iodine, ^{125}I) is used as a marker to follow and measure the course of an immunological reaction. The radioactivity produced by the radionuclide is the end-point signal that is measured. The radioactivity emitted by the radionuclide through the radioactive decay process is not affected by the physicochemical environment. Therefore a separation step for bound and free radiolabeled ligands is always required in RIA.

Other types of labels have been developed as alternative to radionuclides in immunoassays of T_4. Some alternative labels used still require the separation step, but some require no phase separation.

In the enzyme-linked immunosorbent assay (ELISA) for T_4 developed by ICL Scientific (Fountain Valley, Calif.), a serum sample is incubated with peroxidase-labeled T_4 in a test tube coated with T_4-antibody complex (method 3a, Table 57-16). The basic principle involved is the same as that of RIA. The endogenous T_4 competes with the peroxidase-labeled T_4 for the binding sites on the T_4-antibody complex. The test tube is aspirated to remove the incubation mixture and washed. The peroxidase activity remaining in the tube, which is inversely proportional to the T_4 concentration in serum, is determined photometrically by measurement of the ability of the enzyme to oxidize a suitable substrate. As in the case of RIA, ELISA also requires the physical separation of bound and free T_4.

The enzyme-multiplied immunoassay technique (EMIT) for T_4 developed by Syva (Palo Alto, Calif.) is a homogeneous immunoassay system in which the separation step is unnecessary (method 3b, Table 57-16). In this technique, malate dehydrogenase chemically bound to T_4 is used as the tracer. It is postulated that the enzyme with bound T_4 is inactive because of the blocking of the active site of the enzyme by T_4. The enzyme becomes active in the presence of T_4 antibody, since T_4 is displaced from the active site by binding to the antibody. Therefore the degree of the enzyme-labeled T_4 binding to the T_4 antibody is directly proportional to the enzyme activity and can be measured without physical separation of bound and free T_4. One disadvantage of the homogeneous assay over the heterogeneous assay is that the serum sample has to be pretreated with alkali to eliminate serum effects on the enzyme activity measurement. This pretreatment step is not required in ELISA, since the serum is removed from the reaction system before the enzyme activity measurement.

Fluorescent probes are also used in place of the radiolabels in T_4 immunoassays. As in the case of enzyme immunoassays (EIA's), both homogeneous and heterogeneous fluorescence immunoassays have been used to measure the total T_4 concentrations in serum (methods 4a and 4b, Table 57-16). The homogeneous fluorescence immunoassay for T_4 developed by Syva utilizes the fluorescent excitation transfer immunoassay method, in which fluor-labeled T_4 and T_4 in serum samples are allowed to compete for the binding sites on the T_4 antibody labeled with the fluorescent acceptor or quencher, such as rhodamine. The fluorescence intensity of the fluor-labeled T_4 is reduced on complexation with the quencher-labeled antibody, and thus the rate of fluorescence quenching is inversely proportional to the T_4 concentration of serum samples. Since the fluorescence intensity is modified by binding, it can be monitored directly without physical separation of bound and free T_4. The major drawback with the homogeneous fluorescence immunoassay is the presence of the endogenous background fluorescence produced by the proteinaceous material and other species in serum samples, which usually results in a loss of fluorometric sensitivity. The problem of background emission can be circumvented by physical separation of the T_4-antibody complex from other species in the serum sample before fluorometric quantitation. In the heterogeneous immunoassay developed by Bio-Rad Laboratories (Richmond, Calif.), the separation of bound and free T_4 is achieved by the solid-phase antibody technique commonly used in RIA. The fluor-labeled T_4 bound to the solid-phase antibody (T_4 antibody covalently linked to polyacrylamide beads) is separated from the other species in the reaction mixture. The beads are resuspended, and the fluorescence intensity of the suspension is measured using a fluorometer.

Table 57-17 lists reaction conditions for the various methods of T_4 analysis.

All three immunoassay methods—radioimmunoassay, enzyme immunoassay, and fluorescence immunoassay—have been either fully automated or semiautomated. Three fully automated RIA systems are available commercially in the United States at present.

The Concept 4 (Micromedic Systems, Horsham, Penn.) utilizes special 8 × 50 mm tubes coated with antibody to achieve separation of bound and free ligand. The system consists of a pipetting station for sample transfer, reagent addition and mixing, an incubator for incubating the mixture at ambient temperature or at 45° C for up to about 17 hours, an aspirate-wash station for separation of bound and free ligands, a gamma ray–counting station capable of counting two tubes at a time, and a programmable calculator for on-line data reduction. The system is designed to use racks of 10 tubes each and is capable of processing up to 200 antibody-coated tubes per assay.

The Gammaflo system introduced by Squibb (Princeton, N.J.) combines continuous-flow methodology with column chromatographic techniques for separation of bound from free ligand. The system consists of a sipper mechanism for introducing the radioligand, ligand binder, and sample or standard; a peristaltic pump for transporting each reagent to a junction for mixing and for subsequent flow-through processing; a temperature-controlled time-delay coil for incubation of the reaction mixture in an air-segmented stream; a replaceable resin-charcoal column for separating bound from free radioligands; a gamma-ray counter for counting the bound radioligand; a series of solenoid valves to control solution flow; and a microcomputer for data reduction and system control.

The separation of bound from free ligands in the ARIA II introduced by Becton Dickinson (Salt Lake City, Utah) is achieved by sequential pumping of assay mixtures through a reusable antibody chamber containing antibodies covalently bonded to solid support media. The system is composed of a sample carousel for dispensing samples and also for incubation, if needed; a flow-through system consisting of a flexible arrangement of pumps, valves, reservoirs, and tubing; an antibody chamber; a gamma-ray detector for counting both free and bound radioligands; and an on-line microcomputer for data reduction.

Table 57-17. Reaction conditions for thyroxine (T_4) analysis

Conditions	Manual radioimmunoassay (RIA) (Corning)[9]	Automated RIA (ARIA II)[9]	Enzyme-multiplied immunoassay technique (EMIT) (Syva)[4]	Fluorescence immunoassay (FIA) (Syva)[10]	Enzyme-linked immunosorbent assay (ELISA) (ICL Scientific)[11]
Temperature	25° C	25° C	25° C	25° C	25° C
pH	7.4	10.5	9.6	8.4	8.6
Sample volume	25 µL	20 µL	10. µL	25 µL	20 µL
Sample pretreatment	Not required	Not required	0.1 M NaOH at 25° C for 15 min	Anionic detergent at 25° C for 11 sec	Not required
Blocking agent	Thimerosal	ANS*	Not required	Not required	ANS*
Tracer	^{125}I-labeled T_4	^{125}I-labeled T_4	Malate dehydrogenase–labeled T_4	Fluor-labeled T_4	Peroxidase-labeled T_4
Antibody	Rabbit, bound to glass particles	Rabbit, bound to fiber particles	Sheep, liquid	Sheep (quencher dye–labeled), liquid	Rabbit, coated tube
Enzyme substrate	Not required	Not required	NAD†	Not required	H_2O_2, ABTS‡
Reaction time	60 min	Instantaneous	10 min	0.5 min	120 min
Sensitivity§	20 µg/L	10 µg/L	36 µg/L	20 µg/L	25 µg/L
Precision (% CV)	3.3%-10.1%	3.6%-7.8%	2.3%-12.5%	2.9%-5.4%	4.7%-12.5%
Interferences	Abnormal protein concentrations	Abnormal protein concentrations	Abnormal protein concentrations, enzyme inhibitors	Abnormal protein concentrations	Abnormal protein concentrations, enzyme inhibitors

*8-Anilino-1-naphthalenesulfonic acid.
†Nicotinamide adenine dinucleotide.
‡Diammonium 2,2′-azino-bis(2-ethylbenzothiazoline-6-sulfonate), or 2,2′-azino-di-(3-ethyl-benzthiazoline sulfonic acid) diammonium salt.
§Minimum detectable concentration.

REFERENCE AND PREFERRED METHODS

Methods involving the measurement of the iodine content in the partially purified T_4 fraction of serum samples are nonspecific and are seldom used for the evaluation of thyroid status, though they are occasionally used for detecting the presence of iodoprotein abnormality. They are mentioned here only for historical interest. CPBA kits for T_4 are still manufactured by some commercial companies and are used in some clinical laboratories. For example, the 1982 CAP basic ligand assay survey of 1,669 participants lists 26 laboratories still using CPBA T_4 kits manufactured by three companies. However, CPBA requires the tedious extraction step for releasing T_4 bound to T_4-binding proteins, and the extraction efficiency may vary from sample to sample, which might have contributed to the relative imprecision of the assay. This method has now largely been replaced by more sensitive and specific T_4 RIA techniques.

The homogeneous enzyme immunoassay has been developed as an alternative method for T_4. The advantages of this method compared to RIA are that the separation step for bound and free T_4 is not required and that handling of radioactive materials can be avoided. Although enzyme immunoassay is used extensively for therapeutic drug monitoring, it has not gained popularity in T_4 analysis, largely as a result of its relative insensitivity and imprecision, especially at low T_4 concentration. The decreased sensitivity of enzyme immunoassay is believed to be attributed primarily to the steric hindrance introduced into the antigen-antibody reaction by the presence of the enzyme macromolecule.[2]

The homogeneous enzyme-multiplied immunoassay technique for T_4 was developed originally for manual use, but it has also been adapted for use on the Autochemist multichannel analyzer in a fully automated manner, as well as the Abbott Biochromatic Analyzer (ABA-100) in a semiautomated fashion. Recently, Abbott (North Chicago) has introduced a homogeneous enzyme assay kit named A-gent Tetrazyme, designed specifically for the ABA-100. This assay utilizes a phosphonate-T_4 conjugate, which is a potent irreversible inhibitor of acetylcholine esterase. When the conjugate is bound by the T_4 antibody, the inhibitory activity is blocked. Therefore the amount of bound phosphonate-T_4 conjugate is directly proportional to the acetylcholine esterase activity, which can be measured photometrically by monitoring of the change in the optical density without the separation of bound and free T_4 conjugate after the addition of appropriate enzyme substrate.

Fluorescent probes seem to be the more attractive alternative to the radioactive labels, since they are much smaller molecules than enzymes. In addition, one can determine directly the modulation of the signal caused by the antigen-antibody interaction resulting from the binding of the fluor-labeled antigen to the quencher-labeled antibody (homogeneous fluorescence immunoassay) by measuring the change in the fluorescence intensity. In contrast, the determination of the modulation of the enzyme activity observed in the enzyme immunoassay requires the monitoring of the enzyme-substrate interaction, which requires an extra substrate addition step and strict control of the reaction conditions. The development of novel fluorescent probes with greater Stokes' shifts, greater environmental sensitivities, and high-quantum yields of fluorescence will enhance the attractiveness of fluoroimmunoassay. The background interference is commonly encountered in fluoroimmunoassay and is the major cause of a loss of fluorometric sensitivity. The ability to reduce such a high background emission will make heterogeneous fluorescence immunoassays more attractive in the future, though the heterogeneous immunoassays require a separation step for bound and free ligand. The homogeneous fluorescence immunoassay for T_4 introduced by Syva is also used in conjunction with a fully automated system, also manufactured by Syva. Serum samples and the pretreatment solution are dispensed manually into reaction cups in the carousel. After the carousel is loaded onto the instrument, the instrument is activated and it automatically dispenses reagents, aspirates the reaction mixture, measures and stores a 20-second reaction rate, carries out data reduction, and prints out patients' T_4 concentrations.

The use of isotope-dilution mass spectrometry as the definitive method for T_4 has been suggested and is under development.[3]

It appears that RIA is the best method available for the routine measurement of serum T_4 and is the most commonly used method at present. Many commercial T_4 RIA kits with a wide variety of separation techniques are available. Nonisotopic alternative methods for T_4 analysis might be useful in laboratories that do not have dedicated radioactivity counters, radioactive waste disposal facilities, or trained personnel. In addition, many of these nonisotopic procedures are compatible with existing equipment, such as Syva assays on ABA-100.[4]

The precision of manually performed RIA is frequently not so good as that of other clinical laboratory tests because of the labor intensiveness involved in the technique. Therefore automation of RIA is highly desirable. The performance characteristics of three automated RIA systems was evaluated for the T_4 assay.[5] Although the accuracy of all three automated systems compared favorably with established manual systems, we observed the highest precision with the ARIA II system (coefficient of variation, CV: 3.8% to 7.8%), followed by the Gammaflo (CV: 5.8% to 11.5%) and Concept 4 (CV: 6.9% to 11.7%). The Concept 4 gave poor precision, especially at low ligand concentrations, a factor probably resulting from occasional nonuniformity in the antibody coating of the sample tube. The Concept 4 demonstrated the best sensitivity and requires the least amount of serum samples for assay. This was especially advantageous when only a limited amount of sample was available for multiple testing, as with pediatric patients. No appreciable carry-over was observed with the

Concept 4 system, whereas the ARIA II and Gammaflo gave statistically significant carry-over when the difference between ligand concentrations of adjacent samples was great. However, the extent of carry-over observed should not create too much of a problem, especially if the automated system is equipped with a warning system to alert for the presence of adjacent samples with such widespread values. Because of its exceptionally high throughput, the Concept 4 may be more suitable to laboratory personnel faced with large test volumes, whereas the ARIA II and Gammaflo instruments are able to adequately handle the test volumes of laboratories associated with average-sized hospitals. It is obvious that the automated system will not be cost effective if used only for T_4 because of the high cost of the system itself. It should be used with other RIA's available for the automated system.

SPECIMEN

Serum is preferred and should be collected using normal aseptic venipuncture techniques. Plasma may also be used but tends to form fibrin after freezing and thawing, which may mechanically interfere with the assay, especially in an automated system. T_4 in serum is quite stable; it has been shown that storage of serum samples at room temperature up to 14 days resulted in no appreciable loss of T_4. However, it is recommended that serum samples be stored frozen if they will not be tested within 24 hours. Repeated freezing and thawing of the sample should be avoided. Although hemolyzed specimens do not interfere with the assay, use of grossly hemolyzed samples should be avoided, since hemolysis may be sufficient to have diluted the samples. Grossly lipemic specimens should not be used, especially in CPBA, since fatty acids are known to compete with T_4 for the binding sites on TBG.

As in the case of all radioassays, if patients have received diagnostic or therapeutic radionuclides within the 2 weeks immediately before the T_4 determination, the radioactivity of the serum sample from such patients should be checked in a counter set for ^{125}I to determine whether the radioactivity contained in the sample could significantly affect assay results.

Most commercial T_4 assay kits are designed specifically for measurement of T_4 in serum or plasma samples and should not be used for other biological fluids, such as urine or cerebrospinal fluid, because of the extreme sensitivity of the antigen-antibody interaction to the matrix of samples to be measured.

REFERENCE RANGE

As with all diagnostic tests, it is recommended that each laboratory establish its own reference range, thereby allowing for variability resulting from such factors as geography and assay techniques. The reported reference ranges are, in general, within 41 to 120 μg/L. There are conflicting reports regarding the age and sex dependency of serum T_4 concentrations, but most researchers reported no age dependency.[6] Lipson et al.[7] found relatively constant T_4 levels in men of all ages and in women over 60 years of age but significantly high T_4 levels in women under 60 years of age (mean ± standard deviation, 79 ± 13 μg/L, n = 44) compared with men (72 ± 12 μg/L, n = 102) and women over 60 years of age (74 ± 14 μg/L). However, these differences are small and can be ignored for routine testing

Although the age-dependent differences in serum T_4 concentrations are of little significance in adults, they are important during childhood. Serum T_4 concentrations increase from cord blood values of about 127 μg/L to a mean of 165 μg/L by 26 hours after birth and fall gradually thereafter to a mean of 80 μg/L at 17 years of age.[8]

Elevated or decreased levels of serum T_4 may occur in patients with euthyroidism as a result of changes in the circulating concentrations of T_4-binding proteins, especially TBG. Increased serum TBG concentrations will result in an increased serum T_4 level in euthyroid subjects who, for example, are pregnant, are on estrogen therapy (including oral contraceptives), or have viral hepatitis; whereas decreased serum TBG concentrations from androgen therapy, active acromegaly, nephrotic syndrome, malnutrition, and so on will result in a decreased serum T_4 level. The status of T_4-binding proteins should be investigated in these euthyroid subjects (see the discussion of free T_4 on p. 748).

REFERENCES

1. Murphy, B., and Patte, C.J.: Determinatin of thyroxine utilizing the property of protein binding, J. Clin. Endocrinol. Metab. **24:**187, 1964.
2. Yalow, R.S.: Radioimmunoassay: a probe for the fine structure of biologic systems, Science **200:**236, 1978.
3. Tietz, N.W.: A model for a comprehensive measurement system in clinical chemistry, Clin. Chem. **25:**833, 1978.
4. Kaplan, L.A., Chen, I.W., Gau, N., et al.: Evaluation and comparison of radio-, fluorescence, and enzyme-linked immunoassays for serum thyroxine, Clin. Biochem. **14:**182, 1981.
5. Chen, I.W.: Commercially available fully automated systems for radioligand assay. Part I. Overview, Ligand Rev. **2**(2):46, 1980; Part II. Performance characteristics, Ligand Rev. **2**(3):45, 1980.
6. Caplan, R.H., Wickus, G., Glasser, J.E., et al.: Serum concentrations of the iodothyronines in elderly subjects: decreased triiodothyronine (T_3) and free T_3 index, J. Am. Geriatr. Soc. **29:**19, 1981.
7. Lipson A., Nickoloff, E.L., and Hsu, T.H.: A study of age dependent changes in thyroid function tests in adults, J. Nucl. Med. **20:**1124, 1979.
8. Fisher, D.A.: Thyroid physiology and function tests in infancy and childhood. In Werner, S.C., and Ingbard, S.H., editors: The thyroid: a fundamental and clinical text, New York, 1978, Harper & Row, Publishers, Inc., p. 382.
9. Chen, I.W., Maxon, H.R., Heminger, L.A., et al.: Evaluation and comparison of two fully automated radioassay systems with distinctly different modes of analysis, J. Nucl. Med. **21:**1162, 1980.
10. Bellet, N., Winfrey, L., Horton, A., et al.: Development of homogeneous fluorescence rate immunoassays for thyroid function testing, Clin. Chem. **27:**1071, 1981.
11. Enzymune-Test™ T_4, enzyme-linked immunosorbent assay (ELISA), ICL List No. 85020, ICL Scientific, Fountain Valley, Calif.

T_3 uptake

I-WEN CHEN
MATTHEW I. SPERLING

T_3 uptake (3,5,3′-triiodothyronine uptake test; thyroxine-binding globulin, or TBG, unsaturation)
Clinical significance: p. 748

PRINCIPLES OF ANALYSIS AND CURRENT USAGE

The triiodothyronine uptake test (T_3 uptake) is a test for estimating unoccupied binding sites on serum thyroxine–binding (T_4-binding) proteins. It was developed originally as in vitro diagnostic test of thyroid function in 1959 by Hamolsky et al.[1] In the early T_3 uptake tests, a fixed amount of radioactive T_3 was incubated with whole blood, and the incorportion of radioactive T_3 by red blood cells was measured by counting the separated erythrocytes obtained after centrifugation and washing. The radioactivity taken up by red blood cells (the "T_3 uptake") was inversely proportional to the radioactivity bound to T_4-binding serum proteins, especially thyroxine-binding globulins (TBG); that is, a high T_3 uptake meant high TBG saturation and reduced available T_4 binding sites. In hyperthyroid patients serum T_4 concentrations are elevated, TBG is more saturated with T_4, and thus T_3 uptake (by the red blood cells) is increased. On the other hand, in hypothyroid patients serum T_4 concentrations are abnormally low, TBG is less saturated with T_4, and T_3 uptake is reduced. Therefore, although the T_3 uptake test is for indirect estimation of the unoccupied binding sites on T_4-binding serum proteins (primarily TBG), it may be of some value in obtainment of information about the thyrometabolic status of the patient.

In the original T_3 uptake test[1] erythrocytes were used as the secondary binder for estimation of unsaturated T_4-binding capacity of serum proteins. This method, however, was found to be less convenient and less reproducible because of the difficulty in a uniform washing of erythrocytes and the effect of individual hematocrit variations on the test results. Since then, a variety of solid-phase materials have been used to improve the technique and the consistency in measuring T_3 uptake values. For example, ion-exchange resins, charcoal, solid-phase anti-T_3 antibody, organic polymers (Sephadex), silicate, talc, and macroaggregated albumin are used in a variety of commercially available T_3 uptake test kits. The 1982 College of American Pathologists (CAP) survey listed 28 commercial kits manufactured by 23 different companies. Although different secondary binders and variations in technique are used, the principle of the test remains the same: measurement of the distribution of ^{125}I-radiolabeled T_3 between endogenous serum binders (mainly TBG) and the secondary binder (method 1, Table 57-18).

Although radiolabeled T_4 can theoretically be used to assess the number of unoccupied binding sites on serum proteins, radiolabeled T_3 is preferred because of its relatively low affinity for TBG (affinity constant approximately 10^9 compared with 10^{10} mol/L for T_4). The radioactive T_3 will fill the unoccupied binding sites but will not displace bound T_4. Therefore differences between the test results obtained from patients with normal and abnormal numbers of unoccupied binding sites are greater with radioactive T_3 than with radioactive T_4. Furthermore, T_4 binds to all T_4-binding proteins, including TBG, T_4-binding prealbumin, and albumin; thus the T_4 uptake values can be affected by the changes in any of the serum T_4-binding proteins. In the T_3 uptake test the T_3 binding is affected primarily by the TBG concentrations, since T_3 binds to neither prealbumin nor albumin under the conditions of most T_3 uptake tests.

In addition to radioactive T_3, nonradioactive labels are also used in the uptake test. In the A-gent Thyrozyme Uptake Diagnostic Kit developed by Abbott Laboratories (method 2, Table 57-18), phosphonate-T_4 conjugate, a potent inhibitor of acetylcholinesterase that still maintains the ability to bind TBG, is used as the label. When the T_4 conjugate is bound to the TBG, it loses the enzyme-inhibitory activity. Therefore one can assess the extent of the T_4 conjugate binding to TBG and hence the number of available T_4 binding sites on TBG by measuring the change in the enzymatic activity of acetylcholinesterase, which can be estimated photometrically when one monitors the change in the absorbance in the presence of a chromogenic enzyme substrate. The enzyme activity is proportional to the number of available T_4 binding sites. Since separation of free T_4 conjugate from TBG-bound T_4 conjugate is not required, the secondary binder used in the radiolabeled T_3–uptake test is not needed. Phosphonate T_4 conjugate, rather than phosphonate T_3 conjugate is employed in this method because the latter loses almost all of its ability to bind TBG.[2] A somewhat different approach is taken in the Syva Advance T_3 Uptake assay using a fluorescent probe as an alternative to radionuclides (method 3, Table 57-18). Since the fluor-labeled T_3 loses the ability to bind serum proteins, it can no longer be used directly to determine the distribution of T_3 between the serum proteins and the secondary binder as in the radiolabeled T_3–uptake tests.[2] However, the fluor-labeled T_3 still maintains its antigenicity against T_3 antibody. A serum sample is incubated with a known amount of T_3, which binds to the available binding sites on the serum proteins. The amount of T_3 that remains unbound is determined by the fluorescence energy transfer immunoassay technique using fluor-labeled T_3 and quencher-labeled antibody to T_3. The binding of the fluor-labeled T_3 to the quencher-

Table 57-18. Methods of T_3 uptake

Methods (type of tracer used)	Principle	Assay system	Automation	Comments
1. ^{125}I-labled T_3	Direct measurement of distribution of tracer between primary and secondary binder	Heterogeneous	Fully automated	Simplest method for indirect estimation of TBG-binding capacity
2. Enzyme inhibitor-labeled T_4	Indirect measurement of tracer bound to serum T_4-binding proteins (primary binder)	Homogeneous	Semiautomated	Requires no secondary binder Enzyme activity directly proportional to TBG-binding capacity
3. Fluor-labeled T_3	Tracer used to measure unbound T_3 by fluorescence excitation transfer immunoassay technique	Homogeneous	Fully automated	Quencher-labeled T_3 antibody used as specific binder in immunoassay of unbound T_3
4. Enzyme-labeled T_4	Tracer used to measure unbound T_4 by enzyme-linked immunosorbent assay (ELISA)	Heterogeneous	Semiautomated	T_4 antibody used as specific binder in enzyme immunoassay of unbound T_4

labeled antibody will result in the quenching of the fluorescence signal because of the transfer of the emission energy of the fluor to the quencher (method 3, Table 57-18). Since the fluor-labeled T_3 can compete with the unbound T_3 for antibody binding sites, the amount of quenching is reduced in proportion to the amount of unbound T_3 and thus is indirectly related to the T_3 uptake value. This is a homogeneous assay system requiring no separation of free and bound fluor-labeled T_3.

The same approach is also taken in the uptake test utilizing an enzyme-linked immunosorbent assay (ELISA) technique (method 4, Table 57-18). T_4 conjugated with enzymes such as peroxidase (ICL Scientific, Fountain Valley, Calif.) or alkaline phosphatase (Immunotech Corp., Cambridge, Mass.) maintains the reactivity toward T_4 antibody but loses the ability to bind serum proteins. A serum sample is preincubated with a fixed amount of T_4, and the unbound T_4 is determined by the ELISA technique, using enzyme-labeled T_4 and antibody to T_4. The antibody-bound enzyme-labeled T_4 is separated from unbound enzyme-T_4 complex by a solid-phase technique (antibody-coated tube, ICL Scientific) or a double-antibody precipitation technique (Immunotech Corp.) and is quantitated spectrophotometrically at 405 or 420 nm in the presence of H_2O_2 and diammonium 2,2′-azino-bis(3-ethylbenzothiazoline-6-sulfonate) (ICL Scientific) or at 400 to 420 nm in the presence of *p*-nitrophenyl phosphate (Immunotech Corp.) as substrate. As with the radioactive T_3 uptake test, this is a heterogeneous assay system requiring the physical separation of free- and antibody-bound T_4.

Since the development of radioimmunoassays of T_4, the T_3 uptake test by itself is seldom used as a general screening test for thyroid function, since the serum T_4 concentrations show much less overlapping of values among euthyroid, hypothyroid, and hyperthyroid states than do the T_3 uptake values. However, the T_3 uptake test can provide the clinician with information regarding the status of serum binding proteins for thyroid hormones and still is a valuable diagnostic test for thyroid function if used with the T_4 assay.

REFERENCE AND PREFERRED METHODS

Because of its simplicity, the radioactive T_3 uptake test is by far the most widely used method for the indirect estimation of the number of unsaturated T_4 binding sites on serum proteins (Table 57-19). Commercial T_3 uptake test kits with ^{125}I-labeled T_3 as the tracer were used in all 1685 laboratories participating in the 1982 CAP survey. As with all commercial diagnostic test kits, it is strongly recommended that the performance characteristics of any T_3 uptake test kits be evaluated carefully in each laboratory before they are selected for routine use.

Recently, Witherspoon et al.[3] examined eight commercially available T_3 uptake test kits using various secondary binders. They observed that the uptake values obtained from all eight kits exhibited varying degrees of dependencies on incubation time and temperature. The kits using charcoal and antibody-coated tube were found to be quite sensitive to the variation in the incubation time. The charcoal kit also showed pronounced temperature-dependent variation in the uptake values. They also studied the relationship between T_3 uptake and TBG concentrations in serum. It was found that almost all assays exhibited a rather pronounced change in the T_3 uptake value at low TBG concentrations (less than 10 mg/L) but showed very little change in the uptake value at TBG concentrations exceeding about 40 mg/L, indicating that the T_3 uptake test is of little value at very high TBG concentrations. Witherspoon et al. concluded that ". . . the clinically most useful assays exhibited the ability to reflect a wide

Table 57-19. Reaction conditions for T_3 uptake

Conditions	Manual radioimmunoassay[6] (Clinical Assays)	Automated RIA[7] (ARIA-HT)	Fluorescence immunoassay (FIA)[8] (Syva)	Thyrozyme uptake[4] (Abbott)	ELISA[9] (ICL scientific)
Temperature	18°-24° C	30° C	25° C	37° C	20°-25° C
pH	8.6	7.4	8.0	8.5	8.6
Sample volume	25 μL	50 μL	40 μL	2.5 μL	10 μL
Tracer	^{125}I-labeled T_3	^{125}I-labeled T_3	Fluor-labeled T_3	Phosphonate-labeled T_4	Peroxidase-labeled T_4
Secondary binder	T_3-antibody-coated tube	Hydrophobic resin column	Quencher-T_3 antibody	Not required	T_4-antibody-coated tube
Enzyme substrate	Not required	Not required	Not required	AMTCI*	H_2O_2, ABTS†
Reaction time	60 min	Instantaneous	40 sec	5 min	120 min
Precision (coefficient of variation)	1.3%-3.9%	1.6%-3.2%	3%-4%	3.5%-5.5%	2.3%-5.5%
Interference	Radioactive drugs	Radioactive drugs	Hemoglobin	Hemolysis	Hemolyzed serum

*Acetyl-β-(methylthio)choline iodide.
†Diammonium 2,2′-azino-bis(2-ethylbenzothiazoline-6-sulfonate).

range of thyroxine-binding globulin concentrations and demonstrated little or no time or temperature effects . . . ."

Variations in incubation time and temperature can be minimized by automation of the test. All three fully automated radioimmunoassay systems currently available commercially in the United States (ARIA-II or -HT, Gammaflo, Concept 4) are equipped to run the T_3 uptake test. The assay precision (coefficient of variation, CV) of the ARIA-HT evaluated in our laboratory is in the range of 1.6% to 3.2%.

The uptake tests using nonradioactive labels as the tracer have also been either semiautomated or fully automated. Although the nonradioactive alternative methods have not received wide acceptance, they might be useful in laboratories without dedicated radioactivity counters, radioactive waste disposal facilities, or trained personnel. The performance characteristics of the Abbott Thyrozyme Uptake Kit used with the ABA-100 in our laboratories[4] compare favorably with a manual radioactive T_3 uptake method, with CV's in the range of 3.5% to 5.5%.

SPECIMEN

Most kit manufacturers recommend use of serum samples that are collected with use of the routine precautions required in venipuncture for clinical assays. Fresh serum samples may be stored at 2° to 8° C for up to 1 week without appreciable changes in the uptake values. However, the sample should be stored frozen at −20° C if the time between sample collection and analysis is longer than 6 days. Frozen samples should be mixed well by inversion after thawing and before assay. Repeated freezing and thawing and mixing by vortexing or vigorous agitation should be avoided, since this may result in denaturation of proteins. The use of hemolyzed or lipemic samples is not recommended.

REFERENCE RANGE

When the radioactive T_3 uptake tests are used, uptake values are customarily expressed as the percentage of the total radioactivity that is taken up by the secondary binder or that is not bound to TBG. For most assays the reference range is 25% to 35%; values below 25% are suggestive of hypothyroidism, and values greater than 35% are suggestive of hyperthyroidism. However, the actual value depends on various assay conditions employed (the specific activity of the radiolabeled T_3, the avidity of the secondary binder for T_3, the incubation time and temperature, and so on). Variation in the uptake values resulting from imprecise assay conditions can be minimized by use of the T_3 uptake ratio, which is the T_3 uptake of the patient divided by the T_3 uptake of a pool of normal standard reference serum included in the same assay run. If the T_3 uptake value of the normal standard reference serum is 30%, the reference range of the T_3 uptake ratio will be 0.83 to 1.17. The T_3 uptake ratio can be calculated directly from the radioactivity taken up on the secondary binder (absorbent):

$$\text{Patient } T_3 \text{ uptake ratio} = \frac{\text{Absorbent counts for patient serum}}{\text{Absorbent counts for normal reference serum}}$$

The T_3 uptake ratio can then be converted to the percent T_3 uptake by multiplication of the patient T_3 uptake ratio by the percent T_3 uptake of the normal reference standard.

Almost all commercial T_3 uptake kits include a normal reference serum with a known percent T_3 uptake value for the determination of T_3 uptake ratio. However, it is strongly recommended that each laboratory prepare its own pool of normal standard reference serum, stored frozen at −20° C or preferably −70° C in small aliquots, and analyze it in every assay run. This will enable each laboratory to maintain the consistency in the reported T_3 uptake values in case of changes in the assay procedure or the kit components.

The Committee on Nomenclature of the American Thyroid Association (ATA) strongly recommends expression of the T_3 uptake as a ratio and considers it mandatory when one is calculating the free T_4 index (FTI).[5] The ATA committee also recommends that if the T_3 uptake is reported directly, it should always be reported with a serum total T_4 concentration so that the free T_4 index can be calculated. In other words, the T_3 uptake test is useful only as a means of calculating the free T_4 index:

$$\text{FTI} = \text{Total } T_4 \text{ concentration} \times T_3 \text{ uptake ratio}$$

Therefore, if the reference range of the T_3 uptake ratio is 0.83 to 1.17 and that of total T_4 concentration 45 to 115 μg/L, the reference range of the free T_4 index would be 37.4 to 134 μg/L. The ATA committee recommends omission of the units from the free T_4 index to avoid confusion with measurements of T_4 itself. However, as with all diagnostic tests, each laboratory should establish its own reference range, since differences may exist between laboratories and between the populations being studied.

In the case of uptake tests using enzyme-labeled T_4 as the tracer, such as the Thyrozyme and the ELISA, the end point of measurement is the enzyme activity expressed in absorbance units, which may change in the direction opposite to the radioactivity measured in the T_3 uptake test. For example, in the Thyrozyme uptake test, an enzyme inhibitor–T_4 conjugate is used as the tracer, and thus a higher enzyme activity or a higher absorbance reading means a lower free-inhibitor concentration or a greater number of available sites on serum proteins for binding the inhibitor. Therefore the Thyrozyme uptake ratio (absorbance of patient sample divided by absorbance of normal reference serum) used by the kit manufacturer is a reciprocal of the T_3 uptake ratio. It follows that:

$$\text{FTI} = T_3 \text{ uptake ratio} \times T_4 = \frac{T_4}{\text{Thyrozyme uptake ratio}}$$

REFERENCES

1. Hamolsky, M.W., Goledtz, A., and Freedberg, A.S.: The plasma protein–thyroid complex in man. III. Further studies on the use of the *in vitro* red blood cell uptake of ^{131}I L-triiodothyronine as a diagnostic test of thyroid function, J. Clin. Endocrinol. Metab. **19**:103-116, 1959.
2. I.W. Chen: Personal communication, Cincinnati, Ohio, 1983.
3. Witherspoon, L.R., Shuler, S.E., and Garcia, M.M.: The triiodothyronine uptake test: an assessment of methods, Clin. Chem. **27**:1272-1276, 1981.
4. Stein, E.A., Chen, I.W., Gau, N., et al.: Spectrophotometric assays for serum thyroxine (T_4) and unsaturated T_4 binding assays for evaluation of thyroid function, Clin. Chem. **27**:1070, 1981.
5. Solomon, D.H., Benotti, J., DeGroot, L.J., et al.: Revised nomenclature for tests of thyroid hormones in serum, J. Clin. Endocrinol. Metab. **42**:595-598, 1976.
6. GammaCoat (^{125}I) T_3 Uptake Kit, Cat. Nos. CA-539, 559, Clinical Assays, Cambridge, Mass.
7. Instruction manual for determination of T_3 uptake using the ARIA-HT, Becton Dickinson Immunodiagnostics, Salt Lake City, Utah, 1983.
8. Bellet, N., Winfrey, L., Horton, A., et al.: Development of homogeneous fluorescence rate immunoassays for thyroid function testing, Clin. Chem. **27**:1071, 1981.
9. Enzymune-Test™ TBC (T_4 Uptake), ICL List No. 85010, ICL Scientific, Fountain Valley, Calif.

Vanillylmandelic acid

STEVEN J. SOLDIN

Vanillylmandelic acid (VMA, 3-methoxy-4-hydroxymandelic acid)

Clinical significance: p. 503

Molecular formula: $C_9H_{10}O_5$

Molecular weight: 198.17 daltons

Merck Index: 4735

Chemical class: substituted carboxylic acid (catecholamine metabolite)

CH_3O — OH — CHCOOH — HO

PRINCIPLES OF ANALYSIS AND CURRENT USAGE

Von Fraenkel[1] showed in 1886 that the concentration of catecholamines was increased in a patient with "explosive hypertension" (pheochromocytoma) by oxidizing the tumor catechols to *ortho*-quinones, which have a characteristic red-brown color. In 1957 the pioneering work of Armstrong et al.[2] helped to identify the metabolic pathways involved (Fig. 57-4). Currently, the laboratory analysis of urinary catecholamines and metabolites plays a central role in assisting the clinician to arrive at a diagnosis of a neural crest tumor.

The diagnosis of a pheochromocytoma in a patient with hypertension is critical, since the hypertension is potentially curable by resection of the tumor but is often fatal if the tumor escapes detection. Clearly, although neuroblastoma and pheochromocytoma are relatively rare tumors—their prevalences are 1 in 100,000 and 1 in 200,000, respectively[3]—the value of early detection and treatment is substantial, prompting numerous investigations into the most efficient approach to diagnosis.[4-7]

The methods for the measurement of vanillylmandelic acid (VMA) can be divided into those employing spectrophotometric techniques and those using chromatographic techniques. Urine contains a large number of compounds, especially phenols, acid phenols, and other metabolites of aromatic ring compounds, that could potentially interfere with either the colorimetric or chromatographic methods.

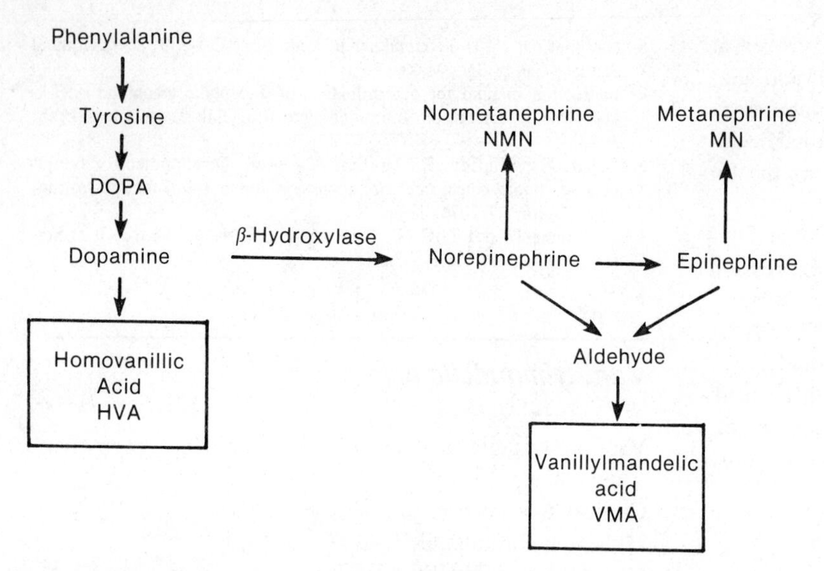

Fig. 57-4. Metabolic pathway of catecholamines.

Table 57-20. Methods of vanillylmandelic acid (VMA) analysis

Method	Principle	Usage	Comment
1. Direct spectrophotometric	VMA + Diazotized *p*-nitroaniline → Purple color (520 nm)	Rare	Used most frequently to develop thin-layer and paper chromatography
2. Direct vanillin method (Pisano)	VMA (3-CH_3O, 4-HO-phenyl–CH(OH)COOH) $\xrightarrow{NaIO_4}$ 3-CH_3O, 4-HO-phenyl–CHO, **Vanillin (360 nm)**	Most common	Need anion-exchange chromatography to minimize interferences
3. Indole condensation	Method 3: VMA (3-CH_3O, 4-HO-phenyl–CH(OH)COOH) $\xrightarrow{+\ 2\ K_3Fe(CN)_6\ +\ 2\ KCl\ +\ 3\ ZnCl_2}$ **Vanillin** (3-CH_3O, 4-HO-phenyl–CHO) $+ K_2Zn_3[Fe(CN)_6]_2\downarrow + 6\ KCl + 2\ HCl + CO_2\uparrow$; Vanillin + **Indole** $\xrightarrow{H_2SO_4,\ H_3PO_4}$ **Carbonium salt (pink) (495 nm)** (3-CH_3O, 4-HO-phenyl–$\overset{\oplus}{C}$H–indol-3-yl)	Common	Needs ion-exchange chromatography to minimize interferences

Table 57-20. Methods of vanillylmandelic acid (VMA) analysis—cont'd

Method	Principle	Usage	Comment
4. Paper chromatography	Separation of VMA from other compounds by differential absorption between paper and mobile phases; VMA spot developed using diazotized *p*-nitroaniline; amount of VMA determined from diameter of spot or by measurement of its absorption at 520-525 nm after extraction from paper	Rare	Requires 12-18 hr for analysis
5. Thin-layer chromatography (TLC)	Separation of VMA from other compounds by differential absorption between silica gel and mobile phases; VMA spot developed by diazotized *p*-nitroaniline; amount of VMA present determined by densitometer or reflectance spectroscopy	Rare	Requires 1-2 hr for analysis. An AACC selected method (1977)
6. High-voltage electrophoresis	Separation on cellulose acetate of VMA from other compounds having different mobilities in high-voltage electrical field (> 200 V); diazotized *p*-nitroaniline used to develop spot, which is quantitated spectrophotometrically (520 nm) after elution	Rare	Requires 4-5 hr for analysis
7. High-performance liquid chromatography (HPLC)	VMA separated from other compounds by differential partitioning between stationary (usually reversed) phase and mobile phase; VMA peak detected by variety of means	Infrequent, but usage increasing	Most specific method, very few compounds known to interfere Single analysis can take 15-20 min Many of these techniques can also be used to measure homovanillic acid (HVA)
a. Ultraviolet	Absorbance at 260 nm		
b. Electrochemical	Oxidation of VMA to quinone forms; amperometric detection of electrons **VMA** (CH_3O, HO, CH(OH)COOH) $\xrightarrow[-H^+]{-2e^-}$ **Quinone** (CH_3O^+, O, CH(OH)COOH) $\xrightarrow{-H_2O}$ **Diquinone** (O, O, CH(OH)COOH) $+ CH_3OH + H^+$ *Or* oxidation of VMA to vanillin and measurement of separated vanillin by its oxidation to its quinone form and amperometric detection of reaction: (CH_3O, HO, CHO) $\xrightarrow{-2e^-}$ (O, O, CHO) $+ CH_3OH$		
c. Post-column reaction	Eluate reacts with sodium periodate to form vanillin, which is measured spectrophotometrically at 360 nm		
8. Gas chromatography (GC)	Silyl and trimethylsilyl derivatives of VMA separated from other compounds by differential absorption on liquid stationary phase (phenylmethylsilicone) and use of a flame ionization detector	Infrequent	Highly specific Method can be adapted to measure HVA Analysis time short

Therefore almost all methods for the quantitation of urinary VMA utilize an extraction step to partially purify the analyte. In addition, a liquid-liquid extraction followed by evaporation and reconstitution in small volume can be used to concentrate the VMA, which improves the sensitivities of all the methods.

The usual procedure employed to extract VMA first converts the VMA to a less soluble, unionized form and then makes this form more insoluble by reducing the amount of water available for solvation. This is accomplished by the addition of concentrated HCl to decrease the pH to less than 2.0 and then by saturation of the solution with excess sodium chloride. The VMA can then be extracted by ethyl acetate with an 80% to 90% efficiency. The alternative method employed to partially purify VMA is anion-exchange, column chromatography.

The oldest quantitative method for VMA analysis is based on the direct method of Armstrong et al.,[2] in which extracted VMA reacts with diazotized *p*-nitroaniline to form a purple chromogen (method 1, Table 57-20). This chromogen is then quantitated at 520 nm.

The other colorimetric methods are based on the conversion of VMA to vanillin. This can be accomplished by autoclaving of the sample or by use of chemical oxidants such as sodium periodate, ferricyanide, Cu^{++} ions, or aluminum oxide. The widely used procedure developed by Pisano et al.[8] measures vanillin directly at 348 to 360 nm (method 2, Table 57-20). The Pisano method uses sodium periodate to convert extracted VMA to vanillin. The vanillin is measured at 360 nm, rather than at its peak at 348 nm because of the presence of other compounds that also absorb at about 348 nm. An alternative procedure was developed by Sunderman et al.[9] based on the observation of Sandler and Ruthven[10] that indole reacts with vanillin to produce a pinkish chromogen (method 3, Table 57-20). This chromogen, which is believed to be a carbonium ion complex, can be quantitated at 495 nm. The Sunderman method employs a number of extraction steps to increase the specificity and accuracy of the procedure: (1) the treatment of urine with magnesium silicate to remove potentially interfering compounds; (2) oxidation of VMA to vanillin with ferricyanide; (3) extraction of vanillin with toluene and then a back-extraction of the vanillin into an alkaline solution; and (4) formation of the pinkish carbonium complex of vanillin and indole. Wybenga and Pileggi[11] also used the indole condensation method for quantitation of VMA but improved on the specificity of the colorimetric reaction by partially purifying the VMA with anion-exchange chromatography before its oxidation by sodium periodate.

The nonspecificity of the colorimetric methods led to the use of chromatographic methods for separating the extracted VMA from potentially interfering compounds. Paper chromatography has been widely used for the analysis of extracted VMA (method 4, Table 57-20). Whatman No. 1 and 3MM papers have been used for both unidirectional[12] and bidirectional[2,13] descending chromatography. All these methods employed diazotized *p*-nitroaniline to develop the chromatogram. The VMA spot was clearly distinguished from other compounds by both its purple color and its mobility (R_f). The amount of VMA present could be determined when the diameter of the spot is related to standards or by elution of the blue color and measurement of its absorbance at 525 nm.

A thin-layer chromatographic (TLC) assay was proposed as a selected method in 1977 by the American Association for Clinical Chemistry (AACC) (method 5, Table 57-20).[14] An ethyl acetate extract was chromatographed on a silica stationary phase, and the chromatogram was developed with a toluene/acetic acid/ethyl acetate solvent phase. The VMA spot, developed with diazotized *p*-nitroaniline, was completely separated from other urinary constituents. The amount of VMA spotted could be determined semiquantitatively by visual comparison with standards or quantitatively by densitometry at 505 nm with the method of standard addition to calibrate the mass of VMA chromatographed. TLC analysis on silica gel has also been reported; reflectance spectroscopy is used to quantitate the yellow color produced by exposing the plate to sunlight for 48 hours.[15]

High-voltage electrophoresis on cellulose acetate has also been employed for VMA analysis (method 6, Table 57-20). Ethyl acetate extracts are applied to cellulose acetate strips, and the components are separated by the influence of an electrical field of 200 to 250 volts.[16] The spots are stained with diazotized *p*-nitroaniline, the purple VMA spot is eluted with an alkaline methanol solution, and VMA is quantitated at 520 nm.

More recently, there has been proposed a number of assays that employ high-performance liquid chromatography (HPLC) to separate VMA from other constituents (method 7, Table 57-20). Although most of these assays used reversed-phase columns (RP-HPLC), adsorption chromatography has also been reported.[16] The VMA can be quantitated by a variety of detectors, including ultraviolet light,[17-19] amperometric detection,[20-23] and post-column reaction.[24,25]

One of the earliest RP-HPLC procedures described was a modification of the Pisano method, in which extracted VMA was converted to vanillin by sodium periodate. The vanillin was isolated by reversed-phase HPLC and quantitated by electrochemical detection.[18] Reverse-phased HPLC assays for VMA using electrochemical detection have also been reported, in which the VMA is partially purified by either liquid extraction[22] or anion-exchange chromatography[20] or with no purification at all.[23] In all these methods the VMA is oxidized to its quinone form by the positive potential (about + 0.75 V) of the electrochemical detector. The magnitude of the current generated by the electrochemical reaction is directly proportional to

Table 57-21. Comparison of methods of VMA analysis

Method parameters	Direct vanillin (Pisano)[8]	Indole condensation[11]	HPLC[21]
Reaction temperature	50° C	50° C	40° C
Time of analysis	1½-2 hr	2 hr	20 min
Sample volume (extracted)	Variable 0.2% of total 24 hr output	5 mL	5 mL
Final concentration of reagents	For formation of vanillin $NaIO_4$: 8.5 mmol/L (for extraction) $Na_2S_2O_5$: 45.8 mmol/L Phosphate buffer: 0.86 mmol/L (pH 7.5)	To 5 mL of column eluate: For vanillin formation: $NaIO_4$: 7.2 mmol/L $Na_2S_2O_5$: 14.3 mmol/L For color formation: Indole: 0.39 mmol/L H_3PO_4: 0.16 mmol/L H_2SO_4: 4.91 mol/L	Mobile phase Phosphate buffer: 100 mmol/L, pH 6.7 Tetrabutylammonium phosphate: 0.5 mmol/L
Detector	Ultraviolet, 360 nm	Visible light: 495 nm	Electrochemical, +0.76 V
Precision ($\bar{X}$, % coefficient of variation)	—	4.8 mg/day, 5%	2.7 mg/day, 9.9%
Interferences	Dietary vanillin, aromatic phenols from various foods (bananas, tea, coffee, chocolate)	α-Methyldopa, *p*-hydroxymandelic acid	

the amount of VMA eluted. Many reversed-phase assays can also be adapted for the simultaneous analysis of homovanillic acid (HVA).[19,21] The post-column reaction used to monitor the HPLC eluate has been a sodium periodate oxidation of VMA to vanillin and measurement of the vanillin at 360 nm.[17,18] One procedure employs direct injection of the urine,[18] whereas the other uses anion-exchange chromatography to purify the VMA partially.[17]

Many gas-chromatographic (GC) methods have been described for VMA quantitation (method 8, Table 57-20).[26,27] These methods require extraction of VMA and subsequent derivatization to form volatile conjugates. Silyl[26] and trimethylsilyl-ether conjugates[27] have been commonly used, though other agents have been reported. A capillary GC method using 3,4-dihydroxybenzoic acid as an internal standard has been reported, with 5% phenylmethylsilicone as the stationary phase and a flame ionization detector.[27] The GC methods have also been used to quantitate HVA simultaneously with VMA. A GC system using mass fragmentography as a detector has also been reported.[28]

According to a recent College of American Pathologists (CAP) Quality Assurance Survey, most participating laboratories report using some form of the Pisano method for VMA analysis. These laboratories most often use anion-exchange chromatography for the partial purification of VMA. Only approximately 7% of the laboratories reported using HPLC or GLC methods, though one would project that the use of these methods would increase as the state of the art changes.

Table 57-21 compares methods of VMA analysis.

REFERENCE AND PREFERRED METHODS

All the methods reported for VMA analysis are subject to wide variety of in vivo interferences. These include a number of drugs (iproniazide, pargyline) that inhibit monoamine oxidase and decrease VMA output.[13,29] L-Dopa, used to treat Parkinson's disease, has been reported to increase VMA output and cause false-positive results.[13] Very serious illnesses such as pulmonary insufficiency, shock, or cancer have also been reported causes of increased VMA output.[29]

The diazotized *p*-nitroaniline method does not have many reported in vitro, methodological interferences, though methocarbamol and *p*-aminosalicylic acid have been reported to produce positive interference. The colorimetric methods that employ vanillin have reported positive interferences by dietary vanillin, as well as by a large number of phenolic acids found in certain foods, such as bananas, coffee, tea, and chocolate. For these methods the diets of the patients must be restricted from use of these foods for several days before the urine collection. If the urine is partially purified by anion-exchange chromatography, most if not all of these potentially interfering compounds, especially vanillin, are removed as sources of interferences.[2,11] However, when patients are being treated with a wide variety of drug combinations, there is always a possibility of interference by drugs—parents or metabolites. Both α-methyldopa and *p*-hydroxymandelic acid, for example, are known to interfere with the procedure of Wybenga and Pileggi.[11] The precision of the colorimetric procedures is reported to be less than 10% coefficient of variation (CV) for values within the normal range. All the colorimetric procedures are very labor intensive, requiring a number of manual steps.

The methods that utilize some chromatographic technique before the quantitation of VMA are greatly improved in their analytical specificity. Few compounds have been shown to interfere with these assays. However, almost all these assays require some purification step before analysis, usually liquid-liquid extraction or perhaps anion-exchange chromatography. Although a liquid-liquid extraction is usually more rapidly performed, there is not a great difference in the time required for these two purification techniques.

Paper chromatography and TLC procedures are considered to be of historical interest only. The time necessary for the analysis of a single sample varies widely among the various chromatographic techniques. Paper chromatography is the longest (12 to 18 hours); whereas high-voltage electrophoresis requires 4 to 5 hours; TLC, 1 to 2 hours; and HPLC, 15 to 20 minutes. The reported precision of the HPLC procedures for VMA concentrations in the normal range tend to be approximately 10% to 12%[20,25] though many report interassay precisions of less than 8% CV.[22,24] The HPLC assays compare well with the Pisano method but tend to be biased slightly lower, suggestive of a better specificity. Although few compounds have been reported to interfere in the HPLC assays, one must carefully review the chromatographs for closely eluting, interfering peaks. With use of 260 nm for the detection of VMA separated by reversed-phase HPLC, positive interferences have been noted for some patients taking certain medications, especially diuretics and propranolol types of beta blockers. Such rare interferences have also been reported to occur in GC analysis.[27]

For overall specificity and sensitivity of analysis, the HPLC procedures are the recommended method of choice. The HPLC procedure listed in references 20 and 21 is reliable and is strongly recommended over the other methods in Table 57-20. The results are free from in vitro interferences and have adequate precision. The major drawback is that the laboratory must invest in HPLC equipment and, of course, must have the suitable personnel to supervise the technique. However, these assays in general are rapidly performed and are suited to the workload of most laboratories. Of the detectors available for monitoring the HPLC eluates, electrochemical detection offers the most sensitive technique. For those laboratories without electrochemical detectors, ultraviolet monitors will suffice.

For the small laboratory with a low volume of VMA requests, a modified Pisano or Sunderman procedure,[11] which employs anion-exchange chromatography to partially purify the VMA, will give adequate precision and specificity.

SPECIMEN

To minimize variations in VMA output occurring throughout the day, a 24-hour urine collection is the recommended specimen of choice. Smaller collection times have also been suggested as being useful, especially when the VMA output is expressed as micrograms of VMA per milligram of creatinine. The urine is best collected in a container with acid preservative added to prevent degradation of VMA. Usually 10 mL of 6 M HCl will suffice for a 24-hour collection, and the urine should be mixed throughout the collection period to ensure an adequately low pH of less than 2.0. The specimen should be stored at 4° C until analyzed for VMA content.

PROCEDURE

Principle

The modified method of Wybenga and Pileggi[11] is a relatively simple procedure for the quantitation of VMA in urine. The procedure employs anion-exchange chromatography to separate VMA from other components in urine. The VMA is eluted from the resin with 3 M sodium chloride and oxidized with sodium periodate to vanillin. VMA is quantified by reacting the vanillin with an indole–phosphoric acid reagent to yield a colored compound with an ultraviolet absorbance maximum at 495 nm.

Reagents

Reagent grade acetic, sulfuric, and phosphoric (85%) acids

1 M acetic acid (1 mol/L). Volumetrically dilute 58.8 mL of concentrated acetic acid to 1 L. This is stable for 12 months at room temperature.

1 M NaOH (1 mol/L). Place 40 g of NaOH pellets in a 1 L volumetric flask and add approximately 800 mL of distilled water to dissolve. When the solution has reached room temperature, bring volume to mark with distilled water. Store in plastic bottle. This solution is stable for 12 months at room temperature.

0.1 M NaOH (0.1 mol/L). Volumetrically dilute 10 mL of 1 M NaOH to 100 mL with distilled water. This is stable at room temperature for 12 months.

1 M potassium carbonate (1 mol/L). Place 138.2 g of K_2CO_3 in a 1 L volumetric flask and add approximately 800 mL of distilled water to dissolve. Bring volume to mark with distilled water. This is stable for 12 months at room temperature in a plastic bottle.

0.2 M sodium acetate (0.2 mol/L). Place 16.41 g of sodium acetate in a 1 L volumetric flask. Add approximately 800 mL of distilled water to dissolve and bring volume to mark with distilled water. This solution is stable for 3 months at room temperature.

3 M sodium chloride (3 mol/L). Place 165.35 g of NaCl in a 1 L volumetric flask. Add approximately 800 mL of distilled water to dissolve and bring volume to mark with distilled water. This is stable for 12 months at room temperature.

Sodium *meta*-periodate, 187 mmol/L (4% w/v). Place 4 g of sodium *meta*-periodate ($NaIO_4$) in a 100 mL volumetric flask. Add approximately 80 mL of distilled water to dissolve. Bring to mark with distilled water. This is stable for 1 month at 4° to 8° C.

Sodium *meta*-bisulfite, 526 mmol/L (10% w/v). Place 10 g of sodium *meta*-bisulfite ($Na_2S_2O_5$) in a 100 mL volumetric flask. Add approximately 80 mL of distilled water to dissolve and bring to mark with additional water. This is stable for 1 week at room temperature.

0.1 M phosphate buffer, pH 6.1 (0.1 mol/L). *Stock buffer:* 0.1 mol/L KH_2PO_4: place 13.61 g of KH_2PO_4 in a 1 L volumetric flask. Add approximately 800 mL of distilled water to dissolve and bring to mark with distilled water. This is stable for 4 months at 4° to 8° C. *Working 0.1 M phosphate buffer:* Add 13.6 mL of 0.1 M NaOH to 100 mL of stock 0.1 M KH_2PO_4 buffer and mix. This is stable for 2 months at 4° to 8° C.

Indole reagent (42.7 mmol/L). Place 500 mg of indole in a 100 mL volumetric flask and dissolve in approximately 80 mL of absolute ethanol. Bring volume to mark with ethanol. This is stable for 2 months at 4° to 8° C.

Indole–phosphoric acid reagent (1.07 mmol/L indole; 43.9 mol/L H_3PO_4). Mix 1 volume of indole reagent with 39 volumes of 85% phosphoric acid. *Prepare just before use and keep in ice bath until used.*

VMA standard (62.5 mg/L, 315.4 μmol/L). Place 62.5 mg of VMA in 1 L volumetric flask. Dissolve in approximately 800 mL of 0.01 mol/L HCl and bring to mark with 0.01 mol/L HCl. This is stable for 4 months at 4° to 8° C.

Preparation of Dowex AG1-X4 resin (Bio-Rad, Richmond, Calif.)

Wash 500 g of the resin with 3 L of 1 M sodium hydroxide. Decant or siphon off the aqueous phase. Repeat three times. Now wash the resin with 3 L of deionized water. Decant. Wash resin with 1 M acetic acid (repeat three times). Finally add 0.2 M sodium acetate and stir. Continue adding 0.2 M sodium acetate until the pH of the solution above the resin is 6.1. The resin is stored in this solution until used.

Assay

Equipment: spectrophotometer with a band pass ≤10 nm; heating bath.

1. With a 1 M NaOH solution adjust the pH of 5 mL of a timed urine collection to 6.1 ± 0.2. Add 5 mL of 0.1 M potassium phosphate buffer (pH 6.1), vortex-mix, and allow to percolate through an ion-exchange column (Dowex AG1-X4, dimensions 1.8 × 0.7 cm). Discard the column effluent. Wash the column with 1 mL of deionized water, which is also discarded. Elute the VMA from the column with 12 mL of 3.0 M NaCl and collect the eluate in a clean glass tube. Mix the eluate. Place the tube in a boiling water bath for approximately 15 minutes.
2. After cooling to room temperature, add 8 mL of 1 M potassium carbonate to the eluate and mix the solution.
3. Three 5 mL aliquots are removed and placed in clean tubes labeled "B" (blank), "U" (unknown), and "U+V" (standard added to unknown). Add 0.1 mL of the VMA standard to the U+V tube and 0.1 mL of 0.01 N HCl to tubes U and B.
4. Add 0.2 mL of 4% sodium *meta*-periodate to tubes U and U+V. Add 0.3 mL of 10% *meta*-bisulfite to B tube to prevent oxidation of VMA to vanillin in the blank.
5. Heat all tubes for 30 minutes at 50° C in a water bath to effect oxidation of VMA to vanillin in tubes U and U+V.
6. At the end of the heating period, add 0.3 mL of 10% *meta*-bisulfite to tubes U and U+V to stop further oxidation and add 0.2 mL of 4% sodium *meta*-periodate to tube B.
7. Chill all tubes in an ice bath. Carefully add 1.5 mL of concentrated H_2SO_4 dropwise to each tube and mix.

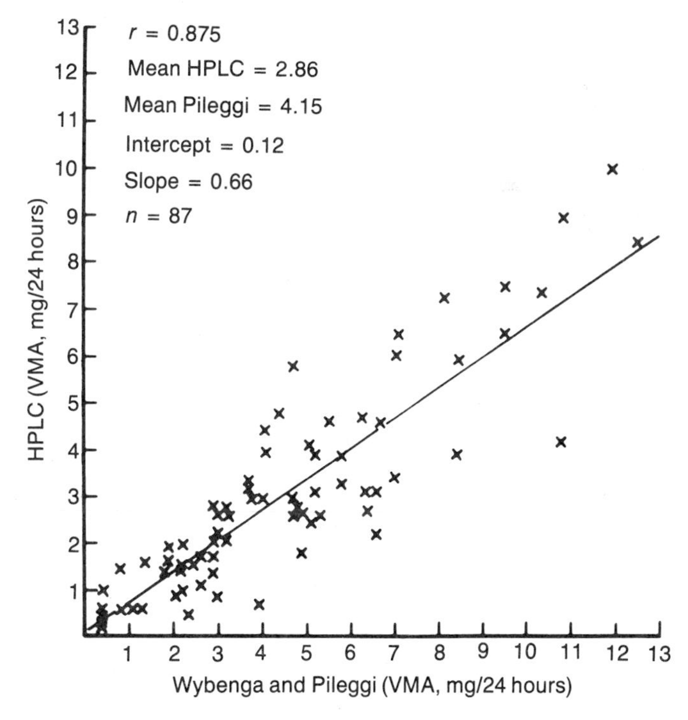

Fig. 57-5. Measurement of urinary vanillylmandelic acid (VMA) (mg/day). Comparison of present and Wybenga-Pileggi methods. (From Soldin, S.J., and Hill, J.G.: Clin. Chem. **26**:291-294, 1980.)

8. Add 4.0 mL of freshly prepared indole–phosphoric acid reagent to tube B, thoroughly, and read absorbance *(A)* at 495 nm. Repeat for tubes U and U+V. The time interval between addition of color reagent and reading of the optical density should be constant for all tubes, since the final color is not stable.

Calculation

Calculate the concentration of VMA using the formula:

$$\text{VMA (mg/24 hr)} = \frac{A_U - A_B}{A_{(U+V)} - A_U} \times \frac{\text{Urine volume (mL/24 hr)}}{200}$$

Interferences

Dietary vanillin does not seem to interfere with this procedure, since it is separated from VMA in the ion-exchange step. Although the procedure is reported to be relatively free of interference,[11] I have found that gross interference (perhaps caused by drugs) occurs in approximately 4% of urine samples tested. In these samples the absorbance of both the blank and the unknown is extremely high. α-Methyldopa and *p*-hydroxymandelic acid both interfere in this procedure.

A comparison of 87 urine samples measured for VMA by both the HPLC and modified Wybenga and Pileggi procedures is shown in Fig. 57-5. In general, the results obtained by HPLC were considerably lower than those obtained by the modified Wybenga-Pileggi method. The normal range for adults is 1.8 to 8.0 mg/24 hr (9.1 to 40.4 μmol/24 hr) for the modified Wybenga-Pileggi procedure.

Table 57-22. Percentile reference values for urinary VMA

Age (years)	Size of population (*n*)	VMA 95th	VMA 100th
Excretion*: mg/24 hr (μmol/24 hr)			
0-1	48	2.3 (11.6)	3.1 (15.6)
2-4	34	3.0 (15.1)	4.0 (20.2)
5-9	20	3.5 (17.7)	8.8 (44.4)
10-19	40	6.0 (30.2)	7.7 (38.9)
>19	56	6.8 (34.3)	8.1 (40.9)
Excretion†: mg/g of creatinine (μmol/μmol of creatinine)			
0-1	37	18.8 (10.7)	59.4 (33.9)
2-4	49	11.0 (6.3)	20.8 (11.9)
5-9	79	8.3 (4.7)	9.4 (5.4)
10-19	55	8.2 6(4.7)	13.9 (7.9)
>19	56	6.0 (3.4)	8.3 (4.7)

*Analyses performed on timed urine samples obtained from patients under investigation for hypertension.
†Analyses performed on random urine specimens obtained from hospital patients not suspected of having a neural crest tumor.

REFERENCE RANGES

Age-related urinary excretion rates for VMA are listed in Table 57-22. These ranges are given as the upper-percentile limits of excretion.

REFERENCES

1. von Fraenkel, F.: Ein Fall von doppelseitigem, völlig latent verlaufenem Nebennierentumor und gleichzeitiger Nephritis mit Veränderungen am Circulationapparat und Retinis, Arch. Pathol. Anat. **103:**244, 1886.
2. Armstrong, M.D., McMillan, A., and Shaw, K.N.F.: 3-Methoxy-4-hydroxy-D-mandelic acid, a urinary metabolite of norepinephrine, Biochim. Biophys. Acta **25:**422-423, 1957.
3. Galen, R.S., and Gambino, S.R.: Beyond normality: the predictive value and efficiency of medical diagnosis, New York, 1975, John Wiley & Sons, Inc., pp. 62-74.
4. Laug, W.E., Siegel, S.E., Shaw, K.N.F., et al.: Initial urinary catecholamine metabolite concentrations and prognosis in neuroblastoma, Pediatrics **62:**77-88, 1978.
5. Gitlow, S.E., Dziedzic, L.B., Strauss, L., et al.: Biochemical and histologic determinants in the prognosis of neuroblastoma, Cancer **32:**898-905, 1973.
6. Labrosse, E.H., Comoy, E., Bohuon, C., et al.: Catecholamine metabolism in neuroblastoma, J. Natl. Cancer Inst. **57:**633-643, 1976.
7. Bravo, E.L., Tarazi, R.C., Gifford, R.W., and Stewart, B.H.: Circulating and urinary catecholamines in pheochromocytoma, N. Engl. J. Med. **301:**682-686, 1979.
8. Pisano, J.J., Crout, R.J., and Abraham, D.: Determination of 3-methoxy-4-hydroxymandelic acid in urine, Clin. Chim. Acta **7:**285-291, 1962.
9. Sunderman, F.W., Jr., Cleveland, P.D., Law, N.C., and Sunderman, F.W.: A method for the determination of 3-methoxy-4-hydroxymandelic acid ("vanilmandelic acid") for the diagnosis of pheochromocytoma, Am. J. Clin. Pathol. **34:**293-312, 1960.
10. Sandler, M., and Ruthven, C.R.J.: Quantitative colorimetric method for estimation of 3-methyl-4-hydroxymandelic acid in urine: value in diagnosis of phaeochromocytoma, Lancet **2:**114-115, 1959.
11. Wybenga, D., and Pileggi, V.J.: Quantitative determination of 3-methoxy-4-hydroxymandelic (VMA) in urine, Clin. Chim. Acta **16:**147-154, 1967.
12. Vahidi, H.R., Roberts, J.S., San Filippo, J., Jr., and Sankar, D.V.S.: Paper chromatographic quantitation of 4-hydroxy-3-methoxymandelic acid (VMA) in urine, Clin. Chem. **17:**903-907, 1971.
13. Gitlow, S.E., Mendlowitz, M., and Bertain, L.: The biochemical techniques for detecting and establishing the presence of a pheochromocytoma, Am. J. Cardiol. **26:**270-279, 1970.
14. Badella, M., Routh, M.W., Gump, B.H., and Gigliotti, H.J.: Thin-layer chromatographic method for urinary 4-hydroxy-3-methoxymandelic acid (vanilmandelic acid). In Cooper, G.R., editor: Selected methods in clinical chemistry, Washington, D.C., 1977, American Association for Clinical Chemistry, pp. 139-145.
15. Huck, H., and Dworzak, E.: Quantitative Analyse von Katecholamin und Serotonin-metaboliten auf der Dünnschichtplatt, mit Remissionmessung nach einer spezifischen photochemischen Reaktion, J. Chromatogr. **74:**303-310, 1972.
16. Hermann, G.A.: The determination of urinary 3-methoxy-4-hydroxymandelic (vanilmandelic) acid by means of electrophoresis with cellulose acetate membrane, Am. J. Clin. Pathol. **41:**373-376, 1964.
17. Yoshida, A., Yoshioka, M., Tanimura, T., and Tamura, Z.: Determination of vanilmandelic acid and homovanillic acid in urine by highspeed liquid chromatography, J. Chromatogr. **116:**240-243, 1976.
18. Felice, L.J., and Kissinger, P.T.: A modification of the Pisano method for vanilmandelic acid using high pressure liquid chromatography, Clin. Chim. Acta **76:**317-320, 1977.
19. Bertani-Dziedzic, L.M., Krstulovic, A.M., Ciriello, S., and Gitlow, S.E.: Routine reversed-phase high performance liquid chromatographic measurement of urinary vanillylmandelic acid in patients with neural crest tumors, J. Chromatogr. **164:**345-353, 1979.
20. Soldin, S.J., and Hill, J.G.: Simultaneous liquid chromatographic

analysis for 4-hydroxy-3-methoxymandelic acid and 4-hydroxy-3-methoxyphenylacetic acid in urine, Clin. Chem. **26:**291-294, 1980.
21. Soldin, S.J., and Hill, J.G.: Liquid chromatographic analysis for urinary 4-hydroxy-3-methoxymandelic acid and 4-hydroxy-3-methoxyphenylacetic acid and its use in investigation of neural crest tumors, Clin. Chem. **27:**502-503, 1981.
22. Moleman, P., and Borstrok, J.J.M.: Determination of urinary vanillylmandelic acid by liquid chromatography with electrochemical detection, Clin. Chem. **29:**878-881, 1983.
23. Fujita, K., Maruta, K., Ito, S., and Nagatsu, T.: Urinary 4-hydroxy-3-methoxymandelic (vanillylmandelic) acid, 4-hydroxy-3-methoxyphenylacetic (homovanillic) acid, and 5-hydroxy-3-indoleacetic acid determined by liquid chromatography with electrochemical detection, Clin. Chem. **29:**876-878, 1983.
24. Rosano, T.G., and Brown, H.H.: Liquid chromatographic assay for urinary 3-methoxy-4-hydroxymandelic acid with use of a periodate oxidative monitor, Clin. Chem. **25:**550-554, 1979.
25. Flood, J.G., Granger, M., and McComb, R.B.: Urinary 3-methoxy-4-hydroxymandelic acid as measured by liquid chromatography, with on-line post–column reaction, Clin. Chem. **25:**1234-1238, 1979.
26. Brewster, M.A., Berry, D.H., and Moriarty, M.: Urinary 3-methoxy-4-hydroxyphenylacetic (homovanillic) and 3-methoxy-4-mandelic (vanillylmandelic) acids: gas-liquid chromatographic methods and experience with 13 cases of neuroblastoma, Clin. Chem. **23:**2247-2248, 1977.
27. Tuchman, M., Crippin, P.J., and Krivit, W.: Capillary gas-chromatographic determination of urinary homovanillic acid and vanillylmandelic acid, Clin. Chem. **29:**828-831, 1983.
28. Takahashi, S., Yoshioka, M., Yoshiue, S., and Tamura, Z.: Mass fragmentographic determination of vanilmandelic acid, homovanillic acid and isohomovanillic acid in human body fluids, J. Chromatogr. **145:**1-9, 1978.
29. Amery, A., and Conway, J.: A critical review of diagnostic tests for pheochromocytoma, Am. Heart. J. **73:**129-133, 1967.

Chapter 58 Lipids

Amniotic fluid phospholipids

PAUL T. RUSSELL

Clinical significance: p. 686

Molecular formula: Depends on fatty acid moieties
- Phosphatidyl choline (lecithin) (L) $C_{40}H_{82}O_9NP$ (dipalmitoyl-)
- Sphingomyelin (S) $C_{39}H_{81}O_7N_2P$ (palmitoyl-)
- Phosphatidyl inositol (PI) $C_{40}H_{80}O_{14}P$ (dipalmitoyl-)
- Phosphatidyl glycerol (PG) $C_{42}H_{83}O_{10}P$ (distearoyl-)

Molecular weight: Depends on fatty acid moieties
- Phosphatidyl choline (lecithin) 752 (dipalmitoyl-)
- Sphingomyelin 721 (palmitoyl-)
- Phosphatidyl inositol about 883
- Phosphatidyl glycerol 779

Merck Index:
- L 5287
- S 8518
- PI —
- PG —

Chemical class: phospholipids

PRINCIPLES OF ANALYSIS AND CURRENT USAGE

Tests for the estimation of pulmonary surfactant in amniotic fluid fall into three general categories: (1) those that measure the chemical constituents of surfactant, (2) those that measure the physical properties of lung surfactant, and (3) those that measure a variety of chemical agents that *correlate* with stages of fetal maturation.

The underlying principle for the analysis of the chemical constituents of surfactant is the ability to separate individual components of phospholipids from a matrix extracted from amniotic fluid. Once the separation is made, quantitation permits the establishment of relationships such as the lecithin-to-sphingomyelin (L/S) ratio, the phospholipid "profile",[1] the fatty acid composition of amniotic fluid lecithin,[2] and many others that provide an index of fetal lung maturation.[3,4] Thin-layer chromatography (TLC) became the most widely used technique for most of these analyses after it was published as the method of choice by Gluck et al.[5-7] in their original work on the L/S ratio.

The thin-layer chromatography systems (method 1, Table 58-1) most extensively used[4] employ silica gel G or H stationary phases and mobile phases consisting of chloroform/methanol/water (65:25:4, by volume) or 25% ammonium hydroxide substituted for the water. Separations by TLC are by adsorption chromatography based on the polarity and charge properties of the phospholipids. Both one-dimensional and two-dimensional TLC systems have been used, the choice determined primarily by the need to effect separations without overlap for the identification of phospholipids in addition to lecithin and sphingomyelin.

Phosphoglyceride structures

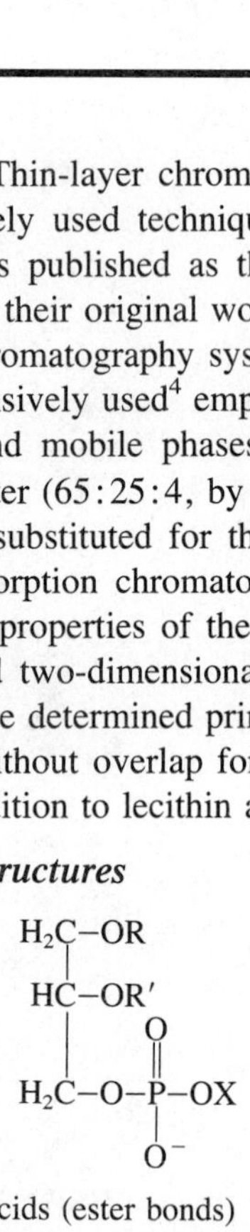

```
H2C—OR
 |
HC—OR'
 |     O
 |     ‖
H2C—O—P—OX
       |
       O⁻
```

R and R′ are fatty acids (ester bonds)

Phosphoglyceride	Component X
Lecithin (phosphatidyl choline)	$HOCH_2CH_2\overset{+}{N}(CH_3)_3$ **Choline**
Phosphatidyl inositol	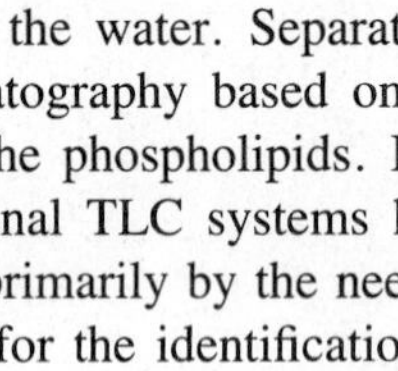 (cyclohexane ring with OH, H substituents) **Inositol**
Phosphatidyl glycerol	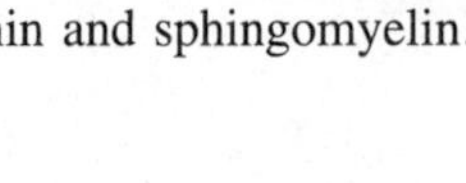 $HOCH_2—CH(OH)—CH_2OH$ **Glycerol**

Table 58-1. Methods of amniotic fluid analysis

Method	Type of analysis	Principle	Usage	Comment
Chromatographic analysis of phospholipids				
1. Thin-layer	Semiquantitative	Silica gel adsorption chromatography; visualization of phospholipids by color-forming sprays, or charring	Most common	Can be run as one dimension or two dimension
2. Gas chromatography	Quantitative	Analysis of fatty acids on polyester columns with flame ionization detector	Rare	Measured as methyl ester derivatives
3. High-performance liquid chromatography	Quantitative	Separation of L, S, PI, PE, PS, and PG by reversed-phase chromatography, detection at 203 nm	Rare	Ability to quantitate multiple components simultaneously may make this an important method in the future
Measurement of surfactant functional activity				
4. Foam stability (shake test, FSI)	Qualitative	Measures presence of surfactant by its ability to support foam bubbles generated by shaking amniotic fluid with ethanol	Common, as stat. procedure	Problem of false-negative results
5. Surface tension	Quantitative	Presence of surfactant related to surface tension properties of amniotic fluid	Rare	Dedicated instrument, but relatively easy to perform
6. Fluorescence polarization	Quantitative	Surfactant increases microviscosity of fluid, which reduces the rotation of dissolved fluorescent probe, resulting in an increase in polarization of incident fluorescent light	Rare	Dedicated instrument, but relatively easy to perform

Sphingomyelin structure

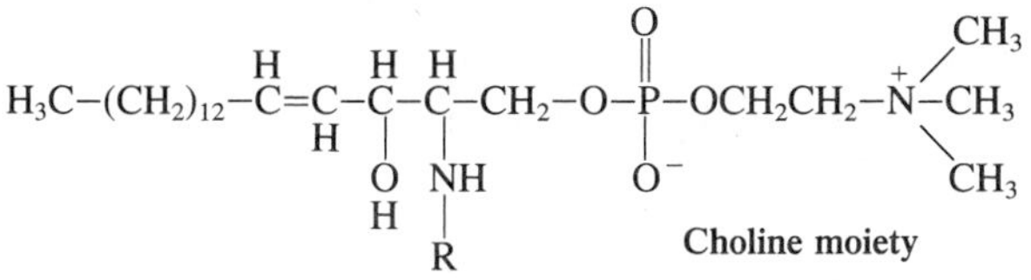

```
                   H    H  H                 O                  CH3
                   |    |  |                 ||              + /
H3C-(CH2)12-C=C-C-C-CH2-O-P-OCH2CH2-N-CH3
                      H  |  |                 |                  \
                         O  NH                O-                  CH3
                         H  |
                            R                 Choline moiety
```

R = fatty acid (amide bond)

Visualization of the separated phospholipids has been performed by any one of a number of techniques. One can char organic materials by spraying the TLC plates with sulfuric or phosphoric acid and then heating the plate to 280° C. Incorporation of ammonium sulfate[6] into the silica gel facilitates charring and eliminates the need for strong acid sprays. Cupric acetate and sulfuric acid with or without dichromate react with the fatty-acid, unsaturated, double bonds, yielding colored products. The intensity of the colored spots depends on the degree of unsaturation of the phospholipid constituent fatty acids. Iodine vapor and molybdate ions also react with the unsaturated bonds in the fatty acid moieties of the phospholipids. Rhodamine B spray viewed under ultraviolet light identifies lipids, and bismuth subnitrate reacts with the choline moiety of the lecithin and sphingomyelin.[8]

After separation of phospholipids by thin-layer chromatography and their identification on the chromatogram by an indicator, evaluation of the L/S is accomplished in a number of ways.

1. The phospholipids can be eluted from the chromatogram and the phospholipid quantitatively estimated by a phosphorus determination.
2. One can evaluate the chromatogram by transmission or reflection densitometry. Compounds isolated on silica gel plates often are charred by strong acid and heat, and the concentration of each "spot" determined densitometrically by use of a scanner and recorder. The L/S ratio is calculated from the recorder densities of the charred lecithin and sphingomyelin.
3. One can estimate the phospholipids on the chromatogram by planimetry. After spraying or visualization of the spots, the product of the length and width of individual spots provides an estimation of the area of the spot. This area in turn is proportional to the quantity of material present. The lecithin and sphingomyelin areas are related to one another by a ratio. Although this is a simple approach to quantitation, it is at best an approximation to the actual value because spots on the chromatogram tend to deviate from a perfect circle or do not appear as rectangles. This inherent

error tends to affect the lecithin component more than the sphingomyelin as the L/S ratio increases.

Gas chromatography (GLC) (method 2, Table 58-1) has provided a means to estimate pulmonary surfactant based on their unique fatty acid composition.[9] These methods include quantification of the palmitic acid concentration of amniotic fluid, establishing a palmitic to stearic acid ratio, and determining the relative content of palmitic acid contained in lecithin isolated from amniotic fluid. Gas chromatographic methods for these fatty acid separations have been performed on polyester columns of diethyleneglycol succinate and ethyleneglycol succinate, with flame ionization detectors. Phosphatidylcholine has been determined as the diacylglycerol trimethylsilyl ether derivative by gas chromatography using a glass capillary column.[10] Preliminary TLC is required for this method, and for some others (method 2, Table 58-1) before GLC analysis.

High-performance liquid chromatography (HPLC) (method 3, Table 58-1) has recently been used to separate the phospholipid components. Diol bonded-phase columns have been developed with gradients (2% to 15% water) of acetonitrile/H_2O as the mobile phases.[11] Detection of the separated phospholipids (PG, PI, PE, PS, L, and S) is accomplished by measurement of fatty acid double-bond absorption at 203 nm. HPLC can be used to measure saturated phospholipid palmitate.[12]

The methods that have been proposed to measure specifically the disaturated phosphatidylcholine eliminate the unwanted unsaturated fatty acid–containing species with osmium tetroxide.[13] In general, some form of chromatography (column, TLC) is then necessary to isolate the unreacted lecithin.

All the procedures that measure phospholipid constituents of amniotic fluid require partial purification of the compounds before chromatography. Purification schemes usually involve one or more of the following steps: centrifugation, extraction, acetone precipitation, evaporation, and reconstitution. However, techniques for the first three steps have prompted considerable controversy and questions about their efficacy. Thus a large number of extraction schemes, using permutations and combinations of these techniques, are in use.

Methods to measure the physical properties of lung surfactant have recently attracted interest because they measure the functional aspects of lung surfactant. Besides multiple phospholipid components, surfactant may also require a protein component for functional integrity; thus circumstances are possible whereby functional properties would give more accurate assessment of potential lung function than the measurement of individual constituents. These include tests for the ability to lower surface tension, to form stable foam (bubbles), to measure microviscosity, and others.[4] Many tests of this type are simpler to perform and quicker to carry out than the chemical tests, and so this approach has strong potential. However, some of these methods require expensive equipment and techniques that are less familiar to most laboratory workers than the chromatographic methods.

One of the physical methods in most common use is the foam stability or "shake" test (method 4, Table 58-1). This test was devised by Clements et al.[14] and calls for serial dilutions of amniotic fluid and equal volumes of 95% ethanol to exclude spurious surface-active compounds. The tubes are inspected for bubbles around the meniscus after shaking. Pulmonary surfactant forms stable surface films that support bubble stability. Therefore the test is positive if bubbles persist.

The foam stability index (FSI)[15] uses a mixture of different volumes of ethanol with 0.5 mL of amniotic fluid and provides semiquantitative evidence of surfactant content. The index is defined as the highest ethanol volume fraction of an amniotic fluid–ethanol mixture that will permit a stable ring of bubbles at the meniscus after vigorous shaking. An FSI value of 0.48 is comparable to an L/S ratio of 2.0 in correlating with fetal pulmonary maturity.[16]

One can also measure surface tension directly[15,17,18] using a surface balance or tensiometer (method 5, Table 58-1). These measurements give good correlations with other tests for lung surfactant.[19]

The fluorescence polarization assay measures the microviscosity of amniotic fluid lipids, which is related to surface tension (method 6, Table 58-1). A fluorescent probe, when mixed into amniotic fluid, dissolves in the hydrocarbon region. Its rotation in this hydrophobic environment depends on the microviscosity of the fluid. The greater the viscosity (that is, the more surfactant present), the more effectively opposed is the rotation of the probe. This results in an increased polarization of the incident fluorescent light, and the extent of polarization can be measured by a specially designed and rather costly instrument.[20]

The nonchromatographic methods of assessing surfactant activity of amniotic fluid have only been recently developed and are rarely used. TLC remains the almost universal technique for amniotic fluid analysis, though there may be an increase in the use of alternative procedures as their clinical utility is demonstrated. The foam stability (shake) test is widely used as a stat. procedure when TLC analysis is unavailable.

REFERENCE AND PREFERRED METHODS

The reference method is that of Gluck et al.[6,7,21] whose method, either in its original form or in some modified form, has been the most widely used. Clinical interpretation and predictions have for the most part used Gluck's criteria.[6]

A large number of variations of the original procedure are in use today because virtually every step has more than one legitimate means for its execution. Other variations have developed because of differences of opinion over the need for specific steps in the procedure, such as the acetone precipitation step. Accurate clinical correlations are the only meaningful criteria for methodological considerations, and so it is incumbent on each laboratory to establish its own criteria that correlate with clinical experience. However, those steps in the generalized procedure that can drastically compromise the method deserve specific mention.

Centrifugation

To remove whole cells and other debris, amniotic fluid is usually centrifuged before further processing. This preliminary step has not been standardized and a wide variety of centrifugal forces have been used.[22]

Amniotic fluid contains phospholipids from many sources. Some of the phospholipid is related to surfactant and much of it is not. When an amniotic fluid specimen is centrifuged, even at relatively low speeds (3000 *g* for 10 minutes), nearly four fifths of the original lecithin is found in the pellet.[23]

Surfactant phospholipid is present in amniotic fluid in membraneous aggregates[24] called "lamellar bodies," and although high-speed centrifugation (33,000 *g* for 60 minutes or longer) is required to sediment these structures, some sedimentation occurs during centrifugation at only 80 *g* for 5 minutes.[25] Centrifugation speeds of 1000 *g* for 5 minutes result in a variable loss of both total phospholipid and lamellar body phospholipid.[26] Often the losses are proportional for lecithin and sphingomyelin so that the ratio remains constant (Fig. 58-1). However, centrifugation speeds can affect lecithin to a larger extent than sphingomyelin since centrifugal forces have been shown to alter

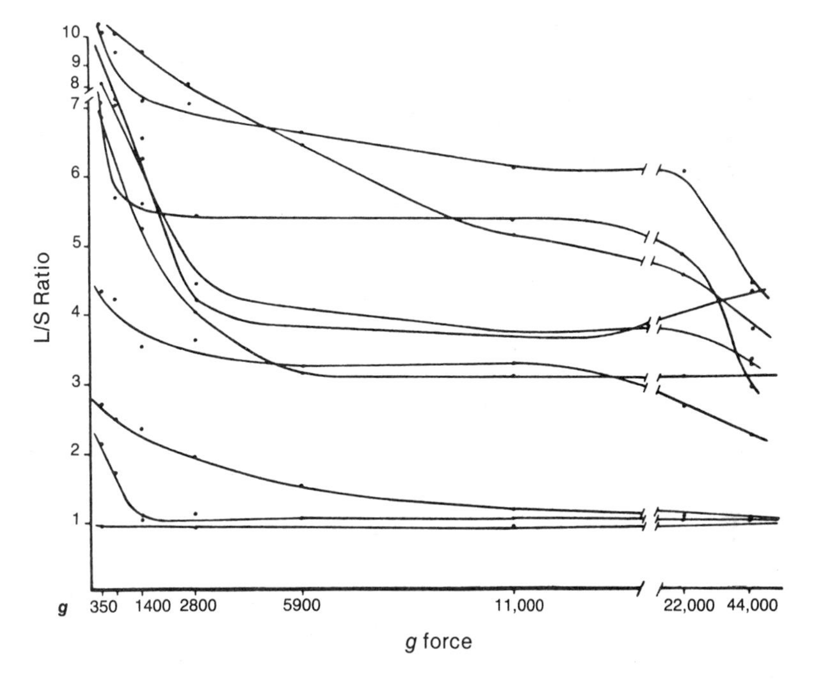

Fig. 58-1. L/S ratio related to centrifugal force. In amniotic fluids with an initial L/S ratio above 2, L/S ratios decrease with increasing *g* force. (From Cherayil, G.D., Wilkinson, E.J., and Borkowf, H.I.: Obstet. Gynecol. **50**:682-688, 1977.)

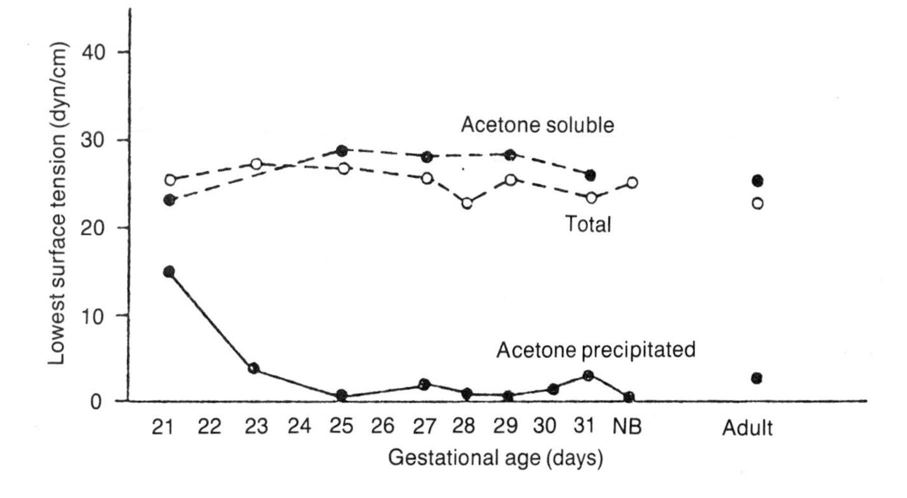

Fig. 58-2. Surface tensions of purified lecithin from lung homogenates at different times of gestation in fetal rabbit. Total lecithin extracted from lung is not surface active. Acetone-precipitated lecithin was surface active. Acetone-precipitated lecithin was surface active and present in lungs from fetuses at all gestational ages examined. (From Gluck, L., Motoyama, E.K., Smits, H.L., and Kulovich, M.V.: Pediatr. Res. **1**:237-246, 1967.)

the L/S ratio.[22,27-29] The effect of centrifugal force on the L/S ratio has been studied by Cherayil et al.[22] and is presented in Fig. 58-1.

Centrifugation also results in the loss of PG, so that when small quantities are present in a sample, losses can influence interpretation of results.

Debris can also be removed by filtration, but losses have been reported[30] of 90% in the total phospholipid content after filtration through Whatman No. 1 filter paper. Filtration also can lower the L/S through disproportionate losses of lecithin compared with sphingomyelin.[28]

Since the acetone precipitation step (see below) provides for the crude isolation of surfactant, I believe that slow centrifugation speeds (500 *g*) are the most suitable.

Extraction

A standardized relationship must exist between the volume of extracting solvent and the volume of the amniotic fluid sample, since the relative proportions of chloroform, methanol, and water significantly affect the extraction of lipid.[31,32] A relationship of one volume of amniotic fluid to one volume of chloroform-methanol (1:1, v/v) is commonly used, although chloroform-methanol (2:1, v/v) also gives good recoveries of phospholipids. After centrifugation to facilitate phase separation, the lower organic phase is removed with a Pasteur pipet and evaporated to moist dryness under a stream of nitrogen. If the sample is taken to complete dryness, the residue will not redissolve easily in chloroform-methanol mixture and losses can occur. This extract is the starting point for the various assays.

Two extractions of the amniotic fluid with equal volume of fluid, methanol, and chloroform are recommended. Although the L/S ratio remains unchanged through a series of extractions,[33] the total amount of lipid obtained from more than one extraction is greater. Since phosphatidylglycerol (PG) is a minor constituent of surfactant and hence extractable lipid, quantitative extraction is an important factor in the detection of PG.

Acetone precipitation

Most naturally occurring lecithins are insoluble in cold acetone[34,35] and, from lung fluid, the acetone-precipitable fraction is enriched in surface activity.[36] Gluck's method used acetone precipitation to effect a crude isolation of surfactant phospholipids,[7,36] and this step has remained controversial ever since. A number of authors claim that the step is unnecessary and time consuming,[3,4] but Gluck is adamant that since surface activity resides solely in the acetone-precipitable material (Fig. 58-2) this step is an essential feature in separating surface-active phospholipids from a total lipid extract.

Lecithins precipitated with cold acetone may range from 60% to 90% of the total surface-active lecithins present in the sample.[37] This means that 10% to 40% of the desired lecithins may be lost. For reliable clinical correlations, it must be assumed that sphingomyelin is lost proportionately, but this may not always be the case. Some authors have reported preferential losses of lecithin[28,38] which falsely reduce the L/S and thus can account for an occasional "immature" L/S when in fact adequate surfactant is present. Other authors have not reported losses as serious with this precipitation step.[4,39] Despite the ongoing controversy, this step remains an integral part of many methods in use today, but strict attention must be paid to the specified ice-cold temperature conditions[7] for this step to be reproducible. Warmer conditions will increase the solubility of surfactant (saturated) lecithin.

Chromatography

The L/S ratio, as introduced by Gluck and co-workers,[5] employed a one-dimensional silica gel thin-layer system. Subsequently, Gluck et al.[1] introduced a two-dimensional system that offered the opportunity to detect other components of the surfactant matrix (notably PI and PG). The change was necessary because the one-dimensional system used initially could not provide sufficient resolution for the phospholipids of interest.

Cochromatography of compounds with lecithin will falsely elevate the L/S. Conversely, other phospholipids cochromatographing with sphingomyelin will falsely decrease the L/S ratio. Painter[40] took advantage of the amphoteric nature of phosphatidylserine and treated his plates with cupric chloride, effecting separations otherwise possible only by two-dimensional TLC. Although many two-dimensional systems provide good precision and chromatographic advantages,[41] one-dimensional chromatography is faster and less expensive. Diffusion of spots during two-dimensional TLC is significantly greater than with one-dimensional systems.[42,43] Thus losses caused by this factor alone can alter results and provide false clinical information.

Methods that use ammonium hydroxide as a constituent in the mobile-phase can have run-to-run variations traceable to the loss of volatile ammonia of the stock solution.[44] The ammonium hydroxide concentrations should be within 15% of the optimum concentration and the working solution should be discarded after 1 week.

Visualization of spots

The literature is replete with methods to visualize spots on thin-layer chromatograms. In the context of the phospholipid methodology for measurement of surfactant, there are some important considerations and the recent paper by Spillman et al.[45] compares seven of the more commonly used techniques for detection of phospholipids after thin-layer chromatography.

It is clear that the visualization method used can distort the L/S ratio. As pointed out by Spillman et al.,[45] the mechanism whereby the visualization reagent interacts with the lipid can often bias the results by underestimation of the detection of saturated lecithin. Reagents that react with double bonds do not give information relating to the saturated lipid of interest. Hence, as the more saturated surfactant is produced by the lung and extracted from amniotic fluid, detection by double-bond reagents deviates more and more from the true saturated lipid content. Also the intensity of the developed spot when treated with reagents that attack double bonds (such as cupric acetate and sulfuric acid) can vary widely (tenfold). One can include ammonium sulfate in this group. This salt is used as part of the stationary phase in certain TLC solvent systems[6] to eliminate the need to spray with strong acid.

Molybdate, ANS (8-anilino-1-naphthalenesulfonic acid), 2′,7′-dichlorofluorescein (DCF), and bromthymol blue stain are independent of the degree of phospholipid saturation. The color intensity of bromthymol blue, however, fades quickly after the plate is removed from the ammonia vapors and for this reason may be less desirable than the other members of this group.

Color development with molybdate and cupric acetate requires heating for development, as do the charring techniques. Background darkening can occur, especially if a plate has been developed in a solvent system that contains an unsaturated component that is not fully removed from the plate before application of the visualization reagent. ANS and DCF are visualized by long-wavelength ultraviolet light so that heating is not required for spot visualization with these reagents. DCF has a higher background than ANS does, but both of these reagents have the advantage that they do not react chemically with lipid so that one can recover spots intact if so desired. Fluorescent indicator is less stable than molybdic stain, and so plates treated in this way should be protected from light if they are kept for extended periods for reference purposes.

If plates are heated for visualization, a preheating period at 110° C before they are placed on the 280° C hot plate is recommended.[41] This avoids the tendency of commercial thin-layer chromatographic plates to shatter.

Lipid staining is facilitated when ethanol is used as solvent for the staining reagent. Reagents dissolved in alcohol more easily penetrate the region containing the phospholipid, resulting in more evenly stained regions. Exposure of the spot to the staining reagent for 2 minutes ensures uniform penetration. I routinely use DCF under ultraviolet light and bromthymol blue for lipid identification.

Detection and quantification

For those laboratories using planimetry for measuring the L/S ratio, lecithin and sphingomyelin spots remain visible longer with phosphomolybdic acid–stannous chloride than with bromthymol blue. Planimetric methods are critically discussed by Whitfield et al.[46]

Other methods have also been developed to measure the L/S ratio. A review of some of these techniques has been published by Olson and Graven.[47] An L/S ratio of 2.0 with reflectance densitometry is about equal to an L/S of 4.0 with gravimetric techniques (Fig. 58-3). It is recommended that a synthetic lecithin and sphingomyelin control

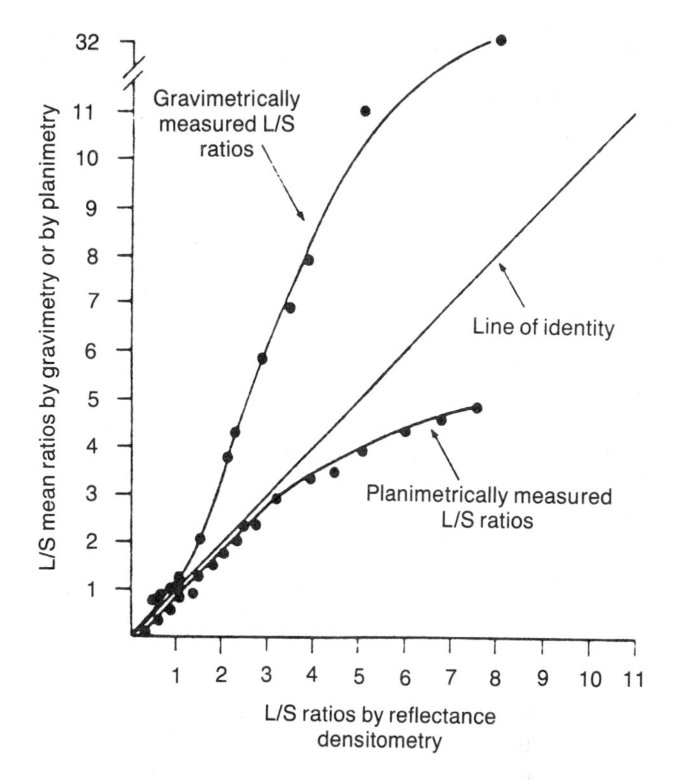

Fig. 58-3. Relationship between densitometry L/S ratios and those with area measurement and by gravimetry. (From Gluck, L., Kulovich, M.V., Borer, R.C., Jr., and Keidel, W.N.: Am. J. Obstet. Gynecol. **120:**142-155, 1974.

be routinely used on plates to monitor spot development to ensure that the end-point has been reached before chromatograms are quantified.[45]

Surfactant that contains PG has better alveolar stabilizing properties than does surfactant that has not yet acquired PG.[49,50] In the absence of PG, the risk of respiratory distress is significant, though development of the disease is not inevitable.[49,51] PG appears when the L/S ratio exceeds 2, and its presence continues to increase in amniotic fluid with continuing gestation (Fig. 58-4). By contrast, PI closely parallels the L/S up to the point of maturity, thereafter falling as the PG continues to increase.[50] The presence of PG appears to provide the final component in the biochemical maturation of the surfactant.

Gluck et al.[6,52] were the first of many to recognize that the L/S ratio was a less reliable predictor of lung maturity in a diabetic pregnancy than in a normal one. They observed a delay in the L/S ratio maturation until 36 weeks of gestation or later. Primarily for this reason, Gluck et al.[1,53] incorporated the pertinent phospholipids into the "lung profile" so that as many factors as possible are considered in predicting fetal lung status. This procedure calls for a quantitative extraction of the phospholipids, acetone precipitation, and two-dimensional thin-layer chromatography followed by quantitation by reflectance densitometry. The unambiguous, simultaneous assessment of PG,

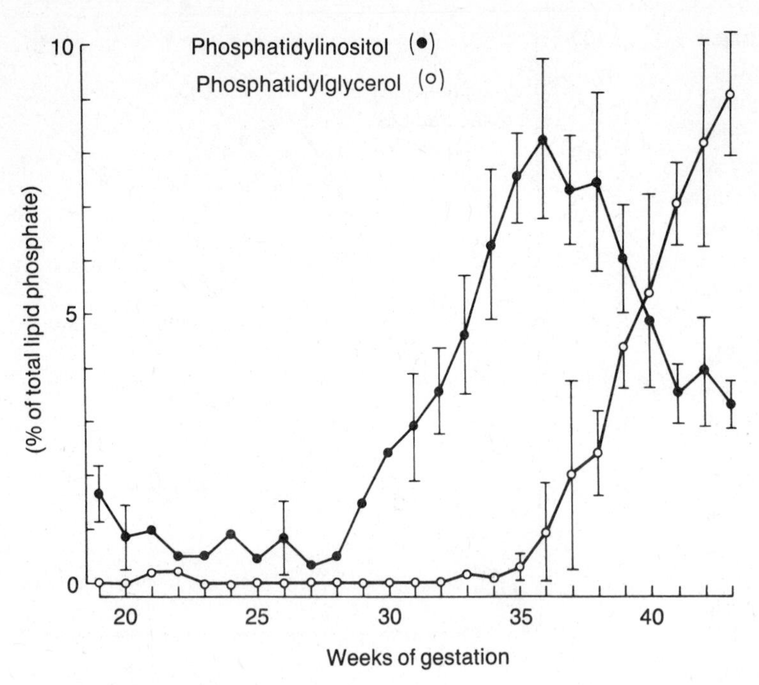

Fig. 58-4. Content of phosphatidyl inositol (PI) *(closed circles)* and phosphatidyl glycerol (PG) *(open circles)* in amniotic fluid during normal gestation. Phospholipids were quantified by measurement of phosphorus (P) content and expressed as percentages of total lipid phosphorus. Means ± standard deviations of three to five samples are shown for each point. (From Hallman, M., Kulovich, M., Kirkpatrick, E., et al.: Am. J. Obstet. Gynecol. **125:**613-617, 1976.)

lecithin, and sphingomyelin is the object of the majority of the chemical methods currently in use, though the need for routine measurement of PG in the nondiabetic may not be needed.

Sources of error in the L/S[48] include excessive amounts of blood because of the spurious phospholipids introduced by this source of contamination. Concentrations in amniotic fluid greater than 0.05 mL of blood per 3 mL of amniotic fluid pose problems, particularly if the L/S is near 2.0. The L/S of blood is near 1.8, and so fluid contaminated with blood would falsely lower the L/S. PG is helpful in cases of blood contamination because PG is not found in blood and its presence therefore indicates mature lung status. Contamination of amniotic fluid from debris, mucus, bacteria, or cellular material from the vagina also cause the L/S to be uninterpretable.

The methods that propose measuring specifically the disaturated phosphatidyl choline[13] eliminate unsaturation with osmium tetroxide. This method is more applicable to the research than the clinical laboratory.[42] Special care must be taken when one is working with osmium tetroxide.

Gas chromatographic methods have been critically reviewed.[3] Although this methodology is more complicated, time consuming, and requires more expensive equipment and higher levels of expertise, useful clinical correlations with surfactant levels have been established.[2,54,55] Correlations of palmitic acid concentrations, palmitic acid–stearic acid ratios, fatty acid compositions of lecithin isolated from amniotic fluid, and lecithin itself[10] have aided in successful prediction of fetal lung status.

High-performance liquid chromatography (HPLC) has many attractive features of potential interest to this area, but the limits imposed by the means of detection (which measures double bonds at 203 nm and not the specific saturated lecithin of interest) and the expense of the equipment relegate this chromatographic approach to status of "promising."

Of the physical methods, the shake test has been the most widely used. The test is very reliable (more than 99%) when the test is positive; that is, less than 1% of infants develop respiratory distress syndrome. However, approximately 50% of infants with a negative test developed respiratory distress syndrome.[56] The test is inexpensive and quick to perform and, when done properly, provides a good adjunct to other methodology. Its reliability requires an uncontaminated sample of amniotic fluid. Bile salts, salts of free fatty acids, and proteins can also exhibit surface activity and contribute to the formation of a stable ring of bubbles. Also, for the test to be reliable all glassware must be free of soap, serum, or biological fluids. The sample must be free of blood contamination and meconium and, if possible, uncentrifuged. Clements[14] suggested a cautious centrifugation at 500 *g* for 5 minutes to remove red cells if necessary, but pointed out that a hematocrit of greater than 3% negated the result. Also of paramount concern is the care with which the ethanol solutions are prepared and maintained. In the original paper, Clements[14] classified the results as negative, intermediate, and positive. In our experience and in the experience of others, a number of results fall in the intermediate range. The shake test is a valuable screening test but does not supersede the necessity for the more specific tests when the shake test is negative. Only when foaming persists at dilutions of 1:2 and over should results be considered positive.[3]

Surface tension measurements[15,17,18] give good correlations with other tests for respiratory distress syndrome and lung surfactant.[13,14,19] Methods that use a tensiometer are relatively rapid but demand meticulous attention to cleaning and maintenance of the equipment. The foam stability index method[15] is simple and rapid and requires no expensive equipment or expertise. This test is sensitive to centrifugation and to contaminants in the amniotic fluid, and patients with hydramnios may have falsely elevated results.[19] Erythrocyte contamination is not a problem if the cells are not hemolyzed, but plasma phospholipids do pose a problem.[15] The final assay mixture is critical to the results, and extreme attention to this detail is required.[4]

Fluorescence polarization methodology is reported to be quicker to perform and less subject to variation than the L/S ratio.[57] It does, however, require dedicated, expensive equipment. The method is sensitive to the effects of centrifugation,[58] and high-density lipoproteins contained in amniotic fluid may hinder the reliability of the method.[59]

The method for amniotic fluid presented below is chosen as a preferred method because of the extensive experience I have had with the method over the past decade. Alternative TLC methods are also reasonable choices.[40,41,60] Basically, the method chosen is only as valid as the clinical correlations that reflect the experience of the individual laboratory.

SPECIMEN

Amniotic fluid obtained by amniocentesis is transported to the laboratory on ice and is promptly centrifuged at 500 *g* for 5 minutes. A 3 mL aliquot of the supernatant is needed for extractions.

Samples should be processed as soon as possible after collection. Untreated amniotic fluid held at room temperature can lose both lecithin and sphingomyelin over a period of 48 hours.[28,61] According to Wagstaff et al., after only 4.5 hours lecithin levels already decrease about 25%. Decreases of a smaller order occur with sphingomyelin over a similar period of time. This disproportionate loss of lecithin relative to sphingomyelin with time results in a consistent fall in the L/S ratio. However, if after collection amniotic fluid is immediately centrifuged at a low speed, such as 500 *g* for 5 minutes, it can be kept at room temperature for at least 4 days with no apparent change in the L/S ratio.[62] Centrifugation of amniotic fluid samples is necessary if storage is contemplated. Samples stored at −20° C are stable for prolonged periods without noticeable alteration of the L/S ratio.

PROCEDURE FOR LECITHIN-SPHINGOMYELIN RATIO (L/S)[40]

Principle

Phospholipids are extracted from the amniotic fluid into chloroform-methanol and then precipitated in cold acetone. The precipitated phospholipids are separated by thin-layer chromatography, visualized, and quantitated. Fetal lung maturity is assessed by an estimation of the amount of lecithin and sphingomyelin present in the amniotic fluid. This is expressed as a ratio of lecithin to sphingomyelin (L/S). Table 58-2 summarizes the important parameters of the procedure.

Table 58-2. Comparison of assay conditions for amniotic fluid analyses

Parameter	Thin-layer chromatography (TLC)	Shake test
Assay temperature	Ambient	Ambient
Sample volume	1 mL	1.0, 0.75, 0.50, 0.25, and 0.20 mL
Fraction of sample volume	0.50 (extraction)	0.50
Final concentration of reagents	(Not applicable)	47.5% ethanol
Interferences	Blood, meconium, high-speed centrifugation before testing	Blood, meconium, soap, serum, biological fluids, surface evaporation from tubes, movement of tubes once they have been shaken, high-speed centrifugation of sample before testing

Reagents

1. Eastman Chromatogram Sheets, 6061 Silica gel without fluorescent indicator, 20 × 20 cm. Cut into 1 × 13 cm strips. Activate by storing in a desiccator over Drierite for 24 hours.
2. Developing solvent: Pipet into a 50 mL round-bottom capped centrifuge tube: 5.0 mL of methanol, 0.8 mL of distilled water, 13.0 mL of chloroform. Vortex. Keep solution capped when not in use. Use fresh daily.
3. Methanol: spectral grade.
4. Chloroform: spectral grade.
5. Acetone: reagent grade.
6. Bromthymol blue (417 mg/L), water soluble (3′3′-dibromothymol sulfone phthalein, sodium salt) available from Aldrich Organic Chemicals (Milwaukee, Wisc.): Add 200 mg of bromthymol blue to 200 mL of distilled water and 32.0 mL of 1 M NaOH. Dilute to 480 mL with distilled water. Add 5.0 gm of boric acid powder (reagent grade). Mix. This solution is stable for 3 months.
7. Nitrogen gas, compressed, 100%.
8. 1 M NaOH (40 g/L): Place 40 g of NaOH in a 1 L volumetric flask and add about 800 mL of water, slowly, to dissolve. When the NaOH is fully dissolved and the temperature of the solution is back to room temperature, add water to mark. Mix thoroughly. Stable for 12 months at room temperature.

9. Standards: The lecithin, sphingomyelin, and phosphatidyl glycerol standards can be purchased from Supelco, Inc. (Bellefonte, Penn.) as chloroform-methanolic solutions. The contents of each ampule are quantitatively transferred to a volumetric flask (with the ampule being rinsed out too) and then diluted with chloroform-methanol (2:1, v/v) to final concentration of 1 mg/mL.

Lecithin (cat. no. 4-6102): 50 mg is diluted to 50 mL.
Sphingomyelin (cat. no. 4-6009): 25 mg is diluted to 25 mL.
Phosphatidyl glycerol (cat. no. 4-6013): 10 mg is diluted to 10 mL.

Assay

Equipment: Centrifuge, 20 × 50 mm test tubes with cork stoppers, forceps, 12 mL conical screw-capped centrifuge tubes, 3 mL conical test tubes.

This procedure is best performed in duplicate.

1. Centrifuge 3 to 5 mL of amniotic fluid at 500 *g* for 5 minutes at ambient temperature.
2. Pipet 1 mL in duplicate into 12 mL conical screw-capped test tubes. Pipet 1 mL in triplicate if PG (phosphatidyl glycerol) is also to be determined.
3. To each 1 mL sample of amniotic fluid, add 1 mL of methanol. Vortex. Add 2 mL of chloroform and vortex vigorously.
4. The tubes are centrifuged (500 *g* for 5 minutes to facilitate phase separations). Three layers should be seen:
   a. Aqueous lay (top)
   b. Protein fluff layer (middle)
   c. Chloroform layer (bottom)
5. Remove as much of the chloroform layer as possible using a disposable Pasteur transfer pipet (go through the top two layers carefully). Place this in a 3 mL conical tube. Evaporate to *moist* dryness under nitrogen. (*Note:* water bath should be below 40° C.)
6. If PG is to be determined, repeat the extraction (steps 3 to 5) for the tube for PG.
7. Wash down the lipid adhering to the sides of each tube with 50 μL of chloroform. Take to moist dryness again.
8. Chill the 3 mL tubes with the lipid extract in ice for 1 minute. With a Pasteur pipet, add 2 drops of ice-cold acetone while swirling in the ice. Eight more drops of ice-cold acetone are added and the tube is iced for another minute. The acetone is decanted. The precipitate is washed with another 10 drops of ice-cold acetone while the tube remains iced. The supernatant is decanted. The precipitate is dried gently under a stream of nitrogen.
9. The cold acetone precipitate is dissolved in 6 mL of chloroform. The entire amount is then spotted 2 cm from the end of the activated TLC strip. (The PG tube is spotted on a plate to be described later.)
10. The strip is placed into 1.5 mL of the developing solvent in the 20 × 150 mm test tube. The tube is stoppered with a cork. The strip is chromatographed to a distance of 2 cm from the top of the strip. (NOTE: Use forceps to handle the strips.)
11. The strips are removed from the tubes with forceps, air dried, and then dipped in the bromthymol blue indicator contained in a 20 × 150 mm test tube. Excess dye is blotted from the strip. The lecithin and sphingomyelin spots are easy to detect; they appear dark yellow in color (Fig. 58-5).

Calculation

1. Circle each spot with a pencil along the outermost edges. Measure maximum width and length of each spot in centimeters. Use the following calculation:

$$\text{L/S ratio} = \frac{(\text{Length of lecithin}) \times (\text{Width of lecithin})}{(\text{Length of sphingomyelin}) \times (\text{Width of sphingomyelin})}$$

2. Average the values from the duplicate samples.
3. Apply L/S standards daily. Spot 30 μL of lecithin standard (10 mg/mL) on top of 15 μL of sphingomyelin standard (10 μg/mL) on a fresh strip and chromatograph.
4. Running time for L/S about 1½ hours.

PROCEDURE FOR PHOSPHATIDYL GLYCEROL (PG)

Reagents

1. Chloroform: spectral grade
2. Methanol: spectral grade
3. Glacial acetic acid: reagent grade
4. Solvent system: chloroform/methanol/acetic acid/water (65:25:8:4, v/v).
5. 2′7′-Dichlorofluorescein (0.2 g/L) in ethanol. Dissolve 2 g into 100 mL of ethanol. This is the stock solution. Dilute 1:100 in ethanol for working solution. When stored at room temperature, the stock is stable for 1 year and the working solution for 3 months.

Assay

Equipment: Silica gel 60 thin-layer plates, 5 × 20 cm, (E. Merck, Darmstadt, Germany). Plates are activated at 110° C for 1 hours before use. Filter paper–lined thin-layer chromatography tank.

1. Triplicate tube (step 6 of L/S ratio method). Lipid contents are dissolved in 10 μL of chloroform and spotted 2 cm from the bottom and a third of the way in from the side of a silical gel (5 × 20 cm) TLC plate (activated).
2. Lecithin (20 μL, of 1 μg/μL) and phosphatidylglycerol

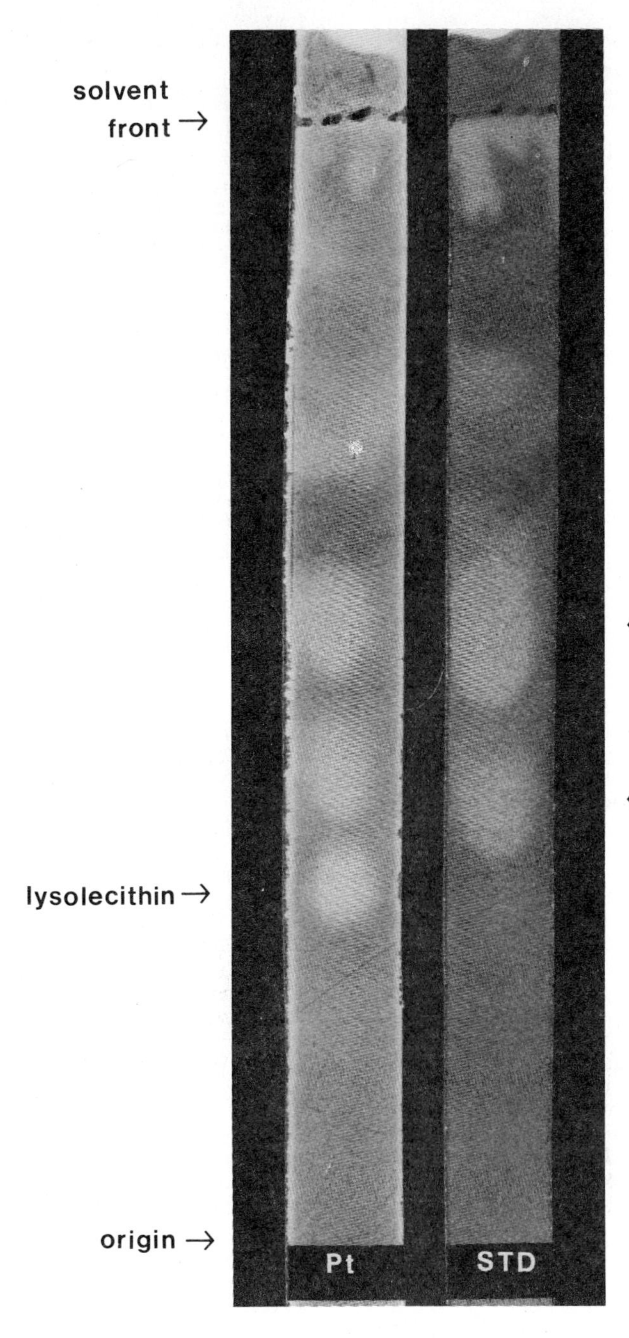

Fig. 58-5. Separation of lecithin and sphingomyelin on thin-layer strips as described for method to determine L/S ratio. *Pt,* Patient sample; *STD,* L and S standards.[7]

(20 μL of 1 μg/μL) standards are spotted 2 cm from the bottom and a third of the way in from the opposite of the same plate.

3. The plate is placed in a filter paper–backed thin-layer tank previously equilibrated with the solvent system for at least 1 hour (replace solvent system weekly or more often if runs become atypical).
4. The solvent system is permitted to run about 10 cm up the plate. The plate is withdrawn from the tank and air dried for 15 to 20 minutes.
5. The plate is sprayed with dichlorofluorescein reagent, allowed to dry, and viewed under ultraviolet light 15 minutes to 1 hour later (Fig. 58-6).
6. The visual identification of a PG spot confirms the presence of PG. This visualization is sensitive to 0.5 μg and indicates the presence of PG in amniotic fluid to be greater than or equal to 0.5 μg/mL.

Interferences

Blood contamination of amniotic fluid has some effect on the L/S ratio, though there is no agreement about whether the blood increases or decreases the L/S ratio.[28,61,63] Samples contaminated with blood are still useful in the prediction of fetal lung maturity if phosphatidylglycerol is found. PG is virtually absent from blood, meconium, and vaginal secretions, all of which contain phospholipids and other components that cause inaccurate L/S ratios.[25,63,64] Another alternative is offered if one isolates the 10,000 *g* pellet from the contaminated fluid and measures fetal pulmonary surfactant from the lambellar bodies.[25,65]

Contamination of amniotic fluid with meconium interferes with the interpretation of chromatograms,[66,67] though Gerbie et al.[68] did not consider that its presence interfered with the estimation of the L/S ratio. Wagstaff et al.[28] demonstrated a consistent rise in the L/S ratio with increasing contamination with meconium. Chromatograms from samples contaminated with meconium are characterized by a lysolecithin spot immediately following the sphingomyelin spot. If chromatographic separation is not good, the lyso-

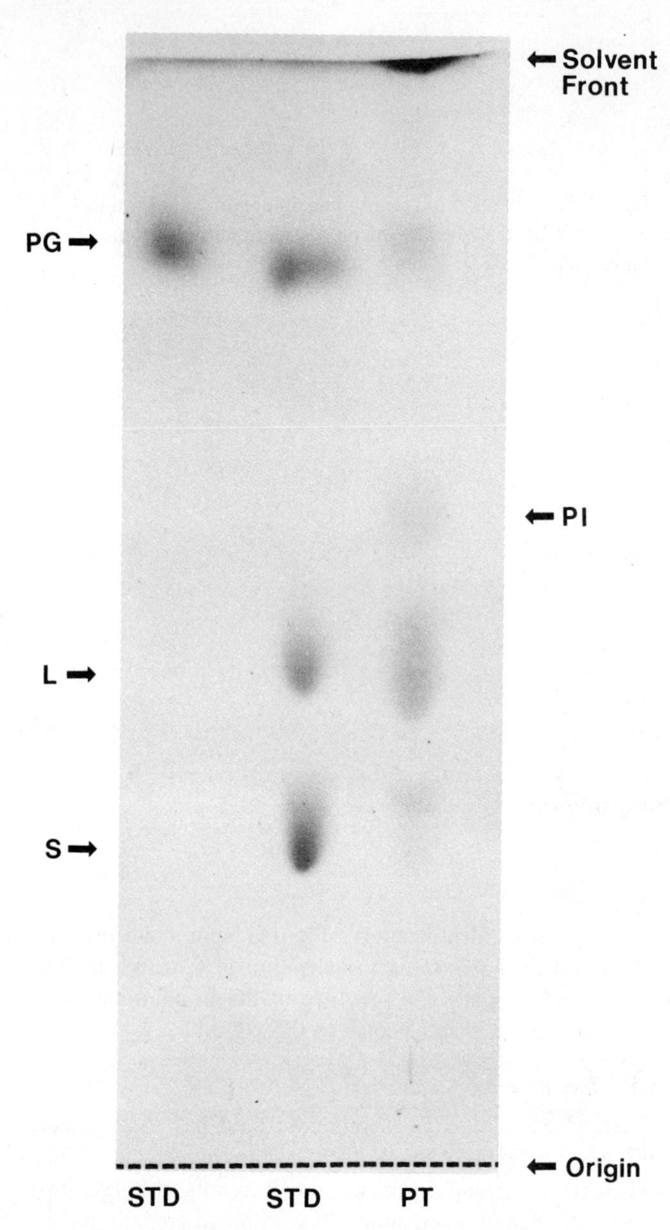

Fig. 58-6. Phosphatidyl glycerol (PG) spot on silica gel thin-layer plate according to method for PG. *L,* Lecithin; *PI,* phosphatidyl inositol, *PT,* patient; *S,* sphingomyelin; *STD,* standard (*left sample,* PG; *middle sample,* PG, L, and S).

lecithin spot will come to lie so close to the sphingomyelin spot that the two may be confused, giving a falsely decreased L/S ratio.[63] Other authors report falsely high L/S ratio from fluid contaminated with meconium.[69] Again, it seems prudent to rely on the presence of PG if samples are contaminated with meconium or to assess surfactant of the 10,000 *g* pellet.[25,65]

REFERENCE RANGE

See Chapter 35 for details. An L/S ratio more than 2.0 is considered indicative of fetal maturity. If a spot is seen in the PG area, the test is considered positive. If no spot is seen, it is considered negative.

REFERENCES

1. Kulovich, M.V., Hallman, M.B., and Gluck, L.: The lung profile. I. Normal pregnancy, Am. J. Obstet. Gynecol.**135:**57-63, 1979.
2. Russell, P.T., Miller, W.J., and McLain, C.R.: Palmitic acid content of amniotic fluid lecithin as an index to fetal lung maturity, Clin. Chem. **20:**1431-1434, 1974.
3. Wagstaff, T.I.: The estimation of pulmonary surfactant in amniotic fluid. In Fairweather, D.V.I., and Eskes, T.K.A.B., editors: Amniotic fluid—research and clinical application, ed. 2, Amsterdam, 1978, Excerpta Medica, pp. 347-391.
4. Freer, D.E., and Statland, B.E.: Measurement of amniotic fluid surfactant, Clin. Chem. **27:**1629-1641, 1981.
5. Gluck, L., Kulovich, M.V., and Borer, R.C., Jr.: The diagnosis of the respiratory distress syndrome (RDS) by amniocentesis, Am. J. Obstet. Gynecol. **109:**440-445, 1971.
6. Gluck, L., and Kulovich, M.V.: Lecithin/sphingomyelin ratios in amniotic fluid in normal and abnormal pregnancy, Am. J. Obstet. Gynecol. **115:**539-546, 1973.
7. Gluck, L., Kulovich, M.V., and Borer, R.C., Jr.: Estimates of fetal lung maturity, Clin. Perinatol. **1:**125-139, 1974.
8. Coch, E.H., Kessler, G., and Meyer, J.S.: Rapid thin-layer chromatographic method for assessing the lecithin/sphingomyelin ratio in amniotic fluid, Select. Methods Clin. Chem. **8:**63-70, 1977.
9. Adams, F.H., Fujiwara, T., Emmanouilides, G.C., and Raiha, N.: Lung phospholipids of human fetuses and infants with and without hyaline membrane disease, J. Pediatr. **77:**833-841, 1970.
10. Lohninger, A., Salzer, H., Simbruner, G., et al.: Relationships among human amniotic fluid dipalmitoyl lecithin, postpartum respiratory compliance, and neonatal respiratory distress syndrome, Clin. Chem. **29:**650-655, 1983.
11. Briand, R.L.: High-performance liquid chromatographic determination of the lecithin/sphingomyelin ratio in amniotic fluid, J. Chromatogr. **223:**227-284, 1981.
12. Sax, S.M., Moore, J.J., Oley, A., et al.: Liquid-chromatographic estimation of saturated phospholipid palmitate in amniotic fluid compared with a thin-layer chromatographic method for acetone-precipitated lecithin, Clin. Chem. **28:**2264-2268, 1982.
13. Mason, R.J., Nellenbogen, J., and Clements, J.A.: Isolation of disaturated phosphatidylcholine with osmium tetroxide, J. Lipid Res. **17:**281-284, 1976.
14. Clements, J.A., Platzker, A.C.G., and Tierney, D.F.: Assessment of risk of respiratory distress syndrome by a rapid test for surfactant in amniotic fluid, N. Engl. J. Med. **286:**1077-1081, 1972.
15. Statland, B.E., and Freer, D.E.: Evaluation of two assays of functional surfactant in amniotic fluid, surface tension lowering ability, and the foam stability index test, Clin. Chem. **25:**1770-1773, 1979.
16. Sher, G., Statland, B.E., and Freer, D.E.: Clinical evaluation of the quantitative foam stability index test, Obstet. Gynecol. **55:**617-620, 1980.
17. Tiwary, C.M., and Goldkrand, J.W.: Assessment of fetal pulmonary maturity by measurement of the surface tension of amniotic fluid lipid extract, Obstet. Gynecol. **48:**191-194, 1976.
18. Goldkrand, J.W., Varki, A., and McClurg, J.E.: Surface tension of amniotic fluid lipid extracts: prediction of pulmonary maturity, Am. J. Obstet. Gynecol. **128:**591-598, 1977.
19. Bichler, A., Daxenbichler, G., Oriner, A., et al.: Amniotic fluid surface tension measurements versus various phospholipid determinations in the assessment of fetal lung maturity, Respiration **37:**114-121, 1979.
20. Shinitsky, M., Goldfisher, A., Bruck, A., et al.: A new method for assessment of fetal lung maturity, Br. J. Obstet. Gynecol. **83:**838-844, 1976.
21. Gluck, L., Landowne, R.A., and Kulovich, M.V.: Biochemical development of surface activity in mammalian lung, Pediatr. Res. **4:**352-364, 1970.
22. Cherayil, G.D., Wilkinson, E.J., and Borkowf, H.I.: Amniotic fluid lecithin/sphingomyelin ratio changes related to centrifugal force, Obstet. Gynecol. **50:**682-688, 1977.

23. Abramovich, D.R., Keeping, J.D., and Thom, H.: The origin of amniotic fluid lecithin, Br. J. Obstset. Gynaecol. **82**:204-207, 1975.
24. Hook, G.E.R., Gilmore, L.B., Tombropoulos, E.G., et al.: Fetal lung lamellar bodies in human amniotic fluid, Annu. Rev. Resp. Dis. **117**:541-550, 1978.
25. Oulton, M.: The role of centrifugation in the measurement of surfactant in amniotic fluid, Am. J. Obstet. Gynecol. **135**:337-343, 1979.
26. Duck-Chong, C.G., Brown, L.M., and Hensley, W.J.: Sedimentation of lung-derived phospholipid during low-speed centrifugation of amniotic fluid, Clin. Chem. **27**:1424-1426, 1971.
27. Wilkinson, E.J., Cherayil, G.D., and Borkowf, H.I.: L/S ratio and the "g-force" factor, N. Engl. J. Med. **296**:286-287, 1977.
28. Wagstaff, T.I., Whyley, G.A., and Freedman, G.: Factors influencing the measurement of the lecithin/sphingomyelin ratio in amniotic fluid, J. Obstet. Gynaecol. Br. Cwlth. **81**:264-277, 1974b.
29. Lindback, T., and Frantz, T.: Effect of centrifugation on amniotic fluid phospholipid recovery, Acta Obstet. Gynecol. Scand. **54**:101-103, 1975.
30. Bedham, L.P., and Worth, H.G.J.: Critical assessment of phospholipid measurement in amniotic fluid, Clin. Chem. **21**:1441-1447, 1975.
31. Bligh, E.G., and Dyer, W.J.: A rapid method of total lipid extraction and purification, Can. J. Biochem. Physiol. **37**:911-917, 1959.
32. Bhagwanani, S.G., Fahmy, D., and Turnbull, A.C.: Quick determination of amniotic fluid lecithin concentration for prediction of neonatal respiratory distress, Lancet **2**:66-67, 1972.
33. Sarkozi, L., Kovacs, H.N., Fox, H.A., and Kerenyi, T.: Modified method for estimating the phosphatidylcholine: sphingomyelin ratio in amniotic fluid, and its use in the assessment of fetal lung maturity, Clin. Chem. **18**:956-960, 1972.
34. Artom, C.: Labeled compounds and phospholipid metabolism. In Colowick. S.P., and Kaplan, N.O., editors: Methods in enzymology, vol. 4, New York, 1957, Academic Press, Inc., p. 823.
35. Dervichian, D.G.: The physical chemistry of phospholipids, Prog. Biophys. Mol. Biol. **14**:26D, 1964.
36. Gluck, L., Motoyama, E.K., Smits, H.L., and Kulovich, M.V.: The biochemical development of surface activity in mammalian lung, Pediatr. Res. **1**:237-246, 1967.
37. Penney, L.L., Hagerman, D.D., and Sei, C.A.: Specificity and reproducibility of acetone precipitation in identifying surface-active phosphatidylcholine in amniotic fluid, Clin. Chem. **22**:681-682, 1976.
38. Armstrong, D., and VanWormer, D.E.: Rapid determination of pulmonary surfactant, Am. J. Obstet. Gynecol. **114**:1083-1086, 1972.
39. Parkinson, C.E., and Harvey, D.R.: A comparison between the lecithin/sphingomyelin ratio and other methods of assessing the presence of fetal lung pulmonary surfactant in amniotic fluid, J. Obstet. Gynaecol. Br. Cwlth. **80**:406-411, 1973.
40. Painter, P.C.: Simultaneous measurement of lecithin, sphingomyelin, phosphatidylglycerol, phosphatidylinositol, phosphatidylethanolamine, and phosphatidylserine in amniotic fluid, Clin. Chem. **26**:1147-1151, 1980.
41. Gross, T.L., Wilson, M.V., Kuhnert, P.M., and Sokol, R.J.: Clinical laboratory determination of phosphatidylglycerol: one- and two-dimensional chromatography compared, Clin. Chem. **24**:486-490, 1981.
42. Tsai, M.Y.: Relative merits of one- and two-dimensional TLC of phospholipids in amniotic fluid, Clin. Chem. **27**:1957-1958, 1981.
43. Cunningham, M.D., McKean, H.E., Gillispie, D.H., and Greene, J.W.: Improved prediction of fetal lung maturity in diabetic pregnancies: a comparison of chromatographic methods, Am. J. Obstet. Gynecol. **142**:197-204, 1982.
44. Glick, J.H., Jr., and Crocker, C.L.: Concentration of ammonium hydroxide is critical in chromatographic solvents for amniotic fluid phospholipids, Clin. Ghem. **28**:1997-1998, 1982.
45. Spillman, T., Cotton, D.B., Lynn, S.C., Jr., and Brétaudière, J.-P.: Influence of phospholipid saturation on classical thin-layer chromatographic detection methods and its effect on amniotic fluid lecithin/sphingomyelin ratio determinations, Clin. Chem. **29**:250-255, 1983.
46. Whitfield, C.R., Chan, W.H., Sproule, W.B., and Stewart, A.D.: Amniotic fluid lecithin: sphingomyelin ratio and fetal lung development, Br. Med. J. **2**:85, 1972.
47. Olson, E.B., and Graven, S.N.: Comparison of visualization methods used to measure the lecithin/sphingomyelin ratio in amniotic fluid, Clin. Chem. **20**:1408-1415, 1974.
48. Gluck, L., Kulovich, M.V., Borer, R.C., Jr., and Keidel, W.N.: The interpretation and significance of the lecithin/sphingomyelin ratio in amniotic fluid, Am. J. Obstet. Gynecol. **120**:142-155, 1974.
49. Hallman, M., Feldman, B.H., Kirkpatrick, E., and Gluck, L.: Absence of phosphatidylglycerol (PG) in respiratory distress syndrome in the newborn, Pediatr. Res. **11**:714-720, 1977.
50. Hallman, M., Kulovich, M., Kirkpatrick, E., et al.: Phosphatidylinositol and phosphatidylglycerol in amniotic fluid: indices of lung maturity, Am. J. Obstet. Gynecol. **125**:613-617, 1976.
51. Hallman, M., Feldman, B., and Gluck, L.: RDS: the absence of phosphatidylglycerol in surfactant, Pediatr. Res. **9**:396, 1975 (Abstract).
52. Donald, I.R., Freeman, R.K., Goebelsmann, U., et al.: Clinical experience with the amniotic fluid lecithin/sphingomyelin ratio, Am. J. Obstet. Gynecol. **115**:547-552, 1973.
53. Kulovich, M.V., and Gluck, L.: The lung profile. II. Complicated pregnancy, Am. J. Obstet. Gynecol. **135**:64-70, 1979.
54. Szalay, J., and Lukacs, E.: Palmitic acid concentration of amniotic fluid in diabetic pregnancy, Arch. Gynaekol. **222**:279-283, 1977.
55. O'Neal, G.J., Davies, I.J., and Siu, J.: Palmitic/stearic ratio of amniotic fluid in diabetic and non-diabetic pregnancies and its relationship to development of respiratory distress syndrome, Am. J. Obstet. Gynecol. **132**:519-523, 1978.
56. Parkinson, G.E., and Harvey, D.: Amniotic fluid and pulmonary maturity. In Sandler, M., editor: Amniotic fluid and its clinical significance, New York, 1981, Marcel Dekker, Inc., pp. 229-252.
57. Gebhardt, D.O.E.: Relationship between the lecithin/sphingomyelin ratio and the P value of amniotic fluid, Am. J. Obstet. Gynecol. **133**:937, 1979.
58. Simon, N.V., Elser, R.C., Levisky, J.S., and Polk, D.T.: Effect of centrifugation on fluorescence polarization of amniotic fluid, Clin. Chem. **27**:930-932, 1981.
59. Gebhardt, D.O.E.: Detection of neonatal respiratory distress on the basis of fluorescence polarization (microviscosity) measurements of amniotic fluid: a word of caution, Clin. Chem. **26**:1629, 1980 (Letter).
60. Pappas, A.A., Mullins, R.E., and Gadsden, R.H.: Improved one-dimensional thin-layer chromatography of phospholipids in amniotic fluid, Clin. Chem. **28**:209-211, 1982.
61. Wagstaff, T.I., Whyley, G.A., and Freedman, G.: The measurement of the lecithin/sphingomyelin ratio of amniotic fluid after thin layer chromatography, Ann. Clin. Biochem. **11**:24-27, 1974a.
62. Whitfield, C.R., and Sproule, W.B.: Fetal lung maturation, Br. J. Hosp. Med. **12**:678-682, 1974.
63. Buhi, W.C., and Spellacy, W.N.: Effects of blood or meconium on the determination of the amniotic fluid lecithin/sphingomyelin ratio, Am. J. Obstet. Gynecol. **121**:321-323, 1975.
64. Torday, J., Carson, L., and Lawson, E.E.: Saturated phosphatidylcholine in amniotic fluid and prediction of the respiratory-distress syndrome, N. Engl. J. Med. **301**:1013-1018, 1979.
65. Oulton, M., Martin, T.R., Faulkner, G.T., et al.: Developmental study of a lamellar body fraction isolated from human amniotic fluid, Pediatr. Res. **14**:722-728, 1980.
66. Bryson, M.J., Gabert, H.A., and Stenchever, M.A.: Amniotic fluid lecithin/sphingomyelin ratio as an assessment of fetal pulmonary maturity, Am. J. Obstet. Gynecol. **114**:208-212, 1972.
67. Hobbins, J.C., Brock, W., Speroff, L., et al.: L/S ratio in predicting pulmonary maturity in utero, Obstet. Gynecol. **39**:660-664, 1972.
68. Gerbie, M.V., Gerbie, A.B., and Boehm, J.: Diagnosis of fetal maturity by amniotic fluid phospholipids, Am. J. Obstet. Gynecol. **114**:1078-1082, 1972.
69. Kulkarni, B.D., Bieniarz, J., Burd, L., and Scommegna, A.: Determination of lecithin-sphingomyelin ratio in amniotic fluid, Obstet. Gynecol. **40**:173-179, 1972.

Cholesterol

HERBERT K. NAITO

Clinical significance: p. 550
Molecular formula: $C_{27}H_{46}O$
Molecular weight: 386.64 daltons
Merck Index: 2192
Chemical class: sterol, lipid

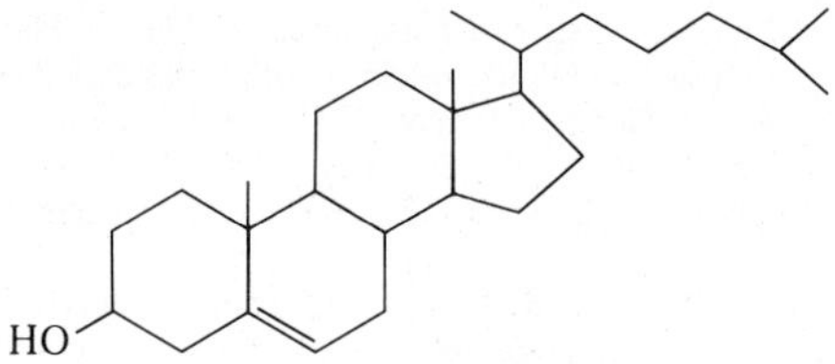

PRINCIPLES OF ANALYSIS AND CURRENT USAGE

Classification of cholesterol methods

Measurement of total cholesterol includes both the ester and free forms of the steroid. In serum or plasma, two thirds of the total cholesterol exists in the esterified form, with the rest in the free form. This has some analytical implications, since in some reactions the color development with the ester cholesterol is greater in intensity than that of free cholesterol. This in turn can lead to a large positive bias. For this reason, understanding the chemistry of the various cholesterol methods and recognizing their limitations are of utmost importance in the selection of the method to be used in the laboratory. Most often there are trade-offs among simplicity, speed, convenience, and accuracy and precision.

Single-step methods. In these direct assays there is no sample preparation, that is, no isolation and purification of the steroid or steroids. Thus direct procedures are those carried out on serum or plasma samples without any prior solvent extraction steps. This advantage allows for simple and rapid analysis, with only one reagent usually needed and little manipulation of the sample, and therefore it is suitable for automation. The colorimetric and fluorometric assays are usually quick and convenient. However, depending on the chemical reaction, these procedures are likely to exhibit both positive and negative errors resulting from the presence of proteins; bilirubin; vitamins A, C, and D; steroid hormones; uric acid; turbidity; and differences in chromogenicity of free cholesterol and ester cholesterol. Many automated procedures are based on direct methods, such as those adapted to the Technicon SMA-12 or SMAC-II instruments. They are usually based on such acceptable methods as Zlatkis et al.,[1] Huang et al.,[2] Pearson et al.,[3] and enzymatic procedures.[4-6]

Two-step methods. In these types of assays an organic-phase extraction step is introduced before measurement of cholesterol and other chemically related steroids. This pretreatment step removes many nonspecific chromogens that might interfere with the assay; the relative chromogenic response of each interfering substance is dependent on the chemistry involved. Since no saponification step is involved in these methods, the deleterious effect of differential color response between free and ester cholesterol still remains, especially in Liebermann-Burchard reactions. However, for most routine clinical work, this two-step procedure has been well accepted; the values obtained correspond closely with that of the Abell et al.[7] method. The correlation is even closer when a calibration factor is added to correct for free and ester cholesterol differential color development. Refer to p. 1206 for details of the calibration factor and a semiautomated procedure that we routinely use in the laboratory. The methods of Zak and Ressler,[8] Carr and Drekter,[9] Bloor,[10] and Chiamori and Henry[11] are examples of two-step procedures.

Three-step methods. These procedures involve, in addition to extraction of cholesterol, a saponification step that hydrolyzes the fatty-acid moiety from the cholesterol ester. Consequently, one measures only free cholesterol. The method of Abell et al.[7] (method 1, Table 58-3) belongs in this classification and has now been considered a reference method. Some modifications have been added to improve the accuracy and precision of the measurement. Since the Liebermann-Burchard method is considered a reference procedure, a modified Abell et al.[7] method is provided on p. 1201.

Four-step methods. These methods go a step further than the three-step method of extraction, saponification, and color development. The total extractable steroids are purified for cholesterol determination by the addition of a saponin, digitonin. The reactive site on the C-3 position on the cyclopentanoperhydrophenanthrene ring in cholesterol is the hydroxyl group that is esterified by the digitonin. This causes the complex to precipitate out. The addition of the digitonin step also eliminates the effect of interfering nonspecific chromogen constituents. Empirically, four-step procedures should be more accurate and precise than any other procedure, with the possible exception of enzymatic cholesterol determinations. However, unless extreme precautions are taken, multiple steps may also mean multiple errors. For instance, whereas digitonin precipitation enhances the accuracy of the method, it must be completely decomposed and removed or it will cause additional color development and thus positive error.

The Schoenheimer and Sperry[12] and Sperry and Webb[13] methods are four-step procedures that have, in the past, been considered reference methods for cholesterol determinations. These two methods also employ Liebermann-Burchard reagents. Although the Abell et al.[7] method may be more tailored to serum or plasma cholesterol analysis, four-step methods may be more suitable for tissue extracts,

Table 58-3. Methods for serum cholesterol analysis

Method	Method classification	Principle	Usage	Comments
1. Abell et al.[7]	Three-step	Cholesterol extracted with zeolite, esters chemically hydrolyzed (saponification), and total cholesterol measured by Liebermann-Burchard reaction	Considered current reference method	Laborious
2. Liebermann-Burchard (L-B)	One-, two-, three-, or four-step	Cholesterol extracted and reacted with strong acid (sulfuric acid) and acetic anhydride to form colored cholestahexaene–sulfonic acid molecule (A_{max}, 410 nm); nonesterified cholesterol precipitated by digitonin, and remaining cholesterol measured and free cholesterol calculated: Total − Esterified = Free	Very common method	Total cholesterol reaction overestimates concentration of esterified cholesterol Unstable color
3. Iron-salt-acid	Two-step	Similar to reaction conditions of method 2, except Fe^{+3} ions are added to yield tetraenylic cation (A_{max}, 563 nm)	Not frequently used	Sevenfold more sensitive than L-B method Free and esterified cholesterol give *same* color; no need to hydrolyze esters
4. *p*-Toluene-sulfonic acid (*p*-TSA)	Three-step	Similar to method 3; *p*-TSA reacts with cholesterol derivative to form chromophore (A_{max}, 550 nm)	Rarely used	Free and esterified cholesterol give same color Bilirubin causes large positive error
5. Enzymatic end-point	One-step	a. Cholesterol-esters $\xrightarrow{\text{Cholesterol esterase}}$ Cholesterol + Fatty acids b.* Cholesterol + O_2 $\xrightarrow{\text{Cholesterol oxidase}}$ Cholest-4-en-3-one + H_2O_2 c. H_2O_2 + 4-Aminophenazone (or other dye) $\xrightarrow{\text{Peroxidase}}$ Oxidized dye (A_{max}, 500 nm) + H_2O	Most common method	Accurate and easily automated Future reference method

*Can monitor reaction by following O_2 consumption with oxygen electrode.

since certain tissues may have appreciable noncholesterol steroids and cholesterol esters that may contribute to the inaccuracy of the final development of color.

Methods of analysis

Literally hundreds of cholesterol methods have been published, usually as modifications of the following reactions: (1) Liebermann-Burchard, (2) iron-salt-acid, (3) *p*-toluene-sulfonic acid, or (4) enzymatic end point. Table 58-4 compares reaction conditions of the candidate reference method and the two most frequently used methods for total cholesterol determination.

Liebermann-Burchard reaction. Among the many nonenzymatic colorimetric reactions for cholesterol, the Liebermann-Burchard (L-B) procedure (method 2, Table 58-3) is perhaps the most widely used. The L-B reaction generally is carried out in a strong acid medium—sulfuric acid, acetic acid, and acetic anhydride. In the chemical reaction cholesterol goes through a stepwise oxidation, with each step yielding a cholestapolyene molecule, which has one more double bond than the compound from which it was derived. The initial L-B step involves protonation of the OH^- group in cholesterol and subsequent loss of water to give the carbonium ion, 3,5-cholestadiene, which is the first step in the color reaction. Sequential oxidation of this allylic carbonium ion by SO_3^- yields a cholestahexaene–sulfonic acid chromophoric compound with an absorbance maximum (A_{max}) of 410 nm. This chemical reaction for cholesterol determination has undergone many modifications through the years and has included the measurements of either free or esterified cholesterol or both. Digitonin, a saponin, reacts with the free OH^- (at the C-3 position) group of the A ring of the sterol to form an insoluble complex, cholesterol digitonide. The measurement of the sterol content in the supernatant solution provides an estimate of esterified cholesterol. To obtain free cholesterol, one subtracts the esterified value from the total (free plus ester) cholesterol value.

Iron-salt-acid reaction. This reaction (method 3, Table 58-3) involves acetic acid–sulfuric acid in the absence of acetic anhydride. In this reaction, however, Fe^{+3} must be added to obtain the desired chromogen. As in the L-B pro-

Table 58-4. Comparison of reaction conditions of methods for total cholesterol analysis

Condition	Abell-Levy-Brodie-Kendall*	Liebermann-Burchard (L-B)†	Enzymatic‡
Temperatures	50° C—hydrolysis 25° C—colorimetric reaction	Ambient—extraction, reaction	37° C
Sample volume	500 μL	500 μL	250 μL
Fraction of sample volume	0.091 for hydrolysis	0.1 for extraction	0.01
Final concentration of reagents	Hydrolysis step KOH: 0.322 mol/L Ethanol: 909 g/L Extraction step Hexane: 50% v/v Colorimetric reaction step Acetic anhydride: 64.5% (v/v) Sulfuric acid: 3.2% (v/v) Acetic acid: 32.3% (v/v)	Insoluble extraction mixture (in 99% isopropanol) 99% isopropanol: 990 g/L Zeolite: 320 g/L $CuSO_4 \cdot H_2O$: 16 g/L Kaolin: 32 g/L $Ca(OH)_2$: 32 g/L Reaction mixture Acetic anhydride: 60% (v/v) Sulfuric acid: 30% (v/v) Acetic acid: 10% (v/v)	Cholesterol ester hydrolase: 40 U/L Cholesterol oxidase: 56 U/L Peroxidase: 1 U/L Sodium cholate: 15 mmol/L 4-Aminophenazone: 2 mmol/L Sodium hydroxybenzoate: 50 mmol/L Phosphate buffer: 0.45 mmol/L (pH 6.75) EDTA: 8 mmol/L
Time of reaction	Hydrolysis, 60 min Reaction, 30 min	Extraction, 20 min AutoAnalyzer-II, 8 min	10 min
Wavelength	620 nm	630 nm	500 nm
Linearity	5 g/L (12.5 mmol/L)	4 g/L	5 g/L
Precision§ ($\bar{X}$, % coefficient of variation)	— —	2879 mg/L, 6.6% 3375 mg/L, 6.5%	2737 mg/L, 6.3% 3246 mg/L, 6.2%

*Candidate reference method from Abell, L.L., et al.: J. Biol. Chem. **195**:357-366, 1952. Modified by Centers for Disease Control Lipid Laboratory, Atlanta (by Duncan et al.: unpublished data).
†Technicon AutoAnalyzer-II method. Modified from Lipid Research Clinics Program method.[19]
‡Sclavo Diagnostics reagents (Wayne, N.J.). Analysis performed on Abbott ABA-100 analyzer.
§From College of American Pathologists Quality Assurance Survey data. L-B without extraction.

cedure, the initial step is the protonation of the OH^- group in the cholesterol molecule and subsequent loss of water to form the carbonium ion (3,5-cholestadiene). Serial oxidation of this allylic carbonium ion by Fe^{+3} yields a tetraenylic cation with an absorbance maximum of 563 nm. The iron-salt-acid procedures are about sevenfold more sensitive than the L-B methods. This increased sensitivity may be attributed largely to the stabilizing effect on enylic cation formation at higher H_2SO_4 concentrations. In general, increasing the H_2SO_4 concentration would be expected to improve the stability of each of the carbonium ions formed in the stepwise oxidation of the sterol, thereby making it much more likely for one to observe carbonium ion formation in the iron-salt-acid reaction than in the L-B reaction. The iron-salt-acid reaction has generated several modifications, including the use of ferrous sulfate, ferric perchlorate, or a mixture of ferric acetate–uranium acetate plus ferrous sulfate.

The original iron-salt method neither included a deproteinization step, nor did it mention the importance of using purified reagents. For example, contaminating glyoxylic acid in glacial acetic acid would react with tryptophan to produce an intense interfering color. The removal of protein from the serum or plasma minimizes the problems of impure acids. Deproteinization can be accomplished by the use of organic solvents or by means of a strongly acid ferric chloride–acetic acid reagent.

***p*-Toluenesulfonic acid reactions.** These methods are based on reactions of cholesterol with *para*-toluenesulfonic acid (*p*-TSA), acetic anhydride, glacial acetic acid, and H_2SO_4 (method 4, Table 58-3).

Enzymatic end-point reactions. The majority of the tests performed in laboratories today are concerned with measurement of total cholesterol. For this reason and because of the urgent demand for speed and ease in processing large volumes of samples, many simple and rapid determinations for total cholesterol have been published. Preliminary treatment of the serum is time consuming, increases the probability of error, and does not readily lend itself to automation. These factors have resulted in a trend to eliminate the preliminary steps and to develop direct serum methods. One simply adds the serum to a solution, incubates the mixture, and measures the color in the presence of all other serum constituents.

Recently, enzymatic techniques (method 5, Table 58-3) for determining cholesterol have emerged to compete with the classical Liebermann-Burchard reaction. The original work used preliminary chemical saponification of the sample to produce only free cholesterol. In the next step, cholesterol oxidase, an enzyme almost entirely specific for cholesterol, is added. This causes the breakdown of cholesterol to cholest-4-en-3-one and hydrogen peroxide, after which several different reaction systems have been used to produce a final chromogen.

The commercial kits on the market now offer a total enzymatic procedure, utilizing the enzyme cholesterol esterase to replace the chemical saponification. Cholesterol esterase is specific for cholesterol esters, splitting the esters into free cholesterol and free fatty acids. The cholesterol oxidase reaction follows, and the same enzyme reaction just described occurs. The amount of color produced is directly proportional to the amount of serum cholesterol.

Because of the trend toward increased use of enzymatic methods in the future, a more extensive discussion is provided here.

The first chemical step in the enzymatic methods for cholesterol uses the enzyme cholesterol esterase to hydrolyze the cholesterol esters present in the serum to free cholesterol and free fatty acids:

$$\text{Cholesterol esters} + H_2O \xrightarrow{\text{Cholesterol esterase}} \text{Cholesterol} + \text{Free fatty acids} \qquad \textit{Eq. 58-1}$$

As discussed previously, the second step uses the enzyme cholesterol oxidase in the presence of oxygen to oxidize the cholesterol (both the free cholesterol found in the serum and the free cholesterol generated in step 1) to cholest-4-en-3-one and hydrogen peroxide:

$$\text{Cholesterol} + O_2 \xrightarrow{\text{Cholesterol oxidase}} \text{Cholest-4-en-3-one} + H_2O_2 \qquad \textit{Eq. 58-2}$$

In this reaction cholesterol concentration can be determined by amperometric measurement of the rate of oxygen depletion. Beckman Instruments (Fullerton, Calif.) produces the Cholesterol Analyzer-2, which measures cholesterol by this method.

Other assays make use of the ability of hydrogen peroxide to oxidize compounds to produce colored species that can be measured spectrophotometrically:

$$2\,H_2O_2 + \text{Phenol} + \text{4-Aminophenazone} \xrightarrow{\text{Peroxidase}} \text{Quinoneimine dye} + 4\,H_2O \qquad \textit{Eq. 58-3}$$

or

$$H_2O_2 + \text{Methanol} \xrightarrow{\text{Catalase}} \text{Formaldehyde} + H_2O \qquad \textit{Eq. 58-4}$$

$$\text{Formaldehyde} + \text{Acetylacetone} \rightarrow \text{3,5-Diacetyl-1,4-dihydrolutidine} \qquad \textit{Eq. 58-5}$$

Reaction 3 (Trinder's reaction), which forms the quinoneimine dye (absorbance maximum 500 to 525 nm), is the basis of the majority of the kits currently on the market.

Reaction 4 has the advantage of not being subject to bilirubin interference and has been developed as a rate procedure to improve on the original time-consuming endpoint assay.

REFERENCE AND PREFERRED METHODS

Liebermann-Burchard methods

One should recognize that the L-B procedure gives a greater color intensity for esterified cholesterol than for free cholesterol. The L-B methods are affected by many variables: the concentration of each reagent, the amount of water in the final reaction mixture, the solvent employed, the time of reaction, the temperature of reaction, the wavelength of light at which the color is measured, the presence of interfering substances such as bilirubin and unreacted digitonin, and the form in which the cholesterol is present (as esters or free or both). Therefore, conditions must be closely adhered to according to rigid specifications. One should note that the L-B reagents are not as stable as other types of cholesterol reagents. The final color is also unstable and can lead to erroneous results.

Iron-salt-acid methods

The iron-salt-acid methods produce the same color intensity for free and esterified cholesterol. Thus saponification is not needed in these chemically based procedures. These methods are also several times more sensitive than the L-B or *p*-TSA reagent systems. Since several times more color development occurs for cholesterol with iron-salt-acid methods than with the other chemical methods, these are more ideally suited for microprocedures.

One of the more important sources of error in the iron–salt–sulfuric acid methods is improper mixing of the reagents after the addition of the H_2SO_4. Variation in the heat of reaction will cause varying results. Thus it is important to give special attention to standardizing a method of mixing the reagents that will produce a maximum temperature in a given time.

Other sources of error are halogens (iodide, bromide), bilirubin, glyoxylic acid, peroxides, hemolysis, and certain drugs and vitamins.

p-Toluenesulfonic acid methods

In general, reactions with *p*-TSA are simpler than those using the L-B methods for three reasons: (1) the reagents are more stable, (2) the final color is more stable (especially at 550 nm), and (3) the same color intensity is produced with free and esterified cholesterol, thereby eliminating the need for saponification. However, bilirubin gives four to five times more color with *p*-TSA than these reagents give with cholesterol; therefore, extensive positive error will occur in blood containing large amounts of bilirubin. Icteric blood always poses a problem for these procedures.

Enzymatic end-point methods

The enzymatic procedure is a one-step method amenable to automated procedures. The harsh reagents involved with the chemical determinations are no longer a problem. The interferences of other constituents are greatly reduced because of the specificity of the enzymes. A few cholesterol analogs may interfere, but the magnitude of interference seen in the chemical determinations is greatly reduced. The option to determine the difference between free and esterified cholesterol is available with these methods.

It has been suggested that with more use and study, enzymatic methods may become the reference methods of choice in cholesterol determinations, since fewer interfering substances affect the reaction and end-point detection. Many enzymatic methods have passed the rigid criteria for accuracy and precision and appear to be even more accurate and precise than the described L-B methods, which are also standardized and certified by the Centers for Disease Control (CDC) in Atlanta.

A properly calibrated enzymatic method is certainly the preferred method for routine laboratory analysis. These methods can be as accurate and precise as the Abell-Kendall reference method. Since these methods are readily available as kits, this technique is not presented in detail. A reference L-B method is presented, since this may be used for the initial calibration of a method.

A few precautions are necessary when one is using the enzymatic kits that have quinoneimine dye as the chromogen. Blanking can cause problems with turbid samples for manual procedures. For automated procedures (as with the Technicon AutoAnalyzer-II), all-glass tubing must be used after the point of the formation of the quinoneimide dye complex, since Tygon tubing will absorb the dye, slowly releasing it in subsequent samples and causing carry-over problems. A perfect bubble pattern and a stable reagent temperature are important for reliable results using the AutoAnalyzer. Standards are another problem with some automated procedures when pumping devices are used. All samples and standards must be of the same viscosity in order for the pump to deliver equivalent aliquots of each sample, reagent, and standards. Since primary cholesterol standards can be dissolved only in organic solutions, a reference serum should be selected. Finding an appropriate calibrating serum and adjusting the set point is not an easy chore. The matrices in calibrators are complex and may not be compatible with a given enzymatic system. In addition, the value or set point suggested by the manufacturer may not necessarily be correct for the equipment and method used. Unfortunately, much care must be taken to find the right set point for a reference serum. Because of different wetting agents, stabilizers, processing, and additives present in both the reference serum and the reagents, the same set points cannot be used for kits obtained from different manufacturers. These interfering substances also make it difficult to attain a set point for an enzymatic method when one is using the reference material calibrated by a reference method. Instead, a comparison of known samples must be used to attain a set point. Some variation in results of different methods may be caused by measurements at different wavelengths. A range of 460 to 560 nm has been proposed, with 500 nm being the maximum absorbance of the dye. Commonly suggested wavelengths for taking measurements are 505, 520, and 525 nm. Despite all these problems, accurate results can be obtained when reliable reference material is used. An enzymatic method is the preferred method for routine analysis of cholesterol.

Reference methods

Reference techniques can be carried out with equipment available in many laboratories and give results statistically coinciding with the definitive method or with a clearly defined bias. The procedures accepted as reference procedures are still based on the Liebermann-Burchard (L-B) reaction (Abell et al.,[7] Schoenheimer-Sperry,[12] or Sperry-Webb[13] methods). The reference procedures are compared to definitive methods, which are discussed on the next page.

The Centers for Disease Control have proposed a modification of the Abell et al.[7] procedure for use as a reference method for cholesterol.[19] This method includes hydrolysis of the cholesterol esters, extraction with petroleum ether, and color development with an acetic acid/acetic anhydride/sulfuric acid reagent. With the proper equipment and good technical skills, this procedure can be used to produce results with both acceptable accuracy and precision. The method can be automated to allow for simultaneous analysis of a greater number of samples than was previously possible. The continuous-flow (AutoAnalyzer-II) method[19] correlates very well with the modified Abell et al. procedure. Both manual and continuous-flow methods are discussed in detail in this chapter.

Since the L-B reagent is so widely used, special emphasis should be placed on its preparation. A common problem associated with methods utilizing the L-B reagent is the lack of uniformity and precision in its laboratory preparation. However, some of the following precautions can help:

1. Try to obtain analytical grade reagents that are certified by the American Chemical Society (ACS).
2. The glacial acetic acid should be aldehyde free. Certain chemical vendors provide this specialized reagent.
3. Moisture is one of the most serious causes of erroneous cholesterol values. Use extreme care when adding the sulfuric acid to the acetic anhydride–gla-

cial acetic acid mixture. To minimize the heat of reaction, keep the sulfuric acid and acetic anhydride–glacial acetic acid mixture in a tightly covered bottle and place it on ice. Add sulfuric acid to the mixture very slowly, each time capping the bottle before mixing the solution.

4. Extra stability can be attained by adding anhydrous sodium sulfate to the L-B reagent.

Definitive methods

Definitive techniques measure the concentration of a substance in a biological sample and give results accepted as the nearest attainable to a true value. Isotopic dilution mass spectrometry has been recommended as a definitive method for cholesterol determination, with a coefficient of variation (CV) no greater than 0.5% and a total uncertainty of no greater than 1% (Cohen et al.[15]). For more information on definitive cholesterol methods, consult Cohen et al.[15] and Wolthers et al.[16]

SPECIMEN CONSIDERATIONS FOR LIPID ANALYSES

Patient preparation

Determinations of lipid constituents in plasma or serum are normally done on blood drawn from patients fasting for 12 to 16 hours. Patient preparation is an integral part of ensuring proper interpretation of lipid analysis. The optimum patient condition at the time of blood drawing is achieved by (1) no change in dietary habits for at least 3 weeks, (2) stable body weight, and (3) fasting for at least 12 hours. Proper fasting means no food of any sort, except for water and possibly black coffee (without sugar) in the morning. The fasting sample is essential for triglyceride analysis, since triglyceride levels increase as soon as 2 hours postprandially and reach a maximum at 4 to 6 hours. Nonfasting samples are not suitable for analysis, since elevated results caused by normal assimilation of the food cannot be distinguished from elevated results resulting from abnormal lipid metabolism or inborn errors of metabolism. Keep in mind that nonfasting may cause hyperchylomicronemia and hyper-pre-β-lipoproteinemia, which may in itself interfere with a given assay. This is especially true for greatly hypertriglyceridemic samples (greater than 20,000 mg/L), which are represented as chylomicrons, whose particle size is greater than 80 nm and can interfere with colorimetric assays. Ethanol consumption causes acute but transient elevations in serum triglyceride concentrations. This is especially evident in carbohydrate-sensitive hypertriglyceridemic persons. Therefore it would be advisable to request the patient to refrain from drinking alcohol for 72 hours before the day of blood drawing. For cholesterol and phospholipid determinations the fasting sample is not so important, though the fasting time may make a difference in cholesterol levels for a small percentage of people.

Blood-drawing techniques

Setting up a standardized procedure for drawing blood is important, especially when long-term comparison studies are being considered. Posture, emotional and physical stress, selected vein, tourniquet use, tube selection, and fasting time are some important factors to be considered. For adults a convenient method is to draw blood from the patient's antecubital vein in a seated position; the person should be subjected to a minimal amount of stress.

It is well recognized that plasma volume increases and the concentrations of nondiffusible plasma components decrease when a standing subject assumes a recumbent position, a result of redistribution of water between the vascular and extravascular compartments. A significant reduction in total plasma cholesterol has been measured after 5 minutes, and decreases of as much as 10% to 15% have been recorded 20 minutes after a recumbent position is assumed. The effect on cholesterol concentration when one changes from a standing position to a sitting position is also significant though somewhat smaller—about 6% after 10 to 20 minutes. These changes should probably not be regarded as causing errors in the lipid analyses, since they occur physiologically; they can nonetheless complicate the interpretation of the measurements made, especially when one is monitoring a patient.

If a tourniquet is used, removal before blood sampling is suggested, since prolonged use of the tourniquet has been reported to produce increased lipid values. Serum cholesterol concentrations were found to increase an average of 10% to 15% after 5 minutes of occlusion. From a practical standpoint errors of this magnitude would not normally be encountered, since the tourniquet is usually removed within 30 to 60 seconds and the changes during this period are insignificant. Increases of 2% to 5% have been observed after about 2 minutes, which could occur if difficulties arise during sampling or when blood is taken by an inexperienced technician.

Choice of plasma or serum

If the preparation is done properly, either plasma or serum is usually suitable for total cholesterol, triglyceride, or phospholipid determinations. Fibrinogen in improperly prepared serum may cause cumbersome problems with continuous-flow systems (such as AutoAnalyzers) by plugging or coating the tubing or joint connectors, causing a change in flow rates. On the other hand, if the plasma is used, spontaneous hydrolysis of triglycerides can occur if the tubes are left at ambient temperature too long. Serum or plasma left at room temperature too long can have increased lecithin-cholesterol acyltransferase (LCAT) activity, which causes altered blood lipid or lipoprotein composition with respect to free/ester cholesterol and lecithin/lysolecithin ratios. Plasma is usually preferred when lipids and lipoproteins are being chemically analyzed. If plasma is chosen, the suggested anticoagulant is solid EDTA

(1 mg/mL of blood), and the blood cells should be separated as soon as possible (within 2 hours).

Certain anticoagulants, such as fluoride, citrate, and oxalate, cause rather large shifts of water from the red blood cells to the plasma, which result in the dilution of plasma components. For example, the apparent total cholesterol concentration of plasma containing 2.5 mg/mL of sodium oxalate was found to be 10% lower than that of serum obtained concurrently without anticoagulant. A recent study by Kuba et al.[17] possibly indicates a similar finding between plasma and serum, except that the difference between the two was about 2% (at levels of 1220 to 2440 mg/L) for cholesterol and 5% (at levels of 380 to 3740 mg/L) for triglycerides. Heparin, on the other hand, causes no detectable change in red cell volume and decreases total cholesterol concentration by only 1% or less. EDTA, which is commonly used to prepare plasma for lipoprotein analysis, causes a slightly greater decrease in lipid concentration. We have found that cholesterol and triglyceride concentrations of plasma containing 1 mg/mL of EDTA are about 3% lower than those of serum. Although EDTA causes slightly greater changes in cholesterol concentration than heparin, it is generally preferred for lipoprotein analysis for several reasons. It retards the auto-oxidation of unsaturated fatty acids and cholesterol by chelating heavy metal ions such as Cu^{++} that promote this change. Oxidation leads to alterations in the physical properties of the lipoproteins, gross denaturation, and apparent degradation of lipoproteins. EDTA retards these changes and can reduce changes that can occur as a result of possible contamination by phospholipase C–producing bacteria. Such contamination is associated with spurious increases in total glyceride concentration that are caused by the production of partial glycerides from some phospholipids. It is also associated with the artifactual appearance of chylomicron-like particles and decreases in α-, β-, and pre-β-lipoproteins. These changes are minimized by the addition of EDTA, which is an antioxidant and an inhibitor of phospholipase C activity.

Sodium fluoride is commonly used when one is collecting plasma for glucose determinations. Its major function is to prevent red blood cell glycolysis, which will alter glucose results. It has been reported by Martinek[18] that colorimetric determination of total cholesterol on samples containing fluoride lowers values by 300 to 500 mg/L. Our data on enzymatic determination of total cholesterol suggest that fluoride produces values about 25% lower than samples without fluoride.

Plasma and serum preparation

For serum, good technique dictates centrifuging the blood samples after an adequate clotting time but without excessive delay; 2 hours clotting at ambient temperature is usually sufficient, with any additional processing done as soon as possible. For plasma, place the sample on ice when drawn and centrifuge as soon as possible. Centrifugation at 4° C is best if a refrigerated centrifuge is available. Centrifuge the samples at 2000 *g* for about 15 minutes. Remove the plasma immediately from cells into a clean tube, since dynamic exchange of free cholesterol between the red blood cell membrane and plasma can occur. Particular attention should be given to the prevention of hemolysis of red blood cells.

It should also be emphasized that when one is centrifuging a blood sample by conventional means for serum or plasma, the tubes should be tightly sealed to prevent evaporation of the sample, which can occur quite readily.

Changes that can occur as a result of delays after the sample is drawn include alteration in lipoprotein composition from the exchange of cholesteryl esters and triglycerides between high-density lipoproteins (HDL) and the other lipoproteins. Thus it is good practice to perform the lipid analysis as soon as possible.

Several of these factors can occur simultaneously, and their cumulative effects might either be additive (they may increase the error of the analysis) or compensatory for another error and therefore not apparent, depending on the particular method used and the procedure's sensitivity to interfering substances.

Storage of samples

Two of the common variables of sample handling that influence measured lipid and lipoprotein values are the length of time and conditions under which samples are stored before analysis. It is generally recommended that plasma be stored in the liquid state when it is to be used for lipid and, in particular, lipoprotein analysis or for lipoprotein electrophoresis studies; the determinations should be performed promptly. In practice, however, delays can occur for a variety of reasons, and it may even be necessary to analyze frozen samples. We have observed that repeated freezing and thawing has a noticeable effect on lipoprotein electrophoretic patterns—that is, the presence of lipid-stainable material at the application point—which suggests lipoprotein degradation. Several groups of workers have measured lipoprotein levels in samples that have been stored at different temperatures for varying periods of time, frozen and unfrozen.[18,19] The measurements were made with the analytical ultracentrifuge, and the results generally indicated that the levels of all the lipoproteins may decrease with storage. No definitive studies have been made of the optimum freezing temperature and the length of storage under these conditions. Cooper[20] recently reported that when specimens are stored in the liquid state at 4° C for longer than 1 week, high-density lipoprotein (HDL) cholesterol values, reproducibility of the test, and resolution of lipoprotein patterns decrease. The HDL cholesterol values were the most stable when the specimens were stored at −60° C, as compared to −20° C or −5°

C, over 10 months. At present we believe that freezing at −60° C provides the longest stable storage and may allow for reproducible results even after a year or more. Freezing at −20° C will keep samples stable for a few months if self-defrosting freezer units are avoided, since samples stored in these units periodically undergo thawing and freezing, causing a breakdown of the lipid and lipoprotein components. If the samples are to be used in a few days, refrigeration at 4° C is sufficient. However, even at refrigerated temperatures, spontaneous hydrolysis of the triglycerides to free glycerol and fatty acids takes place, which effectively reduces the true triglyceride concentration. In addition, the sharpness of the lipoprotein bands decreases with time when electrophoresis is performed. Therefore, if more than short-term storage is anticipated, samples should be frozen to prevent the spontaneous hydrolysis and oxidation of the unsaturated lipids, which occurs during storage, especially at warmer temperatures.

Stored serum or plasma samples must be adequately mixed before use. Lipids will layer by density within the sample when stored and give different results, depending on the layer from which an aliquot is obtained. Simple inversion of the tube that contains the serum or plasma is not sufficient, particularly if the sample was frozen. Instead, the sample should be carefully mixed on a vortex mixer for 5 to 10 seconds. Particular attention should be given to those samples that have appreciable amounts of "standing" chylomicrons. One can obtain accurate lipid results only when the chylomicrons are dispersed evenly through the sample before taking an aliquot for lipid analysis. If pronounced chylomicronemia exists, the sample should be diluted with saline so that positive-displacement errors do not occur with the excessive macromolecules.

If an organic solvent extraction method is being used, the extract can be stored at 4° C for short periods (a few days) when one is concerned only with total cholesterol, triglyceride, or phospholipid analyses. On the other hand, if one is interested in fatty acid profiles, the extract should be frozen immediately to inhibit auto-oxidation. Care must be taken when one is storing the purified lipid extracts. Cork or rubber stoppers or caps with paper, cork, or rubber liners should be avoided when one is storing lipid extracts in organic solvents. Screw-capped tubes with Teflon-lined caps are good storage vials if a good seal is formed. Evaporation of the solvent can be a problem if care is not taken to ensure the selection of a good storage container. A well-sealed container or vial also helps to keep oxygen out of the sample and to prevent oxidation of the unsaturated lipids. If the samples are to be stored in this form for a long period, it is best to overlay the sample extract with nitrogen before sealing the vial.

PROCEDURE

Many methods for cholesterol analysis regularly give inaccurate and imprecise values whereas other methods give incorrect values under various specific conditions. Many methods in current use are not being performed properly, mainly because of the poor understanding and lack of experience with them. Many methods must be further modified to some degree to optimize the procedure to the particular laboratory conditions. Whether in the clinical or in the research laboratory, the need for standardizing and ensuring accurate and reproducible results is being increasingly recognized. Very few laboratories can establish a dependable cholesterol procedure in which the accuracy of the method is no more than a 2% coefficient of variance (CV) from established "target values" with no more than a 2% CV for a within-day variance and no more than a 3% CV on a day-to-day basis over a prolonged period. Although this criterion may appear too stringent, it should be a goal for any laboratory interested in producing valid research or patient care data. These goals are emphasized in the Centers for Disease Control program, and we have adopted this philosophy at The Cleveland Clinic Foundation.

Although cholesterol is one of the most thoroughly investigated biological constituents, it is amazing that its determination still poses a problem in many laboratories. The newest methods of cholesterol analysis by enzymatic systems are starting to make reliable analyses much quicker and more practical. Most methods before this were based on the reaction of strong, concentrated acids with the alcoholic portion of the cholesterol molecule to form a colored product. Many variations on this approach have been proposed to increase the accuracy of the test. The procedures were either performed directly on the blood samples or after various degrees of extraction and purification of the cholesterol by many different organic solvent extraction or precipitation steps. The following section provides a detailed discussion on a candidate reference method that has been adapted as a semiautomated procedure. For the sake of brevity, the automated method includes the simultaneous determinations of cholesterol and triglyceride.

Simultaneous determination of total cholesterol and triglycerides by a semiautomated method using AutoAnalyzer-II* equipment

The following procedure is a modification of the methods validated and standardized by the Lipid Research Clinics Program.[19] They have been chosen as practical and economical methods that have been used successfully for many years in the Lipid Laboratory at The Cleveland Clinic Foundation.

Determinations are made on an isopropanol extract of serum or plasma. For cholesterol analysis, a modified Liebermann-Burchard reagent and the sample extract combine "on stream" and pass through a 60° C heating bath; the

*Technicon Instruments Corp., Tarrytown, NY.

chromogen formed is then read at 630 nm. The triglyceride determination is accomplished by sample saponification with alcoholic potassium hydroxide (mixed "on stream") at 50° C to form glycerol. The glycerol is oxidized by sodium *meta*-periodate to formaldehyde, which is then condensed with acetylacetone to form 3,5-diacetyl-1, 4-dihydrolutidine (DDL). Both reactions take place at 50° C. The DDL is measured fluorometrically with a 400 nm primary filter and a 485 nm secondary filter.

Reagents (See special consideration B.)

A. Extraction
   1. 99% isopropanol, reagent grade (See special consideration B1.)
   2. Zeolite mixture (See special consideration B2.)
      a. Zeolite 200 g, no. 80 mesh (Taylor Chemicals, Inc., product 285)
      b. Cupric sulfate ($CuSO_4 \cdot 5\ H_2O$), 10 g, ACS reagent grade (Fisher Scientific, catalog C-493)
      c. Kaolin, 20 g (Fisher Scientific, catalog K-5)
      d. Calcium hydroxide, 20 g, ACS reagent grade (Fisher Scientific, catalog C-97)

      Place zeolite in a shallow pan or tray and heat in oven overnight at 110° C. Cool and place in a widemouthed jar. Grind cupric sulfate to a fine powder with a mortar and pestle. Add the ground cupric sulfate, kaolin, and calcium hydroxide to the zeolite and mix. Close jar tightly and store desiccated until used.

B. AutoAnalyzer reagents
   1. Liebermann-Burchard reagent
      a. Acetic anhydride, 600 mL, ACS reagent grade (Fisher Scientific, catalog A-10)
      b. Glacial acetic acid, 300 mL, ACS reagent grade, aldehyde free (J.T. Baker Chemical Co., catalog 9511)
      c. Sulfuric acid, 100 mL, ACS reagent grade (J.T. Baker Chemical Co., catalog 9681)

      Place a 2 L Erlenmeyer flask in an ice bucket and surround with ice. Pour acetic anhydride and glacial acetic acid into the flask and swirl to mix. Seal flask tightly to prevent the absorption of moisture and allow to cool until cold—at least 10 minutes. Add sulfuric acid slowly and in several small portions. Seal flask and swirl to mix after each addition. Allow to cool for at least 20 minutes and swirl again before use. Keep reagent on ice during the entire analytical run. Reagent may be stored at 4° C for 1 week.
   2. Potassium hydroxide, 0.8 M, 50.65 g, ACS reagent grade (J.T. Baker Chemical Co., catalog 3140) (See special consideration B3.)

      Add potassium hydroxide to a 1 L volumetric flask containing about 500 mL of reagent-grade water. Dissolve and cool to room temperature, and then dilute to volume with reagent-grade water. Note that poor-grade water can contribute to an elevated background fluorescence.
   3. Periodate reagent
      a. Glacial acetic acid, 115 mL, ACS reagent grade
      b. Sodium *meta*-periodate, 5.4 g, ACS reagent grade (J.T. Baker Chemical Co., catalog 4-3756)

      Place approximately 500 mL of water in a 1 L volumetric flask. Add glacial acetic acid and mix. Add sodium *meta*-periodate and mix until dissolved. Dilute to 1000 mL with reagent-grade water.
   4. Ammonium acetate, 2 M
      a. Ammonium acetate, 154 g, ACS reagent grade (J.T. Baker Chemical Co., catalog 11-0596)
      b. Hydrochloric acid, 2 M

      Add ammonium acetate to approximately 800 mL of reagent-grade water and mix until dissolved. Adjust pH to 6.0 with 2 M HCl and dilute to 1000 mL with reagent-grade water.
   5. Acetylacetone reagent
      a. 99% isopropanol, 12.5 mL, reagent grade
      b. Acetylacetone (2,4-pentanedione), 3.75 mL (Aldrich Chemical Co., catalog P775-4)
      c. Ammonium acetate reagent, 2 M

      Mix 99% isopropanol with acetylacetone in a 500 mL volumetric flask. Dilute to volume with 2 M ammonium acetate reagent. Protect reagent from light. Store in a brown glass bottle. Prepare reagent daily.
   6. 60% isopropanol (volume/volume)

      Add 400 mL of reagent-grade water to 600 mL of 99% isopropanol.
   7. 80% isopropanol (volume/volume)

      Add 200 mL of reagent-grade water to 800 mL of 99% isopropanol.

C. Standards (See special consideration B4.)
   1. Combined cholesterol and triglyceride stock standard, 5000 mg/L of each
      a. Primary cholesterol standard, 500 mg (Eastman Kodak Co., catalog X909)
      b. Triolein, 500 mg (Sigma Chemical Co., catalog T-7502)
      c. 99% isopropanol, reagent grade

      Weigh out exactly 500.0 mg of primary cholesterol standard and transfer quantitatively to a 100 mL volumetric flask. Add approximately 30 mL of isopropanol and mix. This mixture may be heated gently on a hot plate to help the cholesterol dissolve. Under no circumstance should the solution be allowed to boil.

Table 58-5. Preparation of representative working standard concentration

Amount of combined cholesterol and triglyceride stock standard	99% isopropanol	Representative concentration of cholesterol and triglyceride*
1 mL	Dilute to 100 mL	1000 mg/L
2 mL	Dilute to 100 mL	2000 mg/L
3 mL	Dilute to 100 mL	3000 mg/L
4 mL	Dilute to 100 mL	4000 mg/L

*The representative concentrations reflect the dilution factor used for the unknown samples so that one can properly standardize the instrument. The actual concentrations are 20 times less than the representative concentrations. This factor only applies to this AutoAnalyzer-II method. Store working standards in tightly sealed amber glass bottles at 4° C. Working standards are stable for 1 week.

Weigh out exactly 500.0 mg of triolein in a small beaker or glass weighing vial. Add about 30 mL of isopropanol and mix well until the triolein is dissolved. Transfer quantitatively the triolein mixture (with at least 5 rinsings with isopropanol) to the flask containing the cholesterol standard.

After complete transfer of the triolein solution, mix until all cholesterol crystals are dissolved and then dilute to 100 mL with isopropanol. Store in a tightly sealed amber glass bottle. This primary standard is stable for 3 months at 4° C.

2. Combined cholesterol and triglyceride working standards

   Following Table 58-5, measure the specified amounts of stock solution (at room temperature) with a class A volumetric pipet and transfer to a labeled 100 mL volumetric flask.

Assay

Equipment

1. Sample dilution equipment. The Micromedic Automatic Pipette Model 25000 (Micromedic Systems, Inc., Pennsauken, N.J.) is used for aliquoting the samples and standards and dispensing the 99% isopropanol. (See special consideration A at the end of this section.)
2. Centrifuge. This should be capable of 1500 *g*.
3. Automated analytical equipment. Instructions are given for analyses performed with the Technicon AutoAnalyzer-II. Details of the manifold are given in Fig. 58-7. For detailed information on setting up and operating the AutoAnalyzer-II, refer to the Technicon Operating Manual.

A. Extraction
   1. Add 2 g of zeolite mixture to each extraction tube. (See special consideration C.)
   2. Dispense 9.5 mL of 99% isopropanol and 0.5 mL sample into each tube. This is most efficiently and accurately accomplished by use of a Micromedic automatic pipet. Set the 1 mL syringe at 50% and the 5 mL syringe at 90%. Use the following pump sequence:
      a. Aspirate 0.5 mL of sample.
      b. Pipet 0.5 mL of sample followed by 4.5 mL of 99% isopropanol.
      c. Aspirate 0.5 mL of 99% isopropanol.
      d. Pipet 0.5 mL of 99% isopropanol followed by an additional 4.5 mL of 99% isopropanol.

      For blanks, use 0.5 mL of H_2O at step 2a.
   3. Immediately after pipetting, cap each sample tightly and vortex to disperse the zeolite mixture.
   4. Place the tubes in a shaker rack and shake on an automatic shaker for 10 minutes.
   5. After shaking, allow tubes to stand until the zeolite settles. Then, holding the tubes upright, shake so that the isopropanol extract washes the zeolite from the tube walls.
   6. Centrifuge the tubes at 5000 *g* for about 5 minutes. Then directly decant the isopropanol extract into the sample cups, taking care that no zeolite particles are transferred with it. If analysis is to be delayed overnight or longer, the extracts should be transferred to clean, labeled tubes, tightly sealed, and refrigerated. Samples are stable for several days at 4° C.

B. Preparation of the AutoAnalyzer-II system
   1. Prepare all reagents needed for the run.
   2. Change all Solvaflex pump tubes, the L-B reagent pump tube, and the cholesterol flow-cell pull-through pump tube before each run. Change all pump tubes every third run. Pump a 0.05% solution of Brij-35 for a minimum of 30 minutes through all lines except the cholesterol reagent line, the distilled water line, and the 60% isopropanol line.
   3. Turn on:
      a. Colorimeter (minimum warm-up, 30 minutes)
      b. Cholesterol heating bath
      c. Fluoronephelometer (minimum warm-up, 30 minutes)
      d. Triglyceride heating baths
   4. Pump air for 15 minutes through all lines except those which supply the wash reservoirs. Pump distilled water to the cholesterol probe wash reservoir and 60% isopropanol to the triglyceride probe wash reservoir.
   5. Start pumping L-B reagent and lower the cholesterol sample probe into the cholesterol probe

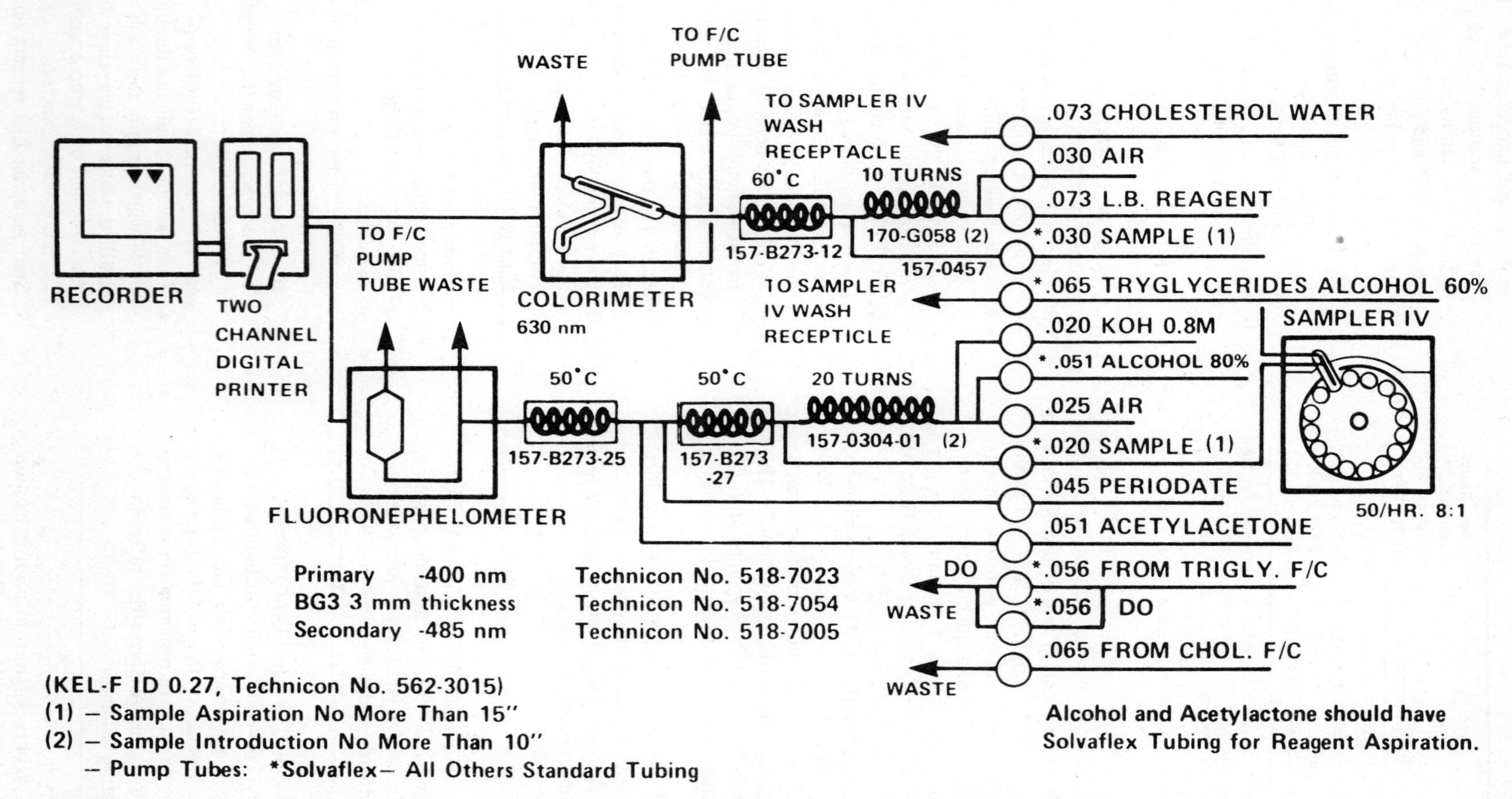

Fig. 58-7. Technicon AutoAnalyzer-II and manifold diagram for simultaneous cholesterol and triglyceride determinations by nonenzymatic procedures.

wash reservoir. Start pumping 80% isopropanol and KOH reagents.

6. When the mixture of 80% isopropanol and KOH reagent emerges from the hydrolysis bath of the triglyceride manifold (5 to 8 minutes), start pumping the sodium *meta*-periodate and acetylacetone reagents. Lower the triglyceride sample probe into the triglyceride wash receptacle.
7. Turn on printer module and channels A and B. Set range for both channels at 500. Turn recorder system switch from off (middle position) to 1+2 position. Turn on chart drive.
8. After all reagents have been pumping for at least 10 minutes, calibrate the colorimeter and fluoronephelometer according to manufacturer's instructions. The damp 2 function settings should not be used during the run, since this excessively reduces the sensitivity of the instruments.
9. Check the heating baths to ensure the proper temperature.
10. Adjust the cholesterol base line to zero on the recorder with the base-line knob on the colorimeter. The visual display should be adjusted to read zero.
11. Adjust the triglyceride base line to zero on the recorder with the base-line knob on the fluoronephelometer. The visual display should be adjusted to read zero.

C. Analytical run
1. Set up sample tray as shown in Table 58-6 and cover the tray to minimize evaporation. (See special consideration D.) Intersperse quality control samples throughout the run if a long run is anticipated.
2. Turn on sampler and release sample hold button on the printer.
3. When the probes enter the first two cups, check and adjust so that the bottom of the probe is just above the bottom of the sample cup.
4. Following sequence in step 1, set base lines and standards, start the printer, and check the linearity curve and quality control samples. If all control samples and the standard curve are within limits, continue running unknowns, setting the base line and standard as necessary throughout the run.
5. If blanking of the triglycerides is desired, at the end of the run modify the triglyceride channel (see special consideration E) and rerun all samples.
6. At the end of the run, turn off sampler and pump air through all tubing for 10 minutes.
7. Wash triglyceride reagent lines and sample lines with 0.05% Brij-35 for 10 minutes and then with distilled water for 10 minutes. *Do not* wash the cholesterol reagent line; air-dry only. After washing, pump air through all lines for 10 minutes. (See special consideration F.)
8. For shutting down, turn off all components and remove pump platen.

Table 58-6. Suggested sample tray setup (see special consideration D)

Tray sample	Instruction	Acceptable range (mg/mL)
1. Blank		
2. Blank		
3. Blank	Set base line	
4. Std 2000 mg/mL	Start printer	
5. Std 2000 mg/mL	Set standard	
6. Blank		
7. Std 1000 mg/mL		980-1020
8. Std 2000 mg/mL	Check linearity	1970-2030
9. Std 3000 mg/mL	Do not adjust	2960-3040
10. Std 4000 mg/mL		3950-4050
11. Blank		
12. Blank	Set base line	
13. Std 2000 mg/mL		
14. Std 2000 mg/mL	Set standard	
15. Quality control		
16. Quality control		
17-30. Unknown samples		
Continue sequence by repeating from 11.		
31. Blank		
32. Blank	Set base line	
33. Std 2000 mg/mL		
34. Std 2000 mg/mL	Set standard	
35-50. Unknown samples		

Std, Standard.

Special considerations

A. Dilution equipment. The Micromedic Automatic Pipette has proved reliable and very efficient for large sample volumes. The precision and accuracy should be checked at regular intervals to ensure reproducible results over time. Alternative pipetting devices may be used but first should be checked for accuracy and precision.

B. Reagents. The manufacturers cited for reagents are those that have been found suitable for use with this method in our laboratory and are listed for the convenience of those trying to set up the procedure. This neither precludes the use of other manufacturers' reagents of equivalent quality, nor does it suggest that these reagents will be adequate from lot to lot. Good quality control practices mandate that each new lot of any reagent be checked out in the system on line.
1. Isopropanol
a. Isopropanol must have a minimal fluorescence. Each new lot number is tested on the AutoAnalyzer using the following procedure.

(1) Plug in all reagents, place the sample probe in water, and obtain a base line.
(2) Switch the probe to the isopropanol to be tested and observe the base line. An increase in the base line greater than 50 mg/L triglyceride equivalents is considered unacceptable.
(3) Perform distillation, which usually removes the excess fluorescence.

b. Isopropanol should also be free of peroxides. Either of the following methods may be used to test for peroxides.
   (1) Method 1: peroxides are detected by the liberation of iodine from a weak acidic solution of potassium iodide.
      (a) Prepare 50 mL of a 2% solution of potassium iodide, KI. Add approximately 3 drops of concentrated HCl.
      (b) Place about 5 mL of organic liquid to be tested (isopropanol) in a test tube (16 × 125 mm).
      (c) Add about 5 mL of KI solution to the tube; stopper and shake.
      (d) Let stand approximately 10 minutes against white background. A brown color indicates the presence of an oxidizing agent (peroxide).
      (e) Blank against the KI solution.
   (2) Method 2
      (a) In a 250 mL beaker place 1.5 g of ferrous sulfate, 1 g of potassium thiocyanate, and approximately 50 mL of distilled water. Add concentrated sulfuric acid until all is in solution.
      (b) Render the pink solution colorless just before use by adding a minimal amount of zinc dust.
      (c) Fill a test tube halfway with the ferrous thiocyanate reagent, and then nearly fill the tube with the organic liquid to be tested so that only a small amount of air is left in contact with the mixture. Stopper tube and shake.
      (d) Check for formation of red color, which indicates an oxidizing agent, probably peroxides.

2. Zeolite mixture. This is stable for several weeks to several months and then may become deactivated. Deactivation will cause a lowering of the expected cholesterol value and an increase in the expected triglyceride values. To reactivate the mixture, heat at 100° C for 1 to 3 hours and then cool in a desiccator until values fall within the expected range. Reactivation overnight may be necessary in some instances.
3. Weigh out the amount of potassium hydroxide equivalent to 45 g of dry potassium hydroxide, taking into account the percentage of water listed on the reagent bottle, for a liter of reagent. For example, 50.65 g is used for KOH at 85% assay.
4. Standards
   a. The quality and purity of the standards is extremely important. If a standard is of unknown quality, comparison to either primary standards from the National Bureau of Standards (NBS) or certified standards from the College of American Pathologists (CAP) is suggested.
   b. Extreme care should be taken when weighing out and transferring the standards, especially the triolein, which is a clear oil and slow to dissolve. Rinse the triolein out of the weighing vial with many small aliquots of isopropanol, making sure none is inadvertently left clinging in droplets to the sides.

C. Zeolite tube filling. It is convenient to use a small scoop marked to contain 2 g of zeolite to fill the extraction tubes. One can make a scoop by bonding an AutoAnalyzer sample cup to a disposable plastic pipet. Tubes prefilled are stable for several days if kept tightly sealed or desiccated.

D. Use of serum calibrator for cholesterol analysis. Plasma cholesterol values determined on the AutoAnalyzer-II are about 6% higher than those determined by a reference method.[7] Plasma samples usually contain a fairly constant proportion of free and esterified cholesterol. However, the total color yield of cholesterol ester is somewhat higher than that of free cholesterol. In the procedure outlined, plasma samples are being compared to a standard that contains only free cholesterol. One can correct the cholesterol values obtained on the AutoAnalyzer-II to the values obtained with a reference method by using a serum calibration sample, which relates the observed value to the reference value. One can set up a calibrator serum by collecting a serum pool and establishing a reference value for the pool. This can be accomplished by analysis of the calibrator serum with a reference method, if one has been used successfully in your laboratory, or by obtainment of a value on the pool from a certified and standardized laboratory. It is also possible to obtain a calibrator from the Centers for Disease Control and use it to obtain a ''corrected'' or reference value for your pool. Prepare and analyze two extracts of the serum calibration pool for each run. The factor used to calculate corrected cholesterol is determined by use of the following equation:

$$\text{Correction factor} = \frac{\text{Reference value of calibrator}}{\text{Mean AutoAnalyzer-II value of calibrator}}$$

Multiplication of cholesterol values by this correction factor gives a "corrected" cholesterol value.

Another possibility is to calculate a factor and then continue to use the same factor for every run; however, one can obtain greater accuracy by using the calibrator and calculating a new factor for each run. The factor for correcting cholesterol values is based on the assumptions that (1) a consistent percentage of the cholesterol is esterified and (2) the distribution of fatty acid esters is consistent for each sample. These assumptions do not hold true for plasma or serum in certain disease states, such as obstructive liver disease. The factor may also not be accurate for tissue samples, nonhuman plasma or serum samples, or other body fluids.

E. The system should be washed weekly or as indicated with a solution of 0.05% Brij-35 in 0.25 N hydrochloric acid for 15 minutes, followed by a 15-minute wash with distilled water. This will help to dissolve small particles of zeolite or other foreign materials, which could cause blockages in the tubing and connectors.

REFERENCES

1. Zlatkis, A., Zak, B., and Boyle, A.J.: A new method for the direct determination of serum cholesterol, J. Lab. Clin. Med. **41**:486-492, 1953.
2. Huang, T.C., Wefler, V., and Raftery, A.: A simplified spectrophotometric method for determination of total and esterified cholesterol with tomatine, Anal. Chem. **35**:1757-1758, 1963.
3. Pearson, S., Stern, S., and McGavack, T., II: A rapid, accurate method for the determination of total cholesterol in serum, Anal. Chem. **25**:813-814, 1953.
4. Allain, C.C., Poon, L.S., Chan, C.S.G., et al.: Enzymatic determination of total serum cholesterol, Clin. Chem. **20**(4):470-475, 1974.
5. Pesce, M.A., and Bodourian, S.H.: Enzymatic rate method for measuring cholesterol in serum, Clin. Chem. **22**:2042-2045, 1976.
6. Kumar, A., and Christian, G.D.: Enzymatic assay of total cholesterol in serum or plasma by amperometric measurement of rate of oxygen depletion following saponification, Clin. Chim. Acta **74**:101-108, 1977.
7. Abell, L.L., Levy, B.B., Brodie, B.B., and Kendall, F.E.: A simplified method for the estimation of total cholesterol in serum and demonstration of its specificity, J. Biol. Chem. **195**:357-366, 1952.
8. Zak, B., and Ressler, N.: Methodology in determination of cholesterol, Am. J. Clin. Pathol. **25**:433-443, 1955.
9. Carr, J.J., and Drekter, I.J.: Simplified rapid technic for the extraction and determination of serum cholesterol without saponification, Clin. Chem. **2**:353-368, 1956.
10. Bloor, W.R.: The determination of small amounts of lipid in blood plasma, J. Biol. Chem. **77**:53-73, 1928.
11. Chiamori, N., and Henry, R.J.: Study of the ferric chloride method for the determination of total cholesterol and cholesterol esters, Am. J. Clin. Pathol. **31**:305-309, 1959.
12. Schoenheimer, R., and Sperry, W.M.: A micromethod for the determination of free and combined cholesterol, J. Biol. Chem. **106**:745-760, 1934.
13. Sperry, W.M., and Webb, M.: A revision of the Schoenheimer and Sperry method for cholesterol determination, J. Biol. Chem. **187**:97-106, 1950.
14. Technicon Methods nos. SF40040F66 and SE0040FC6, Technicon Instruments Corp., Tarrytown, N.Y., March 1976.
15. Cohen, L., Jones, R.L., and Batra, K.V.: Determination of total ester and free cholesterol in serum and serum lipoproteins, Clin. Chim. Acta **6**:613-619, 1961.
16. Wolthers, B.G., Heindriks, F.R., Muskiet, H.J., and Groen, A.: Mass fragmentographic analysis of total cholesterol in serum using heptadeuterated internal standard, Clin. Chim. Acta **103**:305-315, 1980.
17. Kuba, K., Folsom, A.R., Frantz, I.D., Jr., et al: Lipid concentrations in fasting and non-fasting serum and plasma, Clin. Chem. **28**(7):1607, 1982.
18. Martinek, R.G.: Review of methods for determining cholesterol and cholesterol esters in serum, J. Am. Med. Technol. **32**:64-86, 1970.
19. Manual of Laboratory Operations, Lipid Research Clinics Program. Vol. 1. Lipid and lipoprotein analysis, DHEW Publ. No. NIH 75-628, Washington, D.C., May 1974, U.S. Government Printing Office.
20. Cooper, G.R.: High density lipoprotein reference materials. In Lippel, K., editor: Report of the High Density Lipoprotein Methodology Workshop, DHEW Publ. No. NIH 79-1661, Washington, D.C., 1979, U.S. Government Printing Office, p. 178.
21. Folch, J., Lees, M., and Stanley, G.H.S.: A simple method for the isolation and purification of total lipids from animal tissues, J. Biol. Chem. **226**(1):497-509, 1957.
22. Bligh, E.G., and Dyer, W.J.: A rapid method of total lipid extract and purification, Can. J. Biochem. Physiol. **37**(8):911-917, 1959.
23. Nelson, G.J.: Fractionation of phospholipids. In Perkins, E.G., editor: Analysis of lipids and lipoproteins, Champaign, Ill., 1975, American Oil Chemists Society, pp. 70-89.
24. Naito, H.K., and David, J.A.: Laboratory considerations: lipid analyses in blood, tissues and other biological fluids. In Story, J.A., editor: Lipid research methodology, New York, Alan R. Liss, Inc. (In press.)

High-density lipoprotein (HDL) cholesterol

HERBERT K. NAITO

Clinical significance: p. 550
Molecular weight: 1.7 to 3.6 $\times$ 10^5 daltons
Chemical class: lipoprotein

PRINCIPLES OF ANALYSIS AND CURRENT USAGE

The widespread use of high-density lipoprotein (HDL) cholesterol in clinical medicine warrants a critical analysis of the methods currently available for cholesterol and apolipoprotein-lipoprotein quantification. Although HDL cholesterol determinations are fairly simple, it is well established that interlaboratory differences are high, with a (12% to 16% coefficient of variation, CV). However, the coronary heart disease (CHD) risk tables being used by many clinical laboratories for data interpretation are based on the epidemiological studies in Framingham.[1-3] The methods used in the Framingham studies were rigorously standardized by the Centers for Disease Control (CDC). The mean difference in plasma HDL cholesterol concentration between those subjects confronted with coronary artery disease and those apparently normal was about 40 mg/L.[3] This small difference between the two populations highlights the importance of ensuring that the HDL cholesterol method is highly precise so that analytical tests reveal the true sensitivity and specificity of the method. In addition, since HDL cholesterol determination and HDL cho-

Table 58-7. LDL cholesterol–HDL cholesterol risk ratio*

Ratio	Risk
Males	
1.00	One-half average
3.55	Average
6.25	Twice average
7.99	Three times average
Females	
1.47	One-half average
3.22	Average
5.03	Twice average
6.14	Three times average

*This risk ratio is not valid if (1) chylomicrons are present and (2) serum triglyceride is greater than 4000 mg/L. Therefore the serum should be centrifuged in the preparative ultracentrifuge at a density of 1.006 for 20 hours at 30,000 rpm, and the supernatant solution can be used to determine the cholesterol and triglyceride concentration of the *top* fraction. In addition, the *bottom* fraction can be analyzed and LDL cholesterol calculated:
LDL cholesterol = *Bottom* fraction cholesterol − HDL cholesterol

lesterol–total cholesterol ratio and low-density lipoprotein (LDL) cholesterol–HDL cholesterol ratio risk tables (Table 58-7) now being used by physicians are based on the Framingham study, it is essential that HDL cholesterol determinations done in the clinical laboratory be executed with accuracy and precision.

By definition, HDL is the fraction of plasma lipoproteins with a hydrated density of 1.063 to 1.21 g/mL. Electrophoretically, it has a mobility of alpha$_1$-globulins. Being the most dense lipoprotein, it is composed of 50% proteins and 50% lipids.

There are two direct ways to measure HDL: (1) by analytical ultracentrifugation and (2) by isolation of HDL and measurement of the particles gravimetrically. The former method is more accurate and is considered the reference method for quantitation of HDL and other lipoprotein fractions as well. However, there are only a few clinical laboratories that can afford an analytical ultracentrifuge and have the expertise and time to do the laborious and tedious steps. The second method is also impractical, since gravimetric work is time consuming compared to other methods available. To isolate the lipoproteins by column chromatography, electrophoresis, preparative ultracentrifugation, or by polyanion precipitation technique followed by gravimetric determination of the amount of HDL after the salts and water have been removed is also no easy task.

To circumvent some of these problems, the isolated HDL fraction is not measured in its entirety. Instead, either a protein moiety or a lipid moiety is measured as an indirect means of quantitating HDL. HDL is composed of about 50% protein, of which A-I and A-II are the major apoproteins, with A-I being quantitatively more important. The lipids constitute 50% of the HDL molecule; phospholipids, 28%; cholesterol, 18%; and triglycerides, 4%. Thus the phospholipids are quantitatively the larger lipid constituents by mass. However, since cholesterol is simpler to measure than phospholipids are, the HDL cholesterol has prevailed as an indirect means of determining HDL concentrations.

Numerous techniques are available for HDL cholesterol quantitation. They all, however, are based on two steps: (1) isolation of the HDL and (2) quantitation of the cholesterol in the isolated HDL. The various methods differ primarily on how the HDL fraction is isolated. The cholesterol analyses are usually done by one of the many acceptable cholesterol methods available, with the Albers et al.[4] method being recommended and certain compatible enzymatic procedures being preferred (see previous section on cholesterol analysis). The following HDL isolation procedures are used today in the clinical laboratory.[5-16]

Preparative ultracentrifugation (method 1, Table 58-8)

For the simultaneous separation of both very low density lipoproteins (VLDL) and LDL, the plasma or serum is adjusted to a density of 1.063 g/mL (415 mg of KBr per 5 mL) by overlayering of the sample with the potassium bromide solution and centrifuging of the sample at 105,000 *g* for 24 hours at 16° C. After the supernatant solution containing the VLDL and LDL is removed, the infranatant solution can be analyzed for the cholesterol concentration. Technically, the infranatant solution should be adjusted to a density of 1.210 g/mL so that a solution with a density of 1.063 to 1.210 g/mL can be obtained, which truly reflects only HDL. However, there are only minuscule amounts of other lipoproteins besides HDL with a density greater than 1.063. In addition, the extra step of adjusting the solution with a density of 1.063 g/mL to a density of 1.210 g/mL causes greater errors than direct analysis of the fraction with a density greater than 1.063 g/mL, which for the most part represents all the HDL. This method of isolating the HDL and measuring the cholesterol content for HDL cholesterol estimation is a classical procedure and is often regarded as a "reference procedure."

Column chromatography (method 2, Table 58-8)

Ion-exchange chromatography and gel-permeation chromatography have both been used in the isolation of the major lipoprotein groups and subpopulations within each group. These procedures separate HDL subclasses based on differences in charge or molecular size respectively. For instance, using hydroxyapatite packing, the solution with a density of 1.063 to 1.210 g/mL can subfractionated into 12 or 13 HDL subclasses. The significance of each of these subfractions may not seem relevant now, but as more information is gathered about the metabolic role of each of these subclasses of lipoproteins, their distinction will be more important. These methods are not widely used in the

Table 58-8. Methods for high-density lipoprotein (HDL) isolation

Method	Principle	Usage	Comments
1. Ultracentrifugation	Plasma or serum adjusted to density of 1.063 g/mL with potassium bromide and centrifuged at high speeds for 24 hours; all lipoproteins separated by density, with HDL fraction in 1.063 to 1.21 g/mL range	Specialized research laboratories	Reference method Laborious and time consuming
2. Column chromatography	HDL isolated and separated into subclasses on basis of charge (ion-exchange) or molecular size (gel permeation)	Specialized, research laboratories Rarely used	Difficult to properly control chromatographic conditions
3. Starch block electrophoresis	HDL separated from other lipoproteins on basis of charge and size	Rarely used	Used for isolation of large amounts of HDL, not for quantitative purposes
4. Agarose gel electrophoresis	(See method 3)	Frequently used	Precision not adequate for clinical use
5. Precipitation	Polyanions (heparin, dextran sulfate), phosphotungstate, polyethylene glycol, in presence of divalent cations, used to precipitate larger, less dense lipoproteins; HDL quantitated in supernatant as HDL cholesterol	Most frequently used technique	
a. Heparin–manganese chloride		Frequently used reagent	Current method of choice Not compatible with all cholesterol procedures
b. Dextran sulfate		Used occasionally	Use of lower molecular weight dextran can produce biased results; higher molecular weight dextrans produce excellent results Compatible with enzymatic procedures
c. Phosphotungstate		Most commonly used	Underestimates HDL Sensitive to temperature fluctuations
d. Polyethylene glycol		Used infrequently	Poorest accuracy and precision; not recommended

routine clinical laboratory but may be available in specialized laboratories.

Preparative block electrophoresis
(method 3, Table 58-8)

Both starch block and Geon-Pevikon block electrophoresis methods are employed in the isolation of major lipoprotein fractions and their subfractions. These electrophoretic procedures separate the lipoprotein classes on the basis of their net charge and size. The smaller HDL molecules have the highest mobility toward the anode. These methods are mainly used as preparative techniques.

Agarose gel electrophoresis (method 4, Table 58-8)

This method utilizes the standard lipoprotein electrophoresis procedure on agarose medium followed by overlaying the electrophoresed sample with enzymatic cholesterol reagent. A densitometer with an automated integrator is used to scan the agarose strip after color development and quantitate each lipoprotein fraction. (See next section on lipoprotein electrophoresis in this chapter).

Precipitation with polyanion solutions
(method 5, Table 58-8)

Differential precipitation of lipoproteins with various polyanion solutions is common practice and is more suited for the clinical laboratory because of its simplicity, elimination of expensive instrumentation, speed, and low cost. These techniques are all based on the ability of various agents to precipitate selectively the major lipoprotein fractions, except HDL, which is left in the supernatant solution to be quantitated. The agents most frequently employed include heparin–manganese chloride, dextran sulfate–magnesium chloride, sodium phosphotungstate, and polyethylene glycol. Of these, the heparin–manganese chloride and dextran sulfate–magnesium chloride are two of the most frequently used precipitating reagents (Fig. 58-8).

REFERENCE AND PREFERRED METHODS

As already mentioned, the direct gravimetric and ultracentrifugation techniques for HDL quantitation are rarely used for routine work. These methods are too laborious and require highly specialized techniques. However, the

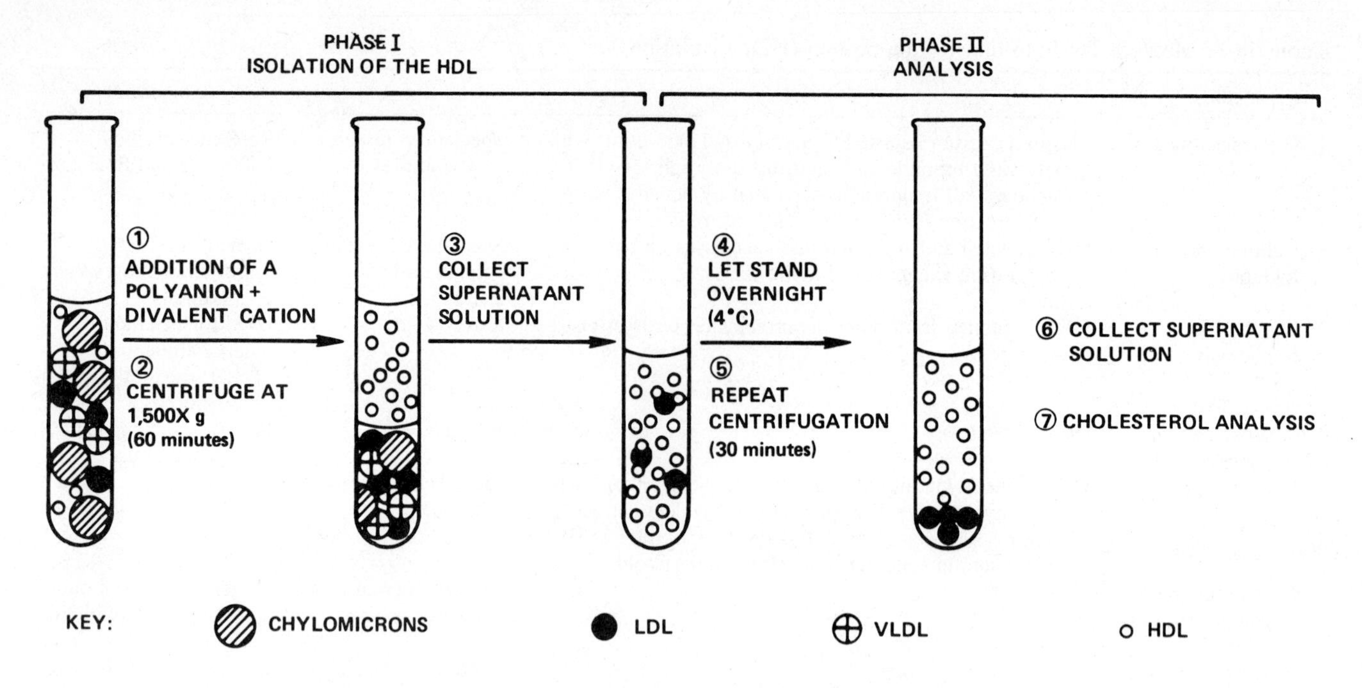

Fig. 58-8. Scheme of a polyanion precipitation method (heparin-$MnCl_2$) for the determination of HDL cholesterol.

ultracentrifugation method is considered the reference method.

The ion-exchange methods are rarely used in clinical laboratories because of the critical care and attention necessary in standardizing the columns, the need to concentrate the eluent for cholesterol analysis, the need for large sample size, and so on. However, it is a commonly used method for research purposes. Similarly, the preparative block electrophoresis methods are almost never used for routine analysis. These methods also require large samples and are very time consuming. Their applications are more suited for the collection of large amounts of lipoproteins.

The agarose gel electrophoresis method is frequently used for routine analysis. Whereas this procedure allows the quantitation of cholesterol in the pre-β and β-lipoproteins as well, the precision of the method is not so good as that in the polyanion precipitation techniques. Also, finding a suitable calibrator for α-lipoprotein cholesterol concentration poses problems for most laboratories utilizing this method.

The most frequently used methods for HDL quantitation are therefore the various precipitation techniques. Of these, the heparin–manganese chloride procedure is one of the most commonly used and is considered one of the methods of choice.

However, the heparin–manganese chloride method is not without faults. This method is not compatible with most cholesterol enzymatic systems, giving a consistent false-positive bias. However, if a reliable and accurate Liebermann-Burchard (L-B) method is used, this becomes an excellent method. Hypertriglyceridemic samples greater than 4000 mg/L tend to cause incomplete precipitation of the β-containing lipoproteins, resulting in turbid solutions that contain substantial quantities of apo-B-lipoproteins. This problem can be minimized by (1) dilution of the samples with 0.9% saline and reprecipitation with heparin–manganese chloride (92 mM of Mn^{+2}); (2) centrifugation of the undiluted sample with heparin–manganese chloride at a higher gravity force, such as 12,000 *g* for 10 minutes; (3) removal of the chylomicron or VLDL fractions first by use of preparative ultracentrifugation at a density of 1.006 g/mL; (4) ultrafiltration of the turbid supernatant solution with a 0.22 μm ultrafilter; or (5) use of undiluted samples but with twice the volume of heparin–manganese chloride solution. In practice, the first and last options are preferred.

Another problem associated with the heparin–manganese chloride method is the interval between sample collection and precipitation. Because the HDL cholesterol values decrease with an increase in this interval, it is recommended that the analysis on the sample be performed soon after blood collection.

A detailed description of the heparin–manganese chloride procedure is included in this chapter. Since the mean difference in HDL cholesterol concentration between persons at normal risk for coronary artery disease and persons at high risk is small (about 40 to 50 mg/L), it is suggested that the minimum precision for this method should be less than 30 mg/L for within-day and day-to-day variance.

Dextran sulfate of various molecular weights with calcium or magnesium salts has also been used for HDL cholesterol quantitation. Most dextran sulfate procedures tend to underestimate HDL cholesterol, though this method does not appear to be sensitive to variations in incubation and centrifugation temperatures. Higher molecular weight (50,000 daltons) dextran sulfates tend to produce HDL cholesterol values closer to the heparin–magnesium chloride methods. Recently, Warnick et al.[17] described a dextran sulfate–magnesium chloride procedure that is compatible with enzymatic cholesterol determination. The method, when compared to the Lipid Research Clinics' method,[5] is considered simple, rapid, accurate, and precise. It seems likely that this procedure will become the method of choice for the clinical laboratory.

Sodium phosphotungstate is another method of precipitating apo-B–containing lipoproteins. This method also underestimates HDL cholesterol, but not so much as polyethylene glycol does. This method is particularly sensitive to temperature fluctuations and reagent concentrations, and it can be a major source of error.

Polyethylene glycol of various molecular weights has been used occasionally for HDL cholesterol determinations. Of all the precipitation techniques, this method has the most serious problem of not being accurate and precise.

Although no ideal method for measuring serum HDL cholesterol is available, it is important to recognize the limitations and varied specificity of all procedures. Particular caution is urged for those setting up this test for clinical evaluation of persons at risk for heart disease. Since the only available data for assigning risk are based on tables generated from the Framingham study, the assay developed in the laboratory must be as accurate as the assay used in the Framingham study. The Centers for Disease Control can play an important role in helping ensure test accuracy. Precision is also important; for example, the mean difference in HDL cholesterol concentration in the Framingham Study population with and without myocardial infarction is only 40 mg/L. As more work is done with different polyanion solutions for the selective precipitation of lipoproteins, a more sensitive, accurate, and precise method will probably evolve.

In conclusion, although no absolute reference or definitive method has been designated for HDL cholesterol determinations, precipitation techniques have been favored, with the heparin–manganese chloride procedure being the method of choice. Included here is a modified Centers for Disease Control–proposed HDL cholesterol procedure[5] for cholesterol analysis by manual or automated Liebermann-Burchardt cholesterol methods.

SPECIMEN

For a detailed review see p. 1199. Briefly, the patient must be fasting for at least 12 hours before the blood is drawn. Serum or plasma can be used as the sample, but since plasma is usually preferred for other lipid analysis, HDL analysis will most frequently be performed on the same plasma specimen. EDTA is again the anticoagulant of choice, and the final concentration should be 1 mg of EDTA per milliliter of blood.

The sample should be removed from the blood clot within 2 hours and stored at 4° C unless analyzed immediately.

PROCEDURE

Principle

The larger, less dense apo-B–containing lipoproteins (low, very low, and intermediate density) are precipitated overnight by heparin–manganese chloride. After centrifugation to separate the precipitated lipoproteins, HDL cholesterol in the supernatant solution is quantitated. (See Fig. 58-8.)

Reagents

Manganese chloride solution, 1.0 mol/L. Dissolve 19.791 g of $MnCl_2 \cdot 4\ H_2O$ in distilled water and bring volume to 100 mL. This is stable for 3 months at 4° C.

Heparin solution, 4000 U/mL. Dilute 1 mL of heparin (A.H. Robins; 10,000 USP units/mL) with 1.5 mL of 0.9% saline solution. This is stable for 1 week at 4° C.

Assay

Equipment: refrigerated centrifuge capable of maintaining temperature of 4° C at 2000 *g*.

1. Pipet 1 mL of serum into a 13 × 100 mm disposable culture tube.
2. Cap tubes with polyethylene stoppers and place in refrigerator for 30 minutes at 4° C.
3. Add 50 μL of heparin solution; mix on a Vortex mixer for 1 minute.
4. Add 50 μL of manganese chloride solution; mix on a Vortex mixer.
5. Cap tubes with polyethylene stoppers and place in refrigerator for 30 minutes at 4° C.
6. Centrifuge the heparin-lipoprotein precipitate at 2000 *g* for 1 hour at 4° C.
7. Using a transfer pipet, transfer supernatant solution into a clean test tube. Store in a refrigerator overnight to ensure complete precipitation of all apo-B–containing lipoproteins.
8. The next morning, sediment any remaining precipitate at 2000 *g* for 30 minutes at 4° C.
9. Using a transfer pipet, transfer the clear supernatant solution into a clean test tube for measurement of cholesterol by a standard method compatible with heparin–manganese chloride. The Liebermann-Burchard method is recommended.

Biological sample

If one uses plasma samples, the 92 mM Mn^{+2} salt concentration should be selected. However, if one uses serum, the 46 mM Mn^{+2} concentration is more ideally suited for HDL cholesterol determinations. If EDTA plasma samples are used with the heparin 46 mM Mn^{+2} method, slightly higher HDL cholesterol values will be obtained because of incomplete precipitation of the apo-B–containing lipoproteins. If one uses the 92 mM Mn^{+2} salt preparation on serum samples, the HDL cholesterol values will be slightly lower than the ultracentrifugal procedure (the reference procedure).

Purity check

If the sample suggests an unusual value (HDL concentrations less than 200 or greater than 800 mg/L) or the α-lipoprotein bands on the electrophoresis medium do not agree with the serum or plasma HDL cholesterol value, perform a purity check or repeat the analysis or do both.

A purity check on the supernatant solution should be done if there is the possibility of incomplete precipitation of apo-B–containing lipoproteins. To do this, electrophorese and stain the supernatant solution with a lipid dye to show whether only one band exists (Fig. 58-9). If there is more than one lipid-staining band and the second band is in the β-globulin area or α_2-globulin area, it probably reflects an apo-B–containing lipoprotein. One can confirm this by doing immunological studies using apo-B antisera.

Double-concentration precipitation

If the serum is turbid or hypertriglyceridemic (greater than 4000 mg/L), a double-strength solution should be used to precipitate the β-containing lipoproteins. To do this, use 1 mL of serum, 100 μL of heparin, and 100 μL manganese chloride solution. Multiply the result by 1.2 to account for the dilution. A double-strength solution should also be performed on a serum sample if, after conventional manganese chloride–heparin precipitation, a floating lipid precipitate is present and the purity check of the supernatant shows contamination.

Pooled human sera for quality control

At present most commercially available quality control materials for HDL cholesterol are not suitable or stable. However, our group has had success with the Hyland Omega Lipid Faction Control Serum (Hyland Diagnostics, Deerfield, IL 60015). Since the matrix of commercial materials may be different from human samples, we also make our own quality control materials by pooling human serum, which has no chylomicrons and low amounts of VLDL and triglyceride (less than 1500 mg/L). This quality control material can be used if it is free of hepatitis antigens. The pooled sample (about 2 L) should be aliquoted into 12 × 75 mm tubes and stored in a freezer at 60° C.

Calculation of VLDL and LDL cholesterol

Although one can isolate VLDL (density less than 1.006 g/mL) and determine the cholesterol content, VLDL cholesterol can be estimated with a fair amount of accuracy by the following calculation:

$$\frac{\text{Triglyceride (mg/L)}}{50} = \text{VLDL cholesterol (mg/L)}$$

This holds true only under the condition that serum triglyceride concentration is less than 4000 mg/L. Furthermore:

$$\text{LDL cholesterol} = \text{Total cholesterol} - (\text{VLDL cholesterol} + \text{HDL cholesterol})$$

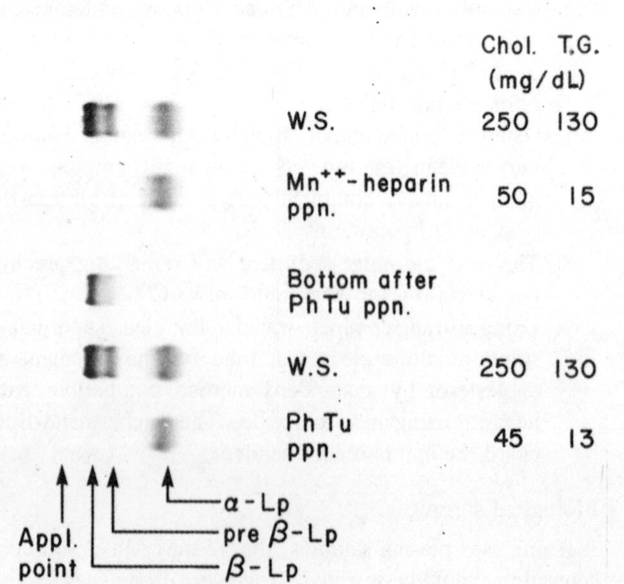

Fig. 58-9. Agarose gel electrophoretic patterns of HDL preparations. Supernatant fractions from phosphotungstic acid *(Ph Tu ppn.)* and manganese-heparin *(Mn^{++}-heparin ppn.)* precipitations contain only an HDL band indicating acceptable purity. The phosphotungstic acid precipitate *(Bottom after Ph Tu ppn.)* contains the LDL and VLDL fractions. *Appl.*, Application; *chol.*, cholesterol; *Lp*, lipoprotein; *Ph Tu*, phosphotungstic acid; *ppn.*, precipitation; *T.G.*, triglyceride; *W.S.*, whole serum.

Use of risk tables for data interpretation

Usually the values for LDL and HDL cholesterol are examined simultaneously as a ratio and compared to coronary heart disease risk assessment tables generated from the Framingham study. These tables assess only one of the risk factors for heart disease, and therefore do not take into account the other primary and secondary risk factors, which ultimately should be included in the overall assessment of a person's risk. Presently, there is no unified system that simultaneously accounts for all of the risk factors.

REFERENCES

1. Gordon, T., Castelli, W.P., Hjortland, M.C., et al.: High-density lipoprotein as a protective factor against coronary heart disease: the Framingham Study, Am. J. Med. **62:**710-713, 1977.
2. Heiss, G., Johnson, N., Reiland, S., et al.: The epidemiology of plasma high-density lipoprotein cholesterol levels, Circulation **62**(suppl.):IV, 116-136, 1980.
3. Castelli, W.P., Doyle, J.T., Gordon, T., et al.: HDL cholesterol and other lipids in coronary heart disease, The Cooperative Lipoprotein Phenotyping Study, Circulation **55:**767-772, 1977.
4. Albers, J.J., Warnick, G.R., Wiebe, D., et al.: Multi-laboratory comparison of three heparin-Mn^{2+} precipitation procedures for estimating cholesterol in high-density lipoprotein, Clin. Chem. **24:**853-856, 1978.
5. Manual of Laboratory Operations, Lipid Research Clinics Program, Lipid and Lipoprotein Analysis, DHEW Pub. No. NIH 751-628, Washington, D.C., May 1974, U.S. Government Printing Office.
6. Bachorik, P.S., Wood, P.D., Albers, J.J., et al.: Plasma high-density lipoprotein cholesterol concentrations determined after removal of other lipoproteins by heparin/manganese precipitation or by ultracentrifugation, Clin. Chem. **22:**1828-1834, 1976.
7. Burstein, M., and Scholnick, H.R.: Lipoprotein-polyanion-metal interactions, Adv. Lipid Res. **11:**67-108, 1973.
8. Burstein, M., Scholnick, H.R., and Morfin, R.: Rapid method for the isolation of lipoproteins from human serum by precipitation with polyanions, J. Lipid Res. **11:**583-595, 1970.
9. Warnick, G.R., and Albers, J.J.: A comprehensive evaluation of the heparin-manganese precipitation procedure for estimating high density lipoprotein cholesterol, J. Lipid Res. **19:**65-76, 1978.
10. Finley, P.R., Schifman, R.B., Williams, R.J., and Lichti, D.A.: Cholesterol in high-density lipoprotein: use of Mg^{2+}/dextran sulfate in its enzymatic measurement, Clin. Chem. **24:**931-933, 1978.
11. Kostner, G.M.: Enzymatic determination of cholesterol in high-density lipoprotein fractions prepared by polyanion precipitation, Clin. Chem. **22:**695, 1976.
12. Hatch, F.T., Lindgren, F.T., Adamson, G., et al.: Quantitative agarose gel electrophoresis of plasma lipoproteins: a sample technique and two methods for standardization, J. Lab. Clin. Med. **81:**946-960, 1973.
13. Viikari, J.: Precipitation of plasma lipoproteins by PEG-6000 and its evaluation with electrophoresis and ultracentrifugation, Scand. J. Clin. Lab. Invest. **36:**265-268, 1976.
14. Allen, J.K., Hensley, W.J., Nicholis, A.V., and Whitfield, J.B.: An enzymic and centrifugal method for estimating high-density lipoprotein cholesterol, Clin. Chem. **25:**325-327, 1979.
15. Lopes-Virella, M.F., Stone, P., Ellis, S., and Colwell, J.A.: Cholesterol determination in high-density lipoproteins separated by three different methods, Clin. Chem. **23:**882-884, 1977.
16. Abell, L.L., Levy, B.B., Brodie, B.B., and Kendall, F.E.: A simplified method for the estimation of total cholesterol in serum and demonstration of its specificity, J. Biol. Chem. **195:**357-366, 1952.
17. Warnick, G.R., Benderson, J., and Albers, J.J.: Dextran sulfate–Mg^{2+} precipitation procedure for quantitation of high-density lipoprotein cholesterol, Clin. Chem. **28:**1379-1387, 1982.

Lipoprotein electrophoresis

HERBERT K. NAITO

Clinical significance: p. 550
Chemical class: lipoproteins

PRINCIPLES OF ANALYSIS AND CURRENT USAGE

The goal of lipoprotein electrophoresis is to separate the major lipoprotein fractions into relatively homogeneous bands of lipoproteins so that one can evaluate, semiquantitatively and qualitatively, the complete lipoprotein profile. Refer to Chapter 8, which discusses electrophoretic theory and technique in detail. In studies using various support media (paper, agarose, agar, starch, polyacrylamide, or cellulose acetate), the same principles apply as in moving boundary procedures.

Thus the major factors generally affecting the migration of molecules in an electrical field must also be controlled for accurate lipoprotein electrophoresis. Briefly, these are:

1. Net charge on the molecule. The pH has to be accurately and reproducibly controlled for proper lipoprotein profiling.
2. Ionic strength and viscosity. For pH determination, the best and most reproducible results require a properly prepared electrophoretic medium. Although high-ionic-strength buffers yield sharper band separations, the resolution may be adversely affected by increased heating, which can denature heat-labile solutes.
3. Strength of the electric field. This must also be reproducibly controlled.
4. Type of support media. This is one of the most important factors in lipoprotein separations. The charge on the support can cause an endosmotic effect; on support media where endosmosis is strong, as in paper cellulose acetate and agar gel, endosmotic effects cause the slowly migrating lipoprotein-X to be swept cathodically behind the application point.

Since the major element in lipoprotein electrophoresis is the support medium, much of this discussion compares the various supports used.

Paper

The majority of paper electrophoretic methods use either Whatman No. 1 or No. 3 paper. This method, once widely used in the clinical laboratory for the separation of serum lipoproteins and proteins, has largely been replaced by agarose gel electrophoresis.

For lipoprotein studies, paper electrophoresis is still considered the conventional and classical method. Two basic types of apparatus for paper electrophoresis have proved most useful. In the first type the amount of moisture on the filter paper is kept uniform throughout the period of electrophoresis. The paper moistened with buffer is laid on a siliconized glass plate, the sample is applied, and then a second siliconized glass plate is laid on top. This system has been referred to as the closed horizontal method.

The second method applies the so-called open strip or inverted V principle. The Beckman-Durrum electrophoretic cell (Beckman Instruments, Inc., Palo Alto, CA 94304) is one of the most popular and widely used cells for this purpose.

In 1953 Swahn introduced the use of filter paper for lipoprotein electrophoresis. The original method of Lees and Hatch[1] was modified to increase the speed of the procedure.[2] In contrast to the method of Jencks and Durrum,[3] Lees and Hatch introduced the use of 1% albumin in barbital buffer and showed that the presence of the albumin provides sharper delineation of the major lipoprotein components and a resolution or partial resolution of the very low density lipoprotein (VLDL). The method of Naito and Lewis[2] used 1% albuminated buffer only for the paper strips, not in the chambers or wells on the cell containing barbital without albumin. This minimized the microbial growth in the buffer, which altered the pH of the buffer and caused unsatisfactory resolution of the lipoprotein bands. In addition, the method increased the ionic strength of the buffer from 0.075 to 0.1 mol/L, resulting in sharper delineation of the lipoprotein bands.

Agarose

Fifteen years after Swahn reported the use of paper as a support medium for electrophoresis of serum lipoproteins, agarose gel was introduced. Agar is composed of at least two major fractions: agaropectin and agarose. Agaropectin contains sulfate and carboxylic acid groups and accounts for the considerable endosmosis and background color observed with most unfractionated agar. Purified agarose is essentially neutral and exhibits few of the problems of unpurified agar. With the advent of premade gel plates by commercial companies, the agarose method came into widespread use. Although the basic knowledge of electrophoretic patterns in health and disease has been acquired from the electrophoresis on filter paper, the agarose method was developed as an analytical, research, and clinical method for studying serum lipoprotein, qualitatively and quantitatively (Fig. 58-10).

The method currently used in most clinical laboratories is agarose electrophoresis, which is described later in this chapter.

Polyacrylamide gel

The technique of disk electrophoresis, the most recent addition to the many variations of electrophoresis, is based on the bands or zones that stack up as a series of concentrated disks at the beginning of the electrophoretic run. The support medium is polyacrylamide gel.

The migration and separation of the components in a given sample are based on (1) the electrophoretic charge of the particle and (2) the molecular size and configuration of the particle. Thus the polyacrylamide support media produces a molecular-sieving effect. As a result of this phenomenon, polyacrylamide gel electrophoresis tends to separate protein and lipoprotein fractions more discretely than other support media (Fig. 58-11).

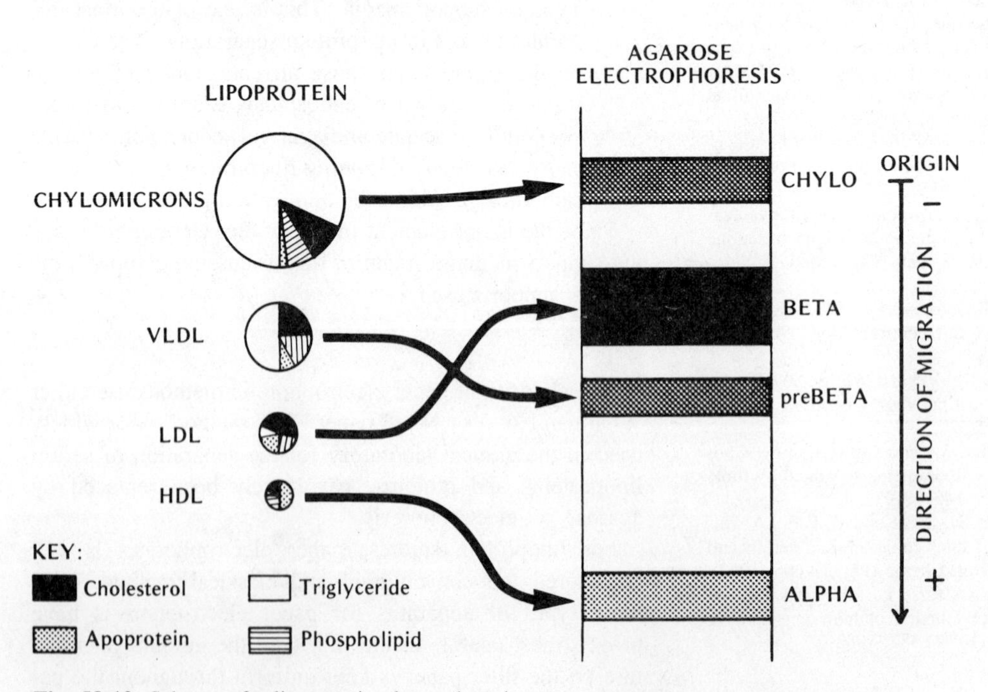

Fig. 58-10. Scheme of a lipoprotein electrophoretic pattern on an agarose support medium.

Cellulose acetate

Before the introduction of agarose electrophoresis, cellulose acetate support medium was once widely used in the clinical laboratory for lipoprotein studies. It is used less frequently now.

Starch gel

For some unknown reason, starch gel electrophoresis has not been utilized in the clinical laboratory as much as other support media. Starch gel has unique properties that permit better electrophoretic separation of proteins and lipoproteins than by other support media. The resolved fractions can be demonstrated by staining, by immunodiffusion, or by radioisotopic or other techniques.

Smithies[4] described the preparation of a starch gel using partially hydrolyzed potato starch as the gel agent. With the use of starch gel in electrophoretic separation of macromolecules, including proteins and lipoproteins, fractionation depends not only on the difference in electrical charge on the protein molecule but also on molecular size. Thus the serum lipoproteins resolved in starch gel, in order of decreasing mobility, are α-lipoproteins, β-lipoproteins, very-low-density lipoproteins (pre-β-fraction by paper electrophoresis), and chylomicrons, which fail to enter the gel during electrophoresis. The molecular-sieving effect of this gel is similar to that of polyacrylamide gel, but with slightly less resolving power.

The original starch gel procedure described by Smithies used a block of starch gel in which a slit was made for insertion of a paper strip moistened with the serum to be tested. The gels were pressed firmly against the filter paper, and electrophoresis was carried out using much the same technique as in horizontal paper electrophoresis. After electrophoresis, the gels were sliced into thin sheets that could be stained for protein and lipoprotein; a third sheet was cut into appropriate sections as determined by the stained sheets and subjected to immunodiffusion studies.

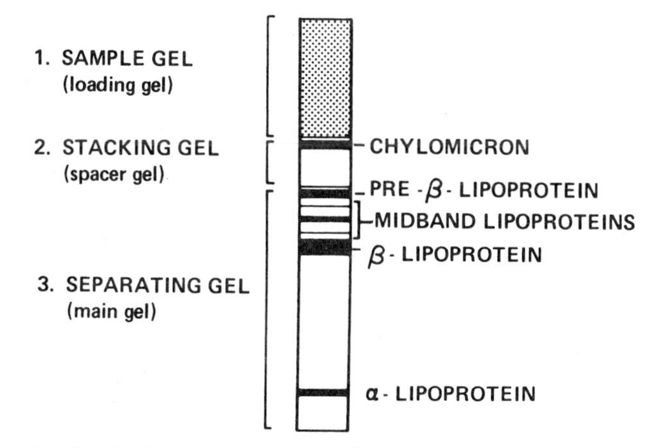

Fig. 58-11. Scheme of a lipoprotein electrophoretic pattern on a polyacrylamide-gel support medium.

A modification of the horizontal technique was subsequently developed in which the starch gel was cast in a vertical holder. It was especially designed for application of larger samples. Because of this arrangement, the filter paper was no longer necessary; at the beginning of electrophoresis a sharpening and concentration of the proteins occurred as they entered the gel. This resulted in sharper resolution of the fractions.

Another modification of the starch gel procedure involves the preparation of a thin-layer starch gel on an inert support film. The sheet coated with gel is placed in a Durrum inverted-V type of electrophoresis cell, and the samples are applied to small depressions made in the gel at the level of the top support rod. After electrophoresis, the sheets are stained with oil red O for lipoprotein, Amido-Schwarz for protein, or other appropriate stains. After staining, they could be scanned in a densitometer to determine the relative percentages of the fractions resolved.

The lipoprotein patterns of young healthy adults obtained from starch gel on serum show the usual components resolved by other techniques such as paper or agarose electrophoresis. α-Lipoprotein is occasionally resolved as double or triple fractions, with one or more possessing mobility slower than normal. Starch gel electrophoresis has proved valuable in the past as a guide in the study, isolation, and purification of lipoproteins and apolipoproteins. It was effectively used as an additional test for evaluation of the whole apolipoprotein or of the polypeptides composing the apoprotein molecule. This electrophoresis system was also used for special lipoprotein studies, as on the serum from multiple myeloma patients in which unusual lipoprotein-immunoglobulin complexes occurred. Since the development of acrylamide gel electrophoresis, the use of starch gel in the evaluation of apolipoproteins and peptides has decreased.

Agar gel

One of the known lipid disturbances in liver disease involves the presence of hypercholesterolemia and an abnormal serum lipoprotein called lipoprotein-X (Lp-X) in patients with cholestasis. Lp-X is characterized by a high content of phospholipid and free cholesterol as well as a very low content of protein. Lp-X migrates in the β-lipoprotein area on most support media. However, in agar gels at the usual pH of 8.4 or 8.6, the Lp-X moves in the opposite direction (cathodically). One can use this peculiar behavior in the identification of Lp-X by electrophoresing the serum on 1% agar support medium and then visualizing it by polyanionic precipitation or by lipid stain. For further details, refer to Seidel et al.[5]

Agarose immunoelectrophoresis and radial immunodiffusion

In the event that the identification of lipoprotein fractions becomes questionable, one is generally advised to perform immunoelectrophoresis and identify the fraction in question by using appropriate antisera, such as anti-α-lipoprotein antisera, anti-β-lipoprotein antisera, or anti–whole serum antisera. One can do further studies on the lipoproteins by performing double immunodiffusion studies. This will provide some indication as to whether there has been alteration in the apoprotein component or structure, that is, incomplete or partial identity as compared to complete identity.

Stains used for visualization of electrophoretic patterns

Many stains are available for visualization of the separated lipoprotein fractions. Sudan black-B was initially used for staining the lipoproteins. Since then, oil red O, fat red 7B, and other specific lipid stains have been used.

REFERENCE AND PREFERRED METHODS

The various electrophoretic methods just listed have a variety of advantages and disadvantages, and the support media chosen must suit the particular needs of the laboratory.

Paper

Advantages

Established technique. This method is still considered the classical method for electrophoresing serum lipoproteins. The classification system of the different types of familial hyperlipoproteinemia developed by Fredrickson et al.[6] was based on the results obtained by the paper method.

Time factor not too critical. After the sample is applied, the sample can be left to electrophorese for 18 to 20 hours, depending on the current used and the temperature of the laboratory. After the electrophoresis the paper strips are dried in the oven. Once dried, one can set aside the paper strips until staining of the strips is desired. Similarly, staining the strips in the lipid stain is an overnight procedure lasting between 16 to 20 hours. Thus this procedure is time dependent, but not to the exact second or even minute. However, to maintain good quality control, every attempt must be made to standardize each phase.

Basically simple. In contrast to starch gel, cellulose acetate, and polyacrylamide gel (PAG) methods, the paper procedure is a relatively straightforward method with fewer steps.

Reveals unusual lipoprotein disorders. The paper electrophoretic method readily shows immunoglobulin-lipoprotein complexes and occasionally high-density lipoprotein bands with two different mobilities. The band of slow-moving α-lipoprotein (in the α_1-globulin area), which carries proportionately higher amounts of cholesterol than the normal α-lipoprotein, has been termed "HDL_C," in contrast to the normal migrating α-lipoprotein, which is called "HDL."

Can apply "macroquantity" of serum sample. This feature is important when there is a need to use the method for semipreparative purposes. One can collect the electrophoresed fractions by cutting the strips and eluting the lipoprotein constituents with saline solution or buffer for further studies.

No destaining necessary. The paper method utilizes either a prestaining or poststaining procedure. In the prestaining procedure, one mixes the lipid-dye solution with the serum sample before electrophoresis. In the poststaining method one stains the paper strip after the drying procedure, which fixes the lipoprotein fractions; this procedure is in wider use. In both methods no destaining procedure is used, unlike that in the agarose and cellulose acetate procedures. The major problem with destaining is twofold: (1) if one is not careful, one can "overdestain" and elute the stain from both the background and the lipoprotein bands; and (2) the destaining technique may selectively elute more stain from one lipoprotein band than from another, thus leaving a false lipoprotein profile.

Disadvantages

Time consuming. From start to finish, the method takes about 3 days. This is long compared to the agarose method, which takes about 1 hour.

Bands not so well resolved. The separation between pre-β- and β-lipoprotein bands is not so clear when compared to media such as agarose and polyacrylamide gels. This problem is particularly aggravated when either of the two lipoprotein bands is in high concentration. This can result in fusion of both bands, giving the impression of a broad β-band or type III hyperlipoproteinemia profile.

In addition, unlike the polyacrylamide gel method, this method cannot demonstrate the intermediate-density lipoprotein.

Appreciable trailing of very-low-density lipoproteins and chylomicrons when concentrations are high. When the pre-β-lipoprotein is elevated, there are some samples in which much VLDL and chylomicron trailing occurs, which makes evaluation of the amount of pre-β-lipoprotein present very difficult. This occurs to a lesser extent in the agarose support medium and seldom in the polyacrylamide gel method.

Sensitive to humidity and temperature fluctuations. Temperature, humidity fluctuations, and drafts of air greatly affect the obtainment of consistent patterns, optimum resolution of the bands, and consistent migration rates of each lipoprotein fraction.

Variability in migration rates. Paper electrophoresis (Durrum cell method) appears to be sensitive to environmental conditions because of the difficulty in obtaining lipoprotein bands with the same mobility within the same cell and between electrophoretic cells. In this respect the agarose method appears to be more suitable.

Semiquantification not ideal with densitometers. Although the paper strips can be examined with a scanning densitometer, there are other electrophoretic methods (agarose, polyacrylamide gel) that are more suitable for this purpose for two major reasons: (1) fewer background interferences and (2) better separation and resolution of the major bands.

Agarose

Advantages

Convenience. Although the agarose gel can be prepared in the laboratory,[7,8] premade gels from Corning Medical (Medfield, MA 02052) make this method very convenient.

Rapid and simple. Within 1 hour the sample is applied on the agarose plate, one can obtain results on the lipoprotein profile. The technique is not only fast, but also fairly simple. It has the further advantage of being a microquantity method, needing less than 2 μL of sample.

Accurate and precise. Lipoprotein profiling with agarose appears to approximate best the pattern observed in the analytical ultracentrifuge, the reference method for quantitative and qualitative lipoprotein studies. In addition, the within-day and day-to-day reproducibility is very good, a feature that makes it an ideal method for the clinical laboratory.

Good separation and resolution of the major lipoprotein bands. The agarose system clearly separates the pre-β- from the β-lipoprotein fraction with good resolution. In addition, the system occasionally shows multiple pre-β-lipoprotein bands; the paper system seldom demonstrates them.

Mobilities. Since this method is highly reproducible, one can use it to measure mobilities of the various lipoprotein fractions. There are many occasions when one will observe major lipoprotein fractions with mobilities that are different from the standard or conventional pattern; the pre-β-lipoprotein and α-lipoprotein bands may also be either fast or slow. Clinically, some investigators have associated the slow pre-β-lipoprotein band (β_1-globulin area) with patients who have coronary artery disease. Similarly, some associations with coronary artery disease have been made with the subfractions of low-density lipoproteins and very-low-density lipoproteins by this method of electrophoresis. However, it must be stressed that more research work is needed in this area to confidently use these additional lipoprotein bands with different mobilities for diagnostic or prognostic purposes.

Can be used for scanning densitometric studies. Because of its relatively clear background, agarose appears to be ideal for scanning studies that provide semiquantitative information, or relative percentage distribution. This should not be confused with the quantitative values achieved by the analytical ultracentrifuge or column chromatographic techniques.

Disadvantages

Need for destaining. The destaining procedure is not particularly desirable because of the potential errors it can cause. This step requires very careful attention when utilized.

Chylomicrons. Sometimes establishing the presence of chylomicrons is difficult. This is especially true of the agarose systems with precast sample wells. When "trace" amounts of chylomicrons are present in the serum, it may be difficult to demonstrate their presence on the gel.

Storage time limited. Because the premade gels are susceptible to drying, the storage time of the gels is limited to about 3 to 6 weeks. One can observe the drying effect by carefully examining the corners of the plates before peeling the Mylar covering off the agarose gel.

Polyacrylamide gel

Advantages

Flexibility. Although the gel can be bought premade, it is recommended that the gels be made in the laboratory and optimized according to the needs and goals of the clinician doing the lipoprotein studies. By this means, the separating-gel concentration, buffer strength, and pH of the gels can be altered to optimize the system. If, for instance, one is interested in the α-lipoproteins and their subclasses, the concentration of the acrylamide in the separating gel should be increased from 37.5 to about 40 g/L. Increasing the concentration of the separating-gel buffer causes the β-lipoprotein to migrate faster, resulting in a wider separation between the pre-β- and β-lipoprotein bands. However, the α-lipoprotein bands are not well resolved in high-separating gel bands. In some cases two runs with different gel concentrations may be helpful. Thus a compromise must be made to optimize the electrophoretic system. The entire procedure takes 5 to 6 hours. Therefore, this method is not so convenient and rapid as the agarose method; the polyacrylamide gel system is primarily used on serum samples that produce complicated or questionable lipoprotein patterns on paper or agarose gels. This is especially true for the broad β-band on paper support medium, which is easier to interpret on polyacrylamide gel.

Good separation and resolution of the lipoprotein bands. In addition to providing extremely good separation and resolution of the major lipoprotein bands once optimized, this method will resolve multiple bands between the pre-β- and β-lipoprotein area, a condition that cannot be matched by other electrophoretic support media. This is one of the major strong points of the gel system. Some midbands (see Fig. 58-11) are believed to be intermediate density lipoproteins. Note that because of the molecular-sieving action of the polyacrylamide gel medium, the pre-β-lipoprotein band migrates slower than the β-lipoprotein band, in contrast to other support media.

Can be used for scanning densitometric studies. Since the polyacrylamide gel method uses a prestaining technique and the background of the gel is transparent, it is an excellent system to semiquantitate the lipoprotein fractions by scanning the gels on a densitometer. The results can be expressed as "relative percent distribution" of the fractions. Sometimes visual inspection of the gel patterns does not reveal as much information as that obtained by study of the densitometric scan of the patterns.

Micromethod. This sensitive technique needs as little as 5 μL of serum, making it useful when sample size is a problem.

Disadvantages

Fairly time consuming. Few commercial companies offer premade polyacrylamide gels. Because polyacrylamide gel is a very sensitive electrophoretic medium, one should always be concerned about storage conditions that could alter the conditions of the premade gels. Making the gels in the laboratory takes about 3 to 4 hours; the electrophoretic process takes about 40 minutes.

Sample-gel background. The polyacrylamide gel procedure uses a prestaining method in which the serum sample and lipid dye are combined with the sample or loading gel. After electrophoresis, some dye may remain in the sample gel, particularly when the serum lipoprotein concentration is low. This makes interpretation of the presence of chylomicrons difficult. One should repeat the analysis and increase the sample size to circumvent this problem.

Resolution of too many bands. This system has such good resolving power that sometimes the lipoprotein patterns cannot be interpreted. This is especially true when the major fractions have different mobilities and more than one representative band. It is well known that each of the major lipoprotein classes are a heterogeneous group of macromolecules. As the electrophoresis systems improve, the complexity of the lipoprotein classes in health and disease may be demonstrated.

Cellulose acetate

Advantages

Micromethod. Depending on the type of electrophoretic cell used, as little as 1 to 2 μL of serum or plasma sample can be used. When this micromethod was introduced to the clinical laboratory, it was considered superior to paper electrophoresis, particularly in pediatric screening for dyslipoproteinemias, in which sample size is often a critical factor.

Comparatively rapid. The cellulose acetate method can be considered fairly rapid compared to the paper method, requiring only a couple of hours to complete the electrophoretic procedure.

Disadvantages

Although this support medium is still used for protein electrophoresis, it is no longer used for the study of lipoproteins because of some major drawbacks. This method is therefore not widely used today in the laboratory.[9,10]

Falsely elevated pre-β-lipoproteins. For some unexplained reason, it has been found that with cellulose acetate the pre-β-lipoprotein fraction is falsely elevated when compared to other established electrophoretic methods and the analytical ultracentrifuge.

High-background staining and need for clearing. This method also has had some problems with a high background stain. The clearing procedure is inconvenient, especially when one uses warm glycerin to "clear" the membrane for easier evaluation of the results.

Storage of membranes. Another disadvantage of this method was the difficulty in storing these membranes after electrophoresis.

Starch gel

The major disadvantage of this system is the time required to prepare the gel. In addition to the special equipment required, an experienced person is needed to consistently produce gels of the same thickness. For information concerning the procedure, refer to Lewis.[11]

• • •

There is little information available as to the comparative accuracy and reproducibilities of the various support media. However, on the basis of speed and convenience, the agarose gel technique can certainly be recommended for most laboratory situations and is the method described in this text.

INTEGRATED APPROACH FOR CHARACTERIZATION OF SERUM LIPOPROTEINS IN THE CLINICAL LABORATORY

In the classification of dyslipoproteinemia, both the serum lipids and the lipoproteins must be evaluated. An integrated approach is shown in Fig. 58-12. The classification system described here is designed to allow the unambiguous assignment of patients to one of several categories if their lipoprotein profiles satisfy certain specific criteria. This algorithm, first described in the Manual of Laboratory Operations,[12] serves as a useful guide to an integrated approach for the characterization of the particular type of lipoprotein abnormality (Fig. 58-12). Lipoprotein profiles of some patients may satisfy some but not all of these criteria; the patients are then arbitrarily assigned to a category designated as "other," pending further consideration when the data are analyzed at the completion of the study. For example, a patient who possesses "floating β-lipoprotein" and who satisfies other clinical criteria for type III hyperlipoproteinemia but who has a plasma VLDL (very-low-density lipoprotein) cholesterol–to–triglyceride ratio of less than 0.3 is classified as "other" rather than as having type III hyperlipoproteinemia. This assignment is made even though the medical management of the patient might follow procedures appropriate to type III hyperlipoproteinemia. The lipoprotein patterns described should not be equated with specific diseases, since the interpretation of these patterns in the conventional medical setting includes the consideration of other clinical tests, physical findings, and family studies. Please refer to Chapter 31 for additional detail.

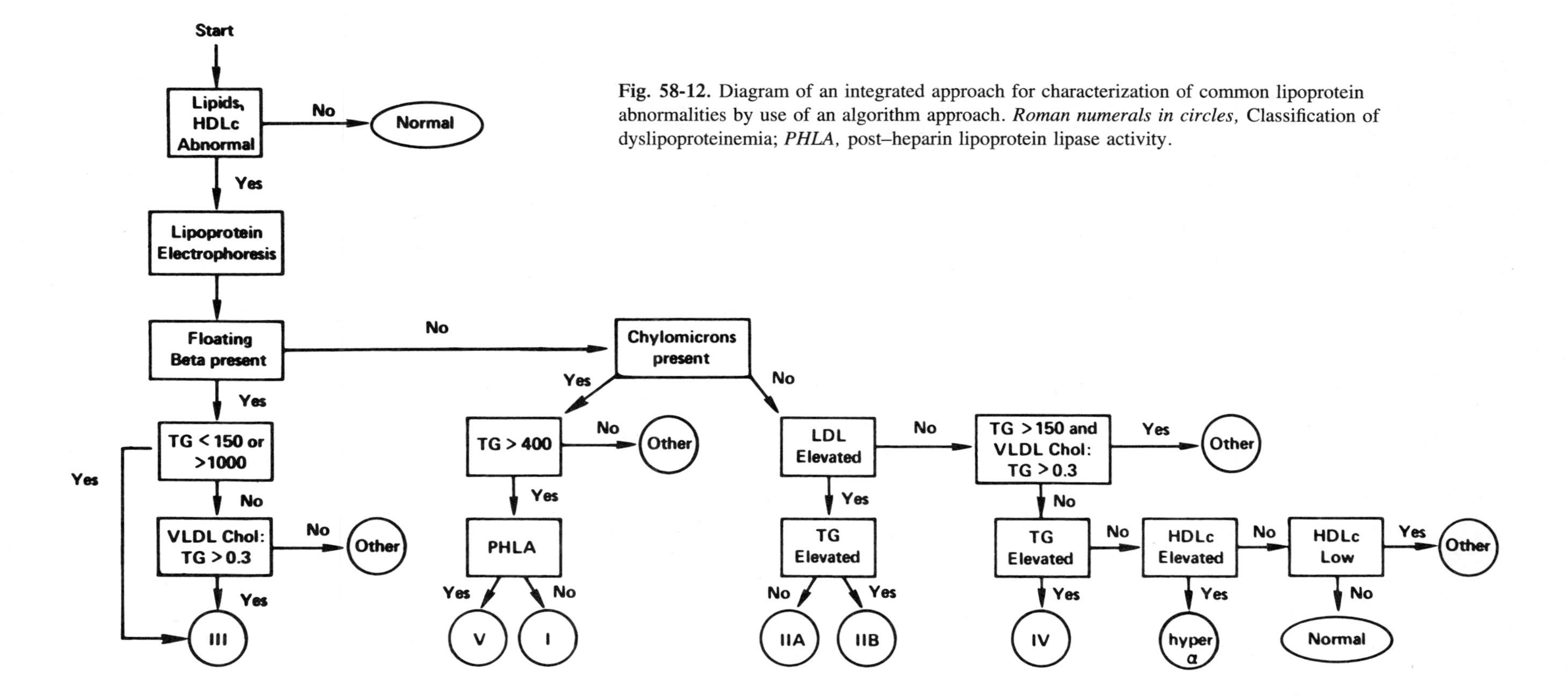

Fig. 58-12. Diagram of an integrated approach for characterization of common lipoprotein abnormalities by use of an algorithm approach. *Roman numerals in circles,* Classification of dyslipoproteinemia; *PHLA,* post–heparin lipoprotein lipase activity.

SPECIMEN

Before one performs any accurate quantitation of lipids and lipoproteins, a series of sampling conditions must be met to allow for a representative analysis. All too often the classification and diagnosis of a hyperlipoproteinemia is based on a single sample that is flawed, or nonrepresentative, because one or more of these sampling conditions for the subject are not met (see p. 1199 for details):

1. Fasting for at least 12 hours before sampling
2. Habitual diet for at least 3 weeks before sampling. Discourage substantial changes in caloric, alcoholic, or type of food intake in advance so that the sample is representative of the person
3. Weight stable or at least no sampling during weight changes, particularly during weight loss
4. No acute illness, trauma, or recent surgery
5. Sampling of postmyocardial infarction patients deferred for a minimum of 3 months
6. Sampling deferred in subjects taking drugs that affect lipid or lipoprotein levels (thyroxine, estrogen, corticosteroids, androgenic compounds, lipid-lowering drugs)

PROCEDURE

Principle

Lipoproteins are separated on thin agarose film, stained with fat red 7B, destained, and scanned.

Reagents

The reagents are most readily obtained primarily from Corning.

Assay

Equipment: electrophoresis apparatus, incubation and drying ovens.

Lipoprotein electrophoresis on agarose gel (Corning) is prepared as follows:

1. Prepare buffer according to manufacturer's instructions.
2. Prepare gel plate and apply samples according to manufacturer's directions.
3. Electrophorese specimens for 30 minutes at 90 ohms. The amperage should remain between 12 to 21 milliamperes (mA) throughout the run.
4. Lift cell cover from cell base, remove agarose film carefully from cell cover, and wipe off *all* buffer condensation from the plastic backing of the agarose film.
5. Dry agarose film by placing it *gel side up* in the oven compartment of the incubator oven (Corning no. 470041), preheated to 60° C. The film should rest flat on the dryer shelf, and the rubber bands should *not* touch the gel.
6. Plug dryer shelf into the oven and set timer for *30 minutes*. NOTE: *Failure to dry the gel adequately will result in excessive background* staining and blurring.
7. Moisten bottom of staining dish with *cold* water and place dried agarose film, agarose side up, into the flat staining dish, which holds the agarose film flat and without air bubbles. Cold water helps to remove excess heat from the stain.
8. Dispense 10 mL of fat red 7B stain and 2.2 mL of type 3 distilled water in the 50 mL beaker. Shake the solution gently until the formed sediment clears up. The ratio of stain to distilled water is *flexible,* depending on the age of the stock stain solution.
9. Pour the stain onto the agarose-gel surface and allow the uniformly flooded agarose film to be stained for 4 to 5 minutes. Staining time is *flexible,* depending on the age of the stock stain solution.
10. Transfer the agarose film into the destaining solution and agitate *gently* until the background clears up.
11. Rinse the agarose film in distilled water for a few seconds.
12. Dry the back of the agarose film with facial tissue; avoid scratching.
13. Place the agarose film in the incubator/oven for *15 minutes*.
14. The electrophoretogram can now be scanned by a densitometer.

Notes

A. Procedure for preparing quality control material for lipoprotein electrophoresis
   1. Obtain one unit of plasma from blood bank.
   2. Determine the lipoprotein profile by doing a lipoprotein electrophoresis on the plasma.
   3. If resolution of the lipoprotein bands is good and distribution is essentially normal, begin the following process of preparation.
   4. Fill the preparative ultracentrifuge tubes (40.3 or 50.3 rotors; Spinco Division of Beckman Instruments, Fullerton, Calif.), cap, and spin at 30,000 rpm for 19½ hours at 16° C.
   5. After spinning, remove the top lipid layer from each tube; this will contain the chylomicrons and the pre-β-lipoprotein layer.
   6. Now check the bottom to see if the resultant lipoprotein pattern is satisfactory. Also, prepare the bottom at a density of 1.21 g/mL for the analytical ultracentrifugation.
   7. Separate the quality control material into 0.5 ml aliquots and freeze at −60° C.
   8. Quality control material is now ready for use; it should be thawed only *once*.
   9. Note that the β- and α-lipoprotein fractions are generally well preserved by freezing at −60° C;

however, if the chylomicrons and pre-β-lipoprotein fractions are not removed from the serum, they will become degraded despite being frozen. This results in poorly resolved bands.

B. Action procedure for lipoprotein-electrophoresis quality control material
   1. The quality control material should be electrophoresed with each plate. The pattern should then be scanned at 500 nm.
   2. When the ratio of the β- and the pre-β-band area to the α-band area of the quality control is outside the 0.9 to 1.3 range, repeat the electrophoresis of the entire run. The percent distribution of β- and α-bands should also be within ±2 standard deviations; otherwise repeat the entire run.
   3. In addition, repeat electrophoresis of all samples that have application artifacts, unusual migration rates, or disagreement between lipid values and the dye intensity of the fractions.
   4. Send reports only after the repeated samples show satisfactory results.

REFERENCES

1. Lees, R.S., and Hatch, F.T.: Sharper separation of lipoprotein species by paper electrophoresis in albumin containing buffer, J. Clin. Med. **61:**518-528, 1963.
2. Naito, H.K., and Lewis, L.A.: Rapid lipid-staining procedure for paper electrophoresis, Clin. Chem. **21:**1454-1456, 1973.
3. Jencks, W.P., and Durrum, E.L.: Paper electrophoresis as a quantitative method: the staining of serum lipoproteins, J. Clin. Invest. **34:**1437-1447, 1955.
4. Smithies, O.: Zone electrophoresis in starch gel: group variations in the serum proteins of normal human adults, Biochem. J. **61:**629-641, 1951.
5. Seidel, D., Wieland, H., and Ruppert, C.: Improved techniques for assessment of plasma lipoprotein patterns. I(1). Precipitation in gels after electrophoresis with polyanionic compounds, Clin. Chem. **19:**737-739, 1973.
6. Fredrickson, D.S., Levy, R.J., and Lees, R.S.: Fat transport in lipoproteins—an integrated approach to mechanisms and disorders, N. Engl. J. Med. **176:**32-44, 94-103, 148-156, 215-224, 273-281; 1967.
7. Noble, R.P., Hatch, F.T., Marrimas, J.A., et al.: Comparison of lipoprotein analysis by agarose gel and paper electrophoresis with analytical ultracentrifugation, Lipids **4:**55-59, 1969.
8. Papadopoulos, N.M., and Kintrios, J.A.: Varieties of human serum lipoprotein pattern: evaluation of agarose gel electrophoresis, Clin. Chem. **17:**427-429, 1971.
9. Chin, H.P., and Blankenhorn, D.H.: On the precision of lipoprotein electrophoresis on cellulose acetate and its use in the diagnosis of hyperlipoproteinemia, Clin. Chim. Acta **23:**239-240, 1969.
10. Beckering, R.E., Jr., and Ellefson, R.D.: A rapid method for lipoprotein electrophoresis using cellulose acetate as support medium, Am. J. Clin. Pathol. **53:**84-88, 1970.
11. Lewis, L.A.: Thin layer starch-gel electrophoresis: a simple accurate method for characterization and quantitation of protein components, Clin. Chem. **12:**596-605, 1966.
12. Manual of Laboratory Operations, Lipid Research Clinics Program. Vol. 1. Lipid and lipoprotein analysis, DHEW Pub. No. NIH 75-628, Washington, D.C., 1974, U.S. Government Printing Office, pp. 9-50.

Triglycerides

HERBERT K. NAITO

Clinical significance: p. 550

Molecular formula: varies with R group of fatty acids

$$
\begin{array}{l}
H_2C-O-\overset{\displaystyle O}{\overset{\|}{C}}-R_1 \\
\;\;| \\
HC-O-\overset{\displaystyle O}{\overset{\|}{C}}-R_2 \\
\;\;| \\
H_2C-O-\overset{\displaystyle O}{\overset{\|}{C}}-R_3
\end{array}
$$

Molecular weight: 800 to 900 daltons (varies with R group of fatty acid); average weight for human samples is 885

Chemical class: neutral lipid

PRINCIPLES OF ANALYSIS AND CURRENT USAGE

Before the 1950s triglycerides were estimated by the subtraction method:

$$\text{Triglycerides} = \text{Total lipids} - (\text{Cholesterol} + \text{Phospholipids})$$

This indirect method was widely used until 1957, when Van Handel and Zilversmit[1] published a direct manual method in which the phospholipids were removed from the lipid extract by an adsorbent and the triglycerides were determined by measurement of the amount of glycerol released by saponification with potassium hydroxide (KOH). This method has been widely adopted and subjected to modification by many investigators. Seven years later a semiautomated adaptation of this procedure for serum triglycerides was established on the Technicon AutoAnalyzer-I by Lofland.[2] In 1966 Kessler and Lederer[3] improved the method by automating the potassium hydroxide saponification step; their method is probably the most widely utilized automated procedure and has been regarded as a potential "reference method."

Most current methods use chemical or enzymatic procedures to determine glyceride glycerol, which are then converted to the equivalent mass concentration of an average triglyceride (in milligrams per liter). Alternatively, concentration may be expressed on a molar basis.

Recently, triglyceride analysis has been simplified by the introduction of enzymatic methods, which have been automated to provide the analyst with quick, easy, and direct procedures.

Since the Hantzsch condensation[1] and automated method[3] is still considered a classical and possibly reference method, the procedure is included in some detail in the section describing the analysis of cholesterol. The classical chromotropic method of Van Handel and Zilversmit[1] is provided under the procedure section. Although triglyceride procedures based on thin-layer, gas-liquid, or column

Table 58-9. Methods of serum triglyceride analysis

Method	Principle	Usage	Comments
Chemical determinations by glycerol	The following three reactions are common to methods 1 to 4: A. Triglycerides are extracted with organic solvents or purified with adsorbents such as zeolite B. Triglycerides $\xrightarrow[OH^-]{ROH}$ Glycerol + 3 fatty acids C. Glycerol + IO_4^- → HCHO (Formaldehyde) + HCOOH (Formic acid) + H_2O + IO_3^-		
1. Egriwe's reaction	HCHO 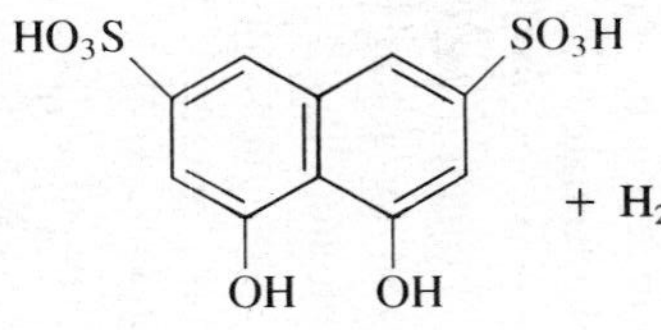 + Chromotropic acid (4,5-dihydroxy-naphthalene with two SO_3H groups: HO_3S, SO_3H, OH, OH) + H_2SO_4 → Chromophore (570 nm)	Rarely used in routine laboratories	Widely used as comparative method
2. Schryver's reaction	HCHO 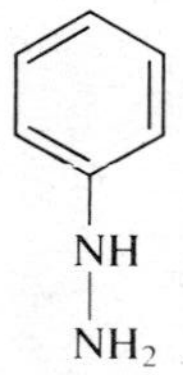 + Phenylhydrazine (C_6H_5–NH–NH_2) $\xrightarrow[HCl]{Ferricyanide}$ Chromophore (540 nm)	Rarely used	
3. Pay's reaction	HCHO + 3-Methyl-benzothiazolin-2-one $\xrightarrow{FeCl_3}$ Chromophore (620 nm)	Rarely used	
4. Hantzsch's reaction	HCHO + NH_4^+ + $2CH_3{-}\overset{O}{\overset{\parallel}{C}}{-}CH_2{-}\overset{O}{\overset{\parallel}{C}}{-}CH_3$ (Acetylacetone) 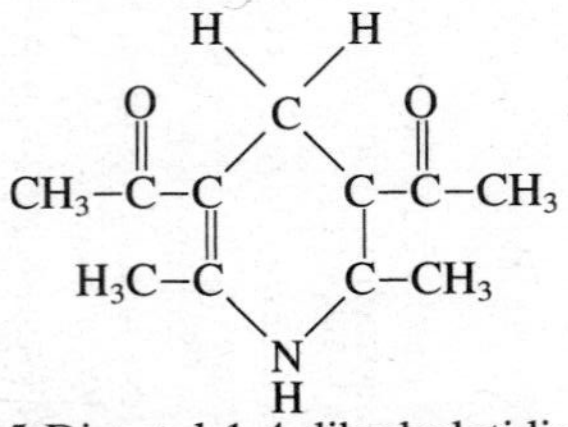 → 3,5-Diacetyl-1,4-dihydrolutidine (A_{max}, 412 nm; $\lambda_{excitation}$, 400 nm; $\lambda_{emission}$, 485 nm)	Most frequently used chemical method in routine laboratories	Fluorescent assay is reference method of Centers for Disease Control
Enzymatic determinations of glycerol		Most commonly used approach	

5. NADH consumption (decreased A_{340})	a. Triglycerides $\xrightarrow[\text{Protease}]{\text{Lipase}}$ Glycerol + 3 fatty acids	
or	b. Glycerol + ATP $\xrightarrow[\text{kinase}]{\text{Glycerol}}$ Glycerol-3 phosphate + ADP	
(decreased fluorescence: λ_{ex}, 355 nm; λ_{em}, 460 nm)	c. ADP + Phospho*enol*pyruvate $\xrightarrow[\text{kinase}]{\text{Pyruvate}}$ ATP + Pyruvate	
	d. Pyruvate + NADH + H^+ $\xrightarrow[\text{dehydrogenase}]{\text{Lactate}}$ Lactate + NAD^+	Very frequently used assay
6. Formazan colorimetric	*Reactions 5a and 5b plus:*	
	Glycerol-3-phosphate + NAD^+ $\xrightarrow[\text{dehydrogenase}]{\text{Glycerol phosphate}}$ Dihydroxyacetone phosphate + NADH + H^+	
	NADH + Oxidized tetrazolium $\xrightarrow{\text{Diaphorase}}$ Reduced tetrazolium (500-590 nm)	
7. Fluorescent	*Reaction 5a plus the following:*	
	Glycerol + NAD^+ $\xrightarrow[\text{dehydrogenase}]{\text{Glycerol}}$ Dihydroxyacetone + NADH + H^+	
	NADH + H^+ + Resazurin $\xrightarrow{\text{Diaphorase}}$ Resorufin + NAD^+ λ_{ex}, 548 λ_{em}, 580 (fluorescent)	

chromatography and infrared spectrophotometry have been published, these are not discussed, since their applications are not for routine use and have very specific and sometimes limited use.

Chemical determination of glyceride glycerol

In reactions based on chemical determination of glyceride glycerol (reactions A to C, Table 58-9), the first step extracts triglycerides and removes interfering substances. The extraction is accomplished using solvents such as methanol, ethanol, isopropanol, or chloroform. These solvents cause the denaturation of the lipoproteins and therefore the dissociation of the bound triglycerides. Interfering substances are removed by either (1) solvent partition with nonane or hexane or (2) use of adsorbents such as zeolite, silicic acid, or Florisil (activated magnesium silicate). The principal interfering substances removed are phospholipids and glucose, along with certain chromogens and sometimes free glycerol.

The second step in these procedures hydrolyzes triglycerides to glycerol and fatty acids and is usually carried out in ethanolic potassium hydroxide at elevated temperatures (saponification):

$$\text{Triglycerides} \rightarrow \text{Glycerol} + 3 \text{ fatty acids}$$

The third step oxidizes glycerol to formaldehyde and is usually accomplished by the following reaction:

$$\text{Glycerol} + \text{Periodate} \rightarrow \text{Formaldehyde} + \text{Formic acid} + \text{Iodate} + \text{Water}$$

An alternative approach is to use an alkoxide transesterification reaction to form glycerol and fatty acids from the triglycerides.

In step four the formaldehyde formed is quantitatively measured by the following various means:

1. Eegriwe's reaction: formaldehyde reaction with a chromotropic acid–sulfuric acid mixture; read at 570 nm (method 1, Table 58-9)
2. Schryver's reaction: formaldehyde reaction with a phenylhydrazine/hydrochloride/ferricyanide/hydrochloric acid mixture; read at 540 nm (method 2, Table 58-9)
3. Pay's reaction: formaldehyde reaction with 3-methyl-2-benzothiazolinone/ferric chloride mixture; read at 620 nm (method 3, Table 58-9)
4. Hantzsch's reaction: formaldehyde reaction with ammonium acetate and acetylacetone; read colorimetrically or fluorometrically (method 4, Table 58-9)

In the Hantzch condensation reaction the following occurs:

$$\text{Acetylacetone} + \text{Formaldehyde} + \text{Ammonium acetate} \rightarrow \text{3,5-Diacetyl-1,4-dihydrolutidine}$$

The yellow end product can be measured colorimetrically at 412 nm or fluorometrically (primary filter at 400 nm, secondary filter at 485 nm). The fluorometric method is more widely used and has been applied to semiautomated methods, one of which is discussed later in this section. The described semiautomated method for the triglyceride determination has been standardized and certified by the Centers for Disease Control for use as a reference method in our laboratory.

In summary, the following four steps are usually used in the chemical determination of triglycerides:

1. Extraction of triglycerides, removal of interfering substances
2. Saponification of the triglycerides to glycerol and fatty acids
3. Oxidation of the glycerol to formaldehyde
4. Measurement of the formaldehyde

Enzymatic determination of glyceride glycerol

The enzymatic methods are based on the determination of the glycerol portion of the triglyceride molecules after hydrolysis (chemical or enzymatic) to remove the fatty acids (see Table 58-9). Methods employing the enzymatic determination of glycerol have been used for many years; however, an alkaline hydrolysis of the triglycerides was used in these methods. The recent development of employing enzymes (lipase, usually combined with a protease) to catalyze hydrolysis has made possible methods that are direct, rapid, and specific. The completely enzymatic systems eliminate the use of caustic reagents, extraction solvents, high-temperature baths, and adsorption mixtures for phospholipid removal. Recent studies have focused on developing lipase reagents that completely hydrolyze triglycerides. The role played by the protease is as yet unknown, since the protease enzymes alone do not hydrolyze triglycerides; however, many methods find a protease enzyme necessary to attain complete hydrolysis. One of the most commonly added proteases is α-chymotrypsin, first reported by Bucolo and David.[4] The critical point of whether phospholipids were hydrolyzed by the enzymes used has also been investigated; it was found that any glycerol from this source was insignificant.

The method of Bucolo and David[4] in common use today is based on the following sequence of reactions (method 5, Table 58-9):

1. $\text{Triglyceride} \xrightarrow[\alpha\text{-Chymotrypsin}]{\text{Lipase}} \text{Glycerol and fatty acids}$

2. $\text{Glycerol} + \text{ATP} \underset{Mg^{++}}{\overset{\text{Glycerol kinase}}{\rightleftharpoons}} \text{Glycerol-3-phosphate} + \text{ADP}$

3. $\text{ADP} + \text{Phospho}\textit{enol}\text{pyruvate} \overset{\text{Pyruvate kinase}}{\rightleftharpoons} \text{ATP} + \text{Pyruvate}$

4. $\text{Pyruvate} + \text{NADH} + H^+ \overset{\text{Lactate dehydrogenase}}{\rightleftharpoons} \text{Lactate} + NAD^+$

The decrease in absorbance of reduced nicotinamide adenine dinucleotide (NADH) is measured at 340 nm. To blank this method, the procedure is repeated with a buffer used in place of the lipase reagent.

Rietz and Guilbault[5] modified the procedure utilizing a fluorometric measurement. The disappearance of NADH fluorescence is read at 460 nm after excitation at 355 nm.

A procedure by Megraw et al.[6] uses reaction steps 1 and 2 and then steps 5 and 6 to form formazan, a highly colored compound that is measured in the 500 to 590 nm range (method 6, Table 58-9):

5. $\text{Glycerophosphate} + NAD^+ \xrightarrow{\text{Glycerol-1-phosphate dehydrogenase}} \text{Dihydroxyacetone phosphate} + \text{NADH} + H^+$

6. $\text{NADH} + H^+ + \text{2-(}p\text{-Iodophenyl)-3-(nitrophenyl)-5-phenyltetrazolium (oxidized)} \xrightarrow{\text{Diaphorase}} \text{2-(}p\text{-Iodophenyl)-3-(}p\text{-nitrophenyl)-5-phenyltetrazolium (reduced) (formazan)} + NAD^+$

On the other hand, the method of Winartasaputra et al.[7] uses glycerol dehydrogenase and diaphorase to form a fluorescent compound from glycerol released by enzymatic hydrolysis (method 7, Table 58-9):

7. $\text{Glycerol} + NAD^+ + \xrightarrow{\text{Glycerol dehydrogenase}} \text{Dihydroxyacetone} + \text{NADH} + H^+$

8. $\text{NADH} + H^+ + \text{Resazurin} \xrightarrow{\text{Diaphorase}} \text{Resorufin (fluorescent)} + NAD^+$

The fluorescence is measured at 580 nm after excitation at 548 nm.

The current enzymatic method that we are using in the laboratory is primarily based on the method of Bucolo and David.[4] It has proved to be as accurate and precise as the chemical method of Kessler and Lederer,[3] which has been considered a reference method for serum triglyceride determinations. (See references 8 to 23 for further reading on enzymatic triglyceride determinations.)

REFERENCE AND PREFERRED METHODS

The newer enzymatic methods are not yet perfected, and several colorimetric and fluorometric methods are still recommended as reference procedures. At present there is no official recognition of a true reference method for triglyceride determinations. The method of Van Handel and Zilversmit or a modification is probably the most widely used comparative method.

Reagent concerns

The major problem one faces when setting up one of the completely enzymatic methods is the inconsistent quality of the enzymes available. A careful check against a reference procedure is suggested to ensure that reliable results are being produced. After a reliable enzymatic procedure is set up, linearity checks and good quality control sera (at high, low, and normal values) will help to determine lot-to-lot variance in purchased kits or raw enzymes. Another important consideration is shelf life and the stability of the reagents; these time periods vary considerably from manufacturer to manufacturer. We have found it necessary to monitor the quality of the reagents during their entire shelf life to ensure optimum reliability, since deterioration before the manufacturer's stated expiration date is not uncommon. Linearity checks are a convenient way to detect reagent deterioration. Since changes are often first detectable at the high or low ends of the concentration range, the linearity checks should cover the entire stated linearity range of the method. One should also take care when running automated equipment for many hours during a single run, since a reagent may deteriorate during the run, producing accurate results at the beginning and inaccurate results at the end of the run. One can monitor reagent stability throughout the day with good quality control practices by running quality control samples (at high, low, and normal values) periodically throughout the run.

Blanking for free glycerol

For an accurate triglyceride level, the free glycerol of the sample should be subtracted from the total glycerol (both free and that produced from saponification of the triglycerides) by a blanking procedure. There has been much discussion in the literature over the years as to whether to blank or not to blank the assay. One commonly does the quantitative analysis for triglycerides in serum and plasma by hydrolyzing the triglycerides and then determining glycerol enzymatically or chemically. When this is done, however, the unesterified "free" glycerol is measured along with the glycerol derived from the triglycerides. Analytically accounting for this free glycerol makes the methods cumbersome and inconvenient. One must either perform a two-part analysis (one before and one after the hydrolysis step) or separate the two serum constituents by extraction or possibly solvent partition.

Another way to account for the free glycerol is to correct the value of total glycerol by a simple calculation. According to Eggstein,[24] a correction by the following equation is acceptable:

$$\text{Triglycerides (mmol/L)} = 0.088 + (0.828 \times \text{Total glycerol [mmol/L]})$$

Van Oers et al.[25] found concentrations of free glycerol ranging from 0.07 mmol/L (when total glycerol was 0 to 0.5 mmol/L) to 0.13 mmol/L (when total glycerol was 5.5 to 6.0 mmol/L). Mourad et al.[13] reported an average concentration of 0.11 mmol of free glycerol per liter. Stinshoff et al.[26] found that the free glycerol followed a log-normal distribution with a mean of 0.12 mmol/L (range: 0.04 to 0.37 mmol/L). They recommended the use of the following correction factor:

$$\text{Triglycerides} = (0.98 \times \text{Total glycerol}) - 0.07\ \text{(mmol/L)}$$

This would be used to subtract from the total glycerol to give an estimated triglyceride value. Whitlow and Gochman[27] found that there were differences in free glycerol values found in patient sera, depending on whether the fluorometric or enzymatic methods were used. Although there was good agreement between the net (total minus blank) results of the fluorometric and enzymatic methods, the fluorometric method consistently had lower and less variable blank values. In about 7% of the analyses the blank values in the enzymatic method were unusually high (as high as 42% of the total glycerol). Our studies on a selected hospital population indicate that the glycerol blank values are not predictable (unpublished results). This is especially true of stressed individuals, such as patients in intensive care wards, patients with liver or kidney disease, or diabetic persons. The net value (true triglycerides), however, was similar whether done enzymatically or fluorometrically. Interestingly, the correlation between triglycerides and the free glycerol (as a percentage of the total) was poor for the diabetic (correlation coefficient r = 0.56) groups. Thus I believe that although one can use "correction factors" to account for the free glycerol, it is probably only reliable for normal, healthy persons.

Any metabolic condition causing the epinephrine effect, in which triglycerides are catabolized rapidly in the adipose tissue to fatty acids and glycerol, will result in elevated free glycerol content in the blood. I have encountered many instances in which the free glycerol (blank) values were higher than the true triglyceride content. Laboratories using nonblanked procedures are thus susceptible to reporting falsely elevated true triglyceride values. We therefore strongly recommend blanking for free glycerol content if one is interested in determining true triglyceride values.

There are instances in which infusates may contain glycerol or glycerol-like products that will cause a high blank value. Lindblad et al.[28] reported the blood glycerol level increasing during total parenteral nutrition (TPN) administration. It has also been found that mannitol (Osmitrol) infusion will cause spuriously high glycerol blank values.[29] An understanding of triglyceride metabolism helps illustrate why blood glycerol can be high on certain occasions (see Chapter 31). Conditions known to contribute to high glycerol levels in the blood are the following:

1. Stress (epinephrine effect)
2. Mannitol infusion
3. Treatment with nitroglycerin
4. Diabetes mellitus
5. Glycerol-coated stopper used in Vacutainer
6. Certain liver disease
7. Hemodialysis for kidney disease

In summary, because there is a risk involved in calculating the free triglyceride concentrations by using factors or percentages, glycerol blanking is necessary for accurate triglyceride analysis.

Standards

Choosing a standard and properly handling it throughout a procedure are critical considerations in producing an accurate triglyceride value.

A pure triglyceride standard is essential to obtain the most accurate results. Highly purified triglycerides are available from commercial sources; however, one should check for impurities by thin-layer chromatography until reliability is verified. This may be accomplished by use of a stationary phase of silicic acid with acetic acid/ethyl ether/light petroleum ether, 1:25:74 (volume ratio) as the mobile phase. Triolein is often the standard of choice, especially when isopropanol is used as a solvent. Witter and Whitner[30] favor the use of a mixture of triolein and tripalmitin, 2:1 (weight ratio), with chloroform extracts, since the degree of unsaturation is similar to that of human serum triglycerides.

Many procedures recommend the use of corn oil, olive oil, or cottonseed oil as a standard. These oils vary considerably in composition and purity as received from the manufacturers. The composition of the oils can also vary from season to season. If they are used, one should first purify them by dissolving them in chloroform and shaking with zeolite. This mixture is centrifuged, and the chloroform layer is removed. The standard is recovered by evaporation of the solvent.

Glycerol and mannitol are suggested standards in some procedures. The main drawback to these compounds is that they do not reflect variations in the extraction, adsorption, and saponification steps. Glycerol has the added drawback of being hygroscopic. Despite this, since glycerol is soluble in water, glycerol standards are often used when an organic solvent cannot be used, as in some enzymatic procedures. If glycerol is used on a blanked two-channel system, each channel must be calibrated separately.

Currently, reference or calibration sera are being used to standardize many of the enzymatic procedures. Determining the correct set point for a reference or calibration serum is difficult but critical for attaining accurate and reliable results. One method is to calibrate the procedure with very carefully prepared glycerol standards that have been checked by refractive index or iodometry. One assays the reference or calibrator serum to attain a set point, which can then be used to standardize the daily runs. This method does not work well with systems that use pumping devices to deliver the sample. The water-based glycerol standards do not have the same viscosity as serum samples, causing a difference in the amount of sample delivered. These methods should be standardized against a reliable reference method using a serum pool. One calibrates the reference method with primary standards, assays the serum pool, uses the pool's value to set the desired method, and then assays the reference or calibrator sera. It is also desirable to compare the results of methods to reference laboratory results or calibrated sera from the Centers for Disease Control to ensure the accuracy.

Another source or error caused by the standard occurs when results are reported in mass/volume units, such as milligrams per liter. The molecular weight of the triglycerides varies; for example, tripalmitin's molecular weight is 807 daltons, whereas triolein's is 885 daltons. In theory nearly 10% disparity would exist in the same method using the two different triglycerides as standards. This disparity indicates that one should report triglyceride values in molar units. The official units designated for triglycerides using the International System of Units (Système International d'Unités, SI units) are millimoles per liter.

A properly calibrated enzymatic method is certainly the preferred method for routine laboratory analysis. Since these methods are readily available in kits, this technique is not presented here in detail. Instead, a method based on the proposed reference method is presented, since this may be used for the initial calibration of a method.

SPECIMEN

To understand blood collection, review the previous section on cholesterol methods in this chapter. Included in that section is a discussion of serum versus plasma and stability of triglycerides.

Briefly, it is imperative that the subjects be fasting for at least 12 hours before blood is drawn. If BD Vacutainer tubes are used, the stoppers should be silicon coated, not glycerin coated. The blood should be rapidly spun down to minimize the spontaneous hydrolysis of triglycerides to glycerol and fatty acids in the blood (see discussion of triglyceride metabolism, p. 550). If the blood sample cannot be analyzed for triglycerides within 25 hours, freezing the samples at $-20°$ C, preferably at $-40°$ or $-60°$ C, is recommended.

PROCEDURE: REFERENCE METHOD (MODIFICATION OF VAN HANDEL AND ZILVERSMIT[1])

Principle

This manual method extracts triglycerides with chloroform while adsorbing phospholipids with zeolite. An aliquot of the chloroform extract is added to alcoholic potassium hydroxide to saponify the triglycerides to glycerol and free fatty acids. Glycerol is then oxidized with periodate to formaldehyde, which is measured colorimetrically after reaction with chromotropic acid. The method is practical for routine testing if large batches of samples are not anticipated.

Reagents

A. Extraction chemicals
   1. Chloroform, reagent grade. Keep in a dark bottle.

2. Zeolite, 80 mesh (W.A. Taylor Co., 7300 York Rd., Baltimore, Md.). Heat for 4 hours at 125° C to activate.

B. Reagents (use American Chemical Society reagent-grade chemicals)
   1. Alcoholic potassium hydroxide
      a. Stock (2%). Dissolve 2 g of potassium hydroxide in 95% ethanol; dilute to 100 mL with 95% ethanol. This is stable for 1 month at ambient temperature.
      b. Working (0.4%). Dilute 10 mL of stock solution to 50 mL with 95% ethanol. Make up fresh daily.
   2. Sulfuric acid (0.1 M). Dilute 3 mL of concentrated sulfuric acid to 500 mL with reagent-grade water. This is stable 6 months at room temperature.
   3. Sodium *meta*periodate (0.05 mol/L). Dissolve 1.07 g of sodium *meta*periodate in reagent-grade water. Dilute to 100 mL. This is stable for 3 months at 4° C.
   4. Sodium arsenite (0.5 mol/L). Dissolve 6.5 g of sodium arsenite in reagent-grade water; dilute to 100 mL. This is stable for 6 months at room temperature.
   5. Chromotropic acid (0.2%). Dissolve 2.00 g of chromotropic acid (or 2.24 g of sodium salt) in 200 mL of reagent-grade water. Add separately 600 mL of concentrated sulfuric acid to 300 mL of reagent-grade water cooled in ice. When it is cool, add the diluted sulfuric acid solution to the chromotropic acid solution. Store in a brown bottle. Prepare fresh every 2 to 3 weeks.

Standards

1. Triglyceride stock standard (5000 mg/L). Dissolve 500 mg of triolein in chloroform. Dilute to 100 mL in a glass-stoppered volumetric flask. Store tightly sealed at 4° C. This is stable for 3 months.
2. Triglyceride working standard (50 mg/L). Dissolve 1 mL of stock standard to 100 mL with chloroform. This is stable for 1 week.

Table 58-10. Comparison of methods for serum triglyceride analysis

Parameter	Egriwe's reaction*	Hantzsch's reaction†	Enzymatic‡
Temperature	Extraction: ambient Saponification: 65° C Colorimetric reaction: 100° C	50° C	37° C
Sample volume	0.5 mL	0.2 mL	0.005 mL
Fraction of sample volume	0.3 (extraction)	Approximately 0.01	0.02
Final concentration of reagents	KOH: 356.5 mmol/L (saponification) $NaIO_4$: 4.5 mmol/L (formaldehyde formation) Colorimetric reaction Chromotropic acid: 1.61 g/L H_2SO_4: 8 mmol/L	KOH: 22.5 mmol/L (saponification) $NaIO_4$: 1.1 mmol/L (formaldehyde formation) Ethanol: 57.5% (v/v) Colorimetric reaction Acetylacetone: 36.6 mmol/L	Glycerol kinase: 0.6×10^3 U/L Lipase: 300×10^3 U/L Lactate dehydrogenase: 0.7×10^3 U/L Pyruvate kinase: 0.8×10^3 U/L Chymotrypsin: 13×10^3 U/L KH_2PO_4: 10 mmol/L K_2HPO_4: 63 mmol/L (pH 7.1) NaOH: 0.3 mmol/L Bovine serum albumin: 1.7 g/L Phospho*enol*pyruvate: 0.8 mmol/L ATP: 0.5 mmol/L Mg^{++}: 6 mmol/L
Time of reaction	Extraction: 30 min Saponification: 15 min Colorimetric reaction: 30 min	Approximately 20 min	10 min
Wavelength	570 nm	Excitation: 400 nm Emission: 485 nm	340 nm
Linearity	—	4000 mg/L	5000 mg/L
Interferences	Free glycerol	Free glycerol	Free glycerol
Precision ($\bar{X}$, % coefficient of variation)§	—	3110 mg/L (11.9%)	1150 mg/L (2.3%) 3360 mg/L (8.0%)

*Modification of Van Handel and Zilversmit.[1]
†Technicon AutoAnalyzer-II method.
‡Eskalab reagent (Smith-Kline, Inc., Sunnyvale, Calif.) adapted to Abbott ABA-100 (Abbott Instruments; Dallas).
§College of American Pathologists Quality Assurance Survey data.

Assay

Equipment

Shaker

Water bath at 65° C

Water bath at 100° C

Spectrophotometer, with band pass ≤10 nm and capable of reading at 570 nm

1. Place 2 g of zeolite in a 15 × 130 mm screw-capped (Teflon) tube.
2. Add 10 mL of chloroform and shake.
3. Add 0.5 mL of serum or plasma and shake for 30 minutes on a mechanical shaker.
4. Filter through coarse fat-free filter paper.
5. Pipet 1 to 3 mL portions of the filtered supernatant (depending on the amount of triglyceride present) into two tubes.
6. Pipet 1 mL portions of the working standard into two tubes.
7. Evaporate all tubes to dryness under nitrogen gas.
8. To one set of standards and unknowns, add 0.5 mL of alcoholic potassium hydroxide (saponified sample). To the other set, add 0.5 mL of 95% ethanol (unsaponified sample).
9. Keep all tubes at 65° C for 15 minutes.
10. Add 0.5 mL of 0.2 N sulfuric acid to each tube. Place uncapped tubes in a 100° C water bath for 15 minutes to remove the alcohol.
11. Remove tubes from water bath, cap, and allow to cool.
12. Add 0.1 mL of periodate solution to each tube. Mix well.
13. After 10 minutes add 0.1 mL of sodium arsenite to stop oxidation. Mix well. A yellowish iodine color appears and then vanishes in a few minutes.
14. Add 5.0 mL of chromotropic acid reagent. Mix well.
15. Heat the tubes in a water bath at 100° C for 30 minutes in the absence of excessive light.
16. Cool in dark to room temperature.
17. Read absorbance *(A)* at 570 nm. The color remains stable for several hours.

Calculation

Triglycerides (mg/L) =

$$\frac{A_{\text{unknown (saponified)}} - A_{\text{unknown (unsaponified)}}}{A_{\text{standard (saponified)}} - A_{\text{standard (unsaponified)}}} \times 0.05 \times 2000$$

Multiply by 0.114 to convert milligrams per liter to millimoles per liter.

REFERENCE RANGE

The reference range varies greatly with age and sex. See Chapter 31 for details on reference ranges.

REFERENCES

1. Van Handel, E., and Zilversmit, B.D.: Micromethod for the direct determination of serum triglycerides, J. Lab. Clin. Med. **50**(1):152-157, 1957.
2. Technicon Methods nos. SF40040F66 and SE0040FC6, Technicon Instruments Corp., Tarrytown, N.Y., March 1976.
3. Kessler, G., and Lederer, H., Fluorometric measurement of triglycerides. In Skeggs, L.T., Jr., et al., editors: Automation in analytical chemistry, Technicon Symposium, Tarrytown, N.Y., 1965, 1966, pp. 341-344.
4. Bucolo, G., and David, H.: Quantitative determination of serum triglycerides by the use of enzymes, Clin. Chem. **19**(5):476-482, 1973.
5. Rietz, E.B., and Guilbault, G.G.: Fluorometric estimation of triglycerides in serum by a modification of the method of Bucolo and David, Clin. Chem. **23**(2 Pt. 1):286-288, 1977.
6. Megraw, R.E., Dunn, D.E., and Biggs, H.G.: Manual and continuous-flow colorimetry of triglycerides by a fully enzymatic method, Clin. Chem. **25**(2):273-278, 1979.
7. Winartasaputra, H., Mallet, V., Kuan, S., and Guilbault, G.: Fluorometric and colorimetric enzymic determination of triglycerides (triacylglycerols) in serum, Clin. Chem. **26**(5):613-617, 1980.
8. Burton, R.M., and Kaplan, N.O.: A PDN specific glycerol dehydrogenase from *Aerobacter aerogenes,* J. Am. Chem. Soc. **75:**1005-1006, 1953.
9. Hagen, J.H., and Hagen, P.B.: An enzymatic method for the estimation of glycerol in blood and its use to determine the effect of noradrenaline on the concentration of glycerol in blood, Can. J. Biochem. **40:**1129-1139, 1962.
10. Strickland, J.E., and Miller, O.N.:Inhibition of glycerol dehydrogenase from *Aerobacter aerogenes* by dihydroxyacetone, high ionic strength, and monovalent cations, Biochem. Biophys. Acta **159:**221-225, 1968.
11. Grossman, S.H., Mollo, E., and Ertingshausen, G.: Simplified totally enzymatic method for determination of serum triglycerides with a centrifugal analyzer, Clin. Chem. **22**(8):1310-1313, 1976.
12. Kreutz, F.H.: Enzymatic determination of glycerol in the measurement of triglycerides, Clin. Chem. **9:**492, 1963.
13. Mourad, N., Zager, R., and Neveu, P.: Semiautomated enzymatic method for determining serum triglycerides by use of the Beckman "DSA 560," Clin. Chem. **19**(1):116-118, 1973.
14. Bucolo, G., Yabat, J., and Chang, T.Y.: Mechanized enzymatic determination of triglycerides, Clin. Chem. **21**(3):420-424, 1975.
15. Wentz, P.W., Cross, R.E., and Savory, J.: An integrated approach to lipid profiling: enzymatic determination of cholesterol and triglycerides with a centrifugal analyzer, Clin. Chem. **22**(2):188-192, 1976.
16. Trocha, P.J.: Improved continuous flow enzymatic determination of total serum cholesterol and triglycerides, Clin. Chem. **23**(1):146-147, 1977 (Letter).
17. DiCesare, J.L.: Optimum kinetic enzymatic procedures for glucose and triglycerides in plasma and serum, Clin. Chem. **21**(10):1448-1453, 1975.
18. Mott, G.E., and Rogers, M.L.: Enzymatic determination of triglycerides in human and baboon serum triglycerides, Clin. Chem. **24:**354-357, 1978.
19. Lehnus, G., and Smith, L.: Automated procedures for kinetic measurement of total triglycerides (as glycerol) in serum, Clin. Chem. **24**(1):27-31, 1978.
20. Bublitz, C., and Kennedy, E.P.: Synthesis of phosphatides in isolated mitochondria. III. The enzymatic phosphorylation of glycerol, J. Biol. Chem. **211:**951-961, 1954.
21. Wieland, O.: Enzymic method for the determination of glycerides, Biochem. Z. **329**(3):313-319, 1957.
22. Barbour, H.M.: Enzymatic determination of cholesterol and triglyceride with the Abbott Bichromatic Analyzer, Am. Clin. Biochem. **14**(1):22-28, 1977.
23. Trappe, J.S., Brockington, P.B., and Vickers, M.F.: Serum triglycerides: a comparison of a nephelometric and an enzymic method, Clin. Biochem. **8**(2):142-144, 1975.
24. Eggstein, M.: Eine neue Bestimmung der Neutralfette in Blutserum und Gewebe-Zuverlässigkeit der Methode andere Neutralfette-Bestimmungen, Normalwette für Triglyceride und Glycerin in menschlichen Blut, Klin. Wochenschr. **44:**267-273, 1966.

25. van Oers, R.J.M., Schiltis, R.J.H., Schmidt, N.A., et al.: Assay of triglycerides using the Perkin Elmer C-4 Automatic Analyzer, Z. Klin. Chem. Klin. Biochem. **9**:516, 1971.
26. Stinshoff, K., Weisshaar, D., Staehler, F., et al.: Relation between concentrations of free glycerol and triglycerides in human sera, Clin. Chem. **23**(6):1029-1032, 1977.
27. Whitlow, K., and Gochman, N.: Continuous-flow enzymic method evaluated for measurement of serum triglycerides with use of an improved lipase reagent, Clin. Chem. **24**(11):2018-2019, 1978.
28. Lindblad, B.S., Settergren, G., Feychting, H., and Persson, B.: Total parenteral nutrition in infants. Blood levels of glucose lactate, pyruvate, free fatty acids, glycerol, D-beta-hydroxybutyrate, triglycerides, free amino acids and insulin, Acta Paediatr. Scand. **66**(4):409-419, 1977.
29. Naito, H.K., Gatautis, V.J., and Popowniak, K.L.: Effect of mannitol (Osmitrol) intoxication on serum "triglyceride" values, Clin. Chem. **22**(6):935-936, 1976 (Letter).
30. Witter, R.F., and Whitner, V.S.: Determination of serum triglycerides. In Nelson, G.J., editor: Blood lipids and lipoproteins: quantitation, composition, and metabolism, New York, 1972, John Wiley & Sons, Inc.–Interscience.

Chapter 59 Nonprotein nitrogenous compounds

Ammonia

NANCY GAU

Ammonia
Clinical significance: pp. 403 and 420
Molecular formula: NH_3
Molecular weight: 17.03 daltons
Merck Index: 506
Chemical class: amine

PRINCIPLES OF ANALYSIS AND CURRENT USAGE

Current methods of analysis for blood ammonia can be generally classified into four groups: (1) diffusion, (2) ion exchange, (3) ammonia-selective electrode, and (4) enzymatic.

Analysis of blood ammonia has traditionally been based on the principle of stoichiometric release of free ammonia base (NH_3) from the specimen by the addition of alkali. Free ammonia diffuses out of the sample into the atmosphere of a closed chamber and is captured by an acid solution. The trapped ammonia can be quantitated by titration with alkali (reaction 1a, Table 59-1)[1] or by nesslerization (reaction 1b, Table 59-1).[2]

In ion-exchange methods, ammonia is isolated after adsorption onto a strongly cationic ion-exchange resin and removal of plasma proteins.[3] Ammonia is then measured colorimetrically with the Berthelot reaction (reaction 2, Table 59-1).[4] These methods are generally more accurate, but a lengthy pretreatment is required, which seems to affect reproducibility.

Ammonia analysis can be performed by use of an ammonia-selective electrode such as that produced by Orion Research (Cambridge, Mass.). Ammonia is liberated from the sample by alkalinization and allowed to diffuse through a hydrophobic gas-permeable membrane into a solution of aqueous ammonium chloride until the partial pressure is the same on both sides of the membrane.[5] Hydrolysis of a small portion of the dissolved ammonia with a glass electrode, using silver–silver chloride as a reference electrode, results in a pH increase, which is then measured. This method is very specific and rapid because of the nature of the electrode, but the stability of the electrode and membrane remains questionable.

Measurement of ammonia by enzymatic reaction has the advantage of increased chemical specificity and greatly decreased time of analysis.[6] The method is based on reaction 4 in Table 59-1.[68] The decrease in absorbance at 340 nm resulting from the consumption of NADPH is measured spectrophotometrically.

REFERENCE AND PREFERRED METHODS

The inherent problems attributed to methods of collection and the limitations of older methods of analysis make it difficult to determine which method most closely measures "true" plasma ammonia.

Diffusion methods, although accurate when tested against standards, are not considered accurate or precise because of the formation of additional ammonia from endogenous compounds in plasma after prolonged periods of alkalinization. Although the nesslerization reaction is more sensitive, the overall accuracy is not greater than that of the titration method. For these reasons, diffusion methods are rarely used.

The cation-exchange method is most often used as the reference assay procedure for ammonia methods.[3] When the microenzymatic method was compared with a modified cation-exchange method, the two showed good correlation. The precision of the modified microenzymatic method (coefficient of variation [CV], 7% to 8%) was slightly better than that of the modified cation exchange method (CV, 8% to 13%) in the normal range.[9]

Similarly, comparison of the ion-selective electrode with a resin method showed excellent correlation in both the normal and abnormal ranges. The results of duplicate analysis by ion-selective electrode show CV's of 4.8% in the "normal" and 3.5% in the "abnormal" range and excellent recovery (99.3%).[5]

Table 59-1. Methods of ammonia analysis

Method	Type of analysis	Principle	Comment	Usage
1. Diffusion				
a. Conway	Quantitative, EP*	Release of NH_3 by alkalinization; titration of released ammonia with acid	Generally poor accuracy, precision, and sensitivity	Historical
b. Seligson	Quantitative, EP	Colorimetric analysis by Nessler's reagent: $2NH_3 + 2HgI_2{\cdot}2KI \rightarrow NH_2Hg_2I_3 + 4KI + NH_4I$ Yellow-orange (450 nm)	Generally poor accuracy, precision, and sensitivity	Historical
2. Ion-exchange	Quantitative, EP	Adsorption of ammonia followed by colorimetric analysis by Berthelot reaction: $NH_3 + NaOCl \rightarrow H_2NCl$ (chloramine) $+ NaOH$; $[Fe(CN)_5H_2O]^{-3} + H_2NCl \rightarrow$ 'Complex'; 'Complex' + Phenol $\xrightarrow{OH}$ Colored indophenol blue (560 nm)	Good sensitivity and accuracy	Manual
3. Ion-selective electrode	Quantitative, EP	Reaction of NH_3 at electrode surface; pH change measured potentiometrically	Good precision and accuracy	Automated
4. Enzymatic	Quantitative EP or K	α-Ketoglutarate $+ NH_4^+ + NADPH \xrightleftharpoons{GLDH}$ Glutamate $+ NADP^+ + H_2O$	Good accuracy and rapid	Manual or semi-automated

**EP*, End-point procedure; *GLDH*, glutamic dehydrogenase; *K*, kinetic.

Most enzymatic methods exhibited excellent overall performance with precision better than 11% and recovery of 95% to 105%.[7-10] Performance of the enzymatic method can be enhanced by the use of adenosine diphosphate to stabilize the glutamate dehydrogenase (GLDH) and increase the rate of reaction. In addition, the use of NADPH as the cofactor has been shown to reduce the incubation time of the reaction.[11] This technique has the advantage of requiring no prior isolation of ammonia and reduces positive biases by avoidance of any alkaline hydrolysis step. The enzymatic methods have been shown to be both precise and accurate, as well as having the additional advantages of small sample volume and adaptability to automated equipment.

Based on accuracy and precision alone, the ion-selective assay would probably be the method of choice. However, the lack of durability and stability of these electrodes has resulted in limited laboratory use. When factors such as cost and ease of automation are considered, the enzymatic method is recommended.

The enzymatic method (see box on p. 1232) involves inexpensive, readily prepared reagents, manual pipetting, and analysis on a spectrophotometer with temperature-controlled cuvette. The method can easily be adapted to other automated spectrophotometric equipment, such as ABA-100 (Abbott Laboratories, Chicago) or CentrifiChem 400 (Baker Instruments, Pleasantville, N.Y.).

SPECIMEN

It has been shown that the accuracy of ammonia determination is highly dependent on sample collection. Blood should be collected with either potassium oxalate, EDTA, or heparin anticoagulants. Obviously ammonium heparin is an unacceptable anticoagulant. Whichever anticoagulant is used, minimum exposure to air is important. Evacuated phlebotomy tubes are recommended for sample collection, and minimum stasis during phlebotomy is necessary. Phlebotomy tubes should be *completely* filled and immediately placed on ice, since the rate of increase of blood ammonia has been shown to be about 0.017 μg/mL of blood/minute at 25° C.[12] If heparin is the anticoagulant selected, each lot of tubes must be corrected for endogenous ammonia.[13] One can correct for small amounts of endogenous ammonia by adding an equivalent amount of heparin to the blank and standard.

After collection, the tube should be centrifuged in the cold and the plasma completely removed from cells and then stored covered at 2° to 4° C until analysis. If analysis cannot be performed within 3 hours, the sample is stable for 24 hours at −18° C.

The presence of noticeable hemolysis is the basis for rejection of a sample for analysis because erythrocytes have ammonia concentration 2.8 times higher than that of plasma.[13]

PROCEDURE

Principle

An enzymatic determination of ammonia is based on the following reaction:

$$\alpha\text{-Ketoglutarate} + NH_4^+ + NADPH \xrightleftharpoons{GLDH} NADP^+ + \text{Glutamate} + H_2O$$

One follows the initial reaction rate by recording the rate of consumption of $NADPH^+$ at 340 nm.

Reaction conditions for ammonia analysis

Condition	Enzymatic procedure
Temperature	37° C
pH	8.0 ± 0.05
Final concentration of reagent	NaK phosphate buffer: 54 mmol/L α-Ketoglutarate: 10 mmol/L NADPH: 120 μmol/L ADP: 0.5 mmol/L Glutamate dehydrogenase: 16,000 U/L
Fraction of sample volume	0.19
Sample	Plasma, urine
Linearity	0-150 μmol/L
Time of reaction	Two-point kinetic, 10 to 70 seconds
Major interferences	Some drugs (such as levodopa and cephalothin) inhibit GLDH activity
Precision* (between day)	Normal range ($\overline{X}$ = 34.8 mmol/L), CV = 15.2%; abnormal range ($\overline{X}$ = 170 mmol/L), CV = 4.8%

*For CentrifiChem procedure see following text.

Reagents

All reagents are prepared with ammonia-free water.

Ammonia-free water. Add 5 g of Dowex 50 (Bio-Rad Laboratories, Richmond, Calif.) or other designated cation-exchange resin to 1 L of distilled H_2O.

Stock KH_2PO_4, 0.066 mol/L. Dissolve 9.12 g of KH_2PO_4 in approximately 500 mL of distilled H_2O. Dilute to 1 L with distilled H_2O. It is stable 6 months at 4° C.

Stock Na_2HPO_4, 0.066 mol/L. Dissolve 9.51 g of Na_2HPO_4 in approximately 500 mL of distilled H_2O. Dilute to 1 L with distilled H_2O. Mixture is stable for 6 months at 4° C.

Working buffer, 0.066 mol/L. Combine 5 mL of stock KH_2PO_4 with 95 mL of stock Na_2HPO_4. Adjust pH to 8.0 ± 0.05. It is stable for 3 weeks at 4° C, but pH should be checked weekly.

α-Ketoglutarate (α-KG), 310 mol/L. Dissolve 0.453 g in 5 mL of ammonia-free water. Carefully adjust pH to 6.8 (± 0.01) with 3 M NaOH. Near the end point (approximately pH 5), switch to 0.1 M NaOH. *Do not overtitrate*. High pH destroys α-KG. Dilute to 10 mL with H_2O. It is stable for 10 days at 4° C.

NADPH (NADPH stock), 13 mmol/L. Weigh approximately 20 mg (stored in desiccator at −20° C) and dissolve in 2 mL of working phosphate buffer. Prepare a one hundredfold dilution by taking 50 μL of stock and diluting to 5 mL with working phosphate buffer. Read the absorbance at 340 nm versus the phosphate buffer. Calculate the *actual* concentration of NADPH in micromoles per milliliter (μmol/mL). NOTE: Millimolar absorptivity of NADPH equals 6.22, and A_{340} should be approximately 0.55.

$$\text{Actual concentration} = \frac{A_{340}}{6.22} \times 100$$

To determine which volume of NADPH *stock* solution is needed to prepare the working GLDH solution (containing 149 μmol/L of NADPH), use this calculation:

$$\text{Required volume of stock NADPH}^+ = \frac{14.9\ \mu\text{mol}}{\text{Actual stock concentration of NADPH}\ \mu\text{mol/mL}}$$

Working glutamate dehydrogenase (GLDH) solution. Weigh out 60 mg of GLDH. GLDH should be stored at −20° C in a desiccator. Weigh 30 mg of ADP. ADP should be stored at −20° C in a desiccator. Add the calculated amount of NADPH stock to a 100 mL volumetric flask along with the ADP and GLDH and dilute to mark with working phosphate buffer. The working solution is stable for 7 days at 4° C.

Standard curve. A standard curve should be obtained with each set of determinations.

Stock standard ammonium sulfate (100 mmol/L). Dissolve 660.7 mg of $(NH_4)SO_4$ in 100 mL of ammonia-free H_2O volumetrically. It is stable for 1 year at 4° C.

1000 μmol/L standard ammonium sulfate. Dilute 2 mL stock to 200 mL volumetrically with ammonia-free H_2O. It remains stable up to 6 months at 4° C.

For the standard curve, run 25, 50, 100, 150 μmol/L standards. Prepare weekly as follows:

μmol/L	Volume of standard	NH_4-free H_2O
25	0.25 ml of 1000 μmol/L standard	Dilute volumetrically to 10 mL
50	0.50 ml of 1000 μmol/L standard	Dilute volumetrically to 10 mL
100	1.0 ml of 1000 μmol/L standard	Dilute volumetrically to 10 mL
150	1.5 ml of 1000 μmol/L standard	Dilute volumetrically to 10 mL

Assay

Equipment: (1) 10 nm band pass spectrophotometer or filter photometer equipped with heated cuvette system; (2) timer; (3) water bath or heating block at 37° C.

1. Set spectrophotometer at 340 nm and cuvette at 37° C.
2. Transfer 1.5 mL of working GLDH reagent into test tube with appropriate label (blank, standard, sample, and so on).
3. Pipet 0.30 mL of water into test tube and mix. Add 0.06 mL of α-KG reagent and mix. Introduce mixture into spectrophotometer and blank the instrument at 100% transmission and absorbance (340 nm) equal to 0.000.
4. Repeat step 2 for each standard and sample. Add 0.30 mL of standard or sample to tube, mix, and incubate for 10 minutes in 37° C incubator.
5. Add 0.06 mL of α-KG, mix, and introduce into spectrophotometer.
6. Allow the reaction to proceed for 10 seconds and record the absorbance (t_0 = 10 seconds).
7. Allow reaction to proceed an additional 60 seconds and record the absorbance (t_f = 70 seconds).
8. Repeat steps 4 through 7 using patient sample, and so on.

Calculation

$$\Delta A = A_{70\ \text{sec}} - A_{10\ \text{sec}}$$

Prepare standard curve by plotting ΔA versus concentration of standards. Calculate ΔA for patient and determine concentration of sample from the curve.

Note

This method has been automated by adaption to a centrifugal analyzer (CentrifiChem, Baker Instruments, Pleasantville, N.Y.). In this procedure the reagents are manually pipetted into the transfer disk. Using the protocol described below, pipet both the GLDH and sample into the larger (center) well. Pipet the α-KG into the shallow (outer) well. Preincubate the transfer disk in the CentrifiChem at 37° C for 10 minutes and analyze as a two-point rate reaction: 340 nm, t_0 = 10 seconds, t_f = 70 seconds, absorbance mode. Plot standard curve and calculate patient values as above.

Manually load the CentrifiChem disk as follows:

Position	Contents	Volume	Well
0-29	α-KG solution	20 μL	Outer
0-29	GLDH solution*	500 μL	Center
2-9	Standards (in duplicate)	100 μL	Center
10-29	Sample (in duplicate)	100 μL	Center
0 (zero)	H_2O	100 μL	Center

*NOTE: Place 600 μl of H_2O in unused cuvettes instead of GLDH for proper balancing.

REFERENCE RANGE

The normal range is 18 to 72 μmol/L. Some studies indicate that ammonia levels in females may be 10% lower than in males.[14]

REFERENCES

1. Conway, E.J., and Cook, R.: Blood ammonia, Biochem. J. **33:**457-478, 1939.
2. Seligson, D., and Seligson H.: A microdiffusion method for the determination of nitrogen liberated as ammonia, J. Lab. Clin. Med. **38:**324-330, 1951.
3. Kingsley, G.R., and Tager, H.S.: Ion exchange method for the determination of plasma ammonia nitrogen with the Berthelot reaction, Stand. Methods Clin. Chem. **6:**115, 1970.
4. Berthelot, M.P.E.: Report, Chim. Appl. **1:**284, 282, 1859.
5. Proelss, H.F., and Wright, B.W.: Rapid determination of ammonia in a perchloric acid supernate from blood, by use of an ammonia-specific electrode, Clin. Chem. **19:**1162-1169, 1973.
6. Humphries, B.A., Melnychuk, M., Donegan, E.J., et al.: Automated enzymatic assay for plasma ammonia, Clin. Chem. **25:**26-30, 1979.
7. Pesh-Iman, M., Kumar, S., and Willis, C.E.: Enzymatic determination of plasma ammonia: evaluation of Sigma and BMC kits, Clin. Chem. **24:**2044-2046, 1978.
8. Bruce, A.W., Leiendecker, C.M., and Freier, E.F.: Two-point determination of plasma ammonia with the centrifugal analyzer, Clin. Chem. **24:**782-787, 1978.
9. Wu, J., Ash, K.O., and Mao, E.: Modified micro-scale enzymatic method for plasma ammonia in newborn and pediatric patients, comparison with a modified cation-exchange procedure, Clin. Chem. **24:**2172-2175, 1978.
10. Li, P.K., and Shull, B.C.: Fixed-time kinetic assay of plasma ammonia, with NaDPH as a cofactor, with a centrifugal analyzer, Clin. Chem. **25:**611-613, 1979.
11. van Anken, H.C., and Schiphorst, M.E.: A kinetic determination of ammonia in plasma, Clin. Chem. Acta **56:**151, 1974.
12. DiGiorgio, J.: Nonprotein nitrogenous constituents. In Henry, R.J., Cannon, D.C., and Winkelman, J., editors: Clinical chemistry: principles and technics, ed. 2, New York, 1974, Harper & Row, Publishers, Inc., pp. 518-526.
13. Seligson, D., and Hirahara, K.: The measurement of ammonia in whole blood, erythrocytes, and plasma, J. Lab. Clin. Med. **49:**962, 1957.
14. Da Fonseca-Wollheim, F.: Direkte Plasmaammoniakbestimmung ohne Enteiweissung. Verbesserter enzymatischen Ammoniaktest, 2, Report, Z. Klin. Chem. Klin. Biochem. **11:**426, 1973.

Bilirubin

JOHN E. SHERWIN
RONALD OVERNOLTE

Clinical significance: pp. 420 and 646
Molecular formula: $C_{33}H_{36}N_4O_6$
Molecular weight: 584.65 daltons
Merck Index: 1236
Chemical class: tetrapyrrole, bile pigment
Isomeric forms: Unconjugated bilirubin is present in several isomeric forms and in serum is circulated as monoglucuronic and diglucuronic acid derivatives

Me, $-CH_3$
R, $-CH{=}CH_2$
P, $=CH_2-CH_2-COOH$

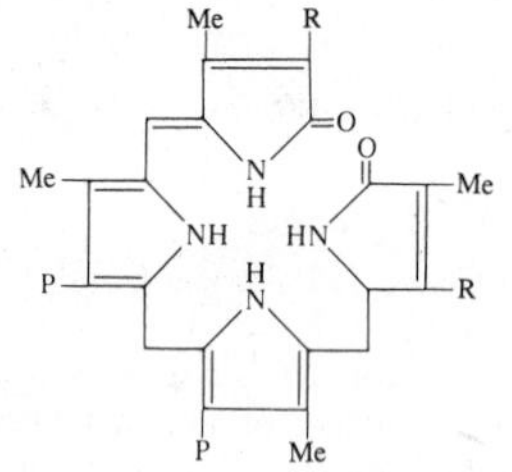

PRINCIPLES OF ANALYSIS AND CURRENT USAGE

Bilirubin was first demonstrated to be present in normal serum by van den Bergh and Snapper.[1] They found that bilirubin in normal serum reacted with Ehrlich's diazo reagent (diazotized sulfanilic acid) only when alcohol was added. It was later observed that pigment in human bile reacted with the diazo reagent without the addition of alcohol.[2] The pigment that reacted in the absence of alcohol was termed "direct." The pigment that required the presence of alcohol was termed the "indirect" bilirubin fraction. The response of serum to van den Bergh's test, with or without the presence of alcohol, has been the basis for several classifications of jaundice.

It is now clear that indirect bilirubin is unconjugated bilirubin bound to albumin en route to the liver from the reticuloendothelial system, where it is formed. The unconjugated bilirubin is a nonpolar molecule and is not soluble in water. Consequently it will react with diazo reagent only in the presence of an agent (historically called an *accelerator*), such as alcohol, in which both it and the diazo reagent are soluble. (The alcohol also enhances the intensity of the formed color.) The nonpolar nature of the unconjugated bilirubin and its strong affinity to albumin are also the bases for the absence of unconjugated bilirubin in urine in more than trace amounts. Unconjugated bilirubin is so tightly bound to albumin that it cannot be filtered at the glomerulus, and there appears to be no known tubular excretion of bilirubin. Therefore unconjugated bilirubin is not normally found in the urine.

In contrast, bilirubin conjugated with glucuronate is a polar and water-soluble compound that exists in plasma in large part not bound to any protein. Therefore conjugated bilirubin reacts *directly* with the diazo reagents to form azobilirubin. It appears in the urine when blood levels of this compound are increased because it is filtered at the glomerulus and not reabsorbed.

Thus the major differentiation of conjugated and unconjugated bilirubin fractions have historically been based on the time of the reaction and solubility of the fractions. Bilirubin that reacts quickly in the absence of solvent is called "direct" or "conjugated bilirubin"; in the presence of a solvent both forms of bilirubin readily react. However, it has been known for some time that a portion of the unconjugated bilirubin is available for reaction in the absence of solvent, and the extent of reaction of unconjugated bilirubin in the absence of solvent is time and temperature dependent as well as dependent on the final concentration of reagents.[3] Decreasing the time of reaction to minimize reactivity of the unconjugated bilirubin in the absence of solvent can increase the specificity of the conjugated (direct) bilirubin assay. The reaction of unconjugated bilirubin in the total reaction is also highly dependent on the final concentration of solvent.

Qualitative analysis of serum bilirubin as indirect or direct according to the type of van den Bergh reaction has long been replaced by quantitative determinations of the amount of conjugated bilirubin and total bilirubin, in which the difference between the total and direct reaction is assumed to represent the indirect or unconjugated bilirubin.

Commonly used procedures for measurement of bilirubin and its fractions are modifications of the method of Malloy and Evelyn.[4,5] All these methods employ some variation of the reaction of bilirubin with diazotized sulfanilic acid to form a colored chromophore (Fig. 59-1). The diazotized sulfanilic acid reacts at the central methylene carbon of bilirubin to split the molecule, forming two molecules of azobilirubin. Modifications of the Malloy-Evelyn procedure primarily differ in the pH at which the reaction is carried out and the reagent used to solubilize the unconjugated (indirect) bilirubin fraction. The Malloy-Evelyn method is typically performed at pH 1.2, at which the azobilirubin is red-purple with an absorption maximum of approximately 560 nm (method 1, Table 59-2). This method has been adapted for microanalysis by the Meites-Hogg technique.[6] The Malloy-Evelyn method most typically uses methanol as a solubilizer of the unconjugated fraction.

The Jendrassik-Grof modification is carried out at a pH near 6.5, but the absorbance of the reaction is measured after alkalization of the reaction solution to pH 13.[7,8] At this pH the absorption spectrum of the azobilirubin is shifted to a more intense blue measured at 600 nm (method 2, Table 59-2). The original Jendrassik-Grof method employed sodium benzoate–caffeine to solubilize unconjugated bilirubin for the measurement of total bilirubin.

Modifications of these two methods vary primarily in

Table 59-2. Methods of bilirubin analysis

Method	Type of analysis	Principle	Usage	Comments
1. Malloy-Evelyn	Kinetic, end point, with or without blank	Reaction shown in Fig. 59-1 performed at pH 1.2; azobilirubin measured at 560 nm	Very frequently used, especially as automated procedure	Susceptible to significant hemoglobin interference
2. Jendrassik-Grof	Kinetic, end point, with or without blank	Reaction shown in Fig. 59-1 performed near neutral pH but chromophore measured at alkaline pH (approximately 13) at 600 nm	Most commonly used method	Has higher molar absorptivity and thus is more sensitive and precise at low bilirubin concentrations than Malloy-Evelyn method
3. Bilirubinometer	Direct spectrophotometric	Bilirubin concentration directly determined by its absorbance at 454 nm; HbO_2 interference corrected by measurement of absorbance at a second wavelength (540 nm)	Not frequently used; primarily for neonatal analysis	Very simple to perform but very strong interference from carotinoids
4. High-performance liquid chromatography	Chromatographic separation	Methyl esters of conjugated and unconjugated bilirubin esters are detected at 430 nm	Research use only	May become future reference method

the solvent used to solubilize the unconjugated bilirubin. Solvents used for the measurement of both the unconjugated and conjugated bilirubin fractions include antipyrine-methanol, urea, and dimethyl sulfoxide (DMSO).[3]

Both the Malloy-Evelyn and Jendrassik-Grof procedures have been successfully automated and are currently the most commonly used methods for bilirubin analysis. Jendrassik-Grof–based procedures are in more frequent use, but the Malloy-Evelyn technique is frequently used as an automated method.

Some laboratories measure total bilirubin by a direct spectrophotometric technique (method 3, Table 59-2).[9] Because the absorption maximum of bilirubin at 454 nm is shared by the absorption of oxyhemoglobin (HbO_2), correction for the contribution of the HbO_2 is accomplished by subtraction of the absorbance at 540 nm, since HbO_2 absorbs equally at each of these wavelengths.[10] Thus either a spectrophotometer or biochromatic photometer capable of measurements at 454 and 540 nm is required. This method does not, however, correct for the presence of carotinoids, which also absorb in the 454 nm region of the spectrum and can cause falsely elevated results. Consequently direct spectrophotometric methods are rarely used for measurement of bilirubin in adults. However, neonates up to 3 months of age do not have sufficient intake of carotene to interfere with this direct spectrophotometric assay. Its rapidity and small sample volume make the direct spectrophotometric assay appealing for use in an immediate response laboratory. Many of these laboratories use a diazo method to measure the concentration of conjugated bilirubin in specimens. Direct spectroanalysis is also employed for the estimation of bilirubin in amniotic fluid as an indicator of a hemolytic disease process in fetuses.

Recently a high-performance liquid chromatography (HPLC) procedure (method 4, Table 59-2) that can separate the methyl esters of unconjugated and conjugated bilirubin isomers has been successfully developed.[11] This technique uses normal-phase (silica) chromatography, and the methylated derivatives are detected when the column eluate is monitored at 430 nm. However, this technique is used only as a research tool at this time.

Other research techniques for measuring bilirubin include derivative spectroscopy,[12] fluorometry,[13] and skin reflectance photometry.[14]

Urine bilirubin measurements are most often made by ''dipsticks'' impregnated with direct-reacting diazo reagent, such as 2,4-dichloroaniline diazonium salts (Ames Co., Elkhart, Ind.). The Ictotest (Ames Co.) uses *p*-nitrobenzene diazonium *p*-toluenesulfonate as the active reagent. With a sensitivity of about 10 mg/L it is two to four times more sensitive than the dipstick procedure. Both methods are highly subject to interferences from endogenous color.

REFERENCE AND PREFERRED METHODS
(Table 59-3)

Because of the limitations in the specificity of the direct and total bilirubin methods, the true concentrations and proportions of conjugated and unconjugated bilirubin in serum have not been known with any great certitude. The recent development of an HPLC procedure[8] which can physically separate and quantitate each fraction has led to a better understanding of the composition of bilirubin in normal blood. The HPLC results suggest that little (less than 5%) conjugated bilirubin actually exists in plasma of healthy persons, and most diazo methods greatly overestimate the concentration of this fraction. In addition, the HPLC analysis indicates the presence in serum of nonbilirubin compounds that are also diazo reacting. Certain compounds present in urine, such as mesobilifuscin and uroerythrin, are known to interfere with urinary bilirubin analysis, and it is presumed that these or similar compounds exist in plasma.

These studies suggest that chemical assays that give a low proportion (at 15% or less) of direct-reacting bilirubin in healthy persons probably give the most accurate results. For a particular diazo method, the amount of direct-reacting bilirubin apparent in normal serum depends on the time and temperature of the reaction.

The primary advantages of the Jendrassik-Grof procedures include the greater sensitivity and precision (especially at low bilirubin concentrations) as a result of the higher molar absorptivity of the alkaline azobilirubin (approximately 7×10^4 L · mol^{-1} · cm^{-1}), relative insensitivity to hemoglobin interference, rapid color development, and a large linear range (up to 1.7 absorbance units). Jendrassik-Grof procedures often include ascorbic acid to further minimize hemoglobin interference, though many Jendrassik-Grof procedures use a sample blank to minimize the effect of endogenous color.

The Malloy-Evelyn method has been readily automated and is used for rapid microanalysis. The interference in this method by hemoglobin, especially at concentrations greater than 1 mg/L, necessitates the use of a serum blank for this method. The Malloy-Evelyn procedure in general tends to give lower results than the Jendrassik-Grof method. However, the discrepancy between the methods has narrowed with the more widespread use of proper calibrators. The precision of both the Jendrassik-Grof and Malloy-Evelyn procedures is similar, tending to have a coefficient of variation (CV) of 14% to 20% at bilirubin levels near normal ranges (approximately 10 mg/L) and 5% to 10% CV at total bilirubin concentrations of 30 to 40 mg/L.

The use of bilirubinometers to measure total bilirubin is useful only in the neonatal population because of the presence of carotinoid-like compounds in adult serum that cause strong positive interference. These instruments are simple to use, however, and have good precision if maintained properly.

There is no acceptable definitive or reference method for bilirubin, though the HPLC procedure may be accepted in the future as a reference procedure. At this time there seems to be no clear-cut choice of either the Jendrassik-Grof or the Malloy-Evelyn procedure. Because both have acceptable precision and are adapted to many automated instruments, the choice will depend on the available instrument or individual laboratory preference. We hope that future work can determine the true specificity of these procedures to allow a clearer choice to be made.

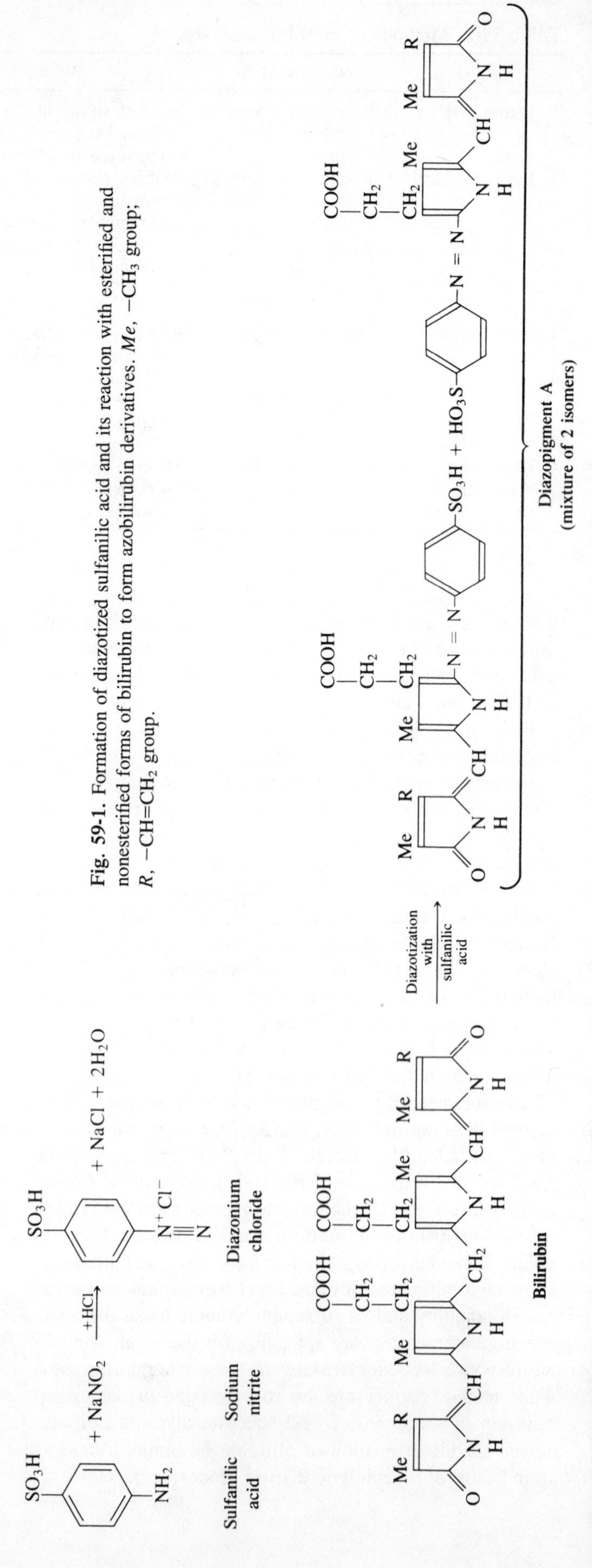

Fig. 59-1. Formation of diazotized sulfanilic acid and its reaction with esterified and nonesterified forms of bilirubin to form azobilirubin derivatives. *Me*, $-CH_3$ group; *R*, $-CH{=}CH_2$ group.

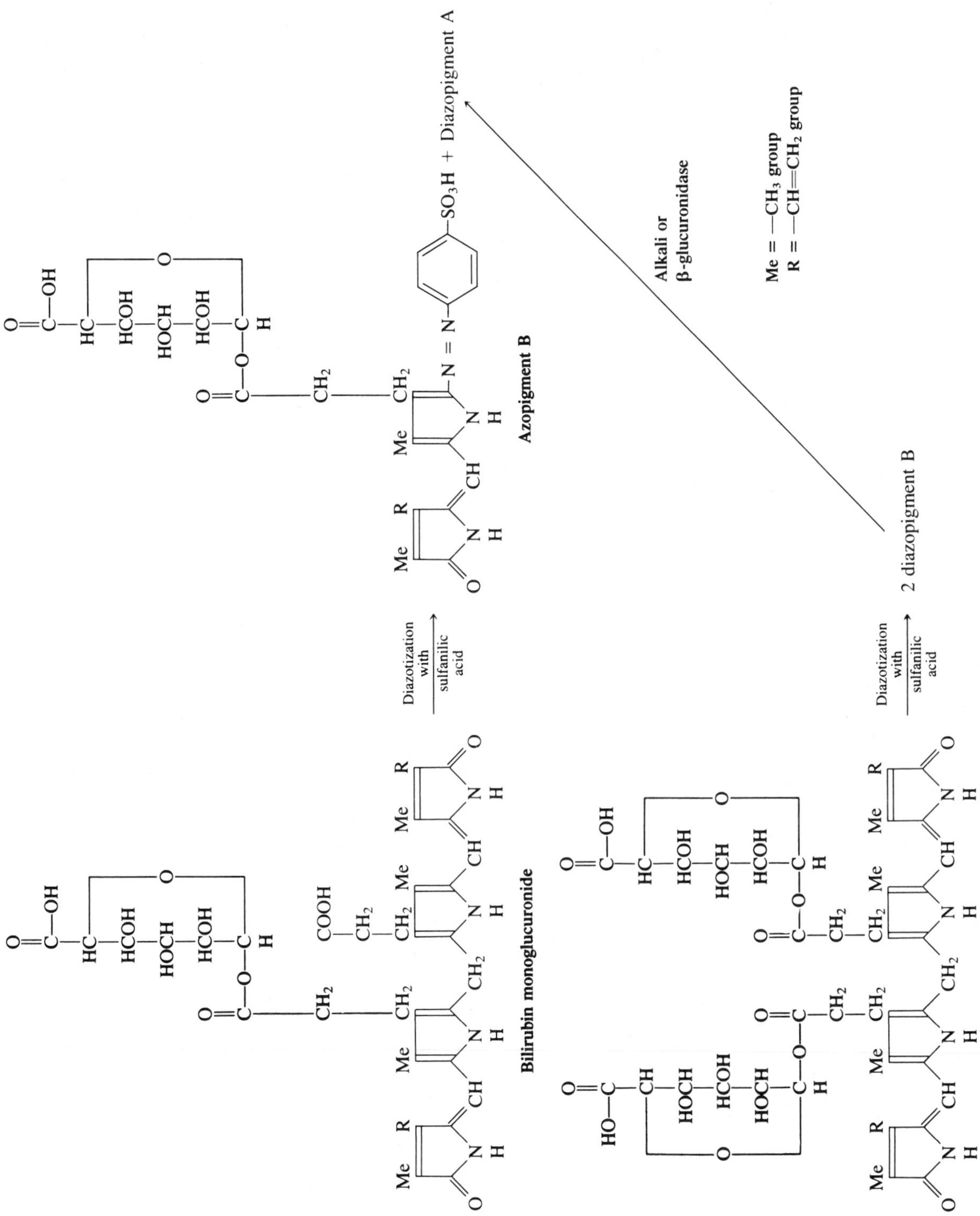

Bilirubin monoglucuronide

Diazotization with sulfanilic acid →

Azopigment B

$-SO_3H$ + Diazopigment A

Alkali or β-glucuronidase

Bilirubin diglucuronide

Diazotization with sulfanilic acid →

2 diazopigment B

Me = $-CH_3$ group
R = $-CH{=}CH_2$ group

Table 59-3. Comparison of reaction conditions for analysis of total and direct bilirubin in serum

Condition	Malloy-Evelyn* (total and direct)		Jendrassik-Grof† (total)	
Temperature	30° or 37° C		30° or 37° C	
Sample volume	0.4 mL of 1:20 diluted sample; uses 100 μL of sample initially		100 μL	
Sample volume fraction	0.4 in reaction mixture, but 0.02 in final dilution		0.0625	
Final concentration of reagents	**Total**	**Direct**		
	Methanol: 50% (v/v) $NaNO_3$: 0.7 mmol/L Sulfanilic acid: 2.5 mmol/L HCl: 0.07 mol/L	$NaNO_3$: 0.7 mmol/L Sulfanilic acid: 2.5 mmol/L HCl: 0.07 mol/L	Sodium acetate: 0.427 mol/L Sodium benzoate: 0.243 mol/L EDTA·2Na: 1.7 mmol/L Caffeine: 0.122 mol/L $NaNO_3$: 0.45 mmol/L Sulfanilic acid: 8.0 mmol/L	In reaction mix
			NaOH: 0.72 mol/L (pH 13) Sodium tartrate: 0.44 mol/L	In final colored solution
Time of reaction	10 minutes	1 minute	10 minutes	
Wavelength	560 nm		600 nm	
Linearity	to 300 mg/L		to 300 mg/L	
Precision‡ (between-day, $\bar{X}$, % CV)	7.2 mg/L, 15.3% 28 mg/L, 6.9% 93 mg/L, 4.9%		15 mg/L, 5.5% 47 mg/L, 3.2% 92 mg/L, 2.2%	
Interference	Hemoglobin		No interference from hemoglobin at concentrations as high as 3320 mg of Hb/L	

*Method as described by Meites et al.[5]
†Method as described Koch et al.[8]
‡Taken from College of American Pathologists (CAP) Quality Assurance Survey data.

Perhaps of greater importance in choosing a method is the problem of choosing a proper calibrator. Many differences and apparent inaccuracies between methods are a result of differences in standardization. Recommendations for a uniform bilirubin standard are available,[15] as are serum-based calibrated bilirubin standards. However, concerned laboratories should check the accuracy of serum-based calibrators by using pure bilirubin standards available from the National Bureau of Standards (Standard Reference Material No. 916). Several methods are available for preparing pure bilirubin standards and are discussed later.

SPECIMEN

Total bilirubin determinations using a diazo method require either serum or plasma, but serum is preferred for the Malloy-Evelyn procedure, since the addition of alcohol in the analysis can precipitate proteins from plasma that interfere with subsequent analysis. The presence of hemolysis falsely depresses the bilirubin result in most assays because of an increase in absorbance of the blank; therefore samples should not have any visible hemolysis for analysis. Red cell contamination should be removed by centrifugation before analysis. Turbid serum creates artifacts in the performance of spectrophotometers and should not be directly analyzed. Bilirubin is sensitive to and destroyed by light and heat; therefore the analysis should be performed promptly in subdued light to avoid falsely low results.

Conjugated bilirubin may be determined by use of either serum or plasma, but serum is the usual sample because total bilirubin is determined concomitantly. Spinal fluid samples can also be analyzed for both total and direct bilirubin. Urine samples can also be analyzed by direct diazo methods, since the polar conjugated bilirubin is, in large part, not protein bound and is filtered at the glomerulus and excreted into urine.

PROCEDURE FOR TOTAL BILIRUBIN

Principle

This method is a variation of the Malloy-Evelyn procedure as described in *Selected Methods of Clinical Chemistry*.[5] The specimen is added to a solution of methanol and diazotized sulfanilic acid in an acid solution. The methanol accelerates the coupling of bilirubin with the diazotized sulfanilic acid. The purple azobilirubin is measured at 560 nm.

Reagents

Sulfanilic acid (26.2 mmol/L in about 0.72 mol/L HCl). Add 5.0 g of sulfanilic acid to 60 mL of concentrated HCl (approximately 12 mol/L); mix and dilute to 1 L with distilled H_2O. The solution is stable indefinitely at ambient temperature.

Sodium nitrite stock (2.35 mmol/L). Place 20 g of sodium nitrate ($NaNO_2$) in a 100 mL volumetric flask, dissolve in 80 mL of distilled H_2O, and dilute to 100 mL. Store at 4° to 8° C in a brown glass-stoppered bottle. Discard when the reagent becomes discolored with yellow nitrate after several months.

Working sodium nitrite solution (0.235 mmol/L). Dilute the stock sodium nitrite tenfold with distilled H_2O to prepare a 0.235 mol/L solution. This reagent is unstable and should be prepared daily.

Diazo blank. Dilute 60 mL of concentrated HCl to 1 L of distilled H_2O. This reagent can be kept at room temperature for at least 1 year.

Diazo reagent. Add 0.3 mL of working sodium nitrite solution to 10 mL of sulfanilic acid reagent and mix. This reagent is unstable and should be used within a few hours of preparation.

Methanol (absolute). Follow ACS (American Chemical Society) specifications for purity, that is, at least 99.5% pure. It must be stored in a glass container.

Serum-based calibrating standard. Bilirubin concentration is approximately 30 to 50 mg/L. Also serum-based controls, preferably one at elevated levels, are available commercially.

Assay

Equipment: spectrophotometer, with band pass ≥ 10 nm, capable of reading at 560 nm

1. Dilute serum samples, standards, and control specimens 1:20 by pipetting 1.9 mL of distilled H_2O into a 12 × 75 mm test tube and adding 100 μL of sample. Mix. This can be accurately done by use of an automatic pipettor (as by Micromedic Systems, Inc., Philadelphia, Pa.).
2. Add 0.5 mL of methanol to cuvettes labeled *T* (for total) and *B* (for blank) for each patient, standard, and control.
3. Add 100 μL of fresh diazo reagent to the cuvette labeled *T* and mix.
4. Add 100 μL of diazo blank reagent to the cuvette labeled *B* and mix.
5. Add 0.5 mL of diluted serum, standard or control, to cuvettes *T* and *B*. Mix thoroughly.
6. Allow reaction to proceed for 10 minutes in covered tubes at room temperature. Read the absorbance of cuvette T at 560 nm against water zero absorbance. Read *B* reaction versus water, and subtract this reading from that of the sample to obtain corrected absorbance.

Calculation

The blank-corrected absorbance of cuvette *T* is compared to the absorbance of the bilirubin calibrator standard to determine the concentration of total bilirubin in the sample. The calculation is as follows:

$$\frac{A_s}{C_s} = \frac{A_u}{C_u} \qquad C_u = \frac{A_u}{A_s} \times C_s$$

where A_s = absorbance of calibrator standard
A_u = absorbance of unknown
C_s = concentration of standard (mg/L)
C_u = concentration of unknown (mg/L)

Notes

1. It is critical to add reagents and diluted samples in the given order to avoid turbidity.
2. This procedure was originally established for pediatric populations. This method can be modified to replace the initial dilution step by an increase in the volumes as follows:
   a. Methanol: 1.0 mL
   b. Diazo reagent: 0.2 mL (or 0.2 mL of diazo blank)
   c. Water: 0.75 mL
   d. Serum: 0.04 mL (using SMI pipettor) Pipet in given order.
3. The linearity of this procedure is up to 300 mg/L. Color is stable for at least 2 hours within this range.
4. Hemolysis causes decreased bilirubin values.
5. To determine the concentration of the unconjugated bilirubin fraction, subtract the concentration of the conjugated bilirubin (see below) from this value for total bilirubin.

Table 59-4. Bilirubin concentrations in various conditions

Condition	Bilirubin concentration (mg/L)		Stool color	Urine bilirubin
	Total	Conjugated		
Normal	5-15	0-2	Brown	—
Prehepatic jaundice				
Infant	150-300	25-75	Brown	Trace
Adult	15-60	1-6	Brown	—
Hepatic jaundice	20-300	10-150	Light brown	2-4+
Posthepatic obstructive jaundice				
Partial	150-300	75-150	Light	4+
Complete	150-300	75-50	Light	4+

PROCEDURE FOR CONJUGATED BILIRUBIN

Principle

The conjugated bilirubin method is also from the *Selected Methods in Clinical Chemistry*[5] and is based on the Malloy-Evelyn procedure. The azobilirubin pigment produced by bilirubin glucuronides and diazotized sulfanilic acid is formed in 1 minute, as opposed to the 10 minutes required in the total bilirubin procedure. The methanol is not added, and the time has been shortened to minimize the reaction of unconjugated bilirubin. To prevent turbidity from occurring, it is critical that the order of addition of reagents and diluted specimen be followed exactly as listed.

Reagents

Same as for total bilirubin on p. 1238.

Assay

Equipment: See the total bilirubin assay on p. 1239.

1. Add 0.5 mL of distilled H_2O to a cuvette labeled *B* (for blank) and 0.5 mL H_2O to a cuvette labeled *C* (for conjugated).
2. Add 100 μL of diazo to each *C* cuvette and 100 μL of diazo blank reagent to each B cuvette.
3. Add 400 μL of diluted serum, standard, or control specimen (see total bilirubin procedure) to their respective *B* (blank) and *C* (conjugated) cuvettes. Mix well.
4. Exactly 1 minute after the addition of diazo reagent in step 3, read the absorbance of cuvettes *B* and *C* at 450 nm against H_2O set at zero absorbance. Subtract the blank absorbance for each sample from the test absorbance to obtain the net reaction absorbance. The addition of samples should be timed every 30 seconds so that one can read the absorbance of each reaction tube *(B* and *C)* within 60 seconds of sample addition.

Note

A commercially available conjugate of bilirubin (Ultimate D; SmithKline-Beckman, Brea, Calif.) can be used to standardize this reaction.

Calculation

The blank-corrected absorbance of cuvette *C* is compared to the absorbance of the standard with respect to concentration to determine the concentration of conjugated bilirubin in the sample (see p. 1239).

PREPARATION OF BILIRUBIN STANDARD

Pure bilirubin can be obtained from several commercial sources (such as Sigma Co., St. Louis, Mo.). However, the best source is the National Bureau of Standards (Standard Reference Material No. 916). It should be used, as described below, to calibrate a serum that is to be used as the reference or standard in the total bilirubin procedure. Thus the commercially available reference pools or laboratory-prepared pools can be checked for the accuracy of the stated bilirubin values. If the procedure is carefully carried out, use the value obtained by the validation method rather than the manufacturer's stated value if there is a substantial difference between the two.

The following procedure must be carried out in a well-darkened room. Weigh out approximately 22 to 24 mg of bilirubin in a tared 50 mL beaker. Add 4 mL of pH 11.3 NaOH (to prepare, dilute 1 mL of 5 N NaOH to 100 mL with 1/15 M pH 7.4 Sørensen buffer*); stir gently with a glass rod but do not form foam or bubbles. The bilirubin will dissolve completely in 2 to 4 minutes to form a clear red-orange solution. Add 40 mL of 40 g/L bovine serum albumin solution dissolved in 1/15 M pH 7.4 Sørensen buffer* and stir. (Albumin is available from a number of commercial suppliers.) An alternative to the 40 g/L albumin solution is to use Plasmonate (Cutter Laboratories, Berkeley, Calif.). Completely transfer the solution to a 100 mL volumetric flask with the aid of additional albumin–7.4 Sørensen buffer and bring to volume with this solution. Prepare a 1:10 dilution of albumin only in the pH 7.4 buffer as an optical reference.

The bilirubin-albumin solution is not stable and must be used within 1 to 2 hours of preparation. Prepare the following standards from the stock solution.

Example: When 24 mg of bilirubin is weighed out (240 mg/L stock):

Concentration (mg/L)	mL of stock	mL of albumin–pH 7.4 Sørensen buffer	
30	1	7	
60	1	3	Use volumetric
120	5	5	pipets
160	2	1	throughout
240	Use as is	—	

When the amount of bilirubin weighed out is less than 240 mg, the concentration of the dilutions must be calculated accordingly.

The standards and the reference solution being calibrated are treated as specimens in the *total* procedure. Analyze all standards and samples in duplicate. The albumin blank is also carried through the procedure and should have 0 mg/L bilirubin. Avoid albumin that is very dark. Calculate the bilirubin concentration of the reference sample from a standard curve drawn from the absorbance values of the standards (A_{560} versus concentration).

An alternate procedure for the preparation of bilirubin standards in a protein matrix is to dissolve approximately 50 mg of pure bilirubin in 25 mL of dimethyl sulfoxide (DMSO) volumetrically. This will give a dark, red-brown stock solution of 2000 mg/L. Dilute this to form a stock serum base standard of 200 mg/L by slowly adding *dropwise* 1 mL of the stock DMSO solution to approximately

*See Appendixes A and B on buffers, pp. 1407 to 1411.

8 mL of Plasmonate solution while swirling the Plasmonate solution. Bring the volume up to exactly 10 mL in a volumetric flask with Plasmonate. The final solution should be clear with no visible precipitate. This protein-based bilirubin standard can be diluted additionally to give standards with lower concentrations.

The Subcommittee on Bilirubin of the American Association for Clinical Chemistry Standards Committee defined acceptable bilirubin as having a molar absorptivity of 60,700 ± 1600 (mean ± 2 SD) at 453 nm in chloroform. If the purity of the bilirubin is in doubt (that is, not from the National Bureau of Standards), it can be checked against the preceding figure in a dual-beam spectrophotometer in 1 cm cells. The cells *must* be capped because of the volatility of chloroform. For this evaluation weigh out approximately 50 mg of bilirubin. The amount does not have to be exactly 50 mg, as long as the actual amount is known to within 2% accuracy. Transfer bilirubin to a 100 mL volumetric flask and dissolve with chloroform to the mark. Mix well. Prepare a 1:100 dilution by volumetrically transferring 1 mL of this solution to a second 100 mL volumetric flask. Dilute to 100 mL mark with chloroform and thoroughly mix. Measure absorbance of diluted bilirubin versus chloroform at 453 nm. The concentration in milligrams per liter can be calculated (from Beer's law) as follows:

By absorbance

$$A = \epsilon c b \qquad \epsilon = 60.7 \text{ L} \cdot \text{cm}^{-1} \cdot \text{mmol}^{-1}$$
$$b = 1 \text{ cm}$$
$$c \text{ (mg/L)} = \frac{A(100)(584.65 \text{ mg/mmol})}{(1)60.7}$$

By weight

$$\frac{\text{mg weighed}}{100 \text{ mL}} = (\text{mg} \times 10) = \text{mg/L}$$

Example: 50.5 mg/100 mL = 505 mg/L

Use the factor of 100 to correct for 1:100 dilution and 584.65 to convert moles to milligrams.

Example: Absorbance = 0.520

$$c \text{ (mg/L)} = \frac{0.520(100)584}{(1)60.7} = 500.9 \text{ mg/L}$$

If the difference between the two calculations is greater than 5%, there is a strong possibility that the bilirubin is not pure. Repeat the purity check. If the discrepancy between the calculations is again excessive, consider an alternative source of bilirubin.

REFERENCE RANGE

The normal value is up to 15 mg/L (median, 7.0 mg/L) total bilirubin in adults. In both sexes the median concentration rises after puberty, falls during the third decade, and thereafter remains stable. After puberty males have both higher median values and a more pronounced skewness to high levels than females do.[16]

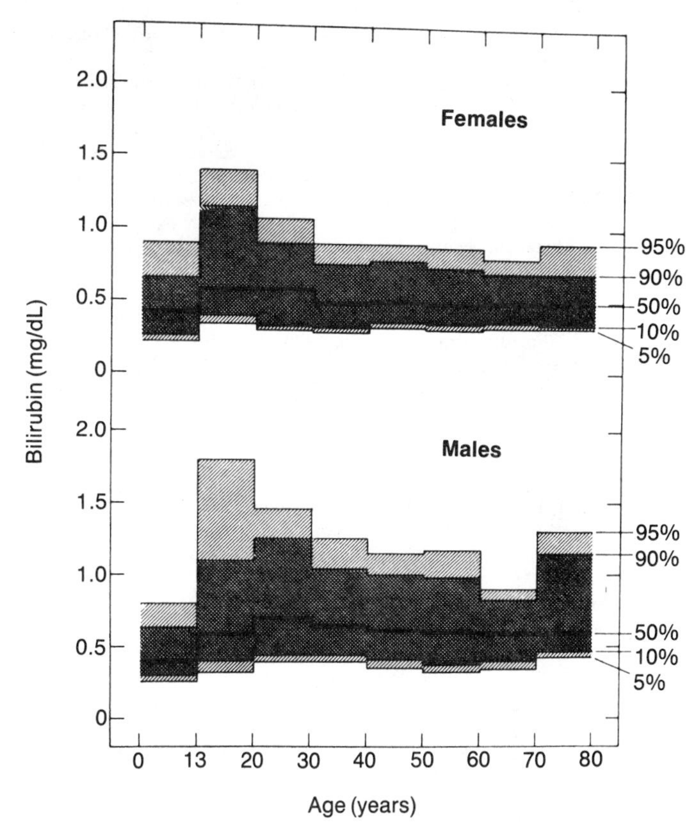

Fig. 59-2. Serum total bilirubin concentrations in percentiles. In both sexes, median concentration rises after puberty, falls during third decade, and thereafter remains stable. Notice that similar shifts are more pronounced at 90th and 95th percentile levels, and 5th and 10th percentile levels hardly change. After puberty, males have both higher median values and a more pronounced skewness at high values than females do. (From Werner, M., Tolls, R.E., Hultin, J.V., and Mellecker, J.: Z. Klin. Chem. **8**:105-115, 1970.)

Conjugated bilirubin levels up to 2 mg/L are found in infants by 1 month of age and remain at this level thereafter.

Fig. 59-2 summarizes serum bilirubin concentrations as a function of age and sex.[16] Elevations resulting from disease states are summarized in Table 59-4.

REFERENCES

1. van den Bergh, A.A.H., and Snapper, J.: Die Farbenstoffe des Blutserums, Dtsch. Arch. Klin. Med. **110**:540-541, 1913.
2. van den Bergh, A.A.H., and Müller, P.: Über eine direkte und eine indirekte Diazoreaktion auf Bilirubin, Biochem. Z. **77**:90-103, 1916.
3. Winkelman, J., Cannon, D.C., and Jacobs, S.L.: Bilirubin. In Henry, R.J., et al., editors: Clinical chemistry: principles and technics, ed. 2, New York, 1974, Harper & Row, Publishers, Inc., pp. 1042-1079.
4. Malloy, H.T., and Evelyn, K.A.: The determination of bilirubin with the photoelectric colorimeter, J. Biol. Chem. **119**:481-490, 1937.
5. Meites, S., Cheng, M.H., Arnold, L.H., et al.: Bilirubin, direct reacting and total, modified Malloy-Evelyn method. In Faulkner, W.R., and Meites, S., editors: Selected methods of clinical chemis-

try, vol. 9, Washington, D.C., 1982, American Association of Clinical Chemists Press, pp. 119-125.

6. Meites, S., and Hogg, C.K.: Studies on the use of the van den Bergh reagent for the determination of serum bilirubin, Clin. Chem. **5**:470-478, 1950.
7. Jendrassik, L., and Grof, P.: Vereinfachte photometrische Methoden zur Bestimmung des Blutbilirubin, Biochem. Z. **297**:81-89, 1938.
8. Koch, T.R., Doumas, D.T., Elser, R.C., et al.: Bilirubin, total and conjugated, modified Jendrassik-Grof method. In Faulkner, W.R., and Meites, S., editors: Selected methods of clinical chemistry, vol. 9, Washington, D.C., 1982, American Association for Clinical Chemistry Press, pp. 113-118.
9. Evans, R.T., and Holton, J.B.: An assessment of a bilirubinometer, Ann. Clin. Biochem. **7**:104-106, 1970.
10. Leukoff, A.H., Westphal, M.C., and Finkler, J.F.: Evaluation of a direct reacting spectrophotometer for neonatal bilirubinometry, Am. J. Clin. Pathol. **54**:562-565, 1970.
11. Blanckaert, N., Kabra, P.M., Fraina, F.A., et al.: Measurement of bilirubin and its monoconjugates and diconjugates in human serum by alkaline methanolysis and high performance liquid chromatography, J. Lab. Clin. Med. **96**:198-212, 1980.
12. Ash, K.O., Holmer, M., and Johnson, C.S.: Bilirubin-protein interactions monitored by difference spectroscopy, Clin. Chem. **24**:1491-1494, 1978.
13. Brown, A.K., Eisinger, J., Blumberg, W.E., et al.: A rapid fluorometric method for determining bilirubin levels and binding in the blood of neonates: comparisons with a diazo method and with 2-(4′-hydroxybenzene) azobenzoic acid dye binding, Pediatrics **65**:767-776, 1980.
14. Yamanouchi, I., Yamauchi, Y., and Igarashi, I.: Transcutaneous bilirubinometry: preliminary studies of noninvasive transcutaneous bilirubin meter in the Okayama National Hospital, Pediatrics **65**:195-202, 1980.
15. Recommendations on a uniform bilirubin standard, Clin. Chem. **8**:405-407, 1962.
16. Werner, M., Toos, R.E., Hultin, J.V., and Mellecker, J.: Influence of sex and age on the normal range of eleven serum constituents, Z. Clin. Chem. **8**:105-115, 1970.

Bilirubin amniotic fluid

PAUL T. RUSSELL

Clinical significance: p. 686
Molecular formula: $C_{33}H_{36}N_4O_6$
Molecular weight: 584.65 daltons
Merck Index: 1236
Chemical class: tetrapyrrole, bile pigment

PRINCIPLES OF ANALYSIS AND CURRENT USAGE

Unconjugated bilirubin is an important constituent of amniotic fluid in the evaluation of Rh-immunized patients.[1] Most laboratories estimate the amount of unconjugated bilirubin indirectly by quantitating the 450 nm absorption peak of the spectrophotometric curve plotted on a semilogarithmic scale as described by Liley.[2] This approach, preceded by the studies of Bevis[3] and Walker,[4] successfully related the magnitude of the peak at 450 nm, the amount of bilirubin in the amniotic fluid, with the severity of Rh-isoimmunization disease in the fetus. From these pioneering studies the magnitude of the optical density difference at 450 nm (ΔOD_{450}) became the basis for prognostications[2,5-8] of delivery of an affected fetus.

Because bilirubin-free amniotic fluid has a high absorbance at 350 nm, which declines steadily toward the longer wavelengths, Liley[2] established an arbitrary line that represents the base line if no bilirubin is present in the sample, connecting points on the curve at 365 and 550 nm. The difference at 450 nm between the peak and this base line represents the absorption in optical density (or absorbance, expressed as ΔOD at 450 nm), a value that is proportional to the bilirubin concentration. Because serial samples were taken, absolute quantitation was not necessary. It was soon realized, however, that it is not always easy to establish satisfactory base lines,[3,5] particularly if the fluid contains either blood or meconium.

In the earlier studies heavy contamination with blood rendered the fluid samples worthless for analysis because oxyhemoglobin peaks at 575, 540, and 412 nm obscured the essential bilirubin reading at 450 nm, and the increase in absorbance from 350 to 400 nm noticeably affected how the base line was established. Blood-contaminated samples gave falsely low ΔOD_{450} values for these reasons. Likewise, because meconium absorbs maximally from 350 to 400 nm and with decreasing absorption up to 550 nm, misleading bilirubin values were associated with the presence of this material.

Modifications of the original spectrophotometric methods attempted to avoid false values resulting from contamination of fluid with blood or meconium. These methods and the direct chemical methods are comprehensibly discussed in recent monographs.[6-9] The most practical of the proposals suggested that contaminated samples first be extracted by chloroform before spectroscopic assay in order to separate bilirubin from other interfering pigments.[10-12] Bilirubin is readily lipid soluble and thus extracts efficiently into chloroform. Potential contaminants from meconium and blood are more water soluble and therefore remain partitioned with the water. Biliverdin, for example, a green oxidation product of bilirubin, is quite water soluble and is separated from bilirubin by chloroform extraction. The same is true of hemoglobin and its degradation products.

Simkins and Worth[13] compared the method of Liley,[2] a modified version of the method of Knox et al.,[14] and the method of Mallikarjuneswara et al.[11] for the estimation of bilirubin in amniotic fluid and for potential effects of various concentrations of hemoglobin, methemoglobin, and methemalbumin on the methods. They found that the methods were in good agreement unless hemoglobin degradation products were present. To circumvent this circumstance, chloroform extraction was necessary.

Many samples of amniotic fluid from affected pregnancies contain oxyhemoglobin as well as methemoglobin and methemalbumin. These pigments most often arise from lysis of erythrocytes that have entered the amniotic fluid from intrauterine hemorrhage associated with partial placental separations or from the amniocentesis itself ("bloody tap"). Intrauterine (fetal) transfusions introduce considerable blood contamination into the fluid, which can take nearly 2 weeks to clear. Oxyhemoglobin is converted first into methemoglobin and then into methemalbumin. The last two compounds may also enter the amniotic fluid directly when the fetus has died or in some cases of fetal distress or impending fetal death. The methemoglobin peak at 620 nm is considered a bad sign.

As mentioned previously, oxyhemoglobin has a large spectral peak at 412 nm (Soret band) that can overlap with absorption at 450 nm if oxyhemoglobin contamination is present. When the ferrous ion of hemoglobin is oxidized to the ferric state for methemoglobin, the absorption spectrum changes. Peaks for oxyhemoglobin at 540 and 575 nm disappear, and a peak appears at 630 nm. This last peak is shifted toward 620 nm when the globin of the compound is replaced by albumin. Alvey[15] found the extinction coefficients for oxyhemoglobin at 574 nm and at 454 nm to be identical. By subtracting the optical density at 574 nm from that at 454 nm, he obtained a "bilirubin index" used for prediction.

Biliverdin, the oxidation product of bilirubin, can also contaminate amniotic fluid and thereby contribute to its spectral character. In the fetus the small part of the bilirubin that is conjugated is excreted into the intestines, where it undergoes conversion into biliverdin. With excretion of meconium the amniotic fluid becomes stained by this pigment. The absorption spectrum of biliverdin is ill-defined, but the amount of biliverdin in amniotic fluid can be estimated by the difference in absorption at 480 and 500 nm.[6] When certain methods[14,16] are used, biliverdin can be mistaken for bilirubin. These methods cannot be used therefore, when even small amounts of meconium contaminate the amniotic fluid.

In amniotic fluid bilirubin forms a complex with the albumin fraction of the proteins. This complex has an absorption spectrum that is different from the spectrum of pure bilirubin. The absorption peak of the bilirubin-albumin complex begins at 520 nm and has a maximum at 460 nm and a minimum at 350 nm.[6] One can easily obtain the spectrum of this compound by measuring a pure solution to which some albumin has been added or by using a diluted bilirubin control serum. The albumin content of amniotic fluid is sufficient to bind 20 to 30 μmol of bilirubin—concentrations that exceed the highest levels of bilirubin observed in severe rhesus (Rh) isoimmunization.

It was recognized early that the content of bilirubin changes in amniotic fluid in a normal pregnancy, depending on time of gestation. Early in pregnancy amniotic fluid bilirubin values are very low. These values reach a peak at approximately 20 to 26 weeks of gestation, after which the concentration of bilirubin in amniotic fluid decreases steadily through each week of gestation, reaching less than ΔOD 0.01 at 36 to 37 weeks of gestation. This decrease with continuing gestation is so consistent that Mandelbaum[17] used it to determine fetal maturity.

Because of the declining presence of bilirubin in amniotic fluid in the final trimester of pregnancy, most methods that allow prediction of the outcome of pregnancy based on bilirubin depend on two or more samples rather than allowing one to attempt to quantitate the bilirubin. These methods include those of Liley,[2,5] Whitfield,[18] Queenan,[19] and Freda.[20] Clinical intrepretations can be found in most perinatal textooks.[21]

REFERENCE AND PREFERRED METHODS
(Table 59-5)

The spectrophotometric method was the first reported for amniotic fluid bilirubin,[2,3] and it prevails as the standard for amniotic fluid bilirubin because it is simple, sensitive, quantitative, and qualitative.

After centrifugation and filtration (if necessary), analysis takes less than 10 minutes on a continuous-recording spectrophotometer scanning between the wavelengths of 350 and 700 nm. If no bilirubin is present, the scan will generally parallel the zero base line with a slightly increased absorbance in the lower wavelengths. When the fluid contains bilirubin, the absorbance starts to increase near 375 nm, reaches a peak at 450 nm, and returns to base line at 525 nm. If only a colorimeter is available, only a few minutes are needed for determination of absorbances at 365, 415, 450, and 550 nm. Absorbances at the wavelengths of 365 and 550 nm establish the base line for the ΔOD when these wavelengths are plotted on semilogarithm paper against wavelength. The absorbance at 450 nm then allows one to assess bilirubin, whereas 415 nm provides information regarding the possibility of oxyhemoglobin contamination. There can be a slight increase in absorbance in

Table 59-5. Amniotic fluid bilirubin

Method	Type of analysis	Principle	Usage	Comments
Spectrophotometry	Differential absorbance	Bilirubin absorbs at 450 nm; background absorption corrected by extrapolation	Common	Accurate except in cases of hemolysis
Extraction spectrophotometry	Differential absorbance	Bilirubin extracted with chloroform; absorbance at 450 nm measured	Less common	Minimizes effects of interferences

the area of 415 nm, which is a common feature of amniotic fluid scan done in early pregnancy; it does not represent contamination by oxyhemoglobin.

The quantitative features of the spectrophotometric analysis of bilirubin (indirect) in amniotic fluid are sensitive and accurate. Errors in analysis are generally a result of contamination of amniotic fluid with oxyhemoglobin or meconium. The scan, unlike chemical bilirubin determinations, provides additional useful information because of the qualitative character of the spectrum. Before intrauterine death, for example, there is a shift in the bilirubin peak at 450 nm toward the shorter wavelengths. This was first described by Freda.[20] This information is valuable in the detection of rapid fetal deterioration.

The limitation to the spectrophotometric analysis of bilirubin in amniotic fluid is directly related to high absorptions that make it difficult or impossible to read a scan with gross contamination of the fluid. If contamination is caused by a single insult of blood (such as an intrauterine transfusion), it can take 2 or more weeks for the amniotic fluid to clear fully. Meanwhile, a method such as that of Brazie et al.[10] or Mallikarjuneswara et al.,[11] using a chloroform extraction, is required to remove bilirubin. A single extraction with chloroform extracts about 90% of the bilirubin present. A spectrophotometer can then be used for the determination.

It is important to remember that differences in optical density, that is, the departure from linearity of the curves, are being measured and that these are not absolute values. The optical density at 450 nm of a specimen can vary appreciably, depending on how vigorously it was centrifuged, but the *departure* from linearity at 450 nm does not vary. Any quantitative estimation of bilirubin in amniotic fluid using the absorption at 450 nm is under the assumption that this pigment obeys the Beer-Lambert law and that there is no interference by other pigments at either 450 nm or the tangential contacts of the base line.

SPECIMEN

The amniotic fluid specimen should be placed in an opaque container to protect it from light. Ultraviolet energy causes a change in the molecules of bilirubin, which can lead to a false low reading.

At the laboratory the specimen is centrifuged at 4000 *g* for 10 minutes in a clinical centrifuge to remove cellular material and vernix. If the fluid still remains turbid, it should be filtered through Whatman no. 42 filter paper. The filtration should be done out of direct sunlight. As long as the sample is protected from light, the spectrophotometric analysis can be performed within the next 24 hours. Storage for longer than 24 hours should be by freezing. Centrifuged amniotic fluid can be stored frozen for months without significant change in spectrophotometric properties.[22]

If blood contaminates the amniotic fluid, the fluid should be centrifuged immediately to remove the red blood cells before they hemolyze. If the amount of blood is small and the fluid is centrifuged immediately, the spectrophotometric absorptions will not be appreciably affected. If the amniotic fluid is blood contaminated and is not centrifuged immediately, the red blood cells hemolyze and the specimen becomes contaminated with oxyhemoglobin, which cannot be removed by centrifugation. Oxyhemoglobin contamination causes peaks at 412, 540, and 575 nm. If the packed cell volume of erythrocytes is greater than 5%, the scan will be noticeably altered even if immediately centrifuged,[4,5,23] since plasma components of the contaminating blood can also distort the spectrophotometric reading.

Meconium in amniotic fluid will give a noticeably abnormal tracing that is easy to identify. There is significant absorbance from 350 to 400 nm, with decreasing absorption up to 550 nm. Small bands in the 412 nm range (Soret bands) may also be seen.

Amniotic fluid is always more or less turbid, depending on the duration of pregnancy. This turbidity is caused by cells and debris derived from the membranes and from meconium and vernix, which are suspended in the amniotic fluid. Centrifugation of the fluid is the most appropriate way to eliminate at least part of the turbidity. Depending on the degree of turbidity, centrifugation can vary from 5 to 15 minutes or more at speeds at or near the maximum for a clinical centrifuge. A further decrease of turbidity can be achieved by means of filtration under pressure through filters with small-sized pores. A disadvantage of filtration, however, is the unwanted loss of sample. A comprehensive discussion of turbidity is given by Van Kessel.[6]

PROCEDURE FOR DIRECT SPECTROPHOTOMETRY[2]

Principle

Amniotic fluid is centrifuged to remove particulate matter and to clear turbidity. The supernatant is then transferred to a cuvette and scanned between 350 and 750 nm. Absorption in the 450 nm range is recorded as optical density.

Assay

Equipment. If a recording spectrophotometer is available, the centrifuged amniotic fluid specimen is scanned between the wavelengths of 350 and 750 nm; 1 cm lightpath glass or silica cuvettes are used.

The scan is normally done on undiluted fluid, but occasionally pigmentation will require dilution of the sample with distilled water.

To establish the base line, a line connecting the optical density points at 365 and 550 nm is drawn. The difference in either absorbance or optical density (ΔOD) is determined by measurement of the deviation between the spectrophotometric tracing at 450 nm and the base line as it intersects at 450 nm (Fig. 59-3).

For spectrophotometers that do not provide the option

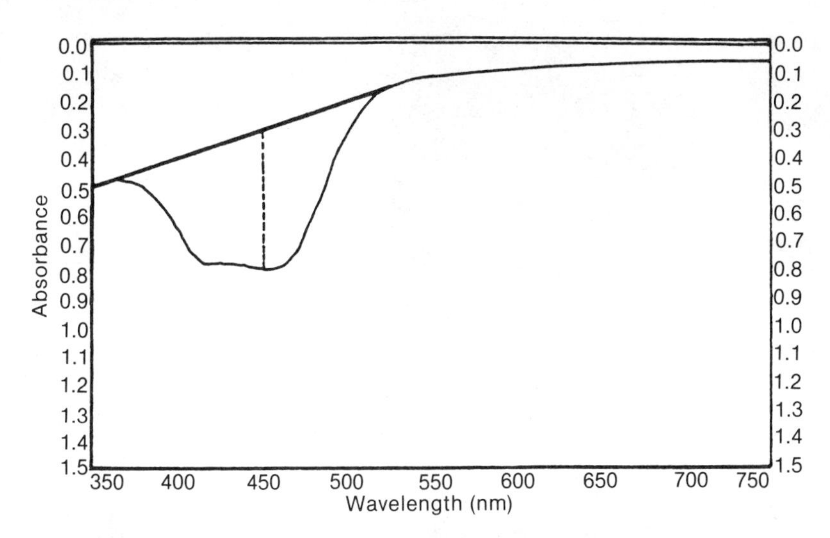

Fig. 59-3. Amniotic fluid scan showing method for determining absorbance at 450 nm. Arbitrary base line drawn from 375 to 525 nm shows where scan would have traced if no bilirubin pigment were present. *Broken line,* Absorbance recorded (ΔOD when ordinate is in optical density units).

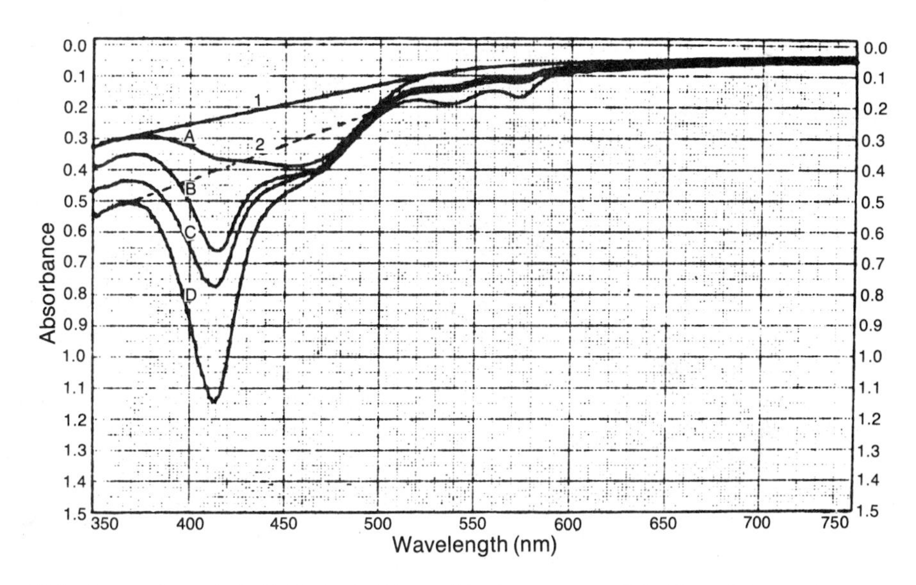

Fig. 59-4. Scans of amniotic fluid, *A,* in Rh-sensitized pregnancy, *B* to *D,* with increasing amounts of hemoglobin added. Line 1 is established base line; line 2 *(dashed)* shows how base line is altered when amniotic fluid is contaminated with hemoglobin. (From Queenan, J.T.: Modern management of the Rh problem, Hagerstown, Md., 1977, Harper & Row, Publishers.)

for a continuous scan, the centrifuged sample is placed in a cuvette and, with distilled water used as a blank, read at 365, 415, 450, and 550 nm. A line is constructed between the points at 365 and 550 nm on semilogarithmic graph paper. The intercepts at 450 and 415 nm on the base line are determined, and these values are subtracted from the optical density readings at 450 and 415 nm, respectively. The ΔOD at 415 indicates possible oxyhemoglobin contamination of the fluid.

Figs. 59-4 and 59-5 show scan patterns of amniotic fluids with varying concentrations of hemoglobin and tracings from clinical situations.

If, from the character of the spectrum, contamination of the amniotic fluid is suspected, chloroform extraction is recommended.

PROCEDURE FOR CHLOROFORM EXTRACTION[10]

Reagents

Chloroform. Chloroform is equilibrated with 0.05 M phosphate buffer, pH 6, and stored in a separatory funnel.

Assay

Equipment: spectrophotometer (≤ 10 nm band pass), glass-stoppered centrifuge tubes (12 mL), laboratory centrifuge in cold room.

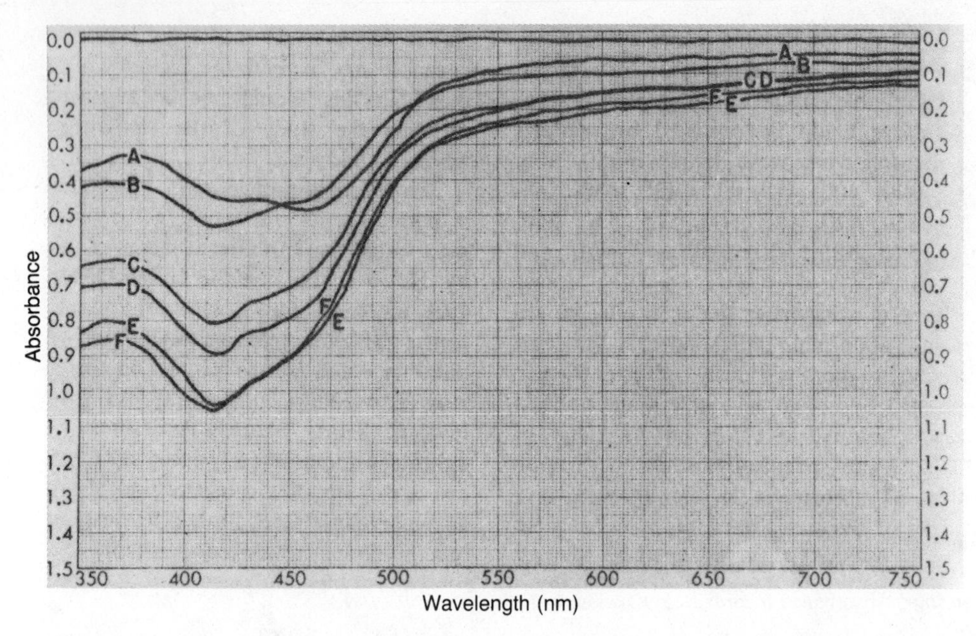

Fig. 59-5. Serial amniotic fluid scans from Rh-sensitized pregnancy eventuating in fetal death at *A*, 25 weeks; *B*, 26 weeks; *C*, 27 weeks; *D* to *F*, 28 weeks of gestation. Notice shape of curves. There is increased absorbance at shorter wavelengths beginning at 26 weeks of gestation (curve *B*). (From Queenan, J.T., and Goetschel, E.E.: Obstet. Gynecol. **32:**120-123, 1968.)

1. Equal quantities of amniotic fluid and chloroform are mixed for 30 seconds in a tight-fitting glass-stoppered tube.
2. The tube is centrifuged in the cold to facilitate separation of the phases. If a protein mesh persists in the chloroform layer after centrifugation, it may be broken up with a glass stirring rod and recentrifuged briefly.
3. The tubes are warmed to room temperature and the chloroform (lower layer) is pipetted into a 1 cm cuvette.
4. The absorbance is read at 450 nm with chloroform being used for the blank.

Interferences

1. Sunlight (ultraviolet light) will cause a false low reading for bilirubin.
2. Oxyhemoglobin interferes because it greatly affects the means of establishing the base line because of increased absorption rendered by the oxyhemoglobin from 350 to 400 nm and because large amounts can magnify the absorption at 450 nm. Oxyhemoglobin interference in this way will give a false low reading for bilirubin. A spectrophotometric method called the "corrected bilirubin method" can correct for bloody contamination.[19]
3. Meconium gives an abnormal tracing that includes noticeable absorbance from 350 to 400 nm. A false low bilirubin value can be expected in the presence of meconium.
4. Hydramnios causes a false lowering of the level of amniotic fluid bilirubin and poses the likelihood of incorrectly predicting the clinical status of the fetus when the amniotic fluid volume is considerably abnormal.[24]
5. If, during the course of amniocentesis, urine is inadvertently aspirated, this circumstance should be readily detectable with the spectrophotometer. The absorption pattern of urine shows increased absorption beginning at 365 nm and decreasing to 550 nm.[19]

REFERENCE RANGE

Values over 0.01 ΔOD indicate possible fetal distress at 36 to 37 weeks of gestation.

REFERENCES

1. Brazie, J.V., Ibbott, F.A., and Bowes, W.A.: Identification of the pigment in amniotic fluid of erythroblastosis as bilirubin, J. Pediatr. **69:**354-358, 1966.
2. Liley, A.W.: Liquor amnii analysis in the management of the pregnancy complicated by rhesus sensitization, Am. J. Obstet. Gynecol. **82:**1359-1370, 1961.
3. Bevis, D.C.A.: Blood pigments in haemolytic disease of the newborn, J. Obstet. Gynaecol. Br. Emp. **63:**68-75, 1956.
4. Walker, A.H.C.: Liquor amnii studies in prediction of haemolytic disease of the newborn, Br. Med. J. **2:**376-378, 1957.
5. Liley, A.W.: Errors in the assessment of hemolytic disease from amniotic fluid, Am. J. Obstet. Gynecol. **86:**485-494, 1963.
6. Van Kessel, H.: Spectrophotometry of amniotic fluid. In Sandler, M., editor: Amniotic fluid and its clinical significance, New York, 1981, Marcel Dekker, Inc., pp. 131-164.
7. Robertson, J.G.: Clinical value and application of measurements of bilirubin and protein levels in amniotic fluid in Rh-isoimmunization. In Sandler, M., editor: Amniotic fluid and its clinical significance, New York, 1981, Marcel Dekker, Inc., pp. 165-186.

8. Koch, T.R.: Bilirubin measurements in neonates, Clin. Lab. Med. **1**:311-327, 1981.
9. Queenan, J.T.: Amniotic fluid analysis in Rh and other blood group immunizations. In Sandler, M., editor: Amniotic fluid and its clinical significance, New York, 1981, Marcel Dekker, Inc., pp. 277-290.
10. Brazie, J.V., Bowes, W.A., and Ibbott, F.A.: An improved, rapid procedure for the determination of amniotic fluid bilirubin and its use in the prediction of the course of Rh-sensitized pregnancies, Am. J. Obstet. Gynecol. **104**:80-86, 1969.
11. Mallikarjuneswara, V.R., Clemetson, C.A.B., and Carr, J.J.: Determination of bilirubin in amniotic fluid, Clin. Chem. **16**:180-184, 1970.
12. Hochberg, C.J., Witheiler, A.P., and Cook, H.: Accurate amniotic fluid bilirubin analysis from the "bloody tap," Am. J. Obstet. Gynecol. **126**:531-534, 1976.
13. Simkins, A., and Worth, H.G.J.: Determination of bilirubin in amniotic fluid: a comparison of some current methods, Ann. Clin. Biochem. **13**:510-515, 1976.
14. Knox, E.G., Fairweather, D.V.I., and Walker, W.: Spectrophotometric measurements on liquor amnii in relation to the severity of haemolytic disease of the newborn, Clin. Sci. **28**:147-156, 1965.
15. Alvey, J.P.: Obstetrical management of Rh-incompatibility based on liquor amnii studies, Am. J. Obstet. Gynecol. **90**:769-775, 1964.
16. Ovenstone, J.A., and Connon, A.F.: Optical density differencing: a new method for the direct measurement of bilirubin in liquor amnii, Clin. Chim. Acta **20**:397-402, 1968.
17. Mandelbaum, B., and Robinson, A.R.: Amniotic fluid pigment in erythroblastosis fetalis, Obstet. Gynecol. **28**:118-120, 1966.
18. Whitfield, C.R.: A three-year assessment of an action line method of timing intervention in rhesus isoimmunization, Am. J. Obstet. Gynecol. **108**:1239-1244, 1970.
19. Queenan, J.T.: Modern management of the Rh problem, New York, 1977, Harper and Row, Publishers, Inc.
20. Freda, V.J.: The Rh problem in obstetrics and a new concept of its management using amniocentesis and spectrophotometric scanning of amniotic fluid, Am. J. Obstet. Gynecol. **92**:341-374, 1965.
21. Queenan, J.T.: Erythroblastosis fetalis. In Quilligan, E.J., and Kretchmer, N., editors: Fetal and maternal medicine, New York, 1980, John Wiley & Sons, Inc., pp. 445-471.
22. Queenan, J.T.: Amniotic fluid analysis, Clin. Obstet. Gynecol. **14**:505-536, 1971.
23. Queenan, J.T., and Goetschel, E.: Amniotic fluid analysis for erythroblastosis fetalis, Obstet. Gynecol. **32**:120-133, 1968.
24. Whitfield, C.R.: Effect of amniotic fluid volume on prediction, Clin. Obstet. Gynecol. **14**:537-547, 1971.

Creatinine

ROBERT L. MURRAY

H
N
HN=C C=O
N — CH_2
CH_3

Clinical significance: pp. 403 and 509
Molecular formula: $C_4H_7N_3O$
Molecular weight: 113.12 daltons
Merck Index: 2557
Chemical class: creatine end-product metabolite

PRINCIPLES OF ANALYSIS AND AND CURRENT USAGE

The Jaffé method for creatinine analysis (method 1, Table 59-6), first described in 1886,[1] has the distinction of being the oldest clinical chemistry method still in common use. This assay is based on the reaction of creatinine (I) with an alkaline solution of sodium picrate (II) to form a red Janovski complex (III). The absorbance of III is measured between 510 and 520 nm, though its maximum is reported as 485 nm.[2] Picrate ion (II, present in excess in the reacting solution) absorbs significantly at wavelengths below 500 nm. When serum or plasma is to be analyzed in an end-point reaction, a protein-free filtrate is used, since the α-ketomethyl or α-ketomethylene groups found in protein are reactive toward alkaline picrate and the resulting complexes are highly colored.

O NH =NH N CH_3

I

O_2N ^-O NO_2 O_2N

II

^-O NH =NH N CH_3 O_2N ^-O NO_2 O_2N

III

The reaction is run at a constant temperature of less than 30° C; at higher temperatures glucose, uric acid, and ascorbic acid can have an unacceptably high reductive reactivity toward picrate, resulting in formation of picramate, which has a maximum absorbance at 482 nm and causes an overestimation of creatinine. *Constant* temperature is also important because both the absorbance of the picrate ion and the creatinine-picrate reaction product increase with increasing temperature. The reaction is usually not buffered but is carried out in about 0.1 mol/L NaOH.

Fuller's earth has been used to increase the specificity of the Jaffé reaction by adsorbing the creatinine present in the protein-free filtrate, thus isolating it from potential interferents (method 2, Table 59-6). Other materials with comparable properties are Lloyd's reagent, bentonite, and floridin (fuller's earth). All of these materials are porous aluminum silicate clays that form colloidal suspensions with high natural adsorptive power. A small amount of the adsorbent is added to a protein-free filtrate at room temperature; adsorption is essentially complete after 1 minute. Creatinine is adsorbed with about 92% efficiency and, after centrifugation and decanting of the interferent-containing supernatant, one can add the alkaline picrate directly to the creatinine-adsorbent pellet. Most potential interferents are not adsorbed: only pyruvate in excess of 0.9 mmol/L and 2-oxoglutarate in excess of 0.5 mmol/L are adsorbed and can thus cause interference.[3] A similar purification has been described using a cation-exchange resin that adsorbs creatinine from an acid solution to the sodium

Table 59-6. Methods of creatinine analysis

Method	Type of analysis	Principle	Usage	Comments
1. Jaffé	Spectrophotometric (520 nm), end point, quantitative	Creatinine + picrate $\xrightarrow{OH^-}$ Janovski complex (red)	Serum, plasma, diluted urine	Described by Jaffé in 1886
2. Jaffé/fuller's earth	Same as 1. Creatinine isolated before analysis. Can be removed with buffer or picrate reagent added directly to creatinine adsorbent suspension	As above	Serum, plasma, diluted urine	Reference method; alternatively one can use cation exchanges as adsorbent
3. Jaffé, kinetic	Spectrophotometric, quantitative, kinetic analysis during early color formation	As above	Serum, plasma, diluted urine	Requires automated equipment for accurate, precise absorbance measurements
4. Jaffé, dual pH	Same as 1, using reaction at pH 9.65 (buffered) as blank	As above	Serum, plasma, diluted urine	Not widely used or automated
5. Creatinine amidohydrolase	Enzymatic hydrolysis to creatine, which reacts in indicator reactions monitored spectrophotometrically at 340 nm	Creatinine $\xrightarrow{\text{Creatinine amidohydrolase}}$ Creatine + H_2O Creatine + ATP $\xrightarrow{CK}$ Creatine phosphate + ADP ADP + Phospho*enol*pyruvate $\xrightarrow{\text{Pyruvate kinase}}$ ATP + Pyruvate Pyruvate + NADH + H^+ $\xrightarrow{LD}$ NAD^+ + Lactate (*CK*, Creatine kinase; *LD*, lactate dehydrogenase)	Serum	Not widely used or automated
6. Creatinine iminohydrolase	Enzymatic hydrolysis of creatinine with formation of ammonia, which can be quantitated spectrophotometrically or electrometrically	Creatinine $\xrightarrow{\text{Creatinine iminohydrolase}}$ *N*-methylhydantoin + NH_3 NH_3 measured by: GLDH reaction, ammonia electrode, or colorimetrically (see p. 1230)	Serum	Requires high enzyme purity, rarely used, may be possible reference method
7. High-performance liquid chromatography (HPLC)	Cation-exchange or reversed-phase chromatographic separation of creatinine from other compounds	Creatinine quantitated by method 1 or absorption at 200 nm	Serum, plasma, urine	Highly specific, not useful for routine analysis, possible reference method

form of the resin. Results from the cation-exchange resin are reported to compare well with those obtained using fuller's earth.[4] With minor modification, either of these isolation techniques can be used with other adaptations of the Jaffé reaction, described below.

The kinetic method for determination of creatinine (method 3, Table 59-6) gained popularity with the availability of instruments capable of making accurate absorbance readings at precise, highly reproducible intervals.[5-7] Sequential readings are made at precise points in the reaction sequence and compared to standards. In addition to centrifugal analyzers, the du Pont ACA and the Beckman Creatinine Analyzer use this technique.

In the kinetic modification, the picrate-creatinine complex formation is monitored after a 10 to 60 second delay after mixing of the reactants; this allows fast-reacting interferents such as acetoacetate to be largely eliminated before the initial absorbance reading is made. Additional measurements are made during or at the end of a period that ranges from 16 to 120 seconds, before the creatinine reaction has gone to completion and before the more slowly reacting interferents have been able to react significantly. Deproteinization is not necessary, since the reaction between alkaline picrate and protein is slow and does not significantly occur during the usual kinetic reaction interval.

The use of a serum "blank" run at pH 9.65 (method 4, Table 59-6) is reported to eliminate interferences caused by protein and other noncreatinine interferents.[8] Phosphate buffers (136 to 173 mmol/L) are used to produce reaction pH's of 9.65 and 11.50; otherwise the reaction conditions are comparable to the manual Jaffé reaction. The reaction interval is 45 minutes at a temperature of 37° C. Under these reaction conditions the commonly encountered pseudocreatinine chromogens are reported to be nonreactive and the interference caused by protein reaction at pH 11.50 is minimized with the pH 9.65 blank.

Separation and quantitation of creatinine has been accomplished with high-performance liquid chromatography (method 7, Table 59-6) by use of cation-exchange stationary phase with citrate buffer (pH 4.25) mobile phase,[9] or with a reversed-phase column and phosphate–sodium lauryl sulfate–methanol (pH 5.1) mobile phase.[10] Quantitation has been reported either with an on-stream Jaffé reaction[9] or with use of creatinine's native absorbance at 200 nm.[10] Both approaches are reported to be rapid, specific, and precise.

Other methods of creatinine analysis involve reaction of creatinine with 3,5-dinitrobenzoic acid or its derivatives,[11-13] with 1-4-naphthoquinone-2-sulfonate,[14] and with *o*-nitrobenzaldehyde.[15] In addition, mass fragmentography has been proposed as a reference method.[16]

Two creatinine-degrading enzymes have been investigated for use in creatinine analysis. First attempted in 1937,[17] the enzymatic approach has only recently become commercially feasible for routine clinical use. The major stumbling block has been a requirement for pure enzyme, and only in recent years has it been possible to obtain sufficiently pure enzyme of constant quality. Creatinine amidohydrolase (EC 3.5.3.10), also referred to as "creatininase" or "creatine hydrolase," has been used to convert creatinine to creatine, coupled to an indicator reaction as shown below (method 5, Table 59-6). This allows the continuous monitoring of NADH disappearance.[18]

$$\text{Creatinine} + H_2O \xrightarrow{\text{Creatinine amidohydrolase}} \text{Creatine}$$

$$\text{Creatine} + \text{ATP} \xrightarrow{\text{Creatine kinase}} \text{Creatine phosphate} + \text{ADP}$$

$$\text{ADP} + \text{PEP} \xrightarrow{\text{Pyruvate kinase}} \text{ATP} + \text{Pyruvate}$$

$$\text{Pyruvate} + \text{NADH} + H^+ \xrightarrow{\text{Lactate dehydrogenase}} \text{Lactate} + \text{NAD}^+$$

The second enzyme system used for creatinine quantitation is creatinine deiminase (EC 3.5.4.21), also known as "creatinine iminohydrolase" (method 6, Table 59-6). This enzyme is responsible for conversion of creatinine to *N*-methyl hydantoin and ammonia. In this method the ammonia is quantitated colorimetrically after its reaction with α-ketoglutarate and the monitoring of NADPH disappearance,[20] or by use of an ammonia electrode.[21] In either of these procedures the reaction mixture must be free of ammonia-producing and ammonia-consuming materials. In addition, a correction for endogenous ammonia must be made. Although the concentration of physiologically produced ammonia is relatively low, substantial amounts of ammonia can appear by protein deamination if blood is left at room temperature.

REFERENCE AND PREFERRED METHODS

The choice of methodology for creatinine analysis in a clinical laboratory involves several considerations. Most important is accuracy, especially for specimens from patients with renal dysfunction, in which a mixture of potential interferents is likely to be present. Furthermore, since analysis is used as the most reliable laboratory indicator of renal function, it must be available with a minimum of delay, 24 hours per day. Particularly when it is ordered on pediatric patients and on dialysis patients, the sample size must be kept to an absolute minimum.

The method that currently best meets these criteria is the Jaffé reaction with pretreatment by fuller's earth. This method is adaptable to standard instrumentation, its resistance to interference has been thoroughly tested over many years, and it has a high level of accuracy and precision. Only pyruvate and oxoglutarate remain potential sources of error when present at high concentrations.[3] The disadvantages of this procedure are that the introduction of fuller's earth necessarily involves a manual procedure; thus the sample requirement (typically 100 μL) is higher than in fully automated methods. This method, listed below, must be considered the reference method at this time. However, for the reasons just discussed it is not used in routine automated analysis.

The major disadvantages of the original Jaffé reaction (without fuller's earth) are related to its lack of specificity. Among substances reported to give a positive reaction are ascorbic acid, pyruvate, acetone, acetoacetic acid, levulose, glucose, aminohippurate, uric acid, protein, and cephalosporin antibiotics.[22-24] Ascorbic acid, levulose, glucose, and uric acid are capable of reducing picrate to picramate, and this product, with its absorbance maximum of 485 nm, causes overestimation of the creatinine. Elevated temperatures (over 30° C) and prolonged reaction times (over 5 minutes) exacerbate this interference. Pyruvate, acetone, acetoacetic acid, amino acids, and protein—or any compound with an active methyl or methylene group—introduce error by virtue of their coupling with picrate by a mechanism analogous to that of creatinine. All these compounds can produce colored compounds, which then increase the absorbance in the region of the creatinine-picrate complex. Bilirubin or other hemoglobin degradation products cause a *negative* bias, probably by their oxidation in strong base to colorless compounds.[25] The resulting decrease in background decreases the measured absorbance and is interpreted as a lower creatinine concentration. This negative bias is frequently seen in kinetic analyses for creatinine.

The disagreement that exists as to the extent of the contribution made by all these interferents is undoubtedly caused by subtle differences in reaction conditions that alter the reactivity of the various noncreatinine compounds toward picrate. Without modifications to the Jaffé reaction one might expect approximately 20% of the "creatinine" in serum or plasma and 5% of the "creatinine" in urine to be caused by noncreatinine materials. (These percentages depend on the concentrations of the interferents and could rise dramatically in specimens from patients with various pathological processes). A minor disadvantage to the Jaffé procedure is the need to work with strong base. With all these drawbacks, it is surprising that this methodology has been as widely used for such a long time, with its strengths being only an adaptability to automation and very low cost.

The kinetic modification of the Jaffé reaction is also highly adaptable to automation; in fact, it virtually *requires* automation. Its drawback is that it is less complete than the fuller's earth modification in its removal of interferences.[6,7] The initial delay reduces error caused by fast-reacting materials, and the final measurement made before the reaction has gone to completion reduces error caused by slow-reacting materials, but no correction is made for noncreatinine reactants that have reaction rates comparable to that of creatinine. Pyruvate, α-ketoglutarate, and oxaloacetate have reaction rates that fall into this range. The major advantages of the kinetic method are the small sample volume required (less than 25 μL), the speed of the analysis, and the adaptability to automation.

Neither high-performance liquid chromatography nor the use of a serum "blank" run at pH 9.65 have been adequately employed in routine operations by enough investigators to firmly establish their ability to eliminate interferents. More importantly, however, both are reported as manual procedures and offer little advantage over the preferred method for routine analysis.

The fact that an enzymatic method is not recommended is not related to lack of specificity or sensitivity but is rather because of the relative unavailability of reagents and because of as-yet-unproved (in routine clinical operations) susceptibility to interfering materials. The recent availability of pure enzyme reagents has allowed enzymatic methods to be adapted for automated analysis.[12,20] These procedures show promise of replacing the chemical creatinine assays now in common use. Although theoretical considerations would predict a significant improvement in specificity, this will be convincingly established only after many patient specimens have been analyzed under widely varying circumstances.*

SPECIMEN

One can analyze serum, plasma, or diluted urine. Urine should be diluted to a final creatinine concentration of approximately 300 to 600 μmol/L (34-68 mg/L); a 1:100 dilution will usually accomplish this. The common anticoagulants (fluoride and heparin) do not cause interference, though heparin, which can be formulated as the ammonium salt, must be avoided in enzymatic methods that measure ammonia production. Such methods also require prompt removal of serum from red cells and prompt analysis to minimize in vitro ammonia production; both of these precautions are necessary because of the imprecision that results from an elevated ammonia background. If analyzed by the Jaffé reaction, specimens are stable for at least 7 days at 4° C.

PROCEDURE: REFERENCE METHOD

Principle

Creatinine purified onto fuller's earth reacts with picrate ions in a highly alkaline solution to form a red complex measured at 509 nm.

Reagents

HCl, 1.0 mol/L. Dilute 85.5 mL of concentrated HCl to 1 liter with distilled water. Stable for 2 years at room temperature.

*Eastman Kodak, for example, has reported a positive bias when serum from patients on 5-fluorocytosine is analyzed for creatinine with their creatinine deiminase–based method on the Ektachem 400.

Tungstate reagent, 0.15 mol/L. Dissolve 50 g of $Na_2WO_4 \cdot 2\,H_2O$ in distilled water; dilute to 1 L. Stable for 1 year at room temperature.

Sulfuric acid, 666 mmol/L. Dilute 18.5 mL of concentrated sulfuric acid to 1 liter with distilled water. Stable for 2 years at room temperature.

Fuller's earth suspension, 6 g/L. Suspend 6 g of fuller's earth (30 to 60 mesh) in 10 mL of HCl (1.0 mol/L) and dilute to 1 liter with distilled water. Stable for 2 years at room temperature.

Picrate solution. Dissolve 2.7 g of picric acid and 6.2 g of NaOH in 1 liter of distilled water. This solution is approximately 12 mmol/L in picrate and 155 mmol/L in NaOH and is stable for at least 3 months at room temperature if stored in a capped, dark bottle. The absolute concentration is not critical; various authors recommend conditions from 50% to 200% of the strengths given here.[5] There are simple alternative methods of preparing bulk quantities of this solution or of purchasing the solutions directly, many of which are acceptable depending on the volumes of solution that are needed and the storage conditions.

Creatinine stock solution, 8.84 mmol/L (1 g/L). Dissolve 500 mg of creatinine in 5 mL of HCl (1.0 mol/L) and dilute to 500 mL. (One can obtain a pure creatinine preparation from the National Bureau of Standards, Washington, DC 20234, as Standard Reference Material No. 914).

Working standard solution, 176.8 μmol/L (20 mg/L). Dilute 2.0 mL of creatinine stock solution to 100 mL with distilled water. This solution is stable for 1 week at room temperature.

Assay

Equipment: spectrophotometer with a $\leq$ 10 nm band pass capable of reading at 509 nm and a 30° ± 0.5° C incubator.

1. Pipet into labeled tubes 200 μL of serum, diluted urine, standard, or water (reagent blank). Add 100 μL of tungstate reagent and 100 μL of sulfuric acid to each tube.
2. Mix well; after 3 minutes at room temperature, centrifuge for 5 minutes at 12,000 *g*, and transfer 300 μL of the supernatant into a second tube. To this add 500 μL of fuller's earth from a vigorously stirred suspension. Mix for 1 minute, centrifuge for 2.5 minutes at 12,000 *g*, and aspirate the supernatant fluid completely. Add to the pellet 500 μL of picrate solution.
3. Mix vigorously until the pellet is thoroughly resuspended. Let the mixture stand in a 30° C incubator for 30 minutes, centrifuge for 1 minute, and measure the absorbance of the supernatant fluid versus the reagent blank at 509 nm (490 to 520 nm). Precision is improved by the use of a constant temperature cuvette (30° or 37° C).

Calculation

The standard solution and the reagent blank are both taken through the same procedure as the unknown samples. The concentration, *C*, of creatinine in the unknown sample is calculated as follows:

$$C = A \times F$$

where *A* is the absorbance of the unknown sample, and *F* is a factor determined in each series as follows:

$$F = \frac{C_{std}}{A_{std}} \times D$$

where A_{std} is the absorbance of the standard; C_{std} is the concentration of the standard (176.8 μmol/L) 20 mg/L; and *D* is the dilution factor (to be used for urines or elevated serum specimens only).

REFERENCE RANGE (Table 59-7)

A person's serum creatinine level is dependent on two factors: the rate of production and the rate of excretion. Because the source of serum creatinine is muscle creatine and creatine phosphate, an increased amount of muscle mass results in an increased serum creatinine. In practice this is noticeable only in studies of large populations when male versus female ranges are studied, or when pediatric subjects are studied (see below). The second and much more significant determinant of serum creatinine levels is the rate of renal excretion. As a measure of renal function, however, the creatinine clearance determination is more sensitive than the serum creatinine level is. When corrected for body surface area (which approximates the total glomerular membrane surface area), reference ranges for the creatinine clearance rate are relatively constant from puberty to midlife, gradually decreasing from the fifth decade on. Infants and children have lower clearance rates

Table 59-7. Reference ranges for creatinine and creatinine clearance

Age	Serum creatinine[3,27] mg/L (μmol/L)	Urine creatinine[20,29] g/day (mmol/day)	Creatinine clearance, height/weight adjusted, mL/min
<12 years	2.5-8.5 (22-75)	0.057 mg (0.5 μmol)/kg of muscle	50-90
Adult male	6.4-10.4 (57-92)	1.0-2.0 (8.8-17.7)	97-137
Adult female	5.7-9.2 (50-81)	0.8-1.8 (7.1-15.9)	88-128

Table 59-8. Comparison of reaction conditions for creatinine

Condition	Reference method Jaffé + fuller's earth*	Kinetic Jaffé†	End-point Jaffé‡
Temperature	Ambient, 30° or 37° C	37° C	Ambient or 37° C
pH	12.4	12	12
Final concentration of reagents	Fuller's earth: 3.7 g/L Picric acid: 12 mmol/L NaOH 155 mmol/L	Picric acid: 12 mmol/L LiOH: 96 mmol/L	Picric acid: 28.4 mmol/L NaOH: 250 mmol/L
Fraction sample volume	0.3 (in protein-free filtrate)	0.056 (serum), 0.0059 (urine)	0.002 after dialysis
Volume of sample	100 μL (serum) (urine diluted 1:100 before use)	40 μL (serum) 4 μL (urine)	378 μL (urine diluted 1:100 before use)
Sample	Serum, plasma, urine	Serum, plasma, urine	Serum, plasma, urine
Time of reaction	30 minutes	16 seconds	9 minutes
Wavelength	509 nm	525 nm	505 nm
Linearity	≤ 100 mg/L serum ≤ 2000 mg/L urine	≤ 150 mg/L serum ≤ 2000 mg/L urine	≤ 2000 mg/L
Interferences	None	Hemoglobin–negative; interference at 1 g/L Hb Bilirubin—negative; interference of 4 mg/L at creatinine concentration, of 14 mg/L (bilirubin, 55 mg/L) Ketones—positive; interference (4 mg/L) at 400 mg/L acetoacetate	Acetoacetic acid, ascorbic acid, glucose, and acetone can give elevated creatinine values
Precision (from CAP quality control surveys)	$\bar{X}$, 10.7 mg/L; CV, 7%-15% $\bar{X}$, 51.1 mg/L; CV, 3%-10%	$\bar{X}$, 10.1 mg/L; CV, 12% $\bar{X}$, 47.6 mg/L; CV, 3%	$\bar{X}$, 10.5 mg/L; CV, 7% $\bar{X}$, 50.9 mg/L; CV, 4%

*Reference method described in text.
†IL919, Instrumentation Laboratory, Inc., Lexington, Mass.
‡Technicon AAII Method No. SE4-0011 FH4, Technicon Instruments Corp., Tarrytown, N.Y.

(even when corrected for body surface area), gradually rising to adult rates by the onset of puberty. Urine determinations of creatinine output are of value only with creatinine clearance determinations or as a check of the completeness of a 24-hour collection. Some, however, have reported a significant inconsistency in day-to-day amounts of creatinine excreted by a person: in one study the intraindividual coefficient of variation averaged 10% among six subjects.[26] Urinary excretion is directly proportional to muscle mass; approximately 0.5 mmol of creatinine is excreted per kilogram of muscle mass. Diet, urine volume, and exercise have little effect on serum creatinine levels; only exercise has an effect on urinary levels, increasing them slightly.

The reference ranges given in Table 59-7 were determined with the fuller's earth modification of the Jaffé reaction (Table 59-8). A less specific method would be expected to result in slightly increased serum levels but would have little effect on the urine levels. Because of this nonuniform effect, a less specific method would result in a lower calculated creatinine-clearance rate.

REFERENCES

1. Jaffé, M.: Ueber den Niederschlag welchen Pikrinsäure in normalen Harn erzeugt und über eine neue Reaction des Kreatinins, Z. Physiol. Chem. **10:**391-400, 1886.
2. Narayanan, S., and Appleton, H.D.: Creatinine: a review, Clin. Chem. **26:**1119-1126, 1980.
3. Haeckel, R.: Assay of creatinine in serum, with use of fuller's earth to remove interferents, Clin. Chem. **27:**179-183, 1981.
4. Polar, E., and Metcoff, J.: "True" creatinine chromogen determination in serum and urine by semi-automated analysis, Clin. Chem. **11:**763-770, 1965.
5. Spencer, K., and Price, C.P.: A review of non–enzyme mediated reactions and their application to centrifugal analysers. In Price, C.P., and Spencer, K., editors: Centrifugal analysers in clinical chemistry, New York, 1980, Praeger Publishers, pp. 231-253.
6. Bowers, L.D.: Kinetic serum creatinine assays. I. The role of various factors in determining specificity, Clin. Chem. **26:**551-554, 1980.
7. Bowers, L.D., and Wong, E.T.: Kinetic serum creatinine assays. II. A critical evaluation and review, Clin. Chem. **26:**555-561, 1980.
8. Yatzidis, H.: New method for direct determination of "true" creatinine, Clin. Chem. **20:**1131-1134, 1974.
9. Brown, N.D., Sing, H.C., Neeley, W.E., and Koetitz, E.S.: Determination of "true" serum creatinine by high-performance liquid chromatography combined with a continuous-flow microanalyzer, Clin. Chem. **23:**1281-1283, 1977.
10. Soldin, S.J., and Hill, G.J.: Micromethod for determination of creatinine in biological fluids by high-performance liquid chromatography, Clin. Chem. **24:**747-750, 1978.

11. Langley, W.D., and Evans, M.: The determination of creatinine with sodium 3,5-dinitrobenzoate, J. Biol. Chem. **115:**333-341, 1936.
12. Parekh, A.C., Cook, S., Sims, C., and Jung, D.H.: A new method for the determination of serum creatinine based on reactions with 3,5-dinitrobenzoyl chloride in an organic medium, Clin. Chim. Acta **73:**221-231, 1976.
13. Sims, C., and Parekh, A.C.: Determination of serum creatinine by reaction with methyl-3,5-dinitrobenzoate in methyl sulfoxide, Ann. Clin. Biochem. **14:**227-232, 1977.
14. Sullivan, M.S., and Irreverre, F.: A highly specific test for creatinine, J. Biol. Chem. **223:**530-533, 1958.
15. Van Pilsum, J.F., Martin, R.P., Kito, E., and Hess, J.: Determination of creatine, creatinine, arginine, guanidinoacetic acid, guanidine, and methyguanidine in biological fluids, J. Biol. Chem. **222:**225-236, 1956.
16. Björkhem, I., Blomstrand, R., and Öhman, G.: Mass fragmentography of creatinine proposed as a reference method, Clin. Chem. **23:**2144-2121, 1977.
17. Miller, B.F., and Dubos, R.: Determination by a specific, enzymatic method of the creatinine content of blood and urine from normal and nephritic individuals, J. Biol. Chem. **121:**457-467, 1937.
18. Moss, G.A., Bondar, R.J.L., and Buzzelli, D.M.: Kinetic enzymatic method for determining serum creatinine, Clin. Chem. **21:**1422-1426, 1975.
19. Slickers, K., Fame, N., Powers, D., and Rand, R.: Performance of Kodak Ektachem clinical chemistry slides for creatinine and ammonia, Clin. Chem. **28:**1570, 1982.
20. Tanganelli, E., Prencipe, L., Bassi, D., et al.: Enzymic assay of creatinine in serum and urine with creatinine iminohydrolase and glutamate dehydrogenase, Clin. Chem. **28:**1461-1464, 1982.
21. Thompson, H., and Rechnitz, G.A.: Ion electrode based enzymatic analysis of creatinine, Anal. Chem. **46:**246-249, 1974.
22. Henry, R.J., Cannon, D.C., and Winkelman, J.W.: Clinical chemistry: principles and technics, New York, 1964, Harper & Row, Publishers, Inc., pp. 287-292.
23. Swain, R.R., and Briggs, S.L.: Positive interference with the Jaffé reaction by cephalosporin antibiotics, Clin. Chem. **23:**1340-1342, 1977.
24. Young, D.S., Pestaner, L.C., and Gibberman, V.: Effects of drugs on clinical laboratory tests, Clin. Chem. **21:**1D-432D, 1975.
25. Soldin, S.J., Henderson, L., and Hill, J.G.: The effect of bilirubin and ketones on reaction rate methods for the measurement of creatinine, Clin. Biochem. **11:**82-86, 1978.
26. Scott, P.J., and Hurley, P.J.: Demonstration of individual variation in constancy of 24-hour urinary creatinine excretion, Clin. Chim. Acta **21:**411-414, 1968.
27. Meites, S., editor: Pediatric clinical chemistry, Washington, D.C., 1981, American Association for Clinical Chemistry, pp. 171-177.
28. Newkirk, R.E., and Rawnsley, H.M.: Creatinine clearance, ASCP Check Sample Clinical Chemistry no. CC-110, Chicago, 1978, American Society of Clinical Pathologists.
29. Faulkner, W.R., and King, J.W.: Renal function. In Tietz, N.W., editor: Fundamentals of clinical chemistry, ed. 2, Philadelphia, 1976, W.B. Saunders Co., pp. 975-1014.

Phenylalanine

HELEN K. BERRY

Phenylalanine
Clinical significance: p. 832
Molecular formula:
$C_6H_5CH_2CH(NH_2)COOH$ ($C_9H_{11}NO_2$)
Molecular weight: 165.2 daltons
Merck Index: 7071
Chemical class: amino acid

CH_2
H—C—$\overset{+}{N}H_3$
C—O^-
O

PRINCIPLES OF ANALYSIS AND CURRENT USAGE

The determination of phenylalanine in serum is most frequently used to confirm a diagnosis of phenylketonuria and to follow changes in blood levels of this amino acid in persons undergoing various types of therapy for the disorder. The most reliable method for measurement of phenylalanine is by automatic ion-exchange column chromatography using an amino acid analyzer (method 1, Table 59-9). The single-column method of Piez and Morris forms the basis for operation of most amino acid analyzers.[1] The ion-exchange resins are highly cross-linked sulfonated styrene copolymers that have an affinity for both the ionic and nonionic portion of the amino acid molecule.[2] Amino acids are eluted from the resin with buffers of different pH and ionic strength together with increasing column temperature. The column effluent is mixed with ninhydrin reagent and heated for 10 to 20 minutes at 100° C. The blue color that develops is measured photometrically, and the absorbance at 570 nm is recorded on a strip chart. The procedure, which measures all amino acids in the sample, requires 4 to 24 hours. A short column method has been described.[3]

Gas-chromatographic procedures are used for separation of amino acids, including phenylalanine, in physiological mixtures (method 2, Table 59-9). Resolution of a mixture by gas chromatography depends on the distribution of the components between a gaseous phase and a solid or liquid stationary phase contained in the separation column. Nonvolatile compounds are converted to derivatives that will vaporize without decomposition. The vaporized mixtures are transported through the column by a carrier gas. The compounds, mixed with the carrier gas, are retained by the stationary phase to different degrees and leave the column in the gaseous state. One can use physical or chemical detectors, the most common of which are flame-ionization and electron-capture detectors. The detector signal is recorded on a linear chart recorder. Substances are identified by their characteristic retention time. A number of derivatizing agents have been used for amino acids. Trimethylsilyl derivatives are frequently used,[4] as are derivatives prepared with bis(trimethylsilyl)-trifluoroacetamide (BSTFA).[5] A method specific for phenylalanine and tyrosine uses the *N*-(*O*-)trifluoroacetyl methyl esters prepared by reaction of the amino acid mixture with acetonitrile and trifluoroacetic anhydride, followed by methylation using diazomethane.[6] One can use a variety of column packings. Conditions of gas flow and column and oven temperatures are adjusted for the particular derivative. Detection is by flame ionization or mass spectrometry with selective ion monitoring.

The Guthrie microbiological inhibition assay is used most commonly for semiquantitative measurement of phenylalanine in blood, particularly in newborn screening programs.[7] The test is based on the inhibition of growth of

Table 59-9. Methods of phenylalanine analysis

Method	Principle	Reaction	Usage	Comment
1. Automatic column chromatography	Cation-exchange chromatography with detection of amino acids by ninhydrin	Triketohydrindene hydrate (ninhydrin) + $RCH(NH_2)CO_2H$ yields diketohydrindylidene-diketohydrindamine (Ruehlman's purple).	Serum, plasma, urine, or other physiological fluids	Most accurate
2. Gas chromatography	Separation of derivatized amino acids on glass columns; packing selected depends on derivative; detection by flame ionization	Formation of *N*-(*O*-)trimethylsilyl-amino acid esters or of amino acid alkyl ester followed by acylation; alkyl group methyl through pentyl have been used	Serum or plasma	Requires removal of interfering substances; not widely used for routine analysis; most valuable when using stable isotopes
3. Guthrie microbiological assay	Inhibition of growth of *Bacillus subtilis* by β-2-thienylalanine; inhibition overcome by phenylalanine		Dried blood spots collected on filter paper	Routinely used for newborn infant screening, semiquantitative
4. Fluorometric	Conversion of phenylalanine to fluorescent derivative	Mechanism of reaction of phenylalanine with ninhydrin to yield fluorescent compound is not known	Serum, plasma, blood eluted from filter paper	Recommended for routine use; has been automated

Bacillus subtilis by β-2-thienylalanine and the ability of phenylalanine to overcome this inhibition (method 3, Table 59-9).

The fluorometric method of McCaman and Robins[8] is preferred for routine use in the clinical laboratory from the standpoint of simplicity, sensitivity, and cost. The procedure is based on the enhanced fluorescence produced when phenylalanine condenses with ninhydrin in the presence of the dipeptide L-leucyl-L-alanine (method 4, Table 59-9). The reaction is carried out at pH 5.9. The fluorescent compound is stabilized by addition of an alkaline copper tartrate reagent. The fluorescence is measured at 489 to 515 nm, produced by excitation at 365 to 390 nm. The procedure described here combines features from modifications of the original method.[9,10]

REFERENCE AND PREFERRED METHODS

The amino acid analyzer yields the most accurate and most precise measurements of phenylalanine. One can measure amino acids in the range of 1.0 μmol/L with a precision of 1% under ideal conditions. In practice one can measure phenylalanine in the normal range of 50 to 100 μmol/L (8 to 16 mg/L) with a precision of 3%. The procedure is relatively time consuming and expensive, both in terms of cost of the equipment and of reagents.

Gas chromatography of amino acids in physiological fluids requires a preliminary separation of amino acids from other substances that may react with the derivatizing agents. Although gas chromatography has great sensitivity and a short analysis time and the apparatus is less expensive than an automatic amino acid analyzer, the problems involved in preparation of the specimen and preparation of the derivative make it undesirable for routine work.

Although the fluorometric method yields less accurate results than the amino acid analyzer, it has the advantage that one can analyze large numbers of samples rapidly with a fair degree of precision. However, reaction of amino acids other than phenylalanine may yield a nonspecific fluorescence that results in overestimation of phenylalanine concentrations in specimens from normal persons.

In a cooperative study, values obtained by several laboratories using the fluorometric procedure were compared with those measured in the same specimens by the amino acid analyzer.[11] At the lowest phenylalanine level tested, 16 mg/L, the mean values from the fluorometric procedure were significantly higher than control values measured on the analyzer. At phenylalanine concentration of 36 mg/L, higher values were obtained by the fluorometric procedure, except in instances in which a serum blank (ninhydrin without peptide) was used to correct for nonspecific fluorescence. At concentrations of phenylalanine above 50 mg/L, none of the mean values from fluorometric procedures was significantly different from the values measured by amino acid analyzer. Dilution of the specimen when one is measuring phenylalanine in patients with phenylketonuria reduces the error from nonspecific fluorescence. However, when phenylalanine concentrations are very low in a nonphenylketonuric patient, the fluorometric proce-

Table 59-10. Reaction conditions for phenylalanine analysis

Condition	Requirements		
Temperature	80° C		
pH	5.9		
Final concentration of reagents (before dilution with copper reagent)*	Succinate buffer: 352 mmol/L Ninhydrin: 7.0 mmol/L L-Leucyl-L-alanine: 0.59 mmol/L		
Fraction of sample volume (protein-free supernatant)	0.0625		
Sample volume	100 μL		
Sample	Protein-free supernatant of serum/plasma or protein-free supernatant of diluted serum/plasma Methanol eluate of dried blood spot may be used		
Time of reaction	25 minutes		
Wavelengths	Excitation: 365-390 nm Emission: 489-515 nm		
Linearity	With most fluorometers the reaction is linear between 5 and 80 mg/L; the range can be extended up to 800 mg/L using the dilutions specified		
Precision: mean (% CV)	**mg/L†**	**mg/L‡**	**mg/L§**
	16 (11.6%)	19 (12.8%)	
	55 (11.8%)	35 (7.1%)	
	110 (6.3%)	140 (10.5%)	167 (4.7%)

*Volume of copper reagent can vary from 3.0 to 4.0 mL.
†From Berry, H.K., and Porter, L.J.: Newborn screening for phenylketonuria, Pediatrics **70**:505-506, 1982.
‡From Sigma Chemical Co., St. Louis, Mo., Technical Bulletin No. 60F.
§My laboratory.

dure may cause overestimation of the serum phenylalanine content, and the phenylalanine deficiency might be falsely diagnosed.

SPECIMEN

Either serum or plasma may be used. The nature of the anticoagulant does not affect the measurement. The specimens are stable at 4° to 8° C for about a week. Blood dried on filter paper is used for regional screening programs and is stable for mailing and handling at room temperature.

PROCEDURE (Table 59-10)

Principle

The chemical basis of the method is the specifically enhanced fluorescence of phenylalanine after reaction with ninhydrin in the presence of a peptide, L-leucyl-L-alanine. A sample blank is included to correct for nonspecific fluorescence.

Reagents

All reagents are prepared in deionized water unless otherwise stated.

Succinate buffer, 0.6 mol/L, pH 5.9. Dissolve 16.2 g of $Na_2C_4H_4O_4 \cdot 6\ H_2O$ (disodium succinate hexahydrate) in 90 mL of water. Adjust pH with 1 N HCl. Dilute to 100 mL. Store at 0° to 5° C. Stable for 2 to 4 months.

Ninhydrin, 30 mmol/L (5.34 g/L). Store in brown bottle at 0° to 5° C. Stable for 7 to 10 days.

L-Leucyl-L-alanine, 5 mmol/L (1.01 g/L). Prepared volumetrically. Store frozen at −20° C in aliquots of 1 mL. Stable for 4 to 6 months.

Ninhydrin-peptide reagent. Prepare *daily* by combining 5 volumes of succinate buffer reagent with 2 volumes of ninhydrin reagent and 1 volume of peptide reagent.

Copper reagent

1. Stock solution:
   a. 16.0 g of Na_2CO_3 (anhydrous sodium carbonate)
   b. 0.650 g of $NaKC_4H_4O_6 \cdot 4\ H_2O$ (sodium potassium tartrate hexahydrate)
   c. 0.600 g of $CuSO_4 \cdot 5\ H_2O$ (copper sulfate pentahydrate)

   Dissolve each in about 300 mL of water; mix together in order listed; dilute to 1 L. Store at room temperature. Stable for 6 to 8 months.
2. Working reagent: For daily use dilute 1 volume of concentrate with 9 volumes of water.

Trichloroacetic acid (TCA). 0.6 mol/L (98 g/L) prepared volumetrically; 0.3 mol/L, 1:1 dilution of 0.6 mol/L. Stable 6 months at refrigerated temperature.

Phenylalanine standard solution

Stock solution, 200 mg/L (1.21 mmol/L). Dissolve 20 mg of L-phenylalanine in 100 mL of water. Store frozen. Stable for 1 to 2 months.

Working standard solution, 20 mg/L (0.121 mmol/L). Combine 1 mL of stock solution, 4 mL of water, and 5 mL of 0.6 mol/L TCA to prepare phenylalanine solution containing 20 mg/L in 0.3 mol/L TCA. Store frozen. Stable for 2 to 4 weeks.

Assay

Equipment

Heating bath. Set at 80° C. Ethylene glycol produces more uniform temperature than water.

Fluorometer. One of the following:

Turner Model 110 or 111 Fluorometer with General Purpose Lamp; Filters: Primary 7-60 (Turner 110-611); Secondary 2A (Turner 110-816) or 8 (Turner 110-817) + 65A (Turner 110-825)

Farrand Model A Fluorometer, filters as above

Aminco Bowman Spectrophotofluorometer; activation 365 nm; emission 515 nm.

Preparation of protein-free supernatant

For analysis of specimens from normal persons: Combine 1 volume (usually 0.100 mL) of serum or plasma with 1 volume of 0.6 mol/L TCA.

For analysis of specimens from untreated patients with phenylketonuria: Dilute 1 volume of serum with 9 volumes of water; combine 1 volume of diluted serum with 1 volume of 0.6 mol/L TCA.

For analysis of specimens from phenylketonuric patients under treatment: Dilute 1 volume of serum with 2 or 5 volumes of water; combine 1 volume of diluted serum with 1 volume of 0.6 mol/L TCA.

The serum-TCA solution is mixed well and allowed to stand at room temperature for 10 minutes. The tubes are centrifuged at 3000 rpm for 10 minutes. The protein-free supernatant is used for analysis.

Blank. 0.040 mL of 0.3 mol/L TCA (in duplicate)

Standard curve. Add 0.010, 0.020, 0.030, and 0.040 mL of phenylalanine working standard (reagent 7b) corresponding to 0.2, 0.4, 0.6, and 0.8 μg per assay. Adjust total volume to 0.040 mL by addition of 0.3 mol/L TCA.

Sample. For each specimen use 0.010, 0.020, and 0.040 mL of protein-free supernatant; bring final volume to 0.04 mL with 0.3 mol/L TCA.

Add 0.60 mL of ninhydrin-peptide mixture (reagent 4).

Mix thoroughly.

Place in heating bath at 80° C for exactly 25 minutes; remove; cool in tap water for 5 minutes.

Add 4.0 mL of dilute copper reagent.

Allow to stand for 10 minutes.

Read in fluorometer.

Calculation

The fluorescence of each standard minus the blank value is plotted versus the respective phenylalanine concentration. The concentration of phenylalanine in the unknown specimen is obtained from interpolation of the standard curve. Average three determinations for each unknown. Dilute if the fluorescence is greater than the highest value on the standard curve.

Example:

Serum diluted 1:6 for preparation of protein-free supernatant:

$$0.020 \text{ mL of serum} + 0.100 \text{ mL of water} + 0.120 \text{ mL of } 0.6 \text{ mol/L TCA} = 0.240 \text{ mL total volume}$$

Calculated volume of serum per assay:

$$\frac{0.020 \text{ mL of serum}}{0.240 \text{ mL (total volume)}} \times 0.010 \text{ mL (sample size)} = 0.000833 \text{ mL}$$

One calculates the concentration of phenylalanine by dividing the micrograms of phenylalanine in each tube (from the standard curve) by the volume of serum per assay as shown in Table 59-11.

REFERENCE RANGE

Serum phenylalanine values measured in several series of normal persons were in milligrams per liter: 15 ± 3,[8] 15.5 ± 3.4,[9] and 21 ± 5.[11] For comparison, mean phenylalanine concentration in 330 specimens from infants and children measured by the amino acid analyzer was 12 mg/L, with values ranging from 8 to 16 mg/L.[12]

Table 59-11. Example of phenylalanine concentration calculation*

Sample size (mL)	μg per assay (from curve)	÷	mL of serum per assay	=	μg per mL serum
0.010	0.133		0.000833		160
0.020	0.264		0.00167		158
0.040	0.551		0.00333		165

$$\text{*Average} = 161\ \mu\text{g/mL} \times \frac{1000 \text{ mL/L}}{1000\ \mu\text{g/mg}} = 161 \text{ mg/L}$$

Phenylalanine concentration in some genetic defects

Defect	Blood phenylalanine concentration (mg/L)
Classical phenylketonuria (phenylalanine hydroxylase deficiency)	
Untreated	300
Treated, good control	30-80
Variant phenylketonuria (partial phenylalanine hydroxylase deficiency) untreated	40-300
Hyperphenylalaninemia	40-200
Dihydropteridine reductase deficiency[14]	160-530
Biopterin synthetase deficiency[14]	210-490

Phenylalanine concentrations greater than 40 mg/L are rarely seen except in patients who have a genetic defect in phenylalanine metabolism. Ranges reported in such patients are shown in the preceding box.[13,14] In a few rare disorders, galactosemia, hereditary fructose intolerance, hereditary tyrosinemia, and transient tyrosinemia of the newborn, abnormal tyrosine metabolism may cause a secondary elevation of phenylalanine.

REFERENCES

1. Piez, K.A., and Morris, L.: An automatic procedure for the automatic analysis of amino acids, Anal. Biochem. **1**:187-201, 1960.
2. Benson, J.V., Jr., and Patterson, J.A.: Accelerated chromatographic analysis of amino acids commonly found in physiological fluids on a spherical resin of specific design, Anal. Biochem. **13**:265-280, 1965.
3. Benson, J.V., Jr., Cormick, J., and Patterson, J.A.: Accelerated chromatography of amino acids associated with phenylketonuria, leucinosis (maple syrup urine disease), and other inborn errors of metabolism, Anal. Biochem. **18**:481-492, 1967.
4. Blau, K.: Biomedical applications of gas chromatography, vol. 2, New York, 1968, Plenum Press.
5. Gehrke, C.W., and Leimer, K.: Trimethylsilylation of amino acids derivatization and chromatography, J. Chromatogr. **58**:219-238, 1971.
6. Zagalak, M.-J., Curtius, H.Ch., Leimbacher, W., and Redweik, U.: Quantitation of deuterated and non-deuterated phenylalanine and tyrosine in human plasma using the selective ion monitoring method with combined gas chromatography-mass spectrometry: application to the *in vivo* measurement of phenylalanine-4-monooxygenase activity, J. Chromatogr. **142**:523-531, 1977.
7. Guthrie, R., and Susi, A.: A simple phenylalanine method for detecting phenylketonuria in large populations of newborn infants, Pediatrics **32**:338-343, 1963.
8. McCaman, M.W., and Robins, E.: Fluorimetric method for the determination of phenylalanine in serum, J. Lab. Clin. Med. **59**:885-890, 1962.
9. Wong, P.W.K., O'Flynn, M.E., and Inouye, T.: Micromethods for measuring phenylalanine and tyrosine in serum, Clin. Chem. **10**:1098-1104, 1964.
10. Ambrose, J.A.: A shortened method for the fluorometric determination of phenylalanine, Clin. Chem. **15**:15-23, 1969.
11. Hsia, D.Y., Berman, J.L., and Slatis, H.M.: Screening newborn infants for phenylketonuria, J.A.M.A. **188**:203-306, 1964.
12. Berry, H.K., and Porter, L.J.: Newborn screening for phenylketonuria, Pediatrics **70**:505-506, 1982.
13. Ambrose, J.A.: Report on a cooperative study of various fluorometric procedures and the Guthrie bacterial inhibition assay in the determination of hyperphenylalaninemia, Centers for Disease Control, Health Services and Mental Health Administration, Public Health Service, Atlanta, 1972, U.S. Government Printing Office.
13. Berry, H.K., The diagnosis of phenylketonuria, Am. J. Dis. Child. **135**:211-213, 1981.
14. Danks, D.M., Bartholome, K., Clayton, B.E., et al.: Malignant hyperphenylalaninemia—current status (June 1977), J. Inherited Metab. Dis. **1**:49-53, 1978.

Urea

LAWRENCE A. KAPLAN

$$NH_2-\overset{\overset{\displaystyle O}{\|}}{C}-NH_2$$

Urea (BUN, carbamide)
Clinical significance: p. 403
Molecular weight: 60.06 daltons
Merck Index: 9525
Chemical class: amino acid metabolite

PRINCIPLES OF ANALYSIS AND CURRENT USAGE

Urea has traditionally been quantitated by either direct chemical analysis or indirectly by a conversion to and subsequent analysis of ammonia (NH_3). Most historical methods involve the measurement of ammonia nitrogen after treatment of samples, either with elevated temperatures (125° C by autoclaving) or by the action of the enzyme urease:

$$2H_2O + O{=}C(NH_2)_2 \xrightarrow[\text{or urease}]{125^\circ\text{ C}} (NH_4)_2CO_3$$

The extraordinarily high specificity of urease for urea makes the enzymatic reaction the preferred method for conversion of urea into ammonia.

Because the analysis of blood urea was originally measured in terms of released nitrogen, concentration was expressed in terms of milligrams of blood urea nitrogen (BUN) per volume (such as deciliters). This term, unfortunately, has survived the years and changes in methodology and is still in current use. The conversion of BUN values to urea concentrations is as follows:

1. Atomic weight of nitrogen = 14 g/mol; molecular weight of urea = 60.06 g/mol.
2. Urea contains two nitrogen atoms per molecule.
3. Urea nitrogen (urea N) is 46.6% by weight of urea (28 divided by 60.06).
4. Therefore:

$$10\text{ mg/L BUN divided by } 0.466 = 21.46\text{ mg/L urea} = 0.36\text{ mmol/L urea}$$

or

$$\text{mg of urea N/L} \times 2.146 = \text{mg of urea/L}$$

$$\text{mg of urea N/L} \times 0.036 = \text{mmol of urea/L}$$

Although it is certainly preferable to report urea concentration in body fluids as mass or moles of urea per liter, the convention of reporting BUN values is used in this text.

The older methods for quantitating ammonia released from urea were performed either by acid titration or by employing the nesslerization or Berthelot reactions to form a colored product.[1,2] The titration method titrated the released gaseous ammonia with dilute sulfuric acid in the presence of a colored pH indicator solution.[1] Both colori-

Table 59-12. Methods of urea analysis

Method	Type of analysis	Principle	Usage	Comments
1. Nesslerization	Quantitative, end-point, spectrophotometric	Urea $\xrightarrow[OH^-]{Urease}$ $(NH_4)_2CO_3$ $2(HgI_2 + 2KI) + NH_4^+ + NaOH \rightarrow$ $NH_2Hg_2I_3 + H_3O^+ + 4KI + NaI$ (yellow-orange colloid)	Rarely used, of historical interest	Nonspecific, long reaction times
2. Berthelot	Quantitative, end-point, spectrophotometric	Urea $\xrightarrow{Urease}$ $(NH_4)_2CO_3$ $NH_4^+ + HOCl \xrightarrow{pH\ 10.5} H_2NCl + H_3O^+$ (hypochlorous acid) (chloramine) H_2NCl + Phenol $\xrightarrow[NP]{OH^-}$ Quinonechloramine ($ClN{=}C_6H_4{=}O$) Quinonechloramine + Phenol $\xrightarrow{OH^-}$ $^-O{-}C_6H_4{-}N{=}C_6H_4{=}O$ **Indolphenol (blue)** (*NP*, Nitroprusside, used as a catalyst)	Rarely used	Nonspecific, relatively long reaction times
3. Coupled enzymatic (urease/glutamate dehydrogenase [GLDH])	Quantitative, end-point, kinetic, spectrophotometric	Urea $\xrightarrow{Urease}$ $(NH_4)_2CO_3$ NH_4^+ + α-Ketoglutaric acid + NADH $\xrightarrow[ADP,\ H^+]{GLDH}$ NAD^+ + Glutamic acid	Most frequently used procedure	Very specific, rapid
4. Diacetyl monoxime	Quantitative, end-point, colorimetric	$CH_3{-}C({=}O){-}C({=}NOH){-}CH_3 \xrightarrow[H^+]{H_2O} CH_3{-}C({=}O){-}C({=}O){-}CH_3 + HNO_2$ Diacetyl monoxime → Diacetyl + Hydroxylamine $NH_2{-}C({=}O){-}NH_2 + CH_3{-}C({=}O){-}C({=}O){-}CH_3 \xrightarrow{H^+} CH_3{-}C{-}C{-}CH_3 + 2\ H_2O$ Urea + Diacetyl → Diazine (yellow) (ring: C=N–C(=O)–N=C)	Very commonly used	Some nonspecificity of reaction, uses noxious, dangerous reagents
5. Conductimetric	Quantitative, kinetic	Urea $\xrightarrow{Urease}$ $(NH_4)_2CO_3 \rightarrow 2NH_4^+ + CO_3$ Increased ions changes conductivity	Used with increased frequency	Very specific, rapid
6. *o*-Phthalaldehyde	Colorimetric, quantitative, end-point	Urea + *o*-Phthalaldehyde $\xrightarrow{H^+}$ Isoindoline Isoindoline + 8-(4-amino-1-methylamino)-6-methoxyquinoline $\xrightarrow{H^+}$ Chromophore (510 nm)	Rarely used	Interferences, often primary amines
7. Ion-selective electrode	Quantitative, enzymatic	Urea $\xrightarrow{Urease}$ $(NH_4)_2CO_3$ NH_4 detected by potentiometric electrode	Of research interest	Very specific

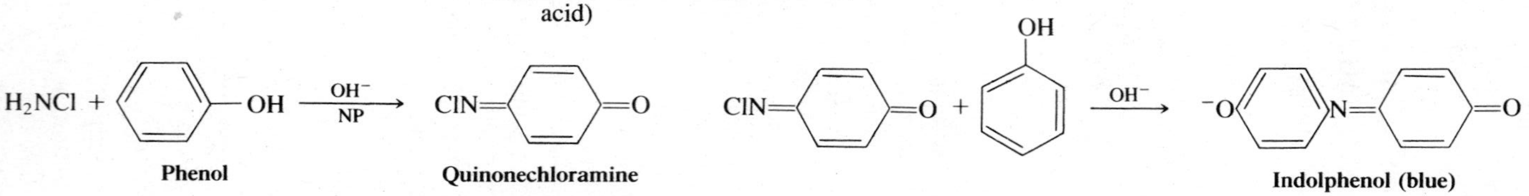

metric reactions could be performed directly on serum or blood, on urine samples, or on a Folin-Wu precipitate.

The nesslerization reaction (reaction 1, Table 59-12) involves the formation of a complex yellow chelate (whose actual composition is not known) between ammonia and HgI_2 under alkaline conditions. The chromogen has an absorption maximum at 410 nm.[2] The Berthelot reaction (reaction 2, Table 59-12) between ammonia and phenol-NaOCl results in a blue indophenol complex whose absorption maximum is at 630 nm.[3,4] This method often is performed with sodium nitroferricyanide [nitroprusside, $Na_2(Fe[CN]_5NO)$] as a catalyst.

The procedures involving the older methods for detection of nitrogen could also be used for urine samples. Since urea is heavily concentrated in urine, the urine samples are usually diluted 10 to 15 times before analysis. There is up to a thousandfold more ammonia in urine than in serum; thus a correction for ammonia levels in urine samples has to be made. The urine samples are analyzed for ammonia before and after urease treatment to correct for the endogenous ammonia levels.

The most common method for measuring urea concentrations in serum or urine by the NH_3 formed by the urease reaction uses a coupled enzyme system employing an NAD/NADH indicator reaction (reaction 3, Table 59-12).[5,6] These reactions are monitored at 340 nm. Inaccuracies of the coupled enzyme system are caused by the indicator reaction (GLDH) where other endogenous enzymes compete to oxidize the NADH. In addition, exogenous ammonia from the reagents may give falsely high values. The urease/GLDH–coupled reaction performed in the kinetic analysis mode is also useful as a measure of urine urea at normal levels of endogenous ammonia. The endogenous urine ammonia is rapidly consumed in the initial seconds of the reaction and the subsequent changes in A_{340} are essentially caused by ammonia generated by the urease reaction with urea.

The only direct chemical analysis for urea has been the diacetyl monoxime reaction[1,2,7] (reaction 4, Table 59-12). Fearon initially observed that diacetyl monoxime will react with several primary and secondary amines containing the general structure $R_1NH\text{-}CO\text{-}NHR_2$, where R_1 is either an H or an aliphatic group and R_2 does not contain an acyl ($-C{=}O$) group.[8] Many other compounds containing urea residues in their structures, such citrulline, alloxan, and allantoin, will also produce a colored product.[1,2] However, the low concentrations of these compounds in serum rarely cause a significant interference. Other compounds present in significant levels in serum, such as creatinine and protein, give chromogens with different absorption maxima so that they do not produce significant interferences. The diacetyl monoxime does not directly react with urea but is first hydrolyzed to form diacetyl and hydroxylamine. The diacetyl condenses with urea in an acid solution to form a yellow diazine product. This chromogen is monitored at 550 nm. The chromogen is also monitored by its flourescence at 415 nm.[9] Several chemicals have been used to enhance and stabilize the color produced in the diacetyl monoxime reaction, either directly (such as thiosemicarbazide, ferric ions, or glucuronolactone) or by elimination of the hydroxylamine formed in the initial hydrolysis (such as potassium persulfate).[2] The diacetyl monoxime reaction has been automated for analysis for both serum and urine samples.

A method for quantitating urea by measurement of the change of conductivity of a sample after the action of urease was reported two decades ago[10] (reaction 5, Table 59-12) This technique has recently been adapted for automated analysis (ASTRA 8, Beckman/Smith-Kline, Sunnyvale, Calif.)[11] In this technique, the reaction is monitored by following the *change* in conductivity with time. The CO_2 and ammonia formed by the urease reaction form ammonium carbonate ($[NH_4]_2CO_2$), which increases the conductivity of the reaction mixture. When performed in a kinetic analysis mode to correct for the endogenous conductivity, one can analyze both serum and urine samples by the conductivity method.

One instrument manufacturer has employed the reaction of *o*-phthalaldehyde with primary amines to quantitate the urea (reaction 6, Table 59-12). The isoindoline product of the reaction is coupled to a complex quinoline to form a chromogen that is monitored at 510 nm.[12] An ammonia ion–selective electrode has also been available for some time but has not been used in the clinical laboratories. This method employs urease, covalently linked to a solid-phase matrix affixed to the electrode, to convert urea to ammonia, which is measured potentiometrically by the electrode[13,14] (method 7, Table 59-12).

Recent CAP quality assurance surveys indicate that the most widely used method for the analysis of serum urea is the coupled urease/GLDH enzymatic assay. Almost one half of the respondents used this method, whereas 21% used the diacetyl monoxime procedure and 17% reported using the conductivity technique. The overwhelming use of all three procedures is by automated analysis procedure. A few laboratories report using nesslerization and Berthelot reactions, usually by nonautomated analysis.

REFERENCE AND PREFERRED METHODS

Methods for quantitating urea by nesslerization and Berthelot reactions are rarely used today. This is most likely because of the inability to readily automate these procedures, technical difficulties inherent in the methods (such as specificity), and their replacemnt by newer, superior methods. Both of these reactions were performed as end-point procedures, which made them sensitive to incomplete conversion of urea to ammonia. Both methods require relatively long incubation times. Although the Berthelot reaction is 10 times more sensitive than nesslerization (molar absorptivity 20,000 versus 3000), it has a relatively narrow range over which it observes Beer's law.

The nesslerization reaction has been reported to have major difficulties.[1,2] The first is the sensitivity to the presence of levels of acetone that can be present in patients with diabetic ketoacidosis. The formation of turbidity as the reaction proceeds is also a major problem of the assay. Attempts have been made to minimize the turbidity by the addition of oxidizing agents or use of protein-free filtrates. At urea concentrations near the upper limit of normal, laboratories using the Berthelot and nesslerization reactions consistently report higher coefficients of variations (7% to 10% CV) on CAP surveys than do laboratories using other methods.

The diacetyl monoxime reaction has been readily adapted for automated analysis with good reported precision (2.5% to 6.5% CV at upper limit of normal concentrations of urea). The major difficulty reported for this method is the photosensitivity and rapid fading of the colored product, which can be partially corrected by the addition of thiosemicarbazide to the reaction. In addition, the method poorly follows Beer's law unless large dilutions are employed, and it uses rather noxious and corrosive reagents. The diacetyl monoxime method is also subject to some interference from other nitrogenous compounds present in serum.

The coupled urease/GLDH procedures and the conductivity method are most frequently employed on automated instruments. These methods probably have the highest degree of specificity of any current urea assay. Other nitrogenous compounds do not interfere and, when run in the kinetic mode with large dilutions, do not show any significant interferences by bilirubin, hemoglobin, or lipemia. The precision of both these methods according to CAP surveys is excellent overall (3.5% to 7.5% CV at upper limits of normal concentrations of urea). However, the precision of the urease/GLDH method can vary considerably with instrument.

The preferred method is certainly either the conductivity method or the urease/GLDH procedure in the kinetic mode because these have the best specificity, speed of analysis, and precision of all current methods. There is no established reference or definitive method for urea at this time.

SPECIMEN

Serum and heparinized plasma are used for the diacetyl monoxime, urease/GLDH, and conductivity methods. Fluoride will inhibit the urease reaction; therefore methods employing urease cannot use serum preserved with fluoride.[3] Ammonium heparin also cannot be used as an anticoagulant for urease methods.

One can analyze urine by all three methods after a 1:20 to 1:50 sample dilution, depending upon the method and instrument employed.

Because of the susceptibiliy of bacterial degradation of urea, serum and urine samples should be kept at 4° to 8° C until analysis. One can also preserve urine samples by maintaining the pH less than 4.

REFERENCE RANGE

The reference ranges for serum BUN will vary with the method; several are listed below. Statistically higher val-

Table 59-13. Comparison of reaction conditions for urea analysis

Reaction component	Coupled enzymatic urease/GLDH*	Conductimetric†	Diacetyl monoxime‡
Temperature	37° C	37° C	37° C
pH	8.1	7.3	Highly acidic
Final concentration of reagents	Urease: ≥ 12 × 10^3 U/L GLDH: ≥ 4 × 10^3 U/L ADP: 2.5 mmol/L NADH: 100 μmol/L α-Ketoglutaric acid: ≥ 6.5 mmol/L Tris buffer: 80 mmol/L	Urease: 2 × 10^5 U/L Tris buffer EDTA	Diacetyl monoxime: 9.4 mmol/L Thiosemicarbazide: 2.1 mmol/L $FeCl_3$: 5.3 μmol/L H_3PO_4: 2.2 mmol/L H_2SO_4: 5.4 mmol/L
Fraction of sample volume	0.0058	0.009	0.111 (before dialysis) (about 0.002 after dialysis)
Linearity (urea nitrogen per liter)	1200 mg/L	1500 mg/L	1500 mg/L
Reaction time	16-second kinetic at 340 nm	11.5-second kinetic	9-minute end-point at 520 nm
Interferences	Fluoride (F^-)	Fluoride (F^-)	
Precision (at interlaboratory precision from CAP surveys)	220 mg of BUN/L—5.2% CV	220 mg of BUN/L—3.4% CV	220 mg of BUN/L—4.1% CV

* IL919, Instrumentation Laboratory, Lexington, Mass.
† Beckman 508, Beckman Instruments, Division of Beckman-SmithKline, Inc., Sunnyvale, Calif. Actual reagent concentrations are proprietary information of Beckman Instruments.
‡Technicon, Inc., Tarrytown, N.Y., Method no. SE4000 IFD4.

ues are seen in men than in women and also for older age groups. For adults, these differences are not clinically significant, and one can combine reference-range data for these groups. Young children have slightly lower serum urea values than older children and adults do. Individual variations in BUN values will also depend on the dietary habits of the person; less affluent people, or those consuming less protein, have lower serum urea concentrations. Similarly urine output of urea will vary with diet.

Method (Table 59-13)	Adult reference range (serum)
Urease/GLDH	<50-170 mg BUN/L (107-365 mg of urea/L, 1.8-6.1 mmol/L)
Urease conductivity	60-200 mg BUN/L (129-429 mg of urea/L, 2.2-7.2 mmol/L)
Diacetyl monoxime	80-260 mg BUN/L (172-558 mg of urea/L, 2.9-9.4 mmol/L)
	Urine urea output (average diet)
Urea/GLDH	7-16 g of BUN/24 hr (0.25-0.57 mol of urea/24 hr)

REFERENCES

1. Natelson, S.: Techniques of clinical chemistry, ed. 3, Springfield, Ill., 1971, Charles C Thomas, Publisher, pp. 728-745.
2. Henry, R.J., Cannon, D.C., and Winkelmann, J.W.: Clinical chemistry principles and technics, ed. 2, New York, 1974, Harper & Row, Publishers, Inc., p. 504-506.
3. Patton, C.J., and Crouch, S.R.: Spectrophotometric and kinetics investigation of the Berthelot reaction for determination of ammonia, Anal. Chem. **49**:464-469, 1977.
4. Faulkner, W.R., and Meites, S., editors: Selected methods of clinical chemistry, vol. 9, Washington, D.C., 1982, American Association for Clinical Chemistry, pp. 357-363.
5. Talke, H., and Schubert, G.E.: Enzymatische Harnstoffbestimmung in Blut und Serum im optischen Test nach Warburg, Klin. Wochenschr. **43**:174-175, 1965.
6. Tiffany, T.O., Jansen, J.M., Burtis, C.A., et al.: Enzymatic kinetic rate and end-point analysis of substrate, by use of a GeMSAEC fast analyzer, Clin. Chem. **18**:829-840, 1972.
7. Faulkner, W.R., and Meites, S., editors: Selected methods of clinical chemistry, vol. 9, Washington, D.C., 1982, American Association for Clinical Chemistry, pp. 365-373.
8. Fearon, W.R.: The carbamido diacetyl reaction: a test for citrulline, Biochem. J. **33**:902-907, 1939.
9. McCleskey, J.E.: Fluorometric method for the determination of urea in blood, Anal. Chem. **36**:1646-1648, 1964.
10. Chin, W.T., and Kroontje, W.: Conductivity method for determination of urea, Anal. Chem. **33**:1757-1760, 1961.
11. Paulson, G., Ray, R., and Sternberg, J.: A rate sensing approach to urea measurement, Clin. Chem. **17**:644, 1971.
12. KDA Application notes, American Monitor Corporation, Indianapolis, Ind.
13. Guilbault, G.G.: Analytical uses of immobilized enzymes, Biotechnol. Bioeng. **3**:361, 1972.
14. Richterich, R., and Colombo, J.P., editors: Clinical chemistry, New York, 1981, John Wiley & Sons, Inc., pp. 190-191.

Uric acid

ARNOLD L. SCHULTZ

Clinical significance: pp. 403 and 710
Molecular formula: $C_5H_4N_4O_3$
Molecular weight: 168.11 daltons
Merck Index: 9538
Chemical class: purine

PRINCIPLES OF ANALYSIS AND CURRENT USAGE

The formation of tungsten blue by reacting an alkaline solution of uric acid with phosphotungstic acid was first applied to the analysis of uric acid in blood in 1912.[1] This early method used protein precipitation and isolation of the uric acid from the filtrate as the silver salt before reaction with phosphotungstic acid in sodium carbonate solution (method 1, Table 59-14). Modifications of this original method isolated uric acid by formation of its magnesium, ammonium, cuprous, or cupric salt. Cyanide was reported to increase the sensitivity of the method, to prevent fading of the color produced, and to dissolve the silver urate.[2] Urea-cyanide was later used as the alkaline reagent.[3,4] This modification did not require isolation of the uric acid from the filtrate. Many other alkaline reagents have been used to enhance the color of the tungsten blue produced (A_{max}, 700 nm). Proteins have been removed by precipitation with tungstic acid, trichloroacetic acid, phosphotungstic acid, heat coagulation, and membrane filtration. Other oxidizing reagents have included arsenotungstic acid, arsenophosphotungstic acid, arsenomolybdic acid, potassium ferricyanide, and uranyl acetate.

To improve the specificity of the colorimetric oxidation procedures, the adsorption of uric acid on an ion-exchange column and the addition of *N*-ethylmaleimide to inactivate the interfering effects of sulfhydryl-containing chromogens have been used.[5] Colorimetric measurements before and after treatment with uricase have also been made to increase the specificity of the phosphotungstic acid oxidation procedure.[6]

Uricase has been used extensively to increase the specificity of the uric acid assay (method 2, Table 59-14). The uricase methods are based on the specificity of the uricase-catalyzed oxidation of uric acid to allantoin and hydrogen peroxide.

Allantoin, unlike uric acid, does not have an absorption peak in the 290 to 293 nm region of the ultraviolet spectrum. Absorption measurements at these wavelengths before and after incubation of uric acid with uricase have been used to quantitate uric acid in serum, plasma, and urine.[7-9]

The hydrogen peroxide produced when uric acid is oxidized in the uricase-catalyzed reaction has been quantified by measurement of the chromogenic response at 530 nm when *o*-dianisidine is oxidized by hydrogen peroxide.[10]

Table 59-14. Methods of uric acid analysis

Method	Type of analysis	Principle	Usage	Comments
1. Phosphotungstic acid	Spectrophotometric	Oxidation of uric acid to allantoin and carbon dioxide, reduction of phosphotungstic acid to tungsten blue (A_{max}, 700 nm)	Serum, urine	Nonspecific, but widely used
2. Uricase	Enzymatic	Oxidation of uric acid to allantoin, hydrogen peroxide, and carbon dioxide	Serum, urine	
	a. Differential absorption	Uric acid absorbs in the 290 to 293 nm (at pH $\geq$ 7) and 283 nm(at pH < 7) region of the ultraviolet spectrum, but allantoin does not		Basis for a candidate reference method, increased specificity
	b. Colorimetric	Quantitation of the hydrogen peroxide produced, especially when coupled to a NAD/NADH indicator reaction		Specificity varies from method to method, NADH reaction widely used
	c. Polarographic	Rate of oxygen consumption		Not widely used, some interferences
	d. Coulometric	Titration with iodine		Instrumentation not readily available
3. High-performance liquid chromatography	Chromatographic			
	a. Spectrophotometric	Reversed-phase chromatography	Serum, urine	Increased specificity and sensitivity
	b. Electrochemical	Ion-exchange separation	Serum, urine	Proposed selected method

Hydrogen peroxide oxidatively reacts with 3-methyl-2-benzothiazolinone and *N,N*-dimethylaniline, in the presence of peroxidase, to produce a blue indamine dye, the absorbance of which is measured at 600 nm,[11] and with 3,5-dichloro-2-hydroxybenzenesulfonic acid and 4-aminophenazone to form a red quinoneimine dye.[12] The absorbance of the quinoneimine dye is measured at 520 nm to avoid spectral interference from hemolysis, bilirubin, and turbidity, though the maximum absorbance occurs at 512 nm and is 3% more intense. 2,4,6-Tribromophenol has also been used in the reaction with 4-aminophenazone.[13] The absorbance of the product from this reaction is read at 492 nm. Similarly, the reaction of formaldehyde (formed in the catalase-catalyzed reaction of hydrogen peroxide and methanol) with acetylacetone and ammonia to produce the yellow dye 3,5-diacetyl-1,4-dihydrolutidine has been measured at 410 nm to quantitate the oxidation of uric acid in the presence of uricase.[14] The hydrogen peroxide produced by the uricase-catalyzed oxidation has been detected through the catalase-catalyzed oxidation of ethanol to acetaldehyde, coupled to the oxidation of acetaldehyde to acetate in the presence of aldehyde dehydrogenase and NAD^+, measurement of the change in absorbance of NADH at 340 nm.[15]

The rate of oxygen consumption, which is a measure of the rate of the uricase-catalyzed oxidation of uric acid and is proportional to the uric acid concentration, has been used to measure uric acid concentrations in serum and urine.[16] A polarographic oxygen sensor is used.

A coulometric method, which employs the uricase differential technique, has been used for the analysis of uric acid in serum and urine.[17] This method is based on the measurement of the total reducing substances produced by coulometric titration with iodine before and after reaction with uricase.

Several high-performance liquid chromatographic (HPLC) procedures for the quantitation of uric acid in serum[18,19] and urine [20] have been introduced (method 3, Table 59-14). These HPLC methods use either reversed-phase chromatography with spectrophotometric detection at 280[18] or 235 nm[20], or ion-exchange separation followed by amperometric detection in a thin-layer flow-through electrochemical cell.[19]

A College of American Pathologists chemistry survey of almost 6000 laboratories showed 34% using phosphotungstate methods, 62% uricase methods, less than 1% oxygen rate methods, and 3% other methods.[21]

REFERENCE AND PREFERRED METHODS

The reaction of 3,5-dichloro-2-hydroxybenzenesulfonic acid with 4-aminophenazone in the presence of peroxidase coupled to the enzymatic oxidation of uric acid has a standard deviation of ±1 mg/L at a level of 53 mg/L.[9] In this procedure, interference from bilirubin is eliminated by the incorporation of potassium ferricyanide into the reagent. The addition of ascorbate oxidase prevents interference by ascorbic acid.

The measurement of the rate of oxygen consumption during the reaction of uric acid and oxygen in the uricase-catalyzed reaction is not widely used. It is reported to have a repeatability of ±2% at 100 mg/L and to be subject to interference by allopurinol, xanthine, and hypoxanthine.[16]

The instrumentation required for the coulometric method employing the uricase differential technique is not readily available, and the method is therefore rarely used.[17] The coulometric method has a standard deviation of ±1 mg/L in the range of 24.1 to 116 mg/L. The recovery for the method ranges from 98.2 to 100.3%.

Over the years, several methods for the determination of uric acid in serum and urine have been suggested as the standard, selected, or reference method.

In 1953, Natelson[22] proposed that the method of Folin[3] as modified by Brown[4] be adopted as the standard method for uric acid determinations. This method incorporates an alkaline oxidation applied directly to a tungstic acid protein-free serum filtrate. The oxidizing agent employed is phosphotungstic acid in the presence of sodium cyanide "buffered" by the urea. However, this method suffers from lack of specificity, with interference resulting from the phosphotungstic acid reagent being reduced by ferrous salts, glutathione, phenols, ascorbic acid, glucose, tyrosine, tryptophan, cystine, and cysteine.

In 1964, a modification of the method of Folin and Denis[1] in which solutions of sodium carbonate and phosphotungstic acid are added directly to a tungstic acid protein-free filtrate of serum was suggested as a standard method.[23] This method was reported to have a standard deviation of ±0.6 mg/L at a level of 57 mg/L. It yields results that are 7% higher than those obtained by the urea-cyanide procedure.

In 1979, an HPLC procedure was proposed as the selected method for uric acid.[19] The advantages of this HPLC procedure using electrochemical detection after chromatography on an ion-exchange material include greater sensitivity and specificity. The relative sensitivity of the amperometric detector is 1 ng/L. Only the metabolites of theophylline (methyluric acids) interfere in the assay of uric acid in urine. The HPLC method has a day-to-day coefficient of variation of about 1% to 2%. The method is linear between 10 and 200 mg/L.

In 1979, the Centers for Disease Control presented a draft of a candidate reference method for uric acid in serum to the American Association for Clinical Chemistry (AACC) Standards Committee, Subcommittee for a Uric Acid Reference Method and to the National Reference Council of the National Committee on Clinical Laboratory Standards.[24] This method includes protein precipitation by trichloroacetic acid to eliminate high serum backgrounds, followed by a manual, equilibrium, ultraviolet, and uricase method. The decrease in absorbance at 283 nm is used to follow the oxidation of uric acid to allantoin catalyzed by high-purity microbial uricase. The coefficient of variation for the method ranges from 2.6% at 30 mg/L to 1.5% at 99 mg/L. The method is linear to 200 mg/L. Recoveries range from 99% to 101%. Of 20 substances tested for interference, including drugs and their metabolites, possible competitive inhibitors, other commonly occurring constituents, thiols, and anticoagulants and preservatives, only xanthine shows significant interference in the manual uricase method. Several other substances, including EDTA and sodium fluoride, exhibit interference that is not considered to be significant.

The preferred method for routine uric acid measurement is an automated coupled-enzymatic procedure available as kits from many commercial sources. The method presented here is based on the Centers for Disease Control manual uricase candidate reference method.[24]

SPECIMEN

Serum or plasma may be used. EDTA and sodium fluoride should be avoided as an anticoagulant and preservative, respectively, because they contribute a positive interference to the method described here. Uric acid is stable at 2° to 6° C for 3 to 5 days and for at least 6 months at −20° C.

Aliquots of a 24-hour urine collection are also useful for uric acid determinations (Table 59-15). To prevent urate precipitation, add 10 mL of 500 g/L sodium hydroxide to the collection bottle before collection of the specimen. Uric acid in urine is usually stable for approximately 3 days at room temperature, provided that there is no bacterial growth to destroy it.

PROCEDURE

Principle

Incubation of uric acid with uricase in Tris buffer, pH 8.5 at 37° C, results in the production of one mole of al-

Table 59-15. Reaction conditions for uric acid analysis

Condition	Candidate reference method
Sample	0.5 mL of serum or plasma or 0.5 mL of a 1:10 dilution of an aliquot of a 24-hour urine collection
Protein precipitant	100 g/L trichloroacetic acid
Reaction temperature	37° C
Reaction time	95 minutes
Linearity	200 mg/L
Precision	2.6% at 30 mg/L, 1.5% at 99 mg/L
Final concentration of reagents	Tris buffer, 0.041 mol/L Uricase, 17 U/L Trichloroacetic acid, 33.3 g/L

$$\text{Uric acid} + O_2 + 2H_2O \xrightarrow{\text{Uricase}} \text{Allantoin} + CO_2 + H_2O_2$$

Uric acid — Allantoin — Hydrogen peroxide

lantoin for every mole of uric acid oxidized. The reaction is quantified by measurement of the decrease in absorbance at 283 nm (see note 1, p. 1265). Uric acid exhibits absorbance at this ultraviolet wavelength, but allantoin does not. Before one takes absorbance readings, proteins are precipitated with trichloroacetic acid to eliminate high serum background absorbances. A blank is also used for each sample to correct for nonprotein, endogenous ultraviolet-absorbing substances that vary from specimen to specimen.

Reagents

Tris hydrochloride, 15.8 g/L (0.1 mol/L). Stable for 2 months at 2° to 6° C if free of bacterial contamination.

Tris base, 12.1 g/L (0.1 mol/L). Stable for 2 months at 2° to 6° C if free of bacterial contamination.

Tris buffer, 0.1 mol/L, pH 8.5 at 37° C. Mix 940 mL of 12.1 g/L Tris base and 224 mL of 15.8 g/L Tris·HCl. Check the pH with a pH meter at 37° C ± 1° C. If necessary, adjust with either Tris·HCl or Tris base solution to pH 8.5 ± 0.1. Filter the buffer through a sterile, 0.45 μm membrane filter and transfer to a sterilized, borosilicate glass, screw-capped bottle. It is stable for 6 months at 2° to 6° C if protected from bacterial contamination.

Trichloroacetic acid, 100 g/L (612 mmol/L). Place 10 g of trichloroacetic acid in a 100 mL volumetric flask and add 50 mL of deionized water to dissolve. Bring volume to mark with water and mix. Filter the solution through a 0.45 μm membrane filter into a sterilized borosilicate glass, screw-capped storage bottle. It is stable for 3 months at 2° to 6° C.

Uricase, 0.1 U/mL. Weigh or measure volumetrically an amount of uricase (from *Bacillus fastidiosus* or *Candida utilis*) estimated to have a total activity of 1000 U at 37° C (see note 2, p. 1265). Transfer this amount to a 1000 mL volumetric flask and dilute to volume with Tris buffer, pH 8.5. One can aliquot the resulting solution in 100 mL portions in borosilicate screw-capped bottles. It is stable for 1 month at −20° C.

"Blank" enzyme reagent. Add 50.0 mL of 100 g/L trichloroacetic acid solution to 25.0 mL of 0.1 U/mL uricase solution. This is enough solution to prepare approximately 25 blanks. Prepare fresh 1 hour before use; prepare an appropriate amount for the number of samples to be analyzed. The solution must be allowed to stand at least 30 minutes to ensure that the uricase has been inactivated.

Uric acid stock standard, 1000 mg/L (5.95 mmol/L). Weigh and transfer 500 mg of anhydrous NBS Certified Uric Acid, Standard Reference Material No. 913, which has been stored in its original capped bottle in a vacuum desiccator over anhydrous calcium sulfate, and 375 mg of ACS-grade lithium carbonate into a 500 mL volumetric flask using 125 mL of sterile, distilled, deionized water that has been heated to 50° C. Mix the uric acid and lithium carbonate into complete solution with swirling. The water should not be heated higher than 50° C. After the solids have completely dissolved, allow the solution to cool to room temperature. Dilute to volume with sterile, distilled, deionized water. Mix well. Divide into five 100 mL portions and transfer each to a screwcapped borosilicate bottle. Stable for 2 months at −20° C. One bottle is used to prepare a complete set of standards.

Uric acid working standards, 20, 40, 60, 80, 100, 150, and 200 mg/L. Dilute 2, 4, 6, 8, 10, 15, and 20 mL of the 1000 mg/L stock uric acid standard to volume with sterile, distilled, deionized water in 100 mL volumetric flasks. Working standards should be dispensed into screw-capped borosilicate vials of about 15 mL capacity and stored tightly capped. Stable for 2 months at −20° C. When a vial of standard is thawed, it is used only that day and discarded.

"Zero" standard, lithium carbonate, 150 mg/L (2 mmol/L). Dissolve and dilute 15 mg of ACS-grade lithium carbonate in sterile, distilled, deionized water to volume in a 100 mL volumetric flask.

Assay

Equipment: ultraviolet spectrophotometer with a ≤10 nm band pass, a 37° C ± 0.5° C water bath.

1. Warm all reagents to room temperature.
2. Measure and record the volume in milliliters of 24-hour urine collections.
3. If the urine is cloudy, warm the specimen to 60° C for 10 minutes to dissolve precipitated urates and uric acid. Cool the specimen to room temperature and centrifuge an aliquot.
4. Prepare a 1:10 dilution of a centrifuged aliquot of urine with deionized water.
5. Label a sufficient number of 16 × 100 mm borosilicate, cappable tubes. Each standard, unknown, and control requires a "test" and a "blank."
6. Transfer 0.5 mL of sample and 2.5 mL of Tris buffer into each "test" and "blank" of the appropriately labeled tubes.

7. Add 1.0 mL of the 0.1 U/mL uricase solution to the tubes marked "test," mix well, and cap.
8. Immerse all the capped "test" and "blank" tubes in a 37° C ± 1° C water bath for 60 minutes.
9. After 60 minutes remove the tubes from the water bath and add 2.0 mL of 100 g/L trichloroacetic acid to all the tubes marked "test." Add 3.0 mL of the "blank" enzyme reagent to all tubes marked "blank." Mix well by inversion, being careful to avoid foaming. (Do *not* shake vigorously and do *not* use a vortex mixer.) Let stand 45 minutes, mixing several times by gentle inversion.
10. Centrifuge all the "test" and "blank" tubes at approximately 1200 *g* for 10 minutes. Remove the tubes from the centrifuge; mix again by gentle inversion. (It is not necessary to remix the precipitate at the bottom of the test tube. This step is helpful in bringing down the fine precipitate that remains at the meniscus.) Centrifuge the tubes for an additional 10 minutes at 1200 *g*. Transfer the supernatants by decanting to clean, dry, borosilicate tubes and centrifuge again for 20 minutes. With Pasteur pipets transfer the supernatants to clean, dry, borosilicate tubes. (Visually inspect all the supernatants and if any particulate matter is noted, centrifuge the tubes again for 20 minutes and transfer the supernatants to clean, dry, borosilicate tubes with Pasteur pipets.) NOTE: *Do not filter*. Uric acid has been shown to adhere to filter paper and to filter membranes.
11. Obtain the absorbance readings at 283 nm for all "tests" and "blanks." Zero the spectrophotometer with deionized water. Read the absorbance of the standards, unknowns, and controls. Timing is not critical, but spectrophotometry is performed in the serial order of addition, so that each sample is read at approximately the same time from the time it was added.

Calculation

1. Subtract the "test" absorbance (A_{test}) from the "blank" absorbance (A_{blank}) for each standard, unknown, and control to obtain the "corrected reading" (A_{corr}),

$$A_{corr} = A_{blank} - A_{test}$$

2. Draw a calibration curve relating the corrected reading to the quantity of uric acid in the standards in mg/L.
3. Read the concentration of uric acid in mg/L in each of the samples and controls from the calibration curve using the corrected absorbance readings.
4. Calculate the amount of uric acid in the 24-hour specimen from the equation:

$$\text{mg of uric acid/24 hour} = \text{mg/L} \times \frac{\text{Total urine volume (mL)}}{1000} \times 10$$

Where 10 is the dilution factor of the aliquot of urine, and 1000 converts mL to L.

5. Samples with readings more than 10% higher than the highest standard accepted for the run, or lying beyond the previously demonstrated linear range, are diluted with equal volumes of deionized water and reanalyzed. Multiply the result obtained from the calibration curve by the dilution factor.

Notes

1. Uric acid in lithium carbonate solution, pH 9.8, exhibits an absorbance maximum at 293 nm. However, in trichloroacetic acid solution, pH 1.1, the absorbance maximum is shifted to 283 nm, as shown in Fig. 59-6.
2. It is not absolutely necessary to measure exactly 1000 U of uricase, but the measurement should be made as accurately as possible. If it is impossible to estimate enzyme activity, prepare only 10 mL of an approximately 1 U/mL uricase solution in Tris buffer, pH 8.5, and assay it for uricase activity on the day of preparation. Store at −20° C.
3. Additional reagents required for the uricase assay are the following:

Albumin, 2 g/L. Dissolve 0.5 g of 96% to 99% pure, fraction V, bovine serum albumin in and dilute to 250 mL with 0.1 mol/L Tris buffer, pH 8.5 at 37° C. Store at 2° to 6° C.

Uric acid, 250 mg/L. Dilute 25.0 mL of the 1000 mg/L uric acid stock standard to volume with deionized water in a 100 mL volumetric flask. Store at 2° to 6° C.

The uricase assay is performed as follows:

1. Warm the albumin, 250 mg/L uric acid standard and the Tris buffer to room temperature.
2. Dilute 1.0 mL of the 1 U/mL uricase solution to volume with the Tris-albumin solution in a 100 mL volumetric flask.
3. Prepare the blanking solution by adding 10.0 mL of the Tris-albumin solution to 1 mL of the 250 mg/L uric acid standard. Mix well.
4. Pipet 1.0 mL of the 250 mg/L uric acid solution and 4.0 mL of Tris-albumin solution into a 16 × 100 mm, screw-capped test tube. Pipet 2.0 mL of diluted uricase solution into another 16 × 100 mm, screw-capped test tube. Place these tubes along with a tube containing the blanking solution into a 37° C ± 1° C water bath for 30 minutes. Transfer the blanking solution to two matched silica cuvettes in a spectrophotometer with the sample compartment controlled at 37° C ± 1° C. Adjust the instrument to zero at 293 nm. Retain the cell in the reference beam. Remove the cell in the sample beam and thoroughly wash and dry it. Replace the cell in the instrument and allow it to equilibrate at 37° C ± 1° C. Prepare the assay solution by pouring the diluted uricase solution into the tube containing the uric acid/Tris-al-

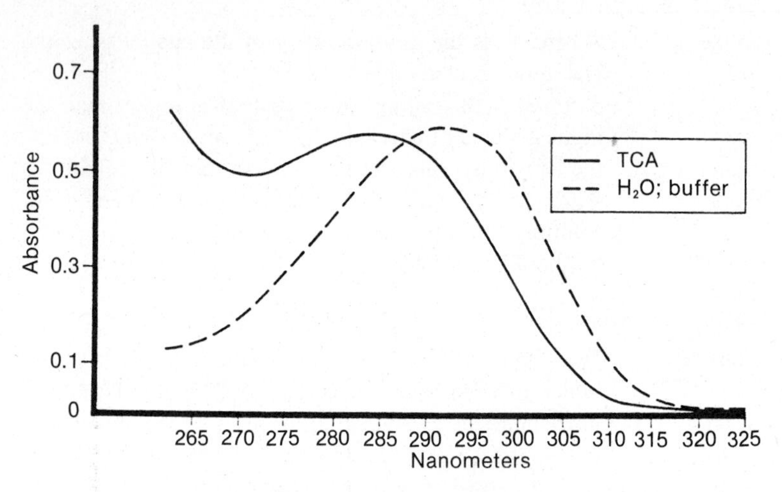

Fig. 59-6. Representative spectra of uric acid (100 mgL) in trichloroacetic acid (TCA), pH 1.1 *(solid line),* and dilutions of stock standard solution *(dashed line).* Stock standard (750 mg of lithium carbonate per liter) was diluted with either deionized, distilled water or Tris buffer, pH 8.5. pH's of resulting solutions were 9.8 and 8.7, respectively. (From Duncan, P., Gochman, N., Cooper, T., et al.: Development and evaluation of a candidate reference method for uric acid in serum, 1979. U.S. Department of Health, Education, and Welfare, Public Health Service, Atlanta, 1980, U.S. Centers for Disease Control.)

bumin solution. Quickly pour back and forth between the two test tubes to thoroughly mix. Immediately fill the cell with the assay solution and place in the sample compartment of the spectrophotometer. Allowing 2 minutes for temperature equilibration, measure the absorbance of the assay solution against the blanking solution at 293 nm continuously for 5 minutes. The absorbance change per minute should be approximately the same for any 1-minute interval over the 5-minute period. If the change in absorbance is so rapid that the intervals are not approximately the same, dilute an aliquot of the dilute uricase solution with a known volume of the Tris-albumin solution and reassay with 2.0 mL of this further dilution. Apply the appropriate dilution factor in the calculation of catalytic activity per milliliter.

5. Calculate the uricase activity from the equation:

$$\text{U/mL in stock uricase} = A_{test(2\ min)} - A_{test(3\ min)} \times 28.2\ \mu mol/min/mL$$

6. With the uricase stock solution prepared and assayed, dilute the uricase stock solution with an appropriate volume of the Tris buffer, pH 8.5 at 37° C, to yield a final concentration of 0.1 U/mL of uricase. The resulting solution can be aliquoted in 100 mL portions in borosilicate screw-capped bottles and stored for up to 1 month at −20° C.

REFERENCE RANGES

Serum or plasma

36 to 77 mg/L (214 to 458 μmol/L) for males
25 to 68 mg/L (149 to 405 μmol/L) for females

Urine

250 to 750 mg (1.49 mmol to 4.46 μmol)/24 hr for average diet
Up to 450 mg (2.68 mmol)/24 hr for low-purine diet
Up to 1 g (5.95 mmol)/24 hr for high-purine diet

REFERENCES

1. Folin, O., and Denis, W.: A new (colorimetric) method for the determination of uric acid in blood, J. Biol. Chem. **13:**469-475, 1912-1913.
2. Benedict, S.R., and Hitchcock, E.H.: On the colorimetric estimation of uric acid in urine, J. Biol. Chem. **20:**619-627, 1915.
3. Folin, O.: An improved method for the determination of uric acid in blood, J. Biol. Chem. **86:**179-187, 1930.
4. Brown, H.: The determination of uric acid in human blood, J. Biol. Chem. **158:**601-608, 1945.
5. Sambhi, M.P., and Grollman, A.: A simplified procedure for the routine determination of uric acid, Clin. Chem. **5:**623-633, 1959.
6. Bulger, H.A., and Johns, H.E.: The determination of uric acid, J. Biol. Chem. **140:**427-440, 1941.
7. Kalckar, H.M.: Differential spectrophotometry of purine compounds by means of specific enzymes. I. Determination of hydroxypurine compounds, J. Biol. Chem. **167:**429-443, 1947.
8. Praetorius, E., and Poulsen, H.: Enzymatic determination of uric acid with detailed directions, Scand. J. Clin. Lab. Invest. **5:**271-280, 1953.
9. Feichtmeir, T.V., and Wrenn, H.T.: Direct determination of uric acid using uricase, Am. J. Clin. Pathol. **25:**833-839, 1955.
10. Marymount, J.H., and London, M.: Analyses performed with heat-coagulated blood and serum. VI. Direct determination of urates by means of *o*-dianisidine oxidation, Am. J. Clin. Pathol. **42:**630-633, 1964.
11. Gochman, N., and Schmitz, J.M.: Automated determination of uric acid, with use of a uricase-peroxidase system, Clin. Chem. **17:**1154-1159, 1971.
12. Fossati, P., Prencipe, L., and Berti, G.: Use of 3,5-dichloro-2-hydroxybenzenesulfonic acid/4-aminophenazone chromogenic system in direct enzymic assay of uric acid in serum and urine, Clin. Chem. **26:**227-231, 1980.

13. Kabasakalian, P., Kalliney, S., and Westcott, A.: Determination of uric acid in serum, with use of uricase and a tribromophenol-aminoantipyrine chromogen, Clin. Chem. **19**:522-524, 1973.
14. Kageyama, N.: A direct colorimetric determination of uric acid in serum and urine with uricase-catalase system, Clin. Chim. Acta **31**:421-426, 1971.
15. Haeckel, R.: The use of aldehyde dehydrogenase to determine H_2O_2-producing reactions. 1. The determination of the uric acid concentration, J. Clin. Chem. Clin. Biochem. **14**:101-107, 1976.
16. Bell, R., and Ray, R.A.: A rate-sensing approach to the measurement of uric acid in serum and urine, Clin. Chem. **17**:644, 1971.
17. Troy, R.J., and Purdy, W.C.: The coulometric determination of uric acid in serum and urine, Clin. Chim. Acta **27**:401-408, 1970.
18. Kiser, E.J., Johnson, G.F., and Witte, D.L.: Serum uric acid determined by reversed-phase liquid chromatography with spectrophotometric determination, Clin. Chem. **24**:536-540, 1978.
19. Pachla, L.A., and Kissinger, P.T.: Measurement of serum uric acid by liquid chromatography, Clin. Chem. **25**:1847-1852, 1979.
20. Hausen, A., Fuchs, D., König, K., and Wachter, H.: Quantitation of urinary uric acid by reversed-phase liquid chromatography, Clin. Chem. **27**:1455-1456, 1981.
21. Comprehensive Chemistry Survey, Set C-A, Skokie, Ill., 1982, College of American Pathologists.
22. Natelson, S.: Uric acid. In Reiner, M., editor: Standard methods of clinical chemistry, vol. 1, New York, 1953, Academic Press, Inc., pp. 123-135.
23. Caraway, W.T.: Uric acid. In Seligson, D., editor: Standard methods of clinical chemistry, vol. 4, New York, 1964, Academic Press, Inc., pp. 239-247.
24. Duncan, P., Gochman, N., Cooper, T., et al.: Development and evaluation of a candidate reference method for uric acid in serum, 1979. U.S. Department of Health, Education, and Welfare, Public Health Service, Atlanta, 1980, U.S. Centers for Disease Control.
25. Fales, F.W.: Recovery of uric acid from serum, Clin. Chem. **14**:449-455, 1968.

Chapter 60 Proteins

Albumin

STEPHEN GENDLER

Clinical interpretation: pp. 403 and 420
Molecular weight: 66,248 daltons
Merck Index: 202
Chemical class: protein, globular

PRINCIPLES OF ANALYSIS AND CURRENT USAGE

Albumin is a globular protein that can be defined by five of its characteristics: (1) it is soluble in 2.03 mol/L ammonium sulfate at 23° C, at pH values greater than 6, and when dialyzed against distilled water; (2) the migration of the protein in an electrophoretic field is −6.0 Tiselius mobility units in barbital buffer (ionic strength 0.1, pH 8.6), in which one mobility unit is $10^{-5}\ cm^2 \cdot V^{-1} \cdot sec^{-1}$; (3) the molecular weight is approximately 66,000 daltons and sediments at a velocity of 4.5 S; (4) the protein is free of carbohydrate; and (5) albumin is the main protein component of normal human serum.[1] Human albumin has been isolated and purified to the extent that is has been sequenced. There are 584 residues, and the molecular weight calculated from this sequence is 66,248 daltons. The protein has diverse functions, playing important roles in the maintenance of the colloid osmotic pressure of the blood; in transport of various ions, amino acids, and hormones; and in nutrition.[2]

The earliest techniques for albumin analysis are based on acid or salt precipitation of the protein (method 1, Table 60-1).[1] In the late nineteenth century albumin was defined as the serum protein that remained in solution at 2.05 mol/L ammonium sulfate at 25° C.[3] The albumin in the supernatant was measured by total nitrogen analysis or by a biuret reaction. Modern salt fractionation techniques are cumbersome because of the many steps required in the procedure and not very specific because of protein interactions.[4] Since they are not easily automated, the precipitation methods are largely of historical interest.

The measurement of albumin can be performed by a direct determination of globulin based on tryptophan content and calculation of the albumin content by subtraction of globulin from total protein (method 2, Table 60-1). The tryptophan content of serum albumin is 7% to 10% that of globulin on a weight basis. A procedure proposed by Goldenberg and Drewes[5] takes advantage of this large difference in tryptophan content between albumin and globulin or globulins. In this one-reagent system, glyoxylic acid, in the presence of Ca^{++} in an acid medium, condenses with the tryptophan residues in globulins to produce a purple color measured at 540 nm. The method needs to be standardized with serum to compensate for the albumin interference in the reaction. Note that in this method, free tryptophan does not interfere. Savory et al.[6] have adapted this globulin technique and a total protein analysis to the AutoAnalyzer (Technicon Instruments Corporation, Tarrytown, NY 10591). Tryptophan content methods have never come into common use because of the ease and specificity of the dye-binding methods for albumin.

Serum albumin can be quantitated by electrophoretic techniques (method 3, Table 60-1). The major classes of serum protein are separated by a serum protein electrophoresis method (such as cellulose acetate or agarose; see p. 1309). The separated fractions are stained and the percentage of each fraction present in the sample is determined by densitometric analysis. The concentration of albumin is calculated by multiplication of the concentration of total protein in the sample by the percentage of albumin. Approximately 1.6% of participants in a recent College of American Physiologists (CAP) Comprehensive Laboratory Proficiency Survey reported using electrophoretic procedures.

Immunological methods (method 4a and 4b, Table 60-1) used for albumin quanitation include radial immunodiffusion (RID) and electroimmunodiffusion (EID) in which albumin either passively diffuses (RID) or is electrophoresed (EID) into a stationary phase (such as agarose) that contains antibodies to albumin. The precipitin lines

Table 60-1. Methods of albumin analysis

Method	Type of analysis	Principle	Usage	Comments
1. Precipitation a. Salt fractionation b. Solvent fractionation c. Acid fractionation	Quantitative	Changes of net charge of protein result in precipitation	Serum, manual	Historical, still used in manufacturing albumin
2. Tryptophan content	Quantitative	Glyoxylic acid + Tryptophan in globulin → Purple chromogen (A_{max}, 540 nm) Total protein − Globulin = Albumin	Serum, manual, and automated	Good correlation with electrophoresis, but requires total protein measurement
3. Electrophoresis a. Moving boundary b. Cellulose acetate c. Cellulose acetate with elution of peak	Quantitative	Albumin separated from other proteins in electrical field Percent staining of albumin fraction multiplied by total protein value	Serum, manual, and automated	Very labor intensive, but if albumin is eluted for measurement, very accurate
4. Immunochemical				
a. Electroimmunoassay	Quantitative	Migration of protein in electrical field through medium containing a specific antibody	Serum, manual	Reference method, somewhat labor intensive
b. Radial immunodiffusion	Quantitative	Diffusion of protein through medium containing specific antibody	Serum, manual	Reference method, very long incubation time
c. Turbidimetry	Quantitative	Antigen-antibody complexes decrease light transmission more than free antigen	Serum, manual, or automated	Reagent cost high
d. Nephelometry	Quantitative	Antigen-antibody complexes scatter light more than free antigen	Serum, manual	Reagent cost high
5. Dye binding				
a. Methyl orange	Quantitative	Albumin binds to dye and changes spectral profile of dye (A_{max}, 550 nm)	Serum, manual, or automated	Nonspecific for albumin
b. HABA(2-[4′-hydroxyazobenzene]-benzoic acid)	Quantitative	Same as above (A_{max}, 485 nm)	Serum, manual, or automated	Specific for albumin; poor sensitivity; many drug interferences
c. BCG (bromcresol green)	Quantitative	Same as above (A_{max}, 628 nm)	Serum, manual, or automated (most often used method)	Nonspecific for albumin if absorbance reading taken after 30 seconds
d. BCP (bromcresol purple)	Quantitative	Same as above (A_{max}, 603 nm)	Serum, manual, or automated	Specific for albumin; albumins from animal sources do not bind equivalently to human albumin

formed by the reaction between albumin and the antibody can be fixed and stained. In RID, the diameter (or square of the diameter) of the precipitin ring formed is proportional to the albumin concentration. For EID, the height of the "rocket" precipitin line is related to albumin concentration.

The reaction between albumin and an antialbumin antibody can be monitored by turbidimetric or nephelometric means (methods 4c and 4d, Table 60-1) (see Chapter 10). The antibody-albumin complexes that form increase the absorption (turbidometry) or the scattering (nephelometry) of incident light, which can be related to the albumin concentration. Automated nephelometers are available, and the turbidimetric technique has been adapted for automated analysis as well.[7-9] However, few laboratories are currently reporting the use of immunological techniques for the quantitation of albumin.

The most widely used methods for the analysis of albumin are dye-binding procedures (method 5, Table 60-1). Albumin has the ability to bind a wide variety of organic anions, including complex dye molecules. The dye-binding techniques are based on a shift in the absorption maximum of the dye when bound to albumin. The shift in the absorption maximum allows the resulting color to be measured in the presence of excess dye, which, in concert with the high-binding affinity to albumin, allows all the albumin molecules to take part in the reaction. A wide variety of dyes has been employed for the measurement of albumin including methyl orange, 2-(4′-hydroxyazobenzene)benzoic acid (HABA), bromcresol green (3,3′,5,5′-tetrabromo-*m*-cresolsulfonphthalein, or BCG) and bromcresol purple (5,5′-dibromo-*o*-cresolsulfonphthalein, or BCP). Introduced by Rodkey in 1965,[10] the bromocresol green reaction is usually performed at pH 4.2 to 4.5 and is monitored at 620 to 630 nm. A bromcresol green method was recommended by the American Association for Clinical Chemistry (AACC) in 1972.[11] In this procedure, the absorbance of bromcresol green when albumin binds the dye is measured at 628 nm in a 0.075 mol/L succinate buffer at pH 4.20. Brij-35, a nonionic detergent, is added to buffer and bromcresol green solutions to reduce the blank absorbance, prevent turbidity, and provide linearity. Human serum albumin Cohn fraction V (HSA-V) is used for standardization in the method published by the AACC, but bovine serum albumin Cohn fraction V has also been used. Bromcresol purple reacts with albumin at pH 5.2 and shows a color shift measured at 603 nm. One can monitor these dye reactions as end-point reactions or as blanked, fast (60-second) reactions. The bromcresol green procedure is the most frequently used dye-binding assay; 98.4% of CAP Comprehensive Laboratory Proficiency Survey participants report using a bromcresol green procedure for the measurement of albumin.

REFERENCE AND PREFERRED METHODS

The coefficient of variation (CV) of the tryptophan assay method is 1.5% at a level of 38 g of globulin per liter and, at a level of 26.3 g of globulin per liter is 4.3%.[5,6] This manual method was compared to a salt fractionation technique. No statistical difference was found between the method means using the paired *t* test at a 95% confidence interval, and recovery of globulin added to serum was quantitative.[5] Comparison of this technique to cellulose acetate electrophoresis shows the slope of the regression line to be 1.003. The standard deviation about the regression line is 3.0 g/L and the correlation coefficient is 0.915. Comparison with the HABA dye technique shows the slope of the regression line to be 0.889. The standard deviation about the regression line is 5.5 g/L, and the correlation coefficient is 0.800. An additional advantage of the tryptophan-globulin method is the excellent reagent stability, reported to be at least 1 year at refrigerated temperatures.[4] The method is linear up to 80 g of globulin per liter.

Electrophoresis is considered by many to be a reference method for albumin determination. However, it is a labor-intensive technique that is difficult to automate. In addition, there is no dye that has been demonstrated to bind to all serum proteins equally or to have a binding strength that is linear with the concentration of all serum proteins.[12] Thus the electrophoretic procedures tend to overestimate albumin, since it tends to bind the stains best. Elution of the proteins from the membrane and use of a standard circumvent these problems but make the method even more labor intensive.[13]

The sensitivity and specificity of EID and RID have been compared and are similar for the analysis of serum.[14] Imprecision of immunological methods measured as within-run coefficients of variation are as follows: RID, 1.4%, EID, 3.8%, laser nephelometry, 3.5%; and turbidimetry, 2.8%. These four methods were compared by analysis of normal human serum pools. The resulting values differ by as much as 9.7% from one method to another. Similar differences are seen with commercially prepared human serum albumin.[15]

Given the availability of automated nephelometers and adaptations of turbidimetric techniques to automated analyzers,[7-9] these two techniques may become more popular. What might limit the use of these techniques for albumin analysis is cost, which would be approximately $0.50 to $1.00 per test for reagents alone.

The dye-binding techniques have variable specificities for albumin because of the ability of other serum proteins to also bind the dyes and cause a color shift. The methyl orange method is nonspecific, with interferences from beta lipoproteins and $alpha_1$ and $alpha_2$ globulins, and it gives falsely elevated results at low albumin concentrations.[16] This technique has less protein and bilirubin interferences than the HABA method does, but albumin measurement by methyl orange is of little value when the patient has nephritis.[1] Although the binding of the HABA dye is fairly specific for albumin, the sensitivity is poor. In addition, salicylate, sulfonamides, penicillin, and conjugated bilirubin interfere with albumin binding to the dye. Heparin is a positive interferent, since it causes turbidity. Human albumin Cohn fraction V or human mercaptoalbumin must be used as a standard, since heat-treated or chemically stabilized albumin does not bind the same as the human product.[16] Correlation between HABA and cellulose acetate is poor, especially in newborns with hemolytic disease and in patients with liver disease.[6] Correlation is also poor with sera from hospitalized patients, in which electrophoretic measurement of albumin is biased higher than HABA in 70% of the cases, with the difference being greater than 4 g/L in 37% of hospitalized patients.[17] The HABA method is also known to underestimate albumin in patients with renal failure by an average of 5.0 g/L.[18]

Specificity of the bromcresol green method is good as shown by correlation with salt fractionation when sera from normal patients were analyzed.[11] Bilirubin, slight to moderate lipemia, and salicylate have no interference with the bromcresol green technique. Hemoglobin decreases the albumin value by 1 g/L for each 100 g/L added. Blanking does not correct this interference, and the negative bias is therefore caused by interference with the dye binding rather than hemoglobin color. The coefficient of variation of this technique in routine use is 2% at a level of 45 g/L.[19]

The question of specificity has become an issue with

the bromcresol green method when used for certain populations. Poor correlation of the BCG method was seen with cellulose acetate electrophoresis, with EID and RID. In the range of 35 to 50 g/L, the results are similar. Specimens having less than 35 g/L are higher by bromcresol green than by electrophoresis.[20] In correlation studies, linear regression equations between electrophoresis (abscissa) and EID is $y = 1.03x - 1.3$, and between electrophoresis and radial immunodiffusion is $y = 0.995x - 0.4$.[13] Yet correlation between electrophoresis and an automated bromcresol green technique yields an equation of $y = 0.70x + 12.2$, and with a semiautomated discrete technique yields $y = 0.77x + 9.9$.[13] This interference has been shown by Webster to be by alpha and beta globulin fractions, and not gamma globulins. A serum from a patient with nephrotic syndrome with high alpha globulin levels was first fractionated by electrophoresis. When "albumin" was measured in all the nonalbumin protein fractions by bromcresol green and by EID, bromcresol green levels of "apparent albumin" were 25 to 55 times greater than EID levels.[20] Speicher et al.[21] showed that alpha globulins produce about one-third, beta globulins produce about one-ninth, fibrinogen one-fifth, and hemoglobin the same color intensity with bromcresol green reagent that an equivalent weight of human serum albumin produces.[21] Proposals to eliminate this interference include adding sodium chloride to a concentration of 0.8 mol/L,[22] reducing reaction temperature from 45° C to room temperature, and reducing detergent concentration.[23]

Probably the most promising adaptation of the bromcresol green reaction for albumin analysis is fast reaction readings. Gustafsson[24] reported that measuring the absorbance of the bromcresol green–protein complex at 629 nm at a time shortly after mixing improves the specificity of the assay. The measurement was made by extrapolation of the reaction-absorbance curve back to zero time. This work showed that albumin changes the A_{629} of bromcresol green at "0 minutes," but "acute phase reactants" such as ceruloplasmin and orosomucoid contribute significantly to A_{629} only at times greater than 5 minutes. The fast reaction for albumin is not affected by pH in the range 3.9 to 4.4 and by temperature in the range 20° to 45° C. The slow-reacting proteins are not affected by the pH 3.9 to 4.4, but as temperature is increased to above 30° C, there is some increase in the A_{629} because of these proteins. Using the method of Doumas and Biggs[11] and varying the time of the absorbance reading, Gustafsson compared EID (abscissa) to bromcresol green at 60 minutes and obtained a regression equation of $y = 0.62x + 23.56$. Comparison of EID to the same bromcresol green method, only with measurement of the A_{629} immediately after mixing, improved this equation to $y = 0.98x + 1.83$. Several such successful modifications have been made to many types of automated instruments. Extrapolation does not appear to be necessary to improve specificity. Nonextrapolated readings taken at 10 to 30 seconds on a spectrophotometer,[25,26] at 10 seconds on a reaction-rate analyzer or discrete multichannel system,[27,28] at 0.5 to 6 seconds on a centrifugal analyzer,[29,30] or at 20 to 30 seconds on a continuous-flow analyzer[31-33] all improve the specificity of the bromcresol green reaction. The concentration of bromcresol green is a factor in the linearity of the reaction and is optimal at 60 μmol/L.[28]

Bromcresol purple seems to have overcome most disadvantages seen in other dye-binding methods. It compares well with EID (abscissa), giving a linear regression equation of $y = 0.95x + 1.72$ when analysis is done on a population selected to provide a maximum range of albumin, bilirubin, and globulin concentrations. Sera with lipemia, hemolysis, and drugs were also part of this tested population. A blank correction is required on icteric sera and on grossly hemolyzed and grossly lipemic sera. About 10 g of hemoglobin per liter is equivalent to 1 g of "albumin" per liter. Bromcresol purple does not bind to nonalbumin proteins, though ceruloplasmin and orosomucoid have not been tested specifically. This specificity is seen without the need of fast reaction readings. Human serum albumin Cohn fraction V must be used as a standard, since bovine and equine serum albumin–dye complexes do not absorb as strongly at 600 nm as human albumin–dye complexes.[34] Heparin has a positive interference with bromcresol purple and with bromcresol green methods. This interference can be eliminated by the addition of hexadimethrine bromide to a concentration of 50 mg/L in the bromcresol purple reagent.[35] The bromcresol purple dye has been adapted for use in continuous-flow analysis,[34] a centrifugal analyzer,[36] and the du Pont ACA clinical analyzer (E.I. du Pont de Nemours & Co., Wilmington, Del.). The procedure on the centrifugal analyzer has a within-day coefficient of variation of 1.1% at 35 g/L and a between-day coefficient of variation of 2.2% at 30 g/L.[36] Given the specificity of the reaction without fast absorbance readings, the lack of many interferences, the excellent correlation with reference methods, and the precision and ease of use, bromcresol purple is the method of choice.

SPECIMEN

Serum is the specimen of choice, but heparinized plasma can also be used if precautions are taken to prevent heparin interference (see above).

PROCEDURE

Principle

Bromcresol purple (BCP) complexes with albumin resulting in the dye having a spectral shift. The presence of albumin increases the absorbance at 603 nm.

Reagents[34,35]

Stock BCP solution (80 mmol/L). Dissolve 1.08 g of BCP (pH indicator grade) in 15 mL of absolute ethanol. When a clear orange solution is obtained, dilute to 25 mL

Men			Women		
Age	Albumin, g/L	(mmol/L)	Age	Albumin, g/L	mmol/L
21-44	33.3-61.2	(0.50-0.92)	20-44	27.8-56.5	(0.42-0.85)
45-54	29.1-61.2	(0.44-0.92)	45-54	24.6-54.4	(0.37-0.82)
55-93	31.6-54.6	(0.48-0.82)	55-81	32.0-52.8	(0.48-0.80)

with absolute ethanol. The reagent is stable at 4° C for at least 3 months.

Brij-35 solution (250 g/L). Dissolve 25 g of "Brij-35" (polyoxyethylene lauryl ether; Sigma Chemical Co., St. Louis, MO 63178) in distilled water, with warming, and dilute to 100 mL. Stable for 1 year at room temperature.

Stock acetic acid solution (1.8 mol/L). Dilute 150 mL of glacial acetic acid (AR grade) to 1 L with distilled water. Stable for 1 year at room temperature.

Working BCP reagent (80 μmol/L). Dissolve 10 g of sodium acetate trihydrate (AR grade) in about 800 mL of distilled water. Add 10 mL of stock acetic acid solution, 1 mL of Brij-35 solution, 1 mL of stock BCP solution, and 50 mg of hexadimethrine bromide (Polybrene; Aldrich Chemical Co., Milwaukee, WI 53233), and dilute to 1 L with distilled water. Check pH and adjust to 5.20 ± 0.03 with stock acetic acid solution or 0.5 mol/L sodium hydroxide solution. The reagent is stable for at least a week at room temperature.

Sodium chloride solution (0.15 mol/L). Dissolve 9 g of sodium chloride in a final volume of 1 L of distilled water. Stable for 1 year at room temperature.

Standardization. At present, there is no adequate human-based albumin standard. Any native serum can be used, as long as it is calibrated against a monomeric albumin solution standardized by a reference method such as EID. For standardization that might be less than optimal, highly purified human serum albumin fraction V might be used, if confirmed for purity by electrophoresis.

Assay

Equipment. GEMSAEC Centrifugal Analyzer (Electro-Nucleonics, Inc., Fairfield, NJ 07006) is used with the following settings:

Sample volume	29 μL
Flush volume (saline)	90 μL
Reagent volume	400 μL
Wavelength	603 nm
Filter	560 to 700 nm
Number of readings	1
Starting absorbance	1.25

Linearity. The method is linear from 5 to 50 g/L.[36]

Interferences. For every 150 mg/L salicylate, 100 mg/L bilirubin, and 4.56 g/L hemoglobin, albumin is decreased by 1 g/L. For severely lipemic sera (10 g/L neutral fat) the "apparent albumin" value is increased by 2 g/L.

REFERENCE RANGE

The reference range by age and sex is shown above.[37]

The normal ranges for children and neonates have been reported to be slightly lower than the ranges seen in adults (newborns, 29-55 g/L [0.44-0.83 mmol/L]; children, 38-55 g/L [0.57-0.80 mmol/L]).[38]

Reaction conditions for albumin analysis

Condition	Requirement
Temperature	Room temperature
pH	5.20 ± 0.03
Final concentration of reagent components in cuvette	Bromcresol purple: 65 μmol/L Brij-35: 204 mg/L Polybrene: 50 mg/L Acetate: 60 mmol/L
Fraction of sample volume	0.41
Sample	Serum, heparinized plasma
Linearity	5-50 g/L
Reaction time	Immediate
Major interferences	Gross hemolysis, gross lipemia, bilirubin greater than 100 mg/L
Precision (between day)	CV: 2.2% at 30 g/L

REFERENCES

1. Peters, T., Jr.: Serum albumin, Adv. Clin. Chem. **13**:37-111, 1970.
2. Peters, T., Jr.: Serum albumin. In Putnam, F.W., editor: The plasma proteins, vol. 1, New York 1975, Academic Press, Inc., pp. 133-181.
3. Watson, D.: Albumin and "total globulin" fractions of blood, Adv. Clin. Chem. **8**:237-303, 1965.
4. Henry, R J.: Proteins. In Henry, R.J.: Clinical chemistry: principles and technics, New York, 1964, Harper & Row, Publishers, Inc., pp. 199-226.
5. Goldenberg, H., and Drewes, P.A.: Direct photometric determination of globulin in serum, Clin. Chem. **17**:358-362, 1971.
6. Savory, J., Heintges, M.G., and Sobel, R.E.: Automated procedure for simultaneously measuring total globulin and total protein in serum, Clin. Chem. **17**:301-306, 1971.
7. Blom, M., and Hjørne, N.: Immunochemical determination of serum albumin with a centrifugal analyzer, Clin. Chem. **21**:195-198, 1975.
8. Kamp, H.H., Luderer, T.K.J., Muller, H.J., and Sopjes-Kruk, A.: Rapid immunoturbidimetric assay of albumin and immunoglobulin G in serum and cerebrospinal fluid with an automatic discrete analyser, Clin. Chim. Acta **114**:195-205, 1981.
9. Dito, W.R.: Rapid immunonephelometric quantitation of eleven serum proteins by centrifugal fast analyzer, Am. J. Clin. Pathol. **71**:301-308, 1979.
10. Rodkey, F.L.: Direct spectrophotometric determination of albumin in human serum, Clin. Chem. **11**:478-487, 1965.

11. Doumas, B.T., and Biggs, H.G.: Determination of serum albumin, Stand. Methods Clin. Chem. **7**:175-188, 1972.
12. Peters, T., Jr.: Serum albumin: recent progress in the understanding of its structure and biosynthesis, Clin. Chem. **23**:5-12, 1977.
13. Webster, D., Bignell, A.H.C., and Attwood, E.C.: An assessment of the suitability of bromocresol green for the determination of serum albumin, Clin. Chim. Acta **53**:101-108, 1974.
14. Laurell, C.B.: Electroimmunoassay, Scand. J. Clin. Lab. Invest. Suppl. **124**:21-37, 1972.
15. Keyser, J.W., Fifield, R., and Watkins, G.L.: Standardization of immunochemical determinations of serum albumin, Clin. Chem. **27**:736-738, 1981.
16. Cannon, D.C., Olitsky, I., and Inkpen, J.A.: Proteins. In Henry, R.J., Cannon, D.C., and Winkelman, J.W., editors: Clinical chemistry: principles and technics, New York, 1974, Harper & Row, Publishers, Inc., pp. 448-502.
17. Arvan, D.A., and Ritz, A.: Measurement of serum albumin by the HABA dye technique: a study of the effect of free and conjugated bilirubin, of bile acids, and of certain drugs, Clin. Chim. Acta **26**:505-516, 1969.
18. Bergeron, L., Talbort, J., Gauvin, R., et al.: Insuffisance rénale et interférence dans le dosage de l'albumine sérique par une méthode de liaison de colorant, Clin. Chim. Acta **75**:49-58, 1977.
19. Doumas, B.T., Watson, W.A., and Biggs, H.G.: Albumin standards and the measurement of serum albumin with bromcresol green, Clin. Chim. Acta **31**:87-96, 1971.
20. Webster, D.: A study of the interaction of bromocresol green with isolated serum globulin fractions, Clin. Chim. Acta **53**:109-115, 1974.
21. Speicher, C.E., Widish, J.R., Gaudot, F.J., and Hepler, B.R.: An evaluation of the overestimation of serum albumin by bromcresol green, Am. J. Clin. Pathol. **69**:347-350, 1978.
22. O'Donnell, N., and Lott, J.A.: Reducing the interference from globulins in the bromocresol green (BCG) determination of serum albumin, Clin. Chem. **24**:1004, 1978 (Abstract).
23. Prior, M.P., O'Leary, T.D., Coles, M.E., et al.: Erroneously high albumin values by SMAC and their correction, Clin. Chem. **22**:2056-2057, 1976 (Letter).
24. Gustafsson, J.E.C.: Improved specificity of serum albumin determination and estimation of "acute phase reactants" by use of the bromocresol green reaction, Clin. Chem. **22**:616-622, 1976.
25. Webster, D.: The immediate reaction between bromcresol green and serum as a measure of albumin content, Clin. Chem. **23**:663-665, 1977.
26. Corcoran, R.M., and Durnan, S.M.: Albumin determination by a modified bromcresol green method, Clin. Chem. **23**:765-766, 1977 (Letter).
27. Gustafsson, J.E.C.: Automated serum albumin determination by use of the immediate reaction with bromcresol green reagent, Clin. Chem. **24**:369-373, 1978.
28. Robertson, W.S.: Optimizing determination of plasma albumin by the bromcresol green dye–binding method, Clin. Chem. **27**:144-146, 1981.
29. Izquierdo, J.M.: Immediate reaction (0.5s) of serum albumin with BCG in a Cobas Bio Analyzer, Clin. Chem. **28**:1629, 1982 (Abstract).
30. King, S.W., Cross, R.E., and Savory, J.: Improved specificity of the bromcresol green (BCG) dye method for serum albumin using an early (6 seconds) absorbance reading, Clin. Chem. **23**:1136, 1977 (Abstract).
31. Cederblad, G., Hickey, B.E., Hollender, A., and Akerlund, G.: Improved continuous-flow (SMAC) determination of serum albumin, Clin. Chem. **24**:1191-1193, 1978.
32. Plesher, C.J.: Continuous-flow analysis for serum albumin, by use of the immediate reaction with bromcresol green, Clin. Chem. **24**:2036-2039, 1978.
33. von Schenck, H., and Rebel, K.: Improved continuous-flow determination of albumin with bromcresol green, Clin. Chem. **28**:1408-1409, 1982 (Letter).
34. Pinnel, A.E., and Northam, B.E.: New automated dye-binding method for serum albumin determination with bromcresol purple, Clin. Chem. **24**:80-86, 1978.
35. Duggan, J., and Duggan, P.F.: Albumin by bromcresol green: a case of laboratory conservatism, Clin. Chem. **28**:1407-1408, 1982 (Letter).
36. Haythorn, P, and Sheehan, M.: Improved centrifugal analyzer assay of albumin, Clin. Chem. **25**:194, 1979 (Letter).
37. Denko, C.W., and Gabriel, P.: Age- and sex-related levels of albumin, ceruloplasmin, α_1-antitrypsin, α_1 acid glycoprotein, and transferrin, Ann. Clin. Lab. Sci. **11**:63-68, 1981.
38. Meites, S., editor: Pediatric clinical chemistry, Washington, D.C., 1977, American Association for Clinical Chemistry, p. 20.

Bence Jones protein

GAYLE BIRKBECK

Clinical significance: p. 882
Molecular weight: 20,000 daltons
Chemical class: protein

PRINCIPLES OF ANALYSIS AND CURRENT USAGE

Bence Jones protein is a monoclonal, free-immunoglobulin light chain synthesized by a clone of plasma cells (usually malignant) in one of the two following patterns: (1) production of only light chains or (2) an unbalanced synthesis resulting in both the production and secretion of intact immunoglobulin molecules and excess light chains.

In 1845, Henry Bence Jones, a physician and chemical pathologist, first described the protein in the urine of a patient now known to have had multiple myeloma. The urinary substance had the peculiar thermal property of precipitating when heated to a range of 45° to 60° C and redissolving on boiling. The thermal property (method 1, Table 60-2) was used as the basis for the Bence Jones heat test, which served as the test for this urinary protein in clinical laboratories until the early 1970s. In 1963, Schwartz and Edelman[1] compared tryptic hydrolysates from the light chain of a myeloma globulin and Bence Jones protein from the same patient. They concluded that Bence Jones protein consists of entire light chains that have not been incorporated into the autologous myeloma protein. Further studies involving amino acid sequencing showed that Bence Jones protein is composed of approximately 214 amino acid residues and has the same structure as normal immunoglobulin light chains.[2] Bence Jones protein can be either a kappa or a lambda light chain.

Heat precipitation, precipitation with acids or salts (method 2, Table 60-2), and electrophoresis and immunoelectrophoresis have been used for the detection of Bence Jones protein. The Bence Jones heat test consisted in heating the urine to 45° to 60° C, at which time a precipitate formed. The precipitate dissolved on boiling, and reprecipitation occurred when it cooled. Tests using precipitation with acids or salts, or both, include the following: (1) layer urine over concentrated HCl or HNO_3, (2) sulfosalicylic acid, (3) trichloroacetic acid, (4) toluenesulfonic acid, and (5) 16% *n*-propanol at 37° C. When Bence Jones proteins are present, a protein precipitate will occur.[3] Protein electrophoresis (method 3, Table 60-2) and immunoelectrophoresis (method 4, Table 60-2; see also discussion of immunoelectrophoresis, p. 1297) are the most commonly used methods for screening for Bence Jones protein. Immunofixation (p. 1298) has been proposed as a more sensitive test for detection of Bence Jones protein in the urine[4] (method 5, Table 60-2).

REFERENCE AND PREFERRED METHODS

The Bence Jones heat test uses the peculiar thermal properties of the protein for detection. Putnam et al.[5] found that the extent of precipitation is dependent on the ionic strength and electrolyte composition of the reaction mixture and on the pH range. A rigid 4.6 to 5.4 pH range is necessary for optimum precipitation. The heat test is relatively insensitive and is unable to detect Bence Jones protein at less than 1450 mg/L.[6] In addition, the heat test is nonspecific. Other proteins, such as transferrin, can give a false-positive heat test.[7] Heterogenous free light chains are found in normal urine in very small amounts, not exceeding 50 mg/24 hours.[8] These polyclonal free light chains (both kappa and lambda) have the thermal properties of Bence Jones proteins. Patients with connective tissue disease and other disorders that have polyclonal hypergammaglobulinemia show an increase in the excretion of free polyclonal light chains, and their urine may produce a false-positive Bence Jones heat test.[9]

All Bence Jones proteins, as defined immunochemically, do not possess the same characteristic thermal properties. Stone and Frenkel[10] found that only six of 18 urine specimens from patients with light-chain disease had a positive heat precipitation test. Deegan[2] reported a patient with light-chain disease who was losing 10 to 14 g of protein every 24 hours in his urine and had several negative heat tests. Therefore, because of the insensitivity and nonspecificity of the heat precipitation test, it should no longer be used as a screening test for Bence Jones protein.

Hobbs[41] reviewed tests using precipitation with acids or salts. On 79 urine specimens containing more than 50 mg/L Bence Jones protein, the following failure rates were observed: (1) layer urine over concentrated HCl, 5%; (2) layer urine over concentrated HNO_3, 15%; (3) 2% to 3% sulfosalicylic acid, 6%, a further 18% incompletely precipitated; (4) 10% trichloroacetic acid, 5%. The failure rates on 66 urine specimens containing more than 150 mg/L Bence Jones protein and using four other less-effective precipitin tests ranged from 31% to 52%. The dipstix test for protein (Ames Division, Miles Laboratories, Elkhart, Ind.) is inconsistent and results in frequent negative results in the presence of significant amounts of Bence

Table 60-2. Methods of Bence-Jones protein analysis

Method	Type of analysis	Principle	Usage	Comments
1. Heat Precipitation	Qualitative	Bence Jones proteins precipitate from acidic solutions heated to 45° to 60° C and redissolve on boiling	Historical	Neither specific nor sensitive
2. Acid Precipitation	Qualitative	Strong acids cause precipitation of Bence Jones proteins at interface between solution and acid	Historical	Not very sensitive
3. Protein Electrophoresis	Qualitative	Bence Jones proteins separate from other proteins and may be identified as an aberrant band	Common	Not sensitive, but useful as a screening procedure Requires confirmation by immunological test (method 4)
4. Immunoelectrophoresis	Qualitative	Bence Jones proteins separated from other proteins by electrophoresis; reacted with antibodies to light chains to form visible precipitate	Common	Sensitive
5. Immunofixation	Qualitative	Similar to method 4	Rare	Most sensitive
6. Total urinary protein analysis	Quantitative	Measurement of total urinary protein excretion; in absence of renal failure is indicative of possible myeloma disease	Rare	Sensitive, but not specific; requires confirmation by immunological procedure

Jones protein.[11] Total urine protein, including Bence Jones protein, may be measured by the biuret method after precipitation from urine with ethanolic phosphotungstic acid.[12] This method has been used in my laboratory for 5 years.

This total protein method has been used with approximately 165 urine protein electrophoresis scans per year. Information on the number of specimens containing Bence Jones protein is unavailable; however, a moderate number of patients with myeloma are diagnosed and followed by our laboratory. No cases have been observed over this 5-year period in which the urine protein electrophoresis has shown the presence of Bence Jones protein and the modified biuret method has failed to detect the protein. Data showing the actual failure rate of this test seem to be unavailable; however, the biuret method seems to be quite effective in my experience.

In summary, none of the precipitation methods by acid or salts seems to be 100% effective in the detection of Bence Jones protein in the urine, and therefore they should not be used as screening tests. The urine dipstick method is an unacceptable screen. However, the chemical precipitation tests are the most practical methods for quantitating Bence Jones protein and are very useful once the presence of Bence Jones protein has been established by serum protein electrophoresis or immunoelectrophoresis.

The preferred methods for screening for Bence Jones protein are protein electrophoresis and immunoelectrophoresis. Urine protein electrophoresis on cellulose acetate or agarose is an appropriate screening test. The urine should be concentrated, usually 50 times. An aliquot from a 24-hour urine collection is the preferred sample, but a sample from a first-voided morning specimen is often adequate. Essentially, any peak other than albumin may be a Bence Jones protein. An immunoelectrophoresis with anti–kappa chain and anti–lambda chain antisera should be performed to ensure that the abnormal peak is not attributable to transferrin, increased amounts of polyclonal light chains, or whole immunoglobulin molecules excreted in elevated amounts because of glomerular disease. The presence of a Bence Jones protein usually results in an abnormal bowing of one light-chain type.

Two types of light-chain antisera are available commercially: light-chain antisera against bound light chains and light-chain antisera against unbound light chains. A whole immunoglobulin molecule is used as the immunogen for the antiserum to the bound chain. Bence Jones protein is used as the immunogen for the unbound chain. One set of light-chain determinants is available only when the light chain is free. Another set of determinants is available when the light chain is either free or bound. Thus, antisera directed against free light chains react with only Bence Jones protein and not bound light chains. Unfortunately, antiserum against free light chains only is not recommended for routine screening for Bence Jones protein.[13,14] A number of subtypes with different antigenic determinants have been found for Bence Jones proteins. Thus antiserum raised against one subtype may react poorly against another. Commercial antiserum usually is prepared against one, two, or at most a few Bence Jones proteins. In general, an antiserum prepared against both free and bound light chains from a variety of sources is preferred.

Immunofixation has been recommended as a more sensitive screen for Bence Jones protein.[4] Whicher found that immunofixation is 10 times more sensitive than electrophoresis and may detect as little as 0.001 g/L Bence Jones protein if the urine is concentrated 200 times. However, one might question the clinical usefulness of such a sensitive test. Monoclonal proteins are present at low levels in the serum of some persons with no apparent disease. These persons may or may not develop a plasma cell malignancy with time. Immunofixation may detect the catabolic light chains from the "benign" monoclonal protein in the urine of these persons. This catabolic product is a free monoclonal light chain but is not the product of a malignant clone that is producing all light chains or more light chains than heavy chains. Dammacco and Waldenström[15] found that 23.8% of 42 patients with benign monoclonal gammapathy had a monoclonal light chain in their urine. However, the amount never exceeded 60 mg/L.

In fact, increased catabolism of a serum monoclonal immunoglobulin may also lead to the excretion of monoclonal light chains into the urine. The standard protein electrophoresis and immunoelectrophoresis will detect these catabolic products. They usually occur at less than 200 mg/L or 300 mg/24 hours.

Therefore, if the serum monoclonal immunoglobulin is greater than 20 g/L, Bence Jones protein should not be reported unless the concentration in the urine is more than 200 mg/L.[9] At levels below 200 mg/L, it is impossible to determine whether the free light chain is a catabolic product or a de novo synthesized monoclonal light chain (Bence Jones).

Kyle and Greipp[54] followed for up to 21 years seven patients with Bence Jones proteinuria (greater than 1.0 g/24 hours) and saw no evidence of malignant disease. Five patients eventually developed multiple myeloma within 7.7 to 21 years. However two patients still had benign Bence Jones proteinuria after 7.7 and 12 years.

Quantitation of Bence Jones protein by immunochemical methods is not effective because of the difference in reactivity of the various subtypes to a particular Bence Jones antiserum. Urine total protein is usually determined by a chemical method and used with the urine protein electrophoresis scan. Sulfosalicylic acid and trichloroacetic acid methods are used for total protein determination. The biuret method[12] is used in my laboratory. Any one of the precipitation methods may not react with a particular Bence Jones protein, and so a urine protein electrophoresis should always accompany the chemical test.

The presence or absence of Bence Jones protein in the urine is dependent on the rate and amount of light-chain synthesis and on the patient's renal status.[2] Labeled Bence Jones protein has a half-life of 4 hours in a normal subject versus 8 to 32 hours in persons with compromised renal status. Thus, if the patient has normal renal function, Bence Jones protein may be seen only in the urine with the protein electrophoresis screen and not in the serum. However, with declining renal status, Bence Jones protein can be detected in the serum. Thus both serum and urine specimens should be screened by electrophoresis before ruling out Bence Jones protein.

Bence Jones protein is most often associated with multiple myeloma; however, the protein has been found in patients with Waldenström's macroglobulinemia, chronic lymphocytic leukemia, amyloidosis, and carcinoma, and in a few patients with no apparent underlying disease.[2]

SPECIMEN

Serum and urine specimens are used to screen for Bence Jones protein. Twenty-four-hour urine specimens are preferred; however, a first-voided morning specimen is often adequate. Serum and urine specimens stored at 4° C for 5 to 7 days remain stable for testing.

PROCEDURE

See sections that discuss immunoelectrophoresis (p. 1297) and urine total protein (p. 1319).

REFERENCES

1. Schwartz, J.H., and Edelman, G.M.: Comparisons of Bence Jones proteins and L-polypeptide chains of myeloma globulins after hydrolysis with trypsin, J. Exp. Med. **118:**41-53, 1963.
2. Deegan, M.J.: Bence Jones proteins: nature, metabolism, detection, and significance, Ann. Clin. Lab. Sci. **6**(1):38-46, 1976.
3. Hobbs, J.R.: The detection of Bence Jones proteins, Biochem. J. **99:**15P, 1966.
4. Whicher, J.T., et al.: Clinical applications of immunofixation: a more sensitive technique for the detection of Bence Jones protein, J. Clin. Pathol. **33:**779-780, 1980.
5. Putnam, F.W., et al.: The heat precipitation of Bence Jones proteins. I. Optimum conditions, Arch. Biochem. Biophys. **83:**115-130, 1959.
6. Lindström, F.D., Williams, R.C., Jr., Swaim, W.R., et al.: Urinary light-chain excretion in myeloma and other disorders: an evaluation of the Bence Jones test, J. Lab. Clin. Med. **71:**812-825, 1968.
7. Bernier, G.M., and Putnam, F.W.: Polymerism, polymorphism, and impurities in Bence Jones proteins, Biochem. Biophys. Acta **86:**295-308, 1964.
8. Dammacco, F., and Waldenström, J.: Bence Jones proteinuria in benign monoclonal gammopathies, Acta Med. Scand. **184:**403-409, 1968.
9. Solomon, A.: Bence Jones proteins and light chains of immunoglobulins. II., N. Engl. J. Med. **294**(2):91-98, 1976.
10. Stone, M.J., and Frenkel, E.P.: The clinical spectrum of light chain myeloma, Am. J. Med. **58:**601-619, 1975.
11. Greenberg, M.S.: Detection of Bence Jones protein in urine, New Engl. J. Med. **289:**806-807, 1973.
12. Savory, J., et al.: A biuret method for determination of protein in normal urine, Clin. Chem. **14:**1160-1171, 1968.
13. Takahashi, M., et al.: Preparation of fluorescent antibody reagents monospecific to light chains of human immunoglobulin, J. Immunol. **102:**1268-1273, 1969.
14. Williams, R.C., et al.: Light chain disease: an abortive variant of multiple myeloma, Ann. Intern. Med. **65:**471-486, 1966.
15. Dammacco, F., and Waldenström, J.: Bence Jones protinuria in benign monoclonal gammapathies, Acta Med. Scand. **184:**403-409, 1968.
16. Kyle, R., and Greipp, P.: ''Idiopathic'' Bence Jones proteinuria, New Engl. J. Med. **306:**564-567, 1982.

Cerebrospinal fluid protein

GAYLE BIRKBECK

Clinical significance: pp. 710 and 718
Chemical class: protein

PRINCIPLES OF ANALYSIS AND CURRENT USAGE

Cerebrospinal fluid (CSF) proteins originate primarily from the ultrafiltration of plasma across the choroidal capillary wall, though some proteins are peculiar to CSF and are synthesized in the central nervous system. The ultrafiltration process removes most plasma proteins so that the total protein concentration of CSF (150 to 450 mg/L) is much lower than that of serum (60 to 78 g/L).[1]

Turbidimetric procedures are most widely used to measure total protein in CSF[1-3] (methods 1 and 2, Table 60-3). These procedures employ either sulfosalicylic acid plus sodium sulfate or trichloroacetic acid to form a protein precipitate in the sample. The turbidity of the precipitate is measured spectrophotometrically. The amount of turbidity produced by albumin differs from that produced by an equal mass of globular proteins. However, it has been reported that these differences are less pronounced when trichloroacetic acid is used.

The Lowry method[1,4] is used in Europe but is infrequently used in this country. The method uses Folin-Ciocalteu reagent and involves the following two steps:

1. Protein reacts with copper in an alkaline solution.
2. Copper-protein complex and any tyrosine and tryptophan present reduce phosphotungstic-phosphomolybdic acids to a colored product, with an A_{max} at 750 nm.

Dye-binding methods have also been reported. A shift in the absorption maximum from 465 to 595 nm is observed when protein is bound to the dye Coomassie brilliant blue G-250. The change in absorbance at 595 nm is used to measure the amount of protein present.[5,6]

REFERENCE AND PREFERRED METHODS
(Table 60-4)

The dye-binding method[5,6] for measurement of total CSF protein is technically simple and requires only 50 to 100 μL of CSF. However, the method is not used widely.[1] The method is difficult to standardize since variability in sensitivity and the standard curve occurs with various proteins.

Table 60-3. Methods of cerebrospinal fluid protein analysis

Method	Type of analysis	Principle	Usage	Comments
1. Sulfosalicylic: 3% sulfosalicylic–7% sodium sulfate	Total protein, turbidimetric	Precipitation of protein with the resultant turbidity measured in a spectrophotometer	CSF (500 μL), commonly used	Sulfosalicylic acid alone results in greater turbidity with albumin than globulin; turbidity with albumin and globulin is the same with sulfosalicylic acid–sodium sulfate combination
2. Trichloroacetic acid	Total protein, turbidimetric	Precipitation of protein with the resultant turbidity measured in a spectrophotometer	CSF (500 μL), commonly used	Less sensitivity and poorer reproducibility than sulfosalicylic acid–sodium sulfate.
3. Lowry method	Total protein, chemical	Uses Folin phenol reagent; (1) protein reacts with copper in an alkaline solution; (2) copper-protein complex and any tyrosine and tryptophan present reduce phosphotungstic-phosphomolybdic acids (A_{max}, 750 nm)	CSF (200 μL), rarely used	Time consuming; influenced by endogenous phenols and drugs
4. Coomassie brilliant blue dye binding	Total protein, chemical, spectrophotometric	When Coomassie brilliant blue G-250 binds to protein, the color changes from brownish orange to an intense blue color, the absorbance of the dye shifts from 465 to 595 nm	CSF (25 to 100 μL), rarely used	Problem with standardization; different proteins show variability in sensitivity and variation in standard curves

Table 60-4. Comparison of reaction conditions for cerebrospinal fluid protein*

Condition	Sulfosalicylic acid	ACA†
Temperature	Ambient	37°
pH	Acid	Acid
Final concentration of reagents	Sulfosalicylic acid, 98.4 mmol/L	Trichloroacetic acid, 0.14 mol/L
Wavelength (nm)	430	340 and 540
Fraction of sample volume	0.166	0.06
Volume of sample	0.2 mL	0.3 mL
Linearity	0-2000 mg/L	0-2000 mg/L
Time of reaction	10 minutes	40 sec
Interferences	Xanthochromia (−) Turbidity (+) Hemolysis (−) Methotrexate (+)	Xanthochromia (−) Turbidity (+) Hemolysis (−)
Precision ($\bar{X}$, % CV)	501 mg/L, 7.3%	500 mg/L, 4%† 200 mg/L, 2%

*(−), Negative interferent; (+), positive interferent.
†From du Pont Instruments, Inc., Wilmington, Del.

The Lowry method[4] is used widely in Europe. It requires only 100 to 200 μL of CSF; however, it is difficult and time consuming. Salicylates, chlorpromazine, tetracyclines, and sulfonamide drugs are among the drugs that interfere with the assay.[1]

CSF total protein is usually measured by turbidimetric methods in the United States. Either the sulfosalicylic acid or the trichloroacetic acid method is acceptable,[2,3] and each agrees well with specific chemical assays. The turbidimetric method is relatively simple and is not affected by most drugs. Xanthochromia and intrathecal methotrexate may interfere.[1] Turbidity can be measured at short (430 nm) or long (650 nm) wavelengths.

SPECIMEN

Cerebrospinal fluid may be stored between 2° and 8° C for at least 5 days if protected from evaporation. Specimens that will not be tested within 5 days should be frozen at 20° C immediately after collection.

PROCEDURE

Principle

Protein is precipitated as a fine white precipitate by the addition of sulfosalicylic acid. The resulting turbidity is determined spectrophotometrically at 430 nm.

Reagents

Sulfosalicylic acid, 30 g/L (118 mmol/L). Dissolve 30 g of sulfosalicylic acid in 800 mL of distilled water and bring to 1 liter. Store in a brown bottle. Stable for 1 year at room temperature.

Control. Ortho Diagnostics CSF.

Assay

Equipment: spectrophotometer capable of reading at 430 nm.

1. Add 1.0 mL of 3% sulfosalicylic acid to "test" tubes; one for an instrument blank, one for the Ortho standard, and one for the patient.
2. Add 0.2 mL of Ortho Diagnostics CSF control to the standard tube (see note 1 at right).
3. Add 0.2 mL of CSF to the patient's tube. Add 0.2 mL of saline solution to "blank" tube.
4. If the spinal fluid is strongly colored, prepare a patient blank using 0.2 mL of CSF and 1.0 mL of saline.
5. Cover all test tubes with plastic caps or Parafilm and gently invert several times.
6. Let stand 10 minutes at room temperature.
7. Set absorbance at zero on the spectrophotometer at 430 nm using the instrument blank of sulfosalicylic acid plus saline.
8. Invert test tube gently to mix, introduce sample into cuvette, and read absorbance *(A)* of all standards and patients.
9. Dilute all samples 1:2 with saline and rerun if concentration is greater than 2000 mg/L.

Calculations

1. Correct absorbance of samples with blanks:

$$A_{\text{patient}} - A_{\text{patient blank}} = A_{\text{corrected patient}}$$

2. Read the corrected absorbance from the standard curve (see below) to obtain the protein concentration.

Controls and standards

1. Since sulfosalicylic acid produces different degrees of turbidity with equal concentrations for different types of proteins, it is necessary to construct a curve using a standard that has the same albumin-globulin ratio as the spinal fluid being tested. Usually spinal fluid and serum have a very close or the same albumin-globulin ratio so that a normal serum control with the proper dilutions can be used to construct the curve. The albumin-globulin ratio of the standard should be between 1.0 and 1.5.
2. To prepare the standard curve dilute a control serum of 70 g/L total protein with saline so that 5 dilutions with protein concentration between 100 and 1500 mg/L are prepared as follows:

Stock volume	Total volume (mL)	Concentration (mg/L)
2.0	100	1400
1.5	100	1050
1.0	100	700
0.5	100	350
0.2	100	140
0	100	0

Do not use a serum control with elevated lipid or bilirubin content.

3. Standard curve. Construct the standard curve, plotting absorbance at 430 nm versus protein concentration. This should be a straight line that passes through the origin.
4. The standard curve needs to be run only when there is a change in the spectrophotometer (that is, bulb change) or lot of quality control sample or the quality control sample does not give the expected value.

Notes

1. The first drop of spinal fluid should be added slowly to the sulfosalicylic acid reagent. If a dense cloud forms immediately, the protein concentration is very high and dilution will be necessary. In this case, the rest of the 0.2 ml sample should not be added to the acid but instead returned to the original tube. The spinal fluid should be diluted with saline.
2. If red blood cells are present in the spinal fluid, the fluid must be centrifuged before protein analysis.

Table 60-5. Cerebrospinal fluid protein in various disease states

Clinical condition	Appearance	Total protein (mg/L)
Normal	Clear, colorless	150-450
Coccal meningitis	Purulent, cells; turbid, opalescent	1000-5000
Tuberculous meningitis	Colorless, fibrin, coagulum, cells	500-3000, occasionally up to 10,000
Benign lymphocytic meningitis	Clear, colorless, lymph cells	300-1000
Encephalitis	Clear, colorless, cells	150-1000
Poliomyelitis	Clear, colorless	100-3000
Neurosyphilis	Clear, colorless	500-1500
Disseminated sclerosis	Clear, colorless	250-500
Spinal cord tumor	Clear, colorless, or xanthochromic	1000-20,000
Brain tumor	Clear	150-2000
Brain abscess	Clear or slightly turbid	300-3000
Cerebral hemorrhage	Colorless, xanthochromic, or bloody	300-1500

Data from Varley, H.: Practical clinical chemistry, ed. 4, New York, 1967, Interscience Publishers, Inc., pp. 708-711; Stewart, C.P., and Dunlop, D.: Clinical chemistry in practical medicine, ed. 6, Edinburgh, 1962, Livingstone, Ltd., pp. 216-217.

REFERENCE RANGE

CSF protein levels usually range between 150 to 450 mg/L. Increases of CSF protein in various disease states can be found in Table 60-5.

REFERENCES

1. Henry, J.B.: Clinical diagnosis and management by laboratory methods, Philadelphia, 1979, The W.B. Saunders Co.
2. Henry, R.J., et al.: Turbidimetric determination of proteins with sulfosalicylic and trichloroacetic acids, Proc. Soc. Exp. Biol. Med. **92:**748-751, 1956.
3. Schriever, H., and Gamblino, S.R.: Protein turbidity produced by trichloroacetic acid and sulfosalicylic acid at varying temperatures and varying ratios of albumin and globulin, Am. J. Clin. Pathol. **44**(6):667-672, 1965.
4. Lowry, O.H., et al.: Protein measurement with the Folin phenol reagent, J. Biol. Chem. **193:**265, 1951.
5. McIntosh, J.C.: Application of a dye-binding method to the determination of protein in urine and cerebrospinal fluid, Clin. Chem. **23**(10):1939-1940, 1977.
6. Spector, T.: Refinement of the Coomassie blue method of protein quantitation, Anal. Biochem. **86:**142-146, 1978.

Ferritin

ROBERT S. FRANCO

Clinical significance: p. 633
Molecular weight (apoferritin): 445,000 daltons
Merck Index: 3957
Chemical class: Protein, ferroprotein

PRINCIPLES OF ANALYSIS AND CURRENT USAGE

Ferritin consists of a protein shell (apoferritin) containing a variable amount of iron. The iron content varies from almost none to 30% and is in the form of colloidal hydrous ferric oxide-phosphate micelles.

Until 1972 no methods with sufficient sensitivity were available to measure ferritin in normal serum. It was therefore assumed that ferritin, known to be a storage form of iron found in many tissues, did not circulate. The advent of assays using radiolabeled antibodies demonstrated not only the presence of ferritin in normal serum but also its usefulness in the diagnosis of disorders resulting in either low or high amounts of storage iron.

The technique introduced by Addison et al.[1] to measure serum ferritin uses an immunoradiometric assay (IRMA). This methodology is similar to radioimmunoassay (RIA) but uses a labeled antibody rather than a labeled antigen.

In the IRMA method (method 1, Table 60-6), serum containing the ferritin antigen to be measured is incubated with an excess of ^{125}I-labeled antibody. Serum ferritin and antibody form a soluble complex that is separated from unreacted antibody by means of an immunoabsorbent consisting of ferritin coupled to cellulose particles. The ferritin-cellulose adsorbs unreacted labeled antibody and removes it from solution after a centrifugation step. The radioactivity remaining in solution represents antibody bound to serum ferritin and is directly related to serum ferritin concentration.

A modification of the IRMA method called the two-site IRMA was introduced for the measurement of serum ferritin in 1974 by Miles et al.[2] In this assay, unlabeled antibody to ferritin is bound to a solid support, usually a test tube or a plastic bead. The antigen (serum ferritin) reacts with the antibody, thus also becoming attached to the solid support. After washing, purified ^{125}I-labeled antibody is added to form an antibody–antigen–^{125}I-antibody complex attached to the solid support. After unreacted ^{125}I antibody is washed away, the amount of radioactivity associated with the solid support is directly related to the serum ferritin concentration (method 2, Table 60-6).

The third type of assay for serum ferritin uses competitive radioimmunoassay.[3] The serum ferritin and radiolabeled ferritin are incubated with antiferritin antibody and compete for antibody binding sites. A higher concentration of ferritin in the serum will result in a larger percentage of the limited number of antibody binding sites being occupied by unlabeled ferritin. The antigen-antibody complexes are precipitated from solution by a second antibody directed against the antiferritin antibody. The radioactivity in the precipitated complex is inversely related to the serum ferritin concentration (method 3, Table 60-6).

The available methods for assay of serum ferritin are listed in Table 60-6.

METHODS OF CHOICE

A disadvantage of the two-site IRMA for ferritin is the high-dose hook effect. This is a paradoxical decrease in radioactive counts at high concentration of antigen, resulting in a maximum in the curve of radioactivity versus concentration. Since the hook effect takes place at very high serum ferritin concentration, this problem does not occur for levels of ferritin in the normal range. However, some patients can have extremely elevated ferritin levels; therefore it is good practice to run two dilutions of each sample and compare the results. If the calculated ferritin concentration of the more dilute sample is significantly higher than the less dilute sample, it could be because of the hook effect.

A number of commercial kits are now available for the assay of serum ferritin using either the 2-site IRMA or RIA. They are listed in Table 60-7 together with some of their important characteristics. All the kits allow the assay of ferritin within 1 working day and require no special skills or equipment beyond a gamma-ray counter. Evaluations of at least one of the kits have been published.[4,5]

REFERENCE RANGE

Expected values for healthy subjects and iron-deficient patients are provided by the kit manufacturer. In general, the normal range is approximately 20 to 200 ng/mL for

Table 60-6. Methods for measurement of serum ferritin

Method	Principle	Usage	Comment
1. IRMA (1-site)	Antibody is labeled, serum ferritin-antibody complex is formed and unreacted antibody removed, remaining radioactivity is directly related to serum ferritin concentration	Not commercially available for clinical use	First method with enough sensitivity to measure normal levels
2. IRMA (2-site)	Serum ferritin bound to solid-phase antibody, labeled antibody then bound to immobilized ferritin; after washing, the radioactivity of the bound labeled antibody is directly related to serum ferritin concentration.	Commercially available for clinical use (Table 60-7)	Hook effect at high concentration (see text)
3. RIA	Serum ferritin and labeled ferritin compete for antibody, antigen-antibody complexes are precipitated and counted, radioactivity is inversely related to serum ferritin concentration	Commercially available for clinical use (Table 60-7)	Require larger sample than 2-site IRMA

Table 60-7. Commercially available ferritin assay kits and some of their characteristics*

Manufacturer	Type of assay	Support material	Sensitivity† (ng/mL)	Precision CV (%)		Sample size (μL)
				Intrarun CV§	Mean CV‖	
Ramco	2-site IRMA	Polystyrene bead	1.4	6.5(68)	14.8	10
Corning	2-site IRMA	Glass particles	1.6	3.9(101)	—	50
American Dade	2-site IRMA	Polypropylene tubes	3.8	7.3(105)	—	10
Clinical Assays	Competitive RIA	None	1.3	10.8(68)	15.8	100
Bio-Rad	2-site IRMA single incubation	Polyacrylamide beads	2.0 0.9‡	5.8(137)	—	50
Becton Dickinson	Competitive RIA	None	6.0 2.0‡	3.5(128)	18.4	200

*This list may not include all available kits.
†From manufacturers' literature.
‡Optional high-sensitivity protocol.
§From manufacturers' literature at concentration shown.
‖Mean between-laboratory coefficient of variation, CAP 1980 survey.

adult males. However, normal values are both age related and sex related, with females and children under 16 exhibiting lower values as a result of physiologically lower iron stores. A serum ferritin of less than 10 ng/mL almost always indicates iron deficiency. Although there is some overlap between the normal and iron-deficient population, this is not usually a diagnostic problem because in anemia resulting from other causes serum ferritin is usually increased.

CHANGES IN DISEASE

Serum ferritin is especially useful in distinguishing iron deficiency from the anemia of chronic disorders because in the latter ferritin levels are increased.[6,7] Serum ferritin is also increased in other anemias[8] including aplastic anemia, sideroblastic anemia, and chronic hemolytic anemias. In idiopathic hemochromatosis and in multiply transfused patients the serum ferritin may be extremely high.

Note that certain diseases can elevate the serum ferritin and obscure its relationship to iron stores.[8] These include liver disease, hematological and nonhematological malignancies, and nonspecific tissue damage. See Chapters 24 and 30 for additional discussion.

REFERENCES

1. Addison, G.M., et al.: An immunoradiometric assay for ferritin in the serum of normal subjects and patients with iron deficiency and iron overload, J. Clin. Pathol. **25:**326-329, 1972.
2. Miles, L.E.M., et al.: Measurement of serum ferritin by a 2-site immunoradiometric assay, Anal. Biochem. **61:**209-224, 1974.
3. Marcus, D.M., and Zinberg, N.: Measurement of serum ferritin by radioimmunoassay: results in normal individuals and patients with breast cancer, J. Natl. Cancer Inst. **55:**791-795, 1975.
4. Sheehan, R.G., Newton, M.J., and Frenkel, E.P.: Evaluation of a packaged kit assay of serum ferritin and application to clinical diagnoses of selected anemias, Am. J. Clin. Pathol. **70:**79-84, 1978.
5. Li, P.K., Humbert, J.R., and Cheng, C.: Evaluation of a commercially obtainable ferritin test kit in relation to the high-dose parabolic phenomenon, Clin. Chem. **24:**650-1651, 1978.
6. Lipschitz, D.A., Cook, J.D., and Finch, C.A.: A clinical evaluation of serum ferritin as an index of iron stores, N. Engl. J. Med. **290:**1213-1216, 1974.
7. Bentley, D.P., and Williams, P.: Serum ferritin concentration as an index of storage iron in rheumatoid arthritis, J. Clin. Pathol. **27:**786-788, 1974.
8. Alfrey, C.P.: Serum ferritin assay, CRC Crit. Rev. Clin. Lab. Sci., pp. 179-208, Nov. 1978.

Glycohemoglobin

MARY ELLEN KING

Clinical significance: p. 520
Molecular weight: 64,500 daltons
Merck Index: 4505 (hemoglobin)
Chemical class: glycosylated protein

HbA–Val–NH_2	HbA–Val–N		HbA–Val–NH
+	‖		|
HCO	HC		CH_2
HCOH	HCOH		C=O
HOCH	HOCH	Amadori	HOCH
HCOH	⇌ HCOH	rearrangement →	HCOH
HCOH	HCOH		HCOH
CH_2OH	CH_2OH		CH_2OH
Glucose	**Unstable Schiff's base**		**Ketoamine**
HbA	**Labile**		**Stable as Hb A_{1c}**

PRINCIPLES OF ANALYSIS AND CURRENT USAGE

Glycohemoglobins (hemoglobins with covalently bound sugars) form nonenzymatically in the red cell in amounts proportional to the cellular glucose level.[1] Hemoglobin A_1, the major, most commonly assayed glycohemoglobin fraction, constitutes 6% to 8% of the total hemoglobin of normal persons but can reach levels of 15% to 18% in diabetics in poor glucose control. Hb A_1 is further separated into subfractions; Hb A_{1c} is the predominant and best characterized component. Hb A_{1c} differs from its parent hemoglobin in that it has a glucose residue covalently linked to the amino group of the N-terminal amino acid of each beta chain.[1] An unstable Schiff's base of sugar and hemoglobin forms initially (also termed the "labile component of glycohemoglobin"[2]) and rearranges to the stable ketoamine product A_{1c}.[1,2] Also present[3] are smaller amounts of glycohemoglobins A_{1a} and A_{1b}, which are products of nonenzymatic glycosylation by other sugars, including fructose and phosphorylated sugars at the N-terminus of the beta chain. Glycosylation can also occur at the N-terminus of the alpha chain and at lysine residues of both alpha and beta chains, but these glycosylated species appear in the hemoglobin A fraction with most current separation procedures and are not quantitated as A_1.[4] A similar variety of glycosylated forms of Hb S and C (Hb S_1 and C_1, respectively) have also been identified.[5]

A wide variety of analytical techniques have been applied to the separation and quantitation of A_1 glycohemoglobins. Two basic approaches are used, either quantitation of the combined A_1 fraction (A_{1a} + A_{1b} + A_{1c}) after its separation from HbA, or separation and quantitation of the subcomponent, A_{1c}. For separations of A_1, one can use either short cation-exchange columns[6,7] (method 2a, Table 60-8) or electrophoresis[8,9] (method 3, Table 60-8). Typically, the sample is applied to a cation-exchange column such as BioRex 70 and the glycosylated subcomponents eluted with a low ionic strength–phosphate buffer. The remaining hemoglobins are then eluted with a high ionic strength–phosphate buffer. Each hemoglobin fraction is quantitated by direct spectrophotometry at 415 nm.

The basis for separation by electrophoresis seems to lie in the ability of the free N-terminus of nonglycosylated hemoglobin to interact with sulfate groups, changing the molecule's electrophoretic mobility. A_1 hemglobins cannot interact with these groups and their mobility is unchanged. In these procedures the hemoglobins are exposed to sulfate groups in an agar-based support medium[8] or to dextran sulfate in the buffer used with a cellulose acetate support[9] and separated by short electrophoretic runs (20 to 45 minutes). Hemoglobin A_1 is separated as one band and measured by densitometry.

To separate A_{1c}, however, the additional resolution of high-performance liquid chromatography (HPLC)[10,11] (method 1, Table 60-8) or isoelectric focusing[12,13] is required. These A_{1c} procedures require more time and more sophisticated skills and equipment and are not generally used as routine procedures.

Separation of a fraction more closely approximating total glycosylated hemoglobin, including alpha-chain and lysine-glycosylated forms, is achieved by boronic acid affinity chromatography columns.[14-16] The glycosylated hemoglobins bind by forming a strong but reversible five-membered ring complex through their *cis*-glycol groups to immobilized boronic acid (method 2b, Table 60-8). The nonglycosylated hemoglobins are eluted first and the glycosylated fraction is then eluted from the column by a sorbitol-containing buffer that competes with the bound glycosylated protein for the boronic acid–binding sites. A_{1c} and some glycosylated non-A_1 hemoglobins bind to the column, but A_{1a} and A_{1b} do not.

Quantitation of separated hemoglobin fractions, either by spectrophotometric techniques (chromatography) or densitometry (electrophoresis, isoelectric focusing), is based on the absorbance of hemoglobin at 415 to 420 nm.

Another analytical approach to determine total glycosylated hemoglobin is based on the release and quantitation of the bound sugar moieties rather than the separation and quantitation of the glycosylated hemoglobins.[17-19] The sample is dialyzed to remove free glucose, fructose, and sucrose, and the hemoglobin-bound sugars are hydrolyzed, converted to 5-hydroxymethylfurfural, and quantitated by a chemical reaction with 2-thiobarbituric acid (TBA), which is monitored at 443 nm (method 4, Table 60-8).

Table 60-8. Methods of glycohemoglobin analysis

Method	Basis of Hb separation	Basis of analysis	Type of glycohemoglobin measured
1. High-performance liquid chromatography	Ion-exchange chromatography	Spectral absorbance of separated Hb components (415 nm)	A_{1c} + Labile A_{1c}
2. Minicolumn	a. Ion-exchange chromatography	Spectral absorbance of separated Hb components (415 nm)	A_1 + Labile A_1
	b. Affinity chromatography	Spectral absorbance of separated Hb components (415 nm)	A_1 + non-A_1 glycohemoglobins Labile glycohemoglobin will not bind
3. Electrophoresis	Charge differences	Spectral absorbance of separated Hb components (415 nm)	A_1 + Labile A_1 S_1, and C_1 can also be measured
4. Colorimetric	No Hb separation required, but interfering sugars must be removed before assay	Hb-bound glucose hydrolyzed, reacted with TBA, reaction monitored at 443 nm*	A_1 + Non-A_1 glycohemoglobins Labile glycohemoglobin will not react

*Footnote:

Ketoamine (Hb A_{1c}) —Heat, $-3H_2O$→ Schiff base of 5-hydroxymethylfurfural (5-HMF) → 5-HMF + Protein (R)

5-HMF + 2-Thiobarbituric acid (TBA) → Colored adduct

REFERENCE AND PREFERRED METHODS

(Table 60-9)

The most generally accepted reference method is HPLC, which allows separation and quantitation of A_1 components.[10,11] Hb A_{1c} is measured separately or summed with a fraction containing A_{1a} + A_{1b} to give a total A_1 value. Interassay coefficients of variation of 3% are reported.[11]

A known interferent is hemoglobin F, which coelutes with A_{1c}, and which can be significantly elevated during pregnancy[20] and the first 6 months of life.[21] Other more rare hemoglobins, such as H, may also coelute with A_{1c}.[22]

Routine procedures using short columns of cation-exchange resin and a discontinuous buffer system similar to that used in HPLC are commercially available. These methods elute total beta-chain N-terminus glycosylated hemoglobins (A_1) together. These procedures, although quick and technically simple to perform, require extremely careful control of sample load and separation temperature.[21,23,24] The user is very dependent on the commercial supplier to provide buffers of precise pH and ionic strength[24] and columns of good quality resin. Fine mesh–sized particles and bead fragments can pass through the column with the A_1 eluting buffer and falsely elevate the A_1 value. Hb F will coelute with A_1. Interassay coefficients of variation of 2% to 16% have been cited.[24-25]

A similar approach is the use of electrophoresis, for which commercially available procedures are also available. Interrun coefficients of variation range from 4% to 10%.[8,9] Hemoglobins S, C, and their glycosylated derivatives can be identified and quantitated with the agar medium, an advantage with some patient populations. However, hemoglobin F interferes in these procedures as in the column procedures.

Boronic acid chromatography using commercially prepared short columns is the most recent approach to the separation of glycosylated hemoglobins. The reference range for a control population is similar to that of A_1 assays,[14,15] probably because, although the small amounts of A_{1a} and A_{1b} are not measured, similarly small amounts of other non-A_1 components are. No simple correlation between A_1 or A_{1c} and boronic acid–separated glycohemoglobin exists for diabetics, however. A comparison of 36 samples from diabetics gave a slightly narrower range (7% to 16%) with affinity chromatography than with an A_1 ion-exchange procedure (7% to 19%).[14] Affinity chromatography gives values 10% greater than A_{1c} values by HPLC ion-exchange assay in the reference range, and up to 20% to 30% greater at levels seen in diabetic patients.[16]

The affinity separation is quick, requiring less than 15 minutes per sample separation time.[15] Labile glycosylated hemoglobin does not interfere[16] and nonglycosylated hemoglobins such as F should not, either.[14,15] Typical between-run coefficients of variation are 1% to 3% at all levels.[15,16] The affinity separation is much less temperature-sensitive than ion-exchange procedures, with a decrease of only 0.1% to 0.3% glycosylated hemoglobin (absolute) per degree Celsius increase.[16] Decreased pH dependence of the separation has also been observed.[14-16]

Colorimetric assays of total glycosylated hemoglobin are becoming more practical for the clinical laboratory, with the reduction of total incubation time to 2 hours[18] and the introduction of automation.[19] The assay can be calibrated with either fructose or 5-hydroxymethylfurfural and results are reported as micromoles per liter (μmol/L) of sugar. This approach is more specific, since neither nonglycosylated hemoglobins such as F nor the labile intermediate of glycosylated hemoglobin will interfere. Disadvantages include length of assay, larger sample requirements, and the use of reporting units that are unfamiliar to most clinicians at this time. (Some authors do convert hemoglobin sugar content to apparent percent A_1 using conversion tables. However, this ignores the differences in specificity of the two types of assays; that is, the colorimetric reaction of glycosylated hemoglobins that are not part of the A_1 fraction separated by other procedures). Interrun coefficients of variation of 4% to 18% are reported.[18,19]

Electrophoresis provides a rapid, reproducible, temperature-insensitive measure of A_1 that allows separate quantitation of S_1 and C_1 if present and avoids problems associated with minicolumns. It is best performed using commercially available kits according to the manufacturer's directions.

Boronic acid chromatography of glycosylated hemoglobin products is a technically simple, rapid separation procedure that avoids pH and temperature control problems of A_1 ion-exchange assays. The procedure does separate a subset of total glycosylated hemoglobin different from those seen with A_1 or A_{1c} methods. Commercially available procedures eliminate the problems of preparing the short columns in the laboratory.[16] Either the affinity or temperature-controlled column assays are recommended for routine use.

SPECIMEN

Blood should be collected with EDTA as anticoagulant. A hemolysate of saline-washed red cells is used. Either whole blood or the hemolysate may be stored for 4 to 7 days at 4° C and up to 30 days at −70° C.[23] Although much attention has been given to the presence of labile glycosylated hemoglobin, the amount present is generally very low. Typically, increases of 1% to 2% in apparent A_1 (expressed here as percent total hemoglobin) were seen both after glucose tolerance tests in normal volunteers and diabetics and after in vitro incubation of glucose with red blood cells.[26-29] Blood glucose values ranged up to 5000 mg/L in these studies. Similarly, small decreases in A_1 values are seen if the labile component is removed by dialysis from samples from diabetic patients and normal persons and the samples reassayed. One report does show a near doubling of normal A_1 levels after incubation with glucose, but only when levels of 10,000 mg/L are used.[30] If removal of the labile component is desired, one can use incubation or dialysis of the erythrocytes with physiological saline[26-29] or a much shorter incubation with semicarbizide plus aniline at pH 5.[30]

REFERENCE RANGE

Reference ranges for glycosylated hemoglobin will vary greatly with the type of procedure employed. The amount of Hb A_1 present in nondiabetic adults, as measured by the temperature-controlled minicolumn assay, is approximately 4.5% to 8.5%.[21] By electrophoretic separation the range is 5% to 7.6%.[8] Hemoglobin A_{1c} levels in nondiabetic persons have been reported to be 4.5% to 5.7% and 4% to 5.2% by HPLC[31] and isoelectric focusing[13] methods, respectively.

Table 60-9. Comparison of methods for glycohemoglobin analysis

Method	Coefficient of variation (%)	Temperature dependence of separation	pH dependence of separation	Assay time of one sample	Interferences
HPLC (ion exchange)	3	Significant	Significant	10-20 min*	Hb F
Minicolumn ion exchange	2-16	Significant	Significant	20-40 min†	Hb F
Minicolumn affinity	1-3	Negligible	Minor	15-30 min†	None
Electrophoresis	4-10	Negligible	Minor	20-45 min†	Hb F
Colorimetric	4-18	‡	‡	hours†	None

*Samples cannot be batched.
†Samples can be batched.
‡No hemoglobin separation step.

Total glycosylated hemoglobin as measured by affinity chromatography is reported to be 5.3% to 7.5%, whereas the colorimetric, thiobarbituric acid procedure has a reference range of 20 to 22 nmol/L of blood.[16]

Since the percent Hb A_1 values by the minicolumn method are reported to be age dependent,[21] this may also be true for other methods. No sex or race dependence for glycosylated hemoglobin has ever been noted.

INTERFERENCES

Hemoglobin F will cause false-positive results for many methods, as listed in Table 60-9. Any increase in turnover of red blood cells, as in certain hemoglobinopathies or anemias, will give low values for glycosylated hemoglobin. Hemoglobin S and C can cause falsely negative results by the minicolumn cation-exchange procedure. Lactescence (from lipemia) will cause falsely elevated results in the minicolumn procedure. The red blood cells in these specimens should be washed with saline to remove the interfering lipemia.

REFERENCES

1. Bunn, H.F., Gabbay, K.H., and Gallop, P.M.: The glycosylation of hemoglobin: relevance to diabetes mellitus, Science **200:**21-27, 1978.
2. Nathan, D.M.: Labile glycosylated hemoglobin contributes to hemoglobin A_1 as measured by liquid chromatography or electrophoresis, Clin. Chem. **27:**1261-1263, 1981.
3. Bunn, H. Franklin: Evaluation of glycosylated hemoglobin in diabetic patients, Diabetes **30:**613-617, 1981.
4. Gabbay, K.H., Sosenko, J.M., Banuchi, G.A., et al.: Glycosylated hemoglobins: increased glycosylation of hemoglobin A in diabetic patients, Diabetes **28:**337-340, 1979.
5. Aleyassine, H., Gardiner, R.J., Blankstein, L.A., and Dempsey, M.E.: Agar gel electrophoretic determination of glycosylated hemoglobin: effect of variant hemoglobins, hyperlipidemia, and temperature, Clin. Chem. **274:**472-475, 1981.
6. Jones, M.B., Koler, R.D., and Jones, R.T.: Micro-column method for the determination of hemoglobin minor fractions A1a+b and A1c, Hemoglobin **2:**53-58, 1978.
7. Abraham, E.C., Huiff, T.A., Cope, J.D., et al.: Determination of the glycosylated (HB A1) hemoglobins with a new microcolumn procedure, Diabetes **27:**931-937, 1978.
8. Menard, L., Dempsey, M.E., Blankstein, L.A., et al.: Quantitative determination of glycosylated hemoglobin A1 by agar gel electrophoresis, Clin. Chem. **26:**1598-1602, 1980.
9. Ambler, J.: The importance of measuring hemoglobin A1 in diabetes mellitus, Electrophoresis Today **3:**1-2, 1982.
10. Davis, J.E., McDonald, J.M., and Jarett, L.: A high performance liquid chromatography method for hemoglobin A1c, Diabetes **27:**102-107, 1978.
11. Dunn, P.J., Cole, R.A., and Soeldner, J.S.: Further development and automation of a high pressure liquid chromatography method for the determination of glycosylated hemoglobins, Metabolism **28:**777-779, 1979.
12. Simon, M., and Cuan, J.: Hemoglobin A1c by isoelectric focusing, Clin. Chem. **28:**9-12, 1982.
13. Spicer, K.M., Allen, R.C., and Buse, M.G.: A simplified assay of hemoglobin A1c in diabetic patients by use of isoelectric focusing and quantitative microdensitometry, Diabetes **27:**384-388, 1978.
14. Bouriotis, V., Stolt, J., Galloway, A., et al.: Measurement of glycosylated hemoglobins using affinity chromatography, Diabetologia **21:**579-580, 1981.
15. Mallia, A.K., Hermanson, G.T., Krohn, R.I., et al.: Preparation and use of a boronic acid affinity support for separation and quantitation of glycosylated hemoglobins, Anal. Lett. **14:**649-661, 1981.
16. Klenk, D.C., Hermanson, G.T., Krohn, R.T., et al.: Determination of glycosylated hemoglobin by affinity chromatography: comparison with colorimetric and ion-exchange methods and effects of common interferences, Clin. Chem. **28:**2088-2094, 1982.
17. Fluckinger, R., and Winterhalter, K.H.: In vitro synthesis of hemoglobin A_{1c}, FEBS Letters **71:**356-360, 1976.
18. Parker, K.M., England, J.D., DaCosta, J., et al.: Improved colorimetric assay for glycosylated hemoglobin, Clin. Chem. **27:**669-672, 1981.
19. Burrin, J., Worth, R., Ashworth, L., et al.: Automated colorimetric estimation of glycosylated hemoglobins, Clin. Chim. Acta **106:**45-50, 1980.
20. McDonald, J.M., and Davis, J.E.: Glycosylated hemoglobins and diabetes mellitus, Hum. Pathol. **10:**279-291, 1979.
21. Kaplan, L.A., Cline, D., Gartside, P., et al.: Hemoglobin A_1 in hemolysates from healthy and insulin-dependent diabetic children, as determined with a temperature-controlled mini-column assay, Clin. Chem. **28:**13-18, 1982.
22. Kruiswijk, T., Diaz, D.P., and Holtkamp, H.C.: Interference of hemoglobin H in the column chromatographic assay of glycosylated hemoglobin A, Clin. Chem. **27:**641-642, (Letter).
23. Worth, R.C., Ashworth, L.A., Burrin, J.M., et al.: Column assay of haemoglobin A1: critical effects of temperature, Clin. Chem. Acta **104:**401-404, 1980.
24. Schellekens, A.P.M., Sanders, G.T.B., Thornton, W., and van Groenestein, T.: Sources of variation in the column-chromatographic determination of glycohemoglobin (Hb A1), Clin. Chem. **27:**94-99, 1981.
25. Hammons, G.T., Junger, K., McDonald, J.M., and Ladenson, J.H.: Evaluation of three minicolumn procedures for measuring hemoglobin A_1, Clin. Chem. **28:**1775-1778, 1982.
26. Svendsen, P.A., Christiansen, J.S., Welinder, B., and Nerup, J.: Fast glycosylation of haemoglobin, Lancet **1:**603, 1979.
27. Botterman, P.: Rapid fluctuations in glycosylated haemoglobin concentration, Diabetologia **20:**159, 1981.
28. Goldstein, D.E., Peth, S.B., England, J.D., et al.: Effects of acute changes in blood glucose on Hb A_{1c}, Diabetes **29:**623-628, 1980.
29. Svendsen, P.A., Christiansen, J.S., Søegaard, U., et al.: Rapid changes in chromatographically determined A_{1c} induced by short-term changes in glucose concentration, Diabetologia **19:**130-136, 1980.
30. Nathan, D.M., Avezzano, E., and Palmer, J.L.: Rapid method for eliminating labile glycosylated hemoglobin from the assay for hemoglobin A_1, Clin. Chem. **28:**512-515, 1982.

Haptoglobin

ROBERT S. FRANCO

Clinical interpretation: p. 611

Molecular weight: 85,000 daltons (type 1-1); other genetically determined types (2-1, 2-2) form a series of higher molecular weight polymers

Chemical class: glycoprotein

PRINCIPLES OF ANALYSIS AND CURRENT USAGE

Haptoglobin is a genetically polymorphic serum protein that binds free hemoglobin and thereby facilitates hemoglobin removal by the reticuloendothelial system. A decreased haptoglobin level is therefore evidence of increased erythrocyte destruction. Earlier methods for quantitation of serum haptoglobin were based on formation of a complex between haptoglobin (Hp) and hemoglobin

Table 60-10. Methods of haptoglobin analysis

Method	Principle	Usage	Comments
1. Hemoglobin-binding			
a. Electrophoresis	Hb separated from Hb-haptoglobin by difference in electrophoretic mobility; peroxidase stain for hemoglobin to visualize bands	Historical	Semiquantitative
b. Gel filtration	Hb separated from Hb-haptoglobin by molecular weight difference; quantitation by measurement of A_{415} of hemoglobin	Not suitable for clinical use	Quantitative
2. Immunochemical			
a. Radial immunodiffusion	Haptoglobin diffuses into gel containing antihaptoglobin antibody; precipitin ring diameter proportional to concentration	Currently in clinical use	Suitable for small number of samples
b. Nephelometry	Antibody reacts with haptoglobin, forming complex that scatters light	Currently in clinical use	Suitable for large number of samples

Table 60-11. Comparison of conditions for haptoglobin analysis

Condition	Radial immunodiffusion*	Nephelometry†
Temperature	23 ± 2° C (RT‡)	22° C (RT‡)
pH	7.4	7.0
Time of reaction	18 hours	30 minutes
Sample	serum	serum
Sample size	5 μL	1 μL
Range (without dilution)	40-370 mg/L	25-2760 mg/L
Precision: Mean (% CV)	40 mg/L (1.9%) 150 mg/L (2.2%-3.5%) 370 mg/L (3.5%-5.3%)	143 mg/L (7.0%) 1323 mg/L (2.5%)

*Quantiplate, Kallestad Laboratories, Chaska, MN 55318.
†"PDQ" Laser nephelometer, Hyland Labs, Inc., Costa Mesa, CA 92626.
‡*RT,* Ambient or room temperature.

(Hb). The addition of a known amount of hemoglobin to serum results in either no formation of complex (no Hp present), all the Hb in the form of Hb-Hp complexes (Hp>Hb), or the presence of complexes and free Hb (Hb>Hp). Separation of the free Hb from the Hb-Hp complex is accomplished by electrophoresis or gel filtration, and the hemoglobin in each component is measured by its peroxidase activity (method 1, Table 60-10). The amount of total Hb complexed with Hp is the hemoglobin-binding capacity and is taken as a measure of the Hp level. Reference ranges for methods based on bonding of Hb are expressed as milligrams of Hb bound per liter.

Modern methods for measuring haptoglobin are immunological, including nephelometry (method 2a, Table 60-10) and radial immunodiffusion (RID) (method 2b, Table 60-10). In the nephelometric method, serum containing an unknown amount of haptoglobin is mixed with antihaptoglobin antibody. The formation of an antigen-antibody complex results in light scattering when a beam of monochromatic light is passed through the sample. The degree of light scattering is proportional to the concentration of haptoglobin.[1] An alternative nephelometric procedure that measures the rate of formation of the antigen-antibody complex is available.[2] This method has been shown to correlate closely with a hemoglobin-binding assay.[2]

In the RID method, an antibody to haptoglobin is evenly dispersed in a thin, two-dimensional gel. Small holes, or wells, are cut into the gel. A small amount of serum (typically 5 μL) is placed in the well and allowed to diffuse radially into the gel surrounding the well forming a precipitin ring. The concentration of haptoglobin in the sample is directly related to the diameter or area of the ring. Reference ranges for immunological methods are expressed in mg/L. Note that both the nephelometric and RID methods tend to overestimate the Hp level for subjects with Hp 1-1. This error appears to be larger with RID and approaches twice the true value.[3]

REFERENCE AND PREFERRED METHODS
(Table 60-11)

Nephelometry is popular in high-volume applications because automated instruments are available (Technicon, Inc., Tarrytown, N.Y.; Hyland Laboratories, Costa Mesa, Calif.). This method has the advantage of a small sample size, a fast turnaround time with results available the same day, and a relatively high precision. Disadvantages are relatively high reagent costs and the need for a specialized piece of equipment. Radial immunodiffusion features a very small sample size, extreme simplicity when commercially prepared plates are used, acceptable precision, and essentially no capital investment. The main disadvantage of RID is the relatively long time before the precipitin rings may be measured, typically 18 to 48 hours. The two methods tend to exhibit extremely good agreement so that the choice of which to use often depends on the number of samples to be run. A high throughput favors nephelometry, whereas RID is advantageous if a relatively small number of samples is to be run. The nephelometric procedure is dependent on the particular instrument used, and the instructions provided by the manufacturer should be followed.

The preparation of RID plates has been described in detail,[4] but it is recommended that plates available from commercial suppliers (such as Behring Diagnostics, Somerville, N.J., and Kallestad Laboratories, Chaska, Minn.) be used. There are two general methods in widespread use. The first is that of Mancini et al.,[4] in which the RID plate is incubated until the precipitin ring ceases to grow. This usually takes about 48 hours. At this time the area (or the square of the diameter) of the ring is proportional to the serum antigen concentration. The actual concentration is read from a standard curve prepared using reference sera supplied by the plate manufacturer. Although the Mancini equilibrium technique is considered most accurate, for clinical use it is often desirable to sacrifice some accuracy in order to obtain the result faster. The method of Fahey and McKelvey[5] is a nonequilibrium technique that is similar to the Mancini procedure except that the precipitin rings are measured before they reach their maximum size, usually after 18 hours. In this case the log of the concentration is proportional to the precipitin-ring diameter. Commercial plates that use either of these methods are available. Typical coefficients of variation for the Fahey technique are 2% to 3.5% within a plate and 2% to 5% between plates.

SPECIMEN

Serum from unhemolyzed blood specimens is used for haptoglobin determination. The serum may be stored for several days at 2° to 8° C or frozen at −20° C for longer storage. Avoid repeated freeze-thaw cycles.

REFERENCE RANGE

An adult normal range is 600 to 2700 mg/L. Newborns have little or no haptoglobin, with levels approaching adult by about 4 months.

REFERENCES

1. Van Lente, F., Marchand, A., and Galen, R.S.: Evaluation of a nephelometric assay for haptoglobin and its clinical usefulness, Clin. Chem. **25**:2007, 1979.
2. Sternberg, J.C.: A rate nephelometer for measuring specific proteins by immunoprecipitin reactions, Clin. Chem. **23**:1456, 1977.
3. Valette, I., Pointis, J., Rondeau, Y., et al.: Is immunochemical determination of haptoglobin phenotype dependent? Clin. Chim. Acta **99**:1-6, 1979.
4. Mancini, G., Carbonara, A.O., and Heremans, J.F.: Immunochemical quantitation of antigens by single radial immunodiffusion, Immunochemistry **2**:235-254, 1965.
5. Fahey, J.L., and McKelvey, E.M.: Quantitative determination of serum immunoglobulins in antibody-agar plates, J. Immunol. **94**:84-90, 1965.

Hemoglobin analysis

GERARDO PERROTTA

Clinical significance: pp. 611 and 832
Molecular weight: 64,000 daltons
Merck Index: 4505
Chemical class: heme protein

PRINCIPLES OF ANALYSIS AND CURRENT USAGE

Three types of hemoglobins are normally found in erythrocytes: hemoglobins A, A_2, and F. They are all composed of two alpha and two nonalpha chains. Hemoglobin A, the most prevalent, is made of two alpha and two beta chains ($\alpha_2\beta_2$); Hb A_2 is present up to 3.5% and is made up of two alpha and two delta chains ($\alpha_2\delta_2$); Hb F may be present normally up to 2% and is made of two alpha and two gamma chains ($\alpha_2\gamma_2$).

This section presents methods that will detect abnormal hemoglobins that are the result of an amino acid substitution, such as sickle cell hemoglobin, or hemoglobinopathies that are the result of aberrant globin-chain synthesis, as in the thalassemias. A combination of methods is usually necessary to separate and identify the presence of normal, abnormal, or variant hemoglobin. These include electrophoretic methods, nonelectrophoretic methods, structural analysis, and others.

Electrophoretic procedures

Electrophoretic methods separate hemoglobins based on differences in electrical charge. Specific methods may vary the strength of the electric field, pH, and type of buffer or the supporting medium used (methods 1 to 3, Table 60-12, and Table 60-13).

The two most common electrophoretic techniques use alkaline and citrate buffer. Cellulose acetate[1] represents by far the most commonly used support medium for the alkaline-buffer procedure, whereas agar gel is preferred for the acid buffer. At alkaline pH, electrophoresis with cellulose acetate, agarose, or starch gel separates HbA, A_2 and F. However, these conditions do not differentiate Hb S from hemoglobins D and G because all three have the same mobility in this system. Likewise, Hb C, E, and O have a mobility similar to the A_2 band at alkaline pH. Citrate agar (acid pH) electrophoresis is used to separate those hemoglobins that are not discriminated from each other in the alkaline cellulose acetate procedure. Thus citrate agar confirms the presence of hemoglobins D or G and differentiates hemoglobinopathies SD from SS and hemoglobins C from E.

After the actual electrophoresis, most procedures require fixation (denaturation) of the hemoglobin, followed by staining with such protein stains as Ponceau S, Paragon, or amido black 10, to render more visible the separated fractions. The position (relative mobility) and intensity of

Table 60-12. Methods of hemoglobin analysis

Method	Type of analysis	Principle	Usage	Comments
Electrophoretic				
1. Cellulose acetate	Quantitative-qualitative alkaline pH (8.5)	Negatively charged Hb migrate anodally **Cathode** – | CA_2 (E,O) S (D,G Lepore) F A | + **Anode**	Whole blood prepared as hemolysate	Simple, fast, and most frequently used
2. Citrate agar	Qualitative acid pH (6.0)	Hb move from anode to cathode – | F A E,D,G (Lepore) S C | +	Whole blood prepared as hemolysate	Not a good screening test; best used to distinguish S from D, A from F, and C from E or O.
3. Globin chain	Qualitative alkaline or acid pH	Under reducing conditions in the presence of 6 mol/L urea, globin chains dissociate and are separated depending on pH of buffer pH 8.9 – | αA βS βA γF | + pH 6.0 | αA βS βA γF |	Whole blood prepared as hemolysate	Separates alpha and nonalpha globin chain variants.
Nonelectrophoretic				
4. Column chromatography	Quantitative for Hb A_2	Hb A binds more strongly to DEAE cellulose than A_2 does, resulting in a faster separation and elution of A_2	Whole blood prepared as hemolysate	For thalassemia states
5. Alkali denaturation	Quantitative for Hb F	Hb F + 1/12 mol/L NaOH → Precipitate	Whole blood prepared as hemolysate	For thalassemia and other hemoglobinopathies
6. Solubility	Qualitative	Hb + $\xrightarrow[\text{Buffer Saponin}]{\text{Dithionite } Na_2S_2O_4}$ Turbid solution if Hb S is present	Whole blood	For Hb S

Table 60-13. Comparison of conditions for hemoglobin analysis by electrophoresis

Technique	CDC	Helena	Beckman	Corning	Gelman
Alkaline					
pH	8.4	8.6	8.6	8.6	Not specified
Support media	Cellulose acetate	Cellulose acetate	Agarose	Agarose	Cellulose acetate
Application (volt/time)	450 V/20 min	450 V/15 min	200 V/25 min	240 V/20 min	400-425 V/20-30 min (membrane dependent)
Stain	Ponceau S	Ponceau S	Paragon blue stain	Amido black 10 colorimetric stain	Ponceau S
Acid					
pH	6.0-6.2	6.3	Not available	6.3	Not available
Support media	Citrate agar	Citrate	Not available	Citrate agar	Not available
Application (volt/time)	70-90 V/70 min	70 V/60 min	Not available	90 V/35 min	Not available
Stain	Benzidine diluted with distilled H_2O	*o*-Dianisidine	Not available	Amido black 10B colorimetric stain	Not available

resultant hemoglobin bands are then compared to known controls. Quantitation is done by a densitometer.

Confirmatory tests for specific hemoglobins

Nonelectrophoretic methods are used to quantitate individual hemoglobin fractions (methods 4 to 6, Table 60-12). The solubility test, column chromatography, and alkali denaturation are the ones most frequently used for hemoglobins S, A_2 and F, respectively.

Hemoglobin S. In 1953, Itano[2] reported that deoxy-Hb S is less soluble than normal Hb A when reduced by dithionite in phosphate buffer solution. In this reduced state a turbid suspension of protein crystals is formed. These crystals disperse transmitted light and the solution becomes turbid. Therefore, when Hb S is present, the lines on a sheet of paper placed behind this solution are invisible. Hb A solutions, on the other hand, remain transparent, and one can easily see the lines. When this test is coupled with alkaline and acid electrophoresis, it confirms or rules out definitively the presence of hemoglobin S. The solubility test does not distinguish among the homozygous SS, the heterozygous AS, SC, and S thalassemia states.

Hemoglobin A_2. Column chromatography provides an accurate method for quantitating hemoglobin A_2. This hemoglobin is separated on a small diethylaminoethyl (DEAE)–cellulose anion-exchange column. Hemoglobin solution is adsorbed onto the column, but Hb A_2, which has a different charge from the other hemoglobins, is not adsorbed and is eluted in the first fraction. A higher ionic strength buffer is then used to elute the remaining hemoglobins from the column. The percentage of Hb A_2 is determined from the absorbances of the eluate at 415 nm (Soret band). Recently, Gottfried et al.[3] reported that A_2 can also be quantitated by densitometry of the electrophoretic patterns when carefully controlled techniques and properly calibrated instruments are used.

Hemoglobin F. Alkali denaturation methods are most frequently used for the determination of Hb F. This technique is based on the principle that fetal hemoglobin is more resistant to denaturation under alkaline conditions. After a strong alkali, such as KOH or NaOH, is added to a hemolysate, the denaturation is stopped by the addition of saturated or half-saturated ammonium sulfate, which results in a lower pH and a precipitated denatured hemoglobin. The amount of unaltered hemoglobin is measured spectrophotometrically and is expressed as the percentage of alkali-resistant fetal hemoglobin.

Other confirmatory tests. Globin-chain electrophoresis, isoelectric focusing (IEF), and amino acid sequencing (fingerprinting) provide additional information for characterizing hemoglobins. These are done in more sophisticated centers and are not part of a routine approach to hemoglobin analysis but are used to confirm the presence of rare hemoglobins. Globin-chain electrophoresis and IEF are the most commonly used methods. Globin-chain electrophoresis separates alpha and nonalpha globin chains by mixing a cell membrane–free hemolysate with 6 M urea in the presence of 0.05 M 2-mercaptoethanol. Fractionation takes place on cellulose acetate plates in Tris-EDTA-borate (TEB) buffer at pH 6 and 8.9. By comparing the unknowns to some known Hb variant, one can make a characterization of the variant globin chain.

IEF is fast becoming a useful tool for identifying Hb mutants. It is based on the principle that all proteins have a zero charge in solution at a specific pH, called the "isoelectric point" (pI). The hemoglobin or hemoglobins being analyzed migrate toward their isoelectric points in a pH-gradient gel or support medium (most commonly a polyacrylamide gel). The band that is formed at this point is referred to as the "focused band." IEF has greater resolving power than either of the two electrophoresis techniques currently in use in most laboratories. The greatest advantage of the IEF technique seems to be in the detection and separation of those mutant hemoglobins that travel with Hb A in cellulose acetate.

Other useful methods employed to differentiate rare abnormal hemoglobins include heat stability tests, the isopropanol test, oxygen affinity determination, and high-pressure liquid chromatography.

Finally, erythrocyte size (MCV) and shape (poikilocytes), serum iron studies, reticulocyte counts, and family studies are necessary and helpful to complete a definitive genetic picture of a suspected hemoglobinopathy or variant.

REFERENCE AND PREFERRED METHODS

Electrophoresis procedures

In studies comparing three methods for quantitating variant Hb fractions, Pearce[4] showed that cellulose acetate electrophoresis with densitometry was better than cellulose acetate with elution or microchromatography.

Schmidt and Holland[5] published results of a survey evaluating three commercially available hemoglobin electrophoretic procedures available from Beckman,* Gelman,† and Helena.‡ All three methods employed cellulose acetate with alkaline buffer (Table 60-13). Their findings showed that both Gelman and Helena assays correctly identified (separated) all the hemoglobin samples tested. The Beckman procedures at that time did not separate Hb A from Hb F for eight of the 36 samples containing both hemoglobins. It was believed that Helena's Zip Zone electrophoresis gave the best separations of hemoglobins A, F, S, and C. No similar studies have since been performed,

*Beckman Instruments, Inc., Clinical Instruments Division, Brea, CA 92621.

†Gelman Sciences, Inc., 600 South Wagner Rd., Ann Arbor, MI 48106.

‡Helena Laboratories, 1530 Lindbergh Dr., P.O. Box 752, Beaumont, TX 77704.

and in the absence of proficiency-testing reports, one cannot say for sure how the same study would fare today.

When Hicks and Hughes[6] compared citrate agar, cellulose acetate, and starch gel electrophoresis, they found that cellulose acetate was simple, rapid, and economical for *initial* identification of hemoglobins S, A, F, C. However, it was not so selective and sensitive as citrate agar and starch gel electrophoresis. Citrate, on the other hand, was more selective than cellulose and starch for identifying Hb S. It was more acceptable than either of the other methods for identifying hemoglobins S, A, F, S–β-thalassemia, and F–β-thalassemia major (Cooley's anemia). Hb F separated well from hemoglobins A and S. Citrate was not so selective as cellulose acetate and starch electrophoresis for identifying Hb AA. It was not an acceptable method for Hb A_2.

Starch gel was found satisfactory for Hb A_2 identification. It was more sensitive than cellulose acetate and citrate for identifying hemoglobins A, F, C, D, or G. However, it was determined to be the most expensive and time-consuming procedure, requiring 24 hours for results. It was not so selective as citrate for Hb S. Hb F in low concentration was not separated clearly from Hb A and Hb S. Vernon[7] found that agarose gel electrophoresis separates commonly encountered abnormal hemoglobins just as well as cellulose acetate does.

Hemoglobin A_2

An interlaboratory study by Schmidt and Brosious on the quantitation of Hb A_2 indicated that methods for quantitating this component of normal hemoglobin may need to be standardized.[8] The 1976 Hemoglobinopathy CDC Proficiency Testing Survey found a great variation among the participant laboratories and the methods used. Standard deviations and coefficients of variation (%CV) for the participant laboratories were more than twice those for the reference laboratories. The methods compared were column chromatography, densitometry, and electrophoresis followed by elution. Of the three, densitometry gave consistently higher coefficients of variation. Although Hb A_2 quantitation using electrophoresis followed by densitometric analysis has been suggested, this technique is generally not recommended because of nonheme protein contamination that has a mobility similar to Hb A_2, thereby producing falsely elevated results.

In 1978, Brosious et al.[9] compared four microchromatographic methods for Hb A_2 quantitation: the Effremov (Tris-HCl) procedure, the Huisman (glycine) method, and two commercially available procedures (Helena and Isolab). The results showed that the Tris method and the two commercial kits had similar precision ($\bar{X}$, SD, %CV) for normal and beta-thalassemia samples. The glycine method, which was considered the faster method, gave coefficients of variation two times higher for the normal and four times higher for the beta-thalassemia specimens. Brosious et al. believed that the Tris-HCl, Helena, and Isolab methods gave a differentiation of normal Hb A_2 levels better and clearer than that of the elevated ones.

One drawback of the two commercial methods mentioned is their inability to measure Hb A_2 in the presence of Hb S or D. Isolab improved its original procedure to provide a better measure of A_2 in people with sickle cell disorders. Baine and Brown[10] compared the modified Isolab procedure, called "Quick-Sep Improved A_2 Test," to the Tris-HCl method. The results showed a similar performance in quantitation of Hb A_2 in the presence of Hb S as the Tris-HCl method. However, the Quick-Sep method gave consistently lower A_2 results than the Tris-HCl method did. When four phenotypic classes were studied (AS, SS, Sβ + thal, $S\beta^0$thal), the means and coefficients of variation obtained showed no statistical difference. The Quick-Sep test was found to be faster than the previous Isolab method, taking 90 minutes to complete the Hb A_2 quantitation. Although the Quick-Sep procedure was able to separate Hb S from Hb A_2, measurement of hemoglobin S by this procedure is not recommended.

In 1980, the National Committee for Clinical Laboratory Standards (NCCLS) published a tentative standard for the quantitation of Hb A_2. They considered the column chromatography technique as simple, precise, and accurate. However, their procedure did not address the issue of measuring Hb A_2 in the presence of Hb S. The NCCLS report did warn that trying to quantitate Hb A_2 from a sample containing hemoglobins S, C, or other slow variants could be a source of error. It remains to be seen, in light of the recent developments by Helena and Isolab, if this tentative proposal will be revised to reflect the advancements made by the two commercially available column procedures.

Hemoglobin S

Several commercially available procedures for the quantitation of Hb S have reached the market based on the principle of decreased solubility in dithionite-phosphate solutions. Most use a single-step test to detect the presence of Hb S; some use a two-step test, not only to identify the presence of Hb S but also to differentiate between the homozygous and heterozygous states. Schmidt and Wilson[11] evaluated several of these commercial procedures and found that no one method was totally accurate. However, they believed at that time that Ortho's Sickledex came closest to being considered acceptable.

Given the ease with which any laboratory can prepare its own reagents for this test, it is the preferred method for use. The procedure described below is essentially that recommended by the Centers for Disease Control (CDC). The reagents used are concentrated phosphate buffer mixed with saponin and sodium dithionite. Regardless of the procedure used, it is recommended that the solubility test be used to confirm the presence of Hb S after presumptive identification by electrophoresis. It should definitely not be

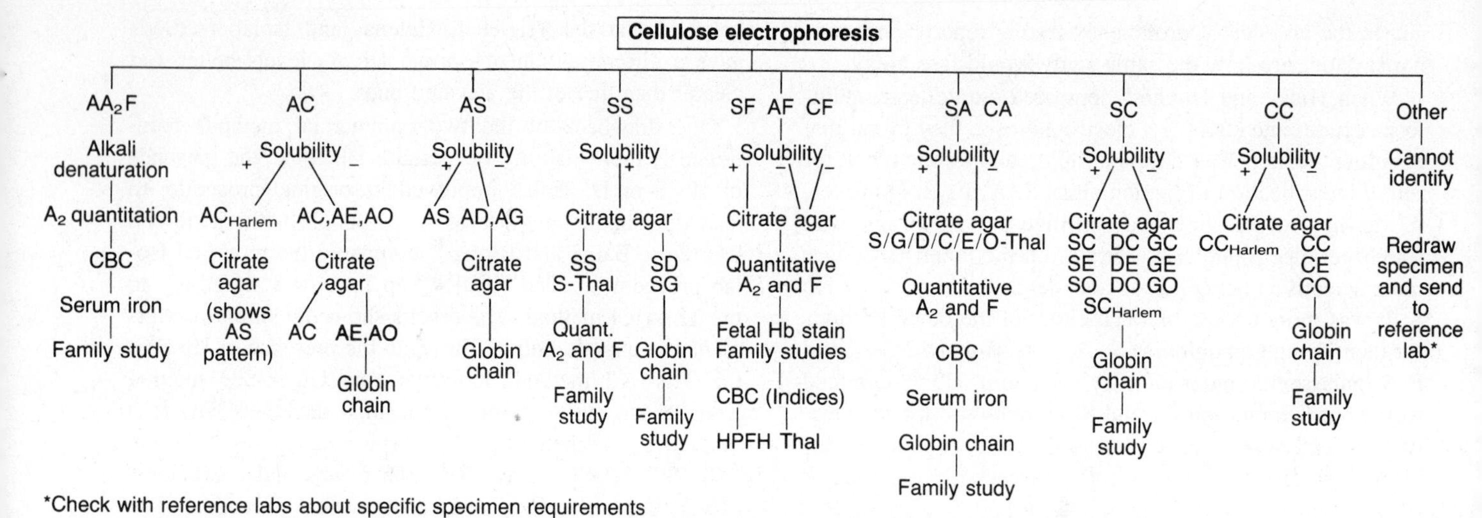

Fig. 60-1. Flow sheet for laboratory diagnosis of common hemoglobinopathies based on cellulose acetate electrophoresis as initial test. *CBC,* Complete blood count; *HPFH,* hereditary persistence of fetal hemoglobin; *Thal,* thalassemia. (Modified from Centers for Disease Control: Hemoglobinopathy manual, HEW publ. no. (CDC) 77-8266, Washington, D.C., 1976, U.S. Government Printing Office.)

used to differentiate the genotype of the sickle cell state because of possible misleading results, as with Hb Sβthal.

Proposed standard procedure

Given this background, the NCCLS has proposed a tentative standard procedure for the detection of abnormal hemoglobins. The Committee report suggests that cellulose acetate can provide accurate and reproducible results as an initial test.[12] The method described below is consistent with these recommendations but uses essentially the procedure outlined by Helena Laboratories.

In summary, Fig. 60-1 describes one of the most established approaches for the detection of hemoglobin variants. This does not mean that every request for Hb electrophoresis goes through this entire process; rather, this sequence of tests is used for those unknowns that cannot be identified at any one particular stage.

Advances in molecular biology may alter this approach in the not-too-distant future. The special or sophisticated procedures of today may be the routine procedures of tomorrow. The goal will always remain the same, that is, to identify disorders and improve the level of care and counseling for the patient concerned.

SPECIMEN

Cellulose acetate

Whole blood is collected in EDTA. Other anticoagulants are also suitable. The sample required for this method should be a hemolysate. This is prepared by centrifugation of an aliquot (1 mL) of whole blood at 2500 rpm (900 *g*) for approximately 5 minutes. The plasma is discarded and the packed red blood cells are washed with 4 mL of 0.9% NaCl. After remixing, centrifuging, and discarding the supernatant, one adds two drops of the washed red blood cells to 12 drops of the hemolysate reagent, consisting of 5 mmol/L EDTA and hemoglobin preservatives. Anticoagulated samples are stable for 2 weeks at 4° C.

Citrate agar

One drop of the hemolysate prepared above is mixed with six drops of distilled H_2O.

Hemoglobin A_2 by column chromatography (Isolab)

The specimen is freshly drawn blood using EDTA or heparin as anticoagulant. Blood should be refrigerated at 4° C and can be stored as whole blood (not as a hemolysate) up to 1 week.

Before testing, one can prepare the hemolysate in two ways:

1. Mix 50 μL of whole blood with 200 μL of distilled H_2O. Allow to stand 5 minutes and mix again.
2. Wash red blood cells three times with saline solution (0.9% NaCl). The cells are then lysed. Fifty microliters of this hemolysate are mixed with 200 μL of distilled H_2O.

Solubility test

The specimen is 20 μL of whole blood, using EDTA or heparin as an anticoagulant.

PROCEDURE FOR CELLULOSE ACETATE ELECTROPHORESIS

Following is the procedure of Helena Laboratories, Inc. (Beaumont, Texas).

Reagents

Hemolysate. As described in specimen section.

Supre Heme Buffer. 14.6 g of Tris-EDTA-boric acid; one packet diluted to 1000 mL of distilled H_2O at pH 8.4.

Titan III cellulose acetate strips. Mylar backed.

Ponceau S stain. Concentration 0.5% (w/v), 5 g/L water. It is stable 6 months at 4° C.

5% acetic acid, 0.88 mol/L. Prepared by diluting 50 mL of glacial acetic acid with 950 mL of distilled water. Stable 1 year at 4° C.

20% acetic acid–80% methanol, 3.52 mol/L. Prepared by dilution of 120 mL of glacial acetic acid with 480 mL of methanol. It is stable 1 year at 4° C.

0.9% saline solution, 0.15 mol/L. 9 g of NaCl dissolved in 1 liter of distilled water. Stable for 3 months at 4° C.

Control. Hemoglobin preparation containing hemoglobins A_1, F, S, and C, commercially available.

Assay

Equipment: electrophoresis power supply capable of delivering 450 V; Salten Hot Tray; 37° C incubator.

1. Follow manufacturer's directions for preparation of cellulose acetate strip and application of patient samples and controls.
2. Electrophoresis is carried out at 59.2 V/cm (450 V) for 15 minutes at ambient temperature.
3. Stain plates for 3 minutes in Ponceau S.
4. Destain in two or more successive washes of 5% acetic acid until background is white.
5. Dehydrate in methanol for 2 minutes.
6. Soak in acetic acid–methanol clearing solution for 6 minutes.
7. Air day plates, cellulose acetate side up, for 2 minutes.
8. With plates lying on a single paper towel, place on Salton Hot Tray to complete clearing. When they are clear, place plates in 37° C incubator until they are dry for at least 5 minutes.
9. After clearing and drying, place plate in a clear plastic envelope to prevent the cellulose acetate from peeling off the Mylar backing.

Calculation

The plate is ready for scanning with a densitometer at 525 nm (green filter). Each fraction results in a peak that can be quantitated. One can determine the percentage of each hemoglobin fraction by dividing the area of each peak by the total area under the densitometric curve.

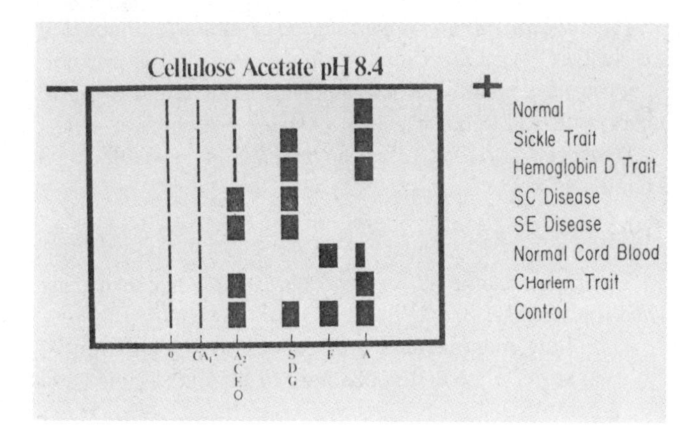

Fig. 60-2. Representation of hemoglobin electrophoresis on cellulose acetate strips, pH 8.4. CA_1, Carbonic anhydrase.

Interpretation

The separated bands are compared to a known control. Migration proceeds from the cathode (−) application point to the anode (+). Very close to the application point, there is at times a faint nonhemoglobin band of carbonic anhydrase (CA). Continuing toward the anode in order of increasing speed of migration, there is Hb A_2 with the possible presence of hemoglobins C, E, and O; next is Hb S (D or G), F, and A (Fig. 60-2).

REFERENCE RANGE

Hb A_1, 96% to 98.5%; Hb A_2, 1.4% to 4.0%; Hb F $<$ 2.0%.

PROCEDURE FOR CITRATE AGAR METHOD

The citrate agar method is for the separation of those hemoglobins that migrate in the same region on cellulose acetate.

Reagents

Patient hemolysate and controls

Titan IV citrate agar plates. Agar plates are 1.5% agar, obtained from Helena.

Citrate agar buffer. Dissolve on package containing 15.5 g of sodium citrate and citric acid in 1000 mL of distilled H_2O.

Remains stable for 1 month at 4° C.

Hemolysate reagent. 5 mmol/L EDTA and preservatives. Stable for 6 months at 4° to 8° C.

5% acetic acid (v/v), 0.88 mol/L. (See above.)

o-Dianisidine stain. Prepare under hood by mixing 10 mL of 5% acetic acid, 5 mL of 0.2% *o*-dianisidine (commercially available), 1 mL of H_2O_2 (3%), and 1 mL of sodium nitroferricyanide. Use gloves. Stable for 3 months at room temperature.

1% sodium nitroferricyanide, 33.6 mmol/L. Place 1 g of $Na_2Fe(CN)_5NO$ (sodium nitroprusside) in a 100 mL volumetric flask. Add distilled water to dissolve and bring to mark with additional water. It is stable 6 months at 4° C.

Hydrogen peroxide (3%), 88 mmol/L. It is stable for 3 months at 4° C.

Assay

Equipment: same as for procedure for cellulose acetate electrophoresis.

1. Follow manufacturer's directions for preparation of agar gel plate and application of patient samples and controls.
2. Electrophoresis is carried out at approximately 1 V/cm (70 V) for 60 minutes. The chambers are placed in a pan containing ice packs.
3. After this time, place the plate gel side up in the stain for 5 minutes.
4. Rinse in 5% acetic acid for several minutes.
5. To preserve the citrate plates, place plate gel side up in pan of water. With a knife, gently lift agar from the plate and slide onto an index card, labeling it appropriately. Overnight drying under the hood leaves a permanent record.

Interpretation

The migration pattern on citrate agar at pH 6.0 shows hemoglobins A, D, E, and G migrating toward the cathode. Hb F is most cathodal, whereas Hb C is most anodal. Hb S is less anodal and migrates closer to the application point (Fig. 60-3).

Notes

The same precautions mentioned in cellulose acetate apply for this method. The cooling is necessary so that one may obtain more dependable separations in a shorter time span.

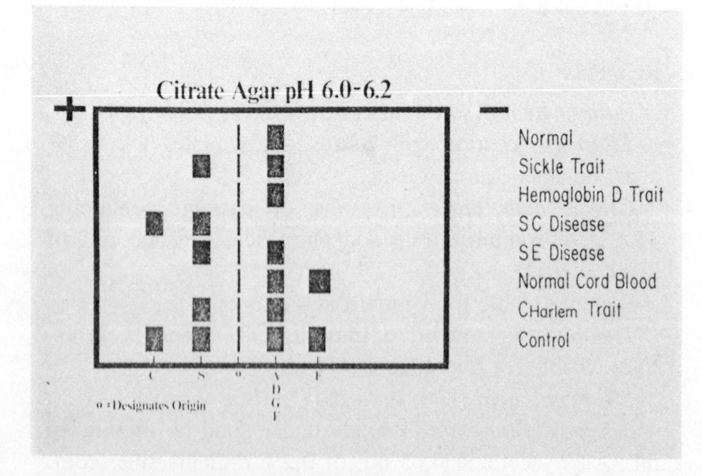

Fig. 60-3. Representation of hemoglobin electrophoresis on citrate agar, pH 6. On some plates the origin may not be as visible or as separated from HbS as shown here.

PROCEDURE FOR HEMOGLOBIN A_2 QUANTITATION BY COLUMN CHROMATOGRAPHY

The method described here is essentially the Isolab Quick-Sep system. It is used to more accurately determine the amount of Hb A_2, which if elevated is an important indicator of beta-thalassemia. A freshly prepared hemolysate is absorbed onto a preconditioned DEAE-cellulose column bed. Alkaline elution agents separate Hb A_2 from the majority of the other hemoglobins. The effluents are collected and are measured as a percentage of the total hemoglobin eluted.

Reagents

1. Patient hemolysate (as described previously) and controls.
2. Columns and accompanying eluting agents are commercially available. The eluting agents for A_2 contain 0.01% KCN buffered at a pH of 7.3. The other hemoglobin eluting agent contains 0.01% KCN in glycine buffer with NaCl.

Assay

Equipment: spectrophotometer with a band pass $\leq$ 10 nm, capable of reading at 415 nm.

1. Follow manufacturer's directions for storage and preparation of columns and for application of patient sample.
2. Elute Hb A_2 with 4 mL of low–ionic strength buffer and the "other hemoglobins" with 4 mL of high–ionic strength buffer; dilute the "other hemoglobin" fraction with 16 mL of distilled water.
3. Read the absorbance of the two eluted Hb fractions collected above at 415 nm against a water blank.

Calculation

Determine the percentage of A_2 by using the following formula:

$$\%\ \text{Hb}\ A_2 = \frac{\text{Absorbance of Hb}\ A_2 \times 100}{\text{Abs of Hb}\ A_2 + 5(\text{Abs of other Hb})}$$

Notes

The procedure described allows the quantitation of hemoglobin A_2, even in the presence of Hb S. If a sample in question is AS or SS, the 50 μL hemolysate is mixed with 50 μL of the Hb A_2 elution reagent. Fifty microliters of this diluted hemolysate is added to the column and analyzed. Hb E and C, however, will be eluted with Hb A_2. In this case, the concentration of A_2 may exceed 10%. When such values are obtained, other tests such as electrophoresis are necessary for confirmation. Other factors that may affect the results include temperature changes, vibration of the column, and improper amount of hemolysate or reagents.

PROCEDURE FOR SOLUBILITY TEST

Principle

Deoxyhemoglobin S is less soluble than Hb A when reduced by dithionite in phosphate buffer solution; thus it forms a turbid suspension of protein crystals. These crystals prevent reading of lines on a paper card.

Reagents

Dithionite. Weigh out the following reagents:

1. Dibasic potassium phosphate, 21.6 g, anhydrous, K_2HPO_4.
2. Monobasic potassium phosphate, 16.9 g, crystals, KH_2PO_4.
3. Sodium hydrosulfite (dithionite), $Na_2S_2O_4$, 0.5 g.
4. Saponin, 0.1 g.

Place the K_2HPO_4 in a 100 mL volumetric flask. To facilitate preparation, first dissolve the K_2HPO_4 in about 60 mL of distilled water. Next, add the KH_2PO_4 and the dithionite and mix until dissolved. It may be necessary to add a little more water to dissolve the crystals. Add the remainder of the water up to the volumetric mark only after everything is dissolved. Finally, add the saponin. Mix well. Date and refrigerate the bottle. It is stable 1 week at 2° to 8° C. Final concentrations are K_2HPO_4 (1.24 mol/L), KH_2PO_4 (1.24 mol/L), $Na_2S_2O_4$ (0.029 mol/L), and saponin (0.1 g/L).

Assay

Equipment

12 × 75 mm glass test tubes and test tube rack.
20 μL micropipet.
5 mL pipets.
White test tube holder with black lines.
Balance and 100 mL volumetric flask.

1. Pipet 2 mL of reagent into the labeled 12 × 75 tubes.
2. Reagents must come to room temperature.
3. Add 20 μL of whole blood, EDTA or heparinized.
4. Mix and wait 10 minutes.
5. Place tube in white test tube holder with black lines (2.5 cm).
6. Read for turbidity and note results.
7. Run positive and negative controls with each group of solubility tests.

Interpretation

1. A *positive test* for sickling hemoglobin is indicated by a very turbid solution. The black lines on the tube holder cannot be seen through the solution (as shown in Fig. 60-4). Other hemoglobins, such as Hb C Harlem and Hb Travis, give a positive result. After electro-

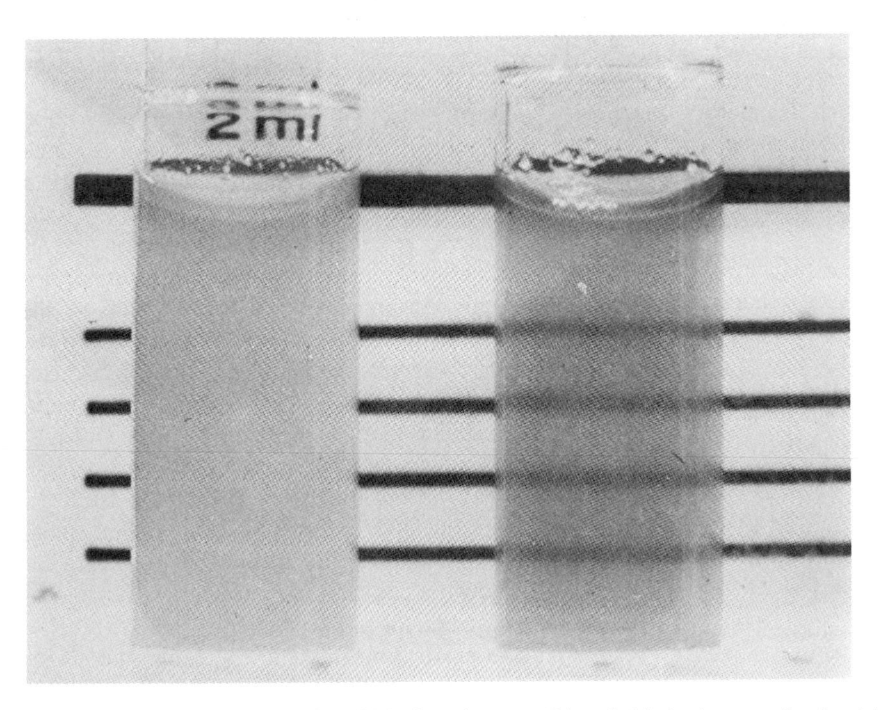

Fig. 60-4. Negative and positive dithionite tube tests. Lines behind tubes are clearly visible for negative test (right tube) and not visible for positive one (left tube).

phoresis, this positive result will distinguish Hb C Harlem or Hb Travis from Hb S.

2. A negative test occurs when one can read the black lines through the solution.

Notes

1. False-positive results may arise from polycythemia, too much blood added to reagent, dysglobulinemia, recent transfusion, and hyperlipidemia.
2. False-negative results are caused by outdated reagents, recent transfusion, tests on newborns and infants less than 3 months old, and anemia with Hb 70 g/L.
3. If Hb is less than 70 g/L, fill a Wintrobe tube with blood. Spin for 5 minutes at velocity 5, pipet off plasma and replace it with an equal volume of normal saline. Remix blood cells in saline with Pasteur pipet, transfer specimen to a clean 12 × 75 mm tube, and add 40 μL blood to 2 mL of reagent. Removal of the plasma eliminates a false-positive result that would otherwise occur because of the increased presence of plasma proteins brought about by the doubling of the blood volume.
4. One can modify the procedure to accommodate specimen collected in microhematocrit tubes. The test is done by use of 10 μL of blood and 1 mL of reagent. The time limit for the reaction remains the same.

Reference range

Hemoglobin electrophoresis. See Figs. 60-2 and 60-3 for the usual hemoglobin patterns.

Hemoglobin A_2. Reference ranges of Hb A_2 will vary between various laboratories. Each laboratory must determine its own range, which must be reported with each test to facilitate interpretation of results. The Isolab Company claims the range of Hb A_2 to be 1.5% to 3.0%. My own experience indicates a range of 1.5% to 4.0%. Values greater than 4.0% are presumptive of the beta-thalassemia trait.

Hemoglobin S. Hemoglobin S is not present in normal blood but will be detected in patients with sickle cell trait or sickle cell anemia.

REFERENCES

1. Brosious, E.M., et al.: Hemoglobinopathy testing, a report of the 1976 and 1977 College of American Pathologists Surveys, Am. J. Clin. Pathol. **70**:563-566, 1978.
2. Nalbandian, R.M., et al.: Dithionite tube test—a rapid inexpensive technique for the detection of hemoglobin S and non-S sickling hemoglobin, Clin. Chem. **17**(17):1028-1032, 1971.
3. Gottfried, E.L., et al.: Reliable estimation of hemoglobin A_2 concentration by electrophoresis with densitometry, Am. J. Clin. Pathol. **72**:415-420, 1979.
4. Pearce, C.: A comparison of three methods for quantitation of variant hemoglobin fractions, Am. J. Med. Technol. **46**(10):698-703, 1980.
5. Schmidt, R.M., and Holland, S.: Standardization in abnormal hemoglobin detection: an evaluation of hemoglobin electrophoresis kit, Clin. Chem. **20**(5):591-594, 1974.
6. Hicks, E.J., and Hughes, B.J.: Comparison of electrophoresis on citrate agar, cellulose acetate, or starch for hemoglobin identification, Clin. Chem. **21**(8):1072-1075, 1975.
7. Vernon, S.E.: Clinical application of agarose-gel electrophoresis in identifying abnormal hemoglobins, Clin. Chem. **26**(8):1230-1231, 1980.
8. Schmidt, R.M, and Brosious, E.M.: Quantitation of hemoglobin A_2, an interlaboratory study, Am. J. Clin. Pathol. **71**:534-539, 1979.
9. Brosious, E.M., et al.: Microchromatographic methods for hemoglobin A_2 quantitation compared, Clin. Chem. **24**(12):2196-2199, 1978.
10. Baine, R.M., and Brown, H.G.: Evaluation of a commercial kit for microchromatographic quantitation of hemoglobin A_2 in the presence of hemoglobin S, Clin. Chem. **27**(7):1244-1247, 1981.
11. Schmidt, R.M., and Wilson, S.M.: Standardization in detection of abnormal hemoglobins: solubility tests for hemoglobin S, J.A.M.A. **225**:1225-1230, 1973.
12. National Committee for Clinical Laboratory Standards (NCCLS): Standard for abnormal hemoglobin detection by cellulose acetate electrophoresis, NCCLS Tentative Standard TSH-8, 1980.

BIBLIOGRAPHY

Schmidt, R.M.: Laboratory diagnosis of hemoglobinopathies, J.A.M.A. **224**:1276-1280, 1973.

Schneider, R.G., et al.: Laboratory indentification of the hemoglobinopathies, Lab. Management **19**:29-43, 1981.

Lubin, B.H., et al.: Sickle cell disease and the thalassemias: diagnostic assays, Lab. Management **18**:38-47, 1980.

Sonnenwirth, A.C., and Jarett, L.: Gradwohl's clinical laboratory methods and diagnosis, ed. 8, vol. 1, St. Louis, 1980, The C.V. Mosby Co.

Hemoglobin F

ROBERT S. FRANCO

Clinical significance: pp. 611 and 833
Molecular weight: 64,000 daltons
Chemical class: hemoprotein

PRINCIPLES OF ANALYSIS AND CURRENT USAGE

The major problem of measurement of hemoglobin F (Hb F) is to quantify it in the presence of large amounts of hemoglobin A (Hb A), which has the structure $\alpha_2\beta_2$ and the same porphyrin groups. Since the porphyrin groups have strong absorptivity properties, many separation procedures use this chromophore for quantitation of the amount of hemoglobin F present. For this to be useful the heme group must be converted into a form that has a quantitatively reproducible color, that is, cyanmethemoglobin. The total hemoglobin may be separated by means of ion-exchange chromatography and each of the hemoglobin subfractions may be quantitated by absorption spectroscopy at 415 nm (method 1, Table 60-14).[1-3] Hemoglobin F consists of two alpha and two gamma chains; hemoglobin A also has two alpha chains; therefore immunological methods are based on the use of antibodies that are specific for the gamma chain. Immunological systems such as radial immunodiffusion (RID),[4] enzyme-linked immunosorbent assay (ELISA),[5] and radioimmunoassay (RIA)[6] are becoming available but are not yet in widespread use

Table 60-14. Method of fetal hemoglobin quantitation

Method	Principle	Usage	Comments
1. Ion-exchange chromatography	Hemoglobins separated according to net ionic charge and charge distribution at a particular pH	Specialized laboratories	Most suitable for specimens containing a high percentage of Hb F
2. Immunological (RIA, ELISA)	Hb F reacts with a specific antibody with detection depending on specific method used	Specialized laboratories	May eventually be method of choice for routine estimation of Hb F
3. Immunological (radial immunodiffusion, RID)	Hb F reacts with a specific antibody with detection as a precipitin ring around the sample well of an antibody-containing gel plate	Rare	See reference 4 for a comparison of RID and alkali denaturation for Hb F
4. Alkali denaturation	At high pH hemoglobins other than Hb F are denatured and may be precipitated with ammonium sulfate	Standard method	Inaccurate for very high levels of Hb F (>40%)
5. Cellulose acetate electrophoresis	Hemoglobins separated electrophoretically and quantitated by densitometric scanning	Rare	Not recommended because of poor correlation with alkali denaturation method (see reference 11)

(method 2, Table 60-14). Assays based on the differential solubility of Hb F and Hb A under alkaline conditions have been in clinical use for many years,[7-10] and that type currently is the most frequently used procedure (method 3, Table 60-14). This assay is based on the relative resistance of Hb F to denaturation at high pH when compared to Hb A or Hb A_2. The denatured adult hemoglobin may be precipitated with ammonium sulfate and removed by filtration (method 4, Table 60-14). Finally, Hb F may be quantitated by densitometric scanning (at 415 nm) of cellulose acetate strips after electrophoretic separation of the hemoglobin subfractions (method 5, Table 60-14).

Table 60-15. Reaction conditions for Hb F quantitation

Conditions	Alkali denaturation
Temperature	Entire procedure performed at room temperature
Hb concentration in hemolysate	80 to 100 g/L
Concentrations during denaturation	Hemoglobin: 4.2 to 5.3 g/L
Time for denaturation	Exactly 2 minutes
Final concentration of reagents	
Cyanmethemoglobin conversion	723 μmol/L KCN; 572 μmol/L $K_3Fe(CN)_6$
Denaturation	0.08 mol/L NaOH
Precipitation	2.23 mol/L $(NH_4)_2SO_4$
Time for precipitation	5 minutes
Linearity	False low values at Hb F levels greater than 40%
% coefficient of variation	8.8% (for normal and elevated Hb F levels)

REFERENCE AND PREFERRED METHODS

The method of choice depends to some extent on the percentage of Hb F in the sample. For Hb F levels higher than 40% of the total hemoglobin concentration, the chromatographic procedures are the most reliable. For very low levels (less than 2%) the sensitive ELISA and RIA methods using anti–human Hb F sera (available commercially, Behring Diagnostics, La Jolla, Calif.) will probably prove to be best. A radial immunodiffusion assay for Hb F is currently available in kit form from Helena Laboratories, Beaumont, Texas. A comparison with alkali denaturation is given in reference 4. It has been found that densitometry of cellulose acetate electrophoretic strips is not an adequate method for quantitation of Hb F at the elevated levels of Hb F seen in hereditary persistence of fetal hemoglobin (Table 60-15).

The recommended method for the routine clinical quantitation of Hb F is alkali denaturation. The unprecipitated hemoglobin is thus "alkali resistant" and is an estimate of the amount of Hb F present. The method of Singer et al.[7] has been widely used but suffers from the fact that oxyhemoglobin is used without conversion to cyanmethemoglobin. Since methemoglobin F is not alkali resistant, false low values may be obtained in specimens that contain significant levels of methemoglobin. Cyanmethemoglobin F is alkali stable; therefore complete conversion to cyanmethemoglobin with Drabkin's solution (see next page) removes the problem.

The method of choice is the Pembrey et al. modification[8] of the Betke et al. procedure.[9] This method incorporates conversion to cyanmethemoglobin as described above and has an increased sensitivity by reading at 415 nm rather than 540 nm. Note that this method is not completely specific for Hb F as a sample of Hb A purified by diethylaminoethyl (DEAE)-Sephadex chromatography was 0.26% alkali resistant.[8] Starch gel electrophoresis of alkali-resistant material shows other compounds in addition to Hb F.[8] None of the alkali-resistant methods are accurate for very high levels of Hb F (greater than 40%), such as those seen in cord blood samples, and a chromatographic method should be used for these samples.

The coefficient of variation of the Pembrey method over the applicable range of Hb F levels is approximately 9%.[8]

SPECIMEN

A variety of anticoagulants may be used, including disodium ethylenediaminetetraacetic acid (EDTA), heparin, and ammonium oxalate. Whole blood may be stored before assay for at least 1 week at 4° C.

PROCEDURE

Principle

Sodium hydroxide is added to cyanmethemoglobin, causing denaturation of nonresistant hemoglobin. Addition of ammonium sulfate precipitates denatured hemoglobins, which are removed by filtration. The resistant hemoglobin in the filtrate is quantitated at 415 nm, and the percentage of resistant hemoglobin is calculated.

Reagents

Sodium hydroxide, 1.2 M. Dissolve 4.8 g of sodium hydroxide in water to make 100 mL. Prepare fresh for each group of analyses.

0.85% saline, 0.144 M. Dissolve 8.5 g of NaCl in 900 mL of distilled water in a 1 L volumetric flask. Bring to mark and mix. The reagent is stable for 3 months at room temperature.

Carbon tetrachloride. Reagent grade. It is stable 1 year at room temperature.

Drabkin's solution, 767 μmol/L KCN and 607 μmol/L $K_3Fe(CN)_6$. Dissolve 0.05 g of potassium cyanide and 0.20 g of potassium ferricyanide with water to make 1 L. CAUTION: Potassium cyanide is a dangerous poison. Gaseous hydrogen cyanide will be released on contact with acids. Store Drabkin's solution in a brown bottle away from light. The solution is stable for 30 days at 4° C.

Ammonium sulfate (saturated), approximately 5.8 mol/L. Place 70 g of ammonium sulfate in 100 mL of distilled water at 80° C. Filter at this temperature and allow to come to room temperature, forming saturated solution. Store at room temperature for up to 12 months.

Assay

Equipment: spectrophotometer of ≤ 10 nm band pass capable of reading at 415 nm; table-top centrifuge and filter paper.

1. Wash red cells three times with 0.85% saline solution.
2. To 1 volume of packed red cells add 2 volumes of distilled water and 0.5 volume of carbon tetrachloride. Mix well and centrifuge 30 minutes at 3000 g. The resulting clear lysate should have a hemoglobin concentration of 80 to 100 g/L. Lysate hemoglobin concentrations below 70 g/L will give false high values for alkali-resistant hemoglobin.[8]
3. Add 0.6 mL of lysate to 10 mL of Drabkin's solution and mix to prepare cyanmethemoglobin solution.
4. Pipet 5.6 mL of the cyanmethemoglobin solution into a test tube.
5. Add 0.40 mL of 1.2 M sodium hydroxide and mix immediately by inversion. Start a stopwatch.
6. Wait exactly 2 minutes, then add 4.0 mL of saturated ammonium sulfate, and mix well.
7. Allow the mixture to stand for 5 minutes and filter through a double layer of Whatman No. 6 filter paper. If the filtrate is cloudy, refilter it. The final filtrate must be completely clear.
8. Prepare an undenatured "total hemoglobin" dilution as follows: 1.4 mL of cyanmethemoglobin solution and 1.6 mL of water and 2 mL of saturated ammonium sulfate. Make a further 10:1 dilution with water.
9. Read the filtrate and the "total hemoglobin" against a water blank at 415 nm.

Calculation

The dilution factor of the cyanmethemoglobin solution in the denaturing (test) solution is (10/5.6). The dilution factor of the cyanmethemoglobin solution in the undenatured (total hemoglobin) solution is $\frac{5.0}{1.4} \times 10$. Therefore the appropriate factor by which the undenatured reading should be multiplied to make it equivalent to the denatured sample is

$$\frac{5.0}{1.4}(10) \div \frac{10}{5.6}$$

which is equal to 20.

The percentage of alkali-resistant hemoglobin may therefore be calculated form the formula:

$$\frac{A_{415,\ \text{filtrate}} \times 100\%}{A_{415,\ \text{undenatured solution}} \times 20}$$

The factor of 100 is to convert to percentage.

The coefficient of variation for this method is 8.8%.[8]

REFERENCE RANGE

Reference values for men and women are less than 0.9%,[8] but a reference range should be established for each laboratory.

REFERENCES

1. Abraham, E.C., et al.: Separation of human hemoglobins by DEAE-cellulose chromatography using glycine-KCN-NaCl developers, Hemoglobin **1**:22-44, 1976.
2. Dozy, A.M., et al.: Studies of the heterogeneity of hemoglobin. XIII. Chromatography of various human and animal hemoglobin types on DEAE-Sephadex, J. Chromatography **32**:723-727, 1968.
3. Huisman, T.H.J., and Jonxis, J.H.P.: The hemoglobinopathies: techniques of identification (clinical and biochemical analysis, 6), New York, 1977, Marcel Dekker, Inc.
4. Chudwin, D.S., and Rucknagel, D.L.: Immunological quantification of hemoglobins F and A_2, Clin. Chim. Acta **50**:413-418, 1974.
5. Makler, M.T., and Pesce, A.J.: ELISA assay for measurement of human hemoglobin A and hemoglobin F, Am. J. Clin. Pathol. **74**:673-676, 1980.
6. Garver, F.A., et al.: Specific radioimmunochemical identification and quantitation of hemoglobins A_2 and F, Am. J. Hematol. **1**:459-469, 1976.
7. Singer, K., et al.: Studies on abnormal hemoglobins. I. Their demonstration in sickle cell anemia and other hematologic disorders by means of alkali denaturation, Blood **6**:413-428, 1951.
8. Pembrey, M.E., et al.: Reliable routine estimation of small amounts of foetal hemoglobin by alkali denaturation, J. Clin. Pathol. **25**:738-740, 1972.
9. Betke, K., et al.: Estimation of small percentages of foetal haemoglobin, Nature **184**:1877-1878, 1959.
10. International Committee for Standardization in Haematology: Recommendations for fetal hemoglobin reference preparations and fetal hemoglobin determination by the alkali denaturation method, Br. J. Haematol. **42**:133-136, 1979.
11. Schmidt, R.M., et al.: Quantitation of fetal hemoglobin by densitometry, J. Lab. Clin. Med. **84**:740, 1974.

Immunoelectrophoresis

GAYLE BIRKBECK

Clinical significance: p. 718

PRINCIPLES OF ANALYSIS AND CURRENT USAGE

Zone electrophoresis and immunodiffusion are commonly used methods in clinical laboratories. Zone electrophoresis employs for an inert solid carrier medium such as cellulose acetate or agarose for the separation of proteins. The technique is routinely used to screen for serum protein abnormalities, hemoglobin variants, and isoenzymes. Specific bands are identified by their mobility when compared to a standard and by staining characteristics. Immunodiffusion techniques are used to quantitate proteins (see immunoglobulin quantitation, Chapter 10 and p. 1304) and also to identify proteins.

Immunoelectrophoresis is a combination of electrophoresis and an antigen-antibody interaction. Immunoelectrophoresis improves zone electrophoresis by enhancing the ability to identify proteins after electrophoretic separation by the use of a specific antibody. Immunoelectrophoresis may also be used to enhance the formation of precipitin lines by use of electrophoresis to amplify the antibody and antigen interaction rather than relying on simple diffusion.

Immunoelectrophoresis may be divided into the following methods: electroimmunodiffusion, counterimmunoelectrophoresis, immunoelectrophoresis (IEP), and immunofixation.

Electroimmunodiffusion or the Laurell rocket technique (method 1, Table 60-16) uses electrophoresis to shorten the diffusion time of the antigen through agarose. The sample is added to a well in an agarose plate with antibody incorporated throughout the agarose. The plate is then placed in an electrophoresis chamber. As each antigen migrates through the agarose under the force of the electric field, precipitin peaks (rockets) form in the agarose. The method is quantitative since the height of the "rocket" is proportional to the immunoglobulin concentration (see methods for immunoglobulin quantitation, p. 1304).

Crossed immunoelectrophoresis (method 2, Table 60-16) may also be used for quantitation purposes. Crossed immunoelectrophoresis or two-dimensional electrophoresis[2] is performed in two steps. In the first step, the sample is placed in a well cut in agarose. The proteins in the sample are separated by conventional zone electrophoresis. The excess agarose above the separated proteins is removed and replaced by an antibody-containing gel. The second-dimensional electrophoresis then causes the separated proteins to migrate perpendicularly to the first separation into the antibody-containing agarose. Antigen-antibody precipitation peaks then form in the second gel. Quantitation is achieved by measurement of the area under the peaks by planimetry or by measurement of the area in square millimeters and comparison of these results to those obtained with standards.[3] Two-dimensional immunoelectrophoresis is very similar to one-dimensional immunoelectrophoresis or electroimmunodiffusion. Both are quantitative; however, only one antibody is used and consequently one antigen is quantitated at a time with electroimmunodiffusion. A polyspecific antiserum is used with two-dimensional immunoelectrophoresis, thus allowing for the quantitation of multiple proteins with each run.

Counterimmunoelectrophoresis[4] (method 3, Table 60-16) again uses electrophoresis to shorten diffusion time. Various terms used for this technique are *countercurrent electrophoresis, electro-osmodiffusion,* and *crossover-electrophoresis.* One performs counterimmunoelectrophoresis by placing an anodically migrating antigen in a well in agarose and a cathodically migrating antibody in another well close by. When current is applied, the antigen and the antibody migrate toward each other and form a precipitin line where they meet. The method is much faster than standard diffusion techniques (that is, Ouchterlony plates) and can be used qualitatively and semiquantitatively (by use of a titer).

Immunoelectrophoresis (method 4, Table 60-16) combines zone electrophoresis and simple antigen-antibody diffusion reactions. This method offers an accurate identification of various proteins by use of specific antibodies at a cost of increased diffusion time over that described in

Table 60-16. Immunoelectrophoresis

Method	Type of analysis	Principle	Usage	Comments
1. Electroimmunodiffusion (EID) (Laurell rocket)	Quantitation by size of immunoprecipitate in gel	Electrophoresis of antigens into antibody-containing agarose gel; height of gel pattern (rocket) proportioned to antigen concentration	Serum	Used for special tests such as factor VIII–related antigen
2. Two-dimensional or crossed immunoelectrophoresis	Quantitation of multiple proteins in one gel by use of polyspecific antiserum; quantitation based on area under precipitin peak	Proteins separated by zone electrophoresis in agarose gel; a second electrophoresis causes separated proteins to migrate perpendicularly to first separation into antibody-containing agarose	Serum, fluids	Too time consuming for the clinical laboratory
3. Counterimmunoelectrophoresis	Qualitative identification of protein by formation of a precipitin band with monospecific antiserum	Anodically migrating antigen and cathodically migrating antibody move through agarose toward each other in an electrophoretic field; a precipitin band forms where they meet	Serum, CSF, other fluids	Used primarily by microbiology laboratories to identify bacterial antigens
4. Immunoelectrophoresis (IEP)	Identification of monoclonal proteins through precipitin arc formation	Proteins separated in agarose by zone electrophoresis; antiserum is placed in a trough horizontal to the electrophoretogram and allowed to diffuse for 24 hours; precipitin arcs form where antigen and antibody meet; monoclonality can be determined from shape and mobility of arc	Serum, urine, body fluids	Should be used as the confirmatory test for the presence and identification of monoclonal proteins
5. Immunofixation	Identification of protein bands on zone electrophoresis by formation of precipitin bands with antisera	Proteins are separated by zone electrophoresis; area containing protein to be identified is overlayed with monospecific antibody and its corresponding antigen	Serum, urine, other body fluids	Can be used with the IEP to help identify monoclonal proteins; more sensitive than IEP

the previous methods. Grabar and Williams[5] first described immunoelectrophoresis in 1953. Scheidegger[6] modified the original technique to a micromethod. Immunoelectrophoresis is usually performed on an agarose plate with wells and troughs precut in the agarose. The plates are available commerically or may be prepared by the laboratory. The patient sample is placed in a well and separated into albumin, $alpha_1$, $alpha_2$, beta, and gamma globulin components by zone electrophoresis. Either a monospecific or a polyspecific antiserum is placed in a trough that is cut parallel to the zone electrophoresis separation. The plate is then incubated in a moist chamber for 24 hours, during which time the antibody and antigen diffuse toward each other and form precipitin arcs where they meet. The arcs can be identified both by the shape and migration rate when compared to a control. Immunoelectrophoresis is both a qualitative and a semiquantitative technique.

Immunofixation[7] (method 5, Table 60-16) also is a combination of zone electrophoresis and an antibody-antigen interaction. Either cellulose acetate or agarose is used as the supporting medium.[8,9] Immunofixation varies from immunoelectrophoresis by elimination of the long diffusion time for antigen-antibody interaction. The proteins first are separated by zone electrophoresis. Strips of cellulose acetate or filter paper, soaked in a monospecific antiserum, are laid over the zone occupied by the protein to be identified and, after approximately 1 hour at room temperature incubation, the strips are removed. The excess protein is washed from the plate and the remaining precipitation band may be stained with a protein stain such as amido black.

REFERENCE AND PREFERRED METHODS

Crossed or two-dimensional immunoelectrophoresis is not used routinely in clinical laboratories. The technique can be used to quantitate a wide variety of proteins. Small

Table 60-17. Comparison of reaction conditions for immunoelectrophoresis

	Electroimmunodiffusion	Counterimmunodiffusion	Immunoelectrophoresis (IEP)	Immunofixation
Temperature	Ambient	Ambient	Ambient	Ambient
pH	Usually pH 8.2	Varies (pH 5 to 10)	Usually pH 8.6	pH 8.6
Buffer	Barbital, 0.2 M	Barbital, 0.05 M	Barbital, 0.2 M	Barbital, 0.2 M
Sample volume	1 μL	20 μL	0.8 μL	100 μL
Time of reaction	2-4 hours	2-4 hours	18-48 hours	8-10 hours

changes in both concentration and mobility can be detected. However, the technique is very time consuming in both the preparation time for the plates and in the actual test-execution time. Commercial plates are not available. Valid quantitative measurements are difficult, since it is necessary to measure a curved area. The measurement techniques are either costly (requiring a computer) or very tedious. In general, the same clinical information can be obtained by other less expensive and less time-consuming methods. (See section on immunoglobulin quantitation for protein quantitation methods, p. 1304.) The immunoelectrophoresis procedures in Table 60-17 offer adequate techniques for protein identification

Electroimmunodiffusion (EID) (Table 60-17) or one-dimensional immunoelectrophoresis is used on a limited basis in some clinical laboratories. Since commercial plates are not available, few laboratories use electroimmunodiffusion for routine protein quantitation such as immunoglobulin or C3 and C4 concentrations. However, the technique is used for the quantitation of factor VIII–related antigen, an important test for the diagnosis of von Willebrand's disease.[10] Calbiochem-Behring (La Jolla, Calif.) offers antiserum for human factor VIII–related antigen, buffer, and a package insert for the electroimmunodiffusion procedure.

Counterimmunoelectrophoresis (Table 60-17) was used by clinical laboratories in the early 1970s to detect hepatitis B surface antigen in both patient and donor blood samples. However, radioimmunoassay and immunoenzyme techniques have replaced counterimmunoelectrophoresis for hepatitis testing. The technique is used in some microbiology laboratories for the detection of bacterial antigens in cerebrospinal fluid (CSF), serum, and urine.[11] The technique is limited by the fact that the antigen and antibody must have opposite charges to migrate toward each other in an electrophoretic field.

Immunofixation (Table 60-17) identifies proteins separated by zone electrophoresis by use of specific antibody. The technique is more sensitive than zone electrophoresis. Chang and Inglis[9] reported a fourfold increase in sensitivity when comparing zone electrophoresis and immunofixation, both performed on cellulose acetate. Using zone electrophoresis and a 0.1% nigrosine stain, they detected a minimum of 500 ng of albumin in a 1 μL sample. With immunofixation and a 0.1% nigrosine stain, the sensitivity was increased to a minimum of 125 ng of protein in a 1 μL sample. Ritchie and Smith[8] reported a minimum sensitivity of 50 ng/μL when using an agarose system for immunofixation. Therefore, bands that are not apparent with zone electrophoresis can be visualized and positively identified by immunofixation by use of a monospecific antibody. Immunofixation also is more sensitive than immunoelectrophoresis, since the protein under study does not diffuse into the agarose before reacting with antibody, as in the latter technique.

Immunofixation has been used to study protein polymorphism[7] and identify specific proteins[8] and in the genetic typing of alpha$_1$-antitrypsin.[12] However, the primary use for immunofixation in the routine clinical laboratory is in the study of monoclonal proteins. Ritchie and Smith[13] demonstrated the use of immunofixation in identifying small amounts of monoclonal protein (minimonoclonals) in serum. Cawley et al.[14] concluded that immunofixation could be used to identify minimonoclonal proteins, biclonal proteins, oligoclonal proteins in cerebrospinal fluid, and verification of mobility shifts in certain proteins such as complement components. Sun et al.[15] and Merlini et al.[16] compared immunofixation and immunoelectrophoresis. Both groups concluded that immunoelectrophoresis should not be used for routine clinical screening but immunofixation could be used for difficult cases as in the presence of biclonal gammopathies or small bands of monoclonal components. Merlini et al.[16] found that the results from the two methods agreed 80% of the time for monoclonal components exceeding 1 to 2 g/L. However, correlation with smaller amounts of monoclonal protein was not so good. Immunofixation appeared more sensitive but less dependable, since technical artifacts interfered more frequently.

The detection of very small amounts of monoclonal protein (oligoclonal bands) may not benefit the clinician. Keshgegian[17] used a high-resolution agarose gel electrophoresis system (Panagel, Worthington Diagnostics Freehold, N.J.) and immunofixation to study the frequency of small homogeneous bands in a hospital population. He found that 10% to 65% of hospitalized patients had small monoclonal bands. The frequency of bands was unrelated to age, sex, race, or medical diagnosis of the patient. However, the bands were found most frequently in patients with a hypergammaglobulinemia (greater than 15 g/L) and in patients with an acute phase reaction. Only one patient had a lymphoproliferative disorder. The bands may reflect a component of the human immune response. Thus the routine screening of patient sera using a sensitive immunofixation technique is not recommended.

Immunoelectrophoresis (IEP) (Table 60-17) is the most commonly used method in clinical laboratories for the identification of monoclonal proteins. However, the detection and identification of monoclonal proteins is accomplished most effectively by a combination of the following tests: protein electrophoresis combined with total protein quantitation, immunoelectrophoresis, immunoglobulin quantitation, and in some cases immunofixation. The first screening tests for serum or urine should be a combination of the quantitation of the total serum proteins and of the protein electrophoresis. If the total protein and protein electrophoresis are normal, no further tests need to be performed at that time. However, if a homogenous band or a suspicious heterogeneous band is detected on the protein electrophoresis, the sample should be evaluated by immunoelectrophoresis. Protein electrophoresis is the screening test and immunoelectrophoresis is the confirmatory test. The appearance of a homogeneous band on the protein electrophoresis does not always indicate the presence of a monoclonal gammopathy. Pseudo-M proteins may be caused by fibrinogen (when left in a serum sample), free hemoglobin, lipoprotein deposits (especially near the point of application), immune complexes, and occasionally hyperlysozymemia.[18] Therefore, when a suspicious band is detected on the protein electrophoresis, the immunoelectrophoresis technique should be used to determine whether a monoclonal gammopathy is present. However, all monoclonal proteins are not always detected on the protein electrophoresis. On occasion the monoclonal protein may be masked by normal proteins, particularly if there is an increase in one of the normal components. Thus, if the clinical evaluation of the patient leads to a strong suspicion of a lymphoproliferative disorder, immunoelectrophoresis should be performed to determine the presence or absence of a monoclonal protein.

When a monoclonal protein is detected by immunoelectrophoresis, immunoglobulin quantitation should be performed. Although the patient's normal immunoglobulins will be quantitated accurately by immunochemical technique, the monoclonal protein will not be accurately measured. Therefore the quantity of monoclonal protein might be best determined from the protein electrophoresis. (See the methods for immunoglobulin quantitation for more details, p. 1304.) Immunofixation might be used for additional information when the immunoelectrophoresis results are dubious or when more than one monoclonal protein is present.

SPECIMEN

The immunoelectrophoresis techniques may be performed on serum, urine, or other body fluids. Plasma is not recommended. The fibrinogen peak or band may interfere with the detection of a monoclonal protein. Urine and body fluids are concentrated 30 to 50 times before analysis by protein electrophoresis, immunoelectrophoresis, or immunofixation. Samples should be frozen if not tested within 1 to 2 days of collection.

PROCEDURE

Methods for electroimmunodiffusion and counterimmunoelectrophoresis will vary somewhat, depending on the specific antigen-antibody system to be assayed. Generally, 1% agarose in barbital buffer, pH 8.6, is used for electroimmunodiffusion. The appropriate antiserum dilution varies according to the specific protein to be assayed. The appropriate initial dilutions of antisera are 1:10, 1:20, and 1:30. Specific procedures will vary for the estimation of the desired antigen and antibody concentrations.

Counterimmunoelectrophoresis is usually performed with 1% agarose in barbital buffer, pH 8.2.[19] At pH 8.2, the antigens generally have a negative charge and migrate toward the anode, whereas the antibody with a slightly positive charge migrates toward the cathode, primarily aided by endosmotic forces. Acetylation and carbamylation can be used to increase the negative charge on the antigen. The ratio of antigen to antibody is critical to obtain a sharp precipitin band between the two wells. Varying dilutions of antibody may be run against varying dilutions of antigen to determine the optimal ratio for a specific antigen-antibody system. In summary, the test can be somewhat difficult to set up, since a number of variables such as antigen-antibody ratio, pH, isoelectric points of the antigen and antibody, buffer, and endosmotic flow must be carefully adjusted and vary depending on the specific antigen-antibody system used. It is easiest to use an established procedure for the specific antigen-antibody system to be assayed.

IMMUNOELECTROPHORESIS

Principle

Serum proteins are separated on the basis of their electrophoretic mobility, and the separated proteins are then allowed to diffuse against the antiserum specific for antigens under investigation. A visible precipitin band is formed where the antigen and its homologous antibody meet (Fig. 60-5).

Reagents

Barbital buffer. Dissolve two vials of powdered buffer (corning ACI electrophoresis universal barbital buffer, Corning Medical and Scientific, Medfield, Mass.) in 4 L of distilled water and pH to 8.6 with 2 N HCl. Refrigerate.

5% glacial acetic acid, 0.88 mol/L. To a 4000 mL volumetric flask, add 200 mL of glacial acetic acid and bring to volume with distilled water. Store at room temperature.

2% glycerol, 21.7 mmol/L. To a 4000 mL volumetric

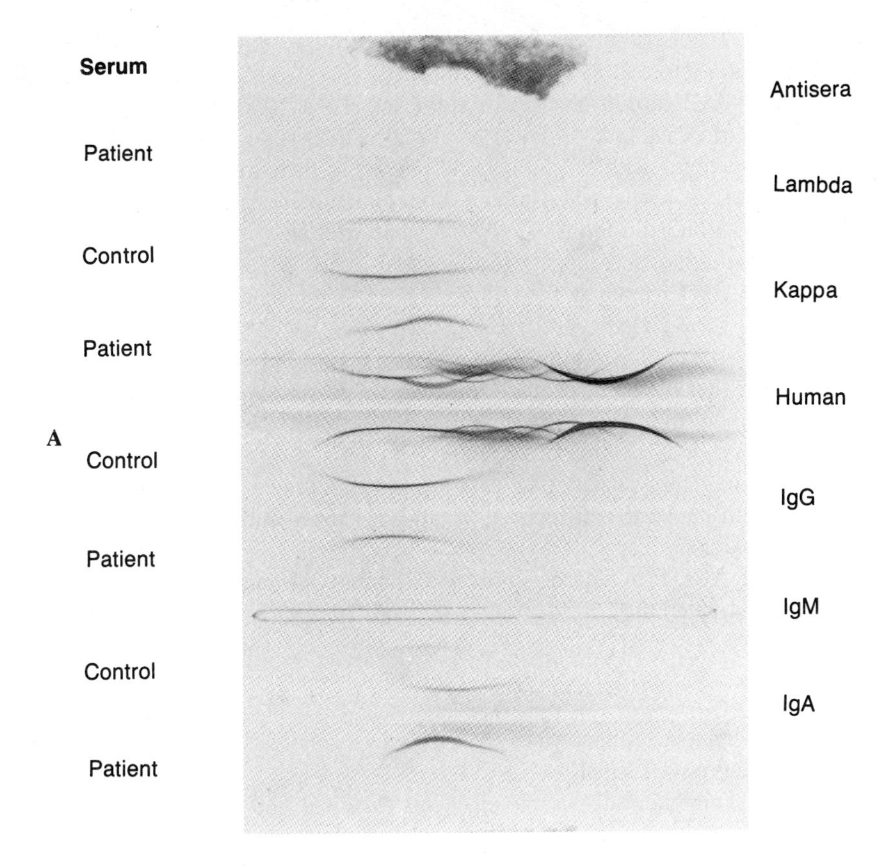

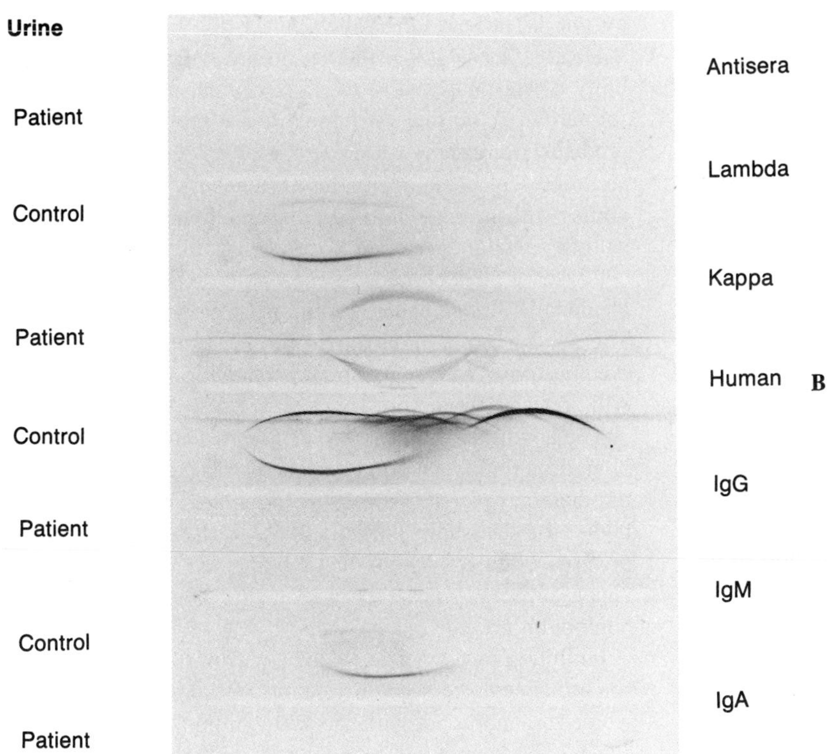

Fig. 60-5. A, Example of serum immunoelectrophoresis analysis. *Left column,* Type of specimen electrophoresed (patient or control); *right column,* antiserum reagent (in this example, anti–human immunoglobulin) placed in troughs running parallel to plane of electrophoresis. **B,** Example of urine immunoelectrophoresis. Columns are as described for **A.**

flask, add 80 mL of glycerol and bring to volume with distilled water (store at room temperature).

Corning ACI amido black 10B stain set. To a 1000 mL volumetric flask, add 1 vial of amido black 10B stain. Rinse vial 3 times with 5% acetic acid and bring flask to volume with 5% acetic acid. Store at room temperature.

Normal saline solution (0.9% NaCl), 0.15 mol/L

Behring diagnostic reagents (Calbiochem-Behring, La Jolla, Calif.): refrigerate at 4° to 6° C.

Anti-IgG (heavy chain specific)
Anti-IgM (heavy chain specific)
Anti-IgA (heavy chain specific)
Anti-lambda (light chain specific)
Anti-kappa (light chain specific)
Anti–human (polyspecific) or anti-TIg (total immunoglobulin contains anti-IgG, -IgM, -IgA, -kappa, and -lambda).

Corning ACI film agarose universal pack (Corning Medical and Scientific). Refrigerate.

Assay

Equipment:

1. ACI agarose film/cassette system (electrophoresis cell and power supply)
2. 20 μL pipet and disposable pipet tips (Medical Laboratory Automation, Inc., Mt. Vernon, N.Y.)
3. Magnetic stirrer and stir bar
4. Oven, 50° to 60° C
5. Pipet

A. Electrophoresis
1. Fill each cell trough with 90 mL of barbital buffer. Level buffer in each cell by tilting cell, and wipe bridge divider free of buffer. Plug into power supply.
2. Set Hamilton pipet to dispense 0.8 μL.
3. Separate plastic plate from agarose plate and discard plastic plate.
4. Pipet 0.8 μL samples of control serum and patient serum to the right (in front) of the preformed well on the agarose plate. Control serum has upper-limit concentrations of IgG, IgM, and IgA (0.1 mL aliquot sample frozen at −70° C).
5. Set agarose plate into cell cover with agarose side up, so that the + on the agarose is on the same side as the + on the cell cover. Electrophorese for 45 minutes.

B. Immunodiffusion
1. Remove agarose plate, carefully blot off excess buffer, and evenly apply 40 μL of antisera to the long antisera trough in two 20 μL applications.
NOTE: Care should be taken not to touch the agarose surface, since this may result in uneven diffusion of antiserum.
2. Place agarose plate into an incubator tray. Place tray in a moisture chamber and allow antiserum to diffuse for 24 hours.

C. Staining immunoelectrophoresis
1. Rinse plate in 0.9% saline overnight.
2. Rinse plate with distilled water.
3. Stain in amido black for 5 minutes. Rinse off excess stain in 5% acetic acid. (Dip 7 to 10 times and wipe back of plate free of moisture.)
4. Place plate in dryer for 1 hour or until completely dry.
5. Decolorize in 5% acetic acid until clear.
6. Rinse in 2% glycerol for 60 seconds for permanent mount.

Interpretation

The immunoelectrophoresis is used to (1) determine a relative increase or decrease in the concentration of antigen and (2) determine the presence or absence of a monoclonal protein, and (3) identify a specific protein. One can determine an increase or decrease in antigen concentration by noting the position of the precipitin arc in relation to the well and to the trough. Since the antibody concentration is constant, an increased antigen concentration will produce an arc that is close to the trough. (The antigen originates in the well and diffuses to the trough.) The arc will also be thicker with an increased antigen concentration. A decreased antigen concentration will result in an arc that is closer to the well.

The presence of a monoclonal protein can be detected by one or more of the following: a distortion of the curvature of the arc, a difference in arc electrophoretic mobility when compared to the control, and inhibition of arc formation. A normal polyclonal arc is the combination of products from many different clones of plasma cells. The arc represents chemically heterogeneous groups of antibodies with no one clonal product predominating. The resulting polyclonal arc is a symmetric, wide, semicircle. In contrast, a monoclonal arc represents a large amount of chemically homogeneous antibody from one cell clone. The presence of a monoclonal protein usually results in a precipitin arc with a humping or bowing effect. The arc will appear asymmetric.

The abnormal protein also may show a different electrophoretic mobility than the normal control. Monoclonal antibodies may vary in size when compared to normal antibodies of the same immunoglobulin class. Bence Jones proteins often produce a fast-moving arc with lambda or kappa antisera, since they are smaller than whole immunoglobulin proteins.

Inhibition of the formation of a part of the precipitin arc sometimes can be seen in relation to light-chain antisera, which may not react with the whole immunoglobulin molecule. The monoclonal protein is in the inhibited area. This phenomenon is seen with IgA_2 proteins in which the light chains are not free to form a cross-linking structure with antibody.[20,21]

Identification of a protein can be performed by use of a monospecific antibody for that protein. The presence is confirmed by the formation of a precipitin arc. The arc should have the same shape and mobility as the arc formed by the known control material.

IMMUNOFIXATION[22]

Principle

Zone electrophoresis is performed on cellulose acetate, agarose, or high-resolution agarose (Panagel, Worthington Diagnostics, Freehold, N.J.). Monospecific antiserum is overlayed on the zone on the electrophoretogram, which contains the unidentified protein. (The entire electrophoretogram may be overlayed if one is screening for a protein). The presence of a precipitin band indicates that the antigen is present for the monospecific antiserum used.

Reagents

Monospecific antisera. (Available commercially)

Saline solution (0.9%) or PBS (phosphate-buffered saline solution) pH 7.0

Protein stain. 1% Coomassie brilliant blue R in 10% acetic acid in 45% absolute methanol. Dissolve 20 g of brilliant blue R in 900 mL of reagent methanol, 200 mL of glacial acetic acid, and 900 mL of distilled water.

Destain solution. 10% acetic acid in 50% reagent alcohol. Mix 2500 mL of 95% reagent alcohol, 500 mL of glacial acetic acid, and 2000 mL of distilled water.

Assay

Equipment

Photographic film tanks with frame and clip hangers
Forced air dryer, 50° to 60° C
Filter paper or cellulose acetate overlay strips, 4 × 20 mm
Filter paper sheets for pressing (Whatman No. 1), 5 × 7 inches

1. Perform zone electrophoresis (see the method for protein electrophoresis, p. 1309).
2. Immunofixation—method for agarose plates.[22]
   a. Overlay the area to be identified on the electrophoretogram with overlay strips moistened with 0.1 mL of monospecific antiserum.
   b. Add an additional 0.1 mL of antiserum to the overlayed strip, avoiding overflow onto the adjacent electrophoretogram.
   c. Incubate the overlaid electrophoretogram in a moist chamber at room temperature for 60 minutes.
   d. Remove the absorbent strips by using a gentle stream of PBS or saline from a squeeze bottle.
   e. Place a 1 cm thickness of filter paper on the gel. Overlay with a Plexiglas plate and press for 15 minutes with a 10 lb weight.
   f. Remove the Plexiglass plate and filter paper and place gel in photographic hanger. Agitate gel in saline or PBS for 15 minutes.
   g. Repeat step *e*, then repeat step *f* and repeat step *e* again.
   h. Dry the plate in the forced air oven at 50° to 60° C.
3. Staining
   a. Place dried gel in 1% Coomassie brilliant blue R for 20 minutes.
   b. Place in destaining solution and agitate for 5 minutes. Repeat destaining with fresh destaining solution.
   c. Rinse in tap water and dry at 50° to 60° C.

Interpretation

A precipitin band will form where the monospecific antiserum and its antigen react. The precipitin band remains trapped in the agarose, and the other free proteins are washed from the gel during the washing process. A comparison is made between the immunofixed electrophoretogram and the zone electrophoresis, which is stained.

REFERENCE RANGE

A decrease or absence of one or more of the immunoglobulin classes may be either congenital or acquired. Acquired hypogammaglobulinemia may be seen in malignant monoclonal B cell disorders as a result of immunosuppressive therapy attributable to an increased loss or increase in immunoglobulin catabolism.

Monoclonal or abnormal proteins are seen in the serum or urine in multiple myeloma, Waldenström's macroglobulinemia, heavy-chain disease, amyloidosis, and monoclonal gammopathy of unknown cause.

REFERENCES

1. Laurell, C.: Quantitative estimation of proteins by electrophoresis in agarose gel containing antibodies, Anal. Biochem. **15:**45-52, 1966.
2. Laurell, C.B.: Antigen-antibody crossed electrophosis, Anal. Biochem. **10:**358-361, 1965.
3. Versey, J.M.B., et al.: Semi-automated two-dimensional immunoelectrophoresis, J. Immunol. Methods **3:**63, 1973.
4. Goecke, D.J., and Howe, C.: Rapid detection of Australia antigen by counterimmunoelectrophoresis, J. Immunol. **104:**1031, 1970.
5. Grabar, P., and Williams, L.A., Jr.: Méthode permettant l'étude conjugée des propriétés électrophorétiques et immunochémiques d'un mélange de protéines. Application au sérum sanguin, Biochim. Biophys. Acta **10:**193, 1953.
6. Scheidegger, J.J.: Une microméthode de l'immunoélectrophorèse, Int. Arch. Allergy Appl. Immunol. **7:**1-3, 1955.
7. Alper, C.A., and Johnson, A.M.: Immunofixation electrophoresis: a technique for the study of protein polymorphism, Vox Sang. **17:**445-452, 1969.
8. Ritchie, R.F., and Smith, R.: Immunofixation. I. General principles and application to agarose gel electrophoresis, Clin. Chem. **22**(4):497-499, 1976.
9. Chang, C., and Inglis, N.: Convenient immunofixation electrophoresis on cellulose acetate membrane, Clin. Chim. Acta **65:**91-97 (1975).
10. Moehring, C.: A rapid factor VII–related antigen electroimmunoassay, Am. J. Clin. Pathol. **72**(5):821-828, 1979.
11. Anhalt, J.P. In Washington, S., editor: Laboratory procedures in clinical microbiology, New York, 1981, Springer-Verlag, pp. 249-277.
12. Ritchie, R.F., and Smith, R.: Immunofixation. II. Application to typing of alpha 1-antitrypsin at acid pH, Clin. Chem. **22:**497-499, 1976.
13. Ritchie, R.F., and Smith, R.: Immunofixation. III. Application to the study of monoclonal proteins, Clin. Chem. **22,**1982-1985, 1976.

14. Cawley, L.P., et al.: Immunofixation electrophoretic techniques applied to identification of proteins in serum and cerebrospinal fluid, Clin. Chem. **22**:1262-1268, 1976.
15. Sun, T., et al.: Study of gammopathies with immunofixation electrophoresis, Am. J. Clin. Pathol. **72**:5-11, 1979.
16. Merlini, G., et al.: Detection and identification of monoclonal components: immunoelectrophoresis on agarose gel and immunofixation on cellulose acetate compared, Clin. Chem. **27**:1862-1865, 1981.
17. Keshgegian, A.A.: Prevalence of small monoclonal proteins in the serum of hospitalized patients, Am. J. Clin. Pathol. **77**:436-441, 1982.
18. Ritzmann, S., and Daniels, J.C.: Serum protein abnormalities, Boston, 1975, Little, Brown & Co., pp. 22-24.
19. Lennette, E.H., et al.: Manual of clinical microbiology, ed. 3, Washington, D.C., 1980, American Society for Microbiology, pp. 525-528.
20. Grey, H.M., et al.: A subclass of IgA globulins (IgM_2), which lacks the disulfide bonds linking heavy and light chains, J. Exp. Med. **128**:1223-1236, 1968.
21. Levin, A.S., and Perin, G.M.: Clinical immunology no. CL-5, alpha heavy chain disease, Chicago, 1977, American Society of Clinical Pathologists.
22. Cawley, L.P., et al.: Electrophoresis and immunochemical reactions in gels, ed. 2, Chicago, 1978, American Society of Clinical Pathologists.

Immunoglobulins

GAYLE BIRKBECK

Clinical significance: p. 718
Molecular formula and weight: See Table 60-18.
Chemical class: protein

PRINCIPLES OF ANALYSIS AND CURRENT USAGE

Historically, measurement of immunoglobulins were included as part of the total globulins present in human serum (Table 60-19). Globulin isolation was achieved by salt or gold precipitation with quantitation by the biuret assay. The introduction of electrophoretic separation techniques allowed fractionation of the globulins into alpha, beta, and gamma components. The gamma fraction consists almost completely of immunoglobulins. Today electrophoretic separation offers an excellent mode for examining serum for total immunoglobulin and for the detection of monoclonal antibodies.

Qualitative and semiquantitative estimates of total immunoglobulin levels may be obtained from immunoelectrophoresis and protein electrophoresis respectively. Knowledge of immunoglobulin structure and availability of antibodies to each immunoglobulin class, that is, specific for the constant regions of the heavy chains, have made it possible to measure each immunoglobulin class. Different immunochemical procedures have been developed for the quantitation of these immunoglobulin classes. In part, the need for a number of different procedures is dictated by the large differences in concentration between each class of immunoglobulin. For example, there is a millionfold difference between the serum concentrations of IgG and IgE.

The most commonly used immunochemical methods for the quantitation of immunoglobulins (G, M, and A) are

Table 60-18. Physical properties of human immunoglobulins

Immunoglobulin class	IgG	IgM	IgA	IgD	IgE
Molecular weight (daltons)	150,000	900,000	160,000 (monomer) 320,000 (dimer)	185,000	200,000
Sedimentation coefficient, S	6.6	18.0-19.0	6.25-10.9	6.2-7.0	7.86-7.92
Heavy chains	γ	μ	α	δ	ϵ
Heavy-chain subclasses	$\gamma_1, \gamma_2, \gamma_3, \gamma_4$	μ_1, μ_2	α_1, α_2	—	—
Light chains	κ or λ	κ or λ	κ or λ	κ or λ	κ or λ
Molecular formula	$IgG(\kappa)2\gamma2\kappa$ $IgG(\lambda)2\gamma2\lambda$	$IgM(\kappa)(2\mu2\kappa)_5$ $IgM(\gamma)(2\mu2\lambda)_5$	$IgA(\kappa)(2\alpha2\kappa)_{1\text{-}3}$ $IgA(\lambda)(2\alpha2\lambda)_{1\text{-}3}$	$IgD(\kappa)2\delta2\kappa$ $IgD(\lambda)2\delta2\lambda$	$IgE(\kappa)2\epsilon2\kappa$ $IgE(\lambda)2\epsilon2\lambda$
Normal serum concentrations (mg/mL) (by age)					
Cord specimen	7.66-16.93	0.04-0.26	0.0004-0.09		
0.5-3 months	2.99-8.52	0.15-1.49	0.03-0.66		
3-6 months	1.42-9.88	0.18-1.18	0.04-0.90		
6-12 months	4.18-11.42	0.43-2.23	0.014-0.95		
1-2 years	3.56-12.04	0.37-2.39	0.13-1.18		116-122 ng/mL
2-3 years	4.92-12.69	0.49-2.04	0.23-1.37		80-122 ng/mL
3-6 years	5.64-13.81	0.51-2.14	0.35-2.09		
4-7 years					140-442 ng/mL
6-9 years	6.58-15.35	0.50-2.28	0.29-3.84		
10-14 years					374-674 ng/mL
12-16 years	6.80-15.48	0.45-2.56	0.81-2.52		
Adult	8.00-16.00	0.50-2.00	1.40-3.50	0-0.14	300 ng/mL

Data from Seligson, O., editor: Handbook series in clinical laboratory science. Section F. Immunology, vol. 1, part 1, Boca Raton, Fla., 1978, CRC Press, Inc., and from Meites, S., editor: Pediatric clinical chemistry, ed. 2, Chicago, 1981, American Association for Clinical Chemistry.

Table 60-19. Methods of immunoglobulin quantitation

Method	Type of analysis	Principle	Usage	Comments
Salt precipitation	Isolation of total immunoglobulin by salt precipitation	Immunoglobulins are less soluble in certain salt solutions and precipitate, amount of immunoglobulins quantified by total protein assay (such as biuret)	Serum	Historical, nonspecific
Gold precipitation	Isolation of total immunoglobulin by gold precipitation	Similar to salt precipitation; gold used as precipitant	Serum	Historical, nonspecific
Electrophoresis	Estimation of total immunoglobulin by physical separation	Proteins separate based on class (albumin, gamma globulin), calculation of each class as percentage of total protein	Serum	Good screening method for detection of monoclonal immunoglobulins; commonly used
Radial immunodiffusion (RID)	Quantitation by immunoprecipitation in gel	Immunoglobulin diffuses into gel containing antibody, forming a ring-shaped immunoprecipitate, diameter of ring proportional to concentration	Serum, body fluids	Accurate, slow, commonly used
Nephelometry	Quantitation by immunoprecipitation in solution	Reaction of immunoglobulin with its specific antibody results in immunoprecipitate, which has light-scattering properties; amount of light scatter proportional to immunoglobulin concentration	Serum, body fluids	Accurate, rapid, commonly used
Immunofluorometry	Quantitation by competition for labeled antibody on a solid phase	Immunoglobulin adsorbed onto solid surface competes for fluorescent-labeled antibody with immunoglobulin sample; fluorescence signal is inversely proportional to immunoglobulin concentration	Serum, some body fluids	Precision slightly less than nephelometer, RID.
Electroimmunodiffusion (Laurell rocket)	Quantitation by size of immunoprecipitate in gel	Immunoglobulin electrophoresis into antibody-containing agarose gel; height of gel pattern (rocket) proportional to immunoglobulin concentration	Serum	Not readily available commercially
Turbidimetry	Quantitation by light absorbance of immune precipitate	Immunoglobulin reacts with antibody or other precipitation agent; turbidity proportional to immunoglobulin concentration	Serum	Requires centrifugal analyzer and high level of expertise to establish procedure
Radioimmunoassay (RIA)	Quantitated by radioisotope	Immunoglobulin reaction with antibody displaces radiolabeled immunoglobulin	Any body fluid	Very sensitive, used for IgE
Radioimmunometric assay (IRMA)	Quantitated by radioisotope	Immunoglobulin reacts with antibody on solid surface and reaction quantitated by second radiolabeled antibody	Any body fluid	Very sensitive, used for IgE (PRIST)
Enzyme immunoassay (EIA)	Quantitated by enzyme	Similar to above radioisotope procedures; substitute enzyme label for radioisotopic ones	Any body fluid	Very sensitive, used for IgE equivalent to paper radioimmunosorbent test (PRIST)

single radial immunodiffusion, rate and end-point nephelometry, immunofluorometric assay, electroimmunodiffusion, and kinetic turbidimetric systems.

Normal serum levels for immunoglobulin D (30 μg/mL) and E (17 to 450 ng/mL) are below the sensitivities of the above immunochemical systems. Serum levels of IgD generally are not of clinical importance, unless a plasma cell malignancy is present with the production of a monoclonal IgD protein. Single radial immunodiffusion may be used to quantitate IgD in this instance, since the serum IgD level is well above normal.

Radioimmunoassay and enzyme immunoassay (ELISA) techniques are used to quantitate serum IgE levels.

The single radial immunodiffusion (RID) test consists of an agar plate with antibody incorporated throughout the agar. Test samples are placed in antigen wells and, on diffusion into the agar, form a precipitin ring around the well. The diameter of the precipitin ring is directly proportional to the amount of antigen placed in the well.

Two radial immunodiffusion methods are commonly used in clinical laboratories. The Mancini method[1] is based on an end-point (equivalence-zone) reading. The plates are incubated for about 48 hours (72 hours for IgM). The diameter of the precipitin ring is measured, and then the antigen level is determined from a standard curve. One prepares the standard curve by plotting on linear graph paper the concentrations of three or four reference sera against the ring diameter squared (mm^2). The Fahey-McKelvey method[2] measures the ring diameter before equivalence (such as 18 ± 2 hours). One constructs the standard curve by plotting ring diameter (mm) against the concentration of the reference standards on two-cycle semilogarithmic graph paper. The Fahey technique has the advantage of more rapid results; however, the accuracy and precision is somewhat less than that obtained with the end-point technique.[3,4] Commercial plates are readily available for radial immunodiffusion quantitation. Many plates can be used with either the Fahey or Mancini technique.

The quantitation of the immunoglobulins by nephelometry is a more recent development. Several commercial nephelometers, with appropriate reagent test kits, are available. The nephelometer measures the light scattering produced by the antibody-antigen complexes formed in the test sample tube. The relative light-scattering values are transformed into reportable units (that is, mg/L or international units) by a data processor that is coupled to the instrument. Both end-point and kinetic methods of measurement are used. The end-point method (such as those by Calbiochem-Behring, Hyland, J.T. Baker) measures the light scattering of the complexes formed after a steady state has been achieved. Thus antibody is added to the patient sample and the sample is incubated for 15 minutes to 1 hour before quantitation. The kinetic or rate method (as by Beckman and J.T. Baker) measures the maximum rate of increase of forward light scatter produced by the formation of antibody-antigen complexes. With this method, a lengthy incubation of the sample is not required before quantitation.

Since light scattering produced by contaminating particles in the antiserum can interfere with quantitation, antiserum, which is free of contamination, must be used. Commercial test kits produced for these systems are readily available. All manufacturers offer IgG, IgM, IgA, C3, and C4 kits as well as kits for a number of other serum proteins. Turbid patient samples may have to be filtered before use on the nephelometer.

The most commonly used immunofluorometric system for the quantitation of IgG, IgM, and IgA by clinical laboratories is the FIAX system (International Diagnostic Technology, Santa Clara, Calif.). With this procedure, the serum sample is incubated with excess fluorescein isothiocyanate (FITC)–labeled monospecific goat antibody to either IgG, IgA, or IgM. The immunoglobulin in the samples competes with a solid-phase bound human immunoglobulin for the FITC-labeled antibody. The solid-phase consists of human IgG, IgA, or IgM antigen immunoadsorbed onto the polymeric surface of a StiQ sampler (a plastic stick resembling a swizzle stick). The coated StiQ sampler is immersed in the reaction mixture of sample and antibody. The remaining unbound fluorescent antibody reacts with the antigen on the StiQ sampler. The fluorescence of the antibody bound on the StiQ is measured at 540 nm when excited by light at 475 nm after the StiQ is inserted into the FIAX fluorometer. The fluorescence is inversely related to the concentration of immunoglobulin in the sample. A microcomputer may be connected to the fluorometer for automatic calculations.

Electroimmunodiffusion or the Laurell rocket technique is performed by addition of a sample to a well in an agarose plate with antibody incorporated throughout the agarose. The plate is then placed in an electrophoresis chamber. As the immunoglobulin migrates through the agarose under the force of the electric field, precipitin peaks (rockets) form in the agarose. The height of the rocket can be plotted as a standard curve.[10] Commercial kits are not readily available for this technique.

Immunoglobulin quantitation can be performed by a turbidimetric method using a centrifugal fast analyzer. With this method, the patient's sample and appropriate reagent antiserum are mixed by the use of centrifugal force and the solution containing an antigen-antibody precipitate is forced into a cuvette at the end of the rotor arm. Light passes perpendicularly through the rotor arm to a photodetector. The increase in absorbance resulting from the presence of the precipitate is used to calculate the antigen concentration. The absorbance reading can be taken at equilibrium or on a kinetic basis (at selected time intervals during the reaction). A standard curve is generated and entered into a computer. The calculation of antigen in the patient sample is generated by use of a computer program.

Finley et al.[11] described a two-point kinetic technique that measured changes in absorbance at 340 nm between 10 and 225 seconds using a 36-place centrifugal analyzer.

Total IgE levels may be quantitated by radioimmunoassay or by enzyme immunoassay (ELISA) technique. Commercial kits are available for both methods.

Standardization and quantitation

The main problem in the accurate quantitation of immunoglobulins is the extreme heterogeneity of the group.[12] The immunoglobulins are divided into five major classes with four immunologically characterized subclasses of IgG, two of IgA, and several probable subclasses of IgM. The ratio of these subclasses may vary with some infectious and autoimmune diseases and certainly with malignant plasma cell dyscrasias. In addition to the immunologically characterized subclasses, the immunoglobulins also vary in size. For instance, IgM is usually present as the stable 19S pentamer of 7S subunits, but occasionally may be found in free 7S form. IgA is 7S in the monomeric form but 11S in the dimeric form. In addition, specificity and binding affinity of the reagent antisera will affect immunochemical systems.

Accurate quantitation of monoclonal proteins is difficult because the reagent antiserum is raised against normal human serum containing a mixture of subclasses. Therefore, in the case of a monoclonal IgG_3 protein, only a portion of the antibodies in the antiserum will react with the IgG_3. In addition, the standard curve is constructed by use of a mixture of all subclasses. Also, variations in size and idiotypical antigens may result in a different reaction between a monoclonal protein and a reagent antiserum.

Daniels et al.[13] found that the serum protein electrophoresis gives a more accurate quantitation of a monoclonal protein (once identified) than immunochemical methods. Although immunochemical methods may either overestimate or underestimate the amount of monoclonal protein; consistency can be obtained if the same immunochemical method is used in following each patient.

True accuracy cannot really be achieved in the quantitation of immunoglobulins. The levels may be reported in international units (arbitrary) or mg/L. The weight/volume method of reporting was used by approximately 96% of the participants in the CAP special diagnostic immunology survey in 1981. Despite its popularity, the true mass of immunoglobin present in a standard may have been incorrectly determined because the preparation was not completely pure or because all immunologically distinct subclasses may not have been represented. The World Health Organization International Reference Preparation of Human Immunoglobulins IgG, IgA, and IgM has been used as a primary standard. Each vial of this freeze-dried pool contains 100 (arbitrary) international units.[12] Rowe et al.[14] determined that 1.00 international unit is equivalent to 80.4 μg of IgG, 14.2 μg of IgA, or 8.47 μg of IgM. The commercial manufacturers have supplied secondary standards calibrated against the World Health Organization preparation. A new *Reference Preparation for Serum Proteins* (RPSP) became available in 1981 through the College of American Pathologists.[15]

REFERENCE AND PREFERRED METHODS

The FIAX immunofluorescent, electroimmunodiffusion, and kinetic-turbidimetric systems are each used by fewer than 20 laboratories participating in CAP surveys. The FIAX system correlates well with radial immunodiffusion and nephelometry methods; although the coefficient of variation seems to be somewhat higher than the other methods.[5,8,9] Blanchard et al.[16] found the technician time for a run of 20 samples for IgG, IgA, and IgM on the FIAX system to be 4 to 5 hours, if the time included in the incubation sequences was included as working time. Kits for electroimmunodiffusion are not readily available through commercial sources.

The kinetic-turbidimetric methods performed on centrifugal analyzers have the advantage of using instrumentation already present in many laboratories. However, turbidimetric measurements are more difficult to perform because a large amount of precipitate must be formed before one observes a significant change in absorbance. Measuring small differences between large absorbances can result in poor precision. In contrast, nephelometric measurements require less precipitate formation to achieve the same differential in signal. With more advanced centrifugal analyzers becoming available (Roche-COBAS; IL-Multistat) this type of analysis may become more widespread.

The two most widely used methods by CAP Diagnostic Immunology Survey participants for 1981 are single radial immunodiffusion and nephelometry. Commercial kits are used by most participants. The coefficients of variation obtained with both methods are comparable.[4-8] Blanchard and Gardner[5] reported the following between-run precision for the radial immunodiffusion method: IgG, 4.4% to 7.1%: coefficients of variation; IgA, 2.3% to 3.7%; IgM, 5.0% to 6.9%. Coefficient of variation ranges between 1% and 11% have been reported for radial immunodiffusion. The best precision can be obtained by use of automated dilutors and by use of only one or two technologists to set up and read the test. The technique variation between technologists is not so significant with the nephelometer as with the radial immunodiffusion method. Alexander[6] reported day-to-day coefficients of variation for the Behring laser nephelometer as follows: IgG, 4.9%; IgA, 5.7% and IgM, 6.7%. The ranges in the literature for nephelometers vary from 0.6% to 10%, with an overall average of around 6%. The correlation coefficients between the various nephelometer and radial immunodiffusion methods are greater than 0.9.[5-7]

The technologist's working time required for radial immunodiffusion plates has been reported as 2 to 3 minutes per sample. Blanchard and Gardner[5] reported 2 hours of technologist time for analysis of a batch of 18 samples for IgG, IgA, and IgM (54 tests in 120 minutes). The technologist's working time required for the quantitation of immunoglobulins with a nephelometer depends on the degree of automation. The working time required for quantitation on a very basic model is approximately the same as with the radial immunodiffusion method. However, all the commercial systems can be expanded from manual methods to ones with a high degree of automation.

The Baker Diagnostics immunology series 420 system is highly automated with features such as automatic antigen excess correction, background rate correction, automatic reference curve generation, and automatic reagent quality control check. The 420 system can operate in both kinetic and end-point modes. Dito et al.[7] reported that the 420 laser nephelometer can be readied to initiate a run in 15 minutes and that the instrument can be left unattended to run 22 to 25 assays per hour.

The main determining factor in choosing between radial immunodiffusion and the nephelometer is the quantity of test specimens. The nephelometer is the method of choice if large batches of specimens are to be run. The test can be almost completely automated, thus saving the technologist's time. In addition, a result is available within several hours from a nephelometer versus 18 to 50 hours with radial immunodiffusion. However, the cost of the instrumentation must be taken into consideration. Radial immunodiffusion may be the method of choice for a laboratory with a small volume. The technologist's time is relatively small if an automated diluter is used in preparing samples and if a computer program is used to determine patient results rather than the manual interpretation of a standard curve. Most of the 18- to 50-hour turnaround time is incubation time. The investment in equipment is minimal (automatic diluter, commercial plate reader). The longer turnaround time may not be of much clinical significance.

The low-level sensitivities of the radial immunodiffusion and nephelometer systems for immunoglobulin quantitation are comparable. The lower limits of detection for the Beckman nephelometer systems for IgG, IgM, and IgA are 17, 14, and 17 mg/L, respectively.

Most commercial manufacturers offer standard-level and low-level radial immunodiffusion plates. The standard-level plates are adequate for most adult serum samples and can be used to quantitate a range from moderately decreased to mildly elevated levels. The plates do not give linear results below the level of the lowest reference standard. Kallestad (Chaska, Minn.) standard plate ranges are approximately 4 to 25 g/L for IgG, 600 to 4000 mg/L for IgA, and 300 to 3000 mg/L for IgM. Sera with concentrations above these limits should be diluted. Concentrations below these limits should be reported as less than the lower limit, or the actual concentration should be determined by use of the low-level plates.

The low-level plates are used for fluids, for children's sera, and for adult sera with low concentrations if the actual concentration is desired. The low-level plates have sensitivity levels equivalent to the nephelometer. The sensitivity of the Kallestad low-level plates for IgG, IgM, and IgA are 18.8, 25.2, and 18.8 mg/L, respectively.

IgE may be quantitated by either radioimmunoassay technique or by solid-phase enzyme immunoassay. The radioimmunoassay method has been used most frequently in the past because of ready availability of commercial kits (Kallestad; Pharmacia, Piscataway, N.J.). However, commercial kits for the enzyme immunoassay technique are now readily available (Calbiochem-Behring, La Jolla, Calif.). The Enzygnost (Calbiochem-Behring) method includes two 1-hour incubation periods at 37° C with a test turnaround time of approximately 3 to 3½ hours. The lowest sensitivity is 0.2 U/mL and the day-to-day coefficients of variation for 16, 113, and 309 U/mL are 5.6%, 4.5%, and 12.5% respectively. The technique will quantitate levels up to 1000 U/mL, which is the highest standard used. The correlation (*r*) with radioimmunoassay is 0.98 (personal communication from Calbiochem-Behring). The Kallestad radioimmunoassay kit has a sensitivity of 2 U/mL and day-to-day coefficients of variation for 9.58, 30.04, and 99.5 U/mL of 5.6%, 6.2%, and 5.6% respectively (personal communication from Kallestad).

In summary, the enzyme immunoassay technique offers the same precision and correlates well with radioimmunoassay. It may offer a wider assay range and a faster turnaround time, depending on the commercial kit used. The enzyme immunoassay does not use radioactive material and requires only a spectrophotometer rather than a gamma-ray counter as with radioimmunoassay. Therefore the immunoassay technique may become the method of choice.

SPECIMEN

Immunoglobulin quantitation, for routine diagnostic purposes, is performed most commonly on serum specimens. IgG is frequently measured in cerebrospinal fluid and combined with the albumin quantitation for a ratio. Normal levels for immunoglobulins in thoracentesis fluid, synovial fluid, and other serous exudates are not well defined, and the clinical usefulness of these fluid quantitations is controversial. Urine specimens generally are concentrated 20 to 50 times before being evaluated by electrophoresis or by immunoelectrophoresis coupled with the total urine protein

Samples for quantitation may be stored for up to 5 days at 2° to 8° C if they are protected from contamination and evaporation. For sample shipment or longer storage periods, samples should be frozen at −20° C or colder. Once thawed, samples generally should not be refrozen, as repeated freezing and thawing may cause deterioration of the proteins.

REFERENCES

1. Mancini, G., Carbonara, A.O., and Heremans, J.F.: Immunochemical quantitation of antigens by single radial immunodiffusion, Immunochemistry **2**:235-254, 1965.
2. Fahey, J.L., and McKelvey, E.M.: Quantitative determination of serum immunoglobulins in antibody-agar plates, J. Immunol. **94**(1):84-90, 1965.
3. Ritzman, S.E., and Daniels, J.C.: Serum protein abnormalities, diagnostic and clinical aspects, Boston, 1975, Little, Brown & Co., pp. 62-64.
4. Berne, B.H.: Differing methodology and equations used in quantitating immunoglobulins by radial immunodiffusion in a comparative evaluation of reported and commercial techniques, Clin. Chem. **20**(1):61-68, 1974.
5. Blanchard, G.C., and Gardner, R.: Two nephelometric methods compared with a radial immunodiffusion method for the measurement of IgG, IgA, and IgM, Clin. Biochem. **13**(2):84-91, 1980.
6. Alexander, R.L.: Comparison of radial immunodiffusion and laser nephelometry for quantitating some serum proteins, Clin. Chem. **26**(2):314-317, 1980.
7. Dito, W.R., et al.: Comparative evaluation of an automated kinetic laser nephelometer with other immunoprecipitin technics for the assay of serum immunoglobulins, Am. J. Clin. Pathol. **76**(6):753-759, 1981.
8. Burdash, N., et al.: Methods for immunoglobulin quantitation compared, Clin. Chem. **27**(2):345-347, 1981.
9. Zaatari, G., et al.: Comparison of nephelometry and immunofluorescence for immunoglobulin quantitation in pathologic sera, Clin. Chim. Acta **103**:357-366, 1980.
10. Laurell, C.: Quantitative estimation of proteins by electrophoresis in

agarose gel containing antibodies, Anal. Biochem. **15:**45-52, 1966.
11. Finley, P.R., et al.: Immunochemical determination of human immunoglobulins: use of kinetic turbidimetry and a 36-place centrifugal analyzer, Clin. Chem. **25**(4):526-530, 1979.
12. Reimer, C.B., and Maddison, S.E.: Standardization of human immunoglobulin quantitation: a review of current status and problems, Clin. Chem. **22**(5):577-582, 1976.
13. Daniels, J.C., et al.: Methodologic differences in values for M-proteins in serum, as measured by three techniques, Clin. Chem. **21**(2):243-248, 1975.
14. Rowe, D.S., Grab, B., and Anderson, S.G.: An international reference preparation for human serum immunoglobulins IgG, IgA, and IgM, Bull. WHO **42:**535, 1970.
15. Reimer, C.B., et al.: Collaborative calibration of the U.S. National and the College of American Pathologists reference preparations for specific serum proteins, Am. J. Clin. Pathol. **77**(1):12-19, 1982.
16. Blanchard G.C., and Gardner, R.: Two immunofluorescent methods compared with a radial immunodiffusion method for measurement of serum immunoglobulins, Clin. Chem. **24**(5):808-814, 1978.

Serum protein electrophoresis

SUMAN PATEL
JOHN A. LOTT

Clinical significance: p. 420

PRINCIPLES OF ANALYSIS AND CURRENT USAGE

Human serum contains more than 125 identified proteins, which perform a number of functions. The protein components of plasma or serum constitute almost all the mass of the serum solutes (approximately 80 g/L). Present are carrier proteins, antibodies, enzyme inhibitors, clotting factors, and so on. Plasma contains fibrinogen (approximately 3 g/L); a protein that is absent from serum. The total serum protein concentrations or the proportions of the individual protein fractions change during a variety of diseases. Quantitation of total serum protein and individual fractions is of considerable value in clinical diagnosis.

One of the simple techniques for quantitation of serum proteins is their separation in an electric field, and this procedure is referred to as "serum protein electrophoresis" (SPE). When an electric field is applied to a medium containing charged particles, the negatively charged particles or molecules migrate toward the positive electrode (anode), while the positively charged particles migrate toward the negative electrode (cathode). This principle is applied to separate the protein fractions in serum. Separation can be followed by permanent fixation of the fractions at the position in the medium to which they migrated.

The separation of proteins in an electric field using a wholly liquid medium was demonstrated by Tiselius[1] and is known as "moving boundary electrophoresis" or "free electrophoresis." Zone electrophoresis, developed later using an inert supporting medium, proved easier to perform and provided better resolution of the bands. (See Chapter 8 for a discussion of the theory of electrophoresis.)

Support

Supporting media fall into two main classes: class I includes paper, cellulose acetate, thin layer materials and agarose gel; class II includes starch gel and acrylamide gel.

Class I supports allow separation of molecules solely on the basis of net molecular charge. Supporting media in class II exert a sieving effect on the compounds being separated. The gel can be considered as a porous medium in which the pore size is of the same order as the protein molecules. The result is that a molecular sieving effect is observed in addition to the electrostatic separation observed with class I supports. Class II supports can separate those large and small molecules that have the same size-to-charge ratio and that cannot be separated on class I supports. Starch and acrylamide gels will act as traps for large molecules, thus impeding their movement through the support. Smaller molecules will migrate without hindrance, and hence the resolution power of these gels is much greater. Five serum protein bands are observed on class I supports, whereas about 25 bands or more are seen on class II supports.

Paper as a support for serum protein electrophoresis is largely of historical interest. Cellulose acetate and agarose gel electrophoresis are now in common use in clinical laboratories, and both material and equipment are commercially available. Other electrophoretic methods called "high-resolution electrophoresis" are also available. High-resolution electrophoresis resolves serum proteins into 13 bands, and the technique is superior to agarose gel electrophoresis for certain diagnostic procedures, such as identification of oligoclonal bands in cerebrospinal fluid for the diagnosis of multiple sclerosis.[2,3]

Buffers and pH for serum protein electrophoresis

Proteins are amphoteric molecules; that is, they are either uncharged or negatively or positively charged particles, depending on the pH of the buffer.

$$\underset{\text{Basic pH, negatively charged}}{H{-}C(NH_2)(R){-}COO^-} \longleftarrow \underset{\text{Isoelectric point, neutral}}{H{-}C(NH_3^+)(R){-}COO^-} \longrightarrow \underset{\text{Acidic pH, positively charged}}{H{-}C(NH_3^+)(R){-}COOH}$$

With knowledge of the value of pI (isoelectric point pH) for the substances to be separated, a buffer for the desired separation can be chosen, such as barbital. Barbital does not precipitate or denature proteins, and it imparts a negative charge to most serum proteins at pH 8.6. In the presence of an electrical field, the negatively charged protein molecules migrate toward the anode.

The buffer helps to maintain a constant pH, and it ensures that each protein will maintain a constant charge throughout the course of the separation. Migration characteristics of proteins depend on the charge; a stable pH is vital if reproducible patterns are to be obtained. The choice of buffer depends on the chemical properties of the components to be separated. The ionic strength of the buffer usually ranges from 0.05 to 0.15 mol/L. Migration becomes slower as the ionic strength increases because of competition for the current by the buffer ions. At higher ionic strengths, the migrated bands are sharper but closer together.

The buffer used in the electrode reservoir is often different from that used to prepare the gel. The net effect is to produce a voltage discontinuity at the interface of the two buffers. As this interface travels through the protein bands, the bands compact to produce narrower zones.

Although barbital buffer with pH 8.6 is commonly used for serum protein electrophoresis, more recently other buffers have been used to improve the separation. The serum protein electrophoresis system of LKB (LKB instruments; Rockville, Md.) uses a Tris-barbital buffer with calcium lactate, pH 8.6 (LKB application Note 310). A buffer solution composed of boric acid, Tris, and EDTA allows the separation of prealbumin, three alpha globulins, three beta globulins, and gamma globulin.[4] Borate buffer ions bind to glycocompounds, which may have produced the improved separation reported by Aronsson and Gronwall.[4] High-resolution capabilities of the relatively new electrophoresis systems like the Panagel (Worthington Diagnostics, Freehold, N.J.) and Gelman (Gelman Sciences, Inc., Ann Arbor, Mich.) systems are believed to result from minor changes in the components of the buffer system and efficient heat dissipation during electrophoresis.

We have replaced the barbital buffer used in the Corning (Medfield, Mass.) serum protein electrophoresis system with Tricine (*N*-[tris(hydroxymethyl)methyl]glycine) (0.05 mol/L, pH 8.6) without affecting the separation of serum proteins. This has eliminated the problem of the carbon-electrode erosion.

Voltage, current, temperature, and time

In practice, it is desirable to have rapid separation of proteins to preserve the sharpness of the bands. The longer the electrophoresis is allowed to proceed, the greater the radial diffusion and the broader the bands. One can decrease the time for a given separation by either decreasing the length of the support or increasing the voltage.

With an increase in voltage, there will be a corresponding current increase, and heating may become a limiting factor because of protein denaturation. One way of reducing heating is by lowering of the ionic strength of the buffer.

Use of constant voltage is a standard practice in serum protein electrophoresis. Because the electrophoretic separation is achieved within about 30 minutes at about 90 to 100 volts, the change in current and temperature that occurs is not a serious problem. However, controlling current and temperature may improve the resolution.

Development of electrophoretogram

Permanent fixing of the zones depends on the type of support medium used. Zones are fixed in the support medium when the medium is dried in an oven to evaporate the solvent. Fixation causes precipitation of the protein zones.

Staining the proteins after electrophoretic separation and subsequent scanning is the most popular method of quantitation of individual protein fractions. Some serum specimens will show diffuse bands between the major electrophoretic fractions. Scanning the pattern gives quantitative information about well-resolved protein fractions only. Observation with an experienced eye is helpful in the detection of minor fractions and unsatisfactory separations. Electrophoresis of a normal control serum on each run is essential.

Different dye-binding characteristics of protein fractions is a factor contributing to the relatively poor precision of serum protein electrophoresis. Analysis of a control serum with the specimens is essential to monitor the variability of this technique.

It is recommended that staining be applied to the dry film rather than to the wet gel. Amido black 10B is widely used for staining agarose gels, because it leaves the background clear and it has adequate sensitivity. Ponceau S stain is preferable for cellulose acetate. Certain methods use bromphenol blue for staining protein bands on agarose, but this stain has a much higher affinity for serum albumin than for the globulins. A combination of amido black 10B and Coomassie brilliant blue R250 has been used to stain proteins that were separated on cellulose acetate.[5] Coomassie brilliant blue R250 gives an intense stain of proteins, making it suitable as a stain for proteins after agarose gel electrophoresis of urine and cerebrospinal fluid. Silver stain has been used for dilute fluids because of its high sensitivity.

REFERENCE AND PREFERRED METHODS

Serum protein electrophoresis can be performed by use of a variety of supporting media (Table 60-20):

1. Polyacrylamide gel
2. Starch gel
3. Paper
4. Cellulose acetate
5. Agarose gel

Electrophoretic techniques using polyacrylamide support medium have served to establish the existence of over 100 serum proteins. The many bands seen with these techniques present a complex picture that is hard to interpret and appears to help little in routine clinical laboratory diagnosis. Except for specialized purposes, polyacrylamide cannot be recommended for routine serum protein electrophoresis.

Table 60-20. Methods of analysis for serum protein electrophoresis

Method	Usages	Comment
Paper[4]	Historical	Has slight denaturing properties causing tailing; difficult to do densitometric analysis
Cellulose acetate[5]	Most widely used	Rapid analysis, separates five major classes of serum proteins, densitometric analysis readily performed; endosmosis effects present; high-resolution electrophoresis resolves serum proteins into 10 to 13 bands
Agarose[7]	Widely used	Rapid analysis, separates five major classes of serum proteins plus prealbumin; densitometry readily performed; endosmosis effects present; high-resolution electrophoresis resolves serum proteins into 10 to 13 bands.
Starch	Rarely used	Sieving properties result in separation based on molecular size and charges: additional protein bands (25) can be separated; opaqueness makes densitometry difficult
Polyacrylamide	Less widely used	Protein separation based on molecular size and charge; up to 100 protein bands can be resolved though of unproved clinical utility; densitometry difficult

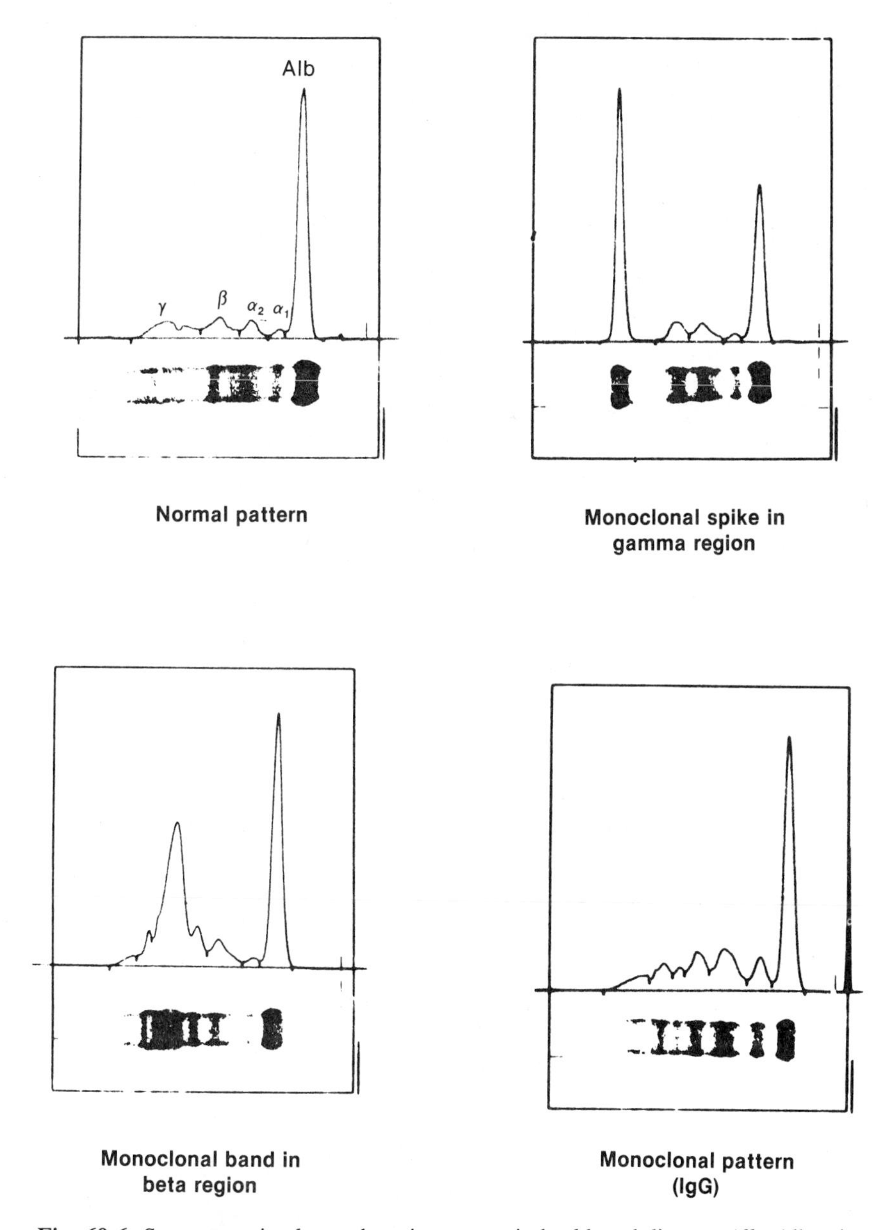

Fig. 60-6. Serum protein electrophoretic patterns in health and disease. *Alb,* Albumin.

Serum protein electrophoresis interpretation

Major proteins found in each fraction

1. Albumin (65,00 daltons)
   Function. Binding and transport of numerous substances (such as amino acids, fatty acids, enzymes, drugs, thyroid hormones, and toxic products); also responsible for controlling the fluid balance between intravascular and extravascular compartments of the body
   Increased. Dehydration
   Decreased. Liver disease, chronic infection, neoplasms, kidney disease, hemorrhage, liver failure, starvation, undernutrition
2. Alpha$_1$ region
   a. Alpha$_1$-antitrypsin (45,000 daltons)
      Function. Neutralization of the proteolytic enzymes trypsin (derived from leukocytes in the lung, pancreas, and other organs) and plasmin
      Increased. Inflammatory reactions
      Decreased. Lung disease, particularly emphysema, patients with alpha$_1$-antitrypsin deficiency tend to develop emphysema
   b. Alpha$_1$-lipoproteins (200,000 daltons)
      Function. Transport of cholesterol and fat-soluble vitamins
      Increased. Hyperlipidemia
      Decreased. Liver disease, particularly Tangier's disease
   c. Alpha$_1$-glycoprotein (44,100 daltons)
      Function. Protein-polysaccharide compounds occurring in tissues and mucus secretions. They have a wide variety of functions, and plasma concentration abnormalities are present in certain inborn errors of metabolism
   d. Prothrombin (72,000 daltons)
      Function. Factor II. Required for blood clotting; converted to thrombin by factor V.
      Decreased. Liver disease
   e. Thyroid-binding globulin (36,500 daltons)
      Function. Carrier of thyroid hormones in blood
      Increased. Pregnancy, use of contraceptive pills
      Decreased. Nephrosis, methyltestosterone treatment
3. Alpha$_2$ region
   a. Alpha$_2$-macroglobulin (820,000 daltons)
      Function. Protease inhibitor, neutralize proteolytic enzymes such as trypsin, plasmin, kallikrein
      Increased. Nephrotic syndrome, emphysema, diabetes mellitus, Down's syndrome, pregnancy
      Decreased. Rheumatoid arthritis, myeloma
   b. Haptoglobin (85,000 to 1 million daltons)
      Function. Hemoglobin (Hb)–binding protein. Hb-haptoglobin complexes preserve the body's stores of iron for reutilization.
      Increased. Acute and chronic inflammation, neoplastic conditions, myocardial infarction, Hodgkin's disease
      Decreased. Liver disease (haptoglobulin is produced in the liver), hemolytic anemia, megaloblastic anemia, liver failure
   c. Ceruloplasmin (132,000 daltons)
      Function. Copper-binding serum protein. Copper bound to ceruloplasmin is nontoxic, but unbound copper is toxic to tissues
      Decreased. Wilson's disease, a congenital disease where ceruloplasmin is not produced
      Increased. Pregnancy, use of contraceptive pills
   d. Alpha$_2$-lipoproteins
      Function. Transport of lipids
      Increased. Hyperlipidemias.
      Decreased. Severe liver disease
   e. Erythropoietin (30,000 daltons)
      Function. A hormone essential for normal erythropoiesis
      Increased. Certain anemias
      Decreased. Kidney disease, certain autoimmune diseases

Serum protein electrophoresis interpretation—cont'd

4. Beta region
   a. Transferrin (80,000 to 90,000 daltons)
      Function. Transferrin functions as a transport protein, in that it shuttles iron between tissues (such as liver) and the bone marrow (site of iron utilization for the production of hemoglobin)
      Increased. Iron-deficiency anemias
      Decreased. Liver diseases, nephrosis, malignant neoplasms
   b. Beta-lipoproteins (3 million daltons)
      Function. Transport of cholesterol, phospholipids, and hormones
      Increased. Nephrosis, hyperlipidemias
      Decreased. Starvation
   c. C3 and C4 (185,000 to 417,000 daltons) (third and fourth component of complement)
      Function. Two of the components of complement (a complex system of nine serum proteins involved in inflammatory reactions)
      Decreased. In active stages of immune disease such as systemic lupus erythematosus, autoimmune hemolytic anemia
   d. C1 esterase inactivator (104,000 daltons)
      Function. Inhibits the activity of C1
      Decreased. Absent or greatly decreased in hereditary angioneurotic edema
   e. Hemopexin (80,000 daltons)
      Function. Specific heme-carrying serum protein
      Increased. Same as haptoglobin
      Decreased. Same conditions that lead to decreased haptoglobin
5. Gamma region
   The gamma region contains most of the immunoglobulins. At present there are five known immunoglobulin classes (Ig): IgG (150,000 daltons), IgA (180,000 daltons), IgM (900,000 daltons), IgD (170,000 daltons), and IgE (190,000 daltons)
   Increased. Hypergammaglobulinemia, liver disease, chronic infections, systemic lupus, monoclonal or polyclonal multiple myeloma, Waldenström's disease, lymphoma
   Decreased. Old age, induced by some drugs, chronic lymphocytic leukemia, light-chain disease, agammaglobulinemia, hypoimmune syndrome

Serum protein electrophoresis with cellulose acetate or agarose yields 5 to 6 bands as shown in Fig. 60-6. Each of these electrophoretic fractions represents a composite of many proteins (see facing box). Serum protein electrophoresis with cellulose acetate or agarose is a simple technique that is helpful in establishing a diagnosis of monoclonal gammopathies, liver cirrhosis, renal failure, hypogammaglobulinemia, multiple myeloma, and so on (Fig. 60-7).

Paper is useful for the separation of small molecules such as amino acids and medium-sized molecules such as proteins. Paper has a slight denaturing effect that results in minor tailing of the bands. Since it is nearly opaque, densitometry is less accurate than with cellulose acetate or agarose. Paper as a support for serum protein electrophoresis is largely of historical interest.

Cellulose acetate has several advantages over paper: adsorption is minimal, resulting in sharp separations and well-defined bands. The separation is rapid (½ to 1 hour), and very small quantities can be separated. It is possible to clear the cellulose acetate to a transparent form after staining, which facilitates densitometry.

Agarose gel gives sharper resolution of the bands than cellulose acetate does, and adsorption is minimal. However, because of the slight negative charge of the agarose, electro-osmosis, or the movement of the buffer salts toward the cathode, occurs. Clearing of the support is possible.

Cellulose acetate and agarose give similar results (Table 60-21); however, agarose gel electrophoresis has found increased use because of more consistent performance and ease of handling, and hence it is considered to be superior to cellulose acetate for electrophoresis.[6]

Serum protein electrophoresis with agarose as the supporting medium is recommended as the preferred method. This method was developed by Wieme,[7] and it has been greatly improved since the original report. Agarose is a neutral mixture of polysaccharides obtained from a certain seaweed. Serum protein electrophoresis does not require a highly purifed agarose, and satisfactory and reproducible separation of proteins from most biological fluids can be achieved with the commercially available agarose despite contamination by some sulfate (charged) groups. Most

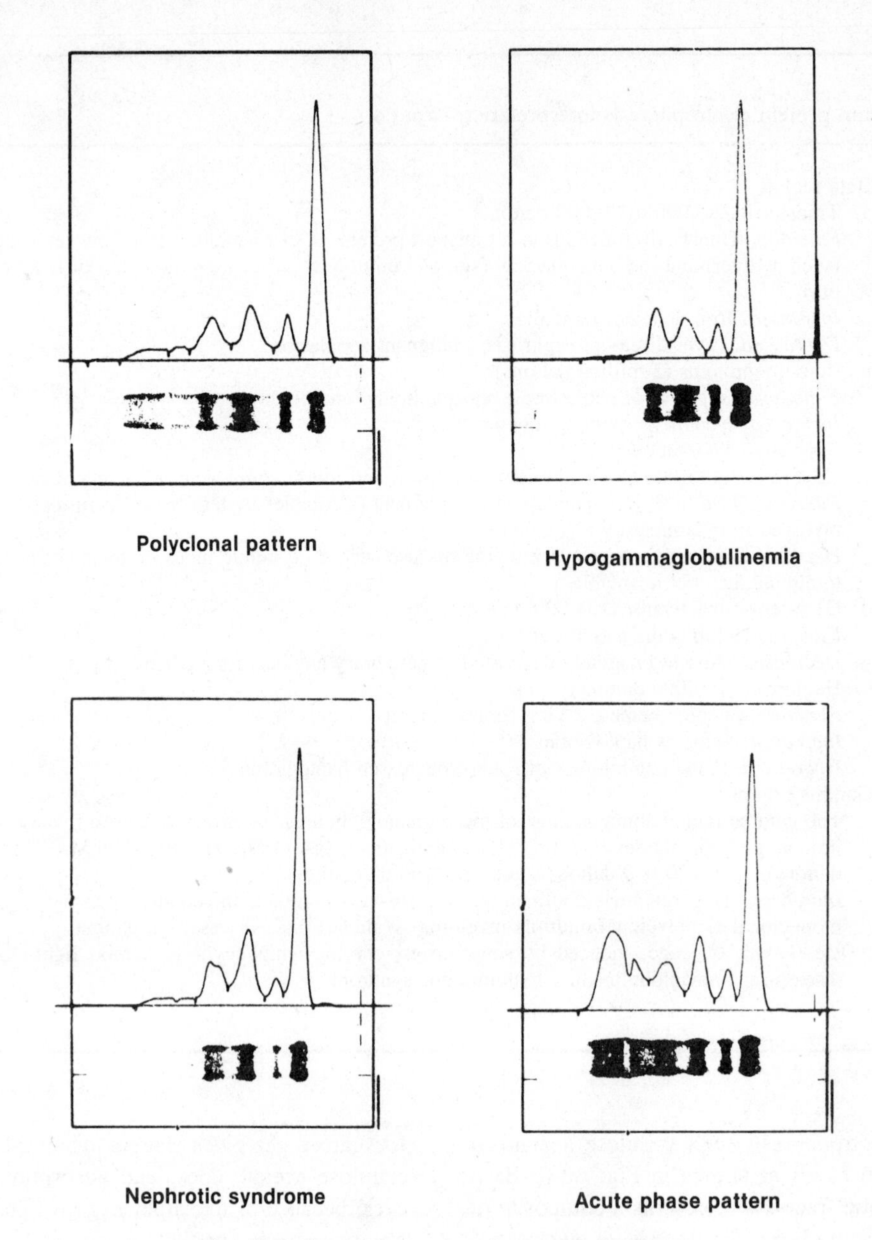

Fig. 60-7. Serum electrophoretic patterns in disease.

Table 60-21. Comparison of results (percent composition) of serum protein electrophoresis

Method	Albumin %	Alpha$_1$ %	Alpha$_2$ %	Beta %	Gamma %
Cellulose acetate electrophoresis*	69.64 ±1.43 SD 2.2 CV	3.22 ±0.37 SD 11.4 CV	7.86 ±0.59 SD 7.7 CV	8.93 ±0.86 SD 9.7 CV	10.34 ±0.82 SD 7.9 CV
Agarose gel electrophoresis†	69.4 ±0.76 SD 1.1 CV	3.78 ±0.35 SD 9.1 CV	9.56 ±0.61 SD 6.4 CV	6.87 ±0.74 SD 10.7 CV	10.23 ±0.66 SD 6.4 CV

*Fisher Pool Lot No. 133-010, n = 107. Ponceau S stain, Beckman Microzone System, Beckman Densitometer at 520 nm.
†Same pool, n = 79. Amido black 10B stain, Corning Electrophoresis System, Beckman Densitometer at 520 nm.
CV, Coefficient of variation; *SD*, standard deviation.

commercially prepared plates are 0.5 mm thick and contain 10 g/L agarose in barbital buffer, pH 8.6.

Many serum protein electrophoresis systems are commercially available. The procedure described here is based on the Corning method.

SPECIMENS

Serum is the specimen of choice. Plasma can be used; however, an extra band produced by fibrinogen will be observed. Urine and cerebrospinal fluid can also be analyzed if their protein content is increased by ultracentrifugation, dialysis, or other concentration techniques.

PROCEDURE

Reagents

Sodium hydroxide, 180 g/L (4.5 mol/L). Dissolve 180 g of NaOH in 700 mL of distilled water. Make up to 1 L of water. Store in plastic bottle. It is stable for 1 year at room temperature.

Tricine buffer, 0.05 mol/L, pH 8.6. Place 8.9 g of Tricine in a liter flask and add 900 mL of distilled water to dissolve. Titrate the pH to 8.6 on a pH meter with 4.5 mol/L NaOH. Make up to 1 L with distilled water. Stable 2 months in a glass bottle at 4° C.

Amido black 10B stain, 2 g/L (3.2 mmol/L) in 5% (v/v) (0.88 mol/L) acetic acid. Transfer 2 g of amido black 10B to a 1 L volumetric flask. Add 5% acetic acid to dissolve the dye, and make up to the 1 L mark. Store the stain solution in an airtight glass container. Stable for 3 months at room temperature.

Five percent acetic acid clearing solution. Add 50 mL of glacial acetic acid to 800 mL of distilled water in a volumetric flask. Allow the solution to cool and then bring to 1 L with distilled water. Stable at least for 6 months at room temperature.

Assay

Equipment

1. Cassette electrophoresis cell and power supply: Several versions of this type of electrophoresis cell are available. They are convenient to use and do not require paper wicks to connect the electrophoresis support medium with the buffer
2. Incubation oven set at 65° C
3. Agarose gel film. As with the electrophoresis cell, several variations of agarose films are available. The agarose film manufactured by Corning Medical comes with sample wells (Agarose Universal Electrophoresis Film). Others use a template for sample application, such as Paragon (Beckman Instruments, Inc., Carlsbad, Calif.), and Panagel
4. Quantitative microliter sample dispenser, such as, Hamilton
5. Disposable sample tips
6. Stir-stain dishes and humidity chambers
7. Disposable liners for stir-stain dishes
8. Densitometer with 520 nm capability

Follow manufacturer's instructions for the electrophoresis procedure.

1. Stain the agarose film in the amido black stain solution for 15 minutes.
2. Transfer the agarose film to the first 5% acetic acid clearing solution and leave there for 30 seconds. Use forceps and disposable gloves to handle the gels.
3. Wipe the moisture from the back of the film and place it in the oven. Dry at 65° C for 15 to 20 minutes or until dry.
4. Remove the agarose film from the oven and allow it to cool.
5. Rinse agarose film in the first 5% acetic acid for 1 minute.
6. Transfer the film to the second 5% acetic acid and rinse until clear.
7. Wipe moisture from the back of the film and dry it in the oven for 10 minutes at 65° C.
8. Scan the plate at 520 nm and obtain the percent value of each fraction.

Calculations

To quantitate each well-defined fraction in g/L, the formula is as follows:

$$g/L = \frac{\text{Fraction (in \%)} \times \text{Total protein (g/L)}}{100}$$

REFERENCE RANGE

Each laboratory performing serum protein electrophoresis should establish reference ranges on the populations it serves. In general, reference values for the five well-defined fractions are as follows:

Protein	g/L
Total protein	60 to 80
Albumin	32 to 50
$Alpha_1$	10 to 40
$Alpha_2$	6 to 10
Beta	6 to 13
Gamma	7 to 15

REFERENCES

1. Tiselius, A.: A new apparatus for electrophoresis: analysis of colloidal mixtures, Trans. Faraday Soc. **33**:524-529, 1937.
2. Laurell, C.B., Jeppsson, J.O., and Tejler, L.: Plasma protein analysis, Bedford, Mass., 1978, Millipore Corp.
3. Sun, T., Lien, Y.Y., and Gross, S.: Clinical applications of a high-resolution electrophoresis system, Ann. Clin. Lab. Sci. **8**:219-227, 1978.
4. Aronsson, T., and Gronwall, A.: Improved separation of serum proteins in paper electrophoresis: a new electrophoresis buffer, Scand. J. Clin. Lab. Invest. **9**:338-345, 1957.
5. Ojala, K., and Weber, T.H.: Some alternatives to the proposed selected method for "agar gel electrophoresis," Clin. Chem. **26**:1754-1755, 1980.
6. Jeppsson, J.O., Laurell, C.B., and Franzen, B.: Agarose gel electrophoresis, Clin. Chem. **25**:629-638 (1979).
7. Wieme, R.J.: Agar gel electrophoresis, New York, 1965, Elsevier/North-Holland, Inc.

Total serum protein

ANTHONY KOLLER

Clinical significance: p. 420
Chemical class: protein

PRINCIPLES OF ANALYSIS AND CURRENT USAGE

The oldest approach to the determination of total protein in serum is the determination of protein nitrogen. The Kjeldahl procedure[1] is a method for determining the total nitrogen content in biologic material (method 1, Table 60-22). The nitrogen-containing compounds in serum are converted to NH_4^+ by oxidation in a digestion mixture of concentrated sulfuric acid, a catalyst, and a salt to increase the boiling point of the mixture. The NH_4^+ is best analyzed by conversion to NH_3 with the addition of alkali. After steam distillation into a boric acid solution, the NH_3 is then titrated with a standardized solution of HCl. Correction for nonprotein nitrogen is performed by use of a protein-free filtrate of serum. The NH_4^+ may also be quantitated photometrically with Nessler's reagent. Based on the assumption that proteins from biological sources contain 16% nitrogen by weight, the total nitrogen content (in g/L) of a sample minus the nonprotein nitrogen is multiplied by 6.25 to obtain the protein content in g/L. The Kjeldahl procedure is rarely used in routine analysis.

The most frequently used method for determining total protein in serum is the biuret reaction. In this reaction, cupric ion reacts with the peptide linkages of protein in a basic solution to form a violet-colored complex with an absorption maximum at 540 nm (method 2, Table 60-22).

The method of Lowry et al.[2] a widely used procedure for the quantitative determination of protein, has also been applied to serum protein analysis (method 3, Table 60-22). In this technique, the protein is pretreated with an alkaline copper solution. On addition of the phenol reagent of Folin and Ciocalteu,[3] the color produced (A_{max}, 745 to 750 nm) results from the reduction of the phosphotungstic and phosphomolybdic acids to molybdenum blue and tungsten blue by the copper-peptide bond complex and by the tyrosine and tryptophan of the protein. Cystine, cysteine, and histidine are also chromogenic, but to a lesser extent.[4]

Serum proteins have also been estimated by ultraviolet absorption (method 4, Table 60-22). Protein solutions show strong absorption in the 270 to 290 nm region and in the 200 to 225 nm region. Virtually all the ultraviolet absorption in serum is attributable to protein. The absorption at the higher wavelengths is attributable to the aromatic rings of tyrosine, tryptophan, and phenylalanine. The absorption at the lower wavelengths is much more intense (about twentyfold more than at 280 nm) and is mostly attributable to the peptide bond.

Another direct approach to the estimation of total protein relies on the measurement of refractive index (method

Table 60-22. Methods of serum total protein analysis

Method	Type of analysis*	Principle	Usage	Comment
1. Kjeldahl	Quantitative, protein nitrogen determination	Oxidation of N-containing compounds to NH_4^+; conversion to NH_3 with alkali; steam distillation into boric acid and titration with standard HCl; correction for nonprotein nitrogen	Historical	Cumbersome and time consuming; good accuracy and precision; used as the reference method in the past
2. Biuret	Quantitative, increased absorption at 540 nm; EP or K	Formation of violet-colored complex between Cu^{++} ions and peptide bonds in an alkaline medium	Usually adapted to automated analysis	Good specificity, accuracy, and precision; has been proposed as the basis for the reference method
3. Lowry	Quantitative, increased absorption at 745 to 750 nm; EP	Pretreatment with alkaline copper solution followed by the addition of Folin and Ciocalteu phenol reagent; reduction of phosphotungstic and phosphomolybdic acids produces the color	Historical for serum, useful for other more dilute biological fluids	Good sensitivity but poor specificity and accuracy
4. Ultraviolet absorption	Quantitative, absorption at 210 nm	Light absorption by the peptide bonds	Manual or semiautomated; not used routinely	Good sensitivity; acceptable accuracy and specificity; rapid
5. Refractometry	Quantitative	Measurement of refractive index	Manual	Acceptable accuracy and precision; rapid; susceptible to false-positive interferences from a variety of compounds

**EP*, end point; *K*, kinetic.

5, Table 60-22). This method is based on the refraction of incident light by total dissolved solids, but for serum this reflects the mass of protein present. Since serum contains a substantial mass of nonprotein-dissolved solids (electrolytes, glucose, and so on), the refractometer must be specifically calibrated with serum of a known protein concentration.

Of the above analytical approaches, the biuret reaction is employed almost exclusively for routine total serum protein determination. The only other approach used routinely to any extent is the measurement of refractive index.

REFERENCE AND PREFERRED METHODS

Although the Kjeldahl procedure is capable of high precision and accuracy,[5] it is too slow and cumbersome for routine analysis of serum protein, even if the ammonia is determined by enzymatic analysis.[6] There is also uncertainty about the correct average factor for conversion of nitrogen into protein. The historical factor of 6.25 is still used, though it is too low and is dependent on the composition of the protein fractions of each patient. The Kjeldahl procedure is still used in method evaluation work, since it has historically been the standard to which other methods were compared.

The Lowry procedure, although a hundred times more sensitive than the biuret method, is not routinely used for serum protein determination. The method lacks specificity, since many substances are known to interfere with it.[4] Also, because the tyrosine levels of the polypeptide portion of serum proteins can vary by a factor of five or more,[7] the amount of color varies with different proteins.[2] For example, gamma globulin produces 23% more color than albumin.[8] Since tyrosine is a principle chromogenic amino acid in the Lowry procedure, results with abnormal serum specimens would not correlate with more specific methods such as the biuret procedure. Other drawbacks of the Lowry procedure are that it is a two-reagent method, and both reagents have short shelf lives.[4]

Since the tyrosine and tryptophan content varies from one serum protein to another,[7] ultraviolet absorbance measurements in the 270 to 290 nm region will also vary with the protein composition of each patient. Thus this method cannot be used for direct measurement of total protein. In contrast, in the far ultraviolet region (200 to 225 nm) the absorbance is attributable mostly to the peptide bond, and all the various serum proteins exhibit similar absorption coefficients.[9] In addition, at the shorter wavelengths, Beer's law is obeyed to 120 g/L.[10] The availability of medium-priced double-beam spectrophotometers capable of accurate measurements in the far-ultraviolet region prompted Ressler et al.[10] to develop a practical method for total protein determination at 210 nm that compared favorably with the biuret and Kjeldahl methods. Nevertheless, determination of total protein by ultraviolet absorption is not routinely used.

An analysis of the 1978 CAP comprehensive chemistry survey of total serum proteins indicated that the biuret and refractometric methods were nearly equivalent in precision.[11] In the determination of the serum total protein by the refractometric method the assumption is made that differences in refractive index are primarily a reflection of differences in protein concentration. This assumption is generally true. Therefore, for clear, nonturbid, and nonpigmented sera, refractometry provides a rapid estimation of total protein with relatively good accuracy.[11,12] The refractometric method is, however, subject to error when the serum is lipemic, hypercholesterolemic, hemolyzed, azotemic, hyperbilirubinemic, hyperglycemic, or hypoproteinemic.[8,11,12] An elevation of serum glucose by 6000 mg/L or BUN of 2800 mg/L would result in an apparent increase in serum protein of 6 g/L when measured by refractometry.[13]

The biuret method is convenient and has good precision. The color produced with the biuret reagent corresponds very closely with the polypeptide portion of the various serum proteins, a distinct advantage over the Lowry method.[8] Interferences are few and most are easily eliminated. The reaction uses a single, very stable reagent and thus has been extensively adapted to a variety of automated equipment.

Kingsley[14] first introduced a single biuret reagent, but it was not very stable. A stable biuret reagent was described by Weichselbaum,[15] who added sodium potassium tartrate as stabilizer and potassium iodide to prevent autoreduction of the alkaline copper tartrate and separation of the cuprous oxide. Gornall et al.[16] introduced a biuret reagent that contained only 1.5 g/L of cupric sulfate ($CuSO_4 \cdot 5H_2O$) and with an optional 1g of potassium iodide. Reinhold[17] described a reagent that was compatible with protein-precipitating salts such as sodium sulfate and sodium sulfite. Reinhold's reagent was essentially the same as the Weichselbaum reagent. Doumas[18] recently introduced a modification of the Reinhold reagent that has a linearity up to 150 g/L. This proposed reagent was incorporated in a candidate reference method for determination of total protein in serum (Table 60-23).

The most widely used biuret reagents currently fall into two groups.[19] One group has a sodium hydroxide concentration of 0.1 to 0.2 mol/L and cupric sulfate concentration of 10 to 30 mmol/L—a version of the Weichselbaum/Reinhold reagent. The other group has a sodium hydroxide concentration of 0.5 to 0.8 mol/L and a cupric sulfate concentration of 4 to 6 mmol/L—a version of the Gornall reagent. The Doumas reagent contains the higher cupric sulfate concentration (12 mmol/L) but also maintains the higher sodium hydroxide concentration (0.6 mol/L).[18]

Some important steps in the standardization of total serum protein analysis have been taken in the last decade. In 1972 the National Committee for Clinical Laboratory Standards (NCCLS) adopted the use of bovine serum albumin as a reference material to be used primarily for as-

Table 60-23. Candidate reference method reaction conditions for total serum protein

Temperature	25° C
Final concentration of reagent components	$CuSO_4 \cdot 5H_2O$: 12 mmol/L KNa tartrate ($KNaC_4H_4O_6 \cdot 4H_2O$): 30 mmol/L KI: 30 mmol/L NaOH: 0.6 mol/L
Fraction of sample volume	0.02
Sample	Serum
Linearity	140 g/L
Time of reaction	End point at 60 minutes
Major interferences	Dextran
Precision	Within-run ($\bar{X}$, 67 g/L), CV, 0.15% Day-to-day ($\bar{X}$, 67 g/L), CV, 0.60%

Doumas, B.T., Bayse, D.D., Carter, R.J., et al.: Clin. Chem. **27**:1642-1650, 1981.

says of total protein by spectrophotometric procedures such as the biuret and the Lowry method. In 1977, the National Bureau of Standards (NBS) released a commercially prepared bovine serum albumin that met NCCLS specifications.[20] The standard is available in lyophilized form (Standard Reference Material No. 926) and as a 70 g/L solution (SRM No. 927). The mean absorptivity of the latter solution was determined as 0.2983 $L \cdot g^{-1} \cdot cm^{-1}$ using the reagent of Doumas.[21] This certified albumin is recommended as the most suitable standard for the biuret procedure. Directions for preparing secondary standards have also been published.[22] Note that since human serum albumin has a lower initial reaction rate with biuret reagent than bovine serum albumin, techniques that measure initial rate *must* be standardized with human albumin.[19,23] Because of the availability of the NBS primary protein standard and a redefinition of a reference method,[24] the biuret reaction was chosen as the basis for a candidate reference method. This choice was based on the relative specificity of the biuret reagent for proteins, the reproducibility of the color development, the similarity of the absorptivity values of the primary serum proteins[7] and the relatively few substances that interfere.[25]

SPECIMEN

Serum and exudates. Plasma may be used and usually yields comparable results.[26] Total protein is stable in serum and plasma for 1 week at room temperature and at least 1 month when refrigerated.[5]

REFERENCE RANGE

The combined male (134 subjects) and female (97 subjects) range established by the Doumas method was 66.6 to 81.4 g/L.[19] The subjects were healthy adults who had fasted for 10 to 12 hours and had been in the upright position for a least 2 hours before blood was collected without anticoagulant. This range is similar to the one established by Reed et al.,[27] which was 66.0 to 83.0 g/L based on 1419 subjects. Note that total serum protein is lower by 4 to 8 g/L with the subject supine than when the subject is in the upright position.[5]

REFERENCES

1. Archibald, R.M.: Nitrogen by the Kjeldahl method, Stand. Methods Clin. Chem. **2**:91-99, 1958.
2. Lowry, O.H., Rosebrough, N.J., Farr, A.L., and Randall, R.J.: Protein measurement with the Folin phenol reagent, J. Biol. Chem. **193**:265-275, 1951.
3. Folin, O., and Ciocalteu, V.: On tyrosine and tryptophane determinations in proteins, J. Biol. Chem. **73**:627-650, 1927.
4. Peterson, G.L.: Review of the Folin phenol protein quantitation method of Lowry, Rosebrough, Farr, and Randall, Anal. Biochem. **100**:201-220, 1979.
5. Cannon, D.C., Olitzky, I., and Inkpen, J.A.: Proteins. In Henry, R.J., Cannon, D.C., and Winkelman, J.W., editors: Clinical chemistry: principles and technics, Hagerstown, Md. 1974, Harper & Row, Publishers, Inc., pp. 405-502.
6. Smit, E.M.: An ultramicro method for the determination of total nitrogen in biological fluids based on Kjeldahl digestion and enzymatic estimation of ammonia, Clin. Chim. Acta **94**:129-135, 1979.
7. Peters, T., Jr.: Proposals for standardization of total protein assays, Clin. Chem. **14**:1147-1159, 1968.
8. Watson, D.: Albumin and "total globulin" fractions of blood, Adv. Clin. Chem. **8**:237-303, 1965.
9. Goldfarb, A.R., Saidel, L.J., and Moscovich, E.: The ultraviolet absorption spectra of proteins, J. Biol. Chem. **193**:397-402, 1951.
10. Ressler, N., Gashkoff, M., and Fischinger, A., Improved method for determinating serum protein concentrations in the far ultraviolet, Clin. Chem. **22**:1355-1360, 1976.
11. Burkhardt, R.T., and Batsakis, J.G.: An interlaboratory comparison of serum total protein analyses, Am. J. Clin. Pathol. **70**:508-510, 1978.
12. Drickman, A., and McKeon, F.A., Jr.: Determination of total protein by means of the refractive index of serum, Am. J. Clin. Pathol. **38**:392-396, 1962.
13. Peters, T., Jr., and Biamonte, G.T.: Protein (total protein) in serum, urine, and cerebrospinal fluid; albumin in serum. In Faulkner, W.R., and Meites, S., editors: Selected methods of clinical chemistry, vol. 9, Washington, D.C., 1982, American Association for Clinical Chemistry, pp. 317-325.
14. Kingsley, G.R.: The direct biuret method for the determination of serum proteins as applied to photoelectric and visual colorimetry, J. Lab. Clin. Med. **27**:840-845, 1942.
15. Weichselbaum, T.E.: An accurate and rapid method for the determination of proteins in small amounts of blood serum and plasma, Am. J. Clin. Pathol. **16**(tech. sect. 10):40-49, 1946.
16. Gornall, A.G., Bardawill, C.J., and David, M.M.: Determination of serum proteins by means of the biuret reagent, J. Biol. Chem. **177**:751-766, 1949.
17. Reinhold, J.G.: Total protein albumin and globulin, Stand. Methods Clin. Chem. **1**:88-97, 1953.
18. Doumas, B.T.: Standards for total serum protein assays—a collaborative study, Clin. Chem. **21**:1159-1166, 1975.
19. Doumas, B.T., Bayse, D.D., Carter, R.J., et al.: A candidate reference method for determination of total protein in serum. I. Development and validation, Clin. Chem. **27**:1642-1650, 1981.
20. Reeder, D.J., and Schaffer, R.: Standard reference material (SRM) for total protein determination—bovine serum albumin, Clin. Chem. **23**:1136, 1977.
21. Doumas, B.T. (Chairman), Bayse, D.D., Borner, K., et al.: A candidate reference method for determination of total protein in serum. II. Test for transferability, Clin. Chem. **27**:1651-1654, 1981.
22. NCCLS Approved Standard: ASC-1, specifications for standardized protein solution (bovine serum albumin), 1979, National Committee for Clinical Laboratory Standards, 771 E. Lancaster Ave., Villanova, PA 19085.

23. Bergkuist, C.E., Whittemore, P., and Trentini, S., A kinetic total protein methology, Clin. Chem. **26:**1057, 1980.
24. Boutwell, J.H., editor: A national understanding for the development of reference materials and methods for clinical chemistry, Washington, D.C., 1978, American Association for Clinical Chemistry, pp. 418-419.
25. Young, D.S., Pestaner, L.C., and Gibberman, V.: Effects of drugs on clinical laboratory tests. Clin. Chem. **21:**351,D-352,D, 1975. (Special issue.)
26. Chorine, V.: Influence des anticoagulants sur le dosage des éléments du sang, Ann. Inst. Pasteur **63:**213-256, 1939.
27. Reed, A.H., Cannon, D.C., Winkelman, J.W., et al.: Estimation of normal ranges from a controlled sample survey. I. Sex and age-related influence on the SMA 12/60 screening group of tests, Clin. Chem. **18:**57-66, 1972.

Total urine protein

ANTHONY KOLLER

Clinical significance: p. 403
Chemical class: protein, glycoprotein

PRINCIPLES OF ANALYSIS AND CURRENT USAGE

The analysis of total protein in urine can be classified into three main approaches: turbidimetric, dye binding, and chemical.

The reagents most commonly used for turbidimetric estimation of protein are trichloroacetic acid, sulfosalicylic acid, and more recently benzethonium chloride in alkali.[1] For dye-binding protein estimation, Coomassie brilliant blue and Ponceau S are used most often. For chemical determination of protein, the biuret reaction, the Folin-Lowry reaction, and the reaction of ferric chloride with tannic acid protein precipitates have been used.

In turbidimetry, a protein precipitant, such as trichloroacetic acid, is added to the sample and the denatured protein precipitates in a fine suspension that is quantitated turbidimetrically (method 1, Table 60-24). The turbidity varies appreciably with the chemical nature of the acid precipitant, the type of protein, the concentration of the acid, the temperature, and the duration of standing after the acid is added. The direct photometric quantitation is dependent on light dispersion rather than on light absorption by the particles in suspension. Any wavelength may be used, but light dispersion will increase as the wavelength is decreased. If nephelometric measurements are made, somewhat greater precision may be achieved.

In 1976, Bradford[2] proposed the use of the dye Coomassie brilliant blue G-250 for the estimation of protein at low concentration (method 2, Table 60-24). The binding of protein causes a shift of the absorption maximum from 465 nm (red form) to 595 nm (blue form). The increase in absorbance at 595 nm is used to monitor the extent of the reaction. The binding is complete in about 2 minutes, and the color is stable for an hour. The method has been adapted for automated analysis.[3,4]

In 1973, Pesce and Strande developed a procedure for the determination of protein in urine based on the formation of a protein-dye complex with Ponceau S (method 3, Table 60-24).[5] Samples are mixed with a trichloroacetic acid–Ponceau S dye solution, which causes the proteins along with bound dye to precipitate. The red precipitate is dissolved in dilute sodium hydroxide, resulting in a violet-colored solution that is measured spectrophotometrically at 560 nm. The color is stable for 6 hours at room temperature. Salo and Honkavaara[6] subsequently introduced a modification because of the difficulty of removing the strongly colored supernatant quantitatively from the protein-dye precipitate in the above procedure. They decreased the concentration of both the trichloroacetic acid and Ponceau S in the reagent and simply measured the decrease in absorbance at 520 nm of the reagent solution after addition of sample followed by centrifugation. A third approach employing Ponceau S dye was introduced by Meola et al.[7] In this procedure for measuring total urinary protein, the protein is adsorbed onto cellulose powder first. Ponceau S dye is then added and allowed to bind to the protein. Excess dye is washed away with dilute acetic acid. The bound dye is then eluted into dilute sodium hydroxide and measured spectrophotometrically at 550 nm.

The biuret reaction involves the reaction of biuret reagent (Cu^{++} ions) in alkaline solution with the peptide bonds in proteins. Since the biuret reaction is too insensitive and suffers from interferences when applied directly to urine, the protein needs to be concentrated first. This is accomplished by precipitation of the protein with either trichloroacetic acid or ethanolic hydrochloride–phosphotungstic acid. The precipitated protein is concentrated by centrifugation. The dissolved protein is then reacted with the biuret reagent (method 4, Table 60-24). An assay for urinary protein developed by Doetsch and Gadsden[8] takes advantage of the biuret reagent specificity for the peptide bond and achieves a sensitivity comparable to the Folin-Lowry reaction. In this approach, interfering substances are removed by gel filtration and cupric ions are stoichiometrically bound to the peptide bonds of the purified protein by the biuret reaction. The protein-copper complex is separated from excess cupric ions by another gel filtration step. Copper bound to peptide bonds is then colorimetrically determined by use of sodium diethyldithiocarbamate.

The Folin-Lowry (method 5, Table 60-24) reaction, which is one hundred times more sensitive than the biuret reaction, has also been applied to urine protein estimations after the separation of interfering materials in urine.

The Folin reagent consists of phosphotungstic and phosphomolybdic acids dissolved in phosphoric and hydrochloric acids. The reaction is an initial interaction of Cu^{++} with protein under alkaline conditions, similar to that in the biuret reaction. The Folin reagent is then added, and causes the oxidation of the Cu^{++} protein complex and

Table 60-24. Methods for total protein in urine

Method	Sensitivity	Principle	Usage	Comments
1. Turbidimetric		A protein-denaturing agent precipitates proteins Resulting turbidity measured photometrically at either 450 or 620 nm	Most common technique	Technically simple, rapid, fairly accurate
a. Sulfosalicylic (SSA) acid	10 to 25 mg/L		Commonly used method	Overestimates albumin producing 4× greater turbidity than for gamma globulins
b. Trichloroacetic (TCA) acid	20 mg/L		Frequently used method	Estimates albumin and gamma globulins equally
c. Benzethonium chloride	10 mg/L		Infrequently used	Most sensitive of turbidimetric techniques
2. Coomassie brilliant blue	2.5 mg/L	Dye binds to NH_3^+ residues in proteins with a resulting absorption at 595 nm	Used as frequently as TCA method	Rapid, highly sensitive; overestimation of albumin
3. Ponceau S	20 mg/L	Precipitation of dye-protein complex, which is redissolved in alkali Color intensity is measured at 560 nm	Infrequently used	Reacts with albumin and gamma globulins equally; aminoglycosides can interfere
4. Biuret	5 to 17 mg/L	Proteins are concentrated by precipitation with TCA or ethanolic-HCl-phosphotungstic acid (Tsuchiya's reagent), and redissolved in biuret reagent (alkaline-Cu^{++}); the Cu^{++} reagent forms a colored complex with peptide bonds that is measured at 540 nm	Used by small percentage of laboratories	With Tsuchiya's reagent this method is very sensitive with a good linear range
5. Folin-Lowry	10 mg/L	Folin reagent (mixture of molybdic and phosphotungstic acids and alkaline copper) reacts with peptide bonds, tyrosine, and tryptophan residues to produce a blue color monitored at 650 nm	Infrequently used	Very sensitive but color varies with amino acid composition of protein Urate can interfere
6. Tannic acid precipitation	5 mg/L	Tannic acid is used to precipitate proteins that are redissolved in triethanolamine-$FeCl_3$ to produce a purple color (510 nm)	Infrequently used	Overestimates albumin; erratic results for urines pH > 6

copper-tyrosine and tryptophan complexes coupled with the reduction of the phosphotungstic and phosphomolybdic acids to the chromogens molybdenum blue and tungsten blue. The blue color has a broad absorption maximum but is most frequently measured at 660 nm.

More recently, Yatzidis[9] proposed a method for total urinary protein determination in which protein is precipitated with tannic acid (method 6, Table 60-24). The precipitate is then dissolved in an aqueous triethanolamine–ferric chloride solution to produce a purple color because of the reaction with tannic acid in the protein complex.

Of all the urine total protein determinations reported in a CAP survey (1982), the sulfosalicylic acid turbidimetric method was used by the largest number of participants (37%). The trichloroacetic acid turbidimetric and Coomassie blue dye binding methods were next (29% each). The remainder of participants (5%) used the biuret reaction.

REFERENCE AND PREFERRED METHODS

Total protein determination in urine is much more difficult than in serum. The concentration of urine protein is normally low (100 to 200 mg/L); large sample-to-sample variation in the amount and composition of proteins are common; the concentration of nonprotein interfering substances is high relative to the protein concentration; and the inorganic ion content is high. All these factors affect the precision and accuracy of the various methods.

Of the various methods and techniques proposed over the years for total protein determination in urine, the turbidimetric procedures have gained the widest acceptance. This is mostly attributable to their sensitivity and simplicity. The sensitivity of benzethonium chloride in alkali is comparable to the Folin-Lowry method,[1] that is, approximately 10 mg/L.[1] Sulfosalicylic acid has a sensitivity range of 10 to 25 mg/L.[7] Trichloroacetic acid is less sensitive, producing about one half the turbidity produced by sulfosalicylic acid.[10] A serious drawback for the sulfosalicylic acid procedure is that is produces four times more turbidity with albumin than it does with gamma globulin.[11] Sulfosalicylic acid also precipitates significant quantities of polypeptides from urine.[12]

The turbidity produced by trichloroacetic acid is not altered by changes in the albumin/globulin ratio in the narrow temperature range of 20° to 25° C. Above 25° C the turbidity produced with albumin is much more than that for globulin.[11] Therefore the temperature has to be controlled to obtain reasonable accuracy with this acid. Although the originators of the benzethonium chloride method claimed that the turbidity produced was relatively insensitive to the albumin/gamma globulin ratio,[1] subsequent evaluations showed that gamma globulin produces 11% to 31% less turbidity than albumin, depending on the total protein concentration (2.4 g/L to 1 g/L).[13,14] The turbidity produced by benzethonium chloride is, however, quite stable between 25° and 40° C.[1] Within-run precision varies greatly for the three turbidimetric methods. The reported coefficient of variation values range anywhere from 5% to 20%.[14]

The benzethonium chloride procedure for urine correlated well with a biuret procedure.[13] The sulfosalicylic acid procedure for urine correlated poorly with two versions of the biuret procedure.[15,16]

Some drugs have been shown to interfere with turbidimetric methods. Miconazole interfered with the trichloroacetic acid procedure in urine when the protein content was above 0.5 g/L.[17] Benzylpenicillin and cloxacillin both interfered with the above procedure applied to urine samples with use of a centrifugal analyzer.[18]

Nakamura et al.[19] postulated an inhibitor in urine that decreases the analytical recovery of protein in the sulfosalicylic acid method.

The protein–Coomassie brilliant blue dye complex has a high extinction coefficient and has four times the sensitivity of the Folin-Lowry method. However, as is true for the Folin-Lowry method, the amount of color development varies with the protein used. The dye-anion interacts with the NH_3^+ groups of proteins. Since not all proteins have the same proportion of NH_3^+ groups and not all NH_3^+ groups react identically, the color development varies with the protein species.[20] Bradford[2] found that interfering color was caused by relatively large amounts of detergents, such as dodecyl sulfate. Thymol, often used as a urine preservative, also was shown to interfere with this procedure, just as salicylates do.[14,21]

McIntosh proposed the application of Coomassie blue to the estimation of protein in urine.[21] The correlation of the Coomassie blue method with the Sephadex-biuret method of Doetsch and Gadsden[8] was acceptable ($r = 0.91$), but the comparison with a turbidimetric method was poor ($r = 0.65$). This was partly ascribed to the low sensitivity of the turbidimetric assay for Tamm-Horsfall mucoprotein and proteins of low molecular weight.

In another study,[14] two modifications of the Bradford method (Pierce Rapid Stat kit and Bio-Rad Protein Assay kit) were compared to a biuret method, using urine samples. The correlations were good, despite the fact that both kits produced a lower absorbance response with gamma globulin than with albumin. Within-day reproducibility ranged from 2.8% to 7.3% for the Bio-Rad kit and 3.0% to 3.3% for the Rapid-Stat kit. Day-to-day reproducibility was 3.2% and 4.6% respectively. Both modifications suffered from lack of linearity. The Rapid Stat kit is only linear to 100 mg/L, and the Bio-Rad kit is linear to 1000 mg/L.

The sensitivity of the Pesce and Strande procedure with Ponceau S is 20 mg/L. The procedure gave identical results at 0°, 24°, and 37° C and was linear between 100 and 1500 mg/L. It is not appreciably affected by the albumin/globulin ratio. The method did not correlate well with a trichloroacetic acid turbidimetric procedure, yielding lower results. However, there was a good correlation with the Kjeldahl method. In a separate study[14] involving urinary proteins, the method of Pesce and Strande correlated quite well with a biuret procedure. The within-run reproducibility in this study was ± 2.0% to 10.2% and the day-to-day reproducibility was ±4.9%. In contrast, Pesce and Strande found the within-run imprecision to be 2% and the day-to-day imprecision 2.5%.

Lievens and Celis[17] found that aminoglycosides (such as gentamicin and kanamycin) at concentrations reached during therapy increased urinary protein results by a factor of 2 to 4 in the Pesce and Strande procedure. To circumvent the interference, they proposed a preliminary precipitation of proteins with trichloracetic acid without Ponceau S. The Meola et al. modification[7] of the Ponceau S procedure, requiring prior absorption of proteins onto cellulose powder, is more cumbersome than the other Ponceau S procedures and cannot be recommended for routine work. In addition, aminoglycosides (such as gentamicin, amikacin, and tobramycin) also cause a positive interference in this procedure.[22]

As already mentioned, the urine protein must be concentrated before the biuret reaction can be applied. Trichloroacetic acid, perchloric acid, or Tsuchiya's reagent (ethanolic HCl–phosphotungstic acid) have been employed as precipitating reagents. More glycoproteins are precipitated in urine with perchloric acid than with trichloroacetic acid.[23] Tamm-Horsfall glycoproteins, the most abundant of the proteins derived from the urinary tract itself, will precipitate with Tsuchiya's reagent.[24]

Piscator[25] employed Tsuchiya's reagent for urinary protein precipitation and allowed the reagent to react for 15 minutes at room temperature before centrifugation. Savory et al.[15] allowed the same reagent to react for 15 minutes at O° C, obtaining somewhat higher (+6.1%) results at this temperature than at 25° C. Rice,[26] on the other hand, proposed the use of 56° C for 15 minutes, which resulted in less absorption of tan-colored pigments. Also the color absorbed during the heating step can more readily be removed by ethanol washing. Rice employed Benedict's qualitative solution for carrying out the biuret reaction. Each sample was set up in duplicate, the blank correcting for precipitated urinary pigments. Rice's procedure was linear to 2000 mg/L with a precision of 2% to 3%.

Lizana et al.,[16] using Tsuchiya's precipitating reagent at room temperature followed by Benedict's solution, found a sensitivity of 17 mg/L. At a protein level of 100 to 200 mg/L, their method showed a recovery of over 96%. Such sensitivity and recovery is important when one is assaying normal urines. This method does require an ultraviolet spectrophotometer, however, since the quantitation was carried out at 330 nm.

Savory et al.,[15] using Tsuchiya's reagent followed by the biuret reaction, found the limit of sensitivity for detection of protein in urine to be 5 mg/L. The coefficient of variation of replicate analyses was 4.2%. The method was linear to 270 mg/L. Each sample required an alkaline tartrate blank.

The method of Doetsch and Gadsden,[8] which was described earlier, has a sensitivity to 10 mg/L. Nonprotein substances in urine should not interfere in this technique unless they simultaneously bind to protein and chelate copper. Radiographic contrast medium has been found to cause positive interference. The method is linear from 200 to 800 mg/L.

When trichloroacetic acid is used to precipitate urinary proteins, the final concentration of acid may vary from 4% to 10%. When the biuret reaction is carried out on the protein precipitate dissolved in sodium hydroxide, a sensitivity of 60 mg/L is attainable.[16] Normally, a correction for residual urine pigments is made by measurement of the absorbance of the protein precipitate dissolved in alkali and subtraction of this value from that obtained after the protein-biuret color is formed.

The Folin-Lowry procedure has also been employed for the determination of urinary proteins after separation from interfering materials by dialysis, precipitation, or ultrafiltration. A precipitation reagent such as perchloric acid–acetone will precipitate mucoproteins along with albumin and globulins.[23] Normal protein content of urine by this approach is in the range of 40 to 400 mg/24 hours. This upper limit is the highest of all the methods discussed.

The tannic acid–ferric chloride method of Yatzidis[9] has some shortcomings. Excess tannic acid must be completely removed by washing or else elevated values will be observed. Gamma globulin, when compared to albumin, resulted in 45% less absorbance in one study.[14] The same study showed that the method was affected by the pH of the urine sample, giving erratic results above pH 6.0.

The turbidimetric methods are the methods of choice because of their simplicity and sensitivity. Although the sulfosalicylic acid methods are the most sensitive, the trichloroacetic acid method is preferred because it measures both albumin and gamma globulins without unacceptable bias for either class of protein. The trichloroacetic acid turbidimetric procedure, listed below, can be performed by equipment readily available in any clinical chemistry laboratory. For analyses that require more accurate values at low protein concentrations, the method using Tsuchiya's reagent followed by the biuret reaction is recommended.

A summary of reaction conditions for common methods of analysis are listed in Table 60-25.

PROCEDURE

Principle

Trichloroacetic acid, 3%, added to a solution of protein in low concentration, precipitates the protein as a fine suspension, which is quantitated photometrically. Trichloroacetic acid produces approximately equal turbidities with both albumin and globulins when one uses the procedure as outlined below.

Reagents

Albustix (Ames, Inc., Divison of Miles Co., Elkart, Ind.)

Stock trichloroacetic acid, 100% (6.12 mol/L)

PRECAUTION: Trichloroacetic acid is extremely corrosive and must be handled with caution. The person preparing this solution must wear gloves and safety glasses during the process. Extreme care must be taken to avoid splashing, and any spills must be carefully cleaned up immediately.

To one 500 g bottle of trichloroacetic acid (Fisher-certified ACS, No. A-322), add 175 mL of deionized water directly into the reagent bottle. Close tightly and allow to dissolve overnight, swirling occasionally. Transfer the solution quantitatively with rinsings to a 500 mL volumetric flask and dilute to 500 mL with deionized water. Mix carefully by inversion 10 times. The 100% trichloroacetic acid should be stored in a brown glass bottle (avoid metal bottle-cap liner), labeled "CORROSIVE," at room temperature in a hood. Ths solution is stable for 12 months at room temperature.

Working trichloroacetic acid, 3% (184 mmol/L)

PRECAUTION: Because of the corrosive nature of trichloroacetic acid, gloves and safety glasses must be worn during preparation of this solution. All spills must be carefully cleaned up.

With a 25 and a 5 mL volumetric pipet, carefully transfer 30 mL of stock 100% trichloroacetic acid into a 1000 mL volumetric flask. Dilute to 1000 mL with deionized water and mix by inversion 10 times. Store in a brown glass bottle (no metal cap liner), labeled "CORROSIVE," at

Table 60-25. Reaction conditions for analysis of total urinary protein

Condition	TCA*	SSA†	Biuret‡	Coomassie brilliant blue§
Temperature	20° to 25° C	Ambient	56° C precipitation Ambient reaction	Ambient
Sample volume	500 μL	500 μL	2000 μL	100 μL
Fraction of sample volume	0.20	0.20	0.5 (for initial precipitation)	0.02
Final concentration of reagents	Trichloroacetic acid: 153 mmol/L (25 g/L) NaCl in blank: 0.12 mol/L	Sulfosalicylic acid: 98 mmol/L (25 g/L) NaCl in blank: 0.12 mol/L	Precipitation step: HCl: 0.34 mol/L Ethanol: 415 mL/L Phosphotungstic acid: 85 g/L Colorimetric reaction: Sodium citrate: 28 mmol/L Sodium carbonate: 44.9 mmol/L Cupric sulfate: 3.3 mmol/L Sodium hydroxide: 714 mmol/L Blank contains all but $CuSO_4$	Coomassie brilliant blue: 100 mg/L Ethanol: 47 g/L Phosphoric acid: 85 g/L
Time of reaction	10 minutes	6 minutes	15-minute precipitation 20-minute reaction	2 minutes
Wavelength	450 nm	620 nm	540 nm	595 nm
Linearity	2400 mg/L	3000 mg/L	2000 mg/L	1000 mg/L
Precision‖ (mean, % CV)	85 mg/L (40%) 132 mg/L (17%)	89 mg/L (47%) 234 mg/L (33%)	115 mg/L (34%)	127 mg/L (25%)
Interferences	Certain penicillin drugs (benzylpenicillin and cloxacillin), miconazole	Most aminoglycosides	None	Detergents, thymol, and salicylates

*Trichloroacetic acid, turbidimetric assay
†Sulfosalicylic acid, turbidimetric assay
‡Biuret using Tsuchiya's reagent to precipitate and concentrate protein; from reference 26.
§From Bradford, M.M.: Anal. Biochem. **72:**248-254, 1976.
‖From College of American Pathologists Quality Assurance Survey data

room temperature. The solution is stable for 1 month at room temperature.

0.9% saline solution, 0.15 mol/L. Place 9 g of NaCl into a 1 L volumetric flask, dissolve with approximately 900 mL of distilled H_2O and dilute to 1 liter with distilled H_2O. The solution is stable for 3 months at room temperature.

Controls. Use commercially available lyophilized serum pools.

Calibration curve

Stock standard. Use pooled serum from routine daily analysis. Pool clear, nonlipemic, nonicteric, nonhemolyzed specimens.

Working standard. With a total protein pool value of approximately 60 g/L *as an example,* the following are saline dilutions that can be used to plot a curve. All dilutions are prepared volumetrically to 100 mL with saline. Refrigerated working standards are stable for several days.

mL of stock (60 g/L)	mL of saline	Working standard (mg/L)
0.5	99.5	300
1.0	99	600
2.0	98	1200
3.0	97	1800
4.0	96	2400

Analyze standards in same manner as urines. Plot %*T* versus concentration in mg/L and make a %*T*–mg/L chart.

Procedure

Equipment: spectrophotometer or photometer capable of reading at 450 nm.

1. Centrifuge 10 mL of urine. Check urine with Albustix and record results. If the dipstick estimate of urine protein concentration is +3 or greater, dilute urine twofold with saline before performing analysis.
2. Set up 2 rows of 12 × 75 mm cuvettes—a test and blank for each urine specimen.
3. Pipet 0.5 mL of urine into both test and blank cuvettes.
4. Add 2 mL of 0.9% saline solution to blanks and 2 mL of 3% trichloroacetic acid to test cuvettes and mix each gently by vortexing on slow speed or inverting with Parafilm. Mix immediately after addition of trichloroacetic acid. Temperature control is critical. Ambient temperature should be between 23° and 27° C for proper results. All reagents and samples must also be at room temperature.
5. Let stand for 10 minutes and invert gently. Avoid

excessive mixing to prevent trapping of air bubbles, which will increase apparent turbidity. Immediately read % *T* at 450 nm against blank. Turbidity is stable for *2 minutes* after reaction period. Run only as many samples at once as can be read in 2 minutes.

Calculations

Read mg/L off %*T*–mg/L table. Do not extend standard curve beyond highest value employed to prepare the curve in present use.

Random urines. Report as mg/L

24-hour urines. To calculate total protein excretion (*TV,* total volume) per 24 hours:

$$\text{mg/L} \times \text{TV (liters)/24 hours} = \text{mg/TV}$$

Dilutions. mg/L × TV/24 hours × Dilution factor = mg/TV

Notes

1. Trichloroacetic acid gives similar turbidities with both albumin and globulin.
2. Any wavelength in the short end of the visible spectrum can be used. The choice of 450 nm is entirely arbitrary.
3. *Interfering colors:* Urine color is usually corrected for by the use of the urine blank.
4. The validity of the preliminary results obtained with the reagent strip tests are dependent on the pH of the urine. In heavily buffered alkaline urine, the buffer on the indicator stick is insufficient to achieve the pH required for adequate indicator function, thus leading to false-positive results. Deeply pigmented urines also tend to give false-positive results.
5. Excretion of protein in the urine is not constant and varies considerably over a 24-hour period with no particular pattern. Testing random samples therefore can be misleading.
6. There can be discrepancies between the Albustix (which measures albumin) and the trichloroacetic acid turbidimetric procedures. A negative or low Albustix and a high trichloroacetic acid result could be caused by the following:
   a. Presence of myeloma protein (globulin, Bence Jones proteins) in urine
   b. Nonhomogeneous suspensions caused by the presence of certain drugs (such as Tolmetrin, a nonsteroid anti-inflammatory drug used in treatment of rheumatoid arthritis). Metabolites of this drug can give false-positive results with acid precipitation procedures.

REFERENCE RANGE

50 to 100 mg/24 hours.

Clinical interpretation[27]

A. Functional: Not associated with easily demonstrable systemic or renal damage
   1. Severe muscular exertion
   2. Pregnancy (eclampsia or toxemia of pregnancy)
   3. Orthostatic proteinuria. This term designates slight to mild proteinuria associated only with the upright position; the exact cause is poorly understood.

B. Organic: Associated with demonstrable systemic disease or renal pathology.
   1. *Prerenal proteinuria:* Not attributable to primary renal disease
      a. Fever, or a variety of toxic conditions—most common cause for organic proteinuria and quite frequent
      b. Venous congestion—chronic passive congestion from heart failure
      c. Relative anoxia—severe dehydration, shock, severe, acidosis
      d. Hypertension
      e. Myxedema
      f. Bence Jones protein of multiple myeloma
   2. *Renal proteinuria* (See Chapter 23.)

REFERENCES

1. Iwata, I., and Nishikaze, O.: New micro-turbidimetric method for determination of protein in cerebrospinal fluid and urine, Clin. Chem. **25:**1317-1319, 1979.
2. Bradford, M.M.: A rapid and sensitive method for the quantitation of microgram quantities of protein utilizing the principle of protein-dye binding, Anal. Biochem. **72:**248-254, 1976.
3. Heick, H.M.C., Begin-Heick, N., Acharya, A., and Mohammed, A.: Automated determinations of urine and cerebrospinal fluid proteins with Coomassie brillant blue and the Abbott ABA-100, Clin. Biochem. **13:**81-85, 1980.
4. Helmer, G.R., Borer, W.Z., and Palmer, J.J.: Urine protein concentration measured by selected method and multistat microprotein method, Clin. Chem. **28:**1628, 1982.
5. Pesce, M.A., and Strande, C.S.: A new micromethod for determination of protein in cerebrospinal fluid and urine, Clin. Chem. **19:**1265-1267, 1973.
6. Salo, E.J., and Honkavaara, E.I.: A linear single reagent method for determination of protein in cerebrospinal fluid, Scand. J. Clin. Lab. Invest. **34:**283-288, 1974.
7. Meola, J.M., Vargas, M.A., and Brown, H.H.: Simple procedure for measuring total protein in urine, Clin. Chem. **23:**975-977, 1977.
8. Doetsch, K., and Gadsden, R.H.: Determination of total urinary protein, combining Lowry sensitivity and biuret specificity, Clin. Chem. **19:**1170-1178, 1973.
9. Yatzidis, H.: New colorimetric method for quantitative determination of protein in urine, Clin. Chem. **23:**811-812, 1977.
10. Cannon, D.C., Olitzky, I., and Inkpen, J.A.: Proteins. In Henry, R.J., Cannon, D.C., and Winkelman, J.W., editors: Clinical chemistry: principles and technics, ed. 2, Hagerstown, Md., 1974, Harper & Row, Publishers, Inc., p. 423.
11. Schriever, H., and Gambino, S.R.: Protein turbidity produced by trichloroacetic acid and sulfosalicylic acid at varying temperatures and varying ratios of albumin and globulin, Am. J. Clin. Pathol. **44:**667-672, 1965.
12. Henry, R.J., Sobel, C., and Segalove, M.: Turbidimetric determination of proteins with sulfosalicylic and trichloroacetic acids, Proc. Soc. Exp. Biol. Med. **92:**748, 1956.
13. Bollin, E., Jr., Schiffreen, R.S., and Ware, E.S.: Quantitation of urinary protein using the DuPont automatic clinical analyzer ('aca'), Clin. Chem. **28:**1579, 1982.
14. McElderry, L.A., Tarbit, I.F., and Cassells-Smith, A.J.: Six methods for urinary protein compared, Clin. Chem **28:**356-360, 1982.
15. Savory, J., Pu, P.H., and Sunderman, F.W., Jr.: A biuret method

for determination of protein in normal urine, Clin. Chem. **14**:1160-1171, 1968.
16. Lizana, J., Brito, M., and Davis, M.R.: Assessment of five quantitative methods for determination of total proteins in urine, Clin. Biochem. **10**:89-93, 1977.
17. Lievens, M.M., and Celis, P.J.: Drug interference in turbidimetry and colorimetry of proteins in urine, Clin. Chem. **28**:2328, 1982.
18. Muir, A., and Hensley, W.J.: Pseudoproteinuria due to penicillins, in the turbidimetric measurement of proteins with trichloroacetic acid, Clin. Chem. **25**:1662-1663, 1979.
19. Nakamura, J., and Yakata, M.: Urine contains an inhibitor for turbidimetric determinations of protein, Clin. Chem. **28**:2294-2296, 1982.
20. Sedmak, J.J., and Grossberg, S.E.: A rapid, sensitive, and versatile assay for protein using Coomassie Brillant Blue G-250, Anal. Biochem. **79**:544-552, 1977.
21. McIntosh, J.C.: Application of a dye-binding method for the determination of protein in urine and cerebrospinal fluid, Clin. Chem. **23**:1939-1940, 1977.
22. Meola, J.M., and Brown, H.H.: Aminoglycoside antibiotic interference with a urinary protein method, Clin. Chem. **25**:1180-1181, 1979.
23. Saifer, A., and Gerstenfeld, S.: Photometric determination of urinary proteins, Clin. Chem. **10**:321-334, 1964.
24. Mazzuchi, N., Pecarovich, R., Ross, N.: Tamm-Horsfall urinary glycoprotein quantitation by radial immunodiffusion: normal patterns, J. Lab. Clin. Med. **84**:771, 1974.
25. Piscator, M.: Proteinuria in chronic cadmium poisoning. II. The applicability of quantitative and qualitative methods of protein determination for the demonstration of cadmium proteinuria, Arch. Environ. Health **5**:325, 1962.
26. Rice, E.W.: Improved biuret procedure for routine determination of urinary total protein in clinical proteinuria, Clin. Chem. **21**:398-401, 1975.
27. Pesce, A.J., and First, M.R.: Proteinuria: an integrated review, New York, 1979, Marcel Dekker, Inc.

Transferrin

ROBERT S. FRANCO

Clinical significance: p. 633
Molecular weight: 90,000 daltons
Chemical class: beta$_2$-glycoprotein

PRINCIPLES OF ANALYSIS AND CURRENT USAGE

Iron in plasma is bound to the transport protein transferrin, and the traditional way of measuring transferrin is as total iron-binding capacity (TIBC). Methods for measuring TIBC are reviewed elsewhere in this volume. Because of the genetic heterogeneity of transferrin, multiple bands can be present in an individual serum as determined by electrophoretic procedures. Therefore, although it can be readily observed on cellulose acetate electrophoresis, this is not a suitable method for quantitative analysis. Transferrin may also be measured directly in serum by immunological means; either nephelometry or radial immunodiffusion (RID) (Table 60-26). In general, the correlation between nonimmunological TIBC and immunological methods for quantitating transferrin is very good.[1-4]

REFERENCE AND PREFERRED METHODS

Nephelometry is popular in high-volume applications, since automated instruments are available (Technicon, Inc., Tarrytown, N.Y.; Hyland Laboratories, Costa Mesa, Calif.) (Table 60-27). This method has the advantage of a small sample size, a fast turnaround time with results available the same day, and a relatively high precision. Disadvantages are relatively high reagent costs and the need for a specialized piece of equipment. Radial immunodiffusion features a very small sample size, extreme simplicity when commercially prepared plates are used, acceptable precision, and essentially no capital investment. The main disadvantage of radial immunodiffusion is the relatively long time before the precipitin rings may be measured, typically 18 to 48 hours (Table 60-27). The two methods exhibit extremely good agreement, and so the choice of which to use is often dependent on the number of samples to be run. A high throughput favors nephelometry, whereas radial immunodiffusion is advantageous if a relatively small number of samples is to be run. The nephelometric procedure is dependent on the particular instrument used, and the instructions provided by the manufacturer should be followed. If commercial plates are used, and they are strongly recommended for the clinical laboratory, the manufacturer's directions should be followed.

An outline for the radial immunodiffusion method of protein determination is given in the haptoglobin method section. Commercial radial immunodiffusion plates for transferring are available (such as Behring Diagnostics, Somerville, N.J., and Kallestad Laboratories, Chaska, Minn.). With the radial immunodiffusion method, serum transferrin may be measured with a coefficient of variation ranging from 3.1% to 6.4%.[3]

Table 60-26. Methods of transferrin analysis

Method	Principle	Usage	Comment
Radial immunodiffusion	Transferrin diffuses into gel containing antitransferrin antibody; precipitin ring diameter is proportional to concentration	Current clinical use	Suitable for small number of samples
Nephelometry	Antibody reacts with transferrin, forming complex that scatters light	Currrent clinical use	Suitable for large number of samples

Table 60-27. Comparison of parameters of transferrin analysis

Parameter	Radial immunodiffusion	Nephelometry*
Temperature	23 ± 2° C (RT)†	22° C (RT)†
pH	7.4	7.0
Time of reaction	18 hours	30 minutes
Sample	Serum	Serum
Sample size	5 μL	10 μL
Range (without dilution)	—	0.44-8 g/L
Precision (% CV)	3.1%-6.4%	2.8%

Data from Buffone, G.J., et al.: Clin. Chem. **24:**1788-1791, 1978.
*"PDQ" Laser Nephelometer (Hyland Laboratories, Inc., Costa Mesa, CA 92626).
†*RT,* Room temperature

SPECIMEN

Blood samples for serum transferrin quantitation should be drawn in the morning. The patient should be in a fasting state.

REFERENCE RANGE

A representative normal range is 1.87 to 3.12 g/L.[2] A discussion of sex and age dependence of normal values is given in reference 4. In general, there is no consistent difference between the normal values for adult men and women, but young children (4 to 10 years) have a normal range shifted slightly (about 0.35 g/L) to higher values.

REFERENCES

1. Burrows S.: Comparison of methods designed to measure transferrin and iron-binding capacity of serum, Am. J. Clin. Pathol. **47:**326-328, 1967.
2. Tsung S.W., Rosenthal, W.A., and Milewski, K.A.: Immunological measurement of transferrin compared with chemical measurement of total iron-binding capacity, Clin. Chem. **21:**1063-1066, 1975.
3. Buffone, G.J., Lewis, S.A., Iosefsohn, M., and Hicks, J.M., Chemical and imunochemical measurement of total iron-binding capacity compared, Clin. Chem. **24:**1788-1791, 1978.
4. Rajamaki, A., Irjala, K., and Aitio, A.: Immunochemical determination of serum transferring, Scand. J. Haematol. **23:**227-231, 1979.

Chapter 61 Toxicology and therapeutic drug monitoring (TDM)

Acetaminophen

JOSEPH SVIRBELY

Acetominophen, *N*-acetyl-*p*-aminophenol, paracetamol
Clinical significance: p. 897
Molecular formula: $C_8H_9NO_2$
Molecular weight: 151.16 daltons
Merck Index: 36
Chemical class: *p*-aminophenol derivative

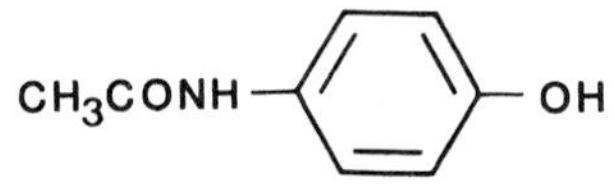

CH_3CONH—⟨benzene ring⟩—OH

PRINCIPLES OF ANALYSIS AND CURRENT USAGE

The first methods for analysis of serum acetaminophen were spectrophotometric. Current methods include spectrophotometric, gas-liquid chromatographic (GLC), and high-performance liquid chromatography (HPLC) procedures.

In 1947, in a paper studying the metabolism of acetanilid, Lester and Greenberg[1] described the analysis of *N*-acetyl-*p*-aminophenol (acetaminophen). Serum was deproteinated with a mixture of 2 N hydrochloric acid, zinc dust, barium sulfate, and capryl alcohol (1-octanol). The filtrate was extracted into ethylene dichloride and back-extracted as the sodium salt by mixture with dilute sodium hydroxide. The alkaline extract was hydrolyzed to *p*-aminophenol by addition of concentrated hydrochloric acid and granulated zinc and was then boiled. Color was developed with α-naphthol and concentrated sodium hydroxide extracted with butanol, and the absorbance was measured with a colorimeter (method 1, Table 61-1).

Brodie and Axelrod[2] reviewed this method and modified it by extracting serum into ether, back-extracting into alkali, and then hydrolyzing the extract as above. The *p*-aminophenol was diazotized with ammonium sulfamate, coupled with α-naphthol, and analyzed spectrophotometrically.

Heirwegh and Fevery[3] further modified the method to make it more reliable. The extraction and hydrolysis steps of Brodie and Axelrod were followed. The *p*-aminophenol solution was then reacted with sodium nitrate to form the nitrous acid derivative. Ammonium sulfamate was added, followed by *N*-(1-naphthyl)ethylenediamine in ethanol. After 4 hours of incubation, absorbances were read at 578 nm.

Glynn and Kendal[4] published further modifications. The hydrolysis step was eliminated, and instead plasma was deproteinated with trichloroacetic acid solution followed by centrifugation. An aliquot of the supernatant was mixed with 6 N hydrochloric acid and sodium nitrite to form the nitrous acid derivative. Sulfamic acid was added, followed by sodium hydroxide; the ensuing yellow color was read at 430 nm. The entire reaction took 15 minutes (method 2, Table 61-1).

Frings and Saloom[5] utilized a different reaction to determine serum acetaminophen. A protein-free filtrate was prepared from serum and then hydrolyzed by boiling with hydrochloric acid to produce *p*-aminophenol. The *p*-aminophenol formed was then reacted with phenol and ammonium hydroxide to form an indophenol blue chromogen, which was read at 620 nm.

Meola[6] developed a procedure that extracted serum with a mixture of isopropanol, methylene chloride, and diethyl ether and then reacted the solvent phase with Folin-Ciocalteu phenol reagent for 25 minutes. The color was read at 660 nm (method 3, Table 61-1).

Routh et al.[7] utilized the differential absorption peak of acetaminophen at 266 nm, which corresponded to the isosbestic point of salicylate and therefore eliminated interference from this analgesic (method 4, Table 61-1). Knepil[8] also used the differential absorption technique but chose the differential absorbance reading at 290 nm to calculate acetaminophen concentration.

More recently, gas-liquid chromatography (GLC) has been used to quantitate acetaminophen (method 5, Table 61-1). Prescott[9] in 1971 published a GLC method utilizing

Table 61-1. Methods for acetaminophen analysis

Method	Type of Analysis	Principle	Usage	Comments
1. Azo dye formation (Lester and Greenberg[1])	Spectrophotometry	$CH_3CONH{-}C_6H_4{-}OH \xrightarrow[H^+]{\Delta} NH_2{-}C_6H_4{-}OH$ Acetaminophen → Aminophenol $NH_2{-}C_6H_4{-}OH + NaNO_2 \rightarrow HN{=}N{-}C_6H_4{-}OH$ Sodium nitrite → Diazophenol $HN{=}N{-}C_6H_4{-}OH$ + Ammonium sulfamate $\xrightarrow[5.5\ N\ NaOH]{\alpha\text{-Naphthol}}$ Azo dye	Hydrolysis time consuming; replaced by gas-liquid (GLC) and high-performance liquid chromatography (HPLC)	Of historical interest
2. Azo dye formation (Glynn and Kendal[4])	Spectrophotometry	$CH_3CONH{-}C_6H_4{-}OH \xrightarrow[HCl]{NaNO_2} CH_3CONH{-}C_6H_3(NO_2){-}OH$ $CH_3CONH{-}C_6H_3(NO_2){-}OH \xrightarrow{OH^-} CH_3CONH{-}C_6H_3(NO_2){-}O^-$ $CH_3CONH{-}C_6H_3(NO_2){-}O^-$ (A_{max} = 430 nm)	Most efficient spectrophotometric method	Linear in range of 100-500 μg/mL; HPLC and GLC more sensitive No known interferences
3. Folin-Ciocalteu reagent (Meola[6])	Spectrophotometry	Indophenol dye complex formed by reaction of acetaminophen with phenol reagent of Folin-Ciocalteu at pH 11.0	Not commonly used because of interferences	Tryptophan, tyrosine, uric acid, and salicylate react to give falsely high readings

4. Spectrophotometry (Routh et al.[7])	Differential ultraviolet spectrophotometry	Measures acetaminophen at differential absorbance peak of 266 nm, avoiding interference with salicylate	Replaced by HPLC and GLC	Reading at isosbestic point of salicylate removes interference from this compound; requires narrow band pass spectrophotometer
5. Gas-liquid chromatography (GLC)	Chromatographic separation	Extracted into ethyl acetate to trimethylsilyl derivatives; injected isothermically onto gas-liquid chromatograph with OV-17 column and flame ionization detector	Method of choice	Sensitive to 1 µg/mL No known interferences Only drawback is time needed for ethyl acetate to evaporate
6. High-performance liquid chromatography (HPLC)				
a. Ion exchange	Chromatographic separation	Separation on cation-exchange column; run time is 50 min	Rarely used	Time-consuming procedure; not to be favored over reversed-phase HPLC
b. Silica adsorption	Chromatographic separation	Separation on Zorbax Sil silica column using mobile phase of chloroform/heptane/ethanol/glacial acetic acid	Alternative to reversed-phase HPLC	Sensitive to 5 µg/mL Recovery of 97% Most drugs do not interfere
c. Reversed phase	Chromatographic separation	Packing octadecylsilane-bonded silica; mobile phase is dilute acetic acid/methanol/ethyl acetate; detection at 254 nm	Method of choice	Most rapid and sensitive (to 1 µg/mL) Interferences with internal standard Future improvements may include use of electrochemical detector

flame ionization detection. The sample was extracted into ethyl acetate, and the extract was then evaporated. After reconstitution in pyridine and derivatization to the trimethylsilyl derivative, the extract was injected onto an OV-17 column at 200° C.

Thomas and Coldwell[10] modified Prescott's method by using a more powerful silylating reagent (Regisil, bis[trimethylsilyl]trifluoroacetamide) and a more selective extractant (diethyl ether). They also substituted an OV-1 column for Prescott's OV-17 column.

Dechtiaruk et al.[11] formed the *o*-heptyl-*N*-methyl derivative and utilized temperature programming from 150° C to 260° C at 16° C per minute with an OV-17 column.

High-performance liquid chromatography (HPLC) has also been used to quantitate acetaminophen using several different chromatographic approaches. In 1977 Blair and Rumack[12] published an HPLC method utilizing a cation-exchange resin column (Aminex A-5) and an isocratic buffer, $NH_4H_2PO_4$ (method 6a, Table 61-1). Detection was at 254 nm, and analysis time was 50 minutes per sample.

Fletterick et al.[13] used a silica-packed column (Zorbax Sil) and a mobile phase of chloroform/heptane/ethanol/glacial acetic acid (method 6b, Table 61-1). Detection was at 254 nm. The sample was mixed with the internal standard, 3-acetaminophenol, and extracted into ethyl acetate. The extract was evaporated, reconstituted, and chromatographed.

Howie et al.[14] have published a method employing reversed-phase HPLC (method 6c, Table 61-1). The column was packed with 10-octadecylsilane–coated silica, and the mobile phase was packed water/acetic acid/ethyl acetate. Detection was at 254 nm. Plasma was deproteinated with trichloroacetic acid containing the internal standard 4-fluorophenol, and the supernatant was injected directly onto the column.

The methods developed by Horvitz and Jatlow,[15] Gotelli et al.,[16] and Lo and Bye[17] also used 10-octadecylsilane–coated silica columns, with detection being at 254 nm. Horvitz and Jatlow and Gotelli et al. used an acetonitrile/phosphate buffer mobile phase, and Lo and Bye used a water/acetonitrile phase. All extracted the sample in ethyl acetate, evaporated the extract, reconstituted it, and injected it onto the column.

Riggin et al.[18] utilized a thin-layer, carbon-paste electrochemical transducer to detect acetaminophen in an HPLC system using a cation-exchange column and a mobile phase of dilute sulfuric acid. Munson et al.[19] used the same electrochemical detector to measure acetaminophen concentrations as low as 0.2 µg/mL.

Recently an enzyme-multiplied immunoassay (Syva Co., Inc., Palo Alto, Calif.) has been introduced to measure acetaminophen. Its use in clinical situations has not been well documented at this time.

REFERENCE AND PREFERRED METHODS

Most spectrophotometric methods require hydrolysis to *p*-aminophenol before analysis. Because the antidote to acetaminophen must be given within 12 hours after ingestion, the time involved in many of these procedures makes them unpractical in the clinical setting. For the laboratory not equipped with a gas-liquid (GLC) or high-performance chromatographic (HPLC) apparatus, the method of Glynn and Kendal[4] is recommended, since analysis time is reported to be 15 minutes. The lower limit of detection is 100 μg/mL, which is a drawback, but few substances interfere. Meola's method,[6] using Folin-Ciocalteu reagent, is not recommended because salicylate, tyrosine, tryptophan, and uric acid all react with this reagent. The differential absorption technique is also not recommended, since acetaminophen has a low molar extinction coefficient ($\epsilon_{249\ nm}$ = 13,500), making the method insensitive.

GLC is a method of choice in any laboratory not yet equipped with HPLC. Prescott's method[9] is sensitive to 1 μg/mL, and the total assay time is about 45 minutes. The limiting factor is the time it takes the ethyl acetate to evaporate. The other GLC methods are modifications of Prescott's method and offer no significant advantages.

HPLC is the preferred method for analysis of acetaminophen. Blair and Rumack's method[12] can quantitate the drug on aliquots as little as 1.5 μL, but analysis time is 50 minutes per sample. The silica absorption method of Fletterick et al.[13] requires 15 minutes per sample and a 1 ml sample size and demonstrates 97% recovery. Sensitivity is 5 μg/mL, and only caffeine and sulfamethoxazole interfere. The reversed-phase method of Howie et al.[14] is favored. There are few interferences, and the limit of sensitivity is 1 μg/mL. The other reversed-phase methods require extraction into ethyl acetate and evaporation, which is time consuming.

A comparison of the conditions for several of these assays is presented in the following chart:

Comparison of reaction conditions for acetaminophen assays

Condition	EMIT*	GLC†	HPLC†
Temperature	30° C	220° C	Ambient
pH	8.0	—	—
Sample volume	50 μL	1 mL	0.1 mL
Fraction of sample volume	0.01	1.0	0.33
Precision	<10%	<10%	5%
Linearity	10-200 μg/mL	20-200 μg/mL	5-300 μg/mL
Interferences	None	None	Cefoxitin Loxitane Mellaril (all interfere with internal standard)

*Enzyme-multiplied immunoassay technique, Syva Co., Inc., Palo Alto, Calif.

†Methods described in text. *GLC*, gas-liquid chromatography; *HPLC*, high-pressure liquid chromatography.

For the future development of HPLC, electrochemical detectors offer promise because of the extreme sensitivity (0.1 μg/mL). At present, this type of detector may be recommended for pharmacokinetic studies.

SPECIMEN

Analysis for acetaminophen using HPLC or GLC techniques can be performed on serum or plasma. No special handling of specimens is required.

PROCEDURE FOR HIGH-PERFORMANCE LIQUID CHROMATOGRAPHY

Principle

This method involves reversed-phase high-performance liquid chromatography of serum after deproteinization with acetonitrile. Detection is by ultraviolet spectroscopy at 254 nm. (Method developed by Waters Associates, Inc., Milford, MA 01757.)

Reagents

Acetic acid reagent (5 mL/L, 88 mmol/L). Dilute 5 mL of glacial acetic acid to 1 L with glass-distilled water. This is stable for 1 year at room temperature.

Acetonitrile and methanol. Ultraviolet grade.

Stock internal standard (1 g/L, 4.8 mmol/L). Place 100 mg of β-8-hydroxyethyltheophylline in a 100 mL volumetric flask and dilute to 100 mL with methanol. This standard is stable for 6 months at 4° C.

Working internal standard (100 mg/L, 480 μmol/L). Dilute 10 mL of stock internal standard to 100 mL with acetonitrile. This is stable for 3 months at 4° C.

Acetaminophen stock standard (1 g/L, 6.61 mmol/L). Place 10 mg of acetaminophen (USP) in a 10 mL volumetric flask and dilute to the mark with methanol. This standard is stable for 1 week at 4° C.

Working acetaminophen calibration standard (50 μg/mL, 330.5 μmol/L). Dilute 5 mL of stock acetaminophen standard to 100 mL with pooled serum.

Acetaminophen standard curve. To 0.5 mL aliquots of pooled serum, add 5, 10, 25, 50, 75, 100, 200, and 300 μL of stock standard and bring to 1 mL with serum. These represent concentrations of 5, 10, 25, 50, 75, 100, 200, and 300 mg/L. Obtain calibration curves with each new column and also periodically to ensure proper quality control.

Mobile phase, acetonitrile–acetic acid (10:90 by volume). Filter 450 mL of acetic acid reagent using a nitrocellulose filter (such as Millipore) of 0.47 μm pore size. Into the acetic acid filter 50 mL of acetonitrile using an organic filter of the same pore size. Degas solution for 15 minutes at room temperature by applying a vacuum.

Assay

Equipment: high-performance liquid chromatography equipment (injection, pump, recorder), including an ultraviolet spectrophotometer capable of reading at 254 nm.

Column is a μBondapak C-18 from Waters Association Inc., Milford, MA 01757.

1. Sample preparation.
   a. Pipet 0.2 mL of working internal standard solution into a 12 × 75 mm glass tube.
   b. Add 0.10 mL of serum.
   c. Vortex and centrifuge 2 minutes at ambient temperature. Save supernatant as sample for high-performance liquid chromatographic analysis.
   d. Prepare calibrator by mixing 0.2 mL of internal standard working solution with 0.10 mL of working acetaminophen calibration standard (50 μg/mL).
2. Chromatographic parameters.
   a. Chart speed: 1.0 cm/min
   b. Ultraviolet sensitivity: 0.02 AUFS (absorbance units at full scale) at 254 nm
   c. Flow rate: 2.0 mL/min at ambient temperature
   d. Run time: Approximately 9 minutes
3. Inject 10 μL of sample or standard.

Calculation

1. Determine the peak height (in millimeters) for the acetaminophen (A) and internal standard (S) peaks on each chromatogram.
2. Calculate peak height ratio, acetaminophen/standard (A/S).
3. Calculate the acetaminophen concentration in the serum samples as follows:

$$\text{Acetaminophen } (\mu g/mL) = \frac{\text{A/S (sample)}}{\text{A/S (calibrator)}} \times 50 \ \mu g/mL$$

NOTE: Because of interference by theobromine, the lower limit of sensitivity is 5 μg/mL.

PROCEDURE FOR GAS-LIQUID CHROMATOGRAPHY

Principle

Acetaminophen is extracted from serum into ethyl acetate, evaporated, derivatized with acetic anhydride, and then analyzed on a gas-liquid chromatograph equipped with a 3% OV-17 column and a flame ionization detector (FID). (Modified from Prescott[9] by F.M. Hassan, Cincinnati).

Reagents

Acetaminophen stock standard (1 g/L, 6.61 mmol/L). Mix 10 mg of acetaminophen (USP) in 10 mL of methanol. This standard is stable for 1 week at 4° C.

Internal standard (1 g/L, 6.0 mmol/L). Mix 20 mg of *N*-butyryl-*p*-aminophenol (from Pfaltz and Bauer, Flushing, N.Y.), in 20 mL of methanol. This is stable for 6 months at 4° C.

Ethyl acetate. Spectroquality.

Pyridine. Silylation grade (from Matheson, Coleman and Bell, Norwood, Ohio). Store at room temperature. This is stable for 2 years.

Acetic anhydride. Chromaquality (from Matheson, Coleman and Bell, Norwood, Ohio). Store at room temperature. This is stable for 2 years.

Phosphate buffer, pH 8.0 (0.1 mol/L). Dilute 13.4 g of Na_2HPO_4 to 500 mL with distilled water (0.1 M Na_2HPO_4). Dilute 6.8 g of KH_2PO_4 to 500 mL with distilled water (0.1 M KH_2PO_4). Both solutions are stable for 1 year at refrigerated temperature. To make a pH 8.0 phosphate buffer, mix 9.5 mL of Na_2HPO_4 solution with 0.5 mL of KH_2PO_4 solution. Make fresh daily.

Acetaminophen standard curve. To 0.5 mL aliquots of pooled serum, add 5, 10, 25, 50, 75, 100, 200, and 300 μL of stock standard and bring to 1 mL with serum. These represent concentrations of 5, 10, 25, 50, 75, 100, 200, and 300 mg/L. Obtain standard curves with each new column and with each batch of internal standard and also periodically to ensure proper quality control.

Assay

Equipment: gas-liquid chromatograph equipped with a flame ionization detector (FID)

1. Sample preparation
   a. To a 10 mL screw-capped tube, add 1 mL of serum, 1 mL of pH 8.0 phosphate buffer, and 100 μL of internal standard and vortex.
   b. Extract with 5 mL of ethyl acetate for 5 minutes. Centrifuge at 1000 *g* (3000 rpm) for 5 minutes at room temperature. Transfer ethyl acetate to an evaporation vial.
   c. Evaporate to dryness in 56° C dry bath under vacuum. Add 10 μL of pyridine and 20 μL of acetic anhydride and vortex.
   d. Inject 1 μL of reaction mixture into the gas chromatograph.
2. GLC parameters
   a. Column: 3% OV-17 column
   b. Detector: flame ionization
   c. Nitrogen flow: 24 mL/min
   d. Column temperature: 220° C
   e. Injector temperature: 250° C
   f. Detector temperature: 300° C
   g. Chart speed: 0.5 cm/min

Calculations and notes

1. Compute peak-height ratio of acetaminophen versus internal standard in standards and unknown.

$$\text{Peak-height ratio} = \frac{\text{Peak height of acetaminophen (A)}}{\text{Peak height of internal standard (S)}}$$

2. Calculate standard curve from peak-height ratio of standards to internal standard and calculate unknown by comparing peak height ratio of unknowns/internal standard to standard curve.
3. Limit of sensitivity is 1 μg/mL.
4. There are no known interfering substances.

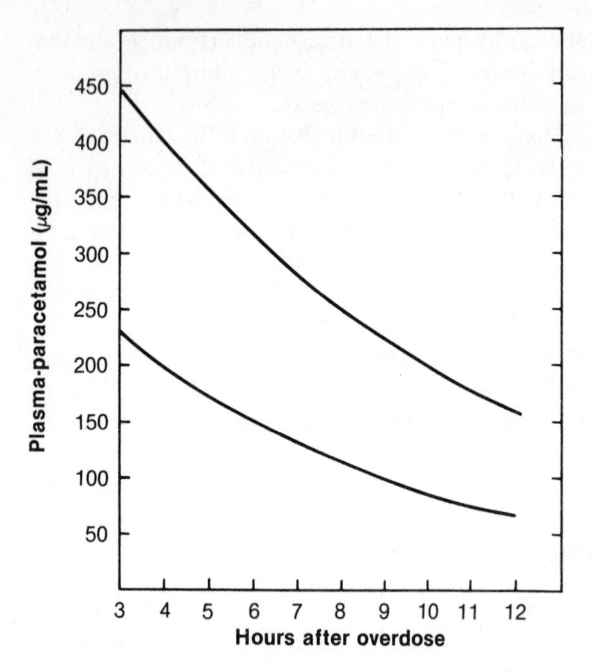

Fig. 61-1. Plasma acetaminophen concentration in relation to time after overdosage. Liver damage is likely to be severe above upper line, severe to mild between lines, and clinically insignificant under lower line. (From Prescott, L.F., Park, J., Sutherland, G.R., et al.: Lancet **2:**109-113, 1976.)

REFERENCE RANGE

Optimal therapeutic concentration of acetaminophen in serum is in the range of 10 to 20 μg/mL. A nomogram of plasma acetaminophen levels versus hours after ingestion is used to estimate probable liver damage (Fig. 61-1).

REFERENCES

1. Lester, L., and Greenberg, L.A.: The metabolic rate of acetanilid and other aniline derivatives, J. Pharmacol. Exp. Ther. **90:**68-75, 1947.
2. Brodie, B.B., and Axelrod, J.: The estimation of acetanilid and its metabolic products, aniline, *N*-acetyl-*p*-aminophenol and *p*-aminophenol (free and total conjugated), in biological fluids and tissues, J. Pharmacol. Exp. Ther. **94:**22-28, 1948.
3. Heirwegh, K.P.M., and Fevery, J.: Determination of unconjugated and total *N*-acetyl-*p*-aminophenol (NAPA) in urine and serum, Clin. Chem. **13:**215-219, 1967.
4. Glynn, J.P., and Kendal, S.E.: Paracetamol measurement, Lancet **1:**1147-1148, 1975.
5. Frings, C.S., and Saloom, J.M.: Colorimetric method for the quantitative determination of acetaminophen in serum, Clin. Toxicol. **15:**67-73, 1979.
6. Meola, J.M.: Emergency determination of acetaminophen, Clin. Chem. **24:**1642-1643, 1978.
7. Routh, J.I., Shane, M.A., Arredondo, E.G., and Paul, W.D.: Determination of *N*-acetyl-*p*-aminophenol in plasma, Clin. Chem. **14:**882-889, 1968.
8. Knepil, J.: A sensitive, specific method for measuring *N*-acetyl-*p*-aminophenol (paracetamol) in blood, Clin. Chem. Acta **52:**369-372, 1974.
9. Prescott, L.F.: The gas-liquid chromatographic estimation of phenacetin and paracetamol in plasma and urine, J. Pharm. Pharmacol. **23:**111-115, 1971.
10. Thomas, B.H., and Coldwell, B.B.: Estimation of phenacetin and paracetamol in plasma and urine by gas-liquid chromatography, J. Pharm. Pharmacol. **24:**243, 1972.
11. Dechtiaruk, W.A., Johnson, G.F., and Solomon, H.M.: Gas-chromatographic method for acetaminophen (*N*-acetyl-*p*-aminophenol) based on sequential alkylation, Clin. Chem. **22:**879-883, 1976.
12. Blair, D., and Rumack, B.H.: Acetaminophen in serum and plasma estimated by high-pressure liquid chromatography: a micro-scale method, Clin. Chem. **23:**743-745, 1977.
13. Fletterick, C.G., Grove, T.H., and Hohnadel, D.C.: Liquid-chromatographic determination of acetaminophen in serum, Clin. Chem. **25:**409-412, 1979.
14. Howie, D., Adriaenssens, P.I., and Prescott, L.F.: Paracetamol metabolism following overdosage: application of high performance liquid chromatography, J. Pharm. Pharmacol. **29:**235-237, 1977.
15. Horvitz, R.A., and Jatlow, P.I.: Determination of acetaminophen concentrations in serum by high-pressure liquid chromatography, Clin. Chem. **23:**1596-1598, 1977.
16. Gotelli, G.R., Kabra, P.M., and Marton, L.J.: Determination of acetaminophen and phenacetin in plasma by high-pressure liquid chromatography, Clin. Chem. **23:**957-959, 1977.
17. Lo, L.Y., and Bye, A.: Rapid determination of paracetamol in plasma by reversed phase high-performance liquid chromatography, J. Chromatogr. **173:**198-201, 1979.
18. Riggin, R.M., Schmidt, A.L., and Kissinger, P.T.: Determination of acetaminophen in pharmaceutical preparations and body fluids by high-performance liquid chromatography with electrochemical detection, J. Pharm. Sci. **64:**680-683, 1975.
19. Munson, J.W., Weierstall, R., and Kostenbauder, H.B.: Determination of acetaminophen in plasma by high-performance liquid chromatography with electrochemical detection, J. Chromatogr. **145:**328-331, 1978.

Alcohol screen

TIMOTHY J. SCHROEDER

Clinical significance: pp. 594 and 897

Molecular formula: C_2H_5OH

```
  H H
HC–C–OH
  H H
```

Molecular weight: 46.07 daltons
Merck Index: 210, 211, 212, 213
Chemical class: carbohydrate

PRINCIPLES OF ANALYSIS AND CURRENT USAGE

Alcohol analysis is frequently requested by the hospital emergency room to establish this agent as a possible cause of a number of pathophysiological changes, including coma. In most cases the analyses must be rapid, and because of medicolegal requirements, they must also be accurate. In addition, in many instances the physician must also know the nature of the alcohol ingested (ethanol, methanol, isopropanol, ethylene glycol), since this can affect care. Techniques for the determination of alcohol include diffusion or distillation, followed by oxidation of the alcohol and osmometric, enzymatic, and gas-chromatographic procedures.

Early techniques for blood alcohol determination used distillation, aeration, or diffusion to separate the alcohol from the plasma matrix. The distilled alcohol was measured by oxidation of the alcohol by strong oxidizing agents. The concomitant reduction of the dichromate, permanganate, or osmic acid oxidizing agents resulted in

Table 61-2. Methods for alcohol analysis

Method	Type of analysis	Principle	Usage	Comments
1. Distillation-oxidation	Colorimetric	Alcohol diffuses into gas phase and reacts with oxidizing agent, changing its color $2K_2Cr_2O_7 + 10H_2SO_4 + 3C_2H_5OH \rightarrow$ (yellow-orange) $2Cr_2(SO_4)_4 + 2K_2SO_4 + 3CH_3COOH + 11H_2O + 4H^+$ (blue-green)	Stat. or routine All body fluids Tissue	Nonspecific Gives reaction with all volatiles
2. Osmometry	Freezing point	Alcohol in high concentration increases serum osmolality; difference between normal and measured value proportional to alcohol levels	Stat. Serum	Nonspecific
3. Enzymatic	Spectrophotometric	$NAD^+ + \text{Alcohol} \xrightarrow{\text{Alcohol dehydrogenase}} NADH + H^+ + \text{Acetaldehyde}$ (NAD, Nicotinamide adenine dinucleotide; NADH, reduced NAD^+)	Stat. or routine Serum	Specific for ethanol; other alcohols not readily measured
4. Gas chromatography	Flame ionization	Alcohol separates on chromatographic column	Stat. or routine All body fluids Tissue	Specific for all alcohols

color changes and was used to monitor the reaction (method 1, Table 61-2). For example, it was observed that when alcohol was placed in a strongly acidic potassium dichromate solution, which was yellow-orange, the dichromate was reduced to a blue-green chromous ion solution, whereas the alcohol was oxidized to products such as acetadehyde, acetic acid, and carbon dioxide and water. The extent of oxidation depended on the reaction conditions.

The Widmark procedure first described in 1922 used a simultaneous distillation and oxidation procedure.[1] The method of Williams et al.[2] used the Conway diffusion dish for distillation of the alcohol. In this procedure a solution of 18 N sulfuric acid–potassium dichromate was placed in the center compartment of the Conway dish. The sample to be tested was placed in the outer ring. The diffusion of alcohol into the acid oxidizing solution was allowed to proceed for 1 hour at 55° to 60° C. The center compartment fluid was made to volume with a 1% brucine buffer solution in 10% sulfuric acid. The blue color of the reduced dichromate was read in a spectrophotometer at 425 nm. The coefficient of variation was reported to be ± 2%.

An osmometric method of analysis for alcohols was first introduced by Redetzki et al. in 1972.[3] This method was based on the increase of blood osmolality by alcohol. Serum osmolality was determined from 2 mL samples by measurement of the increase in the freezing-point depression caused by the significant molal concentration of alcohol in those patients made toxic because of the presence of this agent (method 2, Table 61-2). The contribution of alcohol to the serum osmolality was found to be directly related to its concentration in the blood. These authors found a correlation between true alcohol levels, as determined enzymatically, and those computed from the measurement of the osmolality on the same sample. The alcohol concentration was observed to be proportional to the difference between the osmolality in the presence of alcohol (patient's sample) and the normal serum osmolality occurring in the absence of alcohol, rather than proportional to the absolute osmolality alone. The average normal osmolality of serum (no alcohol) was considered to be 290 mOsm/kg of body water.[4] Conversion of serum osmolality into ethanol concentration was accomplished by use of the equation:

$$([\text{Measured osmolality} - 290] \times 46.07 \text{ g/mol} \times 0.92) = \text{Milligrams of ethanol per liter of serum}$$

where 46.07 g/mol is the molecular weight of ethanol, 290 is the normal serum osmolality, and 0.92 is the correction for the volume fraction of water in serum.

The equation can be reduced to yield:

$$\Delta \text{ osmolality} \times 42.4 = \text{Ethanol (mg/L)}$$

Since the alcohols are volatile and contribute significantly to the vapor pressure above a solution, vapor pressure osmometers cannot be used to estimate blood alcohol.

Enzymatic methods employ alcohol dehydroganse, which reacts primarily with ethanol but not with methanol or acetone. The enzyme does have some reactivity with propanol and butanol. The reaction results in the oxidation of alcohol to acetaldehyde, with consequent reduction of the coenzyme nicotinamide adenine dinucleotide (NAD) to reduced NAD (NADH) (method 3, Table 61-2). The reduced coenzyme may be measured directly at 340 nm or indirectly by a color generated by a coupled-dye system. Roos reported the use of the tetrazolium salt INT (iodonitrotetrazolium) coupled with the hydrogen-ion phenazine methosulfate in forming a colored formazan derivative that is stable between pH 4.0 and 5.0. The formazan derivative has an absorbance maximum at 495 to 500 nm.[5]

Gas-chromatographic analyses (method 4, Table 61-2) of alcohol have used extraction, distillation, and head-space procedures to obtain samples free of the blood matrix. Methods also exist for the direct injection of blood into the chromatograph. In 1958 Cadman and Johns reported a procedure for the separation and quantitation of alcohol and alcohol-related compounds using a gas chromatograph equipped with a thermal conductivity detector.[6] The alcohol was extracted from blood into *n*-propylacetate.

In 1964 Goldbaum et al., using a gas chromatograph equipped with a beta-particle ionization microdetector, introduced the first head-space gas-chromatographic procedure.[7] The method utilized an air sample removed from a confined space above the blood or other biological sample in a closed container. The samples were collected in containers stoppered with a puncture type of rubber cap. Both the unknowns and standards were placed in a water bath at 25° C until the samples were ready. An airtight syringe was filled with 1 mL of air, with the needle of the syringe inserted through the rubber cap and moved up and down several times to obtain an equilibrated air sample. The air sample was then introduced into the chromatograph. Several modifications of this method have been introduced.

In 1962, Machata used a Perkin-Elmer chromatograph equipped with a flame ionization detector for analysis using direct sample injection.[8] In this procedure quantitation was accomplished by pipetting of a known amount of acetone, which served as an internal standard, with the alcohol. A modification of the method of Jain[9] is considered to be the most popular gas-liquid chromatographic procedure for alcohol analysis. The procedure includes the use of a flame ionization detector and a 20% Carbowax column. Injection temperature is at 100° C, and the injection port is kept at 160° C.

REFERENCE AND PREFERRED METHODS

Reference and preferred methods for alcohol determination are intertwined with the medicolegal problems caused by the findings of alcohol levels above certain legal amounts. Therefore in many states there are legal specifications defining the methods by which alcohol levels must be done. Many laboratories engaged in medicolegal analysis must follow these specified methods.

The distillation or diffusion methods for measurement of alcohol take advantage of its volatility. These methods lack specificity, since other oxidizable compounds, including other alcohols, volatiles, and ketones, may also be distilled into and react in the reaction mixture. The sealed diffusion flask of Widmark and the Conway diffusion dish are simple apparatuses that can give rapid, reproducible, quantitative answers. The Conway system, because of its simplicity and low cost, is preferred as a rapid, qualitative stat procedure. The dichromate methods are precise, and equivalent results can be achieved by use of either an iodimetric determination of the residual dichromate or a direct colorimetric measurement of the color change of the dichromate. The Conway diffusion method uses reagent with a long shelf life (more than 1 year) and can be set up in a few minutes using whole blood. In addition, a qualitative answer can be obtained in 20 to 30 minutes. A true quantitative transfer of the alcohol by diffusion process requires at least 18 hours of incubation time at room temperature, which reduces its use as a stat procedure.

One should note that the osmolality method only provides a rapid estimate of blood level alcohols in the range of 1000 to 5000 mg/L. This method is nonspecific; any condition that changes the osmolality changes the estimation of alcohol. Significant error could arise from subjects suffering from diabetic acidosis, uremia, or hypernatremia. It has been reported that values derived from osmolality were approximately 195 mg/L higher than those obtained from the enzymatic method. However, the alcohol concentration calculated from the osmolality was proportional to that determined enzymatically over the same range of 1000 to 5000 mg/L.[4]

The enzymatic methods are fairly specific; however, there is some reaction with other alcohols. This specificity, however, does not allow for the discrimination if other alcohols are present; thus the use of this method is limited to only those patients who have taken ethanol. In terms of simplicity, the preferred enzymatic system monitors only the formation of NADH, as opposed to those systems that monitor the reduction of dyes. The additional steps in these procedures leave open the possibility of additional errors caused by reagent instability and so on.

The enzymatic methods permit the analysis of a large number of samples with rapidity and precision. For example, in a batch mode, 20 samples may be analyzed on an ABA (Abbott, Automatic Bichromic Analysis Instrument) with a coefficient of variation of 5%. In addition, the procedure is similar to that employed for many other enzymatic methods and can be performed on equipment already available in a laboratory, such as centrifugal analyzers and automated spectrophotometers. These procedures offer adequate sensitivity for medically significant measurements. For those laboratories with spectrophotometers capable of accurate measurement at 340 nm, the enzymatic methods may be a practical answer to analysis for ethanol.

The ability to separate, identify, and quantify each type of alcohol that may be present in a patient makes gas-liquid chromatography the preferred technique for the determination of alcohol. The method is very rapid, yielding results within 10 to 15 minutes. However, it is not suited to the rapid measurement of large numbers of samples; since alcohol measurements are rarely performed in large numbers of specimens, this capacity is not a limiting factor. Flame ionization detectors are recommended rather than the less expensive thermal detectors. A most important consideration is that one can use direct sample injection or vapor-phase samples in a flame ionization detector, but thermal detectors usually require that samples be extracted. For the most accurate gas-chromatographic analysis, internal standards must be used.

SPECIMEN

Ethanol swabs should not be used to clean the site of the venipuncture. Specimens must be kept well stoppered and preferably refrigerated to prevent loss of ethanol. For the distillation or diffusion procedure, any body fluid or tissue may be tested. For the enzymatic procedure, plasma, serum, or urine may be used, though serum is preferred. For gas chromatography, any tissue or body fluid may be used.

PROCEDURE

Principle

Volatile compounds such as ethanol and methanol separate from serum in the heated injection port. They are separated by chromatography and are quantitated by peak-height ratio to an internal standard (Table 61-3).

Reagents

Ethanol stock standards. Dilute 100, 200, 300, and 400 μL of 100% ethanol to 100 mL with distilled water. This yields stock standards of 789 μg/mL (17.15 mmol/L), 1578 μg/mL (34.30 mmol/L), 2367 μg/mL (51.45 mmol/L), and 3156 μg/mL (68.60 mmol/L). This is stable for 1 month at room temperature in a stoppered container.

***N*-Propanol internal standard, 3120 μg/mL (51.92 mmol/L).** Dilute 400 μL of *n*-propanol to 100 mL with distilled water. This is stable for 1 year at room temperature in a stoppered container.

Standard serum 789 μg/mL (17.15 mmol/L). Add 100 μL of 100% ethanol to 100 mL of pooled serum and mix well. Store 1 mL aliquots at −20° C. This is stable for 1 year.

Assay

Equipment: gas-liquid chromatograph with flame ionization detector; 6-foot column containing 0.2% Carbowax 1500 on 60/80 Carbopack.

1. Sample preparation. Pipet 200 μL of internal standard into 12 × 75 mm glass tube. Add 200 μL of serum and mix.
2. Chromatographic parameters. Injection temperature, 210° C; column temperature, 100° C detector temperature, 260° C; gas flow at 60 mL/min, with the proportions of nitrogen/air/hydrogen at 40:10:20.
3. Inject 0.5 μL of sample or standard into chromatograph. Chromatographic run time is approximately 6 minutes.

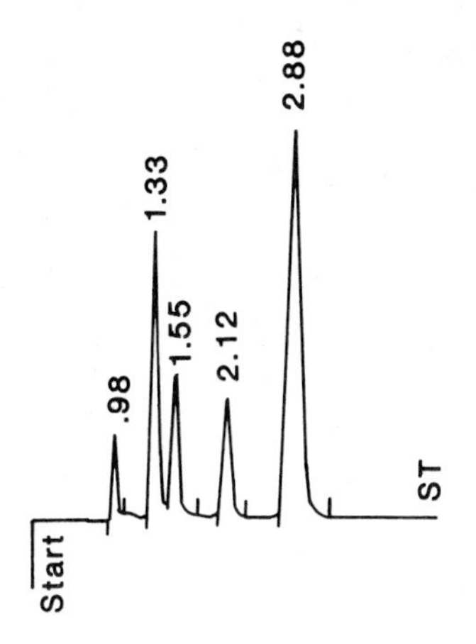

Fig. 61-2. Gas chromatographic tracing of a series of alcohol standards. Retention time in minutes is recorded next to peak height. Alcohols are eluted in following time sequence in minutes: 0.98, methanol; 1.33, ethanol; 1.55, acetone; 2.12, isopropanol; 2.88, *n*-propanol (internal standard). *ST,* Stop.

Calculation

1. Determine the peak height (in millimeters) for the ethanol (E) and internal standard (S) peaks on each chromatogram. (See Fig. 61-2 for sample chromatogram.)
2. Calculate peak-height ratios, E/S.
3. Calculate ethanol in serum samples as follows:

$$\text{Ethanol } (\mu g/mL) = \frac{\text{E/S (sample)}}{\text{E/S (standard serum)}} \times 789\ \mu g/mL$$

Notes

1. Anhydrous alcohols rapidly absorb water. Therefore, once opened, an anhydrous bottle of alcohol cannot be considered as accurate. For this reason, some laboratories use 95% alcohol, which is stable.
2. Isopropanol, methanol, and acetone can also be quantified by this procedure. The same peak-height ratio internal standard and chromatographic conditions can be used.
3. Standards
   a. Isopropanol, 1570 μg/mL (26.12 mmol/L). Add 200 μL of 100% isopropanol to a 100 mL flask and dilute to volume with distilled water.
   b. Acetone, 1576 μg/mL. Prepare in the same manner as isopropanol, using pure acetone.
   c. Methanol, 1582 μg/mL (46.4 mmol/L). Prepare in the same manner as isopropanol, using pure methanol.
4. To calculate serum values, use the formula:

$$\text{Concentration} = \frac{\text{Peak-height ratio (unknown)}}{\text{Peak-height ratio (standard)}} \times \text{Concentration of standard}$$

Table 61-3. Comparison of assay conditions for ethanol

Parameter	Enzyme reaction	Gas chromatography
Temperature	37° C	100°
pH	8.8	7.35
Final concentration of reagent components	0.6 μmol/mL reduced nicotinamide adenine dinucleotide 50 U/mL alcohol dehydrogenase 0.075 mol/L tetrasodium pyrophosphate 0.075 mol/L sodium semicarbazide 0.022 mol/L glycine	1559 μg/mL (25.96 mmol/L) *n*-Propanol
Fraction of sample volume	Approximately 0.008	0.5
Sample volume	0.5 mL	0.1 mL (0.5 μL injection)
Linearity	0-3000 μg/mL (65.1 mmol/L)	0-3000 μg/mL (65.1 mmol/L)
Precision*	5%	9%
Time of reaction	20 min	8 min
Interferences	Other alcohols	None known

*Coefficient of variation; obtained from College of American Pathologists toxicology survey data.

REFERENCE RANGE

A value of 1000 μg/mL (21.7 mmol/L) is considered legally intoxicated in many states.

A value of 3000 μg/mL (65.3 mmol/L) is associated with coma.

REFERENCES

1. Widmark, E.M.P.: Scand. Arch. Physiol. **33**:85, 1918; **35**:128, 1918.
2. Williams, L.A., Linn, R.A., and Zak, B.: Determinations of ethanol in fingertip quantities of blood, Clin. Chim. Acta **3**:169, 1958.
3. Redetzki, H.M., Koerner, T.A., Hughes, J.R., and Smith, A.G.: Osmometry in the evaluation of alcohol intoxicants, Clin. Toxicol. **5**(3):343, 1972.
4. Cravey, R.H., and Jain, N.C.: Current status of blood alcohol methods, J. Chromatogr. Sci. **12**:209, 1974.
5. Roos, K.J.: Rapid, sensitive and inexpensive method for estimating blood and urine alcohol concentrations, Clin. Chim. Acta **31**:285, 1971.
6. Cadman, W.J., and Johns, T.: Paper presented at 9th Annual Conference on Analytical and Chemical Applications in Spectrometry, Pittsburgh, 1958.
7. Goldbaum, L.R., Domanski, T.J., and Schloegel, E.L.: Analysis of biological specimens for volatile compounds by gas chromatography, J. Forensic Sci. **9**:63, 1964.
8. Machata, G.: Diphorase method of blood alcohol determination, Mikrochim. Acta **6**:91, 1962.
9. Jain, N.C.: Direct blood-injection method for gas chromatographic determination of alcohols and other volatile compounds, Clin. Chem. **17**:82, 1971.

Anticonvulsant drugs

STEVEN J. SOLDIN

Clinical significance: pp. 718 and 897

Molecular structure:

Drug:	**Valproic acid**	**Phenobarbital**	**Phenytoin**
Molecular formula:	$C_8H_{16}O_2$	$C_{12}H_2N_2O_3$	$C_{15}H_{12}N_2O_2$
Molecular weight (daltons):	144.21	232.23	252.26
Merck Index:	9574	7032	7130
Chemical class:	Carboxylic acid	Barbiturate	Hydantoin

Drug:	**Carbamazepine**	**Primidone**	**Ethosuximide**
Molecular formula:	$C_{15}H_{12}N_2O$	$C_{12}H_{14}H_2O_2$	$C_7H_{11}NO_2$
Molecular weight (daltons):	236.26	218.25	141.17
Merck Index:	1781	7542	3674
Chemical class:	Dibenzazepine-5-carboxamide	Pyrimidinedione	Succinamide

PRINCIPLES OF ANALYSIS AND CURRENT USAGE

Most anticonvulsant drugs have specific ultraviolet spectral characteristics. These characteristics were exploited in the initial attempts to monitor the therapeutic levels of these drugs.[1,2] The drugs were extracted into an organic solvent; then they were reextracted into an aqueous solution or the absorbance of the solution was recorded (method 1, Table 61-4). By using the appropriate solvents, the drugs could be extracted with relative specificity. By recording the difference spectra of the extract at two pH values, it was also possible to increase specificity.

A second method for measuring anticonvulsant drugs involved extraction and separation by thin-layer chromatography.[3-5] Quantitation was achieved by either ultraviolet scanning of the plate or by elution and recording the ultraviolet absorption (method 2, Table 61-4).

Gas chromatographic techniques (method 3, Table 61-4) were for many years the primary technique used to analyze antiepileptic drugs in biological fluids.[6-11] MacGee[6] developed an on-column methylation technique that resulted in thermally stable and easily volatilized derivatives with quantifiable recovery. Initially, flame ionization detection was the system of choice for quantification of the eluting drug. Current gas chromatographic techniques prepare derivatives of anticonvulsant drugs. Methylation is the most commonly used procedure,[6,12] and in the method described by Solow and Green, seven antiepileptic drugs may be analyzed simultaneously.[12] Considerable interest has been shown for the nitrogen flame ionization detection system. Because almost all antiepileptics except valproic acid consist of compounds containing nitrogen, this approach appears to offer increased specificity and sensitivity.

High-performance liquid chromatography assays (method 4, Table 61-4) involve protein precipitation and resolution of the drug on a column.[13-21] Detection is by

Table 61-4. Methods for anticonvulsant analysis

Method	Principle	Usage	Comment
1. Extraction and ultraviolet spectroscopy	Anticonvulsant extracted free of other ultraviolet-compounds; ultraviolet spectrum or the ultraviolet difference spectra are specific for the drug	Serum	Interference with drugs and endogenous compounds
2. Thin-layer chromatography (TLC)	Extracted anticonvulsant chromatographed and its retardation factor noted; eluted quantitatively and measured by ultraviolet spectroscopy	Serum	Difficult to separate all anticonvulsants Technically difficult to achieve reproducibility
3. Gas chromatography (GC)	Extracted drugs converted to methyl derivatives: $(CH_3)_4N^+OH^-$ + Anticonvulsants → Methyl derivatives Retention time identifies specific drug	Blood, plasma, serum, saliva	Can resolve all anticonvulsants simultaneously Requires extraction and derivatization, therefore technically difficult
4. High-performance liquid chromatography (HPLC)	Protein precipitated from solution and anticonvulsants separated by their affinity for column; monitored by ultraviolet spectroscopy or electrochemical detection	Blood, plasma, serum, saliva	Can resolve all anticonvulsants except valproate simultaneously
5. Competitive-binding assays			
a. Enzyme-multiplied immunoassay technique (EMIT)	Competitive binding with drug attached to enzyme	Plasma, serum, saliva	Can measure all drugs, but each must be done separately
b. Radioimmunoassay (RIA)	Competitive binding of radioactive label; antibody binding to radiolabeled hapten competes for unknown	Plasma, serum, saliva	All problems of radioactive usage; slower than other immunoassays
c. Substrate-labeled fluorescence immunoassay (SLFIA)	Competitive binding substrate label; antibody binding substrate competes for unknown antigen	Plasma, serum, saliva	Fluorescence interferes New method
d. Fluorescence polarization immunoassay (FPIA)	Competitive binding with fluorescent-labeled drug; antibody binding to fluorescent drug competes for unknown	Plasma, serum, saliva	Fluorescence interferes
e. Rate nephelometric inhibition immunoassay	Competitive binding for hapten; antibody binding to hapten-protein competes for unknown	Plasma, serum	Each drug individually assayed

ultraviolet absorption, but electrochemical detectors may also be used. Most methods use reversed-phase liquid chromatography on a C-18 type of column.

Immunological procedures (methods 5a to 5e, Table 61-4) developed for measurement of the anticonvulsants include radioimmunoassay (RIA), enzyme immunoassay (EIA), fluorescence immunoassay (FIA), fluorescence polarization immunoassay (FPIA), and the nephelometric inhibition immunoassay (NIIA). The RIA procedures involve the standard method of displacing the labeled drug with that from the sample.[22]

The introduction of the enzyme-multiplied immunoassay technique (EMIT, Syva Co., Inc., Palo Alto, Calif.) into the clinical chemistry laboratory has been another of the success stories of the 1970s. Homogeneous enzyme immunoassay is based on competitive protein binding, with an enzyme as the label and an antibody as the binding protein. When the enzyme-labeled drug becomes bound to an antibody of the drug, the activity of the enzyme is reduced. Free drug in the sample competes with the enzyme-labeled drug for the antibody and thereby decreases the antibody-induced inactivation of the enzyme. The enzyme activity then correlates with the concentration of the free drug.

The Ames (Elkhart, Ind.) fluorescent immunoassay for anticonvulsant drugs uses the principles of a competitive protein binding to measure levels of these drugs in serum or plasma.[23] The anticonvulsant is labeled with a derivative of the fluorogenic enzyme substrate umbelliferyl-β-D-galactoside. This fluorogenic drug reagent (FDR) is nonfluorescent under the conditions of the assay; however, hydrolysis catalyzed by β-galactosidase yields a fluorescent product. When antibody to the anticonvulsant reacts with FDR, the FDR is virtually inactive as a substrate for the β-galactosidase. Competitive binding reactions are set up with a constant amount of FDR, a limiting amount of antibody, and a patient sample containing the anticonvulsant:

$$\text{FDR} + \text{Antibody} + \text{Anticonvulsant} \rightarrow \text{Antibody-FDR} + \text{Antibody-Anticonvulsant} + \text{FDR} + \text{Anticonvulsant}$$

$$\text{FDR} \xrightarrow{\beta\text{-Galactosidase}} \text{Fluorescent dye} + \text{Galactose}$$

The anticonvulsant in the sample competes with the FDR for the limited number of antibody binding sites. The FDR not bound to antibody is hydrolyzed by β-galactosidase to produce the fluorescent product. Therefore the fluorescence produced is proportional to the anticonvulsant concentration in the sample. The intensity of fluorescence is related to the serum or plasma anticonvulsant concentration by means of a standard curve.

Fluorescence polarization immunoassay (FPIA) uses the principles of a competitive binding assay and measures tracer binding directly by fluorescence polarization. FPIA (Abbott Laboratories, North Chicago) combines the specificity of an immunoassay with the speed and ease of a homogeneous method to provide a reliable procedure for the measurement of anticonvulsants. One makes the fluorescent tracer by covalently attaching fluorescein to the drug being analyzed. When excited with polarized light, this tracer will emit light with a degree of polarization that is inversely proportional to its rate of rotation. The slower the trace molecule rotates, the greater the polarization of the emitted light. When a specific antibody binds to the tracer, its rotation is slowed and the emitted light increases in polarization. The principles of FPIA have been used on the Abbott TDX to provide reliable procedures for the measurement of anticonvulsant drugs.[24]

Rate nephelometric inhibition immunoassay is the procedure employed by the Immunochemistry Systems (ICS) of Beckman Instruments, Inc. (Fullerton, CA 92634). In this assay the anticonvulsant is covalently linked to a protein. Drug from the patient's sample competes with this protein-bound drug for binding to an antibody. The rate of change of light scatter is measured by the instrument.[25]

REFERENCE AND PREFERRED METHODS

The ideal analytical procedure should be:

1. Both accurate and precise, affording the reliable quantitation of the antiepileptic drugs
2. Technically easy, so that little effort is required in training technologists to perform the task
3. Rapid, to minimize delay in reporting the results to the clinician

   Many requests for drug quantitation arise as a result of problems in patient management. In our institution results for about half the antiepileptic drugs quantitated fall outside the therapeutic range and necessitate alterations in dose and dosing schedule. Thus it is important that the interval between specimen receipt and report of the result be short.
4. Cost effective. A key issue here is laboratory work load. It would be inappropriate to spend $30,000 for a piece of equipment (such as a high-performance liquid chromatography apparatus with automated sample injector and data-handling facilities) if the laboratory work load is only 10 samples per week. On the other hand, if the laboratory work load is large (greater than 40 samples per week), the purchase of such equipment would provide a cheaper service than comparable radioimmunoassay and homogeneous enzyme immunoassay procedures, where reagent costs are considerable.
5. Sensitive, to allow analysis for drugs in a microscale sample and permit the simultaneous analysis of the most commonly used drugs. Many patients with epilepsy are on multiple-drug regimens (about half at our institution).

Unfortunately, no method meets all these requirements, and the decision on which procedure to adopt depends

Table 61-5. Comparison of methods for anticonvulsant analysis

Parameter	High-performance liquid chromatography	Gas-liquid chromatography	Enzyme-multiplied immunoassay technique
Temperature	Ambient	Above 200° C	30° C
Sample volume	25 μL	1 mL	50 μL for each determination
Fraction of sample volume	0.50	—	0.01
Linearity μg/mL			
Phenytoin	1-50	1-40	2.5-40
Phenobarbital	1-100	2-80	5-80
Primidone	1-40	2-40	2.5-20
Ethosuximide	1-250	—	10-150
Carbamazepine	1-30	—	1-20
Valproic acid	—	5-180	10-150
Precision (CV)*	5%-9%	5%-9%	8%-10%

*Average coefficient of variation for anticonvulsant drugs within therapeutic dose. From Therapeutic Drug Monitoring Survey of the American Association for Clinical Chemistry, Washington, D.C., 1983.

largely on the environment and on the personal biases of those involved.

Spectrophotometric methods were initially used for the measurement of anticonvulsant drug concentrations in body fluids.[1,2] Although differential extraction procedures were developed to overcome interference by other substances, particularly drugs, the specificity of these and other "wet chemistry" procedures is questionable, and the possibility of drug interference still exists. For these reasons, alternative modes of analysis were developed by use of either chromatographic or immunological procedures (Table 61-5).

Thin-layer chromatography procedures have been described to semiquantitate or quantitate many anticonvulsant drugs.[3-5] This procedure, because of both its lack of specificity and semiquantitative nature, is not a popular mode of analysis for the anticonvulsant drugs in North America.

Gas-liquid chromatography affords reliable analysis for anticonvulsant drugs on microscale samples. The procedure permits the simultaneous analysis of all the anticonvulsant drugs, and these methods are easily automated. Similar to high-performance liquid chromatography (HPLC), GLC is most useful in laboratories with a large work load. Disadvantages of GLC include the requirements of specialized equipment and considerable expertise. GLC is one of the methods of choice for the analysis of valproic acid,[26-29] primarily because of the poor ultraviolet absorption characteristics of this compound.

Although steadily losing ground to both HPLC and enzyme immunoassay (EIA), GLC remains a popular method for anticonvulsant drug analysis. The gas chromatographer must decide between nitrogen flame ionization detection[7-9] and normal flame ionization detection[10,11] and between chromatography of the unchanged drug and a derivative of the drug. Except for valproic acid, the list of antiepileptic drugs shown consists entirely of compounds containing nitrogen. Because the bulk of interfering material is non-nitrogenous, the nitrogen flame ionization detection system affords clear advantages over flame ionization detection.

The advent and application of HPLC to therapeutic drug monitoring has been one of the most significant advances made in the clinical laboratory in recent years. Older procedures, such as gas chromatography, based on lengthy sample extractions and derivatization of the analyte before chromatography, have been largely replaced by HPLC.

Some advantages of HPLC over most other current analytical techniques include the following:

1. Analyte volatility and thermal stability—so essential for gas-liquid chromatography—are not required for this type of chromatography.
2. Relatively little sample work-up is required before analysis.
3. Characteristically, methods involving HPLC require only a short analysis time.
4. There is no need for derivatization of the analyte.
5. For laboratories with a large work load, the cost per analysis is very low because reagent costs are low.
6. HPLC affords good sensitivity. The sensitivity varies with the drug, but reliable quantitation at drug concentrations as low as 1 μg/L can often be achieved.
7. HPLC methods are readily automated. Routine analyses therefore require a minimum of technician time, a major factor in arriving at the low cost per analysis.
8. HPLC allows for the simultaneous analysis of most anticonvulsant drugs in a microsample.

Some limitations of HPLC in drug analysis are as follows:

1. Spectrophotometric, fluorometric, and amperometric detectors are commonly used. Therefore the analyte in question must either absorb light, fluoresce, or be electroactive.
2. Equipment is expensive. Investment of capital (a fully automated system currently costs about $30,000) is only warranted if the laboratory has a

substantial work load (more than 40 samples per week).

3. Although HPLC analysis of valproic acid has been described,[30,31] these procedures require either postcolumn reactions[30] or precolumn derivatization.[31] The methods of choice for the analysis of valproic acid are gas-liquid chromatography, enzyme immunoassay, or fluorescence polarization immunoassay.

Several methods that allow simultaneous analysis for phenobarbital, phenytoin, primidone, ethosuximide, and carbamazepine have been reported in the literature.[13-21] A very recent development, which has simultaneously significantly reduced column costs and increased column life, is the advent of radial compression columns (Radial Pak A, marketed by Waters Associates, Inc., Milford, MA 01757). Using these columns, Soldin[20] and Soldin and Walter[21] have recently described a procedure for the simultaneous analysis of five anticonvulsant drugs and their active metabolites.

Many antiepileptic drugs can be determined by radioimmunoassay.[28] Although extremely sensitive, these techniques have the disadvantage of not allowing for the simultaneous assay of drugs in patients on multiple-drug regimens. They require the availability of a liquid scintillation counter and availability and stability of suitable antisera.

Enzyme immunoassay (EIA), in particular the enzyme-multipled immunoassay technique (EMIT), are very popular. The advantages of the technique are that no separation steps are involved and simple equipment readily available in clinical laboratories is used. This technology requires only a small sample of serum. The technique is easy to learn and does not require highly trained specialists. EIA methods afford good precision and reliable results of phenobarbital, phenytoin, primidone, ethosuximide, carbamazepine, and valproic acid. The EMIT technique has been automated with centrifugal analyzers,[32-35] such as the Roche-COBAS analyzer (Roche-COBAS, Nutley, N.J.), the ABA-100 Biochromatic Analyzer[36] (Abbott Laboratories, North Chicago, Ill.), the Gilford 3500[36] (Gilford Instruments, Oberlin, Ohio), and the du Pont Automatic Clinical Analyzer[37] (du Pont Instruments, Wilmington, Del.). Such approaches to analysis considerably reduce both technician time and reagent cost. Disadvantages include costly reagents required, and EIA does not permit the simultaneous analysis of anticonvulsant drugs.

In general, the chromatographic procedures require greater technical skill to perform than the procedures based on immunologic principles. Chromatographic methods also require expensive equipment; however, they do allow for the simultaneous measurement of many anticonvulsants and their active metabolites. Approximately 50% of our patients receiving anticonvulsants are on multiple-drug regimens, and simultaneous analysis is therefore a considerable advantage over immunological procedures. The reagent cost per analysis is also significantly lower with the chromatographic than with the immunological procedures. Chromatographic methods may well be preferred in university hospital laboratories with large work loads.

The major advantage of the immunological procedures is their ease of operation and 24-hour availability for immediate measurements. This advantage is offset by their greater reagent cost per analysis. These methods are clearly the most popular procedures for measurement of anticonvulsant drugs in North American laboratories.

SPECIMEN

Plasma or serum may be used for the gas chromatography or high-performance liquid chromatography methods. Serum is the preferred specimen for the immunoassays. Icteric or hemolyzed specimens have less effect on the chromatographic methods than on immunoassays. Sera containing the anticonvulsants discussed in this chapter can be stored at room temperature for several hours. If frozen at −20° C, samples containing these drugs are stable for at least 1 year. Saliva may also be used for the measurement of anticonvulsants. The saliva concentration provides a reliable estimate of the free drug concentration for phenytoin and carbamazepine.[38]

PROCEDURE[16-21]

Principle

The five anticonvulsant drugs—primidone, phenobarbital, ethosuximide, carbamazepine, and phenytoin, as well as the primidone metabolite phenylethylmalonamide, and the carbamazepine metabolite carbamazepin-10,11-epoxide—are separated by reversed-phase liquid chromatography and quantitated by measurement of peak areas relative to the peak area of the internal standard, dihydrocarbamazepine. The compounds are identified by their retention times and also by their absorbance ratio at 200 and 254 nm. Each drug has a known absorbance ratio. Any skewing of the ratio alerts the operator to the presence of an interfering compound.

Reagents

Acetonitriles, methanol, and dimethylsulfoxide. May be obtained from Burdick and Jackson (Muskegon, Mich.).

Standards. From Applied Science Laboratories, State College, Penn. (phenobarbital), Parke-Davis, Morris Plains, N.J. (phenytoin and ethosuximide), Ayerst Laboratories, New York, N.Y. (primidone), Aldrich Chemical Co., Milwaukee (10,11-dihydrocarbamazepine), and CIBA-Geigy, Ardsley, N.Y. (carbamazepine).

Potassium phosphate buffer, 10 mmol/L, pH 7.3. This is stable for 3 months at 4° C. Check monthly for microbial growth.

Stock internal standard, 200 mg/mL, 839 μmol/L. Dissolve 20 mg of 10,11-dihydrocarbamazepine in 100 mL of acetonitrile. This is stable for 1 month when stored in a dark bottle.

Mobile phase buffer, acetonitrile-phosphate, 37:63 by volume. Pass 370 mL of acetonitrile through a nylon 66, 0.2 μ filter (D and L Filter Corp., Woburn, MA 01801) or equivalent membrane filter into a 2 L Ehrlenmeyer filtering flask. Replace the filter and pass 630 mL of phosphate buffer into the flask. Degas the solution for 15 minutes at room temperature by applying a vacuum. This is stable for 1 week at room temperature.

Assay

Equipment. A high-performance liquid chromatograph, model ALC/GPC/6000A, with an autosampler (WISP, model 710A), a radial compression system (RC-100) and 10 × 0.8 Radial Pak A columns, all purchased from Waters Associates, Inc. (Milford, MA 01757). Dectectors used were the Waters model 440 with 254 nm kit and the model LC55 variable wavelength spectrophotometer (Perkin-Elmer Corp., Norwalk, CT 06856). The recorder was a Honeywell Electronic, model 196 (Honeywell, Inc., Fort Washington, PA 19034). A Sigma 10 data-handling system (Perkin-Elmer) was employed to receive and compute the spectrophotometer signals.

1. Sample preparation
   a. Add 25 μL of acetonitrile containing approximately 200 mg/L of 10,11-dihydrocarbamazepine to 25 μL of serum or plasma.
   b. Vortex the sample for 30 seconds. Centrifuge for 5 minutes and save the supernatant fraction for analysis.
2. Chromatographic parameters
   a. Chart speed: 1.0 cm/minute
   b. Ultraviolet sensitivity: 254 nm, 0.01 AUFS (absorbance units at full scale)
   c. Ultraviolet sensitivity: 200 nm, 0.05 AUFS
   d. Flow rate: 2.0 mL/min at ambient temperature
3. Inject 4 μL of sample or standard. Sample analysis time is approximately 10 minutes. A typical chromatogram is shown in Fig. 61-3.

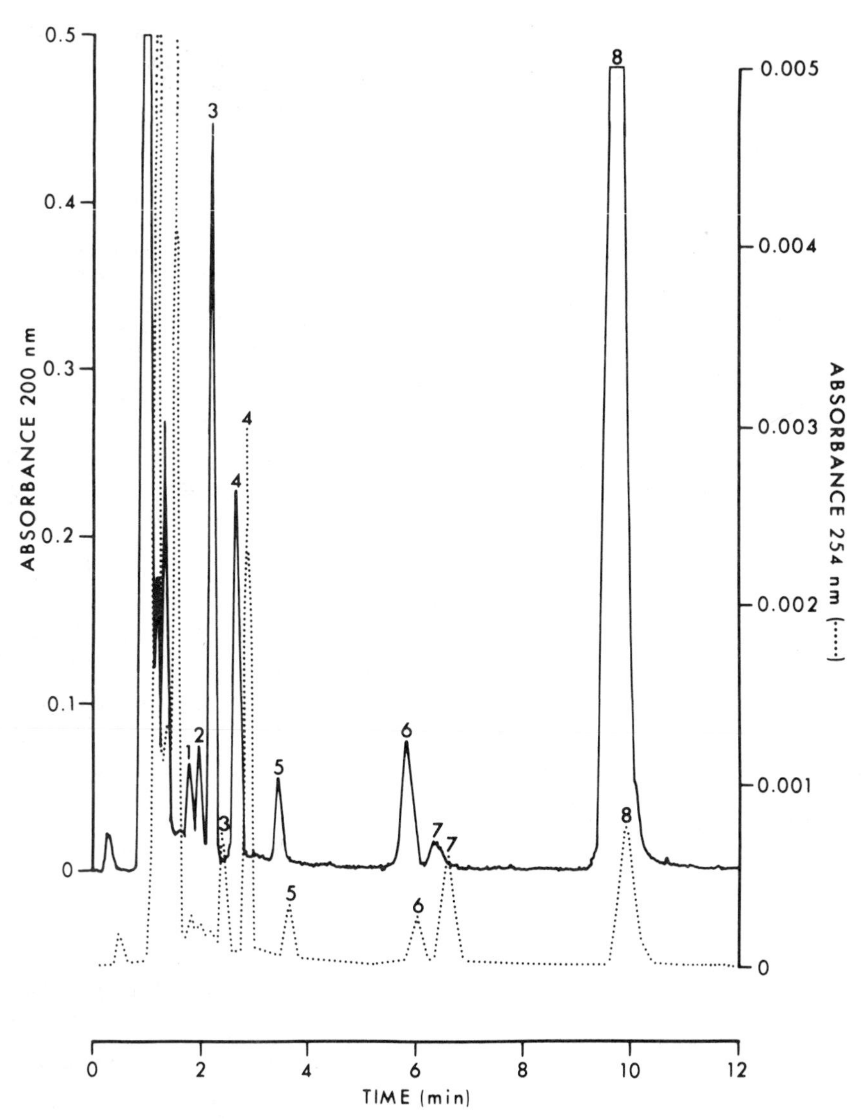

Fig. 61-3. Chromatogram obtained for a plasma specimen containing therapeutic concentrations of primidone, *2*; ethosuximide, *3*; phenobarbital, *4*; phenytoin, *6*; and carbamazepine, *7*. Peaks *1*, *5*, and *8* represent phenylethylmalonamide, carbamazepine-10,11-epoxide, and the internal standard dihydrocarbamazepine, respectively. *Solid tracing,* 200 nm; *dotted tracing,* 254 nm.

Calculations

Standards purchased from the National Bureau of Standards are diluted according to the described directions. Values compared to the internal standard are obtained for control sera (C_S, from Ortho Diagnostics). Peak heights of each standard are compared to the internal standard, and peak-height ratios calculated. Peak heights for the serum calibrators are also calculated, and concentration values for each drug are assigned after comparison of these ratios to the standard ratios. These calibration values for the Ortho sample are used to calculate all patient values. The calculation is as follows:

$$\text{Anticonvulsant drug } (\mu g/mL) = C_S\ (\mu g/mL) \times \frac{\text{Peak height of sample/Peak height of internal standard}}{\text{Peak height of calibrator/Peak height of internal standard}}$$

Notes

1. No interferences are known. The recovery of all the anticonvulsants is greater than 80%.[16]
2. Linearity is achieved for primidone, ethosuximide, phenobarbital, phenytoin, and carbamazepine up to concentrations of at least 40, 250, 100, 50, and 30 μg/L, respectively.
3. The procedure described provides for the simultaneous analysis of the five anticonvulsant drugs, the primidone metabolite phenylethylmalonamide, and the carbamazepine metabolite carbamazepine-10,11-epoxide. It is not recommended for valproic acid.

REFERENCE RANGE

Generally accepted therapeutic ranges for these drugs are as follows:

Drug	μg/mL	μmol/L
Carbamazepine	4-12	17-51
Ethosuximide	40-100	284-459
Phenobarbital	15-40	64-172
Phenytoin	10-20	40-80
Primidone	5-12	23-55
Valproic acid	50-120	347-833

REFERENCES

1. Plaa, G.L., and Hine, C.H.: A method for the simultaneous determination of phenobarbital and diphenylhydantoin in blood, J. Lab. Clin. Med. **47**:649-657, 1956.
2. Svensmark, O., and Kristensen, P.: Determination of diphenylhydantoin and phenobarbital in small amounts of serum, J. Lab. Clin. Med. **61**:501-507, 1963.
3. Gardner-Thorpe, C., Parsonage, M.J.H., Smethurst, P.F., and Toothill, C.: A comprehensive gas chromatographic scheme for the estimation of antiepileptic drugs, Clin. Chim. Acta **36**:223-250, 1971.
4. Pippenger, C.E., Scott, J.E., and Gillen, H.W.: Thin-layer chromatography of anticonvulsant drugs, Clin. Chem. **15**:255-260, 1969.
5. Huisman, J.W.: The estimation of some important anticonvulsant drugs in serum, Clin. Chim. Acta **13**:323-328, 1966.
6. MacGee, J.: Rapid determination of diphenylhydantoin in blood plasma by gas-liquid chromatography, Anal. Chem. **42**:421-442, 1970.
7. Vandemark, F.L., and Adams, R.F.: Ultramicro gas-chromatographic analysis for anticonvulsants, with use of a nitrogen-selective detector, Clin. Chem. **22**:1062-1065, 1976.
8. Sengupta, A., and Peat, M.A.: Gas-liquid chromatography of eight anticonvulsant drugs in plasma, J. Chromatogr. **137**:206-209, 1977.
9. Toseland, P.A.: Gas chromatographic analysis of anticonvulsant drugs, Lablore **7**:433-436, 1977.
10. Least, C.J., Jr., Johnson, G.F., and Solomon, H.M.: Therapeutic monitoring of anticonvulsant drugs: gas-chromatographic simultaneous determination of primidone, phenylethylmalonamide, carbamazepine, and diphenylhydantoin, Clin. Chem. **21**:1658-1662, 1975.
11. Abraham, C.V., and Joslin, H.D.: Simultaneous gas-chromatographic analysis for phenobarbital, diphenylhydantoin, carbamazepine, and primidone in serum, Clin. Chem. **22**:769-771, 1976.
12. Solow, E.B., and Green, J.B.: The simultaneous determination of multiple anticonvulsant drug levels by gas-liquid chromatography, Neurology **22**:540-550, 1972.
13. Adams, R.J., and Vandemark, F.L.: Simultaneous high-pressure liquid-chromatographic determination of some anticonvulsants in serum, Clin. Chem. **22**:25-31, 1976.
14. Kabra, P.M., Stafford, B.E., and Marton, J.L.: Simultaneous measurement of phenobarbital, phenytoin, primidone, ethosuximide, and carbamazepine in serum by high-pressure liquid chromatography, Clin. Chem. **23**:1284-1288, 1971.
15. Kabra, P.M., McDonald, D.M., and Marton, L.J.: A simultaneous high-performance liquid chromatographic analysis of the most common anticonvulsants and their metabolites, J. Anal. Toxicol. **2**:127-133, 1978.
16. Soldin, S.J., and Hill, J.G.: Rapid micromethod for measuring anticonvulsant drugs in serum by high-performance liquid chromatography, Clin. Chem. **22**:856-859, 1976.
17. Soldin, S.J., and Hill, J.G.: Interference with column-chromatographic measurement of primidone, Clin. Chem. **23**:782, 1977.
18. Soldin, S.J., and Hill, J.G.: Routine dual-wavelength analysis of anticonvulsant drugs by high performance liquid chromatography, Clin. Chem. **23**:2352-2353, 1977.
19. Soldin, S.J., and Hill, J.G.: The therapeutic monitoring of anticonvulsant drugs in a 650-bed children's hospital. In Hawk, G.L., editor: Biological/biomedical application of liquid chromatography, New York, 1979, Marcel Dekker, Inc., pp. 559-571.
20. Soldin, S.J.: High performance liquid chromatographic analysis of anticonvulsant drugs using radial compression columns, Clin. Biochem. **13**:99-101, 1980.
21. Soldin, S.J., and Walter, M.: Improved HPLC analysis for anticonvulsant drugs employing radial compression columns, Clin. Biochem. **14**:161, 1981.
22. Cook, C.E., Christensen, H.D., Amerson, E.W., et al.: Radioimmunoassay of anticonvulsant drugs: phenytoin, phenobarbital, and primidone. In Kellaway, P., and Petersen, I.S., editors: Quantitative analytic studies in epilepsy, New York, 1976, Raven Press.
23. Wong, R.C., Burd, J.F., Carrico, R.J., et al.: Substrate-labeled fluoroescent immunoassay for phenytoin in human serum, Clin. Chem. **25**:686-691, 1979.
24. Frings, C.S., and Phillips, G.: Therapeutic drug monitoring of anticonvulsant drugs using fluorescence polarization immunoassay, Clin. Chem. **28**:1611, 1982.
25. Nishikawa, T., Kubo, H., and Saito, M.: Competitive nephelometric immunoassay method for antiepileptic drugs in patient blood, J. Immunol. Methods **29**:85-89, 1979.
26. Willox, S., and Fotte, S.E.: Simple method for measuring valproate in biological fluids, J. Chromatogr. **151**:67-70, 1978.
27. Jakobs, C., Bojasch, M., and Hanefeld, F.: New direct micromethod for determination of valproic acid in serum gas chromatography, J. Chromatogr. **146**:494-497, 1978.
28. Berry, D.J., and Clarke, L.A.: Determination of valproic acid (dipropylacetic acid) in plasma by gas-liquid chromatography, J. Chromatogr. **156**:301-307, 1978.
29. Freeman, D.J., and Rawall, N.: Extraction of underivatized valproic acid from serum before gas chromatography, Clin. Chem. **26**:674-675, 1980.
30. Fairinotte, R., Pfaff, M.C., and Mahuzier, G.: Simultaneous determination of phenobarbital and valproic acid in plasma using high performance liquid chromatography, Ann. Biol. Clin. **36**:347-353, 1978.
31. Gupta, R.N., Keane, P.M., and Gupta, M.L.: Valproic acid in

plasma, as determined by liquid chromatography, Clin. Chem. **25:**1984-1985, 1979.
32. Haven, M.: Phenytoin and phenobarbital measurement by centrifugal analyzer, Clin. Chem. **22:**2057, 1976.
33. London, M., Sanabria, D., and Yau, D.: Enzyme immunoassay of phenytoin and phenobarbital with the centrifugal analyzer, Clin. Chem. **23:**1362, 1977.
34. Long, J.P.: Enzyme immunoassay with use of a centrifugal analyzer of phenytoin, phenobarbital, and theophylline, Clin. Chem. **24:**391, 1978.
35. Lacher, D.A., Valdes, R., Jr., and Savory, J.: Enzyme immunoassay of carbamazepine with centrifugal analyzer, Clin. Chem. **25:**295-298, 1979.
36. Hoelting, C.R., Tieber, V.L., Smith, C.H., and Dietzler, D.N.: EMIT anticonvulsant drug assays with the Gilford 3500 and ABA-100, Clin. Chem. **25:**1096, 1979.
37. Nandedkar, A.K.N.: An evaluation of EMIT reagent system for determination of phenobarbital and phenytoin in du Pont automatic clinical analyzer, Clin. Chem. **25:**1096, 1979.
38. Muckow, J.C.: The use of saliva in therapeutic drug monitoring, Ther. Drug Monit. **4:**29-248, 1982.

Barbiturates

AMADEO J. PESCE

Clinical significance: pp. 718 and 897

Molecular structure:	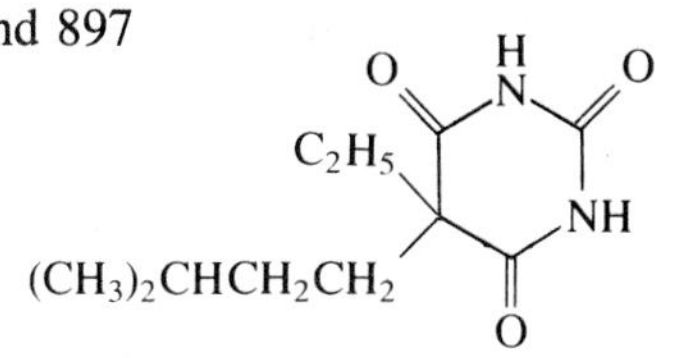	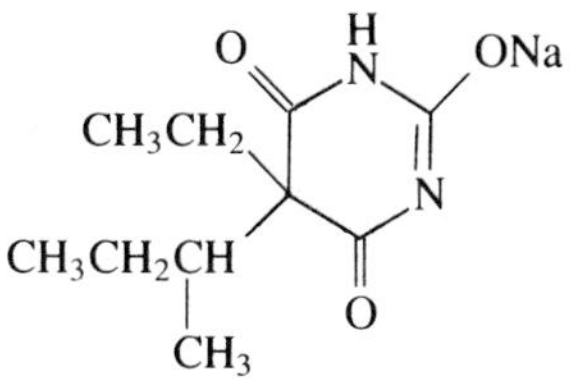	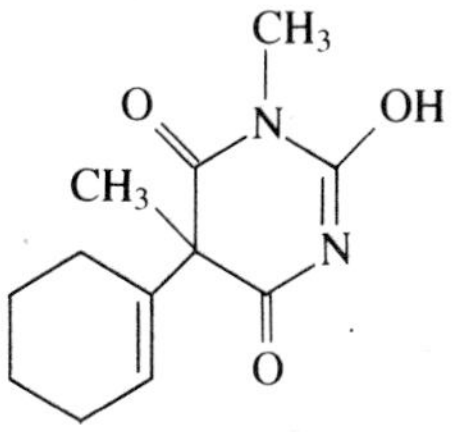
	Amobarbital	**Butabarbital sodium**	**Hexobarbital**
Molecular formula:	$C_{11}H_{18}N_2O_3$	$C_{10}H_{15}N_2NaO_3$	$C_{12}H_{16}N_2O_3$
Molecular weight (daltons):	226.27	234.23	236.26
Merck Index:	605	1487	4576

Molecular structure:	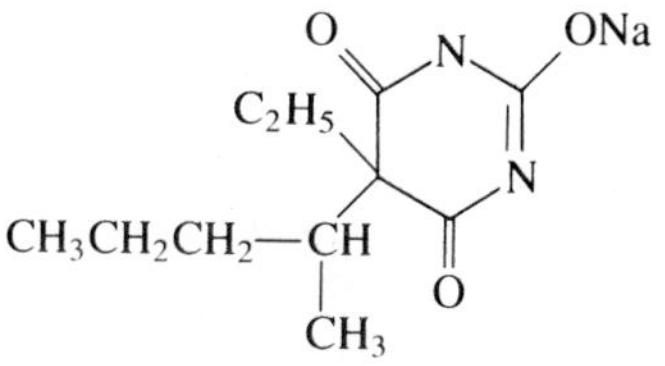	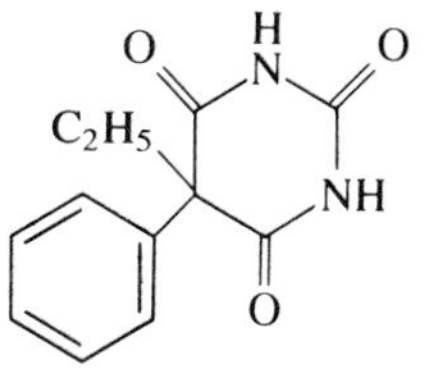	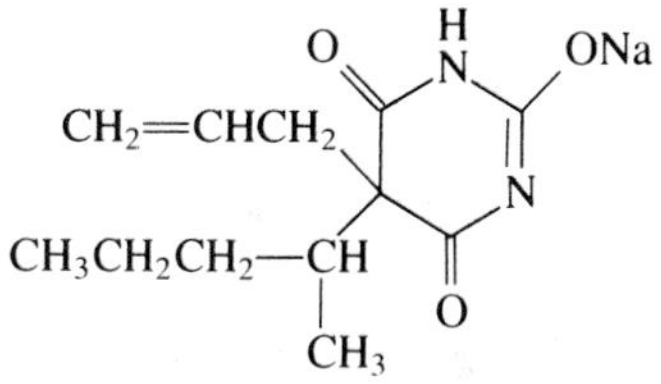
	Pentobarbital sodium	**Phenobarbital**	**Secobarbital sodium**
Molecular formula:	$C_{11}H_{17}N_2NaO_3$	$C_{12}H_{12}N_2O_3$	$C_{12}H_{17}N_2NaO_3$
Molecular weight (daltons):	248.26	232.23	260.27
Merck Index:	6928	7032	8172

Clinical class: barbiturate, organic heterocyclic

PRINCIPLES OF ANALYSIS AND CURRENT USAGE

Barbiturates are a group of drugs classified as hypnotics and are prescribed for a number of conditions. They act as sedatives or analgesics (painkillers) and are used to help control seizure disorders. Most commonly, phenobarbital is taken for the control of epilepsy, and its determination, as part of therapeutic drug monitoring, is usually the most frequently performed analysis for this group of compounds. However, barbiturates are also abused substances; thus tests for their possible presence and amount in the overdosed patient is one of the major components of the toxicological drug screen. Therefore both qualitative and quantitative tests are performed by the laboratory. The qualitative tests are selected for their rapid performance in the drug screen and usually react with any drug of the barbiturate class. Most qualitative analyses are performed on urine or gastric contents, since these body fluids are best for drug screening. In contrast, quantitative analyses are performed on blood, since levels correspond best with pharmacological effect.

Qualitative tests

One performs a qualitative color test for barbiturate by extracting the drug from a slightly acidic solution into an organic solvent such as chloroform or dichloromethane and

Table 61-6. Methods for barbiturate analysis

Method	Principle	Usage	Comments
Qualitative			
1. Extraction and colorimetric reaction	Barbiturate extracted from slightly acidic solution and complexed with Hg^{++}; reacts with diphenylthiocarbazone to form an orange-colored product	Blood, plasma, serum, urine, gastric juice	Nonspecific for barbiturates
2. Extraction and ultraviolet spectroscopy	Barbiturate extracted free of other ultraviolet-absorbing compounds; ultraviolet spectrum is characteristic and specific for each drug	Blood, plasma, serum, urine, gastric contents, tissue	Original specific method Some interferences Can be quantitative *Not* recommended for mixtures
3. Ultraviolet difference spectroscopy	As in method 2, extracted barbiturate is divided into aliquots and difference between spectra at pH greater than 13 and at pH 10.5 recorded	Blood, serum	More specific than method 2 Requires recording dual-beam spectrophotometer
4. Thin-layer chromatography	Extracted barbiturate chromatographed and its R_f factor* and reaction with stain recorded; R_f factor and color reactions presumptive for barbiturates	Blood, plasma, serum, urine, gastric contents, tissue	Used as initial qualitative test or confirmatory Can separate mixtures
5. Gas chromatography	Extracted barbiturate chromatographed and its retention time recorded; parent and possible degradation products have specific retention times	Blood, plasma, serum, urine, gastric contents, tissue	Not quantitative Degradation of some barbiturates, not a gaussian elution pattern Can separate mixtures.
6. Gas chromatography–mass spectrometry	Steps identical as for method 5; mass spectrophotometric detector can identify specific barbiturates and help resolve mixtures	Same as for method 5	Improves identification of method 5
7. EMIT (enzyme-multiplied immunoassay technique)	Competitive binding assay using EMIT technology; antibody has broad specificity and reacts with most barbiturates	Urine	Extent of reaction varies with amount and type of barbiturate
Quantitative—all barbiturates			
Ultraviolet spectroscopy (similar to method 2)	Extracted barbiturate has characteristic absorbance spectrum, with amount of absorbance proportional to concentration; some identification depends on spectral characteristics	Blood, plasma, serum, urine, gastric contents, tissue	Original method, but subject to interference by endogenous compounds and other drugs Does not work well on mixtures
Differential ultraviolet spectroscopy (similar to method 3)	Extracted barbiturate has characteristic differential spectrum when solution at pH > 13 and pH 10.5 are compared	Blood, serum	Less interference than ultraviolet spectroscopy, but specific barbiturate must be known to achieve quantitation
Gas chromatography (variation of method 5)	Extracted barbiturate(s) converted to alkylated derivative(s), such as dimethylphenobarbital $(CH_3)_4N^+OH^-$ + Barbiturates → Dimethylated barbiturates Retention time identifies which barbiturate present	Blood, plasma, serum, urine, gastric contents, tissue	Only a few drugs may interfere Recommended procedure for mixtures.
8. High-performance liquid chromatography	Extracted barbiturates are chromatographed and quantified by refractive index or absorbance	Blood, plasma, serum, urine, gastric contents, tissue	Should be as efficient as gas chromatography, but fewer laboratories have used this technique
Quantitative—phenobarbital in plasma or serum			
Ultraviolet spectroscopy (same as methods 2 or 3)	As for methods 2 and 3		More subject to interference than other procedures
9. Radioimmunoassay	Competitive binding assay		Less used

*A relative mobility factor; see Chapter 4.

Table 61-6. Methods for barbiturate analysis—cont'd

Method	Principle	Usage	Comments
10. Fluorescence immmunoassay	Competitive binding assay		Newly introduced
11. Nephelometric inhibition	Competitive inhibition of precipitation reaction measured by nephelometry		Restricted to laboratories with nephelometers
Enzyme immunoassay (variation of method 7)	Competitive binding assay (EMIT technology)		Most favored technique
Gas-liquid chromatography derivatized or underivatized (variation of method 5)	As for method 5		Requires technical skill Can quantitate several antiepileptics simultaneously
High-performance liquid chromatography (same as method 8)	As for method 8		Requires skill Can quantitate several antiepileptics simultaneously

complexing the barbiturate with mercury. The barbiturate-mercury complex is then allowed to react with diphenylthiocarbazone, and an orange-colored product is formed (method 1, Table 61-6).[1]

One may use ultraviolet absorption spectroscopy to both identify and quantify barbiturates, since the absorption spectrum is fairly specifc for this class of drugs (method 2, Table 61-6). Extraction of the barbiturate is necessary to free it from other analytes that have a similar absorption spectrum.[2] The presumptive barbiturate-containing body fluid is usually acidified and extracted into an organic solvent such as chloroform. It is then reextracted into an alkaline aqueous solution. One scans the spectrum from 230 to 340 nm and records the resulting absorbance[3] (Fig. 61-4). It is compared to the characteristic spectrum of each of the barbiturates. The amount present is proportional to the absorption at specific wavelengths. A refinement of this approach is to use differential spectroscopy of the ultraviolet spectrum for quantitation of blood levels (method 3, Table 61-6). In this test one divides the extracted drug into two aliquots, one with a strongly alkaline pH greater than 13, and the other with a pH of 10.5. The spectral difference between these solutions is obtained by use of a dual-beam recording spectrophotometer.[5] The difference spectrum is definitive for barbiturate, and the amount of the absorbance difference is proportional to concentration (Fig. 61-5).[5]

Chromatographic procedures are also useful for the identification of barbiturates; these primarily include thin-layer chromatography and gas chromatography. Thin-layer chromatography (method 4, Table 61-6) is usually performed on urine that has been acidified and extracted into an organic phase. One places the presumptive drug on a thin-layer chromatographic plate and identifies the position of the barbiturate after chromatography by developing a color reaction with one of several sprays, such as potassium permanganate, mercuric sulfate, and diphenylcarbazone.[6] The R_f values and color are the specific criteria for identification.

Gas chromatography detection (method 5, Table 61-6) is performed by extraction of the presumptive acidified urine with toluene or other selected organic solvents and separation of the extracted analyte.[7] An OV-series column with a flame ionization detector is most commonly used as part of an overall screening procedure (see section on drug screen later in this chapter). The relative retention time is specific for each barbital derivative and is used for presumptive identification. Alternatively, the extract may be analyzed with a gas chromatograph–mass spectrometer (GC-MS); the identification is by the pattern of mass fragmentation (method 6, Table 61-6).

The enzyme-multiplied immunoassay technique (EMIT) for barbiturate in urine (method 7, Table 61-6) is a ho-

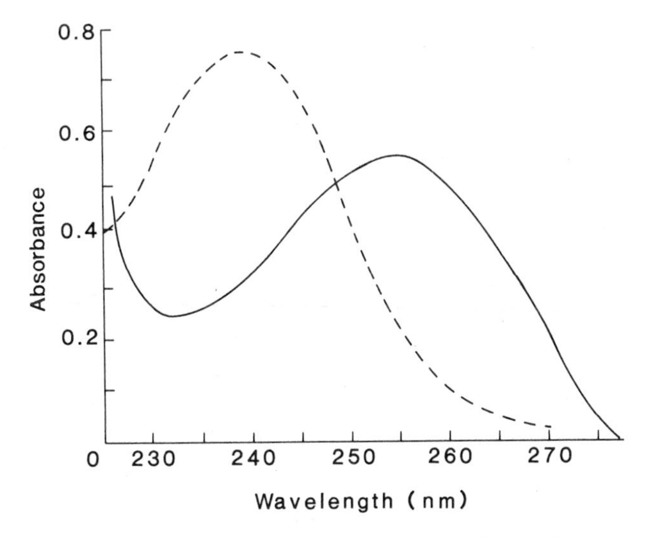

Fig. 61-4. Absorption spectrum of a barbital solution 13.5 mg/mL. In 0.45 M *(solid tracing);* in borate buffer (pH 10.5) *(dashed tracing).* (From Broughton, P.M.G.: Biochemistry **63**:207-213, 1956.)

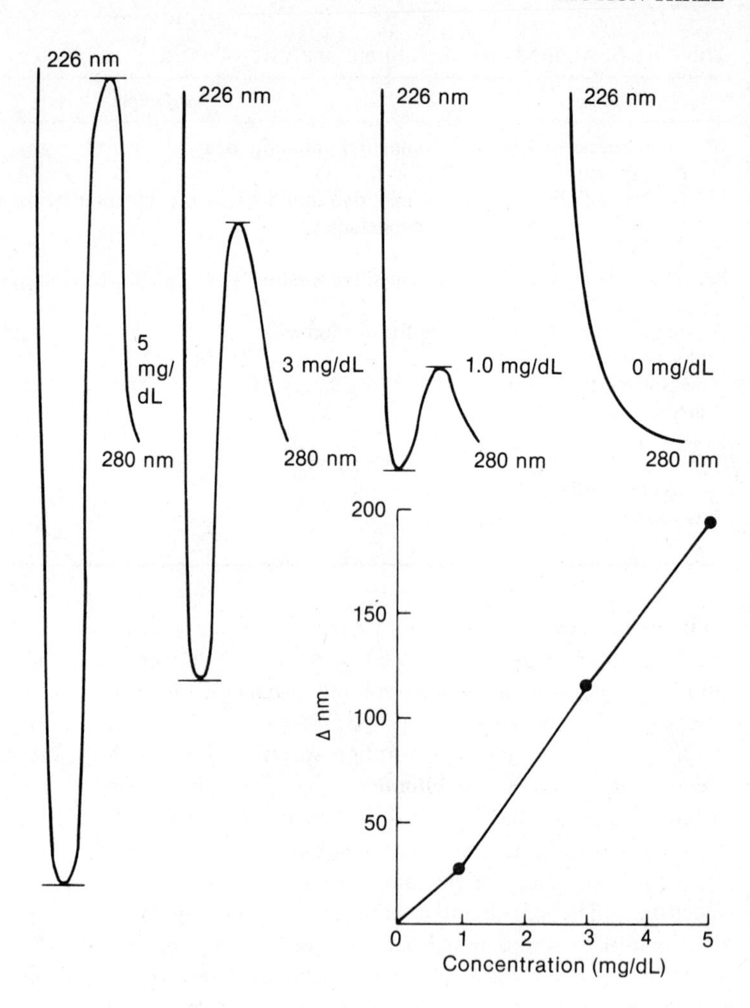

Fig. 61-5. Differential spectra of various concentrations of phenobarbital and a standard calibration curve. Absorbance of sample at pH 13 (sample cuvette) minus absorbance of sample at pH 10 (reference cuvette) is recorded. Difference in absorbance between 240 and 260 nm (Δnm) is plotted versus drug concentration. (From Schumann, G.B., et al.: Am. J. Clin. Pathol. **66:**823-830, 1976.)

mogeneous enzyme immunoassay (EIA). The barbiturate is labeled with an enzyme; the enzyme-labeled drug is then mixed with antibody to barbiturate, which results in inhibition of the reaction. Any drug in the urine sample competes for the antibody; thus the measured enzyme activity is increased. Increased enzyme activity indicates the presence of barbiturates. The enzyme is lysozyme and acts on a bacterial substrate. Following are the lower limits of detection:

Secobarbital	2.0 μg/mL
Pentobarbital	5.0 μg/mL
Butabarbital	10.0 μg/mL
Phenobarbital	3.0 μg/mL
Amobarbital	30 μg/mL

A variation of this assay is used for detection of barbiturates in serum. In this circumstance the enzyme glucose-6-phosphate dehydrogenase is used instead of lysozyme.

Quantitative tests

Techniques of quantitation include enzyme-multiplied immunoassay (EMIT), gas chromatography, high-performance liquid chromatography (HPLC), and other immunoassay procedures. In general, immunoassays are only used when the specific drug is known, such as phenobarbital. These procedures are of the competitive binding type, in which the drug in the unknown sample competes for radiolabeled drug (radioimmunoassay, RIA), enzyme-bound drug (EMIT), antibody-bound drug conjugate (substrate-labeled fluorescence immunoassay, SLFIA), or protein-bound drug (nephelometric inhibition assays). To quantitate several different types of barbiturates, such as phenobarbital versus secobarbital, one must establish the extent of cross-reactivity for each barbiturate.

The ultraviolet absorption analyses are essentially those already described. These have acceptable accuracy and precision, but to quantify mixtures, one must make several calculations and approximations.

Gas chromatographic analysis was the first method to effectively isolate and quantify a mixture of barbiturates with a minimum of possible interference.[8-10] Initial methods required derivatization (methylation) of the barbiturates with trimethylanilinium hydroxide. Current variations use tetramethyl ammonium hydroxide as the derivatizing reagent. Once derivatized, the barbiturate is more stable and is eluted with more symmetrically defined peaks than the underivatized parent. Fig. 61-6 is a comparison of a

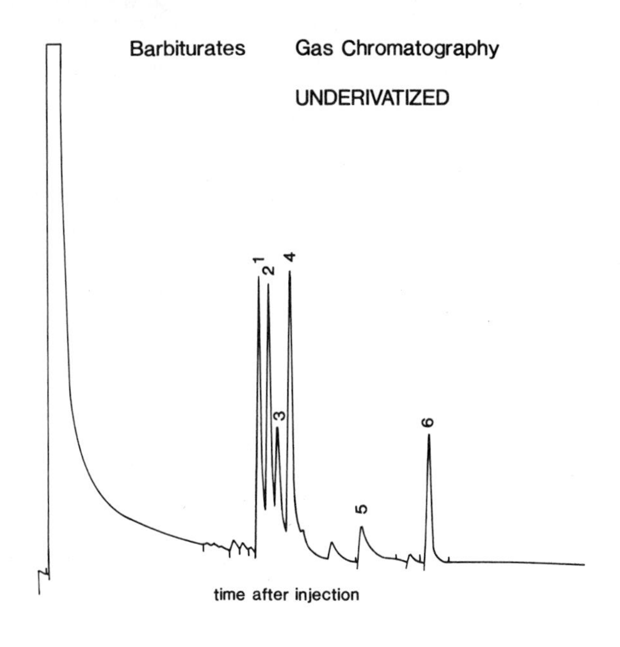

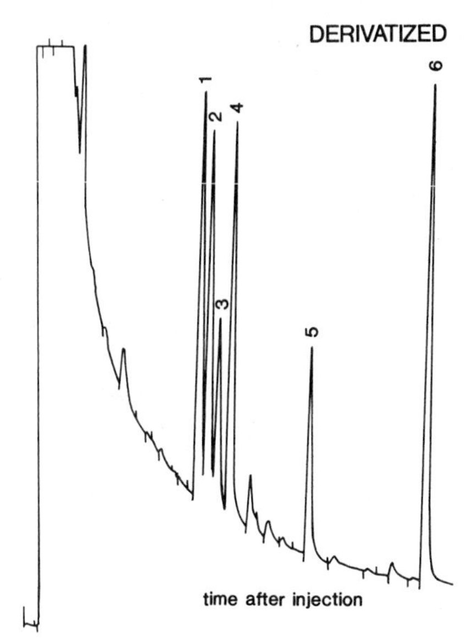

Fig. 61-6. Gas chromatographic tracing of a mixture of barbiturates separated as underivatized *(upper panel)* and derivatized *(lower panel)* form. For derivatization, procedure outlined in manual method was used. Mixed barbiturate standard was 10 mg of each barbiturate dissolved in 10 mL of CH_3OH. *1,* Butabarbital; *2,* amobarbital; *3,* pentobarbital; *4,* secobarbital; *5,* phenobarbital; *6,* phenytoin. Comparative testing was performed with use of Hewlett-Packard 5830A with a flame ionization detector and a 2 mm inner-diameter glass column 6 feet long packed with 3% OV-17 on Chromosorb W(HP) of 80/100 mesh. Nitrogen was used a carrier gas at a flow rate of 30 mL/min. Temperature programming was performed from 115° to 270° at 15° C/min after an initial hold time of 0.5 minutes. (From Hassan, F.M., and Fahey, C.: Unpublished results, University of Cincinnati.)

mixture of five barbiturates and phenytoin separated by gas chromatography in their underivatized and derivatized forms. Note the longer retention times and the more asymmetric breaks eluting from the underivatized mixture. Improved column packings, such as the "SP" (stationary phase support, Supelco, Inc., Bellefonte, Penn.) series, enable analysis of the underivatized drugs without noticeable column effects.

Competitive binding assays have focused primarily on the measurement of phenobarbital. Assays using RIA, EMIT, and SLFIA have all been used. In addition, the nephelometric inhibition assay has been developed for measurement of this specific drug. In these procedures for the measurement of phenobarbital one must keep in mind that there can be cross-reactivity for several or all the other barbitals.

HPLC has been adapted for the measurement of the barbiturates, but the most common procedure quantifies phenobarbital in the presence of other antiepileptic drugs, such as primidone and phenytoin. The chromatographic system used a μPorasil column (Waters Associates, Inc., Milford, MA 01757), 30 × 0.4 cm. The drug was detected using an ultraviolet spectrophotometer, which monitored the absorbance at 254 nm.

Phenobarbital is the most commonly quantified barbiturate because of its use in the treatment of epilepsy. Techniques of quantitation include EMIT, gas chromatography, and HPLC. Other immunoassay procedures are also used, but at present experience is limited.

REFERENCE AND PREFERRED METHODS

Qualitative

The identification of barbiturates, as in all assays of the overdose patient, requires two procedures, one of which must be the confirming procedure. In the case of an overdose, urine is the fluid of choice for testing. In this circumstance, the enzyme-multiplied immunoassay (EMIT) drug abuse test offers the best choice of a rapid screen for the presence of barbiturates. However, one must keep in mind that because there is a cross-reactive response between the barbiturates, this response will vary depending on the nature of the individual barbiturates taken. The presence of barbiturates must be confirmed by thin-layer chromatography, ultraviolet spectroscopy, or another method.[7,12] If thin-layer chromatography is used, one can identify individual drugs by observing the R_f values and the color produced by the staining reagent.

A variation is the use of the EMIT technique for the qualitative assay of barbiturate in serum. This approach is limited by the drug concentrations found in serum lower than those found in urine and the difficulty of obtaining a confirmatory test on serum as opposed to urine. Also, the sensitivity of the test will be poorer for the less common barbiturates, since they are detected as cross-reacting drugs.

Table 61-7. Comparison of assay conditions for serum phenobarbital

Parameter	Colorimetric method	Gas-liquid chromatography	Ultraviolet spectrophotometry (difference)	Enzyme-multiplied immunoassay technique
Sample size	2 mL	1 mL	3 mL	0.05 mL
Sensitivity	10 mg/L	2 mg/L	2 mg/L	5 mg/L
Coefficient of variation	—	8.5%*	5.4%	8.2%
Interference	Glutethimide, meprobamate, tolbutamide, chlorpropamide, diphenylhydantoin, primidone	Phenobarbital and mephobarbital not separated	Endogenous absorption for *p*-aminophenol and sulfonamide drugs	Some barbitals
Time	30 min	1 hour	40 min	15-30 min

Adapted from Schumann, G.B., et al.: Am. J. Clin. Pathol. **66**:823-830, 1976.
*From data of the 1983 Therapeutic Drug Monitoring surveys of the American Association for Clinical Chemistry.

The ultraviolet absorption procedure of Goldbaum, with extraction and measurement of the drug at pH 10.0 and greater than 13.0, also offers a valid assay for analysis; however, it does not work well if there is more than one barbital present. This procedure has been the standard against which others have been compared. It gives consistent results with an estimated coefficient of variation of 5.4% (Table 61-7).

The extraction procedure and complexing with diphenylthiocarbazone is very nonspecific and will react with other drugs commonly present in each patient, such as phenytoin.

Gas chromatography is the procedure of choice for qualitative and quantitative analysis of barbiturates. It yields both the type and the amount of the drug. High-performance liquid chromatography has not been used as extensively for qualitative estimate of barbituates but should yield data comparable to gas chromatography.

Quantitative

For analysis of a mixture of barbiturates, only the gas and high-performance liquid chromatography procedures can be recommended, since the individual barbiturates are separated before quantitation.

When individual barbiturates are known to be present, the ultraviolet absorption procedure also yields accurate results.[13] For phenobarbital any of these procedures may be used.[14] Most often, immunoassays such as the enzyme-multiplied immunoassay technique (EMIT of Syva) or the fluorescence polarization immunoassay of Abbott Diagnostics are performed.

A comparison of the colorimetric, gas-liquid chromatography, ultraviolet absorption, and EMIT procedures is given in Table 61-7. This shows the gas-liquid chromatography method to have the highest variability.

SPECIMEN

The specimen can be blood, plasma, serum, urine, or tissue extract.

PROCEDURE

Principle

The barbiturates are extracted from acidified serum with an organic solvent. They are then back-extracted into tetramethyl ammonium hydroxide. The extracts are introduced into a gas chromatograph, and the methylated derivatives of the drugs are formed on the column. The gas chromatograph detects, separates, and quantitates the drugs of interest, using aprobarbital as an internal standard.

Reagents

Toluene. Chromatoquality. This is stable for 2 years at room temperature.

Methanol (CH_3OH). Chromatoquality. This is stable for 2 years at room temperature.

Tetramethyl ammonium hydroxide. (Eastman). This is stable for 2 years at room temperature.

Barbiturates, obtained in pure form from manufacturer. These are stable for 2 years at room temperature.

1 M H_3PO_4. Add 6 mL of concentrated H_3PO_4 to 80 mL of H_2O. This is stable for 1 year at room temperature.

Working aprobarbital, internal standard, 1000 μg/mL (4.76 μmol/mL). Add 50 mg of aprobarbital to 50 mL of methanol. This is stable for 1 month at 4° C.

Calibration standard. Dissolve 50 mg each of amobarbital, butabarbital, pentobarbital, and secobarbital in 50 mL of methanol. To 99 mL of serum add 1 mL of the dissolved standard. This is mixed and frozen as 1 mL aliquots, which are stable for 1 year. The resultant calibration mixture contains 10 μg/mL (approximately 45 μmol/L) of each of the barbiturates.

Assay

Equipment: gas chromatographic equipment, including a flame ionization detector; a 6-foot glass column, 2 mm inner diameter, packed with 3% OV-1 on 80/100 mesh Chromosorb W (HP).

1. Sample preparation
   a. To a 10 mL screw-capped tube, add 1.0 mL pa-

tient serum (or serum calibrator), 0.5 mL of 1 M H_3PO_4, 100 μL of aprobarbital internal standard, and 5 mL of toluene.
   b. Shake vigorously for 1 minute and centrifuge.
   c. Transfer upper toluene layer to a 5 mL conical centrifuge tube.
   d. Add 12 μL of tetramethyl ammonium hydroxide, shake for 1 minute, and centrifuge. Discard upper layer.
   e. Inject 2 μL of lower layer into gas chromatograph. Start programmer as the solvent peak elutes.
2. Chromatographic parameters
   a. Column program: 90° to 200° C, 8°/min
   b. Injection port temperature: 275° C
   c. Detector temperature: 280° C
   d. Nitrogen flow: 60 mL/min
   e. Attenuation: 16, range 10^{-12}
   f. Recorder: 1 mV, 0.5 inches/min
3. Sample run time is approximately 20 minutes.
4. Run known quality control specimens with each run.

Calculations

1. Measure retention times of each peak and compare to the serum calibrator standards.
2. To calculate each corresponding drug:

$$\frac{\text{Peak height of drug}}{\text{Peak height of internal standard}} = \text{Ratio of drug to aprobarbital}$$

Divide peak height ratios of unknowns by peak-height ratios of the calibration standard. Multiply resulting ratio by 10 μg/mL:

$$\frac{\text{Ratio of unknown}}{\text{Ratio of calibrated material}} \times 10\ \mu\text{g/mL} = \text{Unknown } (\mu\text{g/mL})$$

REFERENCE RANGES

Therapeutic concentrations in serum
1. Amobarbital: up to 8 μg/mL (< 35 μmol/L)
2. Butabarbital: up to 8 μg/mL (< 34 μmol/L)
3. Pentobarbital: up to 4 μg/mL (< 16 μmol/L)
4. Phenobarbital: 15 to 40 μg/mL (64-172 μmol/L)
5. Secobarbital: up to 6 μg/mL (23 μmol/L)

REFERENCES

1. Curry, A.S.: Rapid quantitative barbiturate estimation, Br. Med. J. **1**:354-355, 1964.
2. Walker, J.T., Fisher, R.S., and McHugh J.J.: Qualitative estimation of barbiturates in blood by ultraviolet spectrophotometry, Am. J. Clin. Pathol. **18**:451-461, 1948.
3. Broughton, P.M.G.: A rapid ultraviolet spectrophotometric method for the detection, estimation, and identification of barbiturates in biological material, Biochemistry **63**:207-213, 1956.
4. Goldbaum, L.R.: Determination of barbiturates: ultraviolet spectrophotometric method with differentiation of several barbiturates, Anal. Chem. **24**:1604-1607, 1952.
5. Schumann, G.B., Lauenstein, K., LeFever, D., and Henry, J.B.: Ultraviolet spectrophotometric analysis of barbiturates, Am. J. Clin. Pathol. **66**:823-830, 1976.
6. Ganshirt, H.: Pharmaceutical products. In Stahl, E., editor: Thin-layer chromatography: a laboratory handbook, New York, 1962, Academic Press, Inc., pp. 318-319.
7. Sunshine, I., Maes, R., and Finkle, B.: An evaluation of methods for the determination of barbiturates in biological materials, Clin. Toxicol. **1**:281-296, 1968.
8. MacGee, J.: Rapid determination of diphenylhydantoin in blood plasma by gas liquid chromatography, Anal. Chem. **42**:421-422, 1970.
9. MacGee, J.: Rapid identification in quantitative determination of barbiturates in glutethimide in blood by gas liquid chromatography, Clin. Chem. **17**:587-591, 1971.
10. Solow, E.B., and Green, J.B.: The simultaneous determination of multiple anticonvulsant drug levels by gas liquid chromatography, Neurology **22**:540-550, 1972.
11. Atwell, S., Green, V., and Honey, W.: Development and evaluation of method for simultaneous determination of phenobarbital and diphenylhydantoin in plasma by HPLC, J. Pharm. Sci. **64**:806-809, 1975.
12. Sunshine, I.: Methodology for analytical toxicology, Cleveland, 1975, CRC Press, Inc., pp. 34-44.
13. Spiegel, H.E., Chamberlain, R.T., Dubowski, K.M., et al.: Barbiturates, ultraviolet spectrophotometric method. In Faulkner, W.R., and Meites, S, editors: Selected methods of clinical chemistry: vol. 9, Selected methods for the small clinical laboratory, Washington, D.C., 1982, American Association for Clinical Chemistry, pp. 109-111.
14. Spiehler, V., Sun, L., Miyada, D.S., et al.: Radioimmunoassay, enzyme immunoassay, spectrophotometry and gas-liquid chromatography compared to determination of phenobarbital and diphenylhydantoin, Clin. Chem. **22**:749-753, 1976.

Bromides

JOSEPH SVIRBELY

Clinical significance: p. 897
Atomic formula: Br^-
Atomic weight: 79.90 daltons
Merck Index: potassium bromide, 7395; sodium bromide, 8338
Chemical class: anion (halide)

PRINCIPLES OF ANALYSIS AND CURRENT USAGE

Potassium bromide and sodium bromide are used as sedatives and anticonvulsants. Bromide ion is also an end product of halothane metabolism after halothane anesthesia. These two different sources of bromide ion require different types of analysis, as described in this section.

Most conventional methods for measuring serum bromide are based on the measurement of gold bromide, which is formed by the reaction of this halide with gold chloride solution. The gold chloride reaction was first used to measure bromide ion by Walter[1] in the 1920s. Wuth[2] and Diethelm[3] further modified the procedure and recommended it for diagnosis of bromide intoxication. Both methods made use of color comparison to known standards without optical assistance. Hauptmann's modification[4] involved use of a color comparator and gave better results.

The method of Wuth was reviewed and modified by

Table 61-8. Methods for bromide analysis

Method	Type of analysis	Principle	Usage	Comments
1. Gold chloride	Manual; spectrophotometric at 470 nm	$Au^{+3} + 3Br^{-} \rightarrow$ Gold bromide	Preferred method for monitoring bromide therapy	Simple and relatively sensitive at concentrations above 100 μg/mL
2. Gas-liquid chromatography	Chromatographic; detect with flame ionization detector	$2Br^{-} \rightarrow Br_2$ Br_2 + 2,4-Dimethylphenol → Bromodimethylphenol	Preferred method for low levels of bromide	Sensitive; range of 10-150 μg/mL Uses equipment available in most laboratories Very good accuracy and precision
3. Bromate–rosaniline dye	Spectrophotometric at 570 nm	$Br^{-} + 3ClO^{-} \rightarrow BrO_3 + 3Cl^{-}$ $BrO_3^{-} + 5Br^{-} + 6H^{+} \rightarrow 3Br_2 + 3H_2O$ $2Br_2$ + Rosanilinine dye → Bromorosaniline	Alternative to method 2	Adequate method for measurement of low levels of bromide Sensitive in range of 10-100 μg/mL
4. Neutron activation	Radiochemical	Bromide irradiated with neutrons, forming ^{82}Br; bromine precipitated with silver nitrate and counted	Alternative to method 2	Sensitive, but requires apparatus not commonly used in laboratories Range of 10-80 μg/mL

Gray and Moore,[5] who used a photoelectric comparator that had been standardized with 10 solutions of sodium bromide treated with gold chloride. They measured bromide in protein-free filtrate of the serum. Modifications of this procedure are in common use today and are satisfactory for measuring bromide ion, which is derived from sodium bromide or potassium bromide therapy, when levels are commonly 100 to 2000 μg/mL (method 1, Table 61-8).

Bromide levels are also used to monitor halothane metabolism; the gold chloride method is inadequate for this, since levels from this source are usually less than 100 μg/mL and interference by chloride ion is a problem. Hunter[6] developed a method for microdetermination of bromide in body fluids by using a bromate–rosaniline dye color reaction (method 3, Table 61-8). Organic matter was destroyed by incineration with alkali, a time-consuming step. Goodwin[7] modified the procedure by replacing the incineration step with a deproteination step, using a sodium tungstate–hydrochloric acid solution as the precipitant. The wavelength of maximum absorbance for the chomogen of this reaction is 570 nm, and the range of linearity is 10 to 100 μg/mL.

A sensitive method suitable for measurement of halothane metabolites, described by Corina et al.,[8] uses gas-liquid chromatography (GLC) (method 2, Table 61-8).The technique involves deproteination, followed by oxidation of bromide ion to bromine with hypochlorite. The bromide is derivatized with dimethylphenol and analyzed by GLC on an OV-17 column with a flame ionization detector (FID). Linearity is in the range of 10 to 150 μg/mL, and a rapid analysis time of 6 minutes has been reported.

Atallah and Geddes[9] made use of neutron-activation analysis, a more involved technique than the colorimetric or GLC methods requiring equipment and training not available in most clinical chemistry laboratories (method 4, Table 61-8). Sensitivity is again in the range of 10 μg/mL, but recoveries fall from 100% at 10 μg/mL to 80% at 80 μg/mL.

Measurement of bromide ion in urine by spectrophotometry or GLC requires the addition of charcoal to the urine to remove interfering pigment.[10]

REFERENCE AND PREFERRED METHODS

Excluding the specialized needs of the laboratory studying the pharmacokinetics of halothane anesthesia, the method of choice for measurement of serum bromide is still the gold chloride method using a protein-free filtrate derived by adding 10% trichloroacetic acid to the serum. The sensitivity is adequate for monitoring bromide therapy and interference by chloride ion is not a problem. No specialized equipment is required, and technician training is minimal. The standard curve is stable for each lot of gold chloride solution, eliminating the need to run a standard with the control and sample specimens. The manual method reported below is in common usage and has been used successfully in our laboratory for 5 years.

For measurement of low levels of bromide derived from halothane metabolism the preferred method is the GLC method because of its greater range of linearity (10-150 μg/mL as opposed to 10-100 μg/mL) and because of the rapidity of analysis possible (analysis time of 6 minutes per sample). The colorimetric method is adequate as an alternative to the GLC method, but the neutron activation analysis method requires equipment and expertise not necessarily found in the clinical chemistry laboratory.

SPECIMEN

Serum or plasma is the usual sample.

PROCEDURE

Reaction

The spectrophotometric method used below is based on bromide ion complexing with gold ion to form gold bromide ($AuBr_3$), which is soluble in water and absorbs at 470 nm.[2] See summary chart below:

Summary of reaction conditions for bromide assay

Condition	Colorimetric
Temperature	Ambient
pH	Acid
Sample volume	1 mL
Fraction of sample volume	0.111 (precipitation) 0.044 (reaction)
Precision ($\bar{X}$, %CV)	532 μg/mL, 6.0%
Linearity	150-3000 μg/mL
Interferences	None

Reagents

Gold chloride (3.85 g/L, 12.7 mmol/L). Dissolve 500 mg of gold chloride ($AuCl_3 \cdot HCl \cdot 3\ H_2O$) in 50 mL of distilled water and dilute to 100 mL with distilled water. Stable for 1 year at 4° to 8° C.

Trichloroacetic acid (TCA, 100 g/L, 612 mmol/L). Dissolve 10 g of trichloroacetic acid in distilled water and dilute to 100 mL with distilled water. Stable for 2 years at room temperature.

Stock bromide standard solution (3 g of Br^-/L, 37.6 mmol/L). Dissolve 0.448 g of potassium bromide in distilled water and dilute to 100 mL. This is stable for 1 year at 4° to 8° C.

Working bromide standard solution. Dilute 0.5, 1, 2, 3, 4, 5, 6, 7, 8, 9, and 10 mL of stock bromide standard solution to 10 mL with distilled water. The resulting solutions have bromide concentrations of 150, 300, 600, 900, 1200, 1500, 1800, 2100, 2400, 2700, and 3000 μg/mL, respectively. These standards are stable for 1 year at 4° to 8° C.

Assay

Equipment: ≤10 nm band pass spectrophotometer

1. Add 1.0 mL of serum, control or standard, to a 10 mL screw-capped test tube.
2. Add 4.0 mL distilled water and 4.0 mL of trichloroacetic acid (TCA) reagent.
3. Mix the solution.
4. Let stand 10 minutes at room temperature, centrifuge at 3000 rpm (1000 *g*), and filter through hardened quantitative filter paper (Whatman No. 42).
5. In a 10-inch test tube place 4 mL of clear filtrate.
6. Add 1.0 mL of gold chloride solution.
7. Prepare a blank by adding 1.0 mL of gold chloride solution to a test tube containing 2 mL of distilled water and 2 mL of trichloroacetic acid (TCA) solution.
8. Let stand 10 minutes at room temperature to develop color.
9. Read each specimen against reagent blank in spectrophotometer set at 470 nm.

Calculation

Plot absorbance of each standard versus concentration on linear graph paper. Read absorbance of each control and unknown from the linear curve to determine concentration.

Notes

The standard curve is stable from run to run, requiring that only a control sample be run daily, along with patient samples. A standard curve should be run with each lot of gold chloride or if the control is out of range.

REFERENCE RANGE

The therapeutic range is usually less than 1500 μg/mL.

REFERENCES

1. Walter, F.K.: Studien über die Permeabilität der Meningen, Z. Ges. Neurol. Psychiatr. **95:**522-540, 1925.
2. Wuth, O.: Rational bromide treatment, J.A.M.A. **88:**2013-2017, 1927.
3. Diethelm, O.: On bromide intoxication, J. Nerv. Ment. Dis. **71:**151-165, 278-292, 1930.
4. Hauptmann, A.: Investigations on passage from blood to cerebrospinal fluid in psychoses, Z. Ges. Neurol. Psychiatr. **100:**332-343, 1926.
5. Gray, M.G., and Moore, M.: Blood bromide determination: their use and interpretation, J. Lab. Clin. Med. **27:**680-686, 1941.
6. Hunter, G.: Microdetermination of bromide in body fluids, Biochem. J. **60:**261-264, 1955.
7. Goodwin, J.F.: Colorimetric measurement of serum bromide with a bromate-rosaniline method, Clin. Chem. **17:**544-547, 1971.
8. Corina, D.L., Ballard, K.E., Grice, D., et al.: Bromide measurement in serum and urine by an improved gas chromatographic method, J. Chromatogr. **162:**382-387, 1979.
9. Atallah, M.M., and Geddes, I.C.: Determination of bromides in blood by neutron activation analysis, Br. J. Anaesth. **45:**134-138, 1973.
10. Dunlop, M.: Simple colorimetric method for determination of bromide in urine, J. Clin. Pathol. **20:**300-301, 1967.

Digoxin

I-WEN CHEN
LINDA A. HEMINGER

Clinical significance: p. 488
Molecular formula: $C_{41}H_{64}O_{14}$
Molecular weight: 780.92 daltons
Merck Index: 3144
Chemical class: glycoside

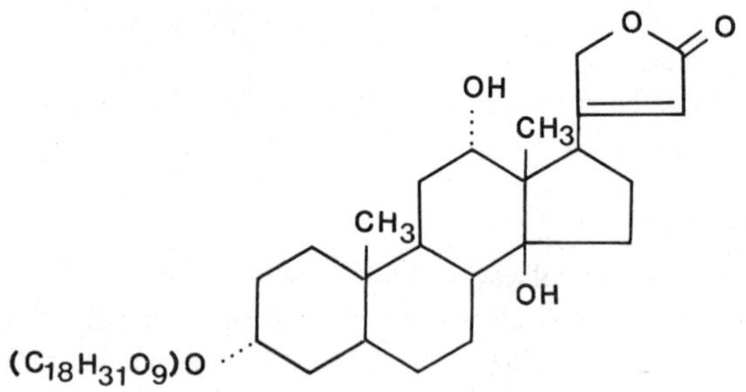

$(C_{18}H_{31}O_9)O$

PRINCIPLES OF ANALYSIS AND CURRENT USAGE

Digoxin is a cardiotonic glycoside obtained in the crystalline form from the leaves of *Digitalis lanata*. It is insoluble in water, chloroform, ether, ethyl acetate, or acetone but soluble in dilute (50%) alcohol or pyridine. It is used to control cardiac arrhythmias through improvement of the strength of myocardial contraction. However, digoxin intoxication is a frequent complication of digoxin therapy and can be life-threatening. Determination of digoxin concentration in blood has been helpful in the diagnosis of digoxin toxicity and the establishment of an optimal dosage because of a reasonably close correlation between blood and tissue concentrations of digoxin.

Many techniques have been used to determine blood levels of digoxin. They include gas chromatography[1] (method 1, Table 61-9), methods based on inhibition by digoxin of sodium-, potassium-activated adenosine triphosphatase (ATPase) in red blood cells[2] (method 2, Table 61-9) or in microsomes,[3] and a sodium-, potassium-activated ATPase isotopic displacement method (method 3, Table 61-9).[4] However, these methods are relatively tedious and have been largely replaced by more practical and sensitive radioimmunoassay (RIA) techniques (method 4, Table 61-9) in routine clinical chemistry laboratories.

Digoxin is one of the first hapten-like small molecules to which the RIA technique was applied.[5] Digoxin antisera were successfully raised in animals immunized with digoxin coupled to a variety of macromolecules, including poly-L-lysine, human serum albumin, and bovine serum albumin. Unlike thyroid hormones, a digoxin molecule contains no iodine as a native atom nor any functional groups that can be iodinated. Tritiated digoxin preparations were initially the only radioligands available for digoxin RIA. Tritiated radioligands as beta-particle emitters had to be measured by means of liquid scintillation counting, which was expensive and inconvenient in comparison with the counting of gamma-ray emitters such as radioactive iodine (^{125}I) in a crystal scintillation well counter. This problem was overcome by the development of techniques that attach to a digoxin molecule a functional group that could be easily iodinated. Tyrosine, histamine, and tyramine have been coupled to a digoxin or digoxigenin molecule for radioiodination. An example of such derivatives is 3-*O*-succinyl digoxigenin tyrosine. When this compound is radioiodinated, a monoiodinated and also a diiodinated derivative can be obtained. It is important that the radioiodinated 3-*O*-succinyl digoxigenin tyrosine used in digoxin RIA be free of the diiodinated derivative, since the diiodinated derivative is known to bind to thyroxine-binding globulins.[6] Nearly every separation technique used in RIA's has been applied successfully to the RIA of digoxin. The dextran-coated charcoal method seems to be the most widely used method, though the solid-phase method is popular among commercial digoxin kits because of its technical simplicity and rapidity.

Other types of labels have been developed as alternatives to radionuclides in digoxin RIA. In the enzyme-linked immunosorbent assay (ELISA) (method 5a, Table 61-9) developed by ICL Scientific (Fountain Valley, Calif.), an enzyme, peroxidase, is used as a label in place of a radionuclide. The peroxidase-labeled digoxin competes with the digoxin in serum for the binding sites on the digoxin antibody coated on the wall of a test tube. After the reaction the test tube is aspirated and washed. The peroxidase activity remaining in the tube, which is inversely proportional to the digoxin concentration in the serum, is determined photometrically, based on the ability of the enzyme to oxidize a suitable substrate. As with RIA, this is a heterogeneous system requiring the physical separation of bound and free digoxin. In the homogeneous enzyme-multiplied immunoassay technique (EMIT, Syva Co., Inc., Palo Alto, Calif.) (method 5b, Table 61-9), an enzyme, glucose-6-phosphate dehydrogenase, chemically coupled to digoxin is used as a label. When the enzyme-digoxin conjugate is bound to digoxin antibody, the enzyme is inactivated. In the presence of digoxin in a sample, the enzyme-digoxin conjugate does not bind to the digoxin antibody and the enzyme is active. Therefore, one can monitor the extent of digoxin-antibody binding and thus the digoxin concentration by measuring the enzyme activity spectrophotometrically without physically separating the bound and free fractions. Although EMIT was developed originally as a manual assay system, it has been successfully adapted to a minidisk centrifugal analyzer[7] and a biochromatic analyzer.[8]

A recently introduced but increasingly popular assay is based upon measurement of the fluorescence polarization

Table 61-9. Methods for digoxin analysis

Method	Principle	Usage	Comments
1. Gas chromatography	Physicochemical separation; tritium (^3H)–digoxin as internal standard	Plasma; may be applied to other biological specimens	Requires prepurification by thin-layer chromatography and derivatization Tedious and time consuming.
2. Adenosine triphosphatase (ATPase) inhibition	Dose-dependent inhibition of sodium- and potassium-activated ATPase	Plasma; may be applied to other biological specimens	Requires extraction Tedious and time consuming Less precise and less specific than method 4
3. Isotope displacement	Displacement of ^3H-ouabain from purified sodium-potassium ATPase preparation by digoxin	Plasma; may be applied to other biological specimen	Requires extraction and beta-particle counting Tedious and nonspecific
4. Radioimmunoassay	Competitive binding of ^3H- or ^{125}I-labeled and nonlabeled digoxin with digoxin antibody	Plasma or serum only	Requires phase separation and beta-particle or gamma-ray counting Fully automated Most sensitive and precise
5. Enzyme immunoassay			
a. Heterogeneous system	Competitive binding of enzyme-labeled digoxin and nonlabeled digoxin with digoxin antibody	Plasma or serum only	Requires no sample pretreatment but requires phase separation System semiautomated
b. Homogeneous system	Competitive binding of enzyme-labeled digoxin and nonlabeled digoxin with digoxin antibody	Plasma or serum only	Requires no phase separation but requires sample pretreatment System semiautomated
c. Fluorescence polarization immunoassay	Competitive binding of fluorescein-labeled digoxin and nonlabeled digoxin with digoxin antibody	Plasma or serum only	Requires deproteinization System semiautomated

in a competitive-binding assay system (Abbott Diagnostics, North Chicago, Ill.) (method 5c, Table 61-9).

REFERENCE AND PREFERRED METHODS

Among several methods developed to determine serum concentrations of digoxin in the recent past, the radioimmunoassay (RIA) technique is used most widely and is considered to be the simplest and most sensitive method. Commercial digoxin RIA kits with ^{125}I-labeled ligands were used by 90% of the laboratories that participated in the 1983 College of American Pathologists survey. Gas chromatography, ATPase inhibition, and isotope displacement methods are described briefly in Table 61-9 only for historical interest or for a specified use. The enzyme and fluorescence immunoassays are used by less than 10% of the participants in the therapeutic drug monitoring survey sponsored by the American Association for Clinical Chemistry. These methods, as currently formulated (1983), have a higher coefficient of variation than RIA methods. For the EMIT system, this is partly a result of incomplete inhibition of the enzyme at low digoxin levels and the need for long incubation times to amplify the enzyme signal.[9]

Digoxin RIA is performed in most laboratories with commercially available kits that contain all the necessary reagents. A variety of commercial digoxin RIA kits with different radiolabeled ligands and different separation techniques are available. The last quarterly report of the 1982 College of American Pathologists survey listed 26 commercial kits manufactured by 22 different companies. Although most commercial kits are well-characterized systems, their performance characteristics should be evaluated carefully in each laboratory before they are selected for routine use. It is strongly recommended that a commercial kit be used according to the exact protocol suggested by the kit manufacturer. This will greatly simplify the quality control problems and troubleshooting in case of assay difficulty. However, it is occasionally desirable to modify the manufacturer's recommended procedure to satisfy the needs of each laboratory. In our laboratory assay conditions of a commercial digoxin RIA kit were modified for use as an emergency procedure.[10]

In the interlaboratory transferability study of digoxin assays sponsored by subcommittees of the Centers for Disease Control, the National Committee for Clinical Standards, and the American Association for Clinical Chemistry, a digoxin RIA method with tritium (^3H)–labeled digoxin as the radioligand and charcoal as the separation agent was used as the candidate reference method, since the ^3H-digoxin method was found to be comparable with the ^{125}I-digoxin method and the ^3H-digoxin provided the advantages of isotope stability and general availability of high-quality labeled digoxin. The preliminary data in-

Table 61-10. Comparison of reaction conditions for digoxin analysis

Conditions	Manual RIA[10] (Becton Dickinson)	Automated RIA[12] (ARIA-II)	EMIT[8] (Syva)	ELISA[15] (ICL Scientific)
Temperature	25° C	25° C	25° C	25° C
pH	7.0	7.2	7.2	7.5
Sample volume	50 μL	50 μL	200 μL	100 μL
Sample pretreatment	Not required	Not required	0.5 M NaOH at 25° C for 5 min	Not required
Tracer	^{125}I	^{125}I	Glucose-6-phosphate dehydrogenase	Peroxidase
Antibody	Rabbit, liquid	Rabbit, liquid	Sheep, liquid	Bovine, coated tube
Enzyme substrate	Not required	Not required	NAD*	H_2O_2, ABTS†
Reaction time	30 min	18 min	45 min	60 min
Sensitivity‡	0.4 μg/L	0.1 μg/L	0.5 μg/L	0.3 μg/L
Precision (coefficient of variation)	5.3%-8.6%	3.6%-5.7%	4.8%-9.9%	5.8%-7.7%
Interferences	Digitoxin, digoxin metabolites	Digitoxin, digoxin metabolites	Digitoxin, digoxin metabolites	Digitoxin, digoxin metabolites

*Nicotinamide adenine dinucleotide.
†Diammonium 2,2-azino-bis(2-ethylbenzothiazoline-6-sulfonate).
‡Minimum detectable concentration.

dicate that a within-run coefficient of variation (CV) of less than 5% is attainable by this reference method.[11]

RIA methods for digoxin have been fully automated. Of the three systems evaluated in our laboratory, the ARIA-II system gave the highest precision (CV, 3.6% to 5.7%), followed by the Gammaflo (CV, 4.0% to 13.5%), and the Concept 4 (CV, 7.6% to 16.3%).[12] Coefficients of variation of the manual system (Becton Dickinson Immunodiagnostics, Orangeburg, N.Y.) used in the study were in the range of 5.3% to 8.6%. Because of the lowest throughput, equivalent to about 30 RIA tubes per hour, the ARIA-II may be less than ideal for laboratories with large test volumes (more than 100 samples assayed in duplicate). The same company recently introduced a second-generation instrument (ARIA-HT), consisting of up to four independent channels equipped with two radioactive detectors each, for simultaneous counting of bound and free fractions. This instrument has a maximum throughput of 170 to 200 analyses per hour.

Enzyme immunoassays have been semiautomated but have not received wide acceptance in clinical laboratories, since they are not fully automated and are not so sensitive or precise as automated RIA methods (Table 61-10). The fluorescence polarization immunoassay is the most rapidly performed commercially available assay (15 minutes) and has a coefficient of variation as good as RIA. This assay is used with increasing frequency.

SPECIMEN

In most instances either serum or plasma may be used for the assay, though some commercial manufacturers specifically indicate not to use plasma in their assay kit. Digoxin in serum is quite stable; no appreciable loss of digoxin is observed in serum stored at room temperature up to 2 weeks. However, if the specimen is not tested within 24 hours, it should preferably be stored frozen.

Time of blood collection is an important factor in determining digoxin toxicity. Serum digoxin levels will rise sharply during the first hour after an oral dose, indicating postabsorption uptake. This is followed by a sharp decrease as digoxin is taken up by the myocardium and other tissues. Usually, 4 to 6 hours are needed for serum and tissue stores of digoxin to reach equilibrium. Thereafter, serum digoxin levels tend to stabilize and decrease very slowly over 24 to 40 hours. One obtains the greatest diagnostic accuracy by drawing the blood sampls at a standard time during the stable phase, when serum digoxin levels reflect the average cardiac concentrations. In our laboratory blood samples are routinely drawn between 5 and 30 hours after the last digoxin dose.

Serum samples obtained from patients who are being switched from digitoxin to digoxin are not suitable for determination of digoxin by immunoassays since both drugs may be present in their sera and since digoxin antisera used in most commercial digoxin immunoassay kits cross-react significiantly with digitoxin (2% to 6% cross-reaction, with some as high as 40%). The problem of cross-reaction is further enhanced by the therapeutic levels of digitoxin, being about 10 times higher than those of digoxin.

REFERENCE RANGE

As with all diagnostic tests, it is advisable for the individual laboratory to accumulate and analyze its own data to establish a range of toxic and nontoxic values; differences in assay techniques used may affect assay results. In our laboratory we found mean digoxin levels of 1.1 ± 0.1 ng/mL for 108 nontoxic patients and 3.8 ± 0.5 ng/mL for 21 toxic patients. Ninety percent of the nontoxic patients had serum digoxin levels below 2 ng/mL, whereas 86% of the toxic patients had levels greater than 2 ng/mL.[13]

It is generally believed that infants are more resistant to cardiac glycosides than adults. O'Malley et al. found that a mean nontoxic digoxin level of 3.8 ± 1.2 ng/mL for eight infants younger than 1 month was significantly higher than 1.4 ± 1.1 ng/mL for 16 adults, but the digoxin level in infants older than 1 month (1.4 ± 0.5, $n = 5$) was no higher on the average than that of the adult controls.[14]

REFERENCES

1. Watson, E., and Kalman, S.M.: Assay of digoxin in plasma by gas chromatography, J. Chromatogr. **56:**209, 1971.
2. Bertler, A., and Redfors, A.: An improved method of estimating digoxin in human plasma, Clin. Pharmacol. Ther. **11:**665, 1970.
3. Burnett, G.H., and Conklin, R.L.: The enzymatic assays of plasma digoxin, J. Lab. Clin. Med. **78:**779, 1971.
4. Brooker, G., and Jelliffe, R.W.: Serum cardiac glycoside assay based upon displacement of ^{3}H-ouabain from Na-K-ATPase, Circulation **45:**20, 1972.
5. Butler, V.P., Jr,. and Chen, J.P.: Digoxin specific antibody, Proc. Natl. Acad. Sci. **57:**71, 1967.
6. Pinter, K., and Vader, C.R.: Interference of iodine-125 ligands in radioimmunoassay: evidence implicating thyroxine-binding globulin, Clin. Chem. **25:**797, 1979.
7. Brunk, S.D., and Malmstadt, H.V.: Adaptation of the EMIT serum digoxin assay to a mini-disc centrifugal analyzer, Clin. Chem. **23:**1054, 1977.
8. Eriksen, P.B., and Andersen, O.: Homogeneous enzyme immunoassay of serum digoxin with use of a biochromatic analyzer, Clin. Chem. **23:**169, 1979.
9. Pesce, A.J., Ford, D.J., and Gaizutis, M.A.: Qualitative and quantitative aspects of immunoassays, Scand. J. Immunol. **8**(suppl. 7):1-6, 1978.
10. Chen, I.W., Sperling, M.T., and Maxon, H.R.: A commercial digoxin radioassay modified for use as an emergency ("stat") procedure, Clin. Chem. **25:**1841, 1979.
11. Myrick, J.E., Hannon, W.H., Powell, M.K., et al.: An interlaboratory transferability study of a candidate reference RIA method for digoxin, Clin. Chem. **26:**971, 1980.
12. Chen, I.W.: Commercially available fully automated systems for radio-ligand assay. Part I. Overview, Ligand Rev. **2**(2):46, 1980; Part II. Performance characteristics, Ligand Rev. **2**(3):45, 1980.
13. Park, H.M., Chen, I.W., Manitasas, G.T., et al.: Clinical evaluation of radioimmunoassay of digoxin, J. Nucl. Med. **14:**531, 1973.
14. O'Malley, K., Coleman, E.N., Doig, W.B., et al.: Plasma digoxin levels in infants, Arch. Dis. Child. **48:**55, 1973.
15. Enzymune-Test™ Digoxin, Enzyme-Linked Immunosorbent Assay (ELISA), ICL list no. 85040, ICL Scientific, Fountain Valley, Calif.

Drug screen

F. MICHAEL HASSAN

Clinical significance: p. 897

PRINCIPLES OF ANALYSIS AND CURRENT USAGE

"Coma panel, toxic screen, complete drug screen, comprehensive screen, drug abuse screen, overdose panel, specific screen for drug identification," and so on—these test names used by various laboratories suggest that the term "drug screen" is nebulous at best. Each laboratory defines an analysis for possible toxic substances in a way that enables its personnel to obtain the information required by the physician ordering the test. There are many different types of qualitative drug screens, each developed with that particular laboratory's interests, finances, and personnel expertise in mind.

To describe a specific drug screen, one must consider the reasons for screening most often encountered by the clinical chemist. Areas in which drug screening is often indicated include (1) clinical toxicology, (2) forensic/medicolegal toxicology, (3) employment screening, (4) drug trials, and (5) drug abuse programs. Screening for forensic purposes is still largely the domain of the coroner's laboratory. Drug screening programs have been adopted by the clinical laboratories, with special emphasis on clinical toxicology (the overdose patient) and drug abuse screening programs.

Defining the test

Qualitative analysis for a drug or a group of drugs can generally be divided into three basic stages: (1) isolation or separation of the compounds of interest from the biological specimen, (2) analysis of its content by a "screening" method, and (3) employment of one or more "confirmatory" methods to validate the findings of the second stage. Screening methods use analytical techniques that give either a positive or negative finding, with a second analytical technique required to confirm any positive findings. Numerous possible analytical permutations exist for the screening-confirmation format.

Most drug screens fall into one of two classes: (1) a specific screen for a particular known or suspected drug (or class of drug), or (2) a complete screen, with analysis for as many drugs as is technically and economically feasible for the laboratory to measure. Of course, searching for one particular compound is much easier than searching for any one or more of a large possible number of compounds. If a single known or suspected drug is of interest, the analyst can immediately optimize the screening test to ensure adequate specificity and sensitivity. If the screen is broadened to include several compounds, a series of specific analyses for each compound can be combined together to form a limited or "panel" type of screen. When arranged in this format, one tests each compound by a specific procedure that optimizes sensitivity. It is evident that if the number of compounds is increased further, the time required for performing specific individual assays for a large group of drugs limits this approach. When it is necessary to expand the screen to include many substances, less specific assays that enable the analyst to look for a vast array of compounds at a single glance must be employed. The actual number of drugs included in this expanded screening process varies from laboratory to laboratory and depends heavily on the experience of the analyst performing the assay.

Once the laboratory has decided which drugs are feasible to include in its qualitative drug identification test list, it must gather the analytical data on these drugs and their metabolites. The necessity of measuring metabolites cannot be overemphasized. Many drugs are present in urine only as metabolites; thus it is essential to have standards of known metabolites and of the parent drugs as well for accurate identification and confirmation.

General analytical procedures

Excellent general references detail not only general screening procedures but also specific individual assay techniques and present fundamental reviews of the technologies involved in toxicological analyses.[1-3] Selected general procedures utilized in drug screening and confirmation procedures are presented here. Table 61-11 presents suggested confirmation procedures for the tests discussed, and Table 61-12 is an overview of each type of test.

Extraction

With the exception of the spot tests, most procedures used by the analyst require preparation of the specimen before actual analysis. For the drug screen, preparation is usually in the form of an extraction procedure. Extraction of urine samples is generally by a two-phase system, in which the drug is separated from the aqueous phase and redistributed into a second, usually organic, phase. This results in the removal of many of the interfering compounds present in the specimen. The degree of complexity of the extraction process depends on the type of analytical method used and the degree of freedom from interfering compounds required of the processed specimen.

Since drugs are extracted from matrices according to their ionization state, partitioning of the specimen constituents into acidic, neutral, and basic fractions is possible. Enhanced extraction efficiencies are obtained by preparation of extractions from specimens made acidic or alkaline to yield higher recoveries of drug. When shortened turnaround times are a necessity, single pH extractions can be performed (see following discussion of thin-layer chromatography). These shortened extractions present a compromise situation, with most drug groups being extracted to some extent but with lower recovery yields.

Of the exchange resins used for extraction of drugs, the nonionic resin XAD-2 (nonionic polystyrene exchange resin, Rohm and Haas Co., Philadelphia) has been most widely used for general drug screening procedures. In this technique, urine is poured over a column bed containing the resin. Drugs are adsorbed onto the resin and then subsequently eluted off with appropriate solvents. As with conventional solvent extractions, the proper selection of sample pH and solvents for elution optimize the selection process.

Table 61-11. Suggested confirmation procedures for screening tests

Screen		Confirmation
SPOT TEST	☞	UV
SPOT TEST	☞	TLC
TLC	☞	UV
TLC	☞	GC
EMIT	☞	TLC
EMIT	☞	GC
EMIT	☞	UV
RIA	☞	TLC
GC	☞	GC/MS

Spot tests

The spot tests are among the simplest and quickest methods of providing an initial screening process or a confirmatory test. Usually the unprocessed specimen (urine or serum) is added directly to test reagents to assay for the presence of a specific drug or class of drugs, which is presumptively indicated by the formation of a colored reaction product. Table 61-13 lists commonly used spot tests routinely employed by the laboratory.[4-11] Although spot tests offer inexpensive testing with good turnaround times and provide an early lead for the analyst, they should be used selectively and with care, since they are relatively nonspecific and often require subjective interpretation.

Ultraviolet spectroscopy

As with the spot tests, ultraviolet analysis is best employed in testing for a specific drug or confirming compounds detected by other techniques. The ultraviolet scan has relatively poor specificity unless the unknown drug is isolated by other test procedures, such as thin-layer chromatography, or extracted by procedures that minimize spectral interferences from the body fluids and from other drugs. Fig. 61-7 emphasizes that compounds may have overlapping or similar ultraviolet spectra, necessitating

Table 61-12. Types of assays used in screening procedures

Assay	Principle	Advantages	Disadvantages
Spot test	Drug or agent in specimen reacts, giving specific color (see Table 61-13)	Very rapid test Indicates possible drug group present	Not specific Different spot test required for each group of drugs Limited to a few drug groups
Enzyme immunoassay	Competitive binding assay between nonlabeled and enzyme-labeled drug	Rapid for single or several assays Most convenient method for detection of certain drugs	Class specific Assays must be done separately for each class of drugs or individual drug Costly
Gas chromatography	Extracted drug chromatographed and identified by retention time	Can separate and detect large variety of drugs	Not possible to detect all drugs Extraction and confirmation necessary
Thin-layer chromatography	Extracted drug chromatographed and detected by specific chemical reaction or properties	Can separate and detect more compounds than any other method	Not possible to detect all drugs with one extraction or one chromatography system Detection is subjective
Ultraviolet spectroscopy	Extracted drug identified by its specific absorbance spectrum	Spectrum often gives specific identification and quantitation of compound	Not all drugs have suitable absorbance spectra Limited sensitivity Some degree of ambiguity among drugs of same class

Table 61-13. Spot tests

Drug	Specimen required	Reaction	Test time (minutes)	Comments	Reference
Acetaminophen	Urine	o-Cresol + Acetaminophen $\xrightarrow{NH_4OH}$ Blue color	15	Highly sensitive and relatively specific	4
Ethanol	Urine, serum	Microdiffusion into dichromate $2K_2CrO_7 + 10H_2SO_4 + 3C_2H_5OH \rightarrow 2Cr_2(SO_4)_4 + 2K_2SO_4 + 3CH_3COOH + 11H_2O + 4H^+$ (Green to blue color)	15-30	Good sensitivity Nonspecific for ethanol	5
Salicylate	Urine, serum	Trinder's solution Salicylate + $FeCl_3$ → Violet-colored complex	2	Good specificity if serum used Good sensitivity	6
Carbamates (meprobamate)	Urine	Furfurol + Meprobamate + Antimony trichloride → Black color on thin-layer chromatography plate	5	Not specific for meprobamate Good sensitivity	7
Imipramine/desipramine	Urine	Forrest reagent K_2CrO_3 (acidic) + Imipramine → Green-colored complex	2	Phenothiazines may interfere	8
Ethchlorvynol	Urine, serum	Diphenylamine + Ethchlorvynol in acid → Red color	10	Good sensitivity and specificity	9
Phenothiazines	Urine	FPN reagent (ferric chloride/perchloric acid/nitric acid) $FeCl_3$ (acidic oxidizing agent) → Red- to violet-colored complex	2	Nonspecific Poor sensitivity for some phenothiazines	10
Iron	Serum	Bathophenanthroline color change with iron (blue)	15	Will not react at normal serum concentrations	11

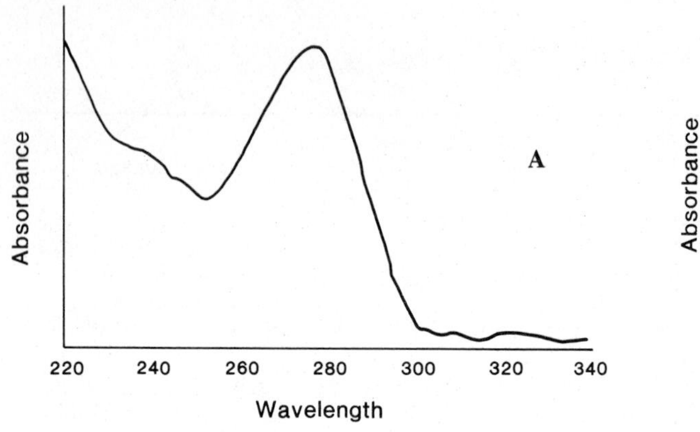

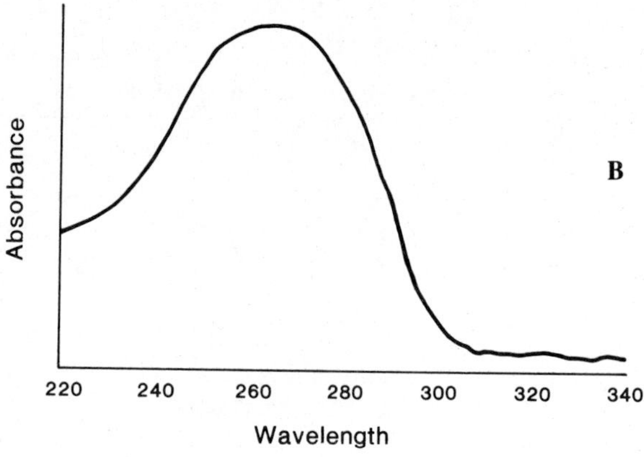

Fig. 61-7. A, Theophylline in 0.1 N NaOH. **B,** Sulfamethoxazole in 0.1 N NaOH.

further confirmatory techniques. Excellent sources have compiled spectra of most drugs encountered by the analyst.[12,13] Ultraviolet spectra are simple to obtain, but the sample preparation time can be lengthy.

Enzyme immunoassay (EMIT)

Enzyme-multiplied imunoassay techniques (EMIT, Syva Co., Inc., Palo Alto, Calif.) offer commercially available kit assays for individual drugs or drug classes. The sample is added to an antibody-substrate reagent, and then an enzyme-labeled drug reagent is introduced, allowing competitive binding of the patient drug and reagent drug. Assays, especially those performed on serum, offer good specificity and adequate sensitivities to detect drugs in both overdose and drug-abuse situations. Following are test assays currently available:

Urine	Serum
Amphetamine	Tricyclic antidepressants
Barbiturates	Barbiturates
Benzodiazepines	Benzodiazepines
Cannabinoids	Selected anticonvulsants
Cocaine	Selected cardioactives
Ethanol	Selected antimicrobials
Methadone	Theophylline
Methaqualone	Acetaminophen
Opiates	Methotrexate
Phencyclidine	
Propoxyphene	

Each of these tests must be performed with appropriate controls. Thus the greater the number of standards and controls tested, the longer the drug screen analysis.

Radioimmunoassay (RIA)

As with enzyme immunoassays, RIA procedures provide the analyst with commercially available assays for a selected number of individual drugs.[14] Specificity of the individual procedures depends completely on the specificity of the assay's antibody. RIA test sensitivities are excellent, and most assays are quantitative, enabling detection of drug concentrations found in most clinical applications. Most assays require more than 30 minutes of incubation time, and few laboratories can offer immediate results for more than one or two of these assays. The cost of these reagents is comparatively high.

Thin-layer chromatography (TLC)

TLC serves as an excellent screening or confirmation technique for both large laboratories and those with more limited capabilities. The method requires no instrumentation and is relatively inexpensive. Additional advantages include the flexibility to perform simultaneous analysis of a greater number of drugs and simultaneous analysis of a greater number of specimens. Sensitivities of most TLC screening procedures are adequate for both overdose and drug abuse cases, and specificity is often acceptable. One inadequancy of the initially described TLC procedures was lengthy turnaround time. Depending on the method utilized, the procedure may take 2 to 4 hours. Newer commercially available methods, based on the same principles of TLC but with accelerated development and detection steps, are now available.[15-17] These have reduced the assay time to 1 to 3 hours.

Gas chromatography (GC) and gas chromatography–mass spectrometry (GC-MS)

GC offers one of the most powerful analytical tools for the separation of complex mixtures, including drug mixtures, in biological fluids. As with thin-layer chromatography, GC enables the analyst to look for a large number of possible constituents simultaneously. Unlike thin-layer chromatography, however, GC techniques cannot screen a large number of patient specimens simultaneously; therefore individual samples must be analyzed consecutively.

Specimens must generally be extracted before GC can be performed, and the drug must either be volatile or derivatized to make it volatile. For screening purposes, samples can be chromatographed on a single column or injected into various compound-selective columns to enhance specificity and sensitivity.[24] The compounds are

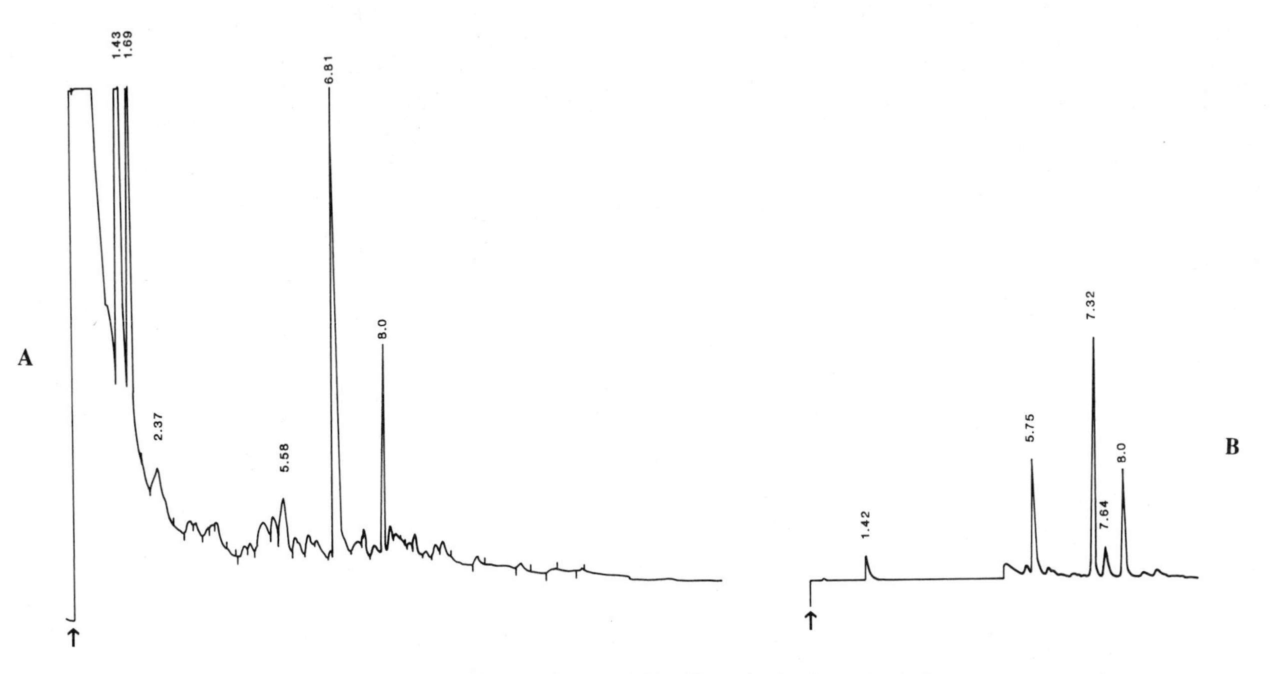

Fig. 61-8. Gas chromatography. **A,** Nonselective detector (FID, flame ionization detector). **B,** Selective detector (NPD, nitrogen-phosphorus detector).

identified by their respective retention times relative to an internal standard (Fig. 61-8, *A*). In addition, specific detectors are available (for nitrogen and phosphorus, or halogens), which can give more specific information. Fig. 61-8, *B,* shows the same sample chromatographed as Fig. 61-8, *A*, but analyzed with a nitrogen detector.

When a gas chromatograph is interfaced to a mass spectrometer (GC-MS), the combination provides the most objective analytical data of existing methods for structural elucidation of an unknown compound. The GC-MS system has been utilized as both a confirmatory procedure and a primary screening method for drug screens when improved turnaround times are necessitated (immediate overdose screening).[25,26] Both conventional GC and GC-MS require considerable technical expertise, and equipment costs, especially for the GC-MS, are high.

Quality control

As with all laboratory tests, one must continuously monitor the performance and reliability of the drug screening procedure for accuracy, documenting adequate sensitivity and specificity of the techniques. One can accomplish this by employing an internal quality control program and by participating in outside proficiency surveys.

Proficiency surveys offer a check of screening accuracy and should tax both the sensitivity and the specificity of the methods utilized. Currently, the programs offered by the College of American Pathologists (PO Box 1234, Traverse City, MI 49684) or by the American Association for Clinical Chemistry (Washington, D.C.) are suitable.

The internal program involves daily analysis of known control materials spiked with an assortment of drugs, which can be provided commercially or can be prepared in-house if the laboratory possesses the proper licenses for handling controlled substances.* Control samples should be processed in the same way as patient specimens and should be submitted to the same rigorous analyses as unknowns. It is important to monitor all three stages of the procedure (extraction or preparation, screening, and confirmation).

REFERENCE AND PREFERRED METHODS

As is so often found with other esoteric laboratory tests, no reference procedure is widely accepted for the qualitative drug screen. The optimal reference method would in theory be difficult to present, since it would entail (1) defining which compounds should be included in various screens, which is impossible, since the drugs of interest are invariably changing; (2) establishing a preferred separation or extraction scheme; (3) describing the screening method; and (4) suggesting various confirmation procedures. The following suggested methods are those that best suit analysts of varying expertises and budgets, encompassing the simplest to the most complex assays. One should note that several analytical techniques—nuclear magnetic resonance, infrared spectroscopy, and high-performance liquid chromatography—have not been considered in the drug screening assays, since these procedures,

*For information regarding the regulations concerning licensing for handling controlled drugs, contact the Drug Enforcement Administration, PO Box 28083, Central Station, Washington, DC 20005.

with the possible exception of high-performance liquid chromatography, have not yet been used extensively and routinely for the qualitative analyses usually encountered by the clinical chemistry laboratory.

SPECIMEN

The submission of proper and timely specimens to the analyst is critical for obtaining useful results. Samples should be collected immediately on the indication that drugs may be involved in a particular case. Depending on the drug screen request, various biological samples are made available to the laboratory (Table 61-14). At present, urine is generally considered to be the optimal specimen for drug screening.

PROCEDURE

Principle

Following is a flow diagram of the suggested method for a comprehensive drug screen:

Urine specimens are extracted with organic solvents at a selected pH, followed by analysis by thin-layer chromatography.[21] Selected ancillary spot tests and enzyme immunoassays are simultaneously performed. Confirmations of positive findings are undertaken using appropriately chosen techniques. A check sheet showing how the data can be recorded is presented in Fig. 61-9.

Reagents (All reagents should be of chromatographic grade)

Trinder's solution ($Fe[NO_3]_3$, 100 mmol/L; $HgCl_2$, 14.7 mmol/L; HCl, 120 mmol/L). Add 4 g of ferric nitrate and 4 g of mercuric chloride to a 100 mL volumetric flask. Dissolve in 12 mL of 1 M hydrochloric acid. Dilute to mark with distilled water. Store at room temperature. This is stable for 6 months.

Potassium dichromate (0.0034 mol/L). To a 1 L volumetric flask add 1 g of potassium dichromate. Dissolve

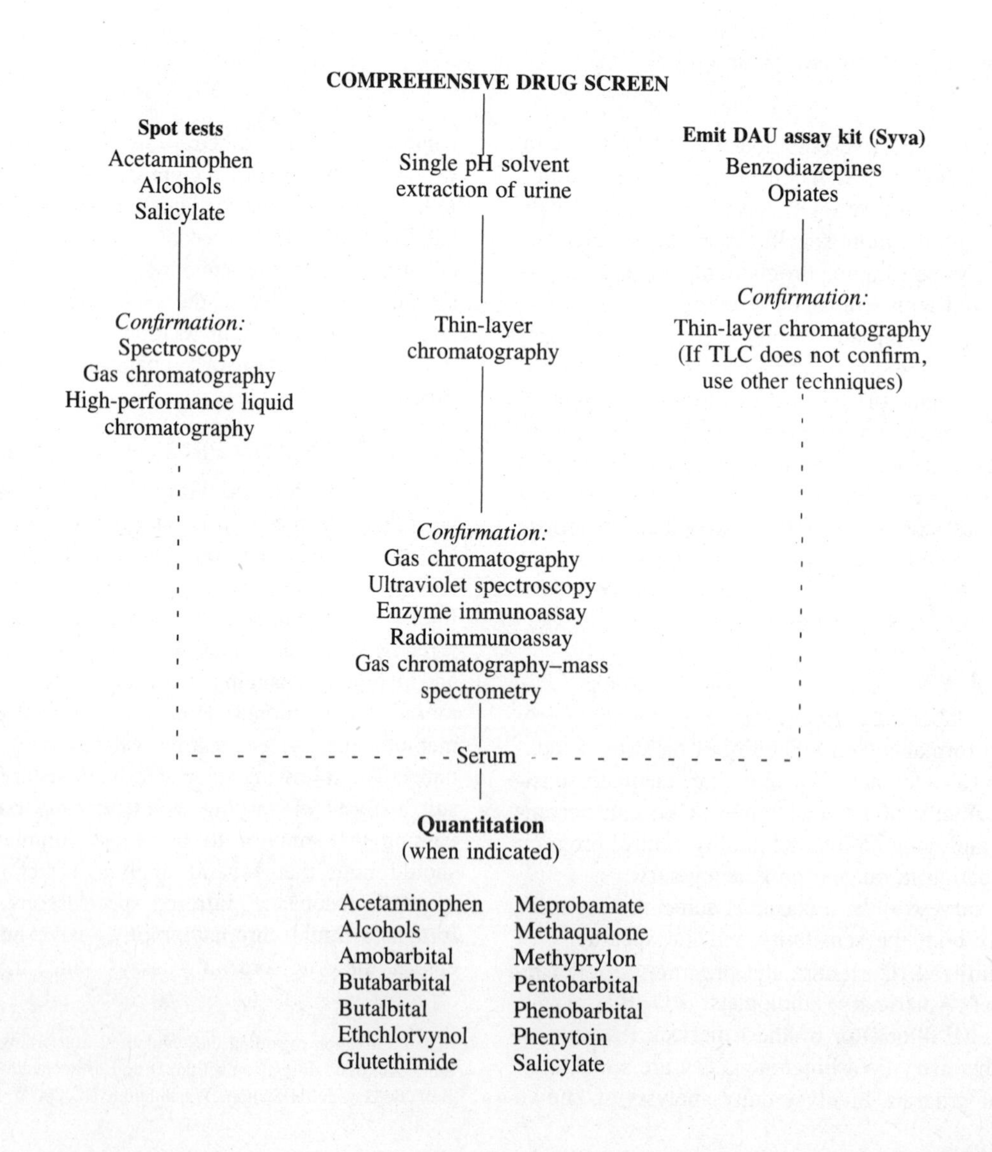

Table 61-14. Specimens of choice in drug screen

Specimen	Volume required	Indication	Advantages	Disadvantages
Urine	30 mL	Drug abuse screening Overdose screening Employment screening	Generally easy to obtain in high volume Most drugs found in sufficient concentration to enable identification	Contains many metabolic products that may interfere with identification Parent drug may not be present Quantitation offers little correlation with clinical effects
Gastric lavage or emesis	30 mL	Overdose screening	Parent drug present	Matrix problems (interference from foodstuffs) Drugs quickly absorbed may be missed by "gastric screen" Drugs not orally ingested will not be detected
Blood, serum, and plasma	10 mL	Overdose screening Therapeutic drug monitoring	Parent drug present Quantitative level may assist with patient management (therapeutic and toxic reference levels often known)	Limitation of sample volume Concentration of select drugs often too low to enable detection (especially in non-overdose situations)

in 500 mL of distilled water. Add 0.1 g of silver nitrate to the flask and then 500 mL of concentrated sulfuric acid. Store in a brown bottle at room temperature. This is stable for 1 year.

***o*-Cresol reagent (0.096 mol/L).** Add 10 mL of *o*-cresol to a 1 L volumetric flask and dilute to mark with distilled water. Store at room temperature. This is stable for 1 year.

Forrest reagent

1. Nitric acid (7.8 mol/L). Add 50 mL of concentrated nitric acid to 50 mL of water.
2. Perchloric acid (1.85 mol/L). Add 20 mL of 70% perchloric acid to 50 mL of water. Allow to cool and bring to 100 mL.
3. Sulfuric acid, 12 N (6 mol/L). Add 30 mL of concentrated sulfuric acid to 50 mL of water. Allow to cool and bring to 100 mL.
4. Potassium dichromate, 2 g/L (0.0068 mol/L). Add 200 mg of potassium dichromate to 80 mL of water. Dissolve and bring to 100 mL.

Mix 100 mL of each of the reagents together. This solution is stable for 1 year at room temperature when stored in an opaque (amber) glass container.

FPN reagent (ferric chloride/perchloric acid/nitric acid)

1. Nitric acid (7.8 mol/L). Add 50 mL of concentrated nitric acid to 50 mL of water.
2. Perchloric acid (1.85 mol/L). Add 20 mL of 70% perchloric acid to 50 mL of water. Allow to cool and bring to 100 mL.
3. Ferric chloride (0.308 mol/L). Add 5 g of ferric chloride to 50 mL of water. Bring to 100 mL.

Mix these solutions in the following proportions: one part ferric chloride, nine parts perchloric acid, and 10 parts nitric acid. This is stable for 1 year at room temperature when stored in an opaque (amber) glass container.

Concentrated hydrochloric acid. This is stable for 2 years at room temperature.

Ammonium hydroxide, 4 mol/L. Dilute 284 mL of concentrated ammonium hydroxide to 1 L with distilled water. This is stable for 2 years at room temperature.

EMIT DAU opiate and benzodiazepine assay kits (Syva Corp., Palo Alto, Calif.). These kits are stable for 1 year before opening. Follow manufacturer's recommendations for storage.

Extraction buffer. Prepare a saturated ammonium chloride solution. Add concentrated ammonium hydroxide to adjust the pH to 9.5. Store at room temperature. This is stable for 6 months.

Extraction solvent. Prepare a mixture of dichloromethane/isopropanol (90:10 by volume). Store at room temperature. Prepare weekly.

Methanolic–0.1 M hydrochloric acid. To a 100 mL volumetric flask add 0.9 mL of concentrated hydrochloric acid. Dilute to mark with methanol. Store at room temperature. This is stable for 1 year.

TOXICOLOGY LABORATORY DATE:________ TECH:________ CHECKED BY________
STAT DRUG SCREEN PATIENT #:________

(1) SPOT TESTS:

"Salicylates"

Serum
Urine (use only if serum is not available)

Sample Pos. Neg.
Std.
Patient

"Acetaminophen"

Use urine only

Sample Pos. Neg.
Std.
Patient

(2) ALCOHOLS: Microdiffusion

Serum
Urine (use only if serum is not available)

Sample Pos. Neg.
Std.
Patient

Gas Chromatography Confirmation

Alcohol detected Ethanol
Methanol
Isopropanol

Quantitation (if serum is available)

________ μg/mL

(3) EMIT DAUS: Use urine only

"Opiates"

Sample OD Pos. Neg.
Neg. Cal. ___
Low Cal. ___
Patient ___
Pat. Blk.

"Benzodiazepines"

Sample OD Pos. Neg.
Neg. Cal. ___
Low Cal. ___
Patient ___
Pat. Blk.

SCREEN: Analyze urine and gastric (if available)

(5) Gas chromatography (confirmation). List relative retention times of note.

(6) GC/MS (Confirmation)
– Utilize GC/MS analysis sheets. List all compounds detected with their respective scan numbers.

– Search all peaks in the "MAP" mode and perform appropriate limited mass analyses in the "CHRO" mode.

(4) TLC

6 Ref.
7 ___
8 ___
9 Std. I
10 Ref.
11 Std. II
12 Std. III
13 ___
14 ___
15 Ref.

COMMENTS:

(7) QUANTITATIONS: (The following are routinely quantitated if detected)

Drug	Concentration (μg/mL)
Acetaminophen	___
Alcohols	(see above)
Barbiturates	___
Specify	___
Ethchlorvynol	___
Glutethimide	___
Meprobamate	___
Methaqualone	___
Methyprylon	___
Phenobarb	___
Phenytoin	___
Salicylate	___

Fig. 61-9. Laboratory check sheet for recording data from a comprehensive drug screen (p. 1360).

Reconstitution solvent. A 1:1 mixture of dichloromethane/methanol. Store at room temperature. This is stable for 1 year.

Thin-layer chromatographic developing solution. Mix 85 mL of ethyl acetate, 10 mL of methanol, and 5 mL of ammonium hydroxide. Prepare fresh for each development.

Thin-layer chromatographic sprays

1. Ninhydrin (5.61 mmol/L). In a 100 mL volumetric flask dissolve 100 mg of 1,2,3-indantrione in 5 mL of acetone and dilute to mark with acetone. Prepare fresh daily.
2. Diphenylcarbazone (2 mmol/L). In a 100 mL volumetric flask dissolve 10 mg of diphenylcarbazone in 5 mL of acetone and dilute to mark with a 1:1 acetone to water solution. This is stable for 1 month at room temperature.
3. Mercuric sulfate (11.5 mmol/L). Add 0.5 g of mer-

curic oxide to 20 mL of concentrated sulfuric acid. Add this acid solution slowly to 150 mL of water, and bring to a final volume of 200 mL with water. This is stable for 6 months at room temperature.

4. Iodoplatinate. *Stock solution* (0.38 mol/L): dissolve 10 g of platinum chloride in 100 mL of distilled water. Refrigerate at 4° C. This is stable for 1 year. *Working solution:* add 5 mL of stock solution to 3 g of potassium iodide in 100 mL of water. Dilute this mixture to 125 mL with water, and then dilute this with an equal volume of methanol (final volume, 250 mL). Refrigerate at 4° C. This is stable for 6 months.
5. Dragendorf reagent. Dissolve 1.3 g of bismuth subnitrate in a solution composed of 60 mL of water and 15 mL of glacial acetic. Dissolve 12 g of potassium iodide in 30 mL of water. Combine the two solutions. Dilute this mixture with 100 mL of water and 25 mL of glacial acetic acid. Refrigerate at 4° C. This is stable for 6 months.

Standards

1. Ethanol (789 mg/L, 17.15 mmol/L). Pipet 0.1 mL of absolute ethanol into a 100 mL volumetric flask. Dilute to mark with distilled water. Store at 4° C. This is stable for 2 months.
2. Salicylate standard (100 mg/L, 72.4 μmol/L). Dissolve 10 mg of salicylic acid (USP grade) in 0.5 mL of methanol in a 100 mL volumetric flask and dilute to 100 mL with distilled water. Store at 4° C. This is stable for 6 months.
3. Acetaminophen urine control (100 μg/mL, 0.66 μmol/mL). To a 100 mL volumetric flask add 10 mg of acetaminophen. Dissolve in 1 mL of methanol. Dilute to mark with urine previously noted to be free of the drug. Refrigerate at 4° C. This is stable for 6 months.
4. Urine drug control. Select a suitable commercially available control or prepare one using the following procedure.
   a. Collect approximately 2 L of drug-free urine (urine previously noted to be negative for all drugs through testing).
   b. Prepare a stock drug standard by adding 4 mg each of phenobarbital, secobarbital, morphine, codeine, meperidine, methyprylon, and oxazepam and 10 mg each of D-amphetamine and methamphetamine to a 10 mL volumetric flask. Dissolve the drugs in methanol and dilute to mark with methanol.
   c. Transfer the entire contents of the drug standard flask to a 2 L volumetric flask. Rinse the standard several times with distilled water. Dilute to the 2 L mark with the urine pool. Mix well.
   d. Divide into 20 mL aliquots and store in plastic vials.
   e. Store at −20° C until used. This is stable for 1 year.
   f. Control contains 2 μg/mL of each drug except the amphetamines, which have concentrations of 5 μg/mL.
5. Thin-layer chromatographic reference standard (1 μg/μL:. To a 10 mL volumetric flask add 10 mg each of propoxyphene, methadone, chlorpromazine, meperidine, quinine, phenobarbital, phenylpropanolamine, and morphine. Dissolve in methanol and dilute to mark with methanol. Refrigerate at 4° C. This is stable for 6 months.
6. Individual drug standards (1 μg/μL). Prepare as for standard 5 with drug of choice. This is stable for 6 months at 4° C.

Assay

Equipment
Porcelain spot test plates
Conway microdiffusion dish
Teflon 125 mL volume separatory funnels (Fisher Scientific)
60 mL glass centrifuge tubes
Sample concentrator/evaporator (Brinkmann, Buchi Rotavap, or other convenient evaporation system)
Thin-layer chromatographic apparatus
  Spotting platen
  20 × 20 cm glass silica gel 60 F-254 plates (EM Science, Gibbstown, N.J.)
  Developing tank to accommodate 20 × 20 cm plates
  Chromatographic spray cans
  Air blower
  Short (254 nm)/long (366 nm) combination ultraviolet light

1. Ancillary tests
   a. Salicylate spot test. In a porcelain dish add three drops of salicylate control or patient specimen, urine or serum. (If urine is used, boil the specimen first to remove diacetic acid, a potentially interfering substance.) Add 3 drops of Trinder's reagent. A violet color indicates presence of salicylate.
   b. Alcohol spot test. To the center well of a microdiffusion dish add 0.5 mL of potassium dichromate reagent. To the outer ring add 0.5 mL of alcohol control or patient specimen (urine or serum). Lightly grease the outer ring of the microdiffusion dish lid with vacuum grease. Cover and allow to stand at room temperature. The center dichromate solution will change from a yellow to light green or blue color in the presence of either methanol, ethanol, or isopropanol.
   c. Acetaminophen spot test. To 1 mL of patient or control urine add 1 mL of concentrated hydrochloric acid. Heat in a boiling water bath for 10 minutes. Dilute 0.1 mL of this sample with 0.9 mL of *o*-cresol reagent. Add 2 mL of 4 M ammonium hydroxide. Blue color indicates presence of acetaminophen.

d. EMIT DAU opiate and benzodiazepine assays (Syva). Use as described by manufacturer.

2. Extraction of urine
   a. Process the drug screen control along with patient specimen.
   b. Into a 125 mL separatory funnel add 20 mL of urine and 2 mL of ammonium chloride–ammonium hydroxide buffer and mix.
   c. Add 30 mL of extraction solvent and extract for 10 minutes.
   d. Allow layers to separate. Collect bottom solvent layer into a 50 mL glass centrifuge tube and centrifuge for 5 minutes.
   e. With a glass stirring rod, break emulsions that may have formed by aggregating and recentrifuge. Aspirate off top aqueous layer and discard. Filter solvent through filter paper into evaporation flask.
   f. Add 2 drops of methanolic–0.1 M hydrochloric acid and evaporate to dryness at 50° C under nitrogen.
   g. Using a Pasteur pipet, reconstitute with 10 drops (approximately 150 μL) of methylene dichloride–methanol solution, being sure to rinse down the sides of the evaporation flask.
3. Thin-layer chromatography (TLC)
   a. TLC spotting. Use a conditioned (preheated for 1 hour at 70° C) TLC plate. Using a spotting platen, mark the spot origin at 2 cm from the bottom of the plate. From this mark measure 15 cm upward and draw the end line. Apply approximately 1 μL of reference standard in the 3, 10, and 17 positions on the plate. Use positions 9, 11, and 12 to spot drug standards of interest (drugs indicated by history and so on). Apply extracted control and patient specimens in open positions. Apply approximately 3 μL at a time and spot 10 times, allowing the applied spots to dry between applications.
   b. Plate development. Prepare the TLC development solvent and pour into a tank lined with 20 × 20 cm filter paper on both sides. Position the plate or plates into the development tank and allow to migrate to end line (approximately 40 minutes).
   c. Detection. Air dry the plate. Observe the plate under long (366 nm) ultraviolet light. Note any fluorescing compounds by lightly tracing the spot with a pencil. Mark subsequent positive spots with a pencil. Place under short (254 nm) ultraviolet light and note any absorbing spots. Spray with ninhydrin, and then heat plate at 70° C for 5 minutes. Amines, such as phenylpropanolamine and D-amphetamine, are noted to stain red. Spray with diphenylcarbazone, followed by the mercuric sulfate spray. Observe the blue-violet colors of barbiturates, phenytoin, glutethimide, and ethchlorvynol. Respray with diphenylcarbazone and heat plate at 70° C for 10 minutes. The pink, red, and violet colors of metabolites of phenothiazine drugs will appear. Observe under long ultraviolet light. Observe the yellow fluorescence of the benzodiazepine drugs and their metabolites. Spray with iodoplatinate and Dragendorf reagents. Most nitrogenous basic drugs will stain dark. Soak the finished plate in water for 30 seconds, remove and dry with a jet of hot air. Observe the chalk-white colors of methyprylon and carbamates, such as meprobamate. R_f values (distance of spot migration divided by distance of developing solvent migration), spray reactions, and ultraviolet detections of unknowns should be compared to known standards.
4. Confirmation. Probable drugs indicated as being positive by the thin-layer chromatographic or ancillary test screening methods must be confirmed by a second analytical procedure (see Table 61-11). Occasionally, when serum is provided and when indicated, as in overdose cases, the quantitation of the suspected drug in serum is the confirmatory step in the method. At other times additional tests must be performed on the urine to confirm the presence of the drug.

REFERENCES

1. Thoma, J.J., editor: Guidelines for analytical toxicology programs, vol. I and II, Cleveland, 1977, CRC Press, Inc.
2. Sunshine, I., editor: Methodology for analytical toxicology, ed. 2, Cleveland, 1975, CRC Press, Inc.
3. Clarke, E.G.C.: Isolation and identification of drugs, London, 1969, Pharmaceutical Press.
4. Berry, D.J., and Grove, J.: Emergency toxicological screening for drugs commonly taken in overdose, J. Chromatogr. **80:**205-220, 1973.
5. Sunshine, I., editor: Methodology for analytical toxicology, ed. 2, Cleveland, 1975, CRC Press, Inc. pp. 145-146.
6. Natelson, S.: Techniques in clinical chemistry, ed. 3, Springfield, Ill., 1971, Charles C Thomas, Publisher, pp. 649-651.
7. Curry, A.: Poison detection in human organs, ed. 2, Springfield, Ill., 1969, Charles C Thomas, Publisher, p. 67.
8. Sunshine, I., editor: Methodology for analytical toxicology, ed. 2, Cleveland, 1975, CRC Press, Inc., p. 192.
9. Haux, P.: Ethchlorvynol estimation in urine and serum, Clin. Chim. Acta **43:**139, 1970.
10. Sunshine, I., editor: Methodology for analytical toxicology, ed. 2, Cleveland, 1975, CRC Press, Inc., p. 301.
11. Fisher, D.S.: A method for rapid detection of acute iron toxicity, Clin. Chem. **13:**6-11, 1967.
12. Jatlow, P.: UV spectrophotometry for sedative drugs frequently involved in overdose emergencies. In Sunshine, I., editor: Methodology for analytical toxicology, ed. 2, Cleveland, 1975, CRC Press, Inc., pp. 414-420.
13. Sunshine, I.: Spectroscopic analysis of drugs including atlas of spectra, Springfield, Ill., 1963, Charles C Thomas, Publisher.
14. Byers, B.J.: Radioimmunoassays. In Thoma, J.J., editor: Guidelines for analytical toxicology programs, VII, Cleveland, 1977, CRC Press, Inc., pp. 207-223.
15. Blass, K.G.: A rapid simple thin-layer chromatography drug screening procedure, J. Chromatogr. **95:**75-79, 1974.
16. Toxi-Lab Applications note, Analytical Systems, Inc., 23162 La-Cadena Dr., Laguna Hills, CA 92653.
17. Michaud, J.D.: Thin layer chromatography for broad spectrum drug detection, Am. Lab. **12:**104, 1980.
18. Mule, S.J.: Routine identification of drugs of abuse in human urine.

II. Development and application of XAD-2 resin column method, J. Chromatogr. **63**:289-301, 1971.
19. Fujimoto, J.M.: A method of identifying narcotic analgesics in human urine after therapeutic doses, Toxicol. Appl. Pharmacol. **16**:186-193, 1970.
20. Weisman, N., and Lowe, M.L.: Screening method for detection of drugs of abuse in human urine, Clin. Chem. **17**:875-881, 1971.
21. Davidow, B.: A thin layer chromatographic screening procedure for detecting drug abuse, Am. J. Clin. Pathol. **50**:714-719, 1968.
22. Tox Elut drugs of abuse of urine, applications note, Analytichem International, Inc., 15620 S. Inglewood Ave., Lawndale, CA 90260.
23. Breiter, J.: Evaluation of column extraction: a new procedure for the analysis of drugs in body fluids, Forensic Sci. **7**:131-140, 1976.
24. Finkle, B.: A GLC based system for the detection of poisons, drugs and human metabolites encountered in forensic toxicology, J. Chromatogr. Sci. **9**:393-419, 1971.
25. Castello, C.E.: Routine use of a flexible gas chromatograph–mass spectrometer–computer system to identify drugs and their metabolites in body fluids of overdose victims, Clin. Chem. **20**:255-265, 1974.
26. Ullucci, P.A.: A comprehensive GC/MS drug screening procedure, J. Anal. Toxicol. **2**:33-38, 1978.

Gentamicin

JOSEPH R. DiPERSIO

Clinical significance: p. 962

Isomeric forms:	C1,	C1a,	C2
Molecular formula:	$C_{21}H_{43}N_5O_7$,	$C_{19}H_{39}N_5O_7$,	$C_{20}H_{41}N_5O_7$
Molecular weight:	477.6	449.6,	463.6

Merck Index: 4224
Chemical class: aminoglycoside

R_1CHNHR_2 ... NH$_2$, NH$_2$, HO, NH$_2$, HO, HNCH$_3$, CH$_3$, OH

C1: $R_1 = R_2 = CH_3$
C1a: $R_1 = R_2 = H$
C2: $R_1 = CH_3, R_2 = H$

PRINCIPLES OF ANALYSIS AND CURRENT USAGE

Gentamicin is a member of the aminoglycoside group of antibiotics and includes a number of structurally related polycationic compounds composed of two or more amino sugars connected by a glycosidic linkage to a six-membered aminocyclitol ring. Other important members of the group include streptomycin, neomycin, kanamycin, tobramycin, and amikacin. All of these are produced by microorganisms, whereas other members such as sisomicin and netilmicin are semisynthetic derivatives. Aminoglycoside antibiotics in general have a narrow margin between effective and toxic concentrations; thus laboratory measurements are often required to ensure adequate body fluid levels and to adjust potentially toxic accumulations.

Before 1970 microbiological assays were commonly used to quantitate aminoglycosides. The method employed most frequently was the agar plate diffusion method (method 1, Table 61-15).[1] In this procedure one adds antibiotic standards (diluted in serum) to blank paper disks in measured amounts bracketing the expected concentration in the unknown serum. Measured amounts of the unknown serum are likewise added to blank disks. All disks are placed on the surface of an agar plate that has been previously seeded with an indicator bacterium susceptible to the antibiotic being measured. The plate is incubated 4 to 18 hours, depending on the antibiotic assayed and the indicator organism used. During this period the indicator organism will grow throughout the agar medium, producing a visual haze of growth. However, zones of growth inhibition will form around the antibiotic-containing disk because of diffusion of drug into the agar. One measures the diameter of each zone and plots it against the appropriate standard disk concentration on semilogarithmic graph paper. The zone of inhibition produced by the unknown serum is also measured, and the concentration of antibiotic is determined from the standard curve. A variation of this method consists in punching small wells in the seeded agar and adding the various dilutions of standards and unknowns to the wells.

If a serum contains other antimicrobial agents in addition to gentamicin, they must be accounted for so that an indicator organism resistant to these antibiotics may be used or techniques that inactivate the interfering drugs can be employed. The various shortcomings of the agar plate method have prompted the development of alternative procedures.

Radioenzymatic procedures were among the first rapid tests to be adapted for the assay of gentamicin and other aminoglycosides (method 2, Table 61-15). In these procedures naturally occurring bacterial enzymes known to inactivate gentamicin are used to form a radioactive derivative by enzymatic transfer of a radiolabeled substrate to the gentamicin. By knowing the specific activity (counts per minute per mole of label) of the labeled substrate, one can determine the gentamicin concentration by counting the amount of radioactivity incorporated into the antibiotic:

$$\text{Labeled group} + \text{Gentamicin} \xrightarrow{\text{Bacterial enzyme}} \text{Labeled gentamicin}$$

In one procedure radiolabeled adenosine triphosphate and an adenylating enzyme of bacterial origin are used to transfer radioactive adenosine monophosphate to gentamicin.[2] The positively charged aminoglycosides are removed from the reaction mixture by adsorption onto negatively charged phosphocellulose paper, which is rinsed, dryed, and counted. The amount of gentamicin is determined by comparison with a standard curve. In another procedure

Table 61-15. Methods for gentamicin analysis

Method	Type of analysis	Principle	Advantages	Disadvantages
1. Microbiological (agar plate diffusion)	Growth inhibition	Zones of bacterial inhibition proportional to antibiotic concentration	Inexpensive, versatile No special equipment needed	Slow (4-18 hr) Variable accuracy
2. Radioenzymatic assay	Radiometry	Enzymatic transfer of radiolabel to antibiotic	Sensitive, specific, accurate	Requires radioisotopes, expensive equipment, and labor-intensive protocol Not appropriate for stat. tests or low work load
3. Radioimmunoassay (RIA)	Competitive binding assay of radioactive drug	Antibody binding to radiolabeled hapten competes for unknown	Sensitive, specific, accurate	Requires radioisotopes and expensive equipment Not appropriate for stat. tests or low work load
4. Enzyme immunoassay (EMIT)	Competitive binding assay, enzyme label	Enzyme-hapten bound to antibody is inhibited in catalytic reactivity	Sensitive, specific, accurate, very rapid (1-3 min) Can handle small work loads and stat. procedures	Moderately expensive equipment May have some problems establishing standard curve
5. Substrate-labeled fluorescence immunoassay (SLFIA)	Competitive binding assay, substrate label	Unbound drug–fluorogenic substrate converted by enzyme to fluorescent product	Sensitive, specific, accurate, rapid (½-1 hr)	Moderately expensive equipment
6. Fluorescence polarization immunoassay (FPIA)	Competitive binding assay, fluorescent-labeled drug	Bound drug has higher fluorescence polarization than free drug	Sensitive, specific, accurate, extremely rapid (1 min) Automated procedure	Specialized equipment Recently introduced
7. Latex agglutination inhibition	Agglutination	Agglutination inhibition of gentamicin-coated latex particles.	Sensitive, specific, accurate, inexpensive, rapid (10-15 min) No expensive equipment needed Can handle small work loads or stat. procedures	Not appropriate for large work loads Rheumatoid factor may interfere Some difficulty in reading end points
8. High-performance liquid chromatography (HPLC)	Chromatography, fluorometry	Chromatographic separation and derivatization	Sensitive, highly specific, accurate, versatile Can handle small work loads	Not appropriate for large work loads Expensive equipment

bacterial acetylating enzymes are used to transfer a labeled acetyl group from radioactive acetyl coenzyme A to the aminoglycoside being measured.[3]

Immunoassays are currently among the most popular methods for measuring aminoglycoside levels. All methods use the principle of competitive ligand binding. Radioimmunoassay (RIA), which was originally employed for the detection of cardiac glycosides,[4] was first used to quantitate gentamicin in 1972 (method 3, Table 61-15).[5] Most RIA procedures now available for aminoglycoside quantitation require the same basic steps. Radiolabeled antigen (aminoglycoside) is bound by specific antibody unless it is blocked by unlabeled antigen present in the unknown serum specimen. The more unlabeled antigen there is in the assay sample, the fewer radioactive counts of labeled antigen complexed with antibody will be in the precipitate or attached to a solid-phase tube.

Homogeneous immunoassays that do not require a separation step before measurement were developed in the late 1970s.[6] The enzyme-multiplied immunoassay technique (EMIT; Syva Co., Inc., Palo Alto, Calif.), which is based on the linkage of the enzyme glucose-6-phosphate dehydrogenase to gentamicin, was the first homogeneous immunoassay to be developed for clinical use (method 4, Table 61-15). When enzyme-labeled gentamicin becomes bound to its specific antibody, the activity of the enzyme is reduced. Patient serum is first mixed with a reagent containing antibody, a coenzyme (nicotinamide adenine dinucleotide, NAD), and an enzyme substrate (glucose-6-phosphate). Binding occurs to any gentamicin in the serum that is recognized by the specific antibody. Next, the enzyme-labeled gentamicin is added, which combines with any remaining unfilled antibody-binding sites, reducing the enzyme activity proportionately. Active enzyme converts NAD^+ to reduced NAD (NADH), resulting in an absorbance change at 340 nm that is measured spectrophotometrically. Enzyme inactivated by bound antibody does not catalyze the reaction. The residual enzyme activity is directly related to the log of the concentration of gentamicin in the serum.

The substrate-labeled fluorescent immunoassay (SLFIA) is another rapid method for quantitating aminoglycosides (method 5, Table 61-15).[7] The principle of this procedure involves competitive binding between gentamicin in the sample and a known amount of sisomicin labeled with a fluorogenic substrate (a derivative of umbelliferyl-β-D-galactoside) for an anti–gentamicin-sisomicin antibody.[8] Binding of the labeled sisomicin to antibody prevents hydrolysis of the substrate to a fluorescent product by the added enzyme β-galactosidase. The greater the amount of gentamicin in the serum, the less antigentamicin antibody will be available to inhibit the hydrolysis to the fluorescent product when specific enzyme (β-galactosidase) is added. The resulting fluorescence is proportional to the level of gentamicin in serum.

The fluorescence polarization immunoassay (FPIA) was first developed by Dandliker et al. (method 6, Table 61-15).[9,10] In this test system gentamicin (fluorescent-labeled and unlabeled unknown) competes for binding sites on a specific antibody.[11] When the labeled drug is bound to the antibody, the fluorophore is constrained from rotating between the time that light is absorbed and emitted. When linearly polarized light is used to excite the labeled drug, the constrained fluorophore allows the emitted fluorescent light to remain highly polarized. On the other hand, when labeled drug is unbound, its rotation is much greater and the emitted light is depolarized. A fluorescence polarization analyzer (FPA) measures the fluorescence generated when two beams of polarized excitation light (horizontally and vertically polarized) are passed through the sample. FPIA provides a direct measure of bound and free labeled drug in a competitive binding immunoassay. Clinical application of the technique has been delayed until recently by the lack of simple, inexpensive, and reliable instrumentation. A bench-top FPA is now available to perform these assays[12] (TDX, Abbott Diagnostics, North Chicago, Ill.).

Another immunological method of measuring gentamicin is by the latex agglutination inhibition card test (method 7, Table 61-15).[8] The basic reagent is composed of latex particles to which gentamicin has been adsorbed. Four dilutions of gentamicin standards ranging from 0.3 to 0.6 μg/mL are pipetted on the test card at their indicated positions. Serial twofold dilutions of patient serum are also pipetted on the card. One adds pretitered antigentamicin antiserum to each standard and each dilution of patient serum. Gentamicin-sensitized latex particles are then added to each dilution, and the card is mechanically rotated for 8 minutes. The latex-bound and free gentamicin compete for the antibody. With no free gentamicin present, agglutination of the latex particles by the antibody will occur. As the levels of free gentamicin increase, agglutination is inhibited. The card is read macroscopically to determine which serum dilutions and standards inhibit the latex agglutination. The gentamicin concentration of the patient's serum is calculated by multiplication of the lowest standard that inhibits agglutination by the reciprocal of the highest serum dilution that also inhibits agglutination.

Thin-layer chromatography and ion-exchange chromatography have also been used to assay gentamicin. These techniques are generally used in research applications but have not proved practical for routine clinical use. High-performance liquid chromatography (HPLC), however, can be used for the rapid and accurate assay of aminoglycoside antibiotics (method 8, Table 61-15).[13] The basic steps of HPLC for the assay of gentamicin include (1) extraction of the antibiotic from serum using a CM-Sephadex column, (2) analysis by reversed-phase ion-pair chromatography, (3) continuous-flow, postcolumn derivatization with *o*-phthalaldehyde to form fluorescent product, and (4) fluorescent detection followed by quantitation using peak-height or peak-area analysis.

REFERENCE AND PREFERRED METHODS

At present there is no one established reference method for the quantitation of gentamicin in body fluids. The various immunoassays are considered as a group to be reference methods. Microbiological assays are still widely used to measure antibiotics, but their use for aminoglycoside determinations has greatly diminished with the development of newer immunoassays. Microbiological assays are inexpensive, easy to set up, and require no instrumentation, but they are relatively slow and the accuracy can vary substantially.[8] It can be argued that the agar plate assay, as opposed to the immunoassays, measures true biological activity of gentamicin in a patient sample. This activity, however, is measured against an indicator organism, not the patient's infecting bacterium. In addition, other factors, some of which are difficult to control in the assay, can significantly affect results. It has also been demonstrated that the recovery of gentamicin from uremic sera can be significantly lower than from normal sera using the agar plate assay.[14] The microbiological assay can also be influenced by the presence of other antibiotics, which can act on the indicator organism if steps are not taken to minimize their effects. The effect of other antibiotics on the assay will go undetected if proper patient information is not provided.

The radioenzymatic procedures are accurate and sufficiently rapid, but the protocols are more labor intensive and require the use of radioisotopes and radioisotopic counting equipment. Only a few commercial kits are available for laboratory use.

Immunoassays are now the most widely used methods for the assay of aminoglycosides. The immunoassays as a whole are sufficiently rapid, sensitive, accurate, and precise. The coefficients of variation for most methods are similar. Radioimmunoassays, as with the radioenzymatic assays, use radioisotopes and counting equipment. Also, these methods are not practical for immediate tests of samples or low work loads, since a standard curve should be generated with each run. Recent therapeutic drug monitoring surveys of the College of American Pathologists have indicated that radioimmunoassay is being replaced by newer, homogeneous immunoassays that do not require radioisotopes. The enzyme-multiplied immunoassay technique (EMIT) is currently the most popular homogeneous immunoassay procedure for measuring aminoglycosides. The EMIT procedure is quite rapid once a satisfactory standard curve has been generated. The curve is usually stable for several days or longer; thus subsequent runs or immediate tests can be performed with minimal effort to check the standard curve.[15] The substrate-labeled fluorescent immunoassay (SLFIA) is also rapid, and as in EMIT, one may not have to perform a standard curve with each run.[16] Both EMIT and SLFIA are semiautomated procedures at present, and the instrument cost for both is comparable. More automated instrumentation to perform these assays will be available soon. SLFIA is not quite so rapid as EMIT, but the reagents may be less expensive.

The fluorescence polarization immunoassay (FPIA) is currently the most automated method available and is probably the fastest. Patient samples and standards are pipetted into tubes that are arranged in a carousel. The carousel is then placed in the fluorescence polarization analyzer (FPA) with a prediluted reagent pack containing one bottle each of diluting buffer, gentamicin antiserum, and gentamicin-fluorescein tracer. The FPA performs all dilutions, analyses, and calculations in minutes. Technician time is minimal, but the FPA is considerably more expensive than the basic equipment needed to perform EMIT or SLFIA. The FPIA may become a popular method for gentamicin quantitation because of its highly accurate results and total automation, but more experience with this procedure is needed.

As with the microbiological assay, the latex agglutination inhibition card test is inexpensive and simple and does not require complex instrumentation. The card test is also sensitive and accurate. Results are rapid (15 minutes per test), but since each test is performed individually, the time involved for a large volume of tests can be considerable.[17] It has been reported that rheumatoid factor may cause falsely low values, and with sera containing less than 2 μg of gentamicin per milliliter, it is sometimes difficult to determine precise agglutination end points.[18] The card test is probably best suited for a low-volume laboratory or as a backup procedure to a more automated assay procedure in a larger laboratory.

High-performance liquid chromatography (HPLC) is a versatile technique, since certain antibiotics other than aminoglycosides can be measured. Also, HPLC is the only method discussed that will quantitate the individual isomers present in solutions of gentamicin. The procedure is accurate and specific, but as in the latex agglutination inhibition card test, large work loads can be time consuming, since each sample is run individually (5 to 10 minutes per sample).[13] HPLC also requires expensive instrumentation.

Table 61-16 compares reaction conditions for the various gentamicin assays.

The final choice of which method to use for gentamicin analysis will depend on many factors, including laboratory work loads, test cost, turnaround time, and available instrumentation. Since most instrumentation described can be used to measure a variety of chemical analytes, versatility may be an important determining factor. Much of the newer instrumentation designed for the clinical chemistry laboratory has the capability of performing homogeneous enzyme assays.

SPECIMEN

The most common sample tested is serum or plasma. However, the microbiological assay may be used to assay gentamicin in other body fluids, including cerebrospinal fluid and joint or any other nonviscous fluid. The antibiotic standards should be prepared in normal fluid from patients not on antibiotics. The immunoassays can be performed on serum or plasma but not on whole blood. There are no concrete data on the general use of these methods with other body fluids.

INTERFERENCES

As stated previously, the microbiological assay will be affected by any additional antibiotic or antibiotics present in the sample to which the indicator organism is sensitive. Unless this activity is counteracted, the results may be invalid. Netilmicin and sisomicin, aminoglycosides structurally related to gentamicin will cross-react with the antibodies to gentamicin in most immunoassays. However, these drugs are not usually administered with gentamicin. The immunoassays may also be affected to varying degrees by hemolytic, lipemic, or icteric serum.

REFERENCE RANGE

The therapeutic range is approximately 2 to 9 μg/mL (4.3 to 19.4 μmol/L).

Table 61-16. Comparison of reaction conditions for gentamicin assays

Condition	Microbiological assay	Enzyme-multiplied immunoassay technique	Substrate-labeled fluorescent immunoassay	Latex agglutination inhibition	High-performance liquid chromatography
Temperature	35° C	30° C	15°-30° C	Room temperature	Ambient
pH	7.9	8.0	—	7.4	—
Sample volume	20 μL	50 μL	100 μL	25 μL	0.4 mL
Linearity	1-20 μg/mL	1-10 μg/mL	1-12 μg/mL	—	—
Precision (coefficient of variation)	5%-50%	≤10%	10.7% (2.5 μg/mL) 3.6% (5.0 μg/mL) 4.0% (12.0 μg/mL)	11% (2.2 μg/mL) 7% (4.4 μg/mL) 9.8% (8.8 μg/mL)	≤5%
Known interferences	Other antimicrobials	Sisomicin; netilmicin; heparin; hemolytic, lipemic, or icteric samples	Sisomicin, netilmicin, lipemic or icteric samples	Sisomicin, rheumatoid factor	—

REFERENCES

1. Sabath, L.D., Casey, J.I., Ruch, R.A., et al.: Rapid microassay of gentamicin, kanamycin, neomycin, streptomycin, and vancomycin in serum or plasma, J. Lab. Clin. Med. **78:**457, 1971.
2. Butcher, R.H.: Rapid serum gentamicin assay by enzymatic adenylation, Am. J. Clin. Pathol. **68:**566, 1977.
3. Stevens, P., Young, L.S., and Hewitt, W.L.: Improved acetylating radioenzymatic assay of amikacin, tobramycin, and sisomicin in serum, Antimicrob. Agents Chemother. **7:**374, 1975.
4. Oliver, G.C., Parker, B.M., Brasfield, D.L., and Parker, C.W.: The measurement of digitoxin in human serum by radioimmunoassay, J. Clin. Invest. **47:**1035, 1968.
5. Lewis, J.E., Nelson, J.C., and Wilson, T.N.: Radioimmunoassay of an antibiotic: gentamicin, Nature (New. Biol.) **239:**214, 1972.
6. Bastiani, R.J.: The EMIT system: a commercially successful innovation, Antibiot. Chemother. **26:**89, 1979.
7. Shaw, E.J., Amina-Watson, R.A., Landon, J., and Smith, D.S.: Estimation of serum gentamicin by quenching fluoroimmunoassay, J. Clin. Pathol. **30:**526, 1977.
8. Selepak, S.T., Witebsky, F.G., Robertson, E.A., and MacLowry, J.D.,: Evaluation of five gentamicin assay procedures for clinical microbiology laboratories, J. Clin. Microbiol. **13:**742, 1981.
9. Dandliker, W.B., Kelly, R.J., Dandliker, J., et ål.: Fluorescence polarization immunoassay: theory and experimental method, Immunochemistry **10:**219, 1973.
10. Dandliker, W.B.: Investigation of immunochemical reactions by fluorescence polarization. In Atassi, M.Z. editor: Immunochemistry of proteins, New York, Pleunum Publishing Co., 1977, p. 231.
11. Jolly, M.D., Stroupe, S.D., Wang, C.J., et al.: Fluorescence polarization immunoassay. I. Monitoring aminoglycoside antibiotics in serum and plasma, Clin. Chem. **17:**1190, 1981.
12. Popelka, S.R., Miller, D.M., Holen, J.T., and Kelso, D.M.: Fluorescence polarization immunoassay. II. Analyzer for rapid, precise measurement of fluorescence polarization with use of disposable cuvettes, Clin. Chem. **27:**1198, 1981.
13. Anhalt, J.P.: Assay of gentamicin in serum by high-pressure liquid chromatography, Antimicrob. Agents Chemother. **11:**651, 1977.
14. Edberg, S.C., and Sabath, L.D.: Determination of antibiotic levels in body fluids: techniques and significance. In Lorian, V., editor: Antibiotics in laboratory medicine, Baltimore, 1980, The Williams & Wilkins Co., pp. 210.
15. Rateliff, R.M., Mirelli, C., Moran, E., et al.: Comparison of five methods for the assay of serum gentamicin, Antimicrob. Agents Chemother. **19:**508, 1981.
16. Burd, J.F., and Clements, H.M.: Standard curve reproducibility in the substrate-labeled fluorescent immunoassay, Lab World **7:**41, 1981.
17. Standiford, H.C., Bernstein, D., Nipper, H.C., et al.: Latex agglutination inhibition card test for gentamicin assay: clinical evaluation and comparison with radioimmunoassay and bioassay, Antimicrob. Agents Chemother. **19:**620, 1981.
18. Johnson, J.E., Crawford, S., and Jorgensen, J.H.: Evaluation of Macro-Vue latex agglutination test for quantitation of gentamicin in human serum, J. Clin. Microbiol. **16:**299, 1982.
19. Anhalt, J.P.: Special tests of antimicrobial activity. In Washington, J.A., editor: Laboratory procedures in clinical microbiology, New York, 1981, Springer-Verlag, New York, Inc., p. 712.

Lead (plumbum)

M. WILSON TABOR

Clinical significance: p. 639
Atomic symbol: Pb
Atomic weight: 207.2 daltons
Merck Index: 5242
Chemical class: metal

PRINCIPLES OF ANALYSIS AND CURRENT USAGE

Lead, a periodic system group 4b metal, occurs in three oxidation states, 0, +2, and +4. The +2 oxidation state is the most widely occurring form of the inorganic salts of lead, whereas Pb (IV) is the most important state in organolead compounds. A complete discussion on lead as to sources of occupational exposure, summary of biological effects, and methods for analysis is contained in the NIOSH (National Institute for Occupational Safety and Health) revised criteria for recommended standards for occupational exposure.[1]

The method historically used to analyze lead in both biological and environmental samples and most commonly used to test the accuracy of other analytical methods is the dithizone compleximetric method[2-8] (method 1, Table 61-17). The sample (10 mL of blood, 25 mL of urine or tissues, particulates, and so on) is ashed by acid oxidation to remove organic material. (This procedure is detailed in the methods section.) After being ashed, the sample is redissolved in acidified water. The pH is adjusted to 9.0 or

Table 61-17. Method of lead analysis

Method	Type of analysis	Principle	Usage	Comments
1. Spectrophotometry with dithizone	Colorometric, quantitative	After ashing, Pb^{++} and dithizone form a complex absorbing at 510 nm	Blood, urine, tissues Environmental testing	Widely used Difficult to obtain good precision
		$Pb^{2+} + 2\,S{=}C(N{=}N{-}C_6H_5)(NH{-}NH{-}C_6H_5)$ (**Dithizone (green)**) $\xrightarrow{pH > 11.0}$ $[\ldots N{-}S^{-}, C{=}N \ldots]_2\,Pb^{2+}$		
2. Atomic absorption spectroscopy (AAS)				
a. Classical AAS	Quantitative	After ashing of lead sample, ionized lead is reduced in flame; absorbs light at 283.3 nm	Blood, urine, tissues Environmental testing	Widely used, straightforward Applicable to variety of macro samples and sizes
b. Delves microscale technique	Quantitative	Same as for method 2a, except sample is oxidized in microcup by H_2O_2	Blood, urine	Same as for method 2a Applicable to microsamples
c. Graphite furnace or carbon rod	Quantitative	Same as for method 2a, except sample is dried, ashed, and vaporized on graphite rod	Blood, urine	Same as for method 2b
3. Electrochemical	Anodic stripping voltammetry	Increased amperage at −430 mV	Blood, urine	Macro- or microsample, Moderately inexpensive Simple, most reliable over wide concentration ranges

10.0, cyanide is added to eliminate reactions with other metals, and then the sample is reacted with diphenylthiocarbazone (dithizone) to form the red lead–dithizonate complex. The complex is extracted into chloroform, and the absorbance of the complex is measured spectrophotometrically at 510 nm.

For atomic absorption spectroscopy (AAS), many methods of sample preparation, sample concentration, and spectrophotometric sampling techniques have been reported.[9-23] However, all have the common features of (1) a digestion or treatment of the sample to allow separation of the lead from the sample matrix; (2) aspiration of the ionic lead into a flame, or heating of the ionic lead solution by electrical means, where it is reduced to the atomic state; and (3) spectrophotometric quantitation by measurement of the light absorbed by the lead at its characteristic resonance frequency of 283.3 nm.

The classical atomic absorption spectroscopy methods[9,10,12-14,18] feature a sample ashing procedure by acid oxidation that is similar to that used for the dithizone method[4-6,8] (method 2a, Table 61-17). This procedure typically requires a 10 mL of blood or 50 mL of urine sample.[12,18] After sample preparation, the solution of ionic lead is aspirated into an air-acetylene flame, where it is reduced to the atomic state. Quantitation is accomplished by measurement of the light absorbed by the lead at 283.3 nm.

Although numerous variations in this atomic absorption spectroscopy procedure have been published,[1] two are noteworthy. The American Association for Clinical Chemistry method[14] features coprecipitation of the lead with bismuth hydroxide after sample digestion. After centrifugation and supernatant fluid removal, the precipitate is dissolved in acid for aspiration into the flame. The second method, a NIOSH atomic absorption spectroscopy procedure, features chelation of the lead with pyrrolidine dithiocarbamate after sample digestion. This lead chelate is extracted into methyl isobutyl ketone (MIBK) and then the MIBK extract is aspirated into the flame.

In addition to these variations, a blood microscale method for flame atomic absorption spectroscopy is now widely used[24-26] for blood lead analysis. This procedure, the Delves microscale technique,[15-17] does not utilize the sample ashing procedure by acid oxidation of the previously described atomic absorption spectroscopy methods (method 2b, Table 61-17). In the Delves cup method 10 μL of blood, obtained by a fingerstick, is added to a special nickel cup that contains hydrogen peroxide to partially oxidize the sample. After the sample preparation procedure, the cup is inserted into the AAS flame for the lead determination.

Other major variations of the classical atomic absorption spectroscopy (AAS) procedures now widely used include the "flameless" AAS procedures, which are also referred to as the graphite-furnace[11,19] and carbon-rod[20-23] techniques (method 2c, Table 61-17). Some procedures[11] feature the sample ashing by acid oxidation, as described for the classical AAS methods, whereas in others[21-23] a microsample of 5 μL is introduced directly into the instrument. The flameless AAS procedures feature a graphite tube or carbon rod held on the optical axis of the atomic absorption unit. The tube is protected from the atmosphere by flowing nitrogen or argon. After sample introduction, the graphite tube or carbon rod is heated by electrical resistance successively to a drying temperature, 100° C, to a charring temperature, 400° C, and then to an atomization (reduction) temperature, 2000° to 2500° C. During atomization a population of atomic lead is formed in the light path, but the atoms quickly escape by diffusion. The instrument output, lead absorbance at 280.2 and 283.3 nm, is a transient signal lasting a few seconds that is subsequently used for quantitation purposes.

The third general category of lead analytical methods is electrochemical. This technique, using anodic stripping voltammetry (ASV) instrumental methods, has become a widely accepted method for lead analysis[27-30] (method 3, Table 61-17). In this methodology ionic lead in blood or urine samples is first reduced to elemental lead by a negative potential applied to a working mercury electrode. After a preselected time the working electrode potential is moved toward more positive values, thereby causing the reoxidation of the lead. The resulting anodic current from this oxidation is measured, which is proportional to the concentration of lead.

The anodic stripping voltammetry (ASV) technique has become a reliable and reproducible method for microliter-sized blood samples because of the design of an efficient mercury-graphite composite electrode by Environmental Science Associates (ESA, Burlington, Mass.). This electrode has been incorporated into equipment marketed by ESA for the semiautomated determination of lead in blood, urine, water, foods, and reagents. Sample preparation for ASV is facilitated by the use of the Metexchange reagent (ESA), a chromate-based digestion solution containing an antifoaming surfactant.

Two anodic stripping voltammetry methods are recommended by NIOSH, one requiring 100 μL of blood and the other requiring 1.0 mL of urine; method no. P&CAM 195[29] is for the analysis of lead in blood, whereas method no. P&CAM 200[30] is for urinary lead measurement. Blood ASV methods are widely applicable to routine monitoring of biological samples during treatment for lead intoxication and for studies of lead exposure in children, and in workers who are occupationally exposed, as in smelting and battery industries.

There are several nonroutine methods for the analysis of lead in a variety of samples matrices.[1,24-26] These methods include electron microprobe, x-ray fluorescence, neutron activation analysis, and mass spectrometry. All require the use of expensive facilities and a high degree of analyst expertise. Although these techniques are widely applicable in industry, both neutron activation analysis and mass spectrometry are used as definitive reference methods. The National Bureau of Standards (NBS) definitive method of isotope dilution utilizing mass spectrometry[31] is employed by the Centers for Disease Control (CDC) for establishing lead values in their samples for the Licensure and Proficiency Testing program.[32]

Of the three main types of methods available for the determination of lead—spectrophotometric, atomic absorption spectroscopy (AAS), and anodic stripping voltammetry (ASV)—two thirds of the laboratories participating in the Centers for Disease Control blood lead proficiency testing for 1982 used AAS-based methods and one third used ASV.[33] The AAS-based methods most commonly used were the extraction/AAS, graphite furnace/AAS, Delves cup/AAS, and carbon rod/AAS.

REFERENCE AND PREFERRED METHODS

The choice of methodology for the determination of lead in a particular laboratory depends on several factors: (1) the availability of equipment, (2) the number of samples to be analyzed per day, (3) the purpose of the analysis, and (4) the background experience and prejudices of the clinicians doing the work.

All the methods discussed appear to have the precision and accuracy to provide acceptable analytical results for the determination of lead in blood and urine. The accuracy and precision reported for the recommended NIOSH methods used in the determination of lead are as follows:

1. Dithizone for lead in blood and urine[5]—an accuracy of 97% ± 2% and coefficient of variation of 6%
2. ASV for lead in blood[29]—a standard devation of 2 ng in a 100 μL sample caused by scatter in the re-

agent blank and an overall sampling plus analytical error of 5% in the determination of 60 ng of lead in a 100 μL sample of blood

3. ASV for lead in urine[30]—a standard deviation of 2 ng in a 1 mL sample and an estimated overall sampling plus analytical error of about 10% in the determination of levels near 200 ng of lead in a 1 mL sample of urine
4. Graphite furnace/AAS for lead in air or blood[11]—although an accuracy has not been reported by NIOSH, an analytical coefficient of variation of 5.7% has been determined
5. Extraction/AAS for lead in blood and urine[12]—an accuracy of 0.99 ± 0.064 μg of lead per gram of blood, with a relative standard deviation of 6.4%
6. Extraction/AAS for lead in air, S341[13]—a coefficient of variation for the total analytical and sampling method in the range of 0.128 to 0.399 mg of lead per cubic meter of air is 7.2%

The atomic absorption method, published as a selected method of the American Association for Clinical Chemistry,[14] has good accuracy, as shown by relative recoveries of 98.8 ± 1.0% for lead in blood, 99% ± 4.5% for lead in urine, and 97% ± 3.6% for lead in hair.

The colorimetric dithizonate method has a class A (recommended) rating as NIOSH method no. P&CAM 102 for the determination of lead in blood and urine.[5] The dithizone procedure is subject to interference from tin (II), bismuth, and thallium, but the method has the advantage of being thoroughly tested and evaluated.[5] Other disadvantages[5] of the method are that (1) large quantities of glassware must be scrupulously cleaned, (2) large quantities of reagents are required, (3) the procedure is highly susceptible to contamination from outside sources, and (4) the procedure is tedious and time consuming and requires a high level of laboratory proficiency from the analyst. However, when care is taken in sample and standard preparation, results can be obtained that are quantitative and precise. Other advantages[7] include (1) the use of simple and relatively inexpensive equipment, (2) linearity of absorbance with lead concentration over a wide range, (3) the need for only a small sample but easy adaptability to large samples, and (4) ready removal of interferences.

An earlier version of the dithizonate method was published as a selected method by the American Association for Clinical Chemistry.[8] However, this association and various government agencies, such as the Centers for Disease Control (CDC) and the National Institute for Occupational Safety and Health (NIOSH), currently prefer the instrumental techniques based on atomic absorption spectroscopy (AAS) and anodic stripping voltammetry (ASV) detection.[27-30] These procedures are summarized in Table 61-17 and are described in more detail in the methods section and in the following paragraphs.

Atomic absorption spectroscopy is recommended by NIOSH in their *General Procedure for Metals,*[10] not only for the determination of lead but also for the determination of many other trace metals in industrial and ambient airborne material. The utility of this method is its wide applicability. A specific AAS method for lead recommended by NIOSH is *Lead in Air or Blood,*[11] which is a graphite furnace (flameless) AAS procedure. The major advantage of this method is a greater sensitivity of one to two orders of magnitude over flame AAS procedures. The main disadvantages of the graphite AAS procedure are that (1) sample matrix effects are more prominent than in flame atomic absorption, requiring the use of the method of standard additions; (2) the analysis is time consuming; (3) the equipment is expensive as compared to flame AAS. Another NIOSH AAS method is *Lead in Blood and Urine,* which is a flame AAS procedure utilizing the methyl isobutyl ketone extraction technique. This proved method offers several advantages, including the speed of analysis, not requiring a high level of technical skill, and no known interferences. The main disadvantage is the requirement for larger sample sizes than those required in some other AAS procedures and the anodic stripping voltammetry procedures. The principal advantage of the bismuth hydroxide precipitation method is its speed, making it practical for routine monitoring of biological samples during treatment for lead intoxication.

The advantages of the Delves cup method are (1) small sample size, (2) speed, and (3) the requirement for only a minimum of technical competence. The disadvantages include (1) interferences from background absorptions and matrix effects, and (2) variation in cup quality from lot to lot, requiring careful attention to the use of each individual cup.

In addition to high sensitivity and wide range of linearity, anodic stripping voltammetry methods are less expensive per analysis than other methods in terms of equipment cost and time required for analysis. The principal disadvantage is that thallium interferes, since the current-voltage peak for this element is not resolved from lead when the two metals are at comparable concentrations. Sensitivity of the anodic stripping voltammetry method for lead is limited by the reagent blank absolute value, which is usually 6 ± 2 ng but is linear to 1000 ng.

All of these methods for lead analysis appear to be reasonably accurate and precise; however, it is the extrinsic factors involved in the analysis of lead that appear to be the major source of error. One is laboratory competence and proficiency of the analyst. For example, the dithizone method is a tedious and time-consuming method that is entirely manual, thus requiring numerous manipulations. Many microscale methods, such as anodic stripping voltammetry, Delves cup/atomic absorption spectroscopy, and graphite furnace/atomic absorption spectoscopy, require a high degree of accuracy in measuring the microliter quantities of sample. In these methods close attention to the use and type of blood capillary sampling tubes is required to prevent contamination.[17] The problem of contamination by the sampling container has been overcome for the macroscale procedures with the availability of lead-free blood-sampling tubes (lead-free Vacutainers, Becton Dickinson Co., Rutherford, N.J.). Another problem is the loss of lead during sample storage. Meranger et al.[35] have reported that it is not only the type of container (glass or plastic) that affects loss by adsorption but also the storage time and temperature that can contribute to significant losses, as much as 80% within 1 week. These results are in agreement with the losses to sample containers reported by Unger and Green.[36] They suggested that the addition of 1% nitric acid or 3% hydrogen peroxide to the sample will minimize losses for storage times of up to 5 days. In the Centers for Disease Control blood lead proficiency testing program, the use of EDTA to a concentration of 1.5 mg per milliliter of blood and the recommendation of storage at $-20°$ C has proved adequate to prevent losses of lead, even for storage for several months.[34]

According to the Centers for Disease Control,[24] most problems associated with obtaining acceptable accuracy and precision in the various methods of analysis for lead appear to be caused by a lack of adherence to method protocols. In terms of accuracy, an overestimation of lead is probably caused by lead contamination of glassware, reagents, or equipment and inadequate blank correction. An underestimation probably results from incomplete analytical recovery in extraction or concentration steps and nonlinearity of calibration curves. If protocols are strictly followed, reliable results can be obtained from all methods discussed.

In a critique[34] of the 1981 Centers for Disease Control blood lead analysis program, it was noted that laboratories using anodic stripping voltammetry had the highest percentage of acceptable responses at lead concentrations, both above and below the reference laboratory mean of 400 μg/L. Those laboratories using graphite furnace/atomic absorption spectroscopy had the highest percentage of unacceptable responses, both above and below the reference laboratory mean. For those using the Delves cup/AAS, the highest percentage of unacceptable responses was for values below the reference laboratory mean, whereas those using extraction/AAS had the highest percentage of unacceptable responses for values above the reference laboratory mean. However, remember that "data reported in an interlaboratory study represent a combination: a laboratory's ability to perform its method, and the accuracy and precision of the method itself."[24]

Based on results of the Centers for Disease Control proficiency testing program, the types of methods for blood lead analysis have been ranked, in decreasing order, for accuracy and precision as follows: anodic stripping voltammetry (ASV) > Delves cup/AAS > extraction/AAS > graphite furnace/AAS > carbon rod/AAS.

In consideration of all these factors, the anodic stripping voltammetry methods appear to be the best methods available for the routine determination of lead in blood and urine.

The dithizone method is recommended for laboratories having only an occasional need for lead analyses or for those on limited budgets. The anodic stripping voltmmetry methods are recommended for laboratories performing lead analyses on a routine basis, particularly for those doing monitoring studies. The costs for equipment are about $5000 and up. The atomic absorption spectroscopy (AAS) methods are recommended for laboratories analyzing lead and also other metals on a routine basis, particularly for those doing analyses on samples in addition to blood and urine. Equipment costs exceed $20,000. The AAS methods have broader applications—they can be used to determine other metals in addition to lead, and many AAS procedures can be used for samples other than blood or urine. The most widely applicable AAS method for various sample types is the extraction/AAS method of Yeager et al.[18] which was adopted as NIOSH method no. P&CAM 262.[12] It is the proved sample preparation aspects of this method that make it adaptable to such a variety of sample types.

SPECIMEN

Whole blood is the sample of choice, although urine can be used for determining lead in humans. Blood must be collected in lead-free heparinized Vacutainers equipped with sterilized stainless-steel needles. Another anticoagulant that can be used is EDTA. Caution should be exercised during sample handling to prevent any external contamination. Heparinized, refrigerated blood samples are stable for 2 weeks, but EDTA blood samples are stable for several months if frozen at $-20°$ C. The sample is suitable for analysis by all methods. If EDTA is used as an anticoagulant, 1.4 mg of calcium chloride per milliliter of blood must be added to enhance lead recovery from the samples.[34] For the extraction/AAS procedure utilizing ammonium pyrrolidine dithiocarbamate as a chelator, the calcium chloride should be added to the system after, not before, the addition of the chelation agent.[37]

Urine samples must be collected in lead-free borosilicate or polyethylene bottles. One should collect at least 50 mL and measure the specific gravity. The sample must be preserved by the addition of 500 mg of thymol per liter of urine. The urine is stable for 1 week if refrigerated. The sample is suitable for analysis by all methods.

PROCEDURE[27] (anodic stripping voltametry)

Principle. Sample is ashed by nitric-perchloric-sulfuric acid mixture. The resultant solution is quantitated by measurement of the current at -430 mV.

Reagents

Double-distilled, concentrated nitric acid (Ultrex grade), double-distilled concentrated sulfuric acid

(Ultrex grade), and double-distilled 70% perchloric acid (G. Frederick Smith Chemical Co., 867 McKinley Ave., Columbus, OH 43223).

Redistilled nitric acid (G. Fredrick Smith Co.).

Acid ashing solution. Combine 32.7 mL of concentrated nitric acid (Ultrex grade) with 65.3 mL of 70% perchloric acid and 2.0 mL of concentrated sulfuric acid. This is stable for 6 months at room temperature.

0.35% perchloric acid (58 mmol/L). Add 1 mL of concentrated (70%), double-distilled perchloric acid to 199 mL of water. This is stable for 1 year at room temperature.

Aqueous standards

Stock (48.20 mmol/L, 1 g/L). Transfer 1.5987 g of lead nitrate to a cleaned 100 mL volumetric flask. Dissolve salt in 30 mL of distilled water, add 3 mL of concentrated, double-distilled nitric acid, dilute to the mark with distilled water, and mix thoroughly. Store in acid-washed polyethylene bottles. Stable for 1 year at room temperature.

Working (48.20 μmol/L, 1 mg/L). Transfer 1 mL of the stock lead solution to a 1-liter volumetric flask containing 1 mL of concentrated, double-distilled nitric acid. Dilute to mark with distilled water and mix thoroughly. Prepare fresh on day of analysis.

Blood working standards. A blood sample is obtained from a healthy person who has not been exposed to lead. Five 100 μL specimens of the blood are placed into each of five Vacutainer tubes. The samples are spiked with 0, 20, 50, 100, and 150 μL of the 1 mg/L working aqueous standard described just above. The samples now correspond to 0, 20, 50, 100, and 150 ng of lead per tube, respectively. The samples are processed as described below and are used to prepare a standard curve. For subsequent analytical runs, only the 100 ng standard needs to analyzed.

Alternatively, a set of standard blood specimens can be obtained from the Centers for Disease Control[32] for use in standard curve and instrument calibration. On receipt of the samples, divide into 250 μl aliquots by adding to clean vials and freeze. On the day an analysis is to be conducted, two aliquots of each standard blood, representing the expected range in the patient samples, are thawed. One set is analyzed to calibrate the instrument, and the second set is analyzed at the end of the day to verify the calibration.

Urine working standards. A urine sample is obtained from a healthy person who has not been exposed to lead. First measure the specific gravity of the sample and discard it if the specific gravity is less than 1.010 g/mL. Six 1.0 mL aliquots are added to six ASV analysis cells. The samples are spiked with 0, 10, 50, 100, 200, and 500 μL of the 1 mg/L working aqueous standard desribed above. This corresponds to 0, 10, 50, 100, 200, and 500 ng of lead per cell. These spiked urine samples are digested and analyzed by the ASV assay described next and are used to prepare a standard curve. For subsequent analytical runs, only the 200 ng standard needs to be analyzed.

Assay

Equipment: Model 301D atomic stripping voltammeter (Environmental Science Associates, or equivalent); Ultrasonics mixer model W-140E scientific (Heat Systems, Ultrasonics Inc., or equivalent).

1. Mix the sample (blood or standard) by rinsing the ultrasonic mixer tip thoroughly in distilled, deionized water, and then place the tip into the sample within 2 cm of the bottom of the sample container and mix at the maximum permissible power for 20 to 30 seconds. If a sample sits for more than 10 minutes after it has been mixed, remix it before taking an aliquot for analysis.
2. Pipet 100 μL of the mixed sample into an analysis cell. Rinse the pipet once with distilled, deionized water and add the rinse to the analysis cell. To prepare the pipet for reuse, rinse it four times with distilled deionized water and discard the rinse water. Since some batches of the micropipets may have an appreciable contamination with lead, it is best to use the same tip throughout a series of samples and to rinse it between each sample. Duplicate 100 μL samples should be taken for analysis.
3. Into the analysis cell containing the sample, pipet 300 μL of acid ashing solution. Place the cell in the digestion rack on a hot plate at 200° to 240° C. The sample will initially turn yellow but will clear in about 10 minutes. A reflux line will be established about halfway up the cell. At the end of 30 minutes there will be approximately 100 to 200 μL of acid left in the bottom of the cell, with most of the nitric acid having evaporated. Continue the digestion for an additional 30 to 60 minutes until the acid is almost completely evaporated. Remove the cell from the hot plate, cool it, and add 4.5 mL of 0.35% perchloric acid. Reconstitute urine digests with 4.5 mL of 0.35% perchloric acid.
4. Following the manufacturer's instructions for the instrument, analyze the sample solution under the following conditions (voltages given with respect to a silver–silver chloride reference electrode): plating potential, −780 mV; plating time, 10 to 30 minutes, making sure that plating time for samples and standards are exactly the same; sweep rate, +60 mV per second; and current range, 0.2 to 1.0 mA, full scale. The stirring rate for the sample solution during plating should not be lower than 170 mL/min, since at lower rates significant variations in the amount of lead plated on the electrode may occur. As the sample is stripped, a cadmium peak may occur first at −600 mV, followed by the lead peak at −430 mV.

Calculations

The amount in nanograms of lead present in each standard solution is plotted versus the peak current value (μA) produced at −430 mV during the reaction. The slope of this line is the response factor in ng/μA that is used in subsequent calculations.

Blood

$$\text{Corrected amount (ng of Pb) in sample} = \text{Peak height current } (\mu A) \times \text{Response factor } (ng/\mu A)$$

$$\text{Concentration of Pb } (\mu g/L) \text{ in blood} = \frac{\text{Corrected amount of Pb (ng)} \times 1000\ (mL/L)}{\text{Sample volume (mL)} \times 1000\ (ng/\mu g)}$$

Urine. Calculate corrected amount as for blood.

$$\text{Concentration of Pb } (\mu g/L) \text{ in urine} = \frac{\text{Corrected amount (ng)} \times \left(\frac{0.024}{\text{Measured spec. grav.} - 1.0}\right) \times 1000\ (mL/L)}{\text{Sample volume (mL)} \times 1000\ (ng/\mu g)}$$

Cleaning of glassware

Reagents

Hot chromic acid. Dissolve 25 g (0.25 mol) of reagent-grade chromic trioxide *(caution—suspected carcinogen)* in 2.5 L of reagent-grade concentrated sulfuric acid. Stir with a Teflon-coated stirring bar. Heat to 50° C before use. This is stable for 6 months at room temperature; discard when the solution changes color from red to green. Dispose according to guidelines of the Environmental Protection Agency (EPA) for hazardous waste materials.

Nitric acid soaking solution. Dilute concentrated nitric acid with an equal volume of glass or double-distilled water. This is stable for 1 year at room temperature.

All glassware must be rendered lead-free before use with the following protocol:

1. Wash in detergent and tap water and follow with tap and distilled water rinses.
2. Soak in hot chromic acid and follow with tap and distilled water rinses.
3. Soak in a 1:1 or concentrated nitric acid bath for 30 minutes and follow with tap, distilled, and then double-distilled water rinses.

Note that the entire procedure must be used when glassware is being initially prepared for use in the lead analysis. For each successive use of equipment previously subjected to the entire cleaning procedure, it is necessary only to use step 3. However, the glassware must be committed to lead analysis to avoid contamination from other sources. Also, polyethylene equipment, used for storage of standards and so on, can be effectively cleaned when one subjects it to this same procedure.

Ashing of urine

1. Add a 25 mL sample of urine to a 125 mL borosilicate glass beaker; then add 7 mL of redistilled nitric acid.
2. Evaporate the sample at 130° C just to the point of dryness; then cover the beaker with a lead-free watch glass.
3. After cooling, add successive portions of the nitric acid, 2.0, 1.5, 1.0, and 0.5 mL, by sliding the watch glass only enough to facilitate each new addition of acid.
4. After each addition and as soon as the residue becomes light colored, heat the sample at 400° C just long enough to blacken the sample.
5. Cool the sample, add an additional portion of the acid, and repeat the heating procedure.
6. Finally, heat the residue for 5 to 10 minutes at 400° C temperature so that it remains pale yellow or light brown.

REFERENCE RANGE

There are two classes of persons exposed to lead: (1) those exposed occupationally and (2) those exposed environmentally. An occupational exposure limit of 100 μg per cubic meter of air, as a time-weighted average exposure for an 8-hour work day, 40 hour work week, has been recommended by NIOSH.[1] The maximum allowable blood level for lead in workers has been set at 600 μg/L,[1] the level at which health effects are observed. The typical range for healthy nonexposed adults is 100 to 200 μg/L,

Table 61-18. Comparison of reaction conditions for methods of lead analysis

Parameter	Dithizone method	Atomic absorption spectroscopy	Anodic stripping voltammetry
Temperature	Ambient	2000° C and above	Ambient
pH	Greater than 11.0	—	Less than 1.0
Final concentration of reagent components	Dithizone: 6.25 mmol/L	Ammonium pyrrolidine dithiocarbamate: 0.12 mol/L	0.35% perchloric acid
Sample volume	10 mL	10 mL	100 μL
Fraction of sample volume	100	100	100
Linearity	0-5 μg/mL	0-1000 ng/mL	6-1000 ng
Precision (coefficient of variation)	6%	5%	5% at 600 ng/mL
Time of reaction	Hours	Minutes	Minutes
Major interferences	Tin, bismuth, thallium	—	Thallium

with values for children being lower.[1,37,38] Blood lead and urine lead concentrations have been reported to be about 10% to 20% higher in males than in females[38]; this has been observed for children and for adults. Other biochemical parameters indicative of lead exposure include δ-aminolevulinic acid, δ-aminolevulinic acid dehydratase, and porphyrins.

REFERENCES

1. Criteria for a recommended standard: occupational exposure to inorganic lead, revised criteria, NIOSH, DHEW Publ. No. 78-158, Washington, D.C., 1978, U.S. Government Printing Office.
2. Bambach, K., and Burkey, R.: Microdetermination of lead by dithizone, Ind. Eng. Chem. Anal. Ed. **14**:904-907, 1942.
3. Cholak, J., Hubbard, D., and Burkey, R.: Microdetermination of lead in biological material with dithizone at high pH, Anal. Chem. **20**:671-672, 1948.
4. Cholak, J.: Analytical methods for determination of lead, Arch. Environ. Health **8**:222-231, 1964.
5. Lead in blood and urine. Method no. P&CAM 102. In Taylor, D.G., manual coordinator: NIOSH manual of analytical methods, vol. I, ed. 2, Washington, D.C., 1977, U.S. Government Printing Office, pp. 102-1 to 102-9.
6. Standard test method for lead in the atmosphere by colorimetric dithizone procedure. In Annual book of ASTM standards. Part 26. Gaseous fuels: coke: atmospheric analyses, method D-3112-77, Philadelphia, 1982, American Society of Testing and Materials, pp. 628-637.
7. Newland, L.W., and Daum, K.A.: Lead. In Hutzinger, O., editor: Anthropogenic compounds, The handbook of environmental compounds, 3, pt. A-B, New York, 1982, Springer-Verlag, New York, Inc., pp. 1-26.
8. Rice, E.W., Fletcher, D.C., and Stumpf, A.: Lead in blood and urine, Stand. Methods Clin. Chem. **5**:121-129, 1965.
9. Christian, G.: Medicine, trace elements and atomic absorption spectroscopy, Anal. Chem. **41**:24A-40A, 1969.
10. General procedure for metals. Methods no. P&CAM 173. In Taylor, D.G., manual coordinator: NIOSH manual of analytical methods, vol. I., ed. 2, Washington, D.C., 1977, U.S. Government Printing Office, pp. 173-1 to 173-10.
11. Lead in air or blood. Method no. P&CAM 214. In Taylor, D.G., manual coordinator: NIOSH manual of analytical methods, vol. I, ed. 2, Washington, D.C., 1977, U.S. Government Printing Office, pp. 214-1 to 214-6.
12. Lead in blood and urine. Method no. P&CAM 262. In Taylor, D.G., manual coordinator: NIOSH manual of analytical methods, vol. I, ed. 2, Washington, D.C., 1977, U.S. Government Printing Office, pp. 262-1 to 262-4.
13. Lead and inorganic lead compounds. Method no. P&CAM S341. In Taylor, D.G., manual coordinator: NIOSH manual of analytical methods, vol. I, ed. 2, Washington, D.C., 1977, U.S. Government Printing Office, pp. S341-1 to S341-7.
14. Kopito, L., and Shwachman, H.: Measurement of lead in blood, urine and scalp hair by atomic absorption spectrometry, Stand. Methods Clin. Chem. **7**:151-162, 1972.
15. Olsen, E.D., and Jatlow, P.I.: An improved Delves cup atomic absorption procedure for determination of lead in blood and urine, Clin. Chem. **18**:1312-1317, 1972.
16. Hicks, J.M., Gutierrez, A.N., and Worthy, B.E.: Evaluation of the Delves micro system for blood lead analysis, Clin. Chem. **19**:322-325, 1973.
17. Marcus, M., Hollander, M., Lucas, R.E., and Pfeiffer, N.C.: Micro-scale blood lead determinations in screening: evaluation of factors affecting results, Clin. Chem. **21**:533-536, 1975.
18. Yeager, D.W., Cholak, J., and Henderson, E.W.: Determination of lead in biological and related material by atomic absorption spectrophotometry, Environ. Sci. Technol. **5**:1020-1022, 1971.
19. Fernandez, F.J.: Micromethod for lead determination in whole blood by atomic absorption with use of the graphite furnace, Clin. Chem. **21**:558-561, 1975.
20. Sunderman, F.W.: Electrothermal atomic absorption spectrometry of trace metals in biological fluids. In Forman, D.T., and Matton, R.W., editors: Clinical chemistry: ACS Symposium Series 36, Washington, D.C., American Chemical Society, 1975, pp. 248-270.
21. Stoeppler, M., and Brandt, K.: Contributions to automated trace analysis. Part II. Rapid method for the automated determination of lead in whole blood by electrothermal atomic-absorption spectrophotometry, Analyst **103**:714-722, 1978.
22. Subramanian, K.S., and Meranger, J.C.: A rapid electrothermal atomic absorption spectrophotometric method for cadmium and lead in human whole blood, Clin. Chem. **27**:1866-1871, 1981.
23. Nise, G., and Vesterberg, O.: Blood lead determination by flameless atomic absorption spectroscopy, Clin. Chem. Acta **84**:129-136, 1978.
24. Boone, J., Hearn, J., and Lewis, S.: Comparison of interlaboratory results for blood lead with results from a definitive method, Clin. Chem. **25**:389-393, 1979.
25. Christian, G.D.: The biochemistry and analysis of lead. In Bodansky, O., and Latner, A.L., editors. Advances in clinical chemistry 18, New York, 1976, Academic Press, Inc., pp. 289-326.
26. Pierce, J.O., Koirtyohann, S.R., Clevenger, T.E., and Lichte, F.E.: The determination of lead in blood: a review and critique of the state of the art 1975, New York, 1976, International Lead Zinc Research Organization, Inc., p. 76.
27. Searle, B., Chan, W., and Davidow, B.: Determination of lead in blood and urine by anodic stripping voltammetry, Clin. Chem. **19**:76-80, 1973.
28. Morrell, G., and Giridhar, G.: Rapid micromethod for blood lead analysis by anodic stripping voltammetry, Clin. Chem. **22**:221-223, 1976.
29. Lead in blood. Method no. P&CAM 195. In Taylor, D.G., manual coordinator: NIOSH manual of analytical methods, ed. 2, vol. I. Washington, D.C., 1977, U.S. Government Printing Office, pp. 195-1 to 195-7.
30. Lead in urine. Method no. PP&CAM 200. In Taylor, D.G., manual coordinator: NIOSH manual of analytical methods, vol. I, ed. 2, Washington, D.C., 1977, U.S. Government Printing Office, pp. 200-1 to 200-8.
31. Barnes, I.L., Murphy, T.J., Gramlich, J.W., and Shields, W.R.: Lead separation by anodic deposition and isotope ratio mass spectrometry of microgram and smaller samples, Anal. Chem. **45**:1881-1884, 1973.
32. Blood lead proficiency testing, Licensure and Proficiency Testing Division, Bureau of Laboratories, Centers for Disease Control, Atlanta, Public Health Service, U.S. Department of Health and Human Services.
33. Blood lead proficiency testing: 1982 data summary, Document No. 44Q421618303, Centers for Disease Control, Atlanta, U.S. Department of Health and Human Services, Jan. 1983.
34. Critique: Blood lead analysis 1981, Document No. 34Q205158208, Centers for Disease Control, Atlanta, U.S. Department of Health and Human Services, May 1982.
35. Meranger, J.C., Hollebone, B.R., and Blanchette, G.A.: The effects of storage times, temperature and container types on the accuracy of atomic absorption determinations of Cd, Cu, Hg, Pb and Zn in whole heparinized blood, J. Anal. Toxicol. **5**:33-41, 1981.
36. Unger, B.C., and Green, V.A.: Blood lead analysis—lead loss to storage containers, Clin. Toxicol. **11**:237-243, 1977.
37. Zinterhofer, L.J.M., Jatlow, P.I., and Fappiano, A.: Atomic absorption determination of lead in blood and urine in the presence of EDTA, J. Lab. Clin. Med. **78**:664-674, 1971.
38. Airborne lead in perspective, report of the Committee on Biological Effects of Atmospheric Pollutants of the National Research Council, Washington, D.C., 1972, National Academy of Sciences, Printing and Publishing Offices.
39. Tsuchiya, K.: Lead. In Friberg, L., Nordberg, G.F., and Vouk, V.B., editors: Handbook on the toxicology of metals, New York, 1979, Elsevier/North-Holland Biomedical Press, pp. 451-484.

Lithium

ROBERT L. MURRAY

Clinical significance: p. 864
Atomic weight: 6.94
Atomic symbol: Li
Merck Index: 5354
Chemical class: element

PRINCIPLES OF ANALYSIS AND CURRENT USAGE

Unlike some clinical laboratory assays in which analytical methodology has developed new directions in recent years, the methods for lithium analysis have remained essentially unchanged from those originally described.[1] Two techniques have been used with almost equal popularity and with equal success: flame emission spectroscopy and atomic absorption spectroscopy (Table 61-19).

In the flame emission technique a 1:50 sample dilution is typically used for serum or plasma. With an exception to be noted later, the diluent is 1.50 millimoles of potassium ion per liter (mmol/L K^+), which serves as the internal standard. Since the serum or plasma adds a small amount of potassium, the diluted sample would contain sightly more potassium than the blank or the standard. This potential inaccuracy is eliminated by the addition to both blank and standard of 5.0 mmol/L potassium chloride, approximating the K^+ concentration of the sample. An aqueous standard of 1.0 mmol/L lithium carbonate is used for calibration. A blank is prepared in the same fashion as the standards, except for the substitution of water in place of the 1.0 mmol/L lithium carbonate.

The principles of flame emission are discussed in detail in Chapter 3. The heat of the air-propane flame (approximately 1925° C) vaporizes the lithium chloride (boiling point, 1350° C). In the presence of the heat and the reducing gases carbon monoxide, hydrogen, and carbon dioxide, the lithium chloride dissoicates into uncharged ground-state atoms, Li^0 and Cl^0. The heat further causes excitation of 2*s* electron to the 2*p* state, which instantly decays to the original 2*s* ground state, with emission of light at 670.8 nm. Although only a very small fraction of the lithium is ultimately raised to the excited state with subsequent emission, this is sufficient for accurate quantitation.

The potassium present as added internal standard and from the sample follows a similar pathway, entering the vapor state at a higher temperature by sublimation at approximately 1500° C. Since the emission intensity depends on conditions subject to minor fluctuations, such as the fuel to oxidant ratio, the sample aspiration rate, the reducing gas composition and concentration, and the flame temperature, these are all potential sources of error. To circumvent this, one measures the ratio of the emission intensity of potassium to the emission intensity of lithium, rather than only that of lithium. In this way instrumental fluctuations that affect both elements are compensated for, and error is minimized.

A modification to this flame emission technique was recently reported.[2] This small but significant change allows the use of cesium instead of potassium as the internal standard. Cesium functions in the same fashion as previously described for potassium, except that it is undetectable in serum or plasma, and therefore (unlike with potassium) no correction or compensation is necessary. In addition, sodium, potassium, and lithium can be analyzed with minimal instrumental modifications. Cesium chloride vaporizes at 2390° C. When thermally excited, its 6*s* electron is raised to the 6*p* orbital; decay of this electon results in light emission at 852 nm.

The theoretical basis of atomic absorption spectroscopy bears a certain similarity to that of flame emission (see Chapter 3). Neutral lithium atoms *absorb* light at 670.8 nm, which is produced by an external source. Because lithium ionizes only to a very small extent (approximately 1%) under the conditions described, little advantage is gained by the use of an ionization-suppression buffer such as tin chloride. In atomic absorption analysis of lithium an aqueous dilution of the serum, plasma, urine, or hemolysate is prepared. For serum, plasma, or hemolysate the typical dilution is 1:10 in physiological solution. For urine the dilution necessary may vary from 1:100 to 1:1000, depending on the lithium concentration in the urine. An estimated value may be calculated based on the patient's dosage and the urine volume, or a range of dilutions may be used, and the appropriate dilution selected for calculation and reporting.

Table 61-19. Methods of lithium analysis

Method	Type of analysis	Principle	Usage
Flame emission spectroscopy	Quantitative	Emission of light at 670.8 nm by Li^0	Serum, plasma
Atomic absorption spectroscopy	Quantitative	Absorption of light at 670.8 nm by Li^0	Serum, plasma urine, red blood cells

REFERENCE AND PREFERRED METHOD

The decision as to the method of choice for lithium analysis in serum or plasma may be made on the basis of criteria other than those of inherent analytical accuracy, precision, cost, and ease of performance.[3,4] The two available techniques compare remarkably well for each of these criteria (Table 61-20). Only when the range of specimens is considered is there a difference: the high and variable potassium content of urine and hemolysate specimens ren-

Table 61-20. Comparison of conditions in methods for lithium analysis

Condition	Flame emission spectroscopy		Atomic absorption spectroscopy	
Sample	Plasma, serum		Plasma, serum, urine, hemolysate	
Dilution (plasma, serum)	1:50		1:10	
Diluent	1.5 mmol/L K^+ or 1.5 mmol/L Cs^+		Water	
Sample consumed	1 mL (approximately)		1 mL (approximately)	
Specimen required	20 μL		100 μL	
Precision*	**SD (mmol/L)**	**Mean (mmol/L)**	**SD (mmol/L)**	**Mean (mmol/L)**
	0.069	0.484	0.037	0.489
	0.093	1.249	0.071	1.237
	0.180	2.730	0.159	2.703
Detection limit	0.01 mmol/L		0.01 mmol/L	
Linearity	0.18-19.1 mmol/L		0-5 mmol/L	

*College of American Pathologists Comprehensive Chemistry Survey, 1982. *SD,* Standard deviation.

ders the flame-emission result unreliable. This can be overcome by prior potassium analysis and addition of equivalent potassium to the standard and blank. However, serum and plasma are overwhelmingly the most common specimens; the choice of methodology can then be based on available instrumentation, degree of automation available, logistic considerations, and so on. Because of these similarities, a detailed procedure for both methods is presented.

SPECIMEN

Serum or plasma drawn 8 to 10 hours after an oral dose of lithium is the specimen of choice for routine monitoring; if repeated levels are drawn, the dose-to-phlebotomy interval should be kept identical for each patient. Some anticoagulants are formulated with lithium as the cation; if plasma is used, such tubes should be avoided. A number of investigators[5,6] have reported an advantage in measuring red blood cell lithium for estimation of tissue levels of lithium and as a method of detecting noncompliance (equilibrium being established only after regular usage). Since the red blood cell lithium is approximately in the same concentration range as the serum or plasma levels, no special manipulations beyond the sample preparation are needed. Urine levels are of extremely limited value, except to verify a patient's intake or to document unusual excretion patterns. When a urinary concentration is compared to an oral intake, it is important to remember that the normal dose of approximately 1200 mg per day is measured in milligrams, not millimoles, and this is measured as lithium carbonate. Thus 1200 mg of lithium carbonate equals 32.5 mmol of lithium ion.

The plasma or serum should be separated from the cells if storage of more than 4 hours is anticipated. In the compliant patient this would not be critical; it is advised only in the event of an unusually large plasma–to–red blood cell concentration disparity. Unseparated anticoagulated blood stored at 4° C shows an increase in red blood cell lithium concentration; storage at 37° C results in a decrease in red blood cell lithium. Once the serum is separated, lithium is stable for at least 24 hours at room temperature, for 7 days at 4° C, and indefinitely if frozen.

PROCEDURE FOR FLAME EMISSION

Reagents

Diluent (1.5 mmol/L potassium chloride). Dissolve 112 mg of potassuim chloride in deionized water, and add water to the mark in a 1 L volumetric flask. This is stable for 1 year at room temperature.

Stock blank (140 mmol/L sodium chloride, 5.0 mmol/L potassium chloride). Dissolve 373 mg of potassium chloride and 9.18 g of sodium chloride in deionized water. Add water to the mark in a 1.0 L volumetric flask. This is stable for 1 year at room temperature.

Stock 1.0 lithium standard (1.0 mmol/L lithium carbonate). Dry analytical-grade lithium carbonate for 4 hours at 200° C; dissolve 73.89 mg of the cooled salt in 50 mL of deionized water containing 20 mL of 0.1 M hydrochloric acid. Add 746 mg of potassium chloride and 16.37 g of sodium chloride directly to the solution. Add water to the mark in a 2 L volumetric flask. This solution is stable for 1 year at room temperature.

Assay

Equipment: any flame emission photometer.

1. Dilute 100 μL of lithium standard, blank, and each unknown serum, plasma, or control with 5.0 mL of diluent. Mix well.
2. The instrument conditions specified by the manufacturer should be followed. In all cases the aspiration chamber and burner should be flushed thoroughly by aspiration of water for 5 minutes before lithium analysis. This is necessary to flush any lithium salts (used as an internal standard for sodium and potassium analysis) from the system.
3. Zero the instrument with the diluted blank, and calibrate with the diluted lithium standard.
4. Analyze the unknown samples and controls, check-

ing the blank and the calibration readings after each unknown. Adjust as necessary.

PROCEDURE FOR ATOMIC ABSORPTION

Reagents

Deionized water should be of such a quality so as to give 10 megaohms per centimeter of specific resistance at 25° C.

Stock blank (140 mmol/L sodium chloride, 5.0 mmol/L potassium chloride). Prepare as in flame emission method.

Stock 1.0 lithium standard (1.0 mmol/L lithium carbonate). Prepare as in flame emission method.

Stock 2.0 lithium standard (2.0 mmol/L lithium carbonate). *See following note 1 before preparing.* Dissolve 73.89 mg of the dried lithium carbonate in 50 mL of deionized water containing 10 mL of 0.1 M hydrochloric acid. Add water to the mark in a 1 L volumetric flask. This is stable for 1 year at room temperature.

Assay

Equipment: any modern atomic absorption spectrophotometer.

1. Dilute 0.50 mL of blank, standard, and each unknown serum, plasma, urine (see note 1), or hemolysate (see note 2) with 9.50 mL of deionized water.
2. Follow the instrument conditions specified by the manufacturer.
3. Zero the instrument with the diluted blank, and calibrate with 1.0 mmol/L lithium standard. Analyze the unknown specimens, checking the blank and calibration readings periodically.
4. Specimens containing more than 2.0 mmol/L lithium should be diluted with diluent until the measured concentration falls below 2.0 mmol/L. The appropriate correction is then made for the dilution.

Notes

1. The 2.0 mmol/L lithium standard is used only if curve correction is not available on the instrument. Since most atomic absorption spectrophotometers can automatically correct for nonlinearity, this standard should be run only initially to verify the accuracy of the correction.
2. Urines are prepared by pipetting 10 mL of urine aliquots to each of the following volumetric flasks: 100 mL, 500 mL, and 1 L. Add water to the mark in each, and then process as described in step 1 of the procedure. Make the appropriate corrections (times 10, times 50, and times 100, respectively), and report the value that reads closest to the 1.0 standard.
3. Hemolysates are made by centrifugation of anticoagulated blood for not less than 15 min at 2000 *g* so as to exclude plasma from the cellular fraction as much as possible. Aspirate completely the plasma and a small portion of the packed red blood cells. Washing the cells is avoided to eliminate the possibility of exchange of lithium with the wash solution. Hemolyze the cells either by freezing and thawing or by immersion of the tube into an ultrasonic bath for 60 seconds. Centrifuge the hemolysate again before removal of the sample.

REFERENCE RANGES

Maintenance therapy	0.4-1.0 mmol/L
Acute therapy	0.9-1.4 mmol/L
Red blood cell concentration	0.2-0.8 mmol/L
Urine content	95%-99% of daily intake (after steady state)

REFERENCES

1. Barrow, G.R.J.: The estimation of lithium in blood, J. Med. Lab. Technol. **17:**236-40, 1960.
2. Bergkuist, C., Kelley, T.F., and Moran, B.L.: Cesium as the internal standard in a new four-element flame photometer, Clin. Chem. **24:**1061, 1978 (Abstract).
3. Prybus, J., and Bowers, G.N., Jr.: Serum lithium determination by atomic absorption spectroscopy. In MacDonald, R.P., editor: Standard Methods of clinical chemistry, vol. 6, New York, 1970, Academic Press, Inc., pp. 189-192.
4. Velapoldi, R.A., Paule, R.C., Schaffer, R., et al.: Standard reference materials: a reference method for the determination of lithium in serum, National Bureau of Standards Special Pub. no. 260-69, Washington, D.C., 1980, U.S. Government Printing Office.
5. Rybarkowski, J., and Strzyzewski, W.: Red blood cell lithium index and long term maintenance treatment, Lancet **1:**1408-1409, 1976.
6. Pandey, G.N., Dorus, E., Lavis, J.M., and Tosteson, D.C.: Lithium transport in human red blood cells, Arch. Gen. Psychiatry **36:**902-908, 1979.

Quinidine

AMADEO J. PESCE

Clinical significance: p. 488
Molecular formula: $C_{20}H_{24}N_2O_2$
Molecular weight: 324.41
Merck Index: 7850, 7852
Chemical class: plant alkaloid

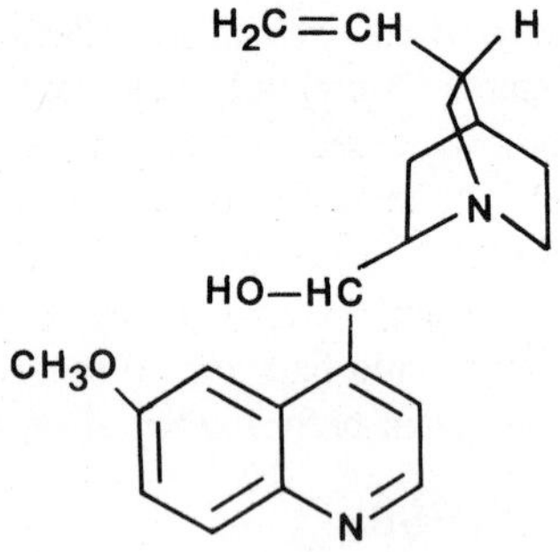

$H_2C{=}CH$ H N HO—HC CH_3O N

PRINCIPLES OF ANALYSIS AND CURRENT USAGE

Serum quinidine is monitored to establish whether or not the drug is present in its appropriate therapeutic range of 2 to 5 μg/mL. Quantitation of quinidine is complicated by the presence of an impurity, dihydroquinidine, a derivative of quinidine that can account for 5% to 25% of the dose. Quinidine is extensively metabolized before excretion, primarily by the hepatic P-450 oxidase mechanism.

The first method for the measurement of quinidine in blood was fluorometric and was adopted from the analysis of quinine (method 1, Table 61-21). The initial method described by Brodie and Udenfreund used dilution, followed by protein precipitation with metaphosphoric acid.[1] The quinidine was strongly fluorescent in the supernatant of this acid solution. To reduce the contribution of non-quinidine fluorescing molecules, Cramer and Isaksson suggested extraction of the quinidine from an alkalinized plasma into an organic solvent (benzene), followed by reextraction into an aqueous acid solvent (method 2, Table 61-21). These authors used a 366 mm mercury line for excitation and recorded the emission at 460 mm.[2] This method is currently used by 25% of the participating laboratories in a recent therapeutic drug monitoring survey.

Immunochemical methods can also be used for the measurement of quinidine (method 3, Table 61-21). Usually, these are competitive binding assays that are commercially available. The most widely used assay is an enzyme-multiplied immunoassay technique (EMIT) sold by Syva Co.[3] Approximately 60% of participants in a therapeutic drug monitoring survey used this method of analysis.

High-performance liquid chromatography resolves the parent compound from the dihydroquinidine derivative and the metabolites of the drug (method 4, Table 61-21). Currently, these methods are used by about 6% of the reporting laboratories. In the procedure of Kates[4] the drug and the internal standard cinchonine are extracted into benzene. The drug is separated from metabolites by a cation-

Table 61-21. Methods for quinidine analysis

Method	Type of analysis	Principle	Usage	Comments
1. Protein precipitation	Fluorometry	Protein is precipitated by metaphosphoric acid; quinidine fluorescent in acid ($\lambda_{excitation}$, 366 nm; $\lambda_{emission}$, 460 nm)	Historical	Other fluorescent compound measured values 2 times higher than actual
2. Double extraction	Fluorometry	Drug extracted into organic phase, then back-extracted into aqueous acid medium; fluorescence proportional to concentration	Common	Gives accurate results of total quinidine compounds
3. EMIT	Competitive binding assay	Antibody binds to drug-enzyme complex; increase in absorbance, A_{340}, proportional to quinidine concentration Does not measure metabolites	Most common	Measures quinidine Some cross-reactivity with other compounds
4. HPLC	Chromatography	Drug separated from metabolites and measured either by absorption (230 or 330 nm) or fluorescence (λ_{ex}, 366 nm; λ_{em}, 460 nm).	Used in small proportion of laboratories	Measures quinidine and metabolites specifically
5. Gas chromatography	Chromatography	Drug derivatized and separated from metabolites; flame ionization or nitrogen detection	Rare	Measures quinidine and its metabolites specifically
6. Thin-layer chromatography	Chromatography	Drug extracted and separated from metabolites and quantified by scraping, eluting, and recording the specific fluorescence	Rare	Measures quinidine and its metabolites Difficult to have good reproducibility

EMIT, Enzyme-multiplied immunoassay technique; *HPLC*, high-performance liquid chromatography.

exchange column (Partisil 10 SCX) using a methanol-water mobile phase buffered at pH 9.0 with potassium hydroxide and trimethylamine hydrochloride. The drug is monitored at 230 nm. The method is sensitive to 0.1 μg/mL and measures serum samples over the range of 1 to 10 μg/mL with a linear response. The method of Drayer et al.[5] is similar in that the drug is extracted into benzene before chromatography. However, these workers used reversed-phase chromatography with a μBondapak C-18 column and a mobile phase of acetonitrile–2.5% acetic acid. The drug was detected using fluorescence with 340 nm excitation, and emission was detected with a KV 418 filter.

Powers and Sadee[6] used a chromatography system similar to that of Drayer et al.—a reversed-phase alkyl-phenyl column with a mobile phase of acetonitrile–sodium acetate buffer with pH 3.6. They chose to precipitate the protein with acetonitrile rather than use an extraction of the serum. They monitored the quinidine at 330 nm and found the method to be linear over the range 0.3 to 10 μg/mL.

Quinidine can also be measured by gas chromatography using column flash methylation to form volatile derivatives[7] (method 5, Table 61-21). The sensitivity of this technique is 50 ng/mL using the hydrogen flame detector. This derivatized form of quinidine can also be measured in a gas chromatograph–mass spectrometer.[7]

Thin-layer chromatography followed by scraping of the plate has also been suggested as a method to quantify quinidine that is free of any contaminants or metabolites[8] (method 6, Table 61-21). The drug was eluted from the scrapings with ethanol-acetone. The fluorescence of the resultant solution was monitored with a 360 primary filter and a Wratten 2A filter for emission.

REFERENCE AND PREFERRED METHODS

The Cramer extraction procedure gives results that are in excellent agreement with the gas chromatographic procedure.[9] The ratio of values for the Brodie/Cramer methods is about 2:1. On this basis, it seems clear that the Cramer extraction procedure gives more accurate results. The interlaboratory precision of the extraction procedure is 11% (coefficient of variation) at quinidine levels of 1.8 μg/mL.[10] Intralaboratory precision is about 4%. The fluorometric procedure is a proposed selected method and is well suited for many laboratories.[11]

Enzyme immunoassays offer a viable alternative for those laboratories not experienced with solvent extraction techniques or without adequate fluorometers. Interlaboratory coefficient of variation of these methods is about 10%. However, when the enzyme immunoassay is compared to high-performance liquid chromatography, a 25% higher value is found for enzyme immunoassay.[12] The reason is that the antibody is not completely specific for quinidine and reacts also with the major metabolite (3*s*)-3-hydroxyquinidine as well as dihydroquinidine. Both these compounds are pharmacologically active, and thus their measurement is acceptable. None of these assays (Cramer, enzyme immunoassay, or HPLC) work well at levels below 1 μg/mL. Survey results indicate variations of 15% by enzyme immunoassay and greater than this by other techniques (fluorometry, HPLC) when the quinidine levels were 0.6 μg/mL.

SPECIMEN

Serum is the preferred body fluid. Serum quinidine is stable when stored in the dark or frozen.

PROCEDURE

Principle

Quinidine is extracted into a benzene solvent from an alkalinized serum followed by reextraction into an aqueous acid solvent. This method separates the quinidine parent compound from its metabolites.

Reagents and materials

0.1 M sodium hydroxide (0.1 mol/L). Dissolve 4 g of sodium hydroxide in 500 mL of water. Bring to 1 L with distilled water. This is stable for 12 months at room temperature.

0.1 M hydrochloric acid (0.1 mol/L). Add 8.35 mL of concentrated hydrochloric acid to 500 mL of water. Bring to 1 L. This is stable for 12 months at room temperature.

Benzene. Reagent grade.

Isoamyl alcohol. Reagent grade.

Quinidine standards. Prepare quinidine sulfate stock solution (3.2 mmol/L, 1 mg/mL) by dissolving 12.1 mg of quinidine sulfate in 10 mL of distilled water. This is stable for 6 months at 4° C.

Quinidine working solution (32 μmol/L). Dilute 1 mL of stock solution to 100 mL of distilled water. This is stable for 1 month at 4° C.

Assay

Equipment: Any fluorometer capable of excitation at 350 nm and reading at 450 nm is satisfactory (Aminco SLM, Turner, and others).

1. To a series of 10 mL screw-capped tubes add 0.5 mL of test sera, controls, and so on. To a second series of tubes add 0.5 mL of pooled sera containing no drug, and then add 0, 50, 100, 250, and 500 μL of working quinidine solution corresponding to concentrations of 0, 1, 2, 5, and 10 μg/mL, respectively.
2. To each tube add 0.5 mL of 0.1 M sodium hydroxide, followed by 5 mL of benzene.
3. Shake for 5 minutes and then centrifuge for 3 minutes at 1000 *g*.
4. Quantitatively pipet off 3 mL of the benzene layer with a serological pipet and place into a second 10 mL screw-capped tube. Add 0.2 mL of isoamyl alcohol and 3 mL of 0.1 M hydrochloric acid.

Table 61-22. Comparison of assay conditions for quinidine

Condition	Cramer (fluorometry)	Gas chromatography	High-performance liquid[4] chromatography	Enzyme-multiplied immunoassay technique
Temperature	Ambient	—	Ambient	30° C
pH	1	14	14	—
Final concentration of reaction compounds	0.1 M H_2SO_4	0.33 M NaOH Benzene 0.2 M trimethylanilinium hydroxide in methanol	0.33 M NaOH Benzene 0.01 M trimethylamine hydrochloride 0.01 KOH pH 9.0 (1:4) methanol	—
Fraction of sample volume	1.00	1.00	1.00	0.03
Sample volume	0.5 mL	1.0 mL	1.0	0.05 mL
Linearity	1-15 μg/mL	0.2-12 μg/mL	0.5-10 μg/mL	0.5-8 μg/mL
Precision*	11%	11%	11%	10%
Time of reaction	> 30 min	> 30 min	> 30 min	1 min
Interferences	Quinine, dihydroxyquinidine	—	—	Dihydroxyquinidine (3S)-3-Hydroxyquinidine

*Coefficient of variation. Estimated from Therapeutic Drug Monitoring Survey data.[10]

5. Shake for 5 minutes and centrifuge for 3 minutes at 1000 *g*.
6. Discard the top benzene layer, taking care not to discard any of the lower acid layer. Transfer at least 1 mL of acid layer to the fluorescence cuvette. Read in the fluorometer.

Calculations

Draw a standard curve on linear graph paper; a straight line should pass through all the points. Interpolate the values of the controls and the patients from the curve.

REFERENCE RANGE

Therapeutic levels are generally 2 to 5 μg/mL (6 to 15 μmol/L).

REFERENCES

1. Brodie, B.B., and Udenfreund, S.: The estimation of quinine in human plasma with a note on the estimation of quinidine, J. Pharmacol. Exp. Ther. **78:**154-158, 1943.
2. Cramer, G., and Isaksson, B.: Quantitative determination of quinidine in plasma, Scand. J. Clin. Lab. Invest. **15:**553-556, 1963.
3. Syva Co., Inc., Palo Alto, CA 94303.
4. Kates, R.E., McKennon, D.W., and Comstock, T.J.: Rapid high pressure liquid chromatographic determination of quinidine and dihydroquinidine in plasma samples, J. Pharm. Sci. **67:**269-270, 1978.
5. Drayer, D.E., Restivo, K., and Reidenberg, M.M.: Specific determination of quinidine and 3-hydroxyquinidine in human serum by high pressure liquid chromatography, J. Lab. Clin. Med. **90:**816-822, 1977.
6. Powers, J.L., and Sadee, W.: Determination of quinidine by high performance liquid chromatography, Clin. Chem. **24:**299-302, 1978.
7. Midha, K.K., and Charette, C.: GLC determination of quinidine from plasma and whole blood, J. Pharm. Sci. **63:**1244-1247, 1974.
8. Hartel, G., and Korhonen, A.: Thin layer chromatography for the quantitative separation of quinidine and quinine metabolites from biological fluids and tissues, J. Chromatogr. **37:**70-75, 1968.
9. Huffman, D.H., and Hignite, C.E.: Serum quinidine concentrations: comparison of fluorescence, gas chromatographic, gas chromatographic/mass spectrophotometric methods, Clin. Chem. **22:**810-812, 1976.
10. Therapeutic Drug Monitoring Survey, American Association for Clinical Chemistry, 1725 K St. NW, Washington, DC 20006.
11. Broussard, L.A., Findley, T.W., Pellegrino, L., et al.: Fluorometric determination of quinidine, Clin. Chem. **27:**1929-1930, 1981.
12. Drayer, D.E., Lorenzo, B., and Reidenberg, M.M.: Liquid chromotography and fluorescence spectroscopy compared with a homogeneous enzyme immunoassay technique for determining quinidine in serum, Clin. Chem. **27:**308-310, 1981.

Salicylate

JOSEPH SVIRBELY

Salicylate, acetylsalicylic acid, aspirin
Clinical significance: p. 897

	Acetylsalicylic acid	Salicylic acid
Molecular formula:	$C_9H_8O_4$	$C_7H_6O_3$
Molecular weight:	180.15	138.12
Merck Index:	874	8093
Chemical class:	Benzoic acid derivative	Hydroxybenzoic acid

COOH
O O
C
CH_3

PRINCIPLES OF ANALYSIS AND CURRENT USAGE

Acetylsalicylic acid (aspirin) is rapidly hydrolized to salicylic acid and acetic acid in the stomach; salicylic acid is rapidly absorbed into the blood. There are a variety of methods for determination of salicylate in serum or plasma. The most widely used methods in the clinical lab-

Table 61-23. Methods for salicylate analysis

Method	Type of analysis	Principle	Usage	Comments
1. Ferric nitrate complex	Spectrophotometry	Salicylates + Fe^{+++} → Ferric complex (A_{max}, 540 nm)	Most commonly used clinical salicylate procedure	Sensitive to 200 mg/L Linear to 1000 mg/L Only interference is salicylate metabolites
a. Brodie et al.		Extracts salicylate into ethylene dichloride		Method 1a uses toxic ethylene dichloride
b. Keller		Eliminates extraction step		Method 1b less complicated but specific enough for clinical needs
c. Trinder		Deproteinates with mercuric chloride		Method 1c uses toxic mercuric chloride
2. Folin-Ciocalteu reagent reduction	Spectrophotometry	Phosphotungstic acid + phosphomolybdic acid reduce salicylates to form blue complex (A_{max}, 660 nm)	Not commonly used because of interferences	Tryptophan, tyrosine, and uric acid react with reagent to give falsely high readings
3. Spectrophotometry	Differential ultraviolet spectrophotometry	Measures salicylate + acetylsalicylate by measuring differences in absorbance at pH 9.0 and pH 13.5	Still in common use, likely to be replaced by gas-liquid chromatography	Has advantage of measuring acetylsalicylate Requires narrow band pass ultraviolet spectrophotometer sensitive to 5 mg/L
4. Fluorometry	Spectrofluorometry	Protein can be removed by precipitation Alkali intensifies fluorescence of salicylate	Not commonly used	Has advantage of greater sensitivity Requires smaller sample than spectrophotometry
5. Gas-liquid chromatography	Chromatographic separation	Acetylsalicylate and major metabolites extracted with ether; derivation with BSTFA (see text) injected into gas-liquid chromatograph using temperature programming and OV-17 column injected with flame ionization detector	Used in research: best candidate for reference method	Sensitive to 1 mg/L Linear to 2000 mg/L No interfering compounds

oratory employ colorimetric techniques, but other methods include gas-liquid chromatography (GLC), ultraviolet spectroscopy, and fluorescence spectrophotometry.

The earliest and still most widely used method for measurement of salicylate is the colorimetric method, which measures the absorbance produced by complexing salicylates with ferric ion (method 1, Table 61-23). The first widely accepted method using this technique was described by Brodie et al.[1] in 1944. In this procedure salicylate is separated from plasma by extraction into acidified ethylene dichloride, followed by back-extraction into an aqueous phase as the ferric complex, which is then measured in a colorimeter at 540 nm.

Keller[2] in 1948 published a modification of Brodie's method in which the extraction steps are eliminated. Instead, total salicylate is measured as the sum of acetylsalicylate and its metabolites.

In another well-known method, published by Trinder[3] in 1950, protein is precipitated with mercuric chloride and hydrochloric acid. Color is produced when salicylate is complexed with ferric ion supplied as ferric nitrate.

A second type of colorimetric method involves reducing salicylate with a mixture of phosphotungstic and phosphomolybdic acids (Folin-Ciocalteu reagent) (method 2, Table 61-23).[4] Weichselbaum and Shapiro[5] published a method in which serum was mixed with this reagent, and the resulting blue color was read at 660 nm. In 1950 Smith and Talbot[6] published a similar method but shortened the color development step from 40 minutes to 3 minutes.

Ultraviolet spectrophotometry has also been employed as an analytical technique to measure acetylsalicylate plus salicylate (method 3, Table 61-23). Routh et al.[7] in 1952 extracted the acetylsalicylate as in the Brodie procedure and quantified the drugs using differential spectrophotometry by measuring the differences in absorbance at pH 9.0 and 13.5.

A third analytical technique frequently employed for measurement of salicylate is spectrofluorometry (method

4, Table 61-23). In 1948 Saltzman[8] published a spectrofluorometric method in which the salicylates were separated from protein by precipitation with dilute tungstic acid. Strong alkali was added to convert the salicylate into a more fluorescent form. The ion was excited at 370 nm, and fluorescent emission was monitored at 460 nm.

Potter and Guy[9] have published a micromethod for plasma that allows determination of protein bound salicylate, free salicylate, acetylsalicylate, and total salicylate in serum. Sephadex-gel columns were used to separate the protein-bound salicylate from the free forms. Salicylate was measured by spectrofluorometry, with excitation at 305 nm and emission at 405.

In a recent fluorometric assay of salicylate described by Lever and Powell[10] in 1973, enhancement and alteration of the salicylate fluorescence is accomplished by the addition of magnesium acetate tetrahydrate and 0.3 mol/L sodium barbitone in ethanediol to serum. Fluorescence is measured with excitation at 340 nm and emission at 390 nm.

Gas-liquid chromatography (GLC) is another attractive analytical technique for measuring salicylates, especially since acetylsalicylate and its metabolites can be separated and quantitated independently by this technique (method 5, Table 61-23). Two recent papers report GLC procedures[11,12] that incorporate a silylation step after extraction of the salicylates from the biological sample. Thomas et al.[11] use a GLC equipped with a flame ionization detector (FID), fitted with a 5% OV-17 column, and run isothermally at a column temperature of 150° C. Plasma is extracted into ether along with *p*-toluic acid, the internal standard, and the ether is evaporated. Bis(trimethylsilyl)trifluoroacetamide (BSTFA) is added to the residue and heated and an aliquot analyzed by GLC. Acetylsalicylate and salicylate are both analyzed.

Rance et al.[12] used a similar procedure but by using temperature programming were able to measure salicylamide in addition to salicylate and acetylsalicylate.

REFERENCE AND PREFERRED METHODS

The simplest method for measuring salicylates on a routine basis is the spectrophotometric method of Keller, which yields a measure of total salicylate and its metabolites. The method of Brodie et al. requires time-consuming extraction steps and involves the use of a toxic solvent, ethylene dichloride. Recovery with the Brodie method is less than 90% because of the double extraction. Likewise, Trinder's method offers no advantages to Keller's method and also makes use of toxic mercuric ion.

A major shortcoming of methods using the Folin-Ciocalteu reagent is the false-positive color produced by reaction with tyrosine, tryptophan, and uric acid.

Ultraviolet spectrophotometry measures salicylate and acetylsalicylate, has a sensitivity of 5 μg/mL, and, because of its lack of specificity, offers no advantages over colorimetric procedures.

The main advantage of spectrofluorometric methods is increased sensitivity as compared with colorimetric methods. Spectrofluorometric methods do not measure acetylsalicylate directly, since acetylsalicylate does not fluoresce. Instead, they derive the value indirectly by subtracting free salicylate from the total salicylate measured after hydrolyzing the sample.

At present, gas-liquid chromatographic techniques (GLC) are methods of choice for separating and quantitating acetylsalicylate and its major metabolites; its use is recommended in studies requiring discrimination between these compounds. GLC methods offer sensitivity in the range of 1 μg/mL and linearity up to 2000 μg/mL for sodium salicylate and 100 μg/mL for acetylsalicylate. Another benefit is that large numbers of samples can be processed simultaneously, provided that the GLC is equipped with an automatic sampler.

SPECIMEN

Either serum or plasma are acceptable samples for analysis. There is no known interference by commonly used anticoagulants. Urine specimens can be analyzed by either method.

PROCEDURE FOR SPECTROPHOTOMETRIC METHOD

Principle

The following spectrophotometric method is based on salicylates complexing with ferric ion, which is supplied by ferric nitrate in dilute nitric acid. The resulting complex is water soluble and absorbs at 540 nm against a serum blank pipetted into dilute nitric acid (Table 61-24).

Reagents

Ferric nitrate solution (6 g/L, 24.8 mmol/L). Dissolve 2.5 of $Fe(NO_3)_3 \cdot 9\,H_2O$ in 1.0 mL of concentrated nitric acid and dilute to 250 mL with distilled water. Filter through Whatman No. 1 filter paper into a glass bottle. This is stable for 1 year at room temperature.

Dilute nitric acid (62 mmol/L). Dilute 1.0 mL of concentrated nitric acid to 250 mL with distilled water. This is stable for 2 years at room temperature.

Stock salicylate standard (1 g/L, 7.24 mmol/L). Dissolved 100 mg of salicylic acid in 100 mL of water. This is stable for 1 year at 4° to 8° C.

Working salicylate standard (100 mg/L, 724 μmol/L). In a 10 mL volumetric flask dilute 1.0 mL of stock standard to 10 mL with water. This is stable for 3 months at 4° to 8° C.

Control. A serum sample containing known amounts of salicylate.

Procedure

Equipment: ≤10 nm band pass spectrophotometer.

1. Label six test tubes: standard, standard blank, control, control blank, unknown(s), and unknown(s) blank

Table 61-24. Comparison of reaction conditions for salicylate analysis

Parameter	Spectrophotometry	Gas-liquid chromatography
Fraction of sample volume	0.077	1.0 mL (sample volume)
Sample	Serum	Serum
Linearity	1000 mg/L	Acetylsalicylate: 100 mg/L Sodium salicylate: 2000 mg/L
Major interferences	Background color caused by iron Interference from salicylate metabolites	None
Precision (coefficient of variation)	9.6%	—

2. To the standard, standard blank, control, and unknown tubes add 3.6 mL of ferric nitrate solution.
3. To unknown and control blank tubes add 3.6 mL of dilute nitric acid.
4. Measure 0.3 mL of standard into standard tube. Mix.
5. Measure 0.3 mL of control into control and control blank tubes. Mix.
6. Measure 0.3 mL of unknown into unknown and unknown blank tubes. Mix.
7. Read the standard, control, and unknown against their respective blanks in the spectrophotometer at a wavelength of 540 nm.

Calculation

$$\frac{\text{Absorbance unknown (or control)}}{\text{Absorbance standard}} \times 100\ \mu g/mL = \text{Salicylate } (\mu g/mL)$$

Notes

1. Because of the presence of iron in the unknown, a slight absorbance reading corresponding to approximately 20 μg/mL will be obtained in the absence of salicylate. A value of 20 μg/mL or less should not be considered significant.[13]
2. The colorimetric reaction is linear from 0 to 1000 μg/mL.
3. Analysis of urine. This procedure can be used for a urine specimen after a heating step to remove volatile interferents (especially ketones). Place a test tube containing 5 mL of centrifuged urine in a 100° C heating block or boiling water for 15 minutes. After cooling, use as a specimen in the ferric nitrate procedure. Phenothiazine, homogentisic acid, and thiocyanates will interfere and cause false-positive results. Salicylate cannot be confirmed in urine in the presence of these compounds.

PROCEDURE FOR GAS-LIQUID CHROMATOGRAPHY

Principle

In the procedure using gas-liquid chromatography, acetylsalicylate, salicylate, and salicylamide are analyzed as their trimethylsilyl derivatives using temperature programming on a gas-liquid chromatograph equipped with a 5% OV-17 column and flame ionization detector (FID).[12] A brief summary is presented in Table 61-24.

Reagents

Bis(trimethylsilyl)trifluoroacetamide (BSTFA). (From Pierce Chemicals, Rockford, Ill.) Store at 4° to 8° C. This is stable for 2 years.

Ether (chromaquality). (From Matheson, Coleman & Bell, Norwood, Ohio.) Store at 4° to 8° C. This is stable for 2 years.

***m*-Toluic acid.** (From Eastman, Rochester, N.Y.) This is stable at room temperature for 2 years.

Potassium bisulfate solution (50 mg/L, 367 μmol/L). To 50 mg of potassium bisulfate in a volumetric flask (1 L) add 800 mL of distilled water and mix. Dilute to 1 L with distilled water. This is stable at room temperature for 2 years.

Phosophate buffer (pH 7.4) (0.1 mol/L). Dilute 13.4 g of Na_2HPO_4 to 500 mL with distilled water (0.1 M sodium phosphate). Dilute 6.8 g of potassium phosphate to 500 mL with distilled water (0.1 M KH_2PO_4). Both solutions are stable for 1 year at refrigerated temperature. To make pH 7.4 phosphate buffer, mix 8 mL of sodium phosphate solution with 2 mL of potassium phosphate solution. Make fresh daily.

Salicylic acid (USP). Store at room temperature. This is stable for 2 years.

Acetylsalicylic acid (USP). Store at room temperature. This is stable for 2 years.

Salicylamide (USP). Store at room temperature. This is stable for 2 years.

***m*-Toluic acid internal standard solution, 734 μmol/L.** Dissolve 10 mg of *m*-toluic acid in 100 mL of phosphate buffer (pH 7.4). This is stable at 4° to 8° C for 6 months.

Salicylic acid stock standard (1000 μg/mL, 7.24 μmol/mL). Dissolve 10 mg of salicylic acid in 1 mL of methanol; then dilute to 10 mL with distilled water. This is stable for 1 year at 4° to 8° C.

Acetylsalicylate stock standard (1000 μg/mL, 5.55 μmol/mL). Dissolve 10 mg of acetylsalicylic acid in 1 mL of methanol and dilute to 10 mL with distilled water. This is stable for 1 year at 4° to 8° C.

Salicylamide stock standard (1000 μg/mL, 7.3 μmol/mL). Dissolve 10 mg of salicylamide in 1 mL of methanol and dilute to 10 mL with distilled water. This is stable for 1 year at 4° to 8° C.

Salicylate standard curve. To tubes containing 1 mL of pooled plasma, add 5, 10, 20, 30, 40, 50, and 75 μL of stock salicylic acid standard with a 50 μL Hamilton syringe. The resulting standards contain concentrations of 5, 10, 20, 30, 40, 50, and 75 μg/mL. If not used immediately, freeze and store. These are stable for 6 months at −20° C.

Acetylsalicylic acid standard curve. To tubes containing 1 mL of pooled plasma, add 1, 2, 5, and 10 μL of stock acetylsalicylic acid standard with a 10 μL Hamilton syringe. The resulting standards contain concentrations of 1, 2, 5, and 10 μg/mL. If not used immediately, freeze and store. These are stable for 6 months at −20° C.

Salicylamide standard curve. To tubes containing 1 mL of pooled plasma, add 5, 10, 20, 30, 40, and 50 μL of stock salicylamide standard with a 50 μL Hamilton syringe. The resulting standards contain concentrations of 5, 10, 20, 30, 40, and 50 μg/mL. If not used immediately, freeze and store. These are stable for 6 months at −20° C.

Procedure

Equipment: gas-liquid chromatograph capable of temperature programming. The GLC should be equipped with a flame ionization detector (FID).

1. To a 10 mL screw-capped tube add 1 mL of plasma or serum, 100 μL of *m*-toluic acid internal standard, and 1 mL of potassium bisulfate solution. Mix.
2. Add 7 mL of ether and extract for 5 minutes. Centrifuge at 1000 *g* (3000 rpm) for 5 minutes at room temperature. Freeze for 30 minutes at −20° C.
3. Pour off ether into Reactivials (Pierce Chemical Co., Rockford, Ill.) and evaporate to dryness at 50° C under a dry nitrogen flow.
4. Add 50 μL of BSTFA, cap and vortex, and incubate for 1 hour at 50° C in a water bath.
5. Inject 1 to 2 μL of BSTFA reacted extract into gas chromatograph.
6. Gas-liquid chromatograhic parameters
   a. Column: 5% OV-17 column
   b. Detector: flame ionization detector
   c. Nitrogen flow: 24 mL/min
   d. Injector temperature: 250° C
   e. Detector temperature: 300° C
   f. Temperature programming:
      Initial, 140° C
      Final, 200° C
      Increase, 5° C/min
   g. Chart speed: 0.5 inches/min

Calculations

Compute peak height ratio of each drug versus internal standard in standards and unknown:

$$\text{Peak height ratio} = \frac{\text{Peak height of compound}}{\text{Peak height internal standard}}$$

Calculate standard curves from the peak-height ratios of the standards/internal standard versus the concentration of the standard and calculate the salicylic acid, acetylsalicylate, and salicylamide concentrations by comparing the peak-height ratios of the unknowns to the standard curves.

Notes

1. Limit of sensitivity is 1 μg/mL for all three drugs.
2. There are no known interfering substances.
3. Because moisture may affect derivatization, it may be necessary to shake the ether extract with sodium sulfate (1 spoonful) before pouring into the Reactivials.

REFERENCE RANGE

A serum concentration of total salicylates greater than 200 μg/mL is considered above the accepted therapeutic range, except for arthritic patients, for whom the therapeutic range is extended to 300 μg/mL.

REFERENCES

1. Brodie, B.B., Udenfriend, S., and Coburn, A.F.: The determination of salicylic acid in plasma, J. Pharmacol. **80:** 114-117, 1944.
2. Keller, W.J.: A rapid method for the determination of salicylates in serum or plasma, Am. J. Clin. Pathol. **17:** 415-417, 1947.
3. Trinder, P.: Rapid determination of salicylate in biological fluid, Biochem. J. **57:**301-303, 1954.
4. Folin, O., and Ciocalteu, V.: On tyrosine and tryptophan determinations in proteins, J. Biol. Chem. **73:**627-650, 1927.
5. Weichselbaum, T.E., and Shapiro, I.: A rapid and simple method for the determination of salicylic acid in small amounts in blood plasma, Am. J. Clin. Pathol. **9:**42-44, 1945.
6. Smith, M.J.H., and Talbot, J.M.: The estimation of plasma salicylate levels, Br. J. Exp. Pathol. **31:**65-69, 1950.
7. Routh, J.I., Shane, N.A., Arredondo, E.G., and Paul, W.D.: Method for the determination of acetylsalicylic acid in the blood Clin. Chem. **13:**734-743, 1967.
8. Saltzmann, A.: Fluorophotometric method for the estimation of salicylate in blood, J. Biol. Chem. **174:**399-404, 1948.
9. Potter, G.D., and Guy, J.L.: A micro method for analysis of plasma salicylate, Proc. Soc. Exp. Biol. Med. **116:**658-660, 1964.
10. Lever, M., and Powell, J.C.: Simplified fluorometric determination of salicylate, Biochem. Med. **7:**203-207, 1973.
11. Thomas, B.J., Solomonraj, G., and Coldwell, B.B.: The estimation of acetylsalicylic acid and salicylate in biological fluids by gas-liquid chromatography, J. Pharm. Pharmacol. **25:**201-204, 1973.
12. Rance, M.L., Jordan, B.I., and Nichols, J.D.: A simultaneous determination of acetylsalicylic acid, salicylic acid, and salicylamide in plasma by gas-liquid chromatography, J. Pharm. Pharmacol. **27:**425-429, 1975.
13. Natelson, S.: Techniques of clinical chemistry, ed. 3, Springfield, Ill., 1971, Charles C Thomas, Publisher, p. 649.

Theophylline

GORDON L. BILLS

Clinical significance: p. 962
Molecular formula: $C_7H_8N_4O_2$
Molecular weight: 180.17
Merck Index: 9004
Chemical class: purine derivative

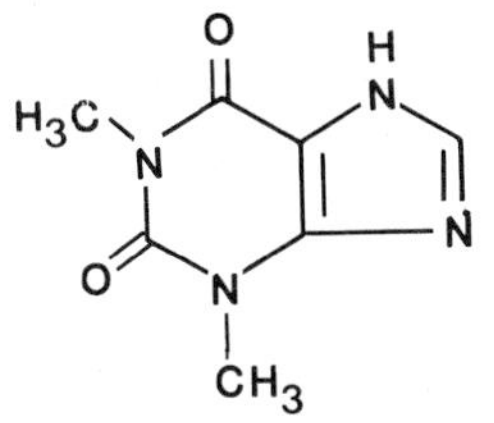

PRINCIPLES OF ANALYSIS AND CURRENT USAGE

The first method used for routine determinations of serum theophylline was the ultraviolet spectrophotometric procedure described by Schack and Waxler in 1949 (method 1, Table 61-25).[1] It involved an initial extraction of serum with a mixture of chloroform and isopropanol, followed by reextraction into dilute sodium hydroxide. The absorbance of the final aqueous extract was measured at 277 and 310 nm, and concentrations were obtained from a standard curve. This procedure was modified by Jatlow in 1975 to eliminate barbiturate interference,[2] which had been a significant problem in the original procedure.

Several procedures have been described using gas-liquid chromatography (GLC) to measure serum theophylline (method 2, Table 61-25). Although there are many differences in the specific details of these various procedures, the majority are basically similar in principle. Most available GLC procedures utilize 3-isobutyl-1-methylxanthine as an internal standard and determine unknown theophylline concentrations from a standard curve. Sample preparation always involves at least one extraction of the serum with an organic solvent, usually chloroform or a mixture of isopropanol and chloroform, and the theophylline is usually derivatized by alkylation. Use of a 3% OV-17 column is fairly standard, though a variety of carrier gases and detection systems have been utilized.[3-6] Initial serum volumes of at least 1 mL are usually required with most GLC procedures; however, use of an organic nitrogen detector has been shown to reduce the volume of serum needed to perform the assay to only 50 μL[5] and a procedure using a flame ionization detector and involving on-column alkylation has been described as requiring only 100 μL of serum.[7] The GLC methods have not been widely used for routine analysis.

In contrast, high-performance liquid chromatography (HPLC) is a widely accepted method for analysis of serum

Table 61-25. Methods for theophylline analysis

Method	Type of analysis	Principle	Usage	Comments
1. Ultraviolet	Spectrophotometry	Extraction of theophylline, absorption measurement at two wavelengths (277 and 310 nm)	Historical	Not specific
2. Gas chromatography	Chromatography, flame ionization detection	Extraction of theophylline and derivatization, chromatography in gas-liquid phase	Limited	Derivatization cumbersome
3. High-performance liquid chromatography	Chromatographic, spectrophotometric detection at 254 nm	Extraction of theophylline, chromatography on reversed phase	Preferred method	Some interferences Specialized equipment required
4. Radioimmunoassay	Competitive binding, radioactive label	Antibody binding to radiolabeled hapten competes for unknown	Limited	All problems of radioactive usage
5. Enzyme-multiplied immunoassay technique	Competitive binding, enzyme label	Antibody binding to enzyme hapten is inhibited in catalytic reactivity	Proposed selected method	Some interferences caused by sample matrix Selected proposed method
6. Substrate-linked fluorescence immunoassays	Competitive binding, substrate label	Unbound drug–fluorescent substrate converted by enzyme to fluorescent product	Recently introduced	Fluorescence interferents New method
7. Fluorescence polarization	Competitive binding, fluorescent-labeled theophylline	Bound drug has higher polarized fluorescence than free drug	Recently introduced but increasingly used	Fluorescence interferents New method

theophylline (method 3, Table 61-25). A variety of procedures are available, essentially all of which utilize reversed-phase chromatography with a C-18 column and an acetonitrile mobile phase.[8-12] Differences between procedures are primarily in the initial steps of sample preparation, which are basically designed to remove serum proteins from the sample. This may be accomplished by either filtration or extraction with acetonitrile. Failure to remove serum proteins before sample injection results in protein binding to the column, leading to the development of significant back pressure after only a short period of use.[8,9] Most HPLC procedures use either 8-chlorotheophylline or 8-hydroxyethyltheophylline as an internal standard and calculate theophylline concentration by comparison of the peak-height ratio of the specimen to that of a plasma standard.[9,11]

Homogeneous enzyme immunoassay (EIA) marketed by the Syva Corporation (Palo Alto, Calif.) as the EMIT-AAD theophylline assay is the most widely employed method of theophylline determination (method 5, Table 61-25). It is a competitive binding immunoassay in which theophylline in the serum sample competes with glucose-6-phosphate dehydrogenase (G6PD)–labeled theophylline to bind with a theophylline specific antibody. Formation of the antibody-G6PD–labeled theophylline complex results in a reduction of enzyme activity. Residual enzyme activity after incubation with the antibody and unlabeled theophylline from the patient's serum is directly proportional to the concentration of unlabeled theophylline in the original serum sample. Enzyme activity is measured spectrophotometrically at 340 nm as the rate of conversion of nicotinamide adenine dinucleotide to reduced NAD.[13]

Radioimmunoassays (RIA's) have also been reported for theophylline analysis (method 4, Table 61-25). These have used the typical competitive binding assay with radiolabeled hapten and antibody. In addition, several new types of assays involving substrate-labeled enzyme immunoassay have also been developed. Although variations exist, these assays may be collectively considered as substrate-linked fluorescence immunoassays (SLFIA's) (method 6, Table 61-25). They are based on the same principles of competitive protein binding as RIA and enzyme immunoassays; however, in the SLFIA assay theophylline is covalently bound to a fluorogenic enzyme substrate that will react with β-galactosidase to produce a fluorescent product. The fluorogenic dye will not fluoresce while conjugated to the theophylline, and hydrolysis by β-galactosidase is blocked by binding of the conjugate to a theophylline-specific antibody. Competition for antibody binding between the theophylline-dye conjugate and free theophylline results in unbound conjugate that will react with β-galactosidase, resulting in a fluorescent product, the intensity of which is directly proportional to the concentration of free theophylline in the specimen. Coefficients of variation of less than 5% have been reported using this method. It has also been shown to demonstrate excellent sensitivity and specificity, uses small (2 μL) serum volumes, and is suitable for automation.[14] SLFIA theophylline assays, however, have had only limited use.

Another recent assay measures the fluorescence polarization of fluorescein-labeled theophylline (method 7, Table 61-25). The drug present in the patient sample competes with labeled drug for a limited amount of antibody. The displaced drug has a lower fluorescence polarization than the bound drug. Curves are drawn, relating fluorescence polarization to the amount of competing drug.

REFERENCE AND PREFERRED METHODS

Currently, there is no established or accepted reference method for the determination of serum theophylline concentration. However, a proposed selected method is an enzyme-multiplied immunoassay technique (EMIT).[13] Based on the data of participating laboratories in the therapeutic drug monitoring program of the American Association for Clinical Chemistry, enzyme immunoassay (EIA) and high-performance liquid chromatography (HPLC) are the most popular methods of those currently available.

Ultraviolet spectrophotometry is the only other method that has ever enjoyed widespread use. Although once the only practical technique for routine determination of serum theophylline, ultraviolet spectrophotometry has been largely replaced in recent years by HPLC and EIA. This has been primarily the result of the increased specificity, low sample volume requirements, and increased throughput of these other methods compared to ultraviolet spectrophotometry.[2,9,13]

Radioimmunoassay (RIA) has not gained the popularity of either HPLC or EIA, despite equal analytical performance and a slightly more rapid throughput.[15] This has most likely been a result of the realization that any potential advantages of RIA are relatively insignificant, whereas the disadvantages of RIA are considerable and include, in addition to increased cost (compared to HPLC),[15] the problems of storing, handling, and disposing of radioactive materials.

Gas-liquid chromatography (GLC) procedures are too complex and time consuming to be useful for routine determination of serum theophylline concentration. Also, most GLC procedures usually require large serum volumes (1 to 3 mL), which make them unsuitable for use with pediatric patients,[3,4] and although sample size can be reduced substantially by use of an organic nitrogen detector,[5] such equipment only further reduces the practicality of GLC for routine use. HPLC in general demonstrates good specificity and sensitivity. Coefficients of variation of less than 10% have been reported in published studies,[15,16] and this level of precision has been shown to be attainable in routine use from review of College of American Pathologists (CAP) survey data. CAP data for HPLC methods of theophylline determination also demonstrate excellent accuracy, with reported mean values ranging from within 1% to 5% of target values throughout the therapeutic toxic ranges. By use of a C-18 column and an acetonitrile mobile phase, theophylline retention time is usually less than 10 minutes, and there is excellent separation of theophylline from metabolites and other xanthine compounds.[8-10] The EMIT procedure can be adapted to any of several available spectrophotometers, including the ABA-100, Gilford Stasar III, Gilford 300 Series Microsample Spectrophotometer, and Roche-COBAS. The precision and accuracy of the EMIT assay are similar to HPLC methods, with coefficients of variation of less than 10% in CAP survey data and reported mean values within 5% of target values over the therapeutic and toxic ranges. There are no reported interferences for metabolities, other xanthines, or concurrently administered drugs using the EMIT method.[13,16]

The preferred method is either HPLC or EIA. Whereas EIA and HPLC can be shown to possess several advantages over other theophylline methods, significant differences between the two are somewhat more subjective. The precision, accuracy, and sensitivity of these two methods are essentially equal, based on published studies[9,13,15,16] and data provided by the therapeutic drug monitoring program of the American Association for Clinical Chemistry. In addition, both utilize small serum samples (less than 50 μL) and can be easily adapted for use with pediatric patients. EIA would appear to be more specific, as there are no reported interferences with EIA caused by other drugs or metabolites. In contrast, several drugs, including ampicillin, methacillin,[12] chloramphenicol, sulfamethoxazole, and some cephalosporins,[17] have been reported to interfere with HPLC. These reports, however, have all been associated with use of the microscale procedure described by Orcutt et al.[9] In the case of the cephalosporins, ampicillin and methacillin, changing the sample preparation to a one-step extraction procedure completely eliminates the interference.[12,17] Although free from drug interferences, the EMIT procedure is subject to interference by severely lipemic, hemolytic, or icteric serum. In contrast, these are not problems with HPLC methods.

HPLC is considerably more economical to perform than EMIT, which has the highest unit cost (several times that of HPLC) of any of the routinely used theophylline assays.[15] Also, EMIT reagents are available only from one source, which users must depend on entirely to provide a product of unvarying quality and performance. The reagents used in HPLC are not only considerably less expensive but are available from a variety of different suppliers. They also possess over twice the shelf life of EMIT reagents.[9] When all these aspects are considered and compared, HPLC would appear to be the best available method for use in routine laboratory determination of serum theophylline, though smaller laboratories with limited equipment and low volumes may find EIA a reasonable alternative. Several of the more common assays are compared in Table 61-26.

SPECIMEN

Analysis using enzyme immunoassay (EIA) or high-performance liquid chromatography (HPLC) may be performed on either plasma or serum. Heparin should be used for anticoagulation in procedures that call for plasma. Otherwise, no special handling of specimens is required. HPLC can also be used for analysis of saliva, which may be collected while the patient chews on paraffin wax and which is then centrifuged to remove sediment.

PROCEDURE

Principle

Reversed-phase high-performance liquid chromatography of serum after deproteinization with acetonitrile. Detection is by ultraviolet spectroscopy at 254 nm.

Reagents

Acetic acid reagent (875 mmol/L). Dilute 5 mL of glacial acetic acid to 1 L with glass-distilled water.

Acetonitrile, ultraviolet grade.

Stock internal standard (100 mg/L, 480 μmol/L). Place 10 mg of β-8-hydroxyethyltheophylline in a 100 mL volumetric flask and dilute to 100 mL with methanol. This is stable for 6 months at 4° C.

Working internal standard (10 mg/L, 48 μmol/L). Dilute 10 mL of stock internal standard to 100 mL with acetonitrile. This is stable for 3 months at 4° C.

Theophylline aqueous standard (20 μg/mL, 111 μmol/L) in filtered glass-distilled water. This is stable for 3 months at 4° C.

Mobile phase (acetonitrile/acetic acid, 10:90 by volume). Filter 450 mL of acetic acid reagent using a nitrocellulose filter (such as Millipore) of 0.47 μm pore size. Into the acetic acid, filter 50 mL of acetonitrile using an organic filter of the same dimension. Degas solution for 15 minutes at room temperature by applying a vacuum.

Assay

Equipment: high-performance liquid chromatography equipment (recorder, pumps, injector), including an ultraviolet spectrophotometer capable of reading at 254 nm. Column is a μBondapak C-18 column from Waters Associates, Milford, Mass.

1. Sample preparation
   a. Pipet 0.10 mL of working internal standard solution into a 12 × 75 mm glass tube.
   b. Add 0.10 mL of serum.
   c. Vortex and centrifuge for 2 minutes at ambient temperature. Save supernatant as sample for HPLC analysis.
   d. Prepare calibrator by mixing 0.10 mL of internal standard working solution with 0.10 mL of theophylline aqueous standard (20 mg/mL).
2. Chromatographic parameters
   a. Chart speed: 1.0 cm/min

Table 61-26. Comparison of reaction conditions for theophylline assays

Condition	EMIT*	FPIA†	HPLC‡
Temperature	30° C	35° C	Ambient
pH	8.0 (55 mmol of Tris)	7.0 (100 mmol of phosphate)	—
Sample volume	50 μL	20 μL	100 μL
Fraction of sample volume	0.01	0.0976	0.5 (extraction)
Precision ($\bar{X}$, %CV)	13.7 μg/mL, 7%	13.7 μg/mL, 4.0%	13.7 μg/mL, 7%
Linearity	2.5-40 μg/mL	2.5-40 μg/mL	2.5-25 μg/mL
Interferences	Caffeine Renal failure§	Caffeine, 9.7% Other xanthine compounds (2%-10%) Renal failure§	Aminoglycosides Cefoxitin Corticosteroids

*Enzyme-multiplied immunoassay technique, Syva Co., Inc., Palo Alto, Calif.
†Fluorescence polarization immunoassay, Abbott Diagnostics, North Chicago, Ill.
‡Assay described in text.
§Data from Elin, R.J., Ruddel, M., Korn, W.R., et al.: Clin. Chem. **29:**1275, 1983.[22]

b. Ultraviolet sensitivity: 0.01 AUFS (absorbance units at full scale) at 254 nm
c. Flow rate: 2.0 mL/min at ambient temperature
3. Inject 10 μL of sample or standard. Sample run time is approximately 9 minutes.

Calculation

1. Determine the peak height (in millimeters) for the theophylline (T) and internal standard (S) peaks on each chromatogram. See Fig. 61-9 for sample chromatogram.
2. Calculate peak-height ratios, T/S.
3. Calcualte theophylline concentrations in the serum samples as follows:

$$\text{Theophylline } (\mu g/mL) = \frac{\text{T/S (sample)}}{\text{T/S (calibrator)}} \times 20\ \mu g/mL$$

REFERENCE RANGE

The therapeutic range for steady-state levels of serum theophylline in children and adults treated prophylactically for asthma has been established as 10 to 20 μg/mL.[18] The range in neonates treated for apnea is 5 to 20 μg/mL.[19] Concentrations greater than 20 μg/mL are associated with signs of toxicity in the majority of patients, regardless of age.[18,21]

REFERENCES

1. Schack, J.A., and Waxler, S.H.: An ultraviolet spectrophotometric method for the determination of theophylline and theobromine in blood and tissues, J. Pharmacol. Exp. Ther. **97:**283, 1949.
2. Jatlow, P.: Ultraviolet spectrophotometry of theophylline in plasma in the presence of barbiturates, Clin. Chem. **21:**1518, 1975.
3. Bailey, D.G., Dais, H.L., and Johnson, G.E.: Improved theophylline serum analysis by an appropriate internal standard for gas chromatography, J. Chromatogr. **121:**263, 1976.
4. Reid, R., Fareed, J., Bermes, E.W., et al.: A rapid gas chromatographic (GLC) procedure for the determination of theophylline in biological fluids using a simple double extraction method, Clin. Chem. **22:**1166, 1976.
5. Lowry, J.D., Williamson, L.J., and Raisys, V.A.: Micromethod for the gas chromatographic determination of serum theophylline utilizing an organic nitrogen sensitive detector, J. Chromatogr. **143:**83, 1977.
6. Dechtiaruk, W., Johnson, G.F., and Solomon, H.M.: Faster gas chromatographic procedure for theophylline, Clin. Chem. **21:**1038, 1975.
7. Perrier, D., and Lear, E.: Gas-chromatographic quantitation of theophylline in small volumes of plasma, Clin. Chem. **22:**898, 1976.
8. Adams, R.F., Vandemark, F.L., and Schmidt, G.J.: More sensitive high pressure liquid chromatography of theophylline in serum, Clin. Chem. **22:**1903, 1976.
9. Orcutt, J.J., Kozak, P.P., Sherwin, A.G., and Cummins, L.H.: Microscale method for theophylline in body fluids by reversed phase, high pressue liquid chromatography, Clin. Chem. **23:**599, 1977.
10. Franconi, L.C., Hawk, G.L., Sandmann, B.J., and Haney, W.B.: Determination of theophylline in plasma ultrafiltrate by reversed phase high pressuure liquid chromatography, Anal. Chem. **48:**372, 1976.
11. Weidner, N., Dietzler, D.N., Ladenson, J.H., et al.: A clinically applicable high pressure liquid chromatographic method for measurement of serum theophylline, with detailed evaluation of interferences, Am. J. Clin. Pathol. **73:**79, 1980.
12. Soldin, S.J., and Hill, J.G.: A rapid micromethod for measuring theophylline and reverse phase high pressure liquid chromatography, Clin. Biochem. **10:**74, 1977.
13. Chang, J., Gotcher, S., and Gashaw, J.B.: Homogeneous enzyme immunoassay for theophylline in serum and plasma, Clin. Chem. **28:**361, 1982.
14. Li, T.M., Benovic, L.J., Buckler, R.T., and Burd, J.F.: Homogeneous substrate-labeled fluorescent immunoassay for theophylline in serum, Clin. Chem. **27:**22, 1981.
15. Kampa, I.S., Dunikoski, L.K., Jarzabek, J.I., and Grubesich, D.: Comparison of three assay procedures for theophylline determination, Ther. Drug Monitor. **1:**249, 1979.
16. Weidner, N., McDonald, J.M., Tieber, V.L., et al.: Assay of theophylline: comparison of EMIT on the ABA-100 to HPLC, GLC and UV procedures with detailed evaluation of interferences, Clin. Chem. Acta **97:**9, 1979.
17. Kelly, R.C., Prentice, D.E., and Hearne, G.M.: Cephalosporin antibiotics interfere with the analysis for theophylline by high pressure liquid chromatography, Clin. Chem. **24:**838, 1978.
18. Jenne, J.W., Wyze, E., Rood, F.S., and McDonald, F.M.: Pharmacokinetics of theophylline: application to adjustment of the clinical dose of aminophylline, Clin. Pharmacol. Ther. **13:**349, 1972.
19. Shannon, D.C., Gotay, F., Stein, I.M., et al.: Prevention of apnea and bradycardia in low birth weight infants, Pediatrics **55:**589, 1975.
20. Zwillich, C.W., Sutton, F., Neff, T.A., et al.: Theophylline induced seizures in adults: correlation with serum concentration, Ann. Intern. Med. **82:**784, 1975.
21. Neese, C.L., and Soyka, L.F.: Development of a radioimmunoassay for theophylline, Clin. Pharmacol. Ther. **21:**633, 1977.
22. Elin, R.J., Ruddel, M., Korn, W.R., et al.: Discrepant results for the determination of theophylline in serum from a patient with renal failure, Clin. Chem. **29:**1275, 1983.

Chapter 62 Urine analysis

Melanins

ARNOLD L. SCHULTZ

Clinical significance: p. 996
Merck Index: 5631
Chemical class: dihydroxyphenylalanine derivatives

PRINCIPLES OF ANALYSIS AND CURRENT USAGE

Melanins are pigments that give rise to the dark color of skin and hair. They can be detected in human urine that shows a pathological condition, where they are an indication of melanotic tumors. These complex polymeric pigments and the melanogens, which are their colorless precursors, are products of the oxidation of dihydroxyphenylalanine (dopa). Dopa itself is a product of tyrosine metabolism.

Melanogens are oxidized to dark brown or black melanins when the urine is exposed to air. This darkening of the urine usually occurs slowly, taking 24 hours or longer to become visible. If the urine is left undisturbed, the darkening occurs from the surface downward.

A number of simple tests[1,2] have been devised to aid in the diagnosis of melanuria and in differentiating it from alkaptonuria, which is associated with an abnormally high concentration of homogentisic acid, and also from the presence of indican in the urine. Indican is indoxyl sulfate, a product of tryptophan that accumulates under conditions of intestinal putrefaction.

Oxidation of melanogens to melanins by agents such as ferric chloride, nitric acid, bromine water, and potassium chlorate in hydrochloric acid is one type of test that has been used (method 1 to 5, Table 62-1). The other primary type of test takes advantage of the reducing properties of the melanogens. In the Thormählen reaction (method 6, Table 62-1), sodium nitroprusside is reduced to ferric ferrocyanide (Prussian blue). Ammoniacal silver nitrate is reduced to colloidal silver by urine that contains melanogens (method 7, Table 62-1). Reactions of melanogens with *p*-dimethylaminobenzaldehyde, sodium nitrate in hydrochloric acid, and diazonium salts have also been reported.

REFERENCE AND PREFERRED METHODS

Urine that contains melanogens or homogentisic acid darkens when exposed to air and sunlight. Testing the urine with ammoniacal silver nitrate will indicate the presence of homogentisic acid if the solution darkens rapidly, whereas slower formation of brown color is indicative of the formation of melanogens. If indican is present, the urine will also have a brown appearance and thus may be mistaken for melanotic urine. However, one can use the Obermayer test for indican to readily differentiate melanins from indican[3] (method 8, Table 62-1).

The oxidation of melanogens to melanins by nitric acid, bromine water, or potassium chlorate in hydrochloric acid does not have sufficient sensitivity to detect small concentrations of melanogens. The reduction of ammoniacal silver nitrate by urine that contains melanogens likewise lacks sensitivity. The most sensitive tests and the ones recommended for routine use are the ferric chloride and the Thormählen tests. The ferric chloride test may be performed either in water or in 10% hydrochloric acid. When the procedure is done in water, the melanins are entrapped on the precipitate of ferric phosphate that is formed, resulting in a gray or black precipitate. If hydrochloric acid is used as the solvent, the precipitate of ferric phosphate does not form and only the color change that results from the oxidation of the melanogens is observed. It has been claimed that the use of hydrochloric acid in the ferric chloride test increases the sensitivity of the reaction.

SPECIMEN

Freshly voided urine must be used. The tests for melanuria are actually reactions involving the presence of melanogens. If the urine is allowed to stand, the melanogens will be converted to melanins, the chromogen will be present at a concentration lower than that in the originally voided urine, and false-negative results may be obtained.

Table 62-1. Methods of melanin analysis

Method	Type of analysis	Principle	Usage	Comments
1. Ferric chloride in water	Qualitative	Oxidation of melanogens to melanins	Urine	Lacks sensitivity
2. Ferric chloride in hydrochloric acid	Qualitative	Oxidation of melanogens to melanins	Urine	Recommended, good sensitivity
3. Nitric acid	Qualitative	Oxidation of melanogens to melanins	Urine	Lacks sensitivity
4. Bromine water	Qualitative	Oxidation of melanogens to melanins	Urine	Lacks sensitivity
5. Potassium chlorate in hydrochloric acid	Qualitative	Oxidation of melanogens to melanins	Urine	Lacks sensitivity
6. Thormählen reaction	Qualitative	Reduction of sodium nitroprusside to ferric ferrocyanide (Prussian blue)	Urine	Recommended, good sensitivity
7. Ammoniacal silver nitrate	Qualitative	Reduction of silver	Urine	Homogentisic acid reacts rapidly; melanogens react slowly Differentiates homogentisic acid from melanins
8. Obermayer test	Qualitative	Conversion of indican to indigo blue	Urine	Differentiates indican from melanins Indican reacts; melanins do not

PROCEDURE

Principle

Melanogens are oxidized to dark brown or black melanins in the presence of ferric chloride in hydrochloric acid.

Prussian blue is formed in the reduction of sodium nitroprusside by urine containing melanogens.

Colloidal silver is rapidly formed in the reduction of silver nitrate by homogentisic acid. Melanogens will also reduce silver nitrate; however, this reaction requires the presence of ammonium hydroxide and proceeds at a slower rate.

The reaction of indican with ferric chloride in the Obermayer test produces indigo blue. The blue pigment is extracted into chloroform.

Reagents (Table 62-2)

Ferric chloride test

1.2 M hydrochloric acid (1.2 mol/L). To approximately 50 ml of deionized water in a 100 ml volumetric flask, add 10.0 ml of concentrated hydrochloric acid. Mix and dilute to volume with deionized water.

Working ferric chloride reagent (100 g/L, 616 mmol/L). Dissolve 10 g of ferric chloride hexahydrate in 1.2 M hydrochloric acid and dilute to 100 mL with 1.2 M hydrochloric acid. Stable for 1 year at room temperature.

Thormählen test

Working sodium nitroferricyanide reagent (50 g/L, 191 mmol/L). Dissolve 0.5 g of sodium nitroferricyanide in 10 mL of deionized water. Prepare fresh.

10 M sodium hydroxide (10 mol/L). Cautiously dissolve 40 g of sodium hydroxide in deionized water and dilute to 100 mL with deionized water. Stable in plastic bottle for 1 year at room temperature.

33% acetic acid (5.61 mol/L). To about 40 mL of deionized water in a 100 mL volumetric flask, add 33 mL of glacial acetic acid. Mix and dilute to volume with deionized water. Stable for 1 year at room temperature.

Ammoniacal silver nitrate test

3% silver nitrate (30 g/L, 176.6 mmol/L). Dissolve 0.3 g of silver nitrate in 10 mL of deionized water. Stable for 6 months at room temperature.

2% ammonium hydroxide (0.3 mol/L). Dilute 0.2 mL of concentrated ammonium hydroxide to 10 mL with deionized water. Stable for 6 months at room temperature.

Obermayer test

Working ferric chloride reagent (4 g/L, 24.6 mmol/L). Dissolve 0.1 g of ferric chloride hexahydrate in 25 mL of concentrated hydrochloric acid. Stable for 6 months at room temperature.

Assay

Ferric chloride test

1. To 5 mL of urine, add 1 mL of 10% ferric chloride in 1.2 M hydrochloric acid.
2. If the urine color rapidly changes to brown, dark brown, or black, melanogens are present.

Thormählen test

1. Add 5 to 6 drops of freshly prepared 5% sodium nitroferricyanide solution to 5.0 mL of urine in a test tube.

Table 62-2. Reaction conditions for melanin analysis

Conditions	Ferric chloride	Thormählen reaction	Ammoniacal silver nitrate	Obermayer reaction
Temperature	Room	Room	Room	Room
Final concentration of reagent components	Ferric chloride: 1.7% Hydrochloric acid: 0.2 M	Sodium nitroferricyanide: 0.4% Sodium hydroxide: 0.2 M Acetic acid: 2.8%	Silver nitrate: 2.5% Ammonium hydroxide: 0.2%	Ferric chloride: 0.2%
Fraction of sample volume	0.83	0.77	0.83	0.50
Reaction color	Brown, dark brown, black	Greenish blue to bluish black	Darkens slowly for melanins, rapidly for homogentisic acid	Clear for melanins, blue for indican

2. Add 0.5 mL of 10 M sodium hydroxide and mix vigorously.
3. Rapidly cool the tube in cold tap water and acidify with 33% acetic acid.
4. With normal urine an amber or pale-brown color is obtained; melanogens give a color that may vary from greenish blue to bluish black, depending on the quantity present. In a positive test, the color also depends on the color of the urine. The deeper the yellow of the urine, the greener the final color. The less pigmented the urine, the bluer the final color.

Ammoniacal silver nitrate test

1. To 0.5 mL of urine, add 5.0 mL of 3% silver nitrate solution and mix.
2. Add 2% ammonium hydroxide until the silver chloride precipitate is almost dissolved.
3. The solution will darken as a result of the formation of both melanins and colloidal silver. The reaction develops slowly. In contrast, homogentisic acid darkens the silver solution rapidly, even before the addition of ammonia.

Obermayer test

1. Mix 6 mL of urine with 6 mL of 0.4% ferric chloride in concentrated hydrochloric acid in an 18 × 200 mm test tube.
2. Add 3 to 4 mL of chloroform.
3. Stopper the tube and mix by inverting the tube vigorously 10 to 12 times.
4. If indican is present, the chloroform (lower layer) will be colored blue (indigo blue).

Note

For quality control a normal urine with about the same pigmentation as the sample to be tested and freshly prepared solutions (1 mg/10 mL of deionized water) of indoxyl-sulfate, homogentisic acid, and melanin (all three available from Sigma Chemical Co., St. Louis, Mo.) should be processed with each test.

REFERENCE RANGE

Negative.

REFERENCES

1. Harrison, G.A.: Chemical methods in clinical medicine: their applications and interpretation with techniques of simple tests, ed. 4, New York, 1957, Grune & Stratton, pp. 252-255.
2. Varley, H.: Practical clinical biochemistry, ed. 2, New York, 1958, Interscience Publishers, Inc., pp. 563-564.
3. Kachmar, J.F.: Proteins and amino acids. In Tietz, N.W., editor: Fundamentals of clinical chemistry, Philadelphia, 1970, W.B. Saunders Co., pp. 257-259.

Porphobilinogen

MICHAEL D.D. McNEELY

Clinical significance: pp. 639 and 996
Molecular formula: $C_{10}H_{14}O_4N_2$
Molecular weight: 226 daltons
Merck Index: 7373
Chemical class: pyrrole

PRINCIPLES OF ANALYSIS AND CURRENT USAGE

Porphobilinogen (PBG) is a monopyrrole precursor of heme that is found in high concentrations in the urine of persons suffering from acute intermittent porphyria (AIP).

Production of this abnormal pigment was the most notable feature of the first recorded case of AIP. It occurred in 1889 in a woman who excreted red urine and died after ingestion of sulfonmethane.[1] In 1931 Sachs discovered that the pigment is not urobilinogen but produces a color reaction with Ehrlich's aldehyde (or diazo) reagent.[2]

The simplest method for porphobilinogen detection is the "window-sill" test, in which fresh urine is acidified and exposed to sunlight. Under these conditions colorless porphobilinogen polymerizes to form porphyrins and other pigments that darken the urine. This crude test is neither sensitive nor specific.

The standard test for clinical detection of porphobilinogen was devised by Watson and Schwartz in 1941.[3] The method, which is qualitative, is designed to indicate the presence or absence of porphobilinogen. The basis of the procedure is the reaction between porphobilinogen and Ehrlich's aldehyde reagent (*p*-dimethylaminobenzaldehyde in strong acid) to form a red condensation compound. The color forms quickly and then fades slowly as the colored product reacts with a second molecule of porphobilinogen to form a series of colorless pyrryl compounds.[4,5] The color production is not specific and is caused by urobilinogen, indole, and indicane.

Sodium acetate is added to the reagent to increase the pH. This maximizes the color produced by urobilinogen and renders it preferentially soluble in chloroform. Thus, if color is produced during the first step of the reaction, a chloroform extraction is carried out. The colored products of urobilinogen and other compounds are then extracted into the chloroform layer. Color that remains in the supernatant aqueous phase is generally caused by porphobilinogen.

This test was evaluated by testing of urine from 1000 hospitalized patients. There were no false-positive results.[6] However, when patients with severe systemic disease were studied, positive reactions were noted in those suffering from Hodgkin's disease, tetanus, hepatic cirrhosis, pancreatic carcinoma, poliomyelitis, carcinomatosis, and intra-abdominal hemorrhage. This study did not confirm the exact nature of the positive pigment.[7]

To increase specificity, the aqueous phase color was further extracted with butanol.[8] Any color not passing into the butanol phase is virtually certain to be porphobilinogen.

Of 1000 hospitalized patients, 59 produced weakly positive reactions when the original Watson-Schwartz method was used. None of these tests remained positive after butanol extraction.[9]

A somewhat simpler method is the Hoesch test,[10,11] which uses Erhlich's aldehyde reagent without sodium acetate. No extractions are required. Test results have been observed to become positive with indoles, indole-3-acetic acid, α-methyl-dopa, phenazopyridine hydrochloride, and end-stage alcoholic malnutrition.[12]

Several variations of the Watson-Schwartz test have been suggested. These include direct application of Ehrlich's aldehyde reagent to urine on a talc pad,[13] preliminary ion-exchange resin isolation before Ehrlich's aldehyde reaction,[14] and dual-wavelength spectrometry.[15]

A popular misconception is that strip tests for urinary urobilinogen that employ Ehrlich's aldehyde reagent will routinely change color in the presence of porphobilinogen. Thus a negative strip test has been suggested to rule out acute intermittent porphyria (AIP). This is inappropriate, since porphobilinogen will not reliably change the test strip's color.[16]

For quantitation the method of Mauzerall and Granick[4] is suitable. It employs Ehrlich's aldehyde reagent to quantitate porphobilinogen after chromatographic isolation.

REFERENCE AND PREFERRED METHODS

The method of choice is the modified Watson and Schwartz technique with butanol extraction. It requires several steps but, if performed and interpreted properly, is very sensitive (more than 99%) and highly specific (at least 99.9%).[8]

The Hoesch test is considerably easier to perform but does not offer the same degree of accuracy. Direct techniques are unacceptably nonspecific and do not offer sufficient sensitivity. Quantitative methods are not needed for routine clinical use.

SPECIMEN

A freshly voided urine specimen collected during or immediately after an attack of acute abdominal pain is recommended. The urine should be cooled to room temperature before being tested to prevent the formation of a "warm aldehyde" reaction. The assay should be conducted within several hours to prevent the degeneration of porphobilinogen.

PROCEDURE

Principle

As described earlier, porphobilinogen in urine reacts with Ehrlich's aldehyde reagent to form a colored chromogen. Sodium acetate is added to adjust the pH, and nonporphobilinogen color is removed by extraction into chloroform. Further specificity is gained by extraction of nonspecific color into a butanol extract.

Reagents

Ehrlich's aldehyde reagent (18.8 mmol/L in 7.2 M HCl). Combine 700 mg of reagent-grade *p*-dimethylaminobenzaldehyde, 150 mL of concentrated HCl, and 100 mL of deionized water. The solution should be colorless or very light yellow. Brown or brownish-red material should not be used. Store in an amber bottle at room temperature. It will remain stable for months. Discard if discolored.

Sodium acetate (saturated). About 200 g/dL. It is stable at room temperature for at least 1 year.

Chloroform

***n*-Butanol**

Control urine. A fresh, otherwise normal, urine is run. Positive controls are not available.

Assay

1. Label two tubes: test and control.
2. To the appropriate tubes, add 1.0 mL control or test urine.

3. Measure 1.0 mL of Ehrlich's reagent into each tube and shake for 30 seconds.
4. Measure 2.0 mL of saturated sodium acetate into each tube and shake for 30 seconds.
5. Check the pH with an indicator paper. The pH should be 4.0 to 5.0. If necessary, further sodium acetate should be added.
6. Measure 2.0 mL of chloroform into each tube and shake for 30 seconds.
7. Observe the tubes. If the aqueous supernatant is colorless or has only a faint trace of color, the test is deemed negative. If color is seen in the supernatant, the test is presumed positive, and step 8 must be undertaken.
8. Transfer the colored aqueous-phase supernatant to a separate tube. Add a half volume of *n*-butanol and shake for 30 seconds.
9. Observe the tube. If the aqueous phase retains color, the specimen has porphobilinogen.

Notes

1. Color identification may be difficult. Usually when acute intermittent porphyria is present, the test color is intense.
2. Particular caution must be exercised if any intense color develops. The color may not be fully extracted into the organic solvent and may lead to a false-positive test. Thus, if the color appears in both phases, a second extraction is warranted.

REFERENCE RANGE

Less than 2 mg/L of urine, which gives a negative result in the screening test.

REFERENCES

1. Magnussen, C.R.: Acute intermittent porphyria: clinical and selected research aspects, Ann. Intern. Med. **83:**851-864, 1975.
2. Sachs, P.: Ein Fall von akuter Porphyrie mit hochgradiger Muskelatrophie, Klin. Wochenschr. **10:**1123-1125, 1931.
3. Watson, C.J., and Schwartz, S.: Simple test for urinary porphobilinogen, Proc. Soc. Exp. Biol. Med. **47:**393-394, 1941.
4. Mauzerall, D., and Granick, S.: Occurrence and determination of δ-amino-levulinic acid and porphobilinogen in urine, J. Biol. Chem. **219:**435-446, 1956.
5. Rimington, C., Krol, S., and Tooth, B.: Detection and determination of porphobilinogen in urine, Scand. J. Clin. Lab. Invest. **8:**251-262, 1956.
6. Hammond, R.L., and Welcker, M.L.: Porphobilinogen tests on 1000 miscellaneous patients in search for false positive reactions, J. Lab. Clin. Med. **33:**1254-1257, 1948.
7. Watson, C.J.: Some studies of nature and clinical significance of porphobilinogen, A.M.A. Arch. Intern. Med. **93:**643-657, 1954.
8. Schwartz, S., Keprios, M., and Schmid, R.: Experimental porphyria: type produced by lead, phenylhydrazine and light, Proc. Soc. Exp. Biol. Med. **79:**463-468, 1952.
9. Townsend, J.D.: Evaluation of recent modification of Watson-Schwartz test for porphobilinogen, Ann. Intern. Med. **60:**306-307, 1964.
10. Hoesch, K.: Über die Pantothensäurebehandlung der Porphyrie, Dtsch. Med. Wochenschr. **72:**252-254, 1947.
11. Lamon, J., With, T.K., and Redeker, A.G.: The Hoesch test: bedside screening for urinary porphobilinogen in patients with suspected porphyria, Clin. Chem. **20:**1438-1440, 1974.
12. Pierach, C.A., Cardinal, R., Bossenmaier, I., and Watson, C.J.: Comparison of the Hoesch and the Watson-Schwartz tests for urinary porphobilinogen, Clin. Chem. **23:**1666-1668, 1977.
13. With, T.K.: Screening test for acute porphyria, Lancet **2:**1187-1188, 1970.
14. Doss, M.O.: Tests for porphyria, Lancet **2:**983-984, 1971.
15. Moore, D.J., and Labbe, R.F.: A quantitative assay for urinary porphobilinogen, Clin. Chem. **10:**1105-1111, 1964.
16. Kanis, J.A.: Detection of urinary porphobilinogen, Lancet **1:**1511, 1973.

Chapter 63 Vitamins

B_{12}

I-WEN CHEN
MATTHEW I. SPERLING
LINDA A. HEMINGER

B_{12}, cyanocobalamin, cobalamin
Clinical significance: p. 656
Molecular formula: $C_{63}H_{88}CoN_{14}O_{14}P$
Molecular weight: 1355.42 daltons
Merck Index: 9670
Chemical class: corrinoid (complexed pyrrole rings)

PRINCIPLES OF ANALYSIS AND CURRENT USAGE

Vitamin B_{12} is a complex corrinoid compound, the central portion of which consists of four reduced and extensively substituted pyrrole rings surrounding a single cobalt atom. Vitamin B_{12} was first isolated in 1948 from liver as a red crystalline compound. It is also known as cyanocobalamin because cyanide occupies one of the coordination positions of the cobalt atom. However, the cyanide is present as an artifact of isolation and does not occur in the vitamin molecule as it exists in natural materials. A vitamin B_{12} molecule without the cyanide group is called *cobalamin*. Substitution of the cyanide group with a hydroxy group forms hydroxycobalamin; with a methyl group, methylcobalamin; and with a 5-deoxyadenosyl group, 5-deoxyadenosylcobalamin. These three cobalamins have been isolated from mammalian tissues. The last two cobalamins are known to act as specific coenzymes in two reactions in mammalian systems: the conversion of methylmalonyl CoA to succinyl CoA and the methylation of homocysteine to methionine.

All three cobalamins are present in plasma bound to plasma-binding proteins called *transcobalamins,* but methylcobalamin is the major form of vitamin B_{12} in the plasma. However, the major form of vitamin B_{12} stored in the liver is 5-deoxyadenosylcobalamin.

Growth-dependent microbiological assays were established in the 1950s for the measurement of vitamin B_{12} concentrations in biological specimens (method 1, Table 63-1). Several microorganisms have been used in such assays: *Lactobacillus lactis, Lactobacillus leichmannii, Euglena gracilis, Ochromonas malhaemensis,* and *Escherichia coli*. *Euglena gracilis* and *Lactobacillus leichmannii* are the two most commonly used at present. The sensitivity of the two organisms to vitamin B_{12} is the same. However, because of strict requirements of temperature and exposure to light that require special and involved apparatus for the growth of *Euglena gracilis, Lactobacillus leichmannii* appears to be the organism of choice at present.[1,2] The growth of the bacteria, as measured by the absorbance at 640 to 700 nm, is related to the concentration of folate in the specimen.

Radioligand assays for serum vitamin B_{12} (method 2, Table 63-1) were developed in the early 1960s.[3] Like other competitive inhibition radioligand assays, the measurement of vitamin B_{12} concentrations is based on the competition between endogenous serum vitamin B_{12} and radiolabeled vitamin B_{12} for the limited number of binding sites on specific vitamin B_{12}–binding proteins. Since vitamin B_{12} is bound to plasma proteins as methylcobalamin before the radioassay, serum is heated in a boiling water bath under acidic or alkaline conditions in the presence of cyanide to denature the endogenous B_{12} binder and convert vitamin B_{12} to the more stable cyanocobalamin derivative.

Commercially available crude hog intrinsic factor had been used as the binder in B_{12} radioassays in the past, but was largely replaced by purified intrinsic factor because the crude intrinsic factor preparation was found to contain more than 50% of R (rapidly migrating) proteins, which could also bind biologically inactive vitamin B_{12} analogs to give spuriously high serum vitamin B_{12} concentrations.[4,5] The presence of vitamin B_{12} analogs in serum has been demonstrated by Kolhouse et al.[4] The National Committee for Clinical Laboratory Standards carried out a thorough investigation of this problem and has established the following guidelines for the intrinsic factor preparations

Table 63-1. Methods of vitamin B_{12} analysis

Method	Type of analysis	Principle	Usage	Comments
1. Microbiological	Turbidimetric	Microorganism growth depends on amount of vitamin B_{12} in serum sample	Original procedure now reference	Tedious, time consuming; not suitable for clinical laboratories
2. Radioligand	Competitive binding	Competitive binding of radiolabeled (^{57}Co) and nonlabeled vitamin B_{12} to limited binding sites on specific binders	Most common	Sensitive and precise; suitable for clinical laboratories; often done with folate assay

used as the binder in vitamin B_{12} radioassays: (1) The intrinsic factor preparations should not measure cobinamide (a vitamin B_{12} analog) up to concentrations of 10 ng/mL in serum. (2) Vitamin B_{12} binding to the binder should be inhibited more than 95% by specific anti–intrinsic factor blocking antibody, or the intrinsic factor–vitamin B_{12} complex should be more than 95% precipitated by specific intrinsic factor precipitating antibody.[6] Affinity chromatography has been used in the purification of intrinsic factor to eliminate essentially all the R-protein binding activity.[7]

Since vitamin B_{12} contains a cobalt atom as an integral part of its molecule, ^{57}Co, a gamma-ray emitting nuclide, is used exclusively for labeling vitamin B_{12} for use in the radioassay. Dextran-coated charcoal seems to be the most widely used agent for separating bound and free ligands in the liquid-phase radioassay of vitamin B_{12}. However, several commercial vitamin B_{12} radioassay kits using the solid-phase technique have become available and are expected to gain popularity because of their technical simplicity and rapidity.

REFERENCE AND PREFERRED METHODS

The microbiological assays require sterile technique and are rather cumbersome and time consuming for routine use by clinical laboratories. Moreover, these assays can be affected by the presence of antibiotics, antimetabolites, and tranquilizers in the blood of the patient, which may prevent the growth of microorganisms.[2] Their use therefore never became widespread in the average clinical setting, especially after the development of competitive radioligand assays.

The microbiological assay may be considered the reference method because it measures only the biologically active vitamin B_{12} derivatives. However, because of the specialized techniques and equipment required, it is used primarily as a research tool by laboratories that have a major commitment to the study of vitamin B_{12} and as a diagnostic service by regional hospitals only. The National Committee for Clinical Laboratory Standards recommends that all commercial radioassay kits be verified by bioassays using either *Euglena gracilis* or *Lactobacillus leichmannii* or by a radioassay that has been previously verified by one of the bioassays.

Radioassays are by far the most commonly used method for the determination of serum vitamin B_{12} levels in routine clinical laboratories mainly because of their technical simplicity and the commercial availability of assay kits. Commercial radioassay kits were used in 1426 of 1436 laboratories participating in the last quarter of the 1982 basic ligand assay survey of the College of American Pathologists. The ''Product Guide for Radioassays and Non-Isotopic Ligand Assays,'' published in the May 1983 issue of *Clinical Chemistry,* [8] listed 15 commercial kits manufactured by nine different companies as of July 1982. It lists various components, assay methods and conditions, performance characteristics, and expected reference ranges of each kit. It is a useful guide for initial screening of commercial kits for clinical use in laboratories, but be reminded that all the information appearing in the guide was provided by each manufacturer and the final decision should be based on the product performance evaluated in each laboratory.

Most commercial radioassay kits for vitamin B_{12} are designed for the simultaneous measurement of serum folate levels because assays for both vitamins are frequently ordered on the same serum sample. For example, about 70% of serum samples sent to our laboratory are for both assays. The reason is that the hematological manifestations of vitamin B_{12} and folate deficiencies are indistinguishable because of close metabolic interrelationships of these two vitamins, and it is important to differentiate between the two deficiencies so that therapy will be appropriate. The simultaneous assay is possible because of the difference in the emission-energy spectrum of ^{57}Co-labeled vitamin B_{12} and ^{125}I-labeled folate used as tracers in the radioassay that allows a certain portion of each radionuclide's energy to be detected almost free from the other. A scintillation detector equipped with at least two pulse-height analyzers is needed for simultaneous counting of two or more radionuclides. All modern gamma-ray scintillation counters are equipped with multichannel analyzers and are suitable for the simultaneous assay. Some commercial kits provide the option for the simultaneous or the separate assay of B_{12} and folate. In our laboratory we find that it is more economical to run the simultaneous assay on all samples because the technical time saved is more than enough to compensate for the extra reagents consumed and the additional information obtained from the simultaneous assay is frequently of clinical value to the clinicians who requested only vitamin B_{12} levels.

In most commercial kits a purified intrinsic factor preparation essentially free of R protein (rapidly migrating protein) is used as the specific binder to determine only the so-called true vitamin B_{12} (biologically active vitamin B_{12} with affinity to intrinsic factor). However, a crude intrinsic factor preparation contaminated with R protein is still used in at least one commercial kit (SimulTRAC Radioassay Kit, Becton Dickinson Immunodiagnostics, Orangeburg, N.Y.). This kit provided the option of measuring either the "true" or the "total" vitamin B_{12} levels (vitamin B_{12} plus its analog) in serum. For the measurement of "true" vitamin B_{12} the assay is carried out in the presence of cobinamide, an analog of vitamin B_{12}, to block binding of endogenous vitamin B_{12} to the R protein, whereas total vitamin B_{12} is measured in the absence of cobinamide. We have found that the "true" vitamin B_{12} levels measured with the blocked crude intrinsic factors are a little higher than those obtained with purified intrinsic factor.[5] Clinical significances of "true" and "total" vitamin B_{12} levels in blood still remain controversial. It has been reported that measurement of "total" vitamin B_{12} with crude intrinsic factor may give an equally good or even better separation between healthy and vitamin B_{12}–deficient patients.[9,10]

Two automated systems (ARIA II by Becton Dickinson Immunodiagnostics and Concept 4 by Micromedic Systems, Inc., Horsham, Penn.) are capable of assaying serum vitamin B_{12} (Table 63-2). In the ARIA II,[11] vitamin B_{12} and folate are assayed simultaneously using purified porcine intrinsic factor and folate binder from bovine milk as the specific binders. The bound and free ligands are separated by passing the incubation mixture through a chamber packed with hydrophobic resins. The Concept 4 instrument measures vitamin B_{12} only and uses a special tube coated with an antibody to vitamin B_{12}. Both systems require off-line denaturation of endogenous serum binders by boiling, which requires manual pipetting of the serum sample and the denaturation buffer, mixing, boiling in a boiling water bath, and cooling. To carry out radioassays of vitamin B_{12} in a fully automated fashion, the denaturation of endogenous binders by processes other than boiling has to be developed. Recently, the use of an alkali treatment has been suggested as an alternative to the heat-denaturation procedure.[12]

The alkali denaturation technique has already been used by commercial kit manufacturers in the so-called no-boil assay kits (Table 63-2)[13] but has not been adapted in the automated radioassay systems. Although the alkali treatment eliminates the need to boil tubes (and hence the necessity to maintain a boiling device such as a water bath), it still requires incubation and an additional pipetting step for the neutralization of the alkaline extractant before the competitive binding reaction. Therefore the "no-boil" technique used manually offers no real advantages in saving technician time. Moreover, recent studies by Zucker, Podell, and Allen[14] indicate that the alkaline denaturation procedures specified by the kit manufacturers fail to denature cobalamin-binding proteins completely in the sera of patients with chronic myelogenous leukemia and fail to denature anti–intrinsic factor blocking antibodies completely in the sera of some patients with vitamin B_{12} defi-

Table 63-2. Comparison of assay conditions for vitamin B_{12} radioassays

	Manual			Automated
Condition	Dualcount (Boil) (Diagnostic Product)	Immo Phase (Corning Medical)	No-Boil Combo Stat II (RIA Products, Inc.)	ARIA II (Becton Dickinson)
Binders	Purified hog IF*	Purified hog IF	Purified hog IF	Purified hog IF
Folate assay	Simultaneous or separate	Simultaneous	Simultaneous	Simultaneous
Sample pretreatment	Boil	Boil	Alkali denaturation	Boil
Separating agent	Charcoal suspension	Solid phase (glass beads)	Solid phase (cellulose particles)	Hydrophobic resin chamber
Sample volume	200 μL	100 μL	200 μL	100 μL
Incubation time	60 minutes	60 minutes	60 minutes	30 minutes
Incubation temperature	25° C	25° C	25° C	25° C
Incubation pH	9.4	9.2	9.3	9.3
Sensitivity	50 pg/mL[11]	17 pg/mL†	24 pg/mL†	100 pg/mL[11]
Precision (coefficient of variation)	3.8%-8.5%[11]	3.0%-12.0%†	2.5%-10.6%†	4.6%-8.2%[11]
Interferences	Ascorbic acid, F^-, radioactive drugs	Ascorbic acid, F^-, radioactive drugs	Ascorbic acid, F^-, radioactive drugs	Ascorbic acid, F^-, radioactive drugs

*Intrinsic factors.
†Reported by manufacturers.

ciency. We have evaluated a recently developed commercial solid-phase no-boil assay kit (Diagnostic Products Corp., Los Angeles) (Table 63-2) and found that denaturation of serum vitamin B_{12}–binding proteins and anti–intrinsic factor blocking antibodies is sufficient to prevent significant interference in the assay, but the precision of the "no-boil" assay (coefficient of variation, CV, about 11%) was not so good as the "boil" assay (CV about 6%) (Table 63-2).[15]

SPECIMEN

Samples should be collected from fasting individuals, since recent food intake may increase vitamin levels in the blood. For patients who are to be tested for malabsorption of B_{12}, blood samples should be drawn before the Schilling test because such patients will receive not only radioactive vitamin B_{12}, but also an injection of a large dose (1 mg) of nonradioactive vitamin B_{12} to block all potential tissue sites of vitamin B_{12} binding.

Either serum or plasma may be used for the assay, but heparin should not be used as an anticoagulant because it shows a modest B_{12}-binding ability. Large amounts of fluoride or ascorbic acid are added in vitro to blood samples for certain procedures; such blood samples should not be used for vitamin B_{12} assay because either of these two agents in high concentrations appears to destroy vitamin B_{12}.

Vitamin B_{12} is susceptible to photolytic degradation; therefore excessive exposure to light should be avoided. Serum samples can be stored overnight in a refrigerator (4° C) or several months in a freezer (−20° C) without appreciable loss of vitamin B_{12}. Repeated freezing and thawing of samples should be avoided.

REFERENCE RANGE

As with all diagnostic tests, the laboratory should establish its own reference ranges to conform with the characteristics of the population being tested and to allow for the variation in assay techniques used. The following reference ranges in picograms per milliliter were obtained in our laboratory using three different intrinsic factor (IF) preparations.[5]

B_{12} status	Crude IF	Pure IF	Crude IF and blocking agent
Deficient	<200	<100	<100
Borderline	200-290	100-180	145-225
Normal	290-900	180-960	225-830
Above normal	>900	>960	>830

A statistically significant negative correlation was found between age and serum levels of vitamin B_{12} measured by microbiological assays.[16] More recently, Magnus, Bache-Wiig, and Anderson reported that more elderly persons in geriatric homes had low B_{12} values when assayed with pure intrinsic factor than when assayed with crude intrinsic factor in radioassay, suggesting that biologically inactive vitamin B_{12} analogs were more frequently present in elderly persons than in the reference sample group.[17] Higher serum vitamin B_{12} levels in the black population have also been reported.[18]

INTERFERENCE

It has been reported that ingestion of megadoses of ascorbic acid may be harmful to human vitamin B_{12} metabolism and may produce artificially low serum vitamin B_{12} levels by radioassay because of destruction of vitamin B_{12} by ascorbic acid.[19] However, we did not find a statistical difference between the mean serum vitamin B_{12} levels (768 ± 60 pg/mL) of 20 children with myelomeningocele receiving daily mean doses of 1.65 g of supplemental ascorbic acid to acidify the urine for prevention of possible urinary tract infection and those of 20 children with myelomeningocele receiving no ascorbic acid therapy (815 ± 55 pg/mL). Their mean serum ascorbic acid levels were 15.1 ± 0.7 and 7.4 ± 0.7 mg/L respectively.[20]

REFERENCES

1. Matthews, D.M.: Observations on the estimation of serum vitamin B_{12} using *Lactobacillus leichmannii,* Clin. Sci. **22:**101-111, 1962.
2. Powell, D.E.B., Thomas, J.H., Mandal, A.R., et al.: Effect of drugs on vitamin B_{12} levels obtained using the *Lactobacillus leichmannii* method, J. Clin. Pathol. **22:**672, 1969.
3. Lau, K.S., Gottlieb, C., Wasserman, L.R., et al.: Measurement of serum vitamin B_{12} level using radioisotope dilution and coated charcoal, Blood **26:**202-214, 1965.
4. Kolhouse, J.F., Kondo, H., Allen, N.C., et al.: Cobalamin analogues are present in human plasma and can mask cobalamin deficiency because current radioisotope dilution assays are not specific for true cobalamin, N. Engl. J. Med. **299:**785-792, 1978.
5. Chen, I.W., Silberstein, E.B., Maxon, H.R., et al.: Clinical significance of serum vitamin B_{12} measured by radioassay using pure intrinsic factor, J. Nucl. Med. **22:**447-451, 1981.
6. National Committee for Clinical Laboratory Standards: Guidelines for evaluating a B_{12} (cobalamin) assay, Villanova, Penn., 1980, National Council on Clinical Laboaratory Standards.
7. Kolhouse, J.F., and Allen, R.H.: Isolation of cobalamin and cobalamin analogs by reverse affinity chromatography, Anal. Biochem. **84:**486-490, 1978.
8. Product guide for radioassays and non-isotopic ligand assays, Clin. Chem. **29:**952-953, 1983.
9. Herbert, V., and Colman, N.: Evidence humans may use some analogues of B_{12} as cobalamins (B_{12}): pure intrinsic factor (IF) radioassay may "diagnose" B_{12} deficiency where it does not exist, Clin. Res. **29:**571A, 1981.
10. Schilling, R.F., Fairbanks, V.F., Miller, R., et al.: "Improved" vitamin B_{12} assays: a report on two commercial kits, Clin. Chem. **29:**582-583, 1983.
11. Chen, I.W., Silberstein, E.B., Maxon, H.R., et al.: Semiautomated system for simultaneous assays of serum vitamin B_{12} and folic acid in serum evaluated, Clin. Chem. **28:**2161-2165, 1982.
12. Ithakissios, D.S., Kubiatowicz, D.O., and Wickes, J.H.: Room temperature radioassay for B_{12} with oyster toadfish *(Opsanus tau)* serum as binder, Clin. Chem. **26:**323-329, 1980.
13. Sturgeon, M.F., Hogle, D.M., Luddy, B.J., et al.: Advantages of solid phase technology over charcoal separation in the determination of B_{12} concentrations by radioassay, Ligand Q. **4:**52-56, 1981.
14. Zucker, R.M., Podell, E.R., and Allen, R.H.: Multiple problems with current no-boil assays for serum cobalamin, Ligand Q. **4:**52-58, 1981.
15. Chen, I.W., Sperling, M.I., Heminger, L.A., et al.: Denaturation of interfering serum proteins under the assay conditions of a solid-phase no-boil vitamin B_{12} (B_{12}) assay kit, Clin. Chem. **29:**1241, 1983.

16. Boger, W.P., Wright, L.D., Strickland, S.C., et al.: Vitamin B_{12}: correlation of serum concentrations and age, Proc. Soc. Exp. Biol. Med. **89:**375-378, 1955.
17. Magnus, E.M., Bache-Wiig, J.E., Anderson, T.R., et al.: Folate and vitamin B_{12} (cobalamin) blood levels in elderly persons in geriatric homes, Scand. J. Haematol. **28:**360-366, 1982.
18. Drum, D.E., and Jankowski, C.B.: A racial influence on serum cobalamin concentration, Syllabus of seventh annual meeting of Clinical Ligand Assay Society, Miami, Fla., 1981, p. 148. (Abstract.)
19. Herbert, V.: Vitamin B_{12}, Am. J. Clin. Nutr. **34:**971-972, 1981.
20. Ekvall, S., Chen, I.W., and Bozian, R.: The effect of supplemental ascorbic acid on serum vitamin levels in myelomeningocele patients, Am. J. Clin. Nutr. **34:**1356-1361, 1981.

Carotenes

LAWRENCE A. KAPLAN

Clinical significance: pp. 468 and 656
Molecular formula: $C_{40}H_{56}$
Molecular weight: 536.85 daltons
Merck Index: 1853 (β-isomer)
Chemical class: unsaturated hydrocarbon, provitamin A
Isomeric forms: 4 known isomers, α, β, γ, and δ

PRINCIPLES OF ANALYSIS AND CURRENT USAGE

β-Carotene of dietary (plant) origin is the primary precursor of vitamin A. β-Carotene can be split in two to yield two molecules of vitamin A, whereas the α and γ isomers of carotene can yield only one molecule of vitamin A. Because of the related structures of the carotenes, older assays had difficulty in specifically measuring vitamin A, or β-carotene. Assays for β-carotene are also hindered by the presence of other carotenoid compounds in serum such as lycopene, xanthophyll, and phytofluene. Most carotene in serum (80% to 90%) is the β-isomer; the remainder is the α-isomer.

The most frequently used β-carotene assay in fact measures *total* carotene, including the variable levels of the other carotenoids that are present in serum (method 1, Table 63-3). This assay extracts serum with organic solvents such as petroleum ether or isooctane.[1] The absorbance at 450 nm is used to quantitate total extractable carotenoids such as β-carotene.

Thin-layer chromatographic separations with alumina and silica have been used to isolate carotenes from the

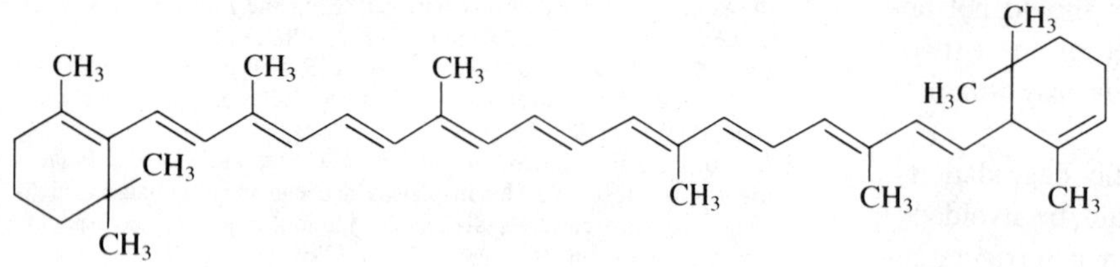

CH3 CH3 CH3 CH3 H3C CH3 CH3 CH3 CH3 CH3

α-Carotene

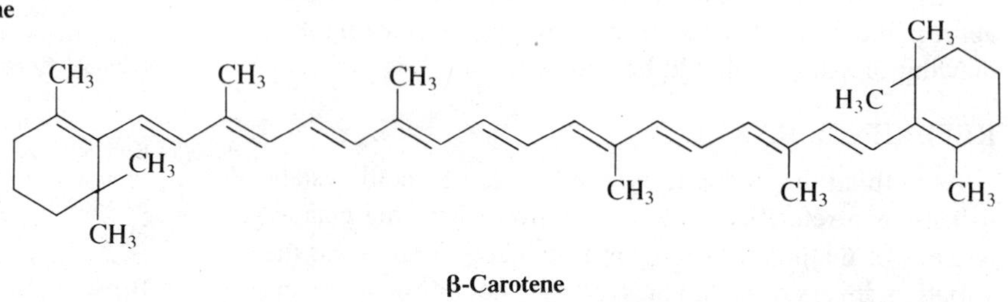

CH3 CH3 CH3 CH3 H3C CH3 CH3 CH3 CH3 CH3

β-Carotene

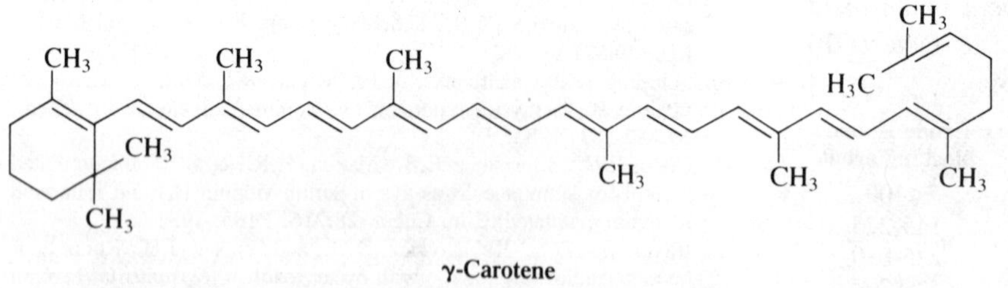

CH3 CH3 CH3 CH3 H3C CH3 CH3 CH3 CH3 CH3

γ-Carotene

CH3 CH3 CH3 CH3 H3C CH3 CH3 CH3 CH3 CH3

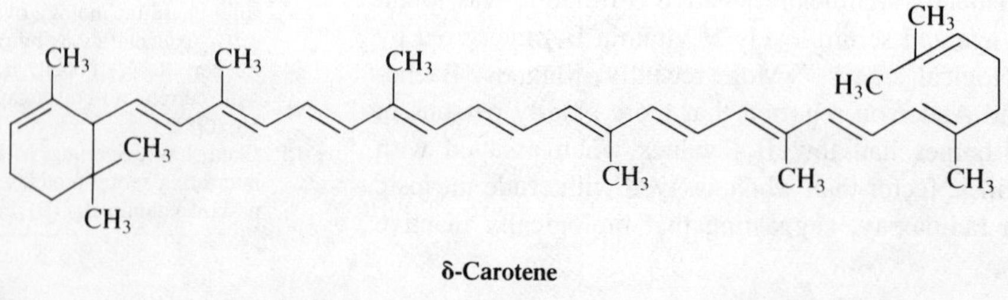

δ-Carotene

Table 63-3. Methods of carotene analysis

Method	Type of analysis	Principle	Usage	Comments
1. Extraction	Spectrophotometric	Carotene extracted into petroleum ether, and the absorbance at 450 mm is related to concentration	Most common, serum or plasma	Very nonspecific, measures entire class of carotenoids
2. High-performance liquid chromatography (HPLC)	Chromatographic	β-Carotene is separated by reversed-phase chromatography; columns monitored at 450 nm	Research only, serum, plasma	Can specifically measure β-carotene

more polar alcohol derivatives such as lycopene.[2,3] These assays have not been widely employed for quantitative analysis. Recently, a high-performance liquid chromatography (HPLC) procedure for quantitative analysis of α- and β-carotene was reported (method 2, Table 63-3). This procedure separates the petroleum ether extract on C-18 reversed-phase columns to give a peak detected at 450 nm containing only the provitamin A compounds, α- and β-carotene.[4]

REFERENCE AND PREFERRED METHODS

The β-carotene assay is most frequently used as a rapid screening procedure for diagnosing fat malabsorption. Since all the carotenoids are most likely absorbed together, the clinical information obtained by measurement of total carotenoids rather than β-carotene alone may be valid.

The HPLC assay for provitamin A (α- and β-carotene) has indicated that the total carotene assays overestimate carotene twofold to tenfold. However, whether the improved analytical assay gives more useful clinical information is not known at this time.

Thus the recommended procedure for serum carotene remains the extraction of total carotenoids with spectrophotometric quantitation at 450 nm.

SPECIMEN

10 mL of blood in a foil-wrapped red-topped tube (with a minimum of 2 mL of serum).

PROCEDURE

Principle

The major precursor of vitamin A in humans is β-carotene. Carotene is extracted into petroleum ether to remove it from interfering substances in serum. Because of its hydrophobic properties, carotene usually binds to serum proteins. Ethanol is added to the serum to break these complexes and release the carotene. The resultant orange-yellow color is read at 450 nm.

Reagents

95% ethanol

Petroleum ether

β-carotene (commercially available from Sigma Co., St. Louis.)

Caution: Decomposes on exposure to light and air; keep refrigerated.

Reagent-grade chloroform

Pooled serum control. Store at −20° C.

Assay (Table 63-4)

Procedure is performed at ambient temperature in subdued light.

Equipment: spectrophotometer, band pass ≤ 10 nm, capable of reading at 450 nm.

1. Label one round-bottomed centrifuge tube each for blank, patient, and control. Wrap tubes in foil. Run duplicates whenever possible.
2. Add 2 mL of water to the blank tube and 2 mL of serum or control to their respective tubes.
3. Add 2 mL of ethanol to each tube. Add dropwise while vortexing.
4. Add 4 mL of petroleum ether to each tube. Cap tubes.
5. Vortex each tube for 2 minutes.
6. Allow layers to separate by standing for 2 to 3 minutes.
7. Carefully pipet off the petroleum ether (top) layer into a cuvette.
8. Read absorbance at 450 nm against the blank.

Calculation

1. Convert absorbance to concentration using the standard curve (see next page).
2. Multiply *all* concentration values by 2 to account for extraction of 2 mL of serum into 4 mL of ether.

Standard curve

1. Transfer 100 mg of β-carotene to a 500 mL volumetric flask. Dissolve in about 20 mL of chloroform and dilute

Table 63-4. Summary of conditions for carotene assay

Condition	Total carotene
Temperature	Ambient
Sample volume	2 mL
Fraction of sample volume	0.5
Linearity	4 μg/mL
Interferences	Other carotenoids

to volume in petroleum ether. This yields a stock standard of 200 mg/L (373 μmol/L).
2. Pipet 10 mL of stock standard into a 100 mL volumetric flask and dilute to volume with petroleum ether; this will yield a working carotene standard of 20 μg/mL.
3. Pipet 2.5, 5, 10, 15, and 20 mL of the 20 μg/mL standard into respective 100 mL volumetric flasks. Dilute to volume with petroleum ether. These yield working standards of 0.5, 1, 2, 3, and 4 μg/mL, respectively.
4. Place standards into cuvettes and read absorbance at 450 nm against a petroleum ether blank.
5. Plot absorbance versus concentration on linear graph paper.
6. Curve should be prepared twice a year.

REFERENCE RANGE

0.50 to 2.5 μg/mL

CLINICAL SIGNIFICANCE

Carotene or provitamin A determination in serum or plasma is the most useful single screening method for the detection of malabsorption of fat. Carotene is a fat-soluble pigment that is slowly absorbed. It is found in leafy vegetables, fruits, and liver. The absorption of carotene requires the presence of dietary lipids; limited amounts of these lipids are stored in the liver. When malabsorption of fats occurs, serum carotene is greatly decreased.

Carotene levels are lowest in primary malabsorption. Levels may also be low in enteritis and many diseases associated with steatorrhea. High fever, hepatic disease, and deficient diet may also cause low carotene levels. Low dietary intake of carotene as a cause of lowered serum carotene values can be determined by repetition of the serum analysis after 15,000 units of carotene in oil is given by mouth for 5 days. If the cause is low dietary intake, a twofold to threefold increase in serum values will occur.

Carotene values are increased in diabetes, hyperlipemia, carotenemia, and hypothyroidism.

REFERENCES

1. Demetriou, J.A.: Vitamins. In Henry, R.J., Cannon, D.C., and Winkelman, J.W., editors: Clinical chemistry: principles and technics, ed. 2, New York, 1974, Harper & Row, Publishers, Inc., pp. 1373-1380.
2. Drujan, B.D., Castillon, R., and Guerrero, R.: Application of fluorometry in determination of vitamin A, Anal. Biochem. **23**:44-52, 1968.
3. Varma, T.N.R., Panalaks, T., and Murray, T.K.: Thin layer chromatography of vitamin A and related compounds, Anal. Chem. **36**:1864-1865, 1964.
4. Katrengi, H., Kaplan, L.A., and Stein, E.A.: Separation and quantitation of serum beta carotenes and other carotenoids by HPLC, J. Lipid Res., 1984. (In press.)

Folate assay

MICHAEL D.D. McNEELY

Clinical significance: p. 656

Folates	*Molecular formula*	*Merck Index*	*Molecular weight (daltons)*
Pteroylglutamic acid (PGA, folic acid)	$C_{19}H_{19}N_7O_6$	4085	441.40
Tetrahydrofolic acid (THFA)	$C_{19}H_{23}N_7O_6$	—	445.40
N-5-Methyltetrahydrofolic acid (MTHFA)	$C_{20}H_{25}N_7O_6$	—	459.40

Chemical class: pteroylglutamic acid

PRINCIPLES OF ANALYSIS AND CURRENT USAGE

Folates are a group of chemically and biologically related compounds[1] that function as coenzymes in the metabolism of one-carbon compounds, in purine and pyrimidine synthesis, and in the degradation of histidine to glutamic acid. The measurement of folate in serum and whole blood is important in the differential diagnosis of megaloblastic anemia.

Folic acid is pteroylglutamic acid (PGA). It can be reduced to tetrahydrofolic acid (THFA), which is the main form found in the bloodstream. *N*-5-Methyltetrahydrofolic acid (MTHFA) is the main storage form and the primary biological coenzyme.

Folate in the blood is 90% unbound or loosely bound to albumin. Ten percent is bound to a specific protein (folic acid–binding protein, FABP). FABP increases as the body store of folate declines.

There are two fundamental approaches to folate measurement: microbiological assays and radiometric competitive protein binding (CPB) techniques. Microbiological assays were developed first, but CPB methods are now the standard assay procedure used in most laboratories.

Table 63-5. Methods of folate analysis

Method	Type of analysis	Principle	Usage	Comments
1. Microbiological	Bioassay	Growth of folate-dependent bacteria related to specimen folate concentration; growth monitored by turbidimetry	Rare	False-negative results because of presence of antibiotics or antibodies to target organism; false-positive results because of turbidity
2. Radiometric	Competitive protein binding assay (CPB)	Folates in sample compete with ^{3}H- or ^{125}I-labeled ligand for binding to milk proteins (such as lactoglobulin); separation of bound and free labeled ligand by dextran-coated charcoal	Most frequently used	Must destroy endogenous folate binders

The microbiological assay (method 1, Table 63-5) depends on the fact that the growth of *Lactobacillus casei* requires (MTHFA).[2] The amount of *L. casei* growth permitted by a patient's serum is compared to the growth permitted by standard solutions of MTHFA. The greater the concentration of folates in the patient's sample, the greater the growth of the indicator organism. Growth is monitored by measurement of the turbidity of the test solutions at 640 to 700 nm, and the absorbance is directly related to the logarithm of folate concentration. This assay requires a 16- to 18-hour incubation period.

Radiometric CPB assays are based on the high affinity of B_{12} or folate for proteins that specifically bind these molecules (method 2, Table 63-5). Using a limited amount of binding protein and radiolabeled vitamin (ligand), one can quantitate the concentration of the molecules (see Chapter 11). The CPB assays are the most widely used methods because they are more efficient, faster, and more precise. Such assays may be arranged to allow the simultaneous measurement of vitamin B_{12} and folate.[3]

The most common binding agent for the CPB assays is derived from milk.[4] Unpurified milk powder or partially purified β-lactoglobulin may be used. Hog kidney[5] and porcine serum[6] contain folate-binding proteins, but these offer no advantage over milk. It is important to recognize that milk binders have different binding affinities for different forms of folate and this binding is pH dependent.[7]

The CPB assay is usually carried out at pH 7.4 to 8.2 as a single or two-stage assay. The one-stage assay is the conventional competitive assay approach in which sample, binder, and tracer are combined simultaneously.

The two-stage assay is noncompetitive and involves sequential incubation. The binder is first incubated with sample (or standard). The temperature is then lowered to 4° C to minimize dissociation. In the second step, tracer is added and occupies the unoccupied binding sites. This approach is more sensitive.

For both assays the separation of bound and free tracer is almost invariably performed using dextran-coated charcoal with centrifugation in the cold.

Red blood cell folate is measured by the same analytical techniques. Folate is much more concentrated in red blood cells than in serum. Its measurement more closely reflects tissue folate stores and is less variable as a consequence of recent folate ingestion.[8,9]

To liberate the intracellular folate, freezing and thawing is considered more complete than incubation with ascorbic acid.[10] A hematocrit should be determined on the sample to express the results as a function of red blood cell mass.

Patient samples contain substances that bind folate and produce a variable blank, which causes false low results.[11] For destruction of these materials the samples may be subjected to severe conditions that inactivate the endogenous binding proteins. The most common approach has been boiling at a high pH. A popular version uses a lysine buffer at pH 10.5 with 2-mercaptoethanol.[12] Since these extraction steps may destroy endogenous folate,[13] a serum blank may be used instead.

Standardization of the assays must be done with care. Impure assay material containing a variety of folate derivatives is a major source of variation between assays.

Methyltetrahydrofolic acid (MTHFA) is the most common standard. It is available as a barium salt. It is important to account for the weight difference between the barium salt and the acid forms when one is preparing standards because a 30% difference may be produced. The purity of MTHFA can be assessed spectrophotometrically at 290 and 245 nm.[8]

Pteroylglutamic acid (PGA) has been advocated as a standard because it is stable, well characterized spectrophotometrically, inexpensive, and easy to purify.[8] If PGA is used, reaction conditions must be established so that PGA and MTHFA react equally with the binding agent.

Tritiated MTHFA cannot be used as an assay tracer because it has a very low specific activity and is inherently unstable. Tritiated PGA has been widely employed. Like all tritiated assays, the specific activity is low and the need to employ liquid scintillation beta-particle counting is inconvenient. Iodinated tracers in which ^{125}I is linked to PGA by tyrosine offer much greater counting activity and the added ease of gamma-ray counting. However, iodinated tracers are not readily available for method preparation within the laboratory.

REFERENCE AND PREFERRED METHODS

The microbiological assay is not suitable for routine work but is appropriate as a reference procedure. False lowering of the results is caused by the presence in serum of antibiotics, folate antagonists, and serum antibodies against *L. casei* as well as exposure of the sample to light. Serum with bacterial contamination or significant turbidity cannot be assayed. These techniques are difficult to perform, require specialized microbiological equipment, and may take up to 3 days to complete.

Radiometric CPB methods using milk binder at pH 8.0, tritiated PGA tracer, MTHFA standard, and dextran-coated charcoal separation provide the most reliable assays. Iodine tracers, if available and proved, are much more convenient than tritiated tracers. A preliminary step should be undertaken to eliminate serum folate binding. Methods that do not use preliminary steps ("no boil") must thoroughly justify themselves. This method is compared to the microbiological assay in Table 63-6.

The recent trend to develop commercial kit assays capable of measuring both vitamin B_{12} and folate must be evaluated with caution. Such methods are certainly more convenient than two individual assays but tend to be optimized for vitamin B_{12}, possibly at the expense of the folate assay.

SPECIMEN

Samples for serum folate are preferentially collected in the fasting state, since blood folate levels vary with meals. Stability is maintained for 24 hours at 4° C; thereafter the serum should be kept frozen at −10° C. Ascorbic acid has been used to stabilize MTHFA but is inconvenient and unnecessary.[14]

Blood can be collected into oxalate or heparin anticoagulants and must be carefully maintained at 4° C; it is stable at this temperature for 3 days.

PROCEDURE[15]

Principle

The folate assay is a competitive protein binding (CPB) method employing tritiated PGA, MTHFA standard, and milk binder in a two-step sequential assay.

Reagents

Milk folate binder. Dissolve 5 g of Carnation brand powdered milk in 50 mL of normal saline and dialyze for 72 hours at 4° C against 100 volumes of saline, changed three times. Dilute the dialyzed milk tenfold with 0.05 M borate-Ringer buffer (B-R buffer), pH 8.0. Before using, prepare a binding curve using various quantities of diluted milk incubated for 30 minutes at 4° C with 0.15 ng of tritiated pteroylglutamic acid (PGA). The dilution that binds 50% to 70% of the labeled folate should be selected. Store in suitable aliquots in polystyrene tubes at −70° C.

Standards

MTHFA. MTHFA (Sigma Co., St. Louis) in 10 mg ampules is made up to 100 mL with 0.05 M borate-Ringer buffer, pH 8.0, containing 200 mg of ascorbic acid (see below). This solution is labeled "stock standard I" (218 μmol/L) and may be stored at −70° C for up to 6 months.

A stock standard II (2.18 μmol/L) solution is prepared by dilution of 1 mL of stock standard I to 100 mL with buffer. This solution may also be stored at −70° C for up to 6 months.

A working standard is prepared by thawing of the stock standard II and dilution of 1 mL to 100 mL with buffer. Each 0.1 mL of the working standard contains 1000 pg.

Finally, prepare a series of doubling dilutions of working standard to produce standards measuring 1000, 500, 250, 125, and 62.5 pg/0.1 mL. These standards are the equivalent of serum samples containing, respectively, 20, 10, 5, 2.5, and 1.25 ng/mL (42.6, 21.8, 10.9, 5.4, and 2.7 nmol/L).

Tritiated PGA. Obtain from Amersham/Searle Corp., Arlington Heights, Ill. The specific activity should be greater than 20 Ci/mmol. The contents of an ampule of tracer containing approximately 250 μCi is dissolved in 25

Table 63-6. Comparison of folate assays

Condition	Bioassay[2]	Radiometric*
Reaction temperature	37° C	25°, 4° C
Sample volume	100-200 μL	50 μL
Fraction of sample volume	0.1	0.086
Time of reaction	16-18 hours	≅3 hours
Final concentration of reagents	—	Milk binder: 0.77 mg/mL Borate buffer (pH 8.0): 0.05 mol/L Ascorbic acid ^{3}H-Folic acid: 115.4 pg/mL (≅3.9 nCi)
Linearity	To 68.2 nmol/L	To 42.6 nmol/L
Interferences	Antibiotics, antibodies to *Lactobacillus casei*	Endogenous folate binders

*Assay given in this chapter.

Tube	B-R buffer	B-R buffer with ascorbate	MTHFA standards	Sample sera	Milk binder	^{3}H-PGA	DCC
NSB	600	200	—	—	—	100	400
Standards	400	200	100	—	100	100	400
Test sera	450	200	—	50	100	100	400

mL of 20% ethanol and stored in 0.1 mL aliquots at −70° C. Before use, thaw and add to 20 mL of borate-Ringer buffer. Each 0.1 mL of this solution should contain about 0.005 μCi and 150 pg of tritiated PGA.

Borate-Ringer buffer. Dissolve 6.2 g of boric acid (100.3 mmol), 8.5 g of $Na_2EDTA \cdot 2\ H_2O$ (22.8 mmol), and 0.5 g of NaCl (8.5 mmol) in 900 mL of distilled water. Adjust pH to 8.0 in 1 M NaOH and adjust the volume to 1 L. Refrigerate at 4° C. This is stable for 1 month.

Borate-Ringer buffer with ascorbic acid. Dissolve 200 mg of ascorbic acid in 50 mL of borate-Ringer buffer and make up to 100 mL with buffer.

Dextran-coated charcoal (DCC). Bring 50 mL of 10% (w/v) Rheomacrodex (Pharmacia Diagnostics, Division of Pharmacia, Inc., Piscataway, NJ 08854) to 1 L with water. Add 50 g of Norit A charcoal and mix well. The suspension is stable at 4° C for at least 2 months.

Assay

Equipment: a liquid scintillation counter.

1. Set up the six standards (including a zero standard), the patient's serum, and control sera in duplicate in 12 × 75 mm polystyrene tubes. A nonspecific binding (NSB) tube is prepared as well. The above box outlines the amounts added in microliters. The sample volume may be varied from 10 to 50 μL.
2. Combine buffer, samples (standards), and milk binder and vortex to mix.
3. Incubate for 30 minutes at 25° C and then incubate for a further 60 minutes at 4° C.
4. Add 100 μL of tracer to each tube, vortex, and incubate for 30 minutes at 4° C.
5. Add 400 μL of dextran-coated charcoal, vortex, and centrifuge 10 minutes at 1000 *g* at 4° C.
6. Transfer the supernatant to scintillation vials and add 10 mL of scintillation cocktail (Unogel, Becton Dickinson Immunodiagnostics, Orangeburg, N.Y.).
7. Determine the number of counts present in each vial with a beta-particle scintillation counter using an appropriate quench correction, and subtract the nonspecific binding tube counts from all other tubes. Plot the results as concentration versus percent of zero binding (B_0).
8. For red blood cell folate a hemolysate may be used as the sample. Prepare the hemolysate from whole blood collected in EDTA and diluted 1⁄10 in deionized water. Allow to stand for 30 minutes; then freeze and thaw twice. Assay 50 μL of the hemolysate. When the assay is complete, compute the concentration of the hemolysate, multiply by 10 to obtain the whole blood folate (WBF), and then apply to the following formula:

$$RCF = \frac{WBF - SF\ (1\ PCV/100)}{(PCV/100)}$$

where *RCF* = Red cell folate
WBF = Whole blood folate
SF = Serum folate
PCV = Packed cell volume

9. If it is desirable to eliminate any possible influence of serum folate–binding substances, a serum blank tube may be prepared. This tube is identical to the serum sample except that no milk binder is added and 500 μL of buffer is employed to compensate for the volume. The value obtained on this tube is subtracted from its serum tube.

REFERENCE RANGE

Serum folate. 1.9 to 14 ng/mL (3.4 to 31.7 pmol/mL)

Red blood cell folate. 200 to 1000 ng/mL (453.1 to 2266 pmol/mL) packed cells

REFERENCES

1. Usdin, E.: Blood folic acid studies. VI. Chromatographic resolution of folic acid–active substances obtained from blood, J. Biol. Chem. **234:**2373-2376, 1959.
2. Herbert, V.: Aseptic addition method for *Lactobacillus casei* assay of folate activity in human serum, J. Clin. Pathol. **19:**12-16, 1966.
3. Gutcho, S., and Mansbach, L.: Simultaneous radioassay of serum vitamin B_{12} and folic acid, Clin. Chem. **23:**1606-1614, 1977.
4. Ghitis, J.: The folate binding in milk, Am. J. Clin. Nutr. **20:**1-4, 1967.
5. Kamen, B.A., and Caston, J.D.: Direct radiochemical assay for serum folate: competition between ^{3}H-folic acid and 5-methyltetrahydrofolic acid for a folate binder, J. Lab. Clin. Med. **83:**164-174, 1974.
6. Mantzos, J.: Radioassay of serum folate with use of pig plasma folate binders, Acta Haematol. **54:**289-296, 1975.
7. Givas, J.K., and Gutcho, S.: pH dependence of the binding of folates to milk binder in radioassay of folates, Clin. Chem. **21:**427-428, 1975.
8. Mortensen, E.: Determination of erythrocyte folate by competitive protein binding assay preceded by extraction, Clin. Chem. **22:**982-992, 1976.
9. Mortensen, E.: Effect of storage on the apparent concentration of folate in erythrocytes, as measured by competitive protein binding radioassay, Clin. Chem. **24:**663-668, 1978.
10. Netteland, B., and Bakke, O.M.: Inadequate sample-preparation technique as a source of error in determination of erythrocyte folate by competitive binding radioassay, Clin. Chem. **23:**1505-1506, 1977.
11. Zettner, A., and Duly, P.E.: New evidence for a binding principle specific for folates as a normal constituent of human serum, Clin. Chem. **20:**1313-1319, 1974.
12. Dunn, R.T., and Foster, L.B.: Radioassay of serum folate, Clin. Chem. **19:**1101-1105, 1973.

13. Mitchell, G.A., Pochron, S.P., Smutny, P.V., and Guity, R.: Decreased radioassay values for folate after serum extraction when pteroylglutamic acid standards are used, Clin. Chem. **22**:647-649, 1976.
14. Waddell, C.C., Domstad, P.A., Pircher, F.J., et al.: Serum folate levels. Comparison of microbiologic assay and radioisotope kit methods, Am. J. Clin. Pathol. **66**:746-752, 1976.
15. Mincey, E.W., Wilcox, E., and Morrison, R.T.: Estimation of serum and red cell folate by a simple radiometric technique, Clin. Biochem. **6**:274-284, 1973.

Appendix A Buffer solutions*

Buffer solutions (or buffers) are solutions whose pH value is to a large degree insensitive to the addition of other substances. It is important to realize, however, that the pH value of a buffer solution does not change only when acids or bases are added or on dilution but also when the temperature changes or neutral salts are added. In accurate work therefore, it is important to check the pH value electrometrically after all the ingredients have been added. The extent to which the pH values of buffer solutions vary when acids or bases are added or the temperature changes is shown in the tables that follow. In general, dilution to half the concentration changes the pH value by only some hundredths of a unit (Buffer No. 1 in the table is an exception in that the change amounts to ca. pH 0.15); addition of 0.1-molar neutral salt solution may change the pH value of ca. 0.1.

In the table opposite the solutions are classified into general buffers (mostly in use for the last 50 years), universal buffers with a low buffering capacity but a wide pH range, and buffers for biological media with a moderate pH range but containing stable ingredients (phosphate and borate, for example, often undergo side reactions with biological media). An important property is often the transparency to ultraviolet light. Occasionally it is desirable to have a volatile buffer, which can be readily removed[1] (examples are buffers Nos. 20 and 21), but the use of very volatile systems makes a close control of the pH essential. Most of the pH data to be found in the literature relate to the SØRENSEN scale, and it should be noted that the values given in the following table of buffers are on the conventional pH scale.

Both stock and buffer solutions should be made up with distilled water free of CO_2. Only standard reagents should be used. If there is any doubt as to the purity or water content of solutions, their molarity must be checked by titration. The amounts x of stock solutions required to make up a buffer solution of the desired pH value are given in the second table in Appendix B.

*This appendix has been compiled by F. KOHLER, Department of Physical Chemistry, University of Vienna, and taken from CIBA-Geigy AG, Basel, Switzerland.

Reference

1. For a list of volatile buffers see MICHL, H., in HEFTMANN, E. (Ed.), *Chromatography*, part 1, Reinhold, New York, 1961, page 250.

No.	Name	pH range	Temperature	pH change per °C
	General buffers			
1	KCl/HCl (CLARK and LUBS)[1]	1.0- 2.2	Room	0
2	Glycine/HCl (SØRENSEN)[2]	1.2- 3.4	Room	0
3	Na citrate/HCl (SØRENSEN)[2]	1.2- 5.0	Room	0
4	K biphthalate/HCl (CLARK and LUBS)[1]	2.4- 4.0	20° C	+0.001
5	K biphthalate/NaOH (Clark and LUBS)[1]	4.2- 6.2	20° C	
6	Na citrate/NaOH (SØRENSEN)[2]	5.2- 6.6	20° C	+0.004
7	Phosphate (SØRENSEN)[2]	5.0- 8.0	20° C	−0.003
8	Barbital-Na/HCl (MICHAELIS)[3]	7.0- 9.0	18° C	
9	Na borate/HCl (SØRENSEN)[2]	7.8- 9.2	20° C	−0.005
10	Glycine/NaOH (SØRENSEN)[2]	8.6-12.8	20° C	−0.025
11	Na borate/NaOH (SØRENSEN)[2]	9.4-10.6	20° C	−0.01
	Universal buffers			
12	Citric acid/phosphate (McILVAINE)[4]	2.2- 7.8	21° C	
13	Citrate-phosphate-borate/HCl (TEORELL and STENHAGEN)[5]	2.0-12.0	20° C	
14	BRITTON-ROBINSON[6]	2.6-11.8	25° C	at low pH 0 at high pH −0.02
	Buffers for biological media			
15	Acetate (WALPOLE)[7-9]	3.8- 5.6	25° C	
16	Dimethylglutaric acid/NaOH[10]	3.2- 7.6	21° C	
17	Piperazine/HCl[11,12]	4.6- 6.4 8.8-10.6	20° C	
18	Tetraethylethylenediamine*[12]	5.0- 6.8 8.2-10.0	20° C	
19	Trismaleate[7,13]	5.2- 8.6	23° C	
20	Dimethylaminoethylamine*[12]	5.6- 7.4 8.6-10.4	20° C	
21	Imidazole/HCl[14]	6.2- 7.8	25° C	
22	Triethanolamine/HCl[15]	7.0- 8.8	25° C	
23	*N*-Dimethylaminoleucylglycine/NaOH[16]	7.0- 8.8	23° C	−0.015
24	Tris/HCl[7]	7.2- 9.0	23° C	−0.02
25	2-Amino-2-methylpropane-1,3-diol/HCl[7,13]	7.8-10.0	23° C	
26	Carbonate (DELORY and KING)[7,17]	9.2-10.8	20° C	

*Can be combined with tris buffer to give a cationic universal buffer (cf. SEMENZA et al.[12]).
From Geigy scientific tables, ed. 8, Basel, Switzerland, 1981, CIBA-Geigy AG.

References

1. CLARK and LUBS, *J. Bact.*, **2,** 1 (1917).
2. SØRENSEN, S.P.L., *Biochem. Z.*, **21,** 131 (1909), and **22,** 352 (1909); *Ergebn. Physiol.*, **12,** 393 (1912); WALBUM, L.E., *Biochem. Z.*, **107,** 219 (1920).
3. MICHAELIS, L., *J. biol. Chem.*, **87,** 33 (1930).
4. McILVAINE, T.C., *J. biol. Chem.*, **49,** 183 (1921).
5. TEORELL and STENHAGEN, *Biochem. Z.*, **299,** 416 (1938).
6. BRITTON and WELFORD, *J. chem. Soc.*, **1937,** 1848.
7. GOMORI, G., in COLOWICK and KAPLAN (Eds.), *Methods in Enzymology,* vol. 1, Academic Press, New York, 1955, page 138.
8. WALPOLE, G.S., *J. chem. Soc.*, **105,** 2501 (1914).
9. GREEN, A.A., *J. Amer. chem. Soc.*, **55,** 2331 (1933).
10. STAFFORD et al., *Biochim. biophys. Acta,* **18,** 319 (1955); KREBS, H.A., unpublished, 1957.
11. SMITH and SMITH, *Biol. Bull.*, **96,** 233 (1949).
12. SEMENZA et al., *Helv. chim. Acta,* **45,** 2306 (1962).
13. GOMORI, G., *Proc. Soc. exp. Biol. (N.Y.),* **68,** 354 (1948).
14. MERTZ and OWEN, *Proc. Soc. exp. Biol. (N.Y.),* **43,** 204 (1940), quoted by RAUEN, H.M. (Ed.), *Biochemisches Taschenbuch,* 2nd ed., part 2, Springer, Berlin, 1964, page 90.
15. BEISENHERZ et al., *Z. Naturforsch.*, **8b**, 555 (1953).
16. LEONIS, J., *C. R. Lab. Carlsberg, Sér. Chim.*, **26,** 357 (1948).
17. DELORY and KING, *Biochem. J.*, **39,** 245 (1945).

Appendix B Preparation of buffer solutions

When not otherwise specified, both stock and buffer solutions should be made up with distilled water free of CO_2. Only standard reagents should be used. If there is any doubt as to the purity or water content of solutions, their molarity must be checked by titration. The amounts x of stock solutions required to make up a buffer solution of the desired pH value are given in the table in the second part of Appendix B.

Buffer No.	Stock solutions		Composition of the buffer
	A	B	
1	KCl 0.2-N (14.91 g/l)	HCl 0.2-N	25 ml A + x ml B made up to 100 ml
2	Glycine 0.1-molar in NaCl 0.1-N (7.507 g glycine + 5.844 g NaCl/l)	HCl 0.1-N	x ml A + (100 − x) ml B
3	Disodium citrate 0.1-molar (21.01 g $C_6H_8O_7 \cdot 1H_2O$ + 200 ml NaOH 1-N per litre)	HCl 0.1-N	x ml A + (100 − x) ml B
4	Potassium biphthalate 0.1-molar (20.42 g $KHC_8H_4O_4$/l)	HCl 0.1-N	50 ml A + x ml B made up to 100 ml
5	As No. 4	NaOH 0.1-N	50 ml A + x ml B made up to 100 ml
6	As No. 3	NaOH 0.1-N	x ml A + (100 − x) ml B
7	Monopotassium phosphate ¹⁄₁₅-molar (9.073 g KH_2PO_4/l)	Disodium phosphate ¹⁄₁₅-molar (11.87 g $Na_2HPO_4 \cdot 2H_2O$/l)	x ml A + (100 − x) ml B
8	Barbital sodium 0.1-molar (20.62 g/l)	HCl 0.1-N	x ml A + (100 − x) ml B
9	Boric acid, half-neutralized, 0.2-molar (corr. to 0.05-molar borax: 12.37 g boric acid + 100 ml NaOH 1-N per litre)	HCl 0.1-N	x ml A + (100 − x) ml B
10	As No. 2	NaOH 0.1-N	x ml A + (100 − x) ml B
11	As No. 9	NaOH 0.1-N	x ml A + (100 − x) ml B
12	Citric acid 0.1-molar (21.01 g $C_6H_8O_7 \cdot 1H_2O$/l)	Disodiumphosphate 0.2-molar (35.60 g $Na_2HPO_4 \cdot 2H_2O$/l)	x ml A + (100 − x) ml B
13	To citric acid and phosphoric acid solutions (ca. 100 ml), each equivalent to 100 ml NaOH 1-N, add 3.54 cryst. orthoboric acid and 343 ml NaOH 1-N, and make up the mixture to 1 litre	HCl 0.1-N	20 ml A + x ml B made up to 100 ml
14	Citric acid, monopotassium phosphate, barbital, boric acid, all 0.02857-molar (6.004 g $C_6H_8O_7 \cdot 1H_2O$, 3.888 g KH_2PO_4, 5.263 g barbital, 1.767 g H_3BO_3/l)	NaOH 0.2-N	100 ml A + x ml B
15	Sodium acetate 0.1-N (8.204 g $C_2H_3O_2Na$ or 13.61 g $C_2H_3O_2Na \cdot 3H_2O$/l)	Acetic acid 0.1-N (6.005 g/l)	x ml A + (100 − x) ml B
16	ββ-Dimethylglutaric acid 0.1-molar (16.02 g/l)	NaOH 0.2-N	(a) 100 ml A + x ml B made up to 1000 ml (b) 100 ml A + x ml B + 5.844 g NaCl made up to 1000 ml (NaCl ≙ 0.1-molar)
17	Piperazine 1-molar (86.14 g/l)	HCl 0.1-N	5 ml A + x ml B made up to 100 ml
18	Tetraethylethylenediamine 1-molar (172.32 g/l)	HCl 0.1-N	5 ml A + x ml B made up to 100 ml
19	Tris acid maleate 0.2-molar (24.23 g tris[hydroxymethyl]-aminomethane + 23.21 g maleic acid or 19.61 g maleic anhydride/l)	NaOH 0.2-N	25 ml A + x ml B made up to 100 ml
20	Dimethylaminoethylamine 1-molar (88 g/l)	HCl 0.1-N	5 ml A + x ml B made up to 100 ml
21	Imidazole 0.2-molar (13.62 g/l)	HCl 0.1-N	25 ml A + x ml B made up to 100 ml
22	Triethanolamine 0.5-molar (76.11 g/l) containing 20 g/l ethylenediaminetetraacetic acid disodium salt ($C_{10}H_{14}O_8N_2Na_2 \cdot 2H_2O$)	HCl 0.05-N	10 ml A + x ml B made up to 100 ml
23	*N*-Dimethylaminoleucylglycine 0.1-molar (24.33 g $C_{10}H_{20}O_3N_2 \cdot \frac{3}{2}H_2O$/l) containing NaCl 0.2-N (11.69 g/l)	NaOH 1-N 100 ml made up to 1 litre with A	x ml A + (100 − x) ml B
24	Tris 0.2-molar (24.23 g tris[hydroxymethyl]aminomethane/l)	HCl 0.1-N	25 ml A + x ml B made up to 100 ml
25	2-Amino-2-methylpropane-1,3-diol 0.1-molar (10.51 g/l)	HCl 0.1-N	50 ml A + x ml B made up to 100 ml
26	Sodium carbonate anhydrous 0.1-molar (10.60 g/l)	Sodium bicarbonate 0.1-molar (8.401 g/l)	x ml A + (100 − x) ml B

From Geigy scientific tables, ed. 8, Basel, Switzerland, 1981, CIBA-Geigy AG.

The table gives the amounts (x ml) of the stock solutions listed in the first part of Appendix B required to make up a buffer solution of the desired pH value.

pH	1	2	3	4	5	6	7	8	9	10	11	12	13	14	15	16a	16b	17	18	19	20	21	22	23	24	25	26	pH
1.0	54.2	...	...	...	...	...	...	...	...	...	...	...	...	...	...	...	...	...	...	...	...	...	...	...	...	...	...	1.0
1.2	36.0	11.1	9.0	...	...	...	...	...	...	...	...	...	...	...	...	...	...	...	...	...	...	...	...	...	...	...	...	1.2
1.4	23.2	26.4	17.9	...	...	...	...	...	...	...	...	...	...	...	...	...	...	...	...	...	...	...	...	...	...	...	...	1.4
1.6	14.7	36.2	23.6	...	...	...	...	...	...	...	...	...	...	...	...	...	...	...	...	...	...	...	...	...	...	...	...	1.6
1.8	9.3	43.9	27.6	...	...	...	...	...	...	...	...	...	...	...	...	...	...	...	...	...	...	...	...	...	...	...	...	1.8
2.0	5.9	50.7	30.2	...	...	...	...	...	...	...	...	...	74.4	...	...	...	...	...	...	...	...	...	...	...	...	...	...	2.0
2.2	3.8	56.5	32.2	...	...	...	...	...	...	...	...	98.8	68.8	...	...	...	...	...	...	...	...	...	...	...	...	...	...	2.2
2.4	...	62.3	34.1	41.0	...	...	...	...	...	...	...	94.5	64.6	...	...	...	...	...	...	...	...	...	...	...	...	...	...	2.4
2.6	...	68.4	36.0	34.3	...	...	...	...	...	...	...	90.0	61.3	1.6	...	...	...	...	...	...	...	...	...	...	...	...	...	2.6
2.8	...	74.7	37.9	27.8	...	...	...	...	...	...	...	85.1	58.9	3.6	...	...	...	...	...	...	...	...	...	...	...	...	...	2.8
3.0	...	81.0	39.9	21.6	...	...	...	...	...	...	...	80.3	56.9	5.7	...	...	...	...	...	...	...	...	...	...	...	...	...	3.0
3.2	...	86.2	42.1	15.9	...	...	...	...	...	...	...	76.0	55.2	7.8	...	7.0	14.4	...	...	...	...	...	...	...	...	...	...	3.2
3.4	...	90.3	44.8	10.9	...	...	...	...	...	...	...	72.0	53.9	9.9	...	13.3	20.9	...	...	...	...	...	...	...	...	...	...	3.4
3.6	...	...	47.8	6.7	...	...	...	...	...	...	...	68.4	52.9	11.7	...	20.7	26.8	...	...	...	...	...	...	...	...	...	...	3.6
3.8	...	...	51.2	3.3	...	...	...	...	...	...	...	65.1	51.8	13.5	10.9	26.3	32.4	...	...	...	...	...	...	...	...	...	...	3.8
4.0	...	...	55.1	0.0	...	...	...	...	...	...	...	62.0	50.7	15.3	16.6	32.4	36.6	...	...	...	...	...	...	...	...	...	...	4.0
4.2	...	...	60.0	...	3.0	...	...	...	...	...	...	59.1	49.7	17.5	23.9	36.2	40.3	...	...	...	...	...	...	...	...	...	...	4.2
4.4	...	...	66.4	...	6.7	...	...	...	...	...	...	56.4	48.6	19.7	33.5	39.3	43.1	...	...	...	...	...	...	...	...	...	...	4.4
4.6	...	...	74.9	...	11.1	...	...	...	...	...	...	53.7	47.5	21.9	44.9	41.3	45.7	94.3	...	...	...	...	...	...	...	...	...	4.6
4.8	...	...	85.6	...	16.5	...	...	...	...	...	...	51.2	46.4	24.1	56.6	43.5	48.3	91.5	...	...	...	...	...	...	...	...	...	4.8
5.0	...	...	100.0	...	22.6	...	99.2	...	...	...	...	49.0	45.4	26.3	67.8	45.7	51.5	87.8	94.3	...	...	...	...	...	...	...	...	5.0
5.2	...	...	...	...	28.8	87.1	98.4	...	...	...	...	46.9	44.3	28.6	76.8	48.4	53.6	83.6	91.5	3.2	...	...	...	...	...	...	...	5.2
5.4	...	...	...	...	34.4	78.0	97.3	...	...	...	...	44.7	43.2	31.0	84.0	51.3	58.2	77.6	87.8	5.0	...	...	...	...	...	...	...	5.4
5.6	...	...	...	...	39.1	70.3	95.5	...	...	...	...	42.4	42.0	33.4	89.3	55.0	63.6	71.8	83.1	7.3	94.3	...	...	...	...	...	...	5.6
5.8	...	...	...	...	42.4	64.5	92.8	...	...	...	...	40.0	40.8	35.8	...	58.8	68.7	66.5	77.6	9.7	91.7	...	...	...	...	...	...	5.8
6.0	...	...	...	...	45.0	60.3	88.9	...	...	...	...	37.4	39.7	38.3	...	63.9	73.6	61.8	71.7	12.4	88.0	...	...	...	...	...	...	6.0
6.2	...	...	...	...	46.7	57.2	83.0	...	...	...	...	34.5	38.4	40.8	...	69.5	78.5	58.2	66.4	15.2	83.3	43.4	...	...	...	...	...	6.2
6.4	...	...	...	...	...	54.8	75.4	...	...	...	...	31.4	37.0	43.3	...	74.1	83.3	55.5	61.7	17.9	77.9	40.4	...	...	...	...	...	6.4
6.6	...	...	...	...	...	53.2	65.3	...	...	...	...	27.9	35.6	45.8	...	83.5	87.4	...	58.0	20.8	72.0	36.5	...	...	...	...	...	6.6
6.8	...	...	...	...	...	...	53.4	...	...	...	...	23.5	34.2	48.3	...	87.4	91.0	...	55.3	22.2	66.6	31.4	...	...	...	...	...	6.8
7.0	...	...	...	...	...	...	41.3	53.3	...	...	...	19.0	32.9	50.9	...	90.0	93.2	...	...	23.7	61.9	25.4	86.2	86.4	...	...	...	7.0
7.2	...	...	...	...	...	...	29.6	55.0	...	...	...	13.8	31.7	53.4	...	91.8	94.9	...	...	25.2	58.1	19.6	79.6	80.6	44.7	...	...	7.2
7.4	...	...	...	...	...	...	19.7	57.6	...	...	...	9.8	30.6	55.8	...	93.0	95.8	...	...	26.7	55.3	14.6	71.3	72.8	42.0	...	...	7.4
7.6	...	...	...	...	...	...	12.8	60.8	...	...	...	6.8	29.6	58.2	...	93.8	96.8	...	...	28.6	...	10.2	62.0	63.2	39.3	...	...	7.6
7.8	...	...	...	...	...	...	7.4	65.2	53.0	...	...	4.6	28.8	60.5	...	...	...	...	...	31.2	...	6.6	52.0	52.1	33.7	43.9	...	7.8
8.0	...	...	...	...	...	...	3.7	70.6	55.4	...	...	...	28.1	62.8	...	...	...	...	...	33.9	...	...	42.0	41.1	27.9	41.6	...	8.0
8.2	...	...	...	...	...	...	...	75.9	58.0	...	...	...	27.6	65.0	...	...	...	...	46.4	36.9	...	...	31.9	31.4	22.9	38.4	...	8.2
8.4	...	...	...	...	...	...	...	81.2	62.1	...	...	...	27.0	67.2	...	...	...	...	43.9	39.9	...	...	22.5	23.0	17.3	34.8	...	8.4
8.6	...	...	...	...	...	...	...	86.2	66.9	94.7	...	...	26.3	69.3	...	...	...	...	40.9	42.7	45.4	...	16.0	15.9	13.0	30.7	...	8.6
8.8	...	...	...	...	...	...	...	90.1	73.6	92.0	...	...	25.2	71.3	...	...	...	45.5	36.8	...	42.8	...	11.7	10.3	8.8	23.3	...	8.8
9.0	...	...	...	...	...	...	...	93.2	83.5	88.4	...	...	24.0	73.2	...	...	...	43.2	31.8	...	39.2	...	...	...	5.3	17.7	...	9.0
9.2	...	...	...	...	...	...	...	...	95.6	84.0	...	...	22.6	75.1	...	...	...	40.0	26.2	...	34.7	...	...	...	...	13.3	10.0	9.2
9.4	...	...	...	...	...	...	...	...	...	78.9	87.0	...	21.4	77.0	...	...	...	35.8	20.4	...	29.3	...	...	...	...	9.2	18.4	9.4
9.6	...	...	...	...	...	...	...	...	...	73.2	75.5	...	20.2	78.8	...	...	...	30.8	15.2	...	23.6	...	...	...	...	5.2	29.3	9.6
9.8	...	...	...	...	...	...	...	...	...	67.2	65.1	...	19.0	80.4	...	...	...	25.0	10.8	...	19.0	...	...	...	...	4.1	42.0	9.8
10.0	...	...	...	...	...	...	...	...	...	62.5	59.6	...	18.1	81.8	...	...	...	19.4	7.4	...	13.1	...	...	...	...	2.3	53.4	10.0
10.2	...	...	...	...	...	...	...	...	...	58.8	56.4	...	17.1	83.1	...	...	...	14.3	...	...	9.2	...	...	...	...	...	63.7	10.2
10.4	...	...	...	...	...	...	...	...	...	55.7	54.1	...	16.5	84.3	...	...	...	10.0	...	...	6.2	...	...	...	...	...	73.1	10.4
10.6	...	...	...	...	...	...	...	...	...	53.6	52.3	...	16.0	85.4	...	...	...	6.9	...	...	...	...	...	...	...	...	81.2	10.6
10.8	...	...	...	...	...	...	...	...	...	52.2	...	...	15.5	86.5	...	...	...	...	...	...	...	...	...	...	...	...	87.9	10.8
11.0	...	...	...	...	...	...	...	...	...	51.2	...	...	14.7	87.8	...	...	...	...	...	...	...	...	...	...	...	...	...	11.0
11.2	...	...	...	...	...	...	...	...	...	50.4	...	...	13.5	89.3	...	...	...	...	...	...	...	...	...	...	...	...	...	11.2
11.4	...	...	...	...	...	...	...	...	...	49.5	...	...	11.7	91.3	...	...	...	...	...	...	...	...	...	...	...	...	...	11.4
11.6	...	...	...	...	...	...	...	...	...	48.7	...	...	9.1	94.5	...	...	...	...	...	...	...	...	...	...	...	...	...	11.6
11.8	...	...	...	...	...	...	...	...	...	47.6	...	...	5.5	99.0	...	...	...	...	...	...	...	...	...	...	...	...	...	11.8
12.0	...	...	...	...	...	...	...	...	...	46.0	...	...	1.3	...	...	...	...	...	...	...	...	...	...	...	...	...	...	12.0
12.2	...	...	...	...	...	...	...	...	...	43.2	...	...	...	...	...	...	...	...	...	...	...	...	...	...	...	...	...	12.2
12.4	...	...	...	...	...	...	...	...	...	39.1	...	...	...	...	...	...	...	...	...	...	...	...	...	...	...	...	...	12.4
12.6	...	...	...	...	...	...	...	...	...	31.8	...	...	...	...	...	...	...	...	...	...	...	...	...	...	...	...	...	12.6
12.8	...	...	...	...	...	...	...	...	...	21.4	...	...	...	...	...	...	...	...	...	...	...	...	...	...	...	...	...	12.8

From Geigy scientific tables, ed. 8, Basel, Switzerland, 1981, CIBA-Geigy AG.

Appendix C Concentrations of common acids and bases

Compound	Molecular weight	Specific gravity	Percent	Normality	ml/liter for 1N* solution
HCl	36.46	1.19	36.0	11.7	85.5
HNO_3	63.02	1.42	69.5	15.6	64.0
H_2SO_4	98.08	1.84	96.0	35.9	28.4
CH_3COOH	60.03	1.06	99.5	17.6	56.9
NH_4OH	35.04	0.90	58.6	15.1	66.5
H_3PO_4	98.00	1.69	85.0	44.1	22.7
Thioglycolic acid	92.12	1.26	80.0	10.9	91.3
HCOOH	46.03	1.21	97.0	25.5	39.2
	46.03	1.19	88.0	22.7	44.1
$HClO_4$	100.5	1.67	70.0	11.65	85.7
Pyridine	79.10	0.98	100.0	12.4	80.6
2-Mercaptoethanol	78.13	1.14	100.0	14.6	68.5

To calculate concentration (c) from the weight percent (w) of a compound, use the formula:

$$\frac{10\,ws}{M} = c$$

M is the molecular weight, and s the specific gravity.

*Remember, the normality (N) is not the same as the molarity (M) for sulfuric and phosphoric acid.
From Brewer, J.M., Pesce, A.J., and Ashworth, H.: Experimental techniques in biochemistry, Engelwood, N.J., 1974, Prentice-Hall, Inc.

Appendix D Gases in common laboratory use—technical information

Product	Formula	Cylinder specifications					Thermophysical properties		
		State	CGA no. valve outlet	Highest purity grade (%)	Cylinder size (cubic feet)	Approximate cylinder pressure (psi)	Molecular weight	Vapor pressure at 21.1° C (psig)	Specific gravity at 21.1° C (1 atm)
Acetylene	C_2H_2	Dissolved gas (in acetone)	510/300	99.6	3-5	250	26.04	635	0.095
Air		Compressed gas	346/677	Mixture	200-500	2200-6000	28.96		1
Carbon dioxide	CO_2	Liquefied gas	320	99.999	2-6	274-838	44.01	839	1.53
Carbon monoxide	CO	Compressed gas	350	99.99	2-6	315-1602	28.01	*	0.97
Helium	He	Compressed gas	580/677	99.9999	200-300	2200-2640	4.003	*	0.138
Hydrogen	H_2	Compressed gas	350	99.9995	2-200	323-2200	2.02	*	0.0695
Methane	CH_4	Compressed gas	350	99.992	12	780-2000	16.04	*	0.555
Nitrogen	N_2	Compressed gas	580/677	99.999	30-200	2200-2640	28.01	*	0.967
Oxygen	O_2	Compressed gas	540	99.995	30-200	2200-2640	32.0	*	1.105
Propane	C_3H_8	Liquefied gas	510	99.98	12	109	44.1	109	1.55

*Above critical temperature at 21.1° C.
SA, Simple asphyxiant.
Data derived from the Airco Industrial Cases catalogue, Murray Hill, N.J., 1977.

Thermophysical properties			Hazardous properties			
			Flammability			
Critical temperature (°C)	Critical pressure (psia)	Specific volume (cf/lb)	Flammable limits in air (vol%)	Ignition temperature (°C)	Physiological properties	Threshold limit value (ppm)
35.1	890	14.7	2.580	305		SA
		13.3			Oxidant	
31.0	1071	8.74			Inert	5000
−140.0	507.4	13.8	12.5-74	651.1	Toxic	50
−267.8	33.2	96.7				SA
−239.98	190.8	192	4.75	585		SA
−82.1	673	23.7	5.15	538		SA
−146.9	492.9	13.8			Inert	SA
−118.4	736.9	12.1			Oxidant	
96.8	617.4	8.5	2.1-9.5	468		SA

Appendix E Major plasma proteins

Protein	Molecular weight	Concentration, mg/100 mL	Electrophoretic† mobility	Biological function
Prealbumins				
Thyroxine binding (TBPA)	55,000	10-40	7.6	Thyroxine transport
Retinol binding (RBP)	21,000	3-6		Vitamin A transport
Albumin	66,300	3500-5500	5.92	Maintain osmotic pressure, transport of bilirubin, free fatty acids, anions, and cations, cell nutrition
α_1 Globulins				
α_1 Acid glycoprotein (α_1S)	40,000	55-140	5.7	Unknown, inactivates progesterone
α_1 Antitrypsin (α_1AT)	54,000	200-400	5.42	Antiserine type of protease
α_1 Glycoprotein (9.5S, α_1M)	308,000	3-8	α_1	Unknown
α_1 Glycoprotein B (α_1B)	50,000	15-30	α_1	Unknown
α_1 Glycoprotein T (α_1T)	60,000	5-12	α_1	Unknown, tryptophan poor
α_1 Antichymotrypsin (α_1X)	68,000	30-60	α_1	Chymotrypsin inhibitor
α_1 Lipoproteins, high density (HDL)	28,000	254-387	α_1	Lipid transport
α_2 Globulins				
G_0 Globulin (Gc)	51,000	40-70	α_2	Vitamin D transport
Ceruloplasmin (Cp)	134,000	15-60	4.6	Copper transport, peroxidase activity
α_2 Glycoprotein, histidine rich (HRG)	58,000	5-15	α_2	Unknown
Zn-α_2-glycoprotein (Znα_2)	41,000	2-15	4.2	Unknown, binds Zn^{2+}
α_2 HS-glycoprotein (α_2HS)	49,000	40-85	4.2	Unknown, binds Ba^{2+}
α_2 Macroglobulin (α_2M)	725,000	150-420	4.2	Inhibitor of thrombin, trypsin, and pepsin
Transcortin (TC)	49,500	<7	α_2	Cortisol transport
Haptoglobins (Hp)				
Type 1-1	100,000	100-200	4.1	Binds hemoglobin, prevents loss of iron
Type 2-1	200,000	160-300	α_2	
Type 2-2	400,000	120-260	α_2	
α_2 Lipoproteins (VLDL)	250,000	150-230	Pre-β	Lipid transport
Thyroxine-binding protein (TBG)	58,000	1-2	α_2	Thyroxine transport
β Globulins				
Hemopexin (Hpx)	57,000	50-100	3.1	Binds heme
Transferrin (Tf)	76,500	200-320	3.1	Iron transport
β Lipoproteins (LDL)	250,000	280-440	3.1	Lipid transport
C4 Complement component (C4)	206,000	40-80	β_1	Complement system
β_2 Microglobulin ($\beta_2\mu$)	11,818	Trace	β_2	Common portion of the HLA transplantation antigen
β_2 Glycoprotein I (β_2 I)	40,000	15-30	1.6	Unknown
β_2 Glycoprotein II (GGG)	63,000	12-30	β_2	C3 activator (activates properidin)
β_2 Glycoprotein III (β_2 III)	35,000	5-15	β_2	Unknown
C-Reactive protein (CRP)	118,000	1	β_2	Opsonin, motivates phagocytosis in inflammatory disease
C3 Complement component (C3)	180,000	55-180	β_2	Complement system
Fibrinogen (ϕ, Fib.)	341,000	200-600	2.1	Blood clotting
γ Globulins				
Immunoglobulin M (IgM)	950,000	60-250	2.1	Antibodies, early response
Immunoglobulin E (IgE)	190,000	0.06	2.1	Reagin of the allergy system
Immunoglobulin A (IgA)	160,000	90-450	2.1	Tissue antibodies
Immunoglobulin D (IgD)	160,000	15	1.9	Cell surface and plasma antibodies
Immunoglobulin G (IgG)	160,000	800-1800	1.2	Antibodies, long range

*Does not include clotting factors, complement factors, or enzymes except fibrinogen, C3 and C4 of complement, which occur in substantial concentrations.

†Tiselius moving boundary electrophoresis in Tiselius units ($cm^2 V^{-1} sec^{-1} \times 10^5$, at 0° C, pH 8.6, and ionic strength 0.15).

From Natelson, S., and Natelson, E.A.: Principles of applied chemistry, vol. 3, New York, 1980, Plenum Pub. Corp.

Appendix F Conversions between conventional and SI units

	Conventional units	× Factor	= SI units
Gram	g/mL	$\frac{10^{15}}{mw}$	pmol/L
	g/100 mL	10	g/L
	g/100 mL	$\frac{10}{mw}$	mol/L
	g/100 mL	$\frac{10^4}{mw}$	mmol/L
	g/d	$\frac{1}{mw}$	mol/d
	g/d	$\frac{10^3}{mw}$	mmol/d
	g/d	$\frac{10^9}{mw}$	nmol/d
Microgram	μg/100 mL	$\frac{10}{mw}$	μmol/L
	μg/d	$\frac{1}{mw}$	μmol/d
	μg/d	$\frac{10^3}{mw}$	nmol/d
Picogram	pg	$\frac{10^3}{mw}$	fmol
	pg/mL	$\frac{10^3}{mw}$	pmol/L
Milliequivalent	mEq/L	$\frac{1}{valence}$	mmol/L
	mEq/kg	$\frac{1}{valence}$	mmol/kg
	mEq/d	$\frac{1}{valence}$	mmol/d
Milligram	mg/100 mL	10^{-2}	g/L
	mg/100 mL	$\frac{10^{-2}}{mw}$	mol/L
	mg/100 mL	$\frac{10}{mw}$	mmol/L
	mg/100 mL	$\frac{10^4}{mw}$	μmol/L
	mg/100 g	10	mg/kg
	mg/100 g	$\frac{10}{mw}$	mmol/kg
	mg/d	$\frac{1}{mw}$	mmol/d
	mg/d	$\frac{10^3}{mw}$	μmol/d
Milliliter	mL/100 g	10	mL/kg
	mL/min	1.667×10^{-2}	mL/s
Millimeters of mercury	mm Hg	1.333	mbar
	mm Hg	0.133	kPa
Minute	min	60	s
	min	0.06	ks
Percent	%	10^{-2}	1 (unit)
	% (g/100 g)	10	g/kg
	% (g/100 g)	10^{-2}	kg/kg
	% (g/100 mL)	10	g/L
	% (g/100 mL)	$\frac{10}{mw}$	mol/L
	% (g/100 mL)	$\frac{10^4}{mw}$	mmol/L
	% (mL/100 mL)	10^{-2}	L/L

Modified from Campbell, J.M., and Campbell, J.B.: Laboratory mathematics, ed. 3, St. Louis, 1983, The C.V. Mosby Co.
d, day; *Eq,* equivalent; *g,* gram; *L,* liter; *min,* minute; *mw,* molecular weight; *Pa,* pascal; *s,* second.
f, Femto (10^{-15}); *p,* pico (10^{-12}); *n,* nano (10^{-9}); *μ,* micro (10^{-6}); *m,* milli (10^{-3}); *k,* kilo (10^3).

Appendix G Conversions between conventional and SI units for specific analytes

Analyte	Conventional units	Multiply by		SI unit
		Conventional to SI	SI to conventional	
Acetominophen	µg/mL	6.61	0.151	µmol/L
Albumin	g/100 mL	144.9	0.0069	µmol/L
Ammonia	µg/100 mL	0.59	1.7	µmol/L
Anticonvulsant drugs				
Carbamazepine	µg/mL	4.32	0.23	µmol/L
Ethosuximide	µg/mL	7.08	0.14	µmol/L
Phenobarbital	µg/mL	4.31	0.23	µmol/L
Phenytoin	µg/mL	3.96	0.25	µmol/L
Primidone	µg/mL	4.58	0.22	µmol/L
Valproic acid	µg/mL	6.93	0.14	µmol/L
Bilirubin	mg/100 mL	17.1	0.059	µmol/L
Bromide	µg/mL	0.0125	80	mmol/L
Calcium	mg/100 mL	0.25	4	mmol/L
Chloride	mEq/L	1	1	mmol/L
Cholesterol	mg/100 mL	0.026	38.7	mmol/L
Cortisol	µg/100 mL	0.0276	36.2	µmol/L
Creatinine	mg/100 mL	88.4	0.0113	µmol/L
Digoxin	ng/mL	1.28	0.781	nmol/L
Estriol	µg/L	3.47	0.288	nmol/L
Ferritin	µg/L	2.2	0.445	pmol/L
Folic acid	µg/100 mL	22.7	0.044	nmol/L
Gentamicin	µg/mL	2.22	0.450	µmol/L
Glucose	mg/100 mL	0.055	18.0	mmol/L
Haptoglobin	mg/100 mL	0.118	8.47	µmol/L
HDL cholesterol	mg/100 mL	0.026	38.7	mmol/L
HCG	U/L	—	—	—
5-HIAA	mg	5.23	0.19	µmol
Ig A	mg/100 mL	0.0625	16	µmol/L
D	mg/100 mL	0.054	18.5	µmol/L
E	ng/mL	0.005	200	nmol/L
G	mg/100 mL	0.067	15	µmol/L
M	mg/100 mL	0.011	91	µmol/L
Insulin	pg/mL	0.174	5.74	nmol/L
	µU/mL	7.25	0.138	nmol/L
Iron	µg/100 mL	0.179	5.58	µmol/L
Ketones (acetoacetate)	mg/L	0.111	9.01	mmol/L
Lead	µg/L	4.83	0.207	nmol/L
Lithium	mEq/L	1	1	mmol/L
LDL cholesterol	mg/100 mL	0.026	38.7	mmol/L
Magnesium	mg/100 mL	0.41	2.43	mmol/L
Phosphorus	mg/100 mL	0.323	3.1	mmol/L
Phenylalanine	mg/L	6.05	0.165	µmol/L
Potassium	mEq/L	1	1	mmol/L
Quinidine	µg/mL	3.09	0.324	µmol/L
Salicylate	mg/100 mL	0.0724	13.8	mmol/L
Sodium	mEq/L	1	1	mmol/L
TIBC	µg/100 mL	0.179	5.58	µmol/L
Theophylline	µg/mL	5.55	0.180	µmol/L
Thyroid-stimulating hormone	mU/L	—	—	—
Thyroxine	µg/100 mL	12.9	0.078	nmol/L
Transferrin	mg/100 mL	0.11	9.09	µmol/L
Triglycerides	mg/100 mL	0.0114	87.5	mmol/L
Urea	mg/100 mL	0.166	6.01	mmol/L
Urea N	mg/100 mL	0.356	2.81	mmol/L
Uric acid	mg/100 mL	59.5	0.0168	µmol/L
Vanillylmandelic acid	mg	5.03	0.20	µmol
VLDL cholesterol	mg/100 mL	0.026	38.7	mmol/L
Vitamin B_{12}	pg/mL	0.738	1.36	pmol/L
Gases	mm Hg	0.133	7.51	kPa
Enzymes	U/L	1.67×10^{-8}	0.6×10^{8}	katal/L

Appendix H

Body Surface of Children

Nomogram for determination of body surface from height and mass[1]

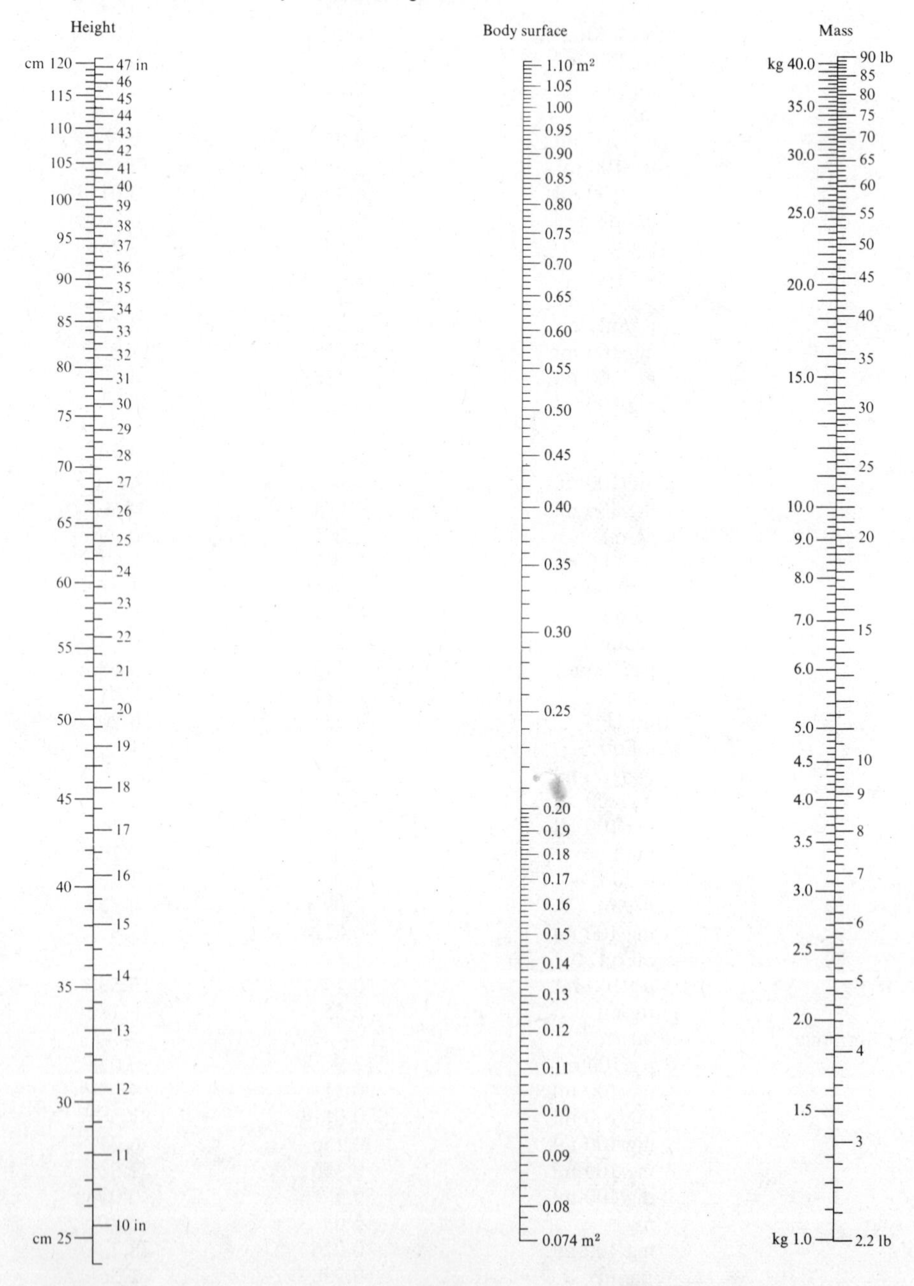

[1] From the formula of Du Bois and Du Bois, *Arch. intern. Med.*, **17**, 863 (1916): $S = M^{0.425} \times H^{0.725} \times 71.84$, or $\log S = \log M \times 0.425 + \log H \times 0.725 + 1.8564$ (S: body surface in cm², M: mass in kg, H: height in cm).

From Geigy scientific tables, ed. 8, Basel, Switzerland, 1981, CIBA-Geigy AG.

Appendix I

Body Surface of Adults

Nomogram for determination of body surface from height and mass[1]

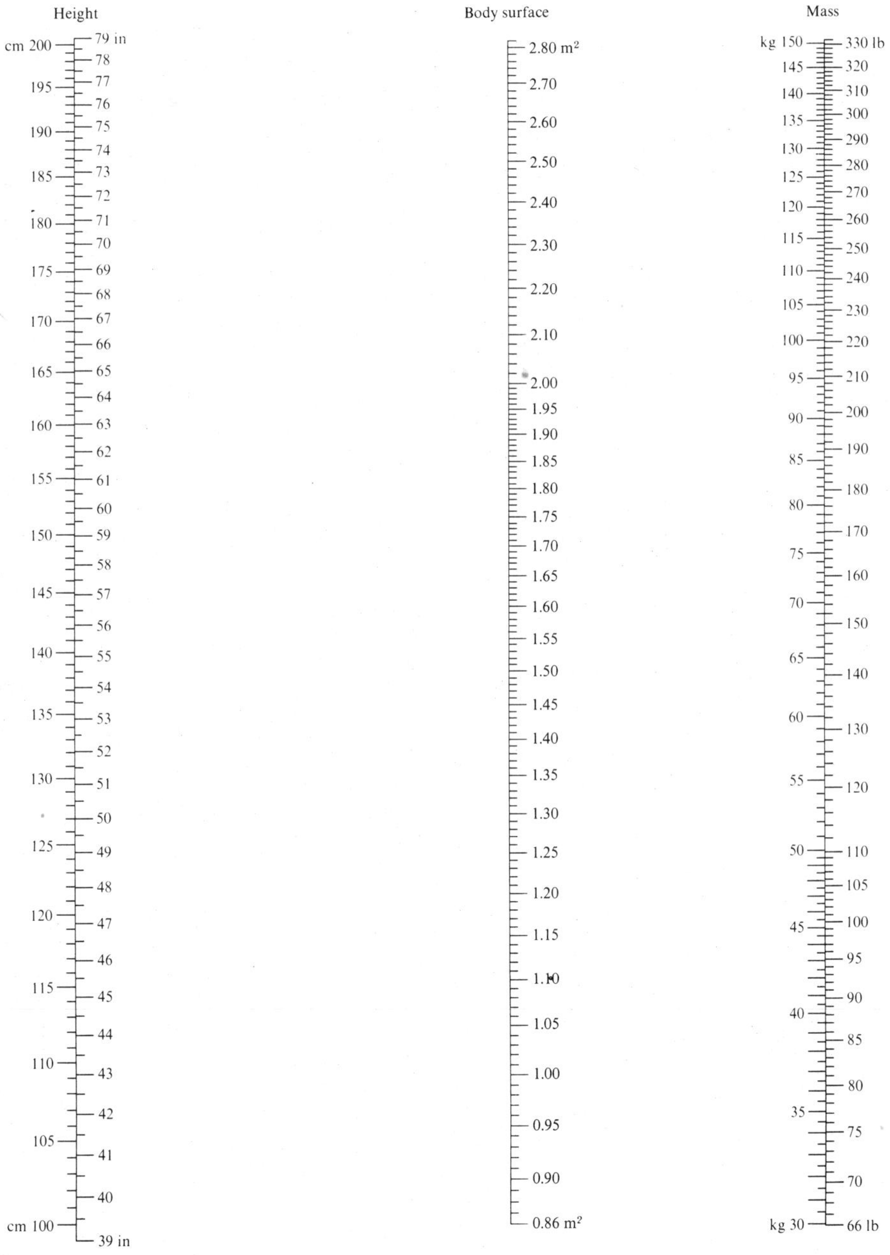

Height

cm 200 — 195 — 190 — 185 — 180 — 175 — 170 — 165 — 160 — 155 — 150 — 145 — 140 — 135 — 130 — 125 — 120 — 115 — 110 — 105 — cm 100

79 in — 78 — 77 — 76 — 75 — 74 — 73 — 72 — 71 — 70 — 69 — 68 — 67 — 66 — 65 — 64 — 63 — 62 — 61 — 60 — 59 — 58 — 57 — 56 — 55 — 54 — 53 — 52 — 51 — 50 — 49 — 48 — 47 — 46 — 45 — 44 — 43 — 42 — 41 — 40 — 39 in

Body surface

2.80 m² — 2.70 — 2.60 — 2.50 — 2.40 — 2.30 — 2.20 — 2.10 — 2.00 — 1.95 — 1.90 — 1.85 — 1.80 — 1.75 — 1.70 — 1.65 — 1.60 — 1.55 — 1.50 — 1.45 — 1.40 — 1.35 — 1.30 — 1.25 — 1.20 — 1.15 — 1.10 — 1.05 — 1.00 — 0.95 — 0.90 — 0.86 m²

Mass

kg 150 — 145 — 140 — 135 — 130 — 125 — 120 — 115 — 110 — 105 — 100 — 95 — 90 — 85 — 80 — 75 — 70 — 65 — 60 — 55 — 50 — 45 — 40 — 35 — kg 30

330 lb — 320 — 310 — 300 — 290 — 280 — 270 — 260 — 250 — 240 — 230 — 220 — 210 — 200 — 190 — 180 — 170 — 160 — 150 — 140 — 130 — 120 — 110 — 105 — 100 — 95 — 90 — 85 — 80 — 75 — 70 — 66 lb

---

[1] From the formula of Du Bois and Du Bois, *Arch. intern. Med.*, **17**, 863 (1916): $S = M^{0.425} \times H^{0.725} \times 71.84$, or $\log S = \log M \times 0.425 + \log H \times 0.725 + 1.8564$ (S: body surface in cm², M: mass in kg, H: height in cm).

From Geigy scientific tables, ed. 8, Basel, Switzerland, 1981, CIBA-Geigy AG.

Index

Page numbers in **boldface** indicate structural formulas.
Page numbers in *italics* indicate illustrations and boxed material.
Page numbers followed by *n* indicate footnotes.
Page numbers followed by *t* indicate tables.

E

M

S

W

X